CHEMICAL PRINCIPLES

13 III IIIA	14 IV IVA	15 V VA	16 VI VIA	17 VII VIIA	18 VIII VIIIA
					2 **He** Helium 4.00 $1s^2$
5 **B** Boron 10.81 $2s^22p^1$	6 **C** Carbon 12.01 $2s^22p^2$	7 **N** Nitrogen 14.01 $2s^22p^3$	8 **O** Oxygen 16.00 $2s^22p^4$	9 **F** Fluorine 19.00 $2s^22p^5$	10 **Ne** Neon 20.18 $2s^22p^6$
13 **Al** Aluminum 26.98 $3s^23p^1$	14 **Si** Silicon 28.09 $3s^23p^2$	15 **P** Phosphorus 30.97 $3s^23p^3$	16 **S** Sulfur 32.06 $3s^23p^4$	17 **Cl** Chlorine 35.45 $3s^23p^5$	18 **Ar** Argon 39.95 $3s^23p^6$

10	11 IB	12 IIB						
28 **Ni** Nickel 58.69 $3d^84s^2$	29 **Cu** Copper 63.55 $3d^{10}4s^1$	30 **Zn** Zinc 65.41 $3d^{10}4s^2$	31 **Ga** Gallium 69.72 $4s^24p^1$	32 **Ge** Germanium 72.64 $4s^24p^2$	33 **As** Arsenic 74.92 $4s^24p^3$	34 **Se** Selenium 78.96 $4s^24p^4$	35 **Br** Bromine 79.90 $4s^24p^5$	36 **Kr** Krypton 83.80 $4s^24p^6$
46 **Pd** Palladium 106.42 $4d^{10}$	47 **Ag** Silver 107.87 $4d^{10}5s^1$	48 **Cd** Cadmium 112.41 $4d^{10}5s^2$	49 **In** Indium 114.82 $5s^25p^1$	50 **Sn** Tin 118.71 $5s^25p^2$	51 **Sb** Antimony 121.76 $5s^25p^3$	52 **Te** Tellurium 127.60 $5s^25p^4$	53 **I** Iodine 126.90 $5s^25p^5$	54 **Xe** Xenon 131.29 $5s^25p^6$
78 **Pt** Platinum 195.08 $5d^96s^1$	79 **Au** Gold 196.97 $5d^{10}6s^1$	80 **Hg** Mercury 200.59 $5d^{10}6s^2$	81 **Tl** Thallium 204.38 $6s^26p^1$	82 **Pb** Lead 207.2 $6s^26p^2$	83 **Bi** Bismuth 208.98 $6s^26p^3$	84 **Po** Polonium (209) $6s^26p^4$	85 **At** Astatine (210) $6s^26p^5$	86 **Rn** Radon (222) $6s^26p^6$
110 **Ds** Darmstadtium $6d^87s^2$ (?)	111 **Rg** Roentgenium $6d^{10}7s^1$ (?)	112	113	114	115	116	117	118

Metals ← | → Nonmetals

Metalloids

62 **Sm** Samarium 150.36 $4f^66s^2$	63 **Eu** Europium 151.96 $4f^76s^2$	64 **Gd** Gadolinium 157.25 $4f^75d^16s^2$	65 **Tb** Terbium 158.93 $4f^96s^2$	66 **Dy** Dysprosium 162.50 $4f^{10}6s^2$	67 **Ho** Holmium 164.93 $4f^{11}6s^2$	68 **Er** Erbium 167.26 $4f^{12}6s^2$	69 **Tm** Thulium 168.93 $4f^{13}6s^2$	70 **Yb** Ytterbium 173.04 $4f^{14}6s^2$
94 **Pu** Plutonium (244) $5f^67s^2$	95 **Am** Americium (243) $5f^77s^2$	96 **Cm** Curium (247) $5f^76d^17s^2$	97 **Bk** Berkelium (247) $5f^97s^2$	98 **Cf** Californium (251) $5f^{10}7s^2$	99 **Es** Einsteinium (252) $5f^{11}7s^2$	100 **Fm** Fermium (257) $5f^{12}7s^2$	101 **Md** Mendelevium (258) $5f^{13}7s^2$	102 **No** Nobelium (259) $5f^{14}7s^2$

FREQUENTLY USED TABLES AND FIGURES

CHEMICAL PRINCIPLES

The Quest for Insight

FOURTH EDITION

PETER ATKINS
Oxford University

LORETTA JONES
University of Northern Colorado

W. H. Freeman and Company / New York

Publisher:	CRAIG BLEYER
Senior Acquisitions Editor:	JESSICA FIORILLO
Marketing Manager:	ANTHONY PALMIOTTO
Senior Developmental Editor:	RANDI ROSSIGNOL
Developmental Assistant:	SUSAN TIMMINS
Media/Supplements Editors:	JEANETTE PICERNO, AMY THORNE
Photo Editor:	BIANCA MOSCATELLI
Cover/Text Designer:	BLAKE LOGAN
Senior Project Editor:	GEORGIA LEE HADLER
Copy Editor:	MARGARET COMASKEY
Illustration Coordinator:	SUSAN TIMMINS
Illustrations:	PETER ATKINS WITH NETWORK GRAPHICS
Production Coordinator:	JULIA DeROSA
Composition:	BLACK DOT GROUP
Printing and Binding:	QW/VERSAILLES

Library of Congress Control Number: 2006936133

ISBN-13: 978-0-7167-7355-9
ISBN-10: 0-7167-7355-4

Printed in the United States of America

Second Printing

W. H. Freeman and Company
41 Madison Avenue
New York, NY 10010
Houndmills, Basingstoke RG21 6XS, England

www.whfreeman.com

Chapter 7 THERMODYNAMICS: THE SECOND AND THIRD LAWS

Chapter 8 PHYSICAL EQUILIBRIA

Chapter 15 THE ELEMENTS: THE LAST FOUR MAIN GROUPS

Chapter 16 THE ELEMENTS: THE d BLOCK

Chapter 17 NUCLEAR CHEMISTRY

LETTER FROM THE AUTHORS

Dear Colleagues,

We are pleased to present this new edition of our textbook, *Chemical Principles: The Quest for Insight*. It is designed for a rigorous course in introductory chemistry and to appeal to those of your students who will focus on engineering, the life sciences and medicine, or environmental sciences, as well as to chemistry majors.

Chemical Principles encourages students to question and to develop insight by emphasizing the nature of chemistry and how chemical knowledge is obtained and developed. We have retained the overall structure of the text, which has been so well received. It begins with a review of the basics in the Fundamentals sections, which instructors have used in a variety of ways. In the main body of the book an "atoms-first" organization introduces students to the most basic structure of matter before building on this knowledge to develop more complex properties and interactions.

With this new edition, we have paid particular attention to three areas: problem solving, the interpretation of mathematical expressions, and encouraging good practice. The most prominent new feature is "Picture the process," the thumbnail illustrations that accompany the steps in most worked Examples and some derivations. These little illustrations serve several purposes. On one hand they will make the text more accessible to a wider range of students; on the other hand, they will encourage students taking higher-level courses to stop and think about what they are doing. They will also help all students make connections between the various representations of chemistry, such as graphs, equations, and molecular structures. In the new "What does this equation mean" feature we walk the reader through the symbols in a mathematical expression, encouraging them to listen to what it is saying. In the new "Notes on good practice" feature we explain common pitfalls and encourage students to express themselves precisely. We have acknowledged—with enthusiasm—the rising interest in nanotechnology and green chemistry and have responded to the request for a wider range of higher-level exercises.

We have paid particular attention to maintaining the length of the text: what growth there has been is due entirely to the additional pedagogical features, such as "Picture the process," that we have introduced to reduce the cognitive load of the material for the reader. A few of the less central topics and some with burdensome mathematics have been moved to the book's Web site.

We are gratified by the response to the first three editions of this book. Our thanks go to all who have provided feedback: we listened carefully to your comments as we prepared this new edition. We hope that you find it an interesting, dynamic, and interactive textbook for the modern student and instructor, with innovative pedagogical features. We look forward very much to hearing from you about it.

Yours sincerely,

Peter Atkins

Loretta Jones

PREFACE

CHEMICAL PRINCIPLES

This text is designed for a rigorous course in introductory chemistry. Its central theme is to challenge students to think and question while providing a sound foundation in the principles of chemistry.

At the same time, students of all levels will benefit from assistance in learning how to think, pose questions, and approach problems. To that end, *Chemical Principles* is organized in a logical way that builds understanding and offers students a wide array of pedagogical supports.

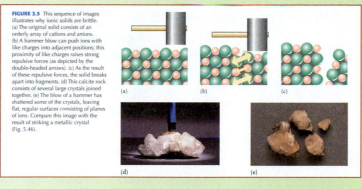

FIGURE 2.5 This sequence of images illustrates why ionic solids are brittle. (a) The original solid consists of an orderly array of cations and anions. (b) A hammer blow can push ions with like charges into adjacent positions; this proximity of like charges raises strong repulsive forces (as depicted by the double-headed arrows). (c) As the result of these repulsive forces, the solid breaks apart into fragments. (d) This calcite rock consists of several large crystals joined together. (e) The blow of a hammer has shattered some of the crystals, leaving flat, regular surfaces consisting of planes of ions. Compare this image with the result of striking a metallic crystal (Fig. 5.46).

ATOMS-FIRST ORGANIZATION

Chemical Principles presents the concepts of chemistry in a logical sequence that enhances student understanding. The **atoms-first sequence** starts with the behavior of atoms and molecules and builds up to more complex properties and interactions.

Chapter 1 has been reorganized in this edition to give readers a gentler introduction to atoms and their structure. Atoms and molecules, including discussions of quantum mechanics and molecular orbitals, provide the foundation for understanding bulk properties and models of gases, liquids, and solids.

Next, an exploration of **thermodynamics and equilibrium,** based on a conceptual understanding of entropy and Gibbs free energy. This integrated presentation lays a common foundation for these concepts and provides a basis for understanding the origin and form of the equilibrium constant and the behavior of equilibrium systems.

Kinetics then shows the dynamic nature of chemistry and the crucial role of insight and model building in identifying reaction mechanisms.

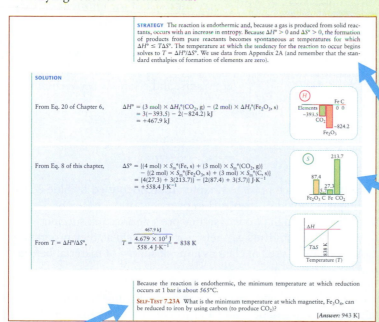

EMPHASIS ON PROBLEM SOLVING

The text features many levels of in-chapter problem-solving support, which build students' confidence in their mastery of the concepts. Problem-solving support has been greatly enhanced in this edition.

- **Strategy** sections in the in-chapter worked Examples show students how to organize information and map out a solution.

- **NEW!** **Picture the process** addresses the idea that learning improves when students can picture their problem-solving steps. **Visual interpretations of the steps in mathematical calculations** appear throughout the text in both worked Examples and derivations. The pictures show exactly what each mathematical step is expressing, helping students to see how the solution is developed. This new three-column format for solutions is designed to enrich the problem-solving experience by helping students to connect the calculation to chemistry concepts and principles, using macroscopic, molecular, and graphical representations.

- **Self-Tests** occur in pairs throughout the book. They enable students to test their understanding of the material covered in the preceding section or worked Example. An answer to the first test gives immediate feedback; the answer to the second is in the back of the book, along with the setup for the calculation.

TOOLBOX 6.1   **HOW TO USE HESS'S LAW**

CONCEPTUAL BASIS

Because enthalpy is a state function, the enthalpy change of a system depends only on its initial and final states. Therefore, we can carry out a reaction in one step or visualize it as the sum of several steps; the reaction enthalpy is the same in each case.

PROCEDURE

To use Hess's law to find the enthalpy of a given reaction, we find a sequence of reactions with known reaction enthalpies that adds up to the reaction of interest. A systematic procedure simplifies that process.

Step 1 Select one of the reactants in the overall reaction and write down a chemical equation in which it also appears as a reactant.

Step 2 Select one of the products in the overall reaction and write down a chemical equation in which it also

appears as a product. Add this equation to the equation written in step 1 and cancel species that appear on both sides of the equation.

Step 3 Cancel unwanted species in the sum obtained in step 2 by adding an equation that has the same substance or substances on the opposite side of the arrow.

Step 4 Once the sequence is complete, combine the standard reaction enthalpies.

In each step, we may need to reverse the equation or multiply it by a factor. Recall from Eq. 16 that, if we want to reverse a chemical equation, we have to change the sign of the reaction enthalpy. If we multiply the stoichiometric coefficients by a factor, we must multiply the reaction enthalpy by the same factor.

This procedure is illustrated in Example 6.9.

• **Toolboxes** summarize *major types* of calculations, demonstrating the connections between concepts and problem solving. They are designed as learning aids and handy summaries of key material. Each Toolbox is immediately followed by a related worked Example.

• **NEW!** **What does this equation tell us?** guides the reader through a qualitative explanation of a mathematical equation and its interpretation.

• **NEW!** **A note on good practice** alerts students to the proper usage of language and of units in calculations.

gravitational field), but it has now been largely displaced by a unit of similar magnitude, the **torr** (Torr). This unit is defined by the exact relation

$$760 \text{ Torr} = 1 \text{ atm}$$

(so 1 Torr = $1.01\ 325 \times 10^5$ Pa/760 = 133.322 Pa). It is normally safe to use the units mmHg and Torr interchangeably, but in very precise work you should note that they are not exactly the same.

A note on good practice: The name of the unit torr, like all names derived from the names of people, has a lower-case initial letter; likewise, the symbol for the unit has an uppercase initial letter (Torr).

We have deduced that expansion through a volume ΔV against a constant external pressure P_{ex} is given by

$$w = -P_{ex}\Delta V \qquad (3)*$$

This expression applies to all systems. A gas is easiest to visualize, but the expression also applies to an expanding liquid or solid. However, Eq. 3 applies *only when the external pressure is constant* during the expansion.

What does this equation tell us? The minus sign in Eq. 3 tells us that the internal energy decreases when the system expands. The factor P_{ex} tells us that more work has to be done when the external pressure (which is responsible for

---

• **Exercise sets** at the end of each chapter present problems at all levels of difficulty, and include questions that require students to use multimedia resources on the book's Web site. In this edition many **new conceptual and higher-level exercises** have been included in each chapter. Problems requiring the use of calculus are marked with a $\boxed{\text{C}}$.

• **NEW!** **Chemistry Connections** are multipart exercises that integrate many different concepts from several chapters. Many of these exercises incorporate illustrations of green chemistry.

Chemistry Connections

5.105 Ethylammonium nitrate, $CH_3CH_2NH_3NO_3$, was the first ionic liquid to be discovered. Its melting point of 12°C was reported in 1914 and it has since been used as a nonpolluting solvent for organic reactions and for facilitating the folding of proteins.
(a) Draw the Lewis structure of each ion in ethylammonium nitrate and indicate the formal charge on each atom (in the cation, the carbon atoms are attached to the N atom in a chain: C—C—N).
(b) Assign a hybridization scheme to each C and N atom.
(c) Ethylammonium nitrate cannot be used as a solvent for some reactions because it can oxidize some compounds. Which ion is more likely to be the oxidizing agent, the cation or anion? Explain your answer.
(d) Ethylammonium nitrate can be prepared by the reaction of gaseous ethylamine, $CH_3CH_2NH_2$, and aqueous nitric acid. Write the chemical equation for the reaction. What type of reaction is this?
(e) 2.00 L of ethylamine at 0.960 atm and 23.2°C was bubbled into 250.0 mL of 0.240 M HNO_3(aq) and 4.10 g of

---

• **NEW!** The **Fact Sheet** at the back of the book provides students with a single source for most of the information they need to solve problems. The fact sheet includes a list of key equations for each chapter; the periodic table; and tables of the elements, SI prefixes, fundamental constants, and relations between units.

FLEXIBLE COVERAGE

• The **Fundamentals** sections, which precede Chapter 1, are identified by blue-edged pages. These 13 minichapters provide a streamlined overview of the basics of chemistry. The sections can be used in two ways: they provide a useful, succinct review of basic material to which students can refer for extra help as they progress through the course, or they can be used in class as a quick survey of material before starting the main text.

• **NEW!** **A diagnostic test for the Fundamentals sections** will allow instructors to determine how much their students understand and where they need additional support. Instructors can then make appropriate assignments from the Fundamentals sections. The test includes 5 to 10 problems for each Fundamentals section. The diagnostic test was created by Cynthia LaBrake of the University of Texas, Austin, and can be found on the Instructor's Resource CD-ROM and in the Instructor section of the Web site.

• **Optional mathematical derivations.** The **How do we do that?** feature sets off derivations of key equations and encourages students to appreciate the power of mathematics by showing how it enables them to make progress and answer questions. All quantitative applications of calculus in the text are confined to this feature. The end-of-chapter exercises that make use of calculus are identified with a $\boxed{C}$.

THE QUEST FOR INSIGHT

Chemical Principles is designed to guide students on the path to a scientific, engineering, or medical career. The text shows students how to think like a scientist, by proposing models based on insight, developing those models with mathematics, and refining them through experimentation. This approach will help students understand key concepts and encourage them to think about how these concepts can be applied to modern research.

HOW DO WE DO THAT?

To calculate the work of reversible, isothermal expansion of a gas, we have to use calculus, starting at Eq. 3 written for an infinitesimal change in volume, dV:

$$dw = -P_{ex}dV$$

Because the external pressure is matched to the pressure, P, of the gas at every stage of a reversible expansion, we write $P_{ex} = P$ and the expression for dw becomes

$$dw = -PdV$$

At each stage of the expansion, the pressure of the gas is related to its volume by the ideal gas law, $PV = nRT$, so we can replace P by nRT/V:

From $dw = -PdV$ and $P = nRT/V$, $\qquad dw = -\dfrac{nRTdV}{V}$

The total work done is the sum (integral) of these infinitesimal contributions as the volume changes from its initial value to its final value. That is, we have to integrate dw from the initial to the final volume with nRT constant (because the change is isothermal):

From $w = \int dw$, with nRT a constant, $\qquad w = -nRT\int_{V_{initial}}^{V_{final}}\dfrac{dV}{V} = -nRT\ln\dfrac{V_{final}}{V_{initial}}$

The final step has made use of the standard integral

$$\int \dfrac{dx}{x} = \ln x + \text{constant}$$

BOX 4.1   HOW DO WE KNOW . . . THE DISTRIBUTION OF MOLECULAR SPEEDS?

The distribution of molecular speeds in a gas can be determined experimentally. To do so, the gas is heated to the required temperature in an oven. The molecules then stream out of the oven through a small hole into an evacuated region. To ensure that the molecules form a narrow beam, they may also pass through a series of slits, and the pressure must be kept very low so that collisions within the beam do not cause spreading.

The molecular beam passes through a series of rotating disks (see diagram). Each disk contains a slit that is offset by a certain angle from its neighbor. A molecule that passes through the first slit will pass through the slit in the next disk only if the time that it takes to pass between the disks is the same as the time required for the slit in the second disk to move into

Diagram of the rotating disks that serve as a velocity selector in a molecular beam apparatus.

The points represent a typ[...] measurement. They are su[...] obtain the fraction of mol[...] Δv, multiply $f(v)$ by Δv.

the orientation origin[...] times must continue t[...] through subsequent d[...] tion of the disks allo[...] having the speed that [...] mine the distribution [...] the intensity of the be[...] for different rotation[...] shown in the graph, a[...] theoretical Maxwell e[...]

• **"How do we know . . . ?"** boxes introduce students to the experiments and techniques used to collect the data underlying some of the fundamental concepts in the text. These boxes reinforce the important role of experimentation in the process of scientific discovery.

MAJOR TECHNIQUE 5   COMPUTATION

Computers have had an enormous impact on the practice of chemistry, where they are used to control apparatus, record and manipulate data, and explore the structures and properties of atoms and molecules. Computer modeling is applied to solids as well as to individual molecules and is useful for predicting the behavior of a material, for example, for indicating which crystal structure of a compound is energetically most favorable, and for predicting phase changes. In this section we concentrate on the use of computers in the exploration of molecular structure. In this context, they are used both to explore practical problems such as predicting the pharmacological activity of compounds and to achieve a deep understanding of the changes that take place during chemical reactions.

In *semi-empirical methods*, complicated integrals are set equal to parameters that provide the best fit to experimental data, such as enthalpies of formation. Semi-empirical methods are applicable to a wide range of molecules with a virtually limitless number of atoms, and are widely popular. The quality of results is very dependent on using a reasonable set of experimental parameters that have the same values across structures, and so this kind of calculation has been very successful in organic chemistry, where there are just a few different elements and molecular geometries.

In the more fundamental *ab initio methods*, an attempt is made to calculate structures from first principles, using only the atomic numbers of the atoms present and their general arrangement in space. Such an approach is intrinsically more reliable than a semi-empirical procedure but it is much more demanding computationally.

A currently popular alternative to the *ab initio* method is *density functional theory*, in which the energy is expressed in terms of the electron density rather than the wavefunction itself. The advantage of this approach is that it is less demanding computationally, requires less computer time, and in some cases—particularly for *d*-metal complexes—gives better agreement with experimental values than other procedures.

The raw output of a molecular structure calculation is a list of the coefficients of the atomic orbitals in each LCAO (linear combination of atomic orbitals) molecular orbital and the energies of the orbitals. The software commonly calculates dipole moments too. Various graphical representations are used to simplify the interpretation of the coefficients. Thus, a typical graphical representation of a molecular orbital uses stylized shapes (spheres for *s*-orbitals, for instance) to represent the basis set and then scales their size to indicate the value of the coefficient in the LCAO. Different signs of the wavefunctions are typically represented by different colors. The total electron density at any point (the sum of the squares of the occupied wavefunctions evaluated at that point) is commonly represented by an isodensity surface, a surface of constant total electron density.

An important aspect of a molecule in addition to its geometrical shape and the energies of its orbitals is the distribution of charge over its surface, because the charge distribution can strongly influence how one molecule (such as a potential drug) can attach to another (and dock into the active site of an enzyme, for instance). A common procedure begins with a calculation of the electrical potential at each point on an isodensity surface. The net potential is determined by subtracting the potential due to the electron density at that point from the potential due to the nuclei. The result is an *electrostatic potential surface* (an "elpot surface") in which net positive potential is shown in one color and net negative potential is shown in another, with intermediate gradations of color. The software can also depict a *solvent-accessible surface*, which represents the shape of the molecule by imagining a sphere representing a solvent molecule rolling across the surface and plotting the locations of its center.

Procedures are now available that allow for the presence of several solvent molecules around a solute molecule. This approach takes into account the effect of molecular interactions with the solvent on properties such as the enthalpy of formation and the shape adopted by a non-rigid molecule, such as a protein or a region of DNA. These studies are important for investigating the structures and reactions of biological molecules in their natural environment.

Related Exercises: Exercises that use software to explore molecular structure are available on the Web site for this text.

(a)   (b)

(c)   (d)

Some graphical representations of fluoroethene, CH_2=CHF: (a) the isodensity surface showing the general shape of the molecule, (b) the electrostatic potential surface showing the relatively negatively charged region (denoted red) close to the fluorine atom, (c) the highest-energy occupied molecular orbital showing the π-bond between the two carbon atoms and the partial involvement of a fluorine orbital, (d) the lowest-energy unoccupied orbital, which is the antibonding counterpart of the π-orbital.

582

• **Major Techniques** introduce students to important experimental methods, connecting the classroom, the laboratory, and the world.

CUTTING EDGE CHEMISTRY . . . FOR ALL YOUR STUDENTS

• **Frontiers of Chemistry** boxed essays explore the achievements of modern chemical research and show students how much research remains to be done. Each Frontiers of Chemistry box asks students, "How might you contribute?"

> **BOX 8.1   FRONTIERS OF CHEMISTRY: DRUG DELIVERY**
>
> The administration of drugs to ease disease and chronic, severe pain or to provide benefits such as hormone replacement therapy is difficult because drugs taken orally may lose much of their potency in the harsh conditions of the digestive system. In addition, they are distributed throughout the entire body, not just where they are needed, and side effects can be significant. Recently, however, techniques have been developed to deliver drugs gradually over time, to the exact location in the body where they are needed, and even at the time when they are needed.
>
> *Transdermal patches* are applied to the skin. The drug is mixed with the adhesive for the patch, and so it lies next to [th]an readily absorb many chemicals and so [s]uch as nitroglycerin (for heart disease), [...]es (for constant, severe pain), estrogen [...]cement therapy), or nicotine (for easing [...]t when a patient stops smoking).
>
> [...]de a means of delivering drugs over a [...]e at a controlled rate inside the body. Sub- [...] skin) implants are used to provide appro- [...]hoactive medications, birth-control drugs, painkillers, and other medications that must be administered
>
> Examples of implants used to insert living cells into the body. The cells continually produce enzymes, hormones, or painkillers needed by the body. Often a long, thin plastic tail is attached as a tether to allow easy retrieval of the implant.
>
> ...ture to the surfactant molecules in detergents: it has a polar head and a nonpolar hydrocarbon tail. Some lipids assemble

> **5.17 Ionic Liquids**
>
> We have seen that molecular substances tend to have low melting points, while network, ionic, and metallic substances tend to have high melting points. Therefore, with a few exceptions, such as mercury, a substance that is liquid at room temperature is likely to be a molecular substance. Liquid solvents are heavily used in industry to extract substances from natural products and to promote the synthesis of desired compounds. Because many of these solvents have high vapor pressures and so give off hazardous fumes, liquids that have low vapor pressures but dissolve

• **NEW! Green chemistry** promotes environmentally sound chemistry. Passages in the text created in consultation with Michael Cann and new end-of-chapter exercises are accompanied by a ⬡. Topics include ionic liquids (Chapter 5), supercritical CO_2 (Chapter 8), yttrium in paint (Chapter 12), chelates as a substitute for chlorine bleach (Chapter 16), and transesterification (Chapter 19).

• **NEW!** Nanotechnology and nanomaterials (Section 15.14)

• **NEW!** Bio-based and biomimetic materials; their properties and uses (Section 8.22)

• **NEW!** Computational chemistry (Major Technique 5)

> **Chemistry Connections**
>
> ⬡ **6.117** Petroleum based fuels create large amounts of pollution, contribute to global warming, and are becoming scarce. Consequently, alternative fuels are being sought (see Box 6.2). Three compounds that could be produced biologically and used as fuels are methane, CH_4, which can be produced from the anaerobic digestion of sewage, dimethyl ether, $H_3C—O—CH_3$, a gas that can be produced from methanol and ethanol, and ethanol, CH_3CH_2OH, a liquid obtained from the fermentation of sugars.
> (a) Draw the Lewis structure of each compound.
> (b) Use bond enthalpies (and, for ethanol, its enthalpy of vaporization) to calculate the enthalpy of combustion of each

OF SPECIAL INTEREST TO ENGINEERING STUDENTS

Impact on Materials sections are devoted to materials chemistry. Topics include:

The main-group elements, pp. 47–48
The transition metals, p. 48
Electronic conduction in solids, p. 126
Real gases: deviations from ideality; liquefaction; equations of state, pp. 165–169
Solids: properties of solids; alloys; liquid crystals, pp. 201–204
Plasma screens, p. 647
Nanotechnology, p. 648
Applications of electrolysis, pp. 514–515
Corrosion, p. 515
Glasses, p. 615
Ceramics, p. 616
Soft materials: colloids, gels, and biomaterials, pp. 342–343
Phosphors and other luminescent materials, p. 647
Steel, p. 690
Nonferrous alloys, p. 692
Magnetic materials, p. 692
Polymers, p. 764
Biology, pp. 771–779

Also:

High-temperature superconductors, p. 192
Fuel cells, p. 519
Nanotubes, p. 608
Self-assembling materials, p. 649
Conducting polymers, p. 772
Paramagnetism and measuring magnetic fields, p. 116
Viscosity and surface tension, pp. 186–187
Crystalline structures, p. 188
STM and AFM, p. 189
Network solids, including ceramics, p. 191
Metallic solids, p. 194

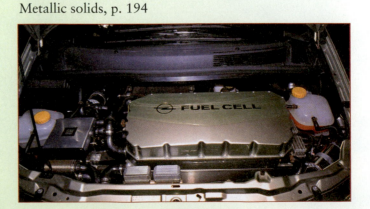

OF SPECIAL INTEREST TO BIOLOGY STUDENTS

Impact on Biology sections are devoted to applications of chemistry to the biological sciences. Topics include:

Gibbs free energy in biological systems, p. 301
Bio-based and biomimetic materials, p. 343
Homeostasis, p. 386
Proteins, p. 771
Carbohydrates, p. 775
Nucleic acids, p. 777

Also:

Chemical self preservation, p. 74
Drugs by design and discovery, p. 96
Drug delivery, p. 344
Physiological buffers,
   p. 453

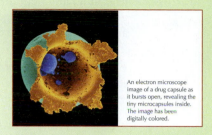

An electron microscope image of a drug capsule as it bursts open, revealing the tiny microcapsules inside. The image has been digitally colored.

Enzymes, p. 569
Self-assembling material,
   p. 649
Why we need to eat *d*-metals, p. 670
Nuclear medicine, p. 708
Biological effects of radiation, p. 709

OF SPECIAL INTEREST TO ENVIRONMENTAL SCIENCE STUDENTS

Alternative fuels, p. 246
Acid rain and the gene pool, p. 430
The ozone layer, p. 568
The greenhouse effect, p. 610
Nuclear power, p. 722
Fossil fuels, p. 745
Green chemistry passages and exercises

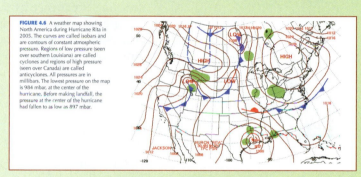

FIGURE 4.6 A weather map showing North America during Hurricane Rita in 2005. The curves are called isobars and are contours of constant atmospheric pressure. Regions of low pressure (seen over southern Louisiana) are called cyclones and regions of high pressure (seen over Canada) are called anticyclones. All pressures are in millibars. The lowest pressure on the map is 984 mbar, at the center of the hurricane. Before making landfall, the pressure at the center of the hurricane had fallen to as low as 897 mbar.

MEDIA INTEGRATION

Selected figures and exercises throughout the book are accompanied by annotated **Media Links** that direct students to Web-based resources (see the more detailed listing under Media-Based Supplements). These integrated links to the book's companion Web site are designed to make the text more dynamic and interactive. *Chemical Principles* contains integrated links to:

• **Living Graphs.**
Selected graphs in the text are available in interactive form. Students can manipulate parameters and see cause-and-effect relationships.

• **Animations.**
Selected art in the text is supported by dynamic media. Students can view motion, three-dimensional effects, and atomic and molecular interactions to learn to visualize as chemists do—at a molecular level.

• **Lab Videos.**
Video clips include many of the reactions described in the book.

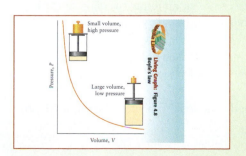

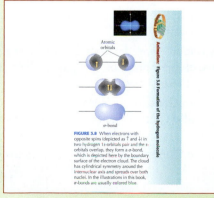

FIGURE 3.8 When electrons with opposite spins (depicted as ↑ and ↓) in two hydrogen 1s-orbitals pair and the s-orbitals overlap, they form a σ-bond, which is depicted here by the boundary surface of the electron cloud. The cloud has cylindrical symmetry around the internuclear axis and spreads over both nuclei. In the illustrations in this book, σ-bonds are usually colored blue.

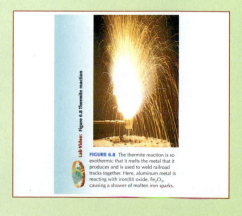

FIGURE 6.8 The thermite reaction is so exothermic that it melts the metal that it produces and is used to weld railroad tracks together. Here, aluminum metal is reacting with iron(III) oxide, Fe_2O_3, causing a shower of molten iron sparks.

- **Tools.** Tools guide the student in carrying out chemical calculations and graphing, and help them to explore periodic properties from different points of view. Flash and a large databank of structures allow students to study structure and function in three dimensions.

- **End-of-Chapter Exercises.** Selected exercises direct students to use media to solve problems.

> **3.72** Use the plotting function on the Web site for this book to plot the amplitude of the bonding and antibonding orbitals in Eqs. 1 and 2, given the information in Exercise 3.71, on a line passing through the two nuclei. Ignore the normalization factors, and take the internuclear distance as (a) a_0, (b) $2a_0$, (c) $5a_0$. Which of these three bond lengths is most likely to be close to the actual bond length?

THE *CHEMICAL PRINCIPLES* e-BOOK

The *Chemical Principles,* fourth edition, **e-book** is a complete online version of the textbook. The e-book offers students substantial savings and provides a rich learning experience by taking full advantage of the electronic medium, integrating all student media resources, and adds features unique to the e-book. The e-book also offers instructors flexibility and customization options not previously possible with any printed textbook. Access to the e-book is included with purchase of the special package of the text (ISBN: 1-4292-0447-8) through use of an activation code card. Individual e-book copies can be purchased online at **www.whfreeman.com**. Key features of the *Chemical Principles* e-book include:

- Easy access from any Internet-connected computer via a standard Web browser.
- Quick, intuitive navigation to any section or subsection, as well as any printed-book page number.
- Integration of all student Web site animations and activities.
- In-text **self-quiz** questions.
- In-text links to all **glossary** entries.
- Interactive **chapter summaries.**
- Text **highlighting,** down to the level of individual phrases.
- **Bookmarking** for quick reference to any page.
- **Notes,** which can be added to any page by students or instructors.
- A full **glossary** and **index.**
- **Full-text search,** including an option to search the glossary and index.
- **Automatic saving** of all notes, highlighting, and bookmarks.

Additional features for instructors:

- Custom Chapter Selection. Instructors can choose the chapters that correspond with their syllabus, and students will get a custom version of the e-book with the selected chapters only.

- Instructor Notes. Instructors can choose to create an annotated version of the e-book with their notes on any page. When students log in, they will see the instructor's version.

- Custom Content. Instructor notes can include text, Web links, and even images, allowing instructors to place any content they choose exactly where they want it.

- Online Quizzing. The online quizzes from the student Web site are integrated into the e-book.

NEW! CHEMISTRY PORTAL: A DIGITAL GATEWAY

ChemPortal is the digital gateway for *Chemical Principles,* fourth edition. Built on a proven learning management system, it has three main components:

- **Online homework server.** The "Assignment Center," allows instructors to assign and automatically grade homework, quizzes, and guided practice. The homework system comes populated with three levels of problems (from easy to challenging) for each chapter. Student progress is tracked in a single, easy-to-use grade book; during assignments students receive immediate feedback so they can prioritize further study. Self-guided tutorials based on all 27 Toolboxes from the text provide student-specific feedback and break down complex problems into their individual steps, offering guidance at each stage to ensure that students fully understand the problem-solving process.

- **Chemistry resource library.** The Resource Center, with links to animations and video, organizes all study materials in one place.

- **Interactive e-book.** The e-book transforms reading into experience through integration of the complete text with animations, simulations, living graphs, practice tests, graphics, and study resources, such as note-taking, highlighting, bookmarking, and a full-text search. Instructors can customize their students' texts through annotations and supplementary links.

For more information about ChemPortal go to www.whfreeman.com/chemportal. (ISBN: 1-4292-0091-X)

SUPPLEMENTS FOR *CHEMICAL PRINCIPLES*

MEDIA-BASED SUPPLEMENTS

The *Chemical Principles* Web site is at www.whfreeman.com/chemicalprinciples4e The Web-based media support for *Chemical Principles* offers students a range of tools for problem solving and chemical exploration, including:

- A calculator adapted for solving equilibrium calculations
- Two- and three-dimensional curve plotters
- "Living Graphs," which allow the user to control the parameters
- Flash for exploring structure and function in 3-D molecular models
- A molecules data bank containing almost all the structures in the book
- An interactive periodic table of the elements
- Animations that allow students to visualize chemical events on a molecular level
- Links that can broaden exposure to up-to-date data and applications

Also available as preformatted PowerPoint slides for the instructor are all the illustrations from the textbook and the new **Diagnostic Test for the Fundamentals sections.** This test will allow instructors to determine what their students understand and where they need additional support. Instructors can then make appropriate assignments from the Fundamentals sections. The test includes 5 to 10 problems for each section.

WEB-BASED ASSESSMENT

Online Quizzing, powered by Question Mark and accessed via the *Chemical Principles* Web site

An excellent online quizzing bank in which instructors can easily quiz students using prewritten, multiple-choice questions for each text chapter (not from the test bank). Students receive instant feedback and can take the quizzes multiple times. Instructors can go to a protected Web site to view results by quiz, student, or question, or they can get weekly results by e-mail. Excellent for practice testing or homework or both.

PRINT SUPPLEMENTS FOR STUDENTS

Student Study Guide and Solutions Manual, John Krenos and Joseph Potenza, Rutgers University; Laurence Lavelle, UCLA; Yinfa Ma, University of Missouri, Rolla; and Carl Hoeger, UCSD (ISBN: 1-4292-0099-5)

This combined manual is designed to help students avoid common mistakes and understand the material better. The solutions manual includes detailed solutions to all odd-numbered exercises in the text, except for the Chemistry Connections exercises.

SUPPLEMENTS FOR INSTRUCTORS

Create a customized lab manual to accompany Atkins/Jones, *Chemical Principles.*

W. H. Freeman and Company is pleased to offer you the ability to create a unique lab manual for your course in just minutes. Drawing from our diverse library of experiments, you can pick and choose the labs you want to use, and even add your own materials. Organize the material as you like, customize the cover, view your newly created book online in PDF format, and then order it online with the click of a button. W. H. Freeman's state-of-the art custom publishing system is easy to use and an excellent alternative to traditional lab manuals. Try it today at http://custompub.whfreeman.com

Instructor's Solutions Manual, Laurence Lavelle, UCLA; Yinfa Ma, University of Missouri, Rolla; and Carl Hoeger, UCSD (ISBN: 1-4292-0100-2)

This manual contains solutions to the even-numbered exercises.

Instructor's Resource CD (ISBN: 1-4292-0088-X)

To help instructors create their own Web sites and prepare dynamic lectures, the CD contains:

- All illustrations from the text in .jpg and preformatted PowerPoint® slides
- All animations, lab videos, and living graphs from the Web site
- All solutions to the end-of-chapter exercises, in editable Word files
- Diagnostic test for the Fundamentals sections

Test Bank CD-ROM (Windows and Mac versions on one disc, ISBN: 1-4292-0089-8; printed version, ISBN: 1-4292-0101-0)

The test bank offers more than 1400 multiple-choice, fill-in-the-blank, and essay questions. The easy-to-use CD includes Windows and Mac versions in a format that lets you add, edit, and resequence questions to suit your needs.

Online Course Materials (WebCT, Blackboard)
As a service for adopters we will provide content files in the appropriate online course format, including the instructor and student resources for this text.

WebAssign Online Homework System (alternative option to ChemPortal)
Developed by instructors at North Carolina State University. WebAssign is filled with end-of-chapter problems, resources, and course management features. WebAssign has content experts on the premises to respond to any question within 24 hours. For more information visit www.webassign.net

FOR THE LABORATORY

Bridging to the Lab, Loretta Jones, University of Northern Colorado, and Roy Tasker, University of Western Sydney (ISBN: 0-7167-4746-4)

The modules are computer-based laboratory simulations with engaging activities that emphasize experimental design and visualization of structures and processes at the molecular level. The modules are designed to help students connect chemical principles from lecture with their practical applications in the lab. Every module has a built-in accountability feature that records section completion for use in setting grades and a workbook for students to record and interpret their work.

Used either as prelaboratory preparation for related laboratory activities or to expose students to additional laboratory activities not available in their program, these modules motivate students to learn by proposing real-life problems in a virtual environment. Students make decisions on experimental design, observe reactions, record data, interpret these data, perform calculations, and draw conclusions from their results. Following a summary of the module, students test their understanding by applying what they have learned to new situations or by analyzing the effect of experimental errors.

Working with Chemistry Lab Manual, Donald J. Wink and Sharon Fetzer Gislason, University of Chicago–Illinois; and Julie Ellefson Kuehn, William Rainey Harper College (ISBN: 0-7167-9607-4)

With this inquiry-based program, students build skills using important chemical concepts and techniques so that they are able to design a solution to a scenario drawn from a professional environment. The scenarios are drawn from the lives of people who work with chemistry every day, ranging from field ecologists to chemical engineers and health professionals. Professors: you can download and class-test sample labs from *Working with Chemistry* at the Web site, www.whfreeman.com/chemistry/wwc2

Chemistry in the Laboratory, sixth edition, James M. Postma, California State University, Chico; and Julian L. Roberts, Jr., and J. Leland Hollenberg, University of Redlands (ISBN: 0-7167-9606-6)

This clearly written, class-tested manual has long given students essential hands-on training with key experiments. Exceptionally compatible with Atkins/Jones, *Chemical Principles*, the manual is known for its clear instructions and illustrations. All experiments are available as lab separates.

HGS Molecular Structure Model Set, Maruzen Company, Ltd. (ISBN: 0-7167-4822-3)

Molecular modeling helps students understand physical and chemical properties by providing a way to visualize the three-dimensional arrangement of atoms. This model set uses polyhedra to represent atoms, and plastic connectors to represent bonds (scaled to correct bond length). Plastic plates representing orbital lobes are included for indicating lone pairs of electrons, radicals, and multiple bonds—a feature unique to this set.

Chemistry Laboratory Student Notebook, second edition, paper (ISBN: 0-7167-3900-3)

The notebook contains 114 pages of graph paper in a convenient $8\frac{1}{2} \times 11$, 3-hole-punched format, carbon included. The new edition adds tables and graphs that make it a handy reference as well.

ACKNOWLEDGMENTS

We are grateful to the many instructors, colleagues, and students who have contributed their expertise to this edition. We would like above all to thank those who carefully evaluated the third edition and commented on drafts of this edition:

Yiyan Bai, *Houston Community College System— Central Campus*
Maria Ballester, *Nova Southeastern University*
Patricia D. Christie, *Massachusetts Institute of Technology*
Henderson J. Cleaves, II, *University of California, San Diego*
Ivan J. Dmochowski, *University of Pennsylvania*
Ronald Drucker, *City College of San Francisco*
Christian Ekberg, *Chalmers University of Technology, Sweden*
Bryan Enderle, *University of California, Davis*
David Erwin, *Rose-Hulman Institute of Technology*
Justin Fermann, *University of Massachusetts*
Regina F. Frey, *Washington University*
P. Shiv Halasyamani, *University of Houston*
Jameica Hill, *Wofford College*
Alan Jircitano, *Penn State, Erie*

Gert Latzel, *Riemerling, Germany*
Nancy E. Lowmaster, *Allegheny College*
Matthew L. Miller, *South Dakota State University*
Clifford B. Murphy, *Boston University*
Maureen Murphy, *Huntingdon College*
Enrique Peacock-López, *Williams College*
LeRoy Peterson, Jr., *Francis Marion University*
Tyler Rencher, *Brigham Young University*
Michael Samide, *Butler University*
Gordy Savela, *Itasca Community College*
Lori Slavin, *College of Saint Catherine*
Mike Solow, *City College of San Francisco*
John E. Straub, *Boston University*
Laura Stultz, *Birmingham-Southern College*
Peter Summer, *Lake Sumter Community College*
David W. Wright, *Vanderbilt University*
Mamudu Yakubu, *Elizabeth City State University*
Meishan Zhao, *University of Chicago*

We offer a special note of gratitude to the participants of a W.H. Freeman workshop on teaching general chemistry. Their insights provided much food for thought:

Keiko Jacobsen, *Tulane University*
Lynn Koplitz, *Loyola University*
Cynthia LaBrake, *University of Texas, Austin*
Paul McCord, *University of Texas, Austin*
Yinfa Ma, *University of Missouri—Rolla*

Joseph Potenza, *Rutgers University*
Sara Sutcliffe, *University of Texas, Austin*
David Vandenbout, *University of Texas, Austin*
Deborah Walker, *University of Texas, Austin*
Zhiping Zheng, *University of Arizona*

The contributions of the reviewers of the first, second, and third editions remain embedded in the text, so we also wish to renew our thanks to:

Thomas Albrecht-Schmidt, *Auburn University*
Matthew Asplund, *Brigham Young University*
Matthew P. Augustine, *University of California, Davis*
David Baker, *Delta College*
Alan L. Balch, *University of California, Davis*
Mario Baur, *University of California, Los Angeles*
Robert K. Bohn, *University of Connecticut*
Paul Braterman, *University of North Texas*
William R. Brennan, *University of Pennsylvania*
Ken Brooks, *New Mexico State University*
Julia R. Burdge, *University of Akron*
Paul Charlesworth, *Michigan Technological University*
Patricia D. Christie, *Massachusetts Institute of Technology*
William Cleaver, *University of Vermont*
David Dalton, *Temple University*
J. M. D'Auria, *Simon Fraser University*
James E. Davis, *Harvard University*
Walter K. Dean, *Lawrence Technological University*
Jimmie Doll, *Brown University*

Ronald Drucker, *City College of San Francisco*
Jetty Duffy-Matzner, *State University of New York, Cortland*
Robert Eierman, *University of Wisconsin*
Kevin L. Evans, *Glenville State College*
Donald D. Fitts, *University of Pennsylvania*
Lawrence Fong, *City College of San Francisco*
Regina F. Frey, *Washington University*
Dennis Gallo, *Augustana College*
David Harris, *University of California, Santa Barbara*
Sheryl Hemkin, *Kenyon College*
Michael Henchman, *Brandeis University*
Geoffrey Herring, *University of British Columbia*
Timothy Hughbanks, *Texas A&M University*
Paul Hunter, *Michigan State University*
Robert C. Kerber, *State University of New York, Stony Brook*
Robert Kolodny, *Armstrong Atlantic State University*
Lynn Vogel Koplitz, *Loyola University*

Petra van Koppen, *University of California, Santa Barbara*
Mariusz Kozik, *Canisius College*
Julie Ellefson Kuehn, *William Rainey Harper College*
Brian B. Laird, *University of Kansas*
Alison McCurdy, *Harvey Mudd College*
Charles W. McLaughlin, *University of Nebraska*
Patricia O'Hara, *Amherst College*
Noel Owen, *Brigham Young University*
Donald Parkhurst, *The Walker School*
Montgomery Pettitt, *University of Houston*
Wallace Pringle, *Wesleyan University*
Philip J. Reid, *University of Washington*
Barbara Sawrey, *University of California, San Diego*
George Schatz, *Northwestern University*
Paula Jean Schlax, *Bates College*
Carl Seliskar, *University of Cincinnati*
Robert Sharp, *University of Michigan, Ann Arbor*
Peter Sheridan, *Colgate University*

Jay Shore, *South Dakota State University*
Herb Silber, *San Jose State University*
Lee G. Sobotka, *Washington University*
Michael Sommer, *Harvard University*
Nanette A. Stevens, *Wake Forest University*
Tim Su, *City College of San Francisco*
Larry Thompson, *University of Minnesota, Duluth*
Dino Tinti, *University of California, Davis*
Sidney Toby, *Rutgers University*
Lindell Ward, *Franklin College*
Thomas R. Webb, *Auburn University*
Peter M. Weber, *Brown University*
David D. Weis, *Skidmore College*
Ken Whitmire, *Rice University*
James Whitten, *University of Massachusetts, Lowell*
Gang Wu, *Queen's University*
Marc Zimmer, *Connecticut College*
Martin Zysmilich, *Massachusetts Institute of Technology*

Some contributed in substantial ways. Roy Tasker, University of Western Sydney, contributed to the Web site for this book, designed related animations, and selected the icons for the animation media links. Michael Cann, University of Scranton, opened our eyes to the world of green chemistry in a way that has greatly enriched this book

We would like particularly to thank Cynthia LaBrake, University of Texas at Austin, for checking and offering very helpful suggestions for improving the worked Examples. We would also like to thank Nathan Barrows, Arizona State University, for checking the accuracy of the solutions manual and the answers at the back of the book. Amelia Lapena and Sanaz Kabehie, UCLA, contributed many improvements to the solutions manual. The supplements authors, especially John Krenos, Joseph Potenza, Laurence Lavelle, Yinfa Ma, and Carl Hoeger, have offered us much useful advice. This book also benefited from suggestions made by Dennis Kohl, University of Texas at Austin; Randall Shirts, Brigham Young University; Catherine Murphy, University of South Carolina; Michael Sailor, University of California at San Diego; Matt Miller and Jay Shore, South Dakota State University; and Peter Garik, Rosina Georgiadis, Mort Hoffman, and Dan Dill, Boston University. As before, students have been our most important critics; they helped us to see the world of learning chemistry through their eyes and made many important suggestions. The students in Rosina Georgiadis's General Chemistry class at Boston University wrote chapter reviews that offered us valuable direction for revising this edition.

We are grateful to the staff members at W. H. Freeman and Company, who understood our vision and helped to bring it to fruition. In particular, we would like to acknowledge Jessica Fiorillo, senior chemistry editor, who organized us and the entire project; Randi Rossignol, our developmental editor, who guided us toward important improvements in this edition; Georgia Lee Hadler, senior project editor, who once again took on the awesome task of overseeing the transformation of our files into paper; Margaret Comaskey, our copyeditor, who organized and coordinated those files with great care; Bianca Moscatelli, who found exactly the right new photographs; and Jeanette Picerno and Amy Thorne, who directed the development and production of the substantial array of print and media supplements. The authors could not have wished for a better or more committed team.

FUNDAMENTALS

WELCOME to chemistry! You are about to embark on an extraordinary voyage that will take you to the center of science. Looking in one direction, toward physics, you will see how the principles of chemistry are based on the behavior of atoms and molecules. Looking in another direction, toward biology, you will see how chemists contribute to an understanding of that most awesome property of matter, life. You will be able to look at an everyday object, see in your mind's eye its composition in terms of atoms, and understand how that composition determines its properties.

INTRODUCTION AND ORIENTATION

Chemistry is the science of matter and the changes it can undergo. The world of chemistry therefore embraces everything material around us—the stones we stand on, the food we eat, the flesh we are made of, and the silicon we build into computers. There is nothing material beyond the reach of chemistry, be it living or dead, vegetable or mineral, on Earth or in a distant star.

Chemistry and Society

We have only to look around us to appreciate the impact of chemistry on technology and society. In the earliest days of civilization, when the Stone Age gave way to the Bronze Age and then to the Iron Age, people did not realize that they were doing chemistry when they changed the material they found as stones—we would now call them *minerals*—into metals (Fig. 1). The possession of metals gave them a new power over their environment, and treacherous nature became less brutal. Civilization emerged as skills in transforming materials grew: glass, jewels, coins, ceramics, and, inevitably, weapons became more varied and effective. Art, agriculture, and warfare became more sophisticated. None of this would have happened without chemistry.

The development of steel accelerated the profound impact of chemistry on society. Better steel led to the Industrial Revolution, when muscles gave way to steam and giant enterprises could be contemplated. With improved transport and greater output from factories came more extensive trade, and the world became simultaneously a smaller but busier place. None of this would have happened without chemistry.

With the twentieth century, and now the twenty-first, came enormous progress in the development of the chemical industry. Chemistry transformed agriculture.

FIGURE 1 Copper is easily extracted from its ores and was one of the first metals worked. The Bronze Age followed the discovery that adding some tin to copper made the metal harder and stronger. These four bronze swords date from 1250 to 850 BCE, the Late Bronze Age, and are from a collection in the Naturhistorisches Museum, Vienna, Austria. From bottom to top, they are a short sword, an antenna-type sword, a tongue-shaped sword, and a Liptau-type sword.

FIGURE 2 Cold weather triggers chemical processes that reduce the amount of the green chlorophyll in leaves, allowing the colors of various other pigments to show.

FIGURE 3 When magnesium burns in air, it gives off a lot of heat and light. The gray-white powdery product looks like smoke.

Synthetic fertilizers provided the means of feeding the enormous, growing population of the world. Chemistry transformed communication and transportation. Today chemistry provides advanced materials, such as polymers for fabrics, ultra-pure silicon for computers, and glass for optical fibers. It is producing more efficient renewable fuels and the tough, light alloys that are needed for modern aircraft and space travel. Chemistry has transformed medicine, substantially extended life expectancy, and provided the foundations of genetic engineering. The deep understanding of life that we are developing through molecular biology is currently one of the most vibrant areas of science. None of this progress would have been achieved without chemistry.

However, the price of all these benefits has been high. The rapid growth of industry and agriculture, for instance, has stressed the Earth and damaged our inheritance. There is now widespread concern about the preservation of our extraordinary planet. It will be up to you and your contemporaries to draw on chemistry—in whatever career you choose—to build on what has already been achieved. Perhaps you will help to start a new phase of civilization based on new materials, just as semiconductors transformed society in the twentieth century. Perhaps you will help to reduce the harshness of the impact of progress on our environment. To do that, you will need chemistry.

Chemistry: A Science at Three Levels

Chemistry operates on three levels. At one level, chemistry is about matter and its transformations. At this level, we can actually see the changes, as when a fuel burns, a leaf changes color in the fall (Fig. 2), or magnesium burns brightly in air (Fig. 3). This level is the **macroscopic level**, the level dealing with the properties of large, visible objects. However, there is an underworld of change, a world that we cannot see directly. At this deeper **microscopic level**, chemistry interprets these phenomena in terms of the rearrangements of atoms (Fig. 4). The third level is the **symbolic level**, the expression of chemical phenomena in terms of chemical symbols and mathematical equations. This level ties the other two levels together. A chemist thinks at the microscopic level, conducts experiments at the macroscopic level, and represents both symbolically. We can map these three aspects of chemistry as a triangle, with the corners labeled macroscopic, microscopic, and symbolic (Fig. 5). As you read further in this text, you will find that sometimes the topics and explanations are close to one vertex of the triangle, sometimes to another. Because it is helpful in understanding chemistry to make connections among these levels, in the worked examples in this book you will find drawings of the molecular level as well as graphical interpretations of equations. As your understanding of chemistry grows so will your ability to travel easily within the triangle.

A sense of scale is important for understanding how chemistry at the macroscopic level is related to the behavior of atoms at the microscopic level. Atoms are extraordinarily small, and there are vast numbers in even very tiny objects. The diameter of a carbon atom is only about 150 trillionths of a meter, and we would have to put 10 million atoms side by side to span the length of this dash: –. Even a small cup of coffee contains more water molecules than there are stars in the visible universe.

How Science Is Done

Scientists pursue ideas in an ill-defined but effective way called the **scientific method**, which takes many forms. There is no strict rule of procedure that leads you from a good idea to a Nobel prize or even to a publishable discovery. Some scientists are meticulously careful, others are highly creative. The best scientists are probably both careful and creative. Although there are various scientific methods in use, a typical approach consists of a series of steps (Fig. 6). The first step is often to collect **data** by making observations and measurements. These measurements are usually made on small **samples** of matter, representative pieces of the material that we want to study.

Scientists are always on the lookout for patterns. When a pattern is observed in the data, it can be stated as a scientific **law**, a succinct summary of a wide range of observations. For example, water was found to have eight times the mass of oxygen as it has of hydrogen, regardless of the source of the water or the size of the sample. One of the earliest laws of chemistry summarized those types of observations as the "law of constant composition," which states that a compound has the same composition regardless of the source of the sample.

Not all observations are summarized by laws. There are many properties of matter (such as superconductivity, the ability of a few cold solids to conduct electricity without any resistance) that are currently at the forefront of research but are not described by grand "laws" that embrace hundreds of different compounds. A major current puzzle, which might be resolved in the future either by finding the appropriate law or by detailed individual computation, is what determines the shapes of big protein molecules. Formulating a law is just one way, not the only way, of summarizing data.

Once they have detected patterns, scientists develop **hypotheses**, possible explanations of the laws—or the observations—in terms of more fundamental concepts. Observation requires careful attention to detail, but the development of a hypothesis requires insight, imagination, and creativity. In 1807, John Dalton interpreted experimental results to propose the hypothesis that matter consists of atoms. Although Dalton could not see individual atoms, he was able to imagine them and formulate his *atomic hypothesis*. Dalton's hypothesis was a monumental insight that helped others understand the world in a new way. The process of scientific discovery never stops. With luck and application, you may acquire that kind of insight as you read through this text, and one day you may make your own extraordinary hypotheses.

After formulating a hypothesis, scientists design further **experiments**—carefully controlled tests—to verify it. Designing and conducting good experiments often requires ingenuity and sometimes good luck. If the results of repeated experiments—often in other laboratories and sometimes by skeptical coworkers—support

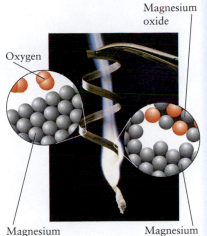

FIGURE 4 When a chemical reaction takes place, atoms exchange partners, as in Fig. 3 where magnesium and oxygen atoms form magnesium oxide. As a result, two forms of matter (left inset) are changed into another form of matter (right inset). Atoms are neither created nor destroyed in chemical reactions.

The postulates of Dalton's atomic hypothesis are described in Section B.

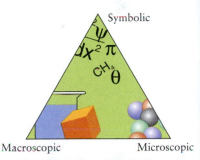

FIGURE 5 This triangle illustrates the three modes of scientific inquiry used in chemistry: macroscopic, microscopic, and symbolic. Sometimes we work more at one corner than at the others, but it is important to be able to move from one approach to another inside the triangle.

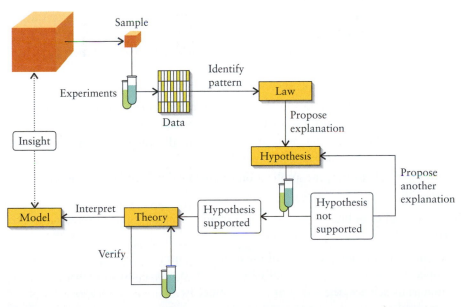

FIGURE 6 A summary of the principal activities that constitute a common version of the scientific method. At each stage, the crucial activity is experiment and its comparison with the ideas proposed.

the hypothesis, scientists may go on to formulate a **theory,** a formal explanation of a law. Quite often the theory is expressed mathematically. A theory originally envisioned as a **qualitative** concept, a concept expressed in words or pictures, is converted into a **quantitative** form, the same concept expressed in terms of mathematics. After a concept has been expressed quantitatively, it can be used to make numerical predictions and subjected to rigorous experimental confirmation. You will have plenty of practice with the quantitative aspect of chemistry while working through this text.

Scientists commonly interpret a theory in terms of a **model,** a simplified version of the object of study. Like hypotheses, theories and models must be subjected to experiment and revised if experimental results do not support them. For example, our current model of the atom has gone through many formulations and progressive revisions, starting from Dalton's vision of an atom as an uncuttable solid sphere to our current much more detailed model, which is described in Chapter 1. One of the main goals of this text is to show you how to build models, turn them into a testable form, and then refine them in the light of additional evidence.

The Branches of Chemistry

Chemistry is more than test tubes and beakers. New technologies have transformed chemistry dramatically in the past 50 years, and new areas of research have emerged (Fig. 7). Traditionally, the field of chemistry has been organized into three main branches:

> **organic chemistry,** the study of compounds of carbon;
>
> **inorganic chemistry,** the study of all the other elements and their compounds; and
>
> **physical chemistry,** the study of the principles of chemistry.

New regions of study have developed as information has been acquired in specialized areas or as a result of the use of particular techniques. It is the nature of a vigorously developing science that the distinctions between its branches are not clear-cut, but nevertheless we still speak of

> **biochemistry,** the study of the chemical compounds, reactions, and other processes in living systems;
>
> **analytical chemistry,** the study of techniques for identifying substances and measuring their amounts;
>
> **theoretical chemistry,** the study of molecular structure and properties in terms of mathematical models;
>
> **computational chemistry,** the computation of molecular properties;
>
> **chemical engineering,** the study and design of industrial chemical processes, including the fabrication of manufacturing plants and their operation;
>
> **medicinal chemistry,** the application of chemical principles to the development of pharmaceuticals; and
>
> **biological chemistry,** the application of chemical principles to biological structures and processes.

Various branches of knowledge emerging from chemistry have adopted names that are rooted in chemistry. They include

> **molecular biology,** the study of the chemical and physical basis of biological function and diversity, especially in relation to genes and proteins;
>
> **materials science,** the study of the chemical structure and composition of materials; and
>
> **nanotechnology,** the study and manipulation of matter at the atomic level (nanometer scale).

FIGURE 7 Scientific research today often requires sophisticated equipment and computers. This chemist is using an Auger electron spectrometer to probe the surface of a crystal. The data collected will allow the chemist to determine which elements are present in the surface.

SOLUTION We rearrange Eq. 1 into

$$V = \frac{m}{d}$$

and then substitute the data. The density of silver is listed in Appendix 2D as 10.50 g·cm^{-3}; so the volume of 5.0 g of solid silver is

From $V = m/d$, $$V = \frac{5.0 \text{ g}}{10.50 \text{ g·cm}^{-3}} = \frac{5.0}{10.50} \text{ cm}^3 = 0.48 \text{ cm}^3$$

0.78 cm

5.0 g

A note on good practice: It is always a good idea to check that the result of a calculation is reasonable. Because the density of silver is about 10 times that of water, the volume 5.0 g of silver should be about 1/10 the volume of 5.0 g of water (5.0 cm^3), or about 0.50 cm^3.

SELF-TEST A.3A The density of selenium is 4.79 g·cm^{-3}. What is the mass of 6.5 cm^3 of selenium?

[*Answer:* 31 g]

SELF-TEST A.3B The density of helium gas at 0°C and 1.00 atm is 0.176 85 g·L^{-1}. What is the volume of a balloon containing 10.0 g of helium under the same conditions?

Notice that in Example A.1 we wrote the result of dividing 5.0 by 10.50 as 0.48, not 0.47619. The number of digits reported in the result of a calculation must reflect the number of digits in the data provided. The number of **significant figures** in a numerical value is the number of digits that can be justified by the data. Thus, the measurement 5.0 g has two significant figures (2 sf) and 10.50 g·cm^{-3} has four (4 sf). The number of significant figures in the result of a calculation cannot exceed the number in the data (you can't generate reliability on a calculator!), so in Example A.2 we limited the result to 2 sf, the lower number of significant figures in the data. The full rules for counting the number of significant figures and determining the number of significant figures in the result of a calculation are given in Appendix 1C, together with the rules for rounding numerical values.

An ambiguity may arise when dealing with a whole number ending in a zero, because the number of significant figures in the number may be less than the number of digits. For example, 400 could have 1, 2, or 3 sf. To avoid this type of ambiguity, in this book when all the digits in numbers ending in zero are significant, the number is followed by a decimal point. Thus, the number 400. has 3 sf.

When scientists measure the properties of a substance, they monitor and report the accuracy and precision of the data. To make sure of their data, scientists usually repeat their measurements several times. The **precision** of a measurement is reflected in the number of significant figures justified by the procedure and depends on how close repeated measurements are to one another. The **accuracy** of a series of measurements is the closeness of their average value to the true value. The illustration in Fig. A.3 distinguishes precision from accuracy. As the illustration suggests, even precise measurements can give inaccurate values. For instance, if there is an unnoticed speck of dust on the pan of a chemical balance that we are using to measure the mass of a sample of silver, then even though we might be justified in reporting our measurements to five significant figures (such as 5.0450 g), the reported mass of the sample will be inaccurate.

More often than not, measurements are accompanied by two kinds of error. A **systematic error** is an error present in every one of a series of repeated measurements. An example is the effect of a speck of dust on a pan, which distorts the mass

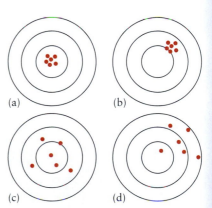

(a)   (b)

(c)   (d)

FIGURE A.3 A representation of measurements that are (a) precise and accurate, (b) precise but inaccurate, (c) imprecise but accurate, and (d) both imprecise and inaccurate.

of each sample in the same direction (the speck makes each sample appear heavier than it is). A **random error** is an error that varies at random and can average to zero over a series of observations. An example is the effect of drafts of air from an open window moving a balance pan either up or down a little, decreasing or increasing the mass measurements randomly. Scientists attempt to minimize random error by making many observations and taking the average of the results. Systematic errors are much harder to identify. However, they can sometimes be eliminated by comparing results obtained independently in different laboratories or by the use of different techniques.

> *Physical properties are those that do not involve changing the identity of a substance. Chemical properties are those that involve changing the identity of a substance. Extensive properties depend on the mass of the sample; intensive properties do not. The precision of a measurement controls the number of significant figures that are justified by the procedure; the accuracy of a measurement is its closeness to the true value.*

A.2 Force

A **force**, F, is an influence that changes the state of motion of an object. For instance, we exert a force to open a door—to start the door swinging open—and we exert a force on a ball when we hit it with a bat. According to Newton's second law of motion, when an object experiences a force, it is accelerated. The **acceleration**, a, of the object, the rate of change of its velocity, is proportional to the force that it experiences:

Acceleration ∝ force, or $a \propto F$

The constant of proportionality between the force and the acceleration it produces is the mass, m:

Force = mass × acceleration, or $F = ma$ $\qquad\qquad$ **(2)**

This expression, in the form $a = F/m$, tells us that a stronger force is required to accelerate a heavy object by a given amount than to accelerate a lighter object by the same amount.

Velocity, the rate of change of position, has both magnitude and direction; so, when a force acts, it can change the magnitude alone, the direction alone, or both simultaneously (Fig. A.4). The magnitude of the velocity of an object—the rate of change of position regardless of the direction of the motion—is called its **speed**, v. When we accelerate a car in a straight line, we change its speed, but not its direction, by applying a force through the rotation of the wheels and their contact with the road. To stop a car, we apply a force that opposes the motion. However, a force can also act without changing the speed: if a body is forced to travel in a different direction at the same speed, it undergoes acceleration because velocity includes direction as well as magnitude. For example, when a ball bounces on the floor, the

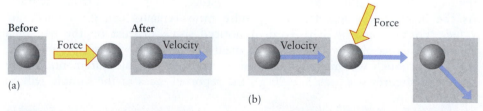

FIGURE A.4 (a) When a force acts along the direction of travel, the speed (the magnitude of the velocity) changes, but the direction of motion does not. (b) The direction of travel can be changed without affecting the speed if the force is applied in an appropriate direction. Both changes in velocity correspond to acceleration.

force exerted by the floor reverses the ball's direction of travel without affecting its speed very much.

Acceleration, the rate of change of velocity, is proportional to applied force.

A.3 Energy

Some chemical changes give off a lot of energy (Fig. A.5); others absorb energy. An understanding of the role of energy is the key to understanding chemical phenomena and the structures of atoms and molecules. But just what is energy?

The word energy is so common in everyday language that most people have a general sense of what it means; however, to get a technical answer to this question, we would have to delve into the theory of relativity, which is far beyond the scope of this book. In chemistry, we use a practical definition of **energy** as the capacity to do work, with **work** defined as motion against an opposing force. Thus, energy is needed to do the work of raising a weight or the work of forcing an electric current through a circuit. The greater the energy of an object, the greater its capacity to do work.

The SI unit for energy is the **joule** (J). As explained in Appendix 1B,

$$1 \text{ J} = 1 \text{ kg} \cdot \text{m}^2 \cdot \text{s}^{-2}$$

Each beat of the human heart uses about 1 J of energy, and to raise this book (of mass close to 1.5 kg) from the floor to a tabletop about 0.97 m above the floor requires about 14 J (Fig. A.6). Because energy changes in chemical reactions tend to be of the order of thousands of joules for the amounts usually studied, it is more common in chemistry to use the kilojoule (kJ, where $1 \text{ kJ} = 10^3 \text{ J}$).

A note on good practice: Names of units derived from the names of people are always lower case (as for joule, named for the scientist J. P. Joule), but their abbreviations are always uppercase (as in J for joule).

There are three contributions to energy: kinetic energy, potential energy, and electromagnetic energy. **Kinetic energy**, E_K, is the energy that a body possesses due to its motion. For a body of mass m traveling at a speed v, the kinetic energy is

$$E_K = \tfrac{1}{2}mv^2 \qquad (3)*$$

A heavy body traveling rapidly has a high kinetic energy. A body at rest (stationary, $v = 0$) has zero kinetic energy.

EXAMPLE A.3   Sample exercise: Calculating kinetic energy

How much energy does it take to accelerate a person and a bicycle of total mass 75 kg to 20. mph ($8.9 \text{ m} \cdot \text{s}^{-1}$), starting from rest and ignoring friction and wind resistance?

SOLUTION To calculate the energy, we substitute the data into Eq. 2:

From $E_K = \tfrac{1}{2}mv^2$,
$$E_K = \tfrac{1}{2} \times \left(75 \text{ kg}\right) \times \left(8.9 \text{ m} \cdot \text{s}^{-1}\right)^2$$
$$= 3.0 \times 10^3 \underbrace{}_{k} \underbrace{\text{kg} \cdot \text{m}^2 \cdot \text{s}^{-2}}_{J} = 3.0 \text{ kJ}$$

You would have to expend more energy to maintain that speed against friction and wind resistance.

SELF-TEST A.4A Calculate the kinetic energy of a ball of mass 0.050 kg traveling at 25 m·s⁻¹.

[*Answer:* 16 J]

SELF-TEST A.4B Calculate the kinetic energy of this book (mass 1.5 kg) just before it lands on your foot after falling off a table, when it is traveling at 3.0 m·s⁻¹.

The **potential energy**, E_P, of an object is the energy that it possesses on account of its position in a field of force. There is no single formula given for the potential

---

FIGURE A.5 When bromine is poured on red phosphorus, a chemical change takes place in which a lot of energy is released as heat and light.

The joule is named for James Joule, the nineteenth-century English scientist who made many contributions to the study of heat.

A star next to an equation number signals that it appears in the list of Key Equations on the Web site for this book: www.whfreeman.com/chemicalprinciples4e.

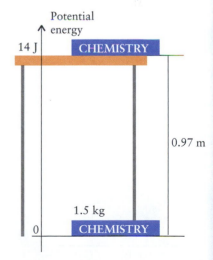

FIGURE A.6 The energy required to raise the book that you are now reading from floor to tabletop is approximately 14 J. The same energy would be released if the book fell from tabletop to floor.

Potential energy is also commonly denoted V.

A field is a region where a force acts.

This formula applies only to objects close to the surface of the Earth.

energy of an object, because the potential energy depends on the nature of the forces that it experiences. However, two simple cases are important in chemistry: gravitational potential energy (for a particle in a gravitational field) and Coulomb potential energy (for a charged particle in an electrostatic field).

A body of mass m at a height h above the surface of the Earth has a gravitational potential energy

$$E_P = mgh \tag{4}*$$

relative to its potential energy on the surface itself (Fig. A.7), where g is the *acceleration of free fall* (and, commonly, the "acceleration of gravity"). The value of g depends on where we are on the Earth, but in most typical locations g has close to its "standard value" of 9.81 m·s^{-2}, and we shall use this value in all calculations. Equation 4 shows that, the greater the altitude of an object, the greater is its gravitational potential energy. For instance, when we raise this book from the floor to the table against the opposing force of gravity, we have done work on the book and its potential energy is increased.

A note on good practice: You will sometimes see kinetic energy denoted KE and potential energy denoted PE. Modern practice is to denote all physical quantities by a single letter (accompanied, if necessary, by subscripts).

EXAMPLE A.4   **Sample exercise: Calculating the gravitational potential energy**

Someone of mass 65 kg walks up a flight of stairs between two floors of a building that are separated by 3.0 m. What is the change in potential energy of the person?

SOLUTION The potential energy of the person on the upper floor relative to that on the lower floor is

From $E_P = mgh$,
$$E_P = (65 \text{ kg}) \times (9.81 \text{ m·s}^{-2}) \times (3.0 \text{ m})$$
$$= 1.9 \times 10^3 \text{ kg·m}^2\text{·s}^{-2} = 1.9 \text{ kJ}$$

+ 1.9 kJ

3.0 m

This energy has to be provided by a chemical process: the digestion of food.

SELF-TEST A.5A What is the gravitational potential energy of this book (mass 1.5 kg) when it is on a table of height 0.82 m relative to its potential energy when it is on the floor?

[*Answer:* 12 J]

SELF-TEST A.5B How much energy has to be expended to raise a can of soda (mass 0.350 kg) to the top of the Sears Tower in Chicago (height 443 m)?

The energy due to attractions and repulsions between electric charges is of great importance in chemistry, which deals with electrons, atomic nuclei, and ions, all of which are charged. The **Coulomb potential energy** of a particle of charge q_1 at a distance r from another particle of charge q_2 is proportional to the two charges and inversely proportional to the distance between them:

$$E_P = \frac{q_1 q_2}{4\pi\varepsilon_0 r} \tag{5}*$$

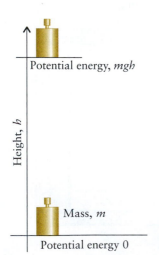

Potential energy, mgh

Height, h

Mass, m

Potential energy 0

FIGURE A.7 The potential energy of a mass m in a gravitational field is proportional to its height h above a point (the "floor"), which is taken to correspond to zero potential energy.

In this expression, which applies when the two charges are separated by a vacuum, ε_0 (epsilon zero) is a fundamental constant called the "vacuum permittivity"; its value is 8.854×10^{-12} J^{-1}·C^2·m^{-1}: the energy is obtained in joules when the charges are in coulombs (C, the SI unit of charge) and their separation is in meters (m). The charge on an electron is $-e$, with $e = 1.602 \times 10^{-19}$ C, the "fundamental charge." The Coulomb potential energy approaches zero as the distance between two particles approaches infinity. If the particles have the same charge—if they are two electrons, for instance—then the numerator $q_1 q_2$, and therefore E_P itself, is positive, and the potential energy *rises* (becomes more strongly positive) as the particles approach each other (*r* decreases). If the particles have opposite charges—an electron and an atomic nucleus, for instance—then the numerator, and therefore E_P, is negative and the potential energy *decreases* (in this case, becomes more negative) as the separation of the particles decreases (Fig. A.8).

What we have termed "electromagnetic energy" is the energy of the **electromagnetic field**, such as the energy carried through space by radio waves, light waves, and x-rays. An electromagnetic field is generated by the acceleration of charged particles and consists of an oscillating **electric field** and an oscillating **magnetic field** (Fig. A.9). The crucial distinction is that an electric field affects charged particles whether they are still or moving, whereas a magnetic field affects only moving charged particles.

The **total energy**, *E*, of a particle is the sum of its kinetic and potential energies:

Total energy = kinetic energy + potential energy, or $E = E_K + E_P$ (6)*

A very important feature of the total energy of an object is that, provided there are no outside influences, it is constant. We summarize this observation by saying that *energy is conserved*. Kinetic energy and potential energy can change into each other, but their sum for a given object is constant. For instance, a ball thrown up into the air initially has high kinetic energy and zero potential energy. At the top of its flight, it has zero kinetic energy and high potential energy. However, as it returns to Earth, its kinetic energy rises and its potential energy approaches zero again. At each stage, its *total* energy is the same as it was initially (Fig. A.10). When it strikes the Earth, the ball is no longer isolated, and its energy is dissipated as **thermal motion**, the chaotic, random motion of atoms and molecules. If we added up all the kinetic and potential energies, we would find that the total energy of the Earth had increased by exactly the same amount as that lost by the ball. No one has ever observed any exception to the **law of conservation of energy**, the observation that energy can be neither created nor destroyed. One region of the universe—an individual atom, for instance—can lose energy, but another region must gain that energy.

Kinetic energy results from motion, potential energy from position. An electromagnetic field carries energy through space; work is motion against an opposing force.

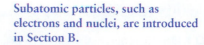

Subatomic particles, such as electrons and nuclei, are introduced in Section B.

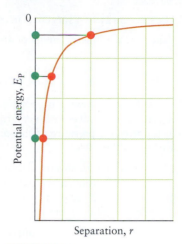

FIGURE A.8 The variation of the Coulomb potential energy of two opposite charges (one represented by the red circle, the other by the green circle) with their separation. Notice that the potential energy decreases as the charges approach each other.

Energy and mass are equivalent and can be interconverted; Einstein deduced that $E = mc^2$, where *c* is the speed of light.

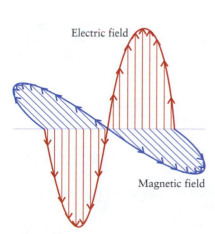

FIGURE A.9 An electromagnetic field oscillates in time and space. The magnetic field is perpendicular to the electric field. The length of an arrow at any point represents the strength of the field at that point, and its orientation denotes its direction. Both fields are perpendicular to the direction of travel of the radiation.

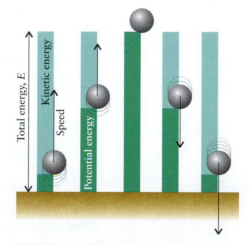

FIGURE A.10 Kinetic energy (represented by the height of the light green bar) and potential energy (the dark green bar) are interconvertible, but their sum (the total height of the bar) is a constant in the absence of external influences, such as air resistance. A ball thrown up from the ground loses kinetic energy as it slows, but gains potential energy. The reverse happens as it falls back to Earth.

FUNDAMENTALS

SKILLS YOU SHOULD HAVE MASTERED

❏ 1 Identify properties as chemical or physical, intensive or extensive.

❏ 2 Use the density of a substance in calculations (Example A.1).

❏ 3 Calculate the kinetic energy of an object (Example A.2).

❏ 4 Calculate the gravitational potential energy of an object (Example A.3).

❏ 5 Distinguish the different forms of energy described in this section.

EXERCISES

A.1 Classify the following properties as chemical or physical: (a) objects made of silver become tarnished; (b) the red color of rubies is due to the presence of chromium ions; (c) the boiling point of ethanol is 78°C.

A.2 A chemist investigates the density, melting point, and flammability of acetone, a component of fingernail polish remover. Which of these properties are physical properties and which are chemical properties?

A.3 Identify all the physical properties and changes in the following statement: "The camp nurse measured the temperature of the injured camper and ignited a propane burner; when the water began to boil some of the water vapor condensed on the cold window."

A.4 Identify all the chemical properties and changes in the following statement: "Copper is a red-brown element obtained from copper sulfide ores by heating them in air, which forms copper oxide. Heating the copper oxide with carbon produces impure copper, which is purified by electrolysis."

A.5 In the containers below, the green spheres represent atoms of one element, the pink spheres the atoms of a second element. In each case, the pictures show either a physical or chemical change; identify the type of change.

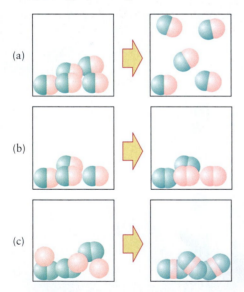

A.6 Which of the containers in Exercise A.5 shows a substance that could be a gas?

A.7 State whether the following properties are extensive or intensive: (a) the temperature at which ice melts; (b) the color of nickel chloride; (c) the energy produced when gasoline burns; (d) the cost of gasoline.

A.8 State whether the following properties are extensive or intensive: (a) the price of platinum; (b) the humidity of the atmosphere; (c) the air pressure in a tire; (d) the hardness of concrete.

A.9 Which value is the largest: (a) 2×10^{-6} km; (b) 2×10^5 μm; (c) 2×10^7 nm?

A.10 Which value is the largest: (a) 5×10^5 pm; (b) 5×10^{-2} cm; (c) 2×10^{-4} dm?

A.11 Express the volume in milliliters of a 1.00-cup sample of milk given that 2 cups = 1 pint, 2 pints = 1 quart.

A.12 The distance for a marathon is 26 miles and 385 yards. Convert this distance to kilometers given that 1 mi = 1760 yd.

A.13 When a piece of metal of mass 112.32 g is dropped into a graduated cylinder containing 23.45 mL of water, the water level rises to 29.27 mL. What is the density of the metal (in grams per cubic centimeter)?

A.14 A flask weighs 43.50 g when it is empty and 105.50 g when filled with water. When the same flask is filled with another liquid, the mass is 96.75 g. What is the density of the second liquid?

A.15 The density of diamond is 3.51 g·cm⁻³. The international (but non-SI) unit for weighing diamonds is the "carat" (1 carat = 200 mg exactly). What is the volume of a 0.750-carat diamond?

A.16 What volume (in cubic centimeters) of lead (of density 11.3 g·cm⁻³) has the same mass as 375 cm³ of a piece of redwood (of density 0.38 g·cm⁻³)?

A.17 Spacecraft are commonly clad with aluminum to provide shielding from radiation. Adequate shielding requires that the cladding provide 20. g of aluminum per square centimeter. How thick does the aluminum cladding need to be to provide adequate shielding?

A.18 To how many significant figures should the result of the following calculation be reported?

$$\frac{604.01 \times 321.81 \times 0.00180}{3.530 \times 10^{-3}}$$

A.19 To how many significant figures should the result of the following calculation be reported?

$$\frac{0.08206 \times (273.15 + 1.2)}{3.25 \times 7.006}$$

A.20 Use the conversion factors in Appendix 1B and inside the back cover to express the following measurements in the designated units: (a) 25 L to m^3; (b) 25 $g \cdot L^{-1}$ to $mg \cdot mL^{-1}$; (c) 1.54 $mm \cdot s^{-1}$ to $pm \cdot \mu s^{-1}$; (d) 7.01 $cm \cdot s^{-1}$ to $km \cdot h^{-1}$; (e) \$2.20/gallon to peso/liter (assume 1 dollar $\approx$ 10.9 peso).

A.21 Use the conversion factors in Appendix 1B and inside the back cover to express the following measurements in the designated units: (a) 4.82 nm to pm; (b) 1.83 $mL \cdot min^{-1}$ to $mm^3 \cdot s^{-1}$; (c) 1.88 ng to kg; (d) 2.66 $g \cdot cm^{-3}$ to $kg \cdot m^{-3}$; (e) 0.044 $g \cdot L^{-1}$ to $mg \cdot cm^{-3}$.

A.22 A chemist determined in a set of four experiments that the density of magnesium metal was 1.68 $g \cdot cm^{-3}$, 1.67 $g \cdot cm^{-3}$, 1.69 $g \cdot cm^{-3}$, 1.69 $g \cdot cm^{-3}$. The accepted value for its density is 1.74 $g \cdot cm^{-3}$. What can you conclude about the precision and accuracy of the chemist's data?

A.23 The density of a metal was measured by two different methods. In each case, calculate the density. Indicate which measurement is more precise. (a) The dimensions of a rectangular block of the metal were measured as 1.10 cm $\times$ 0.531 cm $\times$ 0.212 cm. Its mass was found to be 0.213 g. (b) The mass of a cylinder of water filled to the 19.65-mL mark was found to be 39.753 g. When a piece of the metal was immersed in the water, the level of the water rose to 20.37 mL and the mass of the cylinder with the metal was found to be 41.003 g.

A.24 Assume that the entire mass of an atom is concentrated in its nucleus, a sphere of radius 1.5×10^{-5} pm. (a) If the mass of a carbon atom is 2.0×10^{-23} g, what is the density of a carbon nucleus? The volume of a sphere is $\frac{4}{3}\pi r^3$, where r is its radius. (b) What would the radius of the Earth be if its matter were compressed to the same density as that of a carbon nucleus? (The Earth's average radius is 6.4×10^3 km, and its average density is 5.5 $g \cdot cm^{-3}$.)

A.25 The maximum ground speed of a chicken is 14 $km \cdot h^{-1}$. Calculate the kinetic energy of a chicken of mass 4.2 kg crossing a road at its maximum speed.

A.26 Mars orbits the Sun at 25 $km \cdot s^{-1}$. A spaceship attempting to land on Mars must match its orbital speed. If the mass of the spaceship is 2.5×10^5 kg, what is its kinetic energy when its speed has matched that of Mars?

A.27 A vehicle of mass 2.8 t slows from 100. $km \cdot h^{-1}$ to 50. $km \cdot h^{-1}$ as it enters a city. How much energy could have been recovered instead of being dissipated as heat? To what

height, neglecting friction and other losses, could that energy have been used to drive the vehicle up a hill?

A.28 What is the minimum energy that a football player must expend to kick a football of mass 0.51 kg over a goalpost of height 3.0 m?

A.29 It has been said in jest that the only exercise some people get is raising a fork to their lips. How much energy do we expend to raise a loaded fork of mass 40.0 g a height of 0.50 m 30 times in the course of a meal?

A.30 The expression $E_P = mgh$ applies only close to the surface of the Earth. The general expression for the potential energy of a mass m at a distance R from the center of the Earth (of mass m_E) is $E_P = -Gm_Em/R$. Write $R = R_E + h$, where R_E is the radius of the Earth; and show that, when $h \ll R_E$, this general expression reduces to the special case, and find an expression for g. You will need the expansion $(1 + x)^{-1} = 1 - x + \cdots$.

A.31 The expression for the Coulomb potential energy is very similar to the expression for the general gravitational potential energy given in Exercise A.30. Is there an expression resembling $E_P = mgh$ for the change in potential energy when an electron a long way from a proton is moved through a small distance h? Find the expression of the form $E_P = egh$, with an appropriate expression for g, using the same procedure as in Exercise A.30.

A.32 The weight of an object is the force that it exerts due to the gravitational acceleration of a massive body (such as the Earth) and, for an object of mass m, is equal to mg, where g is the acceleration of free fall. At the surface of the Earth, $g = 9.8$ $m \cdot s^{-2}$. Imagine that, on a planet in a distant galaxy, the acceleration of free fall is 9.8 coburg (where 1 coburg $= 2.78$ $m \cdot s^{-2}$). The inhabitants on this planet discovered an element called yibin, and the weight of a particular sample of yibin is 5.70 lome (1 lome $= 1.3$ N). What would be the weight (in newtons) of this element on Earth?

A.33 Calculate the energy released when an electron is brought from infinity to a distance of 53 pm from a proton. (That distance is the most probable distance of an electron from the nucleus in a hydrogen atom.) The actual energy released when an electron and a proton form a hydrogen atom is 13.6 electronvolts (eV; 1 eV $= 1.602 \times 10^{-19}$ J). Account for the difference.

B   ELEMENTS AND ATOMS

Science is a quest for simplicity. Although the complexity of the world appears boundless, this complexity springs from an underlying simplicity that science attempts to discover. Chemistry's contribution to this quest is to show how everything around us—mountains, trees, people, computers, brains, concrete, oceans— is in fact made up of a handful of simple entities.

   The ancient Greeks had much the same idea. They supposed that there were four elements—earth, air, fire, and water—that could produce all other substances when combined in the right proportions. Their concept of an element is similar to our own; but, on the basis of experiments, we now know that there are actually more than 100 elements, which—in various combinations—make up all the matter on earth (Fig. B.1).

B.1 Atoms

B.2 The Nuclear Model

B.3 Isotopes

B.4 The Organization of the Elements

Appendix 2D lists the names and chemical symbols of the elements and gives the origins of their names.

FIGURE B.1 Samples of common elements. Clockwise from the red-brown liquid bromine are the silvery liquid mercury and the solids iodine, cadmium, red phosphorus, and copper.

B.1 Atoms

The name atom *comes from the Greek word for "not cuttable."*

The Greeks asked what would happen if they continued to cut matter into ever smaller pieces. Is there a point at which they would have to stop because the pieces no longer had the same properties as the whole or could they go on cutting forever? We now know that there is a point at which we have to stop. That is, matter consists of almost unimaginably tiny particles. The smallest particle of an element that can exist is called an **atom.** The story of the development of the modern model of the atom is an excellent illustration of how scientific models are developed.

The first convincing argument for atoms was made in 1807 by the English schoolteacher and chemist John Dalton (Fig. B.2). He made many measurements of the ratios of the masses of elements that combine together to form the substances we call "compounds" (see Section C) and found that the ratios formed patterns. For example, he found that, in every sample of water, there was 8 g of oxygen for every 1 g of hydrogen and that, in another compound of the two elements (hydrogen peroxide), there was 16 g of oxygen for every 1 g of hydrogen. These data led Dalton to develop his **atomic hypothesis:**

FIGURE B.2 John Dalton (1766–1844), the English schoolteacher who used experimental measurements to argue that matter consists of atoms.

1   All the atoms of a given element are identical.

2   The atoms of different elements have different masses.

3   A compound is a specific combination of atoms of more than one element.

4   In a chemical reaction, atoms are neither created nor destroyed; they exchange partners to produce new substances.

The atoms of an element are not all exactly the same, because they can differ slightly in mass. We examine this point in Section B.3.

Today, modern instrumentation provides much more direct evidence of atoms (Fig. B.3). There is no longer any doubt that atoms exist and that they are the units that make up the elements. In fact, chemists use the existence of atoms as the definition of an element: an **element** is a substance composed of only one kind of atom. By 2006, 111 elements had been discovered or created but in some cases in only very small amounts. For instance, when element 110 was made, only two atoms were produced, and even they lasted for only a tiny fraction of a second before disintegrating.

> *All matter is made up of various combinations of the simple forms of matter called the chemical elements. An element is a substance that consists of only one kind of atom.*

B.2 The Nuclear Model

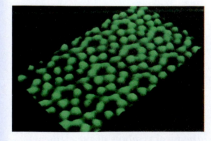

FIGURE B.3 Individual atoms can be seen as bumps on the surface of a solid by the technique called scanning tunneling microscopy (STM). This image is of silicon.

We develop the modern model of an atom in Chapter 1. At this stage, all we need to know is that according to the current **nuclear model** of the atom, an atom consists of a small positively charged **nucleus,** which is responsible for almost all its mass, surrounded by negatively charged **electrons** (denoted e^-). Compared with the size of the nucleus (about 10^{-14} m in diameter), the space occupied by the electrons is enormous (about 10^{-9} m in diameter; a hundred thousand times as great). If the nucleus of an atom were the size of a fly at the center of a baseball field, then the space occupied by the surrounding electrons would be about the size of the entire stadium (Fig. B.4).

The positive charge of the nucleus exactly cancels the negative charge of the surrounding electrons. As a result, an atom is electrically neutral (uncharged). Because each electron has a single negative charge, for each electron in an atom there is a particle inside the nucleus having a single positive charge. These positively charged particles are called **protons** (denoted p); their properties are given in Table B.1. A proton is nearly 2000 times as heavy as an electron.

The number of protons in an element's atomic nucleus is called the **atomic number**, Z, of that element. For example, hydrogen has $Z = 1$; and so we know that the nucleus of a hydrogen atom has one proton; helium has $Z = 2$, and so its nucleus contains two protons. Henry Moseley, a young British scientist, was the first to determine atomic numbers unambiguously, shortly before he was killed in action in World War I. Moseley knew that when elements are bombarded with rapidly moving electrons they emit x-rays. He found that the properties of the x-rays emitted by an element depend on its atomic number; and, by studying the x-rays of many elements, he was able to determine the values of Z for them. Scientists have since determined the atomic numbers of all the known elements (see the list of elements inside the back cover).

Technological advances in electronics early in the twentieth century led to the invention of the **mass spectrometer**, a device for determining the mass of an atom (Fig. B.5). Mass spectrometers are described more fully in Major Technique 6 after Chapter 18. Mass spectrometry has been used to determine the masses of the atoms of all the elements. We now know, for example, that the mass of a hydrogen atom is 1.67×10^{-27} kg and that of a carbon atom is 1.99×10^{-26} kg. Even the heaviest atoms have masses of only about 5×10^{-25} kg. Once we know the mass of an individual atom, we can determine the number of those atoms in a given mass of element simply by dividing the mass of the sample by the mass of one atom.

FIGURE B.4 Think of a fly at the center of this stadium: that is the relative size of the nucleus of an atom if the atom were magnified to the size of the stadium.

EXAMPLE B.1   **Sample exercise: Calculating the number of atoms in a sample**

How many atoms are there in a sample of carbon of mass 10.0 g?

SOLUTION To calculate the number (N) of atoms in the sample, we divide the mass of the sample by the mass of one carbon atom, 1.99×10^{-26} kg, which was given in the text:

From $N = $ (mass of sample)/ (mass of one atom),

$$N = \frac{\overbrace{1.00 \times 10^{-2} \text{ kg}}^{10.0 \text{ g}}}{1.99 \times 10^{-26} \text{ kg}}$$

$$= 5.03 \times 10^{23}$$

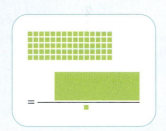

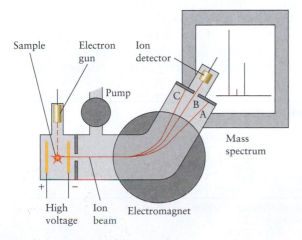

FIGURE B.5 A mass spectrometer is used to measure the masses of atoms. As the strength of the magnetic field is changed, the path of the accelerated ions moves from A to C. When the path is at B, the ion detector sends a signal to the recorder. The mass of the ion is proportional to the strength of the magnetic field needed to move the beam into position.

A note on good practice: Notice how we have converted the units that were given (grams, for the mass of this sample) into units that cancel (here, kilograms). It is often prudent to convert all units to SI base units.

SELF-TEST B.1A The mass of an iron atom is 9.29×10^{-26} kg. How many iron atoms are present in an iron magnet of mass 25.0 g?

[*Answer:* 2.69×10^{23}]

SELF-TEST B.1B A miner panning for gold in an Alaskan creek collects 12.3 g of the fine pieces of gold known as "gold dust." The mass of one gold atom is 3.27×10^{-25} kg. How many gold atoms has the miner collected?

In the nuclear model of the atom, the positive charge and almost all of the mass is concentrated in the tiny nucleus, and the surrounding negatively charged electrons take up most of the space. The atomic number is the number of protons in the nucleus.

B.3 Isotopes

As happens so often in science, a new and more precise technique of measurement led to a major discovery. When scientists first used mass spectrometers they found—much to their surprise—that not all the atoms of a single element have the same mass. In a sample of perfectly pure neon, for example, most of the atoms have mass 3.32×10^{-26} kg, which is about 20 times as great as the mass of a hydrogen atom. Some neon atoms, however, are found to be about 22 times as heavy as hydrogen. Others are about 21 times as heavy (Fig. B.6). All three types of atoms have the same atomic number; so they are definitely atoms of neon.

The observation that atoms of a single element can have different masses helped scientists refine the nuclear model still further. They realized that an atomic nucleus must contain subatomic particles other than protons and proposed that it also contains electrically neutral particles called **neutrons** (denoted n). Because neutrons have no electric charge, their presence does not affect the nuclear charge or the number of electrons in the atom. However, they do add substantially to the mass of the nucleus, so different numbers of neutrons in a nucleus give rise to atoms of different masses, even though the atoms belong to the same element. As we can see from Table B.1, neutrons and protons are very similar apart from their charge; they are jointly known as **nucleons**.

The total number of protons and neutrons in a nucleus is called the **mass number**, A, of the atom. A nucleus of mass number A is about A times as heavy as a hydrogen atom, which has a nucleus that consists of a single proton. Therefore, if we know that an atom is a certain number of times as heavy as a hydrogen atom, then we can infer the mass number of the atom. For example, because mass spectrometry shows that the three varieties of neon atoms are 20, 21, and 22 times as heavy as a hydrogen atom, we know that the mass numbers of the three types of neon atoms are 20, 21, and 22. Because for each of them $Z = 10$, these neon atoms must contain 10, 11, and 12 neutrons, respectively (Fig. B.7).

Atoms with the same atomic number (belonging to the same element) but with different mass numbers are called **isotopes** of the element. All isotopes of an element have exactly the same atomic number; hence they have the same number of protons and electrons. An isotope is named by writing its mass number after the

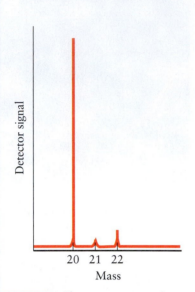

FIGURE B.6 The mass spectrum of neon. The locations of the peaks tell us the relative masses of the atoms, and the intensities tell us the relative numbers of atoms having each mass.

Another, better, name for the mass number is *nucleon number.*

The name isotope comes from the Greek words for "equal place."

TABLE B.1 Properties of Subatomic Particles

Particle	Symbol	Charge*	Mass, kg
electron	e^-	-1	9.109×10^{-31}
proton	p	$+1$	1.673×10^{-27}
neutron	n	0	1.675×10^{-27}

*Charges are given as multiples of the fundamental charge, which in SI units is 1.602×10^{-19} C (see Appendix 1B).

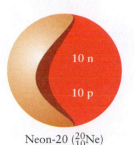

Neon-20 ($^{20}_{10}$Ne)   Neon-21 ($^{21}_{10}$Ne)   Neon-22 ($^{22}_{10}$Ne)

FIGURE B.7 The nuclei of different isotopes of the same element have the same number of protons but different numbers of neutrons. These three diagrams show the composition of the nuclei of the three isotopes of neon. On this scale, the atom itself would be about 1 km in diameter. These diagrams make no attempt to show how the protons and neutrons are arranged inside the nucleus.

FUNDAMENTALS

name of the element, as in neon-20, neon-21, and neon-22. Its symbol is obtained by writing the mass number as a superscript to the left of the chemical symbol of the element, as in ^{20}Ne, ^{21}Ne, and ^{22}Ne. You will occasionally see the atomic number included as a subscript on the lower left, as in the symbol $^{22}_{10}$Ne used in Fig. B.7.

Because isotopes of the same element have the same number of protons and the same number of electrons, they have essentially the same chemical and physical properties. However, the mass differences between isotopes of hydrogen are comparable to the masses themselves, leading to noticeable differences in some physical properties and slight variations in some of their chemical properties. Hydrogen has three isotopes (Table B.2). The most common (^{1}H) has no neutrons; so its nucleus is a lone proton. The other two isotopes are less common but nevertheless so important in chemistry and nuclear physics that they are given special names and symbols. One isotope (^{2}H) is called *deuterium* (D) and the other (^{3}H) is called *tritium* (T).

SELF-TEST B.2A How many protons, neutrons, and electrons are present in (a) an atom of nitrogen-15; (b) an atom of iron-56?

[*Answer:* (a) 7, 8, 7; (b) 26, 30, 26]

SELF-TEST B.2B How many protons, neutrons, and electrons are present in (a) an atom of oxygen-16; (b) an atom of uranium-236?

Isotopes of an element have the same atomic number but different mass numbers. Their nuclei have the same number of protons but different numbers of neutrons.

B.4 The Organization of the Elements

There are 111 known elements, and at first sight the prospect of learning their properties might seem overwhelming. The task is made much easier—and more interesting—by one of the most important discoveries in the history of chemistry. Chemists have found that, when the elements are listed in order of their atomic number and arranged in rows of certain lengths, they form families that show regular trends in properties. The arrangement of elements that shows their family

The story of the discovery of periodic relationships by Dmitri Mendeleev can be found in Box 1.2 in Chapter 1.

TABLE B.2 Some Isotopes of Common Elements

Element	Symbol	Atomic number, Z	Mass number, A	Abundance, %
hydrogen	^{1}H	1	1	99.985
deuterium	^{2}H or D	1	2	0.015
tritium	^{3}H or T	1	3	—*
carbon-12	^{12}C	6	12	98.90
carbon-13	^{13}C	6	13	1.10
oxygen-16	^{16}O	8	16	99.76

*Radioactive, short-lived.

FIGURE B.8 The structure of the periodic table, showing the names of some regions and groups. The groups are the vertical columns, numbered 1 through 18. The periods are the horizontal rows, numbered 1 through 7 (Period 1 is the top row—hydrogen and helium—and is not numbered in the figure). The main-group elements are those in Groups 1, 2, and 13 through 18, together with hydrogen. Some versions of the table use different notations for groups, as in Groups III through VIII shown here. We use both notations for Groups 13 through 18.

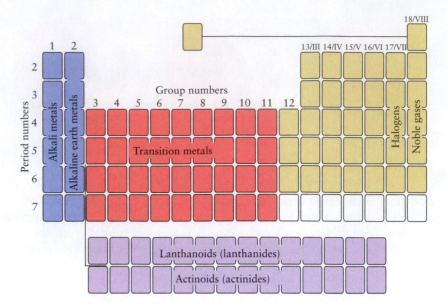

In some versions of the periodic table, you will see a different notation for groups, with the noble gases belonging to Group VIII, VIIIA, or VIIIB. These alternatives are given in the table printed inside the front cover.

In some versions of the periodic table, the lanthanoids begin with cerium (element 58) and the actinoids begin with thorium (element 90).

Lab Video: Figure B.9 Reactions of sodium and water

FIGURE B.9 The alkali metals react with water, producing gaseous hydrogen and heat. Potassium, as shown here, reacts vigorously, producing so much heat that the hydrogen produced is ignited.

relationships is called the **periodic table** (it is printed inside the front cover of this book and repeated schematically in Fig. B.8).

The vertical columns of the periodic table are called **groups**. These groups identify the principal families of elements. The taller columns (Groups 1, 2, and 13/III through 18/VIII) are called the **main groups** of the table. The horizontal rows are called **periods** and are numbered from the top down. The four rectangular regions of the periodic table are called **blocks** and, for reasons related to atomic structure (Section 1.13), are labeled *s*, *p*, *d*, and *f*. The members of the *d* block, with the exception of the elements in Group 12 (the zinc group) are called **transition metals**. As we shall see, these elements are transitional in character between the vigorously reactive metals in the *s* block and the less reactive metals on the left of the *p* block. The members of the *f* block, which is shown below the main table (to save space), are the **inner transition metals**. The upper row of this block, beginning with lanthanum (element 57) in Period 6, consists of the **lanthanoids** (more commonly and traditionally, the "lanthanides"), and the lower row, beginning with actinium (element 89) in Period 7, consists of the **actinoids** (more commonly, the "actinides").

Some of the main groups have special names. The elements in Group 1 are called the **alkali metals.** All of them are soft, lustrous metals that melt at low temperatures. They all produce hydrogen when they come in contact with water (Fig. B.9)—lithium gently, but with increasing violence down the group. The elements calcium (Ca), strontium (Sr), and barium (Ba) in Group 2 are called the **alkaline earth metals,** but the name is often extended to all the members of the group. The Group 2 metals have many properties in common with the Group 1 metals, but their reactions are less vigorous.

On the far right of the table, in Group 18/VIII, are the elements known as the **noble gases.** They are so called because they combine with very few elements—they are chemically aloof. In fact, until the 1960s, they were called the "inert gases" because they were thought not to combine with any elements at all. All the Group 18/VIII elements are colorless, odorless gases. Next to the noble gases are the **halogens** of Group 17/VII. Many of the properties of the halogens vary in a regular fashion from fluorine (F) through chlorine (Cl) and bromine (Br) to iodine (I). Fluorine, for instance, is a very pale yellow, almost colorless gas, chlorine a yellow-green gas, bromine a red-brown liquid, and iodine a purple-black solid (Fig. B.10).

At the head of the periodic table, standing alone, is hydrogen. Some tables place hydrogen in Group 1, others place it in Group 17/VII, and yet others place it in both groups. We treat it as a very special element and place it in none of the groups.

The elements are classified as metals, nonmetals, and metalloids:

A **metal** conducts electricity, has a luster, and is malleable and ductile.

A **nonmetal** does not conduct electricity and is neither malleable nor ductile.

A **metalloid** has the appearance and some properties of a metal but behaves chemically like a nonmetal.

A "malleable" substance (from the Latin word for "hammer") is one that can be hammered into thin sheets (Fig. B.11). A "ductile" substance (from the Latin word for "drawing out") is one that can be drawn out into wires. Copper, for example, is a metal. It conducts electricity, has a luster when polished, and is malleable. It is so ductile that it is readily drawn out to form electrical wires. Sulfur, on the other hand, is a nonmetal. This brittle yellow solid does not conduct electricity, cannot be hammered into thin sheets, and cannot be drawn out into wires. The distinctions between metals and metalloids and between metalloids and nonmetals are not very precise (and not always made), but the metalloids are often taken to be the seven elements shown in Fig. B.12 on a diagonal band between the metals on the left and the nonmetals on the right.

The periodic table is a very useful summary of the properties of the elements. Even if we have never heard of osmium (Os), we can glance at the periodic table and see that it is a metal because it lies to the left of the metalloids (all the *d*-block elements are metals). Similarly, even though chemists might not have investigated the reactions of radon (Rn)—it is a dangerously radioactive gas— we can expect it to resemble the other noble gases, particularly its immediate neighbor in Group 18/VIII, xenon (Xe), and to form very few compounds. As you work through this text and encounter a new element, it is a good idea to refer to the periodic table, identify the neighbors of the element, and try to predict its properties.

The periodic table is an arrangement of the elements that reflects their family relationships; members of the same group typically show a smooth trend in properties.

FIGURE B.10 The halogens are colored elements. From left to right, chlorine is a yellow-green gas, bromine is a red-brown liquid (its vapor fills the flask), and iodine is a dark purple-black solid (note the small crystals).

FIGURE B.11 All metals can be deformed by hammering. Gold can be hammered into a sheet so thin that light can pass through it. Here, it is possible to see the light of a flame through the sheet of gold.

FIGURE B.12 The location of the seven elements commonly regarded as metalloids: these elements have characteristics of both metals and nonmetals. Other elements, notably beryllium and bismuth, are sometimes included in the classification. Boron (B), although not resembling a metal in appearance, is included because it resembles silicon (Si) chemically.

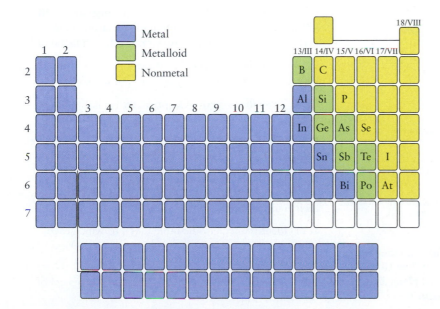

SKILLS YOU SHOULD HAVE MASTERED

❑ 1 Describe the structure of an atom.

❑ 2 Find the number of atoms in a given mass of an element (Example B.1).

❑ 3 Find the numbers of neutrons, protons, and electrons in an isotope (Self-Test B.2).

❑ 4 Describe the organization of the periodic table and the characteristics of elements in different regions of the table.

EXERCISES

B.1 The mass of an atom of beryllium is 1.50×10^{-26} kg. How many beryllium atoms are present in a beryllium film of mass 0.210 g used as a window on an x-ray tube?

B.2 (a) Which part of Dalton's atomic hypothesis has been disproved by experiments? (b) Summarize the evidence disproving this hypothesis.

B.3 Give the number of protons, neutrons, and electrons in an atom of (a) boron-11; (b) ^{10}B; (c) phosphorus-31; (d) ^{238}U.

B.4 Give the number of protons, neutrons, and electrons in an atom of (a) tritium, ^{3}H; (b) ^{27}Al; (c) arsenic-74; (d) ^{237}Np.

B.5 Identify the isotope that has atoms with (a) 117 neutrons, 77 protons, and 77 electrons; (b) 12 neutrons, 10 protons, and 10 electrons; (c) 28 neutrons, 23 protons, and 23 electrons.

B.6 Identify the isotope that has atoms with (a) 106 neutrons, 76 protons, and 76 electrons; (b) 30 neutrons, 28 protons, and 28 electrons; (c) 21 neutrons, 20 protons, and 20 electrons.

B.7 Complete the following table:

Element	Symbol	Protons	Neutrons	Electrons	Mass number
	^{36}Cl				
		30			65
			20	20	
lanthanum			80		

B.8 Complete the following table:

Element	Symbol	Protons	Neutrons	Electrons	Mass number
rhenium					186
	^{208}Pb				
		78			195
			30	28	

B.9 An unstable atomic nucleus gives off nuclear radiation consisting of particles that have a mass of about 1.7×10^{-27} kg. The particles are attracted to a negatively charged plate. The radiation consists of what type of subatomic particle?

B.10 (a) What characteristics do atoms of carbon-12, carbon-13, and carbon-14 have in common? (b) In what ways are they different? (Consider the numbers and types of subatomic particles.)

B.11 (a) What characteristics do atoms of argon-40, potassium-40, and calcium-40 have in common? (b) In what ways are they different? (Consider the numbers and types of subatomic particles.)

B.12 (a) What characteristics do atoms of manganese-55, iron-56, and nickel-58 have in common? (b) In what ways are they different? (Consider the numbers and types of subatomic particles.)

B.13 Determine the fraction of the total mass of a ^{58}Fe atom that is due to (a) neutrons; (b) protons; (c) electrons. (d) What is the mass of neutrons in an automobile of mass 1.000 t? Assume that the total mass of the vehicle is due to ^{56}Fe.

B.14 (a) Determine the total number of protons, neutrons, and electrons in one carbon tetrafluoride molecule, CF_4, assuming that all atoms are the most common isotopes of that element. (b) What is the total mass of protons, of neutrons, and of electrons in one carbon tetrafluoride molecule? Calculate three masses.

B.15 Name each of the following elements: (a) Sc; (b) Sr; (c) S; (d) Sb. List their group numbers in the periodic table. Identify each as a metal, a nonmetal, or a metalloid.

B.16 Name each of the following elements: (a) Al; (b) As; (c) Ag; (d) At. List their group numbers in the periodic table, and identify each as a metal, a nonmetal, or a metalloid.

B.17 Write the symbol of (a) strontium; (b) xenon; (c) silicon. Classify each as a metal, a nonmetal, or a metalloid.

B.18 Write the symbol of (a) iridium; (b) halfnium; (c) mercury. Classify each as a metal, a nonmetal, or a metalloid.

B.19 List the names, symbols, and atomic numbers of the halogens. Identify the normal physical state of each.

B.20 List the names, symbols, and atomic numbers of the alkali metals. Characterize their reactions with water and describe their trend in melting points.

B.21 Identify the periodic table block to which each of the following elements belongs: (a) zirconium; (b) As; (c) Ta; (d) barium; (e) Si; (f) cobalt.

B.22 State three physical properties that are typical of (a) metals; (b) nonmetals.

FUNDAMENTALS

C COMPOUNDS

The small number of elements that make up our world combine to produce matter in a seemingly limitless variety of forms. We have only to look at the vegetation, flesh, landscapes, fabrics, building materials, and other things around us to appreciate the wonderful variety of the material world. A part of chemistry is **analysis**: the discovery of which elements have combined together to form a substance. Another aspect of chemistry is **synthesis**: the process of combining elements to produce compounds or converting one compound into another. If the elements are the alphabet of chemistry, then the compounds are its plays, its poems, and its novels.

C.1 What Are Compounds?

A **compound** is an electrically neutral substance that consists of two or more different elements with their atoms present in a definite ratio. A **binary compound** consists of only two elements. For example, water is a binary compound of hydrogen and oxygen, with two hydrogen atoms for each oxygen atom. Whatever the source of the water, it has exactly the same composition; indeed, a substance with a different ratio of atoms would not be water. Hydrogen peroxide (H_2O_2), for instance, has one hydrogen atom for every oxygen atom.

Compounds are classified as either organic or inorganic. **Organic compounds** contain the element carbon and usually hydrogen, too. There are millions of organic compounds, including fuels such as methane and propane, sugars such as glucose and sucrose, and most medicines. These compounds are called organic because it was once believed, incorrectly, that they could be formed only by living organisms. **Inorganic compounds** are all the other compounds; they include water, calcium sulfate, ammonia, silica, hydrochloric acid, and many, many more. In addition, some very simple carbon compounds, particularly carbon dioxide and the carbonates, which include chalk (calcium carbonate), are treated as inorganic compounds.

The elements in a compound are not just mixed together. Their atoms are actually joined, or *bonded*, to one another in a specific way due to a chemical change (see Section A). The result is a substance with chemical and physical properties different from those of the elements that form it. For example, when sulfur is ignited in air, it combines with oxygen from the air to form the compound sulfur dioxide. Solid yellow sulfur and odorless oxygen gas produce a colorless, pungent, and poisonous gas (Fig. C.1).

Chemists have found that atoms can bond together to form molecules or can be present in compounds as ions:

A **molecule** is a discrete group of atoms bonded together in a specific arrangement.

An **ion** is a positively or negatively charged atom or molecule.

A positively charged ion is called a **cation,** and a negatively charged ion is called an **anion.** For instance, a positively charged sodium atom is a cation and is denoted Na^+; a negatively charged chlorine atom is an anion and is denoted Cl^-. An example of a "polyatomic" (many-atom) cation is the ammonium ion, NH_4^+, and an example of a polyatomic anion is the carbonate ion, CO_3^{2-}; note that the latter has two negative charges. An **ionic compound** consists of ions in a ratio that results in overall electrical neutrality; a **molecular compound** consists of electrically neutral molecules.

In general, binary compounds of two nonmetals are molecular, whereas binary compounds formed by a metal and a nonmetal are ionic. Water (H_2O) is an example of a binary molecular compound, and sodium chloride (NaCl) is an example of a binary ionic compound. As we shall see, these two types of compounds have

FIGURE C.1 Elemental sulfur burns in air with a blue flame and produces the dense gas sulfur dioxide, a compound of sulfur and oxygen.

Bonding is discussed in more detail in Chapters 2 and 3.

The name *ion* comes from the Greek word for "go," because charged particles go either toward or away from a charged electrode.

The prefixes *cat-* and *an-* come from the Greek words for "down" and "up." Oppositely charged ions travel in opposite directions in an electric field.

characteristic properties, and knowing the type of compound that we are studying can be a source of insight into its properties.

> *Compounds are combinations of elements in which the atoms of the different elements are present in a characteristic, constant ratio. A compound is classified as molecular if it consists of molecules and as ionic if it consists of ions.*

C.2 Molecules and Molecular Compounds

The **chemical formula** of a compound represents its composition in terms of chemical symbols. Subscripts show the numbers of atoms of each element present in the smallest unit that is representative of the compound. For molecular compounds, it is common to give the **molecular formula**, a chemical formula that shows how many atoms of each type of element are present in a single molecule of the compound. For instance, the molecular formula for water is H_2O: each molecule contains one O atom and two H atoms. The molecular formula for estrone, a female sex hormone, is $C_{18}H_{22}O_2$, showing that a single molecule of estrone consists of 18 C atoms, 22 H atoms, and 2 O atoms. A molecule of a male sex hormone, testosterone, differs by only a few atoms: its molecular formula is $C_{19}H_{28}O_2$. Think of the consequences of that tiny difference!

Some elements also exist in molecular form. Except for the noble gases, all the elements that are gases at ordinary temperatures are found as **diatomic molecules,** molecules that consist of two atoms. For example, molecules of hydrogen gas contain two hydrogen atoms bonded together and are represented as H_2. The most common form of oxygen consists of diatomic molecules and is known formally as dioxygen, O_2. Solid sulfur exists as S_8 molecules and phosphorus as P_4 molecules. All the halogens exist as diatomic molecules: F_2, Cl_2, Br_2, and I_2.

A **structural formula** indicates how the atoms are linked together, but not their actual three-dimensional arrangement in space. For instance, the molecular formula of methanol (wood spirit) is CH_4O, and its structural formula is shown in (**1**): each line represents a chemical bond (the link between two atoms) and each symbol an atom. Structural formulas contain more information than the chemical formula, but they are cumbersome. So chemists condense them and write, for instance, CH_3OH to represent the structure of methanol. This "condensed" structural formula indicates the groupings of the atoms and summarizes the full structural formula. In most cases, symbols with subscripts represent atoms connected to the preceding element in the formula. A group of atoms attached to another atom in the molecule is set off with parentheses. For example, methylpropane (**2**) has a *methyl* group ($-CH_3$) attached to the central atom in a chain of three carbon atoms, and the condensed version of its structural formula is written $CH_3CH(CH_3)CH_3$ or $HC(CH_3)_3$.

The richness of organic chemistry arises in part from the fact that, although carbon atoms nearly always form four bonds, they can form chains and rings of almost limitless variety. Another source of its richness is that the carbon atoms can be linked together by different kinds of chemical bonds. In Chapter 2, we shall see that atoms can be joined by single bonds, indicated by a single line ($C-C$); double bonds, represented by a double line ($C=C$); and triple bonds, represented by a triple line ($C\equiv C$). A carbon atom can form four single bonds, two double bonds, or any combination that results in four bonds, such as one single and one triple bond.

Organic chemists have found a way to draw complex molecular structures in a very simple way, by not showing the C and H atoms explicitly. A **line structure** represents a chain of carbon atoms by a zigzag line, where each short line indicates a bond and the end of each line represents a carbon atom. Atoms other than C and H are shown by their symbols. Double bonds are represented by a double line and triple bonds by a triple line. Because carbon almost always forms four bonds in organic compounds, there is no need to show the $C-H$ bonds explicitly. We just fill in the correct number of hydrogen atoms mentally: compare the line structure of 2-chlorobutane, $CH_3CHClCH_2CH_3$ (**3a**), with its structural form (**3b**). Line

1   Methanol, CH_3OH

2   Methylpropane, $CH_3CH(CH_3)CH_3$

(a)

(b)

3   2-Chlorobutane, $CH_3CHClCH_2CH_3$

structures are particularly useful for complex molecules, such as testosterone (**4**). They are used occasionally in the early chapters of this text and are explained in more detail in Chapter 18. At this stage, you need only know that a line structure is a display of the "connectivity" of a molecule, a portrayal of which atom is linked to which and by how many bonds.

The next most important aspect of a molecular compound is its shape. The pictorial representations of molecules that most accurately show their shapes are images based on computation or software that represents atoms by spheres of various sizes. An example is the **space-filling model** of an ethanol molecule shown in Fig. C.2a. The atoms are represented by colored spheres (they are not the actual colors of the atoms) that fit into one another. Another representation of the same molecule, called a **ball-and-stick model,** is shown in Fig. C.2b. Each ball represents the location of an atom, and the sticks represent the bonds. Although this kind of model does not represent the actual molecular shape as well as a space-filling model does, it shows bond lengths and angles more clearly. It is also easier to draw and interpret.

Chemists use several other kinds of images to depict molecular structure. The **tube structure** is like a ball-and-stick structure without the balls (Fig. C.2c): different atoms are depicted by different colors at the ends of the tube that represents the link between them. Tube structures and line structures (recall **4**) are useful for showing the structures of complicated molecules. More sophisticated than any of these representations are images that show the distribution of electrons in the molecule. The cloud in Fig. C.2d, for instance, indicates how the electrons are spread throughout an ethanol molecule: it is an example of a **density isosurface** and gives a very real sense of the shape of the molecule (but it is hard to interpret in terms of the location of the atoms). Figure C.2e shows the same diagram with the density isosurface made transparent and a tube structure inside, which helps to clarify the significance of the isosurface.

Chemists also need to know the distribution of electric charge in a molecule, because that distribution affects its physical and chemical properties. To do so, they sometimes use an **electrostatic potential surface** (an "elpot" surface), in which the net electric potential is calculated at each point of the density isosurface and depicted by different colors, as in Fig. C.2f. A blue tint at a point indicates that the positive potential at that point due to the positively charged nuclei outweighs the negative potential due to the negatively charged electrons; a red tint indicates the opposite.

A molecular formula shows the composition of a molecule in terms of the numbers of atoms of each element present. Different styles of molecular models are used to emphasize different molecular characteristics.

C.3 Ions and Ionic Compounds

If we want to imagine what ionic compounds look like, we have to visualize a huge number of cations and anions stacked together in a regular three-dimensional formation. The ions bond together by the attraction between their opposite charges. Each crystal of sodium chloride, for example, consists of an orderly array of a vast number of alternating Na^+ and Cl^- ions (Fig. C.3). When you take a pinch of salt, each crystal that you are picking up consists of more ions than there are stars in the visible universe.

The nuclear model of the atom readily explains the existence of **monatomic ions** (single-atom ions). When an electron is removed from a neutral atom, the charge of the remaining electrons no longer fully cancels the positive charge of the nucleus (Fig. C.4). Because an electron has one unit of negative charge, each electron removed from a neutral atom leaves behind a cation with one additional unit of positive charge. For example, a sodium cation, Na^+, is a sodium atom that has lost one electron. When a calcium atom loses two electrons, it becomes the doubly positively charged calcium ion, Ca^{2+}.

FUNDAMENTALS

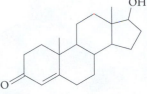

4 Testosterone, $C_{19}H_{28}O_2$

(a) Ethanol, CH_3CH_2OH

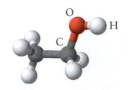

(b) Ethanol, CH_3CH_2OH

(c) Ethanol, CH_3CH_2OH

(d) Density isosurface of C_2H_5OH

(e) Density isosurface of C_2H_5OH

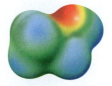

(f) Elpot surface of C_2H_5OH

FIGURE C.2 Representations of an ethanol molecule: (a) space-filling, (b) ball-and-stick, (c) tube, (d) density isosurface, (e) transparent density isosurface, (f) electrostatic potential (elpot) surface.

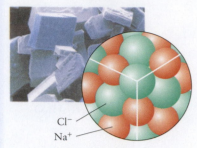

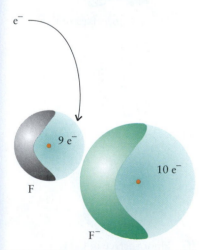

FIGURE C.3 An ionic solid consists of an array of cations and anions stacked together. This illustration shows the arrangement of sodium cations (Na^+) and chlorine anions (chloride ions, Cl^-) in a crystal of sodium chloride (common table salt). The faces of the crystal are where the stacks of ions come to an end.

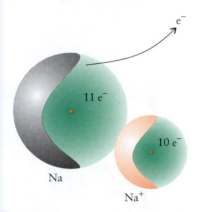

FIGURE C.4 A neutral sodium atom (left) consists of a nucleus that contains 11 protons and is surrounded by 11 electrons. When 1 electron is lost, the remaining 10 electrons cancel only 10 of the proton charges, and the resulting ion (right) has one overall positive charge.

Each electron gained by an atom increases the negative charge by one unit (Fig. C.5). So, when a fluorine atom gains an electron, it becomes the singly negatively charged fluoride ion, F^-. When an oxygen atom gains two electrons, it becomes the doubly charged oxide ion, O^{2-}. When a nitrogen atom gains three electrons, it becomes the triply charged nitride ion, N^{3-}.

The periodic table can help us decide what type of ion an element forms and what charge to expect the ion to have. Fuller details will be given in Chapter 2, but we can begin to see the patterns. One major pattern is that metallic elements—those toward the left of the periodic table—typically form cations by electron loss. Nonmetallic elements—those toward the right of the table—typically form anions by gaining electrons. Thus, the alkali metals form cations, and the halogens form anions.

Figure C.6 shows another pattern in the charges of monatomic cations. For elements in Groups 1 and 2, for instance, the charge of the ion is equal to the group number. Thus, cesium in Group 1 forms Cs^+ ions; barium in Group 2 forms Ba^{2+} ions. Figure C.6 also shows that atoms of the d-block elements and some of the heavier metals of Groups 13/III and 14/IV can form cations with different charges. An iron atom, for instance, can lose two electrons to become Fe^{2+} or three electrons to become Fe^{3+}. Copper can lose either one electron to form Cu^+ or two electrons to become Cu^{2+}.

Some of the anions present in compounds are listed in Fig. C.7. We see another pattern emerging: an element on the right side of the table forms an anion with a negative charge equal to its distance from the noble-gas group. Oxygen is two groups away from the noble gases and forms the oxide ion, O^{2-}; phosphorus, which is three groups away, forms the phosphide ion, P^{3-}. Specifically, if the group number of the element is N (in the 1 to 18 system), the charge of the anions that it forms is $N - 18$.

The pattern of ion formation by main-group elements can be summarized by a single rule: for atoms toward the left or right of the periodic table, *atoms lose or gain electrons until they have the same number of electrons as the nearest noble-gas atom.* Thus, magnesium loses two electrons and becomes Mg^{2+}, which has the same number of electrons as an atom of neon. Selenium gains two electrons and becomes Se^{2+}, which has the same number of electrons as krypton.

Metallic elements typically form cations, and nonmetallic elements typically form anions; the charge of a monatomic ion is related to its group in the periodic table.

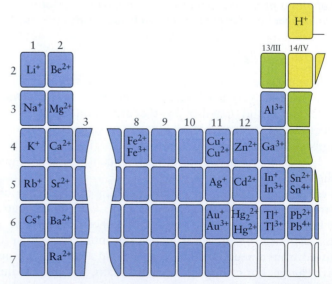

FIGURE C.5 A neutral fluorine atom (left) consists of a nucleus that contains 9 protons and is surrounded by 9 electrons. When the atom gains 1 electron, the 9 proton charges cancel all but 1 of the 10 electron charges, and the resulting ion (right) has one overall negative charge.

FIGURE C.6 The typical cations formed by a selection of elements in the periodic table. The transition metals (Groups 3–11) form a wide variety of cations; we have shown only a few.

FUNDAMENTALS

EXAMPLE C.1 Sample exercise: Identifying the likely charge of a monatomic ion

What ions are (a) nitrogen and (b) calcium likely to form?

SOLUTION

(a) Nitrogen (N) is in Group 15/V and a nonmetal.

$15 - 18 = -3$; therefore, N^{3-}

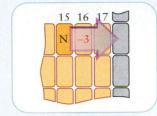

(b) Calcium (Ca) is in Group 2 and a metal.

Ca^{2+}

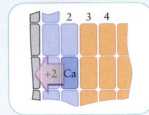

SELF-TEST C.1A What ions are (a) iodine and (b) aluminum likely to form?
[*Answer:* (a) I^-; (b) Al^{3+}]

SELF-TEST C.1B What ions are (a) potassium and (b) sulfur likely to form?

Many ions are **diatomic,** meaning that they consist of two atoms bonded together, or **polyatomic,** meaning that they consist of three or more atoms bonded together. In each case, the ions have an overall positive or negative charge. For example, the cyanide ion, CN^-, is diatomic, and the ammonium ion, NH_4^+, is polyatomic. The most common polyatomic anions are the **oxoanions,** polyatomic anions that contain oxygen. They include the carbonate, CO_3^{2-}; nitrate NO_3^-; phosphate, PO_4^{3-}; and sulfate, SO_4^{2-}, anions. A carbonate ion (**5**), for example, has three O atoms bonded to a C atom and has an additional two electrons. Similarly, a phosphate ion (**6**) consists of four O atoms bonded to a P atom, with three additional electrons. In Chapter 2, we shall see why we have drawn some of these structures with double bonds: for now, we take it as a way of making sure that each atom forms its correct number of bonds.

The formulas of ionic compounds have a different meaning from those of molecular compounds. Each crystal of sodium chloride has a different total number of cations and anions. We cannot simply specify the numbers of ions present as the formula of this ionic compound, because each crystal would have a different formula and the subscripts would be enormous numbers. However, the *ratio* of the number of cations to the number of anions is the same in all the crystals, and the chemical formula shows this ratio. In sodium chloride, there is one Na^+ ion for each Cl^- ion; so its formula is NaCl. Sodium chloride is an example of a **binary ionic compound,** a compound formed from the ions of two elements. Another binary compound, $CaCl_2$, is formed from Ca^{2+} and Cl^- ions in the ratio 1:2, which is required for electrical neutrality.

The formulas of compounds containing polyatomic ions follow similar rules. In sodium carbonate, there are two Na^+ (sodium) ions per CO_3^{2-} (carbonate) ion; so its formula is Na_2CO_3. When a subscript has to be added to a polyatomic ion, the ion is written within parentheses, as in $(NH_4)_2SO_4$, where $(NH_4)_2$ means that there are two NH_4^+ (ammonium) ions for each SO_4^{2-} (sulfate) ion in ammonium sulfate. In each case, the ions combine in such a way that the positive and negative charges cancel: *all compounds are electrically neutral overall.*

We do not usually speak of a "molecule" of an ionic compound. However, it is useful to be able to refer to a group of ions having the number of atoms given by

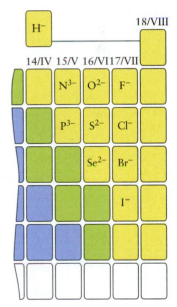

FIGURE C.7 The typical monatomic anions formed by a selection of elements in the periodic table. Notice how the charge on each ion depends on its group number. Only the nonmetals form monatomic anions under common conditions.

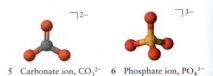

5 Carbonate ion, CO_3^{2-} 6 Phosphate ion, PO_4^{3-}

Clusters of ions, such as Na^+Cl^-, do exist in the gas phase and in concentrated aqueous solutions, in which case they are commonly termed *ion pairs* rather than molecules.

the formula. This group of ions is called a **formula unit.** For example, the formula unit of sodium chloride, NaCl, consists of one Na^+ ion and one Cl^- ion; and the formula unit of ammonium sulfate, $(NH_4)_2SO_4$, consists of two NH_4^+ ions and one SO_4^{2-} ion.

We can often decide whether a substance is an ionic compound or a molecular compound by examining its formula. Binary molecular compounds are typically formed from two nonmetals (such as hydrogen and oxygen, the elements in water). Ionic compounds are typically formed from the combination of a metallic element with nonmetallic elements (such as the combination of potassium with sulfur and oxygen to form potassium sulfate, K_2SO_4). Ionic compounds typically contain one metallic element; the principal exceptions are compounds containing the ammonium ion, such as ammonium nitrate, which are ionic even though all the elements present are nonmetallic.

EXAMPLE C.2 **Sample exercise: Writing the likely formula of a binary ionic compound**

Write the formula of the binary ionic compound formed by magnesium and phosphorus.

SOLUTION

Find the charge on the cation from the periodic table.	Mg is in Group 2; so it forms ions with charge +2.	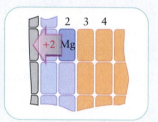
Find the charge on the anion from the periodic table.	P is in Group 15/V and forms ions with charge $15 - 18 = -3$.	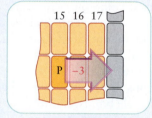
Combine the ions so that their charges cancel.	3 Mg^{2+} ions have a charge of +6. 2 P^{3-} ions have a charge of -6. The formula is Mg_3P_2.	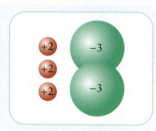

SELF-TEST C.2A Write the formula of the binary ionic compound formed by (a) calcium and chlorine; (b) aluminum and oxygen.

[*Answer:* (a) $CaCl_2$; (b) Al_2O_3]

SELF-TEST C.2B Write the formula of the binary ionic compound formed by (a) lithium and nitrogen; (b) strontium and bromine.

The chemical formula of an ionic compound shows the ratio of the numbers of atoms of each element present in one formula unit. A formula unit of an ionic compound is a group of ions with the same number of atoms of each element as appears in its formula.

SKILLS YOU SHOULD HAVE MASTERED

❏ 1 Distinguish between molecules, ions, and atoms.

❏ 2 Identify compounds as organic or inorganic and as molecular or ionic.

❏ 3 Use various means of representing molecules.

❏ 4 Predict the anion or cation that a main-group element is likely to form (Example C.1).

❏ 5 Interpret chemical formulas in reference to the number of each type of atom present.

❏ 6 Predict the formulas of binary ionic compounds (Example C.2).

EXERCISES

C.1 Each of the containers pictured below holds either a mixture, a single compound, or a single element. The blue spheres represent atoms of one element, the brown spheres the atoms of a second element. In each case, identify the type of contents.

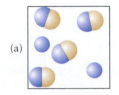
(a)
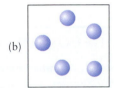
(b)

C.2 Each of the containers pictured below holds either a mixture, a single compound, or a single element. The blue spheres represent atoms of one element, the brown spheres the atoms of a second element. In each case, identify the type of contents.

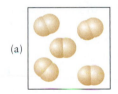
(a)
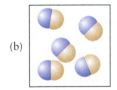
(b)

C.3 The compound xanthophyll is a yellow compound found in bird feathers and flowers. Xanthophyll contains atoms of carbon, hydrogen, and oxygen in the ratio 20:28:1. Its molecules each have six oxygen atoms. Write the chemical formula of xanthophyll.

C.4 Valinomycin is an antibiotic that has been used to treat tuberculosis. Valinomycin contains atoms of carbon, hydrogen, nitrogen, and oxygen in the ratio 9:15:1:3. Its molecules each have 18 oxygen atoms. Write the chemical formula of valinomycin.

C.5 In the following ball-and-stick molecular structures, black indicates carbon, red oxygen, light gray hydrogen, blue nitrogen, and green chlorine. Write the chemical formula of each structure.

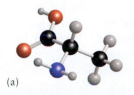
(a)

(b)

C.6 Write the chemical formula of each ball-and-stick structure. See Exercise C.5 for the color code.

(a)

(b)

C.7 State whether each of the following elements is more likely to form a cation or an anion, and write the formula for the ion most likely to be formed: (a) cesium; (b) iodine; (c) selenium; (d) calcium.

C.8 State whether each of the following elements is more likely to form a cation or an anion, and write the formula for the ion most likely to be formed: (a) sulfur; (b) lithium; (c) phosphorus; (d) oxygen.

C.9 How many protons, neutrons, and electrons are present in (a) $^{10}Be^{2+}$; (b) $^{17}O^{2-}$; (c) $^{80}Br^-$; (d) $^{75}As^{3-}$?

C.10 How many protons, neutrons, and electrons are present in (a) $^{64}Cu^{2+}$; (b) $^{138}La^{3+}$; (c) $^{12}C^{2-}$; (d) $^{39}K^+$?

C.11 Write the symbol of the ion that has (a) 9 protons, 10 neutrons, and 10 electrons; (b) 12 protons, 12 neutrons, and 10 electrons; (c) 52 protons, 76 neutrons, and 54 electrons; (d) 37 protons, 49 neutrons, and 36 electrons.

C.12 Write the symbol of the ion that has (a) 11 protons, 13 neutrons, and 10 electrons; (b) 13 protons, 14 neutrons, and 10 electrons; (c) 34 protons, 45 neutrons, and 36 electrons; (d) 24 protons, 28 neutrons, and 22 electrons.

C.13 Write the formula of a compound formed by combining (a) Al and Te; (b) Mg and O; (c) Na and S; (d) Rb and I.

C.14 Write the formula of a compound formed by combining (a) Li and N; (b) Al and S; (c) Sr and Br; (d) Rb and Te.

C.15 Identify the following substances as elements, molecular compounds, or ionic compounds: (a) HCl; (b) S_8; (c) CoS; (d) Ar; (e) CS_2; (f) $SrBr_2$.

C.16 A main-group element in Period 4 forms the molecular compound H_2E and the ionic compound Na_2E. (a) To which group does the element E belong? (b) Write the name and symbol of element E.

C.17 A main-group element in Period 3 forms the following ionic compounds: EBr_3 and E_2O_3. (a) To which group does the element E belong? (b) Write the name and symbol of element E.

C.18 (a) Determine the total number of protons, neutrons, and electrons in one water molecule, H_2O, assuming that only the most common isotopes, ^1H and ^{16}O, are present. (b) What are the total masses of protons, neutrons, and electrons in this water molecule? (c) What fraction of your own mass is due to the neutrons in your body, assuming that you consist primarily of water made from this type of molecule? Note: The masses of free protons and neutrons are slightly higher than the masses of these particles in atoms; so the answer is only an approximation.

C.19 Aluminum oxide, alumina, exists in a variety of crystal structures, some of which are beautiful and rare. Write the formula for aluminum oxide, which is a binary compound of aluminum and oxygen. The mass of a rectangular slab of aluminum oxide of dimensions 2.5 cm × 3.0 cm × 4.0 cm is 102 g. What is the density of aluminum oxide?

C.20 The following compounds contain polyatomic ions. (a) Write the formula for the compound formed from sodium ions and hydrogen phosphite ions (HPO_3^{2-}). (b) Write the formula for ammonium carbonate, which is formed from ammonium ions (NH_4^+) and carbonate ions (CO_3^{2-}). (c) The formula for magnesium sulfate is $MgSO_4$; what is the charge of the cation in $CuSO_4$? (d) The formula for potassium phosphate is K_3PO_4; what is the charge of the cation in $Sn_3(PO_4)_2$?

D.1 Names of Cations
D.2 Names of Anions
D.3 Names of Ionic Compounds
D.4 Names of Inorganic Molecular Compounds
D.5 Names of Some Common Organic Compounds

D   THE NOMENCLATURE OF COMPOUNDS

Many compounds were given informal, **common names** before their compositions were known. Common names include water, salt, sugar, ammonia, and quartz. A **systematic name,** on the other hand, reveals which elements are present and, in some cases, the arrangement of atoms. The systemic naming of compounds, which is called **chemical nomenclature,** follows the simple rules described in this section.

D.1 Names of Cations

The name of a monatomic cation is the same as the name of the element forming it, with the addition of the word ion, as in "sodium ion" for Na^+. When an element can form more than one kind of cation, such as Cu^+ and Cu^{2+} from copper, we use the **oxidation number,** the charge of the cation, written as a Roman numeral in parentheses following the name of the element. Thus, Cu^+ is a copper(I) ion and Cu^{2+} is a copper(II) ion. Similarly, Fe^{2+} is an iron(II) ion and Fe^{3+} is an iron(III) ion. As shown in Fig. C.6, most transition metals form more than one kind of ion; so unless we are given other information we need to include the oxidation number in the names of their compounds.

Some older systems of nomenclature are still in use. For example, some cations were once denoted by the endings *-ous* and *-ic* for the ions with lower and higher charges, respectively. To make matters worse, these endings were in some cases added to the Latin form of the element's name. Thus, iron(II) ions were called ferrous ions and iron(III) ions were called ferric ions (see Appendix 3C). We do not use this system in this text, but you will sometimes come across it and should be aware of it.

In chemical nomenclature, the oxidation number is sometimes called the "Stock number" for the German chemist Alfred Stock, who devised this numbering system. Oxidation numbers are discussed in more detail in Sections K and 2.9.

> The name of a monatomic cation is the name of the element plus the word ion; for elements that can form more than one type of cation, the oxidation number, a Roman numeral indicating the charge, is included.

D.2 Names of Anions

Monatomic anions, such as the Cl^- ions in sodium chloride and the O^{2-} ions in *quicklime* (CaO), are named by adding the suffix *-ide* and the word ion to the first part of the name of the element (the "stem" of its name), as shown in Table D.1; thus, S^{2-} is a sulfide ion and O^{2-} is an oxide ion. There is usually no need to specify the charge, because most elements that form monatomic anions form only one kind of ion. The ions formed by the halogens are collectively called *halide ions* and include fluoride (F^-), chloride (Cl^-), bromide (Br^-), and iodide (I^-) ions.

TABLE D.1 Common Anions and Their Parent Acids

Anion	Parent acid	Anion	Parent acid
fluoride ion, F^-	hydrofluoric acid,* HF (hydrogen fluoride)	nitrite ion, NO_2^-	nitrous acid, HNO_2
		nitrate ion, NO_3^-	nitric acid, HNO_3
chloride ion, Cl^-	hydrochloric acid,* HCl (hydrogen chloride)	phosphate ion, PO_4^{3-}	phosphoric acid, H_3PO_4
		hydrogen phosphate ion, HPO_4^{2-}	
bromide ion, Br^-	hydrobromic acid,* HBr (hydrogen bromide)	dihydrogen phosphate ion, $H_2PO_4^-$	
		sulfite ion, SO_3^{2-}	sulfurous acid, H_2SO_3
iodide ion, I^-	hydroiodic acid,* HI (hydrogen iodide)	hydrogen sulfite ion, HSO_3^-	
		sulfate ion, SO_4^{2-}	sulfuric acid, H_2SO_4
oxide ion, O^{2-}	water, H_2O	hydrogen sulfate ion, HSO_4^-	
hydroxide ion, OH^-		hypochlorite ion, ClO^-	hypochlorous acid, HClO
sulfide ion, S^{2-}	hydrosulfuric acid,* H_2S (hydrogen sulfide)	chlorite ion, ClO_2^-	chlorous acid, $HClO_2$
hydrogen sulfide ion, HS^-		chlorate ion, ClO_3^-	chloric acid, $HClO_3$
cyanide ion, CN^-	hydrocyanic acid,* HCN (hydrogen cyanide)	perchlorate ion, ClO_4^-	perchloric acid, $HClO_4$
acetate ion, $CH_3CO_2^-$	acetic acid, CH_3COOH		
carbonate ion, CO_3^{2-} hydrogen carbonate (bicarbonate) ion, HCO_3^-	carbonic acid, H_2CO_3		

*The name of the aqueous solution of the compound. The name of the compound itself is in parentheses.

Polyatomic ions (see Section C) include the *oxoanions*, which are ions that contain oxygen (see Table D.1). If only one oxoanion of an element exists, its name is formed by adding the suffix *-ate* to the stem of the name of the element, as in the carbonate ion, CO_3^{2-}. Some elements can form two types of oxoanions, with different numbers of oxygen atoms, so we need names that distinguish them. Nitrogen, for example, forms both NO_2^- and NO_3^-. In such cases, the ion with the larger number of oxygen atoms is given the suffix *-ate*, and that with the smaller number of oxygen atoms is given the suffix *-ite*. Thus, NO_3^- is nitrate, and NO_2^- is nitrite. The nitrate ion is an important source of nitrogen for plants and is included in some fertilizers (such as ammonium nitrate, NH_4NO_3).

Some elements—particularly the halogens—form more than two kinds of oxoanions. The name of the oxoanion with the smallest number of oxygen atoms is formed by adding the prefix *hypo-* to the *-ite* form of the name, as in the *hypo*chlorite ion, ClO^-. The oxoanion with the most oxygen atoms is named with the prefix *per-* added to the *-ate* form of the name. An example is the *perchlorate* ion, ClO_4^-. The rules for naming polyatomic ions are summarized in Appendix 3A and common examples are listed in Table D.1.

Hydrogen is present in some anions. Two examples are HS^- and HCO_3^-. The names of these anions begin with "hydrogen." Thus, HS^- is the hydrogen sulfide ion, and HCO_3^- is the hydrogen carbonate ion. You might also see the name of the ion written as a single word, as in hydrogencarbonate ion, which is the more modern style. In an older system of nomenclature, which is still quite widely used, an anion containing hydrogen is named with the prefix *bi-*, as in bicarbonate ion for HCO_3^-. If two hydrogen atoms are present in an anion, as in $H_2PO_4^-$, then the ion is named as a dihydrogen anion, in this case as dihydrogen phosphate.

Hypo comes from the Greek word for "under."

Per is the Latin word for "all over," suggesting that the element's ability to combine with oxygen is finally satisfied.

SELF-TEST D.1A Write (a) the name of the IO^- ion and (b) the formula of the hydrogen sulfite ion.

[*Answer:* (a) Hypoiodite ion; (b) HSO_3^-]

SELF-TEST D.1B Write (a) the name of the $H_2AsO_4^-$ ion and (b) the formula of the chlorate ion.

Lab Video: Figure D.1 Dehydration of copper(II) sulfate pentahydrate

FIGURE D.1 Blue crystals of copper(II) sulfate pentahydrate ($CuSO_4 \cdot 5H_2O$) lose water above 150°C and form the white anhydrous powder ($CuSO_4$) seen in this petri dish. The color is restored when water is added; and, in fact, anhydrous copper sulfate has such a strong attraction for water that it is usually colored a very pale blue from reaction with the water in air.

TABLE D.2 Prefixes Used for Naming Compounds

Prefix	Meaning
mono-	1
di-	2
tri-	3
tetra-	4
penta-	5
hexa-	6
hepta-	7
octa-	8
nona-	9
deca-	10
undeca-	11
dodeca-	12

The distinction can be important: if you buy sodium sulfate decahydrate, $Na_2SO_4 \cdot 10H_2O$, instead of its anhydrous form, Na_2SO_4, you are buying a lot of water!

Names of monatomic anions end in -ide. Oxoanions are anions that contain oxygen. The suffix -ate indicates a greater number of oxygen atoms than the suffixe -ite within the same series of oxoanions.

D.3 Names of Ionic Compounds

An ionic compound is named with the cation name first, followed by the name of the anion; the word *ion* is omitted in each case. The oxidation number of the cation is given if more than one charge is possible. However, if the cation comes from an element that exists in only one charge state (as listed in Fig. C.6), then the oxidation number is omitted. Typical names include potassium chloride (KCl), a compound containing K^+ and Cl^- ions; and ammonium nitrate (NH_4NO_3), which contains NH_4^+ and NO_3^- ions. The cobalt chloride that contains Co^{2+} ions ($CoCl_2$) is called cobalt(II) chloride; $CoCl_3$ contains Co^{3+} ions and is called cobalt(III) chloride.

Some ionic compounds form crystals that incorporate a definite proportion of molecules of water as well as the ions of the compound itself. These compounds are called **hydrates.** For example, copper(II) sulfate normally exists as blue crystals of composition $CuSO_4 \cdot 5H_2O$. The raised dot in this formula separates the waters of hydration from the rest of the formula and the number before the H_2O indicates how many water molecules are present in each formula unit. Hydrates are named by first giving the name of the compound, then adding the word *hydrate* with a Greek prefix indicating how many molecules of water are found in each formula unit (Table D.2). For example, the name of $CuSO_4 \cdot 5H_2O$, the common blue form of this compound, is copper(II) sulfate pentahydrate. When this compound is heated, the water of hydration is lost and the blue crystals crumble to a white powder (Fig. D.1). The white powder is $CuSO_4$ itself. When we wish to emphasize that the compound has lost its water of hydration, we call it **anhydrous.** Thus, $CuSO_4$ is anhydrous copper(II) sulfate.

TOOLBOX D.1   HOW TO NAME IONIC COMPOUNDS

CONCEPTUAL BASIS

The aim of chemical nomenclature is to be simple but unambiguous. Ionic and molecular compounds use different procedures; so it is important first to identify the type of compound. To name an ionic compound, we name the ions present and then combine the names of the ions.

PROCEDURE

Step 1 Identify the cation and the anion (see Table D.1 or Appendix 3A, if necessary). To determine the oxidation number of the cation, decide what cation charge is required to cancel the total negative charge of the anions.

Step 2 Name the cation. If the metal can have more than one oxidation number (most transition metals and some metals in Groups 12 through 15/V), give its charge as a Roman numeral.

Step 3 If the anion is monatomic, change the ending of the element's name to *-ide.*

For an oxoanion:

(a) For elements that form two oxoanions, give the ion with the larger number of oxygen atoms the suffix

-ate and that with the smaller number of oxygen atoms the suffix -ite.

(b) For elements that form a series of four oxoanions, add the prefix *hypo-* to the name of the oxoanion with the smallest number of oxygen atoms. Add the prefix *per-* to the oxoanion with the highest number of oxygen atoms.

For other polyatomic anions, find the name of the ion in Table D.1 or Appendix 3A. If hydrogen is present, add "hydrogen" to the name of the anion. If two hydrogen atoms are present, add "dihydrogen" to the name of the anion.

Step 4 If water molecules appear in the formula, the compound is a hydrate. Add the word *hydrate* with a Greek prefix corresponding to the number of water molecules in front of H_2O.

For examples of how these rules are applied, see Example D.1.

EXAMPLE D.1   Sample exercise: Naming ionic inorganic compounds

Use the rules in Toolbox D.1 to name (a) $CrCl_3 \cdot 6H_2O$ and (b) $Ba(ClO_4)_2$.

SOLUTION

	(a) $CrCl_3 \cdot 6H_2O$	(b) $Ba(ClO_4)_2$
Step 1 Identify the cation and anion.	Cr^{3+}, Cl^-	Ba^{2+}, ClO_4^-
Step 2 Name the cation, giving the charge of the transition-metal cation as a Roman numeral. Note that in (a) there are three Cl^- ions; so the charge on Cr must be +3.	chromium(III)	barium
Step 3 Name the anion and combine the names of the ions with the cation first.	chromium(III) chloride	barium perchlorate
Step 4 If H_2O is present, add hydrate with a Greek prefix.	chromium(III) chloride hexahydrate	

SELF-TEST D.2A Name the compounds (a) $NiCl_2 \cdot 2H_2O$; (b) AlF_3; (c) $Mn(IO_2)_2$.
[*Answer:* (a) Nickel(II) chloride dihydrate; (b) aluminum fluoride; (c) manganese(II) iodite]

SELF-TEST D.2B Name the compounds (a) $AuCl_3$; (b) CaS; (c) Mn_2O_3.

Ionic compounds are named by starting with the name of the cation (with its oxidation number if more than one charge is possible), followed by the name of the anion; hydrates are named by adding the word hydrate, preceded by a Greek prefix indicating the number of water molecules in the formula unit.

D.4 Names of Inorganic Molecular Compounds

Many simple inorganic molecular compounds are named by using the Greek prefixes in Table D.2 to indicate the number of each type of atom present. Usually, we do not use a prefix if only one atom of an element is present; for example, NO_2 is nitrogen dioxide. An important exception to this rule is carbon monoxide, CO. When naming common binary molecular compounds—molecular compounds built from two elements—name the element that occurs further to the right in the periodic table second, with its ending changed to -*ide*:

phosphorus trichloride, PCl_3 dinitrogen oxide, N_2O

sulfur hexafluoride, SF_6 dinitrogen pentoxide, N_2O_5

Some exceptions to these rules are the phosphorus oxides and compounds that are generally known by their common names. The phosphorus oxides are distinguished by the oxidation number of phosphorus, which is written as though phosphorus were a metal and the oxygen present as O^{2-}. Thus, P_4O_6 is named phosphorus(III) oxide as though it were $(P^{3+})_4(O^{2-})_6$, and P_4O_{10} is named phosphorus(V) oxide as though it were $(P^{5+})_4(O^{2-})_{10}$. These compounds, though, are molecular. Certain binary molecular compounds, such as NH_3 and H_2O, have widely used common names (Table D.3).

In both the names and the molecular formulas of compounds formed between hydrogen and nonmetals in Groups 16/VI or 17/VII hydrogen is written first: for example, the formula for hydrogen chloride is HCl and that for hydrogen sulfide is H_2S. Note that, when these compounds are dissolved in water, many act as acids and are named as acids. Binary acids are named by adding the prefix *hydro-* and changing the ending of the name of the second element to -*ic* acid, as in hydrochloric acid for HCl in water and hydrosulfuric acid for H_2S in water. An **aqueous solution**, a solution in water, is indicated by (aq) after the formula. Thus, HCl, the compound itself, is hydrogen chloride, and HCl(aq), its aqueous solution, is hydrochloric acid.

An **oxoacid** is an acidic molecular compound that contains oxygen. Oxoacids are the parents of the oxoanions in the sense that an oxoanion is formed

TABLE D.3 Common Names for Some Simple Molecular Compounds

Formula*	Common name
NH_3	ammonia
N_2H_4	hydrazine
NH_2OH	hydroxylamine
PH_3	phosphine
NO	nitric oxide
N_2O	nitrous oxide
C_2H_4	ethylene
C_2H_2	acetylene

*For historical reasons, the molecular formulas of binary hydrogen compounds of Group 15/V elements are written with the Group 15/V element first.

by removing one or more hydrogen ions from an oxoacid molecule (see Table D.1). In general, *-ic* oxoacids are the parents of *-ate* oxoanions and *-ous* oxoacids are the parents of *-ite* oxoanions. For example, H_2SO_4, sulfuric acid, is the parent acid of the sulfate ion, SO_4^{2-}. Similarly, the parent acid of the sulfite ion, SO_3^{2-}, is the molecular compound H_2SO_3, sulfurous acid.

TOOLBOX D.2   HOW TO NAME SIMPLE INORGANIC MOLECULAR COMPOUNDS

CONCEPTUAL BASIS

The aim of chemical nomenclature is to be simple but unambiguous. A systematic name specifies the elements present in the molecule and the numbers of atoms of each element.

PROCEDURE

Determine the type of compound and then apply the corresponding rules.

Binary molecular compounds other than acids

The compound is generally not an acid if its formula does not begin with H.

Step 1 Write the name of the first element, followed by the name of the second, with its ending changed to *-ide*.

Step 2 Add Greek prefixes to indicate the number of atoms of each element. "Mono-" is usually omitted.

See Example D.2(a) and (b).

Acids

An inorganic acid has a formula that typically begins with H; oxoacids have formulas that begin with H and end in O. We distinguish between binary hydrides, such as HX, which are not named as acids, and their aqueous solutions, HX(aq), which are.

Step 1 If the compound is a binary acid in solution, add "hydro. . .ic acid" to the root of the element's name.

See Example D.2(c).

Step 2 If the compound is an oxoacid, derive the name of the acid from the name of the polyatomic ion that it produces, as in Toolbox D.1. In general,

-ate ions come from *-ic* acids
-ite ions come from *-ous* acids

Retain any prefix, such as *hypo-* or *per-*.

See Example D.2(d).

| EXAMPLE D.2 | Sample exercise: Naming inorganic molecular compounds |

Give the systematic names of (a) N_2O_4, (b) ClO_2, (c) HI(aq), and (d) HNO_2.

SOLUTION

(a) The molecule N_2O_4 has two nitrogen atoms and four oxygen atoms.

Dinitrogen tetroxide

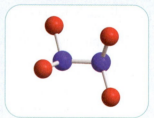

(b) A ClO_2 molecule has one chlorine atom and two oxygen atoms.

Chlorine dioxide

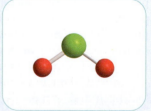

(c) HI(aq) is a binary acid formed when hydrogen iodide dissolves in water.

Hydroiodic acid (The molecular compound HI is hydrogen iodide.)

(d) HNO_2 is an oxoacid, the parent acid of the nitrite ion, NO_2^-.   Nitrous acid

---

SELF-TEST D.3A Name (a) HCN(aq); (b) BCl_3; (c) IF_5.
> [*Answer:* (a) hydrocyanic acid; (b) boron trichloride; (c) iodine pentafluoride]

SELF-TEST D.3B Name (a) PCl_3; (b) SO_3; (c) HBr(aq).

EXAMPLE D.3 **Writing the formula of a binary compound from its name**

Write the formulas of (a) cobalt(II) chloride hexahydrate; (b) diboron trisulfide.

STRATEGY

First check to see whether the compounds are ionic or molecular. Many compounds that contain a metal are ionic. Write the symbol of the metal first, followed by the symbol of the nonmetal. The charges on the ions are determined as shown in Examples C.1 and C.2. Subscripts are chosen to balance charges. Compounds of two nonmetals are normally molecular. Write their formulas by listing the symbols of the elements in the same order as in the name, with subscripts corresponding to the Greek prefixes used.

SOLUTION (a) Determine if the compound is ionic or molecular.	Metal and nonmetal	Ionic
Determine the charge on the cation.	The II in cobalt(II) indicates a +2 charge.	Co^{2+}
Determine the charge on the anion.	Cl is in Group 17/VII, so has the charge $17 - 18 = -1$.	Cl^-
Balance charges.	2 Cl^- ions are required for each Co^{2+} ion.	$CoCl_2$
Add waters of hydration.	Hexahydrate, so has 6 water molecules	$CoCl_2 \cdot 6H_2O$
(b) Determine if the compound is ionic or molecular.	Two nonmetals	Molecular
Convert Greek prefixes into subscripts.	di = 2; tri = 3.	B_2S_3

SELF-TEST D.4A Write the formulas for (a) vanadium(V) oxide; (b) calcium carbide; (c) germanium tetrafluoride; (d) dinitrogen trioxide.
> [*Answer:* (a) V_2O_5; (b) Ca_2C; (c) GeF_4; (d) N_2O_3]

SELF-TEST D.4B Write the formulas for (a) cesium sulfide tetrahydrate; (b) manganese(VII) oxide; (c) hydrogen cyanide (a poisonous gas); (d) disulfur dichloride.

> *Binary molecular compounds are named by using Greek prefixes to indicate the number of atoms of each element present; the element named second has its ending changed to -ide.*

D.5 Names of Some Common Organic Compounds

There are millions of organic compounds; many consist of highly intricate molecules, so their names can be very complicated. You could, for example, find yourself asking for α-D-glucopyranosyl(1→2)-β-D-fructofuranose when all you wanted was sucrose (sugar). However, for most of this text, we will need to know only a few simple organic compounds, and this section will introduce some of them. Chapters 18 and 19 present a more complete introduction to the nomenclature of organic compounds.

At this stage it is helpful to know that compounds of hydrogen and carbon are called **hydrocarbons.** They include methane, CH_4 (**1**); ethane, C_2H_6 (**2**); and benzene, C_6H_6 (**3**). Hydrocarbons that have no carbon–carbon multiple bonds are called **alkanes.** Thus, methane and ethane are both alkanes. The unbranched alkanes with

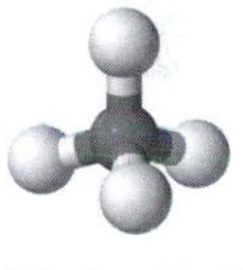

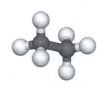

1   Methane, CH_4

2   Ethane, C_2H_6

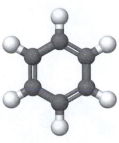

3   Benzene, C_6H_6

TABLE D.4 Alkane Nomenclature

Number of carbon atoms	Formula	Name of alkane	Name of alkyl group
1	CH_4	methane	methyl
2	CH_3CH_3	ethane	ethyl
3	$CH_3CH_2CH_3$	propane	propyl
4	$CH_3(CH_2)_2CH_3$	butane	butyl
5	$CH_3(CH_2)_3CH_3$	pentane	pentyl
6	$CH_3(CH_2)_4CH_3$	hexane	hexyl
7	$CH_3(CH_2)_5CH_3$	heptane	heptyl
8	$CH_3(CH_2)_6CH_3$	octane	octyl
9	$CH_3(CH_2)_7CH_3$	nonane	nonyl
10	$CH_3(CH_2)_8CH_3$	decane	decyl
11	$CH_3(CH_2)_9CH_3$	undecane	undecyl
12	$CH_3(CH_2)_{10}CH_3$	dodecane	dodecyl

up to 12 carbon atoms are listed in Table D.4. Notice that Greek prefixes are used to name all the alkanes with five or more carbon atoms. Hydrocarbons with double bonds are called **alkenes**. Ethene, $CH_2{=}CH_2$, is the simplest example of an alkene. It used to be (and still widely is) called ethylene. Benzene is a hydrocarbon with double bonds, and it has such distinct properties that it is regarded as the parent hydrocarbon of a whole new class of compounds called **aromatic compounds**. The hexagonal benzene ring is exceptionally stable and can be found in many important compounds. Specific groups of atoms derived from hydrocarbons, such as $-CH_3$, methyl, and $-CH_2CH_3$, ethyl, are named by replacing the ending of the parent hydrocarbon's name by *-yl*.

The hydrocarbons are the basic framework for all organic compounds. Different classes of organic compounds have one or more of the hydrogen atoms replaced by other atoms or groups of atoms. All we need to be aware of at this stage are the three classes of compounds known as alcohols, carboxylic acids, and haloalkanes:

An **alcohol** is a type of organic compound that contains an $-OH$ group.

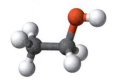

4 Ethanol, CH_3CH_2OH

Ethanol, CH_3CH_2OH (**4**), the "alcohol" of beer and wine, is an ethane molecule in which one H atom has been replaced by an $-OH$ group, and CH_3OH (**5**) is the toxic alcohol called methanol, or wood alcohol.

5 Methanol, CH_3OH

A **carboxylic acid** is a compound that contains the carboxyl group, $-COOH$ (**6**).

6 Carboxyl group, $-COOH$

The most common example is acetic acid, CH_3COOH (**7**), the acid that gives vinegar its sharp taste. Formic acid, HCOOH (**8**), is the acid of ant venom.

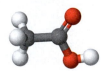

7 Acetic acid, CH_3COOH

A **haloalkane** is an alkane in which one or more H atoms have been replaced by halogen atoms.

8 Formic acid, HCOOH

Haloalkanes include chloromethane, CH_3Cl (**9**), and trichloromethane, $CHCl_3$ (**10**). The latter compound has the common name chloroform and was an early anesthetic. Notice that the names are formed by shortening the name of the halogen to fluoro-, chloro-, bromo-, or iodo- and adding a Greek prefix indicating the number of halogen atoms.

9 Chloromethane, CH_3Cl

SELF-TEST D.5 A(a) Name CH_2BrCl. (b) What type of compound is $CH_3CH(OH)CH_3$?

[*Answer:* (a) Bromochloromethane; (b) alcohol]

SELF-TEST D.5B (a) Name $CH_3CH_2CH_2CH_2CH_3$. (b) What type of compound is CH_3CH_2COOH?

10 Trichloromethane, $CHCl_3$

The names of organic compounds are based on the names of the parent hydrocarbons; alcohols contain $-OH$ groups, carboxylic acids contain $-COOH$ groups, and haloalkanes contain halogen atoms.

SKILLS YOU SHOULD HAVE MASTERED

❏ 1 Name ions, binary inorganic compounds, oxoacids, compounds with common polyatomic ions, and hydrates, and write their formulas (Toolboxes D.1 and D.2, Self-Test D.1, and Examples D.1, D.2, and D.3).

❏ 2 Name simple hydrocarbons and substituted methane (Section D.5).

❏ 3 Identify alcohols and carboxylic acids from their formulas (Self-Test D.5).

EXERCISES

D.1 Write the formula of (a) manganese(II) chloride; (b) calcium phosphate; (c) aluminum sulfite; (d) magnesium nitride.

D.2 Write the formula of (a) strontium hydroxide; (b) cerium(III) phosphate; (c) cadmium chloride; (d) ferrous fluoride.

D.3 Name the following ionic compounds. Write both the old and the modern names wherever appropriate. (a) $Ca_3(PO_4)_2$, the major inorganic component of bones; (b) SnS_2; (c) V_2O_5; (d) Cu_2O.

D.4 The following ionic compounds are commonly found in laboratories. Write the modern name of (a) $NaHCO_3$ (baking soda); (b) Hg_2Cl_2 (calomel): (c) $NaOH$ (lye); (d) ZnO (calamine).

D.5 Write the formula of (a) titanium dioxide; (b) silicon tetrachloride; (c) carbon disulfide; (d) sulfur tetrafluoride; (e) lithium sulfide; (f) antimony pentafluoride; (g) dinitrogen pentoxide; (h) iodine heptafluoride.

D.6 Write the formula of (a) dinitrogen tetroxide: (b) hydrogen sulfide; (c) dichlorine heptoxide; (d) nitrogen triiodide; (e) sulfur dioxide; (f) hydrogen fluoride; (g) diiodine hexachloride.

D.7 Name each of the following binary molecular compounds: (a) SF_6; (b) N_2O_5; (c) NI_3; (d) XeF_4; (e) $AsBr_3$; (f) ClO_2.

D.8 The following compounds are often found in chemical laboratories. Name each compound: (a) SiO_2 (silica); (b) SiC (carborundum); (c) N_2O (a general anesthetic); (d) P_4O_{10} (a drying agent for organic solvents); (e) CS_2 (a solvent); (f) SO_2 (a bleaching agent); (g) NH_3 (a common reagent).

D.9 The following aqueous solutions are common laboratory acids. What are their names? (a) $HCl(aq)$; (b) $H_2SO_4(aq)$; (c) $HNO_3(aq)$; (d) $CH_3COOH(aq)$; (e) $H_2SO_3(aq)$; (f) $H_3PO_4(aq)$.

D.10 The following acids are used in chemical laboratories, although they are less common than those in Exercise D.9. Write the formula of (a) perchloric acid; (b) hypochlorous acid; (c) hypoiodous acid; (d) hydrofluoric acid; (e) phosphorous acid; (f) periodic acid.

D.11 Write the formula for the ionic compound formed from (a) zinc and fluoride ions; (b) barium and nitrate ions; (c) silver and iodide ions; (d) lithium and nitride ions; (e) chromium(III) and sulfide ions.

D.12 Write the formula for the ionic compound formed from (a) strontium and bromide ions; (b) aluminum and sulfate ions; (c) lithium and oxide ions; (d) ammonium and sulfide ions; (e) magnesium and phosphide ions.

D.13 Name each of the following compounds: (a) Na_2SO_3; (b) Fe_2O_3; (c) FeO; (d) $Mg(OH)_2$; (e) $NiSO_4 \cdot 6H_2O$; (f) PCl_5; (g) $Cr(H_2PO_4)_3$; (h) As_2O_3; (i) $RuCl_2$.

D.14 Name each of the following compounds: (a) $CrCl_3 \cdot 6H_2O$; (b) $Co(NO_3)_2 \cdot 6H_2O$; (c) $TlCl$; (d) $BrCl$; (e) MnO_2; (f) $Hg(NO_3)_2$; (g) $Ni(ClO_4)_2$; (h) $Ca(OCl)_2$; (i) Nb_2O_5.

D.15 Each of the following compounds has an incorrect name. Correct each of the mistakes. (a) $CuCO_3$, copper (I) carbonate; (b) K_2SO_3, potassium sulfate; (c) $LiCl$, lithium chlorine.

D.16 Each of the following compounds has an incorrect formula. Correct each of the mistakes. (a) $NaCO_3$, sodium carbonate; (b) $Mg(ClO)_2$, magnesium chlorite; (c) P_2O_6, phosphorus(III) oxide.

D.17 Name each of the following organic compounds: (a) $CH_3CH_2CH_2CH_2CH_2CH_2CH_3$; (b) $CH_3CH_2CH_3$; (c) $CH_3CH_2CH_2CH_2CH_3$; (d) $CH_3CH_2CH_2CH_3$.

D.18 Name each of the following organic compounds: (a) CH_4; (b) CH_3F; (c) CH_3Br; (d) CH_3I.

D.19 You come across some old bottles in a storeroom that are labeled (a) cobaltic oxide monohydrate, (b) cobaltous hydroxide. Using Appendix 3C as a guide, write their modern names and chemical formulas.

D.20 The formal rules of chemical nomenclature result in a certain compound used for electronic components being called barium titanate(IV), in which the oxidation state of titanium is +4. See if you can work out its likely chemical formula. When you have identified the rules for naming oxoanions, suggest a formal name for H_2SO_4.

D.21 A main-group element E in Period 3 forms the molecular compound EH_4 and the ionic compound Na_4E. Identify element E and write the names of the compounds described.

D.22 A main-group element E in Period 6 forms the ionic compounds EBr_2 and EO. Identify element E and write the names of the compounds described.

D.23 The names of some compounds of hydrogen are exceptions to the usual rules of nomenclature. Look up the following compounds, write their names, and identify them as ionic or molecular: (a) $LiAlH_4$; (b) NaH.

D.24 The names of some compounds of oxygen are exceptions to the usual rules of nomenclature. Look up the following compounds, write their names, and identify them as ionic or molecular: (a) KO_2; (b) Na_2O_2; (c) CsO_3.

D.25 Name each of the following compounds, using analogous compounds with phosphorus and sulfur as a guide: (a) H_2SeO_4; (b) Na_3AsO_4; (c) $CaTeO_3$; (d) $Ba_3(AsO_4)_2$; (e) H_3SbO_4; (f) $Ni_2(SeO_4)_3$.

D.26 Name each of the following compounds, using analogous compounds with phosphorus and sulfur as a guide: (a) AsH_3; (b) H_2Se; (c) Cu_2TeO_4; (d) $Ca_3(AsO_3)_2$; (e) NaH_2SbO_4; (f) $BaSeO_3$.

D.27 What type of organic compound is (a) $CH_3CH_2CH_2OH$; (b) $CH_3CH_2CH_2CH_2COOH$; (c) CH_3F?

D.28 Name each of the following compounds, using Appendix 3A: (a) UO_2Cl_2; (b) NaN_3; (c) CaC_2O_4; (d) VOS; (e) $K_2Cr_2O_7$; (f) LiSCN.

E.1 The Mole
E.2 Molar Mass

E   MOLES AND MOLAR MASSES

Because the ratios of atoms of different elements are so important in chemistry, we need to know how to determine the numbers of the different types of atoms, ions, or molecules present in a sample. Knowing the *types* of atoms is fundamental to *qualitative* chemistry—understanding the properties of substances, for instance. Knowing the *numbers* of atoms is fundamental to *quantitative* chemistry—the calculation of the values of these properties.

It is inconvenient to refer to large numbers such as 2.0×10^{25} atoms, just as wholesalers find it inconvenient to report individual items instead of dozens (12) or gross (144). To keep track of the enormous numbers of atoms, ions, or molecules in a sample, we need an efficient way of determining and reporting these numbers.

E.1 The Mole

Chemists report numbers of atoms, ions, and molecules in terms of a unit called a "mole." A mole is the analog of the wholesaler's "dozen." A "dozen" could be defined as the number of soda cans in a "twelve pack" carton supplied by a whole-saler. Even if you could not open the carton to count the number of cans inside, you could find out how many cans are in a dozen by weighing the carton and dividing the mass of the carton by the mass of one can. A similar approach is used to define a **mole** (abbreviated mol):

> 1 mole of objects contains the same number of objects as there are atoms in exactly 12 g of carbon-12.

The name *mole* comes from the Latin word for "massive heap." Coincidentally, the animal of the same name makes massive heaps of soil on lawns.

To tell someone what we mean by 1 mol, we could give them 12 g of carbon-12 and invite them to count the atoms (Fig. E.1). Because counting atoms directly is impractical, we use an indirect route based on the mass of one atom. The mass of a carbon-12 atom has been found by mass spectrometry to be $1.992\,65 \times 10^{-23}$ g. It follows that the number of atoms in exactly 12 g of carbon-12 is

$$\text{Number of C atoms} = \underbrace{\frac{\overbrace{12 \text{ g [exactly]}}^{\text{mass of sample}}}{1.992\,65 \times 10^{-23}\text{g}}}_{\text{mass of one C atom}} = 6.0221 \times 10^{23}$$

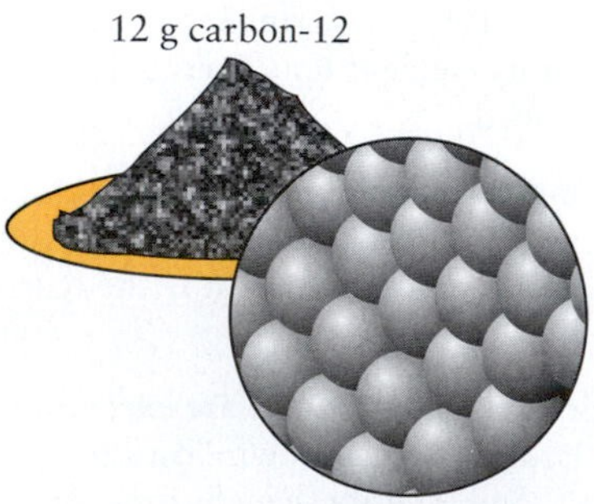

Because the mole tells us the number of atoms in a sample, it follows that 1 mol of atoms of any element is 6.0221×10^{23} atoms of the element (Fig. E.2). The same is true of 1 mol of any objects, including atoms, ions, and molecules:

> 1 mol of objects means 6.0221×10^{23} of those objects.

Just as 1 g and 1 m are units for physical properties, so too is the unit 1 mol. The mole is the unit for the physical property formally called the **amount of substance**, n. However, that name has found little favor among chemists, who colloquially refer to n as "the number of moles." A compromise, which is gaining acceptance, is to refer to n as the "chemical amount" or simply "amount" of entities present in a sample. Thus, 1.0000 mol of hydrogen atoms

FIGURE E.1 The definition of the mole: if we measure out exactly 12 g of carbon-12, then we have exactly 1 mol of carbon-12 atoms. There will be exactly an Avogadro's number of atoms in the pile.

12 g carbon-12

FIGURE E.2 Each sample consists of 1 mol of atoms of the element. Clockwise from the upper right are 32 g of sulfur, 201 g of mercury, 207 g of lead, 64 g of copper, and 12 g of carbon.

$(6.0221 \times 10^{23}$ hydrogen atoms), which is written 1.0000 mol H, is the chemical amount of hydrogen atoms in the sample. Take the advice of your instructor on whether to use the formal term. Like any SI unit, the mole can be used with prefixes. For example, 1 mmol $= 10^{-3}$ mol and 1 μmol $= 10^{-6}$ mol. Chemists encounter such small quantities when dealing with rare or expensive natural products and pharmaceuticals.

The number of objects per mole, 6.0221×10^{23} mol^{-1}, is called **Avogadro's constant**, N_A, in honor of the nineteenth-century Italian scientist Amedeo Avogadro (Fig. E.3), who helped to establish the existence of atoms. Avogadro's constant is used to convert between the chemical amount (the number of moles) and the number of atoms, ions, or molecules in that amount:

Number of objects = amount in moles × number of objects per mole

= amount in moles × Avogadro's constant

If we denote the number of objects by N and the amount of substance (in moles) by n, this relation is written

$$N = nN_A \qquad \qquad (1)^*$$

A note on good practice: Avogadro's constant is a constant with units, not a pure number. You will often hear people referring to *Avogadro's number*: they mean the pure number 6.0221×10^{23}.

To be unambiguous when using moles, we must be specific about which entities (that is, which atoms, molecules, formula units, or ions) we mean. For example, hydrogen occurs naturally as a gas, with each molecule built from two atoms, which is why it is denoted H_2. We write 1 mol H if we mean hydrogen atoms or 1 mol H_2 if we mean hydrogen molecules. Note that 1 mol H_2 corresponds to 2 mol H.

EXAMPLE E.1   Converting number of atoms into an amount in moles

Suppose a sample of vitamin C is known to contain 1.29×10^{24} hydrogen atoms (as well as other kinds of atoms). What is the chemical amount (in moles) of hydrogen atoms in the sample?

STRATEGY Because the number of atoms in the sample is greater than 6×10^{23}, we anticipate that more than 1 mol of atoms is present. Rearrange Eq. 1 into $n = N/N_A$ and then substitute the data.

FIGURE E.3 Lorenzo Romano Amedeo Carlo Avogadro, Count of Quaregna and Cerreto (1776–1856).

SOLUTION

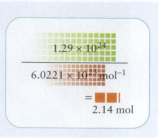

From $n = N/N_A$,

$$n = \frac{1.29 \times 10^{24}\ H}{6.0221 \times 10^{23}\ mol^{-1}} = 2.14\ mol\ H$$

A note on good practice. A good strategy in chemistry is to estimate, as we did here, an approximate answer before doing the calculation; major errors can then be identified quickly.

SELF-TEST E.1A A sample of a drug extracted from a fruit used by the Achuar Jivaro people of Peru to treat fungal infections is found to contain 2.58×10^{24} oxygen atoms. What is the chemical amount (in moles) of O atoms in the sample?

[*Answer:* 4.28 mol O]

SELF-TEST E.1B A small cup of coffee contains 3.14 mol H_2O. What is the number of H atoms present?

> *The amounts of atoms, ions, or molecules in a sample are expressed in moles, and Avogadro's constant, N_A, is used to convert between numbers of these particles and the numbers of moles.*

E.2 Molar Mass

How can we determine the amount of substance present if we can't count the atoms directly? We can find the amount if we know the mass of the sample and the **molar mass**, M, the mass per mole of particles:

> The molar mass of an *element* is the mass per mole of its *atoms*.
> The molar mass of a *molecular compound* is the mass per mole of its *molecules*.
> The molar mass of an *ionic compound* is the mass per mole of its *formula units*.

The units of molar mass in each case are grams per mole ($g \cdot mol^{-1}$). The mass of a sample is the amount (in moles) multiplied by the mass per mole (the molar mass),

> Mass of sample = amount in moles × mass per mole

Therefore, if we denote the total mass of the sample by m, then we can write

$$m = nM \qquad (2)*$$

It follows that $n = m/M$. That is, to find the number of moles, n, we divide the total mass, m, of the sample by the molar mass.

EXAMPLE E.2   **Sample exercise: Calculating the amount of atoms in a sample**

Calculate (a) the amount (n_F, in moles) and (b) the number of F atoms in 22.5 g of fluorine. The molar mass of fluorine atoms is 19.00 $g \cdot mol^{-1}$, or, more specifically, 19.00 $g \cdot (mol\ F)^{-1}$.

SOLUTION

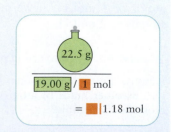

(a) From $n = m/M$,

$$n_F = \frac{22.5\ g}{19.00\ g \cdot (mol\ F)^{-1}}$$

$$= \frac{22.5}{19.00}\ mol\ F = 1.18\ mol\ F$$

A note on good practice: To avoid ambiguity, specify the entities (in this case, fluorine atoms, F, not F_2 molecules) in the units of the calculation.

(b) To calculate the actual number, N, of atoms in the sample, we multiply the amount (in moles) by Avogadro's constant:

From $N = nN_A$,

$$N = (1.18 \text{ mol F}) \times (6.022 \times 10^{23} \text{ mol}^{-1})$$
$$= 7.11 \times 10^{23} \text{ F}$$

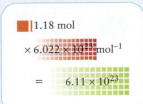

That is, the sample contains 7.11×10^{23} fluorine atoms.

SELF-TEST E.2A The mass of a copper coin is 3.20 g. Suppose it were pure copper. (a) How many moles of Cu atoms would the coin contain, given a molar mass of Cu of 63.55 g·mol^{-1}? (b) How many Cu atoms are present?

[*Answer:* 0.0504 mol Cu; 3.03×10^{22} Cu atoms]

SELF-TEST E.2B In one day, 5.4 kg of aluminum was collected from a recycling bin. (a) How many moles of Al atoms did the bin contain, given that the molar mass of Al is 26.98 g·mol^{-1}? (b) How many Al atoms are present?

The molar masses of elements are determined by using mass spectrometry to measure the masses of the individual isotopes and their abundances. The mass per mole of atoms is the mass of an individual atom multiplied by Avogadro's constant (the number of atoms per mole):

$$M = m_{atom}N_A \tag{3a}$$

The greater the mass of an individual atom, the greater the molar mass of the substance. However, most elements exist in nature as a mixture of isotopes. We saw in Section B, for instance, that neon exists as three isotopes, each with a different mass. In chemistry, we almost always deal with natural samples of elements, which have the natural abundance of isotopes. So, we need the *average* molar mass, the molar mass calculated by taking into account the masses of the isotopes and their relative abundances in typical samples:

$$M = m_{atom, average}N_A \tag{3b}$$

All molar masses quoted in this text refer to these average values. Their values are given in Appendix 2D. They are also included in the periodic table inside the front cover and in the alphabetical list of elements inside the back cover.

EXAMPLE E.3 **Evaluating an average molar mass**

There are two naturally occurring isotopes of chlorine: chlorine-35 and chlorine-37. The mass of an atom of chlorine-35 is 5.807×10^{-23} g and that of an atom of chlorine-37 is 6.139×10^{-23} g. In a typical natural sample of chlorine, 75.77% of the sample is chlorine-35 and 24.23% is chlorine-37. What is the molar mass of a typical sample of chlorine?

STRATEGY First calculate the average atomic mass of the isotopes by adding together the individual masses, each multiplied by the fraction that represents its abundance. Then obtain the molar mass, the mass per mole of atoms, by multiplying the average atomic mass by Avogadro's constant.

SOLUTION The fraction (f) of chlorine-35 atoms in the sample is $75.77/100 = 0.7577$, and the fraction of chlorine-37 is $24.23/100 = 0.2423$. The average mass of an atom of chlorine in a natural sample is

From $m_{atom, average} =$

$$f_{chlorine-35} \times m_{chlorine-35} +$$
$$f_{chlorine-37} \times m_{chlorine-37},$$

$$m_{atom, average} =$$
$$0.7577 \times (5.807 \times 10^{-23} \text{ g}) +$$
$$0.2423 \times (6.139 \times 10^{-23} \text{ g})$$
$$= 5.887 \times 10^{-23} \text{ g}$$

It follows from Eq. 3b that the molar mass of a typical sample of chlorine atoms is

From $M = m_{atom,average}N_A$, $M = (5.887 \times 10^{-23} \text{ g}) \times (6.022 \times 10^{23} \text{ mol}^{-1})$
$$= 35.45 \text{ g·mol}^{-1}$$

SELF-TEST E.3A In a typical sample of magnesium, 78.99% is magnesium-24 (with atomic mass 3.983×10^{-23} g), 10.00% magnesium-25 (4.149×10^{-23} g), and 11.01% magnesium-26 (4.315×10^{-23} g). Calculate the molar mass of a typical sample of magnesium, given the atomic masses (in parentheses).

[*Answer:* 24.31 g·mol^{-1}]

SELF-TEST E.3B Calculate the molar mass of copper, given that a natural sample typically consists of 69.17% copper-63, which has a molar mass of 62.94 g·mol^{-1}, and 30.83% copper-65, which has a molar mass of 64.93 g·mol^{-1}.

To calculate the molar masses of molecular and ionic compounds, we use the molar masses of the elements present: *the molar mass of a compound is the sum of the molar masses of the elements that make up the molecule or the formula unit*. We need only note how many times each atom or ion appears in the molecular formula or the formula unit of the ionic compound. For example, the molar mass of the ionic compound $Al_2(SO_4)_3$ is

$$M_{Al_2(SO_4)_3} = 2M_{Al} + 3M_S + 12M_O$$
$$= 2(26.98 \text{ g·mol}^{-1}) + 3(32.06 \text{ g·mol}^{-1}) + 12(16.00 \text{ g·mol}^{-1})$$
$$= 342.14 \text{ g·mol}^{-1}$$

SELF-TEST E.4A Calculate the molar mass of (a) ethanol, C_2H_5OH; (b) copper(II) sulfate pentahydrate.

[*Answer:* (a) 46.07 g·mol^{-1}; (b) 249.69 g·mol^{-1}]

SELF-TEST E.4B Calculate the molar mass of (a) phenol, C_6H_5OH; (b) sodium carbonate decahydrate.

Two terms used throughout the chemical literature are *atomic weight* and *molecular weight*:

The **atomic weight** of an element is the numerical value of its molar mass.
The **molecular weight** of a molecular compound or the **formula weight** of an ionic compound is the numerical value of its molar mass.

For the most authoritative and up-to-date listing of atomic weights, see the IUPAC link on the Web site for this text.

For instance, the atomic weight of hydrogen (molar mass 1.0079 g·mol^{-1}) is 1.0079, the molecular weight of water (molar mass 18.02 g·mol^{-1}) is 18.02, and the formula weight of sodium chloride (molar mass 58.44 g·mol^{-1}) is 58.44. These three terms are traditional and deeply ingrained in chemical conversations, even though the numbers are not "weights." The mass of an object is a measure of the *quantity of matter* that it contains, whereas the weight of an object is a measure of the *gravitational pull* that it experiences. Mass and weight are proportional to each other, but they are not identical. An astronaut would have the same mass (contain the same quantity of matter) but different weights on Earth and on Mars.

A final point to keep in mind is that the "atomic weight" of an element allows us to make a good estimate of the numbers of protons and neutrons in its nuclei, because each proton and each neutron contribute close to one unit of mass to the atomic weight. Thus, we know that carbon, atomic weight 12.01, contains a total of 12 protons and neutrons. The atomic weight, however, is not exactly the mass number, because the mass of a nucleus is not exactly the sum of the masses of its protons and neutrons, and atomic weights (that is, molar masses) are averages that take into account the isotopic composition of typical samples of the element.

Once the molar mass of a compound has been calculated, we can use the same technique used for elements to determine how many moles of molecules or formula units are in a sample of a given mass.

SELF-TEST E.5A Calculate the amount of urea molecules, $OC(NH_2)_2$, in 2.3×10^5 g of urea, which is used in facial creams and, on a bigger scale, as an agricultural fertilizer.

[*Answer:* 3.8×10^3 mol]

SELF-TEST E.5B Calculate the amount of $Ca(OH)_2$ formula units in 1.00 kg of slaked lime (calcium hydroxide), which is used to adjust the acidity of soils.

Molar mass is important when we need to know the number of atoms in a sample. It would be impossible to count out 6×10^{23} atoms of an element, but it is very easy to measure out a mass equal to the molar mass of the element in grams. Each of the samples shown in Fig. E.2 was obtained in this way: each sample contains the same number of atoms of the element (6.022×10^{23}), but the masses vary because the masses of the atoms are different (Fig. E.4). The same rule applies to compounds. Hence, if we measure out 58.44 g of sodium chloride, we obtain a sample that contains 1.000 mol NaCl formula units (Fig. E.5).

In practice, chemists rarely try to measure out an exact mass. Instead, they estimate the mass required and spoon out that mass approximately. Then they measure the mass of the sample precisely and convert it into moles (by using Eq. 2, $n = m/M$) to find the precise amount that they have obtained.

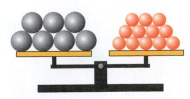

(a)   Equal masses

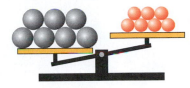

(b)   Equal amounts

FIGURE E.4 (a) The two samples have the same mass but, because the atoms on the right are lighter than those on the left, more of them are required to balance the sample on the left. (b) The two samples contain the same number of atoms but, because the atoms on the right are lighter than those on the left, the mass of the sample on the right is the smaller of the two. Equal amounts (equal numbers of moles) of atoms do not necessarily have the same mass.

EXAMPLE E.4   Sample exercise: Calculating the mass of a sample corresponding to a given amount in moles

Suppose we are preparing a solution of potassium permanganate, $KMnO_4$, and need about 0.10 mol of the compound (that is, 0.10 mol $KMnO_4$). How many grams of the compound do we need?

SOLUTION To find the mass of $KMnO_4$, m_{KMnO_4}, that corresponds to 0.10 mol $KMnO_4$, we note that, because the molar mass of the compound is 158.04 g·mol^{-1},

From $m = nM$, $m_{KMnO_4} = (0.10 \text{ mol}) \times (158.04 \text{ g·mol}^{-1}) = 16$ g

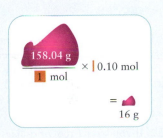

FIGURE E.5 Each sample contains 1 mol of formula units of an ionic compound. From left to right are 58 g of sodium chloride (NaCl), 100 g of calcium carbonate ($CaCO_3$), 278 g of iron(II) sulfate heptahydrate ($FeSO_4 \cdot 7H_2O$), and 78 g of sodium peroxide (Na_2O_2).

So, we need to measure out about 16 g of $KMnO_4$. If, when we weigh the sample, we find that its actual mass is 14.87 g, the amount of $KMnO_4$ that we actually weighed out, n_{KMnO_4}, is

From $n = m/M$,

$$n_{KMnO_4} = \frac{14.87 \text{ g}}{158.04 \text{ g·mol}^{-1}}$$

$$= \frac{14.87}{158.04} \text{ mol} = 0.09409 \text{ mol}$$

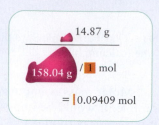

SELF-TEST E.6A What mass of anhydrous sodium hydrogen sulfate should you weigh out to obtain about 0.20 mol $NaHSO_4$?

[*Answer:* About 24 g]

SELF-TEST E.6B> What mass of acetic acid should you weigh out to obtain 1.5 mol CH_3COOH?

The molar mass of a compound, the mass per mole of its molecules or formula units, is used to convert between the mass of a sample and the amount of molecules or formula units that it contains.

SKILLS YOU SHOULD HAVE MASTERED

❑ 1 Use Avogadro's constant to convert between amount in moles and the number of atoms, molecules, or ions in a sample (Examples E.1 and E.2).

❑ 2 Calculate the molar mass of an element, given its isotopic composition (Example E.3).

❑ 3 Calculate the molar mass of a compound, given its chemical formula.

❑ 4 Convert between mass and amount in moles by using the molar mass (Example E.4).

EXERCISES

E.1 You have become very excited about the possibilities of nanotechnology, especially the creation of fibers one atom wide. Suppose you were able to string together 1.00 mole of silver atoms, which have a radius of 144 pm, by encapsulating them in carbon nanotubes (see Box 14.1). How long could the fiber extend?

E.2 If you won 1 mol of dollars in a lottery the day you were born, and spent 1 billion dollars a second for the rest of your life, what percentage of the prize money would remain, if any, when you decide to retire from spending at 90 years of age?

E.3 In your new nanotechnology lab you have the capability to manipulate individual atoms. The atoms on the left are bromine atoms (molar mass 80 g·mol^{-1}), those on the right are atoms of calcium (molar mass 40 g·mol^{-1}). How many calcium atoms would have to be added to the pan on the right for the masses on the two pans to be equal?

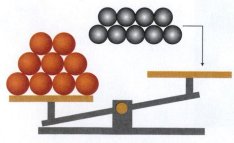

E.4 In your new nanotechnology lab you have the capability to manipulate individual atoms. The atoms on the left are aluminum atoms (molar mass 27 g·mol^{-1}), those on the right are atoms of beryllium (molar mass 9 g·mol^{-1}). How many beryllium atoms would have to be added to the pan on the right for the masses on the two pans to be equal?

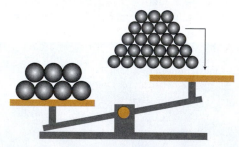

E.5 (a) The approximate population of Earth is 6.0 billion people. How many moles of people inhabit Earth? (b) If all people were pea pickers, then how long would it take for the entire population of Earth to pick 1 mol of peas at the rate of one pea per second, working 24 hours per day, 365 days per year?

E.6 (a) One thousand tonnes (1000 t, 1 t = 10^3 kg) of sand contains about a trillion (10^{12}) grains of sand. How many tonnes of sand are needed to provide 1 mol of grains of sand?

(b) Assuming the volume of a grain of sand is 1 mm^3 and the land area of the continental United States is 3.6×10^6 mi^2, how deep would the sand pile over the United States be if this area were evenly covered with 1 mol of grains of sand?

E.7 The nuclear power industry extracts ^{6}Li but not ^{7}Li from natural samples of lithium. As a result, the molar mass of commercial samples of lithium is increasing. The current abundances of the two isotopes are 7.42% and 92.58%, respectively, and the masses of their atoms are 9.988×10^{-24} g and 1.165×10^{-23} g, respectively. (a) What is the current molar mass of a natural sample of lithium? (b) What will the molar mass be when the abundance of ^{6}Li is reduced to 5.67%?

E.8 Copper metal can be extracted from a copper(II) sulfate solution by electrolysis (as described in Chapter 12). If 29.50 g of copper(II) sulfate pentahydrate, $CuSO_4 \cdot 5H_2O$, is dissolved in 100. mL of water and all the copper is electroplated out, what mass of copper would be obtained?

E.9 Epsom salts consist of magnesium sulfate heptahydrate. Write its formula. (a) How many atoms of oxygen are in 5.15 g of Epsom salts? (b) How many formula units of the compound are present in 5.15 g? (c) How many moles of water molecules are in 5.15 g of Epsom salts?

E.10 Calculate the molar mass of sulfur in a sample that consists of 95.0% ^{32}S (molar mass 31.97 g·mol^{-1}), 0.8% ^{33}S (molar mass 32.97 g·mol^{-1}), and 4.2% ^{34}S (molar mass 33.97 g·mol^{-1}).

E.11 The molar mass of boron atoms in a natural sample is 10.81 g·mol^{-1}. The sample is known to consist of ^{10}B (molar mass 10.013 g·mol^{-1}) and ^{11}B (molar mass 11.093 g·mol^{-1}). What are the percentage abundances of the two isotopes?

E.12 Calculate the molar mass of the noble gas krypton in a natural sample, which is 0.3% ^{78}Kr (molar mass 77.92 g·mol^{-1}), 2.3% ^{80}Kr (molar mass 79.91 g·mol^{-1}), 11.6% ^{82}Kr (molar mass 81.91 g·mol^{-1}), 11.5% ^{83}Kr (molar mass 82.92 g·mol^{-1}), 56.9% ^{84}Kr (molar mass 83.91 g·mol^{-1}), and 17.4% ^{86}Kr (molar mass 85.91 g·mol^{-1}).

E.13 Which sample in each of the following pairs contains the greater number of moles of atoms? (a) 75 g of indium or 80 g of tellurium; (b) 15.0 g of P or 15.0 g of S; (c) 7.36×10^{27} atoms of Ru or 7.36×10^{27} atoms of Fe.

E.14 Calculate the mass, in micrograms, of (a) 5.68×10^{15} Hg atoms; (b) 7.924×10^{-9} mol Hf atoms; (c) 3.49 μmol Gd atoms; (d) 6.29×10^{24} Sb atoms.

E.15 What mass of rhodium contains as many atoms as there are (a) gallium atoms in 36 g of gallium; (b) indium atoms in 36 g of indium?

E.16 Determine the mass of lead that has the same number of atoms as there are in (a) 10.5 mg of silver; (b) 10.5 mg of zinc.

E.17 Calculate the amount (in moles) and the number of molecules and formula units (or atoms, if indicated) in (a) 10.0 g of alumina, Al_2O_3; (b) 25.92 mg of hydrogen fluoride, HF; (c) 1.55 mg of hydrogen peroxide, H_2O_2; (d) 1.25 kg of glucose, $C_6H_{12}O_6$; (e) 4.37 g of nitrogen as N atoms and as N_2 molecules.

E.18 Convert each of the following masses into amount (in moles) and into number of molecules (and atoms, if indicated). (a) 2.40 kg of H_2O; (b) 49 kg of benzene (C_6H_6); (c) 260.0 g of phosphorus, as P atoms and as P_4 molecules; (d) 5.0 g of CO_2; (e) 5.0 g of NO_2.

E.19 Calculate the amount (in moles) of (a) Cu^{2+} ions in 3.00 g of $CuBr_2$; (b) SO_3 molecules in 7.00×10^2 mg of SO_3; (c) F^- ions in 25.2 kg of UF_6; (d) H_2O in 2.00 g of $Na_2CO_3 \cdot 10H_2O$.

E.20 Calculate the amount (in moles) of (a) CN^- in 2.00 g of NaCN; (b) O atoms in 3.00×10^2 ng of H_2O; (c) $CaSO_4$ in 6.00 kg of $CaSO_4$; (d) H_2O in 4.00 g of $Al_2(SO_4)_3 \cdot 8H_2O$.

E.21 (a) Determine the number of formula units in 0.750 mol KNO_3. (b) What is the mass (in milligrams) of 2.39×10^{20} formula units of Ag_2SO_4? (c) Estimate the number of formula units in 3.429 g of $NaHCO_2$, sodium formate, which is used in dyeing and printing fabrics.

E.22 (a) How many CaH_2 formula units are present in 5.294 g of CaH_2? (b) Determine the mass of 6.25×10^{24} formula units of $NaBF_4$, sodium tetrafluoroborate. (c) Calculate the amount (in moles) of 9.54×10^{21} formula units of CeI_3, cerium(III) iodide, a bright yellow, water-soluble solid.

E.23 (a) Calculate the mass, in grams, of one water molecule. (b) Determine the number of H_2O molecules in 1.00 kg of H_2O.

E.24 Octane, C_8H_{18}, is typical of the molecules found in gasoline. (a) Calculate the mass of one octane molecule. (b) Determine the number of C_8H_{18} molecules in 1.00 mL of C_8H_{18}, the mass of which is 0.82 g.

E.25 A chemist measured out 8.61 g of copper(II) chloride tetrahydrate, $CuCl_2 \cdot 4H_2O$. (a) How many moles of $CuCl_2 \cdot 4H_2O$ were measured out? (b) How many moles of Cl^- ions are present in the sample? (c) How many water molecules are present in the sample? (d) What fraction of the total mass of the sample was due to oxygen?

E.26 Anhydrous copper(II) sulfate is difficult to dry completely. What mass of copper(II) sulfate would remain after removing 90% of the water from 250. g of $CuSO_4 \cdot 5H_2O$?

E.27 Suppose you had bought 10. kg of $NaHCO_3 \cdot 10H_2O$ for $72 by mistake, instead of 10. kg of $NaHCO_3$ for $80. (a) How much water did you buy and how much did you pay per liter of water? (The mass of 1 L of water is 1 kg.) (b) What would have been a fair price for the hydrated compound, valuing water at zero cost?

E.28 A chemist wants to extract the gold from 35.25 g of gold(III) chloride dihydrate, $AuCl_3 \cdot 2H_2O$, by electrolysis of an aqueous solution (this technique is described in Chapter 12). What mass of gold could be obtained from the sample?

E.29 The king of Zirconia is naturally fond of the element zirconium and has established an independent definition of the mole based on zirconium. The mass of one zirconium-90 atom is 1.4929×10^{-22} g. If zirconium were the standard used for molar mass (instead of carbon-12), 1 mol would be defined as the amount of substance that contains the same number of entities as there are atoms in exactly 90 g of zirconium-90. In that case, what would be (a) the molar mass of carbon-12; (b) the (average) molar mass of gold?

E.30 The isotope silicon-28 has been proposed as a new standard for the molar masses of elements because it can be prepared to a very high degree of purity. The mass of one silicon-28 atom is 4.64567×10^{-23} g. If silicon were the standard used for molar mass (instead of carbon-12), 1 mol would be defined as the amount of substance that contains the same number of entities as there are atoms in exactly 28 g of silicon-28. In that case, what would be (a) the molar mass of carbon-12; (b) the (average) molar mass of chlorine?

E.31 The density of sodium borohydride is 1.074 g·cm^{-3}. If 3.93 g of sodium borohydride contains 2.50×10^{23} H atoms, how many moles of H atoms are present in 28.0 cm^3 of sodium borohydride?

E.32 Chloroform is produced industrially from dichloroethane ($C_2H_4Cl_2$). Chloroform consists of molecules containing five atoms each and, at 20°C, has a density of 1.492 g·cm^{-3}. Given that 0.250 mol of chloroform molecules occupies 20.0 mL, what mass (in grams) of dichloroethane contains the same number of atoms as 25.5 g of chloroform?

E.33 Suppose that an element has two isotopes, one of mass m_1 and the other of mass m_2, and that the former constitutes $x\%$ of the sample. Plot a graph showing how the mean mass of the atoms varies as x changes from 0% to 100%.

E.34 Now explore a graphical way of expressing the mean mass when three isotopes of masses m_1, m_2, and m_3 contribute. Use the fact that the sum of the three distances shown in the accompanying equilateral triangle diagram satisfies $x_1 + x_2 + x_3 = L$, where L is the length of a side, to plot the mean mass of a sample as its composition varies over the whole range of possibilities, letting x_1, x_2, and x_3 represent the percentages of each isotope.

$$x_1 + x_2 + x_3 = L$$

E.35 A mass spectrum of Br_2 has three peaks, with the mass numbers 158, 160, and 162. (a) Use this information to determine which isotopes of bromine occur in nature. (b) If the relative heights of the peaks, which depend on abundance, are 33.8, 33.3, and 32.9, respectively, which isotope is more abundant?

F   DETERMINATION OF CHEMICAL FORMULAS

Many new drugs are discovered by studying the properties of compounds found in plants or other materials that have been used as medicines for centuries (Fig. F.1). Once chemists have extracted a biologically active compound from a natural product, they identify its molecular structure so that it can be manufactured. This section focuses on the first step in identifying the molecular structure, the determination of the empirical and molecular formulas of the compound.

> The **empirical formula** shows the relative numbers of atoms of each element present in the compound.

For example, the empirical formula of glucose, which is CH_2O, tells us that carbon, hydrogen, and oxygen atoms are present in the ratio 1:2:1. The elements are present in these proportions regardless of the size of the sample. After the empirical formula has been determined, the next step is to determine the molecular formula (Section C): a *molecular formula* tells us the *actual* numbers of atoms of each element in a molecule. The molecular formula for glucose, which is $C_6H_{12}O_6$, tells us that each glucose molecule consists of 6 carbon atoms, 12 hydrogen atoms, and 6 oxygen atoms (**1**, which shows one common configuration of the glucose molecule). In contrast, the empirical formula gives only the *ratios* of the numbers of atoms of each element. Therefore, different compounds with different molecular formulas can have the same empirical formula. For example, formaldehyde, CH_2O (**2**, the preservative in formalin solution), acetic acid, $C_2H_4O_2$ (the acid in vinegar), and lactic acid, $C_3H_6O_3$ (the acid in sour milk), all have the same empirical formula (CH_2O) as that of glucose.

F.1 Mass Percentage Composition

To determine the empirical formula of a compound, we begin by measuring the mass of each element present in a sample. The result is usually reported as the **mass**

F.1 Mass Percentage Composition

F.2 Determining Empirical Formulas

F.3 Determining Molecular Formulas

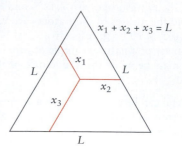

1   α-D-Glucose, $C_6H_{12}O_6$

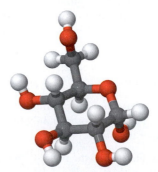

2   Formaldehyde, CH_2O

percentage composition—that is, the mass of each element expressed as a percentage of the total mass:

$$\text{Mass percentage of element} = \frac{\text{mass of element in the sample}}{\text{total mass of sample}} \times 100\% \quad (1)^*$$

Because the mass percentage composition is independent of the size of the sample—in the language of Section A, it is an intensive property—every sample of the substance has that same composition. A principal technique for determining the mass percentage composition of an unknown organic compound is combustion analysis. Chemists commonly send samples to a laboratory or agency for combustion analysis and receive the results as mass percentage composition (see Section M).

SELF-TEST F.1A For centuries, the Australian Aborigines have used the leaves of the eucalyptus tree to alleviate sore throats and other pains. The primary active ingredient has been identified and named eucalyptol. An analysis of a sample of eucalyptol of total mass 3.16 g gave its composition as 2.46 g of carbon, 0.373 g of hydrogen, and 0.329 g of oxygen. Determine the mass percentages of carbon, hydrogen, and oxygen in eucalyptol.

[*Answer:* 77.8% C, 11.8% H, 10.4% O]

SELF-TEST F.1B The compound α-pinene, a natural antiseptic found in the resin of the piñon tree, has been used since ancient times by Zuni healers. A 7.50-g sample of α-pinene contains 6.61 g of carbon and 0.89 g of hydrogen. What are the mass percentages of carbon and hydrogen in α-pinene?

If the chemical formula of a compound is already known, its mass percentage composition can be obtained from the formula.

FIGURE F.1 The research vessel *Alpha Helix* is used by chemists at the University of Illinois at Urbana-Champaign to search for marine organisms that contain compounds of medicinal value. Compounds found to have antifungal or antiviral properties are then subjected to the kinds of analyses described in this section.

EXAMPLE F.1 **Sample exercise: The mass percentage of an element in a compound**

Suppose we are generating hydrogen from water to use as a fuel and need to know how much hydrogen a given mass of water can provide. What is the mass percentage of hydrogen in water?

SOLUTION To calculate the mass percentage of hydrogen in water, we simply find the mass of H atoms in 1 mol H_2O molecules, noting that there are 2 mol H in 1 mol H_2O, divide that mass by the mass of 1 mol H_2O, and multiply by 100%:

From Mass percentage of H

$$= \frac{\text{mass of H atoms}}{\text{mass of } H_2O \text{ molecules}} \times 100\%,$$

Mass percentage of H

$$= \frac{(2 \text{ mol}) \times (1.0079 \text{ g·mol}^{-1})}{(1 \text{ mol}) \times (18.02 \text{ g·mol}^{-1})} \times 100\%$$

$$= 11.19\%$$

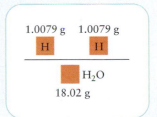

SELF-TEST F.2A Calculate the mass percentage of Cl in NaCl.

[*Answer:* 60.66%]

SELF-TEST F.2B Calculate the mass percentage of Ag in $AgNO_3$.

Mass percentage composition is found by calculating the fraction of the total mass contributed by each element present in a compound and expressing the fraction as a percentage.

F.2 Determining Empirical Formulas

To convert the mass percentage composition obtained from a combustion analysis into an empirical formula, we must convert the mass percentages of each type of atom into the relative numbers of atoms. The simplest procedure is to imagine that we have a sample of mass 100 g exactly. That way, the mass percentage composition tells us the mass in grams of each element. Then we use the molar mass of each element to convert these masses into amounts in moles and go on to find the relative numbers of moles of each type of atom.

EXAMPLE F.2   Sample exercise: Determining the elemental composition from mass percentage composition

Suppose that an analytical laboratory reported a composition of 40.9% carbon, 4.58% hydrogen, and 54.5% oxygen for a sample of vitamin C. In what atom ratios are the elements present in vitamin C?

SOLUTION We consider a sample of exactly 100 g, then convert the masses into amounts in moles by dividing the mass percentage for each element by the element's molar mass. The mass of carbon in a sample of vitamin C of mass 100 g is 40.9 g. Because the molar mass of carbon is 12.01 g·mol^{-1},

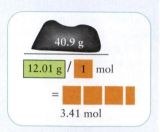

From $n = m/M$,

$$n_C = \frac{40.9\ g}{12.01\ g \cdot (mol\ C)^{-1}} = 3.41\ mol\ C$$

Likewise, from the mass percentages of hydrogen and oxygen, we find

$n_H = 4.54\ mol\ H$ and $n_O = 3.41\ mol\ O$

It follows that, in any sample of vitamin C, the atoms are present in the ratio

3.41 C : 4.54 H : 3.41 O

SELF-TEST F.3A Pyrophosphoric acid has the composition 2.27% H, 34.81% P, and 62.93% O. Find the relative amounts of H, P, and O atoms present in pyrophosphoric acid.

[*Answer:* 100 g contains 2.25 mol H, 1.12 mol P, and 3.93 mol O; 2.25 H : 1.12 P : 3.93 O]

SELF-TEST F.3B Use the information in Self-Test F.1A to find the relative amounts of C, H, and O atoms present in eucalyptol.

To find the empirical formula of vitamin C from the data in Example F.1 we must express the ratios of numbers of atoms as the simplest whole numbers. First, we divide each number by the smallest value (3.41), which gives a ratio of 1.00:1.33:1.00. Molecules contain only whole numbers of atoms, and one of these numbers is still not a whole number. Hence, we must multiply each number by the correct factor so that all numbers can be rounded off to whole numbers. Because 1.33 is ⁴⁄₃ (within experimental error), we multiply all three numbers by 3 to obtain 3.00:3.99:3.00, or approximately 3:4:3. Now we know that the empirical formula of vitamin C is $C_3H_4O_3$.

The calculated values are not exact, because they are derived from experimental data, which are subject to experimental error.

EXAMPLE F.2   Determining the empirical formula from mass percentage composition

The mass percentage composition of a compound that assists in the coagulation of blood is 76.71% C, 7.02% H, and 16.27% N. Determine the empirical formula of the compound.

STRATEGY Divide each mass percentage by the molar mass of the corresponding element to obtain the number of moles of that element found in exactly 100 g of the compound. Divide the number of moles of each element by the smallest number of moles. If fractional numbers result, then multiply by the factor that gives the smallest whole numbers of moles.

SOLUTION

The mass of each element, m_X, in exactly 100 g of the compound is equal to its mass percentage in grams.	$m_C = 76.71$ g $m_H = 7.02$ g $m_N = 16.27$ g	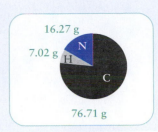
Convert each mass to an amount, n_X, in moles by using the molar mass, M_X, of the element, $n_X = m_X/M_X$.	$n_C = \dfrac{76.71 \text{ g}}{12.01 \text{ g} \cdot (\text{mol C})^{-1}} = 6.387$ mol C $n_H = \dfrac{7.02 \text{ g}}{1.0079 \text{ g} \cdot (\text{mol H})^{-1}} = 6.96$ mol H $n_N = \dfrac{16.27 \text{ g}}{14.01 \text{ g} \cdot (\text{mol N})^{-1}} = 1.161$ mol N	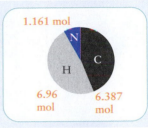
Divide each amount by the smallest amount (1.161 mol).	Carbon: $\dfrac{6.387 \text{ mol}}{1.161 \text{ mol}} = 5.501$ Hydrogen: $\dfrac{6.96 \text{ mol}}{1.161 \text{ mol}} = 5.99$ Nitrogen: $\dfrac{1.161 \text{ mol}}{1.161 \text{ mol}} = 1.000$	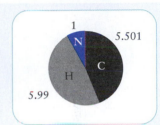

A note on good practice: Note that the number of significant figures in the data controls the number of significant figures in the calculated amounts.

Because 5.501 is approximately 1½, we multiply all the numbers by 2 to get the ratios 11.00:12.0:2.000. The empirical formula is therefore $C_{11}H_{12}N_2$.

SELF-TEST F.4A Use the molar composition of eucalyptol calculated in Self-Test F.3B to determine its empirical formula.

[*Answer:* $C_{10}H_{18}O$]

SELF-TEST F.4B The mass percentage composition of the compound thionyl difluoride is 18.59% O, 37.25% S, and 44.16% F. Calculate its empirical formula.

> *The empirical formula of a compound is determined from the mass percentage composition and the molar masses of the elements present.*

F.3 Determining Molecular Formulas

We have determined that the empirical formula of vitamin C is $C_3H_4O_3$. However, the empirical formula tells us only that the C, H, and O atoms are present in the sample in the ratio 3:4:3, not the number of each type of atom in a molecule. The molecular formula could be $C_3H_4O_3$, $C_6H_8O_6$, $C_9H_{12}O_9$, or any other whole-number multiple of the empirical formula.

To find the molecular formula of a compound, we need one more piece of information—its molar mass. Then all we have to do is to calculate how many empirical formula units are needed to account for the molar mass. One of the best ways of determining the molar mass of an organic compound is by mass spectrometry. We saw this technique applied to atoms in Section B. It can be applied to molecules, too; and, although there are important changes of detail, the technique is essentially the same.

EXAMPLE F.3   Sample exercise: Determining the molecular formula from the empirical formula

Mass spectrometry has been used to show that the molar mass of vitamin C is 176.12 $g \cdot mol^{-1}$. Given its empirical formula of $C_3H_4O_3$, what is the molecular formula of vitamin C?

SOLUTION The molar mass of a $C_3H_4O_3$ formula unit is

Molar mass of $C_3H_4O_3$

$$= 3 \times (12.01 \text{ g} \cdot \text{mol}^{-1}) + 4 \times (1.008 \text{ g} \cdot \text{mol}^{-1}) + 3 \times (16.00 \text{ g} \cdot \text{mol}^{-1})$$

$$= 88.06 \text{ g} \cdot \text{mol}^{-1}$$

To find the number of $C_3H_4O_3$ formula units needed to account for the observed molar mass of vitamin C, we divide the molar mass of the compound by the molar mass of the empirical formula unit:

$$\frac{\text{Molar mass of compound}}{\text{Molar mass of empirical formula unit}} = \frac{176.14 \text{ g} \cdot \text{mol}^{-1}}{88.06 \text{ g} \cdot \text{mol}^{-1}} = 2.000$$

We conclude that the molecular formula of vitamin C is $2 \times (C_3H_4O_3)$, or $C_6H_8O_6$.

SELF-TEST F.5A The molar mass of styrene, which is used in the manufacture of the plastic polystyrene, is 104 $g \cdot mol^{-1}$, and its empirical formula is CH. Deduce its molecular formula.

[*Answer:* C_8H_8]

SELF-TEST F.5B The molar mass of oxalic acid, an acid present in rhubarb, is 90.0 $g \cdot mol^{-1}$, and its empirical formula is CHO_2. What is its molecular formula?

The molecular formula of a compound is found by determining how many empirical formula units are needed to account for the measured molar mass of the compound.

SKILLS YOU SHOULD HAVE MASTERED

❏ 1 Calculate the mass percentage of an element in a compound from a formula (Example F.1).

❏ 2 Calculate the empirical formula of a compound from its mass percentage composition (Example F.2).

❏ 3 Determine the molecular formula of a compound from its empirical formula and its molar mass (Example F.3).

EXERCISES

F.1 Citral is a fragrant component of lemon oil that is used in colognes. It has the molecular structure shown. Calculate the mass percentage composition of citral (black = C, gray = H, red = O).

structure shown. Calculate the mass percentage composition of muscone (black = C, gray = H, red = O).

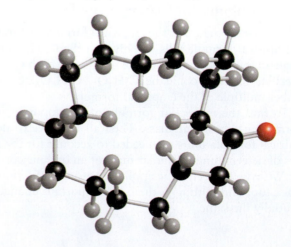

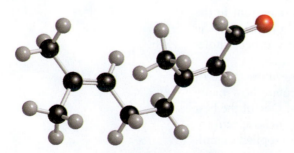

F.2 The compound mainly responsible for the odor of musk produced by musk deer is muscone, which has the molecular

F.3 What is the mass percentage composition of l-carnitine, $C_7H_{15}NO_3$, a compound that is taken as a dietary supplement to reduce muscle fatigue?

F.4 What is the mass percentage composition of aspartame, $C_{14}H_{18}N_2O_5$, an artificial sweetener sold as NutraSweet?

F.5 A metal M forms an oxide with the formula M_2O, for which the mass percentage of the metal is 88.8%. (a) What is the molar mass of the metal? (b) Write the name of the compound.

F.6 A metal M forms an oxide with the formula M_2O_3, for which the mass percentage of the metal is 69.9%. (a) What is the identity of the metal? (b) Write the name of the compound.

F.7 Determine the empirical formulas from the following analyses. (a) The mass percentage composition of cryolite, a compound used in the production of aluminum, is 32.79% Na, 13.02% Al, and 54.19% F. (b) A compound used to generate O_2 gas in the laboratory has mass percentage composition 31.91% K and 28.93% Cl, the remainder being oxygen. (c) A fertilizer is found to have the following mass percentage composition: 12.2% N, 5.26% H, 26.9% P, and 55.6% O.

F.8 Determine the empirical formula of each of the following compounds from the data given. (a) Talc (used in talcum powder) has mass composition 19.2% Mg, 29.6% Si, 42.2% O, and 9.0% H. (b) Saccharin, a sweetening agent, has mass composition 45.89% C, 2.75% H, 7.65% N, 26.20% O, and 17.50% S. (c) Salicylic acid, used in the synthesis of aspirin, has mass composition 60.87% C, 4.38% H, and 34.75% O.

F.9 In an experiment, 4.14 g of phosphorus combined with chlorine to produce 27.8 g of a white solid compound. (a) What is the empirical formula of the compound? (b) Assuming that the empirical and molecular formulas of the compound are the same, what is its name?

F.10 A chemist found that 4.69 g of sulfur combined with fluorine to produce 15.81 g of a gas. (a) What is the empirical formula of the gas? (b) Assuming that the empirical and molecular formulas of the compound are the same, what is its name?

F.11 Diazepam, a drug used to treat anxiety, is 67.49% C, 4.60% H, 12.45% Cl, 9.84% N, and 5.62% O. What is the empirical formula of the compound?

F.12 The compound fluoxetine is sold as Prozac when combined with HCl. Fluoxetine has the composition 66.01% C, 5.87% H, 18.43% F, 4.53% N, and 5.17% O. What is the empirical formula of fluoxetine?

F.13 Osmium forms a number of molecular compounds with carbon monoxide. One light-yellow compound was analyzed to give the following elemental composition: 15.89% C, 21.18% O, and 62.93% Os. (a) What is the empirical formula of this compound? (b) From the mass spectrum of the compound, the molecule was determined to have a molar mass of 907 g·mol^{-1}. What is its molecular formula?

F.14 Paclitaxel, which is extracted from the Pacific yew tree *Taxus brevifolia*, has antitumor activity for ovarian and breast cancer. It is sold under the trade name Taxol. On analysis, its mass percentage composition is 66.11% C, 6.02% H, and 1.64% N, with the balance being oxygen. What is the empirical formula of paclitaxel?

F.15 Caffeine, a stimulant in coffee and tea, has a molar mass of 194.19 g·mol^{-1} and a mass percentage composition of 49.48% C, 5.19% H, 28.85% N, and 16.48% O. What is the molecular formula of caffeine?

F.16 Cacodyl, which has an intolerable garlicky odor and is used in the manufacture of cacodylic acid, a cotton herbicide, has a mass percentage composition of 22.88% C, 5.76% H, and 71.36% As and a molar mass of 209.96 g·mol^{-1}. What is the molecular formula of cacodyl?

F.17 In determining the composition of an unknown compound, elemental analyses are often considered to have an error of ±0.5%. With the use of elemental analyses alone, would it be possible to determine whether a sample was glucose ($C_6H_{12}O_6$) or sucrose ($C_{12}H_{22}O_{11}$)?

F.18 In 1978, scientists extracted a compound with antitumor and antiviral properties from marine animals in the Caribbean Sea. A sample of the compound didemnin-A of mass 1.78 mg was analyzed and found to have the following composition: 1.11 mg C, 0.148 mg H, 0.159 mg N, and 0.363 mg O. The molar mass of didemnin-A was found to be 942 g·mol^{-1}. What is the molecular formula of didemnin-A?

F.19 The CO_2 produced by the combustion of hydrocarbons contributes to global warming. Rank the following fuels according to increasing mass percentage of carbon: (a) ethene, C_2H_4; (b) propanol, C_3H_7OH; (c) heptane, C_7H_{16}.

F.20 Dolomite is a mixed carbonate of calcium and magnesium. Calcium and magnesium carbonates both decompose on heating to produce the metal oxides (MgO and CaO) and carbon dioxide (CO_2). If 4.84 g of residue consisting of MgO and CaO remains when 9.66 g of dolomite is heated until decomposition is complete, what percentage by mass of the original sample was $MgCO_3$?

F.21 In the following ball-and-stick molecular structures, black indicates carbon, gray hydrogen, blue nitrogen, and green chlorine. Write the empirical and molecular formulas of each structure. *Hint:* It may be easier to write the molecular formula first.

(a)

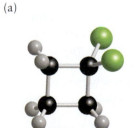

(b)

F.22 Write the empirical and molecular formulas of each ball-and-stick structure. Black indicates carbon, gray hydrogen, red oxygen, and green chlorine. *Hint:* It may be easier to write the molecular formula first.

(a)

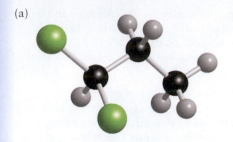

(b)

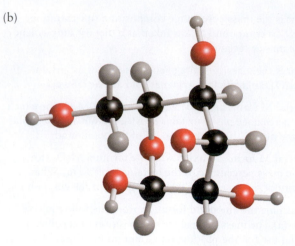

F.23 A mixture of $NaNO_3$ and Na_2SO_4 of mass 5.37 g contains 1.61 g of sodium. What is the percentage by mass of $NaNO_3$ in the mixture?

G   MIXTURES AND SOLUTIONS

So far, we have discussed only pure substances. However, most materials are neither pure elements nor pure compounds, and so they are not "substances" in the technical sense of the term (Section A); they are **mixtures** of these simpler substances, with one substance mingled with another. For example, air, blood, and seawater are mixtures. A medicine, such as a cough syrup, is often a mixture of various ingredients that has been formulated to achieve an overall biological effect. Much the same can be said of a perfume.

Chemists need to be able to specify the composition of mixtures *quantitatively*. For example, a chemist may need to monitor a pollutant, administer a dosage, or transfer a known amount of a solute. In this section we examine the properties and types of mixtures as well as how to use the "molar concentration" of a dissolved substance to analyze solutions quantitatively.

G.1 Classifying Mixtures

A compound has a fixed composition, whereas the composition of a mixture may be varied. There are always two H atoms for each O atom in a sample of the compound water, but sugar and sand, for instance, can be mixed in any proportions. Because the components of a mixture are merely mingled with one another, they retain their own chemical properties in the mixture. In contrast, a compound has chemical properties that differ from those of its component elements. The formation of a mixture is a *physical* change, whereas the formation of a compound requires a *chemical* change. The differences between mixtures and compounds are summarized in Table G.1.

Some mixtures have component particles that are so large we can see them with an optical microscope or even the unaided eye (Fig. G.1). Such a patchwork

FIGURE G.1 This piece of granite is a heterogeneous mixture of several substances.

TABLE G.1   Differences Between Mixtures and Compounds

Mixture	Compound
Components can be separated by using physical techniques	Components cannot be separated by using physical techniques
Composition is variable	Composition is fixed
Properties are related to those of its components	Properties are unlike those of its components

of different substances is called a **heterogeneous mixture**. Many of the rocks that form the landscape are heterogeneous mixtures of crystals of different minerals. Milk, which looks like a pure substance, is in fact a heterogeneous mixture: through a microscope, we can see the individual globules of butterfat floating in a liquid that is largely water. Human bodies are heterogeneous mixtures of thousands of compounds.

In some mixtures, the molecules or ions of the components are so well mixed that the composition is the same throughout, no matter how small the sample. Such a mixture is called a **homogeneous mixture** (Fig. G.2). For example, syrup is a homogeneous mixture of sugar and water. The molecules of sugar are dispersed and mixed so thoroughly with the water that we cannot see separate regions or particles. Even when using a microscope, we cannot distinguish between a pure substance and a homogeneous mixture.

A homogeneous mixture is also called a **solution**. Many of the materials around us are solutions. Root beer is a solution consisting mostly of water, along with sugar, plant extracts, and various additives. Filtered seawater is a solution of common table salt (sodium chloride) and many other substances in water. When we use the everyday term *dissolving*, we mean the process of producing a solution. Usually the component of the solution present in the larger amount (water in these examples) is called the **solvent**, and any dissolved substance is a **solute** (Fig. G.3). However, if we want to emphasize the role of one of the substances as the one that the others are "dissolved in," we may identify that substance as the solvent. For example, in a heavy syrup, the sugar may be present in a much greater amount than the water, but water is considered the solvent. Normally, the solvent determines the physical state of the solution (whether it is a solid, a liquid, or a gas).

Crystallization takes place when the solute slowly comes out of solution as crystals, perhaps as the solvent evaporates. For example, salt crystals that form when water evaporates line the shores of the Great Salt Lake in Utah. In **precipitation**, a solute comes out of solution so rapidly that a single crystal does not have time to form. Instead, the solute forms a finely divided powder called a precipitate. Precipitation is often almost instantaneous (Fig. G.4).

Beverages and seawater are examples of **aqueous solutions**, solutions in which the solvent is water. Aqueous solutions are very common in everyday life and in chemical laboratories; for that reason, most of the solutions mentioned in this text are aqueous. **Nonaqueous solutions** are solutions in which the solvent is not water. Although they are less common than aqueous solutions, nonaqueous solutions have important uses. In "dry cleaning," the grease and dirt on fabrics are dissolved in a nonaqueous liquid solvent such as tetrachloroethene, C_2Cl_4. There are also **solid solutions**, in which the solvent is a solid. An example is a common form of brass that can be regarded as a solution of copper in zinc. Although gaseous

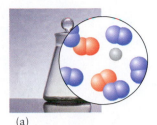

(a)

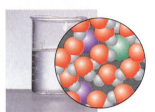

(b)

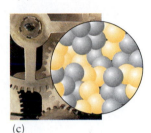

(c)

FIGURE G.2 Three examples of homogeneous mixtures. (a) Air is a homogeneous mixture of many gases, including the nitrogen, oxygen, and argon depicted here. (b) Table salt dissolved in water consists of sodium ions and chloride ions among water molecules. (c) Many alloys are solid homogeneous mixtures of two or more metals. The insets show that the mixing is uniform at the molecular level.

Lab Video: Figure G.4 Formation of lead iodide

FIGURE G.4 Precipitation occurs when an insoluble substance is formed. Here lead(II) iodide, PbI_2, which is an insoluble yellow solid, precipitates when we mix solutions of lead(II) nitrate, $Pb(NO_3)_2$, and potassium iodide, KI.

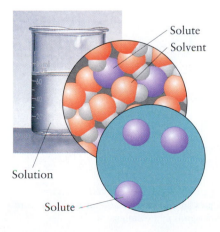

Solute
Solvent

Solution

Solute

FIGURE G.3 Solutions are homogeneous mixtures in which one substance, the solvent (here water), is usually in large excess. A dissolved substance is called a solute.

solutions exist, they are normally referred to as gaseous mixtures. The atmosphere could be regarded as a vast gaseous solution of various substances in nitrogen, its major component, but it is normally termed a gaseous mixture.

Mixtures display the properties of their constituents; they differ from compounds, as summarized in Table G.1. Mixtures are classified as homogeneous or heterogeneous, and solutions are homogeneous mixtures.

G.2 Separation Techniques

Natural products, such as enzymes and vitamins, are almost invariably extracted from mixtures. To analyze the composition of any sample that we suspect is a mixture, we first separate its components by physical means and then identify each individual substance present (Fig. G.5). Common physical separation techniques include decanting, filtration, chromatography, and distillation.

Decanting makes use of differences in density. One liquid floats on another liquid or lies above a solid and is poured off. **Filtration** is used to separate substances when there is a difference in solubility (the ability to dissolve in a particular solvent). The sample is stirred with a liquid and then poured through a fine mesh, the filter. Components of the mixture that dissolve in the liquid pass through the filter, whereas solid components that do not dissolve are captured by the filter. The technique can be used to separate sugar from sand because sugar is soluble in water but sand is not. Filtration is a common first stage in the treatment of domestic water supplies. A related technique, one of the most sensitive techniques available for separating a mixture, is **chromatography**, which relies on the different abilities of substances to **adsorb**, or stick, to surfaces (Fig. G.6). Chromatography is discussed in detail in Major Technique 5, following Chapter 8.

Distillation makes use of differences in boiling points to separate mixtures. When a solution is distilled, the components of the mixture boil away at different temperatures and condense in a cooled tube called a *condenser* (Fig. G.7). Distillation can be used to remove water (which boils at 100°C) from table salt (sodium chloride), which does not even melt until 801°C; the solid salt is left behind when the water evaporates.

Mixtures are separated by making use of the differences in physical properties of the components; common techniques based on physical differences include decanting, filtration, chromatography, and distillation.

The first known distillation apparatus, assembled in the first century CE, is attributed to Mary the Jewess, an alchemist who lived in the Middle East.

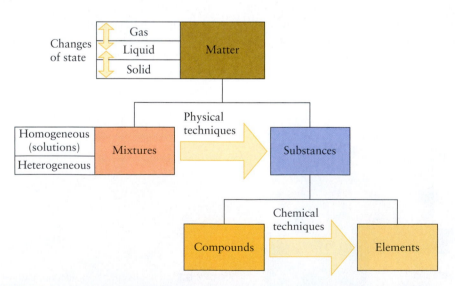

FIGURE G.5 The hierarchy of materials: matter consists of either mixtures or substances; substances consist of either compounds or elements. Physical techniques are used to separate mixtures into pure substances. Chemical techniques are used to separate compounds into elements.

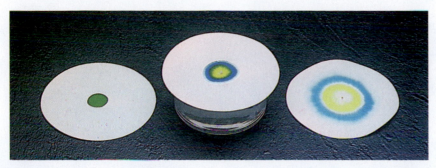

FIGURE G.6 In paper chromatography, the components of a mixture are separated by washing them along a paper—the support—with a solvent. A primitive form of the technique is shown here. On the left is a dry filter paper to which a drop of green food coloring was applied. Solvent was then poured on to the center of the filter paper. The blue and yellow dyes that were combined to make the green color begin to separate. The filter paper on the right was allowed to dry after the solvent had spread out to the edges of the paper, carrying the two dyes to different distances as it spread. The dried support showing the separated components of a mixture is called a chromatogram.

G.3 Molarity

The **molar concentration**, c, of a solute in a solution, which is widely called the "molarity" of the solute, is the amount of solute molecules or formula units (in moles) divided by the volume of the solution (in liters):

The formal name for molarity is the "amount of substance concentration."

$$\text{Molarity} = \frac{\text{amount of solute}}{\text{volume of solution}} \quad \text{or} \quad c = \frac{\overset{\text{mol}}{n}}{\underset{\text{L}}{V}} \qquad (1)^*$$

The units of molarity are moles per liter (mol·L^{-1}), often denoted M:

$$1 \text{ M} = 1 \text{ mol·L}^{-1}$$

The symbol M is often read "molar"; it is not an SI unit. Note that 1 mol·L^{-1} is the same as 1 mmol·mL^{-1}. Chemists working with very low concentrations of solutes also report molar concentrations as millimoles per liter (mmol·L^{-1}) and micromoles per liter ($\mu\text{mol·L}^{-1}$).

> **EXAMPLE G.1** **Sample exercise: Calculating the molarity of a solute**

Suppose we dissolved 10.0 g of cane sugar in enough water to make 200. mL of solution, which we might do (with less precision) if we were making a glass of lemonade. Cane sugar is sucrose ($C_{12}H_{22}O_{11}$), which has a molar mass of 342 g·mol^{-1}. What is the molarity of sucrose molecules in the solution?

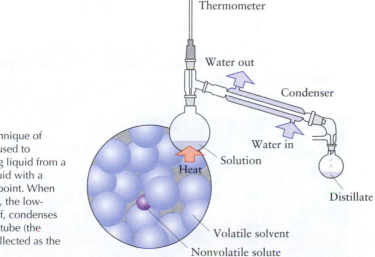

FIGURE G.7 The technique of distillation, which is used to separate a low-boiling liquid from a dissolved solid or liquid with a much higher boiling point. When the solution is heated, the low-boiling liquid boils off, condenses in the water-jacketed tube (the condenser), and is collected as the distillate.

SOLUTION It follows from Eq. 1 that the molarity of the solute in the solution is

From $c = n/V$,

$$c = \frac{\overbrace{(10.0\ \text{g})/(342\ \text{g·mol}^{-1})}^{n=m/M}}{\underbrace{0.200\ \text{L}}_{V}}$$

$$= \frac{10.0\ \text{g}}{(342\ \text{g·mol}^{-1}) \times (0.200\ \text{L})} = 0.146\ \text{mol·L}^{-1}$$

We report this concentration as 0.146 M $C_{12}H_{22}O_{11}$(aq). The (aq) indicates an aqueous solution. If instead of 10.0 g, we were to dissolve 20.0 g of cane sugar in the same volume of solution, the sugar would be twice as concentrated: its molarity would be 0.292 M $C_{12}H_{22}O_{11}$(aq).

A note on good practice: Note that we refer to the molarity of the solute, not the molarity of the solution.

SELF-TEST G.1A What is the molarity of sodium chloride in a solution prepared by dissolving 12.0 g of sodium chloride in enough water to make 250. mL of solution?

[Answer: 0.821 M NaCl(aq)]

SELF-TEST G.1B What is the molarity of sodium sulfate in a solution prepared by dissolving 15.5 g of sodium sulfate in enough water to make 350. mL of solution?

Because molarity is defined in terms of the *volume of the solution,* not the volume of solvent used to prepare the solution, the volume must be measured after the solutes have been added. The usual way to prepare an aqueous solution of a solid substance of given molarity is to transfer a known mass of the solid into a volumetric flask, dissolve it in a little water, fill the flask up to the mark with water, and then mix the solution thoroughly by tipping the flask end over end (Fig. G.8).

To use the molarity of a solution to calculate the amount of solute in a given volume of solution, we use Eq. 1 rearranged into

$$\overset{\text{mol}}{n} = \overset{\text{mol·L}^{-1}}{c} \times \overset{\text{L}}{V} \tag{2}$$

where c is the molarity, V is the volume, and n is the amount. Because we can convert moles into mass by using the molar mass of the solute, we can also use this formula to calculate the mass of solute present in a given volume of solution of known concentration and hence estimate the mass of solute needed to make up that solution.

EXAMPLE G.2 **Sample exercise: Determining the mass of solute required for a given concentration**

Suppose we were asked to prepare 250. mL of a solution that was approximately 0.0380 M $CuSO_4$(aq) from solid copper(II) sulfate pentahydrate, $CuSO_4 \cdot 5H_2O$. What mass of the solid do we need?

FIGURE G.8 The steps in making up a solution of known molarity. A known mass of the solute is dispensed into a volumetric flask (left). Some water is added to dissolve it (center). Finally, water is added up to the mark on the stem of the flask (right). The bottom of the solution's meniscus (the curved top surface of the liquid) should be level with the mark.

SOLUTION First we need to know the amount of $CuSO_4$ in 250. mL (0.250 L) of solution:

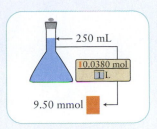

From $n = cV$,

$$n_{CuSO_4} = (0.0380\ mol \cdot L^{-1}) \times (0.250\ L)$$
$$= 0.0380 \times 0.250\ mol = 0.00950\ mol$$

Because 1 mol $CuSO_4 \cdot 5H_2O$ contains 1 mol $CuSO_4$, we need to supply

$$n_{CuSO_4 \cdot 5H_2O} = n_{CuSO_4} = 0.00950\ mol$$

Then, because the molar mass of copper(II) sulfate pentahydrate is 249.6 $g \cdot mol^{-1}$, this amount of the pentahydrate corresponds to the following mass, $m_{CuSO_4 \cdot 5H_2O}$:

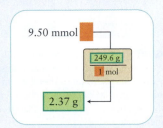

From $m = nM$,

$$m_{CuSO_4 \cdot 5H_2O} = (0.00950\ mol) \times (249.6\ g \cdot mol^{-1})$$
$$= 2.37\ g$$

We conclude that we should measure out about 2.37 g of copper(II) sulfate pentahydrate. In practice, we might find that we had spooned out 2.403 g, in which case the molarity would be 0.0385 M $CuSO_4$(aq).

SELF-TEST G.2A Calculate the mass of glucose needed to prepare 150. mL of 0.442 M $C_6H_{12}O_6$(aq).

[*Answer:* 11.9 g]

SELF-TEST G.2B Calculate the mass of oxalic acid needed to prepare 50.00 mL of 0.125 M $C_2H_2O_4$(aq).

The molarity is also used to calculate the volume of solution, V, that contains a given amount of solute. For this type of calculation, we rearrange Eq. 1 into

$$V = \frac{\overset{mol}{n}}{\underset{mol \cdot L^{-1}}{c}} \quad \overset{L}{} \tag{3}$$

and then substitute the data.

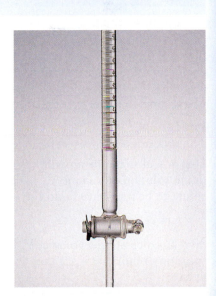

FIGURE G.9 A buret is calibrated so that the volume of liquid delivered can be measured.

EXAMPLE G.3 Sample exercise: Calculating the volume of solution that contains a given amount of solute

Suppose we want to measure out 0.760 mmol CH_3COOH, acetic acid, an acid found in vinegar and often used in the laboratory, and we have available 0.0380 M CH_3COOH(aq). What volume of solution should we use?

SOLUTION To find the volume, V, of solution, we substitute the data into Eq. 3:

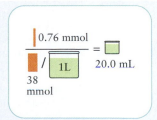

From $V = n/c$,

$$V = \frac{0.760 \times 10^{-3}\ mol}{0.0380\ mol \cdot L^{-1}} = 0.0200\ L$$

We should therefore transfer 20.0 mL of the acetic acid solution into a flask by using a buret or a pipet (Fig. G.9). The flask will then contain 0.760 mmol CH_3COOH.

SELF-TEST G.3A What volume of 1.25×10^{-3} M $C_6H_{12}O_6$(aq) contains 1.44×10^{-6} mol of glucose molecules?

[*Answer:* 1.15 mL]

SELF-TEST G.3B What volume of 0.358 M HCl(aq) contains 2.55 mmol HCl?

The molarity (molar concentration) of a solute in a solution is the amount of solute in moles divided by the volume of the solution in liters.

G.4 Dilution

Before dilution   After dilution

FIGURE G.10 When a solution is diluted, the same number of solute molecules occupy a larger volume. Therefore, the same volume (as depicted by the square) will contain fewer molecules than in the concentrated solution.

A common space-saving practice in chemistry is to store a solution in a concentrated form called a **stock solution** and then to **dilute** it, or reduce its concentration, to whatever concentration is needed. Chemists use solutions and techniques such as dilution when they need very precise control over the amounts of the substances that they are handling, especially very small quantities of substances. For example, pipetting 25.0 mL of 1.50×10^{-3} M NaOH(aq) corresponds to transferring only 37.5 μmol NaOH or 1.50 mg of the compound. A mass this small would be difficult to weigh out accurately.

To dilute a stock solution to a desired concentration, we first use a pipet to transfer the appropriate volume of stock solution to a **volumetric flask**, a flask calibrated to contain a specified volume. Then we add enough solvent to increase the volume of the solution to its final value. Toolbox G.1 shows how to calculate the correct initial volume of stock solution.

TOOLBOX G.1   **HOW TO CALCULATE THE VOLUME OF STOCK SOLUTION REQUIRED FOR A GIVEN DILUTION**

CONCEPTUAL BASIS

This procedure is based on a simple idea: although we may add more solvent to a given volume of solution, we change only the concentration, not the amount of solute (Fig. G.10). After dilution, the same amount of solute simply occupies a larger volume of solution.

PROCEDURE

The procedure has two steps:

Step 1 Calculate the amount of solute, n, in the final, dilute solution of volume V_{final}. (This is the amount of solute to be transferred into the volumetric flask.)

$$n = c_{final} V_{final}$$

Step 2 Calculate the volume, $V_{initial}$, of the initial stock solution of molarity $c_{initial}$ that contains this amount of solute.

$$V_{initial} = \frac{n}{c_{initial}}$$

This procedure is illustrated in Example G.4.

Because the amount of solute, n, is the same in these two expressions, they can be combined into

$$V_{initial} = \frac{c_{final} V_{final}}{c_{initial}} \tag{4a}*$$

and rearranged into an easily remembered form.

$$c_{initial} V_{initial} = c_{final} V_{final} \tag{4b}*$$

In Eq. 4b, the amount of solute in the final solution (the product on the right) is the same as the amount of solute in the initial volume of solution (the product on the left), $n_{final} = n_{initial}$.

EXAMPLE G.4   **Sample Exercise: Calculating the volume of stock solution to dilute**

We need to prepare 250. mL of 1.25×10^{-3} M NaOH(aq) and will use a 0.0380 M NaOH(aq) stock solution. How much stock solution do we need?

SOLUTION We proceed as in Toolbox G.1.

Step 1 From $n = c_{final} V_{final}$,

$n = (1.25 \times 10^{-3} \text{ mol·L}^{-1}) \times (0.250 \text{ L})$

$= (1.25 \times 10^{-3} \times 0.250) \text{ mol}$

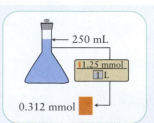

250 mL

1.25 mmol
1 L

0.312 mmol

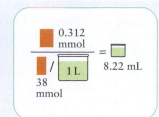

Step 2 From $V_{initial} = n/c_{initial}$,

$$V_{initial} = \frac{(1.25 \times 10^{-3} \times 0.250)\ \text{mol}}{0.0380\ \text{mol·L}^{-1}}$$
$$= 8.22 \times 10^{-3}\ \text{L}$$

We can conclude that 8.22 mL of the stock solution should be measured into a 250.-mL volumetric flask (by using a buret) and water should be added up to the mark (Fig. G.11).

A note on good practice: Note how, to minimize rounding errors, we carry out the calculation in a single step. However, to help guide you through the calculation, we often give intermediate numerical results in the examples.

SELF-TEST G.4A Calculate the volume of 0.0155 M HCl(aq) that we should use to prepare 100. mL of 5.23×10^{-4} M HCl(aq).

[*Answer:* 3.37 mL]

SELF-TEST G.4B Calculate the volume of 0.152 M $C_6H_{12}O_6$(aq) that should be used to prepare 25.00 mL of 1.59×10^{-5} M $C_6H_{12}O_6$(aq).

When a solution is diluted to a larger volume, the total number of moles of solute in the solution does not change, but the concentration of solute is reduced.

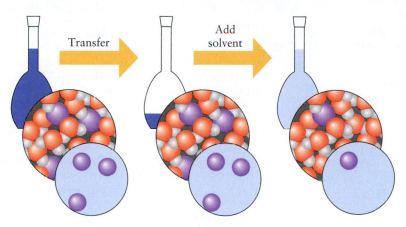

FIGURE G.11 The steps in dilution. A small sample of the original solution is transferred to a volumetric flask, and then solvent is added up to the mark.

SKILLS YOU SHOULD HAVE MASTERED

❏ 1 Distinguish between heterogeneous and homogeneous mixtures and describe methods of separation (Sections G.1 and G.2).

❏ 2 Calculate the molarity of a solute in a solution, volume of solution, and mass of solute, given the other two quantities (Examples G.1–G.3).

❏ 3 Determine the volume of stock solution needed to prepare a dilute solution of a given molarity (Toolbox G.1 and Example G.4).

EXERCISES

G.1 What physical properties are used for separation of the components of a mixture by (a) decanting; (b) chromatography; (c) distillation?

G.2 Water purification, in which impure water is made drinkable, can be achieved by several methods. On the basis of the information in this section, suggest a physical process for converting seawater into pure water.

G.3 Identify the following mixtures as homogeneous or heterogeneous, and suggest a technique for separating their components: (a) oil and vinegar; (b) chalk and table salt; (c) salt water.

G.4 Identify the following mixtures as homogeneous or heterogeneous, and suggest a technique for separating their components: (a) gasoline and sea water; (b) root beer; (c) charcoal and sugar.

G.5 A chemist studying the properties of photographic emulsions needed to prepare 500.0 mL of 0.179 M $AgNO_3$(aq). What mass of silver nitrate must be placed into a 500.0-mL volumetric flask, dissolved, and diluted to the mark with water?

G.6 (a) A chemist prepares a solution by dissolving 4.690 g of $NaNO_3$ in enough water to make 500.0 mL of solution. What molar concentration of sodium nitrate should appear on the label? (b) If the chemist mistakenly uses a 250.0-mL volumetric flask instead of the 500.0-mL flask in part (a), what molar concentration of sodium nitrate will the chemist actually prepare?

G.7 A student prepared a solution of sodium carbonate by adding 2.111 g of the solid to a 250.0-mL volumetric flask and adding water to the mark. Some of this solution was transferred to a buret. What volume of solution should the student transfer into a flask to obtain (a) 2.15 mmol Na^+; (b) 4.98 mmol CO_3^{2-}; (c) 50.0 mg Na_2CO_3?

G.8 A student investigating the properties of solutions containing carbonate ions prepared a solution containing 8.124 g of Na_2CO_3 in a flask of volume 250.0 mL. Some of the solution was transferred to a buret. What volume of solution should be dispensed from the buret to provide (a) 5.124 mmol Na_2CO_3; (b) 8.726 mmol Na^+?

G.9 Explain how you would prepare 0.010 M $KMnO_4$(aq) by starting with (a) solid $KMnO_4$; (b) 0.050 M $KMnO_4$(aq).

G.10 (a) A sample of 1.345 M K_2SO_4(aq) of volume 12.56 mL is diluted to 250.0 mL. What is the molar concentration of K_2SO_4 in the diluted solution? (b) A sample of 0.366 M HCl(aq) of volume 25.00 mL is drawn from a reagent bottle with a pipet. The sample is transferred to a flask of volume 125.00 mL and diluted to the mark with water. What is the molar concentration of the dilute hydrochloric acid solution?

G.11 (a) What volume of 0.778 M Na_2CO_3(aq) should be diluted to 150.0 mL with water to reduce its concentration to 0.0234 M Na_2CO_3(aq)? (b) An experiment requires the use of 60.0 mL of 0.50 M NaOH(aq). The stockroom assistant can only find a reagent bottle of 2.5 M NaOH(aq). How can the 0.50 M NaOH(aq) be prepared?

G.12 A chemist dissolves 0.076 g of $CuSO_4 \cdot 5H_2O$ in water and dilutes the solution to the mark in a 500.0-mL volumetric flask. A 2.00-mL sample of this solution is then transferred to a second 500.0-mL volumetric flask and diluted. (a) What is the molarity of $CuSO_4$ in the final solution? (b) To prepare the final 500.0-mL solution directly, what mass of $CuSO_4 \cdot 5H_2O$ would need to be weighed out?

G.13 (a) Determine the mass of anhydrous copper(II) sulfate that must be used to prepare 250 mL of 0.20 M $CuSO_4$(aq). (b) Determine the mass of $CuSO_4 \cdot 5H_2O$ that must be used to prepare 250 mL of 0.20 M $CuSO_4$(aq).

G.14 The ammonia solution that is purchased for a stockroom has a molarity of 15.0 mol·L^{-1}. (a) Determine the volume of 15.0 M NH_3(aq) that must be diluted to 500. mL to prepare 1.25 M NH_3(aq). (b) An experiment requires 0.32 M NH_3(aq).

The stockroom manager estimates that 12.0 L of the base is needed. What volume of 15.0 M NH_3(aq) must be used for the preparation?

G.15 A solution is prepared by dissolving 0.500 g of KCl, 0.500 g of K_2S, and 0.500 g of K_3PO_4 in 500. mL of water. What is the concentration in the final solution of (a) potassium ions; (b) sulfide ions?

G.16 To prepare a very dilute solution, it is advisable to perform successive dilutions of a single prepared reagent solution, rather than to weigh out a very small mass or to measure a very small volume of stock chemical. A solution was prepared by transferring 0.661 g of $K_2Cr_2O_7$ to a 250.0-mL volumetric flask and adding water to the mark. A sample of this solution of volume 1.000 mL was transferred to a 500.0-mL volumetric flask and diluted to the mark with water. Then 10.0 mL of the diluted solution was transferred to a 250.0-mL flask and diluted to the mark with water. (a) What is the final concentration of $K_2Cr_2O_7$ in solution? (b) What mass of $K_2Cr_2O_7$ is in this final solution? (The answer to the last question gives the amount that would have had to have been weighed out if the solution had been prepared directly.)

G.17 What mass (in grams) of anhydrous solute is needed to prepare each of the following solutions? (a) 1.00 L of 0.125 M K_2SO_4(aq); (b) 375 mL of 0.015 M NaF(aq); (c) 500. mL of 0.35 M $C_{12}H_{22}O_{11}$(aq).

G.18 Describe the preparation of each of the following solutions, starting with the anhydrous solute and water and using the indicated size of volumetric flask: (a) 75.0 mL of 5.0 M NaOH(aq); (b) 5.0 L of 0.21 M $BaCl_2$(aq); (c) 300. mL of 0.0340 M $AgNO_3$(aq).

G.19 In medicine it is sometimes necessary to prepare solutions with a specific concentration of a given ion. A lab technician has made up a 100.0-mL solution containing 0.50 g of NaCl and 0.30 g of KCl, as well as glucose and other sugars. What is the concentration of chloride ions in the solution?

G.20 When a 1.150-g sample of iron ore is treated with 50.0 mL of hydrochloric acid the iron dissolves in the acid to form a solution of $FeCl_3$. The $FeCl_3$ solution is diluted to 100.0 mL and the concentration of Fe^{3+} ions is determined by spectrometry to be 0.095 mol·L^{-1}. What is the mass percentage of iron in the ore?

G.21 Practioners of the branch of alternative medicine known as homeopathy claim that very dilute solutions of substances can have an effect. Is the claim plausible? To explore this question, suppose that you prepare a solution of a supposedly active substance, X, with a molarity of 0.10 mol·L^{-1}. Then you dilute 10. mL of that solution by doubling the volume, doubling it again, and so on, for 90 doublings in all. How many molecules of X will be present in 10. mL of the final solution? Comment on the possible health benefits of the solution.

G.22 Refer to Exercise G.21. How many successive tenfold dilutions of the original solution will result in one molecule of X being left in 10. mL of solution?

G.23 Concentrated hydrochloric acid is 37.50% HCl by mass and has a density of 1.205 g·cm^{-3}. What volume (in milliliters) of concentrated hydrochloric acid must be used to prepare 10.0 L of 0.7436 M HCl(aq)?

G.24 The concentration of toxic chemicals in the environment is often measured in parts per million (ppm) or even parts per billion (ppb). A solution in which the concentration of the solute is 3 ppb by mass has 3 g of the solute for every billion grams (1000 t) of the solution. The World Health Organization has set the acceptable standard for lead in drinking water at 10 ppb. In 1992 the tap water in one-third of the homes in Chicago was found to have a lead concentration of about 10 ppb. If you lived in one of these homes and drank 2 L of tap water at home each day, what is the total mass of lead you would have ingested per year?

H   CHEMICAL EQUATIONS

The growth of a child, the production of polymers from petroleum, and the digestion of food are all the outcome of **chemical reactions,** processes by which one or more substances are converted into other substances. This type of process is a **chemical change.** The starting materials are called the **reactants** and the substances formed are called the **products.** The chemicals available in a laboratory are called **reagents.** In this section, we see how to use the symbolic language of chemistry to describe chemical reactions.

H.1 Symbolizing Chemical Reactions

A chemical reaction is symbolized by an arrow:

Reactants $\longrightarrow$ products

For example, sodium is a soft, shiny metal that reacts vigorously with water. When we drop a small lump of sodium metal into a container of water, hydrogen gas forms rapidly and sodium hydroxide is produced in solution (Fig. H.1). We could describe this reaction in words:

Sodium + water $\longrightarrow$ sodium hydroxide + hydrogen

We can summarize this statement by using chemical formulas:

$Na + H_2O \longrightarrow NaOH + H_2$

This expression is called a **skeletal equation** because it shows the bare bones of the reaction (the identities of the reactants and products) in terms of chemical formulas. A skeletal equation is a *qualitative* summary of a chemical reaction.

To summarize reactions *quantitatively*, we note that atoms are neither created nor destroyed in a chemical reaction: they simply change their partners. The principal evidence for this conclusion is that there is no overall change in mass when a reaction takes place in a sealed container. The observation that the total mass is constant during a chemical reaction is called the **law of conservation of mass.**

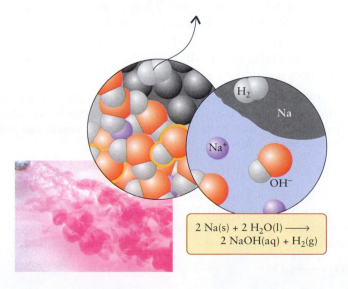

$$2\ Na(s) + 2\ H_2O(l) \longrightarrow$$
$$2\ NaOH(aq) + H_2(g)$$

FIGURE H.1 When a small piece of sodium is dropped into water, a vigorous reaction takes place. Hydrogen gas and sodium hydroxide are formed, and the heat released is enough to melt the sodium into a sphere. The pink color is due to the presence of a dye that changes color in the presence of sodium hydroxide. The chemical equation shows that two sodium atoms give rise to two sodium ions and that two water molecules give rise to one hydrogen molecule (which escapes as a gas) and two hydroxide ions. There is a rearrangement of partners, not a creation or annihilation of atoms. The unreacted water molecules are not shown in the inset on the right.

Because atoms are neither created nor destroyed, chemists regard each elemental symbol as representing one atom of the element (with the subscripts giving the number of each type of atom in a formula) and then multiply formulas by factors to show the same numbers of atoms of each element on both sides of the arrow. The resulting expression is said to be **balanced** and is called a **chemical equation**. For example, there are two H atoms on the left of the preceding skeletal equation but three H atoms on the right. So, we rewrite the expression as

$$2\,Na + 2\,H_2O \longrightarrow 2\,NaOH + H_2$$

Now there are four H atoms, two Na atoms, and two O atoms on each side, and the equation conforms to the law of conservation of mass. The number multiplying an *entire* chemical formula in a chemical equation (for example, the 2 multiplying H_2O) is called the **stoichiometric coefficient** of the substance. A coefficient of 1 (as for H_2) is not written explicitly.

A chemical equation typically also shows the physical state of each reactant and product by using a **state symbol**:

> (s): solid   (l): liquid   (g): gas   (aq): aqueous solution

For the reaction between solid sodium and water, the complete, balanced chemical equation is therefore

$$2\,Na(s) + 2\,H_2O(l) \longrightarrow 2\,NaOH(aq) + H_2(g)$$

When we want to show that a reaction requires a high temperature, we write the Greek letter Δ (delta) over the arrow. For example, the conversion of limestone into quicklime takes place at about 800°C, and we write

$$CaCO_3(s) \xrightarrow{\Delta} CaO(s) + CO_2(g)$$

Sometimes a **catalyst**, a substance that increases the rate of a reaction but is not itself consumed in the reaction, is added. For example, vanadium pentoxide, V_2O_5, is a catalyst in one step of the industrial process for the production of sulfuric acid. The presence of a catalyst is indicated by writing the formula of the catalyst above the reaction arrow:

$$2\,SO_2(g) + O_2(g) \xrightarrow{V_2O_5} 2\,SO_3(g)$$

Now we come to an important interpretation of a chemical equation. First, we note that the equation for the reaction of sodium with water tells us that

- When any 2 *atoms* of sodium react with 2 *molecules* of water, they produce 2 *formula units* of NaOH and one *molecule* of hydrogen.

We can multiply through by Avogadro's number (Section E) and conclude that

- When 2 *moles* of Na atoms react with 2 *moles* of H_2O molecules, they produce 2 *moles* of NaOH formula units and 1 *mole* of H_2 molecules.

In other words, the stoichiometric coefficients multiplying the chemical formulas in any balanced chemical equation tell us the relative number of moles of each substance that reacts or is produced in the reaction.

> *A balanced chemical equation symbolizes both the qualitative and the quantitative changes that take place in a chemical reaction. The stoichiometric coefficients tell us the relative numbers of moles of reactants and products taking part in the reaction.*

H.2 Balancing Chemical Equations

In some cases, the stoichiometric coefficients needed to balance an equation are easy to determine. For example, let's consider the reaction in which hydrogen and oxygen gases combine to form water. We start by summarizing the qualitative information as a skeletal equation:

$$H_2 + O_2 \longrightarrow H_2O$$

The somewhat awkward word "stoichiometric" is derived from the Greek words for "element" and "measure."

We use the international *Hazard!* road sign ⚠ to warn that a skeletal equation is not balanced. Then we balance hydrogen and oxygen atoms:

$$2\,H_2 + O_2 \longrightarrow 2\,H_2O$$

There are four H atoms and two O atoms on each side of the arrow. At this stage, we insert the state symbols:

$$2\,H_2(g) + O_2(g) \longrightarrow 2\,H_2O(l)$$

Figure H.2 is a molecular-level representation of this reaction.

An equation must never be balanced by changing the subscripts in the chemical formulas. That change would imply that different substances were taking part in the reaction. For example, changing H_2O to H_2O_2 in the skeletal equation and writing

$$H_2 + O_2 \longrightarrow H_2O_2$$

certainly results in a balanced equation. However, it is a summary of a different reaction—the formation of hydrogen peroxide (H_2O_2) from its elements. Nor should we write

$$2\,H + O \longrightarrow H_2O$$

Although this equation is balanced, it summarizes the reaction between hydrogen and oxygen atoms, not the molecules that are the actual starting materials.

Although normally the coefficients in a balanced chemical equation are the smallest possible whole numbers, a chemical equation can be multiplied through by a factor and still be a valid equation. At times it is convenient to use fractional coefficients; for example, we could write

$$H_2(g) + \tfrac{1}{2}\,O_2(g) \longrightarrow H_2O(l)$$

if we want the equation to correspond to 1 mol H_2.

Sometimes we need to construct a balanced chemical equation from the description of a reaction. For example, methane, CH_4, is the principal ingredient of natural gas (Fig. H.3). It burns in oxygen to form carbon dioxide and water, both formed initially as gases. To write the balanced equation for the reaction, we first write the skeletal equation:

$$CH_4 + O_2 \longrightarrow CO_2 + H_2O \qquad ⚠$$

Then we balance the equation. A good strategy is to balance one element at a time, starting with one that appears in the fewest formulas, such as carbon and hydrogen. Then specify the states. Because water is produced as a vapor, we write

$$CH_4(g) + 2\,O_2(g) \longrightarrow CO_2(g) + 2\,H_2O(g)$$

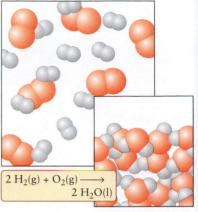

FIGURE H.2 A representation of the reaction between hydrogen and oxygen, with the production of water. No atoms are created or destroyed; they simply change partners. For every two hydrogen molecules that react, one oxygen molecule is consumed and two water molecules are formed.

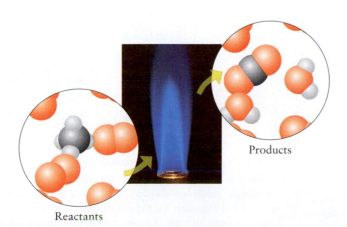

Reactants

Products

FIGURE H.3 When methane burns, it forms carbon dioxide and water. The blue color is due to the presence of C_2 molecules in the flame. If the oxygen supply is inadequate, these carbon molecules can stick together and form soot, thereby producing a smoky flame. Note that one carbon dioxide molecule and two water molecules are produced for each methane molecule that is consumed. The two hydrogen atoms in each water molecule do not necessarily come from the same methane molecule: the illustration depicts the overall outcome, not the specific outcome of the reaction of one molecule. The excess oxygen remains unreacted.

EXAMPLE H.1   **Writing and balancing a chemical equation**

Write and balance the chemical equation for the combustion of hexane, C_6H_{14}, to gaseous carbon dioxide gas and gaseous water.

STRATEGY First write the skeletal equation. Balance first the element that occurs in the fewest formulas. Verify that the coefficients are the smallest whole numbers. Specify the states of each reactant and product.

SOLUTION

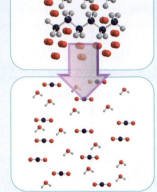

Write the skeletal equation.	$C_6H_{14} + O_2 \longrightarrow CO_2 + H_2O$
Balance carbon and hydrogen.	$C_6H_{14} + O_2 \longrightarrow 6\,CO_2 + 7\,H_2O$
Next balance oxygen. In this case, a fractional stoichiometric coefficient is needed.	$C_6H_{14} + \frac{19}{2}O_2 \longrightarrow 6\,CO_2 + 7\,H_2O$
Multiply by 2 to clear the fraction.	$2\,C_6H_{14} + 19\,O_2 \longrightarrow 12\,CO_2 + 14\,H_2O$
Add the physical states.	$2\,C_6H_{14}(g) + 19\,O_2(g) \longrightarrow 12\,CO_2(g) + 14\,H_2O(g)$

SELF-TEST H.1A When aluminum is melted and heated with solid barium oxide, a vigorous reaction takes place, and elemental molten barium and solid aluminum oxide are formed. Write the balanced chemical equation for the reaction.
[**Answer:** $2\,Al(l) + 3\,BaO(s) \xrightarrow{\Delta} Al_2O_3(s) + 3\,Ba(l)$]

SELF-TEST H.1B Write the balanced chemical equation for the reaction of solid magnesium nitride with aqueous sulfuric acid to form aqueous magnesium sulfate and aqueous ammonium sulfate.

A chemical equation expresses a chemical reaction in terms of chemical formulas; the stoichiometric coefficients are chosen to show that atoms are neither created nor destroyed in the reaction.

SKILLS YOU SHOULD HAVE MASTERED

❑ 1 Explain the role of stoichiometric coefficients (Section H.1).

❑ 2 Write, balance, and label a chemical equation on the basis of information given in sentence form (Example H.1).

EXERCISES

H.1 It appears that balancing the chemical equation $Cu + SO_2 \rightarrow CuO + S$ would be simple if we could just add another O to the product side: $Cu + SO_2 \rightarrow CuO + S + O$. (a) Why is that balancing procedure not allowed? (b) Balance the equation correctly.

H.2 Indicate which of the following are conserved in a chemical reaction: (a) mass; (b) number of atoms; (c) number of molecules; (d) number of electrons.

H.3 The first box below represents the reactants for a chemical reaction and the second box the products that form if all the reactant molecules shown react. Using the key below write a balanced equation for the reaction. Assume that if two atoms

are touching, they are bonded together. Key: ● oxygen; ○ hydrogen; ◆ silicon.

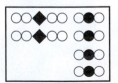
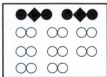

H.4 The first box below represents the reactants for a chemical reaction and the second box the products that form if all the reactant molecules shown react. Using the key below write a balanced equation for the reaction using the smallest whole-number coefficients. Assume that if two atoms are touching, they are bonded together. Key: ● oxygen; ○ hydrogen; ■ nitrogen.

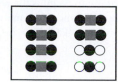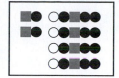

H.5 Balance the following skeletal chemical equations:
(a) $NaBH_4(s) + H_2O(l) \rightarrow NaBO_2(aq) + H_2(g)$
(b) $LiN_3(s) + H_2O(l) \rightarrow LiOH(aq) + HN_3(aq)$
(c) $NaCl(aq) + SO_3(g) + H_2O(l) \rightarrow Na_2SO_4(aq) + HCl(aq)$
(d) $Fe_2P(s) + S(s) \rightarrow P_4S_{10}(s) + FeS(s)$

H.6 Balance the following skeletal chemical equations:
(a) $KClO_3(s) + C_6H_{12}O_6(s) \xrightarrow{\Delta} KCl(s) + CO_2(g) + H_2O(g)$
(b) $P_2S_5(s) + PCl_5(s) \rightarrow PSCl_3(g)$
(c) (in ether) $LiBH_4 + BF_3 \rightarrow B_2H_6 + LiBF_4$
(d) $Ca_3(PO_4)_2(s) + SiO_2(s) + C(s) \xrightarrow{\Delta} CaSiO_3(l) + CO(g) + P_4(g)$

H.7 Write a balanced chemical equation for each of the following reactions. (a) Calcium metal reacts with water to produce hydrogen gas and aqueous calcium hydroxide. (b) The reaction of solid sodium oxide, Na_2O, and water produces aqueous sodium hydroxide. (c) Hot solid magnesium metal reacts in a nitrogen atmosphere to produce solid magnesium nitride, Mg_3N_2. (d) The reaction of ammonia gas with oxygen gas at high temperatures in the presence of a copper metal catalyst produces the gases water and nitrogen dioxide.

H.8 Write a balanced chemical equation for each of the following reactions. (a) In the first step of recovering copper metal from ores containing $CuFeS_2$ the ore is heated in air. During this "roasting" process, molecular oxygen reacts with the $CuFeS_2$ to produce solid copper(II) sulfide, iron(II) oxide, and sulfur dioxide gas. (b) The diamondlike abrasive silicon carbide, SiC, is made by reacting solid silicon dioxide with elemental carbon at 2000°C to produce solid silicon carbide and carbon monoxide gas. (c) The reaction of elemental hydrogen and nitrogen gases is used for the commercial production of gaseous ammonia in the Haber process. (d) Under acidic conditions oxygen gas can react with aqueous hydrobromic acid to form liquid water and liquid bromine.

H.9 In one stage in the commercial production of iron metal in a blast furnace, the iron(III) oxide, Fe_2O_3, reacts with carbon monoxide to form solid Fe_3O_4 and carbon dioxide gas. In a second stage, the Fe_3O_4 reacts further with carbon monoxide to produce solid elemental iron and carbon dioxide. Write the balanced equation for each stage in the process.

H.10 An important role of stratospheric ozone, O_3, is to remove damaging ultraviolet radiation from sunlight. One result is the eventual dissociation of gaseous ozone into molecular oxygen gas. Write a balanced equation for the dissociation reaction.

H.11 When nitrogen and oxygen gases react in the cylinder of an automobile engine, nitric oxide gas, NO, is formed. After it

escapes into the atmosphere with the other exhaust gases, the nitric oxide reacts with oxygen to produce nitrogen dioxide gas, one of the precursors of acid rain. Write the two balanced equations for the reactions leading to the formation of nitrogen dioxide.

H.12 The reaction of boron trifluoride, $BF_3(g)$, with sodium borohydride, $NaBH_4(s)$, leads to the formation of sodium tetrafluoroborate, $NaBF_4(s)$, and diborane gas, $B_2H_6(g)$. The diborane reacts with the oxygen in air, forming boron oxide, $B_2O_3(s)$, and water. Write the two balanced equations leading to the formation of boron oxide.

H.13 Hydrofluoric acid is used to etch grooves in glass because it reacts with the silica, $SiO_2(s)$, in glass. The products of the reaction are aqueous silicon tetrafluoride and water. Write a balanced equation for the reaction.

H.14 The compound $Sb_4O_5Cl_2(s)$, which has been investigated because of its interesting electrical properties, can be prepared by warming a mixture of antimony(III) oxide and antimony(III) chloride, both of which are solids. Write a balanced equation for the reaction.

H.15 Write a balanced equation for the complete combustion (reaction with oxygen) of liquid heptane, C_7H_{16}, a component typical of the hydrocarbons in gasoline, to carbon dioxide gas and water vapor.

H.16 Aspartame, $C_{14}H_{18}N_2O_5$, is a solid used as an artificial sweetener. Write the balanced equation for its combustion to carbon dioxide gas, liquid water, and nitrogen gas.

H.17 The psychoactive drug sold as methamphetamine ("speed"), $C_{10}H_{15}N$, undergoes a series of reactions in the body; the net result of these reactions is the oxidation of solid methamphetamine by oxygen gas to produce carbon dioxide gas, liquid water, and nitrogen gas. Write the balanced equation for this net reaction.

H.18 Aspirin is the analgesic acetylsalicylic acid, $C_9H_8O_4$. Write the balanced equation for the combustion of solid acetylsalicylic acid to carbon dioxide gas and liquid water.

H.19 Sodium thiosulfate, which as the pentahydrate $Na_2S_2O_3 \cdot 5H_2O$ forms the large white crystals used as "photographer's hypo," can be prepared by bubbling oxygen through a solution of sodium polysulfide, Na_2S_5, in alcohol and adding water. Sulfur dioxide gas is formed as a by-product. Sodium polysulfide is made by the action of hydrogen sulfide gas on a solution of sodium sulfide, Na_2S, in alcohol, which, in turn, is made by the reaction of hydrogen sulfide gas, H_2S, with solid sodium hydroxide. Write the three chemical equations that show how hypo is prepared from hydrogen sulfide and sodium hydroxide. Use (alc) to indicate the state of species dissolved in alcohol.

H.20 The first stage in the production of nitric acid by the Ostwald process is the reaction of ammonia gas with oxygen gas, producing nitric oxide gas, NO, and liquid water. The nitric oxide further reacts with oxygen to produce nitrogen dioxide gas, which, when dissolved in water, produces nitric acid and nitric oxide. Write the three balanced equations that lead to the production of nitric acid.

H.21 Phosphorus and oxygen react to form two different phosphorus oxides. The mass percentage of phosphorus in one of these oxides is 43.64%; in the other, it is 56.34%. (a) Write the empirical formula of each phosphorus oxide. (b) The molar mass of the former oxide is 283.33 g·mol^{-1} and that of the latter is 219.88 g·mol^{-1}. Determine the molecular formula and name of each oxide. (c) Write a balanced chemical equation for the formation of each oxide.

H.22 One step in the refining of titanium metal is the reaction of $FeTiO_3$ with chlorine gas and carbon. Balance the equation for the reaction: $FeTiO_3(s) + Cl_2(g) + C(s) \rightarrow TiCl_4(l) + FeCl_3(s) + CO(g)$.

I   AQUEOUS SOLUTIONS AND PRECIPITATION

When we mix two solutions the result is often simply a new solution that contains both solutes. However, in some cases the solutes can react with each other. For instance, when we mix a colorless aqueous solution of silver nitrate with a clear yellow aqueous solution of potassium chromate, a red solid forms, indicating that a chemical reaction has occurred (Fig. I.1). This section and the next two introduce three of the main types of chemical reactions: precipitation reactions, acid–base reactions, and redox reactions, all of which are discussed in more depth in later chapters. (The fourth type of reaction discussed in this text, Lewis acid–base reactions, is introduced in Section 10.2.) Because many chemical reactions take place in solution, particularly in water, in this section we begin by considering the nature of aqueous solutions.

I.1 Electrolytes

A **soluble substance** is one that dissolves to a significant extent in a specified solvent. When we refer simply to solubility without mentioning the solvent, we mean "soluble in water." An **insoluble substance** is one that does not dissolve significantly in a specified solvent; substances are often regarded as "insoluble" if they do not dissolve to more than about 0.1 mol·L^{-1}. Unless otherwise specified in this text, we use the term *insoluble* to mean "insoluble in water." For instance, calcium carbonate, $CaCO_3$, which makes up limestone and chalk, dissolves to form a solution that contains only 0.01 g·L^{-1} (corresponding to 1×10^{-4} mol·L^{-1}), and we regard $CaCO_3$ as insoluble. This insolubility is important for landscapes: chalk hills and limestone buildings do not wash away in natural rainwater.

A solute may be present as ions or as molecules. We can find out if the solute is present as ions by noting whether the solution conducts an electric current. Because a current is a flow of electric charge, only solutions that contain ions conduct electricity. There is such a tiny concentration of ions in pure water (about 10^{-7} mol·L^{-1}) that pure water itself does not conduct electricity significantly.

An **electrolyte** is a substance that, in solution, is present as ions. Ionic solids that are soluble in water are electrolytes because the ions become free to move when the solid dissolves (Fig. I.2). Some electrolytes, however (such as acids), form

FIGURE I.1 When a solution of yellow K_2CrO_4 is mixed with a colorless solution of $AgNO_3$ a red precipitate of silver chromate, Ag_2CrO_4, forms.

The term "electrolyte" is also used to denote a medium, such as a liquid solution or a porous solid, that can conduct electricity by the migration of ions.

FIGURE I.2 Sodium chloride consists of sodium ions and chloride ions. When sodium chloride comes in contact with water (left), the ions are separated by the water molecules, and they spread throughout the solvent (right). The solution consists of water, sodium ions, and chloride ions. There are no NaCl molecules present.

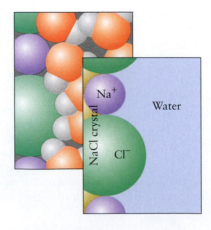

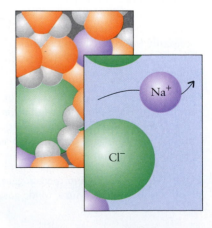

ions only when they dissolve. The term **electrolyte solution** is commonly used to emphasize that the medium is in fact a solution. A **nonelectrolyte** is a substance that does not form ions in solution; it dissolves to give a **nonelectrolyte solution**. Aqueous solutions of acetone (**1**) and glucose (**2**) are nonelectrolyte solutions. If we could see the individual molecules in a nonelectrolyte solution, we would see the intact solute molecules dispersed among the solvent molecules (Fig. I.3).

A **strong electrolyte** is a substance that is present almost entirely as ions in solution. Three types of solutes are strong electrolytes: strong acids and strong bases, which will be described in more detail in Section J, and soluble ionic compounds. Hydrogen chloride is a strong electrolyte; so are sodium hydroxide and the salt sodium chloride. A **weak electrolyte** is a substance that is incompletely ionized in solution; in other words, most of the molecules remain intact. Acetic acid is a weak electrolyte: in aqueous solution at normal concentrations, only a small fraction of CH_3COOH molecules separate into hydrogen ions, H^+, and acetate ions, $CH_3CO_2^-$. One way to distinguish strong and weak electrolytes is to measure the abilities of their solutions to conduct electricity: for the same molar concentration of solute, a solution of a strong electrolyte is a better conductor than a solution of a weak electrolyte (Fig. I.4).

SELF-TEST I.1A Identify each of the following substances as an electrolyte or a nonelectrolyte and predict which will conduct electricity when dissolved in water: (a) NaOH; (b) Br_2.

> [*Answer:* (a) Ionic compound, so a strong electrolyte, conducts electricity; (b) molecular compound and not an acid, so a nonelectrolyte, does not conduct electricity]

SELF-TEST I.1B Identify each of the following substances as an electrolyte or a nonelectrolyte and predict which will conduct electricity when dissolved in water: (a) ethanol, $CH_3CH_2OH(aq)$; (b) $Pb(NO_3)_2(aq)$.

The solute in an aqueous strong electrolyte solution is present as ions that can conduct electricity through the solvent. The solutes in nonelectrolyte solutions are present as molecules. Only a small fraction of the solute molecules in weak electrolyte solutions are present as ions.

I.2 Precipitation Reactions

Some ionic compounds are soluble, others are not. Consider what happens when we pour a solution of sodium chloride (a strong electrolyte) into a solution of silver nitrate (another strong electrolyte). A solution of sodium chloride contains Na^+ cations and Cl^- anions. Similarly, a solution of silver nitrate, $AgNO_3$, contains Ag^+ cations and NO_3^- anions. When we mix these two aqueous solutions, a white **precipitate**, a cloudy, finely divided solid deposit, forms immediately. Analysis shows that the precipitate is silver chloride, AgCl, an insoluble white solid. The

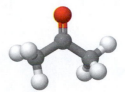

1 Acetone, C_3H_6O

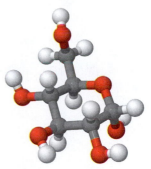

2 Glucose, $C_6H_{12}O_6$

Methanol molecule

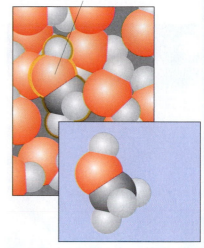

FIGURE I.3 In a nonelectrolyte solution, the solute remains as a molecule and does not break up into ions. Methanol, CH_3OH, is a nonelectrolyte and is present as intact molecules when it is dissolved in water.

(a)   (b)   (c)

FIGURE I.4 Pure water is a poor conductor of electricity, as shown by the very dim glow of the bulb in the circuit on the left (a). However, when ions are present, as in an electrolyte solution, the solution does conduct. The ability of the solution to conduct is low when the solute is a weak electrolyte (b) but significant when the solute is a strong electrolyte (c), even when the solute concentration is the same in each instance.

FUNDAMENTALS

colorless solution remaining above the precipitate contains dissolved Na^+ cations and NO_3^- anions. These ions remain in solution because sodium nitrate, $NaNO_3$, is soluble in water.

In a **precipitation reaction**, an insoluble solid product forms when we mix two electrolyte solutions. When an insoluble substance is formed in water, it immediately precipitates. In the chemical equation for a precipitation reaction, we use (aq) to indicate substances that are dissolved in water and (s) to indicate the solid that has precipitated:

$$AgNO_3(aq) + NaCl(aq) \longrightarrow AgCl(s) + NaNO_3(aq)$$

A precipitation reaction takes place when solutions of two strong electrolytes are mixed and react to form an insoluble solid.

I.3 Ionic and Net Ionic Equations

A **complete ionic equation** for a precipitation reaction shows all the dissolved ions explicitly. For example, the complete ionic equation for the silver chloride precipitation reaction shown in Fig. I.5 is

$$Ag^+(aq) + NO_3^-(aq) + Na^+(aq) + Cl^-(aq) \longrightarrow$$
$$AgCl(s) + Na^+(aq) + NO_3^-(aq)$$

Because the Na^+ and NO_3^- ions appear as both reactants and products, they play no direct role in the reaction. They are **spectator ions**, ions that are present while the reaction takes place but remain unchanged, like spectators at a sports event. Because spectator ions remain unchanged, we can simplify the chemical equation by canceling them on each side of the arrow in the ionic equation:

$$Ag^+(aq) + \cancel{NO_3^-(aq)} + \cancel{Na^+(aq)} + Cl^-(aq) \longrightarrow$$
$$AgCl(s) + \cancel{Na^+(aq)} + \cancel{NO_3^-(aq)}$$

Canceling the spectator ions leaves the **net ionic equation** for the reaction, the chemical equation that displays the net change taking place in the reaction:

$$Ag^+(aq) + Cl^-(aq) \longrightarrow AgCl(s)$$

The net ionic equation shows that Ag^+ ions combine with Cl^- ions to precipitate as solid silver chloride, AgCl (see Fig. I.5). A net ionic equation focuses our attention on the change that results from the chemical reaction.

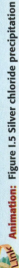

(a)

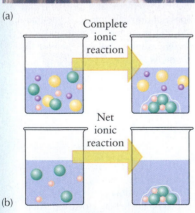

Complete
ionic
reaction

Net
ionic
reaction

(b)

FIGURE I.5 (a) Silver chloride precipitates immediately when sodium chloride solution is added to a solution of silver nitrate. (b) If we imagine the removal of the spectator ions from the complete ionic reaction (top), we can focus on the essential process, the net ionic reaction (bottom).

EXAMPLE I.1   Writing a net ionic equation

When concentrated aqueous solutions of barium nitrate, $Ba(NO_3)_2$, and ammonium iodate, NH_4IO_3, are mixed, insoluble barium iodate, $Ba(IO_3)_2$, forms. The chemical equation for the precipitation reaction is

$$Ba(NO_3)_2(aq) + 2\,NH_4IO_3(aq) \longrightarrow Ba(IO_3)_2(s) + 2\,NH_4NO_3(aq)$$

Write the net ionic equation for the reaction.

STRATEGY First, write and balance the complete ionic equation, showing all the dissolved ions as they actually exist in solution, as separate, charged ions. Insoluble solids are shown as complete compounds. Next, cancel the spectator ions, the ions that remain in solution on both sides of the arrow.

SOLUTION The complete ionic equation, with all the dissolved ions written as they exist in the solutions before and after mixing, is

$$Ba^{2+}(aq) + 2\,\cancel{NO_3^-(aq)} + 2\,\cancel{NH_4^+(aq)} + 2\,IO_3^-(aq) \longrightarrow$$
$$Ba(IO_3)_2(s) + 2\,\cancel{NH_4^+(aq)} + 2\,\cancel{NO_3^-(aq)}$$

We cancel the spectator ions, NH_4^+ and NO_3^-, to obtain the net ionic equation:

$$Ba^{2+}(aq) + 2\,IO_3^-(aq) \longrightarrow Ba(IO_3)_2(s)$$

SELF-TEST I.2A Write the net ionic equation for the reaction in Fig. I.1, in which aqueous solutions of colorless silver nitrate and yellow potassium chromate react to give a precipitate of red silver chromate.

[**Answer:** $2 Ag^+(aq) + CrO_4^{2-}(aq) \rightarrow Ag_2CrO_4(s)$]

SELF-TEST I.2B The mercury(I) ion, Hg_2^{2+}, consists of two Hg^+ ions joined together. Write the net ionic equation for the reaction in which colorless aqueous solutions of mercury(I) nitrate, $Hg_2(NO_3)_2$, and potassium phosphate, K_3PO_4, react to give a white precipitate of mercury(I) phosphate.

A complete ionic equation expresses a reaction in terms of the ions that are present in solution; a net ionic equation is the chemical equation that remains after the cancellation of the spectator ions.

I.4 Putting Precipitation to Work

Precipitation reactions have many applications. One is to make compounds. The strategy is to choose starting solutions that form a precipitate of the desired insoluble compound when they are mixed. Then we can separate the insoluble compound from the reaction mixture by filtration. Another application is in chemical analysis. In **qualitative analysis**—the determination of the substances present in a sample—the formation of a precipitate is used to confirm the identity of certain ions. In **quantitative analysis**, the aim is to determine the amount of each substance or element present. In particular, in **gravimetric analysis**, the amount of substance present is determined by measurements of mass. In this application, an insoluble compound is precipitated, the precipitate is filtered off and weighed, and from its mass the amount of a substance in one of the original solutions is calculated (Fig. I.6). Gravimetric analysis can be used in environmental monitoring to find out how much of a heavy metal ion, such as lead or mercury, is in a sample of water.

Table I.1 summarizes the solubility patterns of common ionic compounds in water. Notice that all nitrates and all common compounds of the Group 1 metals are soluble; so they make useful starting solutions for precipitation reactions. Any spectator ions can be used, provided that they remain in solution and do not otherwise react. For example, Table I.1 shows that mercury(I) iodide, Hg_2I_2, is insoluble. It is formed as a precipitate when solutions containing Hg_2^{2+} ions and I^- ions are mixed:

$$Hg_2^{2+}(aq) + 2 I^-(aq) \longrightarrow Hg_2I_2(s)$$

Because the spectator ions are not shown, the net ionic equation will be the same when any soluble mercury(I) compound is mixed with any soluble iodide.

Section 11.14 describes the use of precipitates in qualitative analysis in more detail.

You will find an example of how to use this technique in Section L.

FIGURE I.6 A step in a gravimetric analysis. An ion has come out of solution as part of a precipitate and is being filtered. The filter paper, which has a known mass, will then be dried and weighed, thereby allowing the mass of the precipitate to be determined.

TABLE I.1   Solubility Rules for Inorganic Compounds

Soluble Compounds	Insoluble Compounds
compounds of Group 1 elements	carbonates (CO_3^{2-}), chromates (CrO_4^{2-}), oxalates ($C_2O_4^{2-}$), and phosphates (PO_4^{3-}), **except** those of the Group 1 elements and NH_4^+
ammonium (NH_4^+) compounds	
chlorides (Cl^-), bromides (Br^-), and iodides (I^-), **except** those of Ag^+, Hg_2^{2+}, and Pb^{2+} *	sulfides (S^{2-}), **except** those of the Group 1 and 2 elements and NH_4^+
nitrates (NO_3^-), acetates ($CH_3CO_2^-$), chlorates (ClO_3^-), and perchlorates (ClO_4^-)	hydroxides (OH^-) and oxides (O^{2-}), **except** those of the Group 1 elements and Group 2 elements below Period 2‡
sulfates (SO_4^{2-}), **except** those of Ca^{2+}, Sr^{2+}, Ba^{2+}, Pb^{2+}, Hg_2^{2+}, and Ag^+†	

* $PbCl_2$ is slightly soluble.
† Ag_2SO_4 is slightly soluble.
‡ $Ca(OH)_2$ and $Sr(OH)_2$ are sparingly (slightly) soluble; $Mg(OH)_2$ is only very slightly soluble.

EXAMPLE I.2   **Predicting the outcome of a precipitation reaction**

Predict the precipitate if any, likely to be formed when aqueous solutions of sodium phosphate, and lead(II) nitrate are mixed. Write the net ionic equation for the reaction.

STRATEGY   Decide which ions are present in the mixed solutions and consider all possible combinations. Use the solubility rules in Table I.1 to decide which combination corresponds to an insoluble compound and write the net ionic equation to match.

SOLUTION   The mixed solution will contain Na^+, PO_4^{3-}, Pb^{2+}, and NO_3^- ions. All nitrates and all compounds of Group 1 metals are soluble, but phosphates of other elements are generally insoluble. Hence, we can predict that Pb^{2+} and PO_4^{3-} ions will form an insoluble compound and that lead(II) phosphate, $Pb_3(PO_4)_2$, will precipitate:

$$3\ Pb^{2+}(aq) + 2\ PO_4^{3-}(aq) \longrightarrow Pb_3(PO_4)_2(s)$$

The Na^+ and NO_3^- ions are spectators, and so they are omitted from the net ionic equation.

SELF-TEST I.3A   Predict the identity of the precipitate that forms, if any, when aqueous solutions of ammonium sulfide and copper(II) sulfate are mixed, and write the net ionic equation for the reaction.

[*Answer:* Copper(II) sulfide; $Cu^{2+}(aq) + S^{2-}(aq) \rightarrow CuS(s)$]

SELF-TEST I.3B   Suggest two solutions that can be mixed to prepare strontium sulfate, and write the net ionic equation for the reaction.

The solubility rules in Table I.1 are used to predict the outcomes of precipitation reactions.

SKILLS YOU SHOULD HAVE MASTERED

❏ 1 Identify electrolytes or nonelectrolytes on the basis of the formulas of the solutes (Self-Test I.1).

❏ 2 Construct balanced complete ionic and net ionic equations for reactions involving ions (Example I.1).

❏ 3 Use the solubility rules to select appropriate solutions that, when mixed, will produce a desired precipitate (Section I.4).

❏ 4 Identify any precipitate that may form on mixing two given solutions (Example I.2).

EXERCISES

I.1 The beaker on the left pictured below contains 0.50 M $CaCl_2(aq)$ and the beaker on the right contains 0.50 M $Na_2SO_4(aq)$. Suppose the contents of the two beakers are mixed. Draw a picture of the resulting products.

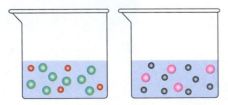

I.2 The beaker on the left pictured below contains 0.50 M $Hg_2(NO_3)_2(aq)$ and the beaker on the right contains 0.50 M $K_3PO_4(aq)$. Suppose the contents of the two beakers are mixed. Draw a picture of the resulting products.

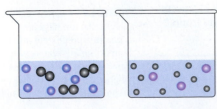

I.3 How would you use the solubility rules in Table I.1 to separate the following pairs of ions? In each case indicate what reagent you would add and write the net ionic equation for the precipitation reaction: (a) lead(II) and copper(II) ions; (b) ammonium and magnesium ions.

I.4 How would you use the solubility rules in Table I.1 to separate the following pairs of ions? In each case indicate what reagent you would add and write the net ionic equation for the precipitation reaction: (a) barium and mercury(I) ions; (b) silver and zinc ions.

I.5 Classify each of the following substances as a strong electrolyte or a nonelectrolyte: (a) CH_3OH; (b) $BaCl_2$; (c) KF.

I.6 Classify each of the following substances as a strong electrolyte, a weak electrolyte, or a nonelectrolyte: (a) HNO_3; (b) NaOH; (c) CH_3COOH.

I.7 Use the information in Table I.1 to classify each of the following ionic compounds as soluble or insoluble in water: (a) potassium phosphate, K_3PO_4; (b) lead(II) chloride, $PbCl_2$; (c) cadmium sulfide, CdS; (d) barium sulfate, $BaSO_4$.

I.8 Use the information in Table I.1 to classify each of the following ionic compounds as soluble or insoluble in water: (a) lead(II) acetate, $Pb(CH_3CO_2)_2$; (b) chromium(III) hydroxide, $Cr(OH)_3$; (c) silver iodide, AgI; (d) copper(II) nitrate, $Cu(NO_3)_2$.

I.9 What are the principal solute species present in an aqueous solution of (a) NaI; (b) Ag_2CO_3; (c) $(NH_4)_3PO_4$; (d) $FeSO_4$?

I.10 What are the principal solute species present in an aqueous solution of (a) $NiSO_4$; (b) Na_2CO_3; (c) $K_2Cr_2O_7$; (d) Hg_2Cl_2?

I.11 (a) When aqueous solutions of iron(III) sulfate and sodium hydroxide were mixed, a precipitate formed. Write the formula of the precipitate. (b) Does a precipitate form when aqueous solutions of silver nitrate, $AgNO_3$, and potassium carbonate are mixed? If so, write the formula of the precipitate. (c) If aqueous solutions of lead(II) nitrate and sodium acetate are mixed, does a precipitate form? If so, write the formula of the precipitate.

I.12 (a) Solid ammonium nitrate and solid calcium chloride were placed in water, and the mixture was stirred. Is the formation of a precipitate expected? If so, write the formula of the precipitate. (b) Solid magnesium carbonate and solid sodium nitrate were mixed, placed in water, and stirred. What is observed? If a precipitate is present, write its formula. (c) Aqueous solutions of sodium sulfate and barium chloride are mixed. What is observed? If a precipitate is present, write its formula.

I.13 When the solution in Beaker 1 is mixed with the solution in Beaker 2, a precipitate forms. Using the following table, write the net ionic equation describing the formation of the precipitate, and then identify the spectator ions.

Beaker 1	Beaker 2
(a) $FeCl_2(aq)$	$Na_2S(aq)$
(b) $Pb(NO_3)_2(aq)$	$KI(aq)$
(c) $Ca(NO_3)_2(aq)$	$K_2SO_4(aq)$
(d) $Na_2CrO_4(aq)$	$Pb(NO_3)_2(aq)$
(e) $Hg_2(NO_3)_2(aq)$	$K_2SO_4(aq)$

I.14 The contents of Beaker 1 are mixed with those of Beaker 2. If a reaction takes place, write the net ionic equation and indicate the spectator ions.

Beaker 1	Beaker 2
(a) $K_2SO_4(aq)$	$LiNO_3(aq)$
(b) $H_3PO_4(aq)$	$CaCl_2(aq)$
(c) $K_2S(aq)$	$NH_4NO_3(aq)$
(d) $CoSO_4(aq)$	$(NH_4)_2CO_3(aq)$
(e) $HNO_3(aq)$	$Ca(OH)_2(aq)$

I.15 Each of the following five procedures results in the formation of a precipitate. For each reaction, write the chemical equations describing the formation of the precipitate: the overall equation, the complete ionic equation, and the net ionic equation. Identify the spectator ions. (a) $(NH_4)_2CrO_4(aq)$ is mixed with $BaCl_2(aq)$. (b) $CuSO_4(aq)$ is mixed with $Na_2S(aq)$. (c) $FeCl_2(aq)$ is mixed with $(NH_4)_3PO_4(aq)$. (d) Potassium oxalate, $K_2C_2O_4(aq)$, is mixed with $Ca(NO_3)_2(aq)$. (e) $NiSO_4(aq)$ is mixed with $Ba(NO_3)_2(aq)$.

I.16 Each of the following five procedures results in the formation of a precipitate. For each reaction, write the chemical equations describing the formation of the precipitate: the overall equation, the complete ionic equation, and the net ionic equation. Identify the spectator ions. (a) $AgNO_3(aq)$ is mixed with $Na_2CO_3(aq)$. (b) $Pb(NO_3)_2(aq)$ is mixed with $KI(aq)$. (c) $Ba(OH)_2(aq)$ is mixed with $H_2SO_4(aq)$. (d) $(NH_4)_2S(aq)$ is mixed with $Cd(NO_3)_2(aq)$. (e) $KOH(aq)$ is mixed with $CuCl_2(aq)$.

I.17 Write the balanced, complete ionic, and net ionic equations corresponding to each of the following reactions:
(a) $Ba(CH_3CO_2)_2(aq) + Li_2CO_3(aq) \rightarrow$
$$BaCO_3(s) + LiCH_3CO_2(aq)$$
(b) $NH_4Cl(aq) + Hg_2(NO_3)_2(aq) \rightarrow NH_4NO_3(aq) + Hg_2Cl_2(s)$
(c) $Cu(NO_3)_2(aq) + Ba(OH)_2(aq) \rightarrow$
$$Cu(OH)_2(s) + Ba(NO_3)_2(aq)$$

I.18 Write the balanced, complete ionic, and net ionic equations corresponding to each of the following reactions:
(a) $Pb(NO_3)_2(aq) + K_3PO_4(aq) \rightarrow Pb_3(PO_4)_2(s) + KNO_3(aq)$
(b) $K_2S(aq) + Bi(NO_3)_3(aq) \rightarrow Bi_2S_3(s) + KNO_3(aq)$
(c) $Na_2C_2O_4(aq) + Mn(CH_3CO_2)_2(aq) \rightarrow$
$$MnC_2O_4(s) + NaCH_3CO_2(aq)$$

I.19 For each of the following reactions, suggest two soluble ionic compounds that, when mixed together in water, result in the net ionic equation given:
(a) $2\,Ag^+(aq) + CrO_4^{2-}(aq) \rightarrow Ag_2CrO_4(s)$
(b) $Ca^{2+}(aq) + CO_3^{2-}(aq) \rightarrow CaCO_3(s)$, the reaction responsible for the deposition of chalk hills and sea urchin spines
(c) $Cd^{2+}(aq) + S^{2-}(aq) \rightarrow CdS(s)$, a yellow substance used to color glass

I.20 For each of the following reactions, suggest two soluble ionic compounds that, when mixed together in water, result in the net ionic equation given:
(a) $2\,Ag^+(aq) + CO_3^{2-}(aq) \rightarrow Ag_2CO_3(s)$
(b) $Mg^{2+}(aq) + 2\,OH^-(aq) \rightarrow Mg(OH)_2(s)$, the suspension present in milk of magnesia
(c) $3\,Ca^{2+}(aq) + 2\,PO_4^{3-}(aq) \rightarrow Ca_3(PO_4)_2(s)$, gypsum, a component of concrete

I.21 Write the net ionic equation for the formation of each of the following insoluble compounds in aqueous solution: (a) silver sulfate, Ag_2SO_4; (b) mercury(II) sulfide, HgS, used as an electrolyte in some primary batteries; (c) calcium phosphate, $Ca_3(PO_4)_2$, a component of bones and teeth. (d) Select two soluble ionic compounds that, when mixed in solution, form each of the insoluble compounds in parts (a), (b), and (c). Identify the spectator ions.

I.22 Write the net ionic equation for the formation of each of the following insoluble compounds in aqueous solution: (a) lead(II) chromate, $PbCrO_4$, the yellow pigment that has been used for centuries in oil paints; (b) aluminum phosphate, $AlPO_4$, used in cements and as an antacid; (c) iron(II) hydroxide, $Fe(OH)_2$; (d) Select two soluble ionic compounds that, when mixed in solution, form each of the insoluble compounds in parts (a), (b), and (c). Identify the spectator ions.

I.23 You are given a solution and asked to analyze it for the cations Ag^+, Ca^{2+}, and Zn^{2+}. You add hydrochloric acid, and a white precipitate forms. You filter out the solid and add sulfuric acid to the solution. Nothing appears to happen. Then you add hydrogen sulfide. A black precipitate forms. Which ions should you report as present in your solution?

I.24 You are given a solution and asked to analyze it for the cations Ag^+, Ca^{2+}, and Hg^{2+}. You add hydrochloric acid. Nothing appears to happen. You then add dilute sulfuric acid, and a white precipitate forms. You filter out the solid and add hydrogen sulfide to the solution that remains. A black precipitate forms. Which ions should you report as present in your solution?

I.25 Suppose that 40.0 mL of 0.100 M NaOH(aq) is added to 10.0 mL of 0.200 M $Cu(NO_3)_2$(aq). (a) Write the chemical equation for the precipitation reaction, the complete ionic

equation, and the net ionic equation. (b) What is the molarity of Na^+ ions in the final solution?

I.26 Suppose that 2.50 g of solid $(NH_4)_3(PO_4)$ is added to 50.0 mL of 0.125 M $CaCl_2$(aq). (a) Write the chemical equation for the precipitation reaction and the net ionic equation. (b) What is the molarity of each spectator ion after reaction is complete? Assume a final volume of 70.0 mL.

I.27 Suppose that 3.50 g of solid potassium chromate is added to 75.0 mL of 0.250 M $Mg(NO_3)_2$(aq). (a) What is the initial molarity of potassium chromate in the solution? (b) What mass of potassium is present in solution? (c) Write the formula for the precipitate that forms. Assume a final volume of 75.0 mL.

J.1 Acids and Bases in Aqueous Solution

J.2 Strong and Weak Acids and Bases

J.3 Neutralization

J ACIDS AND BASES

Early chemists applied the term *acid* to substances that had a sharp or sour taste. Vinegar, for instance, contains acetic acid, CH_3COOH. Aqueous solutions of substances that they called *bases* or *alkalis* were recognized by their soapy feel. Fortunately, there are less hazardous ways of recognizing acids and bases. For instance, acids and bases change the color of certain dyes known as indicators (Fig. J.1). One of the best-known indicators is litmus, a vegetable dye obtained from a lichen. Aqueous solutions of acids turn litmus red; aqueous solutions of bases turn it blue. Later (in Chapter 10), we shall see that the electronic instrument known as a "pH meter" provides a rapid way of identifying a solution as acidic or basic: a pH reading *lower* than 7 denotes an acidic solution and a reading *higher* than 7 denotes a basic solution.

J.1 Acids and Bases in Aqueous Solution

Chemists debated the concepts of acid and base for many years before precise definitions emerged. Among the first useful definitions was the one proposed by the Swedish chemist Svante Arrhenius in about 1884:

> An **acid** is a compound that contains hydrogen and reacts with water to form hydrogen ions.
> A **base** is a compound that produces hydroxide ions in water.

These compounds are called **Arrhenius acids and bases**. For instance, HCl is an Arrhenius acid, because it releases a hydrogen ion, H^+ (a proton), when it dissolves

FIGURE J.1 The acidities of various household products can be demonstrated by adding an indicator (an extract of red cabbage, in this case) and noting the resulting color. Red indicates an acidic solution, blue basic. From left to right, the household products are (a) lemon juice, (b) soda water, (c) 7-Up, (d) vinegar, (e) ammonia, (f) lye, (g) milk of magnesia, and (h) detergent in water. Note that ammonia and lye are such strong bases that they destroy the dye, and a yellow color is obtained instead of the expected blue.

in water; CH_4 is not an Arrhenius acid, because it does not release hydrogen ions in water. Sodium hydroxide is an Arrhenius base, because OH^- ions go into solution when it dissolves; ammonia is an Arrhenius base, because it produces OH^- ions by reacting with water:

$$NH_3(aq) + H_2O(l) \longrightarrow NH_4^+(aq) + OH^-(aq) \tag{1}$$

Sodium metal, although it produces OH^- ions when it reacts with water, is not an Arrhenius base, because it is an element, not a compound, as the definition requires.

The problem with the Arrhenius definitions is that they are specific to one particular solvent, water. When chemists studied nonaqueous solvents, such as liquid ammonia, they found that a number of substances showed the same pattern of acid–base behavior, but plainly the Arrhenius definitions could not be used. A major advance in our understanding of what it means to be an acid or a base came in 1923, when two chemists working independently, Thomas Lowry in England and Johannes Brønsted in Denmark, came up with the same idea. Their insight was to realize that the key process responsible for the properties of acids and bases was the transfer of a proton (a hydrogen ion) from one substance to another. The **Brønsted–Lowry definition** of acids and bases is as follows:

> An **acid** is a proton donor.
>
> A **base** is a proton acceptor.

Acids and bases are described in more detail in Chapter 10.

We call such substances "Brønsted acids and bases" or just plain "acids and bases" because the Brønsted–Lowry definition is the one commonly accepted today and the one used throughout this text.

First consider acids. When a molecule of an acid dissolves in water, it donates a hydrogen ion, H^+, to one of the water molecules and forms a *hydronium ion*, H_3O^+ (**1**). For example, when hydrogen chloride, HCl, dissolves in water, it releases a hydrogen ion to water, and the resulting solution consists of hydronium ions and chloride ions:

$$HCl(aq) + H_2O(l) \longrightarrow H_3O^+(aq) + Cl^-(aq)$$

1 Hydronium ion, H_3O^+

Notice that because H_2O accepts the hydrogen ion to form H_3O^+, water is acting as a Brønsted base in this reaction.

Hydrogen chloride, HCl, and nitric acid, HNO_3, are Brønsted acids. The molecules of both compounds contain hydrogen atoms that they can donate as protons to other substances. Both compounds form hydronium ions in water. Methane, CH_4, is not a Brønsted acid. Although it contains hydrogen, it does not release hydrogen ions to other substances. Acetic acid, CH_3COOH, releases *one* hydrogen ion (from the hydrogen atom of the carboxyl group, $-COOH$ (**2**)) to water and any other Brønsted bases present in the solution. Like HCl and HNO_3, acetic acid is a **monoprotic acid**, an acid that can donate only one proton from each molecule. Sulfuric acid, H_2SO_4, can release both its hydrogen atoms as ions—the first one much more readily than the second—and so it is an example of a **polyprotic acid**, an acid that can donate more than one proton from each molecule.

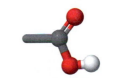

2 Carboxyl group, $-COOH$

The **acidic hydrogen atom** in a compound is the hydrogen atom that can be released as a proton. It is often written as the first element in a molecular formula, as in HCl and HNO_3. However, for organic acids, such as acetic acid, it is more informative to write the formulas to show the carboxyl group, $-COOH$, explicitly, thereby making it easier to remember that the H atom in this group of atoms is the acidic one. Thus, we can immediately recognize from their formulas that HCl, H_2CO_3 (carbonic acid), H_2SO_4 (sulfuric acid), and C_6H_5COOH (benzoic acid) are acids but that CH_4, NH_3 (ammonia), and $CH_3CO_2^-$ (the acetate ion) are not. The common oxoacids, acids containing oxygen, were introduced in Section D and are listed in Table D.1.

Now consider bases, species that accept protons. First, we can see that hydroxide ions are bases, because they accept protons from acids to form molecules of water:

$$OH^-(aq) + CH_3COOH(aq) \longrightarrow H_2O(l) + CH_3CO_2^-(aq)$$

Ammonia is a base because, as we see from Eq. 1, it accepts protons from water and forms NH_4^+ ions. Notice that because water donates a hydrogen ion it is acting as a Brønsted acid.

A note on good practice: Sodium hydroxide is commonly termed a base; however, from the Brønsted point of view it simply *provides* the base OH^-.

SELF-TEST J.1A Which of the following compounds are Brønsted acids or bases in water? (a) HNO_3; (b) C_6H_6; (c) KOH; (d) C_3H_5COOH.
[*Answer:* (a) and (d) are acids; (b) is neither an acid nor a base; (c) *supplies* the base OH^-]

SELF-TEST J.1B Which of the following compounds are Brønsted acids or bases in water? (a) KCl; (b) HClO; (c) HF; (d) $Ca(OH)_2$.

Acids are molecules or ions that are proton donors. Bases are molecules or ions that are proton acceptors.

J.2 Strong and Weak Acids and Bases

We saw in Section I.1 that electrolytes are classified as strong or weak according to the extent to which they are present as ions in solution. We classify acids and bases in a similar fashion. In these definitions the term **deprotonation** means loss of a proton and **protonation** means the gain of a proton:

A **strong acid** is completely deprotonated in solution.

A **weak acid** is incompletely deprotonated in solution.

A **strong base** is completely protonated in solution.

A **weak base** is incompletely protonated in solution.

The terms *ionized* and *dissociated* are commonly used instead of "deprotonated."

By "completely deprotonated," we mean that *each* acid molecule or ion has lost the proton of its acidic hydrogen atom by transferring it as a hydrogen ion to a solvent molecule. "Completely protonated" means that *each* base species has acquired a proton. By "incompletely deprotonated" or "incompletely protonated," we mean that only a fraction (usually a very tiny fraction) of the acid molecules or ions have lost acidic hydrogen atoms as protons or only a tiny fraction of the base species have acquired protons.

Once again, we examine acids first. Hydrogen chloride is a strong acid in water. A solution of hydrogen chloride and water, which we call hydrochloric acid, contains hydronium ions, chloride ions, and virtually no HCl molecules.

$$HCl(g) + H_2O(l) \longrightarrow H_3O^+(aq) + Cl^-(aq)$$

Acetic acid, on the other hand, is a weak acid in water. Only a small fraction of its molecules undergo deprotonation, according to the equation

$$CH_3COOH(aq) + H_2O(l) \longrightarrow H_3O^+(aq) + CH_3CO_2^-(aq)$$

and the solute consists principally of CH_3COOH molecules (Fig. J.2). In fact, 0.1 M $CH_3COOH(aq)$ contains only about one $CH_3CO_2^-$ ion for every hundred acetic acid molecules used to make a solution.

Table J.1 lists all the common strong acids in water. They include three common laboratory reagents—hydrochloric acid, nitric acid, and sulfuric acid (with respect to the loss of one proton from each H_2SO_4 molecule). Most acids are weak in water. All carboxylic acids are weak in water.

Now consider strong and weak bases. The common strong bases are oxide ions and hydroxide ions, which are provided by the alkali metal and alkaline earth metal oxides and hydroxides, such as calcium oxide (see Table J.1). As we have seen,

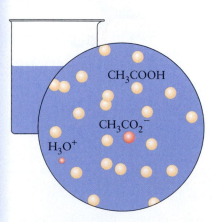

FIGURE J.2 Acetic acid, like all carboxylic acids, is a weak acid in water. This classification means that most of it remains as acetic acid molecules, CH_3COOH; however, a small proportion of these molecules donate a hydrogen ion to a water molecule to form hydronium ions, H_3O^+, and acetate ions, $CH_3CO_2^-$.

TABLE J.1 **The Strong Acids and Bases in Water**

Strong acids	Strong bases
hydrobromic acid, HBr(aq)	Group 1 hydroxides
hydrochloric acid, HCl(aq)	alkaline earth metal hydroxides*
hydroiodic acid, HI(aq)	Group 1 and Group 2 oxides
nitric acid, HNO_3	
perchloric acid, $HClO_4$	
chloric acid, $HClO_3$	
sulfuric acid, H_2SO_4 (to HSO_4^-)	

*$Ca(OH)_2$, $Sr(OH)_2$, $Ba(OH)_2$.

chemists often refer to these oxides and hydroxides as "strong bases." When an oxide dissolves in water, the oxide ions, O^{2-}, accept protons to form hydroxide ions:

$$O^{2-}(aq) + H_2O(l) \longrightarrow 2\,OH^-(aq)$$

Hydroxide ions, such as those provided by sodium hydroxide and calcium hydroxide, are also strong bases in water:

$$H_2O(l) + OH^-(aq) \longrightarrow OH^-(aq) + H_2O(l)$$

Even though a hydroxide ion is a strong base and is protonated in water, it effectively survives, because the H_2O molecule that donates a proton becomes a hydroxide ion!

All other common bases are weak in water. Ammonia is a weak base in water. In its aqueous solutions, it exists almost entirely as NH_3 molecules, with just a small proportion—usually fewer than one in a hundred molecules at normal concentrations—of NH_4^+ cations and OH^- anions. Other common weak bases are the amines, the pungent compounds that are derived from ammonia by replacement of one or more of its hydrogen atoms by an organic group. For example, the replacement of one hydrogen atom in NH_3 by a methyl group, $-CH_3$ (3), results in methylamine, CH_3NH_2 (4). The replacement of all three hydrogen atoms in NH_3 by methyl groups results in trimethylamine, $(CH_3)_3N$ (5), a substance found in decomposing fish and on unwashed dogs.

Strong acids (the acids listed in Table J.1) are completely deprotonated in solution; weak acids (most other acids) are not. Strong bases (the metal oxides and hydroxides listed in Table J.1) are completely protonated in solution. Weak bases (ammonia and its organic derivatives, the amines) are only partially protonated in solution.

J.3 Neutralization

The reaction between an acid and a base is called a **neutralization reaction,** and the ionic compound produced in the reaction is called a **salt.** The general form of a neutralization reaction of a strong acid and a metal hydroxide that provides the hydroxide ion, a strong base, in water is

Acid + metal hydroxide $\longrightarrow$ salt + water

The name *salt* is taken from ordinary table salt, sodium chloride, the ionic product of the reaction between hydrochloric acid and sodium hydroxide:

$$HCl(aq) + NaOH(aq) \longrightarrow NaCl(aq) + H_2O(l)$$

In the neutralization reaction between an acid and a metal hydroxide, the cation of the salt is provided by the metal hydroxide, such as NaOH, and the anion is provided by the acid. Another example is the reaction between nitric acid and barium hydroxide:

$$2\,HNO_3(aq) + Ba(OH)_2(aq) \longrightarrow Ba(NO_3)_2(aq) + 2\,H_2O(l)$$

The barium nitrate remains in solution as Ba^{2+} and NO_3^- ions.

3   Methyl group, $-CH_3$

4   Methylamine, CH_3NH_2

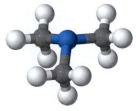

5   Trimethylamine, $(CH_3)_3N$

Although hydrogen ions are always attached to water molecules as hydronium ions, H_3O^+, or more complex species, we write them here as H^+ for simplicity. The (aq) should be taken to imply that the hydrogen ion is actually present as the hydronium ion.

We saw in Section I that the net chemical change in a precipitation reaction is clarified by writing its net ionic equation. The same is true of neutralization reactions. First, we write the complete ionic equation for the neutralization reaction between nitric acid and barium hydroxide in water:

$$2\,H^+(aq) + 2\,NO_3^-(aq) + Ba^{2+}(aq) + 2\,OH^-(aq) \longrightarrow$$
$$Ba^{2+}(aq) + 2\,NO_3^-(aq) + 2\,H_2O(l)$$

The ions common to both sides now cancel, and the net ionic equation of this reaction is therefore

$$2\,H^+(aq) + 2\,OH^-(aq) \longrightarrow 2\,H_2O(l)$$

which simplifies to

$$H^+(aq) + OH^-(aq) \longrightarrow H_2O(l)$$

The net outcome of any neutralization reaction between a strong acid and a strong base in water is the formation of water from hydrogen ions and hydroxide ions.

When we write the net ionic equation for the neutralization of a weak acid or a weak base, we use the molecular form of the weak acid or base, because molecules are the dominant species in solution. For example, we write the net ionic equation for the reaction of the weak acid HCN with the strong base NaOH in water (Fig. J.3) as

$$HCN(aq) + OH^-(aq) \longrightarrow H_2O(l) + CN^-(aq)$$

Similarly, the net ionic equation for the reaction of the weak base ammonia with the strong acid HCl in water is

$$NH_3(aq) + H^+(aq) \longrightarrow NH_4^+(aq)$$

SELF-TEST J.2A What acid and base solutions could you use to prepare rubidium nitrate? Write the chemical equation for the neutralization.
[*Answer:* $HNO_3(aq) + RbOH(aq) \longrightarrow RbNO_3(aq) + H_2O(l)$]

SELF-TEST J.2B Write the chemical equation for a neutralization reaction in which calcium phosphate is produced.

In a neutralization reaction in water, an acid reacts with a base to produce a salt (and water if the base is strong); the net outcome of the reaction between solutions of a strong acid and a strong base is the formation of water from hydrogen ions and hydroxide ions.

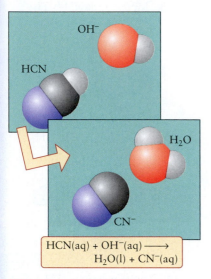

FIGURE J.3 The net ionic equation for the neutralization of HCN, a weak acid, by the strong base NaOH, tells us that the hydroxide ion extracts a hydrogen ion from an acid molecule.

$$HCN(aq) + OH^-(aq) \longrightarrow$$
$$H_2O(l) + CN^-(aq)$$

SKILLS YOU SHOULD HAVE MASTERED

❑ 1 Describe the chemical properties of acids and bases (Section J.1).

❑ 2 Classify substances as acids or bases (Self-Test J.1).

❑ 3 Identify common strong acids and bases (Table J.1).

❑ 4 Predict the outcome of neutralization reactions and write their chemical equations (Self-Test J.2).

EXERCISES

J.1 Identify each compound as either a Brønsted acid or a Brønsted base: (a) NH_3; (b) HBr; (c) KOH; (d) H_2SO_3; (e) $Ca(OH)_2$.

J.2 Classify each compound as either a Brønsted acid or a Brønsted base: (a) H_3AsO_4; (b) CH_3NH_2, a derivative of ammonia; (c) CH_3COOH; (d) LiOH; (e) $HClO_3$.

J.3 Complete the overall equation, and write the complete ionic equation and the net ionic equation for each of

the following acid–base reactions. If the substance is a weak acid or base, leave it in its molecular form in the equations.
(a) $HF(aq) + NaOH(aq) \rightarrow$
(b) $(CH_3)_3N(aq) + HNO_3(aq) \rightarrow$
(c) $LiOH(aq) + HI(aq) \rightarrow$

J.4 Complete the overall equation, and write the complete ionic equation and the net ionic equation for each of the following

acid–base reactions. If the substance is a weak acid or base, leave it in its molecular form in the equations.
(a) $H_3PO_4(aq) + KOH(aq) \rightarrow$
(Phosphoric acid, H_3PO_4, is a triprotic acid. Write the equation for complete reaction with KOH.)
(b) $Ba(OH)_2(aq) + CH_3COOH(aq) \rightarrow$
(c) $Mg(OH)_2(s) + HClO_3(aq) \rightarrow$

J.5 Select an acid and a base for a neutralization reaction that results in the formation of (a) potassium bromide; (b) zinc nitrite; (c) calcium cyanide, $Ca(CN)_2$; (d) potassium phosphate. Write the balanced equation for each reaction.

J.6 Identify the salt that is produced from the acid–base neutralization reaction between (a) potassium hydroxide and acetic acid, CH_3COOH; (b) ammonia and hydroiodic acid; (c) barium hydroxide and sulfuric acid (both H atoms react); (d) sodium hydroxide and hydrocyanic acid, HCN. Write the complete ionic equation for each reaction.

J.7 Identify the acid and the base in each of the following reactions:
(a) $CH_3NH_2(aq) + H_3O^+(aq) \rightarrow CH_3NH_3^+(aq) + H_2O(l)$
(b) $C_2H_5NH_2(aq) + HCl(aq) \rightarrow C_2H_5NH_3^+(aq) + Cl^-(aq)$
(c) $2\ HI(aq) + CaO(s) \rightarrow CaI_2(aq) + H_2O(l)$

J.8 Identify the acid and the base in each of the following reactions:
(a) $CH_3COOH(aq) + NH_3(aq) \rightarrow NH_4^+(aq) + CH_3CO_2^-(aq)$
(b) $(CH_3)_3N(aq) + HCl(aq) \rightarrow (CH_3)_3NH^+(aq) + Cl^-(aq)$
(c) $O^{2-}(aq) + H_2O(l) \rightarrow 2\ OH^-(aq)$

J.9 You are asked to identify compound X, which was extracted from a plant seized by customs inspectors. You run a number of tests and collect the following data. Compound X is a white, crystalline solid. An aqueous solution of X turns litmus red and conducts electricity poorly, even when X is present at appreciable concentrations. When you add sodium hydroxide to the solution a reaction takes place. A solution of the products of the reaction conducts electricity well. An elemental analysis of X shows that the mass percentage composition of the compound is 26.68% C and 2.239% H, with the remainder being oxygen. A mass spectrum of X yields a molar mass of 90.0 g·mol^{-1}. (a) Write the empirical formula of X. (b) Write

the molecular formula of X. (c) Write the balanced chemical equation and the net ionic equation for the reaction of X with sodium hydroxide. (Assume that X has two acidic hydrogen atoms.)

J.10 (a) White phosphorus, which has the formula P_4, burns in air to form compound A, in which the mass percentage of phosphorus is 43.64%, with the remainder oxygen. The mass spectrum of A yields a molar mass of 283.9 g·mol^{-1}. Write the molecular formula of compound A. (b) Compound A reacts with water to form compound B, which turns litmus red and has a mass percentage composition of 3.087% H and 31.60% P, with the remainder oxygen. The mass spectrum of compound B yields a molar mass of 97.99 g·mol^{-1}. Write the molecular formula of compound B. (c) Compound B reacts with an aqueous solution of calcium hydroxide to form compound C, a white precipitate. Write balanced chemical equations for the reactions in parts (a), (b), and (c).

J.11 In each of the following salts, either the cation or the anion is a weak acid or a weak base. Write the chemical equation for the proton transfer reaction of this cation or anion with water: (a) NaC_6H_5O; (b) $KClO$; (c) C_5H_5NHCl; (d) NH_4Br.

J.12 $C_6H_5NH_3Cl$ is a chloride salt with an acidic cation. (a) If 50.0 g of $C_6H_5NH_3Cl$ is dissolved in water to make 150.0 mL of solution, what is the initial molarity of the cation? (b) Write the chemical equation for the proton transfer reaction of the cation with water. Identify the acid and the base in this reaction.

J.13 Na_3AsO_4 is a salt of a weak base that can accept more than one proton. (a) Write the chemical equations for the sequential proton transfer reactions of the anion with water. Identify the acid and the base in each reaction. (b) If 35.0 g of Na_3AsO_4 is dissolved in water to make 250.0 mL of solution, how many moles of sodium cations are in the solution?

J.14 The oxides of nonmetallic elements are called acidic oxides because they form acidic solutions in water. Write the balanced chemical equations for the reaction of one mole of each acidic oxide with one mole of water molecules to form an oxoacid and name the acid formed: (a) CO_2; (b) SO_3.

K   REDOX REACTIONS

Redox reactions constitute the third of the three major classes of chemical reactions treated here. The variety of these reactions is remarkable. Many common reactions, such as combustion, corrosion, photosynthesis, the metabolism of food, and the extraction of metals from their ores, appear to be completely different. However, when we consider these changes at the molecular level with a chemist's eye, we can see that they are all examples of a single type of process.

K.1 Oxidation and Reduction
K.2 Oxidation Numbers: Keeping Track of Electrons
K.3 Oxidizing and Reducing Agents
K.4 Balancing Simple Redox Equations

K.1 Oxidation and Reduction

Let's examine some of these redox reactions and look for their common characteristic. First, consider the reaction between magnesium and oxygen, which produces magnesium oxide (Fig. K.1); this reaction is used in fireworks to produce white sparks. It is also used, less agreeably, in tracer ammunition and incendiary devices. The reaction of magnesium and oxygen is a classic example of an

FIGURE K.1 An example of an oxidation reaction: magnesium burning brightly in air. Magnesium is so easily oxidized that it also burns brightly in water and carbon dioxide; consequently, magnesium fires are very difficult to extinguish.

FIGURE K.2 When chlorine is bubbled through a solution of bromide ions, it oxidizes the ions to bromine, which colors the solution reddish brown.

oxidation reaction, in the original sense of the term, "reaction with oxygen." In the course of the reaction, the Mg atoms in the solid magnesium lose electrons to form Mg^{2+} ions, and the O atoms in the molecular oxygen gain electrons to form O^{2-} ions:

$$2\ Mg(s) + O_2(g) \longrightarrow 2\ Mg^{2+}(s) + 2\ O^{2-}(s),\ as\ 2\ MgO(s)$$

A similar reaction takes place when magnesium reacts with chlorine to produce magnesium chloride:

$$Mg(s) + Cl_2(g) \longrightarrow Mg^{2+}(s) + 2\ Cl^-(s),\ as\ MgCl_2(s)$$

Because the *pattern* of reaction is the same, the second reaction is also regarded as an "oxidation" of magnesium even though no oxygen takes part. In each case, the common feature is the loss of electrons from magnesium and their transfer to another reactant. The loss of electrons from one species to another is now recognized as the essential event in an oxidation, and chemists now define **oxidation** as the loss of electrons, regardless of the species to which the electrons migrate.

We can often recognize loss of electrons by noting the increase in the charge of a species. This rule also applies to anions, as in the oxidation of bromide ions (charge −1) to bromine (charge 0) in a reaction such as the one used commercially to make bromine (Fig. K.2):

$$2\ NaBr(s) + Cl_2(g) \longrightarrow 2\ NaCl(s) + Br_2(l)$$

Here, the bromide ion (as sodium bromide) is oxidized to bromine by the chlorine gas.

The name *reduction* originally referred to the extraction of a metal from its oxide, often by reaction with hydrogen, carbon, or carbon monoxide. One example is the reduction of iron(III) oxide by carbon monoxide in the manufacture of steel:

$$Fe_2O_3(s) + 3\ CO(g) \xrightarrow{\Delta} 2\ Fe(l) + 3\ CO_2(g)$$

In this reaction, an oxide of an element is converted into the free element, the reverse of oxidation. In the reduction of iron(III) oxide, Fe^{3+} ions present in Fe_2O_3 are converted into uncharged Fe atoms when they gain electrons to neutralize their positive charges. That turns out to be the pattern common to all reductions: in a **reduction**, an atom *gains* electrons from another species. Whenever the charge on a species is decreased (as from Fe^{3+} to Fe), we say that reduction has taken place. The same rule applies if the charge is negative. For example, when chlorine is converted into chloride ions in the reaction

$$2\ NaBr(s) + Cl_2(g) \longrightarrow 2\ NaCl(s) + Br_2(l)$$

the charge decreases from 0 (in Cl_2) to −1 (in Cl^-), and we conclude that in this reaction chlorine has been reduced.

SELF-TEST K.1A Identify the species that have been oxidized or reduced in the reaction $3\ Ag^+(aq) + Al(s) \rightarrow 3\ Ag(s) + Al^{3+}(aq)$.

[*Answer:* Al(s) is oxidized, $Ag^+(aq)$ is reduced]

SELF-TEST K.1B Identify the species that have been oxidized or reduced in the reaction $2\ Cu^+(aq) + I_2(s) \rightarrow 2\ Cu^{2+}(aq) + 2\ I^-(aq)$.

We have seen that oxidation is electron loss and reduction is electron gain. Electrons are real particles and cannot just be "lost." Therefore, *whenever a species is oxidized, another species must be reduced.* Oxidation or reduction considered separately is like one hand clapping: one transfer must occur in conjunction with the other for reaction to take place. For instance, in the reaction between chlorine and sodium bromide, the bromide ions are oxidized and the chlorine molecules are reduced. Because oxidation is always accompanied by reduction, chemists speak of **redox reactions**, oxidation–reduction reactions, rather than simply oxidation reactions or reduction reactions.

Oxidation is electron loss; reduction is electron gain. Oxidation and reduction occur together in redox reactions.

K.2 Oxidation Numbers: Keeping Track of Electrons

We recognize redox reactions by noting whether electrons have migrated from one species to another. The loss or gain of electrons is easy to identify for monatomic ions, because we can monitor the charges of the species. Thus, when Br^- ions are converted into bromine atoms (which go on to form Br_2 molecules), we know that each Br^- ion must have lost an electron and hence that it has been oxidized. When O_2 forms oxide ions, O^{2-}, we know that each oxygen atom must have gained two electrons and therefore that it has been reduced. The difficulty arises when the transfer of electrons is accompanied by the transfer of atoms. For example, is chlorine gas, Cl_2, oxidized or reduced when it is converted into hypochlorite ions, ClO^-?

Chemists have found a way to keep track of electrons by assigning an "oxidation number" to each element. The **oxidation number** is defined so that

The concept of oxidation number was first introduced in Section D.

Oxidation corresponds to an *increase* in oxidation number.
Reduction corresponds to a *decrease* in oxidation number.

A redox reaction, therefore, is any reaction in which there are changes in oxidation numbers.

The oxidation number of an element in a monatomic ion is the same as its charge. For example, the oxidation number of magnesium is $+2$ when it is present as Mg^{2+} ions, and the oxidation number of chlorine is -1 when it is present as Cl^- ions. The oxidation number of the elemental form of an element is 0; so magnesium metal has oxidation number 0 and chlorine in the form of Cl_2 molecules also has oxidation number 0. When magnesium combines with chlorine, the oxidation numbers change as follows:

$$\overset{0}{Mg}(s) + \overset{2(0)}{Cl_2}(g) \longrightarrow \overset{+2 \ 2(-1)}{MgCl_2}(s)$$

We see that magnesium has been oxidized and chlorine has been reduced. Similarly, in the reaction between sodium bromide and chlorine,

$$\overset{2(+1 \ -1)}{2\ NaBr}(s) + \overset{2(0)}{Cl_2}(g) \longrightarrow \overset{2(+1 \ -1)}{2\ NaCl}(s) + \overset{2(0)}{Br_2}(l)$$

In this reaction, bromine has been oxidized and chlorine has been reduced, but the sodium ions remain unchanged as Na^+.

When an element is part of a compound or a polyatomic ion, we assign its oxidation number by using the procedure in Toolbox K.1.

A note on good practice: You will hear chemists speaking of both "oxidation number" and "oxidation state." The two terms are commonly used interchangeably, but to be precise, an oxidation *number* is the number assigned according to the rules in Toolbox K.1. An oxidation *state* is the actual condition of a species with a specified oxidation number. Thus, an element *has* a certain oxidation number and *is in* the corresponding oxidation state. For example, Mg^{2+} is the $+2$ oxidation state of magnesium; and, in that state, magnesium has oxidation number $+2$.

TOOLBOX K.1   HOW TO ASSIGN OXIDATION NUMBERS

CONCEPTUAL BASIS

To assign an oxidation number, we imagine that each atom in a molecule, formula unit, or polyatomic ion is present in ionic form (which it might not be). The oxidation number is then taken to be the charge on each "ion." The "anion" is usually oxygen as O^{2-} or the element farthest to the right in the periodic table (actually, the most *electronegative* element; see Section 2.12). We then assign to the other atoms charges that balance the charge on the "anions."

PROCEDURE

To assign an oxidation number to an element, we start with two simple rules:

1   The oxidation number of an element uncombined with other elements is 0.

2   The sum of the oxidation numbers of all the atoms in a species is equal to its total charge.

The oxidation numbers of elements in most of the compounds in this text are assigned by using these two rules along with the following specific values:

• The oxidation number of hydrogen is $+1$ in combination with nonmetals and -1 in combination with metals.

• The oxidation number of elements in Groups 1 and 2 is equal to their group number.

• The oxidation number of all the halogens is -1 unless the halogen is in combination with oxygen or another halogen higher in the group. The oxidation number of fluorine is -1 in all its compounds.

• The oxidation number of oxygen is -2 in most of its compounds. Exceptions are its compounds with fluorine (in which case, the previous statement takes precedence) and its occurrence as peroxides (O_2^{2-}), superoxides (O_2^{-}), and ozonides (O_3^{-}).

This procedure is illustrated in Example K.1.

EXAMPLE K.1   Sample exercise: Assigning oxidation numbers

Is the conversion of SO_2 to SO_4^{2-} oxidation or reduction?

SOLUTION The process is oxidation if the oxidation number of sulfur increases, reduction if it decreases. We need to assign the oxidation numbers of sulfur in SO_2 and SO_4^{2-}, then compare them. To assign an oxidation number to sulfur we represent that number by x and solve for x, by using the rules in Toolbox K.1. The oxidation number of oxygen is -2 in both compounds.

For SO_2: By rule 2, the sum of oxidation numbers of the atoms in the compound must be 0:

[Oxidation number of S]
+ [2 × (oxidation number of O)] = 0

$$x + [2 \times (-2)] = 0$$

S   2 O   zero charge on neutral molecule

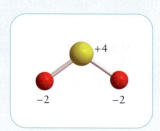

Therefore, the oxidation number of sulfur in SO_2 is $x = +4$.

For SO_4^{2-}: By rule 2, the sum of oxidation numbers of the atoms in the ion is -2; so

[Oxidation number of S]
+ [4 × (oxidation number of O)] = −2

$$x + [4 \times (-2)] = -2$$

S   4 O   total charge on ion

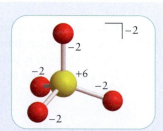

Therefore, the oxidation number of sulfur in SO_4^{2-} is $x = +6$. We can conclude that sulfur is more highly oxidized in the sulfate ion than in sulfur dioxide. Therefore, the conversion of SO_2 to SO_4^{2-} is an oxidation.

SELF-TEST K.2A Find the oxidation numbers of sulfur, phosphorus, and chlorine in (a) H_2S; (b) P_4O_6; (c) ClO^-.

[*Answer:* (a) -2; (b) $+3$; (c) $+1$]

| **SELF-TEST K.2B** Find the oxidation numbers of sulfur, nitrogen, and chlorine in (a) SO_3^{2-}; (b) NO_2^-; (c) $HClO_3$.

> *Oxidation increases the oxidation number of an element; reduction decreases the oxidation number. Oxidation numbers are assigned by using the rules in Toolbox K.1.*

K.3 Oxidizing and Reducing Agents

The species that *causes* oxidation is called the **oxidizing agent** (or "oxidant"). When an oxidizing agent acts, it accepts the electrons released by the species being oxidized. In other words, the oxidizing agent contains an element that undergoes a *decrease* in oxidation number (Fig. K.3). That is,

- The oxidizing agent in a redox reaction is the species that is reduced.

For instance, oxygen removes electrons from magnesium. Because oxygen accepts those electrons, its oxidation number decreases from 0 to -2 (a reduction). Oxygen is therefore the oxidizing agent in this reaction. Oxidizing agents can be elements, ions, or compounds.

The species that brings about reduction is called the **reducing agent** (or "reductant"). Because the reducing agent supplies electrons to the species being reduced, the reducing agent loses electrons. That is, the reducing agent contains an element that undergoes an increase in oxidation number (Fig. K.4). In other words,

- The reducing agent in a redox reaction is the species that is oxidized.

For example, when magnesium metal supplies electrons to oxygen (reducing the oxygen atoms), the magnesium atoms lose electrons and the oxidation number of magnesium increases from 0 to $+2$ (an oxidation). It is the reducing agent in the reaction of magnesium and oxygen.

To identify the oxidizing and reducing agents in a redox reaction, we compare the oxidation numbers of the elements before and after the reaction to see which have changed. The reactant that contains an element that is reduced in the reaction is the oxidizing agent, and the reactant that contains an element that is oxidized is the reducing agent. For example, when a piece of zinc metal is placed in a copper(II) solution (Fig. K.5), the reaction is

$$\overset{0}{Zn}(s) + \overset{+2}{Cu^{2+}}(aq) \longrightarrow \overset{+2}{Zn^{2+}}(aq) + \overset{0}{Cu}(s)$$

The oxidation number of zinc changes from 0 to $+2$ (an oxidation), whereas that of copper decreases from $+2$ to 0 (a reduction). Therefore, because zinc is oxidized, zinc metal is the reducing agent in this reaction. Conversely, because copper is reduced, copper(II) ions are the oxidizing agent.

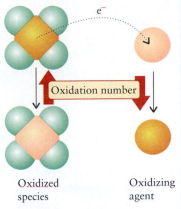

FIGURE K.3 The oxidizing agent (right) is the species containing the element that undergoes a decrease in oxidation number. Here we see how the oxidation number of the species on the left is driven upward as the oxidizing agent gains electrons.

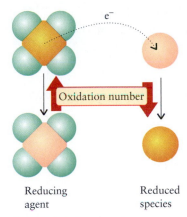

FIGURE K.4 The reducing agent (left) is the species containing the element that undergoes an increase in oxidation number. Here we see how the oxidation number of the species on the right is driven downward as it gains the electrons lost by the reducing agent.

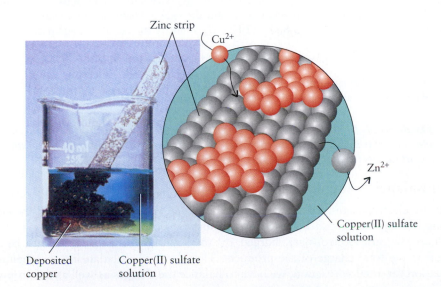

Zinc strip
Cu^{2+}
Zn^{2+}
Copper(II) sulfate solution
Deposited copper
Copper(II) sulfate solution

FIGURE K.5 When a strip of zinc is placed in a solution that contains Cu^{2+} ions, the blue solution slowly becomes colorless and copper metal is deposited on the zinc. The inset shows that, in this redox reaction, the zinc metal is reducing the Cu^{2+} ions to copper metal and the Cu^{2+} ions are oxidizing the zinc metal to Zn^{2+} ions.

Lab Video: Figure K.5 The oxidation of zinc metal by copper ions

FUNDAMENTALS

EXAMPLE K.2   Identifying oxidizing agents and reducing agents

Identify the oxidizing agent and the reducing agent in the following reaction:

$$Cr_2O_7^{2-}(aq) + 6\ Fe^{2+}(aq) + 14\ H^+(aq) \longrightarrow 6\ Fe^{3+}(aq) + 2\ Cr^{3+}(aq) + 7\ H_2O(l)$$

STRATEGY First, determine the oxidation numbers of the elements taking part in the reaction. The oxidizing agent is the species that contains an element that is reduced. Similarly, the reducing agent is the species that contains an element that is oxidized.

SOLUTION The oxidation numbers of H and O have not changed, and so we concentrate on Cr and Fe.

Determine the oxidation numbers of chromium.	As a reactant (in $Cr_2O_7^{2-}$): Let the oxidation number of Cr be x, then $$2x + [7 \times (-2)] = -2, \quad \text{or} \quad 2x - 14 = -2$$ The oxidation number of Cr in $Cr_2O_7^{2-}$ is $x = +6$. As a product (as Cr^{3+}): the oxidation number is +3.	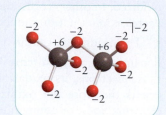
Decide whether Cr is oxidized or reduced.	Because the oxidation number of Cr decreases from +6 to +3, Cr is reduced and the dichromate ion is the oxidizing agent.	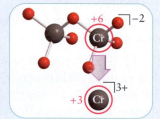
Determine the oxidation numbers of iron.	In Fe^{2+} (reactant): the oxidation number is +2. In Fe^{3+} (product): the oxidation number is +3.	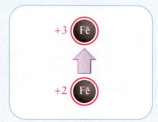
Decide whether Fe is oxidized or reduced.	Because the oxidation number of Fe increases from +2 to +3, Fe is oxidized. Therefore, the Fe^{2+} ion is the reducing agent.	

SELF-TEST K.3A In the Claus process for the recovery of sulfur from natural gas and petroleum, hydrogen sulfide reacts with sulfur dioxide to form elemental sulfur and water: $2\ H_2S(g) + SO_2(g) \rightarrow 3\ S(s) + 2\ H_2O(l)$. Identify the oxidizing agent and the reducing agent.

[*Answer:* H_2S is the reducing agent, SO_2 is the oxidizing agent]

SELF-TEST K.3B When sulfuric acid reacts with sodium iodide, sodium iodate and sulfur dioxide are produced. Identify the oxidizing and reducing agents in this reaction.

Oxidation is brought about by an oxidizing agent, a species that contains an element that undergoes reduction. Reduction is brought about by a reducing agent, a species that contains an element that undergoes oxidation.

K.4 Balancing Simple Redox Equations

Because electrons can be neither lost nor created in a chemical reaction, all the electrons lost by the species being oxidized must be transferred to the species being reduced. Because electrons are charged, the total charge of the reactants must be the same as the total charge of the products. Therefore, when balancing the chemical equation for a redox reaction, we have to balance the charges as well as the atoms.

FIGURE K.6 (a) A silver nitrate solution is colorless. (b) Some time after the insertion of a copper wire, the solution shows the blue color of the copper(II) ion and the formation of crystals of silver metal on the surface of the wire.

For example, consider the net ionic equation for the oxidation of copper metal to copper(II) ions by silver ions (Fig. K.6):

$$Cu(s) + Ag^+(aq) \longrightarrow Cu^{2+}(aq) + Ag(s)$$

At first glance, the equation appears to be balanced, because it has the same number of each kind of atom on each side. However, each copper atom has lost two electrons, whereas each silver atom has gained only one. To balance the electrons, we have to balance the charge. Therefore, we need to write

$$Cu(s) + 2\ Ag^+(aq) \longrightarrow Cu^{2+}(aq) + 2\ Ag(s)$$

SELF-TEST K.4A When tin metal is placed in contact with a solution of Fe^{3+} ions, it reduces the iron to iron(II) and is itself oxidized to tin(II) ions. Write the net ionic equation for the reaction.

[*Answer:* $Sn(s) + 2\ Fe^{3+}(aq) \rightarrow Sn^{2+}(aq) + 2\ Fe^{2+}(aq)$]

SELF-TEST K.4B In aqueous solution, cerium(IV) ions oxidize iodide ions to solid diatomic iodine and are themselves reduced to cerium(III) ions. Write the net ionic equation for the reaction.

Some redox reactions, particularly those involving oxoanions, have complex chemical equations that require special balancing procedures. We meet examples and see how to balance them in Chapter 12.

When balancing the chemical equation for a redox reaction involving ions, the total charge on each side must be balanced.

SKILLS YOU SHOULD HAVE MASTERED

❏ 1 Determine the oxidation number of an element (Toolbox K.1 and Example K.1).

❏ 2 Identify the oxidizing and reducing agents in a reaction (Example K.2).

❏ 3 Write and balance chemical equations for simple redox reactions (Self-Test K.4).

EXERCISES

K.1 Write a balanced equation for each of the following skeletal redox reactions:
(a) $NO_2(g) + O_3(g) \rightarrow N_2O_5(g) + O_2(g)$
(b) $S_8(s) + Na(s) \rightarrow Na_2S(s)$
(c) $Cr^{2+}(aq) + Sn^{4+}(aq) \rightarrow Cr^{3+}(aq) + Sn^{2+}(aq)$
(d) $As(s) + Cl_2(g) \rightarrow AsCl_3(l)$

K.2 Write a balanced equation for each of the following skeletal redox equations:
(a) $Hg^{2+}(aq) + Fe(s) \rightarrow Hg_2^{2+}(aq) + Fe^{3+}(aq)$

(b) $Pt^{4+}(aq) + H_2(g) \rightarrow Pt^{2+}(aq) + H^+(aq)$
(c) $Al(s) + Fe_2O_3(s) \rightarrow Fe(s) + Al_2O_3(s)$
(d) $La(s) + Br_2(l) \rightarrow LaBr_3(s)$

K.3 Write a balanced equation for each of the following redox reactions:
(a) Displacement of copper(II) ion from solution by magnesium metal: $Mg(s) + Cu^{2+}(aq) \rightarrow Mg^{2+}(aq) + Cu(s)$
(b) Formation of iron(III) ion in the following reaction: $Fe^{2+}(aq) + Ce^{4+}(aq) \rightarrow Fe^{3+}(aq) + Ce^{3+}(aq)$

(c) Synthesis of hydrogen chloride from its elements:
$H_2(g) + Cl_2(g) \rightarrow HCl(g)$
(d) Formation of rust (a simplified equation):
$Fe(s) + O_2(g) \rightarrow Fe_2O_3(s)$

K.4 Determine the oxidation number of the italicized element in each of the following compounds: (a) $SOCl_2$; (b) SeO_3; (c) N_2O_5; (d) NO_2; (e) $HBrO_2$; (f) XeF_2.

K.5 Determine the oxidation number of the italicized element in each of the following compounds: (a) H_2CO_3; (b) GeO_2; (c) N_2H_4; (d) P_4O_{10}; (e) S_2Cl_2; (f) P_4.

K.6 Determine the oxidation number of the italicized element in each of the following ions: (a) AlO_2^-; (b) NO_2^-; (c) SO_3^{2-}; (d) $NiCl_4^{2-}$; (e) BrF_3^+.

K.7 Determine the oxidation number of the italicized element in each of the following ions: (a) $Zn(OH)_4^{2-}$; (b) $PdCl_4^{2-}$; (c) UO_2^{2+}; (d) SiF_6^{2-}; (e) IO^-.

K.8 Some compounds of hydrogen and oxygen are exceptions to the common observation that H has oxidation number +1 and O has oxidation number −2. Assuming that each metal has the oxidation number of its most common ion, find the oxidation numbers of H and O in each of the following compounds: (a) KO_2; (b) $LiAlH_4$; (c) Na_2O_2; (d) NaH; (e) KO_3.

K.9 For each of the following redox reactions, identify the substance oxidized and the substance reduced by the change in oxidation numbers.
(a) $CH_3OH(aq) + O_2(g) \rightarrow HCOOH(aq) + H_2O(l)$
(b) $2 MoCl_5(s) + 5 Na_2S(s) \rightarrow 2 MoS_2(s) + 10 NaCl(s) + S(s)$
(c) $3 Tl^+(aq) \rightarrow 2 Tl(s) + Tl^{3+}(aq)$

K.10 In each of the following reactions, use oxidation numbers to identify the substance oxidized and the substance reduced.
(a) Production of iodine from seawater:
$Cl_2(g) + 2 I^-(aq) \rightarrow I_2(aq) + 2 Cl^-(aq)$
(b) Reaction to prepare bleach:
$Cl_2(g) + 2 NaOH(aq) \rightarrow NaCl(aq) + NaOCl(aq) + H_2O(l)$
(c) Reaction that destroys ozone in the stratosphere:
$NO(g) + O_3(g) \rightarrow NO_2(g) + O_2(g)$

K.11 Which do you expect to be the stronger oxidizing agent? Explain your reasoning. (a) Cl_2 or Cl^-; (b) N_2O_5 or N_2O.

K.12 Which do you expect to be the stronger oxidizing agent? Explain your reasoning. (a) KBrO or $KBrO_3$; (b) MnO_4^- or Mn^{2+}.

K.13 Identify the oxidizing agent and the reducing agent in each of the following reactions:
(a) $Zn(s) + 2 HCl(aq) \rightarrow ZnCl_2(aq) + H_2(g)$, a simple means of preparing H_2 gas in the laboratory
(b) $2 H_2S(g) + SO_2(g) \rightarrow 3 S(s) + 2 H_2O(l)$, a reaction used to produce sulfur from hydrogen sulfide, the "sour gas" in natural gas
(c) $B_2O_3(s) + 3 Mg(s) \rightarrow 2 B(s) + 3 MgO(s)$, a preparation of elemental boron

K.14 Identify the oxidizing agent and the reducing agent in each of the following reactions:
(a) $2 Al(l) + Cr_2O_3(s) \xrightarrow{\Delta} Al_2O_3(s) + 2 Cr(l)$, an example of a thermite reaction used to obtain some metals from their ores
(b) $6 Li(s) + N_2(g) \rightarrow 2 Li_3N(s)$, a reaction that shows the similarity of lithium and magnesium

(c) $2 Ca_3(PO_4)_2(s) + 6 SiO_2(s) + 10 C(s) \rightarrow P_4(g) + 6 CaSiO_3(s) + 10 CO(g)$, a reaction for the preparation of elemental phosphorus.

K.15 For each of the following reactions, would you choose an oxidizing agent or a reducing agent to make the conversion?
(a) $ClO_4^-(aq) \rightarrow ClO_2(g)$
(b) $SO_4^{2-}(aq) \rightarrow SO_2(g)$

K.16 For each of the following reactions, would you choose an oxidizing agent or a reducing agent to make the conversion?
(a) $H_3PO_4(aq) \rightarrow P_2O_6(s)$
(b) CH_3OH (methanol) $\rightarrow CH_2O$ (methanal)

K.17 The Sabatier process has been used to remove CO_2 from artificial atmospheres, such as those in submarines and spacecraft. An advantage is that it produces methane, CH_4, which can be burned as a fuel, and water, which can be reused. Balance the equation for the process and identify the type of reaction: $CO_2(g) + H_2(g) \rightarrow CH_4(g) + H_2O(l)$.

K.18 The industrial production of sodium metal and chlorine gas makes use of the Downs process, in which molten sodium chloride is electrolyzed (Chapter 12). Write a balanced equation for the production of the two elements from molten sodium chloride. Which element is produced by oxidation and which by reduction?

K.19 Identify the oxidizing agent and the reducing agent for each of the following reactions:
(a) Production of tungsten metal from its oxide:
$WO_3(s) + 3 H_2(g) \rightarrow W(s) + 3 H_2O(l)$
(b) Generation of hydrogen gas in the laboratory:
$Mg(s) + 2 HCl(aq) \rightarrow H_2(g) + MgCl_2(aq)$
(c) Production of metallic tin from tin(IV) oxide:
$SnO_2(s) + 2 C(s) \xrightarrow{\Delta} Sn(l) + 2 CO(g)$
(d) A reaction used to propel rockets:
$2 N_2H_4(g) + N_2O_4(g) \rightarrow 3 N_2(g) + 4 H_2O(g)$

K.20 Classify each of the following reactions as precipitation, acid–base neutralization, or redox. If a precipitation reaction, write a net ionic equation; if a neutralization reaction, identify the acid and the base; if a redox reaction, identify the oxidizing agent and the reducing agent.
(a) Reaction used to measure the concentration of carbon monoxide in a gas stream:
$5 CO(g) + I_2O_5(s) \rightarrow I_2(s) + 5 CO_2(g)$
(b) Test for the amount of iodine in a sample:
$I_2(aq) + 2 S_2O_3^{2-}(aq) \rightarrow 2 I^-(aq) + S_4O_6^{2-}(aq)$
(c) Test for bromide ions in solution:
$AgNO_3(aq) + Br^-(aq) \rightarrow AgBr(s) + NO_3^-(aq)$
(d) Heating of uranium tetrafluoride with magnesium, one stage in the purification of uranium metal:
$UF_4(g) + 2 Mg(s) \xrightarrow{\Delta} U(s) + 2 MgF_2(s)$

K.21 Nitrogen in the air of spaceships is gradually lost through leakage and must be replaced. One method is to store the nitrogen in the form of hydrazine, $N_2H_4(l)$, from which nitrogen is readily obtained by heating. The ammonia also produced can be processed further to obtain even more nitrogen: $N_2H_4(l) \rightarrow NH_3(g) + N_2(g)$. (a) Balance the equation. (b) Give the oxidation number of nitrogen in each compound. (c) Identify the oxidizing and reducing agent.

(d) Given that 28 g of nitrogen gas occupies 24 L at room temperature and pressure, what volume of nitrogen gas can be obtained from 1.0 L of hydrazine? (The density of hydrazine is 1.004 g·cm^{-3} at room temperature.)

K.22 (a) Determine and tabulate the maximum (most positive) and minimum (most negative) oxidation numbers of the elements in the first seven main groups. *Hint:* Maximum oxidation numbers are often found in the oxoanion with the most oxygen atoms. For minimum oxidation numbers, consult Fig. C.6. (b) Describe any patterns you see in the data.

K.23 A mixture of 5.00 g of $Cr(NO_3)_2$ and 6.00 g of $CuSO_4$ is dissolved in sufficient water to make 250.0 mL of solution, where the cations react. In the reaction, copper metal is formed and each chromium ion loses one electron. (a) Write the net ionic equation. (b) What is the number of electrons transferred in the balanced equation written with the smallest whole-

number coefficients? (c) What are the molar concentrations of the two anions in the final solution?

K.24 Silver tarnish is Ag_2S. (a) When silver metal is tarnished, is it oxidized or reduced? Answer this question by considering the oxidation numbers. (b) If a bar of silver is covered with 5.0 g of tarnish, what amount (in moles) of silver atoms was either oxidized or reduced? (c) What amount (in moles) of electrons was transferred in part (b)?

K.25 The following redox reactions are important in the refining of certain elements. Balance the equations and in each case, write the name of the source compound of the element (in bold face) and the oxidation state in that compound of the element that is being extracted:
(a) $\mathbf{SiCl_4}(l) + H_2(g) \rightarrow Si(s) + HCl(g)$
(b) $\mathbf{SnO_2}(s) + C(s) \xrightarrow{1200^\circ C} Sn(l) + CO_2$
(c) $\mathbf{V_2O_5}(s) + Ca(l) \xrightarrow{\Delta} V(s) + CaO(s)$
(d) $\mathbf{B_2O_3}(s) + Mg(s) \rightarrow B(s) + MgO(s)$

L   REACTION STOICHIOMETRY

L.1 Mole-to-Mole Predictions
L.2 Mass-to-Mass Predictions
L.3 Volumetric Analysis

Sometimes we need to know how much product to expect from a reaction, or how much reactant we need to make a desired amount of product. The quantitative aspect of chemical reactions is the part of chemistry called **reaction stoichiometry**. The key to reaction stoichiometry is the balanced chemical equation. Recall from Section H that a stoichiometric coefficient in a chemical equation tells us the relative amount (number of moles) of a substance that reacts or is produced. Thus, the stoichiometric coefficients in

$$N_2(g) + 3\,H_2(g) \longrightarrow 2\,NH_3(g)$$

tell us that, if 1 mol N_2 reacts, then 3 mol H_2 will be consumed and 2 mol NH_3 will be produced. We summarize this information by writing

$$1\text{ mol }N_2 \simeq 3\text{ mol }H_2 \qquad 1\text{ mol }N_2 \simeq 2\text{ mol }NH_3$$

The sign $\simeq$ is read "is chemically equivalent to," and these expressions are called **stoichiometric relations**.

The term *chemically equivalent* refers only to relative amounts in a specific chemical reaction. Different reactions give rise to different stoichiometric relations.

L.1 Mole-to-Mole Predictions

Stoichiometry has important practical applications, such as predicting how much product can be formed in a reaction. For example, in the space shuttle fuel cell, oxygen reacts with hydrogen to produce water, which is used for life support (Fig. L.1). Let's look at the calculation space shuttle engineers would have to do to find out how much water is formed when 0.25 mol O_2 reacts with hydrogen gas.

First, we write the chemical equation for the reaction:

$$2\,H_2(g) + O_2(g) \rightarrow 2\,H_2O(l)$$

Then we summarize the information that 1 mol O_2 reacts to form 2 mol H_2O by writing the stoichiometric relation between oxygen (the given substance) and water (the required substance):

$$1\text{ mol }O_2 \simeq 2\text{ mol }H_2O$$

Next, we use the stoichiometric relation to set up a conversion factor relating the given substance to the required substance:

$$\frac{\text{Substance required}}{\text{Substance given}} = \frac{2\text{ mol }H_2O}{1\text{ mol }O_2}$$

This ratio, which is commonly called the **mole ratio** for the reaction, allows us to relate the amount of O_2 molecules consumed to the amount of H_2O molecules

FIGURE L.1 One of the three hydrogen–oxygen fuel cells used on the space shuttle to provide life-support electricity and drinking water.

Stoichiometric coefficients are exact numbers; so they do not limit the significant figures of stoichiometric calculations (see Appendix 1C).

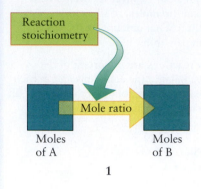

1

produced. We use it in the same way that we use a conversion factor when we are converting units, as illustrated in Section A:

$$\text{Amount of } H_2O \text{ produced (mol)} = (0.25 \text{ mol } O_2) \times \left(\frac{2 \text{ mol } H_2O}{1 \text{ mol } O_2}\right)$$
$$= 0.50 \text{ mol } H_2O$$

Note that the unit *mol* and the species (in this case, O_2 molecules) cancel. The general strategy for this type of calculation is summarized in (**1**).

SELF-TEST L.1A What amount of NH_3 can be produced from 2.0 mol H_2 in the reaction $N_2(g) + 3 H_2(g) \rightarrow 2 NH_3(g)$?

[*Answer:* 1.3 mol NH_3]

SELF-TEST L.1B What amount of Fe atoms can be extracted from 25 mol Fe_2O_3?

The balanced chemical equation for a reaction is used to set up the mole ratio, a factor that is used to convert the amount of one substance into the amount of another.

L.2 Mass-to-Mass Predictions

To find out the mass of a product that can be formed from a known mass of a reactant, we first convert the grams of reactant into moles, use the mole ratio from the balanced equation, and then convert the moles of product formed into grams. Essentially, we go through three steps:

$$\text{Mass (g) of reactant} \longrightarrow \text{amount (mol) of reactant} \longrightarrow$$
$$\text{amount (mol) of product} \longrightarrow \text{mass (g) of product}$$

For example, in Section L.1 we saw that 0.50 mol H_2O can be produced from 0.25 mol O_2. If we want to know the mass of water that can be produced, the additional step is to use the molar mass of water to convert this amount to grams by using $m = nM$:

$$\text{Mass of } H_2O \text{ produced (g)} = 0.50 \text{ mol } H_2O \times 18.02 \text{ g·mol}^{-1}$$
$$= 9.0 \text{ g } H_2O$$

TOOLBOX L.1   HOW TO CARRY OUT MASS-TO-MASS CALCULATIONS

CONCEPTUAL BASIS

A chemical equation tells us the relations between the amounts (in moles) of each reactant and product. By using the molar masses as conversion factors, we can express these relations in terms of masses.

PROCEDURE

The general procedure for mass-to-mass calculations, summarized in diagram (**2**), requires that first we write the

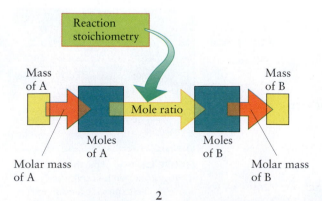

2

balanced chemical equation for the reaction. Then we perform the following calculations:

Step 1 Convert the given mass of one substance (A) in grams into amount in moles by using its molar mass.

$$n_A = \frac{m_A}{M_A}$$

If necessary, first convert the units of mass to grams.

Step 2 Use the mole ratio derived from the stoichiometric coefficients in the balanced chemical equation to convert from the amount of one substance (A) into the amount in moles of the other substance (B).
For $aA \rightarrow bB$ or $aA + bB \rightarrow cC$, use

$$n_B = n_A \times \left(\frac{b \text{ mol B}}{a \text{ mol A}}\right)$$

Step 3 Convert from amount in moles of the second substance to mass (in grams) by using its molar mass.

$$m_B = n_B M_B$$

This procedure is illustrated in Example L.1.

EXAMPLE L.1   **Sample exercise: Calculating the mass of reactant required for a given mass of product**

(a) What mass of iron(III) oxide, Fe_2O_3, present in iron ore is required to produce 10.0 g of iron when it is reduced by carbon monoxide gas to metallic iron and carbon dioxide gas in a blast furnace? (b) The carbon dioxide produced as a by-product must also be monitored to protect the environment. What mass of carbon dioxide is released when 10.0 g of iron is produced?

SOLUTION   The chemical equation is

$$Fe_2O_3(s) + 3\ CO(g) \longrightarrow 2\ Fe(s) + 3\ CO_2(g)$$

which implies that

$$2\ mol\ Fe \simeq 1\ mol\ Fe_2O_3 \quad and \quad 2\ mol\ Fe \simeq 3\ mol\ CO_2$$

(a) The molar mass of iron is 55.85 g·mol^{-1} and that of iron(III) oxide is 159.69 g·mol^{-1}.

Step 1 Convert from mass into amount of product by using its molar mass ($n = m/M$).

$$Amount\ of\ Fe\ (mol) = \frac{10.0\ g}{55.85\ g\cdot(mol\ Fe)^{-1}}$$

$$= \frac{10.0}{55.85}\ mol\ Fe$$

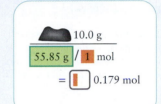

Step 2 Calculate the amount of reactant by using the mole ratio.

$$Amount\ of\ Fe_2O_3\ (mol) = \frac{10.0}{55.85}\ mol\ Fe \times \left(\frac{1\ mol\ Fe_2O_3}{2\ mol\ Fe}\right)$$

$$= \frac{10.0}{55.85 \times 2}\ mol\ Fe_2O_3$$

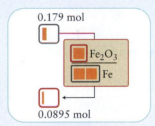

Step 3 Convert from amount into mass of reactant by using its molar mass ($m = nM$).

$$Mass\ of\ Fe_2O_3 = \frac{10.0}{55.85 \times 2}\ mol\ Fe_2O_3 \times 159.69\ g\cdot(mol\ Fe_2O_3)^{-1}$$

$$= \frac{10.0 \times 159.69}{55.85 \times 2}\ g\ =\ 14.3\ g$$

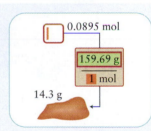

(b) The molar mass of carbon dioxide is 44.01 g·mol^{-1}

Step 1 Convert from mass into amount of product by using its molar mass ($n = m/M$).

$$Amount\ of\ Fe\ (mol) = \frac{10.0\ g\ Fe}{55.85\ g\cdot mol^{-1}}$$

$$= \frac{10.0}{55.85}\ mol\ Fe$$

Step 2 Calculate the amount of product by using the mole ratio.

$$Amount\ of\ CO_2\ (mol) = \frac{10.0}{55.85}\ mol\ Fe \times \left(\frac{3\ mol\ CO_2}{2\ mol\ Fe}\right)$$

$$= \frac{10.0 \times 3}{55.85 \times 2}\ mol\ CO_2$$

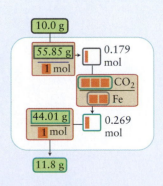

Step 3 Use molar mass to convert from amount into mass (m 5 nM).

$$Mass\ of\ CO_2 = \frac{10.0 \times 3}{55.85 \times 2}\ mol\ CO_2 \times 44.01\ g\cdot(mol\ CO_2)^{-1}$$

$$= \frac{10.0 \times 3 \times 44.01}{55.85 \times 2}\ g\ =\ 11.8\ g$$

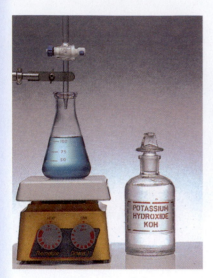

FIGURE L.2 The apparatus typically used for a titration: magnetic stirrer; flask containing the analyte; clamp; buret containing the titrant—in this case, potassium hydroxide.

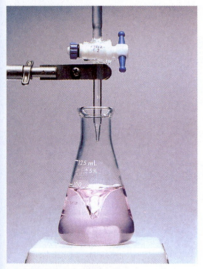

FIGURE L.3 An acid–base titration at the stoichiometric point. The indicator is phenolphthalein.

The stoichiometric point is also called the *equivalence point*.

A note on good practice: Note that although, as usual, we have left the numerical calculation to a single, final step, the same is not true for the units: canceling units does not introduce rounding errors and clarifies each step.

SELF-TEST L.2A Calculate the mass of potassium metal needed to react with 0.450 g of hydrogen gas to produce solid potassium hydride, KH.

[*Answer:* 17.5 g]

SELF-TEST L.2B Carbon dioxide can be removed from power plant exhaust gases by combining it with an aqueous slurry of calcium silicate: $2\ CO_2(g) + H_2O(l) + CaSiO_3(s) \rightarrow SiO_2(s) + Ca(HCO_3)_2(aq)$. What mass of $CaSiO_3$ (having molar mass 116.17 $g \cdot mol^{-1}$) is needed to react completely with 0.300 kg of carbon dioxide?

In a mass-to-mass calculation, convert the given mass into amount in moles, apply the mole-to-mole conversion factor to obtain the amount required, and finally convert amount in moles into mass.

L.3 Volumetric Analysis

A common laboratory technique for determining the concentration of a solute is **titration** (Fig. L.2). Titrations are usually either **acid–base titrations**, in which an acid reacts with a base, or **redox titrations**, in which the reaction is between a reducing agent and an oxidizing agent. Titrations are widely used to monitor water purity and blood composition and for quality control in the food industry.

In a titration, the volume of one solution is known, and we measure the volume of the other solution required for complete reaction. The solution being analyzed is called the **analyte**, and a known volume is transferred into a flask, usually with a pipet. Then a solution containing a known concentration of reactant is measured into the flask from a buret until all the analyte has reacted. The solution in the buret is called the **titrant**, and the difference between the initial and the final volume readings of the buret tells us the volume of titrant that has drained into the flask. The determination of concentration or amount by measuring volume is called **volumetric analysis**.

In a typical acid–base titration, the analyte is a solution of a base and the titrant is a solution of an acid or vice versa. An indicator, a water-soluble dye (Section J), helps us detect the **stoichiometric point**, the stage at which the volume of titrant added is exactly that required by the stoichiometric relation between titrant and analyte. For example, if we titrate hydrochloric acid containing a few drops of the indicator phenolphthalein, the solution is initially colorless. After the stoichiometric point, when excess base is present, the solution in the flask is basic and the indicator is pink. The indicator color change is sudden, so it is easy to detect the stoichiometric point (Fig. L.3). Toolbox L.2 shows how to interpret a titration; the procedure is summarized in diagram (3), where A is the solute in the titrant and B is the solute in the analyte.

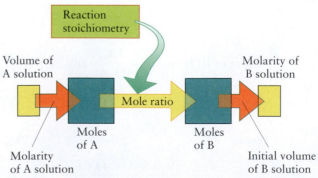

TOOLBOX L.2   HOW TO INTERPRET A TITRATION

CONCEPTUAL BASIS

In a titration one reactant (the titrant) is added gradually in solution to another (the analyte). The goal is to determine the molarity of the reactant in the analyte or its mass. We use the molarity of the titrant to find the amount of reacting titrant species in the volume of titrant supplied.

PROCEDURE

Step 1 Calculate the amount ($n_{titrant}$, in moles) of titrant species added from the volume of titrant ($V_{titrant}$) and its molarity ($c_{titrant}$).

$$\overbrace{n_{titrant}}^{\text{mol titrant}} = \overbrace{c_{titrant}}^{\text{(mol titrant)}\cdot L^{-1}} \times \overbrace{V_{titrant}}^{L}$$

Step 2 (a) Write the chemical equation for the reaction, (b) infer the mole ratio between the titrant species and the analyte species and (c) use it to convert the amount of titrant into amount of analyte ($n_{analyte}$).

$$n_{analyte} = n_{titrant} \times \text{mole ratio}$$

Step 3 Calculate the initial molarity of the analyte ($c_{analyte}$) by dividing the amount of analyte species by the initial volume, $V_{analyte}$, of the analyte.

$$\overbrace{c_{analyte}}^{\text{(mol analyte)}\cdot L^{-1}} = \frac{\overbrace{n_{analyte}}^{\text{mol analyte}}}{\underbrace{V_{analyte}}_{L}}$$

This procedure is illustrated in Example L.2.

If the mass of the analyte is required, instead of step 3, use the molar mass of the analyte to convert moles into grams.

This procedure is demonstrated in Example L.3.

A note on good practice: Because we need to keep track of the substances, we must be sure to indicate the exact species and concentration units, by writing, for example, 1.0 (mol HCl)$\cdot L^{-1}$ or 1.0 M HCl(aq).

EXAMPLE L.2   Sample exercise: Determining the molarity of an acid by titration

Suppose that 25.00 mL of a solution of oxalic acid, $H_2C_2O_4$ (**4**), which has two acidic protons, is titrated with 0.100 M NaOH(aq) and that the stoichiometric point is reached when 38.0 mL of the solution of base is added. Find the molarity of the oxalic acid solution.

4 Oxalic acid, $(COOH)_2$

SOLUTION Because each acid molecule provides two protons, if the concentrations of acid and base were the same, the titration would require a volume of base equal to *twice* the volume of acid. The volume of base added to reach the stoichiometric point is less than twice the volume of acid; so we can expect that the acid is less concentrated than the base. Proceed as in Toolbox L.2.

Step 1 Find the amount of NaOH added.

$$n_{NaOH} = (38.0 \times 10^{-3}\,L) \times 0.100\,(\text{mol NaOH})\cdot L^{-1}$$
$$= 38.0 \times 10^{-3} \times 0.100\ \text{mol NaOH}$$

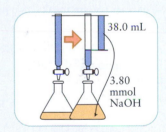

Step 2 (a) Write the chemical equation. (b) Infer the mole ratio. (c) Calculate the amount of acid present.

(a) $H_2C_2O_4(aq) + 2\,NaOH(aq) \longrightarrow Na_2C_2O_4(aq) + 2\,H_2O(l)$
(b) 2 mol NaOH ≏ 1 mol $H_2C_2O_4$
(c) $n_{H_2C_2O_4} = (38.0 \times 10^{-3} \times 0.100\ \text{mol NaOH}) \times \left(\dfrac{1\ \text{mol } H_2C_2O_4}{2\ \text{mol NaOH}}\right)$

$$= \tfrac{1}{2} \times (38.0 \times 10^{-3} \times 0.100)\ \text{mol } H_2C_2O_4$$

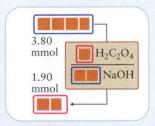

Step 3 Calculate the molarity of the acid.

$$c_{H_2C_2O_4} = \frac{\tfrac{1}{2} \times (38.0 \times 10^3 \times 0.100)\ \text{mol } H_2C_2O_4}{25.00 \times 10^{-3}\ L}$$
$$= 0.0760\ (\text{mol } H_2C_2O_4)\cdot L^{-1}$$

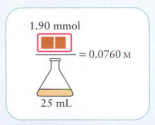

That is, the solution is 0.0760 M $H_2C_2O_4$(aq). As we predicted, the acid is less concentrated than the base.

FUNDAMENTALS

SELF-TEST L.3A A student used a sample of hydrochloric acid that was known to contain 0.72 g of hydrogen chloride in 500.0 mL of solution to titrate 25.0 mL of a solution of calcium hydroxide. The stoichiometric point was reached when 15.1 mL of acid had been added. What was the molarity of the calcium hydroxide solution?

[*Answer:* 0.012 M $Ca(OH)_2(aq)$]

SELF-TEST L.3B Many abandoned mines have exposed nearby communities to the problem of acid mine drainage. Certain minerals, such as pyrite (FeS_2), decompose when exposed to air, forming solutions of sulfuric acid. The acidic mine water then drains into lakes and creeks, killing fish and other animals. At a mine in Colorado, a 16.45-mL sample of mine water was completely neutralized with 25.00 mL of 0.255 M KOH(aq). What is the molar concentration of H_2SO_4 in the water?

EXAMPLE L.3 **Determining the purity of a sample by means of a redox titration**

The iron content of ores can be determined by titrating a sample with a solution of potassium permanganate, $KMnO_4$. The ore is dissolved in hydrochloric acid, forming iron(II) ions, which react with MnO_4^-:

$$5\ Fe^{2+}(aq) + MnO_4^-(aq) + 8\ H^+(aq) \longrightarrow 5\ Fe^{3+}(aq) + Mn^{2+}(aq) + 4\ H_2O(l)$$

The stoichiometric point is reached when all the Fe^{2+} has reacted and is detected when the purple color of the permanganate ion persists. A sample of ore of mass 0.202 g was dissolved in hydrochloric acid, and the resulting solution needed 16.7 mL of 0.0108 M $KMnO_4(aq)$ to reach the stoichiometric point. (a) What mass of iron(II) ions is present? (b) What is the mass percentage of iron in the ore sample?

STRATEGY (a) To obtain the amount of iron(II) in the analyte, we use the volume and concentration of the titrant. We follow the first two steps of the procedure in Toolbox L.2. Then we convert moles of Fe^{2+} ions into mass by using the molar mass of Fe^{2+}: because the mass of electrons is so small, we use the molar mass of elemental iron for the molar mass of iron(II) ions. (b) We divide the mass of iron by the mass of the ore sample and multiply by 100%.

SOLUTION (a) Find the mass of iron present in the sample. From the chemical equation, the stoichiometric relation between the iron and the permanganate ions is

$$5\ mol\ Fe^{2+} \simeq 1\ mol\ MnO_4^-$$

We can set up the mass calculation as follows:

Find the amount of MnO_4^- added from $n = cV$,

$$n_{MnO_4^-} = (1.67 \times 10^{-2}\,L) \times \frac{0.0108\ (mol\ MnO_4^-)}{L}$$

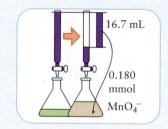

Find the amount of Fe^{2+} present from the mole ratio for the reaction,

$$n_{Fe} = (1.67 \times 10^{-2}\,L) \times \frac{0.0108\ (mol\ MnO_4^-)}{L}$$
$$\times \frac{5\ mol\ Fe^{2+}}{1\ mol\ MnO_4^-}$$

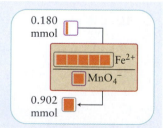

Find the mass of iron from $m = nM$,

$$m_{Fe} = (1.67 \times 10^{-2}\,L) \times \frac{0.0108\ (mol\ MnO_4^-)}{L}$$
$$\times \frac{5\ mol\ Fe^{2+}}{1\ mol\ MnO_4^-} \times 55.85\ g\cdot(mol\ Fe^{2+})^{-1}$$
$$= 0.0504\ g$$

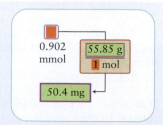

The sample contains 0.0504 g of iron.

(b) The mass percentage of iron in the ore is

From mass percentage = (mass of iron)/(mass of sample),

$$\text{Mass percentage iron} = \frac{\overset{\text{mass of Fe}}{0.0504 \text{ g}}}{\underset{\text{mass of sample}}{0.202 \text{ g}}} \times 100\% = 25.0\%$$

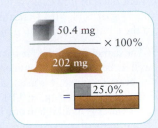

SELF-TEST L.4A A sample of clay of mass 20.750 g for use in making ceramics was analyzed to determine its iron content. The clay was washed with hydrochloric acid, and the iron converted into iron(II) ions. The resulting solution was titrated with cerium(IV) sulfate solution:

$$Fe^{2+}(aq) + Ce^{4+}(aq) \longrightarrow Fe^{3+}(aq) + Ce^{3+}(aq)$$

In the titration, 13.45 mL of 1.340 M $Ce(SO_4)_2$(aq) was needed to reach the stoichiometric point. What is the mass percentage of iron in the clay?

[*Answer:* 4.85%]

SELF-TEST L.4B The amount of arsenic(III) oxide in a mineral can be determined by dissolving the mineral in acid and titrating it with potassium permanganate:

$$24 \text{ H}^+(aq) + 5 \text{ As}_4O_6(s) + 8 \text{ MnO}_4^-(aq) + 18 \text{ H}_2O(l) \longrightarrow$$
$$8 \text{ Mn}^{2+}(aq) + 20 \text{ H}_3AsO_4(aq)$$

A sample of industrial waste was analyzed for the presence of arsenic(III) oxide by titration with 0.0100 M $KMnO_4$. It took 28.15 mL of the titrant to reach the stoichiometric point. How many grams of arsenic(III) oxide did the sample contain?

The stoichiometric relation between analyte and titrant species, together with the molarity of the titrant, is used in titrations to determine the molarity of the analyte.

SKILLS YOU SHOULD HAVE MASTERED

❏ 1 Carry out stoichiometric calculations for any two species taking part in a chemical reaction (Toolbox L.1 and Example L.1).

❏ 2 Calculate the molar concentration (molarity) of a solute from titration data (Toolbox L.2 and Example L.2).

❏ 3 Calculate the mass of a solute from titration data (Toolbox L.2 and Example L.3).

EXERCISES

L.1 Sodium thiosulfate, photographer's hypo, reacts with unexposed silver bromide in film emulsion:

$$2 \text{ Na}_2S_2O_3(aq) + AgBr(s) \longrightarrow$$
$$NaBr(aq) + Na_3[Ag(S_2O_3)_2](aq)$$

(a) How many moles of $Na_2S_2O_3$ are needed to react with 1.0 mg of AgBr? (b) Calculate the mass of silver bromide that will produce 0.033 mol $Na_3[Ag(S_2O_3)_2]$.

L.2 Impure phosphoric acid for use in the manufacture of fertilizers is produced by the reaction of sulfuric acid on phosphate rock, of which a principal component is $Ca_3(PO_4)_2$. The reaction is $Ca_3(PO_4)_2(s) + 3 \text{ H}_2SO_4(aq) \rightarrow 3 \text{ CaSO}_4(s) + 2 \text{ H}_3PO_4(aq)$. (a) How many moles of H_3PO_4 can be produced from the reaction of 120. kg of H_2SO_4? (b) Determine the mass of calcium sulfate that is produced as a by-product of the reaction of 300. mol $Ca_3(PO_4)_2$.

L.3 The solid fuel in the booster stage of the space shuttle is a mixture of ammonium perchlorate and aluminum powder. On ignition, the reaction that takes place is $6 \text{ NH}_4ClO_4(s) + 10 \text{ Al}(s) \rightarrow 5 \text{ Al}_2O_3(s) + 3 \text{ N}_2(g) + 6 \text{ HCl}(g) + 9 \text{ H}_2O(g)$. (a) What mass of aluminum should be mixed with 1.325 kg of NH_4ClO_4 for this reaction? (b) Determine the mass of Al_2O_3 (alumina, a finely divided white powder that is produced as billows of white smoke) formed in the reaction of 3.500×10^3 kg of aluminum.

L.4 The compound diborane, B_2H_6, was at one time considered for use as a rocket fuel. Its combustion reaction is

$$B_2H_6(g) + 3 \text{ O}_2(l) \longrightarrow 2 \text{ HBO}_2(g) + 2 \text{ H}_2O(l)$$

The fact that HBO_2, a reactive compound, was produced rather than the relatively inert B_2O_3 was a factor in the discontinuation of the investigation of diborane as a fuel.

(a) What mass of liquid oxygen (LOX) would be needed to burn 257 g of B_2H_6? (b) Determine the mass of HBO_2 produced from the combustion of 106 g of B_2H_6.

L.5 The camel stores the fat tristearin, $C_{57}H_{110}O_6$, in its hump. As well as being a source of energy, the fat is also a source of water because, when it is used, the reaction $2\ C_{57}H_{110}O_6(s) + 163\ O_2(g) \rightarrow 114\ CO_2(g) + 110\ H_2O(l)$ takes place. (a) What mass of water is available from 1.00 pound (454 g) of this fat? (b) What mass of oxygen is needed to oxidize this amount of tristearin?

L.6 Potassium superoxide, KO_2, is utilized in a closed-system breathing apparatus to remove carbon dioxide and water from exhaled air. The removal of water generates oxygen for breathing by the reaction $4\ KO_2(s) + 2\ H_2O(l) \rightarrow 3\ O_2(g) + 4\ KOH(s)$. The potassium hydroxide removes carbon dioxide from the apparatus by the reaction $KOH(s) + CO_2(g) \rightarrow KHCO_3(s)$. (a) What mass of potassium superoxide generates 115 g of O_2? (b) What mass of CO_2 can be removed from the apparatus by 75.0 g of KO_2?

L.7 When a hydrocarbon burns, water is produced as well as carbon dioxide. The density of gasoline is 0.79 $g \cdot mL^{-1}$. Assume gasoline to be represented by octane, C_8H_{18}, for which the combustion reaction is $2\ C_8H_{18}(l) + 25\ O_2(g) \rightarrow 16\ CO_2(g) + 18\ H_2O(l)$. Calculate the mass of water produced from the combustion of 1.0 gallon (3.8 L) of gasoline.

L.8 The density of oak is 0.72 $g \cdot cm^{-3}$. Assuming oak to have the empirical formula CH_2O, calculate the mass of water produced when a log of dimensions 12 cm $\times$ 14 cm $\times$ 25 cm is completely burned to $CO_2(g)$ and $H_2O(l)$.

L.9 A sample of sodium hydroxide of volume 15.00 mL was titrated to the stoichiometric point with 17.40 mL of 0.234 M $HCl(aq)$. (a) What is the initial molarity of NaOH in the solution? (b) Calculate the mass of NaOH in the solution.

L.10 A sample of oxalic acid, $H_2C_2O_4$ (with two acidic protons), of volume 25.17 mL was titrated to the stoichiometric point with 25.67 mL of 0.327 M $NaOH(aq)$. (a) What is the molarity of the oxalic acid? (b) Determine the mass of oxalic acid in the solution.

L.11 A sample of barium hydroxide of mass 9.670 g was dissolved and diluted to the mark in a 250.0-mL volumetric flask. It was found that 11.56 mL of this solution was needed to reach the stoichiometric point in a titration of 25.0 mL of a nitric acid solution. (a) Calculate the molarity of the HNO_3 solution. (b) What mass of HNO_3 is in the initial sample?

L.12 Suppose that 10.0 mL of 3.0 M $KOH(aq)$ is transferred to a 250.0-mL volumetric flask and diluted to the mark. It was found that 38.5 mL of this diluted solution was needed to reach the stoichiometric point in a titration of 10.0 mL of a phosphoric acid solution according to the reaction $3\ KOH(aq) + H_3PO_4(aq) \rightarrow K_3PO_4(aq) + 3\ H_2O(l)$. (a) Calculate the molarity of H_3PO_4 in the solution. (b) What mass of H_3PO_4 is in the initial sample?

L.13 In a titration, 3.25 g of an acid, HX, requires 68.8 mL of 0.750 M $NaOH(aq)$ for complete reaction. What is the molar mass of the acid?

L.14 Suppose that 14.56 mL of 0.115 M $NaOH(aq)$ was required to titrate 0.2037 g of an unknown acid, HX. What is the molar mass of the acid?

L.15 Excess NaI was added to 50.0 mL of aqueous $CuNO_3$ solution, and 15.75 g of CuI precipitate was formed. What was the molar concentration of $CuNO_3$ in the original solution?

L.16 An excess of $AgNO_3$ reacts with 25.0 mL of 5.0 M $K_2CrO_4(aq)$ to form a precipitate. What is the precipitate, and what mass of precipitate is formed?

L.17 A solution of hydrochloric acid was prepared by measuring 10.00 mL of the concentrated acid into a 1.000-L volumetric flask and adding water up to the mark. Another solution was prepared by adding 0.832 g of anhydrous sodium carbonate to a 100.0-mL volumetric flask and adding water up to the mark. Then, 25.00 mL of the carbonate solution was pipetted into a flask and titrated with the diluted acid. The stoichiometric point was reached after 31.25 mL of acid had been added. (a) Write a balanced equation for the reaction of $HCl(aq)$ with $Na_2CO_3(aq)$. (b) What is the molarity of the original hydrochloric acid?

L.18 A tablet of vitamin C was analyzed to determine whether it did in fact contain, as the manufacturer claimed, 1.0 g of the vitamin. One tablet was dissolved in water to form 100.00 mL of solution, and 10.0-mL of that solution was titrated with iodine (as potassium triiodide). It required 10.1 mL of 0.0521 M $I_3^-(aq)$ to reach the stoichiometric point in the titration. Given that 1 mol I_3^- reacts with 1 mol vitamin C in the reaction, is the manufacturer's claim correct? The molar mass of vitamin C is 176 $g \cdot mol^{-1}$.

L.19 Iodine is a common oxidizing agent, often used as the triiodide ion, I_3^-. Suppose that in the presence of $HCl(aq)$, 25.00 mL of 0.120 M aqueous triiodide solution reacts completely with 30.00 mL of a solution containing 19.0 $g \cdot L^{-1}$ of an ionic compound containing tin and chlorine. The products are iodide ions and another compound of tin and chlorine. The reactant compound is 62.6% tin by mass. Write a balanced equation for the reaction.

L.20 A forensic laboratory is analyzing a mixture of the two solids calcium chloride dihydrate, $CaCl_2 \cdot 2H_2O$, and potassium chloride, KCl. The mixture is heated to drive off the water of hydration: $CaCl_2 \cdot 2H_2O(s) \xrightarrow{\Delta} CaCl_2(s) + 2\ H_2O(g)$. A sample of the mixture weighed 2.543 g before heating. After heating, the resulting mixture of anhydrous $CaCl_2$ and KCl weighed 2.312 g. Calculate the mass percentage of each compound in the original sample.

L.21 Thiosulfate ions ($S_2O_3^{2-}$) "disproportionate" in acidic solution to give solid sulfur (S) and hydrogen sulfite ion (HSO_3^-):

$$2\ S_2O_3^{2-}(aq) + 2\ H_3O^+(aq) \longrightarrow$$
$$2\ HSO_3^-(aq) + 2\ H_2O(l) + 2\ S(s)$$

(a) A disproportionation reaction is a type of oxidation–reduction reaction. Which species is oxidized and which is reduced? (b) If 10.1 mL of 55.0% HSO_3^- by mass is obtained in the reaction, what mass of $S_2O_3^{2-}$ was present initially, assuming the reaction went to completion? The density of the HSO_3^- solution is 1.45 $g \cdot cm^{-3}$.

L.22 Suppose that 25.0 mL of 0.50 M $K_2CrO_4(aq)$ reacts with 15.0 mL of $AgNO_3(aq)$ completely. What mass of NaCl is needed to react completely with 35.0 mL of the same $AgNO_3$ solution?

L.23 The compound $XCl_2(NH_3)_2$ can be formed by reacting XCl_4 with NH_3. Suppose that 3.571 g of XCl_4 reacts with excess NH_3 to give 3.180 g of $XCl_2(NH_3)_2$. What is the element X?

L.24 The reduction of iron(III) oxide to iron metal in a blast furnace is another source of atmospheric carbon dioxide. The reduction takes place in these two steps:

$$2\ C(s) + O_2(g) \longrightarrow 2\ CO(g)$$
$$Fe_2O_3(s) + 3\ CO(g) \longrightarrow 2\ Fe(l) + 3\ CO_2(g)$$

Assume that all the CO generated in the first step reacts in the second. (a) How many C atoms are needed to react with 600. Fe_2O_3 formula units? (b) What is the maximum volume of carbon dioxide (taken to have a density of $1.25\ g \cdot L^{-1}$) that can be generated in the production of 1.0 t of iron (1 t = 10^3 kg)? (c) Assuming a 67.9% yield, what volume of carbon dioxide is released to the atmosphere in the production of 1.0 t of iron? (d) How many kilograms of O_2 are required for the production of 5.00 kg of Fe?

L.25 Barium bromide, $BaBr_x$, can be converted into $BaCl_2$ by treatment with chlorine. It is found that 3.25 g of $BaBr_x$ reacts completely with an excess of chlorine to yield 2.27 g of $BaCl_2$. Determine the value of x and write the balanced chemical equation for the production of $BaCl_2$ from $BaBr_x$.

L.26 Sulfur is an undesirable impurity in coal and petroleum fuels. The mass percentage of sulfur in a fuel can be determined by burning the fuel in oxygen and dissolving the SO_3 produced in water to form aqueous sulfuric acid. In one experiment, 8.54 g of a fuel was burned, and the resulting sulfuric acid was titrated with 17.54 mL of 0.100 M NaOH(aq). (a) Determine the amount (in moles) of H_2SO_4 that was produced. (b) What is the mass percentage of sulfur in the fuel?

L.27 Sodium bromide, NaBr, which is used to produce AgBr for use in photographic film, can itself be prepared as follows.

$$Fe + Br_2 \longrightarrow FeBr_2$$
$$FeBr_2 + Br_2 \longrightarrow Fe_3Br_8$$
$$Fe_3Br_8 + Na_2CO_3 \longrightarrow NaBr + CO_2 + Fe_3O_4$$

How much iron, in kilograms, is needed to produce 2.50 t of NaBr? *Note that these equations must first be balanced!*

L.28 Silver nitrate is an expensive laboratory reagent that is often used for quantitative analysis of chloride ion. A student preparing to conduct a particular analysis needs 100.0 mL of 0.0750 M $AgNO_3$(aq), but finds only about 60 mL of 0.0500 M $AgNO_3$(aq). Instead of making up a fresh solution of the exact concentration desired (0.0750 M), the student decides to pipet 50.0 mL of the existing solution into a 100.0-mL flask, then add enough pure solid $AgNO_3$ to make up the difference and enough water to bring the volume of the resulting solution to exactly 100.0 mL. What mass of solid $AgNO_3$ must be added in the second step?

L.29 (a) How would you prepare 1.00 L of 0.50 M HNO_3(aq) from "concentrated" (16 M) HNO_3(aq)? (b) How many milliliters of 0.20 M NaOH(aq) could be neutralized by 100. mL of the diluted solution?

L.30 You have been given a sample of an unknown diprotic acid. (a) Analysis of the acid shows that a 10.0-g sample contains 0.224 g of hydrogen, 2.67 g of carbon and the rest oxygen. Determine the empirical formula of the acid. (b) A 0.0900-g sample of your unknown acid is dissolved in 30.0 mL of water and titrated to the endpoint with 50.0 mL of 0.040 M NaOH(aq). Determine the molecular formula of the acid. (c) Write a balanced chemical equation for the neutralization of the unknown acid with NaOH(aq).

L.31 A 1.50-g sample of metallic tin was placed in a 26.45-g crucible and heated until all the tin had reacted with the oxygen in air to form an oxide. The crucible and product together were found to weigh 28.35 g. (a) What is the empirical formula of the oxide? (b) Write the name of the oxide.

M   LIMITING REACTANTS

Stoichiometric calculations of the amount of product formed in a reaction are based on an ideal view of the world. They suppose, for instance, that all the reactants react exactly as described in the chemical equation. In practice, that might not be so. Some of the starting materials may be consumed in a **competing reaction,** a reaction taking place at the same time as the one in which we are interested and using some of the same reactants. Another possibility is that the reaction might not be complete at the time we make our measurements. A third possibility—of major importance in chemistry and covered in several chapters of this text—is that many reactions do not go to completion. They appear to stop once a certain proportion of the reactants has been consumed.

M.1 Reaction Yield

The **theoretical yield** of a reaction is the *maximum* quantity (amount, mass, or volume) of product that can be obtained from a given quantity of reactant. The quantities of products calculated from a given mass of reactant in Section L were all theoretical yields. The **percentage yield** is the fraction of the theoretical yield actually produced, expressed as a percentage:

$$\text{Percentage yield} = \frac{\text{actual yield}}{\text{theoretical yield}} \times 100\% \qquad (1)^*$$

EXAMPLE M.1   **Calculating the percentage yield of a product**

Combustion of the octane in a poorly tuned engine can produce toxic carbon monoxide along with the usual carbon dioxide and water. In a test of an automobile engine 1.00 L of octane (702 g) is burned, but only 1.84 kg of carbon dioxide is produced. What is the percentage yield of carbon dioxide?

STRATEGY   Begin by writing the chemical equation for the complete oxidation of octane to carbon dioxide and water. Then calculate the theoretical yield (in grams) of CO_2 by using the procedure in Toolbox L.1. To avoid rounding errors, do all the numerical work at the end of the calculation. To obtain the percentage yield, divide the actual mass produced by the theoretical mass of product and multiply by 100%.

SOLUTION   The molar mass of C_8H_{18} is 114.2 g·mol^{-1} and the molar mass of CO_2 is 44.01 g·mol^{-1}.

Write the chemical equation.

$$2\,C_8H_{18}(l) + 25\,O_2(g) \longrightarrow 16\,CO_2(g) + 18\,H_2O(l)$$

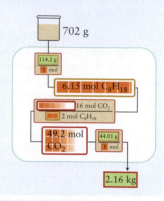

Calculate the theoretical yield of CO_2 for the combustion of 702 g of octane.

$$m_{CO_2} = \frac{702\ \text{g}}{114.2\ \text{g·(mol }C_8H_{18})^{-1}}$$
$$\times \frac{16\ \text{mol }CO_2}{2\ \text{mol }C_8H_{18}} \times 44.01\ \text{g·(mol }CO_2)^{-1}$$
$$= 2.16 \times 10^3\ \text{g}$$

Calculate the percentage yield of carbon dioxide from the fact that only 1.84 kg was produced.

$$\text{Percentage yield of }CO_2 = \frac{1.84\ \text{kg}}{2.16\ \text{kg}} \times 100\% = 85.2\%$$

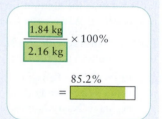

The theoretical yield corresponds to about 8 kg of CO_2 per gallon of fuel. The average car emits about 0.4 kg of CO_2 per mile of travel.

SELF-TEST M.1A   When 24.0 g of potassium nitrate was heated with lead, 13.8 g of potassium nitrite was formed in the reaction $Pb(s) + KNO_3(s) \xrightarrow{\Delta} PbO(s) + KNO_2(s)$. Calculate the percentage yield of potassium nitrite.

[*Answer:* 68.3%]

SELF-TEST M.1B   Reduction of 15 kg of iron(III) oxide in a blast furnace produced 8.8 kg of iron. What is the percentage yield of iron?

The theoretical yield of a product is the maximum quantity that can be expected on the basis of the stoichiometry of a chemical equation. The percentage yield is the percentage of the theoretical yield actually achieved.

M.2 The Limits of Reaction

The **limiting reactant** in a reaction is the reactant that governs the maximum yield of product. A limiting reactant is like a part in short supply in a motorcycle factory. Suppose there are eight wheels and seven motorcycle frames. Because each frame requires two wheels, there are enough wheels for only four motorcycles, so the wheels play the role of the limiting reactant. When all the wheels have been used, three frames remain unused, because they were present in excess.

In some cases, we must determine by calculation which is the limiting reactant. For example, from the equation

$$N_2(g) + 3\,H_2(g) \longrightarrow 2\,NH_3(g)$$

we see that, in the synthesis of ammonia, 1 mol $N_2 \simeq$ 3 mol H_2. To decide which reactant is the limiting one, we compare the number of moles of each reactant supplied with the stoichiometric coefficients. For example, suppose we had available 1 mol N_2 but only 2 mol H_2. Because this amount of hydrogen is less than is required by the stoichiometric relation, hydrogen would be the limiting reactant. Once we have identified the limiting reactant, we can calculate the amount of product that can be formed. We can also calculate the amount of excess reactant that remains at the end of the reaction.

TOOLBOX M.1   HOW TO IDENTIFY THE LIMITING REACTANT

CONCEPTUAL BASIS

The limiting reactant is the reactant that will be completely used up. All other reactants are in excess. Because the limiting reactant is the one that limits the amounts of products that can be formed, the theoretical yield is calculated from the amount of the limiting reactant.

PROCEDURE

There are two ways of determining which reactant is the limiting one.

Method 1:

In this approach, we use the mole ratio from the chemical equation to determine whether there is enough of one reactant to react with another.

Step 1 Calculate the amount of each reactant in moles, converting any masses into amounts by using the molar mass.

Step 2 Choose one of the reactants and use the stoichiometric relation to calculate the theoretical amount of the second reactant needed for complete reaction with the first.

Step 3 If the actual amount of the second reactant is greater than the amount needed (the value calculated in step 2), then the second reactant is present in excess; in this case, the first reactant is the limiting reactant. If the actual amount of the second reactant is less than that calculated, then all of it will react; so it is the limiting reactant and the first reactant is in excess.

This method is used in Example M.2.

Method 2:

Calculate the theoretical molar yield of one of the products for each reactant separately, by using the procedure in Toolbox L.1. This method is a good one to use when there are more than two reactants. The reactant that would produce the smallest amount of product is the limiting reactant.

Step 1 Convert the mass of each reactant into moles, if necessary, by using the molar masses of the substances.

Step 2 Select one of the products. For each reactant, calculate how many moles of the product it can form.

Step 3 The reactant that can produce the least amount of product is the limiting reactant.

This method is used in Example M.3.

EXAMPLE M.2   Sample exercise: Identifying the limiting reactant

Calcium carbide, CaC_2, reacts with water to form calcium hydroxide and the flammable gas ethyne (acetylene). This reaction was once used for lamps on bicycles, because the reactants are easily transported. (a) Which is the limiting reactant when 1.00×10^2 g of water reacts with 1.00×10^2 g of calcium carbide? (b) What mass of ethyne can be produced? (c) What mass of excess reactant remains after reaction is complete? Assume that the calcium carbide is pure and that all the ethyne produced is collected. The chemical equation is

$$CaC_2(s) + 2\,H_2O(l) \longrightarrow Ca(OH)_2(aq) + C_2H_2(g)$$

SOLUTION We follow the procedure in Method 1 of Toolbox M.1. (a) The molar mass of calcium carbide is 64.10 $g \cdot mol^{-1}$ and that of water is 18.02 $g \cdot mol^{-1}$.

Step 1 Determine the amount of each reactant.

$$n_{CaC_2} = \frac{100.\,g}{64.10\,g \cdot (mol\,CaC_2)^{-1}}$$

$$= \frac{100.}{64.10}\,mol\,CaC_2 = 1.56\,mol\,CaC_2$$

$$n_{H_2O} = \frac{100.\,g}{18.02\,g \cdot (mol\,H_2O)^{-1}}$$

$$= \frac{100.}{18.02}\,mol\,H_2O = 5.55\,mol\,H_2O$$

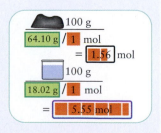

Step 2 Write the stoichiometric relation.

Calculate the amount of H_2O that is needed to react with 1.56 mol CaC_2.

$$1 \text{ mol } CaC_2 \simeq 2 \text{ mol } H_2O$$

$$n_{H_2O} = \left(\frac{100.}{64.10} \text{ mol } CaC_2\right) \times \left(\frac{2 \text{ mol } H_2O}{1 \text{ mol } CaC_2}\right)$$

$$= 3.12 \text{ mol } H_2O$$

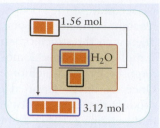

Step 3 Determine which reactant is the limiting reactant.

Because 3.12 mol H_2O is required and 5.55 mol H_2O is supplied, all the calcium carbide can react; so the calcium carbide is the limiting reactant and water is present in excess.

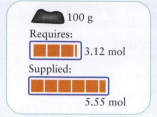

(b) Because CaC_2 is the limiting reactant and 1 mol $CaC_2 \simeq 1$ mol C_2H_2, the mass of ethyne (of molar mass 26.04 g·mol^{-1}) that can be produced is

$$m_{C_2H_2} = \left(\frac{100. \text{ g}}{64.10 \text{ g}\cdot(CaC_2 \text{ mol})^{-1}}\right) \times \left(\frac{1 \text{ mol } C_2H_2}{1 \text{ mol } CaC_2}\right) \times (26.04 \text{ g}\cdot(\text{mol } C_2H_2)^{-1})$$

$$= 40.6 \text{ g}$$

(c) The reactant in excess is water. Because 5.55 mol H_2O was supplied and 3.12 mol H_2O has been consumed, the amount of H_2O remaining is $5.55 - 3.12$ mol $= 2.43$ mol. Therefore, the mass of the excess reactant remaining at the end of the reaction is

$$\text{Mass of } H_2O \text{ remaining} = (2.43 \text{ mol}) \times (18.02 \text{ g}\cdot\text{mol}^{-1}) = 43.8 \text{ g}$$

SELF-TEST M.2A (a) Identify the limiting reactant in the reaction $6 \text{ Na(l)} + Al_2O_3(s) \rightarrow 2 \text{ Al(l)} + 3 \text{ Na}_2O(s)$ when 5.52 g of sodium is heated with 5.10 g of Al_2O_3. (b) What mass of aluminum can be produced? (c) What mass of excess reactant remains at the end of the reaction?

[*Answer:* (a) Sodium; (b) 2.16 g Al; (c) 1.02 g Al_2O_3]

SELF-TEST M.2B (a) What is the limiting reactant for the preparation of urea from ammonia in the reaction $2 \text{ NH}_3(g) + CO_2(g) \rightarrow OC(NH_2)_2(s) + H_2O(l)$ when 14.5 kg of ammonia is available to react with 22.1 kg of carbon dioxide? (b) What mass of urea can be produced? (c) What mass of excess reactant remains at the end of the reaction?

EXAMPLE M.3 Calculating percentage yield from a limiting reactant

Freons, types of chlorofluorocarbons, at one time were used extensively in spray cans and as coolants in refrigerators and air conditioners. Unfortunately, they contribute to global warming and attack the Earth's protective ozone layer. One of the most promising substitutes is $C_2H_2F_4$, which is called HFC-134a in industry. The reaction

$$C_2HF_3(l) + HF(g) \longrightarrow C_2H_2F_4(l)$$

can be used to make HFC-134a. Unfortunately, byproducts can form, reducing the yield. In a certain reaction, 100.0 g of C_2HF_3 is mixed with 30.12 g of HF. The mass of $C_2H_2F_4$ produced was 95.2 g. What was the percentage yield of the reaction?

STRATEGY First, the limiting reactant must be identified (Toolbox M.1). This limiting reactant determines the theoretical yield of the reaction, and so we use it to calculate the theoretical amount of product by Method 2 in Toolbox L.1. The percentage yield is the ratio of the mass produced to the theoretical mass times 100. Molar masses are calculated using the information in the periodic table inside the front cover of this book.

M.12 A compound produced as a by-product in an industrial synthesis of polymers was found to contain carbon, hydrogen, and iodine. A combustion analysis of 1.70 g of the compound produced 1.32 g of CO_2 and 0.631 g of H_2O. The mass percentage of iodine in the compound was determined by converting the iodine in a 0.850-g sample of the compound into 2.31 g of lead(II) iodide. What is the empirical formula of the compound? Could the compound also contain oxygen? Explain your answer.

M.13 When aqueous solutions of calcium nitrate and phosphoric acid are mixed, a white solid precipitates. (a) What is the formula of the solid? (b) How many grams of the solid can be formed from 206 g of calcium nitrate and 150. g of phosphoric acid?

M.14 Small amounts of chlorine gas can be generated in the laboratory from the reaction of manganese (IV) oxide with hydrochloric acid: $4 HCl(aq) + MnO_2(s) \rightarrow 2 H_2O(l) + MnCl_2(s) + Cl_2(g)$. (a) What mass of Cl_2 can be produced from 42.7 g of MnO_2 with an excess of HCl(aq)? (b) What volume of chlorine gas (of density $3.17\ g{\cdot}L^{-1}$) will be produced from the reaction of 300. mL of 0.100 M HCl(aq) with an excess of MnO_2? (c) Suppose only 150. mL of chlorine was produced in the reaction in part (b). What is the percentage yield of the reaction?

M.15 In addition to determining elemental composition of pure unknown compounds, combustion analysis can be used to determine the *purity* of known compounds. A sample of 2-naphthol, $C_{10}H_7OH$, which is used to prepare antioxidants to incorporate into synthetic rubber, was found to be contaminated with a small amount of LiBr. The combustion analysis of this sample gave the following results: 77.48% C and 5.20% H. Assuming that the only species present are 2-naphthol and LiBr, calculate the percentage purity by mass of the sample.

M.16 An organic compound with the formula $C_{14}H_{20}O_2N$ was recrystallized from 1,1,2,2-tetrachloroethane, $C_2H_2Cl_4$. A combustion analysis of the compound gave the following data: 68.50% C, 8.18% H by mass. Because the data were considerably different from that expected for pure $C_{14}H_{20}O_2N$, the sample was examined and found to contain a significant amount of 1,1,2,2-tetrachloroethane. Assuming that only these two compounds are present, what is the percentage purity by mass of the $C_{14}H_{20}O_2N$?

M.17 Tu-jin-pi is a root bark used in traditional Chinese medicines for the treatment of "athlete's foot." One of the active ingredients in tu-jin-pi is pseudolaric acid A, which is known to contain carbon, hydrogen, and oxygen. A chemist wanting to determine the molecular formula of pseudolaric acid A burned 1.000 g of the compound in an elemental analyzer. The products of the combustion were 2.492 g of CO_2 and 0.6495 g of H_2O. (a) Determine the empirical formula of the compound. (b) The molar mass was found to be 388.46 $g{\cdot}mol^{-1}$. What is the molecular formula of pseudolaric acid A?

M.18 A folk medicine used in the Anhui province of China to treat acute dysentery is cha-tiao-qi, a preparation of the leaves of *Acer ginnala*. Reaction of one of the active ingredients in cha-tiao-qi with water yields gallic acid, a powerful antidysenteric agent. Gallic acid is known to contain carbon, hydrogen, and oxygen. A chemist wanting to determine the molecular formula of gallic acid burned 1.000 g of the compound in an elemental analyzer. The products of the combustion were 1.811 g of CO_2 and 0.3172 g of H_2O. (a) Determine the empirical formula of the compound. (b) The molar mass was found to be 170.12 $g{\cdot}mol^{-1}$. What is the molecular formula of gallic acid?

M.19 An industrial by-product consists of C, H, O, and Cl. When 0.100 g of the compound was analyzed by combustion analysis, 0.0682 g of CO_2 and 0.0140 g of H_2O were produced. The mass percentage of Cl in the compound was found to be 55.0%. What are the empirical and molecular formulas of the compound?

M.20 A mixture of 4.94 g of 85.0% pure phosphine, PH_3, and 0.110 kg of $CuSO_4{\cdot}5H_2O$ (of molar mass 249.68 $g{\cdot}mol^{-1}$) is placed in a reaction vessel. (a) Balance the chemical equation for the reaction that takes place, given the skeletal form $CuSO_4{\cdot}5H_2O(s) + PH_3(g) \rightarrow Cu_3P_2(s) + H_2SO_4(aq) + H_2O(l)$. (b) Name each reactant and product. (c) Determine the limiting reactant. (d) Calculate the mass (in grams) of Cu_3P_2 (of molar mass 252.56 $g{\cdot}mol^{-1}$) produced, given that the percentage yield of the reaction is 6.31%.

M.21 The acid HA (where A stands for an unknown group of atoms) has molar mass 231 $g{\cdot}mol^{-1}$. HA reacts with the base XOH (molar mass 125 $g{\cdot}mol^{-1}$) to produce H_2O and the salt XA. In one experiment, 2.45 g of HA react with 1.50 g of XOH to form 2.91 g of XA. What is the percentage yield of the reaction?

M.22 The acid H_2A' (where A' stands for an unknown group of atoms) has molar mass 168 $g{\cdot}mol^{-1}$. H_2A' reacts with the base XOH (molar mass 125 $g{\cdot}mol^{-1}$) to produce H_2O and the salt X_2A'. In one experiment, 1.20 g of H_2A' react with 1.00 g of XOH to form 0.985 g of X_2A'. What is the percentage yield of the reaction?

ATOMS: THE QUANTUM WORLD

What Are the Key Ideas?   Matter is composed of atoms. The structures of atoms can be understood in terms of the theory of matter known as quantum mechanics, in which the properties of particles and waves merge together.

Why Do We Need to Know This Material?   Atoms are the fundamental building blocks of matter. They are the currency of chemistry in the sense that almost all the explanations of chemical phenomena are expressed in terms of atoms. This chapter explores the periodic variation of atomic properties and shows how quantum mechanics is used to account for the structures and therefore the properties of atoms.

What Do We Need to Know Already?   We need to be familiar with the nuclear model of the atom and the general layout of the periodic table (Fundamentals Section B). We also need the concepts of kinetic and potential energy (Section A).

Chapter **1**

We need insight to think like a chemist. Chemical insight means that, when we look at an everyday object or a sample of a chemical, we can imagine the atoms that make it up. Not only that, we need to be able to plunge, in our mind's eye, deep into the center of matter and imagine the internal structure of atoms. To visualize this structure and how it relates to the chemical properties of elements, we need to understand the **electronic structure** of an atom, the description of how its electrons are arranged around the nucleus.

As soon as we start this journey into the atom, we encounter an extraordinary feature of our world. When scientists began to understand the composition of atoms in the early twentieth century (Section B), they expected to be able to use **classical mechanics,** the laws of motion proposed by Newton in the seventeenth century, to describe their structure. After all, classical mechanics had been tremendously successful for describing the motion of visible objects such as balls and planets. However, it soon became clear that classical mechanics fails when applied to electrons in atoms. New laws, which came to be known as **quantum mechanics,** had to be developed.

This chapter builds an understanding of atomic structure in four steps. First, we review the experiments that led to our current nuclear model of the atom and see how spectroscopy reveals information about the arrangement of electrons around the nucleus. Then we describe the experiments that led to the replacement of classical mechanics by quantum mechanics, introduce some of its central features, and illustrate them by considering a very simple system. Next, we apply those ideas to the simplest atom of all, the hydrogen atom. Finally, we extend these concepts to the atoms of all the elements of the periodic table and see the origin of the periodicity of the elements.

INVESTIGATING ATOMS

Dalton pictured atoms as featureless spheres, like billiard balls. Today, we know that atoms have an internal structure: they are built from even smaller **subatomic particles.** In this book, we deal with the three major subatomic particles: the electron, the proton, and the neutron. By investigating the internal structure of atoms, we can come to see how one element differs from another and see how their properties are related to the structures of their atoms.

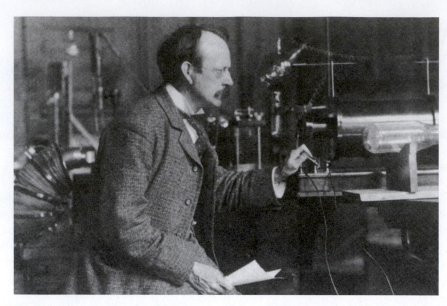

FIGURE 1.1 Joseph John Thomson (1856–1949), with the apparatus that he used to discover the electron.

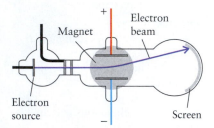

FIGURE 1.2 The apparatus used by Thomson to investigate the properties of electrons. An electric field is set up between the two plates and a magnetic field is applied perpendicular to the electric field.

Recall that C stands for coulomb, the SI unit of electric charge (Section A).

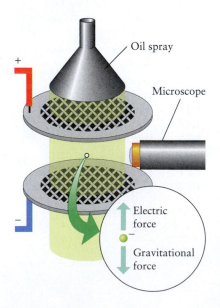

1.1 The Nuclear Atom

The earliest experimental evidence for the internal structure of atoms was the discovery in 1897 of the first subatomic particle, the **electron.** The British physicist J. J. Thomson (Fig. 1.1) was investigating "cathode rays," the rays that are emitted when a high potential difference (a high voltage) is applied between two electrodes (metal contacts) in an evacuated glass tube (Fig. 1.2). Thomson showed that cathode rays are streams of negatively charged particles. They came from inside the atoms that made up the negatively charged electrode, which is called the *cathode.* Thomson found that the charged particles, which came to be called electrons, were the same regardless of the metal he used for the cathode. He concluded that they are part of the makeup of all atoms.

Thomson was able to measure the value of e/m_e, the ratio of the magnitude of the electron's charge e to its mass m_e. Later workers, most notably the American physicist Robert Millikan, carried out experiments that enabled them to determine the value of the charge itself. Millikan designed an ingenious apparatus in which he could observe tiny electrically charged oil droplets (Fig. 1.3). From the strength of the electric field required to overcome the pull of gravity on the droplets, he determined the values of the charges on the particles. Because each oil droplet contained more than one additional electron, he took the charge on one electron to be the smallest increment of charge between droplets. The modern value is $-e$, with $e = 1.602 \times 10^{-19}$ C. This value, $-e$, is considered to be "one unit" of negative charge, and e itself, which is called the **fundamental charge,** is taken to be "one unit" of positive charge. The mass of the electron was calculated by combining this charge with the value of e/m_e measured by Thomson; its modern value is 9.109×10^{-31} kg.

Although electrons have a negative charge, atoms overall have zero charge. Therefore, by the beginning of the twentieth century scientists knew that each atom must contain enough positive charge to cancel the negative charge. But where was the positive charge? Because electrons can be removed easily from atoms but

FIGURE 1.3 A schematic diagram of Millikan's oil-drop experiment. Oil is sprayed as a fine mist into a chamber containing a charged gas, and the location of an oil droplet is monitored by using a microscope. Charged particles (ions) are generated in the gas by exposing it to x-rays. The fall of the charged droplet is balanced by the electric field.

the positive charge cannot, Thomson suggested a model of an atom as a blob of a positively charged, jellylike material, with the electrons suspended in it like raisins in pudding. However, this model was overthrown in 1908 by another experimental observation. Ernest Rutherford (Fig. 1.4) knew that some elements, including radon, emit streams of positively charged particles, which he called **α particles** (alpha particles). He asked two of his students, Hans Geiger and Ernest Marsden, to shoot α particles toward a piece of platinum foil only a few atoms thick (Fig. 1.5). If atoms were indeed like blobs of positively charged jelly, then all the α particles would easily pass through the diffuse positive charge of the foil, with only occasional slight deflections in their paths.

Geiger and Marsden's observations astonished everyone. Although almost all the α particles did pass through and were deflected only very slightly, about 1 in 20 000 was deflected through more than 90°, and a few α particles bounced straight back in the direction from which they had come. "It was almost as incredible," said Rutherford, "as if you had fired a 15-inch shell at a piece of tissue paper and it had come back and hit you."

The results of the Geiger-Marsden experiment suggested the **nuclear model** of the atom, in which there is a dense pointlike center of positive charge, the **nucleus**, surrounded by a large volume of mostly empty space in which the electrons are located. Rutherford reasoned that when a positively charged α particle scored a direct hit on one of the minute but heavy platinum nuclei, the α particle was strongly repelled by the positive charge of the nucleus and deflected through a large angle, like a tennis ball bouncing off a stationary cannonball (Fig. 1.6). Later work showed that the nucleus of an atom contains particles called **protons**, each of which has a charge of $+e$ ("one unit of positive charge"), that are responsible for the positive charge, and **neutrons**, which are uncharged particles. The number of protons in the nucleus is different for each element and is called the **atomic number**, Z, of the element (Section B). The total charge on an atomic nucleus of atomic number Z is $+Ze$ and, for the atoms to be electrically neutral, there must be Z electrons around it.

In the nuclear model of the atom, all the positive charge and almost all the mass is concentrated in the tiny nucleus, and the negatively charged electrons surround the nucleus. The atomic number is the number of protons in the nucleus.

FIGURE 1.4 Ernest Rutherford (1871–1937), who was responsible for many discoveries about the structure of the atom and its nucleus.

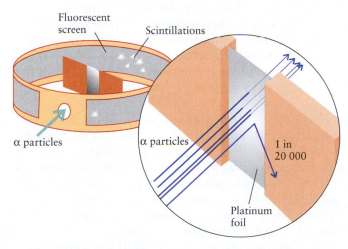

FIGURE 1.5 Part of the experimental arrangement used by Geiger and Marsden. The α particles came from a sample of the radioactive gas radon. They were directed through a hole into a cylindrical chamber with a zinc sulfide coating on the inside. The α particles struck the platinum foil mounted inside the cylinder, and their deflections were measured by observing flashes of light (scintillations) where they struck the screen. About 1 in 20 000 α particles was deflected through very large angles; most went through the thin foil with almost no deflection.

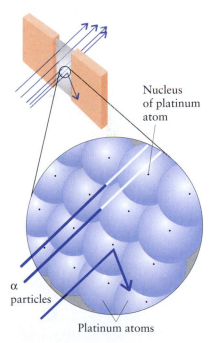

FIGURE 1.6 Rutherford's model of the atom explains why most α particles pass almost straight through the platinum foil, whereas a very few—those scoring a direct hit on the nucleus—undergo very large deflections. Most of the atom is nearly empty space thinly populated by the atom's electrons. The nuclei are much smaller relative to their atoms than shown here.

1.2 The Characteristics of Electromagnetic Radiation

The question that scientists struggled with for years is how those Z electrons are arranged around the nucleus. To investigate the internal structures of objects as small as atoms, scientists observe them indirectly through the properties of the light the atoms emit when stimulated by heat or an electric discharge. The analysis of the light emitted or absorbed by substances is a hugely important branch of chemistry called **spectroscopy.** We shall see how atomic spectroscopy—spectroscopy applied to atoms—allowed scientists to propose a model of the electronic structure of atoms and to test their model against experiment. To do so, we need to understand the nature of light, which is a form of electromagnetic radiation.

Electromagnetic radiation consists of oscillating (time-varying) electric and magnetic fields that travel through empty space at 3.00×10^8 m·s^{-1}, or at just over 670 million miles per hour. This speed is denoted c and called the "speed of light." Visible light is a form of electromagnetic radiation, as are radio waves, microwaves, and x-rays. All these forms of radiation transfer energy from one region of space to another. The warmth we feel from the Sun is a tiny proportion of the total energy it emits, and it is carried to us through space as electromagnetic radiation.

One reason why electromagnetic radiation is a good tool for the study of atoms is that an electric field pushes on charged particles such as electrons. As a light ray passes an electron, its electric field pushes the electron first in one direction and then in the opposite direction, over and over again. That is, the field oscillates in both direction and strength (Fig. 1.7). The number of cycles (complete reversals of direction away from and back to the initial strength and direction) per second is called the **frequency,** ν (the Greek letter nu), of the radiation. The unit of frequency, 1 hertz (1 Hz), is defined as 1 cycle per second:

$$1 \text{ Hz} = 1 \text{ s}^{-1}$$

Electromagnetic radiation of frequency 1 Hz pushes a charge in one direction, then the opposite direction, and returns to the original direction once per second. The frequency of electromagnetic radiation that we see as visible light is close to 10^{15} Hz, and so its electric field changes direction at about a thousand trillion (10^{15}) times a second as it travels past a given point.

An instantaneous snapshot of a wave of electromagnetic radiation spread through space would look like Fig. 1.7. The wave is characterized by its amplitude and wavelength. The **amplitude** is the height of the wave above the center line. The square of the amplitude determines the **intensity,** or brightness, of the radiation. The **wavelength,** λ (the Greek letter lambda), is the peak-to-peak distance. The wavelengths of visible light are close to 500 nm. Although 500 nm is only half of one-thousandth of a millimeter (so you might *just* be able to imagine it), it is much longer than the diameters of atoms, which are typically close to 0.2 nm.

Different wavelengths of electromagnetic radiation correspond to different regions of the spectrum (see Table 1.1). Our eyes detect electromagnetic radiation with wavelengths in the range from 700 nm (red light) to 400 nm (violet light). Radiation in this range is called **visible light,** and the frequency of visible light determines its color. White light, which includes sunlight, is a mixture of all wavelengths of visible light as well as radiation that we cannot see.

Now imagine the wave in Fig. 1.7 zooming along at its actual speed, the speed of light, c. If the wavelength of the light is very short, very many complete oscillations pass a given point in a second (Fig. 1.8a). If the wavelength is long, fewer complete oscillations pass the point in a second (Fig. 1.8b). A short wavelength therefore corresponds to high-frequency radiation and a long wavelength corresponds to low-frequency radiation. The precise relation is

$$\text{Wavelength} \times \text{frequency} = \text{speed of light, or } \lambda\nu = c \qquad \textbf{(1)}^*$$

As we saw in Section A, we can think of a field as a region of influence, like the gravitational field of the Earth.

A radio signal takes about 15 milliseconds to cross the continental USA. Values of the fundamental constants are listed inside the back cover of the book. A more precise value for the speed of light is 2.998×10^8 m·s^{-1}.

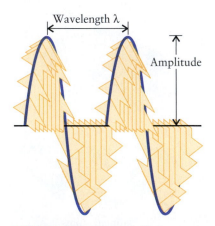

FIGURE 1.7 The electric field of electromagnetic radiation oscillates in space and time. This diagram represents a "snapshot" of an electromagnetic wave at a given instant. The length of an arrow at any point represents the strength of the force that the field exerts on a charged particle at that point. The distance between the peaks is the wavelength of the radiation, and the height of the wave above the center line is the amplitude.

TABLE 1.1 Color, Frequency, and Wavelength of Electromagnetic Radiation

Radiation type	Frequency (10^{14} Hz)	Wavelength (nm, 2 sf)*	Energy per photon (10^{-19} J)
x-rays and γ-rays	$\geq 10^3$	≤ 3	$\geq 10^3$
ultraviolet	8.6	350	5.7
visible light			
violet	7.1	420	4.7
blue	6.4	470	4.2
green	5.7	530	3.8
yellow	5.2	580	3.4
orange	4.8	620	3.2
red	4.3	700	2.8
infrared	3.0	1000	2.0
microwaves and radio waves	$\leq 10^{-3}$	$\geq 3 \times 10^6$	$\leq 10^{-3}$

*The abbreviation sf denotes the number of significant figures in the data. The frequencies, wavelengths, and energies are typical values; they should not be regarded as precise.

EXAMPLE 1.1   Sample Exercise: Calculating the wavelength of light

What is the wavelength of blue light of frequency 6.4×10^{14} Hz?

SOLUTION

From $\lambda \nu = c$ written as $\lambda = c/\nu$,

$$\lambda = \frac{\overbrace{2.998 \times 10^8 \text{ m} \cdot \text{s}^{-1}}^{c}}{\underbrace{6.4 \times 10^{14} \text{ s}^{-1}}_{\text{Hz}}}$$

$$= \frac{2.998 \times 10^8}{6.4 \times 10^{14}} \text{ m} = 4.7 \times 10^{-7} \text{ m}$$

or about 470 nm. Blue light, which has a relatively high frequency, has a shorter wavelength than red light, which has a wavelength near 700 nm.

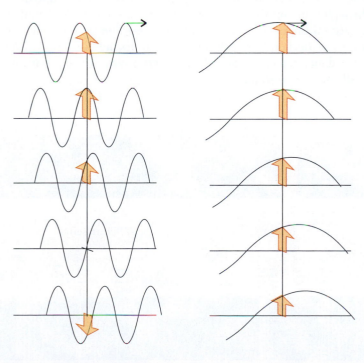

(a) Short wavelength, high frequency   (b) Long wavelength, low frequency

FIGURE 1.8 (a) Short-wavelength radiation: the vertical arrow shows how the electric field changes markedly at five successive instants. (b) For the same five instants, the electric field of the long-wavelength radiation changes much less. The horizontal arrows in the uppermost images show that in each case the wave has traveled the same distance. Short-wavelength radiation has a high frequency, whereas long-wavelength radiation has a low frequency.

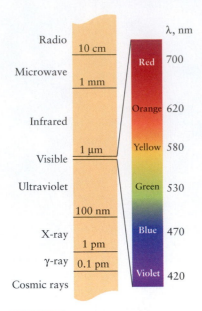

FIGURE 1.9 The electromagnetic spectrum and the names of its regions. The region we call "visible light" occupies a very narrow range of wavelengths. The regions are not drawn to scale.

Modern theories suggest that our concept of space breaks down on a scale of 10^{-34} m, so that may constitute a lower bound to electromagnetic radiation wavelength.

The components of different frequency or wavelength are called lines because, in the early spectroscopic experiments, the radiation from the sample was passed through a slit and then through a prism; the image of the slit was then focused on a photographic plate, where it appeared as a line.

A note on good practice: It is a good idea to test your understanding with the self-tests that follow each worked example. Answers to the B self-tests can be found at the back of this book.

SELF-TEST 1.1A Calculate the wavelengths of the light from traffic signals as they change. Assume that the lights emit the following frequencies: green, 5.75×10^{14} Hz; yellow, 5.15×10^{14} Hz; red, 4.27×10^{14} Hz.

[*Answer:* Green, 521 nm; yellow, 582 nm; red, 702 nm]

SELF-TEST 1.1B What is the wavelength of the signal from a radio station transmitting at 98.4 MHz?

As far as we know, there is neither an upper limit nor a lower limit to the wavelength of electromagnetic radiation (Fig. 1.9). **Ultraviolet radiation** is radiation at higher frequency than violet light; its wavelength is less than about 400 nm. This damaging component of radiation from the Sun is responsible for sunburn and tanning, but it is largely prevented from reaching the surface of the Earth by the ozone layer. **Infrared radiation,** the radiation we experience as heat, has a lower frequency and longer wavelength than red light; its wavelength is greater than about 800 nm. Microwaves, which are used in radar and microwave ovens, have wavelengths in the millimeter-to-centimeter range.

The color of light depends on its frequency and wavelength; long-wavelength radiation has a lower frequency than short-wavelength radiation.

1.3 Atomic Spectra

Now that we know the properties of electromagnetic radiation, we can see how it provides information about atomic structure. In this section we focus on the hydrogen atom, as it contains only a single electron. When an electric current is passed through a low-pressure sample of hydrogen gas, the sample emits light. The electric current, which is like a storm of electrons, breaks up the H_2 molecules and excites the free hydrogen atoms to higher energies. These excited atoms quickly discard their excess energy by giving off electromagnetic radiation; then they combine to form H_2 molecules again.

When white light is passed through a prism, a continuous spectrum of light results (Fig. 1.10a). However, when the light emitted by excited hydrogen atoms is passed through a prism, the radiation is found to consist of a number of components, or **spectral lines** (Fig. 1.10b). The brightest line (at 656 nm) is red, and the excited atoms in the gas glow with this red light. Excited hydrogen atoms also emit ultraviolet and infrared radiation, which are invisible to the eye but can be detected electronically and photographically.

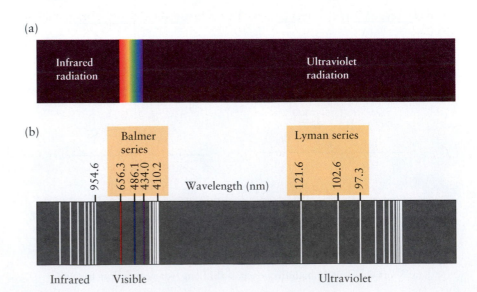

FIGURE 1.10 (a) The visible spectrum. (b) The complete spectrum of atomic hydrogen. The spectral lines have been assigned to various groups called series, two of which are shown with their names.

The first person to identify a pattern in the lines of the visible region of the spectrum was Joseph Balmer, a Swiss schoolteacher. In 1885, he noticed that the frequencies of all the lines then known could be generated by the expression

$$\nu \propto \frac{1}{2^2} - \frac{1}{n^2} \quad n = 3, 4, \ldots$$

As experimental techniques advanced, more lines were discovered, and the Swedish spectroscopist Johann Rydberg noticed that all of them could be predicted from the expression

$$\nu = \mathcal{R}\left\{\frac{1}{n_1^2} - \frac{1}{n_2^2}\right\} \quad n_1 = 1, 2, \ldots, \quad n_2 = n_1 + 1, n_1 + 2, \ldots \quad (2)$$

Here $\mathcal{R}$ is an empirical (experimentally determined) constant now known as the **Rydberg constant**; its value is 3.29×10^{15} Hz. The **Balmer series** is the set of lines with $n_1 = 2$ (and $n_2 = 3, 4, \ldots$). The **Lyman series**, a set of lines in the ultraviolet region of the spectrum, has $n_1 = 1$ (and $n_2 = 2, 3, \ldots$).

EXAMPLE 1.2 Identifying a line in the hydrogen spectrum

Calculate the wavelength of the radiation emitted by a hydrogen atom for $n_1 = 2$ and $n_2 = 3$. Identify the spectral line in Fig. 1.10b.

STRATEGY The frequency is given by Eq. 2. Convert frequency into wavelength by using Eq. 1. The wavelength should match one of the lines in the Balmer series in Fig. 1.10b.

SOLUTION

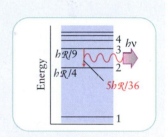

From Eq. 2 with $n_1 = 2$ and $n_2 = 3$,
$$\nu = \mathcal{R}\left(\frac{1}{2^2} - \frac{1}{3^2}\right) = \frac{5}{36}\mathcal{R}$$

From $\lambda \nu = c$,
$$\lambda = \frac{c}{\nu} = \frac{c}{(5\mathcal{R}/36)} = \frac{36c}{5\mathcal{R}}$$

Now substitute the data.
$$\lambda = \frac{36 \times \overbrace{(2.998 \times 10^8 \text{ m·s}^{-1})}^{c}}{5 \times \underbrace{(3.29 \times 10^{15} \text{ s}^{-1})}_{\mathcal{R}}} = \frac{36 \times 2.998 \times 10^8}{5 \times 3.29 \times 10^{15}} \frac{\text{m·s}^{-1}}{\text{s}^{-1}}$$

$$= 6.57 \times 10^{-7} \text{ m}$$

This wavelength, 657 nm, corresponds to the red line in the Balmer series of lines in the spectrum.

SELF-TEST 1.2A Repeat the calculation for $n_1 = 2$ and $n_2 = 4$ and identify the spectral line in Fig. 1.10b.

[*Answer:* 486 nm; blue line]

SELF-TEST 1.2B Repeat the calculation for $n_1 = 2$ and $n_2 = 5$ to and identify the spectral line in Fig. 1.10b.

If we pass white light through a vapor composed of the atoms of an element, we see its *absorption spectrum*, a series of dark lines on an otherwise continuous spectrum (Fig 1.11). The absorption lines have the same frequencies as the lines in the emission spectrum and suggest that an atom can *absorb* radiation only of those same frequencies. Absorption spectra are used by astronomers to identify elements in the outer layers of stars.

We can begin to understand these perplexing features if we suppose that an electron can exist with only certain energies when it is part of a hydrogen atom, and that a spectral line arises from a **transition** between two of the allowed

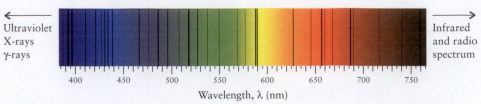

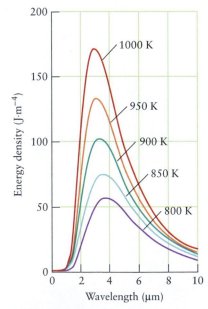

FIGURE 1.11 When white light shines through an atomic vapor, radiation is absorbed at frequencies that correspond to excitation energies of the atoms. Here is a small section of the spectrum of the Sun, in which atoms in its outer layers absorb the radiation from the incandescence below. Many of the lines have been ascribed to hydrogen, showing that hydrogen is present in the cooler outer layers of the Sun.

energies, the difference in energy being discarded as the energy of electromagnetic radiation. If that is the case, then Rydberg's formula begins to suggest that the permitted energies are proportional to $\mathcal{R}/n^2$, for then the difference in energy of the two states involved in the transition would be given by an expression that resembled the right-hand side of Rydberg's formula. But why should the frequency of the emitted radiation be proportional to that energy difference? Furthermore, why does the Rydberg constant have the value that is observed?

> *The observation of discrete spectral lines suggests that an electron in an atom can have only certain energies.*

QUANTUM THEORY

As more experimental information was collected toward the end of the nineteenth century, scientists became increasingly perplexed. They gathered information about electromagnetic radiation that could not be explained by classical mechanics, and the lines in the spectrum of hydrogen remained deeply puzzling. Then, from 1900 on, a series of imaginative suggestions were made, and by 1927 the puzzles had been resolved, only to be replaced by new and even more intriguing puzzles. In this part of the chapter we describe these other puzzles and their resolution in terms of quantum mechanics. We shall then see some of the peculiar, revolutionary results from quantum mechanics that will help us to understand atomic structure, beginning with the hydrogen atom.

1.4 Radiation, Quanta, and Photons

Important clues to the nature of electromagnetic radiation came from observations of objects as they are heated. At high temperatures an object begins to glow—the phenomenon of **incandescence**. As the object is heated to higher temperatures it glows more brightly and the color of light it gives off changes from red through orange and yellow toward white. Those are *qualitative* observations. To understand what the color changes mean, scientists had to study the effect *quantitatively*. They had to measure the intensity of radiation at each wavelength and repeat the measurements at a variety of different temperatures. These experiments caused one of the greatest revolutions that has ever occurred in science. Figure 1.12 shows some of the experimental results. The "hot object" is known as a **black body** (even though it might be glowing white hot!). The name signifies that the object does not favor one wavelength over another in the sense of absorbing a particular wavelength preferentially or emitting one preferentially. The curves in Fig. 1.12 show the intensity of **black-body radiation**, the radiation emitted at different wavelengths by a heated black body, for a series of temperatures. Notice that as the temperature rises the maximum intensity of the radiation emitted occurs at shorter and shorter wavelengths.

Two crucial pieces of experimental information about black-body radiation were discovered in the late nineteenth century. In 1879, Josef Stefan investigated the increasing brightness of a black body as it is heated and discovered that the total intensity of radiation emitted over all wavelengths increases as the fourth

FIGURE 1.12 The intensity of radiation emitted by a heated black body as a function of wavelength. As the temperature increases, the total energy emitted (the area under the curve) increases sharply, and the maximum intensity of emission moves to shorter wavelengths. (To obtain the energy in a volume V and at wavelengths λ and $\lambda + \Delta\lambda$, multiply the energy density by V and $\Delta\lambda$.)

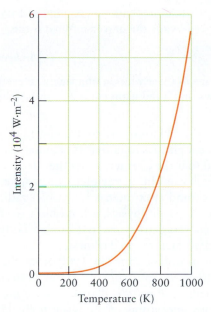

FIGURE 1.13 The total intensity of radiation emitted by a heated black body increases as the fourth power of the temperature, so a body at 1000 K emits more than 120 times as much energy as is emitted by the same body at 300 K.

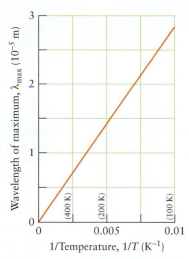

FIGURE 1.14 As the temperature is raised ($1/T$ decreases), the wavelength of maximum emission shifts to smaller values.

power of the temperature (Fig. 1.13). This quantitative relation is now called the **Stefan–Boltzmann law** and is usually written

$$\text{Total intensity} = \text{constant} \times T^4 \qquad (3a)$$

The name Stefan–Boltzmann law recognizes Ludwig Boltzmann's theoretical contribution.

where T signifies an absolute temperature, one reported on the Kelvin scale (Appendix 1B). The experimental value of the constant is 5.67×10^{-8} W·m^{-2}·K^{-4}, where W denotes watts (1 W = 1 J·s^{-1}). A few years later, in 1893, Wilhelm Wien examined the shift in color of black-body radiation as the temperature increases and discovered that the wavelength corresponding to the maximum in the intensity, λ_{max}, is inversely proportional to the temperature, $\lambda_{max} \propto 1/T$ (that is, as T increases the wavelength decreases), and therefore that $\lambda_{max} \times T$ is a constant (Fig. 1.14). This quantitative result is now called **Wien's law** and is normally written

$$T\lambda_{max} = \text{constant} \qquad (3b)$$

The empirical (experimental) value of the constant in this expression is 2.9 K·mm.

EXAMPLE 1.3   Sample exercise: Determining the temperature of the surface of a star

The maximum intensity of solar radiation occurs at 490. nm. What is the temperature of the surface of the Sun?

SOLUTION We can use Wien's law in the form $T = \text{constant}/\lambda_{max}$ to determine the surface temperature of stars treated as hot black bodies:

From $T = \text{constant}/\lambda_{max}$,

$$T = \frac{\overbrace{2.9 \times 10^{-3}}^{2.9\ \text{mm}}\ \text{m·K}}{\underbrace{4.90 \times 10^{-7}\ \text{m}}_{490\ \text{nm}}}$$

$$= \frac{2.9 \times 10^{-3}}{4.90 \times 10^{-7}}\ \text{K} = 5.9 \times 10^3\ \text{K}$$

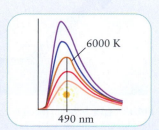

That is, the surface temperature of the Sun is about 6000 K.

SELF-TEST 1.3A In 1965, electromagnetic radiation with a maximum at 1.05 mm (in the microwave region) was discovered to pervade the universe. What is the temperature of "empty" space?

[*Answer:* 2.76 K]

SELF-TEST 1.3B A red giant is a late stage in the evolution of a star. The average wavelength maximum at 700. nm shows that a red giant cools as it dies. What is the surface temperature of a red giant?

For nineteenth-century scientists, the obvious way to account for the laws of black-body radiation was to use classical physics to derive its characteristics. However, much to their dismay, they found that the characteristics they deduced did not match their observations. Worst of all was the **ultraviolet catastrophe:** classical physics predicted that any hot body should emit intense ultraviolet radiation and even x-rays and γ-rays! According to classical physics, a hot object would devastate the countryside with high-frequency radiation. Even a human body at 37°C would glow in the dark. There would, in fact, be no darkness.

The suggestion that resolved the problem came in 1900 from the German physicist Max Planck. He proposed that the exchange of energy between matter and radiation occurs in **quanta,** or packets, of energy. Planck focused his attention on the hot, rapidly oscillating electrons and atoms of the black body. His central idea was that a charged particle oscillating at a frequency v can exchange energy with its surroundings by generating or absorbing electromagnetic radiation only in discrete packets of energy

$$E = hv \tag{4}*$$

The constant h, now called **Planck's constant,** has the value 6.626×10^{-34} J·s. If the oscillating atom releases an energy E into the surroundings, then radiation of frequency $v = E/h$ will be detected.

Planck's hypothesis implies that radiation of frequency v can be generated only if an oscillator of that frequency has acquired the minimum energy required to start oscillating. At low temperatures, there is not enough energy available to stimulate oscillations at very high frequencies, and so the object cannot generate high-frequency, ultraviolet radiation. As a result, the intensity curves in Fig. 1.12 die away at high frequencies (short wavelengths) and the ultraviolet catastrophe is avoided. In contrast, in classical physics it was assumed that an oscillator could oscillate with any energy and therefore, even at low temperatures, high-frequency oscillators could contribute to the emitted radiation. Planck's hypothesis is *quantitatively* successful, too, because not only was he able to use his proposal to derive the Stefan–Boltzmann and Wien laws but he was also able to calculate the variation of intensity with wavelength and to obtain a curve that matched the experimental curve almost exactly.

To achieve this successful theory, Planck had discarded classical physics, which puts no restriction on how small an amount of energy may be transferred from one object to another. He had proposed instead that energy is transferred in discrete packets. To justify such a dramatic revolution, more evidence was needed. That evidence came from the **photoelectric effect,** the ejection of electrons from a metal when its surface is exposed to ultraviolet radiation (Fig. 1.15). The experimental observations were as follows:

1   No electrons are ejected unless the radiation has a frequency above a certain threshold value characteristic of the metal.

2   Electrons are ejected immediately, however low the intensity of the radiation.

3   The kinetic energy of the ejected electrons increases linearly with the frequency of the incident radiation.

A note on good practice: We say that a property y *varies linearly with* x if the relation between y and x can be written $y = b + mx$, where b and m are constants. When $y = mx$ (that is, when $b = 0$), we say that y *is proportional to* x. See Appendix 1E.

The word quantum comes from the Latin for amount—literally, "How much?".

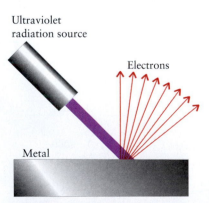

Ultraviolet radiation source

Electrons

Metal

FIGURE 1.15 When a metal is illuminated with ultraviolet radiation, electrons are ejected, provided the frequency is above a threshold frequency that is characteristic of the metal.

Albert Einstein found an explanation of these observations and, in the process, profoundly changed our conception of the electromagnetic field. He proposed that electromagnetic radiation consists of particles, which were later called **photons**. Each photon can be regarded as a packet of energy, and the energy of a single photon is related to the frequency of the radiation by Eq. 4 ($E = h\nu$). For example, ultraviolet photons are more energetic than photons of visible light, which has lower frequencies. According to this photon model of electromagnetic radiation, we can visualize a beam of red light as a stream of photons of one particular energy, yellow light as a stream of photons of higher energy, and green light as a stream of photons of higher energy still. It is important to note that the intensity of radiation is an indication of the *number* of photons present, whereas $E = h\nu$ is a measure of the *energy* of each individual photon.

EXAMPLE 1.4   **Sample exercise: Calculating the energy of a photon**

What is the energy of a single photon of blue light of frequency 6.4×10^{14} Hz?

SOLUTION

From $E = h\nu$, $E = (6.626 \times 10^{-34}\ \text{J·s}) \times (6.4 \times 10^{14}\ \text{Hz}) = 4.2 \times 10^{-19}\ \text{J}$

To derive this value, we have used 1 Hz = 1 s^{-1}, so J·s × Hz = J·s × s^{-1} = J.

SELF-TEST 1.4A What is the energy of a photon of yellow light of frequency 5.2×10^{14} Hz?

[*Answer:* 3.4×10^{-19} J]

SELF-TEST 1.4B What is the energy of a photon of orange light of frequency 4.8×10^{14} Hz?

The characteristics of the photoelectric effect are easy to explain if we visualize electromagnetic radiation as consisting of a stream of photons. If the incident radiation has frequency ν, it consists of photons of energy $h\nu$. These particles collide with the electrons in the metal. The energy required to remove an electron from the surface of a metal is called the **work function** of the metal and denoted Φ (uppercase phi). If the energy of a photon is less than the energy required to remove an electron from the metal, then an electron will not be ejected, regardless of the intensity of the radiation (the rate at which the photons arrive). However, if the energy of the photon, $h\nu$, is greater than Φ, then an electron is ejected with a kinetic energy, $E_K = \frac{1}{2}m_e\nu^2$, equal to the difference between the energy of the incoming photon and the work function: $E_K = h\nu - \Phi$ (Fig. 1.16). It follows that

$$\underbrace{\tfrac{1}{2}m_e\nu^2}_{\substack{\text{kinetic energy} \\ \text{of ejected} \\ \text{electron}}} = \underbrace{h\nu}_{\substack{\text{energy} \\ \text{supplied} \\ \text{by photon}}} - \underbrace{\Phi}_{\substack{\text{energy required} \\ \text{to eject photon}}}$$

(5)

What does this equation tell us? Because the kinetic energy of the ejected electrons varies linearly with frequency, a plot of the kinetic energy against the frequency of the radiation should look like the graph in Fig. 1.17, a straight line of slope h, the same for all metals, and have an extrapolated intercept with the vertical axis at $-\Phi$, different for each metal. The intercept with the horizontal axis (corresponding to zero kinetic energy of the ejected electron) is at Φ/h in each case.

We can now interpret the experimental observations of the photoelectric effect in light of Einstein's theory:

1   An electron can be driven out of the metal only if it receives from the photon during the collision at least a certain minimum energy equal to the work function, Φ. Therefore, the frequency of the radiation must have a certain minimum value if electrons are to be ejected. This minimum frequency depends on the work function and hence on the identity of the metal (as shown in Fig. 1.17).

Be careful to distinguish the symbol for speed, v (for velocity), from the symbol for frequency, ν (the Greek letter nu).

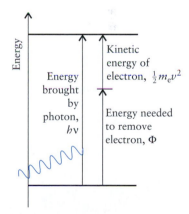

FIGURE 1.16 In the photoelectric effect, a photon with energy $h\nu$ strikes the surface of a metal and its energy is absorbed by an electron. If the energy of the photon is greater than the work function, Φ, of the metal, the electron absorbs enough energy to break away from the metal. The kinetic energy of the ejected electron is the difference between the energy of the photon and the work function: $\frac{1}{2}m_e\nu^2 = h\nu - \Phi$.

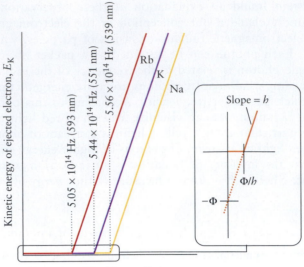

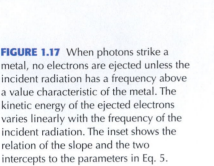

FIGURE 1.17 When photons strike a metal, no electrons are ejected unless the incident radiation has a frequency above a value characteristic of the metal. The kinetic energy of the ejected electrons varies linearly with the frequency of the incident radiation. The inset shows the relation of the slope and the two intercepts to the parameters in Eq. 5.

2   Provided a photon has enough energy, a collision results in the immediate ejection of an electron.

3   The kinetic energy of the electron ejected from the metal increases linearly with the frequency of the incident radiation according to Eq. 5.

EXAMPLE 1.5   **Interpreting a light ray in terms of photons**

A discharge lamp rated with a power of 25 W ($1 \text{ W} = 1 \text{ J} \cdot \text{s}^{-1}$) emits yellow light of wavelength 580 nm. How many photons of yellow light does the lamp generate in 1.0 s?

STRATEGY   We expect a large number of photons because the energy of a single photon is very small. The number of photons generated in a given time interval is the total energy emitted in that interval, the power (P) of the lamp in watts multiplied by the time it is on ($E = P \times t$), divided by the energy of a single photon. We calculate the energy of a photon from its wavelength by combining Eq. 1, which converts the given wavelength into a frequency, with Eq. 4, which converts frequency into energy. Because c is given in meters per second, convert the wavelength from nanometers to meters.

SOLUTION

From $E_{photon} = h\nu$ and $\lambda\nu = c$ in the form $\nu = c/\lambda$,

$$E_{photon} = \frac{hc}{\lambda}$$

From $N = E_{total}/E_{photon}$,

$$N = \frac{E_{total}}{(hc/\lambda)} = \frac{\lambda E_{total}}{hc}$$

From $E_{total} = Pt$,

$$N = \frac{\lambda Pt}{hc}$$

Now substitute the data.

$$N = \frac{\overbrace{(5.80 \times 10^{-7}\text{m})}^{\lambda} \times \overbrace{(25 \text{ J}\cdot\text{s}^{-1})}^{P} \times \overbrace{(1.0 \text{ s})}^{t}}{(6.626 \times 10^{-34} \text{ J s}) \times (3.00 \times 10^{8} \text{ m}\cdot\text{s}^{-1})}$$

$$= \frac{5.80 \times 10^{-7} \times 25 \times 1.0}{6.626 \times 10^{-34} \times 3.00 \times 10^{8}} = 7.3 \times 10^{19}$$

This result means that, when you turn on the lamp, it generates about 10^{20} photons of yellow light each second. As predicted, this number is very large.

A note on good practice: Notice how, in the solution, we went as far as possible algebraically, and introduced numbers only at the end of the calculation. That procedure reduces the possibility of numerical error and produces an equation that can be used for different data.

FIGURE 1.18 When an atom undergoes a transition from a state of higher energy to one of lower energy, it loses energy that is carried away as a photon. The greater the energy loss (A compared with B), the higher the frequency (and the shorter the wavelength) of the radiation emitted.

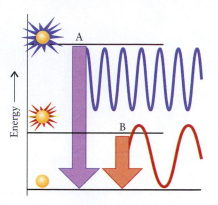

SELF-TEST 1.5A Another discharge lamp produces 5.0 J of energy per second in the blue region of the spectrum. How many photons of blue (470 nm) light would the lamp generate if it were left on for 8.5 s?

[*Answer:* 1.0×10^{20}]

SELF-TEST 1.5B In 1.0 s, a lamp that produces 25 J of energy per second in a certain region of the spectrum emits 5.5×10^{19} photons of light in that region. What is the wavelength of the emitted light?

The existence of photons and the relation between their energy and frequency helps to answer one of the questions posed by the spectrum of atomic hydrogen. At the end of Section 1.3 we started to form the view that a spectral line arises from a transition between two energy levels. Now we can see that if the energy difference is carried away as a photon, then the frequency of an individual line in a spectrum is related to the energy difference between two energy levels involved in the transition (Fig. 1.18):

$$h\nu = E_{\text{upper}} - E_{\text{lower}} \tag{6}*$$

This relation is called the **Bohr frequency condition.** If the energies on the right of this expression are each proportional to $h\mathcal{R}/n^2$, then we have accounted for Rydberg's formula. We still have to explain why the energies have this form, but we have made progress.

Studies of black-body radiation led to Planck's hypothesis of the quantization of electromagnetic radiation. The photoelectric effect provides evidence of the particulate nature of electromagnetic radiation.

1.5 The Wave–Particle Duality of Matter

The photoelectric effect strongly supports the view that electromagnetic radiation consists of photons that behave like particles. However, there is plenty of evidence to show that electromagnetic radiation behaves like waves! The most compelling evidence is the observation of **diffraction,** the pattern of high and low intensities generated by an object in the path of a ray of light (Fig. 1.19). A diffraction pattern results when the peaks and troughs of waves traveling along one path interfere with the peaks and troughs of waves traveling along another path. If the peaks coincide, the amplitude of the wave (its height) is enhanced; we call this enhancement **constructive interference** (Fig. 1.20a). If the peaks of one wave coincide with the troughs of

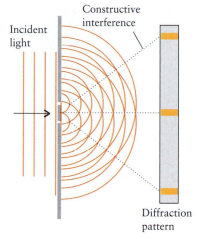

FIGURE 1.19 In this illustration, the peaks of the waves of electromagnetic radiation are represented by orange lines. When radiation coming from the left (the vertical lines) passes through a pair of closely spaced slits, circular waves are generated at each slit. These waves interfere with each other. Where they interfere constructively (as indicated by the positions of the dotted lines), a bright line is seen on the screen behind the slits; where the interference is destructive, the screen is dark.

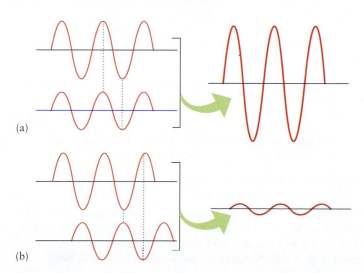

FIGURE 1.20 (a) Constructive interference. The two component waves (left) are "in phase" in the sense that their peaks and troughs coincide. The resultant (right) has an amplitude that is the sum of the amplitudes of the components. The wavelength of the radiation is not changed by interference, only the amplitude is changed. (b) Destructive interference. The two component waves are "out of phase" in the sense that the troughs of one coincide with the peaks of the other. The resultant has a much lower amplitude than either component.

another wave, the amplitude of the wave is diminished by **destructive interference** (Fig. 1.20b). This effect is the basis of a number of useful techniques for studying matter. For example, x-ray diffraction is one of the most important tools for studying the structures of molecules (see Major Technique 4, following Chapter 5).

You can appreciate why scientists were puzzled! The results of some experiments (the photoelectric effect) compelled them to the view that electromagnetic radiation is particlelike. The results of other experiments (diffraction) compelled them equally firmly to the view that electromagnetic radiation is wavelike. Thus we are brought to the heart of modern physics. Experiments oblige us to accept the **wave-particle duality** of electromagnetic radiation, in which the concepts of waves and particles blend together. In the wave model, the intensity of the radiation is proportional to the square of the amplitude of the wave. In the particle model, intensity is proportional to the number of photons present at each instant.

If electromagnetic radiation, which for a long time had been regarded as wave-like, has a dual character, could it be that matter, which since Dalton's day had been regarded as consisting of particles, also has wavelike properties? In 1925, the French scientist Louis de Broglie proposed that *all* particles should be regarded as having wavelike properties. He went on to suggest that the wavelength associated with the "matter wave" is inversely proportional to the particle's mass, m, and speed, v, and that

$$\lambda = \frac{h}{mv} \tag{7a}$$

The product of mass and speed is called the **linear momentum, p,** of a particle, and so this expression is more simply written as the **de Broglie relation:**

$$\lambda = \frac{h}{p} \tag{7b}*$$

EXAMPLE 1.6   **Sample exercise: Calculating the wavelength of a particle**

In order to appreciate why the wavelike properties of particles had not been noticed, calculate the wavelength of a particle of mass 1 g traveling at 1 m·s^{-1}.

SOLUTION

From $\lambda = h/mv$,

$$\lambda = \frac{6.626 \times 10^{-34} \text{ J·s}}{(1 \times 10^{-3} \text{ kg}) \times (1 \text{ m·s}^{-1})} = \frac{6.626 \times 10^{-34}}{1 \times 10^{-3}} \frac{\text{kg·m}^2\text{·s}^{-2}\text{·s}}{\text{kg·m·s}^{-1}}$$
$$= 7 \times 10^{-31} \text{ m}$$

This wavelength is undetectably small; the same is true for any macroscopic (visible) object traveling at normal speeds.

A note on good practice: Notice how, in the calculations, we kept all the units, writing them separately, and then canceling and multiplying them like ordinary numbers. We did not simply "guess" that the wavelength would turn out in meters. This procedure helps you to detect errors and ensures that your answer has the correct units.

SELF-TEST **1.6A** Calculate the wavelength of an electron traveling at 1/1000 the speed of light (see the inside back cover of this book for the mass of an electron).

[*Answer:* 2.43 nm]

SELF-TEST **1.6B** Calculate the wavelength of a rifle bullet of mass 5.0 g traveling at twice the speed of sound (the speed of sound is 331 m·s^{-1}).

The wavelike character of electrons was confirmed by showing that they could be diffracted. The experiment was first performed in 1925 by two American scientists, Clinton Davisson and Lester Germer, who directed a beam of fast electrons at a single crystal of nickel. The regular array of atoms in the crystal, with centers separated by 250 pm, acts as a grid that diffracts waves; and a diffraction pattern was observed (Fig. 1.21). Since then, some molecules have been shown to undergo

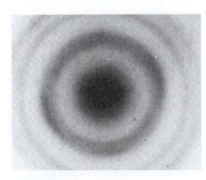

FIGURE 1.21 Davisson and Germer showed that electrons produce a diffraction pattern when reflected from a crystal G. P. Thomson, working in Aberdeen, Scotland, showed that they also produce a diffraction pattern when they pass through a very thin gold foil. The latter is shown here. G. P. Thomson was the son of J. J. Thomson, who identified the electron (Section B). Both received Nobel prizes: J. J. for showing that the electron is a particle and G. P. for showing that it is a wave.

diffraction, and there is no doubt that particles have a wavelike character. Indeed, electron diffraction is now an important technique for determining the structures of molecules and exploring the structures of solid surfaces.

Electrons (and matter in general) have both wavelike and particlelike properties.

1.6 The Uncertainty Principle

The discovery of wave-particle duality not only changes our understanding of electromagnetic radiation and matter, but also sweeps away the foundations of classical physics. In classical mechanics, a particle has a definite **trajectory,** or path on which location and linear momentum are specified at each instant. However, we cannot specify the precise location of a particle if it behaves like a wave: think of a wave in a guitar string, which is spread out all along the string, not localized at a precise point. A particle with a precise linear momentum has a precise wavelength; but, because it is meaningless to speak of the location of a wave, it follows that we cannot specify the location of a particle that has a precise linear momentum. If we keep a hydrogen atom in mind, duality means that however we describe its internal structure, it will not be in terms of an electron orbiting the nucleus in a definite trajectory.

The difficulty will not go away. Wave-particle duality denies the possibility of specifying the location if the linear momentum is known, and so we cannot specify the trajectory of particles. If we know that a particle is *here* at one instant, we can say nothing about where it will be an instant later! The impossibility of knowing the precise position if the linear momentum is known precisely is an aspect of the **complementarity** of location and momentum—if one property is known the other cannot be known simultaneously. The **Heisenberg uncertainty principle**, which was formulated by the German scientist Werner Heisenberg in 1927, expresses this complementarity quantitatively. It states that, if the location of a particle is known to within an uncertainty Δx, then the linear momentum, p, parallel to the x-axis can be known simultaneously only to within an uncertainty Δp, where

$$\Delta p \Delta x \geq \tfrac{1}{2}\hbar \qquad (8)^*$$

The symbol $\hbar$, which is read "h bar," means $h/2\pi$, a useful combination that is found widely in quantum mechanics.

> *What does this equation tell us?* The product of the uncertainties in two simultaneous measurements cannot be less than a certain constant value. Therefore, if the uncertainty in position is very small (Δx very small), then the uncertainty in linear momentum must be large, and vice versa (Fig. 1.22).

The uncertainty principle has negligible practical consequences for macroscopic objects, but it is of profound importance for subatomic particles such as the electrons in atoms and for a scientific understanding of the nature of the world.

EXAMPLE 1.7   Using the uncertainty principle

Estimate the minimum uncertainty in (a) the position of a marble of mass 1.0 g given that its speed is known to within $\pm 1.0 \ \text{mm} \cdot \text{s}^{-1}$ and (b) the speed of an electron confined to within the diameter of a typical atom (200. pm).

STRATEGY We expect the uncertainty in the position of an object as heavy as a marble to be very small but the uncertainty in the speed of an electron, which has a very small mass and is confined to a small region, to be very large. (a) The uncertainty Δp is equal to $m\Delta v$, where Δv is the uncertainty in the speed; we use Eq. 8 to estimate the minimum uncertainty in position, Δx, along the direction of the travel of the marble from $\Delta p \Delta x = \tfrac{1}{2}\hbar$ (the minimum value of the product of uncertainties). (b) We assume Δx to be the diameter of the atom and use Eq. 8 to estimate Δp; by using the mass of the electron inside the back cover, we find Δv from $\Delta p = m\Delta v$.

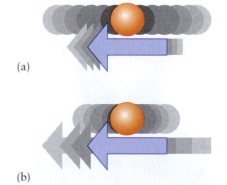

(a)

(b)

FIGURE 1.22 A representation of the uncertainty principle. (a) The location of the particle is ill defined, and so the momentum of the particle (represented by the arrow) can be specified reasonably precisely. (b) The location of the particle is well defined, and so the momentum cannot be specified very precisely.

SOLUTION (a) First we convert mass and speed into SI base units. The mass, m, is 1.0×10^{-3} kg, and the uncertainty in the speed, Δv, is $2 \times (1.0 \times 10^{-3}$ m·s^{-1}). The minimum uncertainty in position, Δx, is then:

From $\Delta p \Delta x = \frac{1}{2}\hbar$ and $\Delta p = m\Delta v$,

$$m\Delta v\Delta x = \frac{\hbar}{2} \text{ or } \Delta x = \frac{\hbar}{2m\Delta v}$$

From $\Delta x = \hbar/2m\Delta v$,

$$\Delta x = \frac{1.05457 \times 10^{-34} \text{ J·s}}{2 \times \underbrace{(1.0 \times 10^{-3} \text{ kg})}_{1.0 \text{ g}} \times \underbrace{(2.0 \times 10^{-3} \text{ m·s}^{-1})}_{2.0 \text{ mm·s}^{-1}}}$$

$$= \frac{1.05457 \times 10^{-34}}{2 \times 1.0 \times 10^{-3} \times 2.0 \times 10^{-3}} \frac{\text{J·s}}{\text{kg·m·s}^{-1}}$$

$$= 2.6 \times 10^{-29} \frac{\overbrace{\text{kg·m}^2\text{·s}^{-2}\text{·s}}^{\text{J}}}{\text{kg·m·s}^{-1}} = 2.6 \times 10^{-29} \text{ m}$$

As predicted, this uncertainty is very small.

A note on good practice: Notice that to manipulate the units we have expressed derived units (J in this case) in terms of base units. Notice too that we are using the more precise values of the fundamental constants given inside the back cover (rather than the less precise values quoted in the text) to ensure reliable results.

(b) The mass of the electron is given inside the back cover; the diameter of the atom is $200. \times 10^{-12}$ m, or 2.00×10^{-10} m. The uncertainty in the speed, Δv, is equal to $\Delta p/m$:

From $\Delta p \Delta x = \frac{1}{2}\hbar$ and $\Delta p = m\Delta v$,

$$\Delta v = \frac{\Delta p}{m} \overset{\Delta p = \hbar/2\Delta x}{=} \frac{\hbar}{2m\Delta x}$$

From $\Delta v = \hbar/2m\Delta x$,

$$\Delta v = \frac{1.05457 \times 10^{-34} \text{ J·s}}{2 \times \underbrace{(9.10939 \times 10^{-31} \text{ kg})}_{m_e} \times \underbrace{(2.00 \times 10^{-10} \text{ m})}_{200. \text{ pm}}}$$

$$= \frac{1.05457 \times 10^{-34}}{2 \times 9.10939 \times 10^{-31} \times 2.00 \times 10^{-10}} \frac{\text{J·s}}{\text{kg·m}}$$

$$= 2.89 \times 10^5 \frac{\overbrace{\text{kg·m}^2\text{·s}^{-2}\text{·s}}^{\text{J}}}{\text{kg·m}} = 2.89 \times 10^5 \text{ m·s}^{-1}$$

As predicted, the uncertainty in the speed of the electron is very large, nearly ± 150 km·s^{-1}.

SELF-TEST 1.7A A proton is accelerated in a cyclotron to a very high speed that is known to within 3.0×10^2 km·s^{-1}. What is the minimum uncertainty in its position?

[*Answer:* 0.10 pm]

SELF-TEST 1.7B The police are monitoring an automobile of mass 2.0 t (1 t = 10^3 kg) speeding along a highway. They are certain of the location of the vehicle only to within 1 m. What is the minimum uncertainty in the speed of the vehicle?

The location and momentum of a particle are complementary; that is, both the location and the momentum cannot be known simultaneously with arbitrary precision. The quantitative relation between the precision of each measurement is described by the Heisenberg uncertainty principle.

1.7 Wavefunctions and Energy Levels

If they were to account for the spectrum of atomic hydrogen and then atoms of the other elements, scientists of the early twentieth century had to revise the nineteenth-century description of matter to take into account wave-particle duality. One of the first people to formulate a successful theory (in 1927) was the Austrian scientist Erwin Schrödinger (Fig. 1.23), who introduced a central concept of quantum theory.

FIGURE 1.23 Erwin Schrödinger (1887–1961).

We shall illustrate the concept with an example (a particle trapped in a box), which though simple, reveals a number of important consequences of his approach. At the end of this section we shall be ready to move on to the hydrogen atom and see how the concepts apply to it and answer the remaining questions about its spectrum.

Because particles have wavelike properties, we cannot expect them to behave like pointlike objects moving along precise trajectories. Schrödinger's approach was to replace the precise trajectory of a particle by a **wavefunction**, ψ (the Greek letter psi), a mathematical function with values that vary with position. Some wavefunctions are very simple: shortly we shall meet one that is simply $\sin x$; when we get to the hydrogen atom, we shall meet one that is like e^{-x}.

The German physicist Max Born suggested how the wavefunction should be interpreted physically. The **Born interpretation** of the wavefunction is that *the probability of finding the particle in a region is proportional to the value of ψ^2* (Fig. 1.24). To be precise, ψ^2 is a **probability density,** the probability that the particle will be found in a small region divided by the volume of the region. "Probability density" is the analog of "mass density" (ordinary "density"), the mass of a region divided by the volume of the region. To calculate the mass of a region, we multiply its mass density by the volume of the region. Likewise, to calculate the probability that a particle is in a region, we multiply the probability density by the volume of the region. For instance, if $\psi^2 = 0.1 \text{ pm}^{-3}$ at a point, then the probability of finding the particle in a region of volume 2 pm^3 located at that point would be $(0.1 \text{ pm}^{-3}) \times (2 \text{ pm}^3) = 0.2$, or 1 chance in 5. Wherever ψ^2 is large, the particle has a high probability density; wherever ψ^2 is small, the particle has only a low probability density.

A note on good practice: Distinguish between *probability* and *probability density*: the former is unitless and lies between 0 (certainly not there) and 1 (certainly there), but the latter has the dimensions of 1/volume. To go from probability density to probability, multiply it by the volume of interest.

Because the square of any number is positive, we don't have to worry about ψ having a negative sign in some regions of space (as a function such as $\sin x$ has): probability density is never negative. Wherever ψ, and hence ψ^2, is zero, the particle has zero probability density. A location where ψ passes *through* zero (not just reaching zero) is called a **node** of the wavefunction; so we can say that a particle has zero probability density wherever the wavefunction has nodes.

To calculate the wavefunction for any particle we use Schrödinger's great contribution, the **Schrödinger equation.** Although we shall not use the equation directly (we shall need to know only the form of some of its solutions, not how those solutions are found), it is appropriate at least to see what it looks like. For a particle of mass m moving in a region where the potential energy is $V(x)$ the equation is

$$-\frac{\hbar^2}{2m}\frac{d^2\psi}{dx^2} + V(x)\psi = E\psi \qquad \text{(9a)}$$

The term $d^2\psi/dx^2$ can be thought of as a measure of how sharply the wavefunction is curved. The left-hand side of the Schrödinger equation is commonly written $H\psi$, where H is called the **hamiltonian** for the system; then the equation takes the deceptively simple form

$$H\psi = E\psi \qquad \text{(9b)}$$

The Schrödinger equation is used to calculate both the wavefunction ψ and the corresponding energy E. To understand what is involved, we shall consider one of the simplest systems, a single particle of mass m confined in a one-dimensional box between two rigid walls a distance L apart, a so-called **particle in a box** (Fig. 1.25). The equation can be solved quite easily for this system (as we show below) and the shapes of the wavefunctions, some of which are shown in the illustration, make sense when we view the particle as a wave. Only certain wavelengths can exist in the box, just as a stretched string can support only certain wavelengths. Think of a

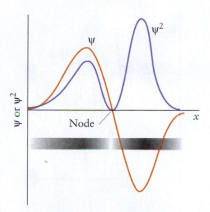

FIGURE 1.24 The Born interpretation of the wavefunction. The probability density (the blue line) is given by the square of the wavefunction and depicted by the density of shading in the band beneath. Note that the probability density is zero at a node. A node is a point where the wavefunction (the orange line) passes through zero, not merely approaches zero.

Note the "passes through"; merely becoming zero is not enough to be considered a node.

The symbol V is commonly used to denote potential energy, rather than E_P, in this context.

The Schrödinger equation is a "differential equation," an equation that relates the "derivatives" of a function (in this case, a second derivative of ψ, $d^2\psi/dx^2$) to the value of the function at each point. Derivatives are reviewed in Appendix 1F.

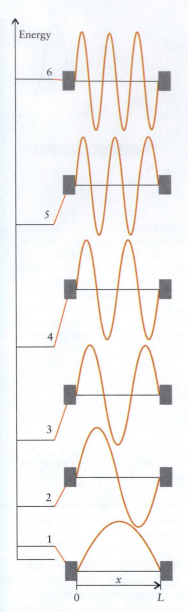

Living Graph: Figure 1.25 Particle-in-a-box wavefunctions

FIGURE 1.25 The arrangement known as "a particle in a box," in which a particle of mass m is confined between two impenetrable walls a distance L apart. The first six wavefunctions and their energies are shown. The numbers on the left are values of the quatum number n. Note that there is a zero-point energy because n cannot be zero.

guitar string: because it is tied down at each end, it can support only shapes like the ones shown in Fig. 1.25. The shapes of the wavefunctions for the particle in the box are identical with the displacements of a stretched string when it vibrates. Their mathematical form is that of a standing wave:

$$\psi_n(x) = \left(\frac{2}{L}\right)^{1/2} \sin\left(\frac{n\pi x}{L}\right) \qquad n = 1, 2, \ldots \tag{10}$$

The integer n labels the wavefunctions and is called a "quantum number." In general, a **quantum number** is an integer (or, in some cases, Section 1.10, a half-integer) that labels a wavefunction, specifies a state, and can be used to calculate the value of a property of the system. For example, we can use n to find an expression for the energy corresponding to each wavefunction.

HOW DO WE DO THAT (LEVEL 1)?

The kinetic energy of a particle of mass m is related to its speed, v, by $E_K = \frac{1}{2}mv^2$. We can relate this energy to the wavelength of the particle by noting that the linear momentum is $p = mv$ and then using the de Broglie relation (Eq. 7):

$$E_K = \frac{1}{2}mv^2 = \frac{1}{2}\frac{(mv)^2}{m} = \frac{p^2}{2m} = \frac{(h/\lambda)^2}{2m} = \frac{h^2}{2m\lambda^2}$$

We assume that the potential energy of the particle is zero everywhere inside the box, and so the total energy, E, is given by the expression for E_K alone.

At this point, we recognize that (like a guitar string) only whole-number multiples of half-wavelengths can fit into the box (see Fig. 1.25; the waves have one bulge, two bulges, three bulges, and so forth, with each "bulge" a half-wavelength). That is, for a box of length L,

$$\tfrac{1}{2}\lambda, \lambda, \tfrac{3}{2}\lambda, \ldots = n \times \tfrac{1}{2}\lambda, \text{ with } n = 1, 2, \ldots$$

Therefore, the allowed wavelengths are

$$2L/n, \text{ with } n = 1, 2, \ldots$$

When this expression for λ is inserted into the expression for the energy, we obtain

$$E_n = \frac{h^2}{2m\lambda^2} = \frac{h^2}{2m(2L/n)^2} = \frac{n^2 h^2}{8mL^2}$$

HOW DO WE DO THAT (LEVEL 2)?

The more sophisticated—and more general—way of finding the energy levels of a particle in a box is to use calculus to solve the Schrödinger equation. First, we note that the potential energy of the particle is zero everywhere inside the box; so $V(x) = 0$, and the equation that we have to solve is

$$-\frac{\hbar^2}{2m}\frac{d^2\psi}{dx^2} = E\psi$$

This equation has the solutions

$$\psi(x) = A \sin kx + B \cos kx$$

with A, B, and k constants. To verify that this is a solution, we substitute it into the differential equation and use $d(\sin kx)/dx = k \cos kx$ and $d(\cos kx)/dx = -k \sin kx$:

$$\frac{d^2\psi}{dx^2} = \frac{d^2}{dx^2}(A \sin kx + B \cos kx) = \frac{d}{dx}(kA \cos kx - kB \sin kx)$$

$$= -k^2 A \sin kx - k^2 B \cos kx = -k^2(A \sin kx + B \cos kx)$$

$$= -k^2\psi$$

This expression has the same form as the Schrödinger equation, and is exactly the same if $k^2 = 2mE/\hbar^2$. It follows that

$$E = \frac{k^2\hbar^2}{2m} = \frac{k^2(h/2\pi)^2}{2m} = \frac{k^2 h^2}{8\pi^2 m}$$

SELF-TEST 1.8A Use the same model for helium but suppose that the box is 100. pm long, because the atom is smaller. Estimate the wavelength of the same transition.

[*Answer*: 11.0 nm]

SELF-TEST 1.8B Use the same model for hydrogen and estimate the wavelength for the transition from the $n = 3$ energy level to the $n = 2$ level.

Another surprising implication of Eq. 11 is that *a particle in a container cannot have zero energy*. Because the lowest value of n is 1 (corresponding to a wave of one-half wavelength fitting into the box), the lowest energy is $E_1 = h^2/8mL^2$. This lowest possible energy is called the **zero-point energy**. The existence of a zero-point energy means that, according to quantum mechanics, a particle can never be perfectly still when it is confined between two walls: it must always possess an energy—in this case, a kinetic energy—of at least $h^2/8mL^2$. This result is consistent with the uncertainty principle. When a particle is confined between two walls, the uncertainty in its position cannot be larger than the distance between the two walls. Because the position is not *completely* uncertain, the linear momentum must be uncertain, too, and so we cannot say that the particle is completely still. The particle must therefore have some kinetic energy. The zero-point energy is a purely quantum mechanical phenomenon and is very small for macroscopic systems. For example, a billiard ball on a pool table has a completely negligible zero-point energy of about 10^{-67} J.

Finally, the shapes of the wavefunctions of a particle in a box also reveal some interesting information. Let's look at the two lowest energy wavefunctions, corresponding to $n = 1$ and $n = 2$. Figure 1.27 shows, by the density of shading, the likelihood of finding a particle: we see that when a particle is described by the wavefunction ψ_1 (and has energy $h^2/8mL^2$), then it is most likely to be found in the center of the box. Conversely, if the particle is described by the wavefunction ψ_2 (and has energy $h^2/2mL^2$), then it is most likely to be found in regions between the center and the walls and is unlikely to be found in the middle of the box. Remember that the wavefunction itself does not have any direct physical significance: we have to take the square of ψ before we can interpret it in terms of the probability of finding a particle somewhere.

> *The probability density for a particle at a location is proportional to the square of the wavefunction at that point; the wavefunction is found by solving the Schrödinger equation for the particle. When the equation is solved subject to the appropriate boundary conditions, it is found that the particle can possess only certain discrete energies.*

THE HYDROGEN ATOM

We are now ready to build a quantum mechanical model of a hydrogen atom. Our task is to combine our knowledge that an electron has wavelike properties and is described by a wavefunction with the nuclear model of the atom, and explain the ladder of energy levels suggested by spectroscopy.

1.8 The Principal Quantum Number

An electron in an atom is like a particle in a box, in the sense that it is confined within the atom by the pull of the nucleus. We can therefore expect the electron's wavefunctions to obey certain boundary conditions, like the constraints we encountered when fitting a wave between the walls of a container. As we saw for a particle in a box, these constraints result in the quantization of energy and the existence of discrete energy levels. Even at this early stage, we can expect the electron to be confined to certain energies, just as spectroscopy requires.

To find the wavefunctions and energy levels of an electron in a hydrogen atom, we must solve the appropriate Schrödinger equation. To set up this equation, which resembles the equation in Eq. 9 but allows for motion in three dimensions, we use the expression for the potential energy of an electron of charge $-e$ at a

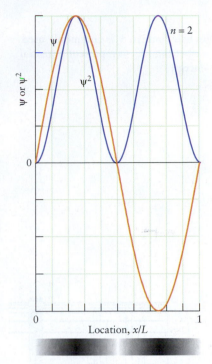

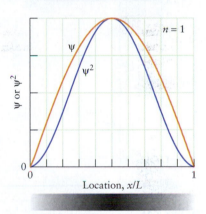

FIGURE 1.27 The two lowest energy wavefunctions (ψ, orange) for a particle in a box and the corresponding probability densities (ψ^2, blue). The probability densities are also shown by the density of shading of the bands beneath each wavefunction.

In this context, V is commonly used rather than E_P to represent potential energy.

distance r from a nucleus of charge $+e$. As we saw in Section A, this "Coulomb" potential energy is

$$V(r) = \frac{(-e)(+e)}{4\pi\varepsilon_0 r} = -\frac{e^2}{4\pi\varepsilon_0 r} \tag{13}$$

Solving the Schrödinger equation for a particle with this potential energy is difficult, but Schrödinger himself achieved it in 1927. He found that the allowed energy levels for an electron in a hydrogen atom are

$$E_n = -\frac{h\mathcal{R}}{n^2} \qquad \mathcal{R} = \frac{m_e e^4}{8h^3\varepsilon_0^2} \qquad n = 1, 2, \ldots \tag{14a}$$

These energy levels have exactly the form suggested spectroscopically, but now we also have an expression for $\mathcal{R}$ in terms of more fundamental constants. When the fundamental constants are inserted into the expression for $\mathcal{R}$, the value obtained is 3.29×10^{15} Hz, the same as the experimental value of the Rydberg constant. This agreement is a triumph for Schrödinger's theory and for quantum mechanics; it is easy to understand the thrill that Schrödinger must have felt when he arrived at this result. A very similar expression applies to other one-electron ions, such as He^+ and even C^{5+}, with atomic number Z:

$$E_n = -\frac{Z^2 h\mathcal{R}}{n^2} \qquad n = 1, 2, \ldots \tag{14b}*$$

What does this equation tell us? All the energies are negative, meaning that the electron has a lower energy in the atom than when it is far from the nucleus. Because Z appears in the numerator, we see that the greater the value of the nuclear charge the more tightly the electron is bound to a nucleus. That n appears in the denominator shows that as n increases, the energy becomes less negative.

The dependence of the energy on Z^2 rather than on Z itself arises from two factors: first, a nucleus of atomic number Z and charge Ze gives rise to a field that is Z times stronger than that of a single proton; second, the electron is drawn in by the higher charge and is Z times closer to the nucleus than it is in hydrogen.

Figure 1.28 shows the energy levels calculated from Eq. 14a. We see that they come closer together as n increases. Each level is labeled by the integer n, which is called the **principal quantum number,** from $n = 1$ for the first (lowest, most negative) level, $n = 2$ for the second, continuing to infinity. The lowest energy possible for an electron in a hydrogen atom, $-h\mathcal{R}$, is obtained when $n = 1$. This lowest energy state is called the **ground state** of the atom. A hydrogen atom is normally found in its ground state, with its electron in the level with $n = 1$. The energy of the bound electron climbs up the ladder of levels as n increases. It reaches the top of the ladder, corresponding to $E = 0$ and freedom, when n reaches infinity. At that point, the electron has left the atom. This process is called **ionization.** The difference in energy between the ground state and the ionized state is the energy required to remove an electron from the neutral atom in its ground state.

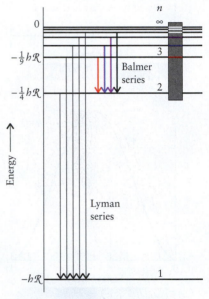

FIGURE 1.28 The permitted energy levels of a hydrogen atom as calculated from Eq. 14. The levels are labeled with the quantum number n, which ranges from 1 (for the lowest state) to infinity (for the separated proton and electron).

The ionization energy is discussed further in Section 1.17.

The energy levels of a hydrogen atom are defined by the principal quantum number, $n = 1, 2, \ldots$, and form a converging series, as shown in Fig. 1.28.

1.9 Atomic Orbitals

We have found the energies and now need to find the corresponding wavefunctions. Once we know the wavefunctions we shall have gone beyond the information provided directly by spectroscopy and know not only the allowed energies of the electron in a hydrogen atom but also how the electron is distributed around the nucleus.

The wavefunctions of electrons in atoms are called **atomic orbitals.** The name was chosen to suggest something less definite than an "orbit" of an electron around a nucleus and to take into account the wave nature of the electron. The mathematical expressions for atomic orbitals—which are obtained as solutions of the Schrödinger equation—are more complicated than the sine functions for the particle in a box, but their essential features are straightforward. Moreover, we must never lose sight of their interpretation, that the *square* of a wavefunction tells us the probability of an electron being at each point. To visualize this probability density, imagine a cloud centered on the nucleus. The density of the cloud at each point represents the probability of finding an electron there. Denser regions of the cloud therefore represent locations where the electron is more likely to be found.

To interpret the information in each atomic orbital, we need a way to identify the location of each point around a nucleus. It is most convenient to describe these locations in terms of **spherical polar coordinates**, in which each point is labeled with three coordinates:

- r, the distance from the center of the atom;
- θ (theta), the angle from the positive z-axis (the "north pole"), which can be thought of as playing the role of the geographical "latitude"; and
- ϕ (phi), the angle about the z-axis, the geographical "longitude."

These coordinates are shown in Fig. 1.29. Each wavefunction, which in general varies from point to point, can be written as a function of the coordinates, $\psi(r,\theta,\phi)$. It turns out, however, that all the wavefunctions can be written as the product of a function that depends only on r and another function that depends only on the angles θ and ϕ. That is,

$$\psi(r,\theta,\phi) = R(r)Y(\theta,\phi) \tag{15}*$$

The function $R(r)$ is called the **radial wavefunction**; it tells us how the wavefunction varies as we move away from the nucleus in any direction. The function $Y(\theta,\phi)$ is called the **angular wavefunction**; it tells us how the wavefunction varies as the angles θ and ϕ change. For example, the wavefunction corresponding to the ground state of the hydrogen atom ($n = 1$) is

$$\psi(r,\theta,\phi) = \overbrace{\frac{2e^{-r/a_0}}{a_0^{3/2}}}^{R(r)} \times \overbrace{\frac{1}{2\pi^{1/2}}}^{Y(\theta,\phi)} = \frac{e^{-r/a_0}}{(\pi a_0^3)^{1/2}} \qquad a_0 = \frac{4\pi\varepsilon_0\hbar^2}{m_e e^2}$$

The quantity a_0 is called the **Bohr radius**; when the values of the fundamental constants are inserted, we find $a_0 = 52.9$ pm.

What does this equation tell us? For this wavefunction, the angular wavefunction Y is a constant, $1/2\pi^{1/2}$, independent of the angles, which means that the wavefunction is the same in all directions. The radial wavefunction $R(r)$ decays exponentially toward zero as r increases, which means that the electron density is highest close to the nucleus ($e^0 = 1$). The Bohr radius tells us how sharply the wavefunction falls away with distance: when $r = a_0$, ψ has fallen to $1/e$ (37%) of its value at the nucleus.

All the higher energy levels have more than one wavefunction for each energy level. One of the wavefunctions for the next higher energy level, with $n = 2$ and $E_2 = -\frac{1}{4}h\mathcal{R}$, is

$$\psi(r,\theta,\phi) = \overbrace{\frac{1}{2\sqrt{6}}\frac{1}{a_0^{5/2}}re^{-r/2a_0}}^{R(r)} \times \overbrace{\left(\frac{3}{4\pi}\right)^{1/2}\sin\theta\cos\phi}^{Y(\theta,\phi)}$$

$$= \frac{1}{4}\left(\frac{1}{2\pi a_0^5}\right)^{1/2}re^{-r/2a_0}\sin\theta\cos\phi$$

Geographical latitudes are measured from the equator, not the north pole.

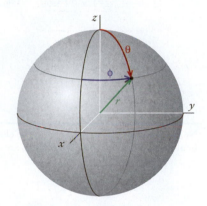

FIGURE 1.29 The spherical polar coordinates: r is the radius, which gives the distance from the center, θ is the colatitude, which gives the angle from the z-axis, and ϕ, the "longitude," is the azimuth, which gives the angle from the x-axis.

In an early model of the hydrogen atom proposed by Niels Bohr, the electron traveled in a circular orbit of radius a_0 around the nucleus. The uncertainty principle rules out this model.

TABLE 1.2 Hydrogenlike Wavefunctions[*] (Atomic Orbitals), $\psi = RY$

(a) Radial wavefunctions, $R_{nl}(r)$			(b) Angular wavefunctions, $Y_{lm_l}(\theta,\phi)$		
n	l	$R_{nl}(r)$	l	"m_l"[†]	$Y_{lm_l}(\theta,\phi)$
1	0	$2\left(\dfrac{Z}{a_0}\right)^{3/2} e^{-Zr/a_0}$	0	0	$\left(\dfrac{1}{4\pi}\right)^{1/2}$
2	0	$\dfrac{1}{2\sqrt{2}}\left(\dfrac{Z}{a_0}\right)^{3/2}\left(2 - \dfrac{Zr}{a_0}\right)e^{-Zr/2a_0}$	1	x	$\left(\dfrac{3}{4\pi}\right)^{1/2}\sin\theta\cos\phi$
	1	$\dfrac{1}{2\sqrt{6}}\left(\dfrac{Z}{a_0}\right)^{3/2}\left(\dfrac{Zr}{a_0}\right)e^{-Zr/2a_0}$		y	$\left(\dfrac{3}{4\pi}\right)^{1/2}\sin\theta\sin\phi$
3	0	$\dfrac{1}{9\sqrt{3}}\left(\dfrac{Z}{a_0}\right)^{3/2}\left(3 - \dfrac{2Zr}{a_0} + \dfrac{2Z^2r^2}{9a_0^2}\right)e^{-Zr/3a_0}$		z	$\left(\dfrac{3}{4\pi}\right)^{1/2}\cos\theta$
	1	$\dfrac{1}{27\sqrt{6}}\left(\dfrac{Z}{a_0}\right)^{3/2}\left(2 - \dfrac{Zr}{3a_0}\right)e^{-Zr/3a_0}$	2	xy	$\left(\dfrac{15}{16\pi}\right)^{1/2}\sin^2\theta\sin 2\phi$
	2	$\dfrac{1}{81\sqrt{30}}\left(\dfrac{Z}{a_0}\right)^{3/2}\left(\dfrac{Zr}{a_0}\right)^2 e^{-Zr/3a_0}$		yz	$\left(\dfrac{15}{4\pi}\right)^{1/2}\cos\theta\sin\theta\sin\phi$
				zx	$\left(\dfrac{15}{4\pi}\right)^{1/2}\cos\theta\sin\theta\cos\phi$
				$x^2 - y^2$	$\left(\dfrac{15}{16\pi}\right)^{1/2}\sin^2\theta\cos 2\phi$
				z^2	$\left(\dfrac{15}{16\pi}\right)^{1/2}(3\cos^2\theta - 1)$

[*]Note: In each case, $a_0 = 4\pi\varepsilon_0\hbar^2/m_ee^2$, or close to 52.9 pm; for hydrogen itself, $Z = 1$.
[†]In all cases except $m_l = 0$, the orbitals are sums and differences of orbitals with specific values of m_l.

What does this equation tell us? This wavefunction also falls exponentially toward zero as r increases. Notice, though, that there is a factor r that multiplies the exponential function, so ψ is zero at the nucleus (at $r = 0$) as well as far away from it. We discuss the angular dependence shortly.

The expressions for a number of other atomic orbitals are shown in Table 1.2a (for R) and Table 1.2b (for Y). To understand these tables, we need to know that each wavefunction is labeled by *three* quantum numbers: n is related to the *size* and *energy* of the orbital, l is related to its *shape*, and m_l is related to its *orientation* in space.

When the Schrödinger equation is solved in detail, it turns out that three quantum numbers are needed to label each wavefunction (because the atom is three-dimensional), and that, for a hydrogen atom, wavefunctions with the same value of n all have the same energy. We have already encountered n, the principal quantum number, which specifies the energy of the orbital in a one-electron atom (through Eq. 14). In a one-electron atom, all atomic orbitals with the same value of the principal quantum number n have the same energy and are said to belong to the same **shell** of the atom. The name "shell" reflects the fact that as n increases, the region of greatest probability density is like a nearly hollow shell of increasing radius. The higher the number of the shell, the further away from the nucleus are the electrons in that shell.

The second quantum number needed to specify an orbital is l, the **orbital angular momentum quantum number.** This quantum number can take the values

$$l = 0, 1, 2, \ldots, n - 1$$

There are n different values of l for a given value of n. For instance, when $n = 3$, l can have any of the three values 0, 1, and 2. The orbitals of a shell with principal quantum number n therefore fall into n **subshells**, groups of orbitals that have the same value of l. There is only one subshell in the $n = 1$ level ($l = 0$), two in the $n = 2$ level ($l = 0$ and 1), three in the $n = 3$ level ($l = 0$, 1, and 2), and so on. All orbitals with $l = 0$ are called *s*-orbitals, those with $l = 1$ are called *p*-**orbitals**, those with $l = 2$ are called *d*-**orbitals**, and those with $l = 3$ are called *f*-**orbitals**:

The names come from the fact that spectroscopic lines were once classified as *s*harp, *p*rincipal, *d*iffuse, and *f*undamental.

Value of l	0	1	2	3
Orbital type	s	p	d	f

Although higher values of l (corresponding to *g*-, *h*-, . . . orbitals) are possible, the lower values of l (0, 1, 2, and 3) are the only ones that chemists need in practice.

Just as the value of n can be used to calculate the energy of an electron, the value of l can be used to calculate another physical property. As its name suggests, l tells us the **orbital angular momentum** of the electron, a measure of the rate at which the electron circulates round the nucleus:

$$\text{Orbital angular momentum} = \{l(l + 1)\}^{1/2}\hbar \tag{16}*$$

An electron in an *s*-orbital (an "*s*-electron"), for which $l = 0$, has zero orbital angular momentum. That means that we should imagine it not as circulating around the nucleus but simply as distributed evenly around it. An electron in a *p*-orbital ($l = 1$) has nonzero orbital angular momentum (of magnitude $2^{1/2}\hbar$); so it can be thought of as circulating around the nucleus. An electron in a *d*-orbital ($l = 2$) has a higher orbital angular momentum ($6^{1/2}\hbar$), one in an *f*-orbital ($l = 3$) has an even higher angular momentum ($12^{1/2}\hbar$), and so on.

An important feature of the hydrogen atom is that all the orbitals of a given shell have the same energy, regardless of the value of their orbital angular momentum (we see from Eq.14 that l does not appear in the expression for the energy). We say that the orbitals of a shell in a hydrogen atom are **degenerate**, which means that they all have the same energy. This degeneracy is true only of the hydrogen atom and one-electron ions (such as He^+ and C^{5+}).

The third quantum number required to specify an orbital is m_l, the **magnetic quantum number**, which distinguishes the individual orbitals within a subshell. This quantum number can take the values

$$m_l = l, l - 1, \ldots, -l$$

There are $2l + 1$ different values of m_l for a given value of l and therefore $2l + 1$ orbitals in a subshell of quantum number l. For example, when $l = 1$, $m_l = +1, 0, -1$; so there are three *p*-orbitals in a given shell. Alternatively, we can say that a subshell with $l = 1$ consists of three orbitals.

A note on good practice: Always write the $+$ sign explicitly for positive values of m_l.

The magnetic quantum number tells us the orientation of the orbital motion of the electron. Specifically, it tells us that the orbital angular momentum around an arbitrary axis is equal to $m_l\hbar$, the rest of the orbital motion (to make up the full amount of $\{l(l + 1)\}^{1/2}\hbar$) being around other axes. For instance, if $m_l = +1$, then the orbital angular momentum of the electron around the arbitrary axis is $+\hbar$, whereas, if $m_l = -1$, then the orbital angular momentum of the electron around the same arbitrary axis is $-\hbar$. The difference in sign simply means that the direction of motion is opposite, the electron in one state circulating clockwise and an electron in the other state circulating counterclockwise. If $m_l = 0$, then the electron is not circulating around the selected axis.

The hierarchy of shells, subshells, and orbitals is summarized in Fig. 1.30 and Table 1.3. Each possible combination of the three quantum numbers specifies an individual orbital. For example, an electron in the ground state of a hydrogen atom has the specification $n = 1$, $l = 0$, $m_l = 0$. Because $l = 0$, the ground-state wavefunction is an example of an *s*-orbital and is denoted $1s$. Each

FIGURE 1.30 A summary of the arrangement of shells, subshells, and orbitals in an atom and the corresponding quantum numbers. Note that the quantum number m_l is an alternative label for the individual orbitals: in chemistry, it is more common to use x, y, and z instead, as shown in Figs. 1.36 through 1.38.

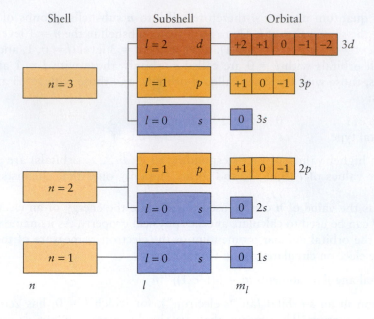

Shell   Subshell   Orbital

$n = 3$ — $l = 2$ d — $+2$ $+1$ 0 -1 -2 $3d$

$l = 1$ p — $+1$ 0 -1 $3p$

$l = 0$ s — 0 $3s$

$n = 2$ — $l = 1$ p — $+1$ 0 -1 $2p$

$l = 0$ s — 0 $2s$

$n = 1$ — $l = 0$ s — 0 $1s$

n l m_l

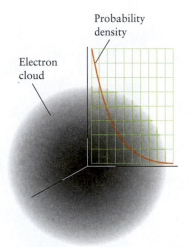

Probability density

Electron cloud

FIGURE 1.31 The three-dimensional electron cloud corresponding to an electron in a 1s-orbital of hydrogen. The density of shading represents the probability of finding the electron at any point. The superimposed graph shows how the probability varies with the distance of the point from the nucleus along any radius.

shell has one s-orbital, and the s-orbital in the shell with quantum number n is called an **ns-orbital.**

All s-orbitals are independent of the angles θ and ϕ, so we say that they are **spherically symmetrical** (Fig. 1.31). The probability density of an electron at the point (r,θ,ϕ) when it is in a 1s-orbital is found from the wavefunction for the ground state of the hydrogen atom. When we square the wavefunction (which was given earlier, but can also be constructed as RY from the entries for R and Y in Tables 1.2a and 1.2b) we find that

$$\psi^2(r,\theta,\phi) = \frac{e^{-2r/a_0}}{\pi a_0^3} \tag{17}$$

In this case, the probability density is independent of angle, and for simplicity is commonly written simply $\psi^2(r)$. In principle, the cloud representing the probability density never thins to exactly zero, no matter how large the value of r. So we could think of an atom as being bigger than the Earth! However, there is virtually no chance of finding an electron farther from the nucleus than about 250 pm, and so atoms are in fact very small. As we can see from the high density of the cloud at the nucleus, an electron in an s-orbital has a nonzero probability of being found right at the nucleus: because $l = 0$, there is no orbital angular momentum to fling the electron away from the nucleus.

EXAMPLE 1.9 **Sample exercise: Calculating the probability of finding an electron at a certain location**

Suppose the electron is in a 1s orbital of a hydrogen atom. What is the probability of finding the electron in a small region a distance a_0 from the nucleus relative to the probability of finding it in the same small region located right at the nucleus?

TABLE 1.3 Quantum Numbers for Electrons in Atoms

Name	Symbol	Values	Specifies	Indicates
principal	n	$1, 2, \ldots$	shell	size
orbital angular momentum*	l	$0, 1, \ldots, n - 1$	subshell: $l = 0, 1, 2, 3, 4, \ldots$ $s, p, d, f, g, \ldots$	shape
magnetic	m_l	$l, l - 1, \ldots, -l$	orbitals of subshell	orientation
spin magnetic	m_s	$-\frac{1}{2}, +\frac{1}{2}$	spin state	spin direction

*Also called the azimuthal quantum number.

SOLUTION The probability density is independent of angle when $l = 0$. We calculate the following ratio of the squares of the wavefunction at the two points

$$\frac{\text{Probability density at } r = a_0}{\text{Probability density at } r = 0} = \frac{\psi^2(a_0)}{\psi^2(0)}$$

From $\psi^2(r, \theta, \phi) = (1/\pi a_0^3)e^{-2r/a_0}$,

$$\frac{\psi^2(a_0)}{\psi^2(0)} = \frac{(1/\pi a_0^3)\overbrace{e^{-2a_0/a_0}}^{e^{-2}}}{(1/\pi a_0^3)\underbrace{e^0}_{1}}$$

$$= e^{-2} = 0.14$$

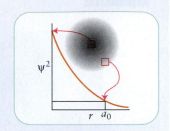

That is, the probability of finding the electron in a small region at a distance a_0 from the nucleus is only 14% of that of finding the electron in a region of the same volume located at the nucleus.

SELF-TEST 1.9A Calculate the same ratio but for the more distant point at $r = 2a_0$, twice as far from the nucleus.

[*Answer:* 0.018]

SELF-TEST 1.9B Calculate the same ratio but for a point at $3a_0$ from the nucleus.

The **radial distribution function**, P, is closely related to the wavefunction $\psi = RY$ and is given by

$$P(r) = r^2 R^2(r) \tag{18a}$$

For *s*-orbitals, $\psi = RY = R/2\pi^{1/2}$, so $R^2 = 4\pi\psi^2$, and this expression is then the same as

$$P(r) = 4\pi r^2 \psi^2(r) \tag{18b}*$$

and this expression is the form that you will normally see used; however, it applies only to *s*-orbitals, whereas Eq. 18a applies to any kind of orbital. The radial distribution function has a very special significance: it tells us *the proba-bility that the electron will be found at a particular distance from the nucleus, regardless of the direction.* Specifically, the probability that the electron will be found anywhere in a thin shell of radius r and thickness δr is given by $P(r)\delta r$ (Fig. 1.32).

A note on good practice: Be careful to distinguish the radial distribution function from the wavefunction and its square, the probability density:
• The wavefunction itself tells us, through $\psi^2(r, \theta, \phi)\delta V$, the probability of finding the electron in the small volume δV at a particular location specified by r, θ, and ϕ.
• The radial distribution function tells us, through $P(r)\delta r$, the probability of finding the electron in the range of radii δr, at a particular value of the radius, summed over all values of θ and ϕ.

The radial distribution function for the population of the Earth, for instance, is zero up to about 6400 km from the center of the Earth, rises sharply, and then falls back to almost zero (the "almost" takes into account the small number of people who are on mountains or flying in airplanes).

Note that for *all* orbitals, not just *s*-orbitals, P is zero at the nucleus, simply because the region in which we are looking for the electron has shrunk to zero size. (The probability density for an *s*-orbital is nonzero at the nucleus, but here we are multiplying it by a volume, $4\pi r^2 \delta r$, that becomes zero at the nucleus, at $r = 0$.) As r increases, the value of $4\pi r^2$ increases (the shell is getting bigger), but, for a 1*s*-orbital, the square of the wavefunction, ψ^2, falls toward zero as r increases; as a result, the product of $4\pi r^2$ and ψ^2 starts off at zero, goes through a maximum, and then declines to zero. The value of P is a maximum at a_0, the Bohr radius. There-fore, the Bohr radius corresponds to the radius at which an electron in a 1*s*-orbital is most likely to be found.

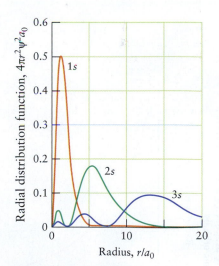

FIGURE 1.32 The radial distribution function tells us the probability density for finding an electron at a given radius summed over all directions. The graph shows the radial distribution function for the 1*s*-, 2*s*-, and 3*s*-orbitals in hydrogen. Note how the most probable radius (corresponding to the greatest maximum) increases as n increases.

Instead of drawing the *s*-orbital as a cloud, chemists usually draw its **boundary surface**, a smooth surface that encloses most of the cloud. However, although the boundary surface is easier to draw, it does not give the best picture of an atom; an atom has fuzzy edges and is not as smooth as the boundary surface might suggest. Despite this limitation, the boundary surface is useful, because an electron is likely to be found only inside the boundary surface of the orbital. An *s*-orbital has a spherical boundary surface (Fig. 1.33), because the electron cloud is spherical. *s*-Orbitals with higher energies have spherical boundary surfaces of greater diameter. They also have a more complicated radial variation, with nodes at locations that can be found by examining the wavefunctions (Fig. 1.34).

The boundary surface of a *p*-orbital has two lobes (Fig. 1.35). These lobes are marked + and − to signify that the wavefunction has different signs in the two regions. For instance, the 2*p*-orbital discussed earlier is proportional to $\sin\theta\cos\phi$, and as ϕ increases from 0 to 2π as we travel around the "equator" of the atom, $\cos\phi$ changes from +1 through 0 (at $\phi = \pi/2$) to −1 (at π) and then through zero and back to +1 again. Thus, in one hemisphere the wavefunction is positive and in the other it is negative, as depicted on the boundary surface in Fig. 1.35. The two lobes of a *p*-orbital are separated by a planar region called a **nodal plane**, which cuts through the nucleus and on which $\phi = 0$. A ***p*-electron**, an electron in a *p*-orbital, will never be found on the nodal plane. Nor will it be found at the nucleus on account of the factor *r* in the wavefunction. This difference from *s*-orbitals stems from the fact that an electron in a *p*-orbital has nonzero orbital angular momentum that flings it away from the nucleus.

There are three *p*-orbitals in each subshell, corresponding to the quantum numbers $m_l = +1$, 0, −1. However, chemists commonly refer to the orbitals according to the axes along which the lobes lie; hence, we refer to p_x, p_y, and p_z orbitals (Fig. 1.36).

A subshell with $l = 2$ consists of five *d*-orbitals. Each *d*-orbital has four lobes, except for the orbital designated d_{z^2}, which has a more complicated shape (Fig. 1.37). A subshell with $l = 3$ consists of seven *f*-orbitals with even more complicated shapes (Fig. 1.38).

The total number of orbitals in a shell with principal quantum number *n* is n^2. To confirm this rule, we need to recall that *l* has *n* integer values from 0 to

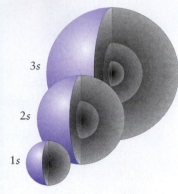

Animation: Figure 1.33 *s*-Orbitals

FIGURE 1.33 The three *s*-orbitals of lowest energy. The simplest way of drawing an atomic orbital is as a boundary surface, a surface within which there is a high probability (typically 90%) of finding the electron. We shall use blue to denote *s*-orbitals, but that color is only an aid to their identification. The shading within the boundary surfaces is an approximate indication of the electron density at each point.

Living Graph: Figure 1.34 Hydrogenic atomic orbitals

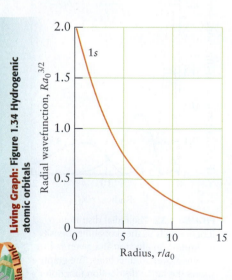

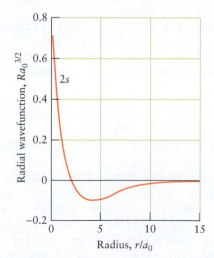

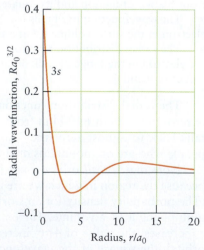

FIGURE 1.34 The radial wavefunctions of the first three *s*-orbitals of a hydrogen atom. Note that the number of radial nodes increases (as $n - 1$), as does the average distance of the electron from the nucleus (compare with Fig. 1.32). Because the probability density is given by ψ^2, all *s*-orbitals correspond to a nonzero probability density at the nucleus.

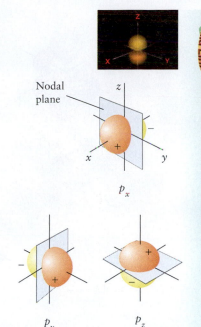

FIGURE 1.35 The boundary surface and the radial variation of a 2p-orbital along the (vertical) z-axis. All p-orbitals have boundary surfaces with similar shapes, including one nodal plane. Note that the orbital has opposite signs (as depicted by the depth of color) on each side of the nodal plane.

FIGURE 1.36 There are three p-orbitals of a given energy, and they lie along three perpendicular axes. We shall use yellow to indicate p-orbitals: dark yellow for the positive lobe and light yellow for the negative lobe.

$n - 1$ and that the number of orbitals in a subshell for a given value of l is $2l + 1$. For instance, for $n = 4$, there are four subshells with $l = 0, 1, 2$, and 3, consisting of one s-orbital, three p-orbitals, five d-orbitals, and seven f-orbitals, respectively. There are therefore $1 + 3 + 5 + 7 = 16$, or 4^2, orbitals in the shell with $n = 4$ (Fig. 1.39).

> *The location of an electron in an atom is described by a wavefunction known as an atomic orbital; atomic orbitals are designated by the quantum numbers n, l, and m_l and fall into shells and subshells as summarized in Fig. 1.30.*

1.10 Electron Spin

Schrödinger's calculation of the energies of the hydrogen orbitals was a milestone in the development of modern atomic theory. Yet the observed spectral lines did not have exactly the frequencies he predicted. In 1925, two Dutch-American physicists, Samuel Goudsmit and George Uhlenbeck, proposed an explanation for the tiny deviations that had been observed. They suggested that an electron behaves in some respects like a spinning sphere, something like a planet rotating on its axis. This property is called **spin**.

According to quantum mechanics, an electron has two spin states, represented by the arrows ↑ (up) and ↓ (down) or the Greek letters α (alpha) and β (beta) We can think of an electron as being able to spin counterclockwise at a certain rate

There is no direct relation between the values of m_l and the x, y, z designation of the orbitals: the orbitals labeled with the axes are combinations of the orbitals labeled with the quantum number m_l.

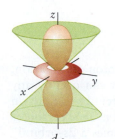

d_{z^2}

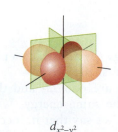

$d_{x^2-y^2}$

d_{zx}

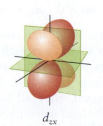

d_{yz}

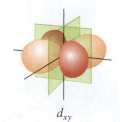

d_{xy}

FIGURE 1.37 The boundary surface of a d-orbital is more complicated than that of an s- or a p-orbital. There are, in fact, five d-orbitals of a given energy; four of them have four lobes, one is slightly different. In each case, an electron that occupies a d-orbital will not be found at the nucleus. We shall use orange to indicate d-orbitals: dark orange for the positive lobes, and light orange for the negative lobes.

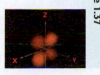

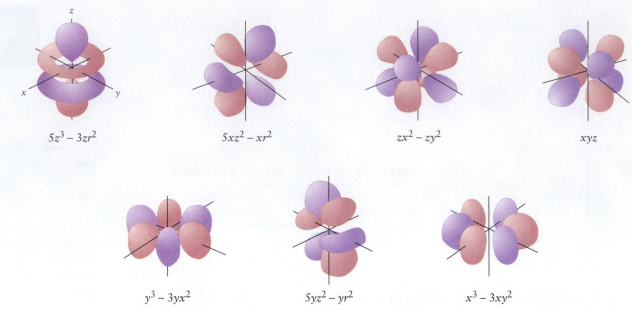

$$5z^3 - 3zr^2 \qquad 5xz^2 - xr^2 \qquad zx^2 - zy^2 \qquad xyz$$

$$y^3 - 3yx^2 \qquad 5yz^2 - yr^2 \qquad x^3 - 3xy^2$$

FIGURE 1.38 The seven *f*-orbitals of a shell (with $n = 3$) have a very complex appearance. Their detailed form will not be used again in this text. However, their existence is important for understanding the periodic table, the presence of the lanthanoids and actinoids, and the properties of the later *d*-block elements. A darker color denotes a positive lobe, a lighter color a negative lobe.

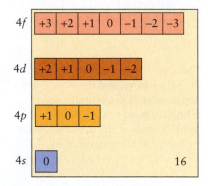

FIGURE 1.39 There are 16 orbitals in the shell with $n = 4$, each of which can hold two electrons (see Section 1.13), for a total of 32 electrons.

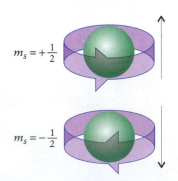

FIGURE 1.40 The two spin states of an electron can be represented as clockwise or counterclockwise rotation around an axis passing through the electron. The two states are identified by the quantum number m_s and depicted by the arrows shown on the right.

(the ↑ state) or clockwise at exactly the same rate (the ↓ state). These two spin states are distinguished by a fourth quantum number, the **spin magnetic quantum number,** m_s. This quantum number can have only two values: $+\frac{1}{2}$ indicates an ↑ electron and $-\frac{1}{2}$ indicates a ↓ electron (Fig. 1.40). Box 1.1 describes an experiment that confirmed these properties of electron spin.

> *An electron has the property of spin; the spin is described by the quantum number m_s, which may have one of two values.*

1.11 The Electronic Structure of Hydrogen

Let's review what we now know about the hydrogen atom by imagining first its ground state and then what happens to its electron as the atom acquires energy. Initially, the electron is in the lowest energy level, the ground state of the atom, with $n = 1$. The only orbital with this energy is the 1*s*-orbital; we say that the electron **occupies** a 1*s*-orbital or that it is a "1*s*-electron." The electron in the ground state of a hydrogen atom is described by the following values of the four quantum numbers:

$$n = 1 \quad l = 0 \quad m_l = 0 \quad m_s = +\tfrac{1}{2} \text{ or } -\tfrac{1}{2}$$

The electron can have either spin state.

When the atom acquires enough energy (by absorbing a photon of radiation, for instance) for its electron to reach the shell with $n = 2$, it can occupy any of the four orbitals in that shell. There are one 2*s*- and three 2*p*-orbitals in this shell; in hydrogen, they all have the same energy. When an electron is described by one of these wavefunctions, we say that it "occupies a 2*s*-orbital" or one of the 2*p*-orbitals or that it is "a 2*s*- or 2*p*-electron." The average distance of an electron from the nucleus when it occupies any of the orbitals in the shell with $n = 2$ is greater than when $n = 1$, and so we can think of the atom as swelling up as it is excited energetically. When the atom acquires even more energy, the electron moves into the shell with $n = 3$; the atom is now even larger. In this shell, the electron can occupy any of nine orbitals (one 3*s*-, three 3*p*-, and five 3*d*-orbitals). More energy moves the electron still farther from

BOX 1.1   HOW DO WE KNOW . . . THAT AN ELECTRON HAS SPIN?

Electron spin was first detected experimentally by two German scientists, Otto Stern and Walter Gerlach, in 1920. Because a moving electric charge generates a magnetic field, they predicted that a spinning electron should behave like a tiny bar magnet.

In their experiment (see the illustration), Stern and Gerlach removed all the air from a container and set up a highly nonuniform magnetic field across it. They then shot a narrow stream of silver atoms through the container toward a detector. As explained in Section 1.13, a silver atom has one unpaired electron, with its remaining 46 electrons paired. The atom therefore behaves like a single unpaired electron riding on a heavy platform, the rest of the atom.

If a spinning electron behaved like a spinning ball, the axis of spin could point in any direction. The electron would behave like a bar magnet that could have any orientation relative to the applied magnetic field. In that case, a broad band of silver atoms should appear at the detector, because the field would push the silver atoms by different amounts according to the orientation of the spin. Indeed, that is exactly what Stern and Gerlach observed when they first carried out the experiment.

However, the first results were misleading. The experiment is difficult because the atoms collide with one another in the beam. An atom moving in one direction might easily be knocked by its neighbors into a different direction. When Stern and Gerlach repeated their experiment, they used a much less dense beam of atoms, thereby reducing the number of collisions between the atoms. They now saw only two narrow bands. One band consisted of atoms flying through the magnetic field with one orientation of their spin; the other band consisted of the atoms with opposite spin. The two narrow bands confirmed not only that an electron has spin but also that it can have only two orientations.

Electron spin is the basis of the experimental technique called *electron paramagnetic resonance* (EPR), which is used to study the structures and motions of molecules and ions that have unpaired electrons. This technique is based on detecting the energy needed to flip an electron between its two spin orientations. Like Stern and Gerlach's experiment, it works only with ions or molecules that have an unpaired electron.

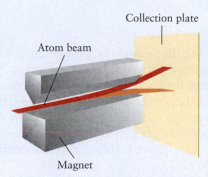

A schematic representation of the apparatus used by Stern and Gerlach. In the experiment, a stream of atoms splits into two as it passes between the poles of a magnet. The atoms in one stream have an odd ↑ electron, and those in the other an odd ↓ electron.

the nucleus to the $n = 4$ shell, where sixteen orbitals are available (one $4s$-, three $4p$-, five $4d$-, and seven $4f$-orbitals). Eventually, enough energy is absorbed so that the electron can escape the pull of the nucleus and leaves the atom.

The state of an electron in a hydrogen atom is defined by the four quantum numbers n, l, m$_l$, and m$_s$; as the value of n increases, the size of the atom increases.

SELF-TEST 1.10A The three quantum numbers for an electron in a hydrogen atom in a certain state are $n = 4$, $l = 2$, and $m_l = -1$. In what type of orbital is the electron located?

[*Answer: 4d*]

SELF-TEST 1.10B The three quantum numbers for an electron in a hydrogen atom in a certain state are $n = 3$, $l = 1$, and $m_l = -1$. In what type of orbital is the electron located?

MANY-ELECTRON ATOMS

The next step in our journey takes us from hydrogen and its single electron to the atoms of all the other elements in the periodic table. A neutral atom other than a hydrogen atom has more than one electron and is known as a **many-electron atom.** In the next three sections, we build on what we have learned about the hydrogen atom to see how the presence of more than one electron affects the energies of

A many-electron atom is also called a *polyelectron atom.*

atomic orbitals. The resulting electronic structures are the key to the periodic properties of the elements and the abilities of atoms to form chemical bonds. This material therefore underlies almost every aspect of chemistry.

1.12 Orbital Energies

The electrons in a many-electron atom occupy orbitals like those of hydrogen. However, the energies of these orbitals are not the same as those for a hydrogen atom. The nucleus of a many-electron atom is more highly charged than the hydrogen nucleus, and the greater charge attracts electrons more strongly and hence lowers their energy. However, the electrons also repel one another; this repulsion opposes the nuclear attraction and raises the energies of the orbitals. In a helium atom, for instance, with two electrons, the charge of the nucleus is $+2e$ and the total potential energy is given by three terms:

$$V = -\overbrace{\frac{2e^2}{4\pi\varepsilon_0 r_1}}^{\substack{\text{Attraction of}\\\text{electron 1 to}\\\text{the nucleus}}} - \overbrace{\frac{2e^2}{4\pi\varepsilon_0 r_2}}^{\substack{\text{Attraction}\\\text{elecron 2 to}\\\text{the nucleus}}} + \overbrace{\frac{e^2}{4\pi\varepsilon_0 r_{12}}}^{\substack{\text{Repulsion}\\\text{between the}\\\text{two electrons}}} \tag{19}$$

where r_1 is the distance of electron 1 from the nucleus, r_2 is the distance of electron 2 from the nucleus, and r_{12} is the distance between the two electrons. The two terms with negative signs (indicating that the energy *falls* as r_1 or r_2 decreases) represent the attractions between the nucleus and the two electrons. The term with a positive sign (indicating an *increase* in energy as r_{12} decreases) represents the repulsion between the two electrons. The Schrödinger equation based on this potential energy is impossibly difficult to solve exactly, but highly accurate numerical solutions can be obtained by using computers.

The number of electrons in an atom affects the properties of the atom. The hydrogen atom, with one electron, has no electron-electron repulsions; therefore, all the orbitals of a given shell in the hydrogen atom are degenerate. For instance, the 2s-orbital and all three 2p-orbitals have the same energy. In many-electron atoms, however, the results of spectroscopic experiments and calculations show

Along with code breakers, weather forecasters, and molecular biologists, chemists are now among the heaviest users of computers, which they use to calculate the detailed electronic structures of atoms and molecules (see Major Technique 5, following Chapter 13).

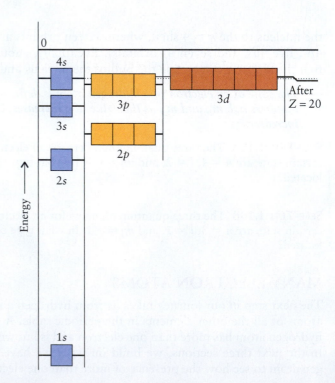

FIGURE 1.41 The relative energies of the shells, subshells, and orbitals in a many-electron atom. Each of the boxes can hold at most two electrons. Note the change in the order of energies of the 3d- and 4s-orbitals after $Z = 20$.

that electron-electron repulsions cause the energy of a 2p-orbital to be higher than that of a 2s-orbital. Similarly, in the $n = 3$ shell, the three 3p-orbitals lie higher than the 3s-orbital, and the five 3d-orbitals lie higher still (Fig. 1.41). How can we explain these energy differences?

As well as being attracted to the nucleus, each electron in a many-electron atom is repelled by the other electrons present. As a result, it is less tightly bound to the nucleus than it would be if those other electrons were absent. We say that each electron is **shielded** from the full attraction of the nucleus by the other electrons in the atom. The shielding effectively reduces the pull of the nucleus on an electron. The **effective nuclear charge**, $Z_{eff}e$, experienced by the electron is always less than the actual nuclear charge, Ze, because the electron-electron repulsions work against the pull of the nucleus. A *very* approximate form of the energy of an electron in a many-electron atom is a version of Eq. 14b in which the true atomic number is replaced by the effective atomic number:

$$E_n = -\frac{Z_{eff}^2 h\mathcal{R}}{n^2} \tag{20}$$

Note that the other electrons do not "block" the influence of the nucleus; they simply provide additional repulsive coulombic interactions that partly counteract the pull of the nucleus. For example, the pull of the nucleus on an electron in the helium atom is less than its charge of $+2e$ would exert but greater than the net charge of $+e$ that we would expect if each electron balanced one positive charge exactly.

An s-electron of any shell can be found very close to the nucleus (remember that ψ^2 for an s-orbital is nonzero at the nucleus), and so we say that it can **penetrate** through the inner shells. A p-electron penetrates much less because its orbital angular momentum prevents it from approaching close to the nucleus (Fig. 1.42). We have seen that its wavefunction vanishes at the nucleus, and so there is zero probability density for finding a p-electron there. Because a p-electron penetrates less than an s-electron through the inner shells of the atom, it is more effectively shielded from the nucleus and hence experiences a smaller effective nuclear charge than an s-electron does. In other words, an s-electron is bound more tightly than a p-electron and has a slightly lower (more negative) energy. A d-electron is bound less tightly than a p-electron of the same shell because its orbital angular momentum is higher and it is therefore even less able to approach the nucleus closely. That is, d-electrons are higher in energy than p-electrons of the same shell, which are in turn higher in energy than s-electrons of that shell.

The effects of penetration and shielding can be large. A 4s-electron generally has a much lower energy than that of a 4p- or 4d-electron; it may even have lower energy than that of a 3d-electron of the same atom (see Fig. 1.41). The precise ordering of orbitals depends on the number of electrons in the atom, as we shall see in the next section.

In a many-electron atom, because of the effects of penetration and shielding, the order of energies of orbitals in a given shell is s < p < d < f.

1.13 The Building-Up Principle

The electronic structure of an atom determines its chemical properties, and so we need to be able to describe that structure. To do so, we write the **electron configuration** of the atom—a list of all its occupied orbitals, with the numbers of electrons that each one contains. In the ground state of a many-electron atom, the electrons occupy atomic orbitals in such a way that the total energy of the atom is a minimum. At first sight, we might expect an atom to have its lowest energy when all its electrons are in the lowest energy orbital (the 1s-orbital), but except for hydrogen and helium, which have no more than two electrons, that can never happen. In 1925, the Austrian scientist Wolfgang Pauli discovered a general and very

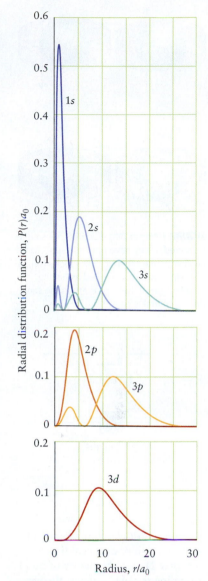

FIGURE 1.42 The radial distribution functions for s-, p-, and d-orbitals in the first three shells of a hydrogen atom. Note that the probability maxima for orbitals of the same shell are close to each other; however, note that an electron in an ns-orbital has a higher probability of being found close to the nucleus than does an electron in an np-orbital or an nd-orbital.

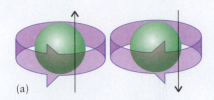

FIGURE 1.43 (a) Two electrons are said to be paired if they have opposite spins (one clockwise, the other counterclockwise). (b) Two electrons are classified as having parallel spins if their spins are in the same direction; in this case, both ↑.

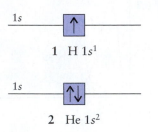

1   H $1s^1$

2   He $1s^2$

The outermost electrons are used in the formation of chemical bonds (Chapter 2), and the theory of bond formation is called *valence theory*; hence the name of these electrons.

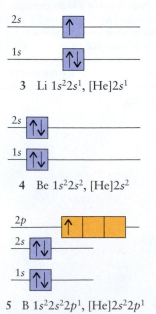

3   Li $1s^2 2s^1$, [He]$2s^1$

4   Be $1s^2 2s^2$, [He]$2s^2$

5   B $1s^2 2s^2 2p^1$, [He]$2s^2 2p^1$

fundamental rule about electrons and orbitals that is now known as the **Pauli exclusion principle**:

No more than two electrons may occupy any given orbital. When two electrons do occupy one orbital, their spins must be paired.

The spins of two electrons are said to be **paired** if one is ↑ and the other ↓ (Fig. 1.43). Paired spins are denoted ↑↓, and electrons with paired spins have spin magnetic quantum numbers of opposite sign. Because an atomic orbital is designated by three quantum numbers (n, l, and m_l) and the two spin states are specified by a fourth quantum number, m_s, another way of expressing the Pauli exclusion principle for atoms is

No two electrons in an atom can have the same set of four quantum numbers.

The exclusion principle implies that each atomic orbital can hold no more than two electrons.

The hydrogen atom in its ground state has one electron in the 1s-orbital. To show this structure, we place a single arrow in the 1s-orbital in a "box diagram," which shows each orbital as a box that can be occupied by no more than two electrons (see diagram **1**, which is a fragment of Fig. 1.41). We then report its **configuration** as $1s^1$ ("one s one"). In the ground state of a helium (He) atom ($Z = 2$), both electrons are in a 1s-orbital, which is reported as $1s^2$ ("one s two"). As we see in (**2**), the two electrons are paired. At this point, the 1s-orbital and the shell with $n = 1$ are fully occupied. We say that the helium atom has a **closed shell**, a shell containing the maximum number of electrons allowed by the exclusion principle.

A note on good practice: When a single electron occupies an orbital we write $1s^1$, for instance, not simply 1s.

Lithium ($Z = 3$) has three electrons. Two electrons occupy the 1s-orbital and complete the $n = 1$ shell. The third electron must occupy the next available orbital up the ladder of energy levels, the 2s-orbital (see Fig. 1.41). The ground state of a lithium (Li) atom is therefore $1s^2 2s^1$ (**3**). We can think of this atom as consisting of a core made up of the inner heliumlike closed shell, the $1s^2$ core, which we denote [He], surrounded by an outer shell containing a higher-energy electron. Therefore, the electron configuration of lithium is [He]$2s^1$. Electrons in the outermost shell are called **valence electrons**. In general, only valence electrons can be lost in chemical reactions, because core electrons (those in lower-energy orbitals) are too tightly bound. Thus, lithium loses only one electron when it forms compounds; it forms Li^+ ions, rather than Li^{2+} or Li^{3+} ions.

The element with $Z = 4$ is beryllium (Be), with four electrons. The first three electrons form the configuration $1s^2 2s^1$, like lithium. The fourth electron pairs with the 2s-electron, giving the configuration $1s^2 2s^2$, or more simply [He]$2s^2$ (**4**). A beryllium atom therefore has a heliumlike core surrounded by a valence shell of two paired electrons. Like lithium—and for the same reason—a Be atom can lose only its valence electrons in chemical reactions. Thus, it loses both 2s-electrons to form a Be^{2+} ion.

Boron ($Z = 5$) has five electrons. Two enter the 1s-orbital and complete the $n = 1$ shell. Two enter the 2s-orbital. The fifth electron occupies an orbital of the next available subshell, which Fig. 1.41 shows is a 2p-orbital. This arrangement of electrons is reported as the configuration $1s^2 2s^2 2p^1$ or [He]$2s^2 2p^1$ (**5**), showing that boron has three valence electrons.

We need to make another decision at carbon ($Z = 6$): does the sixth electron join the one already in the 2p-orbital or does it enter a different 2p-orbital? (Remember, there are three p-orbitals in the subshell, all of the same energy.) To answer this question, we note that electrons are farther from each other and repel each other less when they occupy different p-orbitals than when they occupy the same orbital. So the sixth electron goes into an empty $2p$-orbital, and the ground state of carbon is $1s^2 2s^2 2p_x^1 2p_y^1$ (**6**). We write out the individual orbitals like this only when we need to emphasize that electrons occupy different orbitals within a subshell. In most cases, we can write the shorter form, $[\text{He}]2s^2 2p^2$. Note that in the orbital diagram we have drawn the two $2p$-electrons with **parallel spins** ($\uparrow\uparrow$), indicating that they have the same spin magnetic quantum numbers. For reasons based in quantum mechanics, electrons with parallel spins tend to avoid each other. Therefore, this arrangement has slightly lower energy than that of a paired arrangement. However, it is allowed only when the electrons occupy different orbitals.

The procedure that we have been using is called the **building-up principle**. It can be summarized by two rules. To predict the ground-state configuration of a neutral atom of an element with atomic number Z with its Z electrons:

1   Add Z electrons, one after the other, to the orbitals in the order shown in Fig. 1.44 but with no more than two electrons in any one orbital.

2   If more than one orbital in a subshell is available, add electrons with parallel spins to different orbitals of that subshell rather than pairing two electrons in one of the orbitals.

The first rule takes into account the Pauli exclusion principle. The second rule is called **Hund's rule,** for the German spectroscopist Friedrich Hund, who first proposed it. This procedure gives the configuration of the atom that corresponds to the lowest total energy, which maximizes the attraction of the electrons to the nucleus and minimizes their repulsion by one another. An atom with electrons in energy states higher than predicted by the building-up principle is said to be in an **excited state.** For example, the electron configuration $[\text{He}]2s^1 2p^3$ represents an excited state of a carbon atom. An excited state is unstable and emits a photon as the electron returns to an orbital that restores the atom to a lower energy.

In general, we can think of an atom of any element as having a noble-gas core surrounded by a number of electrons in the valence shell, the outermost occupied shell. The **valence shell** is the occupied shell with the largest value of n.

The underlying organization of the periodic table described in Section B now begins to unfold. All the atoms of the main-group elements in a given period have a valence shell with the same principal quantum number, which is equal to the period number. For example, the valence shell of elements in Period 2 (from lithium to neon) is the shell with $n = 2$. Thus all the atoms in a given *period* have the same type of core and valence electrons with the same principal quantum number. For example, the atoms of Period 2 elements all have a heliumlike $1s^2$ core, and those of Period 3 elements have a neonlike $1s^2 2s^2 2p^6$ core, denoted [Ne]. All the atoms of a given *group* (in the main groups, particularly) have analogous valence electron configurations that differ only in the value of n. For instance, all the members of Group 1 have

6   C $1s^2 2s^2 2p^2$, $[\text{He}]2s^2 2p^2$

The building-up principle is also commonly called the *Aufbau principle*, from the German word for "building up."

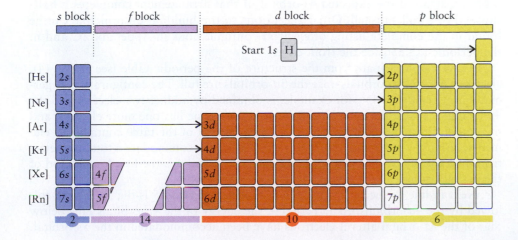

FIGURE 1.44 The order in which atomic orbitals are occupied according to the building-up principle. When we add an electron, we move one place to the right until all the electrons (Z electrons for an element of atomic number Z) have been accommodated. At the end of a row, move to the beginning of the next row down. The names of the blocks of the periodic table indicate the last subshell being occupied according to the building-up principle. The numbers of electrons that each type of orbital can accommodate are shown by the numbers across the bottom of the table. The colors of the blocks match the colors that we are using for the corresponding orbitals.

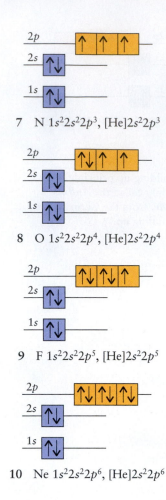

7   N $1s^2 2s^2 2p^3$, [He]$2s^2 2p^3$

8   O $1s^2 2s^2 2p^4$, [He]$2s^2 2p^4$

9   F $1s^2 2s^2 2p^5$, [He]$2s^2 2p^5$

10   Ne $1s^2 2s^2 2p^6$, [He]$2s^2 2p^6$

the valence configuration ns^1; and all the members of Group 14/IV have the valence configuration $ns^2 np^2$. These similar electron configurations give the elements in a group similar chemical properties, as illustrated in Section B.

With these points in mind, let's continue building up the electron configurations across Period 2. Nitrogen has $Z = 7$ and one more electron than carbon, giving [He]$2s^2 2p^3$. Each p-electron occupies a different orbital, and the three have parallel spins (7). Oxygen has $Z = 8$ and one more electron than nitrogen; therefore, its configuration is [He]$2s^2 2p^4$ (8) and two of its $2p$-electrons are paired. Similarly, fluorine, with $Z = 9$ and one more electron than oxygen, has the configuration [He]$2s^2 2p^5$ (9), with only one unpaired electron. Neon, with $Z = 10$, has one more electron than fluorine. This electron completes the $2p$-subshell, giving [He]$2s^2 2p^6$ (10). According to Figs. 1.41 and 1.44, the next electron enters the $3s$-orbital, the lowest-energy orbital of the next shell. The configuration of sodium is therefore [He]$2s^2 2p^6 3s^1$, or more briefly, [Ne]$3s^1$, where [Ne] denotes the neonlike core.

SELF-TEST 1.11A Predict the ground-state configuration of a magnesium atom.

[*Answer:* $1s^2 2s^2 2p^6 3s^2$, or [Ne]$3s^2$]

SELF-TEST 1.11B Predict the ground-state configuration of an aluminum atom.

The s- and p-orbitals of the shell with $n = 3$ are full by the time we get to argon, [Ne]$3s^2 3p^6$, which is a colorless, odorless, unreactive gas resembling neon. Argon completes the third period. From Fig. 1.41, we see that the energy of the $4s$-orbital is slightly lower than that of the $3d$-orbitals. As a result, instead of electrons entering the $3d$-orbitals, the fourth period now begins by filling the $4s$-orbitals (see Fig. 1.41). Hence, the next two electron configurations are [Ar]$4s^1$ for potassium and [Ar]$4s^2$ for calcium, where [Ar] denotes the argonlike core, $1s^2 2s^2 2p^6 3s^2 3p^6$. At this point, however, the $3d$-orbitals begin to be occupied, and there is a change in the rhythm of the periodic table.

According to the pattern of increasing energy of the orbitals (see Fig. 1.41), the next 10 electrons (for scandium, with $Z = 21$, through zinc, with $Z = 30$) enter the $3d$-orbitals. The ground-state electron configuration of scandium, for example, is [Ar]$3d^1 4s^2$, and that of its neighbor titanium is [Ar]$3d^2 4s^2$. Note that, beginning at scandium, we write the $4s$-electrons after the $3d$-electrons: once they contain electrons, the $3d$-orbitals lie lower in energy than the $4s$-orbital (recall Fig. 1.41; the same relation holds true for nd- and $(n + 1)s$-orbitals in subsequent periods). Successive electrons are added to the d-orbitals as Z increases. However, there are two exceptions: the experimental electron configuration of chromium is [Ar]$3d^5 4s^1$ instead of [Ar]$3d^4 4s^2$, and that of copper is [Ar]$3d^{10} 4s^1$ instead of [Ar]$3d^9 4s^2$. This apparent discrepancy occurs because the half-complete subshell configuration d^5 and the complete subshell configuration d^{10} turn out to have a lower energy than simple theory suggests. As a result, a lower total energy may be achieved if an electron enters a $3d$-orbital instead of the expected $4s$-orbital, if that arrangement completes a half-subshell or a full subshell. Other exceptions to the building-up principle can be found in the complete listing of electron configurations in Appendix 2C and in the periodic table inside the front cover.

As we can anticipate from the structure of the periodic table (see Fig. 1.44), electrons occupy $4p$-orbitals once the $3d$-orbitals are full. The configuration of germanium, [Ar]$3d^{10} 4s^2 4p^2$, for example, is obtained by adding two electrons to the $4p$-orbitals outside the completed $3d$-subshell. Arsenic has one more electron and the configuration [Ar]$3d^{10} 4s^2 4p^3$. The fourth period of the table contains 18 elements, because the $4s$- and $4p$-orbitals together can accommodate a total of 8 electrons and the $3d$-orbitals can accommodate 10. Period 4 is the first **long period** of the periodic table.

Next in line for occupation at the beginning of Period 5 is the $5s$-orbital, followed by the $4d$-orbitals. As in Period 4, the energy of the $4d$-orbitals falls below that of the $5s$-orbital after 2 electrons have been accommodated in the $5s$-orbital.

A similar effect is seen in Period 6, but now another set of inner orbitals, the 4*f*-orbitals, begins to be occupied. Cerium, for example, has the configuration $[Xe]4f^15d^16s^2$. Electrons then continue to occupy the seven 4*f*-orbitals, which are complete after 14 electrons have been added, at ytterbium, $[Xe]4f^{14}6s^2$. Next, the 5*d*-orbitals are occupied. The 6*p*-orbitals are occupied only after the 6*s*-, 4*f*-, and 5*d*-orbitals are filled at mercury; thallium, for example, has the configuration $[Xe]4f^{14}5d^{10}6s^26p^1$. You may notice in Appendix 2C that there are several apparent disruptions in the order in which the 4*f* orbitals are filled. The apparent exceptions result because the 4*f* and 5*d* orbitals are very close in energy. Toolbox 1.1 outlines a procedure for writing the electron configuration of a heavy element.

TOOLBOX 1.1   HOW TO PREDICT THE GROUND-STATE ELECTRON CONFIGURATION OF AN ATOM

CONCEPTUAL BASIS

Electrons occupy orbitals in such a way as to minimize the total energy of an atom by maximizing attractions and minimizing repulsions in accord with the Pauli exclusion principle and Hund's rule.

PROCEDURE

We use the following rules of the building-up principle to assign a ground-state configuration to a neutral atom of an element with atomic number Z:

1   Add Z electrons, one after the other, to the orbitals in the order shown in Figs. 1.41 and 1.44 but with no more than two electrons in any one orbital (the Pauli exclusion principle).

2   If more than one orbital in a subshell is available, add electrons to different orbitals of the subshell before doubly occupying any of them (Hund's rule).

3   Write the labels of the orbitals in order of increasing energy, with a superscript that gives the number of electrons in that orbital. The configuration of a filled shell is represented by the symbol of the noble gas having that configuration, as in [He] for $1s^2$.

4   When drawing a box diagram, show the electrons in different orbitals of the same subshell with parallel spins; electrons sharing an orbital have paired spins.

This procedure gives the ground-state electron configuration of an atom. Any other arrangement corresponds to an excited state of the atom. Note that we can use the structure of the periodic table to predict the electron configurations of most elements once we realize which orbitals are being filled in each block of the periodic table (see Fig. 1.44).

A useful shortcut for atoms of elements with large numbers of electrons is to write the electron configuration from the group number, which gives the number of valence electrons in the ground state of the atom, and the period number, which gives the value of the principal quantum number of the valence shell. The core consists of the preceding noble-gas configuration together with any completed *d*- and *f*-subshells.

Example 1.10 shows how these rules (specifically the shortcut) are applied.

EXAMPLE 1.10   Sample exercise: Predicting the ground-state electron configuration of a heavy atom

Predict the ground-state electron configuration of (a) a vanadium atom and (b) a lead atom.

SOLUTION We follow the procedure in Toolbox 1.1. (a) Vanadium is in Period 4, and so it has an argon core. Add two electrons to the 4*s*-orbital, and the last three electrons to three separate 3*d*-orbitals. The electron configuration is $[Ar]3d^34s^2$. (b) Lead belongs to Group 14/IV and Period 6. It therefore has four electrons in its valence shell, two in a 6*s*-orbital and two in different 6*p*-orbitals. The atom has complete 5*d*- and 4*f*-subshells, and the preceding noble gas is xenon. The electron configuration of lead is therefore $[Xe]4f^{14}5d^{10}6s^26p^2$.

SELF-TEST 1.12A Write the ground-state configuration of a bismuth atom.

[*Answer:* $[Xe]4f^{14}5d^{10}6s^26p^3$]

SELF-TEST 1.12B Write the ground-state configuration of an arsenic atom.

We account for the ground-state electron configuration of an atom by using the building-up principle in conjunction with Fig. 1.41, the Pauli exclusion principle, and Hund's rule.

1.14 Electronic Structure and the Periodic Table

The periodic table (Section B) was formulated long before the structures of atoms were known, by noting trends in experimental data (Box 1.2). However, to understand the organization of the periodic table, we need to consider the electron configurations of the elements. The table is divided into s, p, d, and f **blocks**, named for the last subshell that is occupied according to the building-up principle (as shown in Fig. 1.44). Two elements are exceptions. Because it has two $1s$-electrons, we would place helium in the s block, but it is shown in the p block because of its properties. It is a gas with properties matching those of the noble gases in Group 18/VIII, rather than the reactive metals in Group 2. Its place in Group 18/VIII is justified because it has a filled valence shell, like all the other Group 18/VIII

BOX 1.2   THE DEVELOPMENT OF THE PERIODIC TABLE

The periodic table is one of the most notable achievements in chemistry because it helps to organize what would otherwise be a bewildering array of properties of the elements. However, the fact that its structure corresponds to the electronic structure of atoms was unknown to its discoverers. The periodic table was developed solely from a consideration of physical and chemical properties of the elements.

Dmitri Ivanovich Mendeleev (1834–1907).

In 1860, the Congress of Karlsruhe brought together many prominent chemists in an attempt to resolve issues such as the existence of atoms and the correct atomic masses. One of the new ideas presented was Avogadro's principle—that the numbers of molecules in samples of different gases of equal volume, pressure, and temperature are the same (see Section 4.4). This principle allowed the relative atomic masses of the gases to be determined. Two scientists attending the congress were the German Lothar Meyer and the Russian Dmitri Mendeleev, both of whom left with copies of Avogadro's paper. In 1869, Meyer and Mendeleev discovered independ-

ently that the elements fell into families with similar properties when they were arranged in order of increasing atomic mass. Mendeleev called this observation the periodic law.

Mendeleev's chemical insight led him to leave gaps for elements that would be needed to complete the pattern but were unknown at the time. When they were discovered later, he turned out (in most cases) to be strikingly correct. For example, his pattern required an element that he named "eka-silicon" below silicon and between gallium and arsenic. He predicted that the element would have a relative atomic mass of 72 (taking the mass of hydrogen as 1) and properties similar to those of silicon. This prediction spurred the German chemist Clemens Winkler in 1886 to search for eka-silicon, which he eventually discovered and named germanium. It has a relative atomic mass of 72.59 and properties similar to those of silicon, as shown in the accompanying table.

One problem with Mendeleev's table was that some elements seemed to be out of place. For example, when argon was isolated, it did not seem to have the correct mass for its location. Its relative atomic mass of 40 is the same as that of calcium, but argon is an inert gas and calcium a reactive metal. Such anomalies led scientists to question the use of relative atomic mass as the basis for organizing the elements. When Henry Moseley examined x-ray spectra of the elements in the early twentieth century, he realized that he could infer the atomic number itself. It was soon discovered that elements fall into the uniformly repeating pattern of the periodic table if they are organized according to atomic number, rather than atomic mass.

Related Exercises: 1.108 and 1.111.

Mendeleev's Predictions for Eka-Silicon (Germanium)

Property	Eka-silicon, E	Germanium, Ge
molar mass	$72 \ \text{g·mol}^{-1}$	$72.59 \ \text{g·mol}^{-1}$
density	$5.5 \ \text{g·cm}^{-3}$	$5.32 \ \text{g·cm}^{-3}$
melting point	high	937°C
appearance	dark gray	gray-white
oxide	EO_2; white solid; amphoteric; density $4.7 \ \text{g·cm}^{-3}$	GeO_2; white solid; amphoteric; density $4.23 \ \text{g·cm}^{-3}$
chloride	ECl_4; boils below 100°C; density $1.9 \ \text{g·cm}^{-3}$	$GeCl_4$; boils at 84°C; density $1.84 \ \text{g·cm}^{-3}$

elements. Hydrogen occupies a unique position in the periodic table. It has one *s*-electron, and so it belongs in Group 1; but it is also one electron short of a noble-gas configuration, and so it can act like a member of Group 17/VII. Because hydrogen has such a unique character, we do not ascribe it to any group; however, you will often see it placed in Group 1 or Group 17/VII, and sometimes in both.

The *s* and *p* blocks form the **main groups** of the periodic table. The similar valence-shell electron configurations for the elements in the same main group are the reason for the similar properties of these elements. The group number tells us how many valence-shell electrons are present. In the *s* block, the group number (1 or 2) is the same as the number of valence electrons. This relation is also true for all main groups when the Roman numerals (I–VIII) are used to label the groups. However, when the group labels 1–18 are used, in the *p* block we subtract 10 from the group number to find the number of valence electrons. For example, fluorine in Group 17/VII has seven valence electrons.

Each new period corresponds to the occupation of a shell with a higher principal quantum number. This correspondence explains the different lengths of the periods. Period 1 consists of only two elements, H and He, in which the single 1*s*-orbital of the *n* = 1 shell is being filled with its two electrons. Period 2 consists of the eight elements Li through Ne, in which the one 2*s*- and three 2*p*-orbitals are being filled with eight more electrons. In Period 3 (Na through Ar), the 3*s*- and 3*p*-orbitals are being occupied by eight additional electrons. In Period 4, not only are the eight electrons of the 4*s*- and 4*p*-orbitals being added, but so are the ten electrons of the 3*d*-orbitals. Hence there are eighteen elements in Period 4. Period 5 elements add another 18 electrons as the 5*s*-, 4*d*-, and 5*p*-orbitals are filled. In Period 6, a total of 32 electrons are added, because 14 electrons are also being added to the seven 4*f*-orbitals. The *f*-block elements have very similar chemical properties, because their electron configurations differ only in the population of inner *f*-orbitals, and electrons in these orbitals do not participate much in bond formation.

The blocks of the periodic table are named for the last orbital to be occupied according to the building-up principle. The periods are numbered according to the principal quantum number of the valence shell.

THE PERIODICITY OF ATOMIC PROPERTIES

The periodic table can be used to predict a wide range of properties, many of which are crucial for understanding chemistry. The variation of effective nuclear charge through the periodic table plays an important role in the explanation of periodic trends. Figure 1.45 shows the variation for the first three periods. The effective charge increases from left to right across a period and falls back sharply on going to the next period.

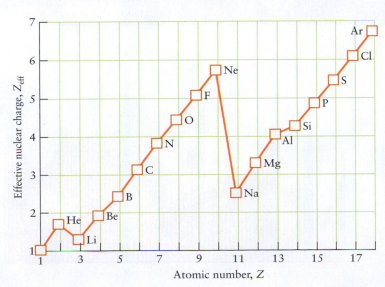

FIGURE 1.45 The variation of the effective nuclear charge for the outermost valence electron with atomic number. Notice that the effective nuclear charge increases from left to right across a period but drops when the outer electrons occupy a new shell. (The effective nuclear charge is actually $Z_{eff}e$, but Z_{eff} itself is commonly referred to as the charge.)

The atomic properties of the elements are listed in Appendix 2D.

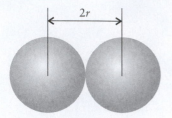

11   Atomic radius

Johannes van der Waals was a Dutch scientist who studied the interactions between molecules; see Chapter 4.

1.15 Atomic Radius

Electron clouds do not have sharp boundaries, and so we cannot measure the exact radius of an atom. However, when atoms pack together in solids and bond together to form molecules, their centers are found at definite distances from one another. The **atomic radius** of an element is defined as half the distance between the centers of neighboring atoms (**11**). If the element is a metal, its atomic radius is taken to be half the distance between the centers of neighboring atoms in a solid sample. For instance, because the distance between neighboring nuclei in solid copper is 256 pm, the atomic radius of copper is 128 pm. If the element is a nonmetal or a metalloid, we use half the distance between the nuclei of atoms joined by a chemical bond; this radius is also called the **covalent radius** of the element. For instance, the distance between the nuclei in a Cl_2 molecule is 198 pm, and so the covalent radius of chlorine is 99 pm. If the element is a noble gas, we use the **van der Waals radius,** which is half the distance between the centers of neighboring atoms in a sample of the solidified gas. The atomic radii of the noble gases listed in Appendix 2D are all van der Waals radii. Because the atoms in a sample of a noble gas are not chemically bonded together, van der Waals radii are generally much larger than covalent radii and are best not included in the discussion of trends.

Figure 1.46 shows some atomic radii, and Fig. 1.47 shows the variation in atomic radius with atomic number. Note the periodic, sawtooth pattern in the latter plot. *Atomic radius generally decreases from left to right across a period and increases down a group.*

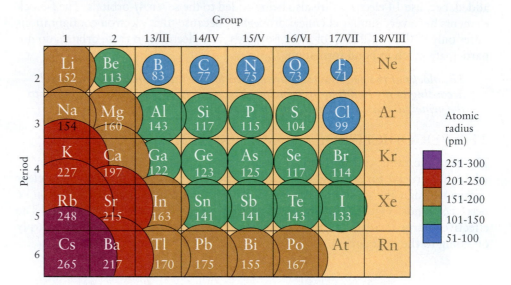

FIGURE 1.46 The atomic radii (in picometers) of the main-group elements. The radii decrease from left to right in a period and increase down a group. The colors used here and in subsequent charts represent the general magnitude of the property, as indicated by the scale on the right. Atomic radii, including those of the *d*-block elements, are listed in Appendix 2D.

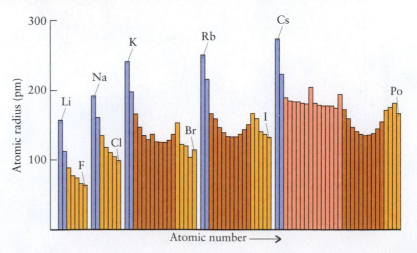

FIGURE 1.47 The periodic variation in the atomic radii of the elements. The variation across a period can be explained in terms of the effect of increasing effective nuclear charge; that down a group by the occupation of shells with increasing principal quantum number.

The increase down a group, such as that from Li to Cs, makes sense: with each new period, the outermost electrons occupy shells with increasing principal quantum number and therefore lie farther from the nucleus. The decrease across a period, such as that from Li to Ne, is surprising at first, because the number of electrons is increasing along with the number of protons. The explanation is that the new electrons are in the same shell of the atom and about as close to the nucleus as other electrons in the same shell. However, because they are spread out, the electrons do not shield one another well from the nuclear charge; so the effective nuclear charge increases across the period. The increasing effective nuclear charge draws the electrons in, and, as a result, the atom is more compact and we see a diagonal trend for atomic radii to increase from the upper right of the periodic table to the lower left.

> *Atomic radii generally decrease from left to right across a period as the effective atomic number increases, and they increase down a group as successive shells are occupied.*

1.16 Ionic Radius

The **ionic radius** of an element is its share of the distance between neighboring ions in an ionic solid (**12**). The distance between the centers of a neighboring cation and anion is the sum of the two ionic radii. In practice, we take the radius of the oxide ion to be 140. pm and calculate the radii of other ions on the basis of that value. For example, because the distance between the centers of neighboring Mg^{2+} and O^{2-} ions in magnesium oxide is 212 pm, the radius of the Mg^{2+} ion is reported as 212 pm − 140 pm = 72 pm.

Figure 1.48 illustrates the trends in ionic radii, and Fig. 1.49 shows the relative sizes of some ions and their parent atoms. All cations are smaller than their parent

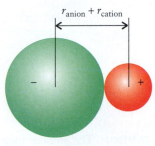

12 Ionic radius

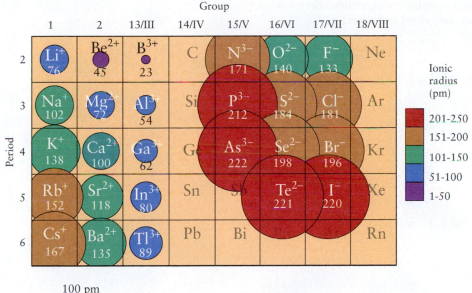

FIGURE 1.48 The ionic radii (in picometers) of the ions of the main-group elements. Note that cations are typically smaller than anions—in some cases, very much smaller.

FIGURE 1.49 The relative sizes of some cations and anions compared with their parent atoms. Note that cations (pink) are smaller than their parent atoms (gray), whereas anions (green) are larger.

atoms, because the atom loses one or more electrons to form the cation and exposes its core, which is generally much smaller than the parent atom. For example, the atomic radius of Li, with the configuration $1s^2 2s^1$, is 152 pm, but the ionic radius of Li^+, the bare heliumlike $1s^2$ core of the parent atom, is only 76 pm. This size difference is comparable to that between a cherry and its pit. Atoms in the same main group tend to form ions with the same charge. Like atomic radii, the radii of these ions increase down each group because electrons are occupying shells with higher principal quantum numbers.

Figure 1.49 shows that anions are larger than their parent atoms. The reason can be traced to the increased number of electrons in the valence shell of the anion and the repulsive effects exerted by electrons on one another. The variation in the radii of anions shows the same diagonal trend as that for atoms and cations, with the smallest at the upper right of the periodic table, close to fluorine.

Atoms and ions with the same number of electrons are called **isoelectronic**. For example, Na^+, F^-, and Mg^{2+} are isoelectronic. All three ions have the same electron configuration, $[He]2s^2 2p^6$, but their radii differ because they have different nuclear charges (see Fig. 1.48). The Mg^{2+} ion has the largest nuclear charge; so it has the strongest attraction for the electrons and therefore the smallest radius. The F^- ion has the lowest nuclear charge of the three isoelectronic ions and, as a result, it has the largest radius.

> **EXAMPLE 1.11   Sample exercise: Deciding the relative sizes of ions**
>
> Arrange each of the following pairs of ions in order of increasing ionic radius: (a) Mg^{2+} and Ca^{2+}; (b) O^{2-} and F^-.

SOLUTION The smaller member of a pair of isoelectronic ions in the same period will be an ion of an element that lies farther to the right in a period, because that ion has the greater effective nuclear charge. If the two ions are in the same group, the smaller ion will be the one that lies higher in the group, because its outermost electrons are closer to the nucleus. Check your answer against the values in Appendix 2C.

(a) Because Mg lies above Ca in Group 2, Mg^{2+} will have the smaller ionic radius.

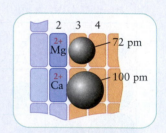

(b) Because F lies to the right of O in Period 2, F^- will have the smaller ionic radius.

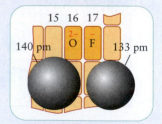

Appendix 2C shows that the actual values are (a) 72 pm for Mg^{2+} and 100 pm for Ca^{2+}; (b) 133 pm for F^- and 140 pm for O^{2-}.

SELF-TEST 1.13A Arrange each of the following pairs of ions in order of increasing ionic radius: (a) Mg^{2+} and Al^{3+}; (b) O^{2-} and S^{2-}.

[*Answer:* (a) $r(Al^{3+}) < r(Mg^{2+})$; (b) $r(O^{2-}) < r(S^{2-})$]

SELF-TEST 1.13B Arrange each of the following pairs of ions in order of increasing ionic radius: (a) Ca^{2+} and K^+; (b) S^{2-} and Cl^-.

Ionic radii generally increase down a group and decrease from left to right across a period. Cations are smaller than their parent atoms and anions are larger.

1.17 Ionization Energy

We shall see in Chapter 2 that the formation of a bond in an ionic compound depends on the removal of one or more electrons from one atom and their transfer to another atom. The energy needed to remove electrons from atoms is therefore of central importance for understanding their chemical properties. The **ionization energy**, I, is the energy needed to remove an electron from an atom in the gas phase:

$$X(g) \longrightarrow X^+(g) + e^-(g) \qquad I = E(X^+) - E(X) \tag{21}*$$

The ionization energy is normally expressed in electronvolts (eV) for a single atom or in kilojoules per mole of atoms ($kJ \cdot mol^{-1}$). The **first ionization energy**, I_1, is the energy needed to remove an electron from a neutral atom in the gas phase. For example, for copper,

$$Cu(g) \longrightarrow Cu^+(g) + e^-(g) \qquad \text{energy required} = I_1 \ (7.73 \text{ eV}, 746 \text{ kJ} \cdot mol^{-1})$$

The **second ionization energy**, I_2, of an element is the energy needed to remove an electron from a singly charged gas-phase cation. For copper,

$$Cu^+(g) \longrightarrow Cu^{2+}(g) + e^-(g) \qquad \text{energy required} = I_2 \ (20.29 \text{ eV}, 1958 \text{ kJ} \cdot mol^{-1})$$

Because ionization energy is a measure of how difficult it is to remove an electron, elements with low ionization energies can be expected to form cations readily and to conduct electricity in their solid forms. Elements with high ionization energies are unlikely to form cations and are unlikely to conduct electricity.

Figure 1.50 shows that first ionization energies generally decrease down a group. The decrease means that it takes less energy to remove an electron from a cesium atom, for instance, than from a sodium atom. Figure 1.51 shows the variation of first ionization energy with atomic number, and we see a periodic sawtooth pattern like that shown by atomic radii. With few exceptions, the first ionization energy rises from left to right across a period. Then it falls back to a lower value at the start of the next period. The lowest values occur at the bottom left of the periodic table (near cesium) and the highest at the upper right (near helium).

Ionization energies typically decrease down a group because, in successive periods, the outermost electron occupies a shell that is farther from the nucleus and is therefore less tightly bound. However, as we go from left to right across a given period, the effective nuclear charge increases. As a result, the outermost electron is gripped more tightly and the ionization energies generally increase. The small departures from these trends can usually be traced to repulsions between electrons, particularly electrons occupying the same orbital.

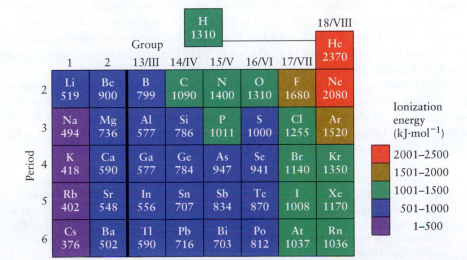

FIGURE 1.50 The first ionization energies of the main-group elements, in kilojoules per mole. In general, low values are found at the lower left of the table and high values are found at the upper right.

FIGURE 1.51 The periodic variation of the first ionization energies of the elements.

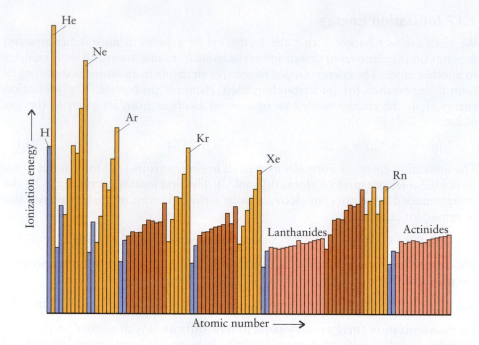

Figure 1.52 shows that the second ionization energy of an element is always higher than its first ionization energy. It takes more energy to remove an electron from a positively charged ion than from a neutral atom. For the Group 1 elements, the second ionization energy is considerably larger than the first; but, in Group 2, the two ionization energies have similar values. This difference makes sense, because the Group 1 elements have an ns^1 valence-shell electron configuration. Although the removal of the first electron requires only a small energy, the second electron must come from the noble-gas core. The core electrons have lower principal quantum numbers and are much closer to the nucleus. They are strongly attracted to it and a lot of energy is needed to remove them.

SELF-TEST 1.14A Account for the slight decrease in first ionization energy between beryllium and boron.

[*Answer:* Boron loses an electron from a higher-energy subshell than beryllium does.]

SELF-TEST 1.14B Account for the large decrease in third ionization energy between beryllium and boron.

The low ionization energies of elements at the lower left of the periodic table account for their metallic character. A block of metal consists of a collection of cations of the element surrounded by a sea of valence electrons that the atoms have lost (Fig. 1.53). Only elements with low ionization energies—the members of the *s* block, the *d* block, the *f* block, and the lower left of the *p* block—can form metallic solids, because only they can lose electrons easily.

The elements at the upper right of the periodic table have high ionization energies; so they do not readily lose electrons and are therefore not metals. Note that our knowledge of electronic structure has helped us to understand a major feature of the periodic table—in this case, why the metals are found toward the lower left and the nonmetals are found toward the upper right.

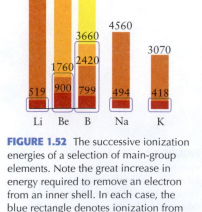

FIGURE 1.52 The successive ionization energies of a selection of main-group elements. Note the great increase in energy required to remove an electron from an inner shell. In each case, the blue rectangle denotes ionization from the valence shell.

The first ionization energy is highest for elements close to helium and is lowest for elements close to cesium. Second ionization energies are higher than first ionization energies (of the same element) and very much higher if the electron is to be removed from a closed shell. Metals are found toward the lower left of the periodic table because these elements have low ionization energies and can readily lose their electrons.

1.18 Electron Affinity

To predict some chemical properties, we need to know how the energy changes when an electron attaches to an atom. The **electron affinity**, E_{ea}, of an element is the energy released when an electron is added to a gas-phase atom. A high electron affinity means that a lot of energy is released when an electron attaches to an atom. A negative electron affinity means that energy must be supplied to push an electron onto an atom. This convention matches the everyday meaning of the term "affinity." More formally, the electron affinity of an element X is defined as

$$X(g) + e^-(g) \longrightarrow X^-(g) \qquad E_{ea}(X) = E(X) - E(X^-) \qquad (22)*$$

where $E(X)$ is the energy of a gas-phase X atom and $E(X^-)$ is the energy of the gas-phase anion. For instance, the electron affinity of chlorine is the energy released in the process

$$Cl(g) + e^-(g) \longrightarrow Cl^-(g) \qquad \text{energy released} = E_{ea} \ (3.62 \text{ eV}, 349 \text{ kJ·mol}^{-1})$$

Because the electron has a lower energy when it occupies one of the atom's orbitals, the difference $E(Cl) - E(Cl^-)$ is positive and the electron affinity of chlorine is positive. Like ionization energies, electron affinities are reported either in electronvolts for a single atom or in joules per mole of atoms.

Figure 1.54 shows the variation in electron affinity in the main groups of the periodic table. It is much less periodic than the variation in radius and ionization energy. However, one broad trend is clearly visible: with the exception of the noble gases, *electron affinities are highest toward the right side of the periodic table*, particularly the upper right, close to oxygen, sulfur, and the halogens. In these atoms, the incoming electron occupies a p-orbital close to a nucleus with a high effective charge and can experience its attraction quite strongly. The noble gases have negative electron affinities because any electron added to them must occupy an orbital outside a closed shell and far from the nucleus: this process requires energy, and so the electron affinity is negative.

Once an electron has entered the single vacancy in the valence shell of a Group 17/VII atom, the shell is complete and any additional electron would have to begin a new shell. In that shell, it not only would be farther from the nucleus but would also feel the repulsion of the negative charge already present. As a result, the second electron affinity of fluorine is strongly negative, meaning that a lot of energy has to be

In some books, you will see electron affinity defined with an opposite-sign convention. Those values are actually the electron-gain enthalpies (Chapter 6).

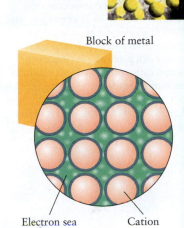

Block of metal

Electron sea   Cation

FIGURE 1.53 A block of metal consists of an array of cations (the spheres) surrounded by a sea of electrons. The charge of the electron sea cancels the charges of the cations. The electrons of the sea are mobile and can move past the cations quite easily and hence conduct an electric current.

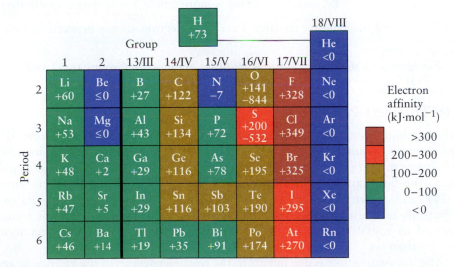

FIGURE 1.54 The variation in electron affinity in kilojoules per mole of the main-group elements. Where two values are given, the first refers to the formation of a singly charged anion and the second is the additional energy needed to produce a doubly charged anion. The negative signs of the second values indicate that energy is required to add an electron to a singly charged anion. The variation is less systematic than that for ionization energy, but high values tend to be found close to fluorine (but not for the noble gases).

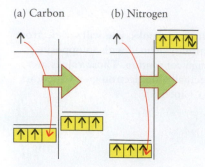

(a) Carbon   (b) Nitrogen

FIGURE 1.55 The energy changes taking place when an electron is added to a carbon atom and a nitrogen atom. (a) A carbon atom can accommodate an additional electron in an empty *p*-orbital. (b) When an electron is added to a nitrogen atom it must pair with an electron in a *p*-orbital. The incoming electron experiences so much repulsion from those already present in the nitrogen atom that the electron affinity of nitrogen is less than that of carbon and is in fact negative.

expended to form F^{2-} from F^-. Ionic compounds of the halogens are therefore built from singly charged ions such as F^- and never from doubly charged ions such as F^{2-}.

A Group 16/VI atom, such as O or S, has two vacancies in its valence-shell *p*-orbitals and can accommodate two additional electrons. The first electron affinity is positive because energy is released when an electron attaches to O or S. However, attachment of the second electron *requires* energy because of the repulsion by the negative charge already present in O^- or S^-. Unlike that of a halogen, however, the valence shell of the anion has room to accommodate the additional electron. Therefore, we expect that less energy will be needed to make O^{2-} from O^- than to make F^{2-} from F^-, where no such vacancy exists (compare diagrams **8** and **9**). In fact, 141 $kJ \cdot mol^{-1}$ is released when the first electron adds to the neutral atom to form O^-, but 844 $kJ \cdot mol^{-1}$ must be supplied to add a second electron to form O^{2-}; so the total energy required to make O^{2-} from O is 703 $kJ \cdot mol^{-1}$. As we shall see in Chapter 2, this energy can be achieved in chemical reactions, and O^{2-} ions are typical of metal oxides. However, no doubly charged anions are stable unless they are surrounded by cations or solvent molecules, as will be shown in Section 2.3.

EXAMPLE 1.12   **Sample exercise: Predicting trends in electron affinity**

The electron affinity of carbon is greater than that of nitrogen; indeed, the latter is negative. Suggest a reason for this observation.

SOLUTION   We expect more energy to be released when an electron enters the N atom, because an N atom is smaller than a C atom and its nucleus is more highly charged: the effective nuclear charges for the outermost electrons *of the neutral atoms* are 3.8 for N and 3.1 for C. However, the opposite is observed, and so we must also consider the effective nuclear charges experienced by the valence electrons in the anions (Fig. 1.55). When C^- forms from C, the additional electron occupies an empty 2*p*-orbital (see **6**). The incoming electron is well separated from the other *p*-electrons, and so it *very approximately* experiences the same effective nuclear charge. When N^- forms from N, the additional electron must occupy a 2*p*-orbital that is already half full (see **7**). The effective nuclear charge for this ion is therefore much less than that in the neutral atom; so energy is actually required when N^- is formed, and the electron affinity of nitrogen is lower than that of carbon.

SELF-TEST 1.15A   Account for the large decrease in electron affinity between lithium and beryllium.

[*Answer:* The additional electron enters a 2*s*-orbital in Li but a 2*p*-orbital in Be, and a 2*s*-electron is more tightly bound than a 2*p*-electron.]

SELF-TEST 1.15B   Account for the large decrease in electron affinity between fluorine and neon.

Elements with the highest electron affinities are those in Groups 16/VI and 17/VII.

1.19 The Inert-Pair Effect

Although both aluminum and indium are in Group 13/III, aluminum forms Al^{3+} ions, whereas indium forms both In^{3+} and In^+ ions. The tendency to form ions two units lower in charge than expected from the group number is called the **inert-pair effect**. Another example of the inert-pair effect is found in Group 14/IV: tin forms tin(IV) oxide when heated in air, but the heavier lead atom loses only its two *p*-electrons and forms lead(II) oxide. Tin(II) oxide can be prepared, but it is readily oxidized to tin(IV) oxide (Fig. 1.56). Lead exhibits the inert-pair effect more strongly than tin.

The inert-pair effect is due in part to the relative energies of the valence *p*- and *s*-electrons. In the later periods of the periodic table, valence *s*-electrons are very low in energy because of their good penetration and the low shielding ability of the *d*-electrons. The valence *s*-electrons may therefore remain attached to the atom during ion formation. The inert-pair effect is most pronounced among the heaviest members of a group, where the difference in energy between *s*- and *p*-electrons is

FIGURE 1.56 When tin(II) oxide is heated in air, it becomes incandescent as it reacts to form tin(IV) oxide. Even without being heated, it smolders and can ignite.

greatest (Fig. 1.57). Even so, the pair of *s*-electrons can be removed from the atom under sufficiently vigorous conditions. An inert pair would be better called a "lazy pair" of electrons.

> *The inert-pair effect is the tendency to form ions two units lower in charge than expected from the group number; it is most pronounced for heavy elements in the p block.*

1.20 Diagonal Relationships

A **diagonal relationship** is a similarity in properties between diagonal neighbors in the main groups of the periodic table (Fig. 1.58). A part of the reason for this similarity can be seen in Figs. 1.46 and 1.50 by concentrating on the colors that show the general trends in atomic radius and ionization energy. The colored bands of similar values lie in diagonal stripes across the table. Because these characteristics affect the chemical properties of an element, it is not surprising to find that the elements within a diagonal band show similar chemical properties (Fig. 1.59). Diagonal relationships are helpful for making predictions about the properties of elements and their compounds.

The diagonal band of metalloids dividing the metals from the nonmetals is one example of a diagonal relationship (Section B). So is the chemical similarity of lithium and magnesium and of beryllium and aluminum. For example, both lithium and magnesium react directly with nitrogen to form nitrides. Like aluminum, beryllium reacts with both acids and bases. We shall see many examples of this diagonal similarity when we look at the main-group elements in detail in Chapters 14 and 15.

> *Diagonally related pairs of elements often show similar chemical properties.*

THE IMPACT ON MATERIALS

We are now at the point where we can begin to use the periodic table as chemists and materials scientists do—to predict the properties of elements and see how they can be used to create the materials around us and to design new materials for tomorrow's technologies.

1.21 The Main-Group Elements

The usefulness of the main-group elements in materials is related to their properties, which can be predicted from periodic trends. For example, an *s*-block element has a low ionization energy, which means that its outermost electrons can easily be lost. An *s*-block element is therefore likely to be a reactive metal with all the characteristics that the name "metal" implies (Table 1.4, Fig. 1.60). Because ionization energies are

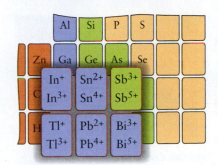

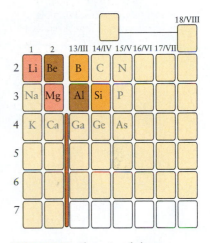

FIGURE 1.57 The typical ions formed by the heavy elements in Groups 13/III through 15/V show the influence of the inert pair—the tendency to form compounds in which the oxidation numbers differ by 2.

FIGURE 1.58 The pairs of elements represented by similarly colored boxes show a strong diagonal relationship to each other.

FIGURE 1.59 Boron (top) and silicon (bottom) have a diagonal relationship. Both are brittle solids with high melting points. They also show a number of chemical similarities.

TABLE 1.4 Characteristics of Metals and Nonmetals

Metals	Nonmetals
Physical properties	
good conductors of electricity	poor conductors of electricity
malleable	not malleable
ductile	not ductile
lustrous	not lustrous
typically:	typically:
solid	solid, liquid, or gas
high melting point	low melting point
good conductors of heat	poor conductors of heat
Chemical properties	
react with acids	do not react with acids
form basic oxides (which react with acids)	form acidic oxides (which react with bases)
form cations	form anions
form ionic halides	form covalent halides

FIGURE 1.60 All the alkali metals are soft, reactive, silvery metals. Sodium is kept under mineral oil to protect it from air, and a freshly cut surface soon becomes covered with the oxide.

lowest at the bottom of each group and the elements there lose their valence electrons most easily, the heavy elements cesium and barium react most vigorously of all *s*-block elements. They have to be kept stored out of contact with air and water. The low melting points of the alkali metals have made them useful as coolants in nuclear reactors that operate above the boiling point of water; they have few other direct uses as elements but their compounds are enormously important.

Elements on the left of the *p* block, especially the heavier elements, have ionization energies that are low enough for these elements to have some of the metallic properties of the members of the *s* block. However, the ionization energies of the *p*-block metals are quite high, and they are less reactive than those in the *s* block. The elements aluminum, tin, and lead, which are important construction materials, all lie in this region of the periodic table (Fig. 1.61).

Elements at the right of the *p* block have characteristically high electron affinities: they tend to gain electrons to complete closed shells. Except for the metalloids tellurium and polonium, the members of Groups 16/VI and 17/VII are nonmetals (Fig. 1.62). They typically form molecular compounds with one another. They react with metals to form the anions in ionic compounds, and hence many of the minerals that surround us, such as limestone and granite, contain anions formed from nonmetals, such as S^{2-}, CO_3^{2-}, and SO_4^{2-}. Much of the metals industry is concerned with the problem of extracting metals from their combinations with nonmetals.

> *All elements in the s block are reactive metals that form basic oxides. The p-block elements tend to gain electrons to complete closed shells; they range from metals through metalloids to nonmetals.*

1.22 The Transition Metals

All *d*-block elements are metals (Fig. 1.63). Their properties are transitional between the *s*- and the *p*-block elements, which (with the exception of the members of Group 12) accounts for their alternative name, the **transition metals.** Because transition metals in the same period differ mainly in the number of *d*-electrons, and these electrons are in inner shells, their properties are very similar.

Variable valence is discussed further in Section 2.1.

When a *d*-metal atom loses electrons to form a cation, it first loses its outer *s*-electrons. However, most transition metals form ions with different oxidation states, because the *d*-electrons have similar energies and a variable number can be lost when they form compounds. Iron, for instance, forms Fe^{2+} and Fe^{3+}; copper forms Cu^+ and Cu^{2+}. Although copper is like potassium in having a single outermost *s*-electron, potassium forms only K^+. The reason for this difference can be understood by comparing their second ionization energies, which are 1958 $kJ \cdot mol^{-1}$ for copper and 3051 $kJ \cdot mol^{-1}$ for potassium. To form Cu^{2+}, an electron is removed

FIGURE 1.61 The Group 14/IV elements. From left to right: carbon (as graphite), silicon, germanium, tin, and lead.

FIGURE 1.62 The Group 16/VI elements. From left to right: oxygen, sulfur, selenium, and tellurium. Note the trend from nonmetal to metalloid.

FIGURE 1.63 The elements in the first row of the *d* block. Top row (left to right): scandium, titanium, vanadium, chromium, and manganese. Bottom row: iron, cobalt, nickel, copper, and zinc.

FIGURE 1.64 A sample of a high-temperature superconductor first produced in 1987, cooled by liquid nitrogen. Here, a small magnet is levitated by the superconductor. If the assembly were turned over, the magnet would hang at about the same distance below the superconductor.

from the *d*-subshell of $[Ar]3d^{10}$; but, to form K^{2+}, the electron would have to be removed from potassium's argonlike core.

The availability of *d*-orbitals and the similarity of the atomic radii of the *d*-block metals have a significant impact on many areas of our lives. The availability of *d*-orbitals is in large measure responsible for the action of transition metals and their compounds as catalysts (substances that accelerate reactions but are not themselves consumed) throughout the chemical industry. Thus, iron is used in the manufacture of ammonia, nickel in the conversion of vegetable oils into shortening, platinum in the manufacture of nitric acid, vanadium(V) oxide in the manufacture of sulfuric acid, and titanium compounds in the manufacture of polyethylene. The ability to form ions with different charges is important for facilitating the very subtle changes that take place in organisms. For instance, iron is present as iron(II) in hemoglobin, the oxygen-transport protein in mammalian blood; copper is present in the proteins responsible for electron transport; and manganese is present in the proteins responsible for photosynthesis. The similarity of their atomic radii is largely responsible for the ability of transition metals to form the mixtures known as *alloys*, especially the wide variety of steels that make modern construction and engineering possible.

Difficulties in separating and isolating the lanthanoids delayed their widespread use in technology. However, today they are studied intensely, because superconducting materials often contain lanthanoids (Fig. 1.64). All the actinoids are radioactive. None of the elements following plutonium occurs naturally on Earth in any significant amount. Because they can be made only in nuclear reactors or particle accelerators, they are available only in small quantities.

All d-block elements are metals with properties between those of s-block and p-block metals. Many d-block elements form cations in more than one oxidation state.

SKILLS YOU SHOULD HAVE MASTERED

❏ 1 Describe the experiments that led to the formulation of the nuclear model of the atom (Section 1.1).

❏ 2 Calculate the wavelength or frequency of light from the relation $\lambda\nu = c$ (Example 1.1).

❏ 3 Use Wien's law to estimate a temperature (Example 1.3).

❏ 4 Use the relation $E = h\nu$ to calculate the energy, frequency, or number of photons emitted from a light source (Examples 1.4 and 1.5).

❏ 5 Estimate the wavelength of a particle (Example 1.6).

❏ 6 Estimate the uncertainty in the location or speed of a particle (Example 1.7).

❏ 7 Calculate the energies and describe the wavefunctions of a particle in a box (Example 1.8).

❏ 8 Explain the origin of the lines in the spectrum of an element and correlate them with specific energy transitions (Example 1.2).

❏ 9 Assess the relative probability of finding an electron at a given distance from the nucleus of an atom (Example 1.9).

❑ 10 Name and explain the relation of each of the four quantum numbers to the properties and relative energies of atomic orbitals (Sections 1.8–1.11).

❑ 11 Describe the factors affecting the energy of an electron in a many-electron atom (Section 1.12).

❑ 12 Write the ground-state electron configuration for an element (Toolbox 1.1 and Example 1.10).

❑ 13 Account for periodic trends in atomic radii, ionization energies, and electron affinities (Examples 1.11 and 1.12).

EXERCISES

Exercises labeled C *require calculus.*

Observing Atoms

1.1 At the time that J. J. Thomson conducted his experiments on cathode rays, the nature of the electron was in doubt. Some considered it to be a form of radiation, like light; others believed the electron to be a particle. Some of the observations made on cathode rays were used to advance one view or the other. Explain how each of the following properties of cathode rays supports either the wave or the particle model of the electron. (a) They pass through metal foils. (b) They travel at speeds slower than that of light. (c) If an object is placed in their path, they cast a shadow. (d) Their path is deflected when they are passed between electrically charged plates.

1.2 J. J. Thomson originally referred to the rays produced in his apparatus (Fig. B.5) as "canal rays." The canal ray is deflected within the region between the poles of a magnet and strikes the phosphor screen. The ratio q/m (where q is the charge and m the mass) of the particles making up the canal rays is found to be 2.410×10^7 C·kg^{-1}. The cathode and anode of the apparatus are made of lithium, and the tube contains helium. Use the information inside the back cover to identify the particles (and their charge) that make up the canal rays. Explain your reasoning.

1.3 Arrange the following types of photons of electromagnetic radiation in order of increasing energy: γ-rays, visible light, ultraviolet radiation, microwaves, x-rays.

1.4 Arrange the following types of photons of electromagnetic radiation in order of increasing frequency: visible light, radio waves, ultraviolet radiation, infrared radiation.

1.5 A university student recently had a busy day. Each of the student's activities on that day (reading, having a dental x-ray, making popcorn in a microwave oven, and getting a suntan) involved radiation from a different part of the electromagnetic spectrum. Complete the following table and match each type of radiation to the appropriate event:

Frequency	Wavelength	Energy of photon	Event
8.7×10^{14} Hz			
		3.3×10^{-19} J	
300 MHz			
	2.5 nm		

1.6 A university professor used a variety of types of electromagnetic radiation when going to a restaurant for lunch (watching a green traffic light change, listening to the car radio, being struck by a cosmic ray while entering the restaurant, and taking food from a serving table heated with an infrared lamp).

Complete the following table and match each type of radiation to the appropriate event:

Frequency	Wavelength	Energy of photon	Event
	50 cm		
	540 nm		
		2.1×10^{-19} J	
3×10^{21} Hz			

Quantum Theory

1.7 The temperature of molten iron can be estimated by using Wien's law. If the melting point of iron is 1540°C, what will be the wavelength (in nanometers) corresponding to maximum intensity when a piece of iron melts?

1.8 An astronomer discovers a new red star and finds that the maximum intensity is at λ = 685 nm. What is the temperature of the surface of the star?

1.9 Sodium vapor lamps, used for public lighting, emit yellow light of wavelength 589 nm. How much energy is emitted by (a) an excited sodium atom when it generates a photon; (b) 5.00 mg of sodium atoms emitting light at this wavelength; (c) 1.00 mol of sodium atoms emitting light at this wavelength?

1.10 When an electron beam strikes a block of copper, x-rays with a frequency of 1.2×10^{17} Hz are emitted. How much energy is emitted at this wavelength by (a) an excited copper atom when it generates an x-ray photon; (b) 2.00 mol of excited copper atoms; (c) 2.00 g of copper atoms?

1.11 The γ-ray photons emitted by the nuclear decay of a technetium-99 atom used in radiopharmaceuticals have an energy of 140.511 keV. Calculate the wavelength of these γ-rays.

1.12 A mixture of argon and mercury vapor used in blue advertising signs emits light of wavelength 470 nm. Calculate the energy change resulting from the emission of 1.00 mol of photons at this wavelength.

1.13 Consider the following statements about electromagnetic radiation and decide whether they are true or false. If they are false, correct them. (a) The total intensity of radiation emitted from a black body at absolute temperature T is directly proportional to the temperature. (b) As the temperature of a black body increases, the wavelength at which the maximum intensity is found decreases. (c) Photons of radio-frequency radiation are higher in energy than photons of ultraviolet radiation.

1.14 Consider the following statements about electromagnetic radiation and decide whether they are true or false. If they are false, correct them. (a) Photons of ultraviolet radiation have less energy than photons of infrared radiation. (b) The kinetic energy of an electron ejected from a metal surface when the metal is irradiated with ultraviolet radiation is independent of the frequency of the radiation. (c) The energy of a photon is inversely proportional to the wavelength of the radiation.

1.15 The velocity of an electron that is emitted from a metallic surface by a photon is 3.6×10^3 km·s^{-1}. (a) What is the wavelength of the ejected electron? (b) No electrons are emitted from the surface of the metal until the frequency of the radiation reaches 2.50×10^{16} Hz. How much energy is required to remove the electron from the metal surface? (c) What is the wavelength of the radiation that caused photoejection of the electron? (d) What kind of electromagnetic radiation was used?

1.16 The work function for chromium metal is 4.37 eV. What wavelength of radiation must be used to eject electrons with a velocity of 1.5×10^3 km·s^{-1}?

1.17 A baseball must weigh between 5.00 and 5.25 ounces (1 ounce = 28.3 g). What is the wavelength of a 5.15-ounce baseball thrown at 92 mph?

1.18 A certain automobile of mass 1645 kg travels on a German autobahn at 162 km·h^{-1}. What is the wavelength of the automobile?

1.19 What is the velocity of a neutron of wavelength 100. pm?

1.20 The average speed of a helium atom at 25°C is 1.23×10^3 m·s^{-1}. What is the average wavelength of a helium atom at this temperature?

1.21 The energy levels of a particle of mass m in a two-dimensional square box of side L are given by $(n_1^2 + n_2^2)h^2/8mL^2$. Are any of the levels degenerate? If so, find the values of the quantum numbers n_1 and n_2 for which these degeneracies arise for the first three cases.

1.22 Refer to Exercise 1.21. If one side of the box is twice that of the other, the energy levels are given by $(n_1^2/L_1^2 + n_2^2/L_2^2) \times h^2/8m$. Are degeneracies allowed? If so, which are the degenerate states of the lowest level that shows degeneracy?

1.23 The energy levels of a particle of mass m confined to a cubix box of side L are

$$E = \frac{h^2}{8m}\left(\frac{n_x^2 + n_y^2 + n_z^2}{L^2}\right)$$

(a) Write expressions for the energies of the three lowest levels. Which of these levels are degenerate? For those levels that are degenerate, give the quantum numbers corresponding to each degenerate level.

1.24 Buckminsterfullerene is form of carbon having nearly spherical molecules that are composed of 60 carbon atoms (see Section 14.16). The interior of a C_{60} molecule is about 0.7 nm in diameter and is being considered as a container for various atoms and molecules. Suppose that buckminsterfullerene were used to transport molecular hydrogen. It can be approximated as a cube with a side of 0.7 nm. Approximating the hydrogen molecule as a point mass, how much energy does a hydrogen molecule inside a C_{60} molecule need to be excited from the lowest energy level to (a) the second energy level; (b) the third energy level? (See Exercise 1.23.)

1.25 (a) Use the Living Graphs on the Web site for this book to plot the particle-in-a-box wavefunction for $n = 2$ and $L = 1$ m. (b) How many nodes does the wavefunction have? Where do these nodes occur? (c) Repeat parts (a) and (b) for $n = 3$. (d) What general conclusion can you draw about the relation between n and the number of nodes present in a wavefunction? (e) Convert the $n = 2$ plot to a probability density distribution: at what values of x is it most likely to find the particle? (f) Repeat part (e) for $n = 3$.

1.26 Verify the conclusion in part (d) of Exercise 1.25 by plotting the wavefunction for $n = 4$ and determining the number of nodes.

[C] **1.27** The wavefunction for a particle in a one-dimensional box is given in Eq. 9. (a) Confirm that the probability of finding the particle in the left half of the box is $\frac{1}{2}$ regardless of the value of n. (b) Does the probability of finding the particle in the left-hand one-third of the box depend on n? If so, find the probability. *Hint:* The indefinite integral of $\sin^2 ax$ is $\frac{1}{2}x - (1/4a)\sin(2ax) + \text{constant}$.

1.28 The *Humphreys series* is set of spectral lines in the emission spectrum of atomic hydrogen that ends in the fifth excited state. If an atom emits a photon of radiation of wavelength 5910 nm, to which spectral line in the Humphreys series does that photon correspond (i.e., the lowest-energy spectral line, the second-lowest-energy spectral line, the third-lowest-energy spectral line, etc.)? Justify your answer with a calculation.

Atomic Spectra

1.29 (a) Use the Rydberg formula for atomic hydrogen to calculate the wavelength for the transition from $n = 4$ to $n = 2$. (b) What is the name given to the spectroscopic series to which this transition belongs? (c) Use Table 1.1 to determine the region of the spectrum in which the transition takes place. If the change takes place in the visible region of the spectrum, what color will be emitted?

1.30 (a) Use the Rydberg formula for atomic hydrogen to calculate the wavelength for the transition from $n = 5$ to $n = 1$. (b) What is the name given to the spectroscopic series to which this transition belongs? (c) Use Table 1.1 to determine the region of the spectrum in which the transition takes place. If the change takes place in the visible region of the spectrum, what color will be emitted?

1.31 The energy levels of hydrogenlike one-electron ions of atomic number Z differ from those of hydrogen by a factor of Z^2. Predict the wavelength of the $2s \rightarrow 1s$ transition in He$^+$.

1.32 Some lasers work by exciting atoms of one element and letting these excited atoms collide with atoms of another element and transfer their excitation energy to those atoms. The transfer is most efficient when the separation of energy levels matches in the two species. Given the information in Exercise 1.31, are there any transitions of He$^+$ (including transitions from its excited states) that could be excited by collision with an excited hydrogen atom with the configuration $2s^1$?

1.33 (a) Using the particle-in-the-box model for the hydrogen atom and treating the atom as an electron in a one-dimensional box of length 150. pm, predict the wavelength of radiation emitted when the electron falls from the level with $n = 3$ to that with $n = 2$. (b) Repeat the calculation for the transition from $n = 4$ to $n = 2$.

1.34 (a) What is the highest energy photon that can be absorbed by a ground-state hydrogen atom without causing ionization? (b) What is the wavelength of this radiation? (c) To what region of the electromagnetic spectrum does this photon belong?

1.35 In the spectrum of atomic hydrogen, several lines are generally classified together as belonging to a series (for example, Balmer series, Lyman series, Paschen series), as shown in Fig. 1.28. What is common to the lines within a series that makes grouping them together logical?

1.36 In the spectrum of atomic hydrogen, a violet line is observed at 434 nm. Determine the beginning and ending energy levels of the electron during the emission of energy that leads to this spectral line.

Models of Atoms

1.37 (a) Sketch the shape of the boundary surfaces corresponding to $1s$-, $2p$-, and $3d$-orbitals. (b) What is meant by a node? (c) How many radial nodes and angular *nodal surfaces* does each orbital have? (d) Predict the number of nodal planes expected for a $4f$-orbital.

1.38 Locate the positions of the radial nodes of (a) a $3s$-orbital; (b) a $4d$-orbital.

1.39 Describe the orientation of the lobes of the p_x-, p_y-, and p_z-orbitals with respect to the reference Cartesian axes.

1.40 Describe the difference in orientation of the d_{xy}- and the $d_{x^2-y^2}$-orbitals with respect to the reference Cartesian axes. You may wish to refer to the animation of the atomic orbitals found on the Web site for this text.

1.41 Evaluate the probability of finding an electron in a small region of a hydrogen $1s$-orbital at a distance $0.55a_0$ from the nucleus relative to finding it in the same small region located at the nucleus.

1.42 Evaluate the probability of finding an electron in a small region of a hydrogen $1s$-orbital at a distance $0.65a_0$ from the nucleus relative to finding it in the same small region located at the nucleus.

1.43 Show that the electron distribution is spherically symmetrical for an atom in which an electron occupies each of the three p-orbitals of a given shell.

1.44 Show that, if the radial distribution function is defined as $P = r^2 R^2$, then the expression for P for an s-orbital is $P = 4\pi r^2 \psi^2$.

1.45 What is the probability of finding an electron anywhere in $\boxed{C}$ a sphere of radius (a) a_0 or (b) $2a_0$ in the ground state of a hydrogen atom?

1.46 At what distance from the nucleus is the electron $\boxed{C}$ most likely to be found if it occupies (a) a $3d$-orbital or (b) a $4s$-orbital in a hydrogen atom?

1.47 How many orbitals are in subshells with l equal to (a) 0; (b) 2; (c) 1; (d) 3?

1.48 (a) How many *subshells* are there for principal quantum

number $n = 5$? (b) Identify the subshells in the form $5s$, etc. (c) How many *orbitals* are there in the shell with $n = 5$?

1.49 (a) How many values of the quantum number l are possible when $n = 7$? (b) How many values of m_l are allowed for an electron in a $6d$-subshell? (c) How many values of m_l are allowed for an electron in a $3p$-subshell? (d) How many subshells are there in the shell with $n = 4$?

1.50 (a) How many values of the quantum number l are possible when $n = 6$? (b) How many values of m_l are allowed for an electron in a $5f$-subshell? (c) How many values of m_l are allowed for an electron in a $2s$-subshell? (d) How many subshells are there in the shell with $n = 3$?

1.51 What are the principal and orbital angular momentum quantum numbers for each of the following orbitals: (a) $6p$; (b) $3d$; (c) $2p$; (d) $5f$?

1.52 What are the principal and orbital angular momentum quantum numbers for each of the following orbitals: (a) $2s$; (b) $6f$; (c) $4d$; (d) $5p$?

1.53 For each orbital listed in Exercise 1.51, give the possible values for the magnetic quantum number.

1.54 For each orbital listed in Exercise 1.52, give the possible values for the magnetic quantum number.

1.55 How many electrons total can occupy (a) the $4p$-orbitals? (b) the $3d$-orbitals? (c) the $1s$-orbital? (d) the $4f$-orbitals?

1.56 How many electrons can occupy a subshell with $l =$ (a) 0; (b) 1; (c) 2; (d) 3?

1.57 Write the subshell notation ($3d$, for instance) and the number of orbitals having the following quantum numbers: (a) $n = 5, l = 2$; (b) $n = 1, l = 0$; (c) $n = 6, l = 3$; (d) $n = 2$, $l = 1$.

1.58 Write the subshell notation ($3d$, for instance) and the number of electrons that can have the following quantum numbers if all the orbitals of that subshell are filled: (a) $n = 3$, $l = 2$; (b) $n = 5, l = 0$; (c) $n = 7, l = 1$; (d) $n = 4, l = 3$.

1.59 How many electrons can have the following quantum numbers in an atom: (a) $n = 2, l = 1$; (b) $n = 4, l = 2$, $m_l = -2$; (c) $n = 2$; (d) $n = 3, l = 2, m_l = +1$?

1.60 How many electrons can have the following quantum numbers in an atom: (a) $n = 3, l = 1$; (b) $n = 5, l = 3$, $m_l = -1$; (c) $n = 2, l = 1, m_l = 0$; (d) $n = 7$?

1.61 Which of the following subshells cannot exist in an atom: (a) $2d$; (b) $4d$; (c) $4g$; (d) $6f$?

1.62 Which of the following subshells cannot exist in an atom: (a) $3d$; (b) $3f$; (c) $5g$; (d) $6i$?

The Structures of Many-Electron Atoms

1.63 (a) Write an expression for the total coulombic potential energy for a lithium atom. (b) What does each individual term represent?

1.64 (a) Write an expression for the total coulombic potential energy for a beryllium atom. (b) If Z denotes the number of electrons present in an atom, write a general expression to represent the total number of terms that will be present in the total coulombic potential energy equation.

1.65 Which of the following statements are true for many-electron atoms? If false, explain why. (a) The effective nuclear charge Z_{eff} is independent of the number of electrons present in an atom. (b) Electrons in an s-orbital are more effective than those in other orbitals at shielding other electrons from the nuclear charge because an electron in an s-orbital can penetrate to the nucleus of the atom. (c) Electrons having $l = 2$ are better at shielding than electrons having $l = 1$. (d) Z_{eff} for an electron in a p-orbital is lower than for an electron in an s-orbital in the same shell.

1.66 For the electrons on a carbon atom in the ground state, decide which of the following statements are true. If false, explain why. (a) Z_{eff} for an electron in a $1s$-orbital is the same as Z_{eff} for an electron in a $2s$-orbital. (b) Z_{eff} for an electron in a $2s$-orbital is the same as Z_{eff} for an electron in a $2p$-orbital. (c) An electron in the $2s$-orbital has the same energy as an electron in the $2p$-orbital. (d) The electrons in the $2p$-orbitals have spin quantum numbers m_s of opposite sign. (e) The electrons in the $2s$-orbital have the same value for the quantum number m_s.

1.67 Determine whether each of the following electron configurations represents the ground state or an excited state of the atom given.

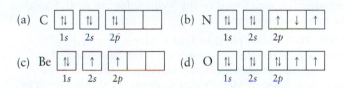

1.68 Each of the following *valence-shell configurations* is possible for a neutral atom for a certain element. What is the element and which configuration represents the ground state?

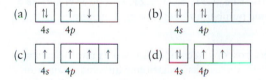

1.69 Of the following sets of four quantum numbers $\{n, l, m_l, m_s\}$, identify the ones that are forbidden for an electron in an atom and explain why they are invalid: (a) $\{4, 2, -1, +\frac{1}{2}\}$; (b) $\{5, 0, -1, +\frac{1}{2}\}$; (c) $\{4, 4, -1, +\frac{1}{2}\}$.

1.70 Of the following sets of four quantum numbers $\{n, l, m_l, m_s\}$, identify the ones that are forbidden for an electron in an atom and explain why they are invalid: (a) $\{2, 2, -1, +\frac{1}{2}\}$; (b) $\{6, 0, 0, +\frac{1}{2}\}$; (c) $\{5, 4, +5, +\frac{1}{2}\}$.

1.71 What is the ground-state electron configuration expected for each of the following elements: (a) silver; (b) beryllium; (c) antimony; (d) gallium; (e) tungsten; (f) iodine?

1.72 What is the ground-state electron configuration expected for each of the following elements: (a) sulfur; (b) cesium; (c) polonium; (d) palladium; (e) rhenium; (f) vanadium?

1.73 Which elements are predicted to have the following ground-state electron configurations: (a) $[Kr]4d^{10}5s^25p^4$; (b) $[Ar]3d^34s^2$; (c) $[He]2s^22p^2$; (d) $[Rn]7s^26d^2$?

1.74 Which elements are predicted to have the following ground-state electron configurations: (a) $[Ar]3d^{10}4s^24p^1$; (b) $[Ne]3s^1$; (c) $[Kr]5s^2$; (d) $[Xe]4f^76s^2$?

1.75 For each of the following ground-state atoms, predict the type of orbital ($1s$, $2p$, $3d$, $4f$, etc.) from which an electron will be removed to form the $+1$ ion: (a) Ge; (b) Mn; (c) Ba; (d) Au.

1.76 For each of the following ground-state atoms, predict the type of orbital ($1s$, $2p$, $3d$, $4f$, etc.) from which an electron will be removed to form the $+1$ ion: (a) Zn; (b) Cl; (c) Al; (d) Cu.

1.77 Predict the number of valence electrons present in each of the following atoms (include the outermost d-electrons: (a) N; (b) Ag; (c) Nb; (d) W.

1.78 Predict the number of valence electrons present in each of the following atoms (include the outermost d-electrons): (a) Bi; (b) Ba; (c) Mn; (d) Zn.

1.79 How many *unpaired* electrons are predicted for the ground-state configuration of each of the following atoms: (a) Bi; (b) Si; (c) Ta; (d) Ni?

1.80 How many *unpaired* electrons are predicted for the ground-state configuration of each of the following atoms: (a) Pb; (b) Ir; (c) Y; (d) Cd?

1.81 The elements Ga, Ge, As, Se, and Br lie in the same period in the periodic table. Write the electron configuration expected for the ground-state atoms of these elements and predict how many unpaired electrons, if any, each atom has.

1.82 The elements N, P, As, Sb, and Bi belong to the same group in the periodic table. Write the electron configuration expected for the ground-state atoms of these elements and predict how many unpaired electrons, if any, each atom has.

1.83 Give the notation for the valence-shell configuration (including the outermost d-electrons) of (a) the alkali metals; (b) Group 15/V elements; (c) Group 5 transition metals; (d) "coinage" metals (Cu, Ag, Au).

1.84 Give the notation for the valence-shell configuration (including the outermost d-electrons) of (a) the halogens; (b) the chalcogens (the Group 16/VI elements); (c) the transition metals in Group 5; (d) the Group 14/IV elements.

The Periodicity of Atomic Properties

1.85 On the basis of your knowledge of periodicity, place each of the following sets of elements in order of *decreasing* ionization energy. Explain your choices. (a) Selenium, oxygen, tellurium; (b) gold, tantalum, osmium; (c) lead, barium, cesium.

1.86 (a) Generally, the first ionization energies for elements in the same *period* increase on going to higher atomic number. Why? (b) Examine the data for the p-block elements given in Fig. 1.50. Note any exceptions to the rule given in part (a). How are these exceptions explained?

1.87 Look up in Appendix 2D the values for the atomic radii for the first row of transition metals. Explain the trends that are observed on going from left to right along the period.

1.88 Account for the fact that the ionization energy of potassium is less than that of sodium despite the latter having the smaller effective nuclear charge.

1.89 Identify which of the following elements experience the inert-pair effect and write the formulas for the ions that they form: (a) Sb; (b) As; (c) Tl; (d) Ba.

1.90 Arrange the elements in each of the following sets in order of *decreasing* atomic radius: (a) sulfur, chlorine, silicon; (b) cobalt, titanium, chromium; (c) zinc, mercury, cadmium; (d) antimony, bismuth, phosphorus.

1.91 Place the following ions in order of *increasing* ionic radius: S^{2-}, Cl^-, P^{3-}.

1.92 Which ion of each of the following pairs has the *larger* radius: (a) Ca^{2+}, Ba^{2+}; (b) As^{3-}, Se^{2-}; (c) Sn^{2+}, Sn^{4+}?

1.93 Which element of each of the following pairs has the *higher* electron affinity: (a) oxygen or fluorine; (b) nitrogen or carbon; (c) chlorine or bromine; (d) lithium or sodium?

1.94 Which element of each of the following pairs has the *higher* electron affinity: (a) aluminum or indium; (b) bismuth or antimony; (c) silicon or lead?

1.95 (a) What is a diagonal relationship? (b) How does it arise? (c) Give two examples to illustrate the concept.

1.96 Use Appendix 2D to find the values for the atomic radii of germanium and antimony as well as the ionic radii for Ge^{2+} and Sb^{3+}. What do these values suggest about the chemical properties of these two ions?

1.97 Which of the following pairs of elements exhibit a diagonal relationship: (a) Li and Mg; (b) Ca and Al; (c) F and S?

1.98 Which of the following pairs of elements do not exhibit a diagonal relationship: (a) Be and Al; (b) As and Sn; (c) Ga and Sn?

The Impact on Materials

1.99 Why are *s*-block metals more reactive than *p*-block metals?

1.100 Which of the following elements are transition metals: (a) radium; (b) radon; (c) hafnium; (d) niobium?

1.101 Identify the following elements as metals, nonmetals, or metalloids: (a) lead; (b) sulfur; (c) zinc; (d) silicon; (e) antimony; (f) cadmium.

1.102 Identify the following elements as metals, nonmetals, or metalloids: (a) aluminum; (b) carbon; (c) germanium; (d) arsenic; (e) selenium; (f) tellurium.

Integrated Exercises

1.103 Infrared spectroscopy is an important tool for studying vibrations of molecules. Just as an atom can absorb a photon of suitable energy to move an electron from one electronic state to another, a molecule can absorb a photon of electromagnetic radiation in the infrared region to move from one *vibrational* energy level to another. In infrared spectroscopy, it is common to express energy in terms of v/c, with the units cm^{-1} (read as *reciprocal centimeters*). (a) If an absorption occurs in the infrared spectrum at 3600 cm^{-1}, what is the frequency of radiation that corresponds to that absorption? (b) What is the energy, in joules (J), of that absorption? (c) How much energy would be absorbed by 1.00 mol of molecules absorbing at 3600 cm^{-1}?

1.104 *Diffraction* of electromagnetic radiation by atoms and molecules occurs when the wavelength of the electromagnetic radiation is similar to the size of the particle that causes the diffraction—in this case, atoms or molecules. (a) Using 2.0×10^2 pm as the diameter of an atom, decide what type(s) of electromagnetic radiation would give rise to diffraction when passed through a sample of atoms or molecules. Beams of electrons and neutrons can also be used in diffraction experiments because of their high speeds and the de Broglie relation. Calculate the speed of (b) an electron and (c) a neutron that would be necessary to generate wavelengths comparable to the diameter of an atom.

1.105 The ground-state electron configurations for Cr and Cu are not those predicted by the building-up principle. Give their actual electron configurations and explain what causes them to deviate from the expected configuration.

1.106 Ionization energies usually increase on going from left to right across the periodic table. The ionization energy for oxygen, however, is lower than that of either nitrogen or fluorine. Explain this anomaly.

1.107 Thallium is the heaviest member of Group 13/III. Aluminum is also a member of that group, and its chemical properties are dominated by the +3 oxidation state. Thallium, however, is found most usually in the +1 oxidation state. Examine this difference by plotting the first, second, and third ionization energies for the Group 13/III elements against atomic number (see Appendix 2D or the periodic-table data found on the Web site for this text). Explain the trends you observe.

1.108 The German physicist Lothar Meyer observed a periodicity in the physical properties of the elements at about the same time as Mendeleev was working on their chemical properties. Some of Meyer's observations can be reproduced by examining the molar volume for the solid element as a function of atomic number. Calculate the molar volumes for the elements in Periods 2 and 3 from the densities of the elements found in Appendix 2D and the following solid densities ($g \cdot cm^{-3}$): nitrogen, 0.88; fluorine, 1.11; neon, 1.21. Plot your results as a function of atomic number and describe any variations that you observe.

1.109 In the spectroscopic technique known as photoelectron spectroscopy (PES), ultraviolet radiation is directed at an atom or a molecule. Electrons are ejected from the valence shell, and their kinetic energies are measured. Because the energy of the incoming ultraviolet photon is known and the kinetic energy of the outgoing electron is measured, the ionization energy, I, can be deduced from the fact that the total energy is conserved. (a) Show that the speed v of the ejected electron and the frequency v of the incomng radiation are related by

$$E = hv = I + \tfrac{1}{2}m_e v^2$$

(b) Use this relation to calculate the ionization energy of a rubidium atom, given that radiation of wavelength 58.4 nm produces electrons with a speed of 2450 $km \cdot s^{-1}$; recall that $1\ J = 1\ kg \cdot m^2 \cdot s^{-2}$.

1.110 In the heavier transition-metal elements, especially the lanthanoids and actinoids, there are numerous exceptions to the regular order of orbital occupation predicted by the building-up principle. Suggest why more exceptions would be noted for these elements.

1.111 Modern periodic tables sometimes differ in which elements are placed immediately to the right of barium and radium. In some cases, the elements are lanthanum and actinium, whereas others place lutetium and lawrencium there. Why? Justify each selection.

[C] **1.112** Wavefunctions corresponding to states of different energy of a particle in a box are mutually "orthogonal" in the sense that, if the two wavefunctions are multiplied together and then integrated over the length of the box, then the outcome is zero. (a) Confirm that the wavefunctions for $n = 1$ and $n = 2$ are orthogonal. (b) Demonstrate, without doing a calculation, that all wavefunctions with even n are orthogonal to all wavefunctions with odd n. *Hint:* Think about the area under the product of any two such functions.

1.113 (a) Evaluate the probability (relative to that of the same small region located at the nucleus) of finding an electron in a small region of a hydrogen 2s-orbital at distances from zero to $3a_0$ in increments of $0.10a_0$. (b) With the help of a computer graphing or a spreadsheet program, such as the *Function Plotter* found on the Web site for this text, plot the data calculated in part (a). (c) Where does the radial node appear?

1.114 The uncertainty principle is negligible for macroscopic objects. Electronic devices, however, are being manufactured on a smaller and smaller scale, and the properties of *nanoparticles,* particles with sizes that range from a few to several hundred nanometers, may be different from those of larger particles as a result of quantum mechanical phenomena. (a) Calculate the minimum uncertainty in the speed of an electron confined in a nanoparticle of diameter 200. nm and compare that uncertainty with the uncertainty in speed of an electron confined to a wire of length 1.00 mm. (b) Calculate the minimum uncertainty in the speed of a Li^+ ion confined in a nanoparticle that has a diameter of 200. nm and is composed of a lithium compound through which the lithium ions can move at elevated temperatures (ionic conductor). (c) Which could be measured more accurately in a nanoparticle, the speed of an electron or the speed of a Li^+ ion?

[C] **1.115** Wavefunctions are "normalized" to 1. This term means that the total probability of finding an electron in the system is 1. Verify this statement for a particle-in-the-box wavefunction (Eq. 10).

[C] **1.116** The intensity of a transition between the states n and n' of a particle in a box is proportional to the square of the integral I, where

$$I_{nn'} = \int_0^L \psi_n x \psi_{n'} \, dx$$

(a) Can there be a transition between states with quantum numbers 3 and 1? (b) Consider the transition between states with quantum numbers 2 and 1. Does the intensity decrease or increase as the box lengthens?

1.117 Millikan measured the charge of the electron in *electrostatic units,* esu. The data that he collected included the following series of charges found on oil drops: 9.60×10^{-10} esu, 1.92×10^{-9} esu, 2.40×10^{-9} esu, 2.88×10^{-9} esu, and 4.80×10^{-9} esu. (a) From this series find the likely charge on the electron in electrostatic units. (b) Predict the number of electrons on an oil drop with the charge 6.72×10^{-9} esu.

1.118 Atomic orbitals may be combined to form molecular orbitals. In such orbitals, there is a nonzero probability of finding an electron on any of the atoms that contribute to that molecular orbital. Consider an electron that is confined in a molecular orbital that extends over two adjacent carbon atoms. The electron can move freely between the two atoms. The C–C distance is 139 pm. (a) Using the one-dimensional particle-in-the-box model, calculate the energy required to promote an electron from the $n = 1$ to the $n = 2$ level, assuming that the length of the box is determined by the distance between two carbon atoms. (b) To what wavelength of radiation does this correspond? (c) Repeat the calculation for a linear chain of 1000 carbon atoms. (d) What can you conclude about the energy separation between energy levels as the size of the atom chain increases?

1.119 The energy required to break a C–C bond in a molecule is 348 $kJ \cdot mol^{-1}$. Will visible light be able to break this bond? If yes, what is the color of that light? If not, what type of electromagnetic radiation will be suitable?

1.120 Apparent anomalies in the filling of electron orbitals in atoms occur in chromium and copper. In these elements an electron expected to fill an s-orbital fills the d-orbitals instead. (a) Explain why these anomalies occurs. (b) Similar anomalies are known to occur in seven other elements. Using Appendix 2C, identify those elements and indicate for which ones the explanation used to rationalize the chromium and copper electron configurations is valid. (c) Explain why there are no elements in which electrons fill $(n + 1)$s-orbitals instead of np-orbitals.

1.121 The electron in a hydrogen atom is excited to a 4d orbital. Calculate the energy of the photon released if the electron were then to move to each of the following orbitals: (a) 1s; (b) 2p; (c) 2s; (d) 4s. (e) If the outermost electron in a potassium atom were excited to a 4d orbital and then fell to the same orbitals, describe qualitatively how the emission spectrum would differ from that of hydrogen (do not do any calculations). Explain your answer.

1.122 The following properties are observed for an unknown element. Identify the element from its properties. (a) The neutral atom has two unpaired electrons. (b) One of the valence electrons in the ground state atom has $m_l = +1$. (c) The most common oxidation state is $+4$. (d) If an electron in a hydrogen atom were excited to the same principal quantum level, n, as the valence electrons in an atom of this element, and fell to the $n - 1$ quantum level, the photon emitted would have an energy of 4.9×10^{-20} J.

1.123 The electron affinity of thulium has been measured by laser photodetachment electron spectroscopy. In this technique a gaseous beam of the anions of an element is bombarded with photons from a laser. The photons eject electrons from some of the anions and the energies of the ejected electrons are detected. The photons had a wavelength of 1064 nm and the ejected electrons had an energy of 0.137 eV. Although the analysis is somewhat more complicated, we can obtain a rough estimate of the electron affinity as the difference in energy between the photons and the energy of the ejected electrons. What is the electron affinity of thulium in $kJ \cdot mol^{-1}$?

1.124 Francium is thought to be the most reactive of the alkali metals. Because it is radioactive and available in only very small amounts it is difficult to study. However, we can predict its properties based on its location in Group 1 of the periodic table. Estimate the following properties of francium: (a) atomic radius; (b) ionic radius of the +1 cation; (c) ionization energy.

1.125 This plot shows the radial distribution function of the 3*s* and 3*p* orbitals of a hydrogen atom. Identify each curve and explain how you made your decision.

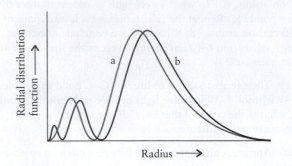

1.126 Below are pictured two reactions: one between atoms of magnesium and atoms of oxygen and one between atoms of sodium and atoms of chlorine. Identify each element and the ions formed.

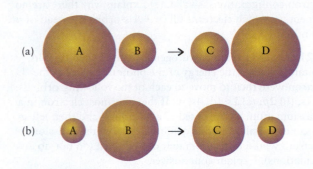

Chemistry Connections

1.127 Green chemistry methods, which use nontoxic chemicals, are replacing elemental chlorine for the bleaching of paper pulp. Chlorine causes problems because it is a strong oxidizing agent that reacts with organic compounds to form toxic byproducts such as furan and dioxins.

(a) Write the electron configuration of a chlorine atom in its ground state. How many unpaired electrons are present in the atom? Write the electron configuration you expect a chloride ion to have. The electron configuration of the chloride ion is identical to that of a neutral atom of what other element?

(b) When a chlorine atom is excited by heat or light, one of its valence electrons may be promoted to a higher energy level. Predict the most likely electron configuration for the lowest possible excited state for an excited chlorine atom.

(c) Estimate the wavelength (in nm) of the energy that needs to be absorbed for the electron to reach the excited state in part (b). To make this estimate, use Eq. 20 and take the effective nuclear charge from Fig. 1.45.

(d) What is the value of the energy required in part (c) in kilojoules per mole and electronvolts?

(e) The proportion of ^{37}Cl in a typical sample is 75.77%, with the remainder being ^{35}Cl. What would the molar mass of a sample of chlorine atoms be if the proportion of ^{37}Cl were reduced to half its current value? The mass of an atom of ^{35}Cl is 5.807×10^{-23} g and that of an atom of ^{37}Cl is 6.139×10^{-23} g.

(f) What are the oxidation numbers of chlorine in the bleaching agents ClO_2 and $NaClO$?

(g) What are the oxidation numbers of chlorine in the oxidizing agents $KClO_3$ and $NaClO_4$?

(h) Write the names of the compounds in parts (f) and (g).

CHEMICAL BONDS

What Are the Key Ideas? Bond formation is accompanied by a lowering of energy. That lowering of energy is due to the attractions between oppositely charged ions or between nuclei and shared electron pairs. The electron configurations of individual atoms control how the atoms combine with one another.

Why Do We Need to Know This Material? The existence of compounds is central to the science of chemistry; and by seeing how bonds form between atoms, we come to see how chemists design new materials. Research into artificial blood, new pharmaceuticals, agricultural chemicals, and the polymers used in materials such as compact discs, cellular phones, and synthetic fibers is based on an understanding of how atoms link together.

What Do We Need to Know Already? This chapter assumes that we know about atomic structure and electron configurations (Chapter 1), the basic features of energy, and the nature of the Coulomb interaction between charges (Section A). It is also helpful to be familiar with the nomenclature of compounds (Section D) and oxidation numbers (Section K).

A chemical bond is the link between atoms. When a chemical bond forms between two atoms the resulting arrangement of the two nuclei and their electrons has a lower energy than the total energy of the separate atoms. If the lowest energy can be achieved by the *complete transfer* of one or more electrons from the atoms of one element to those of another, then ions form and the compound is held together by the electrostatic attraction between them. This attraction is called an **ionic bond**. Sodium and chlorine atoms, for example, bond together as ions because solid sodium chloride, which consists of Na^+ and Cl^- ions, has a lower energy than a collection of widely separated sodium and chlorine atoms. If the lowest energy can be achieved by *sharing* electrons, then the atoms link through a **covalent bond** and discrete molecules are formed. Hydrogen and nitrogen atoms bond together as molecules of ammonia, NH_3, for example, because a gas consisting of NH_3 molecules has a lower energy than a gas consisting of the same number of widely separated nitrogen and hydrogen atoms. A third type of bond is the **metallic bond**, in which large numbers of cations are held together by a sea of electrons. For example, a piece of copper consists of a stack of copper ions held together by a sea of electrons, each of which comes from one of the atoms in the sample (recall Fig. 1.53). We consider the metallic bond in more detail in Section 3.13 and Chapter 5.

The changes in energy responsible for the formation of bonds occur when the **valence electrons** of atoms, the electrons in the outermost shells, move to new locations. Therefore, bond formation depends on the electronic structures of atoms discussed in Chapter 1.

IONIC BONDS

The **ionic model,** the description of bonding in terms of ions, is particularly appropriate for describing binary compounds formed from a metallic element, especially an *s*-block metal, and a nonmetallic element. An **ionic solid** is an assembly of cations and anions stacked together in a regular array. In sodium chloride, sodium ions alternate with chloride ions, and large numbers of oppositely charged ions are lined up in all three dimensions (Fig. 2.1). Ionic solids are examples of **crystalline**

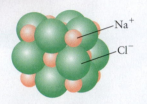
Na⁺
Cl⁻

Animation Figure 2.1 Sodium chloride crystal melting

Media Link

FIGURE 2.1 This tiny fragment of sodium chloride is an example of an ionic solid. The sodium ions are represented by pale red spheres and the chloride ions by green spheres. An ionic solid consists of an array of enormous numbers of cations and anions stacked together to give the lowest energy arrangement. The pattern shown here is repeated throughout the crystal.

Formulas of some common cations are shown in Fig. C.6.

solids, or solids that consist of atoms, molecules, or ions stacked together in a regular pattern. We explore these patterns in Chapter 5; here we concentrate on the changes in the valence electrons that accompany the formation of ions and the energetics of the formation of ionic solids.

2.1 The Ions That Elements Form

When an atom of a metallic element in the s block forms a cation, it loses electrons down to its noble-gas core (Fig. 2.2). In general, that core has an ns^2np^6 outer electron configuration, which is called an **octet** of electrons. For example, sodium ([Ne]$3s^1$) forms Na⁺, which has the same electron configuration as neon, [Ne]. The Na⁺ ions cannot lose more electrons in a chemical reaction, because the ionization energies of core electrons are too high. Hydrogen loses an electron to form a bare proton. Lithium ([He]$2s^1$) and beryllium ([He]$2s^2$) lose their s-electrons, leaving a heliumlike **duplet,** a pair of electrons with the configuration $1s^2$, when they become Li⁺ and Be²⁺. Some typical electron configurations of atoms and the ions they form are shown in Table 2.1.

Likewise, when the atoms of metals on the left of the p block in Periods 2 and 3 lose their valence electrons, they form ions with the electron configuration of the preceding noble gas. Aluminum, [Ne]$3s^23p^1$, for instance, forms Al³⁺ with the same configuration as neon, [Ne] ([He]$2s^22p^6$). However, when the metallic elements in Period 4 and later periods lose their s- and p-electrons, they leave a noble-gas core surrounded by an additional, complete subshell of d-electrons. For instance, gallium forms the ion Ga³⁺ with the configuration [Ar]$3d^{10}$. The d-electrons of the p-block atoms are gripped tightly by the nucleus and, in most cases, cannot be lost.

In the d block, the energies of the $(n-1)d$-orbitals lie below those of the ns-orbitals. Therefore, the ns-electrons are lost first, followed by a variable number of $(n-1)d$-electrons. For example, to obtain the configuration of the Fe³⁺ ion, we start from the configuration of the Fe atom, which is [Ar]$3d^64s^2$, and remove three electrons from it. The first two electrons removed are $4s$-electrons. The third electron comes from the $3d$-subshell, giving [Ar]$3d^5$.

Many metallic elements in the p and d blocks, have atoms that can lose a variable number of electrons. As we saw in Section 1.19, the inert-pair effect implies that the elements listed in Fig. 1.57 can lose either their valence p-electrons alone or all their valence p- and s-electrons. These elements and the d-block metals can form different compounds, such as tin(II) oxide, SnO, and tin(IV) oxide, SnO₂, for tin. The ability of an element to form ions with different charges is called **variable valence.**

> **EXAMPLE 2.1** **Writing the electron configurations of cations**
>
> Write the electron configurations of (a) In⁺ and (b) In³⁺.
>
> **STRATEGY** Determine the configuration of the neutral atom by referring to its position in the periodic table. Remove electrons from the valence-shell p-orbitals first, then

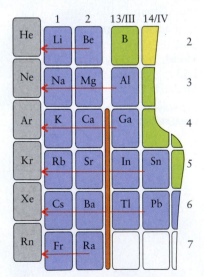

FIGURE 2.2 When a main-group metal atom forms a cation, it loses its valence s- and p-electrons and acquires the electron configuration of the preceding noble-gas atom. The heavier atoms in Groups 13/III and 14/IV retain their complete subshells of d-electrons.

TABLE 2.1 Some Typical Electron Configurations of Atoms and the Ions They Form

Atom	Configuration	Ion	Configuration
Li	[He]$2s^1$	Li⁺	[He] ($1s^2$)
Be	[He]$2s^2$	Be²⁺	[He]
Na	[Ne]$3s^1$	Na⁺	[Ne] ([He]$2s^22p^6$)
Mg	[Ne]$3s^2$	Mg²⁺	[Ne]
Al	[Ne]$3s^23p^1$	Al³⁺	[Ne]
N	[He]$2s^22p^3$	N³⁻	[Ne] ([He]$2s^22p^6$)
O	[He]$2s^22p^4$	O²⁻	[Ne]
F	[He]$2s^22p^5$	F⁻	[Ne]
S	[Ne]$3s^23p^4$	S²⁻	[Ar] ([Ne]$3s^23p^6$)
Cl	[Ne]$3s^23p^5$	Cl⁻	[Ar]

from the *s*-orbitals, and finally, if necessary, from the *d*-orbitals in the next-lower shell, until the number of electrons removed equals the charge on the ion.

SOLUTION

Determine the configuration of the neutral atom.	Indium is in Group 13/III, Period 5. Its ground-state configuration is therefore $[Kr]4d^{10}5s^25p^1$.	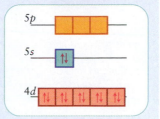
(a) Remove the outermost electron.	In^+ $[Kr]4d^{10}5s^2$	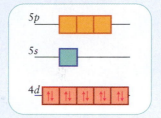
(b) Remove the next two electrons.	In^{3+} $[Kr]4d^{10}$	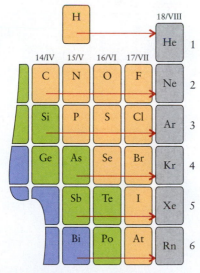

SELF-TEST 2.1A Write the electron configurations of (a) the copper(I) ion and (b) the copper(II) ion.

[***Answer:*** (a) $[Ar]3d^{10}$; (b) $[Ar]3d^9$]

SELF-TEST 2.1B Write the electron configurations of (a) the manganese(II) ion and (b) the lead(IV) ion.

Nonmetals rarely lose electrons in chemical reactions because their ionization energies are too high. However, a nonmetal atom can acquire enough electrons to complete its valence shell and form an anion with an octet corresponding to the configuration of the next noble gas ($1s^2$ in the case of the hydride ion, H^-), Fig. 2.3. It does not gain more electrons, because any additional electrons would have to be accommodated in a higher-energy shell. To form a monatomic anion, we add enough electrons to complete the valence shell. For example, nitrogen has five valence electrons (**1**); so three more electrons are needed to reach a noble-gas configuration, that of neon. Therefore, the ion will be N^{3-} (**2**). Notice that in each case the ion has the electron configuration of the next noble gas.

SELF-TEST 2.2A Predict the chemical formula and electron configuration of the phosphide ion.

[***Answer:*** P^{3-}, $[Ne]3s^23p^6$]

SELF-TEST 2.2B Predict the chemical formula and electron configuration of the iodide ion.

Formulas of some common anions are shown in Fig. C.7.

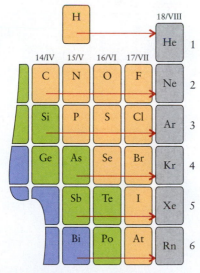

FIGURE 2.3 When nonmetal atoms acquire electrons and form anions, they do so until they have reached the electron configuration of the next noble gas.

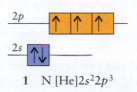

1   N $[He]2s^22p^3$

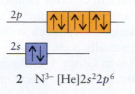

2 N^{3-} $[He]2s^22p^6$

To predict the electron configuration of a monatomic cation, remove outermost electrons in the order np, ns, and (n − 1)d; for a monatomic anion, add electrons until the next noble-gas configuration has been reached. The transfer of electrons results in the formation of an octet (or duplet) of electrons in the valence shell on each of the atoms: metals achieve an octet (or duplet) by electron loss and nonmetals achieve it by electron gain.

2.2 Lewis Symbols

As will become apparent as this chapter progresses, many of our basic ideas on the chemical bond were proposed by G. N. Lewis, one of the greatest of all chemists, in the early years of the twentieth century. Lewis devised a simple way to keep track of valence electrons when atoms form ionic bonds. He represented each valence electron as a dot and arranged the dots around the symbol of the element. A single dot represents an electron alone in an orbital; a pair of dots represents two paired electrons sharing an orbital. Examples of the **Lewis symbols** of atoms are

$$\text{H} \cdot \qquad \text{He:} \qquad :\overset{\cdot}{\underset{\cdot}{\text{N}}} \cdot \qquad \cdot \overset{\cdot \cdot}{\underset{\cdot \cdot}{\text{O}}} \cdot \qquad :\overset{\cdot \cdot}{\underset{\cdot \cdot}{\text{Cl}}} \cdot \qquad \text{K} \cdot \qquad \text{Mg:}$$

The Lewis symbol for nitrogen, for example, represents the valence electron configuration $2s^2 2p_x^1 2p_y^1 2p_z^1$ (see **1**), with two electrons paired in a $2s$-orbital and three unpaired electrons in different $2p$-orbitals. The Lewis symbol is a visual summary of the valence-shell electron configuration of an atom and allows us to see what happens to the electrons when an ion forms.

To work out the formula of an ionic compound by using Lewis symbols, we first represent the cation by removing the dots from the symbol for the metal atom. Then we represent the anion by transferring those dots to the Lewis symbol for the nonmetal atom to complete its valence shell. We might need to adjust the numbers of atoms of each kind so that all the dots removed from the metal atom symbols are accommodated by the nonmetal atom symbols. Finally, we write the charge of each ion as a superscript in the normal way, using brackets to indicate that the charge is the overall charge of the ion. A simple example is the formula of calcium chloride. The calcium atom loses its two valence electrons when it forms the Ca^{2+} ion. Because each chlorine atom has one vacancy, two are required to accommodate the two electrons lost:

$$:\overset{\cdot \cdot}{\underset{\cdot \cdot}{\text{Cl}}} \cdot \ + \ \text{Ca:} \ + \ :\overset{\cdot \cdot}{\underset{\cdot \cdot}{\text{Cl}}} \cdot \ \longrightarrow \ [:\overset{\cdot \cdot}{\underset{\cdot \cdot}{\text{Cl}}}:]^- \ Ca^{2+} \ [:\overset{\cdot \cdot}{\underset{\cdot \cdot}{\text{Cl}}}:]^-$$

The ratio of two chloride ions for each calcium ion results in the formula $CaCl_2$. However, note that this is an empirical formula (Section E). There are no $CaCl_2$ molecules. Crystals of $CaCl_2$ contain many of these ions in three-dimensional arrays.

Formulas of compounds consisting of the monatomic ions of main-group elements can be predicted by assuming that cations have lost all their valence electrons and anions have gained electrons in their valence shells until each ion has an octet of electrons, or a duplet in the case of H, Li, and Be.

2.3 The Energetics of Ionic Bond Formation

To understand why a crystal of sodium chloride, an ionic compound, has a lower energy than widely separated sodium and chlorine atoms, we picture the formation of the solid as taking place in three steps: sodium atoms release electrons, these electrons attach to chlorine atoms, and then the resulting cations and anions clump together as a crystal. Chemists often analyze complex processes by breaking them down into simpler steps such as these, and often consider hypothetical steps (steps that do not actually occur).

Sodium is in Group 1 of the periodic table and can be expected to form a +1 ion. However, the valence electron is tightly held by the effective nuclear charge—

it does not just fall off. In fact, the ionization energy of sodium is 494 kJ·mol^{-1} (see Fig. 1.50), and so we must supply that much energy to form the cations:

$$Na(g) \longrightarrow Na^+(g) + e^-(g) \qquad \text{energy required} = 494 \text{ kJ·mol}^{-1}$$

The electron affinity of chlorine atoms is +349 kJ·mol^{-1} (see Fig. 1.54), and so we know that 349 kJ·mol^{-1} of energy is *released* when electrons attach to chlorine atoms to form anions:

$$Cl(g) + e^-(g) \longrightarrow Cl^-(g) \qquad \text{energy released} = 349 \text{ kJ·mol}^{-1}$$

At this stage, the net change in energy (energy required − energy released) is $494 - 349$ kJ·mol^{-1} = +145 kJ·mol^{-1}, an *increase* in energy. A gas of widely separated Na$^+$ and Cl$^-$ ions has a higher energy than a gas of neutral Na and Cl atoms.

Now consider what happens when gaseous Na$^+$ and Cl$^-$ ions come together to form a crystalline solid. Their mutual attraction releases a lot of energy, and experimentally it is found that

$$Na^+(g) + Cl^-(g) \longrightarrow NaCl(s) \qquad \text{energy released} = 787 \text{ kJ·mol}^{-1}$$

Therefore, the net change in energy for the overall process

$$Na(g) + Cl(g) \longrightarrow NaCl(s)$$

is $145 - 787$ kJ·mol^{-1} = -642 kJ·mol^{-1} (Fig. 2.4), a huge *decrease* in energy. We conclude that a solid composed of Na$^+$ and Cl$^-$ ions has a lower energy than does a collection of widely separated Na and Cl atoms.

We can now begin to understand the formation of ionic bonds and learn when to expect them. There is a net lowering of energy below that of the individual atoms, provided the net attraction between ions is greater than the energy needed to make them. The major contribution to the energy requirement is normally the ionization energy of the element that forms the cation. Although some of this energy may be recovered from the electron affinity of the nonmetal when the anion is formed, in some cases energy is also needed to make the anion. For instance, we saw in Section 1.18 that the formation of an O^{2-} ion from an O atom *requires* 703 kJ·mol^{-1} and adds to the energy that must be recovered from the interactions between ions. Typically, *only metallic elements have ionization energies that are low enough for the formation of cations to be energetically feasible.*

> *The energy required for the formation of ionic bonds is supplied largely by the attraction between oppositely charged ions.*

2.4 Interactions Between Ions

We have seen that a key contribution to the formation of ionic bonds is the strength of the interaction between ions in a solid: it must be strong enough to overcome the energy investment needed to make the ions. However, a very important point to note is that an ionic solid is not held together by bonds between specific pairs of ions: *all* the cations interact to a greater or lesser extent with *all* the anions, *all* the cations repel each other, and *all* the anions repel each other. An ionic bond is a "global" characteristic of the entire crystal, a net lowering of energy of the entire crystal. We need to assess this interaction quantitatively and see what determines the **lattice energy** of the solid, the difference in energy between the ions packed together in a solid and the ions widely separated as a gas. A high lattice energy indicates that the ions interact strongly with one another to give a tightly bonded solid.

The strong attraction between oppositely charged ions accounts for the typical properties of ionic solids, such as their high melting points and their brittleness. A high temperature is required before the ions are able to move past one

A positive electron affinity signifies a release of energy when an electron attaches to a gas-phase atom or ion (Section 1.18).

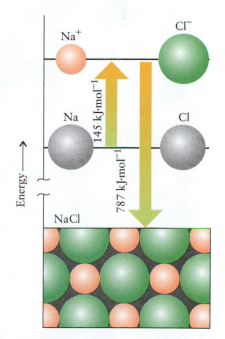

FIGURE 2.4 Considerable energy is needed to produce cations and anions from neutral atoms: the ionization energy of the metal atoms is only partly recovered from the electron affinity of the nonmetal atoms. The overall lowering of energy that drags the ionic solid into existence arises from the strong attraction between cations and anions that occurs in the solid. In Chapter 6, we see how to take into account the chemical reaction between Na(s) and Cl$_2$(g): this illustration is for the relative energies of Na(g) and Cl(g).

FIGURE 2.5 This sequence of images illustrates why ionic solids are brittle. (a) The original solid consists of an orderly array of cations and anions. (b) A hammer blow can push ions with like charges into adjacent positions; this proximity of like charges raises strong repulsive forces (as depicted by the double-headed arrows). (c) As the result of these repulsive forces, the solid breaks apart into fragments. (d) This calcite rock consists of several large crystals joined together. (e) The blow of a hammer has shattered some of the crystals, leaving flat, regular surfaces consisting of planes of ions. Compare this image with the result of striking a metallic crystal (Fig. 5.46).

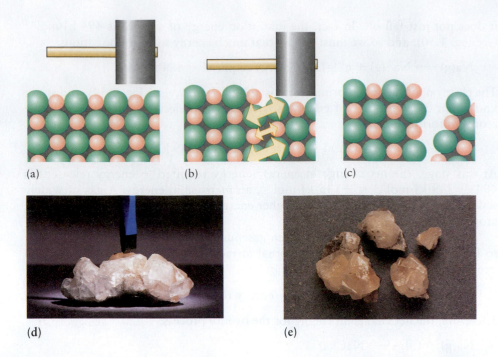

(a) (b) (c)

(d) (e)

another to form a liquid. Ionic solids are brittle because of the same strong attractions and repulsions. We cannot just push a block of ions past another block; instead, when we strike an ionic solid, ions with like charges come into contact and repel one another. The resulting repulsions cause it to shatter into fragments (Fig. 2.5).

Our starting point for understanding the interaction between ions in a solid is the expression for the Coulomb potential energy of the interaction of two individual ions (Section A):

$$E_{P,12} = \frac{(z_1 e) \times (z_2 e)}{4\pi\varepsilon_0 r_{12}} = \frac{z_1 z_2 e^2}{4\pi\varepsilon_0 r_{12}} \tag{1}*$$

In this expression, e is the fundamental charge (the absolute value of the charge of an electron), z_1 and z_2 are the charge numbers of the two ions, r_{12} is the distance between the centers of the ions, and ε_0 is the vacuum permittivity (see inside the back cover for the value of this fundamental constant).

A note on good practice: A charge number, z, is positive for cations and negative for anions and the charge of an ion is ze. However, chemists almost always refer to z itself as the charge, and speak of a charge of $+1$, -1, and so on.

Each ion in a solid experiences attractions from all the other oppositely charged ions and repulsions from all the other like-charged ions. The total potential energy is the sum of all these contributions. Each cation is surrounded by anions, and there is a large negative (energy-lowering) contribution from the attraction of the opposite charges. Beyond those nearest neighbors, there are cations that contribute a positive (repulsive, energy-raising) term to the total potential energy of the central cation. There is also a negative contribution from the anions beyond those cations, a positive contribution from the cations beyond them, and so on, to the edge of the solid. These repulsions and attractions become progressively weaker as the distance from the central ion increases, but because the nearest neighbors of an ion give rise to a strong attraction, the net outcome of all these contributions is a lowering of energy. Our task is to assess how far the energy is lowered by using the Coulomb potential energy expression in Eq. 1.

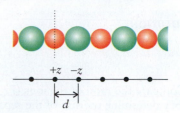

FIGURE 2.6 The arrangement used to calculate the potential energy of an ion in a line of alternating cations (red spheres) and anions (green spheres). We concentrate on one ion, the "central" ion denoted by the vertical dotted line.

HOW DO WE DO THAT?

To calculate the potential energy of an ionic solid we start with a simple model: a single line of uniformly spaced alternating cations and anions, with d (the distance between their centers) the sum of the ionic radii (Fig. 2.6). If the charge numbers of the ions have the same absolute value ($+1$ and -1, or $+2$ and -2, for instance), then $z_1 = +z$, $z_2 = -z$, and $z_1z_2 = -z^2$. The potential energy of the central ion is calculated by summing all the Coulomb potential energy terms, with negative terms representing attractions to oppositely charged ions and positive terms representing repulsions from like-charged ions. For the interaction arising from ions extending in a line to the right of the central ion, the total potential energy of the central ion is

$$E_P = \frac{1}{4\pi\varepsilon_0} \times \left(-\frac{z^2e^2}{d} + \frac{z^2e^2}{2d} - \frac{z^2e^2}{3d} + \frac{z^2e^2}{4d} - \cdots \right)$$

$$= -\frac{z^2e^2}{4\pi\varepsilon_0 d}\left(1 - \frac{1}{2} + \frac{1}{3} - \frac{1}{4} + \cdots \right) = -\frac{z^2e^2}{4\pi\varepsilon_0 d} \times \ln 2$$

In the last step, we have used the relation $1 - \frac{1}{2} + \frac{1}{3} - \frac{1}{4} + \ldots = \ln 2$. Finally, we multiply E_P by 2 to obtain the total energy arising from interactions on each side of the ion and then by Avogadro's constant, N_A, to obtain an expression for the potential energy per mole of ions. The outcome is

$$E_P = -2\ln 2 \times \frac{z^2N_Ae^2}{4\pi\varepsilon_0 d}$$

with $d = r_{cation} + r_{anion}$, the distance between the centers of neighboring ions.

We have found that the potential energy of an ion at the center of a line of alternating cations and anions has the form

$$E_P = -A \times \frac{z^2N_Ae^2}{4\pi\varepsilon_0 d} \tag{2}$$

per mole of ions, with $A = 2\ln 2$ (or 1.386) for this model system.

What does this equation tell us? Because the potential energy is negative, there is a net attraction of the central ion to all the other ions, which means that the attraction between opposite charges overcomes the repulsion between like charges. The potential energy is strongly negative when the ions are highly charged (large values of z) and the separation between them is small (small values of d), which is the case when the ions themselves are small.

The calculation that led to Eq. 2 can be extended to more realistic three-dimensional arrays of ions with different charges, and the result has the same form but with different values of A and $|z_1z_2|$ (that is, the absolute value of z_1z_2, its value without the negative sign) in place of z^2. The factor A is a numerical coefficient called the **Madelung constant**; its value depends on how the ions are arranged about one another (Table 2.2). In all cases, the energy lowering that occurs when an ionic solid forms is greatest for small, highly charged ions. For example, there is a strong interaction between the Mg^{2+} and the O^{2-} ions in magnesium oxide, MgO, because the ions have high charges and small radii. This strong interaction is one reason why magnesium oxide survives at such high temperatures that it can be used for furnace linings. It is an example of a "refractory" material, a substance that can withstand high temperatures.

TABLE 2.2 Madelung Constants

Structural type*	A
cesium chloride	1.763
fluorite	2.519
rock salt	1.748
rutile	2.408

* For information about these structures, see Chapter 5.

EXAMPLE 2.2   Sample exercise: Estimating the relative lattice energies of solids

The ionic solids NaCl and KCl form the same type of crystal structure. In which solid are the ions bound together more strongly by coulombic interactions?

SOLUTION

Compare the radii of the ions (Fig. 1.48), for they determine the size of the separation d.

The radius of Na^+ is 102 pm and that of K^+ is 138 pm; the radius of Cl^- is 181 pm.

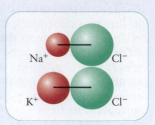

The ions in NaCl should be bound together more strongly than those in KCl because the Na^+ ion has a smaller radius than the K^+ ion and d is smaller in NaCl than in KCl.

SELF-TEST 2.3A The ionic solids CaO and KCl crystallize to form structures of the same type. In which compound are the interactions between the ions stronger?

[*Answer:* CaO, higher charges and smaller radii]

SELF-TEST 2.3B The ionic solids KBr and KCl crystallize to form structures of the same type. In which compound are the interactions between ions stronger?

The potential energy in Eq. 2 becomes more and more negative as the separation d decreases. However, a collection of ions does not collapse to a point because repulsive effects between neighbors become important as soon as they come into contact, and the energy quickly rises again. To take the repulsion effects between close neighbors in an ionic solid into account it is commonly supposed that the repulsive contribution to the potential energy rises exponentially with decreasing separation and therefore has the form

$$E_P{}^* \propto e^{-d/d^*}$$

with d^* a constant (it is commonly taken to be 34.5 pm). The total potential energy is the sum of E_P and $E_P{}^*$ and passes through a minimum as the ion separation decreases and then rises sharply again (Fig. 2.7). The energy at the minimum is given by the **Born–Meyer equation** (which we do not derive here):

$$E_{P,min} = -\frac{N_A |z_1 z_2| e^2}{4\pi\varepsilon_0 d}\left(1 - \frac{d^*}{d}\right) A \qquad (3)^*$$

What does this equation tell us? As before, the negative sign of this potential energy tells us that the ions have a lower potential energy when they are present as a solid rather than widely separated as a gas. The greatest stabilization is expected when the ions are highly charged (so $|z_1 z_2|$ is large) and small (so d is small, but not smaller than d^*).

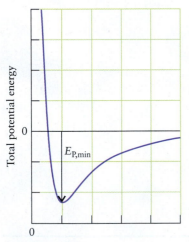

FIGURE 2.7 The potential energy of an ionic solid, taking into account the coulombic interaction of the ions and the exponential increase in their repulsion when they are in contact. The minimum potential energy is given by the Born-Meyer equation, Eq. 3.

We can now see why nature has adopted an ionic solid, calcium phosphate, for our skeletons: the doubly charged small Ca^{2+} ions and the triply charged $PO_4{}^{3-}$ ions attract one another very strongly and clump together tightly to form a rigid, insoluble solid (Fig. 2.8)

Ionic solids typically have high melting points and are brittle. The coulombic interaction between ions in a solid is large when the ions are small and highly charged.

COVALENT BONDS

Because nonmetals do not form monatomic cations, the nature of bonds between atoms of nonmetals puzzled scientists until 1916, when Lewis published his explanation. With brilliant insight, and before anyone knew about quantum mechanics or orbitals, Lewis proposed that a **covalent bond** is a pair of electrons shared between two atoms (**3**). The rest of this chapter and the next develop Lewis's vision of the covalent bond. In this chapter, we consider the types, numbers, and properties of bonds that can be formed by sharing pairs of electrons. In Chapter 3, we revisit Lewis's concept and see how to understand it in terms of orbitals.

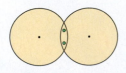

3   Shared electron pair

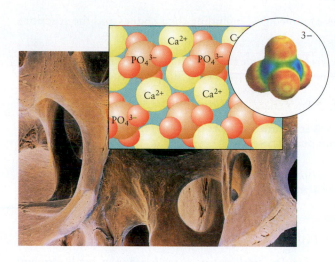

FIGURE 2.8 A micrograph of bone, which owes its rigidity to calcium phosphate. The overlay shows part of the crystal structure of calcium phosphate. Phosphate ions are polyatomic ions; however, as shown in the inset, they are nearly spherical and fit into crystal structures in much the same way as monatomic ions of charge -3.

2.5 Lewis Structures

When ionic bonds form, the atoms of one element lose electrons and the atoms of the second element gain them until both types of atoms have reached a noble-gas configuration. The same idea can be extended to covalent bonds. However, when a covalent bond forms, atoms *share* electrons until they reach a noble-gas configuration. Lewis called this principle the **octet rule**:

In covalent bond formation, atoms go as far as possible toward completing their octets by sharing electron pairs.

For example, nitrogen ($:\!\overset{\cdot}{N}\!\cdot$) has five valence electrons and needs three more electrons to complete its octet. Chlorine ($:\!\overset{\cdot\cdot}{Cl}\!\cdot$) has seven valence electrons and needs one more electron to complete its octet. Argon ($:\!\overset{\cdot\cdot}{Ar}\!:$) already has a complete octet and has no tendency to share any more electrons. Hydrogen (H·) needs one more electron to reach its helium-like duplet. Because hydrogen completes its duplet by sharing one pair of electrons, we say that it has a valence of 1 in all its compounds. In general, the **valence** of an element is the number of bonds that its atoms can form.

We can extend the Lewis symbols introduced in Section 2.2 to describe covalent bonding by using a line (—) to represent a shared pair of electrons. For example, the hydrogen molecule formed when two H· atoms share an electron pair (H:H) is represented by the symbol H—H. A fluorine atom has seven valence electrons and needs one more to complete its octet. It can achieve an octet by accepting a share in an electron supplied by another atom, such as another fluorine atom:

$$:\!\overset{\cdot\cdot}{F}\!\cdot \;+\; \cdot\!\overset{\cdot\cdot}{F}\!: \;\longrightarrow\; \left(:\!\overset{\cdot\cdot}{F}\!\!\left(:\right)\!\!\overset{\cdot\cdot}{F}\!:\right), \quad \text{or} \quad :\!\overset{\cdot\cdot}{F}\!-\!\overset{\cdot\cdot}{F}\!:$$

The circles drawn around each F atom show how each one gets an octet by sharing one electron pair. The valence of fluorine is therefore 1, the same as that of hydrogen.

As well as a bonding pair of electrons, a fluorine molecule also possesses **lone pairs** of electrons; that is, pairs of valence electrons that do not take part in bonding. The lone pairs on one F atom repel the lone pairs on the other F atom, and this repulsion is almost enough to overcome the favorable attractions of the bonding pair that holds the atoms together. This repulsion is one of the reasons why fluorine gas is so reactive: the atoms are bound together as F_2 molecules only very weakly. Among the common diatomic molecules, only H_2 has no lone pairs.

The **Lewis structure** of a molecule shows atoms by their chemical symbols, covalent bonds by lines, and lone pairs by pairs of dots. For example, the Lewis structure of HF is H—$\overset{\cdot\cdot}{\underset{\cdot\cdot}{F}}$:. We shall see that Lewis structures are a great help in

It is sometimes necessary to write a Lewis structure at the end of a sentence or clause: be careful to distinguish electron dots from periods and colons!

understanding the properties of molecules, including their shapes and the reactions that they can undergo.

SELF-TEST 2.4A Write the Lewis structure for the "interhalogen" compound chlorine monofluoride, ClF, and state how many lone pairs each atom possesses in the compound.

[*Answer:* $:\ddot{\mathrm{C}}\mathrm{l}-\ddot{\mathrm{F}}:$; three on each atom]

SELF-TEST 2.4B Write the Lewis structure for the compound HBr and state how many lone pairs each atom in the compound possesses.

> *Nonmetal atoms share electrons until each has completed its octet (or duplet); a Lewis structure shows the arrangement of electrons as lines (bonding pairs) and dots (lone pairs).*

2.6 Lewis Structures for Polyatomic Species

Each atom in a polyatomic molecule completes its octet (or duplet for hydrogen) by sharing pairs of electrons with its immediate neighbors. Each shared pair counts as one covalent bond and is represented by a line between the two atoms. A Lewis structure does not portray the shape of a polyatomic molecule; it simply displays which atoms are bonded together and which atoms have lone pairs.

Let's construct the Lewis structure for the simplest organic molecule, the hydrocarbon methane, CH_4. First, we count the valence electrons available from all the atoms in the molecule. For methane, the Lewis symbols of the atoms are

$$:\dot{\mathrm{C}} \quad \mathrm{H}\cdot \quad \mathrm{H}\cdot \quad \mathrm{H}\cdot \quad \mathrm{H}\cdot$$

and so there are eight valence electrons. The next step is to arrange the dots representing the electrons so that the C atom has an octet and each H atom has a duplet. We draw the arrangement shown on the left in (4); the Lewis structure of methane is then drawn as shown on the right in (4). Because the carbon atom is linked by four bonds to other atoms, we say that carbon is *tetravalent*: it has a valence of 4. We emphasize once again that a Lewis structure does not show the shape of the molecule, just the pattern of bonds, its "connectivity." In Chapter 3, we shall see that the three-dimensional arrangement of bonds in a methane molecule is in fact tetrahedral.

The general procedure for constructing the Lewis structure of any molecule or ion is set out in Toolbox 2.1 at the end of this section, but the following information is essential for applying those rules.

A single shared pair of electrons is called a **single bond**. Two electron pairs shared between two atoms constitute a **double bond**, and three shared electron pairs constitute a **triple bond**. A double bond, such as C::O, is written C=O in a Lewis structure. Similarly, a triple bond, such as C:::C, is written C≡C. Double and triple bonds are collectively called **multiple bonds**. The **bond order** is the number of bonds that link a specific pair of atoms. The bond order in H_2 is 1; in the group C=O, it is 2; and, for C≡C in a molecule such as ethyne, C_2H_2, the bond order is 3.

To write a Lewis structure, we need to know which atoms are linked together in a molecule. A "terminal" atom is bonded to only one other atom; an H atom in methane is an example. Except in the strange compounds called the boranes (Section 14.14), an H atom is always a terminal atom. A "central" atom is bonded to at least two other atoms. Two examples are the O atom in a water molecule, HOH, and the C atom in methane. The arrangement of atoms in a molecule and the identity of the central atom are often known in advance (for instance, it is easy to remember the arrangements of atoms in CH_4, NH_3, and H_2O). If there is doubt, a good rule of thumb for molecules other than compounds of hydrogen is to *choose as the central atom the element with the lowest ionization energy*. This arrangement often results in the lowest energy because an atom in the central

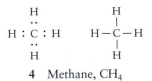

4   Methane, CH_4

In Section 2.12, when we have met the concept of electronegativity, we shall see that another way to express this rule is to say that the central atom is usually the element with lowest electronegativity.

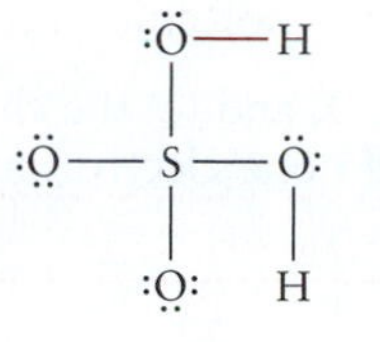

position shares more of its electrons than does a terminal atom. Atoms with higher ionization energies are more reluctant to share and are more likely to hold on to their electrons as lone pairs.

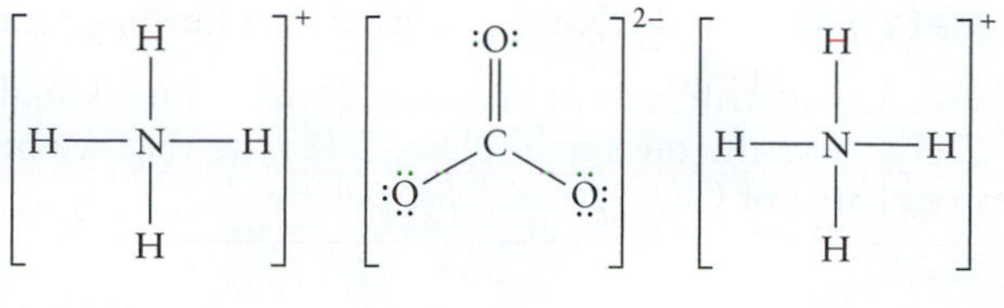

Another rule of thumb for predicting the structure of a molecule is to *arrange the atoms symmetrically around the central atom*. For instance, SO_2 is OSO, not SOO. One common exception to this rule is dinitrogen monoxide, N_2O (nitrous oxide), which has atoms in the asymmetrical arrangement NNO. Another clue for writing the correct arrangement of atoms is that, in simple chemical formulas, the central atom is often written first, followed by the atoms attached to it. For example, in the compound with the chemical formula OF_2, the arrangement of the atoms is actually FOF, not OFF; and, in SF_6, the S atom is surrounded by six F atoms. Acids are exceptions to this rule, because the H atoms are written first, as in H_2S, which has the arrangement HSH. If the compound is an oxoacid, then the acidic hydrogen atoms are attached to oxygen atoms, which in turn are attached to the central atom. For example, the arrangement of atoms in sulfuric acid, H_2SO_4, is $(HO)_2SO_2$ (**5**), and that in hypochlorous acid, HClO, is HOCl.

Finally, the same general procedure is used to determine the Lewis structures of polyatomic ions, except that we add or subtract electrons to account for the charge on the ion. We count the electrons available for bonds and lone pairs, arrange the atoms in the appropriate order, and then construct the Lewis structure. As for neutral molecules, it is essential to know the general arrangement of atoms in the ion. For oxoanions, it is usually the case that (except for H) the atom written first in the chemical formula is the central atom. In CO_3^{2-}, for instance, the C atom is surrounded by three O atoms. Each atom provides the number of dots (electrons) equal to the number of electrons in its valence shell, but we have to adjust the total number of dots to represent the overall charge. For a cation, we subtract one dot for each positive charge. For an anion, we add one dot for each negative charge. The cation and the anion must be treated separately: they are individual ions and are not linked by shared pairs. The Lewis structure of ammonium carbonate, $(NH_4)_2CO_3$, for instance, is written as three bracketed ions (**6**).

| TOOLBOX 2.1 | HOW TO WRITE THE LEWIS STRUCTURE OF A POLYATOMIC SPECIES |

CONCEPTUAL BASIS
We look for ways of using all the valence electrons to complete the octets (or duplets).

PROCEDURE
Step 1 Count the number of valence electrons on each atom; for ions, adjust the number of electrons to account for the charge. Divide the total number of valence electrons in the molecule by 2 to obtain the number of electron pairs.

Step 2 Write down the most likely arrangements of atoms by using common patterns and the clues indicated in the text.

Step 3 Place one electron pair between each pair of bonded atoms.

Step 4 Complete the octet (or duplet, in the case of H) of each atom by placing any remaining electron pairs around the atoms. If there are not enough electron pairs, form multiple bonds.

Step 5 Represent each bonded electron pair by a line.

To check on the validity of a Lewis structure, verify that each atom has an octet or a duplet. As we shall see in Section 2.10, a common exception to this rule arises when the central atom is an atom of an element in Period 3 or higher. Such an atom can accommodate more than eight electrons in its valence shell. Consequently, the most stable Lewis structure may be one in which the central atom has more than eight electrons.

This procedure is illustrated in Examples 2.3 and 2.4.

EXAMPLE 2.3   Sample exercise: Writing the Lewis structure of a molecule or an ion

Write the Lewis structures of (a) water, H_2O; (b) methanal, H_2CO; and (c) the chlorite ion, ClO_2^-. Use the rules in Toolbox 2.1; note that we must add one electron for the negative charge of ClO_2^-.

SOLUTION

	(a) H_2O	(b) H_2CO	(c) ClO_2^-
Step 1 Count the valence electrons and adjust the number for charges on ions.	$1 + 1 + 6 = 8$	$1 + 1 + 4 + 6 = 12$	$7 + 6 + 6 + 1 = 20$
Count the electron pairs.	4	6	10
Step 2 Arrange the atoms.	H O H	H C O H	O Cl O
Step 3 Connect the atoms with bonding electron pairs.	H : O : H	H C : O H	O:Cl:O
Count electron pairs not yet located.	: :(2)	: : :(3)	: : : : : : : :(8)
Step 4 Complete the octets.	H : Ö : H	H C ::Ö H	:Ö:Cl:Ö:
Step 5 Represent the bonds and add any charges.	H—Ö—H	H | C=Ö | H	:Ö—Cl—Ö:⌉⁻

SELF-TEST 2.5A Write a Lewis structure for the cyanate ion, CNO^- (the C atom is in the center).

[*Answer:* N̈=C=Ö⁻ .]

SELF-TEST 2.5B Write a Lewis structure for NH_3.

EXAMPLE 2.4   Sample exercise: writing Lewis structures for molecules with more than one "central" atom

Write the Lewis structure for acetic acid, CH_3COOH, the carboxylic acid in vinegar formed when the ethanol in wine is oxidized. In the —COOH group, both O atoms are attached to the same C atom, and one of them is bonded to the final H atom. The two C atoms are bonded to each other.

SOLUTION The formula for acetic acid suggests that it consists of two groups, with the central C atoms joined together: a CH_3— group and a —COOH group. We can anticipate that the CH_3— group, by analogy with methane, will consist of a C atom joined to three H atoms by single bonds.

CH₃COOH

Step 1 Count the valence electrons. $4 + (3 \times 1) + 4 + 6 + 6 + 1 = 24$

Count the electron pairs.   12

Step 2 Arrange atoms (linked atoms are indicated by the rectangles).

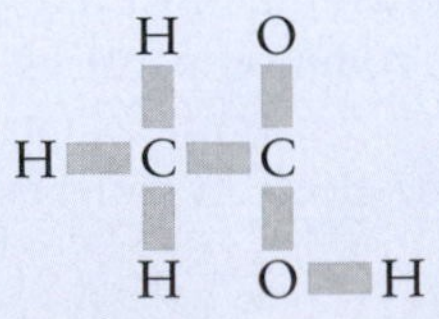

H   O
| |
H — C — C
| |
H   O — H

Step 3 Connect the atoms with bonding electron pairs.

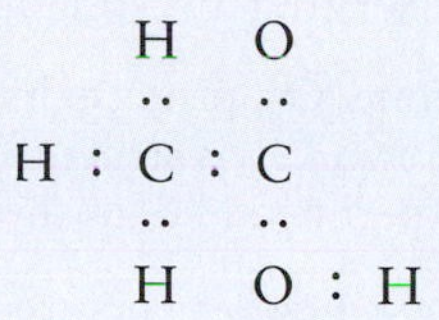

$$
\begin{array}{ccc}
\text{H} & & \text{O} \\
& & \ddot{} \\
\text{H} : \text{C} : \text{C} \\
\ddot{} & & \ddot{} \\
\text{H} & & \text{O} : \text{H}
\end{array}
$$

Count electron pairs not yet located.

$::::: (5\ \text{pairs})$

Step 4 Complete the octets.

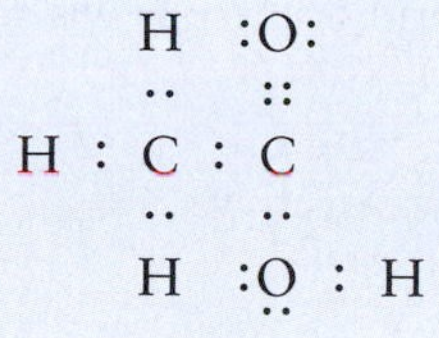

$$
\begin{array}{ccc}
\text{H} & & :\ddot{\text{O}}: \\
\ddot{} & & :: \\
\text{H} : \text{C} : \text{C} \\
\ddot{} & & \ddot{} \\
\text{H} & & :\ddot{\text{O}} : \text{H}
\end{array}
$$

Step 5 Represent the bonds.

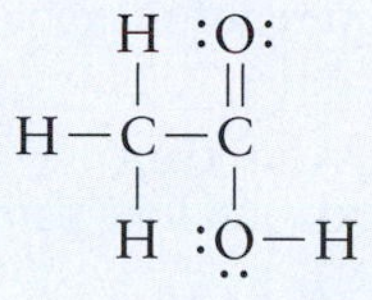

$$
\begin{array}{ccc}
\text{H} & & :\text{O}: \\
| & & \| \\
\text{H}-\text{C}-\text{C} \\
| & & | \\
\text{H} & & :\ddot{\text{O}}-\text{H}
\end{array}
$$

SELF-TEST 2.6A Write a Lewis structure for the urea molecule, $(NH_2)_2CO$.

[*Answer:* See (7).]

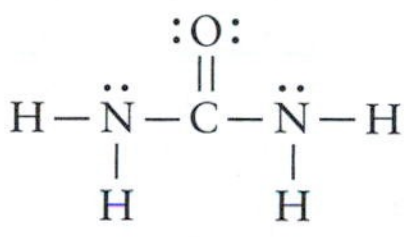

$$
\begin{array}{c}
:\ddot{\text{O}}: \\
\| \\
\text{H}-\ddot{\text{N}}-\text{C}-\ddot{\text{N}}-\text{H} \\
| \qquad\quad | \\
\text{H} \qquad\quad \text{H}
\end{array}
$$

7 Urea, $(NH_2)_2CO$

SELF-TEST 2.6B Write a Lewis structure for hydrazine, H_2NNH_2.

The Lewis structure of a polyatomic species is obtained by using all the valence electrons to complete the octets (or duplets) of the atoms present by forming single or multiple bonds and leaving some electrons as lone pairs.

2.7 Resonance

Some molecules are not represented adequately by a single Lewis structure. For example, consider the nitrate ion, NO_3^-, which, as potassium nitrate, is used in fireworks and fertilizers. The three Lewis structures shown in (8) differ only in the position of the double bond. All are valid structures and have exactly the same energy. If one of the pictured structures were correct, we would expect the nitrate ion to have two long single bonds and one short double bond because a double bond between two atoms is shorter than a single bond between the same types of atoms. However, the experimental evidence is that the bond lengths in a nitrate ion are all the same. At 124 pm, they are longer than a typical $N{=}O$ double bond (120 pm) but shorter than a typical $N{-}O$ single bond (140 pm). The bond order in the nitrate ion lies between 1 (a single bond) and 2 (a double bond).

Because all three bonds are identical, a better model of the nitrate ion is a *blend* of all three Lewis structures with each bond intermediate in properties between a single and a double bond. This blending of structures, which is called **resonance**, is depicted in (9) by double-headed arrows. The blended structure is a **resonance hybrid** of the contributing Lewis structures. A molecule does not flicker between different structures: a resonance hybrid is a blend of structures, just as a mule is a blend of a horse and a donkey, not a creature that flickers between the two.

Electrons that are shown in different positions in a set of resonance structures are said to be **delocalized**. Delocalization means that a shared electron pair is distributed over several pairs of atoms and cannot be identified with just one pair of atoms.

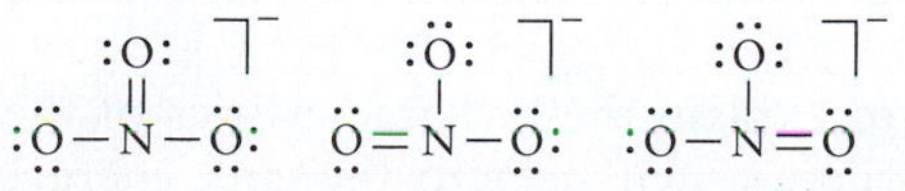

Bond length, the distance between the centers of bonded atoms, is discussed in more detail in Section 2.16.

$$
\left[\ \underset{:\ddot{\text{O}}-\text{N}-\ddot{\text{O}}:}{\overset{:\text{O}:}{\overset{\|}{}}}\ \right]^-
\quad
\left[\ \underset{\ddot{\text{O}}=\text{N}-\ddot{\text{O}}:}{\overset{:\ddot{\text{O}}:}{\overset{|}{}}}\ \right]^-
\quad
\left[\ \underset{:\ddot{\text{O}}-\text{N}=\ddot{\text{O}}}{\overset{:\ddot{\text{O}}:}{\overset{|}{}}}\ \right]^-
$$

8

$$
\left[\ \underset{:\ddot{\text{O}}-\text{N}-\ddot{\text{O}}:}{\overset{:\text{O}:}{\overset{\|}{}}}\ \right]^-
\longleftrightarrow
\left[\ \underset{\ddot{\text{O}}=\text{N}-\ddot{\text{O}}:}{\overset{:\ddot{\text{O}}:}{\overset{|}{}}}\ \right]^-
\longleftrightarrow
\left[\ \underset{:\ddot{\text{O}}-\text{N}=\ddot{\text{O}}}{\overset{:\ddot{\text{O}}:}{\overset{|}{}}}\ \right]^-
$$

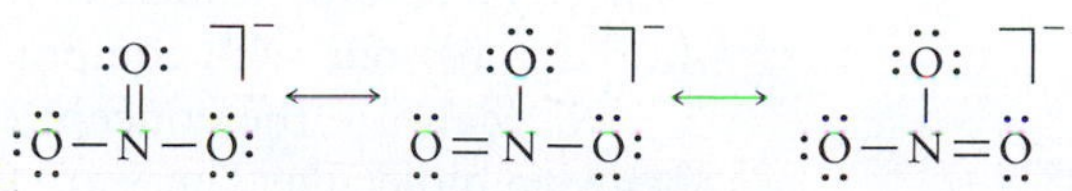

9 Nitrate ion, NO_3^-

FIGURE 2.9 When bromine dissolved in a solvent (the brown liquid) is mixed with an alkene (the colorless liquid), the bromine atoms add to the alkene molecule at the double bond, resulting in colorless products.

The German chemist Friedrich Kekulé first proposed (in 1865) that benzene has a cyclic structure with alternating single and double bonds.

10   Acetate ion, $CH_3CO_2^-$

11   Kekulé structure

12   Kekulé structure, stick form

13   Dichlorobenzene, $C_6H_4Cl_2$

EXAMPLE 2.5   Writing a resonance structure

Stratospheric ozone, O_3, protects life on Earth from harmful ultraviolet radiation from the Sun. Suggest two Lewis structures that contribute to the resonance structure for the O_3 molecule. Experimental data show that the two bond lengths are the same.

STRATEGY Write a Lewis structure for the molecule by using the method outlined in Toolbox 2.1. Decide whether there is another equivalent structure that results from the interchange of a single bond and a double or triple bond. Write the actual structure as a resonance hybrid of these Lewis structures.

SOLUTION

Count the valence electrons.	Oxygen is a member of Group 16/VI; so each atom has six valence electrons: $6 + 6 + 6 = 18$ electrons.
Draw a Lewis structure for the molecule.	$:\ddot{O}-\ddot{O}=\ddot{O}$
Draw a second Lewis structure by rearranging the bonds and lone pairs.	$\ddot{O}=\ddot{O}-\ddot{O}:$
Draw the resonance hybrid.	$:\ddot{O}-\ddot{O}=\ddot{O} \longleftrightarrow \ddot{O}=\ddot{O}-\ddot{O}:$

SELF-TEST 2.7A Write Lewis structures contributing to the resonance hybrid for the acetate ion, $CH_3CO_2^-$. The structure of CH_3COOH is described in Example 2.4; the acetate ion has a similar structure, except that it has lost the final H atom while keeping both electrons from the O—H bond.

[*Answer:* See (10).]

SELF-TEST 2.7B Write Lewis structures contributing to the resonance hybrid for the nitrite ion, NO_2^-.

Benzene, C_6H_6, is another molecule best described as a resonance hybrid. It consists of a planar hexagonal ring of six carbon atoms, each one having a hydrogen atom attached to it. One Lewis structure that contributes to the resonance hybrid is shown in (**11**); it is called a **Kekulé structure**. The structure is normally written as a line structure (see Section C), a simple hexagon with alternating single and double lines (**12**).

The difficulty with a single Kekulé structure is that it does not fit all the experimental evidence:

• *Reactivity*   Benzene does not undergo reactions typical of compounds with double bonds.

For example, when a solution of red-brown bromine is mixed with an alkene such as 1-hexene, CH_2=$CHCH_2CH_2CH_2CH_3$, the color due to bromine is lost as the Br_2 molecules attack the double bond to produce CH_2Br—$CHBrCH_2CH_2CH_2CH_3$ (Fig. 2.9). However, benzene does not react with bromine.

• *Bond lengths*   All the carbon–carbon bonds in benzene are the same length.

A Kekulé structure suggests that benzene should have two different bond lengths: three longer single bonds (154 pm) and three shorter double bonds (134 pm). Instead, all the bonds are found experimentally to have the same intermediate length (139 pm).

• *Structural evidence*   Only one dichlorobenzene in which the two chlorine atoms are attached to adjacent carbon atoms exists.

If the Kekulé structure were correct, there would be two distinct dichlorobenzenes in which the chlorine atoms are attached to adjacent carbon atoms (**13**), one in which the carbon atoms are joined by a single bond and one with a double bond. In fact, only one such compound is known.

We can use the concept of resonance to explain these characteristics of the benzene molecule. There are two Kekulé structures with exactly the same energy: they differ only in the positions of the double bonds. As a result of resonance between these two structures (**14**), the electrons shared in the C=C double bonds are delocalized over the whole molecule, thereby giving each bond a length intermediate between that of a single and that of a double bond. Resonance makes all six C—C bonds identical; this equivalence is implied by representing the double bonds in the resonance hybrid with a circle (**15**). We can see from (**16**) why there can be only one dichlorobenzene with Cl atoms on adjacent C atoms.

An important consequence of resonance is that it stabilizes a molecule by lowering its total energy. This stabilization makes benzene less reactive than expected for a molecule with three carbon–carbon double bonds. Resonance results in the greatest lowering of energy when the contributing structures have equal energies, as for the two Kekulé structures of benzene. However, in general, a molecule is a blend of all reasonable Lewis structures, including those with different energies. In these cases, the lowest energy structures contribute most strongly to the overall structure.

Resonance occurs only between structures with the same arrangement of atoms. For example, although we might be able to write two hypothetical structures for the dinitrogen oxide (nitrous oxide) molecule, NNO and NON, there is no resonance between them, because the atoms lie in different locations.

Resonance is a blending of structures with the same arrangement of atoms but different arrangements of electrons. It spreads multiple bond character over a molecule and results in a lower energy.

2.8 Formal Charge

Different Lewis structures do not in general make the same contribution to a resonance structure. It is possible to decide which structures are likely to make the major contribution by comparing the number of valence electrons distributed around each atom in a structure with the number of valence electrons on each of the free atoms. The smaller these differences for a structure, the greater is its contribution to a resonance hybrid.

The **formal charge** on an atom in a given Lewis structure is the charge it would have if the bonding were perfectly covalent in the sense that the atom had exactly a half-share in the bonding electrons. That is, the formal charge takes into account the number of electrons that an atom can be regarded as "owning" in a molecule. It "owns" all its lone pairs and half of each shared bonding pair. The difference between this number and the number of valence electrons in the free atom is the formal charge:

$$\text{Formal charge} = V - (L + \tfrac{1}{2}B) \qquad (4)*$$

where V is the number of valence electrons in the free atom, L is the number of electrons present on the bonded atom as lone pairs, and B is the number of bonding electrons on the atom. If the atom has more electrons in the molecule than when it is a free, neutral atom, then the atom has a negative formal charge, like a monatomic anion. If the assignment of electrons leaves the atom with fewer electrons than when it is free, then the atom has a positive formal charge, as if it were a monatomic cation.

Formal charge can be used to predict the most favorable arrangement of atoms in a molecule and the most likely Lewis structure for that arrangement: *a Lewis structure in which the formal charges of the individual atoms are closest to zero typically represents the lowest energy arrangement of the atoms and electrons.* A low formal charge indicates that an atom has undergone the least redistribution of electrons relative to the free atom. Such a structure has the lowest energy of all possible structures. For example, the formal charge rule suggests that the structure OCO is more likely for carbon dioxide than COO, as shown in (**17**). Similarly, it also suggests that the structure NNO is more likely than NON for dinitrogen monoxide, as shown in (**18**).

Other arrangements may be drawn, but they differ from the one shown only by a rotation of the molecule.

Resonance stabilization is a quantum mechanical effect: it is discussed further in Section 3.12.

14   Benzene resonance structure

15   Benzene, C_6H_6

16   1,2-Dichlorobenzene, $C_6H_4Cl_2$

17

18

TOOLBOX 2.2   HOW TO USE FORMAL CHARGE TO DETERMINE THE MOST LIKELY LEWIS STRUCTURE

CONCEPTUAL BASIS

To assign a formal charge, we establish the "ownership" of the valence electrons of an atom in a molecule and compare that ownership with the free atom. An atom owns one electron of each bonding pair attached to it and owns its lone pairs completely. The most plausible Lewis structure will be the one in which the formal charges of the atoms are closest to zero.

PROCEDURE

Step 1 Decide on the number of valence electrons (V) possessed by each free atom by noting the number of its group in the periodic table.

Step 2 Draw the Lewis structures.

Step 3 For each bonded atom, count each electron in its lone pairs (L), plus one electron from each of its bonding pairs (B).

Step 4 For each bonded atom, subtract the total number of electrons it "owns" from V.

Each equivalent atom (the same element, the same number of bonds and lone pairs) has the same formal charge. A check on the calculated formal charges is that their sum is equal to the overall charge of the molecule or ion. For an electrically neutral molecule, the sum of the formal charges is zero. Compare the formal charges of each possible structure. The structure with the lowest formal charges represents the least disturbance of the electronic structures of the atoms and is the most plausible (lowest energy) structure.

This procedure is illustrated in Example 2.6.

EXAMPLE 2.6   Sample exercise: Selecting the most likely atom arrangement

One test for the presence of iron(III) ions in solution is to add a solution of potassium thiocyanate, KSCN, and obtain the blood-red color of a compound of iron and the thiocyanate ion. Write three Lewis structures with different atomic arrangements for the thiocyanate ion and select the most likely structure by identifying the structure with formal charges closest to zero. For simplicity, consider only structures with double bonds between neighboring atoms.

SOLUTION We proceed as set out in Toolbox 2.2.

	NCS^-	CNS^-	CSN^-
Step 1 Count the valence electrons V and, for ions, adjust for the charge.	C: 4, N: 5, S: 6 Charge: 1 Total: 16 electrons	C: 4, N: 5, S: 6 Charge: 1 Total: 16 electrons	C: 4, N: 5, S: 6 Charge: 1 Total: 16 electrons
Step 2 Draw Lewis structures.	$\ddot{N}=C=\ddot{S}$	$\ddot{C}=N=\ddot{S}$	$\ddot{C}=S=\ddot{N}$
Step 3 Assign electron ownership, $L + \frac{1}{2}B$.	6 4 6 $\ddot{N}=C=\ddot{S}$	6 4 6 $\ddot{C}=N=\ddot{S}$	6 4 6 $\ddot{C}=S=\ddot{N}$
Step 4 Find formal charge, $V - (L + \frac{1}{2}B)$.	−1 0 0 $\ddot{N}=C=\ddot{S}$	−2 +1 0 $\ddot{C}=N=\ddot{S}$	−2 +2 −1 $\ddot{C}=S=\ddot{N}$

The individual formal charges are closest to zero in the first column; the atom arrangement NCS^- is therefore the most likely one. The same conclusion would be reached if we considered Lewis structures based on the bonding pattern $N\equiv C-S$.

SELF-TEST 2.8A Suggest a likely structure for the poisonous gas phosgene, $COCl_2$. Write its Lewis structure and formal charges.

[*Answer:* See (**19**).]

SELF-TEST 2.8B Suggest a likely structure for the oxygen difluoride molecule. Write its Lewis structure and formal charges.

:O: 0
‖
0 :Cl — C — Cl: 0
0

19 Phosgene, $COCl_2$

Although formal charge and oxidation number both give us information about the number of electrons around an atom in a compound, they are determined by different methods and often have different values. The formal charge exaggerates the covalent character of the bonds by assuming that the electrons are shared equally. Conversely, oxidation number (Sections D and K) is an exaggeration of the ionic character of bonds. It represents the atoms as ions, and all the electrons in a bond are assigned to the more electronegative atom. Thus, although the formal charge of C in structure **17** for CO_2 is 0, its oxidation number is +4 because all the bonding electrons are assigned to the oxygen atoms to give a structure that could be represented $O^{2-}C^{4+}O^{2-}$. Formal charges depend on the particular Lewis structure we write; oxidation numbers do not.

> *The formal charge gives an indication of the extent to which atoms have gained or lost electrons in the process of covalent bond formation; atom arrangements and Lewis structures with lowest formal charges are likely to have the lowest energy.*

EXCEPTIONS TO THE OCTET RULE

The octet rule accounts for the valences of many of the elements and the structures of many compounds. Carbon, nitrogen, oxygen, and fluorine obey the octet rule rigorously, provided there are enough electrons to go around. However, some compounds have an odd number of electrons. In addition, an atom of phosphorus, sulfur, chlorine, or another nonmetal in Period 3 and subsequent periods can accommodate more than eight electrons in its valence shell. The following two sections show how to recognize exceptions to the octet rule.

2.9 Radicals and Biradicals

Some species have an odd number of valence electrons, and so at least one of their atoms cannot have an octet. Species having electrons with unpaired spins are called **radicals**. They are generally highly reactive. One example is the methyl radical, $\cdot CH_3$, which is so reactive that it cannot be stored. It occurs in the flames of burning hydrocarbon fuels. The single unpaired electron is indicated by the dot on the C atom in $\cdot CH_3$.

Radicals are of crucial importance for the chemical reactions that take place in the upper atmosphere, where they contribute to the formation and decomposition of ozone. Radicals also play a role in our daily lives, sometimes a destructive one. They are responsible for the rancidity of foods and the degradation of plastics in sunlight. Damage from radicals can be delayed by an additive called an **antioxidant**, which reacts rapidly with radicals before the radicals have a chance to do their damage. Human aging is believed to be partly due to the action of radicals, and antioxidants such as vitamins C and E may delay the process (see Box 2.1). The nitric oxide molecule, NO, has an unpaired electron and plays an important role as a neurotransmitter. Because it is a radical, NO is very reactive and can be eliminated within a few seconds. Because it is small, the NO molecule can move easily throughout the body. These properties allow NO to play several roles, including controlling the supply of blood to various organs.

A **biradical** is a molecule with two unpaired electrons. The unpaired electrons are usually on different atoms, as depicted in (**20**). In that biradical, one unpaired electron is on one carbon atom of the chain and the second is on another carbon atom several bonds away. In some cases, though, both electrons are on the same atom. One of the most important examples is the oxygen atom itself. Its electron configuration is $[He]2s^2 2p_x^2 2p_y^1 2p_z^1$ and its Lewis symbol is $\cdot \ddot{O} \cdot$. The O atom has two unpaired electrons, and so it can be regarded as a special type of biradical.

SELF-TEST 2.9A Write a Lewis structure for the hydrogenperoxyl radical, HOO·, which plays an important role in atmospheric chemistry and which, in the body, has been implicated in the degeneration of neurons.

[*Answer:* See (**21**).]

SELF-TEST 2.9B Write a Lewis structure for NO_2.

> *A radical is a species with an unpaired electron; a biradical has two unpaired electrons on either the same or different atoms.*

The older, still widely used term for radicals is *free radicals*.

The role of radicals in the depletion of stratospheric ozone is developed further in Box 13.3.

20   A biradical

21   Hydrogenperoxyl, HO_2.

Chemical Self-Preservation

In nearly every pharmacy, supermarket, and health food store, you can find bottles of antioxidants and antioxidant-rich natural products, such as fish oils, *Gingko biloba* leaves, and wheat grass. These dietary supplements are intended to help the body control its population of radicals and, as a result, slow aging and degenerative diseases such as heart failure and cancer.

Radicals occur naturally in the body, partly as a by-product of metabolism. They serve many useful functions but can cause trouble if they are not eliminated when they are no longer needed. They often contain oxygen atoms and oxidize the *lipid* (fat) molecules that make up cell membranes and other vital tissues. This oxidation changes the structures of the lipid molecules and hence affects the function of the membranes. Cell membranes damaged in this way may not be able to protect cells against disease, and heart and nerve cells may lose their function. Recent evidence suggests that radical damage to living cells is the main factor in the aging process. When the nucleic acids such as DNA and RNA are attacked by radicals, they cannot replicate properly. These alterations result in defective cells with diminished function and a lessened ability to protect themselves against diseases such as cancer.

Leaves of the *Gingko biloba* tree, which originated in China. Extracts of these leaves are thought to have antioxidant properties. Some also believe that the extracts improve thinking ability by increasing oxygen flow to the brain.

The body maintains an antioxidant network consisting of vitamins A, C, and E, antioxidant enzymes, and a group of related compounds called coenzyme Q, for which the general formula is shown below. The *n* represents the number of times that a particular group is repeated; it can be 6, 8, or 10. *Antioxidants* are molecules that are easily oxidized, so they react readily with radicals before the radicals can react with other compounds in the body. Many common foods, such as green leafy vegetables, orange juice, and chocolate, contain antioxidants, as do coffee and tea.

n CH$_2$CH = C(CH$_3$)CH$_2$ units (n = 6, 8, 10)

Molecular structure of coenzyme Q, an antioxidant used by the body to control the level of radicals.

Harmful environmental conditions, such as ultraviolet radiation, ozone in the air we breathe, poor nutrition, and tobacco smoke can cause *oxidative stress*, a condition in which the concentration of radicals has become so high that the body's natural antioxidant network cannot cope. Premature aging of skin that has been overexposed to sunlight and lung cancer in smokers are two possible results. Herbal medicines containing certain *phytochemicals*, chemicals derived from plant sources and fish oils, are being investigated as antioxidants that can supplement the diet to protect against damage from free radicals. These same chemicals are also being studied for their ability to prolong life beyond the currently expected lifespan.

Related Exercises: 2.59, 2.60, 2.93, and 2.103.

For Further Reading: D. Harman, "The aging process," *Proceedings of the National Academy of Sciences*, **78** (2004), 7124. C. Goldberg, "The quest for immortality: Science at the frontiers of aging," *Science News*, **162** (August 31, 2002), 143.

2.10 Expanded Valence Shells

The octet rule tells us that eight electrons fill the outer shell of an atom to give a noble-gas ns^2ns^6 valence-shell configuration. However, when the central atom in a molecule has empty *d*-orbitals, it may be able to accommodate 10, 12, or even more electrons. The electrons in such an **expanded valence shell** may be present as lone pairs or may be used by the central atom to form additional bonds.

A note on good practice: Although "expanded valence shell" is the logically precise term, most chemists still use the term *expanded octet*.

Because the additional electrons must be accommodated in valence orbitals, only nonmetal atoms in Period 3 or later periods can expand their valence shells.

Atoms of these elements have empty *d*-orbitals in the valence shell. Another factor—possibly the main factor—in determining whether more atoms than allowed by the octet rule can bond to a central atom is the size of that atom. A P atom is big enough for as many as six Cl atoms to fit comfortably around it, and PCl_5 is a common laboratory chemical. An N atom, though, is too small, and NCl_5 is unknown. A compound that contains an atom with more atoms attached to it than is permitted by the octet rule is called a **hypervalent compound**. This name leaves open the question of whether the additional bonds are due to valence-shell expansion or simply to the size of the central atom.

Elements that can expand their valence shells commonly show **variable covalence**, the ability to form different numbers of covalent bonds. Elements that have variable covalence can form one number of bonds in some compounds and a different number in others. Phosphorus is an example. It reacts directly with a limited supply of chlorine to form the toxic, colorless liquid phosphorus trichloride:

$$P_4(s) + 6\ Cl_2(g) \longrightarrow 4\ PCl_3(l)$$

The Lewis structure of the PCl_3 molecule is shown in (**22**), and we see that it obeys the octet rule. However, when phosphorus trichloride reacts with more chlorine (Fig. 2.10), phosphorus pentachloride, a pale yellow crystalline solid, is produced:

$$PCl_3(l) + Cl_2(g) \longrightarrow PCl_5(s)$$

Phosphorus pentachloride is an ionic solid consisting of PCl_4^+ cations and PCl_6^- anions; but, at 160°C, it vaporizes to a gas of PCl_5 molecules. The Lewis structures of the polyatomic ions and the molecule are shown in (**23**) and (**24**). In the anion, the P atom has expanded its valence shell to 12 electrons, by making use of two of its 3*d*-orbitals. In PCl_5, the P atom has expanded its valence shell to 10 electrons by using one 3*d*-orbital. Phosphorus pentachloride is therefore an example of a hypervalent compound in both its solid and its gaseous forms.

FIGURE 2.10 Phosphorus trichloride is a colorless liquid. When it reacts with chlorine (the pale yellow-green gas in the flask), it forms the very pale yellow solid phosphorus pentachloride (at the bottom of the flask).

22 Phosphorus trichloride, PCl_3

(a) PCl_4^+ (b) PCl_6^-

23 Phosphorus pentachloride, $PCl_5(s)$

24 Phosphorus pentachloride, PCl_5

EXAMPLE 2.7 **Writing a Lewis structure with an expanded valence shell**

The fluoride SF_4 forms when a mixture of fluorine and nitrogen gases is passed over a film of sulfur at 275°C in the absence of oxygen and moisture. Write the Lewis structure of sulfur tetrafluoride and give the number of electrons in the expanded valence shell.

STRATEGY Because sulfur is in Period 3 and has empty 3*d*-orbitals available, it can expand its valence shell to accept additional electrons. After assigning all the valence electrons to bonds and lone pairs to give each atom an octet, assign any remaining electrons to the sulfur atom.

SOLUTION

Count the number of valence electrons.	6 from sulfur ($\cdot \ddot{S} \cdot$)
	7 from each fluorine atom ($:\ddot{F}\cdot$)
Find the number of electron pairs.	There are $6 + (4 \times 7) = 34$ electrons, or 17 electron pairs.
Construct the Lewis structure.	Give each F atom 3 lone pairs and 1 bonding pair shared with the central S atom; place the 2 extra electrons as a lone pair on the S atom.

In this structure sulfur has 10 electrons in its expanded valence shell.

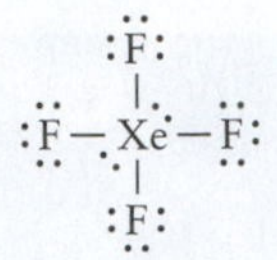

$$:\!\ddot{F}\!:$$

25   Xenon tetrafluoride, XeF₄

26a

26b

26c

27

28

SELF-TEST 2.10A Write the Lewis structure for xenon tetrafluoride, XeF₄, and give the number of electrons in the expanded valence shell.

[*Answer:* See (**25**); 12 electrons]

SELF-TEST 2.10B Write the Lewis structure for the I_3^- ion and give the number of electrons in the expanded valence shell.

When different resonance structures are possible, some giving the central atom in a compound an octet and some an expanded valence shell, the dominant resonance structure is likely to be the one with the lowest formal charges. However, there are many exceptions and the selection of the best structure often depends on a careful analysis of experimental data.

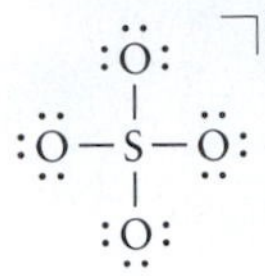

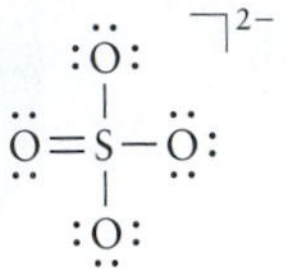

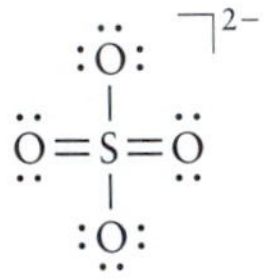

EXAMPLE 2.8   Selecting the dominant resonance structure for a molecule

The sulfate ion, SO_4^{2-}, is present in a number of important minerals, including gypsum ($CaSO_4\cdot 2H_2O$), which is used in cement, and Epsom salts ($MgSO_4\cdot 7H_2O$), which is used as a purgative. Determine the dominant resonance structure of a sulfate ion from the three shown in (**26**) by calculating the formal charges on the atoms in each structure.

STRATEGY Follow the procedure in Toolbox 2.2. We need do only one calculation for equivalent atoms, such as the oxygen atoms in the first diagram, because they all have the same arrangement of electrons and hence the same formal charge.

SOLUTION We draw up the following table, in accord with Toolbox 2.2.

	26a	26b	26c
Step 1 Count the valence electrons (V).	\<center\>O: 6 S: 6\</center\>		
	\<center\>Total: 24 electrons, which provide 12 pairs of electrons\</center\>		
Step 2 Draw the Lewis structure.	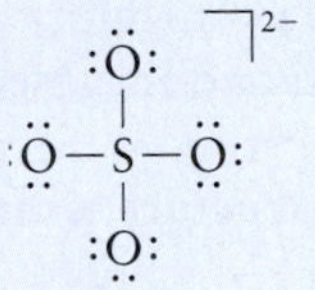	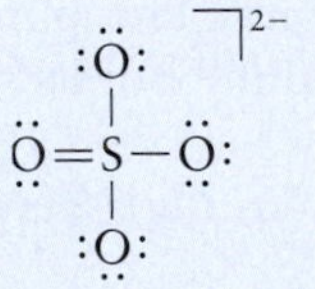	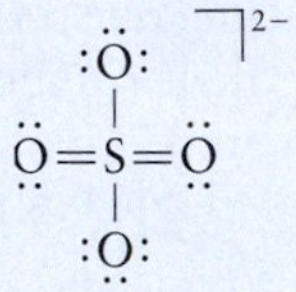
Step 3 Assign electron ownership, $(L + \tfrac{1}{2}B)$.	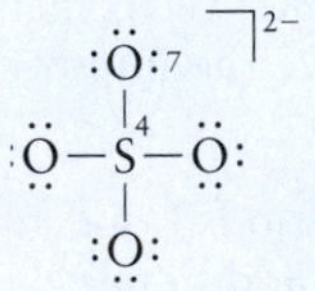	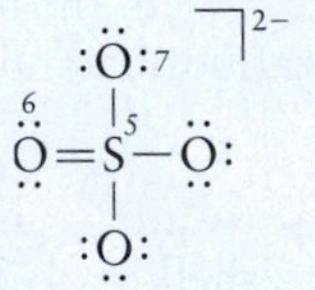	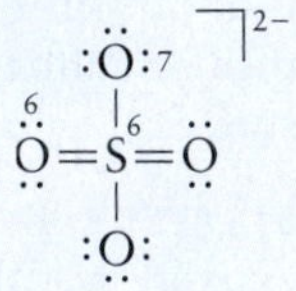
Step 4 Find the formal charge, $V - (L + \tfrac{1}{2}B)$.	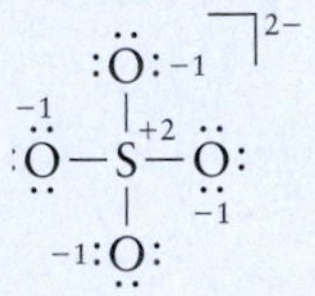	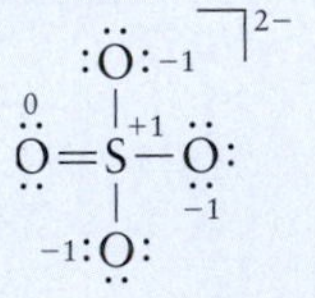	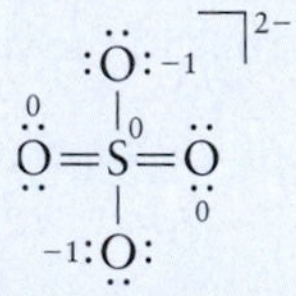

The individual formal charges are closest to zero in structure (**26c**); so that structure is likely to make the biggest contribution to the resonance hybrid, even though the valence shell on the S atom has expanded to hold 12 electrons.

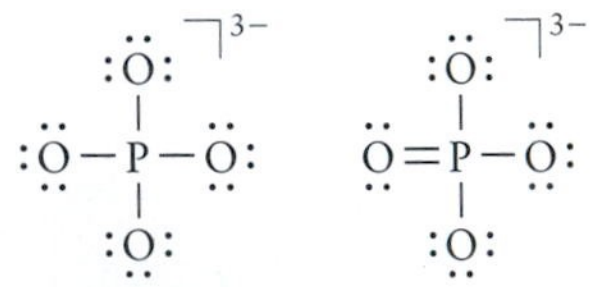

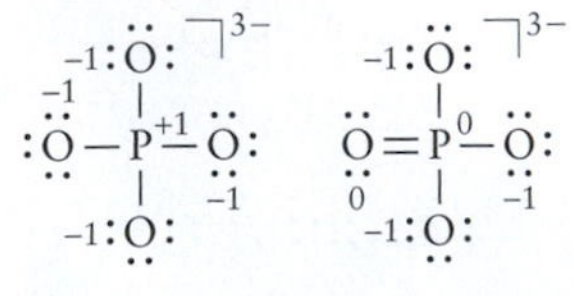

SELF-TEST 2.11A Calculate the formal charges for the two Lewis structures of the phosphate ion shown in (**27**). Which structure is dominant?

[*Answer:* See (**28**); the second structure.]

SELF-TEST 2.11B Calculate the formal charges for the three oxygen atoms in one of the Lewis structures of the ozone resonance structure (Example 2.5).

Octet expansion (expansion of the valence shell to more than eight electrons) can occur in elements of Period 3 and later periods. These elements can exhibit variable covalence and be hypervalent. Formal charge helps to identify the dominant resonance structure.

2.11 The Unusual Structures of Some Group 13/III Compounds

An unusual feature of the Lewis structure of the colorless gas boron trifluoride, BF_3 (**29**), is that the boron atom has an **incomplete octet**: its valence shell consists of only six electrons. We might expect the boron atom to complete its octet by sharing more electrons with fluorine, as depicted in (**30**), but fluorine has such a high ionization energy that it is unlikely to exist with a positive formal charge. Experimental evidence, such as the short B—F bond lengths, suggests that the true structure of BF_3 is a resonance hybrid of both types of Lewis structures, with the singly bonded structure making the major contribution.

   The boron atom in BF_3 can complete its octet if an additional atom or ion with a lone pair of electrons forms a bond by providing *both* electrons. A bond in which both electrons come from one of the atoms is called a **coordinate covalent bond**. For example, the tetrafluoroborate anion, BF_4^- (**31**), forms when boron trifluoride is passed over a metal fluoride. In this anion, the formation of a coordinate covalent bond with a fluoride ion gives the B atom an octet. Another example of a coordinate covalent bond is that formed when boron trifluoride reacts with ammonia:

$$BF_3(g) + NH_3(g) \longrightarrow NH_3BF_3(s)$$

The Lewis structure of the product, a white molecular solid, is shown in (**32**). In this reaction, the lone pair on the nitrogen atom of ammonia, $H_3N:$, can be regarded as completing boron's octet in BF_3 by forming a coordinate covalent bond.

   Boron trichloride, a colorless, reactive gas of BCl_3 molecules, behaves chemically like BF_3. However, the trichloride of aluminum, which is in the same group as boron, forms *dimers*, linked pairs of molecules. Aluminum chloride is a volatile white solid that vaporizes at 180°C to a gas of Al_2Cl_6 molecules. These molecules survive in the gas up to about 200°C and only then fall apart into $AlCl_3$ molecules. The Al_2Cl_6 molecule exists because a Cl atom in one $AlCl_3$ molecule uses one of its lone pairs to form a coordinate covalent bond to the Al atom in a neighboring $AlCl_3$ molecule (**33**). This arrangement can occur in aluminum chloride but not boron trichloride because the atomic radius of Al is bigger than that of B.

Compounds of boron and aluminum may have unusual Lewis structures in which boron and aluminum have incomplete octets or halogen atoms act as bridges.

IONIC VERSUS COVALENT BONDS

Ionic and covalent bonding are two extreme models of the chemical bond. Most actual bonds lie somewhere between purely ionic and purely covalent. When we describe bonds between nonmetals, covalent bonding is a good model. When a metal and nonmetal are present in a simple compound, ionic bonding is a good model. However, the bonds in many compounds seem to have properties between the two extreme models of bonding. Can we describe these bonds more accurately by improving the two basic models?

2.12 Correcting the Covalent Model: Electronegativity

All bonds can be viewed as resonance hybrids of purely covalent and purely ionic structures. For example, the structure of a Cl_2 molecule can be described as

$$:\overset{..}{\underset{..}{Cl}}-\overset{..}{\underset{..}{Cl}}: \longleftrightarrow :\overset{..}{\underset{..}{Cl}}:^- \overset{..}{\underset{..}{Cl}}:^+ \longleftrightarrow :\overset{..}{\underset{..}{Cl}}:^+ :\overset{..}{\underset{..}{Cl}}:^-$$

In this case, the ionic structures make only a small contribution to the resonance hybrid, and we regard the bond as almost purely covalent. Moreover, the two ionic structures have the same energy and make equal contributions to the hybrid; so the average charge on each atom is zero. However, in a molecule composed of different elements, such as HCl, the resonance

$$H-\overset{..}{\underset{..}{Cl}}: \longleftrightarrow H:^- \overset{..}{\underset{..}{Cl}}:^+ \longleftrightarrow H^+ :\overset{..}{\underset{..}{Cl}}:^-$$

29   Boron trifluoride, BF_3

30   Boron trifluoride, BF_3

31   Tetrafluoroborate, BF_4^-

32 NH_3BF_3

33   Aluminum chloride, Al_2Cl_6

has unequal contributions from the two ionic structures. The lower-energy ionic structure is $H^+ : \ddot{\underset{..}{Cl}} :^-$. Because the chlorine atom has a greater attraction for electrons than does the hydrogen atom, the structure with a negative charge on the Cl atom makes a bigger contribution than $H :^- \ddot{\underset{..}{Cl}} :^+$. As a result, there is a small net negative charge on the Cl atom and a small net positive charge on the H atom. Here we see the limitations of using formal charge alone to determine the distribution of electrons in a structure. The formal charge on each atom in HCl is zero.

The charges on the atoms in HCl are called **partial charges**. We show the partial charges on the atoms by writing $^{\delta+}H-Cl^{\delta-}$. A bond in which ionic contributions to the resonance result in partial charges is called a **polar covalent bond**. All bonds between atoms of different elements are polar to some extent. The bonds in homonuclear (same element) diatomic molecules and ions are nonpolar.

The two atoms in a polar covalent bond form an **electric dipole**, a partial positive charge next to an equal but opposite partial negative charge. A dipole is represented by an arrow that points toward the negative partial charge (**34**). The size of an electric dipole—which is a measure of the magnitude of the partial charges—is reported as the **electric dipole moment**, μ (the Greek letter mu), in units called **debye** (D). The debye is defined so that a single negative charge (an electron) separated by 100. pm from a single positive charge (a proton) has a dipole moment of 4.80 D. The dipole moment associated with a Cl—H bond is about 1.1 D. We can think of this dipole as arising from a partial charge of about 23% of an electron's charge on the Cl atom and an equivalent positive charge on the H atom.

A covalent bond is polar if one atom has a greater attraction for electrons than the other atom, because then the electron pair is more likely to be found closer to the former atom. In 1932, the American chemist Linus Pauling proposed a quantitative measure of electron distribution in bonds. The electron-pulling power of an atom when it is part of a molecule is called its **electronegativity**. Electronegativities are denoted χ (the Greek letter chi, pronounced "kye"). The atom of the element with the higher electronegativity has a stronger pulling power on electrons and tends to pull them away from the atom of the element with lower electronegativity (Fig. 2.11). Pauling based his scale on the dissociation energies, D, of the A—A, B—B, and A—B bonds, measured in electronvolts. He defined the difference in electronegativity of the two elements A and B as

$$|\chi_A - \chi_B| = 0.102\{D(A-B) - \tfrac{1}{2}[D(A-A) + D(B-B)]\}^{\frac{1}{2}} \tag{5}$$

A simpler way of setting up a scale of electronegativities was devised by another American chemist, Robert Mulliken. In his approach, the electronegativity is the average of the ionization energy and electron affinity of the element (both expressed in electronvolts):

$$\chi = \tfrac{1}{2}(I + E_a) \tag{6}$$

An atom gives up an electron reluctantly if the ionization energy is high. If the electron affinity is high, attaching electrons to an atom is energetically favorable. Elements with both these properties are reluctant to lose their electrons and tend to gain them; hence, they are classified as highly electronegative. Conversely, if the ionization energy and the electron affinity are both low, then it takes very little energy for the element to give up its electrons and it has little tendency to gain more; hence, the electronegativity is low.

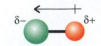

34   Dipole moment

The debye is named for the Dutch chemist Peter Debye, who carried out important studies of dipole moments.

The SI unit of dipole moment is 1 C·m (1 coulomb meter). It is the dipole moment of a charge of 1 C separated from a charge of −1 C by a distance of 1 m; 1 D = 3.336 × 10^{-30} C·m.

Dissociation energies, a measure of the strengths of bonds, are discussed in Section 2.15.

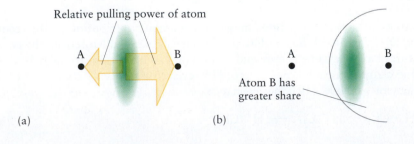

Relative pulling power of atom

Atom B has greater share

(a)   (b)

FIGURE 2.11 The electronegativity of an element is its electron-pulling power when it is part of a compound. (a) An atom with a high electronegativity (B) has a strong pulling power on the electrons (as represented by the larger arrow) that it shares with its neighbor, A. (b) The outcome of the tug-of-war is that the more electronegative atom has a greater share in the electron pair of the covalent bond. Electron density is symbolized by the green cloud in each image.

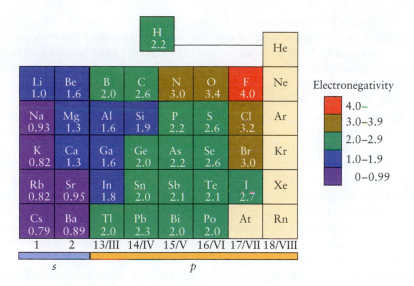

FIGURE 2.12 Variation in the electronegativity of the main-group elements (with the exception of the noble gases). Electronegativity tends to be high toward the upper-right corner of the periodic table and low on the lower left. Elements with low electronegativities (such as the s-block metals) are often called electropositive. These Pauling values are used throughout the text.

Figure 2.12 shows the variation in electronegativity for the main-group elements of the periodic table. Because ionization energies and electron affinities are highest at the top right of the periodic table (close to fluorine), it is not surprising to find that nitrogen, oxygen, bromine, chlorine, and fluorine are the elements with the highest electronegativities.

When the two atoms in a bond have only a small electronegativity difference, the partial charges are very small. As the difference in electronegativities increases, so do the partial charges. If the electronegativities are very different, then one atom can acquire the lion's share of the electron pair, and the corresponding ionic structure makes a large contribution to the resonance. Because it has largely robbed the other atom of its share of the electrons, the highly electronegative element resembles an anion and the other atom resembles a cation. We say that such a bond has considerable **ionic character**. If the difference in electronegativity is large, as in NaCl or KF, the ionic contribution dominates the covalent distribution, and it is better to regard the bond as ionic.

There is no hard-and-fast dividing line between ionic and covalent bonding. However, a good rule of thumb is that an electronegativity difference of about 2 means that there is so much ionic character present that the bond is best regarded as ionic (Fig. 2.13). For electronegativity differences smaller than about 1.5, a covalent description of the bond is probably reasonably reliable. For example, the electronegativities of carbon and oxygen are 2.6 and 3.4, an electronegativity difference of 0.8, and C—O bonds are best regarded as polar covalent. However, the electronegativity of calcium is 1.3, and Ca—O bonds, with an electronegativity difference of 2.1, are considered ionic.

SELF-TEST 2.12A In which of the following compounds do the bonds have greater ionic character: (a) P_4O_{10} or (b) PCl_3?

[*Answer:* (a)]

SELF-TEST 2.12B In which of the following compounds do the bonds have greater ionic character: (a) CO_2 or (b) NO_2?

Electronegativity is a measure of the pulling power of an atom on the electrons in a bond. A polar covalent bond is a bond between two atoms with partial electric charges arising from their difference in electronegativity. The presence of partial charges gives rise to an electric dipole moment.

2.13 Correcting the Ionic Model: Polarizability

All ionic bonds have some covalent character. To see how covalent character can arise, consider a monatomic anion (such as Cl^-) next to a cation (such as Na^+). As the cation's positive charge pulls on the anion's electrons, the spherical electron

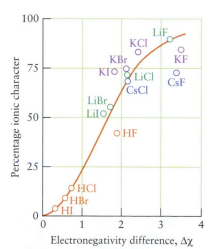

FIGURE 2.13 The dependence of the percentage ionic character of the bond on the difference in electronegativity, $\Delta\chi$, between two bonded atoms for a number of halides.

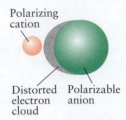

FIGURE 2.14 When a small, highly charged cation is close to a large anion, the electron cloud of the anion is distorted in the process we call polarization. The green sphere represents the shape of anion in the absence of a cation. The gray shadow shows how the shape of the sphere is distorted by the positive charge of the cation.

Polarizing cation

Distorted electron cloud

Polarizable anion

Cations are not significantly polarizable because their electrons are held so tightly.

cloud of the anion becomes distorted in the direction of the cation. We can think of this distortion as the tendency of an electron pair to move into the region between the two nuclei and to form a covalent bond (Fig. 2.14). Ionic bonds acquire more covalent character as the distortion of the electron cloud on the anion increases.

Atoms and ions that readily undergo a large distortion are said to be highly **polarizable**. An anion can be expected to be highly polarizable if it is large, such as an iodide ion, I^-. In such a large, highly polarizable ion, the ion's nucleus exerts only weak control over its outermost electrons because they are so far away. As a result, the electron cloud of the large anion is easily distorted.

Atoms and ions that can *cause* large distortions are said to have a high **polarizing power**. A cation can be expected to have a strong polarizing power if it is small and highly charged, such as an Al^{3+} cation. A small radius means that the center of charge of a highly charged cation can get very close to the anion, where it can exert a strong pull on the anion's electrons. Compounds composed of a small, highly charged cation and a large, polarizable anion tend to have bonds with considerable covalent character.

The polarizabilities of anions and the polarizing powers of cations follow the same diagonal relationships in the periodic table introduced in Section 1.20 and illustrated in Fig. 1.58. Cations become smaller, more highly charged, and hence more strongly polarizing, from left to right across a period. Thus, Be^{2+} is more strongly polarizing than Li^+, and Mg^{2+} is more strongly polarizing than Na^+. On the other hand, cations become larger and hence less strongly polarizing down a group. Thus, Na^+ is less strongly polarizing than Li^+, and Mg^{2+} is less strongly polarizing than Be^{2+}. Now we can see that, because polarizing power increases from Li^+ to Be^{2+} but decreases from Be^{2+} to Mg^{2+}, it follows that the polarizing power of the diagonal neighbors Li^+ and Mg^{2+} should be similar. We can expect such similarities in the properties of other diagonally related neighbors.

SELF-TEST 2.13A In which of the compounds NaBr and $MgBr_2$ do the bonds have greater covalent character?

[*Answer:* $MgBr_2$]

SELF-TEST 2.13B In which of the compounds CaS and CaO do the bonds have greater covalent character?

Compounds composed of highly polarizing cations and highly polarizable anions have a significant covalent character in their bonding.

THE STRENGTHS AND LENGTHS OF COVALENT BONDS

The characteristics of a covalent bond between two atoms are due mainly to the properties of the atoms themselves and vary only a little with the identities of the other atoms present in a molecule. Consequently, we can predict some characteristics of a bond with reasonable certainty once we know the identities of the two bonded atoms. For instance, the length of the bond and its strength are approximately the same regardless of the molecule in which it is found. Thus, to understand the properties of a large molecule, such as how DNA replicates in our cells and transmits genetic information, we can study the character of C=O and N—H bonds in much simpler compounds, such as formaldehyde, H_2C=O, and ammonia, NH_3.

2.14 Bond Strengths

The strength of a chemical bond is measured by its **dissociation energy**, D, the energy required to separate the bonded atoms. On a plot of the potential energy of a diatomic molecule as a function of the internuclear distance, the dissociation energy is the distance between the bottom of the energy well and the energy of the separated atoms (Fig. 2.15). The bond breaking is **homolytic**, which means that each atom retains one of the electrons from the bond. An example is

$$H—Cl(g) \longrightarrow \cdot H(g) + \cdot Cl(g)$$

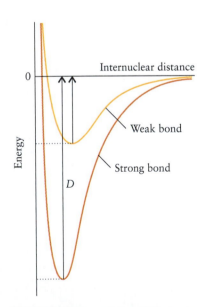

FIGURE 2.15 The variation of the energy of a diatomic molecule with internuclear separation for weak and strong bonds. The dissociation energy is a measure of the depth of the well. (In practice, we have to take into account the small zero-point energy of the vibrating molecule, and so the dissociation energy is slightly less than the depth of the well.)

Internuclear distance

0

Energy

Weak bond

Strong bond

D

A high dissociation energy indicates a deep potential energy well and therefore a strong bond that requires a lot of energy to break. The strongest known bond between two nonmetal atoms is the triple bond in carbon monoxide, for which the dissociation energy is 1062 kJ·mol^{-1}. One of the weakest bonds is that between the iodine atoms in molecular iodine, for which the dissociation energy is only 139 kJ·mol^{-1}.

> *The strength of a bond between two atoms is measured by its dissociation energy: the greater the dissociation energy, the stronger the bond.*

2.15 Variation in Bond Strength

Tables 2.3 and 2.4 list a selection of typical dissociation energies. The values given in Table 2.4 are average dissociation energies for a number of different molecules. For instance, the strength quoted for a C—O single bond is the average strength of such bonds in a selection of organic molecules, such as methanol (CH$_3$—OH), ethanol (CH$_3$CH$_2$—OH), and dimethyl ether (CH$_3$—O—CH$_3$). The values should therefore be regarded as typical rather than as accurate values for a particular molecule.

TABLE 2.3 Bond Dissociation Energies of Diatomic Molecules (kJ·mol^{-1})

Molecule	Bond dissociation energy
H$_2$	424
N$_2$	932
O$_2$	484
CO	1062
F$_2$	146
Cl$_2$	230
Br$_2$	181
I$_2$	139
HF	543
HCl	419
HBr	354
HI	287

TABLE 2.4 Average Bond Dissociation Energies (kJ·mol^{-1})

Bond	Average bond dissociation energy	Bond	Average bond dissociation energy
C—H	412	C—I	238
C—C	348	N—H	388
C=C	612	N—N	163
C⋯C*	518	N=N	409
C≡C	837	N—O	210
C—O	360	N=O	630
C=O	743	N—F	195
C—N	305	N—Cl	381
C—F	484	O—H	463
C—Cl	338	O—O	157
C—Br	276		

*In benzene.

We shall not distinguish between the average dissociation energy and the average dissociation enthalpy, a concept to be introduced in Section 6.20. The two quantities differ by only a few kilojoules per mole.

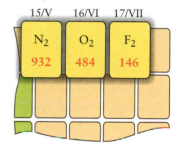

FIGURE 2.16 The bond dissociation energies, in kilojoules per mole of nitrogen, oxygen, and fluorine molecules. Note how the bonds weaken in the change from a triple bond in N$_2$ to a single bond in F$_2$.

The trends in bond strengths shown in Table 2.3 are explained in part by the Lewis structures for the molecules. Consider, for example, the diatomic molecules of nitrogen, oxygen, and fluorine (Fig. 2.16). Note the decline in bond strength as the bond order decreases—from 3 in N$_2$ to 1 in F$_2$. The triple bond in nitrogen is the origin of the inertness mentioned at the beginning of the chapter. A multiple bond is always stronger than a single bond between the same two atoms because more electrons bind the multiply bonded atoms. A triple bond between two atoms is always stronger than a double bond between the same two atoms, and a double bond is always stronger than a single bond between the same two atoms. However, a double bond between two carbon atoms is not twice as strong as a single bond, and a triple bond is a lot less than three times as strong. For example, we see that the average dissociation energy of a C=C double bond is 612 kJ·mol^{-1}, whereas it takes 696 kJ·mol^{-1} to break two C—C single bonds; similarly, the average dissociation energy of a C≡C triple bond is 837 kJ·mol^{-1}, but it takes 1044 kJ·mol^{-1} to break three C—C single bonds (Fig. 2.17). The origin of these differences is in part the repulsions between the electron pairs in a multiple bond, so each pair is not quite as effective at bonding as a pair of electrons in a single bond. Another contributing factor is that, as will be described in Section 3.4, the electrons in double and triple bonds are not as concentrated between the two atoms as they are in a single bond.

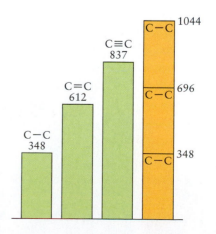

FIGURE 2.17 The strengths (in kilojoules per mole) of single and multiple bonds between two carbon atoms. Note that, for bonds between carbon atoms, a double bond is less than twice as strong as a single bond and a triple bond is less than three times as strong as a single bond, as shown by the fourth column.

FIGURE 2.18 The bond dissociation energies of the hydrogen halide molecules in kilojoules per mole of molecules. Note how the bonds weaken as the halogen atom becomes larger.

FIGURE 2.19 The dissociation energies for bonds between hydrogen and the *p*-block elements. The bond strengths decrease down each group as the atoms increase in size.

The values in Table 2.4 show how resonance affects the strengths of bonds. For example, the strength of a carbon–carbon bond in benzene is intermediate between that of a single and that of a double bond. Resonance spreads multiple bond character over the bonds between atoms: as a result, what were single bonds are strengthened and what were double bonds are weakened. The net effect overall is a stabilization of the molecule.

The presence of lone pairs may influence the strengths of bonds. Lone pairs repel each other; and, if they are on neighboring atoms, that repulsion can weaken the bond. This repulsion between lone pairs helps to explain why the bond in F_2 is weaker than the bond in H_2, because the latter molecule has no lone pairs.

Trends in bond strengths correlate with trends in atomic radii. If the nuclei of the bonded atoms cannot get very close to the electron pair lying between them, the two atoms will be only weakly bonded together. For example, the bond strengths of the hydrogen halides decrease from HF to HI, as shown in Fig. 2.18. The strength of the bond between hydrogen and a Group 14/IV element also decreases down the group (Fig. 2.19). This weakening of the bond correlates with a decrease in the stability of the hydrides down the group. Methane, CH_4, can be kept indefinitely in air at room temperature. Silane, SiH_4, bursts into flame on contact with air. Stannane, SnH_4, decomposes into tin and hydrogen. Plumbane, PbH_4, has never been prepared, except perhaps in trace amounts.

The relative strengths of bonds are important for understanding the way that energy is used in bodies to power our brains and muscles. For instance, adenosine triphosphate, ATP (**35**), is found in every living cell. The triphosphate part of this molecule is a chain of three phosphate groups. One of the phosphate groups is removed in a reaction with water. The P—O bond in ATP requires only 276 $kJ \cdot mol^{-1}$ to break and the new P—O bond formed in $H_2PO_4^-$ releases 350 $kJ \cdot mol^{-1}$ when it forms. As a result, the conversion of ATP to adenosine diphosphate, ADP, in the reaction

35

(where the wiggly line indicates the rest of the molecule) can release energy that is used to power energy-demanding processes in cells.

Closely related to the strength of a bond is its stiffness (its resistance to stretching and compressing), with strong bonds typically being stiffer than weak bonds. The stiffness of bonds is studied by infrared (IR) spectroscopy, as described in Major Technique 1, which follows this chapter, and is used to identify compounds.

The bond strength increases as the multiplicity of a bond increases, decreases as the number of lone pairs on neighboring atoms increases, and decreases as the atomic radii increase. Bonds are strengthened by resonance.

2.16 Bond Lengths

A **bond length** is the distance between the centers of two atoms joined by a covalent bond. It corresponds to the internuclear distance at the potential energy minimum for the two atoms (see Fig. 2.15). Bond lengths affect the overall size and shape of a molecule. The transmission of hereditary information in DNA, for instance, depends on bond lengths because the two strands of the double helix must fit together like pieces of a jigsaw puzzle (Section 19.15). Bond lengths are also crucial to the action of enzymes because only a molecule of the right size and shape will fit into the active site of the enzyme molecule (Section 13.15). As we see from Table 2.5, the lengths of bonds between Period 2 elements typically lie in the range from 100 pm to 150 pm. Bond lengths are determined experimentally by using either spectroscopy or x-ray diffraction (Box 2.2).

Bonds between heavy atoms tend to be longer than those between light atoms because heavier atoms have larger radii than lighter ones (Fig. 2.20). *Multiple bonds are shorter than single bonds* between the same two elements because the additional bonding electrons attract the nuclei more strongly and pull the atoms closer together: compare the lengths of the various carbon–carbon bonds in Table 2.5. We can also see the averaging effect of resonance: the length of the carbon–carbon bond in benzene is intermediate between the lengths of the single and double bonds of a Kekulé structure (but closer to that of a double bond). For bonds between atoms of the same two elements, *the stronger the bond, the shorter it is*. Thus, a C≡C triple bond is both stronger and shorter than a C=C double bond. Similarly, a C=O double bond is both stronger and shorter than a C—O single bond.

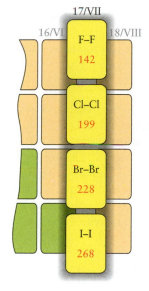

FIGURE 2.20 Bond lengths (in picometers) of the diatomic halogen molecules. Notice how the bond lengths increase down the group as the atomic radii become larger.

TABLE 2.5 Average and Actual Bond Lengths

Bond	Average bond length (pm)	Molecule	Bond length (pm)
C—H	109	H_2	74
C—C	154	N_2	110
C=C	134	O_2	121
C⋯C*	139	F_2	142
C≡C	120	Cl_2	199
C—O	143	Br_2	228
C=O	112	I_2	268
O—H	96		
N—H	101		
N—O	140		
N=O	120		

*In benzene.

BOX 2.2   HOW DO WE KNOW . . . THE LENGTH OF A CHEMICAL BOND?

Because a chemical bond is only about 10^{-10} m long, special techniques have to be used to measure its length. There are two principal techniques: one for solids and the other for gases. The technique used for solids, x-ray diffraction, is described in Major Technique 3, following Chapter 5. *Microwave spectroscopy*, discussed here, is used to determine bond lengths in gas-phase molecules. This branch of spectroscopy makes use of the ability of rotating molecules to absorb microwave radiation, which has a wavelength close to 1 cm.

According to classical physics, a solid body such as a ball can rotate with any energy. According to quantum mechanics, however, rotational energy is quantized, and a body can rotate only with certain energies—that is, only at certain speeds. Let's see what that means for a diatomic molecule AB, with atomic masses m_A and m_B and bond length R. The molecule can rotate only with the following energies:

$$E = \frac{h^2 J(J+1)}{8\pi^2 \mu R^2} \qquad J = 0, 1, 2, \ldots$$

where $\mu = m_A m_B / (m_A + m_B)$, h is Planck's constant, and J is a quantum number. These energy levels are shown in the illustration (right) for two kinds of molecules, one with heavy atoms and a long bond (such as ICl) and the other with light atoms and a short bond (such as HF). We see that the energy levels are much closer together for the heavier molecule than for the lighter molecule. The minimum energy needed to excite a molecule into rotation from rest (corresponding to $J = 0$ and $E = 0$) is

$$\Delta E = \frac{h^2}{4\pi^2 \mu R^2}$$

Perhaps surprisingly, less energy is needed to excite a large heavy molecule than a small light molecule.

Because the energy needed to change the rotational state of a molecule depends on the masses of its atoms and the bond length we can calculate the bond length if we can measure the minimum energy needed to excite the molecule into rotation. Photons corresponding to microwave radiation can supply the energy needed to excite the molecule to higher rotational energies. Therefore, to determine a bond length,

we pass a beam of microwave radiation through a gaseous sample, vary its frequency, and determine the frequency, ν (nu), that results in a strong absorption. The energy of the photons corresponding to this frequency is $h\nu$, and this value can be set equal to the expression for the energy difference given above. We then solve the resulting expression for the value of R, the bond length.

The technique as we have described it works only for polar molecules, because only they can interact with microwave radiation. However, variations of these spectroscopic methods can be used to investigate nonpolar molecules, too. A major limitation of the technique is that only the spectra of simple molecules can be interpreted. For complex molecules, we use solid samples and x-ray diffraction techniques.

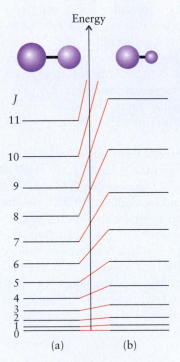

The rotational energy levels of (a) a heavy diatomic molecule and (b) a light diatomic molecule. Note that the energy levels are closer together for the heavy diatomic molecule. Microwaves are absorbed when transitions take place between neighboring energy levels.

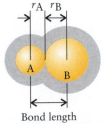

36   Covalent radii

Each atom makes a characteristic contribution, called its **covalent radius**, to the length of a bond (Fig. 2.21). A bond length is approximately the sum of the covalent radii of the two atoms (**36**). The O—H bond length in ethanol, for example, is the sum of the covalent radii of H and O, 37 + 74 pm = 111 pm. We also see from Fig. 2.21 that the covalent radius of an atom taking part in a multiple bond is smaller than that for a single bond of the same atom.

Covalent radii typically decrease from left to right across a period. The reason is the same as for atomic radii (Section 1.15): the increasing effective nuclear

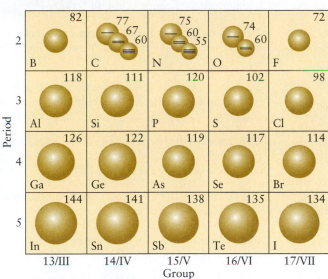

FIGURE 2.21 Covalent radii of hydrogen and the *p*-block elements (in picometers). Where more than one value is given, the values refer to single, double, and triple bonds. Covalent radii tend to become smaller toward fluorine. A bond length is approximately the sum of the covalent radii of the two participating atoms.

charge draws in the electrons and makes the atom more compact. Like atomic radii, covalent radii increase down a group because, in successive periods, the valence electrons occupy shells that are more distant from the nucleus and are better shielded by the inner core of electrons.

The covalent radius of an atom is the contribution it makes to the length of a covalent bond; covalent radii are added together to estimate the lengths of bonds in molecules.

SKILLS YOU SHOULD HAVE MASTERED

❑ 1 Write the electron configuration for an ion (Example 2.1 and Self-Test 2.2).

❑ 2 Compare the relative lattice energies of two ionic compounds (Example 2.2).

❑ 3 Draw the Lewis structures of molecules and ions (Toolbox 2.1 and Examples 2.3, 2.4, and 2.7).

❑ 4 Write the resonance structures for a molecule (Example 2.5).

❑ 5 Use formal charge calculations to evaluate alternative Lewis structures (Toolbox 2.2 and Examples 2.6 and 2.8).

❑ 6 Predict which of two bonds has greater ionic or covalent character (Self-Tests 2.12 and 2.13).

❑ 7 Predict and explain periodic trends in the polarizability of anions and the polarizing power of cations (Section 2.13).

❑ 8 Predict and explain relative bond strengths and lengths (Sections 2.14–2.16).

EXERCISES

Ionic Bonds

2.1 Use data in Appendix 2D to predict which of the following pairs of ions would have the greatest coulombic attraction in a solid compound: (a) K^+, O^{2-}; (b) Ga^{3+}, O^{2-}; (c) Ca^{2+}, O^{2-}.

2.2 Use data from Appendix 2D to predict which of the following pairs of ions would have the greatest coulombic attraction in a solid compound: (a) Mg^{2+}, S^{2-}; (b) Mg^{2+}, Se^{2-}; (c) Mg^{2+}, O^{2-}.

2.3 Explain why the lattice energy of lithium chloride (861 kJ·mol^{-1}) is greater than that of rubidium chloride (695 kJ·mol^{-1}), given that they have similar arrangements of ions in the crystal lattice. See Appendix 2D.

2.4 Explain why the lattice energy of silver bromide (903 kJ·mol^{-1}) is greater than that of silver iodide (887 kJ·mol^{-1}), given that they have similar arrangements of ions in the crystal lattice. See Appendix 2D.

Electron Configurations of Ions

2.5 Give the number of valence electrons for each of the following elements: (a) Sb; (b) Si; (c) Mn; (d) B.

2.6 Give the number of valence electrons for each of the following elements: (a) I; (b) Ni; (c) Re; (d) Sr.

2.7 Give the ground-state electron configuration expected for each of the following ions: (a) S^{2-}; (b) As^{3+}; (c) Ru^{3+}; (d) Ge^{2+}.

2.8 Give the ground-state electron configuration expected for each of the following ions: (a) I^-; (b) Ni^{2+}; (c) Re^{4+}; (d) Sr^{2+}.

2.9 Give the ground-state electron configuration expected for each of the following ions: (a) Cu^+; (b) Bi^{3+}; (c) Ga^{3+}; (d) Tl^{3+}.

2.10 Give the ground-state electron configuration expected for each of the following ions: (a) Al^{3+}; (b) Tc^{4+}; (c) Ra^{2+}; (d) I^-.

2.11 The following species have the same number of electrons: Cd, In^+, and Sn^{2+}. (a) Write the electron configurations for each species. Are they the same or different? Explain. (b) How many unpaired electrons, if any, are present in each species? (c) What neutral atom, if any, has the same electron configuration as that of In^{3+}?

2.12 The following species have the same number of electrons: Ca, Ti^{2+}, and V^{3+}. (a) Write the electron configurations for each ion. Are they the same or different? Explain. (b) How many unpaired electrons, if any, are present in each species? (c) What neutral atom, if any, has the same electron configuration as that of Ti^{3+}?

2.13 Which M^{2+} ions (where M is a metal) are predicted to have the following ground-state electron configurations: (a) $[Ar]3d^7$; (b) $[Ar]3d^6$; (c) $[Kr]4d^4$; (d) $[Kr]4d^3$?

2.14 Which E^{3+} (where E is an element) ions are predicted to have the following ground-state electron configurations: (a) $[Xe]4f^{14}5d^8$; (b) $[Xe]4f^{14}5d^5$; (c) $[Kr]4d^{10}5s^25p^2$; (d) $[Ar]3d^{10}4s^2$?

2.15 Which M^{3+} ions (where M is a metal) are predicted to have the following ground-state electron configurations: (a) $[Ar]3d^6$; (b) $[Ar]3d^5$; (c) $[Kr]4d^5$; (d) $[Kr]4d^3$?

2.16 Which M^{2+} ions (where M is a metal) are predicted to have the following ground-state electron configurations: (a) $[Ar]3d^7$; (b) $[Kr]4d^7$; (c) $[Kr]4d^{10}5s^2$; (d) $[Xe]4f^{14}5d^{10}$?

2.17 For each of the following ground-state atoms, predict the type of orbital (1s, 2p, 3d, 4f, etc.) from which an electron will need to be removed to form the +1 ions: (a) Zn; (b) Cl; (c) Al; (d) Cu.

2.18 For each of the following ground-state ions, predict the type of orbital (1s, 2p, 3d, 4f, etc.) from which an electron will need to be removed to form the ions of one greater positive charge: (a) Ti^{3+}; (b) In^+; (c) Te^{2-}; (d) Ag^+.

2.19 Write the most likely charge for the ions formed by each of the following elements: (a) Br; (b) Te; (c) Cs; (d) Ga; (e) Cd.

2.20 Write the most likely charge for the ions formed by each of the following elements: (a) Sr; (b) Pb; (c) Sc; (d) I; (e) O.

2.21 Predict the number of valence electrons present for each of the following ions: (a) Mn^{4+}; (b) Rh^{3+}; (c) Co^{3+}; (d) P^{3+}.

2.22 Predict the number of valence electrons present for each of the following ions: (a) In^+; (b) Tc^{2+}; (c) Ta^{2+}; (d) Re^+.

2.23 Give the ground-state electron configuration and number of unpaired electrons expected for each of the following ions: (a) Sb^{3+}; (b) Sn^{4+}; (c) W^{2+}; (d) Br^-; (e) Ni^{2+}.

2.24 Give the ground-state electron configuration and number of unpaired electrons expected for each of the following ions: (a) Ga^+; (b) Cu^{2+}; (c) Pb^{2+}; (d) Se^{2-}.

2.25 For each of the following ground-state ions, predict the type of orbital (1s, 2p, 3d, 4f, etc.) that the electrons of highest energy will occupy: (a) Ca^{2+}; (b) In^+; (c) Te^{2-}; (d) Ag^+.

2.26 For each of the following ground-state ions, predict the type of orbital (1s, 2p, 3d, 4f, etc.) that the electrons of highest energy will occupy: (a) Fe^{2+}; (b) Bi^{3+}; (c) Si^{4+}; (d) Br^-.

2.27 Chlorine can exist in both positive and negative oxidation states. What is the maximum (a) positive and (b) negative oxidation number that chlorine can have? (c) Write the electron configuration for each of these states. (d) Explain how you arrived at these values.

2.28 Sulfur can exist in both positive and negative oxidation states. What is the maximum (a) positive and (b) negative oxidation number that sulfur can have? (c) Write the electron configuration for each of these states. (d) Explain how you arrived at these values.

2.29 On the basis of the expected charges on the monatomic ions, give the chemical formula of each of the following compounds: (a) magnesium arsenide; (b) indium(III) sulfide; (c) aluminum hydride; (d) hydrogen telluride; (e) bismuth(III) fluoride.

2.30 On the basis of the expected charges on the monatomic ions, give the chemical formula of each of the following compounds: (a) manganese(II) telluride; (b) barium arsenide; (c) silicon nitride; (d) lithium bismuthide; (e) zirconium(IV) chloride.

2.31 On the basis of the expected charges of the monatomic ions, give the chemical formula of each of the following compounds: (a) bismuth(III) oxide; (b) lead(IV) oxide; (c) thallium(III) oxide.

2.32 On the basis of the expected charges of the monatomic ions, give the chemical formula of each of the following compounds: (a) iron(II) sulfide; (b) cobalt(III) chloride; (c) magnesium phosphide.

Covalent Bonds

2.33 Write the Lewis structure of (a) CCl_4; (b) $COCl_2$; (c) ONF; (d) NF_3.

2.34 Write the Lewis structure of (a) SCl_2; (b) AsH_3; (c) $GeCl_4$; (d) $SnCl_2$.

2.35 Write the Lewis structure of (a) tetrahydridoborate ion, BH_4^-; (b) hypobromite ion, BrO^-; (c) amide ion, NH_2^-.

2.36 Write the Lewis structure of (a) nitronium ion, ONO^+ (b) chlorite ion, ClO_2^-; (c) peroxide ion, O_2^{2-}; (d) formate ion, HCO_2^-.

2.37 Write the complete Lewis structure for each of the following compounds: (a) formaldehyde, HCHO, which as its aqueous solution "formalin" is used to preserve biological specimens; (b) methanol, CH_3OH, the toxic compound also called wood alcohol; (c) glycine, $H_2C(NH_2)COOH$, the simplest of the amino acids, the building blocks of proteins.

2.38 Write the Lewis structure of each of the following organic compounds: (a) ethanol, CH_3CH_2OH, which is also called ethyl alcohol or grain alcohol; (b) methylamine, CH_3NH_2, a putrid-smelling substance formed when flesh decays; (c) formic acid, HCOOH, a component of the venom injected by ants.

2.39 The following Lewis structure was drawn for a Period 3 element. Identify the element.

:O:
||
:Cl—E—Cl:
|
:Cl:

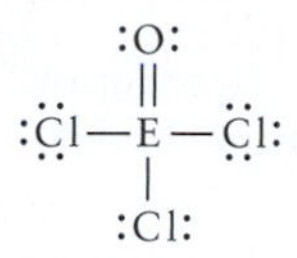

2.40 The following Lewis structure was drawn for a Period 3 element. Identify the element.

:O:
||
:Cl—E=O:
|
:Cl:

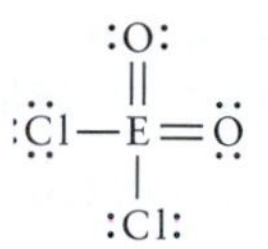

2.41 Write the complete Lewis structure for each of the following compounds: (a) ammonium chloride; (b) potassium phosphide; (c) sodium hypochlorite.

2.42 Write the complete Lewis structure for each of the following compounds: (a) zinc cyanide; (b) potassium tetrafluoroborate; (c) barium peroxide (the peroxide ion is O_2^{2-}).

2.43 Anthracene has the formula $C_{14}H_{10}$. It is similar to benzene but has 3 six-membered rings that share common C—C bonds, as shown below. Complete the structure by drawing in multiple bonds to satisfy the octet rule at each carbon atom. Resonance structures are possible. Draw as many as you can find.

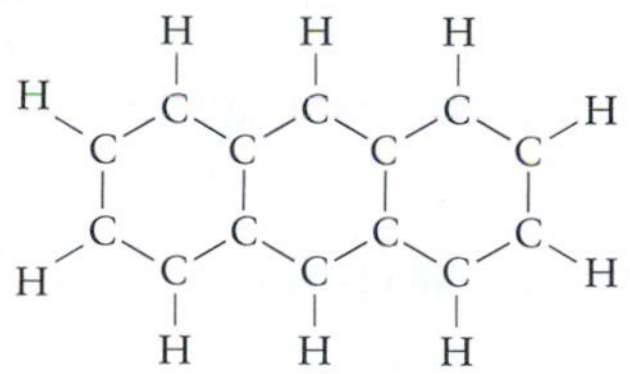

2.44 Write the Lewis structures that contribute to the resonance hybrid of the guanadinium ion, $C(NH_2)_3^+$.

2.45 Draw the Lewis structures that contribute to the resonance hybrid of nitryl chloride, $ClNO_2$ (N is the central atom).

2.46 Do H—$C≡N$ and H—$N≡C$ form a pair of resonance structures? Explain your answer.

2.47 Draw the Lewis structure and determine the formal charge on each atom in (a) NO^+; (b) N_2; (c) CO; (d) C_2^{2-}; (e) CN^-.

2.48 Using only Lewis structures that obey the octet rule, draw the Lewis structures and determine the formal charge on each atom in (a) CH_3^+; (b) OCl^-; (c) BF_4^-.

2.49 Determine the formal charge on each atom in the following molecules. Identify the structure of lower energy in each pair.

(a) 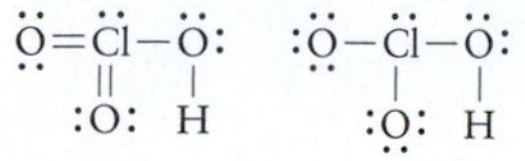   O=Cl—O: :O—Cl—O:
      || | | |
      :O: H   :O: H

(b)    O=C=S   :O—C≡S:

(c)   H—C≡N   H—C=N

2.50 Determine the formal charge on each atom in the following ions. Identify the structure of lowest energy in each case.

(a) 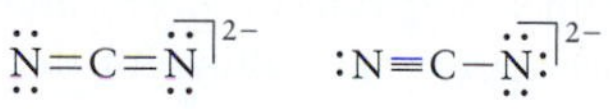   N=C=N²⁻   :N≡C—N:²⁻

(b)   :O—As—O:³⁻   :O—As—O:³⁻   O=As=O³⁻
      | || || ||
      :O: :O: :O:

(c)   O=I=O   O=I=O   :O—I—O:
      || | |
      :O: :O: :O: [each with ⁻ charge]

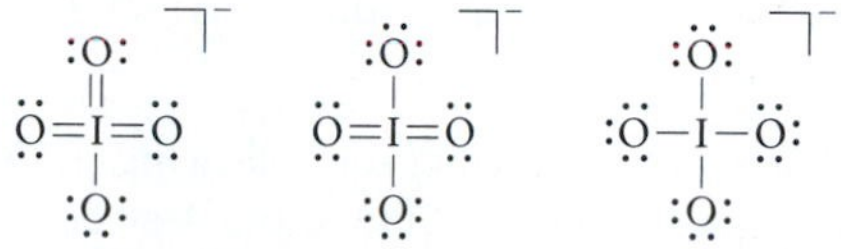

2.51 Two contributions to the resonance structure are shown below for each species. Determine the formal charge on each atom and then, if possible, identify the Lewis structure of lower energy for each species.

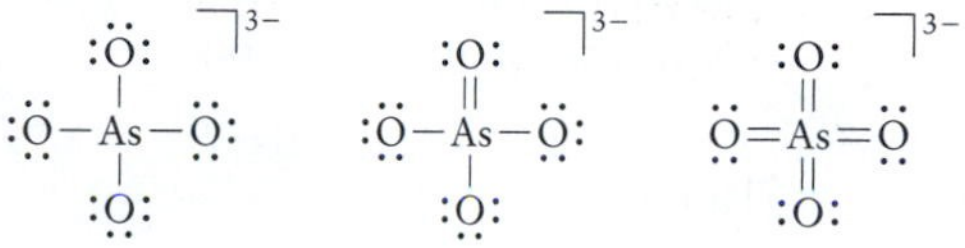

(a) 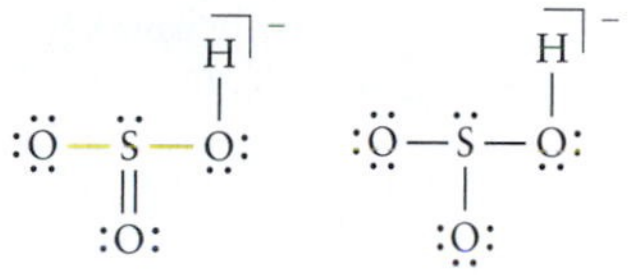   :O—S—O:⁻   :O—S—O:⁻
          | H | H
          || |
          :O: :O: (with H above S)

(b) 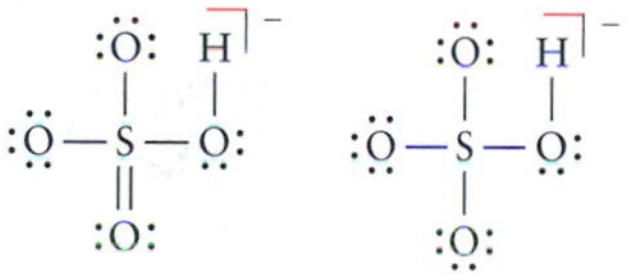   :O—S—O:⁻   :O—S—O:⁻
       | | | |
       | H || H
      :O: :O:

2.52 Two Lewis structures are shown below for each species. Determine the formal charge on each atom and then, if appropriate, identify the Lewis structure of lower energy for each species.

(a)    :O—S=O   O=S=O

(b) 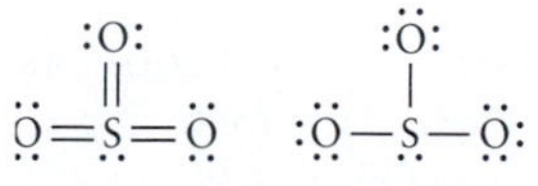   O=S=O   :O—S—O:
          || |
          :O: :O:

2.53 Select from each of the following pairs of Lewis structures the one that is likely to make the dominant contribution to a resonance hybrid. Explain your selection.

(a) 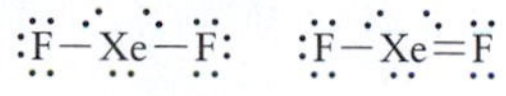   :F—Xe—F: :F—Xe=F

(b) 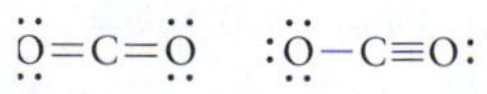   O=C=O   :O—C≡O:

2.54 Select from each of the following pairs of Lewis structures the one that is likely to make the dominant contribution to a resonance hybrid. Explain your selection.

(a) $\ddot{\ddot{N}}=N=\ddot{\ddot{O}}$ $:N\equiv N-\ddot{\ddot{O}}:$

(b) $\ddot{\ddot{O}}=P-\ddot{\ddot{O}}:$ $\left[\begin{array}{c}:\ddot{O}:\\ |\\ \ddot{\ddot{O}}=P-\ddot{\ddot{O}}:\\ |\\ :\ddot{O}:\end{array}\right]^{3-}$ $\ddot{\ddot{O}}=P=\ddot{\ddot{O}}$ $\left[\begin{array}{c}:\ddot{O}:\\ |\\ \ddot{\ddot{O}}=P=\ddot{\ddot{O}}\\ |\\ :\ddot{O}:\end{array}\right]^{3-}$

Exceptions to the Octet Rule

2.55 Write the Lewis structure, including typical contributions to the resonance structure (where appropriate, allow for the possibility of octet expansion, including double bonds in different positions), for (a) sulfite ion; (b) hydrogen sulfite ion; (c) perchlorate ion; (d) nitrite ion.

2.56 Write the Lewis structure, including typical contributions to the resonance structure (where appropriate, allow for the possibility of octet expansion), for (a) dihydrogen phosphate ion; (b) chlorite ion; (c) chlorate ion; (d) nitrate ion.

2.57 Which of the following species are radicals? (a) NO_2^-; (b) CH_3; (c) OH; (d) CH_2O.

2.58 Which of the following species are radicals? (a) ClO_2; (b) Cl_2O; (c) BF_4^-; (d) BrO.

2.59 Write the Lewis structure of each of the following reactive species, all of which are found to contribute to the destruction of the ozone layer, and indicate which are radicals: (a) chlorine monoxide, ClO; (b) dichloroperoxide, Cl—O—O—Cl; (c) chlorine nitrate, $ClONO_2$ (the central O atom is attached to the Cl atom and to the N atom of the NO_2 group).

2.60 Write the Lewis structure of each of the following species and indicate which are radicals: (a) the superoxide ion, O_2^-; (b) the methoxy group, CH_3O; (c) XeO_4; (d) $HXeO_4^-$.

2.61 Determine the numbers of electron pairs (both bonding and lone pairs) on the iodine atom in (a) ICl_2^+; (b) ICl_4^-; (c) ICl_3; (d) ICl_5.

2.62 Determine the numbers of electron pairs (both bonding and lone pairs) on the phosphorus atom in (a) PCl_3; (b) PCl_5; (c) PCl_4^+; (d) PCl_6^-.

2.63 Write the Lewis structure for each of the following molecules or ions and give the number of electrons about the central atom: (a) SF_6; (b) XeF_2; (c) AsF_6^-; (d) $TeCl_4$.

2.64 Write the Lewis structure for each of the following molecules or ions and give the number of electrons about the central atom: (a) SiF_6^{2-}; (b) IF_7; (c) ClF_3; (d) BrF_2^+.

2.65 Write the Lewis structure and state the number of lone pairs on xenon, the central atom of each of the following compounds: (a) $XeOF_2$; (b) XeF_4; (c) $XeOF_4$.

2.66 Write the Lewis structure and state the number of lone pairs on the central atom of each of the following compounds: (a) ClF_3; (b) AsF_5; (c) SF_4.

Ionic Versus Covalent Bonds

2.67 List the halogens in order of increasing electronegativity.

2.68 Just as some elements are electronegative, others can be *electropositive*, a term meaning that the element will readily give up electrons. Elements that are least electronegative are the most electropositive. What trend exists among the alkali metals

and the alkaline earth metals with respect to their electropositive character?

2.69 Place the following elements in order of increasing electronegativity: antimony, tin, selenium, indium.

2.70 Place the following elements in order of increasing electronegativity: oxygen, carbon, nitrogen, fluorine, silicon, phosphorus, sulfur.

2.71 For each pair, determine which compound has bonds with greater ionic character: (a) HCl or HI; (b) CH_4 or CF_4; (c) CO_2 or CS_2.

2.72 For each pair, determine which compound has bonds with greater ionic character: (a) PH_3 or NH_3; (b) SO_2 or NO_2; (c) SF_6 or IF_5.

2.73 Arrange the cations Rb^+, Be^{2+}, and Sr^{2+} in order of increasing polarizing power. Give an explanation of your arrangement.

2.74 Arrange the cations K^+, Mg^{2+}, Al^{3+}, and Cs^+ in order of increasing polarizing power. Give an explanation of your arrangement.

2.75 Arrange the anions Cl^-, Br^-, N^{3-}, and O^{2-} in order of increasing polarizability and give reasons for your decisions.

2.76 Arrange the anions N^{3-}, P^{3-}, I^-, and At^- in order of increasing polarizability and give reasons for your decisions.

The Strengths and Lengths of Covalent Bonds

2.77 Place the following molecules or ions in order of *decreasing* bond length: (a) the CO bond in CO, CO_2, CO_3^{2-}; (b) the SO bond in SO_2, SO_3, SO_3^{2-}; (c) the CN bond in HCN, CH_2NH, CH_3NH_2. Explain your reasoning.

2.78 Place the following molecules or ions in order of *decreasing* bond order: (a) NO bond in NO, NO_2, NO_3^-; (b) CC bond in C_2H_2, C_2H_4, C_2H_6; (c) CO bond in CH_3OH, CH_2O, CH_3OCH_3. Explain your reasoning.

2.79 Use the information in Fig. 2.21 to estimate the bond length of (a) the NN bond in hydrazine, H_2NNH_2; (b) the CO bond in CO_2; (c) the CO and CN bonds in urea, $OC(NH_2)_2$; (d) the NN bond in nitrogen hydride, HNNH.

2.80 Use the information in Fig. 2.21 to estimate the indicated bond length of (a) the CO bond in formaldehyde, H_2CO; (b) the CO bond in dimethyl ether, CH_3OCH_3; (c) the CO bond in methanol, CH_3OH; (d) the CS bond in methanethiol, CH_3SH.

2.81 Use the covalent radii in Fig. 2.21 to calculate the bond lengths in the following molecules. Account for the trends in your calculated values. (a) CF_4; (b) SiF_4; (c) SnF_4.

2.82 Which do you predict to have the strongest CN bond: (a) $NHCH_2$, (b) NH_2CH_3, or (c) HCN? Explain.

Integrated Exercises

2.83 In 1999, Karl Christe synthesized and characterized a salt that contained the N_5^+ cation, in which the five N atoms are connected in a long chain. This cation is the first all-nitrogen species to be isolated in more than 100 years. Draw the most important Lewis structure for this ion, including all equivalent resonance structures. Calculate the formal charges on all atoms.

2.84 Write three Lewis structures that follow the octet rule (including the most important structure) for the isocyanate ion, CNO^-. State which of the three Lewis structures is the most important and explain why it is probably the most important.

2.85 Write the Lewis structure, including resonance structures where appropriate, for (a) the oxalate ion, $C_2O_4^{2-}$ (there is a C—C bond with two oxygen atoms attached to each carbon atom); (b) BrO^+; (c) the acetylide ion, C_2^{2-}. Assign formal charges to each atom.

2.86 Draw resonance structures for the trimethylenemethane anion $C(CH_2)_3^{2-}$ in which a central carbon atom is attached to three CH_2 groups (CH_2 groups are referred to as methylene).

2.87 Show how resonance can occur in the following organic ions: (a) acetate ion, $CH_3CO_2^-$; (b) enolate ion, $CH_2COCH_3^-$, which has one resonance structure with a C=C double bond and an —O^- group on the central carbon atom; (c) allyl cation, $CH_2CHCH_2^+$; (d) amidate ion, CH_3CONH^- (the O and the N atoms are both bonded to the second C atom).

2.88 White phosphorus is composed of tetrahedral molecules of P_4 in which each P atom is connected to three other P atoms. Draw the Lewis structure for this molecule. Does it obey the octet rule?

2.89 Nitrogen, oxygen, phosphorus, and sulfur exist as N_2, O_2, tetrahedral P_4, and cyclic S_8 molecules. Explain this fact in terms of the abilities of the atoms to form different types of bonds with one another.

2.90 Draw the most important Lewis structure for each of the following ring molecules (which have been drawn without showing the locations of the double bonds). Show all lone pairs and nonzero formal charges. If there are equivalent resonance forms, draw them.

| (a) | (b) | (c) | (d) |

2.91 An important principle in chemistry is the *isolobal analogy*. This very simple principle states that chemical fragments with similar valence orbital structures can replace one another in molecules. For example, ·Ċ–H and ·Ṡi–H are isolobal fragments, each having three electrons with which to form bonds in addition to the bond to H. An isolobal series of molecules would be HCCH, HCSiH, HSiSiH. Similarly, a lone pair of electrons can be used to replace a bond so that ·N̈: is isolobal with ·Ċ–H with the lone pair taking the place of the C—H bond. The isolobal set here is HCCH, HCN, NN. (a) Draw the Lewis structures for the molecules HCCH, HCSiH, HSiSiH, HCN, and NN. (b) Using the isolobal principle, draw Lewis structures for molecules based on the structure of benzene, C_6H_6, in which one or more CH groups are replaced with N atoms.

2.92 The cyclopentadienide ion, $C_5H_5^-$, is a common organic anion that forms very stable complexes with metal cations. The anion is derived by removing a proton from cyclopentadiene, C_5H_6, with strong base. The molecule has a five-membered ring of carbon atoms, with four carbon atoms attached to only one proton and one carbon atom bonded to two. Draw the Lewis

structure of cyclopentadiene. Are resonance forms possible for this molecule? The ion $C_5H_5^-$ consists of a planar five-membered ring of carbon atoms, with one hydrogen atom attached to each carbon atom. Draw the Lewis structure for $C_5H_5^-$. How many resonance structures can you draw for this anion?

2.93 Hydrogen peroxide is a powerful oxidizing agent that can damage lung tissue. However, your body is capable of handling small doses of strong oxidizers, because the surface of your lungs is covered with epithelial lining fluid. This thin layer of fluid has several antioxidants dissolved in it, including vitamin C (ascorbic acid, $C_6H_8O_6$). Antioxidants react with oxidizing agents like hydrogen peroxide to form nontoxic products, as shown in the reaction below.

$$C_6H_8O_6 + H_2O_2 \longrightarrow C_6H_6O_6 + 2\,H_2O$$

(a) Verify, using oxidation numbers, whether, in this reaction, H_2O_2 is being reduced or oxidized. Is ascorbic acid being oxidized or reduced? (b) Calculate the formal charges of the atoms in H_2O_2 and H_2O. Which are more useful for determining whether the compound has been oxidized or reduced, oxidation numbers or formal charges? Justify your answer.

2.94 (a) Confirm that lattice energies are inversely proportional to the distance between the ions in MX (M = alkali metal, X = halide ion) by plotting the lattice energies of KF, KCl, and KI against the internuclear distances d_{M-X}. The lattice energies of KF, KCl, and KI are 826, 717, and 645 $kJ \cdot mol^{-1}$, respectively. Use the ionic radii found in Appendix 2D to calculate d_{M-X}. How good is the correlation? You should use a standard graphing program to make the plot that will generate an equation for the line and calculate a correlation coefficient for the fit (see the Web site for this book). (b) Estimate the lattice energy of KBr from your graph. (c) Find an experimental value for the lattice energy of KBr in the chemical literature and compare that value with the value that you calculated in Part (b). How well do they agree?

2.95 (a) Explore whether the lattice energies of the alkali metal iodides are inversely proportional to the distances between the ions in MI (M = alkali metal) by plotting the lattice energies given below against the internuclear distances d_{M-I}.

Alkali metal iodide	Lattice energy ($kJ \cdot mol^{-1}$)
LiI	759
NaI	700
KI	645
RbI	632
CsI	601

Use the ionic radii found in Appendix 2D to calculate d_{M-X}. How good is the correlation? Is a better fit obtained by plotting the lattice energies against $(1 - d^*/d)/d$, as suggested by the Born–Meyer equation? You should use a standard graphing program to make the plot that will generate an equation for the line and calculate a correlation coefficient for the fit (see the Web site for this book). (b) From the ionic radii given in Appendix 2D and the plot given in part (a), estimate the lattice energy for silver iodide. (c) Compare your results from part (b) with the experimentally determined value of 886 $kJ \cdot mol^{-1}$. If they do not agree, provide an explanation for the deviation.

2.96 The framework for the tropylium cation, $C_7H_7^+$, is a seven-membered ring of carbon atoms with a hydrogen atom attached to each carbon atom. Complete the structural drawing by adding the multiple bonds as appropriate. Resonance structures are possible. Draw as many as you can find. Determine the C—C bond order.

2.97 Quinone, $C_6H_4O_2$, is an organic molecule with the structure shown below; it can be reduced to the anion $C_6H_4O_2^{2-}$. (a) Draw the Lewis structure of the reduced product. (b) On the basis of formal charges derived from the Lewis structure, predict which atoms in the molecule are most negatively charged. (c) If two protons are added to the reduced product, where are they most likely to bond?

2.98 In air, the NO radical can react with O_2. What is the most likely product of the reaction? Answer this question by drawing Lewis structures of the reactants and products.

2.99 The atomic numbers (Z), electronic configurations, and numbers of unpaired electrons for five ions are listed in the following table. Assume that all unpaired electrons have parallel spins. Indicate the element symbol, charge, and energy state (that is, ground state or excited state) for each of the five cases.

Z	Configuration	No. of unpaired electrons	Element	Charge	Energy state
26	$[Ar]3d^6$	4			
52	$[Kr]4d^{10}5s^2$ $5p^56s^1$	2			
16	$[Ne]3s^23p^6$	0			
39	$[Kr]4d^1$	1			
30	$[Ar]4s^23d^8$	2			

2.100 The atomic numbers (Z), electronic configurations, and numbers of unpaired electrons for five ions are listed in the following table. Assume that all unpaired electrons have parallel spins. Indicate the element symbol, charge, and energy state (that is ground state, excited state, etc.) for each of the five cases.

Z	Configuration	No. of unpaired electrons	Element	Charge	Energy state
38	$[Kr]5p^1$	1			
45	$[Kr]4d^7$	3			
43	$[Kr]4d^55s^1$	6			
8	$[Ne]$	0			
21	$[Ar]3d^14s^1$	2			

2.101 Draw the most important Lewis structure for each of the following molecules. Show all lone pairs and formal charges. Draw all equivalent resonance forms. (a) HONCO; (b) H_2CSO; (c) H_2CNN; (d) ONCN.

2.102 Determine the formal charges for the atoms in (a) CN^-; (b) CNO^-; (c) N_3^-.

2.103 A common biologically active radical is the pentadienyl radical, RCHCHCHCHCHR′, where the carbons form a long chain, with R and R′, which can be a number of different organic groups, at each end. Draw three resonance structures for this compound that maintain carbon's valence of four.

2.104 Sketch the qualitative molecular potential energy curves for the N—N bond on one graph for N_2H_4, N_2, and N_3^-.

2.105 Thallium and oxygen form two compounds with the following characteristics:

	Compound I	Compound II
Mass percentage Tl	89.49%	96.23%
Melting point	717°C	300°C

(a) Determine the chemical formulas of the two compounds. (b) Determine the oxidation number of thallium in each compound. (c) Assume that the compounds are ionic and write the electron configuration for each thallium ion. (d) Use the melting points to decide which compound has more covalent character in its bonds. Is your finding consistent with what you would predict from the polarizing abilities of the two cations?

2.106 How close are the Mulliken and Pauling electronegativity scales? (a) Use Eq. 6 to calculate the Mulliken electronegativities of C, N, O, and F. Use the values in kJ·mol^{-1} from Figs. 1.50 and 1.54 and divide each value by 230 kJ·mol^{-1} for this comparison. (b) Plot both sets of electronegativities as a function of atomic number on the same graph. (c) Which scale is more periodic (depends more consistently on position in the periodic table)?

2.107 The perchlorate ion, ClO_4^-, is described by resonance structures. (a) Draw the Lewis structures that contribute to the resonance hybrid and identify the most plausible Lewis structures by using formal charge arguments. (b) The average length of a single Cl—O bond is 172 pm and that of a double Cl=O bond can be estimated at 140 pm. The Cl—O bond length in the perchlorate ion is found experimentally to be 144 pm for all four bonds. Identify the most plausible Lewis structures of the perchlorate ion from these experimental data. (c) What is the oxidation number of chlorine in the perchlorate ion? Identify the most plausible Lewis structure by using the oxidation number, assuming that lone pairs belong to the atom to which they are attached but that all electrons shared in a bond belong to the atom of the more negative element. (d) Are these three approaches consistent? Explain why or why not.

2.108 Predict which bond will absorb light of shorter wavelength and explain why: C—H or C—Cl. Refer to Major Technique 1, which follows these exercises.

2.109 Infrared spectra show absorption due to C—H bond stretching at 3.38 μm for a methyl (—CH$_3$) group and at 3.1 μm for an alkyne (—C≡C—H) group. Which C—H bond is stiffer (has the larger force constant k), assuming that the vibrating atoms have the same effective mass? Refer to Major Technique 1, which follows these exercises.

2.110 Vibrational spectra are often so complicated that assignment of a particular absorption to a given bond is difficult. One way to confirm that an assignment is correct is to carry out selective isotopic substitution. For example, we can replace a hydrogen atom with a deuterium atom. If an iron-hydride (Fe—H) stretch occurs at 1950 cm^{-1}, at what energy will this stretch occur, approximately, for a compound that has deuterium in place of the hydrogen? Refer to Major Technique 1, which follows these exercises.

2.111 Certain gases, called greenhouse gases, contribute to global warming by absorbing infrared radiation. Only molecules with dipole moments or nonpolar molecules that undergo distortions that create momentary dipoles (such as CO_2, see Figure 2 in Major Technique 1, which follows these exercises) can absorb infrared radiation. Which of the following gases, all of which occur naturally in air, can function as greenhouse gases? (a) CO; (b) O_2; (b) O_3; (d) SO_2; (e) N_2O; (f) Ar.

2.112 One of the following compounds does not exist. Use Lewis structures to identify that compound. (a) C_2H_2; (b) C_2H_4; (c) C_2H_6; (d) C_2H_8.

2.113 Interhalogen compounds are compounds of two different halogens that have, with few exceptions, the general formula XX'_n. Examples include BrCl, ClF_3 and IF_5. Use Lewis structures to explain why n is always an odd number.

2.114 In the solid state, sulfur is sometimes found in rings of six atoms. (a) Draw a valid Lewis structure for S_6. (b) Is resonance possible in S_6? If so, draw one of the resonance structures.

2.115 Structural isomers are molecules that have the same formula but in which the atoms are connected in a different order. Two isomers of disulfur difluoride, S_2F_2, are known. In each the two S atoms are bonded to each other. In one isomer each of the S atoms is bonded to an F atom. In the other isomer, both F atoms are attached to one of the S atoms. (a) In each isomer the S—S bond length is approximately 190 pm. Are the S—S bonds in these isomers single bonds or do they have some double bond character? (b) Draw two resonance structures for each isomer. (c) Determine for each isomer which structure is favored by formal charge considerations. Are your conclusions consistent with the S—S bond lengths in the compounds?

2.116 Ionic compounds typically have higher boiling points and lower vapor pressures than covalent compounds. Predict which compound in the following pairs has the lower vapor pressure at room temperature: (a) Cl_2O or Na_2O; (b) $InCl_3$ or $SbCl_3$; (c) LiH or HCl; (d) $MgCl_2$ or PCl_3.

Chemistry Connections

2.117 The nitrogen oxides are common pollutants generated by internal combustion engines and power plants. They not only contribute to the respiratory distress caused by smog, but if they reach the stratosphere can also threaten the ozone layer that protects Earth from harmful radiation.

(a) The bond energy in NO is 632 kJ·mol^{-1} and that of each N—O bond in NO_2 is 469 kJ·mol^{-1}. Using Lewis structures and the average bond energies in Table 2.4, explain the difference in bond energies between the two molecules and the fact that the bond energies of the two bonds in NO_2 are the same.

(b) The bond length in NO is 115 pm. Use Fig. 2.21 to predict the length of a single bond and a double bond between nitrogen and oxygen. Use Table 2.5 to estimate the length of a triple bond between nitrogen and oxygen. Predict the bond order in NO from its bond length and explain any difference from the calculated values.

(c) When the NO in smog reacts with NO_2 a bond forms between the two N atoms. Draw the Lewis structure of each reactant and the product and indicate the formal charge on each atom.

(d) The NO_2 in smog also reacts with NO_3 to form a product with an O atom between the two N atoms. Draw the Lewis structure of the most likely product and indicate the formal charge on each atom.

(e) Write the balanced chemical equation for the reaction of the product from part (d) with water to produce an acid. The acid produced acts as a secondary pollutant in the environment. Name the acid.

(f) If 4.05 g of the product from part (d) reacts with water as in part (e) to produce 1.00 L of acidic solution, what will be the concentration of the acid?

(g) Determine the oxidation number of nitrogen in NO, NO_2, and the products in parts (c) and (d). Which of these compounds would you expect to be the most potent oxidizing agent?

When we feel the warmth of the Sun on our faces, we are responding to infrared radiation that has traveled nearly 150 million kilometers through space. That radiation stimulates the molecules in our skin to vibrate, and specially adapted nerve cells detect the vibration and interpret it as "warmth." The same type of radiation is used to identify molecules in samples and to determine the stiffness of bonds in molecules.

The Technique

Infrared radiation is electromagnetic radiation lying at longer wavelengths (lower frequencies) than red light; a typical wavelength is about 1000 nm. A wavelength of 1000 nm corresponds to a frequency of about 3×10^{14} Hz, which is comparable to the frequency at which molecules vibrate. Therefore, molecules can absorb infrared radiation and become vibrationally excited.

Any bond between two atoms vibrates as the atoms move closer to each other and away again. This type of motion is called a "stretching" mode. Polyatomic molecules can also undergo "bending" vibrations in which bond angles periodically increase and decrease. The frequency at which a molecule vibrates depends on the masses of its atoms and the stiffness of its bonds: a molecule made up of light atoms joined by stiff bonds has a higher vibrational frequency than one made up of heavy atoms joined by loose bonds. The former molecule will therefore absorb higher frequency radiation than the latter. Bending motions of molecules tend to be less stiff than stretching motions; so bending vibrations typically absorb at lower frequencies than do stretching vibrations.

The stiffness of a bond is measured by its *force constant, k*. This constant is the same as that in Hooke's law for the restoring force of a spring: Hooke observed that the restoring force is proportional to the displacement of the spring from its resting position, and wrote

Force = $-k \times$ displacement

A stiff bond (like a stiff spring) experiences a strong restoring force, even for quite small displacements, and so in this case k is large. A loose bond (like a weak spring) experiences only a weak restoring force, even for quite large displacements, so its associated k is small. In general, the force constant is larger for stretching displacements of molecules than for bending motions. The stiffness of a bond should not be confused with its strength, which is the energy required to break the bond. Typically, though, the stiffness of a bond increases with the strength of the bond (Fig. 1).

The vibrational frequency, ν (nu), of a bond between two atoms A and B of masses m_A and m_B is given by the expression

$$\nu = \frac{1}{2\pi}\sqrt{\frac{k}{\mu}} \qquad \mu = \frac{m_A m_B}{m_A + m_B}$$

The quantity μ (mu) is called the *effective mass* of the molecular vibration (some people call it the "reduced mass"). As we anticipated, the frequency is higher for stiff bonds (large k) and low atomic masses (low μ). We see that, by measuring

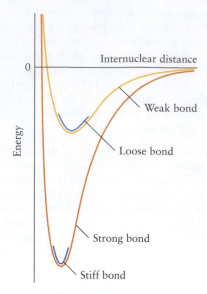

FIGURE 1 The strength of a bond is a measure of the depth of the well in the potential energy curve; the stiffness—which governs the vibrational frequency—is determined by the steepness with which the potential energy rises as the bond is stretched or compressed.

the vibrational frequency of a molecule, which involves measuring the frequency (or wavelength) at which it absorbs infrared radiation, we can measure the stiffness of its bond.

In practice, the vibrational absorption spectrum of a molecule is measured by using an infrared spectrometer. The source of infrared radiation is a hot filament, and the wavelength is selected by diffracting the radiation from a grating. Constructive interference* results in intense radiation being obtained in a given direction for a given angle of the grating; and, as that angle is changed, radiation of varying wavelength is passed through the sample. The beam is divided, one beam passing through the sample and the other beam passing through a blank; the intensities of the two beams are then compared at the detector and the reduction in intensity is monitored. The result is a spectrum in which dips occur at the wavelengths of radiation absorbed by the sample.

Normal Modes and Polyatomic Molecules

A nonlinear molecule consisting of N atoms can vibrate in $3N - 6$ different ways, and a linear molecule can vibrate in $3N - 5$ different ways. The number of ways in which a molecule can vibrate increases rapidly with the number of atoms: a water molecule, with $N = 3$, can vibrate in 3 ways, but a benzene molecule, with $N = 12$, can vibrate in 30 different ways. Some of the vibrations of benzene correspond to expansion and contraction of the ring, others to its elongation, and still others to flexing and bending. Each way in which a molecule can vibrate is called a *normal mode*, and so we say that benzene has 30 normal modes of vibration. Each normal mode has a frequency that depends in a complicated way on the masses of the atoms that move during the vibration and the force constants associated with the motions involved (Fig. 2).

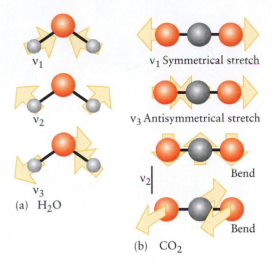

ν_1

ν_2

ν_3

(a) H_2O

ν_1 Symmetrical stretch

ν_3 Antisymmetrical stretch

ν_2| Bend

Bend

(b) CO_2

FIGURE 2 (a) The three normal vibrational modes of H_2O. Two of these modes are principally stretching motions of the bonds, but mode ν_2 is primarily bending. (b) The four normal vibrational modes of CO_2. The first two are symmetrical and antisymmetrical stretching motions, and the last two are perpendicular bending motions.

Except in simple cases, it is very difficult to predict the infrared absorption spectrum of a polyatomic molecule, because each of the modes has its characteristic absorption frequency rather than just the single frequency of a diatomic molecule. However, certain groups, such as a benzene ring or a carbonyl group, have characteristic frequencies, and their presence can often be detected in a spectrum. Thus, an infrared spectrum can be used to identify the species present in a sample by looking for the characteristic absorption bands associated with various groups. An example and its analysis is shown in Fig. 3.

A further aid to identification comes from the fact that a molecule commonly has complex series of absorptions spanning a range of wavelengths. This *fingerprint region* of the spectrum may be too difficult to analyze in detail, but its presence enables us to recognize the substance by comparing the spectrum to an atlas of spectra.

Related Exercises 2.108–2.111.

Interference is discussed more fully in Major Technique 3 following Chapter 5.

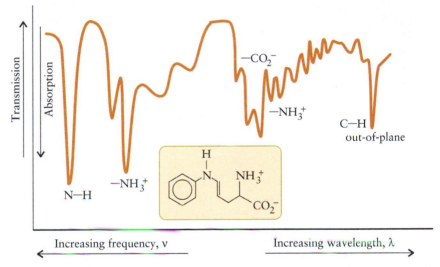

Transmission

Absorption

$-CO_2^-$

$-NH_3^+$

C—H out-of-plane

$-NH_3^+$

N—H

Increasing frequency, ν

Increasing wavelength, λ

FIGURE 3 The infrared spectrum of an amino acid, with the groups contributing to some of the peaks identified. Notice that the spectrum displays the intensity of absorption.

MOLECULAR SHAPE AND STRUCTURE

What Are the Key Ideas?   The central ideas of this chapter are, first, that electrostatic repulsions between electron pairs determine molecular shapes and, second, that chemical bonds can be discussed in terms of two quantum mechanical theories that describe the distribution of electrons in molecules.

Why Do We Need to Know This Material?   The shapes of molecules determine their odors, their tastes, and their actions as drugs. Molecular shape governs the reactions that take place throughout our bodies and are necessary for life. It also affects the properties of the materials around us, including their physical states and their solubilities. Perception, thinking, and learning depend on the shapes of molecules and how they change. Modern theories of the electronic structures of molecules can be extended to describe metals and semiconductors, and so the information in this chapter is also at the root of the development of new technologies.

What Do We Need to Know Already?   This chapter uses atomic orbitals and electron configurations (Chapter 1). It also extends the concept of Lewis structures introduced in Chapter 2. The discussion of polar molecules develops the material on polar bonds described in Section 2.12.

Chapter 3

The bright colors of flowers and the varied hues of autumn leaves have always been a cause for delight, but it was not until the twentieth century that chemists understood how these colors arise from the presence of organic compounds with common structural features. They discovered how small differences in the structures of the molecules of these compounds can enhance photosynthesis, produce important vitamins, and attract pollinating bees. They now know how the shapes of molecules and the orbitals occupied by their electrons explain the properties of these compounds and even the processes taking place in our eyes that allow us to see them.

The impact of modern theories of bonding, however, goes far beyond understanding the colors around us. Increasing knowledge of the electronic structures of the atoms and molecules in polymers and semiconductors has led to the development of new technologies. The pharmaceutical industry is increasingly turning to computer-aided design—in which molecular shape and the distribution of electrons play a crucial role—to discover new, more potent drugs (Box 3.1).

In this chapter we meet three increasingly sophisticated models of molecular shape. The first considers molecular shape to be a consequence merely of the electrostatic (coulombic) interaction between pairs of electrons. The other two models are theories that describe the distribution of electrons and molecular shape in terms of the occupation of orbitals.

THE VSEPR MODEL

In this section, we construct a model of molecular shape **empirically,** which means that we base it on rules suggested by experimental observations rather than on more fundamental principles. We proceed in three steps. First, we set up the basic model for simple molecules without lone pairs on the central atom. Then, we include the effects of lone pairs. Finally, we explore some of the consequences of molecular shape.

BOX 3.1   FRONTIERS OF CHEMISTRY: DRUGS BY DESIGN AND DISCOVERY

The search for new drugs relies not only on the skills of synthetic organic chemists but also on biologists, ethnobotanists, and medical researchers. Because there are so many millions of compounds, it would take too long to start with the elements, combine them in different ways, and then test them. Instead, chemists usually start either by *drug discovery*, the identification and possibly the modification of promising medicines that already exist, or by *rational drug design*, the identification of characteristics of a target enzyme, virus, bacterium, or parasite and the design of new compounds to react with it.

In drug discovery, a chemist usually begins by investigating compounds that have already shown medicinal value. A fruitful path is to find a *natural product*, an organic compound found in nature, that has been shown to have healing characteristics. Nature is the best of all synthetic chemists, with billions of chemicals that fulfill as many different needs. The challenge is to find compounds that have curative powers. These substances are found in different ways: random or "blind" collection of samples that are then tested, or collection of specific samples identified by native healers as medically effective.

Observation of the properties of plants and animals can help to guide a random search. For example, if certain types of fruits remain fresh while others rot, we might expect the former to contain antifungal agents. An example of this type of collection is the gathering of tunicates and sponges in the Caribbean. The chemists harvest the samples by diving from research vessels (see Fig. F.1). The samples are tested for antiviral and antitumor activity in a chemical laboratory on the ship. The antiviral drug didemnin-C and the anticancer drug bryostatin 1 were discovered in marine organisms.

The guided route requires fewer samples for testing because the chemist works with a native healer, the ancient lore guiding the modern chemistry. Often an ethnobotanist, a specialist in plants used for native healing, joins the team. This approach saves time for the scientists and can provide an economic benefit for the healers and their nations as well. Drugs that have been discovered in this way include a variety of anticancer and antimalarial drugs, blood-clotting agents, antibiotics, and medicines for the heart and digestive system.

Once the empirical and molecular formulas of the active compounds are determined, then their structural formulas are sought. At that point, synthetic work can begin. The chemist can identify compounds in the material that have medicinal value and find a way to *synthesize* them, or prepare them in the laboratory, so that they can be made available in large quantities.

In rational drug design, the chemist begins with the tumor or organism that the drug is intended to eradicate. Virtually all processes in living cells depend on specific *enzymes*, types of proteins with very large molecules having specific shapes. Usually there is an *active site* on the enzyme into which only specific molecules can fit and react. If the enzymes that control the growth of parasites or bacteria can be identified and their shapes known, compounds that fit into the active sites and block the reactions can be designed. The chemist taking this route begins by identifying key enzymes in the bacterium or parasite. Then the molecular structure of the enzyme is determined. A computer program is used to design molecules with structures that fit into the active site. The new compounds are synthesized, and their effects and side effects tested.

❓ How Might You Contribute?

Despite all the medicines in a modern pharmacy, there is still a great need for specific chemotherapeutic agents with few side effects. We are beginning to see the investigation of a wide array of natural products and the rational drug design of new therapeutic agents. The selection of good candidates for drug development from these large numbers of compounds and synthetic routes to duplicate them need to be optimized.

Related Exercises: 3.77 and 3.96

For Further Reading: D. Hart, "Designer drugs," *Science Spectra*, no. 8, 1997, p. 52. A. M. Rouhi, "Seeking drugs in natural products," *Chemical and Engineering News*, April 7, 1997, p. 14. J. Staunton and K. Weissman, "Medicines from nature," *The New Chemistry*, edited by N. Hall (Cambridge: Cambridge University Press, 2000), pp. 199–213. R. P. Szuromi, V. Vinson, and E. Marshall, "Rethinking drug discovery," *Science*, vol. 303, p. 1795. March 19, 2004.

A field biologist examines a plant in a South American rainforest. The plant produces chemicals that will be investigated for their medicinal value.

3.1 The Basic VSEPR Model

We begin by looking at molecules that consist of one central atom to which all the other atoms are attached. Many of these molecules have the shapes of the geometrical figures shown in Fig. 3.1; thus, CH_4 (**1**) is tetrahedral, SF_6 (**2**) is octahedral, and PCl_5 (**3**) is trigonal bipyramidal. In a number of these cases the **bond angles,**

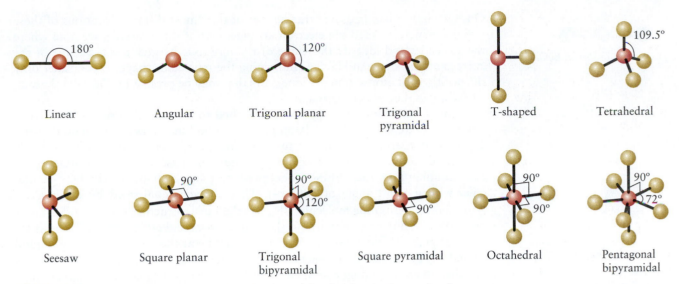

Linear   Angular   Trigonal planar   Trigonal pyramidal   T-shaped   Tetrahedral

Seesaw   Square planar   Trigonal bipyramidal   Square pyramidal   Octahedral   Pentagonal bipyramidal

FIGURE 3.1 The names of the shapes of simple molecules and their bond angles. Lone pairs of electrons are not shown, because they are not included when identifying molecular shapes.

the angles between the bonds (the straight lines that join the atom centers), are fixed by the symmetry of the molecule; these bond angles are indicated in Fig. 3.1. Thus, the HCH angle in CH_4 is 109.5° (the "tetrahedral angle"), the FSF angles in SF_6 are 90° and 180°, and the ClPCl angles in PCl_5 are 90°, 120°, and 180°. The bond angles of molecules that are not fixed by symmetry must be determined experimentally. The HOH bond angle in the angular H_2O molecule, for instance, has been found to be 104.5° and the HNH angle in the trigonal pyramidal NH_3 molecule is measured as 107°. The principal technique for determining bond angles in small molecules is spectroscopy, especially rotational and vibrational spectroscopy; x-ray diffraction is used for larger molecules.

The Lewis structures encountered in Chapter 2 are two-dimensional representations of the links between atoms—their connectivity—and except in the simplest cases do not depict the arrangement of atoms in space. The **valence-shell electron-pair repulsion model** (VSEPR model) extends Lewis's theory of bonding to account for molecular shapes by adding rules that account for bond angles. The model starts from the idea that because electrons repel one another, the shapes of simple molecules correspond to arrangements in which pairs of bonding electrons lie as far apart as possible. Specifically:

Rule 1 Regions of high electron concentration (bonds and lone pairs on the central atom) repel one another and, to minimize their repulsions, these regions move as far apart as possible while maintaining the same distance from the central atom (Fig. 3.2).

1 Methane, CH_4

2 Sulfur hexafluoride, SF_6

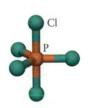

3 Phosphorus pentachloride, PCl_5

The VSEPR model was first explored by the British chemists Nevil Sidgwick and Herbert Powell and has been developed by the Canadian chemist Ronald Gillespie.

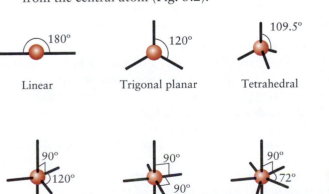

Linear   Trigonal planar   Tetrahedral

Trigonal bipyramidal   Octahedral   Pentagonal bipyramidal

FIGURE 3.2 The positions that two to seven regions of high electron concentration (atoms and lone pairs) take around a central atom. The regions are denoted by the straight lines sticking out of the central atom. Use this diagram to identify the electron arrangement of a molecule, and then use Fig. 3.1 to identify the shape of the molecule from the locations of its atoms.

4   Beryllium chloride, BeCl$_2$

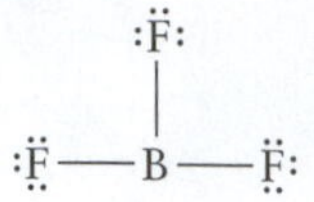

5   Boron trifluoride, BF$_3$

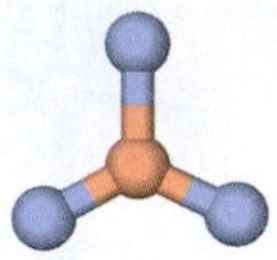

6   Boron trifloride, BF$_3$

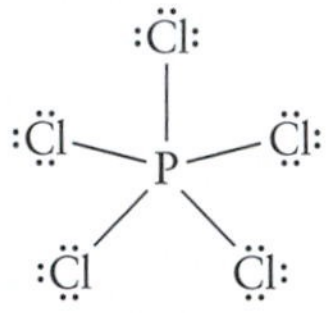

:$\ddot{\text{Cl}}$:

:$\ddot{\text{Cl}}$—P—$\ddot{\text{Cl}}$:

:$\ddot{\text{Cl}}$ $\ddot{\text{Cl}}$:

7   Phosphorus pentachloride, PCl$_5$

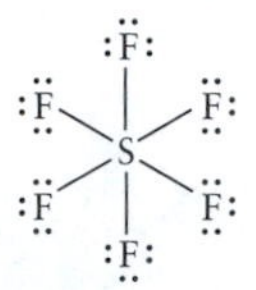

:$\ddot{\text{F}}$:

:$\ddot{\text{F}}$—S—$\ddot{\text{F}}$:

:$\ddot{\text{F}}$ $\ddot{\text{F}}$:

:$\ddot{\text{F}}$:

8   Sulfur hexafluoride, SF$_6$

All six terminal atoms are equivalent in a regular octahedral molecule.

9   Carbon dioxide, CO$_2$

:O: $\rbrack^{2-}$
‖
:$\ddot{\text{O}}$—C—$\ddot{\text{O}}$:

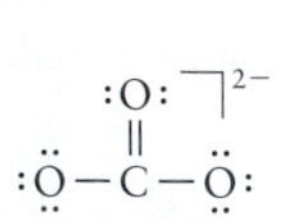

10   Carbonate ion, CO$_3{}^{2-}$

11   Carbonate ion, CO$_3{}^{2-}$

Once we have identified the arrangement of the "most distant" locations of these regions, which is called the **electron arrangement** of the molecule, we note where the atoms lie and identify the shape of the molecule by giving it the name of the corresponding shape in Fig. 3.1. In naming the molecular shape, we consider only the positions of atoms, not any lone pairs that may be present on the central atom, even though they affect the shape.

A molecule with only two atoms attached to the central atom is BeCl$_2$. The Lewis structure is :$\ddot{\text{Cl}}$—Be—$\ddot{\text{Cl}}$:, and there are no lone pairs on the central atom. To be as far apart as possible, the two bonding pairs lie on opposite sides of the Be atom, and so the electron arrangement is linear. Because a Cl atom is attached by each bonding pair, the VSEPR model predicts a linear shape for the BeCl$_2$ molecule, with a bond angle of 180° (**4**). That shape is confirmed by experiment.

A boron trifluoride molecule, BF$_3$, has the Lewis structure shown in (**5**). There are three bonding pairs attached to the central atom and no lone pairs. To be as far apart as possible, the three bonding pairs must lie at the corners of an equilateral triangle. The electron arrangement is trigonal planar. Because an F atom is attached to each bonding pair, the BF$_3$ molecule is trigonal planar (**6**), and all three FBF angles are 120°, as confirmed experimentally.

Methane, CH$_4$, has four bonding pairs on the central atom. To be as far apart as possible, the four pairs must take up a tetrahedral arrangement around the C atom. Because the electron arrangement is tetrahedral and an H atom is attached to each bonding pair, we expect the molecule to be tetrahedral (see **1**), with bond angles of 109.5°. That is the shape found experimentally.

In a phosphorus pentachloride molecule, PCl$_5$ (**7**), there are five bonding pairs and no lone pairs on the central atom. According to the VSEPR model, the five pairs and the atoms that they carry are farthest apart in a trigonal bipyramidal arrangement (see Fig. 3.2). In this arrangement, three atoms lie at the corners of an equilateral triangle and the other two atoms lie above and below the plane of the triangle (see **3**). This structure has three different bond angles: the bond angles in the equatorial plane are 120°; the angle between the axial and the equatorial atoms is 90°; the one axial ClPCl bond angle is 180°. This structure is also confirmed experimentally.

A sulfur hexafluoride molecule, SF$_6$, has six atoms attached to the central S atom and no lone pairs on that atom (**8**). According to the VSEPR model, the electron arrangement is octahedral, with four pairs at the corners of a square on the equator and the remaining two pairs above and below the plane of the square (see Fig. 3.2). An F atom is attached to each electron pair, and so the molecule is predicted to be octahedral. All its bond angles are either 90° or 180°, and all the F atoms are equivalent.

The second rule of the VSEPR model concerns the treatment of multiple bonds:

> **Rule 2** There is no distinction between single and multiple bonds: a multiple bond is treated as a single region of high electron concentration.

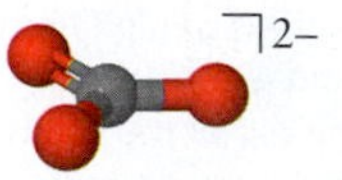

That is, the two electron pairs in a double bond stay together and repel other bonds or lone pairs as a unit. The three electron pairs in a triple bond also stay together and act like a single region of high electron concentration. For instance, a carbon dioxide molecule, $\ddot{\text{O}}$=C=$\ddot{\text{O}}$, has a linear structure similar to that of BeCl$_2$, even though both bonds are double bonds (**9**). One of the Lewis structures of a carbonate ion, CO$_3{}^{2-}$, is shown in (**10**). The two pairs of electrons in the double bond are treated as a unit, and the resulting shape (**11**) is trigonal planar. Because each bond, whether single or multiple, acts as a single unit, to count the number of regions of high electron concentration we simply count the number of atoms attached to the central atom and add the number of lone pairs.

When there is more than one central atom, we consider the bonding about each atom independently. For example, to predict the shape of an ethene (ethylene) molecule, $CH_2=CH_2$, we consider each carbon atom separately. From the Lewis structure (**12**) we note that each carbon atom has three atoms attached but no lone pairs. The arrangement around each carbon atom is therefore trigonal planar. We predict that the HCH and HCC angles will both be 120° (**13**); this prediction is confirmed experimentally.

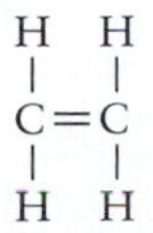

12 Ethene, C_2H_4

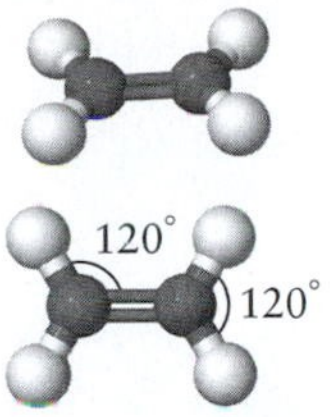

EXAMPLE 3.1	**Predicting the shape of a molecule with no lone pairs on the central atom**

Suggest a shape for the ethyne (acetylene) molecule, HC≡CH.

STRATEGY Write down the Lewis structure and identify the electron arrangement around each "central" atom (each C atom, in this case). Treat each multiple bond as a single unit. Then identify the overall shape of the molecule (refer to Fig. 3.2 if necessary).

13 Ethene, C_2H_4

SOLUTION

Write the Lewis structure of the molecule.	H—C≡C—H
Identify the electron arrangement around each "central" atom.	Linear: Each C atom is attached to two other atoms (one H atom and one C atom), and there are no lone pairs.
Identify the arrangement of atoms around each C atom.	Linear, and the molecule is linear overall.

SELF-TEST 3.1A Predict the shape of an arsenic pentafluoride molecule, AsF_5.
[*Answer:* Trigonal bipyramidal]

SELF-TEST 3.1B Predict the shape of a formaldehyde molecule, CH_2O.

Because we treat single bonds and multiple bonds as equivalent in the VSEPR model, it does not matter which of the Lewis structures contributing to a resonance structure we consider. For example, although we can write several different Lewis structures for a nitrate ion, all of them have three regions of electron concentration around the central N atom—all of them have three atoms attached to the central atom—and in each case we expect a trigonal planar structure (**14**). That the three N—O bonds are all equivalent is confirmed experimentally by the fact that all three have the same length and the three bond angles are identical. It is also confirmed by calculation: an electrostatic potential diagram (Section C) shows the symmetry of the calculated electron distribution (**15**). The equivalence of the three bonds is what we would expect in a resonance structure.

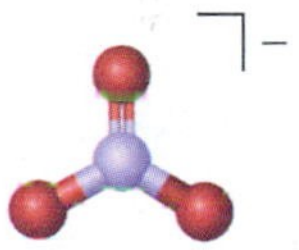

14 Nitrate ion, NO_3^-

Recall that red regions indicate negative potential (an accumulation of electrons) and blue regions indicate positive potential (a deficiency of electrons).

According to the VSEPR model, regions of high electron concentration take up positions that maximize their separations; electron pairs in a multiple bond are treated as a single unit. The shape of the molecule is then identified from the relative locations of its atoms.

3.2 Molecules with Lone Pairs on the Central Atom

To help us predict the shapes of molecules, we use the generic "VSEPR formula" AX_nE_m to identify the different combinations of atoms and lone pairs attached to

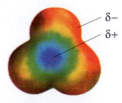

15 Nitrate ion

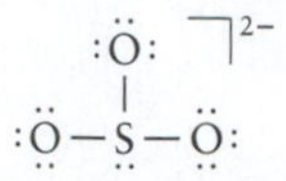

16 Sulfite ion, SO_3^{2-}

17 Sulfite ion, SO_3^{2-}

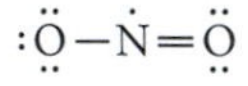

:Ö—N̈=Ö

18 Nitrogen dioxide, NO_2

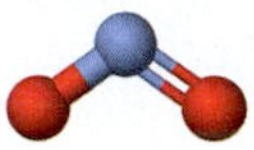

19 Nitrogen dioxide, NO_2

the central atom. We let A represent a central atom, X an attached atom, and E a lone pair. For example, the BF_3 molecule, with three attached fluorine atoms and no lone pairs on B, is an example of an AX_3 species. The sulfite ion, SO_3^{2-} (**16**), which has one lone pair, is an example of an AX_3E species. Molecules with the same VSEPR formula have essentially the same electron arrangement and the same shape; so by recognizing the formula we can immediately predict the shape (but not necessarily the precise numerical values of bond angles that are not governed by symmetry).

If there are no lone pairs on the central atom (an AX_n molecule), each region of high electron concentration corresponds to an atom, and so the molecular shape is the same as the electron arrangement. However, if lone pairs are present, the molecular shape differs from the electron arrangement because only the positions of the atoms are considered when naming the shape. For example, the four regions of high electron concentration in SO_3^{2-} are farthest apart if they adopt a tetrahedral arrangement (see Fig. 3.2). However, the *shape* of the ion is described by the locations of the atoms, not the lone pair. Because only three of the tetrahedral locations are occupied by atoms, the shape of an SO_3^{2-} ion is trigonal pyramidal (**17**). The rule to remember is

> **Rule 3** All regions of high electron density, lone pairs and bonds, are included in a description of the electronic arrangement, but *only the positions of atoms are considered when reporting the shape of a molecule.*

A single unpaired electron on the central atom also is a region of high electron density and is treated like a lone pair when determining molecular shape. For example, radicals such as NO_2 have a single nonbonding electron, a "lone half-pair." Thus, NO_2 (**18**) has a trigonal planar electron arrangement (including the unpaired electron on N), but its shape is angular (**19**).

EXAMPLE 3.2   Predicting the shape of a molecule with lone pairs

Predict the electron arrangement and the shape of a nitrogen trifluoride molecule, NF_3.

STRATEGY For the electron arrangement, draw the Lewis structure and then use the VSEPR model to decide how the bonding pairs and lone pairs are arranged around the central (nitrogen) atom (consult Fig. 3.2 if necessary). Identify the molecular shape from the layout of atoms, as in Fig. 3.1.

SOLUTION

Draw the Lewis structure.	:F̈—N̈—F̈:
	    |
	   :F̈:
		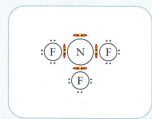

Count the bonds and lone pairs on the central atom.	The central N atom has one electron pair and three bonds, corresponding to four regions of high electron density.	
Assign the electron arrangement.	Tetrahedral	

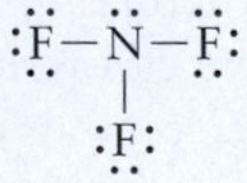

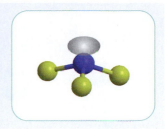

Identify the shape considering only atoms.

The three atoms bonded to N form a trigonal pyramid.

Note that only when using the type of representation on the right do we show the lone pair explicitly. Spectroscopic measurements confirm the prediction of a trigonal pyramidal shape for NF_3.

SELF-TEST 3.2A Predict (a) the electron arrangement and (b) the shape of an IF_5 molecule.

[*Answer:* (a) Octahedral; (b) square pyramidal]

SELF-TEST 3.2B Predict (a) the electron arrangement and (b) the shape of an SO_2 molecule.

So far we have regarded lone pairs as equivalent to atoms, but is that really the case? We have predicted, for instance, that the electron arrangement of an SO_3^{2-} ion is tetrahedral, and so we might expect OSO angles of 109.5°. However, experimental findings have shown that, although the sulfite ion is indeed trigonal pyramidal, its bond angle is only 106° (**20**). Such experimental evidence tells us that the VSEPR model as we have described it is incomplete and needs to be refined.

The final rule of the VSEPR model takes note of the different effects of lone pairs and atoms. Because bond angles in molecules with lone pairs are typically smaller than expected, lone pairs are treated in the VSEPR model as having a more strongly repelling effect than do electrons in bonds. That is, the lone pairs push the atoms bonded to the central atom closer together. One possible rationalization for this effect is that the electron cloud of a lone pair can spread over a larger volume than a bonding pair, because a bonding pair (or several bonding pairs in a multiple bond) is pinned down by two atoms, not one (Fig. 3.3). In summary, the VSEPR model provides reasonably reliable predictions if we adopt the following rule:

> **Rule 4** The strengths of repulsions are in the order lone pair–lone pair > lone pair–atom > atom–atom

Therefore, the lowest energy is achieved when lone pairs are as far from each other as possible. The energy is also lowest if the atoms bonded to the central atom are far from lone pairs, even though that might bring the atoms closer to other atoms.

Our improved model helps to account for the bond angle of the AX_3E sulfite ion. The atoms and the lone pair adopt a tetrahedral arrangement around the S atom. However, the lone pair exerts a strong repulsion on the atoms, forcing them to move together slightly, reducing the OSO angle from the 109.5° of the regular tetrahedron to 106°. Note that, although the VSEPR model can predict the direction of the distortion, it cannot predict its extent. We can predict that, in any AX_3E species, the XAX angle will be less than 109.5°, but we cannot predict its actual value; we must measure it experimentally or calculate it by solving the Schrödinger equation numerically on a computer.

SELF-TEST 3.3A (a) Give the VSEPR formula of an NH_3 molecule. Predict (b) its electron arrangement and (c) its shape.

[*Answer:* (a) AX_3E; (b) tetrahedral; (c) trigonal pyramidal (**21**, LP is lone pair), HNH angle less than 109.5°]

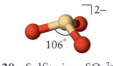

20   Sulfite ion, SO_3^{2-}

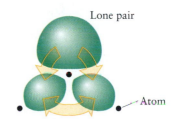

FIGURE 3.3 A possible explanation of why lone pairs have a greater repelling effect than that of bonding electrons. A lone pair is less restrained than bonding pairs and so takes up more space; the bonding pairs (with their atoms) move away from the lone pair in an attempt to lower the repulsion that they experience, thus compressing the bond angle slightly.

21   Ammonia, NH_3

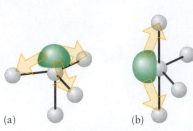

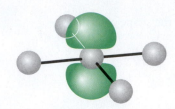

FIGURE 3.4 (a) A lone pair in an axial position is close to three equatorial atoms. (b) In an equatorial position, a lone pair is close to only two atoms, a more favorable arrangement.

(a)   (b)

FIGURE 3.5 Two lone pairs in an AX_3E_2 molecule adopt equatorial positions and move away from each other slightly. As a result, the molecule is approximately T-shaped.

FIGURE 3.6 The square planar arrangement of atoms taken up in AX_4E_2 molecules: the two lone pairs are farthest apart when they are on opposite sides of the central atom.

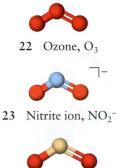

22   Ozone, O_3

23   Nitrite ion, NO_2^-

24   Sulfur dioxide, SO_2

SELF-TEST 3.3B (a) Give the VSEPR formula of a ClO_2^- ion. Predict (b) its electron arrangement and (c) its shape.

Rule 4 allows us to predict the position in which a lone pair will be found. For example, the electron arrangement in an AX_4E molecule or ion, such as IF_4^+, is trigonal bipyramidal, but there are two different possible locations for the lone pair. An **axial lone pair** lies on the axis of the molecule, where it repels three electron pairs strongly; an **equatorial lone pair** lies on the molecule's equator, on the plane perpendicular to the molecular axis, where it repels only two electron pairs strongly (Fig. 3.4). Therefore, the lowest energy is achieved when a lone pair is equatorial, producing a seesaw-shaped molecule. An AX_3E_2 molecule, such as ClF_3, also has a trigonal bipyramidal arrangement of electron pairs, but two of the pairs are lone pairs. These two pairs are farthest apart if they occupy two of the three equatorial positions but move away from each other slightly. The result is a T-shaped molecule (Fig. 3.5). Now consider an AX_4E_2 molecule, which has an octahedral arrangement of electron pairs, two of which are lone pairs. The two lone pairs are farthest apart when they lie opposite each other, and so the molecule is square planar (Fig. 3.6).

Molecules with the same VSEPR formula all have the same general shape, although their bond angles generally differ slightly. For example, O_3 is an AX_2E species (:Ö—Ö=Ö); it has a trigonal planar electron arrangement and an angular molecular shape (**22**). The nitrite ion, NO_2^-, has the same general formula (:Ö—N=Ö⁻) and the same shape (**23**); so too does sulfur dioxide, SO_2 (:Ö—S̈=Ö, **24**).

CONCEPTUAL BASIS

Regions of high electron concentration—bonds to atoms and lone pairs attached to a central atom in a molecule—arrange themselves in such a way as to minimize mutual repulsions.

PROCEDURE

The general procedure for predicting the shape of a molecule is as follows:

Step 1 Decide how many atoms and lone pairs are present on the central atom by writing a Lewis structure for the molecule.

Step 2 Identify the electron arrangement, treating a multiple bond as equivalent to a single bond (see Fig. 3.2).

Step 3 Locate the atoms and identify the molecular shape (according to Fig. 3.1).

Step 4 Allow the molecule to distort so that lone pairs are as far from one another and from bonding pairs as possible. The repulsions are in the order

Lone pair–lone pair > lone pair–atom > atom–atom

Example 3.3 shows how this procedure is used.

EXAMPLE 3.3 **Sample exercise: Predicting a molecular shape**

Predict the shape of a sulfur tetrafluoride molecule, SF_4.

SOLUTION

Step 1 Draw the Lewis structure.	Because we are focusing on the central atom, there is no need to show the lone pairs on the F atoms.	

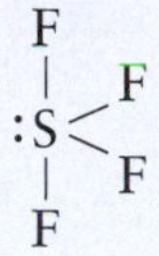

Step 2 Assign the electron arrangement around the central atom.	There are 5 regions of high electron density (4 atoms and 1 lone pair); so trigonal bipyramidal.	

Step 3 Identify the molecular shape.	AX_4E. To minimize electron pair repulsions, the lone pair occupies an equatorial location. SF_4 has a seesaw shape.	

Step 4 Allow for distortions.	The atoms move slightly away from the lone pair.	

This shape (resembling a slightly bent seesaw) is the one found experimentally.

SELF-TEST 3.4A Predict the shape of an I_3^- ion.

[*Answer:* Linear]

SELF-TEST 3.4B Predict the shape of a xenon tetrafluoride molecule, XeF_4.

> *In a molecule that has lone pairs or a single nonbonding electron on the central atom, the valence electrons contribute to the electron arrangement about the central atom but only bonded atoms are considered in the identification of the shape. Lone pairs distort the shape of a molecule so as to reduce lone pair–bonding pair repulsions.*

3.3 Polar Molecules

In Section 2.12, we saw that a polar covalent *bond* in which electrons are not evenly distributed has a nonzero dipole moment. A **polar molecule** is a *molecule* with a nonzero dipole moment. All diatomic molecules are polar if their bonds are polar. An HCl molecule, with its polar covalent bond ($^{\delta+}H-Cl^{\delta-}$), is a polar molecule. Its dipole moment of 1.1 D is typical of polar diatomic molecules (Table 3.1). All diatomic molecules that are composed of atoms of different elements are at least slightly polar. A **nonpolar molecule** is a molecule that has no electric dipole moment. All **homonuclear diatomic molecules**, diatomic molecules containing atoms of only one element, such as O_2, N_2, and Cl_2, are nonpolar, because their bonds are nonpolar.

A polyatomic molecule may be nonpolar even if its bonds are polar. For example, the two $^{\delta+}C-O^{\delta-}$ dipole moments in carbon dioxide, a linear molecule, point in opposite directions, and so they cancel each other (**25**) and CO_2 is a nonpolar

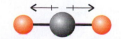

25 Carbon dioxide, CO_2

TABLE 3.1   Dipole Moments of Selected Molecules

Molecule	Dipole moment (D)	Molecule	Dipole moment (D)
HF	1.91	PH_3	0.58
HCl	1.08	AsH_3	0.20
HBr	0.80	SbH_3	0.12
HI	0.42	O_3	0.53
CO	0.12	CO_2	0
ClF	0.88	BF_3	0
NaCl*	9.00	CH_4	0
CsCl*	10.42	*cis*-CHCl=CHCl	1.90
H_2O	1.85	*trans*-CHCl=CHCl	0
NH_3	1.47		

*For pairs of ions in the gas phase, not the bulk ionic solid.

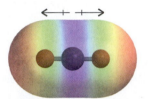

26   Carbon dioxide, CO_2

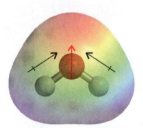

27   Water, H_2O

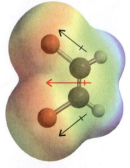

28   *cis*-Dichloroethene, $CH_2CH_2Cl_2$

molecule. Even though there are positive and negative regions of charge within the molecule, the center of positive charge and the center of negative charge coincide, so the molecule itself is nonpolar. The electrostatic potential diagram (**26**) illustrates this conclusion. As noted in Section C, the colors show how electron density is distributed in the molecule. Partially positive regions are blue and partially negative regions red. In contrast, the two $^{\delta+}H\!-\!O^{\delta-}$ dipole moments in H_2O lie at 104.5° to each other and do not cancel, and so H_2O is a polar molecule (**27**). This polarity is part of the reason why water is such a good solvent for ionic compounds.

As we have seen by comparing CO_2 and H_2O, the shape of a polyatomic molecule affects whether or not it is polar. The same is true of more complicated molecules. For instance, the atoms and bonds are the same in *cis*-dichloroethene (**28**) and *trans*-dichloroethene (**29**); but, in the latter, the C—Cl bonds point in opposite directions and the dipoles (which point along the C—Cl bonds) cancel. Thus, whereas *cis*-dichloroethene is polar, *trans*-dichloroethene is nonpolar. Because dipole moments are directional, we can treat each bond dipole moment as a vector. The molecule *as a whole* will be nonpolar if the vector sum of the dipole moments of the bonds is zero.

If the four atoms attached to the central atom in a tetrahedral molecule are the same, as in tetrachloromethane (carbon tetrachloride), CCl_4 (**30**), the dipole moments cancel and the molecule is nonpolar. However, if one or more of the atoms are replaced by different atoms, as in trichloromethane (chloroform), $CHCl_3$, or by lone pairs, as in NH_3, then the dipole moments associated with the bonds are not all the same, so they do not cancel. Thus, the $CHCl_3$ molecule is polar (**31**).

Figure 3.7 summarizes the shapes of simple molecules that result in them being polar or nonpolar.

EXAMPLE 3.4   **Predicting the polar character of a molecule**

Predict whether (a) a boron trifluoride molecule, BF_3, and (b) an ozone molecule, O_3, are polar.

STRATEGY In each case, we must decide on the shape of the molecule by using the VSEPR model and then decide whether the symmetry of the molecule results in the cancellation of the dipole moments associated with the bonds. If necessary, refer to Fig. 3.7.

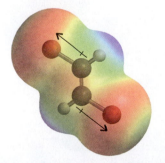

29   *trans*-Dichloroethene, $CH_2CH_2Cl_2$

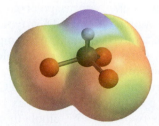

30   Tetrachloromethane, CCl_4

31   Trichloromethane, $CHCl_3$

VSEPR type	Nonpolar	Polar
AX_2	CO_2	N_2O
AX_2E		SO_2, O_3
AX_2E_2		H_2O
AX_2E_3	I_3^-, XeF_2	$BrIF^-$
AX_2E_4	none known	none known
AX_3	BF_3	$COCl_2$
AX_3E		NH_3
AX_3E_2		ClF_3

VSEPR type	Nonpolar	Polar
AX_4	CH_4	CH_3Cl
AX_4E		SF_4
AX_4E_2	XeF_4	
AX_5	PCl_5	PCl_4F
AX_5E		IF_5
AX_6	SF_6	

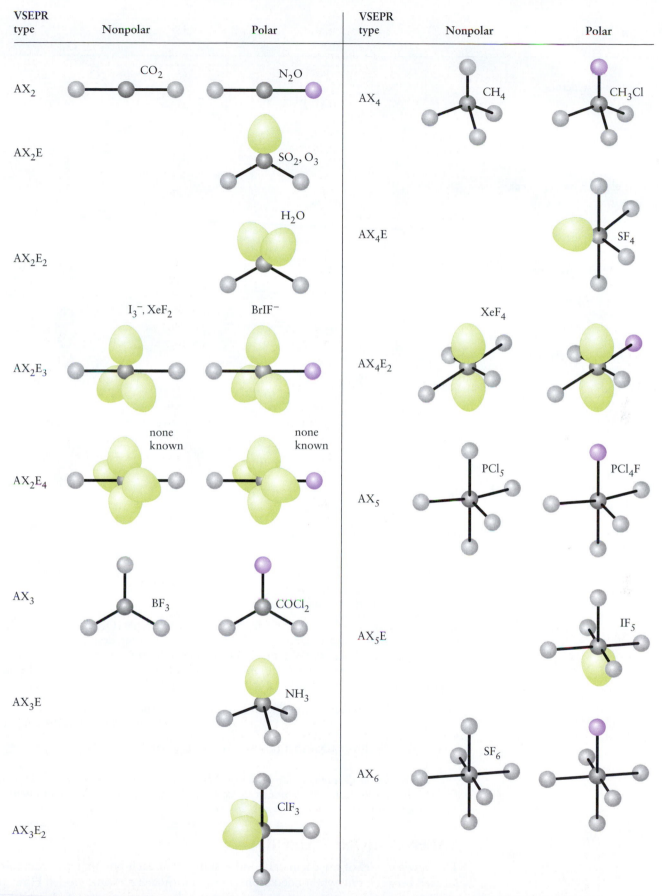

FIGURE 3.7 The arrangements of atoms that give rise to polar and nonpolar molecules. In these VSEPR formulas, A stands for a central atom, X for an attached atom, and E for a lone pair. Identical atoms are the same color; attached atoms colored differently belong to different elements. The green lobes represent lone pairs of electrons.

SOLUTION

	BF_3	O_3
Draw the Lewis structure.		$:\ddot{O}-\ddot{O}=\ddot{O}$
Assign the electron arrangement.		
Identify the VSEPR formula.	AX_3	AX_2E
Name the molecular shape.	Trigonal planar	Angular
Identify the polarity.	Nonpolar: The symmetry allows all three B—F dipoles to c ancel.	Polar: The dipoles do not cancel.

This example shows that a homonuclear polyatomic molecule (O_3) can be polar: shape is more important than differences in atoms and O_3 is polar despite all three atoms being oxygen. In this case, the central O atom has a different electron density associated with it than the outer two O atoms: it is bonded to two O atoms whereas the outer atoms are bonded only to one O atom.

SELF-TEST 3.5A Predict whether (a) SF_4, (b) SF_6 is polar or nonpolar.

[*Answer:* (a) Polar; (b) nonpolar]

SELF-TEST 3.5B Predict whether (a) PCl_5, (b) IF_5 is polar or nonpolar.

A diatomic molecule is polar if its bond is polar. A polyatomic molecule is polar if it has polar bonds arranged in space in such a way that the dipole moments associated with the bonds do not cancel.

VALENCE-BOND THEORY

The Lewis model of the chemical bond assumes that each bonding electron pair is located between the two bonded atoms—it is a *localized electron model*. However, we know from the wave–particle duality of the electron (Sections 1.5–1.7) that the location of an electron in an atom cannot be described in terms of a precise position, but only in terms of the *probability* of finding it somewhere in a region of space defined by its orbital. The same principles apply to electrons in molecules, except that the electrons are distributed over a larger region.

The first description of covalent bonding to be devised in terms of atomic orbitals, by Walter Heitler, Fritz London, John Slater, and Linus Pauling in the late 1920s, is called **valence-bond theory** (VB theory). This theory is a quantum mechanical description of the distribution of electrons in bonds that goes beyond Lewis's theory and the VSEPR model by providing a way of calculating the numerical values of bond angles and bond lengths. We shall not go into the actual calculations, which are very complicated, but shall concentrate instead on some of the qualitative concepts. These bonding concepts are used throughout chemistry, so it is important to know how to apply them.

3.4 Sigma and Pi Bonds

We begin with H_2, the simplest molecule of all, and start by thinking about the two hydrogen atoms from which it is formed. Each hydrogen atom in its ground state has one electron in a 1s-orbital. In valence-bond theory, we suppose that, as two H atoms come together, their 1s-electrons pair (denoted ↑↓, as in the discussion of atomic structure in Section 1.10), and the atomic orbitals merge together (Fig. 3.8). The resulting sausage-shaped distribution of electrons, with an accumulation of electron density between the nuclei, is called a "σ-bond" (a sigma bond). More formally, a **σ-bond** is cylindrically symmetrical (the same in all directions around the long axis of the bond), with no nodal planes containing the internuclear axis. A hydrogen molecule is held together by a σ-bond. The merging of the two atomic orbitals is called the **overlap** of orbitals. A general point to keep in mind throughout this section is that, the greater the extent of orbital overlap, the stronger the bond.

Much the same kind of σ-bond formation ("σ-bonding") occurs in the hydrogen halides. For example, before the H and F atoms combine to form hydrogen fluoride, the unpaired electron on the fluorine atom occupies a $2p_z$-orbital, and the unpaired electron on the hydrogen atom occupies a 1s-orbital. These two electrons are the ones that pair to form a bond (**32**). They pair as the orbitals that they occupy overlap and merge into a cloud that spreads over both atoms (Fig. 3.9). When viewed from the side, the resulting bond has a more complicated shape than that of the σ-bond in H_2; however, the bond looks much the same—it has cyclindrical symmetry and no nodal planes containing the internuclear axis—when viewed along the internuclear (z) axis; hence it too is a σ-bond. All *single* covalent bonds are σ-bonds.

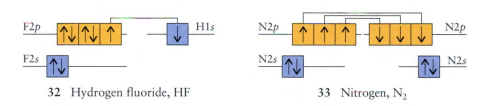

32 Hydrogen fluoride, HF   **33** Nitrogen, N_2

We encounter a different type of bond in a nitrogen molecule, N_2. There is a single electron in each of the three 2p-orbitals on each atom (**33**). However, when we try to pair them and form three bonds, only one of the three orbitals on each atom can overlap end to end to form a σ-bond (Fig. 3.10). Two of the 2p-orbitals on each atom ($2p_x$ and $2p_y$) are perpendicular to the internuclear axis, and each one contains an unpaired electron (Fig. 3.11, top). When the electrons in one of these p-orbitals on each N atom pair, the orbitals can overlap only in a side-by-side arrangement. This overlap results in a "π-bond," a bond in which the two electrons lie in two lobes, one on each side of the internuclear axis (Fig. 3.11, bottom). More formally, a **π-bond** has a single nodal plane containing the internuclear axis. Although a π-bond has electron density on each side of the internuclear axis, it is only one bond, with the electron cloud in the form of two lobes, just as a p-orbital is one orbital with two lobes. In a molecule with two π-bonds, such as N_2, the

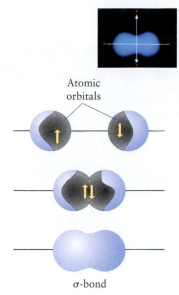

FIGURE 3.8 When electrons with opposite spins (depicted as ↑ and ↓) in two hydrogen 1s-orbitals pair and the s-orbitals overlap, they form a σ-bond, which is depicted here by the boundary surface of the electron cloud. The cloud has cylindrical symmetry around the internuclear axis and spreads over both nuclei. In the illustrations in this book, σ-bonds are usually colored blue.

The Greek letter sigma, σ, is the equivalent of our letter s. It reminds us that, looking along the internuclear axis, the electron distribution resembles that of an s-orbital.

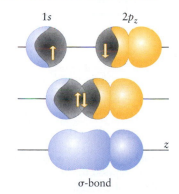

FIGURE 3.9 A σ-bond can also be formed when electrons in 1s- and $2p_z$-orbitals pair (where z is the direction along the internuclear axis). The two electrons in the bond are spread over the entire region of space enclosed by the boundary surface.

By convention, the bond direction defines the z-axis.

The Greek letter pi, π, is the equivalent of our letter p. When we imagine looking along the internuclear axis, a π-bond resembles a pair of electrons in a p-orbital.

FIGURE 3.10 A σ-bond is formed by the pairing of electron spins in two $2p_z$-orbitals on neighboring atoms. At this stage, we are ignoring the interactions of any $2p_x$- (and $2p_y$-) orbitals that also contain unpaired electrons, because they cannot form σ-bonds. The electron pair may be found anywhere within the boundary surface shown in the bottom diagram. Notice that the nodal plane of each p_z-orbital survives in the σ-bond.

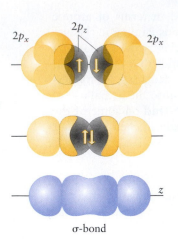

σ-bond

FIGURE 3.11 A π-bond is formed when electrons in two $2p$-orbitals pair and overlap side by side. The middle diagram shows the extent of the overlap, and the bottom diagram shows the corresponding boundary surface. Even though the bond has a complicated shape, with two lobes, it is occupied by one pair of electrons and counts as one bond. In this text, π-bonds are usually colored yellow.

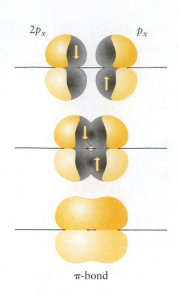

π-bond

electron densities in the two π-bonds merge, and the two atoms appear to be surrounded by a cylinder of electron density (Fig. 3.12).

We can generalize from these examples to the description of a multiply bonded species according to valence-bond theory:

A **single bond** is a σ-bond.

A **double bond** is a σ-bond plus one π-bond.

A **triple bond** is a σ-bond plus two π-bonds.

> There are a few exceptions to the rule about double bonds: in a very few instances both bonds of a double bond are π bonds.

SELF-TEST 3.6A How many σ-bonds and how many π-bonds are there in (a) CO_2 and (b) CO?

[***Answer:*** (a) Two σ, two π; (b) one σ, two π]

SELF-TEST 3.6B How many σ-bonds and how many π-bonds are there in (a) NH_3 and (b) HCN?

In valence-bond theory, we assume that bonds form when unpaired electrons in valence-shell atomic orbitals pair; the atomic orbitals overlap end to end to form σ-bonds or side by side to form π-bonds.

3.5 Electron Promotion and the Hybridization of Orbitals

When we try to apply VB theory to methane we run into difficulties. A carbon atom has the configuration $[He]2s^2 2p_x^{\,1} 2p_y^{\,1}$ with four valence electrons (**34**). However, two valence electrons are already paired and only the two half-filled $2p$-orbitals appear to be available for bonding. It looks as though a carbon atom should have a valence of 2 and form two perpendicular bonds, but in fact it almost always has a valence of 4 (it is commonly "tetravalent") and in CH_4 has a tetrahedral arrangement of bonds.

To overcome this difficulty we note that a carbon atom would have four unpaired electrons available for bonding if an electron is **promoted**—that is, has been relocated to a higher-energy orbital. When we promote a $2s$-electron into an empty $2p$-orbital, we get the configuration $[He]2s^1 2p_x^{\,1} 2p_y^{\,1} 2p_z^{\,1}$ (**35**). Without promotion, a carbon atom can form only two bonds; after promotion, it can form four bonds. Although it takes energy to promote an electron, the overall energy of the CH_4 molecule is lower than if carbon formed only two C—H bonds.

The characteristic tetravalence of carbon is due to the small promotion energy of a carbon atom. The promotion energy is small because a $2s$-electron is transferred from an orbital that it shares with another electron to an empty $2p$-orbital. Although the promoted electron enters an orbital of higher energy, it experiences less repulsion from other electrons than before it was promoted. As a result, only a small energy is needed to promote the electron. Nitrogen, carbon's neighbor, cannot use promotion to increase the number of bonds that it can form, because it has no empty p-orbitals (**36**). The same is true of oxygen and fluorine. Promotion of an

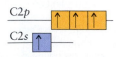

34   Carbon, $[He]2s^2 2p_x^{\,1} 2p_y^{\,1}$

> Carbon monoxide, CO, is the only common exception to the tetravalence of carbon.

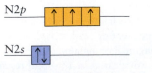

35   Carbon, $[He]2s^1 2p_x^{\,1} 2p_y^{\,1} 2p_z^{\,1}$

36   Nitrogen, $[He]2s^2 2p_x^{\,1} 2p_y^{\,1} 2p_z^{\,1}$

FIGURE 3.12 The bonding pattern in a nitrogen molecule, N_2. (a) The two atoms are held together by one σ-bond (blue) and two perpendicular π-bonds (yellow). (b) When the three bonds are put together, the two π-bonds merge to form a long doughnut-shaped cloud surrounding the σ-bond cloud; so the overall structure resembles a cylindrical hot dog.

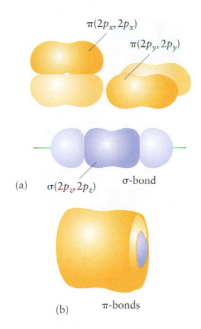

(a)

$\pi(2p_x, 2p_x)$

$\pi(2p_y, 2p_y)$

$\sigma(2p_z, 2p_z)$ σ-bond

(b)   π-bonds

electron is possible if the overall change, taking account of all contributions to the energy and especially the greater number of bonds that can thereby be formed, is toward lower energy. Boron, $[\text{He}]2s^2 2p^1$, like carbon, is an element in which promotion can lead to the formation of more bonds (three in its case), and boron does typically form three bonds.

At this stage, it looks as though electron promotion should result in two different types of bonds in methane, one bond from the overlap of a hydrogen 1s-orbital and a carbon 2s-orbital, and three more bonds from the overlap of hydrogen 1s-orbitals with each of the three carbon 2p-orbitals. The overlap with the 2p-orbitals should result in three σ-bonds at 90° to one another. However, this arrangement is inconsistent with the known tetrahedral structure of methane with four equivalent bonds.

To improve our model we note that s- and p-orbitals are waves of electron density centered on the nucleus of an atom. We imagine that the four orbitals interfere with one another and produce new patterns where they intersect, like waves in water. Where the wavefunctions are all positive or all negative, the amplitudes are increased by this interference; where the wavefunctions have opposite signs, the overall amplitude is reduced and might even be canceled completely. As a result, the interference between the atomic orbitals results in new patterns. These new patterns are called **hybrid orbitals.** Each of the four hybrid orbitals, designated h_n, is formed from a linear combinations of the four atomic orbitals:

$$h_1 = s + p_x + p_y + p_z \qquad h_2 = s - p_x - p_y + p_z$$
$$h_3 = s - p_x + p_y - p_z \qquad h_4 = s + p_x - p_y - p_z$$

For instance, in h_1 the s- and p-orbitals all have their usual signs and their amplitudes add together where they are all positive. In h_2, however, the signs of p_x and p_y are reversed, and so the resulting interference pattern is different.

The four hybrid orbitals that we have constructed differ only in their orientation, with one pointing toward each corner of a tetrahedron (Fig. 3.13); in all other respects, they are identical. These four hybrid orbitals are called sp^3 **hybrids** because they are formed from one s-orbital and three p-orbitals. In an orbital-energy diagram, we represent the hybridization as the formation of four orbitals of equal energy intermediate between the energies of the s- and p-orbitals from which they are constructed (**37**). The hybrids are colored green to remind us that they are a blend of (blue) s-orbitals and (yellow) p-orbitals. An sp^3 hybrid orbital has two lobes, but one lobe extends farther than those of the contributing p-orbitals and the other lobe is shortened. The fact that hybrid orbitals have their amplitudes concentrated on one side of the nucleus allows them to overlap more effectively with other orbitals, and as a result the bonds that they form are stronger than in the absence of hybridization.

We are now ready to account for the bonding in methane. In the promoted, hybridized atom each of the electrons in the four sp^3 hybrid orbitals can pair with an electron in a hydrogen 1s-orbital. Their overlapping orbitals form four σ-bonds that point toward the corners of a tetrahedron (Fig. 3.14). The valence-bond description is now consistent with experimental data on molecular geometry.

When there is more than one "central" atom in a molecule, we concentrate on each atom in turn and match the hybridization of each atom to the shape at that atom predicted by VSEPR. For example, in ethane, C_2H_6 (**38**), the two carbon atoms are both "central" atoms. According to the VSEPR model, the four electron pairs around each carbon atom take up a tetrahedral arrangement. This arrangement suggests sp^3 hybridization of the carbon atoms, as shown in Fig. 3.14. Each

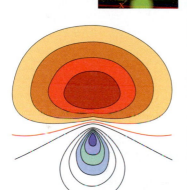

FIGURE 3.13 These contours indicate the amplitude of the sp^3 hybrid orbital wavefunction in a plane that bisects it and passes through the nucleus. Each sp^3 hybrid orbital points toward the corner of a tetrahedron.

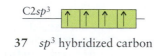

$C2sp^3$

37 sp^3 hybridized carbon

H   H
| |
H—C—C—H
| |
H   H

38 Ethane, CH_3CH_3

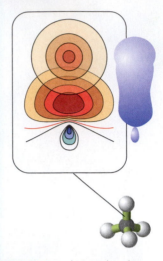

FIGURE 3.14 Each C—H bond in methane is formed by the pairing of an electron in a hydrogen 1s-orbital and an electron in one of the four sp^3 hybrid orbitals of carbon. Therefore, valence-bond theory predicts four equivalent σ-bonds in a tetrahedral arrangement, which is consistent with experimental results.

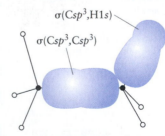

σ(Csp^3,H1s)

σ(Csp^3,Csp^3)

FIGURE 3.15 The valence-bond description of bonding in an ethane molecule, C_2H_6. The boundary surfaces of only two of the bonds are shown. Each pair of neighboring atoms is linked by a σ-bond formed by the pairing of electrons in either H1s-orbitals or C2sp^3 hybrid orbitals. All the bond angles are close to 109.5° (the tetrahedral angle).

C atom has one unpaired electron in each of its four sp^3 hybrid orbitals and can therefore form four σ-bonds that point toward the corners of a regular tetrahedron. The C—C bond is formed by spin-pairing of the electrons in one sp^3 hybrid orbital of each C atom. We label this bond σ(C2sp^3,C2sp^3) to show its composition: C2sp^3 denotes an sp^3 hybrid orbital composed of 2s- and 2p-orbitals on a carbon atom, and the parentheses show which orbitals on each atom overlap (Fig. 3.15). Each C—H bond is formed by spin-pairing of an electron in one of the remaining sp^3 hybrid orbitals with an electron in a 1s-orbital of an H atom (denoted H1s). These bonds are denoted σ(C2sp^3,H1s).

We can extend these ideas to molecules, such as ammonia, that have a lone pair of electrons on the central atom. According to the VSEPR model, the four electron pairs in NH_3 take up a tetrahedral electron arrangement, so we describe the nitrogen atom in terms of four sp^3 hybrid orbitals. Because nitrogen has five valence electrons, one of these hybrid orbitals is already doubly occupied (39). The 1s-electrons of the three hydrogen atoms pair with the three unpaired electrons in the remaining sp^3 hybrid orbitals. This pairing and overlap result in the formation of three N—H σ-bonds. *Whenever an atom in a molecule has a tetrahedral electron arrangement, we say that it is sp^3 hybridized.*

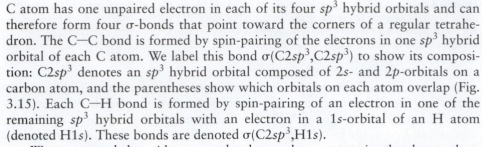

N2sp^3 H1s

39   Ammonia, NH_3

The promotion of electrons will occur if, overall, it leads to a lowering of energy by permitting the formation of more bonds. Hybrid orbitals are constructed on an atom to reproduce the electron arrangement characteristic of the experimentally determined shape of a molecule.

3.6 Other Common Types of Hybridization

We use different hybridization schemes to describe other arrangements of electron pairs (Fig. 3.16). For example, to explain a trigonal planar electron arrangement, like that in BF_3 and each carbon atom in ethene, we mix one s-orbital with two p-orbitals and so produce three sp^2 **hybrid orbitals:**

$$h_1 = s + 2^{1/2}p_y$$
$$h_2 = s + \left(\tfrac{3}{2}\right)^{1/2}p_x - \left(\tfrac{1}{2}\right)^{1/2}p_y$$
$$h_3 = s - \left(\tfrac{3}{2}\right)^{1/2}p_x - \left(\tfrac{1}{2}\right)^{1/2}p_y$$

These identical orbitals all lie in the same plane and point toward the corners of an equilateral triangle.

A linear arrangement of electron pairs requires two hybrid orbitals, and so we mix an s-orbital with a p-orbital to obtain two **sp-hybrid orbitals:**

$$h_1 = s + p$$
$$h_2 = s - p$$

These two sp-hybrid orbitals point away from each other at 180°, and result in bonds that form a straight line. This is the arrangement we see in CO_2.

SELF-TEST 3.7A   Suggest a structure in terms of hybrid orbitals for BF_3.

[*Answer:* Three σ-bonds formed from F2p_z-orbitals and B2sp^2 hybrids in a trigonal planar arrangement]

SELF-TEST 3.7B   Suggest a structure in terms of hybrid orbitals for each carbon atom in ethyne, C_2H_2.

Some of the elements in Period 3 and later periods can accommodate five or more electron pairs, as in PCl_5. We can devise a hybridization scheme to describe

FIGURE 3.16 Three common hybridization schemes shown as outlines of the amplitude of the wavefunction and in terms of the orientations of the hybrid orbitals. (a) An s-orbital and a p-orbital hybridize into two sp hybrid orbitals that point in opposite directions, forming a linear molecular shape. (b) An s-orbital and two p-orbitals can blend together to give three sp^2 hybrid orbitals that point to the corners of an equilateral triangle. (c) An s-orbital and three p-orbitals can blend together to give four sp^3 hybrid orbitals that point to the corners of a tetrahedron.

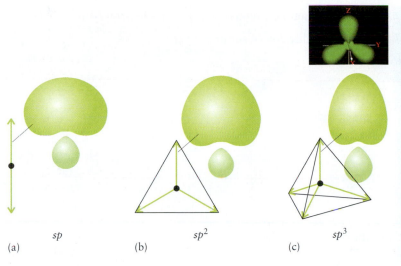

sp sp^2 sp^3

(a)   (b)   (c)

these types of bonds by using the d-orbitals of the central atom. To account for a trigonal bipyramidal arrangement of five electron pairs, we use one d-orbital as well as all the valence s- and p-orbitals of the atom. The resulting five orbitals are called sp^3d **hybrid orbitals** (Fig. 3.17).

We need six orbitals to accommodate six electron pairs around an atom in an octahedral arrangement, as in SF_6 and XeF_4, and so we need to use two d-orbitals in addition to the valence s- and p-orbitals to form six sp^3d^2 **hybrid orbitals** (Fig. 3.18). These identical orbitals point toward the six corners of a regular octahedron.

Table 3.2 summarizes the relation between electron arrangement and hybridization type. No matter how many atomic orbitals we mix together, the number of hybrid orbitals is always the same as the number of atomic orbitals with which we started:

N atomic orbitals always produce N hybrid orbitals.

So far, we have not considered whether terminal atoms, such as the Cl atoms in PCl_5, are hybridized. Because they are bonded to only one other atom, we cannot use bond angles to predict a hybridization scheme. However, spectroscopic data and calculation suggest that both s- and p-orbitals of terminal atoms take part in bond formation, and so it is reasonable to suppose that their orbitals are hybridized. The simplest model is to suppose that the three lone pairs and the bonding pair are arranged tetrahedrally and therefore that the chlorine atoms bond to the phosphorus atom by using sp^3 hybrid orbitals.

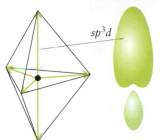

sp^3d

FIGURE 3.17 One of the five sp^3d hybrid orbitals, and their five directions, that account for a trigonal bipyramidal arrangement of electron pairs. The sp^3d hybridization scheme can be applied only when d-orbitals are available on the central atom.

TABLE 3.2 Hybridization and Molecular Shape*

Electron arrangement	Number of atomic orbitals	Hybridization of the central atom	Number of hybrid orbitals
linear	2	sp	2
trigonal planar	3	sp^2	3
tetrahedral	4	sp^3	4
trigonal bipyramidal	5	sp^3d	5
octahedral	6	sp^3d^2	6

*Other combinations of s-, p-, and d-orbitals can give rise to the same or different shapes, but the combinations in the table are the most common.

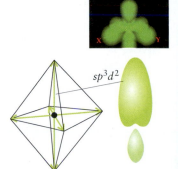

sp^3d^2

FIGURE 3.18 One of the six sp^3d^2 hybrid orbitals, and their six directions, that may be formed when d-orbitals are available and we need to account for an octahedral arrangement of electron pairs.

EXAMPLE 3.5   Sample exercise: Assigning a hybridization scheme

What is the hybridization of sulfur in PF_5?

SOLUTION

Draw the Lewis structure.	[Lewis structure of PF_5 showing P bonded to five F atoms]	
Determine the electron arrangement about the central atom.	Trigonal bipyramidal	
Identify the molecular shape.	Trigonal bipyramidal	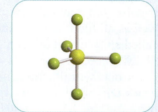
Select the same number of atomic orbitals as there are hybrid orbitals.	5	
Construct the hybrid orbitals, starting with the *s*-orbital, and proceeding to the *p*- and *d*-orbitals.	sp^3d	

SELF-TEST 3.8A   Describe (a) the electron arrangement, (b) the molecular shape, and (c) the hybridization of the central chlorine atom in chlorine trifluoride.

[*Answer:* (a) Trigonal bipyramidal; (b) T-shaped; (c) sp^3d]

SELF-TEST 3.8B   Describe (a) the electron arrangement, (b) the molecular shape, and (c) the hybridization of the central atom in BrF_4^-.

A hybridization scheme is adopted to match the electron arrangement of the molecule. Valence-shell expansion requires the use of d-orbitals.

3.7 Characteristics of Multiple Bonds

Atoms of the Period 2 elements C, N, and O readily form double bonds with one another, with themselves, and (especially for oxygen) with atoms of elements in later periods. However, double bonds are rarely found between atoms of elements in Period 3 and later periods, because the atoms are so large and bond lengths consequently so great that it is difficult for their *p*-orbitals to take part in effective side-by-side overlap.

To describe carbon–carbon double bonds, we use the pattern provided by ethene, $CH_2{=}CH_2$. We know from experimental data that all six atoms in ethene lie in the same plane, with HCH and CCH bond angles of 120°. This angle suggests a trigonal planar electron arrangement and sp^2 hybridization for each C atom (**40**). Each of the three hybrid orbitals on the C atom has one electron available for bonding; the fourth valence electron of each C atom occupies the unhybridized $2p$-orbital, which is perpendicular to the plane formed by the hybrids. The two carbon atoms form a σ-bond by overlap of an sp^2 hybrid orbital on each atom. The H atoms form σ-bonds with the remaining lobes of the sp^2 hybrids. The electrons in the two unhybridized $2p$-orbitals form a π-bond through side-by-side overlap. Figure 3.19 shows that the electron density in the π-bond lies above and below the axis of the C—C σ-bond.

In benzene, the C atoms and their attached H atoms all lie in the same plane, with the C atoms forming a hexagonal ring. To describe the bonding in the Kekulé structures of benzene (Section 2.8) in terms of VB theory, we need hybrid orbitals that match the 120° bond angles of the hexagonal ring. Therefore, we take each carbon atom to be sp^2 hybridized, as in ethene (Fig. 3.20). There is one electron in each of the three hybrid orbitals and one electron in an unhybridized $2p$-orbital perpendicular to the plane of the hybrids. Spin-pairing and overlap of the sp^2 hybrid orbitals on neighboring carbon atoms results in six σ-bonds between them, and spin pairing and overlap between the remaining sp^2 hybrid and hydrogen $1s$-electrons results in six carbon–hydrogen bonds. Finally, spin pairing and side-by-side overlap of the $2p$-orbital on each C atom results in a π-bond between each carbon atom and one of its neighbors (Fig. 3.21). The resulting pattern of π-bonds matches either of the two Kekulé structures, and the overall structure is a resonance hybrid of the two. This resonance ensures that the electrons in the π-bonds are spread around the entire ring (Fig. 3.22).

The presence of a carbon–carbon double bond strongly influences the shape of a molecule because it prevents one part of a molecule from rotating relative to another part. The double bond of ethene, for example, holds the entire molecule flat. Figure 3.19 shows that the two $2p$-orbitals overlap best if the two CH_2 groups lie in the same plane. In order for the molecule to rotate about the double bond, the π-bond would need to break and reform.

Double bonds and their influence on molecular shape are vitally important for living organisms. For instance, they enable you to read these words. Vision depends on the shape of the molecule retinal in the retina of the eye. *cis*-Retinal is held rigid by its double bonds (**41**). When light enters the eye, it excites an electron out of the π-bond marked by the arrow. The double bond is now weaker, and the molecule is free to rotate about the remaining σ-bond. When the excited electron falls back, the molecule has rotated about the double

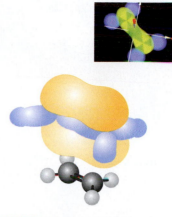

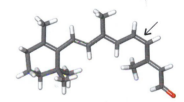

FIGURE 3.19 A view of the bonding pattern in ethene (ethylene), showing the framework of σ-bonds and the single π-bond formed by side-to-side overlap of unhybridized C2p-orbitals. The double bond is resistant to twisting because twisting would reduce the overlap between the two C2p-orbitals and weaken the π-bond.

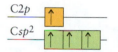

40 sp^2 hybridized carbon

41 *cis*-Retinal

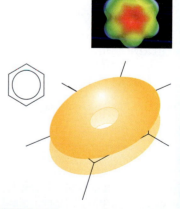

FIGURE 3.20 The framework of σ-bonds in benzene: each carbon atom is sp^2 hybridized, and the array of hybrid orbitals matches the bond angles (of 120°) in the hexagonal molecule. The bonds around only one carbon atom are labeled; all the others are the same.

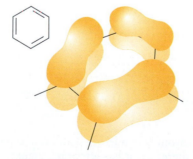

FIGURE 3.21 Unhybridized carbon 2p-orbitals can form a π-bond with either of their immediate neighbors. Two arrangements are possible, each one corresponding to a different Kekulé structure. One Kekulé structure and the corresponding π-bonds are shown here.

FIGURE 3.22 As a result of resonance between two structures like the one shown in Fig. 3.21 (corresponding to resonance of the two Kekulé structures), the π-electrons form a double doughnut-shaped cloud above and below the plane of the ring.

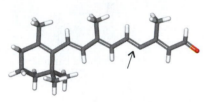

$\pi(C2p_x, C2p_x)$

$\pi(C2p_y, C2p_y)$

z

$\sigma(Csp, Csp)$

(a)

(b)

FIGURE 3.23 The pattern of bonding in ethyne (acetylene). The carbon atoms are *sp* hybridized, and the two remaining *p*-orbitals on each C atom form two π-bonds. (a) The resulting pattern is very similar to that for N_2 (Fig. 3.12), but C—H groups replace the two N atoms. (b) Although the two π orbitals are built from *p*-orbitals, the overall electron density turns out to have cylindrical symmetry.

bond and is now trapped in its trans shape (**42**). This change in shape triggers a signal along the optic nerve and is interpreted by the brain as the sensation of vision.

Now consider the alkynes, hydrocarbons with carbon–carbon triple bonds. The Lewis structure of the linear molecule ethyne (acetylene) is H—C≡C—H. To describe the bonding in a linear molecule, we need a hybridization scheme that produces two equivalent orbitals at 180° from each other: this is *sp* hybridization. Each C atom has one electron in each of its two *sp* hybrid orbitals and one electron in each of its two perpendicular unhybridized 2*p*-orbitals (**43**). The electrons in the *sp* hybrid orbitals on the two carbon atoms pair and form a carbon–carbon σ-bond. The electrons in the remaining *sp* hybrid orbitals pair with hydrogen 1*s*-electrons to form two carbon–hydrogen σ-bonds. The electrons in the two perpendicular sets of 2*p*-orbitals pair with a side-by-side overlap, forming two π-bonds at 90° to each other. As in the N_2 molecule, the electron density in the σ-bonds forms a cylinder about the C—C bond axis. The resulting bonding pattern is shown in Fig. 3.23.

A carbon–carbon double bond is stronger than one carbon–carbon single bond but weaker than the sum of two single bonds (Section 2.15). A carbon–carbon triple bond is weaker than the sum of three carbon–carbon single bonds. Recall that a single C—C bond is a σ-bond, but the additional bonds in a multiple bond are π-bonds. One reason for the difference in strength is that the side-by-side overlap of *p*-orbitals that results in a π-bond is not as great as the end-to-end overlap that results in a σ-bond.

42   *trans*-Retinal

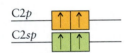

43   *sp* hybridized carbon

C2*p*	↑	↑
C2*sp*	↑	↑

EXAMPLE 3.6   **Accounting for the structure of a molecule with multiple bonds**

Account for the structure of a formic acid molecule, HCOOH, in terms of hybrid orbitals, bond angles, and σ- and π-bonds. The C atom is attached to an H atom, a terminal O atom, and an —OH group.

STRATEGY Use the VSEPR model to identify the shape of the molecule and then assign the hybridization consistent with that shape. All single bonds are σ-bonds and multiple bonds are composed of a σ-bond and one or more π-bonds. Because the C atom is attached to three atoms, we anticipate that its hybridization scheme is sp^2 and that one unhybridized *p*-orbital remains. Finally, we form σ- and π-bonds by allowing the orbitals to overlap.

SOLUTION

Draw the Lewis structure.

$$\begin{array}{c} \ddot{\text{O}} \\ \parallel \\ \text{H}-\text{C} \\ | \\ \ddot{\text{O}}-\text{H} \end{array}$$

Use the VSEPR model to identify the electron arrangements around the central C and O atoms.

The C atom is bonded to 3 atoms and has no lone pairs; therefore, it has a trigonal planar arrangement. The O atom in the —OH group has two single bonds and two lone pairs; thus, it has a tetrahedral electron arrangement.

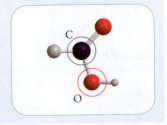

Identify the hybridization and bond angles.	C atom: trigonal planar, so 120° bond angle; sp^2 hybridized. O atom of the —OH group: tetrahedral, so bond angles close to 109.5°; sp^3 hybridized.	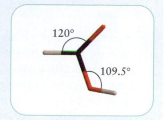
Form the bonds.	A π-bond forms by the overlap of the p-orbital on the C atom with the p-orbital on the terminal O atom.	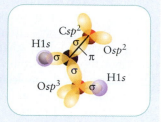

SELF-TEST 3.9A Describe the structure of the carbon suboxide molecule, C_3O_2, in terms of hybrid orbitals, bond angles, and σ- and π-bonds. The atoms lie in the order OCCCO.
[*Answer:* Linear; bond angles all 180°; each C atom is sp hybridized and forms one σ-bond and one π-bond to each adjacent C or O atom.]

SELF-TEST 3.9B Describe the structure of the propene molecule, CH_3—CH=CH_2, in terms of hybrid orbitals, bond angles, and σ- and π-bonds.

Multiple bonds are formed when an atom forms a σ-bond by using an sp or sp^2 hybrid orbital and one or more π-bonds by using unhybridized p-orbitals. The side-by-side overlap that forms a π-bond makes a molecule resistant to twisting, results in bonds weaker than σ-bonds, and prevents atoms with large radii from forming multiple bonds.

MOLECULAR ORBITAL THEORY

Lewis's theory of the chemical bond was brilliant, but it was little more than guesswork inspired by insight. Lewis had no way of knowing why an electron pair was so important for the formation of covalent bonds. Valence-bond theory explained the importance of the electron pair in terms of spin-pairing but it could not explain the properties of some molecules. Molecular orbital theory, which is also based on quantum mechanics and was introduced in the late 1920s by Mulliken and Hund, has proved to be the most successful theory of the chemical bond: it overcomes all the deficiencies of Lewis's theory and is easier to use in calculations than valence-bond theory.

3.8 The Limitations of Lewis's Theory

According to Lewis's approach and valence-bond theory, we should describe the bonding in O_2 as having all the electrons paired. However, oxygen is a paramagnetic gas (Fig. 3.24 and Box 3.2), and paramagnetism is a property of *unpaired* electrons. The paramagnetism of O_2 therefore contradicts both the Lewis structure and the valence-bond description of the molecule.

Lewis's theory also fails to account for the compound diborane, B_2H_6, a colorless gas that bursts into flame on contact with air. The problem is that diborane has only 12 valence electrons (three from each B atom, one from each H atom); but, for a Lewis structure, it needs at least seven bonds, and therefore 14 electrons, to bind the eight atoms together! Diborane is an example of an **electron-deficient compound**, a compound with too few valence electrons to be assigned a valid Lewis structure. Valence-bond theory can account for the structures of electron-deficient compounds in terms of resonance, but the explanation is not straightforward.

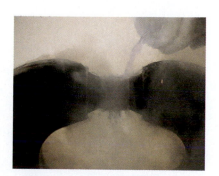

FIGURE 3.24 The paramagnetic properties of oxygen are evident when liquid oxygen is poured between the poles of a magnet. The liquid sticks to the magnet instead of flowing past it.

BOX 3.2   HOW DO WE KNOW . . . THAT ELECTRONS ARE NOT PAIRED?

Most common materials are *diamagnetic*, which means that a sample of the material tends to move out of a magnetic field. The effect is quite small, but it can be detected by hanging a long, thin sample from the pan of a balance and letting it lie between the poles of an electromagnet. This arrangement, which was once the primary technique used to measure the magnetic properties of samples, is called a *Gouy balance*. When the electromagnet is turned on, a diamagnetic sample tends to move upward, out of the field, so it appears to weigh less than in the absence of the field. The diamagnetism arises from the effect of the magnetic field on the electrons in the molecule: the field forces the electrons to circulate through the nuclear framework. Because electrons are charged particles, this circulation corresponds to an electric current circulating within the molecule. That current gives rise to its own magnetic field, which opposes the applied field. The sample tends to move out of the field so as to minimize this opposing field.

Compounds with unpaired electrons are *paramagnetic*. They tend to move into a magnetic field and can be identified because they seem to weigh more in a Gouy balance when a magnetic field is applied than when it is absent. Paramagnetism arises from the electron spins, which behave like tiny bar magnets that tend to line up with the applied field. The more

that can line up in this way, the greater the lowering of energy and the stronger the tendency of the sample to move into the applied field. Oxygen is a paramagnetic substance because it has two unpaired electrons: this property is used to detect the concentration of oxygen in incubators. All radicals are paramagnetic. Many compounds of the *d*-block elements are paramagnetic because they have various numbers of unpaired *d*-electrons.

The modern approach to measuring magnetic properties is to use a *superconducting quantum interference device* (a SQUID), which is highly sensitive to small magnetic fields and can make very precise measurements on small samples.

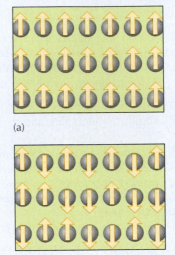

(a)

(b)

A Gouy balance is used to observe the magnetic character of a sample by detecting the extent to which it is drawn into (paramagnetic substances) or driven out of (diamagnetic substances) a magnetic field.

(a) In a magnetic field, the spins of electrons in both paramagnetic and *ferromagnetic* substances are aligned (ferromagnetism is described in Chapter 16). (b) The spins of electrons in a paramagnetic substance return to a random orientation after an applied magnetic field is removed. However, the spins of electrons in a ferromagnetic substance remain aligned after a magnetic field is removed.

The development of **molecular orbital theory** (MO theory) in the late 1920s overcame these difficulties. It explains why the electron pair is so important for bond formation and predicts that oxygen is paramagnetic. It accommodates electron-deficient compounds such as the boranes just as naturally as it deals with methane and water. Furthermore, molecular orbital theory can be extended to account for the structures and properties of metals and semiconductors. It can also be used to account for the electronic spectra of molecules, which arise when an electron makes a transition from an occupied molecular orbital to a vacant molecular orbital.

The VB and MO theories are both procedures for constructing approximations to the wavefunctions of electrons, but they construct these approximations in different ways. The language of valence-bond theory, in which the focus is on bonds between pairs of atoms, pervades the whole of organic chemistry, where chemists speak of σ- and π-bonds between particular pairs of atoms, hybridization, and resonance. However, molecular orbital theory, in which the focus is on electrons that spread throughout the nuclear framework and bind the entire collection of atoms together, has been developed far more extensively than valence-bond theory and is the procedure almost universally used in calculations of molecular structures, like those described in Major Technique 5 following Chapter 13.

> *Unlike Lewis's theory, molecular orbital theory can account for the existence of electron-deficient compounds and the paramagnetism of oxygen.*

3.9 Molecular Orbitals

In molecular orbital theory, electrons occupy orbitals called **molecular orbitals** that spread throughout the entire molecule. In other words, whereas in the Lewis and valence-bond models of molecular structure the electrons are localized on atoms or between pairs of atoms, in molecular orbital theory all valence electrons are **delocalized** over the whole molecule, not confined to individual bonds.

In this section we start, as in valence-bond theory, with a simple molecule, H_2, and in the following sections extend the same principles to more complex molecules and solids. In every case, molecular orbitals are built by adding together—the technical term is **superimposing**—atomic orbitals belonging to the valence shells of the atoms in the molecule. For example, a molecular orbital for H_2 is

$$\psi = \psi_{A1s} + \psi_{B1s} \qquad (1)^*$$

where ψ_{A1s} is a $1s$-orbital centered on one atom (A) and ψ_{B1s} is a $1s$-orbital centered on the other atom (B). The molecular orbital ψ is called a **linear combination of atomic orbitals** (LCAO). Any molecular orbital formed from a linear combination of atomic orbitals on different atoms is called an **LCAO-MO.** Note that at this stage there are no electrons in the molecular orbital: a molecular orbital is just a combination—in this case, a sum—of wavefunctions. Like atomic orbitals, the molecular orbital in Eq. 1 is a well-defined mathematical function that can be evaluated at each point in space and pictured in three dimensions.

The LCAO-MO in Eq. 1 turns out to have a lower energy than either of the atomic orbitals used in its construction. The two atomic orbitals are like waves centered on different nuclei. Between the nuclei, the waves interfere constructively with each other in the sense that the total amplitude of the wavefunction is increased where they overlap (Fig. 3.25). The increased amplitude in the internuclear region means that there is an enhanced probability density between the nuclei. Any electron that occupies that molecular orbital, therefore, is attracted to both nuclei and so has a lower energy than when it is confined to an atomic orbital on one atom. Moreover, because the electron now occupies a greater volume than when it is confined to a single atom, it also has a lower kinetic energy, just like a particle confined to a bigger box (Section 1.7). A combination of atomic orbitals that results in an overall lowering of energy, like that in Eq. 1, is called a **bonding orbital.**

An important feature of MO theory is that

When *N* atomic orbitals overlap, they form *N* molecular orbitals.

In molecular hydrogen, where we are building LCAO-MOs from two atomic orbitals, we expect *two* molecular orbitals. In the second molecular orbital, the two atomic orbitals interfere *destructively* where they overlap. This orbital has the form

$$\psi = \psi_{A1s} - \psi_{B1s} \qquad (2)^*$$

The negative sign indicates that the amplitude of ψ_{B1s} *subtracts* from the amplitude of ψ_{A1s} where they overlap (Fig. 3.26), and there is a nodal surface where the atomic orbitals cancel completely. In a hydrogen molecule, the nodal surface is a plane that lies halfway between the two nuclei. If an electron occupies this orbital, it is largely excluded from the internuclear region and consequently has a higher energy than when it occupies one of the atomic orbitals alone. A combination of atomic orbitals that results in an overall raising of energy, like that in Eq. 2, is called an **antibonding orbital.**

A note on good practice: What the negative sign in Eq. 2 really represents is that we reverse the sign of ψ_{B1s} everywhere (so a peak becomes a trough and vice versa), then superimpose the resulting wavefunction on ψ_{A1s}.

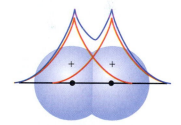

FIGURE 3.25 When two $1s$-orbitals overlap in the same region of space in such a way that their wavefunctions have the same signs in that region, their wavefunctions (red lines) interfere constructively and give rise to a region of enhanced amplitude between the two nuclei (blue line).

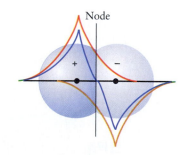

FIGURE 3.26 When two $1s$-orbitals overlap in the same region of space in such a way that their wavefunctions have opposite signs, the wavefunctions (red and orange lines) interfere destructively and give rise to a region of diminished amplitude and a node between the two nuclei (blue line).

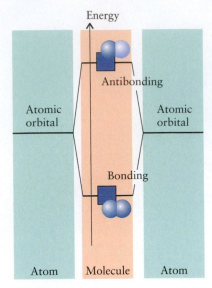

FIGURE 3.27 A molecular orbital energy-level diagram for the bonding and antibonding molecular orbitals that can be built from two s-orbitals. Different signs of the s-orbitals (reflecting how they are combined together to form the molecular orbital) are depicted by the different shades of blue.

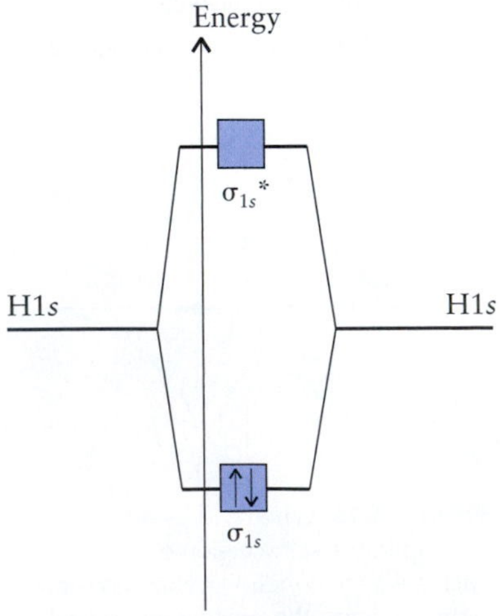

The relative energies of the original atomic orbitals and the bonding and antibonding molecular orbitals are shown in a **molecular orbital energy-level diagram** like that in Fig. 3.27. The increased energy of an antibonding orbital is about equal to or a little greater than the lowering of the energy of the corresponding bonding orbital.

Molecular orbitals are formed by combining atomic orbitals: when atomic orbitals interfere constructively, they give rise to bonding orbitals; when they interfere destructively, they give rise to antibonding orbitals. N atomic orbitals combine to give N molecular orbitals.

3.10 The Electron Configurations of Diatomic Molecules

In the molecular orbital description of homonuclear diatomic molecules, we first build all possible molecular orbitals from the available valence-shell atomic orbitals. Then we accommodate the valence electrons in molecular orbitals by using the same procedure we used in the building-up principle for atoms (Section 1.13). That is,

1   Electrons are accommodated in the lowest-energy molecular orbital, then in orbitals of increasingly higher energy.

2   According to the Pauli exclusion principle, each molecular orbital can accommodate up to two electrons. If two electrons are present in one orbital, they must be paired.

3   If more than one molecular orbital of the same energy is available, the electrons enter them singly and adopt parallel spins (Hund's rule).

We shall illustrate these rules first with H_2 and then with other diatomic molecules. The same principles apply to polyatomic molecules, but their molecular orbitals are more complicated and their energies are harder to predict. Mathematical software for calculating the molecular orbitals and their energies is now widely available, and we shall show some of the results that it provides.

In H_2, two 1s-orbitals (one on each atom) merge to form two molecular orbitals. We denote the bonding orbital σ_{1s} and the antibonding orbital σ_{1s}^{*}. The 1s in the notation shows the atomic orbitals from which the molecular orbitals are formed. The σ indicates that we have built a "σ-orbital," a sausage-shaped orbital. More formally, a **σ-orbital** is a molecular orbital that has cylindrical symmetry and no nodal plane that contains the internuclear axis. Two electrons, one from each H atom, are available. Both occupy the bonding orbital (the lower-energy orbital) and result in the configuration σ_{1s}^{2} (Fig. 3.28). Because only the bonding orbital is occupied, the energy of the molecule is lower than that of the separate atoms, and hydrogen exists as H_2 molecules. Two electrons in a σ-orbital form a σ-bond, like the σ-bond in VB theory. However, even a single electron may be able to hold two atoms together with about half the strength of an electron pair, and so—in contrast to Lewis's theory and VB theory—an electron pair is not required for a bond. A pair is just the maximum number of electrons allowed by the Pauli exclusion principle to occupy any one molecular orbital. Even a single electron can act to bond atoms together.

Now we extend these ideas to other homonuclear diatomic molecules of Period 2 elements. The first step is to build up the molecular orbital energy-level diagram from the valence-shell atomic orbitals provided by the atoms. Because Period 2 atoms have 2s- and 2p-orbitals in their valence shells, we form molecular orbitals from the overlap of these atomic orbitals. There are a total of eight atomic orbitals (one 2s- and three 2p-orbitals on each atom); so we can expect to build eight molecular orbitals. The two 2s-orbitals overlap to form two σ-orbitals, one bonding (the σ_{2s}-orbital) and the other antibonding (the σ_{2s}^{*}-orbital); these orbitals resemble the σ_{1s}- and σ_{1s}^{*}-orbitals in H_2. The six 2p-orbitals (three on each neighboring atom) form the remaining six molecular orbitals. They can overlap in two distinct ways. The two 2p-orbitals that are

FIGURE 3.28 The two electrons in an H_2 molecule occupy the lower-energy (bonding) molecular orbital and result in a stable molecule.

directed toward each other along the internuclear axis form a bonding σ-orbital (σ_{2p}) and an antibonding σ*-orbital (σ_{2p}^*) where they overlap (Fig. 3.29). The two 2p-orbitals on each atom that are perpendicular to the internuclear axis overlap side by side to form bonding and antibonding "π-orbitals" (Fig. 3.30). A **π-orbital** is a molecular orbital with one nodal plane that contains the internuclear axis. There are two perpendicular 2p-orbitals on each atom, and so four molecular orbitals—two bonding π_{2p}-orbitals and two antibonding π_{2p}^*-orbitals—are formed by their overlap.

Detailed calculation shows that there are some small differences in the order of energy levels from molecule to molecule (Box 3.3). Figure 3.31 shows the order for the Period 2 elements with the exception of O_2 and F_2, which lie in the order shown in Fig. 3.32. The order of energy levels is easy to explain for these two molecules. First, because each atom has many electrons that contribute to shielding, the 2s-orbitals lie well below the 2p-orbitals and we can think of building σ-orbitals from the two sets of atomic orbitals separately. However, because the atoms of elements earlier in the period have fewer electrons, their 2s- and 2p-orbitals have more similar energies than in O and F. As a result, it is no longer possible to think of a σ-orbital as being formed from either the 2s-orbitals or the $2p_z$-orbitals separately, and all four of these orbitals must be used to build the four σ-orbitals. It is then hard to predict without detailed calculation where these four orbitals will lie, and it turns out that they in fact lie where we show them in Fig. 3.31.

Once we know what molecular orbitals are available, we can construct the ground-state electron configurations of the molecules by using the building-up principle. For example, consider N_2. Because nitrogen belongs to Group 15/V, each atom supplies five valence electrons. A total of ten electrons must therefore be assigned to the eight molecular orbitals shown in Fig. 3.31. Two fill the σ_{2s}-orbital. The next two fill the σ_{2s}^*-orbital. Next in line for occupation are the two π_{2p}-orbitals, which can hold a total of four electrons. The last two electrons then enter the σ_{2p}-orbital. The ground configuration is therefore

$$N_2: \sigma_{2s}^2 \sigma_{2s}^{*2} \pi_{2p}^4 \sigma_{2p}^2$$

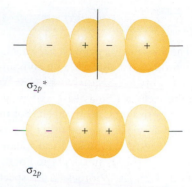

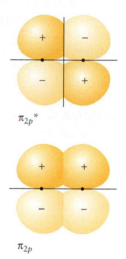

FIGURE 3.29 Two p-orbitals can overlap to give bonding (lower) and antibonding (upper) σ-orbitals. Note that the antibonding combination has a node between the two nuclei. Both σ-orbitals have nodes passing through the nuclei, but no nodes along the axis of the bond.

FIGURE 3.30 Two p-orbitals can overlap side by side to give bonding (below) and antibonding (above) π-orbitals. Note that the latter has a nodal plane between the two nuclei. Both orbitals have a nodal plane through the two nuclei and look like p-orbitals when viewed along the internuclear axis.

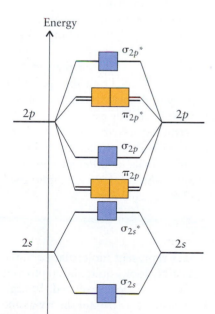

FIGURE 3.31 A typical molecular orbital energy-level diagram for the homonuclear diatomic molecules Li_2 through N_2. Each box represents one molecular orbital and can accommodate up to two electrons.

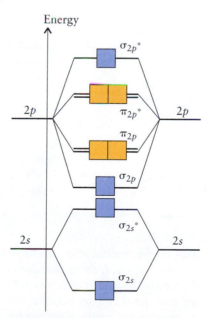

FIGURE 3.32 The molecular orbital energy-level diagram for the homonuclear diatomic molecules on the right-hand side of Period 2, specifically O_2 and F_2.

BOX 3.3   HOW DO WE KNOW . . . THE ENERGIES OF MOLECULAR ORBITALS?

The energies of orbitals are calculated today by solving the Schrödinger equation with computer software. The commercial software available is now so sophisticated that this approach can be as easy as typing in the name of the molecule or drawing it on screen. But these values are theoretical. How do we determine orbital energies experimentally?

One of the most direct methods is photoelectron spectroscopy (PES), an adaptation of the photoelectric effect (Section 1.2). A photoelectron spectrometer (see illustration below) contains a source of high-frequency, short-wavelength radiation. Ultraviolet radiation is used most often for molecules, but x-rays are used to explore orbitals buried deeply inside solids. Photons in both frequency ranges have so much energy that they can eject electrons from the molecular orbitals they occupy.

Let's suppose that the frequency of the radiation is ν (nu), so each photon has an energy $h\nu$. An electron occupying a molecular orbital lies at an energy $E_{orbital}$ below the zero of energy (corresponding to an electron far removed from the molecule). A photon that collides with the electron can eject it from the molecule if the photon can supply at least that much energy. The remaining energy of the photon, $h\nu - E_{orbital}$, then appears as the kinetic energy, E_K, of the ejected electron:

$$h\nu - E_{orbital} = E_K$$

We know ν, the frequency of the radiation being used to bombard the molecules; so, if we could measure the kinetic energy of the ejected electron, E_K, we could solve this expression to find the orbital energy, $E_{orbital}$.

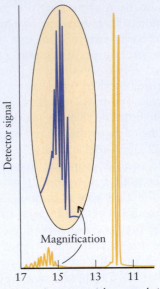

The photoelectron spectrum of nitrogen (N_2) has several peaks, a pattern indicating that electrons can be found in several energy levels in the molecule. Each main group of lines corresponds to the energy of a molecular orbital. The additional "fine structure" on some of the groups of lines is due to the excitation of molecular vibration when an electron is expelled.

The kinetic energy of an ejected electron depends on its speed, v, because $E_K = \frac{1}{2}m_e v^2$ (Section A). A photoelectron spectrometer acts like a mass spectrometer in that it measures v for electrons just as a mass spectrometer can measure v for ions, as we saw in Section B. In this method, the electrons pass through a region of electric or magnetic field, which deflects their path. As the strength of the field is changed, the paths of the electrons change, too, until they fall on a detector and generate a signal. From the strength of the field required to obtain a signal, we can work out the speed of electrons ejected from a given orbital. From the speed, we can calculate the kinetic energy of the electrons and so obtain the energy of the orbital from which they came.

The photoelectron spectrum of nitrogen is shown in the second illustration. There are several peaks, corresponding to electrons being ejected from orbitals of different energy. A detailed analysis shows that the spectrum is a good portrayal of the qualitative structure (as depicted in **44**).

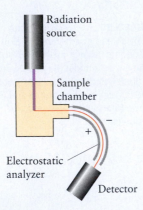

Diagram of a photoelectron spectrometer.

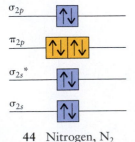

44 Nitrogen, N_2

This configuration is shown as (**44**), where the boxes represent molecular orbitals.

At first sight, the molecular orbital description of N_2 looks quite different from the Lewis description (:N≡N:). However, it is, in fact, very closely related. We can see their similarity by defining the **bond order** (b) in molecular orbital theory as the net number of bonds, allowing for the cancellation of bonds by antibonds:

Bond order = $\frac{1}{2}$ × (number of electrons in bonding orbitals −

number of electrons in antibonding orbitals)

$$b = \tfrac{1}{2} \times (N - N^*) \tag{3}*$$

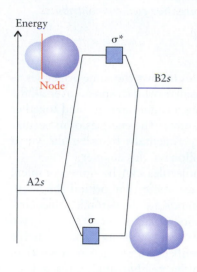

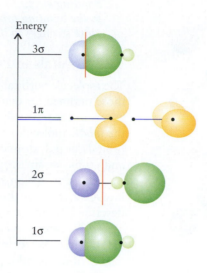

FIGURE 3.33 A typical σ molecular orbital energy-level diagram for a heteronuclear diatomic molecule AB; the relative contributions of the atomic orbitals to the molecular orbitals are represented by the relative sizes of the spheres and the horizontal position of the boxes. In this case, A is the more electronegative of the two elements.

FIGURE 3.34 A schematic diagram of the molecular orbitals of HF. The orbitals are labeled by using a standard convention. The two 1π-orbitals are nonbonding; the three σ-orbitals are progressively more antibonding (1σ, bonding; 3σ, antibonding).

- In an *ionic bond*, the coefficient belonging to one ion is nearly zero because the other ion captures almost all the electron density.

- In a *polar covalent bond*, the atomic orbital belonging to the more electronegative atom has the lower energy, and so it makes the larger contribution to the lowest energy molecular orbital (Fig. 3.33). Conversely, the contribution to the highest-energy (most antibonding) orbital is greater for the higher-energy atomic orbital, which belongs to the less electronegative atom.

To find the ground-state electron configurations of heteronuclear diatomic molecules, we use the same approach that we used for homonuclear diatomic molecules; but first we must modify the energy-level diagrams. For example, consider the HF molecule. The σ-bond in this molecule consists of an electron pair in a σ-orbital built from the $F2p_z$-orbital and the $H1s$-orbital. Because the electronegativity of fluorine is 4.0 and that of hydrogen is 2.2, we can expect the bonding σ-orbital to be mainly $F2p_z$ and the antibonding σ*-orbital to be mainly $H1s$ in character. These expectations are confirmed by calculation (Fig. 3.34). Because the two electrons in the bonding orbital are more likely to be found in the $F2p_z$-orbital than in the $H1s$-orbital, there is a partial negative charge on the F atom and a partial positive charge on the H atom.

The molecular orbital energy-level diagrams of heteronuclear diatomic molecules are much harder to predict qualitatively and we have to calculate each one explicitly because the atomic orbitals contribute differently to each one. Figure 3.35 shows the calculated scheme typically found for CO and NO. We can use this diagram to state the electron configuration by using the same procedure as for homonuclear diatomic molecules.

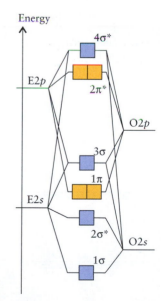

FIGURE 3.35 The molecular orbital schemes typical of those calculated for a diatomic oxide molecule, EO (where E = C for CO and E = N for NO). Note that the σ-orbitals are formed from mixtures of s- and p_z-orbitals on both atoms; accordingly, we label them simply 1σ, 2σ, etc. in order of increasing energy.

EXAMPLE 3.8 **Sample exercise: Writing the configuration of a heteronuclear diatomic molecule or ion**

Write the configuration of the ground state of the carbon monoxide molecule.

SOLUTION There are 4 + 6 = 10 valence electrons to accommodate in the orbitals shown in Fig. 3.35. The resulting configuration is shown in the illustration, and is

CO: $1\sigma^2 2\sigma^{*2} 1\pi^4 3\sigma^2$

SELF-TEST 3.11A Write the configuration of the ground state of the nitric oxide (nitrogen monoxide) molecule.

[*Answer:* $1\sigma^2 2\sigma^{*2} 1\pi^4 3\sigma^2 2\pi^{*1}$]

SELF-TEST 3.11B Write the configuration of the ground state of the cyanide ion, CN^-, assuming that its molecular orbital energy-level diagram is the same as that for CO.

Energy

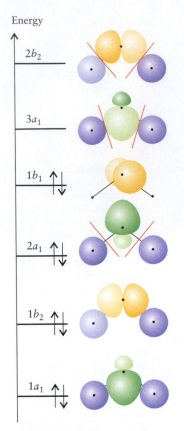

FIGURE 3.36 A schematic diagram of the molecular orbitals of H_2O. The orbitals are labeled according to the conventional notation for angular triatomic molecules. The orbitals become progressively more antibonding ($1a_1$, bonding; $2b_2$, antibonding). The orbital labeled $1b_1$ is nonbonding, because it is composed of an orbital ($O2p_x$) that does not overlap either hydrogen atom.

Energy

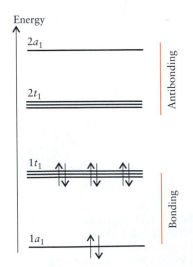

FIGURE 3.37 The molecular orbital energy-level diagram for methane and the occupation of the orbitals by the eight valence electrons of the atoms.

Bonding in heteronuclear diatomic molecules involves an unequal sharing of the bonding electrons. The more electronegative element contributes more strongly to the bonding orbitals, whereas the less electronegative element contributes more strongly to the antibonding orbitals.

3.12 Orbitals in Polyatomic Molecules

The molecular orbital theory of polyatomic molecules follows the same principles as those outlined for diatomic molecules, but the molecular orbitals spread over *all* the atoms in the molecule. An electron pair in a bonding orbital helps to bind together the *whole* molecule, not just an individual pair of atoms. The energies of molecular orbitals in polyatomic molecules can be studied experimentally by using ultraviolet and visible spectroscopy (see Major Technique 2, following this chapter).

The description of bonding in polyatomic molecules can be quite complex. However, we can illustrate qualitatively how to use molecular orbital theory to describe bonding in one of the simplest but most important polyatomic molecules, water. A water molecule has six atomic orbitals (one $O2s$, three $O2p$, and two $H1s$). These six orbitals are used to build six molecular orbitals in which the degree of net bonding character is related to the number of internuclear nodes (Fig. 3.36). The molecular orbital with no nodes between neighboring atoms is fully bonding and, when occupied, contributes to holding all the atoms together; the one with a node between all neighboring pairs of atoms is fully antibonding and, if occupied, contributes to pushing all the atoms apart. There are eight electrons to accommodate: six are provided by the O atom and one comes from each H atom. Two electrons in the lowest energy, most bonding orbital pull all three atoms together. Two electrons in the orbital composed purely of the $O2p_x$-orbital (where x lies perpendicular to the molecular plane) are localized completely on the oxygen atom and do not contribute directly to the bonding. An orbital such as this one that, when occupied, contributes to neither bonding nor antibonding is classified as a **nonbonding orbital.** In some cases, a nonbonding orbital is a single atomic orbital; in others, it may consist of a linear combination of atomic orbitals on atoms that are not neighbors and so overlap to a negligible extent.

When we apply molecular orbital theory to methane we see that valence bond theory misses an interesting point. Valence bond theory predicts that all eight bonding electrons in methane are in identical sp^3-orbitals and should therefore have the same energy. However, although all four bonds are identical, one of the electron pairs is actually slightly lower in energy than the others. This finding is easily explained by molecular orbital theory. In methane eight molecular orbitals are constructed from the four carbon valence orbitals and the four $1s$-orbitals from the four hydrogen atoms. These molecular orbitals have the energies shown in Fig. 3.37: we see that one bonding orbital lies lower in energy than the other three. Therefore, two of the electrons have a lower energy than the others. However, the overall electron density is the same in each bonding region because the molecular orbitals are delocalized and the electron density is spread out uniformly over all C—H regions.

A note on good practice: The concepts of promotion, hybridization, and resonance belong to valence bond theory, not molecular orbital theory. Instead, molecular orbitals are built from all the available atomic orbitals by noting whether or not they have the right shape to overlap with one another.

Another important polyatomic molecule is benzene, C_6H_6, the parent of the aromatic compounds. In the molecular orbital description of benzene, all thirty $C2s$-, $C2p$-, and $H1s$-orbitals contribute to molecular orbitals spreading over all twelve atoms (six C plus six H). The orbitals in the plane of the ring (the $C2s$-, $C2p_x$-, and $C2p_y$-orbitals on each carbon atom and all six $H1s$-orbitals) form delocalized σ-orbitals that bind the C atoms together and link the H atoms to the C atoms. The six $C2p_z$-orbitals, which are perpendicular to the ring, contribute to six delocalized π-orbitals that spread all the way around the ring. However, chemists

FIGURE 3.38 The six π-orbitals of benzene; locations of nodes are shown on the left. Note that the orbitals range from fully bonding (no internuclear nodes) to fully antibonding (six internuclear nodes). The zero of energy corresponds to the total energy of the separated atoms. The three orbitals with negative energies have net bonding character.

(other than those carrying out detailed calculations, who use a pure MO scheme) commonly mix MO and VB descriptions when discussing organic molecules. Because the language of hybridization and orbital overlap is well suited to the description of σ-bonds, chemists typically express the σ-framework of molecules in VB terms. Thus, they think of the σ-framework of benzene as formed by the overlap of sp^2 hybridized orbitals on neighboring atoms. Then, because delocalization is such an important feature of the π-bonding component of **conjugated** double bonds, which alternate as in —C=C—C=C—C=C—, they treat such π-bonds in terms of MO theory. We shall do the same.

First, the VB part of the description of benzene. Each C atom is sp^2 hybridized, with one electron in each hybrid orbital. Each C atom has a p_z-orbital perpendicular to the plane defined by the hybrid orbitals, and it contains one electron. Two sp^2 hybrid orbitals on each C atom overlap and form σ-bonds with similar orbitals on the two neighboring C atoms, forming the 120° internal angle of the benzene hexagon. The third, outward-pointing sp^2 hybrid orbital on each C atom forms a σ-bond with a hydrogen atom. The resulting σ-framework is the same as that illustrated in Fig. 3.20.

Now for the MO part of the description. From the six $C2p_z$-orbitals, we form six delocalized π-orbitals: their shapes are shown in Fig. 3.38 and their energies are shown in Fig. 3.39. The character of the orbitals changes from net bonding to net antibonding as the number of internuclear nodes increases from none (fully bonding) to six (fully antibonding).

Each carbon atom provides one electron for the π-orbitals. Two electrons occupy the lowest energy, most bonding orbital; and the remaining four electrons occupy the next higher energy orbitals (two orbitals of the same energy). Figure 3.39 shows one of the reasons for benzene's great stability: the π-electrons occupy only orbitals with a net bonding effect; none of the destabilizing antibonding orbitals are occupied.

The delocalization of electrons accounts for the existence of electron-deficient molecules. Because the bonding influence of an electron pair is spread over all the atoms in the molecule, there is no need to provide one pair of electrons for each pair of atoms. A smaller number of pairs of electrons spread throughout the molecule may be able to bind all the atoms together, particularly if the nuclei are not highly charged and so do not repel one another strongly. This is the case in diborane, B_2H_6 (Section 14.14), where six electron pairs can hold the eight nuclei together.

Another mystery solved by molecular orbital theory is the existence of hypervalent compounds, compounds in which a central atom forms more bonds than allowed by the octet rule (recall Section 2.10). In valence bond theory, hybridization schemes are required to make sense of these compounds. For example, the expanded valence shell of Period 3 elements in compounds like SF_6 is explained as sp^3d^2 hybridization. However, the d-orbitals of sulfur lie at relatively high energies and may not be accessible for bonding. In molecular orbital theory a bonding scheme can be devised for SF_6 that does not involve d-orbitals. The four valence orbitals provided by the sulfur atom and the six orbitals of the fluorine atoms that point toward the sulfur atom, for a total of 10 atomic orbitals, result in 10 molecular orbitals with the energies shown in Fig. 3.40. The 12 electrons occupy the lowest six orbitals, which are either bonding or nonbonding, and so bind all the atoms together without needing to use d-orbitals. Because four bonding orbitals are occupied, the average bond order of each of the six S—F links is $\frac{2}{3}$.

Finally, we return to the opening remarks of the chapter and see how molecular orbital theory explains the colors of vegetation. The presence of highly delocalized electrons in the large molecules found in the petals of flowers and in fruit and

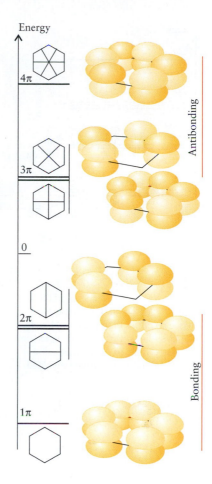

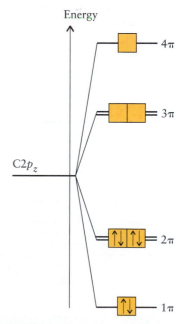

FIGURE 3.39 The molecular orbital energy-level diagram for the π-orbitals of benzene. In the ground state of the molecule, only the net bonding orbitals are occupied.

Energy

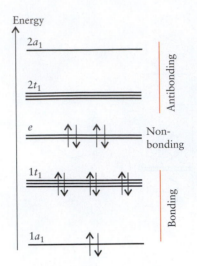

FIGURE 3.40 The molecular orbital energy-level diagram for SF_6 and the occupation of the orbitals by the 12 valence electrons of the atoms. Note that no antibonding orbitals are occupied and that there is a net bonding interaction even though no d-orbitals are involved.

Energy

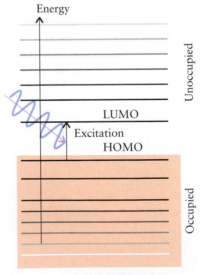

FIGURE 3.41 In large molecules, there are many closely spaced energy levels and the HOMO–LUMO gap is quite small. Such molecules are often colored because photons of visible light can be absorbed when electrons are excited from the HOMO to the LUMO.

vegetables is largely responsible for their colors. Because many carbon atoms contribute p-orbitals to the π system of these molecules, there are many molecular orbitals. An electron in a π system of such a molecule is like a particle in a large one-dimensional box. Because the "box" is very large, the energy levels are very close together. In these large molecules, the **highest occupied molecular orbital** (HOMO) is very close in energy to the **lowest unoccupied molecular orbital** (LUMO). As a result, it takes very little energy to excite an electron from a HOMO to a LUMO (Fig. 3.41). Photons of visible light have enough energy to excite the electrons across this energy gap, and the absorption of these photons results in the colors that we perceive. The coloring agent of carrots and the precursor of vitamin A, β-carotene (**46**), has a highly delocalized π system, as does lycopene (**47**), the red compound that gives tomatoes their color.

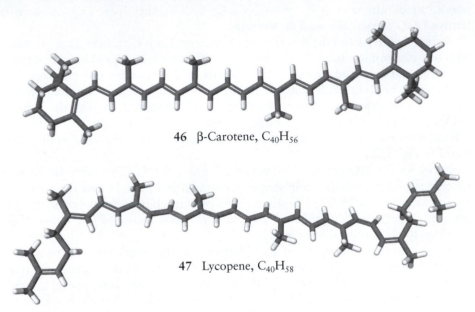

46 β-Carotene, $C_{40}H_{56}$

47 Lycopene, $C_{40}H_{58}$

According to molecular orbital theory, the delocalization of electrons in a polyatomic molecule spreads the bonding effects of electrons over the entire molecule.

IMPACT ON MATERIALS: ELECTRONIC CONDUCTION IN SOLIDS

Molecular orbital theory explains the electrical properties of solids by treating them as one huge molecule and supposing that their valence electrons occupy "molecular orbitals" that spread throughout the solid.

3.13 Bonding in the Solid State

Metals and semiconductors are **electronic conductors** in which an electric current is carried by delocalized electrons. A **metallic conductor** is an electronic conductor in which the electrical conductivity *decreases* as the temperature is raised. A **semiconductor** is an electronic conductor in which the electrical conductivity *increases* as the temperature is raised. In most cases, a metallic conductor has a much higher electrical conductivity than a semiconductor, but it is the temperature dependence of the conductivity that distinguishes the two types of conductors. An **insulator** does not conduct electricity. A **superconductor** is a solid that has zero resistance to an electric current. Some metals become superconductors at very low temperatures, at about 20 K or less, and some compounds also show superconductivity (see Box 5.2). High-temperature superconductors have enormous technological potential because they offer the prospect of more efficient power transmission and the generation of high magnetic fields for use in transport systems (Fig. 3.42).

FIGURE 3.42
Superconductors have the ability to levitate vehicles with embedded magnets. This picture shows an experimental zero-friction train in Japan, built to use helium-cooled metal superconductors.

Electrical conduction in metals can be explained in terms of molecular orbitals that spread throughout the solid. We have already seen that, when N atomic orbitals merge together in a molecule, they form N molecular orbitals. The same is true of a metal; but, for a metal, N is enormous (about 10^{23} for 10 g of copper, for example). Instead of the few molecular orbitals with widely spaced energies typical of small molecules, the huge number of molecular orbitals in a metal are so close together in energy that they form a nearly continuous band (Fig. 3.43).

Consider a metal such as sodium. Each atom contributes one valence orbital (the 3s-orbital in this case) and one valence electron. If there are N atoms in the sample, then the N 3s-orbitals merge to form a band of N molecular orbitals, of which half are net bonding and half are net antibonding. We say "net" bonding or antibonding because, in general and as in benzene, a molecular orbital is bonding between some neighbors and antibonding between others, depending on where its internuclear nodes lie; only the molecular orbital of lowest energy, with no internuclear nodes, is bonding between all neighboring atoms. The N electrons contributed by the N atoms occupy the orbitals according to the building-up principle. Because two electrons can occupy each orbital, the N electrons occupy the lower $\frac{1}{2}N$ bonding orbitals.

An empty or incompletely filled band of molecular orbitals is called a **conduction band.** Because neighboring orbitals lie so close together in energy, it takes very little additional energy to excite the electrons from the topmost filled molecular orbitals to the empty orbitals of the conduction band. Electrons in the conduction band can move freely through the solid, and so they can carry an electric current. The resistance of the metal increases with temperature because, when it is heated, the atoms vibrate more vigorously. The passing electrons collide with the vibrating atoms and do not pass through the solid so easily.

In an insulator, the valence electrons fill all the available molecular orbitals to give a full band called a **valence band.** There is a substantial **band gap**, a range of energies for which there are no orbitals, before the next band, the empty orbitals that form the conduction band, begins (Fig. 3.44). The electrons in the valence band can be excited into the conduction band only by a very large injection of energy. Because the valence band is full and because the conduction band is separated from it by a large energy gap, the electrons are not mobile and the solid does not conduct electricity.

Bonding in solids may be described in terms of bands of molecular orbitals. In metals, the conduction bands are incompletely filled orbitals that allow electrons to flow. In insulators, the valence bands are full and the large band gap prevents the promotion of electrons to empty orbitals.

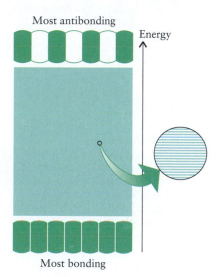

Most antibonding

Energy

Most bonding

FIGURE 3.43 A line of atoms gives rise to an almost continuous band of molecular orbital energies. At the lower edge of the band, the molecular orbitals are fully bonding; at the upper edge, the molecular orbitals are fully antibonding. The enlargement shows that, although the band of allowed energies appears to be continuous, it is in fact composed of discrete, closely spaced levels.

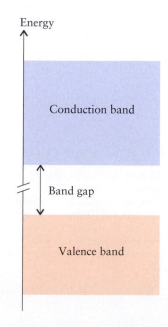

Energy

Conduction band

Band gap

Valence band

FIGURE 3.44 In a typical insulating solid, a full valence band is separated by a substantial energy gap from the empty conduction band. Note the break in the vertical scale.

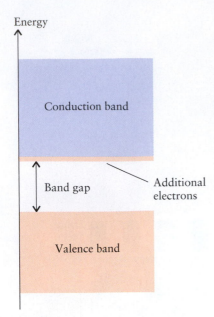

Energy

FIGURE 3.45 In an n-type semiconductor, the additional electrons supplied by the electron-rich dopant atoms enter the conduction band (forming the pink band at the bottom of the conduction band), where they can act as carriers for the current.

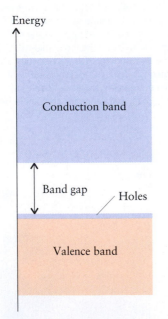

Energy

FIGURE 3.46 In a p-type semiconductor, the electron-poor dopant atoms effectively remove electrons from the valence band, and the "holes" that result (blue band at the top of the valence band) enable the remaining electrons to become mobile and conduct electricity through the valence band.

3.14 Semiconductors

In a semiconductor, an empty conduction band lies close in energy to a full valence band. As a result, as the solid is warmed, electrons can be excited from the valence band into the conduction band, where they can travel throughout the solid. Hence, the resistance of a semiconductor decreases as its temperature is raised.

The ability of a semiconductor to carry an electric current can also be enhanced by adding electrons to the conduction band or by removing some from the valence band. This modification is carried out chemically by **doping** the solid, or spreading small amounts of impurities throughout it. In one example, a minute amount of a Group 15/V element such as arsenic is added to very pure silicon. The arsenic increases the number of electrons in the solid: each Si atom (Group 14/IV) has four valence electrons, whereas each As atom (Group 15/V) has five. The additional electrons enter the upper, normally empty conduction band of silicon and allow the solid to conduct (Fig. 3.45). This type of material is called an **n-type semiconductor** because it contains excess *negatively* charged electrons. When silicon (Group 14/IV) is doped with indium (Group 13/III) instead of arsenic, the solid has fewer valence electrons than does pure silicon; so the valence band is no longer completely full (Fig. 3.46). We say that the valence band now contains "holes." Because the valence band is no longer full, it has been turned into a conduction band, and an electric current can flow. This type of semiconductor is called a **p-type semiconductor,** because the absence of negatively charged electrons is equivalent to the presence of *positively* charged holes. The doped solids are overall electrically neutral, because the nuclei of the doping atoms have a charge that matches the number of their electrons.

Solid-state electronic devices such as diodes, transistors, and integrated circuits contain **p–n junctions** in which a p-type semiconductor is in contact with an n-type semiconductor (Fig. 3.47). The structure of a p–n junction allows an electric current to flow in only one direction. When the electrode attached to the p-type semiconductor has a negative charge, the holes in the p-type semiconductor are attracted to it, the electrons in the n-type semiconductor are attracted to the other (positive) electrode, and current does not flow. When the polarity is reversed, with the negative electrode attached to the n-type semiconductor, electrons flow from the n-type semiconductor through the p-type semiconductor toward the positive electrode.

SELF-TEST 3.12A Which type of semiconductor is germanium doped with arsenic?

[*Answer:* n-type]

SELF-TEST 3.12B Which type of semiconductor is antimony doped with tin?

In semiconductors, empty levels are close in energy to filled levels.

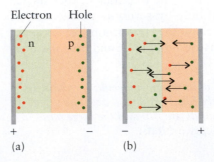

FIGURE 3.47 The structure of a p–n junction allows an electric current to flow in only one direction. (a) Reverse bias: the negative electrode is attached to the p-type semiconductor and current does not flow. (b) Forward bias: the electrodes are reversed to allow charge carriers to be regenerated.

SKILLS YOU SHOULD HAVE MASTERED

❑ 1 Explain the basis of the VSEPR model of bonding in terms of repulsions between electrons (Section 3.1).

❑ 2 Use the VSEPR model to predict the electron arrangement and shape of a molecule or polyatomic ion from its formula (Toolbox 3.1 and Examples 3.1, 3.2, and 3.3).

❑ 3 Predict the polar character of a molecule (Example 3.4).

❑ 4 Account for the structure of a molecule in terms of hybrid orbitals and σ- and π-bonds (Examples 3.5 and 3.6).

❑ 5 Construct and interpret a molecular orbital energy-level diagram for a homonuclear diatomic species (Sections 3.9 and 3.10).

❑ 6 Deduce the ground-state electron configurations of Period 2 diatomic molecules (Toolbox 3.2 and Example 3.7).

❑ 7 Use molecular orbital theory to describe bonding in polyatomic molecules such as the benzene molecule (Section 3.12).

❑ 8 Use molecular orbital theory to account for the differences between metals, insulators, and semiconductors (Sections 3.13 and 3.14).

EXERCISES

Exercises labeled C *require calculus.*

The Shapes of Molecules and Ions

3.1 Below are ball-and-stick models of two molecules. In each case, indicate whether or not there must be, may be, or cannot be one or more lone pairs of electrons on the central atom:

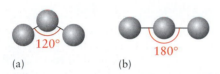

(a)   (b)

3.2 Below are ball-and-stick models of two molecules. In each case, indicate whether or not there must be, may be, or cannot be one or more lone pairs of electrons on the central atom:

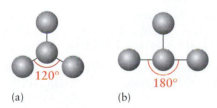

(a)   (b)

3.3 (a) What is the shape of the thionyl chloride molecule, $SOCl_2$? Sulfur is the central atom. (b) How many different OSCl bond angles are there in this molecule? (c) What values are expected for the OSCl and ClSCl bond angles?

3.4 (a) What is the shape of the ClO_3^- ion? (b) What values are expected for the OClO bond angles?

3.5 (a) What is the shape of the ClO_2^+ ion? (b) What is the OClO bond angle?

3.6 (a) What is the shape of the XeF_5^+ ion? (b) How many different FXeF bond angles are there in this molecule? (c) What values are expected for the FXeF bond angles?

3.7 (a) What is the shape of an ICl_2 molecule (iodine is the central atom)? (b) What value is expected for the ClICl angle?

3.8 (a) What is the shape of an SbF_5^{2-} ion? (b) How many different FSbF bond angles are there in this molecule? (c) What values are expected for the FSbF angles?

3.9 Using Lewis structures and VSEPR, give the VSEPR formula for each of the following species and predict its shape: (a) sulfur tetrachloride; (b) iodine trichloride; (c) IF_4^-; (d) xenon trioxide.

3.10 Using Lewis structures and VSEPR, give the VSEPR formula for each of the following species and predict its shape: (a) PF_4^-; (b) ICl_4^+; (c) phosphorus pentafluoride; (d) xenon tetrafluoride.

3.11 Give the VSEPR formula, molecular shape, and bond angles for each of the following species: (a) I_3^-; (b) $SbCl_5$; (c) IO_4^-; (d) NO_2^-.

3.12 Give the VSEPR formula, molecular shape, and bond angles for each of the following species: (a) I_3^+; (b) PCl_3; (c) SeO_3^{2-}; (d) GeH_4.

3.13 Write the Lewis structure and the VSEPR formula, list the shape, and predict the approximate bond angles for (a) CF_3Cl; (b) $TeCl_4$; (c) COF_2; (d) CH_3^-.

3.14 Write the Lewis structure and the VSEPR formula, list the shape, and predict the approximate bond angles for (a) PCl_3F_2; (b) SnF_4; (c) SnF_6^{2-}; (d) IF_5; (e) XeO_4.

3.15 The compound 2,4-pentanedione (also known as acetylacetone and abbreviated to acac) is acidic and can be deprotonated. The anion forms complexes with metals that are used in gasoline additives, lubricants, insecticides, and fungicides. (a) Estimate the bond angles marked with arcs and lowercase letters in 2,4-pentanedione and in the acac ion. (b) What are the differences, if any?

Acetylacetone   Acetylacetonate ion

3.16 Estimate the bond angles marked with arcs and lowercase letters in peroxyacetylnitrate, an eye irritant in smog:

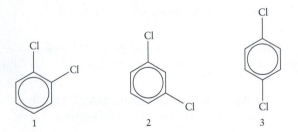

3.17 Predict the bond angles at the central atom of the following molecules and ions: (a) ozone, O_3; (b) azide ion, N_3^-; (c) cyanate ion, CNO^-; (d) hydronium ion, H_3O^+.

3.18 Predict the bond angles at the central atom of the following molecules and ions: (a) OF_2; (b) ClO_2^-; (c) NO_2^-; (d) $SeCl_2$.

3.19 Write its Lewis structure and predict whether each of the following molecules is polar or nonpolar: (a) CH_2Cl_2; (b) CCl_4; (c) CS_2; (d) SF_4.

3.20 Write its Lewis structure and predict whether each of the following molecules is polar or nonpolar: (a) BF_3; (b) PCl_3; (c) SiO_2; (d) H_2S.

3.21 Predict whether each of the following molecules is likely to be polar or nonpolar: (a) C_5H_5N (pyridine, a molecule like benzene except that one —CH— group is replaced by a nitrogen atom); (b) C_2H_6 (ethane); (c) $CHCl_3$ (trichloromethane, also known as chloroform, a common organic solvent and once used as an anesthetic).

3.22 Predict whether each of the following molecules is likely to be polar or nonpolar: (a) CCl_4 (tetrachloromethane); (b) $CH_3CHOHCH_3$ (2-propanol, rubbing alcohol); (c) CH_3COCH_3 (2-propanone, acetone, a common organic solvent used in nail polish remover).

3.23 There are three different dichlorobenzenes, $C_6H_4Cl_2$, which differ in the relative positions of the chlorine atoms on the benzene ring. (a) Which of the three forms are polar? (b) Which has the largest dipole moment?

3.24 There are three different difluoroethenes, $C_2H_2F_2$, which differ in the locations of the fluorine atoms. (a) Which of the forms are polar? (b) Which has the largest dipole moment?

3.25 Acrylonitrile, CH_2CHCN, is used in the synthesis of acrylic fibers (polyacrylonitriles), such as Orlon. Write the Lewis structure of acrylonitrile and describe the hybrid orbitals on each carbon atom. What are the approximate values of the bond angles?

3.26 Xenon forms XeO_3, XeO_4, and XeO_6^{4-}, all of which are powerful oxidizing agents. Give their Lewis structures, their bond angles, and the hybridization of the xenon atom. Which would be expected to have the longest Xe—O distances?

3.27 Predict the shapes and estimate the bond angles of (a) the thiosulfate ion, $S_2O_3^{2-}$; (b) $(CH_3)_2Be$; (c) BH_2^-; (d) $SnCl_2$.

3.28 For each of the following molecules or ions, write the Lewis structure, list the number of lone pairs on the central atom, identify the shape, and estimate the bond angles: (a) PBr_5; (b) $XeOF_2$; (c) SF_5^+; (d) IF_3; (e) BrO_3^-.

3.29 Write the Lewis structure and give the approximate bond angles of (a) C_2H_4; (b) $ClCN$; (c) $OPCl_3$; (d) N_2H_4.

3.30 Write the Lewis structure and predict the shape of (a) TeF_4; (b) NH_2^-; (c) NO_2^+; (d) NH_4^+; (e) SnH_4; (f) OCS.

3.31 Write the Lewis structure and predict the shape of (a) $OSbCl_3$; (b) SO_2Cl_2; (c) $IO_2F_2^-$. The atom in boldface type is the central atom.

3.32 Write the Lewis structure and predict the shape of (a) $OCCl_2$; (b) $OSbCl_2^-$. The atom in boldface type is the central atom.

Valence-Bond Theory

3.33 State the relative orientations of the following hybrid orbitals: (a) sp^3; (b) sp; (c) sp^3d^2; (d) sp^2.

3.34 The orientation of bonds on a central atom of a molecule that has no lone pairs can be any of those listed below. What is the hybridization of the orbitals used by each central atom for its bonding pairs: (a) tetrahedral; (b) trigonal bipyramidal; (c) octahedral; (d) linear?

3.35 State the hybridization of the atom in boldface type in each of the following molecules: (a) $\mathbf{S}F_4$; (b) $\mathbf{B}Cl_3$; (c) $\mathbf{N}H_3$; (d) $(CH_3)_2\mathbf{Be}$.

3.36 State the hybridization of the atom in boldface type in each of the following molecules: (a) $\mathbf{S}F_6$; (b) $O_3\mathbf{Cl}—O—ClO_3$; (c) $H_2\mathbf{N}—CH_2—COOH$ (glycine); (d) $O\mathbf{C}(NH_2)_2$ (urea).

3.37 Identify the hybrid orbitals used by the atom in boldface type in each of the following species: (a) $\mathbf{B}F_3$; (b) $\mathbf{As}F_3$; (c) $\mathbf{Br}F_3$; (d) $\mathbf{Se}F_3^+$.

3.38 Identify the hybrid orbitals used by the atom in boldface type in each of the following molecules: (a) $CH_3\mathbf{C}CCH_3$; (b) $CH_3\mathbf{N}NCH_3$; (c) $(CH_3)_2\mathbf{C}C(CH_3)_2$; (d) $(CH_3)_2\mathbf{N}N(CH_3)_2$.

3.39 Identify the hybrid orbitals used by the phosphorus atom in each of the following species: (a) PCl_4^+; (b) PCl_6^-; (c) PCl_5; (d) PCl_3.

3.40 Identify the hybrid orbitals used by the atom in boldface type in each of the following molecules: (a) $H_2\mathbf{C}CCH_2$; (b) $H_3\mathbf{C}CH_3$; (c) $CH_3\mathbf{N}NN$; (d) $CH_3\mathbf{C}OOH$.

3.41 White phosphorus, P_4, is so reactive that it bursts into flame in air. The four atoms in P_4 form a tetrahedron in which each P atom is connected to three other P atoms. (a) Assign a hybridization scheme to the P_4 molecule. (b) Is the P_4 molecule polar or nonpolar?

3.42 In the vapor phase, phosphorus can exist as P_2 molecules, which are highly reactive, whereas N_2 is relatively inert. Use valence-bond theory to explain this difference.

3.43 Noting that the bond angle of an sp^3 hybridized atom is 109.5° and that of an sp^2 hybridized atom is 120°, do you expect the bond angle between two hybrid orbitals to increase or decrease as the s-character of the hybrids is increased?

3.44 Both NH_2^- and NH_2^+ are angular species, but the bond angle in NH_2^- is less than that in NH_2^+. (a) What is the reason for this difference in bond angles? (b) Take the x-axis as lying perpendicular to the plane of the molecule. Does the $N2p_x$ orbital participate in the hybridization for either species? Briefly explain your answer.

[C] **3.45** Given that the atomic orbitals used to form hybrids are normalized to 1 and mutually orthogonal, (a) show that the two tetrahedral hybrids $h_1 = s + p_x + p_y + p_z$ and $h_3 = s - p_x + p_y - p_z$ are orthogonal. (b) Construct the remaining two tetrahedral hybrids that are orthogonal to these two hybrids.

Hint: Two wavefunctions are orthogonal if $\int \psi_1\psi_2 d\tau = 0$, where $\int \dots d\tau$ means "integrate over all space."

[C] **3.46** The hybrid orbital $h_1 = s + p_x + p_y + p_z$ referred to in Exercise 3.45 is not normalized. Find the normalization factor N, given that all the atomic orbitals are normalized to 1.

3.47 The composition of hybrids can be discussed quantitatively. The outcome is that, if two equivalent hybrids composed of an s-orbital and two p-orbitals make an angle θ to each other, then the hybrids can be regarded as sp^λ, with $\lambda = -\cos\theta/\cos^2(\frac{1}{2}\theta)$. What is the hybridization of the two O—H bonds in H_2O?

3.48 Given the information in Exercise 3.47, plot a graph showing how the hybridization depends on the angle between two hybrids formed from an s-orbital and two p-orbitals, and confirm that it ranges from 90° when no s-orbital is included in the mixture to 120° when the hybridization is sp^2.

Molecular Orbital Theory

3.49 Draw a molecular orbital energy-level diagram and determine the bond order expected for each of the following diatomic species: (a) Li_2; (b) Li_2^+; (c) Li_2^-. State whether each molecule or ion will be paramagnetic or diamagnetic. If it is paramagnetic, give the number of unpaired electrons.

3.50 Draw a molecular orbital energy-level diagram and determine the bond order expected for the following diatomic species: (a) B_2; (b) B_2^-; (c) B_2^+. State whether each molecule or ion will be paramagnetic or diamagnetic. If it is paramagnetic, give the number of unpaired electrons.

3.51 (a) On the basis of the configuration of the neutral molecule F_2, write the molecular orbital configuration of the valence molecular orbitals for (1) F_2^-; (2) F_2^+; (3) F_2^{2-}. (b) For each species, give the expected bond order. (c) Which are paramagnetic, if any? (d) Is the highest-energy orbital that contains an electron σ or π in character?

3.52 (a) On the basis of the configuration for the neutral molecule N_2, write the molecular orbital configuration of the valence molecular orbitals for (1) N_2^+; (2) N_2^{2+}; (3) N_2^{2-}. (b) For each species, give the expected bond order. (c) Which are paramagnetic, if any? (d) Is the highest-energy orbital that contains an electron σ or π in character?

3.53 (a) Draw the molecular orbital energy-level diagram for N_2 and label the energy levels according to the type of orbitals from which they are made, whether they are σ- or π-orbitals and whether they are bonding or antibonding. (b) The orbital structure of the heteronuclear diatomic ion NO^+ is similar to that of N_2. How will the fact that the electronegativity of N differs from that of O affect the molecular orbital energy-level diagram of NO^+ compared with that of N_2? Use this information to draw the energy-level diagram for NO^+. (c) In the molecular orbitals, will the electrons have a higher probability of being at N or at O? Why?

3.54 (a) How does molecular orbital theory make it possible to explain both ionic *and* covalent bonding? (b) The degree of ionic character in bonding was related to electronegativity in Chapter 2. How does electronegativity affect the molecular orbital diagram so that bonds become ionic?

3.55 Write the valence-shell electron configurations and bond orders of (a) B_2; (b) Be_2; (c) F_2.

3.56 Write the valence-shell electron configurations and bond orders of (a) NO; (b) N_2^+; (c) C_2^{2+}.

3.57 Give the valence-shell electron configurations and bond orders for CO and CO^+. Use that information to predict which species has the greater bond enthalpy.

3.58 Give the valence-shell electron configurations and bond orders for CN and CN^-. Use that information to predict which species has the greater bond enthalpy.

3.59 Which of the following species are paramagnetic: (a) B_2; (b) B_2^-; (c) B_2^+? If the species is paramagnetic, how many unpaired electrons does it possess?

3.60 Which of the following species are paramagnetic: (a) N_2^-; (b) F_2^+; (c) O_2^{2+}? If the species is paramagnetic, how many unpaired electrons does it possess?

3.61 Determine the bond orders and use them to predict which species of each of the following pairs has the stronger bond: (a) F_2 or F_2^-; (b) B_2 or B_2^+.

3.62 Determine the bond orders and use them to predict which species of each of the following pairs has the stronger bond: (a) C_2^+ or C_2; (b) O_2 or O_2^+.

3.63 Based on their valence-shell electron configurations which of the following species would you expect to have the lowest ionization energy? (a) C_2^+; (b) C_2; (c) C_2^-.

3.64 Based on their valence-shell electron configurations which of the following species would you expect to have the greatest electron affinity? (a) Be_2; (b) F_2; (c) B_2^+, (d) C_2^+.

3.65 How does the change in conductivity of a semiconductor differ from that of a metal as temperature is increased?

3.66 Normally, in conducting materials, we think of current as being carried by electrons as they move through a solid. In semiconductors, it is also common to talk about the current being carried by the "holes" in the valence band. (a) Explain how holes move through a solid material. (b) If, in a p-type semiconductor device, electric current is moving from left to right, in which direction will the holes be moving?

3.67 Germanium is a semiconductor. If small amounts of the elements In, P, Sb, and Ga are present as impurities, which of them will make germanium into (a) a p-type semiconductor; (b) an n-type semiconductor?

3.68 Gallium arsenide is a semiconducting material. If we wish to modify the sample by replacing a small amount of the arsenic with an element to produce an n-type semiconductor, which element would we choose: selenium, phosphorus, or silicon? Why?

C 3.69 It is usually convenient to deal with wavefunctions that are "normalized," which means that the integral $\int \psi^2 d\tau = 1$.

The bonding orbital in Eq. 1 is not normalized. A wavefunction ψ can be normalized by writing it as $N\psi$ and finding the factor N which ensures that the integral over $(N\psi)^2$ is equal to 1. Find the factor N that normalizes the bonding orbital in Eq. 1, given that the individual atomic orbitals are each normalized. Express your answer in terms of the "overlap integral"

$$S = \int \psi_{A1s}\psi_{B1s}d\tau. \text{ (The notation is the same as in Exercise 3.45.)}$$

C 3.70 The antibonding orbital in Eq. 2 is not normalized (see Exercise 3.69). Find the factor that normalizes it to 1, given that the individual atomic orbitals are each normalized. Express your answer in terms of the overlap integral $S = \int \psi_{A1s}\psi_{B1s}d\tau$.

(The notation is the same as in Exercise 3.45.) Confirm that the bonding and antibonding orbitals are mutually orthogonal—that is, that the integral over the product of the two wavefunctions is zero.

3.71 The two atomic orbitals that contribute to the antibonding orbital in Eq. 2 are each proportional to e^{-r/a_0}, where r is the distance of the point from its parent nucleus. Confirm that there is a nodal plane lying halfway between the two nuclei.

3.72 Use the plotting function on the Web site for this book to plot the amplitude of the bonding and antibonding orbitals in Eqs. 1 and 2, given the information in Exercise 3.71, on a line passing through the two nuclei. Ignore the normalization factors, and take the internuclear distance as (a) a_0, (b) $2a_0$, (c) $5a_0$. Which of these three bond lengths is most likely to be close to the actual bond length?

Integrated Exercises

3.73 For each of the following species, write the Lewis structure, predict the shape and hybridization about each central atom, give the bond angles, and state whether it is polar or nonpolar: (a) $GaCl_4^-$; (b) TeF_4; (c) $SbCl_4^-$; (d) $SiCl_4$.

3.74 For each of the following species, write the Lewis structure, predict the shape and hybridization about each central atom, give the bond angles, and state whether it is polar or nonpolar: (a) $SnCl_3^-$; (b) TeO_3; (c) NO_3^-; (d) ICl_3.

3.75 Polar molecules attract other polar molecules through dipole–dipole intermolecular forces. Polar solutes tend to have higher solubilities in polar solvents than in nonpolar solvents. Which of the following pairs of compounds would be expected to have the higher solubility in hexafluorobenzene, C_6F_6: (a) SiF_4 or PF_3; (b) SF_6 or SF_4; (c) IF_5 or AsF_5?

3.76 The halogens form compounds among themselves. These compounds, called the *interhalogens,* have the formulas XX′, XX′$_3$, and XX′$_5$, where X is the heavier halogen atom. (a) Predict their structures and bond angles. (b) Which of them are polar? (c) Why is the lighter halogen atom not the central atom of such molecules?

3.77 An organic compound distilled from wood was found to have a molar mass of 32.04 g·mol^{-1} and the following composition by mass: 37.5% C, 12.6% H, and 49.9% O. (a) Write the Lewis structure of the compound and determine the bond angles about the carbon and oxygen atoms. (b) Give the hybridization of the carbon and oxygen atoms. (c) Predict whether the molecule is polar or not.

3.78 Draw Lewis structures for each of the following species and predict the hybridization at each carbon atom: (a) H_2CCH^+; (b) $H_2CCH_3^+$; (c) $H_3CCH_2^-$.

3.79 (a) Draw the bonding and antibonding orbitals that correspond to the σ-bond in H_2. (b) Repeat this procedure for HF. (c) How do these orbitals differ?

3.80 (a) Consider a hypothetical species "HeH." What charge (magnitude and sign), if any, should be present on this combination of atoms to produce the most stable molecule or ion possible? (b) What is the maximum bond order that such a molecule or ion could have? (c) If the charge on this species were increased or decreased by one, what would be the effect on the bonding in the molecule?

3.81 (a) Order the following molecules according to increasing C—F bond length: CF^+, CF, CF^-. (b) Which of these molecules will be diamagnetic, if any? Explain your reasoning.

3.82 Describe as completely as you can the structure and bonding in carbamate ion, $CO(NH_2)_2^-$. The C—O bond lengths are both 128 pm, and the C—N bond length is 136 pm.

3.83 Borazine, $B_3N_3H_6$, a compound that has been called "inorganic benzene" because of its similar hexagonal structure (but with alternating B and N atoms in place of C atoms), is the basis of a large class of boron–nitrogen compounds. Write its Lewis structure and predict the composition of the hybrid orbitals used by each B and N atom.

3.84 Given that carbon has a valence of four in nearly all its compounds and can form chains and rings of C atoms, (a) draw any two of the three possible structures for C_3H_4; (b) determine all bond angles in each structure; (c) determine the hybridization of each carbon atom in the two structures; (d) ascertain whether the two structures are resonance structures and explain your reasoning.

3.85 (a) Draw a Lewis structure for each of the following species: CH_3^+; CH_4; CH_3^-; CH_2; CH_2^{2+}; CH_2^{2-}. (b) Identify each as a radical or not. (c) Rank them in order of increasing HCH bond angles. Explain your choices.

3.86 Consider the molecules H_2CCH_2, H_2CCCH_2, and H_2CCCCH_2. (a) Draw Lewis structures for these molecules. (b) What is the hybridization at each C atom? (c) What type of bond connects the carbon atoms (single, double, etc.)? (d) What are the HCH, CCH, and CCC angles in these molecules? (e) Do all the hydrogen atoms lie in the same plane? (f) A generalized formula for molecules of this type is $H_2C(C)_xCH_2$, where x is 0, 1, 2, etc. What can be said, if anything, about the relative orientation of the H atoms at the ends of the chain as a function of x?

3.87 Acetylene (ethyne), C_2H_2, can be polymerized. (a) Draw the Lewis structure for acetylene and draw a Lewis structure for the polymer that results when acetylene is polymerized. The polymer has formula $(CH)_n$, where n is large. (b) Consider the polymers polyacetylene and polyethylene. The latter has the formula $(CH_2)_n$ and is an insulating material (plastic wrap is made of polyethylene), whereas polyacetylene is a darkly colored material that can conduct electricity when properly treated. On the basis of your answer to part (a), suggest an explanation for the difference in the two polymers.

3.88 Dye molecules are of commercial importance because they are very intensely colored. Most dye molecules possess many multiple bonds and are often aromatic. Why is this important to the properties of the dye molecule?

3.89 Treat the π system of a dye molecule composed of a conjugated carbon chain of N carbon atoms as a box of length NR, where R is the average C—C bond length. Given that each C atom contributes one electron and that each state of the box can accommodate two electrons, derive an expression for the wavelength of the light absorbed by the lowest-energy transition. Should you increase or decrease the number of carbon atoms in the chain to shift the wavelength to longer values? Refer to Major Technique 2, which follows these exercises.

3.90 Show that a molecule with configuration π^4 has a cylindrically symmetrical electron distribution. *Hint:* Take the π-orbitals to be equal to xf and yf, where f is a function that is independent of x and y.

3.91 In addition to forming σ and π types of bonds similar to p-orbitals, d-orbitals may overlap to form δ-bonds. (a) Draw overlap diagrams showing three different ways in which d-orbitals can combine to form bonds. (b) Place the three types of d-d bonds—σ, π, and δ—in order of strongest to weakest.

3.92 An s-orbital and a p-orbital on different atoms may overlap to form molecular orbitals. One of these interactions forms a bonding σ-orbital and the other is nonbonding. Draw diagrams to represent the two types of orbital overlap that give rise to σ-bonding and nonbonding orbitals.

3.93 (a) Describe the changes in bonding that would occur in benzene if two electrons were removed from the HOMO (highest occupied molecular orbital). This removal would correspond to an oxidation of benzene to $C_6H_6^{2+}$. (b) Describe the changes in bonding that would occur if two electrons were added to the LUMO (lowest unoccupied molecular orbital). This addition would correspond to a reduction of benzene by two electrons to give $C_6H_6^{2-}$. Do you expect these ions to be diamagnetic or paramagnetic?

3.94 (a) Confirm, by using trigonometry, that the dipoles of the three bonds in a trigonal pyrimidal AB_3 molecule do not cancel, resulting in a polar molecule. (b) Show that the dipoles of the four bonds in a tetrahedral AB_4 molecule cancel and the molecule is nonpolar.

3.95 Benzyne, C_6H_4, is a highly reactive molecule that is detected only at low temperatures. It is related to benzene in that it has a six-membered ring of carbon atoms; but, instead of three double bonds, the structure is normally drawn with two double bonds and a triple bond. (a) Draw a Lewis structure of the benzyne molecule. Indicate on the structure the hybridization at each carbon atom. (b) On the basis of your understanding of bonding, explain why this molecule might be highly reactive.

3.96 The Lewis structure of caffeine, $C_8H_{10}N_4O_2$, a common stimulant, is shown below. (a) Give the hybridization of each atom other than hydrogen and predict the bond angles about that atom. (b) On the basis of your answers in part (a), estimate the bond angles around each carbon and nitrogen atom. (c) Search the chemical literature for the structure of caffeine, and compare the observed structural parameters with your predictions.

3.97 Cyclopropane, C_3H_6, is a hydrocarbon composed of a three-membered ring of carbon atoms. (a) Determine the hybridization of the carbon atoms. (b) Predict the CCC and HCH bond angles at each carbon atom on the basis of your answer to part (a). (c) What must the "real" CCC bond angles in cyclopropane be? (d) What is the defining characteristic of a σ-bond compared with a π-bond, for example? (e) How do the C—C σ-bonds in cyclopropane extend the definition of conventional σ-bonds? (f) Draw a picture depicting the molecular orbitals to illustrate your answer.

3.98 The diameter of a C_{60} molecule (Section 14.16) is approximately 700 pm. (a) Could more than one lanthanum atom occupy the center of a C_{60} molecule? (b) Because it is possible for C_{60} to undergo reduction (six step-by-step reductions to give C_{60}^{6-} have been reported), it is also possible for a lanthanum ion, La^{3+}, to exist inside the C_{60} molecule. Could two La^{3+} ions be placed inside a C_{60} molecule?

3.99 In which of the following molecules could there be an n-to-π^* transition? Explain your choices. (a) Formic acid, HCOOH; (b) ethyne, C_2H_2; (c) methanol, CH_3OH; (d) hydrogen cyanide, HCN. Refer to Major Technique 2, which follows these exercises.

3.100 Aqueous solutions of compounds containing the Cu^{2+} ion are blue as a result of the presence of the $Cu(H_2O)_6^{2+}$ complex. Does this complex absorb in the visible region? Suggest an explanation. Refer to Major Technique 2, which follows these exercises.

3.101 Consider the bonding in $CH_2=CHCHO$. (a) Draw the most important Lewis structure. Include all nonzero formal charges. (b) Identify the composition of the bonds and the hybridization of each lone pair—for example, by writing $\sigma(H1s, C2sp^2)$.

3.102 Light-emitting diodes (LEDs) contain p–n junctions. The circuit in an LED is arranged so that electrons from the power source flow into the conduction band of the n-type side. As electrons continue to flow, they are pushed to the conduction band of the p-type side, which can hold more electrons. These electrons enter the conduction band of the p-type side, because they already occupy the higher-energy band in the n-type side. However, once the electrons are in the higher-energy band of the p-type side, they fall into the lower-energy band unless it is full. As these electrons make transitions to the lower-energy band, energy is released in the form of light. (a) Explain, in terms of the movement of electrons, why an LED cannot be made from the junction of pure silicon and silicon doped with phosphorus. (b) If the direction of the circuit in the LED is reversed, so that the electrons flow from the power source into the p-type side of the p–n junction directly, where would the electrons go once they enter the p–n junction? (That is, indicate which bands would receive the electrons and whether the electrons would then migrate to other bands.) (c) Would you expect the LED to emit light when it is placed in the reverse circuit described in part (b)? Explain your answer.

3.103 The reaction between SbF_3 and CsF produces, among other products, the anion $[Sb_2F_7]^-$. This anion has no F—F bonds and no Sb—Sb bonds. (a) Propose a Lewis structure for the ion. (b) Assign a hybridization scheme to the Sb atoms.

3.104 The following molecules are bases that are part of the nucleic acids involved in the genetic code. Identify (a) the hybridization of each C and N atom, (b) the number of σ- and π-bonds, and (c) the number of lone pairs of electrons in the molecule.

3.105 Just as $AlCl_3$ forms dimers (Section 2.11), in the $[Bi_2Cl_4]^{2-}$ ion two of the Cl atoms form "bridges" between the two Bi atoms. Propose a structure for the $[Bi_2Cl_4]^{2-}$ ion.

3.106 Germanium forms a series of anions called "germides." In the germide ion Ge_4^{n-} the four Ge atoms form a tetrahedron in which each atom is bonded to the other three and each atom has a lone pair of electrons. What is the value of n, the charge on this anion? Explain your reasoning.

3.107 One form of the polyatomic ion I_5^- has an unusual V-shaped structure: one I atom lies at the point of the V, with a linear chain of two I atoms extending on each side. The bond angles are 88° at the central atom and 180° at the two atoms in the side chains. Draw a Lewis structure for I_5^- that explains its shape and indicate the hybridization you would assign to each nonterminal atom.

3.108 Molecules and ions, like atoms, can be isoelectronic. That is, they can have the same number of electrons. For example, CH_4 and NH_4^+ are isoelectronic. Therefore, they have the same molecular shape. Identify a molecule or ion that is isoelectronic with each of the following species and verify that each pair has the same shape: (a) CO_3^{2-}; (b) O_3; (c) OH^-.

3.109 Is the Al_2Cl_6 molecule, described in Section 2.11, polar or nonpolar? Justify your conclusion.

3.110 The molecular structure of benzene, C_6H_6, is planar. Is the molecular structure of cyclohexane, C_6H_{12}, planar as well? (a) Draw a Lewis structure for cyclohexane. (b) Indicate the bond angles about each carbon atom. (c) What is your conclusion about the shape of the cyclohexane molecule?

3.111 Complexes of d- and f-block metals can be described in terms of hybridization schemes, each associated with a particular shape. Bearing in mind that the number of atomic orbitals hybridized must be the same as the number of hybrid orbitals produced, match the hybrid orbitals sp^2d, sp^3d^3, and sp^3d^3f to the following shapes: (a) pentagonal bipyramidal; (b) cubic; (c) square planar.

Chemistry Connections

3.112 Hydrogen peroxide, H_2O_2, is a nontoxic bleaching agent being used as a replacement for chlorine in industry and home laundries. The bleaching process is an oxidation and when hydrogen peroxide acts as a bleaching agent, the only waste it generates is H_2O.

(a) Draw the Lewis structure of hydrogen peroxide and determine the formal charge on each atom. What is the oxidation number of oxygen in hydrogen peroxide? Which is more useful in predicting the ability of H_2O_2 to act as an oxidizing agent, formal charge or oxidation number? Explain your reasoning.

(b) Predict the bond angles about each O atom in H_2O_2. Are all the atoms in the same plane? Is the molecule polar or nonpolar? Explain your reasoning.

(c) Write the valence electron configuration of (1) O_2; (2) O_2^-; (3) O_2^+; (4) O_2^{2-}. For each species, give the expected bond order and indicate which, if any, are paramagnetic.

(d) The following bond lengths have been reported: (1) O_2, 121 pm; (2) O_2^-, 134 pm; (3) O_2^+, 112 pm; (4) O_2^{2-}, 149 pm. Suggest a reason for the differences based on the configurations in part (c).

(e) One reaction in which H_2O_2 acts as an oxidizing agent is $Fe^{2+}(aq) + H_2O_2(aq) + H^+(aq) \rightarrow Fe^{3+}(aq) + H_2O(l)$. Balance the equation and determine what mass of iron(II) can be oxidized to iron(III) by 54.1 mL of 0.200 M $H_2O_2(aq)$. *Hint:* Make sure you balance charge as well as mass.

(f) Hydrogen peroxide can also act as a reducing agent, as in $Fe^{3+}(aq) + H_2O_2(aq) + OH^-(aq) \rightarrow Fe^{2+}(aq) + H_2O(l) + O_2(g)$. Balance the equation and determine what mass of iron(III) can be reduced to iron(II) by 35.6 mL of 0.200 M $H_2O_2(aq)$. *Hint:* Make sure you balance charge as well as mass.

(g) Hydrogen peroxide must be kept in brown bottles because in the presence of light it can *disproportionate*, which means that it oxidizes and reduces itself in the reaction $2 H_2O_2(aq) \rightarrow$

$2 H_2O(l) + O_2(g)$. How many electrons are transferred in the reaction represented by this equation?

(h) Although peroxides do not generate hazardous waste, they can cause problems in the atmosphere. If it makes its way to the stratosphere, a hydrogen peroxide molecule can break into two OH radicals, which threaten the ozone layer that protects Earth from harmful radiation. Use data in Table 2.4 to calculate the minimum frequency of light needed to break the HO—OH bond.

All the colors of vegetation around us arise from the selective absorption and reflection of visible light. If our eyes were sensitive to ultraviolet radiation—the electromagnetic radiation at shorter wavelengths (higher frequencies) than violet light—we would see an even richer tapestry of color, because substances also absorb and reflect selectively in the ultraviolet region. Honeybees do have this ability, and they can "see" radiation in the ultraviolet as well as in the visible range. The selective absorption and transmittance of visible light and ultraviolet radiation are the basis of a spectroscopic technique for identifying compounds and for determining their concentrations in samples.

The Technique

When electromagnetic radiation falls on a molecule, the electrons in the molecule can be excited to a higher energy state. Radiation of the frequency ν (nu) can raise the energy of the molecule by an amount ΔE, where

$$\Delta E = h\nu \tag{1}$$

This is the *Bohr frequency condition*, which was encountered in Section 1.4; h is Planck's constant. For typical ultraviolet wavelengths (of 300 nm and less, corresponding to a frequency of about 10^{15} Hz), each photon brings enough energy to excite the electrons in a molecule into a different distribution. Provided that an empty orbital exists at the right energy, the incoming radiation can excite an electron into it and hence be absorbed. Therefore, the study of visible and ultraviolet absorption gives us information about the electronic energy levels of molecules.

Visible and ultraviolet absorption spectra are measured in an absorption spectrometer. The source gives out intense visible light or ultraviolet radiation. The wavelengths can be selected with a glass prism for visible light and with a quartz prism or a diffraction grating for ultraviolet radiation (which is absorbed by glass). A typical absorption spectrum, that of

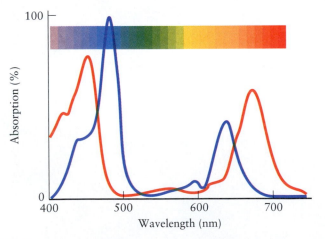

FIGURE 1 The optical absorption spectrum of chlorophyll as a plot of percentage absorption against wavelength. Chlorophyll *a* is shown in red, chlorophyll *b* in blue.

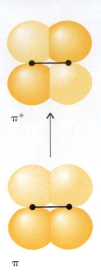

FIGURE 2 In a π-to-π* transition, an electron in a bonding π-orbital is excited into an empty antibonding π*-orbital.

chlorophyll, is shown in Fig. 1. Note that chlorophyll absorbs red and blue light, leaving the green light present in white light to be reflected. That is why most vegetation looks green. The spectrum can help us to assess the absorption quantitatively and to make a precise analysis of the energy-capturing power of the molecule.

Chromophores

The presence of certain absorption bands in visible and ultraviolet spectra can often be traced to the presence of characteristic groups of atoms in the molecules. These groups of atoms are called *chromophores*, from the Greek words for "color bringer."

An important chromophore is a carbon–carbon double bond. In terms of the language of molecular orbital theory, the electronic transition that takes place when energy is absorbed is the excitation of an electron from a bonding π-orbital to the corresponding antibonding orbital, π*. This transition is therefore known as a *π-to-π* (pi to pi star) transition* (Fig. 2); it occurs at about 160 nm, which is in the ultraviolet region. The transition can be brought into the visible region by decreasing the separation between bonding and antibonding orbitals. These orbitals lie close in energy when the molecule has a chain of alternating single and double bonds. Such "conjugated" double bonding occurs in the compound carotene (see structure **46** in Chapter 3), which is partly responsible for the color of carrots, mangoes, and persimmons. Related compounds account for the colors of shrimp and flamingoes. This transition is also responsible for the primary act of vision, as explained in Section 3.7.

Another important chromophore is the carbonyl group, >C=O, which absorbs at about 280 nm. The transition responsible for the absorption is the excitation of a lone-pair electron on the oxygen atom (a "nonbonding" electron, denoted *n*) into the empty antibonding π^*-orbital of the C=O double bond (Fig. 3). This transition is therefore called an *n-to-π* (n to pi star) transition.*

A *d*-metal ion may also be responsible for color, as is apparent from the varied colors of many *d*-metal complexes (see Chapter 16). Two types of transitions may be involved. In one, which is called a *d-to-d transition,* an electron is excited from a *d*-orbital of one energy to a *d*-orbital of higher

FIGURE 4 The perceived color of a complex in white light is the complementary color of the light that it absorbs. In this color wheel, complementary colors are opposite each other. The numbers are approximate wavelengths in nanometers.

energy. Because the energy differences between *d*-orbitals are quite small, visible light brings enough energy to cause this excitation, and so colors are absorbed from white light and the sample takes on colors complementary to those absorbed (Fig. 4). In a second type of transition involving *d*-orbitals, called a *charge-transfer transition,* electrons migrate from the atoms attached to the central metal atom into the latter's *d*-orbitals or vice versa. This transfer of charge can result in very intense absorption; it is responsible, for instance, for the deep purple of permanganate ions, MnO_4^-.

Related Exercises: 3.89, 3.99, and 3.100

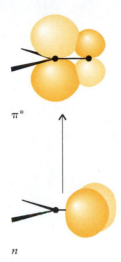

FIGURE 3 In an *n-to-π** transition of a carbonyl group, an electron in a nonbonding orbital (one localized wholly on the oxygen atom) is excited into an antibonding π^*-orbital spread over both atoms.

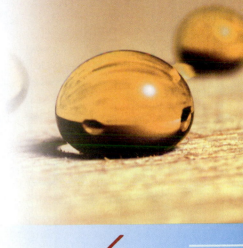

THE PROPERTIES OF GASES

What Are the Key Ideas? We can predict the physical properties of any gas by using the set of equations known as "the gas laws." These equations can be explained in terms of a model of a gas in which the molecules are in ceaseless random motion and so widely separated that they do not interact with one another.

Why Do We Need to Know This Material? In the first three chapters, we investigated the nature of atoms, molecules, and ions. Bulk matter is composed of immense numbers of these particles and its properties emerge from the behavior of the constituent particles. Gases are the simplest state of matter, and so the connections between the properties of individual molecules and those of bulk matter are relatively easy to identify. In later chapters, these concepts will be used to study thermodynamics, equilibrium, and the rates of chemical reactions.

What Do We Need to Know Already? We need to be familiar with SI units (Appendix 1B) and the concepts of force and energy (Section A). This chapter also develops the techniques of reaction stoichiometry (Sections L and M) by extending them to gases.

The most important gas on the planet is the atmosphere, a thin layer of gas held by gravity to the surface of the Earth. Half the mass of the atmosphere lies below an altitude of 5.5 km. If we were to look from a point where the Earth appears to be the size of a basketball, the atmosphere would appear to be only 1 mm thick (Fig. 4.1). Yet this delicate layer is vital to life: it shields us from harmful radiation and supplies substances needed for life, such as oxygen, nitrogen, carbon dioxide, and water.

Eleven elements are gases under normal conditions (Fig. 4.2). So are many compounds with low molar masses, such as carbon dioxide, hydrogen chloride, and organic compounds such as the methane, CH_4, of natural gas and the

FIGURE 4.1 The delicate film of the Earth's atmosphere as seen from space.

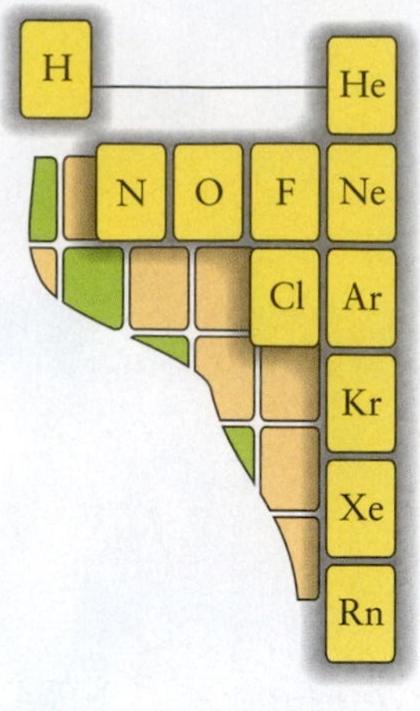

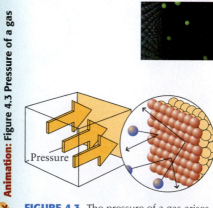

FIGURE 4.2 The 11 elements that are gases under normal conditions. Note how they lie toward the upper right of the periodic table.

propane, C_3H_8, of camping fuel. All substances that are gases at ordinary temperatures are molecular, except the six noble gases, which are monatomic (consist of single atoms).

In this chapter, we follow a typical scientific path. First, we collect experimental observations on the properties of gases and summarize these observations mathematically. We then formulate a simple qualitative molecular model of a gas suggested by these observations and go on to express it quantitatively. Finally, we use more detailed experimental observations to refine the model so that it accounts for the properties of real gases.

THE NATURE OF GASES

A remarkable characteristic of gases is that many of their physical properties are very similar, particularly at low pressures, regardless of the identity of the gas. Therefore, instead of having to describe the properties of each gas individually, we can describe them all simultaneously.

4.1 Observing Gases

Gases are examples of **bulk matter**, forms of matter consisting of large numbers of molecules, and the properties of a gas sample emerge from the collective behavior of vast numbers of its individual particles. For example, when we push on a bicycle pump, we see that air is **compressible**—that is, it can be confined into a smaller volume. The observation that gases are more compressible than solids and liquids suggests that there is a lot of space between the molecules of gases. We also know from everyday experience—by releasing air from an inflated balloon, for instance—that a gas expands rapidly to fill the space available to it. This observation suggests that the molecules are always moving rapidly and hence can respond quickly to changes in the volume available to them. Because the pressure in a balloon is the same in all directions, we can infer that the motion of the molecules is chaotic, not favoring any single direction. Our first primitive picture of a gas could therefore be as a collection of widely spaced molecules in ceaseless rapid chaotic motion. We shall use more quantitative observations to refine this simple model as we proceed through the chapter.

> *The fact that gases are readily compressible and immediately fill the space available to them suggests that molecules of gases are widely separated and in ceaseless chaotic motion.*

4.2 Pressure

If you have ever pumped up a bicycle tire or squeezed an inflated balloon, you have experienced an opposing force arising from the confined air. The **pressure**, P, of a gas is the force, F, exerted by the gas, divided by the area, A, on which the force is exerted:

$$\text{Pressure} = \frac{\text{force}}{\text{area}} \quad \text{or, more simply,} \quad P = \frac{F}{A} \tag{1}$$

The SI unit of pressure, is the **pascal**, Pa:

$$1\ \text{Pa} = 1\ \text{kg·m}^{-1}\text{·s}^{-2}$$

In terms of the model that we are developing, the pressure that a gas exerts on the walls of its container results from the collisions of its molecules with the container's surface (Fig. 4.3). The more vigorous the storm of molecules on a surface, the stronger the force and hence the higher the pressure. Any object on the surface of the Earth stands in an invisible storm of molecules that beat on it incessantly and exert a force all over its surface. Even on an apparently calm day, we are in the midst of a molecular storm.

Animation: Figure 4.3 Pressure of a gas

FIGURE 4.3 The pressure of a gas arises from the collisions that its molecules make with the walls of the container. The storm of collisions, shown in the inset, exerts an almost steady force on the walls.

The pressure exerted by the atmosphere can be measured by various means, most commonly by using a pressure gauge. When we measure the pressure of air in a tire with a pressure gauge, we are actually measuring the "gauge pressure," the difference between the pressure inside the tire and the atmospheric pressure. A flat tire registers a gauge pressure of zero, because the pressure inside the tire is the same as the pressure of the atmosphere. A pressure gauge attached to a piece of scientific apparatus, however, measures the *actual* pressure in the apparatus.

The pressure of the atmosphere can be measured with a **barometer**, an instrument invented in the seventeenth century by Evangelista Torricelli, a student of Galileo. Torricelli (whose name coincidentally means "little tower" in Italian) formed a little tower of liquid mercury. He sealed a long glass tube at one end, filled it with mercury, and inverted it into a beaker (Fig. 4.4). The column of mercury fell until the pressure that it exerted at its base matched the pressure exerted by the atmosphere. To interpret measurements with a barometer, we need to find how the height of the column of mercury depends on the atmospheric pressure.

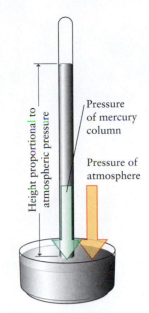

FIGURE 4.4 A barometer is used to measure the pressure of the atmosphere. The pressure of the atmosphere is balanced by the pressure exerted by the column of mercury. The height of the column is proportional to the atmospheric pressure. Therefore, by measuring the height of the column, we can monitor the pressure of the atmosphere.

HOW DO WE DO THAT?

We want to find the relation between the height, h, of the column of mercury in a barometer and the atmospheric pressure, P. Suppose the cross-sectional area of the column is A. The volume of mercury in the column is the height of the cylinder times this area, $V = hA$. The mass, m, of this volume of mercury is the product of mercury's density, d, and the volume; so $m = dV = dhA$. The mercury is pulled down by the force of gravity; and the total force that its mass exerts at its base is the product of the mass and the acceleration of free fall (the acceleration due to gravity), g: $F = mg$. Therefore, the pressure at the base of the column, the force divided by the area, is

From $P = F/A$ and $F = mg$,
$$P = \frac{F}{A} = \frac{mg}{A} = \frac{\overbrace{dhA}^{m}g}{A} = dhg$$

This equation shows that the pressure, P, exerted by a column of mercury is proportional to the height of the column. Mercury inside a tube sealed at one end and inverted in a pool of mercury will fall until the pressure exerted by the mercury balances the atmospheric pressure. Therefore, the height of the column can be used as a measure of atmospheric pressure.

We have shown that the pressure of a gas can be related to the height of a column of liquid and its density by

$$P = dhg \tag{2}$$

The pressure is obtained in pascals when the density, height, and value of g are expressed in SI units.

EXAMPLE 4.1   **Sample exercise: Calculating atmospheric pressure from the height of a column of mercury**

Suppose the height of the column of mercury in a barometer is 760. mm (written 760. mmHg, and read "760 millimeters of mercury") at 15°C. What is the atmospheric pressure in pascals? At 15°C the density of mercury is 13.595 g·cm^{-3} (corresponding to 13 595 kg·m^{-3}) and the standard acceleration of free fall at the surface of the Earth is 9.806 65 m·s^{-2}.

SOLUTION

From $P = dhg$, $P = (13\ 595\ \text{kg·m}^{-3}) \times (0.760\ \text{m}) \times (9.806\ 65\ \text{m·s}^{-2})$

$= 1.01 \times 10^5\ \text{kg·m}^{-1}\text{·s}^{-2} = 1.01 \times 10^5\ \text{Pa}$

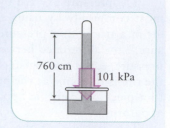

760 cm   101 kPa

This pressure can also be reported as 101 kPa, where 1 kPa = 10^3 Pa. The acceleration of free fall varies over the surface of the Earth and depends on the altitude; in all calculations in this text, we assume that g has the "standard" value used here.

SELF-TEST 4.1A What is the atmospheric pressure when the height of the mercury column in a barometer is 756 mm?

[*Answer:* 100. kPa]

SELF-TEST 4.1B The density of water at 20°C is 0.998 g·cm^{-3}. What height would the column of liquid be in a water barometer when the atmospheric pressure corresponds to 760. mm of mercury?

Although a pressure gauge is more commonly used to measure the pressure inside a laboratory vessel, a **manometer** is sometimes used (Fig. 4.5). It consists of a U-shaped tube connected to the experimental system. The other end of the tube may be either open to the atmosphere or sealed. For an open-tube manometer (like that shown in Fig. 4.5a), the pressure in the system is equal to that of the atmosphere when the levels of the liquid in each arm of the U-tube are the same. If the level of mercury on the system side of an open manometer is above that of the atmosphere side, the pressure in the system is lower than the atmospheric pressure. In a closed-tube manometer (like that shown in Fig. 4.5b), one side is connected to a closed flask (the system) and the other side is vacuum. The difference in heights of the two columns is proportional to the pressure in the system.

EXAMPLE 4.2   **Sample exercise: Calculating the pressure inside an apparatus by using a manometer**

The height of the mercury in the system-side column of an open-tube mercury manometer was 10. mm above that of the open side when the atmospheric pressure corresponded to 756 mm of mercury and the temperature was 15°C. What is the pressure inside the apparatus in millimeters of mercury and in pascals?

FIGURE 4.5 (a) An open-tube manometer. The pressure inside the apparatus to which the narrow horizontal tube is connected pushes against the external pressure. In this instance, the pressure inside the system is lower than the atmospheric pressure by an amount proportional to the difference in heights of the liquid in the two arms. (b) A closed-tube manometer. The pressure in the adjoining apparatus is proportional to the difference in heights of the liquid in the two arms. The space inside the closed end is a vacuum.

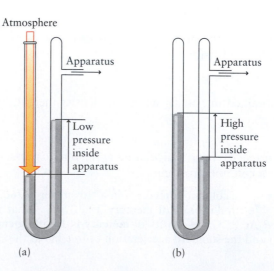

Atmosphere

Apparatus   Apparatus

Low pressure inside apparatus   High pressure inside apparatus

(a)   (b)

SOLUTION The mercury level will be lower on the side with the higher pressure, so we expect that the pressure will be lower inside the system.

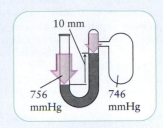

Find the pressure in the system.   756 mm − 10. mm = 746 mm of mercury

From $P = dhg$, $P = (13\,595\ \text{kg·m}^{-3}) \times (0.746\ \text{m}) \times (9.806\,65\ \text{m·s}^{-2})$
$= 9.95 \times 10^4\ \text{kg·m}^{-1}\text{·s}^{-2} = 9.95 \times 10^4\ \text{Pa}$

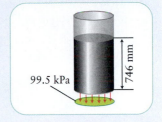

The pressure in the apparatus is 746 mmHg (99.5 kPa).

SELF-TEST 4.2A What is the pressure in a system when the mercury level in the system-side column in an open-tube mercury manometer is 25 mm lower than the mercury level in the atmosphere-side column and the atmospheric pressure corresponds to 760. mmHg at 15°C?

[*Answer:* 785 mmHg, 105 kPa]

SELF-TEST 4.2B What is the pressure in millimeters of mercury inside a system when a *closed* mercury manometer shows a height difference of 10. cm at 15°C?

The pressure of a gas, the force that it exerts divided by the area subjected to the force, arises from the impacts of its molecules.

4.3 Alternative Units of Pressure

Although the SI unit of pressure is the pascal (Pa), there are several other units in common use. Normal atmospheric pressure is close to 100 kPa, and it is useful to have an abbreviation, the **bar**, for exactly 100 kPa:

1 bar = 10^5 Pa

The name *bar* is derived from the Greek word for "heavy."

The **standard pressure** for reporting data is now defined as 1 bar exactly. However, it had long been conventional to report data in terms of a pressure corresponding to 760 millimeters of mercury (written 760 mmHg), corresponding to just over 101 kPa. This convention inspired the use of the unit the **atmosphere** (atm), which is now defined by the exact relation

1 atm = 1.013 25 × 10^5 Pa

Note that 1 atm is slightly greater than 1 bar.

The unit **millimeter of mercury** (mmHg) is defined in terms of the pressure exerted by a column of mercury under certain conditions (15°C and in a standard gravitational field), but it has now been largely displaced by a unit of similar magnitude, the **torr** (Torr). This unit is defined by the exact relation

760 Torr = 1 atm

(so 1 Torr= 1.01 325 × 10^5 Pa/760 = 133.322 Pa). It is normally safe to use the units mmHg and Torr interchangeably, but in very precise work you should note that they are not exactly the same.

A note on good practice: The name of the unit torr, like all names derived from the names of people, has a lower-case initial letter; likewise, the symbol for the unit has an uppercase initial letter (Torr).

FIGURE 4.6 A weather map showing North America during Hurricane Rita in 2005. The curves are called isobars and are contours of constant atmospheric pressure. Regions of low pressure (seen over southern Louisiana) are called cyclones and regions of high pressure (seen over Canada) are called anticyclones. All pressures are in millibars. The lowest pressure on the map is 984 mbar, at the center of the hurricane. Before making landfall, the pressure at the center of the hurricane had fallen to as low as 897 mbar.

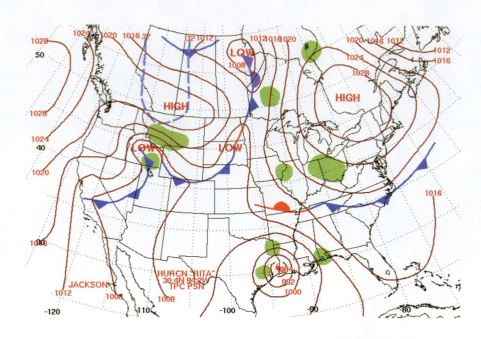

TABLE 4.1 Pressure Units*

SI unit: pascal (Pa)
$1\ Pa = 1\ kg \cdot m^{-1} \cdot s^{-2} = 1\ N \cdot m^{-2}$

Conventional units
$1\ bar = 10^5\ Pa = 100\ kPa$
$1\ atm = \mathbf{1.013\ 25 \times 10^5}\ Pa$
      $= \mathbf{101.325}\ kPa$
$1\ atm = \mathbf{760}\ Torr$
$1\ Torr\ (1\ mmHg) = 133.322\ Pa$
$1\ atm = 14.7\ lb \cdot inch^{-2}\ (psi)$

*Figures in bold are exact. See inside back cover for more relations. N, newton. Note that 1 mmHg is the same as 1 Torr to within 1 part in 10^7.

Pressure units are summarized in Table 4.1. It is important to be familiar with them and to be able to make conversions between them. In Example 4.2, for instance, the pressure in pascals could have been obtained by using a conversion factor derived from Table 4.1:

$$746\ mmHg \times \frac{133.322\ Pa}{1\ mmHg} = 9.95 \times 10^4\ Pa\ or\ 99.5\ kPa$$

The actual pressure exerted by the atmosphere varies with altitude and weather. The pressure of the atmosphere at the cruising height of a commercial jetliner (10 km) is only about 200 Torr (about 0.3 atm), and so airplane cabins must be pressurized. A very low pressure atmospheric region, such as an area of low pressure on the weather chart in Fig. 4.6, typically has a pressure of about 0.98 atm at sea level. A typical region of high pressure is about 1.03 atm.

SELF-TEST 4.3A The U.S. National Hurricane Center reported that the pressure in the eye of hurricane Katrina (2005) fell as low as 902 mbar. What is that pressure in atmospheres?

[*Answer:* 0.890 atm]

SELF-TEST 4.3B The atmospheric pressure in Denver, Colorado, on a certain day was 630. Torr. Express this pressure in pascals.

The principal units for reporting pressure are torr, atmosphere, and (the SI unit) pascal. Pressure units can be interconverted by using the information in Table 4.1.

THE GAS LAWS

Summaries of the properties of gases, particularly the variation of pressure with volume and temperature, are known as the "gas laws." The first reliable measurements of the properties of gases were made by the Anglo-Irish scientist Robert Boyle in 1662 when he examined the effect of pressure on volume. A century and a half later, a new pastime, hot-air ballooning, motivated two French scientists, Jacques Charles and Joseph-Louis Gay-Lussac, to formulate additional gas laws. Charles and

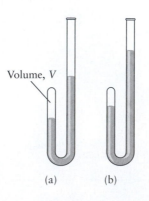
Volume, V

(a)   (b)

FIGURE 4.7 (a) In Boyle's experiment, a gas was trapped by mercury inside the closed end of a J-shaped tube. (b) The volume of the trapped gas decreased as the pressure on it was increased by adding more mercury to the open end of the tube.

FIGURE 4.8 Boyle's law summarizes the effect of pressure on the volume of a fixed amount of gas at constant temperature. As the pressure of a gas sample is increased, the volume of the gas decreases.

Gay-Lussac measured how the temperature of a gas affects its pressure, volume, and density. The Italian scientist Amedeo Avogadro made a further contribution that established the relation between the volume and the amount of molecules in the sample and thereby helped to establish belief in the reality of atoms.

4.4 The Experimental Observations

Boyle took a long tube of glass curved into a J-shape, with the short end sealed (Fig. 4.7). He then poured mercury into the tube, trapping air in the short end of the J. The more mercury he added, the more the air was compressed. He concluded that, at constant temperature the volume of a fixed amount of gas (the air in this case) decreases as the pressure on it increases. Figure 4.8 shows a graph of the dependence. The curve shown there is called an **isotherm**, which is a general term for a constant-temperature plot. Scientists often look for ways of plotting experimental data in a manner that gives straight lines, because such graphs are easier to identify, analyze, and interpret. Boyle's data give a straight line when we plot the pressure against 1/volume (Fig. 4.9). This result implies

> **Boyle's law:** For a fixed amount of gas at constant temperature, volume is inversely proportional to pressure.

Boyle's law is written

$$\text{Volume} \propto \frac{1}{\text{pressure}} \quad \text{or, more simply,} \quad V = \frac{\text{constant}}{P}$$

or equivalently

$$PV = \text{constant (at constant } n \text{ and } T)\tag{3a}$$

It is used in the form

$$P_2 V_2 = P_1 V_1 \tag{3b}$$

when the pressure and volume of a given amount of gas is changed isothermally from P_1 and V_1 to P_2 and V_2. An **isothermal change** is one that takes place at constant temperature.

SELF-TEST 4.4A A sample of neon of volume 10.0 L at 300. Torr is allowed to expand isothermally into an evacuated tube with a volume of 20.0 L. What is the final pressure of the neon in the tube?

[*Answer:* 150. Torr]

SELF-TEST 4.4B In a petroleum refinery a 750.-L container containing ethylene gas at 1.00 bar was compressed isothermally to 5.00 bar. What was the final volume of the container?

Charles and Gay-Lussac carried out a number of experiments with the hope of improving the performance of their balloons. They found that, provided the pressure is kept constant, the volume of a gas increases as its temperature is raised. In this case, a straight-line graph is obtained when the volume is plotted against the temperature (Fig. 4.10). This result implies

> **Charles's law:** For a fixed amount of gas under constant pressure, the volume varies linearly with the temperature.

Charles's law has a very important implication. When the straight lines obtained by plotting volume against temperature for different gases and at various

FIGURE 4.10 When the temperature of a gas is increased and its volume is free to change at constant pressure (as depicted by the constant weight acting on the piston), the volume increases. A graph of volume against temperature is a straight line.

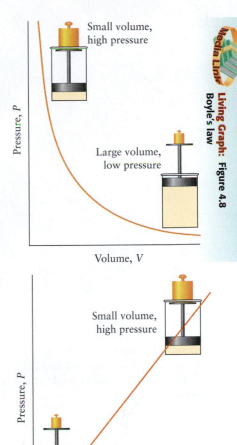

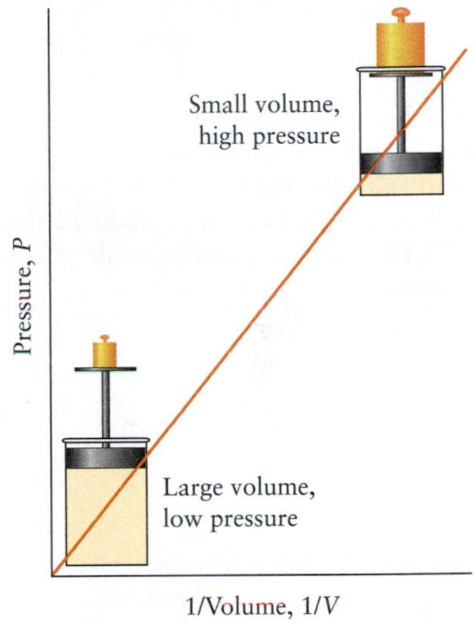

FIGURE 4.9 When the pressure is plotted against 1/volume, a straight line is obtained. Boyle's law breaks down at high pressures, and a straight line is not obtained in these regions (not shown).

Gay-Lussac's name is sometimes associated with the law, but "Charles's law" is now more common.

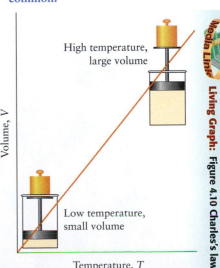

pressure are extrapolated (that is, extended beyond the range of the data), it is found that they all reach zero volume at $-273.15°C$ (Fig. 4.11). This point cannot be reached in practice, because no real gas has zero volume and all real gases condense to a liquid before such low temperatures are reached. Because a volume cannot be negative, $-273.15°C$ must be the lowest possible temperature. It is the value corresponding to zero on the Kelvin scale. It follows that, if we use the absolute temperature, T, then we can write Charles's law as

Volume $\propto$ absolute temperature   or, more simply, $V = \text{constant} \times T$ **(4a)**

A similar expression summarizes the linear variation in pressure of a sample of gas when it is heated in a container of fixed volume. The pressure is found experimentally to extrapolate to zero pressure at $-273.15°C$ (Fig. 4.12). Therefore, an alternative version of Charles's law is

Pressure $\propto$ absolute temperature   or, more simply, $P = \text{constant} \times T$ **(4b)**

It follows that doubling the absolute temperature doubles the pressure of a gas, provided the amount and volume are constant.

A note on good practice: Note that the volume (or the pressure) doubles when the temperature is doubled on the *absolute* (Kelvin) scale, not when it is doubled on the Celsius scale. An increase from 20°C to 40°C corresponds to an increase from 293 K to 313 K, an increase of only 7%.

SELF-TEST 4.5A A rigid oxygen tank stored outside a building has a pressure of 20.00 atm at 6:00 am, when the temperature is 10.°C. What will be the pressure in the tank at 6:00 pm, when the temperature is 30.°C?

[*Answer:* 21.4 atm]

SELF-TEST 4.5B A sample of hydrogen gas at 760. mmHg and 20.°C is heated to 300.°C in a container of constant volume. What is the final pressure of the sample?

A further contribution was made by the Italian scientist Amedeo Avogadro:

Avogadro's principle: *Under the same conditions of temperature and pressure, a given number of gas molecules occupy the same volume regardless of their chemical identity.*

Avogadro's principle is commonly expressed in the terms of the **molar volume**, V_m, the volume occupied per mole of molecules:

$$V_m = \frac{V}{n} \qquad\qquad \text{(5a)*}$$

The Kelvin scale of absolute temperatures is described in Appendix 1B: the temperature in kelvins (K) is obtained by adding 273.15 to the temperature in degrees Celsius (°C).

Avogadro's *principle* rather than law, because it is based not on observation alone but also on a model of matter—namely that matter consists of molecules. Even though there is no longer any doubt that matter consists of atoms and molecules, it remains a principle rather than a law.

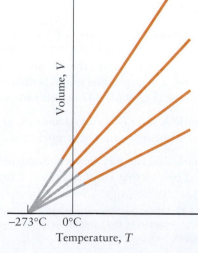

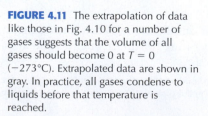

FIGURE 4.11 The extrapolation of data like those in Fig. 4.10 for a number of gases suggests that the volume of all gases should become 0 at $T = 0$ ($-273°C$). Extrapolated data are shown in gray. In practice, all gases condense to liquids before that temperature is reached.

for then we can write

$$V = nV_m \qquad (5b)$$

The molar volume of gases are all close to 22 L·mol^{-1} at 0°C and 1 atm (Fig. 4.13).

SELF-TEST 4.6A A helium weather balloon was filled at −20.°C and a certain pressure to a volume of 2.5×10^4 L with 1.2×10^3 mol He. What is the molar volume of helium under those conditions?

[*Answer:* 21 L·mol^{-1}]

SELF-TEST 4.6B A large natural gas storage tank contains 200. mol $CH_4(g)$ at 1.20 atm. An additional 100. mol $CH_4(g)$ is pumped into the tank at constant temperature. What is the final pressure in the tank?

Each of the properties we have summarized is consistent with our molecular model of a gas as widely spaced molecules in ceaseless motion. Boyle's law is consistent with this model because compression increases the number of molecules in a given region of the sample and hence increases the number of collisions that the molecules make with the walls. As a result, the pressure they exert increases (Fig. 4.14). The effect of temperature on the pressure of a gas in a constant-volume container suggests a new feature: *as the temperature of a gas is raised, the average speed of the molecules increases.* As a result of this increase in average speed, each molecule strikes the walls more often and with greater force. Therefore, the gas exerts a greater pressure as the temperature is increased at constant volume. We can use this same model of a gas to explain the effect of temperature on the volume when the pressure is held constant. To counteract the increase in pressure as the temperature is raised, the space available to the gas must increase so that fewer molecules are available to strike the walls in a given time interval. Finally, Avogadro's principle is consistent with the model because, in order to keep the pressure constant as more molecules are added to a container, the size of the container must increase.

The three properties of a gas expressed by Eqs. 3, 4, and 5 can all be combined into a single expression relating pressure (P), volume (V), temperature (T), and amount (n) of a gas:

$$PV = \text{constant} \times nT$$

When the constant of proportionality for the laws is written as R, this expression becomes the **ideal gas law:**

$$PV = nRT \qquad (6)*$$

The constant R is called the **gas constant** and has the same value for all gases; because R is independent of the identity of the gas, we say that it is a "universal constant." The value of the gas constant can be found by measuring P, V, n, and T and substituting their values into $R = PV/nT$. When we use SI units (pressure in pascals, volume in meters cubed, temperature in kelvins, and amount in moles),

Ideal gas	22.41
Argon	22.09
Carbon dioxide	22.26
Nitrogen	22.40
Oxygen	22.40
Hydrogen	22.43

FIGURE 4.13 The molar volumes (in liters per mole) of various gases at 0°C and 1 atm. The values are all very similar and close to the molar volume of an ideal gas under these conditions, 22.41 L·mol^{-1} (Section 4.5).

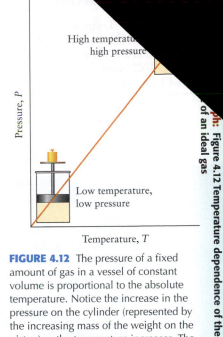

FIGURE 4.12 The pressure of a fixed amount of gas in a vessel of constant volume is proportional to the absolute temperature. Notice the increase in the pressure on the cylinder (represented by the increasing mass of the weight on the piston) as the temperature increases. The pressure extrapolates to 0 at $T = 0$.

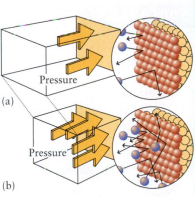

FIGURE 4.14 (a) The pressure of a gas arises from the impact of its molecules on the walls of the container. (b) When the volume of the sample is decreased, there are more molecules in a given volume and so there are more collisions with the same area of the wall in a given time interval. Because the impact on the walls is now greater, so is the pressure.

R is obtained in joules per kelvin per mole: $R = 8.314 \text{ J·K}^{-1}\text{·mol}^{-1}$. Table 4.2 lists the values of R in a variety of different units which are useful if volume or pressure is reported in other than SI units. For instance, it is sometimes convenient to use R in liter-atmospheres per kelvin per mole: $R = 8.206 \times 10^{-2} \text{ L·atm·K}^{-1}\text{·mol}^{-1}$.

A note on good practice: To avoid overwhelming you with data we usually quote the values of fundamental constants to three decimal places. In actual calculations you should use the more precise values given in tables, including those inside the back cover.

The ideal gas law is an example of an **equation of state**, an expression showing how the pressure of a substance—in this case, a gas—is related to its temperature, volume, and amount of substance. A hypothetical gas that obeys the ideal gas law under all conditions is called an **ideal gas**. All real gases are found to obey Eq. 6 with increasing accuracy as the pressure is reduced toward zero (which we write $P \rightarrow 0$). Therefore, the ideal gas law is an example of a **limiting law**, a law that is strictly valid only in some limit—in this case, as $P \rightarrow 0$. Although the ideal gas law is a limiting law, it is in fact reasonably reliable at normal pressures, and so we can use it to describe the behavior of most gases under normal conditions.

> The ideal gas law, $PV = nRT$, is an equation of state that summarizes the relations describing the response of an ideal gas to changes in pressure, volume, temperature, and amount of molecules; it is an example of a limiting law.

4.5 Applications of the Ideal Gas Law

We have seen how to use the individual laws to make predictions when only one variable is changed, such as heating a fixed amount of gas at constant volume. The ideal gas law enables us to make predictions when two or more variables are changed.

To do such calculations, we note that, if the initial conditions of a gas are n_1, P_1, V_1, and T_1, then from Eq. 6 we can write $P_1V_1 = n_1RT_1$, or $P_1V_1/n_1T_1 = R$. After the change, the conditions become n_2, P_2, V_2, and T_2; and, because the ideal gas law still applies, we know that $P_2V_2/n_2T_2 = R$. Because R is a constant, we can equate P_1V_1/n_1T_1 and P_2V_2/n_2T_2 and hence obtain

$$\frac{P_1V_1}{n_1T_1} = \frac{P_2V_2}{n_2T_2} \tag{7}*$$

This expression is called the **combined gas law**. However, it is a direct consequence of the ideal gas law and is not a new law.

TOOLBOX 4.1   HOW TO USE THE IDEAL GAS LAW

CONCEPTUAL BASIS

The ideal gas law, Eq. 6 ($PV = nRT$), summarizes the observation that the pressure is inversely proportional to the volume and directly proportional to the amount of gas and the absolute temperature. It can be used for real gases at low pressures.

PROCEDURE

To calculate the pressure, volume, or temperature of a given sample

Step 1 Express the temperatures in kelvins and the amount in moles.

Step 2 To use Eq. 6, rearrange the equation $PV = nRT$ to give the desired quantity on the left and all other quantities on the right (such as $P = nRT/V$).

Step 3 Substitute the data, if necessary converting from mass of gas to the amount in moles. Select the value of R from Table 4.2 that matches the units of pressure and volume you need to use. Alternatively, convert the pressure units to match the value of R you prefer to use.

The procedure is illustrated in Example 4.3.

To calculate the response of a gas to a change in conditions

The initial and final conditions for a gas undergoing a change are related by Eq. 7 ($P_1V_1/n_1T_1 = P_2V_2/n_2T_2$). To use this relation:

Step 1 Rearrange the relation so that the quantity required is on the left and all other quantities are on the right. Cancel any quantities that are unchanged and, if necessary, express the temperatures in kelvins.

Step 2 Substitute the data and check to see whether the answer agrees with your predictions.

The procedure is illustrated in Examples 4.4 and 4.5.

EXAMPLE 4.3   Sample exercise: Calculating the pressure of a given sample

A television picture tube is a form of cathode-ray tube (Section 1.1) in which the beam of electrons is directed toward a screen that emits light where they strike it. Have you ever wondered what the pressure is inside the tube? Estimate the pressure (in atmospheres), given that the volume of the tube is 5.0 L, its temperature is 23°C, and it contains 0.010 mg of nitrogen gas.

SOLUTION Convert mass into amount of N_2 and convert the temperature into kelvins; then proceed as in Toolbox 4.1.

Step 1 Convert mass to amount ($n = m/M$) and temperature from degrees Celsius to kelvins by adding 273.15.

$$n = \frac{1.0 \times 10^{-5}\text{ g}}{28.02\text{ g·mol}^{-1}} = \frac{1.0 \times 10^{-5}}{28.02}\text{ mol}$$

$$T = (23 + 273.15)\text{ K} = 296\text{ K}$$

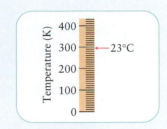

Step 2 From $PV = nRT$,

$$P = \frac{nRT}{V}$$

Step 3 Substitute the data, using R in L·atm·K^{-1}·mol^{-1}.

$$P = \left(\frac{1.0 \times 10^{-5}}{28.02}\text{ mol}\right)$$

$$\times \frac{(8.206 \times 10^{-2}\text{ L·atm·K}^{-1}\text{·mol}^{-1}) \times (296\text{ K})}{5.0\text{ L}}$$

$$= 1.7 \times 10^{-6}\text{ atm}$$

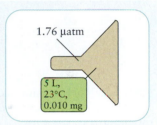

A very low pressure is required to minimize the collisions between the electrons in the beam and the gas molecules; otherwise, deflections of the electrons resulting from collisions would give a blurred, dim picture.

SELF-TEST 4.7A Calculate the pressure (in kilopascals) exerted by 1.0 g of carbon dioxide in a flask of volume 1.0 L at 300.°C.

[Answer: 1.1×10^2 kPa]

SELF-TEST 4.7B An idling, badly tuned automobile engine can release as much as 1.00 mol CO per minute into the atmosphere. At 27°C, what volume of CO, adjusted to 1.00 atm, is emitted per minute?

EXAMPLE 4.4   Sample exercise: Using the combined gas law when one variable is changed

Assume that, when you press in the piston of a bicycle pump, the volume inside the pump is decreased from about 100. cm^3 to 20. cm^3 before the air flows into the tire. Suppose that the compression is isothermal; estimate the final pressure of the compressed air in the pump, given an initial pressure of 1.00 atm.

SOLUTION We expect the final pressure, P_2, to be higher than the initial pressure, P_1, because the volume occupied by the air has been decreased. Follow the second procedure in Toolbox 4.1. Only the pressure and volume change, so all other variables cancel, resulting in Boyle's law.

Step 1 Rearrange $P_1V_1/n_1T_1 = P_2V_2/n_2T_2$ to find P_2; set $n_1 = n_2$ and $T_1 = T_2$ and cancel them.

$$P_2 = \frac{P_1V_1}{n_1T_1} \times \frac{n_2T_2}{V_2} = P_1 \times \frac{V_1}{V_2}$$

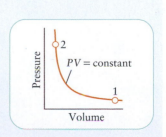

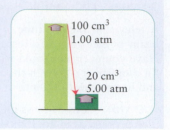

Step 2 Substitute the data:

$$P_2 = \frac{100 \text{ cm}^3}{20 \text{ cm}^3} \times (1.00 \text{ atm}) = 5.0 \text{ atm}$$

The final pressure is higher, as expected.

SELF-TEST 4.8A A sample of argon gas of volume 10.0 mL at 200. Torr is allowed to expand isothermally into an evacuated tube with a volume of 0.200 L. What is the final pressure of the neon in the tube?

[*Answer:* 10.0 Torr]

SELF-TEST 4.8B A sample of dry air in the cylinder of a test engine at 80. cm³ and 1.00 atm is compressed isothermally to 3.20 atm by pushing a piston into the cylinder. What is the final volume of the sample?

EXAMPLE 4.5   **Sample exercise: Using the combined gas law when two variables are changed**

In an investigation of the properties of the coolant gas used in an air-conditioning system, a sample of volume 500. mL at 28.0°C was found to exert a pressure of 92.0 kPa. What pressure will the sample exert when it is compressed to 300. mL and cooled to −5.0°C?

SOLUTION Compression increases the pressure but cooling lowers the pressure, and so it is not easy to predict the outcome without doing a calculation. Follow the procedure set out in Toolbox 4.1.

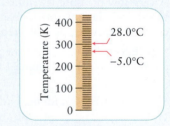

Step 1 Convert the temperatures into kelvins.

$T_1 = (273.15 + 28.0) \text{ K} = 301.2 \text{ K}$
$T_2 = (273.15 − 5.0) \text{ K} = 268.2 \text{ K}$

Step 2 Rearrange $P_1V_1/n_1T_1 = P_2V_2/n_2T_2$ to find P_2 and set $n_1 = n_2$.

$$P_2 = P_1 \times \frac{V_1}{V_2} \times \frac{T_2}{T_1}$$

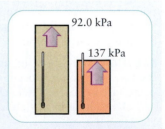

Step 3 Substitute the data:

$$P_2 = (92.0 \text{ kPa}) \times \frac{500. \text{ mL}}{300. \text{ mL}} \times \frac{268.2 \text{ K}}{301.2 \text{ K}} = 137 \text{ kPa}$$

The net outcome is an increase in pressure, so, in this instance, compression has a greater effect than cooling.

SELF-TEST 4.9A A parcel of air (the technical term in meteorology for a small region of the atmosphere) of volume 1.00×10^3 L at 20.°C and 1.00 atm rises up the side of a mountain range. At the summit, where the pressure is 0.750 atm, the parcel of air has cooled to −10.°C. What is the volume of the parcel at that point?

[*Answer:* 1.20×10^3 L]

SELF-TEST 4.9B A weather balloon is filled with helium gas at 20.°C and 1.00 atm. The volume of the balloon is 250. L. When the balloon rises to a layer of air where the temperature is −30.°C, it has expanded to 800. L. What is the pressure of the atmosphere at that point?

The ideal gas law can also be used to predict the molar volume of an ideal gas under any conditions of temperature and pressure. To do so, we combine Eq. 5a and Eq. 6 by writing

$$V_m = \frac{V}{n} = \frac{nRT/P}{n} = \frac{RT}{P} \tag{8}$$

At **standard ambient temperature and pressure** (SATP), which means exactly 25°C (298.15 K) and exactly 1 bar, the conditions commonly used to report data in chemistry, the molar volume of an ideal gas is 24.79 L·mol^{-1}, which is about the volume of a cube 1 ft on a side (Fig. 4.15). The expression **standard temperature and pressure** (STP) means 0°C and 1 atm (both exactly), the conditions formerly used to report data and still widely used in some calculations. At STP, the molar volume of an ideal gas is 22.41 L·mol^{-1}. Note the slightly smaller value: the temperature is lower and the pressure is slightly higher, and so the same amount of gas molecules occupies a smaller volume than at SATP.

Table 4.3 gives values of the molar volume of an ideal gas under a variety of common conditions. To obtain the volume of a known amount of gas at a specified temperature and pressure, we simply multiply the molar volume *at that temperature and pressure* by the amount in moles. Alternatively, we can use the ideal gas law to calculate the volume.

SELF-TEST 4.10A Calculate the volume occupied by 1.0 kg of hydrogen at 25°C and 1.0 atm.

[*Answer:* 1.2 × 10⁴ L]

SELF-TEST 4.10B Calculate the volume occupied by 2.0 g of helium at 25°C and 1.0 atm.

> *The combined gas law describes how a gas responds to changes in conditions. Standard ambient temperature and pressure (SATP) are 25°C (298.15 K) and 1 bar; standard temperature and pressure (STP) are 0°C (273.15 K) and 1 atm.*

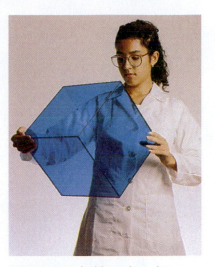

FIGURE 4.15 The blue cube is the volume (25 L) occupied by 1 mol of ideal gas molecules at 25°C and 1 bar.

4.6 Gas Density

As we saw in Section G, the molar concentration of any substance is the amount of molecules (*n*, in moles) divided by the volume that they occupy (*V*). It follows from the ideal gas law that, for a gas behaving ideally (so we can write *n* = *PV/RT*),

$$\text{Molar concentration} = \frac{n}{V} = \frac{PV}{RTV} = \frac{P}{RT} \tag{9}$$

This expression shows that, for a given pressure and temperature, the molar concentration should be the same for every gas. That is, two gas samples of equal volume at the same temperature and pressure should contain the same amount of

TABLE 4.3 Molar Volume of an Ideal Gas

Temperature	Pressure	Molar volume (L·mol^{-1})
0 K	0	0
0°C	1 atm	22.4141
0°C	1 bar	22.7111
25°C	1 atm	24.4655
25°C	1 bar	24.7897

gas molecules, regardless of the identities of the gases (this, of course, is just Avogadro's principle). However, if their molar masses are different, the gas samples will have different masses.

The density, d, of a gas, like that of any substance, is the mass of the sample divided by its volume, $d = m/V$. Because the densities of gases are so low, they are usually expressed in grams per liter ($g·L^{-1}$). The density of air, for instance, is about 1.6 $g·L^{-1}$ at SATP. Because the mass of the sample is equal to the amount in moles times the molar mass, $m = nM$, and $n = PV/RT$, it follows that

$$d = \frac{m}{V} = \frac{nM}{V} = \frac{(PV/RT)M}{V} = \frac{MP}{RT} \tag{10}*$$

We see that, for a given pressure and temperature, the greater the molar mass of the gas, the greater its density. Equation 10 also shows that, at constant temperature, the density of a gas increases with pressure. When a gas is compressed, its density increases because the same number of molecules are confined in a smaller volume. Similarly, heating a gas that is free to expand at constant pressure increases the volume occupied by the gas and therefore reduces its density. The effect of temperature on density is the principle behind hot-air balloons: the hot air inside the envelope of the balloon has a lower density than that of the surrounding cool air. Equation 10 is also the basis for using density measurements to determine the molar mass of a gas or vapor.

EXAMPLE 4.6   **Sample exercise: Calculating the molar mass of a gas from its density**

The volatile organic compound geraniol, a component of oil of roses, is used in perfumery. The density of the vapor at 260.°C and 103 Torr is 0.480 $g·L^{-1}$. What is the molar mass of geraniol?

SOLUTION

List the information given and convert the temperature into kelvins.

$d = 0.480$ $g·L^{-1}$, $P = 103$ Torr, and
$T = (273.15 + 260.)$ K $= 533$ K

Rearrange $d = MP/RT$ into $M = dRT/P$, select a value of R with appropriate units, and substitute the data.

$$M = \frac{(0.480 \text{ g·L}^{-1}) \times (62.364 \text{ L·Torr·K}^{-1}\text{·mol}^{-1}) \times (533 \text{ K})}{103 \text{ Torr}}$$

$$= 155 \text{ g·mol}^{-1}$$

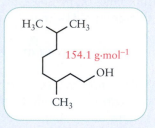

154.1 g·mol^{-1}

The value calculated from the formula of geraniol (shown in the illustration above) is 154.1 g·mol^{-1}.

SELF-TEST 4.11A The oil produced from eucalyptus leaves contains the volatile organic compound eucalyptol. At 190.°C and 60.0 Torr, a sample of eucalyptol vapor had a density of 0.320 g·L^{-1}. Calculate the molar mass of eucalyptol.

[*Answer:* 154 g·mol^{-1}]

SELF-TEST 4.11B The *Codex Ebers*, an Egyptian medical papyrus, describes the use of garlic as an antiseptic. Chemists today have verified that the oxide of diallyl disulfide

TABLE 4.4 Composition of Dry Air at Sea Level

Constituent	Molar mass* (g·mol^{-1})	Composition (%) Volume	Composition (%) Mass
nitrogen, N_2	28.02	78.09	75.52
oxygen, O_2	32.00	20.95	23.14
argon, Ar	39.95	0.93	1.29
carbon dioxide, CO_2	44.01	0.03	0.05

*The average molar mass of molecules in the air, allowing for their different abundances, is 28.97 g·mol^{-1}. The percentage of water vapor in ordinary air varies with the humidity.

(the volatile compound responsible for garlic odor) is a powerful antibacterial agent. At 177°C and 200. Torr, a sample of diallyl disulfide vapor has a density of 1.04 g·L^{-1}. What is the molar mass of diallyl disulfide?

The density of the atmosphere varies greatly from place to place, as does its composition and temperature. The average composition of dry air (air from which water vapor has been removed) is shown in Table 4.4. One reason for the nonuniformity of air is the effect of solar radiation, which causes different chemical reactions at different altitudes. The density of air also varies with altitude. For example, the air outside an airplane cruising at 10 km is only 25% as dense as air at sea level.

The molar concentrations and densities of gases increase as they are compressed but decrease as they are heated. The density of a gas depends on its molar mass.

4.7 The Stoichiometry of Reacting Gases

At this point, we have considered only samples of pure gases. However, some gases react with each other and we can use the ideal gas law to calculate the volume of gas consumed or produced in a reaction. For example, we might need to know the volume of carbon dioxide produced when a fuel burns or the volume of oxygen needed to react with a given mass of hemoglobin in the red cells of our blood. To answer this kind of question, we have to combine the mole-to-mole calculations of the type described in Sections L and M with the conversion of moles of gas molecules into the volume that they occupy. The diagram in (**1**) extends the stoichiometry strategy introduced in Section L to include the volume of a gas.

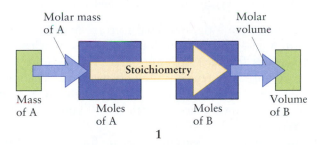

1

EXAMPLE 4.7 Calculating the mass of reagent needed to react with a specified volume of gas

The carbon dioxide generated by the personnel in the artificial atmosphere of submarines and spacecraft must be removed from the air and the oxygen recovered. Submarine design teams have investigated the use of potassium superoxide, KO_2, as an air purifier because this compound reacts with carbon dioxide and releases oxygen (Fig. 4.16):

$$4\ KO_2(s) + 2\ CO_2(g) \longrightarrow 2\ K_2CO_3(s) + 3\ O_2(g)$$

Calculate the mass of KO_2 needed to react with 50. L of carbon dioxide at 25°C and 1.0 atm.

FIGURE 4.16 When carbon dioxide is passed over potassium superoxide (the yellow solid), it reacts to form colorless potassium carbonate (the white solid coating the walls of the tube) and oxygen gas. The reaction is used to remove carbon dioxide from the air in a closed-system breathing environment.

STRATEGY We convert from the given volume of gas into moles of molecules (by using the molar volume), then into moles of reactant molecules or formula units (by using a mole ratio), and then into the mass of reactant (by using its molar mass). If the molar volume at the stated conditions is not available, then use the ideal gas law to calculate the amount of gas molecules.

SOLUTION

Find the molar volume under the stated conditions from a table or calculation.	Table 4.3 gives $V_m = 24.47$ L·mol^{-1}. 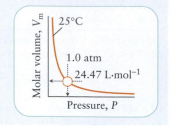
Find the stoichiometric relation between CO_2 and KO_2 from the chemical equation.	2 mol CO_2 ⇌ 4 mol KO_2, or 1 mol CO_2 ⇌ 2 mol KO_2
Find the molar mass of KO_2 (Section E).	$39.10 + 2(16.00)$ g·mol^{-1} = 71.10 g·mol^{-1}
Convert from volume of CO_2 to mass of KO_2.	Mass of KO_2 = $(50 \text{ L}) \times \left(\dfrac{1 \text{ mol } CO_2}{24.47 \text{ L}}\right)$ $\times \left(\dfrac{2 \text{ mol } KO_2}{1 \text{ mol } CO_2}\right) \times \left(\dfrac{71.1 \text{ g}}{1 \text{ mol } KO_2}\right)$ $= 2.9 \times 10^2$ g 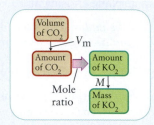

SELF-TEST 4.12A Calculate the volume of carbon dioxide, adjusted to 25°C and 1.0 atm, that plants need to make 1.00 g of glucose, $C_6H_{12}O_6$, by photosynthesis in the reaction $6\ CO_2(g) + 6\ H_2O(l) \rightarrow C_6H_{12}O_6(s) + 6\ O_2(g)$.

[*Answer:* 0.82 L]

SELF-TEST 4.12B The reaction of H_2 and O_2 gases to produce liquid H_2O is used in fuel cells on the space shuttles to provide electricity. What mass of water is produced in the reaction of 100.0 L of oxygen stored at 25°C and 1.00 atm?

There may be more than a thousandfold increase in volume when liquids or solids react to form a gas. The molar volumes of gases are close to 25 L·mol^{-1} under normal conditions (room temperature and pressure), whereas liquids and solids occupy only about a few tens of milliliters per mole. The molar volume of liquid water, for instance, is only 18 mL·mol^{-1}. In other words, 1 mol of gas molecules at 25°C and 1 atm occupies as much as 1000 times the volume of 1 mol of molecules in a typical liquid or solid.

The increase in volume as gaseous products are formed in a chemical reaction is even larger if several gas molecules are produced from each reactant molecule, such as the formation of CO and CO_2 from a solid fuel (Fig. 4.17). Lead azide, $Pb(N_3)_2$, which is used as a detonator for explosives, suddenly releases a large volume of nitrogen gas when it is struck:

$$Pb(N_3)_2(s) \longrightarrow Pb(s) + 3\ N_2(g)$$

An explosion of the same kind, but using sodium azide, NaN_3, is used in air bags in automobiles (Fig. 4.18). The explosive release of nitrogen is detonated electrically when the vehicle decelerates abruptly in a collision.

FIGURE 4.17 An explosion caused by the ignition of coal dust. A shock wave is created by the tremendous expansion of volume as large numbers of gas molecules form.

The molar volume (at the specified temperature and pressure) is used to convert the amount of a reactant or product in a chemical reaction into a volume of gas.

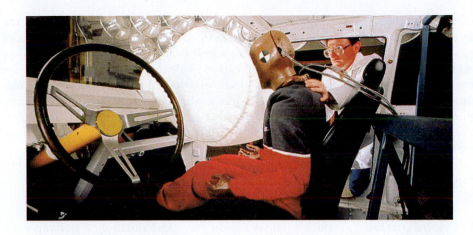

FIGURE 4.18 The rapid decomposition of sodium azide, NaN_3, results in the formation of a large volume of nitrogen gas. The reaction is triggered electrically in this air bag.

4.8 Mixtures of Gases

Many of the gases we meet in chemistry—and in everyday life—are mixtures. The atmosphere, for instance, is a mixture of nitrogen, oxygen, argon, carbon dioxide, and many other gases. The air we inhale is one mixture; the air we exhale is another mixture. Many gaseous anesthetics are carefully controlled mixtures. An important industrial chemical reaction is the synthesis of ammonia, in which a mixture of nitrogen and hydrogen is fed into a reaction vessel. We need to extend the model of a gas that we have been developing to include mixtures of gases.

To extend our model, we should note that, at low pressures at least, all gases respond in the same way to changes in pressure, volume, and temperature. Therefore, for calculations of the type that we are doing in this chapter, it does not matter whether all the molecules in a sample are the same. *A mixture of gases that do not react with one another behaves like a single pure gas.* For instance, we can treat air as a single gas when we want to use the ideal gas law to predict its properties.

John Dalton was the first to see how to calculate the pressure of a mixture of gases. His reasoning was something like this. Imagine that we introduce a certain amount of oxygen into a container and end up with a pressure of 0.60 atm. Then we evacuate the container to empty it of all gas. Now we introduce into the container enough nitrogen gas to give a pressure of 0.40 atm at the same temperature. Dalton wondered what the total pressure would be if these same amounts of the two gases were present in the container simultaneously. From some fairly crude measurements, he concluded that the total pressure resulting from the pressure of both gases in the same container would be 1.00 atm, the sum of the individual pressures.

Dalton summarized his observations in terms of the **partial pressure** of each gas, the pressure that the gas would exert if it occupied the container alone. In our example, the partial pressures of oxygen and nitrogen in the mixture are 0.60 atm and 0.40 atm, respectively, because those are the pressures that the gases exert when each one is in the container alone. Dalton then described the behavior of gaseous mixtures by his **law of partial pressures**:

> The total pressure of a mixture of gases is the sum of the partial pressures of its components.

If we denote the partial pressures of the gases A, B, . . . as P_A, P_B, . . . and the total pressure of the mixture as P, then we can write Dalton's law as follows:

$$P = P_A + P_B + \cdots \qquad (11)^*$$

The law is illustrated in Fig. 4.19. It is exactly true only for gases that behave ideally, but it is a good approximation for nearly all gases under normal conditions.

Dalton's law is consistent with our picture of a gas and adds a little more information. The total pressure of a gas arises, as we have seen, from the battering

This is the same Dalton whose contribution to atomic theory we met in Section B.

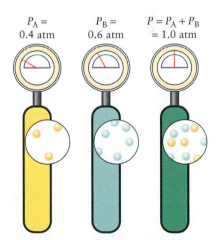

$$P_A = \quad P_B = \quad P = P_A + P_B$$
$$0.4 \text{ atm} \quad 0.6 \text{ atm} \quad = 1.0 \text{ atm}$$

FIGURE 4.19 A representation of the experiment that Dalton performed on a gas mixture. According to Dalton's law, the total pressure, P, of a mixture of gases is the sum of the partial pressures P_A and P_B of gases A and B. Each partial pressure is the pressure that one of the gases would exert if it were the sole gas in the container (at the same temperature).

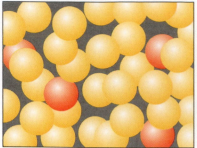

$x_{\text{RED}} = 0.1$

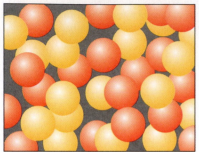

$x_{\text{RED}} = 0.5$

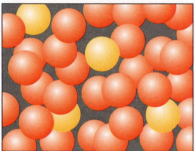

$x_{\text{RED}} = 0.9$

FIGURE 4.20 A mole fraction, x, tells us the fraction of molecules of a particular kind in a mixture of two or more kinds of molecules. In this illustration, the mole fraction of A molecules, which are colored red, is given below each mixture; the B molecules are colored yellow. The mixture can be solid or liquid as well as gaseous.

of the molecules on the walls of the container. That battering is due to all the molecules in a mixture. The molecules of gas A exert a pressure, the molecules of gas B do, too, and the total pressure is the sum of these individual pressures. The additional piece of information that we can use to expand our model is that, for the total pressure to be the sum of the individual contributions, the A molecules must be blind to the presence of the B molecules. That is, there are no interactions—neither attractions nor repulsions—between the two kinds of molecules. Later, we shall see that the absence of any interactions at all is a characteristic feature of an ideal gas.

We use partial pressures to describe the composition of a humid gas. For example, the total pressure of the damp air in our lungs is

$$P = P_{\text{dry air}} + P_{\text{water vapor}}$$

In a closed container, to which a lung is a good first approximation, water vaporizes until its partial pressure has reached a certain value, called its *vapor pressure*. The vapor pressure of water at normal body temperature is 47 Torr. The partial pressure of the air itself in our lungs is therefore

$$P_{\text{dry air}} = P - P_{\text{water vapor}} = P - 47\ \text{Torr}$$

On a typical day, the total pressure at sea level is 760. Torr; so at sea level the pressure in our lungs due to all the gas except the water vapor is 760. − 47 Torr = 713 Torr.

SELF-TEST 4.13A As a sample of oxygen was collected over water at 24°C and 745 Torr, it became saturated with water vapor. At this temperature, the vapor pressure of water is 24.38 Torr. What is the partial pressure of the oxygen?

[*Answer:* 721 Torr]

SELF-TEST 4.13B Students collecting hydrogen and oxygen gases by electrolysis of water failed to separate the two gases. If the total pressure of the dry mixture is 720. Torr, what is the partial pressure of each gas?

The easiest way to express the relation between the total pressure of a mixture and the partial pressures of its components is to introduce the **mole fraction**, x, of each component A, B, . . ., the number of moles of molecules of the gas expressed as a fraction of the total number of moles of molecules in the sample. If the amounts of gas molecules present are n_A, n_B, and so forth, the mole fraction of A is

$$x_A = \frac{n_A}{n_A + n_B + \cdots} \qquad (12)^*$$

and likewise for the mole fractions of the other components. This expression is used to calculate the mole fraction of each component of the mixture. In a binary (two-component) mixture of gases A and B,

$$x_A + x_B = 1 \qquad (13)$$

When $x_A = 1$, the mixture is pure A; when $x_B = 1$, the mixture is pure B. When $x_A = x_B = 0.50$, half the molecules are A and half are B (Fig. 4.20).

Our task is to find the relation between the partial pressure of a gas and its mole fraction.

HOW DO WE DO THAT?

To find the relation between the partial pressure of a gas in a mixture and its mole fraction, we first express the partial pressure, P_A, of a gas A in terms of the amount of A molecules present, n_A, the volume, V, occupied by the mixture, and the temperature, T:

$$P_A = \frac{n_A RT}{V}$$

Next, we express the total pressure in terms of n, the total number of moles of molecules present:

$$P = \frac{nRT}{V} = (n_A + n_B + \cdots)\frac{RT}{V}$$

We can rearrange this relation to

$$\frac{RT}{V} = \frac{P}{n_A + n_B + \cdots}$$

and then substitute this expression for RT/V in the first equation of this derivation ($P_A = n_A RT/V$) to obtain

$$P_A = \frac{n_A P}{n_A + n_B + \cdots} = x_A P$$

The relation that we have derived for the partial pressure of a gas A is

$$P_A = x_A P \tag{14}*$$

where P is the total pressure and x_A is the mole fraction of A in the mixture. In more advanced applications, Eq. 14 is regarded as the *definition* of the partial pressure of a gas, either ideal or real.

EXAMPLE 4.8   **Calculating partial pressures**

Air is a source of reactants for many chemical processes. To determine how much air is needed for these reactions, it is useful to know the partial pressures of the components. A certain sample of dry air of total mass 1.00 g consists almost entirely of 0.76 g of nitrogen and 0.24 g of oxygen. Calculate the partial pressures of these gases when the total pressure is 0.87 atm.

STRATEGY To use Eq. 14, we need the total pressure (given) and the mole fraction of each component. The first step is to calculate the amount (in moles) of each gas present and the total amount (in moles). Then calculate the mole fractions from Eq. 12. To obtain the partial pressures of the gases, multiply the total pressure by the mole fractions of the gases in the mixture (Eq. 14).

SOLUTION

Use their molar masses to find the amounts (in moles) of each gas present.

$$n_{N_2} = \frac{0.76\ \text{g}}{28.02\ \text{g·mol}^{-1}} = \frac{0.76}{28.02}\ \text{mol}\ (= 0.27\ \text{mol})$$

$$n_{O_2} = \frac{0.24\ \text{g}}{32.00\ \text{g·mol}^{-1}} = \frac{0.24}{32.00}\ \text{mol}\ (= 0.0075\ \text{mol})$$

Find the total amount of gas molecules from $n_{\text{Total}} = n_A + n_B$

$$n_{N_2} + n_{O_2} = 0.76/28.02 + 0.24/32.00\ \text{mol}\ (= 0.035\ \text{mol})$$

Calculate the mole fractions from $x_A = n_A/n_{\text{Total}}$.

$$x_{N_2} = \frac{0.76/28.02}{0.76/28.02 + 0.24/32.00} = 0.78$$

$$x_{O_2} = \frac{0.24/32.00}{0.76/28.02 + 0.24/32.00} = 0.22$$

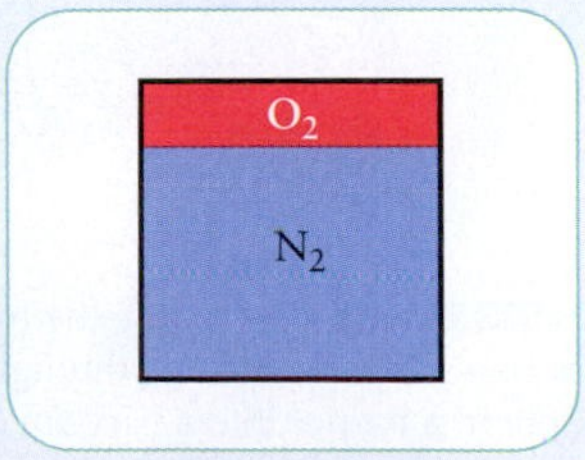

Multiply each mole fraction by the total pressure: $P_X = x_X P$.

$P_{N_2} = 0.78 \times (0.87 \text{ atm}) = 0.68 \text{ atm}$

$P_{O_2} = 0.22 \times (0.87 \text{ atm}) = 0.19 \text{ atm}$

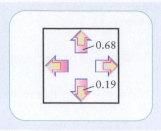

To check this answer, we calculate the total pressure, $0.68 + 0.19 \text{ atm} = 0.87 \text{ atm}$, as in the data.

SELF-TEST 4.14A A baby with a severe bronchial infection is in respiratory distress. The anesthetist administers heliox, a mixture of helium and oxygen with 92.3% by mass O_2. What is the partial pressure of oxygen being administered to the baby if atmospheric pressure is 730 Torr?

[*Answer:* 4.4×10^2 Torr]

SELF-TEST 4.14B Divers exploring a shipwreck and wishing to avoid the narcosis associated with breathing nitrogen under high pressure switch to a neon–oxygen gas mixture containing 141.2 g of oxygen and 335.0 g of neon. The pressure in the gas tanks is 50.0 atm. What is the partial pressure of oxygen in the tanks?

The partial pressure of a gas is the pressure that it would exert if it were alone in the container; the total pressure of a mixture of gases is the sum of the partial pressures of the components; the partial pressure of a gas is related to the total pressure by the mole fraction: $P_A = x_A P$.

MOLECULAR MOTION

The empirical results summarized by the gas laws suggest a model of an ideal gas in which widely spaced (most of the time), noninteracting molecules undergo ceaseless motion, with average speeds that increase with temperature. In the next three sections, we refine our model in two steps. First, we use experimental measurements of the rate at which gases spread from one region to another to discover the *average* speeds of molecules. Then we use these average speeds to express our model of an ideal gas quantitatively, check that it is in agreement with the gas laws, and use it to derive detailed information about the proportion of molecules having any specified speed.

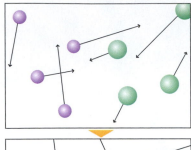

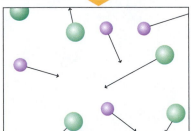

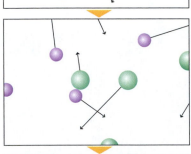

FIGURE 4.21 In diffusion, the molecules of one substance spread into the region occupied by molecules of another in a series of random steps, undergoing collisions as they move.

4.9 Diffusion and Effusion

Two kinds of processes, diffusion and effusion, provide data that show how the average speeds of gas molecules are related to molar mass and temperature. **Diffusion** is the gradual dispersal of one substance through another substance, such as krypton dispersing through a neon atmosphere (Fig. 4.21). Diffusion explains the spread of perfumes and pheromones, the latter being chemical signals exchanged by animals, through air. It also helps to keep the composition of the atmosphere approximately constant, because abnormally high concentrations of one gas diffuse away and disperse. **Effusion** is the escape of a gas through a small hole into a vacuum (Fig. 4.22). Effusion occurs whenever a gas is separated from a vacuum by a porous barrier—a barrier that contains microscopic holes—or a single pinhole. A gas escapes through a pinhole because there are more "collisions" with the hole on the high-pressure side than on the low-pressure side, and so more molecules pass from the high-pressure region into the low-pressure region than pass in the opposite direction. Effusion is easier to treat than diffusion, so we concentrate on it; but similar remarks apply to diffusion too.

The nineteenth-century Scottish chemist Thomas Graham carried out a series of experiments on the rates of effusion of gases. He found that, *at constant tem-*

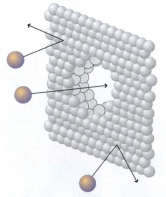

FIGURE 4.22 In effusion, the molecules of one substance escape through a small hole in a barrier into a vacuum or a region of low pressure.

perature, the rate of effusion of a gas is inversely proportional to the square root of its molar mass:

$$\text{Rate of effusion} \propto \frac{1}{\sqrt{\text{molar mass}}} \quad \text{or} \quad \text{Rate of effusion} \propto \frac{1}{\sqrt{M}} \qquad \text{(15a)}$$

This observation is now known as **Graham's law of effusion**. The rate of effusion is proportional to the average speed of the molecules in the gas because the average speed of the molecules determines the rate at which they approach the hole. Therefore, we can conclude that

$$\text{Average speed} \propto \frac{1}{\sqrt{M}} \qquad \text{(15b)}$$

This relation is an important clue about the motion of molecules in a gas, and we shall use it shortly, but first we look at some practical applications of Graham's law.

If we were to write Graham's law for two gases A and B with molar masses M_A and M_B, and divide one equation by the other, we would obtain,

$$\frac{\text{Rate of effusion of A molecules}}{\text{Rate of effusion of B molecules}} = \sqrt{\frac{M_B}{M_A}} \qquad \text{(16a)*}$$

Because the times it takes for the same amount of two substances to effuse through a small hole are inversely proportional to the rates at which they effuse, an equivalent statement is

$$\frac{\text{Time for A to effuse}}{\text{Time for B to effuse}} = \sqrt{\frac{M_A}{M_B}} \qquad \text{(16b)}$$

This relation can be used to estimate the molar mass of a substance by comparing the time required for a given amount of the unknown substance to effuse with that required for the same amount of a substance with a known molar mass.

SELF-TEST 4.15A It takes 30. mL of argon 40. s to effuse through a porous barrier. The same volume of vapor of a volatile compound extracted from Caribbean sponges takes 120. s to effuse through the same barrier under the same conditions. What is the molar mass of this compound?

[***Answer:*** 3.6×10^2 g·mol^{-1}]

SELF-TEST 4.15B It takes a certain amount of helium atoms 10. s to effuse through a porous barrier. How long does it take the same amount of methane, CH_4, molecules to effuse through the same barrier under the same conditions?

We have seen that effusion reveals that the average speed of molecules in a gas is inversely proportional to the square root of their molar mass. In effusion experiments at different temperatures, we find that the rate of effusion increases as the temperature is raised. Specifically, for a given gas, *the rate of effusion increases as the square root of the temperature:*

$$\frac{\text{Rate of effusion at } T_2}{\text{Rate of effusion at } T_1} = \sqrt{\frac{T_2}{T_1}} \qquad \text{(17a)}$$

Because the rate of effusion is proportional to the average speed of the molecules, we can infer that *the average speed of molecules in a gas is proportional to the square root of the temperature:*

$$\text{Average speed} \propto \sqrt{T} \qquad \text{(17b)}$$

This very important relation begins to reveal the significance of one of the most elusive concepts in science: the nature of temperature. We see that, when referring to a gas, the temperature is an indication of the average speed of the molecules, and the higher the temperature, the higher the average speed of the molecules.

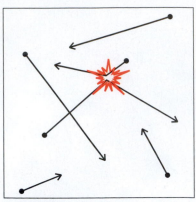

FIGURE 4.23 In the kinetic model of gases, the molecules are regarded as infinitesimal points that travel in straight lines until they undergo instantaneous collisions.

In this context, "molecules" includes all types of particles, whether atoms, ions, or molecules.

We can combine the two relations that we have found. Because the average speed of molecules in a gas is proportional to the square root of the temperature and inversely proportional to the square root of the molar mass, we can write

$$\text{Average speed} \propto \sqrt{\frac{T}{M}} \qquad (18)$$

That is, the higher the temperature and the lower the molar mass, the higher is the average speed of molecules in a gas.

> *The average speed of molecules in a gas is directly proportional to the square root of the temperature and inversely proportional to the square root of the molar mass.*

4.10 The Kinetic Model of Gases

We now have enough information to turn our qualitative ideas about a gas into a quantitative model that can be used to make numerical predictions. The **kinetic model** ("kinetic molecular theory," KMT) of a gas is based on four assumptions (Fig. 4.23):

1   A gas consists of a collection of molecules in continuous random motion.

2   Gas molecules are infinitesimally small points.

3   The molecules move in straight lines until they collide.   •

4   The molecules do not influence one another except during collisions.

The fourth assumption means that we are proposing, as part of the model, that there are no attractive forces between ideal gas molecules and no repulsive forces between them except during collisions.

In the kinetic model of gases, we picture the molecules as widely separated for most of the time and in ceaseless random motion. They zoom from place to place, always in straight lines, changing direction only when they collide with a wall of the container or another molecule. The collisions change the speed and direction of the molecules, just like balls in a three-dimensional cosmic game of pool.

In Section 4.4, we used a molecular model of a gas to explain *qualitatively* why the pressure of a gas rises as the temperature is increased: as a gas is heated, its molecules move faster and strike the walls of their container more often. The kinetic model of a gas allows us to derive the *quantitative* relation between pressure and the speeds of the molecules.

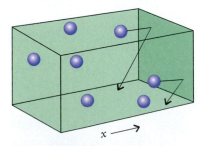

FIGURE 4.24 In the kinetic model of gases, the pressure arises from the force exerted on the walls of the container when the impacting molecules are deflected. We need to know the force of each impact and the number of impacts in a given time interval.

HOW DO WE DO THAT?

The following calculation of the pressure of a gas based on the kinetic model may seem complicated at first, but it breaks down into many small steps.

We begin with the gas sample illustrated in Fig. 4.24 and initially suppose that all the molecules are traveling at the same speed; we remove that constraint later. The molecules strike the wall on the right. If we know how often these impacts occur and what force they exert on the wall, we can calculate the pressure that results.

To calculate the force exerted by a single molecule, we use Newton's second law of motion: force is equal to the rate of change of momentum of a particle (Section A). Momentum is the product of mass and velocity; so, if a molecule of mass m is traveling with a velocity v_x parallel to the side of the box that we are calling x, then its linear momentum before it strikes the wall on the right is mv_x. Immediately after the collision, the momentum of the molecule is $-mv_x$ because the velocity has changed from v_x to $-v_x$.

The change in momentum
of one molecule:

$$2mv_x$$

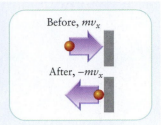

Next, we calculate the number of molecules that can strike the wall in a time interval Δt.

All the molecules within a distance $v_x\Delta t$ of the wall and traveling toward it will strike the wall during the interval Δt.

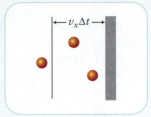

If the wall has area A, all the particles in a volume $Av_x\Delta t$ will reach the wall if they are traveling toward it.

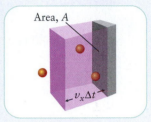

Suppose that the total number of particles in the container is N and that the volume of the container is V. Because a gas fills its container, we know that the molecules are spread evenly throughout the container. Therefore,

The number of molecules in the volume $Av_x\Delta t$ is that fraction of the total volume V, multiplied by the total number of molecules:

$$\text{Number of molecules} = \frac{Av_x\Delta t}{V} \times N$$
$$= \frac{NAv_x\Delta t}{V}$$

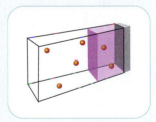

Half the molecules in the box are moving toward the wall on the right and half are moving away from that wall. Therefore,

The average number of collisions with the wall during the interval Δt is half the number in the volume $Av_x\Delta t$:

$$\text{Number of collisions} = \frac{NAv_x\Delta t}{2V}$$

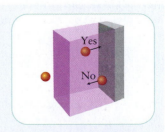

We have calculated the momentum of one molecule and the number of collisions during the interval Δt. Now we bring the parts of the calculation together. The total momentum change in that interval is the change $2mv_x$ that an individual molecule undergoes multiplied by the total number of collisions:

$$\text{Total momentum change} = \frac{NAv_x\Delta t}{2V} \times 2mv_x = \frac{NmAv_x^2\Delta t}{V}$$

At this point, we calculate the rate of change of momentum by dividing this total momentum change by the time interval during which it occurs (Δt) and use Newton's second law that the force is equal to the rate of change of momentum:

From Rate of change of momentum = (total momentum change)/Δt,	$\text{Rate} = \dfrac{NmAv_x^2\Delta t}{V\Delta t} = \dfrac{NmAv_x^2}{V}$	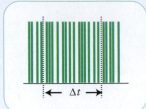
From Newton's second law,	$\text{Force} = \text{rate of change of momentum} = \dfrac{NmAv_x^2}{V}$	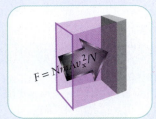
From Pressure = force/area,	$\text{Pressure} = \dfrac{NmAv_x^2}{VA} = \dfrac{Nmv_x^2}{V}$	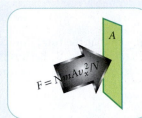

Not all the molecules in the sample are traveling at the same speed. To obtain the detected pressure, P, we need to use the *average* value of v_x^2 in place of v_x^2 for each individual molecule. Averages are commonly denoted by angular brackets; so we write

$$P = \frac{Nm\langle v_x^2\rangle}{V}$$

The importance of the root mean square speed stems from the fact that v_{rms}^2 is proportional to the average kinetic energy of the molecules, $\langle E_K\rangle = \frac{1}{2}m\langle v^2\rangle = \frac{1}{2}mv_{rms}^2$.

where $\langle v_x^2\rangle$ is the average value of v_x^2 for all the molecules in the sample.

At this stage, we relate $\langle v_x^2\rangle$ to the *root mean square speed*, $v_{rms} = \langle v^2\rangle^{1/2}$, the square root of the average of the squares of the molecular speeds (this quantity is explained in more detail in the text following this derivation). First, we note that the speed, v, of a single molecule is related to the velocity parallel to the x, y, and z directions:

From the Pythagorean theorem,	$v^2 = v_x^2 + v_y^2 + v_z^2$	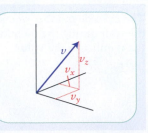

Therefore, the mean square speed is given by

$$v_{rms}^2 = \langle v^2 \rangle = \langle v_x^2 + v_y^2 + v_z^2 \rangle = \langle v_x^2 \rangle + \langle v_y^2 \rangle + \langle v_z^2 \rangle$$

However, because the particles are moving randomly, the average of v_x^2 is the same as the average of v_y^2 and the average of v_z^2, the analogous quantities in the y and z directions. Because $\langle v_x^2 \rangle$, $\langle v_y^2 \rangle$, and $\langle v_z^2 \rangle$ are all equal, we know that $\langle v^2 \rangle = 3\langle v_x^2 \rangle$; therefore, $\langle v_x^2 \rangle = \frac{1}{3} v_{rms}^2$. It follows that

$$P = \frac{N m v_{rms}^2}{3V}$$

The total number of molecules, N, is the product of the amount, n, and Avogadro's constant, N_A; so the last equation becomes

$$P = \frac{n N_A m v_{rms}^2}{3V} = \frac{n M v_{rms}^2}{3V}$$

where m is the mass of one molecule and $M = m N_A$ is the molar mass of the molecules.

We have deduced that the pressure of a gas and the volume are related by

$$PV = \frac{1}{3} n M v_{rms}^2 \qquad (19)$$

where n is the amount (in moles) of gas molecules, M is their molar mass, and v_{rms} is the root mean square speed of the molecules. As remarked in the preceding derivation, the **root mean square speed**, v_{rms}, is the square root of the average value of the squares of the molecular speeds. If there are N molecules in the sample and the speeds of these molecules at some instant are $v_1, v_2, \ldots, v_N$, then the root mean square speed is

$$v_{rms} = \left(\frac{v_1^2 + v_2^2 + \cdots + v_N^2}{N} \right)^{1/2} \qquad (20)$$

We can now do something remarkable: we can use the ideal gas law to calculate the root mean square speed of the molecules of a gas. We know that $PV = nRT$ for an ideal gas; therefore, we can set the right-hand side of Eq. 19 equal to nRT and rearrange the resulting expression ($\frac{1}{3} n M v_{rms}^2 = nRT$) into

$$v_{rms} = \left(\frac{3RT}{M} \right)^{1/2} \qquad (21a)*$$

This important result is used to find the root mean square speeds of the gas-phase molecules at any temperature (Fig. 4.25). We can rewrite this equation to emphasize that, for a gas, the temperature is a measure of mean molecular speed. From $v_{rms}^2 = 3RT/M$, it follows that

$$T = \frac{M v_{rms}^2}{3R} \qquad (21b)$$

That is, *the temperature is proportional to the mean square speed of the molecules.*

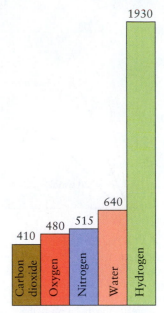

FIGURE 4.25 The root mean square speeds of five gases at 25°C, in meters per second. The gases are some of the components of air; hydrogen is included to show that the root mean square speed of light molecules is much greater than that of heavy molecules.

| EXAMPLE 4.9 | Sample exercise: Calculating the root mean square speed of gas molecules |

What is the root mean square speed of nitrogen molecules in air at 20.°C?

SOLUTION The temperature is 293 K and the molar mass of N_2 is 28.02 g·mol^{-1} (corresponding to 2.802×10^{-2} kg·mol^{-1}).

From $v_{rms} = (3RT/M)^{1/2}$,

$$v_{rms} = \left(\frac{3 \times (8.3145 \text{ J·K}^{-1} \cdot \text{mol}^{-1}) \times (293 \text{K})}{2.802 \times 10^{-2} \text{ kg·mol}^{-1}} \right)^{1/2}$$

$$= 511 \text{ m·s}^{-1}$$

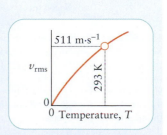

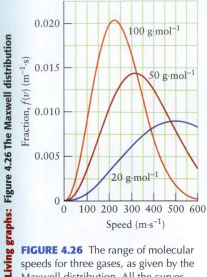

Living graphs: Figure 4.26 The Maxwell distribution

FIGURE 4.26 The range of molecular speeds for three gases, as given by the Maxwell distribution. All the curves correspond to the same temperature (300 K). The greater the molar mass, the lower the average speed and the narrower the spread of speeds. To obtain the fraction of molecules with speeds in the range from v to $v + \Delta v$, multiply $f(v)$ by Δv.

This result means that nitrogen molecules are zooming about your head at about 1140 miles per hour.

A note on good practice: We use SI units throughout: R is in its fundamental SI form and the molar mass is in SI base units to be consistent with the choice for R. The cancellation of the units has made use of the relation $1\ J = 1\ kg \cdot m^2 \cdot s^{-2}$.

SELF-TEST 4.16A Estimate the root mean square speed of water molecules in the vapor above boiling water at 100.°C.

[*Answer:* 719 m·s⁻¹]

SELF-TEST 4.16B Estimate the root mean square speed of CH_4 molecules at 25°C.

> *The kinetic model of gases is consistent with the ideal gas law and provides an expression for the root mean square speed of the molecules:* $v_{rms} = (3RT/M)^{1/2}$. *The molar kinetic energy of a gas is proportional to the temperature.*

4.11 The Maxwell Distribution of Speeds

Useful as it is, Eq. 21a gives only the root mean square speed of gas molecules. Like cars in traffic, individual molecules have speeds that vary over a wide range. Moreover, like a car in a head-on collision, a molecule might be brought almost to a standstill when it collides with another. In the next instant (but now unlike a colliding car), it might be struck by another molecule and move off at the speed of sound. An individual molecule undergoes several billion changes of speed and direction each second.

The formula for calculating the fraction of gas molecules having a given speed, v, at any instant was first derived from the kinetic model by the Scottish scientist James Clerk Maxwell. He derived the expression

$$\Delta N = Nf(v)\,\Delta v \qquad \text{with } f(v) = 4\pi \left(\frac{M}{2\pi RT}\right)^{3/2} v^2 e^{-Mv^2/2RT} \qquad \textbf{(22)}$$

where ΔN is the number of molecules with speeds in the narrow range between v and $v + \Delta v$, N is the total number of molecules in the sample, M is the molar mass, and R is the gas constant. This expression for $f(v)$ is the **Maxwell distribution of speeds** (Box 4.1).

> *What does this equation tell us?* The exponential factor (which falls rapidly toward zero as v increases) means that very few molecules have very high speeds. The factor v^2 that multiplies the exponential factor goes to zero as v goes to zero, so it means that very few molecules have very low speeds. The factor $4\pi(M/2\pi RT)^{3/2}$ simply ensures that the total probability of a molecule having a speed between zero and infinity is 1.

Figure 4.26 shows a plot of the Maxwell distribution against speed for several different gases. We see that heavy molecules (of molar mass 100 g·mol⁻¹, for instance) travel with speeds close to their average values. Light molecules (of 20 g·mol⁻¹, for instance) not only have a higher average speed but also a wider range of speeds. Some molecules of gases with low molar masses have such high speeds that they can escape from the gravitational pull of small planets and go off into space. As a consequence, hydrogen molecules and helium atoms, which are both very light, are rare in the Earth's atmosphere, although they are abundant on massive planets such as Jupiter.

A plot of the Maxwell distribution for the same gas at several different temperatures shows that the average speed increases as the temperature is raised (Fig 4.27). We knew that already (Section 4.9); but the curves also show that the spread of speeds widens as the temperature increases. At low temperatures, most molecules of a gas have speeds close to the average speed. At high temperatures, a high proportion have speeds widely different from their average speed. Because the kinetic energy of a molecule in a gas is proportional to the square of its speed, the distribution of molecular kinetic energies follows the same trends.

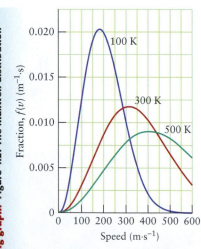

Living graph: Figure 4.27 The Maxwell distribution

FIGURE 4.27 The Maxwell distribution again, but now the curves correspond to the speeds of a single substance (of molar mass 50 g·mol⁻¹) at different temperatures. The higher the temperature, the higher the average speed and the broader the spread of speeds.

BOX 4.1   HOW DO WE KNOW . . . THE DISTRIBUTION OF MOLECULAR SPEEDS?

The distribution of molecular speeds in a gas can be determined experimentally. To do so, the gas is heated to the required temperature in an oven. The molecules then stream out of the oven through a small hole into an evacuated region. To ensure that the molecules form a narrow beam, they may also pass through a series of slits, and the pressure must be kept very low so that collisions within the beam do not cause spreading.

The molecular beam passes through a series of rotating disks (see diagram). Each disk contains a slit that is offset by a certain angle from its neighbor. A molecule that passes through the first slit will pass through the slit in the next disk only if the time that it takes to pass between the disks is the same as the time required for the slit in the second disk to move into

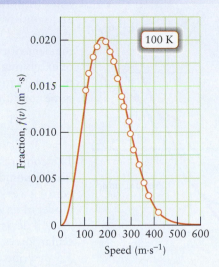

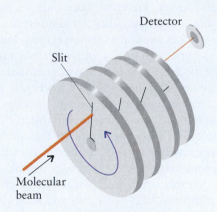

Diagram of the rotating disks that serve as a velocity selector in a molecular beam apparatus.

The points represent a typical result of a molecular speed distribution measurement. They are superimposed on the theoretical curve. To obtain the fraction of molecules with speeds in the range from v to $v + \Delta v$, multiply $f(v)$ by Δv.

the orientation originally occupied by the first disk. The two times must continue to match if the same molecule is to pass through subsequent disks. Therefore, a particular rate of rotation of the disks allows the passage of only those molecules having the speed that carries them through the slits. To determine the distribution of molecular speeds, we would measure the intensity of the beam of molecules arriving at the detector for different rotational rates of the disks. A typical result is shown in the graph, and we see that it is a good match to the theoretical Maxwell expression (see Figs. 4.26 and 4.27).

The molecules of all gases have a wide range of speeds. As the temperature increases, the root mean square speed and the range of speeds both increase. The range of speeds is described by the Maxwell distribution, Eq. 22.

THE IMPACT ON MATERIALS: REAL GASES

In industry and in many research laboratories, gases must be used under high pressures, when the ideal gas law is not followed closely. Recall that the ideal gas law is a limiting law, valid only as $P \to 0$. All actual gases, which are called **real gases**, have properties that differ from those predicted by the ideal gas law. These differences are significant at high pressures and low temperatures. We first explore the experimental evidence for those differences and then relate the behavior of real gases to the properties of their molecules and so come to the final, refined version of our model of gases.

4.12 Deviations from Ideality

Two types of observations show that our model of a gas must be refined. The qualitative observation is that gases can be condensed to liquids when cooled or compressed. This property strongly suggests that, contrary to the assumptions of the kinetic model, gas molecules do attract one another because otherwise they would not cohere (stick together). In addition, liquids are very difficult to compress. This

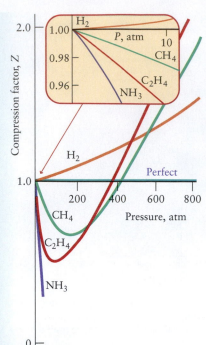

FIGURE 4.28 A plot of the compression factor, Z, as a function of pressure for a variety of gases. An ideal gas has Z = 1 for all pressures. For a few real gases with very weak intermolecular attractions, such as H_2, Z is always greater than 1. For most gases, at low pressures the attractive forces are dominant and Z < 1 (see inset). At high pressures, repulsive forces become dominant and Z > 1 for all gases.

observation strongly suggests that powerful repulsive forces must prevent molecules from being squashed together into a tiny volume. The presence of attractive and repulsive forces implies that once again the kinetic model must be refined.

We can assess the effect of intermolecular forces quantitatively by comparing the behavior of real gases with that expected of an ideal gas. One of the best ways of exhibiting these deviations is to measure the **compression factor**, Z, the ratio of the actual molar volume of the gas to the molar volume of an ideal gas under the same conditions:

$$Z = \frac{V_m}{V_m^{ideal}} \qquad (23)$$

The compression factor of an ideal gas is 1, and so deviations from Z = 1 are a sign of nonideality. Figure 4.28 shows the experimental variation of Z for a number of gases. We see that all gases deviate from Z = 1 as the pressure is raised. Our model of gases must account for these deviations.

All deviations from ideal behavior can be explained by the presence of **intermolecular forces**, or attractions and repulsions between molecules. Sections 5.1 through 5.5 describe the origin of intermolecular forces. Here, we need only know that all molecules attract one another when they are a few molecular diameters apart but (provided they do not react) repel one another as soon as their electron clouds come into contact. Figure 4.29 shows how the potential energy of a molecule varies with its distance from a second molecule. At moderate separations, its potential energy is lower than when it is infinitely far away: attractions always lower the potential energy of an object. As the molecules come into contact, the potential energy starts to rise, because repulsions always increase the potential energy of an object.

The presence of attractive intermolecular forces accounts for the condensation of gases to liquids when they are compressed or cooled. Compression reduces the average separation of the molecules, and neighboring molecules can be captured by one another's attraction, provided that they are traveling slowly enough (that is, the sample is cool enough). The low compressibility of liquids and solids is consistent, as we have seen, with the presence of strong repulsive forces when molecules are pushed together. Another way to describe the repulsive forces between molecules is to say that molecules have definite volumes. When you touch a solid object, you feel its size and shape because your fingers cannot penetrate into it. The resistance to compression offered by the solid is due to the repulsive forces exerted by its atoms on the atoms in your fingers.

The presence of intermolecular forces also accounts for the variation in the compression factor. Thus, for gases under conditions of pressure and temperature such that Z > 1, the repulsions are more important than the attractions. Their molar volumes are greater than expected for an ideal gas because repulsions tend to drive the molecules apart. For example, a hydrogen molecule has so few electrons that the its molecules are only very weakly attracted to one another. For gases under conditions of pressure and temperature such that Z < 1, the attractions are more important than the repulsions, and the molar volume is smaller than for an ideal gas because attractions tend to draw molecules together. To improve our model of a gas, we need to add to it that the molecules of a real gas exert attractive and repulsive forces on one another.

Real gases consist of atoms or molecules with intermolecular attractions and repulsions. Attractions have a longer range than repulsions. The compression factor is a measure of the strength and type of intermolecular forces. When Z > 1, intermolecular repulsions are dominant; when Z < 1, attractions are dominant.

4.13 The Liquefaction of Gases

As described in Section 4.12, at low temperatures, molecules of a real gas move so slowly that intermolecular attractions may result in one molecule being captured by others and sticking to them instead of moving freely. When the tempera-

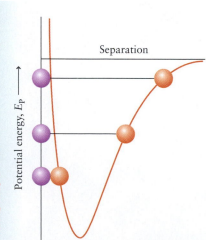

FIGURE 4.29 Variation in the potential energy of a molecule as it approaches another molecule. The potential energy rises sharply once the two molecules come into direct contact.

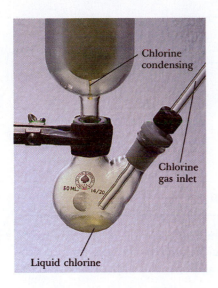

FIGURE 4.30 Chlorine can be condensed to a liquid at atmospheric pressure by cooling it to −35°C or lower. The upper flask contains a "cold finger," a smaller tube filled with dry ice in acetone at −78°C.

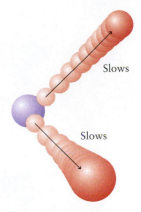

FIGURE 4.31 Cooling by the Joule–Thomson effect can be visualized as a slowing of the molecules as they climb away from each other against the force of attraction between them.

ture falls below the boiling point of the substance, the gas condenses to a liquid (Fig. 4.30).

Gases can also be liquefied by making use of the relation between temperature and molecular speed. Because lower average speed corresponds to lower temperature, slowing the molecules is equivalent to cooling the gas. The molecules of a real gas can be slowed by making use of the attractions between them and allowing the gas to expand: the molecules have to climb away from one another's attractive forces when a gas expands, rather like a ball rising up above the surface of the Earth against the pull of gravity (Fig. 4.31). Therefore, allowing the gas to occupy a larger volume, and hence increasing the average separation of the molecules, results in the molecules having a lower average speed. In other words, provided attractive effects are dominant, a real gas cools as it expands. This observation is called the **Joule–Thomson effect** in honor of the scientists who first studied it, James Joule and William Thomson (later to become Lord Kelvin), the inventor of the absolute temperature scale.

The Joule–Thomson effect is used in some commercial refrigerators to liquefy gases. The gas to be liquefied is compressed and then allowed to expand through a small hole, called the throttle. The gas cools as it expands, and the cooled gas circulates past the incoming compressed gas (Fig. 4.32). This contact cools the incoming gas before it expands and cools still further. As the gas is continually recompressed and recirculated, its temperature progressively falls until finally it condenses to a liquid. If the gas is a mixture, such as air, then the liquid that it forms can later be distilled to separate its components. This technique is used for harvesting nitrogen, oxygen, neon, argon, krypton, and xenon from the atmosphere.

Many gases can be liquefied by making use of the Joule–Thomson effect, cooling brought about by expansion.

4.14 Equations of State of Real Gases

How do we describe the behavior of real gases that do not obey the ideal gas law? A common procedure in chemistry is to suppose that the term on the right-hand side of an equation (such as nRT in $PV = nRT$ of the ideal gas law) is just the leading (and predominant) term of a more complicated expression. Thus, to extend the ideal gas law to real gases, one procedure is to write

$$PV = nRT\left(1 + \frac{B}{V_m} + \frac{C}{V_m^2} + \cdots\right) \qquad (24)$$

Exceptions are helium and hydrogen, which have very weak attractive interactions and relatively strong repulsive interactions; at room temperature, they get warmer as they expand.

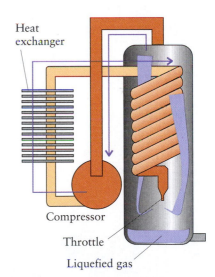

FIGURE 4.32 A Linde refrigerator for liquefying gases. The compressed gas gives up heat to the surroundings in the heat exchanger (left) and passes through the coil (right). The gas is cooled still further by the Joule–Thomson effect as it emerges through the throttle. This gas cools the incoming gas and is recirculated through the system. Eventually, the temperature of the incoming gas is so low that it condenses to a liquid.

TABLE 4.5 Van der Waals Parameters

Gas	a ($L^2 \cdot atm \cdot mol^{-2}$)	b ($10^{-2}\ L \cdot mol^{-1}$)
ammonia	4.225	3.707
argon	1.363	3.219
benzene	18.24	11.54
carbon dioxide	3.640	4.267
chlorine	6.579	5.622
ethane	5.562	6.380
hydrogen	0.2476	2.661
hydrogen sulfide	4.490	4.287
oxygen	1.378	3.183
water	5.536	3.049

The name *virial* comes from the Latin word meaning "force."

This expression is called the **virial equation.** The coefficients B, C, . . . are called the **second virial coefficient, third virial coefficient,** and so on. The virial coefficients, which depend on the temperature, are found by fitting experimental data to the virial equation.

Although the virial equation can be used to make accurate predictions about the properties of a real gas, provided that the virial coefficients are known for the temperature of interest, it is not a source of much insight without a lot of advanced analysis. An equation that is less accurate but easier to interpret was proposed by the Dutch scientist Johannes van der Waals. The **van der Waals equation** is

$$\left(P + a\frac{n^2}{V^2} \right)(V - nb) = nRT \tag{25}$$

The temperature-independent **van der Waals parameters** a and b are unique for each gas and are determined experimentally (Table 4.5). Parameter a represents the role of *attractions*; so it is relatively large for molecules that attract each other strongly and for large molecules with many electrons. Parameter b represents the role of *repulsions*; it can be thought of as representing the volume of an individual molecule (more precisely, the volume per mole of molecules), because it is the repulsive forces between molecules that prevent one molecule from occupying the space already occupied by another molecule.

The roles of the parameters become clearer when we rearrange the van der Waals equation into

$$P = \frac{nRT}{V - nb} - a\frac{n^2}{V^2} \tag{26}*$$

and then write the compression factor (Eq. 23) as

$$Z = \frac{\overbrace{V_m}^{\text{Actual molar volume}}}{\underbrace{RT/P}_{\text{Ideal molar volume}}} = \frac{PV}{nRT} \stackrel{\text{Use Eq. 26 for } P}{=} \frac{V}{V - nb} - \frac{an}{RTV} = \frac{1}{1 - (nb/V)} - \frac{an}{RTV} \tag{27}$$

What does this equation tell us? For an ideal gas, a and b are both zero, and $Z = 1$. We see that $Z > 1$ when the attractive contribution (a) is small and the repulsive contribution (b) is appreciable. Conversely, $Z < 1$ when the repulsive contribution is weak (b is small) and the attractive interaction is strong (a is large).

The values of a and b for a gas are found experimentally by fitting this expression for Z to curves like those in Fig. 4.28. Once the parameters have been determined, they can be used in the van der Waals equation to predict the pressure of the gas under the conditions of interest.

EXAMPLE 4.10 **Sample exercise: Estimating the pressure of a real gas**

Investigators are studying the physical properties of a gas to be used as a refrigerant in an air-conditioning unit. A table of van der Waals parameters shows that for this gas $a = 16.2$ $L^2 \cdot atm \cdot mol^{-2}$ and $b = 8.4 \times 10^{-2}$ $L \cdot mol^{-1}$. Estimate the pressure of the gas when 1.50 mol occupies 5.00 L at 0°C.

SOLUTION To find the pressure of the gas, we substitute the data into Eq. 26 after converting the temperature into the Kelvin scale. We use R in units that match those given (atmospheres and liters).

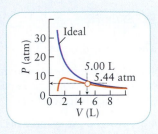

From $P = nRT/(V - nb)$ $- an^2/V^2$,

$$P = \frac{(1.50 \text{ mol}) \times (0.082\,06 \text{ L·atm·K}^{-1}\cdot\text{mol}^{-1}) \times (273 \text{ K})}{5.00 \text{ L} - (1.50 \text{ mol}) \times (8.4 \times 10^{-2} \text{ L·mol}^{-1})}$$

$$-(16.2 \text{ L}^2\cdot\text{atm·mol}^{-2}) \times \frac{(1.50 \text{ mol})^2}{(5.00 \text{ L})^2}$$

$$= \frac{1.50 \times (0.082\,06 \text{ atm}) \times 273}{5.00 - 1.50 \times 8.4 \times 10^{-2}} - (16.2 \text{ atm}) \times \frac{(1.50)^2}{(5.00)^2}$$

$$= 5.44 \text{ atm}$$

An ideal gas under the same circumstances has a pressure of 6.72 atm.

SELF-TEST 4.17A A 10.0-L tank containing 25 mol O_2 is stored in a diving supply shop at 25°C. Use the data in Table 4.5 and the van der Waals equation to estimate the pressure in the tank.

[*Answer:* 58 atm]

SELF-TEST 4.17B The properties of carbon dioxide gas are well known in the bottled beverage industry. In an industrial process, a tank of volume 100. L at 20.°C contains 20. mol CO_2. Use the data in Table 4.5 and the van der Waals equation to estimate the pressure in the tank.

We have arrived at our final model. By drawing on a wide range of experimental information, we now visualize a gas as a large number of molecules in ceaseless, random motion. The average speed—and the spread of speeds—of the molecules increases with temperature and decreases with molar mass. Because the average separation of the molecules is so great, intermolecular forces have only a weak effect on the properties of a gas, and the molecules travel in straight lines until they collide with other molecules or with the walls of the container. However, the intermolecular forces are not completely negligible, and repulsions increase the molar volume (at a given temperature and pressure), whereas attractions decrease the molar volume. Which effect dominates in a real gas depends on the relative strengths of the two kinds of interactions.

The virial equation is a general equation for describing real gases. The van der Waals equation is an approximate equation of state for a real gas; the parameter a represents the role of attractive forces and the parameter b represents the role of repulsive forces.

SKILLS YOU SHOULD HAVE MASTERED

❏ 1 Calculate the pressure at the foot of a column of liquid (Example 4.1).

❏ 2 Interpret a manometer reading (Self-Test 4.2).

❏ 3 Use the gas laws to calculate P, V, T, or n for given conditions or after a change in conditions (Toolbox 4.1 and Examples 4.2–4.5).

❏ 4 Determine molar mass from gas density and vice versa (Example 4.6).

❏ 5 Calculate the mass or volume of reactant needed to react with a specified volume of gas (Example 4.7).

❏ 6 Calculate the partial pressures of gases in a mixture and the total pressure of the mixture (Example 4.8).

❑ 7 Use Graham's law to account for relative rates of effusion (Self-Test 4.15).

❑ 8 Describe the effect of temperature on average speed (Section 4.9).

❑ 9 List and explain the assumptions of the kinetic model of gases (Section 4.10).

❑ 10 Calculate the root mean square speed of molecules in a sample of gas (Example 4.9).

❑ 11 Describe the effect of molar mass and temperature on the Maxwell distribution of molecular speeds (Section 4.11).

❑ 12 Explain how real gases differ from ideal gases (Sections 4.12–4.14).

❑ 13 Use the van der Waals equation to estimate the pressure of a gas (Example 4.10).

EXERCISES

The Nature of Gases

4.1 The pressure needed to make synthetic diamonds from graphite is 8×10^4 atm. Express this pressure in (a) Pa; (b) kbar; (c) Torr; (d) lb·inch^{-2}.

4.2 An argon gas cylinder measures a pressure of 37.2 lb·inch^{-2}. Convert this pressure into (a) kPa; (b) Torr; (c) bar; (d) atm.

4.3 A student attaches a glass bulb containing neon gas to an open-tube manometer (refer to Fig. 4.5) and calculates the pressure of the gas to be 0.890 atm. (a) If the atmospheric pressure is 762 Torr, what height difference between the two sides of the mercury in the manometer did the student find? (b) Which side is higher, the side of the manometer attached to the bulb or the side open to the atmosphere? (c) If the student had mistakenly switched the numbers for the sides of the manometer when recording the data in the laboratory notebook, what would have been the reported pressure in the gas bulb?

4.4 A reaction is performed in a vessel attached to a closed-tube manometer. Before the reaction, the levels of mercury in the two sides of the manometer were at the same height. As the reaction proceeds, a gas is produced. At the end of the reaction, the height of the mercury column on the vacuum side of the manometer has risen 30.74 cm and the height on the side of the manometer connected to the flask has fallen by the same amount. What is the pressure in the apparatus at the end of the reaction expressed in (a) Torr; (b) atm; (c) Pa; (d) bar?

4.5 Suppose you were marooned on a tropical island and had to use seawater (density 1.10 g·cm^{-3}) to make a primitive barometer. What height would the water reach in your barometer when a mercury barometer would reach 73.5 cm? The density of mercury is 13.6 g·cm^{-3}.

4.6 An unknown liquid is used to fill a closed-tube manometer. The atmosphere is found to produce a height difference of 6.14 m in this manometer at the same time that a mercury manometer gives a displacement of 758.7 mm. What is the density of the unknown liquid?

4.7 Assume that the width of your body (across your shoulders) is 20. inches and the depth of your body (chest to back) is 10. inches. If atmospheric pressure is 14.7 lb·inch^{-2}, what mass of air does your body support when you are in an upright position?

4.8 Low-pressure gauges in research laboratories are occasionally calibrated in inches of water (inH$_2$O). Given that the density of mercury at 15°C is 13.5 g·cm^{-3} and the density of water at that temperature is 1.0 g·cm^{-3}, what is the pressure (in Torr) inside a gas cylinder that reads 1.5 inH$_2$O at 15°C?

The Gas Laws

4.9 When Robert Boyle conducted his experiments, he measured pressure in inches of mercury (inHg). On a day when the atmospheric pressure was 29.85 inHg, he trapped some air in the tip of a J-tube (**1**) and measured the difference in height of the mercury in the two arms of the tube (h). When $h = $ 12.0 inches, the height of the gas in the tip of the tube was 32.0 in. Boyle then added additional mercury and the level rose in both arms of the tube so that $h = $ 30.0 inches (**2**). (a) What was the height of the air space (in inches) in the tip of the tube in (**2**)? (b) What was the pressure of the gas in the tube in (**1**) and in (**2**) in inHg?

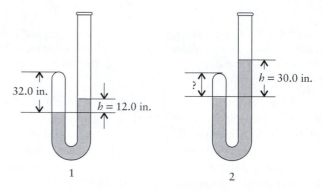

1   2

4.10 Boyle continued to add mercury to the apparatus in Exercise 4.9 until the height of the trapped air had been reduced to 6.85 inches. Assuming that atmospheric pressure had not changed, what was the pressure of the trapped air at that point (in inHg)?

 4.11 (a) Using the Living Graphs on the Web site for this book, prepare and print a single plot of the pressure against temperature for 1.00 mol of gas molecules, showing the curves at volumes ranging from 0.01 L to 0.05 L in increments of 0.01 L for $T = 0$ to 400 K. (b) What is the expression for the slope of each of these lines? (c) What is each intercept? Give your answers to two decimal places.

 4.12 (a) Using the Living Graphs on the Web site for this book, prepare and print a single plot of the volume against temperature graph for 1.00 mol of gas molecules, showing the curves at pressures ranging from 11 000 atm to 15 000 atm in increments of 1000 atm for $T = 0$ to 400 K. (b) What is the expression for the slope of each of these lines? (c) What is each intercept? Give your answers to two decimal places.

4.13 Determine the final pressure when (a) 7.50 mL of krypton at 2.00×10^5 kPa is transferred to a vessel of volume 1.0 L; (b) 54.2 cm^3 of O_2 at 643 Torr is compressed to 7.8 cm^3. Assume constant temperature.

4.14 (a) Suppose that 4.00 L of methane at a pressure of 800. Torr is transferred to a vessel of volume 2.40 L. What is the final pressure of methane if the change occurs at constant temperature? (b) A fluorinated organic gas in a cylinder is compressed from an initial volume of 936 mL at 158 Pa to 468 mL at the same temperature. What is the final pressure?

4.15 A 250.-mL aerosol can at 25°C and 1.10 atm was thrown into an incinerator. When the temperature in the can reached 625°C, it exploded. What was the pressure in the can just before it exploded, assuming it reached the maximum pressure possible at that temperature?

4.16 A helium balloon has a volume of 22.5 L when the pressure is 0.951 atm and the temperature is 18°C. The balloon is cooled at a constant pressure until the temperature is −15°C. What is the volume of the balloon at this stage?

4.17 An outdoor storage vessel for hydrogen gas with a volume of 300. m^3 is at 1.5 atm and 10.°C at 2:00 AM. By 2:00 PM, the temperature has risen to 30.°C. What is the new pressure of the hydrogen in the vessel?

4.18 A chemist prepares a sample of hydrogen bromide and finds that it occupies 250. mL at 65°C and 500. Torr. What volume would it occupy at 0°C at the same pressure?

4.19 A chemist prepares 0.100 mol Ne gas at a certain pressure and temperature in an expandable container. Another 0.010 mol Ne is then added to the same container. How must the volume be changed to keep the pressure and temperature the same?

4.20 A chemist prepares a sample of helium gas at a certain pressure, temperature, and volume and then removes half the gas molecules. How must the temperature be changed to keep the pressure and volume the same?

4.21 A sample of methane gas, CH_4, was slowly heated at a constant pressure of 0.90 bar. The volume of the gas was measured at a series of different temperatures and a plot of volume vs. temperature was constructed. The slope of the line was 2.88×10^{-4} L·K^{-1}. What was the mass of the sample of methane?

4.22 A sample of butane gas, C_4H_{10}, was slowly heated at a constant pressure of 0.80 bar. The volume of the gas was measured at a series of different temperatures and a plot of volume vs. temperature was constructed. The slope of the line was 0.0208 L·K^{-1}. What was the mass of the sample of butane?

4.23 You are told that 35.5 mL of xenon exerts a pressure of 0.255 atm at −45°C. (a) What volume does the sample occupy at 1.00 atm and 298 K? (b) What pressure would it exert if it were transferred to a flask of volume 12.0 mL at 20.°C? (c) Calculate the temperature needed for the xenon to exert a pressure of 5.00×10^2 Torr in the flask.

4.24 A lungful of air (355 cm^3) is exhaled into a machine that measures lung capacity. If the air is exhaled from the lungs at a pressure of 1.08 atm at 37°C but the machine is at ambient conditions of 0.958 atm and 23°C, what is the volume of air measured by the machine?

4.25 (a) A 350.0-mL flask contains 0.1500 mol Ar at 24°C. What is the pressure of the gas in kilopascals? (b) You are told that 23.9 mg of bromine trifluoride exerts a pressure of 10.0 Torr at 100.°C. What is the volume of the container in milliliters? (c) A 100.0-mL flask contains sulfur dioxide at 0.77 atm and 30.°C. What mass of gas is present? (d) A storage tank of volume 6.00×10^3 m^3 contains methane at 129 kPa and 14°C. How many moles of CH_4 are present? (e) A 1.0-μL ampoule of helium has a pressure of 2.00 kPa at −115°C. How many He atoms are present?

4.26 (a) A 125-mL flask contains argon at 1.30 atm and 77°C. What amount of Ar is present (in moles)? (b) A 120.-mL flask contains 2.7 μg of O_2 at 17°C. What is the pressure (in Torr)? (c) A 20.0-L flask at 215 K and 20. Torr contains nitrogen. What mass of nitrogen is present (in grams)? (d) A 16.7-g sample of krypton exerts a pressure of 1.00×10^2 mTorr at 44°C. What is the volume of the container (in liters)? (e) A 2.6-μL ampoule of xenon has a pressure of 2.00 Torr at 15°C. How many Xe atoms are present?

4.27 What is the molar volume of an ideal gas at 1.00 atm and (a) 500.°C; (b) at the boiling point of liquid nitrogen (−196°C)?

4.28 What is the resultant pressure if 1.00 mol of ideal gas molecules at 273 K and 1.00 atm in a closed container of constant volume is heated to (a) 373 K; (b) 500. K?

4.29 Originally at −20.°C and 759 Torr, a 1.00-L sample of air is heated to 235°C. Next, the pressure is increased to 765 Torr. Then it is heated to 1250.°C, and finally the pressure is decreased to 252 Torr. What is the final volume of the air?

4.30 A domestic water-carbonating kit uses steel cylinders of carbon dioxide, each with a volume of 250. mL. Each weighs 1.04 kg when full and 0.74 kg when empty. What is the pressure of gas (in bar) in a full cylinder at 20.°C?

4.31 The effect of high pressure on organisms, including humans, is studied to gain information about deep-sea diving and anesthesia. A sample of air occupied 1.00 L at 25°C and 1.00 atm. What pressure (in atm) is needed to compress it to 239 cm^3 at this temperature?

4.32 To what temperature must a sample of helium gas be cooled from 127.0°C to reduce its volume from 4.60 L to 0.115 L at constant pressure?

4.33 The "air" in the space suit of astronauts is actually pure oxygen supplied at a pressure of 0.30 bar. The two tanks on a space suit each have a volume of 3980. cm^3 and have an initial pressure of 5860. kPa. Assuming a tank temperature of 16°C, what mass of oxygen is contained in the two tanks?

4.34 A balloon vendor has a 20.0 L helium tank under 150 atm pressure at 25°C. How many balloons of 2.50 L each can be filled at 1.0 atm by the helium in this tank?

4.35 Nitrogen monoxide, NO, has been found to act as a neurotransmitter. To prepare to study its effect, a sample was collected in a container of volume 250.0 mL. At 19.5°C, its pressure in this container is found to be 24.5 kPa. What amount (in moles) of NO has been collected?

4.36 Hot-air balloons gain their lift from the reduction in the density of air that occurs when the air in the envelope is heated.

To what temperature should you heat a sample of air, initially at 340. K, to increase its volume by 14%?

4.37 At sea level, where the pressure was 104 kPa and the temperature 21.1°C, a certain mass of air occupied 2.0 m^3. To what volume will the air mass expand when it has risen to an altitude where the pressure and temperature are (a) 52 kPa, −5.0°C; (b) 880. Pa, −52.0°C?

4.38 A meteorological balloon had a radius of 1.0 m when released at sea level at 20.°C. It expanded to a radius of 3.0 m when it had risen to its maximum altitude, where the temperature was −20.°C. What was the pressure inside the balloon at that altitude?

Gas Density

4.39 Order the following gases according to increasing density: CO; CO_2; H_2S. The temperature and pressure are the same for all three samples.

4.40 Order the following gases according to increasing mass density: N_2; NH_3; NO_2. The temperature and pressure are the same for all three samples.

4.41 A 2.00-mg sample of argon is confined to a 0.0500-L vial at 20.°C; 2.00 mg of krypton is confined to a different 0.0500-L vial. What must the temperature of the krypton be if it is to have the same pressure as the argon?

4.42 What mass of ammonia will exert the same pressure as 12 mg of hydrogen sulfide, H_2S, in the same container under the same conditions?

4.43 What is the density (in g·L^{-1}) of chloroform, $CHCl_3$, vapor at (a) 2.00×10^2 Torr and 298 K; (b) 100.°C and 1.00 atm?

4.44 What is the density (in g·L^{-1}) of hydrogen sulfide, H_2S, at (a) 1.00 atm and 298 K; (b) 45.0°C and 0.876 atm?

4.45 A gaseous fluorinated methane compound has a density of 8.0 g·L^{-1} at 2.81 atm and 300. K. (a) What is the molar mass of the compound? (b) What is the formula of the compound if it is composed solely of C, H, and F? (c) What is the density of the gas at 1.00 atm and 298 K?

4.46 The density of a gaseous compound of phosphorus is 0.943 g·L^{-1} at 420. K when its pressure is 727 Torr. (a) What is the molar mass of the compound? (b) If the compound remains gaseous, what would be its density at 1.00 atm and 298 K?

4.47 A compound used in the manufacture of saran is 24.7% C, 2.1% H, and 73.2% Cl by mass. The storage of 3.557 g of the gaseous compound in a 755-mL vessel at 0°C results in a pressure of 1.10 atm. What is the molecular formula of the compound?

4.48 The analysis of a hydrocarbon revealed that it was 85.7% C and 14.3% H by mass. When 1.77 g of the gas was stored in a 1.500-L flask at 17°C, it exerted a pressure of 508 Torr. What is the molecular formula of the hydrocarbon?

4.49 The density of a gaseous compound was found to be 0.943 g·L^{-1} at 298 K and 53.1 kPa. What is the molar mass of the compound?

4.50 A 115-mg sample of eugenol, the compound responsible for the odor of cloves, was placed in an evacuated flask with a volume of 500.0 mL at 280.0°C. The pressure that eugenol

exerted in the flask under those conditions was found to be 48.3 Torr. In a combustion experiment, 18.8 mg of eugenol burned to give 50.0 mg of carbon dioxide and 12.4 mg of water. What is the molecular formula of eugenol?

The Stoichiometry of Reacting Gases

4.51 The Haber process for the synthesis of ammonia is one of the most significant industrial processes for the well-being of humanity. It is used extensively in the production of fertilizers as well as polymers and other products. (a) What volume of hydrogen at 15.00 atm and 350.°C must be supplied to produce 1.0 tonne (1 t = 10^3 kg) of NH_3? (b) What volume of hydrogen is needed in part (a) if it is supplied at 376 atm and 250.°C?

4.52 Nitroglycerin is a shock-sensitive liquid that detonates by the reaction

$$4\ C_3H_5(NO_3)_3(l) \rightarrow$$
$$6\ N_2(g) + 10\ H_2O(g) + 12\ CO_2(g) + O_2(g)$$

Calculate the total volume of product gases at 215 kPa and 275°C from the detonation of 1.00 lb (454 g) of nitroglycerin.

4.53 Which starting condition would produce the larger volume of carbon dioxide by combustion of $CH_4(g)$ with an excess of oxygen gas to produce carbon dioxide and water: (a) 2.00 L of $CH_4(g)$; (b) 2.00 g of $CH_4(g)$? Justify your answer. The system is maintained at a temperature of 75°C and 1.00 atm.

4.54 Which starting condition would produce the larger volume of carbon dioxide by combustion of $C_2H_4(g)$ with an excess of oxygen gas to produce carbon dioxide and water: (a) 1.00 L of $C_2H_4(g)$; (b) 1.20 g of $C_2H_4(g)$? Justify your answer. The system is maintained at a temperature of 45°C and 2.00 atm.

4.55 Through a series of enzymatic steps, carbon dioxide and water undergo photosynthesis to produce glucose and oxygen according to the equation

$$6\ CO_2(g) + 6\ H_2O(l) \rightarrow C_6H_{12}O_6(s) + 6\ O_2(g)$$

Given that the partial pressure of carbon dioxide in the troposphere is 0.26 Torr and that the temperature is 25°C, calculate the volume of air needed to produce 10.0 g of glucose.

4.56 The naturally occurring compound urea, $CO(NH_2)_2$, was first synthesized by Friedrich Wöhler in Germany in 1828 by heating ammonium cyanate. This synthesis was a significant event because it was the first time that an organic compound had been produced from an inorganic substance. Urea may also be made by the reaction of carbon dioxide and ammonia:

$$CO_2(g) + 2\ NH_3(g) \rightarrow CO(NH_2)_2(s) + H_2O(g)$$

What volumes of CO_2 and NH_3 at 200. atm and 450.°C are needed to produce 2.50 kg of urea, assuming the reaction goes to completion?

4.57 A 15.0-mL sample of ammonia gas at 1.00×10^2 Torr and 30.°C is mixed with 25.0 mL of hydrogen chloride gas at 1.50×10^2 Torr and 25°C, and the following reaction takes place:

$$NH_3(g) + HCl(g) \rightarrow NH_4Cl(s)$$

(a) Calculate the mass of NH_4Cl that forms. (b) Identify the gas in excess and determine the pressure of the excess gas at 27°C after the reaction is complete (in the combined volume of the original two flasks).

4.58 A sample of ethene gas, C_2H_4, of volume 1.00 L at 1.00 atm and 298 K is burned in 4.00 L of oxygen gas to form carbon dioxide gas and liquid water at the same pressure and temperature. Ignore the volume of water, and determine the final volume of the reaction mixture at 1.00 atm and 298 K if the reaction goes to completion.

Mixtures of Gases

4.59 A sample of hydrogen chloride gas, HCl, is being collected by bubbling it through liquid benzene into a graduated cylinder. Assume that the molecules pictured as spheres show a representative sample of the mixture of HCl and benzene vapor (● represents an HCl molecule and ○ a benzene molecule). (a) Use the figure to determine the mole fractions of HCl and benzene vapor in the gas inside the container. (b) What are the partial pressures of HCl and benzene in the container when the total pressure inside the container is 0.80 atm?

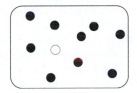

4.60 A piece of lithium metal was added to a flask of water on a day when the atmospheric pressure was 757.5 Torr. The lithium reacted completely with the water to produce 250.0 mL of hydrogen gas, collected over the water at 23°C, at which temperature the vapor pressure of water is 21.07 Torr. (a) What is the partial pressure of hydrogen in the collection flask? (b) What mass of lithium metal reacted?

4.61 A vessel of volume 22.4 L contains 2.0 mol $H_2(g)$ and 1.0 mol $N_2(g)$ at 273.15 K. Calculate (a) their partial pressures and (b) the total pressure.

4.62 An apparatus consists of a 4.0-L flask containing nitrogen gas at 25°C and 803 kPa, joined by a valve to a 10.0-L flask containing argon gas at 25°C and 47.2 kPa. The valve is opened and the gases mix. (a) What is the partial pressure of each gas after mixing? (b) What is the total pressure of the gas mixture?

4.63 A 1.00-L flask contains nitrogen gas at a temperature of 15°C and a pressure of 0.50 bar. 0.10 mol $O_2(g)$ is added to the flask and allowed to mix. Then a stopcock is opened to allow 0.20 mol of molecules to escape. What is the partial pressure of oxygen in the final mixture?

4.64 A gas mixture being used to simulate the atmosphere of another planet consists of 320. mg of methane, 175 mg of argon, and 225 mg of nitrogen. The partial pressure of nitrogen at 300. K is 15.2 kPa. Calculate (a) the total pressure of the mixture and (b) the volume.

4.65 In the course of the electrolysis of water, hydrogen gas was collected at one electrode over water at 20.°C when the external pressure was 756.7 Torr. The vapor pressure of water at 20.°C is 17.54 Torr. The volume of the gas was measured to be 0.220 L. (a) What is the partial pressure of hydrogen? (b) The other product of the electrolysis of water is oxygen gas. Write a balanced equation for the electrolysis of water into H_2 and O_2. (c) What mass of oxygen was also produced in the reaction?

4.66 Dinitrogen oxide, N_2O, gas was generated from the thermal decomposition of ammonium nitrate and collected over water. The wet gas occupied 126 mL at 21°C when the atmospheric pressure was 755 Torr. What volume would the same amount of *dry* dinitrogen oxide have occupied if collected at 755 Torr and 21°C? The vapor pressure of water is 18.65 Torr at 21°C.

Molecular Motion

4.67 Do all the molecules of a gas strike the walls of their container with the same force? Justify your answer on the basis of the kinetic model of gases.

4.68 How does the frequency of collisions of the molecules of a gas with the walls of the container change as the volume of the gas is decreased at constant temperature? Justify your answer on the basis of the kinetic model of gases.

4.69 What is the molecular formula of a compound of empirical formula CH that diffuses 1.24 times more slowly than krypton at the same temperature and pressure?

4.70 What is the molar mass of a compound that takes 2.7 times as long to effuse through a porous plug as it did for the same amount of XeF_2 at the same temperature and pressure?

4.71 A sample of argon gas effuses through a porous plug in 147 s. Calculate the time required for the same number of moles of (a) CO_2, (b) C_2H_4, (c) H_2, and (d) SO_2 to effuse under the same conditions of pressure and temperature.

4.72 Two identical flasks are each filled with a gas at 0°C. One flask contains 1 mol CO_2 and the other 1 mol Ne. In which flask do the particles have more collisions per second with the walls of the container?

4.73 A hydrocarbon of empirical formula C_2H_3 takes 349 s to effuse through a porous plug; under the same conditions of temperature and pressure, it took 210. s for the same number of molecules of argon to effuse. What is the molar mass and molecular formula of the hydrocarbon?

4.74 The composition of a compound used to make polyvinyl chloride (PVC) is 38.4% C, 4.82% H, and 56.8% Cl by mass. It took 7.73 min for a given volume of the compound to effuse through a porous plug, but it took only 6.18 min for the same amount of Ar to diffuse at the same temperature and pressure. What is the molecular formula of the compound?

4.75 Calculate the molar kinetic energy (in joules) of Kr(g) at (a) 55.85°C and (b) 54.85°C. (c) The molar energy difference between the answers to parts (a) and (b) is the energy per mole that it takes to raise the temperature of Kr(g) by 1.00°C. The quantity is known as the molar heat capacity. What is its value?

4.76 Calculate the molar kinetic energy (in joules) of a sample of Ne(g) at (a) 25.00°C and (b) 26.00°C. (c) The energy difference between the answers to parts (a) and (b) is the energy per mole that it takes to raise the temperature of Ne(g) by 1°C. The quantity is known as the molar heat capacity. What is its value?

4.77 Calculate the root mean square speeds of (a) methane, (b) ethane, and (c) propane molecules, all at −20.°C.

4.78 Calculate the root mean square speed of (a) fluorine gas, (b) chlorine gas, and (c) bromine gas molecules, all at 350.°C.

4.79 The root mean square speed of gaseous methane molecules, CH_4, at a certain temperature was found to be 550. $m·s^{-1}$. What is the root mean square speed of krypton atoms at the same temperature?

4.80 In an experiment on gases, you are studying a 1.00-L sample of hydrogen gas at 20°C and 2.40 atm. You heat the gas until the root mean square speed of the molecules of the sample has been doubled. What will be the final pressure of the gas?

4.81 A bottle contains 1.0 mol He(g) and a second bottle contains 1.0 mol Ar(g) at the same temperature. At that temperature, the root mean square speed of He is 1477 $m·s^{-1}$ and that of Ar is 467 $m·s^{-1}$. What is the ratio of the number of He atoms in the first bottle to the number of Ar atoms in the second bottle having these speeds? Assume that both gases behave ideally.

C **4.82** The number of molecules in a gas sample that have the most probable speed (v_{mp}) at a temperature T is one-half the number of the same type of molecules that have the most probable speed at 300. K. What is the temperature?

4.83 Consider the Maxwell distribution of speeds found in Fig. 4.27. (a) From the graph, find the location that represents the most probable speed of the molecules at each temperature. (b) What happens to the percentage of molecules having the most probable speed as the temperature increases?

 4.84 (a) Use the Living Graphs on the Web site for this book to create and print a plot of the Maxwell distribution of molecular speeds for atoms of the noble gases He, Ne, Ar, Xe, and Kr from $s = 0$ to 2000 $m·s^{-1}$. (b) Describe the trends observed in the graph in part (a). (c) Again using the Living Graphs, create graphs for He and for Kr at 100 K and at 300 K. (d) Describe the trends observed in part (c).

Real Gases

4.85 The pressure of a sample of hydrogen fluoride is lower than expected and, as the temperature is increased, rises more quickly than the ideal gas law predicts. Suggest an explanation.

4.86 Under what conditions would you expect a real gas to be (a) more compressible than an ideal gas; (b) less compressible than an ideal gas?

4.87 Using the ideal gas equation, calculate the pressure at 298 K exerted by 1.00 mol CO_2(g) when confined in a volume of (a) 15.0 L; (b) 0.500 L; (c) 50.0 mL. Repeat these calculations, with the use of the van der Waals equation. What do these calculations indicate about how the reliability of the ideal gas law depends on the pressure?

4.88 Using the ideal gas equation, calculate the pressure at 298 K exerted by 1.00 mol H_2(g) when confined in a volume of (a) 30.0 L; (b) 1.00 L; (c) 50.0 mL. Repeat these calculations, with the use of the van der Waals equation. What do these calculations indicate about how the reliability of the ideal gas law depends on the pressure?

4.89 The following table lists the van der Waals a parameters for CO_2, CH_3CN, Ne, and CH_4. Use your knowledge of the factors governing the magnitudes of a to assign each of these four gases to its a value.

a (atm·L^2·mol^{-2})	Substance
17.58	
3.392	
2.253	
0.2107	

4.90 Calculate the pressure exerted by 1.00 mol C_2H_6(g) behaving as (a) an ideal gas; (b) a van der Waals gas when it is confined under the following conditions: (1) at 273.15 K in 22.414 L; (2) at 1.00×10^3 K in 0.100 L.

 4.91 Plot pressure against volume for 1 mol of (a) an ideal gas, (b) ammonia, and (c) oxygen gas, for the range $V = 0.05$ L to 1.0 L at 298 K. Use the van der Waals equation to calculate the pressures of the two real gases. Use a spreadsheet or the graphical plotter on the Web site for this book. Compare the plots and explain the origin of any differences you find.

 4.92 Plot pressure against volume for 1 mol of (a) an ideal gas, (b) carbon dioxide, (c) ammonia, and (d) benzene for the range $V = 0.1$ L to 1.0 L at 298 K. Use the van der Waals equation to determine the pressures of the real gases. Use a spreadsheet or the graphical plotter on the Web site for this book. Compare the shapes of the curves and explain the differences.

Integrated Exercises

4.93 The drawing below shows a tiny section of a flask containing two gases. The orange spheres represent atoms of neon and the blue spheres represent atoms of argon. (a) If the partial pressure of neon in this mixture is 420. Torr, what is the partial pressure of argon? (b) What is the total pressure?

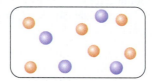

4.94 The following four flasks were prepared with the same volume and temperature. Flask I contains He atoms, Flask II contains Cl_2 molecules, Flask III contains Ar atoms, and Flask IV contains NH_3 molecules. Which flask has (a) the largest number of atoms? (b) The highest pressure? (c) The greatest density? (d) The highest root mean square speed? (e) The highest molar kinetic energy?

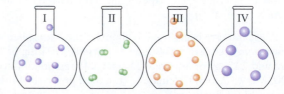

4.95 Photochemical smog is formed, in part, by the action of light on nitrogen dioxide. The wavelength of the radiation absorbed by NO_2 in this reaction is 197 nm.

$$NO_2 + h\nu \rightarrow NO + O$$

(a) Draw the Lewis structure for NO_2 and sketch its π molecular orbitals. (b) When 1.07 mJ of energy is absorbed by 2.5 L of air at 20.°C and 0.85 atm, all the NO_2 molecules in this sample are dissociated by the reaction just shown. Assume that each photon absorbed results in the dissociation (into NO and O) of one NO_2 molecule. What is the proportion, in parts per million, of NO_2 molecules in this sample? Assume that the sample is behaving ideally.

4.96 The reaction of solid dimethylhydrazine, $(CH_3)_2N_2H_2$, and liquefied dinitrogen tetroxide, N_2O_4, has been investigated as a rocket fuel; the reaction produces gaseous carbon dioxide (CO_2), nitrogen (N_2), and water vapor (H_2O), which are ejected as exhaust gases. In a controlled experiment, solid dimethylhydrazine reacted with excess nitrogen tetroxide and the gases were collected in a closed vessel until a pressure of 2.50 atm and a temperature of 400.0 K were reached. What are the partial pressures of CO_2, N_2, and H_2O?

4.97 Suppose that 200. mL of hydrogen chloride at 690. Torr and 20.°C is dissolved in 100. mL of water. The solution was titrated to the stoichiometric point with 15.7 mL of a sodium hydroxide solution. What is the molar concentration of the NaOH in solution?

4.98 Suppose that 2.00 L of propane gas, C_3H_8, at 1.00 atm and 298 K is mixed with 5.00 L of oxygen gas at the same pressure and temperature and burned to form carbon dioxide gas and liquid water. Ignore the volume of water formed, and determine the final volume of the reaction mixture (including products and excess reactant) at 1.00 atm and 298 K if reaction goes to completion.

4.99 A flask of volume 5.00 L is evacuated and 43.78 g of solid dinitrogen tetroxide, N_2O_4, is introduced at $-196°C$. The sample is then warmed to 25°C, during which time the N_2O_4 vaporizes and some of it dissociates to form brown NO_2 gas. The pressure slowly increases until it stabilizes at 2.96 atm. (a) Write a balanced equation for the reaction. (b) If the gas in the flask at 25°C were all N_2O_4, what would the pressure be? (c) If all the gas in the flask converted into NO_2, what would the pressure be? (d) What are the mole fractions of N_2O_4 and NO_2 once the pressure stabilizes at 2.96 atm?

4.100 When 0.40 g of impure zinc reacted with an excess of hydrochloric acid, 127 mL of hydrogen was collected over water at 10.°C. The external pressure was 737.7 Torr. (a) What volume would the dry hydrogen occupy at 1.00 atm and 298 K? (b) What amount (in moles) of H_2 was collected? (c) What is the percentage purity of the zinc, assuming that all the zinc present reacted completely with HCl and that the impurities did not react with HCl to produce hydrogen? The vapor pressure of water at 10.°C is 9.21 Torr.

4.101 Suppose that 0.473 g of an unknown gas that occupies 200. mL at 1.81 atm and 25°C was analyzed and found to contain 0.414 g of nitrogen and 0.0591 g of hydrogen. (a) What is the molecular formula of the compound? (b) Draw the Lewis structure of the molecule. (c) If ammonia effuses through a small opening in a glass apparatus at the rate of 3.5 $\times 10^{-4}$ mol in 15.0 min at 200.°C, how much of the compound will effuse through the same opening in 25.0 min at 200.°C?

4.102 When 2.36 g of phosphorus was burned in chlorine, the product was 10.5 g of a phosphorus chloride. Its vapor took 1.77 times as long to effuse as the same amount of CO_2 under the same conditions of temperature and pressure. What is the molar mass and molecular formula of the phosphorus chloride?

4.103 Determine the ratio of the number of molecules in a gas having a speed ten times as great as the root mean square speed to the number having a speed equal to the root mean square speed. Is this ratio independent of temperature? Why?

4.104 Air bags in automobiles contain crystals of sodium azide, NaN_3, which during a collision decomposes rapidly to give nitrogen gas and sodium metal. The nitrogen gas liberated by this process instantly inflates the air bag. Assume that the nitrogen gas liberated behaves as an ideal gas and that any solid produced has a negligible volume (which may be ignored). (a) Calculate the mass (in grams) of sodium azide required to generate enough nitrogen gas to fill a 57.0-L air bag at 1.37 atm and 25.0°C. (b) What is the root mean square speed of the N_2 gas molecules generated?

4.105 The spreading of smells through still air is due to the diffusion of the gas molecules. In a room that is 5 m long, a vial of ethyl octanoate is opened at the north end of the room, and simultaneously a second vial containing p-anisaldehyde is opened at the south end of the room (5 m away). Ethyl octanoate ($C_{10}H_{20}O_2$) has a fruity smell, and p-anisaldehyde ($C_8H_8O_2$) has a minty smell. How far (in meters) from the north end of the room must a person stand to smell first a minty smell?

4.106 You are told that 2.55 g of a gaseous hydrocarbon occupies a vessel of volume 3.00 L at 0.950 atm and 82.0°C. Draw the Lewis structure of this hydrocarbon.

4.107 A finely powdered solid sample of an osmium oxide (which melts at 40.°C and boils at 130.°C) with a mass of 1.509 g is placed into a cylinder with a movable piston that can expand against the atmospheric pressure of 745 Torr. Assume that the amount of residual air initially present in the cylinder is negligible. When the sample is heated to 200.°C, it is completely vaporized and the volume of the cylinder expands by 235 mL. What is the molar mass of the oxide? Assuming that the oxide is OsO_x, find the value of x.

4.108 How does the root mean square speed of gas molecules vary with temperature? Illustrate this relationship by plotting the root mean square speed of N_2 as a function of temperature from $T = 100$ K to $T = 300$ K.

4.109 A group of chemistry students have injected 46.2 g of a gas into an evacuated, constant-volume container at 27°C and atmospheric pressure. Now the students want to heat the gas at constant pressure by allowing some of the gas to escape during the heating. What mass of gas must be released if the temperature is raised to 327°C?

4.110 The root mean square speed v_{rms} of gas molecules was derived in Section 4.10. Using the Maxwell distribution of speeds, we can also calculate the mean speed and most probable (mp) speed of a collection of molecules. The equations used to calculate these two quantities are $v_{mean} = (8RT/\pi M)^{1/2}$ and $v_{mp} = (2RT/M)^{1/2}$, respectively. These values have a fixed relation to one another. (a) Place these three quantities in order of increasing magnitude. (b) Show that the relative magnitudes are independent of the molar mass of the gas. (c) Using the smallest speed as the reference point, determine the ratio of the larger values to the smallest.

4.111 (a) The van der Waals parameters for helium are $a = 3.412 \times 10^{-2}$ L^2·atm·mol^{-2} and $b = 2.370 \times 10^{-2}$ L·mol^{-1}.

Calculate the apparent volume (in pm^3) and radius (in pm) of a helium atom as determined from the van der Waals parameters. (b) Estimate the volume of a helium atom on the basis of its atomic radius. (c) How do these quantities compare? Should they be same? Discuss.

4.112 Show that the van der Waals parameter b is related to the molecular volume v_{mol} by $b = 4N_A v_{mol}$. Hint. Treat the molecules as spheres of radius r, so that $v_{mol} = \frac{4}{3}\pi r^3$. The closest that the centers of two molecules can approach is $2r$.

4.113 A 1.00-L sample of chlorine gas at 1.00 atm and 298 K reacts completely with 1.00 L of nitrogen gas and 2.00 L of oxygen gas at the same temperature and pressure. There is a single gaseous product, which fills a 2.00-L flask at 1.00 atm and 298 K. Use this information to determine the following characteristics of the product: (a) its empirical formula; (b) its molecular formula; (c) the most favorable Lewis structure based on formal charge arguments (the central atom is an N atom); (d) the molecular shape.

4.114 A sample of gaseous arsine, AsH_3, in a 500.0-mL flask at 300. Torr and 223 K, is heated to 473 K, at which temperature arsine decomposes to solid arsenic and hydrogen gas. The flask is then cooled to 273 K, at which temperature the pressure in the flask is 508 Torr. Has all the arsine decomposed? Calculate the percentage of arsine molecules that have decomposed.

4.115 The van der Waals equation can be arranged into the following cubic equation:

$$V^3 + n\left(\frac{RT + bP}{P}\right)V^2 + \left(\frac{n^2 a}{P}\right)V - \frac{n^3 ab}{P} = 0$$

(a) Using this equation, calculate the volume occupied by 0.505 mol $NH_3(g)$ at 25°C and 95.0 atm. The van der Waals parameters for NH_3 are $a = 4.225$ $L^2 \cdot atm \cdot mol^{-2}$ and $b = 3.707$ $L \cdot mol^{-1}$. (b) Do the attractive or the repulsive forces dominate at this temperature and pressure?

Chemistry Connections

4.116 Iron pyrite, FeS_2, is the form in which much of the sulfur exists in coal. In the combustion of coal, oxygen reacts with iron pyrite to produce iron(III) oxide and sulfur dioxide, which

is a major source of air pollution and a substantial contributor to acid rain.

(a) Write a balanced equation for the burning of FeS_2 in air to produce iron(III) oxide and sulfur dioxide.

(b) Calculate the mass of Fe_2O_3 that is produced from the reaction of 75.0 L of oxygen at 2.33 atm and 150.°C with an excess of iron pyrite.

(c) If the sulfur dioxide that is generated in part (b) is dissolved to form 5.00 L of aqueous solution, what is the molar concentration of H_2SO_3 in the resulting solution?

(d) What mass of SO_2 is produced in the burning of 1.00 tonne (1 t = 10^3 kg) of high-sulfur coal, if the coal is 5% pyrite by mass?

(e) What is the volume of the SO_2 gas generated in part (d) at 1.00 atm and 25°C?

(f) One way to remove SO_2 from exhaust gases is to remove it in the reaction $CaO(s) + SO_2(g) \rightarrow CaSO_3(s)$. In a test of this reaction, a mixture of sulfur dioxide and nitrogen gases is prepared at 25°C in a vessel of volume 500. mL at 1.09 atm. The mixture is passed over warm calcium oxide powder, which removes the sulfur dioxide, and is then transferred to a vessel of volume 150. mL, where the pressure is 1.09 atm at 50.°C. (a) What was the partial pressure of the SO_2 in the initial mixture? (b) What mass of SO_2 was present in the initial mixture?

(g) The van der Waals parameters for SO_2 are $a = 6.865$ $L^2 \cdot atm \cdot mol^{-1}$ and $b = 0.0568$ $L \cdot mol^{-1}$. For $SO_2(g)$ confined in a 1.00-L vessel at 27°C, calculate the pressure of the gas by using the ideal gas law and the van der Waals equation for 0.100 mol to 0.500 mol SO_2 at 0.100-mol increments.

(h) Calculate the percentage deviation of the ideal value from the value calculated from the van der Waals equation at each point in part (g).

(i) Under the conditions in part (g), which term has the larger effect on the pressure of SO_2, the intermolecular attractions or the repulsions?

(j) If we consider to be ideal those gases for which the observed pressure differs by no more than 5% from the ideal value, at what pressure does SO_2 become a "real" gas?

LIQUIDS AND SOLIDS

What Are the Key Ideas? The condensed phases of matter result from attractive intermolecular forces. When atoms, ions, and molecules do not have enough energy to escape from their neighbors they form solids with their atoms in characteristic arrangements. They form liquids when they have enough energy to move past their neighbors but not enough to escape from them entirely.

Why Do We Need to Know This Material? This chapter shows how atomic and molecular properties are related to the structure and properties of bulk solids and liquids, continuing the process begun in Chapter 4 for gases. Because scientists design new materials by considering how the properties and interactions of individual particles give rise to bulk properties, formulating the materials of tomorrow will require an understanding of these relationships.

What Do We Need to Know Already? This chapter uses the concepts of potential energy (Section A), coulombic interactions (Section 2.4), polar molecules and dipoles (Section 3.3), and intermolecular forces in gases (Section 4.12).

Chapter 5

Molecules attract one another. From that simple fact spring fundamentally important consequences. Rivers, lakes, and oceans exist because water molecules attract one another and form a liquid. Without that liquid, there would be no life. Without forces between molecules, our flesh would drip off our bones and the oceans would be gas. Less dramatically, the forces between molecules govern the physical properties of bulk matter and help to account for the differences in the substances around us. They explain why carbon dioxide is a gas that we exhale, why wood is a solid that we can stand on, and why ice floats on water. At very close range, molecules also repel one another. When pressed together, molecules resist further compression.

In Chapter 4 we considered gases, in which intermolecular forces play only a minor role. Here, we deal with liquids and solids, in which the forces that hold molecules together are of crucial importance for determining the physical properties of bulk samples. Individual water molecules, for instance, are not wet, but bulk water is wet because water molecules are attracted to other substances and spread over their surfaces. Individual water molecules neither freeze nor boil, but bulk water does, because in the process of freezing molecules stick together and form a rigid array and in boiling they separate from one another and form a gas.

We have to refine our atomic and molecular model of matter to see how bulk properties can be interpreted in terms of the properties of individual molecules, such as their size, shape, and polarity. We begin by exploring **intermolecular forces**, the forces between molecules, as distinct from the forces responsible for the formation of chemical bonds between atoms. Then we consider how intermolecular forces determine the physical properties of liquids and the structures and physical properties of solids.

INTERMOLECULAR FORCES

The simplest state of matter is a gas. We can understand many of the bulk properties of a gas—its pressure, for instance—in terms of the kinetic model introduced in Chapter 4, in which the molecules do not interact with one another except during collisions. We have also seen that this model can be improved and used to explain the properties of real gases, by taking into account the fact that molecules do in fact attract and repel one another. But what is the origin of these attractive and

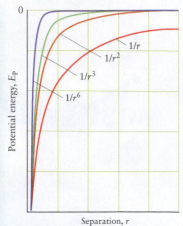

Living Graph: **Figure 5.1 Variation of the coulombic interaction with distance**

FIGURE 5.1 The distance dependence of the potential energy of the interaction between ions (red, lowest line), ions and dipoles (brown), stationary dipoles (green), and rotating dipoles (blue, uppermost line).

TABLE 5.1 **Interionic and Intermolecular Interactions***

Type of interaction	Typical energy (kJ·mol^{-1})	Interacting species
ion–ion	250	ions only
ion–dipole	15	ions and polar molecules
dipole–dipole	2	stationary polar molecules
	0.3	rotating polar molecules
dipole–induced-dipole	2	at least one molecule must be polar
London (dispersion)†	2	all types of molecules
hydrogen bonding	20	molecules containing N, O, F; the link is a shared H atom

*The total interaction experienced by a species is the sum of all the interactions in which it can participate.
†Also known as the induced-dipole–induced-dipole interaction.

repulsive forces? How does their strength relate to the properties of bulk matter? These are some of the questions addressed in the next five sections.

5.1 The Formation of Condensed Phases

Intermolecular forces are responsible for the existence of several different "phases" of matter. A **phase** is a form of matter that is uniform throughout in both chemical composition and physical state. The phases of matter include the three common physical states, solid, liquid, and gas (or vapor), introduced in Section A. Many substances have more than one solid phase, with different arrangements of their atoms or molecules. For instance, carbon has several solid phases; one is the hard, brilliantly transparent diamond we value and treasure and another is the soft, slippery, black graphite we use in common pencil lead. A **condensed phase** means simply a solid or liquid phase. The temperature at which a gas condenses to a liquid or a solid depends on the strength of the attractive forces between its molecules.

All interionic and almost all intermolecular interactions can be traced to the coulombic interaction between two charges (Section 2.4), and throughout the discussion of intermolecular interactions we shall build on Eq. 5 from Section A, the expression for the potential energy E_P of two charges q_1 and q_2 separated by a distance r:

$$E_P = \frac{q_1 q_2}{4\pi\varepsilon_0 r} \tag{1}*$$

Figure 5.1 depicts the distance dependence of this potential energy and that of other interactions described in the following four sections. These interactions are summarized in Table 5.1; notice that the energies of these interactions are much lower than the energies typical of ionic bonds.

Condensed phases form when attractive forces between particles pull them together.

5.2 Ion–Dipole Forces

An ion in water has a number of water molecules attached to it. The attachment of water molecules to solute particles, particularly but not only ions, is called **hydration**. Hydration of ions is due to the polar character of the H_2O molecule (**1**). The partial negative charge on the O atom is attracted to the cation, and the partial positive charges of the H atoms are repelled by it. Therefore, the water molecules can be expected to cluster around the cation, with a lone pair of electrons on each O atom pointing inward and the H atoms pointing outward (Fig. 5.2a). We expect the reverse arrangement around an anion: the H atoms have partial positive charges, and so they are attracted to the anion's negative charge (Fig. 5.2b). Because hydration arises from the interaction between the ion and the partial

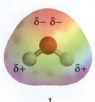

1

charges on the polar water molecule, it is an example of an **ion–dipole interaction.** Neutron diffraction studies of aqueous solutions of LiCl and NaCl show that in dilute solutions individual H atoms are attracted to the anions in a linear hydrogen bond-like arrangement and the partial negative charge on the O atoms is attracted to the cations.

The potential energy of the interaction between the full charge of an ion and the two partial charges of a polar molecule is

$$E_P \propto -\frac{|z|\mu}{r^2} \qquad (2)$$

Here z is the charge number of the ion and μ is the electric dipole moment of the polar molecule. The negative sign means that the potential energy of the ion is *lowered* by its interaction with the polar solvent. The dependence on $1/r^2$ means that the interaction between an ion and a dipole depends more strongly on distance than does the interaction between two ions, which is proportional to $1/r$ (see Eq. 1 and Fig. 5.1); therefore, it acts over a shorter range. Consequently, polar molecules need to be very close to an ion—almost in contact—before the interaction is significant. Even when the molecule and ion are very close, the ion–dipole forces are still smaller than the attraction between two ions, because the dipole moment of a polar molecule represents only partial charges. In addition, an ion attracted to the partial charge at one end of a polar molecule is repelled by the charge at the other end and the two effects partly cancel. At large distances, the two partial charges are at almost the same distance from the ion, and the cancellation is nearly complete: that is why the potential energy of interaction between a point charge and a dipole falls off more quickly with distance (as $1/r^2$) than the interaction of two point charges does (as $1/r$).

When salts crystallize from an aqueous solution, the ions may retain some of the hydrating water molecules and form solid hydrates such as $Na_2CO_3 \cdot 10H_2O$ and $CuSO_4 \cdot 5H_2O$. Both the size of the ion and its charge control the extent of hydration. The strength of the ion–dipole interaction is greater the smaller the value of r in Eq. 2 (that is, the closer the dipole can come to the center of the ion). Because of their stronger ion–dipole interactions, small cations attract the polar H_2O molecules more strongly than large cations do. As a result, small cations are more extensively hydrated than large cations. In fact, lithium and sodium commonly form hydrated salts, whereas the heavier Group 1 elements that have bigger cations—potassium, rubidium, and cesium—do not. Solid ammonium salts are usually **anhydrous,** or water free, for a similar reason: an NH_4^+ ion has about the same radius (143 pm) as an Rb^+ ion (149 pm).

For ions of similar size we should expect hydration to be more important the higher the charge. We can see the effect of charge on the extent of hydration by comparing barium and potassium cations, which have similar radii (136 pm for Ba^{2+} and 138 pm for K^+). In the solid state potassium salts are not hydrated to any appreciable extent, but barium salts are often hydrated. For example, barium chloride is found as $BaCl_2 \cdot 2H_2O$, but potassium chloride is anhydrous. The difference can be traced to the barium ion's higher charge. Lanthanum, barium's neighbor, is both smaller (122 pm) and more highly charged (La^{3+}); so we can expect it to have strong ion–dipole interactions and its compounds to be extensively hydrated. In fact, its salts include $La(NO_3)_3 \cdot 6H_2O$ and $La_2(SO_4)_3 \cdot 9H_2O$.

Ion–dipole interactions are strong for small, highly charged ions; one consequence is that small, highly charged cations are often hydrated in compounds.

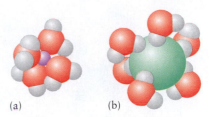

FIGURE 5.2 In water, ions are hydrated. (a) A cation is surrounded by water molecules, oriented with their partially negatively charged oxygen atoms facing the ion. (b) An anion is surrounded by water molecules that direct their partially positively charged hydrogen atoms toward the ion.

5.3 Dipole–Dipole Forces

What would we see if we could observe a polar molecule surrounded by other polar molecules in a solid? An example of a polar molecule is chloromethane, CH_3Cl,

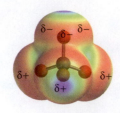

2

FIGURE 5.3 Polar molecules attract each other by the interaction between the partial charges of their electric dipoles (represented by the arrows). Both the relative orientations shown (end to end or side by side) result in a lower energy.

It takes about 1 ps for a small molecule to make a complete revolution in the gas phase.

FIGURE 5.4 A polar molecule rotating near another polar molecule spends more time in the low-energy orientations (shown shaded) that maximize attractions, and so the net interaction is attractive, but not as strong as it would be if the molecules were not rotating.

3   *p*-Dichlorobenzene

4   *o*-Dichlorobenzene

with the partial negative charge on the Cl atom and the partial positive charge spread over the H atoms (**2**). The arrangement of molecules has a low energy if the partial negative charge of a Cl atom on one molecule is closer to the partial positive charge on the H atoms on a neighboring molecule than it is to the partial negative charge. So we would expect to see the surrounding molecules lined up with opposite partial charges on neighboring molecules as close together as possible (Fig. 5.3). The interaction between dipoles is called the **dipole–dipole interaction,** and the resulting potential energy is

$$E_P \propto -\frac{\mu_1\mu_2}{r^3} \tag{3}$$

Here, μ_1 and μ_2 are the dipole moments of the interacting molecules (for a solid composed of identical molecules, $\mu_1 = \mu_2$). The greater the polarity of the molecules, the stronger are the interactions. Equation 3 also shows that doubling the distance between the molecules reduces the strength of their interaction by a factor of $2^3 = 8$. The strong dependence on distance means that the potential energy falls off eight times as fast as the energies of ion–ion interactions and twice as fast as ion–dipole interactions (see Fig. 5.1). The reason for this more rapid falling off is that the opposite partial charges on *each* molecule seem to cancel and merge as the distance between the molecules increases, whereas in the interaction between a point charge and a dipole, only the partial charges on the dipole seem to merge.

Now imagine that we can observe the same chloromethane molecule in a gas. The molecule and its neighbors are all rotating rapidly (Fig. 5.4). For perfectly free rotation, the attractions between opposite partial charges and the repulsions between like partial charges cancel and there is no net interaction. However, in reality the rotating neighbors linger slightly in energetically favorable orientations (their oppositely charged ends adjacent), and so the attractive interactions between opposite partial charges slightly outweigh the repulsive interactions between like partial charges. As a result, there is a weak net attraction between rotating polar molecules in the gas phase. It turns out that the potential energy varies as the *sixth* power of the distance between the molecules (see Fig. 5.1):

$$E_P \propto -\frac{\mu_1^2\mu_2^2}{r^6} \tag{4}$$

Doubling the separation of polar molecules reduces the strength of the interaction by a factor of $2^6 = 64$, and so dipole–dipole interactions between rotating molecules have a significant effect only when the molecules are very close. We can now start to understand why the kinetic model accounts for the properties of gases so well: gas molecules rotate and are far apart for most of the time, so any intermolecular interactions between them are very weak. Equation 4 also describes attractions between rotating molecules in a liquid. However, in the liquid phase, molecules are closer than in the gas phase and therefore the dipole–dipole interactions are much stronger.

EXAMPLE 5.1   **Predicting relative boiling points on the basis of dipole–dipole interactions**

Which would you expect to have the higher boiling point, *p*-dichlorobenzene (**3**) or *o*-dichlorobenzene (**4**)?

STRATEGY   When two compounds have different dipole moments but are otherwise very similar, we expect the molecules with the larger electric dipole moment to interact more strongly. Therefore, assign the higher boiling point to the more strongly polar compound. To decide whether a molecule is polar, determine whether the dipole moments of the bonds cancel each other, as explained in Section 3.3.

SOLUTION

The two C—Cl bonds in *p*-dichlorobenzene lie directly across the ring and their dipole moments cancel, thereby producing a nonpolar molecule.

An *o*-dichlorobenzene molecule is polar because the two C—Cl bond dipoles do not cancel. We therefore predict that *o*-dichlorobenzene will have a higher boiling point than that of *p*-dichlorobenzene.

The experimental values are found to be 180°C for *o*-dichlorobenzene and 174°C for *p*-dichlorobenzene.

SELF-TEST 5.1A Which will have the higher boiling point, *cis*-dichloroethene (**5**) or *trans*-dichloroethene (**6**)?

[*Answer: cis*-Dichloroethene]

SELF-TEST 5.1B Which will have the higher boiling point, 1,1-dichloroethene (**7**) or *trans*-dichloroethene?

> *Polar molecules take part in dipole–dipole interactions, the attraction between the partial charges of their molecules. Dipole–dipole interactions are weaker than forces between ions and fall off rapidly with distance, especially in the liquid and gas phases, where the molecules rotate.*

5.4 London Forces

Attractive interactions are found even between nonpolar molecules. Evidence for these interactions includes the fact that the noble gases—which, because they are monatomic, are necessarily nonpolar—can be liquefied, and many nonpolar compounds, such as the hydrocarbons that make up gasoline, are liquids.

At first thought, there seems to be no means by which two nonpolar molecules could attract each other. To find the explanation, we must refine our model of the electron distribution in a molecule. First, recall that all the representations of electron distributions and charge distributions in the computer graphics that we have seen (for example, structures **1** and **2**) are *average* values. In a nonpolar molecule or a single atom, the electrons appear to be symmetrically distributed. In fact, at any instant, the electron clouds of atoms and molecules are not uniform. If we could take a snapshot of a molecule at a given instant, the electron distribution would look like a swirling fog. Electrons may pile up somewhere in a molecule, leaving a nucleus elsewhere partly exposed. As a result, one region of the molecule will have a fleeting partial negative charge and another region will have a fleeting partial positive charge; in the next moment—about 10^{-16} s later—the charges may be reversed or in other locations. Even a nonpolar molecule may have an *instantaneous*, fleeting dipole moment (Fig. 5.5).

An instantaneous dipole moment on one molecule distorts the electron cloud on a neighboring molecule and gives rise to a dipole moment on that molecule: the two instantaneous dipoles attract each other. An instant later, the swirling electron cloud of the first molecule will give rise to a dipole moment in a different direction,

5   *cis*-Dichloroethene

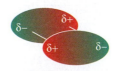

6   *trans*-Dichloroethene

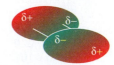

7   1,1-Dichloroethene

FIGURE 5.5 The rapid fluctuations in the electron distribution in two neighboring molecules result in two instantaneous electric dipole moments that attract each other. The fluctuations flicker into different positions, but each new arrangement in one molecule induces an arrangement in the other that results in mutual attraction.

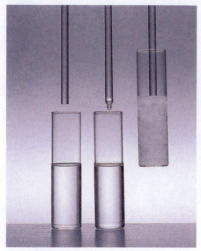

FIGURE 5.6 These hydrocarbons show how London forces increase in strength with molar mass. Pentane is a mobile liquid (left); pentadecane, $C_{15}H_{32}$, a viscous liquid (middle); and octadecane, $C_{18}H_{38}$, a waxy solid (right). To some extent, the effect of increasing intermolecular forces is increased by the ability of long-chain molecules to tangle with one another.

but that new dipole moment induces a dipole moment in the second molecule, and still the two molecules attract each other. That is, although there is a constant flickering of the dipole moment of one molecule from one orientation to another, the induced dipole moment of the second molecule follows it faithfully, and there is a net attraction between the two molecules. This attractive interaction is called the **London interaction.** It acts between *all* molecules and is the only interaction between nonpolar molecules.

The strength of the London interaction depends on the **polarizability,** α (alpha), of the molecules, the ease with which their electron clouds can be distorted. Highly polarizable molecules are those in which the nuclear charges have little control over the surrounding electrons, perhaps because the atoms are large and the distance between electrons and the nuclei is great or because the valence electrons are well shielded by the inner electrons. There can then be big fluctuations in electron density, and hence highly polarizable molecules can have large instantaneous dipole moments and strong London interactions. Detailed calculations show that the potential energy of the London interaction varies as the inverse sixth power of the separation of two molecules:

$$E_P \propto -\frac{\alpha_1\alpha_2}{r^6} \tag{5}$$

Like the potential energy of dipole–dipole interactions between rotating polar molecules, the potential energy of the London interaction decreases very rapidly with distance (as $1/r^6$; see Fig. 5.1). The strength of the interaction increases with the polarizability of the interacting molecules. Because a large molecule with many electrons is typically more polarizable than a small molecule with only a few electrons, a large molecule is likely to have stronger London interactions than a smaller one (Fig. 5.6).

We can now see why the halogens range from gases (F_2 and Cl_2), through a liquid (Br_2), to a solid (I_2) at room temperature: the molecules have an increasing number of electrons, so their polarizabilities and hence their London interactions increase going down the group. The increase in strength of London forces is striking when heavier atoms are substituted for hydrogen atoms. Methane boils at $-161°C$, but tetrachloromethane (carbon tetrachloride, CCl_4) has many more electrons and is a liquid that boils at $77°C$. Tetrabromomethane, CBr_4, has even more electrons: it is a solid at room temperature (Table 5.2).

TABLE 5.2   Melting and Boiling Points of Substances

Substance	Melting point (°C)	Boiling point (°C)	Substance	Melting point (°C)	Boiling point (°C)
Noble gases			Small inorganic species		
He	−270 (3.5 K)*	−269 (4.2 K)	H_2	−259	−253
Ne	−249	−246	N_2	−210	−196
Ar	−189	−186	O_2	−218	−183
Kr	−157	−153	H_2O	0	100
Xe	−112	−108	H_2S	−86	−60
Halogens			NH_3	−78	−33
F_2	−220	−188	CO_2	—	−78s
Cl_2	−101	−34	SO_2	−76	−10
Br_2	−7	59	Organic compounds		
I_2	114	184	CH_4	−182	−162
Hydrogen halides			CF_4	−150	−129
HF	−93	20	CCl_4	−23	77
HCl	−114	−85	C_6H_6	6	80
HBr	−89	−67	CH_3OH	−94	65
HI	−51	−35	glucose	142	d
			sucrose	184d	—

Abbreviations: s, solid sublimes; d, solid decomposes.
*Under pressure.

The effectiveness of London forces also depends on the shapes of molecules. Both pentane (**8**) and 2,2-dimethylpropane (**9**), for instance, have the molecular formula C_5H_{12}, and so they each have the same number of electrons. However, they have different boiling points (36° and 10°C, respectively). Molecules of pentane are relatively long and rod shaped. The instantaneous partial charges on adjacent rod-shaped molecules can be in contact at several points, leading to strong interactions. In contrast, the instantaneous partial charges on more spherical molecules such as 2,2-dimethylpropane cannot get so close to one another, because only a small region of each molecule can be in contact (Fig. 5.7). As a result of the strong dependence on distance ($1/r^6$ in Eq. 5), the London interactions between rod-shaped molecules are more effective than those between spherical molecules with the same number of electrons.

Closely related to the London interaction is the **dipole–induced-dipole interaction,** in which a polar molecule interacts with a nonpolar molecule (for example, when oxygen dissolves in water). Like the London interaction, the dipole–induced-dipole interaction arises from the ability of one molecule to induce a dipole moment in the other. However, in this case, the molecule that induces the dipole moment has a permanent dipole moment. The potential energy of the interaction is

$$E_p \propto -\frac{\mu_1^2 \alpha_2}{r^6} \tag{6}$$

Once again, the potential energy is inversely proportional to the sixth power of the separation. Notice that the potential energies of the dipole–dipole interaction of rotating polar molecules in the gas phase, the London interaction, and the dipole–induced-dipole interaction all have the form

$$E_p \propto -\frac{C}{r^6} \tag{7}*$$

with C depending on the molecules and the type of interaction. Intermolecular interactions that depend on the inverse sixth power of the separation are known collectively as **van der Waals interactions,** after Johannes van der Waals, the Dutch scientist who studied them extensively.

8   Pentane, C_5H_{12}

9   2,2-Dimethylpropane

EXAMPLE 5.2   **Accounting for a trend in boiling points**

Explain the trend in the boiling points of the hydrogen halides: HCl, −85°C; HBr, −67°C; HI, −35°C.

STRATEGY Stronger intermolecular forces result in higher boiling points. The dipole moments, and therefore the strength of the dipole–dipole interactions, increase with the polarity of the H—X bond and therefore with the difference in electronegativity between the hydrogen and the halogen atoms. The strength of London forces increases with the number of electrons. Use the data to decide which is the dominant effect.

SOLUTION The data in Fig. 2.12 show that electronegativity differences decrease from HCl to HI, and so the dipole moments decrease as well. Therefore, dipole–dipole forces decrease, too, a trend suggesting that the boiling points should decrease from HCl to HI. This prediction conflicts with the data; so we examine the London forces. The number of electrons in a molecule increases from HCl to HI, and so the strength of the London interaction increases, too. Therefore, the boiling points should increase from HCl to HI, in accord with the data. This analysis suggests that London forces dominate dipole–dipole interactions for these molecules.

SELF-TEST 5.2A Account for the trend in boiling points of the noble gases, which increase from helium to xenon.

[*Answer:* The strength of the London interaction increases as the number of electrons increases.]

SELF-TEST 5.2B Suggest a reason why trifluoromethane, CHF_3, has a higher boiling point than tetrafluoromethane, CF_4.

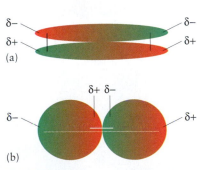

FIGURE 5.7 (a) The instantaneous dipole moments in two neighboring rod-shaped molecules tend to be close together and interact strongly over a relatively broad region of the molecule. (b) Those on neighboring spherical molecules tend to be farther apart and interact weakly over only a small region of the molecule.

FIGURE 5.8 The boiling points of most of the molecular hydrides of the *p*-block elements show a smooth increase with molar mass in each group. However, three compounds—ammonia, water, and hydrogen fluoride—are strikingly out of line.

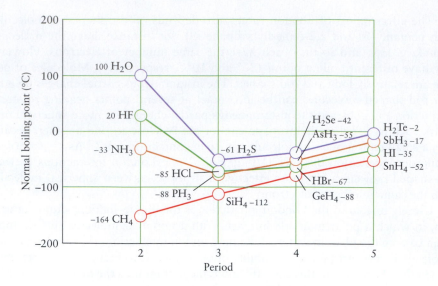

The London interaction arises from the attraction between instantaneous electric dipoles on neighboring molecules and acts between all types of molecules; its strength increases with the number of electrons and occurs in addition to any dipole–dipole interactions. Polar molecules also attract nonpolar molecules by weak dipole–induced-dipole interactions.

5.5 Hydrogen Bonding

The London interaction is "universal" in the sense that it applies to all molecules regardless of their chemical identity. Similarly, the dipole–dipole interaction depends only on the polarity of the molecule, regardless of its chemical identity. However, there is another very strong interaction between molecules that is specific to molecules with certain types of atoms.

We might suspect the existence of a special interaction when we plot the boiling points of the common binary hydrogen compounds of the elements in Groups 14/IV to 17/VII (Fig. 5.8). The trend in Group 14/IV is what we expect for similar compounds that differ in their number of electrons—the boiling points increase down the group because the strength of the London interaction increases. However, ammonia, water, and hydrogen fluoride all show anomalous behavior. Their exceptionally high boiling points suggest that there are unusually strong attractive forces between their molecules.

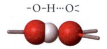

10 Hydrogen bond

The strong interaction responsible for the high boiling points of these substances and certain others is the **hydrogen bond**, an intermolecular attraction in which a hydrogen atom bonded to a small, strongly electronegative atom, specifically N, O, or F, is attracted to a lone pair of electrons on another N, O, or F atom (**10**). To understand how it forms, let's imagine what happens when one water molecule comes close to another. Each O—H bond is polar. The electronegative O atom exerts a strong pull on the electrons in the bond, and the proton of the H atom is almost completely unshielded. Because it is so small, the hydrogen atom, with its partial positive charge, can get very close to one of the lone pairs of electrons on the O atom of another water molecule. The lone pair and the partial positive charge attract each other strongly and form a hydrogen bond. Hydrogen bonding is strongest when the hydrogen atom is on a straight line between the two oxygen atoms. A hydrogen bond is denoted by a dotted line, so the hydrogen bond between two O atoms is denoted O—H···O. The O—H bond length is 101 pm and the H···O distance is somewhat longer; in ice it is 175 pm.

SELF-TEST 5.3A Which of the following intermolecular links can be made by hydrogen bonds: (a) CH_3NH_2 to CH_3NH_2; (b) CH_3OCH_3 to CH_3OCH_3; (c) HBr to HBr?
[*Answer:* Only in (a) is H bonded directly to N, O, or F.]

SELF-TEST 5.3B Which of the following molecules can take part in hydrogen bonding with other molecules of the same compound: (a) CH_3OH; (b) PH_3; (c) HClO (which has the structure Cl—O—H)?

When it can form, a hydrogen bond is so strong—about 10% of a typical covalent bond strength—that it dominates all the other types of intermolecular interactions. Hydrogen bonding is strong enough to survive even in the vapor of some substances. Liquid hydrogen fluoride, for instance, contains zigzag chains of HF molecules (**11**), and the vapor contains short fragments of the chains and $(HF)_6$ rings. The vapor of acetic acid, CH_3COOH, contains **dimers,** pairs of molecules, linked by two hydrogen bonds (**12**).

Hydrogen bonds play a vital role in maintaining the shapes of biological molecules. The shape of a protein molecule is governed largely by hydrogen bonds; once the bonds are broken, the delicately organized protein molecule loses its function. When we cook an egg, the clear albumen becomes milky white, because the heat breaks the hydrogen bonds in its protein molecules, which collapse into a random jumble. Trees are held upright by hydrogen bonds (Fig. 5.9). Cellulose molecules (which have many —OH groups) can form many hydrogen bonds with one another, and the strength of wood is due in large part to the strength of the hydrogen bonds between neighboring ribbonlike cellulose molecules. Hydrogen bonding also binds the two chains of a DNA molecule together, and so hydrogen bonding is a key to understanding reproduction (see Section 19.15). The hydrogen bonds are strong enough to keep the two chains of a DNA molecule together but so much weaker than typical covalent bonds that they readily give way in cell division without affecting the covalent bonds in DNA.

Hydrogen bonding, which occurs when hydrogen atoms are bonded to oxygen, nitrogen, or fluorine atoms, is the strongest type of intermolecular interaction.

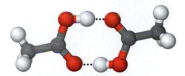

11 Hydrogen Fluoride, $(HF)_n$

12 Acetic acid dimer

FIGURE 5.9 Vegetation, even huge trees such as these, is held upright by the powerful intermolecular hydrogen bonds that exist between the ribbonlike cellulose molecules that form much of its bulk. Without hydrogen bonding, these trees would collapse.

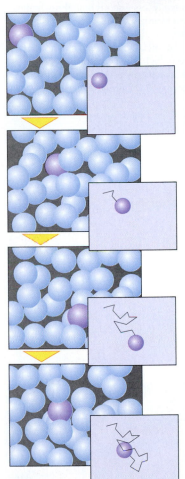

Media Link

Animation **Figure 5.10 Liquid water**

FIGURE 5.10 The structure of a liquid. Although the molecules (represented by the spheres in this series of diagrams) remain in contact with their neighbors, they can move away from one another and have enough energy to push through to a new neighborhood. Consequently, the entire substance is fluid. One sphere is slightly darker so that you can follow its motion. Its path is shown in the insets.

LIQUID STRUCTURE

The molecules in a liquid are in contact with their neighbors, but they can move past and tumble over one another (Fig. 5.10). When we imagine a liquid, we can think of a jostling crowd of molecules, each one constantly changing places with its neighbors. A liquid at rest is like a crowd of people milling about in a stadium; a flowing liquid is like a crowd of people leaving a stadium.

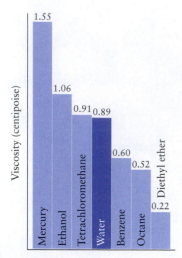

FIGURE 5.11 The viscosities of several liquids. Liquids composed of molecules that cannot form hydrogen bonds are generally less viscous than those that can form hydrogen bonds. Mercury is an exception: its atoms stick together by a kind of metallic bonding, and its viscosity is very high. The centipoise (cP) is the unit commonly used to report viscosity (1 cP = 10^{-3} kg·m·s^{-1}).

5.6 Order in Liquids

The liquid phase of matter is the most difficult to visualize. We have seen that a gas-phase molecule moves with almost complete freedom. The intermolecular forces from other molecules are minimal, and movement is highly disordered. In the solid phase, a molecule is locked in place by intermolecular forces and can only oscillate around an average location. The liquid phase lies between the extremes of the gas and solid phases. The molecules are mobile, but they cannot escape from one another completely.

A crystalline solid has **long-range order.** In other words, its atoms or molecules lie in an orderly arrangement that is repeated over long distances. When the crystal melts, that long-range order is lost. In the liquid, the kinetic energy of the molecules can partly overcome the intermolecular forces, and so the molecules are able to move past one another. However, the molecules still experience strong attractions to one another. In water, for instance, only about 10% of the hydrogen bonds are broken on melting. The rest are continually broken and re-formed with different water molecules. At any given moment, the immediate vicinity of a molecule in the liquid is like that in the solid, but the ordering does not extend very far past the nearest neighbors. This local ordering is called **short-range order.** In water, each molecule is surrounded by an approximately tetrahedral arrangement of other molecules as a result of hydrogen bonds. These interactions are strong enough to affect the local structure of molecules up to the boiling point.

In the liquid phase, molecules have short-range order but not long-range order.

5.7 Viscosity and Surface Tension

The **viscosity** of a liquid is its resistance to flow: the higher the viscosity of the liquid, the more sluggish the flow. A liquid with a high viscosity, such as molasses at room temperature or molten glass, is said to be "viscous." The viscosity of a liquid is an indication of the strength of the forces between its molecules: strong intermolecular forces hold molecules together and do not let them move past one another easily. The strong hydrogen bonds in water give it a viscosity greater than that of benzene: benzene molecules can easily slip past one another in the liquid but, for water molecules to move, hydrogen bonds have to be wrenched apart. Phosphoric acid, H_3PO_4, and glycerol, $HOCH_2CH(OH)CH_2OH$, are very viscous at room temperature because of the numerous hydrogen bonds that their molecules can form. Figure 5.11 compares the relative viscosities of several liquids. The molecules of hydrocarbon oils and greases are nonpolar and so participate only in London forces. However, they have long chains that tangle together like cooked spaghetti (Fig. 5.12), and the molecules move past one another only with difficulty.

Viscosity usually decreases as the temperature rises. Molecules have more energy at high temperatures and can wriggle past their neighbors more readily. The viscosity of water at 100°C, for instance, is only one-sixth of its value at 0°C, and so the same amount of water flows through a tube six times as fast at the higher temperature. In some cases, though, a change in molecular structure takes place in the course of heating, and the viscosity increases. For example, just above 113°C, sulfur is a mobile, straw-colored liquid composed of S_8 rings that can move past one another readily. The viscosity of the liquid increases when it is heated further because the S_8 rings break open into chains that become tangled. The viscosity falls again at still higher temperatures, and the color changes to deep red-brown: now the S_8 chains are breaking up into smaller, more mobile, highly colored S_2 and S_3 molecules.

Another property of a liquid that arises from intermolecular forces is **surface tension.** The surface of a liquid is smooth because intermolecular forces tend to

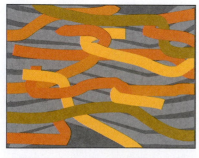

FIGURE 5.12 The long threadlike molecules of heavy hydrocarbon oils tend to get tangled together like a plate of cooked spaghetti. As a result, the molecules do not move past one another very readily and the liquid is very viscous.

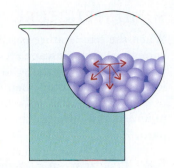

FIGURE 5.13 Surface tension arises from the attractive forces acting on the molecules at the surface. The inset shows that a molecule within the liquid experiences attractive forces from all directions, but a molecule at the surface experiences a net inward force.

TABLE 5.3 Surface Tensions of Liquids at 25°C

Liquid	Surface tension, γ (mN·m^{-1})
benzene	28.88
carbon tetrachloride	27.0
ethanol	22.8
hexane	18.4
mercury	472
methanol	22.6
water	72.75
	58.0 at 100°C

pull the molecules together and inward (Fig. 5.13); the surface tension is the net inward pull. Once again, we can expect liquids composed of molecules with strong intermolecular forces to have high surface tensions, because the inward pull at the surface will be strong. Indeed, the surface tension of water is about three times as high as that of most other common liquids, because of its strong hydrogen bonds (Table 5.3). The surface tension of mercury is higher still—more than six times that of water. Its high surface tension suggests that the bonds between mercury atoms in the liquid are very strong.

Surface tension accounts for a number of everyday phenomena. For example, a droplet of liquid suspended in air or on a waxy surface is spherical because the surface tension pulls the molecules into the most compact shape, a sphere (Fig. 5.14). The attractive forces between water molecules are greater than those between water and wax, which is largely hydrocarbon. Surface tension decreases as the temperature rises and the interactions between molecules are overcome by the increased molecular motion.

Water has strong interactions with paper, wood, and cloth because it can form hydrogen bonds with the molecules in their surfaces. As a result, water maximizes its contact with these materials by spreading over them; in other words, water *wets* them. We now see that water is wet because of the hydrogen bonds that its molecules can form.

Capillary action, the rise of liquids up narrow tubes, occurs when there are favorable attractions between the molecules of the liquid and the tube's inner surface. These attractions are forces of **adhesion,** forces that bind a substance to a surface, as distinct from the forces of **cohesion,** the forces that bind the molecules of a substance together to form a bulk material. An indication of the relative strengths of adhesion and cohesion is the formation of a **meniscus,** the curved surface of a liquid, in a narrow tube (Fig. 5.15). The meniscus of water in a glass capillary is curved upward at the edges (forming a concave shape) because the adhesive forces between water molecules and the oxygen atoms and —OH groups that are present on a typical glass surface are comparable in strength to the cohesive forces between water molecules. The water therefore tends to spread over the greatest possible area of glass. In contrast, the meniscus of mercury curves downward in glass (forming a convex shape). This shape is a sign that the cohesive forces between mercury atoms are stronger than the forces between mercury atoms and the glass, and so the liquid tends to reduce its contact with the glass.

The greater the viscosity of a liquid, the more slowly it flows. Viscosity usually decreases with increasing temperature. Surface tension arises from the imbalance of intermolecular forces at the surface of a liquid. Capillary action arises from the imbalance of adhesive and cohesive forces.

FIGURE 5.14 The nearly spherical shapes of these beads of water on a waxy surface arise from the effect of surface tension.

FIGURE 5.15 When the adhesive forces between a liquid and glass are stronger than the cohesive forces within the liquid, the liquid curves up to maximize contact with the glass, forming the meniscus shown here for water in glass (left). When the cohesive forces are stronger than the adhesive forces (as they are for mercury in glass), the edges of the surface curve downward to minimize contact with the glass (right).

SOLID STRUCTURES

When the temperature is so low that the molecules of a substance do not have enough energy to escape even partially from their neighbors, the substance solidifies. The nature of the solid depends on the types of forces that hold the atoms, ions, or molecules together. An understanding of solids in terms of the properties of their atoms will help us to understand why, for instance, metals can be pounded into different shapes but salt crystals shatter, and why diamonds are so hard.

5.8 Classification of Solids

A **crystalline solid** is a solid in which the atoms, ions, or molecules lie in an orderly array (Fig. 5.16). A crystalline solid has long-range order. An **amorphous solid** is one in which the atoms, ions, or molecules lie in a random jumble, as in butter, rubber, and glass (Fig. 5.17). An amorphous solid has a structure like that of a frozen instant in the life of a liquid, with only short-range order. Crystalline solids typically have flat, well-defined planar surfaces called **crystal faces,** which lie at definite angles to one another. These faces are formed by orderly layers of atoms (Box 5.1). Amorphous solids do not have well-defined faces unless they have been molded or cut.

The arrangement of atoms, ions, and molecules within a crystal is determined by x-ray diffraction (Major Technique 3, which follows this chapter), one of the most useful techniques for determining the structures of solids.

We classify crystalline solids according to the bonds that hold their atoms, ions, or molecules in place:

Molecular solids are assemblies of discrete molecules held in place by intermolecular forces.

Network solids consist of atoms covalently bonded to their neighbors throughout the extent of the solid.

Metallic solids, also called simply *metals,* consist of cations held together by a sea of electrons.

Ionic solids are built from the mutual attractions of cations and anions.

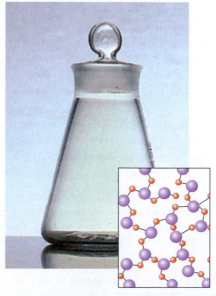

FIGURE 5.17 (Left) Quartz is a crystalline form of silica, SiO_2, with the atoms in an orderly network, represented here in two dimensions. (Right) When molten silica solidifies in an amorphous arrangement, it becomes glass. Now the atoms form a disorderly network.

BOX 5.1   HOW DO WE KNOW ... WHAT A SURFACE LOOKS LIKE?

Even the most powerful optical microscopes cannot reveal the individual atoms of a solid surface. However, the new field of *nanotechnology*, the development and study of devices only nanometers in size, requires the ability to resolve surfaces on an atomic scale. A new technique that allows individual atoms to be visualized, *scanning tunneling microscopy* (STM), is an important tool for nanotechnology. The technique produces images like those on this page. The first image shows individual nickel atoms lined up in rank upon rank to form a clean surface. The other image shows a tiny sodium iodide crystal lying on a copper surface.

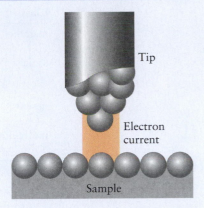

The tip of a scanning tunneling microscope above a surface. Because the tip is too close to the surface for other molecules to interfere, STM devices can be used in a gaseous atmosphere or even in liquids.

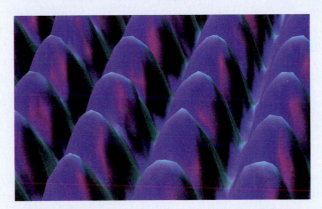

The surface of a single crystal of nickel, showing the regularity of its cubic close-packed structure.

An experimenter attempted to create a two-dimensional sodium iodide crystal on a copper surface, but the ions spontaneously rearranged themselves into a tiny three-dimensional crystal.

The key idea behind STM is that, because electrons have wavelike properties (Section 1.5), they can penetrate into and through regions where classical mechanics would forbid them to be. This penetration is called *tunneling*. The effect is used in STM (hence the "tunneling" in the name) by bringing a fine tip up to a surface and monitoring the current that flows through the gap between the tip and the surface. The magnitude of the current, like the tunneling itself, is very sensitive to the separation of the tip and the surface, and even atomic-scale variations affect it.

To obtain an STM image, a fine tip is moved back and forth across the surface in a series of closely spaced lines (hence the "scanning" in the name). The tip ends in a single atom (see the illustration). As the tip moves across the surface at a constant height, the tunneling ebbs and flows, and the current varies correspondingly through the circuit. The image is a portrayal of the current measured on each scan.

In a modification, the current is maintained at a constant level by varying and monitoring the height of the tip above the surface. That height is controlled by using a *piezoelectric* substance to support the tip, a substance that changes dimensions according to the electrical potential difference applied to it. By measuring the potential difference that must be applied to the piezoelectric support to maintain a constant current through the tip, the height of the tip can be inferred and plotted on a computer screen.

Yet another modification is *atomic force microscopy* (AFM), in which a fine tip attached to a tiny flexible beam is scanned across the surface. The atom at the end of the tip experiences a force that pulls it toward or pushes it away from the atoms on the surface. The deflection of the beam, which shows the shape of the surface, can be monitored by using light from a laser.

The images in this box are computer-generated representations, not true photographs. However, they have opened our eyes to the appearance of surfaces in the most extraordinary ways.

TABLE 5.4   **Typical Characteristics of Solids**

Class	Examples	Characteristics
metallic	*s*- and *d*-block elements	malleable, ductile, lustrous, electrically and thermally conducting
ionic	$NaCl$, KNO_3, $CuSO_4 \cdot 5H_2O$	hard, rigid, brittle; high melting and boiling points; those soluble in water give conducting solutions
network	B, C, black P, BN, SiO_2	hard, rigid, brittle; very high melting points; insoluble in water
molecular	$BeCl_2$, S_8, P_4, I_2, ice, glucose, naphthalene	relatively low melting and boiling points; brittle if pure

Table 5.4 lists examples of each type of solid and their typical characteristics.

In this part of the chapter, we begin with molecular solids and distinguish them from network solids. Then we examine metallic solids, which, if consisting of a single element, are built from identical atoms stacked together in orderly arrays. The structures of ionic solids are based on the same kinds of arrays but are complicated by the need to take into account the presence of ions of opposite charges and different sizes.

Crystalline solids have a regular internal arrangement of atoms or ions; amorphous solids do not. Solids are classified as molecular, network, metallic, or ionic.

5.9 Molecular Solids

Molecular solids consist of molecules held together by intermolecular forces and their physical properties depend on the strengths of those forces. Amorphous molecular solids may be as soft as paraffin wax, which is a mixture of long-chain hydrocarbons. These molecules lie together in a disorderly way, and the forces between them are so weak that they can be pushed past one another very easily. Many other molecular solids have crystalline structures and strong intermolecular forces that make them brittle and hard. For example, sucrose molecules, $C_{12}H_{22}O_{11}$, are held together by hydrogen bonds between their numerous —OH groups. Hydrogen bonding between sucrose molecules is so strong that, by the time the melting point has been reached (at 184°C), the molecules themselves have started to decompose. The mixture of partly decomposed products, called caramel, is used to add flavor and color to food. Some molecular solids are very tough. For example, "ultrahigh-density polyethylene" consists of long hydrocarbon chains that lie densely together like close-packed cylinders: the resulting material is so smooth yet tough that it is used to make bullet-proof vests and joint replacements for the human body.

Because molecules have such widely varying shapes, they stack together in a wide variety of different ways. In ice, for example, each O atom is surrounded by four H atoms in a tetrahedral arrangement. Two of these H atoms are linked covalently to the O atom through σ-bonds. The other two belong to neighboring H_2O molecules and are linked to the O atom by hydrogen bonds. As a result, the structure of ice is an open network of H_2O molecules held in place by hydrogen bonds (Fig. 5.18). Some of the hydrogen bonds break when ice melts; and, as the orderly

Animation: Figure 5.18 Ice

Media Link

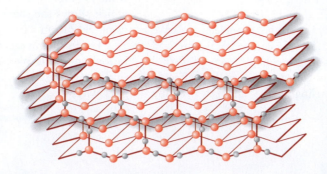

FIGURE 5.18 Ice is made up of water molecules that are held together by hydrogen bonds in an open structure. Each O atom is surrounded tetrahedrally by four hydrogen atoms, two of which are σ-bonded to it and two of which are hydrogen-bonded to it. To show the structure more clearly, only the hydrogen atoms in the front layer are shown.

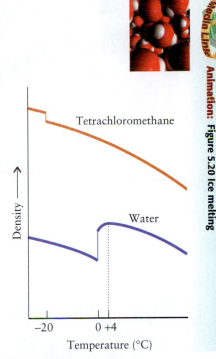

FIGURE 5.19 As a result of its open structure, ice is less dense than liquid water and floats in it (left). Solid benzene is denser than liquid benzene and "benzenebergs" sink in liquid benzene (right).

arrangement collapses, the molecules pack less uniformly but more densely (Fig. 5.19). The openness of the network in ice compared with that in the liquid explains why it has a lower density than liquid water (0.92 g·cm^{-3} and 1.00 g·cm^{-3}, respectively, at 0°C). Solid benzene and solid tetrachloromethane, in contrast, have higher densities than their liquids (Fig. 5.20). Their molecules are held in place by London forces, which are much less directional than hydrogen bonds, and so they can pack together more closely in the solid than in the liquid.

> *Molecular solids are relatively soft and typically melt at relatively low temperatures.*

5.10 Network Solids

Whereas molecular solids consist of molecules held together by relatively weak intermolecular forces, the atoms in network solids are joined to their neighbors by strong covalent bonds that form a framework extending throughout the crystal. In order to break apart a crystal of a network solid, covalent bonds, which are much stronger than intermolecular forces, must be broken. Therefore, network solids are very hard, rigid materials with high melting and boiling points.

Diamond and graphite are elemental network solids. These two forms of carbon are **allotropes,** meaning forms of an element that differ in the way in which the atoms are linked. Each C atom in diamond is covalently bonded to four neighbors through sp^3 hybrid σ-bonds (Fig. 5.21). The tetrahedral framework extends throughout the solid like the steel framework of a large building. This structure accounts for the great hardness of the solid. Diamond is so hard that it is used as a protective coating for drill bits and as a long-lasting abrasive. Because diamond is also one of the best conductors of heat, thin films of diamond are used as a base for some integrated circuits so that they do not overheat. The vigorous vibration of an atom in a hot part of the crystal is rapidly transmitted to distant, cooler parts through the covalent bonds, rather like the effect of slamming a door in a steel-framed building.

Graphite, the most important component of the "lead" of pencils, is a black, lustrous, electrically conducting solid that vaporizes at 3700°C. It consists of flat sheets of sp^2 hybridized carbon atoms bonded covalently into hexagons like chicken wire (Fig. 5.22). There are also weak bonds between the sheets. In the commercially available forms of graphite, there are many impurity atoms trapped between the sheets; these atoms weaken the already weak intersheet bonds and let

FIGURE 5.20 Variation in the densities of water and tetrachloromethane with temperature. Note that ice is less dense than liquid water at its freezing point and that water has its maximum density at 4°C.

FIGURE 5.21 The structure of diamond. Each sphere represents the location of the center of a carbon atom. Each atom is at the center of a tetrahedron formed by the sp^3 hybrid covalent bonds to each of its four neighbors.

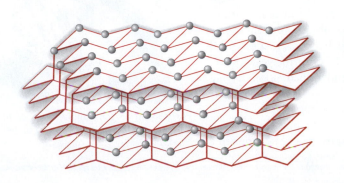

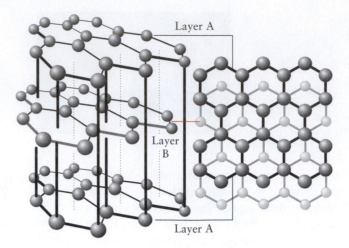

FIGURE 5.22 Graphite consists of layers of hexagonal rings of sp^2 hybridized carbon atoms. The slipperiness of graphite results from the ease with which the layers can slide over one another when there are impurity atoms lying between the planes.

BOX 5.2　FRONTIERS OF CHEMISTRY: HIGH-TEMPERATURE SUPERCONDUCTORS*

Three of the biggest problems facing scientists and engineers in the future will be to provide energy-efficient transportation, improve the capacity and quality of communication systems, and reduce the loss of scarce power resources during transmission. Currently, less than 40% of the energy generated in an electrical power plant actually reaches our homes, most of it being lost during transmission. Scientists and engineers believe that all three of these problems can be resolved with the development of appropriate high-temperature superconductors.

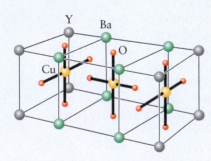

The structure of a "123 superconductor," a ceramic that has the variable formula $YBa_2Cu_3O_{6.5-7.0}$. The numbers 1, 2, and 3 refer to the implied or specified subscripts on the first three elements in the formula.

Superconductivity is the loss of all electrical resistance when a substance is cooled below a certain characteristic transition temperature (T_s). It is thought that the low temperatures are required to reduce the effect of the vibrations of the atoms in their crystalline lattice. Superconductivity was first observed in 1911 in mercury, for which $T_s = 4$ K. Over the years, many other metallic superconductors were identified, some having transition temperatures as high as 23 K. However, low-temperature superconductors need to be cooled with liquid helium, which is very expensive. To use superconducting devices on a large scale, higher transition temperatures would be required.

Research chemists found that they could modify the conducting properties of solids by doping them, a process commonly used to control the properties of semiconductors (see Section 3.14). In 1986, a record-high T_s of 35 K was observed, surprisingly not for a metal, but for a ceramic material, a lanthanum–copper oxide doped with barium. Then early in 1987, a new record T_s of 93 K was set with a yttrium–barium–copper oxide and a series of related compounds. In 1988, two more oxide series of bismuth–strontium–calcium–copper and thallium–barium–calcium–copper exhibited transition temperatures of 110 K and 125 K, respectively. By 2006, the highest transition temperature attained was 130 K. These temperatures can be reached by cooling the materials with liquid nitrogen, at only 0.2% of the cost of liquid helium.

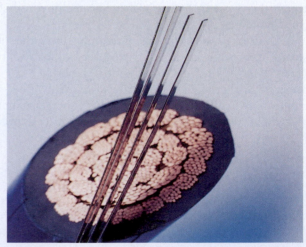

Twenty-five kilograms of this experimental superconducting wire can carry as much current as 1800 kg of the bulky cable shown behind the wire.

the sheets of atoms slide over one another. So, in contrast to the hardness of diamonds, graphite is soft and, when impure, slippery. When we write with a pencil, the mark left on the paper consists of rubbed-off layers of graphite. Electrons can move within the sheets of graphite but much less readily from one sheet to another. Hence, graphite conducts electricity better parallel to the sheets rather than perpendicular to them.

Ceramic materials are typically noncrystalline inorganic oxides prepared by heat-treatment of a powder and have a network structure. They include many silicate minerals, such as quartz (silicon dioxide, which has the empirical formula SiO_2), and high-temperature superconductors (Box 5.2). Ceramic materials have great strength and stability, because covalent bonds must be broken to cause any deformation in the crystal. As a result, ceramic materials under physical stress tend to shatter rather than bend. Section 14.22 contains further information on the properties of ceramic materials.

Network solids are typically hard and rigid; they have high melting and boiling points. Ceramic materials are commonly network solids.

Almost all high-temperature superconductors (HTSCs) are hard, brittle ceramic oxides that have sheets of copper and oxygen atoms sandwiched between layers of either cations or a combination of cations and oxide ions, and all are derived from their respective parent insulators by doping. Because of their layered structure, their electrical and magnetic properties are strongly anisotropic (Section 5.16). The electric current flows easily along the planes of the copper–oxygen sheets, but only weakly perpendicular to them. Thus, the materials must be positioned with the copper–oxygen planes in a favorable orientation. A major challenge to the use of superconductors for transmitting electrical power is the difficulty in making electrical wires from a brittle ceramic material. One solution has been to deposit the superconducting material on the surface of wire or tape made of a metal such as silver. The metal is usually prepared with a textured surface that helps to align the crystal grains in the desired direction.

A better option for the transmission of electricity being considered for use in transformers may be the metallic superconductor magnesium diboride, MgB_2, which superconducts below 39 K by pairing electrons over a relatively large distance of 5 nm. Magnesium diboride is so inexpensive to make that, even with the cost of cooling, it is less expensive than either copper wire or nonmetallic HTSC materials. It requires only inexpensive iron for cladding and can be used to make long, flexible wires. The material is not truly a high-temperature superconductor but, unlike other metallic superconductors, its operating temperature can be reached by the use of a refrigeration unit, and so liquid helium is not required.

❓ HOW MIGHT YOU CONTRIBUTE?

Because superconductors can create a mirror image of a magnetic field within themselves, they trap magnetic fields and can even levitate objects (see Fig.1.64). This property has led to research on exciting possibilities such as levitating railroad trains over magnetic tracks (see Fig. 3.42). However, many technical problems remain to be solved before the promise of high-temperature superconductivity can be realized. Superconducting materials need to be formed into very long, flexible cables that can carry high current without excessive heat loss and that can retain their superconducting properties for a long period of time without chemical or physical degradation. After the discovery of the transistor in 1947, almost 40 years passed before the introduction of the multimegabyte memory chip that is critical to today's computers. If history serves as a guide, the wonderland of high-temperature superconductor applications is destined to become reality in the foreseeable future, and many opportunities to participate in bringing that reality about will be available.

Related Exercises: 5.47, 5.48, and 5.83

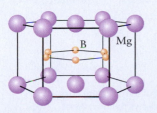

The unit cell of MgB_2 has six boron atoms in the center of a hexagonal array of magnesium atoms. The superconductivity appears to stem from the high-energy vibrational modes of the planes of boron atoms that extend throughout the crystal.

*This box contains contributions from C.W. Chu, Texas Center for Superconductivity, University of Houston.

FIGURE 5.23 The close-packed stacks of apples, oranges, and other produce in a grocery display illustrate how atoms stack together in metals to form crystals with flat faces.

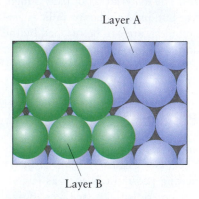

Layer A

Layer B

FIGURE 5.24 A close-packed structure can be built up in stages. The first layer (A) is laid down with minimum waste of space, and the second layer (B) lies in the dips—the depressions—between the spheres of the first layer. Each sphere is touching six other spheres in its layer, as well as three in the layer below and three in the layer above.

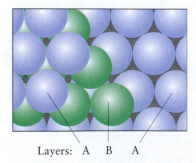

Layers: A   B   A

FIGURE 5.25 When the spheres in the third layer lie directly above the spheres of the first layer, an ABABAB. . . structure is formed.

5.11 Metallic Solids

The cations in a metal are bound together by their interaction with the sea formed by the electrons that they have lost (recall Fig. 1.53). For instance, metallic sodium consists of Na^+ ions held together by electrons that spread throughout the solid, with one electron for each cation. Because the interaction between the ions and the electrons is the same in any direction, the arrangement of the cations can be modeled as hard spheres stacked together. It turns out that we can explain the structures and properties of many metals if we suppose that the spheres representing the cations adopt a **close-packed structure,** in which the spheres stack together with the least waste of space, like oranges in a display (Fig. 5.23).

To see how to stack identical spheres together to give a close-packed structure, look at Fig. 5.24. In the first layer (A) each sphere lies at the center of a hexagon of other spheres. The spheres of the second (upper) layer (B) lie in the dips of the first layer. The third layer of spheres will lie in the dips of the second layer, with the pattern repeating over and over again.

The third layer of spheres may be added in either of two ways. In the first arrangement, the spheres lie in the dips that are directly above the spheres of the first layer (Fig. 5.25). The third layer then duplicates layer A, the next layer duplicates B, and so on. This procedure results in an ABABAB. . . pattern of layers called a **hexagonal close-packed structure** (hcp). The hexagonal pattern in the arrangement of atoms can be seen in Fig. 5.26. Magnesium and zinc are examples of metals that crystallize in this way. As can be seen from Fig. 5.26, each sphere has three nearest neighbors in the plane below, six in its own plane, and three in the plane above, giving twelve in all. This arrangement is reported by saying that the **coordination number,** the number of nearest neighbors of each atom, in the solid is 12. It is impossible to pack identical spheres together with a coordination number greater than 12.

In the second arrangement, the spheres of the third layer lie in the dips of the second layer that do *not* lie directly over the atoms of the first layer (Fig. 5.27). If we call this third layer C, the resulting structure has an ABCABC. . .pattern of layers to give a **cubic close-packed structure** (ccp, Fig. 5.28), because the atoms form cubes when viewed at an angle to the layers. A ccp structure can be thought of as many of these tiny cubes repeating over and over again in all directions. The coordination number is also 12: each sphere has three nearest neighbors in the layer below, six in its own layer, and three in the layer above. Aluminum, copper, silver, and gold are examples of metals that crystallize in this way.

Even in a close-packed structure, hard spheres do not fill all the space in a crystal. The gaps—the interstices—between the atoms are called "holes." To determine just how much space is occupied, we need to calculate the fraction of the total volume occupied by the spheres.

HOW DO WE DO THAT?

To calculate the fraction of occupied space in a close-packed structure, we consider a ccp structure. We can use the radius of the atoms to find the volume of the cube and how much of that volume is taken up by atoms. First, we look at how the cube is built from the atoms. In Fig. 5.29, we see that the corners of the cubes are at the centers of eight atoms. Only 1/8 of each corner atom projects into the cube, so the corner atoms collectively contribute $8 \times 1/8 = 1$ atom to the cube. There is half an atom on each of the six faces, so the atoms on each face contribute $6 \times 1/2 = 3$ atoms, giving four

FIGURE 5.26 A fragment of the structure formed as described in Fig. 5.25 shows the hexagonal symmetry of the arrangement—and the origin of its name, "hexagonal close-packed."

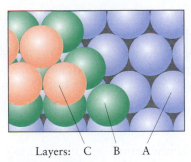

Layers: C B A

FIGURE 5.27 In an alternative to the scheme shown in Fig. 5.25, the spheres of the third layer can lie in the dips in the second layer that are directly above the dips in the first layer, to give an ABCABC. . . arrangement of layers.

FIGURE 5.28 A fragment of the structure constructed as described in Fig. 5.27. This fragment shows the origin of the names "cubic close-packed" or "face-centered cubic" for this arrangement. The layers A, B, and C can be seen along the diagonals of the faces of the cube and are indicated by the different colors of the atoms.

atoms in all within the cube. The length of the diagonal of the face of the cube shown in Fig. 5.29 is $4r$, where r is the atomic radius. Each of the two corner atoms contributes r and the atom at the center of the face contributes $2r$. From the Pythagorean theorem, we know that the length of the side of the face, a, is related to the diagonal by $a^2 + a^2 = (4r)^2$, or $2a^2 = 16r^2$, and so $a = 8^{1/2}r$. The volume of the cube is therefore $a^3 = 8^{3/2}r^3$. The volume of each atom is $\frac{4}{3}\pi r^3$; so the total volume of the atoms inside the cube is $4 \times \frac{4}{3}\pi r^3 = \frac{16}{3}\pi r^3$. The ratio of this occupied volume to the total volume of the cube is therefore

$$\frac{\text{Total volume of spheres}}{\text{Total volume of cube}} = \frac{(16/3)\pi r^3}{8^{3/2}r^3} = \frac{16\pi}{3 \times 8^{3/2}} = 0.74$$

We have shown that 74% of the crystal's space is occupied by atoms and 26% is considered to be empty. Because the hcp structure has the same coordination number of 12, we know that it is packed as densely and must have the same fraction of occupied space.

The holes in the close-packed structure of a metal can be filled with smaller atoms to form alloys (alloys are described in more detail in Section 5.15). If a dip between three atoms is directly covered by another atom, we obtain a **tetrahedral hole**, because it is formed by four atoms at the corners of a regular tetrahedron (Fig. 5.30a). There are two tetrahedral holes per atom in a close-packed lattice. When a dip in a layer coincides with a dip in the next layer, we obtain an **octahedral hole**, because it is formed by six atoms at the corners of a regular octahedron (Fig. 5.30b). There is one octahedral hole for each atom in the lattice. Note that, because holes are formed by two adjacent layers and because neighboring close-packed layers have identical arrangements in hcp and ccp, the numbers of holes are the same for both close-packed structures.

Which close-packed structure—if either—is adopted by a metal depends on which has the lower energy, and that in turn depends on details of its electronic

The Pythagorean theorem states that the square of the hypotenuse of a right-angled triangle is equal to the sums of the squares of the other two sides. That is, if the hypotenuse is c and the other two sides are a and b, then $a^2 + b^2 = c^2$.

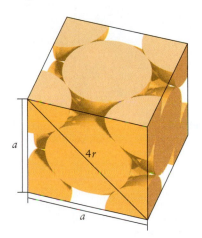

FIGURE 5.29 The relation of the dimensions of a face-centered cubic unit cell to the radius, r, of the spheres. The spheres are in contact along a face diagonal.

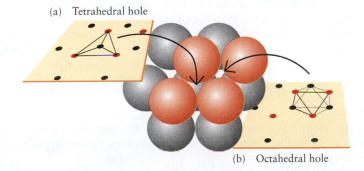

(a) Tetrahedral hole

(b) Octahedral hole

FIGURE 5.30 The locations of (a) tetrahedral and (b) octahedral holes. Note that both types of holes are defined by two neighboring close-packed layers, so they are present with equal abundance in both hcp and ccp structures.

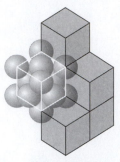

FIGURE 5.31 The entire crystal structure is constructed from a single type of unit cell by stacking the cells together without any gaps.

FIGURE 5.32 The body-centered cubic (bcc) structure. This structure is not packed as closely as the others that we have illustrated. It is less common among metals than the close-packed structures. Some ionic structures are based on this model.

FIGURE 5.33 A primitive cubic unit cell has an atom at each corner. It is rarely found in metals.

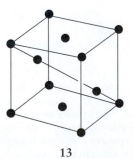

13

structure. In fact, some elements achieve a lower energy by adopting a different arrangement altogether, as we see in the next section.

> *Many metals have close-packed structures, with the atoms stacked in either a hexagonal or a cubic arrangement; close-packed atoms have a coordination number of 12. Close-packed structures have one octahedral and two tetrahedral holes per atom.*

5.12 Unit Cells

We do not have to draw the complete crystal lattice every time we want to convey the structure of a solid. Instead, we can focus our attention on a small region of the crystal that is representative of the entire crystal. The small group illustrated in Fig. 5.28 is an example of a **unit cell,** the smallest unit that, when stacked together repeatedly without any gaps and without rotations, can reproduce the entire crystal (Fig. 5.31). A cubic close-packed unit cell like that illustrated in Fig. 5.31 has an atom at the center of each face of the unit cell; for this reason, it is also called a **face-centered cubic structure** (fcc). In a **body-centered cubic structure** (bcc), a single atom lies at the center of a cube formed by eight other atoms (Fig. 5.32). This structure is not close packed, and metals that have a body-centered cubic structure can often be forced under pressure into a close-packed form. Iron, sodium, and potassium are examples of metals that crystallize with bcc structures. A **primitive cubic structure** has an atom at each corner of a cube. The atoms touch along the edges (Fig. 5.33). This structure is known for only one element, polonium: covalent forces are so strong in this metalloid that they overcome the tendency toward close packing characteristic of metallic bonding. Unit cells are drawn by representing each atom by a dot that marks the location of the atom's center (Fig. 5.34).

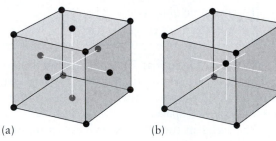

FIGURE 5.34 The unit cells of the (a) ccp (fcc) and (b) bcc structures in which the locations of the centers of the spheres are marked by dots.

(a)   (b)

All crystal structures can be expressed in terms of only 14 basic patterns of unit cell called **Bravais lattices** (Fig. 5.35). We can begin to understand why the number of types of unit cells is limited to such a small number by recognizing that regular pentagonal shapes are absent from the Bravais lattices: regular pentagons cannot cover space without gaps (Fig. 5.36). Similarly, regular heptagonal (seven-sided) and higher polygonal shapes cannot be stacked together to cover all space; hence they do not occur among the Bravais lattices.

The number of atoms in a unit cell is counted by noting how they are shared between neighboring cells. For example, an atom at the center of a cell belongs entirely to that cell, but one on a face is shared between two cells and counts as one-half an atom. As noted earlier, for an fcc structure, the eight corner atoms contribute $8 \times 1/8 = 1$ atom to the cell. The six atoms at the centers of faces contribute $6 \times 1/2 = 3$ atoms (Fig. 5.37). The total number of atoms in an fcc unit cell is therefore $1 + 3 = 4$, and the mass of the unit cell is four times the mass of one atom. For a bcc unit cell (like that in Fig. 5.34b), we count 1 for the atom at the center and 1/8 for each of the eight atoms at the vertices, giving $1 + (8 \times 1/8) = 2$ overall.

SELF-TEST 5.4A How many atoms are there in a primitive cubic cell (see Fig. 5.33)?

[*Answer:* 1]

SELF-TEST 5.4B How many atoms are there in the structure made up of unit cells like the one shown in (**13**), which has an atom at each corner, two on opposite faces, and two inside the cell on a diagonal?

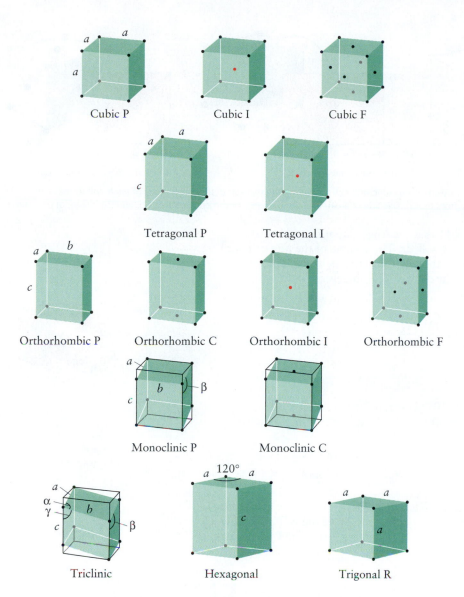

Cubic P   Cubic I   Cubic F

Tetragonal P   Tetragonal I

Orthorhombic P   Orthorhombic C   Orthorhombic I   Orthorhombic F

Monoclinic P   Monoclinic C

Triclinic   Hexagonal   Trigonal R

FIGURE 5.35 The 14 Bravais lattices. P denotes primitive; I, body-centered; F, face-centered; C, with a lattice point on two opposite faces; and R, rhombohedral (a rhomb is an oblique equilateral parallelogram).

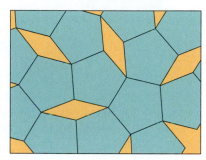

FIGURE 5.36 A plane surface cannot be covered by regular pentagons without leaving gaps. The same is true of regular heptagons.

The best way to determine the type of unit cell adopted by a metal is x-ray diffraction, which gives a characteristic diffraction pattern for each type of unit cell (see Major Technique 3 following his chapter). However, a simpler procedure that can be used to distinguish between close-packed and other structures is to measure the density of the metal; we then calculate the densities of the candidate unit cells and decide which structure accounts for the observed density. Density is an intensive property, which means that it does not depend on the size of the sample (Section A). Therefore, it is the same for a unit cell and a bulk sample. Hexagonal and cubic close packing cannot be distinguished in this way, because they have the same coordination numbers and therefore the same densities (for a given element).

EXAMPLE 5.3 **Deducing the structure of a metal from its density**

The density of copper is 8.93 g·cm^{-3} and its atomic radius is 128 pm. Is the metal (a) close-packed or (b) body-centered cubic?

STRATEGY We calculate the density of the metal by assuming first that its structure is ccp (fcc) and then that it is bcc. The structure with the density closer to the experimental value is more likely to be the actual structure. The mass of a unit cell is the sum of the masses of the atoms that it contains. The mass of each atom is equal to the molar mass of the element divided by Avogadro's constant. The volume of a cubic unit cell is the cube of the length of one of its sides. That length is obtained from the radius of the metal atom, the Pythagorean theorem, and the geometry of the cell.

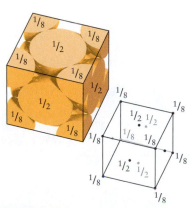

FIGURE 5.37 The calculation of the net number of atoms in a face-centered cubic cell.

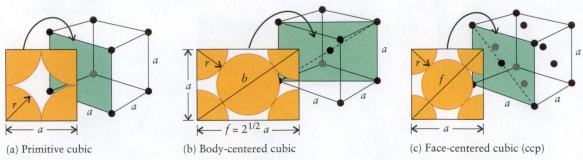

(a) Primitive cubic   (b) Body-centered cubic   (c) Face-centered cubic (ccp)

FIGURE 5.38 The geometries of three cubic unit cells, showing the relation of the dimensions of each cell to the radius, r, of a sphere representing an atom or ion. The side of a cell is a, the diagonal of the body of a cell b, and the diagonal of a face f.

SOLUTION (a) We established in Section 5.11 that the length, a, of the side of an fcc unit cell composed of spheres of radius r is $a = 8^{1/2}r$. The volume of the unit cell is a^3 (Fig. 5.38c). Because there are four atoms in the cell, the mass, m, of one unit cell is four times the mass of one atom (M/N_A). The density, d, is therefore

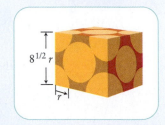

From $d = m/a^3$, $m = 4M/N_A$, and $a = 8^{1/2}r$,
$$d = \frac{4M/N_A}{(8^{1/2}r)^3} = \frac{4M}{8^{3/2}N_A r^3}$$

A radius of 128 pm corresponds to 1.28×10^{-8} cm, and the molar mass of copper (from the periodic table on the inside front cover) is 63.55 g·mol^{-1}. The predicted density is therefore

From $d = 4M/(8^{3/2}N_A r^3)$,
$$d = \frac{4 \times (63.55 \text{ g} \cdot \text{mol}^{-1})}{8^{3/2} \times (6.022 \times 10^{23} \text{ mol}^{-1}) \times (1.28 \times 10^{-8} \text{ cm})^3} = 8.90 \text{ g·cm}^{-3}$$

(b) To calculate the density of a bcc unit cell, we let the length of the diagonal of a face of the cell be f and the length of the diagonal through the body of the cell be b. Then, from Fig. 5.38b and the Pythagorean theorem, $a^2 + f^2 = b^2 = (4r)^2$. The Pythagorean theorem also tells us that $f^2 = 2a^2$, and so

$$a^2 + f^2 = a^2 + 2a^2 = 3a^2$$

It follows that $3a^2 = (4r)^2$ and therefore that $a = 4r/3^{1/2}$. Each unit cell contains one sphere at each of the eight corners and one sphere at the center: $8 \times 1/8 + 1 = 2$ spheres; so the total mass of a body-centered cubic unit cell is $2M/N_A$. Therefore,

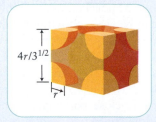

From $d = m/a^3$, $m = 2M/N_A$, and $a = 4r/3^{1/2}$,
$$d = \frac{2M/N_A}{(4r/3^{1/2})^3} = \frac{3^{3/2}M}{32N_A r^3}$$

When we insert numerical values, we obtain

From $d = 3^{3/2}M/(32N_A r^3)$,
$$d = \frac{3^{3/2} \times (63.55 \text{ g} \cdot \text{mol}^{-1})}{32 \times (6.022 \times 10^{23} \text{ mol}^{-1}) \times (1.28 \times 10^{-8} \text{ cm})^3} = 8.17 \text{ g·cm}^{-3}$$

The value is farther from 8.93 g·cm^{-3}, the experimental value, than that for a close-packed structure, 8.90 g·cm^{-3}, and so we conclude that copper has a close-packed structure.

SELF-TEST 5.5A The atomic radius of silver is 144 pm and its density is 10.5 g·cm^{-3}. Is the structure close-packed or body-centered cubic?

[*Answer:* Close-packed cubic]

SELF-TEST 5.5B The atomic radius of iron is 124 pm and its density is 7.87 g·cm^{-3}. Is this density consistent with a close-packed or a body-centered cubic structure?

All crystal structures are derived from the 14 Bravais lattices. The atoms in a unit cell are counted by determining what fraction of each atom resides within the cell. The type of unit cell adopted by a metal can be determined by measuring its density.

5.13 Ionic Structures

As we have seen, we can model elemental metals by spheres of the same radius. However, to model ionic solids, we need to pack together spheres of different radii and opposite charges. Sodium chloride, for instance, is modeled by stacking together positively charged spheres of radius 102 pm, representing the Na^+ ions, and negatively charged spheres of radius 181 pm, representing the Cl^- ions. Because the crystal is electrically neutral overall, each unit cell must reflect the stoichiometry of the compound and itself be electrically neutral.

A helpful starting point is one of the close-packed structures we have already considered. Because anions are usually larger than the accompanying cations, we think of the anions as forming a slightly expanded version of a close-packed structure with the smaller cations occupying some of the enlarged holes in the expanded lattice. A slightly enlarged tetrahedral hole is still relatively small and can accommodate only small cations. Octahedral holes are larger and can accommodate somewhat bigger cations.

The **rock-salt structure** is a common ionic structure that takes its name from the mineral form of sodium chloride. In it, the Cl^- ions lie at the corners and in the centers of the faces of a cube, forming a face-centered cube (Fig. 5.39). This arrangement is like an expanded ccp arrangement: the expansion keeps the anions out of contact with one another, thereby reducing their repulsion, and opens up holes that are big enough to accommodate the Na^+ ions. These ions fit into the octahedral holes between the Cl^- ions. There is one octahedral hole for each anion in the close-packed array, and so all the octahedral holes are occupied. If we look carefully at the structure, we can see that each cation is surrounded by six anions and each anion is surrounded by six cations. The pattern repeats over and over, with each ion surrounded by six other ions of the opposite charge (Fig. 5.40). A crystal of sodium chloride is a three-dimensional array of a vast number of these little cubes.

In an ionic solid, the "coordination number" means the number of ions of *opposite* charge immediately surrounding a specific ion. In the rock-salt structure, the coordination numbers of the cations and the anions are both 6, and the structure overall is described as having *(6,6)-coordination*. In this notation, the first number is the cation coordination number and the second is that of the anion. The rock-salt structure is found for a number of other minerals having ions of the same charge number, including KBr, RbI, MgO, CaO, and AgCl. It is common whenever the cations and anions have very different radii, in which case the smaller cations can fit into the octahedral holes in a face-centered cubic array of anions. The **radius ratio**, ρ (rho), which is defined as

$$\text{Radius ratio} = \frac{\text{radius of smaller ion}}{\text{radius of larger ion}} \quad \text{or} \quad \rho = \frac{r_{\text{smaller}}}{r_{\text{larger}}} \qquad (8)^*$$

is a guide to the type of structure to expect. Although there are many exceptions, a rock-salt structure can be expected when the radius ratio is in the range from 0.4 to 0.7.

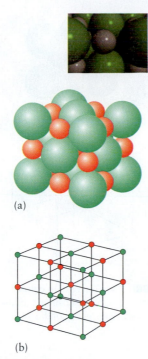

Animation: Figure 5.39 Sodium chloride

(a)

(b)

FIGURE 5.39 The arrangement of ions in the rock-salt structure. (a) The unit cell, showing the packing of the individual ions, and (b) a representation of the same structure by dots that identify the centers of the ions.

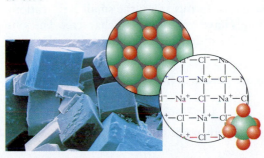

FIGURE 5.40 Billions of unit cells stack together to recreate the smooth faces of the crystal of sodium chloride seen in this micrograph. The first inset shows some of the stacked unit cells. The second inset identifies the individual ions. The third inset (lower right) illustrates the coordination of an anion to its six cation neighbors.

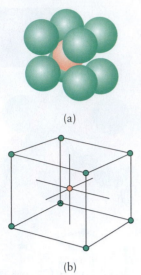

(a)

(b)

FIGURE 5.41 The cesium chloride structure: (a) the unit cell and (b) the location of the centers of the ions.

When the radii of the cations and anions are similar and $\rho > 0.7$, more anions can fit around each cation. Now the ions may adopt the **cesium chloride structure** typified by cesium chloride, CsCl (Fig. 5.41). The radius of a Cs^+ ion is 167 pm and that of a Cl^- ion is 181 pm, giving a radius ratio of 0.923; so the two ions are almost the same size. In this structure, the anions form an expanded primitive cubic array, with a Cl^- ion at the eight corners of each cubic unit cell. There is a large "cubic" hole at the center of the cell, and the Cs^+ ion fits into it. Equivalently, each Cl^- ion can be regarded as being at the center of a cubic unit cell with eight Cs^+ ions at its corners, and so overall the crystal can be thought of as built from two interpenetrating primitive cubic lattices (Fig. 5.42). The coordination number of each type of ion is 8, and overall the structure has *(8,8)-coordination*. The cesium chloride structure is much less common than the rock-salt structure, but it is also found for CsBr, CsI, TlCl, and TlBr.

When the radius ratio of an ionic compound is less than about 0.4, corresponding to cations that are significantly smaller than the anion, the small tetrahedral holes may be occupied. An example is the **zinc-blende structure** (which is also called the *sphalerite structure*), named after a form of the mineral ZnS (Fig. 5.43). This structure is based on an expanded cubic close-packed lattice of the big S^{2-} anions, with the small Zn^{2+} cations occupying half the tetrahedral holes. Each Zn^{2+} ion is surrounded by four S^{2-} ions, and each S^{2-} ion is surrounded by four Zn^{2+} ions; so the zinc-blende structure has *(4,4)-coordination*.

> **EXAMPLE 5.4**   Sample exercise: predicting a structure from the radius ratio rule

Predict the likely structure of solid sodium chloride.

SOLUTION   We base the answer on the radius-ratio rule:

From $\rho = r_{smaller}/r_{larger}$,

$$\rho = \frac{\overbrace{102\ pm}^{radius\ of\ Na^+}}{\underbrace{181\ pm}_{radius\ of\ Cl^-}} = 0.564$$

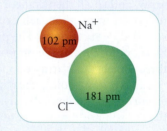

The ratio is consistent with its rock-salt structure.

SELF-TEST 5.6A Predict the likely structure of solid ammonium chloride. Assume that the ammonium ion can be approximated as a sphere with a radius of 151 pm.

[*Answer:* cesium chloride structure]

SELF-TEST 5.6B Predict the likely structure of solid calcium sulfide.

When we discuss the structures of ionic solids, we usually treat ions as spheres of the appropriate radius and stack them together in the arrangement that has the lowest total energy. However, the simple sphere-packing arrangement may break down if the bonding is not purely ionic. In cases where the bonding has a significant degree of covalent character, the ions will lie in specific positions around one another. An example is nickel arsenide, NiAs. In this solid, the small Ni^{3+} cations polarize the big As^{3-} anions, and the bonds have some covalent character. The ions pack together in an arrangement that is quite different from that of a purely ionic, sphere-packing model (Fig. 5.44).

SELF-TEST 5.7A (a) How many Ag^+ ions are present in the AgCl unit cell? (b) What is the coordination type of AgCl?

[*Answer:* (a) 4; (b) (6,6)-coordination]

SELF-TEST 5.7B (a) How many Cl^- ions are there in one unit cell of CsCl? (b) What is the coordination type of CsCl?

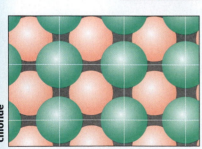

FIGURE 5.42 The repetition of the cesium chloride unit cell recreates the entire crystal. This view is from one side of the crystal and shows several unit cells stacked together.

Ions stack together in the regular crystalline structure corresponding to lowest energy. The structure adopted depends on the radius ratio of cation and anion. Covalent character in an ionic bond imposes a directional character on the bonding.

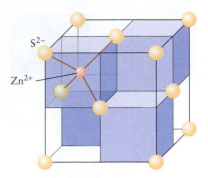

FIGURE 5.43 The zinc-blende (sphalerite) structure. The four zinc ions (pink) form a tetrahedron within a face-centered cubic unit cell composed of sulfide ions (yellow). The zinc ions occupy half the tetrahedral holes between the sulfide ions, and the parts of the unit cell occupied by zinc ions are shaded blue. The detail shows how each zinc ion is surrounded by four sulfide ions; each sulfide ion is similarly surrounded by four zinc ions.

THE IMPACT ON MATERIALS

The properties of materials are consequences of their structures at the molecular level. Solids are the mainstays of technology, and it is hardly surprising that so much effort has gone into the development and understanding of their properties. We dealt with their electrical properties in Sections 3.13 and 3.14. Here we explore some of their other physical properties as well as the properties of the much softer materials known as liquid crystals.

5.14 The Properties of Solids

Solids are dense forms of matter because their atoms, ions, and molecules are packed closely together. Metals are often denser than other kinds of solids because their atoms are packed more closely together. Ionic solids have higher melting points than molecular solids because interionic forces are much stronger than intermolecular forces. Network solids (such as diamond) have very high melting points because they do not melt until the covalent bonds between their atoms have been overcome. The metallic bond is relatively strong. As a result, most metals have high melting points and serve as tough, strong materials for construction.

All metals conduct electricity on account of the mobility of the electrons that bind the atoms together. Ionic, molecular, and network solids are typically electrical insulators or semiconductors (see Sections 3.13 and 3.14), but there are notable exceptions, such as high-temperature superconductors, which are ionic or ceramic solids (see Box 5.2), and there is currently considerable interest in the electrical conductivity of some organic polymers (see Box 19.1).

The characteristic luster of metals is due to the mobility of their electrons (Section 3.13). An incident light wave is an oscillating electromagnetic field. When it strikes the surface, the electric field of the radiation pushes the mobile electrons backward and forward. These oscillating electrons radiate light, and we see it as a luster—essentially a re-emission of the incident light (Fig. 5.45). The electrons oscillate in step with the incident light, and so they generate light of the same frequency. In other words, red light reflected from a metallic surface is red, and blue light is reflected as blue. That is why an image in a mirror—a thin metallic coating on glass—is a faithful portrayal of the reflected object.

The mobility of its electrons explains a metal's **malleability**, the ability to be hammered into shape, and its **ductility**, the ability to be drawn into wires. Because

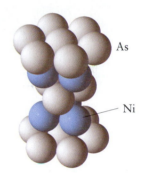

FIGURE 5.44 The structure of nickel arsenide, NiAs. Atypical structures such as this one are often found when the covalent character of the bonding is important and the ions have to take up specific positions relative to one another to maximize their bonding.

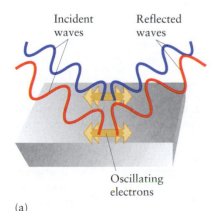

(a)

(b)

FIGURE 5.45 (a) When light of a particular color shines on the surface of a metal, the electrons at the surface oscillate in step. This oscillating motion gives rise to an electromagnetic wave that we perceive as the reflection of the source. (b) Each of these solar mirrors at Sandia National Laboratories in California is positioned at the best angle to reflect sunlight into a collector that uses the incident energy to generate electricity.

Animation: Figure 5.46 Malleability

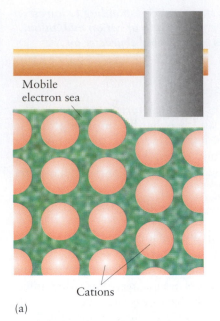

Mobile electron sea

Cations

(a)

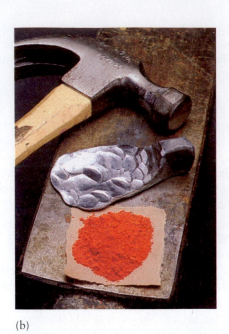

(b)

FIGURE 5.46 (a) When a metal's cations are displaced by a blow from a hammer, the mobile electrons can immediately respond and follow the cations to their new positions, and consequently the metal is malleable. (b) This piece of lead has been flattened by a hammer, whereas crystals of the orange-colored ionic compound lead(II) oxide have shattered.

the cations are surrounded by an electron sea, there is very little directional character in the bonding. As a result, a cation can be pushed past its neighbors without much effort. A blow from a hammer can drive large numbers of cations past their neighbors. The electron sea immediately adjusts, so that the atoms move relatively easily into their new positions (Fig. 5.46). Metals are more malleable than other solids: when groups of atoms are moved from one location to another, the electron sea follows them. Ionic solids are commonly brittle: the interaction between ions is lost when a group of them is moved by a hammer blow (Section 2.4).

The differing malleabilities of metals can be traced to their crystal structures. The crystal structure of a metal typically has **slip planes,** which are planes of atoms that under stress may slip or slide relative to one another. The slip planes of a ccp structure are the close-packed planes, and careful inspection of a unit cell shows that there are eight sets of slip planes in different directions. As a result, metals with cubic close-packed structures, such as copper, are malleable: they can be easily bent, flattened, or pounded into shape. In contrast, a hexagonal close-packed structure has only one set of slip planes, and metals with hexagonal close packing, such as zinc or cadmium, tend to be relatively brittle.

The mobility of the valence electrons in a metal accounts for its electrical conductivity, luster, malleability, and ductility.

5.15 Alloys

Alloys are discussed further in Sections 16.13 and 16.14.

Alloys are metallic materials prepared by mixing two or more molten metals. They are used for many purposes, such as construction, and are central to the transportation and electronics industries. Some common alloys are listed in Table 5.5. In **homogeneous alloys,** atoms of the different elements are distributed uniformly. Examples include brass, bronze, and the coinage alloys. **Heterogeneous alloys,** such as tin–lead solder and the mercury amalgam sometimes used to fill teeth, consist of a mixture of crystalline phases with different compositions.

The structures of alloys are more complicated than those of pure metals because they are built from atoms of two or more elements with different atomic radii. The packing problem is like that of a storekeeper trying to stack oranges and grapefruit in the same pile.

Because the metallic radii of the *d*-block elements are all similar, they can form an extensive range of alloys with one another with little distortion of the original crystal structure. An example is the copper–zinc alloy used for some "copper" coins. Because zinc atoms are nearly the same size as copper atoms and have simi-

TABLE 5.5   Compositions of Typical Alloys

Alloy	Mass percentage composition
brass	up to 40% zinc in copper
bronze	a metal other than zinc or nickel in copper (casting bronze: 10% Sn and 5% Pb)
cupronickel	nickel in copper (coinage cupronickel: 25% Ni)
pewter	6% antimony and 1.5% copper in tin
solder	tin and lead
stainless steel*	more than 12% chromium in iron

*For more detailed information on steels, see Tables 16.2 and 16.6.

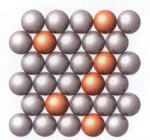

FIGURE 5.47 In a substitutional alloy, the positions of some of the atoms of one metal are taken by atoms of another metal. The two elements must have similar atomic radii.

lar electronic properties (they belong to neighboring groups), they can take the place of some of the copper atoms in the crystal. An alloy in which atoms of one metal are substituted for atoms of another metal is called a **substitutional alloy** (Fig. 5.47). Elements that can form substitutional alloys have atoms with atomic radii that differ by no more than about 15%. Because there are slight differences in size and electronic structure, the less abundant atoms in a substitutional alloy distort the shape of the lattice of the more abundant atoms of the host metal and hinder the flow of electrons. Because the lattice is distorted, it is harder for one plane of atoms to slip past another. Therefore, although a substitutional alloy has lower electrical and thermal conductivity than the pure element, it is harder and stronger.

Steel is an alloy of about 2% or less carbon in iron. Carbon atoms are much smaller than iron atoms, and so they cannot substitute for iron in the crystal lattice. Indeed, they are so small that they can fit into the **interstices** (the holes) in the iron lattice. The resulting material is called an **interstitial alloy** (Fig. 5.48). For two elements to form an interstitial alloy, the atomic radius of the solute element must be less than about 60% of the atomic radius of the host metal. The interstitial atoms interfere with electrical conductivity and with the movement of the atoms forming the lattice. This restricted motion makes the alloy harder and stronger than the pure host metal would be.

Some alloys are softer than the component metals. The presence of big bismuth atoms helps to soften a metal and lower its melting point, much as melons would destabilize a stack of oranges because they just do not fit together well. A low-melting-point alloy of lead, tin, and bismuth is employed to control water sprinklers used in certain fire-extinguishing systems. The heat of the fire melts the alloy, which activates the sprinklers before the fire can spread.

Alloys of metals tend to be stronger and have lower electrical conductivity than pure metals. In substitutional alloys, atoms of the solute metal take the place of some atoms of a metal of similar atomic radius. In interstitial alloys, atoms of the solute element fit into the interstices in a lattice formed by atoms of a metal with a larger atomic radius.

5.16 Liquid Crystals

One type of material that has transformed electronic displays is neither a solid nor a liquid, but something intermediate between the two. **Liquid crystals** are substances that flow like viscous liquids, but their molecules lie in a moderately orderly array, like those in a crystal. They are examples of a **mesophase,** an intermediate state of matter with the fluidity of a liquid and some of the molecular order of a solid. Liquid crystalline materials are finding many applications in the electronics industry because they are responsive to changes in temperature and electric fields.

A typical liquid-crystal molecule, such as *p*-azoxyanisole, is long and rodlike (**14**). Their rodlike shape enables the molecules to stack together like dry, uncooked spaghetti: they lie parallel to one another but are free to slide past one another along their long axes. Liquid crystals are anisotropic because of this ordering. **Anisotropic** materials have properties that depend on the direction of measurement. The viscosity of liquid crystals is least in the direction parallel to the long

Although carbon is not a metal, it is usual to regard steel as an alloy: all commercial steels contain other metals as well as iron and carbon.

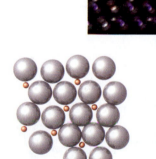

FIGURE 5.48 In an interstitial alloy, the atoms of one metal lie in the gaps between the atoms of another metal. The two elements need to have markedly different atomic radii.

Media Link Animation: Figure 5.48 Stainless steel

14   *p*-Azoxyanisole

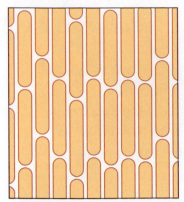

FIGURE 5.49 A representation of the nematic phase of a liquid crystal. The long molecules lie parallel to one another but are staggered along their long axes.

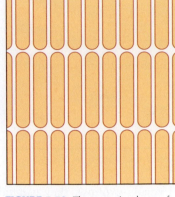

FIGURE 5.50 The smectic phase of a liquid crystal. Not only do the molecules lie parallel to one another, they also line up next to one another to form sheets.

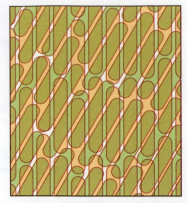

FIGURE 5.51 The cholesteric phase of a liquid crystal. In this phase, sheets of parallel molecules are rotated relative to their neighbors and form a helical structure.

Nematic comes from the Greek word for "thread"; *smectic* comes from the Greek word for "soapy"; *cholesteric* is related to the word *cholesterol,* which comes from the Greek words for "bile solid."

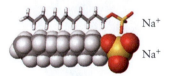

15   Sodium lauryl sulfate

axis of the molecules: it is easier for the long rod-shaped molecules to slip past one another along their axes than to move sideways. **Isotropic materials** have properties that do not depend on the direction of measurement. Ordinary liquids, like water, are isotropic: their viscosities are the same in every direction.

The three classes of liquid crystals differ in the arrangement of their molecules. In the **nematic phase,** the molecules lie together, all in the same direction but staggered, like cars on a busy multilane highway (Fig. 5.49). In the **smectic phase,** the molecules line up like soldiers on parade and form layers (Fig. 5.50). Cell membranes are composed mainly of smectic liquid crystals. In the **cholesteric phase,** the molecules form ordered layers, but neighboring layers have molecules at different angles and so the liquid crystal has a helical arrangement of molecules (Fig. 5.51).

Liquid crystals are also classified by their manner of preparation. **Thermotropic liquid crystals** are made by melting the solid phase. The highly viscous liquid-crystal phase exists over a short temperature range between the solid and the liquid states. Thermotropic liquid crystals become isotropic liquids when they are heated above a characteristic temperature, because then the molecules have enough energy to overcome the attractions that restrict their movement. *p*-Azoxyanisole is a thermotropic liquid crystal. Thermotropic liquid crystals are used in applications such as watches, computer screens, and thermometers. **Lyotropic liquid crystals** are layered structures that result from the action of a solvent on a solid or liquid. Examples are cell membranes and aqueous solutions of detergents and lipids (fats). These molecules, like the detergent sodium lauryl sulfate, have long, nonpolar hydrocarbon chains attached to polar heads (**15**). When the lipids that form cell membranes are mixed with water, they spontaneously form sheets in which the molecules are aligned in rows, forming a double layer, with their polar heads facing outward on each side of the sheet. These sheets form the protective membranes of the cells that make up living tissues (see Fig. 8.44).

Electronic displays make use of the changes in orientation of liquid-crystal molecules caused by an electric field. In an LCD (liquid-crystal display) television or computer monitor, layers of a liquid crystal in a nematic phase lie between the surfaces of two glass or plastic plates. The long axis of the molecules in each layer is parallel to the plates but is oriented in such a way that polarized light (see Section 16.7) is transmitted. However, where a potential difference is applied, the molecules rotate until they are oriented with the electric field and become opaque to the polarized light, hence forming a dark spot on the screen. Cholesteric liquid crystals are also of interest because the helical structure unwinds slightly as the temperature is changed. Because the twist of the helical structure affects the optical properties of the liquid crystal, such as its color, these properties change with temperature. The effect is utilized in liquid-crystal thermometers.

Liquid crystals have a degree of order characteristic of solid crystals, but they can flow like viscous liquids. They are mesophases, intermediate between solids and liquids; their properties can be modified by electric fields and changes in temperature.

5.17 Ionic Liquids

We have seen that molecular substances tend to have low melting points, while network, ionic, and metallic substances tend to have high melting points. Therefore, with a few exceptions, such as mercury, a substance that is liquid at room temperature is likely to be a molecular substance. Liquid solvents are heavily used in industry to extract substances from natural products and to promote the synthesis of desired compounds. Because many of these solvents have high vapor pressures and so give off hazardous fumes, liquids that have low vapor pressures but dissolve organic compounds have been sought.

16   1-Butyl-3-methylimidazolium ion

A new class of solvents called **ionic liquids** has been developed to meet this need. A typical ionic liquid has a relatively small anion, such as BF_4^-, and a relatively large, organic cation, such as 1-butyl-3-methylimidazolium (**16**). Because the cation has a large nonpolar region and is often asymmetrical, the compound does not crystallize easily and so is liquid at room temperature. However, the attractions between the ions reduces the vapor pressure to about the same as that of an ionic solid, thereby reducing air pollution. Because different cations and anions can be used, solvents can be designed for specific uses. For example, one formulation can dissolve the rubber in old tires so that it can be recycled. Other solvents can be used to extract radioactive waste from groundwater.

Ionic liquids are compounds in which one of the ions is a large, organic ion that prevents the liquid from crystallizing at ordinary temperatures. The low vapor pressures of ionic liquids make them desirable solvents that reduce pollution.

SKILLS YOU SHOULD HAVE MASTERED

❏ 1 Predict the relative strengths of ion–dipole interactions (Section 5.2).

❏ 2 Explain how London forces arise and how they vary with the polarizability of an atom and the size and shape of a molecule (Section 5.4).

❏ 3 Predict the relative order of the boiling points of two substances from the strengths of their intermolecular forces (Examples 5.1 and 5.2).

❏ 4 Identify molecules that can take part in hydrogen bonding (Self-Test 5.3).

❏ 5 Describe the structure of a liquid and explain how viscosity and surface tension vary with temperature and the strength of intermolecular forces (Sections 5.6 and 5.7).

❏ 6 Distinguish metallic solids, ionic solids, network solids, and molecular solids by their structures and by their properties (Sections 5.8–5.11 and 5.14).

❏ 7 Determine the fraction of occupied space in a given crystal lattice (Section 5.11).

❏ 8 Give the coordination number of an atom or ion in a given crystal lattice (Section 5.11).

❏ 9 Find the number of atoms or ions in a given unit cell (Self-Test 5.4).

❏ 10 Deduce the crystal structure of a metal from its density (Example 5.3).

❏ 11 Describe the structure of an ionic solid (Section 5.13).

❏ 12 Predict the structure of an ionic solid from the relative radii of the ions (Example 5.4).

❏ 13 Identify the different types of liquid crystals (Section 5.16).

EXERCISES

Exercises labeled C *require calculus.*

Intermolecular Forces

5.1 Identify the kinds of intermolecular forces that might arise between molecules of each of the following substances: (a) HNO_3; (b) N_2H_4; (c) NO; (d) CF_4.

5.2 Identify the kinds of intermolecular forces that might arise between molecules of each of the following substances: (a) CH_3NH_2; (b) PCl_3; (c) PCl_5; (d) H_2Se.

5.3 For which of the following molecules will dipole–dipole interactions be important: (a) CH_4; (b) CH_3Cl; (c) CH_2Cl_2; (d) $CHCl_3$; (e) CCl_4?

5.4 For which of the following molecules will dipole–dipole interactions be important: (a) O_2; (b) O_3; (c) CO_2; (d) SO_2?

5.5 Place the following types of molecular and ion interactions in order of increasing magnitude: (a) ion–dipole; (b) induced-dipole–induced-dipole; (c) dipole–dipole in the gas phase; (d) ion–ion; (e) dipole–dipole in the solid phase.

5.6 Explain why ionic solids such as NaCl have high melting points yet dissolve readily in water, whereas network solids such as diamond have very high melting points and do not dissolve in solvents.

5.7 Which of the following molecules are likely to form hydrogen bonds: (a) H_2S; (b) CH_4; (c) H_2SO_3; (d) PH_3?

5.8 Which of the following molecules are likely to form hydrogen bonds: (a) CH_3OCH_3; (b) CH_3COOH; (c) CH_3CH_2OH; (d) CH_3CHO?

5.9 Identify the arrangement (I, II, or III; all molecules are CH_2Cl_2) that should possess the strongest intermolecular attractions, and justify your selection.

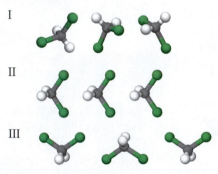

5.10 Identify the arrangement (I, II, or III; all molecules are NH_3) that should possess the strongest intermolecular attractions, and justify your selection.

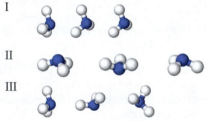

5.11 Suggest, giving reasons, which substance in each of the following pairs is likely to have the higher normal melting point (Lewis structures may help your arguments): (a) HCl or NaCl; (b) $C_2H_5OC_2H_5$ (diethyl ether) or C_4H_9OH (butanol); (c) CHI_3 or CHF_3; (d) C_2H_4 or CH_3OH.

5.12 Suggest, giving reasons, which substance in each of the following pairs is likely to have the higher normal boiling point: (a) H_2S or H_2O; (b) NH_3 or PH_3; (c) KBr or CH_3Br; (d) CH_4 or SiH_4.

5.13 Using the VSEPR model, predict the shapes of each of the following molecules and identify the member of each pair with the higher boiling point: (a) PBr_3 or PF_3; (b) SO_2 or CO_2; (c) BF_3 or BCl_3.

5.14 Using the VSEPR model, predict the shapes of each of the following molecules and identify the member of each pair with the higher boiling point: (a) BF_3 or ClF_3; (b) SF_4 or CF_4; (c) *cis*-CHCl=CHCl or *trans*-CHCl=CHCl. (See structures 5 and 6.)

5.15 Calculate the ratio of the potential energies for the interaction of a water molecule with an Al^{3+} ion and with a Be^{2+} ion. Take the center of the dipole to be located at r_{ion} + 100. pm. Which ion will attract a water molecule more strongly?

5.16 Calculate the ratio of the potential energies for the interaction of a water molecule with a Ca^{2+} ion and with an In^{3+} ion. Take the center of the dipole to be located at r_{ion} + 100. pm. Which ion will attract a water molecule more strongly?

5.17 Account for the following observations in terms of the type and strength of intermolecular forces. (a) The melting point of solid xenon is −112°C and that of solid argon is −189°C. (b) The vapor pressure of diethyl ether ($C_2H_5OC_2H_5$) is greater than that of water. (c) The boiling point of pentane, $CH_3(CH_2)_3CH_3$, is 36.1°C, whereas that of 2,2-dimethylpropane (also known as neopentane), $C(CH_3)_4$, is 9.5°C.

5.18 Two students are assigned two pure compounds to investigate. They observe that compound A boils at 37°C and compound B at 126°C. However, they run out of time to make any additional measurements, so make the following guesses about the two compounds. In each case indicate whether the guess can be justified by the data, is in error, or could be either true or false. Justify your answers. (a) Compound B has the higher molar mass. (b) Compound A is the more viscous. (c) Compound B has stronger intermolecular forces. (d) Compound B has the higher surface tension.

C **5.19** We have been using "intermolecular interaction" and "intermolecular force" almost interchangeably. However, it is important to distinguish the force from the potential energy of interaction. In classical mechanics, the magnitude of the force, F, is related to the distance dependence of the potential energy, E_P, by $F = -dE_P/dr$. How does the intermolecular force depend on separation for a typical intermolecular interaction that varies as $1/r^6$?

5.20 Would you expect the energy of interaction of two rotating polar molecules, as given by Eq. 4, to depend on the temperature? If so, should the interaction increase or decrease as the temperature is raised?

Liquid Structure

5.21 Predict which liquid in each of the following pairs has the greater surface tension: (a) *cis*-dichloroethene or *trans*-dichloroethene (see structures 5 and 6); (b) benzene at 20°C or benzene at 60°C.

5.22 Predict which substance in each of the following pairs has the greater viscosity in its liquid form at 0°C: (a) ethanol, CH_3CH_2OH, or dimethyl ether, CH_3OCH_3; (b) butane, C_4H_{10}, or propanone, CH_3COCH_3.

5.23 Rank the following molecules in order of increasing viscosity at 50°C: C_6H_5SH, C_6H_5OH, C_6H_6.

5.24 Rank the following liquids in order of increasing viscosity at 25°C: C_6H_6, CH_3CH_2OH, $CH_2OHCHOHCH_2OH$, CH_2OHCH_2OH, and H_2O. Explain your ordering.

5.25 The following boiling points correspond to the substances listed. Match the boiling points to the substances by considering the relative strengths of their intermolecular forces. b.p. (°C): −162, −88.5, 28, 36, 64.5, 78.3, 82.5, 140, 205, 290; substance: CH_4, $CH_3CHOHCH_3$, $C_6H_5CH_2OH$ (has a benzene ring), CH_3CH_3, C_5H_9OH (cyclic), $(CH_3)_2CHCH_2CH_3$, CH_3OH, $HOCH_2CHOHCH_2OH$, $CH_3(CH_2)_3CH_3$, CH_3CH_2OH. *Hint:* The boiling point of $(CH_3)_2CHCH_2CH_3$ is 28°C and that of CH_3OH is 64.5°C.

5.26 The following surface tensions (in $mN \cdot m^{-1}$, at 20°C) correspond to the liquids listed. Match the surface tension to the substance. Surface tension: 18.43, 22.75, 27.80, 28.85, 72.75; Compound: H_2O, $CH_3(CH_2)_4CH_3$, C_6H_6, CH_3CH_2OH, CH_3COOH

5.27 Consider Fig. 5.8. (a) Explain the large differences in the boiling points of H_2O and those of the remainder of the series (H_2S—H_2Te). (b) Explain the steady increase in the boiling points in the series CH_4—SnH_4.

5.28 The surface of glass contains many —OH groups that are bonded to the silicon atoms of SiO_2, which is the major component of glass. If the glass is treated with $Si(CH_3)_3Cl$ (chlorotrimethylsilane), a reaction takes place to eliminate HCl and attach the Si atom to the oxygen atom:

$$\text{(surface of glass)}\text{—OH} + Si(CH_3)_3Cl \longrightarrow$$
$$\text{(surface of glass)}\text{—OSi(CH}_3)_3 + HCl$$

How will this rearrangement affect the interaction of liquids with the glass surface?

5.29 The height, h, of a column of liquid in a capillary tube can be estimated by using $h = 2\gamma/gdr$, where γ is the surface tension, d is the density of the liquid, g is the acceleration of free fall, and r is the radius of the tube. Which will rise higher in a tube that is 0.15 mm in diameter at 25°C, water or ethanol? The density of water is $0.997\ g \cdot cm^{-3}$ and that of ethanol is $0.79\ g \cdot cm^{-3}$. See Table 5.3.

5.30 The expression for the capillary rise in Exercise 5.29 assumes that the tube is vertical. How will the expression be modified when the tube is held at an angle θ (theta) to the vertical?

Classification of Solids

5.31 Glucose, benzophenone ($C_6H_5COC_6H_5$), and methane are examples of compounds that form molecular solids. The structures of glucose and benzophenone are given here. (a) What types of forces hold these molecules in a molecular solid? (b) Place the solids in order of increasing melting point.

Glucose

Benzophenone

5.32 Chloromethane (CH_3Cl), methane, and acetic acid (CH_3COOH) form molecular solids. (a) What types of forces hold these molecules in a molecular solid? (b) Place the solids in order of increasing melting point.

5.33 Classify each of the following solids as ionic, network, metallic, or molecular: (a) quartz, SiO_2; (b) limestone, $CaCO_3$; (c) dry ice, CO_2; (d) sucrose, $C_{12}H_{22}O_{11}$; (e) polyethylene, a

polymer with molecules consisting of chains of thousands of repeating —CH_2CH_2— units.

5.34 Classify each of the following solids as ionic, network, metallic, or molecular: (a) iron pyrite (fool's gold), FeS_2; (b) octane (a component of gasoline), C_8H_{18}; (c) cubic boron nitride (a compound with a structure similar to that of diamond, but with alternating boron and nitrogen atoms), BN; (d) calcium sulfate (gypsum), $CaSO_4$; (e) the chromium plate on a motorcycle.

Solid Metallic Structures

5.35 Iron crystallizes in a bcc structure. The atomic radius of iron is 124 pm. Determine (a) the number of atoms per unit cell; (b) the coordination number of the lattice; (c) the length of the side of the unit cell.

5.36 The metal polonium (which was named by Marie Curie after her homeland, Poland) crystallizes in a primitive cubic structure, with an atom at each corner of a cubic unit cell. The atomic radius of polonium is 167 pm. Sketch the unit cell and determine (a) the number of atoms per unit cell; (b) the coordination number of an atom of polonium; (c) the length of the side of the unit cell.

5.37 Calculate the density of each of the following metals from the data given: (a) aluminum, fcc structure, atomic radius 143 pm; (b) potassium, bcc structure, atomic radius 227 pm.

5.38 Calculate the density of each of the following metals from the data given: (a) nickel, fcc structure, atomic radius 125 pm; (b) rubidium, bcc structure, atomic radius 250 pm.

5.39 Calculate the atomic radius of each of the following elements from the data given: (a) platinum, fcc structure, density $21.450\ g \cdot cm^{-3}$; (b) tantalum, bcc structure, density $16.654\ g \cdot cm^{-3}$.

5.40 Calculate the atomic radius of each of the following elements from the data given: (a) silver, fcc structure, density $10.500\ g \cdot cm^{-3}$; (b) chromium, bcc structure, density $7.190\ g \cdot cm^{-3}$.

5.41 One form of silicon has density of $2.33\ g \cdot cm^{-3}$ and crystallizes in a cubic lattice with a unit cell edge of 543 pm. (a) What is the mass of each unit cell? (b) How many silicon atoms does one unit cell contain?

5.42 Krypton crystallizes with a face-centered cubic unit cell of edge 559 pm. (a) What is the density of solid krypton? (b) What is the atomic radius of krypton? (c) What is the volume of one krypton atom? (d) What percentage of the unit cell is empty space if each atom is treated as a hard sphere?

5.43 What percentage of space is occupied by close-packed cylinders of length l and radius r?

5.44 Calculate the radius of the cavity formed by three circular disks of radius r that lie in a close-packed planar arrangement. Check your answer experimentally by using three compact disks and measuring the radius of the cavity that they form.

Nonmetallic Solids

5.45 Calculate the number of cations, anions, and formula units per unit cell in each of the following solids: (a) the cesium chloride unit cell shown in Fig. 5.41; (b) the rutile (TiO_2) unit cell shown here. (c) What are the coordination numbers of the ions in rutile?

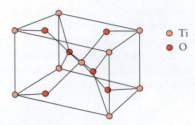

5.46 Calculate the number of cations, anions, and formula units per unit cell in each of the following solids: (a) the rock-salt unit cell shown in Fig. 5.39; (b) the fluorite (CaF_2) unit cell shown here. (c) What are the coordination numbers of the ions in fluorite?

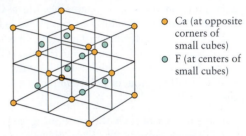

Ca (at opposite corners of small cubes)

F (at centers of small cubes)

5.47 A unit cell of one of the new high-temperature superconductors is shown here. What is its formula?

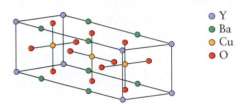

Y
Ba
Cu
O

5.48 A unit cell of the mineral perovskite, which has a structure similar to that of some of the ceramic superconductors, is shown here. What is its formula?

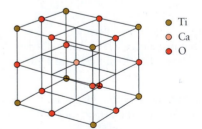

Ti
Ca
O

5.49 Use radius ratios to predict the coordination number of the cation in (a) RbF; (b) MgO; (c) NaBr. (See Fig. 1.48.)

5.50 Use radius ratios to predict the coordination number of the cation in (a) KBr; (b) LiBr; (c) BaO. (See Fig. 1.48.)

5.51 Estimate the density of each of the following solids from the atomic radii of the ions given in Fig. 1.48: (a) calcium oxide (rock-salt structure, Fig. 5.39); (b) cesium bromide (cesium chloride structure, Fig. 5.41).

5.52 Calculate the density of each of the following solids: (a) sodium iodide (rock-salt structure, Fig. 5.39), distance between centers of Na^+ and I^- ions is 322 pm; (b) cesium iodide (cesium chloride structure, Fig. 5.41), distance between the centers of the Cs^+ and I^- ions is 356 pm.

5.53 Graphite forms extended two-dimensional layers (see Fig. 5.22). (a) Draw the smallest possible rectangular unit cell for a layer of graphite. (b) How many carbon atoms are in your unit cell? (c) What is the coordination number of carbon in a single layer of graphite?

5.54 The ammonium ion can be represented by a sphere of radius 151 pm. Use radius ratios to predict the type of lattice structure found in (a) NH_4F; (b) NH_4I.

5.55 If the edge length of a fcc unit cell of RbI is 732.6 pm, how long would the edge of a cubic single crystal of RbI be that contains 1.00 mol RbI?

5.56 The edge length of the fcc unit cell of NaCl is 562.8 pm. (a) How many unit cells are present in a single crystal of NaCl (table salt) that is a cube with edges 1.00 mm in length? (b) What amount (in moles) of NaCl is present in this crystal?

The Impact on Materials

5.57 When iron surfaces are exposed to ammonia at high temperatures, "nitriding"—the incorporation of nitrogen into the iron lattice—occurs. The atomic radius of iron is 124 pm. (a) Is the alloy interstitial or substitutional? Justify your answer. (b) How do you expect nitriding to change the properties of iron?

5.58 Silicon can be doped with small amounts of phosphorus to create a semiconductor used in transistors. (a) Is the alloy interstitial or substitutional? Justify your answer. (b) How do you expect the properties of the doped material to differ from those of pure silicon?

5.59 Calculate the relative number of atoms of each element contained in each of the following alloys: (a) coinage cupronickel, which is 25% Ni by mass in copper; (b) a type of pewter that is about 7% antimony and 3% copper by mass in tin.

5.60 Calculate the relative number of atoms of each element contained in each of the following alloys: (a) Wood's metal, which is a low-melting-point alloy used to trigger automatic sprinkler systems and is 12.5% tin, 12.5% cadmium, and 24% lead by mass in bismuth; (b) a steel that is 1.75% by mass carbon in iron.

5.61 Why do long hydrocarbon molecules that do not have multiple bonds, such as decane, $CH_3(CH_2)_8CH_3$, not form liquid crystals?

5.62 Consider the structure of *p*-azoxyanisole (**14**). (a) Using the VSEPR model, draw a picture that represents the shape of the molecule and predict the CNN bond angles. (b) What features of the bonding of this molecule give rise to its rodlike nature? (c) Use your conclusions in part (b) to design another simple organic molecule that is composed only of C, H, N, and O and might be likely to form rodlike, liquid crystalline materials.

5.63 When long surfactant molecules having a polar headgroup and a nonpolar "tail" are placed into water, *micelles* are formed in which the nonpolar tails aggregate, with the polar headgroups pointing out toward the solvent. *Inverse micelles* are similar but have the nonpolar regions pointing outward. How can inverse micelles be produced?

5.64 The molecule *p*-azoxyanisole (**14**) has a liquid crystalline range of approximately 117° to 137°C. How might this molecule be modified to lower its melting point and make it more suitable for lower-temperature applications (near room temperature, for example)?

5.65 Two solutes were used to study diffusion in liquids, methylbenzene, which is a small molecule that can be approximated as a sphere, and a liquid crystal that is long and rodlike. The two solutes were found to move and rotate in all directions to the same extent in benzene. In a liquid crystal solvent the methylbenzene again moved and rotated to the same extent in all directions, but the liquid crystal solute moved much more rapidly along the long axis of the molecule than it did in a "sideways" mode, perpendicular to the long axis. It also was found to rotate more rapidly around the long axis than perpendicular to it. Explain this behavior.

5.66 The molecular structures of many common liquid crystals are long and rodlike. In addition, they contain polar groups. Explain how both characteristics of liquid crystals contribute to their anisotropic nature.

INTEGRATED EXERCISES

5.67 Draw the Lewis structure of (a) NI_3 and (b) BI_3, name the molecular shape, and indicate whether each can participate in dipole–dipole interactions.

5.68 Draw the Lewis structure of (a) CF_4, (b) SF_4, name the molecular shape, and indicate whether each can participate in dipole–dipole interactions.

5.69 (a) Calculate the surface areas of the isomers 2,2-dimethylpropane and pentane. Assume that 2,2-dimethylpropane is spherical with a radius of 254 pm and that pentane can be approximated by a rectangular prism with dimensions 295 pm $\times$ 692 pm $\times$ 766 pm. (b) Which has the larger surface area? (c) Which do you expect to have the higher boiling point?

5.70 Using the Molecules Database on the Web site for this book, examine the structure of tristearin, formed by the condensation of three molecules of stearic acid, $C_{17}H_{35}COOH$, with glycerol, $C_3H_5(OH)_3$. (a) Draw the Lewis structure of tristearin. (b) What types of forces are responsible for the molecule adopting the shape shown? (c) Do you expect this molecule to be soluble or insoluble in water? For comparison, glycerol is very soluble, but stearic acid is only slightly soluble in water.

5.71 Using the Molecules Database on the Web site for this book, determine the chemical formula and draw the Lewis structure of (a) anthracene; (b) phosgene; (c) glutamic acid. For each molecule, list the types of intermolecular forces expected to be significant.

5.72 (a) Calculate the ratio of potential energies for an ion–ion interaction for Li^+ and for K^+ with the same anion. (b) Repeat the ratio calculation, but for an ion–dipole interaction between Li^+ and K^+ and a water molecule. (c) What do these numbers indicate about the relative importance of the hydration of salts of lithium compared with those of potassium?

5.73 All noble gases except helium crystallize with ccp structures at very low temperatures. Find an equation relating the atomic radius to the density of a ccp solid of given molar mass and apply it to deduce the atomic radius of each of the following noble gases, given the density of each (in $g \cdot cm^{-3}$): Ne, 1.20; Ar, 1.40; Kr, 2.16; Xe, 2.83; Rn, 4.4 (estimated).

5.74 All the alkali metals crystallize with bcc structures. (a) Find a general equation relating the metallic radius to the density of a bcc solid of an element in terms of its molar mass and use it to deduce the atomic radius of each of the following elements, given the density of each (in $g \cdot cm^{-3}$): Li, 0.53; Na, 0.97; K, 0.86; Rb, 1.53; Cs, 1.87. (b) Find a factor for converting the density of a bcc element into the density that it would have if it crystallized in a ccp structure. (c) Calculate what the densities of the alkali metals would be if they were ccp. (d) Which, if any, would float on water?

5.75 Metals with bcc structures, such as tungsten, are not close packed. Therefore, their densities would be greater if they were to change to a ccp structure (under pressure, for instance). What would the density of tungsten be if its structure were ccp rather than bcc? Its actual density is 19.3 $g \cdot cm^{-3}$.

5.76 An oxide of niobium has a unit cell in which there are oxide ions at the middle of each edge and niobium ions at the center of each face. What is the empirical formula of this oxide?

5.77 A number of metal oxides are known to form *nonstoichiometric* compounds, in which the ratios of atoms that make up the compound cannot be expressed in small whole numbers. In the crystal structure of a nonstoichiometric compound, some of the lattice points where one would have expected to find atoms are vacant. Transition metals most easily form nonstoichiometric compounds because of the number of oxidation states that they can have. For example, a titanium oxide with formula $TiO_{1.18}$ is known. (a) Calculate the average oxidation state of titanium in this compound. (b) If the compound has both Ti^{2+} and Ti^{3+} ions present, what fraction of the titanium ions will be in each oxidation state?

5.78 Uranium dioxide, UO_2, can be further oxidized to give a nonstoichiometric compound UO_{2+x}, where $0 < x < 0.25$. See Exercise 5.77 for a description of nonstoichiometric compounds. (a) What is the average oxidation state of uranium in a compound with composition $UO_{2.17}$? (b) If we assume that the uranium exists in either the +4 or the +5 oxidation state, what is the fraction of uranium ions in each?

5.79 Are the following statements true or false? (a) If there is an atom present at the corner of a unit cell, there must be the same type of atom at all the corners of the unit cell. (b) A unit cell must be defined so that there are atoms at the corners. (c) If one face of a unit cell has an atom in its center, then the face opposite that face must also have an atom at its center. (d) If one face of a unit cell has an atom in its center, all the faces of the unit cell must also have atoms at their centers.

5.80 Are the following statements true or false? (a) Because cesium chloride has chloride ions at the corners of the unit cell and a cesium ion at the center of the unit cell, it is classified as having a body-centered unit cell. (b) The density of the unit cell must be the same as the density of the bulk material. (c) When x-rays are passed through a single crystal of a compound, the x-ray beam will be diffracted because it interacts with the electrons in the atoms of the crystal. (d) All the angles of a unit cell must be equal to 90°.

5.81 For each of the two-dimensional arrays shown here, draw a unit cell that, when repeated, generates the entire two-dimensional lattice.

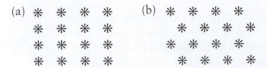

5.82 For each of the two-dimensional arrays shown here, draw a unit cell that, when repeated, generates the entire two-dimensional lattice.

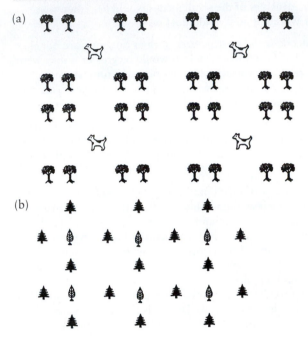

5.83 Buckminsterfullerene is an allotrope of carbon in which the carbon atoms form spheres of 60 atoms each (see Section 14.16). In the pure compound the spheres pack in a cubic close-packed array. (a) The length of a side of the face-centered cubic cell formed by buckminsterfullerene is 142 pm. Use this information to calculate the radius of the buckminsterfullerene molecule treated as a hard sphere. (b) The compound K_3C_{60} is a superconductor at low temperatures. In this compound the K^+ ions lie in holes in the C_{60}^- face-centered cubic lattice. Considering the radius of the K^+ ion and assuming that the radius of C_{60}^- is the same as for the C_{60} molecule, predict in what type of holes the K^+ ions lie (tetrahedral, octahedral, or both) and indicate what percentage of those holes are filled.

5.84 A reflection from a cubic crystal was observed at a glancing angle of 12.1° when x-rays of wavelength 152 pm were used. (a) What is the separation of layers causing this reflection? (b) At what glancing angle would you detect the reflection arising from planes at twice the layer separation? Refer to Major Technique 3 on x-ray diffraction, which follows this set of exercises.

5.85 Glass is composed mainly of silica, SiO_2, with various oxides added to change the properties. The property differences are substantial, and different amounts of added oxide compounds produce glasses with very different applications derived from these properties. Often the added material is B_2O_3 and the products are called borosilicate glasses. From standard reference sources, determine how the properties of glass change as the amount of B_2O_3 is increased.

5.86 In an aqueous solution, solute molecules or ions require a certain amount of time to migrate through the solution. The rate of this migration sets an upper limit on how fast reactions can take place, because no reaction can take place faster than the ions can be supplied. This limit is known as the *diffusion-controlled rate*. It has been found that the diffusion rate for hydrogen ions is about three times as fast as that for other ions in aqueous solution. Explain why this is so.

5.87 Iron corrodes in the presence of oxygen to form rust, which for simplicity can be taken to be iron(III) oxide. If a cubic block of iron of side 1.5 cm reacts with 15.5 L of oxygen at 1.00 atm and 25°C, what is the maximum mass of iron(III) oxide that can be produced? Iron metal has a bcc structure, and the atomic radius of iron is 124 pm. The reaction takes place at 298 K and 1.00 atm.

5.88 Real gases and vapors have intermolecular interactions. Recall that one equation of state for a real gas is the van der Waals equation, which is expressed in terms of two parameters, *a* and *b*. (a) For each of the following pairs of gases, decide which substance has the larger van der Waals *a* parameter: (i) He and Ne; (ii) Ne and O_2; (iii) CO_2 and H_2; (iv) H_2CO and CH_4; (v) C_6H_6 and $CH_3(CH_2)_{10}CH_3$. (b) For each of the following pairs of gases, decide which substance has the larger van der Waals *b* parameter: (i) F_2 and Br_2; (ii) Ne and F_2; (iii) CH_4 and $CH_3CH_2CH_3$; (iv) N_2 and Kr; (v) CO_2 and SO_2.

5.89 A commonly occurring mineral has a cubic unit cell in which the metal cations, M, occupy the corners and face centers. Inside the unit cell, anions, A, occupy all the tetrahedral holes created by the cations. What is the empirical formula of the M_mA_a compound?

5.90 Tetrahedral and octahedral interstitial holes are formed by the vacancies left when anions pack in a ccp array. (a) Which hole can accommodate the larger ions? (b) What is the size ratio of the largest metal cation that can occupy an octahedral hole to the largest that can occupy a tetrahedral hole while maintaining the close-packed nature of the anion lattice? (c) If half the tetrahedral holes are occupied, what will be the empirical formula of the compound M_xA_y, where M represents the cations and A the anions?

5.91 The density of cesium chloride is 3.988 g·cm^{-3}. Calculate the percentage of empty space in a cesium chloride lattice with the ions treated as hard spheres.

5.92 Use information from Exercise 5.46 to calculate the percentage of empty space in a calcium fluoride lattice with the ions treated as hard spheres. The density of CaF_2 is 3.180 g·cm^{-3}.

5.93 (a) If a pure element crystallizes with a primitive cubic lattice, what percentage of the unit cell is empty space? (b) How does this percentage compare with that of empty space in an fcc unit cell?

5.94 Many ionic compounds are considered to pack in such as way that the anions form a close-packed lattice in which the metal cations fill holes or interstitial sites left between the anions. These lattices, however, may not necessarily be as tightly packed as the label "close-packed" implies. The radius of an F^- ion is approximately 133 pm. The edge distances of the cubic unit cells of LiF, NaF, KF, RbF, and CsF, all of which pack in the rock-salt structure, are 568 pm, 652 pm, 754 pm, 796 pm, and 850 pm, respectively. Which of these lattices, if any, can be thought to be based on close-packed arrays of F^- ions? Justify your conclusions.

5.95 Due to its strong hydrogen bonds, in the vapor state hydrogen fluoride is found as short chains and rings. Draw the Lewis structure of an $(HF)_3$ chain and indicate the approximate bond angles.

5.96 Consider a metallic element that crystallizes in a cubic close-packed lattice. The edge length of the unit cell is 408 pm. If close-packed layers are deposited on a flat surface to a depth (of metal) of 0.125 mm, how many close-packed layers are present?

5.97 "Graphite bisulfates" are formed by heating graphite with a mixture of sulfuric and nitric acids. In the reaction, the graphite planes are partially oxidized. There is approximately one positive charge for every 24 carbon atoms, and the HSO_4^- anions are distributed between the planes. (a) What effect is this oxidation likely to have on the electrical conductivity? (b) What effect would you expect it to have on the x-ray diffraction pattern observed for this material? Refer to Major Technique 3 on x-ray diffraction, which follows this set of exercises.

5.98 Using the density of calcium fluoride (CaF_2, found in nature as the mineral fluorite), which is known to be 3.180 g·cm^{-3}, and the information given in Exercise 5.46, calculate (a) the edge length of the unit cell and (b) the Ca—F separation in fluorite. (c) Compare this value to that expected from the ionic radii tabulated in Fig. 1.48.

5.99 The following diagram shows two lattice planes from which two parallel x-rays are diffracted. If the two incoming x-rays are in phase, show that the Bragg equation $2d \sin\theta = n\lambda$ is true when n is an integer. Refer to Major Technique 3 on x-ray diffraction, which follows this set of exercises.

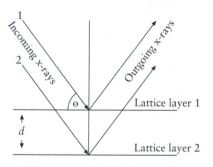

5.100 X-rays generated from a copper target have a wavelength of 154 pm. If an x-ray beam of this type is passed through a single crystal of NaBr, a diffracted beam is observed when the incident x-ray beam is at an angle of 7.42° to the surface of the crystal. What is the minimum spacing between the planes in the crystal that give rise to this diffracted beam? Refer to Major Technique 3 on x-ray diffraction, which follows this set of exercises.

5.101 The unit-cell edge length of lithium fluoride is 401.8 pm. What is the smallest angle at which the x-ray beam generated from a molybdenum source ($\lambda = 71.07$ pm) must strike the planes making up the faces of the unit cell for the beam to be diffracted from those planes? Refer to Major Technique 3 on x-ray diffraction, which follows this set of exercises.

5.102 Could electrons be used for diffraction studies of molecules? The energy of an electron accelerated through a potential difference V (in volts) is eV. What potential difference is needed to accelerate electrons from rest so that they have a wavelength of 100 pm? *Hint:* Use the de Broglie relation and the fact that acceleration of an electron to an energy eV gives it a kinetic energy of $\frac{1}{2}m_e v^2$. Refer to Major Technique 3 on x-ray diffraction, which follows this set of exercises.

5.103 Could neutrons be used for diffraction studies of molecules? The average kinetic energy of neutrons in a beam at a temperature T is $\frac{1}{2}kT$, where k is Boltzmann's constant. What temperature is needed to achieve a wavelength of 100. pm for a neutron? Refer to Major Technique 3 on x-ray diffraction, which follows this set of exercises.

5.104 Calculate the Coulomb potential energy at a point a distance r from a dipole composed of charges q and $-q$ separated by a distance l in the arrangement shown in the diagram. Use the fact that $l < r$ to expand the expression $1/(1 + x)$ into $1 - x + \cdots$ and identify the magnitude of the dipole moment as $\mu = ql$. Show that the potential energy is proportional to $1/r^2$ (as in Eq. 2).

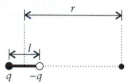

Chemistry Connections

5.105 Ethylammonium nitrate, $CH_3CH_2NH_3NO_3$, was the first ionic liquid to be discovered. Its melting point of 12°C was reported in 1914 and it has since been used as a nonpolluting solvent for organic reactions and for facilitating the folding of proteins.
(a) Draw the Lewis structure of each ion in ethylammonium nitrate and indicate the formal charge on each atom (in the cation, the carbon atoms are attached to the N atom in a chain: C—C—N).
(b) Assign a hybridization scheme to each C and N atom.
(c) Ethylammonium nitrate cannot be used as a solvent for some reactions because it can oxidize some compounds. Which ion is more likely to be the oxidizing agent, the cation or anion? Explain your answer.
(d) Ethylammonium nitrate can be prepared by the reaction of gaseous ethylamine, $CH_3CH_2NH_2$, and aqueous nitric acid. Write the chemical equation for the reaction. What type of reaction is this?
(e) 2.00 L of ethylamine at 0.960 atm and 23.2°C was bubbled into 250.0 mL of 0.240 M HNO_3(aq) and 4.10 g of ethylammonium nitrate was produced. What were the theoretical and percentage yields of the salt?
(f) Suggest ways in which the forces that hold ethylammonium nitrate ions together in the solid state differ from those that hold together salts such as sodium chloride or sodium bromide.
(g) Low-melting salts in which the cation is inorganic and the anion organic have been prepared. Explain the trend in melting point seen in the following series: sodium acetate ($NaCH_3CO_2$), 324°C; sodium propanoate ($NaCH_3CH_2CO_2$), 285°C; sodium butanoate ($NaCH_3CH_2CH_2CO_2$), 76°C; and sodium pentanoate ($NaCH_3CH_2CH_2CH_2CO_2$), 64°C.

A great deal of our knowledge about the interior of solids has come from *x-ray diffraction*. This important technique is used to determine the arrangement of atoms in solid compounds and to measure bond lengths and angles. Almost all recent advances in molecular biology have stemmed from the application of this technique to determine the structures of molecules such as proteins and nucleic acids.

The Technique

When two or more waves pass through the same region of space, the phenomenon of *interference* is observed as an increase or a decrease in the total amplitude of the wave (recall Fig. 1.20). *Constructive interference,* an increase in the total amplitude of the wave, occurs when the peaks of one wave coincide with the peaks of another wave. If the waves are electromagnetic radiation, the increased amplitude corresponds to an increased intensity of the radiation. *Destructive interference,* a decrease in the total amplitude of the waves, occurs when the peaks of one wave coincide with the troughs of the other wave: it results in a reduction in intensity.

The phenomenon of *diffraction* is interference between waves that arises when there is an object in their path. One of the earliest demonstrations of interference was the *Young's slit experiment,* in which light passes through two slits and gives rise to a pattern on a screen (see Fig. 1.19). If we were presented with the pattern and were told the wavelength of the light and the distance of the detection screen from the screen containing the slits, it would be possible to work out the spacing of the two slits. An x-ray diffraction experiment is a more elaborate version of the Young's slit experiment. The regular layers of atoms in a crystal act as a three-dimensional collection of slits and give rise to a diffraction pattern that varies as the crystal is rotated and the "slits" are brought into a new arrangement. The task for the x-ray crystallographer is to use the diffraction pattern to determine the spacing and arrangement of the "slits" that gave rise to it. This enormously complex task is universally carried out on computers.

Why x-rays? Diffraction occurs when the wavelength of the radiation is comparable to characteristic spacings within the object causing the diffraction. Therefore, to obtain diffraction patterns from layers of atoms, we need to use radiation with a wavelength comparable to the spacing of the layers. The separation between layers of atoms in a crystal is about 100 pm, and so electromagnetic radiation of that wavelength, which corresponds to the x-ray region, must be used.

Experimental Techniques

X-rays are generated by accelerating electrons to very high speeds and then letting them plunge into a metal target. This technique generates two types of x-radiation. One type comes from the bombarding electrons themselves. Because accelerating and decelerating charges emit electromagnetic radiation, the electrons generate radiation as they are brought to a standstill in the metal. The radiation covers a wide range of frequencies, including that of x-rays. However, for current versions of x-ray diffraction, we need a well-defined wavelength. Such radiation is generated by a second mechanism. The fast electrons hit electrons occupying orbitals in the inner shells of atoms and drive them out of the atom. That collision leaves a gap in the atom, which is filled when an electron in another shell falls into the vacancy. The difference in energy is given out as a photon. Because the energy difference between shells is so great, the photon has a very high energy, corresponding to the x-ray region. When copper is used as the target, for example, the x-radiation has a wavelength of 154 pm.

In the *powder diffraction technique,* a monochromatic (single-frequency) beam of x-rays is directed at a powdered sample spread on a support, and the diffraction intensity is measured as the detector is moved to different angles (Fig. 1). The pattern obtained is characteristic of the material in the sample, and it can be identified by comparison with a database of patterns. In effect, powder x-ray diffraction takes a fingerprint of the sample. It can also be used to identify the size and shape of the unit cell by measuring the spacing of the lines in the diffraction pattern. The central equation for analyzing the results of a powder diffraction experiment is the *Bragg equation*

$$2d \sin \theta = \lambda$$

which relates the angles θ (theta), at which constructive interference occurs, to the spacing, d, of layers of atoms in the sample for x-rays of wavelength λ (lambda). The illustration in Exercise 5.99 shows how the parameters λ and d relate to the crystal dimensions.

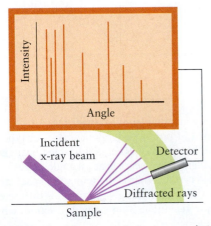

FIGURE 1 In the powder diffraction technique, a sample is spread on a flat plate and exposed to a beam of monochromatic (single-frequency) x-rays. The diffraction pattern (inset) is recorded by moving the detector to different angles.

Sample exercise: Determining the spacing between layers of atoms in a crystal

In an x-ray diffraction experiment on a single crystal of sodium chloride, with the use of radiation from a copper source ($\lambda = 154$ pm), constructive interference was observed at $\theta = 11.2°$. What is the spacing of the layers responsible for the diffraction?

SOLUTION

Rearrange the Bragg equation to give d in terms of θ.

$$d = \frac{\lambda}{2 \sin \theta}$$

Substitute the data.

$$d = \frac{154 \text{ pm}}{2 \sin 11.2°} = 396 \text{ pm}$$

SELF-TEST MT3.1A Constructive interference at 7.23° was observed from a crystal when x-radiation from a molybdenum source ($\lambda = 71.0$ pm) was used. What is the spacing of the layers of atoms responsible for the diffraction?

[*Answer:* 282 pm]

SELF-TEST MT3.1B Constructive interference at 12.1° was observed from a cubic crystal when x-radiation of wavelength 152 pm was used. What is the spacing of the layers of atoms responsible for the diffraction?

The x-ray diffraction pattern from a liquid resembles that from a powdered sample, but the lines are diffuse rather than sharp. This pattern indicates that the molecules of a liquid have a degree of short-range order but that the distances between molecules vary; this variation produces the diffuse lines.

The *single-crystal diffraction* technique is much more elaborate and gives much richer information. The first task is to grow a perfect single crystal of the sample. Whereas that task is usually straightforward for simple inorganic solids, it

is often one of the most difficult parts of the investigation for the huge molecules characteristic of biologically important compounds such as proteins. Only a tiny crystal is needed, about 0.1 mm on a side, but the task of growing one can be very challenging. Once the crystal has been grown, it is placed at the center of a *four-circle diffractometer* (Fig. 2), a device for rotating the crystal and the detector so that the entire diffraction pattern can be recorded under the control of a computer.

The raw data consist of the intensities of the x-rays at all the settings of the angles in the diffractometer. The computational task is to analyze these measurements and convert them into the locations of the atoms. The process of conversion is called *Fourier synthesis* and requires lengthy calculations carried out on a computer that is an integral part of the diffractometer. The end product is a detailed description of the location of all the atoms in the molecule, the bond lengths, and the bond angles. The most spectacular achievement of this kind was the elucidation of the functioning of the genetic messenger, DNA, where x-rays were a source of insight into the workings of life itself. Even at this stage in your studies, you can begin to see the spark that ignited Watson and Crick's vision of the structure of DNA. Figure 3 shows the characteristic X pattern obtained by Rosalind Franklin. The two crossed arms of the X told Watson and Crick that the molecule must be a helix, and from the slopes of the arms and the diffraction spots along them they were able to deduce its pitch (the distance between turns) and radius. The two arcs at the top and bottom of the photograph represent diffraction through a big angle and therefore a narrow spacing: this they could interpret in terms of the spacing of the nucleotide bases along the helix, with 10 bases per turn of helix.

Related Exercises: 5.84, 5.97, and 5.99–5.103.

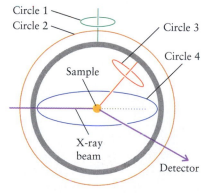

FIGURE 2 A four-circle diffractometer is used to obtain highly detailed information about x-ray diffraction patterns from single crystals. The diffraction pattern is monitored as the orientations around each of the four axes are changed.

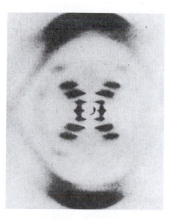

FIGURE 3 The x-ray diffraction pattern that led to the elucidation of the structure of DNA.

THERMODYNAMICS: THE FIRST LAW

Chapter 6

What Are the Key Ideas? Heat and work are equivalent ways of transferring energy between a system and its surroundings. The total energy of an isolated system is constant. The enthalpy change for a process is equal to the heat released at constant pressure.

Why Do We Need to Know This Material? The laws of thermodynamics govern chemistry and life. They explain why reactions take place and let us predict how much heat reactions release and how much work they can do. Thermodynamics plays a role in every part of our lives. For example, the energy released as heat can be used to compare fuels, and the energy resources of food lets us assess its nutritional value. The material in this chapter provides a foundation for the following chapters, in particular Chapter 7, which deals with the driving force of chemical reactions.

What Do We Need to Know Already? This chapter assumes that we are familiar with the concept of energy (Section A), stoichiometry (Sections L and M), and the ideal gas law (Chapter 4). Some of the explanations refer to intermolecular forces (Sections 4.12 and 5.1–5.5). Ionic bonding (Sections 2.3–2.4) and bond strengths (Sections 2.14–2.15) are developed further in this chapter.

Energy is the basis of civilization. Every day we use energy in its various forms to maintain life, to stay sufficiently cool or warm, to move about, and to think. All these processes involve the release, absorption, transfer, or conversion of energy.

The study of the transformations of energy from one form into another is called **thermodynamics.** The *first law of thermodynamics*, which is introduced in this chapter, is concerned with keeping track of energy changes. It allows us to calculate, for instance, how much heat a reaction can produce. The *second law of thermodynamics*, which is the subject of Chapter 7, explains why some chemical reactions take place but others do not. Both laws are summaries of experience with bulk matter and are independent of any model that we might have of its microscopic structure—you can do thermodynamics even if you don't believe in the existence of atoms! However, we can use the models of atoms and molecules we have been developing to elucidate these laws and enrich our understanding of them. The link between the properties of atoms and those of bulk matter is **statistical thermodynamics**, the interpretation of the laws of thermodynamics in terms of the *average* behavior of the large numbers of atoms and molecules that make up a typical sample.

This chapter introduces the first law of thermodynamics and its applications in three main parts. The first part introduces the basic concepts of thermodynamics and the experimental basis of the first law. The second part introduces enthalpy as a measure of the energy transferred as heat during physical changes at constant pressure. The third part shows how the concept of enthalpy is applied to a variety of chemical changes, an important aspect of **bioenergetics**, the use of energy in biological systems.

SYSTEMS, STATES, AND ENERGY

Two of the fundamental concepts of thermodynamics are *heat* and *work*. People once thought that heat was a separate substance, a fluid called *caloric*, which flowed from a hot substance to a cooler one. The French engineer Sadi Carnot

FIGURE 6.1 Nicolas Leonard Sadi Carnot (1796–1832).

FIGURE 6.2 James Prescott Joule (1818–1889).

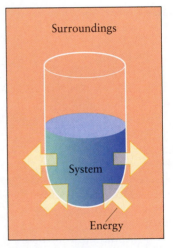

FIGURE 6.3 The system is the sample or reaction mixture in which we are interested. Outside the system are the surroundings. The system plus its surroundings is sometimes called the universe.

(Fig. 6.1), who helped to lay the foundations of thermodynamics, believed that work resulted from the flow of caloric, just as the flow of water turns a water wheel. Some of Carnot's conclusions survive, but we now know that there is no such substance as caloric. About 25 years after Carnot proposed his ideas in the early nineteenth century, the English physicist James Joule showed that both heat and work are forms of energy (Fig. 6.2).

6.1 Systems

In thermodynamics we study how energy is transformed from one form into another and transferred from one place to another. For instance, electricity may be generated in a power station but used a great distance away; food may be digested in your stomach but the energy released may be used in your head. To keep track of changes in energy, we divide the world into two parts. The region in which we are interested, such as a flask of gas, a beaker of acid, a reaction mixture, or a muscle fiber, is called a **system** (Fig. 6.3). Everything else, such as the water bath in which a reaction mixture may be immersed, is called the **surroundings**. The surroundings are where we make observations on the energy transferred into or out of the system. The system and the surroundings jointly make up the **universe**, but often the only part of the actual universe affected by a process consists of the sample, a flask, and a water bath.

A system can be open, closed, or isolated (Fig. 6.4). An **open system** can exchange both matter and energy with the surroundings. Examples of open systems are automobile engines and the human body. A **closed system** has a fixed amount of matter, but it can exchange energy with the surroundings. An example of a closed system is a cold pack used for treating athletic injuries. An **isolated system** has no contact with its surroundings. We can think of an isolated system as sealed inside rigid, thermally insulating walls. A good approximation to an isolated system is a hot liquid inside a sealed vacuum flask.

In thermodynamics, the universe consists of a system and its surroundings. An open system can exchange both matter and energy with the surroundings; a closed system can exchange only energy; an isolated system can exchange nothing.

6.2 Work and Energy

The most fundamental property in thermodynamics—in the sense that it provides a basis for defining the principal concepts—is **work,** or motion against an opposing force (Section A). Work is done when a weight is raised against the pull of

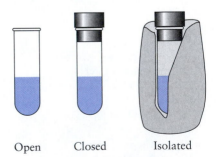

Open   Closed   Isolated

FIGURE 6.4 We can classify any system according to its interactions with its surroundings. An open system can exchange matter and energy with its surroundings. A closed system can exchange energy but not matter. An isolated system can exchange neither matter nor energy.

gravity. The chemical reaction in a battery does work when it pushes an electric current through a circuit. Gas in a cylinder—perhaps the hot gas mixture in the cylinder of an automobile engine—does work when it expands and pushes back a piston. We can identify whether a process can do work by noting whether, in principle at least, it can be harnessed to raise a weight. The expansion of a gas, for example, can be used to raise a weight because the piston can be connected to the weight. An electric current produced by a battery can be used to raise a weight if it is used to power an electric motor.

The work required to move an object a certain distance against an opposing force is calculated by multiplying the opposing force by the distance moved against it:

$$\text{Work} = \text{opposing force} \times \text{distance moved} \tag{1}$$

As pointed out and illustrated in Section A, the unit in which work, and therefore energy, is reported is the *joule*, J, with

$$1\ \text{J} = 1\ \text{kg·m}^2\text{·s}^{-2}$$

In thermodynamics, the capacity of a system to do work—its total store of energy—is called its **internal energy**, U. We cannot measure the absolute value of the internal energy of a system, because that value includes the energies of all the atoms, their electrons, and the components of their nuclei. The best we can do is to measure *changes* in internal energy. For instance, if a system does 15 J of work (and no other changes take place), then it has used up some of its store of energy, and we say that its internal energy has fallen by 15 J and write $\Delta U = -15$ J. Throughout thermodynamics, the symbol ΔX means a difference in a property X:

$$\Delta X = X_{\text{final}} - X_{\text{initial}} \tag{2}$$

A *negative* value of ΔX, as in $\Delta U = -15$ J, signifies that the value of X has decreased.

A note on good practice. Except in the special cases we shall specify, always show the sign of ΔU (and other ΔX) explicitly, even if it is positive. Thus, if the internal energy increases by 15 J during a change, we write $\Delta U = +15$ J, not simply $\Delta U = 15$ J.

When we do work *on* a system (and, once again, there are no other changes taking place), the internal energy of the system is increased. Compressing a gas inside a thermally insulated container increases its internal energy, because the hot compressed gas can do more work than the uncompressed cooler gas. Winding a spring transfers energy to the spring: when the spring is fully wound, it can do work as it unwinds. Passing an electric current through a system, as we do when we charge a battery, is another way to do work on a system.

We use the symbol w to denote the energy transferred to a system by doing work and, *provided that no other type of transfer of energy is taking place*, write $\Delta U = w$. When energy is transferred *to* a system as work, the internal energy of the system increases and w is positive. When energy *leaves* the system as work, the internal energy of the system decreases and w is negative.

Work is the transfer of energy to a system by a process that is equivalent to raising or lowering a weight. The internal energy of a system may be changed by doing work: in the absence of other changes, $\Delta U = w$.

6.3 Expansion Work

A system can do two kinds of work. The first type is **expansion work**, the work needed to change the volume of a system. A gas expanding in a cylinder fitted with a piston pushes out against the atmosphere and thus does expansion work. The second type of work is **nonexpansion work**, work that does not involve a change in volume. A chemical reaction can do nonexpansion work by causing an electrical

The joule is consistent with Eq. 1, because force is measured in newtons ($1\ \text{N} = 1\ \text{kg·m·s}^{-2}$), so the units of force × distance are $\text{kg·m·s}^{-2} \times \text{m} = \text{kg·m}^2\text{·s}^{-2}$, or J.

TABLE 6.1 Varieties of Work

Type of work	w	Comment	Units*
expansion	$-P_{ex}\Delta V$	P_{ex} is the external pressure ΔV is the change in volume	Pa m^3
extension	$f\Delta l$	f is the tension Δl is the change in length	N m
raising a weight	$mg\Delta h$	m is the mass g is the acceleration of free fall Δh is the change in height	kg $m \cdot s^{-2}$ m
electrical	$\phi\Delta q$	ϕ is the electrical potential Δq is the change in charge	V C
surface expansion	$\gamma\Delta A$	γ is the surface tension ΔA is the change in area	$N \cdot m^{-1}$ m^2

*For work in joules (J). Note that 1 N·m = 1 J and 1 V·C = 1 J.

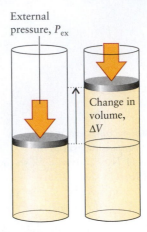

FIGURE 6.5 A system does work when it expands against an external pressure. (a) A gas in a cylinder with a piston held in place. (b) The piston is released and (provided the pressure of the gas is greater than the external pressure, P_{ex}) the gas expands and pushes out against P_{ex}. The work done is proportional to P_{ex} and the change in volume, ΔV, that the system undergoes.

current to flow, and our bodies do nonexpansion work when we move about. Table 6.1 lists some of the kinds of work that a system can do.

First, we consider the expansion work done by a system consisting of a gas in a cylinder. The external pressure acting on the outer face of the piston provides the force opposing expansion. We shall suppose that the external pressure is constant, as when the piston is pressed on by the atmosphere (Fig. 6.5). We need to find how the work done when the system expands through a volume ΔV is related to the external pressure P_{ex}.

HOW DO WE DO THAT?

We can relate pressure to the work of expansion against a constant pressure by using the fact that pressure is the force divided by the area to which it is applied: $P = F/A$ (Section 4.2). Therefore, the force opposing expansion is the product of the pressure acting on the outside of the piston, P_{ex}, and the area of the piston, A ($F = P_{ex}A$). The work needed to drive the piston out through a distance d is therefore

From work = force × distance,

$$\text{Work} = P_{ex}A \times d$$

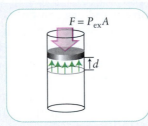

However, the product of area and distance moved is equal to the change in volume of the sample:

From volume = area × height,

$$A \times d = \Delta V$$

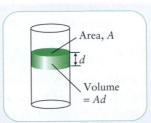

Therefore, the work done by the expanding gas is $P_{ex}\Delta V$. At this point, we match the signs to our convention. When a system expands, it loses energy as work; so, when ΔV is positive (an expansion), w is negative. Therefore,

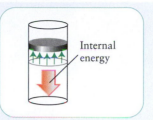

Internal energy

From the sign convention,
$$w = -P_{ex}\Delta V$$

We have deduced that expansion through a volume ΔV against a constant external pressure P_{ex} is given by

$$w = -P_{ex}\Delta V \qquad (3)^*$$

This expression applies to all systems. A gas is easiest to visualize, but the expression also applies to an expanding liquid or solid. However, Eq. 3 applies *only when the external pressure is constant* during the expansion.

> *What does this equation tell us?* The minus sign in Eq. 3 tells us that the internal energy decreases when the system expands. The factor P_{ex} tells us that more work has to be done when the external pressure (which is responsible for the opposing force) is high. The factor ΔV tells us that, for a given external pressure, more work has to be done to expand through a large volume than a smaller volume.

If the external pressure is zero (a vacuum), it follows from Eq. 3 that $w = 0$: that is, *a system does no expansion work when it expands into a vacuum*, because there is no opposing force. You do no work by pushing if there is nothing to push against. Expansion against zero pressure is called **free expansion**.

In SI units, the external pressure is expressed in pascals (1 Pa = 1 $kg \cdot m^{-1} \cdot s^{-2}$, Section 4.2) and the change in volume is expressed in cubic meters (m^3). The product of 1 Pa and 1 m^3 is

$$1\ Pa \cdot m^3 = 1\ kg \cdot m^{-1} \cdot s^{-2} \times 1\ m^3 = 1\ kg \cdot m^2 \cdot s^{-2} = 1\ J$$

Therefore, if we work in pascals and cubic meters, the work is obtained in joules. However, we might have expressed the pressure in atmospheres and the volume in liters. In this case, we may need to convert the answer (in liter-atmospheres) into joules. The conversion factor is obtained by noting that 1 L = 10^{-3} m^3 and 1 atm = 101 325 Pa exactly; therefore

$$1\ L \cdot atm = 10^{-3}\ m^3 \times 101\ 325\ Pa = 101.325\ Pa \cdot m^3 = 101.325\ J\ (exactly)$$

EXAMPLE 6.1 **Sample exercise: Calculating the work done when a gas expands**

Suppose a gas expands by 500. mL (0.500 L) against a pressure of 1.20 atm. How much work is done in the expansion?

SOLUTION We use Eq. 3 to calculate the work, and then convert liter-atmospheres into joules.

From $w = -P_{ex}\Delta V$,
$$w = -(1.20\ atm) \times (0.500\ L) = -0.600\ L \cdot atm$$

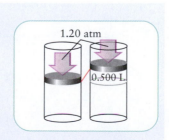

1.20 atm

0.500 L

Convert into joules by using
1 L·atm = 101.325 J.
$$w = -(0.600\ L \cdot atm) \times \left(\frac{101.325\ J}{1\ L \cdot atm}\right) = -60.8\ J$$

The negative sign in $w = -60.8$ J means that the internal energy decreases by 60.8 J when the gas expands and the system does 60.8 kJ of work on its surroundings (and there are no other changes).

SELF-TEST 6.1A Water expands when it freezes. How much work does 100. g of water do when it freezes at 0°C and bursts a water pipe that exerts an opposing pressure of 1070 atm? The densities of water and ice at 0°C are 1.00 g·cm^{-3} and 0.92 g·cm^{-3}, respectively.

[*Answer:* $w = -0.9$ kJ]

SELF-TEST 6.1B The gases in the four cylinders of an automobile engine expand from 0.22 L to 2.2 L during one ignition cycle. Assuming that the gear train maintains a steady pressure of 9.60 atm on the gases, how much work can the engine do in one cycle?

To calculate the work done by a gas that expands against a *changing* external pressure, we need to specify the pressure at each stage of the expansion. In particular, we consider the very important case of "reversible" expansion of an ideal gas. In everyday language, a reversible process is one that can take place in either direction. This common usage is refined in science: in thermodynamics, a **reversible process** is one that can be reversed by an *infinitely small* change in a variable (an "infinitesimal" change). For example, if the external pressure exactly matches the pressure of the gas in the system, then the piston moves in neither direction. If the external pressure is increased infinitesimally, the piston moves in. If, instead, the external pressure is reduced infinitesimally, the piston moves out. Expansion against an external pressure that differs by a finite (measurable) amount from the pressure of the system is an **irreversible process** in the sense that an infinitesimal change in the external pressure does not reverse the direction of travel of the piston. For instance, if the pressure of the system is 2.0 atm at some stage of the expansion and the external pressure is 1.0 atm, then a tiny change in the latter does not convert expansion into compression. Reversible processes are of the greatest importance in thermodynamics, and we shall encounter them many times.

The simplest kind of reversible change to consider is reversible, isothermal (constant temperature) expansion of an ideal gas. We can maintain the same temperature by ensuring that the system is in thermal contact with a constant-temperature water bath at all stages of the expansion. In an isothermal expansion, the pressure of the gas falls as it expands (by Boyle's law); so, to achieve reversible expansion, the external pressure must be reduced in step with the change in volume so that at every stage the external pressure is the same as the pressure of the gas (Fig. 6.6). To calculate the work, we have to take into account the gradual reduction in external pressure and therefore the changing opposing force.

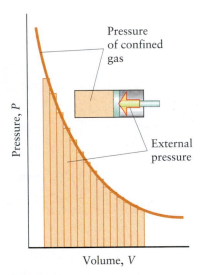

Pressure of confined gas

External pressure

Pressure, P

Volume, V

FIGURE 6.6 When a gas expands reversibly, the external pressure is matched to the pressure of the gas at every stage of the expansion. This arrangement (when the steps corresponding to the increase in volume are infinitesimal) achieves maximum work (the area under the curve).

HOW DO WE DO THAT?

To calculate the work of reversible, isothermal expansion of a gas, we have to use calculus, starting at Eq. 3 written for an infinitesimal change in volume, dV:

$$dw = -P_{ex}dV$$

Because the external pressure is matched to the pressure, P, of the gas at every stage of a reversible expansion, we write $P_{ex} = P$ and the expression for dw becomes

$$dw = -PdV$$

At each stage of the expansion, the pressure of the gas is related to its volume by the ideal gas law, $PV = nRT$, so we can replace P by nRT/V:

From $dw = -PdV$ and $P = nRT/V$,
$$dw = -\frac{nRTdV}{V}$$

The total work done is the sum (integral) of these infinitesimal contributions as the volume changes from its initial value to its final value. That is, we have to integrate dw from the initial to the final volume with nRT a constant (because the change is isothermal):

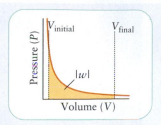

From $w = \int dw$, with nRT a constant,
$$w = -nRT\int_{V_{initial}}^{V_{final}}\frac{dV}{V} = -nRT\ln\frac{V_{final}}{V_{initial}}$$

The final step has made use of the standard integral
$$\int\frac{dx}{x} = \ln x + \text{constant}$$

We have found that the work of reversible, isothermal expansion of an ideal gas from the volume $V_{initial}$ to the volume V_{final} is

$$w = -nRT\ln\frac{V_{final}}{V_{initial}} \tag{4}*$$

where n is the amount of gas molecules (in moles) in the container and T is the temperature. One way to visualize Eq. 4 is to realize that, apart from the sign, the work done by the system as it expands is equal to the *area* beneath the ideal gas isotherm lying between the initial and the final volumes (Fig. 6.7).

What does this equation tell us? For a given initial and final volume, more work is done when the temperature is high than when it is low. For a given volume and amount of molecules, a high temperature corresponds to a high gas pressure, and so the expansion takes place against a stronger opposing force and therefore must do more work. More work is also done if the final volume is very much greater than the initial volume.

FIGURE 6.7 The work done by a system is equal to the area below the curve of the graph of external pressure as a function of volume. The work done *on* a system is equal to the negative of that area.

EXAMPLE 6.2   Sample exercise: Calculating the work of isothermal expansion

A piston confines 0.100 mol Ar(g) in a volume of 1.00 L at 25°C. Two experiments are performed. In one, the piston is allowed to expand through 1.00 L against a constant pressure of 1.00 atm. In the second, it is allowed to expand reversibly and isothermally to the same final volume. Which process does more work?

SOLUTION For expansion against constant external pressure we use Eq. 3 and for reversible, isothermal expansion we use Eq. 4:

From $w = -P_{ex}\Delta V$, and then converting to joules,

$$w = -(1.00\ \text{atm}) \times (1.00\ \text{L})$$
$$= -1.00\ \text{L·atm} \times \frac{101.325\ \text{J}}{1\ \text{L·atm}} = -101\ \text{J}$$

From $w = -nRT\ln(V_{final}/V_{initial})$,

$$w = -(0.100\ \text{mol}) \times (8.3145\ \text{J·K}^{-1}\text{·mol}^{-1})$$
$$\times (298\ \text{K}) \times \ln\left(\frac{2.00\ \text{L}}{1.00\ \text{L}}\right)$$
$$= -172\ \text{J}$$

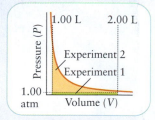

We see that the reversible process does more work.

SELF-TEST 6.2A A cylinder of volume 2.00 L contains 1.00 mol He(g) at 30°C. Which process does more work on the system, compressing the gas isothermally to 1.00 L with a constant external pressure of 5.00 atm or compressing it reversibly and isothermally to the same final volume?

[*Answer:* Reversible compression]

SELF-TEST 6.2B A cylinder of volume 2.00 L contains 1.00 mol He(g) at 30°C. Which process does more work on the surroundings, allowing the gas to expand isothermally to 4.00 L against a constant external pressure of 1.00 atm or allowing it to expand reversibly and isothermally to the same final volume?

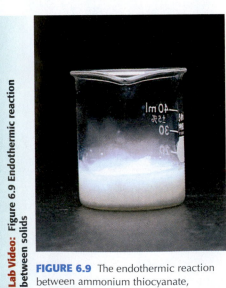

FIGURE 6.8 The thermite reaction is so exothermic that it melts the metal that it produces and is used to weld railroad tracks together. Here, aluminum metal is reacting with iron(III) oxide, Fe_2O_3, causing a shower of molten iron sparks.

The work done by any system on its surroundings during expansion against a constant pressure is calculated from Eq. 3; for a reversible, isothermal expansion of an ideal gas, the work is calculated from Eq. 4. A reversible process is a process that can be reversed by an infinitesimal change in a variable.

6.4 Heat

The internal energy of a system, its capacity to do work, can also be changed by transferring energy to or from the surroundings as heat. *Heat* is a familiar term, but in thermodynamics it has a special meaning. In thermodynamics, **heat** is the energy transferred as a result of a temperature difference. Energy flows as heat from a high-temperature region to a low-temperature region. Therefore, provided that the walls of the system are not thermally insulating, if the system is cooler than its surroundings, energy flows into the system from the surroundings.

In the introduction to this chapter we remarked that heat is not a substance. What, then, does it mean for energy to flow as heat? Molecules in a high temperature region move more vigorously than molecules in a lower temperature region. When the two regions are brought into contact, the energetic molecules in the higher temperature region strike the molecules in the lower temperature region, stimulating them into more vigorous motion. As a result, the internal energy of the cooler system increases as that of the hotter system decreases.

We write the energy transferred *to* a system as heat as q. Therefore, when the internal energy of the system is changed by transferring energy as heat (and no other processes take place), $\Delta U = q$. If energy enters a system as heat, the internal energy of the system increases and q is positive; if energy leaves the system as heat, the internal energy of the system decreases and q is negative. Thus, if 10 J enters the system as heat, we write $q = +10$ J and (provided no work is done on or by the system) $\Delta U = +10$. J. Likewise, if 10 J leaves the system as heat, we write $q = -10$ J and $\Delta U = -10$ J. As these examples show, energy transferred as heat is measured in joules, J. However, a unit still widely used in biochemistry and related fields is the **calorie**, cal. Originally, 1 cal was the energy needed to raise the temperature of 1 g of water by 1°C. The modern definition is

This exact relation defines the calorie in terms of the joule; the joule is the fundamental unit.

$$1 \text{ cal} = 4.184 \text{ J (exactly)}$$

The **nutritional calorie**, Cal, is actually 1 kilocalorie (kcal), and so it is important to note which unit is being used when assessing the energy content of food.

A process that releases heat into the surroundings is called an **exothermic process**. Most common chemical reactions—and all combustions, such as those that power transport and heating—are exothermic (Fig. 6.8). Less familiar are chemical reactions that absorb heat from the surroundings. A process that absorbs heat is called an **endothermic process** (Fig. 6.9). A number of common physical processes are endothermic. For instance, vaporization is endothermic, because heat must be supplied to drive molecules of a liquid apart from one another. The dissolution of ammonium nitrate in water is endothermic; in fact, this process is used in instant cold packs for sports injuries.

Heat is the transfer of energy as a result of a temperature difference. When energy is transferred only by means of heat, $\Delta U = q$.

FIGURE 6.9 The endothermic reaction between ammonium thiocyanate, NH_4SCN, and barium hydroxide octahydrate, $Ba(OH)_2 \cdot 8H_2O$, absorbs a lot of heat and can cause water vapor in the air to freeze on the outside of the beaker.

6.5 The Measurement of Heat

To measure the energy transferred to a system (or surroundings) as heat we use a thermometer to measure the change in temperature the heating normally causes. To interpret the measurement, we need to know how much energy corresponds to a given change in temperature. This correspondence is expressed in terms of the system's **heat capacity**, C, the ratio of the heat supplied to the rise in temperature produced:

$$C = \frac{q}{\Delta T} \tag{5a}$$

A large heat capacity means that a given supply of heat produces only a small rise in temperature. Once we know the heat capacity, we can measure the change in temperature, ΔT, of the system and calculate how much heat has been supplied by using this equation in the form

$$q = C\Delta T \tag{5b}$$

Heat capacity is an extensive property: the larger the sample, the more heat is required to raise its temperature by a given amount and so the greater is its heat capacity (Fig. 6.10). It is therefore common to report either the **specific heat capacity** (often called just "specific heat"), C_s, which is the heat capacity divided by the mass of the sample ($C_s = C/m$), or the **molar heat capacity**, C_m, the heat capacity divided by the amount (in moles) of the sample ($C_m = C/n$). For example, the specific heat capacity of liquid water at room temperature is 4.18 $J \cdot (°C)^{-1} \cdot g^{-1}$, or 4.18 $J \cdot K^{-1} \cdot g^{-1}$, and its molar heat capacity is 75 $J \cdot K^{-1} \cdot mol^{-1}$.

We can calculate the heat capacity of a substance from its mass and its specific heat capacity by using the relation $C = m \times C_s$. If we know the mass of a substance, its specific heat capacity, and the temperature rise it undergoes during an experiment, then the heat supplied to the sample is

$$q = C\Delta T = mC_s\Delta T \tag{6a}$$

A similar expressions is used if we are told the molar heat capacity of a substance and its amount: we use the relation $C = n \times C_m$ and write

$$q = nC_m\Delta T \tag{6b}$$

These expressions may be rearranged to calculate the specific or molar heat capacity from the measured temperature rise caused by a known quantity of heat. The specific heat capacity of a dilute solution is normally taken to be the same as that of the pure solvent (which is commonly water). Table 6.2 lists the specific and molar heat capacities of some common substances.

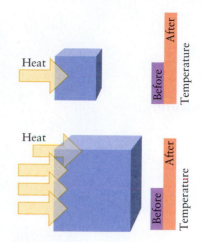

FIGURE 6.10 The heat capacity of an object determines the change in its temperature brought about by the quantity of energy transferred as heat: an object with a large heat capacity requires a lot of heat to bring about a given rise in temperature. Heat capacity is an extensive property; so a large object (bottom) has a larger heat capacity than a small object (top) made of the same material. Heat capacities also, in general, depend on temperature.

TABLE 6.2   Specific and Molar Heat Capacities of Common Materials*

Material	Specific heat capacity ($J \cdot (°C)^{-1} \cdot g^{-1}$)	Molar heat capacity ($J \cdot K^{-1} \cdot mol^{-1}$)
air	1.01	–
benzene	1.05	136
brass	0.37	–
copper	0.38	33
ethanol	2.42	111
glass (Pyrex)	0.78	–
granite	0.80	–
marble	0.84	–
polyethylene	2.3	–
stainless steel	0.51	–
water: solid	2.03	37
liquid	4.184	75
vapor	2.01	34

*More values are available in Appendices 2A and 2D; values assume constant pressure. Specific heat capacities commonly use Celsius degrees in their units, whereas molar heat capacities commonly use kelvins. All values except that for ice are for 25°C.

EXAMPLE 6.3   **Calculating the heat needed to bring about a rise in temperature**

Calculate the heat necessary to increase the temperature of (a) 100. g of water, (b) 2.00 mol $H_2O(l)$ by 20.°C from room temperature.

STRATEGY   The heat required is given by Eq. 6. In each case, $\Delta T = +20.$ K.

SOLUTION

(a) From $q = mC_s\Delta T$,

$$q = (100.\text{ g}) \times (4.18\text{ J·K}^{-1}\text{·g}^{-1}) \times (20.\text{ K}) = +8.4\text{ kJ}$$

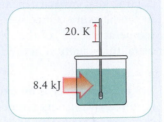

(b) From $q = nC_m\Delta T$,

$$q = (2.00\text{ mol}) \times (75\text{ J·K}^{-1}\text{·mol}^{-1}) \times (20.\text{ K}) = +3.0\text{ kJ}$$

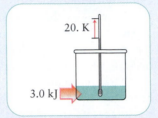

Because 100. g of water is greater than 2.00 mol $H_2O(l)$, which corresponds to 36.02 g, more heat is needed in (a) than in (b). Note that we are assuming that the heat capacity of the beaker itself is zero.

SELF-TEST 6.3A   Potassium perchlorate, $KClO_4$, is used as an oxidizer in fireworks. Calculate the heat required to raise the temperature of 10.0 g of $KClO_4$ from 25°C to an ignition temperature of 900.°C. The specific heat capacity of $KClO_4$ is 0.8111 J·K^{-1}·g^{-1}.

[*Answer:* 7.10 kJ]

SELF-TEST 6.3B   Calculate the heat necessary to increase the temperature of 3.00 mol $CH_3CH_2OH(l)$, ethanol, by 15.0°C from room temperature (see Table 6.2).

Transfers of energy as heat are measured with a **calorimeter**, a device in which heat transfer is monitored by recording the change in temperature that it produces, and then using $q = C_{cal}\Delta T$, where C_{cal} is the heat capacity of the calorimeter (which is sometimes called the "calorimeter constant"). A calorimeter can be as simple as a container immersed in a water bath equipped with a thermometer (Fig. 6.11). A more sophisticated version is a *bomb calorimeter* (Fig. 6.12). The reaction takes place inside the sturdy sealed metal vessel (the bomb), which is immersed in water, and the rise in temperature of the entire assembly is monitored. Its heat capacity is measured by supplying a known quantity of heat and noting the resulting rise in temperature. This process is called "calibrating" the calorimeter.

EXAMPLE 6.4   **Determining the internal energy change accompanying a reaction**

A constant-volume calorimeter was calibrated by carrying out a reaction known to release 1.78 kJ of heat in 0.100 L of solution in the calorimeter, resulting in a temperature rise of 3.65°C. Next, 50. mL of 0.20 M HCl(aq) and 50. mL of 0.20 M NaOH(aq) were mixed in the same calorimeter and the temperature rose by 1.26°C. What is the change in the internal energy of the neutralization reaction?

STRATEGY

There are two steps in the calculation. First, calibrate the calorimeter by calculating its heat capacity from the information on the first reaction, $C_{cal} = q_{cal}/\Delta T$. Second, use that value of C_{cal} to find the energy change of the neutralization reaction. For the second step, use the same equation rearranged to $q_{cal} = C_{cal}\Delta T$, but with ΔT now the change in temperature observed during the reaction. Note that the calorimeter contains the same volume of liquid in both cases. Because dilute aqueous solutions have approximately the same heat capacities as pure water, assume that the heat capacity is the

FIGURE 6.11 The energy released or absorbed as heat by a reaction at constant pressure can be measured in this simple calorimeter. The outer polystyrene cup acts as an extra layer of insulation to ensure that no heat enters or leaves the inner cup. The quantity of energy released or absorbed as heat is proportional to the change in temperature of the calorimeter.

Thermometer

Stirrer

Foamed polystyrene cups

Reaction mixture

same as in the calibration. Finally, relate the heat transferred to the change in internal energy by using $\Delta U = q$.

SOLUTION

Calibration: From $C_{cal} = q_{cal}/\Delta T$, $C_{cal} = \dfrac{1.78 \text{ kJ}}{3.65°C} = \dfrac{1.78}{3.65} \text{ kJ}\cdot(°C)^{-1}$

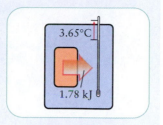

Application: From $q_{cal} = C_{cal}\Delta T$, $q_{cal} = \left(\dfrac{1.78}{3.65} \text{ kJ}\cdot(°C)^{-1}\right) \times (1.26 \text{ °C}) = 0.614 \text{ kJ}$

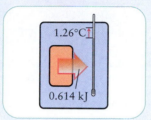

Because the temperature rises, the process is exothermic, energy leaves the system as heat and $\Delta U = -0.614$ kJ.

SELF-TEST 6.4A A small piece of calcium carbonate was placed in the same calorimeter, and 0.100 L of dilute hydrochloric acid was poured over it. The temperature of the calorimeter rose by 3.57°C. What is the value of ΔU?

[*Answer:* −1.74 kJ]

SELF-TEST 6.4B A calorimeter was calibrated by mixing two aqueous solutions together, each of volume 0.100 L. The heat output of the reaction that took place was known to be 4.16 kJ, and the temperature of the calorimeter rose by 3.24°C. Calculate the heat capacity of this calorimeter when it contains 0.200 L of water.

> *The heat capacity of an object is the ratio of the heat supplied to the temperature rise produced. Heat transfers are measured by using a calibrated calorimeter.*

6.6 The First Law

So far, we have considered the transfer of energy as work or heat separately. However, in many processes, the internal energy changes as a result of both work and heat. For example, when a spark ignites the mixture of gasoline vapor and air in the engine of a moving automobile, the vapor burns and expands, transferring energy to its surroundings as both work and heat. In general, the change in the internal energy of a closed system is the net result of both kinds of transfers; so we write

$$\Delta U = q + w \tag{7}*$$

This expression summarizes the experimental fact that both heat and work are means of transferring energy and thereby changing the internal energy of a system. The fundamental molecular difference between work and heat is that when energy is transferred as work the system moves molecules in the surroundings in a definite direction (think of the atoms of a rising weight all moving upward simultaneously), but during the transfer of energy as heat, the molecules of the surroundings are moved in random directions (think of the atoms of a hot object jostling the atoms of the surroundings into more vigorous random motion, Fig. 6.13).

SELF-TEST 6.5A An automobile engine does 520. kJ of work and loses 220. kJ of energy as heat. What is the change in the internal energy of the engine? Treat the engine, fuel, and exhaust gases as a closed system.

[*Answer:* −740. kJ]

SELF-TEST 6.5B A system was heated by using 300. J of heat, yet it was found that its internal energy decreased by 150. J (so $\Delta U = -150.$ J). Calculate w. Was work done on the system or did the system do work?

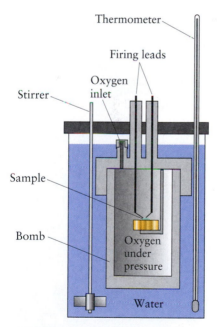

FIGURE 6.12 A bomb calorimeter is used to measure heat transfers at constant volume. The sample in the central rigid container called the bomb is ignited electrically with a fuse wire. Once combustion has begun, energy released as heat spreads through the walls of the bomb into the water. The heat released is proportional to the temperature change of the entire assembly.

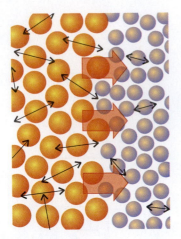

FIGURE 6.13 On an atomic scale, the transfer of energy as heat can be pictured as a process in which the vigorous thermal motion of atoms in the system jostle the less vigorously moving atoms of the surroundings and transfer some of their energy to them. The double-headed arrows represent the motion of the atoms; the large orange arrows represent the direction of heat transfer.

The first law is closely related to the conservation of energy (Section A) but goes beyond it: the concept of heat does not apply to the single particles treated in classical mechanics.

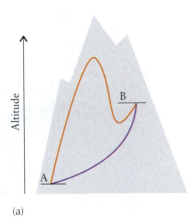

(a)

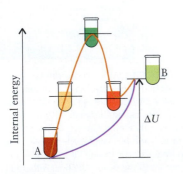

(b)

It is an experimental fact—a fact supported by thousands of experiments—that we cannot use a system to do work, leave it isolated for a while, and then return to it to find its internal energy restored to its original value and ready to provide the same amount of work again. Despite the great amount of effort that has been spent trying to build a "perpetual motion machine," a device that would be an exception to this rule by producing work without using fuel, no one has ever succeeded in building one. In other words, Eq. 7 is a *complete* statement of how changes in internal energy may be achieved in a closed system: the only way to change the internal energy in a closed system is to transfer energy into it as heat or as work. If the system is isolated, then even that ability is eliminated, and the internal energy cannot change at all. This conclusion is known as the **first law of thermodynamics**, which states:

The internal energy of an isolated system is constant.

According to the first law, if an isolated system has a certain internal energy at one instant and we inspect it again later, then we shall find that it still has exactly the same internal energy, no matter how much time has passed. In addition, if we were to prepare a second system consisting of exactly the same amount of substance in exactly the same state as the first, and inspect that system, we would find that the second system has the same internal energy as the first system. We summarize these statements by saying that the internal energy is a **state function,** a property that depends only on the current state of the system and is independent of how that state was prepared. The pressure, volume, temperature, and density of a system are also state functions.

The importance of state functions in thermodynamics is that, *because a state function depends only on the current state of the system, then if the system is changed from one state to another, the change in a state function is independent of how that change was brought about.* A state function is like altitude on a mountain (Fig. 6.14). We may take any number of different paths between two huts on the mountain, but the change in altitude between the two huts is the same regardless of the path. Similarly, if we took 100 g of water at 25°C and raised its temperature to 60°C, its internal energy would change by a certain amount. However, if we took the same mass of water at 25°C and heated it to boiling, vaporized it, condensed the vapor, and then allowed the water to cool to 60°C, the net change in the internal energy of the water would be exactly the same as if we had heated it to 60°C in one step.

The work done by a system is *not* a state function: it depends on how the change is brought about. For example, we could let a gas at 25°C expand at constant temperature (by keeping it in touch with a water bath) through 100 cm³ by two different paths. In the first experiment, the gas might push on a piston and do a certain amount of work against an external force. In the second experiment, the gas might push a piston into a vacuum; it then does no work, because there is no external opposing force (Fig. 6.15). The change in the state of the gas is the same in each case, but the work done by the system is different: in the first case, w is nonzero; in the second case, $w = 0$. Indeed, even everyday language suggests that work is not a state function, because we never speak of a system as possessing a certain amount of "work."

Similarly, heat is not a state function. The energy transferred as heat during a change in the state of a system depends on how the change is brought about. For example, suppose we want to raise the temperature of 100 g of water from 25°C to 30°C. One way to raise the temperature would be to supply energy as heat by

FIGURE 6.14 (a) The altitude of a location on a mountain is like a thermodynamic state function: no matter what route is taken between points A and B, the net change in altitude is the same. (b) Internal energy is a state function: if a system changes from state A to state B (as depicted diagrammatically here), the net change in internal energy is the same whatever the route—the sequence of chemical or physical changes—between the two states.

using an electric heater. The heat required can be calculated from the specific heat capacity of water: $q = (4.18 \text{ J·°C}^{-1}\text{·g}^{-1}) \times (100 \text{ g}) \times (5°\text{C}) = +2 \text{ kJ}$. Another way to raise the temperature would be to stir the water vigorously with paddles until 2 kJ of work has been supplied. In the latter case, all the energy required is transferred as work; none is supplied as heat. So, in the first case, $q = +2 \text{ kJ}$; and, in the second case, $q = 0$. However, the final state of the system is the same in each case. Because heat is not a state function, we should not speak of a system as possessing a certain amount of "heat."

Because internal energy is a state function, we can choose any convenient path between the given initial and final states of a system and calculate ΔU for that path. The result will have the same value of ΔU as the actual path between the same two states, even if the actual path is so complicated that ΔU cannot be calculated directly. In some cases we can use our molecular insight to give the change in internal energy for a process without doing a calculation. For instance, when an ideal gas expands isothermally, its molecules continue to move at the same average speed, so their total kinetic energy remains the same. Because there are no forces between the molecules, their total potential energy also remains the same even though their average separation has increased. Because neither the total kinetic energy nor the total potential energy changes, the internal energy of the gas is unchanged too. That is, $\Delta U = 0$ for the isothermal expansion (or compression) of an ideal gas. It follows that when an ideal gas changes by *any* path between two states, then provided the initial and final states have the same temperature we know at once that $\Delta U = 0$.

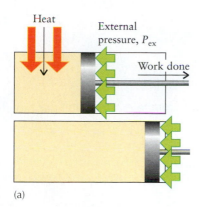
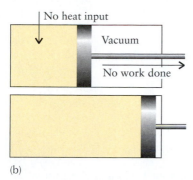

(a)

(b)

FIGURE 6.15 Two different paths between the same initial and final states of an ideal gas. (a) The gas does work as it expands isothermally against an applied, matching pressure. Because energy flows in as heat to restore the energy lost as work, the temperature remains constant. (b) The gas does no work as it expands isothermally into a vacuum. Because internal energy is a state function, the change in internal energy is the same for both processes: $\Delta U = 0$ for the isothermal expansion of an ideal gas on any path. However, the exchange of heat and work is different in each case.

EXAMPLE 6.5 Calculating the work, heat, and change in internal energy for the expansion of an ideal gas

Suppose that 1.00 mol of ideal gas molecules at 292 K and 3.00 atm expands from 8.00 L to 20.00 L and a final pressure of 1.20 atm by two different paths. (a) Path A is an isothermal, reversible expansion. (b) Path B has two parts. In step 1, the gas is cooled at constant volume until its pressure has fallen to 1.20 atm. In step 2, it is heated and allowed to expand against a constant pressure of 1.20 atm until its volume is 20.00 L and $T = 292$ K. Determine for each path the work done, the heat transferred, and the change in internal energy (w, q, and ΔU).

STRATEGY It is a good idea to begin by sketching a plot of each process (Fig. 6.16). (a) For an isothermal, reversible expansion, we use Eq. 4 to calculate w. We expect w to be negative, because energy is lost as work is done. (b) In step 1, the volume does not change, and so no work is done ($w = 0$). Step 2 is a constant-pressure process, so we use Eq. 3 to calculate w. Because internal energy is a state function and because the initial and final states are the same in both paths, ΔU for path B is the same as for path A.

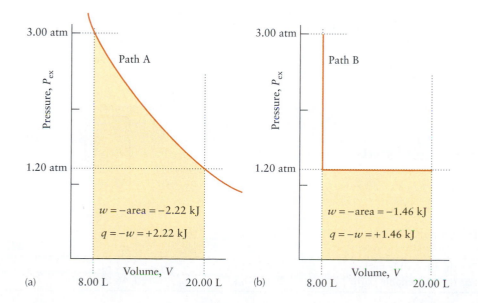

$w = -\text{area} = -2.22 \text{ kJ}$

$q = -w = +2.22 \text{ kJ}$

$w = -\text{area} = -1.46 \text{ kJ}$

$q = -w = +1.46 \text{ kJ}$

(a) 8.00 L   20.00 L   (b) 8.00 L   20.00 L

FIGURE 6.16 (a) On the reversible path, the work done in Example 6.5 is relatively large ($w = -2.22$ kJ); because the change in internal energy is zero, heat flows in to maintain constant temperature and constant internal energy. Therefore, $q = +2.22$ kJ. (b) On the irreversible path, the work done is also equal to the negative of the area beneath the curve and, for this path, is relatively small ($w = -1.46$ kJ). The net heat input, allowing for outflow of heat in the cooling step and influx of heat in the expansion, is $q = +1.46$ kJ.

Living Graph: Figure 6.16 Work of reversible, isothermal expansion of an ideal gas

As remarked in the text, $\Delta U = 0$ for an isothermal expansion of an ideal gas. In each case, we find q for the overall path from $\Delta U = q + w$ with $\Delta U = 0$. Use 1 L·atm = 101.325 J to convert liter-atmospheres into joules.

SOLUTION

(a) From $w = -nRT \ln(V_{final}/V_{initial})$, $w = -(1.00 \text{ mol}) \times (8.3145 \text{ J·K}^{-1}\text{·mol}^{-1}) \times (292 \text{ K})$

$$\times \ln\left(\frac{20.00}{8.00}\right)$$

$$= -2.22 \times 10^3 \text{ J} = -2.22 \text{ kJ}$$

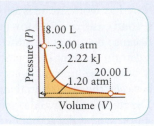

From $\Delta U = q + w = 0$, $q = -w = +2.22 \text{ kJ}$

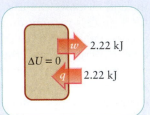

(b) **Step 1** Cool at constant volume, so no work is done. $w = 0$

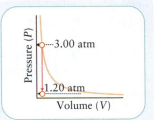

Step 2 From $w = -P_{ex}\Delta V$, $w = -(1.20 \text{ atm}) \times (20.00 - 8.00) \text{ L}$
$$= -14.4 \text{ L·atm}$$

Convert liter-atmospheres to joules. $w = -(14.4 \text{ L·atm}) \times \left(\dfrac{101.325 \text{ J}}{1 \text{ L·atm}}\right)$

$$= -1.46 \times 10^3 \text{ J} = -1.46 \text{ kJ}$$

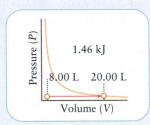

Calculate total work for path B. $w = 0 + (-1.46) \text{ kJ} = -1.46 \text{ kJ}$

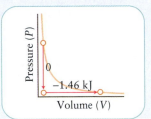

From $\Delta U = q + w = 0$, $q = -w = +1.46 \text{ kJ}$

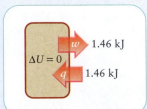

In summary,

	q	w	ΔU
For the reversible path:	+2.22 kJ	−2.22 kJ	0
For the irreversible path:	+1.46 kJ	−1.46 kJ	0

SELF-TEST 6.6A Suppose that 2.00 mol CO_2 at 2.00 atm and 300. K is compressed isothermally and reversibly to half its original volume before being used to produce soda water. Calculate w, q, and ΔU by treating the CO_2 as an ideal gas.

[*Answer:* $w = +3.46$ kJ, $q = -3.46$ kJ, $\Delta U = 0$]

SELF-TEST 6.6B Suppose that 1.00 kJ of energy is transferred as heat to oxygen in a cylinder fitted with a piston; the external pressure is 2.00 atm. The oxygen expands from 1.00 L to 3.00 L against this constant pressure. Calculate w and ΔU for the entire process by treating the O_2 as an ideal gas.

The first law of thermodynamics states that the internal energy of an isolated system is constant. A state function depends only on the current state of a system. The change in a state function between two states is independent of the path between them. Internal energy is a state function; work and heat are not.

6.7 A Molecular Interlude: The Origin of Internal Energy

Internal energy is energy stored in a system as kinetic energy and potential energy. We saw in Section A that kinetic energy is energy due to motion and that the faster a molecule travels, the greater its kinetic energy. When we heat a gas, the average speed of the molecules increases. When we do work on a gas in an insulated container, the molecules are also stimulated to move faster. The increase in the average speed of the gas molecules corresponds to an increase in the total kinetic energy of the molecules and therefore to an increase in the internal energy of the gas. We have already seen (in Section 4.10) that the average speed of gas molecules is an indication of temperature, and so this increase in internal energy also corresponds to an increase in temperature. It is always the case that *a system at high temperature has a greater internal energy than the same system at a lower temperature.*

Molecules in a gas can move in a variety of different ways, and each mode of motion can act as a store of energy (Fig. 6.17). The kinetic energy of an atom or molecule as it moves through space is called its **translational kinetic energy**. Molecules (in contrast to atoms) can also store energy as **rotational kinetic energy**, the kinetic energy arising from their rotational motion. A third store of kinetic energy for molecules is in the oscillation of their atoms relative to one another: this contribution is called **vibrational kinetic energy**; however, most molecules are not vibrationally excited at room temperature, and so we ignore that mode for now.

The contribution of translational and rotational motion to the internal energy can be estimated from the temperature.

HOW DO WE DO THAT?

First, we need to know that a "quadratic contribution" to the energy is an expression that depends on the square of a velocity or a displacement, as in $\frac{1}{2}mv^2$ for translational kinetic energy. The **equipartition theorem** (which we shall not derive here) then states that *the average value of each quadratic contribution to the energy of a molecule in a sample at a temperature T is equal to $\frac{1}{2}kT$.* The name "equipartition" simply means that the available energy is shared (partitioned) equally over all the available modes. In this expression, k is Boltzmann's constant, a fundamental constant with the value 1.381×10^{-23} J·K^{-1}. Boltzmann's constant is related to the gas constant by $R = N_A k$, where N_A is Avogadro's constant. The equipartition theorem is a result from classical mechanics; so we can use it for translational and rotational motion of molecules at room temperature and above, where quantization is unimportant; because large numbers of quantum states are occupied, only averages are observed. We cannot use it reliably for vibrational motion, except at very high temperatures, because at low temperatures only the lowest vibrational

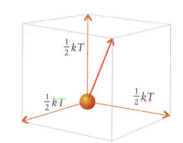

(a)

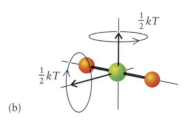

(b)

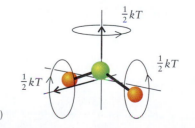

(c)

FIGURE 6.17 The translational and rotational modes of atoms and molecules and the corresponding average energies of each mode at a temperature T. (a) An atom or molecule can undergo translational motion in three dimensions. (b) A linear molecule can also rotate about two axes perpendicular to the line of atoms. (c) A nonlinear molecule can rotate about three perpendicular axes.

quantum state is occupied and the effects of quantization are very important. The following remarks therefore apply only to translational and rotational motion.

A molecule can move through space along any of three dimensions, so it has *three* translational modes of motion, each one giving a quadratic contribution to the energy. It follows from the equipartition theorem that the average translational kinetic energy of a molecule in a sample at a temperature T is $3 \times \frac{1}{2}kT = \frac{3}{2}kT$. The contribution to the molar internal energy (the energy per mole of molecules, U_m) is therefore N_A times this value, or

$$U_m(\text{translation}) = \tfrac{3}{2}N_A kT = \tfrac{3}{2}RT$$

Because $RT = 2.48$ kJ·mol^{-1} at 25°C, the translational motion of gas molecules contributes $\frac{3}{2} \times 2.48$ kJ·mol$^{-1} = 3.72$ kJ·mol^{-1} to the internal energy of the sample at 25°C. Apart from the energy arising from the internal structures of the atoms themselves, this is the only motional contribution to the internal energy of a monatomic gas, such as argon or any other noble gas.

A *linear molecule*, such as any diatomic molecule, carbon dioxide, and ethyne (acetylene, HC≡CH), can rotate about two axes perpendicular to the line of atoms, and so it has two rotational modes of motion. Its average rotational energy is therefore $2 \times \frac{1}{2}kT = kT$, and the contribution to the molar internal energy is N_A times this value:

$$U_m(\text{rotation, linear}) = RT$$

or about 2.48 kJ·mol^{-1} at 25°C. The total motional contribution to the internal energy of a gas of linear molecules is the sum of the translational and rotational contributions, or $\frac{5}{2}RT$, which corresponds to 6.20 kJ·mol^{-1} at 25°C.

A *nonlinear molecule*, such as water, methane, or benzene, can rotate about any of three perpendicular axes, and so it has three rotational modes of motion. The average rotational energy of such a molecule is therefore $3 \times \frac{1}{2}kT = \frac{3}{2}kT$. The contribution of rotation to the molar internal energy of a gas of nonlinear molecules is therefore

$$U_m(\text{rotation, nonlinear}) = \tfrac{3}{2}RT$$

At 25°C, the contribution of rotation to the energy is 3.72 kJ·mol^{-1} (the same as for the translational motion), and the total motional contribution for nonlinear molecules is therefore $3RT$, or about 7.44 kJ·mol^{-1} at 25°C.

We have shown that the contribution to the molar internal energy of a monatomic ideal gas (such as argon) that arises from molecular motion is $\frac{3}{2}RT$. We can conclude that if the gas is heated through ΔT, then the change in its molar internal energy, ΔU_m, is $\Delta U_m = \frac{3}{2}R\Delta T$. For instance, if the gas is heated from 20.°C to 100.°C (so $\Delta T = +80.$ K), then its molar internal energy increases by 1.0 kJ·mol^{-1}.

Internal energy is stored as molecular kinetic and potential energy. The equipartition theorem can be used to estimate the translational and rotational contributions to the internal energy of an ideal gas.

ENTHALPY

In a constant-volume system in which neither expansion work nor nonexpansion work is done, we can set $w = 0$ in Eq. 7 ($\Delta U = w + q$) and obtain

At constant volume: $\Delta U = q$ (8)*

That is, in such a system, the change in the internal energy of the system is equal to the energy supplied to it as heat. In chemistry, though, we are more concerned with heat transfers at constant pressure. Many chemical reactions take place in containers open to the atmosphere and therefore take place at a constant pressure of about 1 atm. Such systems are free to expand or contract. If a gas is evolved, the expanding gas needs to drive back the surrounding atmosphere to make room for itself. Work is done, even though there is no actual piston. For example, if in some process we supply 100 J of heat to a system ($q = +100$ J) and the system does 20 J of work ($w = -20$ J) as it expands at constant pressure, the internal energy of the system rises by only 80 J. Clearly, in this case, the rise in internal energy is not equal to the heat supplied to a system. Is there another state function that will allow us to keep track of energy changes at constant pressure?

6.8 Heat Transfers at Constant Pressure

The state function that allows us to keep track of energy changes at constant pressure is called the **enthalpy**, H:

$$H = U + PV \tag{9}*$$

where U, P, and V are the internal energy, pressure, and volume of the system. We can see that enthalpy is a state function by noting that U (from the first law), P, and V are all state functions, and so $H = U + PV$ must be a state function, too. What we now need to show is that it follows from this definition that *a change in the enthalpy of a system is equal to the heat released or absorbed at constant pressure.*

HOW DO WE DO THAT?

Consider a process at constant pressure for which the change in internal energy is ΔU and the change in volume is ΔV. It then follows from the definition of enthalpy in Eq. 9 that the change in enthalpy is

$$\text{At constant pressure: } \Delta H = \Delta U + P\Delta V \tag{10}$$

Now we use $\Delta U = q + w$, where q is the energy supplied to the system as heat and w is the energy supplied as work. Therefore,

$$\Delta H = q + w + P\Delta V$$

Next, suppose that the system can do no work other than expansion work. In that case, we can use Eq. 3 ($w = -P_{ex}\Delta V$) to write

$$\Delta H = q - P_{ex}\Delta V + P\Delta V$$

Finally, because the system is open to the atmosphere, the pressure of the system is the same as the external pressure; so $P_{ex} = P$, and the last two terms cancel to leave $\Delta H = q$.

We have shown that for a system that can do work only by expansion,

$$\text{At constant pressure: } \Delta H = q \tag{11}*$$

The implication of this equation is that, because chemical reactions typically take place at constant pressure in vessels open to the atmosphere, the heat that they release or require can be equated to the change in enthalpy of the system. It follows that if we study a reaction in a calorimeter that is open to the atmosphere (such as that depicted in Fig. 6.11), then the measurement of its temperature rise gives us the enthalpy change that accompanies the reaction. For instance, if a reaction releases 1.25 kJ of heat in this kind of calorimeter, then we can write $\Delta H = q = -1.25$ kJ.

When we transfer energy to a constant-pressure system as heat, the enthalpy of the system *increases* by that amount. When energy leaves a constant-pressure system as heat, the enthalpy of the system *decreases* by that amount. For example, the formation of zinc iodide from its elements is an exothermic reaction that (at constant pressure) releases 208 kJ of heat to the surroundings for each mole of ZnI_2 formed:

$$Zn(s) + I_2(s) \longrightarrow ZnI_2(s)$$

We can therefore report that $\Delta H = -208$ kJ because the enthalpy of the reaction mixture decreases by 208 kJ in this reaction (Fig. 6.18). An endothermic process absorbs heat, and so when ammonium nitrate dissolves in water the enthalpy of the system increases (Fig. 6.19). Note that $\Delta H < 0$ for exothermic reactions, whereas $\Delta H > 0$ for endothermic reactions.

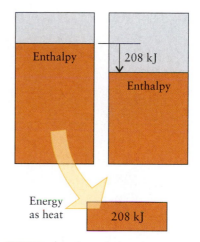

FIGURE 6.18 The enthalpy of a system is like a measure of the height of water in a reservoir. When a reaction releases 208 kJ of heat at constant pressure, the "reservoir" falls by 208 kJ and $\Delta H = -208$ kJ.

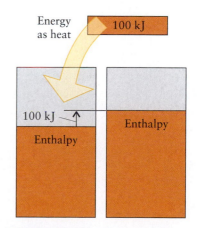

FIGURE 6.19 If an endothermic reaction absorbs 100 kJ of heat at constant pressure, the height of the enthalpy "reservoir" rises by 100 kJ and $\Delta H = +100$ kJ.

SELF-TEST 6.7A   In an exothermic reaction at constant pressure, 50. kJ of energy left the system as heat and 20. kJ of energy left the system as expansion work. What are the values of (a) ΔH and (b) ΔU for this process?

[*Answer:* (a) −50. kJ; (b) −70. kJ]

SELF-TEST 6.7B   In an endothermic reaction at constant pressure, 30. kJ of energy entered the system as heat. The products took up less volume than the reactants, and 40. kJ of energy entered the system as work as the outside atmosphere pressed down on it. What are the values of (a) ΔH and (b) ΔU for this process?

The change in enthalpy of a system is equal to the heat supplied to the system at constant pressure. For an endothermic process, $\Delta H > 0$; for an exothermic process, $\Delta H < 0$.

6.9 Heat Capacities at Constant Volume and Constant Pressure

As explained in Section 6.5, the heat capacity of a substance is the constant of proportionality between the heat supplied to a substance and the temperature rise that results ($q = C\Delta T$). However, the rise in temperature and therefore the heat capacity depend on the conditions under which the heating takes place because, at constant pressure, some of the heat is used to do expansion work rather than to raise the temperature of the system. We need to refine our definition of heat capacity.

Because heat transferred at constant volume can be identified with the change in internal energy, ΔU, we can combine Eq. 8 with $C = q/\Delta T$ and define the **heat capacity at constant volume**, C_V, as

$$C_V = \frac{\Delta U}{\Delta T} \tag{12a}$$

Similarly, because heat transferred at constant pressure can be identified with the change in enthalpy, ΔH, we can define the **heat capacity at constant pressure**, C_P, as

$$C_P = \frac{\Delta H}{\Delta T} \tag{12b}$$

The corresponding molar heat capacities are these quantities divided by the amount of substance and are denoted $C_{V,m}$ and $C_{P,m}$.

The constant-volume and constant-pressure heat capacities of a solid substance are similar; the same is true of a liquid but not of a gas. We can use the definition of enthalpy and the ideal gas law to find a simple quantitative relation between C_P and C_V for an ideal gas.

With rare exceptions, solids and liquids expand when heated, but to a much smaller extent than gases.

HOW DO WE DO THAT?

For an ideal gas, the PV in the definition of enthalpy, $H = U + PV$, can be replaced by nRT, and so

$$H = U + nRT$$

When a sample of an ideal gas is heated, the enthalpy, internal energy, and temperature all change, and it follows that

$$\Delta H = \Delta U + nR\Delta T$$

The heat capacity at constant pressure can therefore be expressed as

$$C_P = \frac{\Delta H}{\Delta T} = \frac{\Delta U + nR\Delta T}{\Delta T} = \frac{\Delta U}{\Delta T} + nR = C_V + nR$$

To obtain the relation between the two molar heat capacities, we divide this expression by n.

We have shown that the two molar heat capacities of an ideal gas are related by

$$C_{P,m} = C_{V,m} + R \tag{13}*$$

As an example, the molar constant-volume heat capacity of argon is 12.8 $\text{J·K}^{-1}\text{·mol}^{-1}$, and so the corresponding constant-pressure value is 12.8 + 8.3

$J \cdot K^{-1} \cdot mol^{-1} = 21.1\ J \cdot K^{-1} \cdot mol^{-1}$, a difference of 65%. The heat capacity at constant pressure is greater than that at constant volume because at constant pressure not all the heat supplied is used to raise the temperature: some returns to the surroundings as expansion work and $C = q/\Delta T$ is larger (because ΔT is smaller) than at constant volume (when all the energy remains inside the system).

> *The molar heat capacity of an ideal gas at constant pressure is greater than that at constant volume; the two quantities are related by Eq. 13.*

6.10 A Molecular Interlude: The Origin of the Heat Capacities of Gases

We can see how the values of heat capacities depend on molecular properties by using the relations in Section 6.7. We start with a simple system, a monatomic ideal gas such as argon. We saw in Section 6.7 that the molar internal energy of a monatomic ideal gas at a temperature T is $\frac{3}{2}RT$ and that the change in molar internal energy when the temperature is changed by ΔT is $\Delta U_m = \frac{3}{2}R\Delta T$. It follows from Eq. 12a that the molar heat capacity at constant volume is

$$\text{For monatomic gases,}\quad C_{V,m} = \frac{\Delta U_m}{\Delta T} = \frac{\frac{3}{2}R\Delta T}{\Delta T} = \frac{3}{2}R$$

or about 12.5 $J \cdot K^{-1} \cdot mol^{-1}$, in agreement with the value found experimentally. Note that the molar heat capacity of a monatomic ideal gas is independent of temperature and pressure. From Eq. 13 ($C_{P,m} = C_{V,m} + R$), the molar heat capacity of an ideal gas at constant pressure is $\frac{3}{2}R + R = \frac{5}{2}R$.

The molar heat capacities of gases composed of molecules (as distinct from atoms) are higher than those of monatomic gases because the molecules can store energy as rotational kinetic energy as well as translational kinetic energy. We saw in Section 6.7 that the rotational motion of linear molecules contributes another RT to the molar internal energy:

$$\text{For linear molecules,}\quad C_{V,m} = \frac{\Delta U_m}{\Delta T} = \frac{\frac{3}{2}R\Delta T + RT}{\Delta T} = \frac{5}{2}R$$

For nonlinear molecules, the rotational contribution is $\frac{3}{2}RT$, for a total contribution of $3RT$. Therefore, we have the following expressions:

	Atoms	Linear molecules	Nonlinear molecules
$C_{V,m}$	$\frac{3}{2}R$	$\frac{5}{2}R$	$3R$
$C_{P,m}$	$\frac{5}{2}R$	$\frac{7}{2}R$	$4R$

(In each case, $C_{P,m}$ has been calculated from $C_{P,m} = C_{V,m} + R$.) Note that the molar heat capacity increases with molecular complexity. The molar heat capacity of nonlinear molecules is higher than that of linear molecules because nonlinear molecules can rotate about three rather than only two axes (recall Fig. 6.17).

The graph in Fig. 6.20 shows how $C_{V,m}$ for iodine vapor, $I_2(g)$, varies with temperature. At very low temperatures $C_{V,m} = \frac{3}{2}R$, but soon rises to $\frac{5}{2}R$ as molecular rotation takes place. At still higher temperatures, molecular vibrations start to absorb energy and the heat capacity rises toward $\frac{7}{2}R$. At 298 K, the experimental value is equivalent to $3.4R$.

EXAMPLE 6.6 Calculating the energy changes when heating an ideal gas

Calculate the final temperature and the change in internal energy when 500. J of energy is transferred as heat to 0.900 mol $O_2(g)$ at 298 K and 1.00 atm at (a) constant volume; (b) constant pressure. Treat the gas as ideal.

STRATEGY We expect the temperature to rise more as a result of heating at constant volume than at constant pressure because at constant pressure some of the energy is used to expand the system. Oxygen is a linear molecule and its heat capacities can be

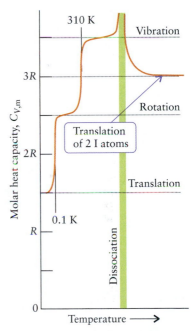

FIGURE 6.20 Variation in the molar heat capacity of iodine vapor at constant volume. The translational motion of the gaseous molecules contributes to the heat capacity even at very low temperatures. As the temperature rises past 0.05 K, rotation makes a significant contribution, but vibrations of the molecule contribute only at high temperatures (above about 310 K; for most other molecules the temperature must be much higher). When the molecules dissociate, the heat capacity becomes very large during dissociation, but then settles down to a value characteristic of 2 mol I atoms undergoing only translational motion.

estimated from the equipartition theorem; then we can use $q = C\Delta T$, with $C = nC_m$, to find the final temperature. The internal energy change at constant volume is equal to the heat supplied. At constant pressure, find the internal energy change by regarding the process as taking place in two steps: heating to the final temperature at constant volume, followed by isothermal expansion.

SOLUTION

From $C_{V,m} = \frac{5}{2}R$ and
$C_{P,m} = C_{V,m} + R = \frac{7}{2}R$,

$C_{V,m} = \frac{5}{2}R = 20.79 \text{ J·K}^{-1}\text{·mol}^{-1}$
$C_{P,m} = \frac{7}{2}R = 29.10 \text{ J·K}^{-1}\text{·mol}^{-1}$

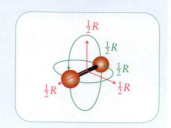

(a) From $\Delta T = q/nC_{V,m}$,

$$\Delta T = \frac{500.\text{ J}}{(0.900 \text{ mol}) \times (20.79 \text{ J·K}^{-1}\text{·mol}^{-1})}$$

$$= +26.7 \text{ K}$$

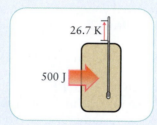

Find the final temperature.

$T = 298 + 26.7 \text{ K} = 325 \text{ K, or } 52°C$

From $\Delta U = q$,

$\Delta U = +500.\text{ J}$

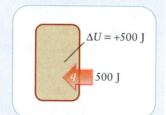

(b) To find ΔU for heating at constant pressure, find the final temperature, then break down the overall process into two steps:

From $\Delta T = q/nC_{P,m}$,

$$\Delta T = \frac{500.\text{ J}}{(0.900 \text{ mol}) \times (29.10 \text{ J·K}^{-1}\text{·mol}^{-1})}$$

$$= +19.1 \text{ K}$$

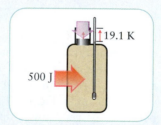

Find the final temperature.

$T = 298 + 19.1 \text{ K} = 317 \text{ K, or } 44°C.$

Heat: Transfer enough energy at constant volume to raise the temperature to its final value (317 K), and use $\Delta U = q$.

$\Delta U = q$
$= (0.900 \text{ mol}) \times (20.79 \text{ J·K}^{-1}\text{·mol}^{-1}) \times (19.1 \text{ K})$
$= +357 \text{ J}$

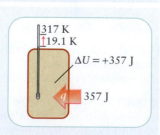

Expand: Allow the sample to expand isothermally to its final volume.

Because U is independent of volume for an ideal gas, $\Delta U_{\text{step 2}} = 0$

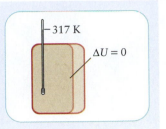

Now add the two changes in internal energy.

$\Delta U = +357\ \text{J}$

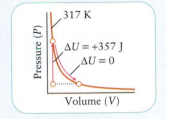

The overall change is smaller than for the constant-volume case, as predicted.

SELF-TEST 6.8A Calculate the final temperature and the change in internal energy when 500. J of energy is transferred as heat to 0.900 mol Ne(g) at 298 K and 1.00 atm (a) at constant volume; (b) at constant pressure. Treat the gas as ideal.

[*Answer:* (a) 343 K, 500. J; (b) 325 K, 300. J]

SELF-TEST 6.8B Calculate the final temperature and the change in internal energy when 1.20 kJ of energy is transferred as heat to 1.00 mol H_2(g) at 298 K and 1.00 atm (a) at constant volume; (b) at constant pressure. Treat the gas as ideal.

> *Rotation requires energy and leads to higher heat capacities for complex molecules; the equipartition theorem can be used to estimate the molar heat capacities of gas-phase molecules.*

6.11 The Enthalpy of Physical Change

A phase change in which the molecules become further separated, such as vaporization, requires energy to break intermolecular attractions and is therefore endothermic. Phase changes that increase molecular contact, such as freezing, are exothermic because energy is given off when attractions form between molecules. When a phase transition takes place at constant pressure, as is most common, the heat transfer due to the phase change is the change in enthalpy of the substance.

The difference in molar enthalpy between the vapor and the liquid states of a substance is called the **enthalpy of vaporization**, ΔH_{vap}:

$$\Delta H_{\text{vap}} = H_{\text{m}}(\text{vapor}) - H_{\text{m}}(\text{liquid}) \tag{14}$$

The enthalpy of vaporization of most substances changes little with temperature. For water at its boiling point, 100°C, $\Delta H_{\text{vap}} = 40.7\ \text{kJ·mol}^{-1}$; but, at 25°C, $\Delta H_{\text{vap}} = 44.0\ \text{kJ·mol}^{-1}$. The latter value means that to vaporize 1.00 mol H_2O(l), corresponding to 18.02 g of water, at 25°C and constant pressure, we have to supply 44.0 kJ of energy as heat. The enthalpy of vaporization of a substance is usually determined by measuring the heat required to vaporize a known mass of the substance.

Why don't we write the signs explicitly? See the next paragraph.

SELF-TEST 6.9A A sample of benzene, C_6H_6, was heated to 80°C, its normal boiling point. The heating was continued until 15.4 kJ had been supplied; as a result, 39.1 g of boiling benzene was vaporized. What is the enthalpy of vaporization of benzene at its boiling point?

[*Answer:* 30.8 kJ·mol^{-1}]

SELF-TEST 6.9B The same heater was used to heat a sample of ethanol, C_2H_5OH, of mass 23 g to its boiling point. It was found that 22 kJ was required to vaporize all the ethanol. What is the enthalpy of vaporization of ethanol at its boiling point?

TABLE 6.3 Standard Enthalpies of Physical Change*

Substance	Formula	Freezing point (K)	$\Delta H_{fus}°$ (kJ·mol^{-1})	Boiling point (K)	$\Delta H_{vap}°$ (kJ·mol^{-1})
acetone	CH_3COCH_3	177.8	5.72	329.4	29.1
ammonia	NH_3	195.4	5.65	239.7	23.4
argon	Ar	83.8	1.2	87.3	6.5
benzene	C_6H_6	278.6	10.59	353.2	30.8
ethanol	C_2H_5OH	158.7	4.60	351.5	43.5
helium	He	3.5	0.021	4.22	0.084
mercury	Hg	234.3	2.292	629.7	59.3
methane	CH_4	90.7	0.94	111.7	8.2
methanol	CH_3OH	175.2	3.16	337.8	35.3
water	H_2O	273.2	6.01	373.2	40.7
					(44.0 at 25°C)

*Values correspond to the temperature of the phase change. The superscript ° signifies that the change takes place at 1 bar and that the substance is pure (that is, the values are for standard states; see Section 6.15).

Table 6.3 lists the enthalpies of vaporization of various substances. All enthalpies of vaporization are positive, and so it is conventional to report them without their sign. Notice that compounds with strong intermolecular forces, such as hydrogen bonds, tend to have the highest enthalpies of vaporization. That correlation is easy to explain, because the enthalpy of vaporization is a measure of the energy needed to separate molecules from their attractions in the liquid state into a free state in the vapor. In plots of the potential energy arising from intermolecular interactions, like that shown in Fig. 6.21, the enthalpy of the substance in the liquid state, where molecular interactions are strong, is related to the depth of the bottom of the well in the curve. The enthalpy of the vapor state, where the interactions are almost insignificant, corresponds to the horizontal part of the curve to the right. A substance with a high molar enthalpy of vaporization has a deep intermolecular potential well, indicating strong intermolecular attractions.

The molar enthalpy change that accompanies melting (fusion) is called the **enthalpy of fusion**, ΔH_{fus}, of the substance:

$$\Delta H_{fus} = H_m(\text{liquid}) - H_m(\text{solid}) \tag{15}$$

Melting is, with only one known exception (helium), endothermic, and so all enthalpies of fusion (with the exception of that special case) are positive and are reported without their sign (see Table 6.3). The enthalpy of fusion of water at 0°C is 6.0 kJ·mol^{-1}: to melt 1.0 mol $H_2O(s)$ (18 g of ice) at 0°C, we have to supply 6.0 kJ of heat. Vaporizing the same amount of water takes much more energy (more than 40 kJ) because, when water is vaporized to a gas, its molecules must be separated completely. In melting, the molecules stay close together, and so the forces of attraction and repulsion are nearly as strong as those experienced in the solid (Fig. 6.22).

The **enthalpy of freezing** is the change in molar enthalpy when a liquid solidifies. Because enthalpy is a state property, the enthalpy of a sample of water must be the same after being frozen and then melted as it was before it was frozen. Therefore, the enthalpy of freezing of a substance is the negative of its enthalpy of fusion. For water at 0°C, the enthalpy of freezing is -6.0 kJ·mol^{-1}, because 6.0 kJ of heat is *released* when 1 mol $H_2O(l)$ freezes. In general, to obtain the enthalpy change for the reverse of any process, we just take the negative of the enthalpy change for the original process:

$$\Delta H_{\text{reverse process}} = -\Delta H_{\text{forward process}} \tag{16}*$$

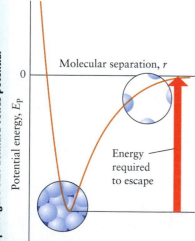

FIGURE 6.21 The potential energy of molecules decreases as they approach one another and experience intermolecular attractions, then rises again as the molecules are pressed closely together. The average intermolecular distance in a liquid is given by the position of the energy minimum. To vaporize a liquid, molecules must be raised from the bottom of the well to the energy of the horizontal part of the curve on the right.

BOX 6.1   HOW DO WE KNOW . . . THE SHAPE OF A HEATING CURVE?

The heating curve of a substance, like that in Fig. 6.26, shows how its temperature changes as heat is supplied at a constant rate, usually at constant pressure. How are these curves produced? Simple laboratory heaters can be used to obtain a crude estimate of a heating curve. However, for accuracy, one of two related techniques is normally used.

In *differential thermal analysis* (DTA), equal masses of a sample and a reference material that will not undergo any phase changes, such as Al_2O_3 (which melts at a very high temperature), are inserted into two separate sample wells in a large steel block that acts as a heat sink (see the following illustration). Because the steel block is so large, it is possible to heat the sample and reference at the same slow and accurately monitored rate. Thermocouples are placed in each well and in the block itself. Then the block is gradually heated and the temperatures of the sample and the reference are compared. An electrical signal is generated if the temperature of the sample suddenly stops rising while that of the reference continues to rise. Such an event signals an endothermic process in the sample, such as a phase change. The output of a DTA analysis is a *thermogram*, which shows the temperatures of phase changes as heat absorption peaks at the transition temperatures.

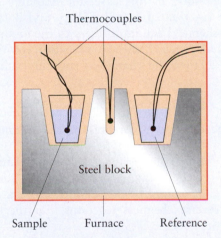

In differential thermal analysis, a sample and reference material are placed in the same large metal heat sink. Changes in the heat capacity of the sample are measured by changes in temperature between the sample and the reference materials as they are heated at the same rate.

In *differential scanning calorimetry* (DSC), higher precision can be obtained and heat capacities can be measured. The apparatus is similar to that for a DTA analysis, with the primary difference being that the sample and reference are in separate heat sinks that are heated by individual heaters (see the following illustration). The temperatures of the two samples are kept the same by differential heating. Even slight

temperature differences between the reference and the sample will trigger a switch that sends more or less power to the sample to maintain the constant temperature. If the heat capacity of the sample is higher than the heat capacity of the reference, energy must be supplied at a greater rate to the sample cell. If a phase transition occurs in the sample, a great deal of energy must be supplied to the sample until the phase transition is complete and the temperature begins to rise again.

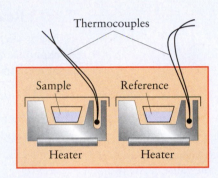

In a differential scanning calorimeter, a sample and reference material are heated in separate, but identical, metal heat sinks. The temperatures of the sample and reference material are kept the same by varying the power supplied to the two heaters. The output is the difference in power as a function of heat added.

The output of a differential scanning calorimeter is a measure of the power (the rate of energy supply) supplied to the sample cell. The thermogram in the third illustration shows a peak that signals a phase change. The thermogram does not look much like a heating curve, but it contains all the necessary information and is easily transformed into the familiar shape.

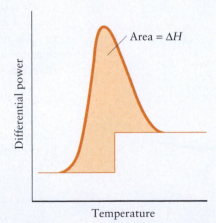

A thermogram from a differential scanning calorimeter. The peak indicates a phase change in the sample, and the difference in base line before and after the phase transition is due to the difference in heat capacities of the two phases.

The temperature of a sample is constant at its melting and boiling points, even though heat is being supplied. The slope of a heating curve is steeper for a phase with a low heat capacity than for one with a high heat capacity.

THE ENTHALPY OF CHEMICAL CHANGE

We have seen that enthalpy changes accompany physical changes, such as vaporization. The same principles apply to the energy and enthalpy changes that accompany chemical changes. Enthalpies of chemical change are important in many areas of chemistry, such as the selection of materials that make good fuels, the design of chemical plants, and the study of biochemical processes.

6.13 Reaction Enthalpies

Every chemical reaction is accompanied by the transfer of energy as heat. For example, complete reaction with oxygen is called **combustion** and the combustion of methane, the major component of natural gas, is the following reaction:

$$CH_4(g) + 2\,O_2(g) \longrightarrow CO_2(g) + 2\,H_2O(l)$$

Calorimetry shows that burning 1.00 mol $CH_4(g)$ produces 890. kJ of heat at 298 K and 1 bar. To report this value, we write

$$CH_4(g) + 2\,O_2(g) \longrightarrow CO_2(g) + 2\,H_2O(l) \qquad \Delta H = -890.\ \text{kJ} \qquad \textbf{(A)}$$

This entire expression is a **thermochemical equation**, a chemical equation together with the **reaction enthalpy**, the corresponding enthalpy change. The stoichiometric coefficients indicate the number of moles that react to give the reported change in enthalpy. Therefore, in this case, the enthalpy change is that resulting from the complete reaction of 1 mol $CH_4(g)$ and 2 mol $O_2(g)$. The enthalpy change in a thermochemical equation refers to the amounts given in the chemical equation exactly as written. If the same reaction is written with the coefficients all multiplied by 2, then the change in enthalpy will be twice as great:

$$2\,CH_4(g) + 4\,O_2(g) \longrightarrow 2\,CO_2(g) + 4\,H_2O(l) \qquad \Delta H = -1780\ \text{kJ}$$

This result makes sense because the equation now represents the burning of twice as much methane.

We saw in Section 6.11 that the first law of thermodynamics implies that, because enthalpy is a state function, the enthalpy change for the reverse of a process is the negative of the enthalpy change of the forward process. The same relation applies to forward and reverse chemical reactions. For the reverse of reaction A, for instance, we can write

$$CO_2(g) + 2\,H_2O(l) \longrightarrow CH_4(g) + 2\,O_2(g) \qquad \Delta H = +890.\ \text{kJ}$$

Once we know the reaction enthalpy, we can calculate the enthalpy change for any amount, mass, or volume of reactant consumed or product formed. As shown in the following example, we carry out a stoichiometry calculation like those described in Section L but with heat treated as a reactant or a product.

> **EXAMPLE 6.7** **Determining a reaction enthalpy from experimental data**
>
> When 0.113 g of benzene, C_6H_6, burns in excess oxygen in a calibrated constant-pressure calorimeter with a heat capacity of 551 $J \cdot (°C)^{-1}$, the temperature of the calorimeter rises by 8.60°C. Write the thermochemical equation for
>
> $$2\,C_6H_6(l) + 15\,O_2(g) \longrightarrow 12\,CO_2(g) + 6\,H_2O(l)$$
>
> **STRATEGY** The heat released by the reaction at constant pressure is calculated from the temperature change multiplied by the heat capacity of the calorimeter. Use the molar mass of one species to convert the heat released into the reaction enthalpy corresponding to the thermochemical equation as written. If the temperature rises, the

Note that, although burning takes place at much higher temperatures than 298 K, the value of ΔH is determined by the difference in enthalpies of the products and reactants, both measured at 298 K.

reaction must be exothermic and ΔH negative; if it falls, the reaction must be endothermic and ΔH positive. From Appendix 2A, the molar mass of benzene is 78.12 g·mol^{-1}.

SOLUTION

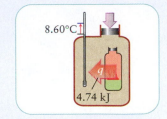

Find the heat transferred to the calorimeter from $q_{cal} = C_{cal}\Delta T$,

$$q_{cal} = [551 \text{ J·(°C)}^{-1}] \times (8.60°\text{C})$$
$$= 8.60 \times 551 \text{ J (4.74 kJ)}$$

Calculate the amount of C_6H_6 that reacts from $n = m/M$.

$$n = \frac{0.113 \text{ g}}{78.12 \text{ g·mol}^{-1}} = \frac{0.113}{78.12} \text{ mol}$$

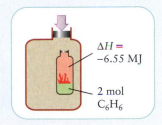

Because the stoichiometric coefficient of C_6H_6 in the chemical equation is 2, calculate ΔH for 2 mol C_6H_6 from $\Delta H = q \times (2 \text{ mol})/n$. Because the reaction is exothermic, ΔH is negative.

$$\Delta H = -\frac{(2 \text{ mol}) \times (8.60 \times 551) \text{ J}}{(0.113/78.12) \text{ mol}}$$

$$= -6.55 \times 10^6 \text{ J} = -6.55 \times 10^3 \text{ kJ}$$

We can write the thermochemical equation as

$$2 \text{ C}_6\text{H}_6(l) + 15 \text{ O}_2(g) \longrightarrow 12 \text{ CO}_2(g) + 6 \text{ H}_2\text{O}(l) \qquad \Delta H = -6.55 \text{ MJ}$$

SELF-TEST 6.11A When 0.231 g of phosphorus reacts with chlorine to form phosphorus trichloride, PCl_3, in a constant-pressure calorimeter of heat capacity 216 J·(°C)$^{-1}$, the temperature of the calorimeter rises by 11.06°C. Write the thermochemical equation for the reaction.

[*Answer:* 2 P(s) + 3 Cl$_2$(g) → 2 PCl$_3$(l), $\Delta H = -641$ kJ]

SELF-TEST 6.11B When 0.338 g of pentane, C_5H_{12}, burns in an excess of oxygen to form carbon dioxide and liquid water in the same calorimeter as that used in Self-Test 6.11A, the temperature rises by 76.7°C. Write the thermochemical equation for the reaction.

A thermochemical equation is a statement of a chemical equation and the corresponding reaction enthalpy, the enthalpy change for the stoichiometric amounts of substances in the chemical equation.

6.14 The Relation Between ΔH and ΔU

We have seen that a constant-pressure calorimeter and a constant-volume bomb calorimeter measure changes in different state functions: at constant volume, the heat transfer is interpreted as ΔU; at constant pressure, it is interpreted as ΔH. However, it is sometimes necessary to convert the measured value of ΔU into ΔH. For example, it is easy to measure the heat released by the combustion of glucose in a bomb calorimeter, but to use that information in assessing energy changes in metabolism, which take place at constant pressure, we need the enthalpy of reaction.

For reactions in which no gas is generated or consumed, little expansion work is done as the reaction proceeds and the difference between ΔH and ΔU is negligible; so we can set $\Delta H = \Delta U$. However, if a gas is formed in the reaction, so much expansion work is done to make room for the gaseous products that the difference can be significant. We can use the ideal gas law to relate the values of ΔH and ΔU for gases that behave ideally.

To calculate the relation between ΔH and ΔU, we suppose that the amount of ideal gas reactant molecules is $n_{initial}$. Because for an ideal gas $PV = nRT$, the initial enthalpy is

$$H_{initial} = U_{initial} + PV_{initial} = U_{initial} + n_{initial}RT$$

After the reaction is complete, the amount of ideal gas product molecules is n_{final}. The enthalpy is then

$$H_{final} = U_{final} + PV_{final} = U_{final} + n_{final}RT$$

The difference is

$$\Delta H = H_{final} - H_{initial} = \Delta U + (n_{final} - n_{initial})RT$$

We have deduced that

$$\Delta H = \Delta U + \Delta n_{gas}RT \qquad\qquad \textbf{(19)}*$$

where $\Delta n_{gas} = n_{final} - n_{initial}$ is the change in the amount of gas molecules in the reaction (positive for net gas formation, negative for net gas consumption). Notice that ΔH is less negative (more positive) than ΔU for reactions that generate gases: less energy can be obtained as heat at constant pressure than at constant volume because the system must use some energy to expand to make room for the reaction products. For reactions with no change in the amount of gas, the two quantities are about the same.

EXAMPLE 6.8   Sample exercise: Relating the enthalpy change and internal energy change for a chemical reaction

A constant-volume calorimeter showed that the heat generated by the combustion of 1.000 mol glucose molecules in the reaction $C_6H_{12}O_6(s) + 6\,O_2(g) \rightarrow 6\,CO_2(g) + 6\,H_2O(g)$ is 2559 kJ at 298 K, and so $\Delta U = -2559$ kJ. What is the change in enthalpy for the same reaction?

SOLUTION

From $\Delta n_{gas} = n_{final} - n_{initial}$, $\Delta n_{gas} = 12 - 6$ mol $= +6$ mol

From $\Delta H = \Delta U + \Delta n_{gas}RT$, $\Delta H = -2559$ kJ $+ [(6\text{ mol}) \times (8.3145\text{ J}\cdot\text{K}^{-1}\cdot\text{mol}^{-1})$
$\times\,(298\text{ K})]$

$= -2559$ kJ $+ 14.9$ kJ $= -2544$ kJ

A note on good practice. Verify that units are consistent when adding numbers. Here we have recognized that 10^3 J $= 1$ kJ.

SELF-TEST 6.12A   The thermochemical equation for the combustion of cyclohexane, C_6H_{12}, is $C_6H_{12}(l) + 9\,O_2(g) \rightarrow 6\,CO_2(g) + 6\,H_2O(l)$, $\Delta H = -3920$ kJ at 298 K. What is the change in internal energy for the combustion of 1.00 mol $C_6H_{12}(l)$ at 298 K?

[*Answer:* -3.91×10^3 kJ]

SELF-TEST 6.12B   The reaction $4\,Al(s) + 3\,O_2(g) \rightarrow 2\,Al_2O_3(s)$ was investigated as part of a study on using aluminum powder as a rocket fuel (Fig. 6.27). It was found that 1.00 mol Al produced 3378 kJ of heat in this reaction under constant-pressure conditions at 1000.°C. What is the change in internal energy for the combustion of 1.00 mol Al at 1000.°C?

The reaction enthalpy is less negative (more positive) than the reaction internal energy for reactions that generate gases; for reactions with no change in the amount of gas, the two quantities are almost the same.

6.15 Standard Reaction Enthalpies

Because the heat released or absorbed by a reaction depends on the physical states of the reactants and products, we need to specify the state of each substance. For

FIGURE 6.27 In this preparation of rocket fuel for the space shuttle, powdered aluminum is mixed with an oxidizing agent in a liquid polymer base that hardens inside the booster rocket shell.

FIGURE 6.28 The enthalpy changes for the reactions in which 1 mol $CH_4(g)$ burns to give carbon dioxide and water in either the gaseous (left) or the liquid (right) state. The difference in enthalpy is equal to 88 kJ, the enthalpy of vaporization of 2 mol $H_2O(l)$.

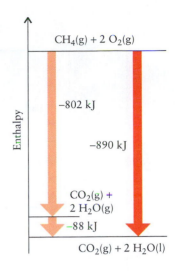

example, when describing the combustion of methane, we can write two different thermochemical equations for two different sets of products:

$$CH_4(g) + 2\,O_2(g) \longrightarrow CO_2(g) + 2\,H_2O(g) \qquad \Delta H = -802\ kJ \qquad \textbf{(B)}$$
$$CH_4(g) + 2\,O_2(g) \longrightarrow CO_2(g) + 2\,H_2O(l) \qquad \Delta H = -890.\ kJ \qquad \textbf{(C)}$$

In the first reaction, the water is produced as a vapor; in the second, it is produced as a liquid. The heat generated is different in each case. We have already seen that the enthalpy of water vapor is 44 kJ·mol^{-1} higher than that of liquid water at 25°C (see Table 6.3). As a result, an additional 88 kJ (for 2 mol H_2O) remains stored in the system if water vapor is formed (Fig. 6.28). If the 2 mol $H_2O(g)$ subsequently condenses, an additional 88 kJ is given off as heat.

An enthalpy of reaction also depends on the conditions (such as the pressure). All the tables in this book list data for reactions in which each reactant and product is in its **standard state**, its pure form at exactly 1 bar. The standard state of liquid water is pure water at 1 bar. The standard state of ice is pure ice at 1 bar. A solute in a liquid solution is in its standard state when its concentration is 1 mol·L^{-1}. The standard value of a property X (that is, the value of X for the standard state of the substance) is denoted $X°$.

The **standard reaction enthalpy**, $\Delta H°$, is the reaction enthalpy when reactants in their standard states change into products in their standard states. For example, for reaction C, the value $\Delta H° = -890.$ kJ signifies that the heat output is 890. kJ when 1 mol $CH_4(g)$ as pure methane gas at 1 bar is allowed to react with pure oxygen gas at 1 bar, giving pure carbon dioxide gas and pure liquid water, both at 1 bar (Fig. 6.29). Reaction enthalpies do not change very much with pressure over the small ranges normally encountered, so the standard value gives a good indication of the change in enthalpy for pressures near 1 bar.

Most thermochemical data are reported for 25°C (more precisely, for 298.15 K). Temperature is not part of the definition of standard states: we can have a standard state at any temperature; 298.15 K is simply the most common temperature used in tables of data. All reaction enthalpies used in this text are for 298.15 K unless another temperature is indicated.

Standard reaction enthalpies refer to reactions in which the reactants and products are in their standard state, the pure form at 1 bar; they are usually reported for a temperature of 298.15 K.

You may see some tables reporting data at 1 atm, the former standard. The small change in standard pressure makes a negligible difference to most numerical values, and so it is normally safe to use data compiled for 1 atm.

6.16 Combining Reaction Enthalpies: Hess's Law

Enthalpy is a state function; therefore, the value of ΔH is independent of the path between given initial and final states. We saw an application of this approach in

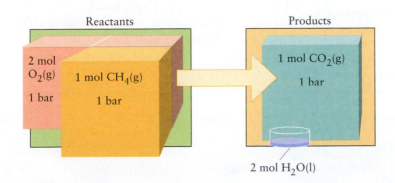

FIGURE 6.29 The standard reaction enthalpy is the difference in enthalpy between the pure products and the pure reactants, each at 1 bar and the specified temperature (which is commonly but not necessarily 298 K). The scheme here is for the combustion of methane gas to carbon dioxide gas and liquid water.

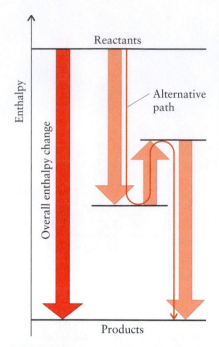

FIGURE 6.30 If an overall reaction can be broken down into a series of steps, then the corresponding overall reaction enthalpy is the sum of the reaction enthalpies of those steps. None of the steps need be a reaction that can actually be carried out in the laboratory.

Section 6.11, when we calculated the enthalpy change for an overall physical process as the sum of the enthalpy changes for a series of two individual steps. The same rule applied to chemical reactions is known as **Hess's law:** *the overall reaction enthalpy is the sum of the reaction enthalpies of the steps into which the reaction can be divided.* Hess's law applies even if the intermediate reactions or the overall reaction cannot actually be carried out. Provided that the equation for each step balances and the individual equations add up to the equation for the reaction of interest, a reaction enthalpy can be calculated from any convenient sequence of reactions (Fig. 6.30).

As an example of Hess's law, let's consider the oxidation of carbon as graphite, denoted $C(gr)$, to carbon dioxide:

$$C(gr) + O_2(g) \longrightarrow CO_2(g)$$

This reaction can be thought of as the outcome of two steps. One step is the oxidation of carbon to carbon monoxide:

$$C(gr) + \tfrac{1}{2}O_2(g) \longrightarrow CO(g) \qquad \Delta H° = -110.5 \text{ kJ}$$

The second step is the oxidation of carbon monoxide to carbon dioxide:

$$CO(g) + \tfrac{1}{2}O_2(g) \longrightarrow CO_2(g) \qquad \Delta H° = -283.0 \text{ kJ}$$

This two-step process is an example of a **reaction sequence,** a series of reactions in which the products of one reaction take part as reactants in another reaction. The equation for the overall reaction, the net outcome of the sequence, is the sum of the equations for the intermediate steps:

$C(gr) + \tfrac{1}{2}O_2(g) \longrightarrow CO(g)$	$\Delta H° = -110.5 \text{ kJ}$	**(a)**
$CO(g) + \tfrac{1}{2}O_2(g) \longrightarrow CO_2(g)$	$\Delta H° = -283.0 \text{ kJ}$	**(b)**
$C(gr) + O_2(g) \longrightarrow CO_2(g)$	$\Delta H° = -393.5 \text{ kJ}$	**(a + b)**

The same procedure is used to predict the enthalpies of reactions that we cannot measure directly in the laboratory. The procedure is described in Toolbox 6.1.

TOOLBOX 6.1   HOW TO USE HESS'S LAW

CONCEPTUAL BASIS

Because enthalpy is a state function, the enthalpy change of a system depends only on its initial and final states. Therefore, we can carry out a reaction in one step or visualize it as the sum of several steps; the reaction enthalpy is the same in each case.

PROCEDURE

To use Hess's law to find the enthalpy of a given reaction, we find a sequence of reactions with known reaction enthalpies that adds up to the reaction of interest. A systematic procedure simplifies that process.

Step 1 Select one of the reactants in the overall reaction and write down a chemical equation in which it also appears as a reactant.

Step 2 Select one of the products in the overall reaction and write down a chemical equation in which it also

appears as a product. Add this equation to the equation written in step 1 and cancel species that appear on both sides of the equation.

Step 3 Cancel unwanted species in the sum obtained in step 2 by adding an equation that has the same substance or substances on the opposite side of the arrow.

Step 4 Once the sequence is complete, combine the standard reaction enthalpies.

In each step, we may need to reverse the equation or multiply it by a factor. Recall from Eq. 16 that, if we want to reverse a chemical equation, we have to change the sign of the reaction enthalpy. If we multiply the stoichiometric coefficients by a factor, we must multiply the reaction enthalpy by the same factor.

This procedure is illustrated in Example 6.9.

EXAMPLE 6.9   **Sample exercise: Using Hess's law**

Consider the synthesis of propane, C_3H_8, a gas used as camping fuel:

$$3\,C(gr) + 4\,H_2(g) \longrightarrow C_3H_8(g)$$

It is difficult to measure the enthalpy change of this reaction. However, standard enthalpies of combustion reactions are easy to measure. Calculate the standard enthalpy of this reaction from the following experimental data:

(a) $C_3H_8(g) + 5\,O_2(g) \longrightarrow 3\,CO_2(g) + 4\,H_2O(l)$ $\Delta H° = -2220.\ kJ$

(b) $C(gr) + O_2(g) \longrightarrow CO_2(g)$ $\Delta H° = -394\ kJ$

(c) $H_2(g) + \frac{1}{2}O_2(g) \longrightarrow H_2O(l)$ $\Delta H° = -286\ kJ$

SOLUTION

Step 1 To treat carbon as a reactant, select (b) and multiply it through by 3.	$3\,C(gr) + 3\,O_2(g) \longrightarrow 3\,CO_2(g)$	$\Delta H° = 3 \times (-394\ kJ)$ $= -1182\ kJ$
Step 2 Reverse equation (a), changing the sign of its reaction enthalpy.	$3\,CO_2(g) + 4\,H_2O(l) \longrightarrow C_3H_8(g) + 5\,O_2(g)$	$\Delta H° = +2220.\ kJ$
Add the two preceding equations and their enthalpies.	$3\,C(gr) + 3\,O_2(g) + 3\,CO_2(g) + 4\,H_2O(l)$ $\longrightarrow C_3H_8(g) + 5\,O_2(g) + 3\,CO_2(g)$	$\Delta H° = (-1182) + (2200)\ kJ$ $= +1038\ kJ$
Simplify the equation.	$3\,C(gr) + 4\,H_2O(l) \longrightarrow C_3H_8(g) + 2\,O_2(g)$	$\Delta H° = +1038\ kJ$
Step 3 To cancel the unwanted H_2O and O_2, multiply equation (c) by 4.	$4\,H_2(g) + 2\,O_2(g) \longrightarrow 4\,H_2O(l)$	$\Delta H° = 4 \times (-286\ kJ)$ $= -1144\ kJ$
Step 4 Add the equations in steps 2 and 3.	$3\,C(gr) + 4\,H_2(g) + 4\,H_2O(l) + 2\,O_2(g)$ $\longrightarrow C_3H_8(g) + 2\,O_2(g) + 4\,H_2O(l)$	$\Delta H° = 1038 + (-1144)\ kJ$ $= -106\ kJ$
Simplify the equation.	$3\,C(gr) + 4\,H_2(g) \longrightarrow C_3H_8(g)$	$\Delta H° = -106\ kJ$

SELF-TEST 6.13A Gasoline, which contains octane, may burn to carbon monoxide if the air supply is restricted. Determine the standard reaction enthalpy for the incomplete combustion of liquid octane in air to carbon monoxide gas and liquid water from the standard reaction enthalpies for the combustions of octane and carbon monoxide:

$2\,C_8H_{18}(l) + 25\,O_2(g) \longrightarrow 16\,CO_2(g) + 18\,H_2O(l)$ $\Delta H° = -10\ 942\ kJ$

$2\,CO(g) + O_2(g) \longrightarrow 2\,CO_2(g)$ $\Delta H° = -566.0\ kJ$

[*Answer:* $2\,C_8H_{18}(l) + 17\,O_2(g) \rightarrow 16\,CO(g) + 18\,H_2O(l)$, $\Delta H° = -6414\ kJ$]

SELF-TEST 6.13B Methanol is a clean-burning liquid fuel proposed as a replacement for gasoline. Suppose it could be produced by the controlled reaction of the oxygen in air with methane. Find the standard reaction enthalpy for the formation of 1 mol $CH_3OH(l)$ from methane and oxygen, given the following information:

$CH_4(g) + H_2O(g) \longrightarrow CO(g) + 3\,H_2(g)$ $\Delta H° = +206.10\ kJ$

$2\,H_2(g) + CO(g) \longrightarrow CH_3OH(l)$ $\Delta H° = -128.33\ kJ$

$2\,H_2(g) + O_2(g) \longrightarrow 2\,H_2O(g)$ $\Delta H° = -483.64\ kJ$

Thermochemical equations for the individual steps of a reaction sequence may be combined to give the thermochemical equation for the overall reaction.

6.17 The Heat Output of Reactions

We could not live without combustion reactions: the oxidation of glucose powers our bodies, and the burning of fossil fuels (coal, petroleum, and natural gas) powers our homes and vehicles. Because fossil fuels reserves are limited, alternatives are being sought (Box 6.2), but even these new fuels will be burned. Consequently, the study of combustion is critically important for our survival.

The **standard enthalpy of combustion**, $\Delta H_c°$, is the change in enthalpy per mole of a substance that is burned in a combustion reaction under standard conditions. The products of the combustion of an organic compound are carbon dioxide

BOX 6.2   WHAT HAS THIS TO DO WITH . . . THE ENVIRONMENT?

Alternative Fuels

Our complex modern life style was made possible by the discovery and refining of fossil fuels, fuels that are the result of the decay of organic matter laid down millions of years ago. The natural gas that heats our homes, the gasoline that powers our automobiles, and the coal that provides much of our electrical power are fossil fuels. Vast reserves of petroleum, the source of liquid hydrocarbon fuels such as gasoline and coal, exist in many areas of the world. However, although large, these reserves are limited, and we are using them up at a much faster rate than they can be replaced.

Alternative and sustainable power generation methods, such as hydroelectric power, wind power, solar power, and alternative fuels are being sought to reduce the demand for fossil fuels. Three of the most promising alternative fuels are hydrogen, ethanol, and methane. Hydrogen is extracted from ocean water by electrolysis. Ethanol is obtained by fermenting biomass, the name given to plant materials that can be burned or reacted to produce fuels. Methane is generated by bacterial digestion of wastes such as sewage and agricultural wastes. In each case, the fuel is renewable. That means the source of the fuel is replenished every year by the Sun. The use of hydrogen as a fuel is discussed in Section 14.5. Here, we look at ethanol and methane.

Ethanol, CH_3CH_2OH, is produced from the biological fermentation of the starches in grains, mainly corn. It currently makes up about 10% by volume of gasoline in the United States, thereby reducing pollution as well as the use of petroleum. The oxygen atom in the ethanol molecule reduces emissions of carbon monoxide and hydrocarbons by helping to ensure complete combustion. One bushel of corn (about 30 L) can produce nearly 10 L of ethanol. A problem with ethanol as a fuel is that the sugars and starches fermented to produce it are expensive. However, the cellulose in straw and cornstalks left behind as stubble when grains are harvested is now attracting attention. Cellulose is the structural material in plants. It is made up of simple sugars, just as starches are, but the bacteria that ferment starches cannot digest cellulose. Research is now being conducted on enzymes that break down cellulose into sugars that can be digested. This process would greatly increase the biomass available for fuel production, because

The cellulose-based waste biomass in this reactor is being digested by special enzymes that break it down into ethanol. These enzymes are being intensively studied to improve the efficiency of conversion.

TABLE 6.4 Standard Enthalpies of Combustion at 25°C (kJ·mol⁻¹)*

Substance	Formula	ΔH_c°
benzene	$C_6H_6(l)$	−3268
carbon	C(s, graphite)	−394
ethanol	$C_2H_5OH(l)$	−1368
ethyne (acetylene)	$C_2H_2(g)$	−1300.
glucose	$C_6H_{12}O_6(s)$	−2808
hydrogen	$H_2(g)$	−286
methane	$CH_4(g)$	−890.
octane	$C_8H_{18}(l)$	−5471
propane	$C_3H_8(g)$	−2220.
urea	$CO(NH_2)_2(s)$	−632

*In a combustion, carbon is converted into carbon dioxide, hydrogen into liquid water, and nitrogen into nitrogen gas. More values are given in Appendix 2A.

These tanks at a water treatment facility are used to generate a mixture of methane and carbon dioxide by anaerobic digestion of sewage. The methane produced provides much of the power needed to run the facility.

straw, wood, grasses, and in fact, nearly any plant materials, could be used for fuel.

Methane, CH_4, is found in underground reserves as the primary component of natural gas, but it is also obtained from biological materials. The "digestion" of the biomaterials by bacteria is anaerobic, which means that it takes place in the absence of oxygen. Currently, many sewage treatment plants have anaerobic digesters that produce the methane used to operate the plants. To generate methane by anaerobic digestion on a large-scale basis, additional materials, such as sugars from the enzymatic breakdown of biomass, would need to be used. Methane would be less useful than ethanol as a transportation fuel, because it is a gas and so has a low *enthalpy density*, the enthalpy of combustion per liter. However, it can be used wherever natural gas is used.

Methane and ethanol do produce carbon dioxide when burned and thus contribute to the greenhouse effect and global warming (Box 14.2). However, they generate less carbon dioxide per gram than gasoline and can be renewed every year, as long as the Sun shines and produces green plants.

Related Exercise: 6.121.

For Further Reading: *Biofuels for Sustainable Transportation*, www.ott.doe.gov/biofuels (U.S. Department of Energy, 2002). Energy Information Administration, *Alternative Fuels Information at a Glance*, www.eia.doe.gov/cneaf/alternate/page/fuelalternate.html (U.S. Department of Energy, 2000). Alternative Fuel Vehicles (AFVs) and High-Efficiency Vehicles, www.energy.ca.gov/afvs/index.html (California Energy Commission, 2000). R. H. Truly, "Ethanol for transportation," *Issues in Science and Technology*. vol. 18, no. 3, Spring 2002, pp. 17–18. *Alternative Fuels Data Center*, http://www.afdc.doe.gov (U.S. Department of Energy, 2002).

gas and liquid water; any nitrogen present is released as N_2, unless other products are specified.

Standard enthalpies of combustion are listed in Table 6.4 and Appendix 2A. We have seen in Toolbox 6.1 how to use enthalpies of combustion to obtain the standard enthalpies of reactions. Here we consider another practical application—the choice of a fuel. For example, suppose we want to know the heat output from the combustion of 150. g of methane. The thermochemical equation allows us to write the following relation

$$1 \text{ mol } CH_4 \mathrel{\widehat{=}} 890. \text{ kJ}$$

and to use it as a conversion factor in the usual way (1).

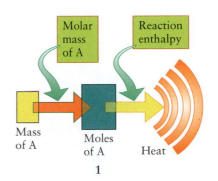

1

EXAMPLE 6.10   Calculating the heat output of a fuel

How much butane should a backpacker carry? Calculate the mass of butane that you would need to burn to obtain 350. kJ of heat, which is just enough energy to heat 1 L of water from 17°C to boiling at sea level (if we ignore all heat losses). The thermochemical equation is

$$2\,C_4H_{10}(g) + 13\,O_2(g) \longrightarrow 8\,CO_2(g) + 10\,H_2O(l) \qquad \Delta H° = -5756\ \text{kJ}$$

STRATEGY The first step is to convert heat output required into moles of fuel molecules by using the thermochemical equation. Then use the molar mass of the fuel to convert from moles of fuel molecules into grams.

SOLUTION Find the relation of the enthalpy change to amount of fuel molecules from the thermochemical equation.

$$5756\ \text{kJ} \simeq 2\ \text{mol}\ C_4H_{10}$$

2 mol C_4H_{10}

Convert heat output into moles of fuel molecules.

$$n_{C_4H_{10}} = (350.\ \text{kJ}) \times \left(\frac{2\ \text{mol}\ C_4H_{10}}{5756\ \text{kJ}}\right)$$

Use the molar mass of butane, 58.12 g·mol^{-1}, to find the mass of product.

$$m_{C_4H_{10}} = (350.\ \text{kJ}) \times \left(\frac{2\ \text{mol}\ C_4H_{10}}{5756\ \text{kJ}}\right) \times (58.12\ \text{g·mol}^{-1})$$

$$= 7.07\ \text{g}\ C_4H_{10}$$

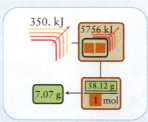

SELF-TEST 6.14A The thermochemical equation for the combustion of propane is

$$C_3H_8(g) + 5\,O_2(g) \longrightarrow 3\,CO_2(g) + 4\,H_2O(l) \qquad \Delta H° = -2220.\ \text{kJ}$$

What mass of propane must be burned to supply 350. kJ as heat? Would it be easier to pack propane rather than butane?

[*Answer:* 6.95 g. Yes, propane would be very slightly lighter to carry.]

SELF-TEST 6.14B Ethanol trapped in a gel is another common camping fuel. What mass of ethanol must be burned to supply 350. kJ as heat? The thermochemical equation for the combustion of ethanol is

$$C_2H_5OH(l) + 3\,O_2(g) \longrightarrow 2\,CO_2(g) + 3\,H_2O(l) \qquad \Delta H° = -1368\ \text{kJ}$$

The heat absorbed or given off by a reaction can be treated like a reactant or product in a stoichiometric calculation.

6.18 Standard Enthalpies of Formation

There are millions of possible reactions, and it is impractical to list every one with its standard reaction enthalpy. However, chemists have devised an ingenious alternative. First, they report the "standard enthalpies of formation" of substances. Then they combine these quantities to obtain the standard enthalpy of reaction needed. Let's look at these two stages in turn.

The **standard enthalpy of formation**, $\Delta H_f°$, of a substance is the standard reaction enthalpy per mole of formula units for the formation of a substance from its elements in their *most stable form*, as in the reaction

> Phosphorus is an exception: white phosphorus is used because it is much easier to obtain pure than the other, more stable allotropes.

$$2\,C(gr) + 3\,H_2(g) + \tfrac{1}{2}O_2(g) \longrightarrow C_2H_5OH(l) \qquad \Delta H° = -277.69\ \text{kJ}$$

where gr denotes graphite, the most stable form of carbon at normal temperatures. The chemical equation that corresponds to the standard enthalpy of formation of a substance has the substance as the sole product with a stoichiometric coefficient of 1 (implying the formation of 1 mol of the substance). Sometimes, as here, fractional coefficients are required for the reactants. Because standard enthalpies of formation are expressed in kilojoules per mole of the substance of interest, we

TABLE 6.5   Standard Enthalpies of Formation at 25°C (kJ·mol^{-1})*

Substance	Formula	$\Delta H_f°$	Substance	Formula	$\Delta H_f°$
Inorganic compounds			**Organic compounds**		
ammonia	$NH_3(g)$	−46.11	benzene	$C_6H_6(l)$	+49.0
carbon dioxide	$CO_2(g)$	−393.51	ethanol	$C_2H_5OH(l)$	−277.69
carbon monoxide	$CO(g)$	−110.53	ethyne	$C_2H_2(g)$	+226.73
dinitrogen tetroxide	$N_2O_4(g)$	+9.16	(acetylene)		
hydrogen chloride	$HCl(g)$	−92.31	glucose	$C_6H_{12}O_6(s)$	−1268
hydrogen fluoride	$HF(g)$	−271.1	methane	$CH_4(g)$	−74.81
nitrogen dioxide	$NO_2(g)$	+33.18			
nitric oxide	$NO(g)$	+90.25			
sodium chloride	$NaCl(s)$	−411.15			
water	$H_2O(l)$	−285.83			
	$H_2O(g)$	−241.82			

*A much longer list is given in Appendix 2A.

report that $\Delta H_f°(C_2H_5OH, l) = -277.69$ kJ·mol^{-1}. Note also how the substance and its state are used to label the enthalpy change so that we know which species and which form of that species (liquid, in this case) we are talking about.

A note on good practice: You should always be alert to the difference between a quantity *per* mole of molecules and the same quantity *for* or *of* a mole of molecules. Standard enthalpies of formation are expressed per mole of molecules, as in −277.69 kJ·mol^{-1}; the standard enthalpy of forming 1 mol $C_2H_5OH(l)$ is −277.69 kJ. The point might seem picky, but it will help you to keep units straight.

It follows from the definition just given that the standard enthalpy of formation of an element in its most stable form is zero. For instance, the standard enthalpy of formation of C(gr) is zero because C(gr) → C(gr) is a "null reaction" (that is, nothing changes). We write, for instance, $\Delta H_f°(C, gr) = 0$. However, the enthalpy of formation of an element in a form other than its most stable one is nonzero. For example, the conversion of carbon from graphite (its most stable form) into diamond is endothermic:

$$C(gr) \longrightarrow C(diamond) \qquad \Delta H° = +1.9 \text{ kJ}$$

The standard enthalpy of formation of diamond is therefore reported as $\Delta H_f°(C, diamond) = +1.9$ kJ·mol^{-1}. Values for a selection of other substances are listed in Table 6.5 and Appendix 2A.

Now let's see how to combine standard enthalpies of formation to calculate a standard reaction enthalpy. To do so, we imagine carrying out the reaction in two steps: we reverse the formation of the reactants from the elements, then combine the elements to form the products. The first step is usually to calculate the reaction enthalpy for the formation of all the products from their elements. For this step, we use the enthalpies of formation of the products. Then, we calculate the reaction enthalpy for the formation of all the reactants from their elements. The difference between these two totals is the standard enthalpy of the reaction (Fig. 6.31):

$$\Delta H° = \sum n\Delta H_f°(\text{products}) - \sum n\Delta H_f°(\text{reactants}) \qquad (20)^*$$

In this expression, the n are the amounts of each substance in the chemical equation and the symbol Σ (sigma) means a sum. The first sum is the total standard enthalpy of formation of the products. The second sum is the similar total for the reactants.

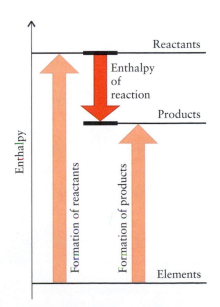

FIGURE 6.31 The reaction enthalpy can be constructed from enthalpies of formation by imagining forming the reactants and products from their elements. Then the reaction enthalpy is the difference between the enthalpies of products and reactants.

EXAMPLE 6.11   **Using standard enthalpies of formation to calculate a standard enthalpy of reaction**

Amino acids are the building blocks of proteins, which have long chainlike molecules. They are oxidized in the body to urea, carbon dioxide, and liquid water. Is this reaction a source of heat for the body? Use the information in Appendix 2A to predict the standard enthalpy of reaction for the oxidation of the simplest amino acid, glycine (NH_2CH_2COOH), a solid, to solid urea (H_2NCONH_2), carbon dioxide gas, and liquid water:

$$2\ NH_2CH_2COOH(s) + 3\ O_2(g) \longrightarrow H_2NCONH_2(s) + 3\ CO_2(g) + 3\ H_2O(l)$$

STRATEGY   We expect a strongly negative value because all combustions are exothermic and this oxidation is like an incomplete combustion. First, add up the individual standard enthalpies of formation of the products, multiplying each value by the appropriate number of moles from the balanced equation. Remember that the standard enthalpy of formation of an element in its most stable form is zero. Then, calculate the total standard enthalpy of formation of the reactants in the same way and use Eq. 20 to calculate the standard reaction enthalpy.

SOLUTION

Calculate the total enthalpy of formation of the products.

$$\sum n\Delta H_f°(\text{products}) = (1\ \text{mol})\Delta H_f°(H_2NCONH_2,\ s)$$
$$+ (3\ \text{mol})\Delta H_f°(CO_2,\ g)$$
$$+ (3\ \text{mol})\Delta H_f°(H_2O,\ l)$$
$$= (1\ \text{mol})(-333.51\ \text{kJ}\cdot\text{mol}^{-1})$$
$$+ (3\ \text{mol})(-393.51\ \text{kJ}\cdot\text{mol}^{-1})$$
$$+ (3\ \text{mol})(-285.83\ \text{kJ}\cdot\text{mol}^{-1})$$
$$= -333.51 + (-1180.53) + (-857.49)\ \text{kJ}$$
$$= -2371.53\ \text{kJ}$$

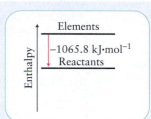

Calculate the total enthalpy of formation of the reactants.

$$\sum nH_f°(\text{reactants}) = (2\ \text{mol})\Delta H_f°(NH_2CH_2COOH,\ s)$$
$$+ (3\ \text{mol})\Delta H_f°(O_2,\ g)$$
$$= (2\ \text{mol})(-532.9\ \text{kJ}\cdot\text{mol}^{-1}) + (3\ \text{mol})(0)$$
$$= -1065.8\ \text{kJ}$$

From $\Delta H° = \sum n\Delta H_f°(\text{products}) - \sum n\Delta H_f°(\text{reactants})$

$$\Delta H° = -2371.53 - (-1065.8)\ \text{kJ}$$
$$= -1305.7\ \text{kJ}$$

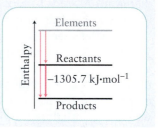

The thermochemical equation is therefore

$$2\ NH_2CH_2COOH(s) + 3\ O_2(g) \longrightarrow$$
$$H_2NCONH_2(s) + 3\ CO_2(g) + 3\ H_2O(l) \qquad \Delta H° = -1305.7\ \text{kJ}$$

Because reactions in the body take place in aqueous solution, this value is not the same as the enthalpy change for the reaction in the body. However, the two values are fairly close. Therefore, the oxidation of glycine, which we have found to be exothermic, is a potential source of energy in the body.

A note on good practice: Enthalpies of formation are expressed in kilojoules per mole and enthalpies of reaction in kilojoules for the reaction as written. Note how the stoichiometric coefficients are interpreted as numbers of moles, and that an unwritten coefficient of 1 for urea is included as 1 mol in the calculation.

SELF-TEST 6.15A Calculate the standard enthalpy of combustion of glucose from the standard enthalpies of formation in Table 6.5 and Appendix 2A.

[*Answer:* -2808 kJ·mol^{-1}]

SELF-TEST 6.15B You have an inspiration: maybe diamonds would make a great fuel! Calculate the standard enthalpy of combustion of diamonds from the information in Appendix 2A.

Standard enthalpies of formation are commonly determined from combustion data by using Eq. 20. The procedure is the same, but the standard reaction enthalpy is known and the unknown value is one of the standard enthalpies of formation.

EXAMPLE 6.12 **Using the enthalpy of combustion to calculate an enthalpy of formation**

Use the information in Table 6.5 and the enthalpy of combustion of propane gas to calculate the enthalpy of formation of propane, a gas commonly used for camping stoves and outdoor barbecues.

STRATEGY Use Eq. 20 and the procedure set out in Example 6.11, but solve for the standard enthalpy of formation of propane.

SOLUTION

Write the thermochemical equation for the combustion of 1 mol C_3H_8.

$$C_3H_8(g) + 5\,O_2(g) \longrightarrow 3\,CO_2(g) + 4\,H_2O(l) \quad \Delta H^\circ = -2220.\ kJ$$

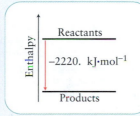

Find the total enthalpy of formation of the products.

$$\sum n\Delta H_f^\circ(\text{products}) = (3\ \text{mol})\Delta H_f^\circ(CO_2,\ g) + (4\ \text{mol})\Delta H_f^\circ(H_2O,\ l)$$
$$= (3\ \text{mol})(-393.51\ \text{kJ·mol}^{-1})$$
$$+ (4\ \text{mol})(-285.83\ \text{kJ·mol}^{-1})$$
$$= -1180.53 - 1143.32\ \text{kJ}$$
$$= -2323.85\ \text{kJ}$$

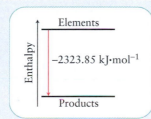

Find the total enthalpy of formation of the reactants.

$$\sum n\Delta H_f^\circ(\text{reactants}) = (1\ \text{mol})\Delta H_f^\circ(C_3H_8,\ g) + (5\ \text{mol})\Delta H_f^\circ(O_2,\ g)$$
$$= (1\ \text{mol})\Delta H_f^\circ(C_3H_8,\ g)$$

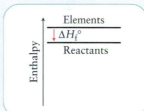

From $\Delta H_r^\circ = \sum n\Delta H_f^\circ\ (\text{products}) - \sum n\Delta H_f^\circ\ (\text{reactants})$

Solve for $\Delta H_f^\circ(C_3H_8,\ g)$.

$$-2220.\ \text{kJ} = -2323.85\ \text{kJ} - (1\ \text{mol})\Delta H_f^\circ(C_3H_8,\ g)$$

$$\Delta H_f^\circ(C_3H_8,\ g) = -2323.85 - (-2220.)\ \text{kJ·mol}^{-1}$$
$$= -104\ \text{kJ·mol}^{-1}$$

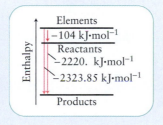

SELF-TEST 6.16A Calculate the standard enthalpy of formation of ethyne, the fuel used in oxyacetylene welding torches, from the information in Tables 6.4 and 6.5.

[*Answer:* $+227$ kJ·mol^{-1}]

SELF-TEST 6.16B Calculate the standard enthalpy of formation of urea, $CO(NH_2)_2$, a by-product of the metabolism of proteins, from the information in Tables 6.4 and 6.5.

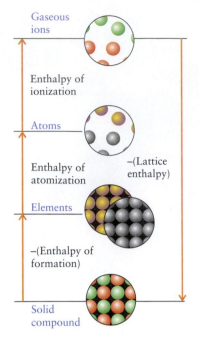

Gaseous ions

Enthalpy of ionization

Atoms

Enthalpy of atomization

–(Lattice enthalpy)

Elements

–(Enthalpy of formation)

Solid compound

FIGURE 6.32 In a Born–Haber cycle, we select a sequence of steps that starts and ends at the same point (the elements, for instance). The lattice enthalpy is the enthalpy change accompanying the reverse of the step in which the solid is formed from a gas of ions. The sum of enthalpy changes around the complete cycle is 0 because enthalpy is a state function.

Standard enthalpies of formation can be combined to obtain the standard enthalpy of any reaction.

6.19 The Born–Haber Cycle

We saw in Section 2.4 that the energy changes accompanying the formation of a solid could be estimated on the basis of a model—the ionic model—in which the principal contribution to the lattice energy was the Coulomb interaction between ions. However, a model can give only an estimate: we need a way to *measure* the energy change. If the measured and calculated energies are similar, we can conclude that the ionic model is reliable for the particular substance. If the two energies are markedly different, then we know that the ionic model must be improved or even discarded.

For a given solid, the difference in molar enthalpy between the solid and a gas of widely separated ions is called the **lattice enthalpy** of the solid, ΔH_L:

$$\Delta H_L = H_m(\text{ions, g}) - H_m(\text{solid}) \tag{21}$$

The lattice enthalpy can be identified with the heat required to vaporize the solid at constant pressure. The greater the lattice enthalpy, the greater is the heat required. Heat equal to the lattice enthalpy is released when the solid forms from gaseous ions. In Section 2.4 we calculated the lattice *energy* and discussed how it depended on the attractions between the ions. The lattice enthalpy differs from the lattice energy by only a few kilojoules per mole and can be interpreted in a similar way.

The lattice enthalpy of a solid cannot be measured directly. However, we can obtain it indirectly by combining other measurements in an application of Hess's law. This approach takes advantage of the first law of thermodynamics and, in particular, the fact that enthalpy is a state function. The procedure uses a **Born–Haber cycle**, a closed path of steps, one of which is the formation of a solid lattice from the gaseous ions. The enthalpy change for this step is the negative of the lattice enthalpy. Table 6.6 lists some lattice enthalpies found in this way.

In a Born–Haber cycle, we imagine that we break apart the bulk elements into atoms, ionize the atoms, combine the gaseous ions to form the ionic solid, then form the elements again from the ionic solid (Fig. 6.32). Only the lattice enthalpy, the enthalpy of the step in which the ionic solid is formed from the gaseous ions, is unknown. The sum of the enthalpy changes for a complete Born–Haber cycle is zero, because the enthalpy of the system must be the same at the start and finish.

TABLE 6.6   Lattice Enthalpies at 25°C (kJ·mol⁻¹)

Halides

LiF	1046	LiCl	861	LiBr	818	LiI	759
NaF	929	NaCl	787	NaBr	751	NaI	700.
KF	826	KCl	717	KBr	689	KI	645
AgF	971	AgCl	916	AgBr	903	AgI	887
$BeCl_2$	3017	$MgCl_2$	2524	$CaCl_2$	2260.	$SrCl_2$	2153
		MgF_2	2961	$CaBr_2$	1984		

Oxides

MgO	3850.	CaO	3461	SrO	3283	BaO	3114

Sulfides

MgS	3406	CaS	3119	SrS	2974	BaS	2832

EXAMPLE 6.13   **Using a Born–Haber cycle to calculate a lattice enthalpy**

Devise and use a Born–Haber cycle to calculate the lattice enthalpy of potassium chloride.

STRATEGY A Born–Haber cycle combines the enthalpy changes that take place when we begin with the pure elements, atomize them to form gaseous atoms, ionize the atoms to form gaseous ions, allow the ions to form an ionic solid, and then convert the solid back into the pure elements. The only unknown enthalpy change is that of the step in which the ionic solid forms from gaseous ions $(-\Delta H_L)$. Because the sum of the enthalpy changes for the complete cycle is zero, we can solve for the lattice enthalpy, ΔH_L.

SOLUTION The Born–Haber cycle for KCl is shown in Fig. 6.33.

Find $\Delta H_f(\text{K, atoms})$ in Appendix 2A.	$+89 \text{ kJ·mol}^{-1}$
Find $\Delta H_f(\text{Cl, atoms})$ in Appendix 2A.	$+122 \text{ kJ·mol}^{-1}$
Find the ionization energy of K in Fig. 1.50 or Appendix 2D.	$+ 418 \text{ kJ·mol}^{-1}$
Write the electron gain enthalpy of Cl as the negative of the electron affinity (Fig. 1.54 or Appendix 2D).	-349 kJ·mol^{-1}
Use Appendix 2A to write the negative of the standard enthalpy of formation of KCl, $-\Delta H_f(\text{KCl})$.	$-(-437 \text{ kJ·mol}^{-1})$
Set up the cycle, with a sum of zero.	$\{89 + 122 + 418 - 349 - (-437)\} \text{ kJ·mol}^{-1} - \Delta H_L = 0$
Solve for ΔH_L.	$\Delta H_L = (89 + 122 + 418 - 349 + 437) \text{ kJ·mol}^{-1}$ $= +717 \text{ kJ·mol}^{-1}$

Therefore, the lattice enthalpy of potassium chloride is 717 kJ·mol^{-1}.

SELF-TEST 6.17A Calculate the lattice enthalpy of calcium chloride, $CaCl_2$, by using the data in Appendices 2A and 2D.

[*Answer:* 2259 kJ·mol^{-1}]

SELF-TEST 6.17B Calculate the lattice enthalpy of magnesium bromide, $MgBr_2$, by using the data in Appendices 2A and 2D.

The strength of interaction between ions in a solid is measured by the lattice enthalpy, which can be determined by using a Born–Haber cycle.

6.20 Bond Enthalpies

In a chemical reaction, old bonds are broken and new ones formed. We can estimate reaction enthalpies if we know the enthalpy changes that accompany the breaking and making of bonds. The strength of a chemical bond is measured by the **bond enthalpy**, ΔH_B, the difference between the standard molar enthalpies of a molecule, X–Y (for instance, H_3C—OH), and its fragments X and Y (such as CH_3 and OH) in the gas phase:

$$\Delta H_B(X-Y) = \{H_m°(X, g) + H_m°(Y, g)\} - H_m°(XY, g) \qquad (22)$$

Whereas a lattice enthalpy is equal to the heat required (at constant pressure) to break up an ionic substance, a bond enthalpy is the heat required to break a specific type of bond at constant pressure. For example, the bond enthalpy of H_2 is derived from the thermochemical equation

$$H_2(g) \longrightarrow 2 H(g) \qquad \Delta H° = +436 \text{ kJ}$$

We write $\Delta H_B(\text{H—H}) = 436 \text{ kJ·mol}^{-1}$ to report this value and conclude that 436 kJ of heat is needed to dissociate 1 mol $H_2(g)$ into atoms.

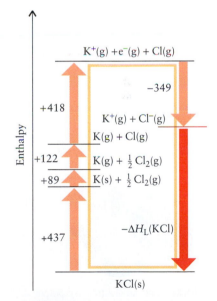

FIGURE 6.33 The Born–Haber cycle used to determine the lattice enthalpy of potassium chloride (see Example 6.13). The enthalpy changes are in kilojoules per mole.

TABLE 6.7 Bond Enthalpies of Diatomic Molecules (kJ·mol^{-1})

Molecule	ΔH_B
H$_2$	436
N$_2$	944
O$_2$	496
CO	1074
F$_2$	158
Cl$_2$	242
Br$_2$	193
I$_2$	151
HF	565
HCl	431
HBr	366
HI	299

A bond dissociation energy (Section 2.14) is strictly the *energy* required to break the bond at $T = 0$; a bond enthalpy is the change in *enthalpy* at the temperature of dissociation (typically 298 K). The two quantities differ by a few kilojoules per mole.

All bond enthalpies are positive because heat must be supplied to break a bond. In other words, *bond breaking is always endothermic and bond formation is always exothermic.* Table 6.7 lists bond enthalpies of some diatomic molecules.

In a polyatomic molecule, all the atoms in the molecule exert a pull—through their electronegativities (recall Section 2.12)—on all the electrons in the molecule (Fig. 6.34). As a result, the bond strength between a given pair of atoms varies to a small extent from one compound to another. For example, the O—H bond enthalpy in water, HO—H (492 kJ·mol^{-1}), is somewhat different from that of the same bond in CH$_3$O—H (437 kJ·mol^{-1}). However, these variations in bond enthalpy are not very great, and so the **mean bond enthalpy**, which we also denote ΔH_B, is a guide to the strength of a bond in any molecule containing the bond (Table 6.8). A bond enthalpy refers to the dissociation of a species in the gas phase; to find the enthalpy change accompanying the dissociation of a liquid or solid sample, we must add the enthalpy of vaporization or sublimation of the sample:

$$X_2(l) \longrightarrow 2\ X(g) \qquad \Delta H° = \Delta H_{vap}°(X_2, l) + \Delta H_B°(X\text{—}X)$$

Reaction enthalpies can be estimated by using mean bond enthalpies to determine the total energy required to break the reactant bonds and form the product bonds. In practice, only the bonds that change are treated. Because bond enthalpies refer to gaseous substances, to use the tabulated values, all substances must be gases or converted into the gas phase.

EXAMPLE 6.14   Using mean bond enthalpies to estimate the enthalpy of a reaction

Estimate the enthalpy change of the reaction between gaseous iodoethane and water vapor:

$$CH_3CH_2I(g) + H_2O(g) \longrightarrow CH_3CH_2OH(g) + HI(g)$$

STRATEGY Decide which bonds are broken and which bonds are formed. Use the mean bond enthalpies in Table 6.8 to estimate the change in enthalpy when the reactant bonds break and the change in enthalpy when the new product bonds form. For diatomic molecules, use the information in Table 6.7 for the specific molecule. Finally, add the enthalpy change required to break the reactant bonds (a positive value) to the enthalpy change that occurs when the product bonds form (a negative value).

Electronegative atom

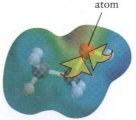

FIGURE 6.34 An electrostatic potential diagram of ethanol, C$_2$H$_5$OH. An electronegative atom (here, the O atom) can pull electrons toward itself from more distant parts of the molecule, as shown by the red tint over the O atom. Hence it can influence the strengths of bonds even between atoms to which it is not directly attached.

TABLE 6.8 Mean Bond Enthalpies (kJ·mol^{-1})

Bond	Mean bond enthalpy	Bond	Mean bond enthalpy
C—H	412	C—I	238
C—C	348	N—H	388
C=C	612	N—N	163
C$\cdots$C*	518	N=N	409
C≡C	837	N—O	210.
C—O	360	N=O	630.
C=O	743	N—F	195
C—N	305	N—Cl	381
C—F	484	O—H	463
C—Cl	338	O—O	157
C—Br	276		

*In benzene.

SOLUTION

Reactants: break a C—I bond in CH_3CH_2I (average value 238 kJ·mol^{-1}) and an O—H bond in H_2O (average value 463 kJ·mol^{-1}).

$\Delta H° = 238 + 463$ kJ $= +701$ kJ

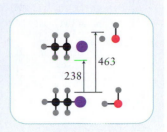

Products: Form a C—O bond (average value 360 kJ·mol^{-1}) and an H—I bond (299 kJ·mol^{-1}).

$\Delta H° = 360 + 299$ kJ $= +659$ kJ

The enthalpy change when the product bonds *form* must be negative.

$\Delta H° = -659$ kJ

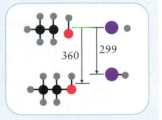

The overall enthalpy change is the sum of the bond enthalpy changes.

$\Delta H° = 701 + (-659)$ kJ $= +42$ kJ

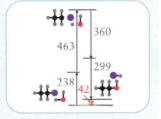

Therefore, the reaction is endothermic, primarily because a relatively large energy is needed to break an O—H bond.

SELF-TEST 6.18A Estimate the standard enthalpy of the reaction $CCl_3CHCl_2(g) + 2 HF(g) \rightarrow CCl_3CHF_2(g) + 2 HCl(g)$.

[***Answer:*** -24 kJ·mol^{-1}]

SELF-TEST 6.18B Estimate the standard enthalpy of the reaction in which 1.00 mol of gaseous CH_4 reacts with gaseous F_2 to form gaseous CH_2F_2 and HF.

A note on good practice. The use of mean bond enthalpies is hazardous because actual bond enthalpies often differ considerably from mean values. The modern procedure for estimating a reaction enthalpy is to use commercial software to calculate the enthalpies of formation of the reactants and products and then to take the difference, as in Section 6.18.

> *A mean bond enthalpy is the average molar enthalpy change accompanying the dissociation of a given type of bond.*

6.21 The Variation of Reaction Enthalpy with Temperature

Suppose we know the reaction enthalpy for one temperature but require it for another temperature. For instance, the temperature of the human body is about 37°C, but the data in Appendix 2A are for 25°C. Does an increase of 12°C make much difference to the reaction enthalpy of a metabolic process?

The enthalpies of both reactants and products increase with temperature. If the total enthalpy of the reactants increases more than that of the products when the temperature is raised, then the reaction enthalpy of an exothermic reaction becomes more negative (Fig. 6.35). On the other hand, if the enthalpy of the products increases more than that of the reactants, then the reaction enthalpy

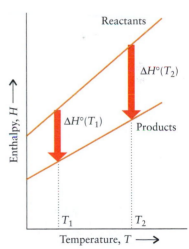

FIGURE 6.35 If the heat capacity of the reactants is larger than that of the products, the enthalpy of the reactants will increase more sharply with increasing temperature. If the reaction is exothermic, the reaction enthalpy will become more negative, as shown here. If the reaction is endothermic, the reaction enthalpy will become less positive and may even become negative.

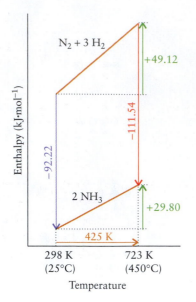

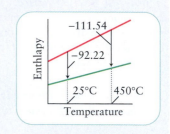

FIGURE 6.36 A depiction of the individual contributions to the change in standard reaction enthalpy for the reaction treated in Example 6.15.

will become less negative. The increase in enthalpy of a substance when the temperature is raised depends on its heat capacity at constant pressure (Eq. 12b), and it is easy to deduce **Kirchhoff's law** (see Exercise 6.91) that

$$\Delta H°(T_2) = \Delta H°(T_1) + (T_2 - T_1)\Delta C_P \qquad (23)*$$

To use this expression, we need to know ΔC_P, the difference between the constant-pressure heat capacities of the products and reactants:

$$\Delta C_P = \sum n C_{P,m}(\text{products}) - \sum n C_{P,m}(\text{reactants}) \qquad (24)$$

These values can be found in Appendix 2A. Because the difference between $\Delta H°(T_2)$ and $\Delta H°(T_1)$ depends on the *difference* in the heat capacities of the reactants and products—a difference that is normally small—in most cases, the reaction enthalpy depends only weakly on the temperature and, for small ranges of temperature, can be treated as a constant.

EXAMPLE 6.15   **Sample exercise: Predicting the reaction enthalpy at a different temperature**

The standard enthalpy of reaction of $N_2(g) + 3 H_2(g) \rightarrow 2 NH_3(g)$ is -92.22 kJ·mol^{-1} at 298 K. The industrial synthesis takes place at 450.°C. What is the standard reaction enthalpy at the latter temperature?

SOLUTION

Calculate the difference in heat capacities.	$\Delta C_P = (2\text{ mol})C_{P,m}(NH_3, g) - \{(1\text{ mol})C_{P,m}(N_2, g)$ $\qquad + (3\text{ mol})C_{P,m}(H_2, g)\}$ $\quad = (2\text{ mol})(35.06\text{ J·K}^{-1}\text{·mol}^{-1}) - \{(1\text{ mol})(29.12\text{ J·K}^{-1}\text{·mol}^{-1})$ $\qquad + (3\text{ mol})(28.82\text{ J·K}^{-1}\text{·mol}^{-1})\}$ $\quad = -45.46\text{ J·K}^{-1}$
Evaluate $T_2 - T_1$.	$T_2 - T_1 = (450 + 273\text{ K}) - 298\text{ K} = 425\text{ K}$
Calculate the enthalpy change at the final temperature from $\Delta H°(T_2) = \Delta H°(T_1) + (T_2 - T_1)\Delta C_P$	$\Delta H°(450\text{ K}) = -92.22\text{ kJ} + (425\text{ K}) \times (-45.46\text{ J·K}^{-1})$ $\qquad = -92.22\text{ kJ} - 1.932 \times 10^4\text{ J}$ $\qquad = -92.22\text{ kJ} - 19.32\text{ kJ}$ $\qquad = -111.54\text{ kJ}$

The individual contributions of the changes in enthalpy of the reactants and products are depicted in Fig. 6.36.

SELF-TEST 6.19A The reaction enthalpy of $4 Al(s) + 3 O_2(g) \rightarrow 2 Al_2O_3(s)$ is -3351 kJ at 298 K. Estimate its value at 1000.°C.

[*Answer:* -3378 kJ]

SELF-TEST 6.19B The standard enthalpy of formation of ammonium nitrate is -365.56 kJ·mol^{-1} at 298 K. Estimate its value at 250.°C.

The temperature variation of the standard reaction enthalpy is given by Kirchhoff's law, Eq. 23, in terms of the difference in molar heat capacities at constant pressure between the products and the reactants.

SKILLS YOU SHOULD HAVE MASTERED

❏ 1 Calculate the work done by a gas due to expansion (Examples 6.1 and 6.2).

❏ 2 Use the heat capacity of a substance to calculate the heat required to raise its temperature by a given amount (Example 6.3).

❏ 3 Calculate energy changes from calorimetry data and write a thermochemical equation (Examples 6.4 and 6.7).

❏ 4 State, and explain the implications of, the first law of thermodynamics (Sections 6.6–6.7).

❏ 5 Calculate the change in internal energy due to heat and work (Self-Test 6.5).

❏ 6 Calculate the work, heat, and change in internal energy when an ideal gas undergoes an isothermal expansion (Example 6.5).

❏ 7 Calculate the change in internal energy when an ideal gas is heated (Example 6.6).

❏ 8 Determine the enthalpy of vaporization (Self-Test 6.9).

❏ 9 Interpret the heating curve of a substance (Section 6.12).

❏ 10 Distinguish between the internal energy and the enthalpy of a process and describe how each is measured (Sections 6.8 and 6.14).

❏ 11 Determine the enthalpy change of a reaction, given the change in internal energy, and vice versa (Example 6.8).

❏ 12 Calculate an overall reaction enthalpy from the enthalpies of the reactions in a reaction sequence by using Hess's law (Toolbox 6.1 and Example 6.9).

❏ 13 Calculate the heat output of a fuel (Example 6.10).

❏ 14 Use standard enthalpies of formation to calculate the standard enthalpy of a reaction, and vice versa (Examples 6.11 and 6.12).

❏ 15 Calculate a lattice enthalpy by using the Born–Haber cycle (Example 6.13).

❏ 16 Use average bond enthalpies to estimate the standard enthalpy of a reaction, (Example 6.14).

❏ 17 Predict the reaction enthalpy at a temperature different from that of the data (Example 6.15).

EXERCISES

Systems

6.1 Identify the following systems as open, closed, or isolated: (a) coffee in a very high quality thermos bottle; (b) coolant in a refrigerator coil; (c) a bomb calorimeter in which benzene is burned; (d) gasoline burning in an automobile engine; (e) mercury in a thermometer; (f) a living plant.

6.2 (a) Describe three ways in which you could increase the internal energy of an open system. (b) Which of these methods could you use to increase the internal energy of a closed system? (c) Which, if any, of these methods could you use to increase the internal energy of an isolated system?

The First Law

6.3 Air in a bicycle pump is compressed by pushing in the handle. If the inner diameter of the pump is 3.0 cm and the pump is depressed 20. cm with a pressure of 2.00 atm, (a) how much work is done in the compression? (b) Is the work positive or negative with respect to the air in the pump?

6.4 The four cylinders of a new type of combustion engine each have a displacement of 2.50 L. (The volume of the cylinder expands 2.50 L each time the fuel is ignited.) (a) If the pistons in the four cylinders are each displaced with a pressure of 1.40 kbar and each cylinder is ignited once per second, how much work can the engine do in 1.00 minutes? (b) Is the work positive or negative with respect to the engine and its contents?

6.5 A gas sample is heated in a cylinder by supplying 524 kJ of heat. At the same time, a piston compresses the gas, doing 340 kJ of work. What is the change in internal energy of the gas during this process?

6.6 A gas sample in a piston assembly expands, doing 536 kJ of work on its surroundings at the same time that 214 kJ of heat is added to the gas. (a) What is the change in internal energy of the gas during this process? (b) Will the pressure of the gas be higher or lower when these changes are completed?

6.7 The internal energy of a system increased by 982 J when it absorbed 492 J of heat. (a) Was work done by or on the system? (b) How much work was done?

6.8 (a) Calculate the work for a system that releases 346 kJ of heat in a process for which the decrease in internal energy is 125 kJ. (b) Is work done on or by the system during this process?

6.9 A gas in a cylinder was placed in a heater and gained 5500 kJ of heat. If the cylinder increased in volume from 345 mL to 1846 mL against an atmospheric pressure of 750. Torr during this process, what is the change in internal energy of the gas in the cylinder?

6.10 A 100.-W electric heater ($1 \text{ W} = 1 \text{ J·s}^{-1}$) operates for 10.0 min to heat the gas in a cylinder. At the same time, the gas expands from 2.00 L to 10.00 L against a constant atmospheric pressure of 0.975 atm. What is the change in internal energy of the gas?

6.11 In a combustion cyclinder, the total internal energy change produced from the burning of a fuel is −2573 kJ. The cooling system that surrounds the cylinder absorbs 947 kJ as heat. How much work can be done by the fuel in the cylinder?

6.12 A laboratory animal exercised on a treadmill that was connected to a weight with a mass of 250 g through a pulley. The work done by the animal raised the weight through 1.32 m. At the same time the animal gave off 8.0 J of energy as heat. Disregarding any other losses and treating the animal as a closed system, what was the change in internal energy of the animal?

6.13 In an adiabatic process, no energy is transferred as heat. Indicate whether each of the following statements about an adiabatic process in a closed system is always true, always false, or true in certain conditions (specify the conditions): (a) $\Delta U = 0$; (b) $q = 0$; (c) $q < 0$; (d) $\Delta U = q$; (e) $\Delta U = w$.

6.14 Indicate whether each of the following statements about a process in a closed system with a constant volume is always true, always false, or true in certain conditions (specify the conditions): (a) $\Delta U = 0$; (b) $w = 0$; (c) $w < 0$; (d) $\Delta U = q$; (e) $\Delta U = w$.

6.15 Each of the pictures below shows a molecular view of a system undergoing a change at constant temperature. In each case, indicate whether heat is absorbed or given off by the system, whether expansion work is done on or by the system, and predict the signs of q and w for the process.

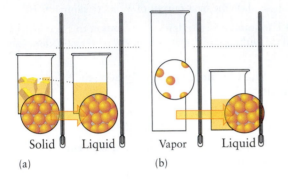

6.16 Each of the pictures below shows a molecular view of a system undergoing a change. In each case, indicate whether heat is absorbed or given off by the system, whether expansion work is done on or by the system, and predict the signs of q and w for the process.

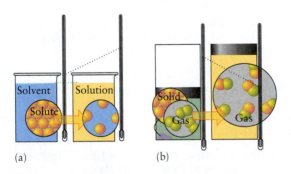

6.17 (a) Calculate the heat that must be supplied to a 500.0-g copper kettle containing 750.0 g of water to raise its temperature from 23.0°C to the boiling point of water, 100.0°C. (b) What percentage of the heat is used to raise the temperature of the water?

6.18 (a) Calculate the heat that must be supplied to a 500.0-g stainless steel vessel containing 450.0 g of water to raise its temperature from 25.0°C to the boiling point of water, 100.0°C. (b) What percentage of the heat is used to raise the temperature of the water? (c) Compare these answers with those of Exercise 6.17.

6.19 A piece of copper of mass 20.0 g at 100.0°C is placed in a vessel of negligible heat capacity but containing 50.7 g of water at 22.0°C. Calculate the final temperature of the water. Assume that there is no energy lost to the surroundings.

6.20 A piece of metal of mass 20.0 g at 100.0°C is placed in a calorimeter containing 50.7 g of water at 22.0°C. The final temperature of the mixture is 25.7°C. What is the specific heat capacity of the metal? Assume that there is no energy lost to the surroundings.

6.21 A calorimeter was calibrated with an electric heater, which supplied 22.5 kJ of energy to the calorimeter and increased the temperature of the calorimeter and its water bath from 22.45°C to 23.97°C. What is the heat capacity of the calorimeter?

6.22 The enthalpy of combustion of benzoic acid, C_6H_5COOH, which is often used to calibrate calorimeters, is -3227 kJ·mol^{-1}. When 1.453 g of benzoic acid was burned in a calorimeter, the temperature increased by 2.265°C. What is the heat capacity of the calorimeter?

6.23 Calculate the work for each of the following processes beginning with a gas sample in a piston assembly with $T = 305$ K, $P = 1.79$ atm, and $V = 4.29$ L: (a) irreversible expansion against a constant external pressure of 1.00 atm to a final volume of 6.52 L; (b) isothermal, reversible expansion to a final volume of 6.52 L.

6.24 A sample of gas in a cylinder of volume 3.42 L at 298 K and 2.57 atm expands to 7.39 L by two different pathways. Path A is an isothermal, reversible expansion. Path B has two steps. In the first step, the gas is cooled at constant volume to 1.19 atm. In the second step, the gas is heated and allowed to expand against a constant external pressure of 1.19 atm until the final volume is 7.39 L. Calculate the work for each path.

Enthalpy

6.25 Which molecular substance do you expect to have the higher molar heat capacity, NO or NO_2? Why?

6.26 Explain why the heat capacities of methane and ethane differ from the values expected for an ideal monatomic gas and from each other. The values are 35.309 J·K^{-1}·mol^{-1} for CH_4 and 52.63 J·K^{-1}·mol^{-1} for C_2H_6.

6.27 Calculate the heat released by 5.025 g of Kr(g) at 0.400 atm as it cools from 97.6°C to 25.0°C at (a) constant pressure and (b) constant volume. Assume that krypton behaves as an ideal gas.

6.28 Calculate the heat that must be supplied to 10.35 g of Ne(g) at 0.150 atm to raise its temperature from 25.0°C to 50.0°C at (a) constant pressure and (b) constant volume. Assume that neon behaves as an ideal gas.

6.29 Predict the contribution to the heat capacity $C_{V,m}$ made by molecular motions for each of the following atoms and molecules: (a) HCN; (b) C_2H_6; (c) Ar; (d) HBr.

6.30 Predict the contribution to the heat capacity $C_{V,m}$ made by molecular motions for each of the following molecules: (a) NO; (b) NH_3; (c) HClO; (d) SO_2.

6.31 In a microwave oven, radiation is absorbed by water in the food and the food is heated. How many photons of wavelength 4.50 mm are required to heat 350. g of water from 25.0°C to 100.0°C, assuming all the energy is used to raise the temperature?

6.32 Samples consisting of 1 mol N_2 and 1 mol CH_4 are in identical but separate containers, with initial temperatures of 500. K. Both gases gain 1200. J of heat at constant volume. Do the gases have the same final temperature? If not, which gas has the higher final temperature? Justify your reasoning.

6.33 The high-temperature contribution of vibrational modes to the molar heat capacity of a solid at constant volume is R for each mode of vibrational motion. Hence, for an atomic solid, the molar heat capacity at constant volume is approximately $3R$. (a) The specific heat capacity of a certain atomic solid is $0.392 \ J \cdot K^{-1} \cdot g^{-1}$. The chloride of this element (XCl_2) is 52.7% chlorine by mass. Identify the element. (b) This element crystallizes in a face-centered cubic unit cell and its atomic radius is 128 pm. What is the density of this atomic solid?

6.34 Estimate the molar heat capacity (at constant volume) of sulfur dioxide gas. In addition to translational and rotational motion, there is vibrational motion. Each vibrational degree of freedom contributes R to the molar heat capacity. The temperature needed for the vibrational modes to be accessible can be approximated by $\theta = h\nu_{vib}/k$, where k is Boltzmann's constant. The vibrational modes have frequencies 3.5×10^{13} Hz, 4.1×10^{13} Hz, and 1.6×10^{13} Hz. (a) What is the high-temperature limit of the molar heat capacity at constant volume? (b) What is the molar heat capacity at constant volume at 1000. K? (c) What is the molar heat capacity at constant volume at room temperature?

6.35 (a) At its boiling point, the vaporization of 0.579 mol $CH_4(l)$ requires 4.76 kJ of heat. What is the enthalpy of vaporization of methane? (b) An electric heater was immersed in a flask of boiling ethanol, C_2H_5OH, and 22.45 g of ethanol was vaporized when 21.2 kJ of energy was supplied. What is the enthalpy of vaporization of ethanol?

6.36 (a) When 25.23 g of methanol, CH_3OH, froze, 4.01 kJ of heat was released. What is the enthalpy of fusion of methanol? (b) A sample of benzene was vaporized at 25°C. When 37.5 kJ of heat was supplied, 95 g of the liquid benzene vaporized. What is the enthalpy of vaporization of benzene at 25°C?

6.37 How much heat is needed to convert 80.0 g of ice at 0.0°C into liquid water at 20.0°C (see Tables 6.2 and 6.3)?

6.38 If we start with 325 g of water at 30.°C, how much heat must we add to convert all the liquid into vapor at 100.°C (see Tables 6.2 and 6.3)?

6.39 A 50.0-g ice cube at 0.0°C is added to a glass containing 400.0 g of water at 45.0°C. What is the final temperature of the system (see Tables 6.2 and 6.3)? Assume that no heat is lost to the surroundings.

6.40 When 25.0 g of a metal at 90.0°C is added to 50.0 g of water at 25.0°C, the temperature of the water rises to 29.8°C. What is the specific heat capacity of the metal (see Table 6.2)?

6.41 In 1750 Joseph Black performed an experiment that eventually led to the discovery of enthalpies of fusion. He placed two 150.-g samples of water at 0.00°C (one ice and one liquid) in a room kept at a constant temperature of 5.00°C. He then observed how long it took for each sample to warm to its final temperature. The liquid sample reached 5.00°C after 30.0 min. However, the ice took 10.5 h to reach 5.00°C. He concluded that the difference in time that the two samples required to reach the same final temperature represented the difference in heat required to raise the temperatures of the samples. Use Black's data to calculate the enthalpy of fusion of ice in $kJ \cdot mol^{-1}$. Use the known heat capacity of liquid water.

6.42 The early notion that heat was a fluid called caloric was disproved in 1798 by Benjamin Thompson (later, Count Rumford). While minister of war in Bavaria and boring cannon, he observed that heat could be produced continuously and endlessly from a given mass of iron, and so it could not be a fluid. In one of his experiments, he used a team of two horses to turn a lathe that bored a hole in a 51-kg piece of cannon iron. The iron was immersed in a wooden box containing 18.77 lb. of water. The assembly was initially at 60.°F and 1.00 atm. After the horses had worked for 2.5 h, the water began to boil. Assuming that all the work done by the horses was converted to heat in the iron and water, use the information in Table 6.2 to determine the work done by the horses per minute.

6.43 The following data were collected for a new compound used in cosmetics: $\Delta H_{fus} = 10.0 \ kJ \cdot mol^{-1}$, $\Delta H_{vap} = 20.0 \ kJ \cdot mol^{-1}$; heat capacities: $30 \ J \cdot mol^{-1}$ for the solid; $60 \ J \cdot mol^{-1}$ for the liquid; $30 \ J \cdot mol^{-1}$ for the gas. Which heating curve below best matches the data for this compound?

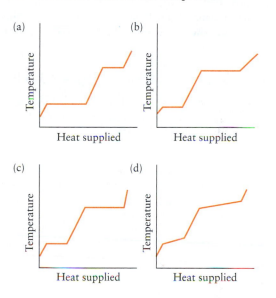

6.44 Use the following information to construct a heating curve for bromine from $-7.2°C$ to $70.0°C$. The molar heat capacity of liquid bromine is $75.69 \ J \cdot K^{-1} \cdot mol^{-1}$ and that of bromine vapor is $36.02 \ J \cdot K^{-1} \cdot mol^{-1}$. The enthalpy of vaporization of liquid bromine is $30.91 \ kJ \cdot mol^{-1}$. Bromine melts at $-7.2°C$ and boils at $58.78°C$.

Reaction Enthalpies

6.45 Carbon disulfide can be prepared from coke (an impure form of carbon) and elemental sulfur:

$$4 \ C(s) + S_8(s) \longrightarrow 4 \ CS_2(l) \quad \Delta H° = +358.8 \ kJ$$

(a) How much heat is absorbed in the reaction of 1.25 mol S_8? (b) Calculate the heat absorbed in the reaction of 197 g of carbon with an excess of sulfur. (c) If the heat absorbed in the reaction was 415 kJ, how much CS_2 was produced?

6.46 The oxidation of nitrogen in the hot exhaust of jet engines and automobiles occurs by the reaction

$$N_2(g) + O_2(g) \longrightarrow 2 \ NO(g) \quad \Delta H° = +180.6 \ kJ$$

(a) How much heat is absorbed in the formation of 1.55 mol NO? (b) How much heat is absorbed in the oxidation of 5.45 L of nitrogen measured at 1.00 atm and 273 K? (c) When the oxidation of N_2 to NO was completed in a bomb calorimeter, the heat absorbed was measured as 492 J. What mass of nitrogen gas was oxidized?

6.47 The combustion of octane is expressed by the thermochemical equation

$$C_8H_{18}(l) + \tfrac{25}{2}O_2(g) \longrightarrow 8\,CO_2(g) + 9\,H_2O(l)$$
$$\Delta H° = -5471\ kJ$$

(a) Estimate the mass of octane that would need to be burned to produce enough heat to raise the temperature of the air in a 12 ft × 12 ft × 8.0 ft room from 40.°F to 78°F on a mild winter's day. Use the normal composition of air to determine its density and assume a pressure of 1.00 atm. (b) How much heat will be evolved from the combustion of 1.0 gal of gasoline (assumed to be exclusively octane)? The density of octane is 0.70 g·mL^{-1}.

6.48 Suppose that coal of density 1.5 g·cm^{-3} is carbon (it is, in fact, much more complicated, but this is a reasonable first approximation). The combustion of carbon is described by the equation

$$C(s) + O_2(g) \longrightarrow CO_2(g) \qquad \Delta H° = -394\ kJ$$

(a) Calculate the heat produced when a lump of coal of size 7.0 cm × 6.0 cm × 5.0 cm is burned. (b) Estimate the mass of water that can be heated from 25°C to 100.°C by burning this piece of coal.

6.49 A student rides a bicycle to class every day, a 10.-mile round trip that takes 30. minutes in each direction. The student burns 1250 kJ·h^{-1} cycling. The same round trip in an automobile would require 0.40 gallons of gasoline. Assume that the student goes to class 150 days per year and that the enthalpy of combustion of gasoline can be approximated by that of octane, which has a density of 0.702 g·cm^{-3} (3.785 L = 1.000 gal). What is the yearly energy requirement of this journey by (a) bicycle and (b) automobile?

6.50 Calculate the heat evolved from a reaction mixture of 13.4 L of sulfur dioxide at 1.00 atm and 273 K and 15.0 g of oxygen in the reaction

$$2\,SO_2(g) + O_2(g) \longrightarrow 2\,SO_3(g) \quad \Delta H° = -198\ kJ$$

6.51 For a certain reaction at constant pressure, $\Delta H = -15$ kJ, and 22 kJ of expansion work is done on the system. What is ΔU for this process?

6.52 For a certain reaction at constant pressure, $\Delta U = -65$ kJ, and 48 kJ of expansion work is done by the system. What is ΔH for this process?

6.53 Hydrochloric acid oxidizes zinc metal in a reaction that produces hydrogen gas and chloride ions. A piece of zinc metal of mass 8.5 g is dropped into an apparatus containing 800.0 mL of 0.500 M HCl(aq). If the initial temperature of the hydrochloric acid solution is 25°C, what is the final temperature of this solution? Assume that the density and molar heat capacity of the hydrochloric acid solution are the same as those of water and that all the heat is used to raise the temperature of the solution.

6.54 "Synthesis gas" is a mixture of carbon monoxide, hydrogen, methane, and some noncombustible gases. A certain synthesis gas is 40.0% by volume carbon monoxide, 25.0% hydrogen gas, 10.0% noncombustible gases, and the rest methane. What volume of this gas must be burned to raise the temperature of 5.5 L of water by 5°C? Assume the gas is at 1.0 atm and 298 K.

6.55 The enthalpy of formation of trinitrotoluene (TNT) is −67 kJ·mol^{-1}, and the density of TNT is 1.65 g·cm^{-3}. In principle, it could be used as a rocket fuel, with the gases resulting from its decomposition streaming out of the rocket to give the required thrust. In practice, of course, it would be extremely dangerous as a fuel because it is sensitive to shock. Explore its potential as a rocket fuel by calculating its enthalpy density (enthalpy released per liter) for the reaction

$$4\,C_7H_5N_3O_6(s) + 21\,O_2(g) \longrightarrow$$
$$28\,CO_2(g) + 10\,H_2O(g) + 6\,N_2(g)$$

6.56 A natural gas mixture is burned in a furnace at a power-generating station at a rate of 13.0 mol per minute. (a) If the fuel consists of 9.3 mol CH_4, 3.1 mol C_2H_6, 0.40 mol C_3H_8, and 0.20 mol C_4H_{10}, what mass of $CO_2(g)$ is produced per minute? (b) How much heat is released per minute?

6.57 The reaction of 1.40 g of carbon monoxide with excess water vapor to produce carbon dioxide and hydrogen gases in a bomb calorimeter causes the temperature of the calorimeter assembly to rise from 22.113°C to 22.799°C. The calorimeter assembly is known to have a total heat capacity of 3.00 kJ·(°C)$^{-1}$. (a) Write a balanced equation for the reaction. (b) Calculate the internal energy change, ΔU, for the combustion of 1.00 mol CO(g).

6.58 Heat is generated in the formation of slaked lime from the reaction $CaO(s) + H_2O(l) \rightarrow Ca(OH)_2(s)$, for which $\Delta H° = -65.17$ kJ. If all this heat could be transferred to 250. g of water, what would be the resulting temperature change of the water?

Hess's Law

6.59 The standard enthalpies of combustion of graphite and diamond are −393.51 kJ·mol^{-1} and −395.41 kJ·mol^{-1}, respectively. Calculate the enthalpy of the graphite → diamond transition.

6.60 Elemental sulfur exists in several forms, with rhombic sulfur the most stable under normal conditions and monoclinic sulfur slightly less stable. The standard enthalpies of combustion of the two forms (to sulfur dioxide) are −296.83 kJ·mol^{-1} and −297.16 kJ·mol^{-1}, respectively. Calculate the enthalpy of the rhombic → monoclinic transition.

6.61 Two successive stages in the industrial manufacture of sulfuric acid are the combustion of sulfur and the oxidation of sulfur dioxide to sulfur trioxide. From the standard reaction enthalpies

$$S(s) + O_2(g) \longrightarrow SO_2(g) \qquad \Delta H° = -296.83\ kJ$$
$$2\,S(s) + 3\,O_2(g) \longrightarrow 2\,SO_3(g) \qquad \Delta H° = -791.44\ kJ$$

calculate the reaction enthalpy for the oxidation of sulfur dioxide to sulfur trioxide in the reaction

$$2\,SO_2(g) + O_2(g) \longrightarrow 2\,SO_3(g)$$

6.62 In the manufacture of nitric acid by the oxidation of ammonia, the first product is nitric oxide, which is then oxidized to nitrogen dioxide. From the standard reaction enthalpies

$$N_2(g) + O_2(g) \longrightarrow 2\,NO(g) \qquad \Delta H° = +180.5\,kJ$$
$$N_2(g) + 2\,O_2(g) \longrightarrow 2\,NO_2(g) \qquad \Delta H° = +66.4\,kJ$$

calculate the standard reaction enthalpy for the oxidation of nitric oxide to nitrogen dioxide:

$$2\,NO(g) + O_2(g) \longrightarrow 2\,NO_2(g)$$

6.63 Determine the reaction enthalpy for the hydrogenation of ethyne to ethane, $C_2H_2(g) + 2\,H_2(g) \rightarrow C_2H_6(g)$, from the following data: $\Delta H_c°(C_2H_2, g) = -1300.\,kJ\cdot mol^{-1}$, $\Delta H_c°(C_2H_6, g) = -1560.\,kJ\cdot mol^{-1}$, $\Delta H_c°(H_2, g) = -286\,kJ\cdot mol^{-1}$.

6.64 Determine the enthalpy for the partial combustion of methane to carbon monoxide, $2\,CH_4(g) + 3\,O_2(g) \rightarrow 2\,CO(g) + 4\,H_2O(l)$, from $\Delta H_c°(CH_4, g) = -890.\,kJ\cdot mol^{-1}$ and $\Delta H_c°(CO, g) = -283.0\,kJ\cdot mol^{-1}$.

6.65 Use the data provided in Appendix 2A to calculate the standard reaction enthalpy for the reaction of pure nitric acid with hydrazine:

$$4\,HNO_3(l) + 5\,N_2H_4(l) \longrightarrow 7\,N_2(g) + 12\,H_2O(l)$$

6.66 Use the data provided in Appendix 2A to calculate the standard reaction enthalpy for the reaction of calcite with hydrochloric acid:

$$CaCO_3(s) + 2\,HCl(aq) \longrightarrow$$
$$CaCl_2(aq) + H_2O(l) + CO_2(g)$$

6.67 Calculate the reaction enthalpy for the synthesis of hydrogen chloride gas, $H_2(g) + Cl_2(g) \rightarrow 2\,HCl(g)$, from the following data:

$$NH_3(g) + HCl(g) \longrightarrow NH_4Cl(s) \quad \Delta H° = -176.0\,kJ$$
$$N_2(g) + 3\,H_2(g) \longrightarrow 2\,NH_3(g) \quad \Delta H° = -92.22\,kJ$$
$$N_2(g) + 4\,H_2(g) + Cl_2(g) \longrightarrow 2\,NH_4Cl(s)$$
$$\Delta H° = -628.86\,kJ$$

6.68 Calculate the reaction enthalpy for the formation of anhydrous aluminum chloride, $2\,Al(s) + 3\,Cl_2(g) \rightarrow 2\,AlCl_3(s)$, from the following data:

$$2\,Al(s) + 6\,HCl(aq) \longrightarrow 2\,AlCl_3(aq) + 3\,H_2(g)$$
$$\Delta H° = -1049\,kJ$$
$$HCl(g) \longrightarrow HCl(aq) \qquad \Delta H° = -74.8\,kJ$$
$$H_2(g) + Cl_2(g) \longrightarrow 2\,HCl(g) \qquad \Delta H° = -185\,kJ$$
$$AlCl_3(s) \longrightarrow AlCl_3(aq) \qquad \Delta H° = -323\,kJ$$

6.69 Consider the following numbered processes:

1. $A \rightarrow 2\,B$

2. $B \rightarrow C + D$

3. $E \rightarrow 2\,D$

Which of the following calculations will give ΔH for the process $A \rightarrow 2\,C + E$?

(a) $\Delta H_1 + \Delta H_2 + \Delta H_3$
(b) $\Delta H_1 + \Delta H_2 - \Delta H_3$
(c) $\Delta H_1 + 2\Delta H_2 - \Delta H_3$
(d) $\Delta H_1 + 2\Delta H_2 + \Delta H_3$

6.70 Consider the following numbered processes:

1. $2\,A + 5\,B \rightarrow 4\,C + 2\,D$

2. $2\,E + 7\,B \rightarrow 4\,C + 6\,D$

3. $2\,F + B \rightarrow 2\,D$

Which of the following calculations will give ΔH for the process $A + 2\,F \rightarrow E$?

(a) $\Delta H_1 - \Delta H_2 + \Delta H_3$
(b) $\frac{1}{2}\Delta H_1 - \frac{1}{2}\Delta H_2 + \Delta H_3$
(c) $\frac{1}{2}\Delta H_1 + \frac{1}{2}\Delta H_2 - \Delta H_3$
(d) $\Delta H_1 + \Delta H_2 + \Delta H_3$

6.71 Calculate the reaction enthalpy for the synthesis of hydrogen bromide gas, $H_2(g) + Br_2(l) \rightarrow 2\,HBr(g)$, from the following data:

$$NH_3(g) + HBr(g) \longrightarrow NH_4Br(s) \quad \Delta H° = -188.32\,kJ$$
$$N_2(g) + 3\,H_2(g) \longrightarrow 2\,NH_3(g) \quad \Delta H° = -92.22\,kJ$$
$$N_2(g) + 4\,H_2(g) + Br_2(l) \longrightarrow 2\,NH_4Br(s)$$
$$\Delta H° = -541.66\,kJ$$

6.72 Calculate the reaction enthalpy for the formation of anhydrous aluminum bromide, $2\,Al(s) + 3\,Br_2(l) \rightarrow 2\,AlBr_3(s)$, from the following data:

$$2\,Al(s) + 6\,HBr(aq) \longrightarrow 2\,AlBr_3(aq) + 3\,H_2(g)$$
$$\Delta H° = -1061\,kJ$$
$$HBr(g) \longrightarrow HBr(aq) \qquad \Delta H° = -81.15\,kJ$$
$$H_2(g) + Br_2(l) \longrightarrow 2\,HBr(g) \qquad \Delta H° = -72.80\,kJ$$
$$AlBr_3(s) \longrightarrow AlBr_3(aq) \qquad \Delta H° = -368\,kJ$$

Standard Enthalpies of Formation

6.73 Calculate the standard enthalpy of formation of dinitrogen pentoxide from the following data

$$2\,NO(g) + O_2(g) \longrightarrow 2\,NO_2(g) \qquad \Delta H° = -114.1\,kJ$$
$$4\,NO_2(g) + O_2(g) \longrightarrow 2\,N_2O_5(g) \quad \Delta H° = -110.2\,kJ$$

and from the standard enthalpy of formation of nitric oxide, NO (see Appendix 2A).

6.74 An important reaction that takes place in the atmosphere is $NO_2(g) \rightarrow NO(g) + O(g)$, which is brought about by sunlight. How much energy must be supplied by the Sun to cause it? Calculate the standard enthalpy of the reaction from the following information

$$O_2(g) \longrightarrow 2\,O(g) \qquad \Delta H° = +498.4\,kJ$$
$$NO(g) + O_3(g) \longrightarrow NO_2(g) + O_2(g)$$
$$\Delta H° = -200.\,kJ$$

and from additional information in Appendix 2A.

6.75 Calculate the standard enthalpy of formation of $PCl_5(s)$ from the standard enthalpy of formation of $PCl_3(l)$ (see Appendix 2A) and $PCl_3(l) + Cl_2(g) \rightarrow PCl_5(s)$, $\Delta H° = -124$ kJ.

6.76 Using standard enthalpies of formation from Appendix 2A, calculate the standard reaction enthalpy for each of the following reactions:

(a) the final stage in the production of nitric acid:

$$3\ NO_2(g) + H_2O(l) \longrightarrow 2\ HNO_3(aq) + NO(g)$$

(b) the industrial synthesis of boron trifluoride:

$$B_2O_3(s) + 3\ CaF_2(s) \longrightarrow 2\ BF_3(g) + 3\ CaO(s)$$

(c) the formation of a sulfide by the action of hydrogen sulfide on an aqueous solution of a base:

$$H_2S(aq) + 2\ KOH(aq) \longrightarrow K_2S(aq) + 2\ H_2O(l)$$

6.77 Use the information in Tables 6.3, 6.7, and 6.8 to estimate the enthalpy of formation of each of the following compounds in the *liquid* state. The standard enthalpy of sublimation of carbon is $+717$ kJ·mol^{-1}. (a) H_2O; (b) methanol, CH_3OH; (c) benzene, C_6H_6 (without resonance); (d) benzene, C_6H_6 (with resonance).

6.78 Use the information in Tables 6.3, 6.7, and 6.8 to estimate the enthalpy of formation of each of the following compounds in the *liquid* state. The standard enthalpy of sublimation of carbon is $+717$ kJ·mol^{-1}. (a) NH_3; (b) ethanol, CH_3CH_2OH; (c) acetone, CH_3COCH_3.

The Born–Haber Cycle

6.79 Use Fig. 1.54, Appendix 2A, Appendix 2D, and the following data to calculate the lattice enthalpy of Na_2O: $\Delta H_f°(Na_2O) = -409$ kJ·mol^{-1}; $\Delta H_f°(O, g) = +249$ kJ·mol^{-1}.

6.80 Use Fig. 1.54, Appendix 2A, Appendix 2D, and the following data to calculate the lattice enthalpy of $AlCl_3$: $\Delta H_f°(Al, g) = +326$ kJ·mol^{-1}.

6.81 Complete the following table (all values are in kilojoules per mole).

Compound MX	$\Delta H_f°$ M(g)	Ionization energy M	$\Delta H_f°$ X(g)	Electron affinity X	ΔH_L MX	$\Delta H_f°$ MX(s)
(a) NaCl	108	494	122	+349	787	?
(b) KBr	89	418	97	+325	?	−394
(c) RbF	?	402	79	+328	774	−558

6.82 By assuming that the lattice enthalpy of $NaCl_2$ is the same as that of $MgCl_2$, use enthalpy arguments based on data in Appendix 2A, Appendix 2D, and Fig. 1.54 to explain why $NaCl_2$ is an unlikely compound.

Bond Enthalpies

6.83 Use the bond enthalpies in Tables 6.7 and 6.8 to estimate the reaction enthalpy for

(a) $3\ C_2H_2(g) \rightarrow C_6H_6(g)$

(b) $CH_4(g) + 4\ Cl_2(g) \rightarrow CCl_4(g) + 4\ HCl(g)$

(c) $CH_4(g) + CCl_4(g) \rightarrow CHCl_3(g) + CH_3Cl(g)$

6.84 Use the bond enthalpies in Tables 6.7 and 6.8 to estimate the reaction enthalpy for

(a) $HCl(g) + F_2(g) \rightarrow HF(g) + ClF(g)$, given that $\Delta H_B(Cl-F) = -256$ kJ·mol^{-1}

(b) $C_2H_4(g) + HCl(g) \rightarrow CH_3CH_2Cl(g)$

(c) $C_2H_4(g) + H_2(g) \rightarrow CH_3CH_3(g)$

6.85 Use the bond enthalpies in Tables 6.7 and 6.8 to estimate the reaction enthalpy for

(a) $N_2(g) + 3\ F_2(g) \rightarrow 2\ NF_3(g)$

(b) $CH_3CH=CH_2(g) + H_2O(g) \rightarrow CH_3CH(OH)CH_3(g)$

(c) $CH_4(g) + Cl_2(g) \rightarrow CH_3Cl(g) + HCl(g)$

6.86 The bond enthalpy in NO is 632 kJ·mol^{-1} and that of each N—O bond in NO_2 is 469 kJ·mol^{-1}. Using Lewis structures and the mean bond enthalpies given in Table 6.8, explain (a) the difference in bond enthalpies between the two molecules; (b) the fact that the bond enthalpies of the two bonds in NO_2 are the same.

6.87 Benzene is more stable and less reactive than would be predicted from its Kekulé structures. Use the mean bond enthalpies in Table 6.8 to calculate the lowering in molar energy when resonance is allowed between the Kekulé structures of benzene.

6.88 Investigate whether the replacement of a carbon-carbon double bond by single bonds is energetically favored by using Tables 6.7 and 6.8 to calculate the reaction enthalpy for the conversion of ethene, C_2H_4, to ethane, C_2H_6. The reaction is $H_2C=CH_2(g) + H_2(g) \rightarrow CH_3-CH_3(g)$.

The Variation of Reaction Enthalpy with Temperature

6.89 (a) From the data in Appendix 2A, calculate the enthalpy of vaporization of benzene (C_6H_6) at 298.2 K. The standard enthalpy of formation of gaseous benzene is $+82.93$ kJ·mol^{-1}. (b) Given that, for liquid benzene, $C_{P,m} = 136.1$ J·mol^{-1}·K^{-1} and that, for gaseous benzene, $C_{P,m} = 81.67$ J·mol^{-1}·K^{-1}, calculate the enthalpy of vaporization of benzene at its boiling point (353.2 K). (c) Compare the value obtained in part (b) with that found in Table 6.3. What is the source of difference between these numbers?

6.90 (a) From the data in Appendix 2A, calculate the enthalpy required to vaporize 1 mol $CH_3OH(l)$ at 298.2 K. (b) Given that the molar heat capacity, $C_{P,m}$, of liquid methanol is 81.6 J·K^{-1}·mol^{-1} and that of gaseous methanol is 43.89 J·K^{-1}·mol^{-1}, calculate the enthalpy of vaporization of methanol at its boiling point (64.7°C). (c) Compare the value obtained in part (b) with that found in Table 6.3. What is the source of difference between these values?

6.91 Derive Kirchhoff's law for a reaction of the form A + 2 B $\rightarrow$ 3 C + D by considering the change in molar enthalpy of each substance when the temperature is increased from T_1 to T_2.

6.92 Use the estimates of molar constant-volume heat capacities given in the text (as multiples of R) to estimate the change in reaction enthalpy of $N_2(g) + 3\ H_2(g) \rightarrow 2\ NH_3(g)$ when the temperature is increased from 300. K to 500. K. Ignore the vibrational contributions to heat capacity. Is the reaction more or less exothermic at the higher temperature?

Integrated Exercises

6.93 How much heat is required to convert a 42.30-g block of ice at $-5.042°C$ into water vapor at $150.35°C$?

6.94 A 155.7-g piece of stainless steel was heated to $475°C$ and quickly added to 25.34 g of ice at $-24°C$ in a well-insulated flask that was immediately sealed. (a) If there is no loss of energy to the surroundings, what will be the final temperature of this system? (b) What phases of water will be present and in what quantities when the system reaches its final temperature?

6.95 The heat capacity of liquid iodine is $80.7\ J\cdot K^{-1}\cdot mol^{-1}$, and the enthalpy of vaporization of iodine is $41.96\ kJ\cdot mol^{-1}$ at its boiling point ($184.3°C$). Using these facts and information in Appendix 2A, calculate the enthalpy of fusion of iodine at $25°C$.

6.96 A typical bathtub can hold 100. gallons of water. (a) Calculate the mass of natural gas that would need to be burned to heat the water for a tub of this size from $65°F$ to $108°F$. Assume that the natural gas is pure methane, CH_4. (b) What volume of natural gas does this correspond to at $25°C$ and 1.00 atm? See Table 6.4.

6.97 (a) Using the Living Graphs, plot the work done by a gas as it expands isothermally and reversibly to 10. times its original volume at 100. K, 200. K, and 300. K. (b) Is more or less work done for this expansion as the temperature is raised? (c) Compare the work done by a gas at any given temperature as it expands reversibly and isothermally from 1.00 L to 5.00 L with that done when it expands reversibly and isothermally from 5.00 L to 9.00 L.

6.98 (a) Calculate the work associated with the isothermal, reversible expansion of 1.000 mol of an ideal gas from 7.00 L to 15.50 L at $25.0°C$. (b) Calculate the work associated with the irreversible adiabatic expansion of the sample of gas described in part (a) against a constant atmospheric pressure of 760. Torr. (c) How will the temperature of the gas in part (b) compare with that in part (a) after the expansion?

6.99 Strong sunshine bombards the Earth with about $1\ kJ\cdot m^{-2}$ in 1 s. Calculate the maximum mass of pure ethanol that can be vaporized in 10. min from a beaker left in strong sunshine, assuming the surface area of the ethanol to be 50. cm^2. Assume that all the heat is used for vaporization, not to increase the temperature.

6.100 Use reactions (a), (b), and (c) to determine the enthalpy change of the following reaction:

$$CH_4(g) + \tfrac{3}{2} O_2(g) \longrightarrow CO(g) + 2\ H_2O(g)$$

(a) $CH_4(g) + 2\ O_2(g) \rightarrow CO_2(g) + 2\ H_2O(g)$ $\Delta H° = -802\ kJ$
(b) $CH_4(g) + CO_2(g) \rightarrow 2\ CO(g) + 2\ H_2(g)$ $\Delta H° = +247\ kJ$
(c) $CH_4(g) + H_2O(g) \rightarrow CO(g) + 3\ H_2(g)$ $\Delta H° = +206\ kJ$

6.101 Aniline, $C_6H_5NH_2(l)$, is a derivative of benzene in which a hydrogen atom has been replaced by an NH_2 group. (a) Write the balanced equation for the combustion of aniline. (b) What is the mass of each product when 0.1754 g of aniline is burned in excess oxygen? (c) If the bomb calorimeter in which this reaction was carried out had a volume of 355 mL, what minimum pressure of oxygen at $23°C$ must have been used to ensure complete combustion? Assume that the volume of the aniline is negligible.

6.102 The standard enthalpy of formation of solid ammonium acetate is $-616.14\ kJ\cdot mol^{-1}$. Calculate its standard enthalpy of combustion.

6.103 (a) Is the production of water gas (a cheap, low-grade industrial fuel) exothermic or endothermic? The reaction is

$$C(s) + H_2O(g) \longrightarrow CO(g) + H_2(g).$$

(b) Calculate the enthalpy change for the production of 200. L of hydrogen at 500. Torr and $65°C$ by this reaction.

6.104 The ABC cereal company is developing a new type of breakfast cereal to compete with a rival product that they call Brand X. You are asked to compare the energy content of the two cereals to see if the new ABC product is lower in calories; so you burn 1.00-g samples of the cereals in oxygen in a calorimeter with a heat capacity of 600. $J\cdot(°C)^{-1}$. When the Brand X cereal sample burned, the temperature rose from 300.2 K to 309.0 K. When the ABC cereal sample burned, the temperature rose from 299.0 K to 307.5 K. (a) What is the heat output of each sample? (b) One serving of each cereal is 30.0 g. How would you label the packages of the two cereals to indicate the fuel value per 30.0-g serving in joules? in nutritional Calories (kilocalories)?

6.105 An experimental automobile burns hydrogen for fuel. At the beginning of a test drive, the rigid 30.0-L tank was filled with hydrogen at 16.0 atm and 298 K. At the end of the drive, the temperature of the tank was still 298 K, but its pressure was 4.0 atm. (a) How many moles of H_2 were burned during the drive? (b) How much heat, in kilojoules, was given off by the combustion of that amount of hydrogen?

6.106 In hot, dry climates an inexpensive alternative to air conditioning is the swamp cooler. In this device water continuously wets porous pads through which fans blow the hot air. The air is cooled as the water evaporates. Use the information in Tables 6.2 and 6.3 to determine how much water must be evaporated to cool the air in a room of dimensions 4.0 m $\times$ 5.0 m $\times$ 3.0 m by $20.°C$. Assume that the enthalpy of vaporization of water is the same as it is at $25°C$.

6.107 (a) Calculate the work that must be done against the atmosphere for the expansion of the gaseous products in the combustion of 1.00 mol $C_6H_6(l)$ at $25°C$ and 1.00 bar. (b) Using data in Appendix 2A, calculate the standard enthalpy of the reaction. (c) Calculate the change in internal energy, $\Delta U°$, of the system.

6.108 A system undergoes a two-step process. In step 1, it absorbs 50. J of heat at constant volume. In step 2, it releases 5 J of heat at 1.00 atm as it is returned to its original internal energy. Find the change in the volume of the system during the second step and identify it as an expansion or compression.

6.109 Consider the hydrogenation of benzene to cyclohexane, which takes place by the step-by-step addition of two H atoms per step:

(1) $C_6H_6(l) + H_2(g) \rightarrow C_6H_8(l)$ $\Delta H° = ?$
(2) $C_6H_8(l) + H_2(g) \rightarrow C_6H_{10}(l)$ $\Delta H° = ?$
(3) $C_6H_{10}(l) + H_2(g) \rightarrow C_6H_{12}(l)$ $\Delta H° = ?$

(a) Draw Lewis structures for the products of the hydrogenation of benzene. If resonance is possible, show only one resonance structure. (b) Use bond enthalpies to estimate the enthalpy changes of each step and the total hydrogenation. Ignore the delocalization of electrons in this calculation and the fact that the compounds are liquids. (c) Use data from Appendix 2A to calculate the enthalpy of the complete hydrogenation of benzene to cyclohexane. (d) Compare the value obtained in part (c) with that obtained in part (b). Explain any difference.

6.110 Draw the Lewis structure for the hypothetical molecule N_6, consisting of a six-membered ring of nitrogen atoms. Using bond enthalpies, calculate the enthalpy of reaction for the decomposition of N_6 to $N_2(g)$. Do you expect N_6 to be a stable molecule?

6.111 Robert Curl, Richard Smalley, and Harold Kroto were awarded the Nobel prize in chemistry in 1996 for the discovery of the soccer-ball-shaped molecule C_{60}. This fundamental molecule was the first of a new series of *molecular* allotropes of carbon. The enthalpy of combustion of C_{60} is $-25\ 937\ kJ \cdot mol^{-1}$, and its enthalpy of sublimation is $+233\ kJ \cdot mol^{-1}$. There are 90 bonds in C_{60}, of which 60 are single bonds and 30 are double bonds. Like benzene, C_{60} has a set of multiple bonds for which resonance structures may be drawn. (a) Determine the enthalpy of formation of C_{60} from its enthalpy of combustion. (b) Calculate the expected enthalpy of formation of C_{60} from bond enthalpies, assuming the bonds to be isolated double and single bonds. (c) Is C_{60} more or less stable than predicted on the basis of the isolated-bond model? (d) Quantify the answer to part (c) by dividing the difference between the enthalpy of formation calculated from the combustion data and that obtained from the bond enthalpy calculation by 60 to obtain a per carbon value. (e) How does the number in part (d) compare with the per carbon resonance stabilization energy of benzene (the total resonance stabilization energy of benzene is approximately $150\ kJ \cdot mol^{-1}$)? (f) Why might these values differ? The enthalpy of atomization of $C(gr)$ is $+717\ kJ \cdot mol^{-1}$.

6.112 Normally, when sulfur is oxidized, the product is SO_2, but SO_3 may also be formed. When 0.6192 g of sulfur was burned, by using ultrapure oxygen in a bomb calorimeter with a heat capacity of $5.270\ kJ \cdot (°C)^{-1}$, the temperature rose $1.140°C$. Assuming that all the sulfur was consumed in the reaction, what was the ratio of sulfur dioxide to sulfur trioxide produced?

6.113 Speculate as to whether the combustion of 1,3,5-trinitrobenzene, $C_6H_3(NO_2)_3$, or the combustion of 1,3,5-triaminobenzene, $C_6H_3(NH_2)_3$, would be more exothermic. Support your answer with appropriate calculations.

6.114 Rank the following compounds in order of increasing enthalpy of vaporization: CH_4, H_2O, N_2, $NaCl$, C_6H_6, and H_2. Do not look up the enthalpy of vaporization. Explain your ordering.

6.115 A technician carries out the reaction $2\ SO_2(g) + O_2(g) \rightarrow 2\ SO_3(g)$ at 25°C and 1.00 atm in a constant-pressure cylinder fitted with a piston. Initially, 0.030 mol SO_2 and 0.030 mol O_2 are present in the cylinder. The technician then adds a catalyst to initiate the reaction. (a) Calculate the volume of the cylinder containing the reactant gases before reaction begins. (b) What is the limiting reactant? (c) Assuming that the reaction goes to completion and that the temperature and pressure of the reaction remain constant, what is the final volume of the cylinder (include any excess reactant)? (d) How much work takes place, and is it done by the system or on the system? (e) How much enthalpy is exchanged, and does it leave or enter the system? (f) From your answers to parts (d) and (e), calculate the change in internal energy for the reaction.

6.116 In Section 6.3 we saw how to calculate the work of reversible, isothermal expansion of a perfect gas. Now suppose that the reversible expansion is not isothermal and that the temperature decreases during expansion. (a) Derive an expression for the work when $T = T_{initial} - c(V - V_{initial})$, with c a positive constant. (b) Is the work in this case greater or smaller than that of isothermal expansion? Explain your observation.

Chemistry Connections

6.117 Petroleum based fuels create large amounts of pollution, contribute to global warming, and are becoming scarce. Consequently, alternative fuels are being sought (see Box 6.2). Three compounds that could be produced biologically and used as fuels are methane, CH_4, which can be produced from the anaerobic digestion of sewage, dimethyl ether, $H_3C-O-CH_3$, a gas that can be produced from methanol and ethanol, and ethanol, CH_3CH_2OH, a liquid obtained from the fermentation of sugars.
(a) Draw the Lewis structure of each compound.
(b) Use bond enthalpies (and, for ethanol, its enthalpy of vaporization) to calculate the enthalpy of combustion of each fuel, assuming that they burn to produce gaseous CO_2 and gaseous H_2O. Explain any differences.
(c) Use the values for the enthalpies of combustion of organic compounds found in Appendix 2A to compare methane and ethanol with octane, a primary constituent of gasoline, as fuels by calculating the heat produced per gram of each fuel. On the basis of this information, which would you choose as a fuel?
(d) What volume of methane gas at 10.00 atm and 298 K would you need to burn at constant pressure to produce the same amount of heat as 10.00 L of octane (the density of octane is $0.70\ g \cdot mL^{-1}$)?
(e) A problem with fuels containing carbon is that they produce carbon dioxide when they burn, and so a consideration governing the selection of a fuel could be the heat per mole of CO_2 produced. Calculate this quantity for methanol, ethanol, and octane. Which process produces more carbon dioxide in the environment for each kilojoule generated?

THERMODYNAMICS: THE SECOND AND THIRD LAWS

What Are the Key Ideas? The direction of natural change corresponds to the increasing disorder of energy and matter. Disorder is measured by the thermodynamic quantity called entropy. A related quantity—the Gibbs free energy—provides a link between thermodynamics and the description of chemical equilibrium.

Why Do We Need to Know This Material? The second law of thermodynamics is the key to understanding why one chemical reaction has a natural tendency to occur but another one does not. We apply the second law by using the very important concepts of entropy and Gibbs free energy. The third law of thermodynamics is the basis of the numerical values of these two quantities. The second and third laws jointly provide a way to predict the effects of changes in temperature and pressure on physical and chemical processes. They also lay the thermodynamic foundations for discussing chemical equilibrium, which the following chapters explore in detail.

What Do We Need to Know Already? The discussion draws on concepts related to the first law of thermodynamics, particularly enthalpy (Section 6.8) and work (Sections 6.2 and 6.3).

Chapter **7**

Some things happen naturally; some things don't. Decay is natural; construction requires work. Water flows downhill naturally; we have to pump it uphill. A spark is enough to initiate a forest fire; the ceaseless input of energy from the Sun is required to rebuild the forest from carbon dioxide and water. Anyone who thinks about the world may wonder what determines the *natural* direction of change. What drives events forward? What drives a reaction toward products, and, on a larger scale, what drives the great and intricate network of reactions in biological systems forward in the extraordinary phenomenon we call life? As this chapter unfolds, we shall see that we can use a single quantity, "entropy," to provide an elegant and quantitative answer to all these questions: an increase in entropy accounts both for physical change and for the more intricate changes that accompany chemical reactions. We shall also see that, although entropy is a thermodynamic concept in the sense that it is a property of bulk samples of matter, it has a very straightforward molecular interpretation.

ENTROPY

The first law of thermodynamics tells us that, if a reaction takes place, then the total energy of the universe (the reaction system and its surroundings) remains unchanged. But the first law does not address the questions that lie behind the "if." Why do some reactions have a tendency to occur, whereas others do not? Why does anything happen at all? To answer these deeply important questions about the world around us, we need to take a further step into thermodynamics and learn more about energy beyond the fact that it is conserved.

7.1 Spontaneous Change

A **spontaneous change** is a change that has a tendency to occur without needing to be driven by an external influence. A simple example is the cooling of a block of hot metal to the temperature of its surroundings (Fig. 7.1). The reverse change, a

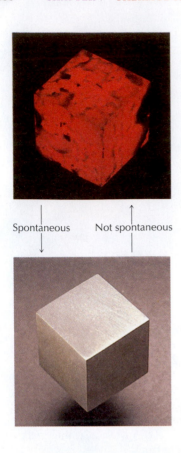

FIGURE 7.1 A hot block of metal (top) spontaneously cools to the temperature of its surroundings (bottom). The reverse process, in which a block at the same temperature as its surroundings spontaneously becomes hotter, does not occur.

Spontaneous   Not spontaneous

block of metal spontaneously growing hotter than its surroundings, has never been observed. The expansion of a gas into a vacuum is spontaneous (Fig. 7.2): a gas has no tendency to contract spontaneously into one part of a container.

A spontaneous change need not be fast and in some cases must be initiated. Molasses has a spontaneous tendency to flow out of an overturned can; but, at low temperatures, that flow may be very slow. Hydrogen and oxygen have a tendency to react to form water—the reaction is spontaneous in the thermodynamic sense—but a mixture of the two gases can be kept safely for centuries, provided it is not ignited with a spark. Diamonds have a natural tendency to turn into graphite, but diamonds last unchanged for countless years—on the human scale, diamonds are, in practice, forever. A spontaneous process has a natural *tendency* to occur; it does not necessarily take place rapidly. Throughout this chapter, and whenever we consider the thermodynamics of change, we must remember that we are exploring only the *tendency* of a process to occur. Whether that tendency is actually realized in practice depends on its rate: rates are outside the reach of thermodynamics and are considered in Chapter 13.

Changes can be made to happen in their "unnatural" direction. For example, we can force an electric current through a block of metal to heat it to a temperature higher than that of its surroundings. We can drive a gas into a smaller volume by pushing in a piston. In every case, though, to bring about a nonspontaneous change in a system, we have to contrive a way of *forcing* it to happen by applying an influence from outside the system. In short, *a nonspontaneous change can be brought about only by doing work.*

> *A process is spontaneous if it has a tendency to occur without being driven by an external influence; spontaneous changes need not be fast.*

7.2 Entropy and Disorder

In science, we look for patterns to discover nature's laws. What is the pattern common to all spontaneous changes? To find a pattern, it is often best to start with very simple examples, because then the pattern is likely to be more obvious. So, let's think about two simple spontaneous changes—the cooling of a hot metal and the expansion of a gas—at a molecular level and search for their common feature.

A hot block of metal cools as the energy of its vigorously vibrating atoms spreads into the surroundings. The vigorously moving atoms of the metal collide with the slower atoms and molecules of the surroundings, transferring some of their energy in the collisions. The reverse change is very improbable, because it would require that energy migrate from the surroundings and concentrate in a small block of metal. Such a process would require that collisions of the less vigorously moving atoms of the surroundings with the more vigorously moving atoms of the metal would cause the latter to move even more vigorously. The randomly moving molecules of a gas spread out all over their container; it is very unlikely that their random motion will bring them all simultaneously back into one corner. The pattern starting to emerge is that *energy and matter tend to disperse in a disorderly fashion.* As this chapter unfolds, we shall see how to use the idea of increasing disorder to account for any spontaneous change. As we shall see, in thermodynamics the meaning of "disorder" encompasses a concept richer than merely "untidy."

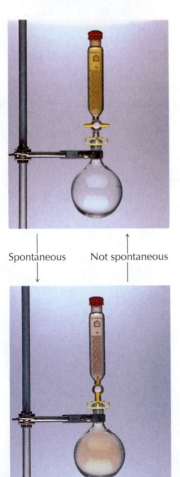

Spontaneous   Not spontaneous

FIGURE 7.2 A gas spontaneously fills its container. A glass cylinder containing the brown gas nitrogen dioxide (upper piece of glassware in the top illustration) is attached to an evacuated flask. When the stopcock between them is opened, the gas spontaneously fills both upper and lower vessels (bottom illustration). The reverse process, in which the gas now in both vessels collects spontaneously back in the upper vessel, does not occur.

In the language of thermodynamics, this simple idea is expressed as the **entropy**, *S*, a measure of disorder. *Low entropy means little disorder; high entropy means great disorder.* It follows that we can express the pattern that we have identified as follows:

The entropy of an isolated system increases in the course of any spontaneous change.

In short, things get worse.

Thus, the cooling of hot metal is accompanied by an increase in entropy as energy spreads into the surroundings. The "isolated system" in this case is taken to be the block and its immediate surroundings. Likewise, the expansion of a gas is accompanied by an increase in entropy as the molecules spread through the container. The pattern we have identified is one version of the **second law of thermodynamics**. The natural progression of a system and its surroundings (which together make up "the universe") is from order to disorder, from lower to higher entropy. For practical measurements, a small isolated region, such as a thermally insulated, sealed flask or a calorimeter, is considered to represent the universe.

We need a quantitative definition of entropy to measure and make precise predictions about disorder. Provided that the temperature is constant, it turns out that a *change* in the entropy of a system can be calculated from the following expression:

We generalize the definition in the next section to changes in which the temperature is not constant.

$$\Delta S = \frac{q_{rev}}{T} \tag{1}*$$

where *q* is the energy transferred as heat and *T* is the (absolute) temperature at which the transfer takes place. The subscript "rev" on *q* signifies that the energy must be transferred reversibly in the sense described in Section 6.3. For a reversible transfer of energy as heat, the temperatures of the surroundings and the system must be only infinitesimally different. With heat measured in joules and temperature in kelvins, the change in entropy (and entropy itself) is measured in joules per kelvin ($J \cdot K^{-1}$).

> *What does this equation tell us?* If a lot of energy is transferred as heat (large q_{rev}), a lot of disorder is stirred up in the system and we expect a correspondingly big increase in entropy. For a given transfer of energy, we expect a greater change in disorder when the temperature is low than when it is high. The arriving energy stirs up the molecules of a cool system, which have little thermal motion, more noticeably than those of a hot system, in which the molecules are already moving vigorously. (Think of how sneezing in a quiet library will attract attention, but sneezing in a noisy street may pass unnoticed.)

EXAMPLE 7.1   **Sample exercise: Calculating the change in entropy when a system is heated**

A large flask of water is placed on a heater and 100. J of energy is transferred reversibly to the water at 25°C. What is the change in entropy of the water?

SOLUTION

From $\Delta S = q_{rev}/T$,
$$\Delta S = \frac{100.\ J}{\underbrace{(273.15 + 25)\ K}_{\text{Convert to kelvins}}} = \frac{1.00 \times 10^2}{298} \frac{J}{K} = +0.336\ J \cdot K^{-1}$$

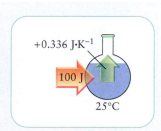

Note that the entropy of the water increases as a result of the flow of heat into it. We refer to a "large flask of water" to ensure that the temperature remains virtually constant as the heat is transferred: Eq. 1 applies only at constant temperature.

A note on good practice: Note the + sign on the answer: always show the sign explicitly for the change in a quantity, even if it is positive.

SELF-TEST 7.1A Calculate the change in entropy of a large block of ice when 50. J of energy is removed reversibly from it as heat at 0°C in a freezer.

[*Answer:* -0.18 J·K^{-1}]

SELF-TEST 7.1B Calculate the change in entropy of a large vat of molten copper when 50. J of energy is removed reversibly from it as heat at 1100.°C.

A very important characteristic of entropy that is not immediately obvious from Eq. 1 but can be proved by using thermodynamics is that *entropy is a state function.* This property is consistent with entropy being a measure of disorder, because the disorder of a system depends only on its current state and is independent of how that state was achieved.

Because entropy is a state function, the change in entropy of a system is independent of the path between its initial and final states. This independence means that, if we want to calculate the entropy difference between a pair of states joined by an *irreversible* path, we can look for a *reversible* path between the same two states and then use Eq. 1 for that path. For example, suppose an ideal gas undergoes free (irreversible) expansion at constant temperature. To calculate the change in entropy, we allow the gas to undergo reversible, isothermal expansion between the same initial and final volumes, calculate the heat absorbed in this process, and use it in Eq.1. Because entropy is a state function, the change in entropy calculated for this reversible path is also the change in entropy for the free expansion between the same two states.

We do the actual calculation in Example 7.12.

> Entropy is a measure of disorder; according to the second law of thermodynamics, the entropy of an isolated system increases in any spontaneous process. Entropy is a state function.

7.3 Changes in Entropy

We can expect disorder to increase when a system is heated because the supply of energy increases the thermal motion of the molecules. Heating increases the **thermal disorder,** the disorder arising from the thermal motion of the molecules. We can also expect the entropy to increase when a given amount of matter spreads into a greater volume or is mixed with another substance. These processes disperse the molecules of the substance over a greater volume and increase the **positional disorder,** the disorder related to the locations of the molecules.

Equation 1 allows us to express these changes quantitatively. First, let's calculate the entropy change that results when a system is heated.

HOW DO WE DO THAT?

These special cases include any isothermal process, such as isothermal expansion of a gas, and transfers of energy at freezing or boiling points (see Section 7.4).

To derive an expression for the change in entropy when a system is heated, we first note that Eq. 1 applies only when the temperature remains constant as heat is supplied to a system. Except in special cases, that can be true only for infinitesimal transfers of heat; so we have to break down the calculation into an infinite number of infinitesimal steps, with each step taking place at a constant but slightly different temperature, and then add together the infinitesimal entropy changes for all the steps. To do this is we use calculus. For an infinitesimal reversible transfer dq_{rev} at the temperature T, the increase in entropy is also infinitesimal; and, instead of Eq. 1, we write

$$dS = \frac{dq_{rev}}{T}$$

The energy supplied as heat is related to the resulting increase in temperature, dT, by the heat capacity, C, of the system:

$$dq_{rev} = CdT$$

(This is the infinitesimal form of Eq. 5a of Chapter 6, which also applies to reversible changes.) If the change in temperature is carried out at constant volume, we use the constant-volume heat capacity, C_V. If the change is carried out at constant pressure, we use the constant-pressure heat capacity, C_P. By combining the preceding two equations, we get

$$dS = \frac{C\,dT}{T}$$

Now suppose that the temperature of a sample is increased from T_1 to T_2. The overall change in entropy is the sum (integral) of all these infinitesimal changes:

$$\Delta S = \int_{T_1}^{T_2} \frac{C\,dT}{T}$$

Provided the heat capacity is independent of temperature in the temperature range of interest, C can be taken outside the integral and we obtain

$$\Delta S = C\int_{T_1}^{T_2} \frac{dT}{T} = C\ln\frac{T_2}{T_1}$$

To evaluate the integral, we have used the standard integral

$$\int \frac{dx}{x} = \ln x + \text{constant}$$

Because an integral can be identified with an area, the expression for ΔS can be identified with the area under the graph of C/T plotted against T enclosed by vertical lines at T_1 and T_2 (see the following Example).

We have shown that, provided the heat capacity can be treated as constant in the temperature range of interest, the entropy change that occurs when a system is heated from T_1 to T_2 is given by

$$\Delta S = C\ln\frac{T_2}{T_1} \qquad (2)*$$

where C is the heat capacity of the system (C_V if the volume is constant, C_P if the pressure is constant).

What does this equation tell us? If T_2 is higher than T_1, then $T_2/T_1 > 1$, the logarithm of this ratio is positive, and therefore ΔS is positive too, corresponding to the expected increase in entropy as the temperature is raised. The greater the heat capacity of the substance, the greater the increase in entropy for a given change in temperature.

Figure 7.3 shows how the entropy of a substance changes as it is heated through a temperature range in which it has a constant heat capacity.

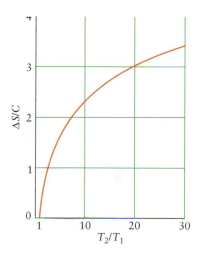

FIGURE 7.3 The change in entropy as a sample is heated for a system with a constant heat capacity (C) in the range of interest. Here we have plotted $\Delta S/C$. The entropy increases logarithmically with temperature.

EXAMPLE 7.2   Calculating the entropy change due to an increase in temperature

A sample of nitrogen gas of volume 20.0 L at 5.00 kPa is heated from 20.°C to 400.°C at constant volume. What is the change in the entropy of the nitrogen? The molar heat capacity of nitrogen at constant volume, $C_{V,m}$, is 20.81 $J \cdot K^{-1} \cdot mol^{-1}$. Assume ideal behavior.

STRATEGY We expect a positive entropy change because the thermal disorder in a system increases as the temperature is raised. We use Eq. 2, with the heat capacity at constant volume, $C_V = nC_{V,m}$. Find the amount (in moles) of gas molecules by using the ideal gas law, $PV = nRT$, and the initial conditions; remember to express temperature in kelvins. Because the data are liters and kilopascals, use R expressed in those units. As always, avoid rounding errors by delaying the numerical calculation to the last possible stage.

SOLUTION

Convert temperatures into kelvins.

$$T_1 = 20. + 273.15\ \text{K} = 293\ \text{K}$$
$$T_2 = 400. + 273.15\ \text{K} = 673\ \text{K}$$

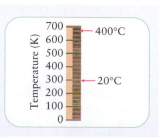

Find the amount of N_2 from $PV = nRT$ in the form $n = PV/RT$.

$$n = \frac{(5.00 \text{ kPa}) \times (20.0 \text{ L})}{(8.3145 \text{ L·kPa·K}^{-1}\text{·mol}^{-1}) \times (293 \text{ K})}$$

$$= \frac{5.00 \times 20.0}{8.3145 \times 293} \text{ mol}$$

Calculate the change in entropy from $\Delta S = C \ln(T_2/T_1)$ with $C = nC_{V,m}$.

$$\Delta S = \left(\frac{5.00 \times 20.0}{8.3145 \times 293} \text{ mol} \right)$$

$$\times (20.81 \text{ J·K}^{-1}\text{·mol}^{-1}) \times \ln\frac{673 \text{ K}}{293 \text{ K}}$$

$$= +0.710 \text{ J·K}^{-1}$$

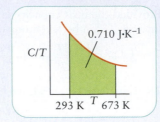

The change is positive (an increase), as expected. Note that, as remarked earlier, the change in entropy is equal to the area beneath the graph of C/T plotted against T.

SELF-TEST 7.2A The temperature of 1.00 mol He(g) is increased from 25°C to 300.°C at constant volume. What is the change in the entropy of the helium? Assume ideal behavior and use $C_{V,m} = \frac{3}{2}R$.

[*Answer:* +8.15 J·K^{-1}]

SELF-TEST 7.2B The temperature of 5.5 g of stainless steel is increased from 20.°C to 100.°C. What is the change in the entropy of the stainless steel? The specific heat capacity of stainless steel is 0.51 J·(°C)$^{-1}$·g^{-1}.

Some changes are accompanied by a change in volume. Because a larger volume provides a greater range of locations for the molecules, we can expect the positional disorder of a gas and therefore its entropy to increase as the volume it occupies is increased. Once again, we can use Eq. 1 to turn this intuitive idea into a quantitative expression of the entropy change for the isothermal expansion of an ideal gas.

HOW DO WE DO THAT?

We choose an isothermal expansion because both temperature and volume affect the entropy of a substance; at this stage, we do not want to have to consider changes in both temperature and volume.

To derive an expression for the dependence of the entropy of an ideal gas on its volume, we note that, when the expansion is isothermal, implying that T is constant, we can use Eq. 1 directly. We then use the first law in the form $\Delta U = q + w$ to relate q for a reversible, isothermal expansion to the change in volume of the gas. We know that $\Delta U = 0$ for the isothermal expansion of an ideal gas (Section 6.7); so we can conclude that $q + w = 0$ and therefore that $q = -w$. Physically, that equality means that the energy the system loses by doing expansion work is replaced by an influx of energy as heat, so the internal energy remains the same. The same relation applies if the change is carried out reversibly, and so we can write $q_{rev} = -w_{rev}$. Therefore, to calculate q_{rev}, we calculate the work done when an ideal gas expands reversibly and isothermally and then change the sign. For that calculation, we can use the following expression, which we derived in Section 6.3:

$$w_{rev} = -nRT\ln\frac{V_2}{V_1}$$

We can substitute this value in the expression for entropy and obtain

$$\Delta S = \frac{q_{rev}}{T} = \frac{(-w_{rev})}{T} = \frac{nRT\ln\dfrac{V_2}{V_1}}{T} = nR\ln\frac{V_2}{V_1}$$

We have shown that the change in entropy of an ideal gas when it expands isothermally from a volume V_1 to a volume V_2 is

$$\Delta S = nR \ln \frac{V_2}{V_1} \qquad \text{(3a)}*$$

where n is the amount of gas molecules (in moles) and R is the gas constant (in joules per kelvin per mole). When the final volume is greater than the initial volume ($V_2 > V_1$), the change in entropy is positive, corresponding to an increase in entropy, as expected (Fig. 7.4).

Notice that here we see an important aspect of thermodynamics: although we have calculated the change in entropy for a reversible path, Eq. 3 is also the change in entropy of the gas when it expands *irreversibly* between the same two states, in this case from V_1 to V_2 at any constant temperature. As we saw in Section 7.2, this conclusion follows from the fact that entropy is a state function, and therefore independent of the path. You should not jump to the conclusion, though, that there is no difference between reversible and irreversible processes: at this point we are considering only the change in entropy of the system and have not yet discussed what is going on in the surroundings. We consider that in Section 7.9.

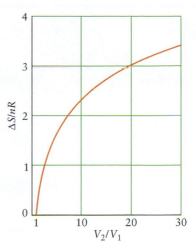

FIGURE 7.4 The change in entropy as a sample of perfect gas expands at constant temperature. Here we have plotted $\Delta S/nR$. The entropy increases logarithmically with volume.

> **EXAMPLE 7.3** Sample exercise: Calculating the change in entropy when an ideal gas expands isothermally

What is the change in entropy of the gas when 1.00 mol $N_2(g)$ expands isothermally from 22.0 L to 44.0 L?

SOLUTION

From $\Delta S = nR \ln (V_2/V_1)$,

$$\Delta S = (1.00 \text{ mol})$$
$$\times (8.3145 \text{ J·K}^{-1}\text{·mol}^{-1}) \times \ln\frac{44.0 \text{ L}}{22.0 \text{ L}}$$
$$= +5.76 \text{ J·K}^{-1}$$

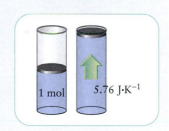

As expected, the entropy increases as the space available to the gas expands.

A note on good practice: Note that the entropy change *for* 1 mol of a substance is reported differently from the entropy change *per* mole: the units of the former are joules per kelvin (J·K^{-1}), those of the latter are joules per kelvin per mole ($\text{J·K}^{-1}\text{·mol}^{-1}$).

SELF-TEST 7.3A Calculate the change in molar entropy of an ideal gas when it is compressed isothermally to one-third its initial volume.

[*Answer:* $-9.13 \text{ J·K}^{-1}\text{·mol}^{-1}$]

SELF-TEST 7.3B Calculate the change in molar entropy of carbon dioxide that is allowed to expand isothermally to 10. times its initial volume (treat carbon dioxide as an ideal gas).

The entropy change accompanying the isothermal compression or expansion of an ideal gas can be expressed in terms of its initial and final pressures. To do so, we use the ideal gas law—specifically, Boyle's law—to express the ratio of volumes in Eq. 3 in terms of the ratio of the initial and final pressures. Because pressure is inversely proportional to volume (Boyle's law), we know that at constant temperature $V_2/V_1 = P_1/P_2$ where P_1 is the initial pressure and P_2 is the final pressure. Therefore,

$$\Delta S = nR \ln \frac{P_1}{P_2} \qquad \text{(3b)}$$

EXAMPLE 7.4   Sample exercise: Calculating the change in entropy with pressure

Calculate the change in entropy when the pressure of 0.321 mol $O_2(g)$ is increased from 0.300 atm to 12.00 atm at constant temperature.

SOLUTION

From $\Delta S = nR \ln(P_1/P_2)$,

$\Delta S = (0.321 \text{ mol})$

$\times (8.3145 \text{ J·K}^{-1}\text{·mol}^{-1}) \times \ln\dfrac{0.300 \text{ atm}}{12.00 \text{ atm}}$

$= -9.85 \text{ J·K}^{-1}$

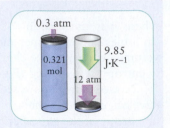

The entropy decreases, as expected for a sample that has been compressed into a smaller volume at constant temperature.

SELF-TEST 7.4A   Calculate the change in entropy when the pressure of 1.50 mol Ne(g) is decreased isothermally from 20.00 bar to 5.00 bar. Assume ideal behavior.

[*Answer:* +17.3 J·K^{-1}]

SELF-TEST 7.4B   Calculate the change in entropy when the pressure of 70.9 g of chlorine gas is increased isothermally from 3.00 kPa to 24.00 kPa. Assume ideal behavior.

To calculate a change in entropy for a process we find a *reversible* path between the initial and final states. It is immaterial whether the actual process is irreversible or reversible. Because entropy is a state function, the change for that path will be the same as that for the irreversible path.

EXAMPLE 7.5   Calculating the change in entropy when both temperature and volume change

In an experiment, 1.00 mol Ar(g) was compressed suddenly (and irreversibly) from 5.00 L to 1.00 L by driving in a piston (like a big bicycle pump), and in the process its temperature increased from 20.0°C to 25.2°C. What is the change in entropy of the gas?

STRATEGY   Because the entropy is a state function, we can calculate the change in entropy by choosing a reversible path that results in the same final state. In this case, consider the following:

Step 1   Reversible isothermal compression at the initial temperature from the initial volume to the final volume (and use Eq. 3a).

Step 2   An increase in temperature of the gas at constant final volume to the final temperature (and use C_V in Eq. 2 with C the heat capacity at constant volume).

SOLUTION

Step 1   From Eq. 3a,
$\Delta S = nR \ln(V_2/V_1)$,

$\Delta S = (1.00 \text{ mol})$

$\times (8.3145 \text{ J·K}^{-1}\text{·mol}^{-1}) \times \ln\dfrac{1.00 \text{ L}}{5.00 \text{ L}}$

$= -13.4 \text{ J·K}^{-1}$

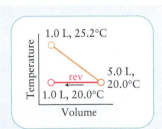

Step 2   From $\Delta S = C \ln(T_2/T_1)$ with $C = nC_{V,m}$,

$\Delta S = (1.00 \text{ mol})$

$\times (12.47 \text{ J·K}^{-1}\text{·mol}^{-1}) \times \ln\dfrac{298.4 \text{ K}}{293.2 \text{ K}}$

$= +0.22 \text{ J·K}^{-1}$

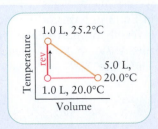

Add the entropy changes for the two steps, $\Delta S = \Delta S$ (Step 1) + ΔS (Step 2).

$$\Delta S = -13.4 + 0.22 \text{ J·K}^{-1} = -13.2 \text{ J·K}^{-1}$$

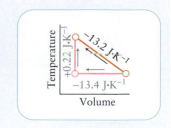

Note that the change in entropy due to the five-fold compression is much greater than that due to the small rise in temperature.

SELF-TEST 7.5A Calculate the change in entropy when the volume of 2.00 mol Ar(g) is increased from 5.00 L to 10.00 L while the temperature increases from 100. K to 300. K. Assume ideal behavior.

[*Answer:* +38.9 J·K^{-1}]

SELF-TEST 7.5B Calculate the change in entropy when the pressure of 23.5 g of oxygen gas is increased from 2.00 kPa to 8.00 kPa while the temperature increases from 240 K to 360 K. Assume ideal behavior.

The entropy of a system increases when its temperature increases and when its volume increases.

7.4 Entropy Changes Accompanying Changes in Physical State

We can expect the entropy to increase when a solid melts and its molecules become more disordered. Similarly, we can expect an even greater increase in entropy when a liquid vaporizes, because its molecules then occupy a much greater volume and their motion is highly chaotic. In this section, we develop expressions for the change in entropy at the transition temperature for the prevailing pressure. For instance, if the pressure is 1 atm, then these expressions are applicable only at the **normal melting point**, T_f (the f stands for fusion), the temperature at which a solid melts when the pressure is 1 atm, or the **normal boiling point**, T_b, the temperature at which a liquid boils when the pressure is 1 atm.

To use Eq. 1 to calculate the entropy change for a substance undergoing a transition from one phase to another at its transition temperature, we need to note three facts:

In most cases, the boiling temperature at 1 bar is negligibly different from the normal boiling point (the boiling temperature at 1 atm).

1   At the transition temperature (such as the boiling point for vaporization), the temperature of the substance remains constant as heat is supplied.

All the energy supplied is used to drive the phase transition, such as the conversion of liquid into vapor, rather than to raise the temperature. The T in the denominator of Eq. 1 is therefore a constant and may be set equal to the transition temperature (in kelvins).

2   At the temperature of a phase transition, the transfer of heat is reversible.

Provided the external pressure is fixed (at 1 atm, for instance), raising the temperature of the surroundings an infinitesimal amount results in complete vaporization and lowering the temperature causes complete condensation.

3   Because the transition takes place at constant pressure (for instance, 1 atm), the heat supplied is equal to the change in enthalpy of the substance (Section 6.8).

It follows that we can replace q_{rev} in the expression for the entropy change by ΔH for the phase change.

The **entropy of vaporization**, ΔS_{vap}, is the change in entropy per mole of molecules when a substance changes from a liquid into a vapor. The heat required per mole to vaporize the liquid at constant pressure is equal to the enthalpy of vaporization (ΔH_{vap}, Section 6.11). It then follows from Eq. 1, by setting $q_{rev} = \Delta H_{vap}$, that the entropy of vaporization at the normal boiling point is

$$\text{At the boiling temperature: } \Delta S_{vap} = \frac{\Delta H_{vap}}{T_b} \qquad (4)*$$

When the liquid and the vapor are in their standard states (both pure, both at 1 bar) and the boiling temperature is for 1 bar, we refer to the **standard entropy of vaporization** and write it $\Delta S_{vap}°$. Because all standard entropies of vaporization are positive, they are normally reported without their positive sign.

> **EXAMPLE 7.6**   **Sample exercise: Calculating the standard entropy of vaporization**
>
> What is the standard entropy of vaporization of acetone at its normal boiling point of 56.2°C?

SOLUTION The boiling point corresponds to 329.4 K, and we note from Table 6.3 that the standard enthalpy of vaporization of acetone at its boiling point is 29.1 kJ·mol^{-1}. Therefore,

From $\Delta S_{vap}° = \Delta H_{vap}°/T_b$,

$$\Delta S_{vap} = \frac{\overbrace{2.91 \times 10^4 \text{ J·mol}^{-1}}^{29.1 \text{ kJ·mol}^{-1}}}{\underbrace{329.4 \text{ K}}_{(273.15 + 56.2) \text{ K}}} = 88.3 \text{ J·K}^{-1}\text{·mol}^{-1}$$

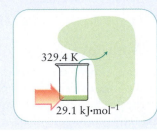

SELF-TEST 7.6A Calculate the standard entropy of vaporization of argon at its boiling point (see Table 6.3).

[*Answer:* 74 J·K^{-1}·mol^{-1}]

SELF-TEST 7.6B Calculate the standard entropy of vaporization of water at its boiling point (see Table 6.3).

Table 7.1 lists the standard entropies of vaporization of a number of liquids. These and other data show a striking pattern: many values are close to 85 J·K^{-1}·mol^{-1}. This observation is called **Trouton's rule**. The explanation of Trouton's rule is that approximately the same increase in positional disorder occurs when any liquid is converted into vapor, and so we can expect the

TABLE 7.1   Standard Entropy of Vaporization at the Normal Boiling Point*

Liquid	Boiling point (K)	$\Delta S_{vap}°$ (J·K^{-1}·mol^{-1})
acetone	329.4	88.3
ammonia	239.7	97.6
argon	87.3	74
benzene	353.2	87.2
ethanol	351.5	124
helium	4.22	20.
mercury	629.7	94.2
methane	111.7	73
methanol	337.8	105
water	373.2	109

*The normal boiling point is the boiling temperature at 1 atm.

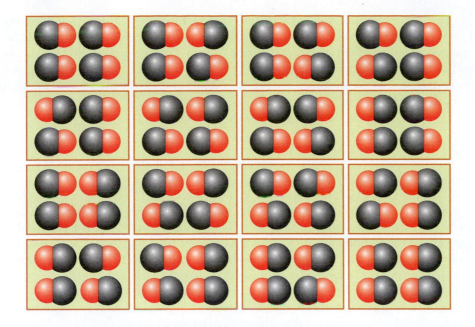

FIGURE 7.7 A tiny sample of solid carbon monoxide consisting of four molecules. Each box represents a different arrangement of the four molecules. When there is only one way of arranging the four molecules so that they all point in one direction (as in the top left image), the entropy of the solid is zero. When all sixteen ways of arranging four CO molecules are accessible, the entropy of the solid is greater than zero.

SOLUTION (a) Because $T = 0$, all motion has ceased. We expect the sample to have zero entropy, because there is no disorder in either location or energy. This value is confirmed by the Boltzmann formula: because there is only one way of arranging the molecules in the perfect crystal, $W = 1$.

From $S = k \ln W$ and $W = 1$ $S = k \ln 1 = 0$

In this case, every member of the ensemble is identical, and we can be sure that any selection results in the same microstate. (b) Because each of the four molecules can take two orientations, the total number of ways of arranging them is

$$W = 2 \times 2 \times 2 \times 2 = 2^4$$

or 16 different possible arrangements corresponding to the same total energy. Now there is only 1 chance in 16 of selecting a given microstate from the ensemble, and so we can be less sure about the actual state of the system. The entropy of this tiny solid is therefore

From $S = k \ln W$ and $W = 2^4$,

$$S = k \ln 2^4 = (1.3807 \times 10^{-23} \text{ J·K}^{-1}) \times \ln 16$$
$$= 3.8281 \times 10^{-23} \text{ J·K}^{-1}$$

As expected, the entropy of the disordered solid is higher than that of the perfectly ordered solid.

SELF-TEST 7.9A Calculate the entropy of a sample of a solid in which the molecules can take any one of three orientations with the same energy. Suppose there are 30 molecules in the sample.

[*Answer:* 4.5×10^{-22} J·K^{-1}]

SELF-TEST 7.9B Calculate the entropy of a sample of a solid in which it is supposed that a substituted benzene molecule, C_6H_5F, can take any one of six orientations with the same energy. Suppose there is 1.0 mol of molecules in the sample.

For a more realistic sample size than that in Example 7.7, one that contains 1.00 mol CO, corresponding to 6.02×10^{23} CO molecules, each of which could be oriented in either of two ways, there are $2^{6.02 \times 10^{23}}$ (an astronomically large number) different microstates, and a chance of only 1 in $2^{6.02 \times 10^{23}}$ of drawing a given microstate in a blind selection. We can expect the entropy of the solid to be high and calculate that

From $S = k \ln W$, $W = 2^{6.02 \times 10^{23}}$ and

$$S = k \ln 2^{6.02 \times 10^{23}}$$
$$= (1.3807 \times 10^{-23} \text{ J·K}^{-1}) \times (6.02 \times 10^{23}) \times \ln 2$$
$$= 5.76 \text{ J·K}^{-1}$$

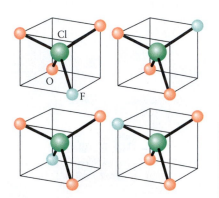

FIGURE 7.8 The four possible orientations of a tetrahedral $FClO_3$ molecule in a solid.

When the entropy of 1.00 mol CO is actually measured at temperatures close to $T = 0$, the value found is 4.6 J·K^{-1}. This value—which is called the **residual entropy** of the sample, the entropy of a sample at $T = 0$ arising from positional disorder surviving at that temperature—is close enough to 5.76 J·K^{-1} to suggest that in the crystal the molecules are indeed arranged nearly randomly. The reason for this randomness is that the electric dipole moment of a CO molecule is very small, and so there is little energy advantage in the molecules lying head-to-tail, head-to-head, or tail-to-tail; thus the molecules lie in either direction. For solid HCl, the same experimental measurements give $S \approx 0$ close to $T = 0$. This value indicates that the bigger dipole moments of the HCl molecules pull them into an orderly arrangement; so they lie strictly head-to-tail and there is no positional disorder at $T = 0$.

EXAMPLE 7.8   **Using the Boltzmann formula to interpret a residual entropy**

The entropy of 1.00 mol $FClO_3$(s) at $T = 0$ is 10.1 J·K^{-1}. Suggest an interpretation.

STRATEGY   The existence of residual entropy at $T = 0$ suggests that the molecules are disordered. From the shape of the molecule (which can be obtained by using VSEPR theory), we need to determine how many orientations, W, it is likely to be able to adopt in a crystal; then we can use the Boltzmann formula to see whether that number of orientations leads to the observed value of S.

SOLUTION   An $FClO_3$(s) molecule is tetrahedral, and so we can expect it to be able to take up any of four orientations in a crystal (Fig. 7.8). The total number of ways of arranging the molecules in a crystal containing N molecules is therefore

$$W = (4 \times 4 \times 4 \times \cdots \times 4)_{N \text{ factors}} = 4^N$$

The entropy is

From $S = k \ln W$, $W = 4^N$ and $N = 6.02 \times 10^{23}$,

$$S = k \ln 4^{6.02 \times 10^{23}}$$
$$= (1.3807 \times 10^{-23} \text{ J·K}^{-1}) \times (6.02 \times 10^{23}) \times \ln 4$$
$$= 11.5 \text{ J·K}^{-1}$$

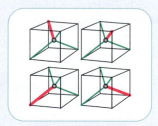

This value is reasonably close to the experimental value of 10.1 J·K^{-1}, which suggests that, at $T = 0$, the molecules lie nearly randomly in any of four possible orientations.

SELF-TEST 7.10A   Explain the observation that the entropy of 1 mol N_2O(s) at $T = 0$ is 6 J·K^{-1}.

[*Answer:* In the crystal, the orientations NNO and ONN are equally likely.]

SELF-TEST 7.10B   Suggest a reason why the entropy of ice is nonzero at $T = 0$; think about how the structure of ice is affected by the hydrogen bonds.

The Boltzmann formula relates the entropy of a substance to the number of arrangements of molecules that result in the same energy: when many energy levels are accessible, this number and the corresponding entropy are large.

7.6 The Equivalence of Statistical and Thermodynamic Entropies

The expressions in Eq. 1 and Eq. 6 are two different definitions of entropy. The first was established by considerations of the behavior of bulk matter and the second by statistical analysis of molecular behavior. To verify that the two definitions are essentially the same we need to show that the entropy changes predicted by Eq. 6 are the same as those deduced from Eq. 1. To do so, we will show that the Boltzmann formula predicts the correct form of the volume dependence of the entropy of an ideal gas (Eq. 3a). More detailed calculations show that the two definitions are consistent with each other in every respect. In the process of developing these ideas, we shall also deepen our understanding of what we mean by "disorder."

An ideal gas consists of a large number of molecules that occupy the energy levels characteristic of a particle in a box. For simplicity, we consider a one-dimensional box (Fig. 7.9a), but the same considerations apply to a real three-dimensional container of any shape. At $T = 0$, only the lowest energy level is occupied; so $W = 1$ and the entropy is zero. There is no "disorder," because we know which state each molecule occupies.

At any temperature above $T = 0$, the molecules occupy large numbers of different energy levels, as shown in Fig. 7.9. Because the molecules can now be found in a larger number of microstates, W is greater than 1. Therefore, the entropy (which is proportional to ln W) is greater than zero. Now the disorder has increased; consequently we are less confident about precisely which state a given molecule occupies. For instance, although at $T = 0$ we can be confident that molecules A and B both occupy the lowest energy level, at a higher temperature we don't know whether it is molecule A that occupies a higher level, with B still in the ground state, or vice versa.

When the length of the box is increased at constant temperature (with $T > 0$), more energy levels become accessible to the molecules because the levels now lie closer together (Fig. 7.9b). The disorder has increased and we are less sure about which energy level any given molecule occupies. Therefore, the value of W increases as the box is lengthened and, by the Boltzmann formula, the entropy increases, too. The same argument applies to a three-dimensional box: as the volume of the box increases, the number of accessible states increases, too.

We can also use the Boltzmann formula to interpret the increase in entropy of a substance as its temperature is raised (Eq. 2 and Table 7.2). We use the same particle-in-a-box model of a gas, but this reasoning also applies to liquids and solids, even though their energy levels are much more complicated. At low temperatures, the molecules of a gas can occupy only a few of the energy levels; so W is small and the entropy is low. As the temperature is raised, the molecules have access to larger numbers of energy levels (Fig. 7.10); so W rises and the entropy increases, too.

We can show that the definition of entropy in Eq. 6 is quantitatively equivalent to that in Eq. 1, even though the equations look totally different.

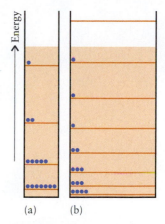

FIGURE 7.9 The energy levels of a particle in a box (a) become closer together as the width of the box is increased. (b) As a result, the number of levels accessible to the particles in the box increases, and the entropy of the system increases accordingly. The range of thermally accessible levels is shown by the tinted band. The change from part (a) to part (b) is a model of the isothermal expansion of an ideal gas. The total energy of the particles is the same in each case.

TABLE 7.2 Standard Molar Entropy of Water at Various Temperatures

Phase	Temperature (°C)	S_{m}° (J·K^{-1}·mol^{-1})
solid	−273 (0 K)	3.4
	0	43.2
liquid	0	65.2
	20	69.6
	50	75.3
	100	86.8
vapor	100	196.9
	200	204.1

HOW DO WE DO THAT?

We can show that the thermodynamic and statistical entropies are equivalent by examining the isothermal expansion of an ideal gas. We have seen that the thermodynamic entropy of an ideal gas increases when it expands isothermally (Eq. 3). If we suppose that the number of microstates available to a single molecule is proportional to the volume available to it, we can write W = constant $\times$ V. For N molecules, the number of microstates is proportional to the Nth power of the volume:

$$W = (\text{constant} \times V) \times (\text{constant} \times V)...\text{\textit{N} times} = (\text{constant} \times V)^N$$

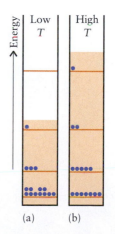

FIGURE 7.10 More energy levels become accessible in a box of fixed width as the temperature is raised. The change from part (a) to part (b) is a model of the effect of heating an ideal gas at constant volume. The thermally accessible levels are shown by the tinted band. The average energy of the molecules also increases as the temperature is raised: that is, both internal energy and entropy increase with temperature.

Therefore, the change in the statistical entropy when a sample expands isothermally from volume V_1 to a volume V_2 is

From $S = k \ln W$,	$\Delta S = k \ln(\text{constant} \times V_2)^N - k \ln(\text{constant} \times V_1)^N$
Use $\ln x^a = a \ln x$.	$\Delta S = Nk \ln(\text{constant} \times V_2) - Nk \ln(\text{constant} \times V_1)$
Use $\ln x - \ln y = \ln(x/y)$ and cancel the constants.	$\Delta S = Nk \ln \dfrac{\text{constant} \times V_2}{\text{constant} \times V_1} = Nk \ln \dfrac{V_2}{V_1}$

This expression is identical to Eq. 3 because $N = nN_A$ (where N_A is Avogadro's constant) and $N_A k = R$, the gas constant; therefore, $Nk = nN_A k = nR$ and $\Delta S = Nk \ln(V_1/V_2)$. In this case at least, the two definitions give identical results. The logarithm of the volume in Eq. 3 can now be seen to stem from the logarithm of W in the Boltzmann formula.

We have shown that the statistical and thermodynamic entropies lead to the same conclusions. We can expect their more general properties to be the same, too:

1   Both are state functions.

Because the number of microstates available to the system depends only on its current state, not on its past history, W depends only on the current state of the system, and therefore the statistical entropy does, too.

2   Both are extensive properties.

Doubling the number of molecules increases the number of microstates from W to W^2, and so the entropy changes from $k \ln W$ to $k \ln W^2$, or $2k \ln W$. Therefore, the statistical entropy, like the thermodynamic entropy, is an extensive property.

3   Both increase in a spontaneous change.

In any irreversible change, the overall disorder of the system and its surroundings increases, which means that the number of microstates increases. If W increases, then so does $\ln W$, and the statistical entropy increases, too.

4   Both increase with temperature.

When the temperature of the system increases, more microstates become accessible, and so the statistical entropy increases.

> *The equations used to calculate changes in the statistical entropy and the thermodynamic entropy lead to the same result.*

7.7 Standard Molar Entropies

To *calculate* the entropy of a substance, we use Boltzmann's formula, but the calculations are sometimes very difficult and require a lot of manipulation of Eq. 6. To *measure* the entropy of a substance, we use the thermodynamic definition, Eq. 1, in combination with the third law of thermodynamics. Because the third law tells us that $S(0) = 0$ and Eq. 2 can be used to calculate the change in entropy as a substance is heated to the temperature of interest, we can write

$$S(T) = S(0) + \Delta S(\text{heating from 0 to } T) = \Delta S(\text{heating from 0 to } T)$$

The change in entropy on heating is calculated from the heat capacity and the temperature, as we did in Example 7.2. However, because we cannot assume that the heat capacity is constant all the way down to $T = 0$, we have to use a more general expression.

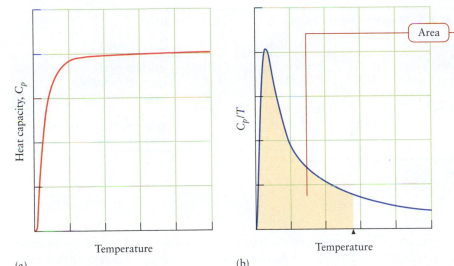

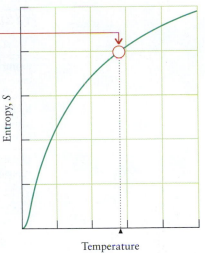

FIGURE 7.11 The experimental determination of entropy. (a) The heat capacity (at constant pressure in this instance) of the substance is determined from close to absolute zero up to the temperature of interest. (b) The area under the plot of C_P/T against T is determined up to the temperature of interest. (c) This area is the entropy of the substance at that temperature.

HOW DO WE DO THAT?

To take into account the possibility that the heat capacity changes with temperature, we go back to the expression

$$\Delta S(\text{heating from } T_1 \text{ to } T_2) = \int_{T_1}^{T_2} \frac{C\,dT}{T}$$

We set $T_1 = 0$ and $T_2 = T$, the temperature of interest, in the limits of the integral. Heating commonly takes place at constant pressure, and so we replace C by C_P:

$$\Delta S(\text{heating from } 0 \text{ to } T) = \int_0^T \frac{C_P\,dT}{T}$$

Therefore, by invoking the third law and setting $S(0) = 0$, we have

$$S(T) = \int_0^T \frac{C_P\,dT}{T}$$

An integral of a function—in this case, the integral of C_P/T—is the area under the graph of the function. Therefore, to measure the entropy of a substance, we need to measure the heat capacity (typically the constant-pressure heat capacity) at all temperatures from $T = 0$ to the temperature of interest. Then the entropy of the substance is obtained by plotting C_P/T against T and measuring the area under the curve (Fig. 7.11).

Note that we have drawn C_P approaching zero as $T \to 0$. This feature is a general phenomenon, and explained by quantum mechanics. At low temperatures, the energy available is so small that there is not enough to stimulate transitions to higher energy states, so the sample cannot take up energy, and its "capacity for heat" is zero.

We have seen that

$$S(T) = \text{area under graph of } C_P/T \text{ against } T \text{ from } 0 \text{ to the}$$
$$\text{temperature of interest} \tag{7}$$

If a phase transition takes place between $T = 0$ and the temperature of interest, then we also have to include the corresponding entropy of transition by using Eq. 5 and possibly Eq. 4 (Fig. 7.12). For instance, if we want to know the entropy of

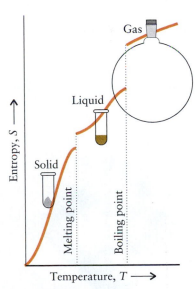

FIGURE 7.12 The entropy of a solid increases as its temperature is raised. The entropy increases sharply when the solid melts to form the more disordered liquid and then gradually increases again up to the boiling point. A second, larger jump in entropy occurs when the liquid vaporizes.

TABLE 7.3 Standard Molar Entropies at 25°C ($J \cdot K^{-1} \cdot mol^{-1}$)*

Gases	$S_m°$	Liquids	$S_m°$	Solids	$S_m°$
ammonia, NH_3	192.4	benzene, C_6H_6	173.3	calcium oxide, CaO	39.8
carbon dioxide, CO_2	213.7	ethanol, C_2H_5OH	160.7	calcium carbonate, $CaCO_3$[†]	92.9
hydrogen, H_2	130.7	water, H_2O	69.9	diamond, C	2.4
nitrogen, N_2	191.6			graphite, C	5.7
oxygen, O_2	205.1			lead, Pb	64.8

*Additional values are given in Appendix 2A.
[†]Calcite.

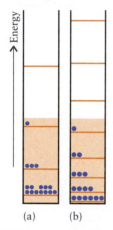

FIGURE 7.13 The energy levels of a particle in a box are more widely spaced for (a) light molecules than for (b) heavy molecules. As a result, the number of thermally accessible levels, as shown by the tinted band, is greater for heavy molecules than for light molecules at the same temperature, and the entropy of the substance with heavy molecules is correspondingly greater.

liquid water at 25°C, we measure the heat capacity of ice from $T = 0$ (or as close to it as we can get) to $T = 273.15$ K, determine the entropy of fusion at that temperature from the enthalpy of fusion, and then measure the heat capacity of liquid water from $T = 273.15$ K to $T = 298.15$ K. Table 7.3 gives selected values of the **standard molar entropy**, $S_m°$, the molar entropy of the pure substance at 1 bar. All the values in Table 7.3 refer to 298.15 K, the conventional temperature for reporting thermodynamic data.

We can use the molecular interpretation of entropy to understand why some substances have high molar entropies and others have low molar entropies. For example, compare the molar entropy of diamond, 2.4 $J \cdot K^{-1} \cdot mol^{-1}$, with the much higher value for lead, 64.8 $J \cdot K^{-1} \cdot mol^{-1}$. The low entropy of diamond is what we should expect for a solid that has rigid bonds: at room temperature, its atoms are not able to jiggle around as much as the atoms of lead. Lead atoms are also much heavier than carbon atoms, and heavier atoms have more vibrational energy levels available than do lighter ones (because their allowed energies are closer together).

The difference in molar entropy between two gases (compare H_2 and N_2 in Table 7.3, for instance) can be understood in terms of the particle-in-a-box model and the fact that the greater the mass of the molecules, the closer together are the energy levels (Fig. 7.13). We also see that large, complex species have higher molar entropies than smaller, simpler ones (compare $CaCO_3$ with CaO or C_2H_5OH with H_2O). Liquids have higher molar entropies than solids because the greater freedom of movement of the molecules in a liquid results in a less ordered state of matter. The molar entropies of gases, in which molecules occupy much larger volumes and undergo almost completely disordered motion, are substantially higher than those of the corresponding liquids.

SELF-TEST 7.11A Which substance in each of the following pairs has the higher molar entropy: (a) 1 mol $CO_2(g)$ at 25°C and 1 bar or 1 mol $CO_2(g)$ at 25°C and 3 bar; (b) 1 mol He(g) at 25°C or 1 mol He(g) at 100°C in the same volume; (c) $Br_2(l)$ or $Br_2(g)$ at the same temperature? Explain your conclusions.

[*Answer:* (a) $CO_2(g)$ at 1 bar, because disorder increases with volume; (b) He(g) at 100°C, because disorder increases with temperature; (c) $Br_2(g)$, because the vapor state has greater disorder than the liquid state.]

SELF-TEST 7.11B Use the information in Table 7.3 or Appendix 2A to determine which allotrope is the more ordered form and predict the sign of ΔS for each transition: (a) white tin (Fig. 7.14) changes into gray tin at 25°C; (b) diamond changes into graphite at 25°C.

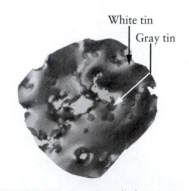

White tin
Gray tin

FIGURE 7.14 Gray tin and white tin are two solid forms of tin. The denser white metallic form is the more stable phase above 13°C, and the powdery gray form is more stable below that temperature.

Standard molar entropies increase as the complexity of a substance increases. The standard molar entropies of gases are higher than those of comparable solids and liquids at the same temperature.

7.8 Standard Reaction Entropies

Entropy is an important concept in chemistry because we can use it to predict the natural direction of a reaction. However, not only does the entropy of the reaction system change as reactants form products, but so too does the entropy of the surroundings as the heat produced or absorbed by the reaction enters or leaves them. Both the entropy change of the system and that of the surroundings affect the direction of a reaction, because both contribute to the entropy of the universe. We explore the contribution of the system in this section and the contribution of the surroundings in the next section.

In some cases, we can predict the sign of the entropy change of the system without doing any calculation. Because the molar entropy of a gas is so much greater than that of solids and liquids, a change in the amount of gas molecules normally dominates any other entropy change in a reaction. A net increase in the amount of gas usually results in a positive change in entropy. Conversely, a net consumption of gas usually results in a negative change. Reactions that produce a large number of small molecules and the dissolving of a substance typically have positive entropies. However, entropy changes are much more finely balanced for reactions in which there is no net change in the amount of gas or in which gases do not take part, and for these reactions we have to use numerical data to predict the sign of a reaction entropy. We also need tables of numerical data (such as Table 7.3 and Appendix 2A) to predict the precise numerical value of any reaction entropy.

SELF-TEST 7.12A Without doing any calculations, estimate the sign of the entropy change for the reaction $N_2(g) + 3\,H_2(g) \rightarrow 2\,NH_3(g)$ and explain your answer.

[**Answer:** negative, because there is a net decrease in the number of moles of gas molecules]

SELF-TEST 7.12B Without doing any calculations, estimate the sign of the entropy change for the reaction $2\,CaCO_3(s) \rightarrow CaO(s) + CO_2(g)$ and explain your answer.

To calculate the change in entropy that accompanies a reaction, we need to know the molar entropies of all the substances taking part; then we calculate the difference between the entropies of the products and those of the reactants. More specifically, the **standard reaction entropy**, $\Delta S°$, is the difference between the standard molar entropies of the products and those of the reactants, taking into account their stoichiometric coefficients:

$$\Delta S° = \sum n S_m°(\text{products}) - \sum n S_m°(\text{reactants}) \qquad (8)*$$

The first term on the right is the total standard entropy of the products, and the second term is that of the reactants; n denotes the amount of each substance in the chemical equation.

EXAMPLE 7.9   Calculating the standard reaction entropy

Calculate the standard reaction entropy of $N_2(g) + 3\,H_2(g) \rightarrow 2\,NH_3(g)$ at 25°C.

STRATEGY We expect a decrease in entropy because there is a net reduction in the amount of gas molecules. To find the numerical value, we use the chemical equation to write an expression for $\Delta S°$, as shown in Eq. 8, and then substitute values from Table 7.3 or Appendix 2A.

SOLUTION From the chemical equation, we can write

$$\Delta S° = \overbrace{(2\ \text{mol})S_m°(NH_3, g)}^{\text{Product}} - \overbrace{\{(1\ \text{mol})S_m°(N_2, g) + (3\ \text{mol})S_m°(H_2, g)\}}^{\text{Reactants}}$$

$$= 2(192.4\ \text{J·K}^{-1}) - \{(191.6\ \text{J·K}^{-1}) + 3(130.7\ \text{J·K}^{-1})\}$$

$$= -198.9\ \text{J·K}^{-1}$$

Because the value of $\Delta S°$ is negative, the product is less disordered than the reactants—in part because it occupies a smaller volume—just as we expected.

A note on good practice: Avoid the error of setting the standard entropies of elements equal to zero, as you would for $\Delta H_f°$: the entropies to use are the absolute values for the given temperature and are zero only at $T = 0$.

SELF-TEST 7.13A Use data from Appendix 2A to calculate the standard entropy of the reaction $N_2O_4(g) \rightarrow 2\ NO_2(g)$ at 25°C.

[*Answer:* +175.83 J·K^{-1}]

SELF-TEST 7.13B Use data from Appendix 2A to calculate the standard entropy of the reaction $C_2H_4(g) + H_2(g) \rightarrow C_2H_6(g)$ at 25°C.

The standard reaction entropy is the difference between the standard molar entropy of the products and that of the reactants weighted by the amounts of each species taking part in the reaction. It is positive (an increase in entropy) if there is a net production of gas in a reaction; it is negative (a decrease) if there is a net consumption of gas.

GLOBAL CHANGES IN ENTROPY

Some processes seem to defy the second law. For example, water spontaneously freezes to ice at low temperatures, and cold packs for athletic injuries spontaneously become ice cold, even on warm days, when the ammonium nitrate inside the pack dissolves in the water that it contains. Life itself appears to go against the second law. Every cell of a living being is organized to an extraordinary extent. Thousands of different compounds, each one having a specific function to perform, move in the intricately organized dance we call life. How can the molecules in our bodies form such highly organized, complex structures from slime, mud, or gas? Our existence seems at first sight to be a contradiction of the second law.

The dilemma is resolved when we realize that the second law refers to an *isolated* system. To interpret the second law correctly, we must consider any system to be part of a larger isolated system that includes the surroundings of the system of interest. The process is spontaneous only if the *total* change in entropy, the sum of the changes in the system and the surroundings, is positive. Our task, therefore, is first to see how to calculate the change in entropy of the surroundings and then to combine that change with the change in entropy of the system. Once we know the total entropy change, we shall be able to predict whether a reaction is spontaneous, whether its reverse is spontaneous, or whether the system is at equilibrium.

7.9 The Surroundings

The system of interest and its surroundings constitute the "isolated system" to which the second law refers (Fig. 7.15). Only if the *total* entropy change,

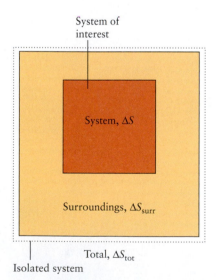

System of interest

System, ΔS

Surroundings, ΔS_{surr}

Total, ΔS_{tot}

Isolated system

FIGURE 7.15 The global isolated system that models events taking place in the world consists of a smaller system together with its immediate surroundings. The only events that can take place spontaneously in the global isolated system are those corresponding to an increase in the total entropy of both the smaller system and its immediate surroundings.

$$\underbrace{\Delta S_{tot}}_{\substack{\text{Total entropy} \\ \text{change}}} = \underbrace{\Delta S}_{\substack{\text{Entropy change} \\ \text{of system}}} + \underbrace{\Delta S_{surr}}_{\substack{\text{Entropy change} \\ \text{of surroundings}}} \qquad (9)*$$

is positive will the process be spontaneous. As is customary, properties of the system are written without subscripts; so here ΔS is the entropy change of the system and ΔS_{surr} that of the surroundings. It is critically important to understand that a process in which ΔS decreases may be spontaneous, provided that the entropy increases so much in the surroundings that ΔS_{tot} increases.

An example of the role of the surroundings in determining the spontaneous direction of a process is the freezing of water. We can see from Table 7.2 that, at 0°C, the molar entropy of liquid water is 22.0 J·K^{-1}·mol^{-1} higher than that of ice

FIGURE 7.16 (a) When a given quantity of heat flows into hot surroundings, it produces very little additional chaos and the increase in entropy is small. (b) When the surroundings are cool, however, the same quantity of heat can make a considerable difference to the disorder, and the change in entropy of the surroundings is correspondingly large.

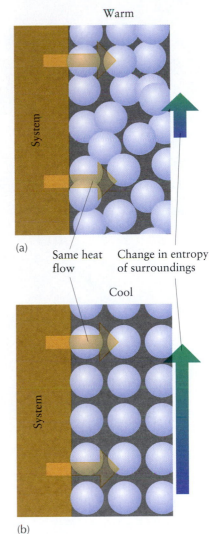

at the same temperature. It follows that, when water freezes at 0°C, its entropy decreases by 22.0 J·K^{-1}·mol^{-1}. Entropy changes do not vary much with temperature; so, just below 0°C, we can expect almost the same decrease. Yet we know from everyday experience that the freezing of water is spontaneous below 0°C. Clearly, the surroundings must be playing a deciding role: if their entropy increases by more than 22.0 J·K^{-1}·mol^{-1} when water freezes, then the total entropy change will be positive.

We can use Eq. 1 to calculate the entropy change of the surroundings, provided that we assume that the surroundings are so large that their temperature and pressure remain constant. If the enthalpy change of the system is ΔH, then, for heat transfers at constant pressure, $q_{surr} = -\Delta H$. We can now use Eq. 1 to write

$$\Delta S_{surr} = -\frac{\Delta H}{T} \qquad \text{at constant temperature and pressure} \qquad (10)^*$$

This important formula, which can be derived more formally from the laws of thermodynamics, applies when any change takes place at constant pressure and temperature. Notice that, for a given enthalpy change of the system (that is, a given output of heat), the entropy of the surroundings increases more if their temperature is low than if it is high (Fig. 7.16). The explanation is the "sneeze in the street" analogy mentioned in Section 7.2. Because ΔH is independent of path, Eq. 10 is applicable whether the process occurs reversibly or irreversibly.

EXAMPLE 7.10 Sample exercise: Calculating the change in entropy of the surroundings

Calculate the change in entropy of the surroundings when water freezes at $-10.°C$; use $\Delta H_{fus}(H_2O) = 6.0$ kJ·mol^{-1} at $-10.°C$.

SOLUTION We can expect the entropy of the surroundings to increase when water freezes because the heat released stirs up the thermal motion of the atoms in the surroundings (Fig. 7.17). To calculate the change in entropy of the surroundings when water freezes we write $\Delta H_{freeze} = -\Delta H_{fus} = -6.0$ kJ·mol^{-1} and $T = (-10. + 273)$ K $= 263$ K. Then

From $\Delta S_{surr} = -\Delta H/T$,
$$\Delta S_{surr} = -\frac{(-6.0 \times 10^3 \text{ J·mol}^{-1})}{263 \text{ K}}$$
$$= +23 \text{ J·K}^{-1}\text{·mol}^{-1}$$

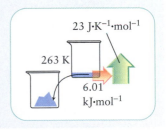

Note that the increase in entropy of the surroundings is indeed greater than the decrease in entropy of the system, -22.0 J·K^{-1}, the value at 0°C (the value at $-10.°C$ is similar), so the overall change is positive and the freezing of water is spontaneous at $-10.°C$.

SELF-TEST 7.14A Calculate the entropy change of the surroundings when 1.00 mol $H_2O(l)$ vaporizes at 90°C and 1 bar. Take the enthalpy of vaporization of water as 40.7 kJ·mol^{-1}.

[*Answer:* -112 J·K^{-1}]

SELF-TEST 7.14B Calculate the entropy change of the surroundings when 2.00 mol $NH_3(g)$ is formed from the elements in their most stable forms at 298 K.

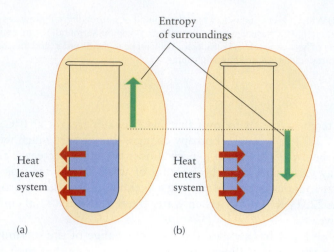

FIGURE 7.17 (a) In an exothermic process, heat escapes into the surroundings and increases their entropy. (b) In an endothermic process, the entropy of the surroundings decreases. The red arrows represent the transfer of heat between system and surroundings, and the green arrows indicate the entropy change of the surroundings.

The entropy change of the surroundings due to a process taking place at constant pressure and temperature is equal to $-\Delta H/T$, *where* ΔH *is the change in enthalpy of the system.*

7.10 The Overall Change in Entropy

As we have already emphasized, *to use the entropy to judge the direction of spontaneous change, we must consider the change in the entropy of the system plus the entropy change in the surroundings*:

If ΔS_{tot} is positive (an increase), the process is spontaneous.

If ΔS_{tot} is negative (a decrease), the reverse process is spontaneous.

If $\Delta S_{tot} = 0$, the process has no tendency to proceed in either direction.

In an exothermic reaction, such as the synthesis of ammonia or a combustion reaction, the heat released by the reaction increases the disorder of the surroundings. In some cases, the entropy of the system may decrease, as when a gaseous reactant is converted into a solid or liquid. However, provided that ΔH is large and negative, the release of energy as heat into the surroundings increases their entropy so much that it dominates the overall change in entropy and the reaction is spontaneous (Fig. 7.18).

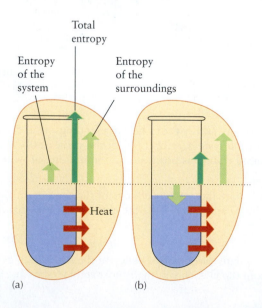

FIGURE 7.18 In an exothermic reaction, (a) the overall entropy change is certainly positive when the entropy of the system increases. (b) The overall entropy change may also be positive when the entropy of the system decreases. The reaction is spontaneous in both cases.

EXAMPLE 7.11 Sample exercise: Calculating the total change in entropy accompanying a reaction

Assess whether the combustion of magnesium is spontaneous at 25°C under standard conditions, given

$$2 \, Mg(s) + O_2(g) \longrightarrow 2 \, MgO(s) \qquad \Delta S° = -217 \, J \cdot K^{-1} \qquad \Delta H° = -1202 \, kJ$$

SOLUTION

Determine the change in entropy of the system (given in this case).

$$\Delta S° = -217 \, J \cdot K^{-1}$$

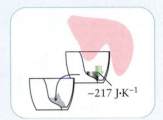

Determine the change in entropy of the surroundings from $\Delta S_{surr}° = -\Delta H°/T$.

$$\Delta S_{surr}° = -\frac{(-1.202 \times 10^6 \, J)}{298 \, K}$$
$$= +4.03 \times 10^3 \, J \cdot K^{-1}$$

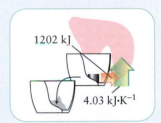

Determine the total change in entropy from $\Delta S_{tot}° = \Delta S° + \Delta S_{surr}$.

$$\Delta S_{tot}° = -217 \, J \cdot K^{-1} + (4.03 \times 10^3 \, J \cdot K^{-1})$$
$$= +3.81 \times 10^3 \, J \cdot K^{-1}$$

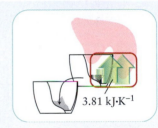

Because $\Delta S_{tot}°$ is positive, the reaction is spontaneous under standard conditions even though the entropy of the system decreases.

SELF-TEST 7.15A Is the formation of hydrogen fluoride from its elements in their most stable forms spontaneous under standard conditions at 25°C? For the reaction $H_2(g) + F_2(g) \rightarrow 2 \, HF(g)$, $\Delta H° = -542.2 \, kJ$ and $\Delta S° = +14.1 \, J \cdot K^{-1}$.
 [*Answer:* $\Delta S_{surr} = +1819 \, J \cdot K^{-1}$; therefore, $\Delta S_{tot} = +1833 \, J \cdot K^{-1}$; spontaneous]

SELF-TEST 7.15B Is the formation of benzene from its elements in their most stable forms spontaneous at 25°C? For the reaction $6 \, C(s) + 3 \, H_2(g) \rightarrow C_6H_6(l)$, $\Delta H° = +49.0 \, kJ$ and $\Delta S° = -253.18 \, J \cdot K^{-1}$.

Spontaneous endothermic reactions were a puzzle for nineteenth-century chemists, who believed that reactions ran only in the direction of decreasing energy of the system. It seemed to them that, in endothermic reactions, reactants were rising spontaneously to higher energies, like a weight suddenly leaping up from the floor onto a table. However, the criterion for spontaneity is an *increase in the total entropy of the system and its surroundings*, not a decrease in the energy of a system. In an endothermic reaction, the entropy of the surroundings certainly decreases as heat flows from them into the system. Nevertheless, there can still be an overall increase in entropy if the disorder of the system increases enough. Every endothermic reaction must be accompanied by increased disorder of the system if it is to be spontaneous (Fig. 7.19): endothermic reactions are driven by the dominating increase in disorder of the system.

By considering the total entropy change, we can draw some far-reaching conclusions about processes going on in the universe. For instance, we saw in Section 6.3 that maximum work is achieved if expansion takes place reversibly, by matching the

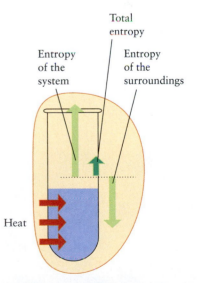

FIGURE 7.19 An endothermic reaction is spontaneous only if the entropy of the system increases enough to overcome the decrease in entropy of the surroundings.

external pressure to the pressure of the system at every stage. That relation is always true: *a process produces maximum work if it takes place reversibly.* That is, w_{rev} is more negative (more energy leaves the system as work) than w_{irrev}. However, because the internal energy is a state function, ΔU is the same for any path between the same two states. Therefore, because $\Delta U = q + w$, it follows that q_{rev}, the heat absorbed along the reversible path, must be greater than q_{irrev}, the heat absorbed along any other path, because only then can the sums of q and w be the same for the two paths. If now we replace q_{rev} in Eq. 1 by the smaller quantity q_{irrev}, we get $\Delta S > q_{irrev}/T$. In general, we can write the **Clausius inequality:**

$$\Delta S \geq \frac{q}{T} \tag{11a}*$$

with the equality valid for a reversible process. For a fully isolated system like that in Fig. 7.15, $q = 0$ for any process taking place inside the system. It follows from Eq. 11a that

$$\Delta S \geq 0 \text{ for any process in an isolated system} \tag{11b}$$

That is, we have shown that *the entropy cannot decrease in an isolated system.* This is another statement of the second law of thermodynamics. It tells us, in effect, that, as a result of all the processes going on around us, the entropy of the universe is steadily increasing.

Now consider an isolated system consisting of both the system that interests us and its surroundings (again like that in Fig. 7.15). For any spontaneous change in this isolated system, we know from Eq. 11b that $\Delta S_{tot} > 0$. If we calculate for a particular hypothetical process that $\Delta S_{tot} < 0$, we can conclude that the reverse of that process is spontaneous.

A final point is that reversible and irreversible paths between two given states of the system leave the surroundings in different states. Because entropy is a state function, the value of ΔS, the change in entropy of the system, is the same for both paths. However, ΔS_{tot} is different for the two paths because the entropy of the surroundings changes by different amounts in each case. For example, the isothermal expansion of an ideal gas always results in the change in the entropy of the system given by Eq. 3. However, the change in entropy of the surroundings is different for reversible and irreversible paths because the surroundings are left in different states in each case (Fig. 7.20). Table 7.4 lists the characteristics of reversible and irreversible processes.

EXAMPLE 7.12   **Calculating the total entropy change for the expansion of an ideal gas**

Calculate ΔS, ΔS_{surr}, and ΔS_{tot} for (a) the isothermal, reversible expansion and (b) the isothermal, free expansion of 1.00 mol of ideal gas molecules from 8.00 L to 20.00 L at 292 K. Explain any differences between the two paths.

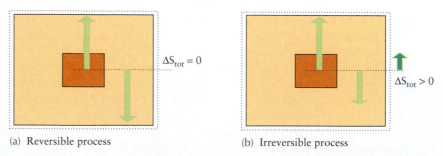

(a) Reversible process   (b) Irreversible process

FIGURE 7.20 (a) When a process takes place reversibly—that is, when the system is at equilibrium with its surroundings—the change in entropy of the surroundings is the negative of the change in entropy of the system and the overall change is zero. (b) For an irreversible change between the same two states of the system, the final state of the surroundings is different from the final state in part (a), and the change in the system is not canceled by the entropy change of the surroundings. Overall, there is an increase in entropy.

TABLE 7.4 Criteria for Spontaneity

ΔS	ΔS_{surr}	ΔS_{tot}	Character
>0	>0	>0	spontaneous
<0	<0	<0	not spontaneous; reverse change is spontaneous
>0	<0		spontaneous if ΔS is greater than $-\Delta S_{surr}$
<0	>0		spontaneous if ΔS_{surr} is greater than $-\Delta S$

STRATEGY Because entropy is a state function, the change in entropy of the system is the same regardless of the path between the two states, so we can use Eq. 3 to calculate ΔS for both part (a) and part (b). For the entropy of the surroundings, we need to find the heat transferred to the surroundings. In each case, we can combine the fact that $\Delta U = 0$ for an isothermal expansion of an ideal gas with $\Delta U = w + q$ and conclude that $q = -w$. We then use Eq. 4 in Chapter 6 to calculate the work done in an isothermal, reversible expansion of an ideal gas and Eq. 9 in this chapter to find the total entropy. The changes that we calculate are summarized in Fig. 7.21.

SOLUTION (a) For the reversible path,

From $\Delta S = nR \ln (V_2/V_1)$,

$$\Delta S = (1.00 \text{ mol})$$
$$\times (8.3145 \text{ J·K}^{-1}\text{·mol}^{-1}) \times \ln\frac{20.00 \text{ L}}{8.00 \text{ L}}$$
$$= +7.6 \text{ J·K}^{-1}$$

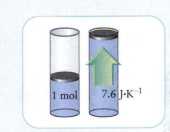

The change in the entropy of the gas is positive, as expected. Because $\Delta U = 0$, $q = -w$. Therefore, because the heat that flows into the surroundings is equal to the heat that flows *out* of the system,

From $q_{surr} = -q$, $q = -w$, and $w = -nRT \ln (V_2/V_1)$,

$$q_{surr} = -nRT \ln\frac{V_2}{V_1}$$

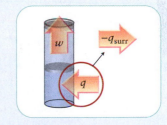

From $\Delta S_{surr} = q_{surr}/T$,

$$\Delta S_{surr} = -nR \ln\frac{V_2}{V_1} = -\Delta S = -7.6 \text{ J·K}^{-1}$$

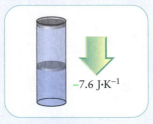

From $\Delta S_{tot} = \Delta S + \Delta S_{surr}$,

$$\Delta S_{tot} = 7.6 \text{ J·K}^{-1} - 7.6 \text{ J·K}^{-1} = 0$$

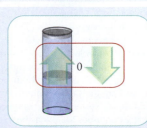

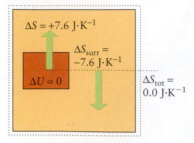

(a) Nonspontaneous process

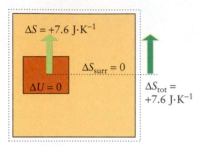

(b) Spontaneous process

FIGURE 7.21 The changes in entropy and internal energy when an ideal gas undergoes (a) reversible and (b) irreversible changes between the same two states, as described in Example 7.12.

Because $\Delta S_{surr} = -\Delta S$, $\Delta S_{tot} = 0$. This value is in accord with the statement that the process is reversible. (b) For the irreversible process, ΔS is the same, at $+7.6$ J·K^{-1}. No work is done in free expansion (Section 6.3), and so $w = 0$. Because $\Delta U = 0$, it follows that $q = 0$. Therefore, no heat is transferred into the surroundings, and their entropy is unchanged: $\Delta S_{surr} = 0$. The total change in entropy is therefore $\Delta S_{tot} = +7.6$ J·K^{-1}. The positive value is consistent with an irreversible expansion.

SELF-TEST 7.16A Determine ΔS, ΔS_{surr}, and ΔS_{tot} for (a) the reversible, isothermal expansion and (b) the isothermal free expansion of 1.00 mol of ideal gas molecules from 10.00 atm and 0.200 L to 1.00 atm and 2.00 L at 298 K.

[*Answer:* (a) $\Delta S = +19.1$ J·K^{-1}; $\Delta S_{surr} = -19.1$ J·K^{-1}; $\Delta S_{tot} = 0$;
(b) $\Delta S = +19.1$ J·K^{-1}; $\Delta S_{surr} = 0$; $\Delta S_{tot} = +19.1$ J·K^{-1}]

SELF-TEST 7.16B Determine ΔS, ΔS_{surr}, and ΔS_{tot} for the reversible, isothermal compression of 2.00 mol of ideal gas molecules from 1.00 atm and 4.00 L to 20.00 atm and 0.200 L at 298 K.

A process is spontaneous if it is accompanied by an increase in the total entropy of the system and the surroundings.

7.11 Equilibrium

Although this section provides only a brief introduction to equilibrium, the principles presented here are critically important, because the tendency of reactions to proceed toward equilibrium is the basis of much of chemistry. The material in this section lays the foundation for the next five chapters, including phase changes, the reactions of acids and bases, and redox reactions.

A system at **equilibrium** has no tendency to change in either direction (forward or reverse). Such a system remains in its current state until it is disturbed by changing the conditions, such as raising the temperature, decreasing the volume, or adding more reactants. The equilibrium state important in chemistry is a **dynamic equilibrium**, in which forward and reverse processes still continue but at matching rates. For example, when a block of metal is at the same temperature as its surroundings, it is in **thermal equilibrium** with its surroundings, and energy has no net tendency to flow into or out of the block as heat. Energy continues to flow in both directions, but there is no *net* flow (Fig. 7.22). When a gas confined to a cylinder by a piston has the same pressure as that of the surroundings, the system is in **mechanical equilibrium** with the surroundings and the gas has no tendency to expand or contract. The pressure inside is causing the piston to move outward, but the pressure outside is causing the piston to move inward to the same extent, and there is no net change.

When a solid, such as ice, is in contact with its liquid form, such as water, at certain conditions of temperature and pressure (at 0°C and 1 atm for water), the two states of matter are in dynamic equilibrium with each other, and there is no tendency for one form of matter to change into the other form. When solid and liquid water are at equilibrium, water molecules continually leave solid ice to form liquid water, and water molecules continually leave the liquid phase to form ice. However, there is no *net* change, because these processes occur at the same rate and so balance each other.

When a chemical reaction mixture reaches a certain composition, the reaction seems to come to a halt. A mixture of substances at **chemical equilibrium** has no tendency either to produce more products or to revert to reactants. At equilibrium, reactants are still forming products, but products are decaying at a matching rate into reactants and there is no *net* change of composition.

The common characteristic of any kind of dynamic equilibrium is the continuation of processes at the microscopic level but no *net* tendency for the system to change in either the forward or the reverse direction. That is, neither the forward nor the reverse process is spontaneous. Expressed thermodynamically,

$$\Delta S_{tot} = 0 \tag{12}*$$

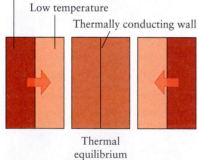

High temperature
Low temperature
Thermally conducting wall

Thermal equilibrium

FIGURE 7.22 Energy flows as heat from a high-temperature region (dark red) to a low-temperature region (light red) through a thermally conducting wall. A system is in thermal equilibrium with its surroundings when the temperatures on both sides of the wall are the same (center). However, because the systems are in contact, energy continues to flow as heat, but at the same rate in each direction.

The total entropy varies with the composition of a reaction mixture and by looking for the composition at which $\Delta S_{tot} = 0$, we can predict the equilibrium composition of the reaction. At this composition, the reaction has no tendency either to form more products or to decompose into reactants.

SELF-TEST 7.17A Confirm that liquid water and water vapor are in equilibrium when the temperature is 100.°C and the pressure is 1 atm. Data are available in Tables 6.3 and 7.2.

[*Answer:* $\Delta S_{tot} = 0$ at 100.°C]

SELF-TEST 7.17B Confirm that liquid benzene and benzene vapor are in equilibrium at the normal boiling point of benzene, 80.1°C, and 1 atm pressure. The enthalpy of vaporization of benzene at its boiling point is 30.8 kJ·mol^{-1} and its entropy of vaporization is 87.2 J·K^{-1}·mol^{-1}.

The general criterion for equilibrium in thermodynamics is $\Delta S_{tot} = 0$.

GIBBS FREE ENERGY

One of the problems with using the second law to judge whether a reaction is spontaneous is that to get the total entropy change we must calculate three quantities: the entropy change of the system, the entropy change of the surroundings, and the sum of the two changes. We could avoid some of this work if a single property combined the entropy calculations for the system and the surroundings. We could then carry out the calculation in one step by using a single table of data. This simplification is achieved by introducing a new state function, the "Gibbs free energy." Not only does the Gibbs free energy allow us to assess whether a reaction is spontaneous, it also tells us how much nonexpansion work we can get from the system—a central consideration in the study of foods and fuels—and how a change in temperature can change the spontaneity of a reaction. In Chapter 9, we shall also see how the Gibbs free energy controls the composition of a reaction mixture at equilibrium and how that composition depends on the conditions. The Gibbs free energy is probably the single most widely used and useful quantity in the application of thermodynamics to chemistry.

The Gibbs free energy is named for Josiah Willard Gibbs (Fig. 7.23), the nineteenth-century American physicist who was responsible for turning thermodynamics from an abstract theory into a subject of great usefulness.

7.12 Focusing on the System

The total entropy change, ΔS_{tot}, is the sum of the changes in the system, ΔS, and its surroundings, ΔS_{surr}, with $\Delta S_{tot} = \Delta S + \Delta S_{surr}$. For a process at constant temperature and pressure, the change in the entropy of the surroundings is given by Eq. 10, $\Delta S_{surr} = -\Delta H/T$. Therefore,

$$\Delta S_{tot} = \Delta S + \underbrace{\Delta S_{surr}}_{-\Delta H/T} = \Delta S - \frac{\Delta H}{T} \quad \text{at constant temperature and pressure} \quad (13)$$

This equation shows how to calculate the total entropy change from information about the system alone. The limitation is that the equation is valid only at constant temperature and pressure.

To take the next step, we introduce the **Gibbs free energy**, G, which is defined as

$$G = H - TS \quad (14)*$$

This quantity, which is commonly known as the *free energy* and more formally as the *Gibbs energy*, is defined solely in terms of state functions, and so G is a state function. Throughout chemistry, we use only changes in Gibbs free energy, not its absolute values; and, for a process occurring at constant temperature, the change in Gibbs free energy is

$$\Delta G = \Delta H - T\Delta S \quad \text{at constant temperature} \quad (15)*$$

FIGURE 7.23 Josiah Willard Gibbs (1839–1903).

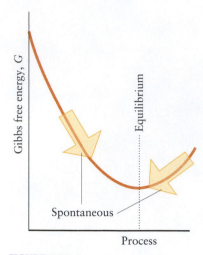

FIGURE 7.24 At constant temperature and pressure, the direction of spontaneous change is toward lower Gibbs free energy. The equilibrium state of a system corresponds to the lowest point on the curve.

By comparing this expression rearranged into

$$\frac{\Delta G}{T} = \frac{\Delta H}{T} - \Delta S$$

with Eq. 13 where there is the additional constraint of constant pressure, we see that $\Delta G/T = -\Delta S_{tot}$ and therefore that

$$\Delta G = -T\Delta S_{tot} \qquad \text{at constant temperature and pressure} \qquad (16)$$

The minus sign in this equation means that, provided the temperature and pressure are constant, an increase in total entropy corresponds to a decrease in Gibbs free energy. Therefore, *at constant temperature and pressure, the direction of spontaneous change is the direction of decreasing Gibbs free energy* (Fig. 7.24). The great importance of the introduction of Gibbs free energy is that, *provided that the temperature and pressure are constant, we can predict the spontaneity of a process solely in terms of the thermodynamic properties of the system.* Note that we have restricted attention to processes at constant temperature and pressure, but by doing so we can express spontaneity under common laboratory conditions in terms of the system alone.

Equation 15 summarizes the factors that determine the direction of spontaneous change at constant temperature and pressure: for a spontaneous change, we look for values of ΔH, ΔS, and T that result in a negative value of ΔG (Table 7.5). One condition that may result in a negative ΔG is a large negative ΔH, such as that for a combustion reaction. As we have seen, a large negative ΔH corresponds to a large increase in the entropy of the surroundings. However, a negative ΔG can occur even if ΔH is positive (an endothermic reaction), provided that $T\Delta S$ is large and positive. In this case, as we saw in Section 7.10, the driving force of the reaction is the increase in entropy of the system.

SELF-TEST 7.18A Can a nonspontaneous process with a negative ΔS become spontaneous if the temperature is increased?

[Answer: No]

SELF-TEST 7.18B Can a nonspontaneous process with a positive ΔS become spontaneous if the temperature is increased?

We have seen that the criterion for equilibrium is $\Delta S_{tot} = 0$. It follows from Eq. 16 that, for a process at constant temperature and pressure, the condition for equilibrium is

$$\Delta G = 0 \qquad \text{at constant temperature and pressure} \qquad (17)$$

If we find that $\Delta G = 0$ for a process, then we know at once that the system is at equilibrium. For example, when ice and water are in equilibrium with each other at a particular temperature and pressure, the Gibbs free energy of 1 mol $H_2O(l)$ must be the same as the Gibbs free energy of 1 mol $H_2O(s)$. In other words, the molar Gibbs free energy, the Gibbs free energy per mole, of water in each phase is the same.

EXAMPLE 7.13 **Sample exercise: Deciding whether a process is spontaneous**

Calculate the change in molar Gibbs free energy, ΔG_m, for the process $H_2O(s) \rightarrow H_2O(l)$ at 1 atm and (a) 10.°C; (b) 0.°C. Decide for each temperature whether melting is spontaneous or not. Treat ΔH_{fus} and ΔS_{fus} as independent of temperature.

TABLE 7.5 **Factors That Favor Spontaneity**

Enthalpy change	Entropy change	Spontaneous?
exothermic ($\Delta H < 0$)	increase ($\Delta S > 0$)	yes, $\Delta G < 0$
exothermic ($\Delta H < 0$)	decrease ($\Delta S < 0$)	yes, if $|T\Delta S| < |\Delta H|$, $\Delta G < 0$
endothermic ($\Delta H > 0$)	increase ($\Delta S > 0$)	yes, if $T\Delta S > \Delta H$, $\Delta G < 0$
endothermic ($\Delta H > 0$)	decrease ($\Delta S < 0$)	no, $\Delta G > 0$

SOLUTION The enthalpy of fusion is 6.01 kJ·mol^{-1} and the entropy of fusion is 22.0 J·K^{-1}·mol^{-1}. These values are almost independent of temperature over the temperature range considered.

(a) At 10.°C

From $\Delta G_m = \Delta H_m - T\Delta S_m$,

$$\Delta G_m = \underbrace{6.01 \text{ kJ·mol}^{-1}}_{\Delta H_m} - \underbrace{(283 \text{ K})}_{(273.15 + 10.) \text{ K}} \times \underbrace{(22.0 \text{ J·K}^{-1}\text{·mol}^{-1})}_{\Delta S_m}$$

$$= 6.01 \text{ kJ·mol}^{-1} - 6.23 \times 10^3 \text{ J·mol}^{-1}$$

$$= 6.01 \text{ kJ·mol}^{-1} - \underbrace{6.23}_{10^3 \text{ J}} \text{ kJ·mol}^{-1}$$

$$= +0.22 \text{ kJ·mol}^{-1}$$

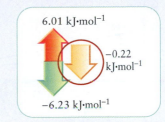

A note on good practice: Be sure to convert joules to kilojoules before subtracting $T\Delta S$ from ΔH.

(b) At 0.°C

From $\Delta G_m = \Delta H_m - T\Delta S_m$,

$$\Delta G_m = 6.01 \text{ kJ·mol}^{-1} - (273 \text{ K}) \times (22.0 \text{ J·K}^{-1}\text{·mol}^{-1}.)$$

$$= 6.01 \text{ kJ·mol}^{-1} - 6.01 \text{ kJ·mol}^{-1} = 0$$

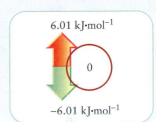

Because the difference in molar Gibbs free energy is negative at 10.°C, melting is spontaneous at that temperature, but at 0.°C, ice and water are in equilibrium.

SELF-TEST 7.19A Calculate the change in molar Gibbs free energy for the process $H_2O(l) \rightarrow H_2O(g)$ at at 1 atm and (a) 95°C; (b) 105°C. The enthalpy of vaporization is 40.7 kJ·mol^{-1}, and the entropy of vaporization is 109.1 J·K^{-1}·mol^{-1}. In each case, indicate whether vaporization would be spontaneous or not.
      [*Answer:* (a) +0.6 kJ·mol^{-1}, not spontaneous; (b) −0.5 kJ·mol^{-1}, spontaneous]

SELF-TEST 7.19B Calculate the change in molar Gibbs free energy for the process $Hg(l) \rightarrow Hg(g)$ at 1 atm and (a) 350.°C; (b) 370.°C. The enthalpy of vaporization is 59.3 kJ·mol^{-1}, and the entropy of vaporization is 94.2 J·K^{-1}·mol^{-1}. In each case, indicate whether or not vaporization would be spontaneous.

The Gibbs free energy of a substance decreases as the temperature is raised at constant pressure. This conclusion follows from the definition $G = H - TS$ and the fact that the entropy of a pure substance is invariably positive; as T increases, TS increases, too, and a larger quantity is subtracted from H. Another important conclusion is that, because the molar entropy of the gas phase of a substance is greater than that of its liquid phase, the Gibbs free energy decreases more sharply with temperature for the gas phase of a substance than for the liquid phase. Similarly, the Gibbs free energy of the liquid phase of a substance decreases more sharply than that of its solid phase (Fig. 7.25).

We can now see the thermodynamic origin of melting, vaporization, and their opposites. At low temperatures, the molar Gibbs free energy of the solid phase lies lowest; so there is a spontaneous tendency for a liquid to freeze and thereby reduce its Gibbs free energy. Above a certain temperature, the molar Gibbs free energy of the liquid has fallen below that of the solid, and the substance has a spontaneous tendency to melt. At an even higher temperature, the molar Gibbs free energy of the gas phase has fallen to below the liquid line, and the substance has a spontaneous tendency to vaporize. The temperature of each phase change corresponds to the point of intersection of the lines for the two phases, as can be seen in Fig. 7.25.

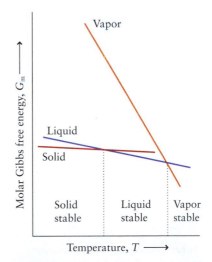

FIGURE 7.25 The variation of the (molar) Gibbs free energy with temperature for three phases of a substance at a given pressure. The most stable phase is the phase with lowest molar Gibbs free energy. We see that, as the temperature is raised, the solid, liquid, and vapor phases in succession become the most stable.

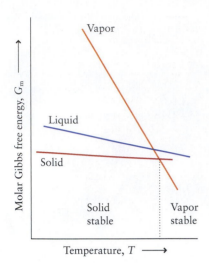

FIGURE 7.26 For some substances and at certain pressures, the molar Gibbs free energy of the liquid phase might never lie lower than those of the other two phases. For such substances, the liquid is never the stable phase and, at constant pressure, the solid sublimes when the temperature is raised to the point of intersection of the solid and vapor lines.

Standard states are defined in Section 6.15.

As remarked previously, in some texts you will see 1 atm used in the definition of standard states. The modern definition is 1 bar. The change in standard pressure has little effect on the tabulated values.

The relative positions of the three lines shown in Fig. 7.25 are different for each substance. One possibility—which depends on the strength of intermolecular interactions in the condensed phases—is for the liquid line to lie in the position shown in Fig. 7.26. In this case, the liquid line is never the lowest line, at any temperature. As soon as the temperature has been raised above the point corresponding to the intersection of the solid and gas lines, the direct transition of the solid to the vapor becomes spontaneous. This plot is the type that we would expect for carbon dioxide, which sublimes at room temperature.

The change in Gibbs free energy for a process is a measure of the change in the total entropy of a system and its surroundings at constant temperature and pressure. Spontaneous processes at constant temperature and pressure are accompanied by a decrease in Gibbs free energy.

7.13 Gibbs Free Energy of Reaction

The decrease in Gibbs free energy as a signpost of spontaneous change and $\Delta G = 0$ as a criterion of equilibrium are applicable to any kind of process, provided that it is occurring at constant temperature and pressure. Because chemical reactions are our principal interest in chemistry, we now concentrate on them and look for a way to calculate ΔG for a reaction.

The thermodynamic function used as the criterion of spontaneity for a chemical reaction is the **Gibbs free energy of reaction, ΔG** (which is commonly referred to as the "reaction free energy"). This quantity is defined as the difference in molar Gibbs free energies, G_m, of the products and the reactants:

$$\Delta G = \sum n G_m(\text{products}) - \sum n G_m(\text{reactants}) \tag{18}$$

The n are the amounts of each substance in the chemical equation. For example, for the formation of ammonia, $N_2(g) + 3\,H_2(g) \rightarrow 2\,NH_3(g)$:

$$\Delta G = \{(2\text{ mol}) \times G_m(NH_3)\} - \{(1\text{ mol}) \times G_m(N_2) + (3\text{ mol}) \times G_m(H_2)\}$$

The molar Gibbs free energy of a substance in a mixture depends on what molecules it has as neighbors, so the molar Gibbs free energies of NH_3, N_2, and H_2 change as the reaction proceeds. For example, at an early stage of the reaction, an NH_3 molecule has mostly N_2 and H_2 molecules around it, but at a later stage of the reaction, most of its neighbors may be NH_3 molecules. Because the individual Gibbs free energies change as the reaction proceeds, the Gibbs free energy of reaction also changes. If $\Delta G < 0$ at a certain composition of the reaction mixture, then the reaction is spontaneous. If $\Delta G > 0$ at a certain composition, then the reverse reaction (the decomposition of ammonia in our example), is spontaneous.

The **standard Gibbs free energy of reaction, $\Delta G°$**, is defined like the Gibbs free energy of reaction but in terms of the *standard* molar Gibbs energies of the reactants and products:

$$\Delta G° = \sum n G_m°(\text{products}) - \sum n G_m°(\text{reactants}) \tag{19}$$

That is, the standard Gibbs free energy of reaction is the difference in Gibbs free energy between the products in their standard states and the pure reactants in their standard states (at the specified temperature). Because the standard state of a substance is its pure form at 1 bar, the *standard* Gibbs free energy of reaction is the difference in Gibbs free energy between the *pure* products and the *pure* reactants: it is a fixed quantity for a given reaction and does not vary as the reaction proceeds. We shall unfold the enormous importance of this quantity shortly. The point to grasp is that $\Delta G°$ is fixed for a given reaction and temperature, but ΔG varies—and might even change sign—as the reaction proceeds.

Equations 18 and 19 are not very useful in practice, because, as remarked earlier, we can know only changes in the Gibbs free energies of substances. However, we can

TABLE 7.6 Examples of the Most Stable Forms of Elements

Element	Most stable form at 25°C and 1 bar
H_2, O_2, Cl_2, Xe	gas
Br_2, Hg	liquid
C	graphite
Na, Fe, I_2	solid

TABLE 7.7 Standard Gibbs Free Energies of Formation at 25°C ($kJ \cdot mol^{-1}$)*

Gases	ΔG_f°	Liquids	ΔG_f°	Solids	ΔG_f°
ammonia, NH_3	−16.45	benzene, C_6H_6	+124.3	calcium carbonate, $CaCO_3^\dagger$	−1128.8
carbon dioxide, CO_2	−394.4	ethanol, C_2H_5OH	−174.8		
nitrogen dioxide, NO_2	+51.3	water, H_2O	−237.1	silver chloride, AgCl	−109.8
water, H_2O	−228.6				

* Additional values are given in Appendix 2A.
† Calcite.

use the same technique that we used to find the standard reaction enthalpy in Section 6.18, where we assigned each compound a standard enthalpy of formation, ΔH_f°. We can also tabulate Gibbs free energies of formation of substances and then use them to calculate ΔG°. The **standard Gibbs free energy of formation**, ΔG_f° (the "standard free energy of formation"), of a substance is *the standard Gibbs free energy of reaction per mole for the formation of a compound from its elements in their most stable form*. The most stable form of an element is the state with the lowest Gibbs free energy (Table 7.6). For example, the standard Gibbs free energy of formation of hydrogen iodide gas at 25°C is $\Delta G_f^\circ(HI, g) = +1.70 \text{ kJ} \cdot mol^{-1}$. It is the standard Gibbs free energy of reaction for $\frac{1}{2}H_2(g) + \frac{1}{2}I_2(s) \rightarrow HI(g)$. The standard Gibbs free energies of formation of elements in their most stable forms are defined as exactly zero, and so $\Delta G_f^\circ(I_2, s) = 0$, but $\Delta G_f^\circ(I_2, g)$ is not zero.

Standard Gibbs free energies of formation can be determined in various ways. One straightforward way is to combine standard enthalpy and entropy data from tables such as Tables 6.5 and 7.3. A list of values for several common substances is given in Table 7.7, and a more extensive one appears in Appendix 2A.

EXAMPLE 7.14 **Calculating a standard Gibbs free energy of formation from enthalpy and entropy data**

Calculate the standard Gibbs free energy of formation of HI(g) at 25°C from its standard molar entropy and standard enthalpy of formation.

STRATEGY We write the chemical equation for the formation of HI(g) and calculate the standard Gibbs free energy of reaction from $\Delta G^\circ = \Delta H^\circ - T\Delta S^\circ$. It is best to write the equation with a stoichiometric coefficient of 1 for the compound of interest, because then $\Delta G^\circ = \Delta G_f^\circ$. The standard enthalpy of formation is found in Appendix 2A. The standard reaction entropy is found as shown in Example 7.9, by using the data from Table 7.3 or Appendix 2A.

SOLUTION The chemical equation is $\frac{1}{2}H_2(g) + \frac{1}{2}I_2(s) \rightarrow HI(g)$.

From data in Appendix 2A, $\Delta H^\circ = (1 \text{ mol}) \times \Delta H_f^\circ(HI, g) = +26.48 \text{ kJ}$

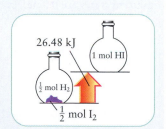

From Eq. 8 of this chapter, $\Delta S^\circ = S_m^\circ(HI, g) - \{\frac{1}{2}S_m^\circ(H_2, g) + \frac{1}{2}S_m^\circ(I_2, s)\}$

$$= \{(1 \text{ mol}) \times (206.6 \text{ J} \cdot K^{-1} \cdot mol^{-1})\}$$

$$- \{(\tfrac{1}{2} \text{ mol}) \times (130.7) + (\tfrac{1}{2} \text{ mol}) \times (116.1)\} \text{ J} \cdot K^{-1} \cdot mol^{-1}$$

$$= +83.2 \text{ J} \cdot K^{-1} = +0.0832 \text{ kJ} \cdot K^{-1}$$

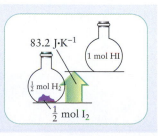

From $\Delta G° = \Delta H° - T\Delta S°$,

$$\Delta G° = (1 \text{ mol}) \times (+26.48 \text{ kJ·mol}^{-1}) - (298 \text{ K})$$
$$\times (0.0832 \text{ kJ·K}^{-1})$$
$$= +1.69 \text{ kJ}$$

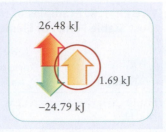

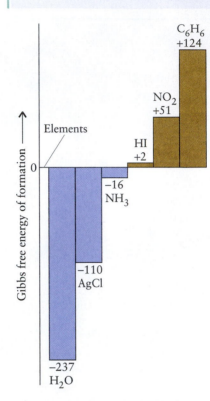

FIGURE 7.27 The standard Gibbs free energy of formation of a compound is defined as the standard reaction Gibbs free energy per mole of formula units of the compound when the compound is formed from its elements. It represents a "thermodynamic altitude" with respect to the elements at "sea level." The numerical values are in kilojoules per mole.

The rates of reactions are discussed in Chapter 13.

The standard Gibbs free energy of formation of HI is therefore +1.69 kJ·mol^{-1}, in good agreement with the value of +1.70 kJ·mol^{-1} quoted in the text. Note that, because this value is positive, the formation of pure HI from the elements at 1 bar is not spontaneous.

SELF-TEST 7.20A Calculate the standard Gibbs free energy of formation of $NH_3(g)$ at 25°C from the enthalpy of formation and the molar entropies of the species taking part in its formation.

[*Answer:* −16.5 kJ·mol^{-1}]

SELF-TEST 7.20B Calculate the standard Gibbs free energy of formation of $C_3H_6(g)$, cyclopropane, at 25°C.

The standard Gibbs free energy of formation at a given temperature is an indication of a compound's stability relative to its elements under standard conditions. The data in Appendix 2A are for 298.15 K, the conventional temperature for reporting thermodynamic data. If $\Delta G_f° < 0$ at a certain temperature, the pure compound has a lower Gibbs free energy than that of its pure elements and the elements are poised to change spontaneously into the compound at that temperature (Fig. 7.27). We say that the compound is "more stable" under standard conditions than the elements. If $\Delta G_f° > 0$, the Gibbs free energy of the compound is higher than that of its elements and the compound is poised to change spontaneously into the pure elements. In this case, we say that the pure elements are "more stable" than the pure compound. For example, the standard Gibbs free energy of formation of benzene is +124 kJ·mol^{-1} at 25°C, and benzene is unstable with respect to its elements under standard conditions at 25°C.

A **thermodynamically stable compound** is a compound with a negative standard Gibbs free energy of formation (water is an example). A **thermodynamically unstable compound** is a compound with a positive standard Gibbs free energy of formation (benzene is an example). Such a compound is poised to decompose spontaneously into its elements. However, that tendency may not be realized in practice, because the decomposition may be very slow. Benzene can, in fact, be kept indefinitely without decomposing at all. Substances that are thermodynamically unstable but survive for long periods are called **nonlabile** or even **inert**. For example, benzene is thermodynamically unstable but nonlabile. Substances that do decompose or react rapidly are called **labile**. Most radicals are labile. It is important to keep in mind the distinction between stability and lability:

Stable and *unstable* are terms that refer to the thermodynamic tendency of a substance to decompose into its elements.

Labile, *nonlabile*, and *inert* are terms that refer to the rate at which a thermodynamic tendency to react is realized.

SELF-TEST 7.21A Is glucose stable relative to its elements at 25°C and under standard conditions?

[*Answer:* Yes; for glucose, $\Delta G_f° = −910.$ kJ·mol^{-1}, a negative value.]

SELF-TEST 7.21B Is methylamine, CH_3NH_2, stable relative to its elements at 25°C and under standard conditions?

Just as we can combine standard enthalpies of formation to obtain standard reaction enthalpies, we can also combine standard Gibbs free energies of formation to obtain standard Gibbs free energies of reaction:

$$\Delta G° = \sum n\Delta G_f°(\text{products}) - \sum n\Delta G_f°(\text{reactants}) \tag{20}*$$

where, as usual, the n are the amounts of each substance in the chemical equation.

EXAMPLE 7.15 **Sample exercise: Calculating the standard Gibbs free energy of reaction**

Calculate the standard Gibbs free energy of the reaction

   $4 NH_3(g) + 5 O_2(g) \rightarrow 4 NO(g) + 6 H_2O(g)$

at 25°C.

SOLUTION

From Appendix 2A and Eq. 20,

$$\Delta G° = \{(4 \text{ mol}) \times \Delta G_f°(\text{NO, g}) + (6 \text{ mol}) \times \Delta G_f°(\text{H}_2\text{O, g})\}$$
$$- \{(4 \text{ mol}) \times \Delta G_f°(\text{NH}_3, \text{g}) + (5 \text{ mol}) \times \Delta G_f°(\text{O}_2, \text{g})\}$$
$$= \{4(86.55) + 6(-228.57)\} - \{4(-16.45) + 0\} \text{ kJ}$$
$$= -959.42 \text{ kJ}$$

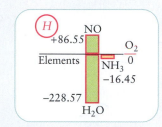

We can conclude that the oxidation of ammonia is spontaneous at 25°C under standard conditions.

A note on good practice: There are limits to this conclusion. By "under standard conditions" we mean that there is a spontaneous tendency for pure reactants, each at 1 bar, to change completely into pure products, also each at 1 bar (at the stated temperature).

SELF-TEST 7.22A Calculate the standard Gibbs free energy of the reaction $2 CO(g) + O_2(g) \rightarrow 2 CO_2(g)$ from standard Gibbs free energies of formation at 25°C.

[*Answer:* $\Delta G° = -514.38$ kJ]

SELF-TEST 7.22B Calculate the standard Gibbs free energy of the reaction $6 CO_2(g) + 6 H_2O(l) \rightarrow C_6H_{12}O_6(s, \text{glucose}) + 6 O_2(g)$ from standard Gibbs free energies of formation at 25°C.

> *The standard Gibbs free energy of formation of a substance is the standard Gibbs free energy of reaction per mole of compound when it is formed from its elements in their most stable forms. The sign of $\Delta G_f°$ tells us whether a compound is stable or unstable with respect to its elements. Standard Gibbs free energies of formation are used to calculate standard Gibbs free energies of reaction by using Eq. 20.*

7.14 The Gibbs Free Energy and Nonexpansion Work

We are now ready to see why the Gibbs free energy is called the *free* energy: we shall see that the change in Gibbs free energy that accompanies a process allows us to predict the maximum *nonexpansion work* that the process can do at constant temperature and pressure. That is, the Gibbs free energy is a measure of the energy *free* to do nonexpansion work. As explained in Section 6.3, nonexpansion work, w_e, is any kind of work other than that due to expansion against an opposing pressure; so it includes electrical work and mechanical work (like stretching a spring or carrying a weight up a slope). The fact that nonexpansion work includes electrical work—the work of pushing electrons through an electrical circuit—is the basis of the chemical generation of electrical power; we pursue that connection in Chapter 12. Nonexpansion work also includes the work of muscular activity, the work of linking amino acids together to form protein molecules, and the work of sending

The subscript e in w_e stands for "extra."

nerve signals through neurons; so a knowledge of changes in Gibbs free energy is central to an understanding of **bioenergetics,** the deployment and utilization of energy in living cells.

The challenge that we now face is to justify these remarks and to derive a *quantitative* relation between the Gibbs free energy and the maximum nonexpansion work that a system can do.

HOW DO WE DO THAT?

To find the relation between the Gibbs free energy and the maximum nonexpansion work, we start with Eq. 15 for an infinitesimal change (denoted d) in G at constant temperature:

$$dG = dH - TdS \qquad \text{at constant temperature}$$

We then use Eq. 9 of Chapter 6 ($H = U + PV$) to express the infinitesimal change in enthalpy at constant pressure in terms of the change in internal energy

$$dH = dU + PdV \qquad \text{at constant pressure}$$

Then we substitute this expression into the first:

$$dG = dU + PdV - TdS \qquad \text{at constant temperature and pressure}$$

Now we use Eq. 7 from Chapter 6 for an infinitesimal change in internal energy ($dU = dw + dq$) and obtain

$$dG = dw + dq + PdV - TdS \qquad \text{at constant temperature and pressure}$$

We are interested in the maximum work that a process can do, which means that the process has to occur reversibly (recall Section 6.3). For a reversible change,

$$dG = dw_{rev} + dq_{rev} + PdV - TdS \qquad \text{at constant temperature and pressure}$$

Now we use the infinitesimal version of Eq. 1 ($dS = dq_{rev}/T$) to replace dq_{rev} by TdS and cancel the two TdS terms:

$$dG = dw_{rev} + TdS + PdV - TdS$$
$$= dw_{rev} + PdV \qquad \text{at constant temperature and pressure}$$

At this point, we recognize that the system may do both expansion work and nonexpansion work:

$$dw_{rev} = dw_{rev,\,e} + dw_{rev,\,expansion}$$

Reversible expansion work (achieved by matching the external to the internal pressure) is given by the infinitesimal version of Eq. 3 of Chapter 6 ($w_{expansion} = -P_{ex}\Delta V$, which becomes $dw_{expansion} = -P_{ex}dV$) and setting the external pressure equal to the pressure of the gas in the system at each stage of the expansion:

$$dw_{rev,\,expansion} = -PdV$$

It follows that

$$dw_{rev} = dw_{rev,\,e} - PdV$$

When we substitute this expression into the last line of the expression for dG, the PdV terms cancel and we are left with

$$dG = dw_{rev,\,e} \qquad \text{at constant temperature and pressure}$$

Because $dw_{rev,\,e}$ is the *maximum* nonexpansion work that the system can do (because it is achieved reversibly), we obtain

$$dG = dw_{e,\,max} \qquad \text{at constant temperature and pressure}$$

When we consider a measurable change in the Gibbs free energy, the equation we have derived becomes

$$\Delta G = w_{e,max} \qquad \text{at constant temperature and pressure} \qquad (21)*$$

This important relation tells us that, if we know the change in Gibbs free energy of a process taking place at constant temperature and pressure, then we immediately know how much nonexpansion work it can do. For instance, for the oxidation of glucose,

$$C_6H_{12}O_6(s) + 6\,O_2(g) \longrightarrow 6\,CO_2(g) + 6\,H_2O(l)$$

the standard Gibbs free energy of reaction is -2879 kJ. Therefore, at 1 bar the maximum nonexpansion work obtainable from 1.000 mol $C_6H_{12}O_6(s)$, corresponding to 180.0 g of glucose, is 2879 kJ. Because about 17 kJ of work must be done to build 1 mol of peptide links (a link between amino acids) in a protein, the oxidation of 180. g of glucose can be used to build about $(2879 \text{ kJ})/(17 \text{ kJ}) = 170$ moles of such links. In other words, the oxidation of one glucose molecule is needed to build about 170 peptide links. In practice, biosynthesis occurs indirectly, there are energy losses, and only about 10 such links can be built. A typical protein has several hundred peptide links, and so several glucose molecules must be sacrificed to build one protein molecule.

The change in Gibbs free energy for a process is equal to the maximum nonexpansion work that the system can do at constant temperature and pressure.

7.15 The Effect of Temperature

The enthalpies of reactants and products are both affected by a temperature rise and the *difference* between their values hardly changes. The same is true of the entropies. As a result, the values of $\Delta H°$ and $\Delta S°$ do not change much with temperature. However, $\Delta G°$ does depend on temperature (remember the T in $\Delta G° = \Delta H° - T\Delta S°$) and might change sign as the temperature is changed. There are four cases to consider:

1   For an exothermic reaction ($\Delta H° < 0$) with a negative reaction entropy ($\Delta S° < 0$), $-T\Delta S°$ contributes a positive term to $\Delta G°$. For such a reaction, $\Delta G°$ is negative (and the pure reactants are poised to form products spontaneously) at low temperatures because $\Delta H°$ dominates $-T\Delta S°$, but it may become positive (and the reverse reaction, the decomposition of pure products, spontaneous) at higher temperatures when $-T\Delta S°$ dominates $\Delta H°$ (Fig. 7.28a).

In this case, the increase in entropy of the surroundings drives the reaction forward, but it is opposed by the decrease in entropy of the system.

2   For an endothermic reaction ($\Delta H° > 0$) with a positive reaction entropy ($\Delta S° > 0$), the reverse is true (Fig. 7.28b). In this case, $\Delta G°$ is positive at low temperatures but may become negative when the temperature is raised to the point that $T\Delta S°$ becomes larger than $\Delta H°$. The formation of products from pure reactants becomes spontaneous when the temperature is high enough.

Here, the increase in entropy of the system is opposed by the decrease in entropy of the surroundings, the effect of which is reduced by raising the temperature.

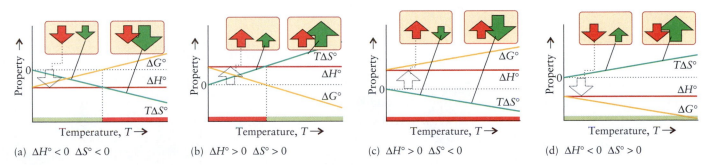

(a) $\Delta H° < 0$ $\Delta S° < 0$ (b) $\Delta H° > 0$ $\Delta S° > 0$ (c) $\Delta H° > 0$ $\Delta S° < 0$ (d) $\Delta H° < 0$ $\Delta S° > 0$

FIGURE 7.28 The effect of an increase in temperature on the spontaneity of a reaction under standard conditions. In each case, "spontaneous" is taken to mean $\Delta G° < 0$ and "nonspontaneous" is taken to mean $\Delta G° > 0$. (a) An exothermic reaction with negative reaction entropy becomes spontaneous *below* the temperature marked by the dotted line. (b) An endothermic reaction with a positive reaction entropy becomes spontaneous *above* the temperature marked by the dotted line. (c) An endothermic reaction with negative reaction entropy is not spontaneous at any temperature. (d) An exothermic reaction with positive reaction entropy is spontaneous at all temperatures.

3   For an endothermic reaction ($\Delta H° > 0$) with a negative reaction entropy ($\Delta S° < 0$), $\Delta G° > 0$ at all temperatures and the reaction is not spontaneous at any temperature (Fig. 7.28c).

In this case, the entropies of the system and the surroundings both decrease whatever the temperature.

4   For an exothermic reaction ($\Delta H° < 0$) with a positive reaction entropy ($\Delta S° > 0$), $\Delta G° < 0$ and the formation of products from pure reactants is spontaneous at all temperatures (Fig. 7.28d).

In this case, the entropies of the system and the surroundings both increase whatever the temperature.

EXAMPLE 7.16   **Calculating the temperature at which an endothermic reaction becomes spontaneous**

Estimate the temperature at which it is thermodynamically possible for carbon to reduce iron(III) oxide to iron under standard conditions by the endothermic reaction

$$2\ Fe_2O_3(s) + 3\ C(s) \longrightarrow 4\ Fe(s) + 3\ CO_2(g)$$

STRATEGY   The reaction is endothermic and, because a gas is produced from solid reactants, occurs with an increase in entropy. Because $\Delta H° > 0$ and $\Delta S° > 0$, the formation of products from pure reactants becomes spontaneous at temperatures for which $\Delta H° \leq T\Delta S°$. The temperature at which the tendency for the reaction to occur begins solves to $T = \Delta H°/\Delta S°$. We use data from Appendix 2A (and remember that the standard enthalpies of formation of elements are zero).

SOLUTION

From Eq. 20 of Chapter 6,	$\Delta H° = (3\ mol) \times \Delta H_f°(CO_2, g) - (2\ mol) \times \Delta H_f°(Fe_2O_3, s)$ $= 3(-393.5) - 2(-824.2)\ kJ$ $= +467.9\ kJ$

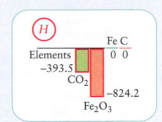

From Eq. 8 of this chapter,	$\Delta S° = \{(4\ mol) \times S_m°(Fe, s) + (3\ mol) \times S_m°(CO_2, g)\}$ $\quad - \{(2\ mol) \times S_m°(Fe_2O_3, s) + (3\ mol) \times S_m°(C, s)\}$ $= \{4(27.3) + 3(213.7)\} - \{2(87.4) + 3(5.7)\}\ J \cdot K^{-1}$ $= +558.4\ J \cdot K^{-1}$

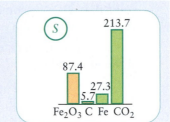

From $T = \Delta H°/\Delta S°$,	$T = \dfrac{\overbrace{4.679 \times 10^5\ J}^{467.9\ kJ}}{558.4\ J \cdot K^{-1}} = 838\ K$

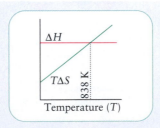

Because the reaction is endothermic, the minimum temperature at which reduction occurs at 1 bar is about 565°C.

SELF-TEST 7.23A   What is the minimum temperature at which magnetite, Fe_3O_4, can be reduced to iron by using carbon (to produce CO_2)?

[*Answer:* 943 K]

SELF-TEST 7.23B Estimate the temperature at which magnesium carbonate can be expected to decompose to magnesium oxide and carbon dioxide.

The Gibbs free energy increases with temperature for reactions with a negative $\Delta S°$ and decreases with temperature for reactions with a positive $\Delta S°$.

7.16 Impact on Biology: Gibbs Free Energy Changes in Biological Systems

Many biological reactions, such as the construction of a protein from amino acids or the construction of a DNA molecule, are not spontaneous and therefore must be driven by an external source of energy. That energy comes from sunlight and the chemicals in food that have stored solar energy. When food is metabolized, the resulting exothermic reaction generates a lot of entropy, and if the reaction is coupled to a biochemical reaction that is nonspontaneous, the *overall* change in entropy may be positive, and the *overall* process spontaneous. In other words, *a reaction that produces a lot of entropy can drive a nonspontaneous reaction forward*. In terms of Gibbs free energy, one biochemical process may be driven uphill in Gibbs free energy by another reaction that rolls downhill. Staying alive is very much like the effect of a heavy weight tied to another weight by a string that passes over a pulley (Fig. 7.29). The lighter weight could never fly up into the air on its own. However, when it is connected to a heavier weight falling downward on the other side of a pulley, the light weight can soar upward.

The hydrolysis of adenosine triphosphate, ATP (**2**), to adenosine diphosphate, ADP (**3**), is the reaction used most frequently by biological organisms to couple with and drive nonspontaneous reactions. This hydrolysis is the key metabolic reaction by which Gibbs free energy is stored and used in living systems. The value of $\Delta G°$ for the hydrolysis of 1 mol ATP is about -30 kJ. To restore ADP back to ATP, which is accompanied by a Gibbs free energy change of $+30$ kJ, an ADP molecule and a phosphate group must be linked by coupling them to another reaction for which the Gibbs free energy of reaction is more negative than -30 kJ. That is one reason why we have to eat. When we eat food containing glucose, we consume a fuel. Like all fuels, it has a spontaneous tendency to form combustion products. If we were simply to burn glucose in an open container, it would do no work other than pushing back the atmosphere, and it would give off a lot of heat. However, in our bodies, the "combustion" is a highly controlled and sophisticated version of burning. In such a controlled reaction, the nonexpansion work that the process can do approaches 2500 kJ per mole of glucose molecules, which is enough to "recharge" about 80 mol ADP molecules.

When living organisms die, they no longer ingest the second-hand sunlight stored in molecules of carbohydrate, protein, and fat. Then the natural direction of change becomes dominant, and the intricate molecules that support life start to decompose. Living organisms are engaged in a constant battle to generate enough

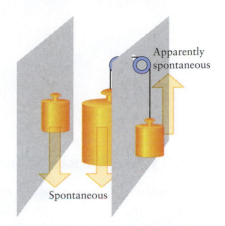

FIGURE 7.29 A natural process can be represented as a falling weight (left). A weight rising spontaneously would be regarded as highly unusual until we recognize that it is actually part of a natural process overall (right). The natural fall of the heavier weight causes the "unnatural" rise of the smaller weight.

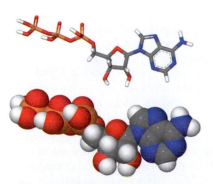

3 Adenosine triphosphate, ATP

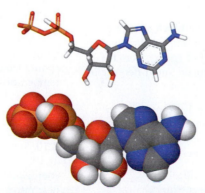

3 Adenosine diphosphate, ADP

entropy in their surroundings to go on building and maintaining their intricate interiors. As soon as they stop the battle, they stop generating that external entropy, and their bodies decay.

Reactions that are nonspontaneous can be made to take place by coupling them with spontaneous reactions. This coupling is used extensively in biological systems.

SKILLS YOU SHOULD HAVE MASTERED

❏ 1 Calculate the entropy change for reversible heat transfer (Example 7.1).

❏ 2 Calculate the entropy change when the temperature of a substance is changed (Example 7.2).

❏ 3 Determine the entropy change for the isothermal expansion or compression of an ideal gas (Examples 7.3, 7.4, and 7.5).

❏ 4 Calculate the standard entropy of a phase change (Example 7.6).

❏ 5 Use the Boltzmann formula to calculate and interpret the entropy of a substance (Examples 7.7 and 7.8).

❏ 6 Predict which of two systems has the greater entropy, given their compositions and conditions (Self-Test 7.11).

❏ 7 Calculate the standard reaction entropy from standard molar entropies (Example 7.9).

❏ 8 Estimate the change in entropy of the surroundings due to heat transfer at constant pressure and temperature (Example 7.10).

❏ 9 Calculate the total change in entropy for a process (Examples 7.11 and 7.12).

❏ 10 Use the change in Gibbs free energy to determine the spontaneity of a process at a given temperature (Example 7.13).

❏ 11 Calculate a standard Gibbs free energy of formation from enthalpy and entropy data (Example 7.14).

❏ 12 Calculate the standard Gibbs free energy of reaction from standard Gibbs free energies of formation (Example 7.15).

❏ 13 Estimate the maximum nonexpansion work that can be done by a process (Section 7.14).

❏ 14 Predict the minimum temperature at which an endothermic reaction can occur spontaneously (Example 7.16).

EXERCISES

Exercises labeled C *require calculus*

Entropy

7.1 A human body generates heat at the rate of about 100. W ($1 \text{ W} = 1 \text{ J·s}^{-1}$). (a) At what rate does your body heat generate entropy in your surroundings, taken to be at 20.°C? (b) How much entropy do you generate each day? (c) Would the entropy generated be greater or less if you were in a room kept at 30.°C? Explain your answer.

7.2 An electric heater is rated at 1.8 kW ($1 \text{ W} = 1 \text{ J·s}^{-1}$). (a) At what rate does it generate entropy in a room maintained at 25°C? (b) How much entropy does it generate in the course of a day? (c) Would the entropy generated be greater or less if the room were maintained at 27°C? Explain your answer.

7.3 (a) Calculate the change in entropy of a block of copper at 25°C that absorbs 65 J of energy from a heater. (b) If the block of copper is at 100.°C and it absorbs 65 J of energy from the heater, what is its entropy change? (c) Explain any difference in entropy change.

7.4 (a) Calculate the change in entropy of 1.0 L of water at 0.°C when it absorbs 470 J of energy from a heater. (b) If the 1.0 L of water is at 99°C, what is its entropy change? (c) Explain any difference in entropy change.

7.5 Assuming that the heat capacity of an ideal gas is independent of temperature, calculate the entropy change associated with raising the temperature of 1.00 mol of ideal gas atoms reversibly from 37.6°C to 157.9°C at (a) constant pressure and (b) constant volume.

7.6 Assuming that the heat capacity of an ideal gas is independent of temperature, calculate the entropy change associated with lowering the temperature of 2.92 mol of ideal gas atoms from 107.35°C to −52.39°C at (a) constant pressure and (b) constant volume.

7.7 Calculate the entropy change associated with the isothermal expansion of 5.25 mol of ideal gas atoms from 24.252 L to 34.058 L.

7.8 Calculate the entropy change associated with the isothermal compression of 6.32 mol of ideal gas atoms from 6.72 atm to 13.44 atm.

7.9 Show that, if two copper blocks with different temperatures are placed in contact, then the direction of spontaneous change is toward the equalization of temperatures. Do so by considering the transfer of 1 J of energy as heat from one to the other and assessing the sign of the entropy change. Assume that the temperatures of the blocks remain constant.

7.10 Suppose that 100. J of energy is taken from a hot source at 300.°C, passes through a turbine that converts some of the energy into work, and then releases the rest of the energy as heat into a cold sink at 20.°C. What is the maximum amount of work that can be produced by this engine if overall it is to operate spontaneously? What is the efficiency of the engine, with work done divided by heat supplied expressed as a percentage? How could the efficiency be increased?

7.11 Use data in Table 6.3 or Appendix 2A to calculate the entropy change for (a) the freezing of 1.00 mol $H_2O(l)$ at 0.00°C; (b) the vaporization of 50.0 g of ethanol, C_2H_5OH, at 351.5 K.

7.12 Use data in Table 6.3 or Table 7.2 to calculate the entropy change for (a) the vaporization of 1.00 mol $H_2O(l)$ at 100.°C and 1 atm; (b) the freezing of 3.33 g of ethanol, C_2H_5OH, at 158.7 K.

7.13 (a) Using data provided in Appendix 2A, estimate the boiling point of ethanal, $CH_3CHO(l)$. (b) From standard reference sources, find the actual boiling point of ethanal. (c) How do these values compare? (d) What gives rise to the differences?

7.14 Using data provided in Appendix 2A, estimate the boiling point of $Br_2(l)$. (b) From standard reference sources, find the actual boiling point of bromine. (c) How do these values compare?

7.15 (a) Using Trouton's rule, estimate the boiling point of dimethyl ether, CH_3OCH_3, given that $\Delta H_{vap}° = 21.51$ kJ·mol^{-1}. (b) Using standard reference sources available in your library, find the actual boiling point of dimethyl ether and compare this value with the value obtained by using Trouton's rule.

7.16 (a) Using Trouton's rule, estimate the boiling point of methylamine, CH_3NH_2, given that $\Delta H_{vap}° = 25.60$ kJ·mol^{-1}. (b) Using standard reference sources available in your library, find the actual boiling point of methylamine and compare this value with the value obtained by using Trouton's rule.

7.17 The standard entropy of vaporization of benzene is approximately 85 J·K^{-1}·mol^{-1} at its boiling point. (a) Estimate the standard enthalpy of vaporization of benzene at its normal boiling point of 80.°C. (b) What is the standard entropy change of the surroundings when 10. g of benzene, C_6H_6, vaporizes at its normal boiling point?

7.18 The standard entropy of vaporization of acetone is approximately 85 J·K^{-1}·mol^{-1} at its boiling point. (a) Estimate the standard enthalpy of vaporization of acetone at its normal boiling point of 56.2°C. (b) What is the entropy change of the surroundings when 10. g of acetone, CH_3COCH_3, condenses at its normal boiling point?

7.19 Which would you expect to have a higher molar entropy at $T = 0$, single crystals of BF_3 or of COF_2? Why?

7.20 On the basis of the structures of each of the following molecules, predict which ones would be most likely to have a residual entropy in their crystal forms at $T = 0$: (a) CO_2; (b) NO; (c) N_2O; (d) Cl_2.

7.21 If SO_2F_2 adopts a disordered arrangement in its crystal form, what would its residual molar entropy be?

7.22 What would you expect the residual molar entropy of PH_2F to be if it were to adopt a disordered arrangement in its crystal form?

7.23 Which substance in each of the following pairs has the higher molar entropy at 298 K: (a) HBr(g) or HF(g); (b) NH_3(g) or Ne(g); (c) I_2(s) or I_2(l); (d) 1.0 mol Ar(g) at 1.00 atm or 1.0 mol Ar(g) at 2.00 atm?

7.24 Which substance in each of the following pairs has the higher molar entropy? (Assume the temperature to be 298 K unless otherwise specified.) (a) CH_4(g) or C_2H_6(g); (b) KCl(aq) or KCl(s); (c) Ne(g) or Kr(g); (d) O_2(g) at 273 K and 1.00 atm or O_2(g) at 450. K and 1.00 atm.

7.25 List the following substances in order of increasing molar entropy at 298 K: $H_2O(l)$, $H_2O(g)$, $H_2O(s)$, C(s, diamond). Explain your reasoning.

7.26 List the following substances in order of increasing molar entropy at 298 K: CO_2(g), Ar(g), $H_2O(l)$, Ne(g). Explain your reasoning.

7.27 Which substance in each of the following pairs would you expect to have the higher standard molar entropy at 298 K? Explain your reasoning. (a) Iodine vapor or bromine vapor; (b) the two liquids cyclopentane and 1-pentene (see structures);

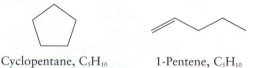

Cyclopentane, C_5H_{10} 1-Pentene, C_5H_{10}

(c) ethene (ethylene) or an equivalent mass of polyethylene, a substance formed by the polymerization of ethylene.

7.28 Predict which of the hydrocarbons below has the greater standard molar entropy at 25°C. Explain your reasoning.

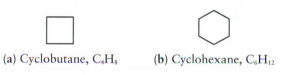

(a) Cyclobutane, C_4H_8 **(b)** Cyclohexane, C_6H_{12}

7.29 Predict which of the organic compounds (a) dimethyl ether, $(CH_3)_2O$, or (b) diethyl ether $(CH_3CH_2)_2O$, has the greater standard molar entropy at 25°C. Explain your reasoning.

7.30 Explain why the standard molar entropy of liquid benzene is less than that of liquid cyclohexane.

7.31 Without performing any calculations, predict whether there is an increase or a decrease in entropy for each of the following processes: (a) Cl_2(g) + $H_2O(l)$ → HCl(aq) + HClO(aq); (b) $Cu_3(PO_4)_2$(s) → 3 Cu^{2+}(aq) + 2 PO_4^{3-}(aq); (c) SO_2(g) + Br_2(g) + 2 $H_2O(l)$ → H_2SO_4(aq) + 2 HBr(aq).

7.32 Without performing any calculations, state whether the entropy of the system increases or decreases in each of the following processes: (a) the dissolution of table salt: NaCl(s) → NaCl(aq); (b) the photosynthesis of glucose: 6 CO_2(g) + 6 $H_2O(l)$ → $C_6H_{12}O_6$(s) + 6 O_2(g); (c) the evaporation of water from damp clothes. Explain your reasoning.

7.33 Container A is filled with 1.0 mol of the atoms of an ideal monatomic gas. Container B has 1.0 mol of atoms bound together as diatomic molecules that are not vibrationally active. Container C has 1.0 mol of atoms bound together as diatomic molecules that are vibrationally active. The containers all start at T_i and the temperature is increased to T_f. Rank the containers in order of increasing change in entropy. Explain your reasoning.

7.34 A closed vessel of volume 2.5 L contains a mixture of neon and fluorine. The total pressure is 3.32 atm at 0.0°C. When the mixture is heated to 15°C, the entropy of the mixture increases by 0.345 J·K^{-1}. What amount (in moles) of each substance (Ne and F$_2$) is present in the mixture?

7.35 Use data in Table 7.3 or Appendix 2A to calculate the standard reaction entropy for each of the following reactions at 25°C. For each reaction, interpret the sign and magnitude of the reaction entropy. (a) The formation of 1.00 mol H$_2$O(l) from the elements in their most stable state at 298 K. (b) The oxidation of 1.00 mol CO(g) to carbon dioxide. (c) The decomposition of 1.00 mol calcite, CaCO$_3$(s), to carbon dioxide gas and solid calcium oxide. (d) The decomposition of potassium chlorate: 4 KClO$_3$(s) → 3 KClO$_4$(s) + KCl(s).

7.36 Use data in Table 7.3 or Appendix 2A to calculate the standard entropy change for each of the following reactions at 25°C. For each reaction, interpret the sign and magnitude of the reaction entropy. (a) The synthesis of carbon disulfide from natural gas (methane): CH$_4$(g) + 4 S(s, rhombic) → CS$_2$(l) + 2 H$_2$S(g). (b) The production of acetylene from calcium carbide and water: CaC$_2$(s) + 2 H$_2$O(l) → C$_2$H$_2$(g) + Ca(OH)$_2$(s). (c) The oxidation of ammonia, which is the first step in the commercial production of nitric acid: 4 NH$_3$(g) + 5 O$_2$(g) → 4 NO(g) + 6 H$_2$O(l). (d) The industrial synthesis of urea, a common fertilizer: CO$_2$(g) + 2 NH$_3$(g) → CO(NH$_2$)$_2$(s) + H$_2$O(l).

C **7.37** The temperature dependence of the heat capacity of a substance is commonly written in the form $C_{P,m} = a + bT + c/T^2$, with a, b, and c constants. Obtain an expression for the entropy change when the substance is heated from T_1 to T_2. Evaluate this change for graphite, for which $a = 16.86$ J·K^{-1}·mol^{-1}, $b = 4.77$ mJ·K^{-2}·mol^{-1}, and $c = -8.54 \times 10^5$ J·K·mol^{-1} heated from 298 K to 400. K. What is the percentage error in assuming that the heat capacity is constant with its mean value in this range?

C **7.38** At low temperatures, heat capacities are proportional to T^3. Show that, near $T = 0$, the entropy of a substance is equal to one-third of its heat capacity at the same temperature.

7.39 Calculate the standard entropy of vaporization of water at 85°C, given that its standard entropy of vaporization at 100.°C is 109.0 J·K^{-1}·mol^{-1} and the molar heat capacities at constant pressure of liquid water and water vapor are 75.3 J·K^{-1}·mol^{-1} and 33.6 J·K^{-1}·mol^{-1}, respectively, in this range.

7.40 Three liquid samples with known masses are heated to their boiling points with the use of a 500.-W heater. Once their boiling points are reached, heating continues for 4.0 min and some of each sample is vaporized. After 4.0 min, the samples are cooled and the masses of the remaining liquids are determined. The process is performed at constant pressure. Using the following data, (a) calculate ΔS_{vap} and ΔH_{vap} for each sample. Assume that all the heat from the heater goes into the sample. (b) What do the values of ΔS_{vap} indicate about the relative degree of order in the liquids?

Liquid	Boiling temperature (°C)	Initial mass (g)	Final mass (g)
C$_2$H$_5$OH	78.3	400.15	271.15
C$_4$H$_{10}$	0.0	398.05	74.95
CH$_3$OH	64.5	395.15	294.25

Global Changes in Entropy

7.41 Suppose that 50.0 g of H$_2$O(l) at 20.0°C is mixed with 65.0 g of H$_2$O(l) at 50.0°C at constant atmospheric pressure in a thermally insulated vessel. Calculate ΔS and ΔS_{tot} for the process.

7.42 Suppose that 150.0 g of ethanol at 22.0°C is mixed with 200.0 g of ethanol at 56.0°C at constant atmospheric pressure in a thermally insulated vessel. Calculate ΔS and ΔS_{tot} for the process.

7.43 Use the information in Table 6.3 to calculate the changes in entropy of the surroundings and of the system for (a) the vaporization of 1.00 mol CH$_4$(l) at its normal boiling point; (b) the melting of 1.00 mol C$_2$H$_5$OH(s) at its normal melting point; (c) the freezing of 1.00 mol C$_2$H$_5$OH(l) at its normal freezing point.

7.44 Use the information in Table 6.3 to calculate the changes in entropy of the surroundings and of the system for (a) the melting of 1.00 mol NH$_3$(s) at its normal melting point; (b) the freezing of 1.00 mol CH$_3$OH(l) at its normal freezing point; (c) the vaporization of 1.00 mol H$_2$O(l) at its normal boiling point.

7.45 Initially a sample of ideal gas at 323 K occupies 1.67 L at 4.95 atm. The gas is allowed to expand to 7.33 L by two pathways: (a) isothermal, reversible expansion; (b) isothermal, irreversible free expansion. Calculate ΔS_{tot}, ΔS, and ΔS_{surr} for each pathway.

7.46 Initially a sample of ideal gas at 412 K occupies 12.62 L at 0.6789 atm. The gas is allowed to expand to 19.44 L by two pathways: (a) isothermal, reversible expansion and (b) isothermal, irreversible free expansion. Calculate ΔS_{tot}, ΔS, and ΔS_{surr} for each pathway.

7.47 The following pictures show a molecular visualization of a system undergoing a spontaneous change. Account for the spontaneity of the process in terms of the entropy changes in the system and the surroundings.

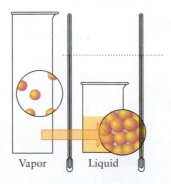

Vapor   Liquid

7.48 The following pictures show a molecular visualization of a system undergoing a spontaneous change. Determine from the picture whether entropy increases or decreases during the process. Account for the spontaneity of the process in terms of the entropy changes in the system and the surroundings.

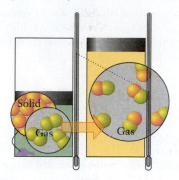

Gibbs Free Energy

7.49 Why are so many exothermic reactions spontaneous?

7.50 Explain how an endothermic reaction can be spontaneous.

7.51 Calculate the standard enthalpy, entropy, and Gibbs free energy at 298 K for each of the following reactions by using data in Appendix 2A. For each case, confirm that the value obtained from the Gibbs free energies of formation is the same as that obtained by using the relation $\Delta G° = \Delta H° - T\Delta S°$.
(a) the oxidation of magnetite to hematite:
$2\ Fe_3O_4(s) + \frac{1}{2}O_2(g) \rightarrow 3\ Fe_2O_3(s)$
(b) the dissolution of CaF_2 in water: $CaF_2(s) \rightarrow CaF_2(aq)$
(c) the dimerization of NO_2: $2\ NO_2(g) \rightarrow N_2O_4(g)$

7.52 Calculate the standard enthalpy, entropy, and Gibbs free energy for each of the following reactions at 298 K by using data in Appendix 2A. For each case, confirm that the value obtained from the Gibbs free energies of formation is the same as that obtained by using the relation $\Delta G° = \Delta H° - T\Delta S°$.
(a) the hydration of copper sulfate:
$CuSO_4(s) + 5\ H_2O(l) \rightarrow CuSO_4\cdot 5H_2O(s)$
(b) the reaction of $H_2SO_4(l)$, an effective dehydrating agent, with water to form aqueous sulfuric acid:
$H_2SO_4(l) \rightarrow H_2SO_4(aq)$
(c) the reaction of calcium oxide with water to form calcium hydroxide: $CaO(s) + H_2O(l) \rightarrow Ca(OH)_2(s)$
(d) Which of the reagents in parts (a)–(c) would you expect to be the most effective at removing water from a substance?

7.53 Write a balanced chemical equation for the formation reaction of (a) $NH_3(g)$; (b) $H_2O(g)$; (c) $CO(g)$; (d) $NO_2(g)$. For each reaction, determine $\Delta H°$, $\Delta S°$, and $\Delta G°$ from data in Appendix 2A.

7.54 Write a balanced chemical equation for the formation reaction of (a) $HCl(g)$; (b) $C_6H_6(l)$; (c) $CuSO_4\cdot 5H_2O(s)$; (d) $CaCO_3(s$, calcite). For each reaction, determine $\Delta H°$, $\Delta S°$, and $\Delta G°$ from data in Appendix 2A.

7.55 Use the standard Gibbs free energies of formation in Appendix 2A to calculate $\Delta G°$ for each of the following reactions at 25°C. Comment on the spontaneity of each reaction under standard conditions at 25°C.
(a) $2\ SO_2(g) + O_2(g) \rightarrow 2\ SO_3(g)$
(b) $CaCO_3(s$, calcite$) \rightarrow CaO(s) + CO_2(g)$
(c) $2\ C_8H_{18}(l) + 25\ O_2(g) \rightarrow 16\ CO_2(g) + 18\ H_2O(l)$

7.56 Use the standard Gibbs free energies of formation in Appendix 2A to calculate $\Delta G°$ for each of the following reactions at 25°C. Comment on the spontaneity of each reaction under standard conditions at 25°C.
(a) $NH_4Cl(s) \rightarrow NH_3(g) + HCl(g)$
(b) $H_2(g) + D_2O(l) \rightarrow D_2(g) + H_2O(l)$
(c) $N_2(g) + NO_2(g) \rightarrow NO(g) + N_2O(g)$
(d) $2\ CH_3OH(g) + 3\ O_2(g) \rightarrow 2\ CO_2(g) + 4\ H_2O(l)$

7.57 Determine which of the following compounds are stable with respect to decomposition into their elements under standard conditions at 25°C (see Appendix 2A): (a) $PCl_5(g)$; (b) $HCN(g)$; (c) $NO(g)$; (d) $SO_2(g)$.

7.58 Determine which of the following compounds are stable with respect to decomposition into their elements under standard conditions at 25°C (see Appendix 2A): (a) $CuO(s)$; (b) $C_6H_{12}(l)$, cyclohexane; (c) $PCl_3(g)$; (d) $N_2H_4(l)$.

7.59 Which of the following compounds become less stable with respect to the elements as the temperature is raised: (a) $PCl_5(g)$; (b) $HCN(g)$; (c) $NO(g)$; (d) $SO_2(g)$?

7.60 Which of the following compounds become less stable with respect to their elements as the temperature is raised: (a) $CuO(s)$; (b) $C_6H_{12}(l)$, cyclohexane; (c) $PCl_3(g)$; (d) $N_2H_4(l)$?

7.61 Calculate the standard reaction entropy, enthalpy, and Gibbs free energy for each of the following reactions from data found in Appendix 2A:
(a) the decomposition of hydrogen peroxide:
$2\ H_2O_2(l) \rightarrow 2\ H_2O(l) + O_2(g)$
(b) the preparation of hydrofluoric acid from fluorine and water: $2\ F_2(g) + 2\ H_2O(l) \rightarrow 4\ HF(aq) + O_2(g)$

7.62 Calculate the standard reaction entropy, enthalpy, and Gibbs free energy for each of the following reactions from data in Appendix 2A:
(a) the production of "synthesis gas," a low-grade industrial fuel: $CH_4(g) + H_2O(g) \rightarrow CO(g) + 3\ H_2(g)$
(b) the thermal decomposition of ammonium nitrate:
$NH_4NO_3(s) \rightarrow N_2O(g) + 2\ H_2O(g)$

7.63 Assume that $\Delta H°$ and $\Delta S°$ are independent of temperature and use data in Appendix 2A to calculate $\Delta G°$ for each of the following reactions at 80.°C. Over what temperature range will each reaction be spontaneous under standard conditions?
(a) $B_2O_3(s) + 6\ HF(g) \rightarrow 2\ BF_3(g) + 3\ H_2O(l)$
(b) $CaC_2(s) + 2\ HCl(aq) \rightarrow CaCl_2(aq) + C_2H_2(g)$
(c) $C(s$, graphite$) \rightarrow C(s$, diamond$)$

7.64 Assume that $\Delta H°$ and $\Delta S°$ are independent of temperature and use data in Appendix 2A to calculate $\Delta G°$ for each of the following reactions at 250.°C. Over what temperature range will each reaction be spontaneous under standard conditions?
(a) $HCN(g) + 2\ H_2(g) \rightarrow CH_3NH_2(g)$
(b) $2\ Cu^+(aq) \rightarrow Cu(s) + Cu^{2+}(aq)$
(c) $CaCl_2(s) + 2\ HF(g) \rightarrow CaF_2(s) + 2\ HCl(g)$

7.65 A scientist proposed the following two reactions to produce ethanol, a liquid fuel:

$C_2H_4(g) + H_2O(g) \longrightarrow CH_3CH_2OH(l)$ (A)

$C_2H_6(g) + H_2O(g) \longrightarrow CH_3CH_2OH(l) + H_2(g)$ (B)

Reaction B is preferred if spontaneous, because $C_2H_6(g)$ is a cheaper starting material than $C_2H_4(g)$. Assume standard-state conditions and determine if either reaction is thermodynamically spontaneous.

7.66 A rocket fuel would be useless if its oxidation were not spontaneous. Although rockets operate under conditions that are far from standard, an initial estimation of the potential of a rocket fuel might assess whether its oxidation at the high temperatures reached in a rocket is spontaneous. A chemist exploring potential fuels for use in space considered using vaporized aluminum chloride in a reaction for which the skeletal equation is

$$AlCl_3(g) + O_2(g) \longrightarrow Al_2O_3(s) + ClO(g)$$

Balance this equation. Then use the following data (which are for 2000 K) to decide whether the fuel is worth further investigation: $\Delta G_f°(AlCl_3, g) = -467$ kJ·mol^{-1}, $\Delta G_f°(Al_2O_3, s) = -1034$ kJ·mol^{-1}, $\Delta G_f°(ClO, g) = +75$ kJ·mol^{-1}.

Impact on Biology

7.67 Coupled reactions are used in organisms to drive important biochemical processes. Individual chemical reactions may be added together to form a net reaction, and the overall reaction Gibbs free energy is the sum of the individual Gibbs free energies of reaction. For example, ATP is the primary molecule that stores and releases energy to drive vital nonspontaneous chemical reactions in our bodies. However, the generation of ATP from ADP is nonspontaneous and must itself be coupled to a spontaneous reaction. This process is called *oxidative phosphorylation* and the reactions are:

(1) $ADP^{3-}(aq) + HPO_4^{2-}(aq) + H^+(aq) \longrightarrow$
$\qquad ATP^{4-}(aq) + H_2O(l) \qquad \Delta G = +30.5$ kJ

(2) $NADH(aq) \longrightarrow$
$\qquad NAD^+(aq) + H^+(aq) + 2\ e^- \qquad \Delta G = -158.3$ kJ

(3) $\frac{1}{2}\ O_2(g) + 2\ H^+(aq) + 2\ e^- \longrightarrow H_2O(l) \qquad \Delta G = -61.9$ kJ

(The reaction Gibbs free energies are for pH = 7 but otherwise standard conditions.) What amount (in moles) of ATP could be formed if all the Gibbs free energy released in the oxidation of 3.00 mol NADH were used to generate ATP?

7.68 For the hydrolysis of acetyl phosphate under the conditions prevailing in the body, $\Delta G = -41$ kJ·mol^{-1}. If the phosphorylation of acetic acid (the reverse of the hydrolysis of acetyl phosphate) was driven by coupling to the hydrolysis of ATP at pH = 7, what is the minimum amount of ATP molecules (in moles) that would have to be hydrolyzed to form 1.0 mol acetyl phosphate molecules by the phosphorylation of acetic acid? (See Exercise 7.67 for more details.)

Integrated Exercises

7.69 (a) Which substance would you expect to have the highest molar entropy in the liquid phase, benzene, methanol (CH_3OH), or 1-propanol ($CH_3CH_2CH_2OH$)? (b) Would your answer be different if these substances were in the gas phase?

7.70 (a) Before checking the numbers, which substance would you expect to have the higher standard molar entropy, $CH_3COOH(l)$ or $CH_3COOH(aq)$? Having made this prediction, examine the values in Appendix 2A and explain your findings.

7.71 Under what conditions, if any, does the sign of each of the following quantities provide a criterion for assessing the spontaneity of a reaction? (a) $\Delta G°$; (b) $\Delta H°$; (c) $\Delta S°$; (d) ΔS_{tot}.

7.72 The normal boiling point of methanol is lower than predicted by Trouton's rule. Provide a molecular interpretation for this observation.

7.73 Assuming statistical disorder, would you expect a crystal of octahedral *cis*-MX_2Y_4 to have the same, higher, or lower residual entropy than the corresponding trans isomer? Explain your conclusion.

cis-MX_2Y_4 *trans*-MX_2Y_4

7.74 There are three different substituted benzene compounds with the formula $C_6H_4F_2$. Assume that the benzene rings pack similarly into their crystal lattices. If the positions of the H and F atoms are statistically disordered in the solid state, which isomer will have the *least* residual molar entropy?

7.75 Suppose that you create two tiny systems consisting of three atoms each, and each atom can accept energy in quanta of the same magnitude. (a) How many distinguishable arrangements are there of two quanta of energy distributed among the three atoms in one of these systems? (b) You now bring the two tiny systems together. How many distinguishable arrangements are there if the two quanta of energy are distributed among the six atoms? (c) In what direction will the energy quanta flow starting from the initial arrangement?

7.76 (a) Which phase of water has the greater standard molar entropy at 0°C, liquid or solid? (b) Which phase of water has the greater molar volume at 0°C, liquid or solid? (c) Are your answers to (a) and (b) consistent? Explain your answer.

7.77 Consider the enthalpies of fusion and the melting points of the following elements: Pb, 5.10 kJ·mol^{-1}, 327°C; Hg, 2.29 kJ·mol^{-1}, −39°C; Na, 2.64 kJ·mol^{-1}, 98°C. Given these data, determine whether a relation similar to Trouton's rule can be obtained for the entropy of fusion of the metallic elements.

7.78 Determine whether titanium dioxide can be reduced by carbon at 1000. K in each of the following reactions:
(a) $TiO_2(s) + 2\ C(s) \longrightarrow Ti(s) + 2\ CO(g)$
(b) $TiO_2(s) + C(s) \longrightarrow Ti(s) + CO_2(g)$

given that, at 1000. K, $\Delta G_f°(CO, g) = -200.$ kJ·mol^{-1}, $\Delta G_f°(CO_2, g) = -396$ kJ·mol^{-1}; and $\Delta G_f°(TiO_2, s) = -762$ kJ·mol^{-1}.

7.79 Which is the thermodynamically more stable iron oxide in air, $Fe_3O_4(s)$ or $Fe_2O_3(s)$? Justify your selection.

7.80 (a) Calculate $\Delta H°$ and $\Delta S°$ at 25°C for the reaction $C_2H_2(g) + 2\ H_2(g) \rightarrow C_2H_6(g)$. (b) Calculate the standard Gibbs free energy of reaction from the equation $\Delta G° = \Delta H° - T\Delta S°$. (c) Interpret the calculated values for $\Delta H°$ and $\Delta S°$. (d) Assuming that $\Delta H°$ and $\Delta S°$ are unaffected by temperature changes, determine the temperature at which $\Delta G° = 0$. (e) What is the significance of this temperature?

7.81 (a) Would you expect pure FeS(s) to convert into Fe(s) and $FeS_2(s)$ if heated? (b) Would you expect pure $FeS_2(s)$ to convert into FeS(s) and S(s) if heated?

7.82 (a) Calculate the work that must be done at 298.15 K and 1.00 bar against the atmosphere for the production of $CO_2(g)$

and $H_2O(g)$ in the combustion of 8.50 mol $C_6H_6(l)$. Using data in Appendix 2A, calculate the standard enthalpy of the reaction. (c) Calculate the change in internal energy, $\Delta U°$, of the system.

7.83 Hydrogen burns in an atmosphere of bromine gas to give hydrogen bromide gas. (a) What is the standard Gibbs free energy of the reaction at 298 K? (b) If 120. mL of H_2 gas at STP combines with a stoichiometric amount of bromine and the resulting hydrogen bromide dissolves to form 150. mL of an aqueous solution, what is the molar concentration of the resulting hydrobromic acid?

7.84 A group of chemists are studying the neutralization reaction $2 HCl(aq) + Ba(OH)_2(aq) \rightarrow BaCl_2(aq) + 2 H_2O(l)$. They mix 300.0 mL of 1.500 M HCl(aq) and 200.0 mL of 1.400 M $Ba(OH)_2(aq)$ in a flask immersed in a water bath kept at 25°C. (a) Use the data in Appendix 2A to determine the entropy change in the system. (b) What is the entropy change in the surroundings?

7.85 Potassium nitrate dissolves readily in water, and its enthalpy of solution is $+34.9$ kJ·mol^{-1}. (a) Does the enthalpy of solution favor the dissolving process? (b) Is the entropy change of the system likely to be positive or negative when the salt dissolves? (c) Is the entropy change of the system primarily a result of changes in positional disorder or thermal disorder? (d) Is the entropy change of the surroundings primarily a result of changes in positional disorder or thermal disorder? (e) What is the driving force for the dissolution of KNO_3?

7.86 Explain why each of the following statements is false. (a) Reactions with negative Gibbs free energies of reaction occur spontaneously and rapidly. (b) Every sample of a pure element, regardless of its physical state, is assigned zero Gibbs free energy of formation. (c) An exothermic reaction producing more moles of gas molecules than are consumed has a positive standard reaction Gibbs free energy.

7.87 A common antiseptic used in first aid for cuts and scrapes is a 3% aqueous solution of hydrogen peroxide. The oxygen that bubbles out of the hydrogen peroxide as it is decomposed by enzymes in blood into oxygen and water helps to clean the wound. Two possible industrial routes for the synthesis of hydrogen peroxide are: (i) $H_2(g) + O_2(g) \xrightarrow{catalyst} H_2O_2(l)$, which uses either a metal such as palladium or the organic compound quinone as catalyst, and (ii) $2 H_2O(l) + O_2(g) \rightarrow 2 H_2O_2(l)$. (a) Which method releases more energy per mole of O_2? (b) Which method has the more negative standard Gibbs free energy? (c) Once hydrogen peroxide has been used at home, can it be regenerated easily from oxygen and water?

7.88 According to current theories of biological evolution, complex amino and nucleic acids were produced from randomly occurring reactions of compounds thought to be present in the Earth's early atmosphere. These simple molecules then assembled into more and more complex molecules, such as DNA and RNA. Is this process consistent with the second law of thermodynamics? Explain your answer.

7.89 Some entries for $S_m°$ in Appendix 2A are negative. What is common about these entries and why would the entropy be negative?

7.90 Acetic acid, $CH_3COOH(l)$, could be produced from (a) the reaction of methanol with carbon monoxide; (b) the oxidation of ethanol; (c) the reaction of carbon dioxide with methane. Write balanced chemical equations for each process. Carry out a thermodynamic analysis of the three possibilities and decide which you would expect to be the easiest to accomplish.

7.91 The enthalpy change accompanying the combustion of 0.50 g of $H_2(g)$ was determined to be -70.9 kJ at 25°C. The standard Gibbs free energy of formation of $H_2O(l)$ is -237.25 kJ·mol^{-1}. What is the reaction entropy for the combustion of this sample at 25°C?

7.92 (a) Write a balanced equation for the production of 1.00 mol $NH_3(g)$ from nitrogen and hydrogen. (b) Calculate the standard enthalpy and entropy changes of this reaction at 298 K. (c) Calculate the standard free-energy change of this reaction at 150. K, 298 K, and 350. K, assuming that $\Delta H°$ and $\Delta S°$ are independent of temperature. (d) The synthesis of ammonia at room temperature is slow. Is raising the temperature to speed up the reaction a good idea? Explain your conclusion.

7.93 Using values in Appendix 2A, calculate the standard Gibbs free energy for the vaporization of water at 25.0°C, 100.0°C, and 150.0°C. (b) What should the value at 100.0°C be? (c) Why is there a discrepancy?

7.94 Propose the argument that, for any liquid at atmospheric pressure (that is, a liquid that boils above room temperature when the external pressure is 1 atm), the numerical value of ΔH_{vap} in joules per mole is greater than the numerical value of ΔS_{vap} in joules per kelvin per mole. (Explain and justify each step and any assumptions.)

7.95 It is helpful in understanding graphs of thermodynamic functions to interpret them in terms of molecular behavior. Consider the plot of the temperature dependence of the standard molar Gibbs free energy of the three phases of a substance in Fig. 7.25. (a) Explain in terms of molecular behavior why the Gibbs free energy of each phase decreases with temperature. (b) Explain in terms of molecular behavior why the Gibbs free energy of the vapor phase decreases more rapidly with temperature than that of the solid or liquid phase.

7.96 Because state functions depend only on the state of the system, when a system undergoes a series of processes that bring it back to the original state, a thermodynamic cycle has been carried out and all the state functions have returned to their original value. However, functions that depend on the path may have changed. (a) Verify that there is no difference in the state function S, for a process in which 1.00 mol of ideal gas molecules in a 3.00 L cylinder at 302 K undergoes the following three steps: (i) cooling at constant volume until $T = 75.6$ K; (ii) heating at constant pressure until $T = 302$ K. (iii) compressing at constant temperature until $V = 3.00$ L. Calculate ΔU and ΔS for this entire cycle. (b) What are the values of q and w for the entire cycle? (c) What are ΔS_{surr} and ΔS_{total} for the cycle? If any values are nonzero, explain how this can be so, despite entropy being a state function. (d) Is the process spontaneous, nonspontaneous, or at equilibrium?

7.97 A technique used to overcome the unfavorable thermodynamics of one reaction is to "couple" that reaction with another process that is thermodynamically favored. For instance, the dehydrogenation of cyclohexane to form benzene and hydrogen gas is not spontaneous. Show that, if another molecule such as ethene is present to act as a hydrogen acceptor (that is, the ethene reacts with the hydrogen produced to form ethane), then the process can be made spontaneous.

7.98 Adenosine triphosphate (ATP) is an extremely important molecule in biological systems. Consult standard reference sources in your library to describe how this molecule is used in energy transfer to facilitate nonspontaneous processes necessary for life.

7.99 Three isomeric alkenes have the formula C_4H_8 (see the following table). (a) Draw Lewis structures of these compounds. (b) Calculate $\Delta G°$, $\Delta H°$, and $\Delta S°$ for the three reactions that interconvert each pair of compounds. (c) Which isomer is the most stable? (d) Rank the isomers in order of decreasing $S_m°$.

Compound	$\Delta H_f°$ (kJ·mol^{-1})	$\Delta G_f°$ (kJ·mol^{-1})
2-methylpropene	−16.90	+58.07
cis-2-butene	−6.99	+65.86
trans-2-butene	−11.17	+62.97

7.100 When a space shuttle re-enters the atmosphere, its temperature rises and it can heat the surrounding air to 1260°C. At that temperature some endothermic reactions take place that at lower temperatures are not spontaneous under standard conditions (pure substances, each at 1 bar). Estimate the temperatures at which the following stratospheric reactions become spontaneous under standard conditions. Which ones could occur during re-entry of the space shuttle? (a) The formation of nitrogen monoxide from nitrogen and oxygen; (b) the formation of ozone from molecular oxygen. *A word of warning*: Because at this stage we have to assume that $\Delta H°$ and $\Delta S°$ are constant over the temperature range and that the conditions are standard (but see Section 7.15), this calculation gives only a very approximate estimate of the temperature at which such reactions are significant in practice.

Chemistry Connections

7.101 Vehicle air bags protect passengers by allowing a chemical reaction to occur that generates gas rapidly. Such a reaction must be both spontaneous and explosively fast. A common reaction is the decomposition of sodium azide, NaN_3, to nitrogen gas and sodium metal.
(a) Write a balanced chemical equation for this reaction using the smallest whole-number coefficients. (b) Predict the sign of the entropy change for this reaction without doing a calculation. Explain your reasoning.
(c) Determine the oxidation number of nitrogen in the azide ion and in nitrogen gas. Is nitrogen oxidized or reduced in the reaction?
(d) Use the data in Appendix 2A and the fact that, for sodium azide, $S_m° = 96.9$ J·K^{-1}·mol^{-1}, to calculate $\Delta S°$ at 298 K for the decomposition of sodium azide.
(e) Use your result from part (d) and the fact that, for sodium azide, $\Delta H_f° = +21.7$ kJ·mol^{-1}, to calculate $\Delta G°$ at 298 K for the decomposition of sodium azide.
(f) Is the reaction spontaneous under standard conditions at 298 K?
(g) Can the reaction become nonspontaneous (under standard conditions) if the temperature is changed? If so, must the temperature be raised or lowered?

PHYSICAL EQUILIBRIA

What Are the Key Ideas? Equilibrium between two phases is reached when the rates of conversion between the two phases are the same in each direction. The rates are equal when the molar Gibbs free energy of the substance is the same in each phase and therefore there is no tendency to change in either direction. The same concepts apply to the dissolving of a solute. The presence of a solute alters the entropy of a solvent and consequently affects its thermodynamic properties.

Why Do We Need to Know This Material? In earlier chapters, we investigated the nature of the solid, liquid, and gaseous states of matter; in this chapter, we extend the discussion to transformations between these states. The discussion introduces the concept of equilibrium between different phases of a substance, a concept that will prove to be of the greatest importance for chemical and biochemical transformations. We also take a deeper look at solutions in this chapter. We shall see how the presence of solutes is used by the body to control the flow of nutrients into and out of living cells and how the properties of solutions are used by oil companies to separate the components of petroleum.

What Do We Need to Know Already? This chapter develops the concepts of physical equilibria introduced in the context of thermodynamics (Chapters 6 and 7) and assumes a knowledge of intermolecular forces (Sections 5.1–5.5). The compositions of some of the solutions discussed are expressed in terms of mole fraction (Section 4.8).

Chapter **8**

Nearly all the gases, liquids, and solids we encounter in our daily life are mixtures or solutions. In Section 4.8 we considered gaseous mixtures. In this chapter we investigate liquid solutions and how the solutes in them affect their physical properties. The largest solutions on Earth are the oceans, which account for 1.4×10^{21} kg of the Earth's surface water. That mass amounts to nearly 2×10^{18} t (1 t = 10^3 kg) of water for each inhabitant, more than enough to supply all the drinking water needed for a thirsty world. However, drinking seawater itself would be fatal, because of the high concentrations of dissolved salts, in particular Na^+ and Cl^- ions (Table 8.1). The transformation of seawater into potable water is a major challenge around the world. The material in this chapter suggests how this purification can be achieved.

Another solution of great importance to our lives is the plasma that carries the erythrocytes (red blood cells) through our bodies. These cells can cease to function if the concentrations of solutes in the plasma drift from their optimal values. This chapter explains why such changes occur and guides us through building a model that accounts for the behavior of solutions. We begin by examining phase changes in pure substances, then we see how the presence of a solute affects the properties of the substance.

PHASES AND PHASE TRANSITIONS

As we saw in Section 5.1, a single substance can exist in different *phases*, or physical forms. The phases of a substance include its solid, liquid, and gaseous forms and its different solid forms, such as the diamond and graphite phases of carbon. In one case—helium—two liquid phases are known to exist. The conversion of a substance from one phase into another, such as the melting of ice, the vaporization of water, and the conversion of graphite into diamond, is called a **phase transition** (recall Section 6.11).

The unifying concept of this chapter is *equilibrium*. For a given pressure, a substance undergoes a phase transition at a specific temperature, such as the

TABLE 8.1 Principal Ions Found in Seawater

Element	Principal form	Concentration $(g \cdot L^{-1})$
Cl	Cl^-	19.0
Na	Na^+	10.5
Mg	Mg^{2+}	1.35
S	SO_4^{2-}	0.89
Ca	Ca^{2+}	0.40
K	K^+	0.38
Br	Br^-	0.065
C	HCO_3^-, H_2CO_3, CO_2	0.028

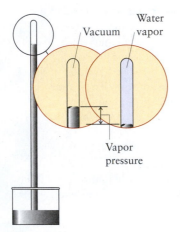

FIGURE 8.1 The apparatus is a mercury barometer. The closeup on the left shows the vacuum above the mercury column. In the right-hand closeup, we see the effect of the addition of a small amount of water. At equilibrium, some of the water has evaporated and the vapor pressure exerted by the water on the mercury has lowered the height of the mercury column. The vapor pressure is the same however much water is present in the column.

TABLE 8.2 Vapor Pressures at 25°C

Substance	Vapor pressure (Torr)
benzene	94.6
ethanol	58.9
mercury	0.0017
methanol	122.7
water*	23.8

* For values at other temperatures, see Table 8.3.

change from solid to liquid at the melting temperature. At this characteristic temperature, the two phases involved in the transformation are in equilibrium. For example, the solid and liquid phases of a substance are in equilibrium at its melting point. To treat these transitions quantitatively, we make use of the fact that at equilibrium there is no difference in Gibbs free energy between the two phases: that is, $\Delta G = 0$, where ΔG is the change in Gibbs free energy when a given amount of one phase changes into the second phase. This chapter explores the consequences of this thermodynamic condition for physical equilibria. In Chapter 9, we shall see how these same ideas can be applied to chemical equilibria.

8.1 Vapor Pressure

A simple experiment can be used to show that, in a closed vessel, the liquid and vapor phases of a substance reach equilibrium with each other. First, we set up a mercury barometer. As we saw in Section 4.2, the mercury inside the tube falls to a height proportional to the external atmospheric pressure, leaving it about 76 cm high at sea level on Earth. The space above the mercury is almost a vacuum (the trace of mercury vapor present is so small that we can ignore it). Now suppose we inject a tiny drop of water into the space above the mercury. All the water added immediately evaporates and fills the space with vapor. The molecules in this vapor push the surface of the mercury down a few millimeters. The pressure exerted by the vapor—as measured by the change in the height of mercury—depends on the amount of water added. However, suppose we add so much water that a little liquid water remains on the surface of the mercury. The pressure of the vapor now remains constant, no matter how much liquid water is present (Fig. 8.1). We conclude that, *at a fixed temperature, as long as some liquid is present, the vapor exerts a characteristic pressure regardless of the amount of liquid water present.* For example, at 20°C, the mercury falls 18 mm, and so the pressure exerted by the water vapor is 18 Torr. The pressure of the water vapor is the same whether there is 0.1 mL or 1 mL of liquid water present. This characteristic pressure is the "vapor pressure" of the liquid at the temperature of the experiment (Table 8.2).

Liquids with high vapor pressures at ordinary temperatures are said to be **volatile**. Methanol (vapor pressure 98 Torr at 20°C) is highly volatile; mercury (1.4 mTorr) is not. Solids also exert a vapor pressure, but their vapor pressures are usually much lower than those of liquids because the molecules are gripped more tightly in a solid than they are in a liquid. Nevertheless, solids vaporize in the process called "sublimation" (Section 6.11), which we can observe in the presence of some pungent solids—such as menthol and mothballs.

To build a molecular model of the equilibrium between a liquid and its vapor we first suppose that the liquid is introduced into an evacuated *closed* container. Vapor forms as molecules leave the surface of the liquid. Most evaporation takes place from the surface of the liquid because the molecules there are least strongly bound to their neighbors and can escape more easily than those in the bulk. However, as the number of molecules in the vapor increases, more of them become available to strike the surface of the liquid, stick to it, and become part of the liquid again. Eventually, the number of molecules returning to the liquid each second matches the number escaping (Fig. 8.2). The vapor is now condensing as fast as the liquid is vaporizing, and so the equilibrium is *dynamic* in the sense introduced in Section 7.11:

Rate of evaporation = rate of condensation

The dynamic equilibrium between liquid water and its vapor is denoted

$$H_2O(l) \rightleftharpoons H_2O(g)$$

Whenever we see the symbol $\rightleftharpoons$, it means that the species on both sides of the symbol are in dynamic equilibrium with each other. Although "products" (water molecules in the gas phase) are being formed from "reactants" (water molecules in the liquid phase), the products are changing back into reactants at a matching rate. With this picture in mind, we can now define the **vapor pressure** of a liquid (or a solid) as the pressure exerted by its vapor when the vapor and the liquid (or the solid) are in dynamic equilibrium with each other.

The vapor pressure of a given phase of a substance is the pressure exerted by its vapor when the vapor is in dynamic equilibrium with the condensed phase.

8.2 Volatility and Intermolecular Forces

We can expect high vapor pressure when the molecules of a liquid are held together by weak intermolecular forces in the liquid and low vapor pressure when the intermolecular forces are strong. We therefore expect compounds capable of forming hydrogen bonds (which are stronger than other types of intermolecular interaction) to be less volatile than others of similar molar mass. We can see the effect of hydrogen bonding clearly when we compare dimethyl ether (**1**) and ethanol (**2**), both of which have the molecular formula C_2H_6O. Because the two compounds have the same numbers of electrons, we can expect them to have similar London interactions and therefore similar vapor pressures. However, an ethanol molecule has an —OH group and can form a hydrogen bond to another ethanol molecule. Ether molecules cannot form hydrogen bonds to one another because all their hydrogen atoms are attached to carbon atoms and the C—H bond is not very polar. As a result of this difference, ethanol is a liquid at room temperature, whereas dimethyl ether is a gas.

SELF-TEST 8.1A Which do you expect to have the higher vapor pressure at room temperature, tetrabromomethane, CBr_4, or tetrachloromethane, CCl_4? Give your reasons.

[*Answer:* CCl_4; weaker London forces]

SELF-TEST 8.1B Which do you expect to have the higher vapor pressure at 25°C, CH_3CHO or $CH_3CH_2CH_3$?

The vapor pressure of a liquid at a given temperature is expected to be low and its enthalpy of vaporization high if the forces acting between its molecules are strong.

8.3 The Variation of Vapor Pressure with Temperature

The vapor pressure of a liquid depends on how readily the molecules in the liquid can escape from the forces that hold them together. More energy to overcome these attractions is available at higher temperatures than at low, and so we can expect the vapor pressure of a liquid to rise with increasing temperature. Table 8.3 shows the temperature dependence of the vapor pressure of water and Fig. 8.3 shows how the vapor pressures of several liquids rise as the temperature increases. We can use the thermodynamic relations introduced in Chapter 7 to find an expression for the temperature dependence of vapor pressure and trace it to the role of intermolecular forces.

HOW DO WE DO THAT?

To find how vapor pressure changes with temperature we make use of the fact that, when a liquid and its vapor are in equilibrium, there is no difference in the molar Gibbs free energies of the two phases:

$$\Delta G_{vap} = G_m(g) - G_m(l) = 0$$

The Gibbs free energy of a liquid is almost independent of pressure, and so we can replace $G_m(l)$ by its standard value (its value at 1 bar), $G_m°(l)$. The Gibbs free energy of an ideal gas does vary with pressure, and thermodynamics can be used to show that, for an ideal gas,

$$G_m(g, P) = G_m°(g) + RT \ln P$$

where P is the numerical value of the pressure in bars and $G_m°$ is the standard molar Gibbs free energy of the gas (its value at 1 bar). This pressure dependence is illustrated in Fig. 8.4. It follows that when the pressure P is the vapor pressure of the liquid,

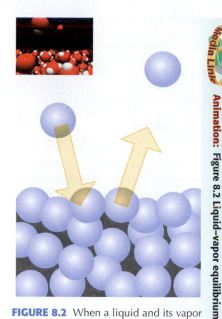

FIGURE 8.2 When a liquid and its vapor are in dynamic equilibrium inside a closed container, the rate of condensation is equal to the rate of evaporation.

1 Dimethyl ether

2 Ethanol

TABLE 8.3 Vapor Pressure of Water

Temperature (°C)	Vapor pressure (Torr)
0	4.58
10	9.21
20	17.54
21	18.65
22	19.83
23	21.07
24	22.38
25	23.76
30	31.83
37*	47.08
40	55.34
60	149.44
80	355.26
100	760.00

*Body temperature.

FIGURE 8.3 The vapor pressures of liquids increase sharply with temperature, as shown here for diethyl ether (red), ethanol (blue), benzene (green), and water (orange). The normal boiling point is the temperature at which the vapor pressure is 1 atm (101.325 kPa). Notice that the curve for ethanol, which has a higher enthalpy of vaporization than benzene, rises more steeply than that of benzene, as predicted by the Clausius–Clapeyron equation. The diagram on the right shows the vapor pressure of water close to its normal boiling point in more detail.

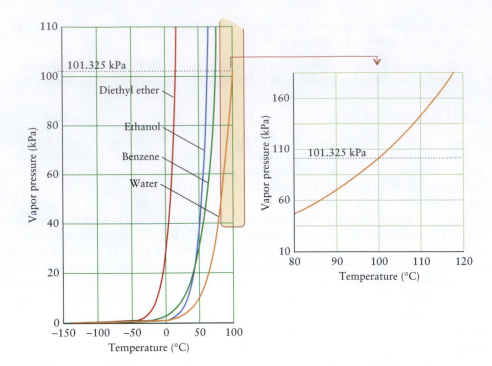

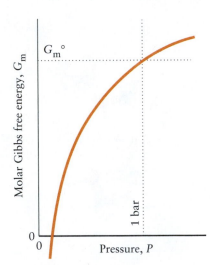

FIGURE 8.4 The variation of the molar Gibbs free energy of an ideal gas with pressure. The Gibbs free energy has its standard value when the pressure of the gas is 1 bar. The value of the Gibbs free energy approaches minus infinity as the pressure falls to zero.

$$\Delta G_{vap} = \{G_m°(g) + RT \ln P\} - G_m°(l) = \{G_m°(g) - G_m°(l)\} + RT \ln P$$
$$= \Delta G_{vap}° + RT \ln P$$

$\Delta G_{vap}°$ is the **standard Gibbs free energy of vaporization**, the change in molar Gibbs free energy when a liquid at 1 bar changes into a vapor at 1 bar. At equilibrium, $\Delta G_{vap} = 0$, and so

$$0 = \Delta G_{vap}° + RT \ln P$$

It follows that

$$\ln P = -\frac{\Delta G_{vap}°}{RT}$$

At this point we can use $\Delta G_{vap}° = \Delta H_{vap}° - T\Delta S_{vap}°$, where $\Delta H_{vap}°$ and $\Delta S_{vap}°$ are the standard enthalpy and entropy of vaporization, respectively, and obtain

$$\ln P = -\frac{\Delta H_{vap}° - T\Delta S_{vap}°}{RT} = -\frac{\Delta H_{vap}°}{RT} + \frac{\Delta S_{vap}°}{R}$$

Neither the enthalpy nor the entropy of vaporization varies much with temperature; so, for a given substance, $\Delta S_{vap}°$ and $\Delta H_{vap}°$ can both be treated as approximately constant. It follows that the vapor pressures P_1 and P_2 at any two temperatures T_1 and T_2 are related by writing this equation for two temperatures and subtracting one from the other. In the process, the entropy terms cancel:

$$\ln P_2 - \ln P_1 = \left(-\frac{\Delta H_{vap}°}{RT_2} + \frac{\Delta S_{vap}°}{R}\right) - \left(-\frac{\Delta H_{vap}°}{RT_1} + \frac{\Delta S_{vap}°}{R}\right)$$
$$= \frac{\Delta H_{vap}°}{R}\left(\frac{1}{T_1} - \frac{1}{T_2}\right)$$

We have deduced the **Clausius–Clapeyron equation** for the vapor pressure of a liquid at two different temperatures:

$$\ln\frac{P_2}{P_1} = \frac{\Delta H_{vap}°}{R}\left(\frac{1}{T_1} - \frac{1}{T_2}\right) \tag{1}$$

What does this equation tell us? When $T_2 > T_1$, the term in parentheses is positive, and because the enthalpy of vaporization is positive, $\ln(P_2/P_1)$ is positive too and we can conclude that P_2 is greater than P_1. In other words, the equation tells us that the vapor pressure increases with increasing temperature. Because $\Delta H_{vap}°$ occurs in the numerator, the increase is greatest for substances with high enthalpies of vaporization (strong intermolecular interactions).

EXAMPLE 8.1 **Estimating the vapor pressure of a liquid from its value at a different temperature**

Tetrachloromethane, CCl_4, which is now known to be carcinogenic, was once used as a dry-cleaning solvent. The enthalpy of vaporization of CCl_4 is 33.05 kJ·mol^{-1} and its vapor pressure at 57.8°C is 405 Torr. What is the vapor pressure of tetrachloromethane at 25.0°C?

STRATEGY We expect the vapor pressure of CCl_4 to be lower at 25.0°C than at 57.8°C. Substitute the temperatures and the enthalpy of vaporization into the Clausius–Clapeyron equation to find the ratio of vapor pressures. Then substitute the known vapor pressure to find the desired one. To use the equation, convert the enthalpy of vaporization into joules per mole and express all temperatures in kelvins.

SOLUTION

Express $\Delta H_{vap}°$ in joules per mole. 33.05 kJ·mol^{-1} corresponds to 3.305×10^4 J·mol^{-1}

Convert temperatures to kelvins.
$T_2 = 25.0 + 273.15$ K $= 298.2$ K
$T_1 = 57.8 + 273.15$ K $= 331.0$ K

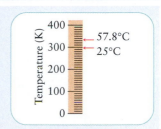

From $\ln (P_2/P_1) =$
$(\Delta H_{vap}°/R) (1/T_1 - 1/T_2)$,

$$\ln\frac{P_2}{P_1} = \frac{3.305 \times 10^4 \text{ J·mol}^{-1}}{8.3145 \text{ J·K}^{-1}\text{·mol}^{-1}}\left(\frac{1}{331.0 \text{ K}} - \frac{1}{298.2 \text{ K}}\right) = \frac{3.305 \times 10^4}{8.3145}\left(\frac{1}{331.0} - \frac{1}{298.2}\right)$$

(This expression evaluates to -1.33.)

Solve for P_2 by taking the exponential (e^x) of both sides and using $P_1 = 405$ Torr.

$$P_2 = (405 \text{ Torr}) \times e^{\frac{3.305 \times 10^4}{8.3145}\left(\frac{1}{331.0} - \frac{1}{298}\right)}$$
$$= 107 \text{ Torr}$$

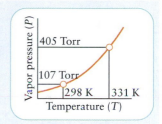

A note on good practice: Exponential functions are very sensitive to rounding errors, so it is important to carry out the numerical calculation in one step. A common error is to forget to express the enthalpy of vaporization in joules (not kilojoules) per mole, but keeping track of units will help you to avoid that mistake.

SELF-TEST 8.2A The vapor pressure of water at 25°C is 23.76 Torr and its standard enthalpy of vaporization at that temperature is 44.0 kJ·mol^{-1}. Estimate the vapor pressure of water at 35°C.

[*Answer:* 42 Torr]

SELF-TEST 8.2B The vapor pressure of benzene at 25°C is 94.6 Torr and its standard enthalpy of vaporization is 30.8 kJ·mol^{-1}. Estimate the vapor pressure of benzene at 35°C.

The vapor pressure of a liquid increases as the temperature increases. The Clausius—Clapeyron equation gives the quantitative dependence of the vapor pressure of a liquid on temperature.

8.4 Boiling

Now consider what happens when we heat a liquid in a container that is open to the atmosphere—water heated in a kettle is an example. When the temperature is raised to the point at which the vapor pressure is equal to the atmospheric pressure (for instance, when water is heated to 100°C and the external pressure is 1 atm), vaporization occurs *throughout* the liquid, not just from its surface. At this temperature, any vapor formed can drive back the atmosphere and make room for itself. Thus, bubbles of vapor form in the liquid and rise to the surface. This rapid vaporization taking place throughout the bulk liquid is called **boiling**. The **normal boiling point**, T_b, of a liquid is the temperature at which a liquid boils when the atmospheric pressure is 1 atm. In other words, *the normal boiling point is the temperature at which the vapor pressure of the liquid is 1 atm*. To find the normal boiling points of the compounds in Fig. 8.3, we draw a horizontal line at $P = 1$ atm (101.325 kPa) and note the temperatures at which it intercepts the curves.

Boiling occurs at a temperature higher than the normal boiling point when the pressure is greater than 1 atm, as it is in a pressure cooker. In this case, a higher temperature is needed to raise the vapor pressure of the liquid to the higher pressure. Boiling occurs at a lower temperature when the pressure is less than 1 atm, because now the vapor pressure matches the external pressure at a lower temperature. At the summit of Mt. Everest—where the pressure is about 240 Torr—water boils at only 70°C.

The lower the vapor pressure, the higher the boiling point. Therefore, a high normal boiling point is a sign of strong intermolecular forces.

EXAMPLE 8.2   **Estimating the boiling point of a liquid**

The vapor pressure of ethanol at 34.9°C is 13.3 kPa. Use the data in Table 6.3 to estimate the normal boiling point of ethanol.

STRATEGY   Use the Clausius–Clapeyron equation to find the temperature at which the vapor pressure has risen to 1 atm (101.325 kPa).

SOLUTION

From Table 6.3,

$$\Delta H_{vap}° = 43.5 \text{ kJ·mol}^{-1} \ (= 4.35 \times 10^4 \text{ J·mol}^{-1})$$

Convert the given temperature to kelvins.

$$T_2 = 34.9 + 273.15 \text{ K} = 308.0 \text{ K}$$

Record the values of the pressure at the two temperatures.

$$P_2 = 13.3 \text{ kPa}$$
$$P_1 = 101.325 \text{ kPa}$$

Rearrange the Clausius–Clapeyron equation to $1/T_1 = 1/T_2 + (R/\Delta H_{vap}°) \ln (P_2/P_1)$ and substitute the data.

$$\frac{1}{T_1} = \frac{1}{308.0 \text{ K}} + \frac{8.3145 \text{ J·K}^{-1}\text{·mol}^{-1}}{4.35 \times 10^4 \text{ J·mol}^{-1}} \ln \frac{13.3 \text{ kPa}}{101.325 \text{ kPa}}$$
$$= \frac{1}{308.0 \text{ K}} + \frac{8.3145}{4.35 \times 10^4 \text{ K}} \ln \frac{13.3}{101.325}$$
$$= \frac{1}{350. \text{ K}}$$

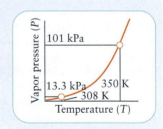

Take the reciprocal of the result to obtain T_1.

$$T_1 = 350. \text{ K}$$

The calculated boiling point corresponds to 77°C; the experimental value is 78°C. The small error probably comes from assuming that the enthalpy of vaporization is a constant over the temperature range of the question and assuming that the vapor behaves like an ideal gas.

SELF-TEST 8.3A   The vapor pressure of acetone, C_3H_6O, at 7.7°C is 13.3 kPa, and its enthalpy of vaporization is 29.1 kJ·mol^{-1}. Estimate the normal boiling point of acetone.

[*Answer:* 62.3°C (actual: 56.2°C)]

SELF-TEST 8.3B   The vapor pressure of methanol, CH_3OH, at 49.9°C is 400. Torr, and its enthalpy of vaporization is 35.3 kJ·mol^{-1}. Estimate the normal boiling point of methanol.

FIGURE 8.5 The structure of ice; notice how the hydrogen bonds hold the water molecules apart from one another in a hexagonal array. The two gray spheres between the oxygen atoms indicate the two possible locations of the hydrogen atom in that region of the structure. Only one of the positions is occupied.

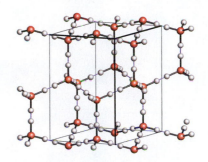

Boiling occurs when the vapor pressure of a liquid is equal to the atmospheric pressure. Strong intermolecular forces usually lead to high normal boiling points.

8.5 Freezing and Melting

A liquid solidifies (freezes) when its molecules have such low energies that they are unable to wriggle past their neighbors. In the solid, the molecules vibrate about their average positions but rarely move from place to place. The **freezing temperature,** the temperature at which the solid and liquid phases are in dynamic equilibrium, varies slightly as the pressure is changed, and the **normal freezing point,** T_f, of a liquid is the temperature at which it freezes at 1 atm. In practice, a liquid sometimes does not freeze until the temperature is a few degrees below its freezing point. A liquid that survives below its freezing point is said to be **supercooled.** The **normal melting point** of a solid is the temperature at which it melts at 1 atm. The **melting temperature** of a solid is the same as the freezing temperature of the liquid.

For most substances, the density of the solid phase of a substance is greater than that of the liquid phase, because the molecules pack together more closely in the solid phase. Applied pressure helps to hold the molecules together, and so a high temperature must be reached before the molecules of a solid at high pressure can break apart. As a result, most solids melt at higher temperatures when subjected to pressure. However, except at very high pressures, the effect of pressure is usually quite small. Iron at 1 atm, for example, melts at 1800 K, but it melts at only a few degrees higher when the pressure is a thousand times as great. At the center of the Earth, however, the pressure is high enough for iron to be solid despite the high temperatures there, so the Earth's inner core is believed to be solid.

Water is highly unusual in that at 0°C the density of the liquid is greater than that of ice. We already know this fact from a common observation: ice floats on water. Therefore, at the melting point, the molar volume of liquid water is less than that of ice. This anomalous behavior is due to the hydrogen bonds in ice, which result in a very open structure (Fig. 8.5; also see Section 5.9). When ice melts, many of the hydrogen bonds collapse, thereby allowing the water molecules to pack together more closely. As a result, high pressure encourages the formation of the denser liquid and the melting point of ice decreases as the pressure is increased.

The melting points of most liquids increase with pressure. Water's hydrogen bonds make it anomalous: its melting point decreases with pressure.

8.6 Phase Diagrams

A **phase diagram** is a map that shows which phase is the most stable at different pressures and temperatures. The phase diagram for water is shown in Fig. 8.6, and that for carbon dioxide is shown in Fig. 8.7. These graphs are examples of phase diagrams for a single substance (water or carbon dioxide) and hence are known as **one-component phase diagrams.** Any point in the region marked "solid" (or, specifically, "ice" for water) corresponds to conditions for which the solid phase of the substance is the most stable; similarly, the regions marked "liquid" and "vapor" (or "gas") indicate the conditions under which liquid and vapor, respectively, are the most stable phase. We can see from the phase diagram for carbon dioxide, for instance, that a sample of carbon dioxide at 10°C and 2 atm will be a gas but, if the pressure is increased at constant temperature to 10 atm, then carbon

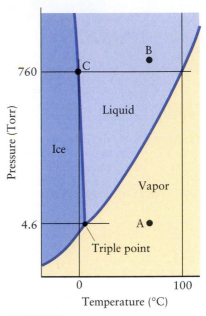

FIGURE 8.6 The phase diagram for water (not to scale). The solid blue lines define the boundaries of the regions of pressure and temperature at which each phase is the most stable. Note that the freezing point decreases slightly with increasing pressure. The triple point is the point at which three phase boundaries meet. The letters A and B are referred to in Example 8.3.

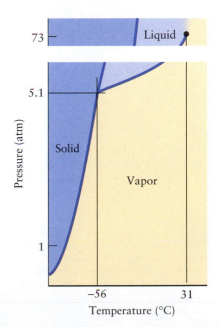

FIGURE 8.7 The phase diagram for carbon dioxide (not to scale). The liquid can exist only at pressures above 5.1 atm. Note the slope of the boundary between the solid and liquid phases; it shows that the freezing point rises as pressure is applied.

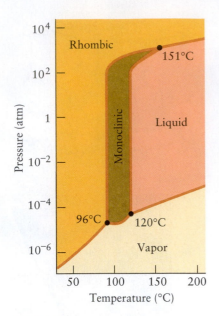

FIGURE 8.8 The phase diagram for sulfur. Notice that there are two solid phases and three triple points. The pressure scale, which is logarithmic, covers a very wide range of values.

dioxide will condense to a liquid. Sulfur has two solid phases (Fig. 8.8), rhombic and monoclinic, corresponding to the two ways in which its crown-like S_8 molecules can stack together. Many substances have several solid phases: even water forms at least ten kinds of ice, depending on how the H_2O molecules fit together, but only one of them is stable at ordinary pressures (Fig. 8.9). Ice-VIII, for instance, is stable only above 20 000 atm and 100°C.

The lines separating the regions in a phase diagram are called **phase boundaries**. At any point on a boundary between two regions, the two neighboring phases coexist in dynamic equilibrium. If one of the phases is a vapor, the pressure corresponding to this equilibrium is just the vapor pressure of the substance. Therefore, the liquid–vapor phase boundary shows how the vapor pressure of the liquid varies with temperature. For example, the point at 80.°C and 0.47 atm in the phase diagram for water lies on the phase boundary between liquid and vapor (Fig. 8.10), and so we know that the vapor pressure of water at 80.°C is 0.47 atm. Similarly, the solid–vapor phase boundary shows how the vapor pressure of the solid varies with temperature (see Fig. 8.6).

The solid–liquid boundary, the almost vertical line in Figs. 8.6 and 8.10, shows the pressures and temperatures at which solid and liquid water coexist in equilibrium. In other words, it shows how the melting temperature of ice (or the freezing temperature of water) varies with pressure. The steepness of the lines indicates that even large changes in pressure result in only small variations in melting temperature. The slope of the solid–liquid boundary reveals the relative densities of solid and liquid. A negative slope, such as that in the phase diagram for water (Fig. 8.6), implies that the liquid is more dense. Notice that as we increase the pressure of ice, it will eventually be converted to liquid. If the slope is positive, as for carbon dioxide (Fig. 8.7), then the solid is the denser phase.

SELF-TEST 8.4A From the phase diagram for carbon dioxide (Fig. 8.7), predict which is more dense, the solid or the liquid phase. Explain your conclusion.

[*Answer:* The solid, because it is the stable phase at higher pressures.]

SELF-TEST 8.4B From the phase diagram for sulfur (Fig. 8.8), predict which phase is more dense, liquid sulfur or monoclinic sulfur. Explain your conclusion.

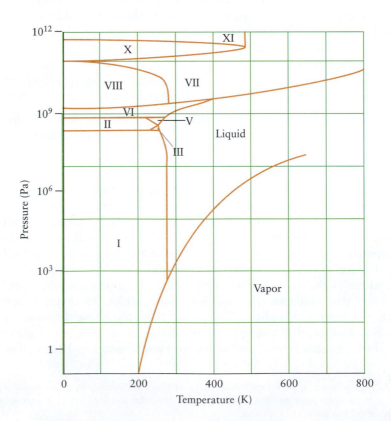

FIGURE 8.9 The phase diagram for water drawn with a logarithmic scale for pressure, in order to show the different solid phases of water in the high-pressure region.

A **triple point** is a point where three phase boundaries meet on a phase diagram. For water, the triple point for the solid, liquid, and vapor phases lies at 4.6 Torr and 0.01°C (see Fig. 8.6). At this triple point, all three phases (ice, liquid, and vapor) coexist in mutual dynamic equilibrium: solid is in equilibrium with liquid, liquid with vapor, and vapor with solid. The location of a triple point of a substance is a fixed property of that substance and cannot be changed by changing the conditions. The triple point of water is used to define the size of the kelvin: by definition, there are exactly 273.16 kelvins between absolute zero and the triple point of water. Because the normal freezing point of water is found to lie 0.01 K below the triple point, 0°C corresponds to 273.15 K.

In Fig. 8.8, we see that sulfur can exist in any of four phases: two solid phases (rhombic and monoclinic sulfur), one liquid phase, and one vapor phase. There are three triple points in the diagram, where various combinations of these phases, such as monoclinic solid, liquid, and vapor or monoclinic solid, rhombic solid, and liquid, coexist. However, four phases in mutual equilibrium (such as the vapor, liquid, and rhombic and monoclinic solid forms of sulfur, all in mutual equilibrium) in a one-component system has never been observed, and thermodynamics can be used to prove that such a "quadruple point" cannot exist.

A phase diagram can be used to explain the changes that take place when the pressure or temperature of a substance is changed. Imagine that we have a sample of water in a cylinder fitted with a piston, that the temperature is constant at 50°C, and that the weights on the piston exert a pressure of 1.0 atm (Fig. 8.11). The piston presses on the surface of the liquid, as in Fig. 8.11a. Now we gradually reduce the pressure by removing some of the weights (Fig. 8.11b). At first, nothing seems to happen. The high pressure is keeping all the water molecules in the liquid state, and the volume of a liquid changes very little with pressure. However, when so many weights have been removed that the pressure has fallen to 0.12 atm (93 Torr, the vapor pressure of water at that temperature), vapor begins to appear (Fig. 8.11c). We are now at the vapor–liquid boundary on the phase diagram. The pressure remains constant so long as the liquid and vapor phases are both present at equilibrium. We are free to pull the piston up by an arbitrary extent (Fig. 8.11d), but enough water will evaporate to maintain the pressure at 0.12 atm. When we pull the piston out far enough, the liquid phase disappears (Fig. 8.11e); we are now free to modify the pressure at will without changing the temperature (Fig. 8.11f).

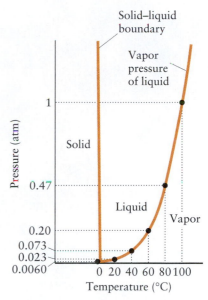

FIGURE 8.10 The liquid–vapor boundary curve is a plot of the vapor pressure of the liquid (in this case, water) as a function of temperature. The liquid and its vapor are in equilibrium at each point on the curve. At each point on the solid–liquid boundary curve (for which the slope is slightly exaggerated), the solid and liquid are in equilibrium.

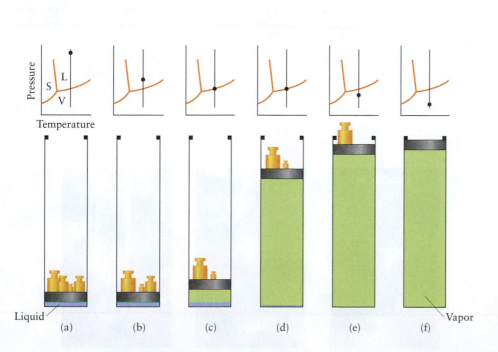

FIGURE 8.11 The changes undergone by a liquid as its pressure is decreased at constant temperature. The large dot on the vertical line in the phase diagram traces the path taken by the system, which is described in the text. The blue region in the container is the liquid and the light green region above it is the vapor.

Use the phase diagram in Fig. 8.6 to describe the physical states and phase changes of water as the pressure on it is increased from 5 Torr to 800 Torr at 70°C.

STRATEGY First, locate the initial and final conditions on the phase diagram. The region in which each of these points lie shows the stable phase of the sample under those conditions. If a point lies on one of the curves, then both phases are present in mutual equilibrium.

SOLUTION Although the phase diagram in Fig. 8.6 is not to scale, we can find the approximate locations of the points. Point A is at 5 Torr and 70°C; so it lies in the vapor region. Increasing the pressure takes the vapor to the liquid–vapor phase boundary, at which point liquid begins to form. At this pressure, liquid and vapor are in equilibrium and the pressure remains constant until all the vapor has condensed. The pressure is increased further to 800 Torr, which takes it to point B, in the liquid region.

SELF-TEST 8.5A The phase diagram for carbon dioxide is shown in Fig. 8.7. Describe the physical states and phase changes of carbon dioxide as it is heated at 2 atm from −155°C to 25°C.

> [*Answer:* Solid CO_2 is heated until it begins to sublime at the solid–vapor boundary. The temperature remains constant until all the CO_2 has vaporized. The vapor is then heated to 25°C.]

SELF-TEST 8.5B Describe what happens when liquid carbon dioxide in a container at 60 atm and 25°C is released into a room at 1 atm and the same temperature.

A phase diagram summarizes the regions of pressure and temperature at which each phase of a substance is most stable. The phase boundaries show the conditions under which two phases can coexist in dynamic equilibrium with each other. Three phases coexist in mutual equilibrium at a triple point.

8.7 Critical Properties

A feature of the phase diagram in Fig. 8.12 is that the liquid–vapor boundary comes to an end at point C. To see what happens at that point, suppose that a vessel like the one shown in Fig. 8.13 contains liquid water and water vapor at 25°C and 24 Torr (the vapor pressure of water at 25°C). The two phases are in equilibrium, and the system lies at point A on the liquid–vapor curve in Fig. 8.12. Now let's raise the temperature, which moves the system from left to right along the phase boundary. At 100.°C, the vapor pressure is 760. Torr; and, at 200.°C, it has reached 11.7 kTorr (15.4 atm, point B). The liquid and vapor are still in dynamic equilibrium, but now the vapor is very dense because it is at such a high pressure.

When the temperature has been increased to 374°C (point C), the vapor pressure has reached 218 atm—the container must be very strong! The density of the vapor is now so great that it is equal to that of the remaining liquid. At this stage, the surface separating the liquid from its vapor vanishes and a single uniform

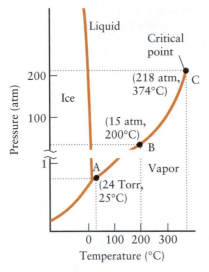

FIGURE 8.12 A quantitative version of the phase diagram for water close to the critical point. Pressures are in atmospheres, except for point A.

FIGURE 8.13 When the temperature of a liquid in a sealed, constant-volume container (left) is raised, the density of the liquid decreases and the density of the vapor increases (center). At the critical temperature, T_c, the density of the vapor becomes the same as the density of the liquid; at and above that temperature, a single uniform phase fills the container (right).

  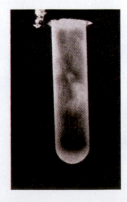

phase fills the container. Because a substance that fills any container that it occupies is by definition a gas, we have to conclude that this single uniform phase is a gas, despite its high density. Moreover, provided the temperature is at or above 374°C, it is found that even if the pressure is increased by compressing the sample, no surface appears. That is, 374°C is the **critical temperature**, T_c, of water, the temperature at and above which there is no transition from vapor to liquid. Similar considerations apply to other substances (Table 8.4); for instance, the critical temperature of carbon dioxide is 31°C, and at and above that temperature CO_2 gas does not undergo a transition to liquid whatever the pressure. The pressure corresponding to the end of the liquid–vapor phase boundary is called the **critical pressure**, P_c, of the substance. The critical pressure of water is 218 atm; that of carbon dioxide is 73 atm. The critical temperature and critical pressure jointly define the **critical point** of the substance.

A gas can be liquefied by applying pressure only if it is below its critical temperature (Fig. 8.14). For example, carbon dioxide can be liquefied by applying pressure only if its temperature is lower than 31°C. According to Table 8.4, the critical temperature of oxygen is −118°C, and so we know that it cannot exist as a liquid at room temperature whatever the pressure.

The dense fluid that exists above the critical temperature and pressure of a substance is called a **supercritical fluid**. It may be so dense that, although it is formally a gas, it is as dense as a liquid phase and can act as a solvent for liquids and solids. Supercritical carbon dioxide, for instance, can dissolve organic compounds. It is used to remove caffeine from coffee beans, to separate drugs from biological fluids for later analysis, and to extract perfumes from flowers and phytochemicals from herbs. The use of supercritical carbon dioxide avoids contamination with potentially harmful solvents and allows rapid extraction on account of the high mobility of the molecules through the fluid. Supercritical hydrocarbons are used to dissolve coal and separate it from ash, and they have been proposed for extracting oil from oil-rich tar sands.

SELF-TEST 8.6A Identify trends in the data in Table 8.4 that indicate the relationship of the strength of London forces to the critical temperature.

> [*Answer:* Noble gases have higher critical temperatures as their atomic number increases; so critical temperature increases with the strength of London forces.]

SELF-TEST 8.6B Identify trends in the data in Table 8.4 that indicate the effect of hydrogen bonding on the critical temperature.

The vapor of a substance can be condensed to a liquid by the application of pressure only if it is below its critical temperature.

SOLUBILITY

The presence of a solute affects the physical properties of the solvent. For instance, when salt is spread on icy sidewalks, a mixture is created with a lower freezing point than that of pure water and the ice melts. In this part of the chapter we explore the molecular nature of these effects and see how to treat them quantitatively.

8.8 The Limits of Solubility

Let's imagine that we could see at the molecular level what happens when we add a crystal of glucose to some water. The water molecules next to the surface of the crystal form hydrogen bonds to the glucose molecules. As a result, the surface glucose molecules are pulled into solution by water molecules but are held back by other glucose molecules. When the interactions with water molecules are comparable to the interactions with the other glucose molecules, the glucose molecules can drift off into the solvent, surrounded by water molecules. A similar process takes place when an ionic solid dissolves. The polar water molecules hydrate the ions

TABLE 8.4 Critical Temperatures and Pressures of Selected Substances

Substance	Critical temperature (°C)	Critical pressure (atm)
He	−268 (5.2 K)	2.3
Ne	−229	27
Ar	−123	48
Kr	−64	54
Xe	17	58
H_2	−240	13
O_2	−118	50
H_2O	374	218
N_2	−147	34
NH_3	132	111
CO_2	31	73
CH_4	−83	46
C_6H_6	289	49

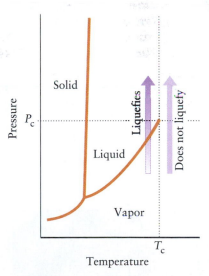

FIGURE 8.14 When pressure is applied to a vapor below its critical temperature, the vapor may be liquefied. However, if the temperature is above the critical temperature (the vertical dotted line), an increase in pressure does not take the vapor into the liquid region. At high pressures above its critical temperature, the substance becomes a supercritical fluid.

FIGURE 8.16 When a little glucose is stirred into 100 mL of water, all the glucose dissolves (left). However, when a large amount is added, some undissolved glucose remains and the solution is saturated with glucose (right).

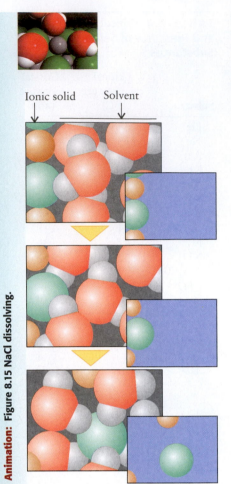

Ionic solid   Solvent

FIGURE 8.15 The events that take place at the interface of a solid ionic solute and a solvent (water). When the ions at the surface of the solid become hydrated, they move off into the solution.

(surround them in a closely held "shell," Section 5.2) and pry them away from the predominately attractive forces within the crystal lattice (Fig. 8.15). Stirring and shaking speed the process because they bring more free water molecules to the surface of the solid and sweep the hydrated ions away.

If we add only a small amount—20 g, for instance—of glucose to 100 mL of water at room temperature, it all dissolves. However, if we add 200 g, some glucose remains undissolved (Fig. 8.16). A solution is said to be **saturated** when the solvent has dissolved all the solute that it can and some undissolved solute remains. The concentration of solid solute in a saturated solution has reached its greatest value and no more can dissolve. The **molar solubility**, s, of a substance is its molar concentration in a saturated solution. In other words, the molar solubility of a substance represents the limit of its ability to dissolve in a given quantity of solvent.

In a saturated solution, any solid solute present still continues to dissolve, but the rate at which it dissolves exactly matches the rate at which the solute returns to the solid (Fig. 8.17). In a saturated solution, the dissolved and undissolved solute are in dynamic equilibrium with each other.

The molar solubility of a substance is its molar concentration in a saturated solution. A saturated solution is one in which the dissolved and undissolved solute are in dynamic equilibrium with each other.

8.9 The Like-Dissolves-Like Rule

Understanding the interplay of forces that takes place when a solute dissolves can help us answer some practical questions. Suppose, for instance, we need to remove some wax that has spilled from a candle onto a tablecloth. How can we select a good solvent for wax? A good guide is the rule that *like dissolves like*. A polar liquid, such as water, is generally the best solvent for ionic and polar compounds. Conversely, nonpolar liquids, including hexane and tetrachloroethene, $Cl_2C{=}CCl_2$, used for dry cleaning, are often better solvents for nonpolar compounds such as wax, which are held together by London forces.

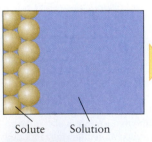   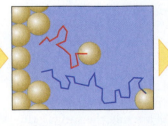   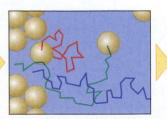   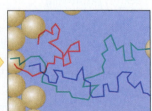

Solute   Solution

FIGURE 8.17 The solute in a saturated solution is in dynamic equilibrium with the undissolved solute. If we could follow the solute particles (yellow spheres), we would sometimes find them in solution and sometimes back within the solid. Red, green, and blue lines represent the paths of individual solute particles. The solvent molecules are not shown.

FIGURE 8.18 A representation of the changes in molecular interactions and energy associated with the formation of a dilute solution. Step 1 is the separation of solute molecules from one another. In step 2, some of the solvent molecules move apart and leave cavities in the solvent. In step 3, the solute molecules occupy the cavities in the solvent, a process that releases energy. The overall energy change for the process is the sum of the energy changes, shown by the dark red arrow. In the actual dissolving process, these steps do not occur independently.

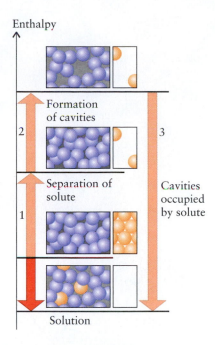

Enthalpy

Formation of cavities

Separation of solute

Cavities occupied by solute

Solution

We can explain the "like-dissolves-like" rule by examining the attractions between solute and solvent molecules. When a solute dissolves in a liquid solvent, the solute–solute attractions must be replaced by solute–solvent attractions, and dissolving can be expected to occur if the new interactions are similar to the original interactions (Fig. 8.18). For example, when the main cohesive forces in a solute are hydrogen bonds, the solute is more likely to dissolve in a hydrogen-bonded solvent than in other solvents. The molecules can slip into solution only if they can replace their solute–solute hydrogen bonds with solute–solvent hydrogen bonds. Glucose, for example, has hydrogen-bond-forming —OH groups and dissolves readily in water but not in hexane.

If the principal cohesive forces between solute molecules are London forces, then the best solvent is likely to be one that can mimic those forces. For example, a good solvent for nonpolar substances is the nonpolar liquid carbon disulfide, CS_2. It is a far better solvent than water for sulfur because solid sulfur is a molecular solid of S_8 molecules held together by London forces (Fig. 8.19). The sulfur molecules cannot penetrate into the strongly hydrogen-bonded structure of water, because they cannot replace those bonds with interactions of similar strength.

The cleaning action of soaps and detergents relies on the like-dissolves-like rule. Soaps are the sodium salts of long-chain carboxylic acids, including sodium stearate (**3**).

3 Sodium stearate,
$NaCH_3(CH_2)_{16}CO_2$

The anions of these acids have a polar carboxylate ($-CO_2^-$) group, called the *head group*, at one end of a nonpolar hydrocarbon chain. The head group is **hydrophilic**, or water attracting, whereas the nonpolar hydrocarbon tails are **hydrophobic**, or water repelling. Because the hydrophilic head group of the anion has a tendency to dissolve in water and the hydrophobic hydrocarbon tails have a tendency to dissolve in grease, soap is very effective at removing grease. The hydrocarbon tails sink into a blob of grease up to the head groups and the hydrophilic head groups remain on the surface of the blob. The casing of soap molecules, which is called a **micelle**, is soluble in water and so carries away the grease (Fig. 8.20).

Soaps are made by heating sodium hydroxide with a fat such as coconut oil, olive oil, or beef fat, which contain *esters* formed between glycerol and fatty acids (see Section 19.7). The sodium hydroxide attacks the esters and forms the soluble soap. In the case of beef fat, stearic acid forms the soap sodium stearate, seen in (**3**). Soaps, however, form a scum in hard water. The scum is an impure precipitate of calcium stearate.

Modern commercial detergents are mixtures. Their most important component is a **surfactant**, or *surface-active agent*, which takes the place of the soap. Surfactant molecules are organic compounds with a structure and action similar to those of soap. A difference is that they typically contain sulfur atoms in their polar groups (**4**).

A general guide to the suitability of a solvent is the rule that like dissolves like. Soaps and detergents contain surfactant molecules that have both hydrophobic and hydrophilic regions.

Hard water is discussed in Section 8.12.

4 A typical surfactant ion

FIGURE 8.19 Sulfur, which is a solid with nonpolar molecules, does not dissolve in water (left), but it does dissolve in carbon disulfide (right), with which the S_8 molecules have favorable London interactions.

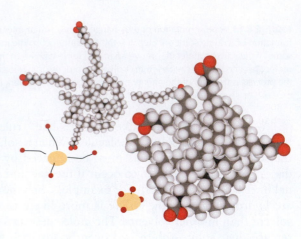

FIGURE 8.20 On the left, the hydrocarbon tails of a soap or surfactant molecule begin to dissolve in grease. (Schematic representations of the molecules, with the grease depicted as a yellow blob and the polar heads of the surfactant molecules shown in red, are shown below the space-filling models.) The water-attracting head groups remain on the surface where they can interact favorably with water (right). As more surfactant molecules dissolve in the grease, the whole blob, called a micelle, dissolves in water and is washed away.

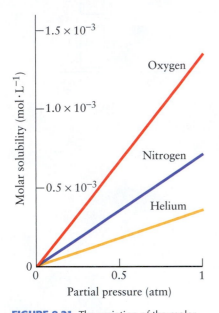

FIGURE 8.21 The variation of the molar solubilities of oxygen, nitrogen, and helium gases with partial pressure. Note that the solubility of each gas doubles when its partial pressure is doubled.

8.10 Pressure and Gas Solubility: Henry's Law

Almost all aquatic organisms rely on the presence of dissolved oxygen for respiration. Although oxygen is nonpolar, it is very slightly soluble in water and the extent to which it dissolves depends on its pressure. We have already seen (in Section 4.2) that the pressure of a gas arises from the impacts of its molecules. When a gas is introduced into the same container as a liquid, the gas molecules can burrow into the liquid like meteorites plunging into the ocean. Because the number of impacts increases as the pressure of a gas increases, we should expect the solubility of the gas—its molar concentration when the dissolved gas is in dynamic equilibrium with the free gas—to increase as its pressure increases. If the gas above the liquid is a mixture (like air), then the solubility of each component depends on that component's partial pressure (Fig. 8.21).

The straight lines in Fig. 8.21 show that *the solubility of a gas is directly proportional to its partial pressure, P.* This observation was first made in 1801 by the English chemist William Henry and is now known as **Henry's law**. The law is normally written

$$s = k_H P \tag{2}*$$

The constant k_H, which is called **Henry's constant**, depends on the gas, the solvent, and the temperature (Table 8.5).

> **EXAMPLE 8.4**   **Sample exercise: Estimating the solubility of a gas in a liquid**
>
> Verify that the concentration of oxygen in lake water is normally adequate to sustain aquatic life, which requires a concentration of at least 0.13 mmol·L^{-1}. The partial pressure of oxygen is 0.21 atm at sea level.

SOLUTION

From $s = k_H P$,

$$s = (1.3 \times 10^{-3}\ \text{mol·L}^{-1}\text{·atm}^{-1}) \times (0.21\ \text{atm})$$
$$= 2.7 \times 10^{-4}\ \text{mol·L}^{-1}$$

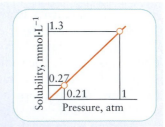

This molar concentration is 0.27 mmol·L^{-1}, more than adequate to sustain life.

SELF-TEST 8.7A At the elevation, 2900 m, of Bear Lake in Rocky Mountain National Park, the partial pressure of oxygen is 0.14 atm. What is the molar solubility of oxygen in Bear Lake at 20.°C?

[*Answer:* 0.18 mmol·L^{-1}]

SELF-TEST 8.7B Use the information in Table 8.5 to calculate the amount (in moles) of CO_2 that will dissolve in enough water to form 900. mL of solution at 20.°C when the partial pressure of CO_2 is 1.00 atm.

TABLE 8.5 Henry's Constants for Gases in Water at 20°C

Gas	k_H $(mol \cdot L^{-1} \cdot atm^{-1})$	Gas	k_H $(mol \cdot L^{-1} \cdot atm^{-1})$
air	7.9×10^{-4}	hydrogen	8.5×10^{-4}
argon	1.5×10^{-3}	neon	5.0×10^{-4}
carbon dioxide	2.3×10^{-2}	nitrogen	7.0×10^{-4}
helium	3.7×10^{-4}	oxygen	1.3×10^{-3}

The solubility of a gas is proportional to its partial pressure, because an increase in pressure corresponds to an increase in the rate at which gas molecules strike the surface of the solvent.

8.11 Temperature and Solubility

Most substances dissolve more quickly at higher temperatures than at low ones. However, that does not necessarily mean that they are more soluble—that is, reach a higher concentration of solute—at higher temperatures. In a number of cases, the solubility turns out to be lower at higher temperatures. We must always distinguish the effect of temperature on the *rate* of a process from its effect on the final outcome.

Near room temperature most gases become less soluble in water as the temperature is raised. The lower solubility of gases in warm water is responsible for the tiny bubbles that appear when cool water from the faucet is left to stand in a warm room. The bubbles consist of air that dissolved when the water was cooler; it comes out of solution as the temperature rises. In contrast, most ionic and molecular solids are more soluble in warm water than in cold (Fig. 8.22). We make use of this characteristic in the laboratory to dissolve a substance and to grow crystals by letting a saturated solution cool slowly. However, a few solids containing ions that are extensively hydrated in water, such as lithium carbonate, are less soluble at high temperatures than at low. A small number of compounds show a mixed behavior. For example, the solubility of sodium sulfate decahydrate increases up to 32°C but then decreases as the temperature is raised further.

Most gases are more soluble in warm water than in cold water; solids show a more varied behavior.

8.12 The Enthalpy of Solution

The change in molar enthalpy when a substance dissolves is called the **enthalpy of solution**, ΔH_{sol}. The change can be measured calorimetrically from the heat released or absorbed when the substance dissolves at constant pressure. However,

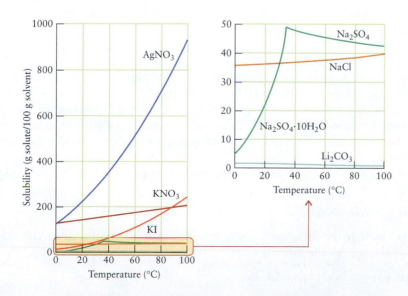

FIGURE 8.22 The variation with temperature of the solubilities of six substances in water.

TABLE 8.6 Limiting Enthalpies of Solution, ΔH_{sol}, at 25°C, in Kilojoules per Mole*

| | Anion | | | | | | | |
Cation	fluoride	chloride	bromide	iodide	hydroxide	carbonate	sulfate	nitrate
lithium	+4.9	−37.0	−48.8	−63.3	−23.6	−18.2	−2.7	−29.8
sodium	+1.9	+3.9	−0.6	−7.5	−44.5	−26.7	+20.4	−2.4
potassium	−17.7	+17.2	+19.9	+20.3	−57.1	−30.9	+34.9	+23.8
ammonium	−1.2	+14.8	+16.0	+13.7	—	—	+25.7	+6.6
silver	−22.5	+65.5	+84.4	+112.2	—	+41.8	+22.6	+17.8
magnesium	−12.6	−160.0	−185.6	−213.2	+2.3	−25.3	−90.9	−91.2
calcium	+11.5	−81.3	−103.1	−119.7	−16.7	−13.1	−19.2	−18.0
aluminum	−27	−329	−368	−385	—	—	—	−350

*The value for silver iodide, for example, is the entry found where the row labeled "silver" intersects the column labeled "iodide."

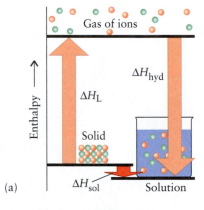

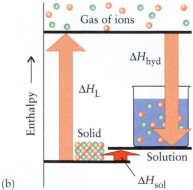

(a)

(b)

FIGURE 8.23 The enthalpy of solution, ΔH_{sol}, is the sum of the enthalpy change required to separate the molecules or ions of the solute, the lattice enthalpy, ΔH_L (step 1 in Fig. 8.18), and the enthalpy change accompanying their hydration, ΔH_{hyd} (steps 2 and 3 in Fig. 8.18). The outcome is finely balanced: (a) in some cases, it is exothermic; (b) in others, it is endothermic. For gaseous solutes, the lattice enthalpy is zero because the molecules are already widely separated.

because solute molecules interact with one another as well as with the solvent, to isolate the solute–solvent interactions we focus on the **limiting enthalpy of solution**, the enthalpy change accompanying the formation of such dilute solutions that the solute–solute interactions are negligible (Table 8.6). The data show that some solids, such as lithium chloride, dissolve exothermically, with a release of heat. Others, such as ammonium nitrate, dissolve endothermically.

To gain insight into the values in Table 8.6, we can think of dissolving as a hypothetical two-step process (Fig. 8.23). In the first step, we imagine the ions separating from the solid to form a gas of ions. The change in molar enthalpy accompanying this highly endothermic step is the lattice enthalpy, ΔH_L, of the solid, which was introduced in Section 6.19 (see Table 6.6 for values). The lattice enthalpy of sodium chloride (787 kJ·mol^{-1}), for instance, is the molar enthalpy change for the process

$$NaCl(s) \longrightarrow Na^+(g) + Cl^-(g)$$

We saw in Section 6.19 that compounds formed from small, highly charged ions (such as Mg^{2+} and O^{2-}) cohere strongly and that a lot of energy is needed to break up the lattice. Such compounds have high lattice enthalpies. Compounds formed from large ions with low charges, such as potassium iodide, typically have weak attractive forces and correspondingly low lattice enthalpies.

In the second hypothetical step, we imagine the gaseous ions plunging into water and forming the final solution. The molar enthalpy of this step is called the **enthalpy of hydration**, ΔH_{hyd}, of the compound (Table 8.7). Enthalpies of hydration are negative and comparable in value to the lattice enthalpies of the compounds. For sodium chloride, for instance, the enthalpy of hydration, the molar enthalpy change for the process

$$Na^+(g) + Cl^-(g) \longrightarrow Na^+(aq) + Cl^-(aq),$$

is −784 kJ·mol^{-1}. That is enough energy for 1 g of NaCl (as a gas of Na$^+$ and Cl$^-$ ions) to raise the temperature of 100 mL of water by nearly 50°C. Hydration is always exothermic for ionic compounds because of the formation of ion–dipole attractions between the water molecules and the ions. It is also exothermic for molecules that can form hydrogen bonds with water, such as sucrose, glucose, acetone, and ethanol.

Now we bring the two steps of the dissolving process together and calculate the enthalpy for the overall change:

$$\Delta H_{sol} = \Delta H_L + \Delta H_{hyd}$$

When we include the data, the limiting enthalpy of solution of sodium chloride, the enthalpy change for the process

$$NaCl(s) \longrightarrow Na^+(aq) + Cl^-(aq)$$

TABLE 8.7 Enthalpies of Hydration, ΔH_{hyd}, at 25°C, of Some Halides, in Kilojoules per Mole*

Cation	Anion			
	F^-	Cl^-	Br^-	I^-
H^+	-1613	-1470	-1439	-1426
Li^+	-1041	-898	-867	-854
Na^+	-927	-784	-753	-740
K^+	-844	-701	-670	-657
Ag^+	-993	-850	-819	-806
Ca^{2+}	—	-2337	—	—

*The entry where the row labeled Na^+ intersects the column labeled Cl^-, for instance, is the enthalpy change, -784 kJ·mol^{-1}, for the process $Na^+(g) + Cl^-(g) \rightarrow Na^+(aq) + Cl^-(aq)$; the values here apply only when the resulting solution is very dilute.

is

$$\Delta H_{sol} = \underset{\text{lattice enthalpy}}{787 \text{ kJ·mol}^{-1}} - \underset{\text{enthalpy of hydration}}{784 \text{ kJ·mol}^{-1}} = \underset{\text{enthalpy of solution}}{+3 \text{ kJ·mol}^{-1}}$$

Because the enthalpy of solution is positive, there is a net inflow of energy as heat when the solid dissolves (recall Fig. 8.23b). Sodium chloride therefore dissolves endothermically, but only to the extent of 3 kJ·mol^{-1}. As this example shows, the overall change in enthalpy depends on a very delicate balance between the lattice enthalpy and the enthalpy of hydration.

High charge and small ionic radius both contribute to a high enthalpy of hydration. However, the same properties also contribute to high lattice enthalpy. It is therefore very difficult to make reliable predictions about solubilities on the basis of ion charge and radius. The best we can do is to use these properties to rationalize what is observed. With that limitation in mind, we can begin to understand the behavior of some everyday substances and the properties of some minerals. For example, nitrates have big, singly charged anions and hence low lattice enthalpies. Their hydration enthalpies, though, are quite large because water can form hydrogen bonds with the nitrate anions. As a result, they are rarely found in mineral deposits because they are soluble in groundwater, the water that trickles through the ground and washes away soluble substances. Carbonate ions are about the same size as nitrate ions, but they are doubly charged. As a result, carbonates commonly have higher lattice enthalpies than do nitrates, and it is much harder to break the ions out of solids such as limestone (calcium carbonate). Hydrogen carbonate ions (bicarbonate ions, HCO_3^-) have a single charge and their salts are more soluble than carbonates.

The difference in solubility between carbonates and hydrogen carbonates is responsible for the behavior of "hard water," which is water that contains dissolved calcium and magnesium salts. Hard water originates as rainwater, which dissolves carbon dioxide from the air and forms a very dilute solution of carbonic acid.

$$CO_2(g) + H_2O(l) \longrightarrow H_2CO_3(aq)$$

As the water runs over and through the ground, the carbonic acid reacts with the calcium carbonate of limestone or chalk and forms the more soluble hydrogen carbonate:

$$CaCO_3(s) + H_2CO_3(aq) \longrightarrow Ca^{2+}(aq) + 2\ HCO_3^-(aq)$$

These two reactions are reversed when water containing $Ca(HCO_3)_2$ is heated in a kettle or furnace:

$$Ca^{2+}(aq) + 2\ HCO_3^-(aq) \longrightarrow CaCO_3(s) + H_2O(l) + CO_2(g)$$

Although we write the reaction as leading to the formation of H_2CO_3, the reaction is an equilibrium that favors reactants. As a result, a high proportion of the dissolved carbon dioxide remains unreacted.

The carbon dioxide is driven off and the calcium carbonate is deposited as the hard, insoluble deposit known as *scale*.

Enthalpies of solution in dilute solutions can be expressed as the sum of the lattice enthalpy and the enthalpy of hydration of the compound.

8.13 The Gibbs Free Energy of Solution

A negative enthalpy of solution tells us that energy is released as heat when a substance dissolves. However, to judge whether dissolving is spontaneous at constant temperature and pressure, we need to consider the change in Gibbs free energy, $\Delta G = \Delta H - T\Delta S$. In other words, we have to consider changes in the entropy of the system, not just the change in enthalpy.

The disorder of a system typically increases when a solid dissolves (Fig. 8.24). Therefore, in most cases, we can expect the entropy of the system to increase when a solution forms. Because $T\Delta S$ is positive, this increase in disorder makes a negative contribution to ΔG. If ΔH is negative, we can be confident that ΔG is negative overall. Therefore, we can expect most substances with negative enthalpies of solution to be soluble.

In some cases, the entropy of the system is lowered when the solution forms because the solvent molecules form cagelike structures around the solute molecules. As a result, ΔS is negative and the term $-T\Delta S$ makes a positive contribution to ΔG. Even if ΔH is negative, ΔG might be positive. In other words, even if energy is released to the surroundings, the increase in the entropy of the surroundings may not be enough to overcome the decrease in entropy of the solution. In such a case ΔG is positive and the substance does not dissolve. For this reason, some hydrocarbons, such as heptane, are insoluble in water even though they have weakly negative enthalpies of solution.

If ΔH is positive, ΔG can be negative only if $T\Delta S$ is positive and larger than ΔH. That is, the solubilities of substances that dissolve endothermically depend on the balance of the entropy change of the system and that of the surroundings. A substance with a strongly positive enthalpy of solution is likely to be insoluble, because the entropy of the surroundings may decrease so much as energy leaves them and enters the solution that dissolving corresponds to an overall decrease in disorder.

Taking into account the Gibbs free energy of the dissolving process helps us to explain the temperature dependence of solubility. Dissolving becomes spontaneous as the temperature increases only when ΔS is positive. For most ionic substances, dissolving does indeed result in an increase in the entropy of the system and these substances are more soluble at higher temperatures (see Fig. 8.22). However, for substances that are extensively hydrated and for gases, which enter a condensed state with much less freedom of movement when they dissolve in a liquid, the entropy of solution is negative and solubility decreases as the temperature rises.

Remember that ΔG is a measure of the overall change in entropy: at constant temperature and pressure, ΔS is the change in entropy of the system and $-\Delta H/T$ is the change in entropy of the surroundings (Section 7.9).

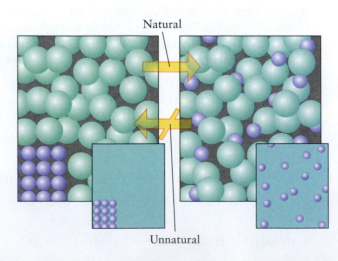

Natural

Unnatural

FIGURE 8.24 The dissolution of a solid is usually accompanied by an increase in disorder. However, to decide whether the dissolution is spontaneous, the accompanying changes in the surroundings must be considered, too. The insets show the solute particles alone.

Because ΔG for the dissolving of a solute depends on the concentration of the solute, even if ΔG is negative at low concentrations, it may become positive at high concentrations (Fig. 8.25). A solute dissolves spontaneously only until $\Delta G = 0$. At that point, dissolved and undissolved solute are in equilibrium—the solution is saturated.

Dissolving depends on the balance between the change in entropy of the solution and the change in entropy of the surroundings.

COLLIGATIVE PROPERTIES

Some properties of a solution are essential to life. For example, *osmosis*, the tendency of a solvent to flow through a membrane into a solution, promotes the flow of nutrients through biological cell walls. When chemists started to study the physical properties of solutions quantitatively, they discovered that some properties, including osmosis, depend only on the relative amounts of solute and solvent and are independent of the chemical identity of the solute. Properties that depend on the relative numbers of solute and solvent molecules and not on the chemical identity of the solute are called **colligative properties**. Four colligative properties of major importance are the lowering of the vapor pressure of the solvent, the raising of its boiling point, the lowering of its freezing point, and osmosis. All four properties involve an equilibrium between two phases of a solvent or (for osmosis) between two solutions with different concentrations.

8.14 Molality

Two measures of concentration that are useful for the study of colligative properties, because they indicate the relative numbers of solute and solvent molecules, are mole fraction and molality. We first met the *mole fraction*, x, in Section 4.8, where we saw that it is the ratio of the number of moles of a species to the total number of moles of all the species present in a mixture. The **molality** of a solute is the amount of solute species (in moles) in a solution divided by the mass of the solvent (in kilograms):

$$\text{Molality of solute} = \frac{\text{amount of solute (mol)}}{\text{mass of solvent (kg)}} \qquad (3)^*$$

Like mole fraction but unlike molarity, the molality is independent of temperature. The units of molality are moles of solute per kilogram of solvent (mol·kg^{-1}); these units are often denoted m (for example, a $1\ m\ NiSO_4(aq)$ solution) and read "molal." Note the emphasis on *solvent* in the definition. To prepare a $1\ m$ $NiSO_4(aq)$ solution, we dissolve 1 mol $NiSO_4$ in 1 kg of water (Fig. 8.26).

A note on good practice: Note that when we refer to the molality (or molarity) of a solute we include the formula of the solute.

EXAMPLE 8.5   **Sample exercise: Calculating the molality of a solute**

What is the molality of NaCl in a solution prepared by dissolving 10.5 g of sodium chloride in 250. g of water?

SOLUTION Convert the mass of sodium chloride into amount of NaCl in moles, express the mass of water in kilograms, and then divide the amount of solute by the mass of solvent.

From molality = $n_{\text{solute}}(\text{mol})/m_{\text{solvent}}(\text{kg})$,   Molality of NaCl = $\dfrac{(10.5\,\text{g})/(58.44\,\text{g·mol}^{-1})}{0.250\,\text{kg}}$

$= 0.719\,\text{mol·kg}^{-1}$

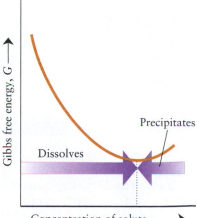

FIGURE 8.25 At low concentrations of solute, dissolving is accompanied by a decrease in Gibbs free energy of the system; so it is spontaneous. At high concentrations, dissolving is accompanied by an increase in Gibbs free energy; so the reverse process, precipitation, is spontaneous. The concentration of a saturated solution corresponds to the state of lowest Gibbs free energy at the temperature of the experiment.

The word *colligative* means "depending on the collection."

FIGURE 8.26 The steps taken to prepare a solution of a given molality. First (left), the required masses of solute and solvent are measured out. Then (right), the solute is dissolved in the solvent. Compare this procedure with that for preparing a solution of given molarity (see Fig. G.8).

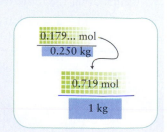

We report this molality as 0.719 m NaCl(aq).

SELF-TEST 8.8A Calculate the molality of $ZnCl_2$ in a solution prepared by dissolving 4.11 g of $ZnCl_2$ in 150. g of water.

[*Answer:* 0.201 mol·kg^{-1}]

SELF-TEST 8.8B Calculate the molality of $KClO_3$ in a solution prepared by dissolving 7.36 g of $KClO_3$ in 200. g of water.

TOOLBOX 8.1   HOW TO USE THE MOLALITY

CONCEPTUAL BASIS

The molality is the concentration of solute in moles per kilogram of solvent. Its value is independent of the temperature and is directly proportional to the numbers of solute and solvent molecules in the solution. To convert molarity to molality, we note that the former is defined in terms of the volume of the solution, so we convert that overall volume to the mass of solvent present.

PROCEDURE

1   Calculating the mass of solute in a given mass of solvent from the molality

Step 1 Calculate the amount of solute molecules, n_{solute}, present in a given mass of solvent, $m_{solvent}$, by rearranging the equation defining molality (Eq. 3) into

$$n_{solute} = \text{molality} \times m_{solvent}$$

Step 2 Use the molar mass of the solute, M_{solute}, to find the mass of the solute from its amount:

$$m_{solute} = n_{solute} M_{solute}$$

2   Calculating the molality from a mole fraction

Step 1 Consider a solution comprised of a total of 1 mol of molecules. If the mole fraction of the solute is x_{solute}, the amount of solute molecules in a total of 1 mol of molecules is

$$n_{solute} = x_{solute} \text{ mol}$$

Step 2 If there is only one solute, the mole fraction of solvent molecules is $1 - x_{solute}$. The amount of solvent molecules in a total of 1 mol of molecules is then $n_{solvent} = (1 - x_{solute})$ mol. Convert this amount into mass in grams by using the molar mass of the solvent, $M_{solvent}$:

$$m_{solvent} = n_{solvent} M_{solvent} = \{(1 - x_{solute}) \text{ mol}\} M_{solvent}$$

and then convert grams into kilograms.

Step 3 Calculate the molality of the solute by dividing the amount of solute molecules (step 1) by the mass of solvent (step 2).

Example 8.6 shows how to carry out this conversion.

3   Calculating the mole fraction from the molality

Step 1 Consider a solution containing exactly 1 kg of solvent. Convert that mass of solvent into an amount of solvent molecules, $n_{solvent}$, by using the molar mass, $M_{solvent}$, of the solvent:

$$n_{solvent} = \frac{1 \text{ kg}}{M_{solvent}} = \frac{10^3 \text{ g}}{M_{solvent}}$$

We already know the amount of solute molecules, n_{solute}, in the solution:

$$n_{solute} = \text{molality} \times (1 \text{ kg})$$

Step 2 Calculate the mole fractions from the amounts of solvent and solute molecules:

$$x_{solute} = \frac{n_{solute}}{n_{solute} + n_{solvent}}$$

4   Calculating the molality, given the molarity

The conversion is more involved because the molality is defined in terms of the mass of *solvent* but the molarity is defined in terms of the volume of *solution*. To carry out the conversion we need to know the density of the solution.

Step 1 Calculate the total mass of exactly 1 L (10^3 mL) of solution by using the density, d, of the solution (not the solvent):

$$m_{solution} = d \times (10^3 \text{ mL})$$

Step 2 The molarity gives the amount of solute in 1 L of solution. Use the molar mass of the solute to convert that amount into the mass of solute present in 1 L of solution:

$$n_{solute} = cV = \text{molality} \times (1 \text{ L})$$
$$m_{solute} = n_{solute} M_{solute} = \text{molality} \times (1 \text{ L}) \times M_{solute}$$

Step 3 Subtract the mass of solute (step 2) from the total mass (step 1) to find the mass of solvent in 1 L of solution,

$$m_{solvent} = m_{solution} - m_{solute}$$

and convert the mass of solvent into kilograms.

Step 4 The molality is the amount of solute (given by the molarity) divided by the mass of the solvent in kilograms (step 3).

See Example 8.7.

EXAMPLE 8.6   Sample exercise: Calculating a molality from a mole fraction

What is the molality of benzene, C_6H_6, dissolved in toluene, $C_6H_5CH_3$, in a solution for which the mole fraction of benzene is 0.150?

SOLUTION From procedure 2 in Toolbox 8.1:

Step 1 Find the amount of solute molecules in a total of exactly 1 mol of solution molecules.	$n_{benzene} = 0.150 \times 1\ mol = 0.150\ mol$

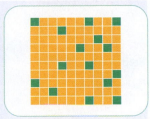

Step 2 Find the mass of solvent (in kg) present.

Mass of toluene (kg)

$$= (1 - 0.150)\ mol \times (92.13\ g\cdot mol^{-1}) \times \left(\frac{1\ kg}{10^3\ g}\right)$$

$$= 0.850 \times 92.13 \times 10^{-3}\ kg\ (=0.0783\ kg)$$

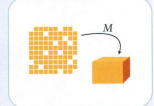

Step 3 From molality $= n_{colute}/m_{solvent}$,

$$\text{Molality of } C_6H_6 = \frac{0.150\ mol}{0.850 \times 92.13 \times 10^{-3}\ kg}$$

$$= 1.92\ mol\cdot kg^{-1}$$

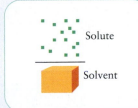

SELF-TEST 8.9A Calculate the molality of toluene dissolved in benzene, given that the mole fraction of toluene is 0.150. The molar mass of benzene is 78.11 g·mol^{-1}.

[*Answer:* 2.26 mol·kg^{-1}]

SELF-TEST 8.9B Calculate the molality of methanol in aqueous solution, given that the mole fraction of methanol is 0.250.

EXAMPLE 8.7 **Sample exercise: Converting molarity into molality**

Find the molality of sucrose, $C_{12}H_{22}O_{11}$, in 1.06 M $C_{12}H_{22}O_{11}$(aq), which is known to have density 1.14 g·mL^{-1}.

SOLUTION From procedure 4 in Toolbox 8.1:

Step 1 Find the mass of exactly 1 L (10^3 mL) of solution from $m = d \times (10^3\ mL)$.

$$m_{solution} = (1.14\ g\cdot mL^{-1}) \times (10^3\ mL) = 1.14 \times 10^3\ g$$

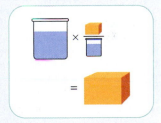

Step 2 Find the mass of solute in exactly 1 L of solution.

$$m_{sucrose} = (1.06\ mol\cdot L^{-1}) \times (1\ L) \times (342.3\ g\cdot mol^{-1})$$

$$= 1.06 \times 342.3\ g$$

$$= 363\ g$$

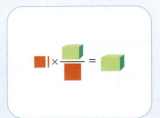

Step 3 Find the mass of water present in exactly 1 L of solution from $m_{water} = m_{solution} - m_{solute}$

$$m_{water} = 1.14 \times 10^3 - 363\ g = 7.8 \times 10^2\ g,\ or\ 0.78\ kg$$

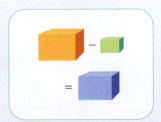

Step 4 From molality $=$ $n_{solute}/m_{solvent}$,

$$\text{Molality of } C_{12}H_{22}O_{11} = \frac{1.06 \text{ mol}}{0.78 \text{ kg}} = 1.4 \text{ mol·kg}^{-1}$$

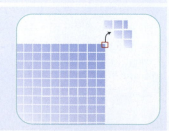

Vapor pressure, P

P_{pure}

0
0   0.25   0.5   0.75   1.0

Mole fraction of solvent, x_A

FIGURE 8.27 Raoult's law predicts that the vapor pressure of a solvent in a solution should be proportional to the mole fraction of the solvent molecules. The horizontal axis shows the mole fraction of the red molecules of substance A in the pure solute, three different solutions, and pure solvent, pictured below the graph. The vapor pressure of pure A is marked as P_{pure}.

SELF-TEST 8.10A Battery acid is 4.27 M H_2SO_4(aq) and has a density of 1.25 g·cm^{-3}. What is the molality of H_2SO_4 in the solution?

[***Answer:*** 5.14 m H_2SO_4(aq)]

SELF-TEST 8.10B The density of 1.83 M NaCl(aq) is 1.07 g·cm^{-3}. What is the molality of NaCl in the solution?

The molality of a solute in a solution is the amount (in moles) of solute divided by the mass (in kilograms) of solvent used to prepare the solution.

8.15 Vapor-Pressure Lowering

The French scientist François-Marie Raoult, who spent much of his life measuring vapor pressures, discovered that *the vapor pressure of a solvent is proportional to its mole fraction in a solution*. This statement, which is called **Raoult's law**, is normally written

$$P = x_{solvent}P_{pure} \tag{4}*$$

where P is the vapor pressure of the solvent in the solution, $x_{solvent}$ is the mole fraction of the solvent, and P_{pure} is the vapor pressure of the pure solvent (Fig. 8.27). According to this equation, at any given temperature, the vapor pressure of the solvent at that temperature is directly proportional to the mole fraction of solvent molecules in the solution. For instance, if 9 in 10 of the molecules present in a solution are solvent molecules, then the vapor pressure of the solvent is nine-tenths that of the pure solvent.

EXAMPLE 8.8   **Using Raoult's law**

Calculate the vapor pressure of water at 20.°C in a solution prepared by dissolving 10.00 g of the nonelectrolyte sucrose, $C_{12}H_{22}O_{11}$, in 100.0 g of water.

STRATEGY Expect a lower vapor pressure when the solute is present. Calculate the mole fraction of the solvent (water) in the solution and then apply Raoult's law. To use Raoult's law, we need the vapor pressure of the pure solvent (Table 8.2 or 8.3).

SOLUTION

Find the amount of each species from the molar masses of solute and solvent.

$$\text{Amount of } C_{12}H_{22}O_{11} = \frac{10.00 \text{ g}}{342.3 \text{ g·mol}^{-1}} = \frac{10.00}{342.3} \text{ mol}$$

$$\text{Amount of } H_2O = \frac{100.0 \text{ g}}{18.02 \text{ g·mol}^{-1}} = \frac{100.0}{18.02} \text{ mol}$$

From $x_{solvent} = n_{solvent}/(n_{solute} + n_{solvent})$,

$$x_{water} = \frac{100.0/18.02 \text{ mol}}{(10.00/342.3 + 100.0/18.02) \text{ mol}}$$

$$= 0.995$$

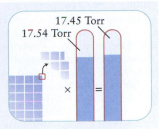

From $P = x_{solvent}P_{pure}$ and the vapor pressure of pure solvent (Table 8.3),

$$P_{water} = \frac{100.0/18.02}{10.00/342.3 + 100.0/18.02} \times (17.54 \text{ Torr})$$

$$= 17.45 \text{ Torr}$$

SELF-TEST 8.11A Calculate the vapor pressure of water at 90.°C for a solution prepared by dissolving 5.00 g of glucose ($C_6H_{12}O_6$) in 100. g of water. The vapor pressure of pure water at 90.°C is 524 Torr.

[*Answer:* 521 Torr]

SELF-TEST 8.11B Calculate the vapor pressure of ethanol in kilopascals (kPa) at 19°C for a solution prepared by dissolving 2.00 g of cinnamaldehyde, C_9H_8O, in 50.0 g of ethanol, C_2H_5OH. The vapor pressure of pure ethanol at that temperature is 5.3 kPa.

A hypothetical solution that obeys Raoult's law exactly at all concentrations is called an **ideal solution**. In an ideal solution, the interactions between solute and solvent molecules are the same as the interactions between solvent molecules in the pure state and between solute molecules in the pure state. Consequently, the solute molecules mingle freely with the solvent molecules. That is, in an ideal solution, the enthalpy of solution is zero. Solutes that form nearly ideal solutions are often similar in composition and structure to the solvent molecules. For instance, methylbenzene (toluene), $C_6H_5CH_3$, forms nearly ideal solutions with benzene, C_6H_6. Real solutions do not obey Raoult's law at all concentrations; but the lower the solute concentration, the more closely they resemble ideal solutions. Raoult's law is another example of a limiting law (Section 4.4), which in this case becomes increasingly valid as the concentration of the solute approaches zero. A solution that does not obey Raoult's law at a particular solute concentration is called a **nonideal solution**. Real solutions are approximately ideal at solute concentrations below about 0.1 M for nonelectrolyte solutions and 0.01 M for electrolyte solutions. The greater departure from ideality in electrolyte solutions arises from the interactions between ions, which occur over a long distance and hence have a pronounced effect. Unless stated otherwise, we shall assume that all the solutions that we meet are ideal.

Electrolyte solutions, which have ionic solutes, and nonelectrolyte solutions, which have molecular solutes, were introduced in Section I.

The lowering of the vapor pressure of a solvent due to the presence of a solute has a thermodynamic basis. We saw in Section 8.2 that at equilibrium, and in the absence of any solute, the molar Gibbs free energy of the vapor is equal to that of the pure liquid solvent. In an ideal solution, a solute increases the disorder and therefore the entropy of the liquid phase: we can no longer be sure that a molecule picked at random from the solution will be a solvent molecule. Because the entropy of the liquid phase is raised by a solute but the enthalpy is left unchanged, overall there is a decrease in the molar Gibbs free energy of the solvent. At equilibrium, the molar Gibbs free energy of the vapor must equal that of the solvent. Because the Gibbs free energy of the solvent has been lowered, for the two phases to remain in equilibrium it follows that the Gibbs free energy of the vapor must be lower, too. Because the Gibbs free energy of a gas depends on its pressure (recall Fig. 8.4), it follows that the vapor pressure also must be lower. In other words, *the vapor pressure of the solvent is lower in the presence of the solute* (Fig. 8.28).

Remember that $G = H - TS$; so if S increases, then G decreases.

The vapor pressure of a solvent is reduced by the presence of a nonvolatile solute: in an ideal solution, the vapor pressure of the solvent is proportional to the mole fraction of solvent.

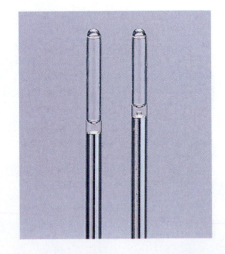

FIGURE 8.28 The vapor pressure of a solvent is lowered by a nonvolatile solute. The barometer tube on the left has a small volume of pure water floating on the mercury. That on the right has a small volume of 10 *m* NaCl(aq) and a lower vapor pressure. Note that the column on the right is depressed less by the vapor in the space above the mercury than the one on the left, showing that the vapor pressure is lower when solute is present.

FIGURE 8.29 (a) The molar Gibbs free energy of a liquid and its vapor both decrease with increasing temperature, but that of the vapor decreases more sharply. The vapor is the stable phase at temperatures higher than the point of intersection of the two lines (the boiling point). (b) When a nonvolatile solute is present, the molar Gibbs free energy of the solvent is lowered (an entropy effect), but that of the vapor is left unchanged. The point of intersection of the lines moves to a slightly higher temperature.

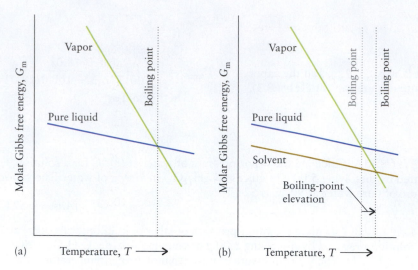

8.16 Boiling-Point Elevation and Freezing-Point Depression

Because the presence of a nonvolatile solute lowers the vapor pressure of the solvent, the boiling point of the solvent rises. This increase is called **boiling-point elevation**. The elevation of the boiling point has the same origin as vapor-pressure lowering and is also due to the effect of the solute on the entropy of the solvent.

Figure 8.29a shows how the molar Gibbs free energies of the liquid and vapor phases of a pure solvent vary with temperature. Above the boiling point, the molar Gibbs free energy of the vapor phase is lower than that of the liquid; therefore, the vapor is the more stable phase. We have already seen that the presence of a solute in the liquid phase of the solvent increases the entropy of the solute and therefore lowers its Gibbs free energy. As shown in Fig. 8.29b, the lines representing the molar Gibbs free energies of the liquid solution and the vapor intersect at a higher temperature than they do for the pure solvent, and so the boiling point is higher in the presence of the solute. The increase is usually quite small and is of little practical importance in science. A 0.1 m aqueous sucrose solution, for instance, boils at 100.05°C.

Of broader practical significance is **freezing-point depression**, the lowering of the freezing point of the solvent caused by a solute. For example, seawater freezes about 1°C lower than fresh water. People living in regions with cold winters make use of the depression of the freezing point when they spread salt on walkways and roads to lower the freezing point of water. In the laboratory, chemists make use of the effect to judge the purity of a solid compound: if impurities are present, the compound's melting point is lower than the accepted value.

Figure 8.30a shows how the molar Gibbs free energies of the liquid and solid phases of a pure solvent vary with temperature. The solid is the more stable phase

FIGURE 8.30 (a) The molar Gibbs free energy of a solid and its liquid phase both decrease with increasing temperature, but that of the liquid decreases slightly more steeply. The liquid is the stable phase at temperatures higher than the point of intersection of the two lines. (b) When a solute is present, the molar Gibbs free energy of the solvent is lowered (an entropy effect), but that of the solid is left unchanged. The point of intersection of the lines moves to a lower temperature.

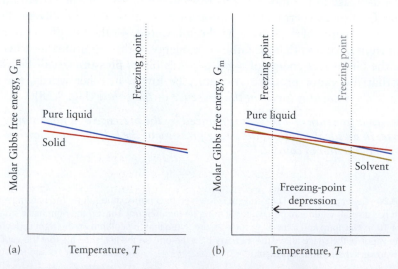

TABLE 8.8 Boiling-Point and Freezing-Point Constants

Solvent	Freezing point (°C)	k_f (K·kg·mol^{-1})	Boiling point (°C)	k_b (K·kg·mol^{-1})
acetone	−95.35	2.40	56.2	1.71
benzene	5.5	5.12	80.1	2.53
camphor	179.8	39.7	204	5.61
carbon tetrachloride	−23	29.8	76.5	4.95
cyclohexane	6.5	20.1	80.7	2.79
naphthalene	80.5	6.94	217.7	5.80
phenol	43	7.27	182	3.04
water	0	1.86	100.0	0.51

below the freezing point, where the molar Gibbs free energy of the solid is lower than that of the liquid. The presence of a solute in the liquid phase of the solvent raises the entropy of the solvent and hence lowers its Gibbs free energy (Fig. 8.30b). However, because the solute is insoluble in the solid solvent, the Gibbs free energy of the solid phase of the solvent is unchanged. The lines representing the molar Gibbs free energies of the liquid and solid phases of the solvent intersect at a lower temperature than they do for the pure solvent, and so the freezing point is lower in the presence of the solute.

It is found empirically and can be justified thermodynamically that the freezing-point depression for an ideal solution is proportional to the molality of the solute. For a nonelectrolyte solution,

$$\text{Freezing-point depression} = k_f \times \text{molality} \qquad (5a)$$

The constant k_f is called the **freezing-point constant** of the solvent; it is different for each solvent and must be determined experimentally (Table 8.8). The effect is quite small: for instance, for a 0.1 m $C_{12}H_{22}O_{11}$(aq) (sucrose) solution,

$$\text{Freezing-point depression} = (1.86 \text{ K·kg·mol}^{-1}) \times (0.1 \text{ mol·kg}^{-1}) = 0.2 \text{ K}$$

Hence the water in the solution freezes at −0.2°C. A similar expression (boiling-point elevation = k_b × molality) is used to relate the elevation of boiling point to the molality of the solute.

SELF-TEST 8.12A Use the data in Table 8.8 to determine at what temperature a 0.20 mol·kg^{-1} solution of the analgesic codeine, $C_{18}H_{21}NO_3$, in benzene will freeze.

[*Answer:* 4.5°C]

SELF-TEST 8.12B Use the data in Table 8.8 to determine at what temperature a 0.050 mol·kg^{-1} solution of the insecticide malathion, $C_{10}H_{19}O_6PS_2$, in camphor will freeze.

In an electrolyte solution, each formula unit contributes two or more ions. Sodium chloride, for instance, dissolves to give Na^+ and Cl^- ions, and both kinds of ions contribute to the depression of the freezing point. The cations and anions contribute nearly independently in very dilute solutions, and so the total solute molality is twice the molality of NaCl formula units. In place of Eq. 5a we write

$$\text{Freezing-point depression} = ik_f \times \text{molality} \qquad (5b)*$$

Here, i, the **van 't Hoff i factor**, is determined experimentally. In a very dilute solution (less than about 10^{-3} mol·L^{-1}), when all ions are independent, $i = 2$ for MX salts such as NaCl, $i = 3$ for MX$_2$ salts such as CaCl$_2$, and so on. For dilute nonelectrolyte solutions, $i = 1$. The i factor is so unreliable, however, that it is best to confine quantitative calculations of freezing-point depression to nonelectrolyte solutions. Even these solutions must be dilute enough to be approximately ideal.

The i factor can be used to help determine the extent to which a substance is dissociated into ions in solution. For example, in dilute solution, HCl has an i factor of 1 in toluene and 2 in water. These values suggest that HCl retains its molecular form in toluene but is fully deprotonated in water. The strength of a weak acid in water can be estimated in this way. In an aqueous solution of a weak acid that is 5% deprotonated (5% of the acid molecules have given up their protons), each deprotonated molecule produces two ions and $i = 0.95 + (0.05 \times 2) = 1.05$.

Cryoscopy is the determination of the molar mass of a solute by measuring the depression of freezing point that it causes when dissolved in a solvent. Camphor has been used as the solvent for organic compounds because it has a large freezing-point constant; consequently, solutes depress its freezing point significantly. However, this procedure is rarely used in modern laboratories because techniques such as mass spectrometry give far more reliable results (see Major Technique 6, following Chapter 18). The procedure is described in Toolbox 8.2 at the end of Section 8.17.

The presence of a solute lowers the freezing point of a solvent; if the solute is nonvolatile, the boiling point is also raised. The freezing-point depression can be used to calculate the molar mass of the solute. If the solute is an electrolyte, the extent of its dissociation, protonation, or deprotonation must also be taken into account.

8.17 Osmosis

The name *osmosis* comes from the Greek word for "push."

Osmosis is the flow of solvent through a membrane into a more concentrated solution. The phenomenon can be demonstrated in the laboratory when a solution and the pure solvent are separated by a **semipermeable membrane**, a membrane that permits only certain types of molecules or ions to pass through (Fig. 8.31). Cellulose acetate, for instance, allows water molecules to pass through it, but not solute molecules or ions with their bulky coating of hydrating water molecules. Initially, the heights of the solution and the pure solvent shown in the illustration are the same. However, the level of the solution inside the tube begins to rise as pure solvent passes through the membrane into the solution. At equilibrium, the pressure exerted by the higher level of the column of solution is sufficiently great that the rate of flow of molecules through the membrane is the same in each direction, and so the net flow is zero. The pressure needed to stop the flow of solvent is called the **osmotic pressure**, Π (the Greek uppercase letter pi). The greater the osmotic pressure, the greater the height of the solution needed to reduce the net flow to zero.

The pressure exerted by a vertical column of liquid is proportional to its height: see Eq. 2 in Section 4.2.

Life depends on osmosis. Biological cell walls act as semipermeable membranes that allow water, small molecules, and hydrated ions to pass (Fig. 8.32). However, they block the passage of the enzymes and proteins that have been synthesized within the cell. The higher concentration of solutes within a cell compared to the solution outside the cell gives rise to an osmotic pressure, and water passes into the more concentrated solution in the interior of the cell, carrying small nutrient molecules with it. This influx of water also keeps the cell turgid (swollen). When the water supply is cut off, the turgidity is lost and the cell becomes dehydrated. In a plant, this dehydration results in wilting. Salted meat is preserved from bacterial attack by osmosis. In this case, the concentrated salt solution dehydrates—and kills—the bacteria by causing water to flow out of them.

FIGURE 8.31 An experiment to illustrate osmosis. Initially, the tube contained a sucrose solution and the beaker contained pure water: the initial heights of the two liquids were the same. At the stage shown here, water has passed into the solution through the membrane by osmosis, and the level of solution in the tube has risen above that of the pure water. The large inset shows the molecules in the pure solvent (below the membrane) tending to join those in the solution (above the membrane) because the presence of solute molecules there has led to increased disorder. The small inset shows just the solute molecules; the yellow arrow shows the direction of flow of solvent molecules.

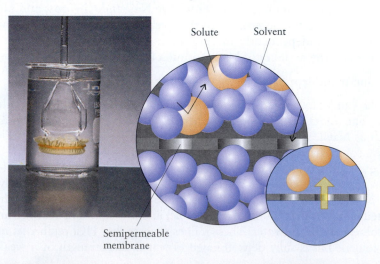

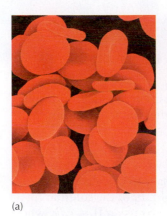
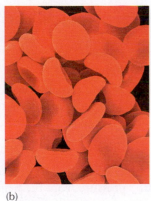

(a)   (b)   (c)

FIGURE 8.32 (a) Red blood cells need to be in a solution of the correct solute concentration if they are to function properly. (b) When the solution is very dilute, too much water passes into them and they burst. (c) When the solution is too concentrated, water flows out of them and they shrivel up.

Because osmosis is a thermodynamic property, we can expect it to be related to the effect of the solute on the enthalpy and entropy of the solution: solvent flows until the molar Gibbs free energy of the solvent is the same on each side of the membrane We have already seen several times that a solute lowers the molar Gibbs free energy of the solution below that of the pure solvent, and solvent therefore has a tendency to pass into the solution (Fig. 8.33).

The same van 't Hoff responsible for the i factor showed that the osmotic pressure of a solution is related to the molarity, c, of the solute in the solution:

$$\Pi = iRTc \tag{6}*$$

where i is the i factor (Section 8.16), R is the gas constant, and T is the temperature. This expression is now known as the **van 't Hoff equation**. Notice that the osmotic pressure depends only on temperature and the total molar concentration of solute. It is independent of the identities of both the solute and the solvent. However, because the densities of solvents differ, the height of the column of solvent depends on its identity (Fig. 8.34).

The van 't Ho'ff equation is used to determine the molar mass of a solute from osmotic pressure measurements. This technique, which is called **osmometry**, involves the determination of the osmotic pressure of a solution prepared by making up a known volume of solution of a known mass of solute with an unknown molar mass. Osmometry is very sensitive, even at low concentrations, and is commonly used to determine very large molar masses, such as those of polymers.

SELF-TEST 8.13A What is the osmotic pressure of 0.0100 M KCl(aq) at 298 K? (Assume $i = 2$.)

[*Answer:* 0.49 atm]

SELF-TEST 8.13B What is the osmotic pressure of a 0.120 M sucrose solution at 298 K?

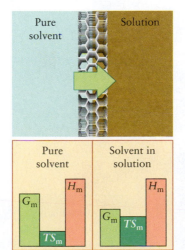

FIGURE 8.33 On the left of the semipermeable membrane is the pure solvent with its characteristic molar enthalpy, entropy, and Gibbs free energy. On the right is the solution. The molar Gibbs free energy of the solvent is lower in the solution (an entropy effect), and so there is a spontaneous tendency for the solvent to flow into the solution.

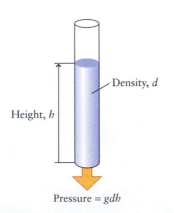

FIGURE 8.34 The pressure at the base of a column of fluid is equal to the product of the acceleration of free fall, g, the density, d, of the liquid, and the height, h, of the column.

TOOLBOX 8.2   HOW TO USE COLLIGATIVE PROPERTIES TO DETERMINE MOLAR MASS

CONCEPTUAL BASIS

The lowering of freezing point and the generation of osmotic pressure both depend on the total concentration of solute particles. Therefore, by using the colligative property to determine the amount of solute present, and knowing its mass, we can infer its molar mass.

PROCEDURE

1   Cryoscopy

Step 1 Convert the observed freezing-point depression into solute molality by writing Eq. 5b in the form

$$\text{Molality} = \frac{\text{freezing-point depression}}{i k_f}$$

Take the freezing-point constant from Table 8.8.

Step 2 Calculate the amount of solute, n_{solute} (in moles), in the sample by multiplying the molality by the mass of solvent, $m_{solvent}$ (in kilograms):

$$n_{solute} = \text{molality} \times m_{solvent}$$

Step 3 Determine the molar mass of the solute by dividing the given mass of solute, m_{solute} (in grams) by the amount in moles (step 2).

$$M_{solute} = \frac{m_{solute}}{n_{solute}}$$

This procedure is illustrated in Example 8.9.

2   Osmometry

Step 1 Convert the observed osmotic pressure into solute molarity (not molality) by writing Eq. 6 in the form

$$c = \frac{\Pi}{iRT}$$

In some cases, it may be necessary to calculate the osmotic pressure from the height, h, of the solution (in an apparatus like that in Fig. 8.31) by using $\Pi = gdh$, where d is the density of the solution and g is the acceleration of free fall (see inside back cover).

Step 2 Use this molarity to calculate the amount of solute, n_{solute} (in moles), in the stated volume, V (in liters), of solution:

$$n_{solute} = cV$$

Step 3 Determine the molar mass of the solute by dividing the given mass of solute, m_{solute} (in grams), by its amount (step 2), as in the procedure for cryoscopy.

This procedure is illustrated in Example 8.10.

EXAMPLE 8.9   **Sample exercise: Determining molar mass cryoscopically**

The addition of 0.24 g of sulfur to 100. g of the solvent carbon tetrachloride lowers the solvent's freezing point by 0.28°C. What is the molar mass and molecular formula of sulfur?

SOLUTION

Use the procedure for cryoscopy in Toolbox 8.2. Sulfur is a nonelectrolyte, so $i = 1$.

Step 1 Assume that $i = 1$. Then find the molality of solute.

$$\text{Molality of solute} = \frac{0.28\ \text{K}}{29.8\ \text{K·kg·mol}^{-1}}$$

$$= \frac{0.28}{29.8}\ \text{mol·kg}^{-1}$$

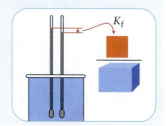

Step 2 Calculate the amount of solute present.

$$n_{S_x} = (0.100\ \text{kg}) \times \left(\frac{0.28}{29.8}\ \text{mol·kg}^{-1}\right)$$

$$= \frac{0.100 \times 0.28}{29.8}\ \text{mol}$$

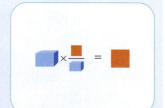

Step 3 Determine the molar mass of the solute.

$$M_{S_x} = \frac{0.24\ \text{g}}{(0.100 \times 0.28/29.8)\ \text{mol}}$$

$$= 2.6 \times 10^2\ \text{g·mol}^{-1}$$

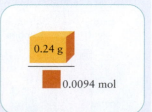

Use the molar mass of atomic sulfur to find the value of x in the molecular formula S_x.

$$x = \frac{2.6 \times 10^2 \text{ g·mol}^{-1}}{32.1 \text{ g·mol}^{-1}} = 8.1$$

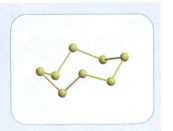

Elemental sulfur is therefore composed of S_8 molecules.

SELF-TEST 8.14A When 250. mg of eugenol, the compound responsible for the odor of oil of cloves, was added to 100. g of camphor, it lowered the freezing point of camphor by 0.62°C. Calculate the molar mass of eugenol.

[*Answer:* 1.6×10^2 g·mol^{-1} (actual: 164.2 g·mol^{-1})]

SELF-TEST 8.14B When 200. mg of linalool, a fragrant compound found in cinnamon oil from Sri Lanka, was added to 100. g of camphor, it lowered the freezing point of camphor by 0.51°C. What is the molar mass of linalool?

EXAMPLE 8.10 Sample exercise: Using osmometry to determine molar mass

The osmotic pressure due to 2.20 g of polyethylene (PE) dissolved in enough benzene to produce 100.0 mL of solution was 1.10×10^{-2} atm at 25°C. Calculate the average molar mass of the polymer, which is a nonelectrolyte.

SOLUTION Polymers commonly have very high molar masses (of the order of kilograms per mole). Use the procedure for osmometry in Toolbox 8.2. Because polyethylene is a nonelectrolyte, $i = 1$. Use R in the units that match the data: in this case, liters and atmospheres.

Step 1 From $c = \Pi/iRT$,

$$c = \frac{1.10 \times 10^{-2} \text{ atm}}{(0.0821 \text{ L·atm·K}^{-1}\text{·mol}^{-1}) \times (298 \text{ K})}$$

$$= \frac{1.10 \times 10^{-2}}{0.0821 \times 298} \text{ mol·L}^{-1}$$

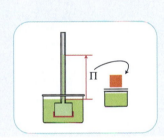

Step 2 Find the amount of solute molecules in the solution from $n = cV$.

$$n_{PE} = (0.100 \text{ L}) \times \left(\frac{1.10 \times 10^{-2}}{0.0821 \times 298} \text{ mol·L}^{-1} \right)$$

$$= \frac{0.100 \times 1.10 \times 10^{-2}}{0.0821 \times 298} \text{ mol } (= 4.50 \times 10^{-5} \text{ mol})$$

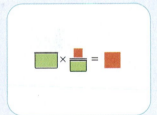

Step 3 Find the molar mass from $M_{solute} = m_{solute}/n_{solute}$.

$$\text{Molar mass of PE} = \frac{2.20 \text{ g}}{\left(\dfrac{0.100 \times 1.10 \times 10^{-2}}{0.0821 \times 298} \right) \text{ mol}}$$

$$= 4.89 \times 10^4 \text{ g·mol}^{-1}$$

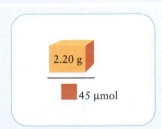

We report this molar mass as 48.9 kg·mol^{-1}.

SELF-TEST 8.15A The osmotic pressure of 3.0 g of polystyrene dissolved in enough benzene to produce 150. mL of solution was 1.21 kPa at 25°C. Calculate the average molar mass of the sample of polystyrene.

[*Answer:* 41 kg·mol^{-1}]

SELF-TEST 8.15B The osmotic pressure of 1.50 g of polymethyl methacrylate dissolved in enough methylbenzene to produce 175 mL of solution was 2.11 kPa at 20°C. Calculate the average molar mass of the sample of polymethyl methacrylate.

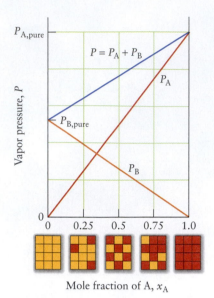

FIGURE 8.35 The vapor pressures of the two components of an ideal binary mixture obey Raoult's law. The total vapor pressure is the sum of the two partial vapor pressures (Dalton's law). The insets below the graph represent the mole fraction of A.

In **reverse osmosis**, a pressure greater than the osmotic pressure is applied to the solution side of the semipermeable membrane. This application of pressure increases the rate at which solvent molecules leave the solution and thus reverses the flow of solvent, forcing it to flow from the solution to pure solvent. Reverse osmosis is used to remove salts from seawater to produce fresh water for drinking and irrigation. The water is almost literally squeezed out of the salt solution through the membrane. The technological challenge is to fabricate new membranes that are strong enough to withstand high pressures and do not easily become clogged. Commercial plants use cellulose acetate membranes at pressures as high as 70 atm.

Osmosis is the flow of solvent through a semipermeable membrane into a solution; the osmotic pressure is proportional to the molar concentration of the solute. Osmometry is used to determine the molar masses of compounds with large molecules, such as polymers; reverse osmosis is used in water purification.

BINARY LIQUID MIXTURES

The following three sections of this chapter examine how the vapor pressure varies with composition when both components of a mixture are volatile and how that information can be used to separate them by distillation. Distillation, which we first encountered in Section G, is used to separate the many compounds that make up petroleum and to purify alternative fuels such as ethanol and methanol.

8.18 The Vapor Pressure of a Binary Liquid Mixture

Consider an ideal binary mixture of the volatile liquids A and B. We could think of A as benzene, C_6H_6, and B as toluene (methylbenzene, $C_6H_5CH_3$), for example, because these two compounds have similar molecular structures and so form nearly ideal solutions. Because the mixture can be treated as ideal, each component has a vapor pressure given by Raoult's law:

$$P_A = x_{A,liquid}P_{A,pure} \qquad P_B = x_{B,liquid}P_{B,pure}$$

In these equations, $x_{A,liquid}$ is the mole fraction of A in the liquid mixture and $P_{A,pure}$ is the vapor pressure of pure A; similarly, $x_{B,liquid}$ is the mole fraction of B in the liquid and $P_{B,pure}$ is the vapor pressure of pure B. According to Dalton's law (Section 4.8), the total pressure of the vapor, P, is the sum of these two partial pressures (Fig. 8.35):

$$P = P_A + P_B = x_{A,liquid}P_{A,pure} + x_{B,liquid}P_{B,pure}$$

EXAMPLE 8.11   **Sample exercise: Predicting the vapor pressure of a mixture of two liquids**

What is the vapor pressure of each component at 25°C and the total vapor pressure of a mixture in which one-third of the molecules are benzene (so $x_{benzene,liquid} = \frac{1}{3}$ and $x_{toluene,liquid} = \frac{2}{3}$)? The vapor pressures of benzene and toluene at 25°C are 94.6 and 29.1 Torr, respectively.

SOLUTION First, we use Raoult's law:

From $P_A = x_{A,liquid}P_{A,pure}$,

From $P_B = x_{B,liquid}P_{B,pure}$,

$P_{benzene} = \frac{1}{3}(94.6 \text{ Torr}) = 31.5 \text{ Torr}$

$P_{toluene} = \frac{2}{3}(29.1 \text{ Torr}) = 19.4 \text{ Torr}$

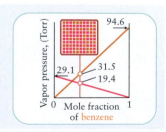

From Dalton's law, $P = P_A + P_B$,

$P_{total} = 31.5 + 19.4 \text{ Torr} = 50.9 \text{ Torr}$

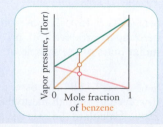

SELF-TEST 8.16A What is the total vapor pressure at 25°C of a mixture of 3.00 mol $C_6H_6(l)$ and 2.00 mol $CH_3C_6H_5(l)$?

[*Answer:* 68.4 Torr]

SELF-TEST 8.16B What is the total vapor pressure at 25°C of a mixture of equal masses of benzene and toluene?

The vapor of the mixture is richer than the liquid in the more volatile component (the component with the greater vapor pressure). Benzene, for instance, is more volatile than toluene, and so we can expect that the vapor in equilibrium with the liquid mixture will be richer in benzene than the liquid is. If we could express the composition of the vapor in terms of the composition of the liquid, we could confirm that the vapor is richer than the liquid in the more volatile component.

HOW DO WE DO THAT?

To express the composition of the vapor in equilibrium with the liquid phase of a binary liquid mixture, we first note that the definition of partial pressure ($P_A = x_A P$ for component A) and Dalton's law ($P = P_A + P_B$) allow us to express the composition of the vapor of a mixture of liquids A and B in terms of the partial pressures of the components:

$$x_{A,vapor} = \frac{P_A}{P} = \frac{P_A}{P_A + P_B}$$

and likewise for $x_{B,vapor}$. We can express the vapor pressure of the components and the total vapor pressure in terms of the composition of the liquid (by using Raoult's law):

$$x_{A,vapor} = \frac{x_{A,liquid}P_{A,pure}}{x_{A,liquid}P_{A,pure} + x_{B,liquid}P_{B,pure}}$$

This expression relates the composition of the vapor (in terms of the mole fraction of A in the vapor) in a binary mixture to the composition of the liquid (in terms of the mole fraction of A in the liquid, remembering that $x_B = 1 - x_A$). It is plotted for benzene and toluene in Fig. 8.36, and we see that $x_{benzene,vapor} > x_{benzene,liquid}$, just as we anticipated.

FIGURE 8.36 The composition of the vapor in equilibrium with a mixture of two volatile liquids (here, benzene and toluene) and its variation with the composition of the liquid. Note that the vapor is richer in benzene than the liquid mixture for each composition of the mixture. For instance, when the mole fraction of benzene is 0.333 in the liquid, in the vapor it is 0.619.

EXAMPLE 8.12 Sample exercise: Predicting the composition of the vapor in equilibrium with a binary liquid mixture

Find the mole fraction of benzene at 25°C in the vapor of a solution of benzene in toluene in which one-third of the molecules in the liquid are benzene. See Example 8.11 for data.

SOLUTION

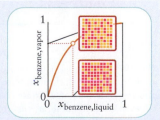

From $x_B = 1 - x_A$,

$$x_{toluene,liquid} = 1 - 0.333 = 0.667$$

From $x_{A,vapor} = x_{A,liquid}P_{A,pure}/(x_{A,liquid}P_{A,pure} + x_{B,liquid}P_{B,pure})$,

$$x_{benzene,vapor} = \frac{0.333 \times 94.6\ \text{Torr}}{0.333 \times 94.6\ \text{Torr} + 0.667 \times 29.1\ \text{Torr}}$$

$$= 0.619$$

Note that the mole fraction of benzene in the vapor is nearly twice that in the liquid, as shown in Fig. 8.36.

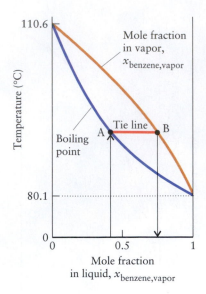

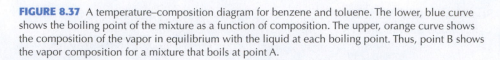

FIGURE 8.37 A temperature–composition diagram for benzene and toluene. The lower, blue curve shows the boiling point of the mixture as a function of composition. The upper, orange curve shows the composition of the vapor in equilibrium with the liquid at each boiling point. Thus, point B shows the vapor composition for a mixture that boils at point A.

Self-Test 8.17A (a) Determine the vapor pressure at 25°C of a solution of toluene in benzene in which the mole fraction of benzene is 0.900. (b) Calculate the mole fractions of benzene and toluene in the vapor.

[*Answer:* (a) 88.0 Torr; (b) 0.967 and 0.033]

Self-Test 8.17B (a) Determine the vapor pressure at 25°C of a solution of benzene in toluene in which the mole fraction of benzene is 0.500. (b) Calculate the mole fractions of benzene and toluene in the vapor.

The vapor of an ideal mixture of two volatile liquids is richer than the liquid in the more volatile component. The contribution of each component to the total vapor pressure and its mole fraction in the vapor can be calculated by combining Raoult's law and Dalton's law.

8.19 Distillation

The normal boiling point of a binary liquid mixture is the temperature at which the total vapor pressure is equal to 1 atm. If we were to heat a sample of pure benzene at a constant pressure of 1 atm, it would boil at 80.1°C. Similarly, pure toluene boils at 110.6°C. Because, at a given temperature, the vapor pressure of a mixture of benzene and toluene is intermediate between that of toluene and benzene, the boiling point of the mixture will be intermediate between that of the two pure liquids. In Fig. 8.37, which is called a **temperature–composition diagram**, the lower curve shows how the normal boiling point of the mixture varies with the composition.

The upper curve in Fig. 8.37 shows the composition of the vapor in equilibrium with the liquid mixture at the boiling point. To find the composition of the vapor, we simply look along the **tie line**, the horizontal line at the boiling point, and see where it cuts through the upper curve. So, if we heat a liquid mixture with the composition given by the vertical line through A in Fig. 8.37 ($x_{benzene,\ liquid}$ = 0.45) at a constant pressure of 1 atm, the mixture boils at the temperature corresponding to point A. At this temperature, the composition of the vapor in equilibrium with the liquid is given by point B ($x_{benzene,\ vapor}$ = 0.73).

When a mixture of benzene and toluene molecules with $x_{benzene}$ = 0.20 begins to boil (point A in Fig. 8.38), the initial composition of the vapor formed is given by point B ($x_{benzene}$ = 0.45). If we were to cool the vapor until it condenses, the first drop of condensed vapor, the **distillate**, would be richer in benzene than the original mixture and the liquid remaining in the pot would be richer in toluene. The separation is not very good: the vapor is still rich in toluene. However, if we take that drop of distillate and reheat it, the condensed liquid will boil at the temperature represented by point C, and the vapor above the boiling solution will have the composition D ($x_{benzene}$ = 0.73), as indicated by the tie line. Note that the distillate from this second stage of distillation is even richer in benzene than the distillate from the first stage. If we continued these steps of boiling, condensation, and boiling again, we would eventually obtain a very tiny amount of nearly pure benzene.

The process called **fractional distillation** uses a method of continuous redistillation to separate mixtures of liquids with similar boiling points, such as benzene and toluene: the mixture is heated and the rising vapor passes up through a tall column packed with material having a high surface area, such as glass beads (Fig. 8.39). The vapor begins to condense on the beads near the bottom of the column. However, as heating continues, the vapor condenses and vaporizes over and over as it rises; the liquid drips back into the boiling mixture. The vapor becomes richer in the component with the lower boiling point as it continues to rise through the column and passes out into the condenser. Therefore, the final distillate is nearly pure benzene, the more volatile of the components, whereas the liquid in the pot is nearly pure toluene.

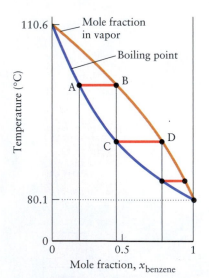

FIGURE 8.38 Some of the steps that represent fractional distillation of a mixture of two volatile liquids (benzene and toluene). The original mixture boils at A and its vapor has composition B. After condensation of the vapor, the resulting liquid boils at C and the vapor has composition D, and so on.

FIGURE 8.39 A schematic illustration of the process of fractional distillation. The temperature in the fractionating column decreases with height. The condensations and reboilings illustrated in Fig. 8.38 occur at increasing heights in the column. The less volatile component returns to the flask beneath the fractionating column, and the more volatile component escapes from the top, to be condensed and collected.

Low-boiling-point component

High-boiling-point component

If the original sample consists of several volatile liquids, the component liquids appear in the distillate in succession in a series of **fractions**, or samples of the distillate that boil in specific temperature ranges. Giant fractionating columns are used in industry to separate complex mixtures such as crude petroleum. We use the volatile fractions as natural gas (boiling below 0°C), gasoline (boiling in the range from 30°C to 200°C), and kerosene (from 180°C to 325°C). The less volatile fractions are used as diesel fuel (above 275°C). The residue that remains behind after distillation is asphalt, which is used for surfacing highways.

Volatile liquids can be separated by fractional distillation. Liquid and vapor are in equilibrium at each point in the fractionating column, but their compositions vary with height. As a result, the lowest-boiling-point component can be removed from the top of the column before the next-higher-boiling-point component distills.

8.20 Azeotropes

Most liquid mixtures are not ideal, and so their vapor pressures do not follow Raoult's law (Fig. 8.40). The direction of a deviation from Raoult's law can be correlated with the **enthalpy of mixing**, ΔH_{mix}, the enthalpy difference between the mixture and the pure components. The enthalpy of mixing of ethanol and benzene is positive—the mixing process is endothermic—and this mixture has a vapor pressure higher than that predicted by Raoult's law (a positive deviation). The enthalpy of mixing of acetone and chloroform is negative—the mixing process is exothermic—and this mixture has a vapor pressure lower than that predicted by the law (a negative deviation).

Deviations from Raoult's law can make it impossible to separate liquids by distillation. The temperature–composition diagrams for mixtures of ethanol and benzene and of acetone and chloroform show why. A positive deviation from Raoult's law means that the attractive forces between solute and solvent are lower than those between the molecules of the pure components. As a result, the boiling point of the mixture is lower than that predicted by Raoult's law. For some pairs of components, the boiling point of the mixture is in fact lower than the boiling point of either constituent (Fig. 8.41). A mixture for which the lowest boiling temperature is below

The enthalpy of mixing is the same as the enthalpy of solution, ΔH_{sol}, but ΔH_{mix} is used more commonly for the mixing of two liquids.

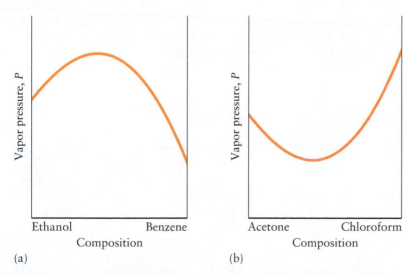

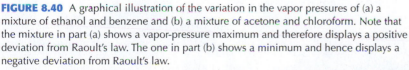

(a) Ethanol — Benzene / Composition

(b) Acetone — Chloroform / Composition

FIGURE 8.40 A graphical illustration of the variation in the vapor pressures of (a) a mixture of ethanol and benzene and (b) a mixture of acetone and chloroform. Note that the mixture in part (a) shows a vapor-pressure maximum and therefore displays a positive deviation from Raoult's law. The one in part (b) shows a minimum and hence displays a negative deviation from Raoult's law.

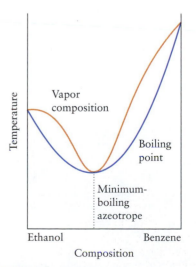

FIGURE 8.41 The temperature–composition diagram of a minimum-boiling azeotrope (such as ethanol and benzene). When this mixture is fractionally distilled, the (more volatile) azeotropic mixture is obtained as the initial distillate.

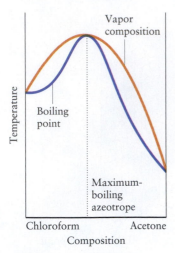

FIGURE 8.42 The temperature–composition diagram showing a maximum-boiling azeotrope (such as acetone and chloroform). When this mixture is fractionally distilled, the (less volatile) azeotropic mixture is left in the flask.

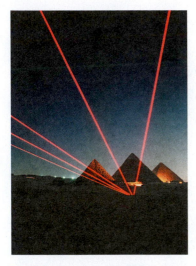

FIGURE 8.43 Laser beams are invisible. However, they can be traced when they pass through smoky or misty environments because the light scatters from the particles suspended in the air.

that of both pure components is called a **minimum-boiling azeotrope,** and the components cannot be separated by distillation: in a fractional distillation, a mixture with the azeotropic composition boils over first, not the more volatile pure liquid.

The opposite behavior is found for a mixture of acetone in chloroform (Fig. 8.42). This **maximum-boiling azeotrope** boils at a higher temperature than either constituent and is the last fraction to be collected, not the less volatile of the pure liquids. The reduction in vapor pressure results from the fact that, although neither molecule can form hydrogen bonds in the pure state, the partially positive hydrogen atom in the highly polar chloroform molecule can form a hydrogen bond to the partially negative oxygen atom in the acetone molecule.

Solutions in which intermolecular forces are stronger in the solution than in the pure components have negative deviations from Raoult's law; some form maximum-boiling azeotropes. Solutions in which intermolecular forces are weaker in the solution than in the pure components have positive deviations from Raoult's law; some form minimum-boiling azeotropes.

THE IMPACT ON BIOLOGY AND MATERIALS

The complexity and versatility of materials made by nature are the envy of scientists. We are only beginning to be able to create materials that have the strong yet porous structure of bone or the strength and flexibility of spider silk (Section 19.13). However, some materials are not strong: they are soft and flexible. These materials, some of which are described in the following two sections, are also important to industry and medicine and some are vital to life.

8.21 Colloids

A **colloid** is a dispersion of large particles (from 1 nm to 1 μm in diameter) in a solvent. Colloidal particles are much larger than most molecules but are too small to be seen with an optical microscope. As a result, colloids have properties between those of a solution and those of a heterogeneous mixture. The small particles give the colloid a homogeneous appearance but are large enough to scatter light. The light scattering explains why milk is white, not transparent, and why searchlight and laser beams are more visible in fog, smoke, and clouds than in clear, dry air (Fig. 8.43). Many foods are colloids, as are clay particles, smoke, and the fluids in living cells. Some self-assembling materials are colloids that form orderly structures spontaneously (see Box 15.1).

Colloids are classified according to the phases of their components (Table 8.9). A colloid that is a suspension of solids in a liquid is called a **sol,** and a suspension of one liquid in another is called an **emulsion.** For example, muddy water is a sol in which tiny flakes of clay are dispersed in water; mayonnaise is an emulsion in which small droplets of water are suspended in vegetable oil. **Foam** is a suspension of a gas in a liquid or solid. Foam rubber, Styrofoam, soapsuds, and aerogels (insu-

TABLE 8.9 The Classification of Colloids*

Dispersed phase	Dispersion medium	Technical name	Examples
solid	gas	aerosol	smoke
liquid	gas	aerosol	hairspray, mist, fog
solid	liquid	sol or gel	printing ink, paint
liquid	liquid	emulsion	milk, mayonnaise
gas	liquid	foam	fire-extinguisher foam
solid	solid	solid dispersion	ruby glass (Au in glass); some alloys
liquid	solid	solid emulsion ice cream	bituminous road paving;
gas	solid	solid foam	insulating foam

*Based on R. J. Hunter, *Foundations of Colloid Science*, Vol. 1 (Oxford: Oxford University Press, 1987).

lating ceramic foams that have densities nearly as low as that of air) are foams. Zeolites (Section 13.14) are a type of solid foam in which the openings in the solid are comparable in size to molecules.

A **solid emulsion** is a suspension of a liquid or solid phase in a solid. For example, opals are solid emulsions formed when partly hydrated silica fills the interstices between close-packed microspheres of silica aggregates. Gelatin desserts are a type of solid emulsion called a **gel**, which is soft but holds its shape. Photographic emulsions are gels that also contain solid colloidal particles of light-sensitive materials such as silver bromide. Many liquid crystalline arrays can be considered colloids. Cell membranes form a two-dimensional colloidal structure (Fig. 8.44).

Aqueous colloids can be classified as hydrophilic or hydrophobic, depending on the strength of the molecular interactions between the suspended substance and water. Suspensions of fat in water (such as milk) and water in fat (such as mayonnaise and hand lotions) are hydrophobic colloids, because fat molecules have little attraction for water molecules. Gels and puddings are examples of hydrophilic colloids. The macromolecules of the proteins in gelatin and the starch in pudding have many hydrophilic groups that attract water. The giant protein molecules in gelatin uncoil in hot water, and their numerous polar groups form hydrogen bonds with the water. When the mixture cools, the protein chains link together again, but now they have entwined to form a three-dimensional web that encloses many water molecules, as well as molecules of sugar, dye, and flavoring agents. The result is a gel: an open network of protein chains that holds the water in a flexible solid structure.

Clusters of metal atoms can form colloidal suspensions. Colloidal clusters of copper, silver, and gold in glass are responsible for some of the vivid colors of stained glass in medieval cathedrals. Even aqueous suspensions of metal clusters are known (Fig. 8.45).

Many precipitates, such as $Fe(OH)_3$, form initially as colloidal suspensions. The tiny particles are kept from settling out by **Brownian motion**, the motion of small particles resulting from constant bombardment by solvent molecules. The sol is further stabilized by the adsorption of ions on the surfaces of the particles. The ions attract a layer of water molecules that prevents the particles from adhering to one another.

Colloids are suspensions of particles too small to be seen with a microscope but large enough to scatter light.

8.22 Bio-based and Biomimetic Materials

Bio-based materials are materials that are taken from or made from natural materials in living things. Examples include packing pellets made from corn and soybeans, polylactic acid (a polymer used to make plastic packaging), and various kinds of pharmaceuticals.

One example of a bio-based pharmaceutical that forms a colloid in water is hyaluronic acid (5), which is found in the body, where it is a major component of the fluid that lubricates joints and plays an important role in the repair of tissues, especially in the skin. The molecules of hyaluronic acid contain many —OH groups with which it can form hydrogen bonds to water. As a result, as it moves through the body large numbers of water molecules move with it. This property makes it useful in regions of the body that require lubrication, such as joints, and in the healing of wounds and connective tissue (Fig. 8.46). Hyaluronic acid is widely used in sports medicine to heal joint injuries by reducing inflammation. Because hyaluronic acid is easily metabolized in the stomach, it must be injected into the region where it is needed.

Biomimetic materials are materials that are modeled after naturally occurring materials. Gels or flexible polymers (Chapter 19) modeled after natural membranes and tissues are biomimetic materials with remarkable properties. Some can be made to crawl on their own like tiny nanometer worms, others pulsate to an internally generated rhythm, and still others respond in seemingly lifelike ways to stimuli.

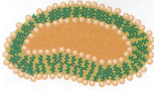

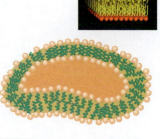

Animation: Figure 8.44 Bilayer membranes

FIGURE 8.44 A cross section of a liposome, a tiny sac enclosed by a bila membrane formed from surfactant-like phospholipid molecules.

FIGURE 8.45 The stability of colloids is illustrated by this violet liquid, which is a colloid of metallic gold that has survived since it was prepared by Michael Faraday in 1857.

FIGURE 8.46 Hyaluronic acid has the consistency of a soft gel. It absorbs water easily and bonds with the proteins in the skin, making it useful in the repair of wounds and joint injuries.

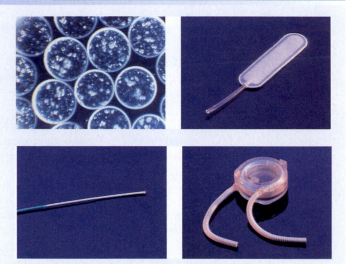

| BOX 8.1 | FRONTIERS OF CHEMISTRY: DRUG DELIVERY |

The administration of drugs to ease disease and chronic, severe pain or to provide benefits such as hormone replacement therapy is difficult because drugs taken orally may lose much of their potency in the harsh conditions of the digestive system. In addition, they are distributed throughout the entire body, not just where they are needed, and side effects can be significant. Recently, however, techniques have been developed to deliver drugs gradually over time, to the exact location in the body where they are needed, and even at the time when they are needed.

Transdermal patches are applied to the skin. The drug is mixed with the adhesive for the patch, and so it lies next to the skin. The skin can readily absorb many chemicals and so can absorb drugs such as nitroglycerin (for heart disease), morphine derivatives (for constant, severe pain), estrogen (for hormone replacement therapy), or nicotine (for easing symptoms that result when a patient stops smoking).

Implants provide a means of delivering drugs over a longer period of time at a controlled rate inside the body. Subcutaneous (under the skin) implants are used to provide appropriate doses of psychoactive medications, birth-control drugs, painkillers, and other medications that must be administered frequently. The implants last for as long as a month and can be easily replaced or renewed. When the location of the drug release is critical, implants can be placed deeper into the body. For example, implants can be introduced into the brain or spinal column to provide effective relief from pain or to protect neurons from degeneration. The implant is contained inside a cylinder of porous foam through which the drug is released. Some implants contain living animal cells that have been engineered to produce natural hormones or painkillers that are released to the body as they are produced. In other types of implants, membranes allow a gradual release of the drug.

Controlled-release drug delivery systems mimic nature. Molecules called lipids are found in fats and also form the membranes of living cells. A lipid molecule is similar in struc-

Examples of implants used to insert living cells into the body. The cells continually produce enzymes, hormones, or painkillers needed by the body. Often a long, thin plastic tail is attached as a tether to allow easy retrieval of the implant.

ture to the surfactant molecules in detergents: it has a polar head and a nonpolar hydrocarbon tail. Some lipids assemble themselves spontaneously into liquid crystal structures in water (Section 8.22); in these structures, sheets consisting of rows of molecules are lined up next to one another. The sheets can be coaxed into forming liposomes (see Fig. 8.44). Liposomes are similar to micelles but are formed from a double layer of molecules, with polar heads forming each surface, like the membrane of a living cell. When a drug is present in the aqueous solution in which liposomes are being formed, some of the drug is encapsulated inside each liposome, which becomes a tiny container for the drug. The liposomes can then be injected into the body, where they stick only to certain types of cells—cancer cells, for instance. Compared with oral or intravenous medicine, a smaller dosage is required and side effects are greatly reduced.

Controlled-release drug delivery systems can be made from liquid crystals (Section 5.16). Surfactant compounds called *phospholipids* are found in fats and form the membranes of living cells. These molecules are liquid crystals similar to detergent molecules (Section 8.9). The membranes of living cells are double layers of phospholipid molecules that line up with their hydrocarbon tails pointing into the membrane and their polar head groups forming the membrane surface. This structure separates the contents of the cell from the intercellular fluid. It is possible to coax phospholipids to form artificial membranes that close up to form tiny bags, called *liposomes*, which are much smaller than living cells (see Fig. 8.44). If the liposomes are formed in a solution containing a drug, some drug molecules become encapsulated. This behavior has led to the study of liposomes as a way of delivering drugs into different organs and regions of the body (Box 8.1).

Biomimetic materials are modeled after naturally occurring biological materials.

This implant containing live hamster cells was inserted into the spinal column of a patient for 17 weeks. After removal, the cells were still alive and secreting the hormone required to keep the patient healthy.

Nanotechnology has led to very efficient versions of liposomes. Tiny hollow spheres only nanometers in diameter hold even tinier capsules of medicine. The spheres are made of silica covered with gold nanoparticles and when they are coated with antibodies they attach to tumor cells. The spheres are sensitive to light of specific wavelengths and when the light is applied, either heat up and destroy the tumor, or burst, releasing the drugs within the capsules directly into the tumor.

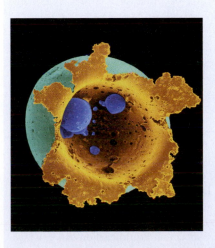

An electron microscope image of a drug capsule as it bursts open, revealing the tiny microcapsules inside. The image has been digitally colored.

Smart gels (see Box 15.1) are being investigated for drug delivery in situations in which the drug dosage must be modified according to conditions in the body. For example, the amount of insulin that a nondiabetic person needs is delivered by the body according to the level of blood sugar. However, a diabetic person must take insulin at specified times of the day and the same amount each time. If the blood sugar is already low, a hypoglycemic reaction, and possibly coma, can result. An insulin delivery system that would be responsive to blood sugar levels and provide the correct dosage when it is needed is now being investigated. The system makes use of a smart gel in which molecules of insulin have been trapped. The gel incorporates into its structure molecules of phenylboronic acid, to which glucose (blood sugar) molecules are attracted. If the blood sugar is high, more and more glucose molecules stick to the gel and cause it to swell. When the level of glucose rises above a certain concentration, the gel swells so much that it becomes porous, releasing insulin into the blood.

❓ HOW MIGHT YOU CONTRIBUTE?

Both basic and applied research are needed for effective drug delivery systems. Basic research on the self-assembly of molecules may allow more innovative solutions in the future. Applied research may have more immediate benefits. For example, the optimal drug delivery system must be designed for each specific drug. Coatings for implants or nanospheres that are nontoxic and are similar in nature to body tissues need to be developed. Both the length of time during which a drug delivery system can remain active inside the body and the stability of the system need to be increased.

Related Exercise: 8.117

For Further Reading: C. M. Henry, "Special delivery," *Chemical and Engineering News* (September 18, 2000), pp. 49–64. M. J. Lysaght and P. Aebischer, "Encapsulated cells as therapy," *Scientific American* (April 1999), pp. 76–82. S. Morrissey, "Nanotech meets medicine," *Chemical and Engineering News* (May 16, 2005), p. 30.

SKILLS YOU SHOULD HAVE MASTERED

❏ 1 Use the Clausius–Clapeyron equation to estimate the vapor pressure or boiling point of a liquid (Examples 8.1 and 8.2).

❏ 2 Interpret a one-component phase diagram (Example 8.3).

❏ 3 Predict relative solubilities from molecular polarity (Section 8.9).

❏ 4 Use Henry's law to calculate the solubility of a gas (Example 8.4).

❏ 5 Interpret enthalpies of solution in terms of lattice enthalpies and enthalpies of hydration (Section 8.12).

❏ 6 Calculate the molality of a solute (Example 8.5).

❏ 7 Convert between molality and mole fraction or molarity (Toolbox 8.1 and Examples 8.6 and 8.7).

❏ 8 Calculate the vapor pressure of a solvent in a solution by using Raoult's law (Example 8.8).

❏ 9 Determine molar mass cryoscopically (Toolbox 8.2 and Example 8.9).

❏ 10 Use osmometry to find the molar mass of a solute (Toolbox 8.2 and Example 8.10).

❏ 11 Calculate the vapor pressure and vapor composition for a solution of two liquids (Examples 8.11 and 8.12).

❏ 12 Interpret a two-component phase diagram and discuss fractional distillation (Sections 8.18, 8.19, and 8.20).

❏ 13 Identify colloids and explain their properties (Section 8.21).

❏ 14 Describe the structure of cell membranes (Section 8.22).

EXERCISES

Phases and Phase Transitions

8.1 Suppose you were to collect 1.0 L of air by passing it slowly through water at 20.°C and into a container. Estimate the mass of water vapor in the collected air, assuming that the air is saturated with it.

8.2 A bottle of mercury at 25°C was left unstoppered in a chemical supply room measuring 3.0 m by 3.0 m by 2.5 m. What mass of mercury vapor would be present if the air became saturated with it? The vapor pressure of mercury at 25°C is 0.227 Pa.

8.3 Use the vapor-pressure curve in Fig. 8.3 to estimate the boiling point of water when the atmospheric pressure is (a) 60. kPa; (b) 160. kPa.

8.4 Use the vapor-pressure curve in Fig. 8.3 to estimate the boiling point of ethanol when the atmospheric pressure is (a) 60. kPa; (b) 85 kPa.

8.5 Arsine, AsH_3, is a highly toxic compound used in the electronics industry for the production of semiconductors. Its vapor pressure is 35 Torr at −111.95°C and 253 Torr at −83.6°C. Using these data, calculate (a) the standard enthalpy of vaporization; (b) the standard entropy of vaporization; (c) the standard Gibbs free energy of vaporization; (d) the normal boiling point of arsine.

8.6 The vapor pressure of chlorine dioxide, ClO_2, is 155 Torr at −22.75°C and 485 Torr at 0.00°C. Calculate (a) the standard enthalpy of vaporization; (b) the standard entropy of vaporization; (c) the standard Gibbs free energy of vaporization; (d) the normal boiling point of ClO_2.

8.7 The normal boiling point of iodomethane, CH_3I, is 42.43°C, and its vapor pressure at 0.00°C is 140. Torr. Calculate (a) the standard enthalpy of vaporization of iodomethane; (b) the standard entropy of vaporization of iodomethane; (c) the vapor pressure of iodomethane at 25.0°C.

8.8 The normal boiling point of trimethylphosphine, $P(CH_3)_3$, is 38.4°C, and its vapor pressure at −45.21°C is 13 Torr.

Calculate (a) the standard enthalpy of vaporization of trimethylphosphine; (b) the standard entropy of vaporization of trimethylphosphine; (c) the vapor pressure of trimethylphosphine at 15.0°C.

8.9 Use data from Table 6.3 to calculate the vapor pressure of methanol at 25.0°C.

8.10 Use data from Table 6.3 to calculate the vapor pressure of ammonia at 235 K.

8.11 Use Fig. 8.6 to predict the state of a sample of water under the following conditions: (a) 1 atm, 200.°C; (b) 100. atm, 50.0°C; (c) 3 Torr, 10.0°C.

8.12 Use Fig. 8.7 to predict the state of a sample of CO_2 under the following conditions: (a) 6 atm, −80.°C; (b) 1 atm, −56°C; (c) 80. atm, 25°C; (c) 5.1 atm, −56°C.

8.13 The phase diagram for helium is shown here. (a) What is the maximum temperature at which superfluid helium-II can exist? (b) What is the minimum pressure at which solid helium can exist? (c) What is the normal boiling point of helium-I? (d) Can solid helium sublime?

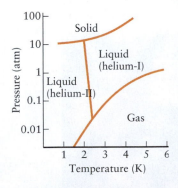

8.14 The phase diagram for carbon, shown here, indicates the extreme conditions that are needed to form diamonds from graphite. (a) At 2000 K, what is the minimum pressure needed before graphite changes into diamond? (b) What is the minimum temperature at which liquid carbon can exist at pressures below 10 000 atm? (c) At what pressure does graphite melt at 3000 K? (d) Are diamonds stable under normal conditions? If not, why is it that people can wear them without having to compress and heat them?

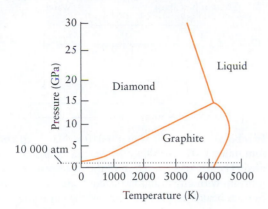

8.15 Use the phase diagram for helium in Exercise 8.13 (a) to describe the phases in equilibrium at each of helium's two triple points; (b) to decide which liquid phase is more dense, helium-I or helium-II.

8.16 Use the phase diagram for carbon in Exercise 8.14 (a) to describe the phase transitions that carbon would undergo if compressed at a constant temperature of 2000 K from 100 atm to 1×10^6 atm; (b) to rank the diamond, graphite, and liquid phases of carbon in order of increasing density.

8.17 Use the phase diagram for carbon dioxide (Fig. 8.7) to predict what would happen to a sample of carbon dioxide gas at $-50°C$ and 1 atm if its pressure were suddenly increased to 73 atm at constant temperature. What would be the final physical state of the carbon dioxide?

8.18 A new substance developed in a laboratory has the following properties: normal melting point, $83.7°C$; normal boiling point, $177°C$; triple point, 200. Torr and $38.6°C$. (a) Sketch the approximate phase diagram and label the solid, liquid, and gaseous phases and the solid–liquid, liquid–gas, and solid–gas phase boundaries. (b) Sketch an approximate heating curve for a sample at constant pressure, beginning at 500. Torr and $25°C$ and ending at $200.°C$.

Solubility

8.19 Which would be the better solvent, water or benzene, for each of the following substances: (a) KCl; (b) CCl_4; (c) CH_3COOH?

8.20 Which would be the better solvent, water or tetrachloromethane, for each of the following substances: (a) CH_4; (b) Br_2; (c) HF?

8.21 The following groups are found in some organic molecules. Which are hydrophilic and which are hydrophobic: (a) $-NH_2$; (b) $-CH_3$; (c) $-Br$; (d) $-COOH$?

8.22 The following groups are found in some organic molecules. Which are hydrophilic and which are hydrophobic: (a) $-OH$; (b) $-CH_2CH_3$; (c) $-CONH_2$; (d) $-Cl$?

8.23 State the molar solubility in water of (a) O_2 at 50. kPa; (b) CO_2 at 500. Torr; (c) CO_2 at 0.10 atm. The temperature in each case is $20.°C$, and the pressures are partial pressures of the gases. Use the information in Table 8.5.

8.24 Calculate the aqueous solubility (in milligrams per liter) of (a) air at 1.0 atm; (b) He at 1.0 atm; (c) He at 25 kPa. The temperature is $20.°C$ in each case, and the pressures are partial pressures of the gases. Use information in Table 8.5.

8.25 The minimum mass concentration of oxygen required for fish life is 4 $mg·L^{-1}$. (a) Assume the density of lake water to be 1.00 $g·mL^{-1}$, and express this concentration in parts per million ($mg·kg^{-1}$; see Table 8.5). (b) What is the minimum partial pressure of O_2 that would supply the minimum mass concentration of oxygen in water to support fish life at $20.°C$? (c) What is the minimum atmospheric pressure that would give this partial pressure, assuming that oxygen exerts about 21% of the atmospheric pressure?

8.26 The volume of blood in the body of a certain deep-sea diver is about 6.00 L. Blood cells make up about 55% of the blood volume, and the remaining 45% is the aqueous solution called plasma. What is the maximum volume of nitrogen measured at 1.00 atm and $37°C$ that could dissolve in the diver's blood plasma at a depth of 93 m, where the pressure is 10.0 atm? (This is the volume that could come out of solution suddenly, causing the painful and dangerous condition called the bends, if the diver were to ascend too quickly.) Assume that Henry's constant for nitrogen at $37°C$ (body temperature) is 5.8×10^{-4} $mol·L^{-1}·atm^{-1}$.

8.27 The carbon dioxide gas dissolved in a sample of water in a partly filled, sealed container has reached equilibrium with its partial pressure in the air above the solution. Explain what happens to the solubility of the CO_2 if (a) the partial pressure of the CO_2 gas is doubled by the addition of more CO_2; (b) the total pressure of the gas above the liquid is doubled by the addition of nitrogen.

8.28 Explain what happens to the solubility of the CO_2 in Exercise 8.27 if (a) the partial pressure of $CO_2(g)$ is increased by compressing the gas to a third of its original volume; (b) the temperature is raised.

8.29 A soft drink is made by dissolving CO_2 gas at 3.60 atm in a flavored solution and sealing the solution in aluminum cans at $20.°C$. What amount of CO_2 is contained in a 420-mL can of the soft drink? At $20.°C$ the Henry's law constant for CO_2 is 2.3×10^{-2} $mol·L^{-1}·atm^{-1}$.

8.30 A soft drink is made by dissolving CO_2 gas at 3.00 atm in a flavored solution and sealing the solution in aluminum cans at $20.°C$. What amount (in moles) of CO_2 is contained in a 240.-mL can of the soft drink? At $20.°C$ the Henry's law constant for CO_2 is 2.3×10^{-2} $mol·L^{-1}·atm^{-1}$.

8.31 Lithium sulfate dissolves exothermically in water. (a) Is the enthalpy of solution for Li_2SO_4 positive or negative? (b) Write the chemical equation for the dissolving process. (c) Which is larger for lithium sulfate, the lattice enthalpy or the enthalpy of hydration?

8.32 The enthalpy of solution of ammonium nitrate in water is positive. (a) Does NH_4NO_3 dissolve endothermically or exothermically? (b) Write the chemical equation for the dissolving process. (c) Which is larger for NH_4NO_3, the lattice enthalpy or the enthalpy of hydration?

8.33 Calculate the heat evolved or absorbed when 10.0 g of (a) NaCl; (b) NaI; (c) $AlCl_3$; (d) NH_4NO_3 is dissolved in 100. g of water. Assume that the enthalpies of solution in Table 8.6 are applicable and that the specific heat capacity of the solution is $4.18 \ \mathrm{J \cdot K^{-1} \cdot g^{-1}}$.

8.34 Determine the temperature change when 10.0 g of (a) KCl; (b) $MgBr_2$; (c) KNO_3; (d) NaOH is dissolved in 100. g of water. Assume that the specific heat capacity of the solution is $4.18 \ \mathrm{J \cdot K^{-1} \cdot g^{-1}}$ and that the enthalpies of solution in Table 8.6 are applicable.

Colligative Properties

8.35 Calculate (a) the molality of sodium chloride in a solution prepared by dissolving 25.0 g of NaCl in 500.0 g of water; (b) the mass (in grams) of NaOH that must be mixed with 345 g of water to prepare 0.18 m NaOH(aq); (c) the molality of urea, $CO(NH_2)_2$, in a solution prepared by dissolving 0.978 g of urea in 285 mL of water.

8.36 Calculate (a) the molality of KOH in a solution prepared from 13.72 g of KOH and 75.0 g of water; (b) the mass (in grams) of ethylene glycol, HOC_2H_4OH, that should be added to 1.5 kg of water to prepare 0.44 m HOC_2H_4OH(aq); (c) the molality of an aqueous 3.89% by mass HCl solution.

8.37 A 5.00% by mass K_3PO_4 aqueous solution has a density of $1.043 \ \mathrm{g \cdot cm^{-3}}$. Determine (a) the molality; (b) the molarity of potassium phosphate in the solution.

8.38 Calculate the concentrations of each of the following solutions: (a) the molality of 13.63 g of sucrose, $C_{12}H_{22}O_{11}$, dissolved in 612 mL of water; (b) the molality of CsCl in a 10.00% by mass aqueous solution; (c) the molality of acetone in an aqueous solution with a mole fraction for acetone of 0.197.

8.39 Calculate the concentrations of each of the following solutions: (a) the molality of chloride ions in an aqueous solution of magnesium chloride in which x_{MgCl_2} is 0.0120; (b) the molality of 6.75 g of sodium hydroxide dissolved in 325 g of water; (c) the molality of 15.00 M HCl(aq) with a density of $1.0745 \ \mathrm{g \cdot cm^{-3}}$.

8.40 Calculate the concentrations of each of the following solutions: (a) the molality of chloride ions in an aqueous solution of iron(III) chloride for which x_{FeCl_3} is 0.0205; (b) the molality of hydroxide ions in a solution prepared from 9.25 g of barium hydroxide dissolved in 183 g of water.; (c) the molality of 12.00 M NH_3(aq) with a density of $0.9519 \ \mathrm{g \cdot cm^{-3}}$.

8.41 (a) Calculate the mass of $CaCl_2 \cdot 6H_2O$ needed to prepare 0.125 m $CaCl_2$(aq) by using 500. g of water. (b) What mass of $NiSO_4 \cdot 6H_2O$ must be dissolved in 500. g of water to produce 0.22 m $NiSO_4$(aq)?

8.42 A 10.0% by mass H_2SO_4(aq) solution has a density of $1.07 \ \mathrm{g \cdot cm^{-3}}$. (a) How many milliliters of solution contain 8.37 g of H_2SO_4? (b) What is the molality of H_2SO_4 in solution? (c) What mass (in grams) of H_2SO_4 is in 250. mL of solution?

8.43 Two beakers, one containing 0.010 m NaCl (aq) and the other containing pure water, are placed inside a bell jar and sealed. The beakers are left until the water vapor has come to equilibrium with any liquid in the container. The levels of the liquid in each beaker at the beginning of the experiment are the

same, as pictured below. Draw the levels of the liquid in each beaker after equilibrium has been reached. Explain your reasoning.

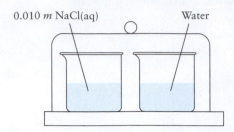

0.010 m NaCl(aq)   Water

8.44 Two beakers, one containing 0.010 m NaCl(aq) and the other containing 0.010 m $AlCl_3$(aq), are placed inside a bell jar and sealed. The beakers are left until the water vapor has come to equilibrium with any liquid in the container. The levels of the liquid in each beaker at the beginning of the experiment are the same, as pictured below. Draw the levels of the liquid in each beaker after equilibrium has been reached. Explain your reasoning.

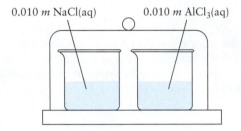

0.010 m NaCl(aq)   0.010 m $AlCl_3$(aq)

8.45 Calculate the vapor pressure of the solvent in each of the following solutions. Use Table 8.3 to find the vapor pressure of water in (a) an aqueous solution at 100.°C in which the mole fraction of sucrose is 0.100; (b) an aqueous solution at 100.°C in which the molality of sucrose is 0.100 $\mathrm{mol \cdot kg^{-1}}$.

8.46 What is the vapor pressure of the solvent in each of the following solutions: (a) the mole fraction of glucose is 0.050 in an aqueous solution at 80.°C; (b) an aqueous solution at 25°C is 0.10 m urea, $CO(NH_2)_2$, a nonelectrolyte? Use the data in Table 8.3 for the vapor pressure of water at various temperatures.

8.47 Benzene has a vapor pressure of 94.6 Torr at 25°C. A nonvolatile compound was added to 0.300 mol benzene at 25°C and the vapor pressure of the benzene in the solution decreased to 75.0 Torr. What amount (in moles) of solute molecules was added to the benzene?

8.48 Benzene has a vapor pressure of 100.0 Torr at 26°C. A nonvolatile compound was added to 0.300 mol benzene at 26°C and the vapor pressure of the benzene in the solution decreased to 60.0 Torr. What amount (in moles) of solute molecules were added to the benzene?

8.49 When 8.05 g of an unknown compound X was dissolved in 100. g of benzene, the vapor pressure of the benzene decreased from 100.0 Torr to 94.8 Torr at 26°C. What is (a) the mole fraction and (b) the molar mass of X?

8.50 The normal boiling point of ethanol is 78.4°C. When 9.15 g of a soluble nonelectrolyte was dissolved in 100. g of ethanol, the vapor pressure of the solution at that temperature was 7.40×10^2 Torr. (a) What are the mole fractions of ethanol and solute? (b) What is the molar mass of the solute?

8.51 (a) What is the normal boiling point of an aqueous solution that has a vapor pressure of 751 Torr at 100.°C? (b) Determine the normal boiling point of a benzene solution that has a vapor pressure of 740. Torr at 80.1°C, the normal boiling point of pure benzene.

8.52 (a) What is the normal boiling point of an aqueous solution that has a freezing point of -1.04°C? (b) The freezing point of a benzene solution is 2.0°C. The normal freezing point of benzene is 5.5°C. What is the expected normal boiling point of the solution? The normal boiling point of benzene is 80.1°C.

8.53 A 1.14-g sample of a molecular substance dissolved in 100. g of camphor freezes at 176.9°C. What is the molar mass of the substance?

8.54 When 1.32 g of a nonpolar solute was dissolved in 50.0 g of phenol, the latter's freezing point was lowered by 1.454°C. Calculate the molar mass of the solute.

8.55 A 1.00% by mass NaCl(aq) solution has a freezing point of -0.593°C. (a) Estimate the van 't Hoff i factor from the data. (b) Determine the total molality of all solute species. (c) Calculate the percentage dissociation of NaCl in this solution. (The molality calculated from the freezing-point depression is the sum of the molalities of the undissociated ion pairs, the Na^+ ions, and the Cl^- ions.)

8.56 A 1.00% by mass $MgSO_4$(aq) solution has a freezing point of -0.192°C. (a) Estimate the van 't Hoff i factor from the data. (b) Determine the total molality of all solute species. (c) Calculate the percentage dissociation of $MgSO_4$ in this solution.

8.57 Two unknown molecular compounds were being studied. A solution containing 5.00 g of compound A in 100. g of water froze at a lower temperature than a solution containing 5.00 g of compound B in 100. g of water. Which compound has the greater molar mass? Explain how you decided your answer.

8.58 Two unknown compounds were being studied. Compound C is molecular and compound D is known to ionize completely in dilute aqueous solutions. A solution containing 0.30 g of compound C in 100. g of water froze at the same temperature as a solution containing 0.30 g of compound D in 100. g of water. Which compound has the greater molar mass? Explain how you decided your answer.

8.59 Determine the freezing point of a 0.10 mol·kg^{-1} aqueous solution of a weak electrolyte that is 7.5% dissociated into two ions.

8.60 A 0.124 m CCl_3COOH(aq) solution has a freezing point of -0.423°C. What is the percentage deprotonation of the acid?

8.61 What is the osmotic pressure at 20.°C of (a) 0.010 M $C_{12}H_{22}O_{11}$(aq); (b) 1.0 M HCl(aq); (c) 0.010 M $CaCl_2$(aq)? Assume complete dissociation of the $CaCl_2$.

8.62 Which of the following solutions has the highest osmotic pressure at 50.°C: (a) 0.10 M KCl(aq); (b) 0.60 M

$CO(NH_2)_2$(aq); (c) 0.30 M K_2SO_4(aq)? Justify your answer by calculating the osmotic pressure of each solution.

8.63 A 0.40-g sample of a polypeptide dissolved in 1.0 L of an aqueous solution at 27°C has an osmotic pressure of 3.74 Torr. What is the molar mass of the polypeptide?

8.64 When 0.10 g of insulin is dissolved in 0.200 L of water, the osmotic pressure is 2.30 Torr at 20.°C. What is the molar mass of insulin?

8.65 A 0.20-g sample of a polymer, dissolved in 0.100 L of toluene, has an osmotic pressure of 6.3 Torr at 20.°C. What is the molar mass of the polymer?

8.66 A solution prepared by adding 0.50 g of a polymer to 0.200 L of toluene (methylbenzene, a common solvent) showed an osmotic pressure of 0.582 Torr at 20.°C. What is the molar mass of the polymer?

8.67 Calculate the osmotic pressure at 20.°C of each of the following solutions; assume complete dissociation for any ionic solutes: (a) 0.050 M $C_{12}H_{22}O_{11}$(aq); (b) 0.0010 M NaCl(aq); (c) a saturated aqueous solution of AgCN of solubility 23 μg/100. g of water.

8.68 Calculate the osmotic pressure at 20°C of each of the following solutions; assume complete dissociation of ionic compounds: (a) 3.0×10^{-3} M $C_6H_{12}O_6$(aq); (b) 2.0×10^{-3} M $CaCl_2$(aq); (c) 0.010 M K_2SO_4(aq).

8.69 Catalase, a liver enzyme, dissolves in water. A 10.0-mL solution containing 0.166 g of catalase exhibits an osmotic pressure of 1.2 Torr at 20.°C. What is the molar mass of catalase?

8.70 Intravenous medications are often administered in 5.0% glucose, $C_6H_{12}O_6$(aq), by mass. What is the osmotic pressure of such solutions at 37°C (body temperature)? Assume that the density of the solution is 1.0 g·mL^{-1}.

Binary Liquid Mixtures

8.71 Benzene, C_6H_6, and toluene, $C_6H_5CH_3$, form an ideal solution. The vapor pressure of benzene is 94.6 Torr and that of toluene is 29.1 Torr at 25°C. Calculate the vapor pressure of each of the following solutions and the mole fraction of each substance in the vapor phase above those solutions at 25°C: (a) 1.50 mol C_6H_6 mixed with 0.50 mol $C_6H_5CH_3$; (b) 15.0 g of benzene mixed with 64.3 g of toluene.

8.72 Hexane, C_6H_{14}, and cyclohexane, C_6H_{12}, form an ideal solution. The vapor pressure of hexane is 151 Torr and that of cyclohexane is 98 Torr at 25.0°C. Calculate the vapor pressure of each of the following solutions and the mole fraction of each substance in the vapor phase above those solutions at 25°C: (a) 0.25 mol C_6H_{14} mixed with 0.65 mol C_6H_{12}; (b) 10.0 g of hexane mixed with 10.0 g of cyclohexane.

8.73 1,1-Dichloroethane, CH_3CHCl_2, has a vapor pressure of 228 Torr at 25°C; at the same temperature, 1,1-dichlorotetrafluoroethane, CF_3CCl_2F, has a vapor pressure of 79 Torr. What mass of 1,1-dichloroethane must be mixed with 100.0 g of 1,1-dichlorotetrafluoroethane to give a solution with vapor pressure 157 Torr at 25°C? Assume ideal behavior.

8.74 Butanone, $CH_3CH_2COCH_3$, has a vapor pressure of 100. Torr at 25°C; at the same temperature, propanone, CH_3COCH_3, has a vapor pressure of 222 Torr. What mass of propanone must be mixed with 350.0 g of butanone to give a solution with a vapor pressure of 135 Torr? Assume ideal behavior.

8.75 Which of the following mixtures would you expect to show a positive deviation, negative deviation, or no deviation (that is, form an ideal solution) from Raoult's law? Explain your conclusion. (a) methanol, CH_3OH, and ethanol, CH_3CH_2OH; (b) HF and H_2O; (c) hexane, C_6H_{14}, and H_2O.

8.76 Which of the following mixtures would you expect to show a positive deviation, negative deviation, or no deviation (that is, form an ideal solution) from Raoult's law? Explain your conclusion. (a) HBr and H_2O; (b) formic acid, HCOOH, and benzene; (c) cyclopentane, C_5H_{10}, and cyclohexane, C_6H_{12}.

Impact on Biology and Materials

8.77 Distinguish between a foam and a sol. Give at least one example of each.

8.78 Distinguish between an emulsion and a gel. Give at least one example of each.

8.79 Some colloidal solutions, such as that in Fig. 8.45, appear at first glance to be solutions. What quick, simple procedure can you use to distinguish colloids from solutions?

8.80 Puddings contain large starch molecules that cause the mixture to thicken by a mechanism similar to that by which gelatin thickens. Which is the best description of how puddings thicken? Explain your choice. (a) The starch molecules in pudding are insoluble in water and precipitate when mixed with water. (b) The strands of the starch molecules connect to each other by forming covalent bonds with each other. (c) The starch molecules form hydrogen bonds with water and encapsulate the water in a network. (d) The water molecules hydrate the starch molecules in pudding and the heat of hydration causes the starch molecules to decompose.

Integrated Exercises

8.81 Are the water molecules oriented in the same way or differently around the positive and negative ions when sodium chloride dissolves? Explain your conclusion.

8.82 Cellulose is a primary structural material in plants. Its molecules consist of long chains of thousands of carbon atoms, most of which have —OH groups attached. Give one reason why, although cellulose can form numerous hydrogen bonds to water, it is insoluble in water.

8.83 Complete the following statements about the effect of intermolecular forces on the physical properties of a substance. (a) The higher the boiling point of a liquid, the (stronger, weaker) are its intermolecular forces. (b) Substances with strong intermolecular forces have (high, low) vapor pressures. (c) Substances with strong intermolecular forces typically have (high, low) surface tensions. (d) The higher the vapor pressure of a liquid, the (stronger, weaker) are its intermolecular forces. (e) Because nitrogen, N_2, has (strong, weak) intermolecular forces, it has a (high, low) critical temperature. (f) Substances with high vapor pressures have

correspondingly (high, low) boiling points. (g) Because water has a high boiling point, it must have (strong, weak) intermolecular forces and a correspondingly (high, low) enthalpy of vaporization.

8.84 Hydrogen peroxide, H_2O_2, is a syrupy liquid with a vapor pressure lower than that of water and a boiling point of 152°C. Account for the differences between these properties and those of water.

8.85 Explain the effect that an increase in temperature has on each of the following properties: (a) viscosity; (b) surface tension; (c) vapor pressure; (d) evaporation rate.

8.86 Explain how the vapor pressure of a liquid is affected by each of the following changes in conditions: (a) an increase in temperature; (b) an increase in surface area of the liquid; (c) an increase in volume above the liquid; (d) the addition of air to the volume above the liquid.

8.87 You have two beakers: one is filled with tetrachloromethane and the other with water. You also have two compounds, butane ($CH_3CH_2CH_2CH_3$) and calcium chloride. (a) In which liquid will butane dissolve? Sketch the local environment of the solute in the solution. (b) In which solvent will calcium chloride dissolve? Sketch the local environment of the solute in the solution.

8.88 Use the phase diagram for compound X below to answer these questions: (a) Is X a solid, liquid, or gas at normal room temperatures? (b) What is the normal melting point of X? (c) What is the vapor pressure of liquid X at −50°C? (d) What is the vapor pressure of solid X at −100°C?

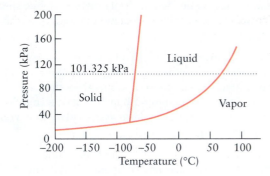

8.89 Relative humidity at a particular temperature is defined as

$$\text{Relative humidity} = \frac{\text{partial pressure of water}}{\text{vapor pressure of water}} \times 100\%$$

The vapor pressure of water at various temperatures is given in Table 8.3. (a) What is the relative humidity at 30.°C when the partial pressure of water is 25.0 Torr? (b) Explain what would be observed if the temperature of the air were to fall to 25°C.

8.90 Suppose that 10.0 g of an organic compound used as a component of mothballs is dissolved in 80.0 g of benzene. The freezing point of the solution is 1.20°C. (a) What is an approximate molar mass of the organic compound? (b) An elemental analysis of that substance indicated that the empirical formula is C_3H_2Cl. What is its molecular formula? (c) Using the atomic molar masses from the periodic table, calculate a more accurate molar mass of the compound.

8.91 When a molar mass is determined from freezing-point depression, it is possible to make each of the following errors (among others). In each case, predict whether the error would cause the reported molar mass to be greater or less than the actual molar mass. (a) There was dust on the balance, causing the mass of solute to appear greater than it actually was. (b) The water was measured by volume, assuming a density of 1.00 g·cm^{-3}, but the water was warmer and less dense than assumed. (c) The thermometer was not calibrated accurately, and so the temperature of the freezing point was actually $0.5°C$ higher than recorded. (d) The solution was not stirred sufficiently, and so not all the solute dissolved.

8.92 An elemental analysis of epinephrine resulted in the following composition: 59.0% carbon, 26.2% oxygen, 7.15% hydrogen, and 7.65% nitrogen by mass. When 0.64 g of epinephrine was dissolved in 36.0 g of benzene, the freezing point decreased by $0.50°C$. (a) Determine the empirical formula of epinephrine. (b) What is the molar mass of epinephrine? (c) Deduce the molecular formula of epinephrine.

8.93 Interpret the following verse from the Coleridge's *Rime of the Ancient Mariner*:

Water, water, every where,
And all the boards did shrink,
Water, water every where,
Nor any drop to drink.

8.94 The vapor pressure of ethanol at $25°C$ is 58.9 Torr. A sample of ethanol vapor at $25°C$ and 58.9 Torr partial pressure is in equilibrium with a very small amount of liquid ethanol in a 10.0-L container also containing dry air, at a total pressure of 750.0 Torr. The volume of the container is then reduced at constant temperature to 5.0 L. (a) What is the partial pressure of ethanol in the smaller volume? Explain your reasoning. (b) What is the total pressure of the mixture?

8.95 The height of a column of liquid that can be supported by a given pressure is inversely proportional to its density. An aqueous solution of 0.010 g of a protein in $10.$ mL of water at $20.°C$ shows a rise of 5.22 cm in the apparatus shown in Fig. 8.31. Assume the density of the solution to be 0.998 g·cm^{-3} and the density of mercury to be 13.6 g·cm^{-3}. (a) What is the molar mass of the protein? (b) What is the freezing point of the solution? (c) Which colligative property is best for measuring the molar mass of these large molecules? Give reasons for your answer.

8.96 A 0.020 M $C_6H_{12}O_6$(aq) solution (glucose) is separated from a 0.050 M $CO(NH_2)_2$(aq) solution (urea) by a semipermeable membrane at $25°C$. For both compounds $i = 1$. (a) Which solution has the higher osmotic pressure? (b) Which solution becomes more dilute with the passage of H_2O molecules through the membrane? (c) To which solution should an external pressure be applied to maintain an equilibrium flow of H_2O molecules across the membrane? (d) What external pressure (in atm) should be applied in part (c)?

8.97 Organic chemists once used freezing-point and boiling-point measurements to determine the molar masses of compounds that they had synthesized. When 0.30 g of a nonvolatile solute is dissolved in 30.0 g of CCl_4, the boiling point of the solution is $77.19°C$. What is the molar mass of the compound?

8.98 A sample consisting of 155 mg of a purified protein is dissolved in 10.0 mL of ethanol. This solution is placed in a device for measuring osmotic pressure and rises to a final height of 32.5 cm above the level of pure ethanol. The experiment is performed at 1.00 atm and 298 K. The density of ethanol at 298 K is 0.79 g·cm^{-3}. What is the molar mass of the protein? Assume that the density of the solution is the same as that of pure ethanol. See Exercise 8.95.

8.99 (a) From data in Appendix 2A, derive a numerical form of the Clausius–Clapeyron equation for methanol. (b) Using this equation, plot the appropriate quantities that should give a straight-line relation between vapor pressure and temperature. (c) Estimate the vapor pressure of methanol at $0.0°C$. (d) Estimate the normal boiling point of methanol.

8.100 (a) From data in Appendix 2A and Table 6.3, derive a numerical form of the Clausius–Clapeyron equation for benzene. (b) Using this equation, plot the appropriate quantities that should give a straight-line relation between vapor pressure and temperature. (c) Estimate the boiling point of benzene when the external pressure is 0.655 atm. (d) Calculate $S_m°$ for gaseous benzene.

8.101 Colligative properties can be sources of insight into not only the properties of solutions, but also the properties of the solute. For example, acetic acid, CH_3COOH, behaves differently in two different solvents. (a) The freezing point of a 5.00% by mass aqueous acetic acid solution is $-1.72°C$. What is the molar mass of the solute? Explain any discrepancy between the experimental and the expected molar mass. (b) The freezing-point depression associated with a 5.00% by mass solution of acetic acid in benzene is $2.32°C$. What is the experimental molar mass of the solute in benzene? What can you conclude about the nature of acetic acid in benzene?

8.102 It is standard practice in chemical laboratories to distill high-boiling-point substances under reduced pressure. Trichloroacetic acid has a standard enthalpy of vaporization of 57.814 kJ·mol^{-1} and a standard entropy of vaporization of 124 J·K^{-1}·mol^{-1}. Use this information to determine the pressure that one would need to achieve to distill trichloroacetic acid at $80.°C$.

8.103 The vapor pressure of phosphoryl chloride difluoride (OPClF$_2$) has been measured as a function of temperature:

Temperature (K)	Vapor pressure (Torr)
190.	3.2
228	68
250.	240.
273	672

(a) Plot $\ln P$ against T^{-1} (this project may be done with the aid of a computer or a graphing calculator that can calculate a least-squares fit to the data). (b) From the plot (or a linear equation derived from it) in part (a), determine the standard enthalpy of vaporization of OPClF$_2$; (c) the standard entropy of vaporization of OPClF$_2$; and (d) the normal boiling point of OPClF$_2$. (e) If the pressure of a sample of OPClF$_2$ is reduced to 15 Torr, at what temperature will the sample boil?

8.104 From literature sources, find the critical pressures and temperatures of methane, methylamine (CH$_3$NH$_2$), ammonia, and tetrafluoromethane. Discuss the suitability of using each of these solvents for a supercritical extraction at room temperature in an autoclave that can withstand pressures of 100. atm.

8.105 From literature sources, find the critical temperatures for the gaseous hydrocarbons methane, ethane, propane, and butane. Explain the trends observed.

8.106 Consider an apparatus in which A and B are two 1.00-L flasks joined by a stopcock C. The volume of the stopcock is negligible. Initially, A and B are evacuated, the stopcock C is closed, and 1.50 g of diethyl ether, $C_2H_5OC_2H_5$, is introduced into flask A. The vapor pressure of diethyl ether is 57 Torr at $-45°C$, 185 Torr at $0.°C$, 534 Torr at $25°C$, and negligible below $-86°C$. (a) If the stopcock is left closed and the flask is brought to equilibrium at $-45°C$, what will be the pressure of diethyl ether in flask A? (b) If the temperature is raised to $25°C$, what will be the pressure of diethyl ether in the flask? (c) If the temperature of the assembly is returned to $-45°C$ and the stopcock C is opened, what will be the pressure of diethyl ether in the apparatus? (d) If flask A is maintained at $-45°C$ and flask B is cooled with liquid nitrogen (boiling point, $-196°C$) with the stopcock open, what changes will take place in the apparatus? Assume ideal behavior.

8.107 The apparatus in Exercise 8.106 is again evacuated. Then 35.0 g of chloroform, $CHCl_3$, is placed in flask A and 35.0 g of acetone, CH_3COCH_3, is placed in flask B. The system is allowed to come to equilibrium at $25°C$ with the stopcock closed. The vapor pressures of chloroform and acetone at $25°C$ are 195 Torr and 222 Torr, respectively. (a) What is the pressure in each flask at equilibrium? (b) The stopcock is now opened. What will be the final composition of the gas and liquid phases in each flask once the system reaches equilibrium, assuming ideal behavior? (c) Acetone and chloroform solutions show negative deviations from Raoult's law. How will this behavior affect the answers given in part (b)?

8.108 Pentane is a liquid with a vapor pressure of 512 Torr at $25°C$; at the same temperature, the vapor pressure of hexane is only 151 Torr. What composition must the liquid phase have if the gas-phase composition is to have equal partial pressures of pentane and hexane?

8.109 Using the Clausius–Clapeyron Equation Living Graph on the Web site for this book, plot on the same set of axes the lines for $\Delta H_{vap} = 15, 20., 25,$ and $30. \text{ kJ·mol}^{-1}$. Is the vapor pressure of a liquid more sensitive to changes in temperature if ΔH_{vap} is small or large? Explain your conclusion.

8.110 A pair of amino acids is separated in a column in which the stationary phase is saturated with water and the carrier solvent is methanol, CH_3OH. The more polar the acid, the more strongly it is adsorbed by the stationary phase. The amino acids that were separated in this column are (a) $HOOCCHNH_2CH_2COOH$ and (b) $HOOCCHNH_2CH(CH_3)_2$. Which amino acid would you expect to be eluted first? Explain your reasoning. Refer to Major Technique 4 on chromatography, which follows these exercises.

8.111 Compounds A and B were extracted from a sample of Martian soil. A mixture of 0.52 mg of A and 2.30 mg of B in 1.00 mL of solution was separated by gas chromatography.

The areas of the two peaks were 5.44 cm² for A and 8.72 cm² for B. A second solution contained an extract with an unknown amount of A. To determine the concentration of A in the solution, 2.00 mg of B was added to 2.0 mL of the solution, which was then introduced into a gas chromatograph. Peak areas of 3.52 cm² for A and 7.58 cm² for B were measured. What is the concentration of A in the second solution? Refer to Major Technique 4, which follows these exercises.

8.112 Pure liquid naphthalene freezes at $80.2°C$ at 1 atm. When 1.00 mol of any nonelectrolyte solute is dissolved in 1.00 kg of naphthalene, the new freezing point of the solution (at 1 atm) is $73.2°C$. When 14.8 g of sulfur is dissolved in 575 g of naphthalene, the new freezing point of the solution is $79.5°C$. What are the molar mass and the molecular formula of the sulfur species in solution?

8.113 Human blood has an osmotic pressure relative to water of approximately 7.7 atm at body temperature ($37°C$). In a hospital, intravenous glucose ($C_6H_{12}O_6$) solutions are often given. If a technician must mix 500. mL of a glucose solution for a patient, what mass of glucose should be used?

8.114 Which of the following liquids is expected to have the highest vapor pressure at $0°C$: octane, $CH_3(CH_2)_6CH_3$; pentane, $CH_3(CH_2)_3CH_3$; or neopentane, 2,2-dimethylpropane, $C(CH_3)_4$? Justify your answer.

8.115 What volume of 0.010 M NaOH(aq) is required to react completely with 30. g of an aqueous acetic acid solution in which the mole fraction of acetic acid is 0.15?

8.116 The combustion analysis of L-carnitine, an organic compound thought to build muscle strength, yielded 52.16% C, 9.38% H, 8.69% N, and 29.78% O. The osmotic pressure of a 100.00-mL solution of 0.322 g of L-carnitine was found to be 0.501 atm at $32°C$. Assuming that L-carnitine does not ionize in methanol, determine (a) the molar mass of L-carnitine; (b) the molecular formula of L-carnitine.

8.117 Pentanol, an organic compound with the formula $CH_3CH_2CH_2CH_2CH_2OH$, has molecules with a nonpolar hydrocarbon chain of medium length attached to a polar —OH group. It is insoluble in water. However, if surfactant is added, the molecules of the three compounds form a layered structure that suspends the pentanol in the water. Propose a structure for the layered structure, describing the arrangement of the water, pentanol, and detergent molecules.

8.118 Hydrogen bonding typically occurs between a highly electronegative atom and a hydrogen atom attached to another electronegative atom. (a) Draw the Lewis structures for $CHCl_3$ and $(CH_3)_2CO$ and orient the structures to show the hydrogen bonding between the pair: (b) Which hydrogen atom is considered to have a low electron density and therefore to be a center of positive partial charge, and why?

Chemistry Connections

8.119 Sports drinks provide water to the body in the form of an isotonic solution (one having the same total molar concentration of solutes as human blood). These drinks contain electrolytes such as NaCl and KCl as well as sugar and flavoring. One of the main flavoring agents in sport drinks is citric acid, shown below.

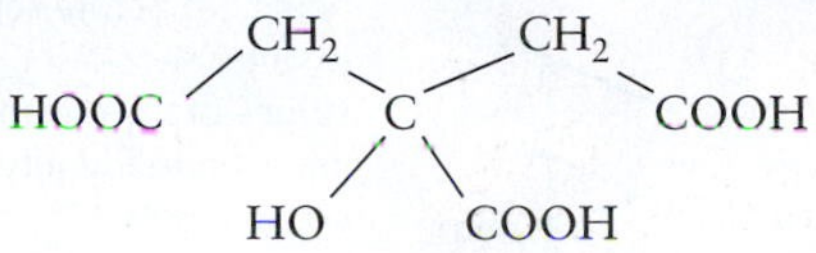

$$HOOC \overset{\displaystyle CH_2}{\diagup} \overset{\displaystyle C}{\underset{\displaystyle \diagdown}{\diagup}} \overset{\displaystyle CH_2}{\diagdown} COOH$$

Citric acid

(a) Indicate the hybridization of each C atom.
(b) Can citric acid take part in hydrogen bonding?
(c) Predict from a consideration of intermolecular forces whether citric acid is a gas, liquid, or solid at 25°C and whether it is soluble in water.

(d) *Normal saline solution* is an isotonic solution containing 0.9% NaCl by mass. Assuming complete dissociation of the NaCl, what is the total molar concentration of all solutes in an isotonic solution? Assume a density of $1.00 \text{ g} \cdot \text{cm}^{-3}$ for the solution.
(e) If you decide to make up 500.0 mL of a sports drink with 1.0 g of NaCl and glucose, what mass of glucose do you need to add to the NaCl and water to make the solution isotonic (see part d)?
(f) A paramedic treating injuries in a remote area has 300.0 mL of a 1.00% by mass solution of boric acid, $B(OH)_3$, that needs to be made isotonic (assume $i = 1$ in this solution and that the density is $1.00 \text{ g} \cdot \text{cm}^{-3}$). What mass of NaCl should be added? Assume that the NaCl is completely dissociated in the solution and that the volume of the solution does not change on addition of the NaCl. Take into account the 0.007% deprotonation of boric acid.

Because the scent of a flower may be due to hundreds of different compounds, it is difficult for perfume manufacturers to duplicate floral scents. Establishing the identities and relative amounts of the components of a fragrance was actually impossible until the development of chromatography. Related techniques are used in forensic laboratories to match samples of fluids, by food manufacturers to test product quality, and to search for evidence of life on other planets. All these techniques depend on subtle differences in intermolecular forces to separate compounds.

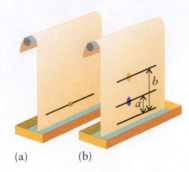

FIGURE 1 Two stages in a paper chromatography separation of a mixture of two components. (a) Before separation; (b) after separation. The relative values of the distances a and b are used to identify the components.

(a)   (b)

Solvent Extraction

If an aqueous solution of a compound is shaken with another liquid that is immiscible (mutually insoluble) with water, some of the compound may dissolve in the other solvent. For example, molecular iodine, I_2, is very slightly soluble in water but is highly soluble in tetrachloromethane, CCl_4, which is immiscible with water. When tetrachloromethane is added to water containing iodine, most of the iodine dissolves in the CCl_4. The solute is said to *partition* itself between the two solvents. *Solvent extraction* is used to obtain plant flavors and aromas from aqueous slurries of the plant that have been crushed in a blender.

In some cases, the solids themselves are subjected to extraction by a solvent. For example, in one process used to decaffeinate coffee, the coffee beans are mixed with activated charcoal and a high-pressure stream of "supercritical carbon dioxide" (carbon dioxide at high pressure and above its critical temperature) is passed over them at approximately 90°C. A supercritical solvent is a highly mobile fluid with a very low viscosity. The carbon dioxide removes the soluble caffeine preferentially without extracting the flavoring agents and evaporates without leaving a harmful residue.

Liquid Chromatography

In the early part of the twentieth century, the Russian botanist M. S. Tsvet found a way to separate the many pigments in flowers and leaves. He ground up the plant materials and dissolved the pigments, then poured the solution over the top of a vertical tube of ground chalk. The different pigments trickled through the chalk at different rates, producing colored bands in the tube, inspiring the name *chromatography* ("color writing"). The separation occurred because the different pigments were absorbed to different extents by the chalk.

Chromatography is one of the most powerful and widely used means for separating mixtures, because it is often inexpensive and it can be used to provide quantitative as well as qualitative information. The simplest method is paper chromatography. A drop of solution is placed near the bottom edge of the *stationary phase*. an absorbant support, such as a strip of paper. The *mobile phase*, a fluid solvent, is added below the spot and the solvent is absorbed on the support. As the mobile phase rises up the stationary phase by capillary action, the materials in the spot are carried upward at a rate that depends on how strongly they are adsorbed on (adhere to) the stationary phase (Fig. 1). The more strongly the solute is adsorbed on the stationary phase, the longer it will take to travel up the support.

The same concepts apply to *column chromatography,* where the stationary phase is normally small particles of silica, SiO_2, or alumina, Al_2O_3. These substances are not very reactive and have specially prepared surfaces to increase their adsorption ability. The column is saturated with solvent, and a small volume of solution containing the solutes is poured onto the top. As soon as it has soaked in, more solvent is added. The solutes travel slowly down the column and are *eluted* (removed as fractions) at the bottom (Fig. 2). If the mobile phase is less polar than the stationary phase, the less polar solutes will be eluted first and the more polar ones last.

To improve the separation of the solutes in a mixture, *high-performance liquid chromatography* (HPLC) was developed. In this technique, the mobile phase is forced under pressure through a long, narrow column, yielding an excellent separation in a relatively short time. HPLC has become the primary means of monitoring the use of therapeutic drugs and of detecting drug abuse; it is also used to separate the compounds contributing to the fragrance of flowers.

SELF-TEST M4.1A A pair of amino acids is separated in a column in which the stationary phase is saturated with water and the mobile phase is methanol, CH_3OH. The more polar the acid, the more strongly it is adsorbed by the

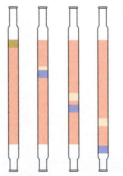

FIGURE 2 A column chromatography experiment. The mixture is placed on top of the column (left). As the solvent passes through the support, the mixture is washed down the column and separates into bands.

stationary phase. The amino acids that were separated in this column are (a) $HOOCCH(NH_2)CH_2OH$ and (b) $HOOCCH(NH_2)CH_3$. Which amino acid will be retained in the column the longest?

[**Answer:** Compound (a) has more polar groups, because it has an —OH group, and so will be retained longer on the column.]

Self-Test M4.1B Inorganic cations can be separated by liquid chromatography according to their ability to form complexes with chloride ions. For the separation, the stationary phase is saturated with water and the mobile phase is a solution of HCl in acetone. The relative solubilities of the following chlorides in concentrated hydrochloric acid are $CuCl_2 > CoCl_2 > NiCl_2$. What is the order of elution of these compounds?

Gas Chromatography

Volatile compounds can be separated in a *gas chromatograph*, in which the mobile phase is usually a relatively unreactive carrier gas such as helium, nitrogen, or hydrogen. The principles are the same as those for liquid chromatography, but the output is more often a *chromatogram*, rather than a series of eluted samples (Fig. 3). The chromatogram shows when each solute was eluted, and the peak areas tell how much of each component is present. The identity of the solute producing each peak can be determined by comparing its location against a database of known compounds.

In *gas–liquid partition chromatography* (GLPC), the stationary phase is a liquid that coats the particles in the tube or the walls of the tube. Often the tube itself is very narrow and long, perhaps 100 m, and has to be coiled (Fig. 4). Solutes are separated, as in liquid chromatography, by their relative solubility in the gas and liquid phases. In

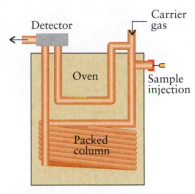

FIGURE 4 A schematic representation of the arrangement in a gas chromatograph. The coiled column, which is packed with the stationary phase, may be as long as 100 m.

gas–solid adsorption chromatography, solid particles coat the inside of the narrow tube. The solute vapors are separated by their relative attraction for the solid particles. In both cases, it is relative polarity that determines the distances between peaks.

As the vapor leaves the tube, the compounds in the sample are detected by a device such as a thermal conductivity detector. This instrument continuously measures the thermal conductivity (the ability to conduct heat) of the carrier gas, which changes when a solute is present. The detection techniques are very sensitive, allowing tiny amounts of solutes to be detected. Many environmental monitoring and forensic applications have been developed.

Gas Chromatography–Mass Spectrometry

Some detectors can give additional information about the *elutes* (the eluted solutes). One example is the *gas chromatograph–mass spectrometer* (GC-MS), which produces a mass spectrum of each compound as well as its mass and location in the chromatogram.* This powerful means of detection can be used when standard samples are not available to help determine the identities of the solutes. A beam of ions bombards each compound as it emerges from the chromatograph. The compound breaks up into ions of different masses, providing a spread of narrow peaks instead of one peak for each compound. The relative amount of each fragment is determined and used to help identify the compound.

The GC-MS instrument can be very compact and used in mobile equipment. A GC-MS device was used on the Viking Lander on the first Mars mission to search for organic compounds that would have provided evidence of life (none was detected).

Related Exercises 8.110 and 8.111

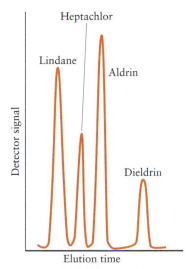

FIGURE 3 A gas chromatogram of a mixture of pesticides from farmland. The relative heights of the peaks indicate the relative abundances of the compounds.

Heptachlor
Lindane
Aldrin
Dieldrin
Detector signal
Elution time

---

* Mass spectrometry is the subject of Major Technique 6, following Chapter 18.

CHEMICAL EQUILIBRIA

What Are the Key Ideas? Instead of going to completion, reactions proceed until the composition of a reaction mixture corresponds to minimum Gibbs free energy. This composition is described by an equilibrium constant that is characteristic of the reaction and depends on the temperature.

Why Do We Need to Know This Material? The dynamic equilibrium toward which every chemical reaction tends is such an important aspect of the study of chemistry that four chapters of this book deal with it. We need to know the composition of a reaction mixture at equilibrium because it tells us how much product we can expect. To control the yield of a reaction, we need to understand the thermodynamic basis of equilibrium and how the position of equilibrium is affected by conditions such as temperature and pressure. The response of equilibria to changes in conditions has considerable economic and biological significance: the regulation of chemical equilibrium affects the yields of products in industrial processes, and living cells struggle to avoid sinking into equilibrium.

What Do We Need to Know Already? The concepts of chemical equilibrium are related to those of physical equilibrium (Sections 8.1–8.3). Because chemical equilibrium depends on the thermodynamics of chemical reactions, we need to know about the Gibbs free energy of reaction (Section 7.13) and standard enthalpies of formation (Section 6.18). Chemical equilibrium calculations require a thorough knowledge of molar concentration (Section G), reaction stoichiometry (Section L), and the gas laws (Chapter 4).

Chapter **9**

Early in the twentieth century, the looming prospect of World War I created a desperate need for nitrogen compounds because nitrates normally used for agriculture were being made into explosives. Almost all the nitrates used for fertilizers and explosives were quarried from deposits in Chile, and the limited supply could not keep up with the demand. Moreover, shipping channels were vulnerable to attack, which threatened to cut off the supply altogether. Although nitrogen is abundant in air, the methods used at the time to convert nitrogen into compounds were too costly to employ on a large scale. Any nation that could develop an economical process to **fix** atmospheric nitrogen—that is, to combine it with other elements—would have all the nitrogen compounds it needed.

Scientists on both sides of the conflict were urgently seeking ways to fix nitrogen. Finally, through determination, application, and—as so often happens in research—a bit of luck, the German chemist Fritz Haber in collaboration with the chemical engineer Carl Bosch found an economical way to harvest the nitrogen of the air and to provide an abundant source of compounds for both agriculture and armaments.

One difficulty Haber faced is that the reactions used to produce compounds from nitrogen do not go to completion, but appear to stop after only some of the reactants have been used up. At this point the mixture of reactants and products has reached **chemical equilibrium**, the stage in a chemical reaction when there is no further tendency for the composition of the reaction mixture—the concentrations or partial pressures of the reactants and products—to change. To achieve the greatest conversion of nitrogen into its compounds, Haber had to understand how a reaction approaches and eventually reaches equilibrium and then use that

FIGURE 9.1 (a) Methane burns in air with a steady flame but, because matter is being added and removed, the reaction is not at equilibrium. (b) This sample of glucose in air is unchanging in composition, but it is not yet in equilibrium with its combustion products; the reaction proceeds far too slowly at room temperature. (c) Nitrogen dioxide (a brown gas) and dinitrogen tetroxide (a colorless volatile solid) are in equilibrium in this vessel. We can tell that they are by changing the temperature slightly and seeing the composition adjust to a new value.

(a)   (b)   (c)

knowledge—as we shall see in Section 9.12—to improve the yield by changing the conditions of the reaction.

Like physical equilibria, all chemical equilibria are dynamic equilibria, with the forward and reverse reactions occurring at the same rate. In Chapter 8, we considered several physical processes, including vaporizing and dissolving, that reach dynamic equilibrium. This chapter shows how to apply the same ideas to chemical changes. It also shows how to use thermodynamics to describe equilibria quantitatively, which puts enormous power into our hands—the power to control the direction of a reaction and the yield of products.

And, we might add, to change the course of history.

REACTIONS AT EQUILIBRIUM

To say that chemical equilibrium is dynamic means that when a reaction has reached equilibrium, the forward and reverse reactions continue to take place, but with reactants being formed as fast as they are consumed. As a result, the composition of the reaction mixture remains constant. We can use these characteristics to decide if the three systems in Fig. 9.1 are at equilibrium. All three systems appear to be unchanging. However, when methane, CH_4, burns with a steady flame to form carbon dioxide gas and water (Fig. 9.1a), the combustion reaction is not at equilibrium because the composition is not constant (reactants continue to be added and the products dissipate instead of reacting). The sample of glucose does not change (Fig. 9.1b), even if we leave it for a very long time. However, the glucose is not at equilibrium with the products of its combustion (carbon dioxide and water); it survives only because the rate of its combustion is immeasurably slow at room temperature. The gas-phase reaction (Fig. 9.1c), however, is at equilibrium because, as well as the composition being constant, additional experiments show that NO_2 is ceaselessly forming N_2O_4 and that N_2O_4 is decomposing into NO_2 at the same rate.

The criteria that identify a dynamic chemical equilibrium are:

1   The forward reaction and its reverse are both taking place.
2   They are doing so at equal rates (so there is no net change).

As we shall see later in the chapter, dynamic equilibria respond to changes in temperature and pressure, and the addition of even a small amount of additional reagent can result in a change in overall composition. A reaction that is simply not taking place (such as a mixture of hydrogen and oxygen at room temperature and pressure) does not respond to small changes in the conditions.

Now we begin to develop these ideas. First, we establish that reactions can take place in their reverse direction as soon as some products have accumulated. Then we see how to relate the equilibrium composition to the thermodynamic properties of the system.

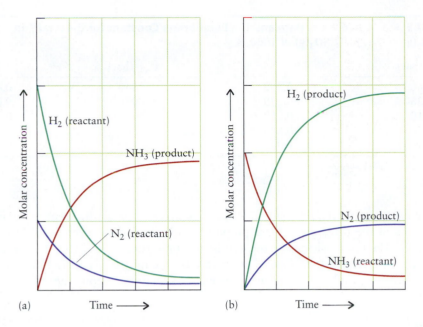

FIGURE 9.2 (a) In the synthesis of ammonia, the concentrations of N_2 and H_2 decrease with time and that of NH_3 increases until they finally settle into values corresponding to a mixture in which all three are present and there is no further net change. (b) If the experiment is repeated with pure ammonia, it decomposes, and the composition settles down into a mixture of ammonia, nitrogen, and hydrogen. (The two graphs correspond to experiments at two different temperatures, and so they correspond to different equilibrium compositions.)

9.1 The Reversibility of Reactions

Some reactions, such as the explosive reaction of hydrogen and oxygen, appear to proceed to completion, but others seem to stop at an early stage. For example, consider the reaction that took place when Haber heated nitrogen and hydrogen under pressure in the presence of a small amount of the metal osmium:

$$N_2(g) + 3 H_2(g) \longrightarrow 2 NH_3(g) \tag{A}$$

The reaction does produce ammonia rapidly at first, but eventually the reaction seems to stop (Fig. 9.2). As the graph shows, no matter how long we wait, no more product forms. The reaction has reached equilibrium.

Like phase changes, chemical reactions tend toward a *dynamic* equilibrium in which, although there is no net change, the forward and reverse reactions are still taking place, but at matching rates. What actually happens when the formation of ammonia *appears* to stop is that the rate of the reverse reaction,

$$2 NH_3(g) \longrightarrow N_2(g) + 3 H_2(g) \tag{B}$$

increases as more ammonia is formed. At equilibrium, the ammonia is decomposing as fast as it is being formed. Just as we did for phase transitions, we express this state of dynamic equilibrium by replacing the arrow in the equation with equilibrium "harpoons":

$$N_2(g) + 3 H_2(g) \rightleftharpoons 2 NH_3(g) \tag{C}$$

All chemical equilibria are dynamic equilibria. Although there is no further net change at equilibrium, the forward and reverse reactions are still taking place.

> *Chemical reactions reach a state of dynamic equilibrium in which the rates of forward and reverse reactions are equal and there is no net change in composition.*

9.2 Equilibrium and the Law of Mass Action

In 1864, before Haber began his work, the Norwegians Cato Guldberg (a mathematician) and Peter Waage (a chemist) had discovered the mathematical relation that summarizes the composition of a reaction mixture at equilibrium. As an example of their approach, look at the data in Table 9.1 for the reaction between SO_2 and O_2:

$$2 SO_2(g) + O_2(g) \rightleftharpoons 2 SO_3(g) \tag{D}$$

The metal acts as a *catalyst* for this reaction, a substance that helps to make a reaction go faster (Section 9.12). Osmium is too expensive to be used commercially: in the industrial process, iron is used instead.

TABLE 9.1 Equilibrium Data and the Equilibrium Constant for the Reaction
$2 SO_2(g) + O_2(g) \rightleftharpoons 2 SO_3(g)$ at 1000. K

P_{SO_2} (bar)	P_{O_2} (bar)	P_{SO_3} (bar)	K^*
0.660	0.390	0.0840	0.0415
0.0380	0.220	0.00360	0.0409
0.110	0.110	0.00750	0.0423
0.950	0.880	0.180	0.0408
1.44	1.98	0.410	0.0409

*Average K: 0.0413

In the experiments reported in Table 9.1, five mixtures with different initial compositions of the three gases were prepared and allowed to reach equilibrium at 1000. K. The compositions of the equilibrium mixtures and the total pressure P were then determined. At first, there seemed to be no pattern in the data. However, Guldberg and Waage noticed an extraordinary relation: the value of the quantity

$$K = \frac{(P_{SO_3}/P^\circ)^2}{(P_{SO_2}/P^\circ)^2(P_{O_2}/P^\circ)}$$

was nearly the same for every experiment, regardless of the initial compositions. Here, P_J is the equilibrium partial pressure of gas J and $P^\circ = 1$ bar, the standard pressure. Note that K is unitless, because the units of P_J are canceled by the units of P° in each term. From now on, though, to keep the notation simple, we shall write simply

$$K = \frac{(P_{SO_3})^2}{(P_{SO_2})^2 P_{O_2}}$$

with the P_J understood to be the numerical value of the pressure in bars.

Within experimental error, Guldberg and Waage obtained the same value of K whatever the initial composition of the reaction mixture. This remarkable result shows that K is characteristic of the composition of the reaction mixture at equilibrium at a given temperature. It is known as the **equilibrium constant** for the reaction. The **law of mass action** summarizes this result: it states that, *at equilibrium, the composition of the reaction mixture can be expressed in terms of an equilibrium constant* where, for any reaction between gases that can be treated as ideal,

$$K = \left\{ \frac{\text{partial pressures of products}}{\text{partial pressures of reactants}} \right\}_{\text{equilibrium}} \tag{1a}$$

with each partial pressure raised to a power equal to the stoichiometric coefficient in the balanced chemical equation for the reaction. We shall not in general write "equilibrium" on expressions for K. Thus, if we are interested in the reaction

$$a\,A(g) + b\,B(g) \rightleftharpoons c\,C(g) + d\,D(g) \tag{E}$$

and if all the gases are treated as ideal, we write

$$K = \frac{(P_C)^c(P_D)^d}{(P_A)^a(P_B)^b} \tag{1b}*$$

with the numerical values of the partial pressures being those at equilibrium (in bar).

EXAMPLE 9.1   **Sample exercise: Writing the expression for an equilibrium constant**

Write the equilibrium constant for the ammonia synthesis reaction, reaction C.

SOLUTION We write the equilibrium constant with the partial pressure of the product, NH_3, in the numerator, raised to a power equal to its coefficient in the balanced equation. We do the same for the reactants, but place their partial pressures in the denominator:

From $K = (P_C)^c(P_D)^d/(P_A)^a(P_B)^b$,

$$K = \frac{(P_{NH_3})^2}{P_{N_2}(P_{H_2})^3}$$

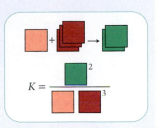

As already explained, each P_J in this expression should be interpreted as $P_J/P°$.

SELF-TEST 9.1A Write the expression for the equilibrium constant for the reaction $4\,NH_3(g) + 5\,O_2(g) \rightleftharpoons 4\,NO(g) + 6\,H_2O(g)$.

[**Answer:** $K = (P_{NO})^4(P_{H_2O})^6/(P_{NH_3})^4(P_{O_2})^5$]

SELF-TEST 9.1B Write the expression for the equilibrium constant for $2\,H_2S(g) + 3\,O_2(g) \rightleftharpoons 2\,SO_2(g) + 2\,H_2O(g)$.

We use a different measure of concentration when writing expressions for the equilibrium constants of reactions that involve species other than gases. Thus, for a species J that forms an ideal solution in a liquid solvent, the partial pressure in the expression for K is replaced by the molarity [J] relative to the standard molarity $c° = 1\ \text{mol·L}^{-1}$. Although K should be written in terms of the dimensionless ratio $[J]/c°$, it is common practice to write K in terms of [J] alone and to interpret each [J] as the molarity with the units struck out. It has been found empirically, and is justified by thermodynamics, that pure liquids or solids should not appear in K. So, even though $CaCO_3(s)$ and $CaO(s)$ occur in the equilibrium

$$CaCO_3(s) \rightleftharpoons CaO(s) + CO_2(g) \tag{F}$$

neither appears in the equilibrium constant, which is $K = P_{CO_2}/P°$ (or, more simply, $K = P_{CO_2}$).

We can summarize these empirical rules by introducing the concept of the **activity**, a_J, of a substance J:

Substance	Activity	Simplified form
Ideal gas	$a_J = P_J/P°$	$a_J = P_J$
Solute in a dilute solution	$a_J = [J]/c°$	$a_J = [J]$
Pure solid or liquid	$a_J = 1$	

Note that all activities are pure numbers and thus are unitless. When using the simplified form, the activity is the numerical value of the pressure in bar or the numerical value of the molarity in moles per liter.

At this stage, activities are simply quantities introduced to facilitate writing the expression for K. In advanced work activities are used to take into account deviations from ideal behavior and to accommodate real gases and real solutions, where intermolecular interactions are important. In those cases the activity of a substance may differ from the calculated pressure or concentration. However, for the low-pressure gases and dilute solutions that concern us here, we can ignore the interactions between molecules and regard activities as providing a simple, uniform way of writing general expressions for equilibrium constants and other quantities that we shall meet shortly. We have to remember, though, that our expressions are approximations.

The use of activities allows us to write a general expression for the equilibrium constant for any reaction:

$$K = \left\{ \frac{\text{activities of products}}{\text{activities of reactants}} \right\}_{\text{equilibrium}} \tag{2a}$$

More specifically, for a generalized version of reaction E, with phases not specified:

$$a\,A + b\,B \rightleftharpoons c\,C + d\,D \qquad K = \frac{(a_C)^c(a_D)^d}{(a_A)^a(a_B)^b} \tag{2b}*$$

Because activities are unitless, so too is K.

A note on good practice: In some cases you will see an equilibrium constant denoted K_P to remind you that it is expressed in terms of partial pressures. However, the subscript P is unnecessary.

Chemical equilibria with reactants and products that are all in the same phase are called **homogeneous equilibria.** Equilibria C, D, and E are homogeneous. Equilibria in systems having more than one phase are called **heterogeneous equilibria.** Equilibrium F is heterogeneous; so too is the equilibrium between water vapor and liquid water in a closed system:

$$H_2O(l) \rightleftharpoons H_2O(g) \tag{G}$$

In this reaction, there is a gas phase and a liquid phase. Likewise, the equilibrium between a solid and its saturated solution is heterogeneous:

$$Ca(OH)_2(s) \rightleftharpoons Ca^{2+}(aq) + 2\,OH^-(aq) \tag{H}$$

The equilibrium constants for heterogeneous reactions are also given by the general expression in Eq. 2: all we have to remember is that the activity of a pure solid or liquid is 1. For instance, for the calcium hydroxide equilibrium (reaction H),

$$K = \frac{a_{Ca^{2+}}(a_{OH^-})^2}{\underbrace{a_{Ca(OH)_2}}_{\text{1 for a pure solid}}} = [Ca^{2+}][OH^-]^2$$

because the calcium hydroxide is a pure solid. Remember that each [J] represents the molar concentration of J with the units struck out. Similarly, in the equilibrium between solid nickel, gaseous carbon monoxide, and gaseous nickel carbonyl used in the purification of nickel,

$$Ni(s) + 4\,CO(g) \rightleftharpoons Ni(CO)_4(g) \qquad K = \frac{a_{Ni(CO)_4}}{\underbrace{a_{Ni}}_{1}(a_{CO})^4} = \frac{P_{Ni(CO)_4}}{(P_{CO})^4}$$

The pure solid nickel must be present for the equilibrium to exist, but it does not appear in the expression for the equilibrium constant.

Some reactions in solution involve the solvent as a reactant or product. When the solution is very dilute, the change in solvent concentration due to the reaction is insignificant. In such cases, the solvent is treated as a pure substance and ignored when writing K. In other words,

for a nearly pure solvent: $a_{\text{solvent}} = 1$

A final point to bear in mind is that, when a reaction involves fully dissociated ionic compounds in solution, then the equilibrium constant should be written for the net ionic equation, by using the activity for each type of ion.

SELF-TEST 9.2A Write the equilibrium constant for the reaction of silver nitrate with sodium hydroxide: $2\,AgNO_3(aq) + 2\,NaOH(aq) \rightleftharpoons Ag_2O(s) + 2\,NaNO_3(aq) + H_2O(l)$. Remember to use the net ionic equation.

[*Answer:* $K = 1/[Ag^+]^2[OH^-]^2$]

SELF-TEST 9.2B Write the equilibrium constant K for $P_4(s) + 5\,O_2(g) \rightleftharpoons P_4O_{10}(s)$.

Each reaction has its own characteristic equilibrium constant, with a value that can be changed only by varying the temperature (Table 9.2). The extraordinary empirical result, which we justify in the next section, is that, *regardless of the initial composition of a reaction mixture, the composition tends to adjust until the*

TABLE 9.2 Equilibrium Constants for Various Reactions

Reaction	T (K)*	K	$K_c^†$
$H_2(g) + Cl_2(g) \rightleftharpoons 2\ HCl(g)$	300	4.0×10^{31}	4.0×10^{31}
	500	4.0×10^{18}	4.0×10^{18}
	1000	5.1×10^{8}	5.1×10^{8}
$H_2(g) + Br_2(g) \rightleftharpoons 2\ HBr(g)$	300	1.9×10^{17}	1.9×10^{17}
	500	1.3×10^{10}	1.3×10^{10}
	1000	3.8×10^{4}	3.8×10^{4}
$H_2(g) + I_2(g) \rightleftharpoons 2\ HI(g)$	298	794	794
	500	160	160
	700	54	54
$2\ BrCl(g) \rightleftharpoons Br_2(g) + Cl_2(g)$	300	377	377
	500	32	32
	1000	5	5
$2\ HD(g) \rightleftharpoons H_2(g) + D_2(g)$	100	0.52	0.52
	500	0.28	0.28
	1000	0.26	0.26
$F_2(g) \rightleftharpoons 2\ F(g)$	500	3.0×10^{-11}	7.3×10^{-13}
	1000	1.0×10^{-2}	1.2×10^{-4}
	1200	0.27	2.7×10^{-3}
$Cl_2(g) \rightleftharpoons 2\ Cl(g)$	1000	1.0×10^{-5}	1.2×10^{-7}
	1200	1.7×10^{-3}	1.7×10^{-5}
$Br_2(g) \rightleftharpoons 2\ Br(g)$	1000	3.4×10^{-5}	4.1×10^{-7}
	1200	1.7×10^{-3}	1.7×10^{-5}
$I_2(g) \rightleftharpoons 2\ I(g)$	800	2.1×10^{-3}	3.1×10^{-5}
	1000	0.26	3.1×10^{-3}
	1200	6.8	6.8×10^{-2}
$N_2(g) + 3\ H_2(g) \rightleftharpoons 2\ NH_3(g)$	298	6.8×10^{5}	4.2×10^{8}
	400	41	4.5×10^{4}
	500	3.6×10^{-2}	62
$2\ SO_2(g) + O_2(g) \rightleftharpoons 2\ SO_3(g)$	298	4.0×10^{24}	9.9×10^{25}
	500	2.5×10^{10}	1.0×10^{12}
	700	3.0×10^{4}	1.7×10^{6}
$N_2O_4(g) \rightleftharpoons 2\ NO_2(g)$	298	0.15	6.1×10^{-3}
	400	47.9	1.44
	500	1.7×10^{3}	41

*Three significant figures
$^†K_c$ is the equilibrium constant in terms of molar concentrations of gases (Section 9.6).

activities give rise to the characteristic value of K for the reaction and temperature (Fig. 9.3).

> *The equilibrium composition of a reaction mixture is described by the equilibrium constant, which is equal to the activities of the products (raised to powers equal to their stoichiometric coefficients in the balanced chemical equation for the reaction) divided by the activities of the reactants (raised to powers equal to their stoichiometric coefficients).*

9.3 The Thermodynamic Origin of Equilibrium Constants

Guldberg and Waage's description of equilibrium was entirely empirical. Now, though, chemists know that the law of mass action is a consequence of thermodynamics and that the apparently mysterious form of K arises from thermodynamic considerations. We begin our exploration of the thermodynamic basis for equilibrium by analyzing how changes in Gibbs free energy determine the tendency of a reaction to go in the forward or the reverse directions. We then explore how the

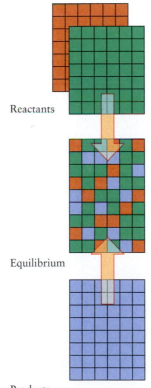

Reactants

Equilibrium

Products

FIGURE 9.3 Whether we start with pure reactants or with pure products, a reaction mixture will always tend toward a mixture of reactants and products that has a composition in accord with the equilibrium constant for the reaction at the temperature of the experiment.

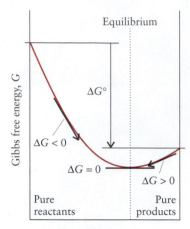

FIGURE 9.4 The variation of Gibbs free energy of a reaction mixture with composition. A reaction mixture has a spontaneous tendency to change in the direction of decreasing Gibbs free energy. Note that ΔG is the *slope* of the line at each composition, whereas $\Delta G°$ is the difference between the standard Gibbs free energies of the pure reactants and the pure products.

Gibbs free energy of reaction depends on the composition of the reaction mixture and how it changes as the reaction approaches equilibrium.

Every chemical reaction has a tendency to proceed spontaneously toward equilibrium. When a reaction mixture has not yet produced enough products to reach equilibrium, the spontaneous direction of change is toward more products. In such a case, $\Delta G < 0$ for the forward reaction at constant temperature and pressure. If, on the other hand, we start with a reaction mixture that has excess products, the *reverse* reaction is spontaneous and in such a case $\Delta G > 0$ for the forward reaction (so that $\Delta G < 0$ for the reverse reaction). For the reaction at equilibrium, with all reactants and products present at their equilibrium concentrations, there is no tendency for spontaneous change in either the forward or the reverse direction and $\Delta G = 0$. Clearly, ΔG changes as the proportions of reactants and products change; the composition of the reaction mixture at equilibrium is the composition at which ΔG is zero (Fig. 9.4). As we can see from the graph, at any point during the reaction the sign of ΔG is indicated by the slope of the curve, which shows how G varies with composition at that point: to the left of the equilibrium composition, the slope is negative and so the forward reaction is spontaneous; to the right of equilibrium, the slope is positive and so the reverse reaction is spontaneous; at equilibrium, at the minimum of the curve, the slope is zero.

The value of ΔG at a particular stage of the reaction is the difference in the molar Gibbs free energies of the products and the reactants *at the partial pressures or concentrations that they have at that stage*, weighted by the stoichiometric coefficients interpreted as amounts in moles:

$$\Delta G = \sum n G_m(\text{products}) - \sum n G_m(\text{reactants}) \qquad \text{Units: kilojoules} \qquad (3a)$$

(This is Eq. 18 of Chapter 7.) We shall sometimes find it useful to interpret the n that appear in Eq. 3a as pure numbers (rather than amounts in moles), so that if $n = 2$ mol, in this convention we would use $n = 2$. To signal that we are using this "molar" convention ("molar" because the units of ΔG then become kilojoules per mole), we attach a subscript r to ΔG (r for reaction) and write

$$\Delta G_r = \sum n G_m(\text{products}) - \sum n G_m(\text{reactants}) \qquad (3b)$$
$$\text{Units: kilojoules per mole}$$

We shall explain later in this section (in the "How Do We Do That?" box) exactly when and why it is necessary to use this "molar" convention.

The molar Gibbs free energies of the reactants and products change during the course of the reaction, so ΔG changes too. It is very important to distinguish this Gibbs free energy of reaction, ΔG, from the *standard* Gibbs free energy of reaction, $\Delta G°$, which does not change during the course of a reaction:

ΔG is the difference in molar Gibbs free energies of the products and the reactants at a specified stage of the reaction.

$\Delta G°$ is the difference in molar Gibbs free energies of the products and the reactants in their standard states and can be calculated from

$$\Delta G° = \sum n \Delta G_f°(\text{products}) - \sum n \Delta G_f°(\text{reactants}) \qquad (3c)$$

(This is Eq. 20 of Chapter 7.) We could write this expression in the molar convention too, as $\Delta G_r°$, in which case the n would be interpreted as pure numbers.

A note on good practice: Recall that the standard state of a pure substance is its pure form at a pressure of 1 bar (Section 6.15). For a solute, the standard state is for a concentration of 1 $\text{mol} \cdot \text{L}^{-1}$. Pure solids and liquids may always be regarded as being in their standard states provided the pressure is close to 1 bar.

To find how ΔG changes with composition, we need to know how the molar Gibbs free energy of each substance varies with its partial pressure, if it is a gas, or with its concentration, if it is a solute. We have already seen (in Section 8.3) that the molar Gibbs free energy of an ideal gas J is related to its partial pressure, P_J, by

$$G_m(J) = G_m°(J) + RT \ln (P_J/P°) \qquad (4a)$$

Note that, because $\ln 1 = 0$, $G_m(J)$ has its standard value, $G_m°(J)$, when $P_J = 1$ bar. Thermodynamic arguments (which we do not reproduce here) show that a similar expression applies to solutes and pure substances. In each case, we can write the molar Gibbs free energy of a substance J as

$$G_m(J) = G_m°(J) + RT \ln a_J \qquad (4b)*$$

with the activity defined as in Section 9.2. For a gas that behaves ideally, Eq. 4b is the same as Eq. 4a. For a solute in an ideal solution, Eq. 4b becomes

$$G_m(J) = G_m°(J) + RT \ln ([J]/c°) \qquad (4c)$$

Note that $G_m(J)$ has its standard value, $G_m°(J)$, when $[J] = 1$ mol·L^{-1}. For a pure solid or liquid,

$$G_m(J) = G_m°(J) \qquad (4d)$$

That is, for a pure solid or liquid, the molar Gibbs free energy always has its standard value (provided the pressure is 1 bar).

We are now ready to find the value of ΔG at any stage of a reaction from the composition of the reaction mixture at that point.

HOW DO WE DO THAT?

To keep the units straight, we need to use the "molar" convention for this calculation and to use the stoichiometric coefficients in reaction E as pure numbers. To find an expression for ΔG_r, we substitute Eq. 4 for each substance into Eq. 3a. For example, for the general reaction E,

$$
\begin{aligned}
\Delta G_r &= \{cG_m(C) + dG_m(D)\} - \{aG_m(A) + bG_m(B)\} \\
&= \{c[G_m°(C) + RT \ln a_C] + d[G_m°(D) + RT \ln a_D]\} \\
&\quad - \{a[G_m°(A) + RT \ln a_A] + b[G_m°(B) + RT \ln a_B]\} \\
&= \{cG_m°(C) + dG_m°(D)\} - \{aG_m°(A) + bG_m°(B)\} \\
&\quad + RT\{(c \ln a_C + d \ln a_D) - (a \ln a_A + b \ln a_B)\}
\end{aligned}
$$

Notice that the combination of the first four terms in the final equation is the standard Gibbs free energy of reaction, $\Delta G_r°$ (Eq. 19 of Chapter 7):

$$\Delta G_r° = \{cG_m°(C) + dG_m°(D)\} - \{aG_m°(A) + bG_m°(B)\}$$

Therefore,

$$\Delta G_r = \Delta G_r° + RT\{(c \ln a_C + d \ln a_D) - (a \ln a_A + b \ln a_B)\}$$

We can also tidy up the four logarithmic terms:

From $s \ln x = \ln x^s$,	$(c \ln a_C + d \ln a_D) - (a \ln a_A + b \ln a_B)$ $= (\ln a_C{}^c + \ln a_D{}^d) - (\ln a_A{}^a + \ln a_B{}^b)$
From $\ln x + \ln y = \ln xy$,	$(\ln a_C{}^c + \ln a_D{}^d) - (\ln a_A{}^a + \ln a_B{}^b)$ $= \ln a_C{}^c a_D{}^d - \ln a_A{}^a a_B{}^b$
From $\ln x - \ln y = \ln (x/y)$,	$\ln a_C{}^c a_D{}^d - \ln a_A{}^a a_B{}^b = \ln (a_C{}^c a_D{}^d / a_A{}^a a_B{}^b)$

Overall, therefore,

$$(c \ln a_C + d \ln a_D) - (a \ln a_A + b \ln a_B) = \ln \frac{(a_C)^c (a_D)^d}{(a_A)^a (a_B)^b}$$

Now we can bring our work together and write:

$$\Delta G_r = \Delta G_r° + RT \ln \frac{(a_C)^c (a_D)^d}{(a_A)^a (a_B)^b}$$

A note on good practice: Notice that by using the "molar" convention, the units match: RT is a molar energy (kilojoules per mole), and so too are the two Gibbs free energy terms.

The expression that we have just derived may be written

$$\Delta G_r = \Delta G_r° + RT \ln Q \tag{5}*$$

with the **reaction quotient** Q defined as

$$Q = \frac{(a_C)^c (a_D)^d}{(a_A)^a (a_B)^b} \tag{6}$$

Equation 5 shows how the Gibbs free energy of reaction varies with the activities (the partial pressures of gases or molarities of solutes) of the reactants and products. The expression for Q has the same form as the expression for K, but the activities refer to *any* stage of the reaction.

EXAMPLE 9.2   Calculating the Gibbs free energy of reaction from the reaction quotient

The standard Gibbs free energy of reaction for $2\ SO_2(g) + O_2(g) \rightarrow 2\ SO_3(g)$ is $\Delta G_r° = -141.74\ \text{kJ·mol}^{-1}$ at 25.00°C. What is the Gibbs free energy of reaction when the partial pressure of each gas is 100. bar? What is the spontaneous direction of the reaction under these conditions?

STRATEGY Calculate the reaction quotient and substitute it and the standard Gibbs free energy of reaction into Eq. 5. If $\Delta G_r < 0$, the forward reaction is spontaneous at the given composition. If $\Delta G_r > 0$, the reverse reaction is spontaneous at the given composition. If $\Delta G_r = 0$, there is no tendency to react in either direction; the reaction is at equilibrium. At 298.15 K, $RT = 2.479\ \text{kJ·mol}^{-1}$.

SOLUTION

From $Q = (a_{SO_3})^2/(a_{SO_2})^2(a_{O_2})$
$= (P_{SO_3})^2/(P_{SO_2})^2(P_{O_2})$,

$$Q = \frac{(100.)^2}{(100.)^2 \times (100.)} = 1.00 \times 10^{-2}$$

From $\Delta G_r = \Delta G_r° + RT \ln Q$,

$$\Delta G_r = -141.74\ \text{kJ·mol}^{-1}$$
$$+ (2.479\ \text{kJ·mol}^{-1}) \ln (1.00 \times 10^{-2})$$
$$= -153.16\ \text{kJ·mol}^{-1}$$

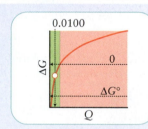

Because the Gibbs free energy of reaction is negative, the formation of products is spontaneous (as indicated by the green region in the diagram) at this composition and temperature.

SELF-TEST 9.3A The standard Gibbs free energy of reaction for $H_2(g) + I_2(g) \rightarrow 2\ HI(g)$ is $\Delta G_r° = -21.1\ \text{kJ·mol}^{-1}$ at 500. K (when $RT = 4.16\ \text{kJ·mol}^{-1}$). What is the value of ΔG_r at 500. K when the partial pressures of the gases are $P_{H_2} = 1.5$ bar, $P_{I_2} = 0.88$ bar, and $P_{HI} = 0.065$ bar? What is the spontaneous direction of the reaction?

[*Answer:* $-45\ \text{kJ·mol}^{-1}$; toward products]

SELF-TEST 9.3B The standard Gibbs free energy of reaction for $N_2O_4(g) \rightarrow 2\ NO_2(g)$ is $\Delta G_r° = +4.73\ \text{kJ·mol}^{-1}$ at 298 K. What is the value of ΔG_r when the partial pressures of the gases are $P_{N_2O_4} = 0.80$ bar and $P_{NO_2} = 2.10$ bar? What is the spontaneous direction of the reaction?

Now we come to the most important point in this chapter. At equilibrium, the activities (the partial pressures or molarities) of all the substances taking part in the reaction have their equilibrium values. At this point the expression for Q (in which the activities have their equilibrium values) has become the equilibrium constant, K, of the reaction. That is,

At equilibrium, $Q = K$ (7)

Thermodynamics has explained the puzzling form of K: it is a direct consequence of Eq. 4b, which shows how the Gibbs free energy of a substance depends on its activity.

We can now take another important step. We know that at equilibrium $\Delta G_r = 0$ and have just seen that at equilibrium $Q = K$. It follows from Eq. 5 that, at equilibrium,

$$0 = \Delta G_r^\circ + RT \ln K$$

and therefore that

$$\Delta G_r^\circ = -RT \ln K \tag{8}*$$

This fundamentally important equation links thermodynamic quantities—which are widely available from tables of thermodynamic data—and the composition of a system at equilibrium.

A note on good practice: Always write Eq. 8 with the standard state sign. Note too that, for the units to match on both sides, we have to use the "molar" convention for the standard Gibbs free energy.

EXAMPLE 9.3   **Sample exercise: Predicting the value of K from the standard Gibbs free energy of reaction**

At 25.00°C, the standard Gibbs free energy of reaction for $\frac{1}{2}H_2(g) + \frac{1}{2}I_2(s) \rightarrow HI(g)$ is $+1.70\ kJ \cdot mol^{-1}$ (see Appendix 2A); calculate the equilibrium constant for this reaction.

SOLUTION

From $\Delta G_r^\circ = -RT \ln K$ in the form $\ln K = -\Delta G_r^\circ/RT$, first converting the units of ΔG_r° to joules per mole,

$$\ln K = -\frac{\overbrace{1.70 \times 10^3\ J \cdot mol^{-1}}^{1.70\ kJ \cdot mol^{-1}}}{(8.3145\ J \cdot K^{-1} \cdot mol^{-1}) \times (298.15\ K)}$$

$$= -\frac{1.70 \times 10^3}{8.3145 \times 298.15} = -0.685$$

Take the inverse logarithm ($e^{\ln x} = x$) $K = 0.50$

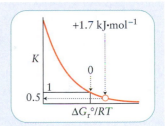

A note on good practice: Exponential functions (inverse logarithms, e^x) are very sensitive to the value of x, so carry out all the arithmetic in one step to avoid rounding errors.

SELF-TEST 9.4A Use the thermodynamic data in Appendix 2A to calculate K from ΔG_r° for $N_2O_4(g) \rightleftharpoons 2\ NO_2(g)$ at 298 K.

[*Answer: $K = 0.15$*]

SELF-TEST 9.4B Use the thermodynamic data in Appendix 2A to calculate K from ΔG_r° for $2\ NO(g) + O_2(g) \rightleftharpoons 2\ NO_2(g)$ at 298 K.

We can now begin to acquire some insight into why some reactions have large equilibrium constants and others have small ones. It follows from $\Delta G_r^\circ = \Delta H_r^\circ - T\Delta S_r^\circ$ and $\Delta G_r^\circ = -RT \ln K$ that

$$\ln K = -\frac{\Delta G_r^\circ}{RT} = -\frac{\Delta H_r^\circ}{RT} + \frac{\Delta S_r^\circ}{R}$$

When we take inverse logarithms of both sides and use $e^{x+y} = e^x e^y$, we obtain

$$K = e^{-\Delta H_r^\circ/RT + \Delta S_r^\circ/R} = e^{-\Delta H_r^\circ/RT}e^{\Delta S_r^\circ/R} \tag{9}$$

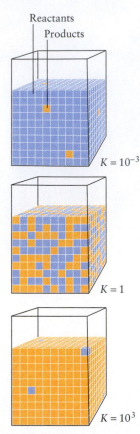

Reactants
Products

$K = 10^{-3}$

$K = 1$

$K = 10^3$

FIGURE 9.5 The size of the equilibrium constant indicates whether the reactants or the products are favored. In this diagram, the reactants are represented by blue cubes and the products by yellow cubes. Note that reactants are favored when K is small (top), products are favored when K is large (bottom), and reactants and products are in equal abundance when $K = 1$.

Because equilibrium constants must be raised to a power when a chemical equation is multiplied by a factor, and therefore change in magnitude, these rules are only general guidelines.

We see that K can be expected to be small if ΔH_r° is positive. An endothermic reaction is therefore likely to have $K < 1$ and is unlikely to form much product. Only if ΔS_r° is large and positive can we expect $K > 1$ for an endothermic reaction. Conversely, if a reaction is strongly exothermic, then ΔH_r° is large and negative. Now we can expect $K > 1$ and the products to be favored. In other words, we can expect strongly exothermic reactions to go to completion.

> *The reaction quotient, Q, has the same form as K, the equilibrium constant, except that Q uses the activities evaluated at an arbitrary stage of the reaction. The equilibrium constant is related to the standard Gibbs free energy of reaction by $\Delta G_r^{\circ} = -RT \ln K$.*

9.4 The Extent of Reaction

When ΔG_r° for a reaction is strongly negative, equilibrium is reached only after a reaction has gone nearly to completion and the equilibrium reaction mixture consists mainly of products. Because the concentrations or partial pressures of products appear in the numerator of K and those of reactants in the denominator, the numerator will be large and the denominator small. Therefore, the value of K is large when the equilibrium mixture consists mostly of products. In contrast, when ΔG_r° for a reaction is strongly positive, equilibrium is reached after very little reaction has taken place, the numerator is small and the denominator large; therefore, K is small.

For instance, consider the reaction

$$H_2(g) + Cl_2(g) \rightleftharpoons 2\, HCl(g) \qquad K = \frac{(P_{HCl})^2}{P_{H_2} P_{Cl_2}}$$

Experiment shows that $K = 4.0 \times 10^{18}$ at 500. K. Such a large value for K tells us that the system does not reach equilibrium until most of the reactants have been converted into HCl. In fact, this reaction essentially goes to completion. Now consider the equilibrium

$$N_2(g) + O_2(g) \rightleftharpoons 2\, NO(g) \qquad K = \frac{(P_{NO})^2}{P_{N_2} P_{O_2}}$$

Experiment gives $K = 3.4 \times 10^{-21}$ at 800. K. The very small value of K tells us that the reactants N_2 and O_2 are the dominant species in the system at equilibrium.

We can summarize these remarks as follows for chemical equations written with the smallest whole-number stoichiometric coefficients (Fig. 9.5):

- Large values of K (larger than about 10^3): the equilibrium favors the products.

- Intermediate values of K (approximately in the range 10^{-3} to 10^3): neither reactants nor products are strongly favored at equilibrium.

- Small values of K (smaller than about 10^{-3}): the equilibrium favors the reactants.

The following example shows how to use the equilibrium constant to calculate an equilibrium concentration.

EXAMPLE 9.4   Calculating an equilibrium composition

Suppose that, in an equilibrium mixture of HCl, Cl_2, and H_2, the partial pressure of H_2 is 4.2 mPa and that of Cl_2 is 8.3 mPa. What is the partial pressure of HCl at 500. K, given $K = 4.0 \times 10^{18}$ for $H_2(g) + Cl_2(g) \rightleftharpoons 2\, HCl(g)$?

STRATEGY   At equilibrium, the partial pressures of the reactants and products (in bar, 1 bar = 10^5 Pa) satisfy the expression for K. Therefore, rearrange the expression to give the one unknown concentration, and substitute the data.

SOLUTION First convert the units of partial pressures to bars using 1 bar = 10^5 Pa: $P_{H_2} = 4.2 \times 10^{-8}$ bar, $P_{Cl_2} = 8.3 \times 10^{-8}$ bar. Write the expression for the equilibrium constant from the chemical equation given.

From $K = (P_{HCl})^2/P_{H_2}P_{Cl_2}$ $P_{HCl} = \{(4.0 \times 10^{18}) \times (4.2 \times 10^{-8}) \times (8.3 \times 10^{-8})\}^{1/2}$
in the form $= 1.2 \times 10^2$
$P_{HCl} = (KP_{H_2}P_{Cl_2})^{1/2},$

This result means that, at equilibrium, the partial pressure of product in the system, 120 bar, is overwhelming relative to the tiny partial pressures of the reactants.

SELF-TEST 9.5A Suppose that the equilibrium partial pressures of H_2 and Cl_2 are both 1.0 μPa. What is the equilibrium partial pressure of HCl at 500. K, given $K = 4.0 \times 10^{18}$?

[**Answer:** P_{HCl} = 20 mbar]

SELF-TEST 9.5B Suppose that the equilibrium partial pressures of N_2 and O_2 in the reaction $N_2(g) + O_2(g) \rightleftharpoons 2\,NO(g)$ at 800. K are both 52 kPa. What is the partial pressure (in pascals) of NO at equilibrium if $K = 3.4 \times 10^{-21}$ at 800. K?

If K is large, products are favored at equilibrium; if K is small, reactants are favored.

9.5 The Direction of Reaction

Example 9.4 deals with a system at equilibrium, but suppose the reaction mixture has arbitrary concentrations. How can we tell whether it will have a tendency to form more products or to decompose into reactants? To answer this question, we first need the equilibrium constant. We may have to determine it experimentally or calculate it from standard Gibbs free energy data. Then we calculate the reaction quotient, Q, from the actual composition of the reaction mixture, as described in Section 9.3. To predict whether a particular mixture of reactants and products will tend to produce more products or more reactants, we compare Q with K:

• If $Q < K$, ΔG is negative; the concentrations or partial pressures of the products are too low relative to those of the reactants for equilibrium. Hence, the reaction has a tendency to proceed toward products.

• If $Q = K$, ΔG is zero; the mixture has its equilibrium composition and has no tendency to change in either direction.

• If $Q > K$, ΔG is positive, the reverse reaction is spontaneous and the products tend to decompose into the reactants.

This pattern is summarized in Fig. 9.6.

The same pattern is seen when we plot the Gibbs free energy of a reaction mixture against its changing composition, as we did in Fig. 9.4. Reaction tends to proceed toward the composition at the lowest point of the curve, because that is the direction of decreasing Gibbs free energy. The composition at the lowest point of the curve—the point of minimum Gibbs free energy—corresponds to equilibrium. For a system at equilibrium, any change—either the forward or the reverse reaction—would lead to a reaction mixture with a greater Gibbs free energy, and so neither change is spontaneous. When the minimum value of the Gibbs free energy lies very close to pure products, the equilibrium composition strongly favors products and "goes to completion" (Fig. 9.7a). When the Gibbs free energy minimum lies very close to the pure reactants, the equilibrium composition strongly favors the reactants and the reaction "does not go" (Fig. 9.7b).

A note on good practice: Whether or not a reaction is spontaneous depends on the composition, so it is better to say that $K > 1$ for a reaction rather than that it is spontaneous. However, for reactions with very large equilibrium constants, it is very unlikely that the mixture of reagents prepared in the laboratory will correspond to $Q > K$, and it is common to refer to such reactions as "spontaneous."

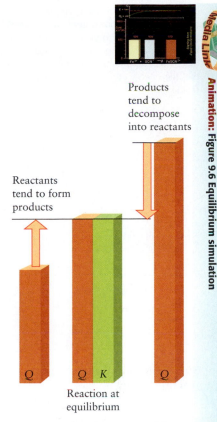

Products tend to decompose into reactants

Reactants tend to form products

Q Q K Q

Reaction at equilibrium

FIGURE 9.6 The relative sizes of the reaction quotient Q and the equilibrium constant K indicate the direction in which a reaction mixture tends to change. The arrows show that, when $Q < K$, reactants form products (left) and when $Q > K$, products form reactants (right). There is no tendency to change once the reaction quotient has become equal to the equilibrium constant.

FIGURE 9.7 (a) A reaction that has the potential to go to completion ($K > 1$) is one in which the minimum of the free-energy curve (the position of equilibrium) lies close to pure products. (b) A reaction that has little tendency to form products ($K < 1$) is one in which the minimum of the free-energy curve lies close to pure reactants.

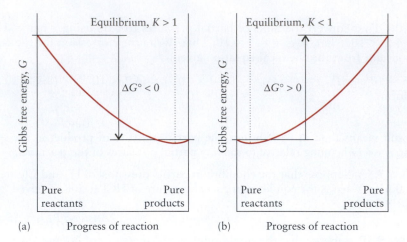

EXAMPLE 9.5   Predicting the direction of reaction

A mixture of hydrogen, iodine, and hydrogen iodide, each at 55 kPa, was introduced into a container heated to 783 K. At this temperature, $K = 46$ for $H_2(g) + I_2(g) \rightleftharpoons 2\, HI(g)$. Predict whether HI has a tendency to form or to decompose into $H_2(g)$ and $I_2(g)$.

STRATEGY We need to calculate Q and compare it with K. If $Q > K$, the products need to decompose into reactants until their concentrations match K. The opposite is true if $Q < K$: in that case, more products need to form.

SOLUTION Substitute the data, noting that 55 kPa is equivalent to 0.55 bar, in the reaction quotient.

From $Q = (P_{HI})^2/P_{H_2}P_{I_2}$,
$$Q = \frac{(0.55)^2}{0.55 \times 0.55} = 1.0$$

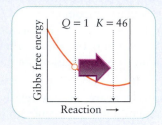

Because $Q < K$, we conclude that the reaction will tend to form more product and consume reactants.

SELF-TEST 9.6A A mixture of H_2, N_2, and NH_3 with partial pressures 22 kPa, 44 kPa, and 18 kPa, respectively, was prepared and heated to 500. K, at which temperature $K = 3.6 \times 10^{-2}$ for reaction C. Decide whether ammonia tends to form or decompose.

[Answer $Q = 6.9$; tends to decompose]

SELF-TEST 9.6B For the reaction $N_2O_4(g) \rightleftharpoons 2\, NO_2(g)$ at 298 K, $K = 0.15$. A mixture of N_2O_4 and NO_2 with initial partial pressures of 2.4 and 1.2 bar, respectively, was prepared at 298 K. Which compound will tend to increase its partial pressure?

A reaction has a tendency to form products if $Q < K$ and to form reactants if $Q > K$.

EQUILIBRIUM CALCULATIONS

We have seen that the value of an equilibrium constant tells us whether we can expect a high or low concentration of product at equilibrium. The constant also allows us to predict the spontaneous direction of reaction in a reaction mixture of any composition. In the following three sections, we see how to express the equilibrium constant in terms of molar concentrations of gases as well as partial pressures and how to predict the equilibrium composition of a reaction mixture, given the value of the equilibrium constant for the reaction. Such information is critical to the success of many industrial processes and is fundamental to the discussion of acids and bases in the following chapters.

9.6 The Equilibrium Constant in Terms of Molar Concentrations of Gases

The equilibrium constant in Eq. 2 is defined in terms of activities, and the activities are interpreted in terms of the partial pressures or concentrations. Gases *always* appear in K as the numerical values of their partial pressures and solutes always appear as the numerical values of their molarities. Often, however, we want to discuss gas-phase equilibria in terms of molar concentrations (the amount of gas molecules in moles divided by the volume of the container, $[J] = n_J/V$), not partial pressures. To do so, we introduce the equilibrium constant K_c, which for reaction E is defined as

$$K_c = \frac{[C]^c[D]^d}{[A]^a[B]^b} \tag{10}$$

with each molar concentration raised to a power equal to the stoichiometric coefficient of the species in the chemical equation. As usual, we have replaced $[J]/c°$ by $[J]$ itself, which represents the numerical value of the molar concentration of the gas J. For example, for the ammonia synthesis equilibrium, reaction C,

$$K_c = \frac{[NH_3]^2}{[N_2][H_2]^3} \tag{11}$$

We are free to choose either K or K_c to report the equilibrium constant of a reaction. However, it is important to remember that calculations of an equilibrium constant from thermodynamic tables of data (standard Gibbs free energies of formation, for instance) and Eq. 8 give K, not K_c. In some cases, we need to know K_c after we have calculated K from thermodynamic data, and so we need to be able to convert between these two constants.

HOW DO WE DO THAT?

The overall strategy for finding the relation between K and K_c is to replace the partial pressures that appear in K by the molar concentrations and thereby generate K_c. For this calculation, we write activities as $P_J/P°$ and $[J]/c°$ and track the units by keeping $P°$ and $c°$ in our expressions.

Our starting point is Eq. 1b in its complete form:

$$K = \frac{(P_C/P°)^c(P_D/P°)^d}{(P_A/P°)^a(P_B/P°)^b} = \frac{(P°)^{a+b}(P_C)^c(P_D)^d}{(P°)^{c+d}(P_A)^a(P_B)^b} = (P°)^{a+b-(c+d)}\frac{(P_C)^c(P_D)^d}{(P_A)^a(P_B)^b}$$

The molar concentration of a gas J is $[J] = n_J/V$. We assume ideal gas behavior and write the ideal gas law $P_J V = n_J RT$ as

$$P_J = \frac{n_J RT}{V} = RT\frac{n_J}{V} = RT[J]$$

When this expression is substituted for each gas in the expression for K, we obtain

$$\frac{(P_C)^c(P_D)^d}{(P_A)^a(P_B)^b} = \frac{(RT[C])^c(RT[D])^d}{(RT[A])^a(RT[B])^b} = (RT)^{c+d-(a+b)}\frac{[C]^c[D]^d}{[A]^a[B]^b}$$

At this point we recognize that K_c, Eq. 10, in its complete form can be written as follows:

$$K_c = \frac{([C]/c°)^c([D]/c°)^d}{([A]/c°)^a([B]/c°)^b} = \frac{(c°)^{a+b}[C]^c[D]^d}{(c°)^{c+d}[A]^a[B]^b} = (c°)^{a+b-(c+d)}\frac{[C]^c[D]^d}{[A]^a[B]^b}$$

and therefore

$$\frac{[C]^c[D]^d}{[A]^a[B]^b} = \frac{K_c}{(c°)^{a+b-(c+d)}}$$

When we substitute this expression into the equation for the ratio of partial pressures we obtain

$$\frac{(P_C)^c(P_D)^d}{(P_A)^a(P_B)^b} = (RT)^{c+d-(a+b)}\frac{K_c}{(c°)^{a+b-(c+d)}} = (c°RT)^{c+d-(a+b)}K_c$$

We conclude that

$$K = (P^\circ)^{a+b-(c+d)}(c^\circ RT)^{c+d-(a+b)}K_c = \left(\frac{c^\circ RT}{P^\circ}\right)^{c+d-(a+b)} K_c$$

A useful way of remembering the general form of the expression that we have derived is to write it as

$$K = \left(\frac{c^\circ RT}{P^\circ}\right)^{\Delta n} K_c \tag{12a}*$$

where Δn is the change in the number of gas molecules between reactants and products in the chemical equation, calculated as $\Delta n = n_{\text{products}} - n_{\text{reactants}}$ (so $\Delta n = 2 - (1 + 3) = -2$ for reaction C). If no gases take part in the reaction or if $\Delta n = 0$, then $K = K_c$. The same relation holds between Q and Q_c, the reaction quotient in terms of concentrations. Equation 12a is commonly written more succinctly as

$$K = (RT)^{\Delta n} K_c \tag{12b}$$

but the full version clarifies the units and should be used in calculations. Some values of K_c are listed in Table 9.2.

EXAMPLE 9.6   **Converting between K and K_c**

At 400.°C, the equilibrium constant K for $2 SO_2(g) + O_2(g) \rightleftharpoons 2 SO_3(g)$ is 3.1×10^4. What is the value of K_c at this temperature?

STRATEGY Because $P^\circ = 1$ bar and $c^\circ = 1$ mol·L^{-1}, it is sensible to use R expressed in bars and liters, $R = 8.3145 \times 10^{-2}$ L·bar·K^{-1}·mol^{-1}, and to note that

$$\frac{P^\circ}{Rc^\circ} = \frac{1 \text{ bar}}{(8.3145 \times 10^{-2} \text{ L·bar·K}^{-1}\text{·mol}^{-1}) \times (1 \text{ mol·L}^{-1})} = 12.03 \text{ K}$$

Then, Eq. 12a can be written as

$$K = \left(\frac{T}{12.03 \text{ K}}\right)^{\Delta n} K_c$$

To use this equation, identify the value of Δn for the reaction, convert the temperature to the Kelvin scale, and rearrange it to solve for K_c.

SOLUTION From the chemical equation $\Delta n = 2 - (2 + 1) = -1$ and $T = 400. + 273.15$ K $= 673$ K.

From $K = (T/12.03 \text{ K})^{\Delta n}K_c$
in the form $K_c = (T/12.03 \text{ K})^{-\Delta n}K$, $K_c = \left(\frac{673 \text{ K}}{12.03 \text{ K}}\right) \times (3.1 \times 10^4) = 1.7 \times 10^6$

SELF-TEST 9.7A The equilibrium constant for the ammonia synthesis (reaction C) is $K = 41$ at 127°C. What is the value of K_c at that temperature?

[Answer: $\Delta n = -2$; so $K_c = 4.5 \times 10^4$]

SELF-TEST 9.7B At 127°C, the equilibrium constant for $N_2O_4(g) \rightleftharpoons 2 NO_2(g)$ is $K = 47.9$. What is the value of K_c at that temperature?

> *For thermodynamic calculations, gas-phase equilibria are expressed in terms of K but, for practical calculations, they may be expressed in terms of molar concentrations by using Eq. 12.*

9.7 Alternative Forms of the Equilibrium Constant

The powers to which the activities are raised in the expression for an equilibrium constant must match the stoichiometric coefficients in the chemical equation, which is normally written with the smallest whole numbers for coefficients. Therefore, if we change the stoichiometric coefficients in a chemical equation (for instance, by multiplying through by a factor), then we must make sure that the equilibrium constant corresponds to that change. For example, at 700 K,

$$H_2(g) + I_2(g) \rightleftharpoons 2\,HI(g) \qquad K_{c1} = \frac{[HI]^2}{[H_2][I_2]} = 54$$

If we rewrite the chemical equation by multiplying through by 2, the equilibrium constant becomes

$$2\,H_2(g) + 2\,I_2(g) \rightleftharpoons 4\,HI(g) \qquad K_{c2} = \frac{[HI]^4}{[H_2]^2[I_2]^2} = (54)^2 = 2.9 \times 10^3$$

In general, *if we multiply a chemical equation by a factor n, we raise K (and K_c) to the nth power.*

Now suppose we reverse the original equation for the reaction:

$$2\,HI(g) \rightleftharpoons H_2(g) + I_2(g)$$

This equation still describes the same equilibrium; but we write the equilibrium constant for this equation as

$$K_{c3} = \frac{[H_2][I_2]}{[HI]^2} = \frac{1}{K_{c1}} = \frac{1}{54} = 0.019$$

In general, *the equilibrium constant for an equilibrium written in one direction is the reciprocal of the equilibrium constant for the equilibrium written in the opposite direction.*

A note on good practice: As these examples have shown, it is important to specify the chemical equation to which the equilibrium constant applies.

If a chemical equation can be expressed as the sum of two or more chemical equations, *the equilibrium constant for the overall reaction is the product of the equilibrium constants for the component reactions.* For example, consider the three gas-phase reactions

$$2\,P(g) + 3\,Cl_2(g) \rightleftharpoons 2\,PCl_3(g) \qquad K_1 = \frac{(P_{PCl_3})^2}{(P_P)^2(P_{Cl_2})^3}$$

$$PCl_3(g) + Cl_2(g) \rightleftharpoons PCl_5(g) \qquad K_2 = \frac{P_{PCl_5}}{P_{PCl_3}P_{Cl_2}}$$

$$2\,P(g) + 5\,Cl_2(g) \rightleftharpoons 2\,PCl_5(g) \qquad K_3 = \frac{(P_{PCl_5})^2}{(P_P)^2(P_{Cl_2})^5}$$

The third reaction is the following sum of the first two reactions (notice that we multiplied the second one by a factor of 2):

$$2\,P(g) + 3\,Cl_2(g) \rightleftharpoons 2\,PCl_3(g)$$
$$2\,PCl_3(g) + 2\,Cl_2(g) \rightleftharpoons 2\,PCl_5(g)$$
$$2\,P(g) + 5\,Cl_2(g) \rightleftharpoons 2\,PCl_5(g)$$

and the equilibrium constant, K_3, of the overall reaction can be written

$$K_3 = \frac{(P_{PCl_5})^2}{(P_P)^2(P_{Cl_2})^5} = \overbrace{\frac{(P_{PCl_3})^2}{(P_P)^2(P_{Cl_2})^3}}^{K_1} \times \overbrace{\frac{(P_{PCl_5})^2}{(P_{PCl_3})^2(P_{Cl_2})^2}}^{K_2^2} = K_1 K_2^2$$

Notice that, because we multiplied the second reaction by two to obtain the sum, its equilibrium constant has been raised to the second power.

SELF-TEST 9.8A At 500. K, K for $H_2(g) + D_2(g) \rightleftharpoons 2\,HD(g)$ is 3.6. What is the value of K for $2\,HD(g) \rightleftharpoons H_2(g) + D_2(g)$?

[Answer: 0.28]

SELF-TEST 9.8B At 500. K, K_c for $F_2(g) \rightleftharpoons 2\,F(g)$ is 7.3×10^{-13}. What is the value of K_c for $\frac{1}{2}F_2(g) \rightleftharpoons F(g)$?

Table 9.3 summarizes the relations between equilibrium constants.

TABLE 9.3 Relations Between Equilibrium Constants

Chemical equation	Equilibrium constant
$a\,A + b\,B \rightleftharpoons$ $\qquad c\,C + d\,D$	K_1
$c\,C + d\,D \rightleftharpoons$ $\qquad a\,A + b\,B$	$K_2 = 1/K_1$ $= K_1^{-1}$
$na\,A + nb\,B \rightleftharpoons$ $\qquad nc\,C + nd\,D$	$K_3 = K_1^n$

For a reaction that can be expressed as the sum of other reactions, the equilibrium constant is the product of the equilibrium constants of the component reactions

9.8 Using Equilibrium Constants

The equilibrium constant of a reaction contains information about the equilibrium composition at the given temperature. However, in many cases, we know only the initial composition of the reaction mixture and are given apparently incomplete information about the equilibrium composition. In fact, the missing information can usually be inferred by using the reaction stoichiometry. The easiest way to proceed is to draw up an **equilibrium table**, a table showing the initial composition, the changes needed to reach equilibrium in terms of some unknown quantity x, and the final equilibrium composition. The procedure is summarized in Toolbox 9.1 and illustrated in the examples that follow.

TOOLBOX 9.1   **HOW TO SET UP AND USE AN EQUILIBRIUM TABLE**

CONCEPTUAL BASIS

The composition of a reaction mixture tends to adjust until the molar concentrations or partial pressures of gases ensure that $Q_c = K_c$ and $Q = K$. A change in the abundance of any one component is linked to changes in the others by the reaction stoichiometry.

PROCEDURE

Step 1 Write the balanced chemical equation for the equilibrium and the corresponding expression for the equilibrium constant. Then set up an equilibrium table as shown here, with columns labeled by the species taking part in the reaction. In the first row, show the initial composition (molar concentration or partial pressure) of each species

	Species 1	Species 2	Species 3	...
Initial composition				...
Change in composition				...
Final composition				...

The first line in the table shows how the reaction system was prepared before reaction began. As always, strike out the units from molar concentrations (in moles per liter) and partial pressures (in bars). Omit pure solids and liquids.

Step 2 In the second row, write the changes in the composition that are needed for the reaction to reach equilibrium.

When we do not know the changes, write one of them as x or a multiple of x and then use the stoichiometry of the balanced equation to express the other changes in terms of that x.

Step 3 In the third row, write the equilibrium compositions by adding the change in composition (from step 2) to the initial value for each substance (from step 1).

Although a change in composition may be positive (an increase) or negative (a decrease), the value of each concentration or partial pressure itself must be positive.

Step 4 Use the equilibrium constant to determine the value of x, the unknown change in molar concentration or partial pressure.

A note on good practice: A good habit to adopt is to check the answer by substituting the equilibrium composition into the expression for K.

An approximation technique can greatly simplify calculations when the change in composition (x) is less than about 5% of the initial value. To use it, assume that x is negligible when added to or subtracted from a number. Thus, we can replace all expressions like $A + x$ or $A - 2x$, by A. When x occurs on its own (not added to or subtracted from another number), it is left unchanged. So, an expression such as $(0.1 - 2x)^2 x$ simplifies to $(0.1)^2 x$, provided that $2x \ll 0.1$ (specifically, if $x < 0.005$). At the end of the calculation, it is important to verify that the calculated value of x is indeed smaller than 5% of the initial values. If it is not, then we must solve the equation without making an approximation.

The approximation procedure is illustrated in Example 9.7.

In some cases, the full equation for x is a quadratic equation of the form

$$ax^2 + bx + c = 0$$

When the approximation procedure is invalid, we use the exact solutions of this equation, which are

$$x = \frac{-b \pm (b^2 - 4ac)^{1/2}}{2a}$$

We have to decide which of the two solutions given by this expression is valid (the one with the $+$ sign or the one with the $-$ sign in front of the square root) by seeing which solution is chemically possible.

See Example 9.8 for an illustration of this procedure.

For some reactions, the equation for x in terms of K may be a higher-order polynomial. If an approximation is not valid, one approach to solving the equation is to use a graphing calculator or mathematical software to find the roots of the equation.

EXAMPLE 9.7   **Sample exercise: Calculating the equilibrium composition by approximation**

Under certain conditions, nitrogen and oxygen react to form dinitrogen oxide, N_2O. Suppose that a mixture of 0.482 mol N_2 and 0.933 mol O_2 is transferred to a reaction vessel of volume 10.0 L and allowed to form N_2O at 800. K; at this temperature $K = 3.2 \times 10^{-28}$ for the reaction $2\,N_2(g) + O_2(g) \rightleftharpoons 2\,N_2O(g)$. Use the procedure in Toolbox 9.1 to calculate the partial pressures of the gases in the equilibrium mixture.

SOLUTION   The equilibrium constant for the reaction is $K = (P_{N_2O})^2/(P_{N_2})^2 P_{O_2}$.

The initial partial pressures are

From $P_J = n_J RT/V$,
$$P_{N_2} = \frac{(0.482\ \text{mol}) \times (8.3145 \times 10^{-2}\ \text{L·bar·K}^{-1}\text{·mol}^{-1}) \times (800.\ \text{K})}{10.0\ \text{L}}$$
$$= 3.21\ \text{bar}$$

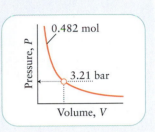

$$P_{O_2} = \frac{(0.933\ \text{mol}) \times (8.3145 \times 10^{-2}\ \text{L·bar·K}^{-1}\text{·mol}^{-1}) \times (800.\ \text{K})}{10.0\ \text{L}}$$
$$= 6.21\ \text{bar}$$

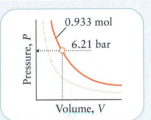

$$P_{N_2O} = 0$$

The stoichiometry of the reaction implies that, if the partial pressure of O_2 decreases by x, then the partial pressure of N_2 decreases by $2x$ and that of N_2O increases by $2x$. Because there is no N_2O present initially, the equilibrium table, with all partial pressures in bar, is

	N_2	O_2	N_2O
Step 1 Initial partial pressure	3.21	6.21	0
Step 2 Changes in partial pressure	$-2x$	$-x$	$+2x$
Step 3 Equilibrium partial pressure	$3.21 - 2x$	$6.21 - x$	$2x$

Step 4 From $K = (P_{N_2O})^2/(P_{N_2})^2(P_{O_2})$ $\quad K = \dfrac{(2x)^2}{(3.21 - 2x)^2 \times (6.21 - x)}$

When rearranged, this equation is a cubic equation (an equation in x^3), which can be solved with a graphing calculator or mathematical software. However, because K is very small, we suppose that x will turn out to be so small that we can use the approximation procedure:

Replace $3.21 - 2x$ by 3.21 and $6.21 - x$ by 6.21.
$$K \approx \frac{(2x)^2}{(3.21)^2 \times (6.21)}$$

Rearrange into an equation for x.
$$x \approx \left\{ \frac{(3.21)^2 \times (6.21) \times K}{4} \right\}^{1/2}$$

Substitute the value of K.
$$x \approx \left\{ \frac{(3.21)^2 \times (6.21) \times (3.2 \times 10^{-28})}{4} \right\}^{1/2} = 7.2 \times 10^{-14}$$

The value of x is very small compared with 3.21 (far smaller than 5% of it), and so the approximation is valid. We conclude that, at equilibrium,

From $P_{N_2} = 3.21 - 2x$ bar,	$P_{N_2} = 3.21$ bar
From $P_{O_2} = 6.21 - x$ bar,	$P_{O_2} = 6.21$ bar
From $P_{N_2O} = 2x$ bar,	$P_{N_2O} = 1.4 \times 10^{-13}$ bar

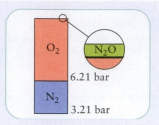

When substituted into the full expression for the equilibrium constant, these values give $K = 3.1 \times 10^{-28}$, in good agreement with the experimental value.

SELF-TEST 9.9A The initial partial pressures of nitrogen and hydrogen in a rigid, sealed vessel are 0.010 and 0.020 bar, respectively. The mixture is heated to a temperature at which $K = 0.11$ for $N_2(g) + 3 H_2(g) \rightleftharpoons 2 NH_3(g)$. What are the equilibrium partial pressures of each substance in the reaction mixture?

[*Answer:* N_2: 0.010 bar; H_2: 0.020 bar; NH_3: 9.4×10^{-5} bar]

SELF-TEST 9.9B Hydrogen chloride gas is added to a reaction vessel containing solid iodine until its partial pressure reaches 0.012 bar. At the temperature of the experiment, $K = 3.5 \times 10^{-32}$ for $2 HCl(g) + I_2(s) \rightleftharpoons 2 HI(g) + Cl_2(g)$. Assume that some I_2 remains at equilibrium. What are the equilibrium partial pressures of each gaseous substance in the reaction mixture?

EXAMPLE 9.8   Sample exercise: Calculating the equilibrium composition by using a quadratic equation

Suppose that we place 3.12 g of PCl_5 in a reaction vessel of volume 500. mL and allow the sample to reach equilibrium with its decomposition products phosphorus trichloride and chlorine at 250.°C, when $K = 78.3$ for the reaction $PCl_5(g) \rightleftharpoons PCl_3(g) + Cl_2(g)$. All three substances are gases at 250°C. Use the general procedure set out in Toolbox 9.1 to find the composition of the equilibrium mixture in moles per liter.

SOLUTION The equilibrium constant for the reaction is $K = P_{PCl_3}P_{Cl_2}/P_{PCl_5}$.

First, we calculate the initial partial pressure of PCl_5 (the initial partial pressures of PCl_3 and Cl_2 are zero):

From $n_J = m_J/M_J$, the amount of PCl_5 added is:	$n_{PCl_5} = \dfrac{3.12 \text{ g}}{208.24 \text{ g·mol}^{-1}} = \dfrac{3.12}{208.24} \text{ mol}$

From $P_J = n_J RT/V$, the initial partial pressure of PCl_5 is therefore	$P_{PCl_5} = \left(\dfrac{3.12}{208.24} \text{ mol}\right)$ $\times \dfrac{(8.3145 \times 10^{-2} \text{ L·bar·K}^{-1}\text{·mol}^{-1}) \times (523 \text{ K})}{0.500 \text{ L}}$ $= 1.30$ bar

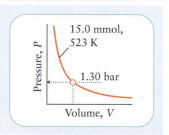

We use $P_{PCl_5} = 1.30$ in the expression for K and let the change in its partial pressure be $-x$. We construct the following table, with all partial pressures in bars.

		PCl_5	PCl_3	Cl_2
Step 1	Initial partial pressure	1.30	0	0
Step 2	Change in partial pressure	$-x$	$+x$	$+x$
Step 3	Equilibrium partial pressure	$1.30 - x$	x	x

From $K = P_{PCl_3}P_{Cl_2}/P_{PCl_5}$,	$K = \dfrac{x \times x}{1.30 - x} = \dfrac{x^2}{1.30 - x}$

Rearrange the equation:	$x^2 + Kx - 1.30K = 0$

Substitute the value of K:	$x^2 + 78.3x - 102 = 0$

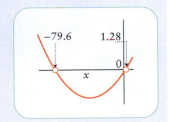

Solve for x by using the quadratic formula $x = \{-b \pm (b^2 - 4ac)^{1/2}\}/2a$.	$x = \dfrac{-78.3 \pm \{(78.3)^2 - 4 \times (1) \times (-102)\}^{1/2}}{2}$ $= -79.6 \text{ or } 1.28$

Because the partial pressures must be positive and because x is the partial pressure of PCl_3, we select 1.28 as the solution. It follows that, at equilibrium,

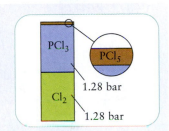

From $P_{PCl_5} = 1.30 - x$ bar,	$P_{PCl_5} = 1.30 - 1.28 \text{ bar} = 0.02 \text{ bar}$
From $P_{PCl_3} = x$ bar,	$P_{PCl_3} = 1.28 \text{ bar}$
From $P_{Cl_2} = x$ bar,	$P_{Cl_2} = 1.28 \text{ bar}$

Figure 9.8 illustrates how the reaction approaches equilibrium.

SELF-TEST 9.10A Bromine monochloride, BrCl, decomposes into bromine and chlorine and reaches the equilibrium $2\,BrCl(g) \rightleftharpoons Br_2(g) + Cl_2(g)$, for which $K = 32$ at 500. K. If initially pure BrCl is present at a concentration of 3.30 mbar, what is its partial pressure in the mixture at equilibrium?

[*Answer:* 0.3 mbar]

SELF-TEST 9.10B Chlorine and fluorine react at 2500. K to produce ClF and reach the equilibrium $Cl_2(g) + F_2(g) \rightleftharpoons 2\,ClF(g)$ with $K = 20$. If a gaseous mixture with $P_{Cl_2} = 0.200$ bar, $P_{F_2} = 0.100$ bar, and $P_{ClF} = 0.100$ bar is allowed to come to equilibrium at 2500. K, what is the partial pressure of ClF in the equilibrium mixture?

To calculate the equilibrium composition of a reaction mixture, set up an equilibrium table in terms of changes in the concentrations of reactants and products, express the equilibrium constant in terms of those changes, and solve the resulting equation.

THE RESPONSE OF EQUILIBRIA TO CHANGES IN CONDITIONS

Because chemical equilibria are dynamic, they respond to changes in the conditions under which a reaction is taking place. When we disturb an equilibrium by adding or removing a reactant, the value of ΔG is changed and the composition shifts until $\Delta G = 0$ again. The composition also might change in response to a change in pressure or temperature. In the following sections, we shall see how Haber used this responsiveness to control the equilibrium between ammonia, nitrogen, and hydrogen to increase the production of ammonia.

We can predict how the composition of a reaction mixture at equilibrium tends to change when the conditions are changed by following the general principle identified by the French chemist Henri Le Chatelier (Fig. 9.9):

Le Chatelier's principle: When a stress is applied to a system in dynamic equilibrium, the equilibrium tends to adjust to minimize the effect of the stress.

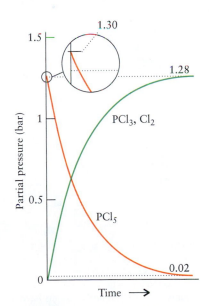

FIGURE 9.8 The approach of the composition of the reaction mixture to equilibrium when PCl_5 decomposes in a closed container. Note that the curves for Cl_2 and PCl_3 are superimposed, as they increase by the same amount.

FIGURE 9.9 Henri Le Chatelier (1850–1936).

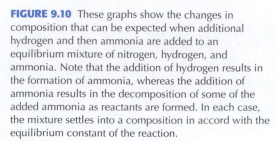

FIGURE 9.10 These graphs show the changes in composition that can be expected when additional hydrogen and then ammonia are added to an equilibrium mixture of nitrogen, hydrogen, and ammonia. Note that the addition of hydrogen results in the formation of ammonia, whereas the addition of ammonia results in the decomposition of some of the added ammonia as reactants are formed. In each case, the mixture settles into a composition in accord with the equilibrium constant of the reaction.

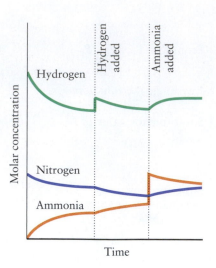

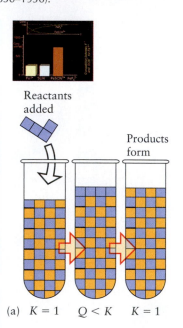

(a) $K = 1$ $Q < K$ $K = 1$

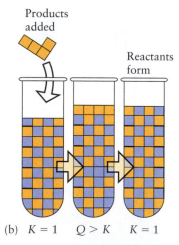

(b) $K = 1$ $Q > K$ $K = 1$

FIGURE 9.11 a) When a reactant (blue) is added to a reaction mixture at equilibrium, $Q < K$ and products have a tendency to form. (b) When a product (yellow) is added instead, $Q > K$ and reactants tend to be formed. For this reaction, we have used $K = 1$ for the equilibrium blue $\rightleftharpoons$ yellow.

Animation: Figure 9.11 Equilibrium simulation

This empirical principle is no more than a rule of thumb. It provides neither an explanation nor a quantitative prediction. However, we shall see that it is consistent with thermodynamics; and, as we develop it, we shall see the powerful *quantitative* conclusions that can be drawn from thermodynamics.

9.9 Adding and Removing Reagents

We start by considering changes in the composition of a system. Let's suppose that the ammonia synthesis reaction (reaction C) has reached equilibrium. Now suppose that we pump in more hydrogen gas. According to Le Chatelier's principle, the increase in the concentration of hydrogen molecules will tend to be minimized by the reaction of hydrogen with nitrogen. As a result, additional ammonia will be formed. If, instead of hydrogen, we were to add some ammonia, then the reaction would tend instead to form reactants at the expense of the added ammonia (Fig. 9.10).

We can explain these responses thermodynamically by considering the relative sizes of Q and K (Fig. 9.11). When reactants are added, the reaction quotient Q falls below K, because the reactant concentrations in the denominator of Q increase. As we have seen, when $Q < K$, the reaction mixture responds by forming products until Q is restored to K. Likewise, when products are added, Q rises above K, because products appear in the numerator. Then, because $Q > K$, the reaction mixture responds by forming reactants at the expense of products until $Q = K$ again. It is important to understand that K is a constant that is not altered by changing concentrations. Only the value of Q changes, and always in a way that brings its value closer to that of K.

EXAMPLE 9.9   **Sample exercise: Predicting the effect of adding or removing reactants and products**

Consider the equilibrium

$$4\,NH_3(g) + 3\,O_2(g) \rightleftharpoons 2\,N_2(g) + 6\,H_2O(g)$$

Predict the effect on each equilibrium concentration of (a) the addition of N_2; (b) the removal of NH_3; (c) the removal of H_2O.

SOLUTION   (a) The addition of N_2 (a product) to the equilibrium mixture causes the reaction to shift toward reactant formation, which increases the concentrations of NH_3 and O_2 while decreasing the concentration of H_2O. The concentration of N_2 remains slightly higher than its original equilibrium value but lower than its concentration immediately after the additional N_2 was supplied. (b) When NH_3 is removed from the system at equilibrium, the reaction shifts to form more reactants. Therefore, the concentration of O_2 will increase and the concentrations of N_2 and H_2O will decrease. The concentration of NH_3 will be somewhat lower than its original equilibrium value but not as low as it was immediately after the removal of NH_3. (c) The removal of H_2O causes the equilibrium to shift in favor of products, which increases the concentration of N_2 while decreasing the concentrations of NH_3 and O_2. The concentration of H_2O is somewhat lower than its original value.

SELF-TEST 9.11A Consider the equilibrium $SO_3(g) + NO(g) \rightleftharpoons SO_2(g) + NO_2(g)$. Predict the effect on the equilibrium of (a) the addition of NO; (b) the addition of NO_2; (c) the removal of SO_2.

> [*Answer:* The equilibrium tends to shift toward (a) products;
> (b) reactants; (c) products.]

SELF-TEST 9.11B Consider the equilibrium $CO(g) + 2\,H_2(g) \rightleftharpoons CH_3OH(g)$. Predict the effect on the equilibrium of (a) the addition of H_2; (b) the removal of CH_3OH; (c) the removal of CO.

Le Chatelier's principle suggests a good way of ensuring that a reaction goes on generating a substance: simply remove products as they are formed. In its continuing hunt for equilibrium, the reaction generates more product. Industrial processes rarely reach equilibrium, for just this reason. In the commercial synthesis of ammonia, for instance, ammonia can be continuously removed by circulating the equilibrium mixture through a refrigeration unit in which only the ammonia condenses. The hydrogen and nitrogen go on to react to form more product.

EXAMPLE 9.10 Calculating the equilibrium composition after addition of a reagent

Suppose that the equilibrium mixture for the reaction $PCl_5(g) \rightleftharpoons PCl_3(g) + Cl_2(g)$ considered in Example 9.8 is perturbed by adding 0.0100 mol $Cl_2(g)$ to the container (of volume 500. mL); then the system is once again allowed to reach equilibrium. Use this information and data from Example 9.8 to calculate the new composition of the equilibrium mixture.

STRATEGY The general procedure is like that set out in Toolbox 9.1. Write the expression for the equilibrium constant, and then set up an equilibrium table. In this case, use as the initial partial pressures those found immediately after addition of the reagent, but before the reaction has responded. Because a product has been added, we expect the reaction to respond by producing more reactants. Obtain the initial equilibrium concentrations and the value of K from Example 9.8.

SOLUTION From Example 9.8 the initial equilibrium partial pressures of the components are $P_{PCl_5} = 0.02$ bar, $P_{PCl_3} = 1.28$ bar, and $P_{Cl_2} = 1.28$ bar. The 0.0100 mol $Cl_2(g)$ added to the container corresponds to an additional partial pressure of chlorine of

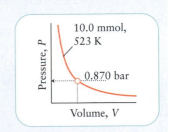

From $P_J = n_J RT/V$, $\quad P_{Cl_2} = \dfrac{(0.0100\ \text{mol}) \times (8.3145 \times 10^{-2}\ \text{L·bar·K}^{-1}\text{·mol}^{-1}) \times (523\ \text{K})}{0.500\ \text{L}}$

$$= 0.870\ \text{bar}$$

The total partial pressure of chlorine immediately after the addition of the chlorine gas is therefore $1.28 + 0.870$ bar $= 2.15$ bar. We draw up the following table, with all partial pressures in bar, noting a decrease for partial pressures of the products and an increase for the partial pressure of the reactant.

	PCl_5	PCl_3	Cl_2
Step 1 Initial partial pressure	0.02	1.28	2.15
Step 2 Change in partial pressure	$+x$	$-x$	$-x$
Step 3 Equilibrium partial pressure	$0.02 + x$	$1.28 - x$	$2.15 - x$

Now we use the equilibrium constant to find x:

From $K = P_{PCl_3}P_{Cl_2}/P_{PCl_5}$,	$K = \dfrac{(1.28 - x) \times (2.15 - x)}{0.02 + x}$
	$= \dfrac{2.75 - 3.43x + x^2}{0.02 + x}$

Rearrange the equation with $K = 78.3$:	$78.3 = \dfrac{2.75 - 3.43x + x^2}{0.02 + x}$
	$78.3(0.02 + x) = 2.75 - 3.43x + x^2$
	$1.57 + 78.3x = 2.75 - 3.43x + x^2$
	Hence, $x^2 - 81.7x + 1.18 = 0$

Solve by using the quadratic formula.	$x = 81.7$ and 0.0144	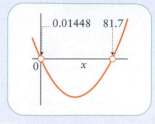

A note on good practice: Rounding to the correct number of significant figures should be carried out only at the end of a calculation to minimize rounding errors at intermediate stages of the calculation. Accordingly, although 78.3×0.02, where 0.02 is a measured quantity with 1 sf, should be rounded to 2, we have retained it as 1.57. Because $1.28 - x$ is a partial pressure, we select 0.0144, rounded to 0.01, as the acceptable solution.

It follows that, at equilibrium,

From $P_{PCl_5} = 0.01 + x$ bar,	$P_{PCl_5} = 0.02 + 0.01$ bar $= 0.03$ bar	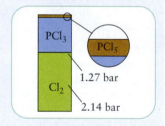
From $P_{PCl_3} = 1.28 - x$ bar,	$P_{PCl_3} = 1.28 - 0.01$ bar $= 1.27$ bar	
From $P_{Cl_2} = 2.15 - x$ bar,	$P_{Cl_2} = 2.15 - 0.01$ bar $= 2.14$ bar	

Notice that, as anticipated, the partial pressure of reactant has increased and the partial pressures of the products have decreased from their initial values (Fig. 9.12).

SELF-TEST 9.12A Suppose that the equilibrium mixture in Example 9.7 is perturbed by adding 3.00 mol $N_2(g)$ and then the system is once again allowed to reach equilibrium. Use this information and data from Example 9.7 to calculate the new composition of the equilibrium mixture.

[*Answer:* 23.2 bar N_2; 6.21 bar O_2; 1.0×10^{-12} bar N_2O]

SELF-TEST 9.12B Nitrogen, hydrogen, and ammonia gases have reached equilibrium in a container of volume 1.00 L at 298 K. Equilibrium partial pressures are 0.080 atm N_2, 0.050 atm H_2, and 2.60 atm NH_3. For $N_2(g) + 3 H_2(g) \rightleftharpoons 2 NH_3(g)$, $K = 6.8 \times 10^5$. Calculate the new equilibrium partial pressures if half the NH_3 is removed from the container and equilibrium is reestablished.

When the equilibrium composition is perturbed by adding or removing a reactant or product, reaction tends to occur in the direction that restores the value of Q to that of K.

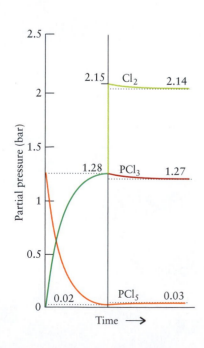

FIGURE 9.12 The response of the equilibrium mixture illustrated in Fig. 9.8 to the addition of chlorine. The curves for Cl_2 and PCl_3 are superimposed until the additional Cl_2 has been added.

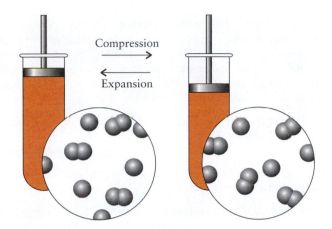

FIGURE 9.13 Le Chatelier's principle predicts that, when a reaction at equilibrium is compressed, the number of molecules in the gas phase will tend to decrease. This diagram illustrates the effect of compression and expansion on the dissociation equilibrium of a diatomic molecule. Note the increase in the relative concentration of diatomic molecules as the system is compressed and the decrease when the system expands.

9.10 Compressing a Reaction Mixture

A gas-phase equilibrium may respond to compression—a reduction in volume—of the reaction vessel. According to Le Chatelier's principle, the composition will tend to change in a way that minimizes the resulting increase in pressure. For instance, in the dissociation of gaseous I_2 to form I atoms, $I_2(g) \rightleftharpoons 2\,I(g)$, 1 mol of gaseous reactant molecules produces 2 mol of gaseous product molecules. The forward reaction increases the number of particles in the container and hence the total pressure of the system and the reverse reaction decreases it. It follows that, when the mixture is compressed, the equilibrium composition will tend to shift in favor of the reactant, I_2, because that response minimizes the increase in pressure (Fig. 9.13). Expansion of the system would result in the opposite response, a tendency for I_2 to dissociate to free atoms. In the formation of ammonia, 2 mol of gas molecules are produced from 4 mol of gas molecules. Therefore, Haber realized that, to increase the yield of ammonia, he needed to carry out the synthesis with highly compressed gases. The actual industrial process uses pressures of 250 atm and more (Fig. 9.14).

> **EXAMPLE 9.11**   **Sample Exercise: Predicting the effect of compression on an equilibrium**
>
> Predict the effect of compression on the equilibrium composition of the reaction mixtures in which the equilibria (a) $2\,NO_2(g) \rightleftharpoons N_2O_4(g)$ and (b) $H_2(g) + I_2(g) \rightleftharpoons 2\,HI(g)$ have been established.

FIGURE 9.14 One of the high-pressure vessels used for the catalytic synthesis of ammonia. The vessel must be able to withstand internal pressures of greater than 250 atm.

SOLUTION Reaction will take place in the direction that reduces the increase in pressure. (a) In the forward reaction, two NO_2 molecules combine to form one N_2O_4 molecule. Hence, compression favors the formation of N_2O_4. (b) Because neither direction corresponds to a reduction of gas-phase molecules, compressing the mixture should have no effect on the composition of the equilibrium mixture. (In practice, there will be a small effect due to the nonideality of the gases.)

SELF-TEST 9.13A Predict the effect of compression on the equilibrium composition of the reaction $CH_4(g) + H_2O(g) \rightleftharpoons CO(g) + 3 H_2(g)$.

[*Answer:* Reactants favored]

SELF-TEST 9.13B Predict the effect of compression on the equilibrium composition of the reaction $CO_2(g) + H_2O(l) \rightleftharpoons H_2CO_3(aq)$.

We can justify the effect of compression on an equilibrium mixture mathematically by showing that compressing a system will change Q, and the reaction tends to adjust in the direction that restores the value of Q to that of K.

HOW DO WE DO THAT?

Suppose we want to discover the effect of compression on the equilibrium in Example 9.11a. We write the equilibrium constant in its complete form as

$$2 NO_2(g) \rightleftharpoons N_2O_4(g) \qquad K = \frac{P_{N_2O_4}/P°}{(P_{NO_2}/P°)^2}$$

Next, we express K in terms of the volume of the system:

From $P_J = n_J RT/V$, $\qquad K = \dfrac{n_{N_2O_4}RT/VP°}{(n_{NO_2}RT/VP°)^2} = \dfrac{n_{N_2O_4}}{(n_{NO_2})^2}\dfrac{P°V}{RT}$

Because $P°/RT$ is a constant, for this expression to remain constant when the volume (V) of the system is reduced, the ratio $n_{N_2O_4}/(n_{NO_2})^2$ must increase. That is, the amount of NO_2 must decrease and the amount of N_2O_4 must increase. Therefore, as we have seen before, as the volume of the system is decreased, the equilibrium shifts to a smaller total number of gas molecules.

Suppose that we were to increase the total pressure inside a reaction vessel by pumping in argon or some other inert gas at constant volume. The reacting gases continue to occupy the same volume, and so their individual molar concentrations and partial pressures remain unchanged despite the presence of an inert gas. In this case, therefore, provided that the gases can be regarded as ideal, the equilibrium composition is unaffected despite the fact that the total pressure has increased.

> *Compression of a reaction mixture at equilibrium tends to drive the reaction in the direction that reduces the number of gas-phase molecules; increasing the pressure by introducing an inert gas has no effect on the equilibrium composition.*

9.11 Temperature and Equilibrium

We can see from Table 9.2 that the equilibrium constant depends on the temperature. For an exothermic reaction, the formation of products is found experimentally to be favored by lowering the temperature. Conversely, for an endothermic reaction, the products are favored by an *increase* in temperature.

Le Chatelier's principle is consistent with these observations. We can imagine the heat generated in an exothermic reaction as helping to offset lowering the temperature. Similarly, we can imagine the heat absorbed in an endothermic reaction as helping to offset an increase in temperature. In other words, *raising the temperature of a reaction mixture at equilibrium will shift the reaction in the endothermic direction*. An example is the decomposition of carbonates. A reaction such as

$$CaCO_3(s) \rightleftharpoons CaO(s) + CO_2(g)$$

is strongly endothermic in the forward direction, and an appreciable partial pressure of carbon dioxide is present at equilibrium only if the temperature is high. For instance, at 800°C, the partial pressure is 0.22 atm at equilibrium. If the heating takes place in an open container, this partial pressure is never reached, because equilibrium is never reached. The gas drifts away, and the calcium carbonate decomposes completely, leaving a solid residue of CaO. However, if the surroundings are already so rich in carbon dioxide that its partial pressure exceeds 0.22 atm, then virtually no decomposition occurs: for every CO_2 molecule that is formed, one is converted back into carbonate. This dynamic process is probably what happens on the hot surface of Venus (Fig. 9.15), where the partial pressure of carbon dioxide is about 87 atm. This high pressure has led to speculation that the planet's surface is rich in carbonates, despite its high temperature (about 500°C).

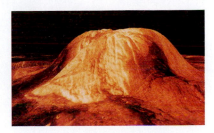

FIGURE 9.15 A radar image of the surface of Venus. Although the rocks are very hot, the partial pressure of carbon dioxide in the atmosphere is so great that carbonates may be abundant.

EXAMPLE 9.12 **Predicting the effect of temperature on an equilibrium**

One stage in the manufacture of sulfuric acid is the formation of sulfur trioxide by the reaction of SO_2 with O_2 in the presence of a vanadium(V) oxide catalyst. Predict how the equilibrium composition for the sulfur trioxide synthesis will tend to change when the temperature is raised.

STRATEGY Raising the temperature of an equilibrium mixture will tend to shift its composition in the endothermic direction of the reaction. A positive reaction enthalpy indicates that the reaction is endothermic in the forward direction. A negative reaction enthalpy indicates that the reaction is endothermic in the reverse direction. To find the standard reaction enthalpy, use the standard enthalpies of formation given in Appendix 2A.

SOLUTION The chemical equation is

$$2\,SO_2(g) + O_2(g) \xrightarrow{V_2O_5} 2\,SO_3(g)$$

The standard reaction enthalpy of the forward reaction is

$$\Delta H° = (2\text{ mol}) \times \Delta H_f°(SO_3, g) - (2\text{ mol}) \times \Delta H_f°(SO_2, g)$$
$$= 2(-395.72\text{ kJ}) - 2(-296.83\text{ kJ}) = -197.78\text{ kJ}$$

Because the formation of SO_3 is exothermic, the reverse reaction is endothermic. Hence, raising the temperature of the equilibrium mixture favors the decomposition of SO_3 to SO_2 and O_2; as a consequence, the pressures of SO_2 and O_2 will increase and that of SO_3 will decrease.

SELF-TEST 9.14A Predict the effect of raising the temperature on the equilibrium composition of the reaction $N_2O_4(g) \rightleftharpoons 2\,NO_2(g)$. See Appendix 2A for data.

[*Answer:* The pressure of NO_2 will increase.]

SELF-TEST 9.14B Predict the effect of lowering the temperature on the equilibrium composition of the $2\,CO(g) + O_2(g) \rightleftharpoons 2\,CO_2(g)$. See Appendix 2A for data.

The effect of temperature on the equilibrium composition arises from the dependence of the equilibrium constant on the temperature. The relation between the equilibrium constant and the standard Gibbs free energy of reaction in Eq. 8 applies to any temperature. Therefore, we ought to be able to use it to relate the equilibrium constant at one temperature to its value at another temperature.

HOW DO WE DO THAT?

To find the quantitative relation between the equilibrium constants for the same reaction at two temperatures T_1 and T_2, we note that Eq. 8 allows us to write the relation between standard Gibbs free energies of reaction and the equilibrium constants K_1 and K_2 at the two temperatures:

$$\Delta G_{r,1}° = -RT_1 \ln K_1 \qquad \Delta G_{r,2}° = -RT_2 \ln K_2$$

These two expressions rearrange to

$$\ln K_1 = -\frac{\Delta G_{r,1}°}{RT_1} \qquad \ln K_2 = -\frac{\Delta G_{r,2}°}{RT_2}$$

Subtraction of the second from the first gives

$$\ln K_1 - \ln K_2 = -\frac{1}{R}\left\{\frac{\Delta G_{r,1}°}{T_1} - \frac{\Delta G_{r,2}°}{T_2}\right\}$$

At this point, we introduce the definition of $\Delta G_r°$ in terms of $\Delta H_r°$ and $\Delta S_r°$:

$$\Delta G_{r,1}° = \Delta H_{r,1}° - T_1 \Delta S_{r,1}° \qquad \Delta G_{r,2}° = \Delta H_{r,2}° - T_2 \Delta S_{r,2}°$$

which gives

$$\ln K_1 - \ln K_2 = -\frac{1}{R}\left\{\frac{\Delta H_{r,1}° - T_1\Delta S_{r,1}°}{T_1} - \frac{\Delta H_{r,2}° - T_1\Delta S_{r,2}°}{T_2}\right\}$$

$$= -\frac{1}{R}\left\{\frac{\Delta H_{r,1}°}{T_1} - \frac{\Delta H_{r,2}°}{T_2} - \Delta S_{r,1}° + \Delta S_{r,2}°\right\}$$

It is usually reasonable to assume that $\Delta H_r°$ and $\Delta S_r°$ are both approximately independent of temperature over the range of temperatures of interest. When we make that approximation, the reaction entropies cancel, and we are left with

$$\ln K_1 - \ln K_2 = -\frac{\Delta H_r°}{R}\left\{\frac{1}{T_1} - \frac{1}{T_2}\right\}$$

A note on good practice: Notice that we have used the "molar" convention for the thermodynamic functions, as this convention is required by Eq. 8.

The expression that we have just derived is a quantitative version of Le Chatelier's principle for the effect of temperature. It is normally rearranged (by multiplying through by -1 and then using $\ln a - \ln b = \ln (a/b)$) into the **van 't Hoff equation:**

> This equation is sometimes called the *van 't Hoff isochore*, to distinguish it from van 't Hoff's osmotic pressure equation (Section 8.17). An isochore is the plot of an equation for a constant-volume process.

$$\ln \frac{K_2}{K_1} = \frac{\Delta H_r°}{R}\left\{\frac{1}{T_1} - \frac{1}{T_2}\right\} \tag{13}*$$

In this expression, K_1 is the equilibrium constant when the temperature is T_1 and K_2 is the equilibrium constant when the temperature is T_2.

> *What does this equation tell us?* Suppose that the reaction is endothermic, then $\Delta H_r°$ is positive. If $T_2 > T_1$, then $1/T_2 < 1/T_1$ and the term in braces is also positive. Therefore, $\ln (K_2/K_1)$ is positive, which implies that $K_2/K_1 > 1$ and therefore that $K_2 > K_1$. In other words, an increase in temperature favors the formation of product if the reaction is endothermic. We predict the opposite effect for an exothermic reaction because $\Delta H_r°$ is then negative. Therefore, the van 't Hoff equation accounts for Le Chatelier's principle for the effect of temperature on an equilibrium.

One word of warning: when using the van 't Hoff equation for reactions involving gases, the equilibrium constants must be K, not K_c. If we want a new value for K_c for a gas-phase reaction, we convert from K_c into K at the initial temperature (by using Eq. 12), use the van 't Hoff equation to calculate the value of K at the new temperature, and then convert that K into the new K_c by using Eq. 12 at the new temperature.

EXAMPLE 9.13 **Predicting the value of an equilibrium constant at a different temperature**

The equilibrium constant K for the synthesis of ammonia (reaction C) is 6.8×10^5 at 298 K. Predict its value at 400. K.

STRATEGY The synthesis of ammonia is exothermic, and so we expect the equilibrium constant to be smaller at the higher temperature. To use the van 't Hoff equation, we need the standard reaction enthalpy, which can be calculated from the standard enthalpies of formation in Appendix 2A. Equation 13 requires us to use the "molar" convention.

SOLUTION The standard reaction enthalpy for the forward reaction in C is

$$\Delta H_r° = 2\Delta H_f°(NH_3, g) = 2(-46.11 \text{ kJ·mol}^{-1})$$
$$= -92.22 \text{ kJ·mol}^{-1} \text{ or } -92.22 \times 10^3 \text{ J·mol}^{-1}$$

Therefore,

From $\ln(K_2/K_1) = (\Delta H_r°/R)\{(1/T_1) - (1/T_2)\}$,

$$\ln\frac{K_2}{K_1} = \overbrace{\frac{-9.222 \times 10^4 \text{ J·mol}^{-1}}{8.3145 \text{ J·K}^{-1}\text{·mol}^{-1}}}^{-92.22 \text{ kJ·mol}^{-1}} \times \left\{\frac{1}{298 \text{ K}} - \frac{1}{400. \text{ K}}\right\}$$
$$= -9.49$$

Take antilogarithms (e^x).

$$K_2 = K_1 e^{-9.49} = (6.8 \times 10^5) \times e^{-9.49} = 51$$

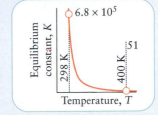

The answer is close to the experimental value of 41 in Table 9.2. It is not exactly the same because $\Delta H_r°$ may change slightly over the temperature range in question.

SELF-TEST 9.15A The equilibrium constant K for $2 SO_3(g) \rightleftharpoons 2 SO_2(g) + O_2(g)$ is 2.5×10^{-25} at 298 K. Predict its value at 500. K.

[*Answer:* 2.5×10^{-11}]

SELF-TEST 9.15B The equilibrium constant K for $PCl_5(g) \rightleftharpoons PCl_3(g) + Cl_2(g)$ is 78.3 at 523 K. Predict its value at 800. K.

Raising the temperature of an exothermic reaction favors the formation of reactants; raising the temperature of an endothermic reaction favors the formation of products.

9.12 Catalysts and Haber's Achievement

A **catalyst** is a substance that increases the rate of a chemical reaction without being consumed itself. We shall see a lot more of catalysts later, when we consider reaction rates in Chapter 13. However, it is important to be aware at this stage that a catalyst has no effect on the equilibrium composition of a reaction mixture. A catalyst can speed up the rate at which a reaction reaches equilibrium, but it does not affect the composition at equilibrium. It acts by providing a faster route to the same destination.

A catalyst speeds up both the forward and the reverse reactions by the same amount. Therefore, the dynamic equilibrium is unaffected. The thermodynamic justification of this observation is based on the fact that the equilibrium constant depends only on the temperature and the value of $\Delta G_r°$. A standard Gibbs free energy of reaction depends only on the identities of the reactants and products and is independent of the rate of the reaction or the presence of any substances that do not appear in the overall chemical equation for the reaction.

We have seen how Haber's understanding of chemical equilibrium enabled him to increase the yield of ammonia (Fig. 9.16). He realized that he had to compress the gases and remove the ammonia as it was formed. As we have seen, compression shifts the equilibrium composition in favor of ammonia, which increases the yield of product. Removing the ammonia also encourages more to be formed. In addition, Haber realized that he ought to run the reaction at as low a temperature as possible: low temperature favors the formation of product because the reaction is exothermic. However, nitrogen and hydrogen combine too slowly at low temperatures, and so Haber solved this problem by developing an appropriate catalyst (he used osmium and later

FIGURE 9.16 Fritz Haber (1868–1934).

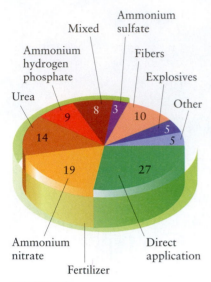

Ammonium
sulfate

Mixed

Ammonium
hydrogen
phosphate

Fibers

Explosives

Urea

Other

8 3

9   10

14

5

5

19   27

Ammonium
nitrate

Direct
application

Fertilizer

FIGURE 9.17 The Haber process is still used to produce almost all the ammonia manufactured in the world. This pie chart shows how the ammonia is used. The figures are percentages. Note that 80%—as shown by the green band—is used as fertilizer, either directly or after conversion into another compound.

uranium). The process was brought to an industrial scale by the chemical engineer Carl Bosch, who identified a cheaper catalyst (iron) and overcame the enormous problems of working at high pressure, when even hydrogen is corrosive. Their process is still in use throughout the world. In the United States alone, it accounts for almost the entire annual production of more than 1.6×10^{10} kg of ammonia (Fig. 9.17).

A catalyst does not affect the equilibrium composition of a reaction mixture.

9.13 The Impact on Biology: Homeostasis

Because a living being is not a closed system, true equilibrium in living systems can be reached only for very fast reactions such as those between acids and bases. In general, however, our bodies maintain certain relatively constant characteristics such as temperature and levels of certain chemicals in the blood. A beneficial environment in the body is maintained through the process of **homeostasis**, the maintenance of constant internal conditions. Homeostasis is not a true equilibrium, because there are usually slight variations above and below the desired point. However, it responds to changes in conditions in the same way as a system in chemical equilibrium and so is governed by Le Chatelier's principle.

An important homeostatic biological process that involves chemical equilibria is the transport of oxygen. Most of the oxygen in the blood is carried by hemoglobin (Hb). When blood flows through lung tissues, about 98% of the hemoglobin molecules pick up oxygen molecules and a small amount of additional oxygen simply dissolves in the blood plasma (the solution in which blood cells are suspended). However, when blood enters the small blood vessels called capillaries in muscle tissue far from the lungs, the hemoglobin molecules are surrounded by tissues that are depleted in oxygen. The equilibrium

$$Hb(aq) + O_2(aq) \rightleftharpoons HbO_2(aq)$$

is disturbed by the lower concentration of oxygen in the blood. Some of the hemoglobin molecules release their oxygen molecules to re-establish the equilibrium compositions. When a person is exercising heavily, muscles use an additional mechanism to encourage the hemoglobin molecules to release oxygen in the muscle tissue. Hydrogen ions from acids can cause the oxygen molecules in hemoglobin to be released; so, when lactic acid is produced by working muscle tissue, additional oxygen molecules are released from the hemoglobin.

Figure 9.18 shows how the O_2 uptake by hemoglobin and myoglobin (Mb), the oxygen storage protein, varies with partial pressure of oxygen. The shape of the Hb saturation curve means that Hb can load O_2 more fully in the lungs than Mb and unload it more completely than Mb in different regions of the organism. In the lungs, where $P_{O_2} < 105$ Torr, 98% of the molecules have taken up O_2, resulting in almost complete saturation. In resting muscular tissue, the concentration of O_2 corresponds to a partial pressure of about 40 Torr, at which value 75% of the Hb molecules are saturated with oxygen. This figure implies that sufficient oxygen is still available should a sudden surge of activity take place. If the local partial pressure falls to 20 Torr, the fraction of Hb molecules saturated falls to about 10%. Note that the steepest part of the curve falls in the range of typical tissue oxygen partial pressure. Myoglobin, on the other hand, begins to release O_2 only when P_{O_2} has fallen below about 20 Torr, and so it acts as a reserve to be drawn on only when the Hb oxygen has been used up.

Overloading the body's coping mechanisms can lead to failure to maintain homeostasis. The result can be sudden and even fatal illness. Mountain climbers encounter low oxygen conditions at high altitudes and, if they climb vigorously, their lungs may not be able to provide sufficient oxygen to maintain homeostasis. For this reason, mountain climbers spend some time in high-altitude base camps before climbing, so that their bodies can adjust by producing more hemoglobin molecules.

Homeostasis, a mechanism similar to chemical equilibrium, allows living organisms to control biological processes at a constant level.

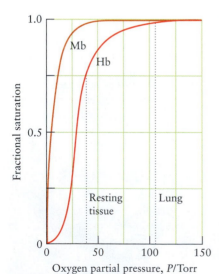

FIGURE 9.18 The variation of the extent of saturation of myoglobin (Mb) and hemoglobin (Hb) with the partial pressure of oxygen. The different shapes of the curves account for the different biological functions of the two proteins.

SKILLS YOU SHOULD HAVE MASTERED

❑ 1 Distinguish homogeneous and heterogeneous equilibria and write equilibrium constants for both types of reaction from a balanced equation (Example 9.1 and Self-Tests 9.2 and 9.5).

❑ 2 Determine the Gibbs free energy of reaction from the reaction quotient (Example 9.2).

❑ 3 Calculate an equilibrium constant from a standard Gibbs free energy (Example 9.3).

❑ 4 Calculate an equilibrium concentration (Example 9.4).

❑ 5 Predict the direction of a reaction, given K and the concentrations of reactants and products (Example 9.5).

❑ 6 Convert between K and K_c (Example 9.6).

❑ 7 Calculate the effect on K of reversing a reaction or multiplying the chemical equation by a factor (Section 9.7).

❑ 8 Use an equilibrium table to carry out equilibrium calculations (Toolbox 9.1 and Examples 9.7, 9.8, and 9.10).

❑ 9 Use Le Chatelier's principle to predict how the equilibrium composition of a reaction mixture is affected by adding or removing reagents, compressing or expanding the mixture, or changing the temperature (Examples 9.9, 9.11, and 9.12).

❑ 10 Predict the value of K at a different temperature (Example 9.13).

EXERCISES

9.1 State whether the following statements are true or false. If false, explain why.
(a) A reaction stops when equilibrium is reached.
(b) An equilibrium reaction is not affected by increasing the concentrations of products.
(c) If one starts with a higher pressure of reactant, the equilibrium constant will be larger.
(d) If one starts with higher concentrations of reactants, the equilibrium concentrations of the products will be larger.

9.2 Determine whether the following statements are true or false. If false, explain why.
(a) In an equilibrium reaction, the reverse reaction will begin as soon as any products are formed.
(b) If we make a reaction go faster, we can increase the amount of product at equilibrium.
(c) The Gibbs free energy of reaction at equilibrium is zero.
(d) The standard Gibbs free energy of reaction at equilibrium is zero.

9.3 Depict the progress of the reaction graphically (as in Fig. 9.2) for the decomposition of PCl_5 to give PCl_3 and Cl_2, using the values given in Example 9.8.

9.4 Depict the progress of the reaction graphically (as in Fig. 9.2) for the reaction $PCl_5(g) \rightleftharpoons PCl_3(g) + Cl_2(g)$, using the conditions described in Example 9.10, starting from the point where additional Cl_2 is added and continuing until equilibrium is restored.

9.5 The following flasks show the dissociation of a diatomic molecule, X_2, over time. (a) Which flask represents the point in time at which the reaction has reached equilibrium? (b) What percentage of the X_2 molecules have decomposed at equilibrium? (c) Assuming that the initial pressure of X_2 was 0.10 bar, calculate the value of K for the decomposition.

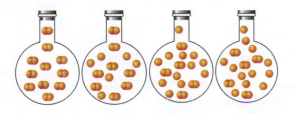

9.6 The flask below contains atoms of A (red) and B (yellow). They react as follows: $2\,A(g) + B(g) \rightleftharpoons A_2B(g)$, $K = 0.25$. Draw a picture of the flask after it has reached equilibrium.

9.7 Write the equilibrium expression K_c for each of the following reactions:
(a) $CO(g) + Cl_2(g) \rightleftharpoons COCl(g) + Cl(g)$
(b) $H_2(g) + Br_2(g) \rightleftharpoons 2\,HBr(g)$
(c) $2\,H_2S(g) + 3\,O_2(g) \rightleftharpoons 2\,SO_2(g) + 2\,H_2O(g)$

9.8 Write the equilibrium expression K for each of the following reactions:
(a) $2\,NO(g) + O_2(g) \rightleftharpoons 2\,NO_2(g)$
(b) $SbCl_5(g) \rightleftharpoons SbCl_3(g) + Cl_2(g)$
(c) $N_2(g) + 2\,H_2(g) \rightleftharpoons N_2H_4(g)$

9.9 A 0.10-mol sample of pure ozone, O_3, is placed in a sealed 1.0-L container and the reaction $2\,O_3(g) \rightleftharpoons 3\,O_2(g)$ is allowed to reach equilibrium. A 0.50-mol sample of pure ozone is placed in a second 1.0-L container at the same temperature and allowed to reach equilibrium. Without doing any calculations, predict which of the following will be different in the two containers at equilibrium. Which will be the same? (a) Amount of O_2; (b) concentration of O_2; (c) the ratio $[O_2]/[O_3]$; (d) the ratio $[O_2]^3/[O_3]^2$; (e) the ratio $[O_3]^2/[O_2]^3$. Explain each of your answers.

9.10 A 0.10-mol sample of $H_2(g)$ and a 0.10-mol sample of $Br_2(g)$ are placed into a 2.0-L container. The reaction $H_2(g) + Br_2(g) \rightleftharpoons 2\,HBr(g)$ is then allowed to come to equilibrium. A 0.20-mol sample of HBr is placed into a second 2.0-L sealed container at the same temperature and allowed to reach equilibrium with H_2 and Br_2. Which of the following will be different in the two containers at equilibrium? Which will be the same? (a) Amount of Br_2; (b) concentration of H_2; (c) the ratio $[HBr]/[H_2][Br_2]$; (d) the ratio $[HBr]/[Br_2]$; (e) the ratio $[HBr]^2/[H_2][Br_2]$; (f) the total pressure in the container. Explain each of your answers.

9.11 Use the following data, which were collected at 460.°C and are equilibrium molar concentrations, to determine K for the reaction $H_2(g) + I_2(g) \rightleftharpoons 2\,HI(g)$:

P_{H_2} (bar)	P_{I_2} (bar)	P_{HI} (bar)
6.47×10^{-3}	0.594×10^{-3}	0.0137
3.84×10^{-3}	1.52×10^{-3}	0.0169
1.43×10^{-3}	1.43×10^{-3}	0.0100

9.12 Determine K from the following equilibrium data collected at 24°C for the reaction $NH_4HS(s) \rightleftharpoons NH_3(g) + H_2S(g)$:

P_{NH_3} (bar)	P_{H_2S} (bar)
0.307	0.307
0.364	0.258
0.539	0.174

9.13 Write the reaction quotients Q for
(a) $4\,Bi(s) + 3\,O_2(g) \rightarrow 2\,Bi_2O_3(s)$
(b) $MgSO_4 \cdot 7H_2O(s) \rightarrow MgSO_4(s) + 7\,H_2O(g)$
(c) $N_2O_3(g) \rightarrow NO(g) + NO_2(g)$

9.14 Write the reaction quotients Q_c for
(a) $3\,ClO^-(aq) \rightarrow 2\,Cl^-(aq) + ClO_3{}^-(aq)$
(b) $Li_2CO_3(s) \rightarrow Li_2O(s) + CO_2(g)$
(c) $CH_3COOH(aq) + H_2O(l) \rightarrow H_3O^+(aq) + CH_3CO_2{}^-(aq)$

9.15 Calculate the equilibrium constant at 25°C for each of the following reactions, using data in Appendix 2A:
(a) the combustion of hydrogen: $2\,H_2(g) + O_2(g) \rightleftharpoons 2\,H_2O(g)$
(b) the oxidation of carbon monoxide:
$2\,CO(g) + O_2(g) \rightleftharpoons 2\,CO_2(g)$
(c) the decomposition of limestone:
$CaCO_3(s) \rightleftharpoons CaO(s) + CO_2(g)$

9.16 Calculate the equilibrium constant at 25°C for each of the following reactions, by using data in Appendix 2A:

(a) The synthesis of carbon disulfide from natural gas (methane):

$$2\,CH_4(g) + S_8(s) \rightleftharpoons 2\,CS_2(l) + 4\,H_2S(g)$$

(b) the production of acetylene from calcium carbide:

$$CaC_2(s) + 2\,H_2O(l) \rightleftharpoons Ca(OH)_2(s) + C_2H_2(g)$$

(c) the oxidation of ammonia, the first step in the production of nitric acid:

$$4\,NH_3(g) + 5\,O_2(g) \rightleftharpoons 4\,NO(g) + 6\,H_2O(l)$$

(d) the industrial synthesis of urea, a common fertilizer:

$$CO_2(g) + 2\,NH_3(g) \rightleftharpoons CO(NH_2)_2(s) + H_2O(l)$$

9.17 Calculate the standard Gibbs free energy of each of the following reactions:
(a) $I_2(g) \rightleftharpoons 2\,I(g)$, $K = 6.8$ at 1200. K
(b) $Ag_2CrO_4(s) \rightleftharpoons 2\,Ag^+(aq) + CrO_4{}^{2-}(aq)$, $K = 1.1 \times 10^{-12}$ at 298 K

9.18 Calculate the standard Gibbs free energy for each of the following reactions:
(a) $H_2(g) + I_2(g) \rightleftharpoons 2\,HI(g)$, $K = 54$ at 700. K
(b) $CCl_3COOH(aq) + H_2O(l) \rightleftharpoons CCl_3CO_2{}^-(aq) + H_3O^+(aq)$, $K = 0.30$ at 298 K

9.19 If $Q = 1.0$ for the reaction $N_2(g) + O_2(g) \rightarrow 2\,NO(g)$ at 25°C, will the reaction have a tendency to form products or reactants or will it be at equilibrium?

9.20 If $Q = 1.0 \times 10^{50}$ for the reaction $C(s) + O_2(g) \rightarrow CO_2(g)$ at 25°C, will the reaction have a tendency to form products or reactants or will it be at equilibrium?

9.21 (a) Calculate the reaction Gibbs free energy of $I_2(g) \rightarrow 2\,I(g)$ at 1200. K ($K = 6.8$) when the partial pressures of I_2 and I are 0.13 bar and 0.98 bar, respectively. (b) What is the spontaneous direction of the reaction? Explain briefly.

9.22 Calculate the reaction Gibbs free energy of $PCl_3(g) + Cl_2(g) \rightarrow PCl_5(g)$ at 230.°C when the partial pressures of PCl_3, Cl_2, and PCl_5 are 0.22 bar, 0.41 bar, and 1.33 bar, respectively. What is the spontaneous direction of change, given that $K = 49$ at 230.°C?

9.23 (a) Calculate the reaction Gibbs free energy of $N_2(g) + 3\,H_2(g) \rightarrow 2\,NH_3(g)$ when the partial pressures of N_2, H_2, and NH_3 are 4.2 bar, 1.8 bar, and 21 bar, respectively, and the temperature is 400. K. For this reaction, $K = 41$ at 400. K. (b) Indicate whether this reaction mixture is likely to form reactants, is likely to form products, or is at equilibrium.

9.24 (a) Calculate the reaction Gibbs free energy of $H_2(g) + I_2(g) \rightarrow 2\,HI(g)$ at 700. K when the partial pressures of H_2, I_2, and HI are 0.16 bar, 0.25 bar, and 2.17 bar, respectively. For this reaction, $K = 54$ at 700. K. (b) Indicate whether this reaction mixture is likely to form reactants, is likely to form products, or is at equilibrium.

9.25 Determine K_c for each of the following equilibria from the value of K:
(a) $2\,NOCl(g) \rightleftharpoons 2\,NO(g) + Cl_2(g)$, $K = 1.8 \times 10^{-2}$ at 500. K
(b) $CaCO_3(s) \rightleftharpoons CaO(s) + CO_2(g)$, $K = 167$ at 1073 K

9.26 Determine K_c for each of the following equilibria from the value of K:
(a) $2\,SO_2(g) + O_2(g) \rightleftharpoons 2\,SO_3(g)$, $K = 3.4$ at 1000. K
(b) $NH_4HS(s) \rightleftharpoons NH_3(g) + H_2S(g)$, $K = 9.4 \times 10^{-2}$ at 24°C

9.27 For the reaction $N_2(g) + 3\,H_2(g) \rightleftharpoons 2\,NH_3(g)$ at 400. K, $K = 41$. Find the value of K for each of the following reactions at the same temperature:
(a) $2\,NH_3(g) \rightleftharpoons N_2(g) + 3\,H_2(g)$
(b) $\frac{1}{2}\,N_2(g) + \frac{3}{2}\,H_2(g) \rightleftharpoons NH_3(g)$
(c) $2\,N_2(g) + 6\,H_2(g) \rightleftharpoons 4\,NH_3(g)$

9.28 The equilibrium constant for the reaction $2\,SO_2(g) + O_2(g) \rightleftharpoons 2\,SO_3(g)$ has the value $K = 2.5 \times 10^{10}$ at 500. K. Find the value of K for each of the following reactions at the same temperature.
(a) $SO_2(g) + \frac{1}{2}\,O_2(g) \rightleftharpoons SO_3(g)$
(b) $SO_3(g) \rightleftharpoons SO_2(g) + \frac{1}{2}\,O_2(g)$
(c) $3\,SO_2(g) + \frac{3}{2}\,O_2(g) \rightleftharpoons 3\,SO_3(g)$

9.29 Use the information in Table 9.2 to determine the value of K at 300 K for the reaction $2\,BrCl(g) + H_2(g) \rightleftharpoons Br_2(g) + 2\,HCl(g)$.

9.30 Use the information in Table 9.2 to determine the value of K at 500 K for the reaction $2\,NH_3(g) + 3\,Br_2(g) \rightleftharpoons N_2(g) + 6\,HBr(g)$.

9.31 In a gas-phase equilibrium mixture of H_2, I_2, and HI at 500. K, $[HI] = 2.21 \times 10^{-3}$ mol·L^{-1} and $[I_2] = 1.46 \times 10^{-3}$ mol·L^{-1}. Given the value of the equilibrium constant in Table 9.2, calculate the concentration of H_2.

9.32 In a gas-phase equilibrium mixture of H_2, Cl_2, and HCl at 1000. K, $[HCl] = 1.45$ mmol·L^{-1} and $[Cl_2] = 2.45$ mmol·L^{-1}. Use the information in Table 9.2 to calculate the concentration of H_2.

9.33 In a gas-phase equilibrium mixture of PCl_5, PCl_3, and Cl_2 at 500. K, $P_{PCl_5} = 1.18$ bar, $P_{Cl_2} = 5.43$ bar. What is the partial pressure of PCl_3, given that $K = 25$ for the reaction $PCl_5(g) \rightleftharpoons PCl_3(g) + Cl_2(g)$?

9.34 In a gas-phase equilibrium mixture of $SbCl_5$, $SbCl_3$, and Cl_2 at 500. K, $P_{SbCl_5} = 0.072$ bar and $P_{SbCl_3} = 5.02$ mbar. Calculate the equilibrium partial pressure of Cl_2, given that $K = 3.5 \times 10^{-4}$ for the reaction $SbCl_5(g) \rightleftharpoons SbCl_3(g) + Cl_2(g)$.

9.35 For the reaction $H_2(g) + I_2(g) \rightleftharpoons 2\,HI(g)$, $K = 160$. at 500 K. An analysis of a reaction mixture at 500. K showed that it had the composition $P_{H_2} = 0.20$ bar, $P_{I_2} = 0.10$ bar, and $P_{HI} = 0.10$ bar. (a) Calculate the reaction quotient. (b) Is the reaction mixture at equilibrium? (c) If not, is there a tendency to form more reactants or more products?

9.36 Analysis of a reaction mixture showed that it had the composition 0.417 mol·L^{-1} N_2, 0.524 mol·L^{-1} H_2, and 0.122 mol·L^{-1} NH_3 at 800. K, at which temperature $K_c = 0.278$ for $N_2(g) + 3\,H_2(g) \rightleftharpoons 2\,NH_3(g)$. (a) Calculate the reaction quotient Q_c. (b) Is the reaction mixture at equilibrium? (c) If not, is there a tendency to form more reactants or more products?

9.37 A 0.500-L reaction vessel at 700. K contains 1.20 mmol $SO_2(g)$, 0.50 mmol $O_2(g)$, and 0.10 mmol $SO_3(g)$. At 700. K, $K_c = 1.7 \times 10^6$ for the equilibrium $2\,SO_2(g) + O_2(g) \rightleftharpoons 2\,SO_3(g)$. (a) Calculate the reaction quotient Q_c. (b) Will more $SO_3(g)$ tend to form?

9.38 Given that $K_c = 62$ for the reaction $N_2(g) + 3\,H_2(g) \rightleftharpoons 2\,NH_3(g)$ at 500. K, calculate whether more ammonia will tend to form when a mixture of composition 2.23 mmol·L^{-1} N_2, 1.24 mmol·L^{-1} H_2, and 0.112 mmol·L^{-1} NH_3 is present in a container at 500. K.

9.39 When 0.0172 mol HI is heated to 500. K in a 2.00-L sealed container, the resulting equilibrium mixture contains 1.90 g of HI. Calculate K for the decomposition reaction $2\,HI(g) \rightleftharpoons H_2(g) + I_2(g)$.

9.40 When 1.00 g of gaseous I_2 is heated to 1000. K in a 1.00-L sealed container, the resulting equilibrium mixture contains 0.830 g of I_2. Calculate K_c for the dissociation equilibrium $I_2(g) \rightleftharpoons 2\,I(g)$.

9.41 A 25.0-g sample of ammonium carbamate, $NH_4(NH_2CO_2)$, was placed in an evacuated 0.250-L flask and kept at 25°C. At equilibrium, 17.4 mg of CO_2 was present. What is the value of K_c for the decomposition of ammonium carbamate into ammonia and carbon dioxide? The reaction is $NH_4(NH_2CO_2)(s) \rightleftharpoons 2\,NH_3(g) + CO_2(g)$.

9.42 Carbon monoxide and water vapor, each at 200. Torr, were introduced into a 250.-mL container. When the mixture reached equilibrium at 700.°C, the partial pressure of $CO_2(g)$ was 88 Torr. Calculate the value of K for the equilibrium $CO(g) + H_2O(g) \rightleftharpoons CO_2(g) + H_2(g)$.

9.43 Consider the reaction $2\,NO(g) \rightleftharpoons N_2(g) + O_2(g)$. If the initial partial pressure of NO(g) is 1.0 bar, and x is the equilibrium concentration of $N_2(g)$, what is the correct equilibrium relation? (a) $K = x^2/(1.0 - x)$; (b) $K = x^2$; (c) $K = x^2/(1.0 - 2x)^2$; (d) $K = 4x^3/(1.0 - 2x)^2$; (e) $K = 2x/(1.0 - x)^2$.

9.44 Consider the reaction $2\,NO_2(g) \rightleftharpoons 2\,NO(g) + O_2(g)$. If the initial partial pressure of $NO_2(g)$ is 3.0 bar, and x is the equilibrium concentration of $O_2(g)$, what is the correct equilibrium relation? (a) $K = x^3$; (b) $K = 2x^2/(3.0 - 2x)^2$; (c) $K = 4x^3/(3.0 - 2x)^2$; (d) $K = x^2/(3.0 - x)$; (e) $K = 2x/(3.0 - x)^2$.

9.45 (a) A sample of 2.0 mmol $Cl_2(g)$ was sealed into a 2.0-L reaction vessel and heated to 1000. K to study its dissociation into Cl atoms. Use the information in Table 9.2 to calculate the equilibrium composition of the mixture. (b) If 2.0 mmol F_2 was placed into the reaction vessel instead of the chlorine, what would be its equilibrium composition at 1000. K? (c) Use your results from parts (a) and (b) to determine which is thermodynamically more stable relative to its atoms at 1000. K, Cl_2 or F_2.

9.46 (a) A sample of 5.0 mmol $Cl_2(g)$ was sealed into a 2.0-L reaction vessel and heated to 1200. K, and the dissociation equilibrium was established. What is the equilibrium composition of the mixture? Use the information in Table 9.2. (b) If 5.0 mol Br_2 is placed into the reaction vessel instead of the chlorine, what would be the equilibrium composition at 1200. K? (c) Use your results from parts (a) and (b) to determine which is thermodynamically more stable relative to its atoms at 1200. K, Cl_2, or Br_2.

9.47 The initial pressure of HBr(g) in a reaction vessel is 1.2 mbar. If the vessel is heated to 500. K, what is the equilibrium composition of the mixture? See Table 9.2 for data on the reaction.

9.48 The initial pressure of BrCl(g) in a reaction vessel is 1.4 mbar. If the vessel is heated to 500. K, what is the equilibrium composition of the mixture? See Table 9.2 for data on the reaction.

9.49 For the reaction $PCl_5(g) \rightleftharpoons PCl_3(g) + Cl_2(g)$ $K_c = 1.1 \times 10^{-2}$ at 400. K. (a) Given that 1.0 g of PCl_5 is placed in a 250.-mL reaction vessel, determine the molar concentrations in the mixture at equilibrium. (b) What percentage of the PCl_5 has decomposed at equilibrium at 400. K?

9.50 For the reaction $PCl_5(g) \rightleftharpoons PCl_3(g) + Cl_2(g)$, $K_c = 0.61$ at 500. K. (a) Calculate the equilibrium molar concentrations of the components in the mixture when 2.0 g of PCl_5 is placed in a 300.-mL reaction vessel and allowed to come to equilibrium. (b) What percentage of the PCl_5 has decomposed at equilibrium at 500. K?

9.51 When solid NH_4HS and 0.400 mol $NH_3(g)$ were placed in a 2.0-L vessel at 24°C, the equilibrium $NH_4HS(s) \rightleftharpoons NH_3(g) + H_2S(g)$, for which $K_c = 1.6 \times 10^{-4}$, was reached. What are the equilibrium concentrations of NH_3 and H_2S?

9.52 When solid NH_4HS and 0.200 mol $NH_3(g)$ were placed into a 2.0-L vessel at 24°C, the equilibrium $NH_4HS(s) \rightleftharpoons NH_3(g) + H_2S(g)$, for which $K_c = 1.6 \times 10^{-4}$, was reached. What are the equilibrium concentrations of NH_3 and H_2S?

9.53 The equilibrium constant K_c for the reaction $N_2(g) + O_2(g) \rightleftharpoons 2\,NO(g)$ at 1200°C is 1.00×10^{-5}. Calculate the equilibrium molar concentrations of NO, N_2, and O_2 in a 1.00-L reaction vessel that initially held 0.114 mol N_2 and 0.114 mol O_2.

9.54 The equilibrium constant K_c for the reaction $N_2(g) + O_2(g) \rightleftharpoons 2\, NO(g)$ at 1200°C is 1.00×10^{-5}. Calculate the equilibrium molar concentrations of NO, N_2, and O_2 in a 10.00-L reaction vessel that initially held 0.0140 mol N_2 and 0.214 mol O_2.

9.55 A reaction mixture that consisted of 0.400 mol H_2 and 1.60 mol I_2 was introduced into a 3.00-L flask and heated. At equilibrium, 60.0% of the hydrogen gas had reacted. What is the equilibrium constant K for the reaction $H_2(g) + I_2(g) \rightleftharpoons 2\, HI(g)$ at this temperature?

9.56 A reaction mixture that consisted of 0.20 mol N_2 and 0.20 mol H_2 was introduced into a 25.0-L reactor and heated. At equilibrium, 5.0% of the nitrogen gas had reacted. What is the value of the equilibrium constant K_c for the reaction $N_2(g) + 3\, H_2(g) \rightleftharpoons 2\, NH_3(g)$ at this temperature?

9.57 The equilibrium constant K_c for the reaction $2\, CO(g) + O_2(g) \rightleftharpoons 2\, CO_2(g)$ is 0.66 at 2000.°C. If 0.28 g of CO and 0.032 g of $O_2(g)$ are placed in a 2.0-L reaction vessel and heated to 2000.°C, what will the equilibrium composition of the system be? (You may wish to use graphing software or a calculator to solve the cubic equation.)

9.58 In the Haber process for ammonia synthesis, $K = 0.036$ for $N_2(g) + 3\, H_2(g) \rightleftharpoons 2\, NH_3(g)$ at 500. K. If a 2.0-L reactor is charged with 0.025 bar of N_2 and 0.015 bar of H_2, what will the equilibrium partial pressures in the mixture be?

9.59 At 500. K, the equilibrium constant for the reaction $Cl_2(g) + Br_2(g) \rightleftharpoons 2\, BrCl(g)$ is $K_c = 0.031$. If the equilibrium composition is 0.495 mol·L^{-1} Cl_2 and 0.145 mol·L^{-1} BrCl, what is the equilibrium concentration of Br_2?

9.60 The equilibrium constant for the reaction $PCl_3(g) + Cl_2(g) \rightleftharpoons PCl_5(g)$ is $K = 3.5 \times 10^4$ at 760.°C. At equilibrium, the partial pressure of PCl_5 was 1.3×10^2 bar and that of PCl_3 was 9.56 bar. What was the equilibrium partial pressure of Cl_2?

9.61 A reaction mixture consisting of 2.00 mol CO and 3.00 mol H_2 is placed in a 10.0-L reaction vessel and heated to 1200. K. At equilibrium, 0.478 mol CH_4 was present in the system. Determine the value of K_c for the reaction $CO(g) + 3\, H_2(g) \rightleftharpoons CH_4(g) + H_2O(g)$ at 1200. K.

9.62 A mixture consisting of 1.000 mol $H_2O(g)$ and 1.000 mol $CO(g)$ is placed in a 10.00-L reaction vessel at 800. K. At equilibrium, 0.665 mol $CO_2(g)$ is present as a result of the reaction $CO(g) + H_2O(g) \rightleftharpoons CO_2(g) + H_2(g)$. What are (a) the equilibrium concentrations for all substances and (b) the value of K_c at 800. K?

9.63 A reaction mixture is prepared by mixing 0.100 mol SO_2, 0.200 mol NO_2, 0.100 mol NO, and 0.150 mol SO_3 in a 5.00-L reaction vessel. The reaction $SO_2(g) + NO_2(g) \rightleftharpoons NO(g) + SO_3(g)$ is allowed to reach equilibrium at 460.°C, when $K_c = 85.0$. What is the equilibrium concentration of each substance?

9.64 A 0.100-mol sample of H_2S is placed in a 10.0-L reaction vessel and heated to 1132°C. At equilibrium, 0.0285 mol H_2 is present. Calculate the value of K_c for the reaction $2\, H_2S(g) \rightleftharpoons 2\, H_2(g) + S_2(g)$ at 1132°C.

9.65 The equilibrium constant $K_c = 0.56$ for the reaction $PCl_3(g) + Cl_2(g) \rightleftharpoons PCl_5(g)$ at 250.°C. On analysis, 1.50 mol PCl_5, 3.00 mol PCl_3, and 0.500 mol Cl_2 were found to be

present in a 0.500-L reaction vessel at 250.°C. (a) Is the reaction at equilibrium? (b) If not, in which direction does it tend to proceed? (c) What is the equilibrium composition of the reaction system?

9.66 An 0.865-mol sample of PCl_5 is placed in a 500.-mL reaction vessel. What is the concentration of each substance when the reaction $PCl_5(g) \rightleftharpoons PCl_3(g) + Cl_2(g)$ has reached equilibrium at 250°C (when $K_c = 1.80$)?

9.67 At 25°C, $K = 3.2 \times 10^{-34}$ for the reaction $2\, HCl(g) \rightleftharpoons H_2(g) + Cl_2(g)$. If a 1.0-L reaction vessel is filled with HCl at 0.22 bar, what are the equilibrium partial pressures of HCl, H_2, and Cl_2?

9.68 If 4.00 L of HCl(g) at 1.00 bar and 273 K and 26.0 g of $I_2(s)$ are transferred to a 12.00-L reaction vessel and heated to 25°C, what will the equilibrium concentrations of HCl, HI, and Cl_2 be? $K_c = 1.6 \times 10^{-34}$ at 25°C for $2\, HCl(g) + I_2(s) \rightleftharpoons 2\, HI(g) + Cl_2(g)$.

9.69 A 3.00-L reaction vessel is filled with 0.342 mol CO, 0.215 mol H_2, and 0.125 mol CH_3OH. Equilibrium is reached in the presence of a zinc oxide–chromium(III) oxide catalyst and, at 300.°C, $K_c = 1.1 \times 10^{-2}$ for the reaction $CO(g) + 2\, H_2(g) \rightleftharpoons CH_3OH(g)$. (a) As the reaction approaches equilibrium, will the concentration of CH_3OH increase, decrease, or remain unchanged? (b) What is the equilibrium composition of the mixture?

9.70 For the reaction $2\, NH_3(g) \rightleftharpoons N_2(g) + 3\, H_2(g)$, $K_c = 0.395$ at 350.°C. A 25.6-g sample of NH_3 is placed in a 5.00-L reaction vessel and heated to 350.°C. What are the equilibrium concentrations of NH_3, N_2, and H_2?

9.71 The reaction $2\, HCl(g) \rightleftharpoons H_2(g) + Cl_2(g)$ has $K = 3.2 \times 10^{-34}$ at 298 K. The initial partial pressures are H_2, 1.0 bar; HCl, 2.0 bar; and Cl_2, 3.0 bar. At equilibrium there is 1.0 mol $H_2(g)$. What is the volume of the container? (*Don't be surprised at the large size of the volume.*)

9.72 Suppose that, in the same reaction as in Exercise 9.71, the total pressure at equilibrium is found to be 3.0 bar and the Cl:H atom ratio is 1:3. What are the partial pressures of the three gases?

The Response of Equilibria to Changes in Conditions

9.73 Consider the equilibrium $CO(g) + H_2O(g) \rightleftharpoons CO_2(g) + H_2(g)$. (a) If the partial pressure of CO_2 is increased, what happens to the partial pressure of H_2? (b) If the partial pressure of CO is decreased, what happens to the partial pressure of CO_2? (c) If the concentration of CO is increased, what happens to the concentration of H_2? (d) If the concentration of H_2O is decreased, what happens to the equilibrium constant for the reaction?

9.74 Consider the equilibrium $CH_4(g) + 2\, O_2(g) \rightleftharpoons CO_2(g) + 2\, H_2O(g)$. (a) If the partial pressure of CO_2 is increased, what happens to the partial pressure of CH_4? (b) If the partial pressure of CH_4 is decreased, what happens to the partial pressure of CO_2? (c) If the concentration of CH_4 is increased, what happens to the equilibrium constant for the reaction? (d) If the concentration of H_2O is decreased, what happens to the concentration of CO_2?

9.75 The four gases NH_3, O_2, NO, and H_2O are mixed in a reaction vessel and allowed to reach equilibrium in the reaction $4 NH_3(g) + 5 O_2(g) \rightleftharpoons 4 NO(g) + 6 H_2O(g)$. Certain changes (see the following table) are then made to this mixture. Considering each change separately, state the effect (increase, decrease, or no change) that the change has on the original equilibrium values of the quantity in the second column (or K, if that is specified). The temperature and volume are constant.

Change	Quantity
(a) add NO	amount of H_2O
(b) add NO	amount of O_2
(c) remove H_2O	amount of NO
(d) remove O_2	amount of NH_3
(e) add NH_3	K
(f) remove NO	amount of NH_3
(g) add NH_3	amount of O_2

9.76 The four substances HCl, I_2, HI, and Cl_2 are mixed in a reaction vessel and allowed to reach equilibrium in the reaction $2 HCl(g) + I_2(s) \rightleftharpoons 2 HI(g) + Cl_2(g)$. Certain changes (which are specified in the first column in the following table) are then made to this mixture. Considering each change separately, state the effect (increase, decrease, or no change) that the change has on the original equilibrium value of the quantity in the second column (or K, if that is specified). The temperature and volume are constant.

Change	Quantity
(a) add HCl	amount of HI
(b) add I_2	amount of Cl_2
(c) remove HI	amount of Cl_2
(d) remove Cl_2	amount of HCl
(e) add HCl	K
(f) remove HCl	amount of I_2
(g) add I_2	K

9.77 State whether reactants or products will be favored by an increase in the total pressure (resulting from compression) on each of the following equilibria. If there is no change, explain why that is so.

(a) $2 O_3(g) \rightleftharpoons 3 O_2(g)$
(b) $H_2O(g) + C(s) \rightleftharpoons H_2(g) + CO(g)$
(c) $4 NH_3(g) + 5 O_2(g) \rightleftharpoons 4 NO(g) + 6 H_2O(g)$
(d) $2 HD(g) \rightleftharpoons H_2(g) + D_2(g)$
(e) $Cl_2(g) \rightleftharpoons 2 Cl(g)$

9.78 State what happens to the concentration of the indicated substance when the total pressure on each of the following equilibria is increased (by compression):
(a) $NO_2(g)$ in $2 Pb(NO_3)_2(s) \rightleftharpoons 2 PbO(s) + 4 NO_2(g) + O_2(g)$
(b) $NO(g)$ in $3 NO_2(g) + H_2O(l) \rightleftharpoons 2 HNO_3(aq) + NO(g)$
(c) $HI(g)$ in $2 HCl(g) + I_2(s) \rightleftharpoons 2 HI(g) + Cl_2(g)$
(d) $SO_2(g)$ in $2 SO_2(g) + O_2(g) \rightleftharpoons 2 SO_3(g)$
(e) $NO_2(g)$ in $2 NO(g) + O_2(g) \rightleftharpoons 2 NO_2(g)$

9.79 Consider the equilibrium $3 NH_3(g) + 5 O_2(g) \rightleftharpoons 4 NO(g) + 6 H_2O(g)$. (a) What happens to the partial pressure of NH_3 when the partial pressure of NO is increased? (b) Does the partial pressure of O_2 decrease when the partial pressure of NH_3 is decreased?

9.80 Consider the equilibrium $2 SO_2(g) + O_2(g) \rightleftharpoons 2 SO_3(g)$. (a) What happens to the partial pressure of SO_3 when the partial pressure of SO_2 is decreased? (b) If the partial pressure of SO_2 is increased, what happens to the partial pressure of O_2?

9.81 Predict whether each of the following equilibria will shift toward products or reactants with a temperature increase:
(a) $N_2O_4(g) \rightleftharpoons 2 NO_2(g)$, $\Delta H° = +57$ kJ
(b) $X_2(g) \rightleftharpoons 2 X(g)$, where X is a halogen
(c) $Ni(s) + 4 CO(g) \rightleftharpoons Ni(CO)_4(g)$, $\Delta H° = -161$ kJ
(d) $CO_2(g) + 2 NH_3(g) \rightleftharpoons CO(NH_2)_2(s) + H_2O(g)$, $\Delta H° = -90$ kJ

9.82 Predict whether each of the following equilibria will shift toward products or reactants with a temperature increase:
(a) $CH_4(g) + H_2O(g) \rightleftharpoons CO(g) + 3 H_2(g)$, $\Delta H° = +206$ kJ
(b) $CO(g) + H_2O(g) \rightleftharpoons CO_2(g) + H_2(g)$, $\Delta H° = -41$ kJ
(c) $2 SO_2(g) + O_2(g) \rightleftharpoons 2 SO_3(g)$, $\Delta H° = -198$ kJ

9.83 A gaseous mixture consisting of 2.23 mmol N_2 and 6.69 mmol H_2 in a 500.-mL container was heated to 600. K and allowed to reach equilibrium. Will more ammonia be formed if that equilibrium mixture is then heated to 700. K? For $N_2(g) + 3 H_2(g) \rightleftharpoons 2 NH_3(g)$, $K = 1.7 \times 10^{-3}$ at 600. K and 7.8×10^{-5} at 700. K.

9.84 A gaseous mixture consisting of 1.1 mmol SO_2 and 2.2 mmol O_2 in a 250-mL container was heated to 500. K and allowed to reach equilibrium. Will more sulfur trioxide be formed if that equilibrium mixture is cooled to 298 K? For the reaction $2 SO_2(g) + O_2(g) \rightleftharpoons 2 SO_3(g)$, $K = 2.5 \times 10^{10}$ at 500. K and 4.0×10^{24} at 298 K.

9.85 At 500.°C, $K_c = 0.061$ for $N_2(g) + 3 H_2(g) \rightleftharpoons 2 NH_3(g)$. If analysis shows that the composition is 3.00 mol·L^{-1} N_2, 2.00 mol·L^{-1} H_2, and 0.500 mol·L^{-1} NH_3, is the reaction at equilibrium? If not, in which direction does the reaction tend to proceed to reach equilibrium?

9.86 At 2500. K, the equilibrium constant is $K_c = 20$. for the reaction $Cl_2(g) + F_2(g) \rightleftharpoons 2 ClF(g)$. An analysis of a reaction vessel at 2500. K revealed the presence of 0.18 mol·L^{-1} Cl_2, 0.31 mol·L^{-1} F_2, and 0.92 mol·L^{-1} ClF. Will ClF tend to form or to decompose as the reaction proceeds toward equilibrium?

9.87 Calculate the equilibrium constant at 25°C and at 150.°C for each of the following reactions, using data available in Appendix 2A:
(a) $NH_4Cl(s) \rightleftharpoons NH_3(g) + HCl(g)$
(b) $H_2(g) + D_2O(l) \rightleftharpoons D_2(g) + H_2O(l)$

9.88 Calculate the equilibrium constant at 25°C and at 100.°C for each of the following reactions, using data available in Appendix 2A:
(a) $HgO(s) \rightleftharpoons Hg(l) + O_2(g)$
(b) propene (C_3H_6, g) $\rightleftharpoons$ cyclopropane (C_3H_6, g)

9.89 By combining the relation for K_c in terms of K with the van 't Hoff equation, find the analog of the van 't Hoff equation for K_c.

9.90 The vaporization of a liquid can be treated as a special case of an equilibrium. How does the vapor pressure of a liquid vary with temperature? *Hint:* Devise a version of the van 't Hoff equation that applies to vapor pressure by first writing the equilibrium constant K for vaporization.

Integrated Exercises

9.91 Dissociation of a diatomic molecule, $X_2(g) \rightleftharpoons 2 X(g)$ occurs at 500 K. Picture (1) shows the equilibrium state of the dissociation and picture (2) shows the equilibrium state in the same container after a change has occurred. Which of the following changes will produce the change shown? (a) Increasing the temperature. (b) Adding X atoms. (c) Decreasing the volume. (d) Adding a catalyst. Explain your selections.

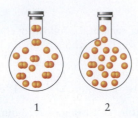

1   2

9.92 At 25°C, $K = 47.9$ for $N_2O_4(g) \rightleftharpoons 2\ NO_2(g)$. (a) If 0.020 mol of N_2O_4 and 0.010 mol of NO_2 are placed in a 2.00-L reaction vessel and the reaction is allowed to reach equilibrium, what are the equilibrium concentrations of N_2O_4 and NO_2? (b) An additional 0.010 mol of NO_2 is added to the flask. How will this change affect the concentration of N_2O_4? (c) Justify your conclusion by calculating the new equilibrium concentrations of NO_2 and N_2O_4.

9.93 The overall photosynthesis reaction is $6\ CO_2(g) + 6\ H_2O(l) \rightleftharpoons C_6H_{12}O_6(aq) + 6\ O_2(g)$, and $\Delta H° = +2802$ kJ. Suppose that the reaction is at equilibrium. State the effect that each of the following changes will have on the equilibrium composition (tends to shift toward the formation of reactants, tends to shift toward the formation of products, or has no effect). (a) The partial pressure of O_2 is increased. (b) The system is compressed. (c) The amount of CO_2 is increased. (d) The temperature is increased. (e) Some of the $C_6H_{12}O_6$ is removed. (f) Water is added. (g) The partial pressure of CO_2 is decreased.

9.94 A certain enzyme-catalyzed reaction in a biochemical cycle has an equilibrium constant that is 10 times the equilibrium constant of the next step in the cycle. If the standard Gibbs free energy of the first reaction is $-200.$ kJ·mol^{-1}, what is the standard Gibbs free energy of the second reaction?

9.95 The following plot shows how the partial pressures of reactant and products vary with time for the decomposition of compound A into compounds B and C. All three compounds are gases. Use this plot to do the following: (a) Write a balanced chemical equation for the reaction. (b) Calculate the equilibrium constant for the reaction. (c) Calculate the value of K_c for the reaction at 25°C.

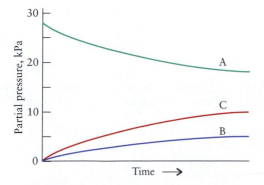

9.96 The following plot shows a system composed of compounds A and B in a rigid, constant-volume flask. The

system was initially at equilibrium, then a change occurred. (a) Describe the change that occurred and how it affected the system. (b) Write the chemical equation for the reaction that occurred. (c) Calculate the value of K for the reaction.

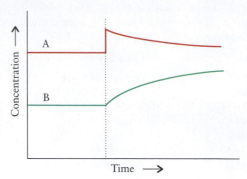

9.97 At 1565 K, the equilibrium constants for the reactions (1) $2\ H_2O(g) \rightleftharpoons 2\ H_2(g) + O_2(g)$ and (2) $2\ CO_2(g) \rightleftharpoons 2\ CO(g) + O_2(g)$ are 1.6×10^{-11} and 1.3×10^{-10}, respectively. (a) What is the equilibrium constant for the reaction (3) $CO_2(g) + H_2(g) \rightleftharpoons H_2O(g) + CO(g)$ at that temperature? (b) Show that the manner in which equilibrium constants are calculated is consistent with the manner in which the $\Delta G_r°$ values are calculated when combining two or more equations by determining $\Delta G_r°$ for reactions (1) and (2) and using those values to calculate $\Delta G_r°$ and K_3 for reaction (3).

9.98 1.50 mol NH_3 decomposed in the reaction $2\ NH_3(g) \rightleftharpoons N_2(g) + 3\ H_2(g)$ in a 1.0-L flask at a certain temperature. At equilibrium 0.20 mol NH_3 remains. (a) Calculate K for the reaction. (b) Use the information in Table 9.2 to estimate the temperature at which the reaction occurred.

9.99 Let α be the fraction of PCl_5 molecules that have decomposed to PCl_3 and Cl_2 in the reaction $PCl_5(g) \rightleftharpoons PCl_3(g) + Cl_2(g)$ in a constant-volume container; then the amount of PCl_5 at equilibrium is $n(1 - \alpha)$, where n is the amount present initially. Derive an equation for K in terms of α and the total pressure P, and solve it for α in terms of P. Calculate the fraction decomposed at 556 K, at which temperature $K = 4.96$, and the total pressure is (a) 0.50 bar; (b) 1.00 bar.

9.100 The three compounds methylpropene, *cis*-2-butene, and *trans*-2-butene are isomers with formula C_4H_8, with $\Delta G_f° = +58.07, +65.86,$ and $+62.97$ kJ·mol^{-1}, respectively. In the presence of a suitable metal catalyst, these three compounds can be interconverted to give an equilibrium gaseous mixture. What will be the percentage of each isomer present at 25°C once equilibrium is established?

9.101 (a) What is the standard Gibbs free energy of the reaction $CO(g) + H_2O(g) \rightleftharpoons CO_2(g) + H_2(g)$ when $K = 1.00$? (b) From data available in Appendix 2A, estimate the temperature at which $K = 1.00$. (c) At this temperature, a cylinder is filled with $CO(g)$ at 10.00 bar, $H_2O(g)$ at 10.00 bar, $H_2(g)$ at 5.00 bar, and $CO_2(g)$ at 5.00 bar. What will be the partial pressure of each of these gases when the system reaches equilibrium? (d) If the cylinder were filled instead with $CO(g)$ at 6.00 bar, $H_2O(g)$ at 4.00 bar, $H_2(g)$ at 5.00 bar, and $CO_2(g)$ at 10.00 bar, what would the partial pressures be at equilibrium?

9.102 Consider the equilibrium $A(g) \rightleftharpoons 2\ B(g) + 3\ C(g)$ at 25°C. When A is loaded into a cylinder at 10.0 atm and the system is allowed to come to equilibrium, the final pressure is found to be 15.76 atm. What is $\Delta G_r°$ for this reaction?

9.103 (a) Calculate the Gibbs free energies of formation of the halogen atoms X(g) at 1000. K from data available in Table 9.2. (b) Show how these data correlate with the X—X bond strength by plotting the Gibbs free energy of formation of the atoms against the bond dissociation energy and atomic number. Rationalize any trends you observe.

9.104 The gas phosphine, PH_3, decomposes by the reaction $2 PH_3(g) \rightarrow 2 P(s) + 3 H_2(g)$. In an experiment, pure phosphine was placed in a rigid, sealed 1.00-L flask at 0.64 bar and 298 K. After equilibrium was attained, the total pressure in the flask was found to be 0.93 bar. (a) Calculate the equilibrium partial pressures of H_2 and PH_3. (b) Calculate the mass (in grams) of P produced once equilibrium was reached. (c) Calculate K for this reaction.

9.105 (a) Calculate K at 25°C for the reaction $Br_2(g) \rightleftharpoons 2 Br(g)$ from the thermodynamic data provided in Appendix 2A. (b) What is the vapor pressure of liquid bromine? (c) What is the partial pressure of Br(g) above the liquid in a bottle of bromine at 25°C? (d) A student wishes to add 0.0100 mol Br_2 to a reaction and will do so by filling an evacuated flask with Br_2 vapor from a reservoir that contains only bromine liquid in equilibrium with its vapor. The flask will be sealed and then transferred to the reaction vessel. What volume container should the student use to deliver 0.010 mol Br_2(g) at 25°C?

9.106 A reaction vessel is filled with Cl_2(g) at 1.00 bar and Br_2(g) at 1.00 bar, which are allowed to react at 1000. K to form BrCl(g) according to the equation $Br_2(g) + Cl_2(g) \rightleftharpoons 2 BrCl(g)$, $K = 0.2$. Construct a plot of the Gibbs free energy of this system as a function of partial pressure of BrCl as the reaction approaches equilibrium.

9.107 A reactor for the production of ammonia by the Haber process is found to be at equilibrium with $P_{N_2} = 3.11$ bar, $P_{H_2} = 1.64$ bar, and $P_{NH_3} = 23.72$ bar. If the partial pressure of N_2 is increased by 1.57 bar, what will be the partial pressure of each gas once equilibrium is re-established?

9.108 Using data from Table 9.2 and standard graphing software, determine the enthalpy and entropy of the equilibrium $N_2O_4(g) \rightleftharpoons 2 NO_2(g)$ and estimate the N—N bond enthalpy in N_2O_4. How does this value compare with the mean N—N bond enthalpy in Table 6.8?

9.109 The distribution of Na^+ ions across a typical biological membrane is 10. $mmol \cdot L^{-1}$ inside the cell and 140 $mmol \cdot L^{-1}$ outside the cell. At equilibrium the concentrations would be equal, but in a living cell the ions are not at equilibrium. What is the Gibbs free energy difference for Na^+ ions across the membrane at 37°C (normal body temperature)? The concentration differential must be maintained by coupling to reactions that have at least that difference of Gibbs free energy.

9.110 ATP is a compound that provides energy for biochemical reactions in the body when it undergoes hydrolysis. For the hydrolysis of ATP at 37°C (normal body temperature) $\Delta H_r° = -20.$ $kJ \cdot mol^{-1}$ and $\Delta S_r° = +34$ $J \cdot K^{-1} \cdot mol^{-1}$. Assuming that these quantities are independent of temperature, calculate the temperature at which the equilibrium constant for the hydrolysis of ATP becomes greater than 1.

9.111 Use the Living Graph "Variation of Equilibrium Constant" on the Web site for this book to construct a plot from 250 K to 350 K for reactions with standard reaction Gibbs free energies of $+11$ $kJ \cdot mol^{-1}$ to $+15$ $kJ \cdot mol^{-1}$ in increments of 1 $kJ \cdot mol^{-1}$. Which equilibrium constant is most sensitive to changes in temperature?

9.112 (a) Use graphing software, such as that found on the Web site for this book, and data at 1000. K and 1200. K from Table 9.2, to plot the van 't Hoff equations for the dissociation of the diatomic halogens into the atoms, $X_2(g) \rightleftharpoons 2 X(g)$. (b) From the graphs, determine the enthalpies and entropies of dissociation. (c) Use these data to calculate the molar entropies of the gaseous halogen atoms X(g).

9.113 The reaction $N_2O_4 \rightleftharpoons 2 NO_2$ is allowed to reach equilibrium in chloroform solution at 25°C. The equilibrium concentrations are 0.405 $mol \cdot L^{-1}$ N_2O_4 and 2.13 $mol \cdot L^{-1}$ NO_2. (a) Calculate K for the reaction. (b) An additional 1.00 mol NO_2 is added to 1.00 L of the solution and the system is allowed to reach equilibrium again at the same temperature. Use Le Chatelier's principle to predict the direction of change (increase, decrease, or no change) for N_2O_4, NO_2, and K_c after the addition of NO_2. (c) Calculate the final equilibrium concentrations after the addition of NO_2 and confirm that your predictions in part (b) were valid. If they do not agree, check your procedure and repeat it if necessary.

9.114 In order to generate the starting material for a polymer that is used in water bottles, hydrogen is removed from the ethane in natural gas to produce ethene in the catalyzed reaction $C_2H_6(g) \rightleftharpoons H_2(g) + C_2H_4(g)$. Use the information in Appendix 2A to calculate the equilibrium constant for the reaction at 298 K. (a) If the reaction is begun by adding the catalyst to a flask containing C_2H_6 at 10.0 bar, what will be the partial pressure of the C_2H_4 at equilibrium? (b) Identify three steps the manufacturer can take to increase the yield of product.

9.115 Determine the vapor pressure of heavy water, D_2O, and of normal water at 25°C by using data in Appendix 2A. How do these values compare with each other? Using your knowledge of intermolecular forces, explain the reason for the difference observed.

9.116 A chemist needs to prepare the compound PH_3BCl_3 by using the reaction $PH_3(g) + BCl_3(g) \rightarrow PH_3BCl_3(s)$ for which $K = 19.2$ at 60.°C. (a) Write the expression for K. (b) What is the value of K_c for this reaction? (c) Some solid PH_3BCl_3 was added to a closed 500.-mL vessel at 60.°C that already contains 0.0128 mol PH_3. What is the equilibrium concentration of PH_3? (d) At 70.°C, $K = 26.2$. Is the reaction endothermic or exothermic? Explain your reasoning. (e) What is the new value of K_c? (f) Can the reactants in the preceding reaction be classified as acids or bases? Explain your answer.

9.117 It is possible to choose a different standard state (after all, molecules don't know what their standard state is) for the pressure or concentration of chemical species. The standard state corresponding to data presented in Appendix 2A is for 1 bar or 1 M, as appropriate. Calculate $\Delta G_f°$ for each of the following species if the standard state condition were changed to that listed (the temperature remains at 298 K):
(a) HI(g), standard state corresponds to 1 atm
(b) CO(g), standard state corresponds to 1 atm
(c) HCN(g), standard state corresponds to 1 Torr
(d) CH_4(g), standard state corresponds to 1 Pa

9.118 Cyclohexane (C) and methylcyclopentane (M) are isomers with the chemical formula C_6H_{12}. The equilibrium constant for the rearrangement $C \rightleftharpoons M$ in solution is 0.140 at 25°C. (a) A solution of 0.0200 mol·L^{-1} cyclohexane and 0.100 mol·L^{-1} methylcyclopentane is prepared. Is the system at equilibrium? If not, will it will form more reactants or more products? (b) What are the concentrations of cyclohexane and methylcyclohexane at equilibrium? (c) If the temperature is raised to 50.°C, the concentration of cyclohexane becomes 0.100 mol·L^{-1} when equilibrium is reestablished. Calculate the new equilibrium constant. (d) Is the reaction exothermic or endothermic at 25°C? Explain your conclusion.

9.119 The two air pollutants SO_3 and NO can react in the atmosphere as follows: $SO_3(g) + NO(g) \rightarrow SO_2(g) + NO_2(g)$. (a) Predict the effect of the following changes to the amount of NO_2 when the reaction has come to equilibrium in a stainless steel bulb equipped with entrants for chemicals: (i) the amount of NO is increased; (ii) the SO_2 is removed by condensation; (iii) the pressure is tripled by pumping in helium. (b) Given that at a certain temperature $K = 6.0 \times 10^3$, calculate the amount (in moles) of NO that must be added to a 1.00-L vessel containing 0.245 mol $SO_3(g)$ to form 0.240 mol $SO_2(g)$ at equilibrium.

Chemistry Connections

9.120 Reactions between gases in the atmosphere are not at equilibrium, but for a thorough understanding of them we need to study both the rates at which they take place and their behavior under equilibrium conditions.
(a) The equilibrium for the depletion of ozone in the stratosphere is summarized by the equation $2\,O_3(g) \rightleftharpoons 3\,O_2(g)$.

From values in Appendix 2A, determine the standard Gibbs free energy and the standard entropy for the reaction.
(b) What is the equilibrium constant of the reaction in part (a) at 25°C? What is the significance of your answer with regard to ozone depletion?
(c) A reaction that destroys ozone in the stratosphere is $O_3(g) + O(g) \rightleftharpoons 2\,O_2(g)$. Calculate the value of the equilibrium constant for this reaction at 25°C given that at that temperature the reaction is catalyzed (accelerated) by NO_2 molecules in a two-step process:

$$NO_2(g) + O(g) \rightleftharpoons NO(g) + O_2(g) \qquad K = 7 \times 10^{103}$$

$$NO(g) + O_3(g) \rightleftharpoons NO_2(g) + O_2(g) \qquad K = 5.8 \times 10^{-34}$$

(d) Use your answer to part (c) to find the standard Gibbs free energy of formation of O atoms.
(e) The temperature dependence of the equilibrium constant of the reaction $N_2(g) + O_2(g) \rightleftharpoons 2\,NO(g)$, which makes an important contribution to the concentration of atmospheric nitrogen oxides, can be expressed as $\ln K = 2.5 - (21\,700\ \text{K})/T$. What is the standard enthalpy of the forward reaction at 298 K? Will this reaction proceed further at the very low temperatures of the stratosphere or at the very high temperatures in an internal combustion engine?
(f) An equimolar mixture of N_2 and O_2 was heated to a certain temperature until the reaction in part (e) came to equilibrium. The equilibrium reaction mixture was found to contain an equal number of moles of each reactant and product. At what temperature was the reaction carried out?
(g) An equimolar mixture of N_2 and O_2 with a total pressure of 4.00 bar was allowed to come to equilibrium in the reaction in part (e) at 1200. K. What will be the partial pressure of each reactant and product at equilibrium?

ACIDS AND BASES

What Are the Key Ideas? Brønsted acids are proton donors; Brønsted bases are proton acceptors. The composition of a solution of an acid or base immediately adjusts to satisfy the values of the equilibrium constants for all the proton transfer reactions taking place.

Why Do We Need to Know This Material? Chapter 9 developed the concepts of chemical equilibria in gaseous systems; this chapter extends those ideas to aqueous systems, which are important throughout chemistry and biology. Equilibria between acids, bases, and water in plant and animal cells are vital for the survival of individual organisms. To sustain human societies and protect our ecosystems, we also need these ideas to understand the acidity of rain, natural waters such as lakes and rivers, and municipal water supplies.

What Do We Need to Know Already? This chapter builds on the introduction to acids and bases in Section J. It also draws on and illustrates the principles of thermodynamics (Chapters 6 and 7) and chemical equilibrium (Chapter 9). To a smaller extent, it uses the concepts of hydrogen bonding (Section 5.5), bond polarity (Section 2.12), and bond strength (Sections 2.14 and 2.15).

I n this chapter, we see what acids and bases are and why they vary in strength. We shall use thermodynamics, particularly equilibrium constants, to discuss the strengths of acids and bases quantitatively and thereby develop our insight into their behavior. We then use our knowledge of equilibria involving acids and bases to examine systems in which more than one equilibrium is taking place simultaneously.

THE NATURE OF ACIDS AND BASES

When chemists see a pattern in the reactions of certain substances, they formulate a definition of a class of substance that captures them all. The reactions of the substances we call "acids" and "bases" are an excellent illustration of this approach. The pattern in these reactions was first identified in aqueous solutions, and led to the Arrhenius definitions of acids and bases (Section J). However, chemists discovered that similar reactions take place in nonaqueous solutions and even in the absence of solvent. The original definitions had to be replaced by more general definitions that encompassed this new knowledge.

10.1 Brønsted–Lowry Acids and Bases

In 1923, the Danish chemist Johannes Brønsted proposed that

An **acid** is a proton donor.

A **base** is a proton acceptor.

The term *proton* in these definitions refers to the hydrogen ion, H^+. An acid is a species containing an **acidic hydrogen atom**, which is a hydrogen atom that can be transferred as its nucleus, a proton, to another species acting as a base. The same definitions were proposed independently by the English chemist Thomas Lowry, and the theory based on them is called the **Brønsted–Lowry theory** of acids and bases.

A substance can act as an acid only if a base is present to accept its acidic protons. An acid does not simply release its acidic proton: the proton is *transferred* to the base through direct contact. For example, HCl is a Brønsted acid. In the gas

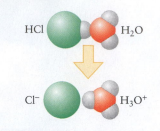

HCl   H₂O

Cl⁻   H₃O⁺

FIGURE 10.1 When an HCl molecule dissolves in water, a hydrogen bond forms between the H atom of HCl (the acid) and the O atom of a neighboring H₂O molecule (the base). The nucleus of the hydrogen atom is pulled out of the HCl molecule to become part of a hydronium ion.

Recall that an Arrhenius acid is a compound that produces hydronium ions in water and an Arrhenius base is a compound that produces hydroxide ions in water.

FIGURE 10.2 Stalactites hang from the roof of a cave and stalagmites grow from the floor. Both are made of insoluble calcium carbonate formed from the soluble hydrogen carbonate ions in groundwater.

A coordinate covalent bond is a bond in which both bonding electrons come from the same atom (Section 2.11).

phase, an HCl molecule remains intact. However, when HCl dissolves in water, each HCl molecule immediately transfers an H^+ ion to a neighboring H_2O molecule, which here is acting as a base (Fig. 10.1). This process is a **proton transfer reaction**, a reaction in which a proton is transferred from one species to another. We say that the HCl molecule becomes **deprotonated**:

$$HCl(aq) + H_2O(l) \longrightarrow H_3O^+(aq) + Cl^-(aq)$$

Because at equilibrium virtually all the HCl molecules have donated their protons to water, HCl is classified as a *strong acid*. The proton transfer reaction essentially goes to completion. The H_3O^+ ion is called the *hydronium ion*. It is strongly hydrated in solution, and there is some evidence that a better representation of the species is $H_9O_4^+$ (or even larger clusters of water molecules attached to a proton). A hydrogen ion in water is sometimes represented as $H^+(aq)$, but we must remember that H^+ does not exist by itself in water and that H_3O^+ is a better representation.

Another example of an acid is hydrogen cyanide, HCN, which transfers its proton to water when it dissolves to form the solution known as hydrocyanic acid, HCN(aq). However, only a small fraction of the HCN molecules donate their protons, and so we classify HCN as a *weak acid* in water. We write the proton transfer reaction with equilibrium half-arrows:

$$HCN(aq) + H_2O(l) \rightleftharpoons H_3O^+(aq) + CN^-(aq)$$

Like all chemical equilibria, this equilibrium is dynamic and we should think of protons as ceaselessly exchanging between HCN and H_2O molecules, with a constant but low concentration of CN^- and H_3O^+ ions. The proton transfer reaction of a strong acid, such as HCl, in water is also dynamic, but the equilibrium lies so strongly in favor of products that we represent it just by its forward reaction with a single arrow.

The Brønsted definition also includes the possibility that an ion is an acid (an option not allowed by the Arrhenius definition). For instance, a hydrogen carbonate ion, HCO_3^-, one of the species present in natural waters, can act as an acid and lose a proton, and the resulting carbonate ion is removed by precipitation if suitable cations are present (Fig. 10.2):

$$Ca^{2+}(aq) + HCO_3^-(aq) + H_2O(l) \rightleftharpoons H_3O^+(aq) + CaCO_3(s)$$

In the Brønsted–Lowry theory, the strength of an acid depends on the extent to which it donates protons to the solvent. We can therefore summarize the distinction between strong and weak acids as follows:

A **strong acid** is fully deprotonated in solution.

A **weak acid** is only partly deprotonated in solution.

The strength of an acid depends on the solvent and an acid that is strong in water may be weak in another solvent or vice versa (see Section 10.8). However, because almost all reactions in living tissues and most reactions in laboratories take place in water, in almost every case the only solvent we consider in this chapter will be water.

A Brønsted base is a proton acceptor, which means that it possesses a lone pair of electrons to which the proton can bond. For example, an oxide ion is a Brønsted base. When CaO dissolves in water, the strong electric field of the small, highly charged O^{2-} ion pulls a proton out of a neighboring H_2O molecule (Fig. 10.3). A coordinate covalent bond then forms between the proton and a lone pair of electrons on the oxide ion. By accepting a proton, the oxide ion has become **protonated**. Every oxide ion present accepts a proton from water, and so O^{2-} is an example of a *strong base* in water, a species that is fully protonated. That is, the following reaction goes essentially to completion:

$$O^{2-}(aq) + H_2O(l) \longrightarrow 2\,OH^-(aq)$$

Another example of a Brønsted base is ammonia. When ammonia dissolves in water, it is protonated when the lone pair of electrons on the N atom accepts a proton from H_2O:

$$NH_3(aq) + H_2O(l) \rightleftharpoons NH_4^+(aq) + OH^-(aq)$$

The fact that the NH_3 molecule is electrically neutral means that it has much less proton-pulling power than the oxide ion, and even less than the hydroxide ion. As a result, only a very small proportion of the NH_3 molecules are converted into NH_4^+ ions (Fig. 10.4). Ammonia is therefore an example of a *weak base*. All amines, organic analogs of ammonia, such as methylamine, CH_3NH_2, are weak bases in water. Because the proton transfer equilibrium in an aqueous solution of ammonia is dynamic, at the molecular level we visualize the protons as ceaselessly exchanging between NH_3 and H_2O molecules, such that there is a constant, low concentration of NH_4^+ and OH^- ions. The proton transfer to the strong base O^{2-} is also dynamic, but the equilibrium lies so strongly in favor of products that, as for a strong acid, we represent it just by its forward reaction with a single arrow.

We can summarize the distinction between strong and weak bases as follows:

A **strong base** is completely protonated in solution.

A **weak base** is only partially protonated in solution.

As for acids, the strength of a base depends on the solvent; a base that is strong in water may be weak in another solvent and vice versa. The common strong bases in aqueous solution are listed in Table J.1.

A note on good practice: The oxides and hydroxides of the alkali and alkaline earth metals are not Brønsted bases: the oxide and hydroxide *ions* they contain are the bases (the cations are spectator ions). However, for convenience, chemists often refer to the compounds themselves as bases.

The products of proton transfer in aqueous solution may also react with water. For example, the CN^- ion produced when HCN loses a proton to water can accept a proton from a water molecule and form HCN again. Therefore, according to the Brønsted definition, CN^- is a base: it is called the "conjugate base" of the acid HCN. In general, a **conjugate base** of an acid is the species left when the acid donates a proton:

$$\text{Acid} \xrightarrow{\text{donates } H^+} \text{conjugate base}$$

Because HCN is the acid that forms when a proton is transferred to a cyanide ion, HCN is the "conjugate acid" of the base CN^-. In general, a **conjugate acid** of a base is the species formed when the base accepts a proton:

$$\text{Base} \xrightarrow{\text{accepts } H^+} \text{conjugate acid}$$

EXAMPLE 10.1   **Sample exercise: Writing the formulas of conjugate acids and bases**

Write the formulas of (a) the conjugate base of HCO_3^- and (b) the conjugate acid of O^{2-}.

SOLUTION

The conjugate base of an acid has one less H^+ ion than the acid; the conjugate acid of a base has one more H^+ ion than that of the base.

(a) The conjugate base of HCO_3^- is CO_3^{2-}.
(b) The conjugate acid of O^{2-} is OH^-.

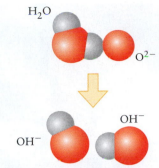

FIGURE 10.3 When an oxide ion is present in water, it exerts such a strong attraction on the nucleus of a hydrogen atom in a neighboring water molecule that the hydrogen ion is pulled out of the molecule as a proton. As a result, the oxide ion forms two hydroxide ions.

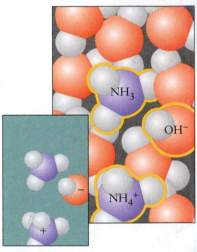

FIGURE 10.4 In this molecular portrayal of the structure of a solution of ammonia in water at equilibrium, we see that NH_3 molecules are still present because only a small percentage of them have been protonated by transfer of hydrogen ions from water. In a typical solution, only about 1 in 100 NH_3 molecules is protonated. The overlay shows only the solute species.

Acid   Base

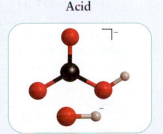
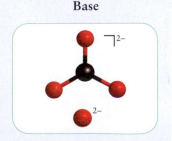

FIGURE 10.5 The white powder is ammonium chloride formed by gaseous ammonia and hydrogen chloride gas escaping from concentrated hydrochloric acid.

SELF-TEST 10.1A What is (a) the conjugate acid of OH^-; (b) the conjugate base of HPO_4^{2-}?

[*Answer:* (a) H_2O; (b) PO_4^{3-}]

SELF-TEST 10.1B What is (a) the conjugate acid of H_2O; (b) the conjugate base of NH_3?

The Brønsted definitions of acids and bases are more general than the Arrhenius definitions: they also apply to species in nonaqueous solvents and even to gas-phase reactions. For example, when pure acetic acid is added to liquid ammonia, proton transfer takes place and the following equilibrium is reached:

$$CH_3COOH(l) + NH_3(l) \rightleftharpoons CH_3CO_2^-(am) + NH_4^+(am)$$

(The label "am" indicates a species dissolved in liquid ammonia.) An example of proton transfer in the gas phase is the reaction of hydrogen chloride and ammonia gases. They produce the fine powder of ammonium chloride often seen coating surfaces in chemical laboratories (Fig. 10.5):

$$HCl(g) + NH_3(g) \longrightarrow NH_4Cl(s)$$

A Brønsted acid is a proton donor and a Brønsted base is a proton acceptor. The conjugate base of an acid is the base formed when the acid has donated a proton. The conjugate acid of a base is the acid that forms when the base has accepted a proton. A strong acid is fully deprotonated in solution; a weak acid is only partially deprotonated in solution. A strong base is completely protonated in solution; a weak base is only partially protonated in solution.

10.2 Lewis Acids and Bases

The Brønsted–Lowry theory focuses on the transfer of a proton from one species to another. However, the concepts of acids and bases have a much wider significance than the transfer of protons. Even more substances can be classified as acids or bases under the definitions developed by G. N. Lewis:

A **Lewis acid** is an electron pair acceptor.

A **Lewis base** is an electron pair donor.

When a Lewis base donates an electron pair to a Lewis acid, the two species share the pair and become joined by a coordinate covalent bond, a bond in which both electrons come from one of the atoms (see Section 2.11).

A proton (H^+) is an electron pair acceptor. It is therefore a Lewis acid because it can attach to ("accept") a lone pair of electrons on a Lewis base. In other words, a Brønsted acid is a supplier of one particular Lewis acid, a proton. The Lewis theory is more general than the Brønsted–Lowry theory. For instance, metal atoms and ions can act as Lewis acids, as in the formation of $Ni(CO)_4$ from nickel atoms (the Lewis acid) and carbon monoxide (the Lewis base), but they are not Brønsted acids. Likewise, a Brønsted base is a special kind of Lewis base, one that can use a lone pair of electrons to form a coordinate covalent bond to a proton. For instance, an oxide ion is a Lewis base. It forms a coordinate covalent bond to a proton, a Lewis acid, by supplying both the electrons for the bond:

(The curved arrows show the direction in which the electrons can be thought to move.) Similarly, when the Lewis base ammonia, NH_3, dissolves in water, some of the molecules accept protons from water molecules:

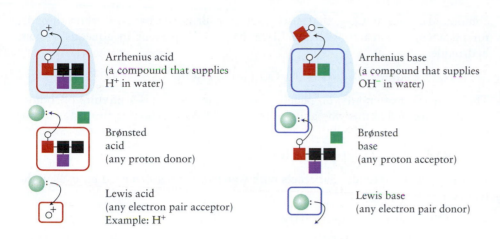

FIGURE 10.6 The left-hand column illustrates the action of acids in the Arrhenius, Brønsted, and Lewis definitions. The right-hand column shows the action of the corresponding bases. In each case the rounded rectangles enclose the acid or base and the small white circles represent hydrogen ions. The dark green square represents an accompanying ion. The Arrhenius definition includes any accompanying ion, whereas the Brønsted definition may apply either to a compound or, as here, to an ion. Notice that only the Arrhenius definition requires the presence of water (the blue background).

The Lewis definition of a base is broader than the Brønsted definition. That is, although every Brønsted base is a Lewis base, not every Lewis base is a Brønsted base. For instance, carbon monoxide is an important Lewis base in its reactions with metals, but it is not a Brønsted base because it does not accept protons.

A note on good practice: The entities that are regarded as acids and bases are different in each theory. In the Lewis theory, the proton is an acid; in the Brønsted theory, the species that *supplies* the proton is the acid. In both the Lewis and Brønsted theories, the species that accepts a proton is a base; in the Arrhenius theory, the species that *supplies* the proton acceptor is the base (Fig. 10.6).

Many nonmetal oxides are Lewis acids that react with water to give Brønsted acids. An example is the reaction of CO_2 with water:

We depict the migration of a proton with a red arrow.

In this reaction, the C atom of CO_2, the Lewis acid, accepts an electron pair from the O atom of a water molecule, the Lewis base, and a proton migrates from an H_2O oxygen atom to a CO_2 oxygen atom. The product, an H_2CO_3 molecule, is a Brønsted acid.

SELF-TEST 10.2A Identify (a) the Brønsted acids and bases in both reactants and products in the proton transfer equilibrium $HNO_2(aq) + HPO_4^{2-}(aq) \rightleftharpoons NO_2^-(aq) + H_2PO_4^-(aq)$. (b) Which species (not necessarily shown explicitly) are Lewis acids and which are Lewis bases?

[*Answer:* (a) Brønsted acids, HNO_2, $H_2PO_4^-$; Brønsted bases, HPO_4^{2-} and NO_2^-; (b) Lewis acid, H^+; Lewis bases, HPO_4^{2-} and NO_2^-]

SELF-TEST 10.2B Identify (a) the Brønsted acids and bases in both reactants and products in the proton transfer equilibrium $HCO_3^-(aq) + NH_4^+(aq) \rightleftharpoons H_2CO_3(aq) + NH_3(aq)$. (b) Which species (not necessarily shown explicitly) are Lewis acids and which are Lewis bases?

A Lewis acid is an electron pair acceptor; a Lewis base is an electron pair donor. A proton is a Lewis acid that attaches to a lone pair provided by a Lewis base.

10.3 Acidic, Basic, and Amphoteric Oxides

An **acidic oxide** is an oxide that reacts with water to form a Brønsted acid: an example is CO_2, as we saw in Section 10.2. Acidic oxides are *molecular*

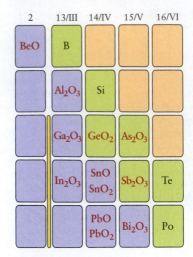

FIGURE 10.7 The elements in and close to the diagonal line of metalloids typically form amphoteric oxides (indicated by the red lettering).

compounds, such as CO_2, that also react with Brønsted bases. Carbon dioxide, for instance, also reacts with the Lewis base OH^- present in aqueous sodium hydroxide:

$$2\ NaOH(aq) + CO_2(g) \longrightarrow Na_2CO_3(aq) + H_2O(l)$$

The white crust often seen on pellets of sodium hydroxide is a mixture of sodium carbonate formed in this way and of sodium hydrogen carbonate formed in a similar reaction:

$$NaOH(aq) + CO_2(g) \longrightarrow NaHCO_3(aq)$$

A **basic oxide** is an oxide that reacts with water to form a Brønsted base, as in the reaction

$$CaO(s) + H_2O(l) \longrightarrow Ca(OH)_2(aq)$$

Basic oxides are *ionic* compounds, such as CaO, that also react with acids to give a salt and water. For instance, magnesium oxide, a basic oxide, reacts with hydrochloric acid:

$$MgO(s) + 2\ HCl(aq) \longrightarrow MgCl_2(aq) + H_2O(l)$$

In this reaction, the base O^{2-} accepts two protons from the hydronium ions present in the hydrochloric acid solution.

Metals typically form basic oxides and nonmetals typically form acidic oxides, but what about the elements that lie on the diagonal frontier between the metals and nonmetals? Along this frontier, from beryllium to polonium, metallic character blends into nonmetallic character, and the oxides of these elements have both acidic and basic character (Fig. 10.7). Substances that react with both acids and bases are classified as **amphoteric**, from the Greek word for "both." For example, aluminum oxide, Al_2O_3, is amphoteric. It reacts with acids:

$$Al_2O_3(s) + 6\ HCl(aq) \longrightarrow 2\ AlCl_3(aq) + 3\ H_2O(l)$$

and with bases:

$$2\ NaOH(aq) + Al_2O_3(s) + 3\ H_2O(l) \longrightarrow 2\ Na[Al(OH)_4](aq)$$

The product of the second reaction is sodium aluminate, which contains the aluminate ion, $Al(OH)_4^-$. Other main-group elements that form amphoteric oxides are shown in Fig.10.7. The acidic, amphoteric, or basic character of the oxides of the *d*-block metals depends on their oxidation state (Fig. 10.8; also see Chapter 16).

Metals form basic oxides, nonmetals form acidic oxides; the elements on a diagonal line from beryllium to polonium and several d-block metals form amphoteric oxides.

10.4 Proton Exchange Between Water Molecules

An important implication of the Brønsted definitions of acids and bases is that the same substance may be able to function as both an acid and a base. For example, we have seen that a water molecule accepts a proton from an acid molecule (such as HCl or HCN) to form an H_3O^+ ion. So water is a base. However, a water molecule can donate a proton to a base (such as O^{2-} or NH_3) and become an OH^- ion. So water is also an acid. We describe water as **amphiprotic**, meaning that an H_2O molecule can act both as a proton donor and as a proton acceptor.

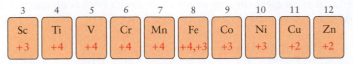

FIGURE 10.8 Certain elements of the *d* block form amphoteric oxides, particularly in oxidation states intermediate in their range (as shown here for the first series).

A note on good practice: Distinguish between *amphoteric* and *amphiprotic*. Aluminum metal is amphoteric (it reacts with both acids and bases), but it has no hydrogen atoms to donate as protons, and is not amphiprotic.

Because water is amphiprotic—because it is both a Brønsted acid and a Brønsted base—proton transfer between water molecules occurs even in pure water, with one molecule acting as a proton donor and a neighboring molecule acting as a base:

$$2\,H_2O(l) \rightleftharpoons H_3O^+(aq) + OH^-(aq) \tag{A}$$

The reaction is very fast in both directions, and so is always at equilibrium in water and in aqueous solutions. In every glass of water, protons from the hydrogen atoms are ceaselessly migrating between the molecules. This type of reaction, in which one molecule transfers a proton to another molecule of the same kind, is called **autoprotolysis** (Fig. 10.9).

The equilibrium constant for reaction A is

$$K = \frac{a_{H_3O^+}\, a_{OH^-}}{(a_{H_2O})^2}$$

In dilute aqueous solutions (the only ones we consider in this chapter), the solvent, water, is very nearly pure, and so its activity may be taken to be 1. The resulting expression is called the **autoprotolysis constant** of water and is written K_w:

$$K_w = a_{H_3O^+}\, a_{OH^-} \tag{1a}$$

As we saw in Section 9.2, the activity of a solute J in a dilute solution is approximately equal to the molar concentration relative to the standard molar concentration, $[J]/c^\circ$, with $c^\circ = 1\ mol\cdot L^{-1}$, and so a practical form of this expression is

$$K_w = [H_3O^+][OH^-] \tag{1b}*$$

where, as in Chapter 9, we have simplified the appearance of the expression by replacing $[J]/c^\circ$ by $[J]$, the value of the molarity with the units struck out.

In pure water at 25°C, the molar concentrations of H_3O^+ and OH^- are equal and are known by experiment to be $1.0 \times 10^{-7}\ mol\cdot L^{-1}$. Therefore,

$$K_w = (1.0 \times 10^{-7}) \times (1.0 \times 10^{-7}) = 1.0 \times 10^{-14}$$

The concentrations of H_3O^+ and OH^- are very low in pure water, which explains why pure water is such a poor conductor of electricity. To imagine the very tiny extent of autoprotolysis, think of each letter in this book as a water molecule. We would need to search through more than 50 books to find one ionized water molecule. The autoprotolysis reaction is endothermic ($\Delta H_r^\circ = +56\ kJ\cdot mol^{-1}$), and so we can expect K_w to increase with temperature, and aqueous solutions to have higher concentrations of both hydronium and hydroxide ions at higher temperatures. Unless otherwise stated, all the calculations in this chapter will be for 25°C.

Now we come to a very important point that will be the basis of much of the discussion in this chapter and the next. *Because K_w is an equilibrium constant, the product of the concentrations of H_3O^+ and OH^- ions is always equal to K_w.* We can increase the concentration of H_3O^+ ions by adding acid, and the concentration of OH^- ions will immediately respond by decreasing to preserve the value of K_w. Alternatively, we can increase the concentration of OH^- ions by adding base, and the concentration of H_3O^+ ions will decrease correspondingly. The autoprotolysis equilibrium links the concentrations of H_3O^+ and OH^- ions rather like a seesaw: when one goes up, the other must go down (Fig. 10.10).

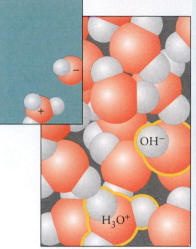

FIGURE 10.9 As a result of autoprotolysis, pure water consists of hydronium ions and hydroxide ions as well as water molecules. The concentration of ions that results from autoprotolysis is only about $10^{-7}\ mol\cdot L^{-1}$, and so only about 1 molecule in 200 million is ionized. The overlay shows only the ions.

K_w is also widely called the *autoionization constant* and sometimes the *ion product constant* of water.

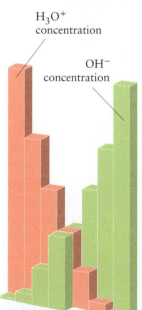

FIGURE 10.10 The product of the concentrations of hydronium and hydroxide ions in water (pure water and aqueous solutions) is a constant. If the concentration of one type of ion increases, the other must decrease to keep the product of the ion concentrations constant.

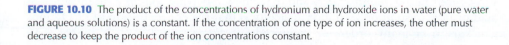

EXAMPLE 10.2 Calculating the concentrations of ions in a solution of a metal hydroxide

What are the concentrations of H_3O^+ and OH^- in 0.0030 M $Ba(OH)_2(aq)$ at 25°C?

STRATEGY When $Ba(OH)_2$ dissolves in water, it provides OH^- ions; most hydroxides of Groups 1 and 2 can be treated as fully dissociated in solution. Decide from the chemical formula how many OH^- ions are provided by each formula unit and calculate the concentrations of these ions in the solution. To find the concentration of H_3O^+ ions, use the water autoprotolysis constant $K_w = [H_3O^+][OH^-]$.

SOLUTION

Decide whether the compound is fully dissociated in solution.	Because barium is an alkaline earth metal, $Ba(OH)_2$ dissociates almost completely in water to provide OH^- ions. 
Find the mole ratio of hydroxide ion concentration to solute concentration.	$Ba(OH)_2(s) \longrightarrow Ba^{2+}(aq) + 2\ OH^-(aq)$ 1 mol $Ba(OH)_2 \simeq 2$ mol OH^-
Calculate the hydroxide ion concentration from the solute concentration.	$[OH^-] = 2 \times 0.0030\ \text{mol·L}^{-1}$ $= 0.0060\ \text{mol·L}^{-1}$
Rearrange $K_w = [H_3O^+][OH^-]$ to find the concentration of H_3O^+ ions.	$\left[H_3O^+\right] = \dfrac{K_w}{[OH^-]} = \dfrac{1.0 \times 10^{-14}}{0.0060}$ $= 1.7 \times 10^{-12}$ 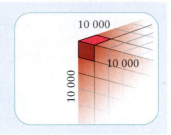

The concentration of H_3O^+ ions is 1.7×10^{-12} mol·L^{-1}, or about 1 in 10^{12}, as illustrated in the graphic above.

SELF-TEST 10.3A Estimate the concentrations of (a) H_3O^+ and (b) OH^- at 25°C in 6.0×10^{-5} M HI(aq).

[*Answer:* (a) 6.0×10^{-5} mol·L^{-1}; (b) 1.7×10^{-10} mol·L^{-1}]

SELF-TEST 10.3B Estimate the concentrations of (a) H_3O^+ and (b) OH^- at 25°C in 2.2×10^{-3} M NaOH(aq).

In aqueous solutions, the concentrations of H_3O^+ and OH^- ions are related by the autoprotolysis equilibrium; if one concentration is increased, then the other must decrease to maintain the value of K_w.

10.5 The pH Scale

So far, we have seen *qualitatively* that hydronium ions are always present in water and that, in an aqueous solution of an acid or base, the concentration of H_3O^+ ions depends on the concentration of solute. The time has come for us to express this concentration *quantitatively* and to see how it depends on the concentration of acid or base in solution. A minor difficulty is that the concentration of H_3O^+ ions can vary over many orders of magnitude: in some solutions, it is higher than

1 mol·L^{-1} and, in others, it is lower than 10^{-14} mol·L^{-1}. Chemists avoid the awkwardness of working with such a wide range of values by reporting the hydronium ion concentration in terms of the **pH** of the solution, the negative logarithm (to the base 10) of the hydronium ion activity:

$$pH = -\log a_{H_3O^+} \tag{2a}$$

where (for solutions so dilute that they may be treated as ideal) $a_{H_3O^+} = [H_3O^+]/c°$. As in Chapter 9, we simplify the appearance of this expression by interpreting $[H_3O^+]$ as the molar concentration of H_3O^+ with the units struck out and write

$$pH = -\log [H_3O^+] \tag{2b}*$$

For example, the pH of pure water, in which the concentration of H_3O^+ ions is 1.0×10^{-7} mol·L^{-1} at 25°C, is

$$pH = -\log(1.0 \times 10^{-7}) = 7.00$$

A note on good practice: The number of digits following the decimal point in a pH value is equal to the number of significant figures in the corresponding molar concentration, because the digits preceding the decimal point simply report the power of 10 in the data (as in log 10^5 = 5).

The negative sign in the definition of pH means that the higher the concentration of H_3O^+ ions, the lower the pH:

- The pH of pure water is 7.
- The pH of an acidic solution is less than 7.
- The pH of a basic solution is greater than 7.

Notice that, as the concentration of hydronium ions increases, the pH decreases. Because pH is a common logarithm (to the base 10), a change of one pH unit means that the concentration of H_3O^+ ions has changed by a factor of 10. For example, when the concentration of H_3O^+ increases by a factor of 10, from 10^{-5} mol·L^{-1} to 10^{-4} mol·L^{-1}, the pH decreases from 5 to 4. Most solutions used in chemistry have a pH ranging from 0 to 14, but values outside this range are possible.

The pH scale was introduced by the Danish chemist Søren Sørensen in 1909 in the course of his work on quality control in the brewing of beer and is now used throughout science, medicine, agriculture, and engineering.

The negative logarithm is used so that most pH values are positive numbers.

EXAMPLE 10.3 Calculating a pH from a concentration

What is the pH of (a) human blood, in which the concentration of H_3O^+ ions is 4.0×10^{-8} mol·L^{-1}; (b) 0.020 M HCl(aq); (c) 0.040 M KOH(aq)?

STRATEGY The pH is calculated from Eq. 2b. For strong acids, the concentration of H_3O^+ is equal to the molarity of the acid; for acids, expect pH < 7. For strong bases, first find the concentration of OH$^-$, then convert that concentration into $[H_3O^+]$ by using $[H_3O^+][OH^-] = K_w$, as described in Example 10.2; for bases, expect pH > 7.

SOLUTION

(a) From pH = $-\log [H_3O^+]$,   pH = $-\log (4.0 \times 10^{-8}) = 7.40$

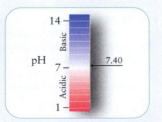

(b) HCl is a strong acid, so it is completely deprotonated in water. $[H_3O^+] = [HCl] = 0.020$ mol·L^{-1}

From pH = $-\log [H_3O^+]$,   pH = $-\log 0.020 = 1.70$

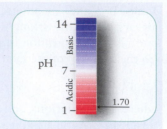

(c) Because KOH provides one OH$^-$ ion per formula unit, $[OH^-] = [KOH] = 0.040$ mol·L^{-1}

Find $[H_3O^+]$ from $[H_3O^+][OH^-] = K_w$ in the form $[H_3O^+] = K_w/[OH^-]$. $[H_3O^+] = \dfrac{K_w}{[OH^-]} = \dfrac{1.0 \times 10^{-14}}{0.040} = 2.5 \times 10^{-13}$

From pH = $-\log [H_3O^+]$,   pH = $-\log (2.5 \times 10^{-13}) = 12.60$

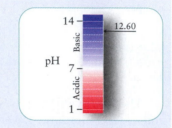

SELF-TEST 10.4A Calculate the pH of (a) household ammonia, in which the OH$^-$ concentration is about 3×10^{-3} mol·L^{-1}; (b) 6.0×10^{-5} M HClO$_4$(aq).

[*Answer:* (a) 11.5; (b) 4.22]

SELF-TEST 10.4B Calculate the pH of 0.077 M NaOH(aq).

We can determine an approximate value of the pH of an aqueous solution very quickly with a strip of *universal indicator paper*, which turns different colors at different pH values. More precise measurements are made with a *pH meter* (Fig. 10.11). This instrument consists of a voltmeter connected to two electrodes that dip into the solution. The difference in electrical potential between the electrodes is proportional to the hydronium ion activity (as will be explained in Section 12.10); so, once the scale on the meter has been calibrated, the pH can be read directly.

To convert a pH into a concentration of H$_3$O$^+$ ions, we reverse the sign of the pH and then take its antilogarithm:

$$[H_3O^+] = 10^{-pH} \text{ mol·L}^{-1} \tag{3}$$

EXAMPLE 10.4 **Sample exercise: Calculating the hydronium ion concentration from the pH**

Find the hydronium ion concentration in a solution with pH = 4.83.

FIGURE 10.11 A pH meter is a voltmeter that measures the pH electrochemically. The two samples are (a) orange juice and (b) lemon juice. Notice that the lemon juice has a lower pH and hence a higher concentration of hydronium ions.

(a)

(b)

SOLUTION Because pH $<$ 7, the solution is acidic and we expect $[H_3O^+] >$ 10^{-7} mol·L^{-1}. To calculate the precise value, we first change the sign of the pH and then take its antilogarithm:

From $[H_3O^+] = 10^{-pH}$ mol·L^{-1}, $[H_3O^+] = 10^{-4.83}$ mol·L$^{-1} = 1.5 \times 10^{-5}$ mol·L^{-1}

SELF-TEST 10.5A The pH of stomach fluids is about 1.7. What is the concentration of H_3O^+ ions in the stomach?

[*Answer:* 2×10^{-2} mol·L^{-1}]

SELF-TEST 10.5B The pH of pancreatic fluids, which help to digest food once it has left the stomach, is about 8.2. What is the approximate concentration of H_3O^+ ions in pancreatic fluids?

Figure 10.12 shows the results of measuring the pH of a selection of liquids and beverages. Fresh lemon juice, as we have seen, has a pH of 2.5, corresponding to an H_3O^+ concentration of 3×10^{-3} mol·L^{-1}. Natural (unpolluted) rain, with an acidity due largely to dissolved carbon dioxide, typically has a pH of about 5.7. In the United States, the Environmental Protection Agency (EPA) defines waste as "corrosive" if its pH is either lower than 3.0 (highly acidic) or higher than 12.5 (highly basic). Lemon juice, with a pH of about 2.5, would be regarded as corrosive; so would solutions of 1 M NaOH(aq), for which the pH is 14.

The pH scale is used to report H_3O^+ concentration: pH = −log $[H_3O^+]$; pH > 7 denotes a basic solution, pH < 7 an acidic solution; a neutral solution has pH = 7.

10.6 The pOH of Solutions

Many quantitative expressions relating to acids and bases are greatly simplified and more easily remembered when we use logarithms. The quantity pX is a generalization of pH:

$$pX = -\log X \tag{4}$$

For example, **pOH** is defined as

$$pOH = -\log a_{OH^-} \tag{5a}$$

which, for the same reasons as those for pH, is normally simplified to

$$pOH = -\log [OH^-] \tag{5b}*$$

The pOH is convenient for reporting the concentration of OH^- ions in solution. For example, in pure water, where $[OH^-] = 1.0 \times 10^{-7}$ mol·L^{-1}, the pOH is 7.00. Similarly, by pK_w we mean

$$pK_w = -\log K_w = -\log(1.0 \times 10^{-14}) = 14.00 \text{ (at 25°C)}$$

The values of pH and pOH are related. To find that relation, we start with the expression for the autoprotolysis constant of water: $K_w = [H_3O^+][OH^-]$. Then we take logarithms of both sides:

$$\log K_w = \log [H_3O^+][OH^-]$$

Now we use log ab = log a + log b and rearrange the equation to obtain:

$$\log [H_3O^+] + \log [OH^-] = \log K_w$$

Multiplication of both sides of the equation by −1 gives

$$-\log [H_3O^+] - \log [OH^-] = -\log K_w$$

which is the same as

$$pH + pOH = pK_w \tag{6a}*$$

Because $pK_w = 14.00$ at 25°C, at that temperature

$$pH + pOH = 14.00 \tag{6b}$$

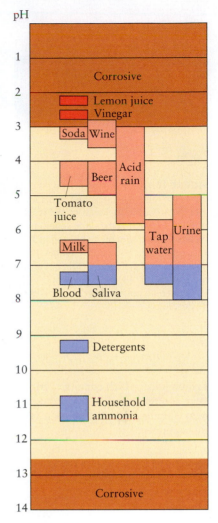

FIGURE 10.12 Typical pH values of some common aqueous solutions. The rust-colored regions indicate the pH ranges for liquids regarded as corrosive.

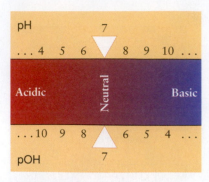

FIGURE 10.13 The numbers along the top of the rectangle are values of pH for a range of solutions. Below each pH value is the pOH value of the same solution. Notice that the sum of the pH and pOH values for a given solution is always 14. Most values of pH and pOH lie in the range from 1 to 14, but pH and pOH values can lie outside this range and even be negative.

Lemon juice, which contains citric acid, is often added to fish dishes to help eliminate the smell of some of these amines.

Equation 6 shows that the pH and pOH of a solution have complementary values: if one increases, the other decreases such that their sum remains constant (Fig. 10.13).

The pH and pOH of a solution are related by the expression $pH + pOH = pK_w$.

WEAK ACIDS AND BASES

Solutions of different acids having the same concentration might not have the same pH. For instance, the pH of 0.10 M $CH_3COOH(aq)$ is close to 3; but that of 0.10 M $HCl(aq)$ is close to 1. We have to conclude that the concentration of H_3O^+ ions in 0.10 M $CH_3COOH(aq)$ is *lower* than that in 0.10 M $HCl(aq)$. Similarly, we find that the concentration of OH^- ions is lower in 0.10 M $NH_3(aq)$ than it is in 0.10 M $NaOH(aq)$. The explanation must be that in water CH_3COOH is not fully deprotonated and NH_3 is not fully protonated. That is, acetic acid and ammonia are, respectively, a weak acid and a weak base. The incomplete deprotonation of CH_3COOH explains why solutions of HCl and CH_3COOH with the same molarity react with a metal at different rates (Fig. 10.14).

Most acids and bases that exist in nature are weak. For example, the natural acidity of river water is generally due to the presence of carbonic acid (H_2CO_3, from dissolved CO_2), hydrogen phosphate ions, HPO_4^{2-}, and dihydrogen phosphate ions, $H_2PO_4^-$ (from fertilizer runoff), and carboxylic acids arising from the degradation of plant tissues. Similarly, most naturally occurring bases are weak. Some arise from the decomposition in the absence of air of compounds containing nitrogen. For example, the odor of dead fish is due to amines, which are weak bases.

In this part of the chapter, we develop a quantitative measure of the strengths of weak acids and bases. We then use this information to explore how acid strength is related to molecular structure.

10.7 Acidity and Basicity Constants

When we visualize the molecular composition of a solution of a weak acid in water, we think of a solution that contains

*   the acid molecules or ions;
*   small concentrations of H_3O^+ ions and the conjugate base of the acid formed by proton transfer to water molecules; and

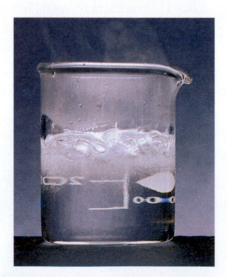
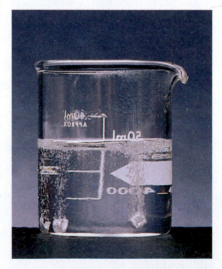

FIGURE 10.14 The same mass of magnesium metal has been added to solutions of HCl, a strong acid (left), and CH_3COOH, a weak acid (right). Although the acid solutions were prepared with the same concentrations, the rate of hydrogen evolution, which depends on the concentration of hydronium ions, is much greater in the strong acid.

- a very, very small concentration of OH⁻ ions, which maintain the autoprotolysis equilibrium.

All these species are in ceaseless dynamic equilibrium. Similarly, for a solution of a weak base, we visualize

- the base molecules or ions;
- small concentrations of OH⁻ ions and the conjugate acid of the base formed by proton transfer from water molecules; and
- a very, very small concentration of H_3O^+ ions that maintain the autoprotolysis equilibrium.

Because conjugate acids and bases are in equilibrium in solution, we use the equilibrium constant for proton transfer between the solute and the solvent as an indicator of the strength of an acid or a base. For example, for acetic acid in water,

$$CH_3COOH(aq) + H_2O(l) \rightleftharpoons H_3O^+(aq) + CH_3CO_2^-(aq) \qquad (B)$$

and the equilibrium constant is

$$K = \frac{a_{H_3O^+}\, a_{CH_3CO_2^-}}{a_{CH_3COOH}\, a_{H_2O}} \qquad (7)$$

Because the solutions that we are considering are dilute and the water is almost pure, the activity of H_2O can be set equal to 1. The resulting expression is called an **acidity constant**, K_a. If we make the further approximation of replacing the activities of the solute species by the numerical values of their molar concentrations, we can write the acidity constant expression for acetic acid as

$$K_a = \frac{[H_3O^+][CH_3CO_2^-]}{[CH_3COOH]}$$

The experimental value of K_a for acetic acid at 25°C is 1.8×10^{-5}. This small value tells us that only a small proportion of CH_3COOH molecules donate their protons when dissolved in water. About 99 of 100 CH_3COOH molecules may remain intact, depending on the concentration. This value is typical of weak acids in water (Fig. 10.15). In general, the acidity constant for an acid HA is

$$HA(aq) + H_2O(l) \rightleftharpoons H_3O^+(aq) + A^-(aq) \qquad K_a = \frac{[H_3O^+][A^-]}{[HA]} \qquad (8)^*$$

Table 10.1 lists the acidity constants of some other weak acids in aqueous solution.
We can also write an equilibrium constant for the proton transfer equilibrium of a base in water. For aqueous ammonia, for instance,

$$NH_3(aq) + H_2O(l) \rightleftharpoons NH_4^+(aq) + OH^-(aq) \qquad (C)$$

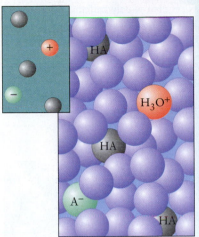

FIGURE 10.15 In a solution of a weak acid, only some of the acidic hydrogen atoms are present as hydronium ions (the red sphere), and the solution contains a high proportion of the original acid molecules (HA, gray spheres). The green sphere represents the conjugate base of the acid and the blue spheres are water molecules. The overlay shows only the solute species.

K_a is also widely called the acid ionization constant or acid dissociation constant.

TABLE 10.1 Acidity Constants at 25°C**

Acid	K_a	pK_a	Acid	K_a	pK_a
trichloroacetic acid, CCl₃COOH	3.0×10^{-1}	0.52	formic acid, HCOOH	1.8×10^{-4}	3.75
benzene sulfonic acid, C₆H₅SO₃H	2.0×10^{-1}	0.70	benzoic acid, C₆H₅COOH	6.5×10^{-5}	4.19
iodic acid, HIO₃	1.7×10^{-1}	0.77	acetic acid, CH₃COOH	1.8×10^{-5}	4.75
sulfurous acid, H₂SO₃	1.5×10^{-2}	1.81	carbonic acid, H₂CO₃	4.3×10^{-7}	6.37
chlorous acid, HClO₂	1.0×10^{-2}	2.00	hypochlorous acid, HClO	3.0×10^{-8}	7.53
phosphoric acid, H₃PO₄	7.6×10^{-3}	2.12	hypobromous acid, HBrO	2.0×10^{-9}	8.69
chloroacetic acid, CH₂ClCOOH	1.4×10^{-3}	2.85	boric acid, B(OH)₃†	7.2×10^{-10}	9.14
lactic acid, CH₃CH(OH)COOH	8.4×10^{-4}	3.08	hydrocyanic acid, HCN	4.9×10^{-10}	9.31
nitrous acid, HNO₂	4.3×10^{-4}	3.37	phenol, C₆H₅OH	1.3×10^{-10}	9.89
hydrofluoric acid, HF	3.5×10^{-4}	3.45	hypoiodous acid, HIO	2.3×10^{-11}	10.64

*The values for K_a listed here have been calculated from pK_a values with more significant figures than shown so as to minimize rounding errors. Values for polyprotic acids—those capable of donating more than one proton—refer to the first deprotonation.
†The proton transfer equilibrium is $B(OH)_3(aq) + 2\,H_2O(l) \rightleftharpoons H_3O^+(aq) + B(OH)_4^-(aq)$.

Animation: Figure 10.15 Acetic acid in water

and the equilibrium constant is

$$K = \frac{a_{NH_4^+} \, a_{OH^-}}{a_{H_2O} \, a_{NH_3}}$$

K_b is also widely called the *base ionization constant.*

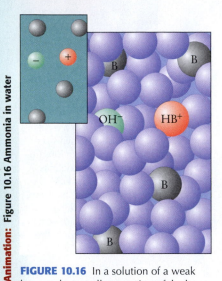

Animation: Figure 10.16 Ammonia in water

FIGURE 10.16 In a solution of a weak base, only a small proportion of the base molecules (B, represented here by gray spheres) have accepted protons from water molecules (the blue spheres) to form HB^+ ions (the red sphere) and OH^- ions (green sphere). The overlay shows only the solute species.

In dilute solutions, the water is almost pure and its activity can be set equal to 1. With this approximation, we obtain the **basicity constant**, K_b. If we make the further approximation of replacing the activities of the solute species by the numerical values of their molar concentrations, we can write the basicity constant expression for ammonia as

$$K_b = \frac{[NH_4^+][OH^-]}{[NH_3]}$$

The experimental value of K_b for ammonia in water at 25°C is 1.8×10^{-5}. This small value tells us that normally only a small proportion of the NH_3 molecules are present as NH_4^+. Equilibrium calculations show that only about 1 in 100 molecules are protonated in a typical solution (Fig. 10.16). In general, the basicity constant for a base B in water is

$$B(aq) + H_2O(l) \rightleftharpoons HB^+(aq) + OH^-(aq) \qquad K_b = \frac{[HB^+][OH^-]}{[B]} \qquad (9)^*$$

The value of K_b tells us how far the reaction proceeds to the right. The smaller the value of K_b the weaker is the ability of the base to accept a proton. Table 10.2 lists the basicity constants of some weak bases in aqueous solution.

Acidity and basicity constants are commonly reported as their negative logarithms, by defining

$$pK_a = -\log K_a \qquad pK_b = -\log K_b \qquad (10)^*$$

We can easily remember the relationship between K_a and acid strength if we recall that:

- The smaller the value of K_a, the greater the value of pK_a and the weaker the acid.

For example, the pK_a of trichloroacetic acid is 0.5, whereas that of acetic acid, a much weaker acid, is nearly 5. Similar remarks apply to bases:

- The smaller the value of K_b, the greater the value of pK_b and the weaker the base.

Values of pK_a and pK_b are included in Tables 10.1 and 10.2.

> *The proton-donating strength of an acid is measured by its acidity constant; the proton-accepting strength of a base is measured by its basicity constant. The smaller the constants, the weaker the respective strengths. The larger the value of pK, the weaker the acid or base.*

TABLE 10.2 Basicity Constants at 25°C*

Base	K_b	pK_b	Base	K_b	pK_b
urea, $CO(NH_2)_2$	1.3×10^{-14}	13.90	ammonia, NH_3	1.8×10^{-5}	4.75
aniline, $C_6H_5NH_2$	4.3×10^{-10}	9.37	trimethylamine, $(CH_3)_3N$	6.5×10^{-5}	4.19
pyridine, C_5H_5N	1.8×10^{-9}	8.75	methylamine, CH_3NH_2	3.6×10^{-4}	3.44
hydroxylamine, NH_2OH	1.1×10^{-8}	7.97	dimethylamine, $(CH_3)_2NH$	5.4×10^{-4}	3.27
nicotine, $C_{10}H_{14}N_2$	1.0×10^{-6}	5.98	ethylamine, $C_2H_5NH_2$	6.5×10^{-4}	3.19
morphine, $C_{17}H_{19}O_3N$	1.6×10^{-6}	5.79	triethylamine, $(C_2H_5)_3N$	1.0×10^{-3}	2.99
hydrazine, NH_2NH_2	1.7×10^{-6}	5.77			

*The values for K_b listed here have been calculated from pK_b values with more significant figures than shown so as to minimize rounding errors.

10.8 The Conjugate Seesaw

Hydrochloric acid is classified as a strong acid because it is fully deprotonated in water. It follows that its conjugate base, Cl^-, must be a very, very weak proton acceptor (weaker, in fact, than H_2O). Conversely, acetic acid is a weak acid. Its conjugate base, the acetate ion, $CH_3CO_2^-$, must be a relatively good proton acceptor, because it readily forms CH_3COOH molecules in water. Similarly, because methylamine, CH_3NH_2, is a stronger base than ammonia (see Table 10.2), the conjugate acid of methylamine—the methylammonium ion, $CH_3NH_3^+$—must be a weaker proton donor (and therefore a weaker acid) than NH_4^+. In general,

- the stronger the acid, the weaker its conjugate base; and
- the stronger the base, the weaker its conjugate acid.

To express the relative strengths of an acid and its conjugate base (a "conjugate acid–base pair"), we consider the special case of the ammonia proton transfer equilibrium, reaction C, for which the basicity constant was given earlier ($K_b = [NH_4^+][OH^-]/[NH_3]$). Now let's consider the proton transfer equilibrium of ammonia's conjugate acid, NH_4^+, in water:

$$NH_4^+(aq) + H_2O(l) \rightleftharpoons H_3O^+(aq) + NH_3(aq) \qquad K_a = \frac{[H_3O^+][NH_3]}{[NH_4^+]}$$

Multiplication of the two equilibrium constants for ammonia gives

$$K_a \times K_b = \frac{[H_3O^+][\cancel{NH_3}]}{[\cancel{NH_4^+}]} \times \frac{[\cancel{NH_4^+}][OH^-]}{[\cancel{NH_3}]} = [H_3O^+][OH^-]$$

We recognize K_w on the right; so we can write

$$K_a \times K_b = K_w \qquad \text{(11a)}*$$

In this expression, K_a is the acidity constant of a weak acid and K_b is the basicity constant of the conjugate base of that acid. The acid and base must form a conjugate acid–base pair (such as $CH_3COOH/CH_3CO_2^-$ or NH_4^+/NH_3). We can express Eq. 11a in another way by taking logarithms of both sides of the equation:

$$\log K_a + \log K_b = \log K_w$$

Multiplication throughout by -1 turns this expression into

$$pK_a + pK_b = pK_w \qquad \text{(11b)}*$$

SELF-TEST 10.6A Write the chemical formula for the conjugate acid of the base pyridine, C_5H_5N, and calculate its pK_a.

[*Answer:* $C_5H_5NH^+$, 5.25]

SELF-TEST 10.6B Write the chemical formula for the conjugate base of the acid HIO_3 and calculate its pK_b.

Equation 11—in either form—confirms the seesaw relation between the strengths of acids and of their conjugate bases. Equation 11a tells us that if an acid has a large K_a, then its conjugate base must have a small K_b. Similarly, if a base has a large K_b, then its conjugate acid must have a small K_a. Equation 11b tells us that if pK_a of an acid is large, then the pK_b of its conjugate base is small, and vice versa. This reciprocal relation is summarized in Fig. 10.17 and Table 10.3. For example, because pK_b for ammonia in water is 4.75, the pK_a of NH_4^+ is

$$pK_a = pK_w - pK_b = 14.00 - 4.75 = 9.25$$

This value shows that NH_4^+ is a weaker acid than boric acid ($pK_a = 9.14$) but stronger than hydrocyanic acid (HCN, $pK_a = 9.31$).

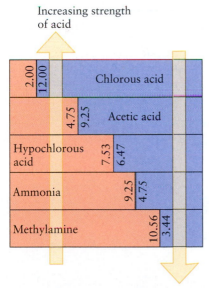

FIGURE 10.17 As shown here for five conjugate acid–base pairs, the sum of the pK_a of an acid (pink) and the pK_b of its conjugate base (blue) is constant and equal to pK_w, which is 14.00 at 25°C.

TABLE 10.3 Conjugate Acid–Base Pairs Arranged by Strength

pK_a	Acid name	Acid formula	Base formula	Base name	pK_b
	Strong acid			*Very weak base*	
	hydroiodic acid	HI	I$^-$	iodide ion	
	perchloric acid	HClO4	ClO$_4^-$	perchlorate ion	
	hydrobromic acid	HBr	Br$^-$	bromide ion	
	hydrochloric acid	HCl	Cl$^-$	chloride ion	
	sulfuric acid	H$_2$SO$_4$	HSO$_4^-$	hydrogen sulfate ion	
	chloric acid	HClO$_3$	ClO$_3^-$	chlorate ion	
	nitric acid	HNO$_3$	NO$_3^-$	nitrate ion	
	hydronium ion	H$_3$O$^+$	H$_2$O	*water*	
1.92	hydrogen sulfate ion	HSO$_4^-$	SO$_4^{2-}$	sulfate ion	12.08
3.37	nitrous acid	HNO$_2$	NO$_2^-$	nitrite ion	10.63
3.45	hydrofluoric acid	HF	F$^-$	fluoride ion	10.55
4.75	acetic acid	CH$_3$COOH	CH$_3$CO$_2^-$	acetate ion	9.25
6.37	carbonic acid	H$_2$CO$_3$	HCO$_3^-$	hydrogen carbonate ion	7.63
6.89	hydrosulfuric acid	H$_2$S	HS$^-$	hydrogen sulfide ion	7.11
9.25	ammonium ion	NH$_4^+$	NH$_3$	ammonia	4.75
9.31	hydrocyanic acid	HCN	CN$^-$	cyanide ion	4.69
10.25	hydrogen carbonate ion	HCO$_3^-$	CO$_3^{2-}$	carbonate ion	3.75
	methylammonium ion	CH$_3$NH$_3^+$	CH$_3$NH$_2$	methylamine	3.44
	water	H$_2$O	OH$^-$	*hydroxide ion*	
	ammonia	NH$_3$	NH$_2^-$	amide ion	
	hydrogen	H$_2$	H$^-$	hydride ion	
	methane	CH$_4$	CH$_3^-$	methide ion	
	hydroxide ion	OH$^-$	O^{2-}	oxide ion	
	Very weak acid			*Strong base*	

Although the proton transfer in a solution of a strong acid is actually an equilibrium, the proton-donating power of a strong acid, HA, is so much greater than that of H$_3$O$^+$ that proton transfer to water effectively goes to completion. As a result, the solution contains only H$_3$O$^+$ ions and A$^-$ ions; there are virtually no HA molecules left. In other words, the only acid species present in an aqueous solution of a strong acid, other than the H$_2$O molecules, is the H$_3$O$^+$ ion. Because all strong acids in water behave as though they were solutions of the acid H$_3$O$^+$, we say that strong acids are **leveled** in water to the strength of the acid H$_3$O$^+$.

Sulfuric acid is a special case, because loss of its first acidic hydrogen leaves a weak acid, the HSO$_4^-$ ion.

EXAMPLE 10.5 Deciding which of two species is the stronger acid or base

Decide which member of each of the following pairs is the stronger acid or base in water: (a) acid: HF or HIO$_3$; (b) base: NO$_2^-$ or CN$^-$.

STRATEGY The greater the K_a of a weak acid, the stronger is the acid and the weaker is its conjugate base. Similarly, the greater the K_b of a weak base, the stronger is the base and the weaker is its conjugate acid.

SOLUTION Compare the relevant K_a and K_b (or pK_a and pK_b) values in Tables 10.1 and 10.2.

(a) Because K_a(HIO$_3$) > K_a(HF), or pK_a(HIO$_3$) < pK_a(HF), HIO$_3$ is a stronger acid than HF.

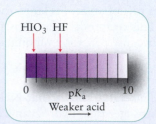

(b) Because $K_a(HNO_2) > K_a(HCN)$, or $pK_a(HNO_2) < pK_a(HCN)$, and because the stronger acid has the weaker conjugate base, NO_2^- is a weaker base than CN^-. Hence, CN^- is the stronger base.

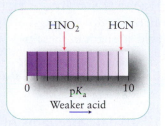

SELF-TEST 10.7A Use Tables 10.1 and 10.2 to decide which species in each of the following pairs is the stronger acid or base: (a) acid: HF or HIO; (b) base: $C_6H_5CO_2^-$ or $CH_2ClCO_2^-$; (c) base: $C_6H_5NH_2$ or $(CH_3)_3N$; (d) acid: $C_6H_5NH_3^+$ or $(CH_3)_3NH^+$.
[***Answer:*** Stronger acids: (a) HF; (d) $C_6H_5NH_3^+$.
Stronger bases: (b) $C_6H_5CO_2^-$; (c) $(CH_3)_3N$]

SELF-TEST 10.7B Use Tables 10.1 and 10.2 to decide which species in each of the following pairs is the stronger acid or base: (a) base: C_5H_5N or NH_2NH_2; (b) acid: $C_5H_5NH^+$ or $NH_2NH_3^+$; (c) acid: HIO_3 or $HClO_2$; (d) base: ClO_2^- or HSO_3^-.

The stronger the acid, the weaker its conjugate base; the stronger the base, the weaker its conjugate acid.

10.9 Molecular Structure and Acid Strength

Chemists commonly interpret trends in the properties of compounds by considering the structures of their molecules. However, there are two reasons why the relative strengths of acids and bases are difficult to predict. First, K_a and K_b are equilibrium constants, and so they are related to the Gibbs free energy of the proton transfer reaction. Their values therefore depend on considerations of entropy as well as energy. Second, the solvent plays a significant role in the proton transfer reactions, and so we cannot expect to relate acid strength solely to the molecule itself. However, although absolute values are difficult to predict, we can look for trends among series of compounds with similar structures. Because acid strength in aqueous solutions depends on the breaking of the H—A bond and the formation of an H—OH$_2^+$ bond, we might suspect that one factor in determining strength is the ease with which these bonds can be broken and formed.

Here is an opportunity for you to come up with a theory.

First, we consider *binary acids*, acids composed of hydrogen and one other element, such as HCl and H$_2$S and, in general, HA. Hydronium ions are generated when an acidic proton initially forms a hydrogen bond to a water molecule and then migrates completely to it. The stronger the hydrogen bond that HA forms with the O atom of an H$_2$O molecule, the more likely it is that the proton will be transferred. We know that the more polar the H—A bond, the greater the partial positive charge on the H atom—therefore the stronger the H$_2$O⋯H—A hydrogen bond and the more readily HA loses its proton. Because the polarity of the H—A bond increases with the electronegativity of A, we can predict that *the greater the electronegativity of A, the stronger the acid HA*. For instance, the electronegativity difference is 0.8 for the N—H bond and 1.8 for F—H (see Fig. 2.12); therefore, the F—H bond is markedly more polar than the N—H bond. This difference is consistent with the observation that HF, but not NH$_3$, is an acid in water. In general, bond polarity dominates the trend of acid strengths for binary acids of elements of the same period.

When we compare the relative strengths of the binary acids within the same group, we find a different pattern. For example, although bond polarities in the hydrogen halides decrease down the group, the acid strengths increase down the group: HF < HCl < HBr < HI. Another factor must be affecting the acidity. In this case, the order of acid strength is consistent with the weakening of the H—A bond down the group (recall Fig. 2.18). The weaker the H—A bond, the easier it is for the proton to leave and the stronger the acid HA. The Group 16/VI acids exhibit the same trend in aqueous solution: the acid strengths lie in the order

Because all the hydrohalic acids except HF are strong acids, they are leveled in water. Therefore, to determine their relative strengths, they must be studied in a solvent that is a poorer proton acceptor than water (such as pure acetic acid).

$H_2O < H_2S < H_2Se < H_2Te$. Because bond strengths decrease and acid strength increases down the group, it appears that bond strength dominates the trend in acid strength in these two sets of binary acids.

> *The more polar the H—A bond, the stronger the acid on going across a row in the periodic table; the weaker the bond, the stronger the acid on going down a group.*

10.10 The Strengths of Oxoacids and Carboxylic Acids

We can learn more about the effect of structure on acidity by considering the oxoacids. These acids form structurally related families, and so we can examine the effect of different central atoms with the same number of O atoms (as in $HClO_3$ and $HBrO_3$). Alternatively, we can look at the influence of different numbers of O atoms attached to the same central atom (as in $HClO_3$ and $HClO_4$).

The high polarity of the O—H bond is one reason why the proton of an —OH group in an oxoacid molecule is acidic. For example, phosphorous acid, H_3PO_3, has the structure $(HO)_2PHO$ (**1**): it can donate the protons from its two —OH groups but not the proton attached directly to the phosphorus atom. The difference in behavior can be traced to the much lower electronegativity of phosphorus relative to that of oxygen. In nearly all common inorganic oxoacids, however, all the hydrogen atoms are attached to oxygen atoms.

Let's consider a family of oxoacids in which the number of O atoms is constant, as in the hypohalous acids HClO, HBrO, and HIO. By referring to Table 10.4, we see that, *the greater the electronegativity of the halogen, the stronger the oxoacid.* A partial explanation of this trend is that electrons are withdrawn slightly from the O—H bond as the electronegativity of the halogen increases. As these bonding electrons move toward the central atom, the O—H bond becomes more polar, and so the molecule becomes a stronger acid. A halogen that has a high electronegativity also weakens the conjugate base by making the electrons in the molecule less accessible to an incoming proton.

Now let's consider a family of oxoacids in which the number of oxygen atoms varies, as in the chlorine oxoacids HClO, $HClO_2$, $HClO_3$, and $HClO_4$ or in the

1 Phosphorous acid, H_3PO_3

TABLE 10.4 Correlation of Acid Strength and Electronegativity

Acid, HXO	Structure*	Electronegativity of atom X	pK_a	
hypochlorous acid, HClO	$:\overset{..}{\underset{..}{Cl}}—\overset{..}{\underset{..}{O}}—H$	3.2	7.53	
hypobromous acid, HBrO	$:\overset{..}{\underset{..}{Br}}—\overset{..}{\underset{..}{O}}—H$	3.0	8.69	
hypoiodous acid, HIO	$:\overset{..}{\underset{..}{I}}—\overset{..}{\underset{..}{O}}—H$	2.7	10.64	

*The red arrows indicate the direction of the shift of electron density away from the O—H bond.

TABLE 10.5 **Correlation of Acid Strength and Oxidation Number**

Acid	Structure*	Oxidation number of chlorine atom	pK_a	
hypochlorous acid, HClO	 $:\ddot{Cl}-\ddot{O}-H$	+1	7.53	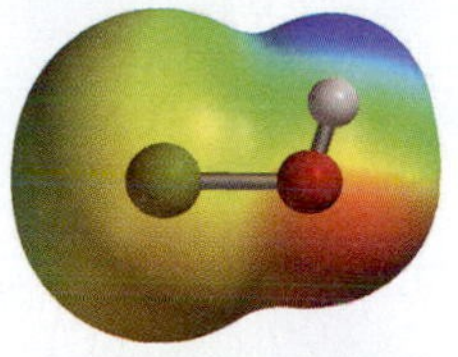
chlorous acid, HClO$_2$	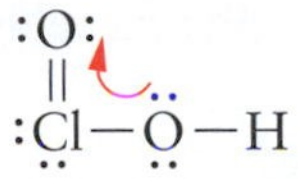 $:\ddot{Cl}-\ddot{O}-H$ with $=O$	+3	2.00	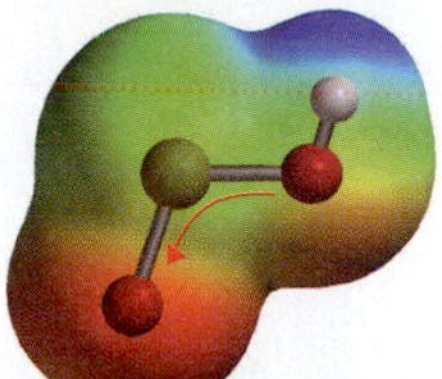
chloric acid, HClO$_3$	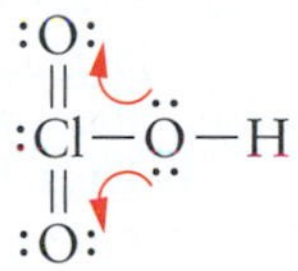 $:\ddot{Cl}-\ddot{O}-H$ with two $=O$	+5	strong	
perchloric acid, HClO$_4$	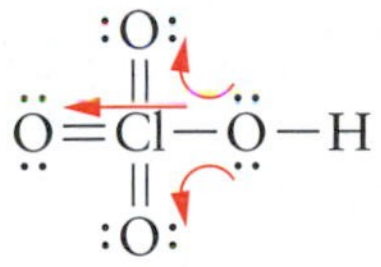 $O=\ddot{Cl}-\ddot{O}-H$ with additional O atoms	+7	strong	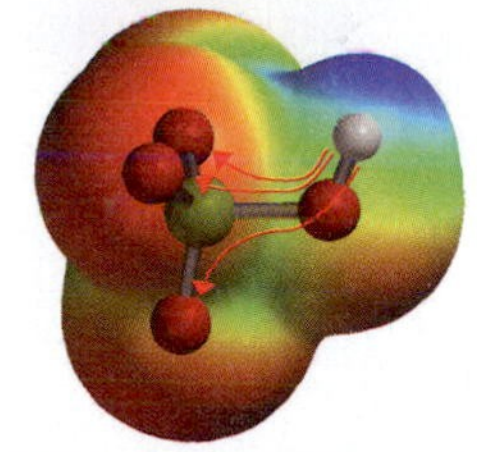 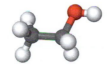

*The red arrows indicate the direction of the shift of electron density away from the O—H bond. The Lewis structures shown are the ones with the most favorable formal charges, but it is unlikely that the bond orders are as high as these structures suggest.

sulfur oxoacids H_2SO_3 and H_2SO_4. When we refer to Table 10.5, we see that, *the greater the number of oxygen atoms attached to the central atom, the stronger the acid.* Because the oxidation number of the central atom increases as the number of O atoms increases, we can also conclude that, *the greater the oxidation number of the central atom, the stronger the acid.*

The effect of the number of O atoms on the strengths of organic acids is similar. For example, we have seen (Section D) that alcohols are organic compounds in which an —OH group is attached to a carbon atom, as in ethanol (**2**). Carboxylic acids, on the other hand, have two O atoms attached to the same carbon atom: one is a doubly bonded terminal O atom and the other is the O atom of an —OH group, as in acetic acid (**3**). Although carboxylic acids are weak acids, they are much stronger acids than alcohols, partly as a result of the electron-withdrawing power of the second O atom. In fact, alcohols have such weak proton-donating power that usually they are not regarded as oxoacids.

The strength of a carboxylic acid is also increased relative to that of an alcohol by electron delocalization in the conjugate base. The second O atom of the carboxyl group, —COOH, provides an additional electronegative atom over which the negative charge of the conjugate base can spread. This electron delocalization stabilizes the carboxylate anion, —CO$_2^-$ (**4**). Moreover, because the charge is spread over several atoms, it is less effective at attracting a proton. A carboxylate ion is therefore a much weaker base than the conjugate base of an alcohol (for example, the ethoxide ion, $CH_3CH_2O^-$).

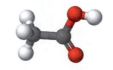

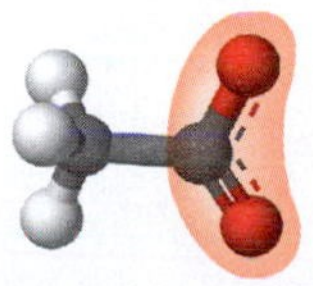

2 Ethanol, CH_3CH_2OH

3 Acetic acid, CH_3COOH

4 Acetate ion, $CH_3CO_2^-$

A note on good practice: The formula of a carboxylic acid is written RCOOH, because the two O atoms are distinct (one is part of an OH group); the formula of a carboxylate ion, however, is written RCO_2^- because the two O atoms are equivalent.

The strengths of carboxylic acids also vary with the total electron-withdrawing power of the atoms bonded to the carboxyl group. Because hydrogen is less electronegative than chlorine, the $-CH_3$ group bonded to $-COOH$ in acetic acid is less electron withdrawing than the $-CCl_3$ group in trichloroacetic acid. Therefore, we expect CCl_3COOH to be a stronger acid than CH_3COOH. In agreement with this prediction, the pK_a of acetic acid is 4.75, whereas that of trichloroacetic acid is 0.52.

EXAMPLE 10.6   **Sample exercise: Predicting relative acid strength from molecular structure**

Predict from their molecular structures which acid in each of the following pairs is the stronger: (a) H_2S and H_2Se; (b) H_2SO_4 and H_2SO_3; (c) H_2SO_4 and H_3PO_4.

SOLUTION   Refer to the summary in Table 10.6. (a) Sulfur and selenium are in the same group, and we expect the H—Se bond to be weaker than the H—S bond. Thus, H_2Se can be expected to be the stronger acid. (b) H_2SO_4 has the greater number of O atoms bonded to the S atom and the oxidation number of sulfur is +6, whereas in H_2SO_3 the sulfur has an oxidation number of only +4. Thus, H_2SO_4 is expected to be the stronger acid. (c) Both acids have four O atoms bonded to the central atom, but the electronegativity of sulfur is greater than that of phosphorus, and so H_2SO_4 is expected to be the stronger acid.

SELF-TEST 10.8A   For each of the following pairs, predict which acid is stronger: (a) H_2S and HCl; (b) HNO_2 and HNO_3; (c) H_2SO_3 and $HClO_3$.

[*Answer:* (a) HCl; (b) HNO_3; (c) $HClO_3$]

SELF-TEST 10.8B   List the following carboxylic acids in order of increasing strength: $CHCl_2COOH$, CH_3COOH, and $CH_2ClCOOH$.

The greater the number of oxygen atoms and the more electronegative the atoms present in the molecules of an acid, the stronger the acid. These trends are summarized in Table 10.6.

THE pH OF SOLUTIONS OF WEAK ACIDS AND BASES

The rest of this chapter is a variation on a theme introduced in Chapter 9: the use of equilibrium constants to calculate the equilibrium composition of solutions of acids, bases, and salts. We shall see how to predict the pH of solutions of weak acids and bases and how to calculate the extent of deprotonation of a weak acid and the extent of protonation of a weak base. We shall also see how to calculate the pH of a solution of a salt in which the cation or anion of the salt may itself be a weak acid or base.

10.11 Solutions of Weak Acids

The initial concentration is sometimes called the *analytical concentration.*

Our first task is to calculate the pH of a solution of a weak acid, such as acetic acid in water. The **initial concentration** of the acid is its concentration as prepared, as if no acid molecules had donated any protons. For a strong acid HA, the H_3O^+ concentration in solution is the same as the initial concentration of the strong acid, because all the HA molecules are deprotonated. However, to find the H_3O^+ concentration in a solution of a weak acid HA, we have to take into account the equilibrium between the acid HA, its conjugate base A^-, and water (Eq. 8). We can expect the pH to lie somewhere between 7, a value indicating no deprotonation, and the value that we would calculate for a strong acid, which undergoes complete deprotonation. The technique, which is based on the use of an equilibrium table like those introduced in Chapter 9, is set out in Toolbox 10.1.

TABLE 10.6   **Correlations of Molecular Structure and Acid Strength**

Acid Type	Trend	
binary	The more polar the H—A bond, the stronger the acid. *This effect is dominant for acids of the same period.*	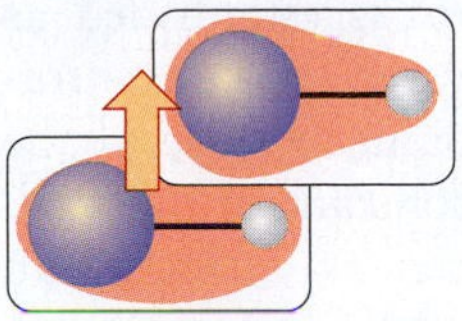
	The weaker the H—A bond, the stronger the acid. *This effect is dominant for acids of the same group.*	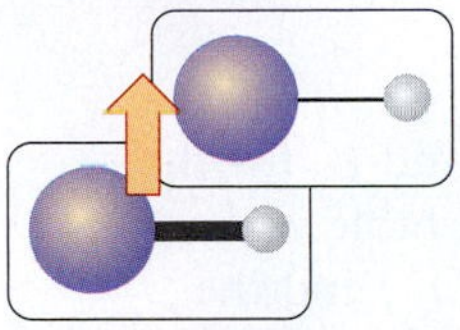
oxoacid	The greater the number of O atoms attached to the central atom (the greater the oxidation number of the central atom), the stronger the acid.	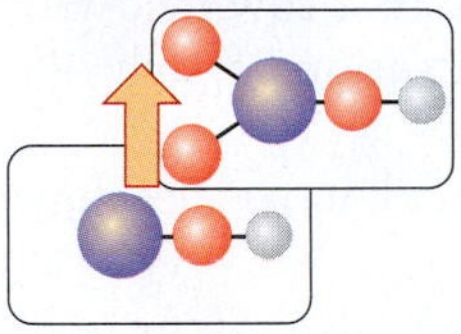
	For the same number of O atoms attached to the central atom, then the greater the electronegativity of the central atom, the stronger the acid.	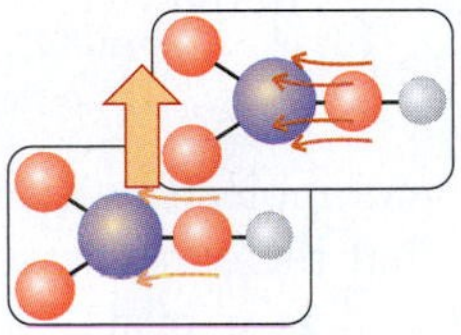
carboxylic	The greater the electronegativities of the groups attached to the carboxyl group, the stronger the acid.	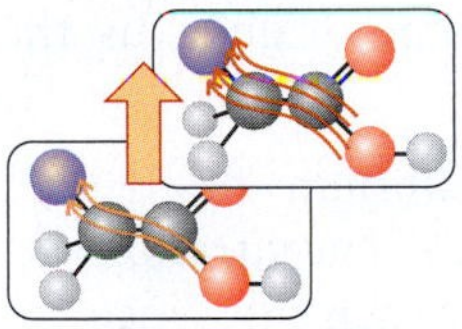

In each diagram, the vertical orange arrow indicates the corresponding increase in acid strength.

The calculation summarized in Toolbox 10.1 also allows us to predict the **percentage deprotonation**, the percentage of HA molecules that are deprotonated in the solution.

$$\text{Percentage deprotonated} = \frac{\text{concentration of A}^-}{\text{initial concentration of HA}} \times 100\% \qquad (12a)^*$$

To express the percentage deprotonation in terms of the pH of the solution, we use the equality $[H_3O^+] = [A^-]$, which follows from the stoichiometric relation 1 mol $A^- \simeq 1$ mol H_3O^+ for the deprotonation reaction in Eq. 8. Then

$$\text{Percentage deprotonated} = \frac{[H_3O^+]}{[HA]_{\text{initial}}} \times 100\% \qquad (12b)^*$$

A small percentage of deprotonated molecules indicates that the acid HA is very weak.

The autoprotolysis of water contributes significantly to the pH when the acid is so dilute or so weak that the calculation predicts an H_3O^+ concentration close to 10^{-7} mol·L^{-1}. In such cases, we must use the procedures described in Sections 10.18 and 10.19. We can ignore the contribution of the autoprotolysis of water in an acidic solution only when the calculated H_3O^+ concentration is substantially (about 10 times) higher than 10^{-7} mol·L^{-1}, corresponding to a pH of 6 or less.

TOOLBOX 10.1   HOW TO CALCULATE THE pH OF A SOLUTION OF A WEAK ACID

CONCEPTUAL BASIS

Because a proton transfer equilibrium is established as soon as a weak acid is dissolved in water, the concentrations of acid, hydronium ion, and conjugate base of the acid must always satisfy the acidity constant of the acid. We can calculate any of these quantities by setting up an equilibrium table like that in Toolbox 9.1.

PROCEDURE

Step 1 Write the chemical equation and K_a for the proton transfer equilibrium. Set up a table with columns labeled by the acid HA, H_3O^+, and the conjugate base of the acid, A^-. In the first row below the headings, show the initial concentration of each species.

For this step, assume that no acid molecules have been deprotonated.

Step 2 In the second row, write the changes in the concentration needed for the reaction to reach equilibrium.

We do not know the number of acid molecules that lose their protons, and so we assume that the concentration of the acid decreases by x mol·L^{-1} as a result of deprotonation. The reaction stoichiometry gives us the other changes in terms of x.

Step 3 In the third row write the equilibrium concentrations by adding the change in concentration (step 2) to the initial concentration for each substance (step 1).

	Acid, HA	H_3O^+	Conjugate base, A^-
Step 1 Initial concentration	$[HA]_{initial}$	0	0
Step 2 Change in concentration	$-x$	$+x$	$+x$
Step 3 Equilibrium concentration	$[HA]_{initial} - x$	x	x

Although a *change* in concentration may be positive (an increase) or negative (a decrease), the value of the concentration must always be positive.

Step 4 Use the value of K_a to calculate the value of x.

The calculation of x can often be simplified, as shown in Toolbox 9.1, by ignoring changes of less than 5% of the initial concentration of the acid. However, at the end of the calculation, we must check that x is consistent with the approximation, by calculating the percentage of acid deprotonated. If this percentage is greater than 5%, then the exact expression for K_a must be solved for x. An exact calculation requires solving a quadratic equation, as explained in Toolbox 9.1. If the pH is greater than 6, the autoprotolysis of water needs to be taken into account. Although we calculate the pH to the number of significant figures appropriate to the data, the answers are often considerably less reliable than that. One reason for this poor reliability is that we are ignoring interactions between the ions in solution.

This procedure is illustrated in Examples 10.7 and 10.8.

EXAMPLE 10.7   Sample exercise: Calculating the pH and percentage deprotonation of a weak acid

Calculate the pH and percentage deprotonation of 0.10 M $CH_3COOH(aq)$, given that K_a for acetic acid is 1.8×10^{-5}.

SOLUTION Following the procedure in Toolbox 10.1, we write the proton transfer equilibrium,

$$CH_3COOH(aq) + H_2O(l) \rightleftharpoons H_3O^+(aq) + CH_3CO_2^-(aq)$$

$$K_a = \frac{[H_3O^+][CH_3CO_2^-]}{[CH_3COOH]}$$

and the equilibrium table, with all the concentrations in moles per liter:

	Species		
	CH_3COOH	H_3O^+	$CH_3CO_2^-$
Step 1 Initial concentration	0.10	0	0
Step 2 Change in concentration	$-x$	$+x$	$+x$
Step 3 Equilibrium concentration	$0.10 - x$	x	x

Step 4 Substitute the equilibrium concentrations into K_a.

$$K_a = 1.8 \times 10^{-5} = \frac{x \times x}{0.10 - x}$$

Assume that $x \ll 0.10$.

$$1.8 \times 10^{-5} \approx \frac{x^2}{0.10}$$

Solve for x (which must be positive).

$$x \approx \sqrt{(0.10) \times (1.8 \times 10^{-5})} = 1.3 \times 10^{-3}$$

From pH = $-\log [H_3O^+]$;
$[H_3O^+] = x$.

$$pH \approx -\log(1.3 \times 10^{-3}) = 2.89$$

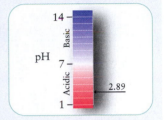

From Percentage deprotonated = $([H_3O^+]/[HA]_{initial}) \times 100\%$ with $[HA]_{initial} = 0.10$,

$$\text{Percentage deprotonated} = \frac{1.3 \times 10^{-3}}{0.10} \times 100\% = 1.3\%$$

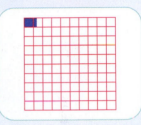

The blue square in the grid in the final box represents the percentage of the acid molecules that are deprotonated. We see that x is less than 5% of 0.10, and the approximation is valid. Because the pH < 6, the assumption that the autoprotolysis of water can be ignored is valid.

SELF-TEST 10.9A Calculate the pH and percentage deprotonation of 0.50 M aqueous lactic acid. See Table 10.1 for K_a. Be sure to check any approximation to see whether it is valid.

[*Answer:* 1.69; 4.1%]

SELF-TEST 10.9B Calculate the pH and percentage deprotonation of 0.22 M aqueous chloroacetic acid. Be sure to check any approximation to see whether it is valid.

EXAMPLE 10.8 **Calculating the K_a of a weak acid from the pH**

The pH of a 0.010 M aqueous solution of mandelic acid, $C_6H_5CH(OH)COOH$, an antiseptic, is 2.95. What is the K_a of mandelic acid?

STRATEGY Calculate the hydronium ion concentration from the pH and then calculate the value of K_a from the initial concentration of the acid and the hydronium ion concentration.

SOLUTION
From $[H_3O^+] = 10^{-pH}$,

$$[H_3O^+] = 10^{-2.95} \text{ mol·L}^{-1} = 0.0011 \text{ mol·L}^{-1}$$

Note relations between equilibrium concentrations.

$$[H_3O^+] = [A^-]$$
$$[HA] = [HA]_{initial} - [H_3O^+]$$

Use K_a in the form
$K_a = [H_3O^+][A^-]/[HA]$
$= [H_3O^+]^2/([HA]_{initial} - [H_3O^+])$
and substitute the data.

$$K_a = \frac{(0.0011)^2}{0.010 - 0.0011} = 1.4 \times 10^{-4}$$

SELF-TEST 10.10A The pH of a 0.20 M aqueous solution of crotonic acid, C_3H_5COOH, which is used in medicinal research and in the manufacture of synthetic vitamin A, is 2.69. What is the K_a of crotonic acid?

[*Answer:* 2.1×10^{-5}]

SELF-TEST 10.10B The pH of a 0.50 M aqueous solution of the metabolic intermediate homogentisic acid is 2.35. What is the K_a of homogentisic acid, $C_7H_5(OH)_2COOH$?

To calculate the pH and percentage deprotonation of a solution of a weak acid, set up an equilibrium table and determine the H_3O^+ concentration by using the acidity constant.

10.12 Solutions of Weak Bases

In the treatment of weak acids, we found that the percentage deprotonated gave an indication of acid strength. Similarly, when we describe the strengths of weak bases, it is useful to know the **percentage protonated**, the percentage of base molecules that have been protonated:

$$\text{Percentage protonated} = \frac{\text{concentration of } HB^+}{\text{initial concentration of B}} \times 100\%$$

$$= \frac{[HB^+]}{[B]_{\text{initial}}} \times 100\% \qquad (13)*$$

Here $[B]_{\text{initial}}$ is the initial molar concentration of base, its concentration based on the assumption that no protonation has occurred.

We calculate the pH of solutions of weak bases in the same way as we calculate the pH of solutions of weak acids—by using an equilibrium table. The protonation equilibrium is given in Eq. 9. To calculate the pH of the solution, we first calculate the concentration of OH^- ions at equilibrium, express that concentration as pOH, and then calculate the pH at 25°C from the relation pH + pOH = 14.00. For very weak or very dilute bases, the autoprotolysis of water must be taken into consideration.

TOOLBOX 10.2   **HOW TO CALCULATE THE pH OF A SOLUTION OF A WEAK BASE**

CONCEPTUAL BASIS

Proton transfer equilibrium is established as soon as a weak base is dissolved in water, and so we can calculate the hydroxide ion concentration from the initial concentration of the base and the value of its basicity constant. Because the hydroxide ions are in equilibrium with the hydronium ions, we can use the pOH and pK_w to calculate the pH.

PROCEDURE

Step 1 Write the proton transfer equilibrium and set up a table with columns labeled by the base B, its conjugate acid BH^+, and OH^-. In the first row, show the initial concentration of each species.

For the initial values, assume that no base molecules have been protonated.

Step 2 Write the changes in the concentrations that are needed for the reaction to reach equilibrium. Assume that the concentration of the base decreases by x mol·L^{-1} as a result of protonation. The reaction stoichiometry gives us the other changes in terms of x.

Step 3 Write the equilibrium concentrations by adding the change in concentration (step 2) to the initial concentration for each substance (step 1).

	B	BH$^+$	OH$^-$
Step 1 Initial concentration	$[B]_{\text{initial}}$	0	0
Step 2 Change in concentration	$-x$	$+x$	$+x$
Step 3 Equilibrium concentration	$[B]_{\text{initial}} - x$	x	x

Although a change in concentration may be positive (an increase) or negative (a decrease), the concentration itself must always be positive.

Step 4 Use the value of K_b to calculate the value of x. The calculation of x can often be simplified, as explained in Toolbox 10.1. We ignore contributions from the autoprotolysis of water to the hydroxide ion concentration if the concentration of hydroxide ions is greater than 10^{-6} mol·L^{-1}.

Step 5 Determine the pOH of the solution and then calculate the pH from the pOH by using Eq. 6b.

This procedure is illustrated in Example 10.9.

As in Toolbox 10.1, although we calculate the pH to the number of significant figures appropriate to the data, the answers are often considerably less reliable than that.

EXAMPLE 10.9   Sample exercise: Calculating the pH and percentage protonation of a weak base

Calculate the pH and percentage protonation of a 0.20 M aqueous solution of methylamine, CH_3NH_2. The K_b for CH_3NH_2 is 3.6×10^{-4}.

SOLUTION   Proceeding as in Toolbox 10.2, we write the proton transfer equilibrium:

$$H_2O(l) + CH_3NH_2(aq) \rightleftharpoons CH_3NH_3^+(aq) + OH^-(aq) \quad K_b = \frac{[CH_3NH_3^+][OH^-]}{[CH_3NH_2]}$$

The equilibrium table, with all concentrations in moles per liter, is

	CH_3NH_2	$CH_3NH_3^+$	OH^-
Step 1 Initial concentration	0.20	0	0
Step 2 Change in concentration	$-x$	$+x$	$+x$
Step 3 Equilibrium concentration	$0.20 - x$	x	x

Step 4 Substitute the equilibrium concentrations into the expression for K_b.	$K_b = 3.6 \times 10^{-4} = \dfrac{x \times x}{0.20 - x}$
Assume that $x \ll 0.20$.	$3.6 \times 10^{-4} \approx \dfrac{x^2}{0.20}$
Solve for x.	$x \approx \sqrt{0.20 \times (3.6 \times 10^{-4})} = 8.5 \times 10^{-3}$
Check the assumption that $x \ll 0.20$.	$8.5 \times 10^{-3} \ll 0.20$, so the approximation is valid
From $pOH = -\log[OH^-]$, with $[OH^-] = x$,	$pOH \approx -\log(8.5 \times 10^{-3}) = 2.07$

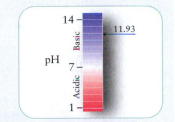

Step 5 From $pH = pK_w - pOH$,	$pH \approx 14.00 - 2.07 = 11.93$

From Eq. 13, with $[HB^+] = x$ and $[B]_{initial} = 0.20$,	Percentage protonated $= \dfrac{8.5 \times 10^{-3}}{0.20} \times 100\%$ $= 4.2\%$

That is, 4.2% of the methylamine is present as the protonated form, $CH_3NH_3^+$. Because the pH is greater than 8, the assumption that this equilibrium dominates the pH and that autoprotolysis can be ignored is valid.

SELF-TEST 10.11A   Estimate the pH and percentage of protonated base in 0.15 M $NH_2OH(aq)$, aqueous hydroxylamine.

[*Answer:* 9.61; 0.027%]

SELF-TEST 10.11B   Estimate the pH and percentage of protonated base in 0.012 M $C_{10}H_{14}N_2(aq)$, nicotine.

To calculate the pH of a solution of a weak base, set up an equilibrium table to calculate pOH from the value of K_b and convert that pOH into pH by using pH + pOH = 14.00.

10.13 The pH of Salt Solutions

We saw in Section J that a salt is produced by the neutralization of an acid by a base. However, if we measure the pH of a solution of a salt, we do not in general find the "neutral" value (pH = 7). For instance, if we neutralize 0.3 M $CH_3COOH(aq)$ with 0.3 M $NaOH(aq)$, the resulting solution of sodium acetate has pH = 9.0. How can this be? The Brønsted–Lowry theory provides the explanation. According to this theory, an ion may be an acid or a base. The acetate ion, for instance, is a base, and the ammonium ion is an acid. The pH of a solution of a salt depends on the relative acidity and basicity of its ions.

Table 10.7 lists some cations that are acidic in water. They fall into four general categories:

- All cations that are the conjugate acids of weak bases produce acidic solutions.

Conjugate acids of weak bases, such as NH_4^+, act as proton donors, and so we can expect them to form acidic solutions.

- Small, highly charged metal cations that can act as Lewis acids in water, such as Al^{3+} and Fe^{3+}, produce acidic solutions, even though the cations themselves have no hydrogen ions to donate (Fig. 10.18).

The protons come from the water molecules that hydrate these metal cations in solution (Fig. 10.19). The water molecules act as Lewis bases and share electrons with the metal cations. This partial loss of electrons weakens the O—H bonds and allows one or more hydrogen ions to be lost from the water molecules. Small, highly charged cations exert the greatest pull on the electrons and so form the most acidic solutions.

- Cations of Group 1 and 2 metals, as well as those of charge +1 from other groups, are such weak Lewis acids that the hydrated ions do not act as acids.

These metal cations are too large or have too low a charge to have an appreciable polarizing effect on the hydrating water molecules that surround them, and so the water molecules do not readily release their protons. These cations are sometimes called "neutral cations," because they have so little effect on the pH.

- Very few anions that contain hydrogen produce acidic solutions.

FIGURE 10.18 These four solutions show that hydrated cations can be significantly acidic. The flasks contain, from left to right, pure water, 0.1 M $Al_2(SO_4)_3(aq)$, 0.1 M $Ti_2(SO_4)_3(aq)$, and 0.1 M $CH_3COOH(aq)$. All four tubes contain a few drops of universal indicator, which changes color from green in a neutral solution through yellow to red with increasing acidity. The superimposed numbers are the pH values of each solution.

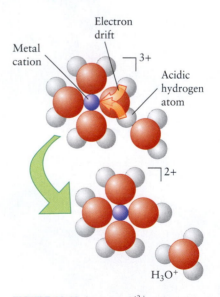

FIGURE 10.19 In water, Al^{3+} cations are hydrated by water molecules that can act as Brønsted acids. Although, for clarity, only four water molecules are shown here, a metal cation typically has six H_2O molecules attached to it.

TABLE 10.7 Acidic Character and K_a Values of Common Cations in Water*

Character	Examples	K_a	pK_a
Acidic			
conjugate acids of weak bases	anilinium ion, $C_6H_5NH_3^+$	2.3×10^{-5}	4.64
	pyridinium ion, $C_5H_5NH^+$	5.6×10^{-6}	5.24
	ammonium ion, NH_4^+	5.6×10^{-10}	9.25
	methylammonium ion, $CH_3NH_3^+$	2.8×10^{-11}	10.56
small, highly charged metal cations	Fe^{3+} as $Fe(H_2O)_6^{3+}$	3.5×10^{-3}	2.46
	Cr^{3+} as $Cr(H_2O)_6^{3+}$	1.3×10^{-4}	3.89
	Al^{3+} as $Al(H_2O)_6^{3+}$	1.4×10^{-5}	4.85
	Cu^{2+} as $Cu(H_2O)_6^{2+}$	3.2×10^{-8}	7.49
	Ni^{2+} as $Ni(H_2O)_6^{2+}$	9.3×10^{-10}	9.03
	Fe^{2+} as $Fe(H_2O)_6^{2+}$	8×10^{-11}	10.1
Neutral			
Group 1 and 2 cations	$Li^+, Na^+, K^+, Mg^{2+}, Ca^{2+}$		
metal cations with charge +1	Ag^+		
Basic	none		

*As in Table 10.1, the experimental pK_a values have more significant figures than shown here, and the K_a values have been calculated from these better data.

TABLE 10.8 Acidic and Basic Character of Common Anions in Water

Character	Examples
Acidic very few	HSO_4^-, $H_2PO_4^-$
Neutral conjugate bases of strong acids	Cl^-, Br^-, I^-, NO_3^-, ClO_4^-
Basic conjugate bases of weak acids	F^-, O^{2-}, OH^-, S^{2-}, HS^-, CN^-, CO_3^{2-}, PO_4^{3-}, NO_2^-, $CH_3CO_2^-$, other carboxylate ions

It is difficult for a positively charged proton to leave a negatively charged anion. The few anions that do act as acids include $H_2PO_4^-$ and HSO_4^-.

Table 10.8 summarizes the properties of common anions in solution. Notice that no cations are listed. Cations cannot readily accept protons, because the positive charge of the cation repels the positive charge of the incoming protons.

- All anions that are the conjugate bases of weak acids produce basic solutions.

For example, formic acid, HCOOH, the acid in ant venom, is a weak acid, and so the formate ion acts as a base in water:

$$H_2O(l) + HCO_2^-(aq) \rightleftharpoons HCOOH(aq) + OH^-(aq)$$

Acetate ions and the other ions listed in last row of Table 10.8 act as bases in water.

- The anions of strong acids—which include Cl^-, Br^-, I^-, NO_3^-, and ClO_4^-—are such weak bases that they have no significant effect on the pH of a solution. These anions are considered "neutral" in water.

To determine whether the solution of a salt will be acidic, basic, or neutral, we must consider both the cation and the anion. First we examine the anion to see whether it is the conjugate base of a weak acid. If the anion is neither acidic nor basic, we examine the cation to see whether it is an acidic metal ion or the conjugate acid of a weak base. If one ion is an acid and the other a base, as in NH_4F, then the pH is affected by the reactions of both ions with water and both equilibria must be considered, as in Section 10.19.

SELF-TEST 10.12A Use Tables 10.7 and 10.8 to decide whether aqueous solutions of the salts (a) $Ba(NO_2)_2$; (b) $CrCl_3$; (c) NH_4NO_3 are acidic, neutral, or basic.
[*Answer:* (a) Basic; (b) acidic; (c) acidic]

SELF-TEST 10.12B Decide whether aqueous solutions of (a) Na_2CO_3; (b) $AlCl_3$; (c) KNO_3 are acidic, neutral, or basic.

To calculate the pH of a salt solution, we can use the equilibrium table procedure described in Toolboxes 10.1 and 10.2—an acidic cation is treated as a weak acid and a basic anion as a weak base. However, often we must first calculate the K_a or K_b for the acidic or basic ion. Examples 10.10 and 10.11 illustrate the procedure.

EXAMPLE 10.10 Calculating the pH of a salt solution with an acidic cation

Estimate the pH of 0.15 M $NH_4Cl(aq)$.

STRATEGY Because NH_4^+ is a weak acid and Cl^- is neutral, we expect pH < 7. We treat the solution as that of a weak acid, using an equilibrium table as in Toolbox 10.1 to calculate the composition and hence the pH. First, write the chemical equation for proton transfer to water and the expression for K_a. Obtain the value of K_a from K_b for the conjugate base by using $K_a = K_w/K_b$ (Eq. 11a). The initial concentration of the acidic cation is equal to the concentration of the cation that the salt would produce if the salt were fully dissociated and the cation retained all its acidic protons. The initial concentrations of its conjugate base and H_3O^+ are assumed to be zero.

SOLUTION The equilibrium to consider is

$$NH_4^+(aq) + H_2O(l) \rightleftharpoons H_3O^+(aq) + NH_3(aq) \qquad K_a = \frac{[H_3O^+][NH_3]}{[NH_4^+]}$$

In Table 10.2, we find $K_b = 1.8 \times 10^{-5}$ for NH_3. We construct the following equilibrium table, with all concentrations in moles per liter:

	Species		
	NH_4^+	H_3O^+	NH_3
Step 1 Initial concentration	0.15	0	0
Step 2 Change in concentration	$-x$	$+x$	$+x$
Step 3 Equilibrium concentration	$0.15 - x$	x	x

Step 4 From $K_a = K_w/K_b$,

$$K_a = \frac{1.0 \times 10^{-14}}{1.8 \times 10^{-5}} = 5.6 \times 10^{-10}$$

Substitute the equilibrium concentrations into the expression for K_a.

$$5.6 \times 10^{-10} = \frac{x \times x}{0.15 - x}$$

Assume that $x \ll 0.15$.

$$5.6 \times 10^{-10} \approx \frac{x^2}{0.15}$$

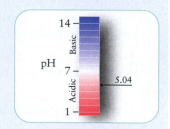

Solve for x.

$$x = \sqrt{0.15 \times (5.6 \times 10^{-10})} = 9.2 \times 10^{-6}$$

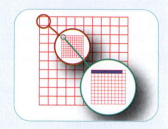

From pH $= -\log [H_3O^+]$ with $[H_3O^+] = x$,

$$pH \approx -\log(9.2 \times 10^{-6}) = 5.04$$

The approximation that x is less than 5% of 0.15 is valid (by a large margin). Moreover, the H_3O^+ concentration (9.2×10^{-6} mol·L^{-1}) is much larger than that generated by the autoprotolysis of water (1.0×10^{-7} mol·L^{-1}), and so ignoring the latter contribution is valid.

SELF-TEST 10.13A Estimate the pH of 0.10 M $CH_3NH_3Cl(aq)$, aqueous methylammonium chloride; the cation is $CH_3NH_3^+$.

[*Answer:* 5.78]

SELF-TEST 10.13B Estimate the pH of 0.10 M $NH_4NO_3(aq)$.

EXAMPLE 10.11 **Calculating the pH of a salt solution with a basic anion**

Estimate the pH of 0.15 M $Ca(CH_3CO_2)_2(aq)$.

STRATEGY The $CH_3CO_2^-$ ion is the conjugate base of a weak acid; so the solution will be basic and we expect pH > 7. Use the procedure in Toolbox 10.2, taking the initial concentration of base from the concentration of added salt. Calculate K_b for the basic anion from K_a for its conjugate acid. Convert pOH into pH by using pH + pOH = 14.00.

SOLUTION The proton transfer equilibrium is

$$H_2O(l) + CH_3CO_2^-(aq) \rightleftharpoons CH_3COOH(aq) + OH^-(aq)$$

$$K_b = \frac{[CH_3COOH][OH^-]}{[CH_3CO_2^-]}$$

The initial concentration of $CH_3CO_2^-$ is 2×0.15 mol·L^{-1} = 0.30 mol·L^{-1}, because each formula unit of salt provides two $CH_3CO_2^-$ ions. Table 10.1 gives the K_a of CH_3COOH as 1.8×10^{-5}.

	$CH_3CO_2^-$	CH_3COOH	OH^-
Step 1 Initial concentration	0.30	0	0
Step 2 Change in concentration	$-x$	$+x$	$+x$
Step 3 Equilibrium concentration	$0.30 - x$	x	x

Step 4 Find the K_b of the $CH_3CO_2^-$ ion from $K_b = K_w/K_a$.	$K_b = \dfrac{1.0 \times 10^{-14}}{1.8 \times 10^{-5}} = 5.6 \times 10^{-10}$
Substitute the equilibrium concentrations into the expression for K_b.	$5.6 \times 10^{-10} = \dfrac{x \times x}{0.30 - x}$
Assume that x is less than 5% of 0.30 (that is, $x < 0.015$).	$5.6 \times 10^{-10} \approx \dfrac{x^2}{0.30}$

Solve for x.	$x = \sqrt{0.30 \times (5.6 \times 10^{-10})} = 1.3 \times 10^{-5}$
Check the assumption that $x < 0.015$.	$1.3 \times 10^{-5} < 0.015$
From pOH = $-\log [OH^-]$ with $[OH^-] = x$,	pOH $\approx -\log (1.3 \times 10^{-5}) = 4.89$
From pH = pK_w − pOH,	pH $\approx 14.00 - 4.89 = 9.11$

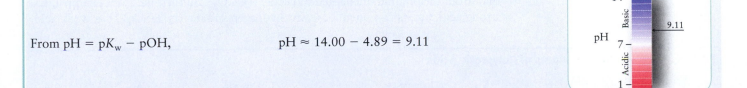

The OH^- concentration arising from the proton transfer equilibrium (1.3×10^{-5} mol·L^{-1}) is much larger than that arising from the autoprotolysis of water (1.0×10^{-7} mol·L^{-1}), and so it is valid to ignore autoprotolysis.

SELF-TEST 10.14A Estimate the pH of 0.10 M $KC_6H_5CO_2$(aq), potassium benzoate; see Table 10.1 for data.

[*Answer:* 8.59]

SELF-TEST 10.14B Estimate the pH of 0.020 M KF(aq); see Table 10.1 for data.

Salts that contain the conjugate acids of weak bases produce acidic aqueous solutions; so do salts that contain small, highly charged metal cations. Salts that contain the conjugate bases of weak acids produce basic aqueous solutions.

POLYPROTIC ACIDS AND BASES

A **polyprotic acid** is a compound that can donate more than one proton. Many common acids are polyprotic, including sulfuric acid, H_2SO_4, and carbonic acid, H_2CO_3, each of which can donate two protons, and phosphoric acid, H_3PO_4, which can donate three protons. Polyprotic acids play a critical role in biological systems, because many enzymes can be regarded as polyprotic acids that carry out their vital functions by donating one proton after another. A **polyprotic base** is a species that can accept more than one proton. Examples include the CO_3^{2-} and SO_3^{2-} anions, both of which can accept two protons, and the PO_4^{3-} anion, which can accept three protons.

The principal difference between a polyprotic acid and a monoprotic acid is that a polyprotic acid donates protons in a succession of deprotonation steps. For example, a carbonic acid molecule can lose one proton to form HCO_3^-, and then that ion can donate the remaining proton to form CO_3^{2-}. We shall see how to take this succession of deprotonations into account when assessing the pH of the solution of a polyprotic acid or one of its salts. In addition, we shall see how the relative concentrations of the ions in solution, such as PO_4^{3-}, HPO_4^{2-}, and $H_2PO_4^-$ depend on the pH of the solution.

10.14 The pH of a Polyprotic Acid Solution

Carbonic acid is an important natural component of the environment because it is formed whenever carbon dioxide dissolves in lake water or seawater. In fact, the oceans provide one of the critical mechanisms for maintaining a constant concentration of carbon dioxide in the atmosphere. Carbonic acid takes part in two successive proton transfer equilibria:

$$H_2CO_3(aq) + H_2O(l) \rightleftharpoons H_3O^+(aq) + HCO_3^-(aq) \qquad K_{a1} = 4.3 \times 10^{-7}$$
$$HCO_3^-(aq) + H_2O(l) \rightleftharpoons H_3O^+(aq) + CO_3^{2-}(aq) \qquad K_{a2} = 5.6 \times 10^{-11}$$

The conjugate base of H_2CO_3 in the first equilibrium, HCO_3^-, acts as an acid in the second equilibrium. That ion produces in turn its own conjugate base, CO_3^{2-}.

Protons are donated successively by polyprotic acids, with the acidity constant decreasing significantly, usually by a factor of about 10^3, with each proton lost (Table 10.9):

$$K_{a1} \gg K_{a2} \gg K_{a3} \gg \cdots$$

The decrease can be traced to the attraction between opposite charges: it is harder to separate a positively charged proton from a negatively charged ion (such as HCO_3^-) than from the original uncharged molecule (H_2CO_3). Sulfuric acid, for example, is a strong acid, but its conjugate base, the hydrogen sulfate ion, HSO_4^-, is a weak acid.

TABLE 10.9 Acidity Constants of Polyprotic Acids

Acid	K_{a1}	pK_{a1}	K_{a2}	pK_{a2}	K_{a3}	pK_{a3}
sulfuric acid, H_2SO_4	strong		1.2×10^{-2}	1.92		
oxalic acid, $(COOH)_2$	5.9×10^{-2}	1.23	6.5×10^{-5}	4.19		
sulfurous acid, H_2SO_3	1.5×10^{-2}	1.81	1.2×10^{-7}	6.91		
phosphorous acid, H_3PO_3	1.0×10^{-2}	2.00	2.6×10^{-7}	6.59		
phosphoric acid, H_3PO_4	7.6×10^{-3}	2.12	6.2×10^{-8}	7.21	2.1×10^{-13}	12.68
tartaric acid, $C_2H_4O_2(COOH)_2$	6.0×10^{-4}	3.22	1.5×10^{-5}	4.82		
carbonic acid, H_2CO_3	4.3×10^{-7}	6.37	5.6×10^{-11}	10.25		
hydrosulfuric acid, H_2S	1.3×10^{-7}	6.89	7.1×10^{-15}	14.15		

Sulfuric acid is the only common polyprotic acid for which the first deprotonation is complete. The second deprotonation adds to the H_3O^+ concentration slightly, and so the overall pH will be slightly less than that due to the first deprotonation alone. For example, in 0.010 M H_2SO_4(aq) the first deprotonation is complete:

$$H_2SO_4(aq) + H_2O(l) \longrightarrow H_3O^+(aq) + HSO_4^-(aq)$$

It results in an H_3O^+ concentration equal to the initial concentration of the acid, 0.010 mol·L^{-1}. This value corresponds to pH = 2.0. However, the conjugate base, HSO_4^-, also contributes protons to the solution, and so the second proton transfer equilibrium needs to be taken into account:

$$HSO_4^-(aq) + H_2O(l) \rightleftharpoons H_3O^+(aq) + SO_4^{2-}(aq) \qquad K_{a2} = 0.012$$

Therefore, to calculate the pH of a solution of sulfuric acid, we set up an equilibrium table in which the initial concentrations are those due to the first deprotonation: $[HSO_4^-] = 0.010$ mol·L^{-1}, $[H_3O^+] = 0.010$ mol·L^{-1}, and $[SO_4^{2-}] = 0$. Then we solve for the new value of the hydronium ion concentration:

$$K_a = \frac{[H_3O^+][SO_4^{2-}]}{[HSO_4^-]}; \qquad 0.012 = \frac{(0.010 + x)(x)}{0.010 - x}$$

leads to $x = 4.5 \times 10^{-3}$, $[H_3O^+] = 0.014$ mol·L^{-1}, and pH = 1.9.

SELF-TEST 10.15A Estimate the pH of 0.050 M H_2SO_4(aq).

[*Answer:* 1.23]

SELF-TEST 10.15B Estimate the pH of 0.10 M H_2SO_4(aq).

The parent acids of common polyprotic acids other than sulfuric are weak and the acidity constants of successive deprotonation steps are normally widely different. As a result, except for sulfuric acid, we can treat a polyprotic acid or the salt of any anion derived from it as the *only significant species* in solution. This approximation leads to a major simplification: to calculate the pH of a polyprotic acid, we just use K_{a1} and take only the first deprotonation into account; that is, we treat the acid as a monoprotic weak acid (see Toolbox 10.1). Subsequent deprotonations do take place, but provided K_{a2} is less than about $K_{a1}/1000$, they do not affect the pH significantly and can be ignored.

Estimate the pH of a polyprotic acid for which all deprotonations are weak by using only the first deprotonation equilibrium and assuming that further deprotonation is insignificant. An exception is sulfuric acid, the only common polyprotic acid that is a strong acid in its first deprotonation.

10.15 Solutions of Salts of Polyprotic Acids

The conjugate base of a polyprotic acid is amphiprotic: it can act as either an acid or a base because it can either donate its remaining acidic hydrogen atom or accept an acidic hydrogen atom and revert to the original acid. For example, a hydrogen sulfide ion, HS^-, in water acts as both an acid and a base:

$$HS^-(aq) + H_2O(l) \rightleftharpoons H_3O^+(aq) + S^{2-}(aq)$$
$$K_{a2} = 7.1 \times 10^{-15} \qquad pK_{a2} = 14.15$$

$$HS^-(aq) + H_2O(l) \rightleftharpoons H_2S(aq) + OH^-(aq)$$
$$K_{b1} = K_w/K_{a1} = 7.7 \times 10^{-8} \qquad pK_{b1} = 7.11$$

Because HS^- is amphiprotic, it is not immediately apparent whether an aqueous solution of NaHS will be acidic or basic. However, we can use the pK_a and pK_b values of the HS^- ion to conclude that:

- HS^- is such a weak acid that the basic character of S^{2-} will dominate, and pH > 7.
- HS^- is a weak base; again pH > 7.

> Exceptions are HSO_4^-, which is an extremely weak base, and HPO_3^{2-}, which does not act as an acid because its proton is not acidic (Section 10.10).

The first conclusion arises from the large pK_{a2} of H_2S. The second conclusion arises from pK_{b1} having an intermediate value and therefore (by the relation in Eq. 11b) to the pK_{a1} of H_2S also having an intermediate value. This reasoning suggests that the pH will be high if both pK_{a1} and pK_{a2} are relatively large. Indeed, if we make some reasonable assumptions we can use

$$pH = \tfrac{1}{2}(pK_{a1} + pK_{a2}) \tag{14}*$$

This formula is reliable provided $S \gg K_w/K_{a2}$ and $S \gg K_{a1}$, where S is the initial (that is, analytical) concentration of the salt. If these criteria are not satisfied, a much more complicated expression must be used: it and its derivation, including the derivation of this simplified version, will be found on the Web site for this text.

EXAMPLE 10.12 **Estimating the pH of a solution of an amphiprotic salt**

Estimate the pH of (a) 0.20 M $NaH_2PO_4(aq)$; (b) 0.20 M $Na_2HC_6H_5O_7(aq)$, a salt of citric acid, $H_3C_6H_5O_7$. For citric acid, $pK_{a2} = 5.95$, and $pK_{a3} = 6.39$.

STRATEGY Verify that Eq. 14 can be used by checking that $S \gg K_w/K_{a2}$ and $S \gg K_{a1}$. If so, we use Eq. 14 to determine the pH of the salts of the diprotic conjugate base (H_2A^-) of a triprotic acid (H_3A) and the monoprotic conjugate base (HA^-) of a diprotic acid (H_2A). However, when the solute is a salt of an anion that has lost two protons, such as $HPO_4{}^{2-}$, we must adjust the expression to use the appropriate neighboring pK_as.

SOLUTION

(a) For H_3PO_4, $K_{a1} = 7.6 \times 10^{-3}$ and $K_{a2} = 6.2 \times 10^{-8}$, and so $pK_{a1} = 2.12$ and $pK_{a2} = 7.21$. Therefore, for the solution of $H_2PO_4^-$,

Check that $S \gg K_w/K_{a2}$ and $S \gg K_{a1}$.

$$0.20 \gg \frac{1.0 \times 10^{-14}}{6.2 \times 10^{-8}} = 1.6 \times 10^{-7}$$

$$0.20 \gg 7.6 \times 10^{-3}$$

Therefore, the use of Eq. 14 is valid.

From $pH = \tfrac{1}{2}(pK_{a1} + pK_{a2})$,

$$pH = \tfrac{1}{2}(2.12 + 7.21) = 4.66$$

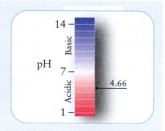

(b) For citric acid, $pK_{a2} = 5.95$ and $pK_{a3} = 6.39$. Therefore, for the solution of $HC_6H_5O_7{}^{2-}$,

Check that $S \gg K_w/K_{a3}$ and $S \gg K_{a2}$.

$$0.20 \gg \frac{1.0 \times 10^{-14}}{4.1 \times 10^{-7}} = 2.4 \times 10^{-8}$$

$$0.20 \gg 1.1 \times 10^{-6}$$

Therefore, the use of Eq. 14 is valid.

From $pH = \tfrac{1}{2}(pK_{a1} + pK_{a2})$,

$$pH = \tfrac{1}{2}(5.95 + 6.39) = 6.17$$

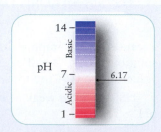

SELF-TEST 10.16A Estimate the pH of 0.10 M $NaHCO_3$(aq).

[*Answer:* 8.31]

SELF-TEST 10.16B Estimate the pH of 0.50 M KH_2PO_4(aq).

Suppose we need to estimate the pH of an aqueous solution of a fully deprotonated polyprotic acid molecule. An example is a solution of sodium sulfide, in which sulfide ions, S^{2-}, are present; another example is a solution of potassium phosphate, which contains PO_4^{3-} ions. In such a solution, the anion acts as a base: it accepts protons from water. For such a solution, we can use the techniques for calculating the pH of a basic anion illustrated in Example 10.11. The K_a to use in the calculation is for the deprotonation that produces the ion being studied. For S^{2-}, we would use K_{a2} for H_2S; and, for PO_4^{3-}, we would use K_{a3} for H_3PO_4.

The pH of the aqueous solution of an amphiprotic salt is equal to the average of the pK_as of the salt and its conjugate acid. The pH of a solution of a salt of the final conjugate base of a polyprotic acid is found from the reaction of the anion with water.

10.16 The Concentrations of Solute Species

Environmental chemists studying the pollution caused by fertilizers in runoff from fields or mineralogists studying the formation of sedimentary rocks as groundwater trickles through rock formations may need to know not only the pH but also the concentrations of each of the ions present in the solution. For example, they may need to know the concentration of sulfite ion in a solution of sulfurous acid or the concentrations of phosphate and hydrogen phosphate ions in a solution of phosphoric acid. The calculations described in Toolbox 10.1 tell us the pH—the hydrogen ion concentration—but they do not give the concentrations of all the solute species in solution, which for H_3PO_4 includes H_3PO_4, $H_2PO_4^-$, HPO_4^{2-}, and PO_4^{3-}. To calculate them, we need to take into account all these simultaneous proton transfer equilibria.

To simplify the calculations, we start by judging the relative concentration of each species in solution and identify terms that can be ignored. To make this judgement, we use the general rule that concentrations of species present in the greater amount are not significantly affected by concentrations of species present in the smaller amount, especially if the difference in concentrations is large.

TOOLBOX 10.3   HOW TO CALCULATE THE CONCENTRATIONS OF ALL SPECIES IN A POLYPROTIC ACID SOLUTION

CONCEPTUAL BASIS

We assume that the polyprotic acid is the solute species present in largest amount. We also assume that only the first deprotonation contributes significantly to $[H_3O^+]$ and that the autoprotolysis of water does not contribute significantly to $[H_3O^+]$ or $[OH^-]$.

PROCEDURE FOR A DIPROTIC ACID

Step 1 From the deprotonation equilibrium of the acid (H_2A), determine the concentrations of the conjugate base (HA^-) and H_3O^+, as illustrated in Example 10.7.

Step 2 Find the concentration of A^{2-} from the second deprotonation equilibrium (that of HA^-) by substituting the concentrations of H_3O^+ and HA^- from step 1 into the expression for K_{a2}.

Step 3 Find the concentration of OH^- by dividing K_w by the concentration of H_3O^+.

PROCEDURE FOR A TRIPROTIC ACID

Step 1 From the deprotonation equilibrium of the acid (H_3A), determine the concentrations of the conjugate base (H_2A^-) and H_3O^+.

Step 2 Find the concentration of HA^{2-} from the second deprotonation equilibrium (that of H_2A^-) by substituting the concentrations of H_3O^+ and H_2A^- from step 1 into the expression for K_{a2}.

Step 3 Find the concentration of A^{3-} from the deprotonation equilibrium of HA^{2-} by substituting the concentrations of H_3O^+ and HA^{2-} from step 2 into the expression for K_{a3}. The concentration of H_3O^+ is the same in all three calculations because only the first deprotonation makes a significant contribution to its value.

Step 4 Find the concentration of OH^- by dividing K_w by the concentration of H_3O^+.

This procedure is illustrated in Example 10.13.

EXAMPLE 10.13 Sample exercise: Calculating the concentrations of all solute species in a polyprotic acid solution

Calculate the concentrations of all solute species in 0.10 M $H_3PO_4(aq)$.

SOLUTION

Step 1 Assume that only H_3PO_4 significantly affects the pH. The primary proton transfer equilibrium is

$$H_3PO_4(aq) + H_2O(l) \rightleftharpoons H_3O^+(aq) + H_2PO_4^-(aq)$$

and the first acidity constant, from Table 10.9, is 7.6×10^{-3}. The equilibrium table, with the concentrations in moles per liter, is

	H_3PO_4	H_3O^+	$H_2PO_4^-$
Initial concentration	0.10	0	0
Change in concentration	$-x$	$+x$	$+x$
Equilibrium concentration	$0.10 - x$	x	x

From $K_{a1} = [H_3O^+][H_2A^-]/[H_3A]$,	$K_{a1} = 7.6 \times 10^{-3} = \dfrac{x^2}{0.10 - x}$
Assume that $x \ll 0.10$ and decide if it is valid to approximate the expression.	$7.6 \times 10^{-3} \approx \dfrac{x^2}{0.10}$ solves to $x \approx 0.028$
Because the value of x is in fact larger than 5% of 0.10, we must use the full quadratic equation.	$(7.6 \times 10^{-3}) \times (0.10 - x) = x^2$
Rearrange the equation.	$x^2 + (7.6 \times 10^{-3})x - 7.6 \times 10^{-4} = 0$
Solve for x with the quadratic formula.	$x = 2.4 \times 10^{-2}$ or -3.2×10^{-2} 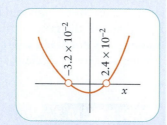
Reject the negative root.	$x = 2.4 \times 10^{-2}$; therefore, $[H_3O^+] = 2.4 \times 10^{-2}$ mol·L^{-1}
From $[H_2PO_4^-] = [H_3O^+]$,	$[H_2PO_4^-] = 2.4 \times 10^{-2}$ mol·L^{-1} 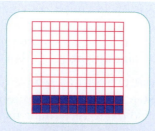
It follows from the equilibrium table that	$[H_3PO_4] \approx 0.10 - 0.024$ mol·L^{-1} $= 0.08$ mol·L^{-1} 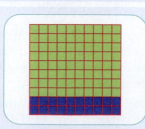

The illustrations display the percentage composition of species in the solution: blue for $H_2PO_4^-$, green for H_3PO_4, and (below) yellow for HPO_4^{2-}.

Step 2 Now we use $K_{a2} = 6.2 \times 10^{-8}$ to find the concentration of HPO_4^{2-}. Because $K_{a2} \ll K_{a1}$, we can safely assume that the H_3O^+ concentration calculated in step 1 is unchanged by the second deprotonation. The proton transfer equilibrium is

$$H_2PO_4^-(aq) + H_2O(l) \rightleftharpoons H_3O^+(aq) + HPO_4^{2-}(aq)$$

We set up the equilibrium table with concentrations in moles per liter, using the results from step 1 for the concentrations of H_3O^+ and $H_2PO_4^-$.

	$H_2PO_4^-$	H_3O^+	HPO_4^{2-}
Initial concentration (from step 1)	2.4×10^{-2}	2.4×10^{-2}	0
Changes in concentration	$-x$	$+x$	$+x$
Equilibrium concentration	$2.4 \times 10^{-2} - x$	$2.4 \times 10^{-2} + x$	x

From $K_{a2} = [H_3O^+][HA^{2-}]/[H_2A^-]$,

$$K_{a2} = 6.2 \times 10^{-8} = \frac{(2.4 \times 10^{-2} + x) \times x}{2.4 \times 10^{-2} - x}$$

Because K_{a2} is very small, we assume that $x \ll 2.4 \times 10^{-2}$.

$$6.2 \times 10^{-8} \approx \frac{(2.4 \times 10^{-2}) \times x}{2.4 \times 10^{-2}} = x$$

Check the assumption that $x \ll 2.4 \times 10^{-2}$.

$$6.2 \times 10^{-8} \ll 2.4 \times 10^{-2}$$

From $[HPO_4^{2-}] = x \approx K_{a2}$,

$$[HPO_4^{2-}] = 6.2 \times 10^{-8} \text{ mol·L}^{-1}$$

We can conclude that the second deprotonation proceeds only to a very small extent.

Step 3 The loss of the final proton from HPO_4^{2-} produces the phosphate ion, PO_4^{3-}:

$$HPO_4^{2-}(aq) + H_2O(l) \rightleftharpoons H_3O^+(aq) + PO_4^{3-}(aq)$$

The equilibrium constant is $K_{a3} = 2.1 \times 10^{-13}$, a very small value. We assume that the concentration of H_3O^+ calculated in step 1 and the concentration of HPO_4^{2-} calculated in step 2 are unaffected by the additional deprotonation.

	HPO_4^{2-}	H_3O^+	PO_4^{3-}
Initial concentration (from step 2)	6.2×10^{-8}	2.4×10^{-2}	0
Changes in concentration	$-x$	$+x$	$+x$
Equilibrium concentration	$6.2 \times 10^{-8} - x$	$2.4 \times 10^{-2} + x$	x

From $K_{a3} = [H_3O^+][A^{3-}]/[HA^{2-}]$,

$$K_{a3} = 2.1 \times 10^{-13} = \frac{(2.4 \times 10^{-2} + x) \times x}{6.2 \times 10^{-8} - x}$$

Because K_{a3} is so small, assume that $x \ll 6.2 \times 10^{-8}$ and simplify the equation.

$$2.1 \times 10^{-13} \approx \frac{(2.4 \times 10^{-2}) \times x}{6.2 \times 10^{-8}}$$

Solve for x.

$$x \approx \frac{(2.1 \times 10^{-13})(6.2 \times 10^{-8})}{2.4 \times 10^{-2}} = 5.4 \times 10^{-19}$$

From $[PO_4^{3-}] = x$,

$$[PO_4^{3-}] \approx 5.4 \times 10^{-19} \text{ mol·L}^{-1}$$

The HPO$_4^{2-}$ ion is deprotonated only to a tiny extent, too small to illustrate with a diagram, and our assumption that H$_3$O$^+$ is unaffected by the third deprotonation is justified.

Step 4 From [OH$^-$] = K_w/[H$_3$O$^+$],

$$[OH^-] = \frac{1.0 \times 10^{-14}}{2.4 \times 10^{-2}} \text{ mol·L}^{-1} = 4.2 \times 10^{-13} \text{ mol·L}^{-1}$$

At this point, we can summarize the concentrations of all the solute species in 0.10 M H$_3$PO$_4$(aq) in order of decreasing concentration:

Species:	H$_3$PO$_4$	H$_3$O$^+$	H$_2$PO$_4^-$	HPO$_4^{2-}$	OH$^-$	PO$_4^{3-}$
Concentration (mol·L^{-1}):	0.08	2.4×10^{-2}	2.4×10^{-2}	6.2×10^{-8}	4.2×10^{-13}	5.4×10^{-19}

SELF-TEST 10.17A Calculate the concentrations of all solute species in 0.20 M H$_2$S(aq).

[*Answer:* With concentrations in mol·L^{-1}: H$_2$S, 0.20; HS$^-$, 1.6×10^{-4}; H$_3$O$^+$, 1.6×10^{-4}; OH$^-$, 6.2×10^{-11}; S^{2-}, 7.1×10^{-15}]

SELF-TEST 10.17B Protonated glycine ($^+$NH$_3$CH$_2$COOH) is a diprotic amino acid with $K_{a1} = 4.5 \times 10^{-3}$ and $K_{a2} = 1.7 \times 10^{-10}$. Calculate the concentrations of all solute species in 0.50 M NH$_3$CH$_2$COOHCl(aq).

BOX 10.1   WHAT HAS THIS TO DO WITH . . . THE ENVIRONMENT?

Acid Rain and the Gene Pool

The human impact on the environment affects many areas of our lives and future. One example is the effect of acid rain on *biodiversity*, the diversity of living things. In the prairies that extend across the heartlands of North America and Asia, native plants have evolved that can survive even nitrogen-poor soil and drought. By studying prairie plants, scientists hope to breed food plants that will be hardy sources of food in times of drought. However, acid rain is making some of these plants extinct.

Acid rain is a regional phenomenon. The areas in different colors on the map to the right indicate constant values of the pH of rain. Notice that the pH of rain decreases down-

Eastern gama grass is a variety of a prairie plant that produces seeds rich in protein. It is the subject of sustainable agriculture research because it produces an abundance of seeds and yet is a perennial plant that is resistant to drought.

wind (generally, east) of heavily populated areas. The low pH in heavily industrialized and populated areas is caused by the acidic oxides sulfur dioxide, SO$_2$, and the nitrogen oxides, NO and NO$_2$.

Rain unaffected by human activity contains mostly weak acids and has a pH of 5.7. The primary acid present is carbonic acid, H$_2$CO$_3$, a weak acid that results when atmospheric carbon dioxide dissolves in water. The major pollutants in acid rain are strong acids that arise from human activities. Atmospheric nitrogen and oxygen can react to form NO, but the endothermic reaction is spontaneous only at the high temperatures of automobile internal combustion engines and electrical power stations:

$$N_2(g) + O_2(g) \rightleftharpoons 2\,NO(g)$$

Nitric oxide, NO, is not very soluble in water, but it can be oxidized further in air to form nitrogen dioxide:

$$2\,NO(g) + O_2(g) \longrightarrow 2\,NO_2(g)$$

The NO$_2$ reacts with water, forming nitric acid and nitric oxide:

$$3\,NO_2(g) + 3\,H_2O(l) \longrightarrow 2\,H_3O^+(aq) + 2\,NO_3^-(aq) + NO(g)$$

The catalytic converters now used in automobiles can reduce NO to harmless N$_2$. They are required in the United States for all new cars and trucks (see Section 13.14).

Sulfur dioxide is produced as a by-product of the burning of fossil fuels. It may combine with water directly to form sulfurous acid, a weak acid:

$$SO_2(g) + H_2O(l) \longrightarrow H_2SO_3(aq)$$

The concentrations of all species in a solution of a polyprotic acid can be calculated by assuming that species present in smaller amounts do not affect the concentrations of species present in larger amounts.

10.17 Composition and pH

Sometimes we need to know how the concentrations of the ions present in a solution of a polyprotic acid vary with pH. This information is particularly important in the study of natural waters, such as rivers and lakes (Box 10.1). For example, if we were examining carbonic acid in rainwater, then, at low pH (when hydronium ions are abundant), we would expect the fully protonated species (H_2CO_3) to be dominant; at high pH (when hydroxide ions are abundant), we expect the fully deprotonated species (CO_3^{2-}) to be dominant; at intermediate pH, we expect the intermediate species (HCO_3^-, in this case) to be dominant (Fig. 10.20). We can verify these expectations quantitatively.

HOW DO WE DO THAT?

To see how the concentrations of species present in a solution vary with pH, we take the carbonic acid system as an example. Consider the following proton transfer equilibria:

$$H_2CO_3(aq) + H_2O(l) \rightleftharpoons H_3O^+(aq) + HCO_3^-(aq) \qquad K_{a1} = \frac{[H_3O^+][HCO_3^-]}{[H_2CO_3]}$$

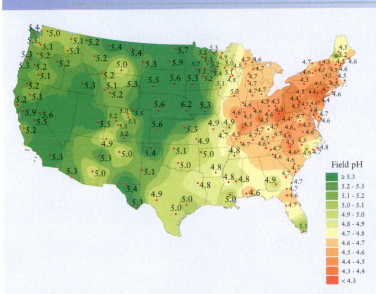

Precipitation over North America gradually becomes more acidic from west to east, especially in industrialized areas of the Northeast. This acid rain may be a result of the release of nitrogen and sulfur oxides into the atmosphere. The colors and numbers (see key) indicate pH measured at field laboratories in 2004. Data from National Atmospheric Deposition Program/National Trends Network http://nadp.sws.uiuc.edu.

Field pH

Acid rain affects plants by changing the conditions in the soil. For example, nitric acid deposits nitrates, which fertilize the land. The nitrates allow fast-growing weeds such as quack grass to replace valuable prairie species. If these species were to become extinct, their genetic material would no longer be available for agricultural research.

Research on air pollution is complex. Forests and prairies cover vast areas, and the interplay of regional air pollutants is so subtle that it may take years to sort out all the environmental stresses. However, controls being put in place are already beginning to reduce the acidity of rain in North America and Europe. Such controls will help us maintain our quality of life without losing our precious heritage of native plants. We can also help by using automobiles less and bicycles more or by taking public transportation when we can.

Related Exercises: 10.135

For Further Reading: J. P. Grime, "Biodiversity and ecosystem function: The debate deepens," *Science*, vol. 277, 1997, pp. 1260–1261. C. K. Lajewski and H. T. Mullins, "Historic calcite record from the Finger Lakes, New York: Impact of acid rain on a buffered terrane," *Geological Society of America Bulletin*, vol. 115, 2003, pp. 373–384. J. Raloff, "Pollution helps weeds take over prairies," *Science News*, vol. 150, 1996, p. 356. Environment Canada, "Acid rain," http://www.ec.gc.ca/acidrain/.

Alternatively, in the presence of particulate matter and aerosols, sulfur dioxide may react with atmospheric oxygen to form sulfur trioxide, which forms sulfuric acid, a strong acid, in water:

$$2\ SO_2(g) + O_2(g) \longrightarrow 2\ SO_3(g)$$

$$SO_3(g) + 2\ H_2O(l) \longrightarrow H_3O^+(aq) + HSO_4^-(aq)$$

FIGURE 10.20 The fractional composition of the species in carbonic acid as a function of pH. Note that the more fully protonated species are dominant at lower pH.

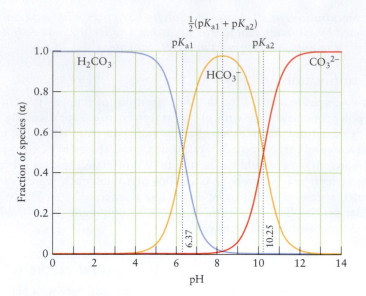

FIGURE 10.20 The fractional composition of the species in carbonic acid as a function of pH. Note that the more fully protonated species are dominant at lower pH.

$$HCO_3^-(aq) + H_2O(l) \rightleftharpoons H_3O^+(aq) + CO_3^{2-}(aq) \qquad K_{a2} = \frac{[H_3O^+][CO_3^{2-}]}{[HCO_3^-]}$$

We can express the composition of the solution in terms of the fraction, $f(X)$, of each species X present, where X could be H_2CO_3, HCO_3^-, or CO_3^{2-}, and

$$f(X) = \frac{[X]}{[H_2CO_3] + [HCO_3^-] + [CO_3^{2-}]}$$

It will turn out to be helpful to express $f(X)$ in terms of the ratio of each species to the intermediate species, HCO_3^-. So we divide the numerator and denominator by $[HCO_3^-]$ and get

$$f(X) = \frac{[X]/[HCO_3^-]}{([H_2CO_3]/[HCO_3^-]) + 1 + ([CO_3^{2-}]/[HCO_3^-])}$$

All three concentration ratios can be written in terms of the hydronium ion concentration. We simply rearrange the expressions for the first and second acidity constants:

$$\frac{[H_2CO_3]}{[HCO_3^-]} = \frac{[H_3O^+]}{K_{a1}}; \qquad \frac{[CO_3^{2-}]}{[HCO_3^-]} = \frac{K_{a2}}{[H_3O^+]}$$

then substitute them into the expression for $f(X)$ and rearrange it to obtain:

$$f(H_2CO_3) = \frac{[H_3O^+]^2}{H}; \quad f(HCO_3^-) = \frac{[H_3O^+]K_{a1}}{H}; \quad f(CO_3^{2-}) = \frac{K_{a1}K_{a2}}{H}$$

where

$$H = [H_3O^+]^2 + [H_3O^+]K_{a1} + K_{a1}K_{a2}$$

We have found expressions for the fractions, f, of species in a solution of carbonic acid. They are easily generalized to any diprotic acid H_2A:

$$f(H_2A) = \frac{[H_3O^+]^2}{H}; \quad f(HA^-) = \frac{[H_3O^+]K_{a1}}{H}; \quad f(A^{2-}) = \frac{K_{a1}K_{a2}}{H} \qquad \text{(15a)}$$

where

$$H = [H_3O^+]^2 + [H_3O^+]K_{a1} + K_{a1}K_{a2} \qquad \text{(15b)}$$

What do these equations tell us? At high pH, the concentration of hydronium ions is very low; therefore the numerators in $f(H_2A)$ and $f(HA^-)$ are very small, and so these species are in very low abundance, as we expected. At low pH, the concentration of hydronium ions is high; therefore the numerator in $f(H_2A)$ is large, and this species dominates.

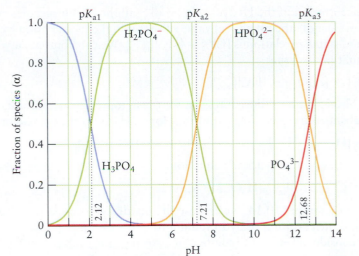

Living Graph: Figure 10.21 Fractional composition of triprotic acids

FIGURE 10.21 The fractional composition of the species in phosphoric acid as a function of pH. As in Fig. 10.20, the more fully protonated the species, the lower the pH at which it is dominant.

The shapes of the curves predicted by Eq. 15 are shown for H_2CO_3 in Fig. 10.20. We see that $f(HCO_3^-) \approx 1$ at intermediate pH. The maximum value of $f(HCO_3^-)$ is at

$$pH = \tfrac{1}{2}(pK_{a1} + pK_{a2})$$

Note that the fully protonated form (H_2CO_3) is dominant when $pH < pK_{a1}$ and the fully deprotonated form (CO_3^{2-}) becomes dominant when $pH > pK_{a2}$. Similar calculations can be carried out for triprotic acids (Fig. 10.21).

> *The fraction of deprotonated species increases as the pH is increased, as summarized in Figs. 10.20 and 10.21.*

AUTOPROTOLYSIS AND pH

Suppose we were asked to estimate the pH of 1.0×10^{-8} M HCl(aq). If we used the techniques of Example 10.3 to calculate the pH from the concentration of the acid itself, we would find pH = 8.00. That value, though, is absurd, because it lies on the basic side of neutrality, whereas HCl is an acid! The error stems from there being two sources of hydronium ions, whereas we have considered only one. At very low acid concentrations, the supply of hydronium ions from the autoprotolysis of water is close to the supply provided by the very low concentration of HCl, and both supplies must be taken into account. The following two sections explain how to take autoprotolysis into account, first for strong acids and bases and then for weak ones.

10.18 Very Dilute Solutions of Strong Acids and Bases

We have to include the contribution of autoprotolysis to pH only when the concentration of strong acid or base is less than about 10^{-6} mol·L^{-1}. To calculate the pH in such a case, we need to consider all species in solution. As an example, consider a solution of HCl, a strong acid. Other than water, the species present are H_3O^+, OH^-, and Cl^-. There are three unknown concentrations. To find them, we need three equations.

The first equation takes **charge balance** into account, the requirement that the solution must be electrically neutral overall. That is, the concentration of cations must equal that of anions. Because there is only one type of cation, H_3O^+, the concentration of H_3O^+ ions must equal the sum of the concentrations of the two types of anions, Cl^- and OH^-. The charge-balance relation $[H_3O^+] = [Cl^-] + [OH^-]$ then tells us that

$$[OH^-] = [H_3O^+] - [Cl^-]$$

The second equation takes **material balance** into account, the requirement that all the added solute must be accounted for even though it is now present as ions. Because HCl is a strong acid, the concentration of Cl^- ions in the solution is equal

If an ion is doubly charged, we multiply its concentration by 2 in the charge balance equation, and likewise for triply charged ions.

to the concentration of the HCl added initially (all the HCl molecules are deprotonated). If we denote the numerical value of that initial concentration by $[HCl]_{initial}$, the material-balance relation is $[Cl^-] = [HCl]_{initial}$. We can combine it with the preceding equation to write

$$[OH^-] = [H_3O^+] - [HCl]_{initial}$$

The third equation is the expression for the autoprotolysis constant, K_w (Eq. 1). We can substitute the preceding expression for $[OH^-]$ into the expression for K_w:

$$K_w = [H_3O^+][OH^-] = [H_3O^+]([H_3O^+] - [HCl]_{initial})$$

and rearrange it into a quadratic equation:

$$[H_3O^+]^2 - [HCl]_{initial}[H_3O^+] - K_w = 0 \tag{16}$$

As shown in Example 10.14, we can use the quadratic formula to solve this equation for the concentration of hydronium ions.

Now consider a very dilute solution of a strong base, such as NaOH. Apart from water, the species present in solution are Na^+, OH^-, and H_3O^+. As we did for HCl, we can write down three equations relating the concentrations of these ions by using charge balance, material balance, and the autoprotolysis constant. Because the cations present are hydronium ions and sodium ions, the charge-balance relation is

$$[OH^-] = [H_3O^+] + [Na^+]$$

The concentration of sodium ions is the same as the initial concentration of the NaOH, $[NaOH]_{initial}$, and so the material-balance relation is $[Na^+] = [NaOH]_{initial}$. It follows that

$$[OH^-] = [H_3O^+] + [NaOH]_{initial}$$

The autoprotolysis constant now becomes

$$K_w = [H_3O^+]([H_3O^+] + [NaOH]_{initial}) \tag{17}$$

The concentration of hydronium ions can be obtained by solving this equation with the quadratic formula.

EXAMPLE 10.14   **Sample exercise: Finding the pH of a very dilute aqueous solution of a strong acid**

What is the pH of 8.0×10^{-8} M HCl(aq)?

SOLUTION   Let $[HCl]_{initial} = 8.0 \times 10^{-8}$ mol·L^{-1} and $[H_3O^+] = x$; substitute them in Eq. 16.

From $[H_3O^+]^2 -$ $[HCl]_{initial}[H_3O^+] -$ $K_w = 0$,

$$x^2 - (8.0 \times 10^{-8})x - (1.0 \times 10^{-14}) = 0$$

From the quadratic formula,

$$x = \frac{(8.0 \times 10^{-8}) \pm \sqrt{(-8.0 \times 10^{-8})^2 - 4(1)(-1.0 \times 10^{-14})}}{2}$$

$$= 1.5 \times 10^{-7} \text{ or } -6.8 \times 10^{-8}$$

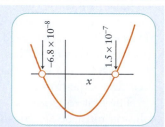

Reject the negative root and use pH = $-\log x$.

$$pH = -\log(1.5 \times 10^{-7}) = 6.82$$

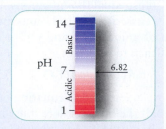

SELF-TEST 10.18A What is the pH of 1.0×10^{-7} M $HNO_3(aq)$?

[*Answer: 6.79*]

SELF-TEST 10.18B What is the pH of 2.0×10^{-7} M $NaOH(aq)$?

> *In very dilute solutions of strong acids and bases, the pH is significantly affected by the autoprotolysis of water. The pH is determined by solving three simultaneous equations: the charge-balance equation, the material-balance equation, and the expression for K_w.*

10.19 Very Dilute Solutions of Weak Acids

Autoprotolysis also contributes to the pH of very dilute solutions of weak acids. In fact, some acids, such as hypoiodous acid, HIO, are so weak and undergo so little deprotonation that the autoprotolysis of water almost always contributes significantly to the pH. To find the pH of these solutions, we must take into account the autoprotolysis of water.

HOW DO WE DO THAT?

The calculation of pH for very dilute solutions of a weak acid HA is similar to that for strong acids in Section 10.18. It is based on the fact that, apart from water, there are four species in solution—namely, HA, A^-, H_3O^+, and OH^-. Because there are four unknowns, we need four equations to find their concentrations. Two relations that we can use are the autoprotolysis constant of water and the acidity constant of the acid HA:

$$K_w = [H_3O^+][OH^-] \qquad K_a = \frac{[H_3O^+][A^-]}{[HA]}$$

Charge balance provides a third equation:

$$[H_3O^+] = [OH^-] + [A^-]$$

Material balance provides the fourth equation: the total concentration of A groups (for instance, F atoms if the acid added is HF) must be equal to the initial concentration of the acid:

$$[HA]_{initial} = [HA] + [A^-]$$

To find an expression for the concentration of hydronium ions in terms of the initial concentration of the acid, we use the charge-balance relation to express the concentration of A^- in terms of $[H_3O^+]$:

$$[A^-] = [H_3O^+] - [OH^-]$$

Then we express $[OH^-]$ in terms of the hydronium ion concentration by using the autoprotolysis expression:

$$[A^-] = [H_3O^+] - \frac{K_w}{[H_3O^+]}$$

When we substitute this expression into the material-balance equation, we obtain

$$[HA] = [HA]_{initial} - [A^-] = [HA]_{initial} - \left([H_3O^+] - \frac{K_w}{[H_3O^+]}\right)$$

$$= [HA]_{initial} - [H_3O^+] + \frac{K_w}{[H_3O^+]}$$

Now we substitute these expressions for [HA] and $[A^-]$ into K_a to obtain

$$K_a = \frac{[H_3O^+]\left([H_3O^+] - \dfrac{K_w}{[H_3O^+]}\right)}{[HA]_{initial} - [H_3O^+] + \dfrac{K_w}{[H_3O^+]}}$$

The expression we just derived is certainly intimidating, but many experimental conditions allow it to be simplified. For example, in many solutions of weak

Remember that $[H_3O^+]$ is really $[H_3O^+]/c°$.

acids, $[H_3O^+] > 10^{-6}$ (that is, pH < 6). Under these conditions, $K_w/[H_3O^+] < 10^{-8}$ and we can ignore this term in both the numerator and the denominator:

$$K_a = \frac{[H_3O^+]^2}{[HA]_{initial} - [H_3O^+]} \qquad (18a)$$

However, when the acid is so dilute or so weak that $[H_3O^+] \leq 10^{-6}$ (that is, when the pH lies between 6 and 7), then we must solve the full expression for K_a, which can be rearranged into

$$[H_3O^+]^3 + K_a[H_3O^+]^2 - (K_w + K_a[HA]_{initial})[H_3O^+] - K_aK_w = 0 \qquad (18b)$$

This is a cubic equation in $[H_3O^+]$. To solve this equation, it is best to use a graphing calculator or mathematical software like that on the Web site for this text.

EXAMPLE 10.15   **Sample exercise: Estimating the pH of a dilute aqueous solution of a weak acid when the autoprotolysis of water must be considered.**

Use Eq. 18b to estimate the pH of a 1.0×10^{-4} M aqueous phenol solution.

SOLUTION Because Eq. 18b is complicated, first find the numerical values of the third and fourth terms. For simplicity, write $x = [H_3O^+]$.

Find K_a for phenol in Table 10.1.	$K_a = 1.3 \times 10^{-10}$
Evaluate $K_w + K_a[HA]_{initial}$.	$K_w + K_a[HA]_{initial} = 1.0 \times 10^{-14} + (1.3 \times 10^{-10}) \times (1.0 \times 10^{-4}) = 2.3 \times 10^{-14}$
Evaluate K_aK_w.	$K_aK_w = (1.3 \times 10^{-10}) \times (1.0 \times 10^{-14}) = 1.3 \times 10^{-24}$
Substitute the values into Eq. 18b.	$x^3 + (1.3 \times 10^{-10})x^2 - (2.3 \times 10^{-14})x - 1.3 \times 10^{-24} = 0$
To simplify the coefficients, write $x = y \times 10^{-7}$ and divide the resulting equation through by 10^{-21}.	$y^3 + 0.0013y^2 - 2.3y - 0.0013 = 0$ 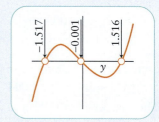
Find the positive root.	The only positive root is $y = 1.516$; so $x = 1.5 \times 10^{-7}$
From pH $= -\log[H_3O^+]$ and $[H_3O^+] = x$,	pH $= -\log(1.5 \times 10^{-7}) = 6.82$ 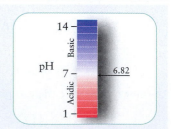

SELF-TEST 10.19A Use Eq. 18b and the information in Table 10.1 to estimate the pH of 2.0×10^{-4} M HCN(aq).

[Answer: 6.48]

SELF-TEST 10.19B Use Eq. 18b and the information in Table 10.1 to estimate the pH of 1.0×10^{-2} M HIO(aq).

In aqueous solutions of very weak acids, the autoprotolysis of water must be taken into account if the hydronium ion concentration is less than 10^{-6} mol·L^{-1}. The expressions for K_w and K_a are combined with the equations for charge balance and material balance to find the pH.

SKILLS YOU SHOULD HAVE MASTERED

❏ 1 Write the formulas for conjugate acids and bases (Example 10.1).

❏ 2 Identify Brønsted and Lewis acids and bases in a chemical reaction (Self-Test 10.2).

❏ 3 Calculate the concentrations of ions in a solution of a strong acid or base (Example 10.2).

❏ 4 Calculate the pH and pOH of a solution of a strong acid or base (Example 10.3 and Section 10.6).

❏ 5 Calculate the H_3O^+ concentration in a solution from its pH (Example 10.4).

❏ 6 Show how the K_a and pK_a of an acid are related to the K_b and pK_b of its conjugate base (Section 10.8).

❏ 7 Use K_a values to predict the relative strengths of two acids or two bases (Example 10.5).

❏ 8 Predict the relative strengths of acids from molecular structures (Example 10.6).

❏ 9 Calculate the pH and percentage deprotonation of a weak acid (Toolbox 10.1 and Example 10.7).

❏ 10 Calculate the value of K_a for a weak acid (Example 10.8).

❏ 11 Calculate the pH and percentage protonation of a weak base (Toolbox 10.2 and Example 10.9).

❏ 12 Calculate the pH of an electrolyte solution (Examples 10.10 and 10.11).

❏ 13 Calculate the pH of a solution of a polyprotic acid (Section 10.14 and Example 10.13).

❏ 14 Estimate the pH of a solution of an amphiprotic salt (Example 10.12).

❏ 15 Calculate the concentrations of all species in a polyprotic acid solution (Toolbox 10.3 and Example 10.13).

❏ 16 Find the pH of an aqueous solution of strong acid or base that is so dilute that the autoprotolysis of water significantly affects the pH (Example 10.14).

❏ 17 Find the pH of an aqueous solution of a weak acid for which the autoprotolysis of water significantly affects the pH (Example 10.15).

EXERCISES

Proton Transfer Reactions

10.1 Write the formulas for the conjugate acids of (a) CH_3NH_2, methylamine; (b) NH_2NH_2, hydrazine; (c) HCO_3^-; and the conjugate bases of (d) HCO_3^-; (e) C_6H_5OH, phenol; (f) CH_3COOH.

10.2 Write the formulas for the conjugate acids of (a) NH_3; (b) S^{2-}; (c) $C_2H_5NH_2$, ethylamine; and the conjugate bases of (d) NH_3; (e) H_2S; (f) $HBrO_4$.

10.3 Write the proton transfer equilibria for the following acids in aqueous solution and identify the conjugate acid–base pairs in each one: (a) H_2SO_4; (b) $C_6H_5NH_3^+$, anilinium ion; (c) $H_2PO_4^-$; (d) HCOOH, formic acid; (e) $NH_2NH_3^+$, hydrazinium ion.

10.4 Write the proton transfer equilibria for the following bases in aqueous solution and identify the conjugate acid–base pairs in each one: (a) CN^-; (b) NH_2NH_2, hydrazine; (c) CO_3^{2-}; (d) HPO_4^{2-}; (e) $CO(NH_2)_2$, urea.

10.5 Identify (a) the Brønsted acid and base in the following reaction, and (b) the conjugate base and acid formed: $HNO_3(aq) + HPO_4^{2-}(aq) \rightleftharpoons NO_3^-(aq) + H_2PO_4^-(aq)$.

10.6 Identify (a) the Brønsted acid and base in the following reaction, and (b) the conjugate base and acid formed: $HSO_3^-(aq) + CH_3NH_3^+(aq) \rightleftharpoons H_2SO_3(aq) + CH_3NH_2(aq)$.

10.7 Below are the molecular models of two oxoacids. Write the name of each acid and then draw the model of its conjugate base. (Red = O, light gray = H, green = Cl, and blue = N.)

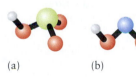

(a)   (b)

10.8 Below are the molecular models of two oxoacids. Write the name of each acid and then draw the model of its conjugate base. (Red = O, light gray = H, green = Cl, and blue = N.)

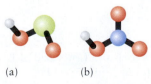

(a)   (b)

10.9 Which of the following reactions can be classified as reactions between Brønsted acids and bases? For those that can be so classified, identify the acid and the base. (*Hint:* It may help to write the net ionic equations.)
(a) $NH_4I(aq) + H_2O(l) \rightarrow NH_3(aq) + H_3O^+(aq) + I^-(aq)$
(b) $NH_4I(s) \xrightarrow{\Delta} NH_3(g) + HI(g)$

(c) $CH_3COOH(aq) + NH_3(aq) \rightarrow CH_3CONH_2(aq) + H_2O(l)$
(d) $NH_4I(am) + KNH_2(am) \rightarrow KI(am) + 2\ NH_3(l)$ (Note: "am" indicates that liquid ammonia is the solvent.)

10.10 Which of the following reactions can be classified as reactions between Brønsted acids and bases? For those that can be so classified, identify the acid and the base. (*Hint:* It may help to write the net ionic equations.)
(a) $KOH(aq) + CH_3I(aq) \rightarrow CH_3OH(aq) + KI(aq)$
(b) $AgNO_3(aq) + HCl(aq) \rightarrow AgCl(s) + HNO_3(aq)$
(c) $2\ NaHCO_3(am) + 2\ NH_3(l) \rightarrow Na_2CO_3(s) + (NH_4)_2CO_3(am)$ (Note: "am" indicates that liquid ammonia is the solvent.)
(d) $H_2S(aq) + Na_2S(s) \rightarrow 2\ NaHS(aq)$

10.11 Write the two proton transfer equilibria that demonstrate the amphiprotic character of (a) HCO_3^-; (b) HPO_4^{2-}. Identify the conjugate acid–base pairs in each equilibrium.

10.12 Write the two proton transfer equilibria that show the amphiprotic character of (a) $H_2PO_4^-$; (b) $HC_2O_4^-$, hydrogen oxalate ion. Identify the conjugate acid–base pairs in each equilibrium.

10.13 The two strands of the nucleic acid DNA are held together by four organic bases. The structure of one of these bases, thymine, is shown below. (a) How many protons can this base accept? (b) Draw the structure of each conjugate acid that can be formed. (c) Mark with an asterisk any structure that can show amphiprotic behavior in aqueous solution.

Thymine

10.14 The two strands of the nucleic acid DNA are held together by four organic bases. The structure of one of these bases, cytosine, is shown below. (a) How many protons can this base accept? (b) Draw the structure of each conjugate acid that can be formed. (c) Mark with an asterisk any structure that can show amphiprotic behavior in aqueous solution.

Cytosine

Lewis Acids and Bases

10.15 Draw the Lewis structure for each of the following species and identify each as a Lewis acid or Lewis base:
(a) NH_3; (b) BF_3; (c) Ag^+; (d) F^-; (e) H^-.
10.16 Draw the Lewis structure for each of the following species and identify each as a Lewis acid or Lewis base: (a) H^+; (b) Al^{3+}; (c) CN^-; (d) NO_2^-; (e) CH_3O^- (the C atom is the central atom).

10.17 Write the Lewis structures of each reactant, identify the Lewis acid and the Lewis base, and then write the Lewis

structure of the product (a complex) for the following Lewis acid–base reactions:
(a) $PF_5 + F^- \rightarrow$
(b) $Cl^- + SO_2 \rightarrow$

10.18 Write the Lewis structures of each reactant, identify the Lewis acid and the Lewis base, and then write the Lewis structure of the product (a complex) for the following Lewis acid–base reactions:
(a) $Cl^- + GaCl_3 \rightarrow$
(b) $SF_4 + F^- \rightarrow$

10.19 State whether the following oxides are acidic, basic, or amphoteric: (a) BaO; (b) SO_3; (c) As_2O_3; (d) Bi_2O_3.

10.20 State whether the following oxides are acidic, basic, or amphoteric: (a) SO_2; (b) CaO; (c) P_4O_{10}; (d) TeO_2.

Strong Acids and Bases

10.21 Calculate the molarity of OH^- in solutions with the following H_3O^+ concentrations: (a) $0.020\ mol \cdot L^{-1}$; (b) $1.0 \times 10^{-5}\ mol \cdot L^{-1}$; (c) $3.1\ mmol \cdot L^{-1}$.

10.22 Estimate the molarity of H_3O^+ in solutions with the following OH^- concentrations: (a) $0.024\ mol \cdot L^{-1}$; (b) $4.5 \times 10^{-5}\ mol \cdot L^{-1}$; (c) $1.60\ mmol \cdot L^{-1}$.

10.23 The value of K_w for water at body temperature (37°C) is 2.1×10^{-14}. (a) What is the molarity of H_3O^+ ions and the pH of neutral water at 37°C? (b) What is the molarity of OH^- in neutral water at 37°C?

10.24 The concentration of H_3O^+ ions at the freezing point of water is $3.9 \times 10^{-8}\ mol \cdot L^{-1}$. (a) Calculate K_w and pK_w at 0.0°C. (b) What is the pH of neutral water at 0.0°C?

10.25 Calculate the initial molarity of $Ba(OH)_2$ and the molarities of Ba^{2+}, OH^-, and H_3O^+ in an aqueous solution that contains 0.25 g of $Ba(OH)_2$ in 0.100 L of solution.

10.26 Calculate the initial molarity of KNH_2 and the molarities of K^+, NH_2^-, OH^-, and H_3O^+ in an aqueous solution that contains 0.50 g of KNH_2 in 0.250 L of solution.

10.27 The pH of several solutions was measured in the research laboratories of a food company; convert each of the following pH values into the molarity of H_3O^+ ions: (a) 3.3 (the pH of sour orange juice); (b) 6.7 (the pH of a saliva sample); (c) 4.4 (the pH of beer); (d) 5.3 (the pH of a coffee sample).

10.28 The pH of several solutions was measured in a hospital laboratory; convert each of the following pH values into the molarity of H_3O^+ ions: (a) 5.0 (the pH of a urine sample); (b) 2.3 (the pH of a sample of lemon juice); (c) 7.4 (the pH of blood); (d) 10.5 (the pH of milk of magnesia).

10.29 A student added solid Na_2O to a 200.0 mL volumetric flask, which was then filled with water, resulting in 200.0 mL of NaOH solution. 5.00 mL of the solution was then transferred to another volumetric flask and diluted to 500.0 mL. The pH of the diluted solution is 13.25. What is the concentration of hydroxide ion in (a) the diluted solution? (b) the original solution? (c) What mass of Na_2O was added to the first flask?

10.30 A student added solid K_2O to a 500.0 mL volumetric flask, which was then filled with water, resulting in 500.0 mL of KOH solution. 10.0 mL of the solution was then transferred to another volumetric flask and diluted to

300.0 mL. The pH of the diluted solution is 14.12. What is the concentration of hydroxide ion in (a) the diluted solution? (b) the original solution? (c) What mass of K_2O was added to the first flask?

10.31 Calculate the pH and pOH of each of the following aqueous solutions of strong acid or base: (a) 0.0146 M $HNO_3(aq)$; (b) 0.11 M HCl(aq); (c) 0.0092 M $Ba(OH)_2(aq)$; (d) 2.00 mL of 0.175 M KOH(aq) after dilution to 0.500 L; (e) 13.6 mg of NaOH dissolved in 0.350 L of solution; (f) 75.0 mL of 3.5×10^{-4} M HBr(aq) after dilution to 0.500 L.

10.32 Calculate the pH and pOH of each of the following aqueous solutions of strong acid or base: (a) 0.0356 M HI(aq); (b) 0.0725 M HCl(aq); (c) 3.46×10^{-3} M $Ba(OH)_2(aq)$; (d) 10.9 mg of KOH dissolved in 10.0 mL of solution; (e) 10.0 mL of 5.00 M NaOH(aq) after dilution to 2.50 L; (f) 5.0 mL of 3.5×10^{-4} M $HClO_4(aq)$ after dilution to 25.0 mL.

Weak Acids and Bases

10.33 Give the K_a value of each of the following acids: (a) phosphoric acid, H_3PO_4, $pK_{a1} = 2.12$; (b) phosphorous acid, H_3PO_3, $pK_{a1} = 2.00$; (c) selenous acid, H_2SeO_3, $pK_{a1} = 2.46$; (d) hydrogen selenate ion, $HSeO_4^-$, $pK_{a2} = 1.92$. (e) List the acids in order of increasing strength.

10.34 Give the pK_b values of the following bases: (a) ammonia, NH_3, $K_b = 1.8 \times 10^{-5}$; (b) deuterated ammonia, ND_3, $K_b = 1.1 \times 10^{-5}$; (c) hydrazine, NH_2NH_2, $K_b = 1.7 \times 10^{-6}$; (d) hydroxylamine, NH_2OH, $K_b = 1.1 \times 10^{-8}$. (e) List the bases in order of increasing strength.

10.35 For each of the following weak acids, write the proton transfer equilibrium equation and the expression for the equilibrium constant K_a. Identify the conjugate base, write the appropriate proton transfer equation, and write the expression for the basicity constant K_b. (a) $HClO_2$; (b) HCN; (c) C_6H_5OH.

10.36 For each of the following weak bases, write the proton transfer equilibrium equation and the expression for the equilibrium constant K_b. Identify the conjugate acid, write the appropriate proton transfer equation, and write the expression for the acidity constant K_a. (a) $(CH_3)_2NH$, dimethylamine; (b) $C_{14}H_{10}N_2$, nicotine; (c) $C_6H_5NH_2$, aniline.

10.37 Using data available in Tables 10.1 and 10.2, place the following acids in order of increasing strength: HNO_2, $HClO_2$, $^+NH_3OH$, $(CH_3)_2NH_2^+$.

10.38 Using data available in Tables 10.1 and 10.2, place the following acids in order of increasing strength: HCOOH, $(CH_3)_3NH^+$, $N_2H_5^+$, HF.

10.39 Using data available in Tables 10.1 and 10.2, place the following bases in order of increasing strength: F^-, NH_3, $CH_3CO_2^-$, C_5H_5N (pyridine).

10.40 Using data available in Tables 10.1 and 10.2, place the following bases in order of increasing strength: CN^-, $(C_2H_5)_3N$, N_2H_4 (hydrazine), BrO^-.

10.41 Label each of the following species as a strong or weak acid. Refer to Tables 10.1, 10.2, and 10.3. (a) $HClO_3$; (b) H_2S; (c) HSO_4^-; (d) $CH_3NH_3^+$; (e) HCO_3^-; (f) HNO_3; (g) CH_4.

10.42 Label each of the following species as a strong or weak base. Refer to Tables 10.1, 10.2, and 10.3. (a) O^{2-}; (b) Br^-; (c) HSO_4^-; (d) HCO_3^-; (e) CH_3NH_2; (f) H_2; (g) CH_3^-.

10.43 The pK_a of HIO, hypoiodous acid, is 10.64 and that of HIO_3, iodic acid, is 0.77. Account for the difference in acid strength.

10.44 The pK_a of HClO, hypochlorous acid, is 7.53 and that of HBrO, hypobromous acid, is 8.69. Account for the difference in acid strength.

10.45 Determine which acid in each of the following pairs is stronger and explain why: (a) HF or HCl; (b) HClO or $HClO_2$; (c) $HBrO_2$ or $HClO_2$; (d) $HClO_4$ or H_3PO_4; (e) HNO_3 or HNO_2; (f) H_2CO_3 or H_2GeO_3.

10.46 Determine which acid in each of the following pairs is stronger and explain why: (a) H_3AsO_4 or H_3PO_4; (b) $HBrO_3$ or HBrO; (c) H_3PO_4 or H_3PO_3; (d) H_2Te or H_2Se; (e) H_2S or HCl; (f) HClO or HIO.

10.47 Suggest an explanation for the different strengths of (a) acetic acid and trichloroacetic acid; (b) acetic acid and formic acid.

10.48 Suggest an explanation for the different strengths of (a) ammonia and methylamine; (b) hydrazine and hydroxylamine.

10.49 The values of K_a for phenol and 2,4,6-trichlorophenol (see following structures) are 1.3×10^{-10} and 1.0×10^{-6}, respectively. Which is the stronger acid? Account for the difference in acid strength.

Phenol   2,4,6-Trichlorophenol

10.50 The value of pK_b for aniline is 9.37 and that for 4-chloroaniline is 9.85 (see the following structures). Which is the stronger base? Account for the difference in base strength.

Aniline   4-Chloroaniline

10.51 Arrange the following bases in order of increasing strength on the basis of the pK_a values of their conjugate acids, which are given in parentheses: (a) ammonia (9.26); (b) methylamine (10.56); (c) ethylamine (10.81); (d) aniline (4.63) (see Exercise 10.50). Is there a simple pattern of strengths?

10.52 Arrange the following bases in order of increasing strength on the basis of the pK_a values of their conjugate acids, which are given in parentheses: (a) aniline (4.63) (see Exercise 10.50); (b) 2-hydroxyaniline (4.72); (c) 3-hydroxyaniline (4.17); (d) 4-hydroxyaniline (5.47). Is there a simple pattern of strengths?

NH$_2$ ... OH

2-Hydroxyaniline   3-Hydroxyaniline   4-Hydroxyaniline

The pH of Solutions of Weak Acids and Bases

Refer to Tables 10.1 and 10.2 for the appropriate K_a and K_b values for the following exercises.

10.53 Calculate the pH and pOH of each of the following aqueous solutions: (a) 0.29 M CH_3COOH(aq); (b) 0.29 M CCl_3COOH(aq); (c) 0.29 M $HCOOH$(aq). (d) Explain any differences in pH on the basis of molecular structure.

10.54 Lactic acid is produced by muscles during exercise. Calculate the pH and pOH of each of the following aqueous solutions of lactic acid, $CH_3CH(OH)COOH$: (a) 0.12 M; (b) 1.2×10^{-3} M; (c) 1.2×10^{-5} M.

10.55 Calculate the pH, pOH, and percentage protonation of solute in each of the following aqueous solutions: (a) 0.057 M NH_3(aq); (b) 0.162 M NH_2OH(aq); (c) 0.35 M $(CH_3)_3N$(aq); (d) 0.0073 M codeine, given that the pK_a of its conjugate acid is 8.21.

10.56 Calculate the pOH, pH, and percentage protonation of solute in each of the following aqueous solutions: (a) 0.075 M C_5H_5N(aq), pyridine; (b) 0.0122 M $C_{10}H_{14}N_2$(aq), nicotine; (c) 0.021 M quinine, given that the pK_a of its conjugate acid is 8.52; (d) 0.059 M strychnine, given that the K_a of its conjugate acid is 5.49×10^{-9}.

10.57 (a) When the pH of 0.10 M $HClO_2$(aq) was measured, it was found to be 1.2. What are the values of K_a and pK_a of chlorous acid? (b) The pH of a 0.10 M propylamine, $C_3H_7NH_2$, aqueous solution was measured as 11.86. What are the values of K_b and pK_b of propylamine?

10.58 (a) The pH of 0.015 M HNO_2(aq) was measured as 2.63. What are the values of K_a and pK_a of nitrous acid? (b) The pH of 0.10 M $C_4H_9NH_2$(aq), butylamine, was measured as 12.04. What are the values of K_b and pK_b of butylamine?

10.59 Find the initial concentration of the weak acid or base in each of the following aqueous solutions: (a) a solution of HClO with pH = 4.60; (b) a solution of hydrazine, NH_2NH_2, with pH = 10.20.

10.60 Find the initial concentration of the weak acid or base in each of the following aqueous solutions: (a) a solution of HCN with pH = 5.3; (b) a solution of pyridine, C_5H_5N, with pH = 8.8.

10.61 The percentage deprotonation of benzoic acid in a 0.110 M solution is 2.4%. What is the pH of the solution and the K_a of benzoic acid?

10.62 The percentage deprotonation of veronal (diethylbarbituric acid) in a 0.020 M aqueous solution is 0.14%. What is the pH of the solution and what is the K_a of veronal?

10.63 The percentage protonation of octylamine (an organic base) in a 0.100 M aqueous solution is 6.7%. What is the pH of the solution and what is the K_b of octylamine?

10.64 An aqueous solution is 35.0% by mass methylamine (CH_3NH_2) and has a density of 0.85 g·cm^{-3}. (a) Draw the Lewis structures of a methylamine molecule and its conjugate acid. (b) If 50.0 mL of this solution is diluted to 1000.0 mL, what is the pH of the resulting solution?

10.65 Write the equilibrium constant for the following reaction and calculate the value of K at 298 K for the reaction HNO_2(aq) + NH_3(aq) $\rightleftharpoons$ NH_4^+(aq) + NO_2^-(aq) using the data in Tables 10.1 and 10.2.

10.66 Write the equilibrium constant for the reaction HIO_3(aq) + NH_2NH_2(aq) $\rightleftharpoons$ $NH_2NH_3^+$(aq) + IO_3^-(aq). and calculate the value of K at 298 K using the data in Tables 10.1 and 10.2.

10.67 Determine whether an aqueous solution of each of the following salts has a pH equal to, greater than, or less than 7. If pH > 7 or pH < 7, write a chemical equation to justify your answer. (a) NH_4Br; (b) Na_2CO_3; (c) KF; (d) KBr; (e) $AlCl_3$; (f) $Cu(NO_3)_2$.

10.68 Determine whether an aqueous solution of each of the following salts has a pH equal to, greater than, or less than 7. If pH > 7 or pH < 7, write a chemical equation to justify your answer. (a) $K_2C_2O_4$, potassium oxalate; (b) $Ca(NO_3)_2$; (c) CH_3NH_3Cl, methylamine hydrochloride; (d) K_3PO_4; (e) $FeCl_3$; (f) C_5H_5NHCl, pyridinium chloride.

10.69 Calculate the pH of each of the following solutions: (a) 0.63 M $NaCH_3CO_2$(aq); (b) 0.19 M NH_4Cl(aq); (c) 0.055 M $AlCl_3$(aq); (d) 0.65 M KCN(aq).

10.70 Calculate the pH of each of the following solutions: (a) 0.25 M CH_3NH_3Cl(aq); (b) 0.13 M Na_2SO_3(aq); (c) 0.071 M $FeCl_3$(aq).

10.71 A 15.5-g sample of CH_3NH_3Cl is dissolved in water to make 450. mL of solution. What is the pH of the solution?

10.72 A 7.8-g sample of $C_6H_5NH_3Cl$ is dissolved in water to make 350. mL of solution. Determine the percentage deprotonation of the cation.

10.73 During the analysis of an unknown acid HA, a 0.010 M solution of the sodium salt of the acid was found to have a pH of 10.35. Use Table 10.1 to write the formula of the acid.

10.74 During the analysis of an unknown weak base B, a 0.10 M solution of the nitrate salt of the base was found to have a pH of 3.13. Use Table 10.2 to write the formula of the base.

10.75 (a) A 150.-mL sample of 0.020 M $NaCH_3CO_2$(aq) is diluted to 0.500 L. What is the concentration of acetic acid at equilibrium? (b) What is the pH of a solution resulting from the dissolution of 2.16 g of NH_4Br in water to form 0.400 L of solution?

10.76 (a) A 25.0-mL sample of 0.250 M KCN(aq) is diluted to 100.0 mL. What is the concentration of hydrocyanic acid at equilibrium? (b) A 1.59-g sample of $NaHCO_3$ is dissolved in water to form 200.0 mL of solution. What is the pH of the solution?

10.77 A 3.38-g sample of the sodium salt of alanine, $NaCH_3CH(NH_2)CO_2$, is dissolved in water, and then the solution is diluted to 50.0 mL. For alanine, $K_{a1} = 4.57 \times 10^{-3}$, $K_{a2} = 1.30 \times 10^{-10}$. What is the pH of the resulting solution?

10.78 The drug amphetamine, $C_6H_5CH_2CH(CH_3)NH_2$, $K_b = 7.8 \times 10^{-4}$, is usually marketed as the hydrogen bromide salt, $C_6H_5CH_2CH(CH_3)NH_3^+Br^-$, because it is much more stable in this solid form. Determine the pH of a solution made from dissolving 6.48 g of the salt to form 200.0 mL of solution.

Polyprotic Acids and Bases

Refer to Table 10.9 for the K_a values needed for the following exercises.

10.79 Write the stepwise proton transfer equilibria for the deprotonation of (a) sulfuric acid, H_2SO_4; (b) arsenic acid, H_3AsO_4; (c) phthalic acid, $C_6H_4(COOH)_2$.

10.80 Write the stepwise proton transfer equilibria for the deprotonation of (a) phosphoric acid, H_3PO_4; (b) adipic acid; $(CH_2)_4(COOH)_2$; (c) succinic acid, $(CH_2)_2(COOH)_2$.

10.81 Calculate the pH of 0.15 M H_2SO_4(aq) at 25°C.

10.82 Calculate the pH of 0.010 M H_2SeO_4(aq), given that K_{a1} is very large and $K_{a2} = 1.2 \times 10^{-2}$.

10.83 Calculate the pH of each of the following diprotic acid solutions at 25°C, ignoring second deprotonations only when the approximation is justified: (a) 0.010 M H_2CO_3(aq); (b) 0.10 M $(COOH)_2$(aq); (c) 0.20 M H_2S(aq).

10.84 Calculate the pH of each of the following diprotic acid solutions at 25°C, ignoring second deprotonations only when that approximation is justified: (a) 0.10 M H_2S(aq); (b) 0.15 M $H_2C_4H_4O_6$(aq), tartaric acid; (c) 1.1×10^{-3} M H_2TeO_4(aq), telluric acid, for which $K_{a1} = 2.1 \times 10^{-8}$ and $K_{a2} = 6.5 \times 10^{-12}$.

10.85 Estimate the pH of (a) 0.15 M $NaHSO_3$(aq); (b) 0.050 M $NaHSO_3$(aq).

10.86 Estimate the pH of (a) 0.125 M $NaHCO_3$(aq); (b) 0.250 M $KHCO_3$(aq).

10.87 Citric acid, which is extracted from citrus fruits and pineapples, undergoes three successive deprotonations with pK_a values of 3.14, 5.95, and 6.39. Estimate the pH of (a) a 0.15 M aqueous solution of the monosodium salt; (b) a 0.075 M aqueous solution of the disodium salt.

10.88 Selenous acid, H_2SeO_3, undergoes two deprotonations with pK_a values of 2.46 and 7.31. Estimate the pH of 0.100 M $NaHSeO_3$(aq).

10.89 Calculate the concentrations of H_2CO_3, HCO_3^-, CO_3^{2-}, H_3O^+, and OH^- present in 0.0456 M H_2CO_3(aq).

10.90 Calculate the concentrations of H_2SO_3, HSO_3^-, SO_3^{2-}, H_3O^+, and OH^- present in 0.125 M H_2SO_3(aq).

10.91 Calculate the concentrations of H_2CO_3, HCO_3^-, CO_3^{2-}, H_3O^+, and OH^- present in 0.0456 M Na_2CO_3(aq).

10.92 Calculate the concentrations of H_2SO_3, HSO_3^-, SO_3^{2-}, H_3O^+, and OH^- present in 0.125 M Na_2SO_3(aq).

10.93 For each of the following polyprotic acids, state which species (H_2A, HA^-, or A^{2-}) you expect to be the form present in highest concentration in aqueous solution at pH = 6.50: (a) phosphorous acid; (b) oxalic acid; (c) hydrosulfuric acid.

10.94 For each of the following polyprotic acids, state which species (H_2A, HA^-, or A^{2-} of the diprotic acid or H_3A, H_2A^-, HA^{2-}, or A^{3-} for triprotic acids) you expect to be the form present in highest concentration in aqueous solution at pH = 5.0: (a) tartaric acid; (b) hydrosulfuric acid; (c) phosphoric acid.

10.95 A large volume of 0.150 M H_2SO_3(aq) is treated with a solid, strong base to adjust the pH to 5.50. Assume that the addition of the base does not significantly affect the volume of the solution. Estimate the concentrations of H_2SO_3, HSO_3^-, and SO_3^{2-} present in the final solution.

10.96 A large volume of 0.250 M H_2S(aq) is treated with a solid, strong base to adjust the pH to 9.35. Assume that the addition of the base does not significantly affect the volume of the solution. Estimate the concentrations of H_2S, HS^-, and S^{2-} present in the final solution.

10.97 Calculate the pH of each of the following acid solutions at 25°C; ignore second deprotonations only when that approximation is justified. (a) 1.0×10^{-4} M H_3BO_3(aq), boric acid acts as a monoprotic acid; (b) 0.015 M H_3PO_4(aq); (c) 0.10 M H_2SO_3(aq).

10.98 Calculate the molarities of $(COOH)_2$, $HOOCCO_2^-$, $(CO_2)_2^{2-}$, H_3O^+, and OH^- in 0.10 M $(COOH)_2$(aq).

10.99 Compute the concentrations of the phosphate species in a H_3PO_4 solution that has pH = 2.25, if the total concentration of all four forms of the dissolved phosphates is 15 mmol·L^{-1}.

10.100 Ammonium acetate is prepared by the reaction of equal amounts of ammonium hydroxide and acetic acid. Determine the concentration of all solute species present in 0.100 M $NH_4CH_3CO_2$(aq).

Autoprotolysis and pH

10.101 Calculate the pH of 6.55×10^{-7} M $HClO_4$(aq).

10.102 Calculate the pH of 7.49×10^{-8} M HI(aq).

10.103 Calculate the pH of 9.78×10^{-8} M KOH(aq).

10.104 Calculate the pH of 8.23×10^{-7} M $NaNH_2$(aq).

For Exercises 10.105 through 10.108, we suggest that you use a graphing calculator to solve the equation or suitable computer software to solve the set of simultaneous equations.

10.105 (a) Calculate the pH of 1.00×10^{-4} M and 1.00×10^{-6} M HBrO(aq), ignoring the effect of the autoprotolysis of water. (b) Repeat the calculations, taking into account the autoprotolysis of water.

10.106 (a) Calculate the pH of 2.50×10^{-4} M and 2.50×10^{-6} M phenol(aq), ignoring the effect of the autoprotolysis of water. (b) Repeat the calculations, taking into account the autoprotolysis of water.

10.107 (a) Calculate the pH of 8.50×10^{-5} M and 7.37×10^{-6} M HCN(aq), ignoring the effect of the autoprotolysis of water. (b) Repeat the calculations, taking into account the autoprotolysis of water.

10.108 (a) Calculate the pH of 1.89×10^{-5} M and 9.64×10^{-7} M HClO(aq), ignoring the effect of the autoprotolysis of water. (b) Repeat the calculations, taking into account the autoprotolysis of water.

Integrated Exercises

10.109 The images below represent the solutes in the solutions of three acids (water molecules are not shown, hydrogen atoms and hydronium ions are represented by small gray spheres, conjugate bases by large colored spheres). (a) Which acid is a strong acid? (b) Which acid has the strongest conjugate base? (c) Which acid has the largest pK_a? Explain each of your answers.

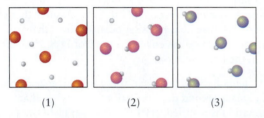

(1)   (2)   (3)

10.110 The images below represent the solutes in the solutions of three salts (water molecules are not shown, hydrogen atoms and ions are represented by small gray spheres, hydroxide ions by red and gray spheres, cations by pink spheres and anions by green spheres). (a) Which salt has a cation that is the conjugate acid of a weak base? (b) Which salt has an anion that is the conjugate base of a weak acid? (c) Which salt has an anion that is the conjugate base of a strong acid? Explain each of your answers.

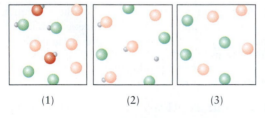

(1)   (2)   (3)

10.111 Hydronium and hydroxide ions appear to move through water much faster than other kinds of ions. Explain this observation.

10.112 In a solution labeled "0.10 M H_2SO_4," which of the following best describes the composition of the solution?
(a) $[H_2SO_4] = 0.10$ M
(b) $[H_3O^+] > 0.10$ M, $[SO_4^{2-}] = 0.10$ M
(c) $[H_3O^+] = 0.20$ M, $[SO_4^{2-}] = 0.10$ M
(d) $[H_3O^+] > 0.10$ M, $[HSO_4^{2-}] < 0.10$ M
(e) $[H_3O^+] = 0.10$ M, $[HSO_4^{2-}] = 0.10$ M

10.113 A combustion analysis of 1.200 g of an anhydrous sodium salt gave 0.942 g of CO_2, 0.0964 g of H_2O, and

0.246 g of Na. The molar mass of the salt is 112.02 g·mol^{-1}. (a) What is the chemical formula of the salt? (b) The salt contains carboxylate groups ($-CO_2^-$), and the carbon atoms are bonded together. Draw the Lewis structure of the anion. (c) Next, 1.50 g of this sodium salt was dissolved in water and diluted to 50.0 mL. Identify the dissolved substance: Is it an acid, a base, or is it amphiprotic? Calculate the pH of the solution.

10.114 Decide on the basis of the information in Table 10.3 whether carbonic acid is a strong or weak acid in liquid ammonia solvent. Explain your answer.

10.115 Draw the Lewis structure for boric acid, $B(OH)_3$. (a) Is resonance important for its description? (b) The proton transfer equilibrium for boric acid is given in a footnote to Table 10.1. In that reaction does boric acid act as a Lewis acid, a Lewis base, or neither? Justify your answer by using Lewis structures of boric acid and its conjugate base.

10.116 Dinitrogen monoxide, N_2O, reacts with water to form hyponitrous acid, $H_2N_2O_2$(aq), in a Lewis acid-base reaction. (a) Write the chemical equation for the reaction. (b) Draw the Lewis structures of N_2O and $H_2N_2O_2$ (the atoms are attached in the order HONNOH). (c) Identify the Lewis acid and Lewis base.

10.117 Acetic acid is used as a solvent for some reactions between acids and bases. (a) Nitrous acid and carbonic acids are both weak acids in water. Will either of them act as a strong acid in acetic acid? Explain your answer. (b) Will ammonia act as a strong or weak base in acetic acid? Explain your answer.

10.118 Sections 10.9 and 10.10 discuss the relationship between molecular structure and the strengths of acids. The same ideas can be applied to bases. (a) Explain the relative strengths of the bases OH^-, NH_2^-, and CH_3^- (see Table 10.3). (b) Explain why NH_3 is a weak base in water, but PH_3 forms essentially neutral solutions. (c) If you were ranking the species in (a) or (b) as Lewis bases, would your rankings be the same or different? Explain your reasoning.

10.119 Using the thermodynamic data available in Appendix 2A, calculate the acidity constant of HF(aq).

10.120 Calculate the pH of (a) 0.095 M NaH_2AsO_4(aq) and (b) 0.148 M Na_2HAsO_4(aq). For H_3AsO_4, $pK_{a1} = 2.25$, $pK_{a2} = 6.77$, and $pK_{a3} = 11.60$.

10.121 The autoprotolysis constant, K_{D_2O}, for heavy water at 25°C is 1.35×10^{-15}. (a) Write the chemical equation for the autoprotolysis of D_2O. (b) Evaluate pK_{D_2O} for D_2O at 25°C. (c) Calculate the molarities of D_3O^+ and OD^- in pure heavy water at 25°C. (d) Evaluate the pD and pOD of heavy water at 25°C. (e) Find the relation between pD, pOD, and pK_{D_2O}.

10.122 Although many chemical reactions take place in water, it is often necessary to use other solvents; liquid ammonia (normal boiling point, $-33°C$) has been used extensively. Many of the reactions that take place in water have analogous reactions in liquid ammonia. (a) Write the chemical equation for the autoprotolysis of NH_3. (b) What are the formulas of the acid and base species that result from the autoprotolysis of liquid ammonia? (c) The autoprotolysis constant, K_{am}, of liquid ammonia has the value 1×10^{-33} at $-35°C$. What is the value of pK_{am} at that temperature? (d) What is the molarity of NH_4^+ ions in liquid ammonia? (e) Evaluate pNH_4 and pNH_2, which are the analogs of pH and pOH, in liquid ammonia at $-35°C$. (f) Determine the relation between pNH_4, pNH_2, and pK_{am}.

10.123 Use Table 10.9 to determine the percentage deprotonation of 1.00 M aqueous lactic acid. At what temperature will the solution freeze? (Assume that the density of the solution is 1.00 g·cm^{-3}.)

10.124 Is the osmotic pressure of 0.10 M H_2SO_4 the same as, less than, or greater than that of 0.10 M HCl(aq)? Calculate the osmotic pressure of each solution to support your conclusion.

10.125 Estimate the enthalpy of deprotonation of formic acid at 25°C, given that $K_a = 1.765 \times 10^{-4}$ at 20°C and 1.768×10^{-4} at 30°C.

10.126 (a) The value of K_w at 40°C is 3.8×10^{-14}. What is the pH of pure water at 40°C? (b) Use data from part (a) and Exercise 10.23 and the values for the autoprotolysis of water at 25°C to determine graphically the enthalpy and entropy of the autoprotolysis of water. (c) Suggest an interpretation of the sign of $\Delta S°$. (d) Write an equation that describes the pH of pure water as a function of temperature.

10.127 Heavy water is deuterium oxide, D_2O. The standard reaction Gibbs free energy for the autoprotolysis of pure deuterium oxide is $+84.8 \text{ kJ·mol}^{-1}$ at 298 K. (a) If pD is defined analogously to pH, what is the pD of pure D_2O at 298 K? (b) Does pD increase or decrease as the temperature is raised?

10.128 The pK for the autoprotolysis (more precisely, the autodeuterolysis, because a deuteron is being transferred) of heavy water (D_2O) is 15.136 at 20.°C and 13.8330 at 30.°C. Assuming $\Delta H_r°$ for this reaction to be independent of temperature, calculate $\Delta S_r°$ for the autoprotolysis reaction. Suggest an interpretation of the sign. Suggest a reason why the autoprotolysis constant of heavy water differs from that of ordinary water.

10.129 (a) Using the Living Graph "Fractional Composition of Triprotic Acids," plot the fraction of solute species for H_3PO_4 in water against pH. (b) On the graph, label the places where the system functions as a buffer solution. (c) For each region in (b), give the major phosphorus-containing species present in solution. (d) Repeat (a) for H_3AsO_4, using data from Exercise 10.120. (e) For pH = 7.00, determine the major species present in solution and compare the relative fractions of these species for H_3PO_4 and H_3AsO_4.

10.130 The amino acid L-histidine has three acid–base equilibria, with pK_a values of 1.78, 5.97, and 8.97. (a) Using the Living Graph "Fractional Composition of Triprotic Acids," plot the fraction of solute species present in solution in the pH range 0 to 14. Using the graph, determine (b) the major species present in solution at pH = 7.5; (c) the pH value at which equal amounts of the singly and doubly deprotonated forms are present; (d) the pH necessary to achieve greater than 99% fully deprotonated

histidine; and (e) the pH range for which three different deprotonated forms are present in significant quantities.

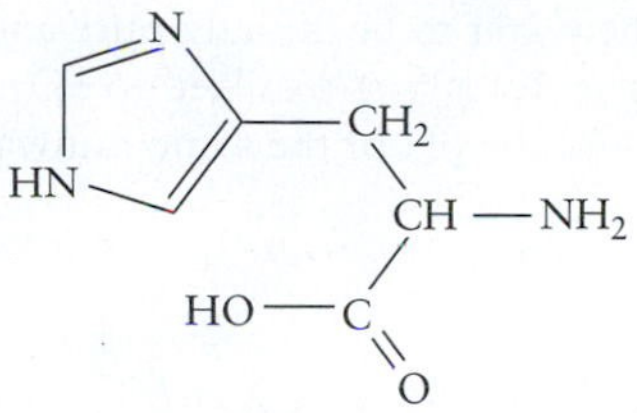

Histidine

10.131 Recall from Section 9.13 that hemoglobin (Hb) molecules in blood carry oxygen molecules from the lungs, where the concentration of oxygen is high, to the tissues where it is low. In the tissues the equilibrium $H_3O^+(aq) + HbO_2^-(aq) \rightleftharpoons HHb(aq) + H_2O(l) + O_2(aq)$ releases oxygen. When muscles work hard, they produce lactic acid as a by-product. (a) What effect will the lactic acid have on the concentration of HbO_2^-? (b) When the hemoglobin returns to the lungs, where oxygen concentration is high, how does the concentration of HbO_2^- change?

10.132 The structures below show a hydrated d-metal ion. Draw the structure of the conjugate base of this complex.

10.133 Like sulfuric acid, a certain diprotic acid, H_2A, is a strong acid in its first deprotonation and a weak acid in its second deprotonation. A solution that is 0.020 M H_2A(aq) has a pH of 1.66. What is the value of K_{a2} for this acid?

10.134 Propanoic acid, CH_3CH_2COOH, has $K_a = 1.3 \times 10^{-5}$. A 50.0-mL sample of 0.250 M CH_3CH_2COOH(aq) is diluted to 850.0 mL. Determine the percentage deprotonation of the acid in the diluted solution.

Chemistry Connections

10.135 Rainwater is naturally slightly acidic due to the dissolved carbon dioxide. Acid rain results when acidic sulfur and nitrogen oxides produced during the combustion of coal and oil react with rainwater (see Box 10.1). (a) The partial pressure of CO_2 in air saturated with water vapor at 25°C and 1.00 atm is 3.04×10^{-4} atm. Henry's constant for CO_2 in water is $2.3 \times 10^{-2} \text{ mol·L}^{-1}\text{·atm}^{-1}$; and, for carbonic acid, $pK_{a1} = 6.37$. Assuming that all the dissolved CO_2 can be thought of as H_2CO_3, verify by calculation that the pH of "normal" rainwater is about 5.7. (b) Scientists investigating acid rain measured the pH of a water sample from a lake and found it to be 4.8. The total concentration of dissolved carbonates in the lake is 4.50 mmol·L^{-1}. Determine the molar concentrations of the carbonate species CO_3^{2-}, HCO_3^-, and H_2CO_3 in the lake. (c) Suppose that 1.00 tonne ($1 \text{ t} = 10^3$ kg) of coal that is 2.5% sulfur by mass is burned in a coal-fired plant. What mass of SO_2 is produced?

(d) What is the pH of rainwater when the SO_2 generated in part (c) dissolves in a volume of water equivalent to 2.0 cm of rainfall over 2.6 km^2? (The pK_{a1} of sulfurous acid is 1.81. Consider the water to be initially pure and at a pH of 7.) (e) If the SO_2 in part (c) is first oxidized to SO_3 before the rainfall, what would the pH of the same rainwater be?

(f) One process used to clean SO_2 from the emissions of coal-fired plants is to pass the stack gases along with air through a wet calcium carbonate slurry, where the following reaction takes place: $CaCO_3(s) + SO_2(g) + O_2(g) \rightarrow CaSO_4(s) + CO_2(g)$. What mass of limestone ($CaCO_3$) is needed to remove 50.0 kg sulfur dioxide from stack gases if the removal process is 90% efficient?

AQUEOUS EQUILIBRIA

What Are the Key Ideas? Ions in solution adopt concentrations that satisfy the equilibrium constants for the reactions in which they participate.

Why Do We Need to Know This Material? The techniques described in this chapter provide the tools that we need to analyze and control the concentrations of ions in solution. A great deal of chemistry is carried out in solution, and so this material is fundamental to understanding chemistry. The ionic compounds released into waterways by individuals, industry, and agriculture can impair the quality of our water supplies. However, these hazardous ions can be identified and removed if we add the right reagents. Aqueous equilibria govern the stabilization of the pH in blood, seawater, and other solutions encountered in biology, medicine, and the environment.

Chapter **11**

What Do We Need to Know Already? This chapter develops the ideas in Chapters 9 and 10 and applies them to equilibria involving ions in aqueous solution. To prepare for the sections on titrations, review Section L. For the discussion of solubility equilibria, review Section I. The discussion of Lewis acids and bases in Section 11.13 is based on Section 10.2.

A primary goal of this chapter is to learn how to achieve control over the pH of solutions of acids, bases, and their salts. The control of pH is crucial for the ability of organisms—including ourselves—to survive, because even minor drifts from the optimum value of the pH can cause enzymes to change their shape and cease to function. The information in this chapter is used in industry to control the pH of reaction mixtures and to purify water. In agriculture it is used to maintain the soil at an optimal pH. In the laboratory it is used to interpret the change in pH of a solution during a titration, one of the most common quantitative analytical technique. It also helps us appreciate the basis of **qualitative analysis,** the identification of the substances and ions present in a sample.

All these applications involve equilibrium between species in solution, especially between species that can exchange protons. The common theme throughout this chapter is that all these equilibria can be treated in a similar way:

1 Identify the solute species present in the solution.

2 Identify the equilibrium relations between the solute species (often by setting up an equilibrium table).

3 Use those relations to calculate the concentrations of the solute species.

MIXED SOLUTIONS AND BUFFERS

We have seen how to estimate the pH of a solution of a weak acid or base (Chapter 10), but suppose that a salt of the acid or base is also present. How does that salt affect the pH of the solution? Suppose we have a dilute hydrochloric acid solution and add to it appreciable concentrations of the conjugate base, the Cl^- ion, as sodium chloride. Because the acid is strong, its conjugate base is extremely weak and so has no measurable effect on pH. The pH of 0.10 M HCl(aq) is about 1.0, even after 0.10 mol NaCl has been added to a liter of the solution. Now suppose instead that the solution contains acetic acid to which sodium acetate has been added (the acetate ion, $CH_3CO_2^-$, is the conjugate base of CH_3COOH). Because the conjugate base of a weak acid is a base, we can predict that adding acetate ions (as sodium acetate) to a solution of acetic acid will increase the pH of the solution. Similarly, suppose we have a solution of ammonia and add ammonium chloride to it. The ammonium ion is a weak acid; therefore, we can predict that adding ammonium ions (as ammonium chloride, for instance) to a solution of ammonia will lower the pH of the solution. Such "mixed solutions"—solutions that contain a weak acid or a weak base and one of its salts—provide a means of stabilizing the pH of aqueous solutions such as blood plasma, seawater, detergents, sap, and reaction mixtures.

11.1 Buffer Action

The most important type of mixed solution is a **buffer,** a solution in which the pH resists change when small amounts of strong acids or bases are added. Buffers are used to calibrate pH meters, to culture bacteria, and to control the pH of solutions in which chemical reactions are taking place. They are also administered intravenously to hospital patients. Human blood plasma is buffered to pH = 7.4; the ocean is buffered to about pH = 8.4 by a complex buffering process that depends on the presence of hydrogen carbonates and silicates. A buffer consists of an aqueous solution of a weak acid and its conjugate base supplied as a salt, or a weak base and its conjugate acid supplied as a salt. Examples are a solution of acetic acid and sodium acetate and a solution of ammonia and ammonium chloride.

When a drop of strong acid is added to water, the pH changes significantly. However, when the same amount is added to a buffer, the pH hardly changes at all. To understand why not, consider the dynamic equilibrium between a weak acid and its conjugate base in water:

$$CH_3COOH(aq) + H_2O(l) \rightleftharpoons H_3O^+(aq) + CH_3CO_2^-(aq)$$

When a few drops of the solution of a strong acid are added to a solution that contains $CH_3CO_2^-$ ions and CH_3COOH molecules in about equal concentrations (and which has pH < 7), the newly arrived H_3O^+ ions transfer protons to the $CH_3CO_2^-$ ions to form CH_3COOH and H_2O molecules (Fig. 11.1). Because the added H_3O^+ ions are removed by the $CH_3CO_2^-$ ions, the pH remains almost unchanged. In effect, the acetate ions act as a "sink" for protons. If a small amount of strong base is added instead, the incoming OH^- ions remove protons from the CH_3COOH molecules to produce $CH_3CO_2^-$ ions and H_2O molecules. In this case, the acetic acid molecules act as a source of protons. Because the added OH^- ions are removed by the CH_3COOH molecules, the concentration of OH^- ions remains nearly unchanged. Consequently, the H_3O^+ concentration (and the pH) is also left nearly unchanged.

Now suppose we add ammonium chloride to aqueous ammonia until the solution contains similar concentrations of $NH_3(aq)$ and $NH_4^+(aq)$. The ammonia equilibrium is

$$NH_3(aq) + H_2O(l) \rightleftharpoons NH_4^+(aq) + OH^-(aq)$$

This solution has pH > 7. When a few drops of a solution of strong base are added, the incoming OH^- ions remove protons from NH_4^+ ions to make NH_3 and H_2O molecules. When instead a few drops of a solution of strong acid are added, the incoming protons attach to NH_3 molecules to make NH_4^+ ions and hence are removed from the solution. In each case, the pH is left almost unchanged.

A buffer is a mixture of a weak conjugate acid–base pair that stabilizes the pH of a solution by providing both a source and a sink for protons (Fig. 11.2).

11.2 Designing a Buffer

Suppose we need to make up a buffer with a particular pH. For instance, we might be culturing bacteria and need to maintain a precise pH to sustain their metabolism. To choose the most appropriate buffer system, we need to know the value of the pH at which a given buffer stabilizes a solution. When doing large numbers of this type of calculation, it is convenient first to rearrange the expression for the

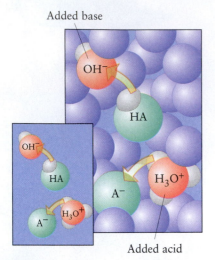

FIGURE 11.1 A solution can act as a buffer if it contains a weak acid, HA, which donates protons when a strong base is added, and the conjugate base, A⁻, which accepts protons when a strong acid is added. In the inset, for clarity water molecules are represented by the blue background.

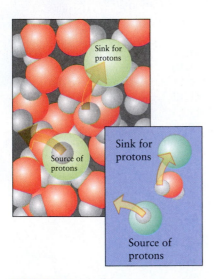

FIGURE 11.2 A buffer solution contains a weak base that acts as a sink for the protons supplied by a strong acid and a weak acid that acts as a source of protons to supply to a strong base being added. The joint action of the source and sink keeps the pH constant when strong acid or base is added. The inset highlights the buffer action by representing the water molecules as a blue background.

acidity constant to give the concentration of hydronium ions in terms of the concentrations of the other species present in the solution.

EXAMPLE 11.1   Calculating the pH of a buffer solution

Suppose we are culturing bacteria that require an acid environment and want to prepare a buffer close to pH = 4. We prepare a buffer solution that is 0.040 M $NaCH_3CO_2(aq)$ and 0.080 M $CH_3COOH(aq)$ at 25°C. What is the pH of the buffer solution?

STRATEGY First, identify the weak acid and its conjugate base. Then, write the proton transfer equilibrium between them, rearrange the expression for K_a to give $[H_3O^+]$, and find the pH, by using the approximation that the equilibrium concentrations of the acid and its conjugate base are essentially the same as the initial concentrations.

SOLUTION The acid is CH_3COOH and its conjugate base is $CH_3CO_2^-$. The equilibrium to consider is

$$CH_3COOH(aq) + H_2O(l) \rightleftharpoons H_3O^+(aq) + CH_3CO_2^-(aq)$$

From Table 10.1, $pK_a = 4.75$ and $K_a = 1.8 \times 10^{-5}$.

Find the equilibrium concentration of H_3O^+ ions from $K_a = [H_3O^+][CH_3CO_2^-]/[CH_3COOH]$.	$$[H_3O^+] = K_a \times \frac{[CH_3COOH]}{[CH_3CO_2^-]}$$
Approximate the conjugate acid and base concentrations by their initial values.	$$[H_3O^+] \approx (1.8 \times 10^{-5}) \times \frac{0.080}{0.040}$$
From pH = $-\log [H_3O^+]$,	$$pH \approx -\log\left\{(1.8 \times 10^{-5}) \times \frac{0.080}{0.040}\right\} = 4.44$$

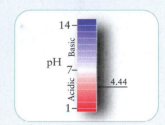

That is, the solution acts as a buffer close to pH = 4.

SELF-TEST 11.1A Calculate the pH of a buffer solution that is 0.15 M $HNO_2(aq)$ and 0.20 M $NaNO_2(aq)$.

[*Answer:* 3.49]

SELF-TEST 11.1B Calculate the pH of a buffer solution that is 0.040 M $NH_4Cl(aq)$ and 0.030 M $NH_3(aq)$.

We can adjust the pH of a buffer solution by adding some acid to lower it or some base to raise it. Another way to adjust the pH of a buffer by adding more salt (which supplies the conjugate acid or base). Example 11.2 shows how to calculate the effect of added acid or base on the pH of a buffer.

EXAMPLE 11.2   Calculating the pH change of a buffered solution

Suppose we dissolve 1.2 g of sodium hydroxide (0.030 mol NaOH) in 500. mL of the buffer solution described in Example 11.1. Calculate the pH of the resulting solution and the change in pH. Assume that the volume of the solution remains unchanged.

STRATEGY The OH⁻ ions added to the buffer solution react with some of the acid of the buffer system, decreasing the amount of acid and increasing the conjugate base by the same amount. We solve this problem in two steps. First, we find the new molar concentrations of the acid and its conjugate base. Then we can rearrange the expression for K_a to obtain the pH of the solution, just as in Example 11.1.

SOLUTION The proton transfer equilibrium is

$$CH_3COOH(aq) + H_2O(l) \rightleftharpoons H_3O^+(aq) + CH_3CO_2^-(aq)$$

$$K_a = \frac{[H_3O^+][CH_3CO_2^-]}{[CH_3COOH]}$$

We use the data in Example 11.1, including $K_a = 1.8 \times 10^{-5}$. The OH^- from the added NaOH reacts with the CH_3COOH according to

$$CH_3COOH(aq) + OH^-(aq) \rightleftharpoons CH_3CO_2^-(aq) + H_2O(l)$$

Step 1 Find the new concentration of acid.

Find the initial amount of CH_3COOH in the solution from $n_J = V[J]$.

$$n_{CH_3COOH}(\text{initial}) = (0.500 \text{ L}) \times (0.080 \text{ mol·L}^{-1})$$
$$= 0.040 \text{ mol}$$

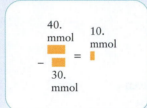

Calculate the amount of CH_3COOH that reacts by using 1 mol CH_3COOH ≏ 1 mol OH^-.

$$n_{CH_3COOH}(\text{reacts}) = (0.030 \text{ mol } OH^-) \times \frac{1 \text{ mol } CH_3COOH}{1 \text{ mol } OH^-}$$
$$= 0.030 \text{ mol } CH_3COOH$$

Calculate the amount of CH_3COOH remaining from $n(\text{final}) = n(\text{initial}) - n(\text{reacts})$.

$$n_{CH_3COOH}(\text{final}) = 0.040 - 0.030 \text{ mol}$$
$$= 0.010 \text{ mol}$$

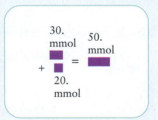

From $[J] = n_J/V$,

$$[CH_3COOH] = \frac{0.010 \text{ mol}}{0.500 \text{ L}} = 0.020 \text{ mol·L}^{-1}$$

Step 2 Find the new concentration of the conjugate base.

Find the initial amount of $CH_3CO_2^-$ in the solution from $n_J = V[J]$.

$$n_{CH_3CO_2^-}(\text{initial}) = (0.500 \text{ L}) \times (0.040 \text{ mol·L}^{-1})$$
$$= 0.020 \text{ mol}$$

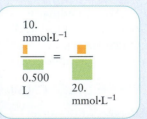

Add the change in the amount of $CH_3CO_2^-$ due to the reaction.

$$n_{CH_3CO_2^-}(\text{final}) = 0.020 + 0.030 \text{ mol} = 0.050 \text{ mol}$$

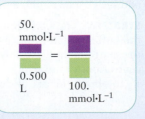

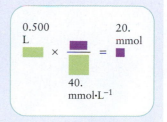

From $[J] = n_J/V$, $$[CH_3CO_2^-] = \frac{0.050 \text{ mol}}{0.500 \text{ L}} = 0.100 \text{ mol·L}^{-1}$$

Step 3 Calculate the pH.

Calculate $[H_3O^+]$ from $[H_3O^+] = K_a[CH_3COOH]/[CH_3CO_2^-]$.

$$[H_3O^+] \approx (1.8 \times 10^{-5}) \times \frac{0.020}{0.100} = 3.6 \times 10^{-6}$$

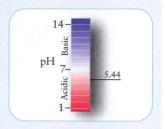

From $pH = -\log[H_3O^+]$, $$pH \approx -\log(3.6 \times 10^{-6}) \approx 5.44$$

We have found that the pH of the solution changes from about 4.4 to about 5.4.

SELF-TEST 11.2A Suppose that 0.0200 mol NaOH(s) is dissolved in 300. mL of the buffer solution of Example 11.1. Calculate the pH of the resulting solution and the change in pH.

[*Answer:* 5.65, an increase of 1.21]

SELF-TEST 11.2B Suppose that 0.0100 mol HCl(g) is dissolved in 500. mL of the buffer solution of Example 11.1. Calculate the pH of the resulting solution and the change in pH.

Buffers are often prepared with equal molar concentrations of the conjugate acid and base. In these "equimolar" solutions,

$$HA(aq) + H_2O(l) \rightleftharpoons H_3O^+(aq) + A^-(aq) \qquad (A)$$

and, when $[HA] = [A^-]$,

$$K_a = \frac{[H_3O^+][A^-]}{[HA]} = \frac{[H_3O^+][A^-]}{[A^-]} = [H_3O^+]$$

It follows from $[H_3O^+] = K_a$ that, when $[HA] = [A^-]$,

$$pH = pK_a \qquad (1)*$$

This very simple result makes it easy to make an initial choice of a buffer: we just select an acid that has a pK_a close to the pH that we require and prepare an equimolar solution with its conjugate base. When we prepare a buffer for pH > 7 we have to remember that the acid is supplied by the salt, that the conjugate base is the base itself, and that the pK_a is that of the *conjugate acid* of the base (and hence related to the pK_b of the base by $pK_a + pK_b = pK_w$). Mixtures in which the conjugate acid and base have unequal concentrations—such as those considered in Examples 11.1 and 11.2—are buffers, but they may be less effective than those in which the concentrations are nearly equal (see Section 11.3). Table 11.1 lists some typical buffer systems.

SELF-TEST 11.3A Which of the buffer systems listed in Table 11.1 would be a good choice to prepare a buffer with pH close to 5?

[*Answer:* CH₃COOH/CH₃CO₂⁻]

TABLE 11.1 Typical Buffer Systems

Composition	pK_a
Acid buffers	
$CH_3COOH/CH_3CO_2^-$	4.75
HNO_2/NO_2^-	3.37
$HClO_2/ClO_2^-$	2.00
Base buffers	
NH_4^+/NH_3	9.25
$(CH_3)_3NH^+/(CH_3)_3N$	9.81
$H_2PO_4^-/HPO_4^{2-}$	7.21

SELF-TEST 11.3B Which of the buffer systems listed in Table 11.1 would be a good choice to prepare a buffer with pH close to 10?

Commercially available buffer solutions can be purchased for virtually any desired pH. A buffer solution commonly used to calibrate pH meters contains 0.025 M $Na_2HPO_4(aq)$ and 0.025 M $KH_2PO_4(aq)$ and has pH = 6.87 at 25°C. However, the method demonstrated in Example 11.1 would give pH = 7.2 for this solution. Because these calculations interpret activities as molarities, not *effective* molarities, they ignore ion–ion interactions; so the values calculated are only approximate.

Because so many chemical reactions in our bodies take place in buffered environments, biochemists commonly need to make quick estimates of pH by employing a form of the expression for K_a that gives the pH directly. For the equilibrium in reaction A, we can rearrange the expression for K_a into

$$[H_3O^+] = K_a \times \frac{[HA]}{[A^-]}$$

from which it follows, by taking the negative logarithms of both sides, that

$$\underbrace{-\log[H_3O^+]}_{pH} = \underbrace{-\log K_a}_{pK_a} - \log\frac{[HA]}{[A^-]}$$

Then, from $\log x = -\log(1/x)$,

$$pH = pK_a - \log\frac{[HA]}{[A^-]} = pK_a + \log\frac{[A^-]}{[HA]}$$

The values of [HA] and $[A^-]$ in this expression are the equilibrium concentrations of acid and base in the solution, not the concentrations added initially. However, a weak acid HA typically loses only a tiny fraction of its protons, and so [HA] is negligibly different from the concentration of the acid used to prepare the buffer, $[HA]_{initial}$. Likewise, only a tiny fraction of the weakly basic anions A^- accept protons, and so $[A^-]$ is negligibly different from the initial concentration of the base used to prepare the buffer. With the approximations $[A^-] \approx [base]_{initial}$ and $[HA] \approx [acid]_{initial}$, we obtain the **Henderson–Hasselbalch equation:**

$$pH = pK_a + \log\frac{[base]_{initial}}{[acid]_{initial}} \qquad (2)*$$

The same expression can be used for a basic buffer, with pK_a that of the conjugate acid of the base (for example, in the case of an ammonia buffer, we would use the pK_a of NH_4^+). If only pK_b is available, calculate pK_a by using Eq. 11b of Chapter 10 ($pK_a + pK_b = pK_w$). For example, for an ammonia/ammonium buffer we would write

$$pH = pK_a(NH_4^+) + \log\frac{[NH_3]_{initial}}{[NH_4^+]_{initial}}$$

A note on good practice: Keep in mind the approximations required for the use of the Henderson–Hasselbalch equation (that the concentrations of both the weak acid and its conjugate base are much greater than the hydronium ion concentration). Because the equation uses molar concentration instead of activities, it also ignores the interactions between ions.

In practice, the Henderson–Hasselbalch equation is used to make rapid estimates of the pH of a mixed solution intended to be used as a buffer, and then the pH is adjusted to the precise value required by adding more acid or base and monitoring the solution with a pH meter.

EXAMPLE 11.3   **Sample exercise: Selecting the composition for a buffer solution with a given pH**

Calculate the ratio of the molarities of CO_3^{2-} and HCO_3^- ions required to achieve buffering at pH = 9.50. The pK_{a2} of H_2CO_3 is 10.25.

SOLUTION The acid is HCO_3^- and its conjugate base is CO_3^{2-}. From Section 10.16, we know that we can ignore any H_2CO_3 formed.

From $pH = pK_a + \log([\text{base}]_{\text{initial}}/[\text{acid}]_{\text{initial}})$,

$$\log\frac{[\text{base}]}{[\text{acid}]} = pH - pK_a$$

Use the initial concentration of CO_3^{2-} for the base and that of HCO_3^- for the acid and substitute the values of pH and pK_a.

$$\log\frac{[CO_3^{2-}]}{[HCO_3^-]} = 9.50 - 10.25 = -0.75$$

Now use $x = 10^{\log x}$:

$$\frac{[CO_3^{2-}]}{[HCO_3^-]} = 10^{-0.75} = 0.18$$

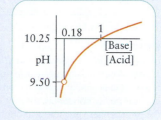

Therefore, the solution acts as a buffer with a pH close to 9.50 if it is prepared by mixing the solutes in the ratio 0.18 mol CO_3^{2-} to 1.0 mol HCO_3^-.

SELF-TEST 11.4A Calculate the ratio of the molarities of acetate ions and acetic acid needed to buffer a solution at pH = 5.25. The pK_a of CH_3COOH is 4.75.

[*Answer:* 3.2:1]

SELF-TEST 11.4B Calculate the ratio of the molarities of benzoate ions and benzoic acid (C_6H_5COOH) needed to buffer a solution at pH = 3.50. The pK_a of C_6H_5COOH is 4.19.

The pH of a buffer solution is close to the pK_a of the weak acid component when the acid and base have similar concentrations.

11.3 Buffer Capacity

Just as a sponge can hold only so much water, a buffer can mop up only so many hydronium ions. Its proton sources and sinks become exhausted if too much strong acid or base is added to the solution. **Buffer capacity** is the maximum amount of acid or base that can be added before the buffer loses its ability to resist large changes in pH. A buffer with a high capacity can maintain its buffering action longer than can one with only a small capacity. The buffer is exhausted when most of the weak base has been converted into its conjugate acid or when most of the weak acid has been converted into its conjugate base. A concentrated buffer has a greater capacity than the same volume of a more dilute solution of the same buffer.

Buffer capacity also depends on the relative concentrations of weak acid and base. Broadly speaking, a buffer is found experimentally to have a high capacity for acid when the amount of base present is at least 10% of the amount of acid. Otherwise, the base is used up quickly as strong acid is added. Similarly, a buffer has a high capacity for base when the amount of acid present is at least 10% of the amount of base, because otherwise the acid is used up quickly as strong base is added.

We can use these numbers to express the range of buffer action in terms of the pH of the solution. The Henderson–Hasselbalch equation shows us that,

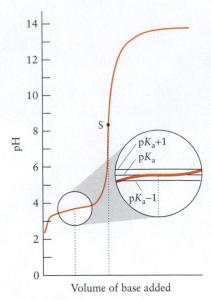

FIGURE 11.3 This plot shows how the pH of a weak acid changes as a strong base is added. When the conjugate acid and base are present at similar concentrations the curve is nearly horizontal, showing that the pH changes very little as more strong base or strong acid is added. As the inset shows, the pH lies between $pK_a \pm 1$ across the buffer region. S denotes the stoichiometric point (Section L).

when the acid is 10 times as abundant as the base ([HA] = 10[A$^-$]), the pH of the solution is

$$pH = pK_a + \log\frac{[A^-]}{10[A^-]} = pK_a + \log\frac{1}{10} = pK_a - 1 \qquad (3a)$$

Likewise, when the base is 10 times as abundant as the acid ([A$^-$] = 10[HA]), the pH is

$$pH = pK_a + \log\frac{10[HA]}{[HA]} = pK_a + \log 10 = pK_a + 1 \qquad (3b)$$

We see that the experimentally determined concentration range converts into a pH range of ± 1. That is, the buffer acts most effectively within a range of ± 1 units of pK_a (Fig. 11.3). For instance, because the pK_a of $H_2PO_4^-$ is 7.21, a KH_2PO_4/K_2HPO_4 buffer can be expected to be most effective between about pH = 6.2 and pH = 8.2.

The composition of blood plasma, in which the concentration of HCO_3^- ions is about 20 times that of H_2CO_3, seems to be outside the range for optimum buffering. However, the principal waste products of living cells are carboxylic acids, such as lactic acid. Plasma, with its relatively high concentration of HCO_3^- ions, can absorb a significant surge of hydrogen ions from these carboxylic acids. The high proportion of HCO_3^- also helps us to withstand disturbances that lead to excess acid, such as disease and shock due to burns (Box 11.1).

The capacity of a buffer is determined by its concentration and pH. A more concentrated buffer can react with more added acid or base than can a less concentrated one. A buffer solution is generally most effective in the range $pK_a \pm 1$.

TITRATIONS

As we saw in Section L, titration involves the addition of a solution, called the *titrant*, from a buret to a flask containing the sample, called the *analyte*. For example, if an environmental chemist is monitoring acid mine drainage and needs to know the concentration of acid in the water, a sample of the effluent from the mine would be the analyte and a solution of base of known concentration would be the titrant. At the *stoichiometric point*, the amount of OH$^-$ (or H$_3$O$^+$) added as titrant is equal to the amount of H$_3$O$^+$ (or OH$^-$) initially present in the analyte. The success of the technique depends on our ability to detect this point. We use the techniques in this chapter to identify the roles of different species in determining the pH and to select the appropriate indicator for a titration.

11.4 Strong Acid–Strong Base Titrations

When a strong acid is mixed with a strong base, a neutralization reaction occurs for which the net ionic equation is

$$H_3O^+(aq) + OH^-(aq) \longrightarrow 2\,H_2O(l)$$

However, it is best to use the full chemical equation when working with titrations to ensure the correct stoichiometry. For example, if hydrochloric acid is used to neutralize $Ca(OH)_2$, we must take into account the fact that each formula unit of $Ca(OH)_2$ provides two OH$^-$ ions:

$$2\,HCl(aq) + Ca(OH)_2(aq) \longrightarrow CaCl_2(aq) + 2\,H_2O(l)$$

A plot of the pH of the analyte solution against the volume of titrant added during a titration is called a **pH curve**. The shape of the pH curve in Fig. 11.4 is typical of titrations in which a strong acid is added to a strong base. Initially, the pH falls slowly. Then, at the stoichiometric point, there is a sudden decrease in pH through 7. At this point, an indicator changes color or an automatic titrator responds electronically to the sudden change in pH. Titrations typically end at this point. However, if we were to continue the titration, we would find that the pH

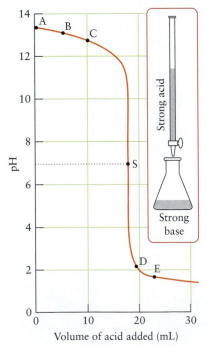

FIGURE 11.4 The variation of pH during the titration of a strong base, 25.00 mL of 0.250 M NaOH(aq), with a strong acid, 0.340 M HCl(aq). The stoichiometric point (S) occurs at pH = 7. The other points on the pH curve are explained in Example 11.4 in the text.

BOX 11.1   WHAT HAS THIS TO DO WITH . . . STAYING ALIVE?

Physiological Buffers

Buffer systems are so vital to the existence of living organisms that the most immediate threat to the survival of a person with severe injury or burns is a change in blood pH. One of a paramedic's first steps in saving a life is to administer intravenous fluids.

Patients who have suffered traumatic injuries must receive an immediate intravenous solution to combat the symptoms of shock and help maintain the pH of blood.

Metabolic processes normally maintain the pH of human blood within a narrow range (7.35–7.45). To control blood pH, the body uses primarily the carbonic acid/hydrogen carbonate (bicarbonate) ion system. The normal ratio of HCO_3^- to H_2CO_3 in the blood is 20:1, with most of the carbonic acid in the form of dissolved CO_2. When the concentration of HCO_3^- increases further relative to that of H_2CO_3, blood pH rises. If the pH rises above the normal range, the condition is called *alkalosis*. Conversely, blood pH decreases when the ratio decreases; and, when blood pH falls below the normal range, the condition is called *acidosis*. Because these conditions are life threatening and death can result within minutes, it is critical that the cause of the pH imbalance be identified and treated quickly.

The body maintains blood pH by two primary mechanisms: respiration and excretion. Carbonic acid concentration is controlled by respiration: as we exhale, we deplete our system of CO_2 and hence deplete it of H_2CO_3, too. This decrease in acid concentration raises the blood pH. Breathing faster and more deeply increases the amount of CO_2 exhaled and hence decreases the carbonic acid concentration in the blood, which in turn raises the blood pH. Hydrogen carbonate ion concentration is controlled by its rate of excretion in urine.

Respiratory acidosis results when decreased respiration raises the concentration of CO_2 in the blood. Asthma, pneumonia, emphysema, or inhaling smoke can all cause respiratory acidosis. So can any condition that reduces a person's ability to breathe. Respiratory acidosis is usually treated with a mechanical ventilator, to assist the victim's breathing. The improved exhalation increases the excretion of CO_2 and raises blood pH. In many cases of asthma, chemicals can facilitate respiration by opening constricted bronchial passages.

Metabolic acidosis is caused by the release into the bloodstream of excessive amounts of lactic acid and other acidic byproducts of metabolism. These acids enter the bloodstream, react with hydrogen carbonate ion to produce H_2CO_3, and shift the ratio HCO_3^-/H_2CO_3 to a lower value. Heavy exercise, diabetes, and fasting can all produce metabolic acidosis. The normal response of the body is to increase the rate of breathing to eliminate some of the CO_2. Thus, we pant heavily when running uphill.

Metabolic acidosis can also result when a person is severely burned. Blood plasma leaks from the circulatory system into the injured area, producing edema (swelling) and reducing the blood volume. If the burned area is large, this loss of blood volume may be sufficient to reduce blood flow and oxygen supply to all the body's tissues. Lack of oxygen, in turn, causes the tissues to produce an excessive amount of lactic acid and leads to metabolic acidosis. To minimize the decrease in pH, the injured person breathes harder to eliminate the excess CO_2. However, if blood volume drops below levels for which the body can compensate, a vicious circle ensues in which blood flow decreases still further, blood pressure falls, CO_2 excretion diminishes, and acidosis becomes more severe. People in this state are said to be in shock and will die if not treated promptly.

The dangers of shock are avoided or treated by intravenous infusion of large volumes of a salt-containing solution that is isotonic with blood (has the same osmotic pressure as blood), usually one known as *lactated Ringer's solution*. The added liquid increases blood volume and blood flow, thereby improving oxygen delivery. The $[HCO_3^-]/[H_2CO_3]$ ratio then increases toward normal and allows the severely injured person to survive.

Respiratory alkalosis is the rise in pH associated with excessive respiration. Hyperventilation, which can result from anxiety or high fever, is a common cause. The body may control blood pH during hyperventilation by fainting, which results in slower respiration. An intervention that may prevent fainting is to have a hyperventilating person breathe into a paper bag, which allows much of the respired CO_2 to be taken up again.

Metabolic alkalosis is the increase in pH resulting from illness or chemical ingestion. Repeated vomiting or the overuse of diuretics can cause metabolic alkalosis. Once again the body compensates, this time by decreasing the rate of respiration.

Related Exercises: 11.107 and 11.108

For Further Reading: J. A. Kraut and N. E. Madias, "Approach to patients with acid–base disorders," *Respiratory Care*, vol. 46, no. 4, April 2001, pp. 392–403. J. Squires, "Artificial blood," *Science*, vol. 295, Feb. 8, 2002, pp. 1002–1005. Lynn Taylor and Norman P. Curthoys, "Glutamine metabolism: Role in Acid–Base Balance, *Biochemistry and Molecular Biology Education*, vol. 32, no. 5, 2004, pp. 291–304.

This box includes contributions from B. A. Pruitt, M.D., and A. D. Mason, M.D., U.S. Army Institute of Surgical Research.

falls slowly toward the value of the acid itself as the dilution due to the original analyte solution becomes less and less important.

Figure 11.5 shows a pH curve for a titration in which the analyte is a strong acid and the titrant is a strong base. This curve is the mirror image of the curve for the titration of a strong base with a strong acid.

TOOLBOX 11.1   HOW TO CALCULATE THE pH DURING A STRONG ACID–STRONG BASE TITRATION

CONCEPTUAL BASIS

The pH during the titration of a strong acid with a strong base is governed by the major species in solution. Because the conjugate base of the strong acid has little effect on the pH, the pH is determined by whichever is in excess, the strong acid or the strong base.

PROCEDURE

First, use the reaction stoichiometry to find the amount of excess acid or base.

Step 1 Calculate the amount of H_3O^+ ions (if the analyte is a strong acid) or OH^- ions (if the analyte is a strong base) in the original analyte solution from the product of the analyte's molarity and its volume (use $n_J = V[J]$, where J is H_3O^+ or OH^-).

Step 2 Calculate the amount of OH^- ions (if the titrant is a strong base) or H_3O^+ ions (if the titrant is a strong acid) in the volume of titrant added from the product of the titrant's molarity and its volume. Use $n_J = V[J]$, where J is OH^- or H_3O^+.

Step 3 Write the chemical equation for the neutralization reaction and use the reaction stoichiometry to find the amount of H_3O^+ ions (or OH^- ions if the analyte is a strong base) that remains in the analyte solution after all the added titrant reacts. Each mole of H_3O^+ ions reacts with 1 mol OH^- ions; therefore, subtract the number of moles of H_3O^+ or OH^- ions that have reacted from the initial number of moles.

Next, determine the concentration.

Step 4 Use the remaining amount of H_3O^+ (or OH^-) and the total volume of the combined solutions, $V = V_{analyte} + V_{titrant}$, to find the molarity of the H_3O^+ (or OH^-) ions in the solution from $[J] = n_J/V$.

Finally, calculate the pH.

Step 5 If acid is in excess, take the negative logarithm of the H_3O^+ molarity to find the pH. If base is in excess, find the pOH, and then convert pOH into pH by using the relation $pH + pOH = pK_w$.

This procedure is illustrated in Example 11.4.

EXAMPLE 11.4   **Sample exercise: Calculating points on the pH curve for a strong acid–strong base titration**

Suppose we are carrying out a titration in which the analyte initially consists of 25.00 mL of 0.250 M NaOH(aq) and the titrant is 0.340 M HCl(aq). After the addition of 5.00 mL of the acid titrant we can expect the pH to decrease slightly from its initial value. Calculate the new pH.

SOLUTION Initially, the pOH of the analyte is $pOH = -\log 0.250 = 0.602$, and so the pH of the solution is $pH = 14.00 - 0.602 = 13.40$. This is point A in Fig. 11.4. We now follow the procedure in Toolbox 11.1.

Step 1 Find the amount of OH^- ions initially present from $n_J = V[J]$.	$n_{OH^-} = (25.00 \times 10^{-3}\ \text{L}) \times (0.250\ \text{mol·L}^{-1})$ $= 6.25 \times 10^{-3}\ \text{mol} = 6.25\ \text{mmol}$	 6.25 mmol OH^-
Step 2 Find the amount of H_3O^+ ions supplied by the titrant from $n_J = V[J]$.	$n_{H_3O^+} = (5.00 \times 10^{-3}\ \text{L}) \times (0.340\ \text{mol·L}^{-1})$ $= 1.70 \times 10^{-3}\ \text{mol} = 1.70\ \text{mmol}$	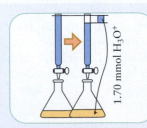 1.70 mmol H_3O^+

Step 3 Write the balanced equation for the neutralization reaction.	$HCl(aq) + NaOH(aq) \longrightarrow NaCl(aq) + H_2O(l)$	
Find the amount of OH^- remaining from 1 mol $NaOH \simeq 1$ mol HCl, after reaction of all the added H_3O^+ ions.	$n_{OH^-}(\text{final}) = 6.25 - 1.70 \text{ mmol} = 4.55 \text{ mmol}$	 4.55 mmol OH^-
Step 4 Find the concentration of OH^- ions from the total volume of the solution and $[J] = n_J/V$.	$[OH^-] = \dfrac{4.55 \times 10^{-3} \text{ mol}}{(25.00 + 5.00) \times 10^{-3} \text{ L}}$ $= 0.152 \text{ mol·L}^{-1}$	
Step 5 Calculate the pH from $pOH = -\log [OH^-]$ and then $pH = pK_w - pOH$.	$pOH = -\log(0.152) = 0.82$ $pH = 14.00 - 0.82 = 13.18$	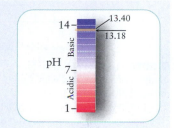

This is point B in Fig. 11.4. Note that the pH has fallen, as expected, but only by a very small amount. This small change is consistent with the shallow slope of the pH curve at the start of the titration.

SELF-TEST 11.5A What is the pH of the solution that results from the addition of a further 5.00 mL of the HCl(aq) titrant to the analyte?

[*Answer:* 12.91, point C]

SELF-TEST 11.5B What is the pH of the solution that results from the addition of yet another 2.00 mL of titrant to the analyte?

Experimentally, we know that the pH changes abruptly close to the stoichiometric point. Suppose we reach the stoichiometric point in the titration described in Example 11.4 and then add an additional 1.00 mL of HCl(aq). To find out by how much the pH changes, we work through the steps in Toolbox 11.1 as in Example 11.4, except that now the acid is in excess. We find that, after the addition, the pH has fallen to 2.1 (point D in Fig. 11.4). This point is well below the pH (of 7) at the stoichiometric point, although only 1 mL more acid has been added.

In the titration of a strong acid with a strong base or of a strong base with a strong acid, the pH changes slowly initially, changes rapidly through pH = 7 at the stoichiometric point, and then changes slowly again.

11.5 Strong Acid–Weak Base and Weak Acid–Strong Base Titrations

In many titrations, one solution—either the analyte or the titrant—contains a weak acid or base and the other solution contains a strong base or acid. For example, if we want to know the concentration of formic acid, the weak acid found in ant venom (**1**), we can titrate it with sodium hydroxide, a strong base. Alternatively, to find the concentration of ammonia, a weak base, in a soil sample, titrate it with hydrochloric acid, a strong acid. Weak acids are not normally titrated with weak bases, because the stoichiometric point is too difficult to locate.

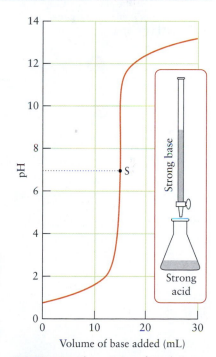

FIGURE 11.5 The variation of pH during a typical titration of a strong acid (the analyte) with a strong base (the titrant). The stoichiometric point (S) occurs at pH = 7.

1 Formic acid, HCOOH

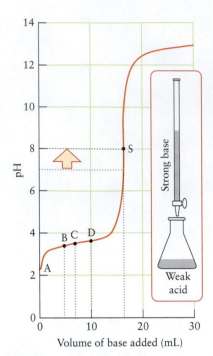

FIGURE 11.6 The pH curve for the titration of a weak acid with a strong base: 25.00 mL of 0.100 M $CH_3COOH(aq)$ with 0.150 M $NaOH(aq)$. The stoichiometric point (S) occurs at pH > 7 because the anion $CH_3CO_2^-$ is a base. The other points on the curve are explained in the text.

2 Formate ion, HCO_2^-

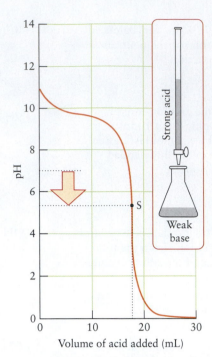

FIGURE 11.7 A typical pH curve for the titration of a weak base with a strong acid. The stoichiometric point (S) occurs at pH < 7 because the salt formed by the neutralization reaction has an acidic cation.

Figures 11.6 and 11.7 show the different pH curves that are found experimentally for these two types of titrations. Notice that the stoichiometric point does not occur at pH = 7. Moreover, although the pH changes reasonably sharply near the stoichiometric point, it does not change as abruptly as it does in a strong acid–strong base titration.

The pH at the stoichiometric point depends on the type of salt produced in the neutralization reaction. At the stoichiometric point of the titration of formic acid, HCOOH, with sodium hydroxide,

$$HCOOH(aq) + NaOH(aq) \longrightarrow NaHCO_2(aq) + H_2O(l)$$

the solution consists of sodium formate, $NaHCO_2$, and water. Because Na^+ ions have virtually no effect on pH and the formate ion, HCO_2^- (**2**), is a base, overall the solution is basic and pH > 7. The same is true for the stoichiometric point of the titration of any weak acid with strong base. At the stoichiometric point of the titration of aqueous ammonia with hydrochloric acid, the solute is ammonium chloride. Because Cl^- ions have virtually no effect on the pH and NH_4^+ is an acid, we expect pH < 7. The same is true for the stoichiometric point of the titration of any weak base with strong acid.

EXAMPLE 11.5   Estimating the pH at the stoichiometric point of the titration of a weak acid with a strong base

Estimate the pH at the stoichiometric point of the titration of 25.00 mL of 0.100 M HCOOH(aq) with 0.150 M NaOH(aq).

STRATEGY Decide whether the salt present at the stoichiometric point provides an ion that acts as a weak base or as a weak acid. If the former, expect pH > 7; if the latter, expect pH < 7. To calculate the pH at the stoichiometric point, proceed as in Example 10.10 or 10.11, noting that the amount of salt at the stoichiometric point is equal to the initial amount of acid and the volume is the total volume of the combined analyte and titrant solutions. The K_b of a weak base is related to the K_a of its conjugate acid by $K_a \times K_b = K_w$; K_a is listed in Table 10.1. Assume that the autoprotolysis of water has no significant effect on the pH, and then check that assumption.

SOLUTION The salt present at the stoichiometric point, sodium formate, provides basic formate ions, and so we expect pH > 7. From Table 10.1, $K_a = 1.8 \times 10^{-4}$ for formic acid; therefore, $K_b = K_w/K_a = 5.6 \times 10^{-11}$.

Find the initial amount of HCOOH in the analyte from $n_J = V[J]$.	$n_{HCOOH} = (2.500 \times 10^{-2}\ L) \times (0.100\ mol\cdot L^{-1})$ $= 2.50 \times 10^{-3}\ mol$, or 2.50 mmol

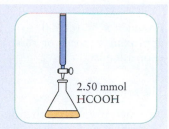
2.50 mmol HCOOH

Find the amount of OH^- required to react with the HCOOH from the reaction stoichiometry, 1 mol $OH^- \simeq$ 1 mol HCOOH.	$n_{OH^-} = (2.50 \times 10^{-3}\ mol\ HCOOH) \times \dfrac{1\ mol\ OH^-}{1\ mol\ HCOOH}$ $= 2.50 \times 10^{-3}\ mol\ OH^-$ (or 2.50 mmol)

The amount of HCO_2^- in the solution at the stoichiometric point is the same as the amount of OH^- added.	$n_{HCO_2^-} = 2.50\ mmol$

Find the volume of titrant containing this amount of OH^- ions, from $V = n_J/[J]$.	$V(added) = \dfrac{2.50 \times 10^{-3}\ mol}{0.150\ mol\cdot L^{-1}} = 1.67 \times 10^{-2}\ L$, or 16.7 mL

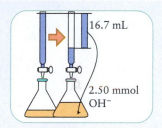
16.7 mL
2.50 mmol OH^-

Find the total volume of the solution at the stoichiometric point from $V(final) = V(initial) + V(added)$.	$V(final) = 25.00 + 16.7\ mL = 41.7\ mL$

2.50 mmol HCO_2^-
41.7 mL

Find the concentration of HCO_2^- ions at the stoichiometric point from $[J] = n_J/V$.	$[HCO_2^-] = \dfrac{2.50 \times 10^{-3}\ mol}{4.17 \times 10^{-2}\ L} = 0.0600\ mol\cdot L^{-1}$

Now that we know the composition of the solution at the stoichiometric point, we can calculate the pH of the solution as described in Toolbox 10.2. The equilibrium to consider is

$$HCO_2^-(aq) + H_2O(l) \rightleftharpoons HCOOH(aq) + OH^-(aq) \qquad K_b = \dfrac{[HCOOH][OH^-]}{[HCO_2^-]}$$

The equilibrium table, with all concentrations in moles per liter, is

	HCO_2^-	HCOOH	OH^-
Step 1 Initial molarity	0.0600	0	0
Step 2 Change in molarity	$-x$	$+x$	$+x$
Step 3 Equilibrium molarity	$0.0600 - x$	x	x

Step 4 From $K_b = [HCOOH][OH^-]/[HCO_2^-]$,	$5.6 \times 10^{-11} = \dfrac{x \times x}{0.0600 - x}$

Provided $x \ll 0.0600$, the approximate form of this expression is:	$5.6 \times 10^{-11} \approx \dfrac{x^2}{0.0600}$

The solution is:	$x \approx (5.6 \times 10^{-11} \times 0.0600)^{1/2} = 1.8 \times 10^{-6}$

It follows from step 3 that the molarity of OH^- is 1.8×10^{-6} mol·L^{-1}, which is about 18 times as great as the molarity of OH^- ions from the autoprotolysis of water (1.0×10^{-7} mol·L^{-1}), and so ignoring the latter is reasonable. Now we can write

$$pOH = -\log(1.8 \times 10^{-6}) = 5.74$$

and therefore,

From $pH + pOH = pK_w$,	$pH = 14.00 - 5.74 = 8.26$ or about 8.3. Notice that, at the stoichiometric point, $pH > 7$ as we expected.	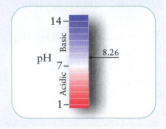

SELF-TEST 11.6A Calculate the pH at the stoichiometric point of the titration of 25.00 mL of 0.010 M HClO(aq) with 0.020 M KOH(aq). See Table 10.1 for K_a.

[*Answer:* 9.67]

SELF-TEST 11.6B Calculate the pH at the stoichiometric point of the titration of 25.00 mL of 0.020 M NH$_3$(aq) with 0.015 M HCl(aq). (For NH$_4^+$, $K_a = 5.6 \times 10^{-10}$.)

Now consider the overall shape of the pH curve. The slow change in pH about halfway to the stoichiometric point indicates that the solution acts as a buffer in that region (see Fig. 11.3). At the halfway point of the titration, [HA] = [A$^-$] and pH = pK_a. In fact, one way to prepare a buffer is to neutralize half the amount of weak acid present with strong base. The flatness of the curve near pH = pK_a illustrates very clearly the ability of a buffer solution to stabilize the pH of the solution. Moreover, we can now see how to determine pK_a: plot the pH curve during a titration, identify the pH halfway to the stoichiometric point, and set pK_a equal to that pH (Fig. 11.8). To obtain the pK_b of a weak base, we find pK_a in the same way but go on to use $pK_a + pK_b = pK_w$. The values recorded in Tables 10.1 and 10.2 were obtained in this way.

Well after the stoichiometric point in the titration of a weak acid with a strong base, the pH depends only on the concentration of excess strong base. For example, suppose we went on to add several liters of strong base from a giant buret. The presence of salt produced by the neutralization reaction would be negligible relative to the concentration of excess base. The pH would be that of the nearly pure titrant (the original base solution).

We have already seen how to estimate the pH of the initial analyte when only weak acid or weak base is present (point A in Fig. 11.6, for instance), as well as the pH at the stoichiometric point (point S). Between these two points lie points corresponding to a mixed solution of some weak acid (or base) and some salt. We can therefore use the techniques described in Toolbox 11.2 and Example 11.6 to account for the shape of the curve.

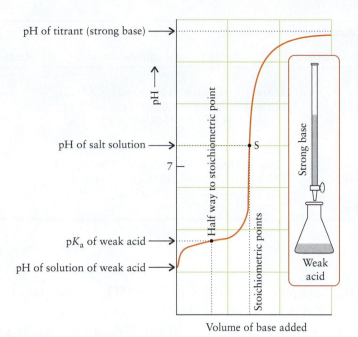

FIGURE 11.8 The pK_a of an acid can be determined by carrying out a titration of the weak acid with a strong base and locating the pH of the solution after the addition of half the volume of base needed to reach the stoichiometric point. The pH at that point is equal to the pK_a of the acid.

TOOLBOX 11.2   **HOW TO CALCULATE THE pH DURING A TITRATION OF A WEAK ACID OR A WEAK BASE**

CONCEPTUAL BASIS

The pH is governed by the major solute species present in solution. As strong base is added to a solution of a weak acid, a salt of the conjugate base of the weak acid is formed. This salt affects the pH and needs to be taken into account, as in a buffer solution. Table 11.2 outlines the regions encountered during a titration and the primary equilibrium to consider in each region.

PROCEDURE

The procedure is like that in Toolbox 11.1, except that an additional step is required to calculate the pH from the proton transfer equilibrium.

Use the reaction stoichiometry to find the amount of excess acid or base.

Begin by writing the chemical equation for the reaction, then:

Step 1 Calculate the amount of weak acid or base in the original analyte solution. Use $n_J = V_{analyte}[J]$.

Step 2 Calculate the amount of OH^- ions (or H_3O^+ ions if the titrant is an acid) in the volume of titrant added. Use $n_J = V_{titrant}[J]$.

Step 3 Use the reaction stoichiometry to calculate the following amounts:

- Weak acid with a strong base: the amount of conjugate base formed in the neutralization reaction, and the amount of weak acid remaining.

- Weak base with a strong acid: the amount of conjugate acid formed in the neutralization reaction, and the amount of weak base remaining.

Determine the concentration.

Step 4 Find the molarities of the conjugate acid and base in solution. Use $[J] = n_J/V$, where V is the total volume of the solution, $V = V_{analyte} + V_{titrant}$.

Determine the pH.

Step 5 Use an equilibrium table to find the H_3O^+ concentration in a weak acid or the OH^- concentration in a weak base. Alternatively, if the concentrations of conjugate acid and base calculated in step 4 are both large relative to the concentration of hydronium ions, use them in the expression for K_a or the Henderson–Hasselbalch equation to determine the pH. In each case, if the pH is less than 6 or greater than 8, assume that the autoprotolysis of water does not significantly affect the pH. If necessary, convert between K_a and K_b by using $K_w = K_a \times K_b$.

This procedure is illustrated in Example 11.6.

TABLE 11.2 **Summary of Weak Acid and Weak Base Titration Equilibria**

Point in titration	Primary species	Proton transfer equilibrium	Related Toolbox
1 Weak acid HA titrated with strong base			
initial	HA	$HA(aq) + H_2O(l) \rightleftharpoons H_3O^+(aq) + A^-(aq)$	10.1
buffer region	HA, A^-	$HA(aq) + H_2O(l) \rightleftharpoons H_3O^+(aq) + A^-(aq)$	11.2
stoichiometric point	A^-	$A^-(aq) + H_2O(l) \rightleftharpoons HA(aq) + OH^-(aq)$	10.2
(Note: This is a solution of a salt with a basic anion.)			
2 Weak base B titrated with strong acid			
initial	B	$B(aq) + H_2O(l) \rightleftharpoons HB^+(aq) + OH^-(aq)$	10.2
buffer region	B, HB^+	$B(aq) + H_2O(l) \rightleftharpoons HB^+(aq) + OH^-(aq)$	11.2
stoichiometric point	HB^+	$HB^+(aq) + H_2O(l) \rightleftharpoons H_3O^+(aq) + B(aq)$	10.1
(Note: This is a solution of a salt with an acidic anion.)			

EXAMPLE 11.6   **Sample exercise: Calculating the pH before the stoichiometric point in a weak acid–strong base titration**

Calculate the pH of (a) 0.100 M HCOOH(aq) and (b) the solution resulting when 5.00 mL of 0.150 M NaOH(aq) is added to 25.00 mL of the acid. Use $K_a = 1.8 \times 10^{-4}$ for HCOOH.

SOLUTION For (a) we use the procedure in Toolbox 10.1. For (b) we expect a pH slightly greater than that of the pure acid, due to the added base, and find the pH using the procedure in Toolbox 11.2.

(a) From $[H_3O^+] = (K_a[HA])^{1/2}$ and $pH = -\log [H_3O^+]$,

$pH = -\log (1.8 \times 10^{-4} \times 0.100)^{1/2}$

$= 2.37$

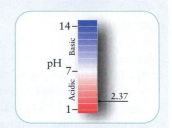

(b) We expect a pH greater than 2.37 after the base has been added. Now proceed as in Toolbox 11.2. The chemical equation is

$$HCOOH(aq) + H_2O(l) \rightleftharpoons H_3O^+(aq) + HCO_2^-(aq)$$

Step 1 Find the initial amount of HCOOH from $n_J = V[J]$.

$n_{HCOOH} = (2.500 \times 10^{-2} \text{ L}) \times (0.100 \text{ mol·L}^{-1})$

$= 2.50 \times 10^{-3}$ mol, or 2.50 mmol

Step 2 Find the amount of OH^- added from $n_J = V[J]$.

$n_{OH^-} = (5.00 \times 10^{-3} \text{ L}) \times (0.150 \text{ mol·L}^{-1})$

$= 7.50 \times 10^{-4}$ mol, or 0.750 mmol

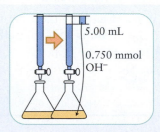

Step 3 Find the amounts of HCOOH and HCO_2^- from 1 mol $OH^- \simeq$ 1 mol HCOOH and 1 mol $OH^- \simeq$ 1 mol HCO_2^-.

0.750 mmol OH^- produces 0.750 mmol HCO_2^- and leaves $2.50 - 0.750$ mmol = 1.75 mmol HCOOH.

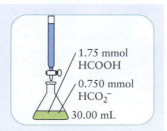

1.75 mmol HCOOH

0.750 mmol HCO_2^-

30.00 mL

Step 4 Find the concentrations of acid and conjugate base from $[J] = n_J/V$, where the total volume is $V = V_{analyte} + V_{titrant}$.

$$[HCOOH] = \frac{1.75 \times 10^{-3} \text{ mol}}{(25.00 + 5.00) \times 10^{-3} \text{ L}} = 0.0583 \text{ mol·L}^{-1}$$

$$[HCO_2^-] = \frac{7.50 \times 10^{-4} \text{ mol}}{(25.00 + 5.00) \times 10^{-3} \text{ L}} = 0.0250 \text{ mol·L}^{-1}$$

Step 5 Now we evaluate the pH. The proton transfer equilibrium for HCOOH in water is

$$HCOOH(aq) + H_2O(l) \rightleftharpoons H_3O^+(aq) + HCO_2^-(aq)$$

From Table 10.1, the pK_a for formic acid is 3.75. Assuming little change in the concentrations of acid and base due to the deprotonation of HCOOH, we write

From $pH = pK_a + \log([HCO_2^-]/[HCOOH])$,

$$pH = 3.75 + \log\frac{0.0250}{0.0583} = 3.38$$

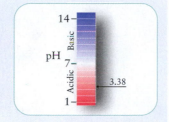

14 —

pH   Basic

7 —

Acidic   3.38

1 —

This pH corresponds to $[H_3O^+] = 4.2 \times 10^{-4}$, point B in Fig. 11.6 and, as expected, the contribution of autoprotolysis is negligible. As predicted, the pH of the mixed solution (3.38) is higher than that of the original acid (2.37).

SELF-TEST 11.7A Calculate the pH of the solution after the addition of another 5.00 mL of 0.150 M NaOH(aq).

[*Answer:* 3.93, point D]

SELF-TEST 11.7B Calculate the pH of the solution after the addition of yet another 5.00 mL of 0.150 M NaOH(aq).

Figure 11.8 summarizes the changes in pH of a solution during a titration of a weak acid by a strong base. Half way to the stoichiometric point, the pH is equal to the pK_a of the acid. The pH is greater than 7 at the stoichiometric point of the titration of a weak acid and strong base. The pH is less than 7 at the stoichiometric point of the titration of a weak base and strong acid.

11.6 Acid–Base Indicators

A simple, reliable, and fast method of determining the pH of a solution and of monitoring a titration is with a **pH meter,** which uses a special electrode to measure H_3O^+ concentration. An automatic titrator monitors the pH of the analyte solution continuously. It detects the stoichiometric point by responding to the characteristic rapid change in pH (Fig. 11.9). Another common technique is to use an indicator to detect the stoichiometric point. An **acid–base indicator** is a water-soluble organic dye with a color that depends on the pH. The sudden change in pH

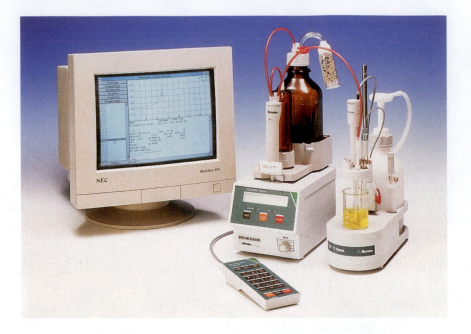

FIGURE 11.10 The stoichiometric point of an acid–base titration may be detected by the color change of an indicator. Here we see the colors of solutions containing a few drops of phenolphthalein at (from left to right) pH of 7.0, 8.5, 9.4 (its end point), 9.8, and 12.0. At the end point, the concentrations of the conjugate acid and base forms of the indicator are equal.

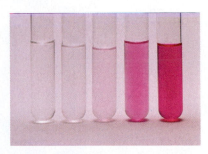

3 Phenolphthalein
(Acid form, colorless)

4 Phenolphthalein
(Base form, pink)

that occurs at the stoichiometric point of a titration is signaled by a sharp change in color of the dye as it responds to the pH.

An acid–base indicator changes color with pH because it is a weak acid that has one color in its acid form (HIn, where In stands for indicator) and another color in its conjugate base form (In$^-$). The color change results because the proton in HIn changes the structure of the molecule in such a way that the light absorption characteristics of HIn are different from those of In$^-$. When the concentration of HIn is much greater than that of In$^-$, the solution has the color of the acid form of the indicator. When the concentration of In$^-$ is much greater than that of HIn, the solution has the color of the base form of the indicator.

Because it is a weak acid, an indicator takes part in a proton transfer equilibrium:

$$HIn(aq) + H_2O(l) \rightleftharpoons H_3O^+(aq) + In^-(aq) \qquad K_{In} = \frac{[H_3O^+][In^-]}{[HIn]} \qquad (4)$$

The **end point** of an indicator is the point at which the concentrations of its acid and base forms are equal: $[HIn] = [In^-]$. When we substitute this equality into the expression for K_{In}, we see that at the end point $[H_3O^+] = K_{In}$. That is, the color change occurs when

$$pH = pK_{In} \qquad\qquad\qquad (5)^*$$

The color starts to change perceptibly about 1 pH unit before pK_{In} and is effectively complete about 1 pH unit after pK_{In}. Table 11.3 gives the values of pK_{In} for a number of common indicators.

One common indicator is phenolphthalein (Fig. 11.10). The acid form of this large molecule (**3**) is colorless; its conjugate base form (**4**) is pink. The structure of the base form of phenolphthalein allows electrons to be delocalized across all three of the benzenelike rings of carbon atoms, and the increase in delocalization is part of the reason for the change in color. The pK_{In} of phenolphthalein is 9.4, and so the end point occurs in slightly basic solution. Litmus, another well-known indicator, has $pK_{In} = 6.5$; it is red for pH < 5 and blue for pH > 8.

There are many naturally occurring indicators. For instance, a single compound is responsible for the colors of red poppies and blue cornflowers: the pH of the sap is different in the two plants. The color of hydrangeas also depends on the acidity of their sap and can be controlled by modifying the acidity of the soil (Fig. 11.11).

TABLE 11.3 Indicator Color Changes*

Indicator	Color of acid form	pH range of color change	pK_{In}	Color of base form	
thymol blue	red	1.2 to 2.8	1.7	yellow	
methyl orange	red	3.2 to 4.4	3.4	yellow	
bromophenol blue	yellow	3.0 to 4.6	3.9	blue	
bromocresol green	yellow	3.8 to 5.4	4.7	blue	
methyl red	red	4.8 to 6.0	5.0	yellow	
litmus	red	5.0 to 8.0	6.5	blue	
bromothymol blue	yellow	6.0 to 7.6	7.1	blue	
phenol red	yellow	6.6 to 8.0	7.9	red	
thymol blue	yellow	8.0 to 9.6	8.9	blue	
phenolphthalein	colorless	8.2 to 10.0	9.4	pink	
alizarin yellow R	yellow	10.1 to 12.0	11.2	red	
alizarin	red	11.0 to 12.4	11.7	purple	

* The colors of the acid and base forms shown on the right are only a symbolic representation of the actual colors.

FIGURE 11.11 The color of these hydrangeas depends on the acidity of the soil in which they are growing: acid soil results in blue flowers, alkaline soil results in pink flowers.

The *end point* is a property of the indicator; the *stoichiometric point* is a property of the chemical reaction taking place during the titration. It is important to select an indicator with an end point close to the stoichiometric point of the titration (Fig. 11.12). In practice, the pK_{In} of the indicator should be within about 1 pH unit of the stoichiometric point:

$$pK_{In} \approx pH(\text{at the stoichiometric point}) \pm 1 \qquad (6)$$

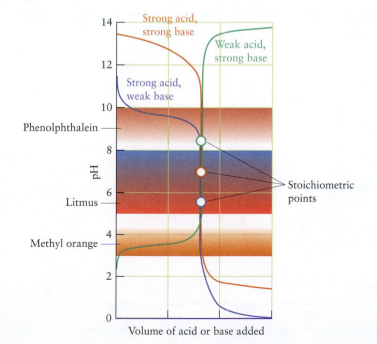

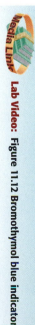

FIGURE 11.12 Ideally, an indicator should have a sharp color change close to the stoichiometric point of the titration, which is at pH = 7 for a strong acid–strong base titration. However, the change in pH is so abrupt that phenolphthalein can be used. Phenolphthalein can also be used to detect the stoichiometric point of a weak acid–strong base titration, but methyl orange can not. However, methyl orange can be used for a weak base–strong acid titration. Phenolphthalein would be inappropriate in this case, because its color change occurs well away from the stoichiometric point.

Phenolphthalein can be used for titrations with a stoichiometric point near pH = 9, such as a titration of a weak acid with a strong base. Methyl orange changes color between pH = 3.2 and pH = 4.4 and can be used in the titration of a weak base with a strong acid. Ideally, indicators for strong acid–strong base titrations should have end points close to pH = 7; however, in strong acid–strong base titrations, the pH changes rapidly over several pH units, and even phenolphthalein can be used. Table 11.3 includes the pH ranges over which several indicators can be used.

Acid–base indicators are weak acids that change color close to pH = pK_{In}; an indicator should be chosen so that its end point is close to the stoichiometric point of the titration.

11.7 Stoichiometry of Polyprotic Acid Titrations

Because many biological systems use polyprotic acids and their anions to control pH, we need to be familiar with pH curves for polyprotic titrations and to be able to calculate the pH during such a titration. The titration of a polyprotic acid proceeds in the same way as that of a monoprotic acid, but there are as many stoichiometric points in the titration as there are acidic hydrogen atoms. We therefore have to keep track of the major species in solution at each stage, as described in Sections 10.16 and 10.17 and summarized in Figs. 10.20 and 10.21.

Suppose we are titrating the triprotic acid H_3PO_4 with a solution of NaOH. The experimentally determined pH curve is shown in Fig. 11.13. Notice that there are three stoichiometric points (B, D, and F) and three buffer regions (A, C, and E). In pH calculations for these systems, we assume that, as we add the hydroxide solution, initially NaOH reacts completely with the acid to form the diprotic conjugate base $H_2PO_4^-$:

$$H_3PO_4(aq) + OH^-(aq) \longrightarrow H_2PO_4^-(aq) + H_2O(l) \tag{B}$$

At point A, the system is in the first buffer region and pH = pK_{a1}. Once all the acid H_3PO_4 molecules have lost their first acidic protons, the system is at B and the primary species in solution are the diprotic conjugate base and sodium ion—we have a solution of $NaH_2PO_4(aq)$. Point B is the first stoichiometric point, and to reach it we need to supply 1 mol NaOH for each mole of H_3PO_4.

As we continue to add base, it reacts with the $H_2PO_4^-$ to form that acid's conjugate base, HPO_4^{2-}:

$$H_2PO_4^-(aq) + OH^-(aq) \longrightarrow HPO_4^{2-}(aq) + H_2O(l) \tag{C}$$

At point C, the system is in the second buffer region and pH = pK_{a2}. Enough base takes us to the second stoichiometric point, D. The primary species in solution are now the monoprotic anions HPO_4^{2-} and sodium ions, which form a solution of

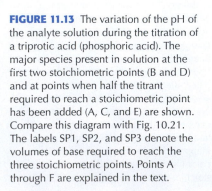

FIGURE 11.13 The variation of the pH of the analyte solution during the titration of a triprotic acid (phosphoric acid). The major species present in solution at the first two stoichiometric points (B and D) and at points when half the titrant required to reach a stoichiometric point has been added (A, C, and E) are shown. Compare this diagram with Fig. 10.21. The labels SP1, SP2, and SP3 denote the volumes of base required to reach the three stoichiometric points. Points A through F are explained in the text.

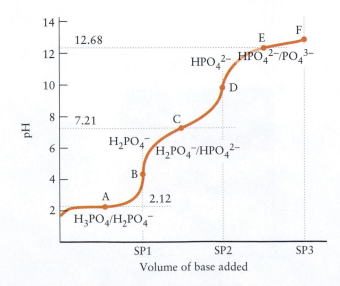

$Na_2HPO_4(aq)$. To reach the second stoichiometric point, an additional mole of NaOH is required for each mole of H_3PO_4 originally present. A total of 2 mol NaOH for each mole of H_3PO_4 has now been added.

Additional base reacts with HPO_4^{2-} to produce phosphate ion, PO_4^{3-}:

$$HPO_4^{2-}(aq) + OH^-(aq) \longrightarrow PO_4^{3-}(aq) + H_2O(l) \qquad \textbf{(D)}$$

At point E, the system is in the third buffer region and $pH = pK_{a3}$. When this reaction is complete, the primary species in solution are phosphate ions and sodium ions, which form a solution of $Na_3PO_4(aq)$. To reach this stoichiometric point (F in the plot), we have to add another mole of OH^- for each mole of H_3PO_4 initially present. At this point, a total of 3 mol OH^- has been added for each mole of H_3PO_4. Notice that the third stoichiometric point (point F) is indistinct, largely because K_{a3} is comparable to K_w. As a result, it is not detected in titrations.

Figure 11.14 shows the pH curve of a diprotic acid, such as oxalic acid, $H_2C_2O_4$. There are two stoichiometric points (B and D) and two buffer regions (A and C). The major species present in solution at each point are indicated. Note that it takes twice as much base to reach the second stoichiometric point as it does to reach the first.

SELF-TEST 11.8A What volume of 0.010 M NaOH(aq) is required to reach (a) the first stoichiometric point and (b) the second stoichiometric point in a titration of 25.00 mL of 0.010 M $H_2SO_3(aq)$?

[*Answer:* (a) 25 mL; (b) 50. mL]

SELF-TEST 11.8B What volume of 0.020 M NaOH(aq) is required to reach (a) the first stoichiometric point and (b) the second stoichiometric point in a titration of 30.00 mL of 0.010 M $H_3PO_4(aq)$?

We can predict the pH at any point in the titration of a polyprotic acid with a strong base by using the reaction stoichiometry to recognize what stage we have reached in the titration. We then identify the principal solute species at that point and the principal proton transfer equilibrium that determines the pH.

For instance, suppose we titrated $H_3PO_4(aq)$ with NaOH(aq). We assume that, up to the first stoichiometric point, each OH^- ion from the base reacts with one molecule of H_3PO_4 until all the latter has been consumed (Eq. B). In this part of the titration, the solution consists of $H_3PO_4(aq)$ and $H_2PO_4^-$, with the latter increasing as more NaOH(aq) is added and with the volume increasing, too. At the first stoichiometric point, the solution contains a salt of $H_2PO_4^-$ with its characteristic pH.

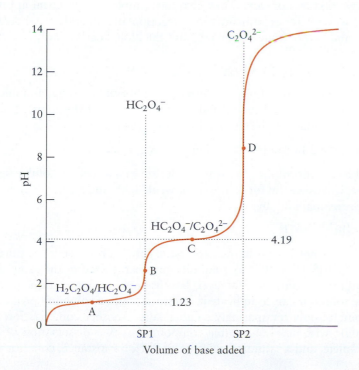

FIGURE 11.14 The variation of the pH of the analyte solution during the titration of a diprotic acid (oxalic acid) and the major species present in solution at the two stoichiometric points (B and D) and at points when half the titrant required to reach a stoichiometric point has been added (A and C). Compare this diagram with Fig. 10.20. The labels SP1 and SP2 denote the volumes of base required to reach the two stoichiometric points.

Between the first and the second stoichiometric points, the $H_2PO_4^-$ anion reacts with additional base (Eq. C). In this region, the solution consists of $H_2PO_4^-$ and HPO_4^{2-} ions with a pH characteristic of that mixture. At the second stoichiometric point, the solution contains a stoichiometric amount of HPO_4^{2-} in an increased volume and the pH is given by its reaction with H_2O. Between the second and the third stoichiometric points, the neutralization reaction is given by Eq. D. The pH of the solution is that of a mixture of HPO_4^{2-} and PO_4^{3-} ions until the third stoichiometric point is reached, at which point a stoichiometric amount of the latter is present. Beyond that point, the pH rises as excess NaOH(aq) is added.

SELF-TEST 11.9A A sample of 0.200 M H_3PO_4(aq) of volume 25.0 mL was titrated with 0.100 M NaOH(aq). Identify the primary species in solution and the principal proton transfer equilibrium in solution after the addition of the following volumes of NaOH solution: (a) 70.0 mL; (b) 100.0 mL.

[*Answer:* (a) Between first and second stoichiometric points, Na^+ from the base, $H_2PO_4^-$, and HPO_4^{2-}; $H_2PO_4^-(aq) + H_2O(l) \rightleftharpoons H_3O^+(aq) + HPO_4^{2-}(aq)$; (b) at the second stoichiometric point, Na_2HPO_4; $HPO_4^{2-}(aq) + H_2O(l) \rightleftharpoons H_2PO_4^-(aq) + OH^-(aq)$.]

SELF-TEST 11.9B A sample of 0.100 M H_2S(aq) of volume 20.0 mL was titrated with 0.300 M NaOH(aq). Identify the primary species in solution and the principal proton transfer equilibrium after the addition of the following volumes of NaOH solution: (a) 5.0 mL; (b) 13.4 mL.

The titration of a polyprotic acid has a stoichiometric point corresponding to the removal of each acidic hydrogen atom. The pH of a solution of a polyprotic acid undergoing a titration is estimated by considering the primary species in solution and the proton transfer equilibrium that determines the pH.

SOLUBILITY EQUILIBRIA

Up to this point, we have focused on aqueous equilibria involving proton transfer. Now we apply the same principles to the equilibrium that exists between a solid salt and its dissolved ions in a saturated solution. We can use the equilibrium constant for the dissolution of a substance to predict the solubility of a salt and to control precipitate formation. These methods are used in the laboratory to separate and analyze mixtures of salts. They also have important practical applications in municipal wastewater treatment, the extraction of minerals from seawater, the formation and loss of bones and teeth, and the global carbon cycle.

11.8 The Solubility Product

K_{sp} is also called the *solubility product constant* and, most simply, the *solubility constant*.

The equilibrium constant for the solubility equilibrium between an ionic solid and its dissolved ions is called the **solubility product**, K_{sp}, of the solute. For example, the solubility product for bismuth sulfide, Bi_2S_3, is defined as

$$Bi_2S_3(s) \rightleftharpoons 2\ Bi^{3+}(aq) + 3\ S^{2-}(aq) \qquad K_{sp} = (a_{Bi^{3+}})^2(a_{S^{2-}})^3$$

Because the concentrations of ions in a solution of a sparingly soluble salt are low, we assume, just as we did for solutions of weak acids and bases (Section 10.7), that we can approximate K_{sp} by

$$K_{sp} = [Bi^{3+}]^2[S^{2-}]^3$$

Solid Bi_2S_3 does not appear in the expression for K_{sp}, because it is a pure solid and its activity is 1 (Section 9.2). A solubility product is used in the same way as any other equilibrium constant. However, because ion–ion interactions in even dilute electrolyte solutions can complicate its interpretation, a solubility product is generally meaningful only for sparingly soluble salts. Another complication that arises when dealing with nearly insoluble compounds is that dissociation of the ions is rarely complete, and a saturated solution of PbI_2, for instance, contains substantial

TABLE 11.4 Solubility Products at 25°C

Compound	Formula	K_{sp}	Compound	Formula	K_{sp}
aluminum hydroxide	$Al(OH)_3$	1.0×10^{-33}	lead(II) fluoride	PbF_2	3.7×10^{-8}
antimony sulfide	Sb_2S_3	1.7×10^{-93}	iodate	$Pb(IO_3)_2$	2.6×10^{-13}
barium carbonate	$BaCO_3$	8.1×10^{-9}	iodide	PbI_2	1.4×10^{-8}
fluoride	BaF_2	1.7×10^{-6}	sulfate	$PbSO_4$	1.6×10^{-8}
sulfate	$BaSO_4$	1.1×10^{-10}	sulfide	PbS	8.8×10^{-29}
bismuth sulfide	Bi_2S_3	1.0×10^{-97}	magnesium ammonium phosphate	$MgNH_4PO_4$	2.5×10^{-13}
calcium carbonate	$CaCO_3$	8.7×10^{-9}	carbonate	$MgCO_3$	1.0×10^{-5}
fluoride	CaF_2	4.0×10^{-11}	fluoride	MgF_2	6.4×10^{-9}
hydroxide	$Ca(OH)_2$	5.5×10^{-6}	hydroxide	$Mg(OH)_2$	1.1×10^{-11}
sulfate	$CaSO_4$	2.4×10^{-5}	mercury(I) chloride	Hg_2Cl_2	2.6×10^{-18}
chromium(III) iodate	$Cr(IO_3)_3$	5.0×10^{-6}	iodide	Hg_2I_2	1.2×10^{-28}
copper(I) bromide	$CuBr$	4.2×10^{-8}	mercury(II) sulfide, black	HgS	1.6×10^{-52}
chloride	$CuCl$	1.0×10^{-6}	sulfide, red	HgS	1.4×10^{-53}
iodide	CuI	5.1×10^{-12}	nickel(II) hydroxide	$Ni(OH)_2$	6.5×10^{-18}
sulfide	Cu_2S	2.0×10^{-47}	silver bromide	$AgBr$	7.7×10^{-13}
copper(II) iodate	$Cu(IO_3)_2$	1.4×10^{-7}	carbonate	Ag_2CO_3	6.2×10^{-12}
oxalate	CuC_2O_4	2.9×10^{-8}	chloride	$AgCl$	1.6×10^{-10}
sulfide	CuS	1.3×10^{-36}	hydroxide	$AgOH$	1.5×10^{-8}
iron(II) hydroxide	$Fe(OH)_2$	1.6×10^{-14}	iodide	AgI	8×10^{-17}
sulfide	FeS	6.3×10^{-18}	sulfide	Ag_2S	6.3×10^{-51}
iron(III) hydroxide	$Fe(OH)_3$	2.0×10^{-39}	zinc hydroxide	$Zn(OH)_2$	2.0×10^{-17}
lead(II) bromide	$PbBr_2$	7.9×10^{-5}	sulfide	ZnS	1.6×10^{-24}
chloride	$PbCl_2$	1.6×10^{-5}			

concentrations of PbI^+ and PbI_2 ion clusters. At best, the calculations that we are about to describe are only estimates.

One of the simplest ways to determine K_{sp} is to measure the molar solubility of the compound, the molar concentration of the compound in a saturated solution, but more advanced and accurate methods are also available. Table 11.4 gives some experimental values. In the following calculations, we use s to denote the numerical value of the molar solubility expressed in moles per liter; for example, if the molar solubility of a compound is 6.5×10^{-5} mol·L^{-1}, then we write $s = 6.5 \times 10^{-5}$.

Electrochemical methods for determining solubility products are discussed in Section 12.8.

EXAMPLE 11.7 Determining the solubility product

The molar solubility of silver chromate, Ag_2CrO_4, is 6.5×10^{-5} mol·L^{-1}. Determine the value of K_{sp} for silver chromate.

STRATEGY First, we write the chemical equation for the equilibrium and the expression for the solubility product. To evaluate K_{sp}, we need to know the molarity of each type of ion formed by the salt. We determine the molarities from the molar solubility, the chemical equation for the equilibrium, and the stoichiometric relations between the species. We assume complete dissociation.

SOLUTION

Write the chemical equation.	$Ag_2CrO_4(s) \rightleftharpoons 2\, Ag^+(aq) + CrO_4^{2-}(aq)$
Write the expression for the solubility product.	$K_{sp} = [Ag^+]^2[CrO_4^{2-}]$
From 2 mol $Ag^+ \simeq 1$ mol Ag_2CrO_4,	$[Ag^+] = 2s = 2 \times (6.5 \times 10^{-5})$
From 1 mol $CrO_4^{2-} \simeq 1$ mol Ag_2CrO_4,	$[CrO_4^{2-}] = s = 6.5 \times 10^{-5}$
From $K_{sp} = [Ag^+]^2[CrO_4^{2-}] = (2s)^2(s) = 4s^3$,	$K_{sp} = 4 \times (6.5 \times 10^{-5})^3 = 1.1 \times 10^{-12}$

SELF-TEST 11.10A The molar solubility of lead(II) iodate, $Pb(IO_3)_2$, at 25°C is 4.0×10^{-5} mol·L^{-1}. What is the value of K_{sp} for lead(II) iodate?

[*Answer:* 2.6×10^{-13}]

SELF-TEST 11.10B The molar solubility of silver bromide, AgBr, at 25°C is 8.8×10^{-7} mol·L^{-1}. What is the value of K_{sp} for silver bromide?

EXAMPLE 11.8   Sample Exercise: Estimating the molar solubility from the solubility product

According to Table 11.4, $K_{sp} = 5.0 \times 10^{-6}$ for chromium(III) iodate in water. Estimate the molar solubility of the compound.

SOLUTION Assume complete dissociation.

Write the chemical equation.	$Cr(IO_3)_3(s) \rightleftharpoons Cr^{3+}(aq) + 3\,IO_3^-(aq)$
Write the expression for the solubility product.	$K_{sp} = [Cr^{3+}][IO_3^-]^3$
From 1 mol $Cr^{3+} \approx 1$ mol $Cr(IO_3)_3$,	$[Cr^{3+}] = s$
From 3 mol $IO_3^- \approx 1$ mol $Cr(IO_3)_3$,	$[IO_3^-] = 3s$
Write the expression for K_{sp} in terms of s.	$K_{sp} = [Cr^{3+}][IO_3^-]^3 = s \times (3s)^3 = 27s^4$
From $s = (K_{sp}/27)^{1/4}$,	$s = \{(5.0 \times 10^{-6})/27\}^{1/4} = 0.021$

A note on good practice: A quick way to evaluate the fourth root is to take the square root twice in succession.

The molar solubility of $Cr(IO_3)_3$ is therefore 0.021 mol·L^{-1}.

SELF-TEST 11.11A The solubility product of silver sulfate, Ag_2SO_4, is 1.4×10^{-5}. Estimate the molar solubility of the salt.

[*Answer:* 1.5×10^{-2} mol·L^{-1}]

SELF-TEST 11.11B The solubility product of lead(II) fluoride, PbF_2, is 3.7×10^{-8}. Estimate the molar solubility of the salt.

The solubility product is the equilibrium constant for the equilibrium between an undissolved salt and its ions in a saturated solution.

11.9 The Common-Ion Effect

Sometimes we have to precipitate one ion of a sparingly soluble salt. For example, heavy metal ions such as lead and mercury can be removed from municipal wastewater by precipitating them as the hydroxides. However, because the ions are in dynamic equilibrium with the solid salt, some heavy metal ions remain in solution. How can we remove more of the ions?

We can use Le Chatelier's principle as a guide. This principle tells us that, if we add a second salt or an acid that supplies one of the same ions—a "common ion"—to a saturated solution of a salt, then the equilibrium will tend to adjust by decreasing the concentration of the added ions (Fig. 11.15). That is, the solubility of the original salt is decreased, and it precipitates. We can conclude that the addition of excess OH$^-$ ions to the water supply should precipitate more of the heavy metal ions as their hydroxides. In other words, the addition of OH$^-$ ions reduces the solubility of the heavy metal hydroxide. The decrease in solubility caused by the addition of a common ion is called the **common-ion effect**.

(a)

(b)

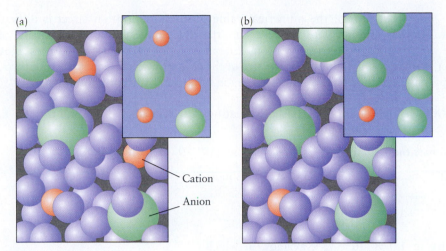

Cation

Anion

FIGURE 11.15 If the concentration of one of the ions of a slightly soluble salt is increased, the concentration of the other decreases to maintain a constant value of K_{sp}. (a) The cations (pink) and anions (green) in solution. (b) When more anions are added (together with their accompanying spectator ions, which are not shown), the concentration of cations decreases. In other words, the solubility of the original compound is reduced by the presence of a common ion. In the insets, the blue background represents the solvent (water).

We can gain a quantitative understanding of the common-ion effect by considering how a change in the concentration of one of the ions affects the solubility product. Suppose we have a saturated solution of silver chloride in water:

$$AgCl(s) \rightleftharpoons Ag^+(aq) + Cl^-(aq) \qquad K_{sp} = [Ag^+][Cl^-]$$

Experimentally, $K_{sp} = 1.6 \times 10^{-10}$ at 25°C, and the molar solubility of AgCl in water is 1.3×10^{-5} mol·L^{-1}. If we add sodium chloride to the solution, the concentration of Cl$^-$ ions increases. For the equilibrium constant to remain constant, the concentration of Ag$^+$ ions must decrease. Because there is now less Ag$^+$ in solution, the solubility of AgCl is lower in a solution of NaCl than it is in pure water. A similar effect occurs whenever two salts having a common ion are mixed (Fig. 11.16).

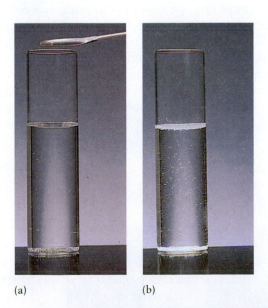

(a)   (b)

FIGURE 11.16 (a) A saturated solution of zinc acetate in water. (b) When additional acetate ions are added (as a single crystal of solid sodium acetate in the spatula shown in part (a)), the solubility of the zinc acetate is significantly reduced and more zinc acetate precipitates.

The prediction of the numerical value of the common-ion effect is difficult. Because ions interact with one another strongly, simple equilibrium calculations are rarely valid: the activities of ions differ markedly from their molarities. However, we can still get an idea of the size of the common-ion effect by solving the K_{sp} expression for the concentration of the ion other than the common ion.

EXAMPLE 11.9 **Sample exercise: Estimating the effect of a common ion on solubility**

Estimate the solubility of silver chloride in 1.0×10^{-4} M NaCl(aq).

SOLUTION For a given concentration of Cl^- ions, the concentration of Ag^+ ions must satisfy K_{sp}.

From $K_{sp} = [Ag^+][Cl^-]$,

$$[Ag^+] = \frac{K_{sp}}{[Cl^-]}$$

It follows that silver chloride will dissolve in 1.0×10^{-4} M NaCl(aq), in which $[Cl^-] = 1.0 \times 10^{-4}$ mol·L^{-1}, until the concentration of Ag^+ ions is given by

With $[Cl^-] = 1.0 \times 10^{-4}$ mol·L^{-1},

$$[Ag^+] = \frac{1.6 \times 10^{-10}}{1.0 \times 10^{-4}} = 1.6 \times 10^{-6}$$

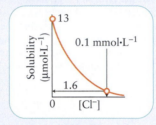

The concentration of Ag^+ ions, and hence the solubility of AgCl formula units, is 1.6×10^{-6} mol·L^{-1}, which is about 10 times less than the solubility of AgCl in pure water.

SELF-TEST 11.12A What is the approximate molar solubility of calcium carbonate in 0.20 M CaCl$_2$(aq)?

[*Answer:* 4.4×10^{-8} mol·L^{-1}]

SELF-TEST 11.12B What is the approximate molar solubility of silver bromide in 0.10 M CaBr$_2$(aq)?

The common-ion effect is the reduction in solubility of a sparingly soluble salt by the addition of a soluble salt that has an ion in common with it.

11.10 Predicting Precipitation

Sometimes it is important to know under what conditions a precipitate will form. For example, if we are analyzing a mixture of ions, we may want to precipitate only one type of ion to separate it from the mixture. In Section 9.5, we saw how to predict the direction in which a reaction will take place by comparing the values of Q, the reaction quotient, and K, the equilibrium constant. Exactly the same techniques can be used to decide whether a precipitate is likely to form when two electrolyte solutions are mixed. In this case, the equilibrium constant is the solubility product, K_{sp}, and the reaction quotient is denoted Q_{sp}. Precipitation occurs when Q_{sp} is greater than K_{sp} (Fig. 11.17).

FIGURE 11.17 The relative magnitudes of the solubility quotient, Q_{sp}, and the solubility product, K_{sp}, are used to decide whether a salt will dissolve (left) or precipitate (right). When the concentrations of the ions are low (left) Q_{sp} is smaller than K_{sp}; when the ion concentrations are high (right), Q_{sp} is larger than K_{sp}.

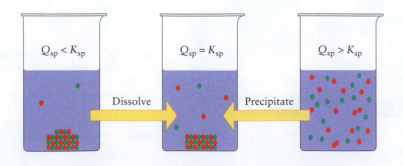

EXAMPLE 11.10   **Predicting whether a precipitate will form when two solutions are mixed**

Suppose we mix equal volumes of 0.2 M $Pb(NO_3)_2$(aq) and KI(aq). Will lead(II) iodide precipitate?

STRATEGY The concentrations of the Pb^{2+} and I^- ions are high, and we can suspect that precipitation will be spontaneous. To verify this prediction we note that, because equal volumes are mixed together, the new volume is twice each initial volume, and so the new molarities are half their original values. We also make the approximation that the salt is fully dissociated.

SOLUTION We know from Table 11.4 that $K_{sp} = 1.4 \times 10^{-8}$ for PbI_2 at 25°C.

Write the chemical equation and its equilibrium constant.	$PbI_2(s) \rightleftharpoons Pb^{2+}(aq) + 2\,I^-(aq), \quad K_{sp} = [Pb^{2+}][I^-]^2$
Calculate the new molarities of the ions.	Pb^{2+}(aq): $(0.2\ \text{mol·L}^{-1})/2 = 0.1\ \text{mol·L}^{-1}$ I^-(aq): $(0.2\ \text{mol·L}^{-1})/2 = 0.1\ \text{mol·L}^{-1}$

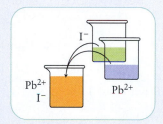

From $Q_{sp} = [Pb^{2+}][I^-]^2$,	$Q_{sp} = 0.1 \times (0.1)^2 = 1 \times 10^{-3}$

This value is considerably higher than K_{sp}, and so a precipitate forms (Fig. 11.18).

SELF-TEST 11.13A Does a precipitate of silver chloride form when 200. mL of 1.0×10^{-4} M $AgNO_3$(aq) and 900. mL of 1.0×10^{-6} M KCl(aq) are mixed? Assume complete dissociation.

[*Answer:* No ($Q_{sp} = 1.5 \times 10^{-11}$; $Q_{sp} < K_{sp}$)]

SELF-TEST 11.13B Does a precipitate of barium fluoride form when 100. mL of 1.0×10^{-3} M $Ba(NO_3)_2$(aq) is mixed with 200. mL of 1.0×10^{-3} M KF(aq)? Ignore possible protonation of F^-.

A salt precipitates if Q_{sp} is greater than K_{sp}.

11.11 Selective Precipitation

It is sometimes possible to separate different cations from a solution by adding a soluble salt containing an anion with which they form insoluble salts. For example, seawater is a mixture of many different ions. It is possible to precipitate magnesium ions from seawater by adding hydroxide ions. However, other cations are also present in seawater. Their individual concentrations and the relative solubilities of their hydroxides determine which will precipitate first if a certain amount of hydroxide is added. Optimum separation of two compounds is achieved when Q_{sp} exceeds the K_{sp} of one species but is less than the K_{sp} of the second species. Example 11.11 illustrates a strategy for predicting the order of precipitation.

EXAMPLE 11.11   **Predicting the order of precipitation**

A sample of seawater contains, among other solutes, the following concentrations of soluble cations: 0.050 mol·L^{-1} Mg^{2+}(aq) and 0.010 mol·L^{-1} Ca^{2+}(aq). (a) Use the information in Table 11.4 to determine the order in which each ion precipitates as solid NaOH is added, and give the concentration of OH^- when precipitation of each begins. Assume no volume change on addition of the NaOH. (b) Calculate the concentration of the first ion to precipitate that remains in solution when the second ion precipitates.

FIGURE 11.18 When a few drops of lead(II) nitrate solution are added to a solution of potassium iodide, yellow lead(II) iodide immediately precipitates.

STRATEGY A salt begins to precipitate when the concentrations of its ions are such that $Q_{sp} > K_{sp}$. (a) Calculate the value of $[OH^-]$ required for each salt to precipitate by writing the expression for K_{sp} for each salt and then substituting the data provided. (b) Calculate the remaining concentration of the first cation to precipitate by substituting the value of $[OH^-]$ into K_{sp} for that hydroxide.

SOLUTION (a) Write the chemical equation and K_{sp} for the dissolving of $Ca(OH)_2$:

$$Ca(OH)_2(s) \rightleftharpoons Ca^{2+}(aq) + 2\ OH^-(aq) \quad K_{sp} = [Ca^{2+}][OH^-]^2$$

and from Table 11.4, $K_{sp} = 5.5 \times 10^{-6}$. Then:

Find $[OH^-]$ from $K_{sp} = [Ca^{2+}][OH^-]^2$ in the form $[OH^-] = \{K_{sp}/[Ca^{2+}]\}^{1/2}$.

$$[OH^-] = \left(\frac{5.5 \times 10^{-6}}{0.010}\right)^{1/2} = 0.023$$

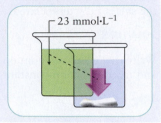

Write the chemical equation and K_{sp} for the dissolving of $Mg(OH)_2$:

$$Mg(OH)_2(s) \rightleftharpoons Mg^{2+}(aq) + 2\ OH^-(aq) \quad K_{sp} = [Mg^{2+}][OH^-]^2$$

From Table 11.4, $K_{sp} = 1.1 \times 10^{-11}$.

Find $[OH^-]$ from $K_{sp} = [Mg^{2+}][OH^-]^2$ in the form $[OH^-] = \{K_{sp}/[Mg^{2+}]\}^{1/2}$.

$$[OH^-] = \left(\frac{1.1 \times 10^{-11}}{0.050}\right)^{1/2} = 1.5 \times 10^{-5}$$

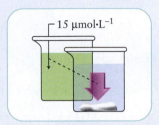

We conclude that the hydroxides precipitate in the order $Mg(OH)_2$ at 1.5×10^{-5} mol·L^{-1} OH$^-$(aq) and $Ca(OH)_2$ at 0.023 mol·L^{-1} OH$^-$(aq).

(b) Find the concentration of magnesium ions when $[OH^-] = 0.023$:

Find $[Mg^{2+}]$ from $K_{sp} = [Mg^{2+}][OH^-]^2$ in the form $[Mg^{2+}] = K_{sp}/[OH^-]^2$.

$$[Mg^{2+}] = \frac{1.1 \times 10^{-11}}{(0.023)^2} = 2.1 \times 10^{-8}$$

The concentration of magnesium ions remaining when $Ca(OH)_2$ begins to precipitate is indeed very small.

SELF-TEST 11.14A Potassium carbonate is added to a solution containing the following concentrations of soluble cations: 0.030 mol·L^{-1} Mg^{2+}(aq) and 0.0010 mol·L^{-1} Ca^{2+}(aq). (a) Use the information in Table 11.4 to determine the order in which each ion precipitates as the concentration of K_2CO_3 is increased and give the concentration of CO_3^{2-} when precipitation of each begins. (b) Calculate the concentration of the first ion to precipitate that remains in solution when the second ion precipitates.

[*Answer:* (a) $CaCO_3$ precipitates first, at 8.7×10^{-6} mol·L^{-1} CO$_3^{2-}$, then $MgCO_3$ at 3.3×10^{-4} mol·L^{-1} CO$_3^{2-}$; (b) 2.6×10^{-5} mol·L^{-1} Ca^{2+}]

SELF-TEST 11.14B Chloride ion is added to a solution containing the following concentrations of soluble salts: 0.020 mol·L^{-1} Pb(NO$_3$)$_2$(aq) and 0.0010 mol·L^{-1} AgNO$_3$(aq). (a) Use the information in Table 11.4 to determine the order in which each ion precipitates as the concentration of chloride ion is increased and give the concentration of Cl$^-$ when precipitation of each begins. (b) Calculate the concentration of the first ion to precipitate that remains in solution when the second ion precipitates.

A mixture of ions in solution can be separated by adding an oppositely charged ion with which they form salts having very different solubilities.

11.12 Dissolving Precipitates

When a precipitate has been formed during the qualitative analysis of the ions present in a solution, it may be necessary to dissolve the precipitate again to identify the cation or anion. One strategy is to remove one of the ions from the solubility equilibrium so that the precipitate will continue to dissolve in a fruitless chase for equilibrium. Suppose, for example, that a solid hydroxide such as iron(III) hydroxide is in equilibrium with its ions in solution:

$$Fe(OH)_3(s) \rightleftharpoons Fe^{3+}(aq) + 3\ OH^-(aq)$$

To dissolve more of the solid, we can add acid. The H_3O^+ ions from the acid remove the OH^- ions by converting them into water, and the $Fe(OH)_3$ dissolves.

Many carbonate, sulfite, and sulfide precipitates can be dissolved by the addition of acid, because the anions react with the acid to form a gas that bubbles out of solution. For example, in a saturated solution of zinc carbonate, solid $ZnCO_3$ is in equilibrium with its ions:

$$ZnCO_3(s) \rightleftharpoons Zn^{2+}(aq) + CO_3^{2-}(aq)$$

The CO_3^{2-} ions react with acid to form CO_2:

$$CO_3^{2-}(aq) + 2\ HNO_3(aq) \longrightarrow CO_2(g) + H_2O(l) + 2\ NO_3^-(aq)$$

The dissolution of carbonates by acid is an undesired result of acid rain, which has damaged the appearance of many historic marble and limestone monuments (Fig. 11.19; marble and limestone are forms of calcium carbonate).

An alternative procedure for removing an ion from solution is to change its identity by changing its oxidation state. The metal ions in very insoluble heavy metal sulfide precipitates can be dissolved by oxidizing the sulfide ion to elemental sulfur. For example, copper(II) sulfide, CuS, takes part in the equilibrium

$$CuS(s) \rightleftharpoons Cu^{2+}(aq) + S^{2-}(aq)$$

However, when nitric acid is added, the sulfide ions are oxidized to elemental sulfur:

$$3\ S^{2-}(aq) + 8\ HNO_3(aq) \longrightarrow 3\ S(s) + 2\ NO(g) + 4\ H_2O(l) + 6\ NO_3^-(aq)$$

This complicated oxidation removes the sulfide ions from the equilibrium, and Cu^{2+} ions dissolve as $Cu(NO_3)_2$.

Some precipitates dissolve when the temperature is changed. This strategy is used to purify precipitates. The mixture is heated to dissolve the solid and filtered to remove insoluble impurities. The solid is then allowed to re-form as the solution cools and is removed from the solution in a second filtration. Complex ion formation (Section 11.13) can also be used to dissolve metal ions.

The solubility of a solid can be increased by removing one of its ions from solution; acid can be used to dissolve a hydroxide, sulfide, sulfite, or carbonate precipitate; and nitric acid can be used to oxidize metal sulfides to sulfur and a soluble salt.

11.13 Complex Ion Formation

To remove an ion, we can use the fact that many metal cations are Lewis acids (Section 10.2). When a Lewis acid and a Lewis base react, they form a coordinate covalent bond and the product is called a **coordination complex**. In this section, we consider complexes in which the Lewis acid is a metal cation, such as Ag^+. An example is the formation of $Ag(NH_3)_2^+$ when an aqueous solution of the Lewis base ammonia is added to a solution of silver ions:

$$Ag^+(aq) + 2\ NH_3(aq) \longrightarrow Ag(NH_3)_2^+(aq)$$

If enough ammonia is added to a precipitate of silver halide, all the precipitate dissolves. A similar procedure is used to remove the silver halide emulsion from an

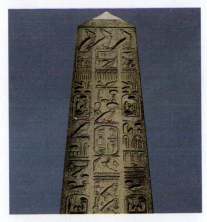

(a)

(b)

FIGURE 11.19 The state of the carving on Cleopatra's Needle has deteriorated as a result of the action of acid rain: (a) after 3500 years in the Egyptian desert; (b) after a further 90 years in Central Park, New York City.

TABLE 11.5 Formation Constants of Complexes in Water at 25°C

Equilibrium	K_f
$Ag^+(aq) + 2\,CN^-(aq) \rightleftharpoons Ag(CN)_2^-(aq)$	5.6×10^8
$Ag^+(aq) + 2\,NH_3(aq) \rightleftharpoons Ag(NH_3)_2^+(aq)$	1.6×10^7
$Au^+(aq) + 2\,CN^-(aq) \rightleftharpoons Au(CN)_2^-(aq)$	2.0×10^{38}
$Cu^{2+}(aq) + 4\,NH_3(aq) \rightleftharpoons Cu(NH_3)_4^{2+}(aq)$	1.2×10^{13}
$Hg^{2+}(aq) + 4\,Cl^-(aq) \rightleftharpoons HgCl_4^{2-}(aq)$	1.2×10^5
$Fe^{2+}(aq) + 6\,CN^-(aq) \rightleftharpoons Fe(CN)_6^{4-}(aq)$	7.7×10^{36}
$Ni^{2+}(aq) + 6\,NH_3(aq) \rightleftharpoons Ni(NH_3)_6^{2+}(aq)$	5.6×10^8

exposed photographic film after it has been developed. In this case, the reagent used to form the complex is the thiosulfate ion, $S_2O_3^{2-}$:

$$Ag^+(aq) + 2\,S_2O_3^{2-}(aq) \longrightarrow Ag(S_2O_3)_2^{3-}(aq)$$

Complex formation removes some of the Ag^+ ions from solution. As a result, to preserve the value of K_{sp}, more silver chloride dissolves. Formation of a complex increases the solubility of a sparingly soluble compound.

To treat complex formation quantitatively, we note that complex formation and the dissolving of the salt are both at equilibrium; so we can write

$$AgCl(s) \rightleftharpoons Ag^+(aq) + Cl^-(aq) \qquad K_{sp} = [Ag^+][Cl^-] \tag{E}$$

$$Ag^+(aq) + 2\,NH_3(aq) \rightleftharpoons Ag(NH_3)_2^+ \qquad K_f = \frac{[Ag(NH_3)_2^+]}{[Ag^+][NH_3]^2} \tag{F}$$

The equilibrium constant for the formation of a complex ion is called the **formation constant**, K_f. At 25°C, $K_f = 1.6 \times 10^7$ for reaction F. Values for other complexes are given in Table 11.5, and Example 11.12 shows how to use them.

EXAMPLE 11.12 Calculating molar solubility in the presence of complex formation

Calculate the molar solubility of silver chloride in 0.10 M $NH_3(aq)$, given that $K_{sp} = 1.6 \times 10^{-10}$ for silver chloride and $K_f = 1.6 \times 10^7$ for the ammonia complex of Ag^+ ions, $Ag(NH_3)_2^+$.

STRATEGY First, we write the chemical equation for the equilibrium between the solid solute and the complex in solution as the sum of the equations for the solubility and complex formation equilibria. The equilibrium constant for the overall equilibrium is therefore the product of the equilibrium constants for the two processes. Then, we set up an equilibrium table and solve for the equilibrium concentrations of ions in solution.

SOLUTION The overall equilibrium is the sum of reactions E and F:

$$AgCl(s) + 2\,NH_3(aq) \rightleftharpoons Ag(NH_3)_2^+(aq) + Cl^-(aq)$$

The equilibrium constant of the overall reaction is the product of the equilibrium constants for the two reactions:

$$K = K_{sp} \times K_f = \frac{[Ag(NH_3)_2^+][Cl^-]}{[NH_3]^2}$$

Because 1 mol AgCl $\rightleftharpoons$ 1 mol Cl^-, the molar solubility of AgCl is given by the equation $s = [Cl^-]$. From the balanced overall equation, we see that $[Cl^-] = [Ag(NH_3)_2^+]$ in the saturated solution. The equilibrium table, with all concentrations in moles per liter, is

	NH_3	$Ag(NH_3)_2^+$	Cl^-
Step 1 Initial molarity	0.10	0	0
Step 2 Change in molarity	$-2x$	$+x$	$+x$
Step 3 Equilibrium molarity	$0.10 - 2x$	x	x

Now determine the value of K and substitute the information from the table.

From $K = K_{sp} \times K_f$,	$K = (1.6 \times 10^{-10}) \times (1.6 \times 10^7) = 2.6 \times 10^{-3}$
Find $[Cl^-]$ from $K = [Ag(NH_3)_2{}^+][Cl^-]/[NH_3]^2$.	$K = \dfrac{x \times x}{(0.10 - 2x)^2} = 2.6 \times 10^{-3}$
Take the square root of each side.	$\dfrac{x}{0.10 - 2x} = 5.1 \times 10^{-2}$
Solve for x.	$x = 5.1 \times 10^{-2} \times (0.10 - 2x)$, hence $x = 4.6 \times 10^{-3}$

From step 3, $x = [Ag(NH_3)_2{}^+] = 4.6 \times 10^{-3}$. Hence, the molar solubility of silver chloride in 0.10 M $NH_3(aq)$ is 4.6×10^{-3} mol·L^{-1}, more than 100 times the molar solubility of silver chloride in pure water (1.3×10^{-5} mol·L^{-1}).

SELF-TEST 11.15A Use data from Tables 11.4 and 11.5 to calculate the molar solubility of silver bromide in 1.0 M $NH_3(aq)$.

[*Answer:* 3.5×10^{-3} mol·L^{-1}]

SELF-TEST 11.15B Use data from Tables 11.4 and 11.5 to calculate the molar solubility of copper(II) sulfide in 1.2 M $NH_3(aq)$.

The solubility of a salt increases if the salt can form a complex ion with other species in the solution.

11.14 Qualitative Analysis

Complex formation, selective precipitation, and control of the pH of a solution all play important roles in the qualitative analysis of the ions present in aqueous solutions. There are many different schemes of analysis, but they follow the same general principles. Let's think through a simple procedure for the identification of five cations by following the steps that might be used in the laboratory. We shall see how each step makes use of solubility equilibria.

Suppose we have a solution that contains lead(II), mercury(I), silver, copper(II), and zinc ions. The method is outlined in Fig. 11.20, which includes additional cations, and is illustrated in Fig. 11.21. Most chlorides are soluble; so, when hydrochloric acid is added to a mixture of salts, only certain chlorides precipitate (see Table 11.4). Silver and mercury(I) chlorides have such small values of K_{sp} that, even with low concentrations of Cl$^-$ ions, the chlorides precipitate. Lead(II) chloride, which is slightly soluble, will precipitate if the chloride ion concentration is

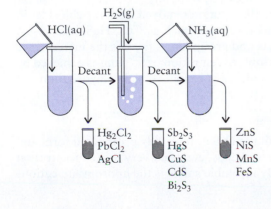

FIGURE 11.20 Part of a simple qualitative analysis scheme used to separate certain cations. In the first step, three cations are separated as insoluble chlorides. In the second step, cations that form highly insoluble sulfides are removed by precipitation at a low pH; and, in the third step, the remaining cations are precipitated as the sulfides at a higher pH.

FIGURE 11.21 Steps in the analysis of five cations by selective precipitation. (a) The original solution contains Pb^{2+}, Hg_2^{2+}, Ag^+, Cu^{2+}, and Zn^{2+} ions (left). Addition of HCl precipitates AgCl, Hg_2Cl_2, and $PbCl_2$, which can be removed by decanting or filtration (right). (b) Addition of H_2S to the solution remaining from the first step (left) precipitates CuS, which can be removed (right). (c) Making the solution from the second step (left) basic by adding ammonia precipitates ZnS (right).

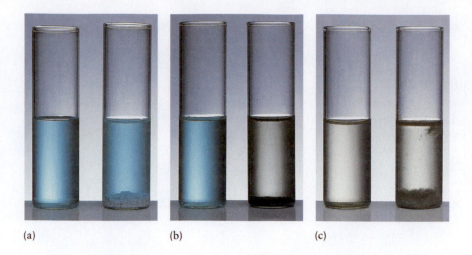

(a)   (b)   (c)

high enough. The hydronium ions provided by the acid play no role in this step; they simply accompany the chloride ions. At this point, the precipitate can be separated from the solution by using a centrifuge to compact the solid and then decanting (pouring off) the solution. The solution now contains copper(II) and zinc ions, whereas the solid consists of $PbCl_2$, Hg_2Cl_2, and AgCl.

Because $PbCl_2$ is slightly soluble, if the precipitate is rinsed with hot water, the lead(II) chloride dissolves and the solution is separated from the precipitate. If sodium chromate is then added to that solution, any lead(II) ions present will precipitate as yellow lead(II) chromate:

$$Pb^{2+}(aq) + CrO_4^{2-}(aq) \longrightarrow PbCrO_4(s)$$

At this point, the silver and mercury(I) chlorides remain as precipitates. When aqueous ammonia is added to the solid mixture, the silver precipitate dissolves as the soluble complex ion $Ag(NH_3)_2^+$ forms:

$$Ag^+(aq) + 2\ NH_3(aq) \longrightarrow Ag(NH_3)_2^+(aq)$$

and mercury(I) reacts with ammonia to form a gray solid consisting of a mixture of mercury(II) ions precipitated as white $HgNH_2Cl(s)$ and black metallic mercury (Fig. 11.22):

$$Hg_2Cl_2(s) + 2\ NH_3(aq) \longrightarrow Hg(l) + HgNH_2Cl(s) + NH_4^+(aq) + Cl^-(aq)$$

At this point, any Hg_2^{2+} has precipitated and any Ag^+ present is in solution. The solution is separated from the solid, and the presence of silver ion in the solution is verified by addition of nitric acid. This acid pulls the ammonia out of the complex as NH_4^+, allowing white silver chloride to precipitate:

$$Ag(NH_3)_2^+(aq) + Cl^-(aq) + 2\ H_3O^+(aq) \longrightarrow$$
$$AgCl(s) + 2\ NH_4^+(aq) + 2\ H_2O(l)$$

FIGURE 11.22 When ammonia is added to a silver chloride precipitate, the precipitate dissolves. However, when ammonia is added to a precipitate of mercury(I) chloride, mercury metal and mercury(II) ions are formed in a redox reaction and the mass turns gray. Left to right: silver chloride in water, silver chloride in aqueous ammonia, mercury(I) chloride in water, and mercury(I) chloride in aqueous ammonia.

Sulfides with widely different solubilities and solubility products can be selectively precipitated by adding S^{2-} ions to the solution removed from the chlorides in the first step (see Fig. 11.20). Some metal sulfides (such as CuS, HgS, and Sb_2S_3) have extremely small solubility products and precipitate if there is the merest trace of S^{2-} ions in the solution. Such a very low concentration of S^{2-} ions is achieved by adding hydrogen sulfide, H_2S, to an acidified solution. A higher hydronium ion concentration shifts the equilibrium

$$H_2S(aq) + 2\ H_2O(l) \rightleftharpoons 2\ H_3O^+(aq) + S^{2-}(aq)$$

to the left, which ensures that almost all the H_2S is in its fully protonated form and hence that very little S^{2-} is present. Nevertheless, even that very low concentration will result in the precipitation of highly insoluble solids if the appropriate cations are present.

To verify the presence of Zn^{2+} ions in the solution remaining after the first two steps, we add H_2S followed by ammonia. The base removes the hydronium ion from the H_2S equilibrium, which shifts the equilibrium in favor of S^{2-} ions. The higher concentration of S^{2-} ions increases the Q_{sp} values of any remaining metal sulfides, such as ZnS or MnS, above their K_{sp} values, and they precipitate.

Qualitative analysis involves the separation and identification of ions by selective precipitation, complex formation, and the control of pH.

SKILLS YOU SHOULD HAVE MASTERED

❏ 1 Calculate the pH of a buffer solution (Example 11.1).

❏ 2 Calculate the pH change when acid or base is added to a buffer solution (Example 11.2).

❏ 3 Specify the composition of a buffer solution with a given pH (Example 11.3).

❏ 4 Interpret the features of the pH curve for the titration of a strong or weak acid with a strong base and a strong or weak base with a strong acid (Sections 11.4 and 11.5).

❏ 5 Calculate the pH at any point in a strong base–strong acid titration (Toolbox 11.1 and Example 11.4).

❏ 6 Calculate the pH at any point in a strong base–weak acid and weak base–strong acid titration (Toolbox 11.2 and Examples 11.5 and 11.6).

❏ 7 Select an appropriate indicator for a given titration (Section 11.6).

❏ 8 Identify the primary species in solution and the principal proton transfer equilibrium at any point in the titration of a polyprotic acid (Self-Test 11.9).

❏ 9 Estimate a solubility product from molar solubility and vice versa (Examples 11.7 and 11.8).

❏ 10 Describe the common-ion effect and assess its magnitude (Example 11.9).

❏ 11 Predict whether a salt will precipitate given the concentrations of its ions in water (Example 11.10).

❏ 12 Predict the order of precipitation of a series of salts (Example 11.11).

❏ 13 Calculate molar solubility in the presence of complex ion formation (Example 11.12).

❏ 14 Use a simple qualitative analysis scheme and justify the steps in terms of solubility equilibria (Section 11.14).

EXERCISES

The values of K_a and K_b for weak acids and bases may be found in Tables 10.1 and 10.2.

Mixed Solutions and Buffers

11.1 Explain what happens to (a) the concentration of H_3O^+ ions in an acetic acid solution when solid sodium acetate is added; (b) the percentage deprotonation of benzoic acid in a benzoic acid solution when hydrochloric acid is added; (c) the pH of the solution when solid ammonium chloride is added to aqueous ammonia.

11.2 Explain what happens to (a) the pH of a solution of phosphoric acid after the addition of solid sodium dihydrogen phosphate; (b) the percentage deprotonation of HCN in a hydrocyanic acid solution after the addition of hydrobromic acid; (c) the concentration of H_3O^+ ions when pyridinium chloride is added to an aqueous solution of the base pyridine.

11.3 A solution of equal concentrations of lactic acid and sodium lactate was found to have pH = 3.08. (a) What are the values of pK_a and K_a of lactic acid? (b) What would the pH be if the acid had twice the concentration of the salt?

11.4 What is the concentration of hydronium ions in (a) a solution that is 0.075 M HCN(aq) and 0.060 M NaCN(aq); (b) a solution that is 0.20 M NH_2NH_2(aq) and 0.30 M NaCl(aq); (c) a solution that is 0.015 M HCN(aq) and 0.030 M NaCN(aq); (d) a solution that is 0.125 M NH_2NH_2(aq) and 0.125 M NH_2NH_3Br(aq)?

11.5 Determine the pH and pOH of (a) a solution that is 0.50 M $NaHSO_4$(aq) and 0.25 M Na_2SO_4(aq); (b) a solution that is 0.50 M $NaHSO_4$(aq) and 0.10 M Na_2SO_4(aq); (c) a solution that is 0.50 M $NaHSO_4$(aq) and 0.50 M Na_2SO_4(aq).

11.6 Calculate the pH and pOH of (a) a solution that is 0.17 M Na_2HPO_4(aq) and 0.25 M Na_3PO_4(aq); (b) a solution that is 0.66 M Na_2HPO_4(aq) and 0.42 M Na_3PO_4(aq); (c) a solution that is 0.12 M Na_2HPO_4(aq) and 0.12 M Na_3PO_4(aq).

11.7 The pH of 0.40 M HF(aq) is 1.93. Calculate the change in pH when 0.356 g of sodium fluoride is added to 50.0 mL of the solution. Ignore any change in volume.

11.8 The pH of 0.50 M HBrO(aq) is 4.50. Calculate the change in pH when 6.80 g of sodium hypobromite is added to 100. mL of the solution. Ignore any change in volume.

11.9 Calculate the pH of the solution that results from mixing (a) 30.0 mL of 0.050 M HCN(aq) with 70.0 mL of 0.030 M NaCN(aq); (b) 40.0 mL of 0.030 M HCN(aq) with 60.0 mL of 0.050 M NaCN(aq); (c) 25.0 mL of 0.105 M HCN(aq) with 25.0 mL of 0.105 M NaCN(aq).

11.10 Calculate the pH of the solution that results from mixing (a) 0.100 L of 0.020 M $(CH_3)_2NH$(aq) with 0.300 L of 0.030 M $(CH_3)_2NH_2Cl$(aq); (b) 65.0 mL of 0.010 M $(CH_3)_2NH$(aq) with 10.0 mL of 0.150 M $(CH_3)_2NH_2Cl$(aq); (c) 50.0 mL of 0.015 M $(CH_3)_2NH$(aq) with 12.5 mL of 0.015 M $(CH_3)_2NH_2Cl$(aq).

11.11 Sodium hypochlorite, $NaClO$, is the active ingredient in many bleaches. Calculate the ratio of the concentrations of ClO^- and $HClO$ in a bleach solution having a pH adjusted to 6.50 by the use of strong acid or strong base.

11.12 Aspirin (shown here as acetylsalicylic acid, $K_a = 3.2 \times 10^{-4}$) is a product of the reaction of salicylic acid with acetic anhydride. Calculate the ratio of the concentrations of the acetylsalicylate ion to acetylsalicylic acid in a solution that has a pH adjusted to 4.13 by the use of strong acid or strong base.

Salicylic acid   Acetylsalicylic acid

11.13 Predict the pH region in which each of the following buffers will be effective, assuming equal molarities of the acid and its conjugate base: (a) sodium lactate and lactic acid; (b) sodium benzoate and benzoic acid; (c) potassium hydrogen phosphate and potassium phosphate; (d) potassium hydrogen phosphate and potassium dihydrogen phosphate; (e) hydroxylamine and hydroxylammonium chloride.

11.14 Predict the pH region in which each of the following buffers will be effective, assuming equal molarities of the acid and its conjugate base: (a) sodium nitrite and nitrous acid; (b) sodium formate and formic acid; (c) sodium carbonate and sodium hydrogen carbonate; (d) ammonia and ammonium chloride; (e) pyridine and pyridinium chloride.

11.15 Use Tables 10.1, 10.2, and 10.9 to suggest a conjugate acid–base system that would be an effective buffer at a pH close to (a) 2; (b) 7; (c) 3; (d) 12.

11.16 Use Tables 10.1, 10.2, and 10.9 to suggest a conjugate acid–base system that would be an effective buffer at a pH close to (a) 4; (b) 9; (c) 5; (d) 11.

11.17 (a) What must be the ratio of the concentrations of CO_3^{2-} and HCO_3^- ions in a buffer solution having a pH of 11.0? (b) What mass of K_2CO_3 must be added to 1.00 L of 0.100 M $KHCO_3(aq)$ to prepare a buffer solution with a pH of 11.0? (c) What mass of $KHCO_3$ must be added to 1.00 L of 0.100 M $K_2CO_3(aq)$ to prepare a buffer solution with a pH of 11.0? (d) What volume of 0.200 M $K_2CO_3(aq)$ must be added to 100. mL of 0.100 M $KHCO_3(aq)$ to prepare a buffer solution with a pH of 11.0?

11.18 (a) What must be the ratio of the molarities of PO_4^{3-} and HPO_4^{2-} ions in a buffer solution having a pH of 12.0? (b) What mass of K_3PO_4 must be added to 1.00 L of 0.100 M $K_2HPO_4(aq)$ to prepare a buffer solution with a pH of 12.0? (c) What mass of K_2HPO_4 must be added to 1.00 L of 0.100 M $K_3PO_4(aq)$ to prepare a buffer solution with a pH of 12.0? (d) What volume of 0.150 M $K_3PO_4(aq)$ must be added to 50.0 mL of 0.100 M $K_2HPO_4(aq)$ to prepare a buffer solution with a pH of 12.0?

11.19 A buffer solution of volume 100.0 mL is 0.100 M $CH_3COOH(aq)$ and 0.100 M $NaCH_3CO_2(aq)$. (a) What are the pH and the pH change resulting from the addition of 10.0 mL of 0.950 M $NaOH(aq)$ to the buffer solution? (b) What are the pH and the pH change resulting from the addition of 20.0 mL of 0.100 M $HNO_3(aq)$ to the initial buffer solution?

11.20 A buffer solution of volume 100.0 mL is 0.150 M $Na_2HPO_4(aq)$ and 0.100 M $KH_2PO_4(aq)$. (a) What are the pH and the pH change resulting from the addition of 80.0 mL of 0.0100 M $NaOH(aq)$ to the buffer solution? (b) What are the pH and the pH change resulting from the addition of 10.0 mL of 1.0 M $HNO_3(aq)$ to the initial buffer solution?

Titrations

11.21 (a) Sketch the titration curve for the titration of 5.00 mL 0.010 M $NaOH(aq)$ with 0.005 M $HCl(aq)$, indicating the pH of the initial and final solutions and the pH at the stoichiometric point. What volume of HCl has been added at (b) the stoichiometric point? (c) the halfway point of the titration?

11.22 (a) Sketch the titration curve for the titration of 5.00 mL 0.010 M $HCl(aq)$ with 0.010 M $Ca(OH)_2(aq)$, indicating the pH of the initial and final solutions and the pH at the stoichiometric point. What volume of titrant has been added at (b) the stoichiometric point? (c) the halfway point of the titration?

11.23 Calculate the volume of 0.150 M $HCl(aq)$ required to neutralize (a) one-half and (b) all the hydroxide ions in 25.0 mL of 0.110 M $NaOH(aq)$. (c) What is the molarity of Na^+ ions at the stoichiometric point? (d) Calculate the pH of the solution after the addition of 20.0 mL of 0.150 M $HCl(aq)$ to 25.0 mL of 0.110 M $NaOH(aq)$.

11.24 Calculate the volume of 0.116 M $HCl(aq)$ required to neutralize (a) one-half and (b) all the hydroxide ions in 25.0 mL of 0.215 M $KOH(aq)$. (c) What is the molarity of Cl^- ions at the stoichiometric point? (d) Calculate the pH of the solution after the addition of 40.0 mL of 0.116 M $HCl(aq)$ to 25.0 mL of 0.215 M $KOH(aq)$.

11.25 Suppose that 4.25 g of an unknown monoprotic weak acid, HA, is dissolved in water. Titration of the solution with 0.350 M $NaOH(aq)$ required 52.0 mL to reach the stoichiometric point. After the addition of 26.0 mL, the pH of the solution was found to be 3.82. (a) What is the molar mass of the acid? (b) What is the value of pK_a for the acid?

11.26 Suppose that 0.483 g of an unknown monoprotic weak acid, HA, is dissolved in water. Titration of the solution with 0.250 M $NaOH(aq)$ required 42.0 mL to reach the stoichiometric point. After the addition of 21.0 mL, the pH of the solution was found to be 3.75. (a) What is the molar mass of the acid? (b) What is the value of pK_a for the acid? Can you identify the acid?

11.27 Suppose that 1.436 g of impure sodium hydroxide is dissolved in 300. mL of aqueous solution and that 25.00 mL of this solution is titrated to the stoichiometric point with 34.20 mL of 0.0695 M $HCl(aq)$. What is the percentage purity of the original sample?

11.28 Suppose that 1.331 g of impure barium hydroxide is dissolved in enough water to produce 250. mL of solution and that 35.0 mL of this solution is titrated to the stoichiometric point with 17.6 mL of 0.0935 M HCl(aq). What is the percentage purity of the original sample?

11.29 Calculate the pH at each stage in the titration for the addition of 0.150 M HCl(aq) to 25.0 mL of 0.110 M NaOH(aq) (a) initially; (b) after the addition of 5.0 mL of acid; (c) after the addition of a further 5.0 mL; (d) at the stoichiometric point; (e) after the addition of 5.0 mL of acid beyond the stoichiometric point; (f) after the addition of 10.0 mL of acid beyond the stoichiometric point.

11.30 Calculate the pH at each stage in the titration in which 0.116 M HCl(aq) is added to 25.0 mL of 0.215 M KOH(aq) (a) initially; (b) after the addition of 5.0 mL of acid; (c) after the addition of a further 5.0 mL; (d) at the stoichiometric point; (e) after the addition of 5.0 mL of acid beyond the stoichiometric point; (f) after the addition of 10.0 mL of acid beyond the stoichiometric point.

11.31 Suppose that 25.0 mL of 0.10 M CH_3COOH(aq) is titrated with 0.10 M NaOH(aq). (a) What is the initial pH of the 0.10 M CH_3COOH(aq) solution? (b) What is the pH after the addition of 10.0 mL of 0.10 M NaOH(aq)? (c) What volume of 0.10 M NaOH(aq) is required to reach halfway to the stoichiometric point? (d) Calculate the pH at that halfway point. (e) What volume of 0.10 M NaOH(aq) is required to reach the stoichiometric point? (f) Calculate the pH at the stoichiometric point.

11.32 Suppose that 30.0 mL of 0.20 M C_6H_5COOH(aq) is titrated with 0.30 M KOH(aq). (a) What is the initial pH of the 0.20 M C_6H_5COOH(aq)? (b) What is the pH after the addition of 15.0 mL of 0.30 M KOH(aq)? (c) What volume of 0.30 M KOH(aq) is required to reach halfway to the stoichiometric point? (d) Calculate the pH at the halfway point. (e) What volume of 0.30 M KOH(aq) is required to reach the stoichiometric point? (f) Calculate the pH at the stoichiometric point.

11.33 Suppose that 15.0 mL of 0.15 M NH_3(aq) is titrated with 0.10 M HCl(aq). (a) What is the initial pH of the 0.15 M NH_3(aq)? (b) What is the pH after the addition of 15.0 mL of 0.10 M HCl(aq)? (c) What volume of 0.10 M HCl(aq) is required to reach halfway to the stoichiometric point? (d) Calculate the pH at the halfway point. (e) What volume of 0.10 M HCl(aq) is required to reach the stoichiometric point? (f) Calculate the pH at the stoichiometric point. (g) Use Table 11.3 to select an indicator for the titration.

11.34 Suppose that 50.0 mL of 0.25 M CH_3NH_2(aq) is titrated with 0.35 M HCl(aq). (a) What is the initial pH of the 0.25 M CH_3NH_2(aq)? (b) What is the pH after the addition of 15.0 mL of 0.35 M HCl(aq)? (c) What volume of 0.35 M HCl(aq) is required to reach halfway to the stoichiometric point? (d) Calculate the pH at the halfway point. (e) What volume of 0.35 M HCl(aq) is required to reach the stoichiometric point? (f) Calculate the pH at the stoichiometric point. (g) Use Table 11.3 to select an indicator for the titration.

11.35 Below is the titration curve for the neutralization of 25 mL of a monoprotic acid with a strong base. Answer the following questions about the reaction and explain your reasoning in each case. (a) Is the acid strong or weak? (b) What is the initial hydronium ion concentration of the acid? (c) What is K_a for the acid? (d) What is the initial concentration of the

acid? (e) What is the concentration of base in the titrant? (f) Use Table 11.3 to select an indicator for the titration.

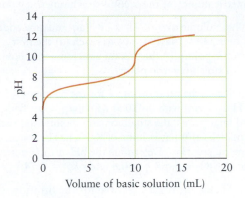

Volume of basic solution (mL)

11.36 Below is the titration curve for the neutralization of 25 mL of a base with a strong monoprotic acid. Answer the following questions about the reaction and explain your reasoning in each case. (a) Is the base strong or weak? (b) What is the initial hydroxide ion concentration of the base? (c) What is K_b for the base? (d) What is the initial concentration of the base? (e) What is the concentration of acid in the titrant? (f) Use Table 11.3 to select an indicator for the titration.

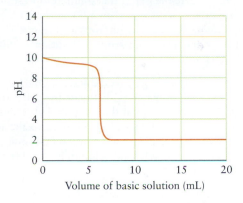

Volume of basic solution (mL)

11.37 Which of the following indicators (see Table 11.3) could you use for a titration of 0.20 M CH_3COOH(aq) with 0.20 M NaOH(aq): (a) methyl orange; (b) litmus; (c) thymol blue; (d) phenolphthalein? Explain your selections.

11.38 Which of the following indicators (see Table 11.3) could you use for a titration of 0.20 M NH_3(aq) with 0.20 M HCl(aq): (a) bromocresol green; (b) methyl red; (c) phenol red; (d) thymol blue? Explain your selections.

11.39 Use Table 11.3 to suggest suitable indicators for the titrations described in Exercises 11.31 and 11.33.

11.40 Use Table 11.3 to suggest suitable indicators for the titrations described in Exercises 11.32 and 11.34.

Polyprotic Acid Titrations

11.41 What volume of 0.275 M KOH(aq) must be added to 75.0 mL of 0.137 M H_3AsO_4(aq) to reach (a) the first stoichiometric point; (b) the second stoichiometric point; (c) the third stoichiometric point?

11.42 What volume of 0.123 M NaOH(aq) must be added to 125 mL of 0.197 M H_2SO_3(aq) to reach (a) the first stoichiometric point; (b) the second stoichiometric point?

11.43 What volume of 0.255 M HNO_3(aq) must be added to 35.5 mL of 0.158 M Na_2HPO_3(aq) to reach (a) the first stoichiometric point; (b) the second stoichiometric point?

11.44 What volume of 0.0848 M HCl(aq) must be added to 88.8 mL of 0.233 M Na_3PO_4(aq) to reach (a) the first stoichiometric point; (b) the second stoichiometric point; (c) the third stoichiometric point?

11.45 (a) Sketch the titration curve for the titration of 5.00 mL 0.010 M $H_2S_2O_3$(aq) with 0.010 M KOH(aq), identifying the initial and final points, the stoichiometric points, and each of the halfway points. (b) What volume of KOH has been added at each stoichiometric point? (c) Determine the pH at each stoichiometric point. For $H_2S_2O_3$(aq), $pK_{a1} = 0.6$ and $pK_{a2} = 1.74$.

11.46 (a) Sketch the titration curve for the titration of 5.00 mL 0.010 M H_3AsO_4(aq) with 0.010 M KOH(aq), identifying the initial and final points, the stoichiometric points, and each of the halfway points. (b) What volume of KOH has been added at each stoichiometric point? (c) Determine the pH at each stoichiometric point. For H_3AsO_4(aq), $pK_{a1} = 2.25$ and $pK_{a2} = 6.77$.

11.47 Suppose that 0.122 g of phosphorous acid, H_3PO_3, is dissolved in water and that the total volume of the solution is 50.0 mL. (a) Estimate the pH of this solution. (b) Estimate the pH of the solution that results when 5.00 mL of 0.175 M NaOH(aq) is added to the phosphorous acid solution. (c) Estimate the pH of the solution if an additional 5.00 mL of 0.175 M NaOH(aq) is added to the solution in part (b).

11.48 Suppose that 0.242 g of oxalic acid, $(COOH)_2$, is dissolved in 50.0 mL of water. (a) Estimate the pH of this solution. (b) Estimate the pH of the solution that results when 15.0 mL of 0.150 M NaOH(aq) is added to the oxalic acid solution. (c) Estimate the pH of the solution that results if an additional 5.00 mL of NaOH(aq) solution is added to the solution from part (b).

11.49 Estimate the pH of the solution that results when each of the following solutions is added to 50.0 mL of 0.275 M Na_2HPO_4(aq): (a) 50.0 mL of 0.275 M HCl(aq); (b) 75.0 mL of 0.275 M HCl(aq); (c) 25.0 mL of 0.275 M HCl(aq).

11.50 Estimate the pH of the solution that results when 75.0 mL of 0.0995 M Na_2CO_3(aq) is mixed with (a) 25.0 mL of 0.130 M HNO_3(aq); (b) 65.0 mL of 0.130 M HNO_3(aq).

Solubility Equilibria

The values for the solubility products of some sparingly soluble salts are listed in Table 11.4.

11.51 Determine K_{sp} for each of the following sparingly soluble substances, given their molar solubilities: (a) AgBr, 8.8×10^{-7} mol·L^{-1}; (b) $PbCrO_4$, 1.3×10^{-7} mol·L^{-1}; (c) $Ba(OH)_2$, 0.11 mol·L^{-1}; (d) MgF_2, 1.2×10^{-3} mol·L^{-1}. For the purpose of this calculation, ignore any reaction of the anion with water and the autoprotolysis of water.

11.52 Determine K_{sp} for each of the following sparingly soluble compounds, given their molar solubilities: (a) AgI, 9.1×10^{-9} mol·L^{-1}; (b) $Ca(OH)_2$, 0.011 mol·L^{-1}; (c) Ag_3PO_4, 2.7×10^{-6} mol·L^{-1}; (d) Hg_2Cl_2, 5.2×10^{-7} mol·L^{-1}. For the purpose of this calculation, ignore any reaction of the anion with water and the autoprotolysis of water.

11.53 Calculate the molar solubility in water of (a) BiI_3 ($K_{sp} = 7.71 \times 10^{-19}$); (b) CuCl; (c) $CaCO_3$. For the purpose of this calculation, ignore any reaction of the anion with water.

11.54 Calculate the molar solubility in water of (a) $PbBr_2$; (b) Ag_2CO_3; (c) $Fe(OH)_2$. For the purpose of this calculation, ignore any reaction of the anion with water and the autoprotolysis of water.

11.55 The molarity of CrO_4^{2-} in a saturated Tl_2CrO_4 solution is 6.3×10^{-5} mol·L^{-1}. What is the K_{sp} of Tl_2CrO_4?

11.56 The molar solubility of cerium(III) hydroxide, $Ce(OH)_3$, is 5.2×10^{-6} mol·L^{-1}. What is the K_{sp} of cerium(III) hydroxide? For the purpose of this calculation, ignore the autoprotolysis of water.

11.57 Use the data in Table 11.4 to calculate the molar solubility of each sparingly soluble substance in its respective solution: (a) silver chloride in 0.20 M NaCl(aq); (b) mercury(I) chloride in 0.150 M NaCl(aq); (c) lead(II) chloride in 0.025 M $CaCl_2$(aq); (d) iron(II) hydroxide in 2.5×10^{-3} M $FeCl_2$(aq). For the purpose of this calculation, ignore the autoprotolysis of water.

11.58 Use the data in Table 11.4 to calculate the solubility of each sparingly soluble substance in its respective solution: (a) silver bromide in 0.050 M NaBr(aq); (b) magnesium carbonate in 1.0×10^{-3} M Na_2CO_3(aq); (c) lead(II) sulfate in 0.25 M Na_2SO_4(aq); (d) nickel(II) hydroxide in 0.125 M $NiSO_4$(aq). For the purpose of this calculation, ignore the autoprotolysis of water.

11.59 (a) What molarity of Ag^+ ions is required for the formation of a precipitate in 1.0×10^{-5} M NaCl(aq)? (b) What mass (in micrograms) of solid $AgNO_3$ needs to be added for the onset of precipitation in 100. mL of the solution in part (a)?

11.60 It is necessary to add iodide ions to precipitate lead(II) ions from 0.0020 M $Pb(NO_3)_2$(aq). (a) What minimum iodide ion concentration is required for the onset of PbI_2 precipitation? (b) What mass (in grams) of KI must be added for PbI_2 to form?

11.61 Determine the pH required for the onset of precipitation of $Ni(OH)_2$ from (a) 0.060 M $NiSO_4$(aq); (b) 0.030 M $NiSO_4$(aq).

11.62 Decide whether a precipitate will form when each of the following pairs of solutions are mixed: (a) 5.0 mL of 0.10 M K_2CO_3(aq) and 1.00 L of 0.010 M $AgNO_3$(aq); (b) 3.3 mL of 1.0 M HCl(aq), 4.9 mL of 0.0030 M $AgNO_3$(aq), and enough water to dilute the solution to 50.0 mL. For the purpose of this calculation, ignore any reaction of the anion with water.

11.63 Suppose that there are typically 20 average-sized drops in 1.0 mL of an aqueous solution. Will a precipitate form when 1 drop of 0.010 M NaCl(aq) is added to 10.0 mL of (a) 0.0040 M $AgNO_3$(aq)? (b) 0.0040 M $Pb(NO_3)_2$(aq)?

11.64 Assume 20 drops per milliliter. Will a precipitate form if (a) 7 drops of 0.0029 M K_2CO_3(aq) are added to 25.0 mL of 0.0018 M $CaCl_2$(aq); (b) 10 drops of 0.010 M Na_2CO_3(aq) are added to 10.0 mL of 0.0040 M $AgNO_3$(aq)? For the purpose of this calculation, ignore any reaction of the anion with water.

11.65 The concentrations of magnesium, calcium, and nickel(II) ions in an aqueous solution are 0.0010 mol·L^{-1}. (a) In what order do they precipitate when solid KOH is added? (b) Determine the pH at which each salt precipitates.

11.66 Suppose that two hydroxides, MOH and $M'(OH)_2$, both have $K_{sp} = 1.0 \times 10^{-12}$ and that initially both cations are present in a solution at concentrations of 0.0010 mol·L^{-1}. Which hydroxide precipitates first, and at what pH, when solid NaOH is added?

11.67 We wish to separate magnesium ions and barium ions by selective precipitation. Which anion, fluoride or carbonate, would be the better choice for achieving this precipitation? Why?

11.68 We wish to separate barium ions from calcium ions by selective precipitation. Which anion, fluoride or carbonate, would be the better choice for achieving this precipitation? Why?

11.69 In the process of separating Pb^{2+} ions from Cu^{2+} ions as sparingly soluble iodates, what is the Pb^{2+} concentration when Cu^{2+} just begins to precipitate as sodium iodate is added to a solution that is initially 0.0010 M $Pb(NO_3)_2$(aq) and 0.0010 M $Cu(NO_3)_2$(aq)?

11.70 A chemist attempts to separate barium ions from lead ions by using the sulfate ion as a precipitating agent. (a) What sulfate ion concentrations are required for the precipitation of $BaSO_4$ and $PbSO_4$ from a solution containing 0.010 M Ba^{2+}(aq) and 0.010 M Pb^{2+}(aq)? (b) What is the concentration of barium ions when the lead sulfate begins to precipitate?

11.71 Use the data in Table 11.4 to calculate the solubility of each of the following sparingly soluble substances in its respective solution: aluminum hydroxide at (a) pH = 7.0; (b) pH = 4.5; zinc hydroxide at (c) pH = 7.0; (d) pH = 6.0.

11.72 Use the data in Table 11.4 to calculate the solubility of each of the following sparingly soluble compounds in its respective solution: iron(III) hydroxide at (a) pH = 11.0; (b) pH = 3.0; iron(II) hydroxide at (c) pH = 8.0; (d) pH = 6.0.

11.73 Consider the two equilibria

$$CaF_2(s) \rightleftharpoons Ca^{2+}(aq) + 2\,F^-(aq) \qquad K_{sp} = 4.0 \times 10^{-11}$$

$$F^-(aq) + H_2O(l) \rightleftharpoons HF(aq) + OH^-(aq)$$
$$K_b(F^-) = 2.9 \times 10^{-11}$$

(a) Write the chemical equation for the overall equilibrium and determine the corresponding equilibrium constant. Determine the solubility of CaF_2 at (b) pH = 7.0; (c) pH = 3.0.

11.74 Consider the two equilibria

$$BaF_2(s) \rightleftharpoons Ba^{2+}(aq) + 2\,F^-(aq) \qquad K_{sp} = 1.7 \times 10^{-6}$$

$$F^-(aq) + H_2O(l) \rightleftharpoons HF(aq) + OH^-(aq)$$
$$K_b(F^-) = 2.9 \times 10^{-11}$$

(a) Write the chemical equation for the overall equilibrium and determine the corresponding equilibrium constant. Determine the solubility of BaF_2 at (b) pH = 7.0; (c) pH = 4.0.

11.75 Calculate the solubility of silver bromide in 0.10 M KCN(aq). Refer to Tables 11.4 and 11.5.

11.76 Precipitated silver chloride dissolves in ammonia solutions as a result of the formation of $Ag(NH_3)_2^+$. What is the solubility of silver chloride in 1.0 M NH_3(aq)?

11.77 You find a bottle of a pure silver halide that could be AgCl or AgI. Develop a simple chemical test that would allow you to distinguish which compound was in the bottle.

11.78 Which of the following compounds, if either, will dissolve in 1.00 M HNO_3(aq): (a) Bi_2S_3(s); (b) FeS(s)? Substantiate your answer by giving an appropriate calculation.

11.79 A metal alloy sample is believed to contain silver, bismuth, and nickel. Explain how it could be determined qualitatively that all three of these metals are present.

11.80 Zinc(II) readily forms the complex ion $Zn(OH)_4^{2-}$. Explain how this fact can be used to distinguish a solution of $ZnCl_2$ from $MgCl_2$.

Integrated Exercises

11.81 A buffer solution containing equal amounts of acetic acid and sodium acetate is prepared. What concentration of the buffer must be prepared to prevent a change in the pH by more than 0.20 pH units after the addition of 1.00 mL of 6.00 M HCl(aq) to 100.0 mL of the buffer solution?

11.82 You require 0.150 L of a buffer with pH = 3.00. On the shelf is a bottle of trichloracetic acid/sodium trichloracetate buffer with pH = 2.95. The label also says [trichloracetate] = 0.200 M. What mass of which substance (trichloracetic acid or sodium trichloracetate) should you add to 0.150 L of the buffer to obtain the desired pH?

11.83 (a) Using the Living Graph "Titration of a Strong Acid with a Strong Base," prepare a plot for the titration of 15.0 mL of 0.0567 M HNO_3(aq) by 0.0296 M KOH(aq). From the graph, determine (b) the volume of titrant needed to reach the stoichiometric point; (c) the starting pH of the acid solution. (d) What is the pH at the stoichiometric point?

11.84 (a) Using the Living Graph "Titration of a Weak Acid with a Strong Base," prepare a plot for the titration of 20.0 mL of 0.0329 M pyridinium chloride, $C_5H_5NH^+Cl^-$(aq), by 0.0257 M KOH(aq). From the graph, determine (b) the volume of titrant needed to reach the stoichiometric point (note that the horizontal axis is given in terms of volume of titrant divided by the volume of the initial acid solution); (c) the starting pH of the acid solution. (d) What is the pH at the stoichiometric point? The K_a for pyridinium chloride is 5.6×10^{-6}.

Pyridinium chloride

11.85 A 60.0-mL sample of 0.10 M $NaHCO_2$(aq) is mixed with 4.0 mL of 0.070 M HCl(aq). Calculate the pH and the molarity of HCOOH in the mixed solution.

11.86 An old bottle labeled "Standardized 6.0 M NaOH" was found at the back of a shelf in the stockroom. Over time, some of the NaOH had reacted with the glass and the solution was no longer 6.0 M. To determine its purity, 5.0 mL of the solution was diluted to 100. mL and titrated to the stoichiometric point with 11.8 mL of 2.05 M HCl(aq). What is the molarity of the sodium hydroxide solution in the bottle?

11.87 Novocaine, which is used by dentists as a local anesthetic, is a weak base with $pK_b = 5.05$. Blood has a pH of 7.4. What is the ratio of the concentration of novocaine to that of its conjugate acid in the bloodstream?

11.88 To simulate blood conditions, a phosphate buffer system with a pH = 7.40 is desired. What mass of Na_2HPO_4 must be added to 0.500 L of 0.10 M NaH_2PO_4(aq) to prepare such a buffer?

11.89 A buffer is prepared by adding 55.0 mL of 0.15 M $HNO_3(aq)$ to 45.0 mL of 0.65 M $NaC_6H_5CO_2(aq)$. Determine the solubility of PbF_2 in this buffer solution.

11.90 A buffer is prepared by adding 25.0 mL of 0.12 M $HCl(aq)$ to 55.0 mL of 0.52 M $NaCH_3CO_2(aq)$. What is the solubility of BaF_2 in this buffer solution?

11.91 Which sulfide precipitates first when sulfide ions are added to a solution containing equal concentrations of Co^{2+}, Cu^{2+}, and Cd^{2+}? For CoS, $K_{sp} = 5 \times 10^{-22}$. Explain your conclusions.

11.92 You decide to use sulfide precipitation to separate the copper(II) ions from the manganese(II) ions in a solution that is 0.20 M $Cu^{2+}(aq)$ and 0.20 M $Mn^{2+}(aq)$. Determine the minimum sulfide ion concentration that will result in the precipitation of one cation (identify the cation) but not the other.

11.93 What is the pH at each stoichiometric point in the titration of 0.20 M $H_2SO_4(aq)$ with 0.20 M $NaOH(aq)$?

11.94 What volume (in liters) of a saturated mercury(II) sulfide, HgS, solution contains an average of one mercury(II) ion, Hg^{2+}?

11.95 In an attempt to determine the amount of sulfur dioxide in the air near a power plant, two students set up a bubbler that passes air through 50.00 mL of 1.0×10^{-4} M $NaOH(aq)$. The temperature is 22°C, and the atmospheric pressure 753 Torr. The air is pumped for 2.5 h at a flow rate of 3.0 L·h^{-1}. The students then returned to the laboratory and titrated the solution with 1.5×10^{-4} M $HCl(aq)$ with phenolphthalein indicator to see how much NaOH was left unreacted. They found that 30.2 mL of $HCl(aq)$ was required to reach the stoichiometric point. (a) Write the balanced chemical equation for the reaction of SO_2 and water. (b) What amount of NaOH (in mol) had reacted with the SO_2? (c) What was the concentration of sulfur dioxide in the air, in parts per million?

11.96 Consider the equilibria

$$ZnS(s) \rightleftharpoons Zn^{2+}(aq) + S^{2-}(aq)$$

$$S^{2-}(aq) + H_2O(l) \rightleftharpoons HS^-(aq) + OH^-(aq)$$

$$HS^-(aq) + H_2O(l) \rightleftharpoons H_2S(aq) + OH^-(aq)$$

(a) Write the chemical equation for the overall equilibrium and determine the corresponding equilibrium constant. (b) Determine the solubility of ZnS in a saturated H_2S solution (0.1 M $H_2S(aq)$) adjusted to pH = 7.0. (c) Determine the solubility of ZnS in a saturated H_2S solution (0.1 M $H_2S(aq)$) adjusted to pH = 10.0.

11.97 Using data available in the tables and appendices, calculate the Gibbs free energy of formation of $PbF_2(s)$.

11.98 Silver iodide is very insoluble in water. A common method for increasing its solubility is to increase the temperature of the solution containing the solid. Estimate the solubility of AgI at 85°C.

11.99 The main buffer in the blood consists primarily of hydrogen carbonate ions (HCO_3^-) and H_3O^+ ions in equilibrium with water and CO_2:

$$H_3O^+(aq) + HCO_3^-(aq) \rightleftharpoons 2 H_2O(l) + CO_2(aq)$$

$$K = 7.9 \times 10^{-7}$$

This reaction assumes that all H_2CO_3 produced decomposes completely to CO_2 and H_2O. Suppose that 1.0 L of blood is

removed from the body and brought to pH = 6.1. (a) If the concentration of HCO_3^- is 5.5 μmol·L^{-1}, calculate the amount (in moles) of CO_2 present in the solution at this pH. (b) Calculate the change in pH that occurs when 0.65 μmol H_3O^+ is added to this sample of blood at this pH (that is, pH = 6.1). See Box 11.1.

11.100 The pH of the blood is maintained by a finely tuned buffering system, consisting primarily of hydrogen carbonate ion (HCO_3^-) and H_3O^+ in equilibrium with water and CO_2.

$$H_3O^+(aq) + HCO_3^-(aq) \rightleftharpoons H_2CO_3(aq) + H_2O(l)$$

$$H_3O^+(aq) + HCO_3^-(aq) \rightleftharpoons 2 H_2O(l) + CO_2(g)$$

(a) During exercise, CO_2 is produced at a rapid rate in muscle tissue. How does exercise affect the pH of blood? (b) Hyperventilation (rapid and deep breathing) can occur during intense exertion. How does hyperventilation affect the pH of the blood? (c) The normal first-aid treatment for hyperventilation is to have the patient breathe into a paper bag. Explain briefly why this treatment works and tell what effect the paper-bag treatment has on the pH of the blood. See Box 11.1.

11.101 A buffer solution of volume 300.0 mL is 0.200 M $CH_3COOH(aq)$ and 0.300 M $NaCH_3CO_2(aq)$. (a) What is the initial pH of this solution? (b) What mass of NaOH would have to be dissolved in this solution to bring the pH to 6.0?

11.102 A buffer solution of volume 300.0 mL is 0.400 M $NH_3(aq)$ and 0.200 M $NH_4Cl(aq)$. (a) What is the initial pH of this solution? (c) What volume of HCl gas at 2.00 atm and 25°C would have to be dissolved in this solution to bring the pH to 8.0?

Chemistry Connections

11.103 Soluble nontoxic salts such as $Fe_2(SO_4)_3$ are often used during water purification to remove soluble toxic contaminants, because they form gelatinous hydroxides that encapsulate the contaminants and allow them to be filtered from the water.
(a) Calculate the molar solubility in water of $Fe(OH)_3$ at 25°C.
(b) What is the concentration of hydroxide ion in a saturated solution of $Fe(OH)_3$? Calculate the pH of the solution.
(c) Discuss whether your result in part (b) is reasonable for a solution of a basic hydroxide. Explain what assumptions were made in your calculations and evaluate the validity of the assumptions.
(d) A simplified equation for the reaction of Fe^{3+} ions with water is $Fe^{3+}(aq) + 6 H_2O(l) \rightleftharpoons Fe(OH)_3(s) + 3 H_3O^+(aq)$. Use the data in Table 11.4 and K_w to calculate the equilibrium constant for this reaction.
(e) If 10.0 g of $Fe_2(SO_4)_3$ is dissolved in enough water to make up 1.00 L of aqueous solution and the pH of the solution is raised to 8.00 by addition of NaOH, what mass of solid $Fe(OH)_3$ will form?
(f) To test for the ability of $Fe_2(SO_4)_3$ to remove chloride ions from water, a standard aqueous solution containing 24.72 g of NaCl in 1.000 L of solution was prepared. A 25.00-mL sample of the NaCl solution was then combined with the mixture described in part (e) and stirred. The $Fe(OH)_3$ precipitate containing encapsulated chloride ion was removed by filtration, then dissolved in acid. An aqueous solution of $AgNO_3$ was then added to the resulting solution and the solid AgCl formed filtered and dried. The mass of AgCl was 0.604 g. What percentage of the chloride ion in the 25.00-mL sample had been removed from the solution?

ELECTROCHEMISTRY

What Are the Key Ideas? The tendency of electrons to be transferred in a chemical reaction depends on the species involved and their concentrations. When the process is spontaneous and reduction and oxidation occur at different locations, the reaction can do work and drive electrons through an external circuit.

Why Do We Need to Know This Material? The topics described in this chapter may one day unlock a virtually inexhaustible supply of clean energy supplied daily by the Sun. The key is electrochemistry, the study of the interaction of electricity and chemical reactions. The transfer of electrons from one species to another is one of the fundamental processes underlying life, photosynthesis, fuel cells, and the refining of metals. An understanding of how electrons are transferred helps us to design ways to use chemical reactions to generate electricity and to use electricity to bring about chemical reactions. Electrochemical measurements also allow us to determine the values of thermodynamic quantities.

What Do We Need to Know Already? This chapter extends the thermodynamic discussion presented in Chapter 7. In particular, it builds on the concept of Gibbs free energy (Section 7.12), its relation to maximum nonexpansion work (Section 7.14), and the dependence of the reaction Gibbs free energy on the reaction quotient (Section 9.3). For a review of redox reactions, see Section K. To prepare for the quantitative treatment of electrolysis, review stoichiometry in Section L.

Chapter **12**

Electricity has been observed since ancient times in the form of lightning, as sparks given off when metals strike one another, and as shocks from static electricity. However, its nature and its usefulness were unknown until the late eighteenth century, when the Italian scientist Luigi Galvani discovered that by touching the muscles of dead animals, mainly frogs, with a rod bearing static electricity, he could make their muscles twitch. He believed that electricity came from the muscles themselves. However, at the end of that century another Italian, Alessandro Volta, suggested that the electricity came from the fact that the muscles were between two different metals when touched by the rod. He proved that the electricity came from metals by constructing a tower of alternating discs of different metals in layers separated by paper strips soaked in a solution of table salt (Fig. 12.1). This apparatus, a "voltaic pile," was the first electrical storage device, a simple battery, but because it opened a door to a new understanding of the structure of matter, it amazed the scientists and even the rulers of the day. Eventually scientists realized that the electricity arose from electrons passing from one metal to another in redox reactions.

These early observations have evolved into the branch of chemistry called **electrochemistry**. This subject deals not only with the use of spontaneous chemical reactions to produce electricity but also with the use of electricity to drive nonspontaneous reactions forward. Electrochemistry also provides techniques for monitoring chemical reactions and measuring properties of solutions such as the pK_a of an acid. Electrochemistry even allows us to monitor the activity of our brain and heart (perhaps while we are trying to master chemistry), the pH of our blood, and the presence of pollutants in our water supply.

REPRESENTING REDOX REACTIONS

We begin with a review of redox reactions, which were introduced in Section K. In this chapter we take a closer look at them and see how they can be used to generate electricity, particularly in aqueous solution.

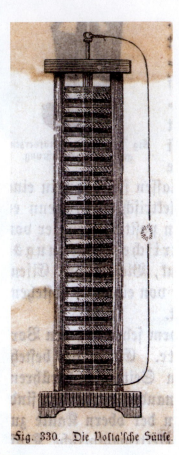

FIGURE 12.1 Volta used this stack of alternating disks of two different metals separated by paper soaked in salt water to produce the first sustainable electric current.

We need to be able to write balanced chemical equations to describe redox reactions. It might seem that this task ought to be simple. However, some redox reactions can be tricky to balance, and special techniques, which we describe in Sections 12.1 and 12.2, have been developed to simplify the procedure.

12.1 Half-Reactions

The key to writing and balancing equations for redox reactions is to think of the reduction and oxidation processes individually. We saw in Section K that oxidation is the loss of electrons and reduction the gain of electrons.

We consider oxidation first. To show the removal of electrons from a species that is being oxidized in a redox reaction, we write the chemical equation for an oxidation half-reaction. A **half-reaction** is the oxidation or reduction part of the reaction considered alone. For example, one battery that Volta built used silver and zinc plates to carry out the reaction

$$Zn(s) + 2\ Ag^+(aq) \longrightarrow Zn^{2+}(aq) + 2\ Ag(s)$$

To show the oxidation of zinc we write

$$Zn(s) \longrightarrow Zn^{2+}(s) + 2\ e^-$$

An oxidation half-reaction is a *conceptual* way of reporting an oxidation: the electrons are never actually free. In an equation for an oxidation half-reaction, the electrons released always appear on the right of the arrow. Their state is not given, because they are in transit and do not have a definite physical state. The reduced and oxidized species in a half-reaction jointly form a **redox couple**. In this example, the redox couple consists of Zn^{2+} and Zn, and is denoted Zn^{2+}/Zn. A redox couple has the form Ox/Red, where Ox is the oxidized form of the species and Red is the reduced form.

A note on good practice: Distinguish a conceptual half-reaction from an actual ionization, where the electron is removed and which we would write, for example, as $Na(g) \rightarrow Na^+(g) + e^-(g)$, with the state of the electron specified.

Now consider reduction. To show the addition of electrons to a species we write the corresponding half-reaction for electron gain. For example, to show the reduction of Ag^+ ions to Ag metal we write

$$Ag^+(aq) + e^- \longrightarrow Ag(s)$$

This half-reaction, too, is conceptual: the electrons are not actually free. In the equation for a reduction half-reaction, the electrons gained always appear on the left of the arrow. In this example, the redox couple is Ag^+/Ag.

> *Half-reactions express the two contributions (oxidation and reduction) to an overall redox reaction.*

12.2 Balancing Redox Equations

Balancing the chemical equation for a redox reaction by inspection can be a real challenge, especially for one taking place in aqueous solution, when water may participate and we must include H_2O and either H^+ or OH^-. In such cases, it is easier to simplify the equation by separating it into its reduction and oxidation half-reactions, balance the half-reactions separately, and then add them together to obtain the balanced equation for the overall reaction. When adding the equations for half-reactions, we match the number of electrons released by oxidation with the number used in reduction, because electrons are neither created nor destroyed in chemical reactions. The procedure is outlined in Toolbox 12.1 and illustrated in Examples 12.1 and 12.2.

In redox reactions, it is conventional to write H^+ rather than H_3O^+.

TOOLBOX 12.1   HOW TO BALANCE COMPLICATED REDOX EQUATIONS

CONCEPTUAL BASIS

When balancing redox equations, we consider the gain of electrons (reduction) separately from the loss of electrons (oxidation), express each of these processes as a half-reaction, and then balance both atoms and charge in each of the two half-reactions. When we combine the half-reactions, the number of electrons released in the oxidation must equal the number used in the reduction.

PROCEDURE

The general procedure for balancing the chemical equation for a redox reaction is first to balance the half-reactions separately.

Step 1 Identify the species being oxidized and the species being reduced from the changes in their oxidation numbers.

Step 2 Write the two skeletal (unbalanced) equations for the oxidation and reduction half-reactions.

Step 3 Balance all elements in the half-reactions except O and H.

Step 4 In acidic solution, balance O by using H_2O and then balance H by using H^+. In basic solution, balance O by using H_2O; then balance H by adding H_2O to the side of each half-reaction that needs H and adding OH^- to the other side.

When we add $\ldots OH^- \ldots \rightarrow \ldots H_2O \ldots$ to a half-reaction, we are effectively adding one H atom to the right. When we add $\ldots H_2O \ldots \rightarrow \ldots OH^- \ldots$, we are effectively adding one H atom to the left. Note that one H_2O molecule is added for each H atom needed.

Step 5 Balance the electric charges by adding electrons to the left for reductions and to the right for oxidations until the charges on the two sides of the arrow are the same.

Step 6 Multiply all species in either one or both half-reactions by factors that result in equal numbers of electrons in the two half-reactions, and then add the two equations and include physical states.

Finally, simplify the appearance of the equation by canceling species that appear on both sides of the arrow and check to make sure that charges as well as numbers of atoms balance. In some cases it is possible to simplify the half-reactions before they are combined.

Examples 12.1 and 12.2 illustrate this procedure.

EXAMPLE 12.1   Sample exercise: Balancing a redox equation in acidic solution

Permanganate ions, MnO_4^-, react with oxalic acid, $H_2C_2O_4$, in acidic aqueous solution, producing manganese(II) ions and carbon dioxide. The skeletal equation is

$$MnO_4^-(aq) + H_2C_2O_4(aq) \longrightarrow Mn^{2+}(aq) + CO_2(g)$$ ⚠

SOLUTION To balance this equation, work through the procedure for an acidic solution set out in Toolbox 12.1.

Reduction half-reaction:

Step 1 Identify the species being reduced.	The oxidation number of Mn decreases from +7 to +2, so the Mn in the MnO_4^- ion is reduced.	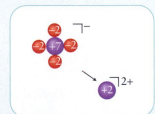
Step 2 Write the skeletal equation for reduction.	$MnO_4^- \longrightarrow Mn^{2+}$	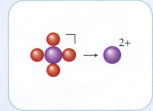

Step 3 Balance all elements except H and O.

$$MnO_4^- \longrightarrow Mn^{2+}$$

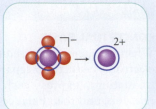

Step 4 Balance the O atoms by adding H_2O.

$$MnO_4^- \longrightarrow Mn^{2+} + 4\,H_2O$$

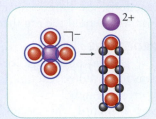

Balance the H atoms by adding H^+.

$$MnO_4^- + 8\,H^+ \longrightarrow Mn^{2+} + 4\,H_2O$$

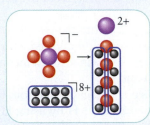

Step 5 Balance the net charges by adding electrons.

Net charge on the left is +7 and on the right it is +2; we need 5 electrons on the left to bring it to +2.

$$MnO_4^- + 8\,H^+ + 5\,e^- \longrightarrow Mn^{2+} + 4\,H_2O$$

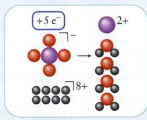

Oxidation half-reaction:

Step 1 Identify the species being oxidized.

The oxidation number of carbon increases from +3 to +4, so the C atoms in the oxalic acid are oxidized.

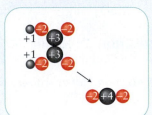

Step 2 Write the skeletal equation for oxidation.

$$H_2C_2O_4 \longrightarrow CO_2$$

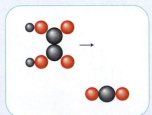

Step 3 Balance all elements except H and O.

$$H_2C_2O_4 \longrightarrow 2\,CO_2$$

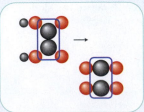

Step 4 Balance the O atoms by adding H_2O (none required).

$$H_2C_2O_4 \longrightarrow 2\ CO_2$$

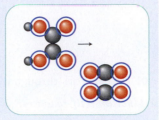

Balance the H atoms by adding H^+.

$$H_2C_2O_4 \longrightarrow 2\ CO_2 + 2\ H^+$$

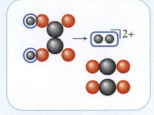

Step 5 Balance charge by adding electrons.

Net charge on the left is 0 and on the right it is +2; we need 2 electrons on the right to bring it to 0.

$$H_2C_2O_4 \longrightarrow 2\ CO_2 + 2\ H^+ + 2\ e^-$$

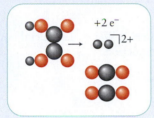

A note on good practice: At this point, check that the number of electrons lost or gained in each half-reaction is the same as the change in oxidation number of the element oxidized or reduced.

Step 6 Write the overall equation. In the half-reactions, 2 electrons are lost, but 5 are gained, so we need 10 in each half-reaction.

Multiply the reduction half-reaction by 2 and multiply the oxidation half-reaction by 5.

$$2\ MnO_4^- + 16\ H^+ + 10\ e^- \longrightarrow 2\ Mn^{2+} + 8\ H_2O$$

$$5\ H_2C_2O_4 \longrightarrow 10\ CO_2 + 10\ H^+ + 10\ e^-$$

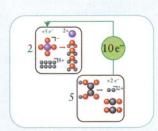

Add the equations and cancel electrons.

$$2\ MnO_4^- + 5\ H_2C_2O_4 + 16\ H^+ \longrightarrow 2\ Mn^{2+} + 8\ H_2O + 10\ CO_2 + 10\ H^+$$

Cancel 10 H^+ ions on each side and include physical states.

$$2\ MnO_4^-(aq) + 5\ H_2C_2O_4(aq) + 6\ H^+(aq) \longrightarrow 2\ Mn^{2+}(aq) + 8\ H_2O(l) + 10\ CO_2(g)$$

Elements and charge are both balanced; so this is the fully balanced net ionic equation.

SELF-TEST 12.1A Copper reacts with dilute nitric acid to form copper(II) nitrate and the gas nitric oxide, NO. Write the net ionic equation for the reaction.
   [*Answer:* $3\ Cu(s) + 2\ NO_3^-(aq) + 8\ H^+(aq) \rightarrow 3\ Cu^{2+}(aq) + 2\ NO(g) + 4\ H_2O(l)$]

SELF-TEST 12.1B Acidified potassium permanganate solution reacts with sulfurous acid, $H_2SO_3(aq)$, to form sulfuric acid and manganese(II) ions. Write the net ionic equation for the reaction. In acidic aqueous solution, H_2SO_3 is present as electrically neutral molecules and sulfuric acid is present as HSO_4^- ions.

EXAMPLE 12.2   **Sample exercise: Balancing a redox equation in basic solution**

The products of the reaction between bromide ions and permanganate ions, MnO_4^-, in basic aqueous solution are solid manganese(IV) oxide, MnO_2, and bromate ions. Balance the net ionic equation for the reaction.

SOLUTION Work through the procedure for a basic solution set out in Toolbox 12.1.

Reduction half-reaction:

Step 1   Identify the species being reduced.

The oxidation number of Mn changes from +7 in MnO_4^- to +4 in MnO_2, so the Mn in MnO_4^- is reduced ("permanganate is reduced").

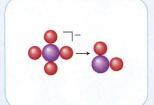

Step 2   Write the skeletal equation for reduction.

$$MnO_4^- \longrightarrow MnO_2$$

Step 3   Mn atoms are balanced.

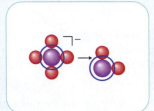

Step 4   Balance the O atoms by adding H_2O.

$$MnO_4^- \longrightarrow MnO_2 + 2\,H_2O$$

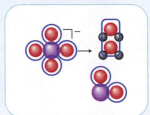

Balance the H atoms by adding (for each H atom needed) one H_2O molecule to the side of each equation that needs hydrogen and one OH^- ion to the opposite side.

$$MnO_4^- + 4\,H_2O \longrightarrow$$
$$MnO_2 + 2\,H_2O + 4\,OH^-$$

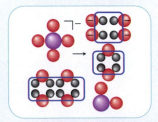

Cancel like species on opposite sides of the arrow (in this case, $2\,H_2O$).

$$MnO_4^- + 2\,H_2O \longrightarrow MnO_2 + 4\,OH^-$$

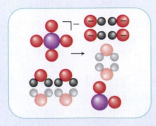

Step 5 Balance charge by adding electrons.

Net charge on the left is -1 and on the right it is -4; we need 3 electrons on the left to bring it to -4.

$$MnO_4^- + 2\,H_2O + 3\,e^- \longrightarrow MnO_2 + 4\,OH^-$$

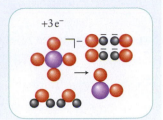

Oxidation half-reaction:

Step 1 Identify the species being oxidized.

The oxidation number of Br increases from -1 in Br^- to $+5$ in BrO_3^-, and so Br^- is oxidized.

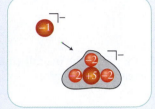

Step 2 Write the skeletal equation for oxidation.

$$Br^- \longrightarrow BrO_3^-$$

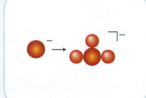

Step 3 The Br atoms are balanced.

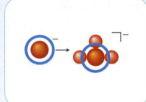

Step 4 Balance the O atoms by adding H_2O.

$$Br^- + 3\,H_2O \longrightarrow BrO_3^-$$

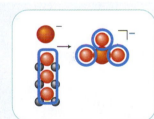

Balance the H atoms by adding (for each H atom needed) one H_2O molecule to the side of each equation that needs hydrogen and one OH^- ion to the opposite side.

$$Br^- + 3\,H_2O + 6\,OH^- \longrightarrow BrO_3^- + 6\,H_2O$$

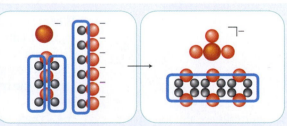

Cancel like species on opposite sides of the arrow (in this case, $3\,H_2O$).

$$Br^- + 6\,OH^- \longrightarrow BrO_3^- + 3\,H_2O$$

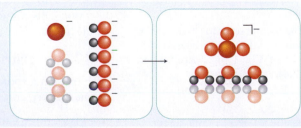

Step 5 Balance charge by adding electrons.

Net charge on the left is -7 and on the right it is -1; we need 6 electrons on the right to bring it to -7.

$$Br^- + 6\,OH^- \longrightarrow BrO_3^- + 3\,H_2O + 6\,e^-$$

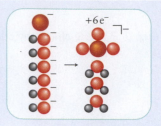

A note on good practice: As before, at this point check that the number of electrons lost or gained in each half-reaction is the same as the change in oxidation number of the element oxidized or reduced.

Step 6 Write the overall equation. Because 6 electrons are lost and 3 gained, we need 6 in each half-reaction.

Multiply the reduction half-reaction by 2.

$$2\,MnO_4^- + 4\,H_2O + 6\,e^- \longrightarrow 2\,MnO_2 + 8\,OH^-$$

The oxidation half-reaction remains unchanged.

$$Br^- + 6\,OH^- \longrightarrow BrO_3^- + 3\,H_2O + 6\,e^-$$

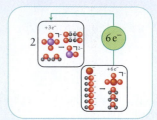

Add the two equations and cancel the electrons.

$$2\,MnO_4^- + Br^- + 6\,OH^- + 4\,H_2O \longrightarrow 2\,MnO_2 + BrO_3^- + 8\,OH^- + 3\,H_2O$$

Cancel 3 H_2O and 6 OH^-, then include the physical states.

$$2\,MnO_4^-(aq) + Br^-(aq) + H_2O(l) \longrightarrow 2\,MnO_2(s) + BrO_3^-(aq) + 2\,OH^-(aq)$$

SELF-TEST 12.2A An alkaline (basic) solution of hypochlorite ions reacts with solid chromium(III) hydroxide to produce aqueous chromate ions and chloride ions. Write the net ionic equation for the reaction.

[Answer: $2\,Cr(OH)_3(s) + 4\,OH^-(aq) + 3\,ClO^-(aq) \rightarrow 2\,CrO_4^{2-}(aq) + 5\,H_2O(l) + 3\,Cl^-(aq)]$

SELF-TEST 12.2B When iodide ions react with iodate ions in basic aqueous solution, triiodide ions, I_3^-, are formed. Write the net ionic equation for the reaction. (Note that the same product is obtained in each half-reaction.)

> *The chemical equation for a reduction half-reaction is added to the equation for an oxidation half-reaction to form the balanced chemical equation for the overall redox reaction.*

GALVANIC CELLS

Whenever we turn on a portable CD player or a laptop computer, we are completing a circuit that allows a chemical reaction to take place in a battery, a direct descendant of Volta's pile of metals and paper. A battery is an example of an electrochemical cell. In general, an **electrochemical cell** is a device in which an electric current—a flow of electrons through a circuit—is either produced by a spontaneous chemical reaction or used to bring about a nonspontaneous reaction. A **galvanic cell** is an electrochemical cell in which a spontaneous chemical reaction is used to generate an electric current. Technically, a **battery** is a collection of galvanic cells joined in series; so the voltage that it produces—its ability to push an electric current through a circuit—is the sum of the voltages of each cell.

Galvanic cells are also known as voltaic cells.

The formal term for "voltage" is **potential difference, measured in volts:** $1\,V = 1\,J \cdot C^{-1}$ *(see Section 12.4).*

12.3 The Structure of Galvanic Cells

How can a spontaneous reaction can be used to generate an electric current? To find the answer, consider the redox reaction between zinc metal and copper(II) ions:

$$Zn(s) + Cu^{2+}(aq) \longrightarrow Zn^{2+}(aq) + Cu(s) \qquad \text{(A)}$$

If we were to place a piece of zinc metal into an aqueous copper(II) sulfate solution, we would see a layer of metallic copper begin to deposit on the surface of the zinc (see Fig. K.5). If we could watch the reaction at the atomic level, we would see that, as the reaction takes place, electrons are transferred from the Zn atoms to adjacent Cu^{2+} ions in the solution. These electrons reduce the Cu^{2+} ions to Cu atoms, which stick to the surface of the zinc or form a finely divided solid deposit in the beaker. The piece of zinc slowly disappears as its atoms give up electrons and form colorless Zn^{2+} ions that drift off into the solution. The Gibbs free energy of the system decreases as electrons are transferred and the reaction approaches equilibrium. However, although energy is released as heat, no electrical work is done.

Now suppose we separate the reactants but provide a pathway for the electrons to travel from the zinc metal to the copper(II) ions. The electrons can then do work—for example, drive an electric motor—as they pass from the species being oxidized to that being reduced. This is what happens when the reaction takes place in a galvanic cell. A galvanic cell consists of two **electrodes**, or metallic conductors, which make electrical contact with the contents of the cell but not with each other, and an **electrolyte**, an ionically conducting medium, inside the cell. In an *ionic conductor* an electric current is carried by the movement of ions. The electrolyte is typically an aqueous solution of an ionic compound. Oxidation takes place at one electrode as the species being oxidized releases electrons into the metallic conductor. Reduction takes place at the other electrode, where the species that is undergoing reduction collects electrons from the metallic conductor (Fig. 12.2). We can think of the overall chemical reaction as pushing electrons onto one electrode and pulling them off the other electrode. This push–pull process sets up a flow of electrons in the external circuit joining the two electrodes, and that current can be used to do electrical work.

The electrode at which oxidation takes place is called the **anode**. The electrode at which reduction takes place is called the **cathode**. Electrons are released by the oxidation half-reaction at the anode, travel through the external circuit, and reenter the cell at the cathode, where they are used in the reduction half-reaction. A commercial galvanic cell has its cathode marked with a + sign and its anode with a − sign.

The *Daniell cell* is an early example of a galvanic cell that makes use of the oxidation of copper by zinc ions, as in reaction A. The British chemist John Daniell invented it in 1836, when the growth of telegraphy created an urgent need for a reliable, steady source of electric current. Daniell set up the arrangement shown in Fig. 12.3, in which the two reactants are separated: the zinc metal is immersed in a solution of zinc sulfate and the copper electrode is immersed in a solution of copper(II) sulfate. For the electrons to travel from Zn atoms to Cu^{2+} ions and bring about the spontaneous reaction, they must pass from the zinc metal through the wire that serves as the external circuit, then through the copper electrode, then to the copper(II) solution. The Cu^{2+} ions are converted into Cu atoms at the cathode by the reduction half-reaction

$$Cu^{2+}(aq) + 2\,e^- \longrightarrow Cu(s)$$

At the same time zinc atoms are converted into Zn^{2+} ions at the anode by the oxidation half-reaction

$$Zn(s) \longrightarrow Zn^{2+}(aq) + 2\,e^-$$

As Cu^{2+} ions are reduced, the solution at the cathode becomes negatively charged and the solution at the anode begins to develop a positive charge as the additional Zn^{2+} ions enter the solution. To prevent this charge buildup, which would quickly stop the flow of electrons, the two solutions are in contact through a porous wall: ions provided by the electrolyte solutions move between the two compartments and complete the electrical circuit.

The electrodes in the Daniell cell are made of the metals involved in the reaction. However, not all electrode reactions include a conducting solid directly. For example, to use the reduction $2\,H^+(aq) + 2\,e^- \rightarrow H_2(g)$ at an electrode, a

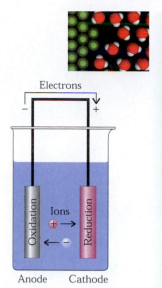

Electrons

Oxidation Ions Reduction

Anode   Cathode

FIGURE 12.2 In an electrochemical cell, a reaction takes place in two separate regions. Oxidation takes place at one electrode (the anode), and the electrons released travel through the external circuit to the other electrode, the cathode, where they cause reduction. The circuit is completed by ions, which carry the electric charge through the solution.

The term "electrolyte" was first introduced in Section I to refer to the solute. In the discussion of electrochemical cells, the term is commonly used to refer to the ionically conducting medium.

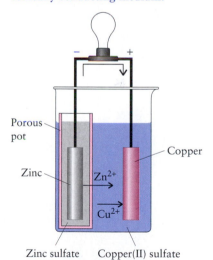

Porous pot

Copper

Zinc

Zn^{2+}

Cu^{2+}

Zinc sulfate   Copper(II) sulfate

FIGURE 12.3 The Daniell cell consists of copper and zinc electrodes dipping into solutions of copper(II) sulfate and zinc sulfate, respectively. The two solutions make contact through the porous barrier, which allows ions to pass through and complete the electric circuit.

chemically inert metallic conductor, such as an unreactive metal or graphite, must carry the electrons into or out of the electrode compartment. Platinum is customarily used as the electrode for hydrogen: the gas is bubbled over the metal immersed in a solution that contains hydrogen ions. This arrangement is called a *hydrogen electrode*. The entire compartment with metal conductor and electrolyte solution is commonly referred to as "the electrode" or as a **half-cell**.

> *In a galvanic cell, a spontaneous chemical reaction draws electrons into the cell through the cathode, the site of reduction, and releases them at the anode, the site of oxidation.*

12.4 Cell Potential and Reaction Gibbs Free Energy

The **cell potential**, E, is a measure of the ability of a cell reaction to force electrons through a circuit. A reaction with a lot of pushing-and-pulling power generates a high cell potential (colloquially, a high voltage). A reaction with little pushing-and-pulling power generates only a small potential (a low voltage). An exhausted battery is a cell in which the reaction is at equilibrium; it has lost its power to move electrons and has a potential of zero. The SI unit of potential is the **volt** (V). A volt is defined so that a charge of one coulomb (1 C) falling through a potential difference of one volt (1 V) releases one joule (1 J) of energy:

$$1 \text{ V·C} = 1 \text{ J}$$

Cell potential is analogous to gravitational potential. The maximum work that a falling weight can do is equal to its mass times the difference in gravitational potential. Similarly, the maximum work that an electron can do is equal to its charge times the difference in electrical potential through which it falls. To interpret this idea in terms of thermodynamics, recall from Section 7.14 that the value of ΔG gives the maximum nonexpansion work that can be obtained from a process at constant temperature and pressure. As explained there, *nonexpansion work* is any work other than that resulting from a change in volume. For example, electrical work and mechanical work, such as muscle contraction (which changes the shape of the muscle, but not its volume), are types of nonexpansion work. Our immediate task, therefore, is to find a relation between the thermodynamic property (the Gibbs free energy of a reaction) and the electrochemical property (the cell potential produced by that reaction).

> One coulomb is the magnitude of the charge delivered by a current of one ampere flowing for one second: $1 \text{ C} = 1 \text{ A·s}$.

HOW DO WE DO THAT?

To find the connection between cell potential and Gibbs free energy, recall that in Section 7.14 (Eq. 21) we saw that the change in Gibbs free energy is the maximum nonexpansion work that a reaction can do at constant pressure and temperature:

$$\Delta G = w_e$$

The work done when an amount n of electrons (in moles) travels through a potential difference E is their total charge times the potential difference. The charge of one electron is $-e$; the charge per mole of electrons is $-eN_A$, where N_A is Avogadro's constant. Therefore, the total charge is $-neN_A$ and the work done is

$$w_e = -neN_A E$$

Faraday's constant, F, is the magnitude of the charge per mole of electrons (the product of the elementary charge e and Avogadro's constant N_A):

$$F = eN_A = (1.602\ 177 \times 10^{-19} \text{ C}) \times \{6.0223 \times 10^{23} \text{ (mol e}^-)^{-1}\}$$
$$= 9.6485 \times 10^4 \text{ C·(mol e}^-)^{-1}$$

Faraday's constant is normally abbreviated to $F = 9.6485 \times 10^4$ C·mol^{-1} (or 96.485 kC·mol^{-1}). We can now write the preceding expression for work as

$$w_e = -nFE$$

When this equation is combined with the thermodynamic equation, we obtain

$$\Delta G = -nFE$$

The units of ΔG are joules (or kilojoules), with a value that depends not only on E, but also on the amount n (in moles) of electrons transferred in the reaction. Thus, in reaction A, $n = 2$ mol. As in the discussion of the relation between Gibbs free energy and equilibrium constants (Section 9.3), we shall sometimes need to use this relation in its "molar" form, with n interpreted as a pure number (its value with the unit "mol" struck out). Then we write

$$\Delta G_r = -nFE$$

The units of ΔG_r are joules (or kilojoules) per mole. The subscript r is always the signal that we are using this "molar" form.

The relation that we have derived,

$$\Delta G = -nFE \qquad\qquad (1)*$$

relates the thermodynamic information that we have been compiling since Chapter 6 to the electrochemical information in this chapter. We see from this equation that E provides an experimental criterion of spontaneity: if the cell potential is positive, then the reaction Gibbs free energy is negative, and the cell reaction has a spontaneous tendency to form products. If the cell potential is negative, then the *reverse* of the cell reaction is spontaneous, and the cell reaction has a spontaneous tendency to form reactants.

There is a hidden assumption in the derivation of Eq. 1. *Maximum* nonexpansion work is obtained when a cell is operating *reversibly*. Therefore, Eq. 1 applies only when the pushing power of the cell is balanced against an external matching source of potential. In practice, that means using a voltmeter with such a high resistance that the potential difference is measured without drawing any current. The cell potential under these conditions is the *maximum* potential that can be produced. It is called the **electromotive force**, emf, of the cell. From now on, E will always be taken to represent this emf. A *working cell*, one actually producing current, such as the cell in a CD player, will produce a potential that is smaller than that predicted by Eq. 1.

> Remember the definition of reversibility in Section 6.3, which requires the pushing force of the system to be balanced against an equal and opposite force.

EXAMPLE 12.3   Sample exercise: Calculating the reaction Gibbs free energy

The emf of the Daniell cell for certain concentrations of copper and zinc ions is 1.04 V. What is the reaction Gibbs free energy under those conditions?

SOLUTION Use Eq. 1 to determine a reaction Gibbs free energy—a thermodynamic quantity—from a cell emf—an electrical quantity. From the chemical equation for the cell reaction (reaction A), we see that $n = 2$ mol.

From $\Delta G = -nFE$,

$$\Delta G = -(2 \text{ mol}) \times (9.6485 \times 10^4 \text{ C·mol}^{-1}) \times (1.04 \text{ V})$$
$$= -2.01 \times 10^5 \text{ C·V}$$

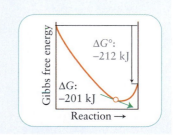

Because 1 C·V = 1 J, we can conclude that the Gibbs free energy of reaction A under these conditions is -201 kJ; because this value is negative, the reaction is spontaneous in the forward direction for this composition of the cell.

A note on good practice: The value of n depends on the balanced equation. Check to ensure that n matches the number of moles of electrons transferred in the balanced equation.

SELF-TEST 12.3A The reaction taking place in a nicad cell is $Cd(s) + 2 Ni(OH)_3(s) \rightarrow Cd(OH)_2(s) + 2 Ni(OH)_2(s)$, and the emf of the cell when fully charged is 1.25 V. What is the reaction Gibbs free energy? (See Table 12.2.)

[*Answer:* -241 kJ]

SELF-TEST 12.3B The reaction taking place in the silver cell used in some cameras and wristwatches is $Ag_2O(s) + Zn(s) \rightarrow 2\ Ag(s) + ZnO(s)$, and the emf of the cell when new is 1.6 V. What is the reaction Gibbs free energy? (See Table 12.2.)

We shall often employ Eq. 1 for the *standard* reaction Gibbs free energy, $\Delta G°$, when it becomes

$$\Delta G° = -nFE° \tag{2}*$$

In this expression, $E°$ is the **standard emf** of the cell, the emf measured when all the species taking part are in their standard states. In practice, this condition means that all gases are at 1 bar, all participating solutes are at $1\ mol \cdot L^{-1}$, and all liquids and solids are pure. For example, to measure the standard emf of the Daniell cell, we use 1 M $CuSO_4$(aq) and a pure copper electrode in one electrode compartment and 1 M $ZnSO_4$(aq) and a pure zinc electrode in the other.

The value of E is the same, regardless of how we write the equation, but the value of $\Delta G°$ depends on the stoichiometric coefficients in the chemical equation. When we multiply all the coefficients by 2 the value of $\Delta G°$ doubles. However, multiplying all the coefficients by 2 also doubles the value of n, and so $E° = -\Delta G°/nF$ remains the same. That is, although the reaction Gibbs free energy changes when the chemical equation is multiplied by a factor, $E°$ does not change:

	$\Delta G°$	$E°$
$Zn(s) + Cu^{2+}(aq) \longrightarrow Zn^{2+}(aq) + Cu(s)$	-212 kJ	$+1.10$ eV
$2\ Zn(s) + 2\ Cu^{2+}(aq) \longrightarrow 2\ Zn^{2+}(aq) + 2\ Cu(s)$	-424 kJ	$+1.10$ eV

A practical implication of this conclusion is that the emf produced by a cell is independent of the size of the cell. To get a higher potential than predicted by Eq. 1, we have to construct a battery by joining cells in series: the potential is then the sum of the potentials of the individual cells (see Section 12.15 for some examples).

> *Cell emf and reaction Gibbs free energy are related by Eq. 1 ($\Delta G = -nFE$), and their standard values are related by Eq. 2 ($\Delta G° = -nFE°$). The magnitude of the emf is independent of how the chemical equation is written.*

12.5 The Notation for Cells

Chemists use a special notation to specify the structure of electrode compartments in a galvanic cell. The two electrodes in the Daniell cell, for instance, are denoted $Zn(s)\,|\,Zn^{2+}(aq)$ and $Cu^{2+}(aq)\,|\,Cu(s)$. Each vertical line represents an interface between phases—in this case, between solid metal and ions in solution in the order reactant $|$ product.

We report the structure of a cell in a symbolic **cell diagram**, by using the conventions specified by IUPAC and used by chemists throughout the world. The diagram for the Daniell cell, for instance, is

$$Zn(s)\,|\,Zn^{2+}(aq)\,|\,Cu^{2+}(aq)\,|\,Cu(s)$$

In the Daniell cell, zinc sulfate and copper(II) sulfate solutions meet inside the porous barrier to complete the circuit. However, when different ions mingle together, they can affect the cell voltage. To keep solutions from mixing, chemists use a **salt bridge** to join the two electrode compartments and complete the electrical circuit. A salt bridge typically consists of a gel containing a concentrated aqueous salt solution in an inverted U-tube (Fig. 12.4). The bridge allows a flow of ions, and so it completes the electrical circuit, but the ions are chosen so that they do not affect the cell reaction (often KCl is used). In a cell diagram, a salt bridge is shown by a double vertical line ($\|$), and so the arrangement in Fig. 12.4 is denoted

$$Zn(s)\,|\,Zn^{2+}(aq)\,\|\,Cu^{2+}(aq)\,|\,Cu(s)$$

More precisely, all the solutes should be at unit activity, not unit molarity. Activities differ appreciably from molarities in electrolyte solution because ions interact over long distances. However, we ignore this complication here.

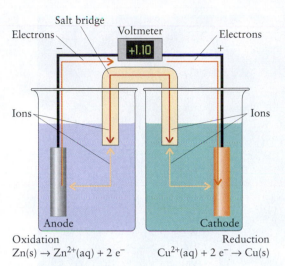

Salt bridge

Electrons

Voltmeter

+1.10

Electrons

−   +

Ions

Ions

Anode   Cathode

Oxidation   Reduction
$Zn(s) \rightarrow Zn^{2+}(aq) + 2\,e^-$ $Cu^{2+}(aq) + 2\,e^- \rightarrow Cu(s)$

FIGURE 12.4 In a galvanic cell electrons produced by oxidation at the anode (−) travel through the external circuit and reenter the cell at the cathode (+), where they cause reduction. The circuit is completed inside the cell by migration of ions through the salt bridge. A salt bridge is not used in a working cell (one used to produce a current). When the emf of a cell is measured, no current actually flows: the voltmeter measures the tendency of the electrons to flow from one electrode to the other.

Any inert metallic component of an electrode is written as the outermost component of that electrode in the cell diagram. For example, a hydrogen electrode constructed with platinum is denoted $H^+(aq)|H_2(g)|Pt(s)$ when it is the right-hand electrode in a cell diagram and $Pt(s)|H_2(g)|H^+(aq)$ when it is the left-hand electrode. An electrode consisting of a platinum wire dipping into a solution of iron(II) and iron(III) ions is denoted either $Fe^{3+}(aq),Fe^{2+}(aq)|Pt(s)$ or $Pt(s)|Fe^{3+}(aq),Fe^{2+}(aq)$. In this case, the oxidized and reduced species are both in the same phase, and so a comma rather than a line is used to separate them. Pairs of ions in solution are normally written in the order Ox,Red.

> When it is important to emphasize the spatial arrangement of an electrode, the order may reflect that arrangement, as in $Cl^-(aq)|Cl_2(g)|Pt(s)$.

SELF-TEST 4A Write the diagram for a cell with a hydrogen electrode on the left and an iron(III)/iron(II) electrode on the right. The two electrode compartments are connected by a salt bridge and platinum is used as the conductor at each electrode.

[*Answer:* $Pt(s)|H_2(g)|H^+(aq)\|Fe^{3+}(aq),Fe^{2+}(aq)|Pt(s)$]

SELF-TEST 4B Write the diagram for a cell that has an electrode consisting of a manganese wire dipping into a solution of manganese(II) ions on the left, a salt bridge, and a copper(II)/copper(I) electrode on the right with a platinum wire.

The cell diagram is written to correspond to how the chemical equation for the reaction is written, not to the way that the cell is arranged in the laboratory. Thus, the Daniell cell can be described as either

$$Zn(s)|Zn^{2+}(aq)\|Cu^{2+}(aq)|Cu(s) \quad \text{or} \quad Cu(s)|Cu^{2+}(aq)\|Zn^{2+}(aq)|Zn(s)$$

However, as we shall shortly see, the two descriptions correspond to the direction in which we write the cell reaction and, therefore, to different signs used to report the cell emf.

As already remarked, the emf of a cell is measured with an electronic voltmeter (Fig. 12.5), and we identify the cathode by noting which terminal is positive. We can think of the + sign on the cathode as indicating where electrons tend to *enter* the cell from the external circuit and the − sign on the anode as showing where electrons tend to leave the cell. If the cathode is the electrode that we have placed on the right *of the cell diagram*, then by convention the emf of the cell described by the cell diagram is reported as positive, as in

$$Zn(s)|Zn^{2+}(aq)\|Cu^{2+}(aq)|Cu(s) \qquad E = +1.10\ \text{V}$$

Anode (−)   Cathode (+)

> Formerly, they were measured with a device called a *potentiometer*, but electronic devices are more reliable and easier to interpret.

(The emf depends on the concentrations of the ions, as we explore later; here we have reported the value observed when the two solutions have equal cation concentrations.) In this case, the electrons can be thought of as tending to travel

FIGURE 12.5 The cell potential is measured with an electronic voltmeter, a device designed to draw negligible current so that the composition of the cell does not change during the measurement. The display shows a positive value when the + terminal of the meter is connected to the cathode of the galvanic cell. The salt bridge completes the electric circuit within the cell.

Lab Video: Figure 12.5 A galvanic cell

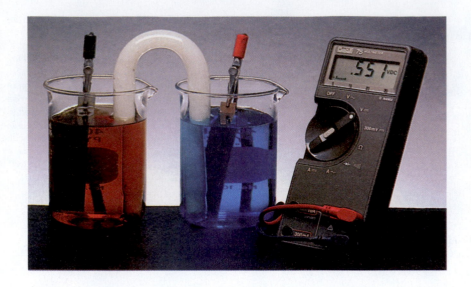

through the external circuit from the left of the cell as written (the anode) to the right (the cathode). However, if the cathode is found to be the electrode that we have placed on the *left* of the cell diagram, then the emf of the cell is reported as negative, as in

$$Cu(s)\,|\,Cu^{2+}(aq)\|Zn^{2+}(aq)\,|\,Zn(s) \qquad E = -1.10\ V$$
$$\quad\text{Cathode (+)} \qquad\qquad \text{Anode (−)}$$

In summary, the sign of the emf reported in conjunction with a cell diagram is the same as the sign of the right-hand electrode in the diagram.

A particular cell diagram corresponds to a specific direction of writing the corresponding cell reaction. *Solely for the purpose of writing the cell reaction corresponding to a given cell diagram*, we suppose that the right-hand electrode in the diagram is the site of reduction and that the left-hand electrode is the site of oxidation, and write the corresponding half-reactions. Thus, for the cell written as $Zn(s)\,|\,Zn^{2+}(aq)\|Cu^{2+}(aq)\,|\,Cu(s)$, we write

Left (L)	$Zn(s)\,|\,Zn^{2+}(aq)\|Cu^{2+}(aq)\,|\,Cu(s)$	Right (R)
$Zn(s) \longrightarrow Zn^{2+}(aq) + 2\ e^-$		$Cu^{2+}(aq) + 2\ e^- \longrightarrow Cu(s)$
(oxidation)		(reduction)
Overall (R + L): $Zn(s) + Cu^{2+}(aq) \longrightarrow Zn^{2+}(aq) + Cu(s)$		$E = +1.10\ V$

Because $E > 0$, and therefore $\Delta G < 0$ for this reaction, the cell reaction as written is spontaneous for the concentrations of ions that we chose. For the alternative way of writing the cell diagram, $Cu(s)\,|\,Cu^{2+}(aq)\|Zn^{2+}(aq)\,|\,Zn(s)$, we would write

Left (L)	$Cu(s)\,|\,Cu^{2+}(aq)\|Zn^{2+}(aq)\,|\,Zn(s)$	Right (R)
$Cu(s) \longrightarrow Cu^{2+}(aq) + 2\ e^-$		$Zn^{2+}(aq) + 2\ e^- \longrightarrow Zn(s)$
(oxidation)		(reduction)
Overall (L + R): $Zn^{2+}(aq) + Cu(s) \longrightarrow Zn(s) + Cu^{2+}(aq)$		$E = -1.10\ V$

Because $E < 0$, and therefore $\Delta G > 0$, the *reverse* of the cell reaction as written is spontaneous for the chosen concentrations of ions.

The general procedure for writing the chemical equation for the reaction corresponding to a given cell diagram is set out in Toolbox 12.2.

TOOLBOX 12.2   HOW TO WRITE A CELL REACTION FOR A CELL DIAGRAM

CONCEPTUAL BASIS

A cell diagram corresponds to a specific cell reaction in which the right-hand electrode in the cell diagram is treated as the site of reduction and the left-hand electrode is treated as the site of oxidation. The sign of the emf then distinguishes whether the resulting reaction is spontaneous in the direction written ($E > 0$) or whether the reverse reaction is spontaneous ($E < 0$).

PROCEDURE

Step 1 Write the equation for the electrode on the right of the cell diagram as a reduction half-reaction (remember: **Right for Reduction**).

Step 2 Write the equation for the electrode on the left of the cell diagram as an oxidation half-reaction.

Step 3 Multiply one or both equations by a factor if necessary to equalize the number of electrons, and then add the two equations.

If the emf of the cell is positive, then the reaction is spontaneous as written. If the emf is negative, then the reverse reaction is spontaneous.

This procedure is illustrated in Example 12.4.

EXAMPLE 12.4   Sample exercise: Writing a cell reaction

Write the cell reaction for the cell $Pt(s)\,|\,H_2(g)\,|\,HCl(aq)\,|\,Hg_2Cl_2(s)\,|\,Hg(l)$.

SOLUTION Follow the procedure set out in Toolbox 12.2:

Step 1 Write the equation for the reduction at the right-hand electrode.	$Hg_2Cl_2(s) + 2\,e^- \longrightarrow 2\,Hg(l) + 2\,Cl^-(aq)$

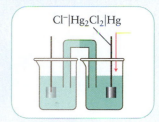

Step 2 Write the equation for the oxidation at the left-hand electrode.	$\tfrac{1}{2}\,H_2(g) \longrightarrow H^+(aq) + e^-$

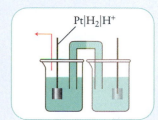

Step 3 To balance electrons multiply the oxidation half-reaction by 2.	$H_2(g) \longrightarrow 2\,H^+(aq) + 2\,e^-$

Add the half-reactions together.	$Hg_2Cl_2(s) + H_2(g) \longrightarrow$ $\quad 2\,Hg(l) + 2\,Cl^-(aq) + 2\,H^+(aq)$

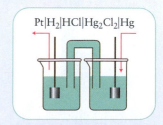

If the concentrations in the cell are such that it is reported as having a positive emf (that is, the mercury/mercury(I) chloride electrode is found to be positive), then the reaction as written is spontaneous. If the concentrations were such that the emf were reported as negative (that is, the hydrogen electrode were found to be positive), then the reverse of the reaction that we have derived would be spontaneous.

SELF-TEST 12.5A (a) Write the chemical equation for the reaction corresponding to the cell $Pt(s)\,|\,H_2(g)\,|\,H^+(aq)\,\|\,Co^{3+}(aq),Co^{2+}(aq)\,|\,Pt(s)$. (b) Given that the cell emf is reported as positive, is the cell reaction spontaneous as written?

[*Answer:* (a) $H_2(g) + 2\,Co^{3+}(aq) \rightarrow 2\,H^+(aq) + 2\,Co^{2+}(aq)$; (b) yes]

Self-Test 12.5B (a) Write the chemical equation for the reaction corresponding to the cell $Hg(l) | Hg_2Cl_2(s) | HCl(aq) \| Hg_2(NO_3)_2(aq) | Hg(l)$. (b) Given that the cell emf is reported as positive, is the cell reaction spontaneous as written?

> *An electrode is designated by representing the interfaces between phases by a vertical line. A cell diagram depicts the physical arrangement of species and interfaces, with any salt bridge denoted by a double vertical line. The sign with which the emf is reported is the same as the measured sign of the right-hand electrode in the cell diagram. A positive emf indicates that the cell reaction is spontaneous as written.*

12.6 Standard Potentials

There are thousands of possible galvanic cells that can be studied. However, instead of having to learn about all these different cells, it is much simpler to learn about the smaller number of electrodes that are combined to form the cells. Under standard conditions (all solutes present at 1 mol·L^{-1}; all gases at 1 bar), we can think of each electrode as making a characteristic contribution to the cell potential called its **standard potential**, $E°$. Each standard potential is a measure of the electron-pulling power of a single electrode. In a galvanic cell, the electrodes pull in opposite directions, so the overall pulling power of the cell, the cell's standard emf, is the *difference* of the standard potentials of the two electrodes (Fig. 12.6). That difference is always written as

$$E° = E°(\text{electrode on right of cell diagram}) -$$
$$E°(\text{electrode on left of cell diagram}) \quad \textbf{(3a)}$$

or, more succinctly,

$$E° = E_R° - E_L° \quad \textbf{(3b)}$$

Note that, because the right side of the cell diagram corresponds to reduction, $E° = E°(\text{for reduction}) - E°(\text{for oxidation})$ where both values of $E°$ are the standard reduction potentials.

If $E° > 0$, then the corresponding cell reaction is spontaneous under standard conditions (that is, as explained in Section 9.3, $K > 1$) and the electrode on the right of the cell diagram serves as the cathode. For example, for the cell

$$Fe(s) | Fe^{2+}(aq) \| Ag^+(aq) | Ag(s) \quad \text{corresponding to}$$
$$2\,Ag^+(aq) + Fe(s) \longrightarrow 2\,Ag(s) + Fe^{2+}(aq)$$

we write

$$E° = E°(Ag^+/Ag) - E°(Fe^{2+}/Fe)$$

and find $E° = +1.24$ V at 25°C. Because $E° > 0$, the cell reaction has $K > 1$, with products dominant at equilibrium, and iron metal can reduce silver ions. If we had written the cell in the opposite order,

$$Ag(s) | Ag^+(aq) \| Fe^{2+}(aq) | Fe(s) \quad \text{corresponding to}$$
$$2\,Ag(s) + Fe^{2+}(aq) \longrightarrow 2\,Ag^+(aq) + Fe(s)$$

we would have written

$$E° = E°(Fe^{2+}/Fe) - E°(Ag^+/Ag)$$

and would have found $E° = -1.24$ V. We would have come to the same conclusion as before: for this reaction $K < 1$, with reactants dominant at equilibrium, and iron has a tendency to reduce silver.

A problem with compiling a list of standard potentials is that we know only the *overall* emf of the cell, not the contribution of a single electrode. A voltmeter placed between the two electrodes of a galvanic cell measures the *difference* of their potentials, not the individual values. To provide numerical values for individual standard potentials, we arbitrarily set the standard potential of one particular electrode, the hydrogen electrode, equal to zero at all temperatures:

$$2\,H^+(g) + 2\,e^- \longrightarrow H_2(g) \qquad E° = 0$$

Standard potentials are also called *standard electrode potentials*. Because they are always written for reduction half-reactions, they are also sometimes called *standard reduction potentials*.

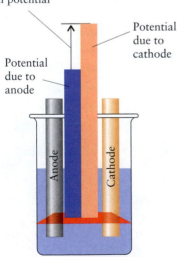

Cell potential

Potential due to cathode

Potential due to anode

Anode

Cathode

FIGURE 12.6 The cell emf can be thought of as the difference of the two potentials produced by the two electrodes.

In redox couple notation, $E°(H^+/H_2) = 0$ at all temperatures. A hydrogen electrode in its standard state, with hydrogen gas at 1 bar and the hydrogen ions present at 1 mol·L^{-1} (strictly, unit activity), is called a **standard hydrogen electrode** (SHE). The standard hydrogen electrode is then used to define the standard potentials of all other electrodes:

> The standard potential of a couple is the standard emf of a cell (including the sign) in which the couple forms the right-hand electrode in the cell diagram and a hydrogen electrode forms the left-hand electrode in the cell diagram.

For example, for the cell

$$Pt(s)\,|\,H_2(g)\,|\,H^+(aq)\,\|\,Cu^{2+}(aq)\,|\,Cu(s)$$

we find that the standard emf is 0.34 V with the copper electrode the cathode, and so $E° = +0.34$ V. Because the hydrogen electrode contributes zero to the standard emf of the cell, the emf is attributed entirely to the copper electrode, and we write

$$Cu^{2+}(aq) + 2\,e^- \longrightarrow Cu(s) \qquad E°(Cu^{2+}/Cu) = +0.34 \text{ V}$$

We can take this analysis a little farther. Because the cell reaction

$$Cu^{2+}(aq) + H_2(g) \longrightarrow Cu(s) + 2\,H^+(aq)$$

is spontaneous in the direction written (under standard conditions), we can consider the oxidizing ability of $Cu^{2+}(aq)$, as represented by the reduction half-reaction $Cu^{2+}(aq) + 2\,e^- \to Cu(s)$, to be greater than the oxidizing ability of $H^+(aq)$, as represented by $2\,H^+(aq) + 2\,e^- \to H_2(g)$. Consequently, Cu^{2+} ions can be reduced to metallic copper by hydrogen gas under standard conditions. In general (Fig. 12.7),

> The more *positive* the potential, the greater the electron-pulling power of the reduction half-reaction and, therefore, the more strongly *oxidizing* the redox couple (the stronger the tendency for the half-reaction to occur as a reduction).

Now consider the cell

$$Pt(s)\,|\,H_2(g)\,|\,H^+(aq)\,\|\,Zn^{2+}(aq)\,|\,Zn(s)$$

and the corresponding cell reaction

$$Zn^{2+}(aq) + H_2(g) \longrightarrow Zn(s) + 2\,H^+(aq)$$

The standard emf is measured as 0.76 V and the hydrogen electrode is found to be the cathode: therefore, this emf is reported as −0.76 V. Because the entire emf is attributed to the zinc electrode, we write

$$Zn^{2+}(aq) + 2\,e^- \longrightarrow Zn(s) \qquad E°(Zn^{2+}/Zn) = -0.76 \text{ V}$$

The negative standard potential means that the Zn^{2+}/Zn electrode is the anode in a cell with H^+/H_2 as the other electrode and, therefore, that the *reverse* of the cell reaction, specifically

$$Zn(s) + 2\,H^+(aq) \longrightarrow Zn^{2+}(aq) + H_2(g)$$

is spontaneous under standard conditions. We can conclude that the reducing ability of $Zn(s)$ in the half-reaction $Zn(s) \to Zn^{2+}(aq) + 2\,e^-$ is greater than the reducing ability of $H_2(g)$ in the half-reaction $H_2(g) \to 2\,H^+(aq) + 2\,e^-$. Consequently, zinc metal can reduce H^+ ions in acidic solution to hydrogen gas under standard conditions. In general (Fig. 12.8),

> The more *negative* the potential, the greater the electron-donating power of the oxidation half-reaction and therefore the more strongly *reducing* the redox couple (that is, the stronger the tendency for the half-reaction to occur as an oxidation).

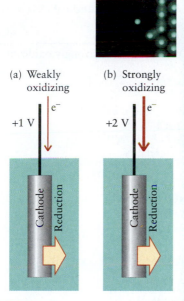

(a) Weakly oxidizing +1 V e⁻ Cathode Reduction

(b) Strongly oxidizing +2 V e⁻ Cathode Reduction

FIGURE 12.7 (a) A couple that has a small positive potential has a weak electron-pulling power (is a weak acceptor of electrons) relative to hydrogen ions and hence is a weak oxidizing agent. (b) A couple with a high positive potential has strong pulling power (is a strong acceptor of electrons) and is a strong oxidizing agent.

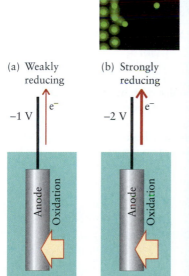

(a) Weakly reducing −1 V e⁻ Anode Oxidation

(b) Strongly reducing −2 V e⁻ Anode Oxidation

FIGURE 12.8 (a) A couple that has a small negative potential has a poor pushing power (is a weak donor of electrons) relative to hydrogen and, hence, is a weak reducing agent. (b) A couple with a large negative potential has strong pushing power (is a strong donor of electrons) and is a strong reducing agent.

Animation: Figure 12.7 Reduction at a cathode

Media Link

Animation: Figure 12.8 Oxidation at an anode

Media Link

TABLE 12.1 Standard Potentials at 25°C*

Species	Reduction half-reaction	$E°$ (V)
Oxidized form is strongly oxidizing		
F_2/F^-	$F_2(g) + 2\,e^- \longrightarrow 2\,F^-(aq)$	+2.87
Au^+/Au	$Au^+(aq) + e^- \longrightarrow Au(s)$	+1.69
Ce^{4+}/Ce^{3+}	$Ce^{4+}(aq) + e^- \longrightarrow Ce^{3+}(aq)$	+1.61
$MnO_4^-,H^+/Mn^{2+},H_2O$	$MnO_4^-(aq) + 8\,H^+(aq) + 5\,e^- \longrightarrow Mn^{2+}(aq) + 4\,H_2O(l)$	+1.51
Cl_2/Cl^-	$Cl_2(g) + 2\,e^- \longrightarrow 2\,Cl^-(aq)$	+1.36
$Cr_2O_7^{2-},H^+/Cr^{3+},H_2O$	$Cr_2O_7^{2-}(aq) + 14\,H^+(aq) + 6\,e^- \longrightarrow 2\,Cr^{3+}(aq) + 7\,H_2O(l)$	+1.33
$O_2,H^+/H_2O$	$O_2(g) + 4\,H^+(aq) + 4\,e^- \longrightarrow 2\,H_2O(l)$	+1.23; +0.82 at pH = 7
Br_2/Br^-	$Br_2(l) + 2\,e^- \longrightarrow 2\,Br^-(aq)$	+1.09
$NO_3^-,H^+/NO,H_2O$	$NO_3^-(aq) + 4\,H^+(aq) + 3\,e^- \longrightarrow NO(g) + 2\,H_2O(l)$	+0.96
Ag^+/Ag	$Ag^+(aq) + e^- \longrightarrow Ag(s)$	+0.80
Fe^{3+}/Fe^{2+}	$Fe^{3+}(aq) + e^- \longrightarrow Fe^{2+}(aq)$	+0.77
I_2/I^-	$I_2(s) + 2\,e^- \longrightarrow 2\,I^-(aq)$	+0.54
$O_2,H_2O/OH^-$	$O_2(g) + 2\,H_2O(l) + 4\,e^- \longrightarrow 4\,OH^-(aq)$	+0.40; +0.82 at pH = 7
Cu^{2+}/Cu	$Cu^{2+}(aq) + 2\,e^- \longrightarrow Cu(s)$	+0.34
$AgCl/Ag,Cl^-$	$AgCl(s) + e^- \longrightarrow Ag(s) + Cl^-(aq)$	+0.22
H^+/H_2	$2\,H^+(aq) + 2\,e^- \longrightarrow H_2(g)$	0, by definition
Fe^{3+}/Fe	$Fe^{3+}(aq) + 3\,e^- \longrightarrow Fe(s)$	−0.04
$O_2,H_2O/HO_2^-,OH^-$	$O_2(g) + H_2O(l) + 2\,e^- \longrightarrow HO_2^-(aq) + OH^-(aq)$	−0.08
Pb^{2+}/Pb	$Pb^{2+}(aq) + 2\,e^- \longrightarrow Pb(s)$	−0.13
Sn^{2+}/Sn	$Sn^{2+}(aq) + 2\,e^- \longrightarrow Sn(s)$	−0.14
Fe^{2+}/Fe	$Fe^{2+}(aq) + 2\,e^- \longrightarrow Fe(s)$	−0.44
Zn^{2+}/Zn	$Zn^{2+}(aq) + 2\,e^- \longrightarrow Zn(s)$	−0.76
$H_2O/H_2\ OH^-$	$2\,H_2O(l) + 2\,e^- \longrightarrow H_2(g) + 2\,OH^-(aq)$	−0.83; −0.42 at pH = 7
Al^{3+}/Al	$Al^{3+}(aq) + 3\,e^- \longrightarrow Al(s)$	−1.66
Mg^{2+}/Mg	$Mg^{2+}(aq) + 2\,e^- \longrightarrow Mg(s)$	−2.36
Na^+/Na	$Na^+(aq) + e^- \longrightarrow Na(s)$	−2.71
K^+/K	$K^+(aq) + e^- \longrightarrow K(s)$	−2.93
Li^+/Li	$Li^+(aq) + e^- \longrightarrow Li(s)$	−3.05
Reduced form is strongly reducing		

* For a more extensive table, see Appendix 2B.

Notice that, in Appendix 2B, standard potentials are listed both by numerical value and alphabetically, to make it easy to find the one you want.

Table 12.1 lists a number of standard electrode potentials measured at 25°C (the only temperature we consider here); a longer list can be found in Appendix 2B. The standard potentials of elements vary in a complicated way through the periodic table (Fig. 12.9). However, the most negative—the most strongly reducing—are usually found toward the left of the periodic table, and the most positive—the most strongly oxidizing—are found toward the upper right.

EXAMPLE 12.5 Sample exercise: Determining the standard potential of an electrode

The standard potential of a zinc electrode is −0.76 V, and the standard emf of the cell

$$Zn(s)\,|\,Zn^{2+}(aq)\,\|\,Sn^{4+}(aq),Sn^{2+}(aq)\,|\,Pt(s)$$

is +0.91 V. What is the standard potential of the Sn^{4+}/Sn^{2+} electrode?

SOLUTION We can determine the standard potential of an electrode by measuring the emf of a standard cell in which the other electrode has a known standard potential and applying Eq. 3.

From Eq. 3, $E° = E_R° − E_L°$,

$$E° = E°(Sn^{4+}/Sn^{2+}) − E°(Zn^{2+}/Zn)$$
$$= +0.91\ V$$

Rearrange Eq. 3 into $E_R° = E° + E_L°$.

$$E°(Sn^{4+}/Sn^{2+}) = E° + E°(Zn^{2+}/Zn)$$
$$= 0.91\ V − 0.76\ V = +0.15\ V$$

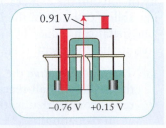
0.91 V

−0.76 V   +0.15 V

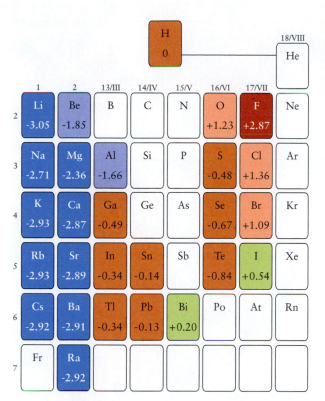

FIGURE 12.9 The variation of standard potentials through the main groups of the periodic table. Note that the most negative values are in the *s* block and that the most positive values are close to fluorine.

SELF-TEST 12.6A The standard potential of the Ag^+/Ag electrode is $+0.80$ V, and the standard emf of the cell $Pt(s)|I_2(s)|I^-(aq)\|Ag^+(aq)|Ag(s)$ is $+0.26$ V at the same temperature. What is the standard potential of the I_2/I^- electrode?

[*Answer:* $+0.54$ V]

SELF-TEST 12.6B The standard potential of the Fe^{2+}/Fe electrode is -0.44 V and the standard emf of the cell $Fe(s)|Fe^{2+}(aq)\|Pb^{2+}(aq)|Pb(s)$ is $+0.31$ V. What is the standard potential of the Pb^{2+}/Pb electrode?

In some cases, we find that available tables of data do not contain the standard potential that we need but do contain closely related values for the same element; for instance, we might require the standard potential of the Ce^{4+}/Ce couple, whereas we know only the values for the Ce^{3+}/Ce and Ce^{4+}/Ce^{3+} couples. In such cases, the potential of a couple cannot be determined by adding or subtracting the standard potentials directly. Instead, we calculate the values of $\Delta G°$ for each half-reaction and combine them into the $\Delta G°$ for the desired half-reaction. We then convert that value of $\Delta G°$ into the corresponding standard potential by using Eq. 2.

EXAMPLE 12.6 Calculating the standard potential of a couple from two related couples

Use the information in Appendix 2B to determine the standard potential for the redox couple Ce^{4+}/Ce, for which the reduction half-reaction is

$$Ce^{4+}(aq) + 4\ e^- \longrightarrow Ce(s) \tag{B}$$

STRATEGY Use the alphabetical listing in Appendix 2B to find half-reactions that can be combined to give the desired half-reaction. Combine these half-reactions and their Gibbs free energies of reaction. Convert the Gibbs free energies into standard potentials by using Eq. 2 and then simplify the resulting expressions.

SOLUTION
From the data in Appendix 2B,

$$Ce^{3+}(aq) + 3\ e^- \longrightarrow Ce(s) \qquad E° = -2.48\ V \tag{C}$$

$$Ce^{4+}(aq) + e^- \longrightarrow Ce^{3+}(aq) \qquad E° = +1.61\ V \tag{D}$$

Add the reaction Gibbs free energies to get the overall reaction Gibbs free energy.	$\Delta G°(B) = \Delta G°(C) + \Delta G°(D)$	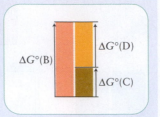
From $\Delta G_r° = -nFE°$,	$-4FE°(B) = -3FE°(C) + \{-FE°(D)\}$	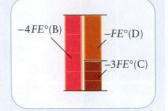
Divide by $-4F$.	$E°(B) = \dfrac{3E°(C) + E°(D)}{4}$	
Insert the data.	$E°(B) = \dfrac{3(-2.48\ V) + 1.61\ V}{4} = -1.46\ V$	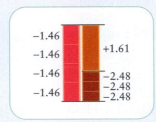

Notice that this value is not the same as the sum of the potentials for half-reactions C and D.

SELF-TEST 12.7A Use the data in Appendix 2B to calculate the standard potential of the couple $Au^{3+}(aq)/Au^{+}(aq)$.

[*Answer:* $+1.26\ V$]

SELF-TEST 12.7B Use the data in Appendix 2B to calculate the standard potential of the couple $Mn^{3+}(aq)/Mn(s)$.

The standard potential of an electrode is the standard emf of a cell in which the electrode on the left in the cell diagram is a hydrogen electrode. A metal with a negative standard potential has a thermodynamic tendency to reduce hydrogen ions in solution; the ions of a metal with a positive standard potential have a tendency to be reduced by hydrogen gas.

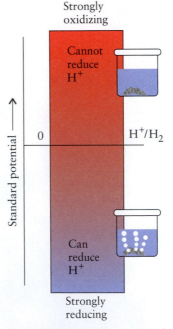

FIGURE 12.10 The significance of standard potentials. Only couples with negative standard potentials (and hence lying below hydrogen in the electrochemical series) can reduce hydrogen ions under standard conditions. The reducing power increases as the standard potential becomes more negative.

12.7 The Electrochemical Series

When redox couples are ordered by their standard potentials, we have a list of oxidizing and reducing agents ordered by their strengths. As we have seen, *the more negative the standard potential of a couple, the greater its reducing strength*. Only a couple with a negative potential can reduce hydrogen ions under standard conditions. A couple with a positive potential, such as Au^{3+}/Au, cannot reduce hydrogen ions under standard conditions (Fig. 12.10).

When Table 12.1 is viewed as a table of relative strengths of oxidizing and reducing agents, it is called the **electrochemical series**. The species on the left of each equation in Table 12.1 are potentially oxidizing agents. They can themselves be reduced. Species on the right of the equations are potentially reducing agents. An oxidized species in the list (on the left of the equation) has a tendency to oxidize a reduced species that lies below it. For example, Cu^{2+} ions oxidize zinc metal. A reduced species (on the right of the equation) has a tendency to reduce an oxidized species that lies above it. For example, zinc metal reduces H^+ ions.

The higher the position of a species on the left side of an equation in Table 12.1, the greater its oxidizing strength. For example, F_2 is a strong oxidizing agent, whereas Li^+ is a very, very poor oxidizing agent. It also follows that, the lower the standard potential, the greater the reducing strength of the reduced species on the right side of an equation in Table 12.1. For example, lithium metal is the strongest reducing agent in the table.

SELF-TEST 12.8A Can lead produce zinc metal from aqueous zinc sulfate under standard conditions?

[***Answer:*** No, because lead lies above zinc in Table 12.1.]

SELF-TEST 12.8B Can chlorine gas oxidize water to oxygen gas under standard conditions in basic solution?

We can use the electrochemical series to predict the thermodynamic tendency for a reaction to take place under standard conditions. A cell reaction that is spontaneous under standard conditions (that is, has $K > 1$) has $\Delta G° < 0$ and therefore the corresponding cell has $E° > 0$. The standard emf is positive when $E_R° > E_L°$; that is, when the standard potential for the reduction half-reaction is more positive than that for the oxidation half-reaction.

EXAMPLE 12.7   **Predicting the spontaneous direction of a redox reaction under standard conditions**

Which is the more powerful oxidizing agent under standard conditions, an acidified aqueous permanganate solution or an acidified aqueous dichromate solution? Specify the cell for the spontaneous reaction of the two couples by writing a cell diagram that under standard conditions has a positive emf. Determine the standard emf of the cell and write the net ionic equation for the spontaneous cell reaction.

STRATEGY Find the standard potentials of the two reduction half-reactions in Appendix 2B. The couple with the more positive potential will act as an oxidizing agent (and be the site of reduction). That couple will be the right-hand electrode in the cell diagram corresponding to the spontaneous cell reaction. To calculate the standard emf of the cell, subtract the standard potential of the oxidation half-reaction (the one with the less-positive standard potential) from that of the reduction half-reaction. To write the cell reaction, follow the procedure in Toolbox 12.2.

SOLUTION

Find the two standard potentials in Appendix 2B.

$$MnO_4^-(aq) + 8\,H^+(aq) + 5\,e^- \longrightarrow$$
$$Mn^{2+}(aq) + 4\,H_2O(l) \qquad E° = +1.51\ V$$

$$Cr_2O_7^{2-}(aq) + 14\,H^+(aq) + 6\,e^- \longrightarrow$$
$$2\,Cr^{3+}(aq) + 7\,H_2O(l) \qquad E° = +1.33\ V$$

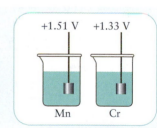

The half-reaction with the more positive value of $E°$ occurs at the cathode in a spontaneous reaction.

$$E°(MnO_4^-,H^+/Mn^{2+},H_2O) >$$
$$E°(Cr_2O_7^{2-},H^+/Cr^{3+},H_2O)$$

Therefore, MnO_4^- is a stronger oxidizing agent than $Cr_2O_7^{2-}$ in acidic solution and so serves as the cathode.

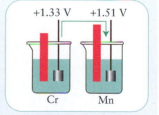

Find the standard emf from $E° = E°(\text{for reduction}) - E°(\text{for oxidation})$.

$$E° = 1.51\ V - 1.33\ V = +0.18\ V$$

Write the cell diagram with the cathode on the right-hand side.

$$Pt(s)\,|\,Cr_2O_7^{2-}(aq),Cr^{3+}(aq),H^+(aq)\,\|$$
$$H^+(aq),MnO_4^-(aq),Mn^{2+}(aq)\,|\,Pt(s)$$

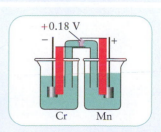

504 CHAPTER 12 ELECTROCHEMISTRY

To construct the spontaneous cell reaction, combine the two half-reactions, leaving the permanganate half-reaction as a reduction and reversing the dichromate half-reaction. To match numbers of electrons, multiply the manganese half-reaction by 6 and the chromium half-reaction by 5:

$$10\ Cr^{3+}(aq) + 35\ H_2O(l) \longrightarrow 5\ Cr_2O_7^{2-}(aq) + 70\ H^+(aq) + 30\ e^-$$
$$6\ MnO_4^-(aq) + 48\ H^+(aq) + 30\ e^- \longrightarrow 6\ Mn^{2+}(aq) + 24\ H_2O(l)$$

Their net sum gives the spontaneous cell reaction:

$$6\ MnO_4^-(aq) + 11\ H_2O(l) + 10\ Cr^{3+}(aq) \longrightarrow$$
$$6\ Mn^{2+}(aq) + 22\ H^+(aq) + 5\ Cr_2O_7^{2-}(aq)$$

SELF-TEST 12.9A Which metal, zinc or nickel, is the stronger reducing agent in aqueous solution under standard conditions? Evaluate the standard emf of the appropriate cell, specify the cell with a cell diagram, and write the net ionic equation for the spontaneous reaction.

[*Answer:* Zinc; +0.53 V; $Zn(s)\,|\,Zn^{2+}(aq)\,\|\,Ni^{2+}(aq)\,|\,Ni(s)$;
$Zn(s) + Ni^{2+}(aq) \rightarrow Zn^{2+}(aq) + Ni(s)$]

SELF-TEST 12.9B Which is the stronger oxidizing agent, Cu^{2+} or Ag^+, in aqueous solution under standard conditions? Evaluate the standard emf of the appropriate cell, specify the cell with a cell diagram, and write the net ionic equation for the corresponding cell reaction.

The oxidizing and reducing power of a redox couple determines its position in the electrochemical series. The strongest oxidizing agents are at the top left of the table; the strongest reducing agents are at the bottom right of the table.

12.8 Standard Potentials and Equilibrium Constants

One of the most useful applications of standard potentials is in the calculation of equilibrium constants from electrochemical data. The techniques that we develop here can be applied to any kind of reaction, including neutralization and precipitation reactions as well as redox reactions, provided that they can be expressed as the difference of two reduction half-reactions.

We saw in Section 9.3 that the standard reaction Gibbs free energy, ΔG_r°, is related to the equilibrium constant of the reaction by $\Delta G_r^\circ = -RT \ln K$. In this chapter, we have seen that the standard reaction Gibbs free energy is related to the standard emf of a galvanic cell by $\Delta G_r^\circ = -nFE^\circ$, with n a pure number. When we combine the two equations, we get

$$nFE^\circ = RT \ln K \tag{4}$$

This expression can be rearranged to allow us to calculate the equilibrium constant from the cell emf:

$$\ln K = \frac{nFE^\circ}{RT} \tag{5}*$$

Because the magnitude of K increases exponentially with E°, a reaction with a large positive E° has $K \gg 1$. A reaction with a large negative E° has $K \ll 1$.

A note on good practice: Equation 5 was derived on the basis of the "molar" convention for writing the reaction Gibbs free energy; that means that the n must be interpreted as a pure number. That convention keeps the units straight: FE° has the units joules per mole, so does RT, so the ratio FE°/RT is a pure number and, with n a pure number, the right hand side is a pure number too (as it must be, if it is to be equal to a logarithm).

The fact that we can calculate E° from standard potentials allows us to calculate equilibrium constants for any reaction that can be expressed as two half-reactions. The reaction does not need to be spontaneous nor does it have to be a redox reaction. Toolbox 12.3 summarizes the steps and Example 12.8 shows the steps in action.

TOOLBOX 12.3   HOW TO CALCULATE EQUILIBRIUM CONSTANTS FROM ELECTROCHEMICAL DATA

CONCEPTUAL BASIS

The equilibrium constant of a reaction is an exponential function of the standard emf of the corresponding cell. We can expect a cell reaction with a large positive emf to have a strong tendency to take place, and therefore to produce a high proportion of products at equilibrium. Therefore, we expect $K > 1$ when $E° > 0$ (and often $K \gg 1$). The opposite is true for a cell reaction with a negative standard emf.

PROCEDURE

The procedure for calculating an equilibrium constant from electrochemical data is as follows.

Step 1 Write the balanced equation for the reaction. Then find two reduction half-reactions in Appendix 2B that combine to give that equation. Reverse one of the half-reactions and add them together.

Step 2 Identify the numerical (unitless) value of n from the change in oxidation numbers or by examining the half-reactions (after multiplication by appropriate factors) for the number of electrons transferred in the balanced equation.

Step 3 To obtain $E°$, subtract the standard potential of the half-reaction that was reversed (oxidation) from the standard potential of the half-reaction that was left as a reduction: $E° = E°(\text{for reduction}) - E°(\text{for oxidation})$. Alternatively, write a cell diagram for the reaction; in that case, $E° = E_R° - E_L°$.

Step 4 Use the relation $\ln K = nFE°/RT$ to calculate the value of K.

At 25.00°C (298.15 K), $RT/F = 0.025\,693$ V; so, at that temperature,

$$\ln K = \frac{nE°}{0.025\,693\ \text{V}}$$

This procedure is illustrated in Example 12.8.

EXAMPLE 12.8   **Sample exercise: Calculating the equilibrium constant for a reaction**

Calculate the equilibrium constant at 25.00°C for the reaction

$$\text{AgCl(s)} \longrightarrow \text{Ag}^+(\text{aq}) + \text{Cl}^-(\text{aq})$$

The equilibrium constant for this reaction is actually the solubility product, $K_{sp} = [\text{Ag}^+][\text{Cl}^-]$, for silver chloride (Section 11.8).

SOLUTION Because silver chloride is almost insoluble, we expect K to be very small (and $E°$ to be negative). Follow the procedure in Toolbox 12.3.

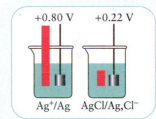

Step 1 Find the two reduction half-reactions required for the cell reaction.	R: $\text{AgCl(s)} + \text{e}^- \longrightarrow \text{Ag(s)} + \text{Cl}^-(\text{aq})$ $E° = +0.22$ V L: $\text{Ag}^+(\text{aq}) + \text{e}^- \longrightarrow \text{Ag(s)}$ $E° = +0.80$ V
Reverse the second half-reaction.	$\text{Ag(s)} \longrightarrow \text{Ag}^+(\text{aq}) + \text{e}^-$
Add this equation to the reduction half-reaction and cancel species that appear on both sides of the equation.	$\text{AgCl(s)} \longrightarrow \text{Ag}^+(\text{aq}) + \text{Cl}^-(\text{aq})$
Step 2 From the half-reactions note the number of electrons transferred.	$n = 1$

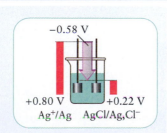

Step 3 Find $E°$ from $E° = E°(\text{for reduction}) - E°(\text{for oxidation})$.	$E° = 0.22$ V $- 0.80$ V $= -0.58$ V

From $\ln K = nFE°/RT =$ $nE°/(RT/F)$,	$\ln K_{sp} = \dfrac{(1) \times (-0.58 \text{ V})}{0.025\,693 \text{ V}} = -\dfrac{0.58}{0.025\,693}$
Take the antilogarithm (e^x) of $\ln K_{sp}$.	$K_{sp} = e^{-0.58/0.025\,693} = 1.6 \times 10^{-10}$

The value of K_{sp} is the same as that listed in Table 11.4. Many of the solubility products listed in tables were determined from emf measurements and calculations like this one.

SELF-TEST 12.10A Use Appendix 2B to calculate the solubility product of mercury(I) chloride, Hg_2Cl_2.

[*Answer:* 2.6×10^{-18}]

SELF-TEST 12.10B Use the tables in Appendix 2B to calculate the solubility product of cadmium hydroxide, $Cd(OH)_2$.

The equilibrium constant of a reaction can be calculated from standard potentials by combining the equations for the half-reactions to give the cell reaction of interest and determining the standard potential of the corresponding cell.

12.9 The Nernst Equation

As a reaction proceeds toward equilibrium, the concentrations of its reactants and products change and ΔG approaches zero. Therefore, as reactants are consumed in a working electrochemical cell, the cell potential also decreases until finally it reaches zero. A "dead" battery is one in which the cell reaction has reached equilibrium. At equilibrium, a cell generates zero potential difference across its electrodes and the reaction can no longer do work. To describe this behavior quantitatively, we need to find how the cell emf varies with the concentrations of species in the cell.

HOW DO WE DO THAT?

To establish how the emf of a cell depends on concentration, we first note that the emf is proportional to the reaction Gibbs free energy (Eq. 2). We already know how ΔG_r varies with composition:

$$\Delta G_r = \Delta G_r° + RT \ln Q$$

where Q is the reaction quotient for the cell reaction (Eq. 5 of Section 9.3). Because $\Delta G_r = -nFE$ and $\Delta G_r° = -nFE°$, it follows at once that

$$-nFE = -nFE° + RT \ln Q$$

At this point, we divide through by $-nF$ to get an expression for E in terms of Q.

The equation for the concentration dependence of the cell emf that we have derived,

$$E = E° - \frac{RT}{nF} \ln Q \qquad (6)*$$

is called the **Nernst equation** for the German electrochemist Walther Nernst, who first derived it. At 298.15 K, $RT/F = 0.025\,693$ V; so at that temperature the Nernst equation takes the form

$$E = E° - \frac{0.025\,693 \text{ V}}{n} \ln Q$$

with n unitless. It is sometimes convenient to use this equation with common logarithms, in which case we make use of the relation $\ln x = 2.303 \log x$. At 298.15 K,

$$E = E° - \frac{2.303\, RT}{nF} \log Q = E° - \frac{0.05917 \text{ V}}{n} \log Q$$

The Nernst equation is widely used to estimate the emf of cells under nonstandard conditions. In biology it is used, among other things, to estimate the potential difference across biological cell membranes, such as those of neurons.

EXAMPLE 12.9   **Using the Nernst equation to predict an emf**

Calculate the emf at 25°C of a Daniell cell in which the concentration of Zn^{2+} ions is 0.10 mol·L^{-1} and that of the Cu^{2+} ions is $0.0010 \text{ mol·L}^{-1}$.

STRATEGY   First, write the balanced equation for the cell reaction and the corresponding expression for Q, and note the value of n. Then determine $E°$ from the standard potentials in Table 12.1 or Appendix 2B. Determine the value of Q for the stated conditions. Calculate the emf by substituting these values into the Nernst equation, Eq. 6. At 25.00°C, $RT/F = 0.025\ 693$ V.

SOLUTION   The Daniell cell and the corresponding cell reaction are

$$Zn(s) \,|\, Zn^{2+}(aq) \,\|\, Cu^{2+}(aq) \,|\, Cu(s) \qquad Cu^{2+}(aq) + Zn(s) \longrightarrow Zn^{2+}(aq) + Cu(s)$$

Set up the reaction quotient.

$$Q = \frac{a_{Zn^{2+}}}{a_{Cu^{2+}}} = \frac{[Zn^{2+}]}{[Cu^{2+}]} = \frac{0.10}{0.0010}$$

From the balanced equation, note the value of n.

$$n = 2$$

Determine the value of $E° = E_R° - E_L°$.

$$E° = 0.34 - (-0.76) \text{ V} = +1.10 \text{ V}$$

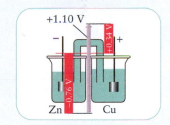

From $E = E° - (RT/nF) \ln Q$,

$$E = 1.10 \text{ V} - \frac{0.025\ 693 \text{ V}}{2} \ln \frac{0.10}{0.0010}$$

$$= 1.10 \text{ V} - 0.059 \text{ V} = +1.04 \text{ V}$$

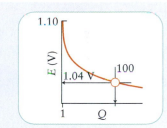

SELF-TEST 12.11A Calculate the emf of the cell $Zn(s) \,|\, Zn^{2+}(aq, 1.50 \text{ mol·L}^{-1}) \,\|\, Fe^{2+}(aq, 0.10 \text{ mol·L}^{-1}) \,|\, Fe(s)$.

[*Answer:* +0.29 V]

SELF-TEST 12.11B Calculate the emf of the concentration cell $Ag(s) \,|\, Ag^{+}(aq, 0.0010 \text{ mol·L}^{-1}) \,\|\, Ag^{+}(aq, 0.010 \text{ mol·L}^{-1}) \,|\, Ag(s)$.

An important application of the Nernst equation is the measurement of concentration. In a **concentration cell,** the two electrodes are identical except for their concentrations. For such a cell, $E° = 0$ and at 25°C the potential corresponding to the cell reaction is related to Q by $E = -(0.025693 \text{ V}/n) \ln Q$. For example, a concentration cell having two Ag^{+}/Ag electrodes is

$$Ag(s) \,|\, Ag^{+}(aq, L) \,\|\, Ag^{+}(aq, R) \,|\, Ag(s) \qquad Ag^{+}(aq, R) \longrightarrow Ag^{+}(aq, L)$$

The cell reaction has $n = 1$ and $Q = [Ag^{+}]_L/[Ag^{+}]_R$. If the concentration of Ag^{+} in the right-hand electrode is 1 mol·L^{-1}, Q is equal to $[Ag^{+}]_L$, and the Nernst equation is

$$E = -(0.025693 \text{ V}) \ln [Ag^{+}]_L$$

Therefore, by measuring E, we can infer the concentration of Ag^+ in the left-hand electrode compartment. If the concentration of Ag^+ ions in the left-hand electrode is less than that in the right, then $E > 0$ for the cell as specified and the right-hand electrode will be found to be the cathode.

EXAMPLE 12.10 Using the Nernst equation to find a concentration

Each electrode compartment of a galvanic cell contains a silver electrode and 10.0 mL of 0.10 M $AgNO_3$(aq); they are connected by a salt bridge. You now add 10.0 mL of 0.10 M NaCl(aq) to the left-hand electrode compartment. Almost all the silver precipitates as silver chloride but a little remains in solution as a saturated solution of AgCl. The measured emf is $E = +0.42$ V. What is the concentration of Ag^+ in the saturated solution?

STRATEGY The cell is a concentration cell in which the concentration of Ag^+ in one compartment is determined by the solubility of AgCl. Use the Nernst equation, Eq. 6, to find the concentration of Ag^+ in the compartment with the precipitate. The standard emf of the cell is 0 (in their standard states the electrodes are identical). At 25.00°C, $RT/F = 0.025\,693$ V.

SOLUTION The cell and the corresponding cell reaction are

$$Ag(s) \mid Ag^+(aq, L) \| Ag^+(aq, R) \mid Ag(s) \qquad Ag^+(aq, R) \longrightarrow Ag^+(aq, L)$$

Set up the reaction quotient, Q.

$$Q = \frac{[Ag^+]_L}{[Ag^+]_R} = \frac{[Ag^+]_L}{0.10}$$

From the balanced equation note the value of n.

$$n = 1$$

From $E = -(RT/nF) \ln Q$ rearranged into $\ln Q = -E/(RT/nF)$,

$$\ln Q = \frac{-0.42 \text{ V}}{0.025\,693 \text{ V}} = -16.34$$

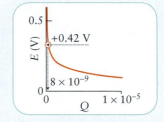

From $Q = e^{\ln Q}$,

$$Q = e^{-16.34}$$

From $[Ag^+]_L = Q[Ag^+]_R$,

$$[Ag^+]_L = e^{-16.34} \times 0.10 = 8.0 \times 10^{-9}$$

That is, the concentration of Ag^+ ions in the saturated solution is 8.0 nmol·L⁻¹.

SELF-TEST 12.12A Calculate the molar concentration of Y^{3+} in a saturated solution of YF_3 by using a cell constructed with two yttrium electrodes. The electrolyte in one compartment is 1.0 M $Y(NO_3)_3$(aq). In the other compartment you have prepared a saturated solution of YF_3. The measured cell potential is +0.34 V at 298 K.

[Answer: 8.3×10^{-18} mol·L⁻¹]

SELF-TEST 12.12B Calculate the emf of a cell constructed with two copper electrodes. The electrolyte in one compartment is 1.0 M $AgNO_3$(aq). In the other compartment NaOH has been added to a $AgNO_3$ solution until the pH = 12.5 at 298 K.

The dependence of emf on composition is expressed by the Nernst equation, Eq. 6.

12.10 Ion-Selective Electrodes

An important application of the Nernst equation is the measurement of pH (and, through pH, acidity constants). The pH of a solution can be measured electrochemically with a device called a pH meter. The technique makes use of a cell in which one electrode is sensitive to the H_3O^+ concentration and another electrode serves as a reference. An electrode sensitive to the concentration of a particular ion is called an **ion-selective electrode**.

One combination of electrodes that could be used to determine pH is a hydrogen electrode connected through a salt bridge to a *calomel electrode*. The reduction half-reaction for the calomel electrode is

$$Hg_2Cl_2(s) + 2\,e^- \longrightarrow 2\,Hg(l) + 2\,Cl^-(aq) \quad E° = +0.27\text{ V}$$

The overall cell reaction is

$$Hg_2Cl_2(s) + H_2(g) \longrightarrow 2\,H^+(aq) + 2\,Hg(l) + 2\,Cl^-(aq) \quad Q = \frac{[H^+]^2[Cl^-]^2}{P_{H_2}}$$

Provided that the pressure of hydrogen is 1 bar, we can write the reaction quotient as $Q = [H^+]^2[Cl^-]^2$. To find the concentration of hydrogen ions, we write the Nernst equation:

$$E = E° - \{\tfrac{1}{2}(0.0257\text{ V}) \times \ln [H^+]^2[Cl^-]^2\}$$

We apply $\ln (ab) = \ln a + \ln b$,

$$E = E° - \{\tfrac{1}{2}(0.0257\text{ V}) \times \ln [Cl^-]^2\} - \{\tfrac{1}{2}(0.0257\text{ V}) \times \ln [H^+]^2\}$$

and then use $\ln a^x = x \ln a$ to obtain

$$E = E° - \{(0.0257\text{ V}) \times \ln [Cl^-]\} - \{(0.0257\text{ V}) \times \ln [H^+]\}$$

The Cl^- concentration of a calomel electrode is fixed at the time of manufacture by saturating the solution with KCl, and so $[Cl^-]$ is a constant. We can therefore combine the first two terms on the right into a single constant, $E' = E° - (0.0257\text{ V}) \times \ln [Cl^-]$. Then, because $\ln x = 2.303 \log x$,

$$E = E' - 2.303 \times (0.0257\text{ V}) \times \log [H^+]$$
$$= E' + (0.0592\text{ V}) \times pH$$

Therefore, by measuring the cell emf, E, we can determine the pH. The value of E' is established by calibrating the cell, which requires measuring E for a solution of known pH.

A **glass electrode**, a thin-walled glass bulb containing an electrolyte, is much easier to use than a hydrogen electrode and has a potential that varies linearly with the pH of the solution outside the glass bulb (Fig. 12.11). Often there is a calomel electrode built into the probe that makes contact with the test solution through a miniature salt bridge. A pH meter therefore usually has only one probe, which forms a complete electrochemical cell once it is dipped into a solution. The meter is calibrated with a buffer of known pH, and the measured cell emf is then automatically converted into the pH of the solution, which is displayed.

Commercially available electrodes used in *pX meters* are sensitive to other ions, such as Na^+, Ca^{2+}, NH_4^+, CN^-, or S^{2-}. They are used to monitor industrial processes and in pollution control.

The pH or concentrations of ions can be measured by using an electrode that responds selectively to only one species of ion.

ELECTROLYTIC CELLS

Redox reactions that have a positive Gibbs free energy of reaction are not spontaneous, but an electric current can be used to make them take place. For example, there is no common spontaneous chemical reaction in which fluorine is a product, and so the element cannot be isolated by any common chemical reaction. It was not until 1886 that the French chemist Henri Moissan found a way to force the

Calomel is the common name for mercury(I) chloride, Hg_2Cl_2.

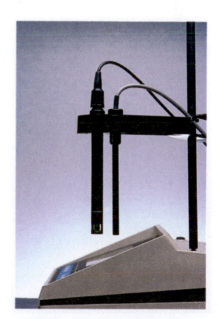

FIGURE 12.11 A glass electrode in a protective plastic sleeve (left) is used to measure pH. It is used in conjunction with a calomel electrode (right) in pH meters such as this one.

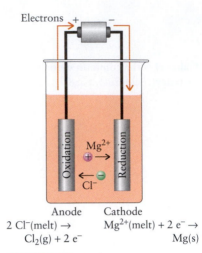

Electrons

Anode	Cathode
$2\,Cl^-(melt) \rightarrow$	$Mg^{2+}(melt) + 2\,e^- \rightarrow$
$Cl_2(g) + 2\,e^-$	$Mg(s)$

FIGURE 12.12 A schematic representation of the electrolytic cell used in the Dow process for magnesium. The electrolyte is molten magnesium chloride. As the current generated by the external source passes through the cell, magnesium ions are reduced to magnesium metal at the cathode and chloride ions are oxidized to chlorine gas at the anode.

The anode of an electrolytic cell is labeled + and the cathode −, the opposite of a galvanic cell.

formation of fluorine by passing an electric current through an anhydrous molten mixture of potassium fluoride and hydrogen fluoride. Fluorine is still prepared commercially by the same process.

In this part of Chapter 12, we study **electrolysis**, the process of driving a reaction in a nonspontaneous direction by using an electric current. First, we see how electrochemical cells are constructed for electrolysis and how to predict the potential needed to bring electrolysis about. Then, we examine the products of electrolysis and see how to predict the amount of products to expect for a given flow of electric current.

12.11 Electrolysis

An electrochemical cell in which electrolysis takes place is called an **electrolytic cell**. The arrangement of components in electrolytic cells is different from that in galvanic cells. Typically, the two electrodes share the same compartment, there is only one electrolyte, and concentrations and pressures are far from standard. As in all electrochemical cells, the current is carried through the electrolyte by the ions present. For example, when copper metal is refined electrolytically, the anode is impure copper, the cathode is pure copper, and the electrolyte is an aqueous solution of $CuSO_4$. As the Cu^{2+} ions in solution are reduced and deposited as Cu atoms at the cathode, more Cu^{2+} ions migrate toward the cathode to take their place, and in turn their concentration is restored by Cu^{2+} produced by oxidation of copper metal at the anode.

Figure 12.12 shows schematically the layout of an electrolytic cell used for the commercial production of magnesium metal from molten magnesium chloride (the *Dow process*). As in a galvanic cell, oxidation takes place at the anode and reduction takes place at the cathode, electrons travel through the external wire from anode to cathode, cations move through the electrolyte toward the cathode, and anions move toward the anode. But unlike the spontaneously generated current in a galvanic cell, a current must be supplied by an external electrical power source for reaction to occur. That power source can be a galvanic cell, which provides the current that drives electrons through the wire in a predetermined direction. The result is to force oxidation at one electrode and reduction at the other. For example, the following half-reactions are made to take place in the Dow process:

Anode reaction: $2\,Cl^-(melt) \longrightarrow Cl_2(g) + 2\,e^-$

Cathode reaction: $Mg^{2+}(melt) + 2\,e^- \longrightarrow Mg(l)$

where "melt" signifies the molten salt. A rechargeable battery functions as a galvanic cell when it is doing work and as an electrolytic cell when it is being recharged.

To drive a reaction in a nonspontaneous direction, the external supply must generate a potential difference greater than the potential difference that would be produced by the reverse reaction. For example,

$2\,H_2(g) + O_2(g) \longrightarrow 2\,H_2O(l)$ $E = +1.23$ V at pH = 7, spontaneous

To achieve the nonspontaneous reaction

$2\,H_2O(l) \longrightarrow 2\,H_2(g) + O_2(g)$ $E = -1.23$ V at pH = 7, nonspontaneous

we must apply at least 1.23 V from the external source to overcome the reaction's natural pushing power in the opposite direction. In practice, the applied potential difference must usually be substantially greater than the cell emf to reverse a spontaneous cell reaction and achieve a significant rate of product formation. The additional potential difference, which varies with the type of electrode, is called the **overpotential**. For platinum electrodes, the overpotential for the production of water from hydrogen and oxygen is about 0.6 V; so about 1.8 V (0.6

+1.23 V) is actually required to electrolyze water when platinum electrodes are used. Much contemporary research on electrochemical cells involves attempts to reduce the overpotential and hence to increase the efficiency of electrolytic processes.

When carrying out an electrolysis in solution, we must consider the possibility that other species present might be oxidized or reduced by the electric current. For example, suppose that we want to electrolyze water to produce hydrogen and oxygen. Because pure water does not carry a current, we must add an ionic solute with ions that are less easily oxidized or reduced than water:

$$O_2(g) + 4\,H^+(aq) + 4\,e^- \longrightarrow 2\,H_2O(l) \qquad E = +0.82\ \text{V at pH} = 7$$

To reverse this half-reaction and bring about the oxidation of water, we need an applied potential difference of at least 0.82 V. Suppose the added salt is sodium chloride. When Cl^- ions are present at $1\ \text{mol·L}^{-1}$ in water, is it possible that they, and not the water, will be oxidized? From Table 12.1, the standard potential for the reduction of chlorine is +1.36 V:

$$Cl_2(g) + 2\,e^- \longrightarrow 2\,Cl^-(aq) \qquad E° = +1.36\ \text{V}$$

To reverse this reaction and oxidize chloride ions, we have to supply at least 1.36 V. Because only 0.82 V is needed to force the oxidation of water but 1.36 V is needed to force the oxidation of Cl^-, it appears that oxygen should be the product at the anode. However, the overpotential for oxygen production can be very high, and in practice chlorine also might be produced. At the cathode, we want the half-reaction $2\,H^+(aq) + 2\,e^- \rightarrow H_2(g)$ to take place. At pH = 7, the potential required for the reduction of hydrogen ions is 0.41 V. Hydrogen, rather than sodium metal, will be produced at the cathode, because the potential required to reduce sodium ions is significantly higher (+2.71 V).

EXAMPLE 12.11 Sample exercise: Deciding which species will be produced at an electrode

Suppose that an aqueous solution with pH = 7 and containing I^- ions at $1\ \text{mol·L}^{-1}$ is being electrolyzed. Will O_2 or I_2 be produced at the anode?

SOLUTION From Table 12.1,

$$I_2(s) + 2\,e^- \longrightarrow 2\,I^-(aq) \qquad E° = +0.54\ \text{V}$$

so we know that at least 0.54 V is needed to oxidize I^-. We have seen that at least 0.82 V is needed to oxidize water:

$$O_2(g) + 4\,H^+(aq) + 4\,e^- \longrightarrow 2\,H_2O(l) \qquad E = +0.82\ \text{V at pH} = 7$$

Therefore, provided that the overpotentials are similar, we expect I^- ions to be oxidized in preference to water.

SELF-TEST 12.13A Predict the products resulting from the electrolysis of 1 M $AgNO_3$(aq).
[*Answer:* Cathode, Ag; anode, O_2]

SELF-TEST 12.13B Predict the products resulting from the electrolysis of 1 M NaBr(aq).

The potential supplied to an electrolytic cell must be at least as great as that of the cell reaction to be reversed. If there is more than one reducible species in solution, the species with the greater potential for reduction is preferentially reduced. The same principle applies to oxidation.

12.12 The Products of Electrolysis

Now we shall see how to calculate the amount of product formed by a given amount of electricity (1). The calculation is based on observations made by Michael Faraday (Fig.12.13) and summarized—in more modern language than he used—as follows:

FIGURE 12.13 Michael Faraday (1791–1867).

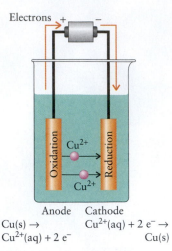

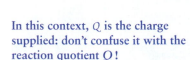

Anode   Cathode
$Cu(s) \rightarrow$ $Cu^{2+}(aq) + 2\,e^- \rightarrow$
$Cu^{2+}(aq) + 2\,e^-$ $Cu(s)$

FIGURE 12.14 A schematic representation showing the electrolytic process for refining copper. The anode is impure copper. The Cu^{2+} ions produced by oxidation of the anode migrate to the cathode, where they are reduced to pure copper metal. A similar arrangement is used for electroplating objects.

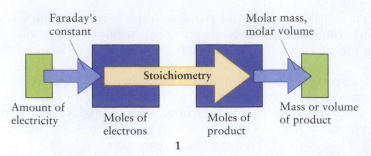

Faraday's law of electrolysis: The amount of product formed or reactant consumed by an electric current is stoichiometrically equivalent to the amount of electrons supplied.

Once we know the amount (in moles) of product formed, we can calculate the masses of the products or, if they are gases, their volumes.

EXAMPLE 12.12   Sample exercise: Calculating how much copper can be produced by electrolysis

Copper is refined electrolytically by using an impure form of copper metal called blister copper as the anode in an electrolytic cell (Fig. 12.14). The current supply drives the oxidation of the blister copper to copper(II) ions, Cu^{2+}, which are then reduced to pure copper metal at the cathode:

$$Cu^{2+}(aq) + 2\,e^- \longrightarrow Cu(s)$$

What amount of copper (in moles) can be produced by using 4.0 mol e^-?

SOLUTION

From 2 mol $e^- \approx$ 1 mol Cu,

$$\text{Amount of Cu (mol)} = (4.0 \text{ mol } e^-) \times \frac{1 \text{ mol Cu}}{2 \text{ mol } e^-}$$
$$= 2.0 \text{ mol Cu}$$

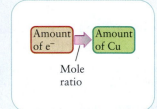

SELF-TEST 12.14A What amount (in moles) of Al(s) can be produced from Al_2O_3 if 5.0 mol e^- is supplied?

[*Answer:* 1.7 mol Al]

SELF-TEST 12.14B What amount (in moles) of Cr(s) can be produced from CrO_3 if 12.0 mol e^- is supplied?

The quantity, Q, of electricity passed through the electrolysis cell is measured in coulombs. It is determined by measuring the current, I, and the time, t, for which the current flows and is calculated from

In this context, Q is the charge supplied: don't confuse it with the reaction quotient Q!

$$\text{Charge supplied (C)} = \text{current (A)} \times \text{time (s)} \qquad \text{or} \qquad Q = It \qquad (7)$$

For example, because 1 A·s = 1 C, if 2.00 A is passed for 125 s, the charge supplied to the cell is

$$Q = (2.00 \text{ A}) \times (125 \text{ s}) = 250. \text{ A·s} = 250. \text{ C}$$

To determine the amount of electrons supplied by a given charge, we use Faraday's constant, F, the magnitude of the charge per mole of electrons (Section 12.4). Because the charge supplied is nF, where n is the amount of electrons (in moles), and $Q = nF$, it follows that

$$n = \frac{Q}{F} = \frac{It}{F} \qquad (8)^*$$

So, by measuring the current and the time for which it flows, we can determine the amount of electrons supplied. By combining the amount of electrons supplied with the mole ratio from the stoichiometry of the electrode reaction, we can deduce the amount of product obtained (see **1**).

TOOLBOX 12.4   HOW TO PREDICT THE RESULT OF ELECTROLYSIS

CONCEPTUAL BASIS

The number of electrons required to reduce a species is related to the stoichiometric coefficients in the reduction half-reaction. The same is true of oxidation. Therefore, we can set up a stoichiometric relation between the reduced or oxidized species and the amount of electrons supplied. The amount of electrons required is calculated from the current and the length of time for which the current flows.

PROCEDURE

To determine the amount of product that can be produced

Step 1 Identify the stoichiometric relation between electrons and the species of interest from the applicable half-reaction.

Step 2 Calculate the amount (in moles) of electrons supplied from Eq. 8, $n = It/F$. Use the stoichiometric relation from step 1 to convert n into the amount of

substance. If required, use the molar mass to convert into mass (or molar volume to convert into volume).

This procedure is illustrated in Example 12.13

To determine the time required for a given amount of product to be produced

Step 1 Identify the stoichiometric relation between electrons and the species of interest from the applicable half-reaction.

Step 2 If required, use the molar mass to convert mass into amount (in moles). Use the stoichiometric relation from step 1 to convert the amount of substance into the amount of electrons passed, n (in moles).

Step 3 Substitute n, the current, and Faraday's constant into Eq. 8 rearranged to $t = Fn/I$ and solve for time.

This procedure is illustrated in Example 12.14.

EXAMPLE 12.13   Sample exercise: Calculating the amount of product formed by electrolysis

Aluminum is produced by electrolysis of its oxide dissolved in molten cryolite (Na_3AlF_6). Find the mass of aluminum that can be produced in 1.00 day (d) in an electrolytic cell operating continuously at 1.00×10^5 A. The cryolite does not react.

SOLUTION   We use the first procedure in Toolbox 12.4.

Step 1 Write the half-reaction for the reduction and find the amount of electrons required to reduce 1 mol metal.

$$Al^{3+}(\text{melt}) + 3\ e^- \longrightarrow Al(l)$$

$$3\ \text{mol } e^- \simeq 1\ \text{mol Al}$$

Step 2 From $n = It/F$, using $M(Al) = 26.98\ \text{g·mol}^{-1}$, 3600 s $= 1$ h, and 24 h $= 1$ d,

$$m_{Al} = n_{e^-} \times \frac{1\ \text{mol Al}}{3\ \text{mol } e^-} \times \frac{26.98\ \text{g Al}}{1\ \text{mol Al}}$$

$$= \frac{(1.00 \times 10^5\ \text{C·s}^{-1}) \times (24.0 \times 3600\ \text{s})}{9.65 \times 10^4\ \text{C·(mol } e^-)^{-1}}$$

$$\times \frac{1\ \text{mol Al}}{3\ \text{mol } e^-} \times \frac{26.98\ \text{g Al}}{1\ \text{mol Al}}$$

$$= 8.05 \times 10^5\ \text{g Al}$$

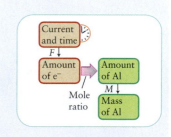

The mass produced corresponds to 805 kg. The fact that the production of 1 mol Al requires 3 mol e^- accounts for the very high consumption of electricity characteristic of aluminum-production plants.

SELF-TEST 12.15A   Determine the mass (in grams) of magnesium metal that can be obtained from molten magnesium chloride, by using a current of 7.30 A for 2.11 h. What volume of chlorine gas at 25°C and 1.00 atm will be produced at the anode?

[*Answer:* 6.98 g; 7.03 L]

SELF-TEST 12.15B   What mass of chromium metal can be obtained from a 1 M solution of CrO_3 in dilute sulfuric acid, by using a current of 6.20 A for 6.00 h?

EXAMPLE 12.14 **Sample exercise: Calculating the time required to produce a given mass of product**

How many hours are required to plate 25.00 g of copper metal from 1.00 M $CuSO_4(aq)$ by using a current of 3.00 A?

SOLUTION We use the second procedure in Toolbox 12.4.

Step 1 Find the stoichiometric relation between electrons and the species of interest from the half-reaction for the electrolysis.

$Cu^{2+}(aq) + 2\,e^- \longrightarrow Cu(s)$;
therefore, $2\text{ mol }e^- \mathrel{\hat=} 1\text{ mol Cu}$

Step 2 To find $n(e^-)$ convert grams of Cu into moles of Cu and moles of Cu into moles of e^-.

$$n_{e^-} = (25.00\text{ g Cu}) \times \left(\frac{1\text{ mol Cu}}{63.54\text{ g Cu}}\right) \times \left(\frac{2\text{ mol }e^-}{1\text{ mol Cu}}\right)$$

$$= \frac{25.00 \times 2}{63.54}\text{ mol }e^-$$

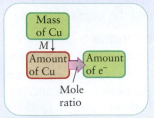

Step 3 Convert seconds to hours and use $t = Fn/I$.

$$t = \frac{9.6485 \times 10^4\text{ C}\cdot(\text{mol }e^-)^{-1}}{3.00\text{ C}\cdot s^{-1}}$$

$$\times \left(\frac{25.00 \times 2}{63.54}\text{ mol }e^-\right) \times \left(\frac{1\text{ h}}{3600\text{ s}}\right)$$

$$= 7.0\text{ h}$$

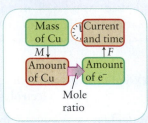

SELF-TEST 12.16A Determine the time, in hours, required to electroplate 7.00 g of magnesium metal from molten magnesium chloride, by using a current of 7.30 A.

[*Answer:* 2.12 h]

SELF-TEST 12.16B How many hours are required to plate 12.00 g of chromium metal from a 1 M solution of CrO_3 in dilute sulfuric acid, by using a current of 6.20 A?

The amount of product in an electrolysis reaction is calculated from the stoichiometry of the half-reaction and the current and time for which the current flows.

THE IMPACT ON MATERIALS

Electrochemical cells play important roles in both the purification and the preservation of metallic materials. Redox reactions are used throughout the chemical industry to extract metals from their ores. However, redox reactions also corrode the artifacts that industry produces. What redox reactions achieve, redox reactions can destroy.

12.13 Applications of Electrolysis

We have already described the refining of copper and the electrolytic extraction of aluminum, magnesium, and fluorine. Another important industrial application of electrolysis is the production of sodium metal by the *Downs process*, the electrolysis of molten rock salt (Fig. 12.15):

   Cathode reaction: $2\text{ Na}^+(\text{melt}) + 2\,e^- \longrightarrow 2\text{ Na}(l)$

   Anode reaction: $2\text{ Cl}^-(\text{melt}) \longrightarrow Cl_2(g) + 2\,e^-$

Sodium chloride is plentiful as the mineral rock salt, but the solid does not conduct electricity, because the ions are locked into place. Sodium chloride must be molten for electrolysis to occur. The electrodes in the cell are made of inert materials such as carbon, and the cell is designed to keep the sodium and chlorine produced by the electrolysis out of contact with each other and away from air. In a modification of the Downs process, the electrolyte is an aqueous solution of sodium chloride (see Section 14.7). The products of this *chloralkali process* are chlorine and aqueous sodium hydroxide.

Electroplating is the electrolytic deposition of a thin film of metal on an object. The object to be electroplated (either metal or graphite-coated plastic) constitutes the cathode, and the electrolyte is an aqueous solution of a salt of the plating metal. Metal is deposited on the cathode by reduction of ions in the electrolyte solution. These cations are supplied either by the added salt or from oxidation of the anode, which is made of the plating metal (Fig. 12.16).

Electrolysis is used industrially to produce aluminum and magnesium; to extract metals from their salts; to prepare chlorine, fluorine, and sodium hydroxide; to refine copper; and in electroplating.

12.14 Corrosion

Corrosion is the unwanted oxidation of a metal. It cuts short the lifetimes of steel products such as bridges and automobiles, and replacing corroded metal parts costs billions of dollars a year. Corrosion is an electrochemical process, and the electrochemical series is a source of insight into why corrosion occurs and how to prevent it.

The main culprit in corrosion is water. One half-reaction that we have to take into account is

$$2\,H_2O(l) + 2\,e^- \longrightarrow H_2(g) + 2\,OH^-(aq) \qquad E° = -0.83\ V$$

This standard potential is for an OH^- concentration of 1 mol·L^{-1}, which corresponds to pH = 14, a strongly basic solution. However, from the Nernst equation, we can calculate that, at pH = 7, this couple has $E = -0.42$ V. Any metal with a standard potential more negative than -0.42 V can therefore reduce water at pH = 7: that is, at this pH, any such metal can be oxidized by water. Because $E° = -0.44$ V for $Fe^{2+}(aq) + 2\,e^- \rightarrow Fe(s)$, iron has only a very slight tendency to be oxidized by water at pH = 7. For this reason, iron can be used for pipes in water supply systems and can be stored in oxygen-free water without rusting (Fig. 12.17).

When iron is exposed to damp air, with both oxygen and water present, the half-reaction

$$O_2(g) + 4\,H^+(aq) + 4\,e^- \longrightarrow 2\,H_2O(l) \qquad E° = +1.23\ V$$

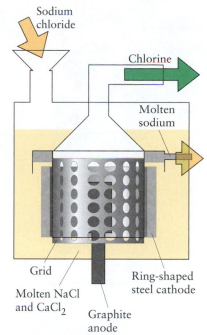

FIGURE 12.15 In the Downs process, molten sodium chloride is electrolyzed with a graphite anode (at which the Cl$^-$ ions are oxidized to chlorine) and a steel cathode (at which the Na$^+$ ions are reduced to sodium). The sodium and chlorine are kept apart by the hoods surrounding the electrodes. Calcium chloride is present to lower the melting point of sodium chloride to an economical temperature.

FIGURE 12.16 Chromium plating lends decorative flair as well as electrochemical protection to the steel of this motorcycle. Large quantities of electricity are needed for chromium plating because six electrons are required to produce each atom of chromium from chromium(VI) oxide.

FIGURE 12.17 Iron nails stored in oxygen-free water (left) do not rust, because the oxidizing power of water itself is weak. When oxygen is present (as a result of air dissolving in the water, right), oxidation is thermodynamically spontaneous and rust soon forms.

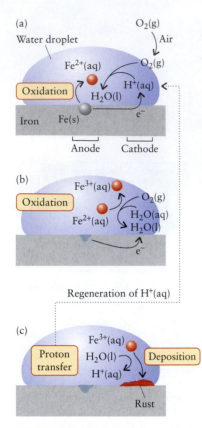

FIGURE 12.18 The mechanism of rust formation in a drop of water. (a) Oxidation of the iron occurs at a location out of contact with the oxygen of the air. The surface of the metal acts as an anode in a tiny galvanic cell, with the metal at the outer edge of the drop serving as the cathode. (b) Further oxidation of Fe^{2+} results in the formation of Fe^{3+} ions. (c) Protons are removed from H_2O as oxide ions combine with Fe^{3+} ions to deposit as rust. These protons are recycled, as indicated by the dotted line.

must be taken into account. The potential of this half-reaction at pH = 7 is +0.82 V, which lies well above the value for iron. Hence, iron can reduce oxygen in aqueous solution at pH = 7. In other words, oxygen and water can jointly oxidize iron metal to iron(II) ions. They can also subsequently work together to oxidize the iron(II) ions to iron(III), because $E° = +0.77$ V for $Fe^{3+}(aq) + e^- \rightarrow Fe^{2+}(aq)$.

Let's look in more detail at the processes. A drop of water on the surface of iron can act as the electrolyte for corrosion in a tiny electrochemical cell (Fig. 12.18). At the edge of the drop, dissolved oxygen oxidizes the iron in the process

$$2\ Fe(s) \longrightarrow 2\ Fe^{2+}(aq) + 4\ e^-$$
$$\underline{O_2(g) + 4\ H^+(aq) + 4\ e^- \longrightarrow 2\ H_2O(l)}$$
Overall: $2\ Fe(s) + O_2(g) + 4\ H^+(aq) \longrightarrow 2\ Fe^{2+}(aq) + 2\ H_2O(l)$ **(E)**

The electrons withdrawn from the metal by this oxidation can be restored from another part of the conducting metal—in particular, from iron lying beneath the oxygen-poor region in the center of the drop. The iron atoms there give up their electrons to the iron atoms at the edge of the drop, form Fe^{2+} ions, and drift away into the surrounding water. This process results in the formation of tiny pits in the surface of the iron. The Fe^{2+} ions are then oxidized further to Fe^{3+} by the dissolved oxygen:

$$2\ Fe^{2+}(aq) \longrightarrow 2\ Fe^{3+}(aq) + 2\ e^-$$
$$\underline{\tfrac{1}{2}\ O_2(g) + 2\ H^+(aq) + 2\ e^- \longrightarrow H_2O(l)}$$
Overall: $2\ Fe^{2+}(aq) + \tfrac{1}{2}\ O_2(g) + 2\ H^+(aq) \longrightarrow 2\ Fe^{3+}(aq) + H_2O(l)$ **(F)**

These ions then precipitate as a hydrated iron(III) oxide, $Fe_2O_3 \cdot H_2O$, the brown, insoluble substance that we call *rust*. The oxide ions can be regarded as coming from deprotonation of water molecules and as immediately forming the hydrated solid by precipitation with the Fe^{3+} ions produced in reaction F:

$$4\ H_2O(l) + 2\ Fe^{3+}(aq) \longrightarrow 6\ H^+(aq) + Fe_2O_3 \cdot H_2O(s)$$ **(G)**

This step restores the $H^+(aq)$ ions needed for reaction E, and so hydrogen ions function as a catalyst. The removal of Fe^{3+} ions from solution drives the reaction forward. The overall process is the sum of reactions E, F, and G:

$$2\ Fe(s) + \tfrac{3}{2}\ O_2(g) + H_2O(l) \longrightarrow Fe_2O_3 \cdot H_2O(s)$$

Water is more highly conducting when it has dissolved ions, and the formation of rust is then accelerated. That is one reason why the salt air of coastal cities and salt used for de-icing highways is so damaging to exposed metal.

Because corrosion is electrochemical, we can use our knowledge of redox reactions to combat it. The simplest way to prevent corrosion is to protect the surface of the metal from exposure to air and water by painting. A method that achieves greater protection is to **galvanize** the metal, coating it with an unbroken film of zinc (Fig. 12.19). Zinc lies below iron in the electrochemical series; so, if a scratch exposes the metal beneath, the more strongly reducing zinc releases electrons to the iron. As a result, the zinc, not the iron, is oxidized. The zinc itself survives exposure on the unbroken surface because it is **passivated**, protected from further reaction, by a protective oxide. In general, the oxide of any metal that takes up more space than does the metal that it replaces acts as a **protective oxide**, an oxide that protects the metal from further oxidation. Zinc and chromium both form low-density protective oxides that can protect iron from oxidation.

It is not possible to galvanize large metal structures, such as ships, underground pipelines, gasoline storage tanks, and bridges, but **cathodic protection** can be used. For example, a block of a more strongly reducing metal than iron, typically zinc or magnesium, can be buried in moist soil and connected to an underground pipeline (Fig. 12.20). The block of magnesium is oxidized preferentially and supplies electrons to the iron for the reduction of oxygen. The block, which is called a **sacrificial anode**, protects the pipeline and is inexpensive to replace. For similar reasons, auto-

FIGURE 12.19 Steel girders are galvanized by immersion in a bath of molten zinc.

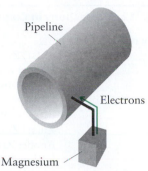

FIGURE 12.20 In the cathodic protection of a buried pipeline or other large metal construction, the artifact is connected to a number of buried blocks of metal, such as magnesium or zinc. The sacrificial anodes (the magnesium block in this illustration) supply electrons to the pipeline (the cathode of the cell), thereby preserving it from oxidation.

mobiles generally have negative ground systems as part of their electrical circuitry, which means that the body of the car is connected to the anode of the battery. The decay of the anode in the battery is the sacrifice that helps preserve the vehicle itself.

A common means of protecting the steel bodies of automobiles and trucks is through *cationic electrodeposition* coatings. In this process, coating containing a sacrificial metal is sprayed on the automobile body as a primer. The coating greatly reduces corrosion of the vehicle body. For many years lead was the only metal that could provide corrosion protection and also be applied in this manner. However, lead is toxic and concerns for environmental pollution stimulated research into alternative metals for cationic electrodeposition coatings. Eventually it was discovered that yttrium also offered corrosion resistance and, in fact, is twice as effective as lead. Yet yttrium is nontoxic and has an oxide that is a ceramic. Thus, the oxide is insoluble in water and cannot be spread through the environment through waterways.

SELF-TEST 12.17A Which of the following procedures helps to prevent the corrosion of an iron rod in water: (a) decreasing the concentration of oxygen in the water; (b) painting the rod?

[*Answer:* Both]

SELF-TEST 12.17B Which of the following elements can act as a sacrificial anode for iron: (a) copper; (b) aluminum; (c) tin?

> *The corrosion of iron is accelerated by the presence of oxygen, moisture, and salt. Corrosion can be inhibited by coating the surface with paint or zinc or by using cathodic protection.*

12.15 Practical Cells

An important application of galvanic cells is their use as the portable power sources called batteries. An ideal battery should be inexpensive, portable, safe to use, and environmentally benign. It should also maintain a potential difference that is stable with the passage of time (Table 12.2). Both the mass and the volume of a battery are critical parameters. The electrolyte in a battery uses as little water as possible, both to reduce leakage of the electrolyte and to keep the mass low. Much of the research on batteries deals with raising the *specific energy*, the Gibbs free energy of reaction per kilogram (typically expressed in kilowatt-hours per kilogram, $kW \cdot h \cdot kg^{-1}$).

A **primary cell** is a galvanic cell with the reactants sealed inside at manufacture. It cannot be recharged; when it runs down, it is discarded. A **fuel cell** is like a primary cell, but the reactants are continuously supplied (Box 12.1).

A *dry cell* is the primary cell used in most common applications, such as portable CD players, remote controls, and flashlights (Fig. 12.21). Its familiar

$1 \text{ kW} \cdot \text{h} = (10^3 \text{ J} \cdot \text{s}^{-1}) \times (3600 \text{ s}) = 3.6 \times 10^6 \text{ J} = 3.6 \text{ MJ exactly.}$

TABLE 12.2 Reactions in Commercial Batteries*

Primary cells

dry	$Zn(s) \mid ZnCl_2(aq), NH_4Cl(aq) \mid MnO(OH)(s) \mid MnO_2(s) \mid graphite, 1.5\ V$

Anode: $Zn(s) \longrightarrow Zn^{2+}(aq) + 2\ e^-$
        followed by $Zn^{2+}(aq) + 4\ NH_3(aq) \longrightarrow [Zn(NH_3)_4]^{2+}(aq)$
Cathode: $MnO_2(s) + H_2O(l) + e^- \longrightarrow MnO(OH)(s) + OH^-(aq)$
        followed by $NH_4^+(aq) + OH^-(aq) \longrightarrow H_2O(l) + NH_3(aq)$

alkaline $Zn(s) \mid ZnO(s) \mid OH^-(aq) \mid Mn(OH)_2(s) \mid MnO_2(s) \mid graphite, 1.5\ V$
Anode: $Zn(s) + 2\ OH^-(aq) \longrightarrow ZnO(s) + H_2O(l) + 2\ e^-$
Cathode: $MnO_2(s) + 2\ H_2O(l) + 2\ e^- \longrightarrow Mn(OH)_2(s) + 2\ OH^-(aq)$

silver $Zn(s) \mid ZnO(s) \mid KOH(aq) \mid Ag_2O(s) \mid Ag(s) \mid steel, 1.6\ V$
Anode: $Zn(s) + 2\ OH^-(aq) \longrightarrow ZnO(s) + H_2O(l) + 2\ e^-$
Cathode: $Ag_2O(s) + H_2O(l) + 2\ e^- \longrightarrow 2\ Ag(s) + 2\ OH^-(aq)$

Secondary cells

lead–acid $Pb(s) \mid PbSO_4(s) \mid H^+(aq), HSO_4^-(aq) \mid PbO_2(s) \mid PbSO_4(s) \mid Pb(s), 2\ V$
Anode: $Pb(s) + HSO_4^-(aq) \longrightarrow PbSO_4(s) + H^+(aq) + 2\ e^-$
Cathode: $PbO_2(s) + 3\ H^+(aq) + HSO_4^-(aq) + 2\ e^- \longrightarrow PbSO_4(s) + 2\ H_2O(l)$

nicad $Cd(s) \mid Cd(OH)_2(s) \mid KOH(aq) \mid Ni(OH)_3(s) \mid Ni(OH)_2(s) \mid Ni(s), 1.25\ V$
Anode: $Cd(s) + 2\ OH^-(aq) \longrightarrow Cd(OH)_2(s) + 2\ e^-$
Cathode: $2\ Ni(OH)_3(s) + 2\ e^- \longrightarrow 2\ Ni(OH)_2(s) + 2\ OH^-(aq)$

NiMH $M(s) \mid MH(s) \mid KOH(aq) \mid NiOOH(s) \mid Ni(OH)_2(s) \mid Ni(s), 1.2\ V$
Anode: $MH(s)^\dagger + OH^-(aq) \longrightarrow M(s) + H_2O(l) + e^-$
Cathode: $NiOOH(s) + H_2O(l) + e^- \longrightarrow Ni(OH)_2(s) + OH^-$

sodium–sulfur $Na(l) \mid Na^+(ceramic\ electrolyte), S^{2-}(ceramic\ electrolyte) \mid S_8(l), 2.2\ V$
Anode: $2\ Na(l) \longrightarrow 2\ Na^+(electrolyte) + 2\ e^-$
Cathode: $S_8(l) + 16\ e^- \longrightarrow 8\ S^{2-}(electrolyte)$

*Cell notation is described in Section 12.5.
$\dagger$The metal in a nickel−metal hydride battery is usually a complex alloy of several metals, such as Cr, Ni, Co, V, Ti, Fe, and Zr.

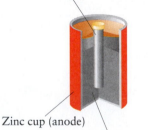

Carbon rod (cathode)

Zinc cup (anode)

MnO_2 + carbon black + NH_4Cl (electrolyte)

FIGURE 12.21 A commercial dry cell. The dry cell is also called the Leclanché cell, for Georges Leclanché, the French engineer who invented it in about 1866. The electrolyte is a moist paste.

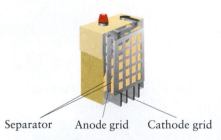

Separator   Anode grid   Cathode grid

FIGURE 12.22 A lead–acid battery consists of a number of cells in series. A series of six cells produces about 12 V.

cylindrical zinc container serves as the anode; in the center is the cathode, a carbon rod. The interior of the container is lined with paper that serves as the porous barrier. The electrolyte is a moist paste of ammonium chloride, manganese(IV) oxide, finely granulated carbon, and an inert filler, usually starch. The ammonia provided by the ammonium ions forms the complex ion $Zn(NH_3)_4^{2+}$ with the Zn^{2+} ions and prevents their buildup and a consequent reduction of the potential.

Two primary cells that provide a more stable and longer-lasting potential than the dry cell are the alkaline cell and silver cell. An *alkaline cell* is similar to a dry cell but uses an alkaline electrolyte, with which the zinc electrode does not readily react when the battery is not in use. As a result, alkaline cells have longer lives than dry cells. They are used in smoke detectors and backup power supplies. A *silver cell* has a cathode made of Ag_2O. The relatively high emf of a silver cell, with its solid reactants and products, is maintained with great reliability over long periods of time and the cell can be manufactured in very small sizes. These features make it desirable for medical implants such as pacemakers, for hearing aids, and for cameras.

Secondary cells are galvanic cells that must be charged before they can be used; this type of cell is normally rechargeable. The batteries used in portable computers and automobiles are secondary cells. In the charging process, an external source of electricity reverses the spontaneous cell reaction and creates a nonequilibrium mixture of reactants. After charging, the cell can again produce electricity.

The *lead–acid cell* of an automobile battery is a secondary cell that contains several grids that act as electrodes (Fig. 12.22). Although it has a low specific energy, because the total surface area of these grids is large, the battery can generate large currents for short periods, such as the time needed for starting an engine. The electrodes are initially a hard lead–antimony alloy covered with a paste of lead(II) sulfate. The electrolyte is dilute sulfuric acid. During the first charging, some of the lead(II) sulfate is reduced to lead on one of the electrodes; that electrode will act as the anode during discharge. Simultaneously, during charging, lead(II) sulfate is

BOX 12.1   FRONTIERS OF CHEMISTRY: FUEL CELLS

Few places are more inhospitable to life than outer space. Any habitat, including a spacecraft, needs a source of electrical power, and astronauts need enough water for drinking and washing. Because the mass of a spacecraft must be kept as low as possible, most batteries—which usually provide energy from the oxidation of a metal—would be too heavy. Electricity can be obtained from combustion reactions by burning a fuel to create heat, which runs a generator. However, producing electricity from burning fuels is very inefficient because energy is wasted as heat.

The problem was solved by Francis Bacon, a British scientist and engineer, who developed an idea proposed by Sir William Grove in 1839. A fuel cell generates electricity directly from a chemical reaction, as in a battery, but uses reactants that are supplied continuously, as in an engine. A fuel cell that runs on hydrogen and oxygen is currently installed on the space shuttle (see Fig. L.1). An advantage of this fuel cell is that the only product of the cell reaction, water, can be used for life support.

In a simple version of a fuel cell, a fuel such as hydrogen gas is passed over a platinum electrode, oxygen is passed over the other, similar electrode, and the electrolyte is aqueous potassium hydroxide. A porous membrane separates the two electrode compartments. Many varieties of fuel cells are possible, and in some the electrolyte is a solid polymer membrane or a ceramic (see Section 14.22). Three of the most promising fuel cells are the alkali fuel cell, the phosphoric acid fuel cell, and the methanol fuel cell.

The hydrogen–oxygen cell used in the space shuttle is called an alkali fuel cell, because it has an alkaline electrolyte:

Anode: $2 H_2(g) + 4 OH^-(aq) \longrightarrow 4 H_2O(l) + 4 e^-$

Electrolyte: $KOH(aq)$

Cathode: $O_2(g) + 4 e^- + 2 H_2O(l) \longrightarrow 4 OH^-(aq)$

Although its cost prohibits its use in many applications, the alkali fuel cell is the primary fuel cell used in the aerospace industry.

If an acid electrolyte is used, water is produced only at the cathode. An example is the phosphoric acid fuel cell:

Anode: $2 H_2(g) \longrightarrow 4 H^+(aq) + 4 e^-$

Electrolyte: $H_3PO_4(aq)$

Cathode: $O_2(g) + 4 H^+(aq) + 4 e^- \longrightarrow 2 H_2O(l)$

This fuel cell has shown promise for combined heat and power systems (CHP systems). In such systems, the waste heat is used to heat buildings or to do work. Efficiency in a CHP plant can reach 80%. These plants could replace heating plants and power sources in colleges and universities, hotels, and apartment buildings.

Although hydrogen gas is an attractive fuel, it has disadvantages for mobile applications: it is difficult to store and dangerous to handle. One possibility for portable fuel cells is to store the hydrogen in carbon nanotubes. Carbon nanofibers in herringbone patterns have been shown to store huge amounts of hydrogen and to result in energy densities twice that of gasoline. Another option is the use of organometallic materials or inorganic hydrides, such as sodium aluminum hydride, $NaAlH_4$, doped with titanium.

Until these materials have been developed, one attractive fuel is methanol, which is easy to handle and is rich in hydrogen atoms:

Anode: $CH_3OH(l) + 6 OH^-(aq) \longrightarrow$
$5 H_2O(l) + CO_2(g) + 6 e^-$

Electrolyte: polymeric materials

Cathode: $O_2(g) + 4 e^- + 2 H_2O(l) \longrightarrow 4 OH^-(aq)$

A disadvantage of methanol, however, is the phenomenon of "electro-osmotic drag" in which protons moving through the

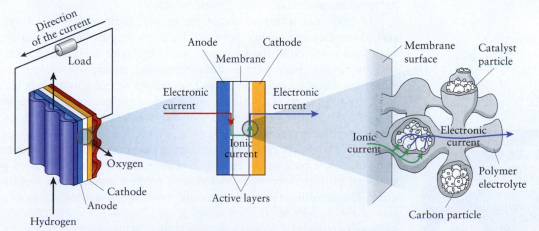

This proton exchange membrane is used in both hydrogen and methanol fuel cells, in which a catalyst at the anode produces hydrogen from the methanol. Because the membrane allows the protons, but not the electrons, to travel through it, the protons flow through the porous membrane to the cathode, where they combine with oxygen to form water, while the electrons flow through an external circuit.

(continued)

BOX 12.1   FRONTIERS OF CHEMISTRY: FUEL CELLS *(continued)*

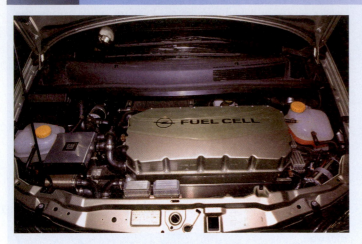

This automobile is powered by a hydrogen fuel cell with a proton exchange membrane. Its operation is pollution free, because the only product of the combustion is water.

polymer proton-exchange electrolyte membrane that separates the anode and cathode carry water and methanol with them into the cathode compartment. In the cathode, the potential is sufficient to oxidize CH_3OH to CO_2, thereby reducing the efficiency of the cell. One solution might be to find a material for the cathode that is selective to oxygen reduction so that any methanol that leaks in is not oxidized.

An exciting emerging technology is the biofuel cell. A biofuel cell is like a conventional fuel cell; however, in place of a platinum catalyst, it uses enzymes or even whole organisms. The electricity will be extracted through organic molecules that can support the transfer of electrons. One application will be as the power source for medical implants, such as pacemakers, perhaps by using the glucose present in the bloodstream as the fuel.

❓ HOW MIGHT YOU CONTRIBUTE?

Fuel cell technologies developed for aerospace have the potential to change the way we live. Automobiles powered by hydrogen fuel cells are already being manufactured (see illustration below). However, there are many obstacles that must be overcome before fuel cells reach their potential to provide us with pollution-free energy. The hydrogen fuel cells are the most attractive, because of their use of a renewable fuel. Hydrogen can be obtained from the water in the oceans. The challenge is to extract it from seawater by using solar energy and to find safe means of transportation and storage. Many practical problems with fuel cells also need to be solved, such as controlling corrosion by electrolytes and reducing the cell size and operating temperature.

Related Exercises: 12.67–12.70.

For Further Reading: Breakthrough Technologies Institute, "The Online Fuel Cell Information Center," http://www.fuelcells.org/. M. Jacoby, "Filling up with Hydrogen," *Chemical and Engineering News*, vol. 83, August 22, 2005, pp. 42–47. P. Weiss, "Pocket sockets," *Science News*, vol. 162, September 7, 2002, pp. 155–156. E. Willcocks, "Fuel cells go mobile," *Chemistry in Britain*, January 2003, p. 27.

oxidized to lead(IV) oxide on the electrode that will later act as the cathode during discharge. The lead–acid cell, which has a potential of 2 V, is used in a series of six cells to provide a 12-V source of power for starting the engine in most vehicles and as the main power source in electric vehicles.

Hybrid vehicles make use of the rechargeable *nickel–metal hydride (NiMH) cell* to supplement energy provided by burning gasoline. In this type of battery, hydrogen is stored in the form of a metal hydride, using a heterogeneous alloy of several metals, commonly including titanium, vanadium, chromium, and nickel. The advantages include low mass, high energy density, long shelf life, high current load capability, rapid charging, and good capacity (long time between charges). Because the materials are nontoxic, disposal of the batteries does not generate environmental problems.

The *lithium-ion cell* is used in laptop computers, because it can be recharged many times. This type of battery has an electrolyte consisting of polypropylene oxide or polyethylene oxide mixed with molten lithium salts and then allowed to cool. The resulting rubbery materials serve as good conductors of Li^+ ions. The low mass density of lithium gives it the highest available energy density, and lithium's very negative electrode potential provides an emf as high as 4 V.

A *sodium–sulfur cell* is one of the more startling batteries (Fig. 12.23). It has liquid reactants (sodium and sulfur) and a solid electrolyte (a porous aluminum oxide ceramic); it must operate at a temperature of about 320°C; and it is highly dangerous in case of breakage. Because sodium has a low density, these cells have a very high specific energy. Their most common application is to power electric

FIGURE 12.23 A sodium–sulfur battery used in an electric vehicle.

vehicles. Once the vehicle is operating, the heat generated by the battery is sufficient to maintain the temperature.

Practical galvanic cells can be classified as primary cells (reactants are sealed inside in a charged state), secondary cells (can be recharged), and fuel cells.

SKILLS YOU SHOULD HAVE MASTERED

❏ 1 Balance chemical equations for redox reactions by the half-reaction method (Toolbox 12.1 and Examples 12.1 and 12.2).

❏ 2 Estimate the reaction Gibbs free energy from a cell emf (Example 12.3).

❏ 3 Write the cell diagram for a redox reaction (Self-Test 12.4).

❏ 4 Write the chemical equation for a cell reaction, given the cell diagram (Toolbox 12.2 and Example 12.4).

❏ 5 Determine the standard potential of an electrode from a cell emf (Example 12.5).

❏ 6 Calculate the standard potential of a redox couple from two others relating to different oxidation states (Example 12.6).

❏ 7 Predict the spontaneous direction of a redox reaction by using the electrochemical series (Example 12.7).

❏ 8 Calculate the equilibrium constant for a reaction from the standard cell emf (Toolbox 12.3 and Example 12.8).

❏ 9 Use the Nernst equation to predict a cell emf (Examples 12.9 and 12.10).

❏ 10 Predict the likely products of electrolysis of an aqueous solution from standard potentials (Example 12.11).

❏ 11 Calculate the amount of product produced from electrolysis or the time required for electrolysis (Toolbox 12.4 and Examples 12.12, 12.13, and 12.14).

❏ 12 Describe corrosion and means for protecting iron from corrosion (Section 12.14).

❏ 13 Describe the operation of commercial practical cells (Section 12.15).

EXERCISES

Assume a temperature of 25°C (298 K) for the following exercises unless instructed otherwise.

Redox Equations

12.1 The following redox reaction is used in acidic solution in the Breathalyzer test to determine the level of alcohol in the blood:

$$H^+(aq) + Cr_2O_7^{2-}(aq) + C_2H_5OH(aq) \longrightarrow$$
$$Cr^{3+}(aq) + C_2H_4O(aq) + H_2O(l)$$

(a) Identify the elements undergoing changes in oxidation state and indicate the initial and final oxidation numbers for these elements. (b) Write and balance the oxidation half-reaction. (c) Write and balance the reduction half-reaction. (d) Combine the half-reactions to produce a balanced redox equation.

12.2 The following redox reaction between persulfate ions and chromium ions is carried out in aqueous acidic solution:

$$O_3SOOSO_3^{2-}(aq) + Cr^{3+}(aq) \longrightarrow$$
$$HSO_4^-(aq) + Cr_2O_7^{2-}(aq)$$

(a) Identify the elements undergoing changes in oxidation state and indicate the initial and final oxidation numbers for these elements. (b) Write and balance the oxidation half-reaction. (c) Write and balance the reduction half-reaction. (d) Combine the half-reactions to produce a balanced redox equation.

12.3 Balance each of the following skeletal equations by using oxidation and reduction half-reactions. All the reactions take place in acidic solution. Identify the oxidizing agent and reducing agent in each reaction.
(a) Reaction of thiosulfate ion with chlorine gas:
$Cl_2(g) + S_2O_3^{2-}(aq) \rightarrow Cl^-(aq) + SO_4^{2-}(aq)$

(b) Action of the permanganate ion on sulfurous acid:
$MnO_4^-(aq) + H_2SO_3(aq) \rightarrow Mn^{2+}(aq) + HSO_4^-(aq)$
(c) Reaction of hydrosulfuric acid with chlorine:
$H_2S(aq) + Cl_2(g) \rightarrow S(s) + Cl^-(aq)$
(d) Reaction of chlorine in water:
$Cl_2(g) \rightarrow HClO(aq) + Cl^-(aq)$

12.4 Balance each of the following skeletal equations by using oxidation and reduction half-reactions. All the reactions take place in acidic solution. Identify the oxidizing agent and reducing agent in each reaction.
(a) Conversion of iron(II) into iron(III) by dichromate ion:
$Fe^{2+}(aq) + Cr_2O_7^{2-}(aq) \rightarrow Fe^{3+}(aq) + Cr^{3+}(aq)$
(b) Formation of acetic acid from ethanol by the action of permanganate ion:
$C_2H_5OH(aq) + MnO_4^-(aq) \rightarrow Mn^{2+}(aq) + CH_3COOH(aq)$
(c) Reaction of iodide with nitric acid:
$I^-(aq) + NO_3^-(aq) \rightarrow I_2(aq) + NO(g)$
(d) Reaction of arsenic(III) sulfide with nitric acid:
$As_2S_3(s) + NO_3^-(aq) \rightarrow H_3AsO_4(aq) + S(s) + NO(g)$

12.5 Balance each of the following skeletal equations by using oxidation and reduction half-reactions. All the reactions take place in basic solution. Identify the oxidizing agent and reducing agent in each reaction.
(a) Action of ozone on bromide ions:
$O_3(aq) + Br^-(aq) \rightarrow O_2(g) + BrO_3^-(aq)$
(b) Reaction of bromine in water:
$Br_2(l) \rightarrow BrO_3^-(aq) + Br^-(aq)$
(c) Formation of chromate ions from chromium(III) ions:
$Cr^{3+}(aq) + MnO_2(s) \rightarrow Mn^{2+}(aq) + CrO_4^{2-}(aq)$
(d) Reaction of elemental phosphorus to form phosphine, PH_3, a poisonous gas with the odor of decaying fish:
$P_4(s) \rightarrow H_2PO_2^-(aq) + PH_3(g)$

12.6 Balance each of the following skeletal equations by using oxidation and reduction half-reactions. All the reactions take place in basic solution. Identify the oxidizing agent and reducing agent in each reaction.
(a) Production of chlorite ions from dichlorine heptoxide:
$Cl_2O_7(g) + H_2O_2(aq) \rightarrow ClO_2^-(aq) + O_2(g)$
(b) Action of permanganate ions on sulfide ions:
$MnO_4^-(aq) + S^{2-}(aq) \rightarrow S(s) + MnO_2(s)$
(c) Reaction of hydrazine with chlorate ions:
$N_2H_4(g) + ClO_3^-(aq) \rightarrow NO(g) + Cl^-(aq)$
(d) Reaction of plumbate ions and hypochlorite ions:
$Pb(OH)_4^{2-}(aq) + ClO^-(aq) \rightarrow PbO_2(s) + Cl^-(aq)$

12.7 The compound P_4S_3 is oxidized by nitrate ions in acid solution to give phosphoric acid, sulfate ions, and nitric oxide, NO. Write the balanced equation for each half-reaction and the overall equation for the reaction.

12.8 Iron(II) hydrogen phosphite, $FeHPO_3$, is oxidized by hypochlorite ions in basic solution. The products are chloride ion, phosphate ion, and iron(III) hydroxide. Write the balanced equation for each half-reaction and the overall equation for the reaction.

Galvanic Cells

12.9 Write the half-reactions and the balanced equation for the cell reaction for each of the following galvanic cells:
(a) $Ni(s) | Ni^{2+}(aq) || Ag^+(aq) | Ag(s)$
(b) $C(gr) | H_2(g) | H^+(aq) || Cl^-(aq) | Cl_2(g) | Pt(s)$
(c) $Cu(s) | Cu^{2+}(aq) || Ce^{4+}(aq), Ce^{3+}(aq) | Pt(s)$
(d) $Pt(s) | O_2(g) | H^+(aq)) || OH^-(aq) | O_2(g) | Pt(s)$
(e) $Pt(s) | Sn^{4+}(aq), Sn^{2+}(aq) || Cl^-(aq) | Hg_2Cl_2(s) | Hg(l)$

12.10 Write the half-reactions and the balanced equation for the cell reaction for each of the following galvanic cells:
(a) $Cu(s) | Cu^{2+}(aq) || Cu^+(aq) | Cu(s)$
(b) $Cr(s) | Cr^{2+}(aq) || Au^{3+}(aq) | Au(s)$
(c) $Ag(s) | AgI(s) | I^-(aq) || Cl^-(aq) | AgCl(s) | Ag(s)$
(d) $Hg(l) | Hg_2Cl_2(s) | Cl^-(aq) || Cl^-(aq) | AgCl(s) | Ag(s)$
(e) $Hg(l) | Hg_2^{2+}(aq) || MnO_4^-(aq), Mn^{2+}(aq), H^+(aq) | Pt(s)$

12.11 Write the half-reactions, the balanced equation for the cell reaction, and the cell diagram for each of the following skeletal equations:
(a) $Ni^{2+}(aq) + Zn(s) \rightarrow Ni(s) + Zn^{2+}(aq)$
(b) $Ce^{4+}(aq) + I^-(aq) \rightarrow I_2(s) + Ce^{3+}(aq)$
(c) $Cl_2(g) + H_2(g) \rightarrow HCl(aq)$
(d) $Au^+(aq) \rightarrow Au(s) + Au^{3+}(aq)$

12.12 Write the half-reactions, the balanced equation for the cell reaction, and the cell diagram for each of the following skeletal equations:
(a) $Mn(s) + Ti^{2+}(aq) \rightarrow Mn^{2+}(aq) + Ti(s)$
(b) $Fe^{3+}(aq) + H_2(g) \rightarrow Fe^{2+}(aq) + H^+(aq)$
(c) $Cu^+(aq) \rightarrow Cu(s) + Cu^{2+}(aq)$
(d) $MnO_4^-(aq) + H^+(aq) + Cl^-(aq) \rightarrow$
$$Cl_2(g) + Mn^{2+}(aq) + H_2O(l)$$

12.13 Write the half-reactions and devise a galvanic cell (write a cell diagram) to study each of the following reactions:
(a) $AgBr(s) \rightleftharpoons Ag^+(aq) + Br^-(aq)$, a solubility equilibrium
(b) $H^+(aq) + OH^-(aq) \rightarrow H_2O(l)$, the Brønsted neutralization reaction
(c) $Cd(s) + 2 Ni(OH)_3(s) \rightarrow Cd(OH)_2(s) + 2 Ni(OH)_2(s)$, the reaction in the nickel–cadmium cell

12.14 Write balanced half-reactions and devise a galvanic cell (write a cell diagram) to study each of the following reactions:
(a) $AgNO_3(aq) + KI(aq) \rightarrow AgI(s) + KNO_3(aq)$, a precipitation reaction
(b) $H_3O^+(aq, concentrated) \rightarrow H_3O^+(aq, dilute)$
(c) $Zn(s) + Ag_2O(s) \rightarrow ZnO(s) + 2 Ag(s)$, the reaction in a silver cell

12.15 (a) Write balanced reduction half-reactions for the redox reaction of an acidified solution of potassium permanganate and iron(II) chloride. (b) Write the balanced equation for the cell reaction and devise a galvanic cell to study the reaction (write its cell diagram).

12.16 (a) Write balanced reduction half-reactions for the redox reaction between sodium dichromate and mercury(I) nitrate in an acidic solution. (b) Write the balanced equation for the cell reaction and devise a galvanic cell to study the reaction (write its cell diagram).

12.17 Predict the standard emf of each of the following galvanic cells:
(a) $Pt(s) | Cr^{3+}(aq), Cr^{2+}(aq) || Cu^{2+}(aq) | Cu(s)$
(b) $Ag(s) | AgI(s) | I^-(aq) || Cl^-(aq) | AgCl(s) | Ag(s)$
(c) $Hg(l) | Hg_2Cl_2(s) | Cl^-(aq) || Hg_2^{2+}(aq) | Hg(l)$
(d) $C(gr) | Sn^{4+}(aq), Sn^{2+}(aq) || Pb^{4+}(aq), Pb^{2+}(aq) | Pt(s)$

12.18 Predict the standard emf of each of the following galvanic cells:
(a) $Pt(s) | Fe^{3+}(aq), Fe^{2+}(aq) || Ag^+(aq) | Ag(s)$
(b) $U(s) | U^{3+}(aq) || V^{2+}(aq) | V(s)$
(c) $Sn(s) | Sn^{2+}(aq) || Sn^{4+}(aq), Sn^{2+}(aq) | Pt(s)$
(d) $Cu(s) | Cu^{2+}(aq) || Au^+(aq) | Au(s)$

12.19 For each reaction that is spontaneous under standard conditions, write a cell diagram, determine the standard cell emf, and calculate $\Delta G°$ for the reaction:
(a) $2 NO_3^-(aq) + 8 H^+(aq) + 6 Hg(l) \rightarrow$
$$3 Hg_2^{2+}(aq) + 2 NO(g) + 4 H_2O(l)$$
(b) $2 Hg^{2+}(aq) + 2 Br^-(aq) \rightarrow Hg_2^{2+}(aq) + Br_2(l)$
(c) $Cr_2O_7^{2-}(aq) + 14 H^+(aq) + 6 Pu^{3+}(aq) \rightarrow$
$$6 Pu^{4+}(aq) + 2 Cr^{3+}(aq) + 7 H_2O(l)$$

12.20 Predict the standard cell emf and calculate the standard reaction Gibbs free energy for galvanic cells having the following cell reactions:
(a) $3 Zn(s) + 2 Bi^{3+}(aq) \rightarrow 3 Zn^{2+}(aq) + 2 Bi(s)$
(b) $2 H_2(g) + O_2(g) \rightarrow 2 H_2O(l)$ in acidic solution
(c) $2 H_2(g) + O_2(g) \rightarrow 2 H_2O(l)$ in basic solution
(d) $3 Au^+(aq) \rightarrow 2 Au(s) + Au^{3+}(aq)$

12.21 A student was given a standard $Cu(s) | Cu^{2+}(aq)$ half-cell and another half-cell containing an unknown metal M immersed in 1.00 M $M(NO_3)_2(aq)$. When the copper was connected as the anode at 25°C, the cell emf was found to be -0.689 V. What is the reduction potential for the unknown M^{2+}/M couple?

12.22 A student was given a standard $Fe(s) | Fe^{2+}(aq)$ half-cell and another half-cell containing an unknown metal M immersed in 1.00 M $MNO_3(aq)$. When these two half-cells were connected at 25°C, the complete cell functioned as a galvanic cell with $E = +1.24$ V. The reaction was allowed to continue overnight and the two electrodes were weighed. The iron electrode was found to be lighter and the unknown metal electrode was heavier. What is the standard potential of the unknown M^+/M couple?

12.23 Arrange the following metals in order of increasing strength as reducing agents for species in aqueous solution: (a) Cu, Zn, Cr, Fe; (b) Li, Na, K, Mg; (c) U, V, Ti, Al; (d) Ni, Sn, Au, Ag.

12.24 Arrange the following species in order of increasing strength as oxidizing agents for species in aqueous solution: (a) Co^{2+}, Cl_2, Ce^{4+}, In^{3+}; (b) NO_3^-, ClO_4^-, $HBrO$, $Cr_2O_7^{2-}$, all in acidic solution; (c) O_2, O_3, $HClO$, and $HBrO$, all in acidic solution; (d) O_2, O_3, ClO^-, BrO^-, all in basic solution.

12.25 Suppose that each of the following pairs of redox couples are joined to form a galvanic cell that generates a current under standard conditions. Identify the oxidizing agent and the reducing agent, write a cell diagram, and calculate the standard cell emf. (a) Co^{2+}/Co and Ti^{3+}/Ti^{2+}; (b) La^{3+}/La and U^{3+}/U; (c) H^+/H_2 and Fe^{3+}/Fe^{2+}; (d) $O_3/O_2,OH^-$ and Ag^+/Ag.

12.26 Suppose that each of the following pairs of redox couples are joined to form a galvanic cell that generates a current under standard conditions. Identify the oxidizing agent and the reducing agent, write a cell diagram, and calculate the standard cell emf. (a) Pt^{2+}/Pt and $AgF/Ag,F^-$; (b) Cr^{3+}/Cr^{2+} and I_3^-/I^-; (c) H^+/H_2 and Ni^{2+}/Ni; (d) $O_3/O_2,OH^-$ and $O_3,H^+/O_2$.

12.27 Identify the reactions with $K > 1$ in the following list and, for each such reaction, identify the oxidizing agent and calculate the standard cell emf.
(a) $Cl_2(g) + 2 Br^-(aq) \rightarrow 2 Cl^-(aq) + Br_2(l)$
(b) $MnO_4^-(aq) + 8 H^+(aq) + 5 Ce^{3+}(aq) \rightarrow$
$\qquad 5 Ce^{4+}(aq) + Mn^{2+}(aq) + 4 H_2O(l)$
(c) $2 Pb^{2+}(aq) \rightarrow Pb(s) + Pb^{4+}(aq)$
(d) $2 NO_3^-(aq) + 4 H^+(aq) + Zn(s) \rightarrow$
$\qquad Zn^{2+}(aq) + 2 NO_2(g) + 2 H_2O(l)$

12.28 Identify the reactions with $K > 1$ among the following reactions and, for each such reaction, write balanced reduction and oxidation half-reactions. For those reactions, show that $K > 1$ by calculating the standard Gibbs free energy of the reaction. Use the smallest whole-number coefficients to balance the equations.
(a) $Mg^{2+}(aq) + Cu(s) \rightarrow ?$
(b) $Al(s) + Pb^{2+}(aq) \rightarrow ?$
(c) $Hg_2^{2+}(aq) + Ce^{3+}(aq) \rightarrow ?$
(d) $Zn(s) + Sn^{2+}(aq) \rightarrow ?$
(e) $O_2(g) + H^+(aq) + Hg(l) \rightarrow ?$

12.29 (a) Using data available in Appendix 2B, write the disproportionation reaction for $Au^+(aq)$. (b) From the appropriate standard potentials, determine whether Au^+ will disproportionate spontaneously in aqueous solution under standard conditions.

12.30 (a) Using data available in Appendix 2B, write the disproportionation reaction for Sn^{2+}. (b) From the appropriate standard potentials, determine whether Sn^{2+} will spontaneously disproportionate in aqueous solution under standard conditions.

12.31 Using data available in Appendix 2B, calculate the standard potential for the half-reaction $U^{4+}(aq) + 4 e^- \rightarrow U(s)$.

12.32 Using data available in Appendix 2B, calculate the standard potential for the half-reaction $Ti^{3+}(aq) + 3 e^- \rightarrow Ti(s)$.

12.33 Determine the equilibrium constants for the following reactions:
(a) $Mn(s) + Ti^{2+}(aq) \rightleftharpoons Mn^{2+}(aq) + Ti(s)$
(b) $In^{3+}(aq) + U^{3+}(aq) \rightleftharpoons In^{2+}(aq) + U^{4+}(aq)$

12.34 Determine the equilibrium constants for the following reactions:
(a) $2 Fe^{3+}(aq) + H_2(g) \rightleftharpoons 2 Fe^{2+}(aq) + 2 H^+(aq)$
(b) $Cr(s) + Zn^{2+}(aq) \rightleftharpoons Cr^{2+}(aq) + Zn(s)$

12.35 Calculate the reaction quotient, Q, for the cell reaction, given the measured values of the cell emf. Balance the chemical equations by using the smallest whole-number coefficients.
(a) $Pt(s)|Sn^{4+}(aq),Sn^{2+}(aq)\|Pb^{4+}(aq),Pb^{2+}(aq)|C(gr)$, $E = +1.33$ V.
(b) $Pt(s)|O_2(g)|H^+(aq)\|Cr_2O_7^{2-}(aq),H^+(aq),Cr^{3+}(aq)|Pt(s)$, $E = +0.10$ V.

12.36 Calculate the reaction quotient, Q, for the cell reaction, given the measured values of the cell emf. Balance the chemical equations by using the smallest whole-number coefficients.
(a) $Ag(s)|Ag^+(aq)\|ClO_4^-(aq),H^+(aq),ClO_3^-(aq)|Pt(s)$, $E = +0.40$ V.
(b) $C(gr)|Cl_2(g)|Cl^-(aq)\|Au^{3+}(aq)|Au(s)$, $E = 0.00$ V.

12.37 Calculate E for each of the following concentration cells:
(a) $Cu(s)|Cu^{2+}(aq, 0.0010\ mol\cdot L^{-1})\|$
$Cu^{2+}(aq, 0.010\ mol\cdot L^{-1})|Cu(s)$
(b) $Pt(s)|H_2(g, 1\ bar)|H^+(aq, pH = 4.0)\|H^+(aq, pH = 3.0)|H_2(g, 1\ bar)|Pt(s)$

12.38 Determine the unknown concentration of the ion in each of the following cells:
(a) $Pb(s)|Pb^{2+}(aq, ?)\|Pb^{2+}(aq, 0.10\ mol\cdot L^{-1})|Pb(s)$, $E = +0.050$ V.
(b) $Pt(s)|Fe^{3+}(aq, 0.10\ mol\cdot L^{-1}),Fe^{2+}(aq, 1.0\ mol\cdot L^{-1})\|$
$Fe^{3+}(aq, ?),Fe^{2+}(aq, 0.0010\ mol\cdot L^{-1})|Pt(s)$, $E = +0.10$ V.

12.39 Determine the emf of each of the following cells:
(a) $Pt(s)|H_2(g, 1.0\ bar)|HCl(aq, 0.075\ M)\|HCl(aq, 1.0\ mol\cdot L^{-1})|H_2(g, 1.0\ bar)|Pt(s)$
(b) $Zn(s)|Zn^{2+}(aq, 0.37\ mol\cdot L^{-1})\|Ni^{2+}(aq, 0.059\ mol\cdot L^{-1})|Ni(s)$
(c) $Pt(s)|Cl_2(g, 250\ Torr)|HCl(aq, 1.0\ M)\|HCl(aq, 0.85\ M)|H_2(g, 125\ Torr)|Pt(s)$
(d) $Sn(s)|Sn^{2+}(aq, 0.277\ mol\cdot L^{-1})\|Sn^{4+}(aq, 0.867\ mol\cdot L^{-1}),Sn^{2+}(aq, 0.55\ mol\cdot L^{-1})|Pt(s)$

12.40 Determine the emf of each of the following cells:
(a) $Cr(s)|Cr^{3+}(aq, 0.37\ mol\cdot L^{-1})\|Pb^{2+}(aq, 9.5 \times 10^{-3}\ mol\cdot L^{-1})|Pb(s)$
(b) $Pt(s)|H_2(g, 2.0\ bar)|H^+(pH = 3.5)\|Cl^-(aq, 0.75\ mol\cdot L^{-1})|Hg_2Cl_2(s)|Hg(l)$
(c) $C(gr)|Sn^{4+}(aq, 0.059\ mol\cdot L^{-1}),Sn^{2+}(aq, 0.059\ mol\cdot L^{-1})\|Fe^{3+}(aq, 0.15\ mol\cdot L^{-1}),Fe^{2+}(aq, 0.15\ mol\cdot L^{-1})|Pt(s)$
(d) $Ag(s)|AgI(s)|I^-(aq, 0.025\ mol\cdot L^{-1})\|Cl^-(aq, 0.67\ mol\cdot L^{-1})|AgCl(s)|Ag(s)$

12.41 Determine the unknown quantity in each of the following cells:
(a) $Pt(s)|H_2(g, 1.0\ bar)|H^+(pH = ?)\|Cl^-(aq, 1.0\ mol\cdot L^{-1})|Hg_2Cl_2(s)|Hg(l)$, $E = +0.33$ V.
(b) $C(gr)|Cl_2(g, 1.0\ bar)|Cl^-; (aq, ?)\|MnO_4^-(aq, 0.010\ mol\cdot L^{-1}),H^+(pH = 4.0),Mn^{2+}(aq, 0.10\ mol\cdot L^{-1})|Pt(s)$, $E = -0.30$ V.

12.42 Determine the unknown quantity in each of the following cells:
(a) $Pt(s)|H_2(g, 1.0 \text{ bar})|H^+(pH = ?)||Cl^-(aq,$
$1.0 \text{ mol}\cdot L^{-1})|AgCl(s)|Ag(s)$, $E = +0.30$ V.
(b) $Pb(s)|Pb^{2+}(aq, ?)||Ni^{2+}(aq, 0.10 \text{ mol}\cdot L^{-1})|Ni(s)$,
$E = +0.040$ V.

12.43 A tin electrode in 0.015 M $Sn(NO_3)_2(aq)$ is connected to a hydrogen electrode in which the pressure of H_2 is 1.0 bar. If the cell potential is 0.061 V at 25°C, what is the pH of the electrolyte at the hydrogen electrode?

12.44 A lead electrode in 0.010 M $Pb(NO_3)_2(aq)$ is connected to a hydrogen electrode in which the pressure of H_2 is 1.0 bar. If the cell potential is 0.057 V at 25°C, what is the pH of the electrolyte at the hydrogen electrode?

12.45 (a) Using data from Appendix 2B, calculate the solubility product of Hg_2Cl_2. (b) Compare this number with the value listed in Table 11.4 and comment on the difference.

12.46 (a) The standard potential of the reduction of Ag_2CrO_4 to $Ag(s)$ and chromate ions is $+0.446$ V. Write the balanced half-reaction for the reduction of silver chromate. (b) Using the data from part (a) and Appendix 2B, calculate the solubility product of $Ag_2CrO_4(s)$.

12.47 Consider the cell $Ag(s)|Ag^+(aq, 5.0 \text{ mmol}\cdot L^{-1})||$
$Ag^+(aq, 0.15 \text{ mol}\cdot L^{-1})|Ag(s)$. Can this cell do work? If so, what is the maximum work that it can perform (per mole of Ag)?

12.48 Consider the cell $Ag(s)|Ag^+(aq, 2.1 \times 10^{-4} \text{ mol}\cdot L^{-1})||$
$Pb^{2+}(aq, 0.10 \text{ mol}\cdot L^{-1})|Pb(s)$. (a) Can this cell do work? If so, what is the maximum work that it can perform (per mole of Pb)? (b) What is the value of ΔH for the cell reaction and what is the sign of ΔS?

12.49 Suppose the reference electrode for Table 12.1 were the standard calomel electrode, $Hg_2Cl_2/Hg,Cl^-([Cl^-] = 1.00 \text{ mol}\cdot L^{-1})$ with its $E°$ set equal to 0. Under this system, what would be the potential for (a) the standard hydrogen electrode; (b) the standard Cu^{2+}/Cu redox couple?

12.50 Consider the question posed in Exercise 12.49 except that a saturated calomel electrode (the solution is saturated with KCl instead of having $[Cl^-] = 1.00 \text{ mol}\cdot L^{-1}$) is used in place of the standard calomel electrode. How will this replacement change the answers given in Exercise 12.49? The solubility of KCl is 35 g/(100 mL H_2O).

Electrolysis

For the exercises in this section, base your answers on the potentials listed in Table 12.1 or Appendix 2B, with the exception of the reduction and oxidation of water at pH = 7:

$$2 H_2O(l) + 2 e^- \longrightarrow H_2(g) + 2 OH^-(aq)$$
$$E = -0.42 \text{ V at pH} = 7$$

$$O_2(g) + 4 H^+(aq) + 4 e^- \longrightarrow 2 H_2O(l)$$
$$E = +0.82 \text{ V at pH} = 7$$

Ignore other factors such as passivation and overpotential.

12.51 A 1.0 M $NiSO_4(aq)$ solution was electrolyzed by using inert electrodes. Write (a) the cathode reaction; (b) the anode reaction. (c) With no overpotential or passivity at the

electrodes, what is the minimum potential that must be supplied to the cell for the onset of electrolysis?

12.52 A 1.0 M KBr(aq) solution was electrolyzed by using inert electrodes. Write (a) the cathode reaction; (b) the anode reaction. (c) With no overpotential or passivity at the electrodes, what is the minimum potential that must be supplied to the cell for the onset of electrolysis?

12.53 Aqueous solutions of (a) Mn^{2+}; (b) Al^{3+}; (c) Ni^{2+}; (d) Au^{3+} with concentrations of 1.0 mol$\cdot L^{-1}$ are electrolyzed at pH = 7. Determine whether the metal ion or water will be reduced at the cathode.

12.54 The anode of an electrolytic cell was constructed from (a) Cr; (b) Pt; (c) Cu; (d) Ni. Determine whether oxidation of the electrode or of water will take place at the anode when the electrolyte consists of a 1.0 M solution of the oxidized metal ions at pH = 7.

12.55 A total charge of 4.5 kC is passed through an electrolytic cell. Determine the quantity of substance produced in each case: (a) the mass (in grams) of bismuth metal from a bismuth nitrate solution; (b) the volume (in liters at 273 K and 1.00 atm) of hydrogen gas from a sulfuric acid solution; (c) the mass of cobalt (in grams) from a cobalt(III) chloride solution.

12.56 A total charge of 96.5 kC is passed through an electrolytic cell. Determine the quantity of substance produced in each case: (a) the mass (in grams) of silver metal from a silver nitrate solution; (b) the volume (in liters at 273 K and 1.00 atm) of chlorine gas from a brine solution (concentrated aqueous sodium chloride solution); (c) the mass of copper (in grams) from a copper(II) chloride solution.

12.57 (a) How much time is required to electroplate 1.50 g of silver from a silver nitrate solution by using a current of 13.6 mA? (b) When the same current is used for the same length of time, what mass of copper can be electroplated from a copper(II) sulfate solution?

12.58 (a) When a current of 324 mA is used for 7.0 h, what volume (measured in liters at 298 K and 1.0 atm) of fluorine gas can be produced from a molten mixture of potassium and hydrogen fluorides? (b) With the same current and time period, how many liters of oxygen gas at 298 K and 1.0 atm can be produced from the electrolysis of water?

12.59 (a) What current is required to produce 2.5 g of chromium metal from chromium(VI) oxide in 12 h? (b) What current is required to produce 2.5 g of sodium metal from molten sodium chloride in the same period?

12.60 What current is required to electroplate 6.66 µg of gold in 30.0 min from a gold(III) chloride aqueous solution? (b) How much time is required to electroplate 6.66 µg of chromium from a potassium dichromate solution, by using a current of 100 mA?

12.61 When a ruthenium chloride solution was electrolyzed for 500 s with a 120-mA current, 31.0 mg of ruthenium was deposited. What is the oxidation number of ruthenium in the ruthenium chloride?

12.62 A 4.9-g sample of manganese was produced from a manganese nitrate aqueous solution when a current of 350 mA was passed for 13.7 h. What is the oxidation number of manganese in the manganese nitrate?

12.63 A current of 15.0 A electroplated 50.0 g of hafnium metal from an aqueous solution in 2.00 h. What is the oxidation number of hafnium in the solution?

12.64 A mass loss of 12.57 g occurred in 6.00 h at a titanium anode when a current of 4.70 A was used in an electrolytic cell. What is the oxidation number of the titanium in solution?

12.65 A metal forms the salt MCl_3. Electrolysis of the molten salt with a current of 0.700 A for 6.63 h produced 3.00 g of the metal. What is the molar mass of the metal?

12.66 Thomas Edison was faced with the problem of measuring the electricity that each of his customers had used. His first solution was to use a zinc "coulometer," an electrolytic cell in which the quantity of electricity is determined by measuring the mass of zinc deposited. Only some of the current used by the customer passed through the coulometer. (a) What mass of zinc would be deposited in 1 month (of 31 days) if 1.0 mA of current passed through the cell continuously? (b) An alternative solution to this problem is to collect the hydrogen produced by electrolysis and measure its volume. What volume would be collected at 298 K and 1.00 bar under the same conditions? (c) Which method would be more practical?

Impact on Materials

For Exercises 12.67–12.70, refer to Box 12.1.

12.67 The "aluminum–air fuel cell" is used as a reserve battery in remote locations. In this cell aluminum reacts with the oxygen in air in basic solution. (a) Write the oxidation and reduction half-reactions for this cell. (b) Calculate the standard cell potential. See Box 12.1.

12.68 A fuel cell in which hydrogen reacts with nitrogen instead of oxygen is proposed. (a) Write the chemical equation for the reaction in water, which produces aqueous ammonia. (b) What would be the maximum free energy output of the cell for the consumption of 28.0 kg nitrogen? (c) Is this type of fuel cell thermodynamically feasible? See Box 12.1.

12.69 The body functions as a kind of fuel cell that uses oxygen from the air to oxidize glucose:

$$C_6H_{12}O_6(aq) + 6\ O_2(g) \longrightarrow 6\ CO_2(g) + 6\ H_2O(l)$$

During normal activity, a person uses the equivalent of about 10 MJ of energy a day. Assume that this value represents ΔG, and estimate the average current through your body in the course of a day, assuming that all the energy that we use arises from the reduction of O_2 in the glucose oxidation reaction. See Box 12.1.

12.70 A photoelectrochemical cell is an electrochemical cell that uses light to carry out a chemical reaction. This type of cell is being considered for the production of hydrogen from water. The silicon electrodes in a photoelectrochemical cell react with water:

$$SiO_2(s) + 4\ H^+(aq) + 4\ e^- \longrightarrow Si(s) + 2\ H_2O(l)$$
$$E° = -0.84\ V$$

Calculate the standard cell potential for the reaction between silicon and water in a cell that also produces hydrogen from water and write the balanced equation for the cell reaction. See Box 12.1.

12.71 What is (a) the electrolyte and (b) the oxidizing agent in the mercury cell shown here? (c) Write the overall cell reaction for a mercury cell.

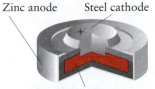

Zinc anode   Steel cathode

HgO in KOH and $Zn(OH)_2$

12.72 What is (a) the electrolyte and (b) the oxidizing agent during discharge in a lead–acid battery? (c) Write the reaction that takes place at the cathode during the charging of the lead–acid battery.

12.73 Explain how a dry cell generates electricity.

12.74 The density of the electrolyte in a lead–acid battery is measured to assess its state of charge. Explain how the density indicates the state of charge of the battery.

12.75 (a) What is the electrolyte in a nickel–cadmium cell? (b) Write the reaction that takes place at the anode when the cell is being charged.

12.76 (a) Why are lead–antimony grids used as electrodes in the lead–acid battery rather than smooth plates? (b) What is the reducing agent in the lead–acid battery? (c) The lead–acid cell potential is about 2 V. How, then, does a car battery produce 12 V for its electrical system?

12.77 A chromium-plated steel bicycle handlebar is scratched. Will rusting of the iron in the steel be encouraged or retarded by the chromium?

12.78 A solution is prepared by dissolving 1 mol each of $Cu(NO_3)_2$, $Ni(NO_3)_2$, and $AgNO_3$ in 1.0 L of water. Using only data from Appendix 2B, identify the metals (if any) that when added to these solutions (a) will leave the Ni^{2+} ions unaffected but will cause Cu and Ag to plate out of solution; (b) will leave the Ni^{2+} and Cu^{2+} ions in solution but will cause Ag to plate out of solution; (c) will leave all three metals ions in solution; (d) will leave Ni^{2+} and Ag^+ ions in solution but will cause Cu to plate out of solution.

12.79 Suppose that 2.69 g of a silver salt (AgX) is dissolved in 550 mL of water. With a current of 3.5 A, 395.0 s was needed to plate out all the silver. (a) What is the mass percentage of silver in the salt? (b) What is the formula of the salt?

12.80 Three electrolytic cells containing solutions of $CuNO_3$, $Sn(NO_3)_2$, and $Fe(NO_3)_3$, respectively, are connected in series. A current of 3.5 A is passed through the cells until 3.05 g of copper has been deposited in the first cell. (a) What masses of tin and iron are deposited? (b) For how long did the current flow?

12.81 (a) What is the approximate chemical formula of rust? (b) What is the oxidizing agent in the formation of rust? (c) How does the presence of salt accelerate the rusting process?

12.82 (a) What is the electrolyte solution in the formation of rust? (b) How are steel (iron) objects protected by galvanizing and by sacrificial anodes? (c) Suggest two metals that could be used in place of zinc for galvanizing iron.

12.83 (a) Suggest two metals that could be used for the cathodic protection of a titanium pipeline. (b) What factors other than relative positions in the electrochemical series need to be considered in practice? (c) Often copper piping is connected to iron pipes in household plumbing systems. What is a possible effect of the copper on the iron pipes?

12.84 (a) Can aluminum be used for the cathodic protection of a steel underground storage container? (b) Which of the metals zinc, silver, copper, and magnesium cannot be used as a sacrificial anode in the protection of a buried iron pipeline? Explain your answer. (c) What is the electrolyte solution for the cathodic protection of an underground pipeline by a sacrificial anode?

Integrated Exercises

12.85 Volta discovered that when he used different metals in his "pile" some combinations had a stronger effect than others. From that information he constructed an electromotive series. How would Volta have ordered the following metals, if he put the most strongly reducing metal first: Fe, Ag, Au, Zn, Cu, Ni, Co, Al?

12.86 One stage in the extraction of gold from rocks involves dissolving the metal from the rock with a basic solution of sodium cyanide that has been thoroughly aerated. This stage results in the formation of soluble $Au(CN)_2^-$ ions. The next stage is to reduce gold to the metal by the addition of zinc dust, forming $Zn(CN)_4^{2-}$. Write the balanced equations for the half-reactions and the overall redox equation for both stages.

12.87 A galvanic cell has the following cell reaction: $M(s) + 2 Zn^{2+}(aq) \rightarrow 2 Zn(s) + M^{4+}(aq)$. The standard emf of the cell is $+0.16$ V. What is the standard potential of the M^{4+}/M redox couple?

12.88 Using data in Appendix 2B, calculate the standard potential for the half-reaction $Ti^{4+}(aq) + 4 e^- \rightarrow Ti(s)$.

12.89 K_{sp} for $Cu(IO_3)_2$ is 1.4×10^{-7}. Using this value and data available in Appendix 2B, calculate $E°$ for the half-reaction $Cu(IO_3)_2(s) + 2 e^- \rightarrow Cu(s) + 2 IO_3^-(aq)$.

12.90 K_{sp} for $Ni(OH)_2$ is 6.5×10^{-18}. Using this value and data from Appendix 2B, calculate $E°$ for the half-reaction $Ni(OH)_2(s) + 2 e^- \rightarrow Ni(s) + 2 OH^-(aq)$.

12.91 A voltaic cell functions only when the electrical circuit is complete. In the external circuit the current is carried by the flow of electrons through a metal wire. Explain how the current is carried through the cell itself.

12.92 A technical handbook contains tables of thermodynamic quantities for common reactions. If you want to know whether a certain cell reaction has a positive standard emf, which of the following properties would give you that information directly (on inspection)? Which would not? Explain. (a) $\Delta G°$; (b) $\Delta H°$; (c) $\Delta S°$; (d) $\Delta U°$; (e) K.

12.93 Answer the following questions and, for each "yes" response, write a balanced cell reaction and calculate the standard cell emf. (a) Can the oxygen present in air oxidize silver metal in acidic solution? (b) Can the oxygen in air oxidize silver metal in basic solution?

12.94 What is the standard potential for the reduction of oxygen to water in (a) an acidic solution? (b) a basic solution?

(c) Is MnO_4^- more stable in an acidic or a basic aerated solution (a solution saturated with oxygen gas at 1 atm)? Explain your conclusion.

12.95 (a) If you were to construct a concentration cell in which one half-cell contains 1.0 M $CrCl_3$ and the other half-cell 0.0010 M $CrCl_3$, and both electrodes were chromium, at which electrode would reduction spontaneously take place? How will each of the following changes affect the cell potential? Justify your answers. (b) Adding 100 mL pure water to the anode compartment. (c) Adding 100 mL 1.0 M $NaOH(aq)$ to the cathode compartment ($Cr(OH)_3$ is insoluble). (d) Increasing the mass of the chromium electrode in the anode compartment.

12.96 Dental amalgam, a solid solution of silver and tin in mercury, is used for filling tooth cavities. Two of the reduction half-reactions that the filling can undergo are

$$3 Hg_2^{2+}(aq) + 4 Ag(s) + 6 e^- \longrightarrow 2 Ag_2Hg_3(s)$$
$$E° = +0.85 \text{ V}$$

$$Sn^{2+}(aq) + 3 Ag(s) + 2 e^- \longrightarrow Ag_3Sn(s)$$
$$E° = -0.05 \text{ V}$$

Suggest a reason why, when you accidentally bite on a piece of aluminum foil with a tooth containing a silver filling, you may feel pain. Write a balanced chemical equation to support your suggestion.

12.97 Suppose that 25.0 mL of a solution of Ag^+ ions of unknown concentration is titrated with 0.015 M $KI(aq)$ at 25°C. A silver electrode is immersed in this solution, and its potential is measured relative to a standard hydrogen electrode. A total of 16.7 mL of $KI(aq)$ was required to reach the stoichiometric point, when the potential was 0.325 V. (a) What is the molar concentration of Ag^+ in the solution? (b) Determine K_{sp} for AgI from these data.

12.98 Suppose that 35.0 mL of 0.012 M $Cu^+(aq)$ is titrated with 0.010 M $KBr(aq)$ at 25°C. A copper electrode is immersed in this solution, and its potential is measured relative to a standard hydrogen electrode. What volume of the KBr solution must be added to reach the stoichiometric point and what will the potential be at that point? $K_{sp}(CuBr) = 5.2 \times 10^{-9}$.

12.99 Use the data in Appendix 2B and the fact that, for the half-reaction $F_2(g) + 2 H^+(aq) + 2 e^- \rightarrow 2 HF(aq)$, $E° = +3.03$ V, to calculate the value of K_a for HF.

12.100 The following items are obtained from the stockroom for the construction of a galvanic cell: two 250-mL beakers and a salt bridge, a voltmeter with attached wires and clips, 200 mL of 0.0080 M $CrCl_3(aq)$, 200 mL of 0.12 M $CuSO_4(aq)$, a piece of copper wire, and a chrome-plated piece of metal. (a) Describe the construction of the galvanic cell. (b) Write the reduction half-reactions. (c) Write the overall cell reaction. (d) Write the cell diagram for the galvanic cell. (e) What is the expected cell emf?

12.101 (a) Considering the dependence of the Gibbs free energy of reaction on emf and on temperature, derive an equation for the temperature dependence of $E°$. (b) Use your equation to predict the standard emf for the formation of water from hydrogen and oxygen in a fuel cell at 80°C. Assume that $\Delta H°$ and $\Delta S°$ are independent of Temperature.

12.102 The standard cell potential ($E°$) for the reaction below is +0.63 V. What is the cell potential for this reaction at 275 K when $[Zn^{2+}] = 2.0 \times 10^{-4}$ mol·L^{-1} and $[Pb^{2+}] = 1.0$ mol·L^{-1}? (See Exercise 12.101.)

$$Pb^{2+}(aq) + Zn(s) \longrightarrow Zn^{2+}(aq) + Pb(s)$$

12.103 In a neuron (a nerve cell), the concentration of K$^+$ ions inside the cell is about 20 to 30 times as great as that outside. What potential difference between the inside and the outside of the cell would you expect to measure if the difference is due only to the imbalance of potassium ions?

12.104 (a) The emf of the cell Zn(s)|Zn^{2+}(aq, ?)||Pb^{2+}(aq, 0.10 mol·L^{-1})|Pb(s) is +0.661 V. What is the molarity of Zn^{2+} ions? (b) Write an equation that gives the Zn^{2+} ion concentration as a function of emf, assuming that all other aspects of the cell remain constant.

12.105 Calculate the standard emf and the actual emf of the galvanic cell Pt(s)|Fe^{3+}(aq, 0.20 mol·L^{-1}),Fe^{2+}(aq, 0.0010 mol·L^{-1})||Ag$^+$(aq, 0.020 mol·L^{-1})|Ag(s). Explain the difference in values.

12.106 Electrochemistry is one of the main methods used to determine equilibrium constants that are either very large or very small. To measure the equilibrium constant for the reaction of Fe(CN)$_6$$^{4-}$ with Na$_2$Cr$_2$O$_7$, the following cell was built:

$$Pt | Fe(CN)_6^{3-}(aq, 0.00010 \text{ mol·L}^{-1}), Fe(CN)_6^{4-}(aq,$$
$$1.0 \text{ mol·L}^{-1}) \| H^+(aq, 1.0 \text{ mol·L}^{-1}), Cr_2O_7^{2-}(aq,$$
$$0.1 \text{ mol·L}^{-1}), Cr^{3+}(aq, 0.010 \text{ mol·L}^{-1}) | Pt$$
$$E = 1.18 \text{ V at 298 K}$$

Write the cell reaction and determine its equilibrium constant.

12.107 The reduction of ClO$_4$$^-$ can be conducted in either basic or acidic solutions. The two half-reactions are

(1, acidic) $ClO_4^-(aq) + 2 H^+(aq) + 2 e^- \longrightarrow$
$ClO_3^-(aq) + H_2O(l)$ $E° = +1.23$ V

(2, basic) $ClO_4^-(aq) + H_2O(l) + 2 e^- \longrightarrow$
$ClO_3^-(aq) + 2 OH^-(l)$ $E° = +0.36$ V

(a) Show how these processes are related by deriving an expression that gives the pH dependence of the emf for each half-reaction. (b) What is the potential of each reaction in neutral solution?

12.108 Calculate the standard potential of the Cu$^+$/Cu couple from those of the Cu^{2+}/Cu$^+$ and Cu^{2+}/Cu couples.

12.109 When a pH meter was standardized with a boric acid–borate buffer with a pH of 9.40, the cell emf was +0.060 V. When the buffer was replaced with a solution of unknown hydronium ion concentration, the cell emf was +0.22 V. What is the pH of the solution?

12.110 Show how a silver–silver chloride electrode (silver in contact with solid AgCl and a solution of Cl$^-$ ions) and a hydrogen electrode can be used to measure (a) pH; (b) pOH.

12.111 What voltage range does a voltmeter need to have in order to measure pH in the range of 1 to 14 at 25°C if the voltage is zero when pH = 7?

12.112 The entropy change of a cell reaction can be determined from the change of the cell potential with temperature. (a) Show that $\Delta S° = nF(E_2° - E_1°)/(T_2 - T_1)$. Assume that $\Delta S°$ and $\Delta H°$ are constant over the temperature range considered. (b) Calculate $\Delta S°$ and $\Delta H°$ for the cell reaction Hg$_2$Cl$_2$(s) + H$_2$(g) → 2 Hg(l) + 2 H$^+$(aq) + 2 Cl$^-$(aq), given that $E° = +0.2699$ V at 293 K and +0.2669 V at 303 K.

12.113 A silver concentration cell is constructed with the electrolyte at both electrodes being initially 0.10 M AgNO(aq) at 25°C. The electrolyte at one electrode is diluted by a factor of 10 five times and the emf measured each time. (a) Plot the emf of this cell on a graph as a function of $\ln [Ag^+]_{anode}$. (b) Calculate the value of the slope of the line. To what term in the Nernst equation does this value correspond? Is the value you determined from the plot consistent with the value you would calculate from the values in that term? If not consistent, calculate your percentage error. (c) What is the value of the y-intercept? To what term in the Nernst equation does this value correspond?

12.114 Consider the electroplating of a metal +1 cation from a solution of unknown concentration according to the half-reaction M$^+$(aq) + e$^-$ → M(s), with a standard potential $E°$. When the half-cell is connected to an appropriate oxidation half-cell and current is passed through it, the M$^+$ cation begins plating out at E_1. To what value (E_2) must the applied potential be adjusted, relative to E_1, if 99.99% of the metal is to be removed from the solution?

12.115 Using only data given in Appendix 2B, calculate the acidity constant for HBrO.

12.116 The absolute magnitudes of the standard potentials of two metals M and X were determined to be

(1) $M^+(aq) + e^- \longrightarrow M(s)$ $|E°| = 0.25$ V

(2) $X^{2+}(aq) + 2 e^- \longrightarrow X(s)$ $|E°| = 0.65$ V

When the two electrodes are connected, current flows from M to X in the external circuit. When the electrode corresponding to half-reaction 1 is connected to the standard hydrogen electrode (SHE), current flows from M to SHE. (a) What are the signs of $E°$ of the two half-reactions? (b) What is the standard cell potential for the cell constructed from these two electrodes?

12.117 An aqueous solution of Na$_2$SO$_4$ was electrolyzed for 30.0 min; 25.0 mL of oxygen was collected at the anode over water at 22°C at a total pressure of 722 Torr. Determine the current that was used to produce the gas. See Table 8.3 for the vapor pressure of water.

12.118 A brine solution is electrolyzed by using a current of 2.0 A. How much time is required to collect 20.0 L of chlorine if the gas is collected over water at 20.0°C and the total pressure is 770. Torr? Assume that the water is already saturated with chlorine, and so no more dissolves. See Table 8.3 for the vapor pressure of water.

Chemistry Connections

12.119 Many important biological reactions involve electron transfer. Because the pH of bodily fluids is close to 7, the "biological standard potential" of an electrode, E^*, is measured at pH = 7.

(a) Calculate the biological standard potential for (i) the reduction of hydrogen ions to hydrogen gas; (ii) the reduction of nitrate ions to NO gas.

(b) Calculate the biological standard potential E^* for the reduction of the biomolecule NAD^+ to NADH in aqueous solution. The reduction half-reaction under thermodynamic standard conditions is $NAD^+(aq) + H^+(aq) + 2 e^- \rightarrow$ NADH(aq), with $E° = -0.099$ V.

(c) The pyruvate ion, $CH_3C(=O)CO_2^-$, is formed during the metabolism of glucose in the body. The ion has a chain of three carbon atoms. The central carbon atom has a double bond to a terminal oxygen atom and one of the end carbon atoms is bonded to two oxygen atoms in a carboxylate group. Draw the Lewis structure of the pyruvate ion and assign a hybridization scheme to each carbon atom.

(d) During exercise the pyruvate ion is converted to lactate ion in the body by coupling to the half reaction for NADH given in part (b). For the half-reaction pyruvate + 2 H^+ + 2 $e^- \rightarrow$ lactate, $E^* = -0.190$ V. Write the cell reaction for the spontaneous reaction that occurs between these two biological couples and calculate E^* and $E°$ for the overall reaction.

(e) Calculate the standard Gibbs free energy of reaction for the overall reaction in part (d).

(f) Calculate the equilibrium constant at 25°C for the overall reaction in part (d).

CHEMICAL KINETICS

What Are the Key Ideas? The rates of chemical reactions are described by simple expressions that enable us to predict the composition of a reaction mixture at any time; these expressions also suggest the steps by which the reaction takes place.

Why Do We Need to Know This Material? Chemical kinetics provides us with tools that we can use to study the rates of chemical reactions on both the macroscopic and the atomic levels. At the atomic level, chemical kinetics is a source of insight into the nature and mechanisms of chemical reactions. At the macroscopic level, information from chemical kinetics allows us to model complex systems, such as the processes taking place in the human body and the atmosphere. The development of catalysts, which are substances that speed up chemical reactions, is a branch of chemical kinetics crucial to the chemical industry, to the solution of major problems such as world hunger, and to the development of new fuels.

What Do We Need to Know Already? Much of this chapter stands alone, but it would be helpful to review the kinetic model of gases (Section 4.10) and equilibrium constants (Section 9.2).

Chapter 13

A lthough thermodynamics can be used to predict the direction and extent of chemical change, it does not tell us how the reaction takes place or how fast. We have seen that some spontaneous reactions—such as the decomposition of benzene into carbon and hydrogen—do not seem to proceed at all, whereas other reactions—such as proton transfer reactions—reach equilibrium very rapidly. In this chapter, we examine the intimate details of how reactions proceed, what determines their rates, and how to control those rates. The study of the rates of chemical reactions is called **chemical kinetics.** When studying thermodynamics, we consider only the initial and final states of a chemical process (its origin and destination) and ignore what happens between them (the journey itself, with all its obstacles). In chemical kinetics, we are interested only in the journey—the changes that take place in the course of reactions.

This chapter begins by explaining how the rates of reactions are determined experimentally and showing how their dependence on concentration can be summarized by succinct expressions known as *rate laws*. We then see how this information gives us insight into how reactions take place at a molecular level. Finally, we see how substances called catalysts accelerate reactions and control biological processes.

REACTION RATES

What do we mean by the "rate" of a chemical reaction? Informally, we consider a reaction to be fast if the products are formed rapidly, as occurs in a precipitation reaction or an explosion (Fig. 13.1). A reaction is slow if the products are formed over a long period of time, as happens in corrosion or the decay of organic material (Fig. 13.2). Our first task is to set up a precise, quantitative definition of the rate of a chemical reaction.

13.1 Concentration and Reaction Rate

In everyday life, a rate is defined as the change in a property divided by the time that it takes for that change to take place. For instance, the speed of an automobile, the rate of change of its position, is defined as distance traveled divided by the time taken. We get the *average speed* if we divide the length of the journey by the total time for the journey; we get the *instantaneous speed* by reading the speedometer at some point on its journey. In chemistry, we express rates in terms of

FIGURE 13.1 Reactions proceed at widely different rates. Some, such as explosions of dynamite, are very fast. Charges have been set off to demolish this old building. The chemical reaction in each explosion is over in a fraction of a second; the gases produced expand more slowly.

FIGURE 13.2 Some reactions are very slow, as in the gradual buildup of corrosion on the prow of the Titanic on the cold floor of the Atlantic Ocean.

how quickly reactants are used up or products are formed. That is, we define the **reaction rate** as the change in concentration of one of the reactants or products divided by the time interval over which the change takes place. Because the rate may change as time passes, we denote the **average reaction rate** as the change in molar concentration of a reactant R, $\Delta[R] = [R]_{t_2} - [R]_{t_1}$, divided by the time interval $\Delta t = t_2 - t_1$:

$$\text{Average rate of consumption of R} = -\frac{\Delta[R]}{\Delta t} \qquad \text{(1a)*}$$

Because reactants are used up in a reaction, the concentration of R decreases as time passes; thus, $\Delta[R]$ is negative. The minus sign in Eq. 1a is included to ensure that the rate is positive, which is the normal convention in chemical kinetics. If we follow the concentration of a product P, we express the rate as

$$\text{Average rate of formation of P} = \frac{\Delta[P]}{\Delta t} \qquad \text{(1b)*}$$

In this expression, $\Delta[P]$ is the change in molar concentration of P during the interval Δt: it is a positive quantity because products are formed as time goes on.

A note on good practice: In chemical kinetics, the square brackets denote molar concentration, with the units $\text{mol} \cdot L^{-1}$ retained.

EXAMPLE 13.1   **Sample exercise: Calculating an average reaction rate**

Suppose we are studying the reaction $2\,HI(g) \rightarrow H_2(g) + I_2(g)$ and find that, in an interval of 100. s, the concentration of HI decreases from 4.00 $\text{mmol} \cdot L^{-1}$ to 3.50 $\text{mmol} \cdot L^{-1}$. What is the average reaction rate?

SOLUTION

From Average rate of consumption of R = $-\Delta[R]/\Delta t$,

Average rate of consumption of HI

$$= -\frac{(3.50 - 4.00)\,(\text{mmol HI}) \cdot L^{-1}}{100.\ \text{s}}$$

$$= 5.0 \times 10^{-3}\,(\text{mmol HI}) \cdot L^{-1} \cdot s^{-1}$$

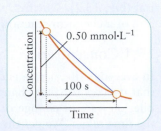

In the plot, the slope of the straight blue line gives the average rate.

A note on good practice: It is best to choose multiples of units that minimize the powers of 10 shown explicitly. In this case, because 10^{-3} mmol = 1 μmol, it would be good practice to report the rate as 5.0 (μmol HI)·L^{-1}·s^{-1}.

SELF-TEST 13.1A When the reaction 2 HI(g) → H$_2$(g) + I$_2$(g) was carried out at a high temperature, the concentration of HI decreased from 6.00 mmol·L^{-1} to 4.20 mmol·L^{-1} in 200. s. What was the average reaction rate?

[*Answer:* 9.00 (μmol HI)·L^{-1}·s^{-1}]

SELF-TEST 13.1B Hemoglobin (Hb) carries oxygen through our bodies by forming a complex with it: Hb(aq) + O$_2$(aq) → HbO$_2$(aq). In a solution of hemoglobin exposed to oxygen, the concentration of hemoglobin fell from 1.2 nmol·L^{-1} (1 nmol = 10^{-9} mol) to 0.80 nmol·L^{-1} in 0.10 μs. What was the average rate at which hemoglobin reacted with oxygen in that solution, in millimoles per liter per microsecond?

In the calculation in Example 13.1, we used the units micromoles per liter per second (μmol·L^{-1}·s^{-1}) to report the reaction rate, but other units for time (such as minutes or even hours) are commonly encountered for slower reactions. Note, too, that, when we report a reaction rate, we must specify the species to which the rate refers. For example, the rate of consumption of HI is twice the rate of formation of H$_2$ in the reaction in Example 13.1, because two HI molecules are used to make one H$_2$ molecule. The various ways of reporting the rate of a given reaction are related by the reaction stoichiometry. Therefore, in our example, we conclude that

$$-\frac{\Delta[H_2]}{\Delta t} = \frac{1}{2}\frac{\Delta[HI]}{\Delta t}$$

To avoid the ambiguity associated with several ways of reporting a reaction rate, we can report a single unique average rate of a reaction without specifying the species. The **unique average rate** of the reaction $a\,A + b\,B \to c\,C + d\,D$ is any of the following four equal quantities:

$$\text{Unique average reaction rate} = -\frac{1}{a}\frac{\Delta[A]}{\Delta t} = -\frac{1}{b}\frac{\Delta[B]}{\Delta t} = \frac{1}{c}\frac{\Delta[C]}{\Delta t} = \frac{1}{d}\frac{\Delta[D]}{\Delta t} \qquad (2)^*$$

Division by the stoichiometric coefficients takes care of the stoichiometric relations between the reactants and products. There is no need to specify the species when reporting the unique average reaction rate, because the value of the rate is the same for each species. However, the unique average rate does depend on the coefficients used in the balanced equation, and so the chemical equation should be specified when reporting the unique rate.

SELF-TEST 13.2A The average rate of the reaction N$_2$(g) + 3 H$_2$(g) → 2 NH$_3$(g) over a certain period is reported as 1.15 (mmol NH$_3$)·L^{-1}·h^{-1}. (a) What is the average rate over the same period in terms of the consumption of H$_2$? (b) What is the unique average rate?

[*Answer:* (a) 1.72 (mmol H$_2$)·L^{-1}·h^{-1}; (b) 0.575 mmol·L^{-1}·h^{-1}]

SELF-TEST 13.2B Consider the reaction in Example 13.1. What is (a) the average rate of formation of H$_2$ in the same reaction and (b) the unique average rate, both over the same period?

The experimental technique that we employ to measure a rate of reaction depends on how rapidly the reaction takes place. Some biologically important reactions, such as the growth of hair, may take weeks to show significant changes, but other reactions, such as the explosion of nitroglycerin, are very rapid. Special techniques have to be used when the reaction is so fast that it is over in seconds. Using lasers, chemists can study reactions that are complete in a picosecond (1 ps = 10^{-12} s). The newest techniques can even monitor reactions that are complete after a few femtoseconds (1 fs = 10^{-15} s, Box 13.1) and research is now starting to investigate processes even on an attosecond timescale (1 as = 10^{-18} s). On that time scale, atoms seem hardly to be moving and are caught red-handed in the act of reaction.

BOX 13.1   HOW DO WE KNOW . . . WHAT HAPPENS TO ATOMS DURING A REACTION?

The events that happen to an atom in a chemical reaction are on a time scale of approximately 1 femtosecond (1 fs = 10^{-15} s), the time that it takes for a bond to stretch or bend and, perhaps, break. If we could follow atoms on that time scale, we could make a movie of the changes in molecules as they take part in a chemical reaction. The new field of *femtochemistry*, the study of very fast chemical processes, is bringing us closer to realizing that dream. Lasers can emit very intense but short pulses of electromagnetic radiation, and so they can be used to study processes on very short time scales.

So far, the technique has been applied only to very simple reactions. For example, it is possible to watch the ion pair Na^+I^- decompose in the gas phase into separate Na and I atoms. At the start, the sodium ion and iodide ion are held together by the coulombic attraction of opposite charges. The pair is then struck by a femtosecond pulse of radiation from a laser. That pulse excites an electron from the I^- ion on to the Na^+ ion, thereby creating an NaI molecule in which the two atoms are held together by a covalent bond. The molecule has a lot of energy, and the bond length varies as the atoms swing in and out. At this point, a second femtosecond pulse is fired at the molecule. The radiation in the second pulse has a frequency that can be absorbed by the molecule only when the atoms have a particular separation. If the pulse is absorbed, we know that the atoms of the vibrating molecule have that special internuclear separation.

The illustration shows a typical result. The absorption reaches a maximum whenever the Na–I bond length returns to the length to which the second pulse is tuned. The peaks show that the sodium atom moves away from the iodine atom (corresponding to the dips in the curve), only to be recaptured

(at the maxima) again. The separation of the maxima is about 1.3 ps, and so the Na atom takes that long to swing out and be recaptured by the I atom. We can see that the peaks progressively decrease in intensity, showing that some Na atoms escape from their I atom partners on each swing. It takes about 10 swings outward before an Na atom can be sure of escaping. When sodium bromide is studied by these methods, the sodium atom escapes after about one swing, showing that an Na atom can escape more readily from a Br atom than from an I atom.

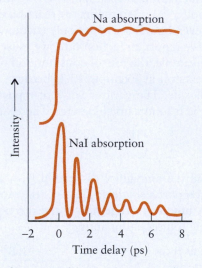

The femtosecond spectrum of the gas-phase NaI molecule as it dissociates into the separate atoms. A peak in the lower spectrum is observed whenever the bond distance in NaI reaches a certain value.

Almost all the techniques used to study fast reactions monitor concentration spectroscopically, as described in Major Technique 2, following Chapter 3. For instance, suppose we were studying the effect of a chlorofluorocarbon on the concentration of ozone, a blue gas that absorbs light of certain wavelengths. We could use a spectrometer to monitor the light absorbed by the ozone at one of those wavelengths and calculate the molar concentration of O_3 molecules from the intensity of absorption. In the **stopped-flow technique,** solutions of the reactants are forced into a mixing chamber very rapidly and the formation of products is observed spectroscopically (Fig. 13.3). This procedure is commonly used to study biologically important reactions.

> *The average rate of a reaction is the change in concentration of a species divided by the time over which the change takes place; the unique average rate is the average rate divided by the stoichiometric coefficient of the species monitored. Spectroscopic techniques are widely used to study reaction rates, particularly for fast reactions.*

13.2 The Instantaneous Rate of Reaction

Most reactions slow down as the reactants are used up. In other words, the average reaction rate decreases as the reaction proceeds. The reaction rate may also change over the time interval during which the change in concentration is being measured.

FIGURE 13.3 In a stopped-flow experiment, the driving syringes on the left push reactant solutions into the mixing chamber, and the stopping syringe on the right stops the flow. The progress of the reaction is then monitored spectroscopically as a function of time.

To determine the reaction rate at a given instant in the course of the reaction, we should make our two concentration measurements as close together in time as possible. In other words, to determine the rate at a single instant we determine the slope of the tangent to the plot of concentration against time at the time of interest (Fig. 13.4). This slope is called the **instantaneous rate** of the reaction. The instantaneous reaction rate changes in the course of the reaction (Fig. 13.5).

From now on, whenever we speak of a reaction rate, we shall always mean an instantaneous rate. The definitions in Eqs.1 and 2 can easily be adapted to refer to the instantaneous rate of a reaction.

HOW DO WE DO THAT?

To set up expressions for the instantaneous rate of a reaction, we consider Δt to be very small so that t and $t + \Delta t$ are close together; we determine the concentration of a reactant or product at those times and find the average rate from Eq. 1. Then we decrease the interval and repeat the calculation. We can imagine continuing the process until the interval Δt has become infinitely small (denoted dt) and the change in molar concentration of a reactant R has become infinitesimal (denoted $d[R]$). Then we define the instantaneous rate as

$$\text{Rate of disappearance of R} = -\frac{d[R]}{dt}$$

For a product P, we write

$$\text{Rate of formation of P} = \frac{d[P]}{dt}$$

The differential coefficients $d[R]/dt$ and $d[P]/dt$ are the mathematical expressions for the slope of the tangent drawn to a curve at the time of interest. Similarly, the unique instantaneous rate of a reaction is defined as in Eq. 2, but with differential coefficients in place of $\Delta[R]/\Delta t$ and $\Delta[P]/\Delta t$:

$$\text{Reaction rate} = -\frac{1}{a}\frac{d[A]}{dt} = -\frac{1}{b}\frac{d[B]}{dt} = \frac{1}{c}\frac{d[C]}{dt} = \frac{1}{d}\frac{d[D]}{dt}$$

Because it is difficult to draw a tangent accurately by eye, it is better to use a computer to analyze graphs of concentration against time. A superior method—which we meet in Section 13.4—is to report rates by using a procedure that, although based on these definitions, avoids the use of tangents altogether.

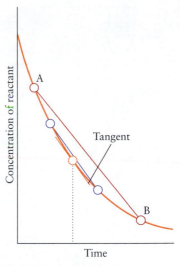

FIGURE 13.4 The rate of reaction is the change in concentration of a reactant (or product) divided by the time interval over which the change occurs (the slope of the line AB, for instance). The *instantaneous* rate is the slope of the tangent to the curve at the time of interest.

This chapter makes extensive use of expressions derived from calculus. For a discussion of derivatives and related functions, see Appendix 1F.

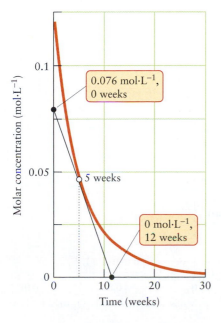

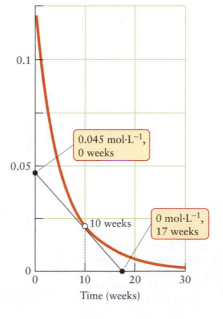

FIGURE 13.5 Determination of the rate of deterioration of penicillin during storage at two different times. Note that the rate (the slope of the tangent to the curve) at 5 weeks is greater than the rate at 10 weeks, when less penicillin is present.

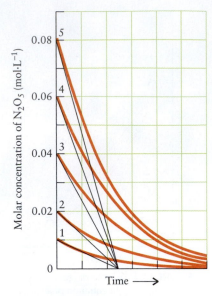

FIGURE 13.6 The definition of the initial rate of reaction. The orange curves show how the concentration of N_2O_5 changes with time for five different initial concentrations. The initial rate of consumption of N_2O_5 can be determined by drawing a tangent (black line) to each curve at the start of the reaction.

The instantaneous reaction rate is the slope of a tangent drawn to the graph of concentration as a function of time; for most reactions, the rate decreases as the reaction proceeds.

13.3 Rate Laws and Reaction Order

Patterns in reaction rate data can often be identified by examining the **initial rate** of reaction, the instantaneous rate of change in concentration of a species at the instant the reaction begins (Fig. 13.6). The advantage of examining the initial rate is that the products present later in the reaction may affect the rate; the interpretation of the rate is then quite complicated. There are no products present at the start of the reaction, and so any pattern due to the reactants is easier to find.

For example, suppose we were to measure different amounts of solid dinitrogen pentoxide, N_2O_5, into five flasks of the same volume, immerse all the flasks in a water bath at 65°C to vaporize all the solid, and then use spectrometry to monitor the concentration of reactant remaining in each flask as the N_2O_5 decomposes:

$$2\ N_2O_5(g) \longrightarrow 4\ NO_2(g) + O_2(g)$$

Each flask has a different initial concentration of N_2O_5. We determine the initial rate of reaction in each flask by plotting the concentration as a function of time for each flask and drawing the tangent to each curve at $t = 0$ (the black lines in Fig. 13.6). We find higher initial rates of decomposition of the vapor—steeper tangents—in the flasks with higher initial concentrations of N_2O_5. In fact, by plotting the initial rate against the initial concentration of N_2O_5 we find that the initial rate is proportional to that initial concentration (Fig. 13.7):

Initial rate of consumption of $N_2O_5 = k \times$ initial concentration of N_2O_5

where k is a constant, called the **rate constant** for the reaction. The experimental value of k for this reaction at 65°C, the slope of the straight line in Fig. 13.7, is $5.2 \times 10^{-3}\ s^{-1}$.

We have found that the initial rate of the reaction is proportional to the initial concentration of N_2O_5. If we were to follow the reaction rate in one of the flasks as the reaction proceeds we would also find that as the concentration of N_2O_5 falls, the rate falls too. More specifically, we would find that the rate at any instant is directly proportional to the concentration of N_2O_5 at that instant, with the same constant of proportionality, k. We therefore conclude that, at *any* stage of the reaction,

Rate of consumption of $N_2O_5 = k[N_2O_5]$

FIGURE 13.7 This graph was obtained by plotting the five initial rates from Fig. 13.6 against the initial concentration of N_2O_5. The initial rate is directly proportional to the initial concentration. This graph also illustrates how we can determine the value of the rate constant k by calculating the slope of the straight line from two points.

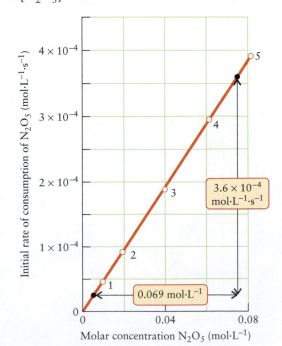

TABLE 13.1   Rate Laws and Rate Constants

Reaction	Rate law*	Temperature (K)†	Rate constant
Gas phase			
$H_2 + I_2 \longrightarrow 2\ HI$	$k[H_2][I_2]$	500	$4.3 \times 10^{-7}\ L \cdot mol^{-1} \cdot s^{-1}$
		600	4.4×10^{-4}
		700	6.3×10^{-2}
		800	2.6
$2\ HI \longrightarrow H_2 + I_2$	$k[HI]^2$	500	$6.4 \times 10^{-9}\ L \cdot mol^{-1} \cdot s^{-1}$
		600	9.7×10^{-6}
		700	1.8×10^{-3}
		800	9.7×10^{-2}
$2\ N_2O_5 \longrightarrow 4\ NO_2 + O_2$	$k[N_2O_5]$	298	$3.7 \times 10^{-5}\ s^{-1}$
		318	5.1×10^{-4}
		328	1.7×10^{-3}
		338	5.2×10^{-3}
$2\ N_2O \longrightarrow 2\ N_2 + O_2$	$k[N_2O]$	1000	$0.76\ s^{-1}$
		1050	3.4
$2\ NO_2 \longrightarrow 2\ NO + O_2$	$k[NO_2]^2$	573	$0.54\ L \cdot mol^{-1} \cdot s^{-1}$
$C_2H_6 \longrightarrow 2\ CH_3$	$k[C_2H_6]$	973	$5.5 \times 10^{-4}\ s^{-1}$
cyclopropane $\longrightarrow$ propene	$k[\text{cyclopropane}]$	773	$6.7 \times 10^{-4}\ s^{-1}$
Aqueous solution			
$H_3O^+ + OH^- \longrightarrow 2\ H_2O$	$k[H^+][OH^-]$	298	$1.5 \times 10^{11}\ L \cdot mol^{-1} \cdot s^{-1}$
$CH_3Br + OH^- \longrightarrow CH_3OH + Br^-$	$k[CH_3Br][OH^-]$	298	$2.8 \times 10^{-4}\ L \cdot mol^{-1} \cdot s^{-1}$
$C_{12}H_{22}O_{11} + H_2O \longrightarrow 2\ C_6H_{12}O_6$	$k[C_{12}H_{22}O_{11}][H^+]$	298	$1.8 \times 10^{-4}\ L \cdot mol^{-1} \cdot s^{-1}$

*For the unique instantaneous rate.
†Three significant figures.

where $[N_2O_5]$ is the molar concentration of N_2O_5. This equation is an example of a **rate law**, an expression for the instantaneous reaction rate in terms of the concentration of a species at any instant. Each reaction has its own characteristic rate law and rate constant k (Table 13.1). The rate constant is independent of the concentrations of the reactants but depends on the temperature. For the decomposition of N_2O_5 at 65°C, $k = 5.2 \times 10^{-3}\ s^{-1}$.

Other rate laws may depend differently on concentration. When we make similar measurements on the reaction

$$2\ NO_2(g) \longrightarrow 2\ NO(g) + O_2(g)$$

we do not get a straight line if we plot rate against concentration of NO_2 (Fig. 13.8a). However, if we plot the rate against the *square* of the concentration of NO_2, then we do get a straight line (Fig. 13.8b). This result shows that the rate is proportional to the square of the concentration:

Rate of consumption of $NO_2 = k[NO_2]^2$

From the slope of the straight line in Fig. 13.8b, $k = 0.54\ L \cdot mol^{-1} \cdot s^{-1}$ at 300.°C.

The rate laws for the decomposition reactions of N_2O_5 and NO_2 are different, but each has the form

Rate = constant × [concentration]^a (3)

with $a = 1$ for the decomposition of N_2O_5 and $a = 2$ for the decomposition of NO_2. The decomposition of N_2O_5 is an example of a **first-order reaction**, because its rate is proportional to the *first* power of the concentration (that is, $a = 1$). The decomposition of NO_2 is an example of a **second-order reaction**, because its rate is proportional to the *second* power of the concentration (that is, $a = 2$). Doubling

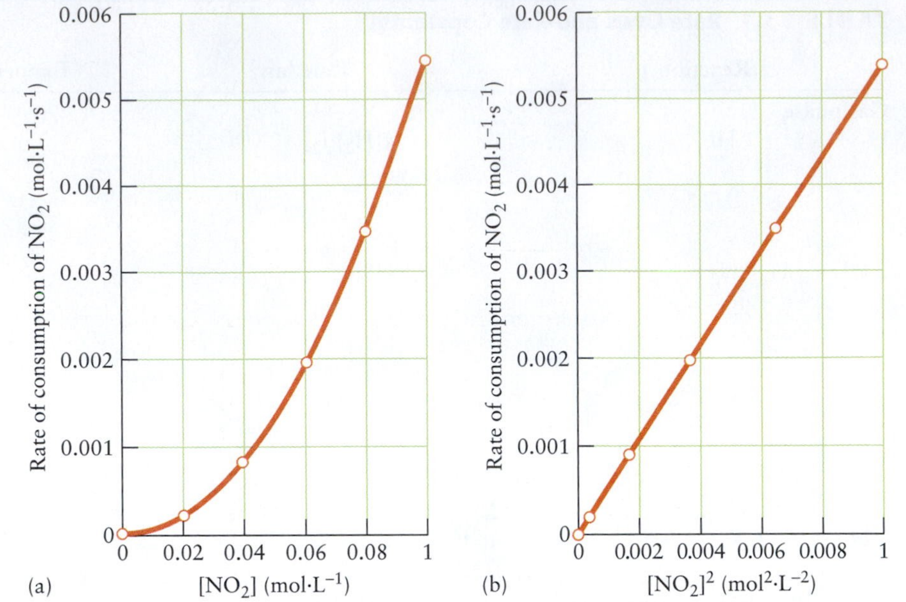

FIGURE 13.8 (a) When the rates of disappearance of NO_2 are plotted against its concentration, a straight line is not obtained. (b) However, a straight line is obtained when the rates are plotted against the square of the concentration, indicating that the rate is directly proportional to the square of the concentration.

the concentration of a reactant in a first-order reaction doubles the reaction rate. Doubling the concentration of the reactant in any second-order reaction increases the reaction rate by a factor of $2^2 = 4$.

Why do many reaction rates decrease over time? In Section 13.8 we see how rate laws provide clues to how a reaction occurs. At this point all we need to know is that reactions occur when molecules disintegrate or when reactant molecules meet. It follows that as the concentrations of reactants decrease, then fewer disintegrations occur in a given time and molecules meet less frequently. As a result, the rate of the reaction decreases.

Most of the reactions that we shall consider are either first or second order in each reactant, but some reactions have other orders (different values of a in Eq. 3). For example, ammonia decomposes into nitrogen and hydrogen on a hot platinum wire:

$$2\, NH_3(g) \longrightarrow N_2(g) + 3\, H_2(g)$$

Experiments show that the decomposition takes place at a constant rate until all the ammonia has been consumed (Fig. 13.9). Its rate law is therefore

$$\text{Rate of consumption of } NH_3 = k$$

Zero-order reactions are so called because Rate $= k \times$ (concentration)$^0 = k$.

This decomposition is an example of a **zero-order reaction,** a reaction for which the rate is independent of concentration. In this case, the rate is controlled by the surface area of the catalyst and so is constant over time.

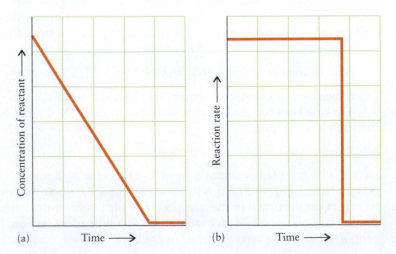

FIGURE 13.9 (a) The concentration of the reactant in a zero-order reaction falls at a constant rate until the reactant is exhausted. (b) The rate of a zero-order reaction is independent of the concentration of the reactant and remains constant until all the reactant has been consumed, when the rate falls abruptly to zero.

We can summarize the rate laws for the three most common orders of reaction as follows:

Order in A	Rate law
0	Rate = k
1	Rate = $k[A]$
2	Rate = $k[A]^2$

The order of a reaction cannot in general be predicted from the chemical equation: a rate law is an *empirical* law:

> The rate law for a reaction is experimentally determined and cannot in general be inferred from the chemical equation for the reaction.

For instance, both the decomposition of N_2O_5 and that of NO_2 have a stoichiometric coefficient of 2 for the reactant, but one reaction is first order and the other is second order.

Many reactions have rate laws that depend on the concentrations of more than one reactant. An example is the redox reaction between persulfate ions and iodide ions:

$$S_2O_8{}^{2-}(aq) + 3\,I^-(aq) \longrightarrow 2\,SO_4{}^{2-}(aq) + I_3{}^-(aq)$$

The rate law of this reaction is found to be

$$\text{Rate of consumption of } S_2O_8{}^{2-} = k[S_2O_8{}^{2-}][I^-]$$

We say that the reaction is first order with respect to $S_2O_8{}^{2-}$ (or "in" $S_2O_8{}^{2-}$) and first order in I^-. Doubling either the $S_2O_8{}^{2-}$ ion concentration or the I^- ion concentration doubles the reaction rate. Doubling both concentrations quadruples the reaction rate. We say that the *overall* order of the reaction is 2. In general, if

$$\text{Rate} = k[A]^a[B]^b\cdots \tag{4}*$$

then the **overall order** is the sum of the powers $a + b + \cdots$.

The units of k depend on the overall order of the reaction and ensure that $k \times$ (concentration)^a has the same units as the rate, namely concentration/time. Thus, when the concentration is expressed in moles per liter and the rate is expressed in $mol \cdot L^{-1} \cdot s^{-1}$, the units of k are as follows:

Overall order:	1	2	3
Units of k:	s^{-1}	$L \cdot mol^{-1} \cdot s^{-1}$	$L^2 \cdot mol^{-2} \cdot s^{-1}$

and so on. If the concentrations are expressed as partial pressures in kilopascals and the rate is reported in $kPa \cdot s^{-1}$, then the units of k are:

Overall order:	1	2	3
Units of k:	s^{-1}	$kPa^{-1} \cdot s^{-1}$	$kPa^{-2} \cdot s^{-1}$

SELF-TEST 13.3A When the NO concentration is doubled, the rate of the reaction $2\,NO(g) + O_2(g) \to 2\,NO_2(g)$ increases by a factor of 4. When both the O_2 and the NO concentrations are doubled, the rate increases by a factor of 8. What are (a) the reactant orders, (b) the overall order of the reaction, and (c) the units of k if the rate is expressed in moles per liter per second?

[*Answer:* (a) Second order in NO, first order in O_2;
(b) third order overall; (c) $L^2 \cdot mol^{-2} \cdot s^{-1}$]

SELF-TEST 13.3B When the concentration of 2-bromo-2-methylpropane, C_4H_9Br, is doubled, the rate of the reaction $C_4H_9Br(aq) + OH^-(aq) \to C_4H_9OH(aq) + Br^-(aq)$ increases by a factor of 2. When both the C_4H_9Br and the OH^- concentrations are doubled, the rate increase is the same, a factor of 2. What are (a) the reactant orders, (b) the overall order of the reaction, and (c) the units of k if the rate is expressed in moles per liter per second?

Orders can be negative, as in (concentration)$^{-1}$, corresponding to order -1. Because $[A]^{-1} = 1/[A]$, a negative order implies that the concentration appears in the denominator of the rate law. Increasing the concentration of that species, usually a product, slows down the reaction because the species participates in a reverse reaction. An example is the decomposition of ozone, O_3, in the upper atmosphere:

$$2 O_3(g) \longrightarrow 3 O_2(g)$$

The experimentally determined rate law for this reaction is

> Note that a rate law may depend on the concentrations of products as well as those of reactants.

$$\text{Rate} = k\frac{[O_3]^2}{[O_2]} = k[O_3]^2[O_2]^{-1}$$

This rate law tells us that the reaction is slower in regions of the atmosphere where O_2 molecules are abundant than where they are not so abundant. In Example 13.7 in Section 13.8, we shall use this rate law as a source of insight into how the reaction takes place.

Some reactions have fractional orders. For example, the oxidation of sulfur dioxide to sulfur trioxide in the presence of platinum,

$$2 SO_2(g) + O_2(g) \longrightarrow 3 SO_3(g)$$

is found to have the rate law

$$\text{Rate} = k\frac{[SO_2]}{[SO_3]^{1/2}} = k[SO_2][SO_3]^{-1/2}$$

and an overall order of $1 - \frac{1}{2} = \frac{1}{2}$. The presence of $[SO_3]$ in the denominator means that the reaction slows down as the concentration of product builds up.

EXAMPLE 13.2   Determining the reaction orders and rate law from experimental data

Four experiments were conducted to discover how the initial rate of consumption of BrO_3^- ions in the reaction $BrO_3^-(aq) + 5 Br^-(aq) + 6 H_3O^+(aq) \rightarrow 3 Br_2(aq) + 9 H_2O(l)$ varies as the concentrations of the reactants are changed. (a) Use the experimental data in the following table to determine the order of the reaction with respect to each reactant and the overall order. (b) Write the rate law for the reaction and determine the value of k.

Experiment	Initial concentration (mol·L^{-1})			Initial rate ((mmol BrO$_3^-$)·L^{-1}·s^{-1})
	BrO_3^-	Br^-	H_3O^+	
1	0.10	0.10	0.10	1.2
2	0.20	0.10	0.10	2.4
3	0.10	0.30	0.10	3.5
4	0.20	0.10	0.15	5.5

STRATEGY Suppose that the concentration of a substance A is increased but no other concentrations change. From the generic rate law, Rate $= k[A]^a[B]^b$, we know that as the concentration of A increases by the factor f, the rate increases by f^a. To isolate the effect of each substance, if possible, compare experiments that differ in the concentration of only one substance at a time.

SOLUTION (a) *Order in BrO_3^-:* By comparing experiments 1 and 2, we see that, when the other concentrations are held constant but the concentration of BrO_3^- is doubled ($f = 2$), the rate also doubles and so $f^a = (2)^a = 2$. Therefore, $a = 1$ and the reaction is first order in BrO_3^-.

Order in Br^-: By comparing experiments 1 and 3, we see that, when the other concentrations are held constant but the concentration of Br^- is changed by a factor of 3.0 ($f = 3.0$), the rate changes by a factor of $3.5/1.2 = 2.9$. Allowing for experimental error, we can deduce that $f^b = (3.0)^b = 3$, and so $b = 1$ and the reaction is first order in Br^-.

Order in H_3O^+: When the concentration of hydronium ions is increased from experiment 2 to experiment 4, by a factor of 1.5 ($f = 1.5$), the rate increases by a factor of $5.5/2.4 = 2.3$ when all other concentrations are held constant. Therefore, $f^c = (1.5)^c = 2.3$. To solve the relation $1.5^c = 2.3$ (and in general $f^c = x$), take logarithms of both sides:

From $f^c = x$, $c \ln f = \ln x$, and hence $c = (\ln x)/(\ln f)$,

$$c = \frac{\ln 2.3}{\ln 1.5} = 2.0$$

The reaction is second order in H_3O^+ and so fourth order overall. We can check our result by verifying that $(1.5)^2 = 2.3$.
(b) The rate law is therefore

$$\text{Rate of consumption of } BrO_3^- = k[BrO_3^-][Br^-][H_3O^+]^2$$

We find k by substituting the values from one of the experiments into the rate law and solving for k. For example, in experiment 4, we note that the rate of reaction is $5.5 \text{ mmol·L}^{-1}\text{·s}^{-1} = 5.5 \times 10^{-3} \text{ mol·L}^{-1}\text{·s}^{-1}$; then we set up the rate law and solve for k:

From $k = (\text{Rate of consumption of } BrO_3^-)/[BrO_3^-][Br^-][H_3O^+]^2$,

$$k = \frac{5.5 \times 10^{-3} \text{ mol·L}^{-1}\text{·s}^{-1}}{(0.20 \text{ mol·L}^{-1}) \times (0.10 \text{ mol·L}^{-1}) \times (0.15 \text{ mol·L}^{-1})^2}$$

$$= 12 \text{ L}^3\text{·mol}^{-3}\text{·s}^{-1}$$

The mean value of k calculated from the four experiments is $12 \text{ L}^3\text{·mol}^{-3}\text{·s}^{-1}$.

SELF-TEST 13.4A The reaction $S_2O_8^{2-}(aq) + 3 I^-(aq) \longrightarrow 2 SO_4^{2-}(aq) + I_3^-(aq)$ is commonly used to produce triiodide ion in a "clock reaction," in which a sudden color change signifies that a reagent has been used up. Write the rate law for the consumption of persulfate ions and determine the value of k, given the following data:

Experiment	Initial concentration (mol·L^{-1})		Initial rate
	$S_2O_8^{2-}$	I^-	$((\text{mol } S_2O_8^{2-})\text{·L}^{-1}\text{·s}^{-1})$
1	0.15	0.21	1.14
2	0.22	0.21	1.70
3	0.22	0.12	0.98

[*Answer:* Rate of consumption of $S_2O_8^{2-} = k[S_2O_8^{2-}][I^-]$, with the use of Experiment 1: $k = 36 \text{ L·mol}^{-1}\text{·s}^{-1}$]

SELF-TEST 13.4B The highly toxic gas carbonyl chloride, $COCl_2$ (phosgene), is used to synthesize many organic compounds. Use the following data to write the rate law and determine the value of k for the reaction used to produce carbonyl chloride: $CO(g) + Cl_2(g) \rightarrow COCl_2(g)$ at a certain temperature:

Experiment	Initial concentration (mol·L^{-1})		Initial rate $((\text{mol } COCl_2)\text{·L}^{-1}\text{·s}^{-1})$
	CO	Cl_2	
1	0.12	0.20	0.121
2	0.24	0.20	0.241
3	0.24	0.40	0.682

Hint: One of the orders is not an integer.

We can study the dependence of the rate on the concentration of one substance by using one experiment even when two or more substances are involved. To see how this is done, consider the rate law for the overall second-order oxidation of iodide ions by persulfate ions:

$$\text{Rate of consumption of } S_2O_8{}^{2-} = k[S_2O_8{}^{2-}][I^-]$$

Suppose we start with persulfate ions in such a high concentration that their concentration barely changes in the course of the reaction. For instance, the concentration of persulfate ions might be 100 times as great as the concentration of iodide ions; so, even when all the iodide ions have been oxidized, the persulfate concentration is almost the same as it was at the beginning of the reaction. Because $[S_2O_8{}^{2-}]$ is virtually constant, we can write the rate law as

$$\text{Rate of consumption of } S_2O_8{}^{2-} = k'[I^-]$$

where $k' = k[S_2O_8{}^{2-}]$, another constant. We have turned the actual second-order reaction into a **pseudo-first-order reaction,** a reaction that is effectively first order. The rate law for a pseudo-first-order reaction is much easier to analyze than the true rate law because it depends on the concentration of only one substance.

The order of a reaction is the power to which the concentration of the species is raised in the rate law; the overall order is the sum of the individual orders.

CONCENTRATION AND TIME

We often want to know how the concentration of a reactant or product varies with time. For example, how long will it take for a pollutant to decay? How much sulfur trioxide can be produced in an hour in an industrial process? How much penicillin will be left after 6 months? These questions can be answered by using formulas derived from the experimental rate law for the reactions. An **integrated rate law** gives the concentration of reactants or products at any time after the start of the reaction. Finding the integrated rate law from a rate law is very much like calculating the distance that a car will have traveled from a knowledge of its speed at each moment of the journey.

The integrated rate law for a zero-order reaction is easy to find. Because the rate is constant (at k), the difference in concentration of a reactant from its initial value, $[A]_0$, is proportional to the time for which the reaction is in progress, and we can write

$$[A]_0 - [A] = kt \quad \text{or} \quad [A] = [A]_0 - kt$$

As shown in Fig. 13.9a, a plot of concentration against time is a straight line of slope $-k$; the reaction comes to an end when $t = [A]_0/k$, because then all the reactant has been consumed ($[A] = 0$).

In the next two sections, we concentrate on first- and second-order rate laws, but the techniques can be used for other reaction orders, too.

13.4 First-Order Integrated Rate Laws

Our aim is to find an expression for the concentration of a reactant A at a time t, given that the initial molar concentration of A is $[A]_0$ and that A is consumed in the first-order reaction A → products.

HOW DO WE DO THAT?

To find the concentration of reactant A in a first-order reaction at any time after it has begun, we begin by writing the rate law for the consumption of A as

$$\text{Rate of disappearance of A} = -\frac{d[A]}{dt} = k[A]$$

Because an instantaneous rate is a derivative of concentration with respect to time, we can use the techniques of integral calculus to find the change in [A] as a function of time. First, we divide both sides by [A] and multiply through by $-dt$:

$$\frac{d[A]}{[A]} = -kdt$$

Now we integrate both sides between the limits $t = 0$ (when $[A] = [A]_0$) and the time of interest, t (when $[A] = [A]_t$):

$$\int_{[A]_0}^{[A]_t} \frac{d[A]}{[A]} = -k\int_0^t dt = -kt$$

To evaluate the integral on the left, we use the standard form

$$\int \frac{dx}{x} = \ln x + \text{constant}$$

and obtain

$$\int_{[A]_0}^{[A]_t} \frac{d[A]}{[A]} = (\ln[A]_t + \text{constant}) - (\ln[A]_0 + \text{constant})$$

$$= \ln[A]_t - \ln[A]_0 = \ln\frac{[A]_t}{[A]_0}$$

We conclude that

$$\ln\frac{[A]_t}{[A]_0} = -kt$$

When we take (natural) antilogarithms of both sides, we obtain $[A]_t/[A]_0 = e^{-kt}$, and therefore

$$[A]_t = [A]_0 e^{-kt}$$

The two equations that we have derived,

$$\ln\frac{[A]_t}{[A]_0} = -kt \qquad (5a)*$$

$$[A]_t = [A]_0 e^{-kt} \qquad (5b)*$$

are two forms of the integrated rate law for a first-order reaction. The variation of concentration with time predicted by Eq. 5b is shown in Fig. 13.10. This behavior is called **exponential decay.** Initially the concentration changes rapidly, but it changes more slowly as time goes on and the reactant is used up.

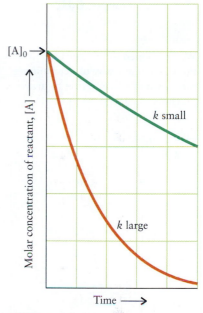

FIGURE 13.10 The graph of the concentration of a reactant in a first-order reaction is an exponential decay, as shown here. The larger the rate constant, the faster the decay from the same initial concentration.

EXAMPLE 13.3 **Sample exercise: Calculating a concentration from a first-order integrated rate law**

What concentration of N_2O_5 remains 10.0 min (600. s) after the start of its decomposition at 65°C when its initial concentration was 0.040 mol·L^{-1}? See Table 13.1 for the rate law.

SOLUTION The reaction and its rate law are

$$2\,N_2O_5(g) \longrightarrow 4\,NO_2(g) + O_2(g) \qquad \text{Rate of decomposition of } N_2O_5 = k[N_2O_5]$$
$$\text{with } k = 5.2 \times 10^{-3}\,s^{-1}.$$

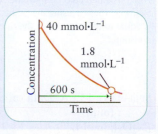

From $[A]_t = [A]_0 e^{-kt}$,
$$[N_2O_5]_t = (0.040\ \text{mol·L}^{-1}) \times e^{-(5.2 \times 10^{-3}\,s^{-1}) \times (600.\,s)}$$
$$= 0.0018\ \text{mol·L}^{-1}$$

1 Cyclopropane, C_3H_6

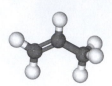

2 Propene, C_3H_6

We mentioned in Section 13.2 that there is a better way of determining rate constants than trying to draw tangents to curves: this is it.

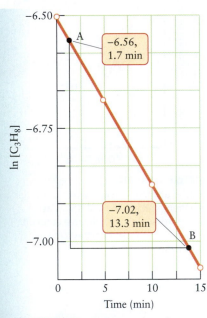

FIGURE 13.11 We can test for a first-order reaction by plotting the natural logarithm of the reactant concentration against time. The graph is linear if the reaction is first order. The slope of the line, which is calculated here for the system in Example 13.4 by using the points A and B, is equal to the negative of the rate constant.

That is, after 600. s, the concentration of N_2O_5 will have fallen from its initial value of 0.040 mol·L^{-1} to 0.0018 mol·L^{-1} (1.8 mmol·L^{-1}).

SELF-TEST 13.5A Calculate the concentration of N_2O remaining after the first-order decomposition

$$2\,N_2O(g) \longrightarrow 2\,N_2(g) + O_2(g) \qquad \text{Rate of decomposition of } N_2O = k[N_2O]$$

has continued at 780°C for 100. ms, if the initial concentration of N_2O is 0.20 mol·L^{-1} and $k = 3.4$ s^{-1}.

[*Answer:* 0.14 mol·L^{-1}]

SELF-TEST 13.5B Calculate the concentration of cyclopropane, C_3H_6 (**1**), remaining after the first-order conversion into its isomer propene (**2**):

$$C_3H_6(g) \longrightarrow CH_3{-}CH{=}CH_2(g) \qquad \text{Rate of conversion of } C_3H_6 = k[C_3H_6]$$

has continued at 773 K for 200. s, if the initial concentration of C_3H_6 is 0.100 mol·L^{-1} and $k = 6.7 \times 10^{-4}$ s^{-1}.

An important application of an integrated rate law is to confirm that a reaction is in fact first order and to measure its rate constant. From Eq. 5a, we can write

$$\ln [A]_t = \ln [A]_0 - kt \tag{6}$$

This equation has the form of an equation for a straight line (see Appendix 1E):

$$y = \text{intercept} + \text{slope} \times x$$

Therefore, if we plot $\ln [A]_t$ as a function of t, we should get a straight line with slope $-k$ and intercept $\ln [A]_0$.

EXAMPLE 13.4 **Measuring a rate constant**

When cyclopropane (C_3H_6, see **1**) is heated to 500.°C (773 K), it changes into an isomer, propene (see **2**). The following data show the concentration of cyclopropane at a series of times after the start of the reaction. Confirm that the reaction is first order in C_3H_6 and calculate the rate constant.

t (min)	0	5	10.	15
$[C_3H_6]_t$ (mol·L^{-1})	1.50×10^{-3}	1.24×10^{-3}	1.00×10^{-3}	0.83×10^{-3}

STRATEGY We need to plot the natural logarithm of the reactant concentration as a function of t. If we get a straight line, the reaction is first order and the slope of the graph is $-k$. We could use a spreadsheet program or the Living Graph "Determination of Rate Constant (first-order rate law)" on the Web site for this book to make the plot.

SOLUTION For the graphical procedure, begin by setting up the following table:

t (min)	0	5	10.	15
$\ln [C_3H_6]_t$	-6.50	-6.69	-6.91	-7.09

The points are plotted in Fig. 13.11. The graph is a straight line, confirming that the reaction is first order in cyclopropane. When we use points A and B on the plot, the slope of the line is

$$\text{Slope} = \frac{(-7.02) - (-6.56)}{(13.3 - 1.7)\,\text{min}} = -0.040\ \text{min}^{-1}$$

Therefore, because $k = -\text{slope}$, $k = 0.040$ min^{-1}. This value is equivalent to $k = 6.7 \times 10^{-4}$ s^{-1}, the value in Table 13.1.

Note on good practice: Be sure to determine and include the appropriate units for k.

SELF-TEST 13.6A Data on the decomposition of N_2O_5 at 25°C are

t (min)	0	200.	400.	600.	800.	1000.
$[N_2O_5]_t$ (mol·L^{-1})	15.0×10^{-3}	9.6×10^{-3}	6.2×10^{-3}	4.0×10^{-3}	2.5×10^{-3}	1.6×10^{-3}

Confirm that the reaction is first order and find the value of k for the reaction $N_2O_5(g) \rightarrow$ products.

[*Answer:* The plot of ln $[N_2O_5]$ against time is linear, and so the reaction is first order; $k = 2.2 \times 10^{-3}$ min^{-1}]

SELF-TEST 13.6B Azomethane, $CH_3N_2CH_3$, decomposes to ethane and nitrogen gas in the reaction $CH_3N_2CH_3(g) \rightarrow CH_3CH_3(g) + N_2(g)$. The reaction was followed at 460. K by measuring the partial pressure of azomethane over time:

t (s)	0.	1000.	2000.	3000.	4000.
$P_{CH_3N_2CH_3}$ (Torr)	8.20×10^{-2}	5.72×10^{-2}	3.99×10^{-2}	2.78×10^{-2}	1.94×10^{-2}

Confirm that the reaction is first order of the form Rate $= kP$, where P is the partial pressure of azomethane, and find the value of k.

EXAMPLE 13.5 **Predicting how long it will take for a concentration to change**

A sample of N_2O_5 is allowed to decompose by the following reaction:

$$2\,N_2O_5(g) \longrightarrow 4\,NO_2(g) + O_2(g)$$

How long will it take for the concentration of N_2O_5 to decrease from 20. mmol·L^{-1} to 2.0 mmol·L^{-1} at 65°C? Use the data in Table 13.1.

STRATEGY First, identify the order of the reaction in N_2O_5 by referring to Table 13.1. If the reaction is first order, rearrange Eq. 5a into an equation for t in terms of the given concentrations. Then substitute the numerical value of the rate constant and the concentration data and evaluate t.

SOLUTION The reaction is first order with $k = 5.2 \times 10^{-3}$ s^{-1} at 65°C (338 K). Therefore,

From ln $([A]_t/[A]_0) = -kt$ rearranged into $t = (1/k)$ ln $([A]_0/[A]_t)$,

$$t = \frac{1}{5.2 \times 10^{-3}\text{ s}}\ln\frac{20.\text{ mmol·L}^{-1}}{2.0\text{ mmol·L}^{-1}}$$

$$= 4.4 \times 10^2\text{ s}$$

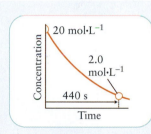

The time required is approximately 7.3 min.

SELF-TEST 13.7A How long does it take for the concentration of A to decrease to 1.0% of its initial value in a first-order reaction of the form A $\rightarrow$ products with $k = 1.0$ s^{-1}?

[*Answer:* 4.6 s]

SELF-TEST 13.7B Cyclopropane isomerizes to propene in a first-order process. How long does it take for the concentration of cyclopropane to decrease from 1.0 mol·L^{-1} to 0.0050 mol·L^{-1} at 500.°C? Use the data in Table 13.1.

In a first-order reaction, the concentration of reactant decays exponentially with time. To verify that a reaction is first order, plot the natural logarithm of its concentration as a function of time and expect a straight line; the slope of the straight line is $-k$.

13.5 Half-Lives for First-Order Reactions

The **half-life**, $t_{1/2}$, of a substance is the time needed for its concentration to fall to one-half its initial value. Knowing the half-lives of pollutants such as chlorofluorocarbons allows us to assess their environmental impact. If their half-lives are short, they may not survive long enough to reach the stratosphere, where they can destroy ozone. Half-lives are also important in planning storage systems for radioactive materials, because the decay of radioactive nuclei is a first-order process.

We already know that the higher the value of k, the more rapid the consumption of a reactant. Therefore, we should be able to deduce a relation for a first-order reaction that shows that, the greater the rate constant, the shorter the half-life.

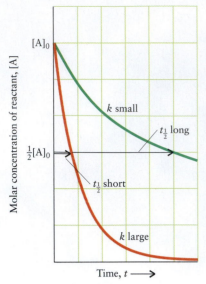

FIGURE 13.12 The change in concentration of the reactant in two first-order reactions plotted on the same graph. When the first-order rate constant is large, the half-life of the reactant is short, because the exponential decay of the concentration of the reactant is then fast.

HOW DO WE DO THAT?

To find the relation between the rate constant and the half-life, we begin by rearranging Eq. 5a into an expression for the time needed to reach a given concentration:

From $\ln ([A]_t/[A]_0) = -kt$,

$$t = \frac{1}{k} \ln \frac{[A]_0}{[A]_t}$$

Now set $t = t_{1/2}$ and $[A]_t = \frac{1}{2}[A]_0$,

$$t_{1/2} = \frac{1}{k} \ln \frac{[A]_0}{\frac{1}{2}[A]_0} = \frac{1}{k} \ln 2$$

Notice that the initial concentration has canceled.

We have shown that, for the rate law Rate of consumption of A = $k[A]$,

$$t_{1/2} = \frac{\ln 2}{k} \tag{7}*$$

with $\ln 2 = 0.693...$. As we anticipated, the greater the value of the rate constant k, the shorter the half-life for the reaction (Fig. 13.12).

EXAMPLE 13.6   Using a half-life to calculate the amount of reactant remaining

In 1989 a teenager in Ohio was poisoned by breathing vapors from spilled mercury. The mercury level in his urine, which is proportional to its concentration in his body, was found to be 1.54 mg·L^{-1}. Mercury(II) is eliminated from the body by a first-order process that has a half-life of 6 days (6 d). What would be the concentration of mercury(II) in the patient's urine in milligrams per liter after 30 d if therapeutic measures were not taken?

STRATEGY The level of mercury(II) in the urine can be predicted by using the integrated first-order rate law, Eq. 5b. To use this equation, we need the rate constant. Therefore, start by calculating the rate constant from the half-life (Eq. 7) and substitute the result into Eq. 5b.

SOLUTION

From $t_{1/2} = (\ln 2)/k$ in the form $k = (\ln 2)/t_{1/2}$,

$$k = \frac{\ln 2}{6 \text{ d}} = \frac{\ln 2}{6} \text{ d}^{-1}$$

From $[A] = [A]_0 e^{-kt}$,

$$[A]_t = [A]_0 e^{-\{(\ln 2)/6 \text{ d}^{-1}\} \times (30 \text{ d})}$$

$$= (1.54 \text{ mg·L}^{-1}) e^{-\{(\ln 2) \times 30/6\}}$$

$$= 0.05 \text{ mg·L}^{-1}$$

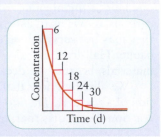

The concentration of mercury in the urine is therefore 0.05 mg·L^{-1}.

A note on good practice: Note that, because exponential functions e^x are so sensitive to the value of x, we avoid rounding errors by leaving the numerical calculation to a single final step.

SELF-TEST 13.8A In 1972 grain treated with methyl mercury was released for human consumption in Iraq, resulting in 459 deaths. The half-life of methyl mercury in body tissues is 70. d. How many days are required for the amount of methyl mercury to drop to 10.% of the original value after ingestion?

[*Answer:* 230 d]

SELF-TEST 13.8B Soil at the Rocky Flats Nuclear Processing Facility in Colorado was found to be contaminated with radioactive plutonium-239, which has a half-life of 24 ka (2.4×10^4 years). The soil was loaded into drums for storage. How many years must pass before the radioactivity drops to 20.% of its initial value?

The concentration of the reactant does not appear in Eq. 7: for a first-order reaction, the half-life is independent of the initial concentration of the reactant. That is, it is constant: regardless of the initial concentration of reactant, half the reactant will have been consumed in the time given by Eq. 7. It follows that we can take the "initial" concentration of A to be its concentration at any stage of the reaction: if at some stage the concentration of A happens to be [A], then after a further time $t_{1/2}$, the concentration of A will have fallen to $\frac{1}{2}$[A], after a further $t_{1/2}$ it will have fallen to $\frac{1}{4}$[A], and so on (Fig. 13.13). In general, the concentration remaining after n half-lives is equal to $(\frac{1}{2})^n$[A]$_0$. For example, in Example 13.6, because 30 days corresponds to 5 half-lives, after that interval [A]$_t$ = $(\frac{1}{2})^5$[A]$_0$, or [A]$_0$/32, which evaluates to 3%, the same as the result obtained in the example.

SELF-TEST 13.9A Calculate (a) the number of half-lives and (b) the time required for the concentration of N$_2$O to fall to one-eighth of its initial value in a first-order decomposition at 1000. K. Consult Table 13.1 for the rate constant.

[*Answer:* (a) 3 half-lives; (b) 2.7 s]

SELF-TEST 13.9B Calculate (a) the number of half-lives and (b) the time required for the concentration of C$_2$H$_6$ to fall to one-sixteenth of its initial value as it dissociates into CH$_3$ radicals at 973 K. Consult Table 13.1 for the rate constant.

The half-life of a first-order reaction is characteristic of the reaction and independent of the initial concentration. A reaction with a large rate constant has a short half-life.

13.6 Second-Order Integrated Rate Laws

Now we derive the integrated rate law for second-order reactions with the rate law

Rate of consumption of A = k[A]^2

HOW DO WE DO THAT?

To obtain the integrated rate law for a second-order reaction, we recognize that the rate law is a differential equation and write it as

$$-\frac{d[A]}{dt} = k[A]^2$$

After division by [A]^2 and multiplication by $-dt$, this equation becomes

$$\frac{d[A]}{[A]^2} = -k\,dt$$

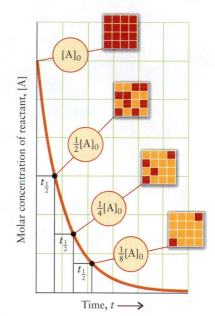

FIGURE 13.13 For first-order reactions, the half-life is the same whatever the concentration at the start of the chosen period. Therefore, it takes one half-life to fall to half the initial concentration, two half-lives to fall to one-fourth the initial concentration, three half-lives to fall to one-eighth, and so on. The boxes portray the composition of the reaction mixture at the end of each half-life; the red squares represent the reactant A and the yellow squares represent the product.

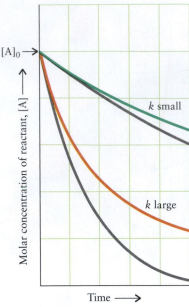

FIGURE 13.14 The characteristic shapes of the time dependence of the concentration of a reactant during a second-order reaction. The larger the rate constant, k, the greater is the dependence of the rate on the concentration of the reactant. The lower gray lines are the curves for first-order reactions with the same initial rates as for the corresponding second-order reactions. Note how the concentrations for second-order reactions fall away much less rapidly at longer times than those for first-order reactions do.

To solve this equation, we integrate it between the same limits used in the first-order case:

$$\int_{[A]_0}^{[A]_t} \frac{d[A]}{[A]^2} = -k \int_0^t dt$$

This time we need the integral

$$\int \frac{dx}{x^2} = -\frac{1}{x} + \text{constant}$$

and write the integral on the left as

$$\int_{[A]_0}^{[A]_t} \frac{d[A]}{[A]^2} = \left(-\frac{1}{[A]_t} + \text{constant}\right) - \left(-\frac{1}{[A]_0} + \text{constant}\right)$$

$$= \frac{1}{[A]_0} - \frac{1}{[A]_t}$$

Thus, we obtain

$$\frac{1}{[A]_t} - \frac{1}{[A]_0} = kt, \text{ which can be rearranged into } [A]_t = \frac{[A]_0}{1 + kt[A]_0}$$

The two equations that we have derived are

$$\frac{1}{[A]_t} - \frac{1}{[A]_0} = kt \tag{8a}*$$

$$[A]_t = \frac{[A]_0}{1 + kt[A]_0} \tag{8b}*$$

Equation 8b is plotted in Fig. 13.14. We see that the concentration of the reactant decreases rapidly at first but then changes more slowly than a first-order reaction with the same initial rate. This slowing down of second-order reactions has important environmental consequences: because many pollutants degrade by second-order reactions, they remain at low concentration in the environment for long periods. Equation 8a can be written in the form of a linear equation:

$$\overbrace{\frac{1}{[A]_t}}^{y} = \overbrace{\frac{1}{[A]_0}}^{\text{intercept}} + \overbrace{k}^{\text{slope}} t \tag{8c}*$$

Therefore, to determine whether a reaction is second order in a reactant, we plot the inverse of the concentration against the time and see whether we get a straight

Living graph: Figure 13.15 Test for a second-order reaction

FIGURE 13.15 The plots that allow the determination of reaction order. (a) If a plot of ln[A] against time is a straight line, the reaction is first order. (b) If a plot of 1/[A] against time is a straight line, the reaction is second order.

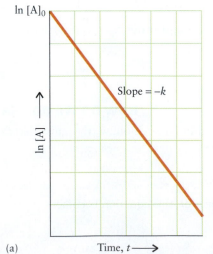

(a)

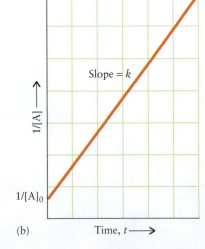

(b)

TABLE 13.2 Rate Law Summary

	Order of reaction		
	0	1	2
Rate law	Rate = k	Rate = $k[A]$	Rate = $k[A]^2$
Integrated rate law	$[A]_t = -k + [A]_0$	$[A]_t = [A]_0 e^{-kt}$	$[A]_t = \dfrac{[A]_0}{1 + [A]_0 t}$ $\dfrac{1}{[A]_t} = \dfrac{1}{[A]_0} + kt$
Plot to determine order	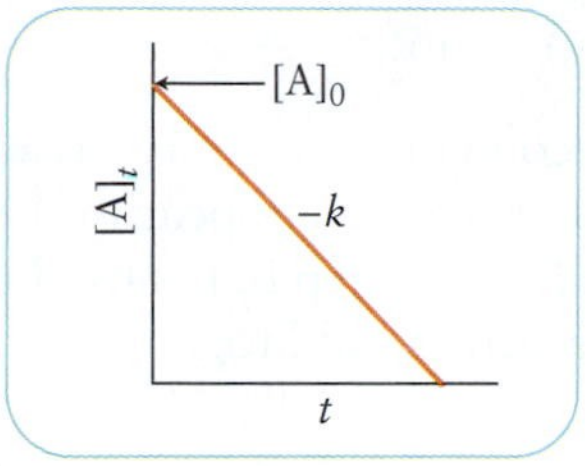	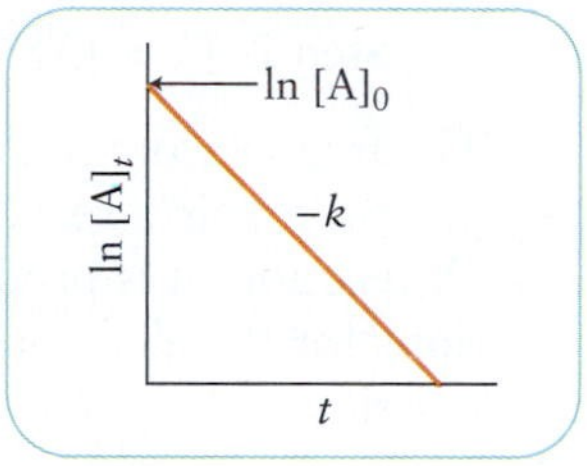	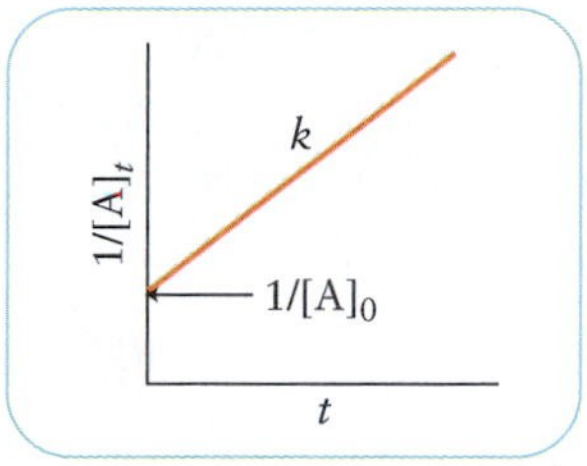
Slope of the line plotted	$-k$	$-k$	k
Half-life	$t_{1/2} = \dfrac{[A]_0}{2k}$ (not used)	$t_{1/2} = \dfrac{\ln 2}{k} \approx \dfrac{0.693}{k}$	$t_{1/2} = \dfrac{1}{k[A]_0}$ (not used)

line. If the line is straight, then the reaction is second order and the slope of the line is equal to k (Fig. 13.15).

As we have seen for first- and second-order rate laws, each integrated rate law can be rearranged into an equation that, when plotted, gives a straight line and the rate constant can then be obtained from the slope of the plot. Table 13.2 summarizes the relationships to use.

A second-order reaction has a long tail of low concentration at long reaction times. The half-life of a second-order reaction is inversely proportional to the concentration of the reactant.

REACTION MECHANISMS

The rate law for a reaction is a window into the changes that take place at the molecular level in the course of the reaction. Knowing how those changes take place provides answers to many important questions. For example, what controls the rate of formation of the DNA double helix from its individual strands? What molecular events convert ozone into oxygen or turn a mixture of fuel and air into carbon dioxide and water when it ignites in an engine?

13.7 Elementary Reactions

We stressed in Section 13.3 that we cannot in general write a rate law from a chemical equation. The reason is that all but the simplest reactions are the outcome of several, and sometimes many, steps called **elementary reactions.** Each elementary reaction describes a distinct event, often a collision of particles. To understand how a reaction takes place, we have to propose a **reaction mechanism,** a sequence of elementary reactions describing the changes that we believe take place as reactants are transformed into products.

Although several different mechanisms might be proposed for a reaction, rate measurements can be used to eliminate some of them. For example, in the decom-

FIGURE 13.16 A representation of a proposed one-step mechanism for the decomposition of ozone in the atmosphere. This reaction takes place in a single bimolecular collision.

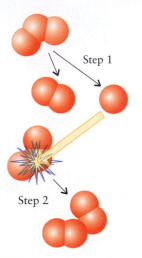

Step 1

Step 2

FIGURE 13.17 An alternative, two-step mechanism for the decomposition of ozone. In the first step, an energized ozone molecule shakes off an oxygen atom. In the second step, that oxygen atom attacks another ozone molecule.

position of ozone, $2\,O_3(g) \longrightarrow 3\,O_2(g)$, we could imagine the reaction taking place two different ways:

One-step mechanism: Two O_3 molecules collide and rearrange into three O_2 molecules (Fig. 13.16):

$$O_3 + O_3 \longrightarrow O_2 + O_2 + O_2$$

Two-step mechanism: In the first step, an O_3 molecule is energized by solar radiation and dissociates into an O atom and an O_2 molecule. In the second step, the O atom attacks another O_3 molecule to produce two more O_2 molecules (Fig. 13.17):

Step 1: $O_3 \longrightarrow O_2 + O$
Step 2: $O + O_3 \longrightarrow O_2 + O_2$

The free O atom in the second mechanism is a **reaction intermediate,** a species that plays a role in a reaction but does not appear in the chemical equation for the overall reaction: it is produced in one step but is used up in a later step. The two equations for the elementary reactions add together to give the equation for the overall reaction.

A note on good practice: The chemical equations for elementary reaction steps are written without the state symbols. They differ from the overall chemical equation, which summarizes bulk behavior, because they show how *individual* atoms and molecules take part in the reaction,. We do not use stoichiometric coefficients for elementary reactions. Instead, to emphasize that we are depicting individual molecules, we write the formula as many times as required.

Elementary reactions are classified according to their **molecularity,** the number of reactant molecules (or atoms and ions) taking part in a specified elementary reaction. The equation for the first step in the two-step mechanism for the decomposition of ozone ($O_3 \rightarrow O_2 + O$) is an example of a **unimolecular reaction,** because only one reactant molecule participates. We can picture an ozone molecule as acquiring energy from sunlight and vibrating so vigorously that it shakes itself apart. In the second step of the two-step mechanism, the O atom produced by the dissociation of O_3 goes on to attack another O_3 molecule ($O + O_3 \rightarrow O_2 + O_2$). This elementary reaction is an example of a **bimolecular reaction,** because two reactant species come together to react. The molecularity of a unimolecular reaction is 1 and that of a bimolecular reaction is 2.

Any forward reaction that can take place is also accompanied, in principle at least, by the corresponding reverse reaction. Therefore, the unimolecular decay of O_3 in step 1 of the two-step mechanism is accompanied by the formation reaction

$$O_2 + O \longrightarrow O_3$$

This step is bimolecular. Similarly, the bimolecular attack of O on O_3 in step 2 of the two-step mechanism is accompanied by its reverse reaction

$$O_2 + O_2 \longrightarrow O + O_3$$

This step also is bimolecular. The last two equations add up to give the reverse of the overall equation, $3\,O_2(g) \rightarrow 2\,O_3(g)$. The forward and reverse reactions jointly provide a mechanism for reaching dynamic equilibrium between the reactants and the products in the overall process.

The reverse of the single elementary step in the one-step mechanism is

$$O_2 + O_2 + O_2 \longrightarrow O_3 + O_3$$

The step is an example of a **termolecular reaction,** an elementary reaction requiring the simultaneous collision of three molecules. Termolecular reactions are uncommon, because it is very unlikely that three molecules will collide simultaneously with one another under normal conditions.

SELF-TEST 13.10A What is the molecularity of the elementary reaction (a) $C_2N_2 \rightarrow$ $CN + CN$; (b) $NO_2 + NO_2 \rightarrow NO + NO_3$?

[*Answer:* (a) Unimolecular; (b) bimolecular]

SELF-TEST 13.10B What is the molecularity of the elementary reaction (a) $C_2H_5Br +$ $OH^- \rightarrow C_2H_5OH + Br^-$; (b) $Br_2 \rightarrow Br + Br$?

Many reactions take place by a series of elementary reactions. The molecularity of an elementary reaction is the number of reactant particles that take part in the step.

13.8 The Rate Laws of Elementary Reactions

To verify that a proposed reaction mechanism agrees with experimental data, we construct the overall rate law implied by the mechanism and check to see whether it is consistent with the experimentally determined rate law. However, although the constructed rate law and the experimental rate law may be the same, the proposed mechanism may still be incorrect because some other mechanism may also lead to the same rate law. Kinetic information can only support a proposed mechanism; it can never prove that a mechanism is correct. The acceptance of a suggested mechanism is more like the process of proof in an ideal court of law than a proof in mathematics, with evidence being assembled to give a convincing, consistent picture.

To construct an overall rate law from a mechanism, write the rate law for each of the elementary reactions that have been proposed; then combine them into an overall rate law. First, it is important to realize that the chemical equation for an elementary reaction is different from the balanced chemical equation for the overall reaction. The overall chemical equation gives the overall stoichiometry of the reaction, but tells us nothing about *how* the reaction occurs; and so we must find the rate law experimentally. In contrast, an elementary step shows explicitly which particles and how many of each we propose come together in that step of the reaction. Because the elementary reaction shows how the reaction occurs, the rate of that step depends on the concentrations of those particles. Therefore, we can write the rate law *for an elementary reaction* (but *not* for the overall reaction) from its chemical equation, with each exponent in the rate law being the same as the number of particles of a given type participating in the reaction, as summarized in Table 13.3.

Let's write the rate laws for the steps in a mechanism proposed for the gas-phase oxidation of NO to NO_2. Its overall rate law has been determined experimentally:

$$2\, NO(g) + O_2(g) \longrightarrow 2\, NO_2(g) \quad \text{Rate of formation of } NO_2 = k[NO]^2[O_2]$$

The following mechanism has been proposed for the reaction.

Step 1 A fast bimolecular dimerization and its reverse:

$$NO + NO \longrightarrow N_2O_2 \qquad \text{Rate of formation of } N_2O_2 = k_1[NO]^2$$
$$N_2O_2 \longrightarrow NO + NO \qquad \text{Rate of consumption of } N_2O_2 = k_1'[N_2O_2]$$

TABLE 13.3 Rate Laws of Elementary Reactions

Molecularity	Elementary reaction		Rate law
1	$A \longrightarrow$	products	rate $= k[A]$
2	$A + B \longrightarrow$	products	rate $= k[A][B]$
	$A + A \longrightarrow$	products	rate $= k[A]^2$
3	$A + B + C \longrightarrow$	products	rate $= k[A][B][C]$
	$A + A + B \longrightarrow$	products	rate $= k[A]^2[B]$
	$A + A + A \longrightarrow$	products	rate $= k[A]^3$

Step 2 A slow bimolecular reaction in which an O_2 molecule collides with the dimer:

$$O_2 + N_2O_2 \longrightarrow NO_2 + NO_2 \quad \text{Rate of consumption of } N_2O_2 = k_2[N_2O_2][O_2]$$

The reverse of step 2 is too slow to include.

To evaluate the plausibility of this mechanism, we need to construct the overall rate law it implies. First, we identify any elementary reaction that results in product and write the equation for the net rate of product formation. In this case, NO_2 is formed only in step 2, and so

$$\text{Rate of formation of } NO_2 = 2k_2[N_2O_2][O_2]$$

The factor of 2 appears in the rate law because two NO_2 molecules are formed from each N_2O_2 molecule consumed. This expression is not yet an acceptable rate law for the overall reaction because it includes the concentration of an intermediate, N_2O_2, and intermediates do not appear in the overall rate law for a reaction.

To eliminate the concentration of the intermediate N_2O_2 we proceed as follows. First, we write the expression for its net rate of formation. According to the mechanism, N_2O_2 is formed in the forward reaction in step 1, removed in the reverse reaction, and removed in step 2. Therefore the net rate of its formation is

$$\text{Net rate of formation of } N_2O_2 = \overset{\text{Step 1}}{k_1[NO]^2} - \overset{\text{Step 1, reverse}}{k_1'[N_2O_2]} - \overset{\text{Step 2}}{k_2[N_2O_2][O_2]}$$

Now we make the **steady-state approximation**, that any intermediate remains at a constant, low concentration. The justification for this approximation is that the intermediate is so reactive that it reacts as soon as it is formed. Because the concentration of the intermediate is constant, its net rate of formation is zero, and the previous equation becomes

$$k_1[NO]^2 - k_1'[N_2O_2] - k_2[N_2O_2][O_2] = 0$$

This equation, which can be written

$$k_1[NO]^2 - (k_1' + k_2[O_2])[N_2O_2] = 0$$

can be rearranged to find the concentration of N_2O_2:

$$[N_2O_2] = \frac{k_1[NO]^2}{k_1' + k_2[O_2]}$$

Now that we have an expression for $[N_2O_2]$, we can substitute that expression into the rate law:

$$\text{Rate of formation of } NO_2 = 2k_2[N_2O_2][O_2] = \frac{2k_1k_2[NO]^2[O_2]}{k_1' + k_2[O_2]} \tag{9}$$

The rate law that we have derived is not the same as the experimental one. We have stressed that a reaction mechanism is plausible only if its predictions are in line with experimental results; so should we discard our proposal? Before doing so, it is always wise to explore whether under certain conditions the predictions do in fact agree with experimental data. In this case, if the rate of step 2 is very slow relative to the rapid equilibrium in step 1—so that $k_1'[N_2O_2] \gg k_2[N_2O_2][O_2]$, which implies that $k_1' \gg k_2[O_2]$ when we cancel the $[N_2O_2]$ on each side of the inequality—we can ignore the term $k_2[O_2]$ in the denominator of Eq. 9. With this approximation, the rate law simplifies to

$$\text{Rate of formation of } NO_2 = \frac{2k_1k_2}{k_1'}[NO]^2[O_2] \tag{10a}$$

which agrees with the experimentally determined rate law if we identify k with

$$k = \frac{2k_1 k_2}{k_1'} \qquad\qquad (10b)$$

The proposed mechanism is therefore consistent with experiment.

The assumption that the rate of consumption of the intermediate in the slow step is insignificant relative to its rates of formation and decomposition in the first step is called a **pre-equilibrium condition**. A pre-equilibrium arises when an intermediate is formed and sustained in a rapid formation reaction and its reverse. The calculation of the rate law is then much simpler. For instance, if we propose that the first step in the NO oxidation gives rise to a pre-equilibrium, we would write

$$NO + NO \rightleftharpoons N_2O_2 \qquad\qquad K = \frac{[N_2O_2]}{[NO]^2}$$

This equilibrium implies that $[N_2O_2] = K[NO]^2$. The rate of formation of NO_2 is therefore

$$\text{Rate of formation of } NO_2 = 2k_2[N_2O_2][O_2] = 2k_2 K[NO]^2[O_2]$$

As before, the mechanism gives rise to an overall third-order rate law, in agreement with experiment. Although this procedure is much simpler than the steady-state approach, it is less flexible: it is more difficult to extend to more complex mechanisms and it is not so easy to establish the conditions under which the approximation is valid.

The slowest elementary step in a sequence of reactions that governs the overall rate of formation of products—in our example, the reaction between O_2 and the intermediate N_2O_2 in step 2—is called the **rate-determining step** of the reaction (Fig. 13.18). A rate-determining step is like a slow ferry on the route between two cities. The rate at which the traffic arrives at its destination is governed by the rate at which it is ferried across the river, because this part of the journey is much slower than any of the others. Steps that follow the rate-determining step take place as soon as the intermediate has been formed and have a negligible effect on the overall rate. Therefore, they can be ignored when writing the overall rate law from the mechanism.

In simple cases, we can identify the rate-determining step by considering the experimental rate law. For example, the reaction

$$(CH_3)_3CBr(sol) + OH^-(sol) \longrightarrow (CH_3)_3COH(sol) + Br^-(sol)$$

takes place in an organic solvent ("sol") with the rate law Rate $= k[(CH_3)_3CBr]$. We can write a two-step mechanism for the reaction:

$$(CH_3)_3CBr \longrightarrow (CH_3)_3C^+ + Br^-$$
$$(CH_3)_3C^+ + OH^- \longrightarrow (CH_3)_3COH$$

In this mechanism, the rate law for the first step is Rate $= k[(CH_3)_3CBr]$, the same as the experimental rate law. Because the reaction is zero order in OH^-, the

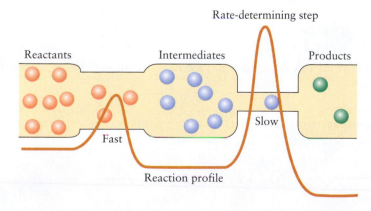

Rate-determining step

Reactants   Intermediates   Products

Fast   Slow

Reaction profile

FIGURE 13.18 The rate-determining step in a reaction is an elementary reaction that governs the rate at which products are formed in a multistep series of reactions. The narrow neck representing the rate-determining step in the drawing is like a ferry that cannot handle the busy traffic on a highway. The superimposed "reaction profile" shows the energy requirements of each step. The step that requires the most energy is the slowest (see Section 13.13).

hydroxide ion concentration does not affect the rate and we can assume that the first step is the slow step. As soon as $(CH_3)_3C^+$ is formed, it is immediately converted into the alcohol. Recall that elementary reactions following the rate-determining step do not contribute to the rate law derived from the mechanism.

EXAMPLE 13.7   **Setting up an overall rate law from a proposed mechanism**

As remarked in Section 13.3, the following rate law has been determined for the decomposition of ozone in the reaction $2\,O_3(g) \rightarrow 3\,O_2(g)$:

$$\text{Rate of decomposition of } O_3 \ = k\,\frac{[O_3]^2}{[O_2]}$$

The following mechanism has been proposed:

Step 1 $O_3 \rightleftharpoons O_2 + O$

Step 2 $O + O_3 \rightleftharpoons O_2 + O_2$

Measurements of the rates of the elementary forward reactions show that the slow step is the second step, the attack of O on O_3. The reverse reaction $O_2 + O_2 \rightarrow O + O_3$ is so slow that it can be ignored. Derive the rate law implied by the mechanism, and confirm that it matches the observed rate law.

A note on good practice: Be careful to distinguish the sign $\rightleftarrows$ (paired regular arrows), which signifies only that both forward and reverse reactions may occur, from the sign $\rightleftharpoons$ (paired single-barbed arrows), which signifies that the reactions are at equilibrium.

STRATEGY   Construct the rate laws for the elementary reactions and combine them into the overall rate law for the decomposition of the reactant. If necessary, use the steady-state approximation for any intermediates and simplify it by using arguments based on rapid pre-equilibria and the existence of a rate-determining step.

SOLUTION   The rate laws for the elementary reactions are

Step 1 (forward) $O_3 \longrightarrow O_2 + O$ Rate of decomposition of $O_3 = k_1[O_3]$ (fast)

(reverse) $O_2 + O \longrightarrow O_3$ Rate of formation of $O_3 = k_1{'}[O_2][O]$ (fast)

Step 2 (forward) $O + O_3 \longrightarrow O_2 + O_2$ Rate of consumption of $O_3 = k_2[O][O_3]$ (slow)

(reverse) $O_2 + O_2 \longrightarrow O + O_3$ Rate of formation of $O_3 = k_2{'}[O_2]^2$ (very slow, ignore)

Write the rate law for the net rate of decomposition of O_3:

$$\text{Net rate of decomposition of } O_3 = k_1[O_3] - k_1{'}[O_2][O] + k_2[O][O_3]$$

To eliminate the O atom concentration from this expression, we consider the net rate of formation of O atoms and use the steady-state approximation to set that net rate equal to zero:

$$k_1[O_3] - k_1{'}[O_2][O] - k_2[O][O_3] = 0$$

This equation solves to

$$[O] = \frac{k_1[O_3]}{k_1{'}[O_2] + k_2[O_3]}$$

When we substitute this expression into the rate law for net rate of decomposition of ozone, we get

$$\text{Net rate of decomposition of } O_3 = k_1[O_3] - \frac{k_1k_1{'}[O_2][O_3]}{k_1{'}[O_2] + k_2[O_3]} + \frac{k_1k_2[O_3]^2}{k_1{'}[O_2] + k_2[O_3]}$$

$$= \frac{2k_1k_2[O_3]^2}{k_1{'}[O_2] + k_2[O_3]}$$

(You should verify for yourself that the three expressions in the first line do combine to give the final expression.) Because step 2 is slow relative to the fast pre-equilibrium, we can make the approximation $k_2[O][O_3] \ll k_1'[O_2][O]$, or equivalently by canceling the $[O]$, $k_2[O_3] \ll k_1'[O_2]$, and simplify this expression to

$$\text{Net rate of decomposition of } O_3 = \frac{2k_1k_2[O_3]^2}{k_1'[O_2]} = k\frac{[O_3]^2}{[O_2]}$$

with $k = 2k_1k_2/k_1'$. This equation has exactly the same form as that of the observed rate law. Because the mechanism is consistent with both the balanced equation and the experimental rate law, we can consider it plausible.

SELF-TEST 13.11A Consider the following mechanism for the formation of a DNA double helix from its strands A and B:

A + B $\rightleftharpoons$ unstable helix (fast, k_1 and k_1' both large)

Unstable helix $\longrightarrow$ stable double helix (slow, k_2 very small)

Derive the rate equation for the formation of the stable double helix and express the rate constant of the overall reaction in terms of the rate constants of the individual steps.

[*Answer:* Rate = $k[A][B]$, $k = k_1k_2/k_1'$]

SELF-TEST 13.11B The proposed two-step mechanism for a reaction is $H_2A + B \rightarrow BH^+ + HA^-$ and its reverse, both of which are fast, followed by $HA^- + B \rightarrow BH^+ + A^{2-}$, which is slow. Find the rate law with HA^- treated as the intermediate and write the equation for the overall reaction.

The rate law of an elementary reaction is written from the equation for the reaction. A rate law is often derived from a proposed mechanism by imposing the steady-state approximation or assuming that there is a pre-equilibrium. To be plausible, a mechanism must be consistent with the experimental rate law.

13.9 Chain Reactions

In the mechanisms considered so far, there have only been one or two intermediates. In a **chain reaction,** a highly reactive intermediate reacts to produce another highly reactive intermediate, which reacts to produce another, and so on (Fig. 13.19). In many cases, the reaction intermediate—which in this context is called a **chain carrier**—is a radical, and the reaction is called a radical chain reaction. In a **radical chain reaction,** one radical reacts with a molecule to produce another radical, that radical goes on to attack another molecule to produce yet another radical, and so on. The ideas presented in the preceding sections apply to chain reactions, too, but they often result in very complex rate laws, which we will not derive.

The formation of HBr in the reaction

$$H_2(g) + Br_2(g) \longrightarrow 2\,HBr(g)$$

takes place by a chain reaction. The chain carriers are hydrogen atoms (H·) and bromine atoms (Br·). The first step in any chain reaction is **initiation,** the formation of chain carriers from a reactant. Often heat (denoted Δ) or light (denoted $h\nu$) is used to generate the chain carriers:

$$Br_2(g) \xrightarrow{\Delta \text{ or } h\nu} Br\cdot + Br\cdot$$

Once chain carriers have been formed, the chain can propagate in a reaction in which one carrier reacts with a reactant molecule to produce another carrier. The elementary reactions for the **propagation** of the chain are

$$Br\cdot + H_2 \longrightarrow HBr + H\cdot$$
$$H\cdot + Br_2 \longrightarrow HBr + Br\cdot$$

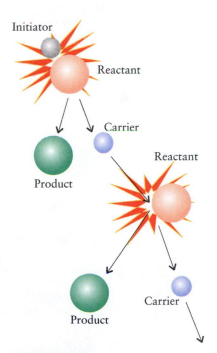

FIGURE 13.19 In a chain reaction, the product of one step in a reaction is a highly reactive reactant in a subsequent step, which in turn produces reactive species that can take part in other reaction steps.

The chain carriers—radicals here—produced in these reactions can go on to attack other reactant molecules (H_2 and Br_2), thereby allowing the chain to continue. The elementary reaction that ends the chain, a process called **termination,** takes place when chain carriers combine to form products. Two examples of termination reactions are:

$$Br\cdot + Br\cdot \longrightarrow Br_2$$
$$H\cdot + Br\cdot \longrightarrow HBr$$

In some cases, a chain propagates explosively. Explosions can be expected when **chain branching** occurs—that is, when more than one chain carrier is formed in a propagation step. Reactions that are explosively fast are often chain reactions. The characteristic pop that occurs when a mixture of hydrogen and oxygen is ignited is a consequence of chain branching. The two gases combine in a radical chain reaction in which the initiation step may be the formation of hydrogen atoms:

$$\text{Initiation: } H_2 \xrightarrow{\Delta} H\cdot + H\cdot$$

After the reaction has been initiated, two new radicals are formed when one hydrogen atom attacks an oxygen molecule:

$$\text{Branching: } H\cdot + O_2 \longrightarrow HO\cdot + \cdot O\cdot$$

The oxygen atom, with valence electron configuration $2s^2 2p_x^2 2p_y^1 2p_z^1$, has two electrons with unpaired spins (its Lewis symbol is $\cdot\ddot{O}\cdot$, which we abbreviate to $\cdot O\cdot$). Two radicals are also produced when the oxygen atom attacks a hydrogen molecule:

$$\text{Branching: } \cdot O\cdot + H_2 \longrightarrow HO\cdot + H\cdot$$

As a result of these branching processes, the chain produces an increasingly larger number of radicals that can take part in even more branching steps. The reaction rate increases rapidly, and an explosion typical of many combustion reactions may occur (Fig. 13.20).

Chain reactions begin with the initiation of a reactive intermediate that propagates the chain and concludes with termination when radicals combine. Branching chain reactions can be explosively fast.

FIGURE 13.20 This flame front was caught during the fast combustion—almost a miniature explosion—that occurs inside an internal combustion engine every time a spark plug ignites the mixture of fuel and air.

13.10 Rates and Equilibrium

At equilibrium, the rates of the forward and reverse reactions are equal. Because the rates depend on rate constants and concentrations, we ought to be able to find a relation between rate constants for elementary reactions and the equilibrium constant for the overall reaction.

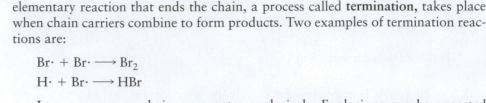

HOW DO WE DO THAT?

To deduce the relation between rate constants and equilibrium constants, we note that the equilibrium constant for a chemical reaction in solution that has the form $A + B \rightleftharpoons C + D$ is

$$K = \frac{[C][D]}{[A][B]}$$

Suppose that experiments show that both the forward reaction and the reverse reaction are elementary second-order reactions. Then we would write the following rate laws:

$$A + B \longrightarrow C + D \qquad \text{Rate} = k[A][B]$$
$$C + D \longrightarrow A + B \qquad \text{Rate} = k'[C][D]$$

At equilibrium, these two rates are equal, and so we can write

$$k[A][B] = k'[C][D]$$

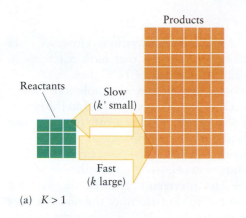
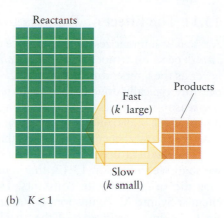

(a) $K > 1$ (b) $K < 1$

FIGURE 13.21 The equilibrium constant
for a reaction is equal to the ratio of the
rate constants for the forward and reverse
reactions. (a) A forward rate constant (k)
that is relatively large compared with the
reverse rate constant means that the
forward rate matches the reverse rate
when the reaction has neared completion
and the concentration of reactants is low.
(b) Conversely, if the reverse rate constant
(k') is larger than the forward rate
constant, then the forward and reverse
rates are equal when little reaction has
taken place and the concentration of
products is low.

It follows that, at equilibrium,

$$\frac{[C][D]}{[A][B]} = \frac{k}{k'}$$

Comparison of this expression with the expression for the equilibrium constant then shows that

$$K = \frac{k}{k'}$$

If the reaction involves gas-phase species and the rate law is expressed in terms of molar concentrations, then instead of K use K_c.

We have deduced that

$$K = \frac{k}{k'} \tag{11}*$$

That is, *the equilibrium constant for a reaction is equal to the ratio of the rate constants for the forward and reverse elementary reactions that contribute to the overall reaction.* We can now see in kinetic terms rather than thermodynamic (Gibbs free energy) terms when to expect a large equilibrium constant: $K \gg 1$ (and products are favored) when k for the forward direction is much larger than k' for the reverse direction. In this case, the fast forward reaction builds up a high concentration of products before reaching equilibrium (Fig. 13.21). In contrast, $K \ll 1$ (and reactants are favored) when k is much smaller than k'. Now the reverse reaction destroys the products rapidly, and so their concentrations are very low.

We derived the relation between the equilibrium constant and the rate constant for a single-step reaction. However, suppose that a reaction has a complex mechanism in which the elementary reactions have rate constants $k_1, k_2, \ldots$ and the reverse elementary reactions have rate constants $k_1', k_2', \ldots$. Then, by an argument similar to that for the single-step reaction, the overall equilibrium constant is related to the rate constants as follows:

$$K = \frac{k_1}{k_1'} \times \frac{k_2}{k_2'} \times \cdots \tag{12}$$

The equilibrium constant for an elementary reaction is equal to the ratio of the forward and reverse rate constants of the reaction.

MODELS OF REACTIONS

Rate laws and rate constants are windows on to the molecular processes of chemical change. We have seen how rate laws reveal details of the mechanisms of reactions; here, we build models of the molecular processes that account for the values of the rate constants that appear in the rate laws.

FIGURE 13.22 Reaction rates almost always increase with temperature. The beaker on the left contains magnesium in cold water and that on the right contains magnesium in hot water. An indicator has been added to show the formation of an alkaline solution as magnesium reacts.

The use of the gas constant does not mean that the Arrhenius equation applies only to gases: it applies to a wide range of reactions in solution, too.

13.11 The Effect of Temperature

A rate law summarizes the dependence of the rate on concentrations. However, rates also depend on temperature. The *qualitative* observation is that most reactions go faster as the temperature is raised (Fig. 13.22). An increase of 10°C from room temperature typically doubles the rate of reaction of organic species in solution. That is one reason why we cook foods: heating accelerates reactions that lead to the breakdown of cell walls and the decomposition of proteins. We refrigerate foods to slow down the natural chemical reactions that lead to their decomposition.

The temperature dependence of a reaction rate lies in the rate constant and, as we shall see in Section 13.12, that temperature dependence gives valuable insight into the origins of rate constants. In the late nineteenth century, the Swedish chemist Svante Arrhenius found that the plot of the logarithm of the rate constant (ln k) against the inverse of the absolute temperature ($1/T$) is a straight line. In other words,

$$\ln k = \text{ intercept } + \text{ slope } \times \frac{1}{T}$$

The intercept is denoted ln A; and, for reasons that will become clear in Section 13.12, the slope is denoted $-E_a/R$, where R is the gas constant. With this notation, the empirical **Arrhenius equation** is

$$\ln k = \ln A - \frac{E_a}{RT} \tag{13a}*$$

An alternative form of this expression, which is obtained by taking antilogarithms of both sides, is

$$k = Ae^{-E_a/RT} \tag{13b}*$$

The two constants, A and E_a, are known as the **Arrhenius parameters** for the reaction and are found from experiment; A is called the **pre-exponential factor**, and E_a is the **activation energy**. Both A and E_a are nearly independent of temperature but have values that depend on the reaction being studied.

EXAMPLE 13.8   Measuring an activation energy

The rate constant for the second-order reaction between bromoethane and hydroxide ions in water, $C_2H_5Br(aq) + OH^-(aq) \rightarrow C_2H_5OH(aq) + Br^-(aq)$, was measured at several temperatures, with the results shown here:

Temperature (°C)	25	30.	35	40.	45	50.
k (L·mol^{-1}·s^{-1})	8.8×10^{-5}	1.6×10^{-4}	2.8×10^{-4}	5.0×10^{-4}	8.5×10^{-4}	1.40×10^{-3}

Find the activation energy of the reaction.

STRATEGY

Activation energies are found from the Arrhenius equation (Eq. 13). We plot ln k against $1/T$, with T in kelvins, and multiply the slope of the graph by $-R$ to find the activation energy, with $R = 8.3145$ J·K^{-1}·mol^{-1}. A spreadsheet, curve-fitting program, or the Living Graph "Determination of Arrhenius Parameters" on the Web site for this book is very useful for this type of calculation.

SOLUTION   The table to use for drawing the graph is

Temperature (°C)	25	30.	35	40.	45	50.
T (K)	298	303	308	313	318	323
$1/T$ (K^{-1})	3.35×10^{-3}	3.30×10^{-3}	3.25×10^{-3}	3.19×10^{-3}	3.14×10^{-3}	3.10×10^{-3}
k (L·mol^{-1}·s^{-1})	8.8×10^{-5}	1.6×10^{-4}	2.8×10^{-4}	5.0×10^{-4}	8.5×10^{-4}	1.40×10^{-3}
ln k	-9.34	-8.74	-8.18	-7.60	-7.07	-6.57

The points are plotted in Fig. 13.23. We calculate the slope from any two points, such as those marked A and B:

$$\text{Slope} = \frac{(-8.74) - (-6.60)}{\{(3.30 \times 10^{-3}) - (3.10 \times 10^{-3})\}K^{-1}} = -\frac{2.14}{2.0 \times 10^{-4} K^{-1}}$$

$$= -1.07 \times 10^4 \text{ K}$$

Because the slope is equal to $-E_a/R$,

From $E_a = -R \times \text{slope}$, $E_a = -(8.3145 \text{ J·K}^{-1}\text{·mol}^{-1}) \times (-1.07 \times 10^4 \text{ K})$

$$= 8.9 \times 10^4 \text{ J·mol}^{-1} = 89 \text{ kJ·mol}^{-1}$$

SELF-TEST 13.12A The rate constant for the second-order gas-phase reaction $HO(g) + H_2(g) \rightarrow H_2O(g) + H(g)$ varies with the temperature as shown here:

Temperature (°C)	100.	200.	300.	400.
k (L·mol^{-1}·s^{-1})	1.1×10^{-9}	1.8×10^{-8}	1.2×10^{-7}	4.4×10^{-7}

Determine the activation energy.

[*Answer:* 42 kJ·mol^{-1}]

SELF-TEST 13.12B The rate of a reaction increased from 3.00 mol·L^{-1}·s^{-1} to 4.35 mol·L^{-1}·s^{-1} when the temperature was raised from 18°C to 30.°C. What is the activation energy of the reaction?

Reactions that give a straight line when ln k is plotted against $1/T$ are said to show **Arrhenius behavior**. A wide range of reactions—not only those in the gas phase—show Arrhenius behavior. Even tropical fireflies flash more quickly on hot nights than on cold nights, and their rate of flashing is Arrhenius-like over a small range of temperatures. We can conclude that the biochemical reactions responsible for the flashing have rate constants that increase with temperature in accord with Eq. 13. Some Arrhenius parameters are listed in Table 13.4.

Because the slope of an Arrhenius plot is proportional to E_a, it follows that *the higher the activation energy E_a, the stronger the temperature dependence of the rate constant.* Reactions with low activation energies (about 10 kJ·mol^{-1}, with not very steep Arrhenius plots) have rates that increase only slightly with temperature. Reactions with high activation energies (above approximately 60 kJ·mol^{-1}, with steep Arrhenius plots) have rates that depend strongly on the temperature (Fig. 13.24).

The Arrhenius equation is used to predict the value of a rate constant at one temperature from its value at another temperature.

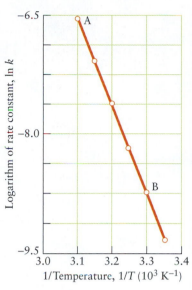

FIGURE 13.23 An Arrhenius plot is a graph of ln k against $1/T$. If, as here, the line is straight, then the reaction is said to show Arrhenius behavior in the temperature range studied. This plot has been constructed from the data in Example 13.8.

TABLE 13.4 Arrhenius Parameters

Reaction	A	E_a (kJ·mol^{-1})
First order, gas phase		
cyclopropane $\longrightarrow$ propene	1.6×10^{15} s^{-1}	272
$CH_3NC \longrightarrow CH_3CN$	4.0×10^{13} s^{-1}	160
$C_2H_6 \longrightarrow 2\ CH_3$	2.5×10^{17} s^{-1}	384
$N_2O \longrightarrow N_2 + O$	8.0×10^{11} s^{-1}	250
$2\ N_2O_5 \longrightarrow 4\ NO_2 + O_2$	4.0×10^{13} s^{-1}	103
Second order, gas phase		
$O + N_2 \longrightarrow NO + N$	1×10^{11} L·mol^{-1}·s^{-1}	315
$OH + H_2 \longrightarrow H_2O + H$	8×10^{10} L·mol^{-1}·s^{-1}	42
$2\ CH_3 \longrightarrow C_2H_6$	2×10^{10} L·mol^{-1}·s^{-1}	0
Second order, in aqueous solution		
$C_2H_5Br + OH^- \longrightarrow C_2H_5OH + Br^-$	4.3×10^{11} L·mol^{-1}·s^{-1}	90
$CO_2 + OH^- \longrightarrow HCO_3^-$	1.5×10^{10} L·mol^{-1}·s^{-1}	38
$C_{12}H_{22}O_{11} + H_2O \longrightarrow 2\ C_6H_{12}O_6$	1.5×10^{15} L·mol^{-1}·s^{-1}	108

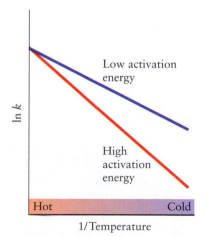

FIGURE 13.24 The dependence of the rate constant on temperature for two reactions with different activation energies. The higher the activation energy, the more strongly the rate constant depends on temperature.

The Arrhenius equations for the two temperatures, which we call T and T', when the rate constant for the reaction has the values k and k', respectively, are

At temperature T': $\ln k' = \ln A - \dfrac{E_a}{RT'}$

At temperature T: $\ln k = \ln A - \dfrac{E_a}{RT}$

We eliminate $\ln A$ by subtracting the second equation from the first:

$$\ln k' - \ln k = -\frac{E_a}{RT'} + \frac{E_a}{RT}$$

The expression that we have derived can easily be rearranged into

$$\ln \frac{k'}{k} = \frac{E_a}{R}\left(\frac{1}{T} - \frac{1}{T'}\right) \tag{14}$$

EXAMPLE 13.9 Sample exercise: Using the activation energy to predict a rate constant

The hydrolysis of sucrose is a part of the digestive process. To investigate how strongly the rate depends on our body temperature, calculate the rate constant for the hydrolysis of sucrose at 35.0°C, given that $k = 1.0$ mL·mol^{-1}·s^{-1} at 37.0°C (normal body temperature) and that the activation energy of the reaction is 108 kJ·mol^{-1}.

SOLUTION We expect a lower rate constant at the lower temperature. We use Eq. 14 with $T = 310.0$ K and $T' = 308.0$ K and express R in kilojoules to cancel the units of E_a:

From $\ln(k'/k) = (E_a/R)(1/T - 1/T')$,

$$\ln \frac{k'}{k} = \frac{108 \text{ kJ·mol}^{-1}}{8.3145 \times 10^{-3} \text{ kJ·K}^{-1}\text{·mol}^{-1}} \times$$

$$\left(\frac{1}{310.0 \text{ K}} - \frac{1}{308.0 \text{ K}}\right)$$

$$= -0.27$$

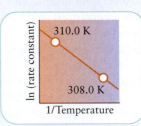

From $x = e^{\ln x}$,

$$\frac{k'}{k} = e^{-0.27}$$

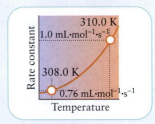

Finally, because $k = 1.0$ mL·mol^{-1}·s^{-1},

$$k' = (1.0 \text{ mL·mol}^{-1}\text{·s}^{-1}) \times e^{-0.27} = 0.76 \text{ mL·mol}^{-1}\text{·s}^{-1}$$

at 35°C. The high activation energy of the reaction means that its rate is very sensitive to temperature.

A note on good practice: We have evaluated the exponent (-0.27) in this example to avoid writing a cumbersome expression for k'/k; our recommended good practice of doing the numerical calculation in one final step does not, in this instance, change the final result (to 2sf).

SELF-TEST 13.13A The rate constant for the second-order reaction between CH_3CH_2Br and OH^- in water is 0.28 mL·mol^{-1}·s^{-1} at 35.0°C. What is the value of the rate constant at 50.0°C? See Table 13.4 for data.

[*Answer:* 1.4 mL·mol^{-1}·s^{-1}]

SELF-TEST 13.13B The rate constant for the first-order isomerization of cyclopropane, C_3H_6, to propene, $CH_3CH=CH_2$, is 6.7×10^{-4} s^{-1} at 500.°C. What is its value at 300.°C? See Table 13.4 for data.

An Arrhenius plot of ln k against 1/T is used to determine the Arrhenius parameters of a reaction; a large activation energy signifies a high sensitivity of the rate constant to changes in temperature.

13.12 Collision Theory

We are now ready to build a model of how chemical reactions take place at the molecular level. Specifically, our model must account for the temperature dependence of rate constants, as expressed by the Arrhenius equation; it should also reveal the significance of the Arrhenius parameters A and E_a. Reactions in the gas phase are conceptually simpler than those in solution, and so we begin with them.

First, we assume that a reaction can take place only if reactants meet. In a gas that meeting is a collision, and so the model that we are building is called the **collision theory** of reactions. In this model, we suppose that molecules behave like defective billiard balls: they bounce apart if they collide at low speed, but they might smash into pieces when the impact is really energetic. If two molecules collide with less than a certain kinetic energy, they simply bounce apart. If they meet with more than that kinetic energy, reactant bonds can break and new bonds can form, resulting in products (Fig. 13.25). We denote the minimum kinetic energy needed for reaction by E_{min}.

To set up a quantitative theory based on this qualitative picture, we need to know the rate at which molecules collide and the fraction of those collisions that have at least the energy E_{min} required for reaction to occur. The collision frequency (the number of collisions per second) between A and B molecules in a gas at a temperature T can be calculated from the kinetic model of a gas (Section 4.10):

$$\text{Collision frequency} = \sigma \bar{v}_{rel} N_A^2 [A][B] \qquad (15)$$

in which N_A is Avogadro's constant and σ (sigma) is the **collision cross section,** the area that a molecule presents as a target during a collision. The bigger the collision cross section, the greater the collision frequency, because big molecules are easier targets than small molecules. The quantity $\bar{v}_{rel}$ is the **mean relative speed,** the mean speed at which the molecules approach each other in a gas. The mean relative speed is calculated by multiplying each possible speed by the fraction of molecules that have that speed and then adding all the products together.

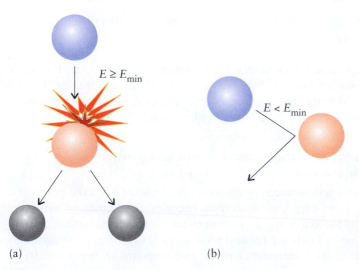

(a)   (b)

FIGURE 13.25 (a) In the collision theory of chemical reactions, reaction may take place only when two molecules collide with a kinetic energy at least equal to a minimum value, E_{min} (which later we identify with the activation energy). (b) Otherwise, they simply bounce apart.

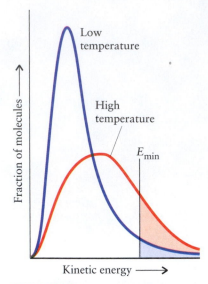

FIGURE 13.26 The fraction of molecules that collide with a kinetic energy that is at least equal to a certain minimum value, E_{min} (which later we show to be the activation energy, E_a), is given by the shaded areas under each curve. Note that this fraction increases rapidly as the temperature is raised.

When the temperature is T and the molar masses are M_A and M_B, this mean relative speed is

$$\bar{v}_{rel} = \left(\frac{8RT}{\pi\mu}\right)^{1/2} \qquad \mu = \frac{M_A M_B}{M_A + M_B} \qquad (16)^*$$

The collision frequency is greater the higher the relative speeds of the molecules, and therefore the higher the temperature.

Although the mean relative speed of the molecules increases with temperature, and the collision frequency therefore increases as well, Eq. 16 shows that the mean relative speed increases only as the square root of the temperature. This dependence is far too weak to account for observation. If we used Eq. 16 to predict the temperature dependence of reaction rates, we would conclude that an increase in temperature of 10°C at about room temperature (from 273 K to 283 K) increases the collision frequency by a factor of only 1.02, whereas experiments show that many reaction rates double over that range. Another factor must be affecting the rate.

To find this factor, we need a measure of the fraction of molecules that collide with a kinetic energy equal to or greater than a certain minimum energy, E_{min}. Because kinetic energy is proportional to the square of the speed, this fraction can be obtained from the Maxwell distribution of speeds (Section 4.11). As indicated for a specific reaction by the shaded area under the blue curve in Fig. 13.26, at a low temperature very few molecules have enough kinetic energy to react. At higher temperatures, a much larger fraction of molecules can react, as represented by the shaded area under the red curve. We need to bring this fraction into our equations.

HOW DO WE DO THAT?

To find how the fraction of molecules that collide with at least an energy E_{min} affects the rate of an elementary reaction, we need to know that, at a temperature T, the fraction of collisions with at least the energy E_{min} is equal to $e^{-E_{min}/RT}$, where R is the gas constant. That result comes from an expression known as the *Boltzmann distribution*, which we do not derive here. The rate of reaction is the product of this factor and the collision frequency:

Rate of reaction = collision frequency × fraction with sufficient energy

$$= \sigma\bar{v}_{rel}N_A^2[A][B] \times e^{-E_{min}/RT}$$

The rate law of an elementary reaction that depends on collisions of A with B is Rate $= k[A][B]$, where k is the rate constant. We can therefore identify the expression for the rate constant as

$$k = \frac{\text{rate of reaction}}{[A][B]} = \sigma\bar{v}_{rel}N_A^2 e^{-E_{min}/RT}$$

We have deduced that, according to collision theory, the rate constant is

$$k = \sigma\bar{v}_{rel}N_A^2 e^{-E_{min}/RT} \qquad (17)$$

This exponential temperature dependence for k is much stronger than the weak temperature dependence of the collision frequency itself. By comparing Eq. 17 with Eq. 13b, we can identify the term $\sigma\bar{v}_{rel}N_A^2$ as the pre-exponential factor, A, and E_{min} as the activation energy, E_a. That is, we can conclude that

- A is a measure of the rate at which molecules collide.

- The activation energy, E_a, is the minimum kinetic energy required for a collision to result in reaction.

We now understand why some spontaneous reactions do not take place at a measurable rate: they have very high activation energies. A mixture of hydrogen and oxygen can survive for years: the activation energy for the production of radicals is very high, and no radicals are formed until a spark or flame is brought into contact with the mixture. The dependence of the rate constant on temperature, its

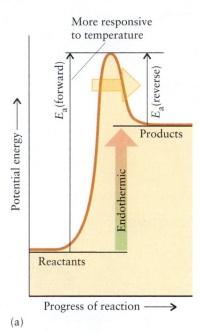

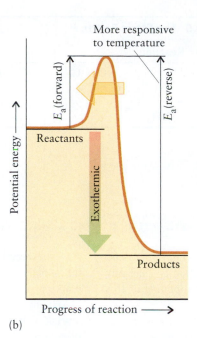

(a)   (b)

FIGURE 13.27 (a) The activation energy for an endothermic reaction is larger in the forward direction than in the reverse, and so the rate of the forward reaction is more sensitive to temperature and the equilibrium shifts toward products as the temperature is raised. (b) The opposite is true for an exothermic reaction, in which case the reverse reaction is more sensitive to temperature and the equilibrium shifts toward reactants as the temperature is raised.

sensitivity to the activation energy, and the fact that the equilibrium constant is equal to the ratio of the forward and reverse rate constants (Eq. 11, $K = k/k'$), provide a kinetic explanation for the temperature dependence of the equilibrium constant.

A diagram called a **reaction profile** is used to represent the energy changes that occur during a reaction. If the forward reaction is endothermic, the activation energy is higher for the forward direction than for the reverse direction (Fig. 13.27). The higher activation energy means that the rate constant of the forward reaction depends more strongly on temperature than does the rate constant of the reverse reaction. Therefore, when the temperature is raised, the rate constant for the forward reaction increases more than that of the reverse reaction. As a result, K will increase and the products will become more favored, just as Le Chatelier's principle predicts.

Now that we have a model, we must check its consistency with various experiments. Sometimes such inconsistencies result in the complete rejection of a model. More often, they indicate that we need to refine the model. In the present case, the results of careful experiments show that the collision model of reactions is not complete, because the experimental rate constant is normally smaller than predicted by collision theory. We can improve the model by realizing that the relative *direction* in which the molecules are moving when they collide also might matter. That is, they need to be oriented a certain way relative to each other. For example, the results of experiments of the kind described in Box 13.2 have shown that, in the gas-phase reaction of chlorine atoms with HI molecules, HI + Cl ⟶ HCl + I, the Cl atom reacts with the HI molecule only if it approaches from a favorable direction (Fig. 13.28). A dependence on direction is called the **steric requirement** of the reaction. It is normally taken into account by introducing an empirical factor, P, called the **steric factor,** and changing Eq. 17 to

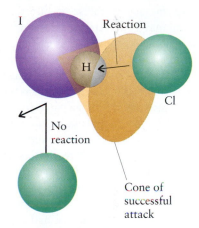

FIGURE 13.28 Whether a reaction takes place when two species collide in the gas phase depends on their relative orientations. In the reaction between a Cl atom and an HI molecule, for example, only those collisions in which the Cl atom approaches the HI molecule from a direction that lies inside the cone indicated here lead to reaction, even though the energy of collisions in other orientations may exceed the activation energy.

$$k = \overbrace{P}^{\substack{\text{Steric}\\\text{requirement}}} \times \overbrace{\sigma\bar{v}_{rel}N_A{}^2}^{\substack{\text{Collision}\\\text{rate}}} \times \overbrace{e^{-E_{min}/RT}}^{\substack{\text{Energy}\\\text{requirement}}} \qquad (18)^*$$

Table 13.5 lists some values of P. All the values shown are less than 1 because the steric requirement reduces the probability of reaction. For collisions between

In certain special cases P is found to exceed 1.

TABLE 13.5 The Steric Factor, *P*

Reaction	P
$NOCl + NOCl \longrightarrow NO + NO + Cl_2$	0.16
$NO_2 + NO_2 \longrightarrow NO + NO + O_2$	5.0×10^{-2}
$ClO + ClO \longrightarrow Cl_2 + O_2$	2.5×10^{-3}
$H_2 + C_2H_4 \longrightarrow C_2H_6$	1.7×10^{-6}

BOX 13.2   HOW DO WE KNOW . . . WHAT HAPPENS DURING A MOLECULAR COLLISION?

When molecules collide, old bonds give way to new ones and atoms in one molecule become part of the other. Is there any way of finding out *experimentally* what is happening at this climactic stage of a reaction?

Molecular beams provide the answer. We first met molecular beams in Box 4.1, where we saw how a velocity selector is constructed. A molecular beam consists of a stream of molecules moving in the same direction with the same speed. A beam may be directed at a gaseous sample or into the path of a second beam, consisting of molecules of a second reactant. The molecules may react when the beams collide; the experimenters can then detect the products of the collision and the direction at which the products emerge from the collision. They also use spectroscopic techniques to determine the vibrational and rotational excitation of the products.

By repeating the experiment with molecules having different speeds and different states of rotational or vibrational excitation, chemists can learn more about the collision itself. For example, experimenters have found that, in the reaction between a Cl atom and an HI molecule, the best direction of attack is within a cone of half-angle 30° surrounding the H atom.

In a "sticky" collision, the reactant molecules orbit around each other for one revolution or more. As a result, the product molecules emerge in random directions because any "memory" of the approach direction is lost. However, a rotation takes time—about 1 ps. If the reaction is over before that, the product molecules will emerge in a specific direction that depends on the direction of the collision. In the collision of K and I_2, for example, most of the products are thrown off in the forward direction. This observation is consistent with the "harpoon mechanism" that had been proposed for this reaction. In this mechanism, an electron flips across from the K atom to the I_2 molecule when they are quite far apart, and the resulting K^+ ion extracts one of the atoms of the negatively charged I_2^- ion. We can think of the electron as a harpoon, the electrostatic attraction as the line attached to the harpoon, and I_2 as a whale. Because the crucial "harpooning" takes place at large distances and there is no actual collision, the products are thrown off in approximately the same direction as the reactants were traveling.

The reaction between a K atom and a CH_3I molecule takes place by a different mechanism. A collision leads to reaction only if the two reactants approach each other very closely. In this mechanism, the K atom effectively bumps into a brick wall, and the KI product bounces out in the backward direction.

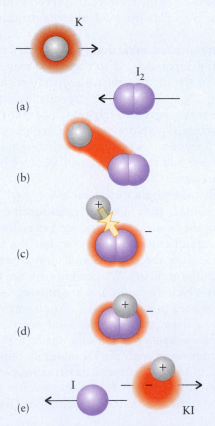

In the harpoon mechanism for the reaction between potassium and iodine to form potassium iodide, as a K atom approaches an I_2 molecule (a), an electron passes from the K atom to the I_2 molecule (b). The charge difference now tethers the two ions together (c and d) until an I^- ion separates and leaves with the K^+ ion (e).

complex species, the steric requirement may be severe, and P is very small. In such cases, the reaction rate is considerably less than the rate at which high-energy collisions occur.

> *According to the collision theory of gas-phase reactions, a reaction takes place only if the reactant molecules collide with a kinetic energy of at least the activation energy, and they do so in the correct orientation.*

13.13 Transition State Theory

Although collision theory applies to gas-phase reactions, we can extend some of its concepts to explain why the Arrhenius equation also applies to reactions in solution. There, molecules do not speed through space and collide, but jostle through the solvent and stay in one another's vicinity for relatively long periods. The more general theory that accounts both for this behavior and for reactions in gases is called **transition state theory** (which is also known as *activated complex theory*). This theory improves on collision theory by suggesting a way of calculating the rate constant even when steric requirements are significant.

In transition state theory we imagine two molecules approaching each other and being distorted as they come close enough to affect each other. In the gas phase, that meeting and distortion are equivalent to the "collision" of collision theory. In solution, the approach is a jostling zigzag walk among solvent molecules, and the distortion might not take place until after the two reactant molecules have met and received a particularly vigorous kick from the solvent molecules around them (Fig. 13.29). In either case, the collision or the kick does not immediately tear the molecules apart. Instead, the encounter results in the formation of an **activated complex,** an arrangement of the two molecules that can either go on to form products or fall apart again into the unchanged reactants. An activated complex is also commonly called a **transition state.**

In the activated complex, the original bonds have lengthened and weakened, and the new bonds are only partly formed. For example, in the proton transfer reaction between the weak acid HCN and water, the activated complex can be pictured as consisting of an HCN molecule with its hydrogen atom in the process of forming a hydrogen bond to the oxygen atom of a water molecule and poised midway between the two molecules. At this point, the hydrogen atom could re-form HCN or leave as the product H_3O^+.

In transition state theory, the activation energy is a measure of the energy of the activated complex (transition state) relative to that of the reactants. The reaction profile in Fig. 13.30 shows how the energy of the reaction mixture changes as the reactants meet, form the activated complex, and then go on to form products. A reaction profile shows the potential energy of the reactants and products, their total energy arising from their relative location, not their relative speed. Now consider the approach of the reactants to each other with a certain kinetic energy. As the reactants approach, they lose kinetic energy as they "climb" up the left side of the barrier (that is, as their potential energies increase due to their closer approach). If the initial kinetic energy of the reactants is less than E_a, they cannot climb to the top of the potential barrier, and they "roll" back down on the left and separate again. If their initial kinetic energy is at least E_a, they can form the activated complex, pass over the top of the barrier, and roll down the other side, where they separate as products.

A **potential energy surface** can help us visualize the energy changes in the course of a reaction as a function of the locations of the atoms. In this three-dimensional

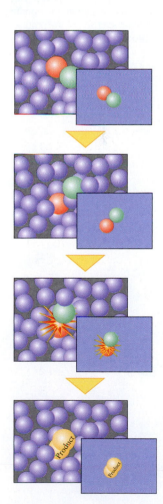

FIGURE 13.29 This sequence of images shows the reactant molecules in solution as they meet, then either move apart, or acquire enough energy by impacts from the solvent molecules to form an activated complex, which may go on to form products.

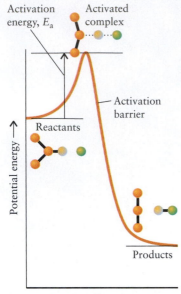

FIGURE 13.30 A reaction profile for an exothermic reaction. In the activated complex theory of reaction rates, it is supposed that the potential energy (the energy due to position) increases as the reactant molecules approach each other and reaches a maximum as they form an activated complex. It then decreases as the atoms rearrange into the bonding pattern characteristic of the products and these products separate. Only molecules with enough energy can cross the activation barrier and react to form products.

plot, the z-axis is a measure of the total potential energy of the reactants and products and the x- and y-axes represent interatomic distances. For example, the plot in Fig. 13.31 shows the potential energy changes that occur in the attack of a bromine atom on a hydrogen molecule and the reverse process, the attack of a hydrogen atom on an HBr molecule:

$$H_2 + Br\cdot \longrightarrow HBr + H\cdot$$
$$HBr + H\cdot \longrightarrow H_2 + Br\cdot$$

The low-energy locations corresponding to the reactants and products are separated by a barrier over which there is a path of minimum potential energy and which the kinetic energy of the approaching molecules must overcome. The actual path of the encounter depends on the total energies of the particles, but we can gain insight into the reaction process by examining potential energy changes alone. For example, suppose the H–H bond remained the same length as the Br atom approached. That would take the system to point A, a state of very high potential energy. In fact, at normal temperatures, the colliding species might not have enough kinetic energy to reach this point. The path that involves lowest potential energy is the one up the foot of the valley, over the "saddle point" (the saddle-shaped region) at the top of the pass, and down the foot of the valley on the other side of the pass. Only the lowest-energy path is available: the H–H bond must lengthen as the new H–Br bond begins to form.

As in collision theory, the rate of the reaction depends on the rate at which reactants can climb to the top of the barrier and form the activated complex. The resulting expression for the rate constant is very similar to the one given in Eq. 15, and so this more general theory also accounts for the form of the Arrhenius equation and the observed dependence of the reaction rate on temperature.

In transition state theory, a reaction takes place only if two molecules acquire enough energy, perhaps from the surrounding solvent, to form an activated complex and cross an energy barrier.

FIGURE 13.31 The contours of a potential energy surface for the reaction between a hydrogen molecule and a bromine atom (on the right). The atoms have been constrained to approach and depart in a straight line. The path of lowest potential energy (blue) is up one valley, across the pass—the saddle-shaped high point—and down the floor of the other valley to the products (on the left). The path shown in red would take the atoms to very high potential energies.

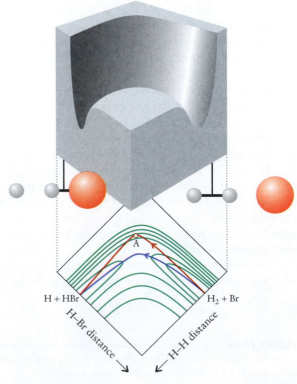
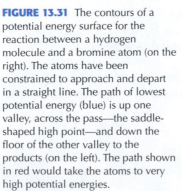

IMPACT ON MATERIALS AND BIOLOGY: ACCELERATING REACTIONS

We have seen that the rates of many reactions increase if we increase the concentration of reactants or the temperature. Similarly, the rate of a heterogeneous reaction can be increased by increasing the surface area of a reactant (Fig. 13.32). But suppose we want to increase the rate for a given concentration or surface area without raising the temperature? These sections describe an alternative.

13.14 Catalysis

One way to increase the rate of a reaction is to use a **catalyst,** a substance that increases the rate without being consumed in the reaction (Fig. 13.33). The name comes from the Greek words meaning "breaking down by coming together." In many cases, only a small amount of catalyst is necessary, because it acts over and over again. This is why small amounts of chlorofluorocarbons can have such a devastating effect on the ozone layer in the stratosphere—they break down into radicals that catalyze the destruction of ozone (Box 13.3).

A catalyst speeds up a reaction by providing an alternative pathway—a different reaction mechanism—between reactants and products. This new pathway has a lower activation energy than the original pathway (Fig. 13.34). At the same temperature, a greater fraction of reactant molecules can cross the lower barrier of the catalyzed path and turn into products than when no catalyst is present. Although the reaction takes place more quickly, a catalyst has no effect on the equilibrium composition. Both forward and reverse reactions are accelerated on the catalyzed path, leaving the equilibrium constant unchanged.

In some cases the original reaction with a slow rate-determining step may continue in parallel with the catalyzed reaction. However, the rate is determined by the faster path, which governs the overall rate of formation of products. A very slow elementary reaction does not control the rate if it can be sidestepped by a faster one on an alternative (usually catalyzed) path (Fig. 13.35).

A **homogeneous catalyst** is a catalyst that is in the same phase as the reactants. For reactants that are gases, a homogeneous catalyst is also a gas. If the reactants are in a liquid solution, a homogeneous catalyst is dissolved in the solution. Dissolved bromine is a homogeneous liquid-phase catalyst for the decomposition of aqueous hydrogen peroxide:

$$2\,H_2O_2(aq) \longrightarrow 2\,H_2O(l) + O_2(g)$$

FIGURE 13.32 Iron pots and pans can be heated in a flame without catching fire. However, a powder of finely divided iron filings oxidizes rapidly in air to form Fe_2O_3, because the powder presents a much greater surface area for reaction.

The Chinese characters for catalyst, which translate as "marriage broker," capture the sense quite well.

(a)   (b)

FIGURE 13.33 A small amount of catalyst—in this case, potassium iodide in aqueous solution—can accelerate the decomposition of hydrogen peroxide to water and oxygen. (a) The slow inflation of the balloon when no catalyst is present. (b) Its rapid inflation when a catalyst is present.

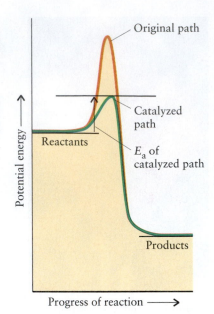

FIGURE 13.34 A catalyst provides a new reaction pathway with a lower activation energy, thereby allowing more reactant molecules to cross the barrier and form products.

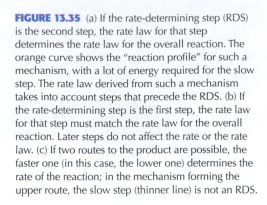

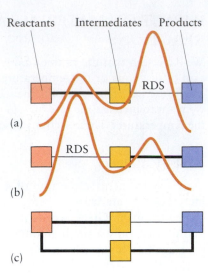

FIGURE 13.35 (a) If the rate-determining step (RDS) is the second step, the rate law for that step determines the rate law for the overall reaction. The orange curve shows the "reaction profile" for such a mechanism, with a lot of energy required for the slow step. The rate law derived from such a mechanism takes into account steps that precede the RDS. (b) If the rate-determining step is the first step, the rate law for that step must match the rate law for the overall reaction. Later steps do not affect the rate or the rate law. (c) If two routes to the product are possible, the faster one (in this case, the lower one) determines the rate of the reaction; in the mechanism forming the upper route, the slow step (thinner line) is not an RDS.

In the absence of bromine or another catalyst, a solution of hydrogen peroxide can be stored for a long time at room temperature; however, bubbles of oxygen form as soon as a drop of bromine is added. Bromine's role in this reaction is believed to be its reduction to Br^- in one step, followed by oxidation back to Br_2 in a second step:

$$Br_2(aq) + H_2O_2(aq) \longrightarrow 2\,Br^-(aq) + 2\,H^+(aq) + O_2(g)$$
$$2\,Br^-(aq) + H_2O_2(aq) + 2\,H^+(aq) \longrightarrow Br_2(aq) + 2\,H_2O(l)$$

When we add the two steps, both the catalyst, Br_2, and the intermediate, Br^-, cancel, leaving the overall equation as $2\,H_2O_2(aq) \rightarrow 2\,H_2O(l) + O_2(g)$. Hence, although Br_2 molecules have participated in the reaction, they are not consumed and can be used over and over again.

Although a catalyst does not appear in the balanced equation for a reaction, the concentration of a homogeneous catalyst does appear in the rate law. For example, the reaction between the triiodide ion and the azide ion is very slow unless a catalyst such as carbon disulfide is present:

$$I_3^-(aq) + 2\,N_3^-(aq) \xrightarrow{CS_2} 3\,I^-(aq) + 3\,N_2(g)$$

The mechanism for the reaction has two steps:

Step 1 Carbon disulfide forms a reactive intermediate, a complex with azide ions:

$$CS_2 + N_3^- \longrightarrow S_2CN_3^- \text{ (slow)}$$

Step 2 The complex then reacts rapidly with triiodide ion in a series of fast elementary reactions that can be summarized as

$$2\,S_2CN_3^- + I_3^- \longrightarrow 2\,CS_2 + 3\,N_2 + 2\,I^- \text{ (fast)}$$

The rate law derived from this mechanism is the same as the experimental rate law:

$$\text{Rate of consumption of } I_3^- = k[CS_2][N_3^-]$$

Notice that the rate law is first order in the catalyst, carbon disulfide, but zero order in triiodide ion, which appears only in the fast step following the slow step.

A **heterogeneous catalyst** is a catalyst present in a phase different from that of the reactants. The most common heterogeneous catalysts are finely divided or porous solids used in gas-phase or liquid-phase reactions. They are finely divided or porous so that they will provide a large surface area for the elementary reactions that provide the catalytic pathway. One example is the iron catalyst used in the

Haber process for ammonia. Another is finely divided vanadium pentoxide, V_2O_5, which is used in the "contact process" for the production of sulfuric acid:

$$2 SO_2(g) + O_2(g) \xrightarrow{V_2O_5} 2 SO_3(g)$$

The reactant is adsorbed on the catalyst's surface. As a reactant molecule attaches to the surface of the catalyst, its bonds are weakened and the reaction can proceed more quickly because the bonds are more easily broken (Fig. 13.36). One important step in the reaction mechanism of the Haber process for the synthesis of ammonia is the adsorption of N_2 molecules on the iron catalyst and the weakening of the strong N≡N triple bond.

The catalytic converters of automobiles use catalysts to bring about the complete and rapid combustion of unburned fuel (Fig. 13.37). The mixture of gases leaving an engine includes carbon monoxide, unburned hydrocarbons, and the nitrogen oxides collectively referred to as NO_x. Air pollution is decreased if the carbon compounds are oxidized to carbon dioxide and the NO_x reduced, by another catalyst, to nitrogen. The challenge is to find a catalyst—or a mixture of catalysts—that will accelerate both the oxidation and the reduction reactions and be active when the car is first started and the engine is cool.

Microporous catalysts are heterogeneous catalysts used in catalytic converters and for many other specialized applications, because of their very large surface areas and reaction specificity. **Zeolites,** for example, are microporous aluminosilicates (see Section 14.19) with three-dimensional structures riddled with hexagonal channels connected by tunnels (Fig. 13.38). The enclosed nature of the active sites in zeolites gives them a special advantage over other heterogeneous catalysts, because an intermediate can be held in place inside the channels until the products form. Moreover, the channels allow products to grow only to a particular size.

Catalysts can be **poisoned,** or inactivated. A common cause of such poisoning is the adsorption of a molecule so tightly to the catalyst that it seals the surface of the catalyst against further reaction. Some heavy metals, especially lead, are very potent poisons for heterogeneous catalysts, which is why lead-free gasoline must be used in engines fitted with catalytic converters. The elimination of

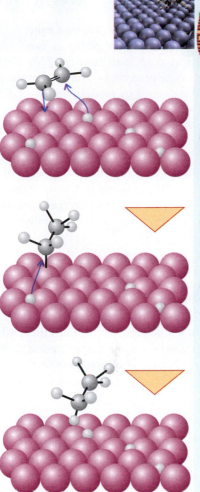

Animation: Figure 13.36 Heterogeneous catalysis

FIGURE 13.36 The reaction between ethene, $CH_2=CH_2$, and hydrogen on a catalytic metal surface. In this sequence of images, we see the ethene molecule approaching the metal surface to which hydrogen molecules have already adsorbed: when they adsorb, they dissociate and stick to the surface as hydrogen atoms. Next, after the ethene molecule also sticks to the surface, it meets a hydrogen atom and forms a bond. At this stage, the ·CH_2CH_3 radical is attached to the surface by one of its carbon atoms. Finally, the radical and another hydrogen atom meet; ethane is formed and escapes from the surface.

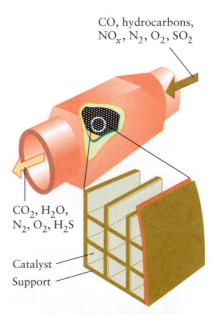

CO, hydrocarbons, NO_x, N_2, O_2, SO_2

CO_2, H_2O, N_2, O_2, H_2S

Catalyst

Support

FIGURE 13.37 The catalytic converter of an automobile is made from a mixture of catalysts bonded to a honeycomb ceramic support.

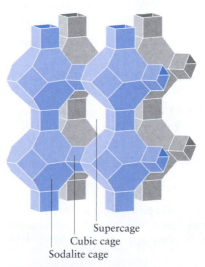

Supercage

Cubic cage

Sodalite cage

FIGURE 13.38 The structure of the ZSM-5 zeolite catalyst. Reactants diffuse through the channels, which are narrow enough to hold intermediates in positions favorable for reaction.

BOX 13.3   WHAT HAS THIS TO DO WITH . . . THE ENVIRONMENT?

Protecting the Ozone Layer

Every year our planet is bombarded with enough energy from the Sun to destroy all life. Only the ozone in the stratosphere protects us from that onslaught. The ozone, though, is threatened by modern life styles. Chemicals used as coolants and propellants, such as chlorofluorocarbons (CFCs), and the nitrogen oxides in jet exhausts, have been found to create holes in Earth's protective ozone layer. Because they act as catalysts, even small amounts of these chemicals can cause large changes in the vast reaches of the stratosphere.

Ozone forms in the stratosphere in two steps. First, O_2 molecules or other oxygen-containing compounds are broken apart into atoms by sunlight, a process called *photodissociation*:

$$O_2 \xrightarrow{\text{sunlight, } \lambda < 340 \text{ nm}} O + O$$

Then the O atoms, which are reactive radicals with two unpaired electrons, react with the more abundant O_2 molecules to form ozone. The ozone molecules are created in such a high-energy state that their vibrational motions would quickly tear them apart unless another molecule, such as O_2 or N_2, collides with them first. The other molecule, indicated as M, carries off some of the energy:

$$O + O_2 \longrightarrow O_3{}^*$$
$$O_3{}^* + M \longrightarrow O_3 + M$$

where * denotes a high-energy state. The net reaction, the sum of these two elementary reactions (after multiplying them by 2) and the reaction presented earlier, is $3\,O_2 \rightarrow 2\,O_3$. Some of the ozone is decomposed by ultraviolet radiation:

$$O_3 \xrightarrow{\text{UV, } \lambda < 340 \text{ nm}} O + O_2$$

The oxygen atom produced in this step can react with oxygen molecules to produce more ozone, and so the ozone concentration in the stratosphere normally remains constant, with seasonal variations. Because the decomposition of ozone

absorbs ultraviolet radiation, ozone helps to shield the Earth from radiation damage.

Warnings of the possibility that *androgenic* (human-generated) chemicals can threaten the ozone in the stratosphere began to appear in 1970 and 1971, when Paul Crutzen determined experimentally that NO and NO_2 molecules catalyze the destruction of ozone. Nitrogen monoxide (nitric oxide), NO, is produced naturally in the atmosphere by lightning, but it is also a by-product of combustion in automobile and airplane engines. The NO is then oxidized by the oxygen in air to nitrogen dioxide, NO_2.

Mario Molina and Sherwood Rowland used Crutzen's work and other data in 1974 to build a model of the stratosphere that explained how chlorofluorocarbons could threaten the ozone layer.[1] In 1985, ozone levels over Antarctica were indeed found to be decreasing and had dropped to the lowest ever observed; by the year 2000, the hole had reached Chile. These losses are now known to be global in extent and it has been postulated that they may be contributing to global warming in the Southern Hemisphere.

The minimum ozone concentration over Antarctica occurs in September or October. This plot shows how the ozone concentration declined rapidly until 1999, when the phasing out of chlorofluorocarbons allowed it to stabilize.

---

lead has the further benefit of decreasing the amount of poisonous lead in the environment.

SELF-TEST 13.14A How does a catalyst affect (a) the rate of the reverse reaction; (b) ΔH for the reaction?

[*Answer:* (a) Increases it; (b) has no effect]

SELF-TEST 13.14B How does a homogeneous catalyst affect (a) the rate law; (b) the equilibrium constant?

Catalysts participate in reactions but are not themselves used up; they provide a reaction pathway with a lower activation energy. Catalysts are classified as either homogeneous or heterogeneous.

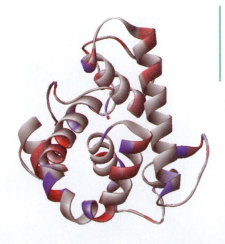

FIGURE 13.39 The lysozyme molecule is a typical enzyme molecule. Lysozyme is present in a number of places in the body, including tears and the mucus in the nose. One of its functions is to attack the cell walls of bacteria and destroy them. This "ribbon" representation shows only the general arrangement of the atoms to emphasize the overall shape of the molecule; the ribbon actually consists of amino acids linked together (Section 19.13).

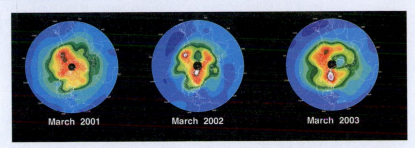

These maps of stratospheric ozone concentration over the North Pole show how the ozone was depleted from 2001 to 2003. Red areas represent ozone concentrations greater than 500 Dobson units (DU); the concentrations decrease through green, yellow, and blue, to purple at less than 270 DU. Normal ozone concentration at temperate latitudes is about 350 DU.

Susan Solomon and James Anderson showed that CFCs produce chlorine atoms and chlorine oxide under the conditions of the ozone layer and identified the CFCs emanating from everyday objects, such as cans of hair spray, refrigerators, and air conditioners, as the primary culprits in the destruction of stratospheric ozone. The CFC molecules are not very polar, and so they do not dissolve in rain or the oceans. Instead, they rise to the stratosphere, where they are exposed to ultraviolet radiation from the Sun. They readily dissociate in the presence of this radiation and form chlorine atoms, which destroy ozone by various mechanisms, one of which is

$$Cl\cdot + O_3 \longrightarrow ClO\cdot + O_2$$
$$ClO\cdot + \cdot O\cdot \longrightarrow Cl\cdot + O_2$$

The O atoms are produced when ozone is decomposed by ultraviolet light, as described previously. Notice that the net reaction, $O_3 + O \rightarrow O_2 + O_2$, does not involve chlorine. Chlorine atoms act as continuously regenerated catalysts, and so even a low abundance can do a lot of damage.

Most nations signed the Montreal Protocol of 1987 and the amendments added in 1992, which required the more dangerous CFCs to be phased out by 1996. The concentration of ozone in the stratosphere is measured by NASA's Total Ozone Mapping Spectrometer (TOMS). Although it takes years for CFC molecules to diffuse into the stratosphere, TOMS data show that the levels of compounds that deplete ozone are gradually beginning to decrease. It is thought that if the Montreal Protocol continues to be observed and there are no major volcanic eruptions (which release dust that accelerates the destruction of ozone), the ozone hole will have begun to decrease in size by the year 2010.

Related Exercises: 13.45, 13.59, and 13.97

For Further Reading: C. Baird and M. Cann, "Stratospheric chemistry: the ozone layer," *Environmental Chemistry* (New York: W. H. Freeman and Company, 2005). P. J. Crutzen, "What is happening to our precious air," *Science Spectra*, vol. 14, 1998, pp. 22–31. D. J. Karoly, "Atmospheric science: Ozone and climate change," *Science*, vol. 302, 2003, pp. 236–237.

[1]Crutzen, Molina, and Rowland were awarded the 1995 Nobel prize in chemistry "for their work in atmospheric chemistry, particularly concerning the formation and decomposition of ozone."

13.15 Living Catalysts: Enzymes

Living cells contain thousands of different kinds of catalysts, each of which is necessary to life. Many of these catalysts are proteins called *enzymes*, large molecules with a slotlike active site, where reaction takes place (Fig. 13.39). The **substrate,** the molecule on which the enzyme acts, fits into the slot as a key fits into a lock (Fig. 13.40). However, unlike an ordinary lock, a protein molecule distorts slightly as the substrate molecule approaches, and its ability to undergo the correct distortion also determines whether the "key" will fit. This refinement of the original lock-and-key model is known as the **induced-fit mechanism** of enzyme action.

Once in the active site, the substrate undergoes reaction. The product is then released for use in the next stage, which is controlled by another enzyme, and the original enzyme molecule is free to receive the next substrate molecule. One example

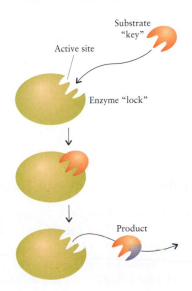

FIGURE 13.40 In the lock-and-key model of enzyme action, the correct substrate is recognized by its ability to fit into the active site like a key into a lock. In a refinement of this model, the enzyme changes its shape slightly as the key enters.

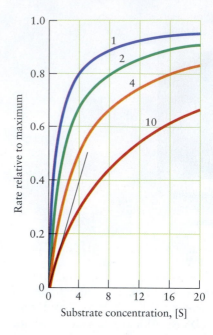

FIGURE 13.41 A plot of the rate of an enzyme-catalyzed reaction (relative to its maximum value, $k_2[E]_0$, when S is in very high concentration) as a function of concentration of substrate for various values of K_M. At low substrate concentrations, the rate of reaction is directly proportional to the substrate concentration (as indicated by the black line for $K_M = 10$). At high substrate concentrations, the rate becomes constant at $k_2[E]_0$ once the enzyme molecules are "saturated" with substrate. The units of [S] are the same as those of K_M.

of an enzyme is the amylase present in your mouth. The amylase in saliva helps to break down starches in food into the more easily digested glucose. If you chew a cracker long enough, you can notice the increased sweetness.

The kinetics of enzyme reactions were first studied by the German chemists Leonor Michaelis and Maud Menten in the early part of the twentieth century. They found that, when the concentration of substrate is low, the rate of an enzyme-catalyzed reaction increases with the concentration of the substrate, as shown in the plot in Fig. 13.41. However, when the concentration of substrate is high, the reaction rate depends only on the concentration of the enzyme. In the **Michaelis–Menten mechanism** of enzyme reaction, the enzyme, E, and substrate, S, reach a rapid pre-equilibrium with the bound enzyme–substrate complex, ES:

$$E + S \rightleftharpoons ES \qquad \text{forward: second order,} \qquad \text{Rate} = k_1[E][S]$$
$$\text{reverse: first order,} \qquad \text{Rate} = k_1'[ES]$$

The complex decays with first-order kinetics, releasing the enzyme to act again:

$$ES \longrightarrow E + \text{product} \qquad \text{forward: first order,} \qquad \text{Rate} = k_2[ES]$$

When the overall rate law is worked out (see Exercise 13.92), we find that

$$\text{Rate of formation of product} = \frac{k_2[E]_0[S]}{K_M + [S]} \qquad (19a)*$$

where $[E]_0$ is the total concentration of enzyme (bound plus unbound) and the **Michaelis constant**, K_M, is

$$K_M = \frac{k_1' + k_2}{k_1} \qquad (19b)*$$

When we plot the rate given by Eq. 19a against the concentration of substrate, we get a curve exactly like the one observed.

One form of biological poisoning mirrors the effect of lead on a catalytic converter. The activity of an enzyme is destroyed if an alien substrate attaches too strongly to the enzyme's active site, because then the site is blocked and made unavailable to the true substrate (Fig. 13.42). As a result, the chain of bio-chemical reactions in the cell stops, and the cell dies. The action of nerve gases is believed to stem from their ability to block the enzyme-controlled reactions that allow impulses to travel through nerves. Arsenic, that favorite of fictional poisoners, acts in a similar way. After ingestion as As(V) in the form of arsenate ions (AsO_4^{3-}), it is reduced to As(III), which binds to enzymes and inhibits their action.

Enzymes are biological catalysts that function by modifying substrate molecules to promote reaction.

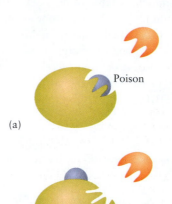

FIGURE 13.42 (a) An enzyme poison (represented by the blue sphere) can act by attaching so strongly to the active site that it blocks the site, thereby taking the enzyme out of action. (b) Alternatively, the poison molecule may attach elsewhere, so distorting the enzyme molecule and its active site that the substrate no longer fits.

SKILLS YOU SHOULD HAVE MASTERED

❏ 1 Determine the order of a reaction, its rate law, and its rate constant from experimental data (Examples 13.1 and 13.2 and Self-Tests 13.2 and 13.3).

❏ 2 Calculate a concentration, time, or rate constant by using an integrated rate law (Examples 13.3, 13.4, and 13.5).

❏ 3 For a first-order process, calculate the rate constant, elapsed time, and amount remaining from the half-life (Example 13.6 and Self-Test 13.9).

❏ 4 Deduce a rate law from a mechanism (Example 13.7).

❑ 5 Show how the equilibrium constant is related to the forward and reverse rate constants of the elementary reactions contributing to an overall reaction (Section 13.10).

❑ 6 Use the Arrhenius equation and temperature dependence of a rate constant to determine an activation energy (Example 13.8).

❑ 7 Use the activation energy to find the rate constant at a given temperature (Example 13.9).

❑ 8 Show how collision theory and transition state theory account for the temperature dependence of reactions (Sections 13.12 and 13.13).

❑ 9 Describe the action of catalysts in terms of a reaction profile (Section 13.14).

❑ 10 Write a rate law for the action of enzymes (Section 13.15).

EXERCISES

Exercises marked $\boxed{C}$ *require calculus.*

All rates are unique reaction rates unless otherwise stated.

Reaction Rates

13.1 Complete the following statements relating to the production of ammonia by the Haber process, for which the overall reaction is $N_2(g) + 3\,H_2(g) \rightarrow 2\,NH_3(g)$. (a) The rate of disappearance of N_2 is _____ times the rate of disappearance of H_2. (b) The rate of formation of NH_3 is _____ times the rate of disappearance of H_2. (c) The rate of formation of NH_3 is _____ times the rate of disappearance of N_2.

13.2 Complete the following statements for the reaction $2\,N_2O_5(g) \rightarrow 4\,NO_2(g) + O_2(g)$. (a) The rate of disappearance of N_2O_5 is _____ times the rate of formation of O_2. (b) The rate of formation of NO_2 is _____ times the rate of disappearance of N_2O_5. (c) The rate of formation of NO_2 is _____ times the rate of formation of O_2.

13.3 (a) The rate of formation of dichromate ions is $0.14\ \text{mol·L}^{-1}\text{·s}^{-1}$ in the reaction $2\,CrO_4^{2-}(aq) + 2\,H^+(aq) \rightarrow Cr_2O_7^{2-}(aq) + H_2O(l)$. What is the rate of reaction of chromate ions in the reaction? (b) What is the unique rate of the reaction?

13.4 (a) In the reaction $3\,ClO^-(aq) \rightarrow 2\,Cl^-(aq) + ClO_3^-(aq)$, the rate of formation of Cl^- is $3.6\ \text{mol·L}^{-1}\text{·min}^{-1}$. What is the rate of reaction of ClO^-? (b) What is the unique rate of the reaction?

13.5 (a) Nitrogen dioxide, NO_2, decomposes at 6.5 $\text{mmol·L}^{-1}\text{·s}^{-1}$ by the reaction $2\,NO_2(g) \rightarrow 2\,NO(g) + O_2(g)$. Determine the rate of formation of O_2. (b) What is the unique rate of the reaction?

13.6 Manganate ions, MnO_4^{2-}, react at $2.0\ \text{mol·L}^{-1}\text{·min}^{-1}$ in acidic solution to form permanganate ions and manganese(IV) oxide: $3\,MnO_4^{2-}(aq) + 4\,H^+(aq) \rightarrow 2\,MnO_4^-(aq) + MnO_2(s) + 2\,H_2O(l)$. (a) What is the rate of formation of permanganate ions? (b) What is the rate of reaction of $H^+(aq)$? (c) What is the unique rate of the reaction?

 13.7 The decomposition of gaseous hydrogen iodide, $2\,HI(g) \rightarrow H_2(g) + I_2(g)$, gives the data shown here for 700. K.

Time (s)	0.	1000.	2000.	3000.	4000.	5000.
[HI] (mmol·L^{-1})	10.0	4.4	2.8	2.1	1.6	1.3

(a) Use standard graphing software, such as that on the Web site for this book, to plot the concentration of HI as a function of time. (b) Estimate the rate of decomposition of HI at each

time. (c) Plot the concentrations of H_2 and I_2 as a function of time on the same graph.

 13.8 The decomposition of gaseous dinitrogen pentoxide in the reaction $2\,N_2O_5(g) \rightarrow 4\,NO_2(g) + O_2(g)$ gives the data shown here at 298 K. (a) Using standard graphing software, such as that on the Web site for this book, plot the concentration of N_2O_5 as a function of time. (b) Estimate the rate of decomposition of N_2O_5 at each time. (c) Plot the concentrations of NO_2 and O_2 as a function of time on the same graph.

Time (s)	0.	1.11	2.22	3.33	4.44
[N$_2$O$_5$] (mmol·L^{-1})	2.15	1.88	1.64	1.43	1.25

Rate Laws and Reaction Order

Rate constants are listed in Table 13.1.

13.9 Express the units for rate constants when the concentrations are in moles per liter and time is in seconds for (a) zero-order reactions; (b) first-order reactions; (c) second-order reactions.

13.10 Because partial pressures are proportional to concentrations, rate laws for gas-phase reactions can also be expressed in terms of partial pressures, for instance, as Rate $= kP_X$ for a first-order reaction of a gas X. What are the units for the rate constants when partial pressures are expressed in torr and time is expressed in seconds for (a) zero-order reactions; (b) first-order reactions; (c) second-order reactions?

13.11 Dinitrogen pentoxide, N_2O_5, decomposes by a first-order reaction. What is the initial rate of decomposition of N_2O_5 when 3.45 g of N_2O_5 is confined in a 0.750-L container and heated to 65°C? For this reaction, $k = 5.2 \times 10^{-3}\ \text{s}^{-1}$ in the rate law (for the rate of decomposition of N_2O_5).

13.12 Ethane, C_2H_6, dissociates into methyl radicals by a first-order reaction at elevated temperatures. If 250. mg of ethane is confined to a 500.-mL reaction vessel and heated to 700°C, what is the initial rate of ethane decomposition if $k = 5.5 \times 10^{-4}\ \text{s}^{-1}$ in the rate law (for the rate of dissociation of C_2H_6)?

13.13 When 0.52 g of H_2 and 0.19 g of I_2 are confined to a 750.-mL reaction vessel and heated to 700. K, they react by a second-order process (first order in each reactant), with $k = 0.063\ \text{L·mol}^{-1}\text{·s}^{-1}$ in the rate law (for the rate of formation of HI). (a) What is the initial reaction rate? (b) By what factor does the reaction rate increase if the concentration of H_2 present in the mixture is doubled?

13.14 When 420. mg of NO_2 is confined to a 150.-mL reaction vessel and heated to 300°C, it decomposes by a second-order process. In the rate law for the decomposition of NO_2, $k = 0.54\ \text{L·mol}^{-1}\text{·s}^{-1}$. (a) What is the initial reaction rate? (b) How does the reaction rate change (and by what factor) if the mass of NO_2 present in the container is increased to 750. mg?

13.15 In the reaction $CH_3Br(aq) + OH^-(aq) \rightarrow CH_3OH(aq) + Br^-(aq)$, when the OH^- concentration alone was doubled, the rate doubled; when the CH_3Br concentration alone was increased by a factor of 1.2, the rate increased by a factor of 1.2. Write the rate law for the reaction.

13.16 In the reaction $2\,NO(g) + O_2(g) \rightarrow 2\,NO_2(g)$, when the NO concentration alone was doubled, the rate increased by a factor of 4; when both the NO and the O_2 concentrations were increased by a factor of 2, the rate increased by a factor of 8. What is the rate law for the reaction?

13.17 The following kinetic data were obtained for the reaction $2\,ICl(g) + H_2(g) \rightarrow I_2(g) + 2\,HCl(g)$.

	Initial concentration ($mmol \cdot L^{-1}$)		
Experiment	$[ICl]_0$	$[H_2]_0$	Initial rate ($mol \cdot L^{-1} \cdot s^{-1}$)
1	1.5	1.5	3.7×10^{-7}
2	3.0	1.5	7.4×10^{-7}
3	3.0	4.5	2.2×10^{-6}
4	4.7	2.7	?

(a) Write the rate law for the reaction. (b) From the data, determine the value of the rate constant. (c) Use the data to predict the reaction rate for Experiment 4.

13.18 The following kinetic data were obtained for the reaction $A(g) + 2\,B(g) \rightarrow$ product.

	Initial concentration ($mol \cdot L^{-1}$)		
Experiment	$[A]_0$	$[B]_0$	Initial rate ($mol \cdot L^{-1} \cdot s^{-1}$)
1	0.60	0.30	12.6
2	0.20	0.30	1.4
3	0.60	0.10	4.2
4	0.17	0.25	?

(a) What is the order with respect to each reactant, and the overall order of the reaction? (b) Write the rate law for the reaction. (c) From the data, determine the value of the rate constant. (d) Use the data to predict the reaction rate for Experiment 4.

13.19 The following data were obtained for the reaction $A + B + C \rightarrow$ products:

	Initial concentration ($mmol \cdot L^{-1}$)			
Experiment	$[A]_0$	$[B]_0$	$[C]_0$	Initial rate (($mmol\ A) \cdot L^{-1} \cdot s^{-1}$)
1	1.25	1.25	1.25	8.7
2	2.5	1.25	1.25	17.4
3	1.25	3.02	1.25	50.8
4	1.25	3.02	3.75	457
5	3.01	1.00	1.15	?

(a) Write the rate law for the reaction. (b) What is the order of the reaction? (c) Determine the value of the rate constant. (d) Use the data to predict the reaction rate for Experiment 5.

13.20 The following kinetic data were obtained for the reaction $3\,A(g) + C(g) \rightarrow$ product:

	Initial concentration ($mol \cdot L^{-1}$)		
Experiment	$[A]_0$	$[C]_0$	Initial rate ($mol \cdot L^{-1} \cdot s^{-1}$)
1	1.72	2.44	0.68
2	3:44	2.44	5.44
3	1.72	0.10	2.8×10^{-2}
4	2.91	1.33	?

(a) What is the order with respect to each reactant, and the overall order of the reaction? (b) Write the rate law for the reaction. (c) From the data, determine the value of the rate constant. (d) Use the data to predict the reaction rate for Experiment 4.

Integrated Rate Laws

13.21 Determine the rate constant for each of the following first-order reactions, in each case expressed for the rate of loss of A: (a) $A \rightarrow B$, given that the concentration of A decreases to one-half its initial value in 1000. s; (b) $A \rightarrow B$, given that the concentration of A decreases from 0.67 $mol \cdot L^{-1}$ to 0.53 $mol \cdot L^{-1}$ in 25 s; (c) $2\,A \rightarrow B + C$, given that $[A]_0 = 0.153$ $mol \cdot L^{-1}$ and that after 115 s the concentration of B rises to 0.034 $mol \cdot L^{-1}$.

13.22 Determine the rate constant for each of the following first-order reactions: (a) $2\,A \rightarrow B + C$, given that the concentration of A decreases to one-fourth its initial value in 38 min; (b) $2\,A \rightarrow B + C$, given that $[A]_0 = 0.039$ $mol \cdot L^{-1}$ and that after 75 s the concentration of B increases to 0.0095 $mol \cdot L^{-1}$; (c) $2\,A \rightarrow 3\,B + C$, given that $[A]_0 = 0.040$ $mol \cdot L^{-1}$ and that after 8.8 min the concentration of B rises to 0.030 $mol \cdot L^{-1}$. In each case, write the rate law for the rate of loss of A.

13.23 Dinitrogen pentoxide, N_2O_5, decomposes by first-order kinetics with a rate constant of 3.7×10^{-5} s^{-1} at 298 K. (a) What is the half-life (in hours) for the decomposition of N_2O_5 at 298 K? (b) If $[N_2O_5]_0 = 0.0567$ $mol \cdot L^{-1}$, what will be the concentration of N_2O_5 after 3.5 h? (c) How much time (in minutes) will elapse before the N_2O_5 concentration decreases from 0.0567 $mol \cdot L^{-1}$ to 0.0135 $mol \cdot L^{-1}$?

13.24 Dinitrogen pentoxide, N_2O_5, decomposes by first-order kinetics with a rate constant of 0.15 s^{-1} at 353 K. (a) What is the half-life (in seconds) for the decomposition of N_2O_5 at 353 K? (b) If $[N_2O_5]_0 = 0.0567$ $mol \cdot L^{-1}$, what will be the concentration of N_2O_5 after 2.0 s? (c) How much time (in minutes) will elapse before the N_2O_5 concentration decreases from 0.0567 $mol \cdot L^{-1}$ to 0.0135 $mol \cdot L^{-1}$?

13.25 The half-life for the first-order decomposition of A is 355 s. How much time must elapse for the concentration of A to decrease to (a) one-fourth; (b) 15% of its original value; (c) one-ninth of its initial concentration?

13.26 The first-order rate constant for the photodissociation of A is 5.74 h^{-1}. Calculate the time needed for the concentration of A to decrease to (a) $\frac{1}{8}[A]_0$; (b) 5% of its original value; (c) one-tenth of its initial concentration.

13.27 Sulfuryl chloride, SO_2Cl_2, decomposes by first-order kinetics, and $k = 2.81 \times 10^{-3}$ min^{-1} at a certain temperature.

(a) Determine the half-life for the reaction. (b) Determine the time needed for the concentration of SO_2Cl_2 to decrease to 10% of its initial concentration. (c) If 14.0 g of SO_2Cl_2 is sealed in a 2500.-L reaction vessel and heated to the specified temperature, what mass will remain after 1.5 h?

13.28 Ethane, C_2H_6, forms $\cdot CH_3$ radicals at 700.°C in a first-order reaction, for which $k = 1.98 \text{ h}^{-1}$. (a) What is the half-life for the reaction? (b) Calculate the time needed for the amount of ethane to fall from 1.15×10^{-3} mol to 2.35×10^{-4} mol in a 500.-mL reaction vessel at 700.°C. (c) How much of a 6.88-mg sample of ethane in a 500.-mL reaction vessel at 700.°C will remain after 45 min?

13.29 For the first-order reaction $A \rightarrow 3 B + C$, when $[A]_0 = 0.015 \text{ mol·L}^{-1}$, the concentration of B increases to 0.018 mol·L^{-1} in 3.0 min. (a) What is the rate constant for the reaction expressed as the rate of loss of A? (b) How much more time would be needed for the concentration of B to increase to 0.030 mol·L^{-1}?

13.30 Pyruvic acid is an intermediate in the fermentation of grains. During fermentation the enzyme pyruvate carboxylase causes the pyruvate ion to release carbon dioxide. In one experiment a 200.-mL aqueous solution of the pyruvate in a sealed, rigid 500.-mL flask at 293 K had an initial concentration of 3.23 mmol·L^{-1}. Because the concentration of the enzyme was kept constant, the reaction was pseudo-first order in pyruvate ion. The elimination of CO_2 by the reaction was monitored by measuring the partial pressure of the CO_2 gas. The pressure of the gas was found to rise from zero to 100. Pa in 522 s. What is the rate constant of the pseudo-first order reaction?

13.31 The data below were collected for the reaction $2 HI(g) \rightarrow H_2(g) + I_2(g)$ at 580 K.

Time (s)	0	1000.	2000.	3000.	4000.
[HI] (mol·L^{-1})	1.0	0.11	0.061	0.041	0.031

(a) Using graphing software, such as that on the Web site for this book, plot the data in an appropriate fashion to determine the order of the reaction. (b) From the graph, determine the rate constant for (i) the rate law for the loss of HI and (ii) the unique rate law.

13.32 The data below were collected for the reaction $H_2(g) + I_2(g) \rightarrow 2 HI(g)$ at 780 K. (a) Using graphing software, such as that on the Web site for this book, plot the data in an appropriate fashion to determine the order of the reaction. (b) From the graph, determine the rate constant for the rate of disappearance of I_2.

Time (s)	0	1.0	2.0	3.0	4.0
[I$_2$] (mmol·L^{-1})	1.00	0.43	0.27	0.20	0.16

13.33 The half-life for the second-order reaction of a substance A is 50.5 s when $[A]_0 = 0.84 \text{ mol·L}^{-1}$. Calculate the time needed for the concentration of A to decrease to (a) one-sixteenth; (b) one-fourth; (c) one-fifth of its original value.

13.34 Determine the rate constant for each of the following second-order reactions: (a) $2 A \rightarrow B + 2 C$, given that the concentration of A decreases from 2.50 mmol·L^{-1} to 1.25

mmol·L^{-1} in 100 s; (b) $A \rightarrow C + 2 D$, given that $[A]_0 = 0.300$ mol·L^{-1} and that the concentration of C increases to 0.010 mol·L^{-1} in 200. s. In each case, express your results in terms of the rate law for the loss of A.

13.35 Determine the time required for each of the following second-order reactions to take place: (a) $2 A \rightarrow B + C$, for the concentration of A to decrease from 0.10 mol·L^{-1} to 0.080 mol·L^{-1}, given that $k = 0.015 \text{ L·mol}^{-1}\cdot\text{min}^{-1}$ for the rate law expressed in terms of the loss of A; (b) $A \rightarrow 2 B + C$, when $[A]_0 = 0.15 \text{ mol·L}^{-1}$, for the concentration of B to increase to 0.19 mol·L^{-1}, given that $k = 0.0035 \text{ L·mol}^{-1}\cdot\text{min}^{-1}$ in the rate law for the loss of A.

13.36 The second-order rate constant for the decomposition of NO_2 (to NO and O_2) at 573 K is $0.54 \text{ L·mol}^{-1}\cdot\text{s}^{-1}$. Calculate the time for an initial NO_2 concentration of 0.20 mol·L^{-1} to decrease to (a) one-half; (b) one-sixteenth; (c) one-ninth of its initial concentration.

[C] **13.37** Suppose a reaction has the form $a A \rightarrow$ products and we consider the unique rate of the reaction and write the rate law Rate $= k[A]$. Derive an expression for the concentration of A at a time t and for the half-life of A in terms of a and k.

[C] **13.38** Suppose a reaction has the form $a A \rightarrow$ products and we consider the unique rate of the reaction and write the rate law Rate $= k[A]^2$. Derive an expression for the concentration of A at a time t in terms of a and k.

[C] **13.39** Derive an expression for the half-life of the reactant A that decays by a third-order reaction with rate constant k.

[C] **13.40** Derive an expression for the half-life of the reactant A that decays by an nth-order reaction (with $n > 1$) with rate constant k.

Reaction Mechanisms

13.41 Write the overall reaction for the mechanism proposed below and identify any reaction intermediates.

Step 1 $AC + B \longrightarrow AB + C$

Step 2 $AC + AB \longrightarrow A_2B + C$

13.42 Write the overall reaction for the mechanism proposed below and identify any reaction intermediates.

Step 1 $Cl_2 \longrightarrow Cl + Cl$

Step 2 $Cl + CO \longrightarrow COCl$

Step 3 $COCl + Cl_2 \longrightarrow COCl_2 + Cl$

13.43 Write the overall reaction for the following proposed mechanism and identify any reaction intermediates:

Step 1
$$\begin{array}{c}H_2C{=}CH\\ \backslash\\ C{-}OH + HCl \longrightarrow\\ /\!/\\ O\end{array}\qquad\begin{array}{c}H_2C{=}CH\\ \backslash\\ C{-}OH + Cl^-\\ /\!/\\ {}^+HO\end{array}$$

Step 2
$$Cl^- + \begin{array}{c}H_2C{=}CH\\ \backslash\\ C{-}OH\\ /\!/\\ {}^+HO\end{array} \longrightarrow \begin{array}{c}H_2C{-}CH\\ \backslash\ \backslash\backslash\\ Cl\quad C{-}OH\\ /\\ HO\end{array}$$

Step 3
$$\begin{array}{c}H_2C{-}CH\\ /\ \backslash\backslash\\ Cl\quad C{-}OH \longrightarrow\\ /\\ HO\end{array}\qquad\begin{array}{c}H_2C{-}CH_2\\ /\ \backslash\\ Cl\quad C{-}OH\\ /\!/\\ O\end{array}$$

13.44 Lead(IV) acetate cleaves organic diols such as 2,3-dihydroxy-2,3-dimethylbutane in aqueous solution to give ketones. Write the overall reaction for the following proposed mechanism and identify any reaction intermediates:

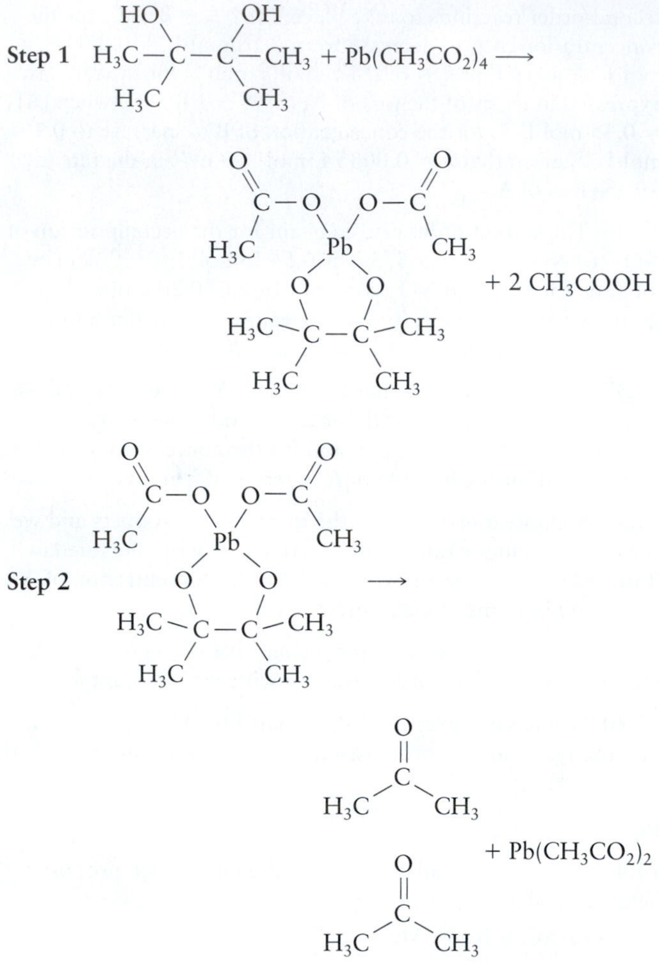

13.45 The contribution to the destruction of the ozone layer caused by high flying aircraft has been attributed to the following mechanism:

> **Step 1** $O_3 + NO \longrightarrow NO_2 + O_2$
>
> **Step 2** $NO_2 + O \longrightarrow NO + O_2$

(a) Write the overall reaction. (b) Write the rate law for each step and indicate its molecularity. (c) What is the reaction intermediate? (d) A catalyst is a substance that accelerates the rate of a reaction and is regenerated in the process. What is the catalyst in the reaction? (See Box 13.3.)

13.46 A reaction was believed to occur by the following mechanism.

> **Step 1** $A_2 \longrightarrow A + A$
>
> **Step 2** $A + A + B \longrightarrow A_2B$
>
> **Step 3** $A_2B + C \longrightarrow A_2 + BC$

(a) Write the overall reaction. (b) Write the rate law for each step and indicate its molecularity. (c) What are the reaction intermediates? (d) A catalyst is a substance that accelerates the

rate of a reaction and is regenerated in the process. What is the catalyst in the reaction?

13.47 The following mechanism has been proposed for the reaction between nitric oxide and bromine:

> **Step 1** $NO + Br_2 \longrightarrow NOBr_2$ (slow)
>
> **Step 2** $NOBr_2 + NO \longrightarrow NOBr + NOBr$ (fast)

Write the rate law for the formation of NOBr implied by this mechanism.

13.48 The mechanism proposed for the oxidation of iodide ion by the hypochlorite ion in aqueous solution is as follows:

> **Step 1** $ClO^- + H_2O \longrightarrow HClO + OH^-$ and its reverse (both fast, equilibrium)
>
> **Step 2** $I^- + HClO \longrightarrow HIO + Cl^-$ (slow)
>
> **Step 3** $HIO + OH^- \longrightarrow IO^- + H_2O$ (fast)

Write the rate law for the formation of HIO implied by this mechanism.

13.49 Three mechanisms for the reaction $NO_2(g) + CO(g) \rightarrow CO_2(g) + NO(g)$ have been proposed:

(a) **Step 1** $NO_2 + CO \rightarrow CO_2 + NO$
(b) **Step 2** $NO_2 + NO_2 \rightarrow NO + NO_3$ (slow)
  Step 3 $NO_3 + CO \rightarrow NO_2 + CO_2$ (fast)
(c) **Step 1** $NO_2 + NO_2 \rightarrow NO + NO_3$ and its reverse (both fast, equilibrium)
  Step 2 $NO_3 + CO \rightarrow NO_2 + CO_2$ (slow)

Which mechanism agrees with the following rate law: Rate = $k[NO_2]^2$? Explain your reasoning.

13.50 When the rate of the reaction $2 NO(g) + O_2(g) \rightarrow 2 NO_2(g)$ was studied, the rate was found to double when the O_2 concentration alone was doubled but to quadruple when the NO concentration alone was doubled. Which of the following mechanisms accounts for these observations? Explain your reasoning.

(a) **Step 1** $NO + O_2 \rightarrow NO_3$ and its reverse (both fast, equilibrium)
  Step 2 $NO + NO_3 \rightarrow NO_2 + NO_2$ (slow)
(b) **Step 1** $NO + NO \rightarrow N_2O_2$ (slow)
  Step 2 $O_2 + N_2O_2 \rightarrow N_2O_4$ (fast)
  Step 3 $N_2O_4 \rightarrow NO_2 + NO_2$ (fast)

Rates and Equilibrium

13.51 Determine whether each of the following statements is true or false. If a statement is false, explain why. (a) For a reaction with a very large equilibrium constant, the rate constant of the forward reaction is much larger than the rate constant of the reverse reaction. (b) At equilibrium, the rate constants of the forward and reverse reactions are equal. (c) Increasing the concentration of a reactant increases the rate of a reaction by increasing the rate constant in the forward direction.

13.52 Determine whether each of the following statements is true or false. If a statement is false, explain why. (a) The equilibrium constant for a reaction equals the rate constant for the forward reaction divided by the rate constant for the reverse reaction. (b) In a reaction that is a series of equilibrium steps, the overall equilibrium constant is equal to the product of all the forward rate constants divided by the product of all the reverse rate constants. (c) Increasing the concentration of a product increases the rate of the reverse reaction, and so the rate of the forward reaction must then increase, too.

[C] **3.53** Consider the reaction A $\rightleftharpoons$ B, which is first order in each direction with rate constants k and k'. Initially, only A is present. Show that the concentrations approach their equilibrium values at a rate that depends on k and k'.

[C] **13.54** Repeat Exercise 13.53 for the same reaction taken to be second-order in each direction.

The Effect of Temperature

13.55 (a) Using standard graphing software, such as that on the Web site for this book, make an appropriate Arrhenius plot of the data shown here for the conversion of cyclopropane into propene and calculate the activation energy for the reaction. (b) What is the value of the rate constant at 600°C?

T (K)	750.	800.	850.	900.
k (s^{-1})	1.8×10^{-4}	2.7×10^{-3}	3.0×10^{-2}	0.26

13.56 (a) Using standard graphing software, such as that on the Web site for this book, make an appropriate Arrhenius plot of the data shown here for the decomposition of iodoethane into ethene and hydrogen iodide, $C_2H_5I(g) \rightarrow C_2H_4(g) + HI(g)$, and determine the activation energy for the reaction. (b) What is the value of the rate constant at 400°C?

T (K)	660	680	720	760
k (s^{-1})	7.2×10^{-4}	2.2×10^{-3}	1.7×10^{-2}	0.11

13.57 The rate constant of the first-order reaction $2 N_2O(g) \rightarrow 2 N_2(g) + O_2(g)$ is 0.76 s^{-1} at 1000. K and 0.87 s^{-1} at 1030. K. Calculate the activation energy of the reaction.

13.58 The rate constant of the second-order reaction $2 HI(g) \rightarrow H_2(g) + I_2(g)$ is 2.4×10^{-6} L·mol^{-1}·s^{-1} at 575 K and 6.0×10^{-5} L·mol^{-1}·s^{-1} at 630. K. Calculate the activation energy of the reaction.

13.59 The rate constant of the reaction $O(g) + N_2(g) \rightarrow NO(g) + N(g)$, which takes place in the stratosphere, is 9.7×10^{10} L·mol^{-1}·s^{-1} at 800.°C. The activation energy of the reaction is 315 kJ·mol^{-1}. Determine the rate constant at 700.°C. (See Box 13.3.)

13.60 The rate constant of the reaction between CO_2 and OH^- in aqueous solution to give the HCO_3^- ion is 1.5×10^{10} L·mol^{-1}·s^{-1} at 25°C. Determine the rate constant at blood temperature (37°C), given that the activation energy for the reaction is 38 kJ·mol^{-1}.

13.61 The rate constant for the decomposition of N_2O_5 at 45°C is $k = 5.1 \times 10^{-4}$ s^{-1}. The activation energy for the reaction is 103 kJ·mol^{-1}. Determine the value of the rate constant at 50°C.

13.62 Ethane, C_2H_6, dissociates into methyl radicals at 700.°C with a rate constant $k = 5.5 \times 10^{-4}$ s^{-1}. Determine the rate constant at 800.°C, given that the activation energy of the reaction is 384 kJ·mol^{-1}.

13.63 For the reversible, one-step reaction $2 A \rightleftharpoons B + C$, the forward rate constant for the formation of B is 265 L·mol^{-1}·min^{-1} and the rate constant for the reverse reaction is 392 L·mol^{-1}·min^{-1}. The activation energy for the forward reaction is 39.7 kJ·mol^{-1} and that of the reverse reaction is 25.4 kJ·mol^{-1}. (a) What is the equilibrium constant for the reaction? (b) Is the reaction exothermic or endothermic? (c) What will be the effect of raising the temperature on the rate constants and the equilibrium constant?

13.64 For the reversible, one-step reaction $A + B \rightleftharpoons C + D$, the forward rate constant is 36.4 L·mol^{-1}·h^{-1} and the rate constant for the reverse reaction is 24.3 L·mol^{-1}·h^{-1}. The activation energy was found to be 33.8 kJ·mol^{-1} for the forward reaction and 45.4 kJ·mol^{-1} for the reverse reaction. (a) What is the equilibrium constant for the reaction? (b) Is the reaction exothermic or endothermic? (c) What will be the effect of raising the temperature on the rate constants and the equilibrium constant?

Catalysis

13.65 The presence of a catalyst provides a reaction pathway in which the activation energy of a certain reaction is reduced from 125 kJ·mol^{-1} to 75 kJ·mol^{-1}. (a) By what factor does the rate of the reaction increase at 298 K, all other factors being equal? (b) By what factor would the rate change if the reaction were carried out at 350. K instead?

13.66 The presence of a catalyst provides a reaction pathway in which the activation energy of a certain reaction is reduced from 88 kJ·mol^{-1} to 62 kJ·mol^{-1}. (a) By what factor does the rate of the reaction increase at 300. K, all other factors being equal? (b) By what factor would the rate change if the reaction were carried out at 500. K instead?

13.67 A reaction rate increases by a factor of 1000. in the presence of a catalyst at 25°C. The activation energy of the original pathway is 98 kJ·mol^{-1}. What is the activation energy of the new pathway, all other factors being equal? In practice, the new pathway also has a different pre-exponential factor.

13.68 A reaction rate increases by a factor of 500. in the presence of a catalyst at 37°C. The activation energy of the original pathway is 106 kJ·mol^{-1}. What is the activation energy of the new pathway, all other factors being equal? In practice, the new pathway also has a different pre-exponential factor.

13.69 The hydrolysis of an organic nitrile, a compound containing a $-C{\equiv}N$ group, in basic solution, is proposed to proceed by the following mechanism. Write a complete balanced equation for the overall reaction, list any intermediates, and identify the catalyst in this reaction.

Step 1 R$-$C$\equiv$N + OH$^-$ $\longrightarrow$ R$-$C(=N$^-$)(OH)

Step 2 R$-$C(=N$^-$)(OH) + H$_2$O $\longrightarrow$ R$-$C(=NH)(OH) + OH$^-$

Step 3 R$-$C(=NH)(OH) $\longrightarrow$ R$-$C(=O)(NH$_2$)

13.70 Consider the following mechanism for the hydrolysis of ethyl acetate. Write a complete balanced equation for the overall reaction, list any intermediates, and identify the catalyst in this reaction.

Step 1 $H_3C - C$ ... (structure with O, $O-CH_2$, CH_3) $+ OH^- \longrightarrow$

$H_3C - C$ (with O and OH) $+ CH_3CH_2O^-$

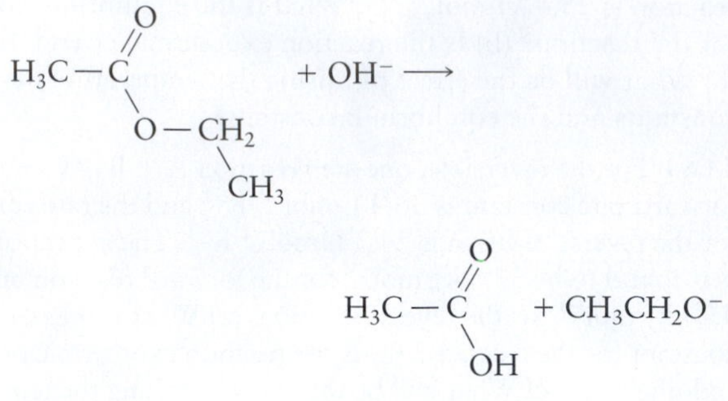

Step 2 $CH_3CH_2O^- + H_2O \longrightarrow CH_3CH_2OH + OH^-$

13.71 Determine which of the following statements about catalysts are true. If the statement is false, explain why. (a) In an equilibrium process, a catalyst increases the rate of the forward reaction but leaves the rate of the reverse reaction unchanged. (b) A catalyst is not consumed in the course of a reaction. (c) The pathway for a reaction is the same in the presence of a catalyst as in its absence, but the rate constants are decreased in both the forward and the reverse directions. (d) A catalyst must be carefully chosen to shift the equilibrium toward the products.

13.72 Determine which of the following statements about catalysts are true. If the statement is false, explain why. (a) A heterogeneous catalyst works by binding one or more of the molecules undergoing reaction to the surface of the catalyst. (b) Enzymes are naturally occurring proteins that serve as catalysts in biological systems. (c) The equilibrium constant for a reaction is greater in the presence of a catalyst. (d) A catalyst changes the pathway of a reaction in such a way that the reaction becomes more exothermic.

13.73 The Michaelis–Menten rate equation for enzyme reactions is typically written as the rate of formation of product (Eq. 19a). This equation implies that 1/Rate (where rate is the rate of formation of product) depends linearly on the inverse of the substrate concentration [S]. This relation allows K_M to be determined. Derive this equation and sketch 1/Rate against 1/[S]. Label the axes, the y-intercept, and the slope with their corresponding functions.

13.74 The Michaelis constant (K_M) is an index of the stability of an enzyme–substrate complex. Does a high Michaelis constant indicate a stable or an unstable enzyme–substrate complex? Explain your reasoning.

Integrated Exercises

13.75 Each of the following steps is an elementary reaction. Write its rate law and indicate its molecularity: (a) $NO + NO \rightarrow N_2O_2$; (b) $Cl_2 \rightarrow Cl + Cl$; (c) $NO_2 + NO_2 \rightarrow NO + NO_3$; (d) Which of these reactions might be radical chain initiating?

13.76 Each of the following is an elementary reaction. Write its rate law and indicate its molecularity. (a) $O + CF_2Cl_2 \rightarrow ClO + CF_2Cl$; (b) $OH + NO_2 + N_2 \rightarrow HNO_3 + N_2$; (c) $ClO^- + H_2O \rightarrow HClO + OH^-$; (d) Which of these reactions might be radical chain propagating?

13.77 The following rate data were collected for the reaction $2\ A(g) + 2\ B(g) + C(g) \rightarrow 3\ G(g) + 4\ F(g)$:

	Initial concentration ($mmol \cdot L^{-1}$)			
Experiment	$[A]_0$	$[B]_0$	$[C]_0$	Initial rate $((mmol\ G) \cdot L^{-1} \cdot s^{-1})$
1	10.	100.	700.	2.0
2	20.	100.	300.	4.0
3	20.	200.	200.	16
4	10.	100.	400.	2.0
5	4.62	0.177	12.4	?

(a) What is the order for each reactant and the overall order of the reaction? (b) Write the rate law for the reaction. (c) Determine the reaction rate constant. (d) Predict the initial rate for Experiment 5.

 13.78 The data shown here were collected for the decomposition of A, $2\ A \rightarrow 4\ B + C$. (a) Using standard graphing software, such as that on the Web site for this book, determine whether this reaction is most likely to be zero, first, or second order with respect to A. (b) Determine the rate constant k for this reaction in the rate law for the decomposition of A.

Time (s)	0	400.	800.	1200.	1600.
[A] ($mmol \cdot L^{-1}$)	2.57	1.5	0.87	0.51	0.30

13.79 The equilibrium constant for the second-order attachment of a substrate to the active site of an enzyme was found to be 326 at 310 K. At the same temperature, the rate constant for the second-order attachment is $7.4 \times 10^7\ L \cdot mol \cdot s^{-1}$. What is the rate constant for the loss of unreacted substrate from the active site (the reverse of the attachment reaction)?

13.80 The decomposition of hydrogen peroxide, $2\ H_2O_2(aq) \rightarrow 2\ H_2O(l) + O_2(g)$, follows first-order kinetics with respect to H_2O_2 and has $k = 0.0410\ min^{-1}$ in the rate law for the decomposition of H_2O_2. (a) If the initial concentration of H_2O_2 is $0.35\ mol \cdot L^{-1}$, what is its concentration after 10 min? (b) How much time will it take for the H_2O_2 concentration to decrease from $0.50\ mol \cdot L^{-1}$ to $0.10\ mol \cdot L^{-1}$? (c) How much time is needed for the H_2O_2 concentration to decrease by one-fourth? (d) Calculate the time needed for the H_2O_2 concentration to decrease by 75%.

13.81 The first-order decomposition of compound X, a gas, is carried out and the data are represented in the following pictures. The green spheres represent the compound; the decomposition products are not shown. The times at which the images were taken are shown below each flask. (a) Determine the half-life of the reaction. (b) Draw the appearance of the molecular image at 8 s.

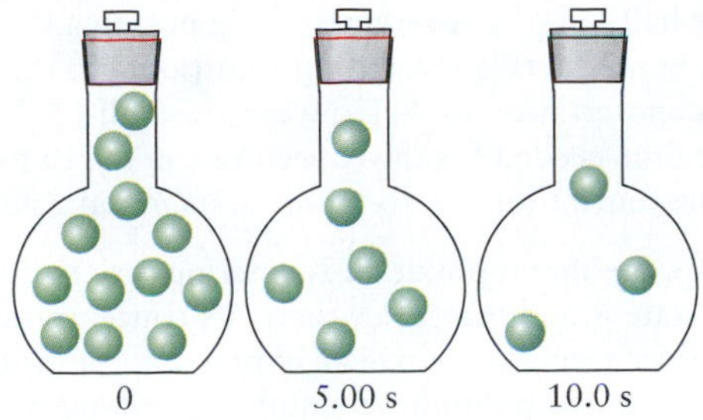

0   5.00 s   10.0 s

13.82 Determine the molecularity of the following elementary reactions:

(a) ○ + ●—● → ○—● + ●

(b) ○—○ → ○ + ○

(c) ○ + ○ + ● → ● + ○—○

13.83 The mechanism of the reaction A → B consists of two steps, with the formation of a reaction intermediate. Overall, the reaction is exothermic. (a) Sketch the reaction profile, labeling the activation energies for each step and the overall enthalpy of reaction. (b) Indicate on the same diagram the effect of a catalyst on the first step of the reaction.

13.84 The half-life for the (first-order) decomposition of azomethane, $CH_3N=NCH_3$, in the reaction $CH_3N=NCH_3(g) →$ $N_2(g) + C_2H_6(g)$ is 1.02 s at 300°C. A 45.0-mg sample of azomethane is placed in a 300.-mL reaction vessel and heated to 300°C. (a) What mass (in milligrams) of azomethane remains after 10. s? (b) Determine the partial pressure exerted by the $N_2(g)$ in the reaction vessel after 3.0 s.

13.85 All radioactive decay processes follow first-order kinetics. The half-life of the radioactive isotope tritium (^3H, or T) is 12.3 years. How much of a 25.0-mg sample of tritium would remain after 10.9 years?

13.86 The following mechanism has been proposed for the formation of hydrazine, $N_2(g) + 2 H_2(g) → N_2H_4(g)$:

Step 1 $N_2 + H_2 \longrightarrow N_2H_2$

Step 2 $H_2 + N_2H_2 \longrightarrow N_2H_4$

The rate law for the reaction is Rate = $k[N_2][H_2]^2$. Which step is the slow step? Show your work.

13.87 The hydrolysis of sucrose ($C_{12}H_{22}O_{11}$) produces fructose and glucose: $C_{12}H_{22}O_{11}(aq) + H_2O(l) → C_6H_{12}O_6(glucose, aq)$ $+ C_6H_{12}O_6(fructose, aq)$. Two mechanisms are proposed for this reaction.

(i) $C_{12}H_{22}O_{11} \longrightarrow C_6H_{12}O_6 + C_6H_{10}O_5$ (slow)
$C_6H_{10}O_5 + H_2O \longrightarrow C_6H_{12}O_6$ (fast)

(ii) $C_{12}H_{22}O_{11} + H_2O \longrightarrow C_6H_{12}O_6 + C_6H_{12}O_6$ (slow)

Under what conditions can these two mechanisms be distinguished by using kinetic data?

13.88 Ketones can react with themselves in a process known as *aldol condensation*. The mechanism for this reaction in acidic solution is shown here. Write the overall reaction, identify any intermediates, and determine the role of the hydrogen ion.

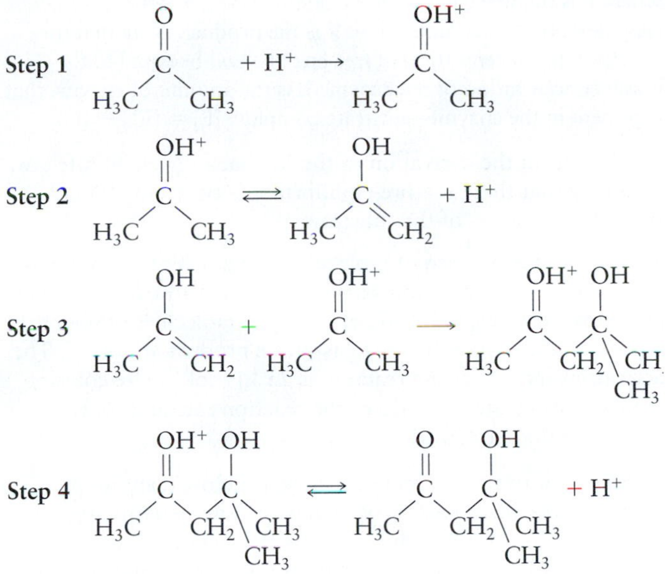

Step 1

Step 2

Step 3

Step 4

13.89 The rate law of the reaction $2 NO(g) + 2 H_2(g) → N_2(g)$ $+ 2 H_2O(g)$ is Rate = $k[NO]^2[H_2]$, and the mechanism that has been proposed is

Step 1 $NO + NO \longrightarrow N_2O_2$

Step 2 $N_2O_2 + H_2 \longrightarrow N_2O + H_2O$

Step 3 $N_2O + H_2 \longrightarrow N_2 + H_2O$

(a) Which step in the mechanism is likely to be rate determining? Explain your answer. (b) Sketch a reaction profile for the overall reaction, which is known to be exothermic. Label the activation energies of each step and the overall reaction enthalpy.

13.90 (a) Using graphing software, such as that on the Web site for this book, calculate the activation energy for the acid hydrolysis of sucrose to give glucose and fructose from an Arrhenius plot of the data shown here. (b) Calculate the rate constant at 37°C (body temperature). (c) From data in Appendix 2A, calculate the enthalpy change for this reaction, assuming that the solvation enthalpies of the sugars are negligible. Draw an energy profile for the overall process.

Temperature (°C)	k (s^{-1})
24	4.8×10^{-3}
28	7.8×10^{-3}
32	13×10^{-3}
36	$20. \times 10^{-3}$
40.	32×10^{-3}

13.91 Raw milk sours in about 4 h at 28°C but in about 48 h in a refrigerator at 5°C. What is the activation energy for the souring of milk?

13.92 (a) From the following mechanism, derive Eq. 19a, which Michaelis and Menten proposed to represent the rate of formation of products in an enzyme-catalyzed reaction. (b) Show that the rate is independent of substrate concentration at high concentrations of substrate.

$E + S \rightleftharpoons ES \quad k_1, k_1'$

$ES \longrightarrow E + P \quad k_2$

where E is the free enzyme, S is the substrate, ES is the enzyme–substrate complex, and P is the product. Note that the steady-state concentration of free enzyme will be equal to the initial concentration of the enzyme less the amount of enzyme that is present in the enzyme–substrate complex: $[E] = [E]_0 - [ES]$.

13.93 Repeat the derivation of the Michaelis–Menten rate law, assuming that there is a pre-equilibrium between the bound and the unbound states of the substrate.

13.94 A gas composed of molecules of diameter 0.5 nm takes part in a chemical reaction at 300. K and 1.0 atm with another gas (present in large excess) consisting of molecules of about the same size and mass to form a gas-phase product at 300. K. The activation energy for the reaction is 25 $kJ \cdot mol^{-1}$. Use collision theory to calculate the ratio of the reaction rate at 320. K relative to that at 300. K.

13.95 Refer to the illustration below. (a) How many steps does this reaction have? (b) Which is the rate-determining step in this reaction? (c) Which step is the fastest? (d) How many intermediates must form in the reaction? (e) A catalyst is added that accelerates the third step only. What effect, if any, will the catalyst have on the rate of the overall reaction?

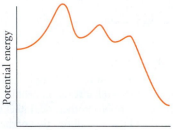

13.96 The second order reactions in Table 13.4 show wide variations in activation energy. In an activated complex, the reactant bonds are lengthened while the product bonds are beginning to form. Consider what bonds need to be stretched to form the activated complex and use bond enthalpies (Sections 2.14, 2.15, and 6.20) to explain the differences in activation energies.

13.97 The following mechanism has been suggested to explain the contribution of chlorofluorocarbons to the destruction of the ozone layer:

Step 1 $O_3 + Cl \longrightarrow ClO + O_2$

Step 2 $ClO + O \longrightarrow Cl + O_2$

(a) What is the reaction intermediate and what is the catalyst? (b) Identify the radicals in the mechanism. (c) Identify the steps as initiating, propagating, or terminating. (d) Write a chain-terminating step for the reaction. (See Box 13.3.)

13.98 Models of population growth are analogous to chemical reaction rate equations. In the model developed by Malthus in 1798, the rate of change of the population N of Earth is dN/dt = births − deaths. The numbers of births and deaths are proportional to the population, with proportionality constants b and d. Derive the integrated rate law for population change. How well does it fit the approximate data for the population of Earth over time given below?

Year	1750	1825	1922	1960	1974	1987	2000
$N/10^9$	0.5	1	2	3	4	5	6

13.99 The half-life of a substance taking part in a third-order reaction A → products is inversely proportional to the square of the initial concentration of A. How can this half-life be used to predict the time needed for the concentration to fall to (a) one-half; (b) one-fourth; (c) one-sixteenth of its initial value?

13.100 Suppose that a pollutant is entering the environment at a steady rate R and that, once there, its concentration decays by a first-order reaction. Derive an expression for (a) the concentration of the pollutant at equilibrium in terms of R and (b) the half-life of the pollutant species when $R = 0$.

13.101 Which of the following plots will be linear? (a) [A] against time for a reaction first order in A; (b) [A] against time for a reaction zero order in A; (c) ln[A] against time for a reaction first order in A; (d) 1/[A] against time for a reaction second order in A; (e) k against temperature; (f) initial rate against [A] for a reaction first order in A; (g) half-life against [A] for a reaction zero order in A; (h) half-life against [A] for a reaction second order in A.

13.102 The pre-equilibrium and the steady-state approximations are two different approaches to deriving a rate law from a proposed mechanism. (a) For the following mechanism, determine the rate law by the steady-state approximation. (c) Under what conditions do the two methods give the same answer? (d) What will the rate law become at high concentrations of Br^-?

$$CH_3OH + H^+ \rightleftharpoons CH_3OH_2^+ \text{ (fast in both directions)}$$

$$CH_3OH_2^- + Br^- \longrightarrow CH_3Br + H_2O \text{ (slow)}$$

13.103 (a) What is the overall reaction for the following mechanism?

$$ClO^- + H_2O \rightleftharpoons HClO + OH^- \text{ (fast equilibrium)}$$

$$HClO + I^- \longrightarrow HIO + Cl^- \text{ (very slow)}$$

$$HIO + OH^- \rightleftharpoons IO^- + H_2O \text{ (fast equilibrium)}$$

(b) Write the rate law based on this mechanism. (c) Will the reaction rate depend on the pH of the solution? (d) How would the rate law differ if the reactions were carried out in an organic solvent?

13.104 The following schematic reaction profile is for the reaction A → D. (a) Is the overall reaction exothermic or endothermic? Explain your answer. (b) How many intermediates are there? Identify them. (c) Identify each activated complex and reaction intermediate. (d) Which step is rate determining? Explain your answer. (e) Which step is the fastest? Explain your answer.

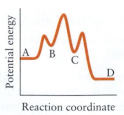

13.105 The decomposition of A has the rate law Rate = $k[A]^a$. Show that for this reaction the ratio $t_{1/2}/t_{3/4}$, where $t_{1/2}$ is the half-life and $t_{3/4}$ is the time for the concentration of A to decrease to $^3/_4$ of its initial concentration, can be written as a function of *a* alone and can therefore be used to make a quick assessment of the order of the reaction in A.

13.106 To prepare a 15-kg dog for surgery, 150 mg of the anesthetic phenobarbitol is administered intravenously. The reaction in which the anesthetic is metabolized (decomposed in the body) is first order in phenobarbitol and has a half-life of 4.5 h. After about 2 hours, the drug begins to lose its effect. However, the surgical procedure requires more time than had been anticipated. What mass of phenobarbitol must be re-injected to restore the original level of the anesthetic in the dog?

Chemistry Connections

13.107 Cyanomethane, commonly known as acetonitrile, CH_3CN, is a toxic volatile liquid that is used as a solvent to purify steriods and to extract fatty acids from fish oils. Acetonitrile can be synthesized from methyl isonitrile by the isomerization reaction $CH_3NC(g) \rightarrow CH_3CN(g)$.

(a) Draw the Lewis structures of methyl isonitrile and cyanomethane; assign a hybridization scheme to each C atom and indicate whether each molecule is polar or nonpolar.

(b) Calculate $\Delta H°$ for the isomerization reaction by using data in Table 6.8. Which isomer is the more stable?

(c) The isomerization reaction obeys the first-order rate law, Rate = $k[CH_3NC]$ in the presence of argon. The activation energy for this reaction is 161 $kJ \cdot mol^{-1}$ and the rate constant at 500. K is $6.6 \times 10^{-4} \ s^{-1}$. Calculate the rate constant at 300. K° and the time (in seconds) needed for the concentration of CH_3NC to decrease to 75% of its initial value at 300. K.

(d) Draw a reaction profile for the isomerization reaction.

(e) Calculate the temperature at which the concentration of CH_3NC will decrease to 75% of its original concentration in 1.0 h.

(f) What purpose does the excess argon in part (c) serve?

(g) At low concentrations of Ar, the rate is no longer first order in $[CH_3NC]$; suggest an explanation.

Computers have had an enormous impact on the practice of chemistry, where they are used to control apparatus, record and manipulate data, and explore the structures and properties of atoms and molecules. Computer modeling is applied to solids as well as to individual molecules and is useful for predicting the behavior of a material, for example, for indicating which crystal structure of a compound is energetically most favorable, and for predicting phase changes. In this section we concentrate on the use of computers in the exploration of molecular structure. In this context, they are used both to explore practical problems such as predicting the pharmacological activity of compounds and to achieve a deep understanding of the changes that take place during chemical reactions.

In *semi-empirical methods*, complicated integrals are set equal to parameters that provide the best fit to experimental data, such as enthalpies of formation. Semi-empirical methods are applicable to a wide range of molecules with a virtually limitless number of atoms, and are widely popular. The quality of results is very dependent on using a reasonable set of experimental parameters that have the same values across structures, and so this kind of calculation has been very successful in organic chemistry, where there are just a few different elements and molecular geometries.

In the more fundamental *ab initio methods*, an attempt is made to calculate structures from first principles, using only the atomic numbers of the atoms present and their general arrangement in space. Such an approach is intrinsically more reliable than a semi-empirical procedure but it is much more demanding computationally.

A currently popular alternative to the *ab initio* method is *density functional theory*, in which the energy is expressed in terms of the electron density rather than the wavefunction itself. The advantage of this approach is that it is less demanding computationally, requires less computer time, and in some cases—particularly for *d*-metal complexes—gives better agreement with experimental values than other procedures.

The raw output of a molecular structure calculation is a list of the coefficients of the atomic orbitals in each LCAO (linear combination of atomic orbitals) molecular orbital and the energies of the orbitals. The software commonly calculates dipole moments too. Various graphical representations are used to simplify the interpretation of the coefficients. Thus, a typical graphical representation of a molecular orbital uses stylized shapes (spheres for *s*-orbitals, for instance) to represent the basis set and then scales their size to indicate the value of the coefficient in the LCAO. Different signs of the wavefunctions are typically represented by different colors. The total electron density at any point (the sum of the squares of the occupied wavefunctions evaluated at that point) is commonly represented by an isodensity surface, a surface of constant total electron density.

An important aspect of a molecule in addition to its geometrical shape and the energies of its orbitals is the distri-bution of charge over its surface, because the charge distribution can strongly influence how one molecule (such as a potential drug) can attach to another (and dock into the active site of an enzyme, for instance). A common procedure begins with a calculation of the electrical potential at each point on an isodensity surface. The net potential is determined by subtracting the potential due to the electron density at that point from the potential due to the nuclei. The result is an *electrostatic potential surface* (an "elpot surface") in which net positive potential is shown in one color and net negative potential is shown in another, with intermediate gradations of color. The software can also depict a *solvent-accessible surface*, which represents the shape of the molecule by imagining a sphere representing a solvent molecule rolling across the surface and plotting the locations of its center.

Procedures are now available that allow for the presence of several solvent molecules around a solute molecule. This approach takes into account the effect of molecular interactions with the solvent on properties such as the enthalpy of formation and the shape adopted by a non-rigid molecule, such as a protein or a region of DNA. These studies are important for investigating the structures and reactions of biological molecules in their natural environment.

Related Exercises: Exercises that use software to explore molecular structure are available on the Web site for this text.

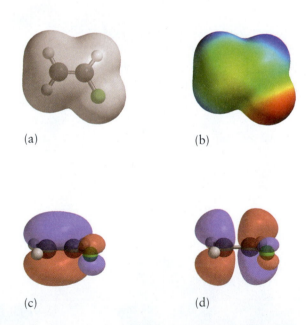

(a)   (b)

(c)   (d)

Some graphical representations of fluoroethene, $CH_2{=}CHF$: (a) the isodensity surface showing the general shape of the molecule, (b) the electrostatic potential surface showing the relatively negatively charged region (denoted red) close to the fluorine atom, (c) the highest-energy occupied molecular orbital showing the π-bond between the two carbon atoms and the partial involvement of a fluorine orbital, (d) the lowest-energy unoccupied orbital, which is the antibonding counterpart of the π-orbital.

THE ELEMENTS: THE FIRST FOUR MAIN GROUPS

Chapter 14

What Are the Key Ideas? The structures of atoms determine their properties; consequently, the behavior of elements is related to their locations in the periodic table.

Why Do We Need to Know This Material? In this chapter, we see how the principles that underlie the properties of atoms account for the behavior and organization of matter. Knowledge of the family relationships of the elements helps us comprehend the nature of matter and allows us to design new materials. The elements in the first four main groups include carbon, the central element of life; silicon and germanium, which are used to create artificial intelligence; calcium, which is responsible for the rigidity of our bones and teeth and the strength of concrete; the space-age metals aluminum and beryllium; the alkali metals; and hydrogen, the fuel of the stars—and perhaps our future energy source on Earth.

What Do We Need to Know Already? The information in this chapter is organized around the principles of atomic structure and specifically the periodic table (Chapter 1). However, the chapter draws on all the preceding chapters, because it uses those principles to account for the properties of the elements.

The twenty-first century demands novel materials of the scientist. New instruments have made possible the field of **nanotechnology,** in which chemists study particles between 1 and 100 nm in diameter, intermediate between the atomic and the bulk levels of matter. Nanotechnology has the promise to provide new materials such as biosensors that monitor and even repair bodily processes, microscopic computers, artificial bone, and lightweight, remarkably strong materials. To conceive and develop such materials, scientists need a thorough knowledge of the elements and their compounds.

This chapter and the following two chapters survey the properties of the elements and their compounds in relation to their locations in the periodic table. To prepare for this journey through the periodic table, we first review the trends in properties discussed in earlier chapters. We then start the journey itself with the unique element hydrogen and move on to the elements of the main groups, working from left to right across the table. The same principles apply to the elements of the *d* and *f* blocks, but these elements have some unique characteristics (mainly their wide variety of oxidation states and their ability to act as Lewis acids), and so they are treated separately in Chapter 16.

Although we concentrate on periodic trends, we also need to know certain details about specific elements, such as where they are found in nature and how they are purified. In addition, some of their compounds have interesting properties and some are of great commercial significance. So, as well as considering the elements in the context of periodic trends, we shall also highlight some of the unique properties and applications that a well-informed chemist is expected to know. These topics are commonly called **descriptive chemistry**—the description of the preparation, properties, and applications of elements and their compounds. In part, the usefulness and availability of an element reflects its abundance on Earth and specifically in the Earth's crust, which may be quite different from its cosmic abundance. Figure 14.1 summarizes the abundances of the elements in three settings: the universe as a whole, the Earth's crust, and the human body.

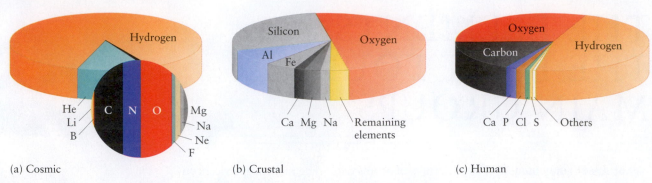

(a) Cosmic   (b) Crustal   (c) Human

FIGURE 14.1 These charts show the relative abundances of the principal elements in (a) the universe (the "cosmic abundances"); (b) the crust of the Earth; and (c) the human body.

PERIODIC TRENDS

In Chapter 1, we saw that the members of each group have analogous valence-shell electron configurations, which give them characteristic properties. However, the element at the head of a group—the lightest element in the group—often has features that distinguish it from its **congeners**, the other members of the group. In this introductory section, we review atomic properties that correlate with the position of an element in the periodic table. Then we investigate periodic trends in bonding by comparing examples of two types of compounds—hydrides and oxides.

14.1 Atomic Properties

The valence-shell electrons directly affect the chemical properties of an element and trends in the effective nuclear charge experienced by these electrons give rise to periodic trends. As we saw in Section 1.12 and Fig. 1.45, valence electrons experience increasing effective nuclear charge on going across a period from left to right. The attraction of a valence electron to the nucleus decreases on going down a group, because valence electrons belong to shells that are successively farther from the nucleus and surround electron-rich cores. These trends in the effective nuclear charge experienced by the valence electrons help account for the diagonal trends in many atomic properties. The valence electrons of elements in the upper right of the periodic table, near fluorine, experience the greatest attraction to their nuclei and those of elements in the lower left, near cesium, experience the least.

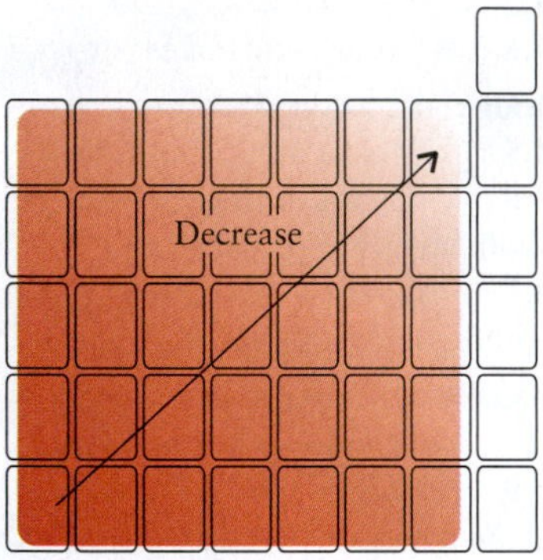

All the elements in a main group have in common a characteristic valence electron configuration. The electron configuration controls the valence of the element (the number of bonds that it can form) and affects its chemical and physical properties. Five atomic properties are principally responsible for the characteristic properties of each element: atomic radius, ionization energy, electron affinity, electronegativity, and polarizability. All five properties are related to trends in the effective nuclear charge experienced by the valence electrons and their distance from the nucleus.

Atomic radii typically decrease from left to right across a period and increase down a group (Fig. 14.2; see also Fig. 1.46). As the nuclear charge experienced by the valence electrons increases across a period, the electrons are pulled closer to the nucleus, so decreasing the atomic radius. Down a group the valence electrons are farther and farther from the nucleus, which increases the atomic radius. Ionic radii follow similar periodic trends (see Fig. 1.48).

First ionization energies typically increase from left to right across a period and decrease down a group (Fig. 14.3; see also Figs. 1.50 and 1.51). The increase from left to right across a period is due to the increasing effective nuclear charge, which grips the electrons more tightly. The decrease in ionization energy down a group shows that it is easier to remove valence electrons from shells that are farther from the nucleus and hence shielded by a greater number of core electrons.

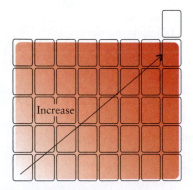

FIGURE 14.2 Atomic radius tends to decrease from left to right across a period and to increase down a group. This diagram and those in Figs. 14.3 through 14.5 are highly schematic representations of periodic trends, with the density of the red tint representing the value of the property: the denser the tint, the higher the value.

FIGURE 14.3 Ionization energy tends to increase from left to right across a period and to decrease down a group.

The highest electron affinities are found at the top right of the periodic table (see Fig. 1.54; the trends are too ill defined to depict simply). The electron affinity of an element is a measure of the energy released when an anion is formed. Except for the noble gases, elements near fluorine have the highest electron affinities, and so we can expect them to be present as anions in compounds with metallic elements. Once again, we see the effect of increasing effective nuclear charge: more energy is released when electrons attach to atoms in which they experience the higher effective nuclear charge.

Electronegativities typically increase from left to right across a period and decrease down a group (Fig. 14.4; see also Fig. 2.12). The greater the effective nuclear charge and the closer the electrons are to the nucleus, the stronger the pull on the electrons in a bond. The electronegativity of an element—a measure of the tendency of an atom to attract electrons to itself when part of a compound—is a useful guide to the type of bond that the element is likely to form. When there is a large difference in electronegativity between two elements, their atoms form ionic bonds with each other; when the difference is small, the bonds are covalent.

Polarizabilities typically decrease from left to right along a period and increase down a group (Fig. 14.5). Polarizability measures the ease with which an electron cloud can be distorted and is greatest for the electron-rich heavier atoms of a group and for negatively charged ions. On the other hand, high polarizing power—the ability to distort the electron cloud of a neighboring atom or ion—is commonly associated with small size and high positive charge. We saw in Section 2.13 that a highly polarizable atom or ion (such as iodine or the iodide ion) in combination with an atom or ion of high polarizing power (such as beryllium) is likely to have considerable covalent character in its bonds.

SELF-TEST 14.1A Which of the elements carbon, aluminum, and germanium has atoms with the greatest polarizability?

[*Answer:* Germanium]

SELF-TEST 14.1B Which of the elements oxygen, gallium, and tellurium has the greatest electron affinity?

> *Atomic radii decrease from left to right across a period and increase down a group; ionization energies increase across a period and decrease down a group; electron affinities and electronegativities are highest near fluorine; polarizabilities decrease from left to right across a period and increase down a group.*

14.2 Bonding Trends

The type and number of bonds an element forms are related to its position in the periodic table. We can usually predict the valence of an element in Period 2 from the number of electrons in the valence shell and the octet rule. For example, carbon, with four valence electrons, commonly forms four bonds; and oxygen, with six valence electrons and therefore needing two more to complete its octet, typically forms two bonds. Elements in Period 3 and higher periods can reach higher oxidation states and have higher valences. These elements have access to empty *d*-orbitals and can use them to expand their valence shells past the usual octet of electrons, or they simply have atoms large enough to bond to more neighbors. Elements at the foot and toward the left of the *p*-block also have variable valence. They can display the inert-pair effect and have an oxidation number 2 less than their group number suggests (Section 1.19).

The radius of an atom helps to determine how many other atoms can bond to it. The small radii of Period 2 atoms, for instance, are largely responsible for the differences between their properties and those of their congeners. As described in Section 2.10, one reason that small atoms typically have low valences is that so few other atoms can pack around them. Nitrogen, for instance, never forms pentahalides, but phosphorus does. With few exceptions, only Period 2 elements form multiple bonds with themselves or other elements in the same period, because only they are small enough for their *p*-orbitals to have substantial π overlap (Fig. 14.6).

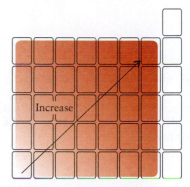

FIGURE 14.4 Electronegativity generally increases from left to right across a period and decreases down a group.

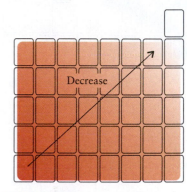

FIGURE 14.5 The polarizability of atoms tends to decrease from left to right across a period and to increase down a group.

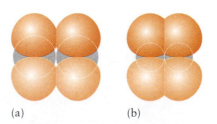

(a)   (b)

FIGURE 14.6 (a) The *p*-orbitals of Period 3 and heavier elements are held apart by the cores of the atoms (shown in gray) and have very little overlap with each other. (b) In contrast, the atoms of elements in Period 2 are small; consequently their *p*-orbitals can overlap effectively with each other and with those of elements in later periods.

Periodic trends in bonding for main-group elements become apparent when we compare the binary compounds that they form with specific elements. All the main-group elements, with the exception of the noble gases and, possibly, indium and thallium, form binary compounds with hydrogen, so we can examine hydrides to look for periodic trends (Section 14.4). The formulas of hydrides of main-group elements are related directly to the group number and reveal the typical valences of the elements (Fig. 14.7). For instance, carbon (Group 14/IV) forms CH_4, nitrogen (Group 15/V) forms NH_3, oxygen (Group 16/VI) forms H_2O, and fluorine (Group 17/VII) forms HF.

The nature of a binary hydride is related to the characteristics of the element bonded to hydrogen (Fig. 14.8). Strongly electropositive metallic elements form ionic compounds with hydrogen in which the latter is present as a hydride ion, H^-. These ionic compounds are called *saline hydrides* (or saltlike hydrides). They are formed by all members of the *s* block, with the exception of beryllium, and are made by heating the metal in hydrogen:

$$2\ K(s) + H_2(g) \xrightarrow{\Delta} 2\ KH(s)$$

The saline hydrides are white, high-melting-point solids with crystal structures that resemble those of the corresponding halides. The alkali metal hydrides, for instance, have the rock-salt structure (Fig. 5.39).

The *metallic hydrides* are black, powdery, electrically conducting solids formed by heating certain of the *d*-block metals in hydrogen (Fig. 14.9):

$$2\ Cu(s) + H_2(g) \xrightarrow{\Delta} 2\ CuH(s)$$

Because the metallic hydrides release their hydrogen (as H_2 gas) when heated or treated with acid, they are being investigated for storing and transporting hydrogen. Both saline and metallic hydrides have the high enthalpy densities desirable in a portable fuel.

Nonmetals form covalent *molecular hydrides*, which consist of discrete molecules. These compounds are volatile and many are Brønsted acids. Some are gases—for example, ammonia, the hydrogen halides (HF, HCl, HBr, HI), and the lighter hydrocarbons such as methane, ethane, ethene, and ethyne. Liquid molecular hydrides include water and hydrocarbons such as octane and benzene.

All the main-group elements except the noble gases also react with oxygen. Like hydrides, the resulting oxides also reveal periodic trends in the elements. The trend in bonding type is from soluble ionic oxides on the left of the periodic table, through insoluble, high-melting-point oxides on the left of the *p* block, to low-melting-point and often gaseous molecular oxides on the right.

Metallic elements with low ionization energies commonly form basic ionic oxides. Elements with intermediate ionization energies, such as beryllium, boron, aluminum, and the metalloids, form amphoteric oxides. These oxides do not react with or dissolve in water, but they do dissolve in both acidic and basic solutions.

Many of these binary compounds have hydrogen in its +1 oxidation state, and so the name hydride is not really appropriate. However, it is the conventional term.

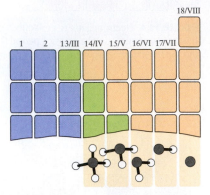

FIGURE 14.7 The chemical formulas of the hydrides of the elements of the main groups display the valence characteristic of each group.

Xenon forms an oxide, but by an indirect route.

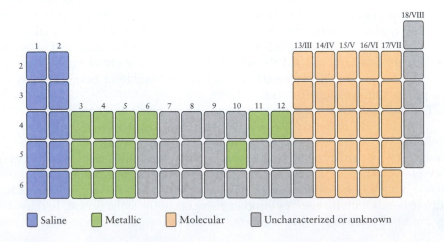

FIGURE 14.8 The different classes of binary hydrogen compounds and their distribution throughout the periodic table.

☐ Saline   ☐ Metallic   ☐ Molecular   ☐ Uncharacterized or unknown

Many oxides of nonmetals are gaseous molecular compounds, such as CO_2, NO, and SO_3. Most can act as Lewis acids, because the electronegative oxygen atoms withdraw electrons from the central atom, enabling it to act as an electron pair acceptor. Many oxides of nonmetals form acidic solutions in water and hence are called **acid anhydrides.** The familiar laboratory acids HNO_3 and H_2SO_4, for instance, are derived from the anhydrides N_2O_5 and SO_3, respectively. Even oxides that do not react with water can be regarded as the *formal* anhydrides of acids. A **formal anhydride** of an acid is the molecule obtained by striking out the elements of water (H, H, and O) from the molecular formula of the acid. Carbon monoxide, for instance, is the formal anhydride of formic acid, HCOOH, although CO does not react with cold water to form the acid.

The oxides of metalloids and some of the less electropositive elements are amphoteric (react with both acids and bases). Aluminum oxide, for instance, reacts with acids and with alkalis (aqueous solutions of strong bases). The oxides reveal a strong diagonal relationship between beryllium and aluminum, because beryllium oxide is also amphoteric.

SELF-TEST 14.2A An element ("E") in Period 4 forms a molecular hydride with the formula HE. Identify the element.

[*Answer:* Bromine]

SELF-TEST 14.2B An element ("E") in Period 3 forms an amphoteric oxide with the formula E_2O_3. Identify the element.

Valence and oxidation state are directly related to the valence-shell electron configuration of a group. Binary hydrides are classified as saline, metallic, or molecular. Oxides of metals tend to be ionic and to form basic solutions in water. Oxides of nonmetals are molecular and many are the anhydrides of acids.

HYDROGEN

Hydrogen is widely considered to be the fuel of our future, and so its properties are of great interest to scientists and engineers. Although hydrogen has the same valence electron configuration as the Group 1 elements, ns^1, and forms +1 ions, hydrogen has few other similarities to the alkali metals. Hydrogen is a nonmetal that resembles the halogens: it needs only one electron to complete its valence electron configuration, it can form -1 ions, and it exists as a diatomic molecule in its most stable form. However, the chemical properties of hydrogen are very different from those of the halogens. Because it does not clearly fit into any group of elements, we do not assign it to a group.

14.3 The Element

Hydrogen is the most abundant element in the universe: it accounts for 89% of all atoms (see Fig. 14.1). These atoms were formed in the first few seconds after the Big Bang, the event that marked the beginning of the universe. However, there is little free hydrogen on Earth because H_2 molecules, being very light, move at such high average speeds that they tend to escape from the Earth's gravity. It takes heavier atoms to anchor hydrogen to the planet in compounds. Most of the Earth's hydrogen is present as water, either in the oceans or trapped inside minerals and clays. Hydrogen is also found in the hydrocarbons that make up the **fossil fuels:** coal, petroleum, and natural gas. It takes energy to release hydrogen from these compounds and one of the challenges to achieving its potential as a fuel is to produce the gas by a means that takes less energy than can be obtained from burning it.

Because water is its only combustion product, hydrogen burns without polluting the air or contributing significantly to the greenhouse effect (see Box 14.2). Petroleum, coal, and natural gas are becoming increasingly rare, but there is enough water in the oceans to generate all the hydrogen fuel we shall ever need. Hydrogen is obtained from water by electrolysis, but that process is economical only where electricity is cheap. Chemists are currently seeking ways of using

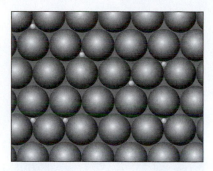

FIGURE 14.9 In a metallic hydride, the tiny hydrogen atoms (the small spheres) occupy gaps—called interstices— between the larger metal atoms (the large spheres).

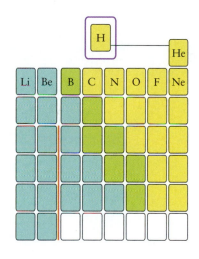

In the United States so many vehicles operate on hydrogen fuel cells (see Box 12.1) that hydrogen refueling stations have opened in many cities, including Washington, DC.

FIGURE 14.10 The two measuring cylinders contain the same mass of liquid. The cylinder on the left holds about 10 mL of water, that on the right liquid hydrogen, which is one-tenth as dense and consequently has a volume of about 100 mL.

sunlight to drive the *water-splitting reaction*, the photochemical decomposition of water into its elements:

$$\text{Water-splitting reaction:}\quad 2\,H_2O(l) \xrightarrow{\text{light}} 2\,H_2(g) + O_2(g) \quad \Delta G° = +474\text{ kJ}$$

At present, most commercial hydrogen is obtained as a by-product of petroleum refining in a sequence of two catalyzed reactions. The first is a *re-forming reaction*, in which a hydrocarbon and steam are converted into carbon monoxide and hydrogen over a nickel catalyst:

$$\text{Reforming reaction:}\quad CH_4(g) + H_2O(g) \xrightarrow{\text{Ni}} CO(g) + 3\,H_2(g)$$

The mixture of products, called *synthesis gas*, is the starting point for the manufacture of many other compounds, including methanol. The re-forming reaction is followed by the *shift reaction*, in which the carbon monoxide in the synthesis gas reacts with more water:

$$\text{Shift reaction:}\quad CO(g) + H_2O(g) \xrightarrow{\text{Fe/Cu}} CO_2(g) + H_2(g)$$

Hydrogen is prepared in small amounts in the laboratory by reducing hydrogen ions from a strong acid (such as hydrochloric acid) with a metal that has a negative standard potential, such as zinc:

$$Zn(s) + 2\,H^+(aq) \longrightarrow Zn^{2+}(aq) + H_2(g)$$

Hydrogen is a colorless, odorless, tasteless gas (Table 14.1). Because H_2 molecules are small and nonpolar, they can attract each other only by very weak London forces. As a result, hydrogen does not condense to a liquid until it is cooled to a very low temperature (20 K). One striking physical property of liquid hydrogen is its very low density (0.070 g·cm^{-3}), which is less than one-tenth that of water (Fig. 14.10). This low density makes hydrogen a very lightweight fuel. Hydrogen has the highest specific enthalpy of any known fuel (the highest enthalpy of combustion per gram), and so liquid hydrogen is used with liquid oxygen to power the space shuttle's main rocket engines.

Each year, about half the 3×10^8 kg of hydrogen used in industry is converted into ammonia by the Haber process (Section 9.12). Through the reactions of ammonia, hydrogen finds its way into numerous other important nitrogen compounds such as hydrazine and sodium amide (see Section 15.2).

Hydrogen is produced as a by-product of the refining of fossil fuels and by electrolysis of water. It has a low density and weak intermolecular forces.

14.4 Compounds of Hydrogen

Hydrogen is unusual because it can form both a cation (H^+) and an anion (H^-). Moreover, its intermediate electronegativity (2.2 on the Pauling scale) means that it can also form covalent bonds with all the nonmetals and metalloids. Because hydrogen forms compounds with so many elements (Table 14.2; also see Section 14.2), we shall meet more of its compounds when we study the other elements.

TABLE 14.1 Physical Properties of Hydrogen

Valence configuration: $1s^1$
Normal form[*]: colorless, odorless gas

Z	Name	Symbol	Molar mass (g·mol^{-1})	Abundance (%)	Melting point (°C)	Boiling point (°C)	Density[†] (g·L^{-1})
1	hydrogen	H	1.008	99.98	−259 (14 K)	−253 (20 K)	0.089
1	deuterium	^{2}H or D	2.014	0.02	−254 (19 K)	−249 (24 K)	0.18
1	tritium	^{3}H or T	3.016	radioactive	−252 (21 K)	−248 (25 K)	0.27

[*]*Normal form* means the state and appearance of the element at 25°C and 1 atm.
[†]At 25°C and 1 atm.

TABLE 14.2 **Chemical Properties of Hydrogen**

Reactant	Reaction with hydrogen
Group 1 metals (M)	$2\,M(s) + H_2(g) \longrightarrow 2\,MH(s)$
Group 2 metals (M, not Be or Mg)	$M(s) + H_2(g) \longrightarrow MH_2(s)$
some d-block metals (M)	$2\,M(s) + x\,H_2(g) \longrightarrow 2\,MH_x(s)$
oxygen	$O_2(g) + 2\,H_2(g) \longrightarrow 2\,H_2O(l)$
nitrogen	$N_2(g) + 3\,H_2(g) \longrightarrow 2\,NH_3(g)$
halogen (X_2)	$X_2(g,l,s) + H_2(g) \longrightarrow 2\,HX(g)$

The hydride ion is large, with a radius of 154 pm (**1**), falling between the fluoride and chloride ions in size. The large radius of the ion makes it highly polarizable and contributes to covalent character in its bonds to cations. However, the single positive charge of the hydrogen atomic nucleus can barely manage to keep control over the two electrons in the H^- ion, and so they are easily lost. The low energy required for a hydride ion to lose an electron causes the ionic hydrides to be very powerful reducing agents, and $E°(H_2/H^-) = -2.25$ V. This value is similar to the standard potential for Na^+/Na ($E° = -2.71$ V) and, like sodium metal, hydride ions react with water as soon as they come into contact with it:

$$NaH(s) + H_2O(l) \longrightarrow NaOH(aq) + H_2(g)$$

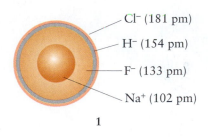

Cl⁻ (181 pm)
H⁻ (154 pm)
F⁻ (133 pm)
Na⁺ (102 pm)

1

Because this reaction produces hydrogen, saline hydrides are potentially useful as transportable sources of hydrogen fuel.

Compounds with a bond between hydrogen and nitrogen, oxygen, or fluorine—elements with small, highly electronegative atoms—participate in the very strong intermolecular force known as *hydrogen bonding* (Section 5.5). A hydrogen bond is about 5% as strong as a covalent bond between the same types of atoms. For example, the H—O bond enthalpy is 463 $kJ \cdot mol^{-1}$, whereas the O···H—O hydrogen bond enthalpy is about 20 $kJ \cdot mol^{-1}$. There are several models of hydrogen bonding (Chapter 5); the simplest is as a coulombic interaction between the partial positive charge on a hydrogen atom and the partial negative charge of another atom, as in $O^{\delta-}\cdots^{\delta+}H$—O.

SELF-TEST 14.3A What common binary compounds of hydrogen are powerful reducing agents?

[*Answer:* The saline (ionic) hydrides]

SELF-TEST 14.3B What are some of the properties of hydrogen that argue against its classification as a Group 1 element?

The hydride ion has a large radius and is highly polarizable. Hydrides are powerful reducing agents. Hydrogen can form hydrogen bonds to lone pairs of electrons on highly electronegative elements.

GROUP 1: THE ALKALI METALS

The members of Group 1 are called the *alkali metals*. The chemical properties of these elements are unique and strikingly similar from one to another. Nevertheless, there are differences, and the subtlety of some of these differences is the basis of the most subtle property of matter: consciousness. Our thinking, which relies on the transmission of signals along neurons, is achieved by the concerted action of sodium and potassium ions and their carefully regulated migration across membranes. So, even to learn about sodium and potassium, we have to make use of them in our brains.

As in the discussion of hydrogen, in this section we examine the properties of the alkali metals in the context of the periodic table and focus on significant applications of the elements and selected compounds. The valence electron configuration of the alkali metals is ns^1, where n is the period number. Their physical and chemical properties are dominated by the ease with which the single valence electron can be removed (Table 14.3).

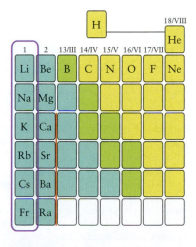

TABLE 14.3 Group 1 Elements: The Alkali Metals

Valence configuration: ns^1
Normal form*: soft, silver-gray metals

Z	Name	Symbol	Molar mass (g·mol^{-1})	Melting point (°C)	Boiling point (°C)	Density (g·cm^{-3})
3	lithium	Li	6.94	181	1347	0.53
11	sodium	Na	22.99	98	883	0.97
19	potassium	K	39.10	64	774	0.86
37	rubidium	Rb	85.47	39	688	1.53
55	cesium	Cs	132.91	28	678	1.87
87	francium	Fr	(223)	27	677	—

*Normal form means the state and appearance of the element at 25°C and 1 atm.

14.5 The Group 1 Elements

The alkali metals are the most violently reactive of all the metals. They are too easily oxidized to be found in the free state in nature and cannot be extracted from their compounds by ordinary chemical reducing agents. The pure metals are obtained by electrolysis of their molten salts, as in the electrolytic Downs process (Section 12.13) or, in the case of potassium, by exposing molten potassium chloride to sodium vapor:

$$KCl(l) + Na(g) \xrightarrow{750°C} NaCl(s) + K(g)$$

Although the equilibrium constant for this reaction is not particularly favorable, the reaction runs to the right because potassium is more volatile than sodium: the potassium vapor is driven off by the heat and condensed in a cooled collecting vessel.

All the Group 1 elements are soft, lustrous metals (Fig. 14.11). Because the valence shell consists of a single electron, bonding in the metals is weak, leading to low melting points, boiling points, and densities. These properties increase down the group, as the size of the atoms increases. For example, the melting points of the

(a)

(b)

(c)

(d)

FIGURE 14.11 The alkali metals of Group 1: (a) lithium; (b) sodium; (c) potassium; (d) rubidium and cesium. Francium has never been isolated in visible quantities. The first three elements were freshly cut and you can see how rapidly they have been corroded in moist air; rubidium and cesium are even more reactive and have to be stored (and photographed) in sealed, airless containers.

alkali metals decrease down the group (Fig. 14.12). Cesium, which melts at 28°C, is barely a solid at room temperature. Lithium is the hardest alkali metal, but even so it is softer than lead. Francium is intensely radioactive, and little is known about the properties of this very rare element.

Lithium metal had few uses until after World War II, when thermonuclear weapons were developed (see Section 17.11). This application has had an effect on the molar mass of lithium. Because only lithium-6 could be used in these weapons, the proportion of lithium-7 and, as a result, the molar mass of commercially available lithium has increased. A growing application of lithium is in the rechargeable lithium-ion battery. Because lithium has the most negative standard potential of all the elements, it can produce a high potential when used in a galvanic cell. Furthermore, because lithium has such a low density, lithium-ion batteries are light.

The alkali metals melt at low temperatures. They are the most reactive metals.

14.6 Chemical Properties of the Alkali Metals

The first ionization energies of the alkali metals are low, and so in their compounds they exist as singly charged cations, such as Na^+. As a result, most of their compounds are ionic. Because their ionization energies are so low, the standard potentials of the alkali metals are all negative. However, $E°$ does not vary as smoothly as the ionization energy, because the lattice energy of the solid and the hydration energy of the ions play a role in determining $E°$; entropy effects also contribute.

Because their standard potentials are so strongly negative, all the alkali metals are strong reducing agents that react with water:

$$2\,Na(s) + 2\,H_2O(l) \longrightarrow 2\,NaOH(aq) + H_2(g)$$

The vigor of this reaction increases uniformly down the group (Fig. 14.13). The reaction of water with sodium and potassium is vigorous enough to ignite the hydrogen; with rubidium and cesium, it is dangerously explosive. Rubidium and cesium are denser than water, and so they sink and react beneath the surface; the rapidly evolved hydrogen gas then forms a shock wave that can shatter the vessel. Lithium is the least active of the Group 1 metals in the reaction with water, but molten lithium is one of the most active metals known. This difference implies that solid lithium reacts more slowly with water than the other alkali metals do mainly because of the greater strength with which its atoms are bound together.

The alkali metals also release their valence electrons when they dissolve in liquid ammonia, but the outcome is different. Instead of reducing the ammonia, the electrons occupy cavities formed by groups of NH_3 molecules and give ink-blue *metal–ammonia solutions* (Fig. 14.14). These solutions of solvated electrons (and cations of the metal) are often used to reduce organic compounds. As the metal concentration is increased, the blue gives way to a metallic bronze, and the solutions begin to conduct electricity like liquid metals.

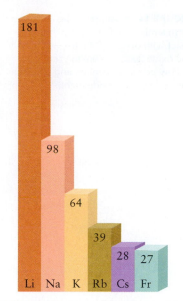

FIGURE 14.12 The melting points of the alkali metals decrease down the group. The numerical values shown here are degrees Celsius.

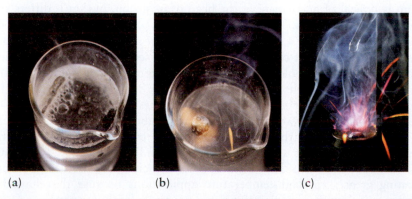

(a)   (b)   (c)

FIGURE 14.13 The alkali metals react with water, producing gaseous hydrogen and a solution of the alkali metal hydroxide. (a) Lithium reacts slowly. (b) Sodium reacts so vigorously that the heat released melts the unreacted metal and ignites the hydrogen. (c) Potassium reacts even more vigorously.

FIGURE 14.14 The blue swirls in this flask of liquid ammonia formed when small pieces of sodium were added and dissolved.

FIGURE 14.15 Although the alkali metals form a mixture of products when they react with oxygen, lithium gives mainly the oxide (left), sodium the very pale yellow peroxide (center), and potassium the yellow superoxide (right).

All alkali metals react directly with almost all nonmetals except the noble gases. However, only lithium reacts with nitrogen, which it reduces to the nitride ion:

$$6\,\text{Li(s)} + \text{N}_2\text{(g)} \longrightarrow 2\,\text{Li}_3\text{N(s)}$$

The principal product of the reaction of the alkali metals with oxygen varies systematically down the group (Fig. 14.15). Ionic compounds formed from cations and anions of similar radius are commonly found to be more stable than those formed from ions with markedly different radii. Such is the case here. Lithium forms mainly the oxide, Li_2O. Sodium, which has a larger cation, forms predominantly the very pale yellow sodium peroxide, Na_2O_2. Potassium, with an even bigger cation, forms mainly the superoxide, KO_2, which contains the superoxide ion, $\text{O}_2{}^-$.

SELF-TEST 14.4A Write the chemical equation for the reaction between lithium and oxygen.

[*Answer:* $4\,\text{Li(s)} + \text{O}_2\text{(g)} \rightarrow 2\,\text{Li}_2\text{O(s)}$]

SELF-TEST 14.4B Write the chemical equation for the reaction between potassium and oxygen.

The alkali metals are usually found as singly charged cations. They react with water with increasing vigor down the group. Cations tend to form their most stable compounds with anions of similar size.

14.7 Compounds of Lithium, Sodium, and Potassium

Lithium is typical of an element at the head of its group, in that it differs significantly from its congeners. The differences stem in part from the small size of the Li^+ cation, which gives it a strong polarizing power and hence a tendency to form bonds with high covalent character. A further consequence of the small size of a Li^+ ion is strong ion–dipole interactions, with the result that many lithium salts form hydrates. Lithium has a diagonal relationship with magnesium and is found in magnesium minerals, where it replaces the magnesium ion.

Lithium compounds are used in ceramics, lubricants, and medicine. Small daily doses of lithium carbonate are an effective treatment for bipolar (manic-depressive) disorder; but scientists still do not fully understand why. Lithium soaps—the lithium salts of long-chain carboxylic acids—are used as thickeners in lubricating greases for high-temperature applications because they have higher melting points than more conventional sodium and potassium soaps.

Two factors that make sodium compounds important are their low cost and their high solubility in water. Sodium chloride is readily mined as rock salt, which

FIGURE 14.16 An evaporation pond. The blue color is due to a dye added to the brine to increase heat absorption and hence speed up evaporation.

is a deposit of sodium chloride left as ancient oceans evaporated; it is also obtained from the evaporation of brine (salt water) from present-day seas and salt lakes (Fig. 14.16). Sodium chloride is used in large quantities in the electrolytic production of chlorine and sodium hydroxide from brine.

Sodium hydroxide, NaOH, is a soft, waxy, white, corrosive solid that is sold commercially as lye. It is an important industrial chemical because it is an inexpensive starting material for the production of other sodium salts. The amount of electricity used to electrolyze brine to produce NaOH in the chloralkali process (Section 12.13) is second only to the amount used to extract aluminum from its ores. The process produces chlorine and hydrogen gases as well as aqueous sodium hydroxide (Fig. 14.17). The net ionic equation for the reaction is

$$2\ Cl^-(aq) + 2\ H_2O(l) \xrightarrow{electrolysis} Cl_2(g) + 2\ OH^-(aq) + H_2(g)$$

Sodium hydrogen carbonate, $NaHCO_3$ (sodium bicarbonate), is commonly called *bicarbonate of soda* or *baking soda*. The rising action of baking soda in batter depends on the reaction of a weak acid, HA, with the hydrogen carbonate ions:

$$HCO_3^-(aq) + HA(aq) \longrightarrow A^-(aq) + H_2O(l) + CO_2(g)$$

The release of gas causes the batter to rise. The weak acids are provided by the recipe, generally in the form of lactic acid from sour milk or buttermilk, citric acid from lemons, or the acetic acid in vinegar. *Baking powder* contains a solid weak acid as well as the hydrogen carbonate, and carbon dioxide is released when water is added.

Sodium carbonate decahydrate, $Na_2CO_3 \cdot 10H_2O$, was once widely used as *washing soda*. It is still sometimes added to water to precipitate Mg^{2+} and Ca^{2+} ions as carbonates, as in

$$Ca^{2+}(aq) + CO_3^{2-}(aq) \longrightarrow CaCO_3(s)$$

and to provide an alkaline environment that helps to remove grease from fabrics. Anhydrous sodium carbonate, or *soda ash*, is used in large amounts in the glass industry as a source of sodium oxide, into which it decomposes when heated (Section 14.21).

Potassium compounds have many similarities to sodium compounds. The principal mineral sources of potassium are *carnallite*, $KCl \cdot MgCl_2 \cdot 6H_2O$, and *sylvite*, KCl, which is incorporated directly into some fertilizers as a source of essential potassium. Potassium compounds are generally more expensive than the corresponding sodium compounds, but in some applications their advantages outweigh their

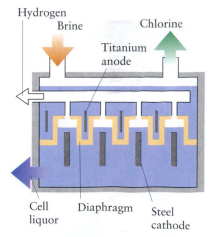

Hydrogen   Brine   Chlorine
Titanium anode
Cell liquor   Diaphragm   Steel cathode

FIGURE 14.17 A diaphragm cell for the electrolytic production of sodium hydroxide from brine (aqueous sodium chloride solution), represented by the blue color. The diaphragm (gold color) prevents the chlorine produced at the titanium anodes from mixing with the hydrogen and the sodium hydroxide formed at the steel cathodes. The liquid (cell liquor) is drawn off and the water is partly evaporated. The unconverted sodium chloride crystallizes, leaving the sodium hydroxide dissolved in the cell liquor.

expense. For example, potassium nitrate, KNO_3, releases oxygen when heated, in the reaction

$$2\ KNO_3(s) \xrightarrow{\Delta} 2\ KNO_2(s) + O_2(g)$$

and is used to facilitate the ignition of matches. It is less hygroscopic (water absorbing) than the corresponding sodium compounds, because the potassium cation is larger and is less strongly hydrated by H_2O molecules.

Self-Test 14.5A Use Table 12.1 to determine the minimum potential difference that must be applied under standard conditions to carry out the chloralkali process.

[*Answer:* 2.19 V]

Self-Test 14.5B Like KO_2, cesium superoxide, CsO_2, can be used to remove exhaled CO_2 and generate oxygen from water. Explain why KO_2 is preferred for this purpose on spacecraft.

Lithium resembles magnesium, and its compounds show covalent character. Sodium compounds are soluble in water, plentiful, and inexpensive. Potassium compounds are generally less hygroscopic than sodium compounds.

GROUP 2: THE ALKALINE EARTH METALS

Calcium, strontium, and barium are called the *alkaline earth metals*, because their "earths"—the old name for oxides—are basic (alkaline). The name alkaline earth metals is often extended to all the members of Group 2 (Table 14.4).

Magnesium and calcium are by far the most important members of the group. Magnesium is, in effect, the doorway to life: it is present in every chlorophyll molecule and hence enables photosynthesis to take place. Calcium is the element of rigidity and construction: it is the cation in the bones of our skeletons, the shells of shellfish, and the concrete, mortar, and limestone of buildings.

14.8 The Group 2 Elements

The valence electron configuration of the atoms of the Group 2 elements is ns^2. The second ionization energy is low enough to be recovered from the lattice enthalpy (Fig. 14.18). Hence, the Group 2 elements occur with an oxidation number of $+2$, as the cation M^{2+}, in all their compounds. Apart from a tendency toward nonmetallic character in beryllium, the elements have all the chemical characteristics of metals, such as forming basic oxides and hydroxides.

All the Group 2 elements are too reactive to occur in the uncombined state in nature (Fig. 14.19). Instead, they are generally found in compounds as doubly charged cations. The element beryllium occurs mainly as *beryl*, $3BeO\cdot Al_2O_3\cdot 6SiO_2$, sometimes in crystals so big that they weigh several tons. The gemstone *emerald* is a form of beryl; its green color is caused by Cr^{3+} ions present as impurities

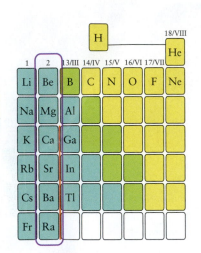

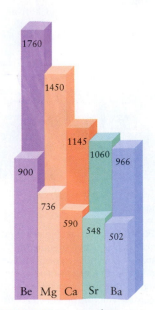

FIGURE 14.18 The first (front row) and second (back row) ionization energies (in kilojoules per mole) of the Group 2 elements. Although the second ionization energies are larger than the first, the are not enormous, and both valence electrons are lost from each atom in all the compounds of these elements.

TABLE 14.4 Group 2 Elements

Valence configuration: ns^2
Normal form*: soft, silver-gray metals

Z	Name	Symbol	Molar mass (g·mol^{-1})	Melting point (°C)	Boiling point (°C)	Density (g·cm^{-3})
4	beryllium	Be	9.01	1285	2470	1.85
12	magnesium	Mg	24.31	650	1100	1.74
20	calcium	Ca	40.08	840	1490	1.53
38	strontium	Sr	87.62	770	1380	2.58
56	barium	Ba	137.33	710	1640	3.59
88	radium	Ra	(226)	700	1500	5.00

*Normal form means the state and appearance of the element at 25°C and 1 atm.

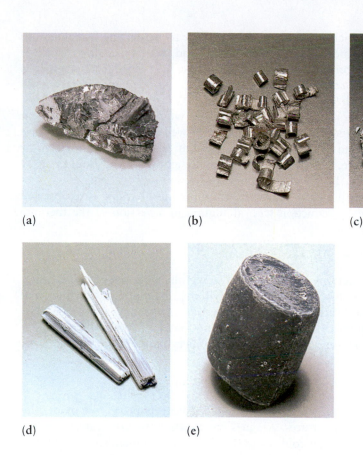

(a)　　　　　(b)　　　　　(c)

(d)　　　　　(e)

FIGURE 14.19 The elements of Group 2:
(a) beryllium; (b) magnesium; (c) calcium;
(d) strontium; and (e) barium. The four
central elements of the group
(magnesium through barium) were
discovered by Humphry Davy in a single
year (1808). The two outer elements were
discovered later: beryllium in 1828 (by
Friedrich Wöhler) and radium (which is
not shown here) in 1898 (by Pierre and
Marie Curie).

(Fig. 14.20). Magnesium occurs in seawater and as the mineral *dolomite*, $CaCO_3 \cdot MgCO_3$. Calcium also occurs as $CaCO_3$ in compressed deposits of the shells of ancient marine organisms and exoskeletons of tiny one-celled organisms; these deposits include *limestone*, *calcite*, and *chalk* (a softer variety of calcium carbonate).

Beryllium is obtained by electrolytic reduction of molten beryllium chloride. The element's low density makes it useful for the construction of missiles and satellites. Beryllium is also used as windows for x-ray tubes: because Be atoms have so few electrons, thin sheets of the metal are transparent to x-rays and allow the rays to escape. Beryllium is added in small amounts to copper: the small Be atoms pin the Cu atoms together in an interstitial alloy that is more rigid than pure copper but still conducts electricity well. These hard, electrically conducting alloys are formed into nonsparking tools for use in oil refineries and grain elevators, where there is a risk of explosion. Beryllium–copper alloys are also used in the electronics industry to form tiny nonmagnetic parts and contacts that resist deformation and corrosion.

Metallic magnesium is produced by either chemical or electrolytic reduction of its compounds. In chemical reduction, first magnesium oxide is obtained from the decomposition of dolomite. Then ferrosilicon, an alloy of iron and silicon, is used to reduce the MgO at about 1200°C. At this temperature, the magnesium produced is immediately vaporized and carried away. The electrolytic method uses seawater as its principal raw material: magnesium hydroxide is precipitated by adding slaked lime ($Ca(OH)_2$, see Section 14.10), the precipitate is filtered off and treated with hydrochloric acid to produce magnesium chloride, and the dried molten salt is electrolyzed.

Magnesium is a silver-white metal that is protected from extensive oxidation in air by a film of white oxide, which gives the metal a dull gray coat. Its density is only two-thirds that of aluminum, and the pure metal is very soft. However, its alloys have great strength, and they are used in applications requiring lightness and toughness—in airplanes, for instance. The use of magnesium alloys in automobiles increased by a factor of 5 between 1989 and 1995 as manufacturers sought to

FIGURE 14.20 An emerald is a crystal of beryl with some Cr^{3+} ions, which are responsible for the green color.

reduce the weight of vehicles. However, don't expect to be driving a magnesium car soon: magnesium is more expensive than steel and more difficult to work. It melts at a low temperature and is deformed by the heat produced when it is machined. Water cannot be used to cool the tools because magnesium reacts with hot water to produce hydrogen.

Magnesium burns vigorously in air with a brilliant white flame, partly because it reacts with the nitrogen and carbon dioxide in air as well as with oxygen. Because the reaction is accelerated when burning magnesium is sprayed with water or exposed to carbon dioxide, neither water nor CO_2 fire extinguishers should ever be used on a magnesium fire. They will only make the fire worse!

The true alkaline earth metals—calcium, strontium, and barium—are obtained either by electrolysis or by reduction with aluminum in a version of the thermite process (see Fig. 6.8):

$$3\ BaO(s) + 2\ Al(s) \xrightarrow{\Delta} Al_2O_3(s) + 3\ Ba(s)$$

As in Group 1, reactions of the Group 2 metals with oxygen and water increase in vigor going down the group. Beryllium, magnesium, calcium, and strontium are partly passivated in air by a protective surface layer of oxide. Barium, however, does not form a protective oxide and may ignite in moist air.

All the Group 2 elements with the exception of beryllium react with water; for example,

$$Ca(s) + 2\ H_2O(l) \longrightarrow Ca^{2+}(aq) + 2\ OH^-(aq) + H_2(g)$$

Beryllium does not react with water, even when red hot: its protective oxide film survives even at high temperatures. Magnesium reacts with hot water (see Fig. 13.22), and calcium reacts with cold water (Fig. 14.21). The metals reduce hydrogen ions to hydrogen gas, but neither beryllium nor magnesium dissolves in nitric acid, because both become passivated by a film of oxide.

The alkaline earth metals can be detected in burning compounds by the colors that they give to flames. Calcium burns orange-red, strontium crimson, and barium yellow-green. Fireworks are often made from their salts (typically nitrates and chlorates, because the anions then provide an additional supply of oxygen) together with magnesium powder.

SELF-TEST 14.6A Write the chemical equation for the reaction that would occur if you tried to put out a magnesium fire with water.

[*Answer:* $Mg(s) + 2\ H_2O(l) \rightarrow Mg^{2+}(aq) + 2\ OH^-(aq) + H_2(g)$]

SELF-TEST 14.6B Write the chemical equation for the reaction of barium with oxygen.

Beryllium shows a hint of nonmetallic character, but the other Group 2 elements are all reactive metals. The vigor of reaction with water and oxygen increases down the group.

14.9 Compounds of Beryllium and Magnesium

Beryllium, at the head of Group 2, resembles its diagonal neighbor aluminum in its chemical properties. It is the least metallic element of the group, and many of its compounds have properties commonly attributed to covalent bonding. Beryllium is amphoteric and reacts with both acids and alkalis. Like aluminum, beryllium reacts with water in the presence of sodium hydroxide; the products are the *beryllate ion*, $Be(OH)_4^{2-}$, and hydrogen:

$$Be(s) + 2\ OH^-(aq) + 2\ H_2O(l) \longrightarrow Be(OH)_4^{2-}(aq) + H_2(g)$$

Beryllium compounds are very toxic and must be handled with great caution. Their properties are dominated by the highly polarizing character of the Be^{2+} ion and its small size. The strong polarizing power results in moderately covalent compounds, and its small size limits to four the number of groups that can attach to the ion. These two features together are responsible for the prominence of the

FIGURE 14.21 Calcium reacts gently with cold water to produce hydrogen and calcium hydroxide, $Ca(OH)_2$.

2 BeX$_4$ unit

3 Beryllium chloride, BeCl$_2$

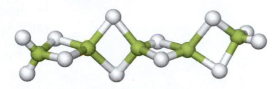

4 Beryllium hydride, BeH$_2$

tetrahedral BeX$_4$ unit (**2**), like that in the beryllate ion. A tetrahedral unit is also found in the solid chloride (**3**) and hydride (**4**). The chloride is made by the action of chlorine on the oxide in the presence of carbon:

$$BeO(s) + C(s) + Cl_2(g) \xrightarrow{600\text{–}800°C} BeCl_2(g) + CO(g)$$

The Be atoms in BeCl$_2$ act as Lewis acids and accept electron pairs from the Cl atoms of the neighboring linear BeCl$_2$ groups, forming a chain of tetrahedral BeCl$_4$ units in the solid.

Magnesium has more pronounced metallic properties than beryllium; its compounds are primarily ionic, with some covalent character. Magnesium oxide, MgO, is formed when magnesium burns in air, but the product is contaminated by magnesium nitride. To prepare the pure oxide, the hydroxide or the carbonate is heated. Magnesium oxide dissolves only very slightly in water. One of its most striking properties is that it is *refractory* (able to withstand high temperatures); it melts at 2800°C. This high stability can be traced to the small ionic radii of the Mg^{2+} and O^{2-} ions and hence to their very strong electrostatic interaction with each other. The oxide has two other useful characteristics: it conducts heat very well, and it conducts electricity poorly. All three properties make it useful as an insulator in electric heaters.

Magnesium hydroxide, Mg(OH)$_2$, is a base. It is not very soluble in water but forms instead a white colloidal suspension, a mist of small particles dispersed through a liquid (see Section 8.21), which is known as *milk of magnesia* and used as a stomach antacid. Because this base is relatively insoluble, it is not absorbed from the stomach but remains to act on whatever acids are present. Hydroxides have an advantage over hydrogen carbonates (which also are used as antacids) in that their neutralization does not lead to the formation of carbon dioxide and its consequence, belching. Milk of magnesia must be used sparingly as an antacid, because the salt produced by the neutralization of milk of magnesia in the stomach is magnesium chloride, which can act as a purgative. Magnesium sulfate, as *Epsom salts*, MgSO$_4$·7H$_2$O, is another common purgative. The magnesium ions inhibit the absorption of water from the intestine. The resulting increased flow of water through the intestine triggers the mechanism that results in defecation.

Arguably the most important compound of magnesium is chlorophyll. This green organic compound consists of large molecules that capture light from the Sun and channel its energy into photosynthesis. One function of the Mg^{2+} ion, which lies just above the plane of the ring (**5**), appears to be to keep the ring rigid. This rigidity helps to ensure that the energy captured from the incoming photon is not lost as heat before it has been used to bring about a chemical reaction. Magnesium also plays an important role in energy generation in living cells. For example, it is involved in the contraction of muscles.

5 Chlorophyll

> *Beryllium compounds have a pronounced covalent character, and the structural unit is commonly tetrahedral. The small size of the magnesium cation results in a thermally stable oxide with low solubility in water.*

14.10 Compounds of Calcium

The most common compound of calcium is calcium carbonate, CaCO$_3$, which occurs naturally in a variety of forms such as chalk and limestone. Marble is a dense form of calcium carbonate that can be given a high polish; it is often colored by impurities, most commonly iron cations. The two most common forms of pure calcium carbonate are calcite and aragonite. All these carbonates are the fossilized remains of marine life.

Calcium carbonate decomposes to calcium oxide, CaO, or *quicklime*, when heated:

$$CaCO_3(s) \xrightarrow{\Delta} CaO(s) + CO_2(g)$$

Calcium oxide is called quicklime because it reacts so exothermically and rapidly with water:

$$CaO(s) + H_2O(l) \longrightarrow Ca^{2+}(aq) + 2\,OH^-(aq)$$

The product, calcium hydroxide, is commonly known as *slaked lime* because, as calcium hydroxide, the thirst of lime for water has been quenched (slaked). Slaked lime is the form in which lime is normally sold because quicklime can set fire to moist wood and paper. In fact, the wooden boats that were once used to transport quicklime sometimes caught fire in the heat of reaction when water seeped into their holds. An aqueous solution of calcium hydroxide, which is slightly soluble in water, is called *lime water*. It is used as a test for carbon dioxide, with which it reacts to form a suspension of the much less soluble calcium carbonate:

$$Ca(OH)_2(aq) + CO_2(g) \longrightarrow CaCO_3(s) + H_2O(l)$$

Quicklime is produced in enormous quantities throughout the world. About 40% of this output is used in metallurgy. In ironmaking (Section 16.13), it is used as a Lewis base; its O^{2-} ion reacts with silica, SiO_2, impurities in the ore to form a liquid *slag*:

$$CaO(s) + SiO_2(s) \xrightarrow{\Delta} CaSiO_3(l)$$

About 50 kg of lime is needed to produce 1 tonne (10^3 kg) of iron.

Slaked lime is used as an inexpensive base in industry, as well as to adjust the pH of soils in agriculture. Perhaps surprisingly, it is also used to *remove* Ca^{2+} ions from hard water containing $Ca(HCO_3)_2$. Its role here is to convert HCO_3^- into CO_3^{2-} by providing OH^- ions:

$$HCO_3^-(aq) + OH^-(aq) \longrightarrow CO_3^{2-}(aq) + H_2O(l)$$

As the concentration of CO_3^{2-} ions rises, Ca^{2+} ions precipitate in the reaction

$$Ca^{2+}(aq) + CO_3^{2-}(aq) \longrightarrow CaCO_3(s)$$

This reaction removes the Ca^{2+} ions that were present initially and those that were added as lime; overall, the Ca^{2+} ion concentration is reduced.

Calcium compounds are often used as structural materials in organisms, buildings, and civil engineering, because of the rigidity of their structures. This rigidity stems from the strength with which the small, highly charged Ca^{2+} cation interacts with its neighbors. *Concrete* is a strong building material that consists of a binder and a filler. The filler is usually gravel, but sometimes polymer or vermiculite pellets are added to lower the density. The binder is *cement*, usually *Portland cement*, which is made by heating a mixture of crushed limestone, clay or shale, sand, and oxides such as iron ore in a kiln. The hard pellets that result, the "clinkers," are a mixture of mainly calcium oxide, calcium silicates, and calcium aluminum silicates. These pellets are ground together with *gypsum*, $CaSO_4 \cdot 2H_2O$, into a powder that sets into a hard mass when mixed with water. The water reacts with the mixture to produce hydrates and hydroxides that bind the salts together into a three-dimensional network. *Mortar* consists of about one part cement and three parts sand (largely silica, SiO_2). It sets to a hard mass as the lime reacts with the carbon dioxide of the air to form the carbonate (Fig. 14.22).

Calcium is also found in the rigid structural components of living organisms, either as the calcium carbonate of the shells of shellfish or the calcium phosphate

FIGURE 14.22 An electron micrograph of the surface of cement, showing the growth of tiny interlocking crystals as carbon dioxide reacts with calcium oxide and silica.

of bone. About a kilogram of calcium is present in an adult human body, mostly in the form of insoluble calcium phosphate, but also as Ca^{2+} ions in other fluids inside our cells. The calcium in newly formed bone is in dynamic equilibrium with the calcium ions in the body fluids, and so calcium must be part of our daily diet to maintain bone strength.

Tooth enamel is a *hydroxyapatite*, $Ca_5(PO_4)_3OH$. Tooth decay begins when acids attack the enamel:

$$Ca_5(PO_4)_3OH(s) + 4 H_3O^+(aq) \longrightarrow 5 Ca^{2+}(aq) + 3 HPO_4^{2-}(aq) + 5 H_2O(l)$$

The principal agents of tooth decay are the carboxylic acids produced when bacteria act on the remains of food. A more resistant coating forms when the OH^- ions in the apatite are replaced by F^- ions. The resulting mineral is called *fluorapatite*:

$$Ca_5(PO_4)_3OH(s) + F^-(aq) \longrightarrow Ca_5(PO_4)_3F(s) + OH^-(aq)$$

The addition of fluoride ions to domestic water supplies (in the form of NaF) is now widespread and has resulted in a dramatic decrease in dental cavities. Fluoridated toothpastes, containing either tin(II) fluoride or sodium monofluorophosphate (MFP, Na_2FPO_3), are also recommended to strengthen tooth enamel.

SELF-TEST 14.7A Explain why beryllium compounds have covalent characteristics.
[*Answer:* The small size and high charge on the beryllium ion make it highly polarizing.]

SELF-TEST 14.7B Is the reaction between CaO and SiO_2 a redox reaction or a Lewis acid–base reaction? If redox, identify the oxidizing and reducing agents. If Lewis acid–base, identify the acid and the base.

Calcium compounds are common in structural materials because the small, highly charged Ca^{2+} ion results in rigid structures.

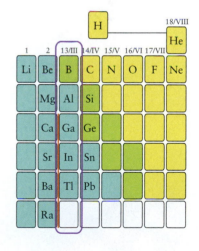

GROUP 13/III: THE BORON FAMILY

Now we move into the *p* block of the periodic table and encounter the complex but fascinating world of the nonmetals. Here, close to the center of the periodic table, we meet strange properties, because the elements are neither so electropositive that they easily lose electrons nor so electronegative that they easily gain them.

Group 13/III is the first group of the *p* block. Its members have an ns^2np^1 electron configuration (Table 14.5), and so we expect a maximum oxidation number of +3. The oxidation numbers of B and Al are +3 in almost all their compounds. However, the heavier elements in the group are more likely to keep their *s*-electrons (the inert-pair effect, Section 1.19); so the oxidation number +1 becomes increasingly important down the group, and thallium(I) compounds are as common as

TABLE 14.5 Group 13/III Elements

Valence configuration: ns^2np^1

Z	Name	Symbol	Molar mass (g·mol^{-1})	Melting point (°C)	Boiling point (°C)	Density (g·cm^{-3})	Normal form*
5	boron	B	10.81	2300	3931	2.47	powdery brown metalloid
13	aluminum	Al	26.98	660	2467	2.70	silver-white metal
31	gallium	Ga	69.72	30	2403	5.91	silver metal
49	indium	In	114.82	156	2080	7.29	silver-white metal
81	thallium	Tl	204.38	304	1457	11.87	soft metal

Normal form means the state and appearance of the element at 25°C and 1 atm.

thallium(III) compounds. We shall concentrate on the two most important members of the group, boron and aluminum.

14.11 The Group 13/III Elements

Boron forms perhaps the most extraordinary structures of all the elements. It has a high ionization energy and is a metalloid that forms covalent bonds, like its diagonal neighbor silicon. However, because it has only three electrons in its valence shell and has a small atomic radius, it tends to form compounds that have incomplete octets (Section 2.11) or are electron deficient (Section 3.8). These unusual bonding characteristics lead to the remarkable properties that have made boron an essential element of modern technology and, in particular, nanotechnology.

Boron is mined as *borax* and *kernite*, $Na_2B_4O_7 \cdot xH_2O$, with $x = 10$ and 4, respectively. Large deposits from ancient hot springs are found in volcanic regions, such as the Mojave Desert region of California. In the extraction process, the ore is converted into boron oxide with acid and then reduced with magnesium to an impure brown, amorphous form of boron:

$$B_2O_3(s) + 3\ Mg(s) \xrightarrow{\Delta} 2\ B(s) + 3\ MgO(s)$$

A product of higher purity is obtained by reducing a volatile boron compound, such as BCl_3 or BBr_3, with hydrogen on a heated filament (tantalum is used for the filament because it has a very high melting point):

$$2\ BBr_3(g) + 3\ H_2(g) \xrightarrow{\Delta} 2\ B(s) + 6\ HBr(g)$$

Boron production remains quite low despite the element's desirable properties of hardness and low density.

Elemental boron exists in a variety of allotropic forms in which each atom strives to find ways to share eight electrons despite being so small and having only three electrons to contribute. It is most commonly found as either a gray-black nonmetallic, high-melting-point solid or a dark brown powder with an icosahedral (20-faced) structure based on clusters of 12 atoms (**6**). Because of the three-dimensional network formed by these bonds, boron is very hard and is chemically unreactive. When boron fibers are incorporated into plastics, the result is a very tough material that is stiffer than steel yet lighter than aluminum; this material is used in aircraft, missiles, and body armor. The element is attacked by only the strongest oxidizing agents.

Aluminum is the most abundant metallic element in the Earth's crust and, after oxygen and silicon, the third most abundant element (see Fig. 14.1). However, the aluminum content in most minerals is low, and the commercial source of aluminum, *bauxite*, is a hydrated, impure oxide, $Al_2O_3 \cdot xH_2O$, where x can range from 1 to 3. Bauxite ore, which is red from the iron oxides that it contains (Fig. 14.23), is processed to obtain alumina, Al_2O_3, in the *Bayer process*. In this process, the ore is first treated with aqueous sodium hydroxide, which dissolves the amphoteric alumina as the aluminate ion, $Al(OH)_4^-(aq)$. Carbon dioxide is then bubbled through the solution to remove OH^- ions as HCO_3^- and to convert some of the aluminate ions into aluminum hydroxide, which precipitates. The aluminum hydroxide is removed and dehydrated to the oxide by heating to 1200°C.

Obtaining aluminum metal from the oxide was a challenge to early scientists and engineers. When it was first isolated, aluminum was a rare and expensive metal. During the nineteenth century, it so symbolized modern technology that the Washington Monument was given an expensive aluminum tip. That rarity and expense were transformed by electrochemistry. Aluminum metal is now obtained on a huge scale by the *Hall process*. In 1886, Charles Hall discovered that, by mixing the mineral cryolite, Na_3AlF_6, with alumina, he got a mixture that melted at a much more economical temperature, 950°C, instead of the 2050°C of pure alu-

Here, perhaps, is an opportunity for another young chemist like Hall to transform the production of boron as Hall did for aluminum (described later in this section).

6 B_{12}

FIGURE 14.23 The iron oxides in bauxite ore give it a red tint.

Napoleon reserved his aluminum plates for his special guests: the rest had to make do with gold.

mina. The melt is electrolyzed in a cell that uses graphite (or carbonized petroleum) anodes and a carbonized steel-lined vat that serves as the cathode (Fig. 14.24). The electrolysis half-reactions are

Cathode reaction: $Al^{3+}(melt) + 3\,e^- \longrightarrow Al(l)$

Anode reaction: $2\,O^{2-}(melt) + C(s,\,gr) \longrightarrow CO_2(g) + 4\,e^-$

The overall reaction is

$$4\,Al^{3+}(melt) + 6\,O^{2-}(melt) + 3\,C(s,\,gr) \longrightarrow 4\,Al(l) + 3\,CO_2(g)$$

Note that the carbon electrode is consumed in the reaction. From the reaction stoichiometry, we can calculate that a current of 1 A must flow for 80 h to produce 1 mol Al (27 g of aluminum, about enough for two soft-drink cans). The very high energy consumption can be greatly reduced by recycling, which requires less than 5% of the electricity needed to extract aluminum from bauxite. Although a 12-ounce beverage can contains less than 14 g of aluminum, the energy that we are throwing away when we discard an aluminum can is equivalent to burning the amount of gasoline that would fill half the can. Note also that the production of 1 tonne of aluminum from alumina is accompanied by the release of more than 1 tonne of carbon dioxide into the atmosphere.

Aluminum has a low density; it is a strong metal and an excellent electrical conductor. Although it is strongly reducing and therefore easily oxidized, aluminum is resistant to corrosion because its surface is passivated in air by a stable oxide film. The thickness of the oxide layer can be increased by making aluminum the anode of an electrolytic cell; the result is called *anodized aluminum*. Dyes may be added to the dilute sulfuric acid electrolyte used in the anodizing process to produce surface layers with different colors.

Aluminum's low density, wide availability, and corrosion resistance make it ideal for construction and for the aerospace industry. Aluminum is a soft metal, and so it is usually alloyed with copper and silicon for greater strength. Its lightness and good electrical conductivity have also led to its use for overhead power lines, and its negative electrode potential has led to its use in fuel cells. Perhaps one day your automobile will not only be made of aluminum but fueled by it, too.

Boron, at the head of Group 13/III, is best regarded as a nonmetal in most of its chemical properties. It has acidic oxides and forms an interesting and extensive range of binary molecular hydrides. Metallic character increases down the group, and even boron's immediate neighbor aluminum is a metal. Nonetheless, aluminum is sufficiently far to the right in the periodic table to show a hint of nonmetallic character. Aluminum is amphoteric, reacting both with nonoxidizing acids (such as hydrochloric acid) to form aluminum ions:

$$2\,Al(s) + 6\,H^+(aq) \longrightarrow 2\,Al^{3+}(aq) + 3\,H_2(g)$$

and with hot aqueous alkali to form aluminate ions:

$$2\,Al(s) + 2\,OH^-(aq) + 6\,H_2O(l) \longrightarrow 2\,Al(OH)_4{}^-(aq) + 3\,H_2(g)$$

Gallium, which is produced as a by-product in the Bayer process, has important uses in the electronics industry. It is a common doping agent for semiconductors and some of its compounds, such as GaAs, are used in light-emitting diodes to convert electricity into light. Thallium, at the bottom of Group 13/III, is a dangerously poisonous heavy metal that has been used as a rat poison. However, concern about its accumulation in the environment has spurred research into alternative pesticides.

Boron is a hard metalloid with pronounced nonmetallic properties. Aluminum is a light, strong, amphoteric, reactive metallic element with a surface that becomes passivated when exposed to air.

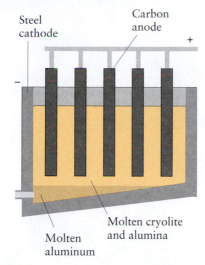

FIGURE 14.24 In the Hall process, aluminum oxide is dissolved in molten cryolite and the mixture is electrolyzed in a cell with carbon anodes and a steel cathode. The molten aluminum flows out of the bottom of the cell.

An oxidizing acid is an oxoacid in which the anion is an oxidizing agent; examples are HNO_3 and $HClO_3$.

14.12 Group 13/III Oxides

Boron, a metalloid with largely nonmetallic properties, has acidic oxides. Aluminum, its metallic neighbor, has amphoteric oxides (like its diagonal neighbor in Group 2, beryllium). The oxides of both elements are important in their own right, as sources of the elements, and as the starting point for the manufacture of other compounds.

Boric acid, $B(OH)_3$, is a white solid that melts at 171°C. It is toxic to bacteria and many insects as well as humans and has long been used as a mild antiseptic and pesticide. Because the boron atom in $B(OH)_3$ has an incomplete octet, it can act as a Lewis acid and form a bond by accepting a lone pair of electrons from an H_2O molecule acting as a Lewis base:

$$(OH)_3B + :OH_2 \longrightarrow (OH)_3B—OH_2$$

The compound so formed is a weak *mono*protic acid:

$$B(OH)_3OH_2(aq) + H_2O(l) \rightleftharpoons H_3O^+(aq) + B(OH)_4^-(aq) \qquad pK_a = 9.14$$

The major use of boric acid is as the starting material for its anhydride, boron oxide, B_2O_3. Because it melts (at 450°C) to a liquid that dissolves many metal oxides, boron oxide (often as the acid) is used as a *flux,* a substance that cleans metals as they are soldered or welded. Boron oxide is also used to make fiberglass and borosilicate glass, a glass with a very low thermal expansion, such as Pyrex (see Section 14.21).

Aluminum oxide, Al_2O_3, is known almost universally as *alumina*. It exists with a variety of crystal structures, many of which form important ceramic materials (see Section 14.22). As α-alumina, it is the very hard, stable, crystalline substance *corundum*; impure microcrystalline corundum is the purple-black abrasive known as *emery*. Some impure forms of alumina are beautiful, rare, and highly prized (Fig. 14.25). A less dense and more reactive form of the oxide is γ-alumina. This form absorbs water and is used as the stationary phase in chromatography.

γ-Alumina is produced by heating aluminum hydroxide. It is moderately reactive and is amphoteric, dissolving readily in bases to produce the aluminate ion and in acids to produce the hydrated Al^{3+} ion:

$$Al_2O_3(s) + 2\ OH^-(aq) + 3\ H_2O(l) \longrightarrow 2\ Al(OH)_4^-(aq)$$
$$Al_2O_3(s) + 6\ H_3O^+(aq) + 3\ H_2O(l) \longrightarrow 2\ Al(H_2O)_6^{3+}(aq)$$

(a)   (b)   (c)

FIGURE 14.25 Some of the impure forms of α-alumina are prized as gems. (a) Ruby is alumina with Cr^{3+} in place of some Al^{3+} ions. (b) Sapphire is alumina with Fe^{3+} and Ti^{4+} impurities. (c) Topaz is alumina with Fe^{3+} impurities.

As described in Section 10.13, the strong polarizing effect of the small, highly charged Al^{3+} ion (a Lewis acid) on the water molecules around it (Lewis bases) gives the $Al(H_2O)_6^{3+}$ ion acidic properties:

$$Al(H_2O)_6^{3+}(aq) + H_2O(l) \rightleftharpoons H_3O^+(aq) + Al(OH)(H_2O)_5^{2+}(aq)$$

One of the most important aluminum salts prepared by the action of an acid on alumina is aluminum sulfate, $Al_2(SO_4)_3$:

$$Al_2O_3(s) + 3 H_2SO_4(aq) \longrightarrow Al_2(SO_4)_3(aq) + 3 H_2O(l)$$

Aluminum sulfate is called *papermaker's alum* and is used in the paper industry to coagulate cellulose fibers into a hard, nonabsorbent surface. True alums (from which aluminum takes its name) are mixed sulfates of formula $M^+M'^{3+}(SO_4)_2 \cdot 12H_2O$. They include potassium alum, $KAl(SO_4)_2 \cdot 12H_2O$ (which is used in water and sewage treatment), and ammonium alum, $NH_4Al(SO_4)_2 \cdot 12H_2O$ (which is used for pickling cucumbers and as the Brønsted acid in baking powders).

Sodium aluminate, $NaAl(OH)_4$, is used along with aluminum sulfate in water purification. When mixed with aluminate ions, the acidic hydrated Al^{3+} cation from the aluminum sulfate produces aluminum hydroxide:

$$Al^{3+}(aq) + 3 Al(OH)_4^-(aq) \longrightarrow 4 Al(OH)_3(s)$$

The aluminum hydroxide is formed as a fluffy, gelatinous network that entraps impurities as it settles, and this precipitate can be removed by filtration (Fig. 14.26).

FIGURE 14.26 Aluminum hydroxide, $Al(OH)_3$, forms as a white, fluffy precipitate. The fluffy form of the solid captures impurities and is used in the purification of water.

SELF-TEST 14.8A Explain how Al^{3+} forms acidic solutions.

> [*Answer:* In water Al^{3+} forms $Al(H_2O)_6^{3+}$; the highly charged Al^{3+} ion polarizes the water molecules, enabling one of them to donate a proton.]

SELF-TEST 14.8B The compound $B(OH)_3$ does not itself lose any protons in water. Why, then, is it called boric acid?

> *Boron oxide is an acid anhydride. Aluminum shows some nonmetallic character in that its oxide is amphoteric.*

14.13 Nitrides and Halides

When boron is heated to white heat in ammonia, boron nitride, BN, is formed as a fluffy, slippery powder:

$$2 B(s) + 2 NH_3(g) \xrightarrow{\Delta} 2 BN(s) + 3 H_2(g)$$

Its structure resembles that of graphite, but the latter's flat planes of carbon hexagons are replaced in boron nitride by planes of hexagons of alternating B and N atoms (Fig. 14.27). Unlike graphite, boron nitride is white and does not conduct

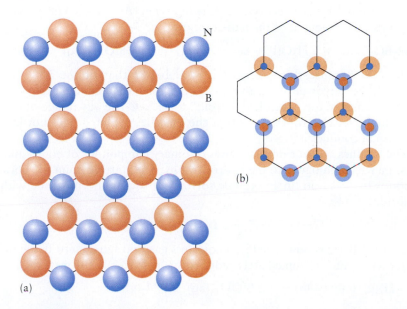

(a)

(b)

FIGURE 14.27 (a) The structure of hexagonal boron nitride, BN, resembles that of graphite, consisting of flat planes of hexagons of alternating B and N atoms (in place of C atoms) but, as shown for two adjacent layers in part (b), the planes are stacked differently, with each B atom directly over an N atom and vice-versa (compare with Fig. 14.29). Note that (to make them distinguishable) the B atoms in the top layer are red and the N atoms blue.

electricity. Under high pressure, boron nitride is converted into a very hard, diamondlike crystalline form called Borazon (a proprietary name). In recent years, boron nitride nanotubes similar to those formed by carbon have been synthesized (Section 14.16), and they have been found to be semiconducting (see Box 14.1).

The boron halides are made either by direct reaction of the elements at a high temperature or from boron oxide. The most important is boron trifluoride, BF_3, an industrial catalyst produced by the reaction between boron oxide, calcium fluoride, and sulfuric acid:

$$B_2O_3(s) + 3\ CaF_2(s) + 3\ H_2SO_4(l) \xrightarrow{\Delta} 2\ BF_3(g) + 3\ CaSO_4(s) + 3\ H_2O(l)$$

Boron trichloride, BCl_3, which is also widely used as a catalyst, is produced commercially by the action of chlorine gas on the oxide in the presence of carbon:

$$B_2O_3(s) + 3\ C(s) + 3\ Cl_2(g) \xrightarrow{500°C} 2\ BCl_3(g) + 3\ CO(g)$$

The B atom has an incomplete octet in all its trihalides. The compounds consist of trigonal planar molecules with an empty $2p$-orbital perpendicular to the molecular plane. The empty orbital allows the molecules to act as Lewis acids, which accounts for the catalytic action of BF_3 and BCl_3.

Aluminum chloride, $AlCl_3$, another major industrial catalyst, is made by the action of chlorine on aluminum or on alumina in the presence of carbon:

$$2\ Al(s) + 3\ Cl_2(g) \longrightarrow 2\ AlCl_3(s)$$
$$Al_2O_3(s) + 3\ C(s) + 3\ Cl_2(g) \longrightarrow 2\ AlCl_3(s) + 3\ CO(g)$$

7   Aluminum chloride dimer, Al_2Cl_6

Aluminum chloride is an ionic solid in which each Al^{3+} ion is coordinated to six Cl^- ions. However, it sublimes at 192°C to a vapor of Al_2Cl_6 molecules (7). In this molecule, the Al atom of one $AlCl_3$ fragment acts as a Lewis acid and accepts an electron pair from a Cl atom of the other $AlCl_3$ fragment, which is acting as a Lewis base.

Boron and aluminum halides have incomplete octets and act as Lewis acids.

14.14 Boranes, Borohydrides, and Borides

Boron forms a remarkable series of binary compounds with hydrogen—the *boranes.* These compounds include diborane, B_2H_6, and more complex compounds such as decaborane, $B_{10}H_{14}$. Anionic versions of these compounds, the *borohydrides,* are known; the most important is BH_4^- as sodium borohydride, $NaBH_4$.

Sodium borohydride is a white crystalline solid produced from the reaction between sodium hydride and boron trichloride dissolved in a nonaqueous solvent:

$$4\ NaH + BCl_3 \longrightarrow NaBH_4 + 3\ NaCl$$

Sodium borohydride is a very useful reducing agent. At pH = 14 (strongly alkaline conditions), the potential of the half-reaction

$$H_2BO_3^-(aq) + 5\ H_2O(l) + 8e^- \longrightarrow BH_4^-(aq) + 8\ OH^-(aq)$$

is −1.24 V. Because this potential is well below that of the Ni^{2+}/Ni couple (−0.23 V), borohydride ions can reduce Ni^{2+} ions to metallic nickel. This reduction is the basis of the "chemical plating" of nickel. The advantage of this chemical plating over electroplating is that the item being plated does not have to be an electrical conductor.

The boranes are an extensive series of binary compounds of boron and hydrogen, somewhat analogous to the hydrocarbons. The starting point for borane production is the reaction (in an organic solvent) of sodium borohydride with boron trifluoride:

$$4\ BF_3 + 3\ BH_4^- \longrightarrow 3\ BF_4^- + 2\ B_2H_6$$

The product B_2H_6 is diborane (8), a colorless gas that bursts into flame in air. On contact with water, it immediately reduces the hydrogen in the water:

8   Diborane, B_2H_6

$$B_2H_6(g) + 6\ H_2O(l) \longrightarrow 2\ B(OH)_3(aq) + 6\ H_2(g)$$

When diborane is heated to a high temperature, it decomposes into hydrogen and pure boron:

$$B_2H_6(g) \xrightarrow{\Delta} 2\ B(s) + 3\ H_2(g)$$

This sequence of reactions is a useful route to the pure element, but more complex boranes form when the heating is less severe. When diborane is heated to 100°C, for instance, it forms decaborane, $B_{10}H_{14}$, a solid that melts at 100°C. Decaborane is stable in air, is oxidized by water only slowly, and is an example of the general rule that heavier boranes are less flammable than boranes of low molar mass.

The boranes are electron-deficient compounds (Section 3.8): we cannot write valid Lewis structures for them, because too few electrons are available. For instance, there are 8 atoms in diborane, so we need at least 7 bonds; however, there are only 12 valence electrons, and so we can form at most 6 electron-pair bonds. In molecular orbital theory, these electron pairs are regarded as delocalized over the entire molecule, and their bonding power is shared by several atoms. In diborane, for instance, a single electron pair is delocalized over a B—H—B unit. It binds all three atoms together with bond order of $\frac{1}{2}$ for each of the B—H bridging bonds. The molecule has two such bridging **three-center bonds** (9).

Numerous borides of the metals and nonmetals are known. Their formulas are typically unrelated to their location in the periodic table and include AlB_2, CaB_6, $B_{13}C_2$, $B_{12}S_2$, Ti_3B_4, TiB, and TiB_2. In some metal borides, the boron atoms are found at the centers of clusters of metal atoms. More commonly, the boron atoms form extended structures such as zigzag chains, branched chains, or networks of hexagonal rings of boron atoms, as in MgB_2 (see Box 5.2).

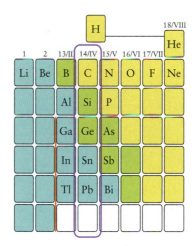

9   Three-center bond

SELF-TEST 14.9A   Compare the reactions for the formation of BF_3 and BCl_3. Which reaction is a redox reaction?

[*Answer:* The formation of BCl_3]

SELF-TEST 14.9B   What is the oxidation number of boron in (a) $NaBH_4$; (b) $H_2BO_3^-$?

The boranes are an extensive series of highly reactive electron-deficient binary compounds of boron and hydrogen. Borohydrides are useful reducing agents.

GROUP 14/IV: THE CARBON FAMILY

Carbon is central to life and natural intelligence. Silicon and germanium are central to electronic technology and artificial intelligence (Fig. 14.28). The unique properties of Group 14/IV elements make both types of intelligence possible. The half-filled valence shell of these elements gives them special properties that straddle

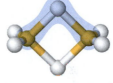

FIGURE 14.28 The elements of Group 14/IV. Back row, from left to right: silicon; tin. Front row: carbon (graphite); germanium; lead.

the line between metals and nonmetals. Carbon, at the head of the group, forms so many compounds that it has its own branch of chemistry, organic chemistry (Chapters 18 and 19).

14.15 The Group 14/IV Elements

The valence electron configuration is ns^2np^2 for all members of the group. All four electrons are approximately equally available for bonding in the lighter elements, and carbon and silicon are characterized by their ability to form four covalent bonds. However, as in Group 13/III, the heavier elements display the inert-pair effect. On descending the group the energy separation between the s- and p-orbitals increases and the s-electrons become progressively less available for bonding; in fact, the most common oxidation number for lead is +2.

The elements show increasing metallic character down the group (Table 14.6). Carbon has definite nonmetallic properties: it forms covalent compounds with nonmetals and ionic compounds with metals. The oxides of carbon and silicon are acidic. Germanium is a typical metalloid in that it exhibits metallic or non-metallic properties according to the other element present in the compound. Tin and, even more so, lead have definite metallic properties. However, even though tin is classified as a metal, it is not far from the metalloids in the periodic table, and it does have some amphoteric properties. For example, tin reacts with both hot concentrated hydrochloric acid and hot alkali:

$$Sn(s) + 2\,H_3O^+(aq) \longrightarrow Sn^{2+}(aq) + H_2(g) + 2\,H_2O(l)$$

$$Sn(s) + 2\,OH^-(aq) + 2\,H_2O(l) \longrightarrow Sn(OH)_4{}^{2-}(aq) + H_2(g)$$

Because carbon stands at the head of its group, we expect it to differ from the other members of the group. In fact, the differences between the element at the head of the group and the other elements are more pronounced in Group 14/IV than anywhere else in the periodic table. Some of the differences between carbon and silicon stem from the smaller atomic radius of carbon, which explains the wide occurrence of C=C and C=O double bonds relative to the rarity of Si=Si and Si=O double bonds. Silicon atoms are too large for the side-by-side overlap of p-orbitals necessary for π-bonds to form between them. Carbon dioxide, which consists of discrete O=C=O molecules, is a gas that we exhale. Silicon dioxide (silica), which consists of networks of —O—Si—O— groups, is a mineral that we stand on.

Silicon compounds can also act as Lewis acids, whereas carbon compounds typically cannot. Because a silicon atom is bigger than a carbon atom and can expand its valence shell by using its d-orbitals, it can accommodate the lone pair of an attacking Lewis base. A carbon atom is smaller and has no available d-orbitals; so in general it cannot act as a Lewis acid. An exception to this behavior is when the carbon atom has multiple bonds, because then a π-bond can give

TABLE 14.6 Group 14/IV Elements

Valence configuration: ns^2np^2

Z	Name	Symbol	Molar mass (g·mol^{-1})	Melting point (°C)	Boiling point (°C)	Density (g·cm^{-3})	Normal form[*]
6	carbon	C	12.01	3370s[†]	—	1.9–2.3	black nonmetal (graphite)
						3.2–3.5	transparent nonmetal (diamond)
							orange nonmetal (fullerite)
14	silicon	Si	28.09	1410	2620	2.33	gray metalloid
32	germanium	Ge	72.61	937	2830	5.32	gray-white metalloid
50	tin	Sn	118.71	232	2720	7.29	white lustrous metal
82	lead	Pb	207.2	328	1760	11.34	blue-white lustrous metal

[*]*Normal form* means the state and appearance of the element at 25°C and 1 atm.
[†]The symbol s denotes that the element sublimes.

way to the formation of a σ-bond, as in the reaction between CO_2 and O^{2-} to form $CO_3{}^{2-}$:

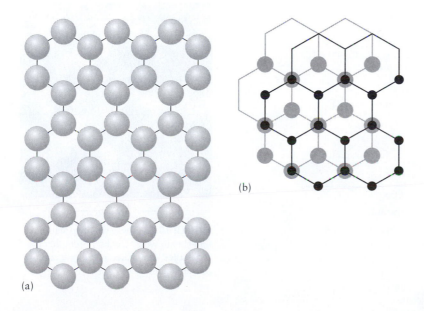

Carbon is the only member of Group 14/IV that commonly forms multiple bonds with itself; singly bonded silicon atoms can act as Lewis acids because a silicon atom can expand its valence shell.

14.16 The Different Forms of Carbon

Solid carbon exists as graphite, diamond, and other phases such as the fullerenes, which have structures related to that of graphite. Graphite is the thermodynamically most stable of these allotropes under ordinary conditions. In this section, we see how the properties of the different allotropes of carbon are related to differences in bonding.

Soot and *carbon black* contain very small crystals of graphite and other forms of carbon. Carbon black, which is produced by heating gaseous hydrocarbons to nearly 1000°C in the absence of air, is used for reinforcing rubber, for pigments, and for printing inks, such as the ink on this page. *Activated carbon*, which is also called *activated charcoal*, consists of granules of microcrystalline carbon. It is produced by heating waste organic matter in the absence of air and then processing it to increase the porosity. The very high specific surface area (as much as about 2000 $m^2 \cdot g^{-1}$) of the porous carbon enables it to remove organic impurities from liquids and gases by adsorption. It is used in air purifiers, gas masks, and aquarium water filters. On a larger scale, activated carbon is used in water purification plants to remove organic compounds from drinking water.

We can understand the differences in properties between the carbon allotropes by comparing their structures. Graphite consists of planar sheets of sp^2 hybridized carbon atoms in a hexagonal network (Fig. 14.29). Electrons are free to move from one carbon atom to another through a delocalized π-network formed by the overlap of unhybridized p-orbitals on each carbon atom. This network spreads across the entire plane. Because of the electron delocalization, graphite is a black, lustrous, electrically conducting solid; indeed, graphite is used as an electrical conductor in industry and as electrodes in electrochemical cells and batteries. Its

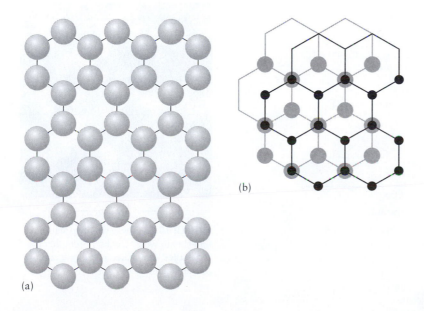

(b)

(a)

FIGURE 14.29 (a) Graphite consists of flat planes of hexagons lying above one another; (b) the ABAB stacking arrangement of adjacent planes. Only the σ-bonding framework is shown: the π-bonds spread above and below the planes. When impurities are present, the planes can slide past one another quite easily. The extensive delocalization through the network of bonding within the planes allows graphite to conduct electricity well along the planes but less well perpendicular to the planes.

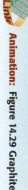

Animation: Figure 14.29 Graphite

Media Link

FIGURE 14.30 Structure of diamond. Each carbon atom is *sp³* hybridized and forms tetrahedral σ-bonds to four neighbors. This pattern is repeated throughout the crystal and accounts for diamond's great hardness.

10   Buckminsterfullerene, C₆₀

slipperiness, which results from the ease with which the flat planes move past one another when impurities are present, leads to its use as a lubricant and, when mixed with clay, as the "lead" in pencils. The graphite found in nature is the result of changes in ancient organic remains. Pure graphite is produced in industry by passing a heavy electric current through rods of *coke* (the solid residue remaining after the destructive distillation of coal).

In diamond, each carbon atom is sp^3 hybridized and linked tetrahedrally to its four neighbors, with all electrons in C—C σ-bonds (Fig. 14.30). Diamond is a rigid, transparent, electrically insulating solid. It is the hardest substance known and the best conductor of heat, being about five times better than copper. These last two properties make it an ideal abrasive, because it can scratch all other substances, yet the heat generated by friction is quickly conducted away.

In nature, diamond is found embedded in a soft rock called *kimberlite*. This rock rises in columns from deep in the Earth, where the diamonds are formed under intense pressure. One method used to make synthetic industrial diamonds recreates the geological conditions that produce natural diamonds by compressing graphite at pressures greater than 80 kbar and temperatures above 1500°C (Fig. 14.31). Small amounts of metals such as chromium and iron are added to the graphite. The molten metals are thought to dissolve the graphite and then, as they cool, to deposit crystals of diamond, which are less soluble than graphite in the molten metal. Most synthetic diamonds are produced by thermal decomposition of methane. In this technique, the carbon atoms settle on a cool surface as both graphite and diamond. However, because hydrogen atoms produced in the decomposition react more quickly with graphite, forming volatile hydrocarbons, more diamond than graphite survives.

Chemists were greatly surprised when soccer-ball-shaped carbon molecules were first identified in 1985, particularly because they might be even more abundant than graphite and diamond! The C₆₀ molecule (**10**) is named buckminsterfullerene after the American architect R. Buckminster Fuller, whose geodesic domes it resembles. Within 2 years, scientists had succeeded in making crystals of buckminsterfullerene; the solid samples are called *fullerite* (Fig. 14.32). The discovery of this molecule and others with similar structures, such as C₇₀, opened up the prospect of a whole new field of chemistry. For instance, the interior of a C₆₀ molecule is big enough to hold an atom of another element, and chemists are now busily preparing a whole new periodic table of these shrink-wrapped atoms.

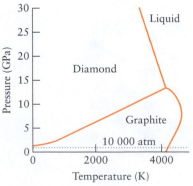

FIGURE 14.31 The phase diagram of carbon, showing the regions of phase stability.

├────100 μm────┤

FIGURE 14.32 These small crystals are fullerite, in which buckminsterfullerene molecules are packed together in a close-packed lattice.

(a)

(b)

(c)

FIGURE 14.33 Three common forms of silica (SiO_2): (a) quartz; (b) quartzite; and (c) cristobalite. The black parts of the sample of cristobalite are obsidian, a volcanic rock that contains silica. Sand consists primarily of small pieces of impure quartz.

The *fullerenes* are members of the family of molecules resembling (and including) buckminsterfullerene. They are formed in smoky flames and by red giants (stars with low surface temperatures and large diameters), and so the universe might contain huge numbers of them. Graphite and diamond are network solids that are insoluble in all liquid solvents except some liquid metals. However, the fullerenes, which are molecular, can be dissolved by suitable solvents (such as benzene); buckminsterfullerene itself forms a red-brown solution. Fullerite currently has few uses, but some of the compounds of the fullerenes have great promise. For example, K_3C_{60} is a superconductor below 18 K, and other compounds appear to be active against cancer and diseases such as AIDS.

A recently discovered form of fibrous carbon consists of concentric tubes with walls like sheets of graphite rolled into cylinders. These tiny structures, called *nanotubes*, form strong, conducting fibers with a large surface area. As a consequence, they have unusually interesting and promising properties that have become a major thrust of nanotechnology research (Box 14.1).

Carbon has an important series of allotropes: diamond, graphite, and the fullerenes.

14.17 Silicon, Germanium, Tin, and Lead

Silicon is the second most abundant element in the Earth's crust. It occurs widely in rocks as *silicates*, compounds containing the silicate ion, SiO_3^{2-}, and as the silica, SiO_2, of sand (Fig. 14.33). Pure silicon is obtained from *quartzite*, a granular form of *quartz* (another solid phase of SiO_2), by reduction with high-purity carbon in an electric arc furnace:

$$SiO_2(s) + 2\,C(s) \xrightarrow{\Delta} Si(s) + 2\,CO(g)$$

The crude product is exposed to chlorine to form silicon tetrachloride, which is then distilled and reduced with hydrogen to a purer form of the element:

$$SiCl_4(l) + 2\,H_2(g) \longrightarrow Si(s) + 4\,HCl(g)$$

Further purification is necessary before silicon can be used in semiconductors. In one process, a large single crystal is grown by pulling a solid rod of the element slowly from the melt. The silicon is then purified by **zone refining**, in which a hot, molten zone is dragged from one end of a cylindrical sample to the other, collecting impurities as it goes (Fig. 14.34). The result is "ultrapure" silicon, which has fewer than one impurity atom per billion Si atoms. An alternative technique is the decomposition of silane, SiH_4, by an electric discharge. This method produces an amorphous form of silicon with a significant hydrogen content. *Amorphous silicon* is used in photovoltaic devices, which produce electricity from sunlight.

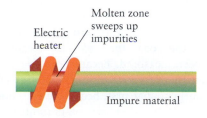
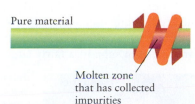

FIGURE 14.34 In the technique of zone refining, a molten zone is passed repeatedly from one end of the solid sample to the other. The impurities collect in the zone and move along the solid with the heater, leaving a pure substance behind.

BOX 14.1   FRONTIERS OF CHEMISTRY: NANOTUBES, NATURE'S SMALLEST PIPES

Imagine a tiny pipet only a few nanometers in diameter that can be used to inject a drug into a single cell, or tiny wires and semiconducting devices in a microprocessor too small to be visible to the naked eye. Then consider huge arrays of these same lilliputian fibers wound into ropes so strong and flexible that they can support bridges and earthquake-proof buildings. These are only a few of the astonishing applications envisioned for *nanotubes*, tiny cylinders made of boron nitride or carbon.

Scientists identified the first carbon nanotubes in 1991. They sealed two graphite rods inside a container of helium gas and sent an electric discharge from one rod to the other. Much of one rod evaporated, but out of the inferno some amazing structures emerged (see illustrations). As well as the tiny 60-atom carbon spheres known as buckminsterfullerene—which had been known since 1985—long, hollow, perfectly straight carbon nanotubes were detected.

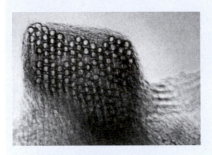

A cluster of carbon nanotubes that has formed a "rope." The surrounding material below the rope in the photograph consists of fullerenes and other carbon structures.

The tubes were found to have a bonding structure similar to that of graphite. A nanotube can be thought of as a narrow sheet of a million or more carbon atoms linked together in six-membered benzenelike rings resembling chicken wire and rolled into a very long cylinder only 1–3 nm in diameter.

Carbon nanotubes conduct electricity because of the extended network of delocalized π-bonds that runs from one end of the tube to the other. Along the long axis of the tube, their conductivity is high enough to be considered metallic. The tubes are very strong, and their tensile strength parallel to the axis of the tube is the greatest of any material that has been measured. Because they have a very low density, their strength-to-mass ratio is 40 times that of steel! This may seem surprising at first, because we do not associate graphite with great strength. However, graphite forms flat sheets, not tubes. Think of how flexible a sheet of paper is. You would have difficulty standing it on edge, for instance. But if you roll it into a tight cylinder, it forms a strong structure that stands on end and can even support a fairly large weight.

This model of a carbon nanotube shows that it consists of several concentric tubes. Such layered structures are very strong.

Nanotubes can be made from other substances, such as WS_2, which forms a solid lubricant. Of great interest are nanotubes formed from boron nitride. The electrical conductivity of carbon nanotubes varies from metallic to semiconducting, depending on factors such as the diameter of the tube. However, boron nitride nanotubes have stable electrical characteristics that do not depend on the diameter of the tube. They have a moderately large band gap, and so they are considered to be only weakly semiconducting. Carbon nanotubes grow until they are capped by bonds to another element or by five-membered rings. The five-membered rings pucker the structure, forming bends along the length of tubes and domes at their ends. In the case of BN nanotubes, however, five-membered rings cannot form, because they would require high-energy boron–boron or nitrogen–nitrogen bonds. Instead, BN nanotubes are capped by the atom of a metal such as tungsten.

Their great strength and conductivity have led to the use of nanotubes in submicroscopic electronic components such as transistors. The rigidity of nanotubes may also allow them to be used as minute molds for other elements. For example, they can be filled with molten lead to create lead wires one atom in diameter and can serve as tiny "test tubes" that hold individual molecules in place. Nanotubes that are filled with biomolecules such as cytochrome *c* hold the promise of acting as nanosensors for medical applications.

The very large surface area provided by the highly porous nanotubes means that atoms of gases are readily adsorbed on the inner surfaces of the tubes. Nanotubes carrying hydrogen molecules could therefore become a high-enthalpy-density storage medium for hydrogen-powered vehicles (see Box 12.1). Such a material would solve a major obstacle to the use of hydrogen fuel cells—the problem of a safe, compact storage medium for hydrogen. Hydrogen absorbed into nanotubes can be stored in a volume much smaller than that required to store the gas.

❓ HOW MIGHT YOU CONTRIBUTE?

The discovery of nanotubes and other nanostructures has opened up an exciting new field of research. But just what other shapes are possible and what other materials will form nanotubes? To find out, we will need to predict the effect of different configurations. There are also many experimental problems to be solved. For example, how would you form an electrical connection to a nanotube? Methods for synthesizing the large amounts of nanotubes needed in large-scale applications of nanotube assemblies also need to be developed.

Related Exercise: 14.115

For Further Reading: S. Borman, "Nanotubes," *Chemical and Engineering News*, December 16, 2002, pp. 45–46. S. Fritz, ed., *Understanding nanotechnology*, New York: Scientific American, 2002. M. Ratner and D. Ratner, *Nanotechnology: The next big idea*, Upper Saddle River: Prentice Hall, 2003. B. I. Yakobson and R. E. Smalley, "Fullerene nanotubes: $C_{1,000,000}$ and beyond," *American Scientist*, July–August, 1997, pp. 324–337.

Germanium was not known until 1886 and has the distinction of having been predicted by Mendeleev before it was discovered (see Box 1.2). It is recovered from the flue dust of industrial plants processing zinc ores (in which it occurs as an impurity). It is used mainly—and increasingly—in the semiconductor industry.

Tin and lead are obtained very easily from their ores and have been known since antiquity. Tin occurs chiefly as the mineral *cassiterite*, SnO_2, and is obtained from it by reduction with carbon at 1200°C:

$$SnO_2(s) + C(s) \xrightarrow{1200°C} Sn(l) + CO_2(g)$$

The principal lead ore is *galena,* PbS. It is roasted in air, which converts it into PbO, and this oxide is then reduced with coke:

$$2\,PbS(s) + 3\,O_2(g) \xrightarrow{\Delta} 2\,PbO(s) + 2\,SO_2(g)$$
$$PbO(s) + C(s) \longrightarrow Pb(s) + CO(g)$$

Tin is expensive and not very strong, but it is resistant to corrosion. Its main use is in tinplating, which accounts for about 40% of its consumption. Tin is also used in alloys such as bronze (with copper) and pewter (with antimony and copper).

Lead's durability (its chemical inertness) and malleability make it useful in the construction industry. The inertness of lead under normal conditions can be traced to the passivation of its surface by oxides, chlorides, and sulfates (see Section 12.14). Passivated lead containers can be used for transporting hot concentrated sulfuric acid but not nitric acid, because lead nitrate is soluble. Another important property of lead is its high density, which makes it useful as a radiation shield because its numerous electrons absorb high-energy radiation. Lead was once commonly used in leaded gasoline (as tetraethyl lead, $Pb(CH_2CH_3)_4$), but lead is a toxic heavy metal and that usage has declined as concern for the amount of lead in the environment has increased. The main use of lead today is for the electrodes of rechargeable storage batteries (see Section 12.15).

Metallic character increases significantly down Group 14/IV.

14.18 Oxides of Carbon

We have already met carbon dioxide, CO_2, many times throughout this book. It is formed when organic matter burns in a plentiful supply of air and during animal respiration. It is normally present in the atmosphere; but there is widespread and well-founded concern that an increase in atmospheric carbon dioxide due to the combustion of fossil fuels is contributing to global warming (Box 14.2).

Carbon dioxide is the acid anhydride of carbonic acid, H_2CO_3, which forms when the gas dissolves in water. However, not all the dissolved molecules react to form the acid, and a solution of carbon dioxide in water is an equilibrium mixture of CO_2, H_2CO_3, HCO_3^-, and a very small amount of CO_3^{2-}. Carbonated beverages are made by using high partial pressures of CO_2 to produce high concentrations of carbon dioxide in water. When the partial pressure of the CO_2 is reduced by removing a bottle cap or the seal, the equilibrium shifts from H_2CO_3 toward CO_2 and the liquid effervesces:

$$H_2CO_3(aq) \rightleftharpoons CO_2(g) + H_2O(l)$$

Carbon monoxide, CO, is produced when carbon or organic compounds burn in a limited supply of air, as happens in cigarettes and badly tuned automobile engines. It is produced commercially as synthesis gas by the re-forming reaction (Section 14.3). Carbon monoxide is the formal anhydride of formic acid, HCOOH, and the gas can be produced in the laboratory by the dehydration of formic acid with hot, concentrated sulfuric acid:

$$HCOOH(l) \xrightarrow{150°C,\ H_2SO_4} CO(g) + H_2O(l)$$

BOX 14.2   WHAT HAS THIS TO DO WITH . . . THE ENVIRONMENT?

The Greenhouse Effect

Every year our planet receives from the Sun more than enough radiant energy to supply all our energy needs. About 55% of solar radiation is reflected away or used in natural processes. The remaining 45% is converted into thermal motion (heat), most of which escapes as infrared radiation with wavelengths between 4 and 50 mm.

The *greenhouse effect* is the trapping of this infrared radiation by certain gases in the atmosphere. This effect warms the Earth, as if the entire planet were enclosed in a huge greenhouse.* Oxygen and nitrogen, which make up roughly 99% of the atmosphere, do not absorb infrared radiation. However, water vapor and CO_2 do. Even though these two gases make up only about 1% of the atmosphere, they trap enough radiation to raise the temperature of the Earth by 33°C. Without this naturally occurring greenhouse effect, the average surface temperature of the Earth would be well below the freezing point of water.

The major greenhouse gases include water vapor, carbon dioxide, methane, dinitrogen oxide (nitrous oxide), ozone, and certain chlorofluorocarbons. Water is the most important greenhouse gas. It absorbs strongly near 6.3 μm and at wavelengths longer than 12 μm, as shown on the plot. The carbon dioxide in the atmosphere absorbs about half the infrared radiation with wavelengths of 14–16 μm (see graph).

The concentration of water vapor in the atmosphere is thought to have remained steady over time, but concentrations of some other greenhouse gases are rising. From the year 1000 or earlier until about 1750, the CO_2 concentration in the atmosphere remained fairly stable at about 280 ± 10 parts per million by volume (ppmv). Since then, the CO_2 concentration has increased by 28% to 360 ppmv (see graph). The concentration of methane, CH_4, has more than doubled during this time and is now at its highest level in 160 000 years. Studies of air pockets in ice cores taken from Antarctica show that changes in the concentrations of

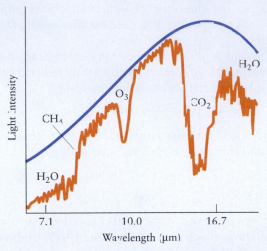

The intensity of infrared radiation at various wavelengths that would be lost from Earth in the absence of greenhouse gases is shown by the smooth line. The jagged line is the intensity of the radiation actually emitted. The maximum wavelength of radiation absorbed by each greenhouse gas is indicated.

both atmospheric carbon dioxide and methane over the past 160 000 years correlate well with changes in the global surface temperature. The rising concentrations of carbon dioxide and methane are therefore of some concern.

Where is the additional CO_2 coming from? Human activities are responsible. Some is generated when limestone, $CaCO_3$, is heated and decomposed in cement making (see Section 14.10). Large amounts of CO_2 are also released to the atmosphere by deforestation, which involves burning large areas of brush and trees. However, most of it comes from the burning of fossil fuels, which began on a large scale after 1850. The additional methane is coming mainly from the petroleum industry and from agriculture.

The temperature of the surface of the Earth is rising about 0.10–0.15°C per decade (see graph). If current trends in population growth and energy use continue, by the middle of the twenty-first century the concentration of CO_2 in the atmosphere will be about twice its value prior to the Industrial Revolution. What are the likely consequences of this doubling of the CO_2 concentration?

---

*The actual insulation mechanism is different in a greenhouse. The glass not only inhibits the escape of infrared radiation but also keeps the warm air inside.

Although the reverse of this reaction cannot be carried out directly, carbon monoxide does react with hydroxide ions in hot alkali to produce formate ions:

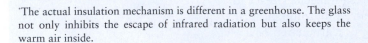

$$CO(g) + OH^-(aq) \longrightarrow HCO_2{}^-(aq)$$

Carbon monoxide is a colorless, odorless, flammable, almost insoluble, very toxic gas that condenses to a colorless liquid at −90°C. It is not very reactive, largely because its bond enthalpy (1074 kJ·mol^{-1}) is higher than that of any other molecule. However, it is a Lewis base, and the lone pair on the carbon atom forms covalent bonds with *d*-block atoms and ions. Carbon monoxide is also a Lewis acid, because its empty antibonding π-orbitals can accept electron density from a

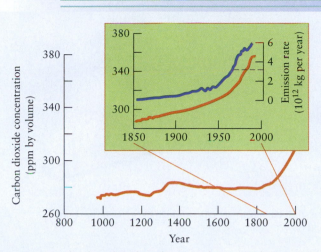

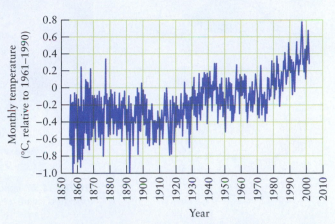

Concentration of carbon dioxide in the atmosphere, over the past 1000 years as determined from ice-core measurements. The blue line shows the emission of CO_2.

The average change in surface temperature of the Earth from 1855 to 2002 compared to the average temperature during 1961–1990.

The Intergovernmental Panel on Climate Change (IPCC) estimated in 2001 that, by the year 2100, the Earth will undergo an increase in temperature of 3°C, with a rise in sea level of about 0.5 m. A rise of 3°C may not sound like much. However, the temperature during the last ice age was only 6°C lower than at present. Furthermore, the *rate* of temperature change is likely to be faster than at any time in the past 10 000 years. Rapid climate changes may bring about droughts and storms and have detrimental effects on many of the Earth's ecosystems.

Computer projections of atmospheric CO_2 concentration for the next 200 years predict escalating increases in CO_2 concentration. Only about half the CO_2 released by humans is absorbed by Earth's natural systems. The other half increases the CO_2 concentration in the atmosphere by about 1.5 ppmv per year. Two conclusions can be drawn from the these facts. First, even if CO_2 emissions were reduced to the amount emitted in 1990 and held constant at that level, the concentration of CO_2 in the atmosphere would continue to increase at about 1.5 ppmv per year for the next century. Second, to maintain CO_2 at its current concentration of 360 ppmv, we would have to reduce fossil fuel consumption by about 50% immediately.

Alternatives to fossil fuels, such as hydrogen, are explored in Box 6.2 and Section 14.3. Coal, which is mostly carbon, can be converted into fuels with a lower proportion of carbon. Its conversion into methane, CH_4, for instance, would reduce CO_2 emissions per unit of energy. We can also work with nature by accelerating the uptake of carbon by the natural processes of the carbon cycle. For example, one proposed solution is to pump CO_2 exhaust deep into the ocean, where it would dissolve to form carbonic acid and bicarbonate ions. Carbon dioxide can also be removed from power plant exhaust gases by passing the exhaust through an aqueous slurry of calcium silicate to produce harmless solid products:

$$2\ CO_2(g) + H_2O(l) + CaSiO_3(s) \longrightarrow$$
$$SiO_2(s) + Ca(HCO_3)_2(s)$$

Related Exercises: 14.107–14.109

For Further Reading: M. Hoffert et al., "Advanced technology paths to global climate stability: Energy for a greenhouse planet," *Science*, vol. 298, 2002, pp. 981–987. R. A. Kerr, "A worrying trend of less ice, higher seas," *Science*, vol. 311, 2006, pp. 1698–1701. F. W. Zwiers and A. J. Weaver, "The causes of 20th century warming," *Science*, vol. 290, 2000, pp. 2081–2083. The United States Environmental Protection Agency Global Warming Web site: http://yosemite.epa.gov/oar/globalwarming.nsf/

metal (Fig. 14.35). This dual character makes carbon monoxide very useful for forming complexes, and numerous metal carbonyls are known (see Section 16.3). Complex formation is also responsible for carbon monoxide's toxicity: it attaches more strongly than oxygen to the iron in hemoglobin and prevents it from accepting oxygen from the air in the lungs. As a result, the victim suffocates.

Because it can be further oxidized, carbon monoxide is a reducing agent. It is used in the production of a number of metals, most notably iron in blast furnaces (Section 16.13):

$$Fe_2O_3(s) + 3\ CO(g) \xrightarrow{\Delta} 2\ Fe(l) + 3\ CO_2(g)$$

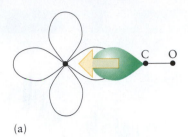

(a)

(b)

FIGURE 14.35 Carbon monoxide can bind to a *d*-metal atom in two ways: (a) by using the lone pair on the C atom to form a σ-bond or (b) by using one of its empty antibonding π-orbitals to accept electrons donated from a *d*-orbital on the metal atom.

FIGURE 14.36 Impure forms of silica: amethyst (left), in which the color is due to Fe^{3+} impurities; agate (center); and onyx (right).

SELF-TEST 14.10A Explain why graphite can conduct electricity but diamond cannot.
[*Answer:* Graphite has a band formed from π-orbitals, through which electrons can be delocalized. In diamond, electrons are confined to σ-bonds.]

SELF-TEST 14.10B What type of reaction is the reaction of carbon monoxide with hydroxide ion?

Carbon has two important oxides: carbon dioxide and carbon monoxide. The former is the acid anhydride of carbonic acid, the parent acid of the hydrogen carbonates and the carbonates.

14.19 Oxides of Silicon: The Silicates

Silica, SiO_2, is a hard, rigid network solid that is insoluble in water. It occurs naturally as quartz and as sand, which consists of small fragments of quartz, usually colored golden brown by iron oxide impurities. Some precious and semi-precious stones are impure silica (Fig. 14.36). *Flint* is silica colored black by carbon impurities.

Silica gets its strength from its covalently bonded network structure. In silica itself, each silicon atom is at the center of a tetrahedron of oxygen atoms, and each corner O atom is shared by two Si atoms (**11**). Hence, each tetrahedron contributes one Si atom and $4 \times \frac{1}{2} = 2$ O atoms to the solid, which has the empirical formula SiO_2. Quartz has a complicated structure; it is built from helical chains of SiO_4 units wound around one another. When it is heated to about 1500°C, it changes to another arrangement, that of the mineral *cristobalite* (Fig. 14.37). This structure is easier to describe: its Si atoms are arranged like the C atoms in diamond; but, in cristobalite, an O atom lies between each pair of neighboring Si atoms.

Metasilicic acid, H_2SiO_3, and *orthosilicic acid*, H_4SiO_4, are weak acids. However, when a solution of sodium orthosilicate is acidified, instead of H_4SiO_4, a gelatinous precipitate of silica is produced:

$$4\ H_3O^+(aq) + SiO_4^{4-}(aq) + x\ H_2O(l) \longrightarrow SiO_2(s) \cdot xH_2O(gel) + 6\ H_2O(l)$$

After it is washed, dried, and granulated, this *silica gel* has a very high specific surface area (about 700 $m^2 \cdot g^{-1}$) and is useful as a drying agent, a support for catalysts, a packing for chromatography columns, and a thermal insulator.

There are a variety of silicates, which can be viewed as various arrangements of tetrahedral oxoanions of silicon in which each Si—O bond has considerable covalent character. The differences in properties between the various silicates are related to the number of negative charges on each tetrahedron, the number of corner O atoms shared with other tetrahedra, and the manner in which chains and sheets of the linked tetrahedra lie together. Differences in the internal structures of

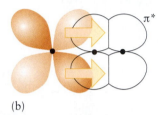

11 An SiO_4 unit

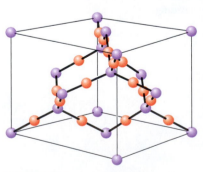

FIGURE 14.37 The structure of cristobalite is like that of diamond except that an O atom (light red) lies between the Si atoms (purple). The arrangement about each Si atom is shown in structure **11**.

FIGURE 14.38 The basic structural unit of the minerals called pyroxenes. Each tetrahedron is an SiO_4 unit (like that in structure **11**), and a shared corner represents a shared O atom (red in the inset where one O atom lies directly over an Si atom). The two unshared O atoms each carry a negative charge.

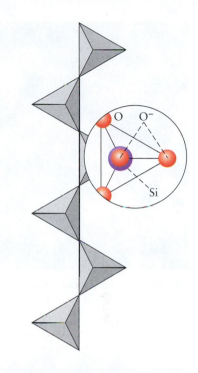

these highly regular network solids lead to a wide array of materials, ranging from gemstones to fibers.

The simplest silicates, the *orthosilicates,* are built from SiO_4^{4-} ions. They are not very common but include the mineral *zircon,* $ZrSiO_4$, which is used as a substitute for diamond in costume jewelry. The *pyroxenes* consist of chains of SiO_4 units in which two corner O atoms are shared by neighboring units (Fig. 14.38); the repeating unit is the metasilicate ion, SiO_3^{2-}. Electrical neutrality is provided by cations regularly spaced along the chain. The pyroxenes include *jade,* $NaAl(SiO_3)_2$.

Chains of silicate units can link together to form ladderlike structures that include *tremolite,* $Ca_2Mg_5(Si_4O_{11})_2(OH)_2$. Tremolite is one of the fibrous minerals called *asbestos,* which can withstand extreme heat (Fig. 14.39). Their fibrous quality is due to the way in which the ladders of SiO_4 units lie together but can easily be torn apart. Because of their great resistance to fire, asbestos fibers were once widely used for heat insulation in buildings. However, these fibers can lodge in lung tissue, where fibrous scar tissue forms around them, giving rise to asbestosis and a susceptibility to lung cancer. In some minerals, the SiO_4 tetrahedra link together to form sheets. An example is *talc,* a hydrated magnesium silicate, $Mg_3(Si_2O_5)_2(OH)_2$. Talc is soft and slippery because the silicate sheets slide past one another.

More complex (and more common) structures result when some of the silicon(IV) in silicates is replaced by aluminum(III) to form the *aluminosilicates.* The missing positive charge is made up by extra cations. These cations account for the difference in properties between the silicate talc and the aluminosilicate *mica.* One form of mica is $KMg_3(Si_3AlO_{10})(OH)_2$. In this mineral, the sheets of tetrahedra are held together by extra K^+ ions. Although it cleaves neatly into transparent layers when the sheets are torn apart, mica is not slippery like talc (Fig. 14.40). Sheets of mica are used for windows in furnaces.

The *feldspars* are aluminosilicates in which as much as half the silicon(IV) has been replaced by aluminum(III). They are the most abundant silicate materials on Earth and are a major component of *granite,* a compressed mixture of

FIGURE 14.39 The minerals commonly called asbestos (from the Greek words meaning "not burning") are fibrous because they consist of long chains based on SiO_4 tetrahedra linked through shared oxygen atoms.

FIGURE 14.40 The aluminosilicate mica cleaves into thin transparent sheets with high melting points. These properties allow it to be used for windows in furnaces.

FIGURE 14.41 The mineral granite is a compressed mixture of mica, quartz, and feldspar.

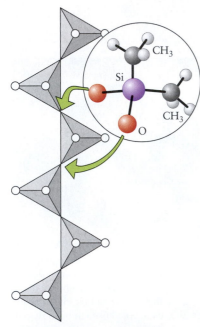

FIGURE 14.42 A typical silicone structure. The hydrocarbon groups give the substance a water-repelling quality. Note the similarity of this structure to that of the purely inorganic pyroxenes in Fig. 14.38.

mica, quartz, and feldspar (Fig. 14.41). When some of the ions between the crystal layers are washed away as these rocks weather, the structure crumbles to clay, one of the main inorganic components of soil. A typical feldspar has the formula $KAlSi_3O_8$. Its weathering by carbon dioxide and water can be described by the equation

$$2\ KAlSi_3O_8(s) + 2\ H_2O(l) + CO_2(g) \longrightarrow$$
$$K_2CO_3(aq) + Al_2Si_2O_5(OH)_4(s) + 4\ SiO_2(s)$$

The potassium carbonate is soluble and washes away, but the aluminosilicate remains as the clay. *Cements* are obtained when aluminosilicates are roasted with limestone and other minerals and then allowed to solidify (see Section 14.10).

Silicones are synthetic materials that consist of long —O—Si—O—Si— chains with the remaining silicon bonding positions occupied by organic groups, such as the methyl group, —CH_3 (Fig. 14.42). Silicones are used to waterproof fabrics because their oxygen atoms attach to the fabric, leaving the hydrophobic (water-repelling) methyl groups like tiny, inside-out umbrellas sticking up out of the fabric's surface.

Silicate structures are based on SiO_4 tetrahedral units with different negative charges and different numbers of shared O atoms.

14.20 Other Important Group 14/IV Compounds

Carbon is the only Group 14/IV element that forms both monatomic and polyatomic anions. There are three classes of carbides: saline carbides (saltlike carbides), covalent carbides, and interstitial carbides. The heavier elements in Group 14/IV form polyatomic anions, such as Si_4^{4-} and Sn_5^{2-}, in which the atoms form a tetrahedron and trigonal bipyramid, respectively.

The *saline carbides* are formed most commonly from the metals of Groups 1 and 2, aluminum, and a few other metals. The s-block metals form saline carbides when their oxides are heated with carbon. The anions present in saline carbides are either C_2^{2-} or C^{4-}. All the C^{4-} carbides, which are called *methides*, produce methane and the corresponding hydroxide in water:

$$Al_4C_3(s) + 12\ H_2O(l) \longrightarrow 4\ Al(OH)_3(s) + 3\ CH_4(g)$$

This reaction shows that the methide ion is a very strong Brønsted base. The species C_2^{2-} is the *acetylide ion*, and the carbides that contain it are called *acetylides*. The acetylide ion is also a strong Brønsted base, and acetylides react with water to produce ethyne (acetylene) and the corresponding hydroxide. Calcium carbide, CaC_2, is the most common saline carbide.

The *covalent carbides* include silicon carbide, SiC, which is sold as carborundum:

$$SiO_2(s) + 3\ C(s) \xrightarrow{2000°C} SiC(s) + 2\ CO(g)$$

Pure silicon carbide is colorless, but iron impurities normally impart an almost black color to the crystals. Carborundum is an excellent abrasive because it is very hard, with a diamondlike structure that fractures into pieces with sharp edges (Fig. 14.43).

The *interstitial carbides* are compounds formed by the direct reaction of a d-block metal and carbon at temperatures above 2000°C. In these compounds, the C atoms occupy the gaps between the metal atoms, as do the H atoms in metallic hydrides (see Fig. 14.9). Here, however, the C atoms pin the metal atoms together into a rigid structure, resulting in very hard substances with melting points often well above 3000°C. Tungsten carbide, WC, is used for the cutting surfaces of drills, and iron carbide, Fe_3C, is an important component of steel.

FIGURE 14.43 Carborundum crystals, showing the sharp fractured edges that give the substance its abrasive power.

All Group 14/IV elements form liquid molecular tetrachlorides. The least stable is $PbCl_4$, which decomposes to solid $PbCl_2$ when it is warmed to about 50°C. Carbon tetrachloride, CCl_4 (tetrachloromethane), was widely used as an industrial solvent; however, now that it is known to be carcinogenic, it is used primarily as the starting point for the manufacture of chlorofluorocarbons. Carbon tetrachloride is formed by the action of chlorine on methane:

$$CH_4(g) + 4\ Cl_2(g) \xrightarrow{650°C} CCl_4(g,\ l\ when\ cooled) + 4\ HCl(g)$$

Silicon reacts directly with chlorine to form silicon tetrachloride, $SiCl_4$ (this reaction was introduced in Section 14.17, as one step in the purification of silicon). This compound differs strikingly from CCl_4 in that it reacts readily with water as a Lewis acid, accepting a lone pair of electrons from H_2O:

$$SiCl_4(l) + 2\ H_2O(l) \longrightarrow SiO_2(s) + 4\ HCl(aq)$$

The cyanide ion, CN^-, is the conjugate base of hydrogen cyanide, HCN. This acid is made by heating ammonia, methane, and air in the presence of a platinum catalyst:

$$2\ CH_4(g) + 2\ NH_3(g) + 3\ O_2(g) \xrightarrow{1100°C,\ Pt} 2\ HCN(g) + 6\ H_2O(g)$$

Cyanides are strong Lewis bases that form a range of complexes with d-block metal ions. They are also famous as poisons. When they are ingested, they combine with certain protein molecules—the cytochromes—involved in the transfer of electrons and the supply of energy in cells, and the victim dies.

Because carbon bonds so readily with itself, there are many hydrocarbons (see Chapter 18). Silicon forms a much smaller number of compounds with hydrogen, called the *silanes*. The simplest silane is silane itself, SiH_4, the analog of methane. Silane is formed by the action of lithium aluminum hydride on silicon halides in ether:

$$SiCl_4 + LiAlH_4 \longrightarrow SiH_4 + LiCl + AlCl_3$$

Silane is much more reactive than methane and bursts into flame on contact with air. Although it survives in pure water, it forms SiO_2 when a trace of alkali is present:

$$SiH_4(g) + 2\ H_2O(l) \xrightarrow{OH^-} SiO_2(s) + 4\ H_2(g)$$

The more complicated silanes, such as $SiH_3\text{—}SiH_2\text{—}SiH_3$, the analog of propane, decompose on standing.

SELF-TEST 14.11A Both carbide ions, C_2^{2-} and C^{4-}, react as bases with water. Predict which carbide ion is the stronger base. Explain your answer.

[*Answer:* The C^{4-} ion, because it has a higher negative charge.]

SELF-TEST 14.11B Explain the fact that SiH_4 reacts with water containing OH^- ions but CH_4 does not.

Carbon forms ionic carbides with the metals of Groups 1 and 2, covalent carbides with nonmetals, and interstitial carbides with d-block metals. Silicon compounds are more reactive than carbon compounds. They can act as Lewis acids.

THE IMPACT ON MATERIALS

Within your lifetime, extraordinary changes will sweep through modern technology. Glasses and ceramics are two of the advanced materials that will make these changes possible. Optical fibers—which are already playing a major role in communication—will control our computers, and our automobiles will be much lighter and more economical.

14.21 Glasses

The growing use of optical fibers to provide wideband telecommunications networks has led to considerable advances in the technology of glassmaking. A **glass**

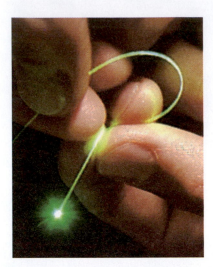

FIGURE 14.44 Glass fibers, such as this one, about the diameter of a human hair, are used in communication networks when large amounts of information must be transmitted in a short time.

is an ionic solid with an amorphous structure resembling that of a liquid. Glass has a network structure based on a nonmetal oxide, usually silica, SiO_2, that has been melted together with metal oxides acting as "network modifiers," which alter the arrangement of bonds in the solid.

To make glass, silica in the form of sand is heated to about 1600°C. Metal oxides, such as MO (where M is a metal cation), are added to the silica. As the mixture melts, many of the Si—O bonds break and the orderly structure of the individual crystals is lost. When the melt cools, the Si—O bonds re-form but are prevented from forming a crystalline lattice because some of the silicon atoms bond with the O^{2-} ions of the metal oxide to give —Si—O⁻—M⁺ groups in place of some of the —Si—O—Si— links present in pure silica. Silicate glasses are generally transparent and durable and can be formed in flat sheets, blown into bottles, or molded into desired shapes. Optical fibers are made by drawing a thin fiber from an optically pure glass rod heated until it softens. The fiber is then coated with plastic (Fig. 14.44).

Almost 90% of all manufactured glass combines sodium and calcium oxides with silica to form *soda-lime glass*. This glass, which is used for windows and bottles, is about 12% Na_2O prepared by the action of heat on sodium carbonate (the soda) and 12% CaO (the lime). When the proportions of soda and lime are reduced and 16% B_2O_3 is added, a *borosilicate glass*, such as Pyrex, is produced. Because borosilicate glasses do not expand much when heated, they survive rapid heating and cooling and are used for ovenware and laboratory beakers.

Glass is resistant to attack by most chemicals. However, the silica in glass reacts with the strong Lewis base F^- from hydrofluoric acid to form fluorosilicate ions:

$$SiO_2(s) + 6\ HF(aq) \longrightarrow SiF_6^{2-}(aq) + 2\ H_3O^+(aq)$$

It is also attacked by the Lewis base OH^- in hot, molten sodium hydroxide and by O^{2-} from the carbonate anion of hot molten sodium carbonate:

$$SiO_2(s) + Na_2CO_3(l) \xrightarrow{1400°C} Na_2SiO_3(s) + CO_2(g)$$

The process by which silica is removed from glass by the ions F^- (from HF), OH^-, and CO_3^{2-} is called *etching*.

SELF-TEST 14.12A Water adheres to glass. Predict the type of intermolecular forces that form between glass and water.

[*Answer:* Ion–dipole forces and hydrogen bonding between H_2O and the partially negative O atoms in SiO_2.]

SELF-TEST 14.12B Predict the products and write the chemical equation for the etching of the SiO_2 in glass by OH^- ions.

Silicate glasses have amorphous structures produced by addition of salts that disrupt the crystalline structure. They can be attacked by strong base and hydrofluoric acid.

14.22 Ceramics

Many of the materials used in the most advanced technologies are made from one of the oldest known materials, common clay. Most clays used commercially are oxides of silicon, aluminum, and magnesium. *China clay* contains primarily *kaolinite*, a form of aluminum aluminosilicate that can be obtained reasonably free of the iron impurities that make many clays look reddish brown, and so it is white. However, other clays contain the iron oxides that cause the orange color of terra cotta tiles and flower pots.

The appearance of a flake of clay reflects its internal structure, which is something like an untidy stack of papers (Fig. 14.45). Sheets of tetrahedral silicate units or octahedral units of aluminum or magnesium oxides are separated by layers of water molecules that serve to bind the layers of the flake together. Each flake of clay is surrounded by a double layer of ions that separates the

FIGURE 14.45 The layers of clay particles can be seen in this micrograph. Because the surfaces of these layers have like charges, they repel one another and easily slide past one another, making clay soft and malleable.

flakes by repelling the like charges on the other flakes. This repulsion allows the flakes to slide past one another and gives the clay some flexibility in response to stress. As a result, clays can be easily molded. When clay is baked in a kiln, it forms the hard, tough *ceramic* materials used in firebricks, tiles, and pots as the water is driven out and strong chemical bonds form between the flakes. Large amounts of china clay, which is used to make ceramics such as porcelain and china, are applied in the coating of paper (such as this page) to give a smooth, nonabsorbent surface.

A **ceramic** is an inorganic material that has been hardened by heating to a high temperature. Typically, a ceramic material is very hard, insoluble in water, and stable to corrosion and high temperatures. These characteristics are the key to their value. Although many ceramics tend to be brittle, they can be used at high temperatures without failing, and they resist deformation. Ceramics are often the oxides of elements that are on the border between metals and nonmetals, but *d*-metal oxides and some compounds of boron and silicon with carbon and nitrogen are ceramic materials. Most ceramics are electrical insulators, but some are semiconductors and some are superconductors (see Box 5.2).

Many aluminosilicate ceramics are created by heating aluminosilicate clays to drive the water out from between the sheets of tetrahedra. This procedure leaves a rigid heterogeneous mass of tiny interlocking crystals bound together by glassy silica. Bone china, which is strong enough to be used for thin, lightweight dishes, is made from china clay with bone ash as a binder that strengthens the ceramic.

Aluminum oxide in the form of corundum makes up about 80% of the advanced ceramics used in high-technology applications. Its hardness, rigidity, thermal conductivity, stability at high temperatures, and electrical insulating ability make it suitable for a wide range of applications, including acting as the substrate for microchips. Corundum is prepared from powdered Al_2O_3 dispersed in liquid. Granules that form in the dispersion are compressed in a mold and *sintered* (heated at a high temperature until hard). Single-crystal forms of aluminum oxide are known and large single-crystal sapphires, which derive their color from iron and titanium impurities, are grown for specialized applications such as acoustic microscopes and as heat-resistant windows on heat-seeking missiles.

The challenge presented by corundum and most other ceramics is to find a way to overcome their brittleness. One route used with some silicon dioxide ceramics is the *sol-gel process*. In this process, an organic silicon compound is dissolved in water and polymerized into a network structure. Many cross-links are formed as the gel forms. These cross-links form a rigid, strong matrix that has few of the tiny cracks that can initiate breakage in brittle ceramics. If the solvent is removed at high temperatures and low pressures, an *aerogel*, a synthetic solid foam that has a density about the same as that of air, but is a good insulator, is formed (Fig. 14.46). Some of the TiO_2 and SiO_2 *nanoparticles* used in ink and coatings can be made by the sol-gel process if care is taken to control particle size and shape during synthesis.

Another means of strengthening ceramics is to create a *composite* material. Flakes of very strong, tough materials such as titanium diboride, TiB_2, are distributed throughout the ceramic material. Even if a crack did form, it would be unable to propagate past the first flake that it encountered.

The stability of ceramic materials at high temperatures has made them useful as furnace liners and has led to interest in ceramic automobile engines, which could endure overheating. Currently, a typical automobile contains about 35 kg of ceramic materials such as spark plugs, pressure and vibration sensors, brake linings, catalytic converters, and thermal and electrical insulation. Some fuel cells make use of a porous solid electrolyte such as zirconia, ZrO_2, that contains a small amount of calcium oxide. It is an electronic insulator, and so electrons do not flow through it, but oxide ions do.

Ceramics are inorganic materials that have been hardened by heating.

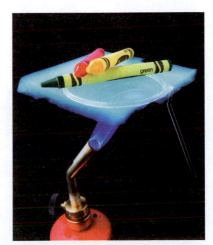

FIGURE 14.46 An aerogel is a ceramic foam. Its low density and low thermal conductivity, combined with great strength, make it an ideal insulating material. Here a thin piece protects three wax crayons from the heat of a flame. Aerogels were used to insulate the Mars Rover.

SKILLS YOU SHOULD HAVE MASTERED

❑ 1 Predict and explain trends in the properties and formulas of the main-group elements.

❑ 2 Describe the names, properties, and reactions of the principal compounds of hydrogen and the Period 1 through 3 elements in Groups 1, 2, 13/III, and 14/IV.

❑ 3 Describe and write balanced equations for the principal reactions used to produce hydrogen and the Period 1 through 3 elements in Groups 1, 2, 13/III, and 14/IV.

❑ 4 Describe the major uses of hydrogen, sodium, potassium, beryllium, magnesium, boron, aluminum, carbon, and silicon.

❑ 5 Describe the reactions of the alkali metals with water and with nonmetals.

❑ 6 Distinguish the allotropes of carbon by their structures and show how their structures affect their properties.

❑ 7 Explain the differences in reactivity between compounds of carbon and those of silicon.

❑ 8 Distinguish the principal silicate structures and describe their properties.

❑ 9 Describe the nature and properties of glasses and ceramics.

EXERCISES

Periodic Trends

These exercises are a review of principles covered in Chapter 1. See also Exercises 1.85–1.98.

14.1 State which atom of each of the following pairs is larger: (a) fluorine, nitrogen; (b) potassium, calcium; (c) gallium, arsenic; (d) chlorine, iodine.

14.2 State which atom of each of the following pairs is larger: (a) tellurium, tin; (b) silicon, lead; (c) calcium, rubidium; (d) germanium, oxygen.

14.3 State which atom of each of the following pairs is more electronegative: (a) sulfur, phosphorus; (b) selenium, tellurium; (c) sodium, cesium; (d) silicon, oxygen.

14.4 State which atom of each of the following pairs is more electropositive: (a) potassium, rubidium; (b) germanium, bromine; (c) oxygen, selenium; (d) aluminum, silicon.

14.5 Arrange the following atoms in order of increasing first ionization energy: oxygen, tellurium, selenium.

14.6 Arrange the following atoms in order of increasing first ionization energy: boron, thallium, gallium.

14.7 (a) Which atom has the greater electron affinity: phosphorus or antimony? (b) Explain your answer.

14.8 (a) Which atom has the greater electron affinity: germanium or selenium? (b) Explain your answer.

14.9 (a) Which of the following species has the greatest polarizability: chloride ions, bromine atoms, bromide ions? (b) Explain your answer.

14.10 (a) Which of the following species has the greatest polarizing power: rubidium ions, potassium atoms, sodium ions?(b) Explain your answer.

14.11 Which bond distance is longer: (a) the Li—Cl distance in lithium chloride or the K—Cl distance in potassium chloride; (b) the K—O distance in potassium oxide or the Ca—O distance in calcium oxide?

14.12 Which of the following bonds do you expect to be longer:(a) the Br—O bond in BrO^- or in BrO_2^-; (b) the C—H bond in CH_4 or the Si—H bond in SiH_4?

Bonding Trends

14.13 Write the balanced equation for the reaction between potassium and hydrogen.

14.14 Write the balanced equation for the reaction between calcium and hydrogen.

14.15 Classify each of the following compounds as a saline, molecular, or metallic hydride: (a) LiH; (b) NH_3; (c) HBr; (d) UH_3.

14.16 Classify each of the following compounds as a saline, molecular, or metallic hydride: (a) B_2H_6; (b) SiH_4; (c) CaH_2; (d) PdH_x, $x < 1$.

14.17 For each of the following oxides, state whether the compound is acidic, basic, or amphoteric: (a) NO_2; (b) Al_2O_3; (c) $B(OH)_3$; (d) MgO.

14.18 For each of the following oxides, state whether the compound is acidic, basic, or amphoteric: (a) Li_2O; (b) Tl_2O_3; (c) Sb_2O_3; (d) SO_3.

14.19 Give the formula for the formal anhydride of each acid: (a) H_2CO_3; (b) $B(OH)_3$.

14.20 Give the formula for the acid corresponding to each of the following formal anhydrides: (a) N_2O_5; (b) P_4O_{10}; (c) SeO_3.

Hydrogen

14.21 Write a balanced chemical equation for (a) the hydrogenation of ethyne (acetylene, C_2H_2) to ethene (C_2H_4) by hydrogen (give the oxidation number of the carbon atoms in the reactant and product); (b) the shift reaction (sometimes called the water gas shift reaction, WGSR); (c) the reaction of barium hydride with water.

14.22 Write a balanced chemical equation for (a) the reaction between sodium hydride and water; (b) the formation of synthesis gas; (c) the hydrogenation of ethene, $H_2C=CH_2$, and give the oxidation number of the carbon atoms in the reactant and product; (d) the reaction of magnesium with hydrochloric acid.

14.23 Use Appendix 2A to determine (a) the standard reaction enthalpy; (b) the standard entropy; (c) the standard Gibbs energy at 25°C for the re-forming reaction of methane.

14.24 Use Appendix 2A to determine (a) the standard reaction enthalpy; (b) the standard entropy; (c) the standard Gibbs energy at 25°C for the shift reaction.

14.25 Identify the products and write a balanced equation for the reaction of hydrogen with (a) chlorine; (b) sodium; (c) phosphorus; (d) copper metal.

14.26 Identify the products and write a balanced equation for the reaction of hydrogen with (a) nitrogen; (b) fluorine; (c) cesium; (d) copper(II) ions.

14.27 The enthalpy of dissociation of hydrogen bonds, ΔH_{HBond}, is a measure of their strength. Explain the trend seen in the data for the following pure substances, which were measured in the gas phase:

Substance	NH_3	H_2O	HF
ΔH_{HBond} (kJ·mol^{-1})	17	25	29

14.28 Methanoic acid, HCOOH, forms dimers in the gas phase. Propose a reason for this behavior.

14.29 (a) Calculate the maximum potential difference that could be produced by a fuel cell that uses the reaction between hydrogen and oxygen in aqueous solution under standard conditions. (b) What are some of the technical problems that would need to be overcome in the design of such a cell?

14.30 (a) What is the maximum potential that could be produced from a cell using the reaction of lithium with Cu^{2+} ions under standard conditions? (b) What technical problems would need to be overcome in the design of such a cell?

Group 1: The Alkali Metals

14.31 Explain why lithium differs from the other Group 1 elements in its chemical and physical properties. Give two examples to support your explanation.

14.32 (a) Write the valence electron configuration for the alkali metal atoms. (b) Explain why the alkali metals are strong reducing agents in terms of electron configurations, ionization energies, and the hydration of their ions.

14.33 Write the chemical equation for the reaction between (a) sodium and oxygen; (b) lithium and nitrogen; (c) sodium and water; (d) potassium superoxide and water.

14.34 Write the chemical equation for the reaction between (a) cesium and oxygen (cesium reacts with oxygen in the same way as potassium); (b) sodium oxide and water; (c) lithium and hydrochloric acid; (d) cesium and iodine.

14.35 Sodium carbonate is often supplied as the decahydrate, $Na_2CO_3 \cdot 10H_2O$. What mass of this solid should be used to prepare 500. mL of 0.135 M Na_2CO_3(aq)?

14.36 Sodium metal is produced from the electrolysis of molten sodium chloride in the Downs process (Section 12.13). Determine (a) the standard Gibbs free energy of the reaction $2\,NaCl(s) \rightarrow 2\,Na(s) + Cl_2(g)$ and (b) the current needed to produce 400. g of sodium in 4.00 h.

Group 2: The Alkaline Earth Metals

14.37 Write the chemical equation for the reaction between magnesium metal and hot water.

14.38 Write the chemical equation for the reaction between barium metal and hydrogen gas.

14.39 CaO and BaO are sometimes used as drying agents for organic solvents such as pyridine, C_5H_5N. The drying agent and the product of the drying reaction are both insoluble in the organic solvent. (a) Write the balanced chemical equation corresponding to the reaction that dries the solvent. (b) For CaO, determine the standard Gibbs free energy of the drying reaction.

14.40 When the mineral dolomite, $CaCO_3 \cdot MgCO_3$, is heated, it gives off carbon dioxide and forms a mixture of a metal oxide and a metal carbonate. Which oxide is formed, CaO or MgO? Which carbonate remains, $CaCO_3$ or $MgCO_3$? Justify your answer.

14.41 Predict and explain the trend in strengths of the Group 2 metals as reducing agents.

14.42 Explain the trend of decreasing lattice enthalpies of the chlorides of the Group 2 metals down the group.

14.43 Aluminum and beryllium have a diagonal relationship. Compare the chemical equations for the reaction of aluminum with aqueous sodium hydroxide to that of beryllium with aqueous sodium hydroxide.

14.44 Write the chemical equation for (a) the industrial preparation of magnesium metal from the magnesium chloride in seawater; (b) the action of water on calcium metal.

14.45 Predict the products of each of the following reactions and then balance each equation:
(a) $Mg(OH)_2 + HCl(aq) \rightarrow$
(b) $Ca(s) + H_2O(l) \rightarrow$
(c) $BaCO_3(s) \xrightarrow{\Delta}$

14.46 Predict the products of each of the following reactions and then balance each equation:
(a) $Mg(s) + Br_2(l) \rightarrow$
(b) $BaO(s) + Al(s) \rightarrow$
(c) $CaO(s) + SiO_2(s) \rightarrow$

14.47 (a) Write the Lewis structures for $BeCl_2$ and $MgCl_2$. How does the Lewis structure of $MgCl_2$ differ from that of $BeCl_2$? (b) Predict the Cl—Be—Cl bond angle. (c) What hybrid orbitals are used in the bonding in $BeCl_2$?

14.48 In the gas phase, $BeCl_2$ forms a dimer by forming chlorine-atom bridges like those in the $AlCl_3$ dimer. Draw the Lewis structure for the $BeCl_2$ dimer and assign formal charges.

14.49 (a) Calculate $\Delta G°$, $\Delta H°$, and $\Delta S°$ for the production of 1.00 mol Ba by the thermite method described in Section 14.8. (b) How would the yield of barium be affected by decreasing the temperature?

14.50 (a) Calculate $\Delta G°$ for the production of 1.00 mol Ba by the electrolysis of molten $BaCl_2$. (b) Why is barium metal not produced from an aqueous solution of the chloride? (c) How would the yield of barium be affected by decreasing the pressure? (d) Compare your value in part (a) with that determined in the previous exercise. Which of the two methods is the more favorable thermodynamically?

14.51 What mass of ethyne is produced when 25.0 g of CaC_2 reacts with 25.0 mL of water?

14.52 (a) Calculate the heat released when 25 kg of quicklime, the amount needed to produce 500. kg of iron, reacts with water. (b) Describe a chemical process that could be used to reconvert the slaked lime into quicklime.

Group 13/III: The Boron Family

14.53 Write a balanced equation for the industrial preparation of aluminum from its oxide.

14.54 Write a balanced equation for the industrial preparation of impure boron.

14.55 Complete and balance the following skeletal equations:
(a) $B_2O_3(s) + Mg(l) \rightarrow$
(b) $Al(s) + Cl_2(g) \rightarrow$
(c) $Al(s) + O_2(g) \rightarrow$

14.56 Complete and balance the following skeletal equations:
(a) $Al_2O_3(s) + OH^-(aq) \rightarrow$
(b) $Al_2O_3(s) + H_3O^+(aq) + H_2O(l) \rightarrow$
(c) $B(s) + NH_3(g) \rightarrow$

14.57 Identify a use for (a) $AlCl_3$; (b) α-alumina; (c) $B(OH)_3$.

14.58 Identify a use for (a) BF_3; (b) $NaBH_4$; (c) $Al_2(SO_4)_3$.

14.59 Suggest a Lewis structure for B_4H_{10} and deduce the formal charges on the atoms. *Hint:* There are four B—H—B bridges.

14.60 BF can be obtained by the reaction between BF_3 and B at a high temperature and low pressure. (a) Determine the electron configuration of the molecule in terms of the occupied molecular orbitals and calculate the bond order. (b) CO is isoelectronic with BF. How do the molecular orbitals in the two molecules differ?

14.61 Many gallium compounds have structures that are similar to those of the corresponding aluminum and boron compounds. Draw the Lewis structure and describe the shape of $[GaBr_4]^-$.

14.62 Many gallium compounds have structures that are similar to those of the corresponding aluminum and boron compounds. Draw the Lewis structure and describe the shape of Ga_2Cl_6.

14.63 What mass of aluminum can be produced by the Hall process in a period of 12.0 h, by using a current of 3.0 MA?

14.64 The annual production of aluminum in the United States in 2000 was 3.6 Mt ($1\ Mt = 1.0 \times 10^9$ kg). What mass of carbon, lost from the anode of the electrolysis cell, was required to produce this amount of aluminum by the Hall process?

14.65 (a) The standard Gibbs free energy of formation of $Tl^{3+}(aq)$ is $+215\ kJ \cdot mol^{-1}$ at 25°C. Calculate the standard potential of the Tl^{3+}/Tl couple. (b) Will Tl^+ disproportionate in aqueous solution?

14.66 The standard potential of the Al^{3+}/Al couple is -1.66 V. Calculate the standard Gibbs free energy of formation for $Al^{3+}(aq)$. Account for any differences between the standard Gibbs free energy of formation of $Tl^{3+}(aq)$ (see Exercise 14.65) and that of $Al^{3+}(aq)$.

Group 14/IV: The Carbon Family

14.67 Describe the sources of silicon and write balanced equations for the three steps in the industrial preparation of silicon.

14.68 Explain why the size of the silicon atom does not permit a silicon analog of the graphite structure.

14.69 Compare the hybridization and structure of carbon in diamond and graphite. How do these features explain the physical properties of the two allotropes?

14.70 Ordinary "hexagonal" graphite has a structure that repeats the ABAB . . . alternation of layers; "rhombohedral graphite" has the repetition ABCABC . . ., with the C layer displaced from the other two. Sketch a structure for rhombohedral graphite.

14.71 Identify the oxidation number of germanium in the following compounds and ions: (a) GeO_4^{4-}; (b) $K_4Ge_4Te_{10}$; (c) Ca_3GeO_5.

14.72 Identify the oxidation number of tin in the following compounds and ions: (a) $Sn_3(OH)_4^{2+}$; (b) K_2SnO_3; (c) $K_2Sn_3O_7$.

14.73 Balance the following skeletal equations and classify them as acid–base or redox:
(a) $MgC_2(s) + H_2O(l) \rightarrow C_2H_2(g) + Mg(OH)_2(s)$
(b) $Pb(NO_3)_2(s) \rightarrow PbO(s) + NO_2(g) + O_2(g)$

14.74 Balance the following skeletal equations and classify them as acid–base or redox:
(a) $CH_4(g) + S_8(s) \rightarrow CS_2(l) + H_2S(g)$
(b) $Sn(s) + KOH(aq) + H_2O(l) \rightarrow K_2Sn(OH)_6(aq) + H_2(g)$

14.75 Write a Lewis structure for the orthosilicate anion, SiO_4^{4-}, and deduce the formal charges and oxidation numbers of the atoms. Use the VSEPR model to predict the shape of the ion.

14.76 Use the VSEPR model to estimate the Si—O—Si bond angle in silica.

14.77 Determine the values of $\Delta H_r°$, $\Delta S_r°$, and $\Delta G_r°$ for the production of high-purity silicon by the reaction $SiO_2(s) + 2\ C(s, graphite) \rightarrow Si(s) + 2\ CO(g)$ at 25°C and estimate the temperature at which the equilibrium constant becomes greater than 1.

14.78 Determine the values of $\Delta H_r°$, $\Delta S_r°$, and $\Delta G_r°$ for the reaction $2\ CO(g) + O_2(g) \rightarrow 2\ CO_2(g)$ at 25°C and estimate the temperature at which the equilibrium constant becomes less than 1.

14.79 Determine the mass of HF as hydrofluoric acid that is required to etch 2.00 mg of SiO_2 from a glass plate in the reaction $SiO_2(s) + 6\ HF(aq) \rightarrow 2\ H_3O^+(aq) + SiF_6^{2-}(aq)$.

14.80 Explain why silicon tetrachloride reacts with water to produce SiO_2 but carbon tetrachloride does not react with water.

14.81 Describe the structures of a silicate in which the silicate tetrahedra share (a) one O atom; (b) two O atoms.

14.82 What is the empirical formula of a potassium silicate in which the silicate tetrahedra share (a) two O atoms and form a chain or (b) three O atoms and form a sheet? In each case, there are single negative charges on the unshared O atoms.

Glasses and Ceramics

14.83 What complex ion is formed when glass is etched by HF?

14.84 In a typical procedure for etching glass, the glass surface is covered by a mask (a protective coating). The mask is then removed from the areas that are to be etched and a paste made of fluorite and sulfuric acid is placed on the surface. From standard reference sources, determine the chemical formula of fluorite and describe the chemical reactions that take place to achieve the etching.

14.85 Why is it unwise to store metal fluorides in glass bottles?

14.86 Solutions of strong bases stored in glass bottles react slowly with their containers. Write the balanced chemical equations for four possible reactions between OH^- and SiO_2.

14.87 Why is it desirable to remove iron from ceramic clays?

14.88 Draw a simple chemical picture to show how the removal of water helps to make aluminosilicates into rigid ceramics.

Integrated Exercises

14.89 State whether the following statements about Group 13/III elements are true or false. If false, explain what is wrong with the statement.
(a) Only two of the elements are metals.
(b) The oxides of all the elements are amphoteric oxides.
(c) The outer s- and p-orbitals become closer in energy going down the group.

14.90 State whether the following statements about Group 15/V elements are true or false. If false, explain what is wrong with the statement.
(a) All the elements are nonmetals.
(b) NH_3 is a stronger Lewis base than PH_3.
(c) The oxides of all the elements are acidic oxides.

14.91 Describe evidence for the statement that hydrogen can act as both a reducing agent and an oxidizing agent. Give chemical equations to support your evidence.

14.92 (a) Reactions of lithium with various oxidizing agents have been examined for use in batteries. A particularly well studied case is that of the lithium–sulfur battery. Determine the potential that is possible for a battery that operates on the reaction of $Li(s)$ with $S(s)$.

 14.93 (a) Using graphing software, such as that on the Web site for this book, plot standard potential against atomic number for the elements of Groups 1 and 2 (refer to Appendix 2B for data). (b) What generalizations can be deduced from the graph?

 14.94 Using graphing software, such as that on the Web site for this book, and data from Fig. 1.50 and Appendix 2B, plot ionization energy against standard potential for the elements of Groups 1 and 2. What generalizations can be drawn from the graph?

14.95 (a) Identify the type of reaction that takes place between calcium oxide and silica in a blast furnace. (b) Write the chemical equation for a related reaction, that between calcium oxide and carbon dioxide.

14.96 Arrange the elements aluminum, gallium, indium, thallium, tin, and germanium in order of increasing electronegativity (see Fig. 2.12) and increasing reducing strength.

14.97 Diborane reacts with ammonia to form an ionic compound (there are no other products). The cation and anion each contain one boron atom. (a) Predict the identity and formula of each ion. (b) Give the hybridization of each boron atom. (c) Identify the type of reaction that has occurred (redox, Lewis acid–base, or Brønsted acid–base).

14.98 Carboranes are compounds in which carbon atoms take some of the positions that would be occupied by boron atoms in boranes. One of the simplest carboranes has the formula $B_4C_2H_6$. (a) The CH group is isoelectronic with the BH^- group. Write the formula of the borane with which this carborane is isoelectronic and indicate its charge. (b) The structure of $B_4C_2H_6$ is an octahedron in which the C atoms take two adjacent position and the B atoms lie in the other positions. Each B and C atom is bonded to three of the other atoms along the edges of the octahedron and is also bonded to an H atom. Draw the Lewis structure of $B_4C_2H_6$. (c) Predict whether this molecule is polar or nonpolar. (d) Calculate the formal charge on each atom. (e) Is this molecule electron deficient? Explain your answer.

14.99 Is there any chemical support for the view that hydrogen should be classified as a member of Group 1? Would it be better to consider hydrogen a member of Group 17/VII? Give evidence that supports each view.

14.100 What justification is there for regarding the ammonium ion as an analog of a Group 1 metal cation? Consider properties such as solubility, charge, and radius. The radius of NH_4^+ is 137 pm.

14.101 Explain the observation that K^+ ions can move more rapidly through water than Li^+ ions can.

14.102 The first ionization energies of the Group 2 elements decrease smoothly down the group, but, in Group 13/III, the values for gallium and thallium are both higher than that for aluminum. Suggest a reason.

14.103 Hydrogen burns in an atmosphere of bromine to give hydrogen bromide. If 135 mL of H_2 gas at 273 K and 1.00 atm combines with a stoichiometric amount of bromine and the resulting hydrogen bromide dissolves to form 225 mL of an aqueous solution, what is the molar concentration of the resulting hydrobromic acid solution?

14.104 The saline hydrides all react rapidly with water. They also react similarly with liquid ammonia. (a) Write a balanced equation for the reaction of CaH_2 with liquid ammonia. (b) Would it be best to classify this reaction as redox, Brønsted acid–base, or Lewis acid–base? Explain your answer.

14.105 Cesium reacts with oxygen to produce a superoxide, CsO_2, which reacts with CO_2 to produce oxygen and Cs_2CO_3. This reaction is used in a self-contained breathing apparatus for firefighters. What volume of oxygen (in liters at 298 K and 1.00 atm) is produced from 30.0 g of CsO_2 with an excess of CO_2?

14.106 Draw simple molecular orbital energy-level diagrams to indicate how the bonding in the saline hydrides, such as NaH or KH, differs from that between hydrogen and a light p-block element such as carbon or nitrogen.

14.107 Infrared radiation is absorbed only when there is a change in the dipole moment of the molecule as the molecule vibrates. Which of the following gases found in the atmosphere can absorb in the infrared region and so function as greenhouse gases: (a) CO; (b) NH_3; (c) CF_4; (d) O_3; (e) Ar? Explain your reasoning. See Box 14.2.

14.108 Carbon dioxide absorbs infrared energy during bending or stretching motions that are accompanied by a change in dipole moment (from zero). Which of the transitions pictured in Fig. 2b of Major Technique 1, following Chapter 2, can absorb infrared radiation? Explain your reasoning. See Box 14.2.

14.109 Methanol, CH_3OH, is a clean-burning liquid fuel being considered as a replacement for gasoline. Calculate the theoretical yield in kilograms of CO_2 produced by the combustion of 1.00 L of methanol (of density 0.791 g·cm^{-3}) and compare it with the 2.16 kg of CO_2 generated by the combustion of 1.00 L of octane. Which fuel contributes more CO_2 per liter to the atmosphere when burned? What other factors would you take into consideration when deciding which of the two fuels to use? See Box 14.2.

14.110 The standard enthalpy of formation of $SiCl_4(g)$ is −662.75 kJ·mol^{-1} and its standard molar entropy is +330.86 J·K^{-1}·mol^{-1}. Calculate the temperature at which the reduction of $SiCl_4(g)$ with hydrogen gas to $Si(s)$ and $HCl(g)$ becomes spontaneous.

  14.111 (a) Examine the structures of diborane, B_2H_6, and $Al_2Cl_6(g)$, which can be found on the Web site for this book. Compare the bonding in these two compounds. How are they similar? (b) What are the differences, if any, in the types of bonds formed? (c) What is the hybridization of the Group 13/III element? (d) Are the molecules planar? If not, describe their shapes.

14.112 In Fig. 14.27 we see that the planes in hexagonal boron nitride take positions in which the B atoms are located directly over N atoms, whereas in graphite (Fig. 14.29), the carbon atoms are offset. Explain this difference in structure between the two substances.

14.113 A unit cell for the calcite structure can be found on the Web site for this book. From this structure, determine (a) the crystal system and (b) the number of formula units present in the unit cell.

14.114 Using the Web site for this book, examine the unit cells of calcite and dolomite. (a) How are these two structures the same? (b) What makes them different? (c) Where are the magnesium and calcium ions located in dolomite?

Chemistry Connections

14.115 The tiny structures such as spheres and tubes formed by carbon atoms are the basis for a large part of the field of nanotechnology. Boron nitride forms similar structures. See Box 14.1.

(a) What is the hybridization of carbon atoms in carbon nanotubes and of boron and nitrogen atoms in boron nitride nanotubes?

(b) A simple carbon nanotube is made up of a sheet resembling a layer of graphite (similar to chicken wire) that has been curled up and bonded to itself. How many hexagons must be strung together (the hexagons denoted by the highlighted bonds in the following diagram) around the circumference of a nanotube to form a tube that is approximately 1.3 nm in diameter? The diagram shows the orientation of hexagons with respect to the curvature of the nanotube. The C—C bond distance in carbon nanotubes is 142 pm.

(c) Buckminsterfullerene, C_{60}, can be hydrogenated, but, to date, a compound with the formula $C_{60}H_{60}$ has not been prepared. The most hydrogenated form of C_{60} known is $C_{60}H_{36}$. Suggest an explanation for why the hydrogenation stops at this point.

(d) Boron nitride can form nanotubes, but does not form spheres that resemble buckminsterfullerene. Suggest a reason why BN cannot form spheres.

(e) In its cubic crystalline form the nitrogen atoms of BN form a face-centered cubic unit cell in which half the tetrahedral interstitial sites are occupied by B atoms (if all B and N atoms were replaced by C atoms, the diamond structure would result). Calculate the density of cubic BN if the edge length of the unit cell is 361.5 pm.

(f) The density of hexagonal BN is 2.29 g·cm^{-3}. Which form of BN is favored at high pressures, hexagonal or cubic?

THE ELEMENTS: THE LAST FOUR MAIN GROUPS

What Are the Key Ideas? As in Chapter 1.4, the key idea is that the elements show a periodicity in their physical and chemical properties, with nonmetallic character becoming more pronounced toward the right of the periodic table.

Why Do We Need to Know This Material? The elements in the last four groups of the periodic table illustrate the rich variety of the properties of the nonmetals and many of the principles of chemistry. These elements include some that are vital to life, such as the nitrogen of proteins, the oxygen of the air, and the phosphorus of our bones, and so a familiarity with their properties helps us to understand living systems. Many of these elements are also central to the materials that provide the backbone of emerging technologies such as the nanosciences, superconductivity, and computer displays.

What Do We Need to Know Already? It would be a good idea to review the information on periodic trends in Sections 1.15–1.22 and 14.1–14.2. Because the nonmetals form molecular compounds, it would also be helpful to review Lewis structures, electronegativity, and covalent bonding in Chapters 2 and 3. The bulk properties of nonmetallic materials are affected by intermolecular forces (Sections 5.1–5.5).

Chapter 15

In the next breath you take, almost all the atoms you inhale will be of elements in the final four groups of the periodic table. Except for the gases containing carbon and hydrogen, air is made up almost entirely of elements from this part of the p block, some as elements and some as compounds. The p-block elements are present in most of the compounds necessary for life and are used to create fascinating and useful modern materials, such as superconductors, plasma screens, and high-performance nanodevices.

We can understand the wide range of properties of the elements in Groups 15/V through 18/VIII by considering their atomic structures. Here, on the far right of the periodic table, the electron configurations are only a few electrons short of a closed-shell, noble-gas configuration, and the effective nuclear charge is high. As a result, the chemistry of the elements is dominated by their high ionization energies and—apart from the noble gases—their high electronegativities. When the elements in this part of the periodic table form compounds with other nonmetals, they do so by sharing electrons to form covalent bonds. Metallic character increases down each group, but the only element in this region considered to be metallic is bismuth, at the bottom of Group 15/V. Arsenic, antimony, tellurium, and polonium are metalloids.

The general structure of this chapter is the same as that of Chapter 14. We work systematically across the remainder of the main-group elements to highlight periodic trends, the production of the elements, and the properties and applications of the elements and their important compounds.

GROUP 15/V: THE NITROGEN FAMILY

Atoms of the Group 15/V elements have the valence electron configuration ns^2np^3 (Table 15.1). The chemical and physical properties of the elements vary sharply in this group, from the almost unreactive gas nitrogen, through the soft nonmetal phosphorus, which is so reactive that it ignites in air, to the important semiconducting

TABLE 15.1 **The Group 15/V Elements**

Valence configuration: ns^2np^3

Z	Name	Symbol	Molar mass (g·mol^{-1})	Melting point (°C)	Boiling point (°C)	Density (g·cm^{-3}) at 25°C	Normal form*
7	nitrogen	N	14.01	−210	−196	1.04†	colorless gas
15	phosphorus	P	30.97	44	280	1.82	white or red nonmetal
33	arsenic	As	74.92	613s‡	—	5.78	gray metalloid
51	antimony	Sb	121.76	631	1750	6.69	blue-white lustrous metalloid
83	bismuth	Bi	208.98	271	1650	8.90	white-pink metal

*Normal form means the appearance and state of the element at 25°C and 1 atm.
†For the liquid at its boiling point.
‡The letter s denotes that the element sublimes.

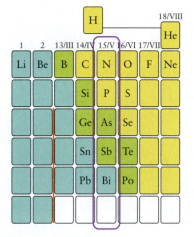

Lavoisier named the element *azote*, which means "lifeless." Ironically, we now know that there would be no life as we know it without nitrogen.

materials arsenic and antimony and the largely metallic bismuth (Fig. 15.1). The range of oxidation states for Group 15/V elements is from −3 to +5, but only nitrogen and phosphorus are found in each oxidation state.

15.1 The Group 15/V Elements

Nitrogen is rare in the Earth's crust, but elemental nitrogen is the principal component of our atmosphere (76% by mass). Pure nitrogen gas is obtained by the fractional distillation of liquid air. Air is cooled to below −196°C by repeated expansion and compression in a refrigerator like that described in Section 4.13. The liquid mixture is then warmed, and the nitrogen (b.p. −196°C) boils off while most of the argon (b.p. −186°C) and oxygen (b.p. −183°C) remain liquid. Nitrogen gas produced industrially is used primarily as a raw material for the synthesis of ammonia in the Haber process.

Plants need nitrogen to grow. However, they cannot use N_2 directly, because the strength of the N≡N (944 kJ·mol^{-1}) makes nitrogen gas almost as inert as the noble gases. To be available for organisms, nitrogen must first be "fixed," or combined with other elements into more useful compounds. Once fixed, nitrogen can be converted into compounds for use as medicines, fertilizers, explosives, and plastics. Lightning converts some nitrogen into its oxides, which dissolve in rain and are then washed into the soil. Some bacteria also fix nitrogen in nodules on the roots of clover, beans, peas, alfalfa, and other legumes (Fig. 15.2). An intensely active field of research is the search for catalysts that can mimic these bacteria and fix nitrogen at ordinary temperatures. At present, the Haber synthesis of ammonia is the main industrial route for fixing nitrogen, but it requires costly high pressures (see Section 9.10).

Like other elements at the head of a group, nitrogen differs in many ways from its congeners. For example, nitrogen is highly electronegative ($\chi = 3.0$, about the same as for chlorine). Because of its high electronegativity, nitrogen is the only

FIGURE 15.1 The elements of Group 15/V. Back row, from left to right: liquid nitrogen, red phosphorus, arsenic. Front row: antimony and bismuth.

element in Group 15/V that forms hydrides capable of hydrogen bonding. Because its atoms are small, nitrogen can form multiple bonds with other Period 2 atoms by using its *p*-orbitals. As we shall see, its small size and the unavailability of *d*-orbitals account for many of the differences between the chemical and physical properties of nitrogen and those of its congeners. Another important chemical property of nitrogen is its wide range of oxidation numbers: compounds are known for each whole-number oxidation state from -3 (in NH_3) to $+5$ (in nitric acid and the nitrates). Nitrogen also occurs with fractional oxidation numbers, such as $-\frac{1}{3}$ in the azide ion, N_3^-.

The properties of phosphorus, in Period 3, differ significantly from those of nitrogen. The atomic radius of phosphorus is nearly 50% larger than that of nitrogen, and so two phosphorus atoms are too big to approach each other closely enough for their $3p$-orbitals to overlap and form π-bonds. Thus, whereas nitrogen can form multiply bonded structures such as N_2O_3 (**1**), phosphorus forms additional single bonds, as in P_4O_6 (**2**). The size of its atoms and the availability of $3d$-orbitals means that phosphorus can form as many as six bonds (as in PCl_6^-), whereas nitrogen can form only four (see Section 2.10).

Phosphorus is obtained from the apatites, which are mineral forms of calcium phosphate, $Ca_3(PO_4)_2$. The rocks are heated in an electric furnace with carbon and sand:

$$2\ Ca_3(PO_4)_2(s) + 6\ SiO_2(s) + 10\ C(s) \xrightarrow{\Delta} P_4(g) + 6\ CaSiO_3(l) + 10\ CO(g)$$

The phosphorus vapor condenses as *white phosphorus*, a soft, white, poisonous molecular solid consisting of tetrahedral P_4 molecules (**3**). This allotrope is highly reactive, in part because of the strain associated with the 60° angles between the bonds. It is very dangerous to handle because it bursts into flame on contact with air and can cause severe burns. White phosphorus is normally stored under water. It changes into *red phosphorus* when heated in the absence of air. Red phosphorus is less reactive than the white allotrope; however, it can be ignited by friction and is used in the striking surfaces of matchbooks. The friction created by rubbing a match across the surface ignites the phosphorus, which, in turn, lights the highly flammable material in the match head. Red phosphorus is thought to consist of chains of linked P_4 tetrahedra.

Vapor rising from white phosphorus in moist air glows with a yellow-green light. The oxides produced by the reaction of phosphorus with the oxygen in air are formed in electronically excited states, and light is given off as electrons fall back into the ground state in the process called **chemiluminescence.**

Arsenic and antimony are metalloids. They have been known in the pure state since ancient times because they are easily obtained from their ores (Fig. 15.3). In the elemental state, they are used primarily in the semiconductor industry and in the lead alloys used as electrodes in storage batteries. Gallium arsenide is used in lasers, including the lasers used in CD players. Metallic bismuth, with its large, weakly bonded atoms, has a low melting point and is used in alloys that serve as fire detectors in sprinkler systems: the alloy melts when a fire breaks out nearby, and the sprinkler system is activated. Like ice, solid bismuth is less dense than the liquid. As a result, molten bismuth does not shrink when it solidifies in molds, and so it is used to make low-temperature castings.

Nitrogen is highly unreactive as an element, largely because of its strong triple bond. White phosphorus is highly reactive. The differences in properties between the nonmetals nitrogen and phosphorus can be traced to the larger atomic radius and lower electronegativity of phosphorus and the availability of d-orbitals in its valence shell. Metallic character increases down the group.

15.2 Compounds with Hydrogen and the Halogens

By far the most important hydrogen compound of a Group 15/V element is ammonia, NH_3, which is prepared in huge amounts by the Haber process. Small quantities of ammonia are present naturally in the atmosphere as a result of the

FIGURE 15.2 The bacteria that inhabit these nodules on the roots of a pea plant are responsible for fixing atmospheric nitrogen and making it available to the plant.

1   Dinitrogen trioxide, N_2O_3

2   Phosphorus(III) oxide, P_4O_6

The name *phosphorus* means "light bringer."

3   Phosphorus, P_4

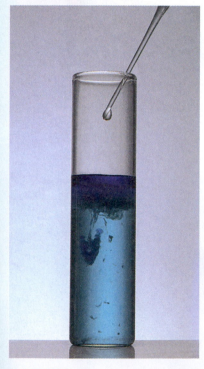

FIGURE 15.3 The minerals (from left to right) orpiment, As_2S_3; stibnite, Sb_2S_3; and realgar, As_4S_4; are all ores that have been used as sources of Group 15/V elements.

The pungency of heated ammonium chloride was known in antiquity to the Ammonians, the worshippers of the Egyptian god Ammon.

Animation: Figure 15.4 Formation of a copper–ammonia complex

FIGURE 15.4 When aqueous ammonia is added to a copper(II) sulfate solution, first a light-blue precipitate of $Cu(OH)_2$ forms (the cloudy region at the top, which appears dark because it is backlit). The precipitate disappears when more ammonia is added to form the dark blue complex $Cu(NH_3)_4^{2+}$ by a Lewis acid–base reaction.

bacterial decomposition of organic matter in the absence of air. This decomposition typically occurs in lake and river beds, in swamps, and in cattle feedlots.

Ammonia is a pungent, toxic gas that condenses to a colorless liquid at $-33°C$. The liquid resembles water in its physical properties, including its ability to act as a solvent for a wide range of substances. Because the dipole moment of the NH_3 molecule (1.47 D) is lower than that of the H_2O molecule (1.85 D), salts with strong ionic character, such as KCl, cannot dissolve in ammonia. Salts with polarizable anions tend to be more soluble in ammonia than are salts with greater ionic character. For example, iodides are more soluble than chlorides in ammonia. Liquid ammonia undergoes much less autoprotolysis than water:

$$2\ NH_3(am) \rightleftharpoons NH_4^+(am) + NH_2^-(am)$$

$$K_{am} = [NH_4^+][NH_2^-] = 1 \times 10^{-33} \text{ at } -35°C$$

Bases that would be fully protonated in water, such as the cyclopentadienide anion, $C_5H_5^-$, behave as very weak bases in ammonia.

Ammonia is very soluble in water because the NH_3 molecules can form hydrogen bonds to H_2O molecules. Ammonia is a weak Brønsted base in water; it is also a reasonably strong Lewis base, particularly toward d-block elements. For example, it reacts with Cu^{2+}(aq) ions to give a deep-blue complex (Fig. 15.4):

$$Cu^{2+}(aq) + 4\ NH_3(aq) \longrightarrow Cu(NH_3)_4^{2+}(aq)$$

Ammonium salts decompose when heated:

$$(NH_4)_2CO_3(s) \xrightarrow{\Delta} 2\ NH_3(g) + CO_2(g) + H_2O(g)$$

The pungent smell of decomposing ammonium carbonate once made it an effective "smelling salt," a stimulant used to revive people who had fainted.

The ammonium cation of an ammonium salt may be oxidized by an anion that has oxidizing character, such as a nitrate. The products depend on the temperature of the reaction:

$$NH_4NO_3(s) \xrightarrow{250°C} N_2O(g) + 2\ H_2O(g)$$
$$2\ NH_4NO_3(s) \xrightarrow{>300°C} 2\ N_2(g) + O_2(g) + 4\ H_2O(g)$$

The explosive violence of the second reaction is the reason that ammonium nitrate is used as a component of dynamite. Ammonium nitrate has a high nitrogen content

(33.5% by mass) and is highly soluble in water. These characteristics make it attractive as a fertilizer, which is its principal use.

Hydrazine, NH_2NH_2, is an oily, colorless liquid. It is prepared by the gentle oxidation of ammonia with alkaline hypochlorite solution:

$$2 NH_3(aq) + ClO^-(aq) \xrightarrow{\text{aqueous alkali}} N_2H_4(aq) + Cl^-(aq) + H_2O(l)$$

Its physical properties are very similar to those of water; for instance, its melting point is 1.5°C and its boiling point is 113°C. However, its chemical properties are very different. It is dangerously explosive and is normally stored and used in aqueous solution. Hydrazine is used as a rocket fuel and to eliminate dissolved, corrosive oxygen from the water used in high-pressure, high-temperature steam furnaces:

$$N_2H_4(aq) + O_2(g) \longrightarrow N_2(g) + 2 H_2O(l)$$

Nitrogen has oxidation number +3 in the highly reactive nitrogen halides. Nitrogen trifluoride, NF_3, is the most stable halide; it does not react with water. However, NCl_3 does react with water, forming ammonia and hypochlorous acid. Nitrogen triiodide, NI_3, which is known only in combination with ammonia as the "ammoniate" $NI_3 \cdot NH_3$ (the analog of a hydrate), is so unstable that it decomposes explosively when lightly touched.

Nitrides are solids that contain the nitride ion, N^{3-}. Nitrides are stable only for small cations such as lithium or magnesium. Magnesium nitride, Mg_3N_2, is formed together with the oxide when magnesium is burned in air (Fig. 15.5):

$$3 Mg(s) + N_2(g) \xrightarrow{\Delta} Mg_3N_2(s)$$

Magnesium nitride, like all nitrides, reacts with water to produce ammonia and the corresponding hydroxide:

$$Mg_3N_2(s) + 6 H_2O(l) \longrightarrow 3 Mg(OH)_2(s) + 2 NH_3(g)$$

In this reaction, the nitride ion acts as a strong base, accepting protons from water to form ammonia.

The azide ion is a highly reactive polyatomic anion of nitrogen, N_3^-. Its most common salt, sodium azide, NaN_3, is prepared from dinitrogen oxide and molten sodium amide:

$$N_2O(g) + 2 NaNH_2(l) \xrightarrow{175°C} NaN_3(s) + NaOH(l) + NH_3(g)$$

Sodium azide, like most azide salts, is shock sensitive. It is used in automobile air bags, where it decomposes to elemental sodium and nitrogen when detonated (see Section 4.7):

$$2 NaN_3(s) \longrightarrow 2 Na(s) + 3 N_2(g)$$

The azide ion is a weak base and accepts a proton to form its conjugate acid, hydrazoic acid, HN_3. Hydrazoic acid is a weak acid similar in strength to acetic acid.

The hydrogen compounds of other members of Group 15/V are much less stable than ammonia and decrease in stability down the group. Phosphine, PH_3, is a poisonous gas that smells faintly of garlic and bursts into flame in air if it is slightly impure. It is much less soluble than ammonia in water because PH_3 cannot form hydrogen bonds to water. Aqueous solutions of phosphine are neutral, because the electronegativity of phosphorus is so low that the lone pair in PH_3 is spread over the hydrogen atoms as well as the phosphorus atom; consequently, the molecule has only a very weak tendency to accept a proton ($pK_b = 27.4$). Because PH_3 is the very weak parent acid of the strong Brønsted base P^{3-}, phosphine can be prepared by protonating phosphide ions with a Brønsted acid. Even water is a sufficiently strong proton donor:

$$2 P^{3-}(s) + 6 H_2O(l) \longrightarrow 2 PH_3(g) + 6 OH^-(aq)$$

Phosphorus trichloride, PCl_3, and phosphorus pentachloride, PCl_5, are the two most important halides of phosphorus. The former is prepared by direct chlorination

FIGURE 15.5 This sample of magnesium nitride was formed by burning magnesium in an atmosphere of nitrogen. When magnesium burns in air, the product is the oxide as well as the nitride.

Lab Video: Figure 15.5 Reaction of magnesium and nitrogen of phosphorus. Phosphorus trichloride, a liquid, is a major intermediate for the production of pesticides, oil additives, and flame retardants. Phosphorus pentachloride, a solid, is made by allowing phosphorus trichloride to react with more chlorine (recall Fig. 2.10).

A typical reaction of the nonmetal halides is their reaction with water to give oxoacids, without a change in oxidation number:

$$PCl_3(l) + 3 H_2O(l) \longrightarrow H_3PO_3(s) + 3 HCl(g)$$

This reaction is an example of a **hydrolysis reaction,** a reaction with water in which new element–oxygen bonds are formed. Another example is the reaction of PCl_5 (phosphorus oxidation state +5) with water to produce phosphoric acid, H_3PO_4 (also phosphorus oxidation state +5):

$$PCl_5(s) + 4 H_2O(l) \longrightarrow H_3PO_4(l) + 5 HCl(g)$$

This reaction is violent and dangerous.

An interesting feature of phosphorus pentachloride is that it is an ionic solid of tetrahedral PCl_4^+ cations and octahedral PCl_6^- anions, but it vaporizes to a gas of trigonal bipyramidal PCl_5 molecules (see Section 2.10). Phosphorus pentabromide is also molecular in the vapor and ionic as the solid; but, in the solid, the anions are simply Br^- ions, presumably because six bulky Br atoms simply do not fit around a central P atom.

SELF-TEST 15.1A Propose an explanation for the fact that the nitrogen trihalides become less stable as the molar mass of the halogen increases.

[*Answer:* The N atom is small; as the molar mass of the halogen increases, fewer can fit around it.]

SELF-TEST 15.1B (a) Write the Lewis structure for the azide ion and assign formal charges to the atoms. (b) You will find it possible to write a number of Lewis structures. Which is likely to make the biggest contribution to the resonance? (c) Predict the shape of the ion and its polarity.

The important compounds of nitrogen with hydrogen are ammonia, hydrazine, and hydrazoic acid, the parent of the shock-sensitive azides. Phosphine forms neutral solutions in water; reaction of phosphorous halides with water produces oxoacids without change in oxidation number.

15.3 Nitrogen Oxides and Oxoacids

Nitrogen forms several oxides, with oxidation numbers ranging from +1 to +5. All nitrogen oxides are acidic oxides and some are the acid anhydrides of the nitrogen oxoacids (Table 15.2). In atmospheric chemistry, where the oxides play an important two-edged role in both maintaining and polluting the atmosphere, they are referred to collectively as NO_x (read "nox").

Dinitrogen oxide, N_2O (oxidation number +1), is commonly called nitrous oxide. It is formed by heating ammonium nitrate gently:

Like many of the reactions in this book, this reaction is extremely hazardous. Don't try it.

$$NH_4NO_3(s) \xrightarrow{250°C} N_2O(g) + 2 H_2O(g)$$

TABLE 15.2 The Oxides and Oxoacids of Nitrogen

Oxidation number	Oxide formula	Oxide name	Oxoacid formula	Oxoacid name
5	N_2O_5	dinitrogen pentoxide	HNO_3	nitric acid
4	NO_2*	nitrogen dioxide	—	
	N_2O_4	dinitrogen tetroxide	—	
3	N_2O_3	dinitrogen trioxide	HNO_2	nitrous acid
2	NO	nitrogen monoxide nitric oxide	—	
1	N_2O	dinitrogen monoxide nitrous oxide	$H_2N_2O_2$	hyponitrous acid

*$2 NO_2 \rightleftharpoons N_2O_4$.

Because it is tasteless, unreactive, nontoxic in small amounts, and soluble in fats, N_2O is sometimes used as a foaming agent and propellant for whipped cream.

Nitrogen oxide (or nitrogen monoxide), NO (oxidation number +2), is commonly called nitric oxide. It is a colorless gas prepared industrially by the catalytic oxidation of ammonia:

$$4 NH_3(g) + 5 O_2(g) \xrightarrow{1000°C, Pt} 4 NO(g) + 6 H_2O(g)$$

In the laboratory, nitrogen oxide can be prepared by reducing a nitrite with a mild reducing agent such as I^-:

$$2 NO_2^-(aq) + 2 I^-(aq) + 4 H^+(aq) \longrightarrow 2 NO(g) + I_2(aq) + 2 H_2O(l)$$

Nitrogen oxide is rapidly oxidized to nitrogen dioxide on exposure to air, a reaction that contributes to acid rain (see Box 10.1):

$$2 NO(g) + O_2(g) \longrightarrow 2 NO_2(g)$$

Nitrogen oxide plays both harmful and beneficial roles in our lives. The conversion of atmospheric nitrogen into NO in hot airplane and automobile engines contributes to the problem of acid rain and the formation of smog as well as to the destruction of the ozone layer (see Box 13.3). However, in small amounts in the body, nitrogen oxide acts as a neurotransmitter, helps to dilate blood vessels, and participates in other physiological changes. It is a subtle neurotransmitter because it is highly mobile, on account of its small size, but is quickly destroyed, on account of being a radical.

Nitrogen dioxide, NO_2 (oxidation number +4), is a choking, poisonous, brown gas that contributes to the color and odor of smog. The molecule has an odd number of electrons, and in the gas phase it exists in equilibrium with its colorless dimer N_2O_4. Only the dimer exists in the solid, and so the brown gas condenses to a colorless solid. When it dissolves in water, NO_2 disproportionates into nitric acid (oxidation number +5) and nitrogen oxide (oxidation number +2):

$$3 NO_2(g) + H_2O(l) \longrightarrow 2 HNO_3(aq) + NO(g)$$

Nitrogen dioxide in the atmosphere undergoes the same reaction and contributes to the formation of acid rain. It also initiates a complex sequence of smog-forming photochemical reactions.

The blue gas dinitrogen trioxide, N_2O_3 (Fig. 15.6, 1), in which the oxidation number of nitrogen is +3, is the anhydride of nitrous acid, HNO_2, and forms that acid when it dissolves in water:

$$N_2O_3(g) + H_2O(l) \longrightarrow 2 HNO_2(aq)$$

Nitrous acid, the parent acid of the nitrites, has not been isolated in pure form but is widely used in aqueous solution. Nitrites are produced by the reduction of nitrates with hot metal:

$$KNO_3(s) + Pb(s) \xrightarrow{350°C} KNO_2(s) + PbO(s)$$

Most nitrites are soluble in water and mildly toxic. Despite their toxicity, nitrites are used in the processing of meat products because they retard bacterial growth and form a pink complex with hemoglobin that inhibits the oxidation of blood (a reaction that would otherwise turn the meat brown). Nitrites are responsible for the pink color of ham, sausages, and other cured meat.

Nitric acid, HNO_3 (oxidation number +5) is used extensively in the production of fertilizers and explosives. It is made by the three-step *Ostwald process*:

Step 1 Oxidation of ammonia; nitrogen's oxidation number increases from -3 to $+2$:

$$4 \overset{-3}{HN_3}(g) + 5 O_2(g) \xrightarrow{850°C, 5 \text{ atm}, Pt/Rh} 4 \overset{+2}{NO}(g) + 6 H_2O(g)$$

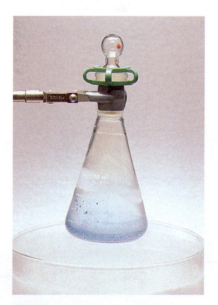

FIGURE 15.6 Dinitrogen trioxide, N_2O_3, condenses to a deep-blue liquid that freezes at $-100°C$ to a pale-blue solid, as shown here. On standing, it turns green as a result of partial decomposition into nitrogen dioxide, a yellow-brown gas.

Step 2 Oxidation of nitrogen oxide; nitrogen's oxidation number increases from +2 to +4:

$$2\ \overset{+2}{N}O(g) + O_2(g) \longrightarrow 2\ \overset{+4}{N}O_2(g)$$

Step 3 Disproportionation in water; nitrogen's oxidation number changes from +4 to +5 and +2:

$$3\ \overset{+4}{N}O_2(g) + H_2O(l) \longrightarrow 2\ H\overset{+5}{N}O_3(aq) + \overset{+2}{N}O(g)$$

Nitric acid, a colorless liquid that boils at 83°C, is normally used in aqueous solution. Concentrated nitric acid is often pale yellow as a result of partial decomposition of the acid to NO_2. Because nitrogen has its highest oxidation number (+5) in HNO_3, nitric acid is an oxidizing agent as well as an acid. It is used to make the explosives nitroglycerin and trinitrotoluene (TNT). Much of its strength as an oxidizing agent—the strongly negative value of the reaction Gibbs free energy when it takes part in an oxidizing reaction—is due to the formation of the strong bond in the N_2 molecule that is produced.

Nitrogen forms oxides in each of its integer oxidation states from +1 to +5; the properties of the oxides and oxoacids are related to the oxidation number of nitrogen in the compound.

15.4 Phosphorus Oxides and Oxoacids

The oxoacids and oxoanions of phosphorus are among the most heavily manufactured chemicals. Phosphate fertilizer production consumes two-thirds of all the sulfuric acid produced in the United States.

4   Phosphorous acid, H_3PO_3

5   Phosphorus(V) oxide, P_5O_{10}

The structures of the phosphorus oxides are based on the tetrahedral PO_4 unit, which is similar to the structural unit in the oxides of its neighbor silicon (see Section 14.19). White phosphorus burns in a limited supply of air to form phosphorus(III) oxide, P_4O_6 (**2**):

$$P_4(s, white) + 3\ O_2(g) \longrightarrow P_4O_6(s)$$

The molecules are tetrahedral, like P_4, but an O atom lies between each pair of P atoms. Phosphorus(III) oxide is the anhydride of phosphorous acid, H_3PO_3 (**4**), and is converted into the acid by cold water:

$$P_4O_6(s) + 6\ H_2O(l) \longrightarrow 4\ H_3PO_3(aq)$$

Although its formula suggests that it should be a triprotic acid, H_3PO_3 is in fact diprotic because one of the H atoms is attached directly to the P atom and the P–H bond is nonpolar (Section 10.10).

When phosphorus burns in an ample supply of air, it forms phosphorus(V) oxide, P_4O_{10} (**5**). This white solid reacts so vigorously with water that it is widely used in the laboratory as a drying agent. Phosphorus(V) oxide is the anhydride of phosphoric acid, H_3PO_4 (**6**):

$$P_4O_{10}(s) + 6\ H_2O(l) \longrightarrow 4\ H_3PO_4(aq)$$

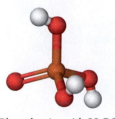

6   Phosphoric acid, H_3PO_4

Phosphoric acid is used primarily for the production of fertilizers, as a food additive, and in detergents. Many soft drinks owe their tart taste to the presence of low concentrations of phosphoric acid. Pure phosphoric acid, H_3PO_4, is a colorless solid with a melting point of 42°C; but, in the laboratory, it is normally a syrupy liquid on account of the water that it has absorbed. In fact, phosphoric acid is usually purchased as 85% H_3PO_4. Its high viscosity can be traced to extensive hydrogen bonding. Although phosphorus has a high oxidation number (+5), the acid shows appreciable oxidizing power only at temperatures above about 350°C; it may be used where nitric acid and sulfuric acid would be too strongly oxidizing.

Phosphoric acid is the parent of the phosphates, which contain the tetrahedral PO_4^{3-} anion and are of great commercial importance. Phosphate rock is mined in huge quantities in Florida and Morocco. After being crushed, it is treated with sulfuric acid to give a mixture of sulfates and phosphates called superphosphate, a major fertilizer:

$$Ca_3(PO_4)_2(s) + 2\ H_2SO_4(l) \longrightarrow 2\ CaSO_4(s) + Ca(H_2PO_4)_2(s)$$

When phosphoric acid is heated, it undergoes a **condensation reaction**, a reaction in which two molecules combine together with the elimination of a small molecule, normally that of water:

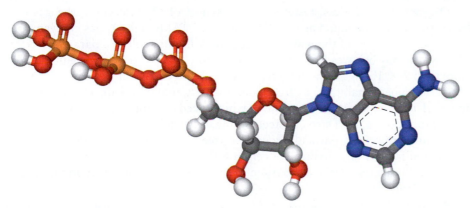

The product, $H_4P_2O_7$, is pyrophosphoric acid. Further heating gives even more complicated products that have chains and rings of PO_4 groups. The products are called polyphosphoric acids. Polyphosphoric acids are far from being of only academic interest: they empower our actions and thoughts. The most important polyphosphate is adenosine triphosphate, ATP (**7**), which is found in every living cell.

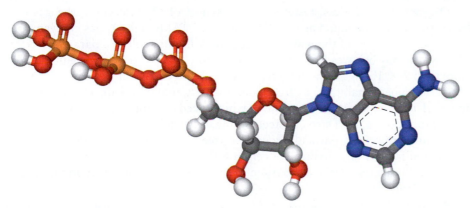

7   Adenosine triphosphate (ATP)

The triphosphate part of this molecule is a chain of three phosphate groups. Its conversion into adenosine diphosphate, ADP, in the reaction

(where the wiggly line indicates the rest of the molecule) releases a large amount of energy ($\Delta G° = -30$ kJ at pH = 7), which is used to power energy-demanding processes in cells.

The oxides of phosphorus have structures based on the tetrahedral PO_4 unit; P_4O_6 and P_4O_{10} are the anhydrides of phosphorous acid and phosphoric acid, respectively. Polyphosphates are extended structures used (as ATP) by living cells to store and transfer energy.

SELF-TEST 15.2A How does (a) the acidity of a nitrogen oxoacid and (b) its strength as an oxidizing agent change as the oxidation number of N increases from +1 to +5?

[*Answer:* (a) Increases; (b) increases]

SELF-TEST 15.2B What is the molarity of phosphoric acid in 85% (by mass) $H_3PO_4(aq)$, which has a density of 1.7 g·mL^{-1}?

GROUP 16/VI: THE OXYGEN FAMILY

Elements become increasingly nonmetallic toward the right-hand side of the periodic table; by Group 16/VI, even polonium, at the foot of the group, is best regarded as a metalloid (Table 15.3). The atoms have the valence electron configuration ns^2np^4, and so they need only two more electrons to complete their valence shells. The members of the group are collectively called the **chalcogens.**

15.5 The Group 16/VI Elements

Oxygen is the most abundant element in the Earth's crust, and the free element accounts for 23% of the mass of the atmosphere. Oxygen is much more reactive than nitrogen, the other major component of our atmosphere. The combustion of all living organisms in oxygen is thermodynamically spontaneous; however, we do not burst into flame at normal temperatures, because combustion has a high activation energy.

Oxygen is a colorless, tasteless, odorless gas of O_2 molecules; the gas condenses to a pale-blue liquid at $-183°C$ (Fig. 15.7). Although O_2 has an even number of electrons, two of them are unpaired, making the molecule paramagnetic; in other words, O_2 behaves like a tiny magnet and is attracted into a magnetic field. More than 2×10^{10} kg of liquid oxygen is produced each year in the United States (about 80 kg per inhabitant) by fractional distillation of liquid air. The biggest consumer of oxygen is the steel industry, which needs about 1 t of oxygen (1 t = 10^3 kg) to produce 1 t of steel. In steelmaking, oxygen is blown into molten iron to oxidize any impurities, particularly carbon (see Section 16.13). Elemental oxygen is also used for welding (to create a very hot flame in oxyacetylene torches) and in medicine. Physicians administer oxygen to relieve strain on the heart and lungs and as a stimulant.

An allotrope of oxygen, ozone, O_3 (**8**), is formed in the stratosphere by the effect of solar radiation on O_2 molecules. Its total abundance in the atmosphere is equivalent to a layer that, at the ordinary conditions of 25°C and 1 bar, would cover the Earth to a thickness of only 3 mm, yet its presence in the stratosphere is vital to the maintenance of life on Earth (see Box 13.3). Ozone can be made in the laboratory by passing an electric discharge through oxygen. It is a blue gas that

Recall Section 3.10, where we saw that in O_2 two unpaired electrons occupy two antibonding π-orbitals of the same energy.

8 Ozone, O_3

TABLE 15.3 The Group 16/VI Elements

Valence configuration: ns^2np^4

Z	Name	Symbol	Molar mass (g·mol^{-1})	Melting point (°C)	Boiling point (°C)	Density (g·cm^{-3}) at 25°C	Normal form*
8	oxygen	O	16.00	−218	−183	1.14†	colorless paramagnetic gas (O_2)
				−192	−112	1.35†	blue gas (ozone, O_3)
16	sulfur	S	32.06	115	445	2.09	yellow nonmetallic solid (S_8)
34	selenium	Se	78.96	220	685	4.79	gray nonmetallic solid
52	tellurium	Te	127.60	450	990	6.25	silver-white metalloid
84	polonium‡	Po	(209)	254	960	9.40	gray metalloid

*Normal form means the appearance and state of the element at 25°C and 1 atm.
†For the liquid at its boiling point.
‡Radioactive.

FIGURE 15.7 Liquid oxygen is pale blue. (The gas itself is colorless.) The inset shows that the gas consists of diatomic molecules known formally as dioxygen.

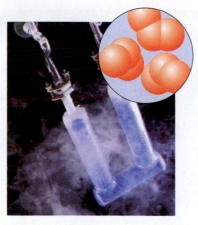

FIGURE 15.8 Ozone is a blue gas that condenses to a dark-blue, highly unstable liquid. The inset shows that the gas consists of triatomic molecules.

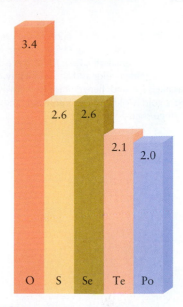

FIGURE 15.9 The electronegativities of the Group 16/VI elements decrease down the group.

condenses at $-112°C$ to an explosive blue liquid that looks like ink (Fig. 15.8). Its pungent smell can often be detected near electrical equipment and after lightning. Ozone is also present in smog, where it is produced by the reaction of oxygen molecules with oxygen atoms:

$$O + O_2 \longrightarrow O_3$$

The oxygen atoms are produced by the photochemical decomposition of NO_2, an emission product of automobile engines:

$$NO_2 \xrightarrow{\text{UV radiation}} NO + O$$

Electronegativities decrease down Group 16 (Fig. 15.9), and atomic and ionic radii increase (Fig. 15.10). The differences between oxygen and sulfur are similar to those between nitrogen and phosphorus, and for similar reasons: sulfur has atoms that are 58% larger than those of oxygen, and it has a lower first ionization energy and electronegativity. The bonds that sulfur forms to hydrogen are much less polar than the bonds between oxygen and hydrogen. Consequently, S—H bonds do not take part in hydrogen bonding; as a result, H_2S is a gas, whereas H_2O is a liquid, despite the smaller number of electrons and hence weaker London forces. Sulfur also has weaker tendencies to form multiple bonds, tending to form additional single bonds to as many as six other atoms and using its d-orbitals to do so.

Sulfur has a striking ability to **catenate,** or form chains of atoms. Oxygen's ability to form chains is very limited, with H_2O_2, O_3, and the anions O_2^-, O_2^{2-}, and O_3^- the only examples. Sulfur's ability is much more pronounced. It appears, for instance, in the existence of S_8 rings, their fragments, and the long strands of "plastic sulfur" that form when sulfur is heated to about 200°C and suddenly

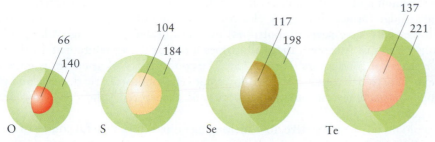

FIGURE 15.10 The atomic and ionic radii of the Group 16/VI elements increase steadily down the group. The values shown are in picometers, and each anion (shown green) is substantially larger than its neutral parent atom.

FIGURE 15.11 A collection of sulfide ores. From left to right: galena, PbS; cinnabar, HgS; pyrite, FeS_2; sphalerite, ZnS. Pyrite has a lustrous golden color and has frequently been mistaken for gold; hence, it is also known as fool's gold. Gold and fool's gold are readily distinguished by their densities.

9 Sulfur, S_8

(a)

(b)

FIGURE 15.12 One of the two most common forms of sulfur is the blocklike rhombic form (a). It differs from the needlelike monoclinic sulfur (b) in the manner in which the S_8 rings are stacked together.

cooled. The —S—S— links that connect different parts of the chains of amino acids in proteins are another example of catenation. These "disulfide links" contribute to the shapes of proteins, including the keratin of our hair; thus, sulfur helps to keep us alive and, perhaps, curly haired.

Sulfur is widely distributed as sulfide ores, which include *galena*, PbS; *cinnabar*, HgS; *iron pyrite*, FeS_2; and *sphalerite*, ZnS (Fig. 15.11). Because these ores are so common, sulfur is a by-product of the extraction of a number of metals, especially copper. Sulfur is also found as deposits of the native element (called brimstone), which are formed by bacterial action on H_2S. The low melting point of sulfur (115°C) is utilized in the **Frasch process**, in which superheated water is used to melt solid sulfur underground and compressed air pushes the resulting slurry to the surface. Sulfur is also commonly found in petroleum, and extracting it chemically has been made inexpensive and safe by the use of heterogeneous catalysts, particularly zeolites (see Section 13.14). One method used to remove sulfur in the form of H_2S from petroleum and natural gas is the **Claus process**, in which some of the H_2S is first oxidized to sulfur dioxide:

$$2\,H_2S(g) + 3\,O_2(g) \longrightarrow 2\,SO_2(g) + 2\,H_2O(l)$$

The SO_2 is then used to oxidize the remainder of the hydrogen sulfide:

$$2\,H_2S(g) + SO_2(g) \xrightarrow{300°C,\ Al_2O_3} 3\,S(s) + 2\,H_2O(l)$$

Sulfur is of major industrial importance. Most of the sulfur that is produced is used to make sulfuric acid, but an appreciable amount is used to vulcanize rubber (Section 19.12).

Elemental sulfur is a yellow, tasteless, almost odorless, insoluble, nonmetallic molecular solid of crownlike S_8 rings (**9**). The two common crystal forms of sulfur are monoclinic sulfur and rhombic sulfur. The more stable form under normal conditions is rhombic sulfur, which forms beautiful yellow crystals (Fig. 15.12). At low temperatures, sulfur vapor consists mainly of S_8 molecules. At temperatures above 720°C, the vapor has a blue tint from the S_2 molecules that form. The latter are paramagnetic, like O_2.

Selenium and tellurium occur in sulfide ores; they are also recovered from the anode sludge formed during the electrolytic refining of copper (see Section 16.4). Both elements have several allotropes, the most stable consisting of long zigzag chains of atoms. Although these allotropes look like silver-white metals, they are poor electrical conductors (Fig. 15.13). The conductivity of selenium is increased by exposure to light, and so it is used in solar cells, photoelectric devices, and photocopying machines. Selenium also occurs as a deep-red solid consisting of Se_8 molecules.

Polonium is a radioactive, low-melting metalloid. It is a useful source of α particles (helium-4 nuclei; they are described in more detail in Section 15.11) and is used in antistatic devices in textile mills: the α particles reduce static by counteracting the negative charges that tend to build up on the fast-moving fabric.

Metallic character increases down Group 16/VI as electronegativity decreases. Oxygen and sulfur occur naturally in the elemental state; sulfur forms chains and rings with itself, but oxygen does not.

15.6 Compounds with Hydrogen

By far the most important compound of oxygen and hydrogen is water, H_2O. Municipal water supplies normally undergo several stages of purification (Fig. 15.14). Raw water is aerated by bubbling air through it to remove foul-smelling dissolved gases such as H_2S, to oxidize some organic compounds to CO_2, and to add oxygen. Adding slaked lime, $Ca(OH)_2$, reduces acidity, precipitates Mg^{2+}, Fe^{3+}, Cu^{2+}, and other metal ions as hydroxides, and softens hard water (see Sections 8.12 and 14.10). After the lime is added, the water is pumped into a primary settling basin. Because the precipitates tend to form as a colloid (a very fine powder that remains suspended in the water, Section 8.21), either $Fe_2(SO_4)_3$ or alum (specifically, $Al_2(SO_4)_3 \cdot 18H_2O$) is added to coagulate and flocculate the precipitate so that it can be filtered. **Coagulation** is the irreversible aggregation of small particles into larger particles; **flocculation** is the reversible, loose aggregation of somewhat larger particles to form a fluffy gel. At this point the water is basic and carbon dioxide is often added to lower the pH, which promotes precipitation of the aluminum as $Al(OH)_3$ so that it can be removed by filtration (Section 14.12).

As the precipitate settles slowly in a secondary basin, it adsorbs any remaining suspended $CaCO_3$, bacteria, and other particles, such as dirt and algae. Precipitates from the primary and secondary basins are combined in a sludge lagoon for disposal. The water is then passed through a sand filter to remove any remaining suspended particles.

The pH of the water is checked again and made slightly basic to reduce acid corrosion of the pipes. At this point, a disinfectant, usually chlorine, is added. In the United States, the chlorine level is required to be greater than 1 g of Cl_2 per 1000 kg (1 ppm by mass) of water at the point of consumption. In water, chlorine forms hypochlorous acid, which is highly toxic to bacteria:

$$Cl_2(g) + 2\,H_2O(l) \longrightarrow H_3O^+(aq) + Cl^-(aq) + HClO(aq)$$

Additional water purification steps, such as reverse osmosis may be required (Section 8.17), depending on the source and condition of the water.

Because water is a common solvent we might think of it merely as a passive medium in which chemical reactions take place. However, water is a reactive compound, and an alien raised in a nonaqueous environment might consider it aggressively corrosive and be surprised at our survival. For instance, water is an oxidizing agent:

$$2\,H_2O(l) + 2\,e^- \longrightarrow 2\,OH^-(aq) + H_2(g) \qquad E = -0.42\,\text{V at pH} = 7$$

One example is its reaction with the alkali metals, as in

$$2\,Na(s) + 2\,H_2O(l) \longrightarrow 2\,NaOH(aq) + H_2(g)$$

FIGURE 15.13 Two of the Group 16/VI elements: selenium, which is a nonmetal, on the left and tellurium, a metalloid, on the right.

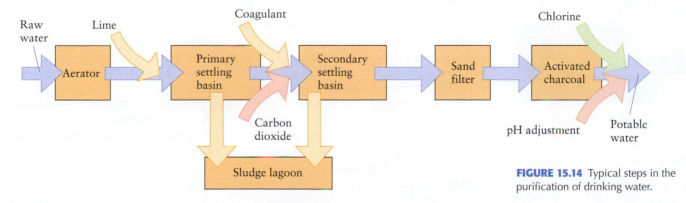

FIGURE 15.14 Typical steps in the purification of drinking water.

10   Hydrogen peroxide, H_2O_2

However, unless the other reactant is a strong reducing agent, water acts as an oxidizing agent only at high temperatures, as in the re-forming reaction (Section 14.3).

Water is a very mild reducing agent:

$$4\,H^+(aq) + O_2(g) + 4\,e^- \longrightarrow 2\,H_2O(l) \qquad E = +0.82\ V\ at\ pH = 7$$

However, few substances besides fluorine are strong enough oxidizing agents to accept the electrons released in the reverse of this half-reaction.

Water is also a Lewis base, because an H_2O molecule can donate one of its lone pairs to a Lewis acid and form complexes such as $Fe(H_2O)_6^{3+}$. Water's ability to act as a Lewis base is also the origin of its ability to hydrolyze substances. The reaction between water and phosphorus pentachloride mentioned in Section 15.2 is an example.

Hydrogen peroxide, H_2O_2 (**10**), is a very pale blue liquid that is appreciably denser than water (1.44 g·mL^{-1} at 25°C) but similar in other physical properties: its melting point is -0.4°C, and its boiling point is 152°C. Chemically, though, hydrogen peroxide and water differ greatly. The presence of the second oxygen atom makes H_2O_2 a very weak acid ($pK_{a1} = 11.75$). Hydrogen peroxide is a stronger oxidizing agent than water. For example, H_2O_2 oxidizes Fe^{2+} and Mn^{2+} in acidic and basic solutions. It can also act as a reducing agent in the presence of more powerful oxidizing agents such as permanganate ions and chlorine (usually in basic solution).

Hydrogen peroxide is normally sold for industrial use as a 30% by mass aqueous solution. When used as a hair bleach (as a 6% solution), it acts by oxidizing the pigments in the hair. A 3% H_2O_2 aqueous solution is used as a mild antiseptic in the home. Contact with blood catalyzes its disproportionation into water and oxygen gas, which cleanses the wound:

$$2\,H_2O_2(aq) \longrightarrow 2\,H_2O(l) + O_2(g)$$

Because it oxidizes unpleasant effluents without producing any harmful by-products, H_2O_2 is also increasingly used as an oxidizing agent to control pollution.

SELF-TEST 15.3A From the Lewis structures of the following molecules, determine which are paramagnetic and explain how you decided: (a) N_2O_3; (b) NO; (c) N_2O.
[*Answer:* (b) NO is paramagnetic, because it has an unpaired electron.]

SELF-TEST 15.3B Write the half-reactions and the overall reaction for the oxidation of water by F_2. Determine the standard potential and $\Delta G°$ for the reaction.

Except for water, all the Group 16/VI binary compounds with hydrogen (the compounds H_2E, where E is a Group 16/VI element) are toxic gases with offensive odors. They are insidious poisons because they paralyze the olfactory nerve and, soon after exposure, the victim cannot smell them. Rotten eggs smell of hydrogen sulfide, H_2S, because egg proteins contain sulfur and give off the gas when they decompose. Another sign of the formation of sulfides in eggs is the pale green discoloration sometimes seen in hard-boiled eggs where the white meets the yolk: this color is a deposit of iron(II) sulfide. Hydrogen sulfide, H_2S, is prepared either by protonation of the sulfide ion, a Brønsted base, in a reaction such as

$$FeS(s) + 2\,HCl(aq) \longrightarrow FeCl_2(aq) + H_2S(g)$$

or by direct reaction of the elements at 600°C.

Hydrogen sulfide dissolves in water to give a solution of hydrosulfuric acid that, as a result of its oxidation by dissolved air, slowly becomes cloudy as S_8 molecules form and then coagulate. Hydrosulfuric acid is a weak diprotic acid and the parent acid of the hydrogen sulfides (which contain the HS^- ion) and the sulfides (which contain the S^{2-} ion). The sulfides of the s-block elements are moderately soluble, whereas the sulfides of the heavy p- and d-block metals are generally very insoluble.

The sulfur analog of hydrogen peroxide also exists and is an example of a *polysulfane*, a catenated molecular compound of composition $HS-S_n-SH$, where n can take on values from 0 through 6. The polysulfide ions obtained from the polysulfanes include two ions found in lapis lazuli (Fig. 15.15).

FIGURE 15.15 The blue stones in this ancient Egyptian ornament are lapis lazuli. This semiprecious stone is an aluminosilicate colored by S_2^- and S_3^- impurities. The blue color is due to S_3^- and the hint of green to S_2^-.

*Water can act as a Lewis base, an oxidizing agent, and a weak reducing
agent; hydrogen peroxide is a stronger oxidizing agent than water. Hydrogen
sulfide is a weak acid; the polysulfanes exhibit the ability of sulfur to
catenate.*

15.7 Sulfur Oxides and Oxoacids

Sulfur forms several oxides that in atmospheric chemistry are referred to collectively as SO_x (read "sox"). The most important oxides and oxoacids of sulfur are the dioxide and trioxide and the corresponding sulfurous and sulfuric acids. Sulfur burns in air to form sulfur dioxide, SO_2 (**11**), a colorless, choking, poisonous gas (recall Fig. C.1). About 7×10^{10} kg of sulfur dioxide is produced annually from the decomposition of vegetation and from volcanic emissions. In addition, approximately 1×10^{11} kg of naturally occurring hydrogen sulfide is oxidized each year to the dioxide by atmospheric oxygen:

$$2\,H_2S(g) + 3\,O_2(g) \longrightarrow 2\,SO_2(g) + 2\,H_2O(g)$$

Industry and transport contribute another 1.5×10^{11} kg of the dioxide, of which about 70% comes from oil and coal combustion—mainly in electricity-generating plants. Because, like many other countries, both the United States and Canada have increased restrictions on emissions of sulfur oxides, emissions of SO_2 into the atmosphere in Canada fell 50% between 1980 and 2000 and in the United States they fell 40% during the same period (see Box 10.1).

Sulfur dioxide is the anhydride of sulfurous acid, H_2SO_3, which is the parent acid of the hydrogen sulfites (or bisulfites) and the sulfites:

$$SO_2(g) + H_2O(l) \longrightarrow H_2SO_3(aq)$$

Sulfurous acid is an equilibrium mixture of two molecules (**12a** and **12b**); in the former, it resembles phosphorous acid, with one of the H atoms attached directly to the S atom. These molecules are also in equilibrium with molecules of SO_2, each of which is surrounded by a cage of water molecules. The evidence for this equilibrium is that crystals of composition $SO_2 \cdot xH_2O$, with x about 7, are obtained when the solution is cooled. Such substances, in which a molecule occupies a cage formed by other molecules, are called **clathrates.** Methane, carbon dioxide, and the noble gases also form clathrates with water.

Sulfur dioxide is easily liquified under pressure and can therefore be used as a refrigerant. It is also a preservative for dried fruit and a bleach for textiles and flour, but its most important use is in the production of sulfuric acid.

The oxidation number of sulfur in sulfur dioxide and the sulfites is +4, an intermediate value in sulfur's range from −2 to +6. Hence, these compounds can act as either oxidizing agents or reducing agents. By far the most important reaction of sulfur dioxide is its slow oxidation to sulfur trioxide, SO_3 (**13**), in which sulfur has the oxidation number +6:

$$2\,SO_2(g) + O_2(g) \longrightarrow 2\,SO_3(g)$$

At normal temperatures, sulfur trioxide is a volatile liquid (its boiling point is 45°C), composed of trigonal planar SO_3 molecules. In the solid and to some extent in the liquid, these molecules form trimers (unions of three molecules) of composition S_3O_9 (**14**) as well as larger clusters.

11 Sulfur dioxide, SO_2

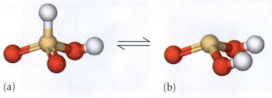

(a)   (b)

12 Sulfurous acid, H_2SO_3

13 Sulfur trioxide, SO_3

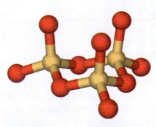

14 Sulfur trioxide trimer, S_3O_9

Sulfuric acid, H_2SO_4, is produced commercially in the *contact process*, in which sulfur is first burned in oxygen and the SO_2 produced is oxidized to SO_3 over a V_2O_5 catalyst:

$$S(s) + O_2(g) \xrightarrow{1000°C} SO_2(g)$$

$$2\,SO_2(g) + O_2(g) \xrightarrow{500°C,\ V_2O_5} 2\,SO_3(g)$$

Because sulfur trioxide forms a corrosive acid mist with water vapor, it is absorbed instead in 98% concentrated sulfuric acid to give the dense, oily liquid called *oleum:*

$$SO_3(g) + H_2SO_4(l) \longrightarrow H_2S_2O_7(l)$$

Oleum is then converted into the acid by reaction with water:

$$H_2S_2O_7(l) + H_2O(l) \longrightarrow 2\,H_2SO_4(l)$$

Sulfuric acid is the most heavily produced inorganic chemical worldwide, the annual production in the United States alone being more than 4×10^{10} kg. The low cost of sulfuric acid leads to its widespread use in industry, particularly for the production of fertilizers, petrochemicals, dyestuffs, and detergents. About two-thirds is used in the manufacture of phosphate and ammonium sulfate fertilizers (see Section 15.4).

Sulfuric acid is a colorless, corrosive, oily liquid that boils (and decomposes) at about 300°C. It has three chemically important properties: it is a strong Brønsted acid, a dehydrating agent, and an oxidizing agent (Fig. 15.16). Sulfuric acid is a strong acid in the sense that its first deprotonation is almost complete at normal concentrations in water. Its conjugate base HSO_4^- is a weak acid, with $pK_a = 1.92$ (Section 10.14). Sulfuric acid forms strong hydrogen bonds with water. The energy released when these bonds form can cause the mixture to boil violently. Sulfuric acid should always be diluted slowly and with great care, making sure to add the acid to the water, instead of the reverse, to avoid splashing concentrated acid.

The powerful dehydrating ability of sulfuric acid is seen when a little concentrated acid is poured on sucrose, $C_{12}H_{22}O_{11}$. A black, frothy mass of carbon forms as a result of the extraction of H_2O (Fig. 15.17):

$$C_{12}H_{22}O_{11}(s) \longrightarrow 12\,C(s) + 11\,H_2O(l)$$

The froth is caused by the generation of the gases CO and CO_2 in side reactions.

Sulfur dioxide is the anhydride of sulfurous acid, and sulfur trioxide is the anhydride of sulfuric acid. Sulfuric acid is a strong acid, a dehydrating agent, and an oxidizing agent.

15.8 Sulfur Halides

Sulfur reacts directly with all the halogens except iodine. It ignites spontaneously in fluorine and burns brightly to give sulfur hexafluoride, SF_6 (**15**), a dense, colorless, odorless, nontoxic, thermally stable, insoluble gas. Despite the high oxidation

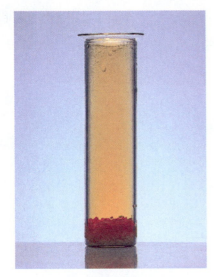

FIGURE 15.16 Sulfuric acid is an oxidizing agent. When some concentrated acid is poured on solid sodium bromide, NaBr, the bromide ions are oxidized to elemental bromine, which colors the mixture red-brown.

Lab Video: Figure 15.17 Dehydration of sucrose by sulfuric acid

FIGURE 15.17 Sulfuric acid is a dehydrating agent. When concentrated sulfuric acid is poured on to sucrose (a), the sucrose, a carbohydrate, is dehydrated (b), leaving a frothy black mass of carbon (c).

(a)

(b)

(c)

number (+6) of the sulfur atom, SF_6 is not a good oxidizing agent. Its low reactivity is in large part due to the F atoms that surround the central S atom like armor and protect it from attack. The ionization energy of SF_6 is very high because any electron that is removed must come from the highly electronegative F atoms. Even quite strong electric fields cannot strip off these electrons, and so SF_6 is a good gas-phase electrical insulator. It is, in fact, a much better insulator than air and is used in switches on high-voltage power lines.

One of the products of the reaction of sulfur with chlorine is disulfur dichloride, S_2Cl_2, a yellow liquid with a nauseating smell; it is used for the vulcanization of rubber. When disulfur dichloride reacts with more chlorine in the presence of iron(III) chloride as a catalyst, the foul-smelling red liquid sulfur dichloride, SCl_2, is produced. Sulfur dichloride reacts with ethene to give mustard gas (**16**), which has been used in chemical warfare. Mustard gas causes blisters, discharges from the nose, and vomiting; it also destroys the cornea of the eye. All in all, it is easy to see why ancient civilizations associated sulfur with the underworld.

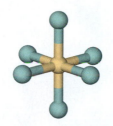

15 Sulfur hexafluoride, SF_6

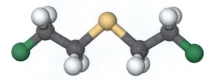

16 Mustard gas, $S(CH_2CH_2Cl)_2$

Sulfur hexafluoride is a thermally stable, nontoxic gas that is a good insulator; sulfur chlorides are toxic liquids.

SELF-TEST 15.4A What is the oxidation number of sulfur in (a) the dithionate ion, $S_2O_6^{2-}$; (b) the thiosulfate ion, $S_2O_3^{2-}$?

[**Answer:** (a) +5; (b) +2]

SELF-TEST 15.4B What is the oxidation number of sulfur in (a) S_2Cl_2; (b) oleum, $H_2S_2O_7$?

GROUP 17/VII: THE HALOGENS

The unique properties of the halogens (Table 15.4), the members of Group 17/VII, can be traced to their valence configuration, ns^2np^5, which needs only one more electron to reach a closed-shell configuration. To complete the octet of valence electrons, in the elemental state all halogen atoms combine to form diatomic molecules, such as F_2 and I_2. With the exception of fluorine, the halogens can also lose valence electrons and their oxidation states can range from -1 to $+7$. The elements form a family showing the smooth trends in physical properties that we would expect when London forces between molecules are the dominant intermolecular force. Because electronegativity decreases down the group (Fig. 15.18) and atomic and ionic radii increase smoothly down the group (Fig. 15.19), chemical properties show smooth trends as well, with the exception of some properties of fluorine.

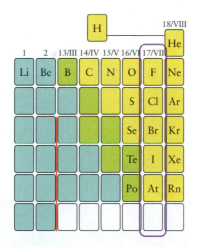

15.9 The Group 17/VII Elements

Fluorine, the first element of the group, is the halogen of greatest abundance in the Earth's crust. It occurs widely in many minerals, including *fluorspar*, CaF_2; *cryolite*, Na_3AlF_6; and the *fluorapatites*, $Ca_5(PO_4)_3F$. Because fluorine is the most strongly oxidizing element ($E° = +2.87$ V), it cannot be obtained from its compounds by

TABLE 15.4 **The Group 17/VII Elements**

Valence configuration: ns^2np^5

Z	Name	Symbol	Molar mass (g·mol⁻¹)	Melting point (°C)	Boiling point (°C)	Density (g·cm⁻³) at 25°C	Normal form*
9	fluorine	F	19.00	−220	−188	1.51†	almost colorless gas
17	chlorine	Cl	35.45	−101	−34	1.66†	yellow-green gas
35	bromine	Br	79.90	27	59	3.12	red-brown liquid
53	iodine	I	126.90	114	184	4.95	purple-black nonmetallic solid
85	astatine‡	At	(210)	300	350	—	nonmetallic solid

*Normal form means the appearance and state of the element at 25°C and 1 atm.
†For the liquid at its boiling point.
‡Radioactive.

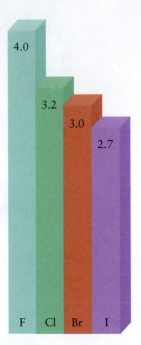

FIGURE 15.18 The electronegativities of the halogens decrease steadily down the group.

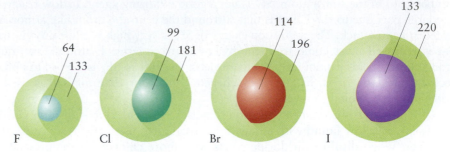

FIGURE 15.19 The atomic and ionic radii of the halogens increase steadily down the group as electrons occupy outer shells of the atoms. The values shown are in picometers. In all cases, the radii of the anions (represented by the green spheres) are larger than the atomic radii.

oxidation with another element. Fluorine is produced by electrolyzing an anhydrous molten mixture of potassium fluoride and hydrogen fluoride at about 75°C with a carbon anode.

Fluorine is a reactive, almost colorless gas of F_2 molecules. Most of the fluorine produced by industry is used to make the volatile solid UF_6 used for processing nuclear fuel (Section 17.12). Much of the rest is used in the production of SF_6 for electrical equipment.

Fluorine has a number of peculiarities that stem from its high electronegativity, small size, and lack of available *d*-orbitals. It is the most electronegative element of all and has an oxidation number of −1 in all its compounds. Its high electronegativity and small size allow it to oxidize other elements to their highest oxidation numbers. The small size helps in this, because it allows several F atoms to pack around a central atom, as in IF_7.

Because the fluoride ion is so small, the lattice enthalpies of its ionic compounds tend to be high (see Table 6.6). As a result, fluorides are less soluble than other halides. This difference in solubility is one of the reasons why the oceans are salty with chlorides rather than fluorides, even though fluorine is more abundant than chlorine in the Earth's crust. Chlorides are more readily dissolved and washed out to sea. There are some exceptions to this trend in solubilities, including AgF, which is soluble; the other silver halides are insoluble. The exception arises because the covalent character of the silver halides increases from AgCl to AgI as the anion becomes larger and more polarizable. Silver fluoride, which contains the small and almost unpolarizable fluoride ion, is freely soluble in water because it is predominantly ionic.

Chlorine is one of the most heavily manufactured chemicals. It is obtained from sodium chloride by electrolysis of molten rock salt or brine (recall Sections 12.13 and 14.7). It is a pale yellow-green gas of Cl_2 molecules that condenses at −34°C. It reacts directly with nearly all the elements (except for carbon, nitrogen, oxygen, and the noble gases). It is a strong oxidizing agent and oxidizes metals to high oxidation states; for example, anhydrous iron(III) chloride, not iron(II) chloride, is formed when chlorine reacts with iron (Fig. 15.20):

$$2\ Fe(s) + 3\ Cl_2(g) \longrightarrow 2\ FeCl_3(s)$$

Chlorine is used in a number of industrial processes, including the manufacture of plastics, solvents, and pesticides. It is used as a bleach in the paper and textile industries and as a disinfectant in water treatment (Section 15.6).

Chlorine is also used to produce bromine from brine wells through the oxidation of Br^- ions (see Fig. K.2):

$$2\ Br^-(aq) + Cl_2(g) \longrightarrow Br_2(l) + 2\ Cl^-(aq)$$

Air is bubbled through the solution to vaporize the bromine and drive it out.

Bromine is a corrosive, red-brown fuming liquid of Br_2 molecules that has a penetrating odor. Bromine is used widely in synthetic organic chemistry because of

FIGURE 15.20 Iron reacts vigorously and exothermically with chlorine to form anhydrous iron(III) chloride.

FIGURE 15.21 Solutions of iodine in a variety of solvents. From left to right, the first three solvents are tetrachloromethane (carbon tetrachloride), water, and potassium iodide solution, in which the brown I_3^- ion forms. In the solution on the far right, some starch has been added to a solution of I_3^-. The intense blue color that results has led to the use of starch as an indicator for the presence of iodine.

the ease with which it can be added to and removed from organic chemicals that are being used to carry out complicated syntheses. Organic bromides are incorporated into textiles as fire retardants and are used as pesticides; inorganic bromides, particularly silver bromide, are used in photographic emulsions.

Iodine occurs as iodide ions in brines and as an impurity in Chile saltpeter. It was once obtained from seaweed, which contains high concentrations accumulated from seawater: 2000 kg of seaweed produce about 1 kg of iodine. The best modern source is the brine from oil wells; the oil itself was produced by the decay of marine organisms that had accumulated the iodine while they were alive. Elemental iodine is produced by oxidation with chlorine:

$$Cl_2(g) + 2\,I^-(aq) \longrightarrow I_2(aq) + 2\,Cl^-(aq) \qquad \Delta G° = -54.54 \text{ kJ}$$

The blue-black lustrous solid sublimes easily and forms a purple vapor.

When iodine dissolves in organic solvents, it produces solutions having a variety of colors. These colors arise from the different interactions between the I_2 molecules and the solvent (Fig. 15.21). The element is only slightly soluble in water, unless I^- ions are present, in which case the soluble, brown triiodide ion, I_3^-, is formed. Iodine itself has few direct uses; but dissolved in alcohol, it is familiar as a mild oxidizing antiseptic. Because it is an essential trace element for living systems but scarce in inland areas, iodides are added to table salt (sold as "iodized salt") in order to prevent an iodine deficiency.

Astatine is a radioactive element that occurs in nature in uranium and thorium ores, but only to a minute extent. Samples are made by bombarding bismuth with α particles in a cyclotron, which accelerates the particles to a very high speed. Astatine isotopes do not exist long enough for its properties to be studied, but it is thought from spectroscopic measurements to have properties similar to those of iodine.

SELF-TEST 15.5A Predict the trend in oxidizing strength of the halogens in aqueous solution.

[*Answer:* Decreases down the group: F > Cl > Br > I]

SELF-TEST 15.5B Explain the trends in the melting points and boiling points of the halogens.

The halogens show smooth trends in chemical properties down the group; fluorine has some anomalous properties, such as its strength as an oxidizing agent and the lower solubilities of most fluorides.

15.10 Compounds of the Halogens

The halogens form compounds among themselves. These **interhalogens** have the formulas XX′, XX′$_3$, XX′$_5$, and XX′$_7$, where X is the heavier (and larger) of the two halogens. Only some of the possible combinations have been synthesized (Table 15.5). They are all prepared by direct reaction of the two halogens, the

TABLE 15.5 Known Interhalogens

Interhalogen	Normal form*
XF$_n$	
ClF	colorless gas
ClF$_3$	colorless gas
ClF$_5$	colorless gas
BrF	pale-brown gas
BrF$_3$	pale-yellow liquid
BrF$_5$	colorless liquid
IF	unstable
IF$_3$	yellow solid
IF$_5$	colorless liquid
IF$_7$	colorless gas
XCl$_n$	
BrCl	red-brown gas
ICl	red solid
I$_2$Cl$_6$	yellow solid
XBr$_n$	
IBr	black solid

*Normal form means the appearance and state of the compound at 25°C and 1 atm.

product formed being determined by the proportions of reactants used. For example,

$$Cl_2(g) + 3\ F_2(g) \longrightarrow 2\ ClF_3(g)$$
$$Cl_2(g) + 5\ F_2(g) \longrightarrow 2\ ClF_5(g)$$

The physical properties of the interhalogens are intermediate between those of their parent halogens. Trends in the chemistry of the interhalogen fluorides can be related to the decrease in bond dissociation energy as the central halogen atom becomes heavier. The fluorides of the heavier halogens are all very reactive: bromine trifluoride gas is so reactive that even asbestos burns in it.

The hydrogen halides, HX, can be prepared by the direct reaction of the elements:

$$H_2(g) + X_2(g) \longrightarrow 2\ HX(g)$$

Fluorine reacts explosively by a radical chain reaction as soon as the gases are mixed. A mixture of hydrogen and chlorine explodes when exposed to light. Bromine and iodine react with hydrogen much more slowly. A less hazardous laboratory source of the hydrogen halides is the action of a nonvolatile acid on a metal halide, as in

$$CaF_2(s) + 2\ H_2SO_4(aq,\ conc) \longrightarrow Ca(HSO_4)_2(aq) + 2\ HF(g)$$

Because Br^- and I^- are oxidized by sulfuric acid, phosphoric acid is used in the preparation of HBr and HI:

$$KI(s) + H_3PO_4(aq) \xrightarrow{\Delta} KH_2PO_4(aq) + HI(g)$$

All the hydrogen halides are colorless, pungent gases except hydrogen fluoride, which is a liquid at temperatures below 20°C. Its low volatility is a sign of extensive hydrogen bonding, and short zigzag chains of hydrogen-bonded molecules, up to about $(HF)_5$, survive to some extent in the vapor. All the hydrogen halides dissolve in water to give acidic solutions. Hydrofluoric acid has the distinctive property of attacking glass and silica. The interiors of lamp bulbs are frosted by the vapors from an aqueous solution of hydrofluoric acid and ammonium fluoride (Fig. 15.22).

 Hydrogen fluoride is used to make fluorinated carbon compounds, such as Teflon (polytetrafluoroethylene) and the refrigerant R134a (CF_3CH_2F).

Most fluoro-substituted hydrocarbons are relatively inert chemically: they are inert to oxidation by air, hot nitric acid, concentrated sulfuric acid, and other strong oxidizing agents. Consequently, Teflon is used to line chemical reactor vessels and to coat the surfaces of nonstick cookware. Because R134a does not contribute to global warming or the destruction of ozone in the stratosphere and is relatively unreactive and nontoxic, it is now the refrigerant most recommended for use in air conditioners in the United States.

The acid strengths and oxidizing abilities of the halogen oxoacids increase with the oxidation number of the halogen. The hypohalous acids, HXO (halogen oxidation number +1), are prepared by direct reaction of the halogen with water. For example, chlorine gas disproportionates in water to produce hypochlorous acid and hydrochloric acid:

$$\overset{0}{Cl_2}(g) + H_2O(aq) \longrightarrow \overset{+1}{H}\overset{}{ClO}(aq) + \overset{-1}{H}Cl(aq)$$

Hypohalite ions, XO^-, are formed when a halogen is added to the aqueous solution of a base. Sodium hypochlorite, NaClO, is produced from the electrolysis of brine when the electrolyte is rapidly stirred, and the chlorine gas produced at the

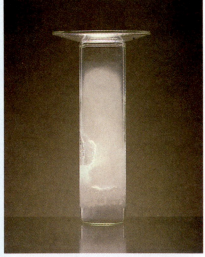

(a)

(b)

FIGURE 15.22 When a mixture of hydrofluoric acid and ammonium fluoride is swirled inside a glass vial (a), reaction with the silica in the glass frosts the surface of the cover glass as well as the walls of the vial (b).

anode reacts with the hydroxide ion generated at the cathode. The chlorine gas disproportionates to produce hypochlorite and chloride ions:

$$\overset{0}{Cl_2}(g) + 2\ OH^-(aq) \longrightarrow \overset{+1}{Cl}O^-(aq) + \overset{-1}{Cl}^-(aq) + H_2O(l)$$

Because hypochlorites oxidize organic material, they are used in liquid household bleaches and as disinfectants. Their action as oxidizing agents stems partly from the decomposition of hypochlorous acid in solution:

$$2\ HClO(aq) \longrightarrow 2\ H^+(aq) + 2\ Cl^-(aq) + O_2(g)$$

The oxygen either bubbles out of the solution or attacks oxidizable material. Calcium hypochlorite is the main component of bleach powders and is used for purifying the water in home swimming pools. It is used for swimming pools in preference to sodium hypochlorite because the Ca^{2+} ions form insoluble calcium carbonate, which is removed by filtration; sodium would remain in solution and make the water too salty.

Chlorate ions, ClO_3^- (oxidation state $+5$), form when chlorine reacts with hot concentrated aqueous alkali:

$$3\ Cl_2(g) + 6\ OH^-(aq) \overset{\Delta}{\longrightarrow} ClO_3^-(aq) + 5\ Cl^-(aq) + 3\ H_2O(l)$$

They decompose when heated, to an extent that depends on whether a catalyst is present:

$$4\ KClO_3(s) \overset{\Delta}{\longrightarrow} 3\ KClO_4(s) + KCl(s)$$
$$2\ KClO_3(s) \xrightarrow{\Delta,\ MnO_2} 2\ KCl(s) + 3\ O_2(g)$$

The latter reaction is a convenient laboratory source of oxygen.

Chlorates are useful oxidizing agents. Potassium chlorate is used as an oxidant in fireworks and in matches. The heads of safety matches consist of a paste of potassium chlorate, antimony sulfide, and sulfur, with powdered glass to create friction when the match is struck; as mentioned in Section 15.1, the striking strip contains red phosphorus, which ignites the match head.

The principal use of sodium chlorate is as a source of chlorine dioxide, ClO_2. The chlorine in ClO_2 has oxidation number $+4$, and so the chlorate must be reduced to form it. Sulfur dioxide is a convenient reducing agent for this reaction:

$$2\ NaClO_3(aq) + SO_2(g) + H_2SO_4(aq, dilute) \longrightarrow 2\ NaHSO_4(aq) + 2\ ClO_2(g)$$

Chlorine dioxide has an odd number of electrons and is a paramagnetic yellow gas. Despite the environmental damage it creates, it is often used to bleach paper pulp, because it can oxidize the various pigments in the pulp without degrading the wood fibers.

The perchlorates, ClO_4^- (oxidation state $+7$), are prepared by electrolytic oxidation of aqueous chlorates:

$$ClO_3^-(aq) + H_2O(l) \longrightarrow ClO_4^-(aq) + 2\ H^+(aq) + 2\ e^-$$

Perchloric acid, $HClO_4$, is prepared by the action of concentrated hydrochloric acid on sodium perchlorate, followed by distillation. It is a colorless liquid and the strongest of all common acids. Because chlorine has its highest oxidation number, $+7$, in these compounds, they are powerful oxidizing agents; contact between perchloric acid and even a small amount of organic material can result in a dangerous explosion.

One spectacular example of the oxidizing ability of perchlorates is their use in the booster rockets of space shuttles. The solid propellant consists of aluminum powder (the fuel), ammonium perchlorate (the oxidizing agent as well as a fuel), and iron(III) oxide (the catalyst). These reactants are mixed into a liquid polymer, which sets to a solid inside the rocket shell. A variety of products can form when the mixture is ignited. One of the reactions is

$$3\ NH_4ClO_4(s) + 3\ Al(s) \xrightarrow{Fe_2O_3} Al_2O_3(s) + AlCl_3(s) + 6\ H_2O(g) + 3\ NO(g)$$

FIGURE 15.23 The white smoke emitted by the shuttle booster rockets consists of powdered aluminum oxide and aluminum chloride.

The solid products form the thick clouds of white powder emitted by the solid rocket boosters during liftoff (Fig. 15.23).

SELF-TEST 15.6A Which oxoacid of bromine is (a) the strongest acid? (b) the strongest oxidizing agent? Explain your answers.

[*Answer:* For both (a) and (b), $HBrO_4$, because it has the most O atoms]

SELF-TEST 15.6B How does pH affect the oxidizing power of $HClO_3$?

> *The interhalogens have properties intermediate between those of the constituent halogens. Nonmetals form covalent halides; metals tend to form ionic halides. The oxoacids of chlorine are all oxidizing agents; both acidity and oxidizing strength of oxoacids increase as the oxidation number of the halogen increases.*

GROUP 18/VIII: THE NOBLE GASES

The elements in Group 18/VIII, the noble gases, get their group name from their very low reactivity (Table 15.6). Experiments on the gases and, later, recognition of their closed-shell electron configurations (ns^2np^6) prompted the belief that these elements were chemically inert. Indeed, no compounds of the noble gases were known until 1962. That year, the British chemist Neil Bartlett synthesized the first noble-gas compound, xenon hexafluoroplatinate, $XePtF_6$. Soon after, chemists at Argonne National Laboratory made xenon tetrafluoride, XeF_4, from a high-temperature mixture of xenon and fluorine.

15.11 The Group 18/VIII Elements

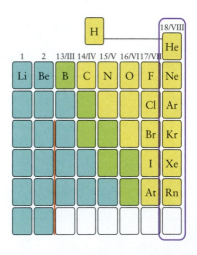

All the Group 18/VIII elements occur in the atmosphere as monatomic gases; together they make up about 1% of its mass. Argon is the third most abundant gas in the atmosphere after nitrogen and oxygen (discounting the variable amount of water vapor). All the noble gases except helium and radon are obtained by the fractional distillation of liquid air.

Helium, the second most abundant element in the universe after hydrogen, is rare on Earth because its atoms are so light that a large proportion of them reach high speeds and escape from the atmosphere; unlike hydrogen, they cannot be anchored to compounds. However, helium is found as a component of natural gases trapped under rock formations (notably in Texas), where it has collected as a result of the emission of α particles by radioactive elements. An α particle is a helium-4 nucleus ($^4He^{2+}$), and an atom of the element forms when the particle picks up two electrons from its surroundings.

Helium gas is twice as dense as hydrogen under the same conditions. Nevertheless, because its density is still very low and it is nonflammable, it is used to

TABLE 15.6 **The Group 18/VIII Elements (the Noble Gases)**

Valence configuration: ns^2np^6
Normal form: colorless monatomic gas

Z	Name	Symbol	Molar mass (g·mol^{-1})	Melting point (°C)	Boiling point (°C)
2	helium	He	4.00	—	−269 (4.2 K)
10	neon	Ne	20.18	−249	−246
18	argon	Ar	39.95	−189	−186
36	krypton	Kr	83.80	−157	−153
54	xenon	Xe	131.29	−112	−108
86	radon*	Rn	(222)	−71	−62

*Radioactive.

provide buoyancy in airships such as blimps. Helium is also used to dilute oxygen for use in hospitals and in deep-sea diving, to pressurize rocket fuels, as a coolant, and in helium–neon lasers. The element has the lowest boiling point of any substance (4.2 K), and it does not freeze to a solid at any temperature unless pressure is applied to hold the light, mobile atoms together. These properties and its chemical inertness make helium useful for **cryogenics**, the study of matter at very low temperatures, such as those used for the study of superconductivity (see Box 5.2). Helium is the only substance known to have more than one liquid phase. Its phase diagram indicates the temperature and pressure ranges over which each phase is stable (Fig. 15.24). Below 2 K, liquid helium-II shows the remarkable property of **superfluidity**, the ability to flow without viscosity.

Neon, which emits an orange-red glow when an electric current flows through it, is widely used in advertising signs and displays (Fig. 15.25). Argon is used to provide an inert atmosphere for welding (to prevent oxidation) and to fill some types of light bulbs, where it conducts heat away from the filament. Krypton gives an intense white light when an electric discharge is passed through it, and so it is used in airport runway lighting. Because krypton is produced by nuclear fission, its atmospheric abundance is one measure of worldwide nuclear activity. Xenon is used in halogen lamps for automobile headlights and in high-speed photographic flash tubes; it is also being investigated as an anesthetic.

The radioactive gas radon seeps out of the ground as a product of radioactive processes deep in the Earth. There is now some concern that its accumulation in buildings and its nuclear decay products can lead to dangerously high levels of radiation.

The noble gases are all found naturally as unreactive monatomic gases. Helium has two liquid phases; the lower-temperature liquid phase exhibits superfluidity.

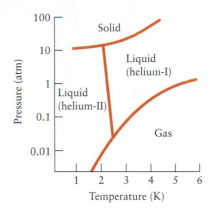

FIGURE 15.24 The phase diagram for helium-4 shows the two liquid phases of helium. Helium-II, the low-temperature liquid phase, is a superfluid.

15.12 Compounds of the Noble Gases

The ionization energies of the noble gases are very high but decrease down the group (Fig. 15.26). Xenon's ionization energy is low enough for electrons to be lost to very electronegative elements. No compounds of helium, neon, and argon exist, except under very special conditions, such as the capture of atoms of He and Ne inside a buckminsterfullerene cage. Krypton forms only one known stable neutral molecule, KrF_2. In 1988, a compound with a Kr—N bond was reported, but it is stable only below $-50°C$. This leaves xenon as the noble gas with the richest chemistry. It forms several compounds with fluorine and oxygen,

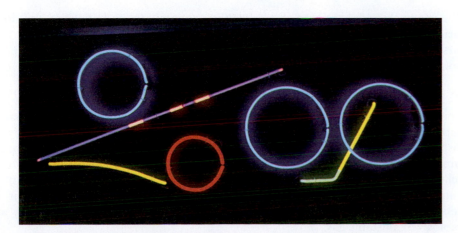

FIGURE 15.25 The colors of this fluorescent lighting art by Tom Anthony are due to emission from noble-gas atoms. Neon is responsible for the red light; when it is mixed with a little argon, the color becomes blue-green. The yellow color is achieved by coating the inside of the glass with phosphors that give off yellow light when excited.

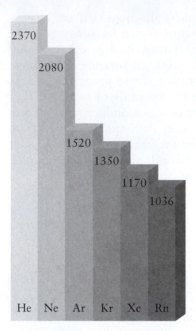

FIGURE 15.26 The ionization energies of the noble gases decrease steadily down the group. The values shown are in kilojoules per mole.

FIGURE 15.27 Crystals of xenon tetrafluoride, XeF_4. This compound was first prepared in 1962 by the reaction of xenon and fluorine at 6 atm and 400°C.

17   Xenon trioxide, XeO_3

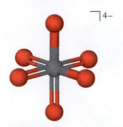

18   Xenic acid, H_2XeO_4

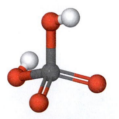

19   Perxenate ion, XeO_6^{4-}

20   Xenon tetroxide, XeO_4

and compounds with Xe—N and Xe—C bonds have been reported, such as $(C_6F_5)_2Xe$.

The starting point for the synthesis of xenon compounds is the preparation of xenon difluoride, XeF_2, and xenon tetrafluoride, XeF_4, by heating a mixture of the elements to 400°C at 6 atm. At higher pressures, fluorination proceeds as far as xenon hexafluoride, XeF_6. All three fluorides are crystalline solids (Fig. 15.27). In the gas phase, all are molecular compounds. Solid xenon hexafluoride, however, is ionic, with a complex structure consisting of XeF_5^+ cations bridged by F^- anions.

The xenon fluorides are used as powerful fluorinating agents (reagents for attaching fluorine atoms to other substances). The tetrafluoride will even fluorinate platinum metal:

$$Pt(s) + XeF_4(s) \longrightarrow Xe(g) + PtF_4(s)$$

The xenon fluorides are used to prepare the xenon oxides and oxoacids and, in a series of disproportionations, to bring the oxidation number of xenon up to +8. First, xenon tetrafluoride is hydrolyzed to xenon trioxide, XeO_3 in a disproportionation reaction:

$$6\ XeF_4(s) + 12\ H_2O(l) \longrightarrow 2\ XeO_3(aq) + 4\ Xe(g) + 3\ O_2(g) + 24\ HF(aq)$$

Xenon trioxide, XeO_3 (**17**), is the anhydride of xenic acid, H_2XeO_4 (**18**). It reacts with aqueous alkali to form a hydrogen xenate ion, $HXeO_4^-$. This ion slowly disproportionates into xenon and the octahedral perxenate ion, XeO_6^{4-} (**19**), in which the oxidation number of xenon is +8. Aqueous perxenate solutions are yellow and are very powerful oxidizing agents as a result of the high oxidation number of xenon. When barium perxenate is treated with sulfuric acid, it is dehydrated to the anhydride of perxenic acid, xenon tetroxide, XeO_4 (**20**), an explosively unstable gas.

Xenon is the only noble gas known to form an extensive series of compounds with fluorine and oxygen; xenon fluorides are powerful fluorinating agents and xenon oxides are powerful oxidizing agents.

THE IMPACT ON MATERIALS

The elements of the *p* block are so rich in properties that it is hardly surprising that they are used to fabricate compounds with unusual and technologically exciting properties. Here we focus on two applications: the generation of light and the emerging world of nanotechnology.

15.13 Luminescent Materials

Incandescence is light emitted by a hot body, such as the filament in a lamp or the particles of hot soot in a candle flame. **Luminescence** is the emission of light by a process other than incandescence. For example, when hydrogen peroxide reacts with chlorine, the O_2 formed by the oxidation of H_2O_2 may be produced in an energetically excited state and emit light as it discards its excess energy. This process is another example of chemiluminescence (Section 15.1), the emission of light by products formed in energetically excited states in a chemical reaction (Fig. 15.28). The light sticks used for emergency lighting glow with light from a chemiluminescent process. *Bioluminescence* is a form of chemiluminescence generated by a living organism. For example, the enzyme *luciferase* produces the luminescent compound *luciferin* in fireflies and some bacteria.

Fluorescence is the emission of light from molecules excited by radiation of higher frequency. For example, fluorescence in the visible region of the spectrum is often observed when a substance is illuminated with ultraviolet radiation. The fluorescence lasts only as long as the illumination lasts. **Phosphorescence** is the emission of light from a molecule that remains excited after the stimulus has ceased. Glow-in-the-dark toys are phosphorescent. **Triboluminescence** is luminescence that results from mechanical shock to a crystal. It is observed in many crystalline substances and can be detected by striking or grinding sugar crystals in a dark room. The luminescence is given off by trapped nitrogen gas escaping from the crystal while in an excited state.

The cathode ray tubes that have been widely used for television and computer displays and the plasma screens that (together with liquid-crystal displays) are becoming more common, both make use of **phosphors**, phosphorescent materials that glow when they are excited by the impact of electrons or ultraviolet radiation. In a cathode ray in a color monitor or television set tube, clusters of three phosphors consisting of zinc sulfide doped by a *d*-block metal are used for each dot in the screen. The red phosphor is often yttrium oxysulfide, Y_2O_2S, activated by europium. Silver-activated ZnS is often used for the blue phosphor and copper-activated ZnS for the green phosphor. Solid zinc sulfide is an insulator, with a large energy difference between the conductance and the valence bands, and the activating metal has an energy level between those two bands. When the zinc sulfide is struck by electrons of sufficient energy, one of its electrons is excited to the conduction band, leaving a positively charged hole in the lattice. The hole "travels" to an atom of the activating metal. The excited electron then drops back from the conduction band into that hole, emitting light of a wavelength corresponding to the energy separation between the activating metal and the excited state. The phosphors must stop glowing as soon as the excitation beam passes on, so that the picture can change rapidly. In a plasma screen the excitation is caused by the ultraviolet radiation emitted by an electric discharge through a low-pressure mixture of noble gases in a cell behind each dot of phosphor.

FIGURE 15.28 Chemiluminescence, the emission of light as the result of a chemical reaction, occurs when hydrogen peroxide is added to a solution of the organic compound perylene. Although hydrogen peroxide itself can fluoresce, in this case the light is emitted by the perylene.

(a)

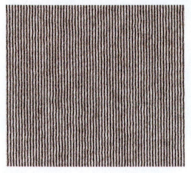

(b)

FIGURE 15.29 These micrographs show that a block copolymer on its own crystallizes in a chaotic pattern (a); nanoscale techniques were used to produce the self-assembled parallel array of the same copolymer in (b). Such regular arrays could be used to make ultra-high density memory chips to store information in miniature computers.

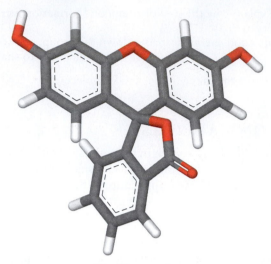

21 Fluorescein

Fluorescent materials are very important in medical research. Dyes such as fluorescein (**21**) can be attached to protein molecules, and the protein can be traced in a biological system by exciting the fluorescein and looking for its emissions. The use of a fluorescent material allows the detection of much smaller concentrations than would otherwise be possible. Because fluorescent materials can be activated by radioactivity, they are also used in scintillation counters to measure radiation (see Box 17.2).

SELF-TEST 15.8A Explain how fluorescent materials can be used to detect radioactivity.

> [*Answer:* The fluorescent materials absorb energy from the radiation and release it as light.]

SELF-TEST 15.8B Explain the difference between chemiluminescence and phosphorescence.

Luminescent materials release energy as light as they return from excited states to lower energy states.

15.14 Nanomaterials

A new area of research with the potential, among other things, for revolutionizing medical diagnosis and treatment and improving our quality of life is **nanoscience.** This field includes the study of materials that are larger than single atoms, but too small to exhibit most bulk properties. In this section we summarize what nanomaterials are, how they are prepared, some of the ways the properties of nanomaterials differ from atomic properties or those of bulk matter, and why they have raised such interest in the scientific and business communities.

Nanoparticles range in size from about 1 to 100 nm and can be manufactured and manipulated at the molecular level (Fig. 15.29). **Nanomaterials** are materials composed of nanoparticles or regular arrays of molecules or atoms such as nanotubes (Box 14.1). These materials have been made possible by advances in **nanotechnology,** such as new imaging technologies, including the scanning tunneling microscope (see Box 5.1), and the discovery of how some nonmetals and metalloids can be manipulated into assembling themselves into regular, extended structures (Box 15.1).

Nanomaterials can be manufactured by one of two groups of methods, one physical and one chemical. In "top-down" approaches, nanoscale materials are carved into shape by the use of physical nanotechnology methods such as lithography (Fig. 15.30). In "bottom-up" approaches, molecules are encouraged to assemble themselves into desired patterns chemically by making use of specific

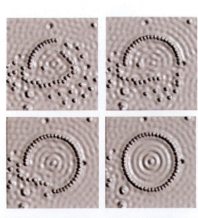

FIGURE 15.30 The images show four stages in the construction of a ring of iron atoms on a copper substrate. The scientists used "top-down" techniques to place 48 iron atoms into a ring. The circular waves seen in the final image show the density of the surface electrons inside the ring, which acts as a "corral" for the electrons.

BOX 15.1   FRONTIERS OF CHEMISTRY: SELF-ASSEMBLING MATERIALS

A major concern of scientists working with nanomaterials is how to coax molecules into organizing themselves into the desired nanoscale arrangements. Many biological processes operate at the nanoscale level. Proteins, for example, fold up into the shape that optimizes their function (see Section 19.13), cell membranes form spontaneously when certain molecules called lipids encounter water (see Box 8.1), and the DNA molecules of our genes replicate themselves every time a cell divides. Although the expression has a variety of definitions, the common meaning of *self-assembly of molecules* is the spontaneous formation of organized structures from individual parts.

Self-assembly can be static or dynamic. In *static self-assembly,* the structure formed is stable and the process does not readily reverse. Two examples are the folding of polypeptide chains into a protein molecule and the formation of the double helix of DNA. *Dynamic self-assembly* involves interactions that dissipate energy and can easily be reversed. Two examples are oscillating chemical reactions and convection patterns. On the macroscopic scale, the coordinated motion of a school of fish or a flock of migrating birds is regarded as an example of dynamic self-assembly. However, the most promising area of self-assembly is in the intermediate realm between molecular and macroscopic, where it may become possible to design nanoscale sensors and machines to carry out specific tasks. In general, we can take advantage of self-assembly only where it already exists in nature. However, designing materials that can assemble themselves is a strategy that is beginning to result in materials that can respond to stimuli and act in seemingly intelligent ways.

You might be using a self-assembled material the next time you use in-line skates. One brand of these skates contains a "smart gel" that is a liquid at normal temperatures but assembles itself into a firm, rubbery gel when exposed to body temperature. The gel fills the space between the lining of the skate boot and the sides and, once the skate has been secured, it sets to conform to the exact shape of the foot. Such a gel generally consists of an aqueous suspension of organic molecules with long chains containing different regions, some of which have strong intermolecular forces and some of which have weak intermolecular forces. The temperature determines how the chains are coiled. At low temperatures, the regions with

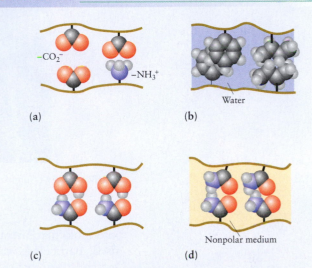

(a)   (b)

(c)   (d)

These four types of forces are responsible for the adaptive behavior of smart gels. The different forces come into play when the network of polymer chains composing a gel is disturbed. (a) Charged ionic regions can attract or repel each other. (b) Nonpolar hydrophobic regions exclude water. (c) Hydrogen bonds may form from one chain to another. (d) Dipole–dipole interactions can attract or repel chains.

strong intermolecular forces are turned inward, and so the molecules remain in liquid solution. However, at higher temperatures, those regions are turned outward, where they can attract other chains to form a flexible, but firm, network.

Another type of gel expands and contracts as its structure changes in response to electrical signals and is being investigated for use in artificial limbs that would respond and feel like real ones. One material being studied for use in artificial muscle contains a mixture of polymers, silicone oil (a polymer with a $-(O-Si-O-Si-)_n-$ backbone and hydrocarbon side chains), and salts. When exposed to an electric field, the molecules of the soft gel rearrange themselves so that the material contracts and stiffens. If struck, the stiffened material can break but, on softening, the gel is reformed. The transition between gel and solid state is therefore reversible.

❓ HOW MIGHT YOU CONTRIBUTE?

The properties of self-assembled materials need to be studied and categorized. For example, how do molecules recognize one another in a mixture? In addition, strategies to promote self-assembly need to be designed. Simple two-dimensional assemblies have been created, and these techniques need to be expanded to three dimensions. Applications are emerging in nanotechnology, in the fields of microelectronics and robotics, and in medicine, where biosensor arrays have shown high selectivity for individual toxins and bioagents.

Related Exercises: 15.67 and 15.68

For Further Reading: R. Dagani, "Intelligent gels," *Chemical and Engineering News,* June 9, 1997, pp. 26–27. R. F. Service, "How far can we push chemical self-assembly?" *Science,* vol. 309, 2005, p. 95. G. Whitesides and B. Grzybowski, "Self-assembly at all scales," *Science,* vol. 295, March 29, 2002, pp. 2418–2421.

In the Belousov–Zhabotinskii reaction, beautiful regular patterns form spontaneously as the result of the oscillating concentrations of reactants and products due to competing reactions.

FIGURE 15.31 A "forest" of carbon nanotubes that grew in vertical alignment in a "bottom-up" fashion.

intermolecular interactions (Fig. 15.31). Chemical approaches are the most widely used, particularly the solution and vapor-phase synthesis methods. In solution methods, tiny crystals are formed in solution and prevented from growing larger by the addition of stabilizers in the form of surfactant molecules. In vapor-phase synthesis a substance is vaporized and then condensed or mixed with a reactant and the product condensed to form the desired tiny crystals. In both methods one-molecule-thick layers of different materials in which the molecules are oriented in a particular direction can be formed, giving the nanomaterial interesting electrical and optical properties.

The attraction of nanotechnology lies in the opportunities for miniaturization. Tiny yet functional electrical circuits, with transistors, diodes, and wires, have been constructed from nanomaterials. Nanotubes filled with atoms of lead or another metal can be used as nanowires with high conductivity. Other nanomaterials function as semiconducting devices. Zinc sulfide nanoparticles serve as microscopic phosphors. Nanoparticles attached to an iron oxide core are used to stick to and destroy tumors. Dendrimers, nano-sized molecules with hollow centers, can be used to carry toxic drugs to specific sites in the body, a strategy that greatly reduces the amount of the toxic drug required. "Nanobots," nanoscale robotic machines, have even been proposed to carry out medical repairs within the body.

Nanomaterials are particles between 1 and 100 nm in diameter. They have properties that differ from the properties of atoms and from those of bulk materials and are used to create miniature circuits and drug delivery systems.

SKILLS YOU SHOULD HAVE MASTERED

❏ 1 Predict and explain periodic trends in the properties of the elements in Groups 15/V through 18/VIII.

❏ 2 Rationalize the properties of the oxides and oxoacids of the nonmetals in terms of the oxidation number of the nonmetal and the identification of acid anhydrides.

❏ 3 Describe how nitrogen is obtained from air and converted into useful compounds.

❏ 4 Describe the structures and properties of the nitrogen, phosphorus, and sulfur oxides.

❏ 5 Describe the acidic and basic character of water and hydrogen peroxide and compare their functions as oxidizing and reducing agents.

❏ 6 Describe the names, properties, and reactions of the principal compounds of the members of Groups 15/V through 18/VIII.

❏ 7 Describe and write balanced equations for the principal reactions used to obtain the elements in Groups 15/V through 17/VII.

❏ 8 Explain the low reactivity of the noble gases and the fact that the ease of compound formation increases down the group.

❏ 9 Explain the properties of luminescent materials

EXERCISES

Group 15/V

15.1 Nitrogen can be found in compounds with oxidation numbers ranging from -3 to $+5$. For each of the possible integral oxidation numbers, give one example of a nitrogen compound or ion.

15.2 Give the oxidation number of phosphorus in (a) white phosphorus; (b) red phosphorus; (c) phosphine, PH_3; (d) diphosphine, P_2H_4; (e) PCl_3; (f) PF_5; (g) phosphate ion, PO_4^{3-}; (h) hydrogen phosphite ion, HPO_3^{2-}.

15.3 Propose an explanation for the fact that the nitrogen trihalides become less stable as the molar mass of the halogen increases.

15.4 A student wrote the formula HN_3 instead of NH_3 on an examination by mistake. (a) Write the name and Lewis structure of the compound corresponding to the formula the student wrote (the N atoms are connected in a chain and the H atom is attached to an N atom at the end of the chain). (b) If HN_3 is added to NH_3 a proton transfer reaction occurs. Write the chemical equation for that reaction.

15.5 Urea, $CO(NH_2)_2$, reacts with water to form ammonium carbonate. Write the chemical equation and calculate the mass of ammonium carbonate that can be obtained from 4.0 kg of urea.

15.6 (a) Nitrous acid reacts with hydrazine in acidic solution to form hydrazoic acid, HN_3. Write the chemical equation and determine the mass of hydrazoic acid that can be produced from 15.0 g of hydrazine. (b) Suggest a method for preparing sodium azide, NaN_3. (c) Is the production of hydrazoic acid an oxidation or a reduction of hydrazine?

15.7 Lead azide, $Pb(N_3)_2$, is used as a detonator. (a) What volume of nitrogen at STP (1 atm, 0°C) does 1.5 g of lead azide produce when it decomposes into lead metal and nitrogen gas? (b) Would 1.5 g of mercury(II) azide, $Hg(N_3)_2$, which is also used as a detonator, produce a larger or smaller volume, given that its decomposition products are elemental mercury and nitrogen gas? (c) Metal azides in general are potent explosives. Why?

15.8 Sodium azide is used to inflate protective air bags in automobiles. What mass of solid sodium azide is needed to provide 100. L of $N_2(g)$ at 1.5 atm and 20.°C?

15.9 The common acid anhydrides of nitrogen are N_2O, N_2O_3, and N_2O_5. Write the formulas of their corresponding acids and chemical equations for the formation of the acids by the reaction of the anhydrides with water.

15.10 The common acid anhydrides of phosphorus are P_4O_6 and P_4O_{10}. Write the formulas of their corresponding acids and chemical equations for the formation of the acids by the reaction of the anhydrides with water.

15.11 The normal boiling point of NH_3 is $-33°C$ and that of NF_3, which has a greater molar mass, is $-129°C$. Explain this difference.

15.12 Explain the observations that NH_3 is a weak Brønsted base in water whereas NF_3 shows no signs of basicity.

15.13 Using the molecule database available on the Web site for this text, compare the structures of phosphorus(III) oxide and phosphorus(V) oxide. Draw Lewis structures for the two molecules. Include in your discussion an analysis of the formal charges, the phosphorus–oxygen bond orders, and the phosphorus–oxygen bond distances. Find the P—O bond distances from a standard library reference source. Do the bond distances agree with your predictions from the Lewis structures?

15.14 Examine the structure of phosphorus(III) oxide found on the Web site of this book. (a) Draw the Lewis structure of this molecule. (b) Does it obey the octet rule? (c) What hydrocarbon fragments are isolobal (see Exercise 2.91) to the phosphorus and oxygen atoms in this structure? (d) Predict the formula of a stable hydrocarbon with the same structure as that of phosphorus(III) oxide.

15.15 (a) Calculate the standard enthalpy for the reaction of nitrogen dioxide with water to produce aqueous nitric acid and nitric oxide. (b) Calculate the standard enthalpy of solution of nitric acid.

15.16 Is the reaction of $PCl_5(g)$ with water to give an aqueous mixture of hydrochloric and phosphoric acids spontaneous (in the sense that $K > 1$) at 298 K?

Group 16/VI

15.17 Write equations for (a) the burning of lithium in oxygen; (b) the reaction of sodium metal with water; (c) the reaction of fluorine gas with water; (d) the oxidation of water at the anode of an electrolytic cell.

15.18 Write equations for the reaction of (a) sodium oxide and water; (b) sodium peroxide and water; (c) sulfur dioxide and water; (d) sulfur dioxide and oxygen, with the use of a vanadium pentoxide catalyst.

15.19 Complete and balance each of the following skeletal equations:
(a) $H_2S(aq) + O_2(g) \rightarrow$
(b) $CaO(s) + H_2O(l) \rightarrow$
(c) $H_2S(g) + SO_2(g) \rightarrow$

15.20 Complete and balance each of the following skeletal equations:
(a) $FeS(s) + HCl(aq) \rightarrow$
(b) $H_2(g) + S_8(s) \rightarrow$
(c) $S_2Cl_2(l) + Cl_2(g) \rightarrow$

15.21 Explain from structural considerations why the dipole moment of the NH_3 molecule is lower than that of the H_2O molecule.

15.22 Explain from structural considerations why water and hydrogen peroxide have similar physical properties but different chemical properties.

15.23 (a) Write the Lewis structure of H_2O_2 and predict the approximate H—O—O bond angle. Which of the following ions will be oxidized by hydrogen peroxide in acidic solution? (b) Cu^+; (c) Mn^{2+}; (d) Ag^+; (e) F^-.

15.24 (a) Draw the Lewis structure for oleum, $(HO)_2OSOSO(OH)_2$, prepared by treating sulfuric acid with SO_3. (b) Determine the formal charges on the sulfur and oxygen atoms. (c) What is the oxidation number for sulfur in this compound?

15.25 If 2.00 g of sodium peroxide is dissolved to form 200. mL of an aqueous solution, what would be the pH of the solution? For H_2O_2, $K_{a1} = 1.8 \times 10^{-12}$ and K_{a2} is negligible.

15.26 What is the pH of 0.040 M NaHS(aq)?

15.27 Describe the trend in acidity of the binary hydrogen compounds of the Group 16/VI elements and account for the trend in terms of bond strength.

15.28 When lead(II) sulfide is treated with hydrogen peroxide, the possible products are either lead(II) sulfate or lead(IV) oxide and sulfur dioxide. (a) Write balanced equations for the two reactions. (b) Using data available in Appendix 2A, determine which possibility is more likely.

15.29 (a) Calculate the standard enthalpy of solution of $H_2SO_4(l)$ in water. (b) Estimate how hot the solution would become if 10.00 g of $H_2SO_4(l)$ were dissolved in 500.0 mL of water at 25°C. Assume that all the heat goes into raising the temperature of the water and that the volume change of the solution is negligible.

15.30 (a) Calculate the standard potential for the reaction of hydrogen peroxide with Fe^{2+} in acidic solution. Will this oxidation be spontaneous? See Appendix 2B. (b) What will the potential for this reaction be in neutral solution? (c) What problem would be encountered in attempting to carry out the oxidation in basic solution?

Group 17/VII

15.31 Describe the preparation of elemental fluorine, chlorine, bromine, and iodine.

15.32 (a) In what form is chlorine found in swimming pools? (b) To what chemical property does this substance owe its disinfecting capabilities?

15.33 Identify the oxidation number of the halogen in (a) hypoiodous acid; (b) ClO_2; (c) dichlorine heptoxide; (d) $NaIO_3$.

15.34 Identify the oxidation number of the halogen atoms in (a) iodine heptafluoride; (b) sodium periodate; (c) hypobromous acid; (d) sodium chlorite.

15.35 Write the balanced chemical equation for (a) the thermal decomposition of potassium chlorate without a catalyst; (b) the reaction of bromine with water; (c) the reaction between sodium chloride and concentrated sulfuric acid. (d) Identify each reaction as a Brønsted acid–base, Lewis acid–base, or redox reaction.

15.36 Write the balanced chemical equation for the reaction of chlorine with water in (a) a neutral aqueous solution; (b) a dilute basic solution; (c) a concentrated basic solution. (d) Verify that each reaction is a disproportionation reaction.

15.37 (a) Arrange the chlorine oxoacids in order of increasing oxidizing strength. (b) Suggest an interpretation of that order in terms of oxidation numbers.

15.38 (a) Arrange the hypohalous acids in order of increasing acid strength. (b) Suggest an interpretation of that order in terms of electronegativities.

15.39 Write the Lewis structure for Cl_2O. Predict the shape of the Cl_2O molecule and estimate the Cl—O—Cl bond angle.

15.40 Write the Lewis structure for BrF_3. What is the hybridization of the bromine atom in the molecule?

15.41 Plot a graph of the standard free energy of formation of the hydrogen halides against the period number of the halogens. What conclusions can be drawn from the graph?

15.42 Consider the plot of the normal boiling points of the hydrogen halides in Fig. 5.8. Account for the trend revealed by the graph.

15.43 Suggest reasons why the following interhalogens are not stable: (a) ICl_5; (b) IF_2; (c) $ClBr_3$.

15.44 Chlorine can be found as the following oxoanions: ClO^-, ClO_2^-, ClO_3^-, and ClO_4^- in many chemical stockrooms. However, fluorine forms no stable oxoanions and iodine is never found as IO_4^-. Explain these observations.

15.45 The interhalogen IF_x can be made only by indirect routes. For example, xenon difluoride gas can react with iodine gas to produce IF_x and xenon gas. In one experiment xenon difluoride is introduced into a rigid container until a pressure of 3.6 atm is reached. Iodine vapor is then introduced until the total pressure is 7.2 atm. Reaction is then allowed to proceed at constant temperature until completion by solidifying the IF_x as it is produced. The final pressure in the flask due to the xenon and excess iodine vapor is 6.0 atm. (a) What is the formula of the interhalogen? (b) Write the chemical equation for its formation.

15.46 The interhalogen ClF_x has been used as a rocket fuel. It reacts with hydrazine to form the gases hydrogen fluoride, nitrogen, and chlorine. In one study of this reaction, ClF_x gas is introduced into a rigid container until a pressure of 1.2 atm is reached. Gaseous hydrazine is then introduced until the total pressure is 3.0 atm. Reaction is then allowed to proceed at constant temperature until completion. The final pressure in the flask is 6.0 atm, which includes 0.9 atm due to excess hydrazine. (a) What is the formula of the interhalogen? (b) Write the chemical equation for its reaction with hydrazine.

15.47 Use data from Appendix 2B to determine whether chlorine gas will oxidize Mn^{2+} to form the permanganate ion in an acidic solution.

15.48 (a) Use the data in Appendix 2B to determine which is the stronger oxidizing agent, ozone or fluorine. (b) Does your answer depend on whether the reaction is carried out in acidic or basic solution?

15.49 The concentration of F^- ions can be measured by adding an excess of lead(II) chloride solution and weighing the lead(II) chlorofluoride (PbClF) precipitate. Calculate the molarity of F^- ions in 25.00 mL of a solution that gave a lead chlorofluoride precipitate of mass 0.765 g.

15.50 Suppose 25.00 mL of an aqueous solution of iodine was titrated with 0.0250 M $Na_2S_2O_3(aq)$, with starch as the indicator. The blue color of the starch–iodine complex disappeared when 27.65 mL of the thiosulfate solution had been added. What was the molar concentration of I_2 in the original solution? The titration reaction is $I_2(aq) + 2 S_2O_3{}^{2-}(aq) \rightarrow 2 I^-(aq) + S_4O_6{}^{2-}(aq)$.

15.51 Calculate the heat released when ammonium perchlorate (1.00 kg) reacts with aluminum powder (1.00 kg) as rocket fuel: $3 NH_4ClO_4(s) + 3 Al(s) \rightarrow Al_2O_3(s) + AlCl_3(s) + 6 H_2O(g) + 3 NO(g)$.

15.52 (a) Using data available in the Appendices, calculate the standard Gibbs free energy of the reaction $Cl_2(g) + H_2O(l) \rightarrow HClO(aq) + HCl(aq)$. (b) What is the equilibrium constant for this process?

Group 18/VIII

15.53 What are the sources for the production of helium and argon?

15.54 What are the sources for the production of krypton and xenon?

15.55 Determine the oxidation number of the noble gas in (a) KrF_2; (b) XeF_6; (c) KrF_4; (d) $XeO_4{}^{2-}$.

15.56 Determine the oxidation number of the noble gas in (a) XeO_3; (b) $XeO_6{}^{4-}$; (c) XeF_2; (d) $HXeO_4{}^-$.

15.57 Xenon tetrafluoride is a powerful oxidizing agent. In an acidic solution, it is reduced to xenon. Write the corresponding half-reaction.

15.58 Complete and balance each of the following reactions:
(a) $XeF_6(s) + H_2O(l) \rightarrow XeO_3(aq) + HF(aq)$
(b) $Pt(s) + XeF_4(s) \rightarrow$
(c) $Kr(g) + F_2(g) \xrightarrow{\text{electric discharge}}$

15.59 Predict the relative acid strengths of H_2XeO_4 and H_4XeO_6. Explain your conclusions.

15.60 Predict the relative oxidizing strengths of H_2XeO_4 and H_4XeO_6. Explain your conclusions.

Impact on Materials

15.61 Distinguish between fluorescence and phosphorescence.

15.62 Describe how triboluminescence and fluorescence are produced.

15.63 What is the main advantage of using a fluorescent dye to trace the function of biomolecules in a living cell?

15.64 In fluorescence, how does the energy of the emitted radiation compare with the energy of the exciting radiation?

15.65 Distinguish between the "top down" and "bottom up" approaches to manufacturing nanomaterials.

15.66 Describe the structure of a nanowire and how it conducts electricity.

15.67 (a) Consider the substances shown in parts (a), (c), and (d) in the second illustration in Box 15.1. (b) In which of these three parts of the figure will the interactions be the strongest? (c) In which will they be the weakest?

15.68 A silicone oil being studied for use in artificial limbs has a structure in which each silicon atom in the $-(O-Si-O-Si-)_n-$ backbone is attached to two methyl groups (see Box 15.1 and Fig. 14.42). Draw the molecular structure of three repeating units of this oil.

Integrated Exercises

15.69 (a) Draw the Lewis structure and assign a hybridization scheme to the atoms in the P_4 molecule. (b) Explain why the bonds in this molecule are thought to be strained.

15.70 Like elements, molecules have ionization energies. (a) Define the ionization energy of a molecule. (b) Predict which of these compounds has a higher ionization energy and justify your prediction: $SiCl_4$ or SiI_4.

15.71 Isoelectronic species have the same number of electrons. (a) Divide the following species into two isoelectronic groups: CN^-, N_2, $NO_2{}^-$, $C_2{}^{2-}$, O_3. Which species in each group is likely to be (b) the strongest Lewis base; (c) the strongest reducing agent?

15.72 Isoelectronic species have the same number of electrons. (a) Divide the following species into three isoelectronic groups: $CO_2{}^-$, NH_3, NO, $NO_2{}^+$, N_2O, H_3O^+, $O_2{}^+$. (b) Which species in each group is likely to be (b) the strongest Lewis acid; (c) the strongest oxidizing agent?

15.73 Refer to Appendix 2B and arrange the halogens in order of increasing oxidizing strength in water.

15.74 Refer to Appendix 2B and arrange S, O_2 (to H_2O), Cl_2, and H_2O_2 (to H_2O) in order of increasing oxidizing strength in water.

15.75 Balance each of the following skeletal equations and classify the reaction as Brønsted–Lowry acid–base, Lewis acid–base, or redox. For both types of acid–base reactions, identify the acid and the base. For redox reactions, identify the reducing agent and the oxidizing agent.

(a) $NH_3(aq) + ClO^-(aq) \xrightarrow{\text{aqueous alkali}} N_2H_4(aq) + Cl^-(aq) + H_2O(l)$

(b) $Mg_3N_2(s) + H_2O(l) \rightarrow Mg(OH)_2(s) + NH_3(g)$

(c) $P^{3-}(s) + H_2O(l) \rightarrow PH_3(g) + OH^-(aq)$

(d) $Cl_2(g) + H_2O(aq) \rightarrow HClO(aq) + HCl(aq)$

15.76 Balance each of the following skeletal equations and classify the reaction as Brønsted–Lowry acid–base, Lewis acid–base, or redox. For both types of acid–base reactions, identify the acid and the base. For redox reactions, identify the reducing agent and the oxidizing agent.

(a) $PCl_5(s) + H_2O(l) \rightarrow H_3PO_4(l) + HCl(g)$

(b) $CaF_2(s) + H_2SO_4(aq, conc) \rightarrow Ca(HSO_4)_2(aq) + HF(g)$

(c) $KI(s) + H_3PO_4(aq) \xrightarrow{\Delta} KH_2PO_4(aq) + HI(g)$

(d) $H_2S(g) + O_2(g) \rightarrow SO_2(g) + H_2O(g)$

15.77 The ground state of O_2 has two unpaired π^* electrons with parallel spins. There are two known low-lying excited states of O_2. State A has the two π^* electrons with the spins antiparallel but in different orbitals. State B has the two π^* electrons paired in the same orbital. (a) The energies of the excited states are 94.72 $kJ \cdot mol^{-1}$ and 157.85 $kJ \cdot mol^{-1}$ above the ground state. Which state corresponds to which excitation energy? (b) What is the wavelength of the radiation absorbed in the transition from the ground state to the first excited state?

15.78 Predict the formula and Lewis structure of the anhydride that corresponds to each of the following acids: (a) HClO; (b) HIO_3; (c) $HClO_4$.

15.79 Thiosulfuric acid, $H_2S_2O_3$, has a structure similar to that of sulfuric acid, except that the terminal oxygen atom has been replaced by a sulfur atom. How would you expect the physical and chemical properties of thiosulfuric acid and sulfuric acid to differ?

15.80 The concentration of Cl^- ions can be measured gravimetrically by precipitating silver chloride, with silver nitrate as the precipitating reagent in the presence of dilute nitric acid. The white precipitate is filtered off and its mass is determined. (a) Calculate the Cl^- ion concentration in 50.00 mL of a solution that gave a silver chloride precipitate of mass 1.972 g. (b) Why is the method inappropriate for measuring the concentration of fluoride ions?

15.81 Xenon fluorides are used as fluorinating agents for organic and inorganic compounds. Xenon tetrafluoride reacts with sulfur tetrafluoride to produce sulfur hexafluoride and xenon. What mass of sulfur hexafluoride can be produced from 330.0 g of xenon tetrafluoride and 250.0 g of sulfur tetrafluoride?

15.82 (a) When Xe and Pt are treated with fluorine gas under pressure at 200°C, the ionic solid $XeF_5^+PtF_6^-$ is formed. Write the Lewis structure for XeF_5^+ and, from VSEPR theory, predict its shape. (b) When two molecules of XeF_2 react with one molecule of AsF_5, the resulting compound has the formula $Xe_2F_3^+AsF_6^-$. Predict the Lewis structure of the $Xe_2F_3^+$ ion.

15.83 From standard reference sources, determine the chemical composition and color of the minerals orpiment and realgar. Mention a use of these substances.

15.84 Chapters 14 and 15 have been organized by group. Discuss whether it would be helpful to organize the elements according to period and give examples of trends that could be displayed helpfully in that system.

15.85 The azide ion has an ionic radius of 148 pm and forms many ionic and covalent compounds that are similar to those of the halides. (a) Write the Lewis formula for the azide ion and predict the N—N—N bond angle. (b) On the basis of its ionic radius, where in Group 17/VII would you place the azide ion? (c) Compare the acidity of hydrazoic acid with those of the hydrohalic acids and explain any differences (for HN_3, $K_a = 1.7 \times 10^{-5}$). (d) Write the formulas of three ionic or covalent azides.

15.86 The concentration of nitrate ion can be determined by a multistep process. A 25.00-mL sample of water from a rural well contaminated with $NO_3^-(aq)$ was made basic and treated with an excess of zinc metal, which reduces the nitrate ion to aqueous ammonia. The evolved ammonia gas was passed into 50.00 mL of 2.50×10^{-3} M HCl(aq). The unreacted HCl(aq) was titrated to the stoichiometric point with 28.22 mL of 1.50×10^{-3} M NaOH(aq). (a) Write balanced chemical equations for the three reactions. (b) What is the molar concentration of nitrate ion in the well water?

15.87 The concentration of hypochlorite ions in a solution can be determined by adding a sample of known volume to a solution containing excess I^- ions, which are oxidized to iodine:

$$ClO^-(aq) + 2\ I^-(aq) + H_2O(l) \longrightarrow$$
$$I_2(aq) + Cl^-(aq) + 2\ OH^-(aq)$$

The iodine concentration is then measured by titration with sodium thiosulfate:

$$I_2(aq) + 2\ S_2O_3^{2-}(aq) \longrightarrow 2\ I^-(aq) + S_4O_6^{2-}(aq)$$

In one experiment, 10.00 mL of ClO^- solution was added to a KI solution, which in turn required 28.34 mL of 0.110 M $Na_2S_2O_3(aq)$ to reach the stoichiometric point. Calculate the molar concentration of ClO^- ions in the original solution.

15.88 The triiodide ion (I_3^-), which forms a brown solution in water, is obtained from the reaction $I_2(s) + I^-(aq) \rightarrow I_3^-(aq)$. The equilibrium constant for the formation of I_3^- at 298 K is 698. If 35.0 g of solid I_2 is added to 250. mL of 0.50 M KI(aq), what is the molar concentration of I_3^- at equilibrium? How many grams of I_2 are present at equilibrium? Assume that the volume of the solution does not change when the I_2 is added.

15.89 Account for the observation that solubility in water generally increases from chloride to iodide for ionic halides with low covalent character (such as the potassium halides) but decreases from chloride to iodide for ionic halides in which the bonds are significantly covalent (such as the silver halides).

15.90 Account for the observation that melting and boiling points generally decrease from fluoride to iodide for ionic halides but increase from fluoride to iodide for molecular halides.

15.91 (a) Using data available in Appendix 2B, determine the standard potential for the redox reaction $2\ Ag^+(aq) + 2\ I^-(aq) \rightarrow 2\ Ag(s) + I_2(s)$. Is this process spontaneous under standard conditions? (b) When equal volumes of 2.0 M $AgNO_3(aq)$ and 2.0 M NaI(aq) are mixed, the redox reaction described in part (a) does not take place. Instead, a precipitate of AgI is formed. Why doesn't the redox reaction take place? Support your answer with an appropriate calculation.

15.92 (a) Draw the molecular orbital energy-level diagram for O_2. Using the diagram, determine the bond order and magnetic properties of O_2. (b) What molecular property of oxygen is explained by its molecular orbital diagram but not by its Lewis structure? (c) Describe the nature of the highest occupied molecular orbital—is it bonding, antibonding, or nonbonding? (d) Predict the bond order and magnetic properties of the peroxide ion and the superoxide ion on the basis of molecular orbital theory.

15.93 (a) The nitrosyl ion, NO^+, is isoelectronic with N_2 and has two fewer electrons than O_2. The ion is stable and can be purchased as its hexafluorophosphate (PF_6^-) or tetrafluoroborate (BF_4^-) salt. Draw its molecular orbital diagram and compare it with those of N_2 and O_2. (b) NO^+ is diamagnetic. Can we tell which orbital, the σ_{2p} or the π_{2p}, is higher in energy from this information?

15.94 (a) Draw the Lewis structure of N_2O_3. (b) What is the N—N bond order? (c) Explain why N_2O_3 adopts this structure, whereas the analogous compound for phosphorus is P_4O_6.

15.95 Pyrite is a sulfur-based mineral with the formula FeS_2. View the unit cell of pyrite on the Web site for this text. (a) What is the type of unit cell adopted by pyrite? (b) Describe the locations of the iron and sulfur atoms in this unit cell (unit-cell faces, corners, edges, etc.). (c) What is the coordination number of iron? (d) To what is each sulfur atom bonded? (e) On the basis of these observations, how would you best describe the oxidation numbers of iron and sulfur in pyrite?

15.96 The structure of apatite, $Ca_5(PO_4)_3X$ (where X is a monovalent anion such as OH^-, F^-, or Cl^-) can be viewed on the Web site for this text. From this model, determine (a) the type of unit cell adopted by apatite; (b) the number of Ca^{2+} ions and PO_4^{3-} ions in the unit cell; (c) the locations of the Ca^{2+} and PO_4^{3-} ions (unit-cell corners, faces, edges, etc.). (d) What is missing from the Web model for the structure of apatite?

Chemistry Connections

15.97 The gas NO is released to the stratosphere by jet engines. Because it can contribute to the destruction of stratospheric ozone, its concentration is monitored closely. One monitoring technique measures the chemiluminescence given off by the reaction $NO(g) + O_3(g) \rightarrow NO_2^*(g) + O_2(g)$. The asterisk indicates that NO_2 is produced in an excited state. The excited NO_2 molecule gives off red light and infrared radiation as it returns to the ground state. Because both NO and NO_2 may be present in the air sample, part of the sample is exposed to a reducing agent, which converts all of the NO_2 to NO. Two samples are then analyzed, first air from the unreduced original sample, which gives the concentration of NO, then air from the reduced sample. The concentration of NO_2 in the air is the difference of the two measurements. In one experiment, two samples were collected, one from a cloud and one from clear air. The following measurements were made at 9 km above the Pacific Ocean (ppt denotes parts per trillion by molecules):

	Unreduced sample	Reduced sample
Cloud:	860 ppt	1110 ppt
Clear air:	480 ppt	740 ppt

(a) Determine the concentrations (in ppt) of NO and NO_2 in each air sample.
(b) In which part of the atmosphere is the concentration of NO greater?
(c) Explain the difference, taking into account other sources of NO in the atmosphere besides airplanes.
(d) Some of the NO molecules released by jet engines react with the hydroxyl radical, $\cdot OH$. Write the Lewis structure of the most likely product and state its name.
(e) Is the product of the reaction in part (d) likely to be more stable than NO? Explain your reasoning.
(f) In the stratosphere NO catalyzes the conversion of O_3 to O_2 in a two-step reaction with the intermediate NO_2. The overall equation for the reaction is $O_3(g) + O(g) \rightarrow 2\,O_2(g)$ and the rate law for the reaction is Rate = $k[NO][O_3]$. Write an acceptable two-step mechanism for the reaction, indicating which step is the slow step.
(g) At $-30°C$, the temperature of the air from which the samples were taken, the rate constant for the reaction in part (f) is 6×10^{-15} $cm^3 \cdot molecule^{-1} \cdot s^{-1}$. If the concentration of ozone in the air from which the NO clear air sample was taken was 4×10^{-12} molecules·cm^3 and the total pressure of the air was 220 mbar, what was the rate at which ozone was being destroyed in the air at $-30°C$?

THE ELEMENTS: THE *d* BLOCK

What Are the Key Ideas?   The properties of the *d*-block metals are governed by the availability of *d*-orbitals, their variable valence, and their ability to act as Lewis acids.

Why Do We Need to Know This Material?   The *d*-block metals are the workhorse elements of the periodic table. Iron and copper helped civilization rise from the Stone Age and are still our most important industrial metals. Other members of the block include the metals of new technologies, such as titanium for the aerospace industry and vanadium for catalysts in the petrochemical industry. The precious metals—silver, platinum, and gold—are prized as much for their appearance, rarity, and durability as for their usefulness. Compounds of *d*-block metals give color to paint, turn sunlight into electricity, serve as powerful oxidizing agents, and form the basis of some cancer treatments.

What Do We Need to Know Already?   This chapter draws on many of the principles introduced in the preceding chapters. In particular, it makes use of the electron configurations of atoms and ions (Sections 1.13 and 2.1) and the classification of species as Lewis acids and bases (Section 10.2). Molecular orbital theory (Sections 3.8 through 3.12) plays an important role in Section 16.12.

Chapter **16**

The Industrial Revolution was made possible by iron in the form of steel, an alloy used for construction and transportation. Other *d*-block metals, both as the elements and in compounds, are transforming our present. Copper, for instance, is an essential component of some superconductors. Vanadium and platinum are used in the development of catalysts to reduce pollution and in the continuing effort to make hydrogen the fuel of our future.

The compounds of the *d*-block elements show a wide range of interesting properties. Some are vital to life. Iron is an essential component of mammalian blood. Compounds of cobalt, molybdenum, and zinc are found in vitamins and essential enzymes. Other compounds simply make life more interesting and colorful. The beautiful color of cobalt blue glass, the brilliant greens and blues of kiln-baked pottery, and many pigments used by artists make use of *d*-block compounds.

We begin this chapter by summarizing the major periodic trends exhibited by the *d*-block elements and their compounds. Then we describe some of the properties and key reactions of selected elements. The *d*-block metals form a wide variety of complexes and, in the second half of the chapter, we describe their structures and the two principal theories of their bonding. We end the chapter by examining the contribution of *d*-block elements to some important modern materials.

THE *d*-BLOCK ELEMENTS AND THEIR COMPOUNDS

The elements in Groups 3 through 11 are called the *transition metals* because they represent a transition from the highly reactive metals of the *s* block to the much less reactive metals of Group 12 and the *p* block (Fig. 16.1). Note that the transition metals do not extend all the way across the *d* block: the Group 12 elements (zinc, cadmium, and mercury) are not normally considered to be transition elements. Because their *d*-orbitals are full, the Group 12 elements have properties that are more like those of main-group metals than those of transition metals. Just after

This distinction is not always made: take the advice of your instructor.

657

FIGURE 16.1 The orange rectangles identify the elements in the *d* block of the periodic table. Note that the *f* block, consisting of the inner transition metals, intervenes in Periods 6 and 7 as indicated by the purple bar. The *f* block may begin at Ce and Th or at La and Ac, as shown here, but always contains 14 elements. The horizontal gray bar designates the groups containing transition metals, Groups 3–11.

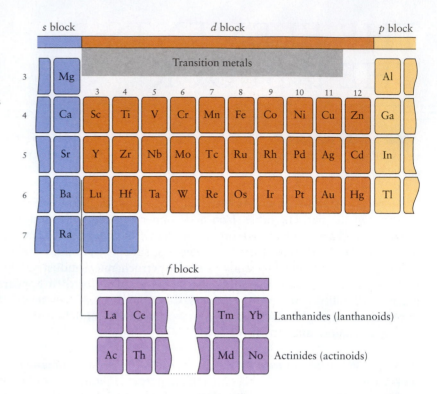

the third row of the *d* block has been started, following lanthanum, electrons begin to occupy the seven 4*f*-orbitals and the lanthanoids (the "rare earths"; commonly known as the lanthanides) delay the completion of Period 6. These elements, together with the actinoids (commonly called the actinides), the analogous series in Period 7, are sometimes referred to as the *inner transition metals*.

The incompletely filled *d*-subshell is responsible for the wide range of colors shown by compounds of the *d*-block elements. Furthermore, many *d*-metal compounds are paramagnetic (see Box 3.2). One of the challenges that we face in this chapter is to build a model of bonding that accounts for color and magnetism in a unified way. First, though, we consider the physical and chemical properties of the elements themselves.

16.1 Trends in Physical Properties

In Chapters 14 and 15, we saw that the main-group elements have physical properties related to their valence electron configurations, resulting in distinctive group characteristics. The ground-state electron configurations of the atoms of *d*-block elements, however, differ primarily in the occupation of the $(n - 1)d$-orbitals. According to the rules of the building-up principle, these orbitals are the last orbitals to be occupied. However, once they are occupied, they lie slightly lower in energy than the outer *ns*-orbitals. Because there are five *d*-orbitals in a given shell, and each one can accommodate up to two electrons, there are 10 elements in each row of the *d* block. The electron configurations of the *d*-block metals differ primarily in the occupation of these inner *d*-orbitals, and so the physical properties of the elements tend to be similar.

All the *d*-block elements are metals. Most of these "*d*-metals" are good electrical conductors. In fact, at room temperature silver is the best electrical conductor of all elements. Most of the *d*-metals are malleable, ductile, lustrous, and silver-white in color. Generally, their melting and boiling points are higher than those of the main-group elements. There are a few notable exceptions: copper is red-brown, gold is yellow, and mercury has such a low melting point that it is a liquid at room temperature.

The shapes of the *d*-orbitals affect the properties of the *d*-block elements in two ways (see Fig. 1.37):

- The lobes of two *d*-orbitals on the same atom occupy markedly different regions of space. As a result, electrons in different *d*-orbitals are relatively far apart and repel one another only weakly.

- The electron density in *d*-orbitals is low near the nucleus, and so they are not very effective at shielding other electrons from the positive nuclear charge.

One consequence of these two characteristics is the trend in atomic radii of the *d*-block metals, which decrease gradually across a period and then rise again (Fig. 16.2). Nuclear charge and number of *d*-electrons both increase from left to right across each row (from scandium to zinc, for instance). Because the repulsion between *d*-electrons is weak, initially the increasing nuclear charge can draw them inward, and so the atoms become smaller. Farther across the block, there are so many *d*-electrons that electron–electron repulsion increases more rapidly than the nuclear charge, and the radii begin to increase again. Because these attractions and repulsions are finely balanced, the range of *d*-metal atomic radii is not very great. In fact, some of the atoms of one *d*-metal can easily replace atoms of another *d*-metal in a crystal lattice (Fig. 16.3). The *d*-metals can therefore form a wide range of alloys, including the many varieties of steel (see Sections 16.13 and 16.14).

The atomic radii of the second row of *d*-metals (Period 5) are typically greater than those in the first row (Period 4). The atomic radii in the third row (Period 6), however, are about the same as those in the second row and smaller than expected. This effect is due to the **lanthanide contraction,** the decrease in radius along the first row of the *f* block (Fig. 16.4). This decrease is due to the increasing nuclear charge along the period coupled with the poor shielding ability of *f*-electrons. When the *d* block resumes (at lutetium), the atomic radius has fallen from 217 pm for barium to 173 pm for lutetium.

A technologically important effect of the lanthanide contraction is the high density of the Period 6 elements (Fig. 16.5). The atomic radii of these elements are comparable to those of the Period 5 elements, but their atomic masses are about twice as large; so more mass is packed into the same volume. A block of iridium, for example, contains about as many atoms as a block of rhodium of the same volume. However, each iridium atom is nearly twice as heavy as a rhodium atom, and so the density of the sample is nearly twice as great. In fact, iridium is one of the two densest elements; its neighbor osmium is the other. Another effect of the contraction is the low reactivity—the "nobility"—of gold and platinum. Because their valence electrons are relatively close to the nucleus, they are tightly bound and not readily available for chemical reactions.

The atomic radii of the d-block metals are similar but tend to decrease across a series. The lanthanide contraction accounts for the smaller than expected radii and higher densities of the d-block atoms in Period 6.

FIGURE 16.2 The atomic radii (in picometers) of the elements of the first row of the *d* block.

Some periodic charts designate the elements Ce–Lu as the *f* block. In this book, the elements La–Yb are designated as the *f* block, because Lu matches the properties of the elements in Group 3 better than La does.

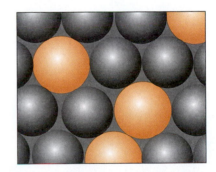

FIGURE 16.3 Because the atomic radii of the *d*-block elements are so similar, the atoms of one element can replace the atoms of another element with minor modification of the atomic locations; consequently, *d*-block metals form a wide range of alloys.

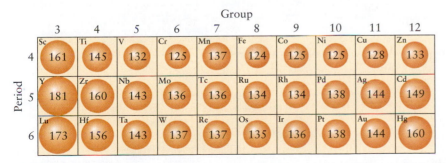

FIGURE 16.4 The atomic radii of the *d*-block elements (in picometers). Notice the similarity of all the values and, in particular, the close similarity between the second and the third rows as a result of the lanthanide contraction.

FIGURE 16.5 The densities (in grams per centimeter cubed, g·cm^{-3}) of the *d*-metals at 25°C. The lanthanide contraction has a pronounced effect on the densities of the elements in Period 6 (front row in this illustration), which are among the densest of all the elements.

16.2 Trends in Chemical Properties

The *d*-block elements tend to lose their valence *s*-electrons when they form compounds. Most of them can also lose a variable number of *d*-electrons and exist in a variety of oxidation states. The only elements of the block that do not use their *d*-electrons in compound formation are the members of Group 12 (zinc, cadmium, and mercury). The ability to exist in different oxidation states is responsible for many of the special chemical properties of these elements and plays a role in the action of many biomolecules.

Most *d*-block elements have more than one common oxidation state (Fig. 16.6). The distribution of oxidation states looks daunting at first sight, but if we examine the figure, we can see two patterns:

1   Elements close to the center of each row have the widest range of oxidation states.

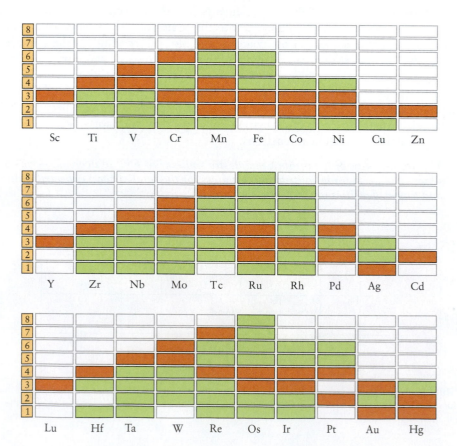

FIGURE 16.6 The oxidation numbers of the *d*-block elements. The orange blocks mark the common oxidation numbers for each element; the green blocks mark the other known states.

Except for mercury, the elements at the ends of each row of the *d* block occur in only one oxidation state other than 0. All the other elements of each row have at least two oxidation states and manganese, at the center of its row, has seven oxidation states.

2 Elements in the second and third rows of the block are more likely to reach higher oxidation states than are those in the first row.

Notice that ruthenium and osmium have all possible oxidation states and that even gold and mercury, which lie near the end of the block, can be found in three oxidation states.

The pattern of oxidation states underlies trends in the chemical properties of the *d*-block elements. An element in a high oxidation state is easily reduced; therefore, the compound is likely to be a good oxidizing agent. For example, manganese has oxidation number $+7$ in the permanganate ion, MnO_4^-, and this ion is a good oxidizing agent in acidic solution:

$$MnO_4^-(aq) + 8\,H^+(aq) + 5\,e^- \longrightarrow Mn^{2+}(aq) + 4\,H_2O(l) \qquad E° = +1.51\,V$$

Compounds that contain the element in a low oxidation state, such as Cr^{2+}, are often good reducing agents:

$$Cr^{3+}(aq) + e^- \longrightarrow Cr^{2+}(s) \qquad E° = -0.41\,V$$

The pattern of oxidation states correlates with the pattern of acid–base behavior of *d*-metal oxides. Although most *d*-metal oxides are basic, the oxides of a given element show a shift toward acidic character with increasing oxidation number, just as the oxoacids do (recall Section 10.10). The family of chromium oxides is a good example:

CrO	$+2$	basic
Cr_2O_3	$+3$	amphoteric
CrO_3	$+6$	acidic

Chromium(VI) oxide, CrO_3, is the anhydride of chromic acid, H_2CrO_4, the parent acid of the chromates. In this highly oxidized state, chromium is electron poor and the oxygen atoms attached to it are less likely to share electrons with a proton.

The elements on the left of the *d* block resemble the *s*-block metals in being much more difficult to extract from their ores than the metals on the right. Indeed, if we go to the right side of the *d* block and move back across it from right to left, we encounter the elements in very roughly the order in which they were put to use as metals. On the far right are copper and zinc, which jointly were responsible for the Bronze Age. As metal workers discovered how to achieve higher temperatures, they found out how to reduce iron oxides, and the Bronze Age was succeeded by the Iron Age. The metals at the left of the block—for example, titanium—require such extreme conditions for their extraction—including the use of other active metals or electrolysis—that they became widely available only in the twentieth century, when these techniques were developed.

SELF-TEST 16.1A Predict trends in ionization energies of the *d*-block metals.

[*Answer:* Ionization energy increases from left to right across a row and decreases down a group.]

SELF-TEST 16.1B Six of the *d*-block metals in Period 4 form $+1$ ions. Predict trends in the radii of those ions.

The range of oxidation states of a d-block element increases toward the center of the block. Compounds in which the d-block element has a high oxidation state tend to be oxidizing; those in which it has a low oxidation state tend to be reducing. The acidic character of oxides increases with the oxidation state of the element.

SELECTED ELEMENTS: A SURVEY

Although the physical properties of the *d*-block elements are similar, the chemical properties of these elements are so diverse that it is impossible to summarize them fully. We can, however, observe some of the major trends in properties within the *d* block by considering the properties of certain representative elements, particularly those in the first row of the block.

16.3 Scandium Through Nickel

Table 16.1 summarizes the physical properties of the elements from scandium through nickel. Notice the similarities in their melting and boiling points, but the gradual increase in density.

Scandium, Sc, which was first isolated in 1937, is a reactive metal: it reacts with water about as vigorously as calcium does. It has few uses and is not thought to be essential to life. The small, highly charged Sc^{3+} ion is strongly hydrated in water (like Al^{3+}), and the resulting $[Sc(H_2O)_6]^{3+}$ complex is about as strong a Brønsted acid as acetic acid.

Titanium, Ti, a light, strong metal, is used where these properties are critical—in widely diverse applications such as jet engines and dental fixtures such as partial plates. Although titanium is relatively reactive, unlike scandium it is resistant to corrosion because it is passivated by a protective skin of oxide on its surface. The principal sources of the metal are the ores *ilmenite*, $FeTiO_3$, and *rutile*, TiO_2.

Titanium requires strong reducing agents for extraction from its ores. It was not widely exploited commercially until demand from the aerospace industry increased in the last part of the twentieth century. The metal is obtained by first treating the ores with chlorine in the presence of coke (impure carbon obtained by heating coal in the absence of air) to form titanium(IV) chloride; the volatile chloride is then reduced by passing it through liquid magnesium:

$$TiCl_4(g) + 2\ Mg(l) \xrightarrow{700°C} Ti(s) + 2\ MgCl_2(s)$$

The most common oxidation state of titanium is +4, in which the atom has lost both its 4*s*-electrons and its two 3*d*-electrons. Its most important compound is titanium(IV) oxide, TiO_2, which is almost universally known as titanium dioxide. This oxide is a brilliantly white (when finely powdered), nontoxic, stable solid used as the white pigment in paints and paper. It acts as a semiconductor in the presence of light, and so it is used to convert solar radiation into electrical energy in solar cells.

Titanium forms a series of oxoanions called *titanates,* which are prepared by heating TiO_2 with a stoichiometric amount of the oxide or carbonate of a second metal. One of these compounds, barium titanate, $BaTiO_3$, is **piezoelectric**, which means that it becomes electrically charged when it is mechanically distorted. The ability to convert mechanical vibration into an electrical signal makes barium titanate useful for underwater sound detection.

Vanadium, V, a soft silver-gray metal, is produced by reducing its oxide or chloride. For example, vanadium(V) oxide is reduced by calcium:

$$V_2O_5(s) + 5\ Ca(l) \xrightarrow{\Delta} 2\ V(s) + 5\ CaO(s)$$

*Square brackets are commonly used to indicate the presence of a *d*-metal complex.*

TABLE 16.1 **Properties of the *d*-Block Elements Scandium Through Nickel**

Z	Name	Symbol	Valence electron configuration	Melting point (°C)	Boiling point (°C)	Density, (g·cm^{-3})
21	scandium	Sc	$3d^1 4s^2$	1540	2800	2.99
22	titanium	Ti	$3d^2 4s^2$	1660	3300	4.55
23	vanadium	V	$3d^3 4s^2$	1920	3400	6.11
24	chromium	Cr	$3d^5 4s^1$	1860	2600	7.19
25	manganese	Mn	$3d^5 4s^2$	1250	2120	7.47
26	iron	Fe	$3d^6 4s^2$	1540	2760	7.87
27	cobalt	Co	$3d^7 4s^2$	1494	2900	8.80
28	nickel	Ni	$3d^8 4s^2$	1455	2150	8.91

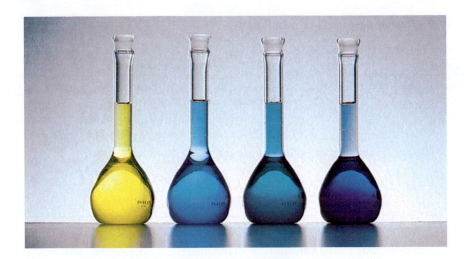

FIGURE 16.7 Many vanadium compounds form vividly colored solutions in water. They are also used in pottery glazes. The blue colors here are due to the vanadyl ion, VO^{2+}.

Vanadium is used to make tough steels for automobile and truck springs. Because it is not economical to add the pure metal to iron, a **ferroalloy** of the metal, an alloy with iron and carbon, that is less expensive to produce, is used instead.

Vanadium(V) oxide, V_2O_5, commonly known as vanadium pentoxide, is the most important compound of vanadium. This orange-yellow solid is used as an oxidizing agent and as an oxidizing catalyst in the contact process for the manufacture of sulfuric acid (Section 15.7). The wide range of colors of vanadium compounds, including the blue of the vanadyl ion, VO^{2+} (Fig. 16.7), has led to their use as glazes in the ceramics industry.

Chromium, Cr, is a bright, lustrous, corrosion-resistant metal. Its name, which comes from the Greek word for "color," was inspired by its colorful compounds. The metal is obtained from the mineral *chromite*, $FeCr_2O_4$, by reduction with carbon in an electric arc furnace:

$$FeCr_2O_4(s) + 4\ C(s) \xrightarrow{\Delta} Fe(l) + 2\ Cr(l) + 4\ CO(g)$$

Chromium metal is also reduced by aluminum in the thermite process:

$$Cr_2O_3(s) + 2\ Al(s) \xrightarrow{\Delta} Al_2O_3(s) + 2\ Cr(l)$$

Chromium metal is important in metallurgy, because it is used to make stainless steel and for chromium plating (Section 12.13). Chromium(IV) oxide, CrO_2, is a ferromagnetic material that is used for coating "chrome" recording tapes.

Sodium chromate, Na_2CrO_4, a yellow solid, is the source of most other chromium compounds, including fungicides, pigments, and ceramic glazes. The chromate ion changes into the orange dichromate ion, $Cr_2O_7^{2-}$, in the presence of acid (Fig. 16.8):

$$2\ CrO_4^{2-}(aq) + 2\ H^+(aq) \longrightarrow Cr_2O_7^{2-}(aq) + H_2O(l)$$

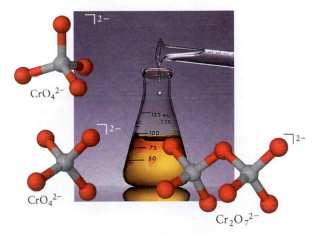

FIGURE 16.8 The chromate ion, CrO_4^{2-}, is yellow. When acid is added to a chromate solution, the ions form orange dichromate ions, $Cr_2O_7^{2-}$.

TABLE 16.2 **Composition of Different Steels**

Element blended into iron	Typical amount (%)	Effect
manganese	0.5 to 1.0	increases strength and hardness but lowers ductility
	13	increases wear resistance
nickel	< 5	increases strength and shock resistance
	> 5	increases corrosion resistance (stainless) and hardness
chromium	variable	increases hardness and wear resistance
	> 12	increases corrosion resistance (stainless)
vanadium	variable	increases hardness
tungsten	< 20	increases hardness, especially at high temperatures

FIGURE 16.9 Manganese nodules that litter the ocean floor are potentially a valuable source of the element.

In the laboratory, acidified solutions of dichromates, in which the oxidation number of chromium is +6, are useful oxidizing agents:

$$Cr_2O_7{}^{2-}(aq) + 14\ H^+(aq) + 6\ e^- \longrightarrow 2\ Cr^{3+}(aq) + 7\ H_2O(l)$$

$$E° = +1.33\ V$$

Manganese, Mn, is a gray metal that resembles iron. It is much less resistant to corrosion than chromium is and becomes coated with a thin brown layer of oxide when exposed to air. The metal is rarely used alone, but it is an important component of alloys. When it is added to iron as ferromanganese to make steel, it removes sulfur by forming a sulfide. It also increases iron's hardness, toughness, and resistance to abrasion (Table 16.2). Another useful alloy is *manganese bronze* (39% by mass Zn, 1% Mn, a small amount of iron and aluminum, and the rest copper), which is very resistant to corrosion and is used for the propellers of ships. Manganese is also alloyed with aluminum to increase the stiffness of beverage cans, allowing them to be manufactured with thinner walls.

A rich supply of manganese lies in nodules of ore that litter the ocean floors (Fig. 16.9). These nodules range in diameter from millimeters to meters and are lumps of the oxides of iron, manganese, and other elements. However, because this source is technically difficult to exploit, manganese is currently obtained by the thermite process from *pyrolusite*, a mineral form of manganese dioxide:

$$3\ MnO_2(s) + 4\ Al(s) \xrightarrow{\Delta} 3\ Mn(l) + 2\ Al_2O_3(s)$$

Manganese lies near the center of its row (in Group 7) and occurs in a wide variety of oxidation states. The most stable state is +2, but +4, +7, and, to a lesser extent, +3 are common in manganese compounds. Its most important compound is manganese(IV) oxide, MnO_2, commonly called manganese dioxide. This compound is a brown-black solid used in dry cells, as a decolorizer to conceal the green tint of glass, and as the starting point for the production of other manganese compounds.

Potassium permanganate is a strong oxidizing agent in acidic solution and is used to oxidize organic compounds and as a mild disinfectant. Its usefulness stems not only from its thermodynamic tendency to oxidize other species but also from its ability to act by a variety of mechanisms; hence, it is likely to be able to find a path with low activation energy and act rapidly.

Iron, Fe, the most widely used of all the *d*-metals, is the most abundant element on Earth and the second most abundant metal in the Earth's crust (after aluminum). Its principal ores are the oxides *hematite*, Fe_2O_3, and *magnetite*, Fe_3O_4. The sulfide mineral *pyrite*, FeS_2 (see Fig. 15.11), is widely available, but it is not used in steelmaking because the sulfur is difficult to remove.

Steelmaking is described in Section 16.13.

Iron is quite reactive and corrodes in moist air. It is passivated by oxidizing acids such as HNO_3 (Section 12.14) but reacts with nonoxidizing acids, evolving hydrogen and forming iron(II) salts. The colors of these salts vary from pale yellow to dark green-brown. Iron(II) salts are readily oxidized to iron(III) salts.

The oxidation is slow in acidic solution but rapid in basic solution, where insoluble iron(III) hydroxide, $Fe(OH)_3$, is precipitated. Although $[Fe(H_2O)_6]^{3+}$ ions are pale purple and Fe^{3+} ions give amethyst its purple color, the colors of aqueous solutions of iron(III) salts are dominated by the conjugate base of $[Fe(H_2O)_6]^{3+}$, the yellow $[FeOH(H_2O)_5]^{2+}$ ion:

$$[Fe(H_2O)_6]^{3+}(aq) + H_2O(l) \rightleftharpoons H_3O^+(aq) + [FeOH(H_2O)_5]^{2+}(aq)$$

Like some other d-block metals, such as nickel, iron can form compounds in which its oxidation number is zero. For example, when iron is heated in carbon monoxide, it reacts to form *iron pentacarbonyl*, $Fe(CO)_5$, a yellow molecular liquid that boils at 103°C.

A healthy adult human body contains about 3 g of iron, mostly as hemoglobin. Because about 1 mg is lost daily (in sweat, feces, and hair), and women lose about 20 mg in each menstrual cycle, iron must be ingested daily to maintain the balance. Iron deficiency, or anemia, results in reduced transport of oxygen to the brain and muscles, and an early symptom is chronic tiredness.

Cobalt ores are often found in association with copper(II) sulfide. Cobalt is a silver-gray metal and is used mainly for alloying with iron. *Alnico steel,* an alloy of iron, nickel, cobalt, and aluminum, is used to make permanent magnets such as those in loudspeakers. Cobalt steels are hard enough to be used as surgical steels, drill bits, and lathe tools. The color of cobalt glass is due to a blue pigment that forms when cobalt(II) oxide is heated with silica and alumina.

Nickel, Ni, is also used in alloys. It is a hard, silver-white metal used mainly for the production of stainless steel and for alloying with copper to produce *cupronickels,* the alloys used for nickel coins (which are about 25% Ni and 75% Cu). Nickel is also used in nicad batteries and as a catalyst, especially for the addition of hydrogen to organic compounds, as in the hydrogenation of vegetable oils (Section 18.6).

The slightly yellow hue of cupronickels is removed by the addition of small amounts of cobalt.

About 70% of the western world's supply of nickel comes from iron and nickel sulfide ores that were brought close to the surface nearly 2 billion years ago by the violent impact of a huge meteor at Sudbury, Ontario. The ore is first *roasted* (heated in air) to form nickel(II) oxide, which is reduced to the metal either electrolytically or by reaction with hydrogen gas in the first step of the *Mond process:*

$$NiO(s) + H_2(g) \xrightarrow{\Delta} Ni(s) + H_2O(g)$$

The impure nickel is then refined by first exposing it to carbon monoxide, with which it forms nickel tetracarbonyl, $Ni(CO)_4$:

$$Ni(s) + 4\ CO(g) \longrightarrow Ni(CO)_4(g)$$

Nickel tetracarbonyl is a volatile, poisonous liquid that boils at 43°C and can therefore be removed from impurities. Nickel metal is then obtained by heating the pure nickel tetracarbonyl to about 200°C, when it decomposes.

Nickel's most common oxidation state is +2, and the green color of aqueous solutions of nickel salts is due to the presence of $[Ni(H_2O)_6]^{2+}$ ions.

The Period 4 d-block elements titanium through nickel are obtained chemically from their ores, with the ease of reduction increasing to the right in the periodic table. They have many industrial uses, particularly as alloys.

16.4 Groups 11 and 12

The elements close to the right-hand edge of the d block have full d-orbitals. Group 11 contains the **coinage metals**—copper, silver, and gold—which have $(n-1)d^{10}ns^1$ valence electron configurations (Table 16.3). Group 12 contains zinc, cadmium, and mercury, with valence configurations $(n-1)d^{10}ns^2$. The low reactivity of the coinage metals is partly due to the poor shielding abilities of the d-electrons and hence the tight grip that the nucleus can exert on the outermost electrons. This effect is enhanced in Period 6 by the lanthanide contraction, which helps to account for the inertness of gold.

TABLE 16.3 Properties of Elements in Groups 11 and 12

Z	Name	Symbol	Valence electron configuration	Melting point (°C)	Boiling point (°C)	Density, (g·cm^{-3})
29	copper	Cu	$3d^{1}4s^{1}$	1083	2567	8.93
47	silver	Ag	$4d^{10}5s^{1}$	962	2212	10.50
79	gold	Au	$5d^{10}6s^{2}$	1064	2807	19.28
30	zinc	Zn	$3d^{10}4s^{2}$	420	907	7.14
48	cadmium	Cd	$4d^{10}5s^{2}$	321	765	8.65
80	mercury	Hg	$5d^{10}6s^{2}$	−39	357	13.55

Copper, Cu, is unreactive enough for some to be found as the metal, but most is produced from its sulfides, particularly the ore *chalcopyrite*, CuFeS$_2$ (Fig. 16.10). The crushed and ground ore is separated from excess rock by *froth flotation*, a process that depends on the ability of sulfide ores to be wetted by oils but not by water. In this process, the powdered ore is combined with oil, water, and detergents (Fig. 16.11). Then air is blown through the mixture; the oil-coated sulfide mineral floats to the surface with the froth, and the unwanted copper-poor residue, which is called *gangue*, sinks to the bottom.

Processes for extracting metals from their ores are generally classified as **pyrometallurgical,** when high temperatures are used (Fig. 16.12), or **hydrometallurgical,** when aqueous solutions are used. Copper is extracted by both methods. In pyrometallurgical extraction, the enriched ore is roasted:

$$2\ CuFeS_2(s) + 3\ O_2(g) \xrightarrow{\Delta} 2\ CuS(s) + 2\ FeO(s) + 2\ SO_2(g)$$

The CuS is then *smelted*, a process in which metal ions are reduced by heating the ore with a reducing agent such as carbon (in the form of coke). At the same time, the sulfur is oxidized to SO$_2$ by blowing compressed air through the mixture of ore, limestone, and sand:

$$CuS(s) + O_2(g) \xrightarrow{\Delta} Cu(l) + SO_2(g)$$

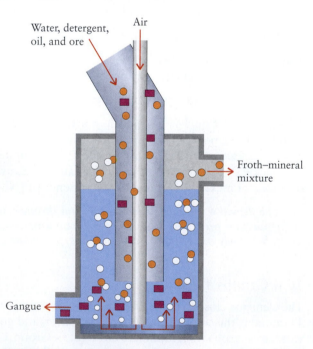

FIGURE 16.11 In the froth flotation process, a stream of bubbles (white circles) is passed through a mixture of ore (orange circles), rock (brown rectangles), and detergent. The ore is buoyed up by the froth of bubbles and is removed from the top of the chamber. The unwanted gangue is washed away through the bottom of the container.

FIGURE 16.10 Three important copper ores (from left to right): chalcopyrite, CuFeS$_2$; malachite, CuCO$_3$·Cu(OH)$_2$; and chalcocite, Cu$_2$S.

FIGURE 16.12 In this industrial-scale copper refinery, the molten impure copper produced by smelting is poured into molds. Next, the copper will be purified by electrolysis.

The copper product is known as *blister copper* because of the appearance of air bubbles in the solidified metal. In the hydrometallurgical process, soluble Cu^{2+} ions are formed by the action of sulfuric acid on the ores. Then the metal is obtained by reducing these ions in aqueous solution either electrolytically or chemically, by using an inexpensive reducing agent that has a more negative standard potential than that of copper, such as hydrogen or iron (see Section 14.3):

$$Cu^{2+}(aq) + H_2(g) \longrightarrow Cu(s) + 2 H^+(aq) \qquad \Delta G° = -65 \text{ kJ}$$

The reduction is thermodynamically favored, because the standard potential of the couple Cu^{2+}/Cu is positive ($E° = +0.34$ V). Metals with negative standard potentials, such as zinc ($E° = -0.76$ V) and nickel ($E° = -0.23$ V), cannot be extracted hydrometallurgically.

The impure copper from either process is refined electrolytically: it is made into anodes and plated onto cathodes of pure copper. Other metals may be present in the impure copper and those with highly positive electrode potentials also are reduced. The rare metals—most notably, platinum, silver, and gold—obtained from the anode sludge are sold to recover much of the cost of the electricity used in the electrolysis.

Copper alloys such as brass and bronze, which are harder and more resistant to corrosion than is copper, are important construction materials. Copper corrodes in moist air in the presence of oxygen and carbon dioxide:

$$2 Cu(s) + H_2O(l) + O_2(g) + CO_2(g) \longrightarrow Cu_2(OH)_2CO_3(s)$$

The pale green product is called *basic copper carbonate* and is responsible for the green patina of copper and bronze objects (Fig. 16.13). The patina adheres to the surface, protects the metal, and has a pleasing appearance.

Like all the coinage metals, copper forms compounds with oxidation number +1. However, in water, copper(I) salts disproportionate into metallic copper and copper(II) ions. The latter exist as pale blue $[Cu(H_2O)_6]^{2+}$ ions in water.

Copper is essential in animal metabolism. In some animals, such as the octopus and certain arthropods, it transports oxygen through the blood, a role performed by iron in mammals. As a result, the blood of these animals is green rather than red. In mammals, copper-bearing enzymes are necessary for healthy nerves and connective tissue.

Silver, Ag, is rarely found as the metal. Most is obtained as a by-product of the refining of copper and lead, and a considerable amount is recycled through the photographic industry. Silver has a positive standard potential, and so it does not

FIGURE 16.13 Copper corrodes in air to form an attractive pale green layer of basic copper carbonate. This patina, or incrustation, passivates the surface, which helps to protect it from further corrosion.

FIGURE 16.14 The color of commercial gold depends on its composition. Left to right: 8-carat gold, 14-carat gold, white gold, 18-carat gold, and 24-carat gold. White gold consists of 6 parts Au and 18 parts Ag by mass.

reduce $H^+(aq)$ to hydrogen. Silver reacts readily with sulfur and sulfur compounds, producing the familiar black tarnish on silver dishes and cutlery.

Silver(I) does not disproportionate in aqueous solution; and, in almost all its compounds, silver has oxidation number +1. Apart from silver nitrate, $AgNO_3$, and silver fluoride, silver salts are generally only sparingly soluble in water. Silver nitrate is the most important compound of silver and the starting point for the manufacture of silver halides for use in photography.

Gold, Au, is so inert that most of it is found in nature as the metal. Pure gold is classified as 24-carat gold. Its alloys with silver and copper, which differ in hardness and hue, are classified according to the proportion of gold that they contain (Fig. 16.14). For example, 10- and 14-carat golds contain, respectively, $\frac{10}{24}$ and $\frac{14}{24}$ parts by mass of gold. Gold is a highly malleable metal, and 1 g of gold can be worked into a leaf covering an area of about 1 m^2 or pulled out into a wire more than 2 km long. Gold leaf provides a valuable decorative protection, such as that on dishes and books.

Gold is too noble to react even with strong oxidizing agents such as nitric acid. Both the gold couples

$$Au^+(aq) + e^- \longrightarrow Au(s) \qquad E^\circ = +1.69 \text{ V}$$
$$Au^{3+}(aq) + 3 e^- \longrightarrow Au(s) \qquad E^\circ = +1.40 \text{ V}$$

lie above H^+/H_2 and $NO_3^-,H^+/NO,H_2O$:

$$NO_3^-(aq) + 4 H^+(aq) + 3 e^- \longrightarrow NO(g) + 2 H_2O(l) \qquad E^\circ = +0.96 \text{ V}$$

However, gold does react with *aqua regia*, a mixture of concentrated nitric and hydrochloric acids, because the complex ion $[AuCl_4]^-$ forms:

$$Au(s) + 6 H^+(aq) + 3 NO_3^-(aq) + 4 Cl^-(aq) \longrightarrow$$
$$[AuCl_4]^-(aq) + 3 NO_2(g) + 3 H_2O(l)$$

Even though the equilibrium constant for the formation of Au^{3+} from gold is very unfavorable, the reaction proceeds because any Au^{3+} ions formed are immediately complexed by Cl^- ions and removed from the equilibrium. In a process widely used in the refining of the metal, gold also reacts with sodium cyanide in an aerated aqueous solution to form the complex ion $[Au(CN)_2]^-$:

$$4 Au(s) + 8 NaCN(aq) + O_2(aq) + 2 H_2O(l) \longrightarrow$$
$$4 Na[Au(CN)_2](aq) + 4 NaOH(aq)$$

Zinc, Zn, is found mainly as its sulfide, ZnS, in *sphalerite*, often in association with lead ores (see Fig. 15.11). The ore is concentrated by froth flotation, and the metal is extracted by roasting and then smelting with coke:

$$2 ZnS(s) + 3 O_2(g) \xrightarrow{\Delta} 2 ZnO(s) + 2 SO_2(g)$$
$$ZnO(s) + C(s) \xrightarrow{\Delta} Zn(l) + CO(g)$$

Zinc is used mainly for galvanizing iron; like copper, it is protected by a hard film of basic carbonate, $Zn_2(OH)_2CO_3$, which forms on contact with air.

Zinc and cadmium are both silvery, reactive metals that are similar to each other but differ sharply from mercury. Zinc is amphoteric (like its main-group neighbor aluminum). It reacts with acids to form Zn^{2+} ions and with alkalis to form the zincate ion, $[Zn(OH)_4]^{2-}$:

$$Zn(s) + 2 OH^-(aq) + 2 H_2O(l) \longrightarrow [Zn(OH)_4]^{2-}(aq) + H_2(g)$$

Galvanized containers should therefore not be used for transporting alkalis. Cadmium, which is lower down the group and is more metallic, has a more basic oxide.

Zinc and cadmium have an oxidation number of +2 in all their compounds. Zinc is an essential element for human health. It is present in many enzymes and plays a role in the expression of DNA and in growth. Zinc is toxic only in very high amounts. However, cadmium is a deadly poison that disrupts metabolism by substituting for other essential metals in the body such as zinc and calcium, leading to soft bones and to kidney and lung disorders.

Mercury, Hg, occurs mainly as HgS in the mineral *cinnabar* (see Fig. 15.11), from which it is separated by froth flotation and then roasting in air:

$$HgS(s) + O_2(g) \xrightarrow{\Delta} Hg(g) + SO_2(g)$$

The volatile metal is separated by distillation and condensed. Mercury is the only metallic element that is liquid at room temperature (gallium and cesium are liquids on warm days). It has a long liquid range, from its melting point of $-39°C$ to its boiling point of $357°C$, and so it is well suited for its use in thermometers, silent electrical switches, and high-vacuum pumps.

Because mercury lies above hydrogen in the electrochemical series, it is not oxidized by hydrogen ions. However, it does react with nitric acid:

$$3\ Hg(l) + 8\ H^+(aq) + 2\ NO_3^-(aq) \longrightarrow 3\ Hg^{2+}(aq) + 2\ NO(g) + 4\ H_2O(l)$$

In compounds, mercury has the oxidation number $+1$ or $+2$. Its compounds with oxidation number $+1$ are unusual in that the mercury(I) cation is the covalently bonded diatomic ion $(Hg—Hg)^{2+}$, written Hg_2^{2+}.

Compounds containing mercury, particularly its organic compounds, are acutely poisonous. Mercury vapor is an insidious poison because its effect is cumulative. Frequent exposure to low levels of mercury vapor can allow mercury to accumulate in the body. The effects include impaired neurological function, hearing loss, and other ailments.

SELF-TEST 16.2A Use standard Gibbs free energies of formation to calculate $\Delta G°$ at 298 K for the reaction $CuS(s) + O_2(g) \rightarrow Cu(s) + SO_2(g)$. ($\Delta G_f°(CuS) = -49.0\ kJ \cdot mol^{-1}$.)

[*Answer:* $\Delta G° = -251.2$ kJ]

SELF-TEST 16.2B Calculate $E°$ for a cell powered by the reaction of mercury metal and nitric acid to form mercury(I) and NO.

Metals in Groups 11 and 12 are easily reduced from their compounds and have low reactivity as a result of poor shielding of the nuclear charge by the d-electrons. Copper is extracted from its ores by either pyrometallurgical or hydrometallurgical processes.

COORDINATION COMPOUNDS

Many of the *d*-block elements form characteristically colored solutions in water. For example, although solid copper(II) chloride is brown and copper(II) bromide is black, their aqueous solutions are both light blue. The blue color is due to the hydrated copper(II) ions, $[Cu(H_2O)_6]^{2+}$, that form when the solids dissolve. As the formula suggests, these hydrated ions have a specific composition; they also have definite shapes and properties. They can be regarded as the outcome of a reaction in which the water molecules act as Lewis bases (electron pair donors, Section 10.2) and the Cu^{2+} ion acts as a Lewis acid (an electron pair acceptor). This type of Lewis acid–base reaction is characteristic of many cations of *d*-block elements.

The hydrated ion $[Cu(H_2O)_6]^{2+}$ is an example of a **complex**, a species consisting of a central metal atom or ion to which a number of molecules or ions are attached by coordinate covalent bonds. A **coordination compound** is an electrically neutral compound in which at least one of the ions present is a complex. However, the terms *coordination compound* (the overall neutral compound) and *complex* (one or more of the ions or neutral species present in the compound) are often used interchangeably. Coordination compounds include complexes in which the central metal atom is electrically neutral, such as $Ni(CO)_4$, and ionic compounds, such as $K_4[Fe(CN)_6]$.

Much research focuses on the structures, properties, and uses of the complexes formed between *d*-metal ions acting as Lewis acids and a variety of Lewis bases, partly because they participate in many biological reactions. Hemoglobin and vitamin B_{12}, for example, are both complexes—the former of iron and the latter of cobalt (Box 16.1). Complexes of the *d*-metals are often brightly colored and magnetic and are used in chemistry for analysis, to dissolve ions (Section 11.13), in the

The formation of coordinate covalent bonds is described in Sections 2.11 and 10.2.

BOX 16.1   WHAT HAS THIS TO DO WITH . . . STAYING ALIVE?

Why We Need to Eat *d*-Metals

Some of the critical enzymes in our cells are *metalloproteins*, large organic molecules made up of folded polymerized chains of amino acids that also include at least one metal atom. These metalloproteins are intensely studied by biochemists, because they control life and protect against disease. They have also been used to trace evolutionary paths. The *d*-block metals catalyze redox reactions, form components of membrane, muscle, skin, and bone, catalyze acid–base reactions, control the flow of energy and oxygen, and carry out nitrogen fixation.

Hemoglobin, in which an iron atom lies as iron(II) at the center of a heme group, is the most familiar metalloprotein. Four nitrogen atoms from amine groups in the heme serve as the ligands in a square planar arrangement. The oxygen molecule serves as a fifth ligand, attaching itself directly to the iron atom and producing a modified square pyramidal shape about the iron atom. Carbon monoxide forms a much stronger bond to the iron atom than oxygen does—hence its high toxicity: the attached CO ligand prevents an O_2 molecule from attaching to the iron, and the victim suffocates.

Cobalt is a *d*-metal needed to prevent pernicious anemia and some kinds of mental illness. It is an essential part of a coenzyme required for the activity of vitamin B_{12} (which is also called cobalamin) and gives the vitamin its red color. The cobalt atom is found in an octahedral complex in which five of the ligands are attached through nitrogen atoms from organic amine groups and one ligand attaches through a —CH_2— group. Cobalamin is the only biomolecule known to have a metal–carbon bond. The ease

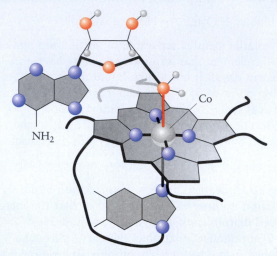

In cobalamin, vitamin B_{12}, one of the six ligands forming an octahedral structure around a cobalt atom is an organic molecule attached through a carbon–cobalt bond (red). The bond is weak and easily broken.

with which this bond is broken and the ability of the cobalt ion to change from one oxidation state to another are responsible for the importance of cobalamin as a biological catalyst.

Zinc enzymes play several important roles in metabolism, including the expression of our genes, the digestion of food, the storage of insulin, and the building of collagen. In fact, zinc has so many functions in our systems that it has been called a "master hormone." Its concentration in our bodies is about the same as that of iron, but the intracellular concentration of zinc declines with aging. It has even been suggested that aging is actually a result of the decline in intracellular zinc concentration. However, this decline cannot be slowed by eating more zinc; intracellular zinc concentrations are controlled by enzymes, not by the lack of zinc in the diet.

Other *d*-metals are also vital to health. For example, chromium(III) plays a role in the regulation of glucose metabolism. Copper(I) is an essential nutrient for healthy cells and is the only biologically available Lewis acid with a +1 charge.

Related Exercise: 16.109.

For Further Reading: J. J. R. Fraústo da Silva and R. J. P. Williams, *The Biological Chemistry of the Elements: The Inorganic Chemistry of Life* (Oxford: Oxford University Press, 1991). M. E. Wastney, W. A. House, R. M. Barnes, and K. N. S. Subramanian, "Kinetics of zinc metabolism: variation with diet, genetics and disease," *Journal of Nutrition*, vol. 130, 2000, pp. 1355S–1359S.

Animation: Box 16.1 Oxygenation of hemoglobin

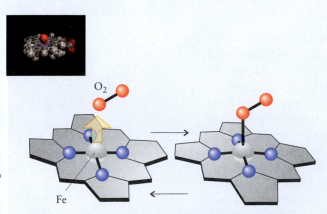

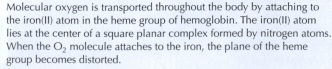

Molecular oxygen is transported throughout the body by attaching to the iron(II) atom in the heme group of hemoglobin. The iron(II) atom lies at the center of a square planar complex formed by nitrogen atoms. When the O_2 molecule attaches to the iron, the plane of the heme group becomes distorted.

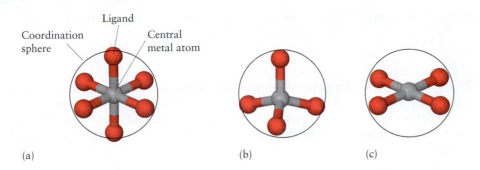

Coordination sphere
Ligand
Central metal atom

(a)   (b)   (c)

FIGURE 16.15 (a) Almost all six-coordinate complexes are octahedral. Four-coordinate complexes are either (b) tetrahedral or (c) square planar.

electroplating of metals, and for catalysis. They are also the target of current research in solar-energy conversion, in atmospheric nitrogen fixation, and in pharmaceuticals.

16.5 Coordination Complexes

The Lewis bases attached to the central metal atom or ion in a *d*-metal complex are known as **ligands**; they can be either ions or molecules. An example of an ionic ligand is the cyanide ion. In the hexacyanoferrate(II) ion, $[Fe(CN)_6]^{4-}$, the CN^- ions provide the electron pairs that form bonds to the Lewis acid Fe^{2+}. In the neutral complex $Ni(CO)_4$, the Ni atom acts as the Lewis acid and the ligands are the CO molecules.

The name "ligand" comes from the Latin word meaning "to bind."

Each ligand in a complex has at least one lone pair of electrons with which it bonds to the central atom or ion by forming a coordinate covalent bond. We say that the ligands **coordinate** to the metal when they form the complex in this way. In the chemical formula for a complex, we usually represent the ligands directly attached to the central ion by enclosing them and the central ion within rectangular brackets (the exceptions are neutral complexes). These ligands make up the **coordination sphere** of the central ion. The number of points at which ligands are attached to the central metal atom is called the **coordination number** of the complex (Fig. 16.15). The coordination number is 4 in $Ni(CO)_4$ and 6 in $[Fe(CN)_6]^{4-}$.

Because water is a Lewis base, it forms complexes with most *d*-block ions when they dissolve in it. Aqueous solutions of *d*-metal ions are usually solutions of their H_2O complexes: $Fe^{2+}(aq)$, for instance, is more accurately $[Fe(H_2O)_6]^{2+}$. Many complexes are prepared simply by mixing aqueous solutions of a *d*-metal ion with the appropriate Lewis base (Fig. 16.16); for example,

$$[Fe(H_2O)_6]^{2+}(aq) + 6\,CN^-(aq) \longrightarrow [Fe(CN)_6]^{4-}(aq) + 6\,H_2O(l)$$

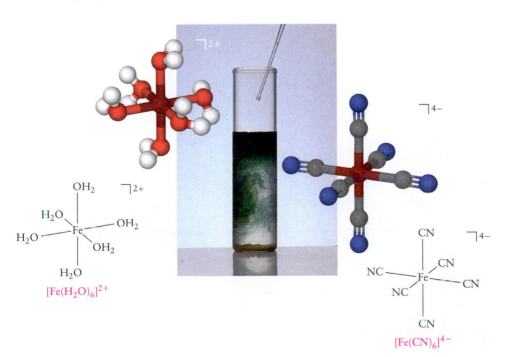

$[Fe(H_2O)_6]^{2+}$

$[Fe(CN)_6]^{4-}$

FIGURE 16.16 When potassium cyanide is added to a solution of iron(II) sulfate, the cyanide ions replace the H_2O ligands of the $[Fe(H_2O)_6]^{2+}$ complex (left) and produce a new complex, the hexacyanoferrate(II) ion, $[Fe(CN)_6]^{4-}$ (right). The blue color is due to the polymeric compound called *Prussian blue*, which forms from the cyanoferrate ion.

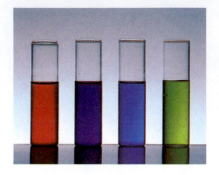

FIGURE 16.17 Some of the highly colored *d*-metal complexes. Left to right: aqueous solutions of $[Fe(SCN)(H_2O)_5]^{2+}$, $[Co(SCN)_4(H_2O)_2]^{2-}$, $[Cu(NH_3)_4(H_2O)_2]^{2+}$, and $[CuBr_4]^{2-}$.

This is an example of a **substitution reaction,** a reaction in which one Lewis base takes the place of another. Here the CN^- ions drive out H_2O molecules from the coordination sphere of the $[Fe(H_2O)_6]^{2+}$ complex and take their place. Replacement is less complete when certain other ions, such as Cl^-, are added to an iron(II) solution:

$$[Fe(H_2O)_6]^{2+}(aq) + Cl^-(aq) \longrightarrow [FeCl(H_2O)_5]^+(aq) + H_2O(l)$$

Because the color of a *d*-metal complex depends on the identity of the ligands as well as that of the metal, impressive changes in color often accompany substitution reactions (Fig. 16.17).

The names of coordination compounds can become awesomely long because the identity and number of each type of ligand must be included. In most cases, chemists avoid the problem by using the chemical formula rather than the name itself. For instance, it is much easier to refer to $[FeCl(H_2O)_5]^+$ than to pentaaquachloroiron(II) ion, its formal name. However, names are sometimes needed, and they can be constructed and interpreted, in simple cases at least, by using the rules set out in Toolbox 16.1. Table 16.4 gives the names of common ligands and their abbreviations, which are used in the formulas of complexes.

TABLE 16.4 Common Ligands

Formula*	Name
Neutral ligands	
H_2O	aqua
NH_3	ammine
NO	nitrosyl
CO	carbonyl
$NH_2CH_2CH_2NH_2$	ethylenediamine (en)[†]
$NH_2CH_2CH_2NHCH_2CH_2NH_2$	diethylenetriamine (dien)[‡]
Anionic ligands	
F^-	fluoro
Cl^-	chloro
Br^-	bromo
I^-	iodo
OH^-	hydroxo
O^{2-}	oxo
CN^-	cyano
CN^-	isocyano
NCS^-	thiocyanato
NCS^-	isothiocyanato
NO_2^- as ONO^-	nitrito
NO_2^- as NO_2^-	nitro
CO_3^{2-} as OCO_2^{2-}	carbonato
$C_2O_4^{2-}$ as $^-OOC{-}COO^-$	oxalato (ox)[†]
	ethylenediaminetetraacetato (edta)[§]
SO_4^{2-} as OSO_3^{2-}	sulfato

*Ligand atoms that bond to the metal atom are in red.
[†]Bidentate (attaches to two sites).
[‡]Tridentate (attaches to three sites).
[§]Hexadentate (attaches to six sites).

TOOLBOX 16.1   HOW TO NAME *d*-METAL COMPLEXES AND COORDINATION COMPOUNDS

CONCEPTUAL BASIS

The aim of nomenclature is to be succinct but unambiguous. *d*-Metal complexes are identified by giving the names and numbers of the individual ligands. Because some names can be quite long, interpreting them is rather like eating a large bun: nibble it bite by bite, don't try to swallow it in one gulp.

PROCEDURE

The following rules are adequate for most common complexes; more elaborate rules are needed if the complex contains more than one metal atom.

1 Name the ligands first and then the metal atom or ion.

2 Neutral ligands, such as $H_2NCH_2CH_2NH_2$ (ethylenediamine), have the same name as the molecule, except for H_2O (aqua), NH_3 (ammine), CO (carbonyl), and NO (nitrosyl).

3 Anionic ligands end in -o; for anions that end in -ide (such as chloride), -ate (such as sulfate), and -ite (such as nitrite), change the endings as follows:

-ide ⟶ -o   -ate ⟶ -ato   -ite ⟶ -ito

Examples: chloro, sulfato, and nitrito.

4 Greek prefixes indicate the number of each type of ligand in the complex ion:

2	3	4	5	6	. . .
di-	tri-	tetra-	penta-	hexa-	. . .

If the ligand already contains a Greek prefix (as in ethylenediamine) or if it is polydentate (able to attach at more than one binding site), then the following prefixes are used instead:

2	3	4	. . .
bis-	tris-	tetrakis-	. . .

5 Ligands are named in alphabetical order, ignoring the Greek prefix that indicates the number of each one present.

6 The chemical symbols of anionic ligands precede those of neutral ligands in the chemical formula of the complex (but not necessarily in its name). Thus, Cl^- precedes H_2O and NH_3.

7 A Roman numeral denotes the oxidation number of the central metal ion:

$[FeCl(H_2O)_5]^+$ pentaaquachloroiron(II) ion

$[CrCl_2(NH_3)_4]^+$ tetraamminedichlorochromium(III) ion

$[Co(en)_3]^{3+}$ tris(ethylenediamine)cobalt(III) ion

(Note that the Cl_2 in the second formula denotes two Cl^- ligands, not one Cl_2 ligand.)

8 If the complex has an overall negative charge (an anionic complex), the suffix -ate is added to the stem of the metal's name. If the symbol of the metal originates from a Latin name (as listed in Appendix 2D), then the Latin stem is used. For example, the symbol for iron is Fe, from the Latin *ferrum*. Therefore, any anionic complex of iron ends with -ferrate followed by the oxidation number of the metal in Roman numerals:

$[Fe(CN)_6]^{4-}$ hexacyanoferrate(II) ion

$[Ni(CN)_4]^{2-}$ tetracyanonickelate(II) ion

9 The name of a coordination compound (as distinct from a complex cation or anion) is built in the same way as that of a simple compound, with the cation named before the anion:

$NH_4[PtCl_3(NH_3)]$
            ammonium amminetrichloroplatinate(II)

$[Cr(OH)_2(NH_3)_4]Br$
            tetraamminedihydroxochromium(III) bromide

This procedure is illustrated in Example 16.1.

EXAMPLE 16.1   Sample exercise: Naming complexes and coordination compounds

(a) Name the coordination compound $[Co(NH_3)_3(H_2O)_3]_2(SO_4)_3$. (b) Write the formula of sodium dichlorobis(oxalato)platinate(IV).

SOLUTION   We apply the rules in Toolbox 16.1.

(a) There are three SO_4^{2-} ions for every two complex ions.

The complex cation must have a charge of +3: $[Co(NH_3)_3(H_2O)_3]^{3+}$.

All the ligands are neutral.

Co is present as cobalt(III).

There are three NH_3 molecules (ammine) and three H_2O molecules (aqua).

The cation is triamminetriaquacobalt(III), and the compound is triamminetriaquacobalt(III) sulfate.

(b) Two Cl^- ligands and two $C_2O_4^{2-}$ ions are attached to Pt^{4+}.

From Table 16.4, the symbol for oxalate is ox.

The charge on the complex is −2.

The complex anion is $[PtCl_2(ox)_2]^{2-}$.

The compound is $Na_2[PtCl_2(ox)_2]$.

SELF-TEST 16.3A (a) Name the compound $[FeOH(H_2O)_5]Cl_2$. (b) Write the formula of potassium diaquabis(oxalato)chromate(II).

[*Answer:* (a) Pentaaquahydroxoiron(III) chloride; (b) $K_2[Cr(ox)_2(H_2O)_2]$]

SELF-TEST 16.3B (a) Name the compound $[CoBr(NH_3)_5]SO_4$. (b) Write the formula of tetraamminediaquachromium(III) bromide.

A complex is formed between a Lewis acid (the metal atom or ion) and a number of Lewis bases (the ligands).

16.6 The Shapes of Complexes

The richness of coordination chemistry is enhanced by the variety of shapes that complexes can adopt. The most common complexes have coordination number 6. Almost all these species have their ligands at the vertices of a regular octahedron, with the metal ion at the center, and are called **octahedral complexes** (**1**). An example of an octahedral complex is the hexacyanoferrate(II) ion, $[Fe(CN)_6]^{4-}$.

The next most common coordination number is 4. Two shapes are typically found for this coordination number. In a **tetrahedral complex,** the four ligands are found at the vertices of a tetrahedron, as in the tetrachlorocobaltate(II) ion, $[CoCl_4]^{2-}$ (**2**). An alternative arrangement, most notably for atoms and ions with d^8 electron configurations such as Pt^{2+} and Au^{3+}, is for the ligands to lie at the corners of a square, giving a **square planar complex** (**3**).

Many other shapes are possible for complexes. The simplest are linear, with coordination number 2. An example is dimethylmercury(0), $Hg(CH_3)_2$ (**4**), which is a toxic compound formed by bacterial action on aqueous solutions of Hg^{2+} ions. Coordination numbers as high as 12 are found for members of the *f* block, but they are rare in the *d* block. One interesting type of *d*-metal compound in which there are 10 links between the ligands and the central metal ion is ferrocene, dicyclopentadienyliron(0), $[Fe(C_5H_5)_2]$ (**5**). Ferrocene is an aptly named "sandwich compound," with the two planar cyclopentadienyl ligands the "bread" and the metal atom the "filling." The formal name for a sandwich compound is a **metallocene.**

Complexes of molybdenum and tungsten with eight ligands are known. These complexes have antiprismatic (**6**) or dodecahedral shapes (**7**). However, complexes with more than six ligands are rare.

Some ligands are **polydentate** ("many toothed") and can occupy more than one binding site simultaneously. Each end of the two-toothed (that is, bidentate)

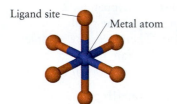

Ligand site ——— Metal atom

1   An octahedral complex

Ligand

2   Tetrahedral complex

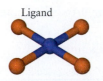

Ligand

3   Square planar complex

4   Dimethylmercury(0), $Hg(CH_3)_2$

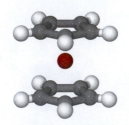

5   Ferrocene, $Fe(C_5H_5)_2$

Ligand

6   Square antiprism

7   Dodecahedral complex

ethylenediamine molecule, $NH_2CH_2CH_2NH_2$ (**8**), has a nitrogen atom with a lone pair of electrons. This ligand is widely used in coordination chemistry and is abbreviated to en, as in tris(ethylenediamine)cobalt(III), $[Co(en)_3]^{3+}$ (**9**). The metal ion in $[Co(en)_3]^{3+}$ lies at the center of the three ligands as though pinched by three molecular claws. It is an example of a **chelate** (from the Greek word for "claw"), a complex containing one or more ligands that form a ring of atoms that includes the central metal atom. There are few hexadentate ligands, but a common example is the ethylenediaminetetraacetate ion, edta (the fully protonated acid is shown as **10**; the green arrows show the points of attachment). This ligand forms complexes with many metal ions, including Pb^{2+} (**11**); hence, it is used as an antidote to lead poisoning.

The production of some chelates releases toxic chemicals such as cyanides to the environment. However, new types of chelates that **sequester** d-block metals, bind with them and remove them from solution, may provide solutions to some of society's trickiest environmental problems. For example, the chelating agent sodium iminodisuccinate, which contains the hexadentate iminodisuccinate ion (**12**), can scavenge metal ions from solutions used to develop photographs and from drinking water and serves as a nontoxic additive to detergents. It is rapidly degraded to nontoxic products in the environment. Other environmentally friendly chelates accelerate the action of hydrogen peroxide and the combination is replacing chlorine bleaches in the production of paper, greatly reducing the release of toxic pollutants to the environment.

Chelating ligands are quite common in nature. Mosses and lichens secrete chelating ligands to extract essential metal ions from the rocks on which they dwell. Chelate formation also lies behind the body's strategy of producing a fever when infected by bacteria. The higher temperature kills bacteria by reducing their ability to synthesize a particular iron-chelating ligand.

Complexes with coordination number 6 tend to be octahedral; those with coordination number 4 are either tetrahedral or square planar. Polydentate ligands can form chelates.

16.7 Isomers

Many complexes and coordination compounds exist as **isomers,** compounds that contain the same numbers of the same atoms but in different arrangements. For example, the ions shown in (**13a**) and (**13b**) differ only in the positions of the Cl^- ligands, but they are distinct species, because they have different physical and chemical properties. Isomerism is of more than academic interest: for example, anticancer drugs based on complexes of platinum are active only if they are the correct isomer. The complex needs to have a particular shape to interact with DNA molecules.

8   Ethylenediamine, $NH_2CH_2CH_2NH_2$

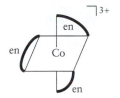

9 $[Co(en)_3]^{3+}$

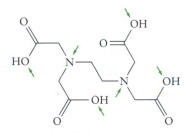

10   Ethylenediaminetetraacetic acid

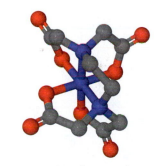

11   An edta complex

12   Iminodisuccinate ion

Isomerism is very important in organic chemistry, and the topic is developed again in Section 18.2.

(a)   *trans*-$[CoCl_2(NH_3)_4]^+$ **(b)**   *cis*-$[CoCl_2(NH_3)_4]^+$

13

Figure 16.18 summarizes the types of isomerism found in coordination complexes. The two major classes of isomers are **structural isomers,** in which the atoms are connected to different partners, and **stereoisomers,** in which the atoms have the same partners but are arranged differently in space. Structural isomers of coordination compounds are subdivided into ionization, hydrate, linkage, and coordination isomers.

FIGURE 16.18 The various types of isomerism in coordination compounds.

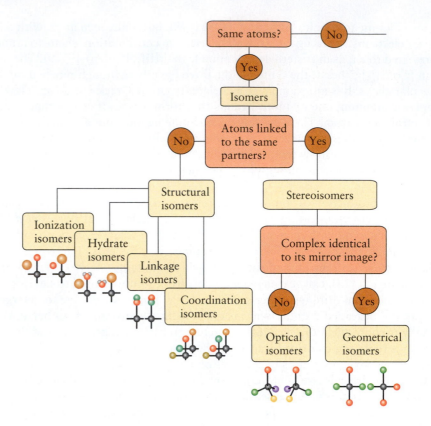

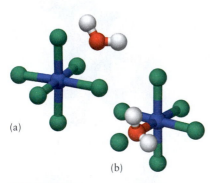

(a)

(b)

FIGURE 16.19 Hydrate isomers. In part (a), the water molecule is simply part of the surrounding solvent; in part (b), the water molecule is present in the coordination sphere and a ligand (green sphere) is now present in the solution.

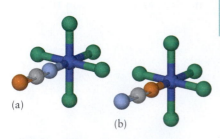

(a)

(b)

FIGURE 16.20 Linkage isomers. In part (a), the ligand (here NCS⁻) is attached through its N atom; but in part (b), it is attached through its S atom.

Ionization isomers differ by the exchange of a ligand with an anion or a neutral molecule outside the coordination sphere. For instance, $[CoBr(NH_3)_5]SO_4$ and $[CoSO_4(NH_3)_5]Br$ are ionization isomers because the Br^- ion is a ligand of the cobalt in the former but an accompanying anion in the latter. The isomers can be distinguished by their chemical properties, because an ion inside the complex is not available for reaction. Thus, the addition of a barium salt will result in the precipitation of barium sulfate from a solution of $[CoBr(NH_3)_5]SO_4$, but not from a solution of $[CoSO_4(NH_3)_5]Br$.

Hydrate isomers differ by the exchange of an H_2O molecule with another ligand in the coordination sphere (Fig. 16.19). For example, the solid hexahydrate of chromium(III) chloride, $CrCl_3 \cdot 6H_2O$, may also be any of the three compounds $[Cr(H_2O)_6]Cl_3$, $[CrCl(H_2O)_5]Cl_2 \cdot H_2O$, or $[CrCl_2(H_2O)_4]Cl \cdot 2H_2O$. Hydrate isomers can often be distinguished by the stoichiometry of reactions in which the ion is exchanged with water. For example, 2 mol AgCl can be precipitated from 1 mol $[CrCl(H_2O)_5]Cl_2 \cdot H_2O$, but only 1 mol AgCl can be precipitated from 1 mol $[CrCl_2(H_2O)_4]Cl \cdot 2H_2O$.

SELF-TEST 16.4A When excess silver nitrate is added to 0.0010 mol $CrCl_3 \cdot 6H_2O$ in aqueous solution, 0.0010 mol AgCl is formed. Which hydrate isomer is present?

[Answer: $[CrCl_2(H_2O)_4]Cl \cdot 2H_2O]$

SELF-TEST 16.4B When excess silver nitrate is added to 0.0010 mol $CrCl_3 \cdot 6H_2O$ in aqueous solution, 0.0030 mol AgCl is formed. Which hydrate isomer is present?

Linkage isomers differ in the identity of the atom used by a given ligand to attach to the metal ion (Fig. 16.20). Common ligands that show linkage isomerism are SCN^- versus NCS^-, NO_2^- versus ONO^-, and CN^- versus NC^-, where the coordinating atom is written first in each pair. For example, NO_2^- can form $[CoCl(NO_2)(NH_3)_4]^+$ and $[CoCl(ONO)(NH_3)_4]^+$. The name used to specify the ligand is different in each case. For instance, nitro signifies that the ligand is linked through the N atom and nitrito that it is linked through an O atom. Table 16.4 lists the names to use for these so-called **ambidentate ligands** that can attach through atoms of different elements.

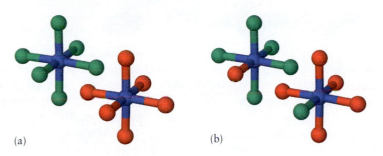

FIGURE 16.21 The compounds in parts (a) and (b) are coordination isomers. In these compounds, a ligand has been exchanged between the cationic and anionic complexes.

(a)   (b)

Coordination isomers differ by the exchange of one or more ligands between a cationic complex and an anionic complex (Fig. 16.21). Thus, $[Cr(NH_3)_6][Fe(CN)_6]$ and $[Fe(NH_3)_6][Cr(CN)_6]$ are coordination isomers.

SELF-TEST 16.5A Identify the type of isomer represented by each of the following pairs: (a) $[Cu(NH_3)_4][PtCl_4]$ and $[Pt(NH_3)_4][CuCl_4]$; (b) $[Cr(OH)_2(NH_3)_4]Br$ and $[CrBr(OH)(NH_3)_4]OH$.

[*Answer:* (a) Coordination; (b) ionization]

SELF-TEST 16.5B Identify the type of isomer represented by each of the following pairs: (a) $[Co(NCS)(NH_3)_5]Cl_2$ and $[Co(SCN)(NH_3)_5]Cl_2$; (b) $[CrCl(H_2O)_5]Cl_2 \cdot H_2O$ and $[CrCl_2(H_2O)_4]Cl \cdot 2H_2O$.

Although they are built from the same numbers and kinds of atoms, *structural isomers* have different chemical formulas, because the formulas show how the atoms are grouped in or outside the coordination sphere. *Stereoisomers*, on the other hand, have the same formulas, because their atoms have the same partners in the coordination spheres; only the spatial arrangement of the ligands differs. There are two types of stereoisomerism, geometrical and optical.

In **geometrical isomers**, atoms are bonded to the same neighbors but have different locations relative to each other, as in (**13a**) and (**13b**): the complex with the Cl^- ligands on opposite sides of the central atom is called the *trans isomer,* and the complex with the ligands on the same side is called the *cis isomer.* Geometrical isomers can occur for square planar and octahedral complexes but not for tetrahedral complexes, because in the latter any pair of vertices is equivalent to any other pair. The chemical and physiological properties of geometrical isomers can differ greatly. For example, *cis*-$[PtCl_2(NH_3)_2]$ (**14a**) is used for chemotherapy treatment of cancer patients, but *trans*-$[PtCl_2(NH_3)_2]$ (**14b**) is therapeutically inactive. **Optical isomers** are nonsuperimposable mirror images of each other (Fig. 16.22). Both geometrical and optical isomerism can occur in an octahedral complex, as in $[CoCl_2(en)_2]^+$: the trans isomer (**15a**) is green, and the two alternative cis isomers (**15b**) and (**15c**), which are optical isomers of one another, are violet. Optical isomers can also occur whenever four different groups form a tetrahedral complex, but not if they form a square planar complex.

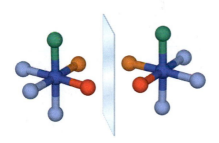

(a) *cis*-$[PtCl_2(NH_3)_2]$

(b) *trans*-$[PtCl_2(NH_3)_2]$

14

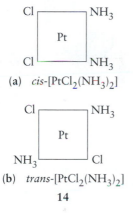

FIGURE 16.22 Optical isomers. The two complexes are each other's mirror image; no matter how we rotate them, one complex cannot be superimposed on the other.

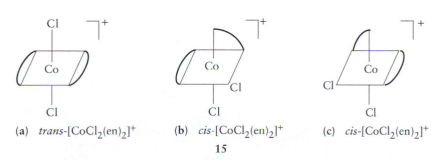

(a) *trans*-$[CoCl_2(en)_2]^+$ (b) *cis*-$[CoCl_2(en)_2]^+$ (c) *cis*-$[CoCl_2(en)_2]^+$

15

A **chiral complex** is one that is not identical to its mirror image. Thus, all optical isomers are chiral. The cis isomers of $[CoCl_2(en)_2]^+$ are chiral, and a chiral complex and its mirror image form a pair of **enantiomers**. The trans isomer is superimposable on its mirror image; complexes with this property are called **achiral**. Enantiomers differ in one physical property: chiral molecules display

The name *enantiomer* comes from the Greek words meaning "both parts."

BOX 16.2 HOW DO WE KNOW . . . THAT A COMPLEX IS OPTICALLY ACTIVE?

The electric field of plane-polarized light oscillates in a single plane. It can be prepared by passing ordinary, unpolarized light through a polarizer, which consists of a material that allows the light to pass only if the electric field is aligned in a certain direction.

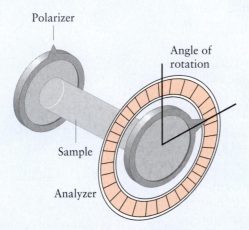

A polarimeter is used to determine the optical activity of a substance by measuring the angle through which plane-polarized light is rotated by a sample.

An optically active substance, such as a chiral complex, rotates the plane of polarization of a beam of light. The light is passed through a sample cell about 10 cm long. To detect chirality, a solution of the chiral complex is placed in the cell. When the light emerges from the far end of the cell, the angle of its plane of polarization may have rotated from its original angle. To determine the angle, the light is passed through an analyzer that contains another polarizing filter. The filter is rotated until the intensity of the light that has passed through the polarizer, sample, and filter reaches its maximum. The angle of the plane of polarization is determined from the angle through which the filter was rotated to achieve this maximum setting. If the sample is not optically active, the light is not rotated by the sample and the maximum intensity is observed at an angle of 0°. The sample is optically active if the angle of rotation is different from 0°. The actual value depends on the identity of the complex, its concentration, the wavelength of the light, and the length of the sample cell.

The determination of the angle of rotation is called *polarimetry*. In some cases, it can help a chemist follow a reaction. For example, if a reaction destroys the chirality of a complex, then the angle of optical rotation decreases with time as the concentration of the complex falls.

optical activity, the ability to rotate the plane of polarization of light (Box 16.2). In ordinary light, the plane of wave motion lies at random orientations around the direction of travel. In plane-polarized light, the wave lies in a single plane (Fig. 16.23). Plane-polarized light can be prepared by passing ordinary light through a special filter, such as the material used to make polarized sunglasses. One enantiomer of a chiral complex rotates the plane of polarization clockwise; its mirror-image partner rotates it by the same amount the other way. Achiral complexes are not optically active: they do not rotate the plane of polarization of polarized light.

Some complexes are synthesized in the laboratory as **racemic mixtures,** or mixtures of enantiomers in equal proportions. Because enantiomers rotate the

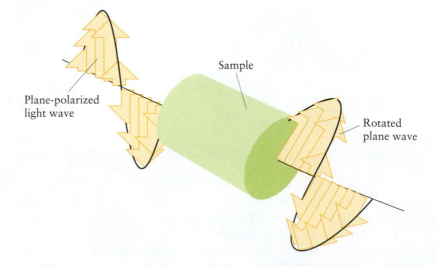

FIGURE 16.23 Plane-polarized light consists of radiation in which all the wave motion lies in one plane (as represented by the arrows on the left). When such light passes through a solution of an optically active substance, the plane of the polarization is rotated through a characteristic angle that depends on the identity and concentration of the solute and the length of the path through the sample (right).

plane of polarization of light in opposite directions, a racemic sample is not optically active.

EXAMPLE 16.2 Identifying optical isomerism

Which of the following complexes are chiral, and which form enantiomeric pairs?

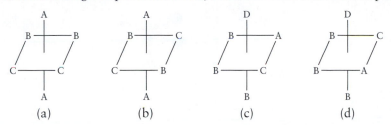

(a)   (b)   (c)   (d)

STRATEGY Draw the mirror image of each complex and mentally rotate it; judge whether any rotation will allow the mirror image to be superimposed on the original molecule. If not, then the complex is chiral. Determine which complexes form enantiomeric pairs by finding pairs in which the two complexes are the nonsuperimposable mirror images of each other. If imagining the three-dimensional structure is difficult, build simple paper models of the complexes.

SOLUTION The mirror image of each complex is shown on the right of each pair.

(a) Rotating the mirror image about A—A gives a structure identical to the original, and so superimposable on it.

Not chiral

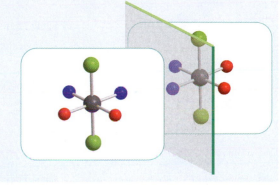

(b) The mirror image is superimposable on the original

Not chiral

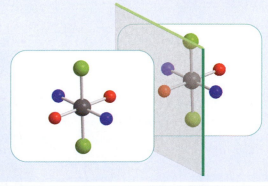

(c) No rotation will allow the complex to be superimposed on its mirror image.

Chiral

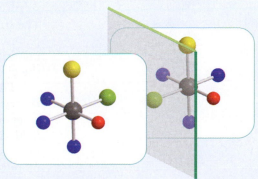

(d) No rotation will allow the complex to be superimposed on its mirror image.

Chiral

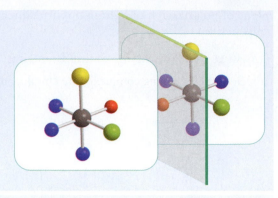

When the mirror image of (c) is rotated by 90° around the vertical B–D axis, it becomes the complex (d).

(c) and (d) are a pair of enantiomers

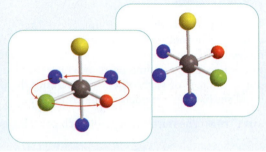

SELF-TEST 16.6A Repeat Example 16.2 for the following complexes:

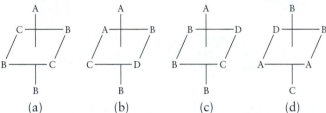

(a)   (b)   (c)   (d)

[*Answer:* (a) Not chiral; (c) chiral; (b, d) chiral and enantiomeric]

SELF-TEST 16.6B Repeat Example 16.2 for the following complexes:

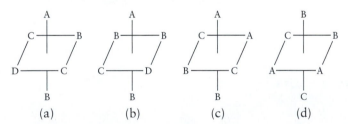

(a)   (b)   (c)   (d)

The varieties of isomerism are summarized in Fig. 16.18. Enantiomeric pairs of optical isomers rotate the plane of polarization of light in opposite directions.

THE ELECTRONIC STRUCTURES OF COMPLEXES

The striking feature of many coordination compounds is that they are colored or paramagnetic or both. How do these properties arise? To find out, we need to understand the electronic structures of complexes, the details of the bonding, and the distribution of their electrons.

There are two major theories of bonding in *d*-metal complexes. *Crystal field theory* was first devised to explain the colors of solids, particularly ruby, which owes its color to Cr^{3+} ions, and then adapted to individual complexes. Crystal field theory is simple to apply and enables us to make useful predictions with very little labor. However, it does not account for all the properties of complexes. A more sophisticated approach, *ligand field theory* (Section 16.12), is based on molecular orbital theory.

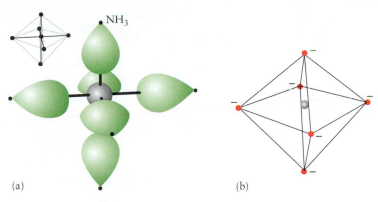

(a)   (b)

FIGURE 16.24 In the crystal field theory of complexes, the lone pairs of electrons that serve as the Lewis base sites on the ligands (a) are treated as equivalent to point negative charges (b).

16.8 Crystal Field Theory

Crystal field theory takes a very simple view of the environment of the central metal atom (or ion): it supposes that each ligand can be represented by a point negative charge. These negative charges represent the ligand lone pairs directed toward the central metal atom (Fig. 16.24). Because the metal atom at the center of a complex is usually a positively charged ion, the negative charges representing the ligands are attracted to it. This attraction results in the formation of the complex. However, in most cases, there are still d-electrons on the central metal ion and the point charges representing the ligands interact with each electron to different extents that depend on the orientation of the d-orbital that it occupies. Crystal field theory explores these differences and uses them to account for the optical and magnetic properties of the complex.

As an example, let's consider an octahedral d^1 complex, such as one containing a Ti^{3+} ion. In a free Ti^{3+} ion, all five $3d$-orbitals have the same energy and the d-electron is equally likely to occupy any one of them. However, when a Ti^{3+} is dissolved in water, six H_2O molecules surround it and form a $[Ti(H_2O)_6]^{3+}$ complex. The six point charges representing the ligands lie on opposite sides of the central metal ion along the x-, y-, and z-axes. From Fig. 16.25, we can see that three of the orbitals (d_{xy}, d_{yz}, and d_{zx}) have their lobes directed *between* the point charges. These three d-orbitals are called t_{2g}-**orbitals**. The other two d-orbitals (d_{z^2} and $d_{x^2-y^2}$), have lobes

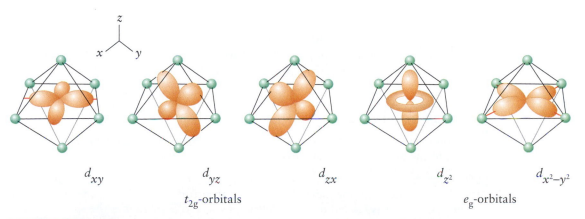

d_{xy} d_{yz} d_{zx} d_{z^2} $d_{x^2-y^2}$

t_{2g}-orbitals e_g-orbitals

FIGURE 16.25 In an octahedral complex with a central d-metal atom or ion, a d_{xy}-orbital is directed between the ligand sites, and an electron that occupies it has a relatively low energy. The same lowering of energy occurs for d_{yz}- and d_{zx}-orbitals. A d_{z^2}-orbital points directly toward two ligands, and an electron that occupies it has a relatively high energy. The same rise in energy occurs for a $d_{x^2-y^2}$-electron.

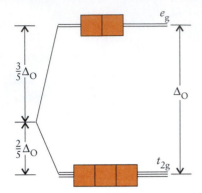

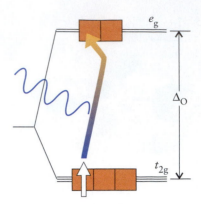

FIGURE 16.26 The energy levels of the *d*-orbitals in an octahedral complex with the ligand field splitting Δ_O. The horizontal line on the far left represents the average energy of the *d*-orbitals once the complex has formed; the lines on the right show the modification of their energies due to their different interactions with the ligands. Each orbital (represented by a box) can hold two electrons.

FIGURE 16.27 When a complex is exposed to light with the correct frequency, an electron can be excited to a higher energy orbital and a photon of light is absorbed.

The labels t_{2g} and e_g are derived from group theory, the mathematical theory of symmetry. The letter g indicates that the orbital does not change sign when we start from any point, pass through the nucleus, and end at the corresponding point on the other side of the nucleus.

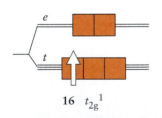

16 t_{2g}^{1}

that point *directly* toward the point charges, and are called e_g-**orbitals.** Because the negative point charges representing the ligands repel electrons, the energy of the single *d*-electron is increased, but it is raised more if it occupies an e_g-orbital than if it occupies a t_{2g}-orbital.

To express the fact that an electron in an octahedral complex has a different energy depending on which of the orbitals it occupies, we say that the three t_{2g}-orbitals have a lower energy than the two e_g-orbitals. The stability of a complex depends on the interaction energy between the central positive ion and its ligands. Typically the difference in energies of electrons in the t_{2g}- and the e_g-orbitals account only for about 10% of that energy, but it plays a major role in the optical and magnetic properties of the complex.

The energy-level diagram in Fig. 16.26 helps to summarize these ideas. The energy separation between the two sets of orbitals is called the **ligand field splitting,** Δ_O (the O denotes octahedral). The three t_{2g}-orbitals lie at an energy that is $\frac{2}{5}\Delta_O$ below the average *d*-orbital energy in the complex, and the two e_g-orbitals lie at an energy $\frac{3}{5}\Delta_O$ above the average. Because the t_{2g}-orbitals have the lower energy, we can predict that, in the ground state of the $[\mathrm{Ti(H_2O)_6}]^{3+}$ complex, the electron occupies one of them in preference to an e_g-orbital and hence that the lowest energy electron configuration of the complex is t_{2g}^{1}. This ground-state configuration is represented by the box diagram in (**16**).

The single *d*-electron of an octahedral $[\mathrm{Ti(H_2O)_6}]^{3+}$ complex can be excited from the t_{2g}-orbital into an e_g-orbital if it absorbs a photon of energy Δ_O (Fig. 16.27). The greater the splitting, the shorter the wavelength of the light that is absorbed by the complex. Therefore, we can use the wavelength of the electromagnetic radiation absorbed by a complex to determine the ligand field splitting.

EXAMPLE 16.3 **Determining the ligand field splitting**

The complex $[\mathrm{Ti(H_2O)_6}]^{3+}$ absorbs light of wavelength 510. nm. What is the ligand field splitting in the complex in kilojoules per mole ($\mathrm{kJ \cdot mol^{-1}}$)?

STRATEGY Because a photon carries an energy $h\nu$, where h is Planck's constant and ν (nu) is the frequency of the radiation, it can be absorbed if $h\nu = \Delta_O$. The wavelength, λ (lambda), of light is related to the frequency by $\lambda = c/\nu$, where c is the speed of light (Section 1.2). Therefore, the wavelength of light absorbed and the ligand field splitting are related by

$$\Delta_O = \frac{hc}{\lambda}$$

Ligand field splittings are normally reported as a molar energy, and so we need to multiply this expression by Avogadro's constant:

$$\Delta_O = \frac{N_A hc}{\lambda}$$

We can then substitute the data.

SOLUTION Because the wavelength absorbed is 510. nm (corresponding to 5.10×10^{-7} m), it follows that the ligand field splitting is:

From
$\Delta_O = N_A hc/\lambda$,

$$\Delta_O = \frac{(6.022 \times 10^{23} \text{ mol}^{-1}) \times (6.626 \times 10^{-34} \text{ J·s}) \times (2.998 \times 10^8 \text{ m·s}^{-1})}{5.10 \times 10^{-7} \text{ m}}$$

$$= 2.35 \times 10^5 \text{ J·mol}^{-1} = 235 \text{ kJ·mol}^{-1}$$

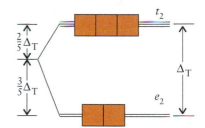

SELF-TEST 16.7A The complex $[Fe(H_2O)_6]^{3+}$ absorbs light of wavelength 700. nm. What is the value (in kilojoules per mole) of the ligand field splitting?

[*Answer:* 171 kJ·mol^{-1}]

SELF-TEST 16.7B The complex $[Fe(CN)_6]^{4-}$ absorbs light of wavelength 305 nm. What is the value (in kilojoules per mole) of the ligand field splitting?

The relative energies of the *d*-orbitals are different in complexes with different shapes. For example, in a tetrahedral complex, the three t_2-orbitals point more directly at the ligands than the two *e*-orbitals do. As a result, in a tetrahedral complex, the t_2-orbitals have a higher energy than the *e*-orbitals (Fig. 16.28). The ligand field splitting, Δ_T (where the T denotes tetrahedral), is generally smaller than in octahedral complexes, in part because there are fewer repelling ligands.

> *In octahedral complexes, the e_g-orbitals (d_{z^2} and $d_{x^2-y^2}$) lie higher in energy than the t_{2g}-orbitals (d_{xy}, d_{yz}, and d_{zx}). The opposite is true in a tetrahedral complex, for which the ligand field splitting is smaller.*

16.9 The Spectrochemical Series

Different ligands affect the *d*-orbitals of a given metal atom or ion to different degrees and thus produce different values of the ligand field splitting. For example, the ligand field splitting is much greater in $[Fe(CN)_6]^{4-}$ than it is in $[Fe(H_2O)_6]^{2+}$. The relative strengths of the splitting produced by a given ligand are much the same regardless of the identity of the *d*-metal in the complex. Thus, ligands can be arranged in a **spectrochemical series** according to the relative magnitudes of the ligand field splittings that they produce (Fig. 16.29). Ligands below the horizontal line in Fig. 16.29 produce only a small ligand field splitting, and so they are called **weak-field ligands**; ligands above the line produce a larger splitting and are called **strong-field ligands**. A CN$^-$ ion is therefore a strong-field ligand, whereas an H$_2$O molecule is a weak-field ligand.

Knowledge of relative ligand strengths allows us to understand the color of a complex ion and, as we shall see in Section 16.11, its magnetism as well. The replacement of one ligand by another gives us chemical control over color. Similarly, the substitution of weak-field ligands for strong-field ligands (or vice versa) acts like a chemical switch for turning paramagnetism on and off. The key idea is that the ligand field splitting affects the electron configuration of the complex. Because all five *d*-orbitals of an isolated metal atom or ion have the same energy, electrons occupy each orbital separately (Hund's rule, Section 1.13) until five electrons have been accommodated. When the atom is part of a complex, we also have to take into account the difference in energies of the two sets of *d*-orbitals, because the order in which the orbitals are occupied depends on that difference.

The subscript g is not used to label the orbitals in a tetrahedral complex because there is no center of symmetry.

FIGURE 16.28 The energy levels of the *d*-orbitals in a tetrahedral complex with the ligand field splitting Δ_T. Each box (that is, orbital) can hold two electrons. The subscript g is not used to label the orbitals in a tetrahedral complex.

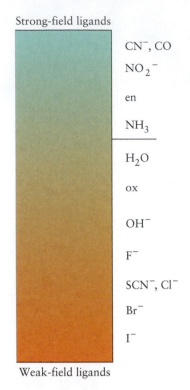

Strong-field ligands

CN⁻, CO

NO₂⁻

en

NH₃

H₂O

ox

OH⁻

F⁻

SCN⁻, Cl⁻

Br⁻

I⁻

Weak-field ligands

FIGURE 16.29 The spectrochemical series. Strong-field ligands give rise to a large splitting between the *t*- and *e*-orbitals, whereas weak-field ligands give rise to only a small splitting. The horizontal line marks the approximate frontier between the two kinds of ligands. The changing color represents the increasing energy of light absorbed as the field strength increases.

First, consider the metal atom or ion at the center of an octahedral complex. The energies of its *d*-orbitals are split by the ligands as shown in Fig. 16.26. The three t_{2g}-orbitals all have the same energy and lie below the two e_g-orbitals. The single *d*-electron of a d^1 complex occupies one of the t_{2g}-orbitals, and so the ground-state configuration is t_{2g}^1 (see **16**). The two *d*-electrons of a d^2 complex occupy separate t_{2g}-orbitals and give rise to the configuration t_{2g}^2 (**17**). Similarly, a d^3 complex will have the ground-state configuration t_{2g}^3 (**18**). According to Hund's rule, all these electrons have parallel spins as that arrangement corresponds to the lowest energy.

A d^4 octahedral complex presents a problem. The fourth electron could enter a t_{2g}-orbital, resulting in a t_{2g}^4 configuration. However, to do so, it would have to enter an orbital that is already half full; hence, it would experience a strong repulsion from the electron already there (**19**). To avoid this repulsion, it could occupy an empty e_g-orbital to give a $t_{2g}^3 e_g^1$ configuration (**20**), but now it experiences a strong repulsion from the ligands. Which configuration has the lower energy depends on the ligands present. If Δ_O is large (as it is for strong-field ligands), the energy difference between the t_{2g}- and e_g-orbitals will be large and the configuration t_{2g}^4 will have a lower energy than $t_{2g}^3 e_g^1$. If Δ_O is small (as it is for weak-field ligands), $t_{2g}^3 e_g^1$ will be the lower energy configuration and hence the one adopted by the complex.

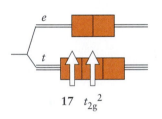

17 t_{2g}^2

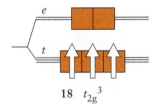

18 t_{2g}^3

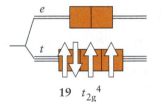

19 t_{2g}^4

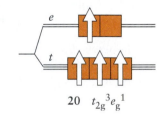

20 $t_{2g}^3 e_g^1$

A note on good practice: Note that a configuration with a single electron in an orbital is written with a superscript 1, as in $t_{2g}^3 e_g^1$, not $t_{2g}^3 e_g$.

EXAMPLE 16.4 **Predicting the electron configuration of a complex**

Predict the electron configuration of an octahedral d^5 complex with (a) strong-field ligands and (b) weak-field ligands, and state the number of unpaired electrons in each case.

STRATEGY Electrons occupy the orbitals that result in the lowest energy configuration. (a) For strong-field ligands, Δ_O is large and electrons are likely to pair in the lower-energy orbitals and fill them completely before occupying any of the higher-energy orbitals. (b) For weak-field ligands, Δ_O is small. In this case, electrons are likely to occupy all the vacant orbitals, even those at the higher energy, before pairing.

SOLUTION

(a) All 5 electrons enter the t_{2g}-orbitals and 4 electrons must pair.

t_{2g}^5; 1 unpaired electron

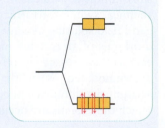

(b) The 5 electrons occupy all 5 orbitals without pairing. $t_{2g}^{3}e_g^{2}$; 5 unpaired electrons

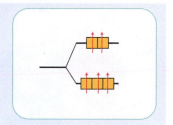

SELF-TEST 16.8A Predict the electron configurations and the number of unpaired electrons of an octahedral d^6 complex with (a) strong-field ligands and (b) weak-field ligands.

[*Answer:* (a) t_{2g}^{6}, 0; (b) $t_{2g}^{4}e_g^{2}$, 4]

SELF-TEST 16.8B Predict the electron configurations and the number of unpaired electrons of an octahedral d^7 complex with (a) strong-field ligands and (b) weak-field ligands.

Table 16.5 lists the configurations for d^1 through d^{10} octahedral complexes, including the alternative configurations for d^4 through d^7 octahedral complexes. A d^n complex with the maximum number of unpaired spins is called a **high-spin complex**. High-spin complexes are expected for weak-field ligands because the electrons can easily occupy both the t_{2g}- and the e_g-orbitals, and the greatest number of electrons then have parallel spins. A d^n complex with the minimum number of unpaired spins is called a **low-spin complex**. We should expect a low-spin complex when the ligands are strong-field, because electrons enter the t_{2g}-orbitals until they are completely full, even though they have to pair their spins. We can predict whether an octahedral complex is likely to be a high-spin or a low-spin complex by noting where the ligands lie in the spectrochemical series. If they are strong-field ligands, we expect a low-spin complex; if they are weak-field ligands, we expect a high-spin complex.

Tetrahedral complexes are almost always high spin. We saw that ligand field splittings are smaller for tetrahedral than for octahedral complexes. Therefore, even if the ligands are classified as strong-field ligands for octahedral complexes, the splitting is so small in the corresponding tetrahedral complex that the t_2-orbitals are energetically accessible.

The electron configurations of d-block metal atoms and ions in complexes are obtained by applying the building-up principle to the d-orbitals, taking into account the strength of the ligand field splitting. Relative field-splitting strengths are summarized by the spectrochemical series.

TABLE 16.5 Electron Configurations of d^n Complexes

Number of d-electrons	Configuration			
	Octahedral complexes			Tetrahedral complexes
d^1	t_{2g}^{1}			e^1
d^2	t_{2g}^{2}			e^2
d^3	t_{2g}^{3}			$e^2 t_2^{1}$
	Low spin	High spin		
d^4	t_{2g}^{4}	$t_{2g}^{3}e_g^{1}$		$e^2 t_2^{2}$
d^5	t_{2g}^{5}	$t_{2g}^{3}e_g^{2}$		$e^2 t_2^{3}$
d^6	t_{2g}^{6}	$t_{2g}^{4}e_g^{2}$		$e^3 t_2^{3}$
d^7	$t_{2g}^{6}e_g^{1}$	$t_{2g}^{5}e_g^{2}$		$e^4 t_2^{3}$
d^8		$t_{2g}^{6}e_g^{2}$		$e^4 t_2^{4}$
d^9		$t_{2g}^{6}e_g^{3}$		$e^4 t_2^{5}$
d^{10}		$t_{2g}^{6}e_g^{4}$		$e^4 t_2^{6}$

FIGURE 16.30 In a color wheel the color of light absorbed is opposite the color perceived. For example, a complex that absorbs orange light appears blue to the eye.

FIGURE 16.31 Because $[Ti(H_2O)_6]^{3+}$ absorbs yellow-green light (of wavelengths close to 510 nm), it looks violet in white light.

FIGURE 16.32 The effect on the color of the complex of substituting ligands with different ligand field strengths in octahedral cobalt(III) complexes in aqueous solution. The ligand field strengths increase from left to right.

16.10 The Colors of Complexes

White light is a mixture of all wavelengths of electromagnetic radiation from about 400 nm (violet) to about 800 nm (red). When some of these wavelengths are removed from a beam of white light by passing the light through a sample, the emerging light is no longer white to the eye. For example, if red light is absorbed from white light, then the light that remains appears green. Conversely, if green is removed, then the light appears red. We say that red and green are each other's **complementary color**—each is the color that white light appears when the other is removed (Fig. 16.30).

We can see from the color wheel that, for example, if a substance looks blue (as does a copper(II) sulfate solution, for instance), then it is absorbing orange (580–620 nm) light. Conversely, if we know the wavelength (and therefore the color) of the light that a substance absorbs, then we can predict the color of the substance by noting the complementary color on the color wheel. Because $[Ti(H_2O)_6]^{3+}$ absorbs 510-nm light, which is yellow-green light, the complex looks violet (Fig. 16.31).

Because weak-field ligands generate small splittings, the complexes that they form absorb low-energy, long-wavelength radiation. The long wavelengths correspond to red light, and so these complexes exhibit colors near green. Because strong-field ligands give large splittings, the complexes that they form should absorb high-energy, short-wavelength radiation, corresponding to the violet end of the visible spectrum. Such complexes can therefore be expected to have colors near orange and yellow (Fig. 16.32).

The colors that we have described arise from *d–d* **transitions**, in which an electron is excited from one *d*-orbital into another. In a **charge-transfer transition** an electron is excited from a ligand onto the metal atom or vice versa. Charge-transfer transitions are often very intense and are the most common cause of the familiar colors of *d*-metal complexes, such as the transition responsible for the deep purple of permanganate ions, MnO_4^- (Fig. 16.33).

> *Transitions between d-orbitals or between the ligands and the metal atom in complexes give rise to color; the wavelength of d–d transitions can be correlated with the magnitude of the ligand field splitting.*

16.11 Magnetic Properties of Complexes

As we saw in Box 3.2, a substance with unpaired electrons is paramagnetic and is pulled into a magnetic field. A substance without unpaired electrons is diamagnetic and is pushed out of a magnetic field. The apparatus shown in Fig. 16.34 allows us to distinguish paramagnetism and diamagnetism experimentally. A sample is hung from a balance so that it lies between the poles of an electromagnet. When the magnet is turned on, a paramagnetic substance is pulled into the field and appears

FIGURE 16.33 In a ligand-to-metal charge-transfer transition, an energetically excited electron migrates from a ligand to the central metal ion. This type of transition is responsible for the intense purple of the permanganate ion, MnO_4^-.

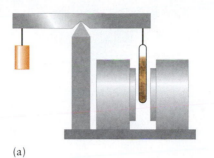 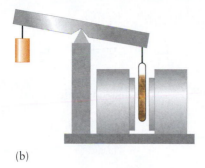 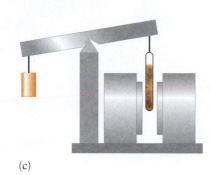

(a)   (b)   (c)

FIGURE 16.34 The magnetic character of a complex can be studied with a Gouy balance. (a) A sample is hung from a balance so that it lies partly between the poles of an electromagnet. (b) When the magnetic field is turned on, a paramagnetic sample is drawn into it, and so the sample seems to weigh more. (c) In contrast, a diamagnetic sample is pushed out of the field when the field is turned on and seems to weigh less.

to weigh more than when the magnet is off. A diamagnetic substance is pushed out of the field and appears to weigh less.

Many *d*-metal complexes have unpaired *d*-electrons and are therefore paramagnetic. We have just seen that a high-spin d^n complex has more unpaired electrons than does a low-spin d^n complex. The high-spin complex is therefore more strongly paramagnetic and drawn more strongly into a magnetic field. Whether a complex is high spin or low spin depends on the ligands present. The d^4 through d^7 complexes of strong-field ligands have large energy gaps and so tend to be low spin and diamagnetic or only weakly paramagnetic (Fig. 16.35). The d^4 through d^7 complexes of weak-field ligands create a small energy gap and therefore tend to be high spin and strongly paramagnetic.

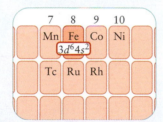

FIGURE 16.35 (a) A strong-field ligand is likely to lead to a low-spin complex (in this case, the configuration is that of Fe^{3+}). (b) Substituting weak-field ligands is likely to result in a high-spin complex.

EXAMPLE 16.5 Predicting the magnetic properties of a complex

Compare the magnetic properties of (a) $[Fe(H_2O)_6]^{2+}$; (b) $[Fe(CN)_6]^{4-}$.

STRATEGY Decide from their positions in the spectrochemical series whether the ligands are weak-field or strong-field. Then judge whether the complex is high spin or low spin.

SOLUTION

(a) Determine the number of *d*-electrons.

The Fe^{2+} ion is a d^6 ion.

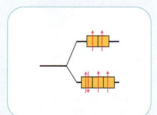

Classify the ligand strength.

H_2O is a weak-field ligand.

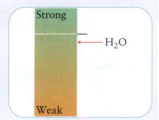

For a weak-field ligand, predict a high-spin $t_{2g}^4 e_g^2$ configuration.

$[Fe(H_2O)_6]^{2+}$ will have 4 unpaired electrons; therefore, paramagnetic.

(b) Determine the number of *d*-electrons.

The Fe^{2+} ion is a d^6 ion.

Classify the strength of the ligand.

CN^- is a strong-field ligand.

Strong ←— CN^-

Weak

For a strong-field ligand, predict a low-spin t_{2g}^6 configuration.

$[Fe(CN)_6]^{4-}$ will have no unpaired electrons; therefore, diamagnetic.

Self-Test 16.9A What change in magnetic properties can be expected when NO_2^- ligands in an octahedral complex are replaced by Cl^- ligands in (a) a d^6 complex; (b) a d^3 complex?

[*Answer:* (a) The complex becomes paramagnetic; (b) there is no change in magnetic properties.]

Self-Test 16.9B Compare the magnetic properties of $[Ni(en)_3]^{2+}$ with those of $[Ni(H_2O)_6]^{2+}$.

The magnetic properties of a complex depend on the magnitude of the ligand field splitting. Strong-field ligands tend to form low-spin, weakly paramagnetic complexes; weak-field ligands tend to form high-spin, strongly paramagnetic complexes.

16.12 Ligand Field Theory

Crystal field theory is based on a very primitive model of bonding. For example, ligands are not point charges: they are actual molecules or ions. The theory also leaves several questions unanswered. Why, for instance, is the electrically neutral molecule CO a strong-field ligand but the negatively charged ion Cl^- a weak-field ligand?

To improve the model of bonding in complexes, chemists have turned to molecular orbital theory (Section 3.9). **Ligand field theory** describes bonding in complexes in terms of molecular orbitals built from the metal atom *d*-orbitals and ligand orbitals. In contrast to crystal field theory, which models the structure of the complex in terms of point charges, ligand field theory assumes, more realistically, that ligands are attached to the central metal atom or ion by covalent bonds. As we shall see, much of the work that we have done in connection with crystal field theory can be transferred into ligand field theory: the principal difference is the origin of the ligand field splitting.

To describe the electronic structure of a complex, first we set up molecular orbitals from the available atomic orbitals in the complex, just as we would for a molecule. Consider an octahedral complex of a *d*-metal in Period 4, such as iron, cobalt, or copper. We need to consider the nine 4*s*-, 4*p*-, and 3*d*-orbitals of the central metal ion, because all these orbitals have similar energies. To simplify the discussion, we use only one atomic orbital on each of the ligands. For instance, for a Cl^- ligand, we use the $Cl3p$-orbital directed toward the metal atom; for an NH_3 ligand, we use the sp^3 lone-pair orbital of the nitrogen atom. The six orbitals provided by the six ligands in an octahedral complex are represented by the tear-shaped lobes in Fig. 16.36. Each of these orbitals has cylindrical symmetry around the metal–ligand axis, and so each can form a σ-orbital.

There are nine valence orbitals on the metal atom and six on the ligands, giving 15 in all. We can therefore expect to find 15 molecular orbitals: six are bonding, three are nonbonding, and six are antibonding. The energies of all 15 are displayed in Fig. 16.37, together with the labels that they are commonly given. Notice in Fig.

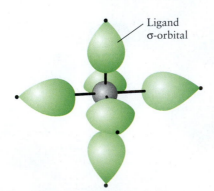

Ligand σ-orbital

FIGURE 16.36 The tear-shaped objects are representations of the six ligand atomic orbitals that are used to build the molecular orbitals of an octahedral complex in ligand field theory. They might represent *s*- or *p*-orbitals on the ligands or hybrids of the two.

16.37 that the t_{2g}-orbitals on the metal atom have no partners among the ligands. There are simply no ligand orbitals to match them. Therefore, these three orbitals are nonbonding orbitals in the complex.

Now we proceed exactly as we did in the discussion of diatomic molecules (Section 3.10): we use the building-up principle to work out the ground-state electron configuration. First, we need to count the available electrons. In a d^n complex, n electrons are supplied by the metal; each ligand orbital supplies two electrons, and so 12 electrons are supplied by the ligands, giving $12 + n$ in all. The first 12 electrons fill the six bonding orbitals. That leaves n electrons to be accommodated in the nonbonding and antibonding orbitals. At this point, notice that the next available orbitals (those in the blue box in Fig. 16.37) lie in exactly the same pattern as that encountered in crystal field theory. The only difference is that, in ligand field theory, we recognize the three t_{2g}-orbitals as nonbonding orbitals and the two e_g-orbitals as antibonding between the metal and the ligands. That is, the ligand field splitting can be identified with the energy separation between nonbonding and antibonding orbitals. The remaining four orbitals are high-energy antibonding orbitals that are ordinarily unavailable to the electrons.

From this point on, the analysis is the same as in crystal field theory. The order of filling these two sets of orbitals is exactly the same as in crystal field theory, and so is the discussion of optical and magnetic properties. If the ligand field splitting is large, then the t_{2g}-orbitals are filled first, and we can expect a low-spin complex. If the ligand field splitting is small, then the e_g-orbitals are occupied before spin-pairing begins in the t_{2g}-orbitals, and we expect a high-spin complex.

Although ligand field theory puts the discussion of bonding on a firmer basis, it has not yet explained all the puzzling features of crystal field theory. In particular, why is CO a strong-field ligand? Why is Cl^- a weak-field ligand despite its negative charge?

We need to develop the model by considering the effects of other ligand orbitals. When we constructed the molecular orbitals, we considered only ligand orbitals that pointed directly at the central metal atom. Ligands also have orbitals perpendicular to the metal–ligand axis, which we can expect to contribute to bonding and antibonding π-orbitals. As Fig. 16.38 shows, a p-orbital on the ligand

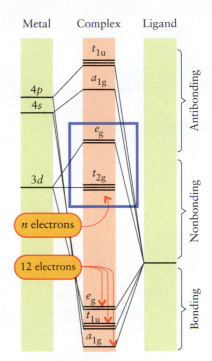

FIGURE 16.37 The molecular orbital energy-level diagram for an octahedral complex. The 12 electrons provided by the six ligands fill the lowest six orbitals, which are all bonding orbitals. The n d-electrons provided by the central metal atom or ion are accommodated in the orbitals inside the blue box. The ligand field splitting is the energy separation of the nonbonding (t_{2g}) and antibonding (e_g) orbitals in the box.

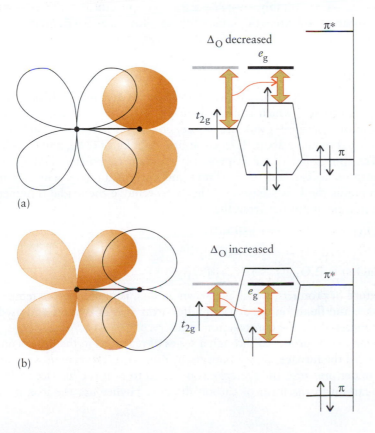

FIGURE 16.38 The effect of π-bonding on ligand field splitting. (a) In this case, the occupied π-orbital of the ligand is close in energy to the ligand t_{2g} orbitals and they overlap to form bonding and antibonding combinations. The ligand-field splitting is reduced. (b) In this case, the unoccupied antibonding π^*-orbital of the ligand is close in energy to the ligand t_{2g} orbitals and they overlap to form bonding and antibonding combinations. In this case, the ligand-field splitting is increased.

perpendicular to the axis of the metal–ligand bond can overlap with one of the t_{2g}-orbitals to produce two new molecular orbitals, one bonding and one antibonding. The resulting bonding orbital lies below the energy of the original t_{2g}-orbitals; the antibonding orbital lies above them.

Now we count the number of electrons to accommodate. If the ligand is Cl^-, the $Cl3p$-orbital that we have used to build the metal–ligand π-orbital is full. It provides two electrons, which occupy a new molecular orbital, the bonding metal–ligand combination shown in Fig. 16.38a. The *n* *d*-electrons provided by the metal must therefore occupy the *antibonding* metal–ligand orbital. Because this molecular orbital is higher in energy than the original t_{2g}-orbitals, the ligand field splitting is decreased by π-bonding. We see how Cl^- can be a weak-field ligand despite its negative charge.

Now suppose the ligand is CO. The orbital that overlaps with the metal t_{2g}-orbitals in this case is either the full bonding π-orbital or the empty antibonding π^*-orbital of the CO molecule. It turns out that the latter orbital is closer in energy to the metal orbitals, and so it plays the dominant role in bond formation to the metal, as in Fig. 16.38b. There are no electrons from the ligand to accommodate because its π^*-orbital is empty. The *n* *d*-electrons therefore enter the *bonding* metal–ligand orbital. Because this new molecular orbital is lower in energy than the original t_{2g}-orbitals, the ligand field splitting is increased by π-bonding and CO is a strong-field ligand despite being electrically neutral.

According to ligand field theory, the ligand field splitting is the energy separation between nonbonding and antibonding molecular orbitals built principally from d-orbitals. When π-bonding is possible, the ligand field splitting is decreased if the ligand supplies π-electrons and is increased if the ligand does not supply π-electrons.

THE IMPACT ON MATERIALS

We have already mentioned some of the important roles that the *d*-block metals play in virtually every aspect of our lives. Steel, an alloy based on iron, is important in construction and transportation and the "nonferrous" alloys, those based on other metals—most notably, copper—are also important in industry, for their corrosion resistance and strength. Some of these alloys are also desired for their magnetic properties.

16.13 Steel

The production of steel begins when iron ore is fed into a blast furnace (Fig. 16.39). The furnace, which is approximately 40 m high, is continuously replenished from the top with a mixture of ore, coke, and limestone. Each kilogram of iron produced requires about 1.75 kg of ore, 0.75 kg of coke, and 0.25 kg of limestone. The limestone, which is primarily calcium carbonate, undergoes thermal decomposition to calcium oxide (lime) and carbon dioxide. The calcium oxide, which contains the Lewis base O^{2-}, helps to remove the acidic (nonmetal oxide) and amphoteric impurities from the ore:

$$CaO(s) + SiO_2(s) \xrightarrow{\Delta} CaSiO_3(l)$$
$$CaO(s) + Al_2O_3(s) \xrightarrow{\Delta} Ca(AlO_2)_2(l)$$
$$6\ CaO(s) + P_4O_{10}(s) \xrightarrow{\Delta} 2\ Ca_3(PO_4)_2(l)$$

The mixture of products, which is known as *slag,* is molten at the temperatures in the furnace and floats on the denser molten iron. It is drawn off and used to make rocklike material for the construction industry.

Molten iron is produced through a series of reactions in the four main temperature zones of the furnace. At the bottom, in Zone A, preheated air is blown into the furnace under pressure, and the coke is oxidized to heat the furnace to 1900°C and provide carbon in the form of carbon dioxide. Higher up, the iron is reduced in

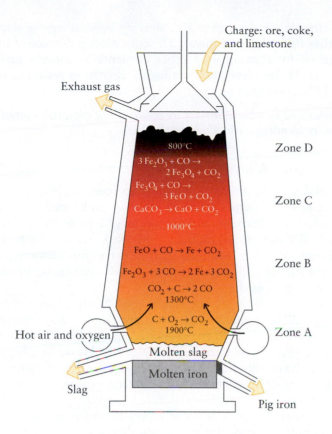

Charge: ore, coke, and limestone

Exhaust gas

800°C

$3\,Fe_2O_3 + CO \rightarrow$
$\quad 2\,Fe_3O_4 + CO_2$

$Fe_3O_4 + CO \rightarrow$
$\quad 3\,FeO + CO_2$

$CaCO_3 \rightarrow CaO + CO_2$

1000°C

$FeO + CO \rightarrow Fe + CO_2$

$Fe_2O_3 + 3\,CO \rightarrow 2\,Fe + 3\,CO_2$

$CO_2 + C \rightarrow 2\,CO$
1300°C

$C + O_2 \rightarrow CO_2$
1900°C

Zone D

Zone C

Zone B

Zone A

Hot air and oxygen

Molten slag

Molten iron

Slag

Pig iron

FIGURE 16.39 The reduction of iron ore takes place in a blast furnace containing a mixture of the ore with coke and limestone. Different reactions take place in different zones when the blast of air and oxygen is admitted. The ore, an oxide, is reduced to the metal by reduction with carbon monoxide produced in the furnace.

stages to the metal, which melts and flows from Zone C to Zone A. Although the melting point of pure iron is 1540°C, iron mixed with 4% carbon melts at about 1015°C. As the carbon dioxide moves up through the furnace to Zone B, it reacts with some of the carbon provided by the coke, producing carbon monoxide. This reaction is endothermic and lowers the temperature to 1300°C. The carbon monoxide produced in this reaction rises to Zones C and D, where it reduces iron ore in a series of reactions, some of which are shown in Fig. 16.39. Molten iron runs off as *pig iron*, which is 90–95% iron, 3–5% carbon, 2% silicon, and trace amounts of other elements found in the original ore. *Cast iron* is similar to pig iron, but with fewer impurities and a carbon content that is usually greater than 2%.

Pure iron is relatively flexible and malleable, but the carbon atoms make cast iron very hard and brittle. Cast iron is used for objects that experience little mechanical and thermal shock, such as ornamental railings, engine blocks, brake drums, and transmission housings.

Pig iron is processed further to produce steel. The first stage is to lower the carbon content of the iron and to remove the remaining impurities, which include silicon, phosphorus, and sulfur. In the *basic oxygen process,* oxygen and powdered limestone are forced through the molten metal (Fig. 16.40). In the second stage, steels are produced by adding the appropriate metals, often as ferroalloys, to the molten iron. The steel that results is a homogeneous alloy, a solid solution of 2% or less carbon in iron. Different formulations of steels have various hardnesses, tensile strengths, and ductilities; the higher the carbon content, the harder and more brittle the steel (Table 16.6). Heat treatment can greatly

FIGURE 16.40 In the basic oxygen process, a blast of oxygen and powdered limestone is used to purify molten iron by oxidizing and combining with the impurities present in it.

TABLE 16.6 The Effect of Carbon Content on Steel

Type of steel	Carbon content (%)	Properties and applications
low-carbon steel	< 0.15	ductility and low hardness, iron wire
mild-carbon steel	0.15 to 0.25	cables, nails, chains, and horseshoes
medium-carbon steel	0.20 to 0.60	nails, girders, rails, and structural purposes
high-carbon steel	0.61 to 1.5	knives, razors, cutting tools, drill bits

Radii:
128 pm   133 pm

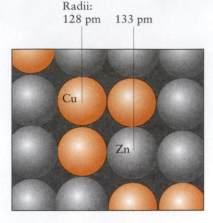

FIGURE 16.41 The atomic radii of copper and zinc are similar, and atoms of one element fit reasonably comfortably into the lattice of the other element to give the range of alloys that we know as brass.

(a)

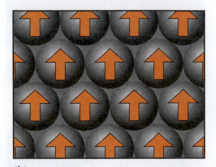

(b)

FIGURE 16.42 Ferromagnetic materials include iron, cobalt, and the iron oxide mineral magnetite. They consist of crystals in which the electrons of many atoms spin in the same direction and give rise to a strong magnetic field. (a) Before magnetization, when the spins are almost randomly aligned; (b) after magnetization. The orange arrows represent the electron spins.

increase the strength and toughness of steel by controlling the sizes of the tiny crystallites that form the solid metal. The corrosion resistance of iron is significantly improved by alloying with other elements to form a variety of steels (recall Table 16.2). Stainless steels are highly corrosion resistant; they are typically about 15% chromium by mass.

Iron is produced in a blast furnace by reduction of ores; its properties are modified by blending with other metals.

16.14 Nonferrous Alloys

Alloys are solid metallic mixtures designed to meet specific needs (see Section 5.15). For example, the frames of racing bicycles can be made of a steel that contains manganese, molybdenum, and carbon to give them the stiffness needed to resist mechanical shock. Titanium frames are used, but not the pure metal. Titanium metal stretches easily, so much so that it becomes deformed under stress. However, when alloyed with metals such as tin and aluminum, titanium maintains its flexibility but keeps its shape.

The properties of alloys are affected by their composition and structure. Not only is the crystalline structure important, but the size and texture of the individual grains also contribute to the properties of an alloy. Some metal alloys are one-phase homogeneous solutions. Examples are brass, bronze, and the gold coinage alloys. Other alloys are heterogeneous mixtures of different crystalline phases, such as tin–lead solder and the mercury–silver amalgams used to fill teeth.

Homogeneous alloys of metals with atoms of similar radius are substitutional alloys. For example, in brass, zinc atoms readily replace copper atoms in the crystalline lattice, because they are nearly the same size (Fig. 16.41). However, the presence of the substituted atoms changes the lattice parameters and distorts the local electronic structure. This distortion lowers the electrical and thermal conductivity of the host metal, but it also increases hardness and strength. Coinage alloys are usually substitutional alloys. They are selected for durability—a coin must last for at least 3 years—and electrical resistance so that genuine coins can be identified by vending machines.

Alloys are mixtures of metals and other elements formulated to achieve desired properties. Generally, an alloy is harder and stronger than the pure metal but has lower electrical conductivity.

16.15 Magnetic Materials

The presence of unpaired *d*-electrons in the ground states of the *d*-block elements explains why some of these metals—notably, iron, cobalt, and nickel—make good permanent magnets. We need to distinguish between two types of magnetism. **Paramagnetism** is the tendency of a substance to move into a magnetic field (see Box 3.2). It arises when an atom or molecule has at least one unpaired electron, which aligns with the applied field. However, because the spins on neighboring atoms or molecules are aligned almost randomly, paramagnetism is very weak and dissipates when the magnetic field is removed. In some *d*-metals, however, when the unpaired electrons of many neighboring atoms align with one another in an applied magnetic field, the much stronger effect of **ferromagnetism** arises. The regions of aligned spins, which are called **domains** (Fig. 16.42), survive even after the applied field is turned off. Ferromagnetism also occurs in some compounds of the *d*-metals, such as the oxides of iron and chromium.

Ferromagnetism is much stronger than paramagnetism, and ferromagnetic materials are used to make permanent magnets. They are also used in the coatings of cassette tapes and computer disks. The electromagnetic recording heads align large numbers of spins as the tape passes underneath, and the spin alignment in the

FIGURE 16.43 When a magnet is pulled up from this viscous ferrofluid, the particles of iron(III) oxide align themselves with the magnetic field. Because strong attractions exist between the particles and the detergent molecules in the oil, the liquid is pulled into the field along with the particles.

domains remains for years. In an **antiferromagnetic** material, neighboring spins are locked into an *antiparallel* arrangement; so the magnetic moments cancel. Manganese is antiferromagnetic.

Ceramic magnets, which are used in refrigerator magnets, are made of barium ferrite ($BaO \cdot nFe_2O_3$) or strontium ferrite ($SrO \cdot nFe_2O_3$). They are made by compressing the powdered ferrite in a magnetic field and heating it until it hardens. Because these magnets are ceramic, they are hard and brittle, with low densities. However, they are the most widely used magnets because of their low cost.

Ferrofluids are ferromagnetic liquids. They are suspensions of finely powdered magnetite, Fe_3O_4, in a viscous, oily liquid (such as mineral oil) that contains a detergent (such as oleic acid, a long-chain carboxylic acid). The iron oxide particles do not settle out, because they are attracted to the polar ends of the detergent molecules, which form inverted micelles, or compact clusters (Section 8.9), around the particles. The nonpolar ends of the detergent molecules point outward, allowing the micelles to be suspended in the oil. When a magnet is brought near a ferrofluid, the particles in the liquid try to align themselves with the magnetic field, but they are kept in place by the oil (Fig. 16.43). As a result, it is possible to control the flow and position of the ferrofluid by means of an applied magnetic field. One application of ferrofluids is in the braking systems of exercise machines. The stronger the magnetic field, the greater the resistance to motion.

Recently interest in single molecule magnets has grown. These molecules contain several *d*-block metal atoms bonded to groups of nonmetal atoms such as carbon, hydrogen, and oxygen. Individual molecular magnets can respond to a magnetic field as if they were nano-size compass needles and thus have great potential for miniaturizing electronic storage media such as computer disks. In these molecules typically the *d*-block metal atoms are imbedded in cage-like structures and some of the nano-size molecular magnets have interesting shapes, such as rings, tubes, or spheres (Fig. 16.44).

Magnetic materials can be paramagnetic, ferromagnetic, or antiferromagnetic. In ferromagnetic materials, large domains of electrons have the same magnetic moment.

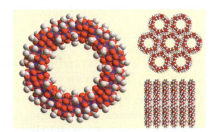

FIGURE 16.44 The structure of a molecular magnet. The nano-size molecular torus contains 84 manganese atoms and is approximately 4 nm in diameter. The manganese atoms are bonded to groups of carbon atoms in the form of acetate ions, water molecules, and chlorine atoms. In this molecule the manganese atoms act as ferromagnets.

SKILLS YOU SHOULD HAVE MASTERED

❑ 1 Explain trends in chemical and physical properties among the *d*-block elements (Sections 16.1 and 16.2).

❑ 2 Describe and write balanced equations for the principal reactions used to produce the elements in the first row (Period 4) of the *d* block and in Groups 11 and 12 (Sections 16.3 and 16.4).

❑ 3 Describe the names, properties, and reactions of some of the principal compounds of the elements in the first row of the *d* block (Sections 16.3 and 16.4).

❑ 4 Name and write formulas for *d*-metal complexes (Toolbox 16.1 and Example 16.1).

❑ 5 Identify pairs of ionization, linkage, hydrate, coordination, geometrical, and optical isomers (Self-Tests 16.4 and 16.5 and Example 16.2).

❑ 6 Determine the ligand field splitting from the wavelength of light absorbed by a complex (Example 16.3).

❑ 7 Use the spectrochemical series to predict the effect of a ligand on the color, electron configuration, and magnetic properties of a *d*-metal complex (Examples 16.4 and 16.5).

❑ 8 Describe bonding in *d*-metal complexes in terms of ligand field theory (Section 16.12).

❑ 9 Describe the operation of a blast furnace and how steel is made by the basic oxygen process (Section 16.13).

❑ 10 Distinguish the types of magnetism (Section 16.15).

EXERCISES

Trends in Properties

You may find Appendix 2 helpful in answering these questions.

16.1 Which members of the *d* block, those at the left or at the right of block, are likely to have the more strongly negative standard potentials? Explain this observation.

16.2 Name five elements in the *d* block that have positive standard potentials.

16.3 Identify the element with the larger atomic radius in each of the following pairs: (a) vanadium and titanium; (b) silver and gold; (c) vanadium and tantalum; (d) rhodium and iridium.

16.4 Identify the element with the larger atomic radius in each of the following pairs: (a) iron and nickel; (b) copper and silver; (c) iridium and gold; (d) titanium and zirconium.

16.5 Identify the element with the higher first ionization energy in each of the following pairs: (a) iron and nickel; (b) nickel and copper; (c) osmium and platinum; (d) nickel and palladium; (e) hafnium and tantalum.

16.6 Identify the element with the higher first ionization energy in each of the following pairs: (a) manganese and cobalt; (b) manganese and rhenium; (c) chromium and zinc; (d) chromium and molybdenum; (e) palladium and platinum.

16.7 Explain why the density of mercury (13.55 g·cm^{-3}) is significantly higher than that of cadmium (8.65 g·cm^{-3}), whereas the density of cadmium is only slightly greater than that of zinc (7.14 g·cm^{-3}).

16.8 Explain why the density of vanadium (6.11 g·cm^{-3}) is significantly less than that of chromium (7.19 g·cm^{-3}). Both vanadium and chromium crystallize in a body-centered cubic lattice.

16.9 (a) Describe the trend in the stability of oxidation states moving down a group in the *d* block (for example, from chromium to molybdenum to tungsten). (b) How does this trend compare with the trend in the stabilities of oxidation states observed for the *p*-block elements on moving down a group?

16.10 Which oxoanion, CrO_4^{2-} or WO_4^{2-}, is expected to be the stronger oxidizing agent? Explain your choice.

16.11 Which of the elements vanadium, chromium, and manganese is most likely to form an oxide with the formula MO_3? Explain your answer.

16.12 Which of the elements zirconium, chromium, and iron is most likely to form a chloride with the formula MCl_4? Explain your answer.

Selected Elements: A Survey

16.13 Predict the major products of each of the following reactions and then balance the skeletal equations:
(a) $TiCl_4(s) + Mg(s) \xrightarrow{\Delta}$
(b) $CoCO_3(s) + HNO_3(aq) \rightarrow$
(c) $V_2O_5(s) + Ca(l) \xrightarrow{\Delta}$

16.14 Predict the major products of each of the following reactions and then balance the skeletal equations:
(a) $FeCr_2O_4(s) + C(s) \xrightarrow{\Delta}$
(b) $CrO_4^{2-}(s) + H_3O^+(aq) \rightarrow$
(c) $MnO_2(s) + Al(s) \xrightarrow{\Delta}$

16.15 Give the systematic name and chemical formula of the principal component of (a) rutile; (b) hematite; (c) pyrolusite; (d) chromite.

16.16 Give the systematic name and chemical formula of the principal component of (a) magnetite; (b) malachite; (c) ilmenite; (d) chalcocite.

16.17 Use Appendix 2B to determine whether an acidic sodium dichromate solution can oxidize (a) bromide ions to bromine and (b) silver(I) ions to silver(II) ions under standard conditions.

16.18 Use Appendix 2B to determine whether an acidic potassium permanganate solution can oxidize (a) chloride ions to chlorine and (b) mercury metal to mercury(I) ions under standard conditions.

16.19 Write the chemical equation that describes each of the following processes: (a) solid V_2O_5 reacts with acid to form the VO^{2+} ion; (b) solid V_2O_5 reacts with base to form the VO_4^{3-} ion.

16.20 Write the chemical equation that describes each of the following processes: (a) the production of chromium by the thermite reaction; (b) the corrosion of copper metal by carbon dioxide in moist air; (c) the purification of nickel by using carbon monoxide.

16.21 By considering electron configurations, explain why gold and silver are less reactive than copper.

16.22 By considering electron configurations, suggest a reason why iron(III) is readily prepared from iron(II) but the conversion of nickel(II) and cobalt(II) into nickel(III) and cobalt(III), respectively, is much more difficult.

16.23 (a) Explain why the dissolution of a chromium(III) salt produces an acidic solution. (b) Explain why the slow addition of hydroxide ions to a solution containing chromium(III) ions first produces a gelatinous precipitate that subsequently dissolves with further addition of hydroxide ions. Write chemical equations showing these aspects of the behavior of chromium(III) ions.

16.24 Some of the atomic properties of manganese differ markedly from its neighbors. For example, at constant pressure it takes 400 kJ (2 sf) to atomize 1.0 mol Cr(s) and 420 kJ to atomize 1.0 mol Fe(s), but only 280 kJ to atomize 1.0 mol Mn(s). Propose an explanation, using the electron configurations of the gaseous atoms, for the lower enthalpy of atomization of manganese.

Coordination Compounds
16.25 Name each of the following complex ions and determine the oxidation number of the metal: (a) $[Fe(CN)_6]^{4-}$; (b) $[Co(NH_3)_6]^{3+}$; (c) $[Co(CN)_5(H_2O)]^{2-}$; (d) $[Co(SO_4)(NH_3)_5]^+$.

16.26 Name each of the following complex ions and determine the oxidation number of the metal: (a) $[Fe(ox)(Cl)_4]^{3-}$; (b) $[Fe(OH)(H_2O)_5]^{2+}$; (c) $[CoBr_2(NH_3)_3(H_2O)]^+$; (d) $[Ir(en)_3]^{3+}$.

16.27 Use the information in Table 16.4 to write the formula for each of the following coordination compounds:
(a) potassium hexacyanochromate(III);
(b) pentaamminesulfatocobalt(III) chloride;

(c) tetraamminediaquacobalt(III) bromide;

(d) sodium bisoxalato(diaqua)ferrate(III).

16.28 Use the information in Table 16.4 to write the formula for each of the following coordination compounds:
(a) tetraamminedihydroxochromium(III) chloride;
(b) sodium tetrachloroplatinate(II);
(c) triamminetrichloronickel(IV) sulfate;
(d) sodium tris(oxalato)rhodium(III);
(e) lithium chlorohydroxobis(oxalato)rhodate(III) octahydrate.

16.29 Which of the following ligands can be polydentate? If the ligand can be polydentate, give the maximum number of places on the ligand that can bind simultaneously to a single metal center. (a) $HN(CH_2CH_2NH_2)_2$; (b) CO_3^{2-}; (c) H_2O; (d) oxalate.

16.30 Which of the following ligands can be polydentate? If the ligand can be polydentate, give the maximum number of places on the ligand that can bind simultaneously to a single metal center. (a) Chloride ion; (b) cyanide ion; (c) ethylenediaminetetraacetate; (d) $N(CH_2CH_2NH_2)_3$.

16.31 Which of the following isomers of diaminobenzene can form chelating complexes? Explain.

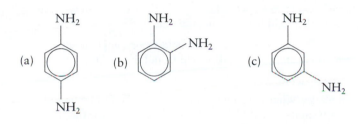

16.32 Which of the following ligands do you expect to form chelating complexes? Explain.

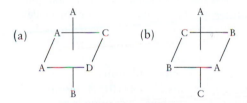

  (a) Bipyridine   (b) *ortho*-Phenanthroline   (c) Pyrimidine

16.33 With the help of Table 16.4, determine the coordination number of the metal ion in each of the following complexes:
(a) $[NiCl_4]^{2-}$; (b) $[Ag(NH_3)_2]^+$; (c) $[PtCl_2(en)_2]^{2+}$;
(d) $[Cr(edta)]^-$.

16.34 With the help of Table 16.4, determine the coordination number of the metal ion in each of the following complexes:
(a) $[Ir(en)_3]^{3+}$; (b) $[Fe(ox)_3]^{3-}$ (the oxalato ligand is bidentate);
(c) $[PtCl_2(NH_3)_2]$; (d) $Fe(CO)_5$.

16.35 Identify the type of structural isomerism that exists in each of the following pairs of compounds:
(a) $[Co(NO_2)(NH_3)_5]Br_2$ and $[Co(ONO)(NH_3)_5]Br_2$
(b) $[Pt(SO_4)(NH_3)_4](OH)_2$ and $[Pt(OH)_2(NH_3)_4]SO_4$
(c) $[CoCl(SCN)(NH_3)_4]Cl$ and $[CoCl(NCS)(NH_3)_4]Cl$
(d) $[CrCl(NH_3)_5]Br$ and $[CrBr(NH_3)_5]Cl$

16.36 Identify the type of structural isomerism that exists in each of the following pairs of compounds:
(a) $[Cr(en)_3][Co(ox)_3]$ and $[Co(en)_3][Cr(ox)_3]$
(b) $[CoCl_2(NH_3)_4]Cl \cdot H_2O$ and $[CoCl(NH_3)_4(H_2O)]Cl_2$

(c) $[Co(CN)_5(NCS)]^{3-}$ and $[Co(CN)_5(SCN)]^{3-}$
(d) $[Pt(NH_3)_4][PtCl_6]$ and $[PtCl_2(NH_3)_4][PtCl_4]$

16.37 Which of the following coordination compounds can have cis and trans isomers? If such isomerism exists, draw the two structures and name the compound.
(a) $[CoCl_2(NH_3)_4]Cl \cdot H_2O$; (b) $[CoCl(NH_3)_5]Br$;
(c) $[PtCl_2(NH_3)_2]$, a square planar complex.

16.38 Which of the following coordination compounds can have cis and trans isomers? If such isomerism exists, draw the two structures and label the compounds.
(a) $[Fe(OH)(H_2O)_5]^{2+}$; (b) $[RuBr_2(NH_3)_4]^+$;
(c) $[Co(NH_3)_4(H_2O)_2]^{3+}$.

16.39 Is either of the following complexes chiral? If both complexes are chiral, do they form an enantiomeric pair?

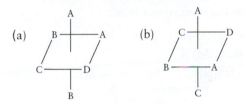

16.40 Is either of the following complexes chiral? If both complexes are chiral, do they form an enantiomeric pair?

(a) and (b) diagrams

The Electronic Structures of Complexes

16.41 Determine the number of valence electrons present in each of the following metal ions: (a) Ti^{3+}; (b) Fe^{2+}; (c) Mn^{2+}; (d) Cr^{3+}; (e) Os^{2+}; (f) Pd^{4+}.

16.42 Determine the number of valence electrons present in each of the following metal ions: (a) V^{2+}; (b) Ru^{2+}; (c) Rh^{3+}; (d) Cu^{2+}; (e) Os^{4+}; (f) Co^{3+}.

16.43 Determine the number of valence electrons present in each of the following metal ions: (a) Ti^{2+}; (b) Tc^{2+}; (c) Ir^+; (d) Ag^+; (e) Y^{3+}; (f) Zn^{2+}.

16.44 Determine the number of valence electrons present in each of the following metal ions: (a) Nb^{3+}; (b) Mo^{3+}; (c) Rh^+; (d) Ni^{2+}; (e) Re^{4+}; (f) Au^{3+}.

16.45 Draw an orbital energy-level diagram (like those in Figs. 16.26 and 16.28) showing the configuration of *d*-electrons on the metal ion in each of the following complexes:
(a) $[Co(NH_3)_6]^{3+}$; (b) $[NiCl_4]^{2-}$ (tetrahedral);
(c) $[Fe(H_2O)_6]^{3+}$; (d) $[Fe(CN)_6]^{3-}$. Predict the number of unpaired electrons for each complex.

16.46 Draw an orbital energy-level diagram (like those in Figs. 16.26 and 16.28) showing the configuration of *d*-electrons on the metal ion in each of the following complexes:
(a) $[Zn(H_2O)_6]^{2+}$; (b) $[CoCl_4]^{2-}$ (tetrahedral); (c) $[Co(CN)_6]^{3-}$;
(d) $[CoF_6]^{3-}$. Predict the number of unpaired electrons for each complex.

16.47 The complexes (a) $[Co(en)_3]^{3+}$ and (b) $[Mn(CN)_6]^{3-}$ have low-spin electron configurations. How many unpaired electrons, if any, are there in each complex?

16.48 The complexes (a) $[FeF_6]^{3-}$ and (b) $[Co(ox)_3]^{4-}$ have high-spin electron configurations. How many unpaired electrons are there in each complex?

16.49 Explain the difference between a weak-field ligand and a strong-field ligand. What measurements can be used to classify them as such?

16.50 Describe the changes that may take place in a compound's properties when weak-field ligands are replaced by strong-field ligands.

16.51 Of the two complexes (a) $[CoF_6]^{3-}$ and (b) $[Co(en)_3]^{3+}$, one appears yellow and the other appears blue. Match the complex to the color and explain your choice.

16.52 Which of the complexes $[Co(NH_3)_6]^{3+}$ or $[CoCl_6]^{3-}$ absorbs at the longer wavelength?

16.53 The complex $[Ni(NH_3)_6]^{2+}$ has a ligand field splitting of 209 kJ·mol^{-1} and forms a purple solution. What is the wavelength and color of the absorbed light?

16.54 In aqueous solution, water competes effectively with bromide ions for coordination to Cu^{2+} ions. The hexaaquacopper(II) ion is the predominant species in solution. However, in the presence of a large concentration of bromide ions, the solution becomes deep violet. This violet color is due to the presence of the tetrabromocuprate(II) ions, which are tetrahedral. This process is reversible, and so the solution becomes light blue again on dilution with water. (a) Write the formulas of the two complex ions of copper(II) that form. (b) Is the change in color from violet to blue on dilution expected? Explain your reasoning.

16.55 Suggest a reason why $Zn^{2+}(aq)$ ions are colorless. Would you expect zinc compounds to be paramagnetic? Explain your answer.

16.56 Suggest a reason why copper(II) compounds are often colored but copper(I) compounds are colorless. Which oxidation number results in paramagnetic compounds?

16.57 Estimate the ligand field splitting for (a) $[CrCl_6]^{3-}$ ($\lambda_{max} = 740$ nm), (b) $[Cr(NH_3)_6]^{3+}$ ($\lambda_{max} = 460$ nm), and (c) $[Cr(H_2O)_6]^{3+}$ ($\lambda_{max} = 575$ nm), where λ_{max} is the wavelength of the most intensely absorbed light. (d) Arrange the ligands in order of increasing ligand field strength.

16.58 Estimate the ligand field splitting for (a) $[Co(CN)_6]^{3-}$ ($\lambda_{max} = 295$ nm), (b) $[Co(NH_3)_6]^{3+}$ ($\lambda_{max} = 435$ nm), and (c) $[Co(H_2O)_6]^{3+}$ ($\lambda_{max} = 540$ nm), where λ_{max} is the wavelength of the most intensely absorbed light. (d) Arrange the ligands in order of increasing ligand field strength.

16.59 Which *d*-orbitals on the metal ion are used to form σ-bonds between octahedral metal ions and ligands?

16.60 Which *d*-orbitals on the metal ion are used to form π-bonds between octahedral metal ions and ligands?

16.61 Ligands that can interact with a metal center by forming a π-type bond are commonly called π-*acids* and π-*bases*. The definitions of π-acid and π-base are similar to those used for

Lewis acidity. A π-acid is a ligand that can accept electrons by using π-orbitals, and a π-base donates electrons by using π-orbitals. On the basis of these definitions, describe each of the following ligands as either a π-acid, a π-base, or neither: (a) CN^-; (b) Cl^-; (c) H_2O; (d) en. (e) Place them in order of increasing ligand field splitting.

16.62 Describe each of the following ligands as either a π-acid, a π-base, or neither (see Exercise 16.61): (a) NH_3; (b) ox; (c) F^-; (d) CO. (e) Place them in order of increasing ligand field splitting.

16.63 Is the best description of the t_{2g}-orbitals in $[Co(H_2O)_6]^{3+}$ (a) bonding, (b) antibonding, or (c) nonbonding? Explain how you reached that conclusion.

16.64 Is the best description of the t_{2g}-orbitals in $[Fe(CN)_6]^{3-}$ (a) bonding, (b) antibonding, or (c) nonbonding? Explain how you reached that conclusion.

16.65 Is the best description of the e_g-orbitals in $[Fe(H_2O)_6]^{3+}$ (a) bonding, (b) antibonding, or (c) nonbonding? Explain how you reached that conclusion.

16.66 Is the best description of the t_{2g}-orbitals in $[CoF_6]^{3-}$ (a) bonding, (b) antibonding, or (c) nonbonding? Explain how you reached that conclusion.

16.67 Using concepts of ligand field theory, explain why water is a weaker field ligand than ammonia.

16.68 Using concepts of ligand field theory, explain why ethylenediamine is a weaker field ligand than CO.

The Impact on Materials

16.69 (a) What reducing agent is used in the production of iron from its ore? (b) Write chemical equations for the production of iron in a blast furnace. (c) What is the major impurity in the product of the blast furnace?

16.70 (a) What is the purpose of adding limestone to a blast furnace? (b) Write chemical equations that show the reactions of lime in a blast furnace.

16.71 How is the impurity present in pig iron removed to create high-quality steel?

16.72 What metal is generally added to steel to make it corrosion resistant?

16.73 What metals are the primary components of bronze? (You might want to refer to Table 5.5.)

16.74 What metals are the primary components of brass? (You might want to refer to Table 5.5.)

16.75 How do alloys differ from the pure metals from which they are made?

16.76 What is the oxidation state of iron in the alkaline earth ferrites?

16.77 The *magnetic susceptibility*, χ (chi), is an indication of the number of unpaired electrons in a compound; it varies with temperature, *T*. A graph of magnetic susceptibility as a function of temperature for a typical ferromagnetic material is shown here. The temperature labeled T_C is known as the *Curie temperature*. On one side of the Curie temperature, the compound is a simple paramagnetic material and, on the other side, the substance is ferromagnetic. From the graph, determine whether this compound is ferromagnetic above or below the Curie temperature. Explain your reasoning.

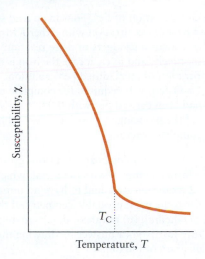

16.78 Describe the differences between diamagnetism, paramagnetism, ferromagnetism, and antiferromagnetism.

Integrated Exercises

16.79 An analytical chemistry team analyzed a green mineral salt found in the glaze of an ancient pot. When they heated the mineral it gave off a colorless gas that turned limewater milky white. When they dissolved the mineral in sulfuric acid the same colorless gas was released and a blue solution formed. Suggest a possible formula for the compound and justify your conclusion.

16.80 An analytical chemistry team analyzed a yellow mineral salt found in the pigments in the studio of a Renaissance artist. When hydrochloric acid was added the mineral dissolved and formed an orange solution. The original salt was dissolved in water to form a yellow solution. When barium chloride was added to the solution a yellow solid formed. The solid was filtered and the water evaporated from the filtrate. The colorless chloride remaining was subjected to a flame test. It gave a bright yellow color to the flame. Suggest a possible formula for the compound and justify your conclusion.

16.81 Vanadium is used in steelmaking because it forms V_4C_3, which increases the strength of the steel and its resistance to wear. It is formed in the steel by adding V_2O_5 to iron ore as it is reduced by carbon during the refining process. (a) What is the likely oxidation state of vanadium in V_4C_3? (b) Write a balanced equation for the reaction of V_2O_5 with carbon to form V_4C_3 and CO_2.

16.82 When 5.25 g of $[Ni(H_2O)_x]Cl_2$ is heated, 2.387 g of water and 1.57 g of chlorine gas are collected. How many water molecules are present in the complex?

16.83 The compound $Cr(OH)_3$ is very insoluble in water; therefore, electrochemical methods must be used to determine its K_{sp}. Given that the reduction of $Cr(OH)_3(s)$ to $Cr(s)$ and hydroxide ions has a standard potential of -1.34 V, calculate the solubility product for $Cr(OH)_3$.

16.84 The complex ion $[Ni(NH_3)_6]^{2+}$ forms in a solution containing 0.75 M $NH_3(aq)$ and 0.10 M $Ni^{2+}(aq)$. If the formation constant for $[Ni(NH_3)_6]^{2+}$ is 1.0×10^9, what are the equilibrium concentrations?

16.85 (a) Draw all the possible isomers of the square planar complex $[PtBrCl(NH_3)_2]$ and name each isomer. (b) How can the existence of these isomers be used to show that the complex is square planar rather than tetrahedral?

16.86 Draw the structures of the isomeric forms of $[CrClBrI(NH_3)_3]$. Which isomers are chiral?

16.87 Suggest a chemical test for distinguishing between (a) $[Ni(SO_4)(en)_2]Cl_2$ and $[NiCl_2(en)_2]SO_4$; (b) $[NiI_2(en)_2]Cl_2$ and $[NiCl_2(en)_2]I_2$.

16.88 Trigonal bipyramidal complexes in which a metal ion is surrounded by five ligands are more rare than octahedral or tetrahedral complexes, but many are known. Will trigonal bipyramidal compounds of the formula MX_3Y_2 exhibit isomerism? If so what types of isomerism are possible?

16.89 (a) Sketch the orbital energy-level diagrams for $[MnCl_6]^{4-}$ and $[Mn(CN)_6]^{4-}$. (b) How many unpaired electrons are present in each complex? (c) Which complex absorbs the longer wavelengths of incident electromagnetic radiation? Explain your reasoning.

16.90 (a) Which types of ligands are π-acid ligands in general (see Exercise 16.61), strong field or weak field? (b) Which types of ligands are π-base ligands in general, strong field or weak field?

16.91 Explain why high-spin Mn^{2+} ions form complexes that are only faintly colored.

16.92 The hydride ion, H^-, can function as a ligand to d-block metals. Compare the differences in bonding to metal ions that you would expect between H^- and the halide ions.

16.93 The trigonal prismatic structure shown in the illustration was once proposed for the $[CoCl_2(NH_3)_4]$ complex. Use the fact that only two isomers of the complex are known to rule out the prismatic structure.

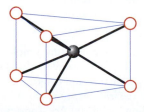

16.94 A hexagonal planar structure was once proposed for the $[CrCl_3(NH_3)_3]$ complex. Use the fact that only two isomers of the complex are known to rule out the hexagonal planar structure.

16.95 Before the structures of octahedral complexes were determined, various means were used to explain the fact that d-metal ions could bind to a greater number of ligands than expected on the basis of their charges. For example, Co^{3+} can bind to six ligands, not just three. An early theory attempted to explain this behavior by postulating that once three ligands had attached to a metal ion of charge +3, the others bonded to the attached ligands, forming chains of ligands. Thus, the compound $[Co(NH_3)_6]Cl_3$, which we now know to have an octahedral structure, would have been described as $Co(NH_3—NH_3—Cl)_3$. Show how the chain theory is not consistent with at least two properties of coordination compounds.

16.96 Suggest the form that the orbital energy-level diagram would take for a square planar complex with the ligands in the *xy* plane, and discuss how the building-up principle applies. *Hint:* The d_{z^2}-orbital has more electron density in the *xy* plane than the d_{zx}- or d_{yz}-orbitals but less than the d_{xy}-orbital.

16.97 Is there a correlation between the ligand field strength of the halide ions F^-, Cl^-, Br^-, and I^- and the electronegativity of the halogen? If so, can this correlation be explained by ligand field theory? Justify your answer.

16.98 The complexes $[Co(NH_3)_6]^{2+}$, $[Co(H_2O)_6]^{2+}$, and $[CoCl_4]^{2-}$ form colored solutions. One is pink, one yellow, and the third blue. Use the spectrochemical series and the relative magnitudes of Δ_O and Δ_T to match each color to a complex. Explain your reasoning.

16.99 Ligand field theory predicts that different types of metal ions will form more stable complexes with certain types of ligands. From your understanding of ligand field theory, predict what types of ligands (weak-field or strong-field) would form the more stable complexes with the early transition metals in their highest oxidation states. Likewise, predict the types of ligands that would form the more stable complexes of the late transition metals (those to the right of the *d* block) in their lowest oxidation states. Explain the reasoning behind your choices.

16.100 (a) A solid pink salt has the formula $CoCl_3 \cdot 5NH_3 \cdot H_2O$. When $AgNO_3$ is added to the pink solution containing 0.0010 mol of the salt, 0.43 g of AgCl precipitates. Write the correct formula and name of the salt. (b) Heating the salt produces a purple solid with the formula $CoCl_3 \cdot 5NH_3$. When $AgNO_3$ is added to a solution containing 0.0010 mol of this salt, 0.29 g of AgCl precipitates. Write the correct formula and name of the purple salt.

16.101 *cis*-Platin is an anticancer drug with a structure that can be viewed on the Web site. (a) What is the formula and systematic name for the compound *cis*-Platin? (b) Draw any isomers that are possible for this compound. Label any isomers that are optically active. (c) What is the coordination geometry of the platinum atom?

16.102 The Web site contains a molecular model of an iridium complex with formula $IrCl(CO)_2\{P(C_6H_5)_3\}_2$. (a) View this structure and determine the coordination geometry of the iridium atom. (b) What is the oxidation number of the iridium atom? (c) Are isomers of this compound possible? If so, draw them. Label any that are optical isomers.

16.103 Two chemists prepared a complex and determined its formula, which they wrote as $[CrNH_3Cl_3] \cdot 2H_2O$. However, when they dissolved 2.11 g of the compound in water and added an excess of silver nitrate, 2.87 g of AgCl precipitated and they realized that the formula was incorrect. Write the correct formula of the compound and draw its structure, including all possible isomers.

16.104 Use the information in Appendix 2B to determine the equilibrium constant for the disproportionation of copper(I) ions in aqueous solution at 25°C to copper metal and copper(II) ions.

16.105 Use the information in Appendix 2B to determine the equilibrium constant for the disproportionation of mercury(I) ions in aqueous solution at 25°C to mercury metal and mercury(II) ions.

16.106 The concentration of Fe^{2+} ions in an acid solution can be determined by a redox titration with either $KMnO_4$ or $K_2Cr_2O_7$. The reduction products of these reactions are Mn^{2+} and Cr^{3+}, respectively, and in each case the iron is oxidized to Fe^{3+}. In one titration of an acidified Fe^{2+} solution, 25.20 mL of 0.0210 M $K_2Cr_2O_7$(aq) was required for complete reaction. If the titration had been carried out with 0.0420 M $KMnO_4$(aq), what volume of the permanganate solution would have been required for complete reaction?

16.107 In a nickel-containing enzyme various groups of atoms in the enzyme form a complex with the metal, which was found to be in the +2 oxidation state and to have no unpaired electrons. What is the most probable geometry of the Ni^{2+} complex: (a) octahedral; (b) tetrahedral; (c) square planar (see Exercise 16.96)? Justify your answer by drawing the orbital energy-level diagram of the ion.

16.108 How many chelate rings are present in: (a) $[Ru(ox)_3]^{3-}$; (b) $[Fe(trien)]^{3+}$; (c) $[Cu(dien)_2]^{2+}$?

Chemistry Connections

16.109 Hemoglobin contains one heme group per subunit. The heme group is an Fe^{2+} complex that is coordinated to the four N atoms in a porphyrin ligand in a square planar arrangement and to an N atom on the histidine residue (see Table 19.4) on the subunit. The hemoglobin molecule transports O_2 through the body by using a bond between O_2 and the Fe^{2+} ion at the center of the heme group. The Fe^{2+} ion can be viewed as having octahedral coordination, with one position unoccupied and available for bonding to oxygen. (See Box 16.1 and Section 9.13.)
(a) Draw the structure of the heme group with the sixth site not coordinated.
(b) The deoxygenated heme is a high-spin complex of the Fe^{2+} ion. When the oxygen molecule binds to the Fe^{2+} ion as the sixth ligand the resulting octahedral complex is low-spin. Predict the number of unpaired electrons in (i) deoxygenated and (ii) oxygenated heme.
(c) Other compounds can bond to the iron atom, displacing oxygen. Identify which of the following species cannot bond to the iron in a heme group and explain your reasoning: CO; Cl^-; BF_3; NO_2^-.
(d) Hemoglobin in both its oxygenated (HbO$_2$) and deoxygenated (Hb) forms helps to maintain the pH of blood at an optimal level. Hemoglobin has several acidic protons, but the most important deprotonation is the first. At 25°C, for the deoxygenated form, $HbH(aq) + H_2O(l) \rightleftharpoons Hb^-(aq) + H_3O^+(aq)$ and $pK_{a1} = 6.62$. For the oxygenated form, $HbO_2H(aq) + H_2O(l) \rightleftharpoons HbO_2^-(aq) + H_3O^+(aq)$ and $pK_{a1} = 8.18$. Calculate the percentage of each form of hemoglobin that is deprotonated at the pH of blood, 7.4.
(e) Use your answer to part (d) to determine how the oxygenation of hemoglobin affects the pH of blood. That is, will the pH increase, decrease, or stay the same as more hemoglobin molecules become oxygenated? Explain your reasoning.

NUCLEAR CHEMISTRY

What Are the Key Ideas? Changes in the nucleus of an atom can result in the ejection of particles, the transformation of the atom into another element, and the release of energy.

Why Do We Need to Know This Material? Nuclear chemistry is central to the development of nuclear energy. Chemistry provides techniques for preparing and recycling nuclear fuels and for the disposal of hazardous radioactive waste. Nuclear chemistry is used in medicine to treat cancer and to produce images of organs inside living bodies. It is used in chemistry to investigate reaction mechanisms and in archaeology to date ancient objects; nuclear chemistry is also part of the military strategies of many nations. Furthermore, this material informs the complex political, environmental, and economic debate about the deployment of nuclear energy.

What Do We Need to Know Already? Nuclear processes can be understood in terms of atomic structure (Section B and Chapter 1) and energy changes (Chapter 6). The section on rates of radioactive decay builds on chemical kinetics (particularly Sections 13.4 and 13.5).

Chapter **17**

So far, we have regarded the atomic nucleus as an unchanging passenger in chemical reactions and have concentrated on the changes that electrons undergo. However, nuclei can undergo changes, and **nuclear chemistry** explores the chemical consequences of those changes. One kind of change is **nuclear fission,** the fragmentation of large nuclei into smaller nuclei. Although fission is a source of energy, the resulting radioactive waste presents serious hazards to life. Another kind of change is **nuclear fusion,** the merging of small nuclei into larger ones. Fusion, too, is a source of energy, but the technique is technologically challenging and therefore expensive to develop. Nuclear processes are critical for the future of humanity, because they might prove to be the answer to one of our greatest challenges—the development of adequate energy supplies. The issues involved and chemistry's contributions to their solution are the subjects of this chapter.

This chapter has three main themes. First, we consider the pattern of nuclear stability and the manner in which unstable nuclei spontaneously undergo change. Changes in the composition of nuclei are the source of all the elements except hydrogen, and so we also discuss the origin of the elements. Next, we consider the effect of radiation on living beings and the uses to which nuclear radiation is put in chemistry. Finally, we consider nuclear energy and how chemistry is used to resolve the problems associated with its use.

NUCLEAR DECAY

Atomic nuclei are extraordinary particles. They contain all the protons in the atom crammed together in a tiny volume, despite their positive charges (Fig. 17.1). Yet most nuclei survive indefinitely despite the immense repulsive forces between their protons. In some nuclei, though, the repulsions that protons exert on one another overcome the force that holds the nucleons together. Fragments of the nucleus are then ejected, and the nucleus is said to "decay."

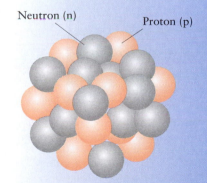

FIGURE 17.1 A nucleus can be visualized as a collection of tightly bonded protons (pink) and neutrons (gray). The diameter of a nucleus is about 10 fm (1 fm = 10^{-15} m).

FIGURE 17.2 Henri Becquerel discovered radioactivity when he noticed that an unexposed photographic plate left near some uranium oxide became fogged. This photograph shows one of his original plates annotated with his record of the event.

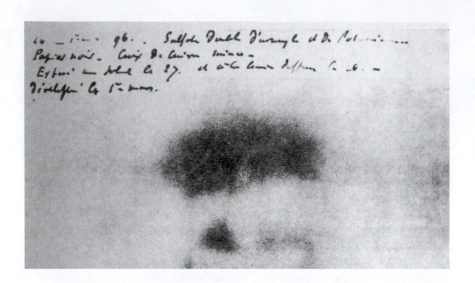

FIGURE 17.3 Marie Sklodowska Curie (1867–1934).

17.1 The Evidence for Spontaneous Nuclear Decay

In 1896, the French scientist Henri Becquerel happened to store a sample of uranium oxide in a drawer that contained some photographic plates (Fig. 17.2). He was astonished to find that the uranium compound darkened the plates even though they were covered with an opaque material. Becquerel realized that the uranium compound must give off some kind of radiation. Marie Sklodowska Curie (Fig. 17.3), a young Polish doctoral student, showed that the radiation, which she called **radioactivity,** was emitted by uranium regardless of the compound in which it was found. She concluded that the source must be the uranium atoms themselves. Together with her husband, Pierre, she went on to show that thorium, radium, and polonium are also radioactive.

The origin of the rays was initially a mystery, because the existence of the atomic nucleus was unknown at the time. However, in 1898, Ernest Rutherford took the first step to discover their origin when he identified three different types of radioactivity by observing the effect of electric fields on radioactive emissions (Fig. 17.4). Rutherford called the three types α (alpha), β (beta), and γ (gamma) radiation.

When Rutherford allowed the radiation to pass between two electrically charged electrodes, he found that one type was attracted to the negatively charged electrode. He proposed that the radiation attracted to the negative electrode consists of positively charged particles, which he called **α particles.** From the charge and mass of the particles, he was able to identify them as helium atoms that had lost their two electrons. Once Rutherford had identified the atomic nucleus (in 1908, Section B), he realized that an α particle must be a helium nucleus, He^{2+}. An α particle is denoted $^4_2\alpha$, or simply α. We can think of it as a tightly bound cluster of two protons and two neutrons (Fig. 17.5).

Rutherford found that a second type of radiation was attracted to the positively charged electrode. He proposed that this type of radiation consists of a stream of negatively charged particles. By measuring the charge and mass of these particles, he showed that they are electrons. The rapidly moving electrons emitted by nuclei are called **β particles** and denoted β^-. Because a β particle has no protons or neutrons, its mass number is 0 and it can be written $^0_{-1}e$.

The third common type of radiation that Rutherford identified, **γ radiation,** is not affected by an electrical field. Like light, γ radiation is electromagnetic radiation but of much higher frequency—greater than about 10^{20} Hz and corresponding to

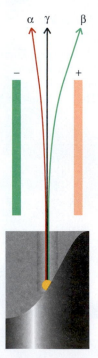

FIGURE 17.4 The effects of an electric field on nuclear radiation. The direction of deflection shows that α-rays are positively charged, β-rays negatively charged, and γ-rays uncharged.

FIGURE 17.5 An α particle has two positive charges and a mass number of 4. It consists of two protons and two neutrons and is the same as the nucleus of a helium-4 atom.

wavelengths less than about 1 pm. It can be regarded as a stream of very high energy photons, each photon being emitted by a single nucleus as that nucleus discards energy. The frequencies, ν, of γ rays are given by the relation $\nu = \Delta E/h$ (see Section 1.4), where ΔE is the energy discarded by the nucleus. The frequency is very high because the energy difference between the excited and the ground nuclear states is very large. Both α and β radiation are often accompanied by γ radiation: the new nucleus may be formed with its nucleons in a high-energy arrangement, and a γ-ray photon is emitted when nucleons settle down into a state of lower energy (Fig. 17.6).

Since Rutherford's work, scientists have identified other types of nuclear radiation. Some consist of rapidly moving particles, such as neutrons or protons. Others consist of rapidly moving **antiparticles**, particles with a mass equal to that of one of the subatomic particles but with an opposite charge. For example, the **positron** has the same mass as an electron but a positive charge; it is denoted β^+ or $^0_{+1}e$. When an antiparticle encounters its corresponding particle, both particles are annihilated and completely converted into energy. Table 17.1 summarizes the properties of particles commonly found in nuclear radiation.

> *The most common types of radiation emitted by radioactive nuclei are α particles (the nuclei of helium atoms), β particles (fast electrons ejected from the nucleus), and γ rays (high-frequency electromagnetic radiation).*

17.2 Nuclear Reactions

The discoveries of Becquerel, Curie, and Rutherford and Rutherford's later development of the nuclear model of the atom (Section B) showed that radioactivity is produced by **nuclear decay**, the partial breakup of a nucleus. The change in the composition of a nucleus is called a **nuclear reaction**. Recall from Section B that nuclei are composed of protons and neutrons that are collectively called *nucleons*; a specific nucleus with a given atomic number and mass number is called a **nuclide**. Thus, ^{1}H, ^{2}H, and ^{16}O are three different nuclides; the first two being isotopes of the same element. Nuclei that change their structure spontaneously and emit radiation are called **radioactive**. Often the result is a different nuclide.

Nuclear reactions differ in some important ways from chemical reactions. First, different isotopes of the same element undergo essentially the same chemical reactions, but their nuclei undergo very different nuclear reactions. Second, when α or β particles are emitted from the nucleus, they leave behind a nucleus with a different number of protons. The product, which is called the **daughter nucleus**

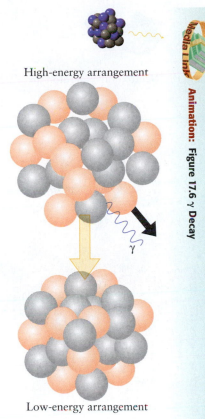

High-energy arrangement

Low-energy arrangement

FIGURE 17.6 After a nucleus decays, the nucleons remaining in the nucleus may be left in a high-energy state, as shown by the expanded arrangement in the upper part of the illustration. As the nucleons adjust to a lower-energy arrangement (bottom), the excess energy is released as a γ-ray photon.

Media Link

Animation: Figure 17.6 γ Decay

TABLE 17.1 Nuclear Radiation

Type	Degree of penetration	Speed*	Particle[†]	Mass number	Charge	Example
α	not penetrating but damaging	10% of c	helium-4 nucleus, $^4_2\text{He}^{2+}$, $^4_2\alpha$, α	4	+2	$^{226}_{88}\text{Ra} \longrightarrow {}^{222}_{86}\text{Rn} + \alpha$ (Fig. 17.7)
β	moderately penetrating	<90% of c	electron, $^0_{-1}e$, β^-, β, e^-	0	−1	$^3_1\text{H} \longrightarrow {}^3_2\text{He} + {}^0_{-1}e$ (Fig. 17.8)
electron capture[‡]	—	—	electron, $^0_{-1}e$, e^-	0	−1	$^{44}_{22}\text{Ti} + {}^0_{-1}e \longrightarrow {}^{44}_{21}\text{Sc}$ (Fig. 17.9)
γ	very penetrating; often accompanies other radiation	c	photon, γ	0	0	$^{60}_{27}\text{Co}^{*\S} \longrightarrow {}^{60}_{27}\text{Co} + \gamma$ (Fig. 17.6)
β^+	moderately penetrating	<90% of c	positron, $^0_{+1}e$, β^+	0	+1	$^{22}_{11}\text{Na} \longrightarrow {}^{22}_{10}\text{Ne} + {}^0_{+1}e$ (Fig. 17.10)
p	moderate or low penetration	10% of c	proton, $^1_1\text{H}^+$, $^1_1$p, p	1	+1	$^{53}_{27}\text{Co} \longrightarrow {}^{52}_{26}\text{Fe} + {}^1_1\text{p}$
n	very penetrating	<10% of c	neutron, $^1_0$n, n	1	0	$^{137}_{53}\text{I} \longrightarrow {}^{136}_{53}\text{I} + {}^1_0\text{n}$

*c is the speed of light.
[†]Alternative symbols are given for the particles; often it is sufficient to use the simplest (the one on the right).
[‡]Electron capture is not nuclear radiation but is included for completeness.
[§]An energetically excited state of a nucleus is usually denoted by an asterisk (*).

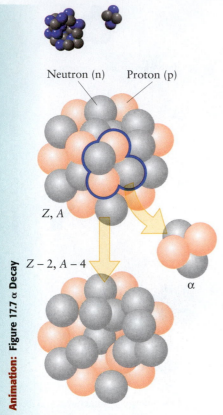

Neutron (n)   Proton (p)

Z, A

$Z - 2, A - 4$

α

FIGURE 17.7 When a nucleus ejects an α particle, the atomic number of the atom decreases by 2 and the mass number decreases by 4. The nucleons ejected from the upper nucleus are indicated by the blue boundary.

(Fig. 17.7), is therefore the nucleus of an atom of a different element. For example, when a radon-222 nucleus emits an α particle, a polonium-218 nucleus is formed. In this case, a **nuclear transmutation**, the conversion of one element into another, has taken place. Another important difference between nuclear and chemical reactions is that energy changes are very much greater for nuclear reactions than for chemical reactions. For example, the combustion of 1.0 g of methane produces about 52 kJ of energy as heat. In contrast, a nuclear reaction of 1.0 g of uranium-235 produces about 8.2×10^7 kJ of energy, more than a million times as much.

To predict the identity of a daughter nucleus, we note how the atomic number and mass number change when the parent nucleus ejects a particle. For example, when a radium-226 nucleus, with $Z = 88$, undergoes α decay, it emits an α particle, which has a nuclear charge of $+2$ and a mass number of 4. Because both total mass number and total nuclear charge are conserved in a nuclear reaction, the fragment remaining must be a nucleus of atomic number 86 (radon) and mass number 222; so the daughter nucleus is radon-222:

$$^{222}_{88}\text{Ra} \longrightarrow ^{222}_{86}\text{Rn} + ^4_2\alpha$$

The expression for these changes is called a **nuclear equation**. The following examples show how to use nuclear equations to identify daughter nuclei.

EXAMPLE 17.1   **Predicting the outcome of α and β decay**

What nuclide is produced by (a) α decay of polonium-211; (b) β decay of sodium-24?

STRATEGY Write the nuclear equation for each reaction, representing the daughter nuclide as E, with atomic number Z and mass number A. Then find Z and A from the requirement that both mass number and atomic number are conserved in a nuclear reaction. (a) In α decay, two protons and two neutrons are lost. As a result, the mass number decreases by 4 and the atomic number decreases by 2 (see Fig. 17.7). (b) The loss of one negative charge when an electron is ejected from the nucleus (Fig. 17.8) can be interpreted as the conversion of a neutron into a proton within the nucleus:

$$^1_0\text{n} \longrightarrow ^1_1\text{p} + ^{\;0}_{-1}\text{e}, \quad \text{or, more simply,} \quad \text{n} \longrightarrow \text{p} + \beta$$

The atomic number of the daughter nucleus is greater by 1 than that of the parent nucleus, because it has one more proton, but the mass number is unchanged, because the total number of nucleons in the nucleus is the same.

SOLUTION

(a) Write the equation for the nuclear reaction.	$^{211}_{84}\text{Po} \longrightarrow ^A_Z\text{E} + ^4_2\alpha$
Express mass and charge conservation.	$211 = A + 4$, or $A = 207$ $84 = Z + 2$, $Z = 82$
Identify the element.	$Z = 82$ corresponds to Pb
Write the nuclear equation.	$^{211}_{84}\text{Po} \longrightarrow ^{207}_{82}\text{Pb} + ^4_2\alpha$

(b) Write the equation for the nuclear reaction.	$^{24}_{11}\text{Na} \longrightarrow ^A_Z\text{E} + ^{\;0}_{-1}\text{e}$
Express mass and charge conservation.	$24 = A + 0$, or $A = 24$, unchanged $11 = Z - 1$, $Z = 12$

Identify the element.	$Z = 12$ corresponds to Mg

Write the equation for the nuclear reaction.	$^{24}_{11}\text{Na} \longrightarrow \, ^{24}_{12}\text{Mg} + \, ^{0}_{-1}\text{e}$

SELF-TEST 17.1A Identify the nuclide produced by (a) α decay of uranium-235, (b) β decay of lithium-9.

[**Answer:** (a) Thorium-231; (b) beryllium-9]

SELF-TEST 17.1B Identify the nuclide produced by (a) α decay of thorium-232, (b) β decay of radium-228.

EXAMPLE 17.2 Predicting the outcome of electron capture and positron emission

What nuclide is produced when (a) calcium-41 undergoes electron capture; (b) oxygen-15 undergoes positron emission?

STRATEGY (a) In electron capture, a nucleus captures one of the surrounding electrons; a proton is turned into a neutron and, although there is no change in mass number, the atomic number is reduced by 1 (Fig. 17.9). (b) A positron has the same tiny mass as an electron but a single positive charge. Positron emission can be thought of as the positive charge shrugged off by a proton as it is converted into a neutron. As a result, the atomic number decreases by 1 but there is no change in mass number (Fig. 17.10).

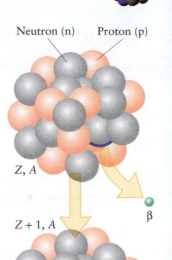

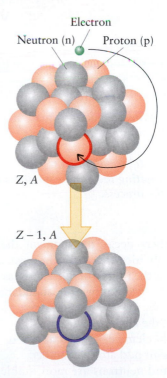

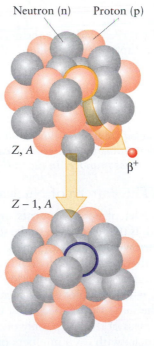

FIGURE 17.8 When a nucleus ejects a β particle, the atomic number of the atom increases by 1 and the mass number remains unchanged. The neutron that we can regard as the source of the electron is indicated by the blue boundary in the upper nucleus.

FIGURE 17.9 In electron capture, a nucleus captures one of the surrounding electrons. The effect is to convert a proton (outlined in red) into a neutron (outlined in blue). As a result, the atomic number decreases by 1 but the mass number remains the same.

FIGURE 17.10 In positron (β⁺) emission, the nucleus ejects a positron. The effect is to convert a proton into a neutron (outlined in blue). As a result, the atomic number decreases by 1 but the mass number remains the same.

SOLUTION

(a) Write the equation for the nuclear reaction.	$^{41}_{20}\text{Ca} + ^{0}_{-1}\text{e} \longrightarrow ^{A}_{Z}\text{E}$
Express mass and charge conservation.	$41 + 0 = A$, or $A = 41$, unchanged $20 - 1 = Z$, $Z = 19$
Identify the element.	$Z = 19$ corresponds to K.
Write the nuclear equation.	$^{41}_{20}\text{Ca} + ^{0}_{-1}\text{e} \longrightarrow ^{41}_{19}\text{K}$

	3 Li	4 Be	
	11 Na	12 Mg	e⁻
19 K	20 Ca	21 Sc	22 Ti

(b) Write the equation for the nuclear reaction.	$^{15}_{8}\text{O} \longrightarrow ^{A}_{Z}\text{E} + ^{0}_{+1}\text{e}$
Express mass and charge conservation.	$15 = A + 0$, or $A = 15$, unchanged $8 = Z - 1$, $Z = 7$
Identify the element.	$Z = 7$ corresponds to N.
Write the nuclear equation.	$^{15}_{8}\text{O} \longrightarrow ^{15}_{7}\text{N} + ^{0}_{+1}\text{e}$

		e⁺	2 He
7 N	8 O	9 F	10 Ne
15 P	16 S	17 Cl	18 Ar

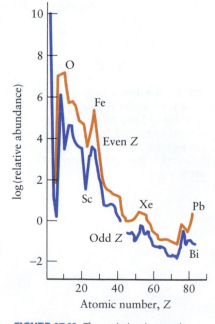

FIGURE 17.11 The variation in cosmic nuclear abundance with atomic number. Note that elements with even atomic numbers (brown curve) are consistently more abundant than neighboring elements with odd atomic numbers (blue curve).

SELF-TEST 17.2A Identify the nuclide produced and write the nuclear equation for (a) electron capture by beryllium-7; (b) positron emission by sodium-22.

[*Answer:* (a) $^{7}_{4}\text{Be} + ^{0}_{-1}\text{e} \longrightarrow ^{7}_{3}\text{Li}$; (b) $^{22}_{11}\text{Na} \longrightarrow ^{0}_{+1}\text{e} + ^{22}_{10}\text{Ne}$]

SELF-TEST 17.2B Identify the nuclide produced and write the nuclear equation for (a) electron capture by iron-55; (b) positron emission by carbon-11.

Nuclear reactions may result in the formation of different elements. The transmutation of a nucleus can be predicted by noting the atomic numbers and the mass numbers in the nuclear equation for the process.

17.3 The Pattern of Nuclear Stability

The nuclei of some elements are stable, but others decay the moment they are formed. Is there a pattern to the stabilities and instabilities of nuclei? The existence of a pattern would allow us to make predictions about the modes of nuclear decay. One clue is that elements with even atomic numbers are consistently more abundant than neighboring elements with odd atomic numbers. We can see this difference in Fig. 17.11, which is a plot of the cosmic abundance of the elements against atomic number. The same pattern occurs on Earth. Of the eight elements present as 1% or more of the mass of the Earth, only one, aluminum, has an odd atomic number.

Nuclei with even numbers of both protons and neutrons are more stable than those with any other combination. Conversely, nuclei with odd numbers of both protons and neutrons are the least stable (Fig. 17.12). Nuclei are more likely to be stable if they are built from certain numbers of either kind of nucleons. These numbers—namely, 2, 8, 20, 50, 82, 114, 126, and 184—are called **magic numbers**. For example, there are ten stable isotopes of tin ($Z = 50$), the most of any element, but only two stable isotopes of its neighbor antimony ($Z = 51$). The α particle itself is a "doubly magic" nucleus, with two protons and two neutrons. We shall see later that a number of actinides decay through a series of steps until they reach Pb, another doubly magic nuclide, with 126 neutrons and 82 protons. This pattern of nuclear stability is similar to the pattern of electron stability in atoms: the noble gas atoms have 2, 10, 18, 36, 54, and 86 electrons.

Figure 17.13 is a plot of mass number against atomic number for known nuclides. Stable nuclei are found in a **band of stability** surrounded by a **sea of instability,** the region of unstable nuclides that decay with the emission of radiation. For atomic numbers up to about 20, the stable nuclides have approximately equal numbers of neutrons and protons, and so A is close to $2Z$. For higher atomic numbers, all known nuclides—both stable and unstable—have more neutrons than protons, and so $A > 2Z$.

The increase in the ratio of neutrons to protons with increasing atomic number can be explained by considering the role of the neutrons in helping to overcome the electrical repulsions between protons. The **strong force** that holds the protons and neutrons together in a nucleus is powerful enough to overcome the electrostatic repulsion between the protons, but it can act only over a very short distance—about the diameter of a nucleus (Fig. 17.14). Because neutrons are uncharged, they can contribute to the strong force without adding to the electrostatic repulsion. A lot of neutrons are needed to overcome the mutual repulsion of the protons in a nucleus of high atomic number, which explains why the band of stability curves upward.

Nuclei with certain even numbers of protons and of neutrons are the most stable.

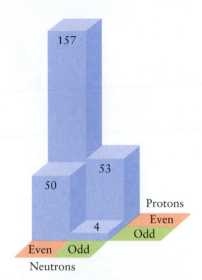

FIGURE 17.12 The numbers of stable nuclides having even or odd numbers of neutrons and protons. With the exception of hydrogen, by far the greatest number of stable nuclides (157) have even numbers of both protons and neutrons. Only four stable nuclides have odd numbers of both protons and neutrons.

17.4 Predicting the Type of Nuclear Decay

We can use Fig. 17.13 to predict the type of disintegration that a radioactive nuclide is likely to undergo. Nuclei that lie above the band of stability are **neutron rich:** they have a high proportion of neutrons. These nuclei tend to decay in such a way that the final n/p ratio is closer to that found in the band of stability. For example, a $^{14}_{6}\text{C}$ nucleus can reach a more stable state by ejecting a β particle, which reduces the n/p ratio as a result of the conversion of a neutron into a proton (Fig. 17.15):

$$^{14}_{6}\text{C} \longrightarrow {}^{14}_{7}\text{N} + {}^{0}_{-1}\text{e}$$

Nuclides that lie below the band of stability have a low proportion of neutrons and are classified as **proton rich.** These isotopes tend to decay in such a way that the

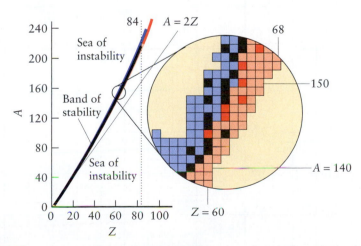

FIGURE 17.13 The manner in which nuclear stability depends on the atomic number and the mass number. Nuclides along the narrow black band (the band of stability) are generally stable. Nuclides in the blue region are likely to emit a β particle, and those in the red region are likely to emit an α particle. Nuclei in the pink region are likely either to emit positrons or to undergo electron capture. The straight line indicates the position that nuclides would have if the numbers of neutrons and protons were equal ($A = 2Z$). The inset shows a magnified view of the diagram near $Z = 60$.

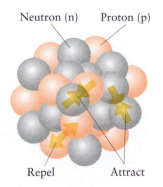

FIGURE 17.14 The protons in a nucleus repel one another electrically, but the strong force, which acts between all nucleons, holds the nucleus together.

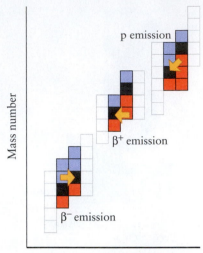

FIGURE 17.15 Three different ways of reaching the band of stability (black). Nuclei that are neutron rich (blue region) tend to convert neutrons into protons by β emission; nuclei that are proton rich (red) tend to reach stability (black) by emitting a positron, capturing an electron, or emitting a proton.

Bismuth ($Z = 83$) was thought to have a stable isotope, but in 2005 even that isotope was found to decay extremely slowly.

atomic number is reduced. For example, proton-rich $^{29}_{15}\text{P}$ is radioactive and decays by emitting a positron, a process that converts a proton into a neutron and raises the final n/p ratio:

$$^{29}_{15}\text{P} \longrightarrow {}^{29}_{14}\text{S} + {}^{0}_{+1}\text{e}$$

As shown in Example 17.2, electron capture and proton emission also decrease the proton count of proton-rich nuclides.

Very few nuclides with $Z < 60$ emit α particles. All nuclei with $Z > 82$ are unstable and decay mainly by α-particle emission. They must discard protons to reduce their atomic number and generally need to lose neutrons, too. These nuclei decay in a step-by-step manner and give rise to a **radioactive series,** a characteristic sequence of nuclides (Fig. 17.16). First, one α particle is ejected, then another α particle or a β-particle is ejected, and so on, until a stable nucleus, such as an isotope of lead (with the magic atomic number 82) is formed. For example, the uranium-238 series ends at lead-206, the uranium-235 series ends at lead-207, and the thorium-232 series ends at lead-208.

Self-Test 17.3A Which of the following processes, (a) electron capture, (b) proton emission, (c) β⁻ emission, (d) β⁺ emission, might a $^{145}_{64}\text{Gd}$ nucleus undergo to begin to reach stability? Refer to Figs. 17.13 and 17.15.

[*Answer:* a, b, d]

Self-Test 17.3B Which of the set of processes listed in Self-Test 17.3A might a $^{148}_{58}\text{Ce}$ nucleus undergo to begin to reach stability?

The pattern of nuclear stability can be used to predict the likely mode of radioactive decay: neutron-rich nuclei tend to reduce their neutron count; proton-rich nuclei tend to reduce their proton count. In general, only heavy nuclides emit α particles.

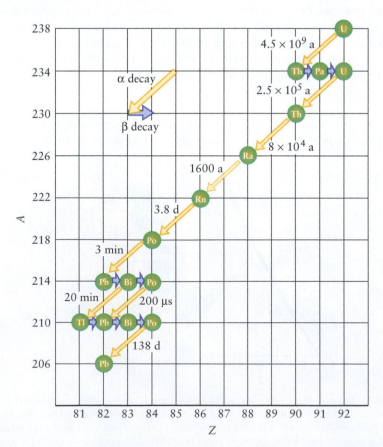

FIGURE 17.16 The uranium-238 decay series. The times are the half-lives of the nuclides (see Section 17.7). The unit a, for annum, is the SI abbreviation for year.

17.5 Nucleosynthesis

Nucleosynthesis is the formation of elements. Hydrogen and helium were produced in the Big Bang; all other elements are descended from these two, as a result of nuclear reactions taking place either in stars or in space. Some elements—among them technetium and promethium—are found in only trace amounts on Earth. Although these elements were made in stars, their short lifetimes did not allow them to survive long enough to contribute to the formation of our planet. However, nuclides that are too unstable to be found on Earth can be made by artificial techniques, and scientists have added about 2200 different nuclides to the 300 or so that occur naturally.

To overcome the energy barriers to nuclear synthesis, particles must collide vigorously with one another (Fig. 17.17), as they do in stars. Therefore, to make elements artificially, we need to simulate the conditions found inside a star. If it is traveling fast enough, a proton, an α particle, or another positively charged nucleus has enough kinetic energy to overcome the electrostatic repulsion from the nucleus. The incoming particle penetrates into the nucleus, where it is captured by the strong force. The high speeds required can be achieved in a particle accelerator.

The **transmutation** of elements—the conversion of one element into another, particularly of lead into gold—was the dream of the alchemists and one of the roots of modern chemistry. However, the alchemists had access only to chemical techniques, which proved ineffective because they involved energy changes far too weak to force nucleons into nuclei. Transmutation has now been recognized in nature and achieved in the laboratory, but by using methods undreamed of by the early alchemists. Rutherford achieved the first artificial nuclear transmutation in 1919. He bombarded nitrogen-14 nuclei with high-speed α particles. The products of the transmutation were oxygen-17 and a proton:

$$^{14}_{7}N + ^{4}_{2}\alpha \longrightarrow ^{17}_{8}O + ^{1}_{1}p$$

Such transmutation processes are commonly written simply as

$$^{4}_{7}N(\alpha, p)^{17}_{8}O$$

or, in general,

Target(incoming particle, ejected particle)product

A large number of nuclides have been synthesized on Earth. For instance, technetium was prepared (as technetium-97) for the first time on Earth in 1937 by the reaction between molybdenum and deuterium nuclei:

$$^{97}_{42}Mo + ^{2}_{1}H \longrightarrow ^{97}_{43}Tc + 2\,^{1}_{0}n, \quad \text{or} \quad ^{97}_{42}Mo(d, 2n)^{97}_{43}Tc$$

where d denotes a deuteron, $^{2}H^{+}$. Technetium is now moderately abundant because it accumulates in the decay products of nuclear power plants. Another isotope, technetium-99, has pharmaceutical applications, particularly for bone scans (Box 17.1).

A neutron can get close to a target nucleus more easily than a proton can. Because a neutron has no charge and hence is not repelled by the nuclear charge, it need not be accelerated to such high speeds. An example of **neutron-induced transmutation** is the formation of cobalt-60, which is used in the radiation treatment of cancer. The three-step process starts from iron-58. First, iron-59 is produced:

$$^{58}_{26}Fe + ^{1}_{0}n \longrightarrow ^{59}_{26}Fe$$

The second step is the β decay of iron-59 to cobalt-59:

$$^{59}_{26}Fe \longrightarrow ^{59}_{27}Co + ^{0}_{-1}e$$

In the final step, cobalt-59 absorbs another neutron from the incident neutron beam and is converted into cobalt-60:

$$^{59}_{27}Co + ^{1}_{0}n \longrightarrow ^{60}_{27}Co$$

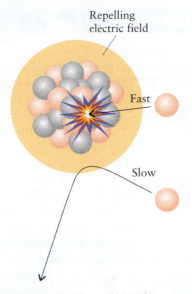

FIGURE 17.17 When a positively charged particle approaches a nucleus, it is repelled strongly. However, a very fast particle can reach the nucleus before the repulsion turns it aside, and a nuclear reaction may result.

BOX 17.1   WHAT HAS THIS TO DO WITH . . . STAYING ALIVE?

Nuclear Medicine

Nuclear chemistry has transformed medical diagnosis, treatment, and research. Radioactive tracers are used to measure organ function, sodium-24 to monitor blood flow, and strontium-87 to study bone growth. However, the most dramatic effect of radioisotopes on diagnosis has been in the field of imaging. Technetium-99m (see Exercise 17.46) is the most widely used radioactive nuclide in medicine, especially for bone scans. This isotope is highly active and gives off γ rays that pass through the body. The γ rays cause much less damage than α particles would, and the isotope is so short-lived that the risk to the patient is minimal.

Positron emission tomography (PET) makes use of a short-lived positron emitter such as fluorine-18 to image human tissue with a degree of detail not possible with x-rays. It has been used extensively to study brain function (see illustration) and in medical diagnosis. For example, when the hormone estrogen is labelled with fluorine-18 and injected into a cancer patient, the fluorine-bearing compound is preferentially absorbed by the tumor. The positrons given off by the fluorine atoms are quickly annihilated when they meet

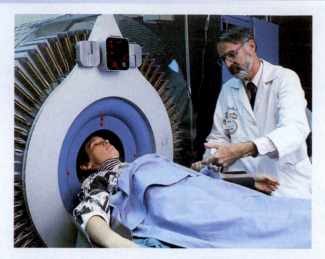

This patient is about to undergo a PET scan of brain function.

electrons. The resulting γ rays are detected by a scanner that moves slowly over the part of the body containing the tumor. The growth of the tumor can be estimated quickly and accurately with this technique. A PET imaging facility must be located near a cyclotron so that the positron emitters can be incorporated into the desired compounds as soon as they are created.

Several kinds of cancer therapy use radiation to destroy malignant cells. *Boron neutron capture therapy* is unusual in that boron-10, the isotope injected, is not radioactive. However, when boron-10 is bombarded with neutrons, it gives off highly destructive α particles. In boron neutron capture therapy, the boron-10 is incorporated into a compound that is preferentially absorbed by tumors. The patient is then exposed to brief periods of neutron bombardment. As soon as the bombardment ceases, the boron-10 stops generating α particles.

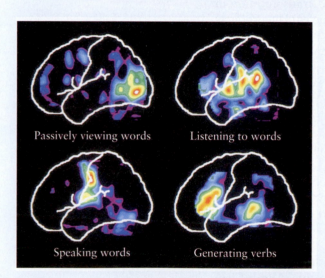

These four PET scans show how blood flow to different parts of the brain is affected by various activities. In this case, an oxygen isotope that is taken up by the hemoglobin in blood is used as a source of positrons.

Passively viewing words   Listening to words

Speaking words   Generating verbs

Related exercises: 17.45, 17.46, 17.79, 17.80, 17.85, and 17.86.

For Further Reading: V. Marx, "Molecular imaging," *Chemical and Engineering News,* vol. 83(30), 2005, pp. 25–34. P. Weiss, "Neutrons may spotlight cancers," *Science News,* vol. 166(8), Aug. 21, 2004, p. 125.

The overall reaction is

$$^{58}_{26}\text{Fe} + 2\,^{1}_{0}\text{n} \longrightarrow\, ^{60}_{27}\text{Co} + \,^{0}_{-1}\text{e}, \quad \text{or} \quad ^{58}_{26}\text{Fe}(2\text{n}, \beta^-)^{60}_{27}\text{Co}$$

SELF-TEST 17.4A Complete the following nuclear reactions: (a) ? $+\,^{4}_{2}\alpha \rightarrow\, ^{243}_{96}\text{Cm} + \,^{1}_{0}\text{n}$; (b) $^{242}_{96}\text{Cm} + \,^{4}_{2}\alpha \rightarrow\, ^{245}_{98}\text{Cf} + ?$.

[*Answer:* (a) $^{240}_{94}\text{Pu}$; (b) $^{1}_{0}\text{n}$]

SELF-TEST 17.4B Complete the following nuclear reactions: (a) $^{250}_{98}\text{Cf} + ? \rightarrow\, ^{257}_{103}\text{Lr} + 4\,^{1}_{0}\text{n}$; (b) ? $+\,^{12}_{6}\text{C} \rightarrow\, ^{254}_{102}\text{No} + 4\,^{1}_{0}\text{n}$.

The **transuranium elements** are the elements following uranium in the periodic table. The elements from rutherfordium (Rf, Z = 104) through meitnerium

(Mt, $Z = 109$) were formally named in 1997. The **transmeitnerium elements,** the elements beyond meitnerium (including hypothetical nuclides that have not yet been made) are named systematically, at least until they have been identified and there is international agreement on a permanent name. Their systematic names use the prefixes in Table 17.2, which identify their atomic numbers, with the ending -*ium*. Thus, element 110 was known as ununnilium until it was named darmstadtium (Ds) in 2003.

> *New elements and isotopes of known elements are made by nucleosynthesis; the repulsive electrical forces of like-charged particles are overcome when very fast particles collide.*

NUCLEAR RADIATION

Nuclear radiation is sometimes called **ionizing radiation,** because it is energetic enough to eject electrons from atoms. Hospitals use nuclear radiation to kill unwanted tissue, such as cancerous cells (see Box 17.1). Yet the same powerful effects that help to diagnose and cure illness can also damage healthy tissue. The damage depends on the strength of the source, the type of radiation, and the length of exposure. The three principal types of nuclear radiation have different abilities to penetrate matter (Table 17.3).

17.6 The Biological Effects of Radiation

Relatively massive, highly charged α particles interact so strongly with matter that they slow down, capture electrons from surrounding matter, and change into bulky helium atoms before traveling very far. They penetrate only the first layer of skin and can be stopped even by a sheet of paper. Most α radiation is absorbed by the surface layer of dead skin, where it can do little harm. However, α particles can be extremely dangerous if inhaled or ingested. The energy of their impact can knock atoms out of molecules, which can lead to serious illness and death. For example, plutonium, considered to be one of the most toxic radioactive materials, is an α emitter and can be handled safely with minimal shielding. However, it is easily oxidized to Pu^{4+}, which has chemical properties similar to those of Fe^{3+}. Plutonium can take the place of iron in the body and become absorbed into bone, where it destroys the body's ability to produce red blood cells. The result is radiation sickness, cancer, and death.

Second in penetrating power is β radiation. These fast electrons can penetrate about 1 cm into flesh before their electrostatic interactions with the electrons and nuclei of molecules bring them to a standstill.

Most penetrating of all is γ radiation. The high-energy γ-ray photons can pass right through buildings and bodies and can cause damage by ionizing the molecules in their path. Protein molecules and DNA that have been damaged in this way can no longer function, and the result can be radiation sickness and cancer. Intense sources of γ rays must be surrounded by shields built from lead bricks or thick concrete to absorb this penetrating radiation.

The **absorbed dose** of radiation is the energy deposited in a sample (in particular, the human body) when it is exposed to radiation. The SI unit of absorbed dose is the **gray,** Gy, which corresponds to an energy deposit of 1 J·kg^{-1}. The original unit used for reporting dose was the **radiation absorbed dose** (rad), the amount of

TABLE 17.2 Notation for the Systematic Nomenclature of Elements*

Digit	Prefix	Abbreviation
0	nil	n
1	un	u
2	bi	b
3	tri	t
4	quad	q
5	pent	p
6	hex	h
7	sep	s
8	oct	o
9	enn	e

*For instance, element 123 would be named unbitriium, Ubt.

TABLE 17.3 Shielding Requirements of α, β, and γ Radiation

Radiation	Relative penetrating power	Shielding required
α	1	paper, skin
β	100	3-mm aluminum
γ	10 000	concrete, lead

radiation that deposits 10^{-2} J of energy per kilogram of tissue, and so 1 rad = 10^{-2} Gy. A dose of 1 rad corresponds to a 65-kg person absorbing a total of 0.65 J, which is not very much energy: it is enough to boil only 0.2 mg of water. However, the energy of a particle of nuclear radiation is highly localized, like the impact of a subatomic bullet. As a result, the incoming particles can break individual bonds as they collide with molecules in their path.

The extent of radiation damage to living tissue depends on the type of radiation and the type of tissue. We must therefore include the **relative biological effectiveness,** Q, when assessing the damage that a given dose of each type of radiation may cause. For β and γ radiation, Q is set arbitrarily at about 1; but, for α radiation, Q is close to 20. A dose of 1 Gy of γ radiation causes about the same amount of damage as 1 Gy of β radiation, but 1 Gy of α particles is about 20 times as damaging (even though these particles are the least penetrating). The precise figures depend on the total dose, the rate at which the dose accumulates, and the type of tissue, but these values are typical.

The **dose equivalent** is the actual dose modified to take into account the different destructive powers of the various types of radiation in combination with various types of tissue. It is obtained by multiplying the actual dose (in gray) by the value of Q for the radiation type. The result is expressed in the SI unit called a **sievert** (Sv):

$$\text{Dose equivalent (Sv)} = Q \times \text{absorbed dose (Gy)} \tag{1}$$

The former (non-SI) unit of dose equivalent was the **roentgen equivalent man** (rem), which was defined in the same way as the sievert but with the absorbed dose in rad; thus, 1 rem = 10^{-2} Sv.

A dose of 0.3 Gy (30 rad) of γ radiation corresponds to a dose equivalent of 0.3 Sv (30 rem), enough to cause a reduction in the number of white blood cells (the cells that fight infection), but 0.3 Gy of α radiation corresponds to 6 Sv (600 rem), which is enough to cause death. A typical average annual dose equivalent that we each receive from natural sources, called **background radiation,** is about 2 mSv·a^{-1} (where a, for annum, is the SI abbreviation for year), but this figure varies, depending on our life style and where we live. About 20% of background radiation comes from our own bodies. About 30% comes from cosmic rays (a mix of γ rays and high-energy subatomic particles from outer space) that continuously bombard the Earth, and 40% comes from radon seeping out of the ground. The remaining 10% is a result largely of medical diagnoses (a typical chest x-ray gives a dose equivalent of about 0.07 mSv). Emissions from nuclear power plants and other nuclear facilities contribute about 0.1% in countries where they are widely used.

> *Human exposure to radiation is monitored by reporting the absorbed dose and the dose equivalent; the latter takes into account the effects of different types of radiation on tissues.*

17.7 Measuring the Rate of Nuclear Decay

Geiger counters and scintillation counters are used to measure the rate at which a radioactive nucleus decays (Box 17.2). Each click of a Geiger counter or flash of a phosphor in a scintillation counter indicates that one nuclear disintegration has been detected. The **activity** of a sample is the number of nuclear disintegrations in a given time interval divided by the length of the interval. The SI unit of activity is the **becquerel** (Bq): 1 Bq is equal to one nuclear disintegration per second. Another commonly used (non-SI) unit of radioactivity is the curie (Ci). It is equal to 3.7×10^{10} nuclear disintegrations per second, the radioactive output of 1 g of radium-226. Because the curie is a very large unit, most activities are expressed in millicuries (mCi) or microcuries (μCi). Table 17.4 summarizes these units.

The principal source of radioactivity in the human body is potassium-40. About 35 000 potassium-40 nuclei disintegrated in your body while you were reading this sentence.

BOX 17.2   HOW DO WE KNOW . . . HOW RADIOACTIVE A MATERIAL IS?

The ability of nuclear radiation to eject electrons from atoms and ions can be used to measure its intensity. Becquerel first gauged the intensity of radiation by determining the degree to which it blackened a photographic film. The blackening results from the same redox processes as those of ordinary photography, such as

$$Ag^+ + Br^- \xrightarrow{\text{light}} Ag + Br$$

except that the initial oxidation of the bromide ions is caused by nuclear radiation instead of light. Becquerel's technique is still used in film badges, which monitor the exposure of workers to radiation.

A *Geiger counter* monitors radiation by detecting the ionization of a low-pressure gas, as shown in the illustration. The radiation ionizes atoms of the gas inside a cylinder and allows a brief flow of current between the electrodes. The resulting electrical signal can be recorded directly or converted into an audible click. The frequency of the clicks indicates the intensity of the radiation. A limitation of Geiger counters is that they do not respond well to γ rays. Only about 1% of the γ-ray photons are detected, whereas all the β particles incident on the counter are detected. Because the efficiency of a Geiger counter depends on the size of the tube, a counter used to monitor a wide range of activities usually contains two tubes of different sizes.

A *scintillation counter* makes use of the fact that phosphors—phosphorescent substances such as sodium iodide and zinc sulfide (see Section 15.14)—give a flash of light—a scintillation—when exposed to radiation. The counter also contains a photomultiplier tube, which converts light into an electrical signal. The intensity of the radiation is determined from the strength of the electronic signal.

A *dosimeter* is used to collect cumulative evidence of exposure to radiation and is worn as a badge. Dosimeters contain a thermoluminescent material such as lithium fluoride. Any incident radiation knocks electrons out of the flu-

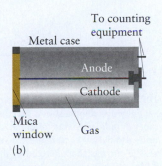

(a) A Geiger counter with a piece of uranium ore. (b) The detector in a Geiger counter contains a gas (often either argon and a little ethanol vapor or neon with some bromine vapor) in a cylinder with a high potential difference (500–1200 V) between the central wire and the walls. When radiation ionizes the gas, the ions allow current to flow briefly, resulting in the characteristic click.

oride ions. The electrons migrate away from the fluorine atoms but are trapped in the crystal. When the crystal is heated, they fall back into the fluorine atoms and give off the difference in energy as light. The dose of radiation received is determined by the intensity of the light. Dosimeters can be used over a period of time ranging from one day to several weeks, because the excited electrons accumulate with continued exposure, allowing a long-term dosage to be evaluated.

TABLE 17.4   Radiation Units*

Property	Unit name	Symbol	Definition
activity	becquerel	Bq	1 disintegration per second
	curie	Ci	3.7×10^{10} disintegrations per second
absorbed dose	gray	Gy	1 J·kg^{-1}
	radiation absorbed dose	rad	$10^{-2} \text{ J·kg}^{-1}$
dose equivalent	sievert	Sv	$Q \times$ absorbed dose[†]
	roentgen equivalent man	rem	$Q \times$ absorbed dose[†]

*Former units are in red.
[†] Q is the relative biological effectiveness of the radiation. Normally, $Q < 1 \text{ Sv·Gy}^{-1}$ for γ, β, and most other radiation, but $Q < 20 \text{ Sv·Gy}^{-1}$ for α radiation and fast neutrons. A further factor of 5 (that is, $5Q$) is used for bone under certain circumstances.

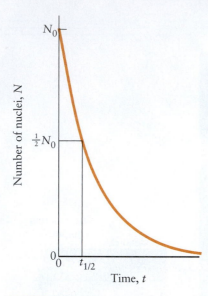

FIGURE 17.18 The exponential decay of the activity of a sample shows that the number of radioactive nuclei in the sample also decays exponentially with time. The curve is characterized by the half-life, $t_{1/2}$.

The equation for the decay of a nucleus (parent nucleus → daughter nucleus + radiation) has exactly the same form as a unimolecular elementary reaction (Section 13.7), with an unstable nucleus taking the place of a reactant molecule. This type of decay is expected for a process that does not depend on any external factors but only on the instability of the nucleus. The rate of nuclear decay depends only on the identity of the isotope, not on its chemical form or temperature.

As in a unimolecular chemical reaction, the rate law for nuclear decay is first order. That is, the relation between the rate of decay and the number N of radioactive nuclei present is given by the **law of radioactive decay:**

$$\text{Activity} = \text{rate of decay} = k \times N \tag{2}*$$

In this context, k is called the **decay constant.** The law tells us that the activity of a radioactive sample is proportional to the number of atoms in the sample. As we saw in Section 13.4, a first-order rate law implies an exponential decay. It follows that the number N of nuclei remaining after a time t is given by

$$N = N_0 e^{-kt} \tag{3}*$$

where N_0 is the number of radioactive nuclei present initially (at $t = 0$); this expression is plotted in Fig. 17.18.

EXAMPLE 17.3   Using the law of radioactive decay

One of the reasons why thermonuclear weapons have to be serviced regularly is the nuclear decay of the tritium that they contain. Suppose a tritium sample of mass 1.00 g is stored. What mass of that isotope will remain after 5.0 a (1 a = 1 year)? The decay constant of tritium is 0.0564 a^{-1}.

STRATEGY The total mass of isotope in a sample is proportional to the number of nuclei of that isotope that the sample contains; therefore, the time dependence of the mass follows the radioactive decay law, Eq. 3. We let m denote the total mass of the radioactive isotope at time t and m_0 its initial mass.

SOLUTION

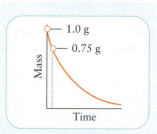

From $m = m_0 e^{-kt}$, $m = (1.00 \text{ g}) \times e^{-(0.0564 \text{ a}^{-1}) \times (5.0 \text{ a})}$

$$= 0.75 \text{ g}$$

That is, 0.75 g of the isotope remains after 5.0 years.

SELF-TEST 17.5A The decay constant for fermium-254 is 210 s^{-1}. What mass of the isotope will be present if a sample of mass 1.00 μg is kept for 10. ms?

[*Answer:* 0.12 μg]

SELF-TEST 17.5B The decay constant for the nuclide neptunium-237 is 3.3×10^{-7} a^{-1}. What mass of the isotope will be present if a sample of mass 5.0 μg survives for 1.0 Ma (1.0 million years)?

Radioactive decay is normally discussed in terms of the **half-life,** $t_{1/2}$, the time needed for half the initial number of nuclei to disintegrate. Just as we did in Section 13.5, we find the relation between $t_{1/2}$ and k by setting $N = \frac{1}{2}N_0$ and $t = t_{1/2}$ in Eq. 3:

$$t_{1/2} = \frac{\ln 2}{k} \tag{4}*$$

This equation shows that, the larger the value of k, the shorter the half-life of the nuclide. Nuclides with short half-lives are less stable than nuclides with long half-lives. They are more likely to decay in a given period of time and are "hotter" (more intensely radioactive) than nuclides with long half-lives.

Half-lives span a very wide range (Table 17.5). Consider strontium-90, for which the half-life is 28 a. This nuclide is present in nuclear **fallout,** the fine dust that settles from clouds of airborne particles after the explosion of a nuclear bomb, and may also be present in the accidental release of radioactive materials into the air. Because it is chemically very similar to calcium, strontium may accompany that element through the environment and become incorporated into bones; once there, it continues to emit radiation for many years. About 10 half-lives (for strontium-90, 280 a) must pass before the activity of a sample has fallen to 1/1000 of its initial value. Iodine-131, which was released in the accidental fire at the Chernobyl nuclear power plant, has a half-life of only 8.05 d, but it accumulates in the thyroid gland. Several cases of thyroid cancer have been linked to iodine-131 exposure from the accident. Plutonium-239 has a half-life of 24 ka (24 000 years). Consequently, very long term storage facilities are required for plutonium waste, and land contaminated with plutonium cannot be inhabited again for thousands of years without expensive remediation efforts.

The constant half-life of a nuclide is used to determine the ages of archaeological artifacts. In **isotopic dating,** we measure the activity of the radioactive isotopes that they contain. Isotopes used for dating objects include uranium-238, potassium-40, and tritium. However, the most important example is **radiocarbon dating,** which uses the β decay of carbon-14, for which the half-life is 5730 a.

Carbon-12 is the principal isotope of carbon, but a small proportion of carbon-14 is present in all living organisms. Its nuclei are produced when nitrogen nuclei in the atmosphere are bombarded by neutrons formed in the collisions of cosmic rays with other nuclei:

$$^{14}_{7}\text{N} + {}^{1}_{0}\text{n} \longrightarrow {}^{14}_{6}\text{C} + {}^{1}_{1}\text{p}$$

The carbon-14 atoms are produced in the atmosphere at an almost constant rate and, as a result, the proportion of carbon-14 to carbon-12 atoms in the atmosphere tends to be constant over time. The carbon-14 atoms enter living organisms as $^{14}\text{CO}_2$ through photosynthesis and digestion. They leave the organisms by the normal processes of excretion and respiration and also because the nuclei decay at a steady rate. As a result, all living organisms have a fixed ratio (of about 1 to 10^{12}) of carbon-14 atoms to carbon-12 atoms, and 1.0 g of naturally occurring carbon has an activity of 15 disintegrations per minute.

When an organism dies, it no longer exchanges carbon with its surroundings. However, carbon-14 nuclei already inside the organism continue to decay with a constant half-life, and so the ratio of carbon-14 to carbon-12 decreases. The ratio observed in a sample of dead tissue can therefore be used to estimate the time since death.

In the technique developed by Willard Libby in Chicago in the late 1940s, the proportion of carbon-14 in a sample is determined by monitoring the β radiation from CO_2 obtained by burning the sample. This procedure is illustrated in Example 17.4. In the modern version of the technique, which requires only a few milligrams of sample, the carbon atoms are converted into C^- ions by bombardment of the sample with cesium atoms. The C^- ions are then accelerated with electric fields, and the carbon isotopes are separated and counted with a mass spectrometer (Fig. 17.19).

TABLE 17.5 Half-Lives of Radioactive Isotopes*

Nuclide	Half-life, $t_{1/2}$
tritium	12.3 a
carbon-14	5.73 ka
carbon-15	2.4 s
potassium-40	1.26 Ga
cobalt-60	5.26 a
strontium-90	28.1 a
iodine-131	8.05 d
cesium-137	30.17 a
radium-226	1.60 ka
uranium-235	0.71 Ga
uranium-238	4.5 Ga
fermium-244	3.3 ms

*d = day, a = year.

Nuclear testing has increased the amount of carbon-14 in the air, and sensitive radiocarbon dating techniques take this increase into account.

EXAMPLE 17.4 Interpreting carbon-14 dating

A carbon sample of mass 1.00 g from wood found in an archaeological site in Arizona underwent 7.90×10^3 carbon-14 disintegrations in a period of 20.0 h. In the same period, 1.00 g of carbon from a modern source underwent 1.84×10^4 disintegrations. Calculate the age of the sample given that the half-life of ^{14}C is 5.73 ka.

STRATEGY First, we rearrange Eq. 3 to give an expression for the time and then express k in terms of $t_{1/2}$ by using Eq. 4. We assume that the carbon-14 activity of the

FIGURE 17.19 In a modern carbon-14 dating measurement, a mass spectrometer is used to determine the proportion of carbon-14 nuclei in the sample relative to the number of carbon-12 nuclei.

modern sample is the same as the *original* activity in the ancient sample. Therefore, N/N_0 is equal to the ratio of the number of disintegrations in the ancient and modern samples.

SOLUTION

From $N = N_0 e^{-kt}$,

$$t = -\frac{1}{k} \ln\left(\frac{N}{N_0}\right)$$

From $t_{1/2} = (\ln 2)/k$,

$$t = -\frac{t_{1/2}}{\ln 2} \ln\left(\frac{N}{N_0}\right)$$

Substitute the data.

$$t = -\frac{5.73 \text{ ka}}{\ln 2} \times \ln\left(\frac{7900}{18400}\right)$$

$$= 6.99 \text{ ka}$$

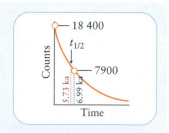

We conclude that approximately 7000 years have elapsed since the piece of wood was part of a living tree.

SELF-TEST 17.6A A sample of carbon of mass 250. mg from wood found in a tomb in Israel underwent 2480 carbon-14 disintegrations in 20. h. Estimate the time since death, assuming the same activity for a modern sample as in Example 17.4.

[*Answer:* 5.1 ka]

SELF-TEST 17.6B A sample of carbon of mass 1.00 g from scrolls found near the Dead Sea underwent 1.4×10^4 carbon-14 disintegrations in 20. h. Estimate the approximate time since the sheepskins composing the scrolls were removed from the sheep, assuming the same activity for a modern sample as in Example 17.4.

The law of radioactive decay implies that the number of radioactive nuclei decreases exponentially with time with a characteristic half-life. Radioactive isotopes are used to determine the ages of objects.

17.8 Uses of Radioisotopes

Radioisotopes are radioactive isotopes. They are used not only to cure disease (as described in Box 17.1), but also to preserve food, to trace mechanisms of reactions, and to power spacecraft.

Radioactive **tracers** are isotopes that are used to track changes and locations. For example, a sample of sugar can be *labeled* with carbon-14; that is, some carbon-12 atoms in sugar molecules are replaced by carbon-14 atoms, which can be detected by radiation counters. In this way, the progress of even tiny amounts of sugar molecules through the body can be monitored. Fertilizers labeled with radioactive nitrogen, phosphorus, or potassium are used to follow the mechanism of plant growth and the passage of these elements through the environment. Chemists and biochemists use tracers to help determine the mechanisms of reactions. For example, if water containing oxygen-18 is used in photosynthesis, the oxygen produced is oxygen-18:

$$6\ CO_2(g) + 6\ H_2{}^{18}O(l) \longrightarrow C_6H_{12}O_6(s,\ glucose) + 6\ {}^{18}O_2(g)$$

This result shows that the oxygen produced in photosynthesis comes from the water molecules, not the carbon dioxide molecules.

Radioisotopes have important commercial applications. For example, americium-241 is used in smoke detectors. Its role is to ionize any smoke particles, which then allow a current to flow and set off the alarm. Exposure to radiation is also used to sterilize food and inhibit the sprouting of potatoes. Radioisotopes that give off a lot of energy as heat are also used to provide power in remote locations, where refueling of generators is not possible. Unmanned spacecraft, such as *Voyager 2*, are powered by radiation from plutonium.

Isotopes are also used to determine properties of the environment. Just as carbon-14 is used to date organic materials, geologists can determine the age of very old substances such as rocks by measuring the abundance in rocks of radioisotopes with longer half-lives. Uranium-238 ($t_{1/2} = 4.5$ Ga, 1 Ga $= 10^9$ years) and potassium-40 ($t_{1/2} = 1.26$ Ga) are used to date very old rocks. For example, potassium-40 decays by electron capture to form argon-40. The rock is placed under vacuum and crushed, and a mass spectrometer is used to measure the amount of argon gas that escapes. This technique was used to determine the age of rocks collected on the surface of the Moon; they were found to be 3.5–4.0 billion years old, about the same age as the Earth.

Radioisotopes are used as long-lasting power sources, to study the environment, and to track movement. They are used in biology to trace metabolic pathways, in chemistry to trace reaction mechanisms, and in geology to determine the ages of rocks.

NUCLEAR ENERGY

Nuclear reactions can release huge amounts of energy, and nuclear reactors are widely used to provide power. The benefits of nuclear power include the large quantity of energy that can be obtained from a small mass of fuel and the absence of chemical pollution of the kind associated with fossil fuels. However, like other powerful resources, nuclear energy presents us with great technical challenges and hazards. The following sections outline some of the principles.

17.9 Mass–Energy Conversion

The energy released when the nucleons in a nucleus adopt a more stable arrangement can be calculated by comparing the masses of the nuclear reactants and the nuclear products. Einstein's theory of relativity tells us that the mass of an object is a measure of its energy content: the greater the mass of an object, the greater its energy. Specifically, the total energy, E, and the mass, m, are related by Einstein's famous equation

$$E = mc^2 \tag{5}*$$

where c is the speed of light (3.00×10^8 m·s^{-1}). It follows from this relation that loss of energy is always accompanied by loss of mass.

The mass loss that always accompanies the loss of energy is normally far too small to detect. Even in a strongly exothermic chemical reaction, such as one that releases 10^3 kJ of energy, the mass of the products and the reactants differ by only 10^{-8} g. In a nuclear reaction, the energy changes are very large, the mass loss is measurable, and we can calculate the energy released from the observed change in mass.

The **nuclear binding energy,** E_{bind}, is the energy *released* when protons and neutrons come together to form a nucleus. All binding energies are positive, which means that a nucleus has a lower energy than its constituent nucleons; the greater the binding energy, the lower the energy of the nuclide. Binding energy is usually reported as the energy per nucleon, and another way to think of binding energy is as the average energy that must be supplied to separate a nucleon from a nucleus. We can use Einstein's equation to calculate the nuclear binding energy from the difference in mass, Δm, between the nucleus and the separated nucleons. For example, iron-56 has 26 protons, each of mass m_p, and 30 neutrons, each of mass m_n. The difference in mass between the nucleus and the separate nucleons is

$$\Delta m = \sum m(\text{products}) - \sum m(\text{reactants})$$
$$= m(^{56}_{26}\text{Fe}) - m(26m_p + 30m_n)$$

We can calculate the binding energy from the difference in mass:

$$E_{bind} = |\Delta m| \times c^2 \tag{6}*$$

The absolute value of the mass difference is used because all binding energies are positive. We report binding energies in electronvolts (eV) or, more specifically, millions of electronvolts (1 MeV = 10^6 eV):

$$1 \text{ eV} = 1.602\ 18 \times 10^{-19} \text{ J}$$

Because the masses of nuclides are so small, they are normally reported as a multiple of the **atomic mass constant,** m_u (formerly: atomic mass unit, amu). The atomic mass constant is defined as exactly $^1/_{12}$ the mass of one atom of carbon-12:

$$m_u = \frac{m(^{12}\text{C})}{12}$$

Current determinations of the mass of carbon-12 give

$$m_u = 1.6605 \times 10^{-27} \text{ kg}$$

Note that because m_u is a constant, not a unit, there is no space between it and the preceding number.

The mass of an atom in terms of the atomic mass constant is numerically the same as the molar mass in grams per mole. For instance, the molar mass of carbon-12 is exactly 12 g·mol^{-1}, and the mass of a carbon-12 atom is exactly $12m_u$.

When calculating $|\Delta m|$ we use the mass of an atom of ^{1}H instead of the mass of a proton. This strategy allows us to use readily available isotope masses instead of the masses of bare atomic nuclei to calculate $|\Delta m|$, because the number of electrons in the isotope will be the same as the total number of electrons in the hydrogen atoms on the other side of the equation and the masses of the electrons cancel. The electron–nucleus binding energy, which contributes to the mass of an atom, is only about $10^{-6}\ m_u$ per proton, and so it can be ignored in elementary calculations.

EXAMPLE 17.5   Calculating the nuclear binding energy

Calculate the nuclear binding energy in electronvolts for a helium-4 nucleus, given the following masses: ^{4}He, $4.0026m_u$; ^{1}H, $1.0078m_u$; n, $1.0087m_u$.

STRATEGY The nuclear binding energy is the energy released in the formation of the nucleus from its nucleons. Use H atoms instead of protons to account for the masses of the electrons in the He atom produced. Write the nuclear equation for the formation of the nuclide from hydrogen atoms and neutrons, and calculate the difference in masses between the products and the reactants; convert the result from a multiple of the atomic mass constant into kilograms. Then, use the Einstein relation to calculate the energy corresponding to this loss of mass and convert the units to electronvolts.

SOLUTION

Write the nuclear equation.	$2\ ^1H + 2\ ^1n \longrightarrow {}^4He$
Calculate the change in mass.	$\Delta m = m(^4He) - (2m_p + 2m_n)$
	$= 4.0026m_u - \{2(1.0078) + 2(1.0087)\}m_u$
	$= -0.0304m_u$

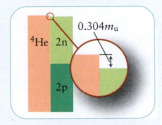

Express in kilograms.	$\Delta m = -0.0304 \times 1.6605 \times 10^{-27}\ kg$

From $E_{bind} = |\Delta m| \times c^2$,

$$E_{bind} = |-0.0304 \times 1.6605 \times 10^{-27}\ kg| \times (3.00 \times 10^8\ m \cdot s^{-1})^2$$

$$= 4.54 \times 10^{-12}\ kg \cdot m^2 \cdot s^{-2} = 4.54 \times 10^{-12}\ J$$

(In the final step we used $1\ kg \cdot m^2 \cdot s^{-2} = 1\ J$.) The value of the binding energy shows that 4.54 pJ ($1\ pJ = 10^{-12}\ J$) is released when one ^4He nucleus forms from its nucleons.

Convert to electronvolts.	$4.54 \times 10^{-12}\ J \times \dfrac{1\ eV}{1.602 \times 10^{-19}\ J} \times \dfrac{1\ MeV}{10^6\ eV} = 28.3\ MeV$

SELF-TEST 17.7A Calculate the binding energy of a carbon-12 nucleus in electronvolts.

[*Answer:* 92.3 MeV]

SELF-TEST 17.7B Calculate the binding energy of a uranium-235 nucleus in electronvolts. The mass of one uranium-235 atom is $235.0439m_u$.

Figure 17.20 shows the binding energy per nucleon, E_{bind}/A, for the elements. The graph shows that nucleons are bonded together most strongly in the elements near iron and nickel. This high binding energy is one of the reasons why iron and nickel are so abundant in meteorites and on a rocky planet such as Earth. The binding energy per nucleon is less for all other nuclides. We can infer that nuclei of light atoms become more stable when they "fuse" together and that heavy nuclei become more stable when they undergo "fission" and split into lighter nuclei.

Nuclear binding energies are determined by applying Einstein's formula to the mass difference between the nucleus and its components. Iron and nickel have the highest binding energy per nucleon.

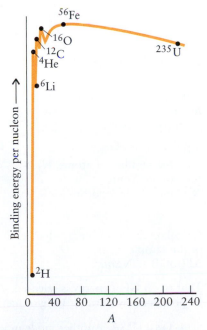

FIGURE 17.20 The variation of the nuclear binding energy per nucleon. The maximum binding energy per nucleon occurs near iron and nickel. Their nuclei have the lowest energies of all because their nucleons are most tightly bound. (The vertical axis is E_{bind}/A.)

17.10 Nuclear Fission

In 1938, Lise Meitner, Otto Hahn, and Fritz Strassmann realized that, by bombarding heavy atoms such as uranium with neutrons, they could "split" the atoms into smaller fragments in fission reactions, releasing huge amounts of energy. We can estimate the energy that would be released by using Einstein's equation, as we did in Example 17.5.

EXAMPLE 17.6 **Calculating the energy released during fission**

When uranium-235 nuclei are bombarded with neutrons, they can split apart in a variety of ways, like glass balls that shatter into pieces of different sizes. In one process, uranium-235 forms barium-142 and krypton-92:

$$^{235}_{92}\text{U} + ^{1}_{0}\text{n} \longrightarrow ^{142}_{56}\text{Ba} + ^{92}_{36}\text{Kr} + 2\ ^{1}_{0}\text{n}$$

Calculate the energy (in joules) released when 1.0 g of uranium-235 undergoes this fission reaction. The masses of the particles are $^{235}_{92}\text{U}$, $235.04m_u$; $^{142}_{56}\text{Ba}$, $141.92m_u$; $^{92}_{36}\text{Kr}$, $91.92m_u$; n, $1.0087m_u$.

Notice that we do not cancel neutrons, although they appear on both sides of the equation. Like equations for elementary chemical reactions, nuclear equations show the specific process.

STRATEGY If we know the mass loss, we can find the energy released by using Einstein's equation. Therefore, we must calculate the total mass of the particles on each side of the nuclear equation, take the difference, and substitute the mass difference into Eq. 6. Then we determine the number of nuclei in the sample from $N = m(\text{sample})/m(\text{atom})$ and, finally, multiply the energy released from the fission of one nucleus by that number to find the energy released by the sample.

SOLUTION

Calculate mass of products.

$$m(\text{products}) = m(\text{Ba}) + m(\text{Kr}) + 2m_n$$
$$= \{141.92 + 91.92 + 2(1.0087)\}m_u$$
$$= 235.86m_u$$

Calculate mass of reactants.

$$m(\text{reactants}) = m(\text{U}) + m_n$$
$$= 235.04m_u + 1.0087m_u = 236.05m_u$$

Calculate the change in mass.

$$\Delta m = 235.86m_u - 236.05m_u = -0.19m_u$$

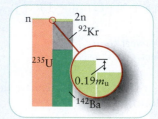

Express this change in kilograms.

$$\Delta m = -0.19 \times 1.6605 \times 10^{-27}\ \text{kg}$$

Calculate the change in energy for the fission of one nucleus from $\Delta E = \Delta mc^2$.

$$\Delta E = (-0.19 \times 1.6605 \times 10^{-27}\ \text{kg}) \times (3.00 \times 10^8\ \text{m·s}^{-1})^2$$
$$= (-0.19 \times 1.6605 \times 10^{-27}) \times (3.00 \times 10^8)^2\ \text{J}$$

Find the number of atoms, N, in the sample from $N = m(\text{sample})/m(\text{atom})$.

$$N = \frac{1.0 \times 10^{-3}\ \text{kg}}{235.04m_u} = \frac{1.0 \times 10^{-3}}{235.04 \times 1.6605 \times 10^{-27}}$$

Calculate the total energy change of the sample from $\Delta E(\text{total}) = N\Delta mc^2$.

$$\Delta E = \frac{1.0 \times 10^{-3}}{235.04 \times 1.6605 \times 10^{-27}} \times \{-0.19 \times 1.6605 \times 10^{-27} \times (3.00 \times 10^8)^2\}\ \text{J}$$
$$= -7.3 \times 10^{10}\ \text{J}$$

That is, 73 GJ of energy is released, which is 1.3 million times as much energy as would be produced by burning 1.0 g of methane, the main component of natural gas.

SELF-TEST 17.8A Another mode in which uranium-235 can undergo fission is

$$^{235}_{52}\text{U} + ^{1}_{0}\text{n} \longrightarrow ^{135}_{52}\text{Te} + ^{100}_{40}\text{Zr} + ^{1}_{0}\text{n}$$

Calculate the energy released when 1.0 g of uranium-235 undergoes fission in this way. The masses needed are $^{235}_{92}\text{U}$, $235.04m_u$; n, $1.0087m_u$; $^{135}_{52}\text{Te}$, $134.92m_u$; $^{100}_{40}\text{Zr}$, $99.92m_u$.

[*Answer:* 77 GJ]

(a)   (b)

FIGURE 17.21 In spontaneous nuclear fission, the oscillations of the heavy nucleus (a) in effect tear the nucleus apart, thereby forming two or more smaller nuclei of similar mass (b). Here, two neutrons are released.

SELF-TEST 17.8B A nuclear reaction that can cause great destruction is one of many that take place in the ^{235}U atomic bomb:

$$^{235}_{92}U + ^{1}_{0}n \longrightarrow ^{138}_{56}Ba + ^{86}_{36}Kr + 12\,^{1}_{0}n$$

How much energy is released when 1.0 g of uranium-235 undergoes fission in this manner? The additional masses needed are $^{138}_{56}Ba$, $137.91m_u$; $^{86}_{36}Kr$, $85.91m_u$.

Spontaneous nuclear fission takes place when the natural oscillations of a heavy nucleus cause it to break into two nuclei of similar mass (Fig. 17.21). We can think of the nucleus as distorting into a dumbbell shape and then breaking into two smaller nuclei. An example is the spontaneous disintegration of americium-244 into iodine and molybdenum:

$$^{244}_{95}Am \longrightarrow ^{134}_{53}I + ^{107}_{42}Mo + 3\,^{1}_{0}n$$

Fission does not occur in precisely the same way in every instance. For example, more than 200 isotopes of 35 different elements have been identified among the fission products of uranium-235, with most products having mass numbers close to 90 or 130 (Fig. 17.22).

Induced nuclear fission is fission caused by bombarding a heavy nucleus with neutrons (Fig. 17.23). The nucleus breaks into two fragments when struck by a projectile. Nuclei that can undergo induced fission are called **fissionable.** For most nuclei, fission takes place only if the impinging neutrons travel so rapidly that they can smash into the nucleus and drive it apart with the shock of impact; uranium-238 undergoes fission in this way. **Fissile nuclei,** however, are nuclei that can be nudged into breaking apart even by slow neutrons. They include uranium-235, uranium-233, and plutonium-239—the fuels of nuclear power plants.

Once nuclear fission has been induced, it can continue, even if the supply of neutrons from outside is discontinued, provided that the fission produces more neutrons. Such self-sustaining fission takes place in uranium-235, which undergoes numerous fission processes, including

$$^{235}_{92}U + ^{1}_{0}n \longrightarrow ^{141}_{56}Ba + ^{92}_{36}Kr + 3\,^{1}_{0}n$$

If the three product neutrons strike three other fissile nuclei, then after the next round of fission there will be nine neutrons, which can induce fission in nine more nuclei. In the language of Section 13.9, neutrons are **carriers** in a branched chain reaction (Fig. 17.24).

Neutrons produced in a chain reaction are moving very fast, and most escape into the surroundings without colliding with another fissionable nucleus. However, if a large enough number of uranium nuclei are present in the sample, enough neutrons can be captured to sustain the chain reaction. In that case, there is a **critical mass,** a mass of fissionable material above which so few neutrons escape from the sample that the fission chain reaction is sustained. If a sample is **supercritical,**

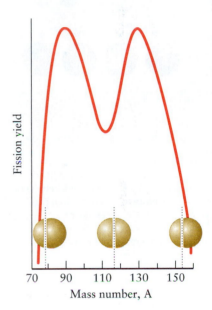

FIGURE 17.22 The fission yield of uranium-235. Note that the majority of fission products lie in the regions close to $A = 90$ and 130 and that relatively few nuclides corresponding to symmetrical fission (A close to 117) are formed.

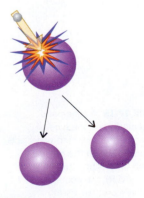

FIGURE 17.23 In induced nuclear fission, the impact of an incoming neutron causes the nucleus to break apart.

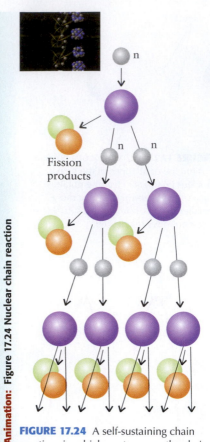

Animation: Figure 17.24 Nuclear chain reaction

n

Fission
products

n   n

FIGURE 17.24 A self-sustaining chain reaction, in which neutrons are the chain carriers, takes place when induced fission produces more than one neutron per fission event. These newly produced neutrons can stimulate fission in increasingly greater numbers of other nuclei.

with a mass in excess of the critical value, then the reaction is not only self-sustaining but difficult to control and may result in an explosion. The critical mass for a solid sphere of pure plutonium of normal density is about 15 kg, about the size of a grapefruit. The critical mass is smaller if the metal is compressed by detonating a conventional explosive that surrounds it. Then the nuclei are pressed closer together and become more effective at blocking the escape of neutrons. The critical mass can be as low as 5 kg for highly compressed plutonium. A sample that has less than the critical mass for its density is said to be **subcritical**.

In a nuclear weapon, the fissile material is initially subcritical. The challenge is to produce a supercritical mass so rapidly that the chain reaction takes place uniformly throughout the metal. Supercriticality can be achieved by shooting two subcritical blocks toward each other (as was done in the bomb that fell on Hiroshima) or by implosion of a single subcritical mass (the technique used in the bomb that destroyed Nagasaki). A strong neutron emitter, typically polonium, helps to initiate the chain reaction.

Explosive fission cannot occur in a nuclear reactor, because the fuel is not dense enough. Instead, reactors sustain a much slower, controlled chain reaction by making efficient use of a limited supply of neutrons and slowing the neutrons down. The fuel is shaped into long rods and inserted into a **moderator,** a material that slows down the neutrons as they pass between fuel rods; the slower neutrons have a greater probability of collision with a nucleus (Fig. 17.25). Slow neutrons have three significant roles: they do not induce the fission of fissionable (as distinct from fissile) material; they are most effectively absorbed by the fissile uranium-235; and they are more easily controlled. The first moderator used was graphite. Heavy water, D_2O, is also an effective neutron moderator, but light-water reactors (LWRs) use ordinary water as a moderator. They are the most common type of nuclear reactor in the United States (Fig. 17.26).

If the rate of the chain reaction exceeds a certain level, the reactor will become too hot and begin to melt. Control rods—rods made from neutron-absorbing elements, such as boron or cadmium, and inserted between the fuel rods—help to control the number of available neutrons and the rate of nuclear reaction.

One of the many problems of nuclear power is the availability of fuel: uranium-235 reserves are only about 0.7% those of the nonfissile uranium-238, and the separation of the isotopes is costly (Section 17.12). One solution is to synthesize fissile nuclides from other elements. In a **breeder reactor,** a reactor that is used to create nuclear fuel, the neutrons are not moderated. Their high speeds result in

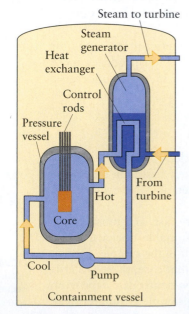

Steam to turbine

Steam generator

Heat exchanger

Control rods

Pressure vessel

From turbine

Hot

Core

Cool

Pump

Containment vessel

FIGURE 17.25 A schematic representation of one type of nuclear reactor in which water acts as a moderator for the nuclear reaction. In this pressurized water reactor (PWR), the coolant is water under pressure. The fission reactions produce heat, which boils water in the steam generator; the resulting steam turns the turbines that generate electricity.

FIGURE 17.26 The reactor core of a typical light-water reactor (LWR) nuclear power plant is immersed in water.

the formation of not only uranium-235 but also some fissile plutonium-239, which can be used as fuel (or for warheads). However, breeder reactors are more hazardous to operate than nuclear power plants. They run very hot, and the fast reactions require more careful control than reactors used for nuclear power generation.

Nuclear energy can be extracted by arranging for a nuclear chain reaction to take place in a critical mass of fissionable material, with neutrons as the chain carriers. A moderator is used to reduce the speeds of the neutrons in a reactor that uses fissile material.

17.11 Nuclear Fusion

Although fission reactors do not generate chemical pollution, they do produce highly hazardous radioactive waste. However, another kind of nuclear reaction being investigated for power generation is essentially free of long-lived radioactive waste products, and its abundant fuel is readily extracted from seawater. The reaction is the fusion of hydrogen nuclei to form helium nuclei.

From the plot in Fig. 17.20, we can see that there is a large increase in nuclear binding energy on going from light elements such as deuterium to heavier elements. Consequently, energy is released when hydrogen nuclei fuse together to form nuclei of those elements. Unfortunately, the strong electrical repulsion between protons makes it difficult for them to approach each other closely enough to fuse together. The nuclei of the heavier isotopes of hydrogen fuse together more readily, because the additional neutrons contribute to the strong force and help to capture the approaching protons. To achieve the high kinetic energies needed for successful collisions, fusion reactors need to operate at temperatures above 10^8 K.

One fusion scheme uses deuterium (D) and tritium (T) in the following sequence of nuclear reactions:

$$D + D \longrightarrow {}^3He + n$$
$$D + D \longrightarrow T + p$$
$$D + T \longrightarrow {}^4He + n$$
$$D + {}^3He \longrightarrow {}^4He + p$$

Overall reaction: $6 \, D \longrightarrow 2 \, {}^4He + 2 \, p + 2 \, n$

The overall reaction releases 3×10^8 kJ for each gram of deuterium consumed. That energy corresponds to the energy generated when the Hoover Dam operates at full capacity for about an hour. Additional tritium is supplied to facilitate the process. Because tritium has a very low natural abundance and is radioactive, it is generated by bombarding lithium-6 with neutrons in the immediate surroundings of the reaction zone:

$${}^6Li + n \longrightarrow T + {}^4He$$

Nuclear fusion is very difficult to achieve because the charged nuclei must be hurled at each other with enormous kinetic energies. One way of accelerating nuclei to sufficiently high speeds is to heat them with a fission explosion: this method is used to produce a **thermonuclear explosion,** an explosion due to nuclear fusion. In "hydrogen bombs," a fission bomb (using uranium or plutonium) ignites a lithium-6 fusion bomb. Fusion bombs can be of varying destructive capacities, because there is no critical mass to maintain. Typical fusion bombs have a destructive capacity greater than 200 times that of the fission bombs that were dropped on Hiroshima and Nagasaki in World War II.

A more constructive approach to nuclear fusion—one that achieves a controlled release of nuclear energy—is to heat a **plasma,** or ionized gas, by passing an electric current through it. The very fast ions in the plasma are kept away from the walls of the container with magnetic fields. This method of achieving fusion is the subject of intense research and is beginning to show signs of success (Fig. 17.27).

Nuclear fusion makes use of the energy released when light nuclei fuse together to form heavier nuclei.

FIGURE 17.27 Research into controlled nuclear fusion is being carried out in several countries. Here we see the Tokomak fusion test reactor at the Princeton Plasma Physics Laboratory.

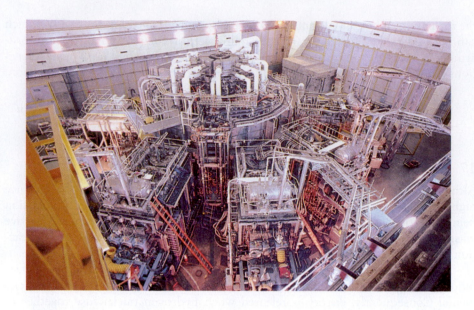

17.12 The Chemistry of Nuclear Power

Chemistry is the key to the safe use of nuclear power. It is used in the preparation of the fuel itself, the recovery of important fission products, and the safe disposal or utilization of nuclear waste.

Uranium is the fuel of nuclear reactors. The most important of its minerals is *pitchblende*, UO_2 (Fig. 17.28), much of which is obtained from strip mines in New Mexico and Wyoming. Uranium is refined to reduce the ore to the metal and to enrich it; that is, to increase the abundance of a specific isotope—in this case, uranium-235. The natural abundance of uranium-235 is about 0.7%; for use in a nuclear reactor, this fraction must be increased to about 3%.

The enrichment procedure uses the small mass difference between the hexafluorides of uranium-235 and uranium-238 to separate them. The first procedure to be developed converts the uranium into uranium hexafluoride, UF_6, which can be vaporized readily. The different effusion rates of the two isotopic fluorides are then used to separate them. From Graham's law of effusion (rate of effusion $\propto 1/(\text{molar mass})^{1/2}$; Section 4.9), the rates of effusion of $^{235}UF_6$ (molar mass, 349.0 $g \cdot mol^{-1}$) and $^{238}UF_6$ (molar mass, 352.1 $g \cdot mol^{-1}$) should be in the ratio

$$\frac{\text{Rate of effusion of } {}^{235}UF_6}{\text{Rate of effusion of } {}^{238}UF_6} = \sqrt{\frac{352.1}{349.0}} = 1.004$$

Because the ratio is so close to 1, the vapor must be allowed to effuse repeatedly through porous barriers consisting of screens with large numbers of minute holes. In practice, it is allowed to do so thousands of times.

Because the effusion process is technically demanding and uses a lot of energy, scientists and engineers continue to look for alternative enrichment procedures. One of these approaches uses a centrifuge that rotates samples of uranium hexafluoride vapor at very high speed. This rotation causes the heavier $^{238}UF_6$ molecules to be thrown outward and collected as a solid on the outer parts of the rotor, leaving a high proportion of $^{235}UF_6$ closer to the axis of the rotor, from where it can be removed.

Spent nuclear fuel remains radioactive and consists of a mixture of uranium and fission products. Nuclear reactor waste can be processed and some of it reused, but the percentage processed depends on the price of uranium. When the price is high, as it is in the early twenty-first century, much of the nuclear waste is processed for reuse.

FIGURE 17.28 Pitchblende is a common uranium ore. It is a variety of uranite, UO_2.

FIGURE 17.29 Containers of high-level waste products, including cesium-137 and strontium-90, glow under a protective layer of water. If the canisters were unshielded, the radiation that they emit would be great enough to cause death within about 4 s.

FIGURE 17.30 This 35-year-old drum of radioactive waste has corroded and leaked radioactive materials into the soil. The drum was located in one of the nuclear waste disposal sites at the U.S. Department of Energy's Hanford, Washington, nuclear manufacturing and research facility. Several storage sites at this facility have become seriously contaminated.

The processing of nuclear waste is complex. Any remaining uranium-235 must be recovered, any plutonium produced must be extracted, and the largely useless but radioactive fission products must be stored safely (Fig. 17.29). The highly radioactive fission (HRF) products from used nuclear fuel rods must be stored until their level of radioactivity is no longer dangerous (about 10 half-lives). Generally they are buried underground, but even burial of radioactive wastes is not without problems. Metal storage drums can corrode and allow liquid radioactive waste to seep into aquifers that supply drinking water (Fig. 17.30). Leakage can be minimized by incorporating the HRF products into a glass—a solid, complex network of silicon and oxygen atoms. Most of the fission products are oxides of the type that form one of the components of glass—they are network formers (see Section 14.21); that is, they promote the formation of a relatively disorderly Si—O network rather than inducing crystallization into an orderly array of atoms. Crystallization is dangerous because crystalline regions easily fracture, exposing the incorporated radioactive materials to moisture. Water could dissolve them and carry them away from the storage area. An alternative is to incorporate radioactive waste into hard ceramic materials. An example is Synroc, a fracture-resistant titanate-based ceramic that can accept radioactive waste products into its lattice.

Uranium is extracted by a series of reactions that lead to uranium hexafluoride; the isotopes are then separated by a variety of procedures. Some radioactive waste is currently converted into glass or ceramic materials for storage underground.

SKILLS YOU SHOULD HAVE MASTERED

❏ 1 Write, complete, and balance nuclear equations (Examples 17.1 and 17.2).

❏ 2 Use the band of stability to predict the types of decay that a given radioactive nucleus is likely to undergo (Self-Test 17.3).

❏ 3 Distinguish α, β, and γ radiation by their response to an electric field, penetrating power, and relative biological effectiveness (Sections 17.1 and 17.6).

❏ 4 Predict the amount of a radioactive sample that will remain after a given time period, given the decay constant or half-life of the sample (Example 17.3).

❏ 5 Use the half-life of an isotope to determine the age of an object (Example 17.4).

❏ 6 Calculate the nuclear binding energy of a given nuclide (Example 17.5).

❏ 7 Calculate the energy released during a nuclear reaction (Example 17.6).

❏ 8 Distinguish nuclear fission from nuclear fusion and predict which nuclides undergo each type of process (Sections 17.10 and 17.11).

❏ 9 Describe some ways of storing radioactive waste (Section 17.12).

724 CHAPTER 17 NUCLEAR CHEMISTRY

EXERCISES

Note that the SI symbol for 1 year is 1 a and that it takes the usual prefixes, as in 1 ka = 10^3 a and 1 Ga = 10^9 a.

Exercises labeled $\boxed{C}$ *require calculus*

Radioactive Decay

17.1 When the nucleons rearrange in the following daughter nuclei, the energy changes by the amount shown and a γ ray is emitted. Determine the frequency and wavelength of the γ ray in each case: (a) cobalt-60, 1.33 MeV; (b) arsenic-80, 1.64 MeV; (c) iron-59, 1.10 MeV. (1 MeV = 1.602×10^{-13} J.)

17.2 Determine the frequency and wavelength of the γ ray emitted in the decay of each of following nuclides: (a) carbon-15, 5.30 MeV; (b) scandium-50, 0.26 MeV; (c) bromine-87, 5.4 MeV. (1 MeV = 1.602×10^{-13} J.)

17.3 *Isotones* are nuclides that have the same number of neutrons. Which isotopes of argon and calcium are isotones of potassium-40?

17.4 Which isotopes of nickel and chromium are isotones (see Exercise 17.3) of iron-56?

17.5 Write the balanced nuclear equation for each of the following decays: (a) β^- decay of tritium; (b) β^+ decay of yttrium-83; (c) β^- decay of krypton-87; (d) α decay of protactinium-225.

17.6 Write the balanced nuclear equation for each of the following decays: (a) β^- decay of actinium-228; (b) α decay of radon-212; (c) α decay of francium-221; (d) electron capture by protactinium-230.

17.7 Write the balanced nuclear equation for each of the following radioactive decays: (a) β^+ decay of boron-8; (b) β^- decay of nickel-63; (c) α decay of gold-185; (d) electron capture by beryllium-7.

17.8 Write the balanced nuclear equation for each of the following radioactive decays: (a) β^- decay of uranium-233; (b) proton emission of cobalt-56; (c) β^+ decay of holmium-158; (d) α decay of polonium-212.

17.9 Determine the particle emitted and write the balanced nuclear equation for each of the following nuclear transformations: (a) sodium-24 to magnesium-24; (b) ^{128}Sn to ^{128}Sb; (c) lanthanum-140 to barium-140; (d) ^{228}Th to ^{224}Ra.

17.10 Determine the particle emitted and write the balanced nuclear equation for each of the following nuclear transformations: (a) carbon-14 to nitrogen-14; (b) neon-19 to fluorine-19; (c) gold-188 to platinum-188; (d) uranium-229 to thorium-225.

17.11 Complete the following equations for nuclear reactions:
(a) ^{11}B + ? $\rightarrow$ 2 n + ^{13}N
(b) ? + D $\rightarrow$ n + ^{36}Ar
(c) ^{96}Mo + D $\rightarrow$? + ^{97}Tc
(d) ^{45}Sc + n $\rightarrow$ α + ?

17.12 Write balanced nuclear equations for the radioactive decay of each of the following nuclides: (a) ^{74}Kr, β^+ emission; (b) ^{174}Hf, α emission; (c) ^{98}Tc, β^- emission; (d) ^{41}Ca, electron capture.

The Pattern of Nuclear Stability

17.13 The following nuclides lie outside the band of stability. Predict whether each is most likely to undergo β^- decay, β^+ decay, or α decay, and identify the daughter nucleus: (a) copper-68; (b) cadmium-103; (c) berkelium-243; (d) dubnium-260.

17.14 The following nuclides lie outside the band of stability. Predict whether each is most likely to undergo β^- decay, β^+ decay, or α decay, and identify the daughter nucleus: (a) copper-60; (b) xenon-140; (c) americium-246; (d) neptunium-240.

17.15 Identify the daughter nuclides in each step of the radioactive decay of uranium-235, if the string of particle emissions is α, β, α, β, α, α, α, β, α, β, α. Write a balanced nuclear equation for each step.

17.16 Neptunium-237 undergoes an α, β, α, α, β, α, α, α, β, α, β sequence of radioactive decays. Write a balanced nuclear equation for each step.

Nucleosynthesis

17.17 Complete the following nuclear equations:
(a) $^{14}_{7}$N + ? $\rightarrow$ $^{17}_{8}$O + $^{1}_{1}$p
(b) ? + $^{1}_{0}$n $\rightarrow$ $^{249}_{97}$Bk + $^{0}_{-1}$e
(c) $^{243}_{95}$Am + $^{1}_{0}$n $\rightarrow$ $^{244}_{96}$Cm + ? + γ
(d) $^{13}_{6}$C + $^{1}_{0}$n $\rightarrow$? + γ

17.18 Complete the following nuclear equations:
(a) ? + $^{1}_{1}$p $\rightarrow$ $^{21}_{11}$Na + γ
(b) $^{1}_{1}$H + $^{1}_{1}$p $\rightarrow$ $^{2}_{1}$H + ?
(c) $^{15}_{7}$N + $^{1}_{1}$p $\rightarrow$ $^{12}_{6}$C + ?
(d) $^{20}_{10}$Ne + ? $\rightarrow$ $^{24}_{12}$Mg + γ

17.19 Complete the following equations for nuclear transmutation:
(a) $^{20}_{10}$Ne + $^{4}_{2}\alpha$ $\rightarrow$? + $^{16}_{8}$O
(b) $^{20}_{10}$Ne + $^{20}_{10}$Ne $\rightarrow$ $^{16}_{8}$O + ?
(c) $^{44}_{20}$Ca + ? $\rightarrow$ γ + $^{48}_{22}$Ti
(d) $^{27}_{13}$Al + $^{2}_{1}$H $\rightarrow$? + $^{28}_{13}$Al

17.20 Complete the following equations for nuclear transmutation:
(a) ? + γ $\rightarrow$ $^{0}_{-1}$e + $^{20}_{10}$Ne
(b) $^{44}_{22}$Ti + $^{0}_{-1}$e $\rightarrow$ $^{0}_{+1}$e + ?
(c) $^{241}_{95}$Am + ? $\rightarrow$ 4 $^{1}_{0}$n + $^{248}_{100}$Fm
(d) ? + $^{1}_{0}$n $\rightarrow$ $^{0}_{-1}$e + $^{244}_{96}$Cm

17.21 Write a nuclear equation for each of the following processes: (a) oxygen-17 produced by α particle bombardment of nitrogen-14; (b) americium-240 produced by neutron bombardment of plutonium-239.

17.22 Write a nuclear equation for each of the following transformations: (a) ^{257}Rf produced by the bombardment of californium-245 with carbon-12 nuclei; (b) the first synthesis of ^{266}Mt by the bombardment of bismuth-209 with iron-58 nuclei. Given that the first decay of meitnerium is by α emission, what is the daughter nucleus?

17.23 What will be the systematic name and atomic symbol given to (a) element 126; (b) element 136; (c) element 200?

17.24 What will be the systematic name and atomic symbol given to (a) element 118; (b) element 127; (c) element 202?

Nuclear Radiation

17.25 The activity of a certain radioactive source is 5.3×10^5 Bq. Express this activity in curies.

17.26 The activity of a sample containing carbon-14 is 21.4 Bq. Express this activity in microcuries.

17.27 Determine the number of disintegrations per second for radioactive sources of each of the following activities: (a) 2.5 μCi; (b) 142 Ci; (c) 7.2 mCi.

17.28 A certain Geiger counter is known to respond to only 1 of every 1000 radiation events from a sample. Calculate the activity of each radioactive source in curies, given the following data: (a) 591 clicks in 100. s; (b) 2.7×10^4 clicks in 1.5 h; (c) 159 clicks in 1.0 min.

17.29 A 2.0-kg sample absorbs an energy of 1.5 J as a result of its exposure to β radiation. Calculate the dose in rads and the dose equivalent in rems and in sieverts.

17.30 A 1.5-g sample of muscle tissue absorbs 2.6 J of energy as a result of its exposure to α radiation. Calculate the dose in rads and the dose equivalent in rems and in sieverts.

17.31 Someone is exposed to a source of β radiation that results in a dose rate of 1.0 rad·d^{-1}. Given that nausea begins after a dose equivalent of about 100 rem, after what period will that symptom of radiation sickness be apparent?

17.32 Someone is exposed to a source of α radiation that results in a dose rate of 2.0 mrad·d^{-1}. If nausea begins after a dose equivalent of about 100 rem, after what period will nausea become apparent?

The Rate of Nuclear Decay

17.33 Determine the decay constant for (a) tritium, $t_{1/2} = 12.3$ a; (b) lithium-8, $t_{1/2} = 0.84$ s; (c) nitrogen-13, $t_{1/2} = 10.0$ min.

17.34 Determine the half-life of (a) potassium-40, $k = 5.3 \times 10^{-10}$ a^{-1}; (b) cobalt-60, $k = 0.132$ a^{-1}; (c) nobelium-255, $k = 3.85 \times 10^{-3}$ s^{-1}.

17.35 The activity of a sample of a radioisotope was found to be 2150 disintegrations per minute. After 6.0 hours the activity was found to be 1324 disintegrations per minute. What is the half-life of the radioisotope?

17.36 (a) A sample of pure cobalt-60 has an activity of 1 μCi. (a) How many atoms of cobalt-60 are present in the sample? (b) What is the mass in grams of the sample?

17.37 (a) What percentage of a carbon-14 sample remains after 3.00 ka? (b) Determine the percentage of a tritium sample that remains after 12.0 a.

17.38 (a) What percentage of a strontium-90 sample remains after 8.5 a? (b) Determine the percentage of an iodine-131 sample that remains after 6.0 d.

17.39 (a) What fraction of the original activity of ^{238}U remains after 3.9 Ga (1 Ga = 10^9 a)? (b) Potassium-40, which is presumed to have existed at the formation of the Earth, is used for dating minerals. If three-fifths of the original potassium-40 exists in a rock, how old is the rock?

17.40 (a) A sample of krypton-85 ($t_{1/2} = 10.8$ a) is released into the atmosphere. What fraction of the krypton-85 remains after 35.0 a? (b) A piece of wood, found in an archaeological dig, has a carbon-14 activity that is 62% of the current carbon-14 activity. How old is the piece of wood?

17.41 A 250.-mg sample of carbon from a piece of cloth excavated from an ancient tomb in Nubia undergoes 1.50×10^3 disintegrations in 10.0 h. If a current 1.00-g sample of carbon shows 921 disintegrations per hour, how old is the piece of cloth?

17.42 A current 1.00-g sample of carbon shows 921 disintegrations per hour. If 1.00 g of charcoal from an archaeological dig in a limestone cave in Slovenia shows 5.50×10^3 disintegrations in 24.0 h, what is the age of the charcoal sample?

17.43 Use the law of radioactive decay to determine the activity of (a) a 1.0-mg sample of radium-226 ($t_{1/2} = 1.60$ ka); (b) a 2.0-mg sample of strontium-90 ($t_{1/2} = 28.1$ a); (c) a 0.43-mg sample of promethium-147 ($t_{1/2} = 2.6$ a). The mass of each nuclide as a multiple of the atomic mass constant (m_u) is equal to its mass number, within two significant figures.

17.44 Use the law of radioactive decay to determine the activity of (a) a 1.0-g sample of ^{235}UO$_2$ ($t_{1/2} = 7.1 \times 10^8$ a); (b) a 1.0-g sample of cobalt that is 1.0% ^{60}Co ($t_{1/2} = 5.26$ a); (c) a 5.0-mg sample of thallium-200 ($t_{1/2} = 26.1$ h). The mass of each nuclide as a multiple of the atomic mass constant (m_u) is equal to its mass number, within two significant figures.

17.45 Deoxyglucose labeled with fluorine-18 is commonly used in PET scans to locate tumors. Fluorine-18 has a half-life of 109 min. How long will it take for the level of fluorine-18 in the body to drop to 10% of its initial value?

17.46 Technetium-99m (the m signifies a "metastable," or moderately stable, species) is generated in nuclear reactors and shipped to hospitals for use in medical imaging. The radioisotope has a half-life of 6.01 h. If a 165-mg sample of technetium-99m is shipped from a nuclear reactor to a hospital 125 kilometers away in a truck that averages 50.0 km·h^{-1}, what mass of technetium-99m will remain when it arrives at the hospital?

17.47 A 1.40-g sample containing radioactive cobalt was kept for 2.50 a, at which time it was found to contain 0.266 g of ^{67}Co. The half-life of ^{67}Co is 5.27 a. What percentage (by mass) of the original sample was ^{67}Co?

17.48 A radioactive sample contains 3.25×10^{18} atoms of a nuclide that decays at a rate of 3.4×10^{13} disintegrations per 15 min. (a) What percentage of the nuclide will have decayed after 150 d? (b) How many atoms of the nuclide will remain in the sample? (c) What is the half-life of the nuclide?

C **17.49** A radioactive isotope X with a half-life of 27.4 d decays into another radioactive isotope Y with a half-life of 18.7 d, which decays into the stable isotope Z. Set up and solve the rate laws for the amounts of the three nuclides as a function of time, and plot your results as a graph.

C **17.50** Suppose that the nuclide Y in Exercise 17.49 is needed for medical research and that 2.00 g of nuclide X was supplied at $t = 0$. At what time will Y be most abundant in the sample?

Uses of Radioisotopes

17.51 A chemist is studying the mechanism of the following hydrolysis reaction of the organic ester methyl acetate: $CH_3COOCH_3 + H_2O \rightarrow CH_3COOH + CH_3OH$. The question arises whether the O atom in the product methanol comes from the methyl acetate or from the water. Propose an experiment using isotopes that would allow the chemist to determine the origin of the oxygen atom.

17.52 The blood volume in a cancer patient was measured by injecting 5.0 mL of $Na_2SO_4(aq)$ labeled with ^{35}S ($t_{1/2} = 87.4$ d). The activity of the sample was 300 μCi. After 30. min, 10.0 mL of blood was withdrawn from the man and the activity of that sample was found to be 0.025 μCi. Report the blood volume of the patient.

17.53 What would you expect to happen to the vibrational frequency of the C—H bond of methane if the hydrogen atoms, which are normally present as ^1H, are replaced by ^2H? See Major Technique 1, Infrared Spectroscopy.

17.54 Normal water, H_2O, is necessary for life, but heavy water, D_2O, is toxic. Suggest a reason for this difference. The pK_a of deuterium oxide is 14.955 at 25°C.

17.55 The biological half-life of a radioisotope is the time required for the body to excrete half of the radioisotope. The *effective half-life* is the time required for the amount of a radioisotope in the body to be reduced to half its original amount, as a result of both the decay of the radioisotope and its excretion. Sulfur-35 ($t_{1/2} = 87.4$ d) is used in cancer research. The biological half-life of sulfur-35 in the human body is 90. d. What is the effective half-life of sulfur-35?

17.56 Barium-140 ($t_{1/2} = 12.8$ d) released in the fire at the Chernobyl nuclear plant has been found in some agricultural products in the region. The biological half-life of barium-140 in the human body is 65 d. What is the effective half-life (see Exercise 17.55) of barium-140?

Nuclear Energy

17.57 Calculate the energy in joules that is equivalent to (a) 1.0 g of matter; (b) one electron; (c) 1.0 pg of matter; (d) one proton.

17.58 Calculate the energy in joules that is equivalent to (a) 1.00 kg of matter; (b) 1.00 lb of matter (1 lb = 454 g); (c) one neutron; (d) one hydrogen atom.

17.59 The Sun emits radiant energy at the rate of 3.9×10^{26} J·s^{-1}. What is the rate of mass loss (in kilograms per second) of the Sun?

17.60 (a) For the fusion reaction $6 D \rightarrow 2\ ^4He + 2\ ^1H + 2$ n, 3×10^8 kJ of energy is released by a certain sample of deuterium. What is the mass loss (in grams) for the reaction? (b) What was the mass of deuterium converted? The molar mass of deuterium is 2.014 g·mol^{-1}.

17.61 Calculate the binding energy per nucleon (J·nucleon^{-1}) for (a) ^{62}Ni, 61.928 346m_u; (b) ^{239}Pu, 239.0522m_u; (c) ^2H,

2.0141m_u; (d) ^3H, 3.016 05m_u. (e) Which nuclide is the most stable?

17.62 Calculate the binding energy per nucleon (J·nucleon^{-1}) for (a) ^{98}Mo, 97.9055m_u; (b) ^{151}Eu, 150.9196m_u; (c) ^{56}Fe, 55.9349m_u; (d) ^{232}Th, 232.0382m_u. (e) Which nuclide is the most stable?

17.63 Calculate the energy released per gram of starting material in the fusion reaction represented by each of the following equations.
(a) $D + D \rightarrow\ ^3He + n$ (D, 2.0141m_u; ^3He, 3.0160m_u)
(b) $^3He + D \rightarrow\ ^4He +\ ^1H$ (^1H, 1.0078m_u; ^4He, 4.0026m_u)
(c) $^7Li +\ ^1H \rightarrow 2\ ^4He$ (^7Li, 7.0160m_u)
(d) $D + T \rightarrow\ ^4He + n$ (T, 3.0160m_u)

17.64 Calculate the energy released per gram of starting material in the nuclear reaction represented by each of the following equations.
(a) $^7Li +\ ^1H \rightarrow n +\ ^7Be$ (^1H, 1.0078m_u; ^7Li, 7.0160m_u; ^7Be, 7.0169m_u)
(b) $^{59}Co + D \rightarrow\ ^1H +\ ^{60}Co$ (^{59}Co, 58.9332m_u; ^{60}Co, 59.9529m_u; D, 2.0141m_u)
(c) $^{40}K + \beta \rightarrow\ ^{40}Ar$ (^{40}K, 39.9640m_u; ^{40}Ar, 39.9624m_u; β, 0.0005m_u)
(d) $^{10}B + n \rightarrow\ ^4He +\ ^7Li$ (^{10}B, 10.0129m_u; ^4He, 4.0026m_u)

17.65 Sodium-24 (23.990 96m_u) decays to magnesium-24 (23.985 04m_u). (a) Write a nuclear equation for the decay. (b) Determine the change in energy that accompanies the decay. (c) Calculate the change in the binding energy per nucleon.

17.66 (a) How much energy is emitted in each α decay of plutonium-234? (^{234}Pu, 234.0433m_u; ^{230}U, 230.0339m_u). (b) The half-life of plutonium-234 is 8.8 h. How much heat is released by the α decay of a 1.00-μg sample of plutonium-234 in a 24-h period?

17.67 Each of the following equations represents a fission reaction. Complete and balance the equations:
(a) $^{244}_{95}Am \rightarrow\ ^{134}_{53}I +\ ^{107}_{42}Mo + 3$?
(b) $^{235}_{92}U +\ ^1_0n \rightarrow\ ? +\ ^{138}_{52}Te + 2\ ^1_0n$
(c) $^{235}_{92}U +\ ^1_0n \rightarrow\ ^{101}_{42}Mo +\ ^{132}_{50}Sn + ?$

17.68 Complete each of the following nuclear equations for fission reactions:
(a) $^{239}_{94}Pu +\ ^1_0n \rightarrow\ ^{98}_{42}Mo +\ ^{138}_{52}Te + ?$
(b) $^{239}_{94}Pu +\ ^1_0n \rightarrow\ ^{100}_{43}Tc + ? + 4\ ^1_0n$
(c) $^{239}_{94}Pu +\ ^1_0n \rightarrow\ ? +\ ^{139}_{49}In + 3\ ^1_0n$

Integrated Exercises

17.69 State whether the following statements are true or false. If false, explain why. (a) The dose equivalent is lower than the actual dose of radiation because it takes into account the differential effects of different types of radiation. (b) Exposure to 1×10^8 Bq of radiation would be much more hazardous than exposure to 10 Ci of radiation. (c) Spontaneous radioactive decay follows first-order kinetics. (d) Fissile nuclei can undergo fission when struck with slow neutrons, whereas fast neutrons are required to split fissionable nuclei.

17.70 State whether the following statements are true or false. If false, explain why. (a) A subcritical mass of fissionable material is unstable and likely to explode. (b) In order for fusion to occur, the colliding particles must have high kinetic energy. (c) Highly reactive fission products are considered to be no longer dangerous after two half-lives. (d) The larger the binding energy per nucleon, the more stable is the nucleus.

17.71 (a) How many radon-222 nuclei ($t_{1/2} = 3.82$ d) decay per minute to produce an activity of 4 pCi? (b) A bathroom in the basement of a home measures 2.0 m $\times$ 3.0 m $\times$ 2.5 m. If the activity of radon-222 in the room is 4.0 pCi·L^{-1}, how many nuclei decay during a shower lasting 5.0 min?

17.72 Tritium undergoes β decay, and the emitted β particle has an energy of 0.0186 MeV (1 MeV = 1.602×10^{-13} J). If a 1.0-g sample of tissue absorbs 10% of the decay products of 1.0 mg of tritium, what dose equivalent does the tissue absorb?

17.73 It is found that 2.0×10^{-5} mol ^{222}Rn ($t_{1/2} = 3.82$ d) has seeped into a closed basement with a volume of 2.0×10^3 m^3. (a) What is the initial activity of the radon in picocuries per liter (pCi·L^{-1})? (b) How many atoms of ^{222}Rn will remain after 1 day (24 h)? (c) How long will it take for the radon to decay to below the EPA recommended level of 4 pCi·L^{-1}?

17.74 Explain why containers of radioactive waste are more susceptible to corrosion than containers of nonradioactive waste with the same chemical reactivity.

17.75 Uranium-238 decays through a series of α and β emissions to lead-206, with an overall half-life for the entire process of 4.5 Ga. How old is a uranium-bearing ore that is found to have a ^{238}U/^{206}Pb ratio of (a) 1.00 and (b) 1.25?

17.76 The age of a bottle of wine was determined by monitoring the tritium level in the wine. The activity of the tritium is determined to be 9.1% that of a sample of fresh grape juice from the same region in which the wine was bottled. How old is the wine?

17.77 Sodium-24 is used for monitoring blood circulation. (a) If 2.0 mg of sodium-24 has an activity of 17.3 Ci, what is its decay constant and its half-life? (b) What mass of the sodium-24 sample remains after 2.0 days? The mass of a sodium-24 atom is $24m_u$.

17.78 A radioactive sample contains ^{32}P (half-life, 14.28 d), ^{33}S (half-life, 87.2 d), and ^{59}Fe (half-life, 44.6 d). After 90 days, a sample that originally weighed 8.00 g contains 0.0254 g of ^{32}P, 1.466 g of ^{33}S, and 0.744 g of ^{59}Fe. What was the percentage composition (by mass) of the original sample?

17.79 Radiopharmaceuticals have one of two general functions: (1) they may be used to *detect* or *image* biological problems such as tumors and (2) they may be used to *treat* an illness. Which type of radiation (α, β, or γ) would be the most suitable for (a) detection and (b) therapy? Justify your selections. (c) From standard literature sources, find at least two radionuclides that have been used for imaging body tissues. (d) What are the half-lives of these radionuclides? See Box 17.1.

17.80 Radioactive metal ions that have short half-lives are being intensely studied as pharmaceuticals. The strategy is to attach a well-designed ligand to the metal ion so that the complex very selectively aggregates in one particular type of body tissue. What properties of the ligand are important to

the design of effective radiopharmaceutical therapeutic agents? See Box 17.1.

17.81 Technetium-99m is produced by a sequence of reactions in which molybdenum-98 is bombarded with neutrons to form molybdenum-99, which undergoes β decay to technetium-99m. (a) Write the balanced nuclear equations for this sequence. (b) Compare the neutron-to-proton ratio of the final daughter product with that of technetium-99m. Which is closer to the band of stability?

17.82 Actinium-225 decays by successive emission of three α particles. (a) Write the nuclear equations for the three decay processes. (b) Compare the neutron-to-proton ratio of the final daughter product with that of actinium-225. Which is closer to the band of stability?

17.83 What volume of helium at 1.0 atm and 298 K will be collected if 2.5 g of ^{222}Rn is stored in a container able to expand to maintain constant pressure for 15 days? (^{222}Rn decays to ^{218}Po with a half-life of 3.824 d.)

17.84 Why is the boiling point of heavy water greater than that of ordinary water? (a) Propose a model that allows you to estimate the difference or, from the known difference, to estimate any unknown parameters in the model. (b) Use your model to estimate the boiling point of "extraheavy water," T$_2$O.

17.85 A positron has the same mass as that of an electron, but the opposite charge. When a positron emitted in a PET scan encounters an electron, annihilation occurs in the body: electromagnetic energy is produced and no matter remains. How much energy (in joules) is produced in the encounter? See Box 17.1.

17.86 Do nuclei that are positron emitters lie above or below the band of stability? Which of the following isotopes might be suitable for PET scans? Explain your reasoning and write the equation for the decay: (a) ^{18}O; (b) ^{13}N; (c) ^{11}C; (d) ^{20}F; (e) ^{15}O. See Box 17.1.

17.87 The radioactivity from a sample of Na$_2$^{14}CO$_3$ was measured by noting the time required for the count of disintegrations to reach 8000. Five times were recorded: 21.25, 23.46, 20.97, 22.54, and 23.01 min. Background radiation was measured by noting the time required for 500 counts to be registered. Three times were recorded: 5.26, 5.12, and 4.95 min. What is the average level of radioactivity in the Na$_2$^{14}CO$_3$ sample, corrected for background radiation, in (a) disintegrations per minute; (b) microcuries?

17.88 (a) Using a standard graphing program such as that found on the Web site for this text, plot the fraction of ^{14}C remaining in a 4.00×10^4-year-old archaeological sample as a function of time. For convenience, use intervals of 1000 a. (b) Plot the natural logarithm of the fraction of ^{14}C remaining as a function of time. (c) After what period of time will less than 1.00% of the original ^{14}C in the sample remain?

Chemistry Connections

17.89 A person with pernicious anemia lacks "intrinsic factor," a compound required for the absorption of vitamin B$_{12}$ and its storage in the liver. The diagnosis is confirmed with the Schilling test. In this test the patient is given a small dose of vitamin B_{12} labeled with radioactive ^{57}Co or ^{58}Co, followed by a saturating dose of nonlabeled B_{12}, which releases the stored B_{12}. If the patient has intrinsic factor, a 24-hour urine sample will contain 13–15% of the labeled B_{12}. If intrinsic factor is absent, less than 6% will be excreted. The patient is then given instrinsic factor and the test is repeated for comparison. In one administration of the Schilling test, the patient was given a capsule containing 0.5 μCi of $^{58}CoB_{12}$, followed by 1.0 mg of nonlabeled B_{12}. 1200 mL of urine was collected over the next 24 hours. A 3.0-mL sample of the urine gave 83 cpm (counts per minute) and a 3.0-mL standard sample containing 0.4 nCi per mL gave 910 cpm. The test was repeated one week later with the administration of 30. mg of intrinsic factor. The 3.0-mL sample of urine from the second test gave 120 cpm. The half-life of ^{58}Co is 72 days.

(a) Calculate the percentage of ^{58}Co excreted with and without the intrinsic factor. Assume that the first test did not contaminate the second test.
(b) What would be the activity of the $^{58}CoB_{12}$ in the original capsule if it were stored for 7 days?
(c) If the biological half-life of B_{12} is 180 days, what is the effective half-life of $^{58}CoB_{12}$ in the body? See Exercise 17.55.
(d) Use the effective half life of $^{58}CoB_{12}$ to determine what fraction of the counts in the second test would be due to the dose given in the first test.
(e) The radioisotope ^{58}Co decays to another radioisotope, ^{59}Fe. The total mass of iron in the patient was 2.5 g. If all the ^{59}Fe produced in the first test were incorporated into hemoglobin, what percentage of the total iron would be ^{59}Fe the day after the first test? Ignore the decay of ^{59}Fe.

ORGANIC CHEMISTRY I: THE HYDROCARBONS

What Are the Key Ideas? The large numbers of different hydrocarbons arise from the ability of carbon atoms to form long chains and rings with one another; the types of carbon–carbon bonds that are present give the hydrocarbons characteristic properties.

Why Do We Need to Know This Material? All life on Earth is based on carbon; so is the fuel we burn, our food, and the clothes we wear. Therefore, to understand a major part of the everyday world, we need to be familiar with the chemistry of this extraordinary element. Compounds of carbon and hydrogen are the foundation of the petrochemical industry; petroleum products are used to generate electricity and to heat our homes. They are also used to make the flexible, strong polymeric and composite materials that make modern communication and transportation possible.

What Do We Need to Know Already? This chapter draws on the introduction to organic formulas and nomenclature in Sections C and D, the structure of molecules (Chapters 2 and 3), intermolecular forces (Sections 5.3–5.5), reaction enthalpy (Section 6.13), reaction mechanisms (Sections 13.7–13.9), and isomers (Section 16.7).

Chapter **18**

Carbon forms such an amazing variety of compounds that an entire field of chemistry, organic chemistry, is devoted to its study. Carbon atoms are so adaptable because they can string together to form chains and rings of endless variety. This versatility allows the element to form thousands of complicated biomolecules. Chemists use carbon's properties to design materials with properties that range from the softness of artificial skin and cartilage to the tough composite materials used in experimental aircraft.

This chapter deals with **hydrocarbons**, compounds built from only carbon and hydrogen. There are two broad classes of hydrocarbons: an **aromatic hydrocarbon** has a benzene ring as a part of its molecular structure; an **aliphatic hydrocarbon** does not. The compound shown as (**1**) is aromatic; compound (**2**) is aliphatic. We can describe more complex molecules as having an "aromatic region" consisting of benzenelike rings and an "aliphatic region" consisting of chains of carbon atoms. In Chapter 19, we shall see that aromatic and aliphatic hydrocarbons can be regarded as the basic framework for all organic compounds.

1 Ethylbenzene, $C_6H_5CH_2CH_3$

2 Pentane, C_5H_{12}

ALIPHATIC HYDROCARBONS

The different types of hydrocarbons are distinguished by the type of bonding between the carbon atoms, such as whether all the bonds are single or whether some are multiple. The types of bonds determine the types of reactions that the hydrocarbon can undergo.

18.1 Types of Aliphatic Hydrocarbons

A **saturated hydrocarbon** is an aliphatic hydrocarbon with no multiple carbon–carbon bonds; an **unsaturated hydrocarbon** has one or more double or triple carbon–carbon bonds. More hydrogen atoms can be added to compounds in which there are multiple bonds, but compounds with only single bonds are "saturated" with hydrogen. Compound (**3**) is saturated; compounds (**4**) and (**5**) are unsaturated.

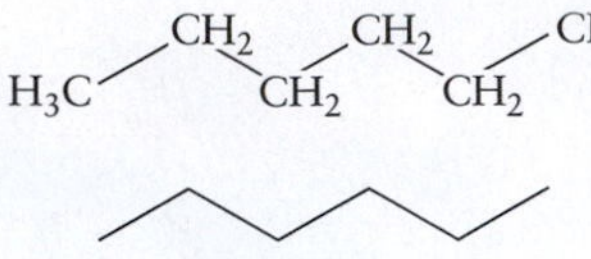

3   Hexane, C_6H_{14}

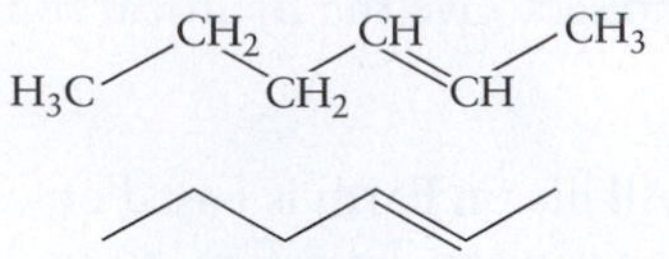

4   2-Hexene, C_6H_{12}

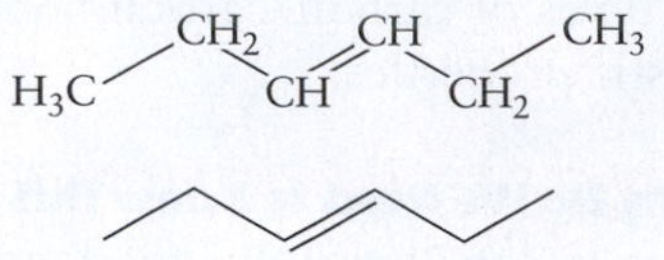

5   3-Hexene, C_6H_{12}

Because many organic molecules are very complicated, we need a simple way to represent their structures. It is often sufficient to give a **condensed structural formula,** which shows how the atoms are grouped together (see Section C). For instance, we write $CH_3CH_2CH_2CH_3$ for butane and $CH_3CH(CH_3)CH_3$ for methylpropane. The parentheses around the CH_3 group show that it is attached to the carbon atom on its left (or, if the formula begins with the parenthetical group, the carbon atom on its right). When several groups of atoms are repeated, we collect them together; so we can write butane as $CH_3(CH_2)_2CH_3$ and methylpropane as $(CH_3)_3CH$.

A line structure (introduced in Section C) represents a chain of carbon atoms as a zigzag line. The end of each short line in the zigzag represents a carbon atom. Because carbon nearly always has a valence of 4 in organic compounds, we do not need to show the C—H bonds. We just fill in the correct number of hydrogen atoms mentally, as we see for methylbutane (**6**), isoprene (**7**), and propyne (**8**). As explained in Section 2.7, a benzene ring is represented by a circle inside a hexagon, and we need to remember that one hydrogen atom is attached to each carbon atom.

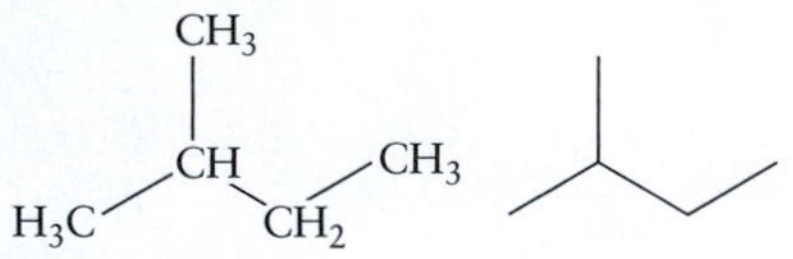

6   Methylbutane, $(CH_3)_2CHCH_2CH_3$

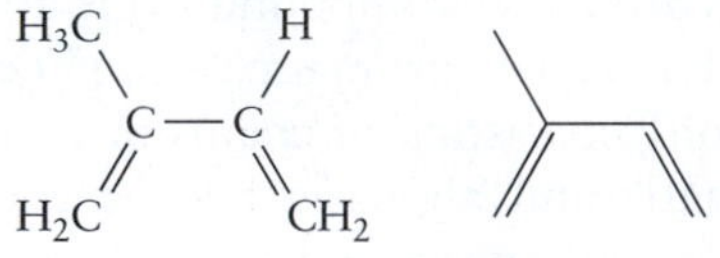

7   Isoprene, CH_2=$C(CH_3)CH$=CH_2

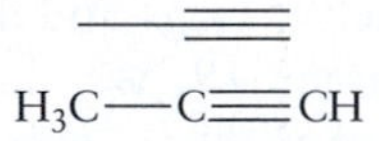

8   Propyne, CH_3C=CH

SELF-TEST 18.1A Draw (a) the line structure of aspirin (**9a**) and (b) the structural formula for $CH_3(CH_2)_2C(CH_3)_2CH_2C(CH_3)_3$.

[*Answer:* (a) (**9b**); (b) (**10**)]

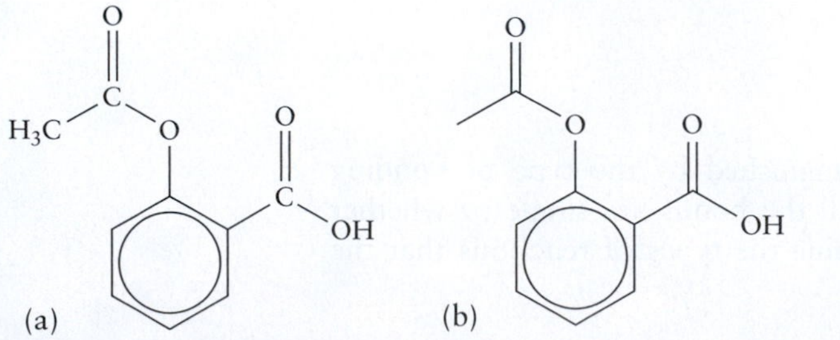

(a)

(b)

9   Acetylsalicylic acid (aspirin)

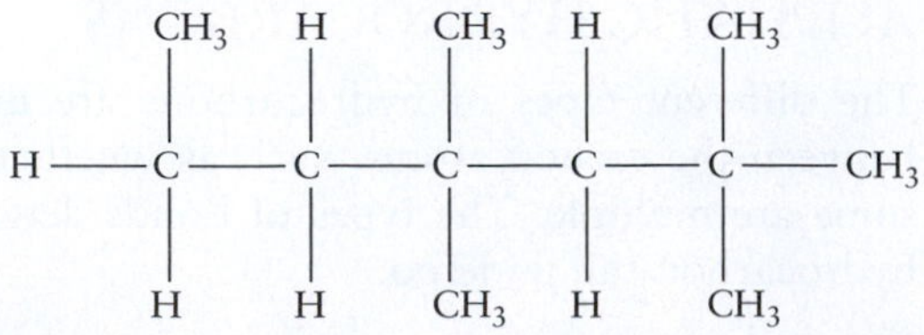

10

SELF-TEST 18.1B Write (a) the structural formula of carvone (**11**) and (b) the condensed structural formula for

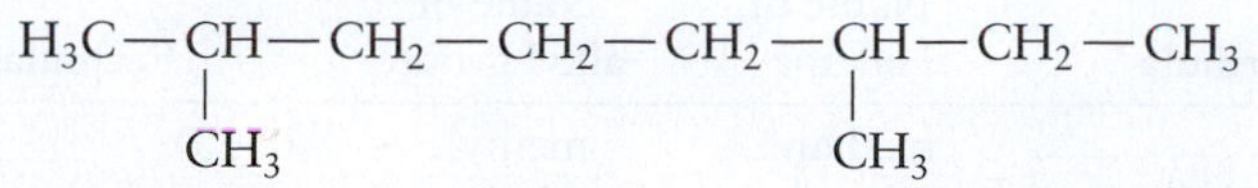

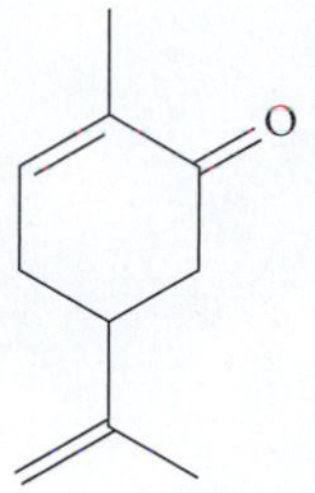

$$H_3C-CH-CH_2-CH_2-CH_2-CH-CH_2-CH_3$$
$$\qquad\ \ |\qquad\qquad\qquad\qquad\ \ |$$
$$\qquad\ \ CH_3\qquad\qquad\qquad\quad\ CH_3$$

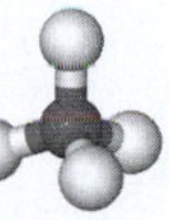

Saturated hydrocarbons are called **alkanes.** As we saw in Section 3.5, the bonds to each carbon atom in an alkane are single bonds that lie in a tetrahedral arrangement, with sp^3 hybridization. The simplest alkane is methane, CH_4 (**12**). The formulas of other alkanes are derived from CH_4 by inserting CH_2 groups between pairs of atoms. Although we write alkanes as linear structures, they really consist of tetrahedral arrangements of bonds at each carbon atom. Moreover, because all the C—C bonds are single, the different parts of an alkane molecule can rotate relative to each other. In liquids and gases, the chains of atoms in an alkane molecule are in constant motion, at some instants rolled up into a ball (**13**) and at others stretched out into a zigzag (**14**).

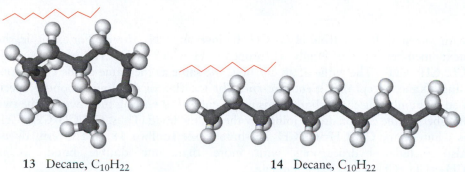

13   Decane, $C_{10}H_{22}$ **14**   Decane, $C_{10}H_{22}$

To name an alkane in which the carbon atoms form a single chain, we combine a prefix denoting the number of carbon atoms with the suffix *-ane* (Table 18.1). For example, CH_3-CH_3 (more simply, CH_3CH_3) is ethane and $CH_3-CH_2-CH_3$ (that is, $CH_3CH_2CH_3$) is propane. Cyclopropane, C_3H_6 (**15**), and cyclohexane, C_6H_{12} (**16**), are **cycloalkanes,** alkanes that contain rings of carbon atoms.

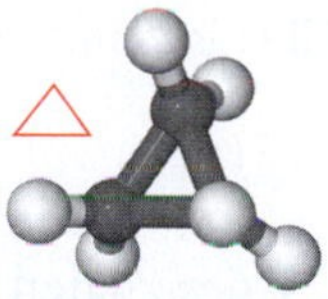

15   Cyclopropane, C_3H_6 **16**   Cyclohexane, C_6H_{12}

With C_4H_{10}, we find another reason for the variety of compounds that carbon can form: the same atoms can bond together in different arrangements. Four carbon atoms can link together in a chain to form butane (**17**) or in a Y-shape to form methylpropane (**18**). As we saw in Section 16.7, different compounds with the same molecular formula are called **isomers.** Thus, butane and methylpropane are isomers with the same molecular formula, C_4H_{10}.

The simplest unsaturated hydrocarbon is ethene, C_2H_4 or $H_2C=CH_2$, commonly called ethylene (**19**). It is the parent of the **alkenes,** a series of compounds

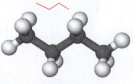

17   Butane, C_4H_{10}

18   Methylpropane, C_4H_{10}

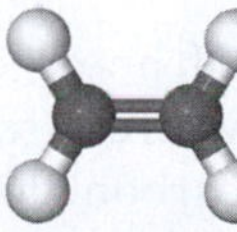

19   Ethene, C_2H_4

(top right)

11   Carvone

(below)

12   Methane, CH_4

(sidebar) The name *isomer* comes from the Greek words for "equal parts," suggesting that isomers are built from the same kit of parts.

TABLE 18.1   Alkane Nomenclature

Number of carbon atoms	Formula	Name of alkane	Name of alkyl group	Formula
1	CH_4	methane	methyl	CH_3-
2	CH_3CH_3	ethane	ethyl	CH_3CH_2-
3	$CH_3CH_2CH_3$	propane	propyl	$CH_3CH_2CH_2-$
4	$CH_3(CH_2)_2CH_3$	butane	butyl	$CH_3(CH_2)_2CH_2-$
5	$CH_3(CH_2)_3CH_3$	pentane	pentyl	$CH_3(CH_2)_3CH_2-$
6	$CH_3(CH_2)_4CH_3$	hexane	hexyl	$CH_3(CH_2)_4CH_2-$
7	$CH_3(CH_2)_5CH_3$	heptane	heptyl	$CH_3(CH_2)_5CH_2-$
8	$CH_3(CH_2)_6CH_3$	octane	octyl	$CH_3(CH_2)_6CH_2-$
9	$CH_3(CH_2)_7CH_3$	nonane	nonyl	$CH_3(CH_2)_7CH_2-$
10	$CH_3(CH_2)_8CH_3$	decane	decyl	$CH_3(CH_2)_8CH_2-$
11	$CH_3(CH_2)_9CH_3$	undecane	undecyl	$CH_3(CH_2)_9CH_2-$
12	$CH_3(CH_2)_{10}CH_3$	dodecane	dodecyl	$CH_3(CH_2)_{10}CH_2-$

with formulas derived from $H_2C=CH_2$ by inserting CH_2 groups. For example, the next member of the family is propene, $H-CH_2-CH=CH_2$ (more simply, $CH_3CH=CH_2$). The name of an alkene is the same as the name of the corresponding alkane, except that it ends in -ene. The location of the double bond is specified by numbering the carbon atoms in the chain and writing the lower of the two numbers of the carbon atoms joined by the double bond. Thus, $CH_3CH_2CH=CH_2$ is 1-butene and $CH_3CH=CHCH_3$ is 2-butene (see Toolbox 18.1). The term alkene also includes hydrocarbons with more than one double bond, as in $CH_2=CH-CH=CH_2$, 1,3-butadiene.

The **alkynes** are hydrocarbons that have at least one carbon–carbon triple bond. The simplest is ethyne, $HC\equiv CH$, which is commonly called acetylene (**20**). Alkynes are named like the alkenes but with the suffix -yne.

20 Ethyne, C_2H_2

TOOLBOX 18.1   **HOW TO NAME ALIPHATIC HYDROCARBONS**

CONCEPTUAL BASIS

Because the types of carbon–carbon bonds present in the molecule tend to dominate its properties, an aliphatic hydrocarbon is first classified as an alkane, alkene, or alkyne. Then the longest chain of carbon atoms is used to form the "root" of the name. Other hydrocarbon groups attached to the longest chain are named as side chains.

PROCEDURE

The following rules have been adopted by the International Union for Pure and Applied Chemistry (IUPAC).

Alkanes

The names of straight-chain alkanes are given in Table 18.1: they all end in -ane. Name a hydrocarbon side chain as a substituent by changing the ending -ane to -yl (as in the last two columns in Table 18.1).

Example: CH_3CH_2- is the ethyl group

The names of branched-chain hydrocarbons and hydrocarbon derivatives are based on the name of the *longest* continuous carbon chain in the molecule (which may not be shown in a horizontal line).

Example: $CH_3-CH-CH_3$ is 2-methylbutane
  $|$
  CH_2-CH_3

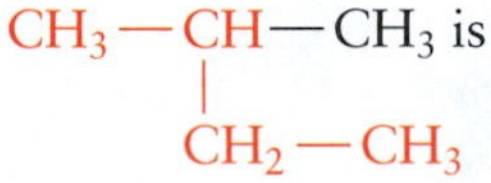

A cyclic (ring) hydrocarbon is designated by the prefix *cyclo-*.

Example: CH_2 is cyclopropane
  $/\backslash$
  H_2C-CH_2

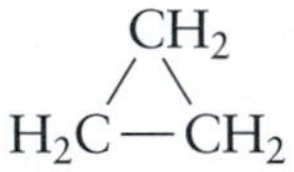

To indicate the position of a branch or substituent, the carbons in the longest chain are numbered consecutively from one end to the other, starting at the end that will give the *lower* number(s) to the substituent(s).

Examples:

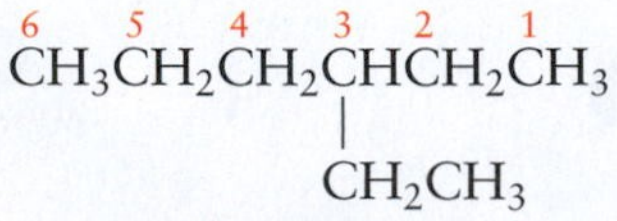

$\overset{6}{C}H_3\overset{5}{C}H_2\overset{4}{C}H_2\overset{3}{C}H\overset{2}{C}H_2\overset{1}{C}H_3$ $\overset{1}{C}H_3\overset{2}{C}(CH_3)_2\overset{3}{C}H_2\overset{4}{C}H_3$
  $|$
  CH_2CH_3

3-Ethylhexane   2,2-Dimethylbutane

The prefixes *di-*, *tri-*, *tetra-*, *penta-*, *hexa-*, and so on, indicate how many of each substituent are in the molecule. Numbers set off by hyphens specify to which carbon atoms the groups are attached.

Examples:

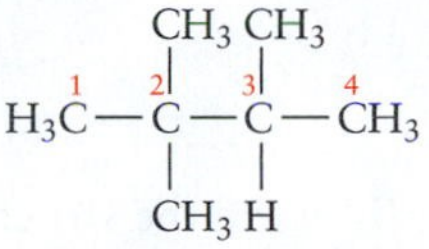

2,2,3-Trimethylbutane 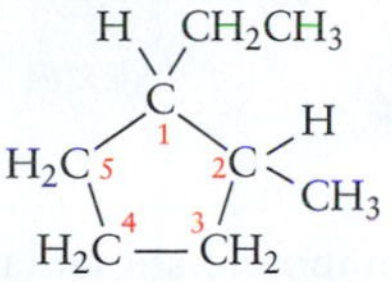   1-Ethyl-2-methylcyclopentane

The substituents are listed in alphabetical order (disregarding the Greek prefixes) and attached to the root name. The names of substituents other than hydrocarbon groups are discussed more fully in Chapter 19 (see Toolbox 19.1).

Example 18.1 demonstrates how to name alkanes.

Alkenes and alkynes

Double bonds in hydrocarbons are indicated by changing the suffix *-ane* to *-ene*, and triple bonds are indicated by changing the suffix to *-yne*. The position of the multiple bond is given by the number of the first (lower-numbered) carbon atom involved in the multiple bond. If more than one multiple bond of the same type is present, the number of these bonds is indicated by a Greek prefix. Then follow the rules for naming alkanes.

Examples:

$$CH_3-CH_2-CH_2-C{\equiv}CH$$
1-Pentyne

$$CH_3-CH_2-CH{=}CH-CH_3$$
2-Pentene

$$CH_2{=}CH-CH_2-CH{=}CH_2$$
1,4-Pentadiene

When numbering atoms in the chain, the lowest numbers are given preferentially to (a) functional groups named by suffixes (see Toolbox 19.1), (b) double bonds, (c) triple bonds, and (d) groups named by prefixes.

Example 18.2 illustrates how to name alkenes.

EXAMPLE 18.1 **Sample exercise: Naming alkanes and cycloalkanes**

(a) Name the compound shown as (**21**) and (b) write the structural formula of 2-ethyl-1,1-dimethylcyclohexane.

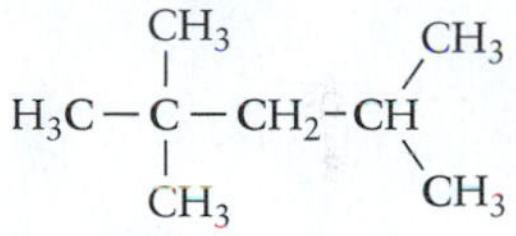

21

SOLUTION

(a) Count carbon atoms in the longest chain.	The longest carbon chain (in red) of (**21**) has 5 carbon atoms. The molecule is a substituted pentane.	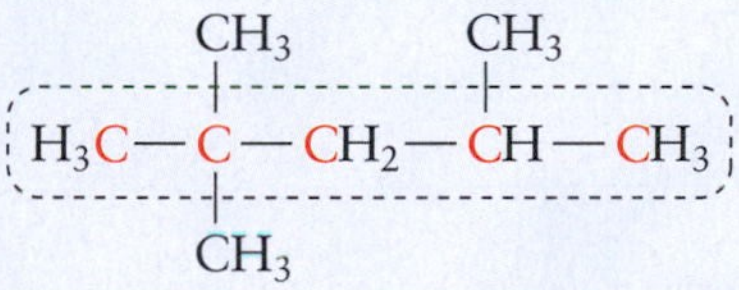
Identify and count substituents.	There are 3 methyl groups ($CH_3{-}$) on the longest chain. The molecule is a trimethylpentane.	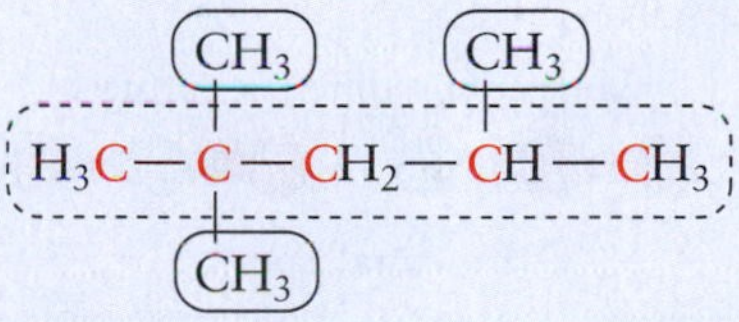
Number the backbone carbon atoms to give the lowest numbers to the substituents.	2,2,4-trimethylpentane.	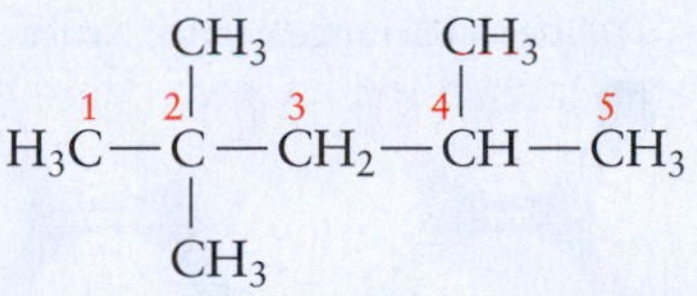
(b) Draw the longest chain of carbon atoms first.	"Cyclohexane" tells us that the molecule has a ring of 6 carbon atoms.	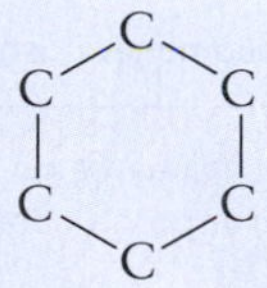
Number the carbon atoms and add the substituents according to the number given in the name.	Add 2 methyl groups to one carbon atom, which becomes carbon atom 1. Then add an ethyl group to carbon atom 2.	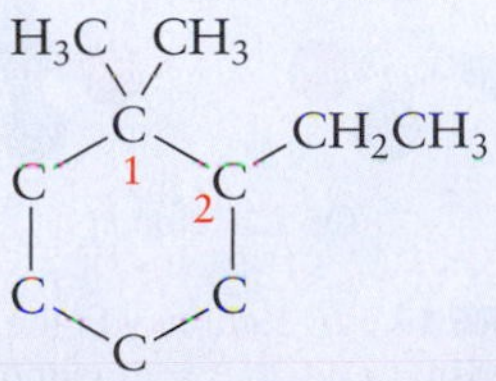

Add hydrogen atoms as needed to give each carbon atom a valence of four.

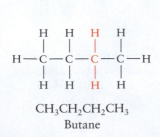

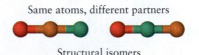

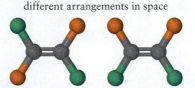

22

SELF-TEST 18.2A (a) Name the compound shown as (**22**) and (b) write the structural formula of 5-ethyl-2,2-dimethyloctane.

[*Answer:* (a) 4-Ethyl-3-methyloctane; (b) (**23**)]

$$CH_3 \quad\quad CH_2CH_3$$
$$H_3C-\overset{|}{\underset{|}{C}}-CH_2-CH_2-\overset{|}{CH}-CH_2-CH_2-CH_3$$
$$CH_3$$

23

SELF-TEST 18.2B Name the compound $(CH_3)_2CHCH_2CH(CH_2CH_3)_2$ and write the structural formula of 3,3,5-triethylheptane.

EXAMPLE 18.2 Sample exercise: Naming alkenes

(a) Name the alkene $CH_3CH_2CH{=}CH_2$ and (b) write the condensed structural formula for 5-methyl-1,3-hexadiene.

SOLUTION
(a) From Toolbox 18.1, $CH_3CH_2CH{=}CH_2$ is 1-butene (not 3-butene). (b) 5-Methyl-1,3-hexadiene has two double bonds, at the first and third carbon atoms, with a methyl group on the fifth carbon atom: $CH_2{=}CHCH{=}CHCH(CH_3)CH_3$.

SELF-TEST 18.3A (a) Name the alkene $(CH_3)_2CHCH{=}CH_2$ and (b) write the condensed structural formula for 2-methylpropene.

[*Answer:* (a) 3-Methyl-1-butene; (b) $CH_2{=}C(CH_3)_2$]

SELF-TEST 18.3B (a) Name the alkene $(CH_3CH_2)_2CHCH{=}CHCH_3$ and (b) write the structural formula for cyclopropene.

Saturated hydrocarbons have only single bonds; unsaturated hydrocarbons have one or more multiple bonds. Alkanes are saturated hydrocarbons. Alkenes and alkynes are unsaturated hydrocarbons: the former have carbon–carbon double bonds and the latter have triple bonds.

18.2 Isomers

Figure 18.1 summarizes the types of isomerism that are found in organic compounds. Molecules that are **structural isomers** are built from the same atoms, but the atoms are connected differently; that is, the molecules have a different **connectivity**. For example, we can insert a —CH_2— group into the C_3H_8 molecule in two different ways to give two different compounds with the formula C_4H_{10}:

Same atoms, different partners

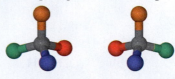

Structural isomers

Same atoms, same partners
different arrangements in space

Geometrical isomers

Same atoms, same partners
nonsuperimposable mirror images

Optical isomers

FIGURE 18.1 A summary of the types of isomerism in organic compounds.

$CH_3CH_2CH_2CH_3$
Butane

$CH(CH_3)_3$
Methylpropane

Although the —CH_2— group could be inserted in other places, the free rotation about single C—C bonds in hydrocarbons allows the resulting molecules to be twisted into one or the other of these two isomers. Both compounds are gases, but butane (**24**) condenses at −1°C, whereas methylpropane (**25**) condenses at −12°C. Two molecules that differ only by rotation about one or more bonds may look different on paper, but they are not isomers of each other; they are different **conformations** of the same molecule. Example 18.3 illustrates how to tell if two molecules are different isomers or different conformations of the same isomer.

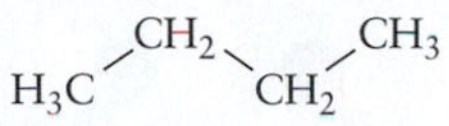

24 Butane, C_4H_{10}

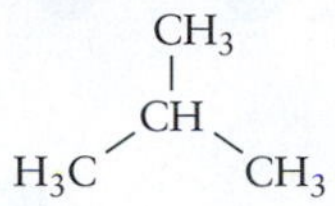

25 Methylpropane, C_4H_{10}

EXAMPLE 18.3 **Writing the formulas of structural isomers**

Draw two-dimensional structural formulas for all the isomeric alkanes of formula C_5H_{12}.

STRATEGY Isomers cannot be changed into one another simply by rotating either the entire formula or parts of the formula on the page. One approach is to insert —CH_2— groups into different parts of the formulas of the two C_4H_8 molecules given earlier, then discard formulas that repeat those already obtained. It is often easier to identify different conformations of the same isomer by building molecular models that allow rotation about single bonds.

SOLUTION

From butane, we can form

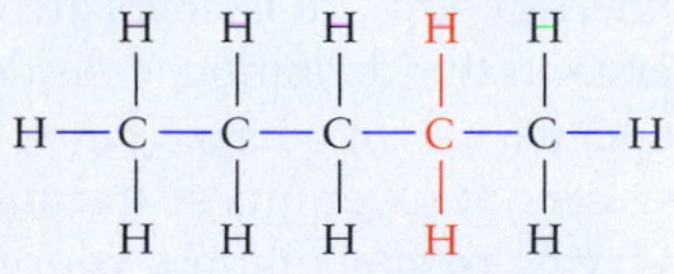

(a) 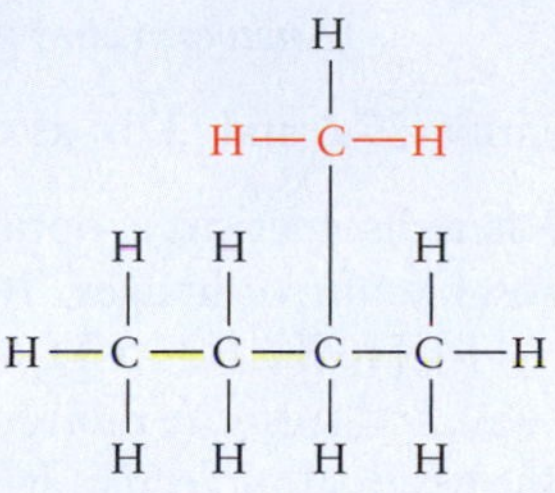   (b)

From methylpropane, we can form

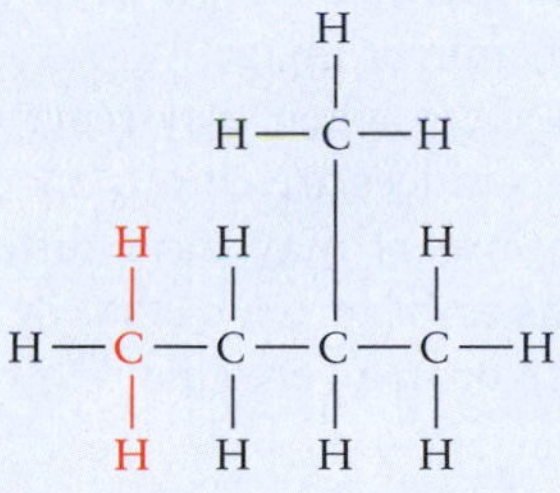

(c) 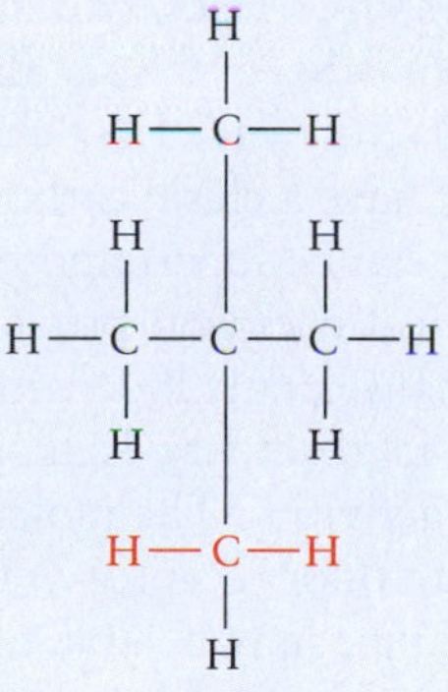   (d) 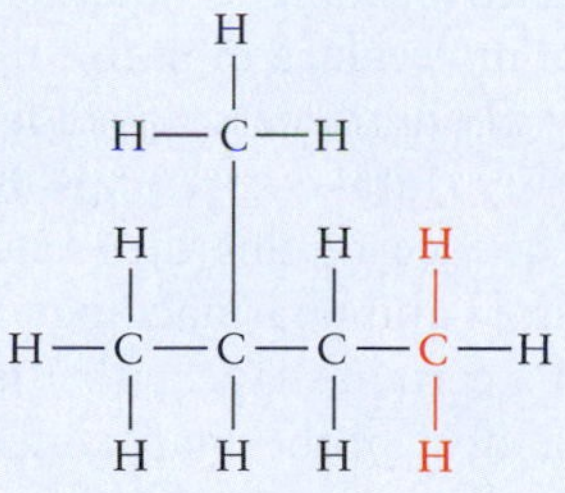   (e)

Molecules (b) and (c) are identical. The atoms of molecule (e) are joined together in the same arrangement as that of (b) and (c), but the structure looks different because it has been rotated 180° and twisted; so (b), (c), and (e) are the same. There are therefore only three distinct isomers with formula C_5H_{12}: (a), (b), and (d).

SELF-TEST 18.4A Write the condensed structural formulas for the five isomeric alkanes of molecular formula C_6H_{14}.

[*Answer:* $CH_3(CH_2)_4CH_3$; $CH_3(CH_2)_2CH(CH_3)CH_3$; $CH_3CH_2CH(CH_3)CH_2CH_3$; $CH_3CH_2C(CH_3)_2CH_3$; $CH_3CH(CH_3)CH(CH_3)CH_3$]

SELF-TEST 18.4B Halogen atoms can substitute for hydrogen atoms in hydrocarbons. Write the condensed structural formulas for the four isomers with the molecular formula C_4H_9Br.

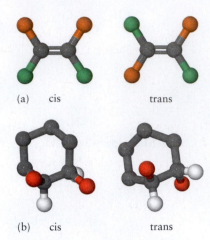

(a)   cis   trans

(b)   cis   trans

FIGURE 18.2 Two pairs of geometrical isomers. In geometrical isomerism two groups take different positions across a double bond, as in part (a); or different positions above and below a ring, as in part (b). Note that the neighbors of each atom in a pair of isomers are the same, but the arrangement of the atoms in space is different. (The compound in part (a) is 2-butene when the green sphere is CH_3 and the orange sphere is H.)

Recall from Section 16.7 that the cis isomer has similar groups on the same side of the molecule, whereas the trans isomer has similar groups on opposite sides of the molecule.

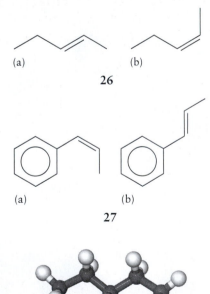

(a)   (b)

26

(a)   (b)

27

28 3-Methylpentane, C_6H_{14}

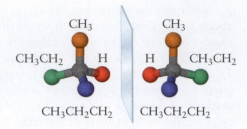

FIGURE 18.3 The molecule of 3-methylhexane on the right is the mirror image of the one on the left. Each group is represented by a sphere of a different color. The molecule on the left cannot be superimposed on the one on the right, and so these two molecules are distinct optical isomers.

In **stereoisomers**, the molecules have the same connectivity but the atoms are arranged differently in space. One class of stereoisomers are **geometrical isomers**, in which atoms have different arrangements on either side of a double bond or above and below the ring of a cycloalkane (Fig. 18.2). Geometrical isomers of organic molecules are distinguished by the prefixes *cis-* and *trans-*. For example, in the upper part of the illustration, we can see that there are two different 2-butenes: in the cis isomer both methyl groups are on the same side of the double bond and in the trans isomer the methyl groups are across the double bond. Geometrical isomers have the same molecular formula and structural formula but different properties.

SELF-TEST 18.5A Identify (**26a**) and (**26b**) as cis or trans.

[*Answer:* (**26a**) is *trans*-2-pentene; (**26b**) is *cis*-2-pentene]

SELF-TEST 18.5B Identify (**27a**) and (**27b**) as cis or trans.

Another type of stereoisomerism is optical isomerism. **Optical isomers** are each other's nonsuperimposable mirror images. To understand this definition, consider 3-methylhexane, $CH_3CH_2CH(CH_3)CH_2CH_2CH_3$, and its mirror image (Fig. 18.3). No matter how we twist and turn the molecules, we cannot superimpose the mirror-image molecule on the original molecule. It is like trying to superimpose your right hand on your left hand. A **chiral molecule**, such as 3-methylhexane, is a molecule that is not identical to its mirror image. A chiral molecule and its mirror image form a pair of **enantiomers,** or mirror-image isomers. Although they have the same composition, the two enantiomers are, in fact, two distinct compounds. In organic compounds, optical isomers occur whenever four different groups are attached to the same carbon atom, which is then called a "chiral carbon atom." The alkane 3-methylpentane (**28**), does not have a chiral carbon atom. It is an example of an **achiral molecule,** a molecule that can be superimposed on its mirror image.

Enantiomers have identical chemical properties, except when they react with other chiral compounds. Because many biochemical substances are chiral, one consequence of this difference in reactivity is that enantiomers may have different odors and pharmacological activities. The molecule has to fit into a cavity, or slot, of a certain shape, either in an odor receptor in the nose or in an enzyme. Only one member of the enantiomeric pair may be able to fit.

Enantiomers differ in one physical property: chiral molecules display optical activity, the ability to rotate the plane of polarization of light (Section 16.7 and Box 16.2). If a chiral molecule rotates the plane of polarization clockwise, then its mirror-image partner rotates it through the same angle in the opposite direction.

Often organic compounds synthesized in the laboratory are racemic mixtures, or mixtures of enantiomers in equal proportions (Section 16.7). In contrast, reactions in living cells commonly lead to only one enantiomer. It is a remarkable feature of nature that almost all naturally occurring amino acids in animals have the same handedness.

EXAMPLE 18.4   **Sample exercise: Deciding whether a compound is chiral**

A bromoalkane is formed from an alkane when a bromine atom replaces a hydrogen atom. Decide whether the bromoalkanes (a) $CH_3CH_2CHBrCH_3$ and (b) $CH_3CHBrCH_3$ are chiral.

SOLUTION

Draw the compounds.	(a) $CH_3CH_2CHBrCH_3$	(b) $CH_3CHBrCH_3$
	H | Br⸌⸌⸌C*⸌CH₂CH₃ CH₃	H | Br⸌⸌⸌C⸌CH₃ CH₃
Mark carbon atoms as chiral if they have four different groups attached.	The C atom marked with a * is chiral; therefore the molecule is nonsuperimposable on its mirror image.	No atom is chiral. The molecule is superimposable on its mirror image.
	Chiral.	Not chiral.

A note on good practice: Dashed and solid wedge-shaped bonds are commonly used when displaying organic structures to convey a sense of the three-dimensional shapes. The dashed wedge-shaped bonds go into the page and the solid wedge-shaped bonds come toward us. The thin lines are in the plane of the paper.

SELF-TEST 18.6A A chlorofluorocarbon molecule contains Cl and F atoms as well as C and H. Which of the following chlorofluorocarbons is chiral: (a) CH_3CF_2Cl; (b) CH_3CHFCl; (c) CH_2FCl?

[*Answer:* (b)]

SELF-TEST 18.6B In an alcohol, an —OH group is attached to a C atom. Which of the following alcohols is chiral: (a) CH_3CH_2OH; (b) $CH_3CH(OH)CH_3$; (c) $CH_3CH(OH)CH_2CH_3$?

Structural isomers have identical molecular formulas, but their atoms are linked to different neighbors. Geometrical isomers have the same molecular and structural formulas but different arrangements in space. Molecules with four different groups attached to a single carbon atom are chiral; they are optical isomers.

18.3 Properties of Alkanes

The electronegativities of carbon and hydrogen (2.6 and 2.2, respectively) are so similar and rotation about bonds so free that hydrocarbon molecules are best regarded as nonpolar. The dominant interaction between alkane molecules is therefore the London force (Section 5.4). Because the strength of this interaction increases with the number of electrons in the molecule, we can expect the alkanes in petroleum, their major source, to become less volatile with increasing molar mass and therefore to be separable by fractional distillation, as described in Section 8.19 (Fig. 18.4). The lightest members, methane through butane, are gases at room temperature. Pentane is a volatile liquid, and hexane through undecane ($C_{11}H_{24}$) are moderately volatile liquids that are present in gasoline, which is discussed further in Section 18.9. All alkanes are insoluble in water. Because their densities are lower than that of water, they float on the surface of water. Even a small oil spill at sea can form an organic layer on the surface of the ocean and spread over a huge area.

Alkanes with long, unbranched chains tend to have higher melting points, boiling points, and enthalpies of vaporization than those of their branched isomers. The difference arises because, compared with unbranched molecules, the atoms of neighboring branched molecules cannot get as close together (Fig. 18.5). As a result, molecules with branched chains have weaker intermolecular forces than their unbranched isomers.

SELF-TEST 18.7A Which compound has the higher boiling point: (a) $CH_3CH_2CH_2CH_2CH_3$ or (b) $(CH_3)_2CHCH_2CH_3$? Why?

[*Answer:* (a), because it is not branched and so has stronger London forces.]

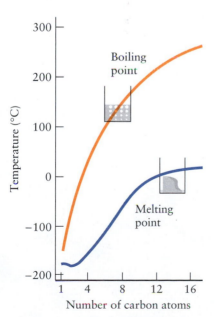

FIGURE 18.4 The melting and boiling points of the unbranched alkanes from CH_4 to $C_{16}H_{34}$.

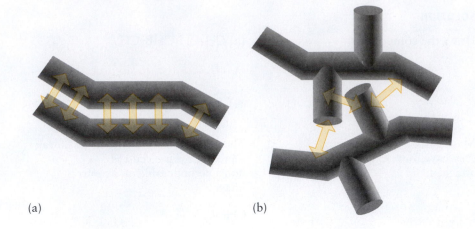

FIGURE 18.5 (a) The atoms in neighboring straight-chain alkanes, represented by the tubelike structures, can lie close together. (b) Fewer of the atoms of neighboring branched alkane molecules can get so close together overall, and so the London forces (represented by double-headed arrows) are weaker and branched alkanes are more volatile.

(a)   (b)

SELF-TEST 18.7B Which compound has the higher boiling point:
(a) $CH_3CH_2CH_2CH_2CH_3$ or (b) $CH_3CH_2CH_2CH_3$? Why?

The alkanes were once called the *paraffins,* from Latin words meaning "little affinity." As this name suggests, they are not very reactive. They are unaffected by concentrated sulfuric acid, by boiling nitric acid, by strong oxidizing agents such as potassium permanganate, and by boiling aqueous sodium hydroxide. One reason for their resistance to chemical onslaught is thermodynamic: the C—C and C—H bonds are strong (their mean bond enthalpies are 348 kJ·mol^{-1} and 412 kJ·mol^{-1}, respectively), and so there is little energy advantage in replacing them with most other bonds. The most notable exceptions are C=O (743 kJ·mol^{-1}), C—O (360 kJ·mol^{-1}), and C—F (484 kJ·mol^{-1}) bonds.

In addition to the catalyzed reactions used in the refining of petroleum (see Sections 18.5 and 18.9), alkanes undergo two common reactions: oxidation and substitution (Section 18.4). The alkanes are commonly used as fuels, because their combustion to carbon dioxide and water is highly exothermic (Section 6.17):

$$CH_4(g) + 2\,O_2(g) \longrightarrow CO_2(g) + 2\,H_2O(g) \qquad \Delta H° = -890 \text{ kJ}$$

In this reaction, the strong carbon–hydrogen bonds are replaced by the even stronger O—H bonds (463 kJ·mol^{-1}), and the oxygen–oxygen bond (496 kJ·mol^{-1}) is replaced by two very strong C=O bonds. The excess energy is given off as heat.

The strength of the London forces between alkane molecules increases as the molar mass of the molecules increases; hydrocarbons with unbranched chains pack together more closely than their branched isomers. Alkanes are not very reactive, but they do undergo oxidation (combustion) and substitution reactions.

18.4 Alkane Substitution Reactions

Alkanes are used as the raw materials for the synthesis of many more reactive compounds. Starting with alkanes obtained from the refining of petroleum, organic chemists introduce reactive groups of atoms into the molecules in the process called **functionalization.** The functionalization of alkanes can be achieved by a **substitution reaction,** a reaction in which an atom or group of atoms replaces an atom (in alkanes, a hydrogen atom) in the original molecule (Fig. 18.6). An example of a substitution reaction is that between methane and chlorine. A mixture of these two gases survives indefinitely in the dark but, when exposed to ultraviolet radiation or heated to more than 300°C, the gases react explosively:

$$CH_4(g) + Cl_2(g) \xrightarrow[\text{light or heat}]{} CH_3Cl(g) + HCl(g)$$

Chloromethane (CH_3Cl) is only one of the products; dichloromethane (CH_2Cl_2), trichloromethane ($CHCl_3$), and tetrachloromethane (CCl_4) also form, especially at high concentrations of chlorine.

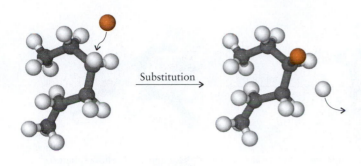

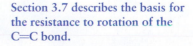

FIGURE 18.6 In an alkane substitution reaction, an incoming atom or group of atoms (represented by the orange sphere) replaces a hydrogen atom in the alkane molecule.

The results of kinetic studies suggest that alkane substitution reactions typically proceed by a radical chain mechanism (Section 13.9). The *initiation* step in the chlorination of methane is the dissociation of chlorine:

$$Cl_2 \xrightarrow{\text{light or heat}} Cl\cdot + \cdot Cl$$

Chlorine atoms (each of which has one unpaired electron) are highly reactive; they attack methane molecules and extract a hydrogen atom, leaving a methyl radical behind:

$$Cl\cdot + CH_4 \longrightarrow HCl + \cdot CH_3$$

Recall from Section 2.9 that most radicals are very reactive. Because one of the products is another radical, this reaction is a *propagation* step (a step in which one reactive radical intermediate produces another). In a second propagation step, the methyl radical may react with a chlorine molecule:

$$Cl_2 + \cdot CH_3 \longrightarrow CH_3Cl + Cl\cdot$$

A *termination* step is reached when two radicals combine to form a nonradical product, as in

$$Cl\cdot + \cdot Cl \longrightarrow Cl_2$$

The product of substitution reactions of alkanes with the halogens is typically a complex mixture of *haloalkanes* (halogenated alkanes). One way to limit the production of the more highly substituted alkanes is to use a large excess of the alkane; then most reactions take place with the original hydrocarbon rather than with any haloalkanes produced in the reaction.

Alkane substitution takes place by a radical chain mechanism.

18.5 Properties of Alkenes

The carbon–carbon double bond in alkenes is more reactive than carbon–carbon single bonds and gives alkenes their characteristic properties. As we saw in Section 3.4, a double bond consists of a σ-bond and a π-bond. Each carbon atom in a double bond is sp^2 hybridized and uses the three hybrid orbitals to form three σ-bonds. The unhybridized p-orbitals on each carbon atom overlap each other and form a π-bond. As we saw in Section 3.7, the carbon–carbon π-bond is relatively weak because the overlap responsible for the formation of the π-bond is less extensive than that responsible for the formation of the σ-bond and the enhanced electron density does not lie directly between the two nuclei. A consequence of this weakness is the reaction most characteristic of alkenes, the replacement of the π-bond by two new σ-bonds, which is discussed in Section 18.6.

The C=C group and all four atoms attached to it lie in the same plane and are locked into that arrangement by the resistance to twisting of the π-bond (Fig. 18.7). Because alkene molecules cannot roll up into a ball as compactly as alkanes or rotate into favorable positions, they cannot pack together as closely as alkanes; so alkenes have lower melting points than alkanes of similar molar mass.

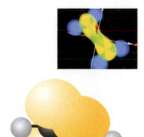

FIGURE 18.7 The π-bond (represented by the yellow electron clouds) in an alkene molecule makes the molecule resistant to twisting around a double bond. Consequently, all six atoms (the two C atoms that form the bond and the four atoms attached to them) lie in the same plane.

Section 3.7 describes the basis for the resistance to rotation of the C=C bond.

Animation: Figure 18.7 Ethene bonding

Media Link

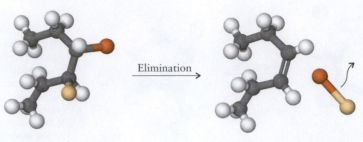

FIGURE 18.8 In an elimination reaction, two atoms or groups (the orange and yellow spheres) attached to neighboring carbon atoms are eliminated from the molecule, leaving a double bond between the two carbon atoms.

Most alkenes used in industry are produced during the refining of petroleum. One of the first refining steps is a reaction that converts some of the abundant alkanes into the more reactive alkenes:

$$CH_3CH_3(g) \xrightarrow{Cr_2O_3} CH_2{=}CH_2(g) + H_2(g)$$

This step is an example of an **elimination reaction,** a reaction in which two groups or atoms on neighboring carbon atoms are removed from a molecule, leaving a multiple bond (Fig. 18.8).

Another route commonly used in the laboratory to produce alkenes is the **dehydrohalogenation** of haloalkanes, the removal of a hydrogen atom and a halogen atom from neighboring carbon atoms:

$$CH_3CH_2CHBrCH_3 + CH_3CH_2O^- \xrightarrow{\text{ethanol at 70°C}}$$
$$CH_3CH{=}CHCH_3 + CH_3CH_2OH + Br^-$$

The states of reactants and products are often not given for organic reactions, because the reaction may take place at the surface of a catalyst or it may take place in a nonaqueous solvent, as here. The reaction is another example of an elimination reaction and is carried out in hot ethanol, with sodium ethoxide, $NaCH_3CH_2O$, as the reagent. Some $CH_3CH_2CH{=}CH_2$ is also formed in this reaction.

The results of kinetic studies suggest that dehydrohalogenation occurs by the attack of the ethoxide ion ($CH_3CH_2O^-$) on a hydrogen atom attached to the carbon atom next to the carbon atom carrying the bromine atom. The ethoxide ion, acting as a very strong base, pulls an H^+ ion out of the molecule to form $CH_3CH_2O{-}H$, leaving the carbon atom with a lone pair of electrons and a negative charge. The carbon atom uses those electrons to form a second bond to its neighbor as the bromine atom is effectively squeezed out of the molecule as a Br^- ion:

$$CH_3CH_2O^- \qquad\qquad CH_3CH_2OH$$
$$H \qquad\qquad\qquad \longrightarrow \qquad {=} \quad + \; Br^-$$
$$Br$$

Curved arrows like the ones here are often used to illustrate organic reaction mechanisms. They show the direction in which electron pairs move as they form new bonds.

The double bonds in alkenes can be generated by elimination reactions.

18.6 Electrophilic Addition

The most characteristic chemical reaction of an alkene is an **addition reaction,** in which atoms supplied by the reactant form σ-bonds to the two atoms originally joined by the double bond (Fig. 18.9). In the process, the π-bond is lost. An example is **halogenation,** the addition of two halogen atoms at a double bond, as in the formation of 1,2-dichloroethane:

$$CH_2{=}CH_2 + Cl_2 \longrightarrow CH_2Cl{-}CH_2Cl$$

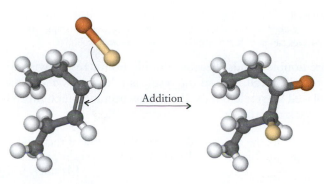

FIGURE 18.9 In an addition reaction, the atoms provided by an incoming molecule form bonds to the carbon atoms originally joined by a multiple bond.

The addition of hydrogen chloride to give chloroethane is an example of a **hydrohalogenation** reaction:

$$CH_2=CH_2 + HCl \longrightarrow CH_3-CH_2Cl$$

The model that organic chemists have built of the mechanism of alkene addition reactions is quite different from that of alkane substitution reactions. The crucial difference between the two types of molecules is the high density of electrons in the region of the double bond (Fig. 18.10). Because electrons are negatively charged, this region represents an accumulation of negative charge that can attract a positively charged reactant. A reactant that is attracted to a region of high electron density is called an **electrophile.** The mechanism of alkene addition is by electrophilic attack on the carbon atoms that make up the double bond. An electrophile may be a positively charged species or it may be a species that has a partial positive charge or can acquire one in the course of the reaction.

An example is the bromination of ethene. When ethene (or any other alkene) is bubbled through a solution of bromine, the solution is decolorized as the bromine reacts to form dibromoethane (this reaction is illustrated in Fig. 2.9). Bromine molecules are polarizable; and, as a Br_2 molecule approaches the region of high electron density of an alkene double bond, a partial positive charge is induced on the Br atom closer to the double bond (Fig. 18.11). This separation of charges means that a Br_2 molecule can act as an electrophile. As it moves in for the attack, the partially positively charged Br atom becomes more and more like Br^+, and its partner becomes more like Br^-. The bond between the two Br atoms snaps, and the Br^+ ion forms a bridge between the two carbon atoms of the alkene, giving a cyclic "bromonium ion":

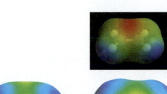

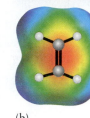

FIGURE 18.10 (a) The electrostatic potential diagram for an ethane molecule: the blue regions show where the positive charge of the nuclei outweighs the negative charge of the electrons, and the red regions show where the opposite is true. (b) The electron distribution in an ethene molecule shows the negative region of charge associated with the accumulation of electrons in the region of the double bond.

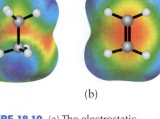

Almost immediately, a Br^- ion swoops in for the attack, attracted by the positive charge of the bromonium ion. It forms a bond to one carbon atom, and the bromine atom already present forms another bond to the second carbon atom, giving 1,2-dibromoethane:

With the use of a solid-state catalyst, hydrogen can be added to carbon–carbon double bonds in a **hydrogenation** reaction:

$$H_2(g) + \cdots C=C\cdots \xrightarrow{\text{catalyst}} \cdots CH-CH\cdots$$

This reaction is used in the food industry to convert vegetable oils into shortening (Fig. 18.12). Oil and solid fat molecules both have long hydrocarbon chains, but

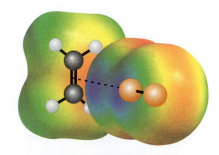

FIGURE 18.11 As a bromine molecule approaches a double bond in an alkene, the atom closer to the ethene molecule acquires a partial positive charge (the blue region). The computation that produced this image was carried out for the point at which the bromine molecule is so close to the double bond that a carbon–bromine bond is starting to form.

FIGURE 18.12 When the runny oil (top) is hydrogenated, it is converted into a solid fat (bottom). Hydrogen adds to the carbon–carbon double bonds, converting them into single bonds. The resulting more flexible molecules can pack together more closely and form a solid.

oils have more double bonds. Because double bonds resist twisting, oil molecules do not pack together well, and so the result is a liquid. When some of the double bonds are replaced by single bonds, the chains become much more flexible, and so the molecules pack together better and form a solid.

SELF-TEST 18.8A Write the condensed structural formula of the product of the addition of hydrogen to 2-butene: $CH_3CH{=}CHCH_3 + H_2 \rightarrow$ product.

[*Answer:* $CH_3CH_2CH_2CH_3$]

SELF-TEST 18.8B Write the condensed structural formula of the compound formed by the addition of hydrogen chloride to 2-butene.

The mechanism of addition to alkenes is electrophilic attack.

29   Benzene, C_6H_6

AROMATIC COMPOUNDS

Aromatic hydrocarbons, which originally got their name from the distinctive odors many of them have, are called **arenes.** They all contain an aromatic ring, usually the six-membered ring of benzene, which was introduced in Sections 2.7, 3.7, and 3.12. An abundant source of arenes is coal, which is a very complex mixture of compounds, many of which consist of extensive networks containing aromatic rings (Section 18.10).

30   Naphthalene, $C_{10}H_8$

18.7 Nomenclature of Arenes

The parent compound of aromatic hydrocarbons is benzene, C_6H_6 (**29**). When the benzene ring is named as a substituent, it is called the *phenyl* group, as in 2-phenylbutane, $CH_3CH(C_6H_5)CH_2CH_3$. In general, aromatic hydrocarbon groups are called *aryl* groups. Aromatic compounds also include the fused-ring analogs of benzene, such as naphthalene, $C_{10}H_8$ (**30**), and anthracene, $C_{14}H_{10}$ (**31**), obtained by the distillation of coal.

In an older but still widely used system of nomenclature, when there is a substituent in location 1 of a benzene ring, locations 2, 3, and 4 are denoted *ortho-* (abbreviated *o-*), *meta-* (*m-*), and *para-* (*p-*), respectively. Thus, (**32**) is *ortho-*dinitrobenzene. In systematic nomenclature, the locations of substituents on the aromatic ring are identified by numbering the carbon atoms from 1 to 6 around the ring, selecting the direction that corresponds to the lower substituent numbers. In this system, (**32**) is 1,2-dinitrobenzene ($-NO_2$ is the nitro group). However, (**33**) is 2,4,6-trinitrophenol because the compound C_6H_5OH is phenol, and so the carbon atom attached to the $-OH$ group is numbered 1.

31   Anthracene, $C_{14}H_{10}$

32   1,2-Dinitrobenzene

33   2,4,6-Trinitrophenol

34

EXAMPLE 18.5   Sample Exercise: Naming an aromatic compound

Name (a) compound (**34**) and (b) the three dimethylbenzenes (xylenes).

SOLUTION   Name the compounds by counting around the ring in the direction that gives the smallest numbers to the substituents.

(a)

1-ethyl-3-methylbenzene

(b)

1,2-dimethylbenzene
(*o*-xylene)

1,3-dimethylbenzene
(*m*-xylene)

1,4-dimethylbenzene
(*p*-xylene)

SELF-TEST 18.9A Name compound (35).

[*Answer:* 1-Ethyl-3-propylbenzene]

SELF-TEST 18.9B Name compound (36).

Aromatic compounds are named by giving the substituents on the benzene ring the lowest numbers. When the benzene ring is itself named as a substituent, it is called a phenyl group.

18.8 Electrophilic Substitution

Arenes are unsaturated but, unlike the alkenes, they are not very reactive. Whereas alkenes commonly take part in addition reactions, arenes undergo predominantly *substitution* reactions, with the π-bonds of the ring left intact. For example, bromine immediately adds to a double bond of an alkene but reacts with benzene only in the presence of a catalyst—typically, iron(III) bromide—and it does not affect the bonding in the ring. Instead, one of the bromine atoms replaces a hydrogen atom to give bromobenzene, C_6H_5Br:

$$C_6H_6 + Br_2 \xrightarrow{FeBr_3} C_6H_5Br + HBr$$

The mechanism of substitution on an electron-rich benzene ring is **electrophilic substitution**, electrophilic attack on an atom and the replacement of one atom by another or by a group of atoms. The fact that substitution occurs rather than addition to the double bonds can be traced to the stability of the delocalized π-electrons in the ring. Delocalization gives the electrons such low energy—that is, they are bound so tightly—that they are unavailable for forming new σ-bonds (see Sections 2.7 and 3.12).

The bromination of benzene illustrates the difference between addition to alkenes and substitution of arenes. First, to achieve the bromination of benzene it is necessary to use a catalyst, such as iron(III) bromide. The catalyst acts as a Lewis acid, binding to the bromine molecule (a Lewis base) and ensuring that the outer bromine atom has a pronounced partial positive charge:

$$:Br-Br: + FeBr_3 \longrightarrow {}^{\delta+}:Br-Br-FeBr_3{}^{\delta-}$$

(For simplicity, we show only one lone pair on each bromine atom.) The outer bromine atom of the complex is now primed to act as a strong electrophile (Fig. 18.13).

Electrophilic substitution begins like electrophilic addition, with an attack on a region of high electron density to form a positively charged intermediate:

However, because the delocalized π-electrons constitute such a stable arrangement, the hydrogen atom shown can be pulled from the ring as H^+ by a bromine atom in the $FeBr_4^-$ complex; in that way, delocalization is regained:

The iron(III) bromide is released in this step and is free to activate another bromine molecule.

CH₂CH₂CH₃ / **CH₂CH₃**

35

CH₂CH₃ / **CH₂CH₂CH₃** / **H₃C** / **CH₃**

36

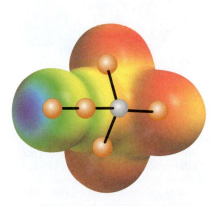

FIGURE 18.13 The catalyst $FeBr_3$ acts by forming a complex with a bromine molecule. As a result, the bromine atom not directly attached to the iron atom acquires a partial positive charge (the blue region). This partial charge enhances the ability of the bromine molecule to act as an electrophile.

One of the most extensively studied examples of electrophilic substitution is the nitration of benzene. A mixture of nitric acid and concentrated sulfuric acid converts benzene slowly into nitrobenzene. The actual nitrating agent is the electrophile NO_2^+ (the nitronium ion, ONO^+), a linear triatomic ion:

$$HNO_3 + H_2SO_4 \longrightarrow NO_2^+ + HSO_4^- + H_2O$$

The accepted mechanism of the reaction is

37   o-Nitrophenol

38   p-Nitrophenol

In the second step, the hydrogen ion is pulled out of the ring by the HSO_4^- ion acting as a Brønsted base. As in the bromination reaction, the restoration of the delocalization of the π-electrons facilitates the removal of the hydrogen ion.

Certain groups attached to an aromatic ring can donate electrons into its delocalized molecular orbitals. Examples of these electron-donating substituents include —NH_2 and —OH. Electrophilic substitution of benzene is much faster when an electron-donating substituent is present. For example, the nitration of phenol, C_6H_5OH, proceeds so quickly that it requires no catalyst. Moreover, when the products are analyzed, the only products are found to be 2-nitrophenol (*ortho*-nitrophenol, 37) and 4-nitrophenol (*para*-nitrophenol, 38).

Why is the *meta* position such an unattractive location for substitution, and why does phenol react much faster than benzene? An electrophile is attracted to regions of high electron density. Therefore, to account for the fast reaction of phenol, the electron density must be greater in the ring when the electron-donating —OH substituent is present. To account for the dominance of the *ortho* and *para* products, electron density must be relatively high at the *ortho* and *para* positions. A molecular orbital calculation, such as those described in Major Technique 5, shows that there is indeed a higher concentration of electrons in the ring in phenol, especially at the ortho and para positions, than in benzene, largely because the O atom has lone pairs of electrons that can participate in π-bonding with the carbon atoms (Fig. 18.14).

Long before their theories were supported by computations, organic chemists found a way to use resonance structures to explain the product distribution in electrophilic substitution. Thus, the Lewis structure for phenol is regarded as a resonance hybrid of the following structures:

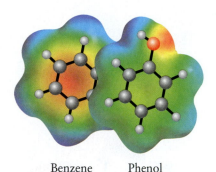

Benzene   Phenol

FIGURE 18.14 The presence of an —OH group in phenol (foreground) alters the distribution of electrons in the benzene ring (background). The blue regions denote relatively positively charged parts of the molecule and the green, yellow, and red regions denote progressively more negatively charged regions. Note how the electron-rich green regions spread over the carbon and hydrogen atoms much more in phenol than in benzene; so these atoms are more susceptible to electrophilic attack. A more detailed analysis shows that the ortho and para carbon atoms have the greatest accumulation of negative charge.

The curved arrows show how one resonance structure relates to another. Notice that the formal negative charge is located on the ortho and para positions, exactly where reaction takes place most quickly. Other ortho- and para-directing groups include —NH_2, —Cl, and —Br. All have an atom with a lone pair of electrons next to the ring, and all accelerate reaction.

We have seen how an —OH group can accelerate a reaction. Are there substituents that can slow down the electrophilic substitution of benzene? One way to reduce the electron density in the benzene ring and to make it less attractive to electrophiles is to substitute the benzene ring with a highly electronegative substituent, which can withdraw electrons partially. Alternatively, we can use a substituent that removes electrons by resonance. A number of substituents, such as the carboxyl group (—COOH), have both effects. For instance, the nitration of benzoic acid, C_6H_5COOH, is found to be much slower than the nitration of benzene; moreover, most of the product is the meta nitro compound:

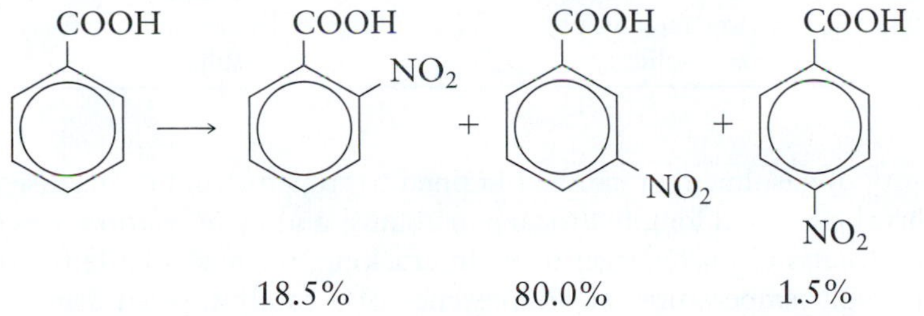

18.5%   80.0%   1.5%

The electronegative O atoms of the carboxylic acid group withdraw electrons from the whole ring, thereby reducing its overall electron density. Moreover, resonance preferentially removes electrons from the ortho and para positions. To focus on the essentials, only the lone pairs of electrons involved in resonance are shown:

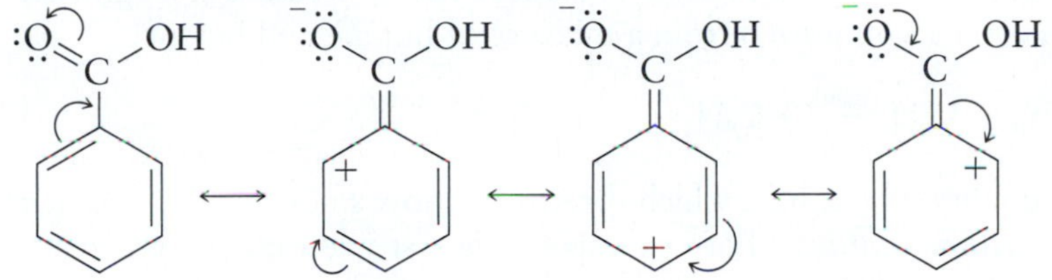

As a result of both effects, the reaction rate is lowered, especially at the ortho and para positions, and the meta position is left as the most likely place for attack. Other meta-directing electron-withdrawing substituents include —NO_2, —CF_3, and —C≡N. Notice that none of these substituents has a lone pair of electrons on the atom next to the ring.

> *Aromatic rings are much less reactive than their double-bond character would suggest; they commonly undergo substitution rather than addition. Electrophilic substitution of benzene with electron-donating substituents is accelerated and takes place at the ortho and para positions preferentially. Electrophilic substitution of benzene with electron-withdrawing substituents takes place at a reduced rate and primarily at the meta positions.*

IMPACT ON MATERIALS: FUELS

As we have seen, the primary sources of hydrocarbons are the fossil fuels petroleum and coal. Aliphatic hydrocarbons are obtained primarily from petroleum, which is a mixture of aliphatic and aromatic hydrocarbons, together with some organic compounds containing sulfur and nitrogen (Fig. 18.15). Coal is another major source of aromatic hydrocarbons.

18.9 Gasoline

The hydrocarbons in petroleum are separated by fractional distillation (Table 18.2). *Kerosene*, a fuel used in jet and diesel engines, contains a number of alkanes in the range C_{10} to C_{16}. Lubricating oils are mixtures in the range C_{17} to C_{22}. The heavier members of the series include the *paraffin waxes* and *asphalt*. However, the primary use of petroleum is for gasoline production, and the gasoline fraction (from C_5 to C_{11} hydrocarbons) is too small to meet the demand. In addition, the straight-chain alkanes make poor gasoline. Therefore, petroleum is refined to increase both the quantity and the quality of gasoline.

Fractional distillation is discussed in more detail in Section 8.19.

FIGURE 18.15 Because fossil fuel reserves are limited, they must be extracted from wherever they are found. This platform is used to extract petroleum from beneath the ocean. The natural gas accompanying it cannot easily be transported, and so it is burned off.

TABLE 18.2 **Hydrocarbon Constituents of Petroleum**

Hydrocarbons	Boiling range (°C)	Fraction
C_1 to C_4	−160 to 0	natural gas and propane
C_5 to C_{11}	30 to 200	gasoline
C_{10} to C_{16}	180 to 400	kerosene, fuel oil
C_{17} to C_{22}	350 and above	lubricants
C_{23} to C_{34}	low-melting-point solids	paraffin wax
C_{35} upward	soft solids	asphalt

The quantity of gasoline that can be obtained from petroleum is increased by *cracking,* or breaking down long hydrocarbon chains, and by *alkylation,* or combining small molecules to make larger ones. In cracking, the less volatile fractions are exposed to high temperatures in the presence of a catalyst, often a modified zeolite (see Section 13.14). For example, fuel oil can be converted into a mixture of octene and octane isomers:

$$C_{16}H_{34} \xrightarrow{\Delta,\ \text{catalyst}} C_8H_{16} + C_8H_{18}$$

Alkylation also requires a catalyst to achieve the desired chain length. Octane, for instance, can be synthesized from a mixture of butane and butene:

$$C_4H_{10} + C_4H_8 \xrightarrow{\text{catalyst}} C_8H_{18}$$

The quality of gasoline, which determines how smoothly it burns, is measured by the *octane rating.* For example, the straight-chain molecule octane, $CH_3(CH_2)_6CH_3$, burns so poorly that it has an octane rating of −19, but its isomer 2,2,4-trimethylpentane, which is commonly called isooctane, has an octane rating of 100. The octane rating is improved by increasing the branching of the molecules and by introducing unsaturation and rings. In *isomerization,* straight-chain hydrocarbons are converted into their branched-chain isomers:

$$CH_3(CH_2)_6CH_3 \xrightarrow{AlCl_3} CH_3C(CH_3)_2CH_2CH(CH_3)CH_3$$

Aromatization is the conversion of an alkane into an arene:

$$CH_3(CH_2)_5CH_3 \xrightarrow{Al_2O_3,\ Cr_2O_3} CH_3C_6H_5 + 4\ H_2$$

The product of this reaction, toluene (methylbenzene), has an octane rating of 120.

The quality of gasoline is also improved by the addition of ethanol, which has an octane rating of 120. The use of ethanol helps to reduce the demand for petroleum. Unlike petroleum, ethanol is a renewable fuel that can be regenerated every year (see Box 6.2).

Petroleum consists mainly of hydrocarbons, which are separated by molar mass in fractional distillation during the refining process.

18.10 Coal

As reserves of petroleum decline worldwide, interest in making better use of coal is growing. The thought of an automobile that runs on coal is strange, but the use of gasoline substitutes derived from coal is a real possibility. Unfortunately, the increased usage of coal raises environmental concerns. Coal contains a much lower ratio of hydrogen to carbon than petroleum and is harder to purify, to work, and to transport. Although both coal and petroleum contribute to the greenhouse effect, the use of coal is potentially more damaging. When it burns, coal also releases a large amount of pollution in the form of particulate matter (primarily ash) and sulfur and nitrogen oxides. Much of the research on coal is aimed at converting it into more useful fuels.

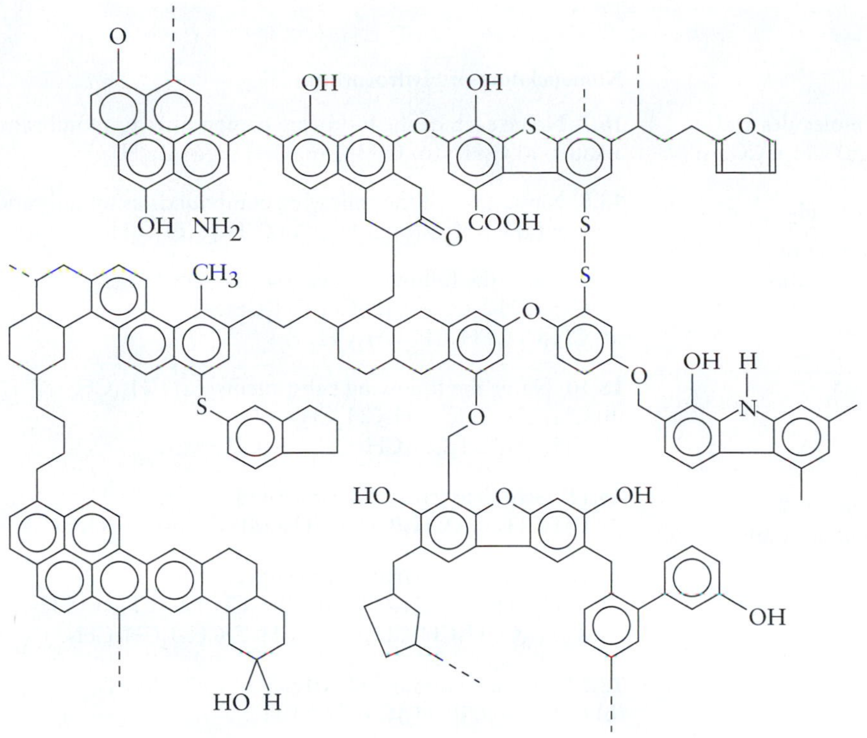

FIGURE 18.16 A highly schematic representation of a part of the structure of coal. When coal is heated in the absence of oxygen, the structure breaks up and a complex mixture of products—many of them aromatic—is obtained.

Coal contains many aromatic rings (Fig. 18.16). Coal is the end product of the decay of swamp vegetation in *anaerobic* conditions (with a very low concentration of oxygen). Oxygen and hydrogen are gradually lost through the *coalification* process. As coalification proceeds, hydrogen is released and the aromatic structures increase.

When coal is destructively distilled—heated in the absence of oxygen so that it decomposes and vaporizes—its sheetlike molecules break up, and the fragments include the aromatic hydrocarbons and their derivatives. *Coal gas,* which is given off first, contains carbon monoxide, hydrogen, methane, and small amounts of other gases. The complex liquid mixture that remains is called *coal tar.* A large number of pharmaceuticals, dyes, and fertilizers come from coal tar. Benzene is the raw material for many plastics, detergents, and pesticides. Naphthalene is used to make synthetic indigo (the dye in blue jeans), ammonia is used as fertilizer, and pitch, which contains the heaviest tars, is used for waterproofing and rustproofing.

Coal is primarily aromatic in nature.

SKILLS YOU SHOULD HAVE MASTERED

❑ 1 Distinguish alkanes, alkenes, alkynes, and arenes by differences in bonding, structure, and reactivity.

❑ 2 Name simple hydrocarbons (Toolbox 18.1 and Examples 18.1, 18.2, and 18.5).

❑ 3 Identify two molecules, given their structural formulas, as structural, geometrical, or optical isomers (Section 18.2).

❑ 4 Write the formulas of isomeric molecules (Example 18.3).

❑ 5 Judge whether a compound is chiral (Example 18.4).

❑ 6 Describe general trends in the physical properties of alkanes (Section 18.3).

❑ 7 Predict the products of given elimination, addition, and substitution reactions (Sections 18.4, 18.6, and 18.8).

❑ 8 Explain why ortho- and para-directing groups accelerate electrophilic substitution on the benzene ring and identify such groups (Section 18.8).

❑ 9 Describe the primary processes used in the refining of petroleum (Section 18.9).

EXERCISES

Types of Hydrocarbons

18.1 Draw Lewis structures of the following molecules and identify each as an alkane, alkene, or alkyne: (a) CH_3CCCH_3; (b) $CH_3CH_2CH_2CH_3$; (c) $CH_2CHCH_2CH_3$; (d) $CH_3CHCHCH_2CCCH_3$; (e) $CH_2CHCH_2CHCH_2$.

18.2 Give the molecular formula and identify each of the following molecules as being an alkane, alkene, or alkyne:

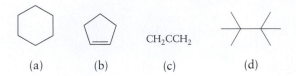

(a)   (b)   (c)   (d)

18.3 Give the molecular formulas corresponding to the following stick figures and identify each as an alkane, an alkene, an alkyne, or an aromatic hydrocarbon:

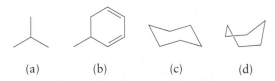

(a)   (b)   (c)   (d)

18.4 Give the molecular formulas corresponding to the following stick figures and identify each as an alkane, an alkene, an alkyne, or an aromatic hydrocarbon:

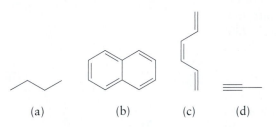

(a)   (b)   (c)   (d)

18.5 Give the molecular formulas corresponding to the following stick figures and identify each as an alkane, an alkene, an alkyne, or an aromatic hydrocarbon:

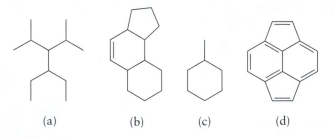

(a)   (b)   (c)   (d)

18.6 Give the molecular formulas corresponding to the following stick figures and identify each as an alkane, an alkene, an alkyne, or an aromatic hydrocarbon:

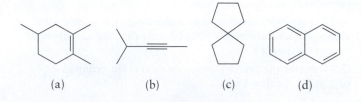

(a)   (b)   (c)   (d)

Nomenclature of Hydrocarbons

18.7 Name each of the following compounds as an unbranched alkane: (a) C_3H_8; (b) C_4H_{10}; (c) C_7H_{16}; (d) $C_{10}H_{22}$.

18.8 Name each of the following compounds as an unbranched alkane: (a) $C_{11}H_{24}$; (b) C_5H_{12}; (c) C_7H_{16}; (d) C_8H_{18}.

18.9 Name the following substituents: (a) CH_3—; (b) $CH_3(CH_2)_3CH_2$—; (c) $CH_3CH_2CH_2$—; (d) $CH_3CH_2CH_2CH_2CH_2CH_2$—.

18.10 Name the following substituents: (a) $CH_3(CH_2)_8CH_2$—; (b) CH_3CH_2—; (c) $CH_3CH_2CH_2CH_2$—; (d) $CH_3CH_2CH_2CH_2CH_2CH_2CH_2CH_2$—.

18.11 Give the systematic name of (a) $CH_3CH_2CH_3$; (b) CH_3CH_3; (c) $CH_3(CH_2)_3CH_3$; (d) $(CH_3)_2CHCH(CH_3)_2$.

18.12 Give the systematic name of (a) $CH_3CH_2CH_2CH_2CH_2CH_3$; (b) $CH_3CH(CH_2CH_3)C(CH_3)_3$; (c) $(CH_3)_3CCH_2CH(CH_3)_2$; (d) $CH_3C(CH_3)_2CH_2CH_3$.

18.13 Name compound (a) $CH_3CH{=}CHCH(CH_3)_2$; (b) $CH_3CH_2CH(CH_3)C(C_6H_5)(CH_3)_2$.

18.14 Name compound (a) $CH_2{=}CHCH(C_6H_5)(CH_2)_4CH_3$; (b) $(CH_3)_2CHCH(CH_3)CHClC{\equiv}CH$.

18.15 Write the shortened (condensed) structural formula of (a) 3-methyl-1-pentene; (b) 4-ethyl-3,3-dimethylheptane; (c) 5,5-dimethyl-1-hexyne; (d) 3-ethyl-2,4-dimethylpentane.

18.16 Write the shortened (condensed) structural formula of (a) 4-ethyl-2-methylhexane; (b) 3,3-dimethyl-1-pentene; (c) *cis*-4-methyl-2-hexene; (d) *trans*-2-butene.

18.17 Write the structural formula of (a) 4,4-dimethylnonane; (b) 4-propyl-5,5-diethyl-1-decyne; (c) 2,2,4-trimethylpentane; (d) *trans*-3-hexene.

18.18 Write the structural formula of (a) 3,4,9-trimethyldecane; (b) 2,3,5-trimethyl-4-propylheptane; (c) 1,3-dimethylcyclohexane; (d) 5-ethyl-3-heptene.

18.19 Draw stick structures to represent each of the following molecules: (a) nonane, $CH_3(CH_2)_7CH_3$; (b) cyclopropane, C_3H_6; (c) cyclohexene, C_6H_{10}.

18.20 Draw stick structures to represent each of the following species: (a) 2,2,3,3-tetramethylhexane, $CH_3C(CH_3)_2C(CH_3)_2CH_2CH_2CH_3$; (b) the trityl cation, $(C_6H_5)_3C^+$; (c) butadiene, $CH_2CHCHCH_2$.

Isomers

18.21 Write the structural formulas for and name (a) at least 10 *alkene* isomers having the formula C_6H_{12}; (b) at least 10 *cycloalkane* isomers have the formula C_6H_{12}.

18.22 Write the structural formulas for and name all the isomers (including geometrical isomers) of the alkenes (a) C_4H_8; (b) C_5H_{10}.

18.23 Identify each of the following pairs as structural isomers, geometrical isomers, or not isomers: (a) butane and cyclobutane; (b) cyclopentane and pentene;

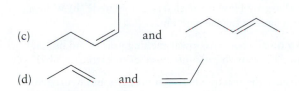

(c) and

(d) and

18.24 Identify each of the following pairs as structural isomers, geometrical isomers, or not isomers:
(a) 1-chlorohexane and chlorocyclohexane;

(b) and

(c) and

18.25 A branched hydrocarbon C_4H_{10} reacts with chlorine in the presence of light to give two branched structural isomers with the formula C_4H_9Cl. Write the structural formulas of (a) the hydrocarbon; (b) the isomeric products.

18.26 A branched hydrocarbon C_6H_{14} reacts with chlorine in the presence of light to give only two structural isomers with the formula $C_6H_{13}Cl$. Write the structural formulas of (a) the hydrocarbon; (b) the two isomeric products.

18.27 Indicate which of the following molecules are optical isomers and identify the chiral carbon atoms in those that are: (a) $CH_3CHBrCH_2CH_3$; (b) $CH_3CH_2CHCl_2$; (c) 1-bromo-2-chloropropane; (d) 1,2-dichloropentane.

18.28 Indicate which of the following molecules are optical isomers and identify the chiral carbon atoms in those that are: (a) $CH_3CHBrCl$; (b) $CH_3CH_2CHClCH_3$; (c) 2-bromo-2-chloropropane; (d) 1,2-dibromobutane.

Alkanes

18.29 Why do branched-chain alkanes have lower melting points and boiling points than unbranched alkanes with the same number of carbon atoms?

18.30 Arrange the following compounds in order of increasing boiling point: $CH_3CH_2CH_2CH_2CH_3$, $CH_3CH_2CH(CH_3)_2$, $C(CH_3)_4$. Explain your reasoning.

18.31 Using data available in Appendix 2A, write balanced equations and calculate the heat released when (a) 1.00 mol and (b) 1.00 g of each of the following compounds is burned in excess oxygen: propane, butane, and pentane. Is there a trend in the amount of heat released per mole of molecules or per gram of compound? If so, what is it?

18.32 Write the balanced chemical equation for the complete fluorination of methane to tetrafluoromethane. Using bond enthalpies, estimate the enthalpy of this reaction. The corresponding reaction using chlorine is much less exothermic. To what can this difference be attributed?

18.33 How many different products containing two carbon atoms are possible in the reaction of chlorine with ethane? Do any of the products exist as optical isomers?

18.34 How many different products that retain the three-membered ring are possible in the reaction of chlorine with cyclopropane? Do any of the products exist as stereoisomers?

Alkenes

18.35 Draw the structures of *cis*-1,2-dichloropropene and *trans*-1,2-dichloropropene. Which of these molecules is polar?

18.36 Draw the structure of 1,3-pentadiene. Use valence-bond and molecular orbital pictures to describe the bonding for the σ-framework and π-orbitals, respectively.

18.37 Two structural isomers can result when hydrogen bromide reacts with 2-pentene. (a) Write their structural formulas. (b) What name is given to this type of reaction?

18.38 (a) Write a balanced equation for the production of 2,3-dibromobutane from 2-butyne. (b) What name is given to this type of reaction?

18.39 (a) Write a balanced equation for the reaction of bromocyclohexane with sodium ethoxide in ethanol. (b) Draw structural formulas of the cyclic reactant and product. (c) What name is given to this type of reaction?

18.40 Draw stick structures of the possible products for the reaction of sodium ethoxide with (a) 1-bromobutane; (b) 2-bromobutane.

18.41 Compare the reaction enthalpies for the halogenation of ethene by chlorine, bromine, and iodine. What trend, if any, exists in these numbers? Use bond enthalpies to estimate the enthalpies of reaction.

18.42 Compare the reaction enthalpies for the hydrohalogenation of ethene by HX, where X = Cl, Br, I. What trend, if any, exists in these numbers? Use bond enthalpies to estimate the enthalpies of reaction.

Aromatic Compounds

18.43 Name the following compounds:

(a) (b)

18.44 Name the following compounds:

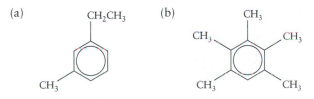

(a) (b)

18.45 Write the structural formula of (a) methylbenzene, more commonly known as toluene; (b) *p*-chlorotoluene; (c) 1,3-dimethylbenzene; (d) 4-chloromethylbenzene.

18.46 Write the structural formula of (a) *p*-xylene; (b) 1,2-dibromobenzene; (c) 3-phenylpropene; (d) 2-ethyl-1,4-dimethylbenzene.

18.47 (a) Draw all the isomeric dichloromethylbenzenes. (b) Name each one and indicate which are polar and which are nonpolar.

18.48 (a) Draw all the isomeric diaminodichlorobenzenes. (b) Name each one and indicate which are polar and which are nonpolar.

18.49 Draw resonance structures of cyanobenzene (C_6H_5CN) that show how it functions as a meta-directing substituent.

18.50 One can prepare bromonitrobenzenes either by nitrating bromobenzene or by brominating nitrobenzene. Will these two reactions give the same product (or product distribution)? If not, how will the products differ?

18.51 How many different compounds can be produced when naphthalene, $C_{10}H_8$ (30), undergoes electrophilic aromatic substitution by a single electrophile (denoted E)? Draw stick figures to represent them.

18.52 The structure of benzaldehyde is

Do you expect the aldehyde functional group (CHO) to act as a meta-directing or as an ortho, para-directing group? Explain.

Fuels

18.53 Why are hydrocarbons having between one and four carbon atoms not suitable for gasoline?

18.54 What are the major problems associated with using coal as a fuel?

18.55 In the production of gasoline, both *cracking* and *alkylation* are employed, although they are chemical opposites of each other. Define cracking and alkylation as they apply to the production of gasoline.

18.56 What reasons are there for using ethanol as a part of gasoline mixtures?

Integrated Exercises

18.57 Identify the type and number of bonds on carbon atom 2 in (a) pentane; (b) 2-pentene; (c) 2-pentyne.

18.58 Predict the geometry and hybridization of the orbitals used in bonding on carbon atom 2 in (a) pentane; (b) 2-pentene; (c) 2-pentyne.

18.59 Classify each of the following reactions as addition or substitution and write its chemical equation: (a) chlorine reacts with methane when exposed to light; (b) bromine reacts with ethene in the absence of light.

18.60 Classify each of the following reactions as addition or substitution and write its chemical equation: (a) hydrogen reacts with 2-pentene in the presence of a nickel catalyst; (b) hydrogen chloride reacts with propene.

18.61 The dehydrohalogenation of haloalkanes to produce alkenes is always carried out in a nonaqueous solvent, usually ethanol. Give two reasons why water cannot be used as the solvent for the reaction.

18.62 View the animation of the alkene addition mechanism associated with Fig. 18.9 on the Web site for this book. (a) In the addition of HCl to propene an intermediate is formed. Is that intermediate positively or negatively charged? How is the charge eliminated? (b) Write a two-step mechanism for the reaction.

18.63 Why do alkenes have lower melting points and boiling points than alkanes with the same number of carbon atoms?

18.64 Consider the reaction between propene and hydrogen. (a) Calculate the equilibrium constant at 298 K. (b) Does K increase or decrease as the temperature is raised?

18.65 Give the systematic name of each of the following compounds. If geometrical isomers are possible, write the names of each one: (a) $CH_2{=}C(CH_3)_2$; (b) $CH_3CH{=}C(CH_3)CH_2CH_3$; (c) $HC{\equiv}CCH_2CH_2CH_2CH_3$; (d) $CH_3CH_2C{\equiv}CCH_2CH_3$; (e) $CH_3CH_2CH_2C{\equiv}CCH_3$.

18.66 Give the systematic name of each of the following compounds. If geometrical isomers are possible, write the names of each one: (a) $CH_3CH_2CH_2CH{=}CH_2$; (b) $CH_3CH_2C(CH_3){=}CHCH_3$; (c) $(CH_3)_2C{=}CHCH(CH_3)_2$; (d) $CH_3C{\equiv}CCH_2CH(CH_2CH_3)CH_3$; (e) $CH_3CH_2C{\equiv}CCH(CH_2CH_3)CH_3$.

18.67 The structure of decalin can be found on the Web site for this book. (a) From an examination of this structure, determine the chemical formula of decalin. (b) From what aromatic hydrocarbon may decalin be produced by complete hydrogenation? (c) Are isomers possible for decalin? If so, draw the appropriate Lewis structures.

18.68 Nonsystematic names for organic compounds may still be found in the chemical literature and chemical supply catalogs, and so it is important to be somewhat familiar with these as well as with the IUPAC rules. Give the systematic name for (a) isobutane and (b) isopentane. (c) Formulate a rule for the usage of the prefix *iso-* and predict the structure of isohexane. Structures for these compounds can be found on the Web site for this book.

18.69 1-Butene reacts with hydrogen chloride in a hydrohalogenation reaction. Estimate the standard reaction enthalpy from bond enthalpies.

18.70 How many liters of hydrogen at 1.00 atm and 298 K are needed to hydrogenate (a) 1.00 mol C_6H_{10}, cyclohexene; (b) 1.00 mol C_6H_6, benzene, completely? (c) Estimate the reaction enthalpy of each hydrogenation from the average bond enthalpies in Tables 6.7 and 6.8. (d) The Lewis structures imply that the enthalpy of hydrogenation of benzene is three times that of cyclohexene. Do your calculations support this implication? Explain any differences.

18.71 In a combustion analysis, 3.21 g of a hydrocarbon formed 4.48 g of water and 9.72 g of carbon dioxide. Deduce its empirical formula and state whether it is likely to be an alkane, an alkene, or an alkyne. Explain your reasoning.

18.72 In a combustion analysis, 5.535 g of a hydrocarbon formed 6.750 g of water and 17.550 g of carbon dioxide. Deduce the empirical formula of the hydrocarbon and state whether it is likely to be an alkane, an alkene, or an alkyne. Explain your reasoning.

18.73 The following names of organic molecules are incorrect. Draw line structures and determine the correct systematic name for (a) 4-methyl-3-propylheptane; (b) 4,6-dimethyloctane; (c) 2,2-dimethyl-4-propylhexane; (d) 2,2-dimethyl-3-ethylhexane.

18.74 Alkanes undergo substitution of hydrogen atoms when treated with halogens. Bromination of which of the following compounds could give rise to chiral monosubstituted products: (a) ethane; (b) propane; (c) butane; (d) pentane? Include in your answer the names of the chiral compounds that would be formed.

18.75 A hydrocarbon with the formula C_3H_6 does not react with bromine in the absence of light but does react in the presence of light, producing C_3H_5Br. What is the name of this hydrocarbon?

18.76 A hydrocarbon is 90% carbon by mass and 10% hydrogen by mass and has a molar mass of 40 $g \cdot mol^{-1}$. It decolorizes bromine water, and 1.46 g of the hydrocarbon reacts with 1.60 L of hydrogen (measured at STP) in the presence of a nickel catalyst. Write the molecular formula of the hydrocarbon and the structural formulas of two possible isomers.

18.77 The compound 1-bromo-4-nitrobenzene (*p*-bromonitrobenzene) can be brominated. What do you expect to be the major product and why? Draw appropriate Lewis structures to support your answer.

18.78 Consider the nitration of the compound

If the reaction can be controlled so that one NO_2 group replaces one H atom of the molecule, where do you expect the nitro group to end up in the product?

18.79 (a) Draw the structure of the hydrocarbon 3,4,6-trimethyl-1-heptene. (b) Identify the chiral carbon atoms in the structure with stars. (c) Are cis and trans isomers possible for this molecule?

18.80 Trinitrotoluene, TNT, is a well-known explosive. (a) Using the structure available on the Web site, determine the systematic name for TNT. (b) TNT is made by nitrating toluene (methylbenzene) with a mixture of concentrated nitric and sulfuric acids. Explain why this particular trisubstituted isomer is the one that is formed during the nitration.

18.81 A chemist obtains the mass spectrum of 1,2-dichloro-4-ethylbenzene. Give at least four possible fragments and the masses at which you would expect them to occur. Chlorine has two naturally occurring isotopes: ^{35}Cl, $34.969m_u$, 75.53%, and ^{37}Cl, $36.966m_u$, 24.47%. The mass of ^1H is $1.0078m_u$. See Major Technique 6, Mass Spectrometry, which follows this set of exercises.

18.82 How will the mass spectrum of D_2O differ from that described in the text for H_2O? See Major Technique 6, Mass Spectrometry, which follows this set of exercises.

18.83 The occurrence of isotopes of some elements allows the presence of these elements to be readily identified in a mass spectrum of a compound. For example, Br has two naturally occurring isotopes with masses $79m_u$ and $81m_u$, in the relative abundances 50.5% and 49.5%. How will this information make the determination of the number of bromine atoms in an organic molecule relatively easy? For example, in the bromination of an alkane, it is possible that a number of products with more than one bromine atom might be produced. See Major Technique 6, Mass Spectrometry, which follows this set of exercises.

18.84 A chemist treats toluene (methylbenzene) with chlorine in the presence of a metal catalyst. A sample of the reaction mixture is injected into a GC-MS unit (gas chromatograph–mass spectrometer). One of the compounds that is separated by using gas chromatography gives a mass spectrum with peaks at $128m_u$, $126m_u$, $113m_u$, $111m_u$, and $91m_u$, among others. The peak at $128m_u$ is approximately one-third the size of the peak at $126m_u$, and the peak at $113m_u$ is approximately one-third the size of the peak at $111m_u$. What is the likely composition of this compound, and what fragments may be assigned to these peaks? See Major Technique 6, Mass Spectrometry, which follows this set of exercises.

18.85 Conjugated polyenes are hydrocarbons with alternating single and double bonds. They are commonly used as dyes because they absorb in the visible range. Two such molecules have the formulas C_6H_8 and C_8H_{10}. Of the two, which has its maximum absorption at the longer wavelength? Justify your answer. See Major Technique 2, following Chapter 3.

Chemistry Connections

18.86 A gaseous compound used to make both bubble gum and automobile tires was analyzed to determine its properties and toxicity.
(a) When 0.108 g of the compound was analyzed by combustion analysis, 0.352 g of CO_2 and 0.109 g of H_2O were produced. What is the empirical formula of the compound?
(b) The molar mass of the compound was found to be 54.09 $g \cdot mol^{-1}$. What is the molecular formula of the compound?
(c) Draw the structural formulas of at least six possible structural isomers of the compound.
(d) When the compound is exposed to hydrogen gas over a catalyst, the molar mass of the hydrogenated product is 58.12 $g \cdot mol^{-1}$. Which of the isomers in part (c) does this information help to eliminate?
(e) A 2.45-L flask was filled with the compound at 25°C and 1.0 atm. Hydrogen bromide gas was pumped into the flask until the total pressure in the flask reached 5.0 atm. Reaction then proceeded to completion because the single product was removed as fast as it was formed. After completion of the reaction the pressure of the excess HBr in the flask was 2.0 atm. What is the molecular formula of the product?
(f) The product has three structural isomers, two of which are optical isomers. Draw the structural formulas of the original compound and each of the product isomers. Place an asterisk on each chiral carbon atom.
(g) Assign a hybridization scheme to each C atom in the original compound.
(h) Write the name of the original compound.

In spectroscopy, we use a diffraction grating to separate photons of light by their wavelengths. In *mass spectrometry,* we use a magnetic field to separate molecular ions by their masses. This technique is one of our most powerful analytical methods because it provides both quantitative and qualitative information about the substance being analyzed, it requires only tiny samples, and it can be conducted in the field, by using lightweight, portable units. Mass spectrometry is used to determine isotopic abundance (Section B), to study the composition of bone and other body tissues, to analyze the blood of newborns for congenital diseases, to detect vanishingly small concentrations of drugs in urine, and to determine the structure of the human genome.

The Technique

We met a classic type of mass spectrometer in Section B. Another common type of mass spectrometer is the *large magnet mass spectrometer* (Fig. 1). In each case, the spectrometer uses the degree to which charged particles are deflected by a magnetic field to determine the relative masses of the particles. In a mass spectrometer, the sample is first vaporized and then ionized. The resulting ions are accelerated by an electric field into a narrow beam that, as it passes between a set of magnets, is deflected toward a detector. The heavier the particle, the less it is deflected, and the degree of deflection allows the relative masses of the particles to be determined. The results are displayed as a series of peaks, with the height of each peak being proportional to the relative number of particles with a given mass.

The entire spectrometer must first be evacuated (have all the air pumped out) to ensure that no gas molecules can collide with the ions produced from the sample and deflect them in unpredictable ways. Then the sample is introduced in the form of a vapor into the sample inlet chamber and allowed into the ionization chamber. In this chamber, rapidly moving electrons collide violently with the molecules of the vapor.

When one of the accelerated electrons collides with a molecule, the colliding electron knocks another electron out of the molecule, leaving a molecular cation. The molecular ion may split apart into two or more charged fragments. The positive ions—the original ("parent") molecular ions and their fragments—are accelerated out of the chamber by a strong electric field applied across a series of metal grids. The speeds reached by the ions depend on their masses: light ions reach higher speeds than heavier ones.

The rapidly moving ions pass between the poles of an electromagnet. The magnetic field bends the paths of the ions to an extent that depends on their speed and on the strength of the field. In the large magnet mass spectrometer, the magnetic field is kept constant but the strength of the electric field is varied, and so the molecular ions are accelerated to different speeds. At a given electric field strength, ions of only one particular mass reach the detector. Any ions with different masses collide with the walls of the chamber. The strength of the electric field is varied and the result is a mass spectrum—a plot of the detector signal against the strength of the electric field. The positions of the peaks in the mass spectrum give the mass-to-charge ratios of the ions. If all the ions have charges of +1, then the peaks give the mass ratios and their relative heights give the proportions of ions of various masses.

Analyzing a Mass Spectrum

If the sample consists of atoms of one element, the mass spectrum gives the isotopic distribution of the sample. The relative molar masses of the isotopes can be determined by comparison with atoms of carbon-12. If the sample is a compound, the formula and structure of the compound can be determined by studying the fragments. For example, the +1 ions that CH_4 could produce are CH_4^+, CH_3^+, CH_2^+, CH^+, C^+, and H^+. Some of the particles that strike the detector are those that result when the molecule simply loses an electron (for example, to produce CH_4^+ from methane):

$$\text{Molecule} + e^- \longrightarrow \text{molecule}^+ + 2\,e^-$$

The mass of this molecular ion, the *parent ion,* is called the *parent mass.* The parent ion has essentially the same molar mass as the compound itself. However, if the electron beam is moving at a very high speed, only fragments of the molecule may survive. Some fragments are almost always produced, and other information must be used to help with the interpretation. For example, a sample of a known compound can be analyzed and its spectrum compared to that of an unknown sample to look for similar fragments. The sample can also be passed through a gas chromatograph first (see Major Technique 4). The chromatogram allows comparison with a large number of possible compounds, and the mass spectrum verifies the identification.

The identification of a molecular structure from a mass spectrum requires good chemical detective work. Let's see how that is done by trying to identify a simple compound,

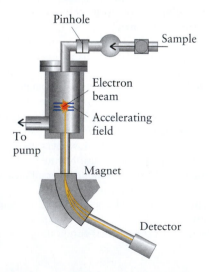

FIGURE 1 A schematic drawing of a large magnet mass spectrometer.

Mass Spectrum of an Unknown Compound

m/e*	Peak intensity as a percentage of largest peak
1	< 0.1
16	1.0
17	21.0
18	100.0
19	0.08
20	0.22

* The quantity m/e is the conventional notation for the mass-to-charge ratio of the molecular ions, with the mass a multiple of the atomic mass constant m_u and the charge a multiple of the fundamental charge e.

the one that produced the mass spectrum summarized in the table above. The parent mass is 18, 19, or 20. If the parent mass is 20, the compound could be HF. However, fluorine has only one stable isotope, fluorine-19; so we could not explain the large peaks at 17 and 18. Because of the very small abundances of the peaks at 19 and 20, we can also guess that they represent the less abundant isotopes of one or more of the elements in the compound. We can conclude that the parent mass is 18 and the compound is probably water. The peaks at 19 and 20 would be due to the mass of parent molecules containing naturally occurring isotopes such as oxygen-18 and deuterium. The peak at 17 represents the OH^+ fragment, the peak at 16 would be O^+ ions, and the peak at 1 the few H^+ ions that form in the ionization process.

More complex detective work is required to analyze large biomolecules and drugs. However, fragmentation generally follows predictable patterns, and one compound can be identified by comparing its mass spectrum with those of other known compounds with similar structures. In Fig. 2, we see the spectrum of a sample of blood from a newborn infant. The blood is being analyzed to determine whether the child has phenylketonuria. The presence of the compound phenylalanine is a positive indication of the condition. Some

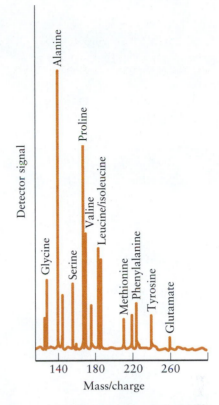

FIGURE 2 The two phenylalanine peaks include the radioactive tracer (the right-hand peak) and the phenylalanine in the blood (the peak $5m_u$ to the left). The presence of the latter peak shows that the baby from whom the blood was drawn has phenylketonuria.

phenylalanine labeled with a radioactive isotope has been injected along with the sample to identify the parent mass. The radioactive compound, which contains an isotope making it $5m_u$ heavier than the phenylalanine, appears in the spectrum $5m_u$ above it, thereby allowing a positive identification of the presence of the chemical.

Related Exercises 18.83–18.86

ORGANIC CHEMISTRY II: POLYMERS AND BIOLOGICAL COMPOUNDS

Chapter **19**

What Are the Key Ideas? The properties and reactions of many organic compounds are due to the presence of characteristic groups of atoms known as functional groups.

Why Do We Need to Know This Material? Organic chemistry is the foundation of the pharmaceutical industry and, consequently, of medicine. Organic chemistry is also the basis of biochemistry and molecular biology; through those disciplines, it provides an understanding of life. The great diversity of organic molecules found in living systems results from the presence of groups of atoms with characteristic functions. Once we understand the properties of these groups, we can predict the properties of other organic molecules, even those of the giant biomolecules of proteins, carbohydrates, and genetic material, and the synthetic polymeric materials that have revolutionized our lives.

What Do We Need to Know Already? This chapter builds on Chapter 18 and requires the same preparation, as well as the concepts of acid–base and redox reactions (Sections J and K). It also makes use of the concepts of molecular polarity (Section 3.3) and solubility (Section 8.9).

Despite their immense variety, organic compounds can be understood in terms of **functional groups,** which are small groups of atoms with characteristic properties. Whereas the hydrocarbons are built only from carbon and hydrogen, functional groups may include atoms of other elements and accordingly bring strikingly different properties to organic compounds. Moreover, because functional groups have characteristic chemical properties, they help to make organic chemistry a highly systematic subject: we do not have to learn all the properties of each different compound—once we identify the functional groups that are present, we have gone a long way toward understanding and predicting the likely chemical properties of the compound.

This chapter has three parts. In the first part, we look at the structure and properties of some of the common functional groups and describe some of the characteristic mechanisms by which the groups react. Then we examine how functional groups are used to create modern polymers. As in so many spheres, nature has preceded chemists' explorations. In the final part of the chapter, we see some examples of how functional groups in nature sustain us, feed us, and replicate our genetic material.

COMMON FUNCTIONAL GROUPS

Functional groups are either attached to the carbon backbone of a molecule or form part of that chain. Examples are the chlorine atom in chloroethane, CH_3CH_2Cl, and the —OH group in ethanol, CH_3CH_2OH. Carbon–carbon multiple bonds, such as the double bond in 2-butene, are also often considered functional groups. Table 19.1 lists the most common functional groups. Double and triple carbon–carbon bonds were considered in Chapter 18. In the following eight

TABLE 19.1 Common Functional Groups

Group	Class of compound	Group	Class of compound
—X	halide (X = F, Cl, Br, or I)	—COOH	carboxylic acid
—OH	alcohol, phenol	—COOR	ester
—O—	ether	—N<	amine
—CHO	aldehyde	—CO—N<	amide
—CO—	ketone		

sections, we examine some other important functional groups, look at their characteristic chemical properties, and examine some of the reaction mechanisms that account for these properties.

19.1 Haloalkanes

The **haloalkanes** (also called alkyl halides) are alkanes in which at least one hydrogen atom has been replaced by a halogen atom. Although they have important uses, many haloalkanes are highly toxic and a threat to the environment. The haloalkane 1,2-dichlorofluoroethane, $CHClFCH_2Cl$, is an example of a chlorofluorocarbon (CFC), one of the compounds held responsible for the depletion of the ozone layer (see Box 13.3). Many pesticides are aromatic compounds with several halogen atoms.

Carbon–halogen bonds are polar (**1**). The carbon atom to which the halogen atom is attached is partially positive, making it susceptible to **nucleophilic substitution**, in which a nucleophile replaces the halogen atom. A **nucleophile** is a reactant that seeks out centers of positive charge in a molecule. A typical example is the hydroxide ion, OH^-: its negative charge is attracted to regions of a molecule that have a partial positive charge. Another example of a nucleophile is H_2O: a lone pair on the oxygen atom is attracted to regions of partial positive charge. Water acts as a nucleophile in a **hydrolysis reaction**, a reaction with water in which a carbon–element bond is replaced by a carbon–oxygen bond. For example, bromomethane undergoes hydrolysis by hydroxide ions in aqueous ethanol to methanol and bromide ions:

$$CH_3Br + OH^- \longrightarrow CH_3OH + Br^-$$

Haloalkanes are alkanes in which at least one hydrogen atom has been replaced by a halogen atom; they undergo nucleophilic substitution.

19.2 Alcohols

The **hydroxyl group** is an —OH group covalently bonded to a carbon atom. An **alcohol** is an organic compound that contains a hydroxyl group not connected directly to a benzene ring or to a >C=O group. One of the best-known organic compounds is ethanol, CH_3CH_2OH, which is also called *ethyl alcohol* and *grain alcohol*. The —OH group in an alcohol molecule is polar and alcohols can give up their hydroxyl protons in certain solvents (Fig. 19.1), but their conjugate bases are so strong that they are not acids in water.

Alcohols are named by adding the suffix *-ol* to the stem of the parent hydrocarbon, as in methanol and ethanol. When the location of the —OH group needs to be specified, the number of the carbon atom to which it is attached is given, as in 1-propanol for $CH_3CH_2CH_2OH$ and 2-propanol for $CH_3CH(OH)CH_3$. Sometimes the hydrocarbon chain of an alcohol is named as a group, as in methyl alcohol, CH_3OH, and ethyl alcohol, CH_3CH_2OH. The —OH group may also be named as a substituent, in which case the name *hydroxy* is used, as in 2-hydroxybutane for $CH_3CH(OH)CH_2CH_3$.

1 Chloromethane, CH_3Cl

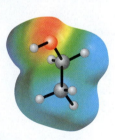

FIGURE 19.1 The charge distribution in an ethanol molecule. The red color denotes the region of net negative charge around the oxygen atom and the blue color denotes regions of net positive charge. Because the molecule is polar and can form hydrogen bonds, it is highly soluble in water.

2-Propanol is often known by its common name, isopropyl alcohol.

Alcohols are divided into three classes according to the number of organic groups attached to the carbon atom connected to the —OH group:

A **primary alcohol** has the form RCH_2—OH.

A **secondary alcohol** has the form R_2CH—OH.

A **tertiary alcohol** has the form R_3C—OH.

Each R represents an organic group, such as methyl or ethyl. They need not all be the same. Ethanol is a primary alcohol and 2-butanol is a secondary alcohol. An example of a tertiary alcohol is 2-methyl-2-propanol, $(CH_3)_3COH$.

Methanol is usually prepared industrially from synthesis gas, which is obtained from coal (Section 14.3):

$$CO(g) + 2 H_2(g) \xrightarrow{\text{catalyst, 250°C, 50–100 atm}} CH_3OH(g)$$

The catalyst is a mixture of copper, zinc oxide, and chromium(III) oxide. Ethanol is produced in large quantities throughout the world by the fermentation of carbohydrates. It is also prepared by the hydration of ethene in an addition reaction:

$$CH_2{=}CH_2(g) + H_2O(g) \xrightarrow{\text{catalyst, 300°C}} CH_3CH_2OH(g)$$

The catalyst is phosphoric acid. The laboratory synthesis of alcohols is by nucleophilic substitution of haloalkanes.

Ethylene glycol, or 1,2-ethanediol, $HOCH_2CH_2OH$ (**2**), is an example of a **diol,** which is a compound with two hydroxyl groups. Ethylene glycol is used as a component of antifreeze mixtures and in the manufacture of some synthetic fibers.

Alcohols with low molar masses are liquids, and alcohols have much lower vapor pressures than do hydrocarbons with approximately the same molar mass. For example, ethanol is a liquid at room temperature, but butane, which has a higher molar mass than ethanol, is a gas. The relatively low volatility of alcohols is a sign of the strength of hydrogen bonds. The ability of alcohols to form hydrogen bonds also accounts for the solubility in water of alcohols with low molar mass.

The formulas of alcohols are derived from that of water by replacing one of the hydrogen atoms with an organic group. Like water, they form intermolecular hydrogen bonds.

19.3 Ethers

An **ether** is an organic compound of the form R—O—R, where R is an alkyl group (the two R groups need not be the same). We can think of an ether as an HOH molecule in which both H atoms have been replaced by alkyl groups:

H—O—H	CH_3CH_2—O—H	CH_3CH_2—O—CH_2CH_3
Water	Ethanol (an alcohol)	Diethyl ether (an ether)

Ethers are more volatile than alcohols of the same molar mass because their molecules do not form hydrogen bonds to one another (Fig. 19.2). They are also less soluble in water because they have a lower ability to form hydrogen bonds to water molecules. Because ethers are not very reactive and have low molecular polarity, they are useful solvents for other organic compounds. However, ethers are flammable; diethyl ether is easily ignited and must be used with great care.

Cyclic ethers (ethers in the form of a ring) with alternating —CH_2CH_2—O— units are called **crown ethers,** owing to the crownlike shape of the molecules (**3**). Crown ethers bind very strongly to some metal cations; for example, a crown ether effectively encases K^+ in a shell with the O atoms pointing inward toward the K^+ ion and the hydrocarbon chain wrapped around the outside. Thus, the crown ether acts as a sort of "Trojan horse" for the K^+ ions, which in this form are soluble in nonpolar solvents. Potassium permanganate, for example, dissolves readily in benzene when a crown ether with six O atoms is added to it. The resulting solution can be used to carry out oxidation reactions in organic solvents.

The common names of 2-methyl-2-propanol are tertiary-butanol and tertiary-butyl alcohol (commonly shortened to *tert*-butanol and *tert*-butyl alcohol or even *t*-butanol and *t*-butyl alcohol).

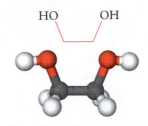

2 Ethanediol, $HOCH_2CH_2OH$

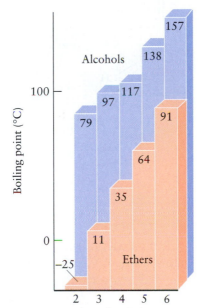

FIGURE 19.2 The boiling points of ethers are lower than those of isomeric alcohols, because there is hydrogen bonding in alcohols but not in ethers. All the molecules represented here are unbranched.

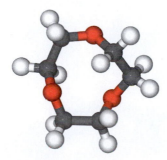

3 A crown ether

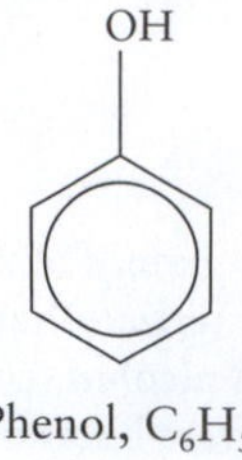

4   Phenol, C_6H_5OH

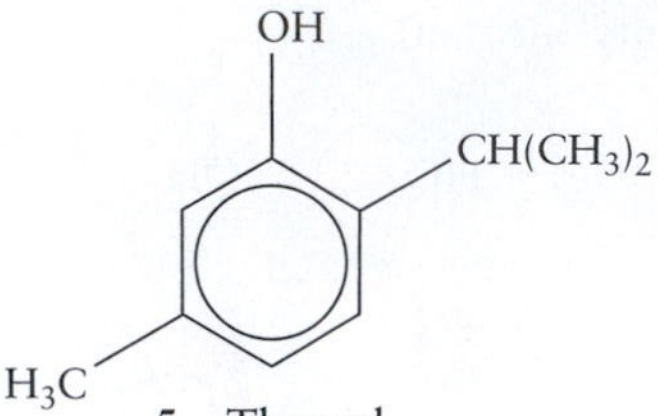

5   Thymol

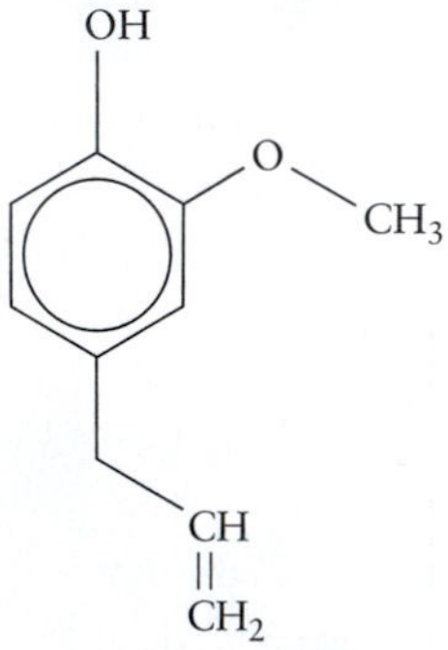

6   Eugenol

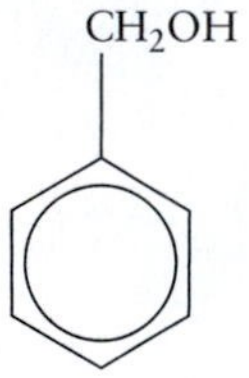

7   Benzyl alcohol,
phenylmethanol, $C_6H_5CH_2OH$

Ethers are not very reactive. They are more volatile than alcohols with similar molar masses because their molecules cannot form hydrogen bonds with one another. Crown ethers adopt shapes that can enclose ions and carry them into nonpolar solvents.

19.4 Phenols

In a **phenol**, a hydroxyl group is attached directly to an aromatic ring. The parent compound, phenol itself, C_6H_5OH (**4**), is a white, crystalline, molecular solid. It was once obtained from the distillation of coal tar, but now it is mainly synthesized from benzene. Many substituted phenols occur naturally, some being responsible for the fragrances of plants. They are often components of *essential oils,* the oils that can be distilled from flowers and leaves. Thymol (**5**), for instance, is the active ingredient of oil of thyme, and eugenol (**6**) provides most of the scent and flavor of oil of cloves.

Phenols differ from alcohols in that they are weak acids. Similar to the resonance in phenol itself (Section 18.8), the resonance in the anion

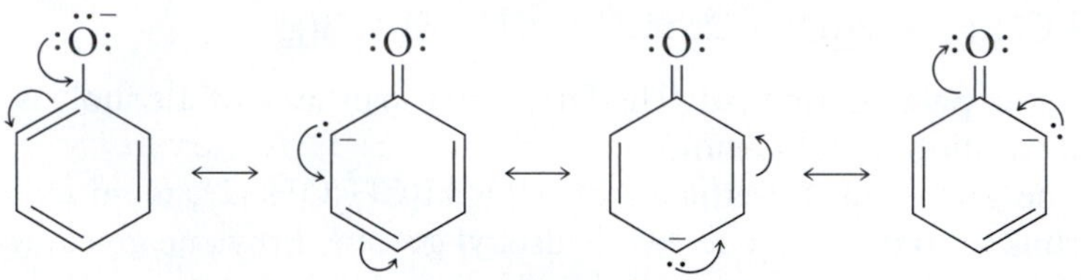

delocalizes the negative charge of the conjugate base of phenol and hence stabilizes the anion. As a result, $C_6H_5O^-$ is a weaker conjugate base than the corresponding conjugate base of an alcohol, such as $CH_3CH_2O^-$, the *ethoxide ion,* which is formed from ethanol. Consequently, phenol, C_6H_5OH, is a stronger acid than ethanol, CH_3CH_2OH, and phenols that are otherwise insoluble in water are soluble in basic solutions. However, even a $-CH_2-$ group can insulate the O atom from the benzene ring; and phenylmethanol, $C_6H_5CH_2OH$ (benzyl alcohol, **7**), is an alcohol, not a phenol.

Phenols are weak acids as a result of delocalization and stabilization of the conjugate base.

19.5 Aldehydes and Ketones

The **carbonyl group,** $>C=O$, occurs in two closely related families of compounds:

Aldehydes are compounds of the form $R-\overset{\displaystyle O}{\underset{\displaystyle H}{\overset{\|}{C}}}$

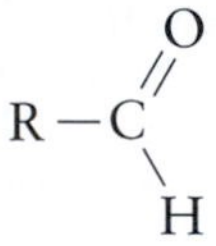

Ketones are compounds of the form $R-\overset{\displaystyle O}{\underset{\displaystyle R}{\overset{\|}{C}}}$

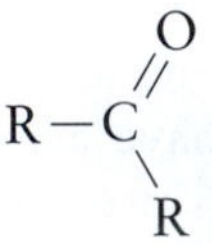

In an aldehyde, the carbonyl group is found at the end of the carbon chain, whereas, in a ketone, it is at an intermediate position. The R groups may be either aliphatic or aromatic and the two R groups of a ketone need not be the same. In condensed structural formulas, the carbonyl group of a ketone is written $-CO-$, as in CH_3COCH_3, propanone (acetone), a common laboratory solvent. The carbonyl group in an aldehyde is normally written $-CHO$, as in HCHO (formaldehyde), the simplest aldehyde. *Formalin,* the liquid used to preserve biological specimens, is an aqueous solution of formaldehyde. Wood smoke contains formaldehyde, and formaldehyde's destructive effect on bacteria is one reason why smoking food helps to preserve it.

The systematic names of aldehydes are obtained by replacing the ending -*e* by -*al,* as in methanal for HCHO and ethanal for CH_3CHO. Note that the carbon atom of the carbonyl group is included in the count of carbon atoms when determining the alkane from which the aldehyde is derived. Ketones are given the suffix -*one,* as in propanone, CH_3COCH_3. To avoid ambiguity, a number is used to denote the carbon atom that has become the carbonyl group; thus, $CH_3CH_2CH_2COCH_3$ is 2-pentanone and $CH_3CH_2COCH_2CH_3$ is 3-pentanone.

Aldehydes occur naturally in essential oils and contribute to the flavors of fruits and the odors of plants. Benzaldehyde, C_6H_5CHO (**8**), contributes to the characteristic aroma of cherries and almonds. Cinnamaldehyde (**9**) is found in cinnamon, and vanilla extract contains vanillin (**10**), which is present in oil of vanilla. Ketones can also be fragrant. For example, carvone (Section 18.1) is the essential oil of spearmint.

Formaldehyde is prepared industrially (for the manufacture of phenol–formaldehyde resins) by the catalytic oxidation of methanol:

$$2\ CH_3OH(g) + O_2(g) \xrightarrow{\ 600°C,\ Ag\ } 2\ HCHO(g) + 2\ H_2O(g)$$

Further oxidation of the product to a carboxylic acid (Section 19.6) can be avoided by using a mild oxidizing agent. There is less risk of further oxidation for ketones than for aldehydes, because a C—C bond would have to be broken for a carboxylic acid to form.

Dichromate oxidation of secondary alcohols produces ketones in good yield, with little additional oxidation. For example, $CH_3CH_2CH(OH)CH_3$ can be oxidized to $CH_3CH_2COCH_3$. The difference between the ease of oxidation of aldehydes and that of ketones is used to distinguish them. Aldehydes can reduce silver ions to form a silver mirror—a coating of silver on test-tube walls—with *Tollens reagent*, a solution of Ag^+ ions in aqueous ammonia (Fig. 19.3):

Aldehydes: $CH_3CH_2CHO + Ag^+$ (in Tollens reagent) gives CH_3CH_2COOH and Ag(s)

Ketones: $CH_3COCH_3 + Ag^+$ (in Tollens reagent) gives no reaction
Aldehydes and ketones can be prepared by the oxidation of alcohols. Aldehydes can be more easily oxidized than ketones can.

8 Benzaldehyde, C_6H_5CHO

9 Cinnamaldehyde

10 Vanillin

19.6 Carboxylic Acids

The **carboxyl group,** $-C\!\!\begin{smallmatrix}O\\ \\OH\end{smallmatrix}$, normally abbreviated —COOH, is the functional group found in **carboxylic acids,** which are weak acids of the form R—COOH. The strengths of carboxylic acids are related to their structures, as described in Section 10.10. The simplest carboxylic acid is formic acid, HCOOH, the acid in ant venom. Another common carboxylic acid is acetic acid, CH_3COOH, the acid of vinegar. It forms when the ethanol in wine is oxidized by air:

$$CH_3CH_2OH \xrightarrow{\ O_2\ } H_3C-C\!\!\begin{smallmatrix}O\\ \\OH\end{smallmatrix} + H_2O$$

Carboxylic acids are named systematically by replacing the -*e* of the parent hydrocarbon by the suffix -*oic acid;* the carbon atom of the carboxyl group is included in the count of atoms to determine the parent hydrocarbon molecule. Thus, formic acid is formally methanoic acid, and acetic acid is ethanoic acid.

FIGURE 19.3 An aldehyde (left) produces a silver mirror with Tollens reagent, but a ketone (right) does not.

Lab Video: Figure 19.3
Formation of a silver mirror

Carboxylic acids can be prepared by oxidizing primary alcohols and aldehydes with a strong oxidizing agent, such as acidified aqueous potassium permanganate solution. In some cases, an alkyl group can be oxidized directly to a carboxyl group. This process is very important industrially.

Carboxylic acids have an —OH group attached to a carbonyl group to form the carboxyl group, —COOH.

19.7 Esters

The product of the reaction between a carboxylic acid and an alcohol is called an **ester (11)**. Acetic acid and ethanol, for example, react when heated to about 100°C in the presence of a strong acid. The products of this **esterification** are ethyl acetate and water:

$$H_3C-C\overset{\displaystyle O}{\underset{\boxed{OH}}{\Big\langle}} \quad \boxed{H}-O-CH_2CH_3 \longrightarrow H_3C-C\overset{\displaystyle O}{\underset{O-CH_2CH_3}{\Big\langle}} + H_2O$$

11   Ethyl acetate, $CH_3COOC_2H_5$

$$H_3C-C\overset{\displaystyle O}{\underset{O-CH_2CH_3}{\Big\langle}}$$

Many esters have fragrant odors and contribute to the flavors of fruits. For example, benzyl acetate, $CH_3COOCH_2C_6H_5$, is an active component of oil of jasmine. Other naturally occurring esters include fats and oils. For example, the animal fat tristearin **(12)**, which is a component of beef fat, is an ester formed from glycerol and stearic acid.

12   Tristearin, $C_{57}H_{110}O_6$

Ester formation is an example of a **condensation reaction** in which two molecules combine to form a larger one and a small molecule is eliminated (Fig. 19.4). The reaction is catalyzed by a small amount of strong acid, such as sulfuric acid. In an esterification of a carboxylic acid and an alcohol, the eliminated molecule is H_2O.

Condensation reactions provide a solution to a problem resulting from the hydrogenation of oils (Section 18.6). During the hydrogenation process the remaining double bonds are converted from their naturally occurring cis forms to the trans isomer. These fats, which are called "trans fats," can contribute to health problems such as arteriosclerosis. *Transesterification* is a technique in which enzymes are used to promote switching of the long carboxylic acid chains from one alcohol location to another in a fatty acid (see Section 8.9). The resulting fat has the desired higher melting point yet is still unsaturated and has no trans fats. The new process also reduces the need for cooling water and does not use toxic or flammable solvents.

SELF-TEST 19.1A (a) Write the condensed structural formula of the ester formed from the reaction between propanoic acid, CH_3CH_2COOH, and methanol, CH_3OH. (b) Write the condensed structural formulas of the acid and alcohol that react to form pentyl ethanoate, $CH_3COOC_5H_{11}$, a contributor to the flavor of bananas.
[*Answer:* (a) $CH_3CH_2COOCH_3$; (b) CH_3COOH and $CH_3(CH_2)_3CH_2OH$]

SELF-TEST 19.1B (a) Write the condensed structural formula of the ester formed from the reaction between formic acid, HCOOH, and ethanol, CH_3CH_2OH. (b) Write the condensed structural formulas of the acid and alcohol that react to form methyl butanoate, $CH_3(CH_2)_2COOCH_3$, a contributor to the flavor of apples.

Alcohols condense with carboxylic acids to form esters.

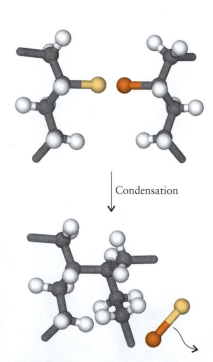

Condensation

FIGURE 19.4 In a condensation reaction, two molecules are linked as a result of removing two atoms or groups (the orange and yellow spheres) as a small molecule.

19.8 Amines, Amino Acids, and Amides

An **amine** is a compound with a formula derived from NH_3 in which various numbers of H atoms are replaced by organic groups. Amines are classified as primary, secondary, or tertiary according to the number of R groups on the nitrogen atom:

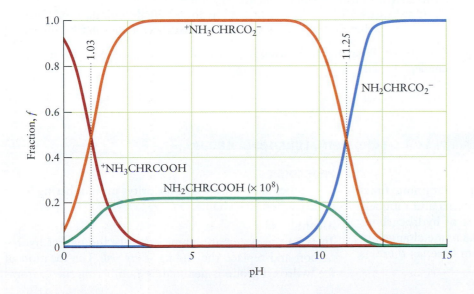

| Amine | Methylamine (primary) | Dimethylamine (secondary) | Trimethylamine (tertiary) |

In each case, the N atom is sp^3 hybridized, with one lone pair of electrons and three σ-bonds. A *quaternary ammonium ion* is a tetrahedral ion of the form R_4N^+, where as many as three of the R groups (which may all be different) can be replaced by H atoms. For example, the tetramethylammonium ion, $(CH_3)_4N^+$, and the trimethylammonium ion, $(CH_3)_3NH^+$, are quaternary ammonium ions. The **amino group**, the parent functional group of amines, is $-NH_2$.

Amines are widespread in nature. Many of them have a pungent, often unpleasant odor. Because proteins are organic polymers containing nitrogen, amines are present in the decomposing remains of living matter and, together with sulfur compounds, are responsible for the stench of decaying flesh. The common names of two diamines—putrescine, $NH_2(CH_2)_4NH_2$, and cadaverine, $NH_2(CH_2)_5NH_2$—speak for themselves. Like ammonia, amines are weak bases (Section 10.7), but quaternary ammonium ions with at least one H atom attached to the N atom are generally acidic.

An **amino acid** is a carboxylic acid that contains an amino group as well as a carboxyl group. The simplest example is glycine, NH_2CH_2COOH (**13**). Notice that an amino acid has both a basic group ($-NH_2$) and an acidic group ($-COOH$) in the same molecule. In aqueous solution with pH close to 7, amino acids are present as the **zwitterion**, such as $^+H_3NCH_2CO_2^-$, in which the amino group is protonated and the carboxyl group is deprotonated. There are four possible forms of an amino acid, which differ in the extent of protonation of the two functional groups. In very acidic solution, alanine exists as $^+H_3NCH(CH_3)COOH$, but, as base is added, the proton on the N is removed first, leaving $NH_2CH(CH_3)COOH$ (mainly in its zwitterion form), and then the proton from the carboxylic acid, to form $NH_2CH(CH_3)CO_2^-$. The concentrations of the various species in a solution of an amino acid can be calculated by the same methods used for polyprotic acids in Section 10.16. Figure 19.5 shows how the concentrations of alanine species

13   Glycine, NH_2CH_2COOH

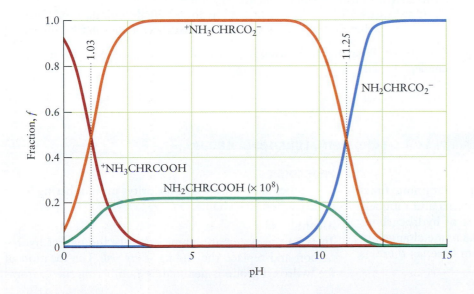

FIGURE 19.5 The fractional composition of a solution of alanine, a typical amino acid, as a function of pH. Notice that the concentration of the molecular form is extremely low at all pH values; its concentration had to be multiplied by a factor of 10^8 for it to be visible on the graph. Amino acids are present almost entirely in ionic form in aqueous solution.

present in a solution vary with pH. Notice that the molecular form of alanine is almost nonexistent at any pH: its concentration had to be multiplied by a factor of 100 million for it to be visible on the plot.

By far the most important amino acids are the α-amino acids, in which the —NH_2 group is attached to the carbon atom next to the carboxyl group, as in glycine. However, other types of amino acids are common and play an important biological role. For instance, the γ-amino acid $NH_2CH_2CH_2CH_2COOH$ is the neurotransmitter GABA.

Like alcohols, amines condense with carboxylic acids:

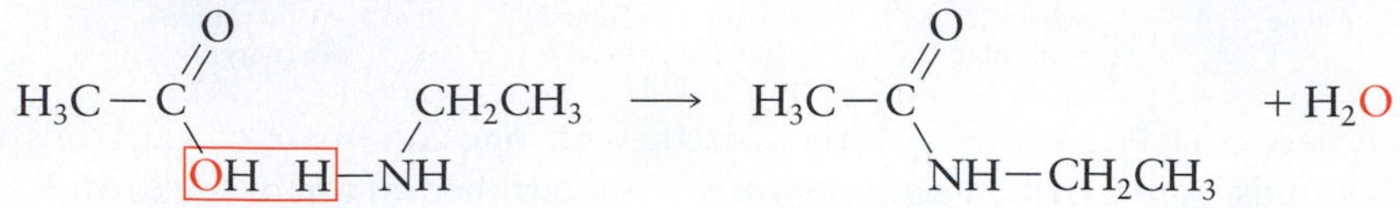

The product is an **amide**. When the reactant is a primary amine, RNH_2, the product is a molecule of the form R—(CO)—NHR.

The mechanism of amide formation is a source of insight into the properties of carboxylic acids and amines. Initially, we might expect an amine to act as a base and simply accept a proton from the carboxylic acid. Indeed, that does happen, and a quaternary ammonium salt is formed when the reagents are mixed in the absence of a solvent. For example,

$$CH_3COOH + CH_3NH_2 \longrightarrow CH_3CO_2^- + CH_3NH_3^+$$

However, on heating to about 200°C, a thermodynamically more favorable reaction takes place. The proton transfer is reversed, and the amine acts as a nucleophile as it attacks the carbon atom of the carboxyl group in a condensation reaction:

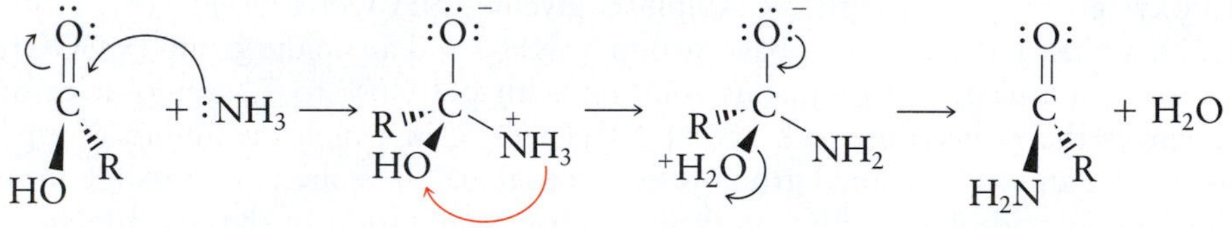

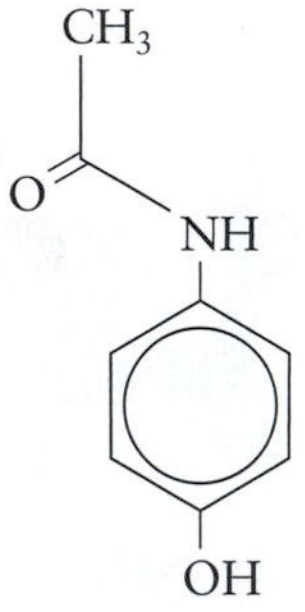

One example of a familiar amide is the pain-relieving drug sold as Tylenol (**14**); we shall see another important example when we consider the polymer known as nylon in Section 19.10. Many amides have N—H bonds that can take part in hydrogen bonding, and so the intermolecular forces between their molecules are relatively strong.

The red arrow on the second structure shows the migration of a proton.

14 Tylenol

SELF-TEST 19.2A Predict whether an ester or a primary amine of the same molar mass is likely to have the higher boiling point and explain why.

[*Answer:* The amine; it can form hydrogen bonds with the —NH_2 group, but the ester group cannot form hydrogen bonds in the pure state.]

SELF-TEST 19.2B What are the hybridization schemes of the C and N atoms in formamide, $HCONH_2$?

TOOLBOX 19.1   **HOW TO NAME SIMPLE COMPOUNDS WITH FUNCTIONAL GROUPS**

CONCEPTUAL BASIS

In general, the names of compounds containing functional groups follow the same conventions and numbering system as those used in the names of hydrocarbons (Toolbox 18.1), with the ending of the name changed to indicate the functional group. The aim here, as there, is to be succinct but unambiguous.

PROCEDURE

The functional group has priority in numbering over hydrocarbon side chains.

Alcohols

To form the systematic name, identify the parent hydrocarbon and replace the ending *-e* by *-ol*. The location of the hydroxyl group is denoted by numbering the C atoms of the backbone, starting at the end of the chain that results in the lowest number for the —OH group location. When —OH is named as a substituent, it is called *hydroxy*. For example, $CH_3CH(OH)CH_2CH_2CH_3$ is 2-pentanol or 2-hydroxypentane.

Aldehydes and ketones For aldehydes, identify the parent hydrocarbon: include the C of —CHO in the count of carbon atoms. Then change the final *-e* of the hydrocarbon name to *-al*. The C in the —CHO group is always carbon 1, at the end of a carbon chain, and is not explicitly numbered. For ketones, change the *-e* of the parent hydrocarbon to *-one* and number the chain in the order that gives the carbonyl group the lower number. Thus, $CH_3CH_2CH_2COCH_3$ is 2-pentanone.

Carboxylic acids Change the *-e* of the parent hydrocarbon to *-oic acid*. To identify the parent acid, include the C atom of the —COOH group when counting carbon atoms. Thus, $CH_3CH_2CH_2COOH$ is butanoic acid.

Esters Change the *-anol* of the alcohol to *-yl* and the *-oic acid* of the parent acid to *-oate*. Thus, CH_3COOCH_3 (from meth*anol* and ethan*oic acid*) is meth*yl* ethan*oate*.

Amines Amines are named systematically by specifying the groups attached to the nitrogen atom in alphabetical order (disregarding prefixes), followed by the suffix *-amine*. Amines with two amino groups are called *diamines*. The —NH_2 group is called *amino-* when it is a substituent. Thus, $(CH_3CH_2)_2CH_3N$ is diethylmethylamine and $CH_3CH(NH_2)CH_3$ is 2-aminopropane.

Halides Name the halogen atom as a substituent by changing the *-ine* part of its name to *-o*, as in 2-chlorobutane, $CH_3CHClCH_2CH_3$.

These rules are illustrated in Example 19.1.

EXAMPLE 19.1 **Sample exercise: Naming compounds with functional groups**

Name the compounds (a) $CH_3CH(CH_3)CH(OH)CH_3$; (b) $CH_3CHClCH_2COCH_3$; and (c) $(CH_3CH_2)_2NCH_2CH_2CH_3$.

SOLUTION
(a) We determine the number of carbon atoms in the longest chain and replace the ending of the name of that chain with *-ol*:

$$\underset{4}{H_3C}-\underset{3}{CH}-\underset{2}{\overset{\displaystyle CH_3}{CH}}-\underset{1}{\overset{\displaystyle OH}{CH}}-CH_3$$

The alcohol is 3-methyl-2-butanol. (b) The compound is both a ketone and a haloalkane. Identify the hydrocarbon chain and number the chain in the direction that gives the ketone group the lower number. The chain has five carbon atoms; the ketone group is on the second carbon atom from one end, and the chlorine atom is on the fourth:

$$\underset{5}{H_3C}-\underset{4}{\overset{\displaystyle Cl}{CH}}-\underset{3}{CH_2}-\underset{2}{\overset{\displaystyle O}{\overset{\displaystyle \|}{C}}}-\underset{1}{CH_3}$$

The compound is 4-chloro-2-pentanone. (c) The compound $(CH_3CH_2)_2NCH_2CH_2CH_3$ has two ethyl groups and one propyl group attached to a nitrogen atom. Give the ethyl groups the prefix *di-* to indicate their number and list the groups in alphabetical order. The compound is diethylpropylamine.

SELF-TEST 19.3A Name (a) $CH_3CH(CH_2CH_2OH)CH_3$; (b) $CH_3CH(CHO)CH_2CH_3$; (c) $(C_6H_5)_3N$.
    [*Answer:* (a) 3-Methyl-1-butanol; (b) 2-methylbutanal; (c) triphenylamine]

SELF-TEST 19.3B Name (a) $CH_3CH_2CH(OH)CH_2CH_3$; (b) $CH_3CH_2COCH_2CH_3$; (c) $CH_3CH_2NHCH_3$.

Amines are derived from ammonia by the replacement of hydrogen atoms with organic groups. Amides result from the condensation of amines with carboxylic acids. Amines and many amides take part in hydrogen bonding.

THE IMPACT ON MATERIALS

The chains of carbon atoms in organic compounds can reach enormous lengths and give rise to *macromolecules*. **Polymers** are compounds in which chains or networks of small repeating units form giant molecules, such as polypropylene and polytetrafluoroethylene (sold as Teflon). Plastics are simply polymers that can be molded into shape. Polymers are made by two main types of reactions: *addition reactions* and *condensation reactions*. Which type of reaction is used depends on the functional groups present in the starting materials. We begin by describing the two main types of reactions used to prepare polymers and then examine the characteristic physical properties of polymers.

19.9 Addition Polymerization

Alkenes can react with themselves to form long chains in a process called **addition polymerization.** For example, an ethene molecule may form a bond to another ethene molecule; another ethene molecule may add to that, and so on, forming a long hydrocarbon chain. The original alkene, such as ethene, is a small molecule called the **monomer.** Each monomer becomes one **repeating unit,** the structure that repeats over and over to produce the polymer chain. The product, the chain of covalently linked repeating units, is the polymer. The simplest addition polymer is polyethylene, $-(CH_2CH_2)_n-$, which is made by polymerizing ethene and so consists of long chains of thousands of repeating $-CH_2CH_2-$ units. Many addition polymer molecules also have a number of branches, which are generated as new chains sprout from intermediate points along the "backbone" chain.

The plastics industry has developed polymers from a number of monomers with the formula $CHX=CH_2$, where X is a single atom (such as the Cl in vinyl chloride, $CHCl=CH_2$) or a group of atoms (such as the CH_3 in propene). These substituted ethenes are used to make polymers of formula $-(CHXCH_2)_n-$, including polyvinyl chloride (PVC), $-(CHClCH_2)_n-$, and polypropylene, $-(CH(CH_3)CH_2)_n-$ (Table 19.2). They differ in appearance, rigidity, transparency, and resistance to weathering. Many plastic materials made of addition polymers can be recycled by melting and reprocessing (Table 19.3).

A widely used synthetic procedure is **radical polymerization,** polymerization by a radical chain reaction (Section 13.9). In a typical procedure, a monomer (such as ethene) is compressed to about 1000 atm and heated to 100°C in the presence of a small amount of an organic peroxide (a compound of formula $R-O-O-R$,

TABLE 19.2 Common Addition Polymers

Monomer name	Formula	Polymer formula	Common name
ethene[*]	$CH_2=CH_2$	$-(CH_2-CH_2)_n-$	polyethylene
vinyl chloride	$CHCl=CH_2$	$-(CHCl-CH_2)_n-$	polyvinyl chloride
styrene	$CH(C_6H_5)=CH_2$	$-(CH(C_6H_5)-CH_2)_n-$	polystyrene
acrylonitrile	$CH(CN)=CH_2$	$-(CH(CN)-CH_2)_n-$	Orlon, Acrilan
propene[*]	$CH(CH_3)=CH_2$	$-(CH(CH_3)-CH_2)_n-$	polypropylene
methyl methacrylate	$CH_3OOCC(CH_3)C=CH_2$	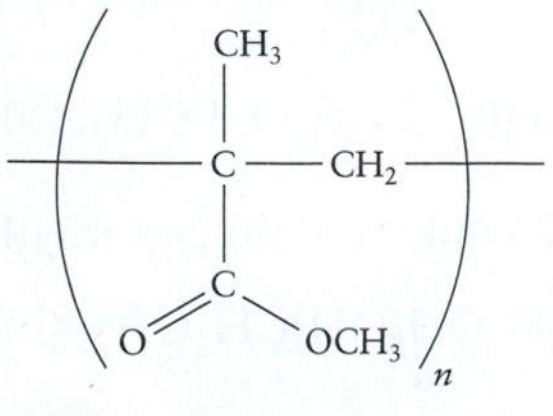 (structure shown)	Plexiglas, Lucite
tetrafluoroethene[*]	$CF_2=CF_2$	$-(CF_2-CF_2)_n-$	Teflon, PTFE[†]

[*]The suffix *-ene* is replaced by *-ylene* in the common names of these compounds, hence the names of the corresponding polymers.
[†]PTFE, polytetrafluoroethylene.

TABLE 19.3 Recycling Codes

Recycling code	Polymer	Recycling code	Polymer
1 PET(E)	polyethylene terephthalate	5 PP	polypropylene
2 HDPE	high-density polyethylene	6 PS	polystyrene
3 PVC or VC	polyvinyl chloride	7	other
4 LDPE	low-density polyethylene		

where R is an organic group). The reaction is initiated by dissociation of the O—O bond, giving two radicals:

$$R—O—O—R \longrightarrow R—O\cdot + \cdot O—R$$

Once it has been initiated, the chain reaction can propagate as the radicals attack the monomer molecules $CHX{=}CH_2$ (with X = H for ethene itself) and form a new, highly reactive radical:

$$R—O\cdot + H_2C{=}\overset{H}{\underset{X}{C}} \longrightarrow R—O—CH_2—\overset{H}{\underset{X}{C}}\cdot$$

This radical attacks another monomer molecule, and the chain grows longer:

$$R—O—CH_2—\overset{H}{\underset{X}{C}}\cdot + H_2C{=}\overset{H}{\underset{X}{C}} \longrightarrow R—O—CH_2—\overset{H}{\underset{X}{C}}—CH_2—\overset{H}{\underset{X}{C}}\cdot$$

The reaction continues until all the monomer has been used up or until it terminates when pairs of chains have linked together into single nonradical species. The product consists of molecules with many repeating units. Polyethylene (with X = H), for instance, consists of long chains of formula $—(CH_2CH_2)_n—$ in which n can reach many thousands.

Strong, tough polymers have chains that pack together well. One problem with early attempts to make polypropylene was that the orientations of the H and CH_3 groups on each C atom were random, thereby preventing the chains from packing well. The resulting material was amorphous, sticky, and nearly useless. Now, however, the stereochemistry of the chains can be controlled by using a **Ziegler–Natta catalyst,** a catalyst consisting of the compounds titanium tetrachloride, $TiCl_4$, and triethylaluminum, $(CH_3CH_2)_3Al$. A polymer in which each unit or pair of repeating units has the same relative orientation is described as being **stereoregular;** the stereoregularity arises from the manner in which the chains grow on the catalyst (Fig. 19.6). The chains of stereoregular polymers produced by Ziegler–Natta catalysts pack together well to form highly crystalline, dense materials (Fig. 19.7).

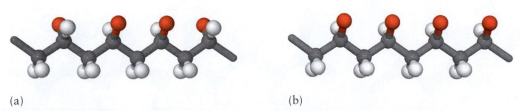

(a)   (b)

FIGURE 19.6 (a) A polymer in which the substituents lie on random sides of the chain. (b) A stereoregular polymer produced by using a Ziegler–Natta catalyst. In this case, the substituents all lie on the same side of the chain.

FIGURE 19.7 The two samples of polyethylene in the test tube were produced by different processes. The floating (low-density) one was produced by high-pressure polymerization. The one at the bottom (high-density) was produced with a Ziegler–Natta catalyst. As the insets show, the latter has a higher density because the stereoregular polymer chains pack together better.

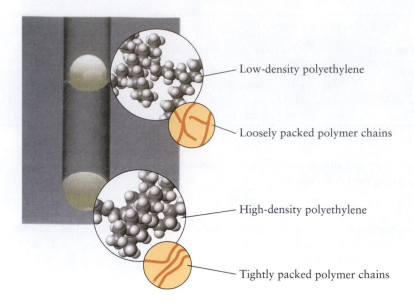

Low-density polyethylene

Loosely packed polymer chains

High-density polyethylene

Tightly packed polymer chains

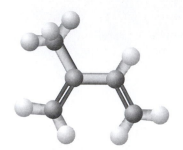

15 Isoprene

FIGURE 19.8 Collecting latex from a rubber tree in Malaysia, a principal producer of latex.

Rubber is a polymer of isoprene (**15**). Natural rubber is obtained from the bark of the rubber tree as a milky white liquid, which is called *latex* (Fig. 19.8) and consists of a suspension of rubber particles in water. The rubber itself is a soft white solid that becomes even softer when warm. It is used for pencil erasers and was once used as crepe rubber for the soles of shoes.

Chemists were unable to synthesize rubber for a long time, even though they knew it was a polymer of isoprene. The enzymes in the rubber tree produce a stereoregular polymer in which all the links between monomers are in a cis arrangement (Fig. 19.9); straightforward radical polymerization, however, produces a random mixture of cis and trans links and a sticky, useless product. The stereoregular polymer was achieved with a Ziegler–Natta catalyst, and almost pure, rubbery *cis*-polyisoprene can now be produced. *trans*-Polyisoprene, in which all the links are trans, is the hard, naturally occurring material *gutta-percha,* once used inside golf balls and still used to fill root canals in teeth.

> *Alkenes undergo addition polymerization. When a Ziegler–Natta catalyst is used, the polymer is stereoregular and has a high density.*

19.10 Condensation Polymerization

In **condensation polymers,** the monomers are linked together by condensation reactions, like those used to form ester or amide links. Polymers formed by linking together monomers that have carboxylic acid groups with those that have alcohol groups are called **polyesters.** Polymers of this type are widely used to make artificial fibers. A typical polyester is Dacron, or Terylene, a polymer produced from the

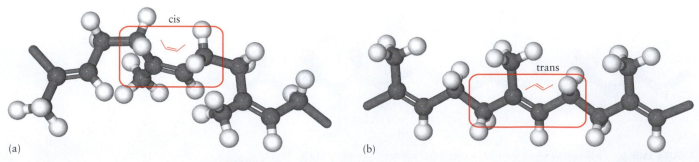

(a)   cis   (b)   trans

FIGURE 19.9 (a) In natural rubber, the isoprene units are polymerized to be all cis. (b) The harder material, gutta-percha, is the all-trans polymer.

esterification of terephthalic acid with ethylene glycol (see **2**, 1,2-ethanediol): its technical name is polyethylene terephthalate. The first condensation is

FIGURE 19.10 Synthetic fibers are made by extruding liquid polymer from small holes in an industrial version of a spider's spinneret.

A new ethylene glycol molecule can condense with the carboxyl group on one end of the product, and another terephthalic acid molecule can condense with the hydroxyl group on the other end. As a result, the polymer grows at both ends and becomes

In the radical addition polymerization of alkenes, side chains can grow out from the main chain. However, in condensation polymerization, growth can occur only at the functional groups, and so chain branching is much less likely. As a result, polyester molecules make good fibers, because the unbranched chains can be made to lie side by side by stretching the heated product and forcing it through small holes (Fig. 19.10). The fibers produced can then be spun into yarn (Fig. 19.11). Polyesters can also be molded and used in surgical implants, such as artificial hearts, or made into thin films for cassette tapes.

Condensation polymerization of amines with carboxylic acids leads to the **polyamides**, substances more commonly known as *nylons*. A common polyamide is nylon-66, which is a polymer of 1,6-diaminohexane, $H_2N(CH_2)_6NH_2$, and adipic acid, $HOOC(CH_2)_4COOH$. The 66 in the name indicates the numbers of carbon atoms in the two monomers.

For condensation polymerization, it is necessary to have two functional groups on each monomer and to mix stoichiometric amounts of the reactants. In polyamide production, the starting materials first form "nylon salt" by proton transfer:

$$HOOC(CH_2)_4COOH + H_2N(CH_2)_6NH_2 \longrightarrow$$

$$^-O_2C(CH_2)_4CO_2^- + {}^+H_3N(CH_2)_6NH_3^+$$

At this point, the excess acid or amine can be removed. Then, when the nylon salt is heated, the condensation begins, just as in the preparation of simple amides. The first step is

FIGURE 19.11 A scanning electron micrograph of Dacron polyester and cotton fibers in a blended shirt fabric. The cotton fibers have been colored green. Compare the smooth cylinders of polyester (orange) with the irregular surface of the cotton (green). The smooth polyester fibers resist wrinkles, and the irregular cotton fibers produce a more comfortable and absorbent texture.

FIGURE 19.12 A rather crude nylon fiber can be made by dissolving the salt of an amine in water and dissolving the acid in a layer of hexane, which floats on the water. The polymer forms at the interface of the two layers, and a long string can be slowly pulled out.

The amide grows at both ends by further condensations (Fig. 19.12), and the final product is

$$\text{}^-\text{O}-\overset{\displaystyle \text{O}}{\underset{\displaystyle \text{O}}{\text{C}}}-(CH_2)_4-C \left(\overset{\displaystyle \text{O}}{\underset{\displaystyle \text{NH}(CH_2)_6\text{NH}}{\text{C}}}-(CH_2)_4-\overset{\displaystyle \text{O}}{\text{C}} \right)_n \text{NH}(CH_2)_6\text{NH}_3{}^+$$

The long polyamide (nylon) chains can be spun into fibers (like polyesters) or molded. The N—H···O=C hydrogen bonding that takes place between neighboring chains accounts for much of the strength of nylon fibers (Fig. 19.13). This ability to form hydrogen bonds also accounts for the tendency of nylon to absorb moisture, because H_2O molecules can hydrogen bond to the polymer and worm their way in among the chains.

EXAMPLE 19.2   Determining the formulas of polymers and monomers

Write the formulas of (a) the monomers of Kevlar, a strong fiber used to make bullet-proof vests:

$$\left(\text{NH}-\overset{\displaystyle \text{O}}{\text{C}}-\text{C}_6\text{H}_4-\overset{\displaystyle \text{O}}{\text{C}}-\text{NH}-\text{C}_6\text{H}_4 \right)_n$$

(b) two repeating units of the polymer that is formed when peroxides are added to $CH_3CH_2CH=CH_2$ at a high temperature and pressure.

STRATEGY (a) Look at the backbone of the polymer, the long chain to which the other groups are attached. If the atoms are all carbon atoms, then the compound is an addition polymer. If ester groups are present in the backbone, then the polymer is a polyester and the monomers will be an acid and an alcohol. If the backbone contains amide groups, then the polymer is a polyamide and the monomers will be an acid and an amine. (b) If the monomer is an alkene or alkyne, then the monomers will add to one another; a π-bond will be replaced by new σ-bonds between the monomers. If the monomers are an acid and an alcohol or amine, then a condensation polymer forms with the loss of a molecule of water.

FIGURE 19.13 The strength of nylon fibers is an indication of the strength of the hydrogen bonds between neighboring polyamide chains.

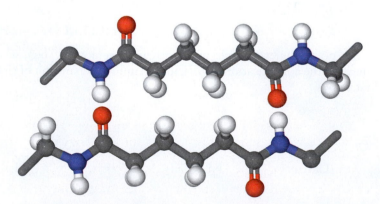

SOLUTION (a) Amide groups are present in the backbone, and so the polymer is a polyamide. Because the amide groups face in opposite directions, there are two different monomers, one with two acid groups and the other with two amine groups. We split each amide group apart and add a molecule of water to each amide link.

(b) The monomer is an alkene; so it forms an addition polymer. Replace each π-bond by an additional σ-bond to an adjacent monomer:

SELF-TEST 19.4A (a) Write the formula for the monomer of the polymer sold as Teflon, $-(CF_2CF_2)_n-$. (b) The polymer of lactic acid (**16**) is a biodegradable polymer made from renewable resources. It is used in surgical sutures that dissolve in the body. Write the formula for a repeating unit of this polymer.

[*Answer:* (a) $CF_2=CF_2$; (b) $(OCH(CH_3)CO)_n$]

SELF-TEST 19.4B Write the formula for (a) the monomer of poly(methyl methacrylate), used in contact lenses (**17**); (b) two repeating units of polyalanine, the polymer of the amino acid alanine, $CH_3CH(NH_2)COOH$.

Most condensation polymers are formed by the reaction of a carboxylic acid with an alcohol to form a polyester or with an amine to form a polyamide.

19.11 Copolymers and Composites

Copolymers are polymers made up of more than one type of repeating unit (Fig. 19.14). One example is nylon-66, in which the repeating units are formed from 1,6-diaminohexane, $H_2N(CH_2)_6NH_2$, and adipic acid, $HOOC(CH_2)_4COOH$. They form an **alternating copolymer,** in which acid and amine monomers alternate.

In a **block copolymer,** a long segment made from one monomer is followed by a segment formed from the other monomer. One example is the block copolymer formed from styrene and butadiene. Pure polystyrene is a transparent, brittle material that is easily broken; polybutadiene is a synthetic rubber that is very resilient, but soft and opaque. A block copolymer of the two monomers produces *high-impact polystyrene,* a material that is a durable, strong, yet transparent plastic. A different formulation of the two polymers produces *styrene–butadiene rubber* (SBR), which is used mainly for automobile tires and running shoes, but also in chewing gum.

In a **random copolymer,** different monomers are linked in no particular order. A **graft copolymer** consists of long chains of one monomer with shorter chains of the other monomer attached as side groups. For example, the polymer used to make hard contact lenses is a nonpolar hydrocarbon that repels water. The polymer used to make soft contact lenses is a graft copolymer that has a backbone of

16   Lactic acid, $CH_2CH(OH)COOH$

17   Poly(methyl methacrylate)

(a) Simple polymer

(b) Alternating copolymer

(c) Block copolymer

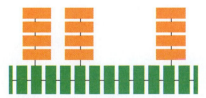

(d) Graft copolymer

FIGURE 19.14 The classification of copolymers. (a) A simple polymer made from a single monomer represented by the green rectangles. (b) An alternating copolymer of two monomers, represented by the green and orange rectangles. (c) A block copolymer. (d) A graft copolymer.

FIGURE 19.15 A photomicrograph of the cross section of the mother-of-pearl lining a mollusk shell. The composite material making up mother-of-pearl consists of flat crystals of calcium carbonate embedded in a tough, flexible organic matrix that resists cracking.

FIGURE 19.16 The material used to make this high-performance car protects the driver by making use of a design like that of the mollusk shell in Fig. 19.15. The car is made of composite materials that are stronger than steel.

nonpolar monomers but side groups of a different water-absorbing monomer. So much water is absorbed by the side chains that 50% of the volume of the contact lens is water, which makes it pliable, soft, and more comfortable to wear than a hard contact lens.

SELF-TEST 19.5A Use Fig. 19.14 to identify the type of copolymer formed by monomers A and B: —AAAABBBBB—.

[*Answer:* Block copolymer]

SELF-TEST 19.5B Identify the type of copolymer formed by monomers A and B: —ABABABAB—.

A **composite material** consists of two or more materials that have been solidified together. Composite materials are found in nature. For instance, seashells owe their strength to a tough organic matrix and their hardness to the calcium carbonate crystals imbedded in the matrix (Fig. 19.15). Bones have low density, but they are strong because of their composite nature. In bone tissue, crystals of phosphate salts are embedded in fibers of a natural polymer called *collagen* that hold them in place but allow a small amount of flexibility. Artificial bone and cartilage are now becoming realities through the preparation of synthetic composite materials such as fiberglass. These materials mimic bone in that they have inorganic solids mixed into a crack-resistant polymer matrix. The result is a material that has great strength yet remains flexible. Some lightweight composites, such as the graphite composite used for tennis rackets and the body of the space shuttle, can have three times the strength-to-density ratio of steel (Fig. 19.16).

Composite materials and copolymers combine the advantages of more than one component material. Copolymers contain more than one type of monomer.

19.12 Physical Properties of Polymers

A polymer can be designed to have the desired properties for an application. The first consideration is chain length. Because synthetic polymers consist of molecules with different lengths, they do not have definite molar masses. We can speak only of the *average* molar mass and the *average* chain length of a polymer. Nor do polymers have definite melting points; rather, they soften gradually as the temperature is raised. The viscosity of a polymer, its ability to flow when molten (Section 5.7), depends on its chain length. The longer the chains, the more tangled together they may be, and hence the slower their flow.

The mechanical strength of a polymer increases as the strength of the interactions between chains increases. Therefore, the longer the chains, the greater the strength of the polymer. For chains of the same length, stronger intermolecular forces result in greater mechanical strength. The properties of the functional groups in a polymer affect the strength of intermolecular forces and contribute to the mechanical strength of the polymer. For example, nylon is a polyamide, and its —NH— and —(CO)— groups can take part in hydrogen bonding. Consequently, nylon is a strong polymer.

The chain-packing arrangements that maximize intermolecular contact result in greater strength as well as greater density. Long, unbranched chains can line up next to one another like uncooked spaghetti and form crystalline regions that result in strong, dense materials. Branched polymer chains cannot fit together as closely and form weaker, less dense materials (recall Fig. 19.7).

The **elasticity** of a polymer is its ability to return to its original shape after being stretched. Natural rubber has low elasticity and is easily softened by heating. However, the "vulcanization" of rubber increases its elasticity. In vulcanization, rubber is heated with sulfur. The sulfur atoms form cross-links between the polyisoprene chains and produce a three-dimensional network of atoms (Fig. 19.17). Because the chains are covalently linked together, vulcanized rubber does not soften as much as natural rubber when the temperature is raised. Vulcanized rubber is also much more resistant to deformation when stretched, because the cross-

Animation: Figure 19.17 Rubber

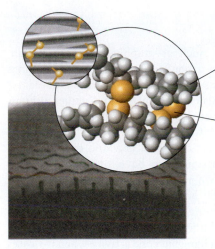

Polyisoprene molecule

Disulfide link

FIGURE 19.17 The gray cylinders in the small inset represent polyisoprene molecules, and the beaded yellow strings represent disulfide (—S—S—) links that are introduced when the rubber is vulcanized, or heated with sulfur. These cross-links increase the resilience of the rubber and make it more useful than natural rubber. Automobile tires are made of vulcanized rubber and a number of additives, including carbon.

links pull it back. Polymeric materials that return to their original shapes after stretching are called **elastomers.** However, extensive cross-linking can produce a rigid network that resists stretching. For example, high concentrations of sulfur result in very extensive cross-linking and the hard material called *ebonite*.

Because they are molecular compounds, most polymers are electrical insulators. However, polymers that have alternating double bonds along the chain can be used to conduct electricity (Box 19.1). These conducting polymers tend to have long chains with few branches.

Polymers melt over a range of temperatures, and polymers consisting of long chains tend to have high viscosities. Polymer strength increases with increasing chain length and the extent of crystallization.

THE IMPACT ON BIOLOGY

At one level, life can be regarded as a collection of hugely complex reactions taking place between organic compounds in oddly shaped containers. Many of these organic compounds are polymers, including the cellulose of wood, natural fibers such as cotton and silk, the proteins and carbohydrates in our food, and the nucleic acids of our genes.

19.13 Proteins

Protein molecules are condensation copolymers of up to twenty different naturally occurring amino acids that differ in their side chains (Table 19.4). Our bodies can synthesize eleven amino acids in sufficient amounts for our needs. But we cannot produce the proteins necessary for life unless we ingest the other nine, which are known as the **essential amino acids.**

A molecule formed by the condensation of two or more amino acids is called a **peptide.** An example is the combination of glycine and alanine, denoted Gly-Ala:

$$\text{H}_2\text{N}-\text{CH}_2-\overset{\displaystyle \text{O}}{\overset{\|}{\text{C}}}-\text{NH}-\overset{\displaystyle \text{CH}_3}{\underset{\displaystyle \text{COOH}}{\text{CH}}}$$

The —CO—NH— link shown in the red box is called a **peptide bond,** and each monomer used to form a peptide is called a **residue.** A typical protein is a polypeptide chain of more than a hundred residues joined through peptide bonds and arranged in a strict order. When only a few amino acid residues are present, we call the molecule an **oligopeptide.** The artificial sweetening agent aspartame is a type of oligopeptide called a **dipeptide** because it has two residues.

BOX 19.1   FRONTIERS OF CHEMISTRY: CONDUCTING POLYMERS*

One day, if you want to check your e-mail in a remote location, you might unroll a plastic sheet with a tiny microprocessor embedded in it. When you activate the microprocessor, your messages will appear and you might reply by writing on the screen with a special pen or by speaking to it. The remarkable material of this flat computer already exists: one form of it was discovered by accident in the early 1970s when a chemist who was polymerizing ethyne (acetylene) added a thousand times too much catalyst. Instead of a synthetic rubber, he made a thin, flexible film. It looked like metal foil tinted pink (see the photograph), and—very much like a metal—it conducted electricity.

Metals conduct electricity because their valence electrons move easily from atom to atom. Most covalently bonded solids do not conduct electricity, because their valence electrons are locked into individual bonds and are not free to

This flexible polyacetylene sheet was peeled from the walls of the reaction flask in which it was made.

move. The exceptions, such as graphite and carbon nanotubes, have delocalized π-bonds in connected aromatic rings through which electrons can move freely because there are empty molecular orbitals close in energy to occupied orbitals (Section 14.16). However, a disadvantage is that commercial graphite is fragile and brittle.

Conducting polymers provide a new and exciting alternative. They do not rust and they have low densities. They can be molded or drawn into shells, fibers, or thin plastic sheets, but they can still act like metallic conductors. They can be made to glow with almost any color and to change conductivity with conditions. Imagine cases of food labeled with polymer tags that change in conductivity when the cases are left unrefrigerated for too long.

All conducting polymers have a common feature: a long chain of sp^2 hybridized carbon atoms, often with nitrogen or sulfur atoms included in the chains. Polyacetylene, the first conducting polymer, is also the simplest, consisting of thousands of —CH=CH—units:

The alternation of double and single bonds means that each C atom has an unhybridized p-orbital that can overlap with the p-orbital on either side. This arrangement allows electrons to be delocalized along the entire chain like a one-dimensional version of graphite.

One conducting polymer, polypyrrole,

The **primary structure** of a protein is the sequence of residues in the peptide chain. Aspartame consists of phenylalanine (Phe) and aspartic acid (Asp), and so its primary structure is Phe-Asp. Three fragments of the primary structure of human hemoglobin are

Leu-Ser-Pro-Ala-Asp-Lys-Thr-Asn-Val-Lys-...

...-Val-Lys-Gly-Trp-Ala-Ala-...

...-Ser-Thr-Val-Leu-Thr-Ser-Lys-Ser-Lys-Tyr-Arg

The determination of the primary structure of a protein is a very demanding analytical task, but, thanks to automated procedures, many of these structures are now known. Any modification of the primary structure of a protein—the replacement of one amino acid residue by another—may lead to a congenital disease. Even one wrong amino acid in the chain can disrupt the normal function of the molecule (Fig. 19.18).

The **secondary structure** of a protein is the shape adopted by the polypeptide chain—in particular, how it coils or forms sheets. The order of the amino acids in the chain controls the secondary structure, because their intermolecular forces hold the chains together. The most common secondary structure in animal proteins is the **α helix,** a helical conformation of a polypeptide chain held in place by hydrogen bonds between residues (Fig. 19.19). One alternative secondary structure is the **β sheet,** which is characteristic of the protein that we know as silk. In silk, protein

FIGURE 19.18 Sickle-shaped red blood cells form when only one amino acid (glutamic acid) in a polypeptide chain is replaced by another amino acid (valine). These cells are less able to take up oxygen than normal cells.

has been used in smart windows, which darken from a transparent yellow-green to a nearly opaque blue-black in bright sunlight. Fibers of polypyrrole are also woven into radar camouflage cloth that absorbs microwaves. Because it does not reflect microwaves back to their source, the cloth appears on radar as a patch of empty space.

Polyaniline, which has the structure

is being used in flexible coaxial cable; rechargeable, flat, buttonlike batteries; and laminated, rolled films that could be used as flexible computer or television displays. Thin films of poly-*p*-phenylenevinylene, PPV,

give off light when exposed to an electric field, a process called *electroluminescence*. By varying the composition of the polymer, scientists have coaxed it to glow in a wide range of colors. Such multicolor light-emitting diode (LED) displays could be as bright as fluorescent ones.

HOW MIGHT YOU CONTRIBUTE?

Current PPV displays last only about 10% as long as fluorescent displays. If their longevity grows, however, they could begin to replace the computer and television screens that we use today. Polyaniline could one day serve both as a nonmetallic solder for the printed circuit boards inside a computer and as electrical shielding for the case. All-plastic transistors and other electronic components can be miniaturized to an amazing degree, raising the possibility of microscopic microprocessors and computers that could survive highly corrosive conditions, such as those at marine research sites. Because conducting polymers can also be designed to change shape with the level of electrical current, they could serve as artificial muscles or give flexibility of movement to robots. For these goals to be achieved, the characteristics of these polymers must be known and their responses to various conditions have to be studied.

Related Exercises: 19.85, 19.86

For Further Reading: R. E. Gleason, "How far will circuits shrink?" *Science Spectra*, vol. 20, 2000, pp. 32–40. B. Halford, "Better electrochromics," *Chemical and Engineering News*, November 30, 2005, http://pubs.acs.org/cen/news/83/i49/8349 electrochromics.html. P. Yam, "Plastics get wired," *Scientific American*, July 1995, pp. 83–89. Information (advanced) on the Nobel Prize 2000, at http://nobelprize.org/nobel_prizes/chemistry/laureates/2000/adv.html

---

*The 2000 Nobel Prize in chemistry was awarded to A. J. Heeger, A. G. MacDiarmid, and H. Shirakawa for their discovery of conducting polymers.

---

molecules lie side by side to form nearly flat sheets. The molecules of many other proteins consist of alternating α-helical and β-sheet regions (Fig. 19.20).

The **tertiary structure** of a protein is the shape into which its secondary structure is folded as a result of interactions between residues. The globular form of each chain in hemoglobin is an example. One important type of link responsible for tertiary structure is the covalent **disulfide link**, —S—S—, between residues containing sulfur. Other links are formed by the various types of intermolecular forces. In most cases, a given protein will fold into a precise three-dimensional conformation determined by the locations of the hydrophobic and hydrophilic groups on the chain (Fig. 19.21). Sometimes, however, the proteins do not fold properly. When this incorrect folding happens in the body, it can lead to illnesses such as Alzheimer's disease. In Alzheimer's disease, brain function is constricted by deposits of proteins that have misfolded and can no longer perform their functions. The so-called "prion diseases," such as variant Creutzfeld–Jakob disease ("mad cow disease") are also due to protein misfolding. If scientists can solve the puzzle of why proteins misfold and discover how to correct that, then diseases now considered untreatable can be cured.

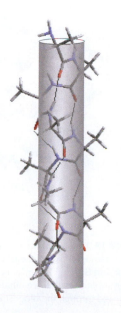

FIGURE 19.19 A representation of part of an α helix, one of the secondary structures adopted by polypeptide chains. The cylinder encloses the "backbone" of the polypeptide chain, and the side groups project outward from it. The thin lines represent the hydrogen bonds that maintain the helical shape.

TABLE 19.4   The Naturally Occurring Amino Acids, X—CH(NH₂)COOH

X	Name	Abbreviation	X	Name	Abbreviation
—H	glycine	Gly	—CH₂(CH₂)₃NH₂	lysine*	Lys
—CH₃	alanine	Ala			
—CH₂—C₆H₅	phenylalanine*	Phe	—CH₂(CH₂)₂NH—C(NH₂)=NH	arginine	Arg
—CH(CH₃)₂	valine*	Val			
—CH₂CH(CH₃)₂	leucine*	Leu	—CH₂—(imidazole)	histidine*	His
—CH(CH₃)CH₂CH₃	isoleucine*	Ile			
—CH₂OH	serine	Ser			
—CH(OH)CH₃	threonine*	Thr			
—CH₂—C₆H₄—OH	tyrosine	Tyr	—CH₂—(indole)	tryptophan*	Trp
—CH₂COOH	aspartic acid	Asp	—CH₂CONH₂	asparagine	Asn
—CH₂CH₂COOH	glutamic acid	Glu	—CH₂CH₂CONH₂	glutamine	Gln
—CH₂SH	cysteine	Cys	(proline ring structure)	proline†	Pro
—CH₂CH₂SCH₃	methionine*	Met			

*Essential amino acids for humans.
†The entire amino acid is shown.

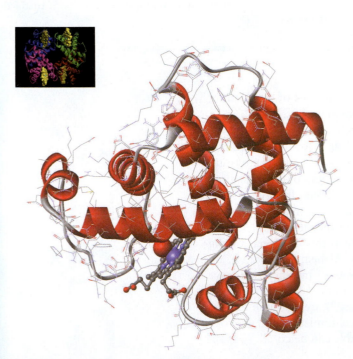

FIGURE 19.20 One of the four polypeptide chains that make up the human hemoglobin molecule. The chains consist of alternating regions of α helix and β sheet. The α-helix regions are represented by red helices. The oxygen molecules that we inhale attach to the iron atom (blue sphere) and are carried through the bloodstream.

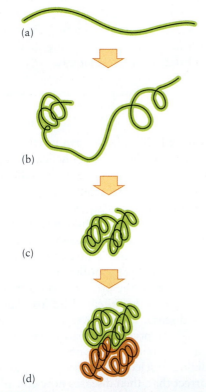

(a)

(b)

(c)

(d)

FIGURE 19.21 These structures show how a protein first forms α helices and β sheets and then how the coils and sheets fold together to form the shape of a protein. Finally, if the protein has a quaternary structure, the protein subunits stack together. (a) Newly formed polypeptide; (b) intermediate; (c) subunit; (d) mature (in this case, dimeric) protein.

FIGURE 19.22 The protein produced by spiders to make a web is a form of silk that can be exceptionally strong.

FIGURE 19.23 Artificial spider silk can now be made in bulk. It can be spun into thin, tough thread, like that on the spools shown here, or wound into cables strong enough to support suspension bridges.

Proteins may also have a **quaternary structure,** in which neighboring polypeptide units stack together in a specific arrangement. The hemoglobin molecule, for example, has a quaternary structure of four polypeptide units, one of which is shown in Fig. 19.20.

The structures of some natural protein-based materials, such as silk and wool, result in strong, tough fibers. Spiders and silkworms use proteins as a structural material of remarkable strength (Fig. 19.22). Chemists are duplicating nature by making artificial spider silk (Fig. 19.23), which is one of the strongest fibers known.

The loss of structure by a protein is called **denaturation.** This structural change may be a loss of quaternary, tertiary, or secondary structure; it may also be degradation of the primary structure by cleavage of the peptide bonds. Even mild heating can cause irreversible denaturation. When we cook an egg, the protein called albumen denatures into a white mass. The "permanent waving" of hair, which consists primarily of long α helices of the protein keratin, is a result of partial denaturation.

Proteins are polymers made of amino acid units. The primary structure of a polypeptide is the sequence of amino acid residues; secondary structure is the formation of helices and sheets; tertiary structure is the folding into a compact unit; quaternary structure is the packing of individual protein units together.

19.14 Carbohydrates

The **carbohydrates** are so called because many have the empirical formula CH_2O, which suggests a hydrate of carbon. They include starches, cellulose, and sugars such as glucose, $C_6H_{12}O_6$ (**18**), which contains an aldehyde group, and fructose (fruit sugar), a structural isomer of glucose that is a ketone (**19**). Carbohydrates have many —OH groups, and so they may also be regarded as alcohols. The presence of these —OH groups allows them to form numerous hydrogen bonds with one another and with water.

18 Glucose, $C_6H_{12}O_6$ **19** Fructose, $C_6H_{12}O_6$

The **polysaccharides** are polymers of glucose. They include starch, which we can digest, and cellulose, which we cannot. Starch is made up of two components:

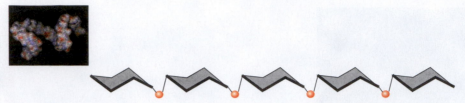

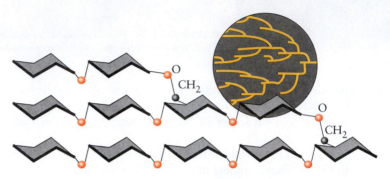

FIGURE 19.24 The amylose molecule, one component of starch, is a polysaccharide, a polymer of glucose. It consists of glucose units linked together to give a structure like the one shown but with a moderate degree of branching.

FIGURE 19.25 The amylopectin molecule is another component of starch. It has a more highly branched structure than amylose, as emphasized in the inset.

amylose and amylopectin. Amylose, which makes up about 20% to 25% of most starches, consists of chains made up of several thousand glucose units linked together (Fig. 19.24). Amylopectin is also made up of glucose chains (Fig. 19.25), but its chains are linked into a branched structure and its molecules are much larger. Each molecule consists of about a million glucose units.

Cellulose is the structural material of plants. It is a polymer with the same monomer (glucose) that starch has, but the units in cellulose are linked differently, forming flat, ribbonlike strands (Fig. 19.26). Hydrogen bonds between these strands lock them together into a rigid structure that for us (but not for termites) is indigestible. Cellulose is the most abundant organic chemical in the world, and billions of tons of it are produced annually by photosynthesis. In research on alternative fuels, enzymes are used to break down the cellulose in waste biomass into glucose, which is then fermented to make ethanol for use as a fuel (see Box 6.2).

Carbohydrates include sugars, starches, and cellulose. Glucose is an alcohol and an aldehyde that polymerizes to form starch and cellulose.

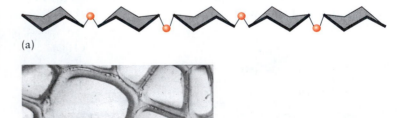

(a)

FIGURE 19.26 (a) Cellulose is yet another polysaccharide constructed from glucose units. The units in cellulose link together in such a way that they form long, flat ribbons that can produce a fibrous material through hydrogen bonding. (b) These long tubes of cellulose (shown in cross section) form the structural material of an aspen tree.

(b)

19.15 Nucleic Acids

The nucleus of every living cell contains at least one deoxyribonucleic acid (DNA) molecule to control the production of proteins and to carry genetic information from one generation of cells to the next. Human DNA molecules are immense: if a molecule of DNA could be extracted without damage from a cell nucleus and drawn out straight from its highly coiled natural shape, it would be about 2 m long (Fig. 19.27). The ribonucleic acid (RNA) molecule is closely related to DNA. One of its functions is to carry information stored by DNA to a region of the cell where that information can be used in protein synthesis.

DNA is a polymer with a backbone built of repeating units derived from the sugar ribose (**20**). For DNA, the ribose molecule has been modified by removing the oxygen atom at carbon atom 2, the second carbon atom clockwise from the ether oxygen atom in the five-membered ring. Therefore, the repeating unit—the monomer—is called deoxyribose (**21**).

FIGURE 19.27 A DNA molecule is very large, even in bacteria. In this micrograph, a DNA molecule has spilled out through the damaged cell wall of a bacterium.

20 Ribose, $C_5H_{10}O_5$ **21** Deoxyribose, $C_5H_{10}O_4$

Attached by a covalent bond to carbon atom 1 of the deoxyribose ring is an amine (and therefore a base), which may be adenine, A (**22**); guanine, G (**23**); cytosine, C (**24**); or thymine, T (**25**). In RNA, uracil, U (**26**), replaces thymine. The base bonds to carbon atom 1 of deoxyribose through the nitrogen of the —NH— group (printed in red); and the compound so formed is called a **nucleoside**. All nucleosides have a similar structure, which we can summarize as the shape shown in (**27**); the lens-shaped object represents the attached amine.

22 Adenine   **23** Guanine   **24** Cytosine   **25** Thymine

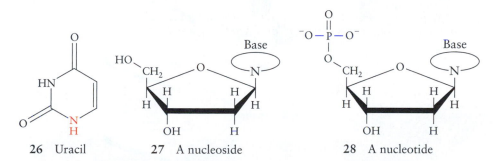

26 Uracil   **27** A nucleoside   **28** A nucleotide

The DNA monomers are each completed by a phosphate group, —O—PO_3^{2-}, covalently bonded to carbon atom 5 of the ribose unit to give a compound called a **nucleotide** (**28**). Because there are four possible nucleoside monomers (one for each base), there are four possible nucleotides in each type of nucleic acid.

Molecules of DNA and RNA are **polynucleotides,** polymeric species built from nucleotide units. Polymerization takes place when the phosphate group of one nucleotide (which is the conjugate base of an organic phosphoric acid) condenses

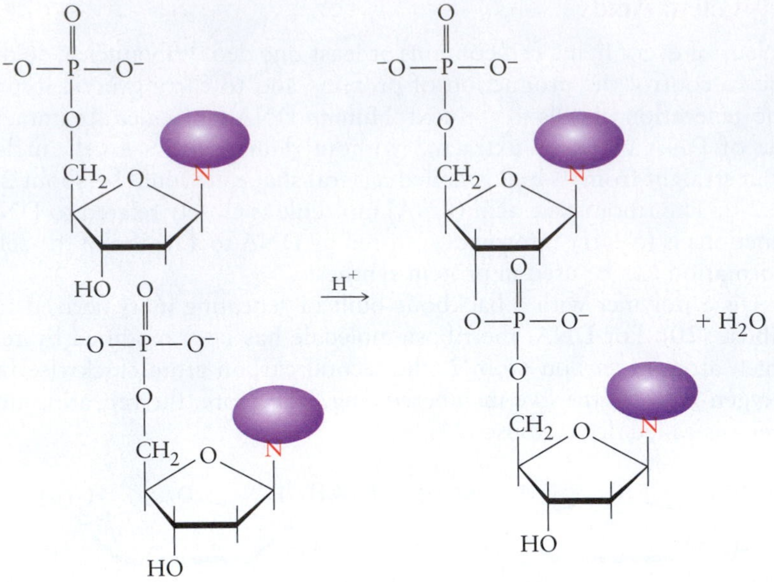

FIGURE 19.28 The condensation of nucleotides that leads to the formation of a nucleic acid, a polynucleotide.

with the —OH group on carbon atom 3 of the ribose unit of another nucleotide, thereby forming an ester link and releasing a molecule of water. As this condensation continues, it results in a structure like that shown in Fig. 19.28, a compound known as a **nucleic acid.** The DNA molecule itself is a double helix in which two long nucleic acid strands are wound around each other.

The ability of DNA to replicate lies in its double-helical structure. There is a precise correspondence between the bases in the two strands. Adenine in one strand always forms two hydrogen bonds to thymine in the other, and guanine always forms three hydrogen bonds to cytosine; so, across the helix, the base pairs are always AT and GC (Fig. 19.29). Any other combination would not be held together as well. During replication of the DNA, the hydrogen bonds, which are

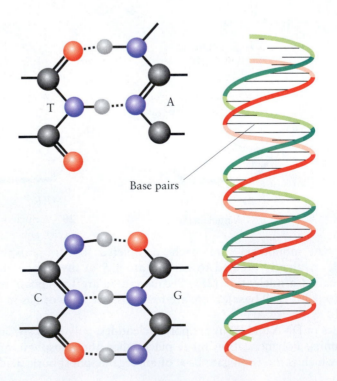

Base pairs

FIGURE 19.29 The bases in the DNA double helix fit together by virtue of the hydrogen bonds that they can form, as shown on the left. Once formed, the AT and GC pairs are almost identical in size and shape. As a result, the turns of the helix shown on the right are regular and consistent.

relatively weak compared with the covalent bonds in the strands, can be broken by an enzyme while the strands themselves remain intact. Nucleotides in the cellular fluid then attach in the appropriate places on each strand to form the double helices of two new molecules of DNA.

As well as replication—the production of copies of itself for reproduction and cell division—DNA governs the production of proteins by serving as the template during the synthesis of molecules of RNA. These new molecules, with U in place of T, carry information about segments of the genetic message out of the cell nucleus to the locations where protein synthesis takes place. In this way, the chemical reactions of functional groups, and in a broader sense, the principles of chemistry, bring matter to life.

Nucleic acids are copolymers of four nucleotides joined by phosphate ester links. The nucleotide sequence stores all genetic information.

SKILLS YOU SHOULD HAVE MASTERED

❏ 1 Recognize a simple haloalkane, alcohol, ether, phenol, aldehyde, ketone, carboxylic acid, amine, amide, or ester, given a molecular structure.

❏ 2 Predict the oxidation products of aldehydes and ketones (Section 19.5).

❏ 3 Write the structural formula of an ester or an amide formed from the condensation reaction of a given carboxylic acid with a given alcohol or amine (Sections 19.7 and 19.8).

❏ 4 Predict the influence of hydrogen bonding on the physical properties of organic compounds (Self-Test 19.2).

❏ 5 Name simple functional groups (Toolbox 19.1 and Example 19.1).

❏ 6 Predict the type of polymer that a given monomer can form and identify monomers, given the repeating unit of a polymer (Example 19.2).

❏ 7 Distinguish the various types of copolymers (Self-Test 19.5).

❏ 8 Explain the role of chain length, crystallinity, network formation, cross-linking, and intermolecular forces in determining the physical properties of polymers (Section 19.12).

❏ 9 Describe the composition of proteins and distinguish their primary, secondary, tertiary, and quaternary structures (Section 19.13).

❏ 10 Describe the composition of carbohydrates (Section 19.14).

❏ 11 Describe the structures and functions of nucleic acids (Section 19.15).

EXERCISES

Identifying Functional Groups and Naming Their Compounds

19.1 Write the general formula of each of the following types of compounds, using R to denote an organic group: (a) amine; (b) alcohol; (c) carboxylic acid; (d) aldehyde.

19.2 Write the general formula of each of the following types of compounds, using R to denote an organic group: (a) ether; (b) ketone; (c) ester; (d) amide.

19.3 Identify each type of compound: (a) R—O—R; (b) R—CO—R; (c) R—NH$_2$; (d) R—COOR.

19.4 Identify each type of compound: (a) R—CHO; (b) R—COOH; (c) R—CONHR; (d) R—OH.

19.5 Name the following compounds: (a) $CH_3CHClCH_3$; (b) $CH_3CH_2C(CH_3)ClCH_2CHClCH_3$; (c) CH_3CI_3; (d) CH_2Cl_2.

19.6 Name the following compounds: (a) CH_2CHCH_2Br; (b) $CF_2CHCHCH_2$; (c) $CH_3CH_2CCCH_2Br$; (d) $Cl_3CCOCH_2CH_3$.

19.7 Write the formulas of the following compounds and state whether each one is a primary, secondary, or tertiary alcohol or a phenol: (a) 1-chloro-2-hydroxybenzene;

(b) 2-methyl-3-pentanol; (c) 2,4-dimethyl-1-hexanol; (d) 2-methyl-2-butanol.

19.8 Write the formulas of the following compounds and state whether each one is a primary, secondary, or tertiary alcohol or a phenol: (a) 2-methyl-1-propanol; (b) 2-propanol; (c) *p*-hydroxytoluene; (d) 1-bromo-2-(hydroxymethyl)benzene.

19.9 Write the formula of (a) ethyl methyl ether, (b) ethyl propyl ether; (c) dimethyl ether.

19.10 Write the formula of (a) diethyl ether; (b) dibutyl ether; (c) methyl propyl ether.

19.11 Name each compound: (a) $CH_3CH_2CH_2OCH_2CH_2CH_2CH_3$; (b) $C_6H_5OCH_3$; (c) $CH_3CH_2CH_2OCH_2CH_2CH_2CH_2CH_3$.

19.12 Name each compound: (a) $CH_3CH_2CH_2OCH_2CH_3$; (b) $CH_3OCH_2CH_2CH_2CH_3$; (c) $CH_3CH_2CH_2OCH_2CH_2CH_3$.

19.13 Identify each compound as an aldehyde or a ketone and give its systematic name: (a) CH_3CHO; (b) CH_3COCH_3; (c) $(CH_3CH_2)_2CO$.

19.14 Identify each compound as an aldehyde or a ketone and give its systematic name: (a) CH_3CH_2CHO;

(b)

(c) $(CH_3CH_2)_2CHCH_2COCH_3$

19.15 Write the structural formula of (a) butanal; (b) 3-hexanone; (c) 2-heptanone.

19.16 Write the structural formula of (a) 2-ethyl-2-methylpentanal; (b) 3,5-dihydroxy-4-octanone; (c) 4,5-dimethyl-3-hexanone.

19.17 Give the systematic name of (a) CH_3COOH; (b) $CH_3CH_2CH_2COOH$; (c) $CH_2(NH_2)COOH$.

19.18 Give the systematic name of (a) $CH_3CH_2CH_2COOH$; (b) $CHCl_2COOH$; (c) $CH_3(CH_2)_8COOH$.

19.19 Draw the structure of (a) benzoic acid, C_6H_5COOH; (b) 2-chloro-3-methylpentanoic acid; (c) hexanoic acid; (d) propenoic acid.

19.20 Draw the structure of (a) 2-methylpropanoic acid; (b) 2,2-dichlorobutanoic acid; (c) 2,2,2-trifluoroethanoic acid; (d) 4,4-dimethylpentanoic acid.

19.21 Give the systematic name of each of the following amines: (a) CH_3NH_2; (b) $(CH_3CH_2)_2NH$; (c) o-$CH_3C_6H_4NH_2$.

19.22 Give the systematic name of each of the following amines: (a) $CH_3CH_2CH_2NH_2$; (b) $(CH_3CH_2)_4N^+$; (c) p-$ClC_6H_4NH_2$.

19.23 Write the structural formula of each of the following amines: (a) o-methylphenylamine; (b) triethylamine; (c) tetramethylammonium ion.

19.24 Write the structural formula of each of the following amines: (a) methylpropylamine; (b) dimethylamine; (c) m-methylphenylamine.

Reactivity Patterns and Functional Groups

19.25 Which of the following molecules or ions may function as a nucleophile in a nucleophilic substitution reaction: (a) NH_3; (b) CO_2; (c) Br^-; (d) SiH_4?

19.26 Which of the following molecules or ions may function as a nucleophile in a nucleophilic substitution reaction: (a) CH_3O^-; (b) CN^-; (c) NH_4^+; (d) PH_3?

19.27 Suggest an alcohol that could be used for the preparation of each of the following compounds and indicate how the reaction would be carried out; (a) ethanal; (b) 2-octanone; (c) 5-methyloctanal.

19.28 Suggest an alcohol that could be used for the preparation of each of the following compounds and indicate how the reaction would be carried out: (a) methanal; (b) propanone; (c) 6-methyl-5-decanone.

19.29 Draw the structure of the principal product formed from each of the following condensation reactions: (a) butanoic acid

with 2-propanol; (b) ethanoic acid with 1-pentanol; (c) hexanoic acid with methylethyl amine; (d) ethanoic acid with propylamine.

19.30 Draw the structure of the principal product formed from each of the following condensation reactions: (a) octanoic acid with methanol; (b) propanoic acid with ethanol; (c) propanoic acid with methylamine; (d) methanoic acid with diethylamine.

19.31 You are given samples of propanal, 2-propanone, and ethanoic acid. Describe how you would use chemical tests, such as acid–base indicators and oxidizing agents, to distinguish among the three compounds.

19.32 You are given samples of 1-propanol, pentane, and ethanoic acid. Describe how you would use chemical tests, such as aqueous solubility and acid–base indicators, to distinguish among the three compounds.

19.33 Rank the following acids in order of strength: $ClCH_2COOH$, Cl_3CCOOH, CH_3COOH, and CH_3CH_2COOH. Justify your answer.

19.34 Rank methylamine, dimethylamine, and diethylamine in order of increasing base strength. Explain your rankings in relation to molecular structure.

Polymers

19.35 Sketch three repeating units of the polymer formed from (a) $CH_2=C(CH_3)_2$; (b) $CH_2=CHCN$; (c) isoprene,

19.36 Sketch three repeating units of the polymer formed from (a) tetrafluoroethene; (b) phenylethene; (c) $CH_3CH=CHCH_3$.

19.37 Write the structural formulas of the monomers of each of the following polymers, for which one repeating unit is shown: (a) polyvinylchloride (PVC), $-(CHClCH_2)_n-$; (b) Kel-F, $-(CFClCF_2)_n-$.

19.38 Write the structural formulas of the monomers of each of the following polymers, for which one repeating unit is shown: (a) a polymer used to make carpets, $-(OC(CH_3)_2-CO)_n-$; (b) $-(CH(CH_3)CH_2)_n-$; (c) a polypeptide, $-(NHCH_2CO)_n-$.

19.39 Write the structural formula of two units of the polymer formed from (a) the reaction of oxalic acid (ethanedioic acid), $HOOCCOOH$, with 1,4-diaminobutane, $H_2NCH_2CH_2CH_2CH_2NH_2$; (b) the polymerization of the amino acid alanine (2-aminopropanoic acid).

19.40 Write the structural formula of two units of the polymer formed from (a) the reaction of terephthalic acid with 1,2-diaminoethane, $H_2NCH_2CH_2NH_2$; (b) the polymerization of 4-hydroxybenzoic acid.

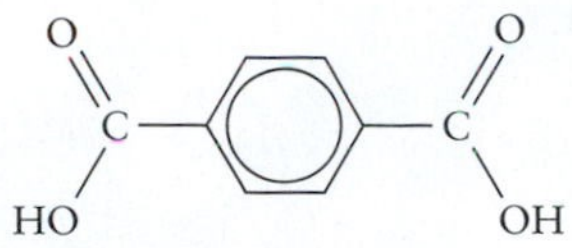

Terephthalic acid 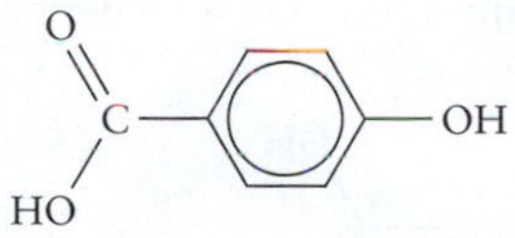   4-Hydroxybenzoic acid

19.41 Identify the type of copolymer formed by monomers A and B: —BBBBAA—.

19.42 Identify the type of copolymer formed by monomers A and B: —AABABBAA—.

19.43 How does average molar mass affect each of the following polymer characteristics: (a) softening point; (b) viscosity; (c) strength?

19.44 How does the polarity of the functional groups in a polymer affect each of the following characteristics of the polymer: (a) softening point; (b) viscosity; (c) strength?

19.45 Describe how the linearity of the polymer chain affects polymer strength.

19.46 Describe how cross-linking affects the elasticity and rigidity of a polymer.

19.47 Why do polymers not have definite molar masses? How does the fact that polymers have average molar masses affect their melting points?

19.48 Rank the following polymers according to increasing value as fibers: polyesters, polyamides, polyalkenes. Explain your reasoning.

Biological Compounds

19.49 (a) Draw the structure of the peptide bond that links the amino acids in proteins. (b) Identify the functional group formed. (c) Identify the type of polymer formed (addition or condensation).

19.50 (a) Draw the structure of the link between glucose units that creates amylose. (b) Identify the functional group formed. (c) Identify the type of polymer formed (addition or condensation).

19.51 Name the amino acids in Table 19.4 that contain side groups capable of forming hydrogen bonds. This interaction contributes to the tertiary structure of a protein.

19.52 Name the amino acids in Table 19.4 that contain nonpolar side groups. These groups contribute to the tertiary structure of a protein by preventing contact with water.

19.53 Draw the structure of the peptide formed from the reaction of the acid group of tyrosine with the amine group of glycine.

19.54 Draw the structure of the peptide formed from the reaction of the acid group of glycine with the amine group of tyrosine.

19.55 Identify (a) the functional groups and (b) the chiral carbon atoms in the mannose molecule shown here.

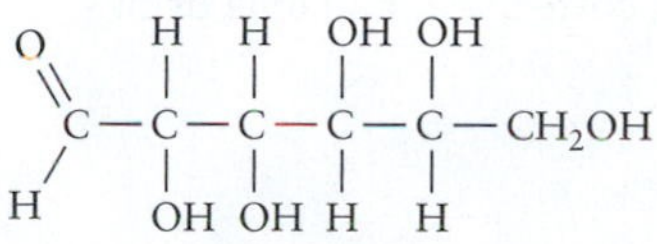

19.56 Identify (a) the functional groups and (b) the chiral carbon atoms in the histidine molecule shown here.

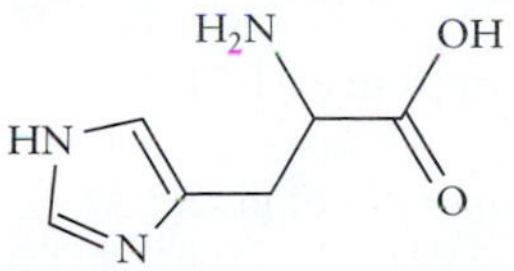

19.57 Write the complementary nucleic acid sequence that would pair with each of the following DNA sequences: (a) CATGAGTTA; (b) TGAATTGCA.

19.58 Write the complementary nucleic acid sequence that would pair with each of the following DNA sequences: (a) GGATCTCAG; (b) CTAGCCTGT.

Integrated Exercises

19.59 Write the chemical formula of the compound represented by each of the following line structures:

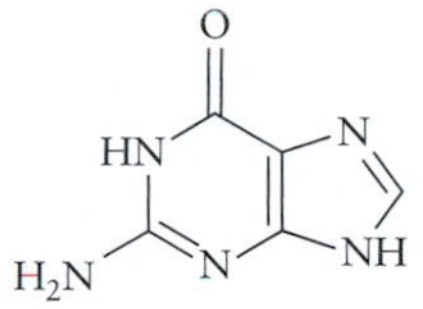

(a) Guanine 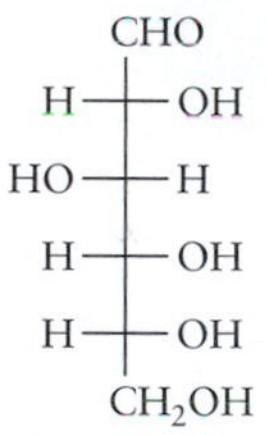   (b) D-Glucose 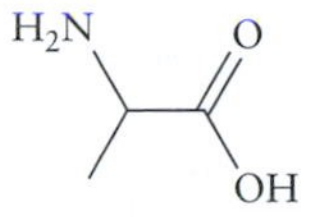   (c) Alanine

19.60 Write the chemical formula of the compound represented by each of the following line structures:

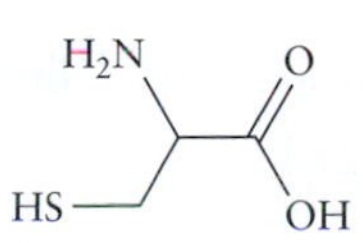

(a) Cysteine 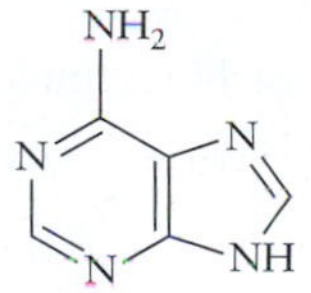   (b) Adenine 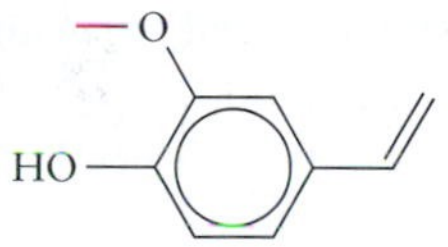   (c) Eugenol

19.61 Identify all the functional groups in each of the following compounds:
(a) vanillin, the compound responsible for vanilla flavor,

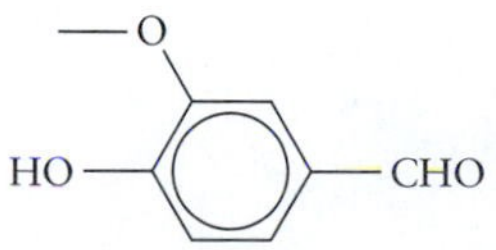

(b) carvone, the compound responsible for spearmint flavor,

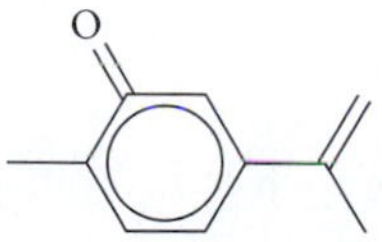

(c) caffeine, the stimulant in coffee, tea, and cola drinks,

19.62 Identify all the functional groups in each of the following compounds:
(a) zingerone, the pungent, hot component of ginger,

(b) Tylenol, an analgesic,

(c) procaine, a local anesthetic,

19.63 Identify the chiral carbon atoms in each of the following compounds:
(a) camphor, used in cooling salves,

(b) testosterone, a male sex hormone,

19.64 Identify the chiral carbon atoms in each of the following compounds:

(a) menthol, the flavor of peppermint,

(b) estradiol, a female sex hormone,

19.65 The structures of the following molecules can be found on the Web site for this text. Draw the structure of each and identify its chiral carbon atoms:
(a) cocaine; (b) aflatoxin B2, a toxin and carcinogen that occurs naturally in peanuts as a by-product of the growth of fungi.

19.66 The structures of the following molecules can be found on the Web site for this text. Draw the line structure of each and identify its chiral carbon atoms:
(a) cephalosporin C, toxic to penicillin-resistant staphylococci; (b) thromboxane A2, a substance that promotes the clotting of blood.

19.67 (a) Write the structural formulas of diethyl ether and 1-butanol (note that they are isomers). (b) The boiling point of 1-butanol is 117°C, higher than that of diethyl ether (35°C), yet the solubility of both compounds in water is about 8 g per 100 mL. Account for these observations.

19.68 Consider the following organic molecules, which have approximately the same molar masses but contain different functional groups: $CH_3CH_2CH_2CH_2CH_2CH_3$, $CH_3CH_2CH_2CH_2CHCH_2$, $CH_3CH_2CH_2CH_2CH_2OH$, $CH_3CH_2CH_2CH_2CHO$, $CH_3CH_2CH_2COOH$, $CH_3CH_2COOCH_3$, $CH_3CH_2COCH_2CH_3$, $CH_3CH_2CH_2CH_2OCH_3$. (a) Draw a Lewis structure for each molecule, name it, and classify it by functional group. (b) Which molecules are isomers of each other? Are any chiral? If so, which ones? (c) For each molecule, list the types of intermolecular forces that are present. (d) Use your answers to parts (a) and (b) to predict the relative boiling points, from lowest to highest.

19.69 Write the structural formula for the product of (a) the reaction of glycerol (1,2,3-trihydroxypropane) with stearic acid, $CH_3(CH_2)_{16}COOH$, to produce a saturated fat; (b) the oxidation of 4-hydroxybenzyl alcohol by sodium dichromate in an acidic organic solvent.

19.70 Pheromones are commonly called sex attractants, although they have other signaling functions, too. A pheromone of the queen bee is *trans*-$CH_3CO(CH_2)_5CH{=}CHCOOH$. (a) Write the structural formula of the pheromone. (b) Identify and name the functional groups in the molecule.

19.71 Classify each of the following reactions as (1) an addition reaction, (2) a nucleophilic substitution reaction, (3) an electrophilic substitution reaction, or (4) a condensation reaction: (a) the reaction of butene with chlorine; (b) the polymerization of the amino acid glycine; (c) the hydrogenation of butyne; (d) the polymerization of styrene, $CH_2CHC_6H_5$, by tertiary butyl hydroperoxide, $(CH_3)_3COOH$; (e) the reaction of methylamine with butanoic acid.

19.72 Classify each of the following reactions as (1) an addition reaction, (2) a nucleophilic substitution reaction, (3) an electrophilic substitution reaction, or (4) a condensation reaction: (a) the reaction of terephthalic acid with 1,2-ethanediol; (b) the reaction of 3-chlorohexane with concentrated sodium hydroxide; (c) the reaction of water with 2-iodo-2-methylpropane; (d) the reaction of propanoic acid with ethanol; (e) the reaction of toluene with bromine in the presence of $FeBr_3$.

19.73 Write the condensed structural formulas of the principal products of the reaction that takes place when (a) ethylene glycol, 1,2-ethanediol, is heated with stearic acid, $CH_3(CH_2)_{16}COOH$; (b) ethanol is heated with oxalic acid, $HOOCCOOH$; (c) 1-butanol is heated with propanoic acid.

19.74 A certain polyester has the repeating unit $-(OCH_2-C_6H_4-CH_2OOC-C_6H_4-CO)_n-$. Identify the monomers of the polyester.

19.75 Acrylic resins are polymeric materials used to make warm yet lightweight garments. The osmotic pressure of a solution prepared by dissolving 47.7 g of an acrylic resin in enough water to make 500. mL of solution is 0.325 atm at 25°C. (a) What is the average molar mass of the polymer? (b) How many monomers constitute the "average" molecule? The repeating unit of this acrylic resin has the formula $-CH_2CH(CN)-$. (c) What would be the vapor pressure of the solution if the vapor pressure of pure water at 25°C is 0.0313 atm? (Assume that the density of the solution is 1.00 g·cm^{-3}.) (d) Which approach (osmometry or the lowering of vapor pressure) would you prefer for the determination of very high molar masses such as those of acrylic resins? Why?

19.76 The average molar mass of a sample of polypropylene was determined by measuring the osmotic pressure of 500. mL of a solution of 3.16 g of polypropylene in benzene. A pressure of 0.0112 atm was observed at 25°C. (a) What is the average molar mass of the polymer? (b) How many propene monomer units with formula $-CH(CH_3)CH_2-$ were combined on average to create each chain? (c) If this sample consists only of linear chains and the carbon–carbon bond lengths in the polymer are close to their average value, what is the average chain length?

19.77 (a) Explain the differences among the primary, secondary, tertiary, and quaternary structures of a protein. (b) Identify the forces holding each structure together as covalent bonds or primarily intermolecular forces.

19.78 Haloalkanes may react with hydroxide ions, undergoing nucleophilic displacement of the halide ion to form an alcohol. A complication of such reactions is competition from elimination reactions rather than substitution (see Section 18.5). (a) Predict the possible products from the reaction of 2-bromopentane with sodium hydroxide. (b) What can be done to favor the

substitution reaction over the elimination pathway or vice versa?

19.79 The protonated form of glycine ($^+H_3NCH_2COOH$) has $K_{a1} = 4.47 \times 10^{-3}$, and $K_{a2} = 1.66 \times 10^{-10}$. (a) Write the chemical equations for the proton transfer equilibria. (b) What is the dominant form of glycine in solution at pH = 2, pH = 5, and pH = 12?

19.80 The pK_a values for phenol, o-nitrophenol, m-nitrophenol, and p-nitrophenol are 9.89, 7.17, 8.28, and 7.15, respectively. Explain the origin of these differences in pK_a.

19.81 Explain the process of condensation polymerization. How might the polymer obtained from benzene-1,2-dicarboxylic acid and ethylene glycol differ from Dacron?

19.82 The average molar mass of a hydrogen-bonded pair of nucleotides in a DNA molecule is 625 g·mol^{-1}. Each successive pair is found at a distance of 340 pm along the chain. If the total length of one strand of a DNA molecule is 0.299 m, what is the molar mass of the molecule?

19.83 A segment of a protein is analyzed and found to contain the amino acid sequence Glu-Leu-Asp. Draw the Lewis structure of this segment, showing the peptide bonds.

19.84 Ferritin is a globular iron-storage protein that stores iron as Fe^{3+}. To leave the ferritin, Fe^{3+} must first be reduced to Fe^{2+}. Ferritin has two types of channels through which the Fe^{2+} could leave: a "three-fold channel" and a "four-fold channel." The three-fold channel is lined with the amino acids aspartate (Asp) and glutamate (Glu) and the four-fold channel is lined with the amino acid leucine (Leu). Through which channel is the Fe^{2+} more likely to leave the ferritin protein? Explain your answer.

19.85 The monomer of the conducting polymer polyaniline is the compound aniline (aminobenzene). (a) Draw the structural formula of the aniline monomer. What is the hybridization of the N atom in (b) aniline? (c) polyaniline? (d) Indicate the locations of lone pairs, if any, in polyaniline. (e) Do the N atoms help to carry the current? Explain your reasoning. See Box 19.1.

19.86 Fibers of the conducting polymer polypyrrole are woven into radar camouflage cloth. Because it absorbs microwaves, rather than reflecting them back to their source, the cloth appears to be a patch of empty space on radar. (a) What is the hybridization of the N atom in polypyrrole? (b) Explain why polypyrrole absorbs microwave radiation, but small organic molecules do not. See Box 19.1 for the structure of polypyrrole.

For the following nine exercises, see Major Technique 7, Nuclear Magnetic Resonance, following this chapter.

19.87 Predict the features of the ^1H NMR spectrum of ethanal, CH_3CHO.

19.88 Predict the features of the ^1H NMR spectrum of propane.

19.89 Explain the features of the following ^{1}H NMR spectrum of ethyl 3,3-dimethylbutanoate.

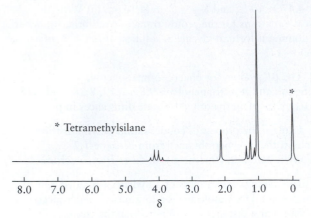

19.90 The ^{1}H NMR spectra of (a) 2-butanone and (b) ethyl acetate are shown here. The spectra are very similar yet have some important differences. Explain the observed similarities and differences.

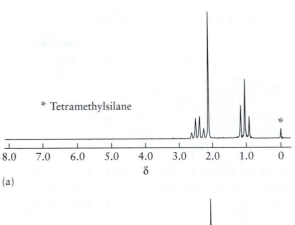

(a)

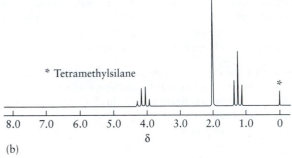

(b)

19.91 When propane is treated with chlorine gas, a mixture of compounds results. The mixture is separated and analysis of one of the components gives the ^{1}H NMR spectrum shown here. What is this product?

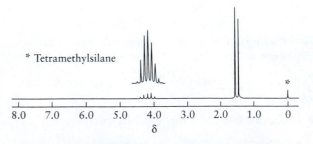

19.92 Other nuclei besides the proton have a nuclear spin of $\frac{1}{2}$ and, in principle at least, may be suitable for nuclear magnetic

resonance spectroscopy. From standard literature sources, find at least five other naturally occurring, nonradioactive elements that have spin of $\frac{1}{2}$ and that can be examined by NMR spectroscopy.

19.93 One naturally occurring isotope of carbon is suitable for NMR spectroscopy. (a) From literature sources, determine which isotope of carbon has a spin of $\frac{1}{2}$. (b) What is the natural abundance of this isotope of carbon? (c) Given that only nuclei having a nuclear spin may interact to produce fine structure in an NMR spectrum, do you expect to see the splitting due to interactions between adjacent carbon atoms in an organic molecule such as propane in its carbon NMR spectrum? (d) Do you expect to see splitting of the proton signal in the ^{1}H NMR spectrum of propane due to the interactions between the carbon and the proton nuclei? (e) Do you expect to see splitting of the carbon signal in the carbon NMR spectrum of propane due to the interactions between the carbon and the proton nuclei?

19.94 Predict the features of the ^{1}H NMR spectra of benzene and the three isomeric dichlorobenzenes. Is it possible to distinguish the isomeric dichlorobenzenes on the basis of their NMR spectra alone?

19.95 The NMR spectrum of a compound can reveal whether the atoms within the molecule are equivalent or different. For example, the hydrogen atoms in $CH_2{=}CH_2$ are equivalent: they all exist in an identical environment. However, in $CH_2{=}CHCl$, the two hydrogen atoms attached to the same carbon atom are equivalent, but different from the hydrogen atom on the other carbon atom. Predict how many different types of hydrogen atoms can be identified in the ^{1}H NMR spectrum of each of the following molecules: (a) C_2H_2; (b) $cis\text{-}C_2H_2Cl_2$; (c) $trans\text{-}C_2H_2Cl_2$; (d) CH_3OH; (e) In the molecule C_2H_5Cl, the hydrogen atoms would all be different if the molecule existed in only one conformation. However, only two types of hydrogen atoms are detected. Propose an explanation.

Chemistry Connections

19.96 Waste reduction is an important goal of the green chemistry movement. In many chemical syntheses in industry, not all the atoms required for the reaction appear in the product. Some end up in by-products and are wasted. "Atom economy" is the use of as few atoms as possible to reach an end product and is calculated as a percentage, using atom economy = (mass of desired product obtained)/(mass of all reactants consumed) $\times$ 100%.
(a) Consider the following synthesis of $CH_3CH{=}CHCH_3$:

$$CH_3CH_2CHBrCH_3 + CH_3CH_2O^- \xrightarrow{\text{ethanol}}$$
$$CH_3CH{=}CHCH_3 + CH_3CH_2OH + Br^-$$

Identify the type of reaction (substitution, elimination, addition).
(b) Name each organic reactant and product.
(c) Does the $CH_3CH_2O^-$ ion function as a nucleophile, an electrophile, or neither?
(d) Calculate the atom economy of the reaction, assuming 100% yield.
(e) An alternative synthesis of $CH_3CH{=}CHCH_3$ is

$$CH_3CH_2CHBrCH_3 + CH_3O^- \xrightarrow{\text{methanol}}$$
$$CH_3CH{=}CHCH_3 + CH_3OH + Br^-$$

Calculate the atom economy for the reaction, assuming 100% yield.

(f) Another alternative synthesis of $CH_3CH=CHCH_3$ is

$$CH_3CH_2CHBrCH_3 + CH_3S^- \xrightarrow{\text{methanethiol}}$$
$$CH_3CH=CHCH_3 + CH_3SH + Br^-$$

Calculate the atom economy for the reaction, assuming 100% yield.

(g) Which of the three syntheses produces the lowest mass of waste? Which produces the highest?

(h) Assume that you carry out each of the three syntheses, starting with 50.0 g of $CH_3CH_2CHBrCH_3$ and with the second reagent in excess in each case. Your yields of $CH_3CH=CHCH_3$ for the three reactions are as follows: (a) 16.2 g; (e) 15.4 g; (f) 13.1 g. Calculate the percentage yield and experimental atom economy for each reaction.

(i) Which reaction would you recommend to a manufacturer? Explain your reasoning.

Nuclear magnetic resonance (NMR) is the principal technique for the identification of organic compounds and is among the leading techniques for the determination of their structures. The technique has also been developed, as *magnetic resonance imaging* (MRI), as a diagnostic procedure in medicine.

The Technique

Many atomic nuclei behave like small bar magnets, with energies that depend on their orientation in a magnetic field. An NMR spectrometer detects transitions between these energy levels. The nucleus most widely used for NMR is the proton, and we shall concentrate on it. Two other very common nuclei, those of carbon-12 and oxygen-16, are nonmagnetic, so they are invisible in NMR.

Like an electron, a proton possesses the property of spin, which for elementary purposes can be thought of as an actual spinning motion. Because a proton has an electric charge and because a moving charge generates a magnetic field, a proton acts as a tiny bar magnet that can have either of two orientations, which are denoted ↑ (or α) and ↓ (or β). In the presence of a magnetic field these two orientations have different energies.

If the sample is exposed to electromagnetic radiation, the nuclei are flipped from one orientation to another when the energy of the incoming photons (which is $h\nu$, where ν is the frequency of the radiation) matches the energy separation of the two spin orientations. The strong coupling between nuclei and radiation that occurs when this condition is satisfied is called *resonance*. At resonance, the radiation is strongly absorbed and a sharp peak appears in the detector output. Superconducting magnets are used to generate very high magnetic fields, and resonance requires radio-frequency radiation at about 500 MHz.

Each proton in a compound comes into resonance at a frequency related to its location in the molecule. Figure 1, for instance, shows the NMR spectrum of ethanol; and we see that there are three groups of peaks and a characteristic pattern of splitting within the groups. Because all compounds that contain hydrogen give a characteristic NMR fingerprint, or pattern of peaks, many can be recognized by comparing the observed pattern with a library of patterns from known substances or by calculating the expected pattern of lines.

The Chemical Shift

The separation of the absorption into groups of lines is due to the different proton environments within the molecule. Thus, in ethanol, CH_3CH_2OH, three protons are present in the methyl group (CH_3), two are present in the methylene group (CH_2), and one is present in the hydroxyl group (OH). The applied magnetic field acts on the electrons in these three groups and causes them to circulate throughout the nuclear

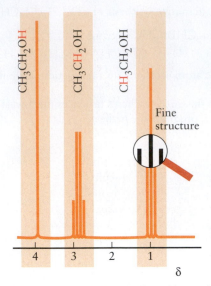

FIGURE 1 The NMR spectrum of ethanol. The red letters denote the protons that give rise to the associated peaks.

framework. These circulating charged particles give rise to an additional magnetic field, and so the protons in each of the three groups experience a local magnetic field that is not the same as the applied magnetic field. Because the electronic structure in each of the groups is different, the protons in each group experience slightly different local magnetic fields. As a result, slightly different radio-frequency fields are needed to bring them into resonance, and three groups of peaks are observed in the NMR spectrum.

We say that each group of protons has a characteristic *chemical shift*. The measurement of the chemical shift helps to identify the type of group responsible for the absorption and indicates what groups are present in the molecule. The chemical shift of a group of lines is expressed in terms of the δ *scale* (delta scale), which measures the difference in absorption frequency between the sample (ν) and a standard ($\nu°$):

$$\delta = \frac{\nu - \nu°}{\nu°} \times 10^6$$

The standard is typically tetramethylsilane, $Si(CH_3)_4$, which has a lot of protons and dissolves in many samples without reaction. Each group has a characteristic chemical shift, although the precise value depends on the other groups attached to the group of interest. For instance, if we observe a resonance at $\delta = 1$, we can be reasonably confident that it arises from a methyl group in an alcohol.*

The intensities (heights) of the peaks are proportional to the numbers of protons that they represent. The three peaks in the ethanol spectrum, for example, have overall intensities in the ratio 3:2:1, which is what we would expect for the three methyl, two methylene, and one hydroxyl protons.

The Fine Structure

The *fine structure* of the spectrum is the splitting of the resonance into sharp peaks. Note that the methyl resonance in ethanol at $\delta = 1$ consists of three peaks with intensities in the ratio 1:2:1. The fine structure arises from the presence of other magnetic nuclei close to the protons undergoing resonance. The fine structure of the methyl group in ethanol, for instance, arises from the presence of the protons in the neighboring methylene group.

Let's imagine that we are a proton in the methyl group in ethanol and can sense the spins of the two neighboring methylene ($-CH_2-$) group protons. There are four possible orientations of these two spins: $\alpha\alpha$, $\alpha\beta$, $\beta\alpha$, and $\beta\beta$. Suppose our neighbors have $\alpha\alpha$ spins. This arrangement gives rise to a magnetic field that we experience in addition to the applied magnetic field; so we (the methyl protons) come into resonance at the corresponding frequency. If we are in a molecule in which our neighbors are $\beta\beta$, we experience a different local magnetic field, and we come into resonance at a different frequency. If our neighbors happen to be either $\alpha\beta$ or $\beta\alpha$, the magnetic field arising from the α spin cancels the field arising from the β spin; so the local field that we experience is the same as it would be in the absence of any neighbors, and we resonate at the characteristic frequency. Because there are two arrangements that give this resonance ($\alpha\beta$ and $\beta\alpha$), the central line of the resonance will be twice as intense as the two outer lines (which arise when our neighbors are $\alpha\alpha$ and $\beta\beta$, respectively). Thus, we expect a 1:2:1 fine structure, just as observed. If there are three equivalent protons in a nearby group (as there are for ethanol's methylene group resonance, because the methylene group has a methyl group as a neighbor), we expect four lines in the intensity ratio 1:3:3:1, just as observed. Four equivalent neighboring protons would yield lines in the intensity ratio 1:4:6:4:1, five would produce intensities in the ratios 1:5:10:10:5:1, and so on.

The hydroxyl resonance is not split by the other protons in the molecule because its proton is very mobile: it can jump from one ethanol molecule to another ethanol molecule or to any water molecules present. As a result, the proton does not stay on one molecule long enough to show any characteristic splitting or to give rise to splitting in other groups.

Magnetic Resonance Imaging

Magnetic resonance imaging is a noninvasive structural technique for complex systems of molecules, such as people. In its simplest form, MRI portrays the concentration of protons in a sample. If the sample—which may be a living human body—is exposed to a uniform magnetic field in an NMR spectrometer and if we work at a resolution that does not show any chemical shifts or fine structure, then the protons

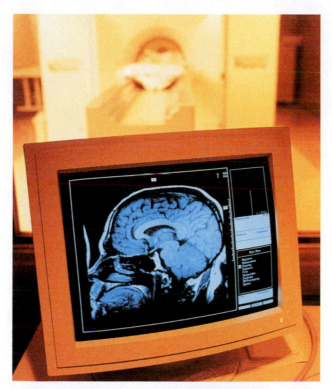

FIGURE 2 An MRI image of a human brain. The patient must lie within the strong magnetic field. The detectors can be rotated around the patient's head, thereby allowing many different views to be recorded.

give rise to a single resonance line. However, if the magnetic field varies linearly across the sample, the protons will resonate at different frequencies according to their location in the field. Moreover, the intensity of the resonance at a given field will be proportional to the number of protons at the spatial location corresponding to that particular field value. If the field gradient is rotated into different orientations, another portrayal of the concentration of protons through the sample is obtained. After several such measurements of the absorption intensity, the data can be analyzed on a computer, which constructs a two-dimensional image of any section, or "slice" through the sample. Figure 2 shows the distribution of protons—in large measure, the distribution of water in a brain—and the different regions can be identified. A major advantage of MRI over x-rays is that, if a series of slices are obtained, they can be assembled into a 3-D image that yields a much more accurate image of soft tissues than is possible with x-rays.

Related Exercises 19.87–19.95

---

* Chemical shifts are sometimes expressed in parts per million, such as $\delta = 1$ ppm for the methyl resonance; ignore the ppm in calculations.

APPENDIX 1: Symbols, Units, and Mathematical Techniques

1A   SYMBOLS

Each physical quantity is represented by an Italic or Greek symbol (thus, m for mass, not m). Table 1 lists most of the symbols used in this textbook together with their units (see also Appendix 1B). The symbols may be modified by attaching subscripts, as set out in Table 2. Fundamental constants are not included in the lists but can be found inside the back cover of the book.

TABLE 1   Common Symbols and Units

Symbol	Physical quantity	SI unit
α (alpha)	polarizability	$C^2 \cdot m^2 \cdot J^{-1}$
γ (gamma)	surface tension	$N \cdot m^{-1}$
δ (delta)	chemical shift	—
θ (theta)	colatitude	degree, rad
λ (lambda)	wavelength	m
μ (mu)	dipole moment	$C \cdot m$
ν (nu)	frequency	Hz
Π (pi)	osmotic pressure	Pa
σ (sigma)	cross section	m^2
ϕ (phi)	azimuth	degree (°), rad
χ (chi)	electronegativity	—
ψ (psi)	wavefunction	$m^{-n/2}$ (in n dimensions)
a	activity	—
	van der Waals parameter	$L^2 \cdot atm \cdot mol^{-2}$
	unit-cell parameter	m
A	area	m^2
	mass number	—
	Madelung constant	—
b	van der Waals parameter	$L \cdot mol^{-1}$
B	second virial coefficient	$L \cdot mol^{-1}$
C	heat capacity	$J \cdot K^{-1}$
	third virial coefficient	$L^2 \cdot mol^{-2}$
c	molar concentration, molarity	$mol \cdot L^{-1}$, M
c_2	second radiation constant	$K \cdot m$
d	density	$kg \cdot m^{-3}$ ($g \cdot cm^{-3}$)
	length of unit-cell diagonal	m
E	energy	J
	electrode potential	V
	electromotive force	V
E_a	activation energy	$J \cdot mol^{-1}$ ($kJ \cdot mol^{-1}$)
E_{bind}	nuclear binding energy	J
E_{ea}	electron affinity	$J \cdot mol^{-1}$ ($kJ \cdot mol^{-1}$)
E_K	kinetic energy	J
E_P	potential energy	J
e	elementary charge	C
F	force	N
G	Gibbs free energy	J
H	enthalpy	J
h	height	m
I	ionization energy	$J \cdot mol^{-1}$ ($kJ \cdot mol^{-1}$)
	electric current	A ($C \cdot s^{-1}$)
i	i factor	—

(continued)

TABLE 1 Common Symbols and Units   *(continued)*

Symbol	Physical quantity	SI unit
[J]	molarity	$mol \cdot L^{-1}$, M
k	rate constant	(depends on order)
	decay constant	s^{-1}
k_b	boiling-point constant	$K \cdot kg \cdot mol^{-1}$
k_f	freezing-point constant	$K \cdot kg \cdot mol^{-1}$
k_H	Henry's law constant	$mol \cdot L^{-1} \cdot atm^{-1}$
K	equilibrium constant	—
K_a	acidity constant	—
K_b	basicity constant	—
K_c	equilibrium constant	—
K_f	formation constant	—
K_M	Michaelis constant	—
K_p	equilibrium constant	—
K_{sp}	solubility product	—
K_w	water autoprotolysis constant	—
l, L	length	m
m	mass	kg
	molality	$mol \cdot kg^{-1}$, m
M	molar mass	$kg \cdot mol^{-1}$ $(g \cdot mol^{-1})$
N	number of entities	—
n	amount of substance	mol
p	linear momentum	$kg \cdot m \cdot s^{-1}$
P	pressure	Pa
P_A	partial pressure	Pa
q	heat	J
	electric charge	C
Q	reaction quotient	—
	relative biological effectiveness	—
r	radius	m
R	radial wavefunction	$m^{-3/2}$
S	entropy	$J \cdot K^{-1}$
	molar solubility	$mol \cdot L^{-1}$
s	dimensionless molar solubility	—
t	time	s
$t_{1/2}$	half-life	s
T	absolute temperature	K
U	internal energy	J
v	velocity	$m \cdot s^{-1}$
V	volume	m^3, L
w	work	J
x_A	mole fraction	—
Y	angular wavefunction	—
Z	compression factor	—
	atomic number	—

TABLE 2 Subscripts for Symbols

Subscript	Meaning	Example (units)
a	acid	acidity constant, K_a
b	base	basicity constant, K_b
	boiling	boiling temperature, T_b (K)
B	bond	bond enthalpy, ΔH_B $(kJ \cdot mol^{-1})$
bind	binding	binding energy, E_{bind} (eV)

(continued)

TABLE 2 Subscripts for Symbols   *(continued)*

Subscript	Meaning	Example (units)
c	concentration	equilibrium constant, K_c
	combustion	enthalpy of combustion, ΔH_c (kJ·mol^{-1})
	critical	critical temperature, T_c (K)
e	nonexpansion (extra) work	electrical work, w_e (J)
f	formation	enthalpy of formation, ΔH_f (kJ·mol^{-1})
		formation constant, K_f
	freezing	freezing temperature, T_f (K)
fus	fusion	enthalpy of fusion, ΔH_{fus} (kJ·mol^{-1})
H	Henry	Henry's law constant, k_H
In	indicator	indicator constant, K_{In}
K	kinetic	kinetic energy, E_K (J)
L	lattice	lattice enthalpy, ΔH_L (kJ·mol^{-1})
m	molar	molar volume, $V_m = V/n$ (L·mol^{-1})
M	Michaelis	Michaelis constant, K_M
mix	mixing	enthalpy of mixing, ΔH_{mix} (kJ·mol^{-1})
P	potential	potential energy, E_P (J)
P	constant pressure	heat capacity at constant pressure, C_P (J·K^{-1})
r	reaction	reaction enthalpy, ΔH_r (kJ·mol^{-1})
s	specific	specific heat capacity, $C_s = C/m$ (J·K^{-1}·g^{-1})
sol	solution	enthalpy of solution, ΔH_{sol} (kJ·mol^{-1})
sp	solubility product	solubility product, K_{sp}
sub	sublimation	enthalpy of sublimation, ΔH_{sub} (kJ·mol^{-1})
surr	surroundings	entropy of surroundings, S_{surr} (J·K^{-1})
tot	total	total entropy, S_{tot} (J·K^{-1})
V	constant volume	heat capacity at constant volume, C_V (J·K^{-1})
vap	vaporization	enthalpy of vaporization, ΔH_{vap} (kJ·mol^{-1})
w	water	water autoprotolysis constant, K_w
0	initial	initial concentration, $[A]_0$
	ground state	wavefunction, ψ_0

1B UNITS AND UNIT CONVERSIONS

Each physical quantity is reported as a multiple of a defined unit:

Physical quantity = numerical value × unit

For instance, a length may be expressed as a multiple of the unit of length, the meter, m; so we write $l = 2.0$ m. All units are denoted by roman letters, such as m for meter and s for second.

The Système International (SI) is the internationally accepted form of the metric system. It defines seven **base units** in terms of which all physical quantities can be expressed:

meter, m The meter, the unit of length, is the length of the path traveled by light during a time interval of 1/299 792 458 of a second.

kilogram, kg The kilogram, the unit of mass, is the mass of a standard cylinder maintained at a laboratory in France.

second, s The second, the unit of time, is 9 192 631 770 periods of a certain spectroscopic transition in a cesium-133 atom.

ampere, A The ampere, the unit of electric current, is defined in terms of the force exerted between two parallel wires carrying the current.

kelvin, K The kelvin, the unit of temperature, is 1/273.16 of the absolute temperature of the triple point of water.

mole, mol The mole, the unit of chemical amount, is the amount of substance that contains as many specified entities as there are atoms in exactly 12 g of carbon-12.

candela, cd The candela, the unit of luminous intensity, is defined in terms of a carefully specified source. We do not use the candela in this book.

Any unit may be modified by one of the prefixes given in Table 3, which denote multiplication or division by a power of 10 of the unit. Thus, 1 mm = 10^{-3} m and 1 MK = 10^6 K. Note that all the prefixes are upright (roman) rather than Italic.

TABLE 3 **Typical SI Prefixes**

Prefix:	deca-	kilo-	mega-	giga-	tera-
Abbreviation:	da	k	M	G	T
Factor:	10	10^3	10^6	10^9	10^{12}

Prefix:	deci-	centi-	milli-	micro-	nano-	pico-	femto-	atto-	zepto-
Abbreviation:	d	c	m	μ (mu)	n	p	f	a	z
Factor:	10^{-1}	10^{-2}	10^{-3}	10^{-6}	10^{-9}	10^{-12}	10^{-15}	10^{-18}	10^{-21}

Derived units are combinations of the base units (Section A). Table 4 lists some derived units. Note that the names of units derived from the names of people all begin with a lowercase letter but their abbreviations are uppercase.

TABLE 4 **Derived Units with Special Names**

Physical quantity	Name of unit	Abbreviation	Definition
absorbed dose	gray	Gy	$J \cdot kg^{-1}$
dose equivalent	sievert	Sv	$J \cdot kg^{-1}$
electric charge	coulomb	C	$A \cdot s$
electric potential	volt	V	$J \cdot C^{-1}$
energy	joule	J	$N \cdot m$, $kg \cdot m^2 \cdot s^{-2}$
force	newton	N	$kg \cdot m \cdot s^{-2}$
frequency	hertz	Hz	s^{-1}
power	watt	W	$J \cdot s^{-1}$
pressure	pascal	Pa	$N \cdot m^{-2}$, $kg \cdot m^{-1} \cdot s^{-2}$
volume	liter	L	dm^3

It is often necessary to convert one set of units (for instance, calories for energy and inches for length) into SI units. Table 5 lists some common conversions; the values in boldface type are exact.

TABLE 5 **Relations Between Units**

Physical quantity	Common unit	Abbreviation	SI equivalent
mass	pound	lb	**0.453 592 37 kg**
	tonne	t	**10^3 kg (1 Mg)**
	ton (short, U.S.)	ton	**907.184 74 kg**
	ton (long, U.K.)	ton	1016.046 kg
length	inch	in.	**2.54 cm**
	foot	ft	**30.48 cm**
volume	U.S. quart	qt	**0.946 352 5 L**
	U.S. gallon	gal	**3.785 41 L**
	Imperial quart	qt	**1.136 522 5 L**
	Imperial gallon	gal	**4.546 09 L**
time	minute	min	**60 s**
	hour	h	**3600 s**
energy	calorie (thermochemical)	cal	**4.184 J**
	electronvolt	eV	$1.602\ 177 \times 10^{-19}$ J
	kilowatt-hour	kWh	**3.6×10^6 J**
	liter-atmosphere	L·atm	**101.325 J**
pressure	torr	Torr	133.322 Pa
	atmosphere	atm	**101 325 Pa (760 Torr)**
	bar	bar	**10^5 Pa**
	pounds/square inch	psi	6894.76 Pa
power	horsepower	hp	745.7 W
dipole moment	debye	D	$3.335\ 64 \times 10^{-30}$ C·m

As explained in Section A, to convert between units, we use a **conversion factor** of the form

$$\text{Conversion factor} = \frac{\text{units required}}{\text{units given}}$$

When using a conversion factor, we treat the units just like algebraic quantities: they are multiplied or canceled in the normal way. Thus, the units in the denominator of the conversion factor cancel the units in the original data, leaving the units in the numerator of the conversion factor.

The conversion of temperatures is carried out slightly differently. Because the Fahrenheit degree (°F) is smaller than a Celsius degree by a factor of $\frac{5}{9}$ (because there are 180 Fahrenheit degrees between the freezing point and boiling point of water but only 100 Celsius degrees between the same two points) and because 0°C coincides with 32°F, we use

$$\text{Temperature (°F)} = \{\tfrac{9}{5} \times \text{temperature (°C)}\} + 32$$

(The 32 is exact.) For example, to convert 37°C (blood temperature) into degrees Fahrenheit, we write

$$\text{Temperature (°F)} = \{\tfrac{9}{5} \times 37\} + 32 = 99$$

and the temperature is reported as 99°F. A more sophisticated way of expressing the same relation is to write

$$\text{Temperature/°F} = \{\tfrac{9}{5} \times \text{temperature/°C}\} + 32$$

In this expression, we treat temperature units like numbers and cancel them when it is appropriate. The same conversion then becomes

$$\text{Temperature/°F} = \{\tfrac{9}{5} \times (37°C)/°C\} + 32$$
$$= \{\tfrac{9}{5} \times 37\} + 32 = 99$$

and multiplication through by °F gives

$$\text{Temperature} = 99°F$$

The corresponding expression for conversion between the Celsius and Kelvin scales is

$$\text{Temperature/°C} = \text{temperature/K} + 273.15$$

(The 273.15 is exact.) Note that the size of the degree Celsius is the same as that of the kelvin.

1C SCIENTIFIC NOTATION

In **scientific notation**, a number is written as $A \times 10^a$. Here A is a decimal number with one nonzero digit in front of the decimal point and a is a whole number. For example, 333 is written 3.33×10^2 in scientific notation, because $10^2 = 10 \times 10 = 100$:

$$333 = 3.33 \times 100 = 3.33 \times 10^2$$

We use

$$10^1 = 10$$
$$10^2 = 10 \times 10 = 100$$
$$10^3 = 10 \times 10 \times 10 = 1000$$
$$10^4 = 10 \times 10 \times 10 \times 10 = 10\,000$$

and so on. Note that the number of zeros following 1 is equal to the power of 10.

Numbers between 0 and 1 are expressed in the same way, but with a negative power of 10; they have the form $A \times 10^{-a}$, with $10^{-1} = 0.1$, and so on. Thus, 0.0333 in decimal notation is 3.33×10^{-2} because

$$10^{-2} = \frac{1}{10} \times \frac{1}{10} = \frac{1}{100}$$

and therefore

$$0.0333 = 3.33 \times \frac{1}{100} = 3.33 \times 10^{-2}$$

We use

$$10^{-2} = 10^{-1} \times 10^{-1} = 0.01$$
$$10^{-3} = 10^{-1} \times 10^{-1} \times 10^{-1} = 0.001$$
$$10^{-4} = 10^{-1} \times 10^{-1} \times 10^{-1} \times 10^{-1} = 0.0001$$

When a negative power of 10 is written out as a decimal number, the number of zeros following the decimal point is one less than the number (disregarding the sign) to which 10 is raised. Thus, 10^{-5} is written as a decimal point followed by $5 - 1 = 4$ zeros and then a 1:

$$10^{-5} = 10^{-1} \times 10^{-1} \times 10^{-1} \times 10^{-1} \times 10^{-1}$$
$$= 0.000\,01$$

The digits in a reported measurement are called the **significant figures**. There are two significant figures (written 2 sf) in 1.2 cm³ and 3 sf in 1.78 g. Section A describes how to find the number of significant figures in a measurement.

Some zeros are legitimately measured digits, but other zeros serve only to mark the place of the decimal point. Trailing zeros (the last ones after a decimal point), as in 22.0 mL, are significant, because they were measured. Thus, 22.0 mL has 3 sf. The "captive" zero in 80.1 kg is a measured digit, and so 80.1 kg has 3 sf. However, the leading digits in 0.0025 g are not significant; they are only placeholders used to indicate powers of 10, not measured numbers. We can see that they are only placeholders by reporting the mass as 2.5×10^{-3} g, which has 2 sf.

We distinguish between the results of measurements, which are always uncertain, and the results of *counting*, which are *exact*. For example, the report "12 eggs" means that there are exactly 12 eggs present, not a number somewhere between 11.5 and 12.5.

Some ambiguity arises with whole numbers ending in zero. Does a length reported as 400 m have 3 sf (4.00×10^2), 2 sf (4.0×10^2), or only 1 sf (4×10^2)? In such cases, the use of scientific notation removes any ambiguity. If it is not convenient to use scientific notation, a final decimal point can be used to indicate that every digit to the left of the decimal is significant. Thus, 400 m is ambiguous and cannot be taken to have more than 1 sf unless other information is given. However, 400. m unambiguously has 3 sf.

Different rounding-off rules are needed for addition (and its reverse, subtraction) and multiplication (and its reverse, division). In both procedures, we round off the answers to the correct number of significant figures.

Rounding off In calculations, round *up* if the last digit is above 5 and round *down* if it is below 5. For numbers ending in 5, always round to the nearest even number. For example, 2.35 rounds to 2.4 and 2.65 rounds to 2.6. In a calculation with multiple steps, round off only in the final step; if possible, carry all digits in the memory of the calculator until that stage.

Addition and subtraction When adding or subtracting, make sure that the number of decimal places in the result is the same as the *smallest number of decimal places* in the data. For example, 0.10 g + 0.024 g = 0.12 g.

Multiplication and division When multiplying or dividing, make sure that the number of significant figures in the result is the same as the *smallest number of significant figures* in the data. For example, $(8.62 \text{ g})/(2.0 \text{ cm}^3) = 4.3 \text{ g·cm}^{-3}$.

Integers and exact numbers In multiplication or division by an integer or an exact number, the uncertainty of the result is determined by the measured value. Some unit conversion factors are defined exactly, even though they are not whole numbers. For example, 1 in. is defined as *exactly* 2.54 cm and the 273.15 in the conversion between Celsius and Kelvin temperatures is exact; so 100.000°C converts into 373.150 K.

Logarithms and exponentials The mantissa of a common logarithm (the digits following the decimal point, see Appendix 1D) has the same number of significant figures as the original number. Thus, log 2.45 = 0.389. A common antilogarithm of a number has the same number of significant figures as the mantissa of the original number. Thus, $10^{0.389} = 2.45$ and $10^{12.389} = 2.45 \times 10^{12}$. There is no simple rule for assessing the correct number of significant figures when natural logarithms are used: one way is to convert natural logarithms into common logarithms and then to use the rules just specified.

1D EXPONENTS AND LOGARITHMS

To multiply numbers in scientific notation, multiply the decimal parts of the numbers and add the powers of 10:

$$(A \times 10^a) \times (B \times 10^b) = (A \times B) \times 10^{a+b}$$

An example is

$$(1.23 \times 10^2) \times (4.56 \times 10^3) = 1.23 \times 4.56 \times 10^{2+3}$$
$$= 5.61 \times 10^5$$

This rule also applies if the powers of 10 are negative:

$$(1.23 \times 10^{-2}) \times (4.56 \times 10^{-3}) = 1.23 \times 4.56 \times 10^{-2-3}$$
$$= 5.61 \times 10^{-5}$$

The results of such calculations are then adjusted so that one digit precedes the decimal point:

$$(4.56 \times 10^{-3}) \times (7.65 \times 10^6) = 34.88 \times 10^3$$
$$= 3.488 \times 10^4$$

When dividing two numbers in scientific notation, divide the decimal parts of the numbers and subtract the powers of 10:

$$\frac{A \times 10^a}{B \times 10^b} = \frac{A}{B} \times 10^{a-b}$$

An example is

$$\frac{4.31 \times 10^5}{9.87 \times 10^{-8}} = \frac{4.31}{9.87} \times 10^{5-(-8)} = 0.437 \times 10^{13}$$
$$= 4.37 \times 10^{12}$$

Before adding and subtracting numbers in scientific notation, we rewrite the numbers as decimal numbers multiplied by the same power of 10:

$$1.00 \times 10^3 + 2.00 \times 10^2 = 1.00 \times 10^3 + 0.200 \times 10^3$$
$$= 1.20 \times 10^3$$

When raising a number in scientific notation to a particular power, raise the decimal part of the number to the power and multiply the power of 10 by the power:

$$(A \times 10^a)^b = A^b \times 10^{a \times b}$$

For example, 2.88×10^4 raised to the third power is

$$(2.88 \times 10^4)^3 = 2.88^3 \times (10^4)^3 = 2.88^3 \times 10^{3 \times 4}$$
$$= 23.9 \times 10^{12} = 2.39 \times 10^{13}$$

The rule follows from the fact that

$$(10^4)^3 = 10^4 \times 10^4 \times 10^4 = 10^{4+4+4} = 10^{3 \times 4}$$

The **common logarithm** of a number x, denoted log x, is the power to which 10 must be raised to equal x. Thus, the logarithm of 100 is 2, written log 100 = 2, because 10^2 = 100. The logarithm of 1.5×10^2 is 2.18 because

$$10^{2.18} = 10^{0.18+2} = 10^{0.18} \times 10^2 = 1.5 \times 10^2$$

The number to the left of the decimal point in the logarithm (the 2 in log (1.5×10^2) = 2.18) is called the **characteristic** of the logarithm: it is the power of 10 in the original number (the power 2 in 1.5×10^2). The decimal fraction (the numbers to the right of the decimal point; such as 0.18 in our example) is called the **mantissa**. It is the logarithm of the decimal number written with one nonzero digit to the left of the decimal point (the 1.5 in the example).

The distinction between the characteristic and the mantissa is important when we have to decide how many significant figures to retain in a calculation that includes logarithms (as in the calculation of pH). Just as the power of 10 in a decimal number indicates only the location of the decimal point and plays no role in the determination of significant figures, so the characteristic of a logarithm is not included in the count of significant figures in a logarithm (see Appendix 1C). The number of significant figures in the mantissa is equal to the number of significant figures in the decimal number.

The **common antilogarithm** of a number x is the number that has x as its common logarithm. In practice, the common antilogarithm of x is simply another name for 10^x, and so the common antilogarithm of 2 is $10^2 = 100$ and that of 2.18 is

$$10^{2.18} = 10^{0.18 + 2} = 10^{0.18} \times 10^2 = 1.5 \times 10^2$$

The logarithm of a number greater than 1 is positive, and the logarithm of a number less than 1 (but greater than 0) is negative:

If $x > 1$, $\log x > 0$
If $x = 1$, $\log x = 0$
If $x < 1$, $\log x < 0$

Logarithms are not defined either for 0 or for negative numbers.

The **natural logarithm** of a number x, denoted $\ln x$, is the power to which the number $e = 2.718. . .$ must be raised to equal x. Thus, $\ln 10.0 = 2.303$, signifying that $e^{2.303} = 10.0$. The value of e may seem a peculiar choice, but it occurs naturally in a number of mathematical expressions, and its use simplifies many formulas. Common and natural logarithms are related by the expression

$$\ln x = \ln 10 \times \log x$$

In practice, a convenient approximation is

$$\ln x \approx 2.303 \times \log x$$

The **natural antilogarithm** of x is normally called the **exponential** of e; it is the value of e raised to the power x. Thus, the natural antilogarithm of 2.303 is $e^{2.303} = 10.0$.

The following relations between logarithms are useful. Written here mainly for common logarithms, they also apply to natural logarithms.

Relation	Example
$\log 10^x = x$	$\log 10^{-7} = -7$
$\ln e^x = x$	$\ln e^{-kt} = -kt$
$\log x + \log y = \log xy$	$\log [Ag^+] + \log [Cl^-] = \log [Ag^+][Cl^-]$
$\log x - \log y = \log (x/y)$	$\log A_0 - \log A = \log (A_0/A)$
$x \log y = \log y^x$	$2 \log [H^+] = \log ([H^+]^2)$
$\log(1/x) = -\log x$	$\log (1/[H^+]) = -\log [H^+]$

Logarithms are useful for solving expressions of the form

$$a^x = b$$

for the unknown x. (This type of calculation can arise in the study of chemical kinetics to determine the order of a reaction.) We take logarithms of both sides

$$\log a^x = \log b$$

and, from a relation given in the preceding table, we write the expression as

$$x \log a = \log b$$

Therefore,

$$x = \frac{\log b}{\log a}$$

1E EQUATIONS AND GRAPHS

A **quadratic equation** is an equation of the form

$$ax^2 + bx + c = 0$$

The two **roots** of the equation (the solutions) are given by the expression

$$x = \frac{-b \pm \sqrt{b^2 - 4ac}}{2a}$$

The roots can also be determined graphically (by using a graphing calculator, for instance) by noting where the graph of $y(x) = ax^2 + bx + c$ against x passes through $y = 0$ (Fig. 1). When a quadratic equation arises in connection with a chemical calculation, we accept only the root that leads to a physically plausible result. For example, if x is a concentration, then it must be a positive number, and a negative root can be ignored. A calculation may result in a cubic equation:

$$ax^3 + bx^2 + cx + d = 0$$

Cubic equations are often very tedious to solve exactly, and so it is better to use mathematical software, a graphing calculator, or a plotter, such as the one on the Web site for this book, and to identify the locations where the graph of $y(x)$ against x passes through $y = 0$ (Fig. 2).

Experimental data can often be analyzed most effectively from a graph. In many

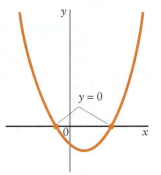

FIGURE 1 A graph of a function of the form $y(x) = ax^2 + bx + c$ passes through $y = 0$ at two points, which are the two roots of the quadratic equation $ax^2 + bx + c = 0$.

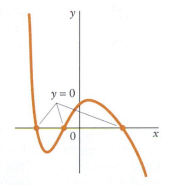

FIGURE 2 A graph of a function of the form $y(x) = ax^3 + bx^2 + cx + d$ passes through $y = 0$ at three points, which are the three roots of the cubic equation $ax^3 + bx^2 + cx + d = 0$.

cases, the best procedure is to find a way of plotting the data as a straight line. It is easier to tell whether the data in fact fall on the line, whereas small deviations from a curve are harder to detect. Moreover, it is also easy to calculate the slope of a straight line, to **extrapolate** (extend) a straight line beyond the range of the data, and to **interpolate** between the data points (that is, find a value between two measured values).

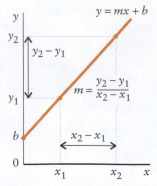

FIGURE 3 The straight line $y(x) = mx + b$; its intercept with the vertical axis at $x = 0$ is b and its slope is m.

The formula of a straight-line graph of y (the vertical axis) plotted against x (the horizontal axis) is

$$y = mx + b$$

Here b is the **intercept** of the line with the y-axis (Fig. 3), the value of y when $x = 0$. The **slope** of the graph, its gradient, is m. The slope can be calculated by choosing two points, x_1 and x_2, and their corresponding values on the y-axis, y_1 and y_2, and substituting the values into the formula

$$m = \frac{y_2 - y_1}{x_2 - x_1}$$

Because b is the intercept and m is the slope, the equation of the straight line is equivalent to

$$y = (\text{slope} \times x) + \text{intercept}$$

Many of the equations that we meet in the text can be rearranged to give a straight-line graph when plotted as shown in the following table.

Application	y	=	slope $\times x$	+	intercept
temperature scale conversions	temp./°C	=	$1 \times T/K$		-273.15
	temp./°F	=	$\frac{9}{5} \times$ temp./°C		$+32$
ideal gas law	P	=	$nRT \times (1/V)$		
first-order integrated rate law	$\ln[A]$	=	$-k \times t$		$+\ln [A]_0$
second-order integrated rate law	$1/[A]$	=	$k \times t$		$+1/[A]_0$
Arrhenius's equation	$\ln k$	=	$(-E_a/R) \times (1/T)$		$+\ln A$

The slope of a straight line is the same at all points. On a curve, however, the slope changes from point to point. The slope at a specified point is given by the slope of the line tangent to the curve at that point. The tangent can be found by a series of approximations, as shown in Fig. 4. We might start (approximation 1) by drawing a point on the curve on each side of the point of interest (corresponding to equal distances along the x-axis) and joining them by a straight line. A better approximation

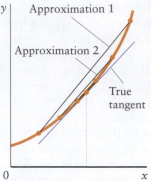

FIGURE 4 Successive approximations to the true tangent are obtained as the two points defining the straight line come closer together and finally coincide.

(approximation 2) is obtained by moving the points an equal distance closer to the point of interest and drawing a new line. The exact tangent is obtained when the two points virtually coincide with the point of interest. Its slope is then equal to the slope of the curve at the point of interest. This technique can be used to measure the rate of a chemical reaction at a specified time.

1F CALCULUS

Differential calculus is the part of mathematics that deals with the slopes of curves and with infinitesimal quantities. Suppose we are studying a function $y(x)$. As explained in Appendix 1E, the slope of its graph at a point can be calculated by considering the straight line joining two points x and $x + \delta x$, where δx is small. The slope of this line is

$$\text{Slope} = \frac{y(x + \delta x) - y(x)}{\delta x}$$

In differential calculus, the slope of the curve is found by letting the separation of the points become infinitesimally small. The **first derivative** of the function y with respect to x is then defined as

$$\frac{dy}{dx} = \lim_{\delta x \to 0} \frac{y(x + \delta x) - y(x)}{\delta x}$$

where "lim" means the limit of whatever follows—in this case, as x approaches zero. For example, if $y(x) = x^2$,

$$\frac{dy}{dx} = \lim_{\delta x \to 0} \frac{(x + \delta x)^2 - x^2}{\delta x}$$

$$= \lim_{\delta x \to 0} \frac{x^2 + 2x\delta x + (\delta x)^2 - x^2}{\delta x}$$

$$= \lim_{\delta x \to 0} \frac{2x\delta x + (\delta x)^2}{\delta x} = \lim_{\delta x \to 0} (2x + \delta x) = 2x$$

Therefore, the slope of the graph of the function $y = x^2$ at any point x is $2x$. The same procedure can be applied to other functions. However, in practice, it is usually more convenient to consult tables of first derivatives that have already been worked out. A selection of common functions and their first derivatives is given here.

Function, $y(x)$	Derivative, dy/dx
x^n	nx^{n-1}
$\ln x$	$1/x$
e^{ax}	ae^{ax}
$\sin ax$	$a \cos ax$
$\cos ax$	$-a \sin ax$

The **second derivative** of a function, denoted d^2y/dx^2, is defined like the first derivative but is applied to the function obtained by taking the first derivative. For example, the second derivative of the function x^2 is the derivative of the function $2x$, which is the constant 2. Likewise, the second derivative of $\sin ax$ is $-a^2 \sin ax$. The second derivative is an indication of the curvature of the function. Where d^2y/dx^2 is positive, the graph has a $\cup$ shape; where it is negative, the graph has a $\cap$ shape. The greater the magnitude of d^2y/dx^2, the sharper the curvature of the graph.

Integral calculus provides a way to determine the original function, given its first derivative. Thus, if we know that the first derivative is $2x$, then the integral calculus allows us to deduce that the function itself is $y = x^2$. Formally, we write

$$\int (2x)\,dx = x^2$$

It follows that the functions in the left-hand column of the preceding table are the integrals of the functions in the right-hand column. More formally, they are the **indefinite** integrals of the function, in contrast with the "definite" integrals described next. Tables of indefinite integrals may be consulted for more complex examples, and mathematical software or a graphing calculator can be used to evaluate them.

An integral has a further important interpretation: the integral of a function evaluated between two points is the *area* beneath the graph of the function between the two points (Fig. 5). For example, the area beneath the curve $y(x) = \sin x$ between $x = 0$ and $x = \pi$ is

$$\text{Area} = \int_0^\pi \sin x\,dx = \left. (-\cos x) \right|_0^\pi$$

$$= -\overbrace{\cos \pi}^{-1} - (-\overbrace{\cos 0}^{1}) = 1 + 1 = 2$$

An integral with limits attached, as in this example, is called a **definite integral**.

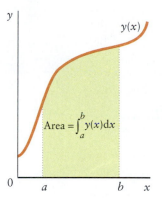

FIGURE 5 The definite integral of the function $y(x)$ between $x = a$ and $x = b$ is equal to the area bounded by the curve, the x-axis, and the two vertical lines through a and b.

APPENDIX 2: Experimental Data

2A   THERMODYNAMIC DATA AT 25°C

Inorganic Substances

Substance	Molar mass, M (g·mol^{-1})	Enthalpy of formation, $\Delta H_f°$ (kJ·mol^{-1})	Gibbs free energy of formation, $\Delta G_f°$ (kJ·mol^{-1})	Molar heat capacity, $C_{P,m}$ (J·K^{-1}·mol^{-1})	Molar entropy,* $S_m°$ (J·K^{-1}·mol^{-1})
Aluminum					
Al(s)	26.98	0	0	24.35	28.33
Al^{3+}(aq)	26.98	−524.7	−481.2	—	−321.7
Al$_2$O$_3$(s)	101.96	−1675.7	−1582.35	79.04	50.92
Al(OH)$_3$(s)	78.00	−1276	—	—	—
AlCl$_3$(s)	133.33	−704.2	−628.8	91.84	110.67
Antimony					
Sb(s)	121.76	0	0	25.23	45.69
SbH$_3$(g)	124.78	+145.11	+147.75	41.05	232.78
SbCl$_3$(g)	228.11	−313.8	−301.2	76.69	337.80
SbCl$_5$(g)	299.01	−394.34	−334.29	121.13	401.94
Arsenic					
As(s), gray	74.92	0	0	24.64	35.1
As$_2$S$_3$(s)	246.05	−169.0	−168.6	116.3	163.6
AsO$_4$$^{3-}$(aq)	138.92	−888.14	−648.41	—	−162.8
Barium					
Ba(s)	137.33	0	0	28.07	62.8
Ba^{2+}(aq)	137.33	−537.64	−560.77	—	+9.6
BaO(s)	153.33	−553.5	−525.1	47.78	70.42
BaCO$_3$(s)	197.34	−1216.3	−1137.6	85.35	112.1
BaCO$_3$(aq)	197.34	−1214.78	−1088.59	—	−47.3
Boron					
B(s)	10.81	0	0	11.09	5.86
B$_2$O$_3$(s)	69.62	−1272.8	−1193.7	62.93	53.97
BF$_3$(g)	67.81	−1137.0	−1120.3	50.46	254.12
Bromine					
Br$_2$(l)	159.80	0	0	75.69	152.23
Br$_2$(g)	159.80	+30.91	+3.11	36.02	245.46
Br(g)	79.90	+111.88	+82.40	20.79	175.02
Br$^-$(aq)	79.90	−121.55	−103.96	—	+82.4
HBr(g)	80.91	−36.40	−53.45	29.14	198.70
Calcium					
Ca(s)	40.08	0	0	25.31	41.42
Ca(g)	40.08	+178.2	+144.3	20.79	154.88
Ca^{2+}(aq)	40.08	−542.83	−553.58	—	−53.1

(continued)

Inorganic Substances (continued)

Substance	Molar mass, M (g·mol^{-1})	Enthalpy of formation, $\Delta H_f°$ (kJ·mol^{-1})	Gibbs free energy of formation, $\Delta G_f°$ (kJ·mol^{-1})	Molar heat capacity, $C_{P,m}$ (J·K^{-1}·mol^{-1})	Molar entropy,* $S_m°$ (J·K^{-1}·mol^{-1})
CaO(s)	56.08	−635.09	−604.03	42.80	39.75
Ca(OH)$_2$(s)	74.10	−986.09	−898.49	87.49	83.39
Ca(OH)$_2$(aq)	74.10	−1002.82	−868.07	—	−74.5
CaCO$_3$(s), calcite	100.09	−1206.9	−1128.8	81.88	92.9
CaCO$_3$(s), aragonite	100.09	−1207.1	−1127.8	81.25	88.7
CaCO$_3$(aq)	100.09	−1219.97	−1081.39	—	−110.0
CaF$_2$(s)	78.08	−1219.6	−1167.3	67.03	68.87
CaF$_2$(aq)	78.08	−1208.09	−1111.15	—	−80.8
CaCl$_2$(s)	110.98	−795.8	−748.1	72.59	104.6
CaCl$_2$(aq)	110.98	−877.1	−816.0	—	59.8
CaBr$_2$(s)	199.88	−682.8	−663.6	72.59	130
CaC$_2$(s)	64.10	−59.8	−64.9	62.72	69.96
CaSO$_4$(s)	136.14	−1434.11	−1321.79	99.66	106.7
CaSO$_4$(aq)	136.14	−1452.10	−1298.10	—	−33.1
Carbon (for organic compounds, see the next table)					
C(s), graphite	12.01	0	0	8.53	5.740
C(s), diamond	12.01	+1.895	+2.900	6.11	2.377
C(g)	12.01	+716.68	+671.26	20.84	158.10
CO(g)	28.01	−110.53	−137.17	29.14	197.67
CO$_2$(g)	44.01	−393.51	−394.36	37.11	213.74
CO$_3^{2-}$(aq)	60.01	−677.14	−527.81	—	−56.9
CCl$_4$(l)	153.81	−135.44	−65.21	131.75	216.40
CS$_2$(l)	76.15	+89.70	+65.27	75.7	151.34
HCN(g)	27.03	+135.1	+124.7	35.86	201.78
HCN(l)	27.03	+108.87	+124.97	70.63	112.84
HCN(aq)	27.03	+107.1	+119.7	—	124.7
Cerium					
Ce(s)	140.12	0	0	26.94	72.0
Ce^{3+}(aq)	140.12	−696.2	−672.0	—	−205
Ce^{4+}(aq)	140.12	−537.2	−503.8	—	−301
Chlorine					
Cl$_2$(g)	70.90	0	0	33.91	223.07
Cl(g)	35.45	+121.68	+105.68	21.84	165.20
Cl$^-$(aq)	35.45	−167.16	−131.23	—	+56.5
HCl(g)	36.46	−92.31	−95.30	29.12	186.91
HCl(aq)	36.46	−167.16	−131.23	—	56.5
Copper					
Cu(s)	63.55	0	0	24.44	33.15
Cu$^+$(aq)	63.55	+71.67	+49.98	—	+40.6
Cu^{2+}(aq)	63.55	+64.77	+65.49	—	−99.6
Cu$_2$O(s)	143.10	−168.6	−146.0	63.64	93.14
CuO(s)	79.55	−157.3	−129.7	42.30	42.63
CuSO$_4$(s)	159.61	−771.36	−661.8	100.0	109
CuSO$_4$·5H$_2$O(s)	249.69	−2279.7	−1879.7	280	300.4
Deuterium					
D$_2$(g)	4.028	0	0	29.20	144.96
D$_2$O(g)	20.028	−249.20	−234.54	34.27	198.34
D$_2$O(l)	20.028	−294.60	−243.44	34.27	75.94

Substance	Molar mass, M (g·mol^{-1})	Enthalpy of formation, $\Delta H_f°$ (kJ·mol^{-1})	Gibbs free energy of formation, $\Delta G_f°$ (kJ·mol^{-1})	Molar heat capacity, $C_{P,m}$ (J·K^{-1}·mol^{-1})	Molar entropy,* $S_m°$ (J·K^{-1}·mol^{-1})
Fluorine					
$F_2(g)$	38.00	0	0	31.30	202.78
$F^-(aq)$	19.00	−332.63	−278.79	—	−13.8
$HF(g)$	20.01	−271.1	−273.2	29.13	173.78
$HF(aq)$	20.01	−330.08	−296.82	—	88.7
Hydrogen (see also Deuterium)					
$H_2(g)$	2.0158	0	0	28.82	130.68
$H(g)$	1.0079	+217.97	+203.25	20.78	114.71
$H^+(aq)$	1.0079	0	0	0	0
$H_2O(l)$	18.02	−285.83	−237.13	75.29	69.91
$H_2O(g)$	18.02	−241.82	−228.57	33.58	188.83
$H_2O_2(l)$	34.02	−187.78	−120.35	89.1	109.6
$H_2O_2(aq)$	34.02	−191.17	−134.03	—	143.9
$H_3O^+(aq)$	19.02	−285.83	−237.13	75.29	+69.91
Iodine					
$I_2(s)$	253.80	0	0	54.44	116.14
$I_2(g)$	253.80	+62.44	+19.33	36.90	260.69
$I^-(aq)$	126.90	−55.19	−51.57	—	+111.3
$HI(g)$	127.91	+26.48	−1.70	29.16	206.59
Iron					
$Fe(s)$	55.84	0	0	25.10	27.28
$Fe^{2+}(aq)$	55.84	−89.1	−78.90	—	−137.7
$Fe^{3+}(aq)$	55.84	−48.5	−4.7	—	−315.9
$Fe_3O_4(s)$, magnetite	231.52	−1118.4	−1015.4	143.43	146.4
$Fe_2O_3(s)$, hematite	159.68	−824.2	−742.2	103.85	87.40
$FeS(s, \alpha)$	87.90	−100.0	−100.4	50.54	60.29
$FeS(aq)$	87.90	—	+6.9	—	—
$FeS_2(s)$	119.96	−178.2	−166.9	62.17	52.93
Lead					
$Pb(s)$	207.2	0	0	26.44	64.81
$Pb^{2+}(aq)$	207.2	−1.7	−24.43	—	+10.5
$PbO_2(s)$	239.2	−277.4	−217.33	64.64	68.6
$PbSO_4(s)$	303.3	−919.94	−813.14	103.21	148.57
$PbBr_2(s)$	367.0	−278.7	−261.92	80.12	161.5
$PbBr_2(aq)$	367.0	−244.8	−232.34	—	175.3
Magnesium					
$Mg(s)$	24.31	0	0	24.89	32.68
$Mg(g)$	24.31	+147.70	−113.10	20.79	148.65
$Mg^{2+}(aq)$	24.31	−466.85	−454.8	—	−138.1
$MgO(s)$	40.31	−601.70	−569.43	37.15	26.94
$MgCO_3(s)$	84.32	−1095.8	−1012.1	75.52	65.7
$MgBr_2(s)$	184.11	−524.3	−503.8	—	117.2
Mercury					
$Hg(l)$	200.59	0	0	27.98	76.02
$Hg(g)$	200.59	+61.32	+31.82	20.79	174.96
$HgO(s)$	216.59	−90.83	−58.54	44.06	70.29
$Hg_2Cl_2(s)$	472.08	−265.22	−210.75	102	192.5

(continued)

Inorganic Substances *(continued)*

Substance	Molar mass, M (g·mol^{-1})	Enthalpy of formation, $\Delta H_f°$ (kJ·mol^{-1})	Gibbs free energy of formation, $\Delta G_f°$ (kJ·mol^{-1})	Molar heat capacity, $C_{P,m}$ (J·K^{-1}·mol^{-1})	Molar entropy,* $S_m°$ (J·K^{-1}·mol^{-1})
Nitrogen					
$N_2(g)$	28.02	0	0	29.12	191.61
$NO(g)$	30.01	+90.25	+86.55	29.84	210.76
$N_2O(g)$	44.02	+82.05	+104.20	38.45	219.85
$NO_2(g)$	46.01	+33.18	+51.31	37.20	240.06
$N_2O_4(g)$	92.02	+9.16	+97.89	77.28	304.29
$HNO_3(l)$	63.02	−174.10	−80.71	109.87	155.60
$HNO_3(aq)$	63.02	−207.36	−111.25	—	146.4
$NO_3^-(aq)$	62.02	−205.0	−108.74	—	+146.4
$NH_3(g)$	17.03	−46.11	−16.45	35.06	192.45
$NH_3(aq)$	17.03	−80.29	−26.50	—	111.3
$NH_4^+(aq)$	18.04	−132.51	−79.31	—	+113.4
$NH_2OH(s)$	33.03	−114.2	—	—	—
$HN_3(g)$	43.04	+294.1	+328.1	98.87	238.97
$N_2H_4(l)$	32.05	+50.63	+149.34	139.3	121.21
$NH_4NO_3(s)$	80.05	−365.56	−183.87	84.1	151.08
$NH_4Cl(s)$	53.49	−314.43	−202.87	—	94.6
$NH_4ClO_4(s)$	117.49	−295.31	−88.75	—	186.2
Oxygen					
$O_2(g)$	32.00	0	0	29.36	205.14
$O_3(g)$	48.00	+142.7	+163.2	39.29	238.93
$OH^-(aq)$	17.01	−229.99	−157.24	—	−10.75
Phosphorus					
$P(s)$, white	30.97	0	0	23.84	41.09
$P_4(g)$	123.88	+58.91	+24.44	67.15	279.98
$PH_3(g)$	33.99	+5.4	+13.4	37.11	210.23
$P_4O_{10}(s)$	283.88	−2984.0	−2697.0	—	228.86
$H_3PO_3(aq)$	81.99	−964.8	—	—	—
$H_3PO_4(l)$	97.99	−1266.9	—	—	—
$H_3PO_4(aq)$	97.99	−1288.34	−1142.54	—	158.2
$PCl_3(l)$	137.32	−319.7	−272.3	—	217.18
$PCl_3(g)$	137.32	−287.0	−267.8	71.84	311.78
$PCl_5(g)$	208.22	−374.9	−305.0	112.8	364.6
$PCl_5(s)$	208.22	−443.5	—	—	—
Potassium					
$K(s)$	39.10	0	0	29.58	64.18
$K(g)$	39.10	+89.24	+60.59	20.79	160.34
$K^+(aq)$	39.10	−252.38	−283.27	—	+102.5
$KOH(s)$	56.11	−424.76	−379.08	64.9	78.9
$KOH(aq)$	56.11	−482.37	−440.50	—	91.6
$KF(s)$	58.10	−567.27	−537.75	49.04	66.57
$KCl(s)$	74.55	−436.75	−409.14	51.30	82.59
$KBr(s)$	119.00	−393.80	−380.66	52.30	95.90
$KI(s)$	166.00	−327.90	−324.89	52.93	106.32
$KClO_3(s)$	122.55	−397.73	−296.25	100.25	143.1
$KClO_4(s)$	138.55	−432.75	−303.09	112.38	151.0
$K_2S(s)$	110.26	−380.7	−364.0	—	105
$K_2S(aq)$	110.26	−471.5	−480.7	—	190.4
Silicon					
$Si(s)$	28.09	0	0	20.00	18.83
$SiO_2(s, \alpha)$	60.09	−910.94	−856.64	44.43	41.84

Substance	Molar mass, M (g·mol^{-1})	Enthalpy of formation, ΔH_f° (kJ·mol^{-1})	Gibbs free energy of formation, ΔG_f° (kJ·mol^{-1})	Molar heat capacity, $C_{P,m}$ (J·K^{-1}·mol^{-1})	Molar entropy,* S_m° (J·K^{-1}·mol^{-1})
Silver					
Ag(s)	107.87	0	0	25.35	42.55
Ag$^+$(aq)	107.87	+105.58	+77.11	—	+72.68
Ag$_2$O(s)	231.74	−31.05	−11.20	65.86	121.3
AgBr(s)	187.77	−100.37	−96.90	52.38	107.1
AgBr(aq)	187.77	−15.98	−26.86	—	155.2
AgCl(s)	143.32	−127.07	−109.79	50.79	96.2
AgCl(aq)	143.32	−61.58	−54.12	—	129.3
AgI(s)	234.77	−61.84	−66.19	56.82	115.5
AgI(aq)	234.77	+50.38	+25.52	—	184.1
AgNO$_3$(s)	169.88	−124.39	−33.41	93.05	140.92
Sodium					
Na(s)	22.99	0	0	28.24	51.21
Na(g)	22.99	+107.32	+76.76	20.79	153.71
Na$^+$(aq)	22.99	−240.12	−261.91	—	+59.0
NaOH(s)	40.00	−425.61	−379.49	59.54	64.46
NaOH(aq)	40.00	−470.11	−419.15	—	48.1
NaCl(s)	58.44	−411.15	−384.14	50.50	72.13
NaBr(s)	102.89	−361.06	−348.98	51.38	86.82
NaI(s)	149.89	−287.78	−286.06	52.09	98.53
Sulfur					
S(s), rhombic	32.06	0	0	22.64	31.80
S(s), monoclinic	32.06	+0.33	+0.1	23.6	32.6
S^{2-}(aq)	32.06	+33.1	+85.8	—	−14.6
SO$_2$(g)	64.06	−296.83	−300.19	39.87	248.22
SO$_3$(g)	80.06	−395.72	−371.06	50.67	256.76
H$_2$SO$_4$(l)	98.08	−813.99	−690.00	138.9	156.90
SO$_4^{2-}$(aq)	96.06	−909.27	−744.53	—	+20.1
HSO$_4^-$(aq)	97.07	−887.34	−755.91	—	+131.8
H$_2$S(g)	34.08	−20.63	−33.56	34.23	205.79
H$_2$S(aq)	34.08	−39.7	−27.83	—	121
SF$_6$(g)	146.06	−1209	−1105.3	97.28	291.82
Tin					
Sn(s), white	118.71	0	0	26.99	51.55
Sn(s), gray	118.71	−2.09	+0.13	25.77	44.14
SnO(s)	134.71	−285.8	−256.9	44.31	56.5
SnO$_2$(s)	150.71	−580.7	−519.6	52.59	52.3
Zinc					
Zn(s)	65.41	0	0	25.40	41.63
Zn^{2+}(aq)	65.41	−153.89	−147.06	—	−112.1
ZnO(s)	81.41	−348.28	−318.30	40.25	43.64

*The entropies of individual ions in solution are determined by setting the entropy of H$^+$ in water equal to 0 and then defining the entropies of all other ions relative to this value; hence a negative entropy is one that is lower than the entropy of H$^+$ in water. All *absolute* entropies are positive, and no sign need be given; all entropies of ions are relative to that of H$^+$ and are listed here with a sign (either + or −).

Organic Compounds

Substance	Molar mass, M (g·mol^{-1})	Enthalpy of combustion, $\Delta H_c°$ (kJ·mol^{-1})	Enthalpy of formation, $\Delta H_f°$ (kJ·mol^{-1})	Gibbs free energy of formation, $\Delta G_f°$ (kJ·mol^{-1})	Molar heat capacity, $C_{P,m}$ (J·K^{-1}·mol^{-1})	Molar entropy, $S_m°$ (J·K^{-1}·mol^{-1})
Hydrocarbons						
CH$_4$(g), methane	16.04	−890	−74.81	−50.72	35.31	186.26
C$_2$H$_2$(g), ethyne (acetylene)	26.04	−1300	+226.73	+209.20	43.93	200.94
C$_2$H$_4$(g), ethene (ethylene)	28.05	−1411	+52.26	+68.15	43.56	219.56
C$_2$H$_6$(g), ethane	30.07	−1560	−84.68	−32.82	52.63	229.60
C$_3$H$_6$(g), propene (propylene)	42.08	−2058	+20.42	+62.78	63.89	266.6
C$_3$H$_6$(g), cyclopropane	42.08	−2091	+53.30	+104.45	55.94	237.4
C$_3$H$_8$(g), propane	44.09	−2220	−103.85	−23.49	73.5	270.2
C$_4$H$_{10}$(g), butane	58.12	−2878	−126.15	−17.03	97.45	310.1
C$_5$H$_{12}$(g), pentane	72.14	−3537	−146.44	−8.20	120.2	349
C$_6$H$_6$(l), benzene	78.11	−3268	+49.0	+124.3	136.1	173.3
C$_6$H$_6$(g)	78.11	−3302	+82.9	+129.72	81.67	269.31
C$_7$H$_8$(l), toluene	92.13	−3910	+12.0	+113.8	—	221.0
C$_7$H$_8$(g)	92.13	−3953	+50.0	+122.0	103.6	320.7
C$_6$H$_{12}$(l), cyclohexane	84.15	−3920	−156.4	+26.7	156.5	204.4
C$_6$H$_{12}$(g)	84.15	−3953	—	—	—	—
C$_8$H$_{18}$(l), octane	114.22	−5471	−249.9	+6.4	—	358
Alcohols and phenols						
CH$_3$OH(l), methanol	32.04	−726	−238.86	−166.27	81.6	126.8
CH$_3$OH(g)	32.04	−764	−200.66	−161.96	43.89	239.81
C$_2$H$_5$OH(l), ethanol	46.07	−1368	−277.69	−174.78	111.46	160.7
C$_2$H$_5$OH(g)	46.07	−1409	−235.10	−168.49	65.44	282.70
C$_6$H$_5$OH(s), phenol	94.11	−3054	−164.6	−50.42	—	144.0
Carboxylic acids						
HCOOH(l), formic acid	46.02	−255	−424.72	−361.35	99.04	128.95
CH$_3$COOH(l), acetic acid	60.05	−875	−484.5	−389.9	124.3	159.8
CH$_3$COOH(aq)	60.05	—	−485.76	−396.46	—	86.6
(COOH)$_2$(s), oxalic acid	90.04	−254	−827.2	−697.9	117	120
C$_6$H$_5$COOH(s), benzoic acid	122.12	−3227	−385.1	−245.3	146.8	167.6
Aldehydes and ketones						
HCHO(g), methanal (formaldehyde)	30.03	−571	−108.57	−102.53	35.40	218.77
CH$_3$CHO(l), ethanal (acetaldehyde)	44.05	−1166	−192.30	−128.12	—	160.2
CH$_3$CHO(g)	44.05	−1192	−166.19	−128.86	57.3	250.3
CH$_3$COCH$_3$(l), propanone (acetone)	58.08	−1790	−248.1	−155.4	124.7	200

Potentials in Alphabetical Order

Reduction half-reaction	$E°$ (V)	Reduction half-reaction	$E°$ (V)
$Ag^+ + e^- \rightarrow Ag$	+0.80	$In^{2+} + e^- \rightarrow In^+$	−0.40
$Ag^{2+} + e^- \rightarrow Ag^+$	+1.98	$In^{3+} + e^- \rightarrow In^{2+}$	−0.49
$AgBr + e^- \rightarrow Ag + Br^-$	+0.07	$In^{3+} + 2\,e^- \rightarrow In^+$	−0.44
$AgCl + e^- \rightarrow Ag + Cl^-$	+0.22	$In^{3+} + 3\,e^- \rightarrow In$	−0.34
$AgF + e^- \rightarrow Ag + F^-$	+0.78	$K^+ + e^- \rightarrow K$	−2.93
$AgI + e^- \rightarrow Ag + I^-$	−0.15	$La^{3+} + 3\,e^- \rightarrow La$	−2.52
$Al^{3+} + 3\,e^- \rightarrow Al$	−1.66	$Li^+ + e^- \rightarrow Li$	−3.05
$Au^+ + e^- \rightarrow Au$	+1.69	$Mg^{2+} + 2\,e^- \rightarrow Mg$	−2.36
$Au^{3+} + 3\,e^- \rightarrow Au$	+1.40	$Mn^{2+} + 2\,e^- \rightarrow Mn$	−1.18
$Ba^{2+} + 2\,e^- \rightarrow Ba$	−2.91	$Mn^{3+} + e^- \rightarrow Mn^{2+}$	+1.51
$Be^{2+} + 2\,e^- \rightarrow Be$	−1.85	$MnO_2 + 4\,H^+ + 2\,e^- \rightarrow Mn^{2+} + 2\,H_2O$	+1.23
$Bi^{3+} + 3\,e^- \rightarrow Bi$	+0.20	$MnO_4^- + e^- \rightarrow MnO_4^{2-}$	+0.56
$Br_2 + 2\,e^- \rightarrow 2\,Br^-$	+1.09	$MnO_4^- + 8\,H^+ + 5\,e^- \rightarrow Mn^{2+} + 4\,H_2O$	+1.51
$BrO^- + H_2O + 2\,e^- \rightarrow Br^- + 2\,OH^-$	+0.76	$MnO_4^{2-} + 2\,H_2O + 2\,e^- \rightarrow MnO_2 + 4\,OH^-$	+0.60
$Ca^{2+} + 2\,e^- \rightarrow Ca$	−2.87	$NO_3^- + 2\,H^+ + e^- \rightarrow NO_2 + H_2O$	+0.80
$Cd^{2+} + 2\,e^- \rightarrow Cd$	−0.40	$NO_3^- + 4\,H^+ + 3\,e^- \rightarrow NO + 2\,H_2O$	+0.96
$Cd(OH)_2 + 2\,e^- \rightarrow Cd + 2\,OH^-$	−0.81	$NO_3^- + H_2O + 2\,e^- \rightarrow NO_2^- + 2\,OH^-$	+0.01
$Ce^{3+} + 3\,e^- \rightarrow Ce$	−2.48	$Na^+ + e^- \rightarrow Na$	−2.71
$Ce^{4+} + e^- \rightarrow Ce^{3+}$	+1.61	$Ni^{2+} + 2\,e^- \rightarrow Ni$	−0.23
$Cl_2 + 2\,e^- \rightarrow 2\,Cl^-$	+1.36	$Ni(OH)_3 + e^- \rightarrow Ni(OH)_2 + OH^-$	+0.49
$ClO^- + H_2O + 2\,e^- \rightarrow Cl^- + 2\,OH^-$	+0.89	$O_2 + e^- \rightarrow O_2^-$	−0.56
$ClO_4^- + 2\,H^+ + 2\,e^- \rightarrow ClO_3^- + H_2O$	+1.23	$O_2 + 4\,H^+ + 4\,e^- \rightarrow 2\,H_2O$	+1.23
$ClO_4^- + H_2O + 2\,e^- \rightarrow ClO_3^- + 2\,OH^-$	+0.36	$O_2 + H_2O + 2\,e^- \rightarrow HO_2^- + OH^-$	−0.08
$Co^{2+} + 2\,e^- \rightarrow Co$	−0.28	$O_2 + 2\,H_2O + 4\,e^- \rightarrow 4\,OH^-$	+0.40
$Co^{3+} + e^- \rightarrow Co^{2+}$	+1.81	$O_3 + 2\,H^+ + 2\,e^- \rightarrow O_2 + H_2O$	+2.07
$Cr^{2+} + 2\,e^- \rightarrow Cr$	−0.91	$O_3 + H_2O + 2\,e^- \rightarrow O_2 + 2\,OH^-$	+1.24
$Cr_2O_7^{2-} + 14\,H^+ + 6\,e^- \rightarrow 2\,Cr^{3+} + 7\,H_2O$	+1.33	$Pb^{2+} + 2\,e^- \rightarrow Pb$	−0.13
$Cr^{3+} + 3\,e^- \rightarrow Cr$	−0.74	$Pb^{4+} + 2\,e^- \rightarrow Pb^{2+}$	+1.67
$Cr^{3+} + e^- \rightarrow Cr^{2+}$	−0.41	$PbSO_4 + 2\,e^- \rightarrow Pb + SO_4^{2-}$	−0.36
$Cs^+ + e^- \rightarrow Cs$	−2.92	$Pt^{2+} + 2\,e^- \rightarrow Pt$	+1.20
$Cu^+ + e^- \rightarrow Cu$	+0.52	$Pu^{4+} + e^- \rightarrow Pu^{3+}$	+0.97
$Cu^{2+} + 2\,e^- \rightarrow Cu$	+0.34	$Ra^{2+} + 2\,e^- \rightarrow Ra$	−2.92
$Cu^{2+} + e^- \rightarrow Cu^+$	+0.15	$Rb^+ + e^- \rightarrow Rb$	−2.93
$F_2 + 2\,e^- \rightarrow 2\,F^-$	+2.87	$S + 2\,e^- \rightarrow S^{2-}$	−0.48
$Fe^{2+} + 2\,e^- \rightarrow Fe$	−0.44	$SO_4^{2-} + 4\,H^+ + 2\,e^- \rightarrow H_2SO_3 + H_2O$	+0.17
$Fe^{3+} + 3\,e^- \rightarrow Fe$	−0.04	$S_2O_8^{2-} + 2\,e^- \rightarrow 2\,SO_4^{2-}$	+2.05
$Fe^{2+} + e^- \rightarrow Fe^{2+}$	+0.77	$Se + 2\,e^- \rightarrow Se^{2-}$	−0.67
$Ga^+ + e^- \rightarrow Ga$	−0.53	$Sn^{2+} + 2\,e^- \rightarrow Sn$	−0.14
$2\,H^+ + 2\,e^- \rightarrow H_2$	0, by definition	$Sn^{4+} + 2\,e^- \rightarrow Sn^{2+}$	+0.15
$2\,HBrO + 2\,H^+ + 2\,e^- \rightarrow Br_2 + 2\,H_2O$	+1.60	$Sr^{2+} + 2\,e^- \rightarrow Sr$	−2.89
$2\,HClO + 2\,H^+ + 2\,e^- \rightarrow Cl_2 + 2\,H_2O$	+1.63	$Te + 2\,e^- \rightarrow Te^{2-}$	−0.84
$2\,H_2O + 2\,e^- \rightarrow H_2 + 2\,OH^-$	−0.83	$Ti^{2+} + 2\,e^- \rightarrow Ti$	−1.63
$H_2O_2 + 2\,H^+ + 2\,e^- \rightarrow 2\,H_2O$	+1.78	$Ti^{3+} + e^- \rightarrow Ti^{2+}$	−0.37
$H_4XeO_6 + 2\,H^+ + 2\,e^- \rightarrow XeO_3 + 3\,H_2O$	+3.0	$Ti^{4+} + e^- \rightarrow Ti^{3+}$	0.00
$Hg_2^{2+} + 2\,e^- \rightarrow 2\,Hg$	+0.79	$Tl^+ + e^- \rightarrow Tl$	−0.34
$Hg^{2+} + 2\,e^- \rightarrow Hg$	+0.85	$U^{3+} + 3\,e^- \rightarrow U$	−1.79
$2\,Hg^{2+} + 2\,e^- \rightarrow Hg_2^{2+}$	+0.92	$U^{4+} + e^- \rightarrow U^{3+}$	−0.61
$Hg_2Cl_2 + 2\,e^- \rightarrow 2\,Hg + 2\,Cl^-$	+0.27	$V^{2+} + 2\,e^- \rightarrow V$	−1.19
$I_2 + 2\,e^- \rightarrow 2\,I^-$	+0.54	$V^{3+} + e^- \rightarrow V^{2+}$	−0.26
$I_3^- + 2\,e^- \rightarrow 3\,I^-$	+0.53	$Zn^{2+} + 2\,e^- \rightarrow Zn$	−0.76
$In^+ + e^- \rightarrow In$	−0.14		

2C GROUND-STATE ELECTRON CONFIGURATIONS*

Z	Symbol	Configuration	Z	Symbol	Configuration
1	H	$1s^1$	29	Cu	$[Ar]3d^{10}4s^1$
2	He	$1s^2$	30	Zn	$[Ar]3d^{10}4s^2$
3	Li	$[He]2s^1$	31	Ga	$[Ar]3d^{10}4s^24p^1$
4	Be	$[He]2s^2$	32	Ge	$[Ar]3d^{10}4s^24p^2$
5	B	$[He]2s^22p^1$	33	As	$[Ar]3d^{10}4s^24p^3$
6	C	$[He]2s^22p^2$	34	Se	$[Ar]3d^{10}4s^24p^4$
7	N	$[He]2s^22p^3$	35	Br	$[Ar]3d^{10}4s^24p^5$
8	O	$[He]2s^22p^4$	36	Kr	$[Ar]3d^{10}4s^24p^6$
9	F	$[He]2s^22p^5$	37	Rb	$[Kr]5s^1$
10	Ne	$[He]2s^22p^6$	38	Sr	$[Kr]5s^2$
11	Na	$[Ne]3s^1$	39	Y	$[Kr]4d^15s^2$
12	Mg	$[Ne]3s^2$	40	Zr	$[Kr]4d^25s^2$
13	Al	$[Ne]3s^23p^1$	41	Nb	$[Kr]4d^45s^1$
14	Si	$[Ne]3s^23p^2$	42	Mo	$[Kr]4d^55s^1$
15	P	$[Ne]3s^23p^3$	43	Tc	$[Kr]4d^55s^2$
16	S	$[Ne]3s^23p^4$	44	Ru	$[Kr]4d^75s^1$
17	Cl	$[Ne]3s^23p^5$	45	Rh	$[Kr]4d^85s^1$
18	Ar	$[Ne]3s^23p^6$	46	Pd	$[Kr]4d^{10}$
19	K	$[Ar]4s^1$	47	Ag	$[Kr]4d^{10}5s^1$
20	Ca	$[Ar]4s^2$	48	Cd	$[Kr]4d^{10}5s^2$
21	Sc	$[Ar]3d^14s^2$	49	In	$[Kr]4d^{10}5s^25p^1$
22	Ti	$[Ar]3d^24s^2$	50	Sn	$[Kr]4d^{10}5s^25p^2$
23	V	$[Ar]3d^34s^2$	51	Sb	$[Kr]4d^{10}5s^25p^3$
24	Cr	$[Ar]3d^54s^1$	52	Te	$[Kr]4d^{10}5s^25p^4$
25	Mn	$[Ar]3d^54s^2$	53	I	$[Kr]4d^{10}5s^25p^5$
26	Fe	$[Ar]3d^64s^2$	54	Xe	$[Kr]4d^{10}5s^25p^6$
27	Co	$[Ar]3d^74s^2$	55	Cs	$[Xe]6s^1$
28	Ni	$[Ar]3d^84s^2$	56	Ba	$[Xe]6s^2$

Z	Symbol	Configuration	Z	Symbol	Configuration
57	La	$[Xe]5d^16s^2$	85	At	$[Xe]4f^{14}5d^{10}6s^26p^5$
58	Ce	$[Xe]4f^15d^16s^2$	86	Rn	$[Xe]4f^{14}5d^{10}6s^26p^6$
59	Pr	$[Xe]4f^36s^2$	87	Fr	$[Rn]7s^1$
60	Nd	$[Xe]4f^46s^2$	88	Ra	$[Rn]7s^2$
61	Pm	$[Xe]4f^56s^2$	89	Ac	$[Rn]6d^17s^2$
62	Sm	$[Xe]4f^66s^2$	90	Th	$[Rn]6d^27s^2$
63	Eu	$[Xe]4f^76s^2$	91	Pa	$[Rn]5f^26d^17s^2$
64	Gd	$[Xe]4f^75d^16s^2$	92	U	$[Rn]5f^36d^17s^2$
65	Tb	$[Xe]4f^96s^2$	93	Np	$[Rn]5f^46d^17s^2$
66	Dy	$[Xe]4f^{10}6s^2$	94	Pu	$[Rn]5f^67s^2$
67	Ho	$[Xe]4f^{11}6s^2$	95	Am	$[Rn]5f^77s^2$
68	Er	$[Xe]4f^{12}6s^2$	96	Cm	$[Rn]5f^76d^17s^2$
69	Tm	$[Xe]4f^{13}6s^2$	97	Bk	$[Rn]5f^97s^2$
70	Yb	$[Xe]4f^{14}6s^2$	98	Cf	$[Rn]5f^{10}7s^2$
71	Lu	$[Xe]4f^{14}5d^16s^2$	99	Es	$[Rn]5f^{11}7s^2$
72	Hf	$[Xe]4f^{14}5d^26s^2$	100	Fm	$[Rn]5f^{12}7s^2$
73	Ta	$[Xe]4f^{14}5d^36s^2$	101	Md	$[Rn]5f^{13}7s^2$
74	W	$[Xe]4f^{14}5d^46s^2$	102	No	$[Rn]5f^{14}7s^2$
75	Re	$[Xe]4f^{14}5d^56s^2$	103	Lr	$[Rn]5f^{14}6d^17s^2$
76	Os	$[Xe]4f^{14}5d^66s^2$	104	Rf	$[Rn]5f^{14}6d^27s^2$ (?)
77	Ir	$[Xe]4f^{14}5d^76s^2$	105	Db	$[Rn]5f^{14}6d^37s^2$ (?)
78	Pt	$[Xe]4f^{14}5d^96s^1$	106	Sg	$[Rn]5f^{14}6d^47s^2$ (?)
79	Au	$[Xe]4f^{14}5d^{10}6s^1$	107	Bh	$[Rn]5f^{14}6d^57s^2$ (?)
80	Hg	$[Xe]4f^{14}5d^{10}6s^2$	108	Hs	$[Rn]5f^{14}6d^67s^2$ (?)
81	Tl	$[Xe]4f^{14}5d^{10}6s^26p^1$	109	Mt	$[Rn]5f^{14}6d^77s^2$ (?)
82	Pb	$[Xe]4f^{14}5d^{10}6s^26p^2$	110	Ds	$[Rn]5f^{14}6d^87s^2$ (?)
83	Bi	$[Xe]4f^{14}5d^{10}6s^26p^3$	111	Rg	$[Rn]5f^{14}6d^{10}7s^1$ (?)
84	Po	$[Xe]4f^{14}5d^{10}6s^26p^4$			

*The electron configurations followed by a question mark are speculative.

2D   THE ELEMENTS

Element	Symbol	Atomic number	Molar mass* (g·mol^{-1})	Normal state†	Density (g·cm^{-3})	Melting point (°C)
actinium (Greek *aktis*, ray)	Ac	89	(227)	s, m	10.07	1230
aluminum (from alum, salts of the form KAl(SO$_4$)$_2$·12H$_2$O)	Al	13	26.98	s, m	2.70	660
americium (the Americas)	Am	95	(243)	s, m	13.67	990
antimony (probably a corruption of an old Arabic word; Latin *stibium*)	Sb	51	121.76	s, md	6.69	631
argon (Greek *argos*, inactive)	Ar	18	39.95	g, nm	1.66‡	−189
arsenic (Greek *arsenikos*, male)	As	33	74.92	s, md	5.78	613§
astatine (Greek *astatos*, unstable)	At	85	(210)	s, nm	—	300
barium (Greek *barys*, heavy)	Ba	56	137.33	s, m	3.59	710
berkelium (Berkeley, California)	Bk	97	(247)	s, m	14.79	986
beryllium (from the mineral beryl, Be$_3$Al$_2$SiO$_{18}$)	Be	4	9.01	s, m	1.85	1285
bismuth (German *weisse Masse*, white mass)	Bi	83	208.98	s, m	8.90	271
bohrium (Niels Bohr)	Bh	107	(264)	—	—	—
boron (Arabic *buraq*, borax, Na$_2$B$_4$O$_7$·10H$_2$O; *bor*(ax) + (carb)*on*	B	5	10.81	s, md	2.47	2300
bromine (Greek *bromos*, stench)	Br	35	79.90	l, nm	3.12	−7
cadmium (Greek *Cadmus*, founder of Thebes)	Cd	48	112.41	s, m	8.65	321
calcium (Latin *calx*, lime)	Ca	20	40.08	s, m	1.53	840
californium (California)	Cf	98	(251)	s, m	—	—
carbon (Latin *carbo*, coal or charcoal)	C	6	12.01	s, nm	2.27	3700§
cerium (the asteroid Ceres, discovered 2 days earlier)	Ce	58	140.12	s, m	6.71	800
cesium (Latin *caesius*, sky blue)	Cs	55	132.91	s, m	1.87	28
chlorine (Greek *chloros*, yellowish green)	Cl	17	35.45	g, nm	1.66‡	−101
chromium (Greek *chroma*, color)	Cr	24	52.00	s, m	7.19	1860§
cobalt (German *Kobold*, evil spirit; Greek *kobalos*, goblin)	Co	27	58.93	s, m	8.80	1494

Boiling point (°C)	Ionization energies (kJ·mol⁻¹)	Electron affinity (kJ·mol⁻¹)	Electronegativity	Principal oxidation states	Atomic radius (pm)	Ionic radius (pm)
3200	499, 1170, 1900	—	1.1	+3	188	118(3+)
2467	577, 1817, 2744	+43	1.6	+3	143	54(3+)
2600	578	—	1.3	+3	173	107(3+)
1750	834, 1794, 2443	+103	2.1	−3, +3, +5	141	89(3+)
−186	1520	<0	—	0	174	—
—	947, 1798	+78	2.2	−3, +3, +5	125	222(3−)
350	1037, 1600	+270	2.0	−1	—	227(1−)
1640	502, 965	+14	0.89	+2	217	135(2+)
—	601	—	1.3	+3	—	87(4+)
2470	900, 1757	<0	1.6	+2	113	34(2+)
1650	703, 1610, 2466	+91	2.0	+3, +5	155	96(3+)
—	660	—	—	+5	128#	83(5+)#
3931	799, 2427, 3660	+27	2.0	+3	83	23(3+)
59	1140, 2104	+325	3.0	−1, +1, +3, +4, +5, +7	114	196(1−)
765	868, 1631	<0	1.7	+2	149	103(2+)
1490	590, 1145, 4910	+2	1.3	+2	197	100(2+)
—	608	—	1.3	+3	169	117(2+)
—	1090, 2352, 4620	+122	2.6	−4, −1, +2, +4	77	260(4−)
3000	527, 1047, 1949	<50	1.1	+3, +4	183	107(3+)
678	376, 2420	146	0.79	+1	265	167(1+)
234	1255, 2297	+349	3.2	−1, +1, +3, +4, +5, +6, +7	99	181(1−)
2600	653, 1592, 2987	+64	1.7	+2, +3	125	84(2+)
2900	760, 1646, 3232	+64	1.9	+3, +6	125	64(3+)

(continued)

Element	Symbol	Atomic number	Molar mass* (g·mol⁻¹)	Normal state†	Density (g·cm⁻³)	Melting point (°C)
copper (Latin *cuprum*, from Cyprus)	Cu	29	63.55	s, m	8.93	1083
curium (Marie Curie)	Cm	96	(247)	s, m	13.30	1340
darmstadtium (town in Germany)	Ds	110	—	—	—	—
dubnium (Dubna)	Db	105	(262)	s, m	29	—
dysprosium (Greek *dysprositos*, hard to get at)	Dy	66	162.50	s, m	8.53	1410
einsteinium (Albert Einstein)	Es	99	(252)	s, m	—	—
erbium (Ytterby, a town in Sweden)	Er	68	167.26	s, m	9.04	1520
europium (Europe)	Eu	63	151.96	s, m	5.25	820
fermium (Enrico Fermi, an Italian physicist)	Fm	100	(257)	s, m	—	—
fluorine (Latin *fluere*, to flow)	F	9	19.00	g, nm	1.51‡	−220
francium (France)	Fr	87	(223)	s, m	—	27
gadolinium (Johann Gadolin, a Finnish chemist)	Gd	64	157.25	s, m	7.87	1310
gallium (Latin *Gallia*, France; also a pun on the discoverer's forename, Le Coq)	Ga	31	69.72	s, m	5.91	30
germanium (Latin *Germania*, Germany)	Ge	32	72.64	s, md	5.32	937
gold (Anglo-Saxon *gold*; Latin *aurum*, gold)	Au	79	196.97	s, m	19.28	1064
hafnium (Latin *Hafnia*, Copenhagen)	Hf	72	178.49	s, m	13.28	2230
hassium (Hesse, the German state)	Hs	108	(277)	—	—	—
helium (Greek *helios*, the sun)	He	2	4.00	g, nm	0.12‡	—
holmium (Latin *Holmia*, Stockholm)	Ho	67	164.93	s, m	8.80	1470
hydrogen (Greek *hydro* + *genes*, water-forming)	H	1	1.0079	g, nm	0.070‡	−259
indium (from the bright indigo line in its spectrum)	In	49	114.82	s, m	7.29	156
iodine (Greek *ioeidēs*, violet)	I	53	126.90	s, nm	4.95	114
iridium (Greek and Latin *iris*, rainbow)	Ir	77	192.22	s, m	22.56	2447
iron (Anglo-Saxon *iron*; Latin *ferrum*)	Fe	26	55.84	s, m	7.87	1540

Boiling point (°C)	Ionization energies (kJ·mol^{-1})	Electron affinity (kJ·mol^{-1})	Electronegativity	Principal oxidation states	Atomic radius (pm)	Ionic radius$^{\|}$ (pm)
2567	785, 1958, 3554	+118	1.9	+1, +2	128	72(2+)
—	581	—	1.3	+3	174	99(3+)
—	—	—	—	—	—	—
—	640	—	—	+5	139$^{\#}$	68(5+)$^{\#}$
2600	572, 1126, 2200	—	1.2	+3	177	91(3+)
—	619	<50	1.3	+3	203	98(3+)
2600	589, 1151, 2194	<50	1.2	+3	176	89(3+)
1450	547, 1085, 2404	<50	—	+3	204	98(3+)
—	627	—	1.3	+3	—	91(3+)
−188	1680, 3374	+328	4.0	−1	71	133(1−)
677	400	+44	0.7	+1	270	180(1+)
3000	592, 1167, 1990	<50	1.2	+2, +3	180	97(3+)
2403	577, 1979, 2963	+29	1.6	+1, +3	122	62(3+)
2830	784, 1557, 3302	+116	2.0	+2, +4	123	90(2+)
2807	890, 1980	+223	2.5	+1, +3	144	91(3+)
5300	642, 1440, 2250	0	1.3	+4	156	84(3+)
—	750	—	—	+3	126$^{\#}$	80(4+)$^{\#}$
−269	2370, 5250	<0	—	0	128	—
2300	581, 1139	<50	1.2	+3	177	89(3+)
−253	1310	+73	2.2	−1, +1	78	154(1−)
2080	556, 1821	+29	1.8	+1, +3	163	80(3+)
184	1008, 1846	+295	2.7	−1, +1, +3, +5, +7	133	220(1−)
4550	880	+151	2.2	+3, +4	136	75(3+)
2760	759, 1561, 2957	+16	1.8	+2, +3	124	82(2+)

(continued)

Element	Symbol	Atomic number	Molar mass* (g·mol^{-1})	Normal state†	Density (g·cm^{-3})	Melting point (°C)
krypton (Greek *kryptos,* hidden)	Kr	36	83.80	g, nm	3.00‡	−157
lanthanum (Greek *lanthanein,* to lie hidden)	La	57	138.91	s, m	6.17	920
lawrencium (Ernest Lawrence, an American physicist)	Lr	103	(262)	s, m	—	—
lead (Anglo-Saxon *lead;* Latin *plumbum*)	Pb	82	207.2	s, m	11.34	328
lithium (Greek *lithos,* stone)	Li	3	6.94	s, m	0.53	181
lutetium (*Lutetia,* ancient name of Paris)	Lu	71	174.97	s, m	9.84	1700
magnesium (Magnesia, a district in Thessaly, Greece)	Mg	12	24.31	s, m	1.74	650
manganese (Greek and Latin *magnes,* magnet)	Mn	25	54.94	s, m	7.47	1250
meitnerium (Lise Meitner)	Mt	109	(268)	—	—	—
mendelevium (D. Mendeleev)	Md	101	(258)	—	—	—
mercury (the planet Mercury; Latin *hydrargyrum,* liquid silver)	Hg	80	200.59	l, m	13.55	−39
molybdenum (Greek *molybdos,* lead)	Mo	42	95.94	s, m	10.22	2620
neodymium (Greek *neos + didymos,* new twin)	Nd	60	144.24	s, m	7.00	1024
neon (Greek *neos,* new)	Ne	10	20.18	g, nm	1.44‡	−249
neptunium (the planet Neptune)	Np	93	(237)	s, m	20.45	640
nickel (German *Nickel,* Old Nick, Satan)	Ni	28	58.69	s, m	8.91	1455
niobium (Niobe, daughter of Tantalus; see tantalum)	Nb	41	92.91	s, m	8.57	2425
nitrogen (Greek *nitron + genes,* soda-forming)	N	7	14.01	g, nm	1.04‡	−210
nobelium (Alfred Nobel, the founder of the Nobel prizes)	No	102	(259)	s, m	—	—
osmium (Greek *osme,* a smell)	Os	76	190.23	s, m	22.58	3030
oxygen (Greek *oxys + genes,* acid forming)	O	8	16.00	g, nm	1.14‡	−218
palladium (the asteroid Pallas, discovered at about the same time)	Pd	46	106.42	s, m	12.00	1554
phosphorus (Greek *phosphoros,* light bearing)	P	15	30.97	s, nm	1.82	44

Boiling point (°C)	Ionization energies (kJ·mol^{-1})	Electron affinity (kJ·mol^{-1})	Electronegativity	Principal oxidation states	Atomic radius (pm)	Ionic radius$^{\parallel}$ (pm)
−153	1350, 2350	<0	—	+2	189	169(1+)
3450	538, 1067, 1850	+50	1.1	+3	188	122(3+)
—	—	—	1.3	+3	—	88(3+)
1760	716, 1450	+35	2.3	+2, +4	175	132(2+)
1347	519, 7298	+60	1.0	+1	152	76(1+)
3400	524, 1340, 2022	<50	1.3	+3	173	85(3+)
1100	736, 1451	<0	1.3	+2	160	72(2+)
2120	717, 1509	<0	1.6	+2, +3, +4, +7	137	91(2+)
—	840	—	—	+2	—	83(2+)
—	635	—	1.3	+3	—	90(3+)
357	1007, 1810	−18	2.0	+1, +2	160	112(2+)
4830	685, 1558, 2621	+72	2.2	+4, +5, +6	136	92(2+)
3100	530, 1035	<0	1.1	+3	182	104(3+)
−246	2080, 3952	0	—	0	—	—
—	597	—	1.4	+5	150	88(5+)
2150	737, 1753	+156	1.9	+2, +3	125	78(2+)
5000	664, 1382	+86	1.6	+5	143	69(5+)
−196	1400, 2856	−7	3.0	−3, +3, +5	75	171(3−)
—	642	—	1.3	+2	—	113(2+)
5000	840	+106	2.2	+3, +4	135	81(3+)
−183	1310, 3388	+141, −844	3.4	−2	73	140(2−)
3000	805, 1875	+54	2.2	+2, +4	138	86(2+)
280	1011, 1903, 2912	+72	2.2	−3, +3, +5	115	212(3−)

(continued)

Element	Symbol	Atomic number	Molar mass* (g·mol^{-1})	Normal state†	Density (g·cm^{-3})	Melting point (°C)
platinum (Spanish *plata*, silver)	Pt	78	195.08	s, m	21.45	1772
plutonium (the planetlike Pluto)	Pu	94	(244)	s, m	19.81	640
polonium (Poland)	Po	84	(209)	s, md	9.40	254
potassium (from potash; Latin *kalium* and Arabic *qali*, alkali)	K	19	39.10	s, m	0.86	64
praseodymium (Greek *prasios* + *didymos*, green twin)	Pr	59	140.91	s, m	6.78	935
promethium (Prometheus, the Greek god)	Pm	61	(145)	s, m	7.22	1168
protactinium (Greek *protos* + *aktis*, first ray)	Pa	91	231.04	s, m	15.37	1200
radium (Latin *radius*, ray)	Ra	88	(226)	s, m	5.00	700
radon (from radium)	Rn	86	(222)	g, nm	4.40‡	−71
rhenium (Latin *Rhenus*, Rhine)	Re	75	186.21	s, m	21.02	3180
rhodium (Greek *rhodon*, rose; its aqueous solutions are often rose-colored)	Rh	45	102.90	s, m	12.42	1963
roentgenium (W. Roentgen, discoverer of x-rays)	Rg	111	—	—	—	—
rubidium (Latin *rubidus*, deep red, "flushed")	Rb	37	85.47	s, m	1.53	39
ruthenium (Latin *Ruthenia*, Russia)	Ru	44	101.07	s, m	12.36	2310
rutherfordium (Ernest Rutherford)	Rf	104	(261)	—	—	—
samarium (from samarskite, a mineral)	Sm	62	150.36	s, m	7.54	1060
scandium (Latin *Scandia*, Scandinavia)	Sc	21	44.96	s, m	2.99	1540
seaborgium (Glenn Seaborg)	Sg	106	(266)	—	—	—
selenium (Greek *sēlēnē*, the moon)	Se	34	78.96	s, nm	4.79	220
silicon (Latin *silex*, flint)	Si	14	28.09	s, md	2.33	1410
silver (Anglo-Saxon *seolfor*; Latin *argentum*)	Ag	47	107.87	s, m	10.50	962
sodium (English *soda*; Latin *natrium*)	Na	11	22.99	s, m	0.97	98
strontium (Strontian, Scotland)	Sr	38	87.62	s, m	2.58	770
sulfur (Sanskrit *sulvere*)	S	16	32.06	s, nm	2.09	115
tantalum (Tantalos, Greek mythological figure)	Ta	73	180.95	s, m	16.65	3000

Boiling point (°C)	Ionization energies (kJ·mol^{-1})	Electron affinity (kJ·mol^{-1})	Electronegativity	Principal oxidation states	Atomic radius (pm)	Ionic radius$^{\|}$ (pm)
3720	870, 1791	+205	2.3	+2, +4	138	85(2+)
3200	585	—	1.3	+3, +4	151	108(3+)
960	812	+174	2.0	+2, +4	167	65(4+)
774	418, 3051	+48	0.82	+1	227	138(1+)
3000	523, 1018	<50	1.1	+3	183	106(3+)
3300	536, 1052	<50	—	+3	181	106(3+)
4000	568	—	1.5	+5	161	89(5+)
1500	509, 979	—	0.9	+2	223	152(2+)
−62	1036, 1930	<0	—	+2	—	—
5600	760, 1260	+14	1.9	+4, +7	137	72(4+)
3700	720, 1744	+110	2.3	+3	134	75(3+)
—	—	—	—	—	—	—
688	402, 2632	+47	0.82	+1	248	152(1+)
4100	711, 1617	+101	2.2	+2, +3, +4	134	77(3+)
—	490	—	—	+4	150$^{\#}$	67(4+)$^{\#}$
1600	543, 1068	<50	1.2	+3	180	100(3+)
2800	631, 1235	+18	1.4	+3	161	83(3+)
—	730	—	—	+6	132$^{\#}$	86(5+)$^{\#}$
685	941, 2044	+195	2.6	−2, +4, +6	117	198(2−)
2620	786, 1577	+134	1.9	+4	117	26(4+)
2212	731, 2073	+126	1.9	+1	144	113(1+)
883	494, 4562	+53	0.93	+1	154	102(1+)
1380	548, 1064	+5	0.95	+2	215	118(2+)
445	1000, 2251	+200, −532	2.6	−2, +4, +6	104	184(2−)
5400	761	+14	1.5	+5	143	72(3+)

(continued)

Element	Symbol	Atomic number	Molar mass* (g·mol^{-1})	Normal state†	Density (g·cm^{-3})	Melting point (°C)
technetium (Greek *technētos*, artificial)	Tc	43	(98)	s, m	11.50	2200
tellurium (Latin *tellus*, earth)	Te	52	127.60	s, md	6.25	450
terbium (Ytterby, a town in Sweden)	Tb	65	158.93	s, m	8.27	1360
thallium (Greek *thallos*, a green shoot)	Tl	81	204.38	s, m	11.87	304
thorium (Thor, Norse god of thunder, weather, and crops)	Th	90	232.04	s, m	11.73	1700
thulium (Thule, early name for Scandinavia)	Tm	69	168.93	s, m	9.33	1550
tin (Anglo-Saxon *tin*; Latin *stannum*)	Sn	50	118.71	s, m	7.29	232
titanium (Titans, Greek mythological figures, sons of the Earth)	Ti	22	47.87	s, m	4.55	1660
tungsten (Swedish *tung + sten*, heavy stone; from wolframite)	W	74	183.84	s, m	19.30	3387
uranium (the planet Uranus)	U	92	238.03	s, m	18.95	1135
vanadium (Vanadis, Scandinavian mythological figure)	V	23	50.94	s, m	6.11	1920
xenon (Greek *xenos*, stranger)	Xe	54	131.29	g, nm	3.56‡	−112
ytterbium (Ytterby, a town in Sweden)	Yb	70	173.04	s, m	6.97	824
yttrium (Ytterby, a town in Sweden)	Y	39	88.91	s, m	4.48	1510
zinc (Anglo-Saxon *zinc*)	Zn	30	65.41	s, m	7.14	420
zirconium (Arabic *zargun*, gold color)	Zr	40	91.22	s, m	6.51	1850

Boiling point (°C)	Ionization energies (kJ·mol^{-1})	Electron affinity (kJ·mol^{-1})	Electronegativity	Principal oxidation states	Atomic radius (pm)	Ionic radius$^{\parallel}$ (pm)
4600	702, 1472	+96	1.9	+4, +7	136	72(4+)
990	870, 1775	+190	2.1	−2, +4	143	221(2−)
2500	565, 1112	<50	—	+3	178	97(3+)
1457	590, 1971	+19	2.0	+1, +3	170	105(3+)
4500	587, 1110	—	1.3	+4	180	99(4+)
2000	597, 1163	<50	1.2	+3	175	94(3+)
2720	707, 1412	+116	2.0	+2, +4	141	93(2+)
3300	658, 1310	+7.6	1.5	+4	145	69(4+)
5420	770	+79	2.4	+5, +6	137	62(6+)
4000	584, 1420	—	1.4	+6	154	80(6+)
3400	650, 1414	+51	1.6	+4, +5	132	61(4+)
−108	1170, 2046	<0	2.6	+2, +4, +6	218	190(1+)
1500	603, 1176	<50	—	+3	194	86(3+)
3300	616, 1181	+30	1.2	+3	181	106(3+)
907	906, 1733	+9	1.6	+2	133	83(2+)
4400	660, 1267	+41	1.3	+4	160	87(4+)

*Parentheses around molar mass indicate the most stable isotope of a radioactive element.
†The normal state is the state of the element at normal temperature and pressure (20°C and 1 atm).
s denotes solid, l, liquid, and g, gas; m denotes metal, nm, nonmetal, and md, metalloid.
‡The density quoted is for the liquid.
§The solid sublimes.
$^{\parallel}$Charge in parentheses.
#Atomic and ionic radii are estimated.

2E   THE TOP 23 CHEMICALS BY INDUSTRIAL PRODUCTION IN THE UNITED STATES IN 2005

Production data are compiled annually by the American Chemical Society and published in *Chemical and Engineering News*. This table is based on the information about production in 2005 that was published in the July 10, 2006, issue. Water, sodium chloride, and steel traditionally are not included and would outrank the rest if they were. Hydrogen is heavily used but almost always "on site" as soon as it has been prepared.

Rank	Name	Annual production (10^9 kg)	Comment on source
1	sulfuric acid	36.5	contact process
2	ethene (ethylene)	24.0	thermal cracking of petroleum
3	polyethylene	16.3	polymerization of ethene
4	propene (propylene)	15.3	thermal cracking of petroleum
5	phosphoric acid	11.6	from phosphate rocks
6	dichloroethane (ethylene dichloride)	11.3	chlorination of ethene
7	chlorine	10.2	electrolysis
8	diammonium hydrogen phosphate	10.0	processing of phosphate rocks
9	ammonia	9.8	Haber process
10	sodium hydroxide	8.4	electrolysis of brine
11	polypropylene	8.1	polymerization of propylene
12	polyvinyl chloride and copolymers	6.9	polymerization of vinyl chloride
13	nitric acid	6.3	Ostwald process
14	ammonium nitrate	6.4	ammonia + nitric acid
15	urea	5.8	ammonia + carbon dioxide
16	ethylbenzene	5.3	Friedel–Crafts alkylation of benzene
17	ammonium dihydrogen phosphate	5.2	processing of phosphate rocks
18	styrene	5.0	dehydration of ethylbenzene
19	hydrogen chloride	4.4	by-product of hydrocarbon chlorination
20	cumene (isopropylbenzene)	3.5	alkylation of benzene
21	ethylene oxide	3.1	addition of O_2 to ethene
22	polystyrene	2.9	polymerization of styrene
23	ammonium sulfate	2.6	ammonia + sulfuric acid

APPENDIX 3: Nomenclature

3A   THE NOMENCLATURE OF POLYATOMIC IONS

Charge number	Chemical formula	Name	Oxidation number of central element	Charge number	Chemical formula	Name	Oxidation number of central element
+2	Hg_2^{2+}	mercury(I)	+1		O_3^-	ozonide	$-\frac{1}{3}$
	UO_2^{2+}	uranyl	+6		OH^-	hydroxide	$-2(O)$
	VO^{2+}	vanadyl	+4		SCN^-	thiocyanate	—
+1	NH_4^+	ammonium	−3	−2	C_2^{2-}	carbide (acetylide)	−1
	PH_4^+	phosphonium	−3				
−1	$CH_3CO_2^-$	acetate (ethanoate)	0(C)		CO_3^{2-}	carbonate	+4
					$C_2O_4^{2-}$	oxalate	+3
	HCO_2^-	formate (methanoate)	+2(C)		CrO_4^{2-}	chromate	+6
					$Cr_2O_7^{2-}$	dichromate	+6
	CN^-	cyanide	+2(C), −3(N)		O_2^{2-}	peroxide	−1
	ClO_4^-	perchlorate*	+7		S_2^{2-}	disulfide	−1
	ClO_3^-	chlorate*	+5		SiO_3^{2-}	metasilicate	+4
	ClO_2^-	chlorite*	+3		SO_4^{2-}	sulfate	+6
	ClO^-	hypochlorite*	+1(Cl)		SO_3^{2-}	sulfite	+4
	MnO_4^-	permanganate	+7		$S_2O_3^{2-}$	thiosulfate	+2
	NO_3^-	nitrate	+5	−3	AsO_4^{3-}	arsenate	+5
	NO_2^-	nitrite	+3		BO_3^{3-}	borate	+3
	N_3^-	azide	$-\frac{1}{3}$		PO_4^{3-}	phosphate	+5

*These names are representative of the halogen oxoanions.

When a hydrogen ion bonds to a −2 or −3 anion, add "hydrogen" before the name of the anion. For example, HSO_3^- is hydrogen sulfite (or hydrogensulfite). If two hydrogen ions bond to a −3 anion, add "dihydrogen" before the name of the anion. For example, $H_2PO_4^-$ is dihydrogen phosphate (or dihydrogenphosphate).

Oxoacids and Oxoanions

The names of oxoanions and their parent acids can be determined by noting the oxidation number of the central atom and then referring to the table on the right. For example, the nitrogen in $N_2O_2^{2-}$ has an oxidation number of +1; because nitrogen belongs to Group 15/V, the ion is a hyponitrite ion.

	Group number					
14/IV	15/V	16/VI	17/VII	Oxoanion	Oxoacid
—	—	—	+7	per . . . ate	per . . . ic acid
+4	+5	+6	+5	. . . ate	. . . ic acid
—	+3	+4	+3	. . . ite	. . . ous acid
—	+1	+2	+1	hypo . . . ite	hypo . . . ous acid

3B   COMMON NAMES OF CHEMICALS

Many chemicals are often referred to by their common names, sometimes as a result of their use over hundreds of years and sometimes because they appear on the labels of consumer products, such as detergents, beverages, and antacids. The following names are just a few that have found their way into the language of everyday life.

Common name	Formula	Chemical name
baking soda	$NaHCO_3$	sodium hydrogen carbonate (sodium bicarbonate)
bleach, laundry	$NaClO$	sodium hypochlorite
borax	$Na_2B_4O_7 \cdot 10H_2O$	sodium tetraborate decahydrate
brimstone	S_8	sulfur
calamine	$ZnCO_3$	zinc carbonate
chalk	$CaCO_3$	calcium carbonate
Epsom salts	$MgSO_4 \cdot 7H_2O$	magnesium sulfate heptahydrate
fool's gold	FeS_2	iron(II) disulfide
gypsum	$CaSO_4 \cdot 2H_2O$	calcium sulfate dihydrate
lime (quicklime)	CaO	calcium oxide
lime (slaked lime)	$Ca(OH)_2$	calcium hydroxide
limestone	$CaCO_3$	calcium carbonate
lye, caustic soda	$NaOH$	sodium hydroxide
marble	$CaCO_3$	calcium carbonate
milk of magnesia	$Mg(OH)_2$	magnesium hydroxide
plaster of Paris	$CaSO_4 \cdot \frac{1}{2}H_2O$	calcium sulfate hemihydrate
potash*	K_2CO_3	potassium carbonate
quartz	SiO_2	silcon dioxide
table salt	$NaCl$	sodium chloride
vinegar	CH_3COOH	acetic acid (ethanoic acid)
washing soda	$Na_2CO_3 \cdot 10H_2O$	sodium carbonate decahydrate

*Potash also refers collectively to K_2CO_3, KOH, K_2SO_4, KCl, and KNO_3.

3C   NAMES OF SOME COMMON CATIONS WITH VARIABLE CHARGE NUMBERS

Modern nomenclature includes the oxidation number of elements with variable oxidation states in the names of their compounds, as in cobalt(II) chloride. However, the traditional nomenclature, in which the suffixes -ous and -ic are used, is still encountered. The table on the right translates from one system into the other for some common elements.

Element	Cation	Old-style name	Modern name
cobalt	Co^{2+}	cobaltous	cobalt(II)
	Co^{3+}	cobaltic	cobalt(III)
copper	Cu^{+}	cuprous	copper(I)
	Cu^{2+}	cupric	copper(II)
iron	Fe^{2+}	ferrous	iron(II)
	Fe^{3+}	ferric	iron(III)
lead	Pb^{2+}	plumbous	lead(II)
	Pb^{4+}	plumbic	lead(IV)
manganese	Mn^{2+}	manganous	manganese(II)
	Mn^{3+}	manganic	manganese(III)
mercury	Hg_2^{2+}	mercurous	mercury(I)
	Hg^{2+}	mercuric	mercury(II)
tin	Sn^{2+}	stannous	tin(II)
	Sn^{4+}	stannic	tin(IV)

GLOSSARY

ab initio method The calculation of molecular structure by solving the *Schrödinger equation* numerically. Compare with *semiempirical method*.

absolute zero ($T = 0$; that is, 0 on the *Kelvin scale*) The lowest possible temperature ($-273.15°C$).

absorb To accept one substance into and throughout the bulk of another substance. Compare with *adsorb*.

absorbance (A) A measure of the extent of absorption of radiation by a sample: $A = \log(I_0/I)$.

absorbed dose (of radiation) The energy deposited in a given mass of sample when it is exposed to radiation (particularly but not exclusively nuclear radiation). Absorbed dose is measured in *rad* or *gray*.

absorption spectrum The wavelength dependence of the absorption of a sample, determined by measuring the extent to which the sample absorbs electromagnetic radiation as the wavelength is varied over a range.

abundance (of an isotope) The percentage (in terms of the numbers of atoms) of the isotope present in a sample of the element. See also *natural abundance*.

acceleration The rate of change of velocity (either its direction or its magnitude).

acceleration of free fall (g) The acceleration experienced by a body owing to the gravitational field at the surface of the Earth.

accuracy Freedom from systematic error. Compare with *precision*.

accurate measurement A measurement that has small systematic error and gives a result close to the accepted value of the property.

achiral Not chiral: identical with its mirror image. See also *chiral*.

acid See *Arrhenius acid*; *Brønsted acid*; *Lewis acid*. Used alone, "acid" normally means a Brønsted acid.

acid anhydride A compound that forms an oxoacid when it reacts with water. See also *formal anhydride*. *Example*: SO_3, the anhydride of sulfuric acid.

acid–base indicator See *indicator*.

acid–base titration See *titration*.

acid buffer See *buffer*.

acidic hydrogen atom A hydrogen atom (more exactly, the proton of that hydrogen atom) that can be donated to a base.

acidic ion An ion that acts as a Brønsted acid. *Examples*: NH_4^+; $[Al(H_2O)_6]^{3+}$.

acidic oxide An oxide that reacts with water to give an acid; the oxides of nonmetallic elements generally are acidic oxides. *Examples*: CO_2; SO_3.

acidic solution A solution with pH < 7.

acid ionization (dissociation) constant (K_a) See *acidity constant*.

acidity The strength of the tendency to donate a proton.

acidity constant (K_a) The equilibrium constant for proton transfer to water; for an acid HA, $K_a = [H_3O^+][A^-]/[HA]$ at equilibrium.

actinide Former (and still very common) term for *actinoid*.

actinoid A member of the second row of the *f* block (actinium through nobelium).

activated complex An unstable combination of reactant molecules that can either go on to form products or fall apart into the unchanged reactants.

activated complex theory See *transition-state theory*.

activation energy (E_a) (1) The minimum energy needed for reaction. (2) The height of the activation barrier. (3) An empirical parameter that describes the temperature dependence of the rate constant of a reaction.

activity (1) In thermodynamics, a_J, the effective concentration or pressure of a species J expressed as the partial pressure or concentration of the species relative to its standard value. (2) In radioactivity, the number of nuclear disintegrations per second.

activity coefficient (γ_J) The proportionality factor between the activity of a species and its molar concentration or partial pressure (divided by its standard value). *Example*: for a solute J, $a_J = \gamma_J[J]/c°$.

addition polymerization The polymerization, usually of alkenes, by an addition reaction propagated by radical or ionic intermediates.

addition reaction A chemical reaction in which atoms or groups bond to two atoms joined by a multiple bond. The product of the reaction is a single molecule that contains all the reactant atoms. *Example*: $CH_3CH{=}CH_2 + HBr \rightarrow CH_3CH_2CH_2Br$.

adhesion Binding to a surface.

adhesive forces Forces that bind a substance to a surface.

adsorb To bind a substance to a surface; the surface *adsorbs* the substance. Distinguish from *absorb*.

aerosol A fine mist of solid particles or droplets of liquid suspended in a gas.

alcohol An organic molecule containing an —OH group attached to a carbon atom that is not part of a carbonyl group or an aromatic ring. Alcohols are classified as *primary, secondary*, and *tertiary* according to the number of carbon atoms attached to the C—OH carbon atom. *Examples*: CH_3CH_2OH (primary); $(CH_3)_2CHOH$ (secondary); $(CH_3)_3COH$ (tertiary).

aldehyde   An organic compound containing the —CHO group. *Examples:* CH_3CHO, ethanal (acetaldehyde); C_6H_5CHO, benzaldehyde.

aliphatic hydrocarbon   A hydrocarbon that does not have benzene rings in its structure.

alkali   An aqueous solution of a strong base. *Example:* aqueous NaOH.

alkali metal   A member of Group 1 of the periodic table (the lithium family).

alkaline earth metal   Calcium, strontium, and barium; more informally, a member of Group 2 of the periodic table (the beryllium family).

alkaline solution   An aqueous solution with pH > 7.

alkane   (1) A hydrocarbon with no carbon–carbon multiple bonds. (2) A saturated hydrocarbon. (3) A member of a series of hydrocarbons derived from methane by the repetitive insertion of —CH_2— groups; alkanes have the molecular formula C_nH_{2n+2}. *Examples:* CH_4; CH_3CH_3; $CH_3(CH_2)_6CH_3$.

alkene   (1) A hydrocarbon with at least one carbon–carbon double bond. (2) A member of a series of hydrocarbons derived from ethene by the repetitive insertion of —CH_2— groups; alkenes with one double bond have the molecular formula C_nH_{2n}. *Examples:* $CH_2=CH_2$; $CH_3CH=CH_2$; $CH_3CH=CHCH_2CH_3$.

alkyne   (1) A hydrocarbon with at least one carbon–carbon triple bond. (2) A member of a series of hydrocarbons derived from ethyne by the repetitive insertion of —CH_2— groups; alkynes with one triple bond have the molecular formula C_nH_{2n-2}. *Examples:* $CH\equiv CH$; $CH_3C\equiv CCH_3$.

allotropes   Alternative forms of an element that differ in the way in which the atoms are linked. *Examples:* O_2 and O_3; white and gray tin.

alloy   A mixture of two or more metals formed by melting, mixing, and then cooling. A *substitutional alloy* is an alloy in which atoms of one metal are substituted for atoms of another metal. An *interstitial alloy* is an alloy in which atoms of one metal lie in the gaps in the lattice formed by atoms of another metal. A *homogeneous alloy* is an alloy in which the atoms of the elements are distributed uniformly. A *heterogeneous alloy* is an alloy that consists of (micro)crystalline phases with different compositions.

alpha (α) helix   One type of secondary structure adopted by a polypeptide chain, in the form of a right-handed helix.

alpha (α) particle   Positively charged, subatomic particle emitted from some radioactive nuclei; nucleus of a helium atom ($^4_2He^{2+}$).

alternating copolymer   See *copolymer*.

ambidentate ligand   A ligand that can coordinate to a metal atom by using atoms of different elements. *Example:* SCN^-, which can coordinate through S or N.

amide   An organic compound formed by the reaction of an amine and a carboxylic acid in which the acidic —OH group has been replaced by an amino group or a substituted amino group. An amide contains the group —$CONR_2$. *Example:* CH_3CONH_2, acetamide.

amine   A compound derived from ammonia by replacing various numbers of H atoms with organic groups; the number of hydrogen atoms replaced determines the classification as *primary, secondary,* or *tertiary. Examples:* CH_3NH_2 (primary); $(CH_3)_2NH$ (secondary); $(CH_3)_3N$ (tertiary). See also *quaternary ammonium ion.*

amino acid   A carboxylic acid that also contains an amino group. The *essential amino acids* are amino acids that must be ingested as a part of the diet. *Example:* NH_2CH_2COOH, glycine. See Table 19.4.

amino group   The functional group —NH_2 characteristic of *amines.*

amorphous solid   A solid in which the atoms, ions, or molecules lie in a random jumble with no long-range order. *Examples:* glass; butter. Compare with *crystalline solid.*

amount of substance (n)   The number of entities in a sample divided by Avogadro's constant. Also referred to as *chemical amount.* See *mole.*

ampere (A)   The SI unit of electric current. See also Appendix 1B.

amphiprotic   Having the ability both to donate and to accept protons. See *amphoteric. Examples:* H_2O; HCO_3^-.

amphoteric   Having the ability to react with both acids and bases. See *amphiprotic. Examples:* Al; Al_2O_3.

amplitude   The height of a mathematical function above 0. On a graph depicting a wave, the height of the wave above the center line.

analysis   See *chemical analysis.*

analyte   The solution of unknown concentration in a titration. Normally, the analyte is in the flask, not the buret.

analytical chemistry   The study and application of the techniques for identifying substances and measuring their amounts.

angular wavefunction ($Y(\theta,\phi)$)   The angular part of a wavefunction, particularly the angular component of the wavefunctions of the hydrogen atom; the probability amplitude of an electron as a function of orientation around the nucleus.

anhydride   See *acid anhydride.*

anhydrous   Lacking water. *Example:* $CuSO_4$, the anhydrous form of copper(II) sulfate. Compare with *hydrate.*

anion   A negatively charged ion. *Examples:* F^-; SO_4^{2-}.

anisotropic   Depending on orientation.

anode   The electrode at which oxidation takes place.

antibonding orbital   A molecular orbital that, when occupied, contributes to an overall raising of the energy of a molecule.

antiferromagnetic material   A substance in which electron spins on neighboring atoms are locked into an antiparallel arrangement over large regions. *Example:* manganese.

antilogarithm If the logarithm to the base B is x, then the antilogarithm of x is B^x. The *common antilogarithm* of x is 10^x. The *natural antilogarithm* of x is the exponential e^x.

antioxidant A substance that reacts with radicals and so prevents the oxidation of another substance.

antiparticle A particle with the same mass as a subatomic particle but with opposite charge. *Example:* positron, the antiparticle of an electron.

applied research Investigations directed toward the solution of real-world problems. See also *basic research*.

aqueous solution A solution in which the solvent is water.

arene An aromatic hydrocarbon.

aromatic compound An organic compound that includes a benzene ring as part of its structure. *Examples:* C_6H_6 (benzene); C_6H_5Cl (chlorobenzene); $C_{10}H_8$ (naphthalene).

aromatic hydrocarbon See *aromatic compound* (the more general term).

Arrhenius acid A compound that contains hydrogen and releases hydrogen ions (H^+) in water. *Examples:* HCl; CH_3COOH; but not CH_4.

Arrhenius base A compound that produces hydroxide ions (OH^-) in water. *Examples:* NaOH; NH_3; but not Na, because it is not a compound.

Arrhenius behavior A reaction shows Arrhenius behavior if a plot of $\ln k$ against $1/T$ is a straight line. See *Arrhenius equation*.

Arrhenius equation The equation $\ln k = \ln A - E_a/RT$ for the commonly observed temperature dependence of a rate constant k. An *Arrhenius plot* is a graph of $\ln k$ against $1/T$.

Arrhenius parameters The *pre-exponential factor A* (also called the *frequency factor*) and the *activation energy E_a*. See also *Arrhenius equation*.

aryl group An aromatic group. *Example:* $-C_6H_5$, phenyl.

atmosphere (1) The layer of gases surrounding a planet (specifically, the air for planet Earth). (2) A unit of pressure (1 atm = $1.013\ 25 \times 10^5$ Pa).

atom (1) The smallest particle of an element that has the chemical properties of that element. (2) An electrically neutral species consisting of a nucleus and its surrounding electrons.

atomic hypothesis The proposal advanced by John Dalton that matter is composed of atoms.

atomic mass constant (m_u, formerly amu) One-twelfth the mass of one atom of carbon-12.

atomic nucleus See *nucleus*.

atomic number (Z) The number of protons in the nucleus of an atom; this number determines the identity of the element and the number of electrons in the neutral atom.

atomic orbital A region of space in which there is a high probability of finding an electron in an atom. An *s-orbital* is a spherical region; a *p-orbital* has two lobes, on opposite sides of the nucleus; a *d-orbital* typically has four lobes, with the nucleus at the center; an *f-orbital* has a more complicated arrangement of lobes.

atomic radius Half the distance between the centers of neighboring atoms in a solid or a homonuclear molecule.

atomic structure The arrangement of electrons around the nucleus of an atom.

atomic weight See *molar mass*.

***Aufbau* principle** See *building-up principle*.

autoionization See *autoprotolysis*.

autoprotolysis A reaction in which a proton is transferred between two molecules of the same substance. The products are the conjugate acid and conjugate base of the substance. *Example:* $2\ H_2O(l) \rightleftharpoons H_3O^+(aq) + OH^-(aq)$.

autoprotolysis constant The equilibrium constant for an autoprotolysis reaction. *Example:* for water, K_w, with $K_w = [H_3O^+][OH^-]$.

average bond enthalpy ($\Delta H_B(A-B)$) See *mean bond enthalpy*.

average reaction rate The reaction rate calculated by measuring the change in concentration of a reactant or product over a finite time interval (and hence an average of the changing rate within that interval). See also *unique average reaction rate*.

Avogadro's constant The number of objects per mole of objects ($N_A = 6.022\ 14 \times 10^{23}$ mol^{-1}). *Avogadro's number* is the number of objects in one mole of objects (that is, the dimensionless number $6.022\ 14 \times 10^{23}$).

Avogadro's principle The volume of a sample of gas at a given temperature and pressure is proportional to the amount of gas molecules in the sample: $V \propto n$.

axial bond A bond that is perpendicular to the molecular plane in a bipyramidal molecule.

axial lone pair A lone pair lying on the axis of a bipyramidal molecule.

azeotrope A mixture of liquids that boils without change of composition. A *minimum-boiling azeotrope* has a boiling point lower than that of either component; a *maximum-boiling azeotrope* has a boiling point higher than that of either component.

azimuthal quantum number (l) See *orbital angular momentum quantum number*.

background radiation The average nuclear radiation to which the Earth's inhabitants are exposed daily.

balanced equation See *chemical equation*.

ball-and-stick model A depiction of a molecule in which atoms are represented by balls and bonds are represented by sticks.

Balmer series A family of spectral lines (some of which lie in the visible region) in the spectrum of atomic hydrogen.

band gap   A range of energies for which there are no orbitals in a solid. Unless stated otherwise, the band gap refers to the gap between the valence band and the conduction band.

band of stability   A region of a plot of mass number against atomic number corresponding to the existence of stable nuclei.

bar   A unit of pressure: 1 bar = 10^5 Pa.

barometer   An instrument for measuring the atmospheric pressure.

base   See *Arrhenius base; Brønsted base; Lewis base.* Used alone, "base" normally means a Brønsted base.

base buffer   See *buffer.*

base ionization constant   See *basicity constant.*

base pair   Two specific nucleotides that link one complementary strand of a DNA molecule to the other by means of hydrogen bonding: adenine pairs with thymine and guanine pairs with cytosine.

base units   The units of measurement in the International System (SI) in terms of which all other units are defined. *Examples: kilogram* for mass; *meter* for length; *second* for time; *kelvin* for temperature; *ampere* for electric current.

basic ion   An ion that acts as a Brønsted base. *Example:* $CH_3CO_2^-$.

basicity constant (K_b)   The equilibrium constant for proton transfer from water to a base; for a base B, $K_b = [BH^+][OH^-]/[B]$.

basic oxide   An oxide that is a Brønsted base. The oxides of metallic elements are generally basic. *Examples:* Na_2O; MgO.

basic research   Investigations aimed at discovering the reasons for phenomena and the fundamental principles of chemistry, the synthesis of new materials, and the investigation of their properties. See also *applied research.*

basic solution   A solution with pH > 7.

battery   A collection of galvanic cells joined in series; the voltage that the battery produces is the sum of the voltages of each cell.

becquerel (Bq)   The SI unit of radioactivity (one disintegration per second).

Beer's law   The absorbance of electromagnetic radiation by a sample is proportional to the molar concentration of the absorbing species and the length of the sample through which the radiation passes.

beta (β) decay   Nuclear decay due to *β-particle* emission.

beta (β) particle   A fast electron emitted from a nucleus in a radioactive decay.

beta (β) sheet   One type of planar secondary structure adopted by a polypeptide, in the form of a pleated sheet.

bimolecular reaction   An elementary reaction in which two molecules, atoms, or ions come together and form a product. *Example:* $O + O_3 \rightarrow O_2 + O_2$.

binary   Consisting of two components, as in *binary mixture* and *binary (ionic or molecular) compound.*

Examples: acetone and water (a binary mixture); HCl, $CaCl_2$, C_6H_6 (binary compounds; $CaCl_2$ is ionic, HCl and C_6H_6 are molecular).

bio-based material   A material taken from or made from natural materials in living things.

biochemistry   The study of biologically important substances, processes, and reactions.

bioenergetics   The deployment and utilization of energy in living cells.

biomass   The organic material of the planet produced annually by photosynthesis.

biological chemistry   The application of chemical principles to biological structures and processes.

biomimetic material   A material modeled after a naturally occurring material.

biradical   A species with two unpaired electrons. *Example:* $\cdot CH_2CH_2CH_2\cdot$.

black body   An object that absorbs and emits all frequencies of radiation without favor.

black-body radiation   The electromagnetic radiation emitted by a *black body.*

block (*s* block, *p* block, *d* block, *f* block)   The region of the periodic table containing elements for which, according to the building-up principle, the corresponding subshell is currently being filled.

block copolymer   See *copolymer.*

body-centered cubic structure (bcc)   A crystal structure with a unit cell in which a central atom lies at the center of a cube formed by eight others.

Bohr frequency condition   The relation between the change in energy of an atom or molecule and the frequency of radiation emitted or absorbed: $\Delta E = h\nu$.

Bohr radius (a_0)   In an early model of the hydrogen atom, the radius of the lowest energy orbit; now a specific combination of fundamental constants ($a_0 = 4\pi\varepsilon_0\hbar^2/m_e e^2 \approx 52.9$ pm) in the description of the wavefunctions of hydrogen.

boiling   Rapid vaporization taking place throughout a liquid. See *boiling temperature.*

boiling point (b.p.)   See *boiling temperature; normal boiling point.*

boiling-point constant (k_b)   The constant of proportionality between the boiling-point elevation and the molality of a solute.

boiling-point elevation   The increase in normal boiling point of a solvent caused by the presence of a solute (a *colligative property*).

boiling temperature   (1) The temperature at which a liquid boils. (2) The temperature at which a liquid is in equilibrium with its vapor at the pressure of the surroundings; vaporization then occurs throughout the liquid, not only at the liquid's surface.

Boltzmann formula (for the entropy)   The formula $S = k \ln W$, where k is *Boltzmann's constant* and W is the number of atomic arrangements that correspond to the same energy.

Boltzmann's constant (k)   A fundamental constant; $k = 1.380\,66 \times 10^{-23}$ J·K^{-1}. Note that $R = N_A k$.

bond   A link between atoms. See also *covalent bond; double bond; ionic bond; triple bond.*

bond angle   In an A—B—C molecule or part of a molecule, the angle between the B—A and B—C bonds.

bond enthalpy ($\Delta H_B(X—Y)$)   The enthalpy change accompanying the dissociation of a bond. *Example:* $H_2(g) \rightarrow 2\,H(g)$, $\Delta H_B(H—H) = +436$ kJ·mol^{-1}.

bonding orbital   A molecular orbital that, when occupied, results in an overall lowering of the energy of a molecule.

bond length   The distance between the centers of two atoms joined by a bond.

bond order   The number of electron pair bonds that link a specific pair of atoms.

Born–Haber cycle   A closed series of reactions used to express the enthalpy of formation of an ionic solid in terms of contributions that include the lattice enthalpy.

Born interpretation   The interpretation of the square of the wavefunction, ψ, of a particle as the probability density for finding the particle in a region of space.

Born–Meyer equation   The formula for the minimum energy of an ionic solid.

boundary condition   A constraint on the value of the wavefunction of a particle.

boundary surface   The surface showing the region of space within which there is about 90% probability of finding an electron when it occupies a specific orbital in an atom or molecule.

Boyle's law   At constant temperature, and for a given sample of gas, the volume is inversely proportional to the pressure: $P \propto 1/V$.

Bragg equation   An equation relating the angle of diffraction of x-rays to the spacing of layers of atoms in a crystal ($\lambda = 2d \sin \theta$).

branched alkane   An alkane with hydrocarbon side chains.

branching   Description of a step in a chain reaction in which more than one chain carrier is formed in a propagation step. *Example:* $\cdot O\cdot + H_2 \rightarrow \cdot OH + \cdot H$. See also *propagation.*

Bravais lattices   The 14 basic patterns of unit cells from which a crystal can be built.

breeder reactor   A reactor used to generate nuclear fuel by making use of neutrons that are not moderated.

Brønsted acid   A proton donor (a source of hydrogen ions, H$^+$). *Examples:* HCl; CH_3COOH; HCO_3^-; NH_4^+.

Brønsted base   A proton acceptor (a species to which hydrogen ions, H$^+$, can bond). *Examples:* OH$^-$; Cl$^-$; $CH_3CO_2^-$; HCO_3^-; NH_3.

Brønsted–Lowry definition   A definition of acids and bases in terms of the ability of molecules and ions to participate in proton transfer.

Brønsted–Lowry theory   A theory of acids and bases involving proton transfer from one species to another.

Brownian motion   The ceaseless jittering motion of colloidal particles caused by the impact of solvent molecules.

buffer   A solution that resists any change in pH when small amounts of acid or base are added. An *acid buffer* stabilizes solutions at pH < 7 and a *base buffer* stabilizes solutions at pH > 7. *Examples:* a solution containing CH_3COOH and $CH_3CO_2^-$ (acid buffer); a solution containing NH_3 and NH_4^+ (base buffer).

buffer capacity   An indication of the amount of acid or base that can be added before a buffer loses its ability to resist the change in pH.

building-up principle   The procedure for arriving at the ground-state electron configurations of atoms and molecules.

bulk matter   Matter composed of large numbers of atoms. See *bulk property.*

bulk property   A property that depends on the collective behavior of large numbers of atoms. *Examples:* melting point; vapor pressure; internal energy.

buret   A narrow, graduated tube fitted with a stopcock, used to measure the volume of liquid delivered into another vessel.

calibration   Interpretation of an observation by comparison with known information.

calorie (cal)   A unit of energy. The unit is now defined in terms of the joule by 1 cal = 4.184 J exactly. The *nutritional calorie* is 1 kcal.

calorimeter   An apparatus used to determine the heat released or absorbed in a process by measuring the temperature change.

candela (cd)   The SI unit of luminous intensity. See also Appendix 1B.

capillary action   The rise of liquids up narrow tubes.

carbohydrate   A compound of general formula $C_m(H_2O)_n$, although small deviations from this general formula are often encountered. Carbohydrates include cellulose, starches, and sugars. *Examples:* $C_6H_{12}O_6$, glucose; $C_{12}H_{22}O_{11}$, sucrose.

carbonyl group   A >CO group in an inorganic or organic compound.

carboxyl group   The functional group —COOH. See *carboxylic acid.*

carboxylic acid   An organic compound containing the carboxyl group, —COOH. *Examples:* CH_3COOH, acetic acid; C_6H_5COOH, benzoic acid.

carrier   See *chain carrier.*

catalyst   A substance that increases the rate of a reaction without being consumed in the reaction. A catalyst is *homogeneous* if it is present in the same phase as the reactants and *heterogeneous* if it is in a different phase from the reactants. *Examples:* homogeneous, Br$^-$(aq) for the decomposition of H_2O_2(aq); heterogeneous, Pt in the Ostwald process.

catenate To form chains or rings of atoms. *Examples:* O_3; S_8.

cathode The electrode at which reduction takes place.

cathodic protection Protection of a metal object by connecting it to a more strongly reducing metal.

cation A positively charged ion. *Examples:* Na^+; NH_4^+; Al^{3+}.

cell diagram A description of an electrochemical cell that corresponds to a given cell reaction. *Example:* $Zn(s) \mid Zn^{2+}(aq) \parallel Cu^{2+}(aq) \mid Cu(s)$.

cell potential See *electromotive force.*

Celsius scale A temperature scale on which the freezing point of water is at 0 degrees and its normal boiling point is at 100 degrees. Units on this scale are degrees Celsius, °C.

ceramic (1) A solid obtained by the action of heat on clay. (2) A noncrystalline inorganic solid usually containing oxides, borides, and carbides.

cesium chloride structure A crystal structure the same as that of solid cesium chloride.

chain branching A propagation step in a chain reaction when more than one chain carrier is formed.

chain carrier An intermediate in a chain reaction.

chain reaction A reaction that is propagated when an intermediate reacts to produce another intermediate in a series of elementary reactions. *Example:* $Br\cdot + H_2 \rightarrow HBr + H\cdot$ followed by $H\cdot + Br_2 \rightarrow HBr + Br\cdot$.

chalcogens Oxygen, sulfur, selenium, and tellurium in Group 16/VI of the periodic table.

change of state The change of a substance from one of its physical states to another of its physical states. *Example:* melting, solid → liquid.

characteristic (of a logarithm) The number preceding the decimal point.

charge A measure of the strength with which a particle can interact electrostatically with another particle.

charge balance The requirement that, because a solution is neutral overall, the concentration of positive charge due to cations must equal the concentration of negative charge due to anions.

charge-transfer transition A transition in which an electron is excited from the ligands of a complex to the metal atom or vice versa.

Charles's law The volume of a given sample of gas at constant pressure is directly proportional to its absolute temperature: $V \propto T$.

chelate A complex containing at least one polydentate ligand that forms a ring of atoms including the central metal atom. *Example:* $[Co(en)_3]^{3+}$.

chemical analysis The determination of the chemical composition of a sample. See also *qualitative, quantitative.*

chemical bond See *bond.*

chemical change The conversion of one or more substances into different substances.

chemical element See *element.*

chemical engineering The study of industrial chemical processes.

chemical equation A statement in terms of chemical formulas summarizing the qualitative information about the chemical changes taking place in a reaction and the quantitative information that atoms are neither created nor destroyed in a chemical reaction. In a *balanced chemical equation* (commonly called a "chemical equation"), the same number of atoms of each element appears on both sides of the equation.

chemical equilibrium A dynamic equilibrium between reactants and products in a chemical reaction.

chemical formula A collection of chemical symbols and subscripts that shows the composition of a substance. See also *condensed structural formula; empirical formula; molecular formula; structural formula.*

chemical kinetics The study of the rates of chemical reactions and the steps by which they take place.

chemical nomenclature The systematic naming of compounds.

chemical plating The deposition of a metal surface on an object by making use of a chemical reduction reaction.

chemical property The ability of a substance to participate in a chemical reaction.

chemical reaction A chemical change in which one substance responds to the presence of another, to a change of temperature, or to some other influence.

chemical symbol The abbreviation of the name of an element.

chemiluminescence The emission of light by products formed in energetically excited states during a chemical reaction.

chemistry The branch of science concerned with the study of matter and the changes that matter can undergo.

chiral (molecule or complex) Not able to be superimposed on its own mirror image. *Examples:* $CH_3CH(NH_2)COOH$; $CHBrClF$; $[Co(en)_3]^{3+}$.

chloralkali process The production of chlorine and sodium hydroxide by the electrolysis of aqueous sodium chloride.

cholesteric phase A liquid crystal phase in which layers of parallel molecules are twisted relative to one another in such a way that the orientations of the molecules form a helical structure.

chromatogram A record of the signal from the detector (or the paper record) obtained in a chromatographic analysis of a mixture.

chromatography A separation technique that relies on the ability of different phases to adsorb substances to different extents.

cis–trans isomerization The conversion of a cis isomer into a trans isomer and vice versa. *Example:* *cis*-butene ⇌ *trans*-butene.

classical mechanics   The laws of motion proposed by Isaac Newton in which particles travel in definite paths in response to forces.

clathrate   A structure in which a molecule of one substance sits in a cage made up of molecules of another substance, typically water. *Example:* SO_2 in water.

Clausius–Clapeyron equation   An equation that gives the quantitative dependence of the vapor pressure of a substance on the temperature.

Clausius inequality   The relation $\Delta S \geq q/T$.

Claus process   A process for obtaining sulfur from the H_2S in oil wells by the oxidation of H_2S with SO_2; the latter is formed by the oxidation of H_2S with oxygen.

closed shell (or subshell)   A shell (or subshell) containing the maximum number of electrons allowed by the exclusion principle. *Example:* the neonlike core $1s^2 2s^2 2p^6$.

closed system   See *system*.

close-packed structure   A crystal structure in which atoms occupy the smallest total volume with the least empty space. *Examples:* hexagonal close packing and cubic close packing of identical spheres.

coagulation   The formation of aggregates from colloidal particles.

cohesion   The act or state in which the particles of a substance stick to one another.

cohesive forces   The forces that bind the molecules of a substance together to form a bulk material and are responsible for condensation.

coinage metals   The elements copper, silver, and gold.

colligative property   A property that depends only on the relative number of solute and solvent particles present in a solution and not on the chemical identity of the solute. *Examples:* elevation of boiling point; depression of freezing point; osmosis.

collision cross section   The area that a molecule presents as a target during a collision.

collision theory   The theory of elementary gas-phase bimolecular reactions in which molecules are assumed to react only if they collide with a characteristic minimum kinetic energy.

colloid (or colloidal suspension)   A dispersion of tiny particles with diameters between 1 nm and 1 μm in a gas, liquid, or solid. *Example:* milk.

combined gas law   A combination of Boyle's law and Charles's law that allows the pressure, volume, or temperature of a sample of an ideal gas to be predicted after a change in state. $P_1 V_1/n_1 T_1 = P_2 V_2/n_2 T_2$.

combustion   A reaction in which an element or compound burns in oxygen. *Example:* $CH_4(g) + 2\,O_2(g) \rightarrow CO_2(g) + 2\,H_2O(l)$.

combustion analysis   The determination of the composition of a sample by the measurement of the masses of the products of its combustion.

common antilogarithm   See *antilogarithm*.

common-ion effect   Reduction of the solubility of one salt by the presence of another salt with one ion in common. *Example:* the lower solubility of AgCl in NaCl(aq) than in pure water.

common logarithm   See *logarithm*.

common name   An informal name for a compound that may give little or no clue to the compound's composition. *Examples:* water; aspirin; acetic acid.

competing reaction   A reaction taking place at the same time as the reaction of interest and using some of the same reactants.

complementarity   The impossibility of knowing the position of a particle with arbitrarily great precision if its linear momentum is known precisely.

complementary color   The color that white light becomes when one of the colors present in it is removed.

complete ionic equation   A balanced chemical equation expressed in terms of the cations and anions present in solution. *Example:* $Ag^+(aq) + NO_3^-(aq) + Na^+(aq) + Cl^-(aq) \rightarrow AgCl(s) + NO_3^-(aq) + Na^+(aq)$.

complex   (1) The combination of a Lewis acid and a Lewis base linked by a coordinate covalent bond. (2) A species consisting of several ligands (the Lewis bases) that have an independent existence bonded to a single central metal atom or ion (the Lewis acid). *Examples:* (1) $H_3N{-}BF_3$; (2) $[Fe(H_2O)_6]^{3+}$; $[PtCl_4]^-$.

composite material   A synthetic material composed of a polymer and one or more other substances that have been solidified together.

compound   (1) A specific combination of elements that can be separated into its elements by chemical techniques but not physical techniques. (2) A substance consisting of atoms of two or more elements in a definite, unchangeable ratio.

compress   To reduce the volume of a sample.

compressible   Able to be compressed into a smaller volume.

compression factor (Z)   The ratio of the actual molar volume of a gas to the molar volume of an ideal gas under the same conditions.

computational chemistry   The application of computational techniques to the determination of the structures and properties of molecules and bulk matter.

concentration   The quantity of a substance in a given volume. See also *molar concentration*.

concentration cell   A galvanic cell in which the electrodes have the same composition but are at different concentrations.

condensation   The formation of a liquid or solid phase from the gas phase of the substance.

condensation polymer   A polymer formed by a chain of condensation reactions. *Examples:* polyesters; polyamides (nylon).

condensation reaction   A reaction in which two molecules combine to form a larger one and a small molecule is eliminated. *Example:* $CH_3COOH + C_2H_5OH \rightarrow CH_3COOC_2H_5 + H_2O$.

condensed phase   A solid or liquid phase; not a gas.

condensed structural formula   A compact version of the structural formula, showing how the atoms are grouped together. *Example:* $CH_3CH(CH_3)CH_3$ for methylpropane.

conduction band   An incompletely occupied band of energy levels in a solid.

configuration   See *electron configuration.*

conformations   Molecular shapes that can be interchanged by rotation about bonds, without bond breakage and reformation.

congeners   Elements in the same group of the periodic table.

conjugate acid   The Brønsted acid formed when a Brønsted base has accepted a proton. *Example:* NH_4^+ is the conjugate acid of NH_3.

conjugate acid–base pair   A Brønsted acid and its conjugate base. *Examples:* HCl and Cl^-; NH_4^+ and NH_3.

conjugate base   The Brønsted base formed when a Brønsted acid has donated a proton. *Example:* NH_3 is the conjugate base of NH_4^+.

conjugated double bonds   A sequence of alternating single and double bonds, as in $-C=C-C=C-$.

connectivity (of atoms in a molecule)   The pattern in which the atoms in a molecule are bonded to one another.

constructive interference   Interference that results in an increased amplitude of a wave. Compare with *destructive interference.*

contact process   The production of sulfuric acid by the combustion of sulfur and the catalyzed oxidation of sulfur dioxide to sulfur trioxide.

conversion factor   A factor that is used to convert a measurement from one unit into another.

cooling curve   A graph of the variation of the temperature of a sample over time as it loses energy at a constant rate. Compare with *heating curve.*

coordinate   Use of a lone pair to form a coordinate covalent bond. *Examples:* $F_3B + :NH_3 \rightarrow F_3B-NH_3$; $Ni + 4\,CO \rightarrow Ni(CO)_4$.

coordinate covalent bond   A bond formed between a Lewis base and a Lewis acid by sharing an electron pair originally belonging to the Lewis base.

coordination compound   A neutral complex or an ionic compound in which at least one of the ions is a complex. *Examples:* $Ni(CO)_4$; $K_3[Fe(CN)_6]$.

coordination isomers   Isomers that differ by the exchange of one or more ligands between a cationic complex and an anionic complex.

coordination number   (1) The number of nearest neighbors of an atom in a solid. (2) For ionic solids, the coordination number of an ion is the number of nearest neighbors of opposite charge. (3) For complexes, the number of points at which ligands are attached to the central metal ion.

coordination sphere   The ligands directly attached to the central ion in a complex.

copolymer   A polymer formed from a mixture of different monomers. In *random copolymers,* the sequence of monomers has no particular order; in *alternating copolymers,* two monomers alternate; in *block copolymers,* regions of one monomer alternate with regions of another; in *graft copolymers,* chains of one monomer are attached to a backbone chain of a second monomer.

core   The inner closed shells of an atom.

core electrons   The electrons that belong to an atom's core.

corrosion   The unwanted reaction of a material that results in the dissolution or consumption of the material. *Example:* the unwanted oxidation of a metal.

corrosive   (1) A reagent that can cause corrosion. (2) Having a high reactivity, such as the reactivity of a strong oxidizing agent or a concentrated acid or base.

coulombic attraction   The attraction between opposite electric charges.

Coulomb potential energy   The potential energy of an electric charge in the vicinity of another electric charge; the potential energy is inversely proportional to the separation of the charges.

Coulomb's law   The potential energy of a pair of electric charges is inversely proportional to the distance between them and proportional to the product of the charges.

couple   See *redox couple.*

covalent bond   A pair of electrons shared between two atoms.

covalent radius   The contribution of an atom to the length of a covalent bond.

cracking   The process of converting petroleum fractions into smaller molecules with more double bonds. *Example:* $CH_3(CH_2)_6CH_3 \rightarrow CH_3(CH_2)_3CH_3 + CH_3CH=CH_2$.

critical mass   The mass of fissionable material above which so few neutrons escape from a sample of nuclear fuel that the fission chain reaction is sustained; a greater mass is *supercritical* and a smaller mass is *subcritical.*

critical point   The point in a phase diagram at the critical pressure and critical temperature.

critical pressure (P_c)   The vapor pressure of a liquid at its critical temperature.

critical temperature (T_c)   The temperature at and above which a substance cannot exist as a liquid.

crown ether   A cyclic ether with alternating $-CH_2CH_2O-$ units.

cryogenics   The study of matter at very low temperatures.

cryoscopy   The measurement of molar mass by using the depression of freezing point.

crystal face   A flat plane forming a side of a crystal.

crystal field   The electrostatic influence of the ligands (modeled as point negative charges) on the central ion of a complex. *Crystal field theory* is a rationalization of the optical, magnetic, and thermodynamic properties of complexes in terms of the crystal field of their ligands.

crystalline solid   A solid in which the atoms, ions, or molecules lie in an orderly array. *Examples:* NaCl; diamond; graphite. Compare with *amorphous solid*.

crystallization   The process in which a solute comes out of solution as crystals.

cubic close-packed structure (ccp)   A close-packed structure with an ABCABC . . . pattern of layers.

curie (Ci)   A unit of activity (for radioactivity).

current (I)   The rate of supply of charge; current is measured in *amperes* (A), with $1\ A = 1\ C \cdot s^{-1}$.

cycle   (1) In thermodynamics, a sequence of changes that begins and ends at the same state. (2) In spectroscopy, one complete reversal of the direction of the electromagnetic field and its return to the original direction.

cycloalkane   A saturated aliphatic hydrocarbon in which the carbon atoms form a ring. *Example:* C_6H_{12}, cyclohexane.

Dalton's law of partial pressures   The total pressure of a mixture of gases is the sum of the partial pressures of its components.

data   The information provided or obtained from experiments.

daughter nucleus   A nucleus that is the product of a nuclear decay.

d–d transition   A transition in which an electron is excited from one *d*-orbital to another.

de Broglie relation   The proposal that every particle has wavelike properties and that its wavelength, λ, is related to its mass by $\lambda = h/(\text{mass} \times \text{velocity})$.

debye (D)   The unit used to report electric dipole moments: $1\ D = 3.336 \times 10^{-30}\ C \cdot m$.

decant   To pour off a liquid from on top of another, denser liquid or from a solid.

decay constant (k)   The rate constant for radioactive decay.

decomposition   A reaction in which a substance is broken down into simpler substances; *thermal decomposition* is decomposition brought about by heat. *Example:* $CaCO_3(s) \xrightarrow{\Delta} CaO(s) + CO_2(g)$.

definite integral   An *integral* with limits attached. See also Appendix 1F.

degenerate   Having the same energy. *Example:* atomic orbitals in the same subshell.

dehydrating agent   A reagent that removes water or the elements of water from a compound. *Example:* H_2SO_4.

dehydrogenation   The removal of a hydrogen atom from each of two neighboring carbon atoms, resulting in the formation of a carbon–carbon multiple bond.

dehydrohalogenation   The removal of a hydrogen atom and a halogen atom from neighboring carbon atoms in a haloalkane.

delocalized   Spread over a region. In particular, *delocalized electrons* are electrons that spread over several atoms in a molecule.

delta (Δ, in a chemical equation)   A symbol that signifies that the reaction takes place at elevated temperatures.

delta X (ΔX)   The difference between the final and initial values of a property, $\Delta X = X_{final} - X_{initial}$. *Examples:* ΔT; ΔE.

denaturation   The loss of structure of a large molecule, such as a protein.

density (d)   The mass of a sample of a substance divided by its volume: $d = m/V$.

density isosurface   A graphic image that represents a molecular structure as a surface and shows the distribution of electrons in a molecule; the surface corresponds to locations with the same electron density.

deposition   The condensation of a vapor directly to a solid. Deposition is the reverse of *sublimation*.

deprotonation   Loss of a proton from a Brønsted acid. *Example:* $NH_4^+(aq) + H_2O(l) \rightarrow H_3O^+(aq) + NH_3(aq)$.

derived unit   A combination of base units. *Examples:* centimeters cubed (cm^3); joules ($kg \cdot m^2 \cdot s^{-2}$).

descriptive chemistry   The description of the preparation, properties, and applications of the elements and their compounds.

destructive interference   Interference that results in a reduced amplitude of a wave. Compare with *constructive interference*.

deuteron   The nucleus of a deuterium atom, $^2H^+$, consisting of a proton and a neutron.

diagonal relationship   A similarity in properties between diagonal neighbors in the periodic table, especially for main-group elements in Periods 2 and 3 at the left-hand side of the table. *Examples:* Li and Mg; Be and Al.

diamagnetic (substance)   A substance that tends to be pushed out of a magnetic field; consisting of atoms, ions, or molecules with no unpaired electrons. *Examples:* most common substances.

diamine   An organic compound that contains two —NH_2 groups.

diatomic ion   An ion that consists of two atoms with a net charge.

diatomic molecule   A molecule that consists of two atoms. *Examples:* H_2; CO.

differential calculus   The part of mathematics that deals with the slopes of curves and with infinitesimal quantities. See also Appendix 1F.

diffraction   The deflection of waves and the resulting interference caused by an object in their path. See also *x-ray diffraction*.

diffraction pattern   The pattern of bright spots against a dark background resulting from diffraction.

diffusion   The spreading of one substance through another substance.

dilute   (1) verb: To reduce the concentration of a solute by adding more solvent. (2) adjective: Describes a solution in which the solute has a low concentration.

dimer   The union of two identical molecules. *Example:* Al_2Cl_6 formed from two $AlCl_3$ molecules.

diol   An organic compound with two —OH groups.

dipeptide   A *peptide* formed by the condensation of two amino acids.

dipole   See *electric dipole; instantaneous dipole moment.*

dipole–dipole interaction   The interaction between two electric dipoles: like partial charges repel and opposite partial charges attract.

dipole–induced dipole interaction   The interaction between an electric dipole and the instantaneous dipole that it induces in a nonpolar molecule.

dipole moment   See *electric dipole moment.*

diprotic   An acid with two acidic hydrogen atoms. See also *polyprotic acid or base.*

disaccharide   A carbohydrate molecule that is composed of two saccharide units. *Example:* $C_{12}H_{22}O_{11}$, sucrose.

dispersion   (1) The spatial separation of light into its component colors (as by a prism). (2) See *suspension.*

dispersion force   See *London force.*

disproportionation   A redox reaction in which a single element is simultaneously oxidized and reduced. *Example:* $2\,Cu^+(aq) \rightarrow Cu(s) + Cu^{2+}(aq)$.

dissociation   (1) The breaking of a bond. (2) The separation of ions that occurs when an ionic solid dissolves.

dissociation constant   See *acidity constant.*

dissociation energy (*D*)   The energy required to separate bonded atoms.

distillate   A liquid obtained by distillation.

distillation   The separation of the components of a mixture by making use of their different volatilities.

distribution (of molecular speeds)   The fraction of gas molecules moving at each speed at any instant.

disulfide link   An —S—S— link that contributes to the secondary and tertiary structures of polypeptides.

domain   A region of a metal in which the electron spins of the atoms are aligned, resulting in *ferromagnetism.*

doping   The addition of a known, small amount of a second substance to an otherwise pure solid substance.

d-orbital   See *atomic orbital.*

dose equivalent   The actual dose of radiation experienced by a sample, modified to take into account the *relative biological effectiveness* of the radiation. The dose equivalent is measured in sievert (and formerly rem).

double bond   (1) Two electron pairs shared by neighboring atoms. (2) One σ-bond and one π-bond between neighboring atoms.

drying agent   A substance that absorbs water and thus maintains a dry atmosphere. *Example:* phosphorus(V) oxide.

ductility   The ability to be drawn out into a wire (as for a metal).

duplet   The $1s^2$ electron pair of the heliumlike electron configuration.

dynamic equilibrium   The condition in which a forward process and its reverse are taking place simultaneously at equal rates. *Examples:* vaporizing and condensing; chemical reactions at equilibrium.

effective nuclear charge (Z_{eff})   The net nuclear charge after taking into account the shielding caused by other electrons in the atom.

effervesce   To bubble out of solution as a gas.

effusion   The escape of a substance (particularly a gas) through a small hole.

elasticity   The ability to return to the original shape after distortion.

elastomer   An elastic polymer. *Example:* rubber (polyisoprene).

electrical conduction   The conduction of electric charge through matter. See also *electronic conductor; ionic conduction.*

electric current   See *current.*

electric dipole   A positive charge next to an equal but opposite negative charge.

electric dipole moment (μ)   The magnitude of the electric dipole (in debye).

electric field   A region of influence that affects charged particles.

electrochemical cell   A system consisting of two electrodes in contact with an electrolyte. A *galvanic cell* (*voltaic cell*) is an electrochemical cell used to produce electricity, and an *electrolytic cell* is an electrochemical cell in which an electric current is used to cause chemical change.

electrochemical series   Redox couples arranged in order of oxidizing and reducing strengths; usually arranged with strong oxidizing agents at the top of the list and strong reducing agents at the bottom.

electrochemistry   The branch of chemistry that deals with the use of chemical reactions to produce electricity, the relative strengths of oxidizing and reducing agents, and the use of electricity to produce chemical change.

electrode   A metallic conductor that makes contact with an electrolyte in an electrochemical cell.

electrolysis   A process in which a chemical change is produced by passing an electric current through a liquid.

electrolyte   (1) An ionically conducting medium. (2) A substance that dissolves to give an electrically conducting solution. A *strong electrolyte* is a substance that is fully ionized in solution. A *weak electrolyte* is a molecular substance that is only partially ionized in solution. A *nonelectrolyte* does not ionize in solution.

Examples: NaCl is a strong electrolyte; CH_3COOH is a weak electrolyte; $C_6H_{12}O_6$ is a nonelectrolyte.

electrolyte solution   A solution of an electrolyte.

electrolytic cell   See *electrochemical cell.*

electromagnetic field   The region of influence generated by accelerated charged particles.

electromagnetic radiation   A wave of oscillating electric and magnetic fields; includes light, x-rays, and γ rays.

electromotive force (emf, *E*)   (1) The pushing and pulling power of a reaction in an electrochemical cell. (2) The potential difference between the electrodes of an electrochemical cell when it is producing no current.

electron (e^-)   A negatively charged subatomic particle found outside the nucleus of an atom.

electron affinity (E_{ea})   The energy released when an electron is added to a gas-phase atom or monatomic ion.

electron arrangement (VSEPR model)   The three-dimensional geometry of the arrangement of bonds and lone pairs about a central atom in a molecule or ion.

electron capture   The capture by a nucleus of one of its own atom's *s*-electrons.

electron configuration   The occupancy of orbitals in an atom or molecule. *Example:* N, $1s^2 2s^2 2p^3$.

electron-deficient compound   A compound with too few valence electrons for it to be assigned a valid Lewis structure. *Example:* B_2H_6.

electronegative element   An element with a high electronegativity. *Examples:* O; F.

electronegativity (χ, chi)   The ability of an atom to attract electrons to itself when it is part of a compound.

electronic conductor   A substance that conducts electricity by the transfer of electrons.

electronic structure   The details of the distribution of the electrons that surround the nuclei in atoms and molecules.

electronvolt (eV)   A unit of energy; the change in potential energy of an electron when it moves through a potential difference of 1 V; 1 eV = 1.602 18 × 10^{-19} J.

electrophile   A reactant that is attracted to a region of high electron density. *Examples:* Br_2; NO_2^+.

electrophilic substitution   Substitution that takes place as a result of an attack by an *electrophile. Example:* nitration of benzene.

electroplating   The deposition of a thin film of metal on an object by electrolysis.

electropositive element   An element that has a low electronegativity and is likely to give up electrons to another element on compound formation. *Examples:* Cs; Mg.

electrostatic potential surface   A molecular structure in which the net charge is calculated at each point of the density isosurface and depicted by different colors; an "elpot" surface.

element   (1) A substance that cannot be separated into simpler components by chemical techniques. (2) A substance consisting of atoms of the same atomic number. *Examples:* hydrogen; gold; uranium.

elementary reaction   An individual step in a proposed reaction mechanism.

elimination reaction   A reaction in which two groups or atoms on neighboring carbon atoms are removed from a molecule, thereby leaving a multiple bond between the carbon atoms. *Example:* $CH_3CHBrCH_3 + OH^- \rightarrow CH_3CH{=}CH_2 + H_2O + Br^-$.

empirical   Determined by experiment.

empirical formula   A chemical formula that shows the relative numbers of atoms of each element in a compound by using the simplest whole-number subscripts. *Examples:* P_2O_5; CH for benzene.

emulsion   A suspension of droplets of one liquid dispersed throughout another liquid.

enantiomers   A pair of optical isomers that are mirror images of, but not superimposable on, each other.

endothermic process   A process that absorbs heat ($\Delta H > 0$). *Examples:* vaporization; $N_2O_4(g) \rightarrow 2\ NO_2(g)$.

end point   The stage in a titration at which enough titrant has been added to bring the indicator to a color halfway between its initial and its final colors.

energy (*E*)   The capacity of a system to do work or supply heat. *Kinetic energy* is the energy of motion, and *potential energy* is the energy arising from position. The *total energy* is the sum of the kinetic and potential energies.

energy level   A permitted value of the energy in a quantized system such as an atom or a molecule.

enrich   In nuclear chemistry, to increase the abundance of a specific isotope.

ensemble   A collection of hypothetical replications of a system.

enthalpy (*H*)   A state property; $H = U + PV$. A change in enthalpy is equal to the heat transferred at constant pressure.

enthalpy density (of a fuel)   The enthalpy of combustion per liter (without the negative sign).

enthalpy of freezing   The enthalpy change per mole accompanying freezing; the negative of the *enthalpy of fusion.*

enthalpy of fusion (ΔH_{fus})   The enthalpy change per mole accompanying fusion (melting).

enthalpy of hydration (ΔH_{hyd})   The enthalpy change accompanying the hydration of gas-phase ions.

enthalpy of ionization   The change in enthalpy for the process $E(g) \rightarrow E^+(g) + e^-(g)$.

enthalpy of melting (ΔH_{melt})   See *enthalpy of fusion.*

enthalpy of mixing (ΔH_{mix})   The change in enthalpy when two fluids (liquids or gases) mix.

enthalpy of solution (ΔH_{sol})   The change in enthalpy when a substance dissolves. The *limiting enthalpy of solution* is the enthalpy of solution for the formation of an infinitely dilute solution.

enthalpy of sublimation (ΔH_{sub})   The enthalpy change per mole accompanying sublimation (the direct conversion of a solid into a vapor).

enthalpy of vaporization (ΔH_{vap})   The enthalpy change per mole accompanying vaporization (the conversion of a substance from the liquid state into the vapor state).

entropy (S)   (1) A measure of the disorder of a system. (2) A change in entropy is equal to the heat supplied reversibly to a system divided by the temperature at which the transfer takes place.

entropy of vaporization (ΔS_{vap})   The entropy change per mole accompanying vaporization (the conversion of a substance from the liquid state into the vapor state).

enzyme   A biological catalyst.

e-orbital   One of the orbitals d_{z^2} or $d_{x^2-y^2}$, in an octahedral or tetrahedral complex. In an octahedral complex, the orbitals are designated e_g.

equation of state   A mathematical expression relating the pressure, volume, temperature, and amount of substance present in a sample. *Example:* ideal gas law, $PV = nRT$.

equatorial bond   A bond perpendicular to the axis of a molecule (particularly trigonal bipyramidal and octahedral molecules).

equatorial lone pair   A *lone pair* in the plane perpendicular to the molecular axis.

equilibrium   See *chemical equilibrium; dynamic equilibrium.*

equilibrium constant (K)   An expression characteristic of the equilibrium composition of the reaction mixture, with a form given by the law of mass action. *Example:* $N_2(g) + 3 H_2(g) \rightleftharpoons 2 NH_3(g)$, $K = (P_{NH_3})^2/P_{N_2}(P_{H_2})^3$.

equilibrium table   A table used to calculate the composition of a reaction mixture at equilibrium, given the initial composition. The columns are headed by the species and the rows are, successively, the initial composition, the change to reach equilibrium, and the equilibrium composition.

equimolar   Having the same molar concentration or the same number of moles.

equipartition theorem   The average energy of each quadratic contribution to the energy of a molecule in a sample at a temperature T is equal to $\frac{1}{2}kT$ (where k is *Boltzmann's constant*).

equivalence point   See *stoichiometric point.*

essential amino acid   An amino acid that is an essential component of the diet because it cannot be synthesized in the body.

ester   The product (other than water) of the reaction between a carboxylic acid and an alcohol and having the formula RCOOR'. *Example:* $CH_3COOC_2H_5$, ethyl acetate.

esterification   The formation of an *ester.*

ether   An organic compound of the form R—O—R'. *Examples:* $CH_3OC_2H_5$, ethyl methyl ether; $C_2H_5OC_2H_5$, diethyl ether.

evaporate   To vaporize completely.

excited state   A state other than the state of lowest energy.

exclusion principle   No more than two electrons can occupy any given orbital; and, when two electrons do occupy one orbital, their spins must be paired.

exothermic process   A process that releases heat ($\Delta H < 0$). *Examples:* freezing; $N_2(g) + 3 H_2(g) \rightarrow 2 NH_3(g)$.

expanded valence shell   A valence shell containing more than eight electrons. Also called an *expanded octet.* *Examples:* the valence shells of P and S in PCl_5 and SF_6.

expansion work   See *work.*

experiment   A test carried out under carefully controlled conditions.

exponential   The exponential of x is the natural antilogarithm of x, namely, e^x.

exponential decay   A variation with time of the form e^{-kt}. *Example:* $[A] = [A]_0e^{-kt}$.

extensive property   A physical property of a substance that depends on the mass of the sample. *Examples:* volume; internal energy; enthalpy; entropy.

extrapolate   To extend a graph outside the region covered by the data.

face   See *crystal face.*

face-centered cubic structure (fcc)   A crystal structure built from a cubic unit cell in which there is an atom at the center of each face and one at each corner.

Fahrenheit scale   A temperature scale on which the freezing point of water is at 32 degrees and the normal boiling point is at 212 degrees. Units on this scale are degrees Fahrenheit, °F.

fallout   The fine dust that settles from clouds of airborne particles after a nuclear bomb test.

Faraday's constant (F)   The magnitude of the charge per mole of electrons; $F = N_Ae = 96.485$ kC·mol^{-1}.

Faraday's law of electrolysis   The amount of product formed by an electric current is chemically equivalent to the amount of electrons supplied.

fat   An ester of glycerol and carboxylic acids with long hydrocarbon chains; fats act as long-term energy storage in living systems.

fatty acid   A carboxylic acid with a long hydrocarbon chain. *Example:* $CH_3(CH_2)_{16}COOH$, stearic acid.

ferroalloy   An alloy of a metal with iron and, often, carbon. *Example:* ferrovanadium.

ferrofluid   A magnetic liquid that is a suspension of a finely powdered magnetic material such as magnetite, Fe_3O_4, in a viscous, oily liquid (such as mineral oil) that contains a detergent.

ferromagnetism   The ability of some substances to be permanently magnetized. *Examples:* iron; magnetite, Fe_3O_4.

field   An influence spreading over a region of space. *Examples:* an *electric field* from a charge; a *magnetic field* from a magnet or a moving charge.

filtration   The separation of a heterogeneous mixture of a solid and liquid by passing the mixture through a fine mesh.

first derivative (dy/dx)   A measure of the slope of a curve. See also Appendix 1F.

first ionization energy (I_1)   See *ionization energy.*

first law of thermodynamics   The internal energy of an isolated system is constant.

first-order reaction   A reaction in which the rate is proportional to the first power of the concentration of a substance.

fissile nucleus   A nucleus having the ability to undergo fission induced by slow neutrons. *Example:* ^{235}U is fissile.

fission (nuclear)   The breakup of a nucleus into two smaller nuclei of similar mass; fission may be *spontaneous* or *induced* (particularly by the impact of neutrons).

fissionable   Having the ability to undergo induced nuclear fission.

fixation of nitrogen   Conversion of elemental nitrogen to its compounds, particularly ammonia.

flammability   The ability of a substance to burn in air.

flocculation   The reversible aggregation of colloidal particles into larger particles that can be filtered.

fluorescence   The emission of light from molecules excited by radiation of higher frequency.

foam   (1) A frothy collection of bubbles formed by a liquid. (2) A type of *colloid* formed by a gas of tiny bubbles in a liquid or solid.

***f*-orbital**   See *atomic orbital.*

force (*F*)   An influence that changes the state of motion of an object. *Examples:* an electrostatic force from an electric charge; a mechanical force from an impact.

formal anhydride   An *acid anhydride* that does not necessarily react with water to form the corresponding acid. *Example:* CO, the formal anhydride of formic acid, HCOOH.

formal charge   (1) The electric charge of an atom in a molecule assigned on the assumption that the bonding is nonpolar covalent. (2) Formal charge (*FC*) = number of valence electrons in the free atom − (number of lone-pair electrons + $\frac{1}{2}$ × number of shared electrons).

formation constant (K_f)   The equilibrium constant for complex formation. The *overall formation constant* is the product of *step-by-step formation constants*. The inverse of the formation constant ($1/K_f$) is called the *stability constant.*

formula unit   The group of ions that matches the formula of the smallest unit of an ionic compound. *Example:* NaCl, one Na^+ ion and one Cl^- ion.

formula weight   See *molar mass.*

fossil fuel   The partly decomposed remains of vegetable and marine life (mainly coal, oil, and natural gas).

fraction   Samples of distillate obtained in different ranges of boiling temperatures.

fractional distillation   Separation of the components of a liquid mixture by repeated distillation, making use of their differing volatilities.

Frasch process   A process for mining sulfur that uses superheated water to melt the sulfur and compressed air to force it to the surface.

free energy   See *Gibbs free energy.*

free expansion   Expansion against zero opposing pressure.

freezing-point constant (k_f)   The constant of proportionality between the freezing-point depression and the solute molality.

freezing-point depression   The lowering of the freezing point of a solvent caused by the presence of a solute (a colligative property).

freezing temperature   The temperature at which a liquid freezes. The *normal freezing point* is the freezing temperature under a pressure of 1 atm.

frequency (of radiation) (v, nu)   The number of cycles (repeats of the waveform) per second (unit: *hertz*, Hz).

fuel cell   A primary electrochemical cell in which the reactants are supplied continuously from outside while the cell is in use.

functional group   A group of atoms that brings a characteristic set of chemical properties to an organic molecule. *Examples:* —OH ; —Br ; —COOH.

functionalization   The introduction of functional groups into an alkane molecule.

fundamental charge (*e*)   The magnitude of the charge of an electron.

fusion   (1) Melting. (2) The merging of nuclei to form the nucleus of a heavier element.

galvanic cell   See *electrochemical cell.*

galvanize   Coat a metal with an unbroken film of zinc.

gamma (γ) radiation   Very high frequency, short wavelength electromagnetic radiation emitted by nuclei.

gas   A fluid form of matter that fills the container that it occupies and can easily be compressed into a much smaller volume. (A gas differs from a vapor in that a gas is a substance at a higher temperature than its critical temperature; a vapor is a gaseous form of matter at a temperature below its critical temperature.)

gas constant (*R*)   (1) The constant that appears in the ideal gas law. See inside the back cover of the book for values. (2) $R = N_A k$, where *k* is *Boltzmann's constant.*

gas–liquid chromatography   A version of chromatography in which a gas carries the sample over a stationary liquid phase.

gauge pressure   The pressure inside a container less that outside the container.

Geiger counter   A device that is used to detect and measure radioactivity by relying on ionization caused by incident radiation.

gel   A soft, solid colloid that typically consists of a liquid trapped within a solid network.

geometrical isomers   Stereoisomers that differ in the spatial arrangement of the atoms. Geometrical isomers exhibit *cis–trans* isomerism.

Gibbs free energy ($G = H - TS$)   The energy of a system that is free to do work at constant temperature and pressure. The direction of spontaneous change at constant pressure and temperature is the direction of decreasing Gibbs free energy.

Gibbs free energy of reaction   The difference in molar Gibbs free energies of the products and reactants, weighted by the stoichiometric coefficients in the chemical equation.

glass   An ionic solid with an amorphous structure resembling that of a liquid.

glass electrode   A thin-walled glass bulb containing an electrolyte solution and a metallic contact; used for measuring pH.

graft copolymer   See *copolymer.*

Graham's law of effusion   The rate of effusion of a gas is inversely proportional to the square root of its molar mass.

gravimetric analysis   An analytical method using measurements of mass.

gray (Gy)   The SI unit of *absorbed dose*; 1 Gy corresponds to an energy deposit of 1 J·kg^{-1}. See also *rad.*

green chemistry   The practice of chemistry that conserves resources and minimizes impact on the environment.

greenhouse effect   The blocking by some atmospheric gases (notably carbon dioxide) of the radiation of heat from the surface of the Earth back into space, leading to the possibility of a worldwide rise in temperature.

greenhouse gas   A gas that contributes to the *greenhouse effect.*

ground state   The state of lowest energy.

group   A vertical column in the periodic table.

Haber process (Haber–Bosch process)   The catalyzed synthesis of ammonia at high pressure.

half-cell   One compartment of an electrochemical cell consisting of an electrode and an electrolyte.

half-life ($t_{1/2}$)   (1) In chemical kinetics, the time needed for the concentration of a substance to fall to half its initial value. (2) In radioactivity, the time needed for half the initial number of radioactive nuclei to disintegrate.

half-reaction   A hypothetical oxidation or reduction reaction showing either electron loss or electron gain. *Examples:* $\text{Na(s)} \rightarrow \text{Na}^+\text{(aq)} + \text{e}^-$; $\text{Cl}_2\text{(g)} + 2 \text{ e}^- \rightarrow 2 \text{ Cl}^-\text{(aq)}$.

halide ion   An anion formed from a halogen atom. *Examples:* F^-; I^-.

Hall process (Hall–Hérault process)   The production of aluminum by the electrolysis of aluminum oxide dissolved in molten cryolite.

haloalkane   An alkane with a halogen substituent. *Example:* CH_3Cl, chloromethane.

halogen   A member of Group 17/VII.

halogenation   The incorporation of a halogen into a compound (particularly, into an organic compound).

hamiltonian   The operator H in the Schrödinger equation; $H\psi = E\psi$.

hard water   Water that contains dissolved calcium and magnesium salts.

heat (q)   The energy that is transferred as the result of a temperature difference between a system and its surroundings.

heat capacity (C)   The ratio of heat supplied to the temperature rise produced. The *heat capacity at constant pressure,* C_P, and the *heat capacity at constant volume,* C_V, are normally distinguished. See also *molar heat capacity; specific heat capacity.*

heating   The act of transferring energy as heat.

heating curve   A graph of the variation of the temperature of a sample as it is heated at a constant rate.

Heisenberg uncertainty principle   If the location of a particle is known to within an uncertainty Δx, then the linear momentum parallel to the x-axis can be known only to within an uncertainty Δp, where $\Delta p \Delta x \geq \hbar/2$.

Henderson–Hasselbalch equation   An approximate equation for estimating the pH of a solution containing a conjugate acid and base. See also Section 11.2.

Henry's constant   The constant k_H that appears in *Henry's law.*

Henry's law   The solubility of a gas in a liquid is proportional to its partial pressure above the liquid: solubility = $k_H \times$ *partial pressure.*

hertz (Hz)   The SI unit of frequency: 1 Hz is one complete cycle per second; $1 \text{ Hz} = 1 \text{ s}^{-1}$.

Hess's law   A reaction enthalpy is the sum of the enthalpies of any sequence of reactions (at the same temperature and pressure) into which the overall reaction can be divided.

heterogenous alloy   See *alloy.*

heterogeneous catalyst   See *catalyst.*

heterogeneous equilibrium   An equilibrium in which at least one substance is in a different phase from the others. *Example:* $\text{AgCl(s)} \rightleftharpoons \text{Ag}^+\text{(aq)} + \text{Cl}^-\text{(aq)}$.

heterogeneous mixture   A mixture in which the individual components, although mixed together, lie in distinct regions that can be distinguished with an optical microscope. *Example:* a mixture of sand and sugar.

heterolytic dissociation   Dissociation into ions. *Example:* $\text{CH}_3\text{I} \rightarrow \text{CH}_3^+ + \text{I}^-$. Compare with *homolytic dissociation.*

heteronuclear diatomic molecule   A molecule consisting of two atoms of different elements. *Examples:* HCl; CO.

hexagonal close-packed structure (hcp)   A close-packed structure with an ABABAB . . . pattern of layers.

high-boiling azeotrope   See *azeotrope*.

highest occupied molecular orbital (HOMO)   The highest-energy molecular orbital in the ground state of a molecule occupied by at least one electron.

high-spin complex   A d^n complex with the maximum number of unpaired electron spins.

high-temperature superconductor   A material that becomes superconducting at temperatures well above the transition temperature for the first generation of superconductors, typically 100 K and above.

homeostasis   The maintenance of constant physiological conditions.

homogeneous alloy   See *alloy*.

homogeneous catalyst   See *catalyst*.

homogeneous equilibrium   A chemical equilibrium in which all the substances taking part are in the same phase. *Example:* $H_2(g) + I_2(g) \rightleftharpoons 2\ HI(g)$.

homogeneous mixture   A mixture in which the individual components are uniformly mixed, even on a molecular scale. *Examples:* air; solutions.

homolytic dissociation   Dissociation into radicals. *Example:* $CH_3I \rightarrow \cdot CH_3 + \cdot I$. Compare with *heterolytic dissociation*.

homonuclear diatomic molecule   A molecule consisting of two atoms of the same element. *Examples:* H_2; N_2.

Hund's rule   If more than one orbital in a subshell is available, add electrons with parallel spins to different orbitals of that subshell.

hybridization   The formation of *hybrid orbitals*.

hybrid orbital   A mixed orbital formed by blending together atomic orbitals on the same atom. *Example:* an sp^3 hybrid orbital.

hydrate   A solid compound containing H_2O molecules. *Example:* $CuSO_4 \cdot 5H_2O$.

hydrated   Having water molecules attached. See *hydration*.

hydrate isomers   Isomers that differ by an exchange of an H_2O molecule and a ligand in the coordination sphere.

hydration   (1) of ions: The attachment of water molecules to a central ion. (2) of organic compounds: The addition of water across a multiple bond (H to one carbon atom, OH to the other). *Example:* $CH_2{=}CH_2 + H_2O \rightarrow CH_3CH_2OH$.

hydride   A binary compound of a metal or metalloid with hydrogen; the term is often extended to include all binary compounds of hydrogen. A *saline* or *saltlike hydride* is a compound of hydrogen and a strongly electropositive metal; a *molecular hydride* is a compound of hydrogen and a nonmetal; a *metallic hydride* is a compound of certain *d*-block metals and hydrogen.

hydrocarbon   A binary compound of carbon and hydrogen. *Examples:* CH_4; C_6H_6.

hydrogenation   The addition of hydrogen to multiple bonds. *Example:* $CH_3CH{=}CH_2 + H_2 \rightarrow CH_3CH_2CH_3$.

hydrogen bond   A link formed by a hydrogen atom lying between two strongly electronegative atoms (O, N, or F). The electronegative atoms may be located on different molecules or in different regions of the same molecule.

hydrohalogenation   The addition of a hydrogen halide to an alkene to form a haloalkane. *Example:* $CH_3CH{=}CH_2 + HCl \rightarrow CH_3CHClCH_3$.

hydrolysis reaction   The reaction of water with a substance, resulting in the formation of a new element–oxygen bond. *Example:* $PCl_5(s) + 4\ H_2O(l) \rightarrow H_3PO_4(aq) + 5\ HCl(aq)$.

hydrolyze   To undergo hydrolysis. See also *hydrolysis reaction*.

hydrometallurgical process   The extraction of metals by reduction of their ions in aqueous solution. *Example:* $Cu^{2+}(aq) + Fe(s) \rightarrow Cu(s) + Fe^{2+}(aq)$.

hydronium ion   The ion H_3O^+.

hydrophilic   Water attracting. *Example:* hydroxyl groups are hydrophilic.

hydrophobic   Water repelling. *Example:* hydrocarbon chains are hydrophobic.

hydrostatic pressure   The pressure exerted by a column of fluid.

hydroxyl group   An —OH group in an organic compound.

hypervalent compound   A compound that contains an atom with more atoms attached than allowed by the octet rule. *Example:* SF_6.

hypothesis   A suggestion put forward to account for a series of observations. *Example:* Dalton's atomic hypothesis.

ideal gas   A gas that satisfies the ideal gas law and is described by the *kinetic model*.

ideal gas law ($PV = nRT$)   All gases obey the law more and more closely as the pressure is reduced to very low values.

ideal solution   A solution that obeys *Raoult's law* at any concentration; all solutions behave ideally as the concentration approaches zero. *Example:* benzene and toluene form an almost ideal system.

i **factor**   A factor that takes into account the existence of ions in an electrolyte solution, particularly for the interpretation of colligative properties. It indicates the number of particles formed from one formula unit of the solute. *Example:* $i \approx 2$ for very dilute NaCl(aq).

incandescence   Light emitted by a hot body.

incomplete octet   A valence shell of an atom that has fewer than eight electrons. *Example:* the valence shell of B in BF_3.

indefinite integral   An *integral* without limits attached. See also Appendix 1F.

indicator   A substance that changes color when it goes from its acidic to its basic form (an *acid–base indicator*) or from its oxidized to its reduced form (a *redox indicator*).

induced dipole moment   An electric dipole moment produced in a polarizable molecule by a neighboring charge or partial charge.

induced-fit mechanism   A model of the action of an enzyme in which the enzyme molecule adjusts its shape to accommodate the incoming substrate molecule. A modification of the *lock-and-key mechanism* of enzyme action.

induced nuclear fission   See *fission*.

inert   (1) Unreactive. (2) Thermodynamically unstable but survives for long periods (*nonlabile*).

inert-pair effect   The observation that an element displays a valence lower than expected from its group number. An *inert pair* is a pair of valence shell *s*-electrons that are tightly bound to the atom and might not participate in bond formation.

infrared radiation   Electromagnetic radiation with a lower frequency (longer wavelength) than that of red light but a higher frequency (shorter wavelength) than microwave radiation.

initial concentration (of a weak acid or base)   The concentration as prepared, as if no deprotonation or protonation had taken place.

initial rate   The rate at the start of the reaction when products are present in concentrations too low to affect the rate.

initiation   The formation of reactive intermediates that serve as chain carriers from a reactant at the start of a chain reaction. *Example:* $Br_2 \rightarrow Br\cdot + Br\cdot$.

inner transition metal   A member of the *f* block of the periodic table (the *lanthanoids* and *actinoids*).

inorganic chemistry   The study of the elements other than carbon and their compounds.

inorganic compound   A compound that is not organic. See also *organic compound*.

insoluble substance   A substance that does not dissolve in a specified solvent. When the solvent is not specified, water is generally meant.

instantaneous dipole moment   A dipole moment that arises from a transient redistribution of charge and is responsible for the *London force*.

instantaneous rate   The slope of the tangent of a graph of concentration against time.

insulator (electrical)   A substance that does not conduct electricity. *Examples:* nonmetallic elements; molecular solids.

integral   (1) The sum of infinitesimal quantities. (2) The inverse of a derivative in the sense that the integral of the first derivative of a function is the original function. See *definite integral, indefinite integral*.

integral calculus   The part of mathematics that deals with the combination of infinitesimal quantities and the areas under curves. See also Appendix 1F.

integrated rate law   An expression for the concentration of a reactant or product in terms of the time, obtained from the rate law of the reaction. *Example:* $[A] = [A]_0 e^{-kt}$.

intensity   The brightness of electromagnetic radiation. The intensity of a wave of electromagnetic radiation is proportional to the square of its *amplitude*.

intensive property   A physical property of a substance that is independent of the mass of the sample. *Examples:* density; molar volume; temperature.

intercept (of a graph)   The value at which the line cuts the specified (usually vertical) axis. See also Appendix 1E.

interference   Interaction between waves, leading to a greater amplitude (*constructive interference*) or to a smaller one (*destructive interference*).

interhalogen   A binary compound of two halogens. *Example:* IF_3.

intermediate   See *reaction intermediate*.

intermolecular   Between molecules.

intermolecular forces   The forces of attraction and repulsion between molecules. *Examples:* hydrogen bonding; dipole–dipole force; London force. See also *van der Waals interactions*.

internal energy (*U*)   In thermodynamics, the total energy of a system.

International System   See *SI*.

internuclear axis   The straight line between the nuclei of two bonded atoms.

interpolate   To find a value between two measured values.

interstice   A hole or gap in a crystal lattice.

interstitial alloy   See *alloy*.

interstitial compound   A compound in which one type of atom occupies the gaps in a lattice of other atoms. *Example:* an interstitial carbide.

intramolecular   Within a molecule.

ion   An electrically charged atom or group of atoms. *Examples:* Al^{3+}; SO_4^{2-}. See also *anion; cation*.

ion–dipole interaction   The attraction between an ion and the opposite partial charge of the electric dipole of a polar molecule.

ion exchange   The exchange of one type of ion in solution for another.

ionic bond   The attraction between the opposite charges of cations and anions.

ionic character   The extent to which ionic structures contribute to the resonance of a molecule or ion.

ionic compound   A compound that consists of ions. *Examples:* $NaCl$; KNO_3.

ionic conduction   Electrical conduction in which the charge is carried by ions. See also *electrical conduction; electronic conductor*.

ionic equation   See *complete ionic equation*.

ionic liquid   An ionic compound that is liquid at ordinary temperatures because one of its ions is a relatively large, organic ion. Ionic liquids are used as nontoxic, nonvolatile solvents.

ionic model   The description of bonding in terms of ions.

ionic radius   The contribution of an ion to the distance between neighboring ions in a solid ionic compound. In practice, the radius of an ion is defined as its contribution to the distance between the centers of neighboring ions, with the radius of the O^{2-} ion set equal to 140. pm.

ionic solid   A solid built from cations and anions. *Examples:* NaCl; KNO_3.

ionization   (1) of atoms and molecules: Conversion into ions by the transfer of electrons. *Example:* $K(g) \rightarrow K^+(g) + e^-(g)$. (2) of acids and bases: See *protonation* and *deprotonation*.

ionization constant   See *acidity constant*.

ionization energy (*I*)   The minimum energy required to remove an electron from the ground state of a gaseous atom, molecule, or ion. The *first ionization energy* is the ionization energy for the removal of the first electron from an atom. The *second ionization energy* is the ionization energy for removal of a second electron, and so on.

ionization isomers   Isomers that differ by the exchange of a ligand with an anion or neutral molecule outside the coordination sphere.

ionizing radiation   High-energy radiation (typically but not necessarily nuclear radiation) that can cause ionization.

ion pair   A cation and anion in close proximity.

ion-selective electrode   An electrode sensitive to the concentration of a particular ion.

irreversible process   A process that is not reversed by an infinitesimal change in a variable.

isoelectronic species   Species with the same number of atoms and the same number of valence electrons. *Examples:* F^- and Ne; SO_2 and O_3; CN^- and CO.

isolated system   See *system*.

isomer   One of two or more compounds that contain the same number of the same atoms in different arrangements. In *structural isomers*, the atoms have different partners or lie in a different order; in *stereoisomers*, the atoms have the same partners but are in different arrangements in space. *Optical isomers* are related like an object and its mirror image; they are types of stereoisomers. *Examples:* CH_3OCH_3 and CH_3CH_2OH (structural isomers); *cis*- and *trans*-2-butene (stereoisomers).

isomerization   A reaction in which a compound is converted into one of its isomers. *Example: cis*-butene $\rightarrow$ *trans*-butene.

isotherm   A line on a graph depicting the variation of a property at constant temperature.

isothermal change   A change that occurs at constant temperature.

isotope   One of two or more atoms that have the same atomic number but different mass number. *Example:* ^1H, ^2H, and ^3H are all isotopes of hydrogen.

isotopic abundance   See *abundance*.

isotopic dating   The determination of the age of objects by measuring the activity of the radioactive isotopes that they contain, particularly ^{14}C.

isotopic label   See *tracer*.

isotopomers   Molecules that differ in their isotopic composition. *Examples:* $^{12}CH_4$ and $^{13}CH_4$.

isotropic   Not depending on orientation.

joule (J)   The SI unit of energy (1 J = 1 $kg \cdot m^2 \cdot s^{-2}$).

Joule–Thomson effect   The cooling of a gas as it expands.

Kekulé structures   Two Lewis structures of benzene, consisting of alternating single and double bonds.

kelvin (K)   The SI unit of temperature. See also Appendix 1B.

Kelvin scale   A fundamental scale of temperature on which the triple point of water lies at 273.16 K and the lowest attainable temperature is at 0. The unit on the Kelvin scale is the *kelvin*, K.

ketone   An organic compound containing a carbonyl group between two carbon atoms, having the form R—CO—R′. *Example:* CH_3—CO—CH_2CH_3, butanone.

kilogram (kg)   The SI unit of mass. See also Appendix 1B.

kinetic energy (E_K)   The energy of a particle due to its motion. Kinetic energy may be *translational* (arising from motion through space), *rotational* (arising from rotation about a center of mass), or *vibrational* (arising from the oscillating motion of the atoms in a molecule). *Example:* the translational kinetic energy of a particle of mass *m* and speed *v* is $\frac{1}{2}mv^2$.

kinetic model   A model of the properties of an ideal gas in which pointlike molecules are in continuous random motion in straight lines until collisions occur between them.

kinetic molecular theory (KMT)   The mathematical version of the *kinetic model* of gases.

Kirchhoff's law   The relation between the standard reaction enthalpies at two temperatures in terms of the temperature difference and the difference in heat capacities (at constant pressure) of the products and reactants.

labile   Refers to species that survive only for short periods.

lanthanide   Former (and still widely used) term for *lanthanoid*.

lanthanoid   A member of the first row of the *f* block (lanthanum through ytterbium).

lanthanide contraction   The reduction of atomic radius of the elements following the lanthanides below the values that would be expected by extrapolation of the trend down a group (and arising from the poor shielding ability of *f*-electrons).

lattice   An orderly array of atoms, molecules, or ions in a crystal.

lattice energy The difference between the potential energy of ions in a crystal and that of the same widely separated ions in a gas.

lattice enthalpy The standard enthalpy change for the conversion of an ionic solid into a gas of ions.

law A summary of an extensive series of observations.

law of conservation of energy Energy can be neither created nor destroyed.

law of conservation of mass Matter (and specifically atoms) is neither created nor destroyed in a chemical reaction.

law of constant composition A compound has the same composition whatever its source.

law of mass action For an equilibrium of the form a A $+ b$ B $\rightleftharpoons c$ C $+ d$ D, the ratio $a_C{}^c a_D{}^d / a_A{}^a a_B{}^b$ evaluated at equilibrium is equal to a constant K, which has a specific value for a given chemical equation and temperature.

law of partial pressures See *Dalton's law of partial pressures*.

law of radioactive decay The rate of decay is proportional to the number of radioactive nuclides in the sample.

LCAO See *linear combination of atomic orbitals*.

Le Chatelier's principle When a stress is applied to a system in dynamic equilibrium, the equilibrium adjusts to minimize the effect of the stress. *Example:* a reaction at equilibrium tends to proceed in the endothermic direction when the temperature is raised.

leveling The observation that strong acids all have the same strength in water, and all behave as though they were solutions of H_3O^+ ions.

Lewis acid An electron pair acceptor. *Examples:* H^+; Fe^{3+}; BF_3.

Lewis base An electron pair donor. *Examples:* OH^-; H_2O; NH_3.

Lewis formula (for an ionic compound) A representation of the structure of an ionic compound showing the formula unit of ions in terms of their Lewis diagrams.

Lewis structure A diagram showing how electron pairs are shared between atoms in a molecule. *Examples:* H—C̈l̈: and Ö=C=Ö.

Lewis symbol (for atoms and ions) The chemical symbol of an element, with a dot for each valence electron.

ligand A group attached to the central metal ion in a complex; a *polydentate ligand* occupies more than one binding site. See also *ambidentate ligand*.

ligand field splitting (Δ) The energy separation of the e- and t-orbitals in a complex induced by the presence of ligands.

ligand field theory The theory of bonding in d-metal complexes, a more complete version of *crystal field theory*. See also *crystal field*.

light See *visible radiation*.

limiting enthalpy of solution See *enthalpy of solution*.

limiting law A law that is accurately obeyed only at the limit of a property, such as when a property (the pressure of a gas, for example) is made very small.

limiting reactant The reactant that governs the theoretical yield of product in a given reaction.

linear combination of atomic orbitals (LCAO) A molecular orbital formed by superimposing atomic orbitals.

linear momentum (p) The product of mass and velocity.

line structure A representation of the structure of an organic molecule in which lines represent bonds; carbon atoms and the hydrogen atoms attached to them are not usually shown explicitly.

linkage isomers Isomers that differ in the identity of the atom that a ligand uses to attach to a metal ion.

lipid A naturally occurring organic compound that dissolves in hydrocarbons but not in water. *Examples:* fats; steroids; terpenes; the molecules that form cell membranes.

liquid A fluid form of matter that has a well-defined surface and takes the shape of the part of the container that it occupies.

liquid crystal A substance that flows like a liquid but has molecules that lie in a moderately orderly array. Liquid crystals may be *nematic, smectic,* or *cholesteric,* depending on the arrangement of the molecules.

lock-and-key mechanism A model of enzyme action in which the enzyme is thought of as a lock and its substrate as a matching key.

logarithm If a number x is written as B^y, then y is the logarithm of x to the base B. For *common logarithms* (denoted log x), $B = 10$. For *natural logarithms* (denoted ln x), $B = $ e. See also Appendix 1D.

London force The force of attraction that arises from the interaction between instantaneous electric dipoles on neighboring molecules.

lone pair A pair of valence electrons that is not taking part in bonding.

long period A period of the periodic table with more than eight members.

long-range order An orderly arrangement of atoms or molecules that is repeated over long distances.

lowest unoccupied molecular orbital (LUMO) The lowest-energy molecular orbital that is unoccupied in the ground state.

low-spin complex A d^n complex with the minimum number of unpaired electron spins.

luminescence The emission of light from a process, other than incandescence, that results from the formation of an excited state.

Lyman series A series of lines in the spectrum of atomic hydrogen in which the transitions are to orbitals with $n = 1$.

lyotropic liquid crystal A liquid crystal that results from the action of a solvent on a solute.

macroscopic level   The level at which visible objects can be observed directly.

Madelung constant (*A*)   A number that appears in the expression for the lattice energy and depends on the type of crystal lattice. *Example: A* = 1.748 for the rock-salt structure.

magic numbers   The numbers of protons or neutrons that correlate with enhanced nuclear stability. *Examples:* 2, 8, 20, 50, 82, and 126.

magnetic field   A region of influence that affects the motion of moving charged particles

magnetic quantum number (m_l)   The quantum number that identifies the individual orbitals of a subshell of an atom and determines their orientation in space.

main group   Any one of the groups forming the *s* and *p* blocks of the periodic table (Groups 1, 2, and 13/III through 18/VIII).

malleability   The ability to be deformed by being struck by a hammer (as a metal).

manometer   An instrument used for measuring the pressure of a gas confined inside a container.

mantissa (of a logarithm)   The numbers to the right of the decimal point.

many-electron atom   An atom with more than one electron.

mass (*m*)   The quantity of matter in a sample.

mass number (*A*)   The total number of nucleons (protons plus neutrons) in the nucleus of an atom. *Example:* ^{14}C, with mass number 14, has 14 nucleons (6 protons and 8 neutrons).

mass percentage composition   The mass of a substance present in a sample, expressed as a percentage of the total mass of the sample.

mass spectrometer   An instrument used in *mass spectrometry.*

mass spectrometry   Technique for measuring the masses and abundances of atoms and molecules by passing a beam of ions through a magnetic field.

mass spectrum   The display of the relative number of particles with each specified mass; the output generated by the detector of a mass spectrometer. See also *mass spectrometry.*

material balance   The requirement that the sum of the concentrations of all forms of a solute in a solution be equal to the initial concentration of the solute. *Example:* the sum of the concentrations of HCN and CN^- in an aqueous solution of HCN is equal to the initial concentration of HCN.

materials science   The study of the chemical structures, compositions, and properties of materials.

matter   Anything that has mass and takes up space.

maximum-boiling azeotrope   See *azeotrope.*

Maxwell distribution of molecular speeds   The formula for calculating the percentage of molecules that move at any given speed in a gas at a specified temperature.

mean bond enthalpy (ΔH_B)   The mean of A—B bond enthalpies for a number of different molecules containing the A—B bond. See also *bond enthalpy.*

mean free path   The average distance that a molecule travels between collisions.

mean relative speed   The mean speed at which two molecules approach each other in a gas.

mechanical equilibrium   The state in which the pressure of a system is equal to that of its surroundings.

mechanism   See *reaction mechanism.*

medicinal chemistry   The application of chemical principles to the development of pharmaceuticals.

melting temperature   The temperature at which a substance melts. The *normal melting point* is the melting point under a pressure of 1 atm.

meniscus   The curved surface that a liquid forms in a narrow tube.

mesophase   A state of matter showing some of the properties of both a liquid and a solid (a liquid crystal).

metal   (1) A substance that conducts electricity, has a metallic luster, is malleable and ductile, forms cations, and has basic oxides. (2) A metal consists of cations held together by a sea of electrons. *Examples:* iron; copper; uranium.

metallic conductor   An electronic conductor with a resistance that increases as the temperature is raised.

metallic hydride   See *hydride.*

metallic bond   The form of bonding characteristic of metals in which cations are held together by a sea of electrons.

metallic solid   See *metal.*

metallocene   A compound in which a metal atom lies between two cyclic ligands, thus resembling a sandwich. *Example:* ferrocene (dicyclopentadienyliron(0), [Fe(C_5H_5)$_2$]).

metalloid   An element that has the physical appearance and properties of a metal but behaves chemically like a nonmetal. *Examples:* arsenic; polonium.

meter (m)   The SI unit of length. See also Appendix 1B.

micelle   A compact, often nearly spherical cluster of oriented detergent (surfactant) molecules.

Michaelis constant (K_M)   A constant in the rate law for the *Michaelis–Menten mechanism.*

Michaelis–Menten mechanism   A model of enzyme catalysis in which the enzyme and its substrate reach a rapid *pre-equilibrium* with the bound substrate–enzyme complex.

microporous catalyst   A catalyst with an open, porous structure. *Example:* A zeolite.

microscopic level   A level of description that refers to the very small, such as atoms.

microstate   A permissible arrangement of molecules in a sample (in the context of statistical thermodynamics and the statistical definition of entropy).

microwaves   Electromagnetic radiation with wavelengths close to 1 cm.

millimeter of mercury (mmHg)   The pressure exerted by a column of mercury of height 1 mm (at 15°C and in a standard gravitational field).

minerals   Substances that are mined; more generally, inorganic substances.

minimum-boiling azeotrope   See *azeotrope*.

mixture   A type of matter that consists of more than one substance and can be separated into its components by making use of the different physical properties of the substances present.

model   A simplified description of nature.

moderator   A substance that slows neutrons. *Examples:* graphite; heavy water.

molality   The amount of solute (in moles) per kilogram of solvent.

molar   Refers to the quantity per mole. *Examples: molar mass,* the mass per mole; *molar volume,* the volume per mole. (*Molar concentration* and some related quantities are exceptions.)

molar absorption coefficient   The constant of proportionality between the absorbance of a sample and the product of its molar concentration and path length. See also *Beer's law*.

molar concentration ([J])   The amount (in moles) of solute per liter of solution.

molar heat capacity (C_m)   The heat capacity per mole of substance.

molarity ([J])   Molar concentration.

molar mass   (1) The mass per mole of atoms of an element (formerly, *atomic weight*). (2) The mass per mole of molecules of a compound (formerly, *molecular weight*). (3) The mass per mole of formula units of an ionic compound (formerly, *formula weight*).

molar solubility (*s*)   The numerical value of the molar concentration of a saturated solution of a substance.

molar volume   The volume of a sample divided by the amount (in moles) of atoms or molecules that it contains.

mole (mol)   The SI unit of chemical amount. See also Appendix 1B.

molecular biology   The study of the functions of living organisms in terms of their molecular composition.

molecular compound   A compound that consists of molecules. *Examples:* water; sulfur hexafluoride; benzoic acid.

molecular formula   A combination of chemical symbols and subscripts showing the actual numbers of atoms of each element present in a molecule. *Examples:* H_2O; SF_6; C_6H_5COOH.

molecular hydride   See *hydride*.

molecularity   The number of reactant molecules (or free atoms) taking part in an *elementary reaction*. See also *bimolecular reaction; termolecular reaction; unimolecular reaction*.

molecular orbital   A one-electron wavefunction that spreads throughout a molecule and gives the probability (through its square) that an electron will be found at each location.

molecular orbital energy-level diagram   A portrayal of the relative energies of the molecular orbitals in a molecule.

molecular orbital theory   The description of molecular structure in which electrons occupy orbitals that spread throughout a molecule.

molecular solid   A solid consisting of a collection of individual molecules held together by intermolecular forces. *Examples:* glucose; aspirin; sulfur.

molecular weight   See *molar mass*.

molecule   (1) The smallest particle of a compound that possesses the chemical properties of the compound. (2) A definite, distinct, electrically neutral group of bonded atoms. *Examples:* H_2; NH_3; CH_3COOH.

mole fraction (*x*)   The amount of particles (molecules, atoms, or ions) of a substance in a mixture expressed as a fraction of the total amount of particles in the mixture.

mole ratio   The stoichiometric relation between two species in a chemical reaction written as a conversion factor. *Example:* (2 mol H_2)/(1 mol O_2) in the reaction 2 H_2(g) + O_2(g) → 2 H_2O(l).

momentum (*p*)   See *linear momentum*.

monatomic ion   An ion formed from a single atom. *Examples:* Na^+; Cl^-.

Mond process   The purification of nickel by the formation and decomposition of nickel carbonyl.

monomer   A small molecule from which a polymer is formed. *Examples:* $CH_2{=}CH_2$ for polyethylene; $NH_2(CH_2)_6NH_2$ for nylon.

monoprotic acid   A Brønsted acid with one acidic hydrogen atom. *Example:* CH_3COOH.

monosaccharide   An individual unit from which carbohydrates are considered to be composed. *Example:* $C_6H_{12}O_6$, glucose.

multiple bond   A double or triple bond between two atoms.

nanomaterials   Materials composed of nanoparticles or regular arrays of molecules or atoms such as nanotubes.

nanoparticles   Particles that range in size from about 1 to 100 nm and can be manufactured and manipulated at the molecular level.

nanoscience   The study of materials at the nanometer scale. These materials are larger than single atoms, but too small to exhibit most bulk properties.

nanotechnology   The study and manipulation of matter at the atomic level (nanometer scale).

natural abundance (of an isotope)   The abundance of an isotope in a sample of a naturally occurring material.

natural antilogarithm   See *exponential*.

natural logarithm   See *logarithm*.

naturally occurring   Found in nature without needing to be synthesized.

natural product   An organic substance that occurs naturally in the environment.

negative deviation (from Raoult's law)   The tendency of a nonideal solution to have a vapor pressure lower than predicted by Raoult's law.

nematic phase   A liquid-crystal phase in which rod-shaped molecules have axes arranged parallel to each other but staggered with respect to one another in other directions.

Nernst equation   The equation expressing the emf of an electrochemical cell in terms of the concentrations of the reagents taking part in the cell reaction; $E = E° - (RT/nF) \ln Q$.

net ionic equation   The equation showing the net change in a chemical reaction, obtained by canceling the spectator ions in a complete ionic equation. *Example*: $Ag^+(aq) + Cl^-(aq) \rightarrow AgCl(s)$.

network solid   A solid consisting of atoms linked together covalently throughout its extent. *Examples*: diamond; silica.

neutralization reaction   The reaction of an acid with a base to form a salt and water or another molecular compound. *Example*: $HCl(aq) + NaOH(aq) \rightarrow NaCl(aq) + H_2O(l)$.

neutron (n)   An electrically neutral subatomic particle found in the nucleus of an atom; it has approximately the same mass as a proton.

neutron emission   A nuclear decay process in which a neutron is emitted. In neutron emission, the mass number decreases by 1, but the charge number remains the same.

neutron-induced transmutation   The conversion of one nucleus into another by the impact of a neutron. *Example*: $^{58}_{26}Fe + 2\,^{1}_{0}n \rightarrow\,^{60}_{27}Co +\,^{0}_{-1}e$.

neutron-rich nucleus   A nucleus with such a high proportion of neutrons that it lies above the *band of stability*.

NO$_x$ An oxide, or mixture of oxides, of nitrogen, typically in atmospheric chemistry.

noble gas   A member of Group 18/VIII of the periodic table (the helium family).

nodal plane   A plane on which an electron will not be found.

node   A point or surface on which a wavefunction passes through zero.

nomenclature   See *chemical nomenclature*.

nonaqueous solution   A solution in which the solvent is not water. *Example*: sulfur in carbon disulfide.

nonbonding orbital   A valence-shell atomic orbital that has not been used to form a bond to another atom.

nonelectrolyte   A substance that dissolves to give a solution that does not conduct electricity. *Example*: sucrose.

nonelectrolyte solution   A solution of a nonelectrolyte.

nonexpansion work (w_e)   See *work*.

nonideal solution   A solution that does not obey Raoult's law. Compare with *ideal solution*.

nonlabile   Thermodynamically unstable but survives for a long period.

nonmetal   A substance that does not conduct electricity and is neither malleable nor ductile. *Examples*: all gases; phosphorus; sodium chloride.

nonpolar bond   (1) A covalent bond between two atoms that have zero partial charges. (2) A covalent bond between two atoms with the same or nearly the same electronegativity.

nonpolar molecule   A molecule with zero electric dipole moment.

normal boiling point (T_b)   (1) The boiling temperature when the pressure is 1 atm. (2) The temperature at which the vapor pressure of a liquid is 1 atm.

normal form   The form of a substance under typical everyday conditions (for instance, close to 1 atm, 25°C).

normal freezing point (T_f)   The temperature at which a liquid freezes at 1 atm.

normal melting point (T_f)   The melting point of a substance at a pressure of 1 atm.

***ns*-orbital**   An atomic orbital with principal quantum number n and $l = 0$.

n-type semiconductor   See *semiconductor*.

nuclear atom   The structure of the atom proposed by Rutherford: a central, small, very dense, positively charged nucleus surrounded by electrons.

nuclear binding energy (E_{bind})   The energy released when Z protons and $A - Z$ neutrons come together to form a nucleus. The greater the binding energy per nucleon, the lower the energy of the nucleus.

nuclear chemistry   The study of the chemical consequences of nuclear reactions.

nuclear decay   The spontaneous partial breakup of a nucleus (including its fission). Nuclear decay is also referred to as *nuclear disintegration*. Example: $^{226}_{88}Ra \rightarrow\,^{222}_{86}Rn +\,^{4}_{2}\alpha$.

nuclear equation   A summary of the changes in a nuclear reaction, written in a form resembling a chemical equation.

nuclear fission   See *fission*.

nuclear fusion   See *fusion*.

nuclear model   A model of the atom in which the electrons surround a minute central nucleus.

nuclear reaction   A change that a nucleus undergoes (such as a nuclear transmutation).

nuclear reactor   A device for achieving controlled self-sustaining nuclear fission.

nuclear transmutation   The conversion of one element into another. *Example*: $^{12}_{6}C +\,^{4}_{2}\alpha \rightarrow\,^{16}_{8}O + \gamma$.

nucleic acid   (1) The product of a condensation of nucleotides. (2) A molecule containing an organism's genetic information.

nucleon   A proton or a neutron; thus, either of the two principal components of a nucleus.

nucleophile   A reactant that seeks out centers of positive charge in a molecule. *Examples:* H_2O, OH^-.

nucleophilic substitition   Substitution that results from attack by a nucleophile. *Examples:* hydrolysis of haloalkanes, $CH_3Br + H_2O \rightarrow CH_3OH + HBr$.

nucleoside   A combination of an organic base and a ribose or deoxyribose molecule.

nucleosynthesis   The formation of an element.

nucleotide   A nucleoside with a phosphate group attached to the carbohydrate ring; one of the units from which nucleic acids are made.

nucleus   The small, positively charged particle at the center of an atom that is responsible for most of its mass.

nuclide   A specific nucleus. *Examples:* ^1_1H; $^{16}_8O$.

nutritional calorie (Cal)   A unit used in food science to measure the caloric value of food; 1 nutritional calorie = 1 kilocalorie.

occupy   Have the characteristics of the wavefunction of a specified state; to be in a specified state.

octahedral complex   A complex in which six ligands are arranged at the corners of a regular octahedron, with a metal atom at the center. *Example:* $[Fe(CN)_6]^{4-}$.

octahedral hole   A cavity in a (usually close packed) crystal lattice formed from six spheres at the corners of a regular octahedron.

octet   An ns^2np^6 valence electron configuration.

octet rule   When atoms form bonds, they proceed as far as possible toward completing their octets by sharing electron pairs.

oligopeptide   A short chain of amino acids connected by amide (peptide) bonds.

one-component phase diagram   See *phase diagram*.

open system   See *system*.

optical activity   The ability of a substance to rotate the plane of polarized light passing through it.

optical isomers   Isomers that are related like an object and its mirror image. *Optical isomerism* is the existence of optical isomers. *Optical isomerism* is a type of *stereoisomerism*.

orbital   See *atomic orbital* and *molecular orbital*.

orbital angular momentum   A measure of the rate of rotation.

orbital angular momentum quantum number (*l*)   The quantum number that specifies the subshell of a given shell in an atom and determines the shapes of the orbitals in the subshell; $l = 0, 1, 2, \ldots, n-1$. *Examples:* $l = 0$ for the *s*-subshell; $l = 1$ for the *p*-subshell. (The quantum number *l* also specifies the magnitude of the angular momentum of the electron around the nucleus.)

order of reaction   The power to which the concentration of a single substance is raised in a rate law. *Example:* if the rate = $k[SO_2][SO_3]^{-1/2}$, then the reaction is first order in SO_2, and of order $-\frac{1}{2}$ in SO_3.

ore   The natural mineral source of a metal. *Example:* Fe_2O_3, hematite, an iron ore.

organic chemistry   The branch of chemistry that deals with organic compounds.

organic compound   A compound containing the element carbon and usually hydrogen. (The carbonates are normally excluded.)

organometallic compound   A compound that contains a metal–carbon bond. *Example:* $Ni(CO)_4$ (complexes of CN^- are normally excluded).

oscillating   Varying in a periodic manner with time.

osmosis   The tendency of a solvent to flow through a semipermeable membrane into a more concentrated solution (a colligative property).

osmotic pressure (Π, pi)   The pressure needed to stop the flow of solvent through a semipermeable membrane. See also *osmosis*.

overall formation constant   See *formation constant*.

overall order   The sum of the powers to which individual concentrations are raised in the rate law of a reaction. *Example:* If rate = $k[SO_2][SO_3]^{-1/2}$, then the overall order is $\frac{1}{2}$.

overall reaction   The net outcome of a sequence of reactions.

overlap   The merging of orbitals belonging to different atoms of a molecule.

overpotential   The additional potential difference that must be applied beyond the cell potential to cause appreciable electrolysis.

oxidation   (1) Combination with oxygen. (2) A reaction in which an atom, ion, or molecule loses an electron. (3) A half-reaction in which the oxidation number of an element is increased. *Examples:* $2Mg(s) + O_2(g) \rightarrow 2MgO(s)$; (2, 3) $Mg(s) \rightarrow Mg^{2+}(s) + 2e^-$.

oxidation number   The effective charge on an atom in a compound, calculated according to a set of rules (see Toolbox K.1). An increase in oxidation number corresponds to oxidation; a decrease corresponds to reduction.

oxidation–reduction reaction   See *redox reaction*.

oxidation state   The actual condition of a species with a specified oxidation number.

oxidizing agent   A species that removes electrons from a species being oxidized (and is itself reduced) in a redox reaction. *Examples:* O_2; O_3; MnO_4^-; Fe^{3+}.

oxoacid   An acid that contains oxygen. *Examples:* H_2CO_3; HNO_3; HNO_2; $HClO$.

oxoanion   An anion of an oxoacid. *Examples:* HCO_3^-; CO_3^{2-}.

paired electrons   Two electrons with opposed spins ($\uparrow\downarrow$).

parallel spins   Electrons with spin in the same direction ($\uparrow\uparrow$).

paramagnetism   The tendency to be pulled into a magnetic field; a paramagnetic substance is composed of atoms or molecules with unpaired electrons. *Examples:* O_2; $[Fe(CN)_6]^{3-}$.

parent nucleus   In a nuclear reaction, the nucleus that undergoes disintegration or transmutation.

partial charge   A charge arising from small shifts in the distributions of electrons. A partial charge can be either *positive* ($\delta+$) or *negative* ($\delta-$).

partial pressure (P_X)   The pressure that a gas (X) in a mixture would exert if it alone occupied the container.

particle in a box   A particle confined between rigid walls.

parts per million (ppm)   (1) The ratio of the mass of a solute to the mass of the solution, multiplied by 10^6. (2) The *mass percentage composition* multiplied by 10^4. (*Parts per billion*, ppb, the mass ratio multiplied by 10^9, also may be used.)

pascal (Pa)   The SI unit of pressure: $1\ Pa = 1\ kg \cdot m^{-1} \cdot s^{-2}$. See also Appendix 1B.

passivation   Protection from further reaction by a surface film. *Example:* aluminum in air.

Pauli exclusion principle   See *exclusion principle*.

p-electron   An electron in a p-orbital.

penetration   The possibility that an s-electron may be found inside the inner shells of an atom and hence close to the nucleus.

peptide   A molecule formed by a condensation reaction between amino acids; often described in terms of the number of units, for example, *dipeptide, oligopeptide, polypeptide*.

peptide bond   The —CONH— group.

percentage composition   See *mass percentage composition; volume percentage composition*.

percentage deprotonation   The fraction of a weak acid, expressed as a percentage, that is present as its conjugate base in a solution.

percentage ionization   The fraction of molecules of a substance, expressed as a percentage, that is present as ions.

percentage protonated   The fraction of a base, expressed as a percentage, that is present as its conjugate acid in a solution.

percentage yield   The percentage of the theoretical yield of a product achieved in practice.

period   A horizontal row in the periodic table; the number of the period is equal to the principal quantum number of the valence shell of the atoms.

periodic table   A chart in which the elements are arranged in order of atomic number and divided into groups and periods in a manner that shows the relationships between the properties of the elements.

pH   The negative logarithm of the hydronium ion molarity in a solution: $pH = -\log[H_3O^+]$. A $pH < 7$ indicates an acidic solution; $pH = 7$, a neutral solution; and $pH > 7$, a basic solution.

phase   A specific physical state of matter. A substance may exist in solid, liquid, and gaseous phases and, in certain cases, in more than one solid or liquid phase. *Examples:* white and gray tin are two solid phases of tin; ice, liquid, and vapor are three phases of water.

phase boundary   A line separating two areas in a phase diagram; the points on a phase boundary represent the conditions at which the two adjoining phases are in dynamic equilibrium.

phase diagram   A summary in graphical form of the conditions of temperature and pressure at which the various solid, liquid, and gaseous phases of a substance exist. A *one-component phase diagram* is a phase diagram for a single substance.

phase transition   The conversion of a substance from one phase into another phase. *Examples:* vaporization; white tin $\rightarrow$ gray tin.

pH curve   A graph of the pH of a reaction mixture against volume of titrant added in an acid–base titration.

phenol   An organic compound in which a hydroxyl group is attached directly to an aromatic ring (Ar—OH). *Example:* C_6H_5OH, phenol.

phosphor   A phosphorescent material that emits light after excitation to higher energy states.

phosphorescence   Long-lasting *luminescence*.

photochemical reaction   A reaction caused by light. *Example:* $H_2(g) + Cl_2(g) \xrightarrow{h\nu} 2\ HCl(g)$.

photoelectric effect   The emission of electrons from the surface of a metal when electromagnetic radiation strikes it.

photon   A particle-like packet of electromagnetic radiation. The energy of a photon of frequency ν is $E = h\nu$.

physical chemistry   The study of the principles of chemistry.

physical equilibrium   A state in which two or more phases of a substance coexist without a tendency to change. *Example:* ice and water at 0°C and 1 atm.

physical property   A characteristic that we observe or measure without changing the identity of the substance.

physical state   The condition of being a solid, a liquid, or a gas at a particular temperature.

pi (π) bond   A bond formed by the side-to-side overlap of two p-orbitals.

piezoelectric   Having the property of becoming electrically charged when mechanically distorted. *Example:* $BaTiO_3$.

pi (π) orbital   A molecular orbital that has one nodal plane cutting through the internuclear axis.

pipet   A narrow tube, sometimes with a central bulb, calibrated to deliver a specified volume.

pK_a and pK_b   The negative logarithms of the acidity and basicity constants: $pK = -\log K$. The larger the value of pK_a or pK_b, the weaker the acid or base, respectively.

Planck's constant (h)   A fundamental constant of nature with the value $6.626\ 08 \times 10^{-34}\ J \cdot s$.

plasma   (1) An ionized gas. (2) In biology, the colorless component of blood in which the red and white blood cells are dispersed.

p–n junction   An interface between a p-type and an n-type semiconductor.

pOH   The negative logarithm of the hydroxide ion molarity in a solution; $pOH = -\log[OH^-]$.

poison   Verb: To inactivate a catalyst.

polar covalent bond   A covalent bond between atoms that have partial electric charges. *Examples:* H—Cl; O—S.

polarizability (α)   The ease with which the electron cloud of a molecule can be distorted.

polarizable   An easily polarized species.

polarize   To distort the electron cloud of an atom or ion.

polarized light   Plane-polarized light is light in which the wave motion occurs in a single plane.

polarizing power   The ability of an ion to polarize a neighboring atom or ion.

polar molecule   A molecule with a nonzero electric dipole moment. *Examples:* HCl; NH_3.

polyamide   A polymer in which the monomers are linked by amide bonds formed by condensation.

polymerization   The formation of a *polymer* from its *monomers*.

polymer   A compound in which chains or networks of small repeating units form giant molecules. *Examples:* polypropylene; polytetrafluoroethylene.

polyatomic ion   An ion in which more than two atoms are linked by covalent bonds. *Examples:* NH_4^+; NO_3^-; SiF_6^{2-}.

polyatomic molecule   A molecule that consists of more than two atoms. *Examples:* O_3; $C_{12}H_{22}O_{11}$.

polycyclic compound   An aromatic compound containing two or more benzene rings that share two neighboring carbon atoms. *Example:* naphthalene.

polydentate ligand   A ligand that can attach at several binding sites.

polyester   A polymer in which the monomers are linked by ester groups formed by condensation polymerization.

polymer   A substance with large molecules consisting of chains of covalently linked repeating units formed from small molecules known as *monomers*. *Examples:* polyethylene; nylon. See also *copolymer*.

polynucleotide   A polymer built from nucleotide units. *Examples:* DNA; RNA.

polypeptide   A polymer formed by the condensation of amino acids.

polyprotic acid or base   A Brønsted acid or base that can donate or accept more than one proton. (A polyprotic acid is sometimes called a polybasic acid.) *Examples:* H_3PO_4, triprotic acid; N_2H_4, diprotic base.

polysaccharide   A chain of many saccharide units, such as glucose, linked together. *Examples:* cellulose; amylose.

p-orbital   See *atomic orbital*.

positional disorder   Disorderly locations of molecules; a contribution to the entropy.

positive ion   An ion formed by loss of one or more electrons from an atom or molecule; a *cation*.

positron   A fundamental particle with the same mass as an electron but with opposite charge.

positron emission   A mode of radioactive decay in which a nucleus emits a positron.

potential difference   An electrical potential difference between two points is a measure of the work that must be done to move an electric charge from one point to the other. Potential difference is measured in volts, V, and is commonly called the *voltage*.

potential energy (E_P)   The energy arising from position. *Example:* the Coulomb potential energy of a charge is inversely proportional to its distance from another charge.

potential energy surface   A surface showing the variation of the potential energy with the relative location of the atoms in a polyatomic cluster of atoms (as in a collision between a diatomic molecule and an atom).

power   The rate of supply of energy. The SI unit of power is the watt, W ($1\ W = 1\ J \cdot s^{-1}$). See also Appendix 1B.

precipitate   The solid formed in a precipitation reaction.

precipitation   The process in which a solute comes out of solution rapidly as a finely divided powder, called a *precipitate*.

precipitation reaction   A reaction in which a solid product is formed when two solutions are mixed. *Example:* $KBr(aq) + AgNO_3(aq) \rightarrow KNO_3(aq) + AgBr(s)$.

precise measurements   (1) Measurements with a large number of significant figures. (2) A series of measurements with small random error and hence in close mutual agreement.

precision   Freedom from random error. Compare with *accuracy*.

pre-equilibrium condition   A pre-equilibrium arises (or is assumed) when an intermediate is formed in a rapid equilibrium reaction prior to a slow step in a reaction mechanism.

pre-exponential factor (A)   The constant obtained from the y intercept in an Arrhenius plot.

pressure (P)   Force divided by the area to which it is applied.

primary alcohol   See *alcohol*.

primary cell   A *galvanic cell* that produces electricity from chemicals sealed within it at the time of manufacture. It cannot be recharged.

primary pollutant   A pollutant directly introduced into the environment. *Example:* SO_2.

primary structure   The sequence of amino acids in the polypeptide chain of a protein.

primitive cubic structure   A structure in which the unit cell consists of spheres (representing atoms or ions) at the corners of a cube.

principal quantum number (n) The quantum number that specifies the energy of an electron in a hydrogen atom and labels the shells of the atom.

probability density (of a particle) A function that, when multiplied by the volume of the region, gives the probability that the particle will be found in that region of space. See also *Born interpretation*.

product A species formed in a chemical reaction.

promotion (of an electron) The conceptual excitation of an electron to an orbital of higher energy in the description of bond formation.

propagation A series of steps in a chain reaction in which one chain carrier reacts with a reactant molecule to produce another carrier. *Examples:* $Br\cdot + H_2 \rightarrow HBr + H\cdot$; $H\cdot + Br_2 \rightarrow HBr + Br\cdot$. See also *chain reaction*.

property A characteristic of matter. *Examples:* vapor pressure; color; density; temperature. See also *chemical property; physical property*.

protective oxide An oxide that protects a metal from oxidation. *Example:* aluminum oxide.

proton (p) A positively charged subatomic particle found in the nucleus of an atom.

protonation Proton transfer to a Brønsted base. *Example:* $H_3O^+(aq) + HS^-(s) \rightarrow H_2S(g) + H_2O(l)$.

proton emission A nuclear decay process in which a proton is emitted. In proton emission, the mass and charge numbers of the nucleus both decrease by 1.

proton-rich nucleus A nucleus that has a low proportion of neutrons and lies below the *band of stability*.

proton transfer equilibrium The equilibrium involving the transfer of a hydrogen ion between an acid and a base.

proton transfer reaction See *proton transfer equilibrium*.

pseudo-first-order reaction A reaction with a rate law that is effectively first order because all but one species have virtually constant concentrations.

p-type semiconductor See *semiconductor*.

pX The quantity $-\log X$. *Example:* $pOH = -\log [OH^-]$.

pyrometallurgical process The extraction of metals by using reactions at high temperatures. *Example:* $Fe_2O_3(s) + 3\ CO(g) \xrightarrow{\Delta} 2\ Fe(l) + 3\ CO_2(g)$.

quadratic equation An equation of the form $ax^2 + bx + c = 0$. See also Appendix 1E.

qualitative A nonnumerical description of the properties of a substance, system, or process. *Example:* qualitative analysis, the identification of the substances present in a sample.

qualitative analysis The determination of substances present in a sample.

quanta The plural of *quantum*.

quantitative A numerical description of the properties of a substance, system, or process. *Example:* quantitative analysis, the determination of the amounts or concentrations of substances present in a sample.

quantitative analysis The determination of the amount of each substance present in a sample.

quantization The restriction of a property to certain values. *Examples:* the quantization of energy and angular momentum.

quantum A packet of energy.

quantum mechanics The description of matter that takes into account the wave–particle duality of matter and the fact that the energy of an object may be changed only in discrete steps.

quantum number An integer (sometimes, a half-integer) that labels a wavefunction and specifies the value of a property. *Example:* principal quantum number, n.

quaternary ammonium ion An ion of the form NR_4^+, where R denotes hydrogen or an alkyl group (the four groups may be different).

quaternary structure The manner in which neighboring polypeptide units stack together to form a protein molecule.

racemic mixture A mixture containing equal amounts of two enantiomers.

rad A (non-SI) unit of *absorbed dose* of radiation; 1 rad corresponds to an energy deposit of 10^{-2} J·kg^{-1}. See also *gray*.

radial distribution function The function that gives the probability that an electron in an atom will be found at a particular radius regardless of the direction.

radial wavefunction ($R(r)$) The radial part of a wavefunction, particularly the radial component of the wavefunctions of the hydrogen atom; the probability amplitude of an electron as a function of distance from the nucleus.

radiation absorbed dose (rad) The amount of radiation that deposits 10^{-2} J of energy per kilogram of tissue.

radical An atom, molecule, or ion with at least one unpaired electron. *Examples:* $\cdot NO$; $\cdot O\cdot$; $\cdot CH_3$.

radical chain reaction A chain reaction propagated by radicals.

radical polymerization A polymerization procedure that utilizes a radical chain reaction.

radical scavenger A substance that reacts with radicals and inhibits a chain reaction.

radioactive A nucleus is radioactive if it can change its structure spontaneously and emit radiation.

radioactive series A step-by-step nuclear decay path in which α and β particles are successively ejected and which terminates at a stable nuclide (often a nuclide of lead).

radioactivity The spontaneous emission of radiation by nuclei. Such nuclei are radioactive.

radiocarbon dating *Isotopic dating* based specifically on the use of carbon-14.

radioisotope A radioactive isotope.

radius ratio The ratio of the radius of the smaller ion in an ionic solid to the radius of the larger ion. The radius ratio controls which crystal structure is adopted by a simple ionic solid.

random copolymer See *copolymer*.

random error An error that varies randomly from measurement to measurement, sometimes giving a high value and sometimes a low one.

Raoult's law The vapor pressure of a liquid solution of a nonvolatile solute is directly proportional to the mole fraction of the solvent in the solution: $P = x_{solvent}P_{pure}$, where P_{pure} is the vapor pressure of the pure solvent.

rate The change in a property divided by the time interval.

rate constant (k) The constant of proportionality in a rate law.

rate-determining step The elementary reaction that governs the rate of the overall reaction. *Example:* the step $O + O_3 \rightarrow O_2 + O_2$ in the decomposition of ozone.

rate law An equation expressing the instantaneous reaction rate in terms of the concentrations, at any instant, of the substances taking part in the reaction. *Example:* rate = $k[NO_2]^2$.

rate of reaction See *instantaneous rate; reaction rate*.

reactant A species acting as a starting material in a chemical reaction; a reagent taking part in a specified reaction.

reaction enthalpy (ΔH) The change of enthalpy for the reaction as expressed by the chemical equation, with stoichiometric coefficients interpreted as the amounts in moles. *Example:* $CH_4(g) + 2 O_2(g) \rightarrow CO_2(g) + 2 H_2O(l)$, $\Delta H = -890$ kJ.

reaction Gibbs free energy (ΔG) The change in Gibbs free energy for the reaction as expressed by the chemical equation with stoichiometric coefficients interpreted as amounts in moles.

reaction intermediate A species that is produced and consumed during a reaction but does not appear in the overall chemical equation.

reaction mechanism The pathway that is proposed for an overall reaction and accounts for the experimental rate law.

reaction order See *order of reaction*.

reaction profile The variation in potential energy as two reactants meet, form an activated complex, and separate as products.

reaction quotient (Q) The ratio of the activities of the products to those of the reactants, each raised to a power equal to the stoichiometric coefficient (as in the definition of the equilibrium constant, but at an arbitrary stage of a reaction). *Example:* for $N_2(g) + 3 H_2(g) \rightarrow 2 NH_3(g)$, $Q = (P_{NH_3})^2/P_{N_2}(P_{H_2})^3$.

reaction rate The unique rate of a chemical reaction calculated by dividing the change in concentration of a substance by the interval during which the change takes place and by taking into account the stoichiometric coefficient of the substance. See also *unique average reaction rate*.

reaction sequence A series of reactions in which the products of one reaction take part as reactants in the next. *Example:* $2 C(s) + O_2(g) \rightarrow 2 CO(g)$, followed by $2 CO(g) + O_2(g) \rightarrow 2 CO_2(g)$.

reaction stoichiometry The quantitative relation between the amounts of reactants consumed and those of products formed in chemical reactions as expressed by the balanced chemical equation for the reaction.

reagent A substance or a solution that reacts with other substances.

real gas An actual gas; a gas that differs from an ideal gas in its behavior.

recrystallization Purification by repeated dissolving and crystallization.

redox couple The oxidized and reduced forms of a substance taking part in a reduction or oxidation half-reaction. The notation is oxidized species/reduced species. *Example:* H^+/H_2.

redox indicator See *indicator*.

redox reaction A reaction in which oxidation and reduction take place. *Example:* $S(s) + 3 F_2(g) \rightarrow SF_6(g)$.

redox titration See *titration*.

reducing agent The species that supplies electrons to a substance being reduced (and is itself oxidized) in a redox reaction. *Examples:* H_2; H_2S; SO_3^{2-}.

reduction (1) The removal of oxygen from, or the addition of hydrogen to, a compound. (2) A reaction in which an atom, ion, or molecule gains an electron. (3) A half-reaction in which the oxidation number of an element is decreased. *Example:* $Cl_2(g) + 2 e^- \rightarrow 2 Cl^-(aq)$.

re-forming reaction A reaction in which a hydrocarbon is converted into carbon monoxide and hydrogen over a nickel catalyst.

refractory Able to withstand high temperatures.

relative biological effectiveness (Q) A factor used when assessing the damage caused by a given dose of radiation.

rem See *roentgen equivalent man*.

repeating unit The combination of atoms in a polymer that repeats over and over again in the polymer chain.

residual entropy The nonzero entropy at $T = 0$ in certain systems, which is due to a surviving disorder in the orientation of molecules.

residue (biochemical) An amino acid in a polypeptide chain.

resistance (electrical) A measure of the ability of matter to conduct electricity: the lower the resistance, the better it conducts.

resonance A blending of Lewis structures into a single composite, hybrid structure. *Example:* $:\ddot{O}-\ddot{S}=\ddot{O} \longleftrightarrow \ddot{O}=\ddot{S}-\ddot{O}:$.

resonance hybrid The composite structure that results from resonance.

reverse osmosis   The passage of solvent out of a solution when a pressure greater than the osmotic pressure is applied on the solution side of a semipermeable membrane.

reversible isothermal expansion   Expansion at constant temperature against an external pressure matched to the pressure of the system.

reversible process   A process that can be reversed by an infinitesimal change in a variable.

rock-salt structure   A crystal structure the same as that of a mineral form of sodium chloride.

roentgen equivalent man (rem)   The (non-SI) unit for reporting *dose equivalent*. See also *sievert*.

root mean square speed (v_{rms})   The square root of the average value of the squares of the speeds of the molecules in a sample.

roots (of an equation)   The solutions of the equation $f(x) = 0$. See also Appendix 1E.

rotational kinetic energy   See *kinetic energy*.

rounding off   The adjustment of a numerical result to the correct number of significant figures.

Rydberg constant ($\mathcal{R}$)   The constant in the formula for the frequencies of the lines in the spectrum of atomic hydrogen; $\mathcal{R} = 3.289\ 84 \times 10^{15}$ Hz.

sacrificial anode   A metal electrode that is allowed to decay in order to protect a metallic artifact. See *cathodic protection*.

saline hydride   See *hydride*.

salt   (1) An ionic compound. (2) The ionic product of the reaction between an acid and a base. *Examples:* $NaCl$; K_2SO_4.

salt bridge   An inverted U-shaped tube containing a concentrated salt (potassium chloride or potassium nitrate) in a gel that acts as an electrolyte and provides a conducting path between two compartments of an electrochemical cell.

sample   A representative part of a whole.

saturated   Unable to take up further material.

saturated hydrocarbon   A hydrocarbon with no carbon–carbon multiple bonds. *Example:* CH_3CH_3.

saturated solution   A solution in which the dissolved and undissolved solute are in dynamic equilibrium.

Schrödinger equation   An equation for calculating the wavefunction of a particle, especially for an electron in an atom or molecule. See also *wavefunction*.

science   The systematically collected and organized body of knowledge based on experiment, observation, and careful reasoning.

scientific method   A set of procedures employed to develop a scientific understanding of nature.

scientific notation   The expression of numbers in the form $n.nnn \ldots \times 10^a$.

scintillation counter   A device for detecting and measuring radioactivity that makes use of the fact that certain substances give a flash of light when they are exposed to radiation.

sea of instability   A region in a graph of mass number against atomic number corresponding to unstable nuclei that decay spontaneously with the emission of radiation. See also *band of stability*.

second (s)   The SI unit of time. See also Appendix 1B.

secondary alcohol   See *alcohol*.

secondary cell   A *galvanic cell* that must be charged (or recharged) by using an externally supplied current before it can be used.

secondary pollutant   A pollutant formed by the chemical reaction of another species in the environment. *Example:* SO_3 from the oxidation of SO_2.

secondary structure   The manner in which a polypeptide chain is coiled. *Examples:* α helix; β sheet.

second derivative (d^2y/dx^2)   A measure of the curvature of a function. See also Appendix 1F.

second ionization energy (I_2)   The energy needed to remove an electron from a singly charged gas-phase cation. *Example:* $Cu^+(g) \rightarrow Cu^{2+}(g) + e^-(g)$, $I_2 = 1958$ kJ·mol^{-1}.

second law of thermodynamics   A spontaneous change is accompanied by an increase in the total entropy of the system and its surroundings.

second-order reaction   (1) A reaction with a rate that is proportional to the square of the molar concentration of a reactant. (2) A reaction with an overall order of 2.

second virial coefficient (B)   See *virial equation*.

selective precipitation   The precipitation of one compound in the presence of other, more soluble compounds.

self-sustaining fission   Induced nuclear fission that, once it is initiated, can continue even if the supply of neutrons from outside is discontinued.

semiconductor   An *electronic conductor* with a resistance that decreases as the temperature is raised. In an *n-type semiconductor,* the current is carried by electrons in a largely empty band; in a *p-type semiconductor,* the conduction is a result of electrons missing from otherwise filled bands.

semiempirical method   The calculation of molecular structure that draws on experimental information to simplify the procedure. Compare with *ab initio method*.

semipermeable membrane   A membrane that allows only certain types of molecules or ions to pass.

sequester   Form a complex between a cation and a bulky molecule or ion that enwraps the central ion. *Example:* Ca^{2+} and $O_3POPO_2OPO_3^{3-}$.

series (in spectroscopy)   A family of spectral lines arising from transitions that have one state in common. *Example:* the Balmer series in the spectrum of atomic hydrogen.

shell   All the orbitals of a given principal quantum number. *Example:* the single 2s- and three 2p-orbitals of the shell with $n = 2$.

shielding   The repulsion experienced by an electron in an atom that arises from the other electrons present and opposes the attraction exerted by the nucleus.

shift reaction   A reaction between carbon monoxide and water: $CO(g) + H_2O(g) \rightarrow CO_2(g) + H_2(g)$; the reaction is used in the manufacture of hydrogen.

short-range order   Atoms or molecules lying in a regular arrangement that does not extend very far past their nearest neighbors.

SI (Système International)   The International System of units; a collection of definitions of units and symbols and their deployment. It is an extension and rational ization of the metric system. See also Appendix 1B.

side chain   A hydrocarbon substituent on a hydrocarbon chain.

sievert (Sv)   The SI unit of dose equivalent: 1 Sv = 1 J·kg^{-1}.

sigma (σ) bond   Two electrons in a cylindrically symmetrical cloud between two atoms.

sigma (σ) orbital   A molecular orbital that has no nodal plane containing the internuclear axis.

significant figures (sf, in a measurement)   The digits in the measurement, up to and including the first uncertain digit in scientific notation. *Example:* 0.0260 mL (that is, 2.60×10^{-2} mL), a measurement with three significant figures (3 sf).

single bond   A shared electron pair.

skeletal equation   An unbalanced equation that summarizes the qualitative information about the reaction. *Example:* $H_2 + O_2 \rightarrow H_2O$.

slip plane   A layer of atoms in a solid that may slide relatively easily across a neighboring layer.

slope (of graph)   The gradient of a graph. See also Appendix 1E.

smectic phase   A liquid-crystal phase in which molecules lie parallel to one another and form layers.

sol   A colloidal dispersion of solid particles in a liquid.

solid   A rigid form of matter that maintains the same shape whatever the shape of its container.

solid emulsion   A colloidal dispersion of a liquid in a solid. *Example:* butter, an emulsion of water in butterfat.

solid solution   A solid homogeneous mixture of two or more substances.

solubility   The concentration of a saturated solution of a substance.

solubility constant   See *solubility product*.

solubility product (K_{sp})   The product of ionic molar concentrations in a saturated solution; the dissolution equilibrium constant. *Example:* $Hg_2Cl_2(s) \rightleftharpoons Hg_2^{2+}(aq) + 2 Cl^-(aq)$, $K_{sp} = [Hg_2^{2+}][Cl^-]^2$.

solubility rules   A summary of the solubility pattern of a range of common compounds in water. See also Table I.1.

soluble substance   A substance that dissolves to a significant extent in a specified solvent. When the solvent is not specified, water is generally meant.

solute   A dissolved substance.

solution   A homogeneous mixture. See also *solute; solvent.*

solvated   Surrounded by and linked to solvent molecules. *Hydration* is the special case in which the solvent is water.

solvent   The most abundant component of a solution.

solvent extraction   A process for separating a mixture of substances that makes use of their differing solubilities in various solvents.

s-orbital   See *atomic orbital.*

space-filling model   A depiction of a molecule in which atoms are represented by spheres that indicate the space occupied by each atom.

sp^3d^n hybrid orbital   A hybrid orbital formed from one s-orbital, three p-orbitals, and n d-orbitals.

species   An atom, ion, or molecule.

specific enthalpy (of a fuel)   The enthalpy of combustion per gram (without the negative sign).

specific heat capacity   The heat capacity of a sample divided by its mass.

spectator ion   An ion that is present but remains unchanged during a reaction. *Examples:* Na^+ and NO_3^- in $NaCl(aq) + AgNO_3(aq) \rightarrow NaNO_3(aq) + AgCl(s)$.

spectral line   Radiation of a single wavelength emitted or absorbed by an atom or molecule.

spectrochemical series   Ligands ordered according to the strength of the ligand field splitting that they produce.

spectrometer   An instrument for recording the spectrum of a sample.

spectrophotometer   An instrument for measuring and recording electronically the intensity of radiation passing through a sample as the wavelength of the radiation is changed and hence recording the spectrum of the sample.

spectroscopy   The analysis of the electromagnetic radiation emitted, absorbed, or scattered by substances.

spectrum   The frequencies or wavelengths of the electromagnetic radiation emitted, absorbed, or scattered by substances.

speed   The magnitude of the velocity; the rate of change of position.

sphalerite structure   See *zinc-blende structure.*

spherically symmetrical   Independent of orientation about a central point.

spherical polar coordinates   The coordinates of a point expressed in terms of the *radius r*, the *colatitude θ*, and the *azimuth ϕ*.

sp^n hybrid   A hybrid orbital constructed from an s-orbital and n p-orbitals. There are two *sp hybrids,* three *sp² hybrids,* and four *sp³ hybrids.*

sp^n **hybrid orbital**   A hybrid orbital formed from one *s*-orbital and *n p*-orbitals.

spin   The intrinsic angular momentum of an electron; the spin cannot be eliminated and can occur in only two senses, denoted ↑and ↓ or α and β.

spin magnetic quantum number (m_s)   The quantum number that distinguishes the two spin states of an electron: $m_s = +\frac{1}{2}$ (↑) and $m_s = -\frac{1}{2}$ (↓).

spontaneous change   A natural change, one that has a tendency to occur without needing to be driven by an external influence. *Examples:* a gas expanding into a vacuum; a hot object cooling; methane burning.

spontaneous neutron emission   The decay of *neutron-rich nuclei* by the emission of neutrons without an external stimulus to do so.

spontaneous nuclear fission   See *fission.*

square planar complex   A complex in which four ligands lie at the corners of a square with the metal atom at the center.

stability constant   See *formation constant.*

stable   See *thermodynamically stable compound.*

standard ambient temperature and pressure (SATP)   25°C (298.15 K) and 1 bar.

standard cell potential   See *standard emf.*

standard emf ($E°$)   The emf when the concentration of each solute taking part in the cell reaction is 1 mol·L^{-1} (strictly, unit activity) and all the gases are at 1 bar. The *standard emf* of a galvanic cell is the difference between its two standard potentials: $E° = E°$(cathode) $- E°$(anode).

standard enthalpy of combustion ($\Delta H_c°$)   The change of enthalpy per mole of substance when it burns (reacts with oxygen) completely under standard conditions.

standard enthalpy of formation ($\Delta H_f°$)   The standard reaction enthalpy per mole of compound for the compound's synthesis from its elements in their most stable form at 1 bar and the specified temperature.

standard entropy of fusion ($\Delta S_{fus}°$)   The standard entropy change per mole accompanying fusion (the conversion of a substance from the solid state into the liquid state).

standard entropy of vaporization ($\Delta S_{vap}°$)   The standard entropy change per mole accompanying vaporization (the conversion of a substance from the liquid state into the vapor state).

standard Gibbs free energy of formation ($\Delta G_f°$)   The standard reaction Gibbs free energy per mole for the formation of a compound from its elements in their most stable form.

standard Gibbs free energy of fusion ($\Delta G_{fus}°$)   The standard Gibbs free energy change per mole accompanying fusion (the conversion of a substance from the solid state into the liquid state).

standard Gibbs free energy of reaction ($\Delta G°$)   The Gibbs free energy of reaction under standard conditions.

standard Gibbs free energy of vaporization ($\Delta G_{vap}°$)   The standard Gibbs free energy change per mole accompanying vaporization (the conversion of a substance from the liquid state into the vapor state).

standard hydrogen electrode (SHE)   A hydrogen electrode that is in its standard state (hydrogen ions at concentration 1 mol·L^{-1} (strictly, unit activity) and hydrogen pressure 1 bar) and is defined as having $E° = 0$ at all temperatures.

standard molar entropy ($S_m°$)   The entropy per mole of a pure substance at 1 bar.

standard potential ($E°$)   (1) The contribution of an electrode to the standard emf of a cell. (2) The standard emf of a cell when the left-hand electrode is a standard hydrogen electrode and the right-hand electrode is the electrode of interest.

standard pressure ($P°$)   A pressure of 1 bar exactly.

standard reaction enthalpy ($\Delta H°$)   The reaction enthalpy under standard conditions.

standard reaction entropy ($\Delta S°$)   The reaction entropy under standard conditions.

standard state   The pure form of a substance at 1 bar; for a solute, the concentration 1 mol·L^{-1}.

standard temperature and pressure (STP)   0°C (273.15 K) and 1 atm (101.325 kPa).

standing wave   A wave stable over time; a wave with peaks and troughs that do not migrate with time.

state function   A property of a substance that is independent of how the sample was prepared. *Examples:* pressure; enthalpy; entropy; color.

state of matter   The physical condition of a sample. The most common states of a pure substance are solid, liquid, or gas (vapor).

state property   See *state function.*

state symbol   A symbol (abbreviation) denoting the state of a species. *Examples:* s (solid); l (liquid); g (gas); aq (aqueous solution).

statistical entropy   The entropy calculated from statistical thermodynamics; $S = k \ln W$.

statistical thermodynamics   The interpretation of the laws of thermodynamics in terms of the behavior of large numbers of atoms and molecules.

steady-state approximation   The assumption that the net rate of formation of reaction intermediates is 0.

Stefan–Boltzmann law   The total intensity of radiation emitted by a heated black body is proportional to the fourth power of the absolute temperature.

stereoisomers   Isomers in which atoms have the same partners arranged differently in space.

stereoregular polymer   A polymer in which each unit or pair of repeating units has the same relative orientation.

steric factor (P)   An empirical factor that takes into account the steric requirement of a reaction.

steric requirement   A constraint on an elementary reaction in which the successful collision of two molecules depends on their relative orientation.

Stern–Gerlach experiment   The demonstration of the quantization of electron spin by passing a beam of atoms through a magnetic field.

stick structure   See *line structure*.

stock solution   A solution stored in concentrated form.

stoichiometric coefficients   The numbers multiplying chemical formulas in a chemical equation. *Examples:* 1, 1, and 2 in $H_2 + Br_2 \rightarrow 2\ HBr$.

stoichiometric point   The stage in a titration when exactly the right volume of solution needed to complete the reaction has been added.

stoichiometric proportions   Reactants in the same proportions as their coefficients in the chemical equation. *Example:* equal amounts of H_2 and Br_2 in the formation of HBr.

stoichiometric relation   An expression that equates the relative amounts of reactants and products that participate in a reaction. *Example:* $1\ \text{mol}\ H_2 \triangleq 2\ \text{mol}\ HBr$ for $H_2 + Br_2 \rightarrow 2\ HBr$.

stoichiometry   See *reaction stoichiometry*.

stopped-flow technique   A procedure for observing fast reactions, involving the spectrometric analysis of a reaction mixture immediately after the rapid injection of reactants into a mixing chamber.

STP   See *standard temperature and pressure*.

strong acids and bases   Acids and bases that are fully deprotonated or protonated, respectively, in solution. *Examples:* HCl, $HClO_4$ (strong acids); NaOH, $Ca(OH)_2$ (strong bases).

strong electrolyte   See *electrolyte*.

strong-field ligand   A ligand that produces a large *ligand field splitting* and lies above H_2O in the spectrochemical series.

strong force   A short-range but very strong force that acts between nucleons and binds them together to form a nucleus.

structural formula   A chemical formula that shows how atoms in a compound are attached to one another. See also *condensed structural formula*.

structural isomers   Isomers in which the atoms have different partners.

subatomic particle   A particle smaller than an atom. *Examples:* electron; proton; neutron.

subcritical   Having a mass less than the *critical mass*.

sublimation   The direct conversion of a solid into a vapor without first forming a liquid.

sublimation vapor pressure   The vapor pressure of a solid.

subshell   All the atomic orbitals of a given shell of an atom that have the same value of the quantum number l. *Example:* the five $3d$-orbitals of an atom.

substance   A single, pure type of matter; either a compound or an element.

substituent   An atom or group that has replaced a hydrogen atom in an organic molecule.

substitutional alloy   See *alloy*.

substitution reaction   (1) A reaction in which an atom (or a group of atoms) replaces an atom in the original molecule. (2) In complexes, a reaction in which one Lewis base expels another and takes its place. *Examples:* (1) $C_6H_5OH + Br_2 \rightarrow BrC_6H_4OH + HBr$; (2) $[Fe(H_2O)_6]^{3+}(aq) + 6\ CN^-(aq) \rightarrow [Fe(CN)_6]^{3-}(aq) + 6\ H_2O(l)$.

substrate   The chemical species on which an enzyme acts.

superconductor   An *electronic conductor* that conducts electricity with zero resistance. See also *high-temperature superconductor*.

supercooled   Refers to a liquid cooled to below its freezing point but not yet frozen.

supercritical fluid   A fluid phase of a substance above its *critical temperature* and *critical pressure*.

supercritical   Having a mass greater than the *critical mass*.

superfluidity   The ability to flow without viscosity.

superimpose   The combination of atomic orbitals to form a molecular orbital.

surface-active agent   See *surfactant*.

surface tension (γ)   The tendency of molecules at the surface of a liquid to be pulled inward, resulting in a smooth surface.

surfactant   A substance that accumulates at the surface of a solution and affects the surface tension of the solvent; a component of detergents. *Example:* the stearate ion of soaps.

surroundings   The region outside a system, where observations are made.

suspension   A mist of small particles in a fluid medium.

sustainable development   The economical utilization and renewal of resources coupled with hazardous waste reduction and concern for the environment.

symbolic level   The discussion of chemical phenomena in terms of chemical symbols and mathematical equations.

synthesis   A reaction in which a substance is formed from simpler starting materials. *Example:* $N_2(g) + 3\ H_2(g) \rightarrow 2\ NH_3(g)$.

synthesis gas   A mixture of carbon monoxide and hydrogen produced by the catalyzed reaction of a hydrocarbon and water.

system   The object of study, usually a reaction vessel and its contents. An *open system* can exchange both matter and energy with the surroundings. A *closed system* has a fixed amount of matter but can exchange energy with the surroundings. An *isolated system* has no contact with its surroundings.

systematic error   An error that persists in a series of measurements and does not average out. See also *accuracy*.

systematic name   The name of a compound that reveals which elements are present (and, in its most complete form, how the atoms are arranged). *Example:* methylbenzene is the systematic name for toluene.

Système International d'Unités   See *SI*.

temperature (T)   (1) How hot or cold a sample is. (2) The intensive property that determines the direction in which heat will flow between two objects in contact.

temperature–composition diagram   A *phase diagram* showing how the normal boiling point of a liquid mixture varies with composition.

termination   A step in a *chain reaction* in which chain carriers combine to form products. *Example:* Br· + Br· → Br$_2$.

termolecular reaction   An *elementary reaction* in which three species collide simultaneously.

tertiary alcohol   See *alcohol*.

tertiary structure   The shape into which the α-helical and β-sheet sections of a polypeptide are twisted as a result of interactions between peptide groups lying in different parts of the primary structure.

tetrahedral complex   A complex in which four ligands lie at the corners of a regular tetrahedron with a metal atom at the center. *Example:* [Cu(NH$_3$)$_4$]$^{2+}$.

tetrahedral hole   A cavity in a (usually close-packed) crystal structure formed by one sphere lying in the dip formed by three others.

theoretical chemistry   The study of molecular structure and properties in terms of mathematical models.

theoretical yield   The maximum quantity of product that can be obtained, according to the reaction stoichiometry, from a given quantity of a specified reactant.

theory   A collection of ideas and concepts used to account for a scientific law.

thermal decomposition   See *decomposition*.

thermal disorder   The disorder arising from the thermal motion of molecules.

thermal equilibrium   The state in which a system is at the same temperature as its surroundings.

thermal motion   The random, chaotic motion of atoms.

thermal pollution   The damage caused to the environment by the waste heat of an industrial process.

thermite reaction (thermite process)   The reduction of a metal oxide by aluminum. *Example:* 2 Al(s) + Fe$_2$O$_3$(s) → Al$_2$O$_3$(s) + 2 Fe(l).

thermochemical equation   An expression consisting of the balanced chemical equation and the corresponding reaction enthalpy.

thermochemistry   The study of the heat released or absorbed by chemical reactions; a branch of thermodynamics.

thermodynamically stable compound   (1) A compound with no thermodynamic tendency to decompose into its elements. (2) A compound with a negative Gibbs free energy of formation.

thermodynamically unstable compound   (1) A compound with a thermodynamic tendency to decompose into its elements. (2) A compound with a positive Gibbs free energy of formation.

thermodynamics   The study of the transformations of energy from one form to another. See also *first law of thermodynamics; second law of thermodynamics, third law of thermodynamics*.

thermonuclear explosion   An explosion resulting from uncontrolled nuclear fusion.

thermotropic liquid crystal   A liquid crystal prepared by melting the solid phase.

third law of thermodynamics   The entropies of all perfect crystals are the same at the absolute zero of temperature.

third virial coefficient   See *virial equation*.

three-center bond   A chemical bond in which a hydrogen atom lies between two other atoms (typically boron atoms) and one electron pair binds all three atoms together.

tie line (in a temperature-composition phase diagram)   A line joining the point indicating the boiling point of a mixture of given composition with the corresponding composition of the vapor at that temperature.

titrant   The solution of known concentration added from a buret in a titration.

titration   The analysis of composition by measuring the volume of one solution needed to react with a given volume of another solution. In an *acid–base titration*, an acid is titrated with a base; in a *redox titration*, an oxidizing agent is titrated with a reducing agent.

t-orbital   One of the orbitals d_{xy}, d_{yz}, and d_{zx} in an octahedral or tetrahedral complex. In an octahedral complex, these orbitals are designated t_{2g}.

torr (symbol: Torr)   A unit of pressure: 760 Torr = 1 atm exactly.

total energy   See *energy*.

tracer   An isotope that can be tracked from compound to compound in the course of a sequence of reactions.

trajectory   The path of a particle on which location and linear momentum are specified at each instant.

transition   A change of state. (1) In thermodynamics, a change of physical state. (2) In spectroscopy, a change of quantum state.

transition metal   An element that belongs to Groups 3 through 11. *Examples:* vanadium; iron; gold.

translational kinetic energy   See *kinetic energy*.

transition state   See *activated complex*.

transition-state theory   A theory of reaction rates in which the reactants form an activated complex.

transmeitnerium elements   The elements beyond meitnerium; those with $Z > 109$.

transmutation   See *nuclear transmutation*.

transuranium elements   The elements beyond uranium; those with $Z > 92$.

triboluminescence   Luminescence that results from mechanical shock to a crystal.

triple bond   (1) Three electron pairs shared by two neighboring atoms. (2) One σ-bond and two π-bonds between neighboring atoms.

triple point   The point where three phase boundaries meet in a phase diagram. Under the conditions represented by the triple point, all three adjoining phases coexist in dynamic equilibrium.

triprotic   See *polyprotic acid or base*.

Trouton's rule   The empirical observation that the entropy of vaporization at the boiling point (the enthalpy of vaporization divided by the boiling temperature) is approximately 85 $J \cdot K^{-1} \cdot mol^{-1}$ for many liquids.

tube structure   A representation of molecular structure that uses tubes to indicate the lengths and distances of bonds. The ends of each tube are colored to indicate the identities of the elements forming the bond.

ultraviolet catastrophe   The classical prediction that any black body at any temperature should emit intense ultraviolet radiation.

ultraviolet radiation   Electromagnetic radiation with a higher frequency (shorter wavelength) than that of violet light.

unbranched alkane   An alkane with no side chains, in which all the carbon atoms lie in a linear chain.

uncertainty principle   See *Heisenberg uncertainty principle*.

unimolecular reaction   An elementary reaction in which a single reactant molecule changes into products. *Example:* $O_3 \rightarrow O_2 + O$.

unique average reaction rate   The rate of change in the concentration of a reactant or product divided by its stoichiometric coefficient in the balanced equation. All unique average rates are reported as positive quantities. See also *average reaction rate*.

unit   See *base units*.

unit cell   The smallest unit that, when stacked together repeatedly without any gaps, can reproduce an entire crystal.

unsaturated hydrocarbon   A hydrocarbon with at least one carbon–carbon multiple bond. *Examples:* $CH_2{=}CH_2$; C_6H_6.

valence   The number of bonds that an atom can form.

valence band   In the theory of solids, a band of energy levels fully occupied by electrons.

valence-bond theory   The description of bond formation in terms of the pairing of spins in the atomic orbitals of neighboring atoms.

valence electrons   The electrons that belong to the valence shell.

valence shell   The outermost shell of an atom. *Example:* the $n = 2$ shell of Period 2 atoms.

valence-shell electron-pair repulsion model (VSEPR model)   A model for predicting the shapes of molecules, using the fact that electron pairs repel one another.

van der Waals equation   An approximate equation of state for a real gas in which two parameters represent the effects of intermolecular forces.

van der Waals interactions   Intermolecular interactions that depend on the inverse sixth power of the separation. See *intermolecular forces*.

van der Waals parameters   The experimentally determined coefficients that appear in the van der Waals equation and are unique for each real gas. The parameter *a* is an indication of the strength of attractive intermolecular forces, and the parameter *b* is an indication of the strength of repulsive intermolecular forces. See also *van der Waals equation*.

van der Waals radius   Half the distance between the centers of nonbonded, touching atoms in a solid.

van 't Hoff equation   (1) The equation for the osmotic pressure in terms of the molarity, $\Pi = i[J]RT$. (2) An equation that shows how the equilibrium constant varies with temperature.

van 't Hoff *i* factor   See *i factor*.

vapor   The gaseous phase of a substance (specifically, of a substance that is a liquid or solid at the temperature in question). See also *gas*.

vaporization   The formation of a gas or a vapor from a liquid.

vapor pressure   The pressure exerted by the vapor of a liquid (or a solid) when the vapor and the liquid (or solid) are in dynamic equilibrium.

variable covalence   The ability of an element to form different numbers of covalent bonds. *Example:* S in SO_2 and SO_3.

variable valence   The ability of an element to form ions with different charges. *Example:* In^+ and In^{3+}.

velocity   The rate of change of position.

vibrational kinetic energy   See *kinetic energy*.

virial equation   An equation of state expressed in powers of $1/V_m$: specifically, $PV = nRT(1 + B/V_m + C/V_m^2 + \cdots)$, where B is the *second virial coefficient* and C is the *third virial coefficient*.

viscosity   The resistance of a fluid (a gas or a liquid) to flow: the higher the viscosity, the slower the flow.

visible light   See *visible radiation*.

visible radiation   Electromagnetic radiation that can be detected by the human eye, with wavelengths in the range from 700 nm to 400 nm. Visible radiation is also called *visible light* or simply *light*.

volatile   Having a high vapor pressure at ordinary temperatures. A substance is typically regarded as volatile if its boiling point is below 100°C.

volatility   The readiness with which a substance vaporizes. See also *volatile*.

volt (V)   The SI unit of electrical potential. See also Appendix 1B.

voltaic cell   See *electrochemical cell*.

volume (V)   The amount of space that a sample occupies.

volume percentage composition   The volume of a substance present in a mixture, expressed as a percentage of the total volume.

volumetric analysis   An analytical method using measurement of volume, as in a *titration*.

volumetric flask   A flask calibrated to contain a specified volume.

water autoprotolysis constant (K_w)   The equilibrium constant for the autoprotolysis (autoionization) of water, $2 H_2O(l) \rightleftharpoons H_3O^+(aq) + OH^-(aq)$, $K_w = [H_3O^+][OH^-]$.

water of hydration   See *hydration*.

water-splitting reaction   The photochemical decomposition of water into hydrogen and oxygen.

wavefunction (ψ)   A solution of the Schrödinger equation; the probability amplitude.

wavelength (λ)   The peak-to-peak distance of a wave.

wave–particle duality   The combined wavelike and particle-like character of both radiation and matter.

weak acids and bases   Acids and bases that are only incompletely deprotonated or protonated, respectively (commonly, in aqueous solution) at normal concentrations. *Examples:* HF, CH_3COOH (weak acids); NH_3, CH_3NH_2 (weak bases).

weak electrolyte   See *electrolyte*.

weak-field ligand   A ligand that produces a small *ligand field splitting* and that lies below NH_3 in the *spectrochemical series*.

Wien's law   The wavelength corresponding to the maximum in the radiation emitted by a heated *black body* is inversely proportional to the absolute temperature.

work (w)   The energy expended during the act of moving an object against an opposing force. In *expansion work*, the system expands against an opposing pressure; *nonexpansion work* is work that does not arise from a change in volume.

work function (Φ)   The energy required to remove an electron from a metal.

x-ray   Electromagnetic radiation with wavelengths from about 10 pm to about 1000 pm.

x-ray diffraction   The analysis of crystal structures by studying the interference pattern in a beam of x-rays.

yield   See *percentage yield; theoretical yield*.

zeolite   A microporous aluminosilicate.

zero-order reaction   A reaction with a rate that is independent of the concentration of the reactant. *Example:* the catalyzed decomposition of ammonia.

zero-point energy   The lowest possible energy of a system. *Example:* $E = h^2/8mL^2$ for a particle of mass m in a box of length L.

Ziegler–Natta catalyst   A stereospecific catalyst for polymerization reactions, consisting of titanium tetrachloride and triethylaluminum.

zinc-blende structure   A crystal structure in which the cations occupy half the tetrahedral holes in a nearly close packed cubic lattice of anions; also known as *sphalerite structure*.

zone refining   A method for purifying a solid by repeatedly passing a molten zone along the length of a sample.

zwitterion   A form of an amino acid in which in which the amino group is protonated and the carboxyl group is deprotonated. *Example:* $^+H_3NCH_2CO_2^-$.

ANSWERS

Self-Tests B

Fundamentals

A.1B 250. g $\times$ 1 lb/453.59 g $\times$ 16 oz/1 lb = 8.82 oz

A.2B 9.81 m·s^2 $\times$ 1 km/10^3 m $\times$ (3600 s/1 h)^2 = 1.27 $\times$ 10^5 km·h^{-2}

A.3B $V = m/d$ = (10.0 g)/(0.17685 g·L^{-1}) = 56.5 L

A.4B $E_K = mv^2/2 = \frac{1}{2} \times$ (1.5 kg) $\times$ (3.0 m·s^{-1})^2 = 6.8 J

A.5B $E_K = mgh$ = (0.350 kg) $\times$ (9.81 m·s^{-2}) $\times$ (443 m) $\times$ (10^{-3} kJ/J) = 1.52 kJ

B.1B number of Au atoms = mass (sample)/mass (one atom) = (0.0123 kg)/(3.27 $\times$ 10^{-25} kg) = 3.76 $\times$ 10^{22} Au atoms

B.2B (a) 8, 8, 8; (b) 92, 144, 92

C.1B (a) Potassium is a Group 1 metal. Cation, +1, so K$^+$. (b) Sulfur is a nonmetal in Group 16/VI. Anion, 16 − 18 = −2, so S^{2-}.

C.2B (a) Li$_3$N; (b) SrBr$_2$

D.1B (a) dihydrogen arsenate; (b) ClO$_3^-$

D.2B (a) gold (III) chloride; (b) calcium sulfide; (c) manganese (III) oxide

D.3B (a) phosphorus trichloride; (b) sulfur trioxide; (c) hydrobromic acid

D.4B (a) Cs$_2$S·4H$_2$O; (b) Mn$_2$O$_7$; (c) HCN; (d) S$_2$Cl$_2$

D.5B (a) pentane; (b) a carboxylic acid

E.1B atoms of H = (3.14 mol H$_2$O) $\times$ (2 mol H)/ (1 mol H$_2$O) $\times$ (6.022 $\times$ 10^{23} atoms/mol) = 3.78 $\times$ 10^{24} H atoms

E.2B (a) $n = m/M$ = (5.4 $\times$ 10^3 g)/(26.98 g·mol^{-1}) = 2.0 $\times$ 10^2 mol; (b) $N = N_A \times n$ = (6.022 $\times$ 10^{23} Al atoms/mol) $\times$ (2.0 $\times$ 10^2 mol) = 1.2 $\times$ 10^{26} Al atoms

E.3B copper-63: (62.94 g·mol^{-1}) $\times$ 0.6917 = 43.536 g·mol^{-1}; copper-65: (64.93 g·mol^{-1}) $\times$ 0.3083 = 20.018 g·mol^{-1}; 43.536 g·mol^{-1} + 20.018 g·mol^{-1} = 63.55 g·mol^{-1}

E.4B (a) phenol: 6 C, 6 H, 1 O; 6(12.01 g·mol^{-1}) + 6(1.008 g·mol^{-1}) + (16.00 g·mol^{-1}) = 94.11 g·mol^{-1}; (b) Na$_2$CO$_3$·10H$_2$O: 2 Na, 1 C, 13 O, 20 H; 2(22.99 g·mol^{-1}) + (12.01 g·mol^{-1}) + 13(16.00 g·mol^{-1}) + 20(1.008 g·mol^{-1}) = 286.15 g·mol^{-1}

E.5B Ca(OH)$_2$: 1 Ca, 2 O, 2 H; (40.08 g·mol^{-1}) + 2(16.00 g·mol^{-1}) + 2(1.008 g·mol^{-1}) = 74.10 g·mol^{-1}; (1.0 $\times$ 10^3 g lime)/(74.10 g·mol^{-1}) = 13.5 lime molecules

E.6B CH$_3$COOH: 2 C, 4 H, 2 O; 2(12.01 g·mol^{-1}) + 4(1.008 g·mol^{-1}) + 2(16.00 g·mol^{-1}) = 60.05 g·mol^{-1}; (60.05 g·mol^{-1})(1.5 mol) = 90. g

F.1B % C = (6.61 g/7.50 g) $\times$ 100% = 88.1%; % H = (0.89 g/7.50 g) $\times$ 100% = 11.9%

F.2B AgNO$_3$: (107.87 g·mol^{-1}) + (14.01 g·mol^{-1}) + 3(16.00 g·mol^{-1}) = 169.88 g·mol^{-1}; % Ag = (107.87 g·mol^{-1})/(169.88 g·mol^{-1}) $\times$ 100% = 63.50%

F.3B carbon: 0.778 $\times$ (100 g eucalyptol) = 77.8 g C; (77.8 g C)/(12.01 g·mol^{-1}) = 6.48 mol C; hydrogen: (11.8 g H)/(1.0079 g·mol^{-1}) = 11.7 mol H; oxygen: (10.4 g O)/(16.00 g·mol^{-1}) = 0.650 mol O; 6.48 C:11.7 H:0.650 O

F.4B mol O = (18.59 g)/(16.00 g·mol^{-1}) = 1.162 mol O; mol S = (37.25 g)/(32.07 g·mol^{-1}) = 1.162 mol; F molecules = (44.16 g)/(19.00 g·mol^{-1}) = 2.324 mol. 1:1:2 ratio, formula is SOF$_2$.

F.5B Molar mass of CHO$_2$ is 45.012 g·mol^{-1}. (90.0 g·mol^{-1})/(45.012 g·mol^{-1}) = 2.00; 2 $\times$ (CHO$_2$) = C$_2$H$_2$O$_4$

G.1B Molar mass of Na$_2$SO$_4$ is 142.05 g·mol^{-1}. (15.5 g)/(142.05 g·mol^{-1}) = 0.109 mol; (0.109 mol)/ (0.350 L) = 0.312 M Na$_2$SO$_4$(aq)

G.2B (0.125 mol/L) $\times$ (0.050 L) = 0.00625 mol oxalic acid. Molar mass of oxalic acid is 90.036 g·mol^{-1}. (0.00625 mol) $\times$ (90.036 g·mol^{-1}) = 0.563 g oxalic acid

G.3B (2.55 $\times$ 10^{-3} mol HCl)/(0.358 mol HCl/L) = 7.12 mL

G.4B $V_{init} = (c_{final} \times V_{final})/c_{init}$ = (1.59 $\times$ 10^{-5} mol/L) $\times$ (0.02500 L)/(0.152 mol/L) = 2.62 $\times$ 10^{-3} mL

H.1B Mg$_3$N$_2$(s) + 4 H$_2$SO$_4$(aq) → 3 MgSO$_4$(aq) + (NH$_4$)$_2$SO$_4$(aq)

I.1B (a) nonelectrolyte, does not conduct electricity; (b) strong electrolyte, conducts electricity

I.2B 3 Hg$_2^{2+}$(aq) + 2 PO$_4^{3-}$ (aq) → (Hg$_2$)$_3$(PO$_4$)$_2$(s)

I.3B SrCl$_2$ and Na$_2$SO$_4$; Sr^{2+}(aq) + SO$_4^{2-}$ (aq) → SrSO$_4$(s).

J.1B (a) neither; (b) acid; (c) acid; (d) base

J.2B 3 Ca(OH)$_2$(aq) + 2 H$_3$PO$_4$(aq) → Ca$_3$(PO$_4$)$_2$(s) + 6 H$_2$O

K.1B Cu$^+$ is oxidized, I$_2$ is reduced.

K.2B (a) $x + 3(-2) = -2$; $x = +4$ for S; (b) $x + 2(-2) = -1$; $x = +3$ for N; (c) $x + 1 + 3(-2) = 0$; $x = +5$ for Cl

K.3B H$_2$SO$_4$ is the oxidizing agent (S is reduced from +6 to +4); NaI is the reducing agent (I is oxidized from −1 to +5).

K.4B 2 Ce^{4+}(aq) + 2 I$^-$(aq) → 2 Ce^{3+}(aq) + I$_2$(s)

L.1B (2 mol Fe)/(1 mol Fe$_2$O$_3$) $\times$ 25 mol Fe$_2$O$_3$ = 50. mol Fe

L.2B 2 mol CO$_2$/1 mol CaSiO$_3$; mol CO$_2$ = (3.00 $\times$10^2 g)/ (44.01 g·mol^{-1}) = 6.82 mol; (1 mol CaSiO$_3$/2 mol CO$_2$) $\times$ (6.82 mol CO$_2$) = 3.41 mol CaSiO$_3$; (3.41 mol CaSiO$_3$) $\times$ (116.17 g·mol^{-1} CaSiO$_3$) = 396 g CaSiO$_3$

L.3B 2 KOH + H$_2$SO$_4$ → K$_2$SO$_4$ + 2 H$_2$O; 2 mol KOH ≏ 1 mol H$_2$SO$_4$; (0.255 mol KOH/L) $\times$ (0.025 L) = 6.375 $\times$ 10^{-3} mol KOH. (6.375 $\times$ 10^{-3} mol KOH) $\times$(1 mol H$_2$SO$_4$)/ (2 mol KOH) = 3.19 $\times$ 10^{-3} mol H$_2$SO$_4$; (3.19 $\times$ 10^{-3} mol H$_2$SO$_4$)/ (0.01645 L) = 0.194 M H$_2$SO$_4$

L.4B (0.100 $\times$ 0.02815) mol KMnO$_4$ $\times$ (5 mol As$_4$O$_6$)/ (8 mol KMnO$_4$) $\times$ 395.28 g·mol^{-1} = 6.96 $\times$ 10^{-2} g As$_4$O$_6$

M.1B 2 Fe$_2$O$_3$(s) → 4 Fe(s) + 3 O$_2$ (g); (15 kg Fe$_2$O$_3$)/ 159.69 g·mol^{-1}) $\times$ (2 mol Fe)/(1 mol Fe$_2$O$_3$) $\times$(55.85 g·mol^{-1}) = 10.5 kg Fe; 8.8 kg/10.5 kg $\times$ 100% = 84% yield

M.2B 2 NH$_3$ + CO$_2$ → OC(NH$_2$)$_2$ + H$_2$O; n_{NH_3} = (14.5 $\times$ 10^3 g)/(17.034 g·mol^{-1}) = 851 mol NH$_3$; n_{CO_2} = (22.1 $\times$ 10^3 g)/(44.01 g·mol^{-1}) = 502 mol CO$_2$; 2 mol NH$_3$ ≏ 1 mol CO$_2$; (a) NH$_3$ is the limiting reagent. (851 mol NH$_3$/2) < 502 mol CO$_2$).

(b) 2 mol NH_3/1 mol urea. 426 mol urea can be produced = 25.6 kg urea.
(c) $(502 - 426)$ mol = 76 mol CO_2 excess = 3.3 kg CO_2
M.3B $3 NO_2(g) + H_2O(l) \rightarrow 2 HNO_3(l) + NO(g)$. There is 0.61 mol NO_2 and 1.0 mol H_2O. 1 mol $H_2O \simeq 3$ mol NO_2; therefore, there is not enough NO_2, so NO_2 is the limiting reactant. 22 g, or 0.35 mol HNO_3, were produced. Theoretical yield is $(0.61$ mol $NO_2) \times (2$ mol $HNO_3)/(3$ mol $NO_2) = 0.407$ mol HNO_3. Percentage yield = $(0.35$ mol$)/(0.407$ mol$) \times 100\% = 86\%$.
M.4B   The sample contains 0.0118 mol C (0.142 g C) and 0.0105 mol H (0.0106 g H). Mass of O = $0.236 - (0.142 + 0.0105)$g $= 0.0834$ g O (0.00521 mol O). The C:H:O mole ratios are 0.0118:0.0105:0.00521, or 2.26:2.02:1. Multiplying these numbers by 4 gives 9:8:4 and the empirical formula $C_9H_8O_4$.

Chapter 1

1.1B $\lambda = c/\nu = (2.998 \times 10^8$ m·s$^{-1})/(98.4 \times 10^6$ Hz$)$ $= 3.05$ m
1.2B $\nu = \mathcal{R}(1/2^2 - 1/5^2) = 21\mathcal{R}/100$; $\lambda = c/\nu = 100c/21\mathcal{R}$ $= (100 \times 2.998 \times 10^8$ m·s$^{-1})/(21 \times 3.29 \times 10^{15}$ s$^{-1}) =$ 434 nm; violet line
1.3B $T = $ constant$/\lambda_{max} = (2.9 \times 10^{-3}$ m·K$)/(700. \times 10^{-9}$ m$) = 4.1 \times 10^3$ K
1.4B $E = h\nu = (6.626 \times 10^{-34}$ J·s$) \times (4.8 \times 10^{14}$ Hz$) =$ 3.2×10^{-19} J
1.5B $\lambda = (hc) \times$ (number of photons)$/E_{tot} = (6.626 \times 10^{-34}$ J·s$) \times (3.00 \times 10^8$ m·s$^{-1}) \times (5.5 \times 10^{19}$ photons$)/25$ J $= 440$ nm
1.6B $\lambda = h/mv = (6.626 \times 10^{-34}$ J·s$)/(0.0050$ kg $\times 2 \times 331$ m·s$^{-1}) = 2.0 \times 10^{-34}$ m
1.7B $\Delta v = \hbar/2m\Delta x = (1.05457 \times 10^{-34}$ J$)/(2 \times 2.0 \times 10^3$ kg $\times 1$ m$) = 2.6 \times 10^{-38}$ m·s^{-1}
1.8B $E_3 - E_2 = 5h^2/8m_eL^2 = h\nu$; $\nu = 5h/8m_eL^2$; $\lambda = c/\nu = 8m_ecL^2/5h = [8 \times (9.109 \times 10^{-31}$ kg$) \times (2.998 \times 10^8$ m·s$^{-1}) \times (1.50 \times 10^{-10}$ m$)^2]/(5 \times 6.626 \times 10^{-34}$ J·s$) = 14.8 \times 10^{-9}$ m, or 14.8 nm
1.9B   ratio = $(e^{-6a_0/a_0}/\pi a_0^3)/(1/\pi a_0^3) = e^{-6} = 2.5 \times 10^{-3}$ (0.25%)
1.10B $3p$
1.11B $1s^22s^22p^63s^23p^1$ or [Ne]$3s^23p^1$
1.12B   [Ar]$3d^{10}4s^24p^3$
1.13B   (a) $r(Ca^{2+}) < r(K^+)$; (b) $r(Cl^-) < r(S^{2-})$
1.14B   Formation of Be^{3+} requires removing an electron from an ion with a filled shell, whereas an electron is removed only from the outer shell to form B^{3+}.
1.15B   In fluorine (Group 17), an additional electron fills the single vacancy in the valence shell; the shell now has the noble-gas configuration of neon and is complete. In neon, an additional electron would have to enter a new shell, where it would be farther from the attraction of the nucleus.

Chapter 2

2.1B   (a) [Ar]$3d^5$; (b) [Xe]$4f^{14}5d^{10}$
2.2B I^-, [Kr]$4d^{10}5s^25p^6$
2.3B   KCl, because Cl^- has a smaller radius than Br^-
2.4B   H—B̈r: ; H has no lone pairs, Br has 3 lone pairs.
2.5B   H—N̈—H
　　　　　|
　　　　　H

2.6B   H—N̈—N̈—H
　　　　　|　　|
　　　　　H　　H

2.7B $\left[:\ddot{O}-\ddot{N}=\ddot{O} \right]^- \longleftrightarrow \left[\ddot{O}=\ddot{N}-\ddot{O}: \right]^-$

2.8B   :F̈—Ö—F̈:
　　　　　0　　0　　0

2.9B $\left[\ddot{O}=\ddot{N}-\ddot{O}: \right] \longleftrightarrow \left[:\ddot{O}-\ddot{N}=\ddot{O} \right]$

2.10B   :Ï—Ï—Ï:⁻ ; 10 electrons

2.11B   Ö=Ö—Ö:
　　　　　0　+1　−1

2.12B CO_2
2.13B   CaS

Chapter 3

3.1B   trigonal planar
3.2B   (a) trigonal planar; (b) angular
3.3B   (a) AX_2E_2; (b) tetrahedral; (c) angular
3.4B   square planar
3.5B   (a) nonpolar; (b) polar
3.6B   (a) 3 σ, no π; (b) 2 σ, 2 π
3.7B   Three σ bonds formed from two C2sp hybrids, one bond between the two C atoms and two connecting each C atom to an H atom in a linear arrangement; two π bonds, one between the two C2p_x-orbitals and the other between the two C2p_y-orbitals.
3.8B   (a) octahedral; (b) square planar; (c) sp^3d^2
3.9B   Carbon atom of CH_3 group is sp^3 hybridized and forms four σ bonds at 109.5°. The other two carbon atoms are both sp^2 hybridized and each forms three σ bonds and one π bond; bond angles are about 120°.
3.10B O_2^+: $\sigma_{2s}^2 \sigma_{2s}^{*2} \sigma_{2p}^2 \pi_{2p}^4 \pi_{2p}^{*1}$; BO = $(8-3)/2 = 2.5$
3.11B CN^-: $1\sigma^22\sigma^{*2}1\pi^43\sigma^2$
3.12B   p-type

Chapter 4

4.1B $h = P/dg = (1.01 \times 10^5$ kg·m^{-1}·s$^{-2})/[(998$ kg·m$^{-3}) \times (9.80665$ m·s$^{-2})] = 10.3$ m
4.2B $P = (10.$ cmHg$) \times (10$ mm$)/(1$ cm$) = 1.0 \times 10^2$ mmHg
4.3B $(630$ Torr$) \times (1$ atm$/760$ Torr$) \times (1.01325 \times 10^5$ Pa/atm$) = 8.4 \times 10^4$ Pa or 84 kPa
4.4B $V_2 = P_1V_1/P_2 = (1.00$ bar$) \times (750.$ L$)/(5.00$ bar$) = 150.$ L
4.5B $P_2 = P_1T_2/T_1 = (760.$ mmHg$) \times (573$ K$)/(293$ K$) = 1.49 \times 10^3$ mmHg
4.6B $P_2 = P_1n_2/n_1 = (1.20$ atm$) \times (300.$ mol$)/(200.$ mol$) = 1.80$ atm
4.7B $V/\text{min} = (n/\text{min}) \times (RT)/P = (1.00$ mol/min$) \times (8.206 \times 10^{-2}$ L·atm·K^{-1}·mol$^{-1}) \times (300.$ K$)/(1.00$ atm$) = 24.6$ L·min^{-1}
4.8B $V_2 = P_1V_1/P_2 = (1.00$ atm$) \times (80.$ cm$^3)/(3.20$ atm$) = 25$ cm^3
4.9B $P_2 = P_1V_1T_2/V_2T_1 = [(1.00$ atm$) \times (250.$ L$) \times (243$ K$)]/[(800.$ L$) \times (293$ K$)] = 0.259$ atm
4.10B $n = (1$ mol He$/4.003$ g He$) \times (2.0$ g He$) = 0.50$ mol He; $V = nRT/P = (0.50$ mol$) \times (24.47$ L mol$^{-1}) = 12$ L

4.11B $M = dRT/P = (1.04 \text{ g·L}^{-1}) \times$ (62.364 L·Torr·K^{-1}·mol^{-1}) $\times$ (450. K)/(200. Torr) = 146 g·mol^{-1}

4.12B $2 H_2(g) + O_2(g) \rightarrow 2 H_2O(l)$, and so 2 mol $H_2O \simeq$ 1 mol O_2; $n_{O_2} = PV/RT = [(1.00 \text{ atm}) \times (100.0 \text{ L})]/[(8.206 \times 10^{-2} \text{ L·atm·K}^{-1}\text{·mol}^{-1}) \times (298 \text{ K})] = 4.09 \text{ mol } O_2$.

$n_{H_2O} = 2(4.09 \text{ mol } O_2) = 8.18 \text{ mol } H_2O$; $m_{H_2O} =$ (8.18 mol) $\times$ (18.02 g·mol^{-1}) = 147 g H_2O.

4.13B $2 H_2O(l) \rightarrow 2 H_2(g) + O_2(g)$; (2 mol H_2/3 gas molecules) $\times$ (720. Torr) = 480. Torr H_2; (1 mol O_2/3 gas molecules) $\times$ (720. Torr) = 240. Torr H_2

4.14B $n_{O_2} = (141.2 \text{ g } O_2)/(32.00 \text{ g·mol}^{-1}) = 4.412 \text{ mol } O_2$; $n_{Ne} = (335.0 \text{ g Ne})/(20.179 \text{ g·mol}^{-1}) = 16.60 \text{ mol Ne}$; $P_{O_2} =$ (4.412 mol O_2/21.01 mol total) $\times$ (50.0 atm) = 10.5 atm.

4.15B $t_{CH_4} = t_{He} \times (M_{CH_4}/M_{He})^{1/2} = (10. \text{ s}) \times$ [(16.04 g·mol^{-1})/(4.00 g·mol^{-1})]$^{1/2}$ = 20. s

4.16B $v_{rms} = (3RT/M)^{1/2} = [3 \times (8.3145 \text{ J·K}^{-1}\text{·mol}^{-1}) \times$ (298 K)/(16.04 $\times$ 10^{-3} kg·mol^{-1})]$^{1/2}$ = 681 m·s^{-1}

4.17B $P = [nRT/(V - nb)] - an^2/V^2 = \{20. \times (8.206 \times 10^{-2} \text{ L·atm·K}^{-1}\text{·mol}^{-1}) \times (293 \text{ K})\}/\{100. - (20. \times 4.267 \times 10^{-2} \text{ L·mol}^{-1})\} - (3.640 \text{ L}^2\text{·atm·mol}^{-2} \times 20.^2)/100.^2 = 4.7 \text{ atm}$

Chapter 5

5.1B 1,1-dichloroethane, because it has a dipole moment

5.2B Unlike CF_4, CHF_3 has a net dipole moment. The resulting dipole–dipole interactions account for the higher boiling point of CHF_3 even though one might expect the CF_4 molecule (with more electrons) to exhibit stronger London interactions.

5.3B (a) CH_3OH and (c) $HClO$ (in solution)

5.4B 8(1/8) + 2(1/2) + 2(1) = 4 atoms

5.5B $d_{fcc} = (4M)/(8^{3/2}N_A r^3) = (4 \times 55.85 \text{ g·mol}^{-1})/\{8^{3/2} \times 6.022 \times 10^{23} \text{ mol}^{-1} \times (1.24 \times 10^{-8} \text{ cm})^3\} = 8.60 \text{ g·cm}^{-3}$; $d_{bcc} = (3^{3/2}M)/(32N_A r^3) = (3^{3/2} \times 55.85 \text{ g·mol}^{-1})/\{32 \times 6.022 \times 10^{23} \text{ mol}^{-1} \times (1.24 \times 10^{-8} \text{ cm})^3\} = 7.90 \text{ g·cm}^{-3}$. The observed density is closer to that predicted for a body-centered cubic structure.

5.6B ρ = (100 pm)/(184 pm) = 0.54, rock salt structure

5.7B (a) 8 corner ions = 8(1/8) = 1 chloride ion per unit cell of CsCl. (b) Each Cs^+ ion is surrounded by 8 Cl^- ions, and each Cl^- ion is surrounded by 8 Cs^+ ions, so it has (8,8)-coordination.

Major Technique 3

MT3.1B d = (152 pm)/(2 sin 12.1°) = 363 pm

Chapter 6

6.1B $w = -P\delta V = -(9.60 \text{ atm})(2.2 \text{ L} - 0.22 \text{ L}) \times$ 101.325 J/(1 L·atm) = -1926 J = -1.9 kJ

6.2B $w = -P\Delta V = -(1.00 \text{ atm})(4.00 \text{ L} - 2.00 \text{ L}) \times$ 101.325 J/1 L·atm = -202 J
$w = -nRT \ln(V_{final}/V_{initial}) = -(1.00 \text{ mol})$ (8.3145 J·K^{-1}·mol^{-1})(303 K) $\times \ln(4.00 \text{ L}/2.00 \text{ L})$ = -1.75 kJ; isothermal reversible expansion does more work

6.3B $q = nC_m\Delta T = (3.00 \text{ mol}) \times (111 \text{ J·K}^{-1}\text{·mol}^{-1}) \times$ (15 K) = 5.00 kJ

6.4B $q_{cal} = -q_{rxn} = +4.16 \text{ kJ}$; $C_{cal} = 4.16 \text{ kJ}/\Delta T$ = 4.16 kJ/3.24°C = 1.28 kJ·(°C)$^{-1}$

6.5B $w = \Delta U - q = -150 \text{ J} - (+300 \text{ J}) = -450 \text{ J}$; $w < 0$ (i.e., the system does the work)

6.6B $q = +1.00$ kJ (heat is absorbed); $w = -(2.00 \text{ atm})(3.00 \text{ L} - 1.00 \text{ L}) = -(4.00 \text{ L·atm})$ $\times \dfrac{101.325 \text{ J}}{1 \text{ L·atm}} = -405 \text{ J}$;
$\Delta U = q + w = 1.00 \text{ kJ} + (-0.405 \text{ kJ}) = +0.60 \text{ kJ}$

6.7B (a) $\Delta H = +30$ kJ; (b) $\Delta U = q + w = 30 \text{ kJ} + 40 \text{ kJ} = +70 \text{ kJ}$

6.8B (a) $w = 0$, $\Delta U = q = +1.20$ kJ;
$\Delta T = \dfrac{q}{nC_V} = \dfrac{(1200 \text{ J})}{(1 \text{ mol}) \times (5/2) \times (8.3145 \text{ J·K}^{-1}\text{·mol}^{-1})}$ = 57.7 K; $T_f = 298 + 57.7 = 356$ K
(b) (step 1) constant volume, find final temperature: $q = nC_P\Delta T = 1.20$ kJ;
$\Delta T = \dfrac{1200 \text{ J}}{(1.0 \text{ mol}) \times (7/2) \times (8.3145 \text{ J·K}^{-1}\text{·mol}^{-1})} = 41.2 \text{ K}$;
T_f = 339 K;
$w = 0$, so $\Delta U = q = nC_V\Delta T = (1 \text{ mol})(2.5)$ (8.3145 J·mol^{-1}·K^{-1})(41.2 K) = 856 J
(step 2) isothermal, $\Delta T = 0$, so $\Delta U = 0$
Therefore, after both steps, $\Delta U = +856$ J, $T_f = 339$ K

6.9B $\Delta H = 22 \text{ kJ}/23 \text{ g} \times 46.07 \text{ g·mol}^{-1} = 44 \text{ kJ·mol}^{-1}$

6.10B $\Delta H_{sub} = \Delta H_{vap} + \Delta H_{fus} = (38 + 3) \text{ kJ·mol}^{-1} =$ 41 kJ·mol^{-1}

6.11B $q_r = -q_{cal} = -(216 \text{ J·°C}^{-1})(76.7 \text{ °C}) =$ -1.66×10^3 J;
$\Delta H_r = \dfrac{-1.66 \times 10^4 \text{ J}}{0.338 \text{ g}} \times (72.15 \text{ g·mol}^{-1}) \times \left(\dfrac{1 \text{ kJ}}{10^3 \text{ J}}\right) =$ -3.54×10^3 kJ;
$C_5H_{12}(l) + 8 O_2(g) \rightarrow 5 CO_2(g) + 6 H_2O(l)$, $\Delta H = -3.54 \times 10^3$ kJ

6.12B $\Delta U = \Delta H - n_{(gas)}RT = -3378 \text{ kJ} - (-\frac{3}{4} \text{ mol}) \times$ (8.3145 J·mol^{-1}·K^{-1})(1273 K)(1 kJ/10^3 J) = -3.37×10^3 kJ

6.13B

$CH_4(g) + H_2O(g) \rightarrow CO(g) + 3 H_2(g)$	$\Delta H = +206.10$ kJ
$2 H_2(g) + CO(g) \rightarrow CH_3OH(l)$	$\Delta H = -128.33$ kJ
$H_2(g) + \frac{1}{2} O_2(g) \rightarrow H_2O(g)$	$\Delta H = \frac{1}{2}(-483.64)$ kJ

$CH_4(g) + \frac{1}{2} O_2(g) \rightarrow CH_3OH(l)$ $\Delta H = -164.05$ kJ

6.14B $m(C_2H_5OH) = 350 \text{ kJ} \times \dfrac{1 \text{ mol } C_2H_5OH}{1368 \text{ kJ}} \times$ 46.07 g·mol^{-1} = 11.8 g C_2H_5OH

6.15B $C(\text{diamond}) + O_2(g) \rightarrow CO_2(g)$;
$\Delta H_r° = \Delta H_f°(CO_2) - \Delta H_f°(C,\text{diamond}) - \Delta H_f°(O_2,g) =$ $-393.51 \text{ kJ·mol}^{-1} - (+1.895 \text{ kJ·mol}^{-1}) - 0 \text{ kJ·mol}^{-1} =$ $-395.41 \text{ kJ·mol}^{-1}$

6.16B $CO(NH_2)_2(s) + \frac{3}{2} O_2(g) \rightarrow CO_2(g) + 2 H_2O(l) + N_2(g)$; $\Delta H_r° = \Delta H_f°(CO_2) + 2 \Delta H_f°(H_2O) + \Delta H_f°(N_2) - \Delta H_f°(CO(NH_2)_2)$; $-632 \text{ kJ} = -393.51 \text{ kJ} + 2(-285.83 \text{ kJ}) + 0 \text{ kJ} - \Delta H_f°(CO(NH_2)_2)$; $\Delta H_f°(CO(NH_2)_2) = -333 \text{ kJ·mol}^{-1}$

6.17B $[524.3 + 147.70 + 2 (111.88) + 736 + 1451 - 2 (325)] \text{ kJ} - \Delta H_L = 0$; $\Delta H_L = +2433$ kJ

6.18B $CH_4(g) + 2 F_2(g) \rightarrow CH_2F_2(g) + 2 HF(g)$; bonds broken [2(C−H), 2(F−F)]: 2(412 kJ·mol^{-1}) + 2(158 kJ·mol^{-1}) = 1140 kJ·mol^{-1}; bonds formed [2(C−F), 2(H−F)]: 2(484 kJ·mol^{-1}) + 2(565 kJ·mol^{-1}) = 2098 kJ·mol^{-1}; H = 1140 kJ·mol^{-1} − 2098 kJ·mol^{-1} = -958 kJ·mol^{-1}

6.19B $\Delta H_{523\,K} = \Delta H_{298\,K} + \Delta C_p \Delta T = -365.56 \text{ kJ·mol}^{-1} +$
$[84.1 - 29.12 - 2(28.82) - (\frac{3}{2})(29.36)] \text{ J·K}^{-1}\text{·mol}^{-1} \times$
$(523 \text{ K} - 298 \text{ K}) \times \dfrac{1 \text{ kJ}}{10^3 \text{ J}} = -365.56 + (-46.7)(225)\,(10^{-3})$
$= -376.1 \text{ kJ·mol}^{-1}$

Chapter 7

7.1B $\Delta S = -50. \text{ J}/1373 \text{ K} = -0.036 \text{ J·K}^{-1}$

7.2B $\Delta S = (5.5 \text{ g})(0.51 \text{ J·K}^{-1}\text{·g}^{-1}) \ln(373/293)$
$= +0.68 \text{ J·K}^{-1}$

7.3B $\Delta S = nR \ln(V_2/V_1) = (8.314 \text{ J·K}^{-1}\text{·mol}^{-1}) \ln(10)$
$= +19 \text{ J·K}^{-1}\text{·mol}^{-1}$

7.4B $\Delta S = nR \ln(P_1/P_2) = \left(70.9 \text{ g} \times \dfrac{1 \text{ mol}}{70.9 \text{ g}}\right) \times$
$(8.3145 \text{ J·mol}^{-1}\text{·K}^{-1}) \ln(3.00 \text{ kPa}/24.00 \text{ kPa}) =$
-17.3 J·K^{-1}

7.5B   (1) $\Delta S = (23.5 \text{ g})(1 \text{ mol}/32.00 \text{ g})(8.3145 \text{ J·K}^{-1}\text{·mol}^{-1})$
$\times \ln(2.00 \text{ kPa}/8.00 \text{ kPa}) = -8.46 \text{ J·K}^{-1}$;
(2) $\Delta S = (23.5 \text{ g})(1 \text{ mol}/32.00 \text{ g})(20.786 \text{ J·K}^{-1}\text{·mol}^{-1})$
$\times \ln(360 \text{ K}/240 \text{ K}) = +6.19 \text{ J·K}^{-1}$;
$\Delta S = -8.46 + 6.19 \text{ J·K}^{-1} = -2.27 \text{ J·K}^{-1}$

7.6B $\Delta S_{vap} = \dfrac{\Delta H_{vap}}{T_b} = \dfrac{40.7 \text{ kJ · mol}^{-1}}{373.2 \text{ K}} \times \dfrac{10^3 \text{ J}}{1 \text{ kJ}}$
$= 109 \text{ J·K}^{-1}\text{·mol}^{-1}$

7.7B $\Delta H_{vap}^{\circ} = (85 \text{ J·K}^{-1}\text{·mol}^{-1})(307.5 \text{ K}) \times 1 \text{ kJ}/10^3 \text{ J}$
$= 26.1 \text{ kJ·mol}^{-1}$

7.8B $\Delta S_{fus}^{\circ} = \Delta H_{fus}^{\circ}/T_f = (10.59 \times 10^3 \text{ J·mol}^{-1})/(278.6 \text{ K})$
$= 38.01 \text{ J·K}^{-1}\text{·mol}^{-1}$

7.9B $\Delta S = (1.38066 \times 10^{-23} \text{ J·K}^{-1}) \ln (6)^{6.022 \times 10^{23}} =$
$(1.38066 \times 10^{-23} \text{ J·K}^{-1})(6.022 \times 10^{23}) \ln(6) = +15 \text{ J·K}^{-1}$

7.10B   Nonzero entropy at $T = 0$ indicates disorder. This disorder results when a molecule can orient itself more than one way in the crystal. In ice, each O atom is surrounded by four H atoms, of which there are two types. Two of the H atoms are covalently bonded to the O atom and the other two H atoms, which belong to neighboring water molecules, are interacting with the central O atom through hydrogen bonds. Thus, more than one orientation is possible in the crystal and entropy will not be zero at $T = 0$.

7.11B   (a) $\Delta S = S_{gray} - S_{white} = (44.14 - 51.55) \text{ J·K}^{-1}\text{·mol}^{-1}$
$= -7.41 \text{ J·K}^{-1}\text{·mol}^{-1}$; the gray form;
(b) $\Delta S = S_{graphite} - S_{diamond} = (5.7 - 2.4) \text{ J·K}^{-1}\text{·mol}^{-1}$
$= +3.3 \text{ J·K}^{-1}\text{·mol}^{-1}$; diamond

7.12B   positive, because a gas is produced from a solid

7.13B $\Delta S_r^{\circ} = S_{C_2H_6\,(g)} - S_{C_2H_4\,(g)} - S_{H_2(g)}$
$= 229.60 \text{ J·K}^{-1}\text{·mol}^{-1} - 219.56 \text{ J·K}^{-1}\text{·mol}^{-1} -$
$130.68 \text{ J·K}^{-1}\text{·mol}^{-1} = -120.64 \text{ J·K}^{-1}\text{·mol}^{-1}$

7.14B $\Delta S_{surr} = -\dfrac{\Delta H}{T} = \dfrac{-(2.00 \text{ mol})(-46.11 \text{ kJ·mol}^{-1})}{298 \text{ K}}$
$\times \dfrac{10^3 \text{ J}}{1 \text{ kJ}} = +309 \text{ J·K}^{-1}$

7.15B $\Delta S_{surr} = -\dfrac{\Delta H}{T} = -\dfrac{49.0 \text{ kJ}}{298 \text{ K}} \times \dfrac{10^3 \text{ J}}{1 \text{ kJ}} = -164 \text{ J·K}^{-1}$;
$\Delta S_{tot} = -164 \text{ J·K}^{-1} + (-253.18 \text{ J·K}^{-1})$
$= -417 \text{ J·K}^{-1}$; no

7.16B $\Delta S = nR \ln\left(\dfrac{V_2}{V_1}\right) = (2.00 \text{ mol}) \times$
$(8.314 \text{ J·K}^{-1}\text{·mol}^{-1}) \ln\left(\dfrac{0.200 \text{ L}}{4.00 \text{ L}}\right)$
$= -49.8 \text{ J · K}^{-1} \cdot \text{mol}^{-1}; \Delta S_{surr}$
$= +49.8 \text{ J·K}^{-1}\text{·mol}^{-1}; \Delta S_{tot} = 0$

7.17B $\Delta S_{surr} = \dfrac{-\Delta H_{vap}}{T} = \dfrac{-30.8 \times 10^3 \text{ kJ·mol}^{-1}}{353.3 \text{ K}}$
$= -87.2 \text{ J·K}^{-1}\text{·mol}^{-1}$;
$\Delta S = +87.2 \text{ J·K}^{-1}; \Delta S_{tot} = \Delta S + \Delta S_{surr}$
$= 87.2 \text{ J·K}^{-1} + (-87.2 \text{ J·K}^{-1}) = 0$

7.18B   Yes. $\Delta G = \Delta H - T\Delta S$. When $\Delta S > 0$, $T\Delta S > 0$. Hence, as temperature increases, $-T\Delta S$ becomes more negative; eventually $\Delta G < 0$ and the process becomes spontaneous.

7.19B   (a) $\Delta G = \Delta H - T\Delta S = 59.3 \text{ kJ·mol}^{-1} - (623 \text{ K}) \times$
$(0.0942 \text{ kJ·mol}^{-1}) = +0.6 \text{ kJ·mol}^{-1}$; vaporization is nonspontaneous; (b) $\Delta G = \Delta H - T\Delta S = 59.3 \text{ kJ·mol}^{-1} -$
$(643 \text{ K})(0.0942 \text{ kJ·mol}^{-1}) = -1.3 \text{ kJ·mol}^{-1}$; vaporization is spontaneous.

7.20B $3 \text{ H}_2 \text{ (g)} + 3 \text{ C (s, graphite)} \rightarrow \text{C}_3\text{H}_6 \text{ (g)}$
$\Delta S_r = 237.4 \text{ J·K}^{-1}\text{·mol}^{-1} - [3(130.68 \text{ J·K}^{-1}\text{·mol}^{-1})$
$+ 3(5.740 \text{ J·K}^{-1}\text{·mol}^{-1})]$
$= -171.86 \text{ J·K}^{-1}\text{·mol}^{-1}$
$\Delta G_r = \Delta H_r - T\Delta S_r = +53.30 \text{ kJ·mol}^{-1}$
$- (298 \text{ K})(-171.86 \text{ J·K}^{-1}\text{·mol}^{-1}) \times \dfrac{1 \text{ kJ}}{10^3 \text{ J}} = +104.5 \text{ kJ·mol}^{-1}$

7.21B   From Appendix 2A, $\Delta G_f^{\circ}(\text{CH}_3\text{NH}_2, \text{g})$
$= +32.16 \text{ kJ·mol}^{-1}$ at 298 K. Because $\Delta G_f^{\circ} > 0$, CH_3NH_2 is less stable than its elements at the stated conditions.

7.22B $\Delta G = [-910 + 6(0)] - [6(-394.36) + 6(-237.13)] =$
$+2879 \text{ kJ·mol}^{-1}$

7.23B $\text{MgCO}_3(s) \rightarrow \text{MgO}(s) + \text{CO}_2(g)$;
$\Delta H^{\circ} = -601.70 + (-393.51) - (-1095.8) =$
$+100.6 \text{ kJ·mol}^{-1}$;
$\Delta S^{\circ} = 26.94 + 213.74 - 65.7 \text{ J·K}^{-1} =$
$+175.0 \text{ J·K}^{-1}\text{·mol}^{-1}$;
$$T = \dfrac{\Delta H^{\circ}}{\Delta S^{\circ}} = \dfrac{100.6 \text{ kJ·mol}^{-1}}{175.0 \text{ J·mol}^{-1}\text{·K}^{-1}} \times \dfrac{10^3 \text{ J}}{1 \text{ kJ}} = 574.9 \text{ K}$$

Chapter 8

8.1B $\text{CH}_3\text{CH}_2\text{CH}_3$; molecules of the two substances have the same molar mass and thus the same number of electrons and comparable London forces. However, CH_3CHO is polar and also experiences dipole–dipole forces.

8.2B

$$\ln\left(\dfrac{P_2}{94.6 \text{ Torr}}\right) = \dfrac{30.8 \times 10^3 \text{ J·mol}^{-1}}{8.3145 \text{ J·mol}^{-1}\text{·K}^{-1}}\left(\dfrac{1}{298 \text{ K}} - \dfrac{1}{308 \text{ K}}\right);$$
$P_2 = 142 \text{ Torr}$

8.3B $\ln\left(\dfrac{760 \text{ Torr}}{400 \text{ Torr}}\right) = \dfrac{35.3 \times 10^3 \text{ J·mol}^{-1}}{8.3145 \text{ J·mol}^{-1}\text{·K}^{-1}}\left(\dfrac{1}{323 \text{ K}} - \dfrac{1}{T_2}\right)$;
$T_2 = 340 \text{ K}$

8.4B   The positive slope of the solid–liquid boundary shows that monoclinic sulfur is more dense than liquid sulfur over the temperature range in which monoclinic sulfur is stable; the solid is more stable at high pressure.

8.5B   Carbon dioxide is liquid at 60 atm and 25° C. When it is released into a room at 1 atm and 25° C, as the pressure lowers, the system reaches the liquid–vapor boundary, at which pressure the liquid is changed to vapor. The vaporization absorbs sufficient heat to cool the CO_2 to below its sublimation temperature at 1 atm. As a result, fine particles of solid CO_2 "snow" are produced.

8.6B   Critical temperature increases with the strength of intermolecular forces. For example, CH_4 cannot form hydrogen bonds; so it has a lower critical temperature than either NH_3 or H_2O, which can form hydrogen bonds.

8.7B $s = (2.3 \times 10^{-2} \text{ mol·L}^{-1}\text{·atm}^{-1})(1.00 \text{ atm})$
$= 2.3 \times 10^{-2} \text{ mol·L}^{-1}$; $n_{CO_2} = 0.900 \text{ L} \times 2.3 \times 10^{-2} \text{ mol·L}^{-1}$
$= 0.0207 \text{ mol}$

8.8B molality $= \dfrac{7.36 \text{ g KClO}_3}{0.200 \text{ kg H}_2\text{O}} \times \dfrac{1 \text{ mol KClO}_3}{122.55 \text{ g KClO}_3}$

$= 0.300 \text{ mol·kg}^{-1}$

8.9B $n_{H_2O} = 1 - n_{CH_3OH} = 1 - 0.250 = 0.750 \text{ mol}$;
$m_{H_2O} = 0.750 \text{ mol H}_2\text{O} \times 18.02 \text{ g·mol}^{-1} \times 1 \text{ kg}/10^3 \text{ g}$

$= 0.0135 \text{ kg H}_2\text{O}$; molality $= \dfrac{0.250 \text{ mol CH}_3\text{OH}}{0.0135 \text{ kg H}_2\text{O}}$

$= 18.5 \text{ mol·kg}^{-1}$

8.10B Assume a 1 L solution; $m_{NaCl} = 1 \text{ L}$

$\times \dfrac{1.83 \text{ mol NaCl}}{1 \text{ L soln}} \times 58.44 \text{ g·mol}^{-1} = 106.9 \text{ g NaCl}$;

$m_{solution} = 1 \text{ L soln} \times \dfrac{10^3 \text{ mL}}{1 \text{ L}} \times \dfrac{1.07 \text{ g}}{1 \text{ mL}} = 1.07 \times 10^3 \text{ g}$;

$m_{H_2O} = 1070 \text{ g} - 107 \text{ g} = 9.6 \times 10^3 \text{ g}$; $1.83 \text{ mol}/0.96 \text{ kg} =$
1.9 mol·kg^{-1}

8.11B $n_{C_2H_5OH} = 50.0 \text{ g}/46.07 \text{ g·mol}^{-1} = 1.09 \text{ mol}$;

$n_{C_9H_8O} = 2.00 \text{ g} \times \dfrac{1 \text{ mol}}{132.16 \text{ g}} = 0.0151 \text{ mol}$;

$x = 1.09/(1.09 + 0.0151) = 0.986$; $P = (0.986)(5.3 \text{ k Pa})$
$= 5.2 \text{ kPa}$

8.12B $\Delta T_f = k_f m = (39.7 \text{ K·kg·mol}^{-1})(0.050 \text{ mol·kg}^{-1}) =$
$1.99 \text{ K} = 1.99°\text{C}$; $T_f = 179.8°\text{C} - 1.99°\text{C} = 177.8°\text{C}$

8.13B $i = 1$ since sucrose is a nonelectrolyte and does not
dissociate; $\Pi = (1)(0.08206 \text{ L·atm·mol}^{-1}\text{·K}^{-1})(298 \text{ K})$
$(0.120 \text{ mol·L}^{-1}) = 2.93 \text{ atm}$

8.14B molality $= 0.51 \text{ K}/39.7 \text{ kg·K·mol}^{-1}$
$= 0.0128 \text{ mol·kg}^{-1}$; $n_{linalool} = 0.100 \text{ kg} \times (0.0128 \text{ mol}/$
$1 \text{ kg}) = 1.28 \times 10^{-3} \text{ mol}$; $M = 0.200 \text{ g}/(1.28 \times 10^{-3} \text{ mol})$
$= 156 \text{ g·mol}^{-1}$

8.15B $n = cV = \dfrac{2.11 \text{ kPa}}{(1)(8.3145 \text{ L·kPa·K}^{-1}\text{·mol}^{-1})(293 \text{ K})}$

$\times 0.175 \text{ L} = 1.52 \times 10^{-4} \text{ mol}$;

$M = \dfrac{1.50 \text{ g}}{1.52 \times 10^{-4}\text{mol}} = 9.90 \times 10^3 \text{ g·mol}^{-1}$

or 9.90 kg·mol^{-1}

8.16B Equal masses; therefore, assume 50 g of each.
$n_{C_6H_6} = 50 \text{ g}/(78.11 \text{ g·mol}^{-1}) = 0.640 \text{ mol}$;
$n_{CH_3C_6H_5} = 50 \text{ g}/(92.13 \text{ g·mol}^{-1}) = 0.543 \text{ mol}$;

$x_{C_6H_6} = \dfrac{0.640}{0.640 + 0.543} = 0.541$;

$x_{CH_3C_6H_5} = \dfrac{0.543}{0.640 + 0.543} = 0.459$; $P_{total} = (0.541) \times$
$(94.6 \text{ Torr}) + (0.459)(29.1 \text{ Torr}) = 64.5 \text{ Torr}$

8.17B (a) $P = (0.500)(94.6 \text{ Torr}) + (0.500)(29.1 \text{ Torr}) =$
61.8 Torr;

(b) $x_{C_6H_6,vap} = \dfrac{(0.500)(94.6 \text{ Torr})}{61.8 \text{ Torr}} = 0.765$;

$x_{CH_3C_6H_5,vap} = 1 - 0.765 = 0.235$

Major Technique 4

MT4.1B HCl is in the mobile phase; the order of elution of
the compounds will be the same as the order of their solubilities in HCl: $CuCl_2 > CoCl_2 > NiCl_2$.

Chapter 9

9.1B $K = (P_{SO_2})^2(P_{H_2O})^2/(P_{H_2S})^2(P_{O_2})^3$

9.2B $K = 1/(P_{O_2})^5$

9.3B $\Delta G_r = 4.73 \text{ kJ·mol}^{-1} + (8.3145 \text{ J·mol}^{-1}\text{·K}^{-1})$
$(298 \text{ K}) \times \ln\{(2.10)^2/0.80\} \times (1 \text{ kJ}/10^3 \text{ J}) = +8.96 \text{ kJ·mol}^{-1}$.
Because $\Delta G > 0$, reaction proceeds toward reactants.

9.4B $\Delta G_r° = 2\Delta G_f° (\text{NO}_2 (g)) - \Delta G_f° (\text{O}_2 (g))$
$- 2\Delta G_f° (\text{NO}(g)) = 2(51.31) - [0 + 2(86.55)] \text{ kJ·mol}^{-1}$
$= -70.48 \text{ kJ·mol}^{-1}$; $\ln K = -(-70.48 \times 10^3 \text{ J·mol}^{-1})/$
$(8.3145 \text{ kJ·K}^{-1}\text{·mol}^{-1} \times 298 \text{ K}) = 28.45$; $K = 2.3 \times 10^{12}$

9.5B $52 \text{ kPa} \times (1\text{bar})/(10^2 \text{ kPa}) = 0.52 \text{ bar}$;
$K = 3.4 \times 10^{-21} = (P_{NO})^2/(0.52)^2$; $P_{NO} = 3.0 \times 10^{-11} \text{ bar}$,
or $3.0 \times 10^{-6} \text{ Pa}$

9.6B $Q = (1.2)^2/2.4 = 0.60$. $K = 0.15$, and so $Q > K$ and
$P_{N_2O_4}$ will increase.

9.7B $K_c = K(P°/RTc°)^{\Delta n}$, $\Delta n = 2 - 1 = +1$;
$K_c = K(12.03 \text{ K}/T) = (47.9)[(12.03 \text{ K})/(400 \text{ K})] = 1.44$

9.8B $K_c = (7.3 \times 10^{-13})^{1/2} = 8.5 \times 10^{-7}$

9.9B $K = (2x)^2(x)/(0.012 - 2x) \approx 4x^2/(0.012)^2$
$= 3.5 \times 10^{-32}$; approximate to $x = 1.1 \times 10^{-12}$. At equilib-
rium: $P_{HCl} = 0.012 \text{ bar}$; $P_{HI} = 2.2 \times 10^{-12} \text{ bar}$; $P_{Cl_2} = 1.1 \times$
10^{-12} bar; some solid I_2 remains as well.

9.10B $Q = (0.100)^2/(0.100)(0.200) = 0.5 < K$; $K = 20 =$
$(0.100 + 2x)^2/\{(0.200 - x)(0.100 - x)\}$; $x = 0.0750$; $P_{ClF} =$
$0.100 + 2x = 0.250 \text{ bar}$

9.11B The equilibrium tends to shift toward (a) products;
(b) products; (c) reactants.

9.12B $Q = (1.30)^2/(0.080)(0.050)^3 = 1.7 \times 10^5$; $Q < K$.
Initially, $P_{NH_3} = \frac{1}{2}(2.60 \text{ bar}) = 1.30 \text{ bar}$;
$K = (1.30 + 2x)^2/(0.080 - x)(0.050 - x) = 6.8 \times 10^5$;
$x = 5.86 \times 10^{-3} \text{ bar}$; $P_{N_2} = 0.074 \text{ bar}$; $P_{NH_3} = 1.31 \text{ bar}$; P_{H_2}
$= 0.032 \text{ bar}$.

9.13B Compression affects gaseous species only. In the forward reaction, 1 mol of $CO_2(g)$ reacts to form an aqueous
species. Therefore, compression favors the formation of
$H_2CO_3(aq)$.

9.14B $\Delta H_r° = 2(-393.51 \text{ kJ·mol}^{-1})$
$- 2(-110.53 \text{ kJ·mol}^{-1}) - 0 = -565.96 \text{ kJ·mol}^{-1}$. The reaction is exothermic; therefore, lowering the temperature shifts
the reaction to the products. CO_2 is favored.

9.15B $\Delta H_r° = 0 + (-287.0 \text{ kJ·mol}^{-1})$
$- (-374.9 \text{ kJ·mol}^{-1}) = +87.9 \text{ kJ·mol}^{-1}$;

$$\ln\left(\frac{K_2}{K_1}\right) = \frac{87.9 \times 10^3 \text{ J·mol}^{-1}}{8.3145 \text{ J·mol}^{-1}\text{·K}^{-1}}\left[\frac{1}{523 \text{ K}} - \frac{1}{800 \text{ K}}\right]$$

$$= 7.0;$$
$$K_2 = K_1 e^{7.0} = (78.3)e^{7.0} = 8.6 \times 10^4$$

Chapter 10

10.1B (a) H_3O^+ (it has one more H^+ than H_2O);
(b) NH_2^- (it has one less H^+ than NH_3)

10.2B (a) Brønsted acids: $NH_4^+(aq)$, $H_2CO_3(aq)$; Brønsted
bases: $HCO_3^-(aq)$, $NH_3(aq)$;
(b) Lewis acid: $H^+ (aq)$; Lewis bases: $NH_3(aq)$, $HCO_3^- (aq)$

10.3B (a) $[H_3O^+] = (1.0 \times 10^{-14})/(2.2 \times 10^{-3})$
$= 4.5 \times 10^{-12} \text{ mol·L}^{-1}$; (b) $[OH^-] = [NaOH]$
$= 2.2 \times 10^{-3} \text{ mol·L}^{-1}$

10.4B $[OH^-] = [NaOH] = 0.077 \text{ mol·L}^{-1}$;
$[H_3O^+] = (1.0 \times 10^{-14})/(0.077) = 1.30 \times 10^{-13} \text{ mol·L}^{-1}$;
pH $= -\log (1.3 \times 10^{-13}) = 12.89$

10.5B $[H_3O^+] = 10^{-8.2} = 6 \times 10^{-9} \text{ mol·L}^{-1}$

10.6B The conjugate base of HIO_3 is IO_3^-;
$pK_b = 14.00 - pK_a = 14.00 - 0.77 = 13.23$

10.7B (a) $1.8 \times 10^{-9} = K_b(C_5H_5N) < K_b(NH_2NH_2)$ $= 1.7 \times 10^{-6}$. Therefore, NH_2NH_2 is the stronger base. (b) From part a, C_5H_5N is the weaker base; therefore, $C_5H_5NH^+$ is the stronger acid. (c) 1.7×10^{-1} $= K_a(HIO_3) > K_a(HClO_2) = 1.0 \times 10^{-2}$; therefore, HIO_3 is the stronger acid. (d) $K_b(HSO_3^-) = (1.0 \times 10^{-14})/(1.5 \times 10^{-2}) =$ 6.7×10^{-13}; $K_b(ClO_2^-) = \dfrac{1.0 \times 10^{-14}}{1.0 \times 10^{-2}}$ $= 1.0 \times 10^{-12}$; $K_b(HSO_3^-) < K_b(ClO_2^-)$; therefore, ClO_2^- is the stronger base.

10.8B $CH_3COOH < CH_2ClCOOH < CHCl_2COOH$ (The electronegativity of Cl is greater than that of H. Therefore, the acidity increases as the number of attached chlorine atoms increases.)

10.9B $K_a = 1.4 \times 10^{-3} = x^2/(0.22 - x)$; we must use a quadratic equation: $x^2 + (1.4 \times 10^{-3})x - 3.1 \times 10^{-4}$ $= 0$; $x = 1.7 \times 10^{-2} = [H_3O^+]$; pH = 1.77; percentage deprotonation $= \{(1.7 \times 10^{-2})/0.22\} \times 100\% = 7.7\%$

10.10B $[H_3O^+] = 10^{-2.35} = 4.5 \times 10^{-3}$; $K_a =$ $(4.5 \times 10^{-3})^2/(0.50 - 4.5 \times 10^{-3}) = 4.1 \times 10^{-5}$

10.11B $K_b = 1.0 \times 10^{-6} = x^2/(0.012 - x) \approx x^2/.012$; $x = 1.1 \times 10^{-4} = [OH^-]$; pOH = 3.96; pH = 14.00 $- 3.96 = 10.04$; percentage protonated $= \dfrac{1.1 \times 10^{-4}}{0.012}$ $\times 100 = 0.92\%$

10.12B (a) CO_3^{2-} is the conjugate base of HCO_3^-, which is a weak acid; therefore, the solution is basic.

(b) $K_a(Al(H_2O)_6{}^{3+}) = 1.4 \times 10^{-5}$. Therefore, the solution is acidic. (c) K^+ is a "neutral cation" and NO_3^- is a conjugate base of a strong acid; therefore, the aqueous solution is neutral.

10.13B $NH_4^+(aq) + H_2O(l) \rightleftharpoons NH_3(aq) + H_3O^+(aq)$; $K_b(NH_3) = 1.8 \times 10^{-5}$; $K_a(NH_4^+) = (1.0 \times 10^{-14})/$ $(1.8 \times 10^{-5}) = 5.6 \times 10^{-10}$; $K_a = 5.6 \times 10^{-10} = x^2/$ $(0.10 - x) \approx x^2/0.10$; $x = 7.5 \times 10^{-6}$; pH = $-\log(7.5 \times 10^{-6}) = 5.12$

10.14B $F^- + H_2O \rightleftharpoons HF + OH^-$; $K_a(HF) =$ 3.5×10^{-4}; $K_b(F^-) = (1.0 \times 10^{-14})/(3.5 \times 10^{-4}) =$ 2.9×10^{-11}; $K_b = 2.9 \times 10^{-11} = x^2/(0.020 - x) \approx x^2/0.020$; $x = 7.56 \times 10^{-7}$; pOH = 6.12; pH = 14.00 - 6.12 $= 7.88$

10.15B $0.012 = (0.10 + x)(x)/(0.10 - x)$; $x^2 + 0.11x - 0.0012 = 0$; from the quadratic formula, $x = 0.0098$; $[H_3O^+] = 0.10 + 0.0098 = 0.11$; pH = 0.96

10.16B $pK_{a1}(H_3PO_4) = 2.12$; $pK_{a2}(H_3PO_4) = 7.21$; pH = $\frac{1}{2}(2.12 + 7.21) = 4.66$

10.17B $[Cl^-] = 0.50$ mol·L^{-1}; $^+NH_3CH_2COOH(aq)$ $+ H_2O(l) \rightleftharpoons NH_2CH_2COOH(aq) + H_3O^+(aq)$, $K_{a1} = 4.5 \times 10^{-3}$; $NH_2CH_2COOH(aq) + H_2O(l)$ $\rightleftharpoons NH_2CH_2CO_2^-(aq) + H_3O^+(aq)$, $K_{a2} = 1.7 \times 10^{-10}$; $K_{a1} =$ $4.5 \times 10^{-3} = x^2/(0.50 - x)$; we must solve the quadratic equation: $x^2 + (4.5 \times 10^{-3})x - 0.00225 = 0$; $x = 0.045$ mol·L$^{-1} = [H_3O^+] = [NH_2CH_2COOH]$; $[^+NH_3CH_2COOH] = 0.50 - 0.045$ mol·L^{-1} $= 0.45$ mol·L^{-1}; $K_{a2} = 1.7 \times 10^{-10} = (0.045)(x)/0.045$; $x =$ $1.7 \times 10^{-10} = [NH_2CH_2CO_2^-]$; $[OH^-]$ $= K_w/[H_3O^+] = (1.0 \times 10^{-14})/(0.045)$ $= 2.2 \times 10^{-13}$ mol·L^{-1}

10.18B $[H_3O^+]^2 + [NaOH]_{initial}[H_3O^+] - K_w = 0$; $x^2 + (2.0 \times 10^{-7})x - (1.0 \times 10^{-14}) = 0$; $[H_3O^+] =$ $x = 4.1 \times 10^{-8}$ mol·L^{-1}; pH = $-\log(4.1 \times 10^{-8}) = 7.38$

10.19B $K_a(HIO) = 2.3 \times 10^{-11}$; $[H_3O^+] = [(1.0 \times 10^{-14}) +$ $(2.3 \times 10^{-11}) \times (1.0 \times 10^{-2})]^{1/2}$; $[H_3O^+] = 4.9 \times 10^{-7}$; pH = 6.31

Chapter 11

11.1B $K_a = (1.0 \times 10^{-14})/(1.8 \times 10^{-5}) = 5.6 \times 10^{-10}$; pH = $-\log(5.6 \times 10^{-10}) + \log(0.030/0.040) = 9.13$

11.2B Assume that 0.020 mol HCl has been added to 1.00 L of buffer. $[CH_3CO_2^-] = 0.040 - 0.020$ $= 0.020$ mol·L^{-1}; $[CH_3COOH] = 0.080 + 0.020$ $= 0.100$ mol·L^{-1}; pH = $4.75 + \log(0.020/0.100)$ $= 4.05$; $pH_{final} - pH_{initial} = 4.05 - 4.45 = -0.4$ (a decrease of 0.4)

11.3B $(CH_3)_3NH^+/(CH_3)_3N$, because the $pK_a = 9.81$, which is close to 10

11.4B pH $- pK_a = 3.50 - 4.19 = -0.69$; $[C_6H_5COO^-]/$ $[C_6H_5COOH] = 10^{-0.69} = 0.20$

11.5B Amount of H_3O^+ added = 0.012 L $\times$ 0.340 mol/1 L $= 0.00408$ mol; amount of OH^- remaining = 0.00625 $-$ 0.00408 mol = 0.00217 mol. $[OH^-] = 0.00217$ mol/(0.0120 + 0.0250) L = 0.0586 mol·L^{-1}; pOH = 1.232; pH = $14.00 - 1.232 = 12.77$

11.6B initial amount of $NH_3 = 0.025$ L $\times$ (0.020 mol·L^{-1}) $= 5.0 \times 10^{-4}$ mol; volume of HCl added $= (5.0 \times 10^{-4}$ mol)/(0.015 mol·L^{-1}) = 0.0333 L; $[NH_4^+] = \dfrac{(5.0 \times 10^{-4}$ mol$)}{(0.03333 + 0.02500)$ L$} = 0.0086$ mol·L^{-1}; $K_a = \dfrac{[H_3O^+][NH_3]}{[NH_4^+]} = 5.6 \times 10^{-10} = (x^2)/(0.0086 - x)$ $\approx x^2/0.0086$; $[H_3O^+] = x = 2.2 \times 10^{-6}$ mol·L^{-1}; pH = 5.66

11.7B final volume of solution = 25.00 + 15.00 mL $= 40.00$ mL; amount of HCO_2^- formed = amount of OH^- added = (0.150 mol·L^{-1}) $\times$ (15.00 mL) $= 2.25$ mmol; amount of HCOOH remaining $= 2.50 - 2.25$ mmol = 0.25 mmol; [HCOOH] $= (0.25 \times 10^{-3}$ mol)/(0.04000 L) = 0.062 mol·L^{-1}; $[HCO_2^-]$ $= (2.25 \times 10^{-3}$ mol)/(0.04000 L) = 0.0562 mol·L^{-1}; pH $= 3.75 + \log(0.0562/0.0062) = 4.71$

11.8B (a) amount of $H_3PO_4 = (0.030$ L$)(0.010$ mol·L$^{-1}) =$ 3.0×10^{-4} mol; at first stoichiometric point, initial amount of $H_3PO_4 =$ amount of NaOH added; volume of NaOH = 3.0×10^{-4} mol/ 0.020 mol·L^{-1} = 0.015 L, or 15 mL. (b) 2(15 mL) = 30 mL

11.9B $(0.020$ L$)(0.100$ mol·L$^{-1}) = 0.0020$ mol H_2S; 0.0020 mol/ (0.300 mol·L^{-1}) = 0.0067 L, or 6.7 mL NaOH for first stoichiometric point and 2×6.7 mL = 13.4 mL for second stoichiometric point; (a) before first stoichiometric point; $H_2S(aq) + H_2O(l) \rightleftharpoons HS^-(aq) + H_3O^+(aq)$; primary species present are Na^+, H_2S, and HS^-; (b) at second stoichiometric point, primary equilibrium is $S^{2-}(aq) + H_2O(l)$ $\rightleftharpoons HS^-(aq) + OH^-(aq)$; species present are Na^+ and S^{2-} (also HS^- and OH^- because S^{2-} has a relatively large K_b)

11.10B $K_{sp} = [Ag^+][Br^-] = (8.8 \times 10^{-7})^2 = 7.7 \times 10^{-13}$

11.11B $K_{sp} = [Pb^{2+}][F^-]^2 = (s)(2s)^2 = 4s^3$; $3.7 \times 10^{-8} =$ $4s^3$; $s = 2.1 \times 10^{-3}$ mol·L^{-1}

11.12B in 0.10 M $CaBr_2(aq)$, $[Br^-] = 2(0.10$ mol·L$^{-1})$ $= 0.20$ mol·L^{-1}; $s = [Ag^+] = K_{sp}/[Br^-] = (7.7 \times 10^{-13})/$ $(0.20) = 3.8 \times 10^{-12}$ mol·L^{-1}

11.13B $[Ba^{2+}] = \{(1.0 \times 10^{-3}$ mol·L$^{-1}) \times (100$ mL$)\}/$ 300 mL = 3.3×10^{-4} mol·L^{-1}.

$[F^-] = \{(1.0 \times 10^{-3}\ mol\cdot L^{-1}) \times (200\ mL)\}/300\ mL$
$= 6.7 \times 10^{-4}\ mol\cdot L^{-1}$. $Q_{sp} = (3.3 \times 10^{-4})(6.7 \times 10^{-4})^2$
$= 1.5 \times 10^{-10} < K_{sp} = 1.7 \times 10^{-6}$; therefore, no precipitate of BaF_2 forms.

11.14B (a) For $PbCl_2$ to precipitate, $[Cl^-] = (K_{sp}/[Pb^{2+}])^{1/2}$
$= (1.6 \times 10^{-5}/0.020)^{1/2} = 2.8 \times 10^{-2}\ mol\cdot L^{-1}$. For AgCl to precipitate, $[Cl^-] = (K_{sp}/[Ag^+]) = (1.6 \times 10^{-10}/0.0010)$
$= 1.6 \times 10^{-7}\ mol\cdot L^{-1}$. Therefore, AgCl precipitates first at $[Cl^-] = 1.6 \times 10^{-7}\ mol\cdot L^{-1}$, then $PbCl_2$ precipitates at $[Cl^-]$
$= 2.8 \times 10^{-2}\ mol\cdot L^{-1}$. (b) When the $PbCl_2$ precipitates, $[Ag^+]$
$= (K_{sp}/[Cl^-]) = (1.6 \times 10^{-10})/(2.8 \times 10^{-2}) = 5.7 \times 10^{-9}\ mol\cdot L^{-1}$.

11.15B $CuS(s) + 4\ NH_3(aq) \rightleftharpoons Cu(NH_3)_4^{2+}(aq) + S^{2-}(aq)$;
$K = (1.3 \times 10^{-36}) \times (1.2 \times 10^{13}) = 1.6 \times 10^{-23}$; $K =$
$[Cu(NH_3)_4^{2+}][S^{2-}]/[NH_3]^4 = 1.6 \times 10^{-23} = x^2/$
$(1.2 - 4x)^4 \approx x^2/(1.2)^4$; $x/(1.2)^2 = 4.0 \times 10^{-12}$;
$x = [S^{2-}] = 5.8 \times 10^{-12}\ mol\cdot L^{-1}$

Chapter 12

12.1B reduction: $(8\ H^+ + MnO_4^- + 5\ e^- \rightarrow$
$Mn^{2+} + 4\ H_2O) \times 2$; oxidation: $(H_2O + H_2SO_3 \rightarrow$
$HSO_4^- + 2\ e^- + 3\ H^+) \times 5$; net reaction: $H^+(aq) + 2MnO_4^-$
$(aq) + 5\ H_2SO_3(aq) \rightarrow 2\ Mn^{2+}(aq) + 5\ HSO_4^-(aq)$
$+ 3\ H_2O(l)$

12.2B oxidation: $(3\ I^- \rightarrow I_3^- + 2\ e^-) \times 8$; reduction: $9\ H_2O$
$+ 16\ e^- + 3\ IO_3^- \rightarrow I_3^- + 18\ OH^-$; net reaction: $3\ H_2O(l) +$
$IO_3^-(aq) + 8\ I^-(aq) \rightarrow 3\ I_3^-(aq) + 6\ OH^-(aq)$

12.3B Half reactions for the silver battery in Table 12.2 show that $n = 2$; $\Delta G = -(2)(9.6485 \times 10^4\ C\cdot mol^{-1})(1.6V) =$
$-3.09 \times 10^5\ C\cdot V\cdot mol^{-1} = -310\ kJ\cdot mol^{-1}$.

12.4B $Mn(s)|Mn^{2+}(aq)||Cu^{2+}(aq),Cu^+(aq)|Pt(s)$

12.5B (a) left: $2\ Hg(l) + 2\ HCl(aq) \rightarrow Hg_2Cl_2(s) +$
$2\ e^-$; right: $2\ e^- + Hg_2(NO_3)_2(aq) \rightarrow 2\ Hg(l) + 2\ NO_3^-(aq)$;
$2\ HCl(aq) + Hg_2(NO_3)_2(aq) \rightarrow Hg_2Cl_2(s) +$
$2\ HNO_3(aq)$. (b) yes

12.6B $E(Pb^{2+}/Pb) - E(Fe^{2+}/Fe) = 0.31\ V + (-0.44\ V) =$
$-0.13\ V$

12.7B $Mn^{3+} + e^- \rightarrow Mn^{2+}$, $E_1 = +1.51\ V$;
$Mn^{2+} + 2\ e^- \rightarrow Mn(s)$, $E_2 = -1.18\ V$; net half reaction: Mn^{3+}
$+ 3\ e^- \rightarrow Mn(s)$, $E_3 = ?$

$$E_3 = \frac{n_1 E_1 + n_2 E_2}{n_3}$$
$$= \frac{(1\ mol)(1.51\ V) + (2\ mol)(-1.18\ V)}{3\ mol}$$
$$= -0.28\ V$$

12.8B $O_2 + 2\ H_2O + 4e^- \rightarrow 4\ OH^-$, $E° = +0.40\ V$; $Cl_2 +$
$2\ e^- \rightarrow 2\ Cl^-$, $E° = +1.36\ V$. Yes, $Cl_2(g)$ can oxidize H_2O to $O_2(g)$ in basic solution under standard conditions, because the Cl_2 reduction half-reaction has a more positive standard potential than the O_2 reduction half-reaction.

12.9B oxidation half-reaction: $Cu^{2+} + 2\ e^- \rightarrow$
Cu, $E° = +0.34\ V$; reduction half-reaction: $Ag^+ + e^- \rightarrow$
Ag, $E° = +0.80\ V$; net reaction: $Cu(s) + 2\ Ag^{2+}(aq) \rightarrow$
$Cu^{2+}(aq) + 2\ Ag(s)$, $E° = +0.46\ V$. Therefore, Ag^+ is the stronger oxidizing agent. Cell diagram:
$Cu(s)|Cu^{2+}(aq)||Ag^+(aq)|Ag(s)$

12.10B $Cd(OH)_2 + 2\ e^- \rightarrow Cd + 2\ OH^-$, $E° = -0.81\ V$;
$Cd^{2+} + 2\ e^- \rightarrow Cd$, $E° = -0.40\ V$; net: $Cd(OH)_2(s) \rightarrow$
$Cd^{2+}(aq) + 2\ OH^-(aq)$, $E° = -0.41\ V$;
$\ln K_{sp} = (2)(-0.41\ V)/0.025693\ V = -31.92$;
$K_{sp} = 1.4 \times 10^{-14}$

12.11B cell reaction: $Ag^+(aq, 0.010\ mol\cdot L^{-1}) \rightarrow$
$Ag^+(aq, 0.0010\ mol\cdot L^{-1})$, $E° = 0.0\ V$, $n = 1$;

$$E = 0.0\ V - 0.025693\ V\ln\left(\frac{0.0010}{0.010}\right) = +0.059\ V$$

12.12B At pH = 12.5, pOH = 1.5 and $[OH^-] = 10^{-1.5} = 0.032$; $[Ag^+] = (K_{sp}/[OH^-]) = (1.5 \times 10^{-8}/(0.032)$
$= 4.7 \times 10^{-7}\ mol\cdot L^{-1}$. $E =$
$-(0.025693\ V/1)\ln(4.7 \times 10^{-7}/1.0) = 0.37\ V$

12.13B The reduction half-reactions are:
$Br_2(l) + 2\ e^- \rightarrow 2\ Br^-(aq)$, $E° = +1.09\ V$;
$O_2(g) + 4\ H^+(aq) + 4\ e^- \rightarrow 2H_2O(l)$, $E = +0.82\ V$ at pH =
7; $2\ H^+(aq) + 2\ e^- \rightarrow 2\ H_2(g)$, $E° = 0.00\ V$. Product at the cathode: H_2; product at the anode: O_2 and Br_2. (We predict that H_2O will be oxidized rather than Br^-. However, because of the high overpotential for oxygen production, bromine may also be produced.)

12.14B $12.0\ mol\ e^- \times (1\ mol\ Cr)/(6\ mol\ e^-) = 2\ mol\ Cr$

12.15B $m_{Cr} = \dfrac{(6.20\ C\cdot s^{-1})(6.00 \times 3600s)}{9.6485 \times 10^4\ C\cdot(mole^-)^{-1}} \times$

$\dfrac{1\ mol\ Cr}{6\ mole^-} \times \dfrac{52.00g\ Cr}{1\ mol\ Cr} = 12.0\ g\ Cr$

12.16B $12.00\ g\ Cr \times \dfrac{1\ mol\ Cr}{52.00\ g\ Cr} \times \dfrac{6\ mol\ e^-}{1\ mol\ Cr} = 1.385\ mol\ e^-$;

$t = \dfrac{(1.385\ mol\ e^-)(9.6485 \times 10^4\ C\cdot mol^{-1})}{6.20\ C\cdot s^{-1}} \times \dfrac{1\ h}{3600\ s} = 5.99h$

12.17B (b) Aluminum. It is the only one of the metals with a standard potential $(-1.66\ V)$ lower than that of iron $(+0.44\ V)$, so it is the only one more easily oxidized than iron.

Chapter 13

13.1B average rate of disappearance of Hb
$$= \frac{-[(8.0 \times 10^{-7}) - (1.2 \times 10^{-6})]\ mmol\cdot L^{-1}}{0.10\ \mu s}$$
$$= 4 \times 10^{-6}\ mmol\cdot L^{-1}\cdot \mu s^{-1}$$

13.2B (a) $\frac{1}{2}(5.0 \times 10^{-3})(mmol\ HI)\cdot L^{-1}\cdot s^{-1} =$
$2.5 \times 10^{-3}\ (mmol\ H_2)\cdot L^{-1}\cdot s^{-1}$; (b) unique average rate =
$\Delta[H_2]/\Delta t = \Delta[I_2]/\Delta t = -\frac{1}{2}(\Delta[HI]/\Delta t) = 2.5 \times 10^{-3}\ mmol\cdot L^{-1}\cdot s^{-1}$

13.3B (a) first order in C_4H_9Br; zero order in OH^-; (b) first order overall; (c) s^{-1}

13.4B rate $= k\ [CO]^m\ [Cl_2]^n$

$\dfrac{rate\ (2)}{rate\ (1)} = \dfrac{0.241}{0.121} = \left(\dfrac{0.24}{0.12}\right)^m \left(\dfrac{0.20}{0.20}\right)^n$; $2 = (2)^m$; $m = 1$

$\dfrac{rate\ (3)}{rate\ (2)} = \dfrac{0.682}{0.241} = \left(\dfrac{0.24}{0.24}\right)^m \left(\dfrac{0.40}{0.20}\right)^n$; $2.8 = (2)^n$;

$n = \dfrac{\log 2.8}{\log 2} = 1.5$

Therefore, rate $= k[CO][Cl_2]^{3/2}$. From experiment 1, $k =$
$\dfrac{0.121\ mol\cdot L^{-1}\cdot s^{-1}}{(0.12\ mol\cdot L^{-1})(0.20\ mol\cdot L^{-1})^{3/2}} = 11\ L^{3/2}\cdot mol^{-3/2}\cdot s^{-1}$

13.5B $[C_3H_6]_t = [C_3H_6]_0\ e^{-kt} = (0.100\ mol\cdot L^{-1})$
$e^{-(6.7 \times 10^{-4}\ s^{-1})(200\ s)} = 0.087\ mol\cdot L^{-1}$

13.6B A plot of $\ln(CH_3N_2CH_3)$ against time is linear, showing that the reaction must be first order; $k = 3.60 \times 10^{-4}\ s^{-1}$.

13.7B Reaction is first order with $k = 6.7 \times 10^{-4}\ s^{-1}$ at 500° C;

$$t = \frac{1}{k}\ln\frac{[C_3H_6]_0}{[C_3H_6]_t} = \frac{1}{6.7 \times 10^{-4}s^{-1}}\ln\left(\frac{1.0\ mol\cdot L^{-1}}{0.005\ mol\cdot L^{-1}}\right)$$
$$= 7.9 \times 10^3\ s = 2.2\ h$$

13.8B Reaction is first order with $k = 5.5 \times 10^{-4}$ s^{-1} at 973 K. (a) $t_{1/2} = (\ln 2)/k = (\ln 2)/(5.5 \times 10^{-4}$ s$^{-1}) = 1.3 \times 10^3$ s = 21 min. (b) For the concentration to fall to one-sixteenth of its initial value requires four half-lives, and so $t = 4t_{1/2} = 4 \times 21$ min = 84 min.

13.9B $k = \dfrac{\ln 2}{2.4 \times 10^4 \text{ y}}$; $t = \dfrac{2.4 \times 10^4 \text{ y}}{\ln 2} \ln\left(\dfrac{1.0}{0.20}\right)$
$= 5.6 \times 10^4$ y

13.10B (a) bimolecular (two reactants); (b) unimolecular (one reactant)

13.11B net rate$_{\text{disappearance of B}}$ =
$k_1[\text{H}_2\text{A}][\text{B}] + k_2[\text{HA}^-][\text{B}] - k_1'[\text{HA}^-][\text{BH}^+]$;

net rate$_{\text{formation of HA}^-}$ =

$k_1[\text{H}_2\text{A}][\text{B}] - k_1'[\text{HA}^-][\text{BH}^+] - k_2[\text{HA}^-][\text{B}] = 0$;

$[\text{HA}^-] = \dfrac{k_1[\text{H}_2\text{A}][\text{B}]}{k_1'[\text{BH}^+] + k_2[\text{B}]}$;

substitute for $[\text{HA}^-]$:
net rate$_{\text{disappearance of B}}$ =

$k_1[\text{H}_2\text{A}][\text{B}] + \dfrac{k_2k_1[\text{H}_2\text{A}][\text{B}]^2}{k_1'[\text{BH}^+] + k_2[\text{B}]} - \dfrac{k_1'k_1[\text{H}_2\text{A}][\text{B}][\text{BH}^+]}{k_1'[\text{BH}^+] + k_2[\text{B}]} =$

$\dfrac{k_2k_1[\text{H}_2\text{A}][\text{B}]^2}{k_1'[\text{BH}^+]}$; assume $k_2[\text{B}] \ll k_1'[\text{BH}^+]$;

net rate$_{\text{disappearance of B}}$ = $k[\text{H}_2\text{A}][\text{B}]^2[\text{BH}^+]^{-1}$; $k = \dfrac{2k_2k_1}{k_1'}$

13.12B $\ln\left(\dfrac{k'}{k}\right) = \dfrac{E_a}{R}\left[\dfrac{1}{T} - \dfrac{1}{T'}\right]$;

$\ln\left(\dfrac{4.35}{3.00}\right) = \dfrac{E_a}{8.3145 \text{ J·mol}^{-1}\cdot\text{K}^{-1}}\left[\dfrac{1}{291 \text{ K}} - \dfrac{1}{303 \text{ K}}\right]$;

$E_a = 2.3 \times 10^5$ J·mol^{-1} = 23 kJ·mol^{-1}

13.13B $\ln\left(\dfrac{k'}{k}\right) = \dfrac{E_a}{R}\left[\dfrac{1}{T} - \dfrac{1}{T'}\right]$; $\ln\left(\dfrac{k'}{k}\right)$

$= \dfrac{272 \text{ kJ·mol}^{-1}}{8.314 \times 10^{-3} \text{ kJ·mol}^{-1}\cdot\text{K}^{-1}}\left[\dfrac{1}{773 \text{ K}} - \dfrac{1}{573 \text{ K}}\right] =$

-14.77; $\dfrac{k'}{k} = e^{-14.77}$; $k' = (6.7 \times 10^{-4}$ s$^{-1})(e^{-14.77}) =$

2.6×10^{-10} s^{-1}

13.14B (a) The homogeneous catalyst changes the reaction pathway, and therefore changes the rate law. (b) Since a catalyst does not change the thermodynamics of the reaction, a homogeneous catalyst does not change the equilibrium constant.

Chapter 14

14.1B Oxygen. It is above both gallium and tellurium in the periodic table and to the right of gallium.

14.2B Aluminum forms an amphoteric oxide in which it has the oxidation state +3; therefore, aluminum is the element.

14.3B Hydrogen is a nonmetal and a diatomic gas at room temperature. It has an intermediate electronegativity ($\chi = 2.2$), so it forms covalent bonds with nonmetals and forms anions in combination with metals. In contrast, Group 1 elements are solid metals that have low electronegativities and form cations in combination with nonmetals.

14.4B $\text{K}(s) + \text{O}_2(g) \rightarrow \text{KO}_2(s)$

14.5B The molar mass of potassium is smaller than that of cesium. Thus the same amount of O_2 can be obtained from a smaller mass of KO_2 than can be obtained from a larger mass of CsO_2.

14.6B $2 \text{ Ba}(s) + \text{O}_2(g) \rightarrow 2 \text{ BaO}(s)$

14.7B Lewis acid–base reaction; CaO is the base and SiO_2 is the acid.

14.8B Boric acid acts as a Lewis acid. The boron atom in B(OH)_3 has an incomplete octet and forms a bond by accepting a lone pair of electrons from a water molecule, which is acting as a Lewis base. The complex formed is a weak Brønsted acid in which an acidic proton can be lost from the H_2O molecule in the complex.

14.9B (a) +3, because the oxidation number of H in this hydride is taken to be -1; (b) +3

14.10B Lewis acid–base reaction; CO is the Lewis acid and OH$^-$ is the Lewis base.

14.11B SiH_4 is a Lewis acid that can react with the Lewis base OH$^-$. CH_4 is not a Lewis acid, because the C atom is much smaller than the Si atom and has no accessible d-orbitals to accommodate added electron pairs.

14.12B $\text{SiO}_2(s) + 2 \text{ OH}^-(aq) \rightarrow \text{SiO}_3^{2-}(aq) + \text{H}_2\text{O}(l)$

Chapter 15

15.1B (a) $\left[\overset{-1}{\text{N}}=\overset{+1}{\text{N}}=\overset{-1}{\text{N}}\right]^-$. (b) The structure shown minimizes the formal charges on each atom and so is the most stable. (c) Linear and nonpolar

15.2B assume 1 L solution;

$c = 1 \text{ L soln} \times \dfrac{10^3 \text{mL}}{1 \text{ L}} \times \dfrac{1.7 \text{ g soln}}{1 \text{ mL soln}} \times \dfrac{85 \text{ g H}_3\text{PO}_4}{100 \text{ g soln}} \times$

$\dfrac{1 \text{ mol H}_3\text{PO}_4}{97.99 \text{ g H}_3\text{PO}_4} = 15 \text{ M H}_3\text{PO}_4(aq)$

15.3B $\underline{\begin{array}{l} 2\,(2 \text{ e}^- + \text{F}_2 \rightarrow 2 \text{ F}^-) \\ 2 \text{ H}_2\text{O} \rightarrow \text{O}_2 + 4 \text{ H}^+ + 4 \text{ e}^- \end{array}}$
$2 \text{ H}_2\text{O}(l) + 2 \text{ F}_2(g) \rightarrow \text{O}_2(g) + 4 \text{ H}^+(aq) + 4 \text{ F}^-(aq)$
$E° = 2.87 - 1.23 \text{ V} = +1.64 \text{ V}$
$\Delta G° = -(4 \text{ mol})(96.485 \text{ kC·mol}^{-1})(1.64 \text{ V}) = -633 \text{ kJ}$

15.4B (a) +1; (b) +6

15.5B The melting and boiling points of the halogens increase down the group because the intermolecular attractions become stronger; London forces become stronger as the number of electrons present increase.

15.6B In the reduction of HClO_3 an acid is a reactant. Therefore, increasing the pH reduces the oxidizing ability of HClO_3.

15.7B

Square pyramid

15.8B In chemiluminescence light is emitted as the result of molecules being excited during a chemical reaction. In phosphorescence light is emitted when molecules remain excited after the stimulus has ceased.

Chapter 16

16.1B Each metal atom will lose one $4s$-electron. V$^+$, Mn$^+$, Co$^+$, and Ni$^+$ will have one electron in the $4s$-orbital, but Cr$^+$ and Cu$^+$ will have zero electrons in the $4s$-orbital. Because their outermost valence electrons will be in $3d$-orbitals and not the $4s$-orbital, their radii should be smaller than the other cations.

16.2B
oxidation: $\qquad\qquad\qquad\qquad 2 \text{ Hg}(l) \rightarrow \text{Hg}_2^{2+}(aq) + 2 \text{ e}^-$
reduction: $\quad 4 \text{ H}^+(aq) + \text{NO}_3^-(aq) + 3\text{e}^- \rightarrow \text{NO}(aq) + 2 \text{ H}_2\text{O}(l)$

net: $\quad 6 \text{ Hg}(l) + 8 \text{ H}^+(aq) + 2 \text{ NO}_3^-(aq) \rightarrow$
$\qquad\qquad\qquad 3 \text{ Hg}_2^{2+}(aq) + 2 \text{ NO}(aq) + 4 \text{ H}_2\text{O}(l)$
$E° = 0.96 \text{ V} - 0.79 \text{ V} = +0.17 \text{ V}$

16.3B   (a) pentaaminebromocobalt(III) sulfate; (b) $[Cr(NH_3)_4(H_2O)_2]Br_3$

16.4B   The ratio of $CrCl_3 \cdot 6H_2O$ to Cl is 1:3; therefore, the isomer $[Cr(H_2O)_6]Cl_3$ is present.

16.5B   (a) NCS^- is replaced by SCN^-; linkage isomers; (b) H_2O is exchanged for Cl^-; hydrate isomers.

16.6B   (a) not chiral; (b) chiral; (c) not chiral; (d) chiral; no entantiomeric pairs

16.7B $\Delta_O = \dfrac{hc}{\lambda}$

$$= \frac{(6.022 \times 10^{23}\ mol^{-1}) \times (6.626 \times 10^{-34}\ J \cdot s) \times (2.998 \times 10^8\ m \cdot s^{-1})}{305 \times 10^{-9}\ nm}$$

$$\times 1\ kJ/10^3\ J$$

$$= 392\ kJ \cdot mol^{-1}$$

16.8B   (a) Strong-field ligands have large Δ_O. The configuration is $t_{2g}^6 e_g^1$ and there is one unpaired electron. (b) Weak-field ligands have small Δ_O. The configuration is $t_{2g}^5 e_g^2$ and there are 3 unpaired electrons.

16.9B   The ligand en is strong-field, and H_2O is a weak-field ligand. However, because Ni^{2+} has eight d-electrons it is a d^8 complex and there is no empty e_g orbital. Therefore, both $[Ni(en)_3]^{2+}$ and $[Ni(H_2O)_6]^{2+}$ have a $t_{2g}^6 e_g^2$ configuration, and are paramagnetic.

Chapter 17

17.1B   (a) $^{232}_{90}Th \rightarrow\ ^A_Z E +\ ^4_2\alpha$; $A = 232 - 4 = 228$; $Z = 90 - 2 = 88$; therefore, $^{228}_{88}Ra$ is produced.

(b) $^{228}_{887}Ra \rightarrow\ ^A_Z E +\ ^0_{-1}e$; $A = 228 - 0 = 228$; $Z = 88 - (-1) = 89$; therefore, $^{228}_{89}Ac$ is produced.

17.2B   (a) $^{55}_{26}Fe +\ ^0_{-1}e \rightarrow\ ^A_Z E$; $A = 55 + 0 = 55$; $Z = 26 + (-1) = 25$; therefore, $^{55}_{25}Mn$ is produced.

(b) $^{11}_6C \rightarrow\ ^A_Z E +\ ^0_{+1}e$; $A = 11 - 0 = 11$; $Z = 6 - 1 = 5$; therefore, $^{11}_5B$ is produced.

17.3B   (c) $^{148}_{58}Cs$ is neutron rich (in the blue band) and therefore might undergo β emission, which converts a neutron into a proton, to reach stability.

17.4B   (a) Balance mass number: $257 + 4\ (1) - 250 = 11$; balance atomic number: $103 + 4\ (0) - 98 = 5$; the missing nuclide is $^{11}_5B$. (b) Balance mass number: $254 + 4\ (1) - 12 = 246$; balance atomic number: $102 + 4\ (0) - 6 = 96$; the missing nuclide is $^{246}_{96}Cm$.

17.5B $m = (5.0\ \mu g)\ e^{-(3.3 \times 10^{-7}y^{-1}) \times (1.0 \times 10^6 y)} = 3.6\ \mu g$

17.6B $t = -\left(\dfrac{5.73 \times 10^3\ y}{\ln 2}\right) \times \ln\left(\dfrac{1.4 \times 10^4}{18400}\right) = 2.3 \times 10^3\ y$

17.7B $\Delta m = \{(235.0439 m_u) - 92(1.0078 m_u) - 143(1.0087 m_u)\} \times 1.6605 \times 10^{-27}\ kg = -3.1845 \times 10^{-27}\ kg$; $E_{bind} = |3.1845 \times 10^{-27}\ kg| \times (3.00 \times 10^8\ m\ s^{-1})^2$

$\times \dfrac{1\ eV}{1.60218 \times 10^{-19}J} = 1.79 \times 10^9\ eV$

17.8B $\Delta m = \{137.91 m_u + 85.91 m_u + 12(1.0087 m_u) - (235.04 m_u + 1.0087 m_u)\} \times 1.6605 \times 10^{-27}\ kg = -2.06 \times 10^{-28}\ kg$ per nucleus;

$\Delta E = \left(\dfrac{1.0 \times 10^{-3} kg}{235.0 \times 1.6605 \times 10^{-27} kg}\right) \times (-2.06 \times 10^{-28} kg)$

$\times (3.00 \times 10^8\ m \cdot s^{-1})^2 = -4.8 \times 10^{10}\ J$

Chapter 18

18.1B   (a)

(b) $(CH_3)_2CH(CH_2)_3CH(CH_3)CH_2CH_3$

18.2B   (a) 4-ethyl-2-methylhexane

(b)

18.3B   (a) 4-ethyl-2-hexene; (b)

18.4B $CH_3(CH_2)_2CH_2Br$; $CH_3CH_2CHBrCH_3$; $(CH_3)_2CHCH_2Br$; $(CH_3)_3CBr$

18.5B   27a is cis; 27b is trans.

18.6B   (c) is chiral, because the second C atom is attached to four different groups.

18.7B   (a) has the higher boiling point because it has a longer chain than (b), and therefore has stronger London forces.

18.8B $CH_3CHClCH_2CH_3$

18.9B   1-ethyl-3,5-dimethyl-2-propylbenzene

Chapter 19

19.1B   (a) $HCOOH + CH_3CH_2OH \rightarrow HCOOCH_2CH_3$; ethyl methanoate; (b) $CH_3OH + CH_3(CH_2)_2COOH \rightarrow CH_3(CH_2)_2 COOCH_3$

19.2B

C: sp^2 hybridization ; N: sp^3 hybridization

19.3B   (a)

3-Pentanol

(b)

3-Pentanone

(c)

Ethylmethylamine

19.4B   (a) $CH_2{=}C(CH_3)COOCH_3$

(b)

19.5B   The A and B monomers alternate; therefore, —ABABAB— copolymer is an alternating copolymer.

Odd-Numbered Exercises

Fundamentals

A.1   (a) chemical; (b) physical; (c) physical

A.3   The temperature of the injured camper, the evaporation and condensation of water are physical properties. The ignition of propane is a chemical change.

A.5   (a) physical; (b) chemical; (c) chemical

A.7   (a) intensive; (b) intensive; (c) extensive; (d) extensive

A.9   (b) 2×10^5 μm

A.11   236 mL

A.13   19.3 $g \cdot cm^{-3}$

A.15   0.0427 cm^3

A.17   7.41 cm

A.19   3 (0.989)

A.21   (a) 4.82×10^3 pm; (b) 30.5 $mm^3 \cdot s^{-1}$; (c) 1.88×10^{-12} kg; (d) 2.66×10^3 $kg \cdot m^{-3}$; (e) 0.044 $mg \cdot cm^{-3}$

A.23   (a) 1.72 $g \cdot cm^{-3}$ (more precise); (b) 1.7 $g \cdot cm^{-3}$

A.25   32 J

A.27 8.1×10^2 kJ; 29 m

A.29   6.0 J

A.31 $E_p = egh$; $g = e/4\pi\varepsilon_0 r^2$

A.33   27 eV. Some energy is retained as the kinetic energy of the electron.

B.1 1.40×10^{22} atoms

B.3   (a) 5 p, 6 n, 5 e; (b) 5 p, 5 n, 5 e; (c) 15 p, 16 n, 15 e; (d) 92 p, 146 n, 92 e

B.5   (a) ^{194}Ir; (b) ^{22}Ne; (c) ^{51}V

B.7   chlorine, ^{36}Cl, 17, 19, 17, 36; zinc, ^{65}Zn, 30, 35, 30, 65; calcium, ^{40}Ca, 20, 20, 20, 40; lanthanum, ^{137}La, 57, 80, 57, 137

B.9   protons

B.11   (a) same mass; (b) different numbers of protons, neutrons, and electrons

B.13   (a) 0.5519; (b) 0.4479; (c) 2.439×10^{-4}; (d) 535.9 kg

B.15   (a) scandium, Group 3/IIIB metal; (b) strontium, Group 2/IIA metal; (c) sulfur, Group 16/VIA nonmetal; (d) antimony, Group 15/VA metalloid

B.17   (a) Sr, metal; (b) Xe, nonmetal; (c) Si, metalloid

B.19   fluorine, F, $Z = 9$, gas; chlorine, Cl, $Z = 17$, gas; bromine, Br, $Z = 35$, liquid; iodine, I, $Z = 53$, solid

B.21   (a) d block; (b) p block; (c) d block; (d) s block; (e) p block; (f) d block

C.1   (a) mixture, one element and one compound; (b) single element

C.3 $C_{40}H_{56}O_2$

C.5   (a) $C_3H_7O_2N$; (b) C_2H_7N

C.7   (a) cation, Cs^+; (b) anion, I^-; (c) Se^{2-}; (d) Ca^{2+}

C.9   (a) 4 p, 6 n, 2 e; (b) 8 p, 9 n, 10 e; (c) 35 p, 45 n, 36 e; (d) 33 p, 42 n, 36 e

C.11   (a) $^{19}F^-$; (b) $^{24}Mg^{2+}$; (c) $^{128}Te^{2-}$; (d) $^{86}Rb^+$

C.13   (a) Al_2Te_3; (b) MgO; (c) Na_2S; (d) RbI

C.15   (a) molecular compound; (b) element; (c) ionic compound; (d) element; (e) molecular compound; (f) ionic compound

C.17   (a) Group 13/IIIA; (b) aluminum, Al

C.19 Al_2O_3; 3.4 $g \cdot cm^{-3}$

D.1   (a) $MnCl_2$; (b) $Ca_3(PO_4)_2$; (c) $Al_2(SO_3)_3$; (d) Mg_3N_2

D.3   (a) calcium phosphate; (b) tin(IV) sulfide, stannic sulfide; (c) vanadium(V) oxide; (d) copper (I) oxide, cuprous oxide

D.5   (a) TiO_2; (b) $SiCl_4$; (c) CS_2; (d) SF_4; (e) Li_2S; (f) SbF_5; (g) N_2O_5; (h) IF_7

D.7   (a) sulfur hexafluoride; (b) dinitrogen pentoxide; (c) nitrogen triiodide; (d) xenon tetrafluoride; (e) arsenic tribromide; (f) chlorine dioxide

D.9   (a) hydrochloric acid; (b) sulfuric acid; (c) nitric acid; (d) acetic acid; (e) sulfurous acid; (f) phosphoric acid

D.11   (a) ZnF_2; (b) $Ba(NO_3)_2$; (c) AgI; (d) Li_3N; (e) Cr_2S_3

D.13   (a) sodium sulfite; (b) iron(III) oxide, ferric oxide; (c) iron(II) oxide, ferrous oxide; (d) magnesium hydroxide; (e) nickel(II) sulfate hexahydrate; (f) phosphorus pentachloride; (g) chromium(III) dihydrogen phosphate; (h) diarsenic trioxide; (i) ruthenium(II) chloride

D.15   (a) copper(II) carbonate; (b) potassium sulfite; (c) lithium chloride

D.17   (a) heptane; (b) propane; (c) pentane; (d) butane

D.19   (a) cobalt(III) oxide monohydrate, $Co_2O_3 \cdot H_2O$; (b) cobalt(II) hydroxide, $Co(OH)_2$

D.21   E = Si; silicon tetrahydride; sodium silicide

D.23   (a) lithium aluminum hydride, ionic; (b) sodium hydride, ionic

D.25   (a) selenic acid; (b) sodium arsenate; (c) calcium tellurite; (d) barium arsenate; (e) antimonic acid; (f) nickel(III) selenate

D.27   (a) alcohol; (b) carboxylic acid; (c) haloalkane

E.1 8.67×10^{10} km

E.3   18 Ca atoms

E.5   (a) 1.0×10^{-14} mol; (b) 3.2×10^6 years

E.7   (a) 6.94 $g \cdot mol^{-1}$; (b) 6.96 $g \cdot mol^{-1}$

E.9   (a) 1.38×10^{23} O atoms; (b) 1.26×10^{22} formula units; (c) 0.146 mol H_2O

E.11   % ^{11}B = 73.8%; % ^{10}B = 26.2%

E.13   (a) 75 g of In. (b) 15.0 g of P. (c) Neither, they have the same number of moles.

E.15   (a) 53 g of Rh; (b) 32 g of Rh

E.17   (a) 0.0981 mol, 5.91×10^{22} molecules; (b) 1.30×10^{-3} mol, 7.83×10^{20} molecules; (c) 4.56×10^{-5} mol, 2.75×10^{19} molecules; (d) 6.94 mol, 4.18×10^{24} molecules; (e) 0.312 mol N atoms, 1.88×10^{23} N atoms; 0.156 mol N_2 molecules, 9.39×10^{22} N_2 molecules

E.19   (a) 0.0134 mol; (b) 8.74×10^{-3} mol; (c) 430. mol; (d) 0.0699 mol

E.21   (a) 4.52×10^{23} formula units; (b) 124 mg; (c) 3.036×10^{22} formula units

E.23   (a) 2.992×10^{-23} $g \cdot molecule^{-1}$; (b) 3.34×10^{25} molecules

E.25   (a) 0.0417 mol; (b) 0.0834 mol; (c) 1.00×10^{23} molecules; (d) 0.3099

E.27   (a) 6.82 kg, $10.60 L^{-1}; (b) $25.44

E.29   (a) 12.013 $g \cdot mol^{-1}$; (b) 197.18 $g \cdot mol^{-1}$

E.31   3.18 mol

E.33 For example, Br has $m_1 = 80.9163$ g·mol^{-1}, $m_2 = 78.9183$ g·mol^{-1}.

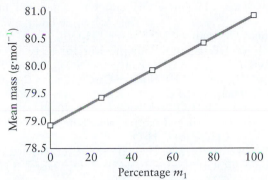

E.35 (a) ($Br_{isotope1} + Br_{isotope1}$) = 158, $Br_{isotope1}$ = 79; ($Br_{isotope1} + Br_{isotope2}$) = 160, $Br_{(isotope2)}$ = 81; (b) bromine-79

F.1 78.90% C, 10.59% H, 10.51% O

F.3 52.15% C, 9.3987% H, 8.691% N, 29.78% O

F.5 (a) 63.4 g·mol^{-1}; (b) copper(I) oxide

F.7 (a) Na_3AlF_6; (b) $KClO_3$; (c) NH_6PO_4 or $[NH_4][H_2PO_4]$

F.9 (a) PCl_5; (b) phosphorus pentachloride

F.11 $C_{16}H_{13}ClN_2O$

F.13 (a) OsC_4O_4; (b) $Os_3C_{12}O_{12}$

F.15 $C_8H_{10}N_4O_2$

F.17 The elemental compositions are close, but the C and O values differ sufficiently to make them distinguishable.

F.19 ethene (85.63%) > heptane (83.91%) > propanol (59.96%)

F.21 (a) empirical formula: C_2H_3Cl, molecular formula: $C_4H_6Cl_2$; (b) empirical formula: CH_4N, molecular formula: $C_2H_8N_2$

F.23 45.1%

G.1 (a) density; (b) the abilities of the components to adsorb; (c) boiling point

G.3 (a) heterogeneous, decanting; (b) heterogeneous, dissolving followed by filtration and distillation; (c) homogeneous, distillation

G.5 15.2 g

G.7 (a) 13.5 mL; (b) 62.5 mL; (c) 5.92 mL

G.9 (a) Weigh 1.6 g (0.010 mol, molar mass of $KMnO_4$ = 158.04 g·mol^{-1}) into a 1.0-L volumetric flask and add water to give a total volume of 1.0 L. Smaller (or larger) volumes could also be prepared by using a proportionally smaller (or larger) mass of $KMnO_4$. (b) Starting with 0.050 mol·L^{-1} $KMnO_4$, add four volumes of water to one volume of starting solution, because the concentration desired is one-fifth of the starting solution. This relation can be derived from the expression $M_i \times V_i = M_f \times V_f$, where i represents the initial solution and f the final solution. For example, to prepare 50 mL of solution, you would add 40 mL of water to 10 mL of 0.050 mol·L^{-1} $KMnO_4$.

G.11 (a) 4.51 mL. (b) Add 48.0 mL of water to 12.0 mL of 2.5 mol·L^{-1} NaOH.

G.13 (a) 8.0 g; (b) 12 g

G.15 (a) 4.58×10^{-2} mol·L^{-1}; (b) 9.07×10^{-3} mol·L^{-1}

G.17 (a) 21.8 g; (b) 0.24 g; (c) 60. g

G.19 0.13 mol·L^{-1}

G.21 Less than one molecule will be left after 69 doublings; no health benefits.

G.23 600. mL

H.1 (a) A compound or element which is not involved in the chemical reaction cannot be added to the chemical equation. (b) $2\,Cu + SO_2 \rightarrow 2\,CuO + S$

H.3 $2\,SiH_4 + 4\,H_2O \rightarrow 2\,SiO_2 + 8\,H_2$

H.5 (a) $NaBH_4(s) + 2\,H_2O(l) \rightarrow NaBO_2(aq) + 4\,H_2(g)$
(b) $LiN_3(s) + H_2O(l) \rightarrow LiOH(aq) \rightarrow HN_3(aq)$
(c) $2\,NaCl(aq) + SO_3(g) + H_2O(l) \rightarrow Na_2SO_4(aq) + 2\,HCl(aq)$
(d) $4\,Fe_2P(s) + 18\,S(s) \rightarrow P_4S_{10}(s) + 8\,FeS(s)$

H.7 (a) $Ca(s) + 2\,H_2O(l) \rightarrow H_2(g) + Ca(OH)_2(aq)$
(b) $Na_2O(s) + H_2O(l) \rightarrow 2\,NaOH(aq)$
(c) $3\,Mg(s) + N_2(g) \rightarrow Mg_3N_2(s)$
(d) $4\,NH_3(g) + 7\,O_2(g) \rightarrow 6\,H_2O(g) + 4\,NO_2(g)$

H.9 (I) $3\,Fe_2O_3(s) + CO(g) \rightarrow 2\,Fe_3O_4(s) + CO_2\,(g)$
(II) $Fe_3O_4(s) + 4\,CO(g) \rightarrow 3\,Fe(s) + 4\,CO_2(g)$

H.11 (I) $N_2(g) + O_2(g) \rightarrow 2\,NO(g)$
(II) $2\,NO(g) + O_2(g) \rightarrow 2\,NO_2(g)$

H.13 $4\,HF(aq) + SiO_2(s) \rightarrow SiF_4(aq) + 2\,H_2O(l)$

H.15 $C_7H_{16}(l) + 11\,O_2(g) \rightarrow 7\,CO_2(g) + 8\,H_2O(g)$

H.17 $4\,C_{10}H_{15}N(s) + 55\,O_2\,(g) \rightarrow$
$40\,CO_2(g) + 30\,H_2O(l) + 2\,N_2(g)$

H.19 (I) $H_2S(g) + 2\,NaOH(s) \rightarrow Na_2S(aq) + 2\,H_2O(l)$
(II) $4\,H_2S(g) + Na_2S(alc) \rightarrow Na_2S_5(alc) + 4\,H_2(g)$
(III) $2\,Na_2S_5(alc) + 9\,O_2(g) + 10\,H_2O(l) \rightarrow$
$2\,Na_2S_2O_3 \cdot 5H_2O(s) + 6\,SO_2(g)$

H.21 (a) P_2O_5, P_2O_3; (b) P_4O_{10}: phosphorus(V) oxide; P_4O_6: phosphorus(III) oxide;
(c) $P_4(s) + 3\,O_2(g) \rightarrow P_4O_6(s)$, $P_4(s) + 5\,O_2(g) \rightarrow P_4O_{10}(s)$

I.1 The picture would show a precipitate, $CaSO_4(s)$, at the bottom of the flask composed of red and pink particles in a 1:1 ratio. Sodium (black) and chloride (blue) ions, NaCl(aq), would remain throughout the solution.

I.3 (a) dilute H_2SO_4, $Pb^{2+}(aq) + SO_4^{2-}(aq) \rightarrow PbSO_4(s)$
(b) H_2S solution, $Mg^{2+}(aq) + S^{2-}(aq) \rightarrow MgS(s)$

I.5 (a) nonelectrolyte; (b) strong electrolyte; (c) strong electrolyte

I.7 (a) soluble; (b) slightly soluble; (c) insoluble; (d) insoluble

I.9 (a) $Na^+(aq)$ and $I^-(aq)$; (b) $Ag^+(aq)$ and CO_3^{2-}, Ag_2CO_3 is insoluble; (c) $NH_4^+(aq)$ and $PO_4^{3-}(aq)$; (d) $Fe^{2+}(aq)$ and $SO_4^{2-}(aq)$

I.11 (a) $Fe(OH)_3$, precipitate; (b) Ag_2CO_3, precipitate; (c) no precipitate

I.13 (a) $Fe^{2+}(aq) + S^{2-}(aq) \rightarrow FeS(s)$; Na^+ and Cl^-
(b) $Pb^{2+}(aq) + 2\,I^-(aq) \rightarrow PbI_2(s)$; K^+ and NO_3^-
(c) $Ca^{2+}(aq) + SO_4^{2-}(aq) \rightarrow CaSO_4(s)$; NO_3^- and K^+
(d) $Pb^{2+}(aq) + CrO_4^{2-}(aq) \rightarrow PbCrO_4(s)$; Na^+ and NO_3^-
(e) $Hg_2^{2+}(aq) + SO_4^{2-}(aq) \rightarrow Hg_2SO_4(s)$; K^+ and NO_3^-

I.15 (a) $(NH_4)_2CrO_4(aq) + BaCl_2(aq) \rightarrow$
$$BaCrO_4(s) + 2\,NH_4Cl(aq)$$
complete ionic:
$2\,NH_4^+(aq) + CrO_4^{2-}(aq) + Ba^{2+}(aq) + 2\,Cl^-(aq) \rightarrow$
$$BaCrO_4(s) + 2\,NH_4^+(aq) + 2\,Cl^-(aq)$$
net ionic: $Ba^{2+}(aq) + CrO_4^{2-}(aq) \rightarrow BaCrO_4(s)$
spectator ions: NH_4^+ and Cl^-
(b) $CuSO_4(aq) + Na_2S(aq) \rightarrow CuS(s) + Na_2SO_4(aq)$
complete ionic:
$Cu^{2+}(aq) + SO_4^{2-}(aq) + 2\,Na^+(aq) + S^{2-}(aq) \rightarrow$
$$CuS(s) + 2\,Na^+(aq) + SO_4^{2-}(aq)$$
net ionic: $Cu^{2+}(aq) + S^{2-}(aq) \rightarrow CuS(s)$
spectator ions: Na^+ and SO_4^{2-}

(c) $3 FeCl_2(aq) + 2 (NH_4)_3PO_4(aq) \rightarrow$
$$Fe_3(PO_4)_2(s) + 6 NH_4Cl(aq)$$
complete ionic:
$3 Fe^{2+}(aq) + 6 Cl^-(aq) + 6 NH_4^+(aq) + 2 PO_4^{3-}(aq) \rightarrow$
$$Fe_3(PO_4)_2(s) + 6 NH_4^+(aq) + 6 Cl^-(aq)$$
net ionic: $3 Fe^{2+}(aq) + 2 PO_4^{3-}(aq) \rightarrow Fe_3(PO_4)_2(s)$
spectator ions: NH_4^+ and Cl^-
(d) $K_2C_2O_4(aq) + Ca(NO_3)_2(aq) \rightarrow CaC_2O_4(s) + 2 KNO_3(aq)$
complete ionic:
$2 K^+(aq) + C_2O_4^{2-}(aq) + Ca^{2+}(aq) + 2 NO_3^-(aq) \rightarrow$
$$CaC_2O_4(s) + 2 K^+ + 2 NO_3^-(aq)$$
net ionic: $Ca^{2+}(aq) + C_2O_4^{2-}(aq) \rightarrow CaC_2O_4(s)$
spectator ions: K^+ and NO_3^-
(e) $NiSO_4(aq) + Ba(NO_3)_2(aq) \rightarrow BaSO_4(s) + Ni(NO_3)_2(aq)$
complete ionic:
$Ni^{2+}(aq) + SO_4^{2-}(aq) + Ba^{2+}(aq) + 2 NO_3^-(aq) \rightarrow$
$$BaSO_4(s) + Ni^{2+}(aq) + 2 NO_3^-(aq)$$
net ionic: $Ba^{2+}(aq) + SO_4^{2-} \rightarrow BaSO_4(s)$
spectator ions: Ni^{2+} and NO_3^-
I.17 (a) $Ba(CH_3CO_2)_2(aq) + Li_2CO_3(aq) \rightarrow$
$$BaCO_3(s) + 2 LiCH_3CO_2(aq)$$
complete ionic:
$Ba^{2+}(aq) + 2 CH_3CO_2^-(aq) + Li^+(aq) + CO_3^{2-}(aq) \rightarrow$
$$BaCO_3(s) + 2 Li^+(aq) + 2 CH_3CO_2^-(aq)$$
net ionic: $Ba^{2+}(aq) + CO_3^{2-}(aq) \rightarrow BaCO_3(s)$
(b) $2 NH_4Cl(aq) + Hg_2(NO_3)_2(aq) \rightarrow$
$$2 NH_4NO_3(aq) + Hg_2Cl_2(s)$$
complete ionic:
$2 NH_4^+(aq) + 2 Cl^-(aq) + Hg_2^{2+}(aq) + 2 NO_3^-(aq) \rightarrow$
$$Hg_2Cl_2(s) + 2 NH_4^+(aq) + 2 NO_3^-(aq)$$
net ionic: $Hg_2^{2+}(aq) + 2 Cl^-(aq) \rightarrow Hg_2Cl_2(s)$
(c) $Cu(NO_3)_2(aq) + Ba(OH)_2(aq) \rightarrow$
$$Cu(OH)_2(s) + Ba(NO_3)_2(aq)$$
complete ionic:
$Cu^{2+}(aq) + 2 NO_3^-(aq) + Ba^{2+}(aq) + 2 OH^-(aq) \rightarrow$
$$Cu(OH)_2(s) + Ba^{2+}(aq) + 2 NO_3^-(aq)$$
net ionic: $Cu^{2+}(aq) + 2 OH^-(aq) \rightarrow Cu(OH)_2(s)$
I.19 (a) $AgNO_3$ and Na_2CrO_4; (b) $CaCl_2$ and Na_2CO_3;
(c) $Cd(ClO_4)_2$ and $(NH_4)_2S$
I.21 (a) $2 Ag^+(aq) + SO_4^{2-}(aq) \rightarrow Ag_2SO_4(s)$
(b) $Hg^{2+}(aq) + S^{2-}(aq) \rightarrow HgS(s)$
(c) $3 Ca^{2+}(aq) + 2 PO_4^{3-}(aq) \rightarrow Ca_3(PO_4)_2(s)$
(d) $AgNO_3$ and Na_2SO_4, Na^+ and NO_3^-; $Hg(CH_3CO_2)_2$ and
Li_2S, Li^+ and $CH_3CO_2^-$; $CaCl_2$ and K_3PO_4, K^+ and Cl^-
I.23 white precipitate $= AgCl(s)$, Ag^+; no precipitate with
$H_2SO_4 =$ no Ca^{2+}; black precipitate $= ZnS(s)$, Zn^{2+}
I.25 (a) $2 NaOH(aq) + Cu(NO_3)_2(aq) \rightarrow$
$$Cu(OH)_2(s) + 2 NaNO_3(aq)$$
complete ionic:
$2 Na^+(aq) + 2 OH^-(aq) + Cu^{2+}(aq) + 2 NO_3^-(aq) \rightarrow$
$$Cu(OH)_2(s) + 2 Na^+(aq) + 2 NO_3^-(aq)$$
net ionic: $Cu^{2+}(aq) + 2 OH^-(aq) \rightarrow Cu(OH)_2(s)$
(b) 0.0800 mol·L^{-1}
I.27 (a) 0.240 mol·L^{-1}; (b) 1.41 g; (c) $MgCrO_4$
J.1 (a) base; (b) acid; (c) base; (d) acid; (e) base
J.3 (a) $HF(aq) + NaOH(aq) \rightarrow NaF(aq) + H_2O(l)$
complete ionic: $HF(aq) + Na^+(aq) + OH^-(aq) \rightarrow$
$$Na^+(aq) + F^-(aq) + H_2O(l)$$
net ionic: $HF(aq) + OH^-(aq) \rightarrow F^-(aq) + H_2O(l)$
(b) $(CH_3)_3N(aq) + HNO_3(aq) \rightarrow (CH_3)_3NHNO_3(aq)$
complete ionic: $(CH_3)_3N(aq) + H_3O^+ + NO_3^-(aq) \rightarrow$
$$(CH_3)_3NH^+(aq) + NO_3^-(aq) + H_2O(l)$$

net ionic: $(CH_3)_3N(aq) + H_3O^+ \rightarrow (CH_3)_3NH^+(aq) + H_2O(l)$
(c) $LiOH(aq) + HI(aq) \rightarrow LiI(aq) + H_2O(l)$
complete ionic: $Li^+(aq) + OH^-(aq) + H_3O^+(aq) + I^-(aq) \rightarrow$
$$Li^+(aq) + I^-(aq) + 2 H_2O(l)$$
net ionic: $OH^-(aq) + H_3O^+(aq) \rightarrow 2 H_2O(l)$
J.5 (a) $HBr(aq) + KOH(aq) \rightarrow KBr(aq) + H_2O(l)$
(b) $Zn(OH)_2(aq) + 2 HNO_2(aq) \rightarrow Zn(NO_2)_2(aq) + 2 H_2O(l)$
(c) $Ca(OH)_2(aq) + 2 HCN(aq) \rightarrow Ca(CN)_2(aq) + 2 H_2O(l)$
(d) $3 KOH(aq) + H_3PO_4(aq) \rightarrow K_3PO_4(aq) + 3 H_2O(l)$
J.7 (a) acid: $H_3O^+(aq)$, base: $CH_3NH_2(aq)$; (b) acid:
$HCl(aq)$, base: $C_2H_5NH_2(aq)$; (c) acid: $HI(aq)$, base: $CaO(s)$
J.9 (a) CHO_2; (b) $C_2H_2O_4$;
(c) $(COOH)_2(aq) + 2 NaOH(aq) \rightarrow Na_2C_2O_4(aq) + 2 H_2O(l)$
net ionic: $(COOH)_2(aq) + 2 OH^- \rightarrow C_2O_4^{2-}(aq) + 2 H_2O(l)$
J.11 (a) $C_6H_5O^-(aq) + H_2O(l) \rightarrow$
$$C_6H_5OH(aq) + OH^-(aq)$$
(b) $ClO^-(aq) + H_2O(l) \rightarrow HClO(aq) + OH^-(aq)$
(c) $C_5H_5NH^+(aq) + H_2O(l) \rightarrow C_5H_5N(aq) + H_3O^+(aq)$
(d) $NH_4^+(aq) + H_2O(l) \rightarrow NH_3(aq) + H_3O^+(aq)$
J.13 (a) (I) $AsO_4^{3-}(aq) + H_2O(l) \rightarrow$
$$HAsO_4^{2-}(aq) + OH^-(aq)$$
(II) $HAsO_4^{2-}(aq) + H_2O(l) \rightarrow H_2AsO_4^-(aq) + OH^-(aq)$
(III) $H_2AsO_4^-(aq) + H_2O(l) \rightarrow H_3AsO_4(aq) + OH^-(aq)$
In each equation, H_2O is the acid. (b) 0.505 mol
K.1 (a) $2 NO_2(g) + O_3(g) \rightarrow N_2O_5(g) + O_2(g)$
(b) $S_8(s) + 16 Na(s) \rightarrow 8 Na_2S(s)$
(c) $2 Cr^{2+}(aq) + Sn^{4+}(aq) \rightarrow 2 Cr^{3+}(aq) + Sn^{2+}(aq)$
(d) $2 As(s) + 3 Cl_2(g) \rightarrow 2 AsCl_3(l)$
K.3 (a) $Mg(s) + Cu^{2+}(aq) \rightarrow Mg^{2+}(aq) + Cu(s)$
(b) $Fe^{2+}(aq) + Ce^{4+}(aq) \rightarrow Fe^{3+}(aq) + Ce^{3+}(aq)$
(c) $H_2(g) + Cl_2(g) \rightarrow 2 HCl(g)$
(d) $4 Fe(s) + 3 O_2(g) \rightarrow 2 Fe_2O_3(s)$
K.5 (a) $+4$; (b) $+4$; (c) -2; (d) $+5$; (e) $+1$; (f) 0
K.7 (a) $+2$; (b) $+2$; (c) $+6$; (d) $+4$; (e) $+1$
K.9 (a) oxidized: CH_3OH; reduced: O_2. (b) oxidized: *some*
sulfur; reduced: Mo. (c) Tl^+ is both oxidized and reduced.
K.11 (a) Cl_2 will be reduced more easily and is therefore a
stronger oxidizing agent than Cl^-. (b) N_2O_5 will be a stronger
oxidizing agent because it will be readily reduced. N^{5+} will
accept e^- more readily than will N^+.
K.13 (a) oxidizing agent: H^+ in $HCl(aq)$; reducing agent:
$Zn(s)$; (b) oxidizing agent: $SO_2(g)$; reducing agent: $H_2S(g)$;
(c) oxidizing agent: $B_2O_3(s)$; reducing agent: $Mg(s)$
K.15 (a) oxidizing agent, Cl goes from $+7$ to $+4$; reducing
agent, S goes from $+6$ to $+4$
K.17 $CO_2(g) + 4 H_2(g) \rightarrow CH_4(g) + 2 H_2O(l)$;
oxidation–reduction reaction
K.19 (a) oxidizing agent: $WO_3(s)$; reducing agent: $H_2(g)$;
(b) oxidizing agent: $HCl(aq)$; reducing agent: $Mg(s)$;
(c) oxidizing agent: $SnO_2(s)$; reducing agent: $C(s)$;
(d) oxidizing agent: $N_2O_4(g)$; reducing agent: $N_2H_4(g)$
K.21 (a) $3 N_2H_4(l) \rightarrow 4 NH_3(g) + N_2(g)$. (b) -2 in N_2H_4, -3
in NH_3, 0 in N_2. (c) N_2H_4 is both oxidizing and reducing agent.
(d) 2.5×10^2 L
K.23 (a) $2 Cr^{2+} + Cu^{2+} \rightarrow 2 Cr^{3+} + Cu(s)$; (b) $2 e^-$;
(c) 2.27 mol·L^{-1} NO_3^-, 1.50 mol·L^{-1} SO_4^{2-}
K.25 (a) $SiCl_4(l) + 2 H_2(g) \rightarrow Si(s) + 4 HCl(g)$,
silicon tetrachloride, Si^{4+};
(b) $SnO_2(s) + C(s) \rightarrow Sn(l) + CO_2(g)$, tin(IV) oxide, Sn^{4+};
(c) $V_2O_5(s) + 5 Ca(l) \rightarrow 2 V(s) + 5 CaO(s)$, vanadium(V)
oxide, V^{5+}; (d) $B_2O_3(s) + 3 Mg(s) \rightarrow 2 B(s) + 3 MgO(s)$,
boron oxide, B^{3+}

L.1 1.1×10^{-5} mol; (b) 6.2 g

L.3 (a) 507.1 g; (b) 6.613×10^3 kg

L.5 (a) 505 g; (b) 1.33×10^3 g

L.7 4.2 kg

L.9 (a) 0.271 mol·L^{-1}; (b) 0.163 g

L.11 (a) 0.209 mol·L^{-1}; (b) 0.329 g

L.13 63.0 g·mol^{-1}

L.15 1.65 mol·L^{-1}

L.17 (a) $Na_2CO_3(aq) + 2 HCl(aq) \rightarrow$ $2 NaCl(aq) + H_2CO_3(aq)$; (b) 12.6 mol·L^{-1}

L.19 $I_3^-(aq) + Sn^{2+}(aq) \rightarrow 3 I^- + Sn^{4+}(aq)$

L.21 (a) $S_2O_3^{2-}$ is both oxidized and reduced. (b) 11.1 g

L.23 Pt

L.25 $x = 2$, $BaBr_2 + Cl_2 \rightarrow BaCl_2 + Br_2$

L.27 509 kg

L.29 (a) Pipette 31 mL of 16 M HNO_3 into a 1.00-L volumetric flask which contains about 800 mL of H_2O. Dilute to the mark with H_2O. Shake the flask to mix the solution thoroughly. (b) 2.5×10^2 mL

L.31 (a) SnO_2; (b) tin(IV) oxide

M.1 93.1%

M.3 6 Cl atoms

M.5 (a) O_2; (b) 5.77 g P_4O_{10}; (c) 5.7 g P_4O_6

M.7 (a) $Cu^{2+}(aq) + 2 OH^-(aq) \rightarrow Cu(OH)_2(s)$; (b) 2.44 g

M.9 $C_4H_5N_2O$, $C_8H_{10}N_4O_2$, $2 C_8H_{10}N_4O_2(s) + 19 O_2(g) \rightarrow 16 CO_2(g) + 10 H_2O(l) + 4 N_2(g)$

M.11 $C_8H_{16}N_4O_3$

M.13 (a) $Ca_3(PO_4)_2$; (b) 130. g

M.15 93.0%

M.17 (a) $C_{11}H_{14}O_3$; (b) $C_{22}H_{28}O_6$

M.19 The empirical and molecular formulas are both CHOCl.

M.21 81.2%

Chapter 1

1.1 (a) Radiation may pass through a metal foil. (b) All light (electromagnetic radiation) travels at the same speed; the slower speed supports the particle model. (c) This observation supports the radiation model. (d) This observation supports the particle model; electromagnetic radiation has no mass and no charge.

1.3 microwaves < visible light < ultraviolet light < x-rays < γ-rays

1.5 8.7×10^{14} Hz, 340 nm, 5.8×10^{-19} J, suntan; 5.0×10^{14} Hz, 600 nm, 3.3×10^{-19} J reading; 300 MHz, 1 m, 2×10^{-25} J, microwave popcorn; 1.2×10^{17} Hz, 2.5 nm, 7.9×10^{-17} J, dental x-ray

1.7 1590 nm

1.9 (a) 3.37×10^{-19} J; (b) 44.1 J; (c) 2.03×10^5 J

1.11 8.8237×10^{-12} m or 8.8237 pm

1.13 (a) False. The total intensity is proportional to T^4. (b) True. (c) False. Photons of radio frequency radiation are lower in energy than photons of ultraviolet radiation.

1.15 (a) 2.0×10^{-10} m; (b) 1.66×10^{-17} J; (c) 8.8×10^{-9} m or 8.8 nm; (d) 8.8 nm: x-ray/gamma ray region

1.17 1.1×10^{-34} m

1.19 3.96×10^3 m·s^{-1}

1.21 Yes, $n_1 = 1$, $n_2 = 2$ is degenerate with $n_1 = 2$, $n_2 = 1$; $n_1 = 1$, $n_2 = 3$ is degenerate with $n_1 = 3$, $n_2 = 1$; and $n_1 = 2$, $n_2 = 3$ is degenerate with $n_1 = 3$, $n_2 = 2$

1.23 $\dfrac{3h^2}{8mL^2}$; $\dfrac{3h^2}{4mL^2}$ (211, 121, 112); $\dfrac{9h^2}{8mL^2}$ (221, 212, 122)

1.25 (a)

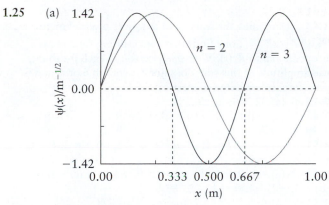

(b) for $n = 2$ there is one node at $x = 0.500$ m; (c) for $n = 3$ there are two nodes, one at $x = 0.333$ m and 0.667 m; (d) the number of nodes is equal to $n - 1$;

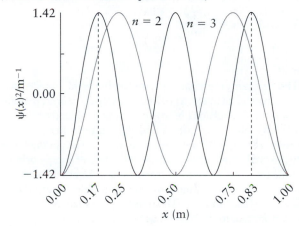

(e) for $n = 2$, $x = 0.25$ m and 0.75 m; (f) for $n = 3$, $x = 0.17$ m, 0.50m, and 0.83 m

1.27 (a) $P = l\left(\dfrac{1}{L} - \dfrac{1}{2n\pi} \sin 2 \dfrac{\pi n l}{L}\right)$; when $l = L/2$,

$$P = \dfrac{1}{2} - \dfrac{L}{4n\pi} \sin n\pi = \dfrac{1}{2} - 0, \text{ and so } P = \dfrac{1}{2}$$

(b) $P = \dfrac{1}{3} - \dfrac{1}{6n\pi} \sin \dfrac{2}{3} n\pi$. Probability is a function of n for small n. However, $P = \dfrac{1}{3}$ if n is a multiple of 3 and P approaches $\dfrac{1}{3}$ for all large n.

1.29 (a) 486 nm; (b) Balmer; (c) visible, blue

1.31 30.4 nm

1.33 (a) 1.48×10^{-8} m; (b) 6.18×10^{-9} m

1.35 Within each series, the principal quantum number for the lower energy level involved is the same.

1.37 (a)

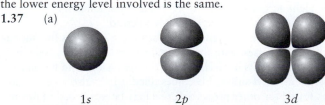

1s   2p   3d

(b) A node is a region where the probability of finding an electron is zero ($\psi^2 = 0$). (c) 1s: no radial nodes or nodal surfaces; 2p: one radial node and one nodal surface; 3d: two radial nodes and two nodal surfaces; (d) three nodal planes

1.39 The lobes of the p_x orbital lie along the x-axis, whereas the lobes of the p_y orbital and p_z orbitals lie along the y- and z-axes, respectively.

1.41   0.33

1.43   Summing the squares of the three p wavefunctions, $\psi_{p_x}^2 + \psi_{p_y}^2 + \psi_{p_z}^2$, one finds that the resulting function is proportional to $R^2(r)$. With one electron in each p-orbital, the amplitude of this sum does not depend on θ or ϕ and is, therefore, spherically symmetrical.

1.45   (a) 32.3%; (b) 76.1%

1.47   (a) 1; (b) 5; (c) 3; (d) 7

1.49   (a) 7; (b) 5; (c) 3; (d) 4

1.51   (a) $n = 6, l = 1$; (b) $n = 3, l = 2$; (c) $n = 2, l = 1$; (d) $n = 5, l = 3$

1.53   (a) $-1, 0, +1$; (b) $-2, -1, 0, +1, +2$; (c) $-1, 0, +1$; (d) $-3, -2, -1, 0, +1, +2, +3$

1.55   (a) 6; (b) 10; (c) 2; (d) 14

1.57   (a) $5d$, 5; (b) $1s$, 1; (c) $6f$, 7; (d) $2p$, 3

1.59   (a) 6; (b) 2; (c) 8; (d) 2

1.61   (a) cannot exist; (b) exists; (c) cannot exist; (d) exists

1.63   (a) $V(r) = \left(\dfrac{-3e^2}{4\pi\varepsilon_0}\right)\left(\dfrac{1}{r_2} + \dfrac{1}{r_2} + \dfrac{1}{r_3}\right) + \dfrac{e^2}{4\pi\varepsilon_0}\left(\dfrac{1}{r_{12}} + \dfrac{1}{r_{13}} + \dfrac{1}{r_{23}}\right)$

(b) The first term represents the coulombic attractions between the nucleus and each electron, and the second term represents the coulombic repulsions between each pair of electrons.

1.65   (a) False. Z_{eff} is affected by the total number of electrons present in the atom because the electrons in the lower energy orbitals shield the electrons in the higher energy orbitals from the nucleus. This effect arises because the electron–electron repulsions tend to offset the attraction of the electron to the nucleus. (b) True. (c) False. The electrons are increasingly less able to penetrate to the nucleus as l increases. (d) True.

1.67   (a) excited state; (b) excited state; (c) excited state; (d) ground state

1.69   (a) allowed; (b) forbidden, m_l cannot be less than $-l$; (c) forbidden, l must always be less than n

1.71   (a) $[Kr]4d^{10}5s^1$; (b) $[He]2s^2$; (c) $[Kr]4d^{10}5s^25p^3$; (d) $[Ar]3d^{10}4s^24p^1$; (e) $[Xe]4f^{14}5d^46s^2$; (f) $[Kr]4d^{10}5s^25p^5$

1.73   (a) tellurium; (b) vanadium; (c) carbon; (d) thorium

1.75   (a) $4p$; (b) $4s$; (c) $6s$; (d) $6s$

1.77   (a) 5; (b) 11; (c) 5; (d) 20

1.79   (a) 3; (b) 2; (c) 3; (d) 2

1.81   Ga: $[Ar]3d^{10}4s^24p^1$, 1; Ge: $[Ar]3d^{10}4s^24p^2$, 2; As: $[Ar]3d^{10}4s^24p^3$, 3; Se: $[Ar]3d^{10}4s^24p^4$, 2; Br: $[Ar]3d^{10}4s^24p^5$, 1

1.83   (a) ns^1; (b) ns^2np^3; (c) $(n-1)d^5ns^2$; (d) $(n-1)d^{10}ns^1$

1.85   (a) O > Se > Te, ionization energy decreases down a group; (b) Au > Os > Ta, ionization energy increases from left to right across a period; (c) Pb > Ba > Cs, ionization energy increases from left to right across a period

1.87   The major trend is for decreasing radius as the nuclear charge increases, with the exception that Cu and Zn begin to show the effects of electron-electron repulsions and become larger as the d-subshell becomes filled. Mn is also an exception as found for other properties; this may be attributed to having the d-shell half-filled.

1.89   (a) Sb^{3+}, Sb^{5+}; (c) Tl^+, Tl^{3+}; (b) and (d) only form one positive ion each

1.91 $Cl^- < S^{2-} < P^{3-}$

1.93   (a) fluorine; (b) carbon; (c) chlorine; (d) lithium

1.95   (a) A diagonal relationship is a similarity in chemical properties between an element in the periodic table and an element lying one period lower and one group to the right.

(b) It is caused by the similarity in size of the ions. The lower-right element in the pair would generally be larger because it lies in a higher period, but it also will have a higher oxidation state, which will cause the ion to be smaller. (c) Al and Ge, Li and Mg

1.97   (b) Li and Mg

1.99   The s-block metals have low ionization energies, which enables them to easily lose electrons in chemical reactions.

1.101   (a) metal; (b) nonmetal; (c) metal; (d) metalloid; (e) metalloid; (f) metal

1.103   (a) 1.1×10^{14} s^{-1}; (b) 7.2×10^{-20} J; (c) 4.3×10^4 J

1.105   Cu: $[Ar]3d^{10}4s^1$ and Cr: $[Ar]3d^54s^1$. In copper it is energetically favorable for an electron to be promoted from the $4s$-orbital to a $3d$-orbital, giving a completely filled $3d$-subshell. In the case of Cr, it is energetically favorable for an electron to be promoted from the $4s$-orbital to a $3d$-orbital to half fill the $3d$-subshell.

1.107   This trend is attributed to the inert-pair effect, which states that the s-electrons are less available for bonding in the heavier elements. Thus, there is an increasing trend as we descend the periodic table for the preferred oxidation number to be 2 units lower than the maximum one. As one descends the periodic table, ionization energies tend to decrease. For Tl, however, the values are slightly higher than those of its lighter analogs.

1.109   (a) The energy of the photon entering, E_{total}, must be equal to the energy required to eject the electron, $E_{ejection}$, plus the energy that ends up as kinetic energy, E_K, in the movement of the electron: $E_{total} = E_{ejection} + E_K$, but E_{total} for the photon $= h\nu$ and $E_K = \frac{1}{2}m\upsilon^2$, where m is the mass of the object and υ is its speed. $E_{ejection}$ corresponds to the ionization energy, I; so we arrive at the final relation desired. (b) 6.67×10^{-19} J

1.111   La, Ac, Lu, and Lr have one electron in a d-orbital, so placement in the third column of the periodic table could be considered appropriate for each, depending on what aspects of the chemistry of these elements we are comparing.

1.113   (a) The points fit on the line described by the equation $f(x) = \dfrac{(2-x)^2 e^{-x}}{4}$ where x is the multiplication factor of a_0 ($r = xa_0$).

(b)

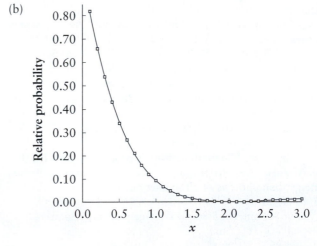

(c) The node occurs when $x = 2$ ($r = 2a_0$).

1.115   The probability of finding the particle somewhere in the box is given by the integral $\displaystyle\int_0^L \left(\dfrac{2}{L}\right)\sin^2\left(\dfrac{n\pi x}{L}\right)dx$, which can be shown to be equal to 1.

1.117 (a) 4.8×10^{-10} esu; (b) 14 electrons
1.119 No, ultraviolet is required.
1.121 (a) -2.04×10^{-18} J. (b) -4.09×10^{-19} J.
(c) -4.09×10^{-19} J. (d) no photon emitted. (e) Whereas hydrogen shows only two lines, potassium should show four lines. Due to its larger numbers of electrons and protons, the orbitals within a given shell will have different energies in a potassium atom. Therefore, in a K atom the $4d$–$2s$ and $4d$–$2p$ transitions will emit photons of different energies and the $4d$–$4s$ transition is no longer between degenerate orbitals.
1.123 99.2 $kJ\cdot mol^{-1}$
1.125 The plot with the most electron density closest to the origin (0, 0—the nucleus) arises from the s-orbital. Curve (b) corresponds to the $3s$-orbital; curve (a) corresponds to the $3p$-orbital.
1.127 (a) $[Ne]3s^23p^5$, 1, $[Ne]3s^23p^6$, argon; (b) $[Ne]3s^23p^44s^1$; (c) 52.0 nm; (d) 2.30×10^3 $kJ\cdot mol^{-1}$ or 23.8 eV; (e) 35.73 $g\cdot mol^{-1}$; (f)–(h) ClO_2, +4, chlorine dioxide; $NaClO$, +1, sodium hypochlorite; $KClO_3$, +5, potassium chlorate; $NaClO_4$, +7, sodium perchlorate

Chapter 2

2.1 (b) Ca^{3+}, O^{2-}
2.3 Li^+ has a smaller radius. Because lattice energy is related to the coulombic attraction of ions, it is inversely proportional to the distance between the ions.
2.5 (a) 5; (b) 4; (c) 7; (d) 3
2.7 (a) $[Ar]$; (b) $[Ar]3d^{10}4s^2$; (c) $[Kr]4d^5$; (d) $[Ar]3d^{10}4s^2$
2.9 (a) $[Ar]3d^{10}$; (b) $[Xe]4f^{14}5d^{10}6s^2$; (c) $[Ar]3d^{10}$; (d) $[Xe]4f^{14}5d^{10}$
2.11 (a) $[Kr]4d^{10}5s^2$, same, In^+ and Sn^{2+} each lose $5p$ valence electrons; (b) none; (c) Pd
2.13 (a) Co^{2+}; (b) Fe^{2+}; (c) Mo^{2+}; (d) Nb^{2+}
2.15 (a) Co^{3+}; (b) Fe^{3+}; (c) Ru^{3+}; (d) Mo^{3+}
2.17 (a) $4s$; (b) $3p$; (c) $3p$; (d) $4s$
2.19 (a) -1; (b) -2; (c) $+1$;
(d) $+3$ ($+1$ sometimes observed); (e) $+2$
2.21 (a) 3; (b) 6; (c) 6; (d) 2
2.23 (a) $[Kr]4d^{10}5s^2$, no unpaired electrons;
(b) $[Kr]4d^{10}$, no unpaired electrons; (c) $[Xe]4f^{14}5d^4$, four unpaired electrons; (d) $[Kr]$, no unpaired electrons;
(e) $[Ar]3d^8$, two unpaired electrons
2.25 (a) $3p$; (b) $5s$; (c) $5p$; (d) $4d$
2.27 (a) $+7$. (b) -1. (c) $[Ne]$ for $+7$, $[Ar]$ for -1.
(d) Electrons are lost or added to give noble-gas configuration.
2.29 (a) Mg_3As_2; (b) In_2S_3; (c) AlH_3; (d) H_2Te; (e) BiF_3
2.31 (a) Bi_2O_3; (b) PbO_2; (c) Tl_2O_3

2.33 (a) Cl_3C structure   (b) Cl_2CO structure
(c) $O=N-F$ structure   (d) F_3N structure

2.35 (a) BH_4^- (b) $[Br-O]^-$ (c) $[NH_2]^-$

2.37 (a) H_2CO (H–C(=O)–H)   (b) CH_3OH (H–C(H)(H)–O–H)
(c) glycine-type structure: $H-\ddot{N}(H)-C(H)(H)-C(=O)-\ddot{O}-H$

2.39 P

2.41 (a) $[NH_4]^+$ $[\ddot{C}l:]^-$ (b) $K^+[\ddot{P}:]^{3-}$, K^+, K^+
(c) $Na^+[\ddot{C}l-\ddot{O}:]^-$

2.43 (polycyclic aromatic hydrocarbon structures)

2.45 $\ddot{O}=N-\ddot{O}$ $\ddot{O}-N=\ddot{O}$

2.47 (a) $[:N\equiv O:]^+$ (b) $:N\equiv N:$ (c) $:C\equiv O:$
(d) $[:C\equiv C:]^{2-}$ (e) $[:C\equiv N:]^-$

2.49 (a)

$$0 \ \ddot{O}=Cl-\ddot{O}: 0 \qquad\qquad :\ddot{O}-\overset{..}{Cl}-\ddot{O}: 0$$
$$\qquad \underset{0}{\overset{\|}{\underset{:O:}{\,}}} \quad | \qquad\qquad \overset{-1}{\,}\quad |\ \overset{+2}{\,} |$$
$$\qquad\qquad :O:\ \ H \qquad\qquad\qquad :\ddot{O}:\ \ H$$
$$\qquad\qquad\quad 0 \qquad\qquad\qquad\qquad\quad -1$$

lower energy

(b)
$$\overset{0}{\,}$$
$$0 \ \ddot{O}=C=\ddot{S}\ 0 \qquad\qquad \overset{-1}{:}\ddot{O}-C\equiv \overset{+1}{S} \quad \overset{0}{\,}$$

lower energy

(c)
$$H-C\equiv N: \qquad\qquad H-\ddot{C}=N:$$
$$\ 0 \quad\ \ 0 \quad\ \ 0 \qquad\qquad\quad 0 \ \ -1 \ +1$$

lower energy

2.51 (a)

$$\left[\ \overset{H}{\underset{:\ddot{O}:_0}{\overset{-1}{:}\ddot{O}-\overset{..}{\underset{\|}{S}}\overset{0}{-}\ddot{O}:}}\ \right]^{-} \qquad \left[\ \overset{H}{\underset{:\ddot{O}:^{-1}}{\overset{-1}{:}\ddot{O}-\overset{+1}{\underset{\|}{S}}-\ddot{O}:}}\ \right]^{-}$$

lower energy

(b)

$$\left[\ \overset{\overset{0}{:}\ddot{O}:\ \ H}{\overset{..}{:}\ddot{O}-\overset{+1}{\underset{\underset{-1}{:}\ddot{O}:}{\|}}S-\ddot{O}:}\ \right]^{-} \qquad \left[\ \overset{\overset{-1}{:}\ddot{O}:\ \ H}{:\ddot{O}-\overset{+2}{\underset{\underset{-1}{:}\ddot{O}:}{\|}}S-\ddot{O}:}\ \right]^{-}$$

lower energy

2.53 (a) In the first structure, the formal charges at Xe and F are 0, whereas, in the second structure, Xe is −1, one F is 0, and the other F is +1. The first structure is favored on the basis of formal charges. (b) In the first structure, all of the atoms have formal charges of 0, whereas, in the second structure, one O atom has a formal charge of +1 and the other O has a formal charge of −1. The first structure is thus preferred.

2.55 (a) The sulfite ion has one Lewis structure that obeys the octet rule:

$$\left[\ :\ddot{O}-\overset{..}{\underset{:\ddot{O}:}{S}}-\ddot{O}:\ \right]^{2-}$$

and three with an expanded octet:

$$\left[\ \ddot{O}=\overset{..}{\underset{:\ddot{O}:}{S}}-\ddot{O}:\ \right]^{2-} \qquad \left[\ :\ddot{O}-\overset{..}{\underset{:\ddot{O}:}{S}}=\ddot{O}\ \right]^{2-} \qquad \left[\ :\ddot{O}-\overset{..}{\underset{:O:}{S}}-\ddot{O}:\ \right]^{2-}$$

The structures with expanded octets have lower formal charges.
(b) There is one Lewis structure that obeys the octet rule:

$$\left[\ :\ddot{O}-\overset{..}{\underset{:O-H}{S}}-\ddot{O}:\ \right]^{-}$$

The formal charge at sulfur can be reduced to 0 by including one double bond contribution. This change gives rise to two expanded octet structures:

$$\left[\ :\ddot{O}-S=\ddot{O}\ \right]^{-} \qquad \left[\ \ddot{O}=S-\ddot{O}:\ \right]^{-}$$
$$\quad\ \ |\qquad\qquad\qquad\quad |$$
$$\quad :O-H\qquad\qquad\quad :O-H$$

Notice that, unlike the sulfite ion, which has three resonance forms, the presence of the hydrogen ion restricts the electrons to the oxygen atom to which it is attached. Because H is electropositive, its placement near an oxygen atom makes it less likely for that oxygen atom to donate a lone pair to an adjacent atom.
(c) The perchlorate ion has one Lewis structure that obeys the octet rule:

$$\left[\ :\ddot{O}-\overset{\overset{:\ddot{O}:}{|}}{\underset{:\ddot{O}:}{Cl}}-\ddot{O}:\ \right]^{-}$$

The formal charge at Cl can be reduced to 0 by including three double-bond contributions, thereby giving rise to four resonance forms:

$$\left[\ \ddot{O}=\overset{\overset{:O:}{\|}}{\underset{:O:}{Cl}}=\ddot{O}\ \right]^{-} \qquad \left[\ :\ddot{O}-\overset{\overset{:O:}{\|}}{\underset{:O:}{Cl}}=\ddot{O}\ \right]^{-}$$

$$\left[\ \ddot{O}=\overset{\overset{:O:}{\|}}{\underset{:O:}{Cl}}-\ddot{O}:\ \right]^{-} \qquad \left[\ \ddot{O}=\overset{\overset{:\ddot{O}:}{|}}{\underset{:O:}{Cl}}=\ddot{O}\ \right]^{-}$$

For the nitrite ion, there are two resonance forms, both of which obey the octet rule:

$$\ddot{O}=\ddot{N}-\ddot{O}:^{-} \qquad\qquad :\ddot{O}-\ddot{N}=\ddot{O}^{-}$$

2.57 (b) and (c)
2.59 The Lewis structures are

(a) $:\dot{\underset{..}{C}l}-\ddot{O}:;$ (b) $:\ddot{\underset{..}{C}l}-\ddot{O}-\ddot{O}-\ddot{\underset{..}{C}l}:;$

(c)
$$\overset{:\ddot{O}:}{\underset{:\ddot{O}=\ddot{N}-\ddot{O}-\ddot{\underset{..}{C}l}:}{|}}$$

Radicals are species with an unpaired electron, therefore (a) and (c) are radicals.
2.61 (a) 4; (b) 6; (c) 5; (d) 6

2.63 (a)
$$\overset{:\ddot{F}:}{\underset{:\ddot{F}\ \ :\ddot{F}:}{\overset{:\ddot{F}\diagdown\ |\ \diagup\ddot{F}:}{S}}}$$
12 electrons

(b)
$$\overset{:\ddot{F}:}{\underset{:\ddot{F}:}{\overset{|}{:\overset{..}{X}e:}}}$$
10 electrons

(c)
$$\left[\ \overset{:\ddot{F}:}{\underset{:\ddot{F}\ \ :\ddot{F}:}{\overset{:\ddot{F}\diagdown\ |\ \diagup\ddot{F}:}{As}}}\ \right]^{-}$$
12 electrons

(d)
$$\overset{:\ddot{Cl}:}{\underset{:\ddot{Cl}:}{\overset{|}{:\ddot{\underset{..}{C}l}\diagdown_{Te:}}}}$$
10 electrons

2.65 (a) 2 lone pairs

$$:O:$$
$$\|$$
$$:F-Xe-F:$$

(b) 2 lone pairs

$$:\ddot{F}:$$
$$\|$$
$$:F-\dot{Xe}-\ddot{F}:$$
$$:\ddot{F}:$$

(c) 1 lone pair

structure with Xe and four F and O

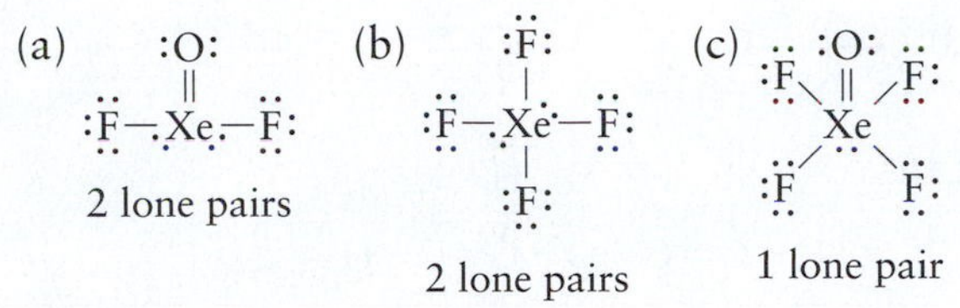

2.67 I (2.7) < Br (3.0) < Cl (3.2) < F (4.0)

2.69 In (1.8) < Sn (2.0) < Sb (2.1) < Se (2.6)

2.71 (a) HCl; (b) CF_4; (c) CO_2

2.73 $Rb^+ < Sr^{2+} < Be^{2+}$. Smaller, more highly charged cations have greater polarizing power.

2.75 $O^{2-} < N^{3-} < Cl^- < Br^-$. Polarizability increases with increasing radius.

2.77 (a) $CO_3^{2-} > CO_2 > CO$; (b) $SO_3^{2-} > SO_2 \approx SO_3$; (c) $CH_3NH_2 > CH_2NH > HCN$

2.79 (a) 150 pm; (b) 127 pm; (c) CO: 127 pm, CN: 142 pm; (d) 120 pm

2.81 (a) 149 pm; (b) 183 pm; (c) 213 pm. Bond length increases with size down Group 14/IV.

2.83
$$:N\equiv N-\overset{..}{\underset{}{N}}-N\equiv N:$$
(+1 −1 +1)

$$\ddot{N}=N=\overset{..}{N}-N\equiv N:$$
(−1 +1 +1)

$$:N\equiv N-\overset{..}{N}=N=\ddot{N}$$
(+1 +1 −1)

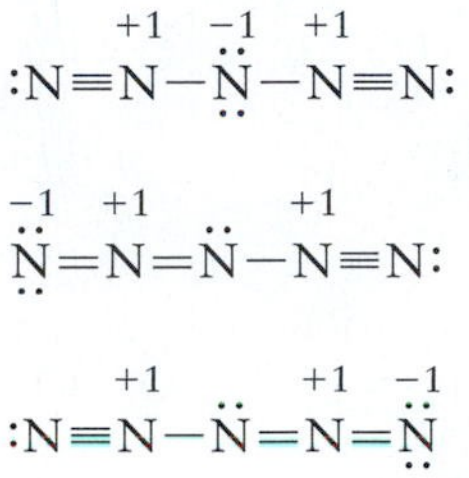

2.85 (a)
structures of $C_2O_4^{2-}$ resonance forms

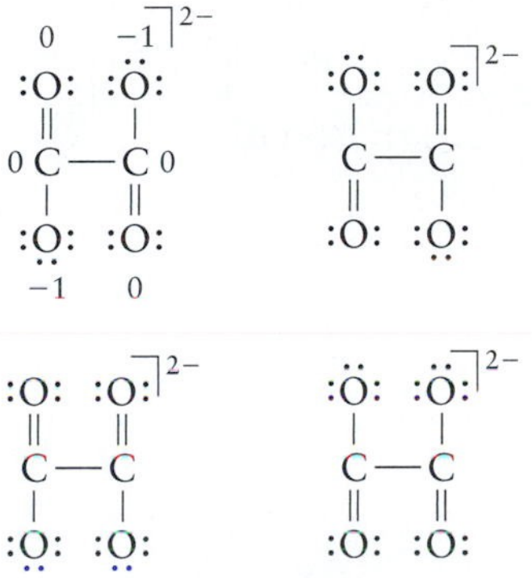

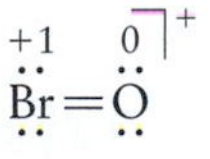

(b) $\overset{..}{Br}=\overset{..}{O}$ (+1 0)$^+$

(c) $:C\equiv C:$ (−1 −1)$^{2-}$

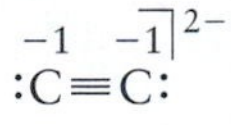

2.87 (a)
$$:O:$$
$$\|$$
$$CH_3-C-\ddot{O}:$$
and
$$CH_3-C=\ddot{O}$$
with :O:

(b)
$$H-C=C-\ddot{O}:$$ with H and CH_3
and
$$H-C-C=\ddot{O}$$ with H and CH_3

(c)
$$H-C=C-C-H$$ (H H H)$^+$
and
$$H-C-C=C-H$$ (H H H)$^+$

(d)
$$:O:$$
$$\|$$
$$CH_3-C-N:$$
and
$$CH_3-C=N:$$ with :Ö:

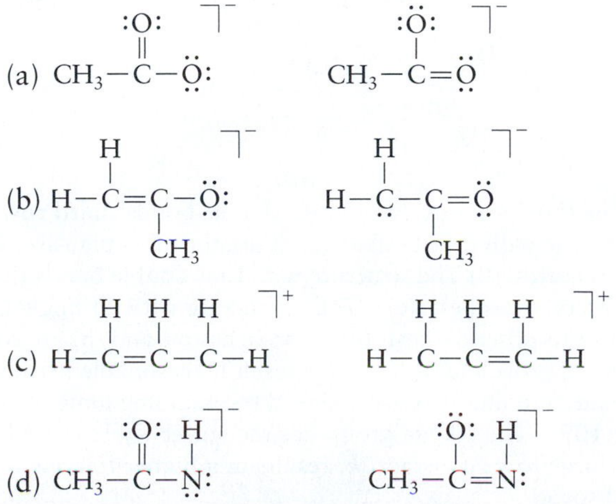

2.89 P and S are larger atoms that are less able to form multiple bonds to themselves, unlike the small N and O atoms. All bonds in P_4 and S_8 are single bonds, whereas N_2 has a triple bond and O_2 a double bond.

2.91 (a) $H-C\equiv C-H$ $H-C\equiv Si-H$

$H-Si\equiv Si-H$ $H-C\equiv N:$ $:N\equiv N:$

(b)
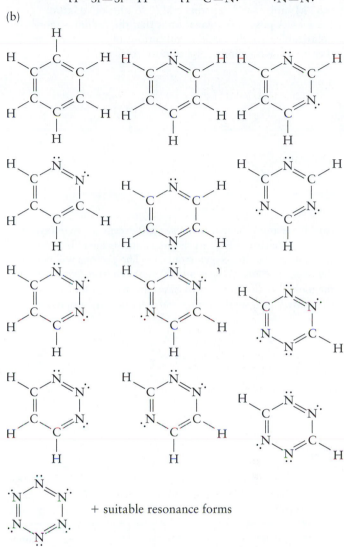

+ suitable resonance forms

2.93 (a) H_2O_2 is reduced because the oxidation numbers of its oxygen atoms change from −1 to −2. $C_6H_8O_6$ is oxidized because the oxidation numbers of its carbon atoms change from $+\frac{2}{3}$ to +1. (b) All atoms have formal charges of zero. Oxidation states are more useful because they indicate that H_2O_2 can undergo an oxidation–reduction reaction to form H_2O, which it does. Formal charges do not allow the same prediction.

2.95 (a) Alkali metal iodide lattice energies versus d_{M-I}:

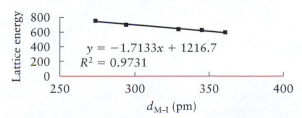

A high correlation ($R^2 = 0.9731$) exists between lattice energy and $d_{(M–I)}$. A better fit ($R^2 = 0.9847$) is obtained between lattice energy and $(1 - d*/d)/d$.
(b) Using the equation (lattice energy) = $218331(1 - d*/d)/d$ + 54.887 and the Ag–I distance (309 pm), the estimated AgI lattice energy is 683 kJ·mol^{-1}.
(c) There is not close agreement between the estimated (683 kJ·mol^{-1}) and experimental (886 kJ·mol^{-1}) lattice energies. A possible explanation is that the Ag ion is more polarizable than the alkali metal cations of similar size and therefore the bonding in AgI is more covalent.

2.97 (a)

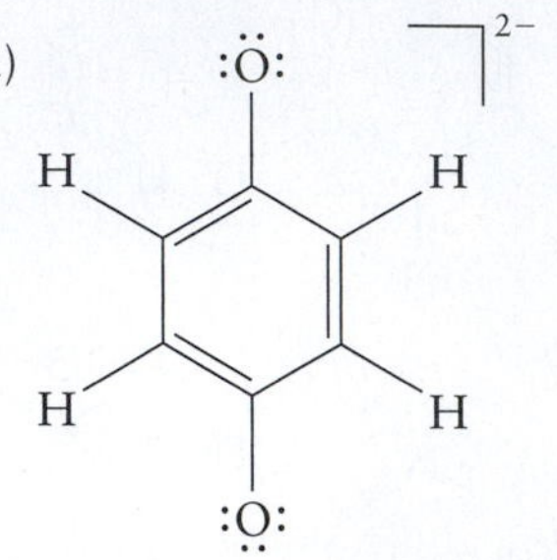

(b) All the atoms have formal charge 0 except the two oxygen atoms, which are -1. The negative charge is most likely to be concentrated at the oxygen atoms. (c) The protons will bond to the oxygen atoms. Oxygen atoms are the most negative sites in the molecule and act as Lewis bases due to their lone pairs of electrons. The resulting compound is named hydroquinone.

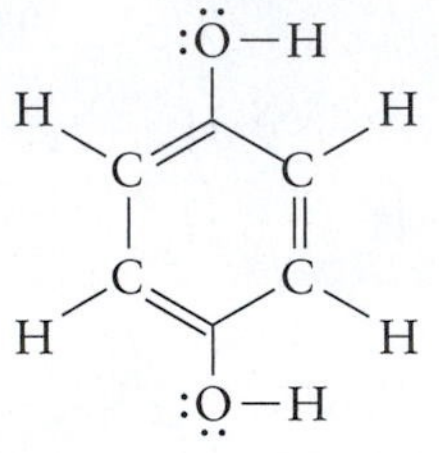

Hydroquinone

2.99

Z	Configuration	Number of unpaired e$^-$	Element	Charge	Energy state
26	[Ar]$3d^6$	4	Fe	+2	ground
52	[Kr]$5s^2 4d^{10} 5p^5 6s^1$	2	Te	-2	excited
16	[Ne]$3s^2 3p^6$	0	S	-2	ground
39	[Kr]$4d^1$	1	Y	+2	ground
30	[Ar]$4s^2 3d^8$	2	Zn	+2	excited

2.101 (a)

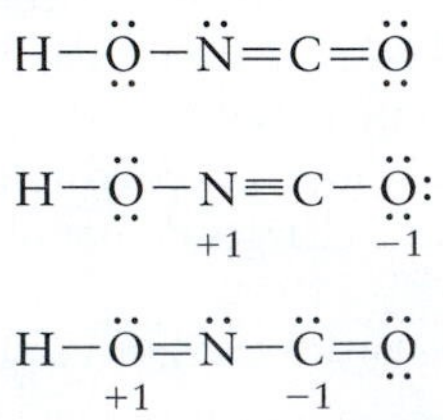

(b)

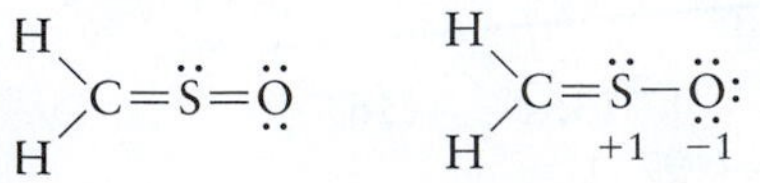

(c)

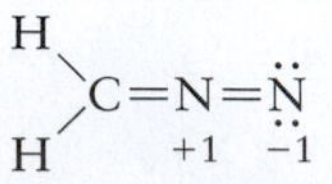

(d)

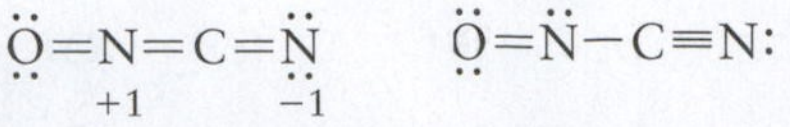

2.103

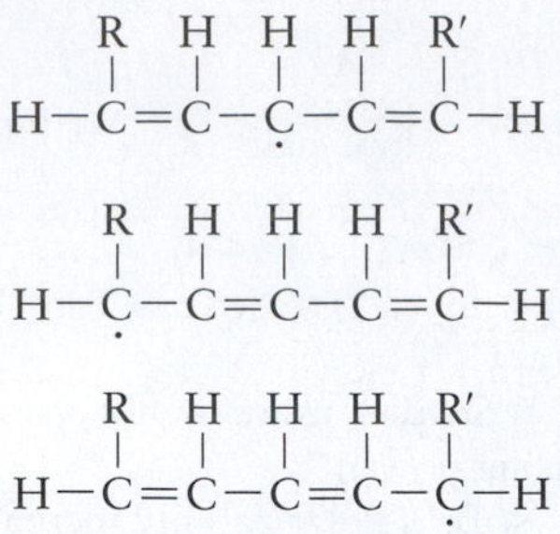

2.105 (a) I: Tl_2O_3; II: Tl_2O. (b) +3; +1. (c) [Xe]$4f^{14}5d^{10}$, [Xe] $4f^{14}5d^{10}6s^2$. (d) Because compound II has a lower melting point, it is probably more covalent, which is consistent with the fact that the +3 ion is more polarizing.

2.107 (a)

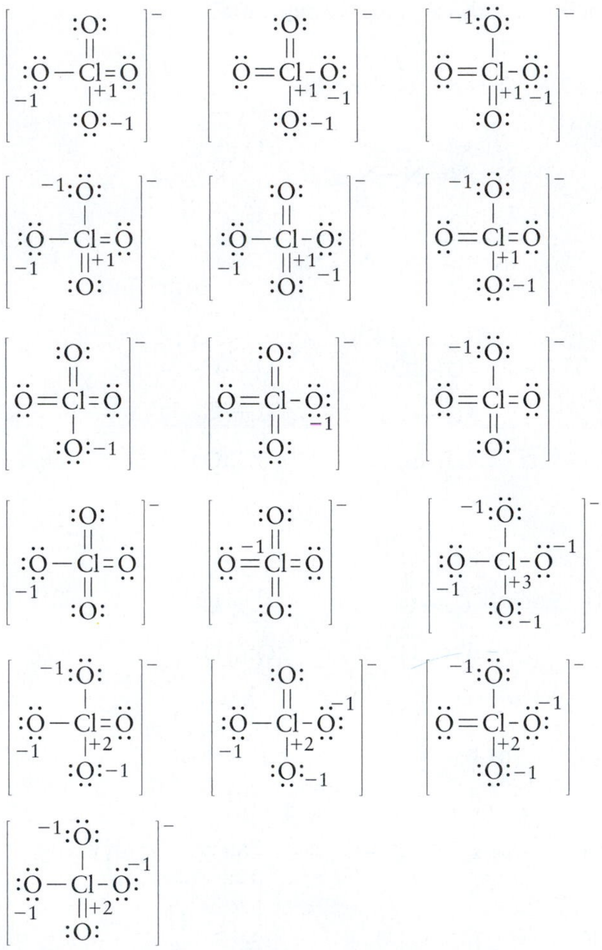

The four structures with three double bonds (third row) and the one with four double bonds are the most plausible Lewis structures. (b) The structure with four double bonds fits these observations best. (c) +7; the structure with all single bonds fits this criterion best. (d) Approaches (a) and (b) are consistent but approach (c) is not. This result is reasonable because oxidation numbers are assigned by assuming ionic bonding.

2.109 The alkyne group has the stiffer C—H bond because a large force constant, k, results in a higher-frequency absorption.

2.111 Look at the Lewis structures for the molecules:

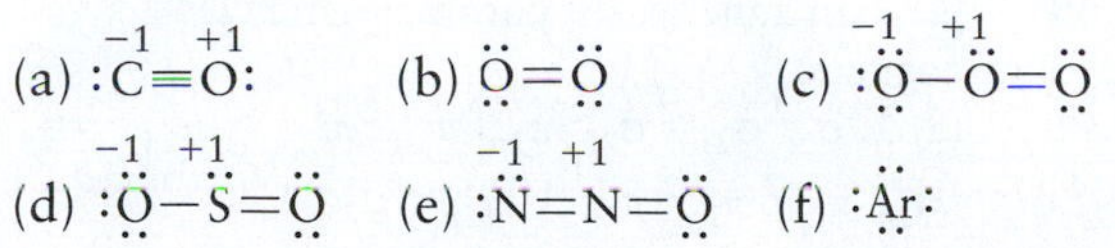

(a) $:C≡O:$ with charges -1 on C, $+1$ on O

(b) $\ddot{O}=\ddot{O}$

(c) $:\ddot{O}-\ddot{O}=\ddot{O}$ with charges -1, $+1$

(d) $:\ddot{O}-\ddot{S}=\ddot{O}$ with charges -1, $+1$

(e) $:N=N=\ddot{O}$ with charges -1, $+1$

(f) $:\ddot{A}r:$

Of these, (a), (c), (d), and (e) can all function as greenhouse gases.

2.113 All halogens have an odd number of valence electrons (7). As a consequence interhalogen compounds of the type XX'_n will be extremely reactive radicals unless the total number of halogens is an even number, which can only be achieved if n is odd. Look at ICl_2 vs. ICl_3 as examples:

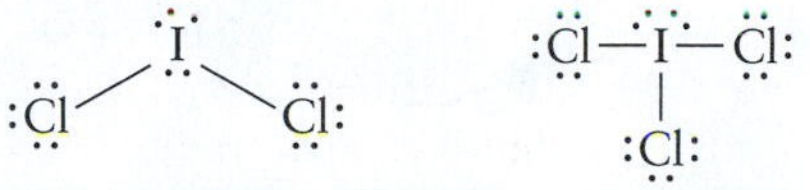

2.115 (a) They have some double bond character. (b) and (c) The Lewis structures for the two possible S_2F_2 are:

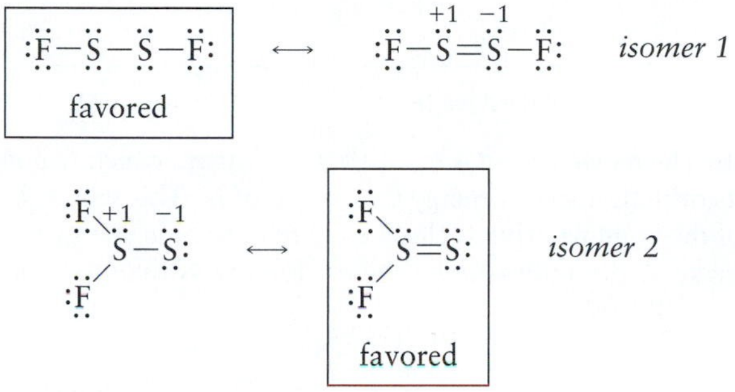

isomer 1

isomer 2

If resonance is occurring, then one would expect that the S—S bond length is indeed between a single and a double bond in length.

2.117 (a) The Lewis structures of NO and NO_2 are:

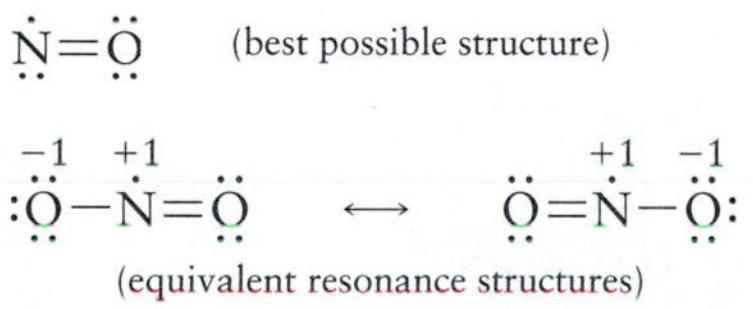

$$\dot{N}=\ddot{O} \quad \text{(best possible structure)}$$

$$:\ddot{O}-\dot{N}=\ddot{O} \longleftrightarrow \ddot{O}=\dot{N}-\ddot{O}: \quad \text{charges } -1, +1 \text{ and } +1, -1$$

(equivalent resonance structures)

From Table 2.4, the average bond dissociation energy of an N=O bond is 630 kJ mol^{-1}, which is right in line with the Lewis structure of NO. The bond dissociation energy of each NO bond in NO_2 is 469 kJ·mol^{-1}, which is about half-way between an N—O double and an N—O single bond, suggesting that the resonance picture of NO_2 is reasonable. (b) An N—O single bond should have a bond length of 149 pm while an N—O double bond should have a bond length of 120 pm (from covalent radii provided in Figure 2.21). From Table 2.5 the length of an N—O triple bond can be estimated to be between 105 and 110 pm. Since N—O has a bond length of 115 pm, this suggests that its actual bond order is somewhere between that of a double and a triple bond.

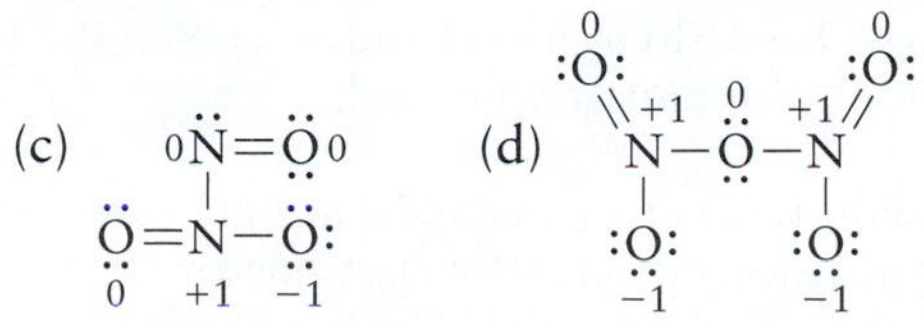

(c)

(d)

(e) $N_2O_5(g) + H_2O(l) \rightarrow 2\ HNO_3(aq)$, nitric acid;
(f) 7.50×10^{-2} mol·L^{-1}
(g) NO: $+2$, NO_2: $+4$, N_2O_3: $+3$, N_2O_5: $+5$; N_2O_5

Chapter 3

3.1 (a) must have lone pairs; (b) may have lone pairs

3.3 (a) trigonal pyramidal; (b) all O–S–Cl are identical; (c) slightly less than 109.5°

3.5 (a) angular; (b) slightly less than 120°

3.7 (a) linear; (b) slightly less than 180°

3.9

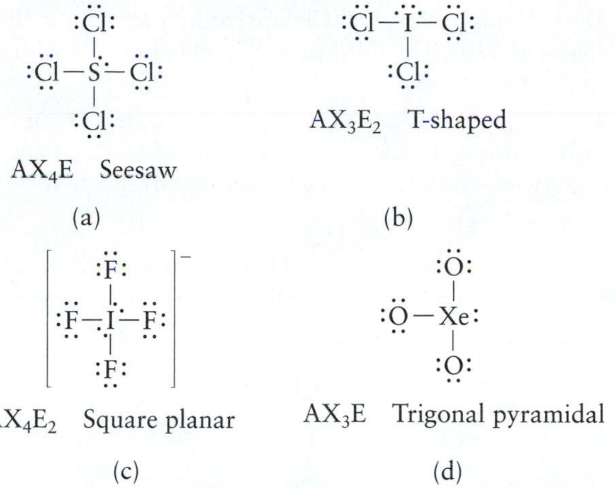

AX_4E Seesaw

(a)

AX_3E_2 T-shaped

(b)

AX_4E_2 Square planar

(c)

AX_3E Trigonal pyramidal

(d)

3.11 (a) AX_2E_3, linear, 180°; (b) AX_5, trigonal bipyramidal, three of 120° and two of 90°; (c) AX_4, tetrahedral, 109.5°; (d) AX_2E, angular, slightly less than 120°

3.13

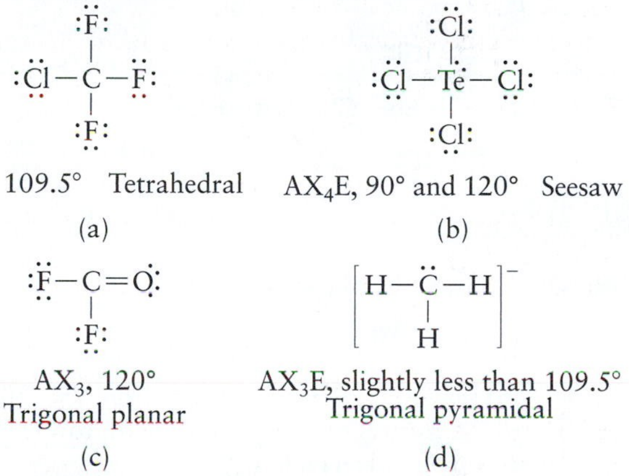

AX_4, 109.5° Tetrahedral

(a)

AX_4E, 90° and 120° Seesaw

(b)

AX_3, 120°
Trigonal planar

(c)

AX_3E, slightly less than 109.5°
Trigonal pyramidal

(d)

3.15 (a) Angles a and b are 120° and c is 109.5° in 2,4-pentanedione. All angles are about 120° in the actylacetonate ion. (b) The major differences arise in the hybridization of the originally sp^3 hybridized CH_2 group, which becomes sp^2 when it loses the hydrogen ion.

3.17 (a) slightly less than 120°; (b) 180°; (c) 180°; (d) slightly less than 109.5°

3.19 The Lewis structures are

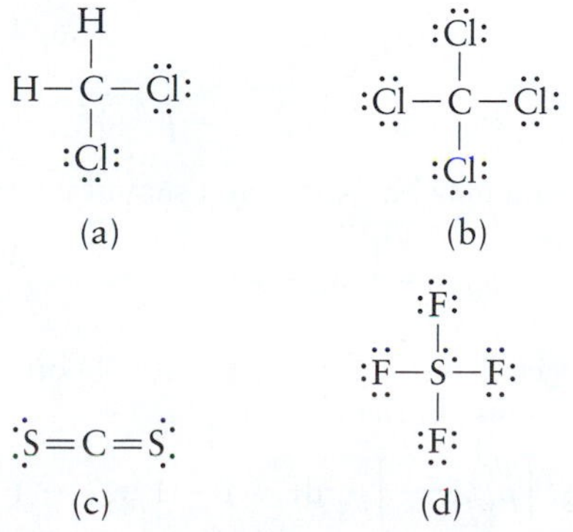

(a)

(b)

(c)

(d)

Molecules (a) and (d) are polar; (b) and (c) are nonpolar.

3.21   (a) polar; (b) nonpolar; (c) polar

3.23   (a) 1 and 2; (b) 1

3.25

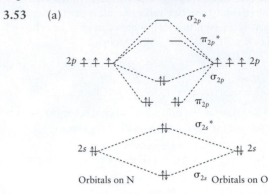

The carbons in CH_2 and CH are sp^2 hybridized with bond angles of 120°. The third carbon is sp hybridized with a bond angle of 180°.

3.27   (a) tetrahedral, 109.5°; (b) tetrahedral about the carbon atoms, 109.5°, C—Be—C angle of 180°; (c) angular, slightly less than 120°; (d) angular, slightly less than 120°

3.29   (a)

H—C—H and H—C—C angles of 120°

(b) $Cl—C≡N:$ linear, 180°

(c) tetrahedral, 109.5°

(d) H—N—N—H   arrangement about each N is trigonal pyramidal, 107°

3.31   (a) tetrahedral

(b) tetrahedral   (c) seesaw

3.33   (a) toward corners of a tetrahedron; (b) 180° apart; (c) toward the corners of an octahedron; (d) toward the corners of an equilateral triangle

3.35   (a) sp^3d; (b) sp^2; (c) sp^3; (d) sp

3.37   (a) sp^2; (b) sp^3; (c) sp^3d; (d) sp^3

3.39   (a) sp^3; (b) sp^3d^2; (c) sp^3d; (d) sp^3

3.41   (a) sp^3; (b) nonpolar

3.43   increase

3.45 $\int h_1 h_2 \, d\tau = \int (s + p_x + p_y + p_z)(s - p_x + p_y - p_z) d\tau =$

$\int (s^2 - sp_x + sp_y - sp_z + sp_x - p_x^2 + p_x p_y - p_x p_z + sp_y -$

$p_x p_y + p_y^2 - p_y p_z + sp_z - p_z p_x + p_z p_y - p_z^2) d\tau$

This integral of a sum may be written as a sum of integrals:

$\int s^2 d\tau - \int sp_x d\tau + \int sp_y d\tau - \int sp_z d\tau + \cdots$

Because the hydrogen wavefunctions are mutually orthogonal, this sum of integrals simplifies to

$\int s^2 d\tau - \int p_x^2 d\tau + \int p_y^2 d\tau - \int p_z^2 d\tau = 1 - 1 + 1 - 1 = 0$

3.47 $\lambda = 0.67$, $sp^{0.67}$

3.49   (a) diamagnetic; (b) paramagnetic, 1; (c) paramagnetic, 1

3.51   (a) 1: $\sigma_{2s}^2 \sigma_{2s}^{*2} \sigma_{2p}^2 \pi_{2p_x}^2 \pi_{2p_y}^2 \pi_{2p_x}^{*2} \pi_{2p_y}^{*2} \sigma_{2p}^{*1}$, 1: $\sigma_{2s}^2 \sigma_{2s}^{*2} \sigma_{2p}^2 \pi_{2p_x}^2 \pi_{2p_y}^2 \pi_{2p_x}^{*2} \pi_{2p_y}^{*1}$,
3: $\sigma_{2s}^2 \sigma_{2s}^{*2} \sigma_{2p}^2 \pi_{2p_x}^2 \pi_{2p_y}^2 \pi_{2p_x}^{*2} \pi_{2p_y}^{*2} \sigma_{2p}^{*2}$.
(b) 1: 0.5; 2: 1.5; 3: 0. (c) 1 and 2 are paramagnetic with one unpaired electron each. (d) σ for 1 and 3, π for 2

3.53   (a)

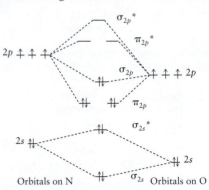

(b) The oxygen atom is more electronegative, which will make its orbitals lower in energy than those of N. This will make all of the bonding orbitals closer to O than to N in energy and will make all the antibonding orbitals closer to N than to O in energy.

(c) The electrons in the bonding orbitals will have a higher probability of being at O because O is more electronegative and its orbitals are lower in energy.

3.55   (a) $(\sigma_{2s})^2 (\sigma_{2s}^*)^2 (\pi_{2p_x})^1 (\pi_{2p_y})^1$, 1; (b) $(\sigma_{2s})^2 (\sigma_{2s}^*)^2$, 0; (c) $(\sigma_{2s})^2 (\sigma_{2s}^*)^2 (\sigma_{2p})^2 (\pi_{2p_x})^2 (\pi_{2p_y})^2 (\pi_{2p_x}^*)^2 (\pi_{2p_y}^*)^2$, 1

3.57   CO: $\sigma_{2s}^2 \sigma_{2s}^{*2} \pi_{2p_x}^2 \pi_{2p_y}^2 \sigma_{2p}^2$, bond order = 3.
CO^+: $\sigma_{2s}^2 \sigma_{2s}^{*2} \pi_{2p_x}^2 \pi_{2p_y}^2 \sigma_{2p}^1$, bond order = 2.5. Due to the higher bond order for CO, it should form a stronger bond and therefore have the higher bond enthalpy.

3.59   All are paramagnetic: B_2^- and B_2^+ have one unpaired electron whereas B_2 has two.

3.61   (a) F_2 has a bond order of 1; F_2^- has a bond order of $\frac{1}{2}$. F_2 has the stronger bond. (b) B_2 has a bond order of 1; B_2^+ has a bond order of $\frac{1}{2}$. B_2 has the stronger bond.

3.63 C_2^-

3.65   The conductivity of a semiconductor increases with temperature as increasing numbers of electrons are promoted into the conduction band, whereas the conductivity of a metal will decrease as the motion of the atoms will slow down the migration of electrons.

3.67 (a) In and Ga; (b) P and Sb

3.69 Given the overlap integral $S = \int \psi_{A1s}\psi_{B1s}d\tau$, the bonding orbital $\psi = \psi_{A1s} + \psi_{B1s}$, and the fact that the individual atomic orbitals are normalized, we are asked to find the normalization constant N which will normalize the bonding orbital ψ such that

$$\int N^2\psi^2 d\tau = N^2 \int (\psi_{A1s} + \psi_{B1s})^2 dt = 1$$

$$N^2 \int (\psi_{A1s} + \psi_{B1s})^2 d\tau$$

$$= N^2 \int (\psi_{A1s}^2 + 2\psi_{A1s}\psi_{B1s} + \psi_{B1s}^2)d\tau$$

$$= N^2 \int (\psi_{A1s}^2 d\tau + 2\int \psi_{A1s}\psi_{B1s}d\tau + \int \psi_{B1s}^2 d\tau)$$

Given the definition of the overlap integral above and the fact that the individual orbitals are normalized, this expression simplifies to $N^2(1 + 2S + 1) = 1$ and

$$N = \sqrt{\frac{1}{2 + 2S}}$$

3.71 The antibonding molecular orbital is obtained by taking the difference between two atomic orbitals that are proportional to e^{-r/a_o}. Halfway between the two nuclei, the distance from the first nucleus, r_1, is equal to the distance to the second nucleus, r_2, and the antibonding orbital is proportional to $\psi \propto e^{-r/a_o} - e^{-r/a_o} = 0$.

3.73

$$\left[\begin{array}{c} :\ddot{Cl}: \\ | \\ :\ddot{Cl}-Ga-\ddot{Cl}: \\ | \\ :\ddot{Cl}: \end{array} \right]^{-}$$

Shape: tetrahedral
Hybridization: sp^3
Bond angles: 109.5°
Nonpolar

(b)

$$\begin{array}{c} :\ddot{F}: \\ :\ddot{F}\diagdown \\ \quad Te-\ddot{F}: \\ :\ddot{F}: \end{array}$$

Shape: seesaw
Hybridization: dsp^3
Bond angles: > 90° and >120°
Polar

(c)

$$\left[\begin{array}{c} :\ddot{Cl}: \\ :\ddot{Cl}\diagdown \\ \quad Sb-\ddot{Cl}: \\ :\ddot{Cl}: \end{array} \right]^{-}$$

Shape: seesaw
Hybridization: dsp^3
Bond angles: > 90° and >120°
Polar

(d)

$$\begin{array}{c} :\ddot{Cl}: \\ | \\ :\ddot{Cl}-Si-\ddot{Cl}: \\ | \\ :\ddot{Cl}: \end{array}$$

Shape: tetrahedral
Hybridization: sp^3
Bond angles: 109.5°
Nonpolar

3.75 (a) SiF_4; (b) SF_6; (c) AsF_5

3.77 (a)

$$\begin{array}{c} H \\ | \\ H-C-\ddot{O}-H \\ | \\ H \end{array}$$

(b) H—C—O and H—C—H angles are 109.5°, C—O—H angle is slightly less than 109.5°, and therefore the hybridization of both C and O is sp^3. (c) polar

3.79

Bonding   Antibonding

(a)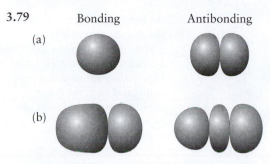

(b)

(c) The bonding and antibonding orbitals for HF appear different due to the fact that a p orbital from the F atom is used to construct bonding and antibonding orbitals, whereas in the H_2 molecule s orbitals on each atom are used.

3.81 (a) $CF^- > CF > CF^+$; (b) CF^+ because it has no unpaired electrons; CF has one unpaired electron and CF^- has two.

3.83 Orbitals at each B and N atom are sp^2 hybridized.

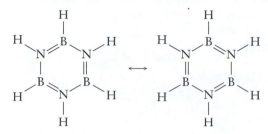

3.85 (a) The Lewis structures are:

$$\left[\begin{array}{c} H \\ | \\ H-C \\ | \\ H \end{array} \right]^{+} \quad \begin{array}{c} H \\ | \\ H-C-H \\ | \\ H \end{array} \quad \left[\begin{array}{c} H \\ | \\ H-C: \\ | \\ H \end{array} \right]^{-} \quad \begin{array}{c} H \\ | \\ C: \\ | \\ H \end{array} \quad \left[\begin{array}{c} H \\ | \\ C \\ | \\ H \end{array} \right]^{2+} \quad \left[\begin{array}{c} H \\ | \\ :C: \\ | \\ H \end{array} \right]^{2-}$$

(b) None are radicals. (c) $CH_2^{2-} < CH_3^- < CH_4 < CH_2 < CH_3^+ < CH_2^{2+}$

3.87 (a) Acetylene: $H-C\equiv C-H$

Polymer:

$$\left(\begin{array}{cc} H & H \\ | & | \\ -C=C-C=C- \\ | & | \\ H & H \end{array} \right)$$

(b) Resonance forms of the Lewis structure of polyacetylene can be drawn showing that the electrons may be delocalized along the carbon chain. No such resonance form is possible for polyethylene.

3.89 $\Delta E = \dfrac{h^2}{8mN^2R^2}\left[\left(\dfrac{N}{2}+1\right)^2 - \left(\dfrac{N}{2}\right)^2 \right] = \dfrac{h^2(N+1)}{8mN^2R^2}$

To shift the wavelength of the absorption to longer wavelengths (lower energies), the length of the carbon chain, N, must increase.

3.91

(a)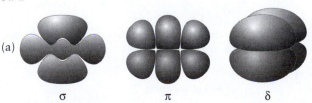

σ π δ

(b) $\sigma > \pi > \delta$

3.93 (a) Net π-bond order would drop from three to two. (b) Net π-bond order would drop from three to two. Both are paramagnetic.

3.95 (a)

H
| sp²
H sp² C sp
\ ‖‖
sp² C sp
H sp²
|
H

(b) Benzyne would be highly reactive because the two carbon atoms that are *sp* hybridized are constrained to have a very strained structure compared with the linear structure associated with their hybridization.

3.97 (a) The carbon atoms are all sp^3 hybridized. (b) The C—C—C, H—C—H, and H—C—C bond angles should be 109.5°. (c) Because of the ring structure, however, the C–C–C bond angles must be 60°. (d) The σ-bond will have the electron density of the bond located on a line between the two atoms that it joins. (e) If the C atoms are truly sp^3 hybridized, then the bonding orbitals will not necessarily point directly between the C atoms. (f) The sp^3 hybridized orbitals can still overlap even if they do not point directly between the atoms.

3.99 (a) and (d) have the possibility of *n*-to-π* transitions because these molecules possess both an atom with a lone pair of electrons and a π-bond to that atom.

3.101 (a) and (b)

σ(C2sp², C2sp²)   O sp²
σ(H1s, C2sp²)   H   Ö: σ(C2sp², O2sp²),
                  | ‖   π(C2p, O2p)
            H—C=C—C
                  | \   σ(H1s, C2sp²)
σ(H1s, C2sp²)   H   H
         σ(C2sp², C2sp²),
         π(C2p, C2p)

All atoms in this molecule have a formal charge of zero.

3.103 (a)

$$\left[\begin{array}{c} :\ddot{F}: \quad +1 \quad :\ddot{F}: \\ :\ddot{F}\diagdown \quad \ddot{F} \quad \diagup\ddot{F}: \\ \cdot\cdot\text{Sb} \quad \quad \text{Sb}\cdot\cdot \\ \cdot\cdot|{-1} \quad {-1}|\cdot\cdot \\ :\ddot{F}: \quad \quad :\ddot{F}: \end{array} \right]^{-}$$

(b) sp^3d

3.105

$$\left[\begin{array}{c} +1 \\ \ddot{C}l \\ :\ddot{C}l—Bi \quad\diagup\diagup\diagdown\diagdown\quad Bi—\ddot{C}l: \\ {-2} \quad \ddot{C}l \quad {-2} \\ +1 \end{array} \right]^{2-}$$

3.107

$$\left[\begin{array}{c} a \\ \quad b\diagup \ddot{I}: \\ :\cdot\cdot c \quad \ddot{I}: \\ :\ddot{I} \quad \ddot{I}: \\ \quad d \quad \ddot{I}: \\ \quad e \end{array} \right]^{-}$$

I_b and I_d are sp^3d hybridized with the attached atoms occupying axial positions, which results in a bond angle of 180° around each axial atom. I_c is sp^3 hybridized.

3.109 Al_2Cl_6 is nonpolar; all dipoles in the molecule (due to the Al—Cl bonds) will cancel:

These cancel

↑ ←These cancel

Cl   Cl   Cl
⟷   Al   Al   ⟷
Cl   Cl   Cl
↓

3.111 (a) sp^3d^3; (b) sp^3d^3f; (c) sp^2d

Chapter 4

4.1 (a) 8×10^9 Pa; (b) 80 kbar; (c) 6×10^7 Torr; (d) 1×10^6 lb·in^{-2}

4.3 (a) 86 mmHg. (b) The side attached to the bulb is higher. (c) 848 Torr

4.5 909 cm

4.7 2.9×10^3 lb

4.9 (a) 22.4 in.; (b) tube (1): 41.85 inHg, tube (2): 59.85 inHg

4.11 (a)

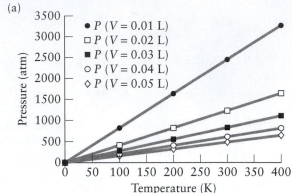

(b) Slope = $\dfrac{nR}{V}$; (c) 0.00

4.13 (a) 1.5×10^3 kPa; (b) 4.5×10^3 Torr

4.15 3.31 atm

4.17 1.6 atm

4.19 The volume must be increased by 10.% to keep *P* and *T* constant.

4.21 0.50 g

4.23 (a) 11.8 mL; (b) 0.969 atm; (c) 199 K

4.25 (a) 1.06×10^3 kPa; (b) 4.06×10^2 mL; (c) 0.20 g; (d) 3.24×10^5 mol; (e) 9.2×10^{14} atoms

4.27 (a) 63.4 L; (b) 6.3 L

4.29 18.1 L

4.31 4.18 atm

4.33 621 g

4.35 2.52×10^{-3} mol

4.37 (a) 3.6 m^3; (b) 1.8×10^2 m^3

4.39 CO < H$_2$S < CO$_2$

4.41 342° C

4.43 (a) 1.28 g·L^{-1}; (b) 3.90 g·L^{-1}

4.45 (a) 70. g·mol^{-1}; (b) CHF$_3$; (c) 2.9 g·L^{-1}

4.47 C$_2$H$_2$Cl$_2$

4.49 44.0 g·mol^{-1}

4.51 (a) 3.0×10^5 L; (b) 1.0×10^4 L

4.53 2.00 g of CH$_4$

4.55 2.4×10^4 L

4.57 (a) 4.25×10^{-3} g; (b) HCl; (c) 0.0757 atm

4.59 (a) $x_{HCl} = 0.9$, $x_{benzene} = 0.1$; (b) $P_{HCl} = 0.72$ atm, $P_{benzene} = 0.08$ atm

4.61 (a) $P_{N_2} = 1.0$ atm, $P_{H_2} = 2.0$ atm; (b) 3.0 atm

4.63 2.0 atm

4.65 (a) 739.2 Torr; (b) 2 H$_2$O(l) → 2 H$_2$(g) + O$_2$(g); (c) 0.142 g

4.67 No. Force is proportional to ν_x and because molecules move at different velocities, they will collide with the walls of their container with different forces.

4.69 C$_{10}$H$_{10}$

4.71 (a) 154 s; (b) 123 s; (c) 33.0 s; (d) 186 s

4.73 110 g·mol^{-1}; C$_8$H$_{12}$

4.75   (a) 4103.2 J·mol^{-1}; (b) 4090.7 J·mol^{-1}; (c) 12.5 J·mol^{-1}

4.77   (a) 627 m·s^{-1}; (b) 458 m·s^{-1}; (c) 378 m·s^{-1}

4.79   241 m.s^{-1}

4.81   0.316

4.83   (a) The most probable speed is the one that corresponds to the maximum on the distribution curve. (b) The percentage decreases.

4.85   At low temperatures, this hydrogen bonding causes the molecules of HF to be attracted to each other more strongly, thus lowering the pressure. As the temperature is increased, the hydrogen bonds are broken and the pressure rises more quickly than for an ideal gas.

4.87   (a) 1.63 atm, 1.62 atm. (b) 48.9 atm, 38.9 atm. (c) 489 atm, 1.88×10^3 atm. At low pressures, the ideal gas law gives essentially the same value as the van der Waals equation does; but, at high pressures, there is a very significant difference.

4.89   17.58, CH$_3$CN; 3.392, CO$_2$; 2.253, CH$_4$; 0.2107

4.91   Clearly, the greater deviation from the ideal gas law values occurs at low volumes or higher pressures. Ammonia deviates more strongly and its van der Waals constants are larger than those for oxygen. This may likely arise because ammonia is more polar and will have stronger intermolecular interactions.

4.93   (a) 280. Torr; (b) 700. Torr

4.95   (a) $\ddot{O}=\dot{N}-\ddot{O}:\ \longleftrightarrow\ :\ddot{O}-\dot{N}=\ddot{O}$

(b) 0.020 ppm

4.97   0.481 mol·L^{-1}

4.99   (a) N$_2$O$_4$(g) → 2 NO$_2$(g); (b) 2.33 atm; (c) 4.65 atm; (d) $x_{NO_2} = 0.426$, $x_{N_2O_4} = 0.574$

4.101   (a) N$_2$H$_4$; (b) $H-\ddot{N}-\ddot{N}-H$ (c) 4.2×10^{-4} mol
   with H, H below the two N atoms

4.103 $\dfrac{f(10\nu_{rms})}{f(\nu_{rms})} = 100e^{\frac{-99M\nu_{rms}^2}{2RT}}$

No, the distribution changes with temperature.

4.105   >2.35 M

4.107   254 g·mol^{-1}; $x = 4$

4.109   23.1 g

4.111   (a) $V = 3.936 \times 10^7$ pm^3, $r = 211$ pm. (b) $r = 128$ pm, $V = 8.78 \times 10^6$ pm^3. (c) The difference in these values illustrates that there is no easy definition for the boundaries of an atom. The van der Waals value obtained from the correction for molar volume is considerably larger than the atomic radius, owing perhaps to longer-range and weak interactions between atoms. One should also bear in mind that the value for the van der Waals b is a parameter used to obtain a good fit to a curve, and its interpretation is more complicated than that of a simple molar volume.

4.113   (a) ClNO$_2$; (b) ClNO$_2$; (c) $:\ddot{C}l:$ $:\ddot{C}l:$

(d) trigonal planar $\ddot{O}=\dot{N}-\ddot{O}:$ $:\ddot{O}-\dot{N}=\ddot{O}$

4.115   (a) 0.0979 L; (b) $V_{ideal} = 0.130$ L $> V_{vdW}$, attractive forces dominate

Chapter 5

5.1   (a) London forces, dipole–dipole, hydrogen bonding; (b) London forces, dipole–dipole, hydrogen bonding; (c) London forces, dipole–dipole; (d) London forces

5.3   (b), (c), and (d)

5.5   (b) ≈ (c) < (e) < (a) < (d)

5.7   (c)

5.9   II, because the dipole–dipole interactions are maximized along the length of the chain

5.11   (a) NaCl, ionic compound; (b) butanol, hydrogen bonding; (c) CH$_3$I, London forces; (d) CH$_3$OH, hydrogen bonding

5.13   (a) both trigonal pyramidal, PBr$_3$; (b) SO$_2$ is bent and CO$_2$ is linear, SO$_2$; (c) both trigonal planar, BCl$_3$

5.15   1.03; Al^{3+}

5.17   (a) Xenon has stronger London forces. (b) Hydrogen bonding occurs in water. (c) 2,2-dimethylpropane is more compact.

5.19 $1/r^7$

5.21   (a) *cis*-dichloroethene, it is polar; (b) benzene at 20°, at higher temperatures intermolecular forces are disrupted

5.23   C$_6$H$_6$ < C$_6$H$_5$SH < C$_6$H$_5$OH

5.25   CH$_4$, −162°C; CH$_3$CH$_3$, −88.5°C; (CH$_3$)$_2$CHCH$_2$CH$_3$, 28°C; CH$_3$(CH$_2$)$_3$CH$_3$, 36°C; CH$_3$OH, 64.5°C; CH$_3$CH$_2$OH, 78.3°C; CH$_3$CHOHCH$_3$, 82.5°C; C$_5$H$_9$OH (cyclic), 140°C; C$_6$H$_5$CH$_3$OH (aromatic ring), 205°C; HOCH$_2$CHOHCH$_2$OH, 290°C

5.27   (a) Hydrogen bonding is present in H$_2$O. (b) London dispersion forces increase.

5.29   water

5.31   (a) glucose: London forces, dipole–dipole interactions, and hydrogen bonds; benzophenone: London forces and dipole–dipole interactions; methane: London forces; (b) methane (m.p. = −182°C) < benzophenone (m.p. = 48°C) < glucose (m.p. = 148°C–155°C)

5.33   (a) network; (b) ionic; (c) molecular; (d) molecular; (e) network

5.35   (a) 2; (b) 8; (c) 286 pm

5.37   (a) 2.72 g·cm^{-3}; (b) 0.813 g·cm^{-3}

5.39   (a) 139 pm; (b) 143 pm

5.41   (a) 3.73×10^{-22} g; (b) 8

5.43   90.7%

5.45   (a) 1 Cs$^+$, 1 Cl$^-$, 1 formula unit. (b) 2 Ti^{4+}, 4 O^{2-}, 2 formula units. (c) The Ti atoms are 6-coordinate and the O atoms are 3-coordinate.

5.47   YBa$_2$Cu$_3$O$_7$

5.49   (a) 8; (b) 6; (c) 6

5.51   (a) 3.36 g·cm^{-1}; (g) 4.67 g·cm^{-1}

5.53   (a) the smallest possible rectangular unit cell is

(b) 4 carbon atoms; (c) 3

5.55   3.897 pm

5.57   (a) interstitial, atomic radius of nitrogen is much smaller than that of iron; (b) becomes harder and stronger, with a lower electrical conductivity

5.59   (a) 2.8 Cu per Ni; (b) 15 Sn:1.2 Sb:1Cu

5.61   Because there are so many ways that these molecules can rotate and twist, they do not remain rodlike. The molecules have only single bonds, which allow rotation about the bonds, and so each molecule can adopt many configurations.

5.63   Use of a nonpolar solvent such as hexane or benzene in place of water should give rise to the formation of inverse micelles.

5.65   Benzene is an isotropic solvent; its viscosity is the same in every direction. However, a liquid crystal solvent is an anisotropic solvent; its viscosity is smaller in the direction parallel to the long axis of the molecule than the perpendicular direction. Methylbenzene is a small, spherical molecule, so its interactions with either solvent are similar in all directions.

5.67   (a) NI_3: trigonal pyramidal, dipole–dipole interactions are possible

(b) BI_3: trigonal planar, dipole–dipole interactions are not possible

5.69   (a) 8.11×10^5 pm^2 for 2,2-dimethylpropane, 9.91×10^5 pm^2; (b) pentane; (c) pentane

5.71

(a) anthracene, $C_{14}H_{10}$, London forces

(b) phosgene, $COCl_2$, dipole–dipole forces, London forces

(c) glutamic acid, $C_5H_9NO_4$, hydrogen bonding, dipole–dipole forces, London forces

5.73 $r = [(2.936 \times 10^5)M/d]^{1/3}$; Ne, 170 pm; Ar, 203 pm; Kr, 225 pm; Xe, 239 pm; Rn, 246 pm

5.75   21.3 g·cm^{-3}

5.77   (a) +2.36; (b) fraction of Ti^{2+} = 0.64, fraction of Ti^{3+} = 0.36

5.79   (a) true; (b) false; (c) true; (d) false

5.81   There are several ways to draw unit cells that will repeat to generate the entire lattice. The choice of unit cell is determined by conventions that are beyond the scope of this text (the smallest unit cell that indicates all of the symmetry present in the lattice is typically the one of choice).

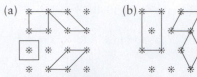

5.83   (a) 50.2 pm; (b) octahedral, 75%

5.85   Glasses melt at lower temperatures as the amount of non-SiO_2 materials increases. Fused silica (fused quartz) is predominantly SiO_2 with very few impurities; it is the most refractory, so it can be used at the highest temperatures. Quartz vessels are routinely used for reactions that must be carried out at temperatures up to 1000°C. Vycor (96% SiO_2) can be used at up to about 900°C, and normal borosilicate glasses at up to approximately 550°C. The borosilicate glasses (Pyrex or Kimax) are commonly used because their lower softening points allow them to be more easily molded and shaped into different types of glass objects, such as reaction flasks, beakers, and other types of laboratory and technical glassware.

5.87   38.1 g

5.89 MA_2

5.91   35%

5.93   (a) 48%; (b) fcc = 26%, fcc cell is packed more efficiently

5.95   The Lewis structure of $(HF)_3$ chain is $H-\ddot{F}:\cdots H-\ddot{F}:\cdots H-\ddot{F}:$; the FHF bond angle is close to 180° and the HFH bond is close to 109°.

5.97   (a) increases the electrical conductivity. (b) Spacing will increase and the angle of the x-ray beam will decrease.

5.99   The solution lies in showing that the extra distance traveled by x-ray 2 must be equal to an integral number of wavelengths in order for x-rays 1 and 2 to be in phase as they leave the crystal.

5.101   5.074°

5.103   633 K

5.105   (a)–(b) The Lewis structures of ethylammonium nitrate (EAN) and the formal charge of each atom are shown as

(c) NO_3^-, N^{5+} can be reduced to a lower oxidation state by a reducing agent;
(d) $CH_3CH_2NH_2 + HNO_3 \rightarrow CH_3CH_2NH_3^+NO_3^-$, acid–base;
(e) 6.49 g, 63.2%; (f) EAN experiences ion–ion, hydrogen-bonding, and London forces, whereas NaCl and NaBr only have ion–ion and London forces holding them together. However, the ion–ion forces in NaCl and NaBr are much stronger that those of EAN because of the smaller ion sizes of NaCl and NaBr. (g) As the ion size increases, the distance between ions will increase, the ion–ion forces will decrease, and the melting point will decrease.

Chapter 6

6.1   (a) isolated; (b) closed; (c) isolated; (d) open; (e) closed; (f) open

6.3   (a) 28 J; (b) positive

6.5   864 kJ

6.7   (a) on; (b) 4.90×10^2 J

6.9   5500 kJ

6.11   −1626 kJ

6.13 (a) true if no work is done; (b) always true; (c) always false; (d) true only if $w = 0$; (e) always true

6.15 (a) absorbed, by the system, q is positive, w is negative; (b) released, on the system, q is negative, w is positive

6.17 (a) 2.6×10^2 kJ; (b) 94.3%

6.19 25°C

6.21 14.8 kJ·(°C)$^{-1}$

6.23 (a) -226 J; (b) -326 J

6.25 NO_2. The heat capacity increases with molecular complexity: as more atoms are present in the molecule, there are more possible bond vibrations that can absorb added energy.

6.27 (a) -90.6 J; (b) -54.4 J

6.29 (a) $\frac{5}{2}R$; (b) $3R$; (c) $\frac{3}{2}R$; (d) $\frac{5}{2}R$

6.31 2.49×10^{27} photons

6.33 (a) Cu; (b) 8.90 g·cm^{-3}

6.35 (a) 8.22 kJ·mol^{-1}; (b) 43.5 kJ·mol^{-1}

6.37 33.4 kJ

6.39 31°C

6.41 7.53 kJ·mol^{-1}

6.43 (b)

6.45 (a) 448 kJ; (b) 1.47×10^3 kJ; (c) 352 g

6.47 (a) 2.2×10^1 g; (b) -1.3×10^5 kJ

6.49 (a) 1.9×10^5 kJ·yr^{-1}; (b) 7.6×10^6 kJ·yr^{-1}

6.51 7 kJ

6.53 31°C

6.55 2.39×10^4 kJ·L^{-1}

6.57 (a) $CO(g) + H_2O(g) \rightarrow CO_2(g) + H_2(g)$; (b) -41.2 kJ·mol^{-1}

6.59 1.90 kJ·mol^{-1}

6.61 -197.78 kJ

6.63 -312 kJ·mol^{-1}

6.65 -2987 kJ·mol^{-1}

6.67 -184.6 kJ

6.69 (c)

6.71 -72.80 kJ

6.73 11.3 kJ

6.75 -444 kJ·mol^{-1}

6.77 (a) -286 kJ·mol^{-1}; (b) -257 kJ·mol^{-1}; (c) 227 kJ·mol^{-1}; (d) -1 kJ·mol^{-1}

6.79 2564 kJ·mol^{-1}

6.81 (a) -412 kJ·mol^{-1}; (b) 673 kJ·mol^{-1}; 63 kJ·mol^{-1}

6.83 (a) -597 kJ·mol^{-1}; (b) -460 kJ·mol^{-1}; (c) 0 kJ·mol^{-1}

6.85 (a) 248 kJ·mol^{-1}; (b) -45 kJ·mol^{-1}; (c) -115 kJ·mol^{-1}

6.87 approximately 228 kJ

6.89 (a) 33.93 kJ·mol^{-1}. (b) 30.94 kJ·mol^{-1}. (c) The values are similar (table value = 30.8 kJ·mol^{-1}). Heat capacities vary with temperature.

6.91 For the reaction $A + 2B \rightarrow 3C + D$ the molar enthalpy of reaction at temperature 2 is given by

$$\Delta H_{r,2}° = H_{m,2}°(\text{products}) - H_{m,2}°(\text{reactants})$$

$$= 3H_{m,2}°(C) + H_{m,2}°(D) - H_{m,2}°(A) - 2H_{m,2}°(B)$$

$$= 3[H_{m,1}°(C) + C_{p,m}(C)(T_2 - T_1)] +$$

$$[H_{m,1}°(D) + C_{p,m}(D)(T_2 - T_1)]$$

$$- [H_{m,1}°(A) + C_{p,m}(A)(T_2 - T_1)]$$

$$- 2[H_{m,1}°(B) + C_{p,m}(B)(T_2 - T_1)]$$

$$= 3H_{m,1}°(C) + H_{m,1}°(D) - H_{m,1}(A) - 2H_{m,1}°(B)$$

$$+ [3C_{p,m}(C) + C_{p,m}(D) - C_{p,m}(A)$$

$$- 2C_{p,m}(B)](T_2 - T_1)$$

$$= \Delta H_{r,1}° + [3C_{p,m}(C) + C_{p,m}(D) - C_{p,m}(A)$$

$$- 2C_{p,m}(B)](T_2 - T_1)$$

Finally, $\Delta H_{r,2}° = \Delta H_{r,1}° + \Delta C_p(T_2 - T_1)$, which is Kirchhoff's law.

6.93 132.0 kJ

6.95 12.71 kJ·mol^{-1}

6.97 (a)

$n = \boxed{1.0}$ $T = \boxed{300}$ $V_f/V_i = \boxed{1}$ to $\boxed{10}$

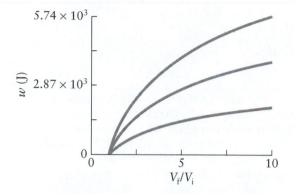

(b) more. (c) Second expansion produces less work.

6.99 3 g

6.101 (a) $C_6H_5NH_2(l) + \frac{31}{4}O_2(g) \rightarrow$

$$6\,CO_2(g) + \frac{7}{2}H_2O(l) + \frac{1}{2}N_2(g);$$

(b) $m_{CO_2} = 0.4873$ g, $m_{H_2O} = 0.1188$ g, $m_{N_2} = 0.02639$ g; (c) 0.999 atm

6.103 (a) 131.29 kJ·mol^{-1}, endothermic; (b) 623 kJ

6.105 (a) 14.7 mol; (b) -4.20×10^3 kJ

6.107 (a) 3.72 kJ; (b) -3267.5 kJ; (c) -3263.8 kJ

6.109 (a)

(b) -372 kJ. (c) -205.4 kJ·mol^{-1}. (d) The hydrogenation of benzene is much less exothermic than predicted by bond enthalpy estimations. Part of this difference can be due to the inherent inaccuracy of using average values, but the difference is so large that this cannot be the complete explanation. As may be expected, the resonance energy of benzene makes it more stable than would be expected by treating it as a set of three isolated double and three isolated single bonds. The difference in these two values $[-205\text{ kJ} - (-372\text{ kJ}) = 167\text{ kJ}]$ is a measure of how much more stable benzene is than the Kekulé structure would predict.

6.111 (a) 2326 kJ·mol^{-1}. (b) 3547 kJ. (c) more stable. (d) 20 kJ per carbon atom. (e) 25 kJ per carbon atom. (f) There is slightly less stabilization per carbon atom in C_{60} than in benzene. This fits with expectations, as the C_{60} molecule is forced by its geometry to be curved.

6.113 From the combustion equations we can see that the differences are (1) the consumption of $\frac{18}{4}$ more moles of $O_2(g)$ and

(2) the production of three more moles of $H_2O(l)$ for the combustion of aniline. Because the ΔH_f° of $O_2(g)$ is 0, the net difference will be the production of 3 more moles of $H_2O(l)$ or $3 \times (-285.83 \text{ kJ·mol}^{-1}) = -857.49 \text{ kJ}$

6.115 (a) 1.5 L; (b) $SO_2(g)$; (c) 1.1 L; (d) 40 J, on the system; (e) -3.0 kJ, heat leaves the system; (f) -2960 J

6.117 (a)

(b) CH_4: -698 kJ·mol^{-1}; CH_3OCH_3: $-1070 \text{ kJ·mol}^{-1}$; CH_3CH_2OH: -988 kJ·mol^{-1}. The burning of 1 mol CH_3OCH_3 releases the most heat. (c) Methane as it releases the most heat per gram. (d) 1.1×10^3 L. (e) methanol: -890 kJ·mol^{-1} CO_2 (less CO_2); ethanol: -684 kJ·mol^{-1} CO_2 (more CO_2); octane: -684 kJ·mol^{-1} CO_2 (more CO_2)

Chapter 7

7.1 (a) 0.341 J·K^{-1}·s^{-1}. (b) 29.5 kJ·K^{-1}·day^{-1}.

(c) Less, because in the equation $\Delta S = \dfrac{-\Delta H}{T}$, if T is larger,

ΔS is smaller.

7.3 (a) 0.22 J·K^{-1}. (b) 0.17 J·K^{-1}. (c) The entropy change is smaller at higher temperatures because the matter is already more chaotic.

7.5 (a) 6.80 J·K^{-1}; (b) 4.08 J·K^{-1}

7.7 14.8 J·K^{-1}

7.9 Let block 1 gain 1 J of energy ($q_1 = 1$ J) and block 2 lose 1 J ($q_2 = -1$ J). The total change in entropy will then be

$\Delta S = \Delta S_{\text{block 1}} + \Delta S_{\text{block 2}} = \dfrac{1\text{J}}{T_1} + \dfrac{-1\text{J}}{T_2}$, where T_1 and T_2 are the temperatures of block 1 and block 2, respectively. For the process to be spontaneous, ΔS must be greater than zero and, therefore, T_2 must be greater than T_1.

7.11 (a) -22.0 J·K^{-1}; (b) 134 J·K^{-1}

7.13 (a) 290 K; (b) 293.9 K; (c) good agreement; (d) Because ethanal does not boil at 298 K, the enthalpy and entropy of vaporization are slightly different from the values calculated at 298 K.

7.15 (a) 253 K. (b) 248 K. Trouton's rule gives a reasonable estimation of the boiling temperature.

7.17 (a) 30. kJ·mol^{-1}; (b) -11 J·K^{-1}

7.19 COF_2. COF_2 and BF_3 are both trigonal planar molecules, but the COF_2 molecules could be disordered, with the fluorine and oxygen atoms occupying different locations in the crystal. BF_3 is a symmetrical molecule and such disorder is not possible.

7.21 14.9 J·K^{-1}

7.23 (a) HBr(g); (b) $NH_3(g)$; (c) $I_2(l)$; (d) 1.0 mol Ar(g) at 1.00 atm

7.25 C(s, diamond) < $H_2O(s)$ < $H_2O(l)$ < $H_2O(g)$. Diamond, being a covalently bound monatomic solid, has less disorder than $H_2O(s)$, which is held together by weaker intermolecular forces. The three phases of water increase in entropy in changing from a solid to a liquid to a gas.

7.27 (a) Iodine, due to its larger mass and larger number of

fundamental particles. (b) 1-Pentene, due its more flexible framework. (c) Ethene, because it is a gas and for the same mass, a sample of ethene will be composed of many small molecules, whereas polyethylene will be made up of fewer but larger molecules.

7.29 Diethyl ether, $(CH_3CH_2)_2O$, has the greater standard molar entropy because its molecular structure consists of more atoms than dimethyl ether, $(CH_3)_2O$.

7.31 (a) decrease; (b) increase; (c) decrease

7.33 $\Delta S_B < \Delta S_C < \Delta S_A$. The change in entropy for container A is greater than that for container B or C due to the greater number of particles. The change in entropy in container C is greater than that of container B because of the disorder due to the vibrational motion of the molecules in container C.

7.35 (a) -163.34 J·K^{-1}; the entropy change is negative because 1 mole of a liquid is formed from 1.5 moles of gas. (b) -86.5 J·K^{-1}; the entropy change is negative because the number of moles of gas decreases. (c) 160.6 J·K^{-1}; the entropy is positive because gas is produced. (d) -36.8 J·K^{-1}; the 4 moles of solid products have less disorder (have more simple structures) than the 4 moles of reactant.

7.37 $dS = \dfrac{dq_{\text{rev}}}{T} = \dfrac{C_{p,m}\,dT}{T}$ and, therefore,

$$\Delta S = \int_{T_1}^{T_2} \frac{C_{p,m}}{T}\,dT$$

If $C_{p,m} = a + bT + c/T^2$, then

$$\Delta S = \int_{T_1}^{T_2} \frac{a + bT + c/T^2}{T}\,dT$$

$$= \int_{T_1}^{T_2} \left(\frac{a}{T} + b + \frac{c}{T^3}\right)dT$$

$$= \left(a\ln(T) + bT - \frac{c}{2T^2}\right)\Bigg|_{T_1}^{T_2}$$

$$= a\ln\left(\frac{T_2}{T_1}\right) + b(T_2 - T_1) - \frac{c}{2}\left(\frac{1}{T_2^2} - \frac{1}{T_1^2}\right)$$

$\Delta S(\text{mean}) = 3.41$ J·mol^{-1}·K^{-1}; $\Delta S(\text{true}) = 3.31$ J·mol^{-1}·K^{-1}; 3.0% error

7.39 111 J·mol^{-1}·K^{-1}

7.41 $\Delta S_{\text{cold}} = 11.9$ J·K^{-1}, $\Delta S_{\text{hot}} = -11.1$ J·K^{-1}, $\Delta S_{\text{tot}} = 0.8$ J·K^{-1}

7.43 (a) $\Delta S_{\text{sys}} = 73$ J·K^{-1}, $\Delta S_{\text{surr}} = -73$ J·K^{-1}; (b) $\Delta S_{\text{sys}} = 29.0$ J·K^{-1}, $\Delta S_{\text{surr}} = -29.0$ J·K^{-1}; (c) $\Delta S_{\text{sys}} = -29.0$ J·K^{-1}, $\Delta S_{\text{surr}} = 29.0$ J·K^{-1}

7.45 (a) $\Delta S_{\text{tot}} = 0$ J·K^{-1}, $\Delta S_{\text{surr}} = -\Delta S = -3.84$ J·K^{-1}; (b) $\Delta S_{\text{tot}} = 3.84$ J·K^{-1}, $\Delta S_{\text{surr}} = 0$, $\Delta S = 3.84$ J·K^{-1}

7.47 Spontaneous change means $\Delta S_{\text{tot}} > 0$. As a vapor condenses to a liquid, it releases heat but as the temperature of the system did not change, the heat must have left the system and entered the surroundings. Since heat flows from hot to cold, the temperature of the surroundings (T_{surr}) must be lower than the temperature of the system (T_{sys}), and since

$\Delta S_{\text{sys}} = \dfrac{q_{\text{rev}}}{T_{\text{sys}}}$ and $\Delta S_{\text{surr}} = \dfrac{q_{\text{rev}}}{T_{\text{surr}}}$, then $\Delta S_{\text{tot}} > 0$.

7.49 Exothermic reactions increase disorder by releasing heat into the surroundings. As a result, strongly exothermic reactions are spontaneous even if the change in entropy for the reacting system is itself negative.

7.51 (a) $\Delta H_r^\circ = -235.8$ kJ·mol^{-1},

$\Delta S_r° = -133.17$ J·mol⁻¹·K⁻¹, $\Delta G_r° = -196.1$ kJ·mol⁻¹;
(b) $\Delta H_r° = -11.5$ kJ·mol⁻¹, $\Delta S_r° = -149.7$ J·mol⁻¹·K⁻¹,
$\Delta G_r° = 56.1$ kJ·mol⁻¹; (c) $\Delta H_r° = -57.20$ kJ·mol⁻¹,
$\Delta S_r° = -175.83$ J·mol⁻¹·K⁻¹, $\Delta G_r° = -4.80$ kJ·mol⁻¹

7.53 (a) $\tfrac{1}{2}N_2(g) + \tfrac{3}{2}H_2(g) \rightarrow NH_3(g)$,
$\Delta H_r° = -46.11$ kJ·mol⁻¹, $\Delta S_r° = -99.38$ J·mol⁻¹·K⁻¹,
$\Delta G_r° = -16.49$ kJ·mol⁻¹; (b) $H_2(g) + \tfrac{1}{2}O_2(g) \rightarrow H_2O(g)$,
$\Delta H_r° = -241.82$ kJ·mol⁻¹, $\Delta S_r° = -44.42$ J·mol⁻¹·K⁻¹,
$\Delta G_r° = -228.58$ kJ·mol⁻¹;
(c) $C(s,graphite) + \tfrac{1}{2}O_2(g) \rightarrow CO(g)$,
$\Delta H_r° = -110.53$ kJ·mol⁻¹, $\Delta S_r° = 89.36$ J·mol⁻¹·K⁻¹,
$\Delta G_r° = -137.2$ kJ·mol⁻¹; (d) $\tfrac{1}{2}N_2(g) + O_2(g) \rightarrow NO_2(g)$,
$\Delta H_r° = 33.18$ kJ·mol⁻¹, $\Delta S_r° = -60.89$ J·mol⁻¹·K⁻¹,
$\Delta G_r° = 51.33$ kJ·mol⁻¹

7.55 (a) -141.74 kJ·mol⁻¹, spontaneous; (b) 130.4
kJ·mol⁻¹, not spontaneous;
(c) -1.05909×10^4 kJ·mol⁻¹, spontaneous

7.57 (a) and (d) are stable.

7.59 (a) $PCl_5(g)$

7.61 (a) $\Delta S_r° = 125.8$ J·mol⁻¹·K⁻¹,
$\Delta H_r° = -196.10$ kJ·mol⁻¹, $\Delta G_r° = -233.56$ kJ·mol⁻¹;
(b) $\Delta S_r° = 14.6$ J·mol⁻¹·K⁻¹, $\Delta H_r° = -748.66$ kJ·mol⁻¹,
$\Delta G_r° = -713.02$ kJ·mol⁻¹

7.63 (a) $\Delta G_r° = -98.42$ kJ·mol⁻¹, spontaneous below
612.9 K; (b) $\Delta G_r° = -283.7$ kJ·mol⁻¹, spontaneous at all
temperatures; (c) $\Delta G_r° = 3.082$ kJ·mol⁻¹, nonspontaneous at
all temperatures

7.65 Reaction A is spontaneous. Reaction B is not spontaneous at any temperature.

7.67 21.6 mol

7.69 (a) 1-propanol; (b) no

7.71 (a) $\Delta G° < 0$; (b) $\Delta H°$ unable to tell; (c) $\Delta S°$ unable to
tell; (d) $\Delta S_{total} > 0$

7.73 Higher, the cis compound will have 12 different orientations and the trans compound will have only 3 different
orientations.

7.75 (a) $W = 3$. (b) $W = 12$. (c) Initially one of the three-
atom systems had two atoms in higher energy states. In part
(b) the system will be at equilibrium when each three-atom system has one quantum of energy. Therefore energy will flow
from the system with two quanta to the system with none.

7.77 According to Trouton's rule, the entropy of vaporization of an organic liquid is a constant of approximately
85 J·mol⁻¹·K⁻¹. The relationship between entropy of fusion,
enthalpy of fusion, and melting point is given by
$$\Delta S°_{fus} = \frac{\Delta H°_{fus}}{T_{fus}}.$$ For Pb: $\Delta S_{fus}° = 8.50$ J·K⁻¹; for Hg:
$\Delta S_{fus}° = 9.79$ J·K⁻¹; for Na: $\Delta S_{fus}° = 7.12$ J·K⁻¹. These numbers are reasonably close but clearly much smaller than the
value associated with Trouton's rule.

7.79 Fe_2O_3 is thermodynamically more stable because $\Delta G_r°$
for the interconversion reaction is negative.

7.81 (a) no, not spontaneous; (b) no, not spontaneous

7.83 (a) -110.0 kJ·mol⁻¹; (b) 7.14×10^{-2} mol·L⁻¹

7.85 (a) no; (b) positive; (c) positional disorder; (d) thermal
disorder; (e) dispersal of matter

7.87 (a) method (i); (b) method (i); (c) no

7.89 The entries all correspond to aqueous ions. The fact
that they are negative is due to the reference point that has
been established. Because ions cannot actually be separated
and measured independently, a reference point that defines
$S_m°(H^+, aq) = 0$ has been established. This definition is then
used to calculate the standard entropies for the other ions. The
fact that they are negative will arise in part because the solvated ion $M(H_2O)_x^{n+}$ will be more ordered than the isolated
ion and solvent molecules (M^{n+} and xH_2O).

7.91 -40.6 J·K⁻¹

7.93 (a) 8.57 kJ at 298 K, -0.35 kJ at 373K, -6.29 kJ at
423 K. (b) 0 kJ. (c) The discrepancy arises because the enthalpy
and entropy values calculated from the tables are not rigorously constant with temperature.

7.95 (a) Since standard molar entropies increase with temperature (more translational, vibrational, and rotational
motion), the $-T\Delta S_m°$ term becomes more negative at higher
temperatures. (b) For gases with translational motion, their
increase in standard molar entropy, on heating, is much larger
than for solids and liquids. Therefore, the $-T\Delta S_m°$ terms of the
gases are more negative.

7.97 The dehydrogenation of cyclohexane to form benzene
is as follows:
$$C_6H_{12}(l) \rightarrow C_6H_6(l) + 3H_2(g) \quad \Delta G_r° = 97.6 \text{ kJ·mol}^{-1}$$
The reaction of ethane with hydrogen can be exam-
ined similarly:
$$C_2H_2(g) + H_2(g) \rightarrow C_2H_6(g)$$
$$\Delta G_r° = -100.97 \text{ kJ·mol}^{-1}$$
We can now combine these two reactions:
$$C_6H_{12}(l) \rightarrow C_6H_6(l) + 3H_2(g)$$
$$\Delta G_r° = 97.6 \text{ kJ·mol}^{-1}$$
$$+ \ 3[C_2H_2(g) + H_2(g) \rightarrow C_2H_6(g)]$$
$$\Delta G_r° = 3(-100.97 \text{ kJ·mol}^{-1})$$
$$\overline{C_6H_{12}(l) + 3 C_2H_2(g) \rightarrow C_6H_6(l) + 3 C_2H_6(g)}$$
$$\Delta G_r° = -205.13 \text{ kJ·mol}^{-1}$$
We can see that by combining these two reactions, the over-
all process becomes spontaneous.

7.99 (a)

cis-2-Butene   trans-2-Butene   2-Methylpropene

(b) cis-2-butene ⇌ trans-2-butene:
$\Delta G_r° = -2.89$ kJ·mol⁻¹, $\Delta H_r° = -4.18$ kJ·mol⁻¹,
$\Delta S_r° = -4.33$ J·mol⁻¹·K⁻¹; cis-2-butene ⇌ 2-methylpropene:
$\Delta G_r° = -7.79$ kJ·mol⁻¹, $\Delta H_r° = -9.91$ kJ·mol⁻¹,
$\Delta S_r° = -7.11$ J·mol⁻¹·K⁻¹; trans-2-butene ⇌ 2-methylpropene:
$\Delta G_r° = -4.90$ kJ·mol⁻¹, $\Delta H_r° = -5.73$ kJ·mol⁻¹,
$\Delta S_r° = -2.78$ J·mol⁻¹·K⁻¹; (c) The most stable is
2-methylpropene. (d) $S_m°(cis$-2-butene$) >$
$S_m°(trans$-2-butene$) > S_m°(2$-methylpropene$)$

7.101 (a) $2 NaN_3(s) \rightarrow 3 N_2(g) + 2 Na(g)$;
(b) $\Delta S_r°$ is positive because a gas is produced from a solid.
(c) $Na^+\overline{N}_3 = \overset{+}{N} = \overset{0}{\overline{N}}, \overset{0}{N} \equiv N$, nitrogen is oxidized.
(d) 688.4 J·mol⁻¹·K⁻¹; (e) -3.39×10^4 J·mol⁻¹; (f) yes;
(g) yes, lowered

Chapter 8

8.1 0.017 g

8.3 (a) 87°C; (b) 113°C

8.5 (a) 28.3 kJ·mol⁻¹; (b) 91.3 J·K⁻¹·mol⁻¹;
(c) 1.1 kJ·mol⁻¹; (d) 309 K

8.7 (a) 28.6 kJ·mol⁻¹; (b) 90.6 J·K⁻¹mol⁻¹; (c) 0.53 atm

8.9   0.19 atm

8.11   (a) vapor; (b) liquid; (c) vapor; (d) vapor

8.13   (a) 2.4 K; (b) about 10 atm; (c) 5.5 K; (d) no

8.15   (a) At the lower-pressure triple point, liquid helium-I and -II are in equilibrium with helium gas; at the higher-pressure triple point, liquid helium-I and -II are in equilibrium with solid helium. (b) helium-I

8.17   It would become a solid.

8.19   (a) water; (b) benzene; (c) water

8.21   (a) hydrophilic; (b) hydrophobic; (c) hydrophobic; (d) hydrophilic

8.23   (a) 6.4×10^{-4} mol·L^{-1}; (b) 1.5×10^{-3} mol·L^{-1}; (c) 2.3×10^{-3} mol·L^{-1}

8.25   (a) 4 mg·L^{-1} or 4 ppm; (b) 0.1 atm; (c) 0.5 atm

8.27   (a) By Henry's law, the concentration of CO_2 in solution will double. (b) No change in the equilibrium will occur; the partial pressure of CO_2 is unchanged and the concentration is unchanged.

8.29   1.5 g

8.31   (a) negative; (b) $Li_2SO_4(s) \rightarrow 2\ Li^+(aq) + SO_4^{2-}(aq)$ + heat; (c) enthalpy of hydration

8.33   (a) 6.7×10^2 J; (b) -5.0×10^2 J; (c) -24.7 kJ; (d) 8.2×10^2 J

8.35   (a) 0.856 *m*; (b) 2.5 g; (c) 0.0571 *m*

8.37   (a) 0.248 *m*; (b) 0.246 mol·L^{-1}

8.39   (a) 1.35 *m*; (b) 0.519 *m*; (c) 28.43 *m*

8.41   (a) 13.9 g; (b) 29 g

8.43   The free energy of the solvent in the NaCl solution will always be lower than the free energy of pure water; if given enough time, all of the water from the "pure water" beaker will become part of the NaCl solution, leaving an empty beaker. An equilibrium will be established between the solvent and the water vapor.

8.45   (a) 684 Torr; (b) 758 Torr

8.47 7.80×10^{-3} mol

8.49   (a) 0.052; (b) 115 g·mol^{-1}

8.51   (a) 100.34°C; (b) 81.0°C

8.53 1.6×10^2 g·mol^{-1}

8.55   (a) 1.84; (b) 0.318 mol·kg^{-1}; (c) 83.8%

8.57   Compound B has the larger molar mass. The compound that freezes at the lower temperature (Compound A) will have the larger concentration of solute particles, assuming $i = 1$ for each solute. As the same mass of each solute was dissolved in the same mass of solvent, the compound with the smaller molar mass will have the larger number of moles and the larger concentration of solute particles (Compound A).

8.59 -0.20°C

8.61   (a) 0.24 atm; (b) 48 atm; (c) 0.72 atm

8.63 2.0×10^3 g·mol^{-1}

8.65 5.8×10^3 g·mol^{-1}

8.67   (a) 1.2 atm; (b) 0.048 atm; (c) 8.3×10^{-5} atm

8.69 2.5×10^5 g·mol^{-1}

8.71   (a) $P = 78.2$ Torr, $x_{\text{benzene}} = 0.91$, $x_{\text{toluene}} = 0.09$; (b) $P = 43.0$ Torr, $x_{\text{benzene}} = 0.469$, $x_{\text{toluene}} = 0.531$

8.73   63 g

8.75   Raoult's law applies to the vapor pressure of the mixture, so positive deviation means that the vapor pressure is higher than expected for an ideal solution. Negative deviation means that the vapor pressure is lower than expected for an ideal solution. Negative deviation will occur when the interactions between the different molecules are somewhat stronger than the interactions between molecules of the same

kind. (a) For methanol and ethanol, we expect the types of intermolecular attractions in the mixture to be similar to those in the component liquids, so that an ideal solution is predicted.
(b) For HF and H_2O, the possibility of intermolecular hydrogen bonding between water and HF would suggest that negative deviation would be observed, which is the case. HF and H_2O form an azeotrope that boils at 111°C, a temperature higher than the boiling point of either HF (19.4°C) or water.
(c) Because hexane is nonpolar and water is polar with hydrogen bonding, we would expect a mixture of these two to exhibit positive deviation (the interactions between the different molecules would be weaker than the intermolecular forces between like molecules). Hexane and water do form an azeotrope that boils at 61.6°C, a temperature below the boiling point of either hexane or water.

8.77   A foam is a colloid formed by suspending a gas in a liquid or a solid matrix, while a sol is a suspension of a solid in a liquid. Some examples of foams are styrofoam and soapsuds; examples of sols are muddy water and mayonnaise.

8.79   Colloids will reflect or scatter light while true solutions do not; this is known as the Tyndall effect.

8.81   Water is a polar molecule and as a result will orient itself differently around cations and anions. It will align its dipole to present the most favorable interaction possible; hydrogen atoms will be closer to an anion whereas oxygen atoms will be closer to a cation.

8.83   (a) stronger; (b) low; (c) high; (d) weaker; (e) weak, low; (f) low; (g) strong, high

8.85   (a, b) Viscosity and surface tension decrease with increasing temperature; at high temperatures the molecules readily move away from their neighbors because of increased kinetic energy. (c, d) Evaporation rate and vapor pressure increase with increasing temperature because the kinetic energy of the molecules increases with temperature, and the molecules are more likely to escape into the gas phase.

8.87   (a) Butane will dissolve in tetrachloromethane.

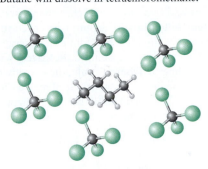

(b) Calcium chloride will dissolve in water.

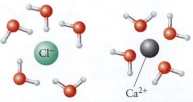

8.89   (a) 78.5%. (b) Some of the water vapor in the air would condense as dew or fog.

8.91   (a) The apparent mass$_{\text{solute}}$ is greater than the true mass$_{\text{solute}}$, so M_{solute} will appear greater than the actual M_{solute}. Also, the ΔT measured will be smaller because less solute will actually be dissolved. This has the same effect as increasing the apparent M_{solute}. (b) Because the true $mass_{solvent} = d \times V$, if $d_{solvent}$ is less than 1.00 g·cm^{-3}, then the true $mass_{solvent}$ will be less than the assumed mass. M_{solute} is inversely proportional to $mass_{solvent}$, so an artificially high $mass_{solvent}$ will lead to an artificially low M_{solute}. (c) If the true freezing point is higher than the recorded freezing point, the assumed ΔT is greater than the true ΔT and M_{solute} appears smaller than the actual M_{solute} because ΔT is in the denominator. (d) If not all the solute dissolved, the assumed $mass_{solute}$ is greater than the true $mass_{solute}$, and M_{solute} appears greater than the actual M_{solute} because $mass_{solute}$ occurs in the numerator.

8.93 The water Coleridge referred to was seawater. The boards shrank due to osmosis (a net movement of water from the cells of the wood to the saline water). The same happens to the cells of your body when you drink seawater, which causes serious health problems and even death.

8.95 (a) 4.8×10^3 g·mol^{-1}; (b) -3.9×10^{-4}°C; (c) osmotic pressure, because the change in freezing point is too small to measure accurately

8.97 76 g·mol^{-1}

8.99 (a)

$$\ln P = -\frac{38\,200 \text{ J·mol}^{-1}}{8.314 \text{ J · K}^{-1}\text{·mol}^{-1}} \cdot \frac{1}{T} + \frac{113.0 \text{ J·K}^{-1}\text{·mol}^{-1}}{8.314 \text{ J·K}^{-1}\text{·mol}^{-1}}$$

$$= -\frac{4595 \text{ K}}{T} + 13.59$$

(b) The relation to plot is $\ln P$ versus $1/T$, which should result in a straight line whose slope is $-\Delta H_{vap}°/R$ and whose intercept is $\Delta S_{vap}°/R$. (c) 30. Torr; (d) 338.1 K

8.101 (a) 56.9 g·mol^{-1}; this experimental molar mass of acetic acid is less than the calculated molar mass (60.0 g·mol^{-1}) because the van 't Hoff i factor is greater than 1 owing to the partial dissociation of acetic acid in aqueous solution. (b) 116 g·mol^{-1}; this experimental molar mass is significantly higher than 60.0 g·mol^{-1} because i is less than 1, which indicates that acetic acid is incompletely dissolved in the benzene or that the acetic acid molecules are dimerizing in solution.

8.103 (a) $y = -3358.714x + 12.247$; (b) 28 kJ·mol^{-1}; (c) 1.0×10^2 J·K^{-1}·mol^{-1}; (d) 2.7×10^2 K; (e) 2.1×10^2 K

8.105 The critical temperatures are: CH_4, -82.1°C; C_2H_6, 32.2°C; C_3H_8, 96.8°C; C_4H_{10}, 152°C. The critical temperatures increase with increasing mass, showing the influence of the stronger London forces.

8.107 (a) 195 Torr in flask A, 222 Torr in flask B. (b) $x_{acetone}$, gas $= 0.70$, $x_{chloroform}$, gas $= 0.30$, $x_{acetone} = 0.67$, $x_{chloroform} = 0.33$. (c) Negative deviation from Raoult's law means that solute and solvent molecules attract each other slightly more than molecules of the same kind. This results in a vapor pressure that is lower than expected from the ideal calculation and gives rise to a high-boiling azeotrope. The gas phase composition will be slightly different from that calculated from the ideal state, but whether acetone or chloroform would be richer in the gas phase depends on which side of the azeotrope the composition lies. Because we are not given the composition of the azeotrope, we cannot state which way the values will vary.

8.109 The vapor pressure is more sensitive if ΔH_{vap} is small. The fact that ΔH_{vap} is small indicates that it takes little energy to volatilize the sample, which means that the intermolecular forces are weaker. Hence we expect the vapor pressure to be more dramatically affected by small changes in temperature.

8.111 0.32 mg A

8.113 27.3 g

8.115 18.5 L

8.117 The nonpolar chains of both the surfactant and pentanol will interact to form a hydrophobic region with the heads of the two molecules pointing away from this region toward the aqueous solution. To prevent the heads of the shorter pentanol molecules from winding up in the hydrophobic region, the layered structure might be comprised of a water region, a surfactant layer (heads pointing toward the water), a pentanol layer (with tails pointing toward the hydrophobic tails of the surfactant), and back to an aqueous region.

8.119 (a) The hybridization of the carbon atoms in citric acid are:

(b) yes; (c) solid, soluble; (d) 0.3 mol·L^{-1}; (e) 2×10^1 g; (f) 1 g

Chapter 9

9.1 (a) False, at equilibrium the concentrations of reactants and products will not change, but the reaction will continue to proceed in both directions. (b) False, equilibrium reactions are affected by the presence of both products and reactants. (c) False, the value of the equilibrium constant is not affected by the amounts of reactants or products added as long as the temperature is constant. (d) True.

9.3

9.5 (a) Flask 3; (b) 54.5%; (c) 0.16

9.7 (a) $K_c = \dfrac{[COCl][Cl]}{[CO][Cl_2]}$; (b) $K_c = \dfrac{[HBr]^2}{[H_2][Br_2]}$;

(c) $K_c = \dfrac{[SO_2]^2[H_2O]^2}{[H_2S]^2[O_2]^3}$

9.9 (a) Because the volume is the same, the number of moles of O_2 is larger in the second experiment. (b) Because K_c is a constant and the denominator is larger in the second case, the numerator must also be larger; so the concentration of O_2 is larger in the second case. (c) Although $[O_2]^3/[O_3]^2$ is the same, $[O_2]/[O_3]$ will be different, a result seen by solving for K_C in each case. (d) Because K_C is a constant, $[O_2]^3/[O_3]^2$ is the same. (e) Because $[O_2]^3/[O_3]^2$ is the same, its reciprocal must be the same.

9.11 48.8 for condition 1, 48.9 for conditions 2 and 3

9.13 (a) $\dfrac{1}{P_{O_2}^3}$; (b) $P_{H_2O}^7$; (c) $\dfrac{P_{NO}P_{NO_2}}{P_{N_2O_3}}$

9.15 (a) 1×10^{80}; (b) 1×10^{90}; (c) 1×10^{-23}

9.17 (a) $-19\ kJ\cdot mol^{-1}$; (b) $68\ kJ\cdot mol^{-1}$

9.19 form reactants

9.21 (a) $8.3 \times 10^{-1}\ kJ\cdot mol^{-1}$. (b) Because ΔG_r is positive, the reaction is spontaneous toward reactants.

9.23 (a) $-27\ kJ\cdot mol^{-1}$; (b) form products

9.25 (a) 4.3×10^{-4}; (b) 1.87

9.27 (a) 0.024; (b) 6.4; (c) 1.7×10^3

9.29 1.5×10^{34}

9.31 $2.1 \times 10^{-5}\ mol\cdot L^{-1}$

9.33 5.4 bar

9.35 (a) 0.50; (b) no; (c) form products

9.37 (a) 6.9; (b) yes

9.39 6.6×10^{-3}

9.41 1.58×10^{-8}

9.43 (c) $K = x^2/(1.0 - 2x)^2$

9.45 (a) $[Cl] = 1.1 \times 10^{-5}\ mol\cdot L^{-1}$, $[Cl_2] = 0.0010\ mol\cdot L^{-1}$; (b) $[F] = 3.2 \times 10^{-4}\ mol\cdot L^{-1}$, $[F_2] = 8 \times 10^{-4}\ mol\cdot L^{-1}$; (c) chlorine

9.47 $P_{Br_2} = P_{H_2} = 1.1 \times 10^{-5}\ mbar$, $P_{HBr} = 1.2\ mbar$

9.49 (a) $[PCl_3] = [Cl_2] = 0.010\ mol\cdot L^{-1}$, $[PCl_5] = 0.009\ mol\cdot L^{-1}$; (b) 53%

9.51 $[NH_3] = 0.200\ mol\cdot L^{-1}$, $[H_2S] = 8 \times 10^{-4}\ mol\cdot L^{-1}$

9.53 $[NO] = 3.6 \times 10^{-4}\ mol\cdot L^{-1}$, $[N_2]$ and $[O_2]$ remain essentially unchanged at $0.114\ mol\cdot L^{-1}$

9.55 1.1

9.57 $[CO_2] = 8.6 \times 10^{-5}\ mol\cdot L^{-1}$, $[CO] = 4.9 \times 10^{-3}\ mol\cdot L^{-1}$, $[O_2] = 4.6 \times 10^{-4}\ mol\cdot L^{-1}$

9.59 $1.4\ mol\cdot L^{-1}$

9.61 3.88

9.63 $[SO_2] = 0.0011\ mol\cdot L^{-1}$, $[NO_2] = 0.0211\ mol\cdot L^{-1}$, $[NO] = 0.0389\ mol\cdot L^{-1}$, $[SO_3] = 0.0489\ mol\cdot L^{-1}$

9.65 (a) no; (b) form products; (c) $[PCl_5] = 3.07\ mol\cdot L^{-1}$, $[PCl_3] = 5.93\ mol\cdot L^{-1}$; $[Cl_2] = 0.93\ mol\cdot L^{-1}$

9.67 $P_{HCl} = 0.22\ bar, P_{H_2} = P_{Cl_2} = 3.9 \times 10^{-18}\ bar$

9.69 (a) decrease; (b) $[CO] = 0.156\ mol\cdot L^{-1}$, $[H_2] = 0.155\ mol\cdot L^{-1}$, $[CH_3OH] = 2.6 \times 10^{-4}\ mol\cdot L^{-1}$

9.71 $9.7 \times 10^{33}\ L$

9.73 (a) decreases; (b) decreases; (c) increases; (d) no change

9.75 (a) decreases; (b) increases; (c) increases; (d) increases; (e) no change; (f) decreases; (g) decreases

9.77 (a) reactants; (b) reactants; (c) reactants; (d) no change; (e) reactants

9.79 (a) increases; (b) no

9.81 (a) products; (b) products; (c) reactants; (d) reactants

9.83 No, less ammonia will be formed at 700. K.

9.85 No, the reaction proceeds to form more products

9.87 (a) 1×10^{-16} at 25°C, 1×10^{-7} at 150.°C; (b) 7.8×10^{-2} at 25°C, 0.22 at 150.°C

9.89 $\ln\left(\dfrac{K_{c2}}{K_{c1}}\right) = -\dfrac{\Delta H_r^\circ}{R}\left(\dfrac{1}{T_1} - \dfrac{1}{T_2}\right) - \Delta n \ln\left(\dfrac{T_2}{T_1}\right)$

9.91 Increasing the temperature (a) would increase the formation of X(g) because the reaction is endothermic (bond breaking).

9.93 (a) reactants; (b) no effect; (c) products; (d) products; (e) products; (f) no effect for a dilute glucose solution; (g) reactants

9.95 (a) $2\ A(g) \rightarrow B(g) + 2\ C(g)$; (b) 1.54×10^{-2}; (c) 6.21×10^{-4}

9.97 (a) 2.9; (b) $\Delta G_{r(3)}^\circ = -10\ kJ\cdot mol^{-1}$, $K_3 = 2$

9.99 $\alpha = \sqrt{K/(P + K)}$; (a) 0.953; (b) 0.912

9.101 (a) 0; (b) 978 K; (c) 7.50 bar; (d) $P_{CO(g)} = 7.04\ bar$, $P_{H_2O(g)} = 5.04\ bar$, $P_{CO_2(g)} = 8.96\ bar$, $P_{H_2(g)} = 3.96\ bar$

9.103 (a) F(g), $19.1\ kJ\cdot mol^{-1}$; Cl(g), $47.9\ kJ\cdot mol^{-1}$; Br(g), $42.8\ kJ\cdot mol^{-1}$; I(g), $5.6\ kJ\cdot mol^{-1}$. (b) There is a correlation between the bond dissociation energy and the free energy of formation of the atomic species, but the relationship is clearly not linear. For the heavier three halogens, there is a trend to decreasing free energy of formation of the atoms as the element becomes heavier, but fluorine is anomalous. The F—F bond energy is lower than expected, owing to repulsions of the lone pairs of electrons on the adjacent F atoms because the F—F bond distance is so short.

9.105 (a) 4.5×10^{-29}; (b) 0.285 bar or 0.289 atm; (c) 3.6×10^{-15} bar or 3.6×10^{-15} atm; (d) 846 mL

9.107 $P_{N_2} = 4.61\ bar$, $P_{H_2} = 1.44\ bar$, $P_{NH_3} = 23.85\ bar$

9.109 $6.8\ kJ\cdot mol^{-1}$

9.111 Reactions with larger values of ΔG_r° have equilibrium constants that are more sensitive to temperature changes.

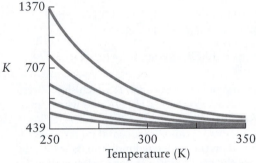

9.113 (a) 11.2. (b) If NO_2 is added, the equilibrium will shift to produce more N_2O_4. The amount of NO_2 will be greater than initially present, but less than the $3.13\ mol\cdot L^{-1}$ present immediately upon making the addition. K_c will not be affected. (c) $[N_2O_4] = 0.64\ mol\cdot L^{-1}$, $[NO_2] = 2.67\ mol\cdot L^{-1}$

9.115 H_2O, 24 Torr; D_2O, 21 Torr. Because D is heavier than H, it has a lower zero-point vibrational energy; so the O—D···O bond is stronger than the O—H···O bond, and hence the liquid is less volatile.

9.117 (a) $1.70\ kJ\cdot mol^{-1}$; (b) $-137.15\ kJ\cdot mol^{-1}$; (c) $124.7\ kJ\cdot mol^{-1}$; (d) $-22.2\ kJ\cdot mol^{-1}$

9.119 (a) (i) increases, (ii) increases, (iii) no change; (b) 0.242 mol

Chapter 10

10.1 (a) $CH_3NH_3^+$; (b) $CH_3NH_3^+$; (c) H_2CO_3; (d) CO_3^{2-}; (e) $C_6H_5O^-$; (f) $CH_3CO_2^-$

10.3 For all parts, H_2O and H_3O^+ form an acid–base pair in which H_2O is the conjugate base and H_3O^+ is the acid.
(a) $H_2SO_4(aq) + H_2O(l) \rightleftharpoons H_3O^+(aq) + HSO_4^-(aq)$, $H_2SO_4(aq)$ (acid) and $HSO_4^-(aq)$ (base);
(b) $C_6H_5NH_3^+(aq) + H_2O(l) \rightleftharpoons H_3O^+(aq) + C_6H_5NH_2(aq)$, $C_6H_5NH_3^+(aq)$ (acid) and $C_6H_5NH_2(aq)$ (base);
(c) $H_2PO_4^-(aq) + H_2O(l) \rightleftharpoons H_3O^+(aq) + HPO_4^{2-}(aq)$, $H_2PO_4^-(aq)$ (acid) and $HPO_4^{2-}(aq)$ (base);
(d) $HCOOH(aq) + H_2O(l) \rightleftharpoons H_3O^+(aq) + HCO_2^-(aq)$, $HCOOH(aq)$ (acid) and $HCO_2^-(aq)$ (base);
(e) $NH_2NH_3^+(aq) + H_2O(l) \rightleftharpoons H_3O^+(aq) + NH_2NH_2(aq)$, $NH_2NH_3^+(aq)$ (acid) and $NH_2NH_2(aq)$ (base)

10.5 (a) Brønsted acid: HNO_3, Brønsted base: HPO_4^{2-}; (b) conjugate base of HNO_3 is NO_3^-, conjugate acid of HPO_4^{2-} is $H_2PO_4^-$

10.7 (a) $HClO_3$, chloric acid, conjugate base is ClO_3^-

(b) HNO_2, nitrous acid, conjugate base: NO_2^-

10.9 (a) yes, NH_4^+ (acid), H_2O (base); (b) yes, NH_4^+ (acid), I^- (base); (c) no; (d) yes, NH_4^+ (acid), NH_2^- (base)

10.11 (a) HCO_3^- as an acid: $HCO_3^-(aq) + H_2O(l) \rightleftharpoons H_3O^+(aq) + CO_3^{2-}(aq)$, HCO_3^- (acid) and CO_3^{2-} (base), H_2O (base) and H_3O^+ (acid);
HCO_3^- as a base: $H_2O(l) + HCO_3^-(aq) \rightleftharpoons H_2CO_3(aq) + OH^-(aq)$, HCO_3^- (base) and H_2CO_3 (acid), H_2O (acid) and OH^- (base).
(b) HPO_4^{2-} as an acid: $HPO_4^{2-}(aq) + H_2O(l) \rightleftharpoons H_3O^+(aq) + PO_4^{3-}(aq)$, HPO_4^{2-} (acid) and PO_4^{3-} (base), H_2O (base) and H_3O^+ (acid);
HPO_4^{2-} as a base: $HPO_4^{2-}(aq) + H_2O(l) \rightleftharpoons H_2PO_4^-(aq) + OH^-(aq)$, HPO_4^{2-} (base) and $H_2PO_4^-$ (acid), H_2O (acid) and OH^- (base).

10.13 (a) Two protons can be accepted (one on each N);
(b)

(c) Each of the two nitro- groups will show amphiprotic behavior in aqueous solution (can either accept a proton and donate a proton).

10.15 The Lewis structures of (a) to (e) are as follows:

(a) Lewis base (b) Lewis acid (c) Lewis acid (d) Lewis base (e) Lewis base

10.17 (a)

Lewis acid   Lewis base   product

(b)

Lewis acid   Lewis base   product

10.19 (a) basic; (b) acidic; (c) amphoteric; (d) basic
10.21 (a) 5.0×10^{-13} mol·L^{-1}; (b) 1.0×10^{-9} mol·L^{-1}; (c) 3.2×10^{-12} mol·L^{-1}
10.23 (a) 1.4×10^{-7} mol·L^{-1}, 6.80; (b) 1.4×10^{-7} mol·L^{-1}
10.25 1.5×10^{-2} mol·L^{-1} Ba^{2+}, 2.9×10^{-2} mol·L^{-1} OH^-, 3.4×10^{-13} mol·L^{-1} H_3O^+
10.27 (a) 5×10^{-4} mol·L^{-1}; (b) 2×10^{-7} mol·L^{-1}; (c) 4×10^{-5} mol·L^{-1}; (d) 5×10^{-6} mol·L^{-1}
10.29 (a) 0.18 mol·L^{-1}; (b) 18 mol·L^{-1}; (c) 1.1×10^2 g
10.31 (a) pH = 1.84, pOH = 12.16; (b) pH = 0.96, pOH = 13.04; (c) pH = 12.26, pOH = 1.74; (d) pH = 10.85, pOH = 3.15; (e) pH = 10.99, pOH = 3.01; (f) pH = 4.28, pOH = 9.72
10.33 (a) 7.6×10^{-3}; (b) 1.0×10^{-2}; (c) 3.5×10^{-3}; (d) 1.2×10^{-3}; (e) $H_2SeO_3 < H_3PO_4 < H_3PO_3 < HSeO_4^-$
10.35 (a) $HClO_2(aq) + H_2O(l) \rightleftharpoons H_3O^+(aq) + ClO_2^-(aq)$

$$K_a = \frac{[H_3O^+][ClO_2^-]}{[HClO_2]}$$

$$ClO_2^-(aq) + H_2O(l) \rightleftharpoons HClO_2(aq) + OH^-(aq)$$

$$K_b = \frac{[HClO_2][OH^-]}{[ClO_2^-]}$$

(b) $HCN(aq) + H_2O(l) \rightleftharpoons H_3O^+(aq) + CN^-(aq)$

$$K_a = \frac{[H_3O^+][CN^-]}{[HCN]}$$

$$CN^-(aq) + H_2O(l) \rightleftharpoons HCN(aq) + OH^-(aq)$$

$$K_b = \frac{[HCN][OH^-]}{[CN^-]}$$

(c) $C_6H_5OH(aq) + H_2O(l) \rightleftharpoons H_3O^+(aq) + C_6H_5O^-(aq)$

$$K_a = \frac{[H_3O^+][C_6H_5O^-]}{[C_6H_5OH]}$$

$$C_6H_5O^-(aq) + H_2O(l) \rightleftharpoons C_6H_5OH(aq) + OH^-(aq)$$

$$K_b = \frac{[C_6H_5OH][OH^-]}{[C_6H_5O^-]}$$

10.37 $(CH_3)_2NH_2^+$ ($pK_a = 10.73$) < $^+NH_3OH$ (6.03) < HNO_2 (3.37) < $HClO_2$ (2.00)
10.39 F^- ($pK_b = 10.55$) < CH_3CO^{2-} (9.25) < C_5H_5N (8.75) < NH_3 (4.75)
10.41 (a) strong; (b) weak; (c) weak; (d) weak; (e) weak; (f) strong; (g) weak
10.43 For oxoacids, the greater the number of highly electronegative O atoms attached to the central atom, the stronger the acid. This effect is related to the increased oxidation number of the central atom as the number of O atoms increases. Therefore, HIO_3 is the stronger acid, with the lower pK_a.
10.45 (a) HCl is the stronger acid because its bond strength is much weaker than the bond in HF. (b) $HClO_2$ is stronger because one more oxygen is attached to the Cl atom, which helps pull electrons out of the H—O bond. (c) $HClO_2$ is stronger because Cl is more electronegative than Br. (d) $HClO_4$ is stronger because Cl is more electronegative than P. (e) HNO_3 is stronger for the same reason as given in part (b). (f) H_2CO_3 is stronger because C is more electronegative than Ge.
10.47 (a) The —CCl_3 group is more electron withdrawing than the —CH_3 group, thus making trichloroacetic acid the stronger acid. (b) The —CH_3 group has electron-donating properties, which means that it is less electron withdrawing than the —H attached to the carboxyl group in formic acid. Thus, formic acid is slightly stronger than acetic acid.

10.49   The larger the K_a, the stronger the corresponding acid. 2,4,6-Trichlorophenol is the stronger acid because the chlorine atoms have a greater electron-withdrawing power than the hydrogen atoms present in the unsubstituted phenol.

10.51   (d) < (a) < (b) < (c). Amines in which the nitrogen atom of the amine group is attached to an aromatic ring such as benzene appear to be weaker than ammonia, which is weaker than alkylamines.

10.53   (a) pH = 2.64, pOH = 11.36. (b) pH = 0.74, pOH = 13.26. (c) pH = 2.14, pOH = 11.86. (d) Acidity increases when the hydrogen atoms in the methyl group of acetic acid are replaced by atoms that have a higher electronegativity, such as chlorine.

10.55   (a) pOH = 3.00, pH = 11.00, 1.8%; (b) pOH = 4.38, pH = 9.62, 0.026%; (c) pOH = 2.32, pH = 11.68, 1.4%; (d) pOH = 3.96, pH = 10.04, 2.5%

10.57   (a) K_a = 0.09, pK_a = 1.0; (b) K_b = 5.6×10^{-4}, pK_b = 3.25

10.59   (a) 0.021 mol·L^{-1}; (b) 0.015 mol·L^{-1}

10.61   pH = 2.58, K_a = 6.5×10^{-5}

10.63   pH = 11.83, K_b = 4.8×10^{-4}

10.65 7.7×10^{-9}

10.67   (a) less than 7, $NH_4^+(aq) + H_2O(l) \rightleftharpoons H_3O^+(aq) + NH_3(aq)$; (b) greater than 7, $H_2O(l) + CO_3^{2-}(aq) \rightleftharpoons HCO_3^-(aq) + OH^-(aq)$; (c) greater than 7, $H_2O(l) + F^-(aq) \rightleftharpoons HF(aq) + OH^-(aq)$; (d) neutral; (e) less than 7, $Al(H_2O)_6^{3+}(aq) + H_2O(l) \rightleftharpoons H_3O^+(aq) + Al(H_2O)_5OH^{2+}(aq)$; (f) less than 7, $Cu(H_2O)_6^{2+}(aq) + H_2O(l) \rightleftharpoons H_3O^+(aq) + Cu(H_2O)_5OH^+(aq)$

10.69   (a) 9.28; (b) 5.00; (c) 3.06; (d) 11.04

10.71   5.42

10.73   HBrO

10.75   (a) 1.8×10^{-6} mol·L^{-1}; (b) 5.26

10.77   6.113

10.79   (a) $H_2SO_4(aq) + H_2O(l) \rightleftharpoons H_3O^+(aq) + HSO_4^-(aq)$ $HSO_4^-(aq) + H_2O(l) \rightleftharpoons H_3O^+(aq) + SO_4^{2-}(aq)$; (b) $H_3AsO_4(aq) + H_2O(l) \rightleftharpoons H_3O^+(aq) + H_2AsO_4^-(aq)$, $H_2AsO_4^-(aq) + H_2O(l) \rightleftharpoons H_3O^+(aq) + HAsO_4^{2-}(aq)$, $HAsO_4^{2-}(aq) + H_2O(l) \rightleftharpoons H_3O^+(aq) + AsO_4^{3-}(aq)$; (c) $C_6H_4(COOH)_2(aq) + H_2O(l) \rightleftharpoons H_3O^+(aq) + C_6H_4(COOH)CO_2^-(aq)$, $C_6H_4(COOH)CO_2^-(aq) + H_2O(l) \rightleftharpoons H_3O^+(aq) + C_6H_4(CO_2)_2^{2-}(aq)$

10.81   0.80

10.83   (a) 4.18; (b) 1.28; (c) 3.80

10.85   (a) 4.37; (b) 4.37

10.87   (a) 4.55; (b) 6.17

10.89   [H_2CO_3] = 0.0455 mol·L^{-1}; [H_3O^+] = [HCO_3^-] = 1.4×10^{-4} mol·L^{-1}; [CO_3^{2-}] = 5.6×10^{-11} mol·L^{-1}; [OH^-] = 7.1×10^{-11} mol·L^{-1}

10.91   [H_2CO_3] = 2.3×10^{-8} mol·L^{-1}; [OH^-] = [HCO_3^-] = 0.0028 mol·L^{-1}; [CO_3^{2-}] = 0.0428 mol·L^{-1}; [H_3O^+] = 3.6×10^{-12} mol·L^{-1}

10.93   (a) HA$^-$; (b) A^{2-}; (c) H$_2$A

10.95   [HSO_3^-] = 0.14 mol·L^{-1}; [H_2SO_3] = 3.2×10^{-5} mol·L^{-1}; [SO_3^{2-}] = 0.0054 mol·L^{-1}

10.97   (a) 6.54; (b) 2.12; (c) 1.49

10.99   [H_3PO_4] = 6.4×10^{-3} mol·L^{-1}, [$H_2PO_4^-$] = 8.6×10^{-3} mol·L^{-1}, [HPO_4^{2-}] = 9.5×10^{-8} mol·L^{-1}, [PO_4^{3-}] = 3.5×10^{-18} mol·L^{-1}

10.101   6.174

10.103   7.205

10.105   (a) 6.35 for 1.00×10^{-4} mol·L^{-1}, 7.35 for 1.00×10^{-6} mol·L^{-1}; (b) 6.34 for 1.00×10^{-4} mol·L^{-1}, 6.96 for 1.00×10^{-6} mol·L^{-1}

10.107   (a) 6.69 for 8.50×10^{-5} mol·L^{-1}, 7.22 for 7.37×10^{-6} mol·L^{-1}; (b) 6.64 for 8.50×10^{-5} mol·L^{-1}, 6.92 for 7.37×10^{-6} mol·L^{-1}

10.109   (a) Acid (1) is a strong acid because it totally dissociates. (b) Acid (3) has the strongest conjugate base because it is the weakest acid. (c) Acid (3) has the largest pK_a because it is the weakest acid (smallest K_a results in largest pK_a).

10.111   The apparent motion of hydronium and hydroxide ions is not as dependent on the diffusion of individual ions as is that of other ions; they are formed and reformed as protons are transferred from and to water molecules in solution. Autoprotolysis allows rapid proton transfer between water molecules.

10.113   (a) NaC$_2$HO$_4$; (b) [structure]

(c) sodium oxalate, amphiprotic, 2.71

10.115   The Lewis structure of boric acid is:

boric acid   conjugate base

(a) no. (b) Boric acid acts as a Lewis acid because it accepts electron pairs (OH$^-$) from water. Its acidity is not due to its dissociation.

10.117   (a) Nitrous acid will act like a strong acid because its conjugate base, the nitrite ion, is a weaker base than the acetate ion. The presence of acetate ions will shift the proton transfer equilibrium of nitrous acid toward the products (NO$_2^-$ and H_3O^+) by consuming H_3O^+, which will increase the apparent K_a of nitrous acid. Carbonic acid is a weaker acid than acetic acid, so no shift in equilibrium occurs. (b) Ammonia will act like a strong base because its conjugate acid, the ammonium ion, is a weaker acid than acetic acid. The presence of acetic acid will shift the proton transfer equilibrium of ammonia with water toward the products (NH$_4^+$ and OH$^-$) by consuming OH$^-$, which will increase the apparent K_b of ammonia.

10.119 6.9×10^{-4}

10.121   (a) $D_2O + D_2O \rightleftharpoons D_3O^+ + OD^-$; (b) 14.870; (c) 3.67×10^{-8} mol·L^{-1}; (d) pD = pOD = 7.435; (e) pD + pOD = pK_{D_2O} = 14.870

10.123   2.9% deprotonated; −2.11°C

10.125 1.3×10^2 J·mol^{-1}

10.127   (a) 7.40; (b) decrease

10.129 (a) and (b) Buffer regions are marked A, B, and C. (c) Region A: H_3PO_4 and $H_2PO_4^-$, Region B: $H_2PO_4^-$ and HPO_4^{2-}, Region C: HPO_4^{2-} and PO_4^{3-}

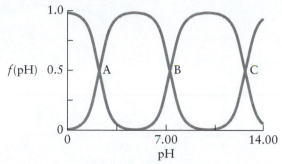

(d) plot is similar to (a). (e) The major species present are similar for both H_3PO_4 and H_3AsO_4: $H_2EO_4^-$ and HEO_4^{2-} where E = P or As. For As, there is more $HAsO_4^{2-}$ than $H_2AsO_4^-$, with a ratio of approximately 0.63 to 0.37, or 1.7:1. For P, the situation is reversed, with more $H_2PO_4^-$ than HPO_4^{2-} in a ratio of about 0.61 to 0.39, or 2.2:1.

10.131 (a) HbO_2^- concentration will be lower in the tissues. (b) HbO_2^- concentration will increase.

10.133 2.2×10^{-4}

10.135 (a) CO_2 will react with water to form carbonic acid, H_2CO_3. Determine the $[H_3O^+]$ due to the deprotonation of H_2CO_3. Under the specified conditions, pH = 5.75. (b) $[H_2CO_3] = 4.38 \times 10^{-3}$ mol·L^{-1}, $[HCO_3^-] = 1.19 \times 10^{-4}$ mol·L^{-1}, $[CO_3^{2-}] = 4.23 \times 10^{-10}$ mol·L^{-1}; (c) 50. kg; (d) 4.82; (e) 4.52; (f) 86.9 kg

Chapter 11

11.1 (a) When solid sodium acetate is added to an acetic acid solution, the concentration of H_3O^+ decreases because the equilibrium $HC_2H_3O_2(aq) + H_2O(l) \rightleftharpoons H_3O^+ + C_2H_3O_2^-(aq)$ shifts to the left to relieve the stress imposed by the increase of $[C_2H_3O_2^-]$. (b) When HCl is added to a benzoic acid solution, the percentage of benzoic acid that is deprotonated decreases because the equilibrium $C_6H_5COOH(aq) + H_2O(l) \rightleftharpoons H_3O^+(aq) + C_6H_5CO_2^-(aq)$ shifts to the left to relieve the stress imposed by the increased $[H_3O^+]$. (c) When solid NH_4Cl is added to an ammonia solution, the concentration of OH^- decreases because the equilibrium $NH_3(aq) + H_2O(l) \rightleftharpoons NH_4^+(aq) + OH^-(aq)$ shifts to the left to relieve the stress imposed by the increased $[NH_4^+]$. As $[OH^-]$ decreases, $[H_3O^+]$ increases and pH decreases.

11.3 (a) $pK_a = 3.08$, $K_a = 8.3 \times 10^{-4}$; (b) 2.77

11.5 (a) pH = 1.62, pOH = 12.38; (b) pH = 1.22, pOH = 12.78; (c) pH = 1.92, pOH = 12.08

11.7 increases by 1.16

11.9 (a) 9.46; (b) 9.71; (c) 9.31

11.11 $[ClO^-]/[HClO] = 9.3 \times 10^{-2}$

11.13 (a) 2–4; (b) 3–5; (c) 11.5–13.5; (d) 6–8; (e) 5–7

11.15 (a) $HClO_2$ and $NaClO_2$; (b) NaH_2PO_4 and Na_2HPO_4; (c) $CH_2ClCOOH$ and $NaCH_2ClCO_2$; (d) Na_2HPO_4 and Na_3PO_4

11.17 (a) $[CO_3^{2-}]/[HCO_3^-] = 5.6$; (b) 77 g; (c) 1.8 g; (d) 2.8×10^2 mL

11.19 (a) pH = 6.3, change of 1.5; (b) pH = 4.58, change of -0.17

11.21 (a)

(b) pH at stoichiometric point = 7.0

(b) pH at halfway titration = 11.4

(b) 10 mL; (c) 5 mL

11.23 (a) 9.17×10^{-3} L; (b) 1.83×10^{-2} L; (c) 6.35×10^{-2} mol·L^{-1}; (d) 2.25

11.25 (a) 234 g·mol^{-1}; (b) 3.82

11.27 79.4%

11.29 (a) 13.04; (b) 12.82; (c) 12.55; (d) 7.0; (e) 1.80; (f) 1.55

11.31 (a) 2.89; (b) 4.56; (c) 12.5 mL; (d) 4.75; (e) 25.0 mL; (f) 8.72

11.33 (a) 11.20; (b) 8.99; (c) 11 mL; (d) 9.25; (e) 22 mL; (f) 5.24; (g) methyl red

11.35 (a) weak, because the initial pH is near 5.0 and pH at the stoichiometric point is greater than 7; (b) 1×10^{-5} mol·L^{-1}; (c) $K_a = 10^{-7.5} = 3 \times 10^{-8}$; (d) 3×10^{-3} mol·L^{-1}; (e) 8×10^{-3} mol·L^{-1}; (f) phenolphthalein

11.37 The pH at the stoichiometric point is 8.8, which lies within the range of (c) thymol blue and (d) phenolphthalein.

11.39 Exercise 11.31: thymol blue or phenolphthalein; Exercise 11.33: methyl red or bromocresol green

11.41 (a) 37.4 mL; (b) 74.8 mL; (c) 112 mL

11.43 (a) 22.0 mL; (b) 44.0 mL

11.45 (a)

a: initial, b: first halfway point, c: first stoichiometric point, d: second halfway point, e: second stoichiometric point. Note: (i) the initial pH is higher than points b and c because we assume that the initial $[H^+]$ is from the first deprotonation only (actually, the second deprotonation contributes also); (ii) due to the close pK_a values of thiosulfuric acid, only one obvious titration curve can be observed. In an actual aqueous titration, two protons will be titrated at the same time for this acid.
(b) Point C: 5.0 mL, Point E: 10. mL; (c) first stoichiometric point: 1.2, second stoichiometric point: 7.0
11.47   (a) 1.89; (b) 1.96; (c) 11.6
11.49   (a) 4.66; (b) 2.80; (c) 7.21
11.51   (a) 7.7×10^{-13}; (b) 1.7×10^{-14}; (c) 5.3×10^{-3}; (d) 6.9×10^{-9}
11.53   (a) 1.30×10^{-5} mol·L^{-1}; (b) 1.0×10^{-3} mol·L^{-1}; (c) 9.3×10^{-5} mol·L^{-1}
11.55 1.0×10^{-12}
11.57   (a) 8.0×10^{-10} mol·L^{-1}; (b) 1.2×10^{-16} mol·L^{-1}; (c) 4.6×10^{-3} mol·L^{-1}; (d) 1.3×10^{-6} mol·L^{-1}
11.59   (a) 1.6×10^{-5} mol·L^{-1}; (b) 2.7×10^2 μg
11.61   (a) 6.00; (b) 6.18
11.63   (a) yes; (b) no
11.65   (a) first to last: Ni(OH)$_2$, Mg(OH)$_2$, Ca(OH)$_2$; (b) Ni(OH)$_2$ at pH $\approx$ 7, Mg(OH)$_2$ at pH $\approx$ 10, Ca(OH)$_2$ at pH $\approx$ 13
11.67   carbonate, because the solubility difference is larger
11.69 1.8×10^{-9} mol·L^{-1}
11.71   (a) 1.0×10^{-12} mol·L^{-1}; (b) 3.1×10^{-5} mol·L^{-1}; (c) 2.0×10^{-3} mol·L^{-1}; (d) 2.0×10^{-1} mol·L^{-1}
11.73   (a) CaF$_2$(s) + 2 H$_2$O $\rightleftharpoons$ Ca^{2+}(aq) + 2 HF(aq) + 2 OH$^-$(aq), $K = 3.4 \times 10^{-32}$; (b) 2.2×10^{-4} mol·L^{-1}; (c) 4×10^{-4} mol·L^{-1}
11.75 2.0×10^{-3} mol·L^{-1}
11.77   The two salts can be distinguished by their solubility in 1.0 mol·L^{-1} NH$_3$(aq). The AgCl will dissolve, but most of the AgI will not.
11.79   To use qualitative analyses, the sample must first be dissolved, which can be accomplished by heating the sample with concentrated HNO$_3$ and then diluting the resulting solution. HCl or H$_2$SO$_4$ cannot not be used because some of the metal compounds formed will be insoluble, whereas all of the nitrates will dissolve. After the sample is dissolved and diluted, an aqueous solution containing chloride ions can be introduced, which should precipitate the Ag$^+$ as AgCl but leave the bismuth and nickel in solution, as long as the solution is acidic. The remaining solution can then be treated with H$_2$S. In acidic solution, Bi$_2$S$_3$ will precipitate but NiS will not. Once the Bi$_2$S$_3$ has precipitated, the pH of the solution can be raised by the addition of base, after which NiS will precipitate.
11.81   0.260 mol·L^{-1} in both acetic acid and sodium acetate
11.83   (a) $M_a = 0.0567$, $M_b = 0.0296$, $V_a = 15.0$, $V_b = 0.0$ to 50.0

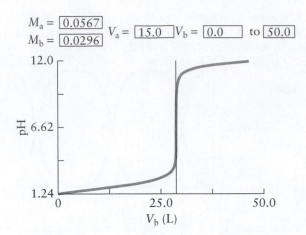

(b) 28.6 mL; (c) 1.24; (d) 7.00, strong acid with strong base titration
11.85   pH = 5.05, 4.4×10^{-3} mol·L^{-1}
11.87 2.8×10^{-2}
11.89 2.2×10^{-3} mol·L^{-1}
11.91   The K_{sp} values are: Cu^{2+}, 1.3×10^{-36}; Co^{2+}, 5×10^{-22}; Cd^{2+}, 4×10^{-29}. All the salts have the same expression for K_{sp}: $K_{sp} = [M^{2+}][S^{2-}]$, so the compound with the smallest K_{sp} value will precipitate first—in this case, CuS.
11.93   At the first stoichiometric point, pH = 1.54; at the second stoichiometric point, pH = 7.46.
11.95   (a) SO$_2$(g) + H$_2$O(l) $\rightleftharpoons$ H$_2$SO$_3$(aq); (b) 1.9 ppm
11.97 -345.65 kJ·mol^{-1}
11.99   (a) 3×10^{-18} mol; (b) ph = 5.8; change = 0.3
11.101   (a) 4.93; (b) 2.08 g
11.103   (a) 9.3×10^{-11} mol·L^{-1}. (b) [OH$^-$] = 2.8×10^{-10}, pH = 4.45. (c) The results in (b) are not reasonable because the OH$^-$ from water is ignored. The contribution of OH$^-$ from water is much greater than from Fe(OH)$_3$, so the [OH$^-$] of the solution is actually 1×10^{-7} mol·L^{-1} and the pH is 7.0. (d) 5.0×10^{-4}; (e) 5.3 g; (f) 60.1%

Chapter 12

12.1   (a) Cr reduced from +6 to +3; C oxidized from -2 to -1; (b) C$_2$H$_5$OH(aq) $\rightarrow$ C$_2$H$_4$O(aq) + 2 H$^+$(aq) + 2 e$^-$; (c) Cr$_2$O$_7{}^{2-}$(aq) + 14 H$^+$(aq) + 6 e$^-$ $\rightarrow$ 2 Cr^{3+}(aq) + 7 H$_2$O(l); (d) 8 H$^+$(aq) + Cr$_2$O$_4{}^{2-}$(aq) + 3 C$_2$H$_5$OH(aq) $\rightarrow$ 2 Cr^{3+}(aq) + 3 C$_2$H$_4$O(aq) + 7 H$_2$O(l)
12.3   (a) 4 Cl(g) + S$_2$O$_3{}^{2-}$(aq) + 5 H$_2$O(l) $\rightarrow$ 8 Cl$^-$(aq) + 2 SO$_4{}^{2-}$(aq) + 10 H$^+$(aq); Cl$_2$ is the oxidizing agent, and S$_2$O$_3{}^{2-}$ is the reducing agent. (b) 2 MnO$_4{}^-$(aq) + H$^+$(aq) + 5 H$_2$SO$_3$(aq) $\rightarrow$ 2 Mn^{2+}(aq) + 3 H$_2$O(l) + 5 HSO$_4{}^-$(aq); MnO$_4{}^-$ is the oxidizing agent and H$_2$SO$_3$ is the reducing agent. (c) Cl$_2$(g) + H$_2$S(aq) $\rightarrow$ 2 Cl$^-$(aq) + S(s) + 2 H$^+$(aq); Cl$_2$ is the oxidizing agent and H$_2$S is the reducing agent. (d) H$_2$O(l) + Cl$_2$(g) $\rightarrow$ HOCl(aq) + H$^+$(aq) + Cl$^-$(aq); Cl$_2$ is both the oxidizing and the reducing agent.
12.5   (a) 3 O$_3$(g) + Br$^-$(aq) $\rightarrow$ 3 O$_2$(g) + BrO$_3{}^-$(aq); O$_3$ is the oxidizing agent and Br$^-$ is the reducing agent.

(b) $3\,Br_2(l)\ +\ 6\,OH^-(aq) \rightarrow 5\,Br^-(aq)\ +\ BrO_3{}^-(aq)\ +$
$3\,H_2O(l)$; Br_2 is both the oxidizing agent and the reducing
agent. (c) $2\,Cr^{3+}(aq)\ +\ 4\,OH^-(aq)\ +\ 3\,MnO_2(s) \rightarrow$
$2\,CrO_4{}^{2-}(aq)\ +\ 2\,H_2O(l)\ +\ 3\,Mn^{2+}(aq)$; Cr^{3+} is the reducing agent and MnO_2 is the oxidizing agent.
(d) $P_4(s)\ +\ 3\,H_2O(l)\ +$
$3\,OH^-(aq) \rightarrow 3\,H_2PO_2{}^-(aq)\ +\ PH_3(g)$; $P_4(s)$ is both the
oxidizing and the reducing agent.

12.7 half-reactions: $P_4S_3(aq)\ +\ 28\,H_2O(l) \rightarrow$
$4\,H_3PO_4(aq)\ +\ 3\,SO_4{}^{2-}(aq)\ +\ 44\,H^+(aq)\ +\ 38\,e^-$;
$NO_3{}^-(aq)\ +\ 4\,H^+(aq)\ +\ 3\,e^- \rightarrow NO(g)\ +\ 2\,H_2O(l)$;
overall reaction: $3\,P_4S_3(aq)\ +\ 38\,NO_3{}^-(aq)\ +\ 20\,H^+(aq)\ +$
$8\,H_2O(l) \rightarrow 12\,H_3PO_4(aq)\ +\ 9\,SO_4{}^{2-}(aq)\ +\ 38\,NO(g)$

12.9 (a) anode: $Ni(s) \rightarrow Ni^{2+}(aq)\ +\ 2\,e^-$, cathode:
$Ag^+(aq)\ +\ e^- \rightarrow Ag(s)$, overall: $2\,Ag^+(aq)\ +\ Ni(s) \rightarrow 2\,Ag(s)$
$Ni^{2+}(aq)$; (b) anode: $H_2(g) \rightarrow 2\,H^+(aq)\ +\ 2\,e^-$, cathode:
$Cl_2(g)\ +\ 2\,e^- \rightarrow 2\,Cl^-(aq)$, overall: $Cl_2(g)\ +\ H_2(g) \rightarrow$
$2\,H^+(aq)\ +\ 2\,Cl^-(aq)$; (c) anode: $Cu(s) \rightarrow Cu^{2+}(aq)\ +\ 2\,e^-$,
cathode: $Ce^{4+}(aq)\ +\ e^- \rightarrow Ce^{3+}(aq)$, overall: $2\,Ce^{4+}(aq)\ +$
$Cu(s) \rightarrow Cu^{2+}(aq)\ +\ 2\,Ce^{3+}(aq)$; (d) anode: $2\,H_2O(l) \rightarrow$
$O_2(g)\ +\ 4\,H^+(aq)\ +\ 4\,e^-$, cathode: $O_2(g)\ +\ 2\,H_2O(l)\ +$
$4\,e^- \rightarrow 4\,OH^-(aq)$, overall: $H_2O(l) \rightarrow H^+(aq)\ +\ OH^-(aq)$;
(e) anode: $Sn^{2+}(aq) \rightarrow Sn^{4+}(aq)\ +\ 2\,e^-$, cathode:
$Hg_2Cl_2(s)\ +\ 2\,e^- \rightarrow 2\,Hg(l)\ +\ 2\,Cl^-(aq)$ overall:
$Sn^{2+}(aq)\ +Hg_2Cl_2(s) \rightarrow 2\,Hg(l)\ +\ 2\,Cl^-(aq)\ +\ Sn^{4+}(aq)$

12.11 (a) anode: $Zn(s) \rightarrow Zn^{2+}(aq)\ +\ 2\,e^-$, cathode:
$Ni^{2+}(aq)\ +\ 2\,e^- \rightarrow Ni(s)$, overall: $Ni^{2+}(aq)\ +\ Zn(s) \rightarrow$
$Ni(s)\ +\ Zn^{2+}(aq)$, $Zn(s)|Zn^{2+}(aq)||Ni^{2+}(aq)|Ni(s)$;
(b) anode: $2\,I^-(aq) \rightarrow 2\,e^-\ +\ I_2(s)$, cathode: $Ce^{4+}(aq)\ +$
$e^- \rightarrow Ce^{3+}(aq)$, overall: $2\,I^-(aq)\ +\ 2\,Ce^{4+}(aq) \rightarrow$
$2\,Ce^{3+}(aq)\ +\ I_2(s)$, $Pt(s)|I^-(aq)|I_2(s)||Ce^{4+}(aq)$,
$Ce^{3+}(aq)|Pt(s)$; (c) anode: $H_2(g) \rightarrow 2\,H^+(aq)\ +\ 2\,e^-$,
cathode: $Cl_2(g)\ +\ 2\,e^- \rightarrow 2\,Cl^-(aq)$, overall: $H_2(g)\ +$
$Cl_2(g) \rightarrow 2\,HCl(aq)$, $Pt(s)|H_2(g)|H^+(aq)||Cl^-(aq)|$
$Cl_2(g)|Pt(s)$; (d) anode: $Au(s) \rightarrow Au^{3+}(aq)\ +\ 3\,e^-$, cathode:
$Au^+(aq)\ +\ e^- \rightarrow Au(s)$, overall: $3\,Au^+(aq) \rightarrow 2\,Au(s)\ +$
$Au^{3+}(aq)$, $Au(s)|Au^{3+}(aq)||Au^+(aq)|Au(s)$;

12.13 (a) anode: $Ag(s)\ +\ Br^-(aq) \rightarrow AgBr(s)\ +\ e^-$,
cathode: $Ag^+(aq)\ +\ e^- \rightarrow Ag(s)$,
$Ag(s)|AgBr(s)|Br^-(aq)||Ag^+(aq)|Ag(s)$;
(b) anode: $4\,OH^-(aq) \rightarrow O_2(g)\ +\ 2\,H_2O(l)\ +\ 4\,e^-$,
cathode: $O_2(g)\ +\ 4\,H^+(aq)\ +\ 4\,e^- \rightarrow 2\,H_2O(l)$,
$Pt(s)|O_2(g)|OH^-(aq)||H^+(aq)|O_2(g)|Pt(s)$;
(c) anode: $Cd(s)\ +\ 2\,OH^-(aq) \rightarrow Cd(OH)_2(s)\ +\ 2\,e^-$,
cathode: $Ni(OH)_3(s)\ +\ e^- \rightarrow Ni(OH)_2(s)\ +\ OH^-(aq)$,
$Cd(s)|Cd(OH)_2(s)|KOH(aq)||Ni(OH)_3(s)|$
$Ni(OH)_2(s)|Ni(s)$;

12.15 (a) anode: $Fe^{2+}(aq) \rightarrow Fe^{3+}(aq)\ +\ e^-$, cathode:
$MnO_4{}^-(aq)\ +\ 8\,H^+(aq)\ +\ 5\,e^- \rightarrow Mn^{2+}(aq)\ +\ 4\,H_2O(l)$;
(b)$MnO_4{}^-(aq)\ +\ 5\,Fe^{2+}(aq)\ +\ 8\,H^+(aq) \rightarrow$
$Mn^{2+}(aq)\ +\ 5\,Fe^{3+}(aq)\ +\ 4\,H_2O(l)$,
$Pt(s)|Fe^{3+}(aq),Fe^{2+}(aq)|H^+(aq),MnO_4{}^-(aq)$,
$Mn^{2+}(aq)|Pt(s)$

12.17 (a) $+0.75$ V; (b) $+0.37$ V; (c) $+0.52$ V; (d) $+1.52$ V

12.19 (a) $Hg(l)|Hg_2{}^{2+}(aq)||NO_3{}^-(aq),H^+(aq)|NO(g)|$
$Pt(s)$, $+0.17$ V, -98 kJ·mol^{-1}; (b) not spontaneous;
(c) $Pt(s)|Pu^{3+}(aq),Pu^{4+}(aq)||Cr_2O_7{}^{2-}(aq),Cr^{3+}(aq)$,
$H^+(aq)|Pt(s)$, $+0.36$ V, -208 kJ·mol^{-1}

12.21 -0.349 V

12.23 (a) Cu < Fe < Zn < Cr; (b) Mg < Na < K < Li;
(c) V < Ti < Al < U; (d) Au < Ag < Sn < Ni

12.25 (a) oxidizing agent: Co^{2+}, reducing agent: Ti^{2+},
$Pt(s)|Ti^{2+}(aq),Ti^{3+}(aq)||Co^{2+}(aq)|Co(s)$, $+0.09$ V;
(b) oxidizing agent: U^{3+}, reducing agent: La,
$La(s)|La^{3+}(aq)||U^{3+}(aq)|U(s)$, $+0.73$ V; (c) oxidizing agent:
Fe^{3+}, reducing agent: H_2,
$Pt(s)|H_2(g)|H^+(aq)||Fe^{2+}(aq),Fe^{3+}(aq)|Pt(s)$, $+0.77$ V;
(d) oxidizing agent: O_3, reducing agent: Ag,
$Ag(s)|Ag^+(aq)||OH^-(aq)|O_3(g),O_2(g)|Pt(s)$, $+0.44$ V

12.27 (a) $Cl_2(g)$, $+0.27$. (b) and (c) do not favor products.
(d) $NO_3{}^-$, $+1.56$ V

12.29 (a) $3\,Au^+(aq) \rightarrow 2\,Au(s)\ +\ Au^{3+}(aq)$; (b) yes

12.31 -1.50 V

12.33 (a) 6×10^{-16}; (b) 1×10^4

12.35 (a) $Pb^{4+}(aq)\ +\ Sn^{2+}(aq) \rightarrow Pb^{2+}(aq)\ +\ Sn^{4+}(aq)$,
$Q = 10^6$; (b) $2\,Cr_2O_7{}^{2-}(aq)\ +\ 16\,H^+(aq) \rightarrow$
$4\,Cr^{3+}(aq)\ +\ 8\,H_2O(l)\ +\ 3\,O_2(g)$, $Q = 1.0$

12.37 (a) $+0.030$ V; (b) $+0.06$ V

12.39 (a) $+0.067$ V; (b) $+0.51$ V; (c) -1.33 V; (d) $+0.31$ V

12.41 (a) pH $= 1.0$; $[Cl^-] = 0.1$ mol·L^{-1}

12.43 2.25

12.45 (a) 1.2×10^{-17}. (b) This value is a factor of 10 greater
than the tabular value.

12.47 yes, 8.4 kJ per mole of Ag

12.49 (a) -0.27 V; (b) $+0.07$ V

12.51 (a) cathode: $Ni^{2+}(aq)\ +\ 2\,e^- \rightarrow Ni(s)$; (b) anode:
$2\,H_2O(l) \rightarrow O_2(g)\ +\ 4\,H^+(aq)\ +\ 4\,e^-$; (c) $+1.46$ V

12.53 (a) water; (b) water; (c) Ni^{2+}; (d) Au^{3+}

12.55 (a) 3.3 g; (b) 0.59 L; (c) 0.94 g

12.57 (a) 27 h; (b) 0.44 g

12.59 (a) 0.64 A; (b) 0.24 A

12.61 $+2$

12.63 $+4$

12.65 52.0 g·mol^{-1}

12.67 (a) oxidation: $2\,Al(s) \rightarrow 2\,Al^{3+}(aq)$
$+\ 6\,e^-\ +\ 3\,H_2O(l)$, reduction: $3\,H_2O(l)\ +$
$1\tfrac{1}{2}\,O_2(g)\ +\ 6\,e^- \rightarrow 6\,OH^-(aq)$; (b) $+2.90$ V

12.69 100 A

12.71 (a) KOH(aq)/HgO(s); (b) HgO(s);
(c) $HgO(s)\ +\ Zn(s) \rightarrow Hg(l)\ +\ ZnO(s)$

12.73 The anode reaction is $Zn(s) \rightarrow Zn^{2+}(aq)\ +\ 2\,e^-$; this
reaction supplies the electrons to the external circuit. The cathode reaction is $MnO_2(s)\ +\ H_2O(l)\ +\ e^- \rightarrow MnO(OH)_2(s)\ +$
$OH^-(aq)$. The $OH^-(aq)$ produced reacts with $NH_4{}^+(aq)$ from
the $NH_4Cl(aq)$ present: $NH_4{}^+(aq)\ +\ OH^-(aq) \rightarrow H_2O(l)\ +$
$NH_3(g)$. The $NH_3(g)$ produced complexes with the $Zn^{2+}(aq)$
produced in the anode reaction $Zn^{2+}(aq)\ +\ 4\,NH_3(g) \rightarrow$
$[Zn(NH_3)_4]^{2+}(aq)$.

12.75 (a) KOH(aq); (b) $2\,Ni(OH)_2(s) + 2\,OH^-(aq) \rightarrow$ $2\,Ni(OH)_3 + 2\,e^-$

12.77 Comparison of the reduction potentials shows that Cr is more easily oxidized than Fe, so the presence of Cr retards the rusting of Fe.

12.79 (a) 57.4%; (b) AgBr

12.81 (a) $Fe_2O_3 \cdot H_2O$; (b) H_2O and O_2 jointly oxidize iron. (c) Water is more highly conducting if it contains dissolved ions, so the rate of rusting is increased.

12.83 (a) aluminum or magnesium. (b) cost, availability, and toxicity of products in the environment. (c) Fe could act as the anode of an electrochemical cell if Cu^{2+} or Cu^+ were present; therefore, it could be oxidized at the point of contact. Water with dissolved ions would act as the electrolyte.

12.85 Al, Zn, Fe, Co, Ni, Cu, Ag, Au

12.87 -0.92 V

12.89 $+0.14$ V

12.91 A negatively charged electrolyte flows from the cathode to the anode.

12.93 (a) yes, $O_2(g) + 4\,H^+(aq) + 4\,Ag(s) \rightarrow$ $4\,Ag^+(aq) + 2\,H_2O(l)$, $+0.42$ V; (b) yes, $O_2(g) +$ $2\,H_2O(l) + 4\,Ag(s) \rightarrow 4\,Ag^+(aq) + 4\,OH^-(aq)$, $+0.05$ V

12.95 (a) Reduction takes place at the electrode with the higher concentration, which would be the chromium electrode in contact with the 1.0 M $CrCl_3$. (b) yes; (c) yes; (d) no

12.97 (a) 1.0×10^{-2} mol·L^{-1}; (b) 8.5×10^{-17}

12.99 1×10^{-3}

12.101 (a) $\Delta G_{r1}° = -nFE_1° = \Delta H_{r1}° - T_1\Delta S_{r1}°$, where r_1 represents the reaction at T_1. $\Delta G_{r2}° = -nFE_2° = \Delta H_{r2}° - T_2\Delta S_{r2}°$, where r_2 represents the reaction at T_2. On subtracting the first reaction from the second, we obtain $-nFE_2° + nFE_1° = \Delta H_{r2}° - T_2\Delta S_{r2}° - [\Delta H_{r1}° - T_1\Delta S_{r1}°]$. Since $\Delta H_{r1}° = \Delta H_{r2}°$, we obtain $nFE_1° - nFE_2° = -T_2\Delta S_{r2}° + T_1\Delta S_{r1}°$, which can be rewritten as $-nFE_1° + nFE_2° = T_2\Delta S_{r2}° - T_1\Delta S_{r1}°$. Since $\Delta S_{r1}° = \Delta S_{r2}° = \Delta S_r°$, we obtain $-nFE_1° + nFE_2° = \Delta S_r°(T_2 - T_1)$; $nFE_2° = nFE_1° + \Delta S_r°(T_2 - T_1)$; $E_2° = E_1° + \Delta S_r°(T_2 - T_1)/nF$; (b) $+1.18$ V

12.103 $+0.08$ V to $+0.09$ V

12.105 $E_{cell}° = +0.03$ V, $E_{cell} = -0.21$ V. The cell changes from spontaneous to nonspontaneous as a function of concentration.

12.107 (a) $E'(1) = 1.23\,V - (0.05916\,V)\,pH$, $E'(2) = +0.36\,V - (0.05916\,V)\,pOH$; $pOH = 14.00 - pH$
$E'(2) = 0.36\,V + (0.05916\,V)(14.00 - pH)$
$\quad\quad = 0.36\,V + 0.83\,V - (0.05916\,V)\,pH$
$\quad\quad = 1.19\,V - (0.05916\,V)\,pH$ (b) $E'(1) = +0.82\,V$,
$E'(2) = +0.77\,V$, potentials should be identical but differ because of the limited number of significant figures used to derive and evaluate the equations.

12.109 12

12.111 -0.828 V to $+0.828$ V

12.113 (a) A plot of cell voltage, E, versus $\ln[Ag^+]_{anode}$ would be a linear increase with positive slope. (b) slope = 0.025693, corresponds to RT/nF, consistent. (c) $E°$

12.115 4×10^{-9}

12.117 0.205 A

12.119 (a) (i) -0.41 V, (ii) $+0.41$ V; (b) -0.31 V;

(c)

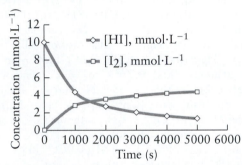

(d) $NADH(aq) + pyruvate + H^+(aq) \rightarrow NAD^+(aq) + lactate$, $E° = +0.12$ V, $E° = +0.33$ V; (e) -64 kJ·mol^{-1}; (f) 1.43×10^{11}

Chapter 13

13.1 (a) one-third; (b) two-thirds; (c) two

13.3 (a) 0.28 mol·L·s^{-1}; (b) 0.14 mol·L^{-1}·s^{-1}

13.5 (a) 3.3×10^{-3} (mol O_2)·L^{-1}·s^{-1}; (b) 3.3×10^{-3} mol·L^{-1}·s^{-1}

13.7 (a) and (c)

[graph: Concentration (mmol·L^{-1}) vs Time (s), showing [HI], mmol·L^{-1} and [I₂], mmol·L^{-1} curves]

Note that the curves for the $[I_2]$ and $[H_2]$ are identical and only the $[I_2]$ curve is shown.

(b)

Time (s)	Rate (mmol·L^{-1}·s^{-1})
0	0.0060
1000	0.003
2000	0.000 98
3000	0.000 61
4000	0.000 40
5000	0.000 31

13.9 (a) mol·L^{-1}·s^{-1}; (b) s^{-1}; (c) L·mol^{-1}·s^{-1}

13.11 2.2×10^{-4} (mol N_2O_5)·L^{-1}·s^{-1}

13.13 (a) 2.2×10^{-5} mol·L^{-1}·s^{-1}; (b) factor of 2

13.15 rate = $k[CH_3Br][OH^-]$

13.17 (a) rate = $k[ICl][H_2]$; (b) 0.16 L·mol^{-1}·s^{-1}; (c) 2.0×10^{-6} mol·L^{-1}·s^{-1}

13.19 (a) rate = $k[A][B]^2[C]^2$; (b) 5; (c) 2.85×10^{12} L^4·mol^{-4}·s^{-1}; (d) 1.13×10^{-2} mol·L^{-1}·s^{-1}

13.21 (a) 6.93×10^{-4} s^{-1}; (b) 9.4×10^{-3} s^{-1}; (c) 5.1×10^{-3} s^{-1}

13.23 (a) 5.2 h; (b) 3.5×10^{-2} mol·L^{-1}; (c) 6.5×10^2 min

13.25 (a) 710 s; (b) 9.7×10^2 s; (c) 1.1×10^3 s

13.27 (a) 247 min; (b) 819 min; (c) 10.9 g

13.29 (a) 0.17 min^{-1}; (b) an additional 3.5 min

13.31 (a)

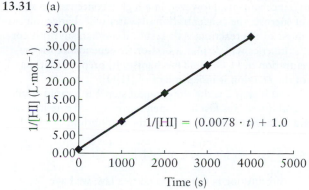

$1/[HI] = (0.0078 \cdot t) + 1.0$

(b) (i) 7.8×10^{-3} L·mol^{-1}·s^{-1}, (ii) 3.9×10^{-3} L·mol^{-1}·s^{-1}
13.33 (a) 7.4×10^2 s; (b) 1.5×10^2 s; (c) 2.0×10^2 s
13.35 (a) 1.7×10^2 min; (b) 3.3×10^3 min
13.37 $[A]_t = [A]_0\, e^{-akt}$, $t_{1/2} = \ln 2/ak$
13.39 $t_{1/2} = 3/(2k[A]_0^2)$
13.41 $2\,AC + B \rightarrow A_2B + 2\,C$. Intermediate is AB.
13.43 $CH_2=CHCOOH + HCl \rightarrow ClCH_2CH_2COOH$;
intermediates: Cl^-, $CH_2=CHC(OH)_2^+$, and
$ClCH_2CHC(OH)_2$
13.45 (a) $O_3 + O \rightarrow 2\,O_2$; (b) Step 1: rate = $k[O_3][NO]$,
bimolecular; Step 2: rate = $k[NO_2][O]$, bimolecular; (c) NO_2;
(d) NO
13.47 rate = $k[NO][Br_2]$
13.49 If mechanism (a) were correct, the rate law would be
rate = $k[NO_2][CO]$. But this expression does not agree with
the experimental result and can be eliminated as a possibility.
Mechanism (b) has rate = $k[NO_2]^2$ from the slow step. Step 2
does not influence the overall rate, but it is necessary to
achieve the correct overall reaction; thus this mechanism agrees
with the experimental data. Mechanism (c) is not correct,
which can be seen from the rate expression for the slow step,
rate = $k[NO_3][CO]$. [CO] cannot be eliminated from this
expression to yield the experimental result, which does not
contain [CO].
13.51 (a) True. (b) False. At equilibrium, the rates of the forward and reverse reactions are equal, not the rate constants.
(c) False. Increasing the concentration of a reactant causes the
rate to increase by providing more reacting molecules. It does
not affect the rate constant of the reaction.
13.53 The overall rate of formation of A is
rate = $-k[A] + k'[B]$. The first term accounts for the forward reaction and is negative as this reaction reduces [A]. The
second term, which is positive, accounts for the back reaction
which increases [A]. Given the 1:1 stoichiometry of the reaction, if no B was present at the beginning of the reaction, [A]
and [B] at any time are related by the equation:
$[A] + [B] = [A]_0$ where $[A]_0$ is the initial concentration of A.
Therefore,

$$\frac{d[A]}{dt} = -k[A] + k'([A]_0 - [A])$$

$$= -(k + k')[A] + k'[A]_0$$

The solution of this first-order differential equation is

$$[A] = \frac{k' + ke^{-(k'+k)t}}{k' + k}[A]_0$$

As $t \rightarrow \infty$ the exponential term in the numerator goes to zero
and the concentrations reach their equilibrium values given by:

$$[A]_{eq} = \frac{k'[A]_0}{k' + k} \quad \text{and} \quad [B]_{eq} = [A]_0 - [A]_\infty = \frac{k[A]_0}{k + k'}$$

taking the ratio of products over reactants we see that

$$\frac{[B]_{eq}}{[A]_{eq}} = \frac{k}{k'} = K \quad \text{where } K \text{ is the equilibrium constant}$$

for the reaction.

13.55 (a)

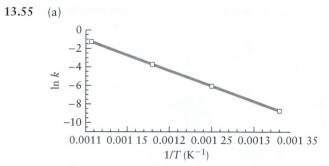

2.72×10^2 kJ·mol^{-1}; (b) 8.8×10^{-2} s^{-1}
13.57 39 kJ·mol^{-1}
13.59 2.5×10^9 L·mol^{-1}·s^{-1}
13.61 9.2×10^{-4} s^{-1}
13.63 (a) 0.676. (b) endothermic. (c) Raising the temperature
will increase the rate constant of the reaction with the higher
activation barrier more than it will the rate constant of the
reaction with the lower energy barrier. We expect the rate of
the forward reaction to go up substantially more than for the
reverse reaction in this case. k will increase more than k' and
consequently the equilibrium constant K will increase.
13.65 (a) 6×10^8; (b) 3×10^7
13.67 81 kJ·mol^{-1}
13.69 $RCN + H_2O \rightarrow RC(=O)NH_2$, $RC(=N^-)OH$ and
$RC(=NH)OH$ are intermediates, OH^- is the catalyst
13.71 (a) False. A catalyst increases the rate of both the forward and reverse reactions by providing a completely different
pathway. (b) True, although a catalyst may be poisoned and
lose activity. (c) False. There is a completely different pathway
provided for the reaction in the presence of a catalyst. (d)
False. The position of the equilibrium is unaffected by the presence of a catalyst.
13.73 (a) To obtain the Michaelis−Menten rate equation, we
will begin by employing the steady-state approximation, setting
the rate of change in the concentration of the ES intermediate
equal to zero: $d[ES]/dt = k_1[E][S] - k_1'[ES] - k_2[ES] = 0$.
Rearranging gives

$$[E][S] = \left(\frac{k_2 + k_1'}{k_1}\right)[ES] = K_M[ES]$$

The total bound and unbound enzyme concentration, $[E]_0$, is given by $[E]_0 = [E] + [ES]$, and, therefore, $[E] = [ES] - [E]_0$. Substituting this expression for $[E]$ into the preceding equation, we obtain $([ES] - [E]_0)[S] = K_M [ES]$. Rearranging to obtain $[ES]$ gives

$$[ES] = \frac{[E]_0[S]}{K_M + [S]}$$

From the mechanism, the rate of appearance of the product is given by rate $= k_2[ES]$. Substituting the preceding equation for $[ES]$, we obtain

$$\text{rate} = \frac{k_2[E]_0[S]}{K_M + [S]},$$

the Michaelis–Menten rate equation, which can be rearranged to obtain

$$\frac{1}{\text{rate}} = \frac{K_M}{k_2[E]_0[S]} + \frac{1}{k_2[E]_0}.$$

If one plots 1/rate versus 1/[S], the slope will be $K_M/k_2[E]_0$ and the y-intercept will be $1/k_2[E]_0$.

(b)

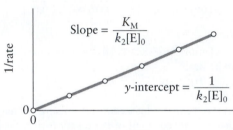

Slope $= \dfrac{K_M}{k_2[E]_0}$

y-intercept $= \dfrac{1}{k_2[E]_0}$

1/[S]

13.75   (a) rate $= k[NO]^2$, bimolecular; (b) rate $= k[Cl_2]$, unimolecular; (c) rate $= k[NO_2]^2$, bimolecular; (d) b and c
13.77   (a) A is 1, B is 2, C is 0; overall is 3; (b) rate $= k[A][B]^2$; (c) 2.0×10^{-5} $L^2 \cdot mmol^{-2} \cdot s^{-1}$; (d) 2.9×10^{-6} mmol $L^{-1} \cdot s^{-1}$
13.79 2.3×10^5 $L \cdot mol \cdot s^{-1}$
13.81   (a) 5.00 s; (b) There should be four molecules at $t = 8$ s.
13.83

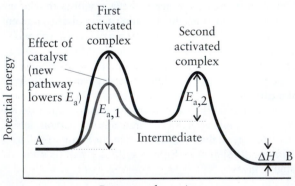

Progress of reaction

13.85   13.5 mg
13.87   The anticipated rate for mechanism (i) is: rate $= k[C_{12}H_{22}O_{11}]$, while the expected rate for mechanism (ii) is: rate $= k[C_{12}H_{22}O_{11}][H_2O]$. The rate for mechanism (ii) will be pseudo-first-order in dilute solutions of sucrose because the concentration of water will not change. Therefore, in dilute solutions kinetic data can not be used to distinguish between

the two mechanisms. However, in a highly concentrated solution of sucrose, the concentration of water will change during the course of the reaction. As a result, if mechanism (ii) is correct the kinetics will display a first-order dependence on the concentration of H_2O while mechanism (i) predicts that the rate of the reaction is independent of $[H_2O]$.
13.89   (a) If step 2 is the slow step, if step 1 is a rapid equilibrium, and if step 3 is fast also, then our proposed rate law will be rate $= k_2[N_2O_2][H_2]$. Consider the equilibrium of step 1:

$$k_1[NO]^2 = k_1'[N_2O_2]; \quad [N_2O_2] = \frac{k}{k_1'}[NO]^2$$

Substituting in our proposed rate law, we have rate $= k_2(k_1/k_1')[NO]^2[H_2] = k[NO]^2[H_2]$ where $k = k_2(k_1/k_1')$. The assumptions made above reproduce the observed rate law; therefore, step 2 is the slow step.

(b)

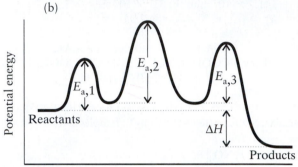

Progress of reaction

13.91   75 $kJ \cdot mol^{-1}$
13.93   (a) To obtain the Michaelis–Menten rate equation, assuming a pre-equilibrium between the bound and unbound states of the substratewe begin with the expression for the equilibrium constant of the fast equilibrium between the bound and unbound substrate: $K = [ES]/[E][S]$. Solving for $[ES]$ we obtain $[ES] = K[E][S]$.

As before in problem 13.69, the total bound and unbound enzyme concentration, $[E]_0$, is given by $[E]_0 = [E] + [ES]$, and, therefore, $[E] = [E]_0 - [ES]$. Substituting this expression for $[E]$ into the equation above we obtain: $K([E]_0 - [ES])[S] = [ES]$. Rearraining to obtain $[ES]$:

$$[ES] = \frac{K[E]_0[S]}{1 + K[S]} = \frac{[E]_0[S]}{K^{-1} + [S]}$$

From the mechanism, the rate of appearance of the product is given by rate $= k_2[ES]$. Substituting the equation above for $[ES]$ one obtains

$$\text{rate} = \frac{k_2[E]_0[S]}{K^{-1} + [S]},$$ the Michaelis–Menten rate equation.

13.95   (a) three; (b) first; (c) third; (d) two; (e) none
13.97   (a) ClO is the reaction intermediate and Cl is the catalyst. (b) Cl, ClO, O, O_2; (c) Step 1 and step 2 are propagating. (d) $Cl + Cl \rightarrow Cl_2$
13.99   For a third-order reaction,

$$t_{1/2} \propto \frac{1}{[A_0]^2} \text{ or } t_{1/2} = \frac{\text{constant}}{[A_0]^2};$$ (a) first half-life $= t_1 = t_{1/2}$

$$= \frac{\text{constant}}{[A_0]^2};$$ (b) second half-life $= t_2$

$$= \frac{\text{constant}}{(\frac{1}{2}[A_0])^2} = \frac{4(\text{constant})}{[A_0]^2} = 4t_1,$$ total time $=$

$$= t_1 + t_2 = t_1 + 4t_1 = 5t_1 = t_{1/4};\text{ (c) third half-life}$$

$$= t_3 = \frac{\text{constant}}{(\frac{1}{4}[A_0])^2} = \frac{16(\text{constant})}{[A_0]^2} = 16t_1;\text{ fourth half-life}$$

$$= t_4 = \frac{\text{constant}}{(\frac{1}{8}[A_0])^2} = \frac{64(\text{constant})}{[A_0]^2} = 64t_1,\text{ total time}$$

$$= t_1 + t_2 + t_3 + t_4 = t_1 + 4t_1 + 16t_1 + 64t_1$$

$$= 85t_1 = t_{1/16}$$

13.101 The linear plots are (b), (c), (d), (f), and (g).
13.103 (a) The overall reaction is: $OC^- + I^- \rightarrow OI^- + CI^-$.

(b) rate $= \dfrac{k_2 k_1}{k_1'}\dfrac{[OCl^-][I^-]}{[OH^-]}$

(c) yes. (d) If the reaction is carried out in an organic solvent, then H_2O is no longer the solvent and its concentration must be included in calculating the equilibrium concentration of HOCl:

$$\text{rate} = \frac{k_2 k_1}{k_1'}\frac{[OCl^-][I^-][H_2O]}{[OH^-]}$$

13.105 To get an expression for $t_{1/2}$ in terms of n, we need to evaluate an integral such as:

$$\int_{[A]_0}^{[A]}\frac{d[A]}{[A]^n} = -k\int_0^t dt = -kt$$

$$\frac{1}{n-1}\left(\frac{1}{[A]^{n-1}} - \frac{1}{[A]_0^{n-1}}\right) = kt$$

An expression for $t_{1/2}$ is then:

$$\frac{1}{n-1}\left(\frac{2^{n-1}}{[A]_0^{n-1}} - \frac{1}{[A]_0^{n-1}}\right) = kt_{1/2}$$

$$\frac{1}{n-1}\left(\frac{2^{n-1} - 1}{[A]_0^{n-1}}\right) = kt_{1/2}$$

An expression for $t_{3/4}$ could be found by setting up $[A] = \frac{3}{4}[A]_0$

$$\frac{1}{n-1}\left(\frac{4^{n-1}}{3^{n-1}[A]_0^{n-1}} - \frac{1}{[A]_0^{n-1}}\right) = kt_{3/4}$$

$$\frac{1}{n-1}\left(\frac{(4/3)^n - 1}{[A]_0^{n-1}}\right) = kt_{3/4}$$

The ratio $t_{1/2}/t_{3/4}$ is then

$$t_{1/2}/t_{3/4} = \left(\frac{2^{n-1} - 1}{(4/3)^{n-1} - 1}\right)$$

13.107 (a) methyl isonitrile

cyanomethane

(b) -43 kJ·mol^{-1}, cyanomethane. (c) $k' = 4.04 \times 10^{-15}$ s^{-1}; $t = 7.12 \times 10^{13}$ s. (d) As the reaction is exothermic (negative $\Delta H°$), the reaction profile has the product lower than the reactant. (e) 474 K. (f) Argon serves as a collision partner. Collisions between the $CH_3NC(g)$ and argon atoms provide the energy needed to overcome the activation energy and allow the isomerization reaction to proceed toward products. The argon atoms also serve as an energy sink, accepting the energy released during the isomerization reaction. (g) At high argon concentrations, collisions resulting in reaction are between the $CH_3NC(g)$ and argon atoms: $CH_3NC(g) + Ar(g) \rightarrow CH_3CN(g) + Ar(g)$. Given this elementary reaction one would predict that the reaction was first order with respect to $CH_3NC(g)$ concentration and would appear first order overall if the concentration of Ar(g) was large and unchanging. If, however, the concentration of Ar(g) was greatly reduced, the isomerization reaction would also proceed through collisions between $CH_3NC(g)$ molecules, making the reaction second order with respect to $CH_3NC(g)$.

Chapter 14

14.1   (a) nitrogen; (b) potassium; (c) gallium; (d) iodine
14.3   (a) sulfur; (b) selenium; (c) sodium; (d) oxygen
14.5   tellurium < selenium < oxygen
14.7   (a) antimony. (b) Antimony has a higher effective nuclear charge.
14.9   (a) bromide ion. (b) Bromide is the largest.
14.11 (a) KCl; (b) K—O
14.13 $2 K(s) + H_2(g) \rightarrow 2 KH(s)$
14.15 (a) saline; (b) molecular; (c) molecular; (d) metallic
14.17 (a) acidic; (b) amphoteric; (c) acidic; (d) basic
14.19 (a) CO_2; (b) B_2O_3
14.21 (a) $C_2H_2(g) + H_2(g) \rightarrow H_2C=CH_2(g)$, oxidation number of C in C_2H_2 is -1; of C in $H_2C=CH_2$ is -2, carbon has been reduced; (b) $CO(g) + H_2O(g) \rightarrow CO_2(g) + H_2(g)$; (c) $BaH_2(s) + 2 H_2O(l) \rightarrow Ba(OH)_2 + 2 H_2(g)$
14.23 (a) 206.10 kJ·mol^{-1}; (b) 214.62 J·K^{-1}·mol^{-1}; (c) 142.12 kJ·mol^{-1}
14.25 (a) $H_2(g) + Cl_2(g) \rightarrow 2 HCl(g)$; (b) $H_2(g) + 2 Na(l) \rightarrow 2 NaH(s)$; (c) $P_4(s) + 6 H_2(g) \rightarrow 4 PH_3(g)$; (d) $2 Cu(s) + H_2(g) \rightarrow 2 CuH(s)$
14.27 This trend is mainly due to the electronegativity of the central atom (N < O < F), which makes the partial charge on F more negative and better able to attract H from another HF atom.
14.29 (a) 1.23 V. (b) The difficulty is isolating the two half-cells but still maintaining electrical contact. Ions need to flow through the system to maintain charge balance in the reaction. In this case, a material that allows hydrogen ions but not hydrogen gas or oxygen gas to pass through would be necessary.
14.31 Lithium is the only Group 1 element that reacts directly with nitrogen to form lithium nitride; it reacts with oxygen to form mainly the oxide, whereas the other members of the group form mainly the peroxide or superoxide. Lithium exhibits the diagonal relationship that is common to many first members of a group. Li is similar in many of its compounds to the compounds of Mg. This behavior is related to the small ionic radius of Li$^+$, 58 pm, which is closer to the ionic radius of Mg^{2+}, 72 pm, but substantially less than that of Na$^+$, 102 pm.
14.33 (a) $4 Na(s) + O_2(g) \rightarrow 2 Na_2O(s)$; (b) $6 Li(s) + N_2(g) \rightarrow 2 Li_3N(s)$; (c) $2 Na(s) + 2 H_2O(l) \rightarrow 2 NaOH(aq) + H_2(g)$; (d) $4 KO_2(s) + 2 H_2O(g) \rightarrow 4 KOH(s) + 3 O_2(g)$

14.35   19.3 g

14.37 $Mg(s) + 2 H_2O(l) \rightarrow Mg(OH)_2 + H_2(g)$

14.39   (a) $CaO(s) + H_2O(l) \rightarrow Ca(OH)_2(s)$; (b) $-57.33 \text{ kJ·mol}^{-1}$

14.41   Be is the weakest reducing agent; Mg is stronger than Be but weaker than the remaining members of the group, all of which have approximately the same reducing strength. This effect is related to the very small radius of the Be^{2+} ion, 27 pm; its strong polarizing power introduces much covalent character into its compounds. Thus, Be attracts electrons more strongly and does not release them as readily as do other members of the group. Mg^{2+} also is a small ion, 58 pm, and so the same reasoning applies to it as well but to a lesser extent. The remaining ions of the group are considerably larger, release electrons more readily, and are better reducing agents.

14.43 $2 Al(s) + 2 OH^-(aq) + 6 H_2O(l) \rightarrow 2[Al(OH)_4]^-(aq) + 3 H_2(g)$. $Be(s) + 2 OH^-(aq) + 2 H_2O(l) \rightarrow [Be(OH)_4]^{2-}(aq) + H_2(g)$. Be and Al are diagonal neighbors in the periodic table and exhibit similar chemical behavior.

14.45   (a) $Mg(OH)_2(s) + 2 HCl(aq) \rightarrow MgCl_2(aq) + 2 H_2O(l)$; (b) $Ca(s) + 2 H_2O(l) \rightarrow Ca(OH)_2(aq) + H_2(g)$; (c) $BaCO_3(s) \rightarrow BaO(s) + CO_2(g)$

14.47   (a)

$$:\!\ddot{C}l\!-\!Be\!-\!\ddot{C}l\!: \qquad Mg^{2+} \quad 2\left[:\!\ddot{\underset{\displaystyle ..}{C}}l\!:\right]^- \ ;$$

$MgCl_2$ is ionic; $BeCl_2$ is a molecular compound. (b) 180°; (c) *sp*

14.49   (a) $\Delta G_r° = -2.35 \text{ kJ}$, $\Delta H_r° = -5.07 \text{ kJ}$; $\Delta S_r° = -9.53 \text{ J}$. (b) Decreasing the temperature will increase the yield of barium but doing so may slow the reaction rate.

14.51   10.2 g

14.53 $4 Al^{3+}(melt) + 6 O^{2-}(melt) + 3 C(s, gr) \rightarrow 4 Al(s) + 3 CO_2(g)$

14.55   (a) $B_2O_3(s) + 3 Mg(l) \rightarrow 2 B(s) + 3 MgO(s)$; (b) $2 Al(s) + 3 Cl_2(g) \rightarrow 2 AlCl_3(s)$; (c) $4 Al(s) + 3 O_2(g) \rightarrow 2 Al_2O_3(s)$

14.57   (a) The hydrate of $AlCl_3$ ($AlCl_3 \cdot 6H_2O$) functions as a deodorant and antiperspirant. (b) corundum, α-alumina, is used as an abrasive in sandpaper. (c) $B(OH)_3$ is an antiseptic and insecticide.

14.59

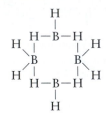

Bridging H atoms and four-coordinate B atoms have formal charge of −1; terminal H atoms and three-coordinate B atoms have formal charge of 0.

14.61   The shape of $GaBr_4^-$ is tetrahedral.

$$\left[\begin{array}{c} :\!\ddot{B}r\!: \\ | \\ :\!\ddot{B}r\!-\!Ga\!-\!\ddot{B}r\!: \\ | \\ :\!\ddot{B}r\!: \end{array}\right]^-$$

14.63 1.2×10^7 g

14.65   (a) 0.743 V; (b) no

14.67   Silicon occurs widely in the Earth's crust in the form of silicates in rocks and as silicon dioxide in sand. It is obtained from quartzite, a form of quartz (SiO_2), by the following processes: (1) reduction in an electric arc furnace, $SiO_2(s) + 2 C(s) \rightarrow Si(s, crude) + 2 CO(g)$; (2) purification of the crude product in two steps, $Si(s, crude) + 2 Cl_2(g) \rightarrow SiCl_4(l)$; followed by reduction with hydrogen to the pure element, $SiCl_4(l) + 2 H_2(g) \rightarrow Si(s, pure) + 4 HCl(g)$

14.69   In diamond, carbon is sp^3 hybridized and forms a tetrahedral, three-dimensional network structure, which is extremely rigid. Graphite carbon is sp^2 hybridized and planar. Its application as a lubricant results from the fact that the two-dimensional sheets can "slide" across one another, thereby reducing friction. In graphite, the unhybridized *p*-electrons are free to move from one carbon atom to another, which results in its high electrical conductivity. In diamond, all electrons are localized in sp^3 hybridized C—C σ-bonds, so diamond is a poor conductor of electricity.

14.71   (a) 4; (b) 4; (c) 4

14.73   (a) $MgC_2(s) + 2 H_2O(l) \rightarrow C_2H_2(g) + Mg(OH)_2(s)$, acid−base; (b) $2 Pb(NO_3)_2(s) \rightarrow 2 PbO(s) + 4 NO_2(g) + O_2(g)$, redox

14.75

$$\left[\begin{array}{c} :\!\ddot{O}\!: \\ | \\ :\!\ddot{O}\!-\!Si\!-\!\ddot{O}\!: \\ | \\ :\!\ddot{O}\!: \end{array}\right]^{4-}$$

Formal charges: Si = 0, O = −1
Oxidation numbers: Si = 4, O = −2
This is an AX_4 VSEPR structure; therefore, the shape is tetrahedral.

14.77 $\Delta H_r° = 689.88 \text{ kJ·mol}^{-1}$, $\Delta S_r° = 360.85 \text{ J·K}^{-1}\text{·mol}^{-1}$, $\Delta G_r° = 582.29 \text{ kJ·mol}^{-1}$, 1912 K

14.79   5.99 mg

14.81   The $Si_2O_7^{6-}$ ion is built from two SiO_4^{4-} tetrahedral ions in which the silicate tetrahedra share one O atom. This is the only case in which one O is shared. (b) The pyroxenes—for example, jade, $NaAl(SiO_3)_2$—consist of chains of SiO_4 units in which two O atoms are shared by neighboring units. The repeating unit has the formula SiO_3^{2-}.

14.83 SiF_6^{2-}

14.85   Ionic fluorides react with water to liberate HF, which then reacts with the glass. Glass bottles used to store metal fluorides become brittle and may disintegrate while standing on the shelf.

14.87   The iron ions impart a deep red color to the clay, which is not desirable for the manufacture of fine china; a white base is aesthetically more pleasing.

14.89   (a) False. Four elements are metals. (b) False. B_2O_3 is an acidic anhydride. (c) False. The outer *s*- and *p*-orbitals become further apart in energy going down the group (the inert-pair effect).

14.91   In most of its reactions, hydrogen acts as a reducing agent. Examples are $2 H_2(g) + O_2(g) \rightarrow 2 H_2O(l)$ and various ore reduction processes, such as $NiO(s) + H_2(g) \rightarrow Ni(s) + H_2O(g)$. With highly electropositive elements, such as the alkali metals, $H_2(g)$ acts as an oxidizing agent and forms metal hydrides—for example, $2 K(s) + H_2(g) \rightarrow 2 KH(s)$.

14.93 (a)

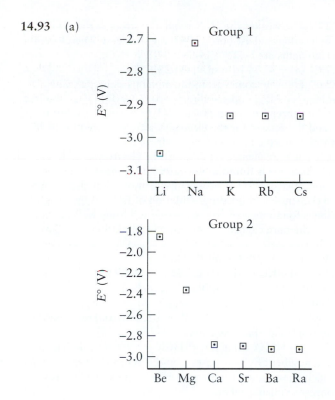

(b) For both groups, the trend in standard potentials with increasing atomic number is overall downward (they become more negative), but lithium is anomalous. This overall downward trend makes sense because we expect that it is easier to remove electrons that are farther away from the nuclei. However, because there are several factors that influence ease of removal, the trend is not smooth. The potentials are a net composite of the free energies of sublimation of solids, dissociation of gaseous molecules, ionization enthalpies, and enthalpies of hydration of gaseous ions. The origin of the anomalously strong reducing power of Li is the strongly exothermic energy of hydration of the very small Li$^+$ ion, which favors the ionization of the element in aqueous solution.

14.95 (a) acid−base; (b) CaO(s) + CO$_2$(g) → CaCO$_3$(s)

14.97 (a) The +1 cation produced is a tetrahedral B atom with two H and two NH$_3$ groups, the −1 anion is the boron terahydride ion.

(b) Hybridization is sp^3 for both B atoms.

(c) The reaction is Lewis acid–base because each N atom provides the electrons for a coordinate covalent bond. The oxidation numbers do not change (+3 for B and −1 for H).

14.99 In most of its reactions, hydrogen acts as a reducing agent, like the alkali metals. However, it may also act as an oxidizing agent, like the halogens. Consequently, H$_2$ will oxidize elements with standard reduction potentials more negative than −2.25 V, such as the alkali and alkaline earth metals (except Be). The compounds formed are hydrides and contain the H$^-$ ion; the singly charged negative ion is reminiscent of the halide ions. Hydrogen also forms diatomic molecules and covalent bonds like the halogens. Consequently, hydrogen could be placed in either Group 1 or Group 17. But it is best to think of hydrogen as a unique element that has properties in common with both metals and nonmetals; therefore, it should probably be centered in the periodic table, as it is shown in the table in the text.

14.101 Li$^+$ is smaller than K$^+$, so it has stronger ion-dipole interactions with H$_2$O than K$^+$, which will make it move more slowly than K$^+$.

14.103 0.0538 mol·L^{-1}

14.105 3.34 L

14.107 Gases (a), (b), (c), and (d) can all function as greenhouse gases, whereas gas (e) cannot; argon is monoatomic.

14.109 1.09 kg. Octane contributes more CO$_2$ per liter burned. We also need to consider how much energy is produced per liter of fuel burned.

14.111 (a) Diborane (B$_2$H$_6$) and Al$_2$Cl$_6$(g) have the same basic structure in the way in which the atoms are arranged in space. (b) The bonding between the boron atoms and the bridging hydrogen atoms is electron deficient. There are three atoms and only two electrons to hold them together in a three-center−two-electron bond. The bonding in Al$_2$Cl$_6$ is conventional in that all the bonds entail two atoms and two electrons. Here, the lone pair of a Cl atom is donated to an adjacent Al atom. (c) The hybridization is sp^3 at the B and Al atoms. (d) The molecules are not planar. The Group 13/III element and the terminal atoms to which it is bound lie in a plane that is perpendicular to the plane that contains the bridging atoms.

14.113 (a) hexagonal. (b) The total number of carbonate ions in the unit cell is 6. The total number of calcium ions in the cell is 6, agreeing with the overall stoichiometry of calcite, CaCO$_3$.

14.115 (a) C: sp^2, B: sp^2, N: sp^2. (b) 10. (c) In C$_{60}$ the carbon atoms are sp^2 hybridized and are nearly planar. However, the curvature of the molecule introduces some strain at the carbon atoms so that there is some tendency for some of the carbon atoms to undergo conversion to sp^3 hybridization. However, to make every carbon sp^3 hybridized would introduce much more strain on the carbon cage and, after a certain point, further addition of hydrogen becomes unfavorable. (d) The spherical structures required the formation of five-member rings. Boron nitride cannot form these rings because they would require high-energy boron–boron or nitrogen–nitrogen bonds. (e) 3.491 g·cm^{-1}. (f) cubic

Chapter 15

15.1 −3: NH$_3$, Li$_3$N, LiNH$_2$, NH$_2^-$; −2: H$_2$NNH$_2$; −1: N$_2$H$_2$, NH$_2$OH; 0: N$_2$; +1: N$_2$O, N$_2$F$_2$; +2: NO; +3: NF$_3$, NO$_2^-$, NO$^+$; +4: NO$_2$, N$_2$O$_4$; +5: HNO$_3$, NO$_3^-$, NO$_2$F

15.3 As the molar mass of the halogen increases, its size increases and it becomes increasingly difficult for nitrogen to arrange three halogen atoms around itself in a trigonal pyramid due to the lone-pair−lone-pair repulsions on the halogens.

15.5 CO(NH$_2$)$_2$(aq) + 2 H$_2$O(l) → (NH$_4$)$_2$CO$_3$(aq); 6.4 kg

15.7 (a) 0.35 L. (b) larger volume. (c) Metal azides are potent explosives because the azide ion is thermodynamically unstable with respect to the production of N$_2$(g). The azide is unstable because the N≡N triple bond is very strong and because the production of a gas is favored by the increase in entropy.

15.9 N_2O: $H_2N_2O_2$; $N_2O(g) + H_2O(l) \rightarrow H_2N_2O_2(aq)$; N_2O_3: HNO_2; $N_2O_3(g) + H_2O(l) \rightarrow 2\ HNO_2(aq)$; N_2O_5: HNO_3; $N_2O_5(g) + H_2O(l) \rightarrow 2\ HNO_3(aq)$

15.11   Ammonia can form hydrogen bonds with itself, whereas NF_3 cannot.

15.13   The Lewis structures of phosphorous (III) oxide (P_4O_6) and phosphorous (V) oxide (P_4O_{10}) are:

Phosphorus(III) oxide   Phosphorus(V) oxide

The basic structure of the two molecules is the same. The phosphorus atoms lie in a tetrahedral arrangement in which there are bridging oxygen atoms to the other phosphorus atoms. In phosphorus(V) oxide, there is an additional terminal oxygen atom bonded to each phosphorus atom. In phosphorus(III) oxide, each oxygen atom has a formal charge of 0, as does each phosphorus atom. In phosphorus(V) oxide, this is also true. According to the Lewis structures, all P–O bonds in phosphorus(III) oxide have a bond order of 1, while in phosphorus(V) oxide, the terminal oxygen atoms have a bond order of 2 between them and the phosphorus atoms to which they are attached. If one examines the molecular parameters, one sees that all of the P–O$_{bridging}$ distances for phosphorus(III) oxide are slightly longer than those of phosphorus(V) oxide (163.8 pm versus 160.4 pm). This is expected because the radius of phosphorus(V) should be smaller than that of phosphorus(III). The terminal P=O distances for phosphorus(V) oxide are considerably shorter (142.9 pm); this agrees with the higher bond order between phosphorus and these atoms.

15.15   (a) $-138.18\ kJ\cdot mol^{-1}$; (b) $-33.26\ kJ\cdot mol^{-1}$

15.17   (a) $4\ Li(s) + O_2(g) \overset{\Delta}{\rightarrow} 2\ Li_2O(s)$;
(b) $2\ Na(s) + 2\ H_2O(l) \rightarrow 2\ NaOH(aq) + H_2(g)$;
(c) $2\ F_2(g) + 2\ H_2O(l) \rightarrow 4\ HF(aq) + O_2(g)$;
(d) $2\ H_2O(l) \rightarrow O_2(g) + 4\ H^+(aq) + 4\ e^-$

15.19   (a) $2\ H_2S(g) + 3\ O_2(g) \rightarrow 2\ SO_2(g) + 2\ H_2O(g)$;
(b) $CaO(s) + H_2O(l) \rightarrow Ca(OH)_2(aq)$;
(c) $2\ H_2S(g) + SO_2(g) \rightarrow 3\ S(s) + 2\ H_2O(l)$

15.1   Both water and ammonia have four groups attached to their central atom and therefore both possess a tetrahedral electronic (or VSEPR) geometry. However, H_2O has two unshared electron pairs while NH_3 only has one, producing a larger dipole moment for H_2O.

15.23   (a)

H
|
:O—O:
|
H

The bond angle is predicted to be less than 109.5°. The experimental value is 97°.

(b) Cu^+ and (c) Mn^{2+} will be oxidized.

15.25   13.11 (13.12 if K_{a2} is not ignored)

15.27   The weaker the H–X bond, the stronger the acid. H_2Te has the weakest bond; H_2O, the strongest. Therefore, the acid strengths are $H_2Te > H_2Se > H_2S > H_2O$.

15.29   (a) $-73.35\ kJ\cdot mol^{-1}$; (b) 28.6°C

15.31   Fluorine comes from the minerals fluorspar, CaF_2; cryolite, Na_3AlF_6; and the fluorapatites, $Ca_5F(PO_4)_3$. The free element is prepared from HF and KF by electrolysis, but the HF and KF needed for the electrolysis are prepared in the laboratory. Chlorine primarily comes from the mineral rock salt, NaCl. The pure element is obtained by electrolysis of liquid NaCl. Bromine is found in seawater and brine wells as the Br^- ion; it is also found as a component of saline deposits; the pure element is obtained by oxidation of $Br^-(aq)$ by $Cl_2(g)$. Iodine is found in seawater, seaweed, and brine wells as the I^- ion; the pure element is obtained by oxidation of $I^-(aq)$ by $Cl_2(g)$.

15.33   (a) +1; (b) +4; (c) +7; (d) +5

15.35   (a) $4\ KClO_3(l) \rightarrow 3\ KClO_4(s) + KCl(s)$
(b) $Br_2(l) + H_2O(l) \overset{\Delta}{\rightarrow} HBrO(aq) + HBr(aq)$
(c) $NaCl(s) + H_2SO_4(aq) \rightarrow NaHSO_4(aq) + HCl(g)$
(d) Reactions (a) and (b) are redox reactions and reaction (c) is a Brønsted acid−base reaction.

15.37   (a) $HClO < HClO_2 < HClO_3 < HClO_4$. (b) The oxidation number of Cl increases from $HClO$ to $HClO_4$. Chlorine has its highest oxidation number of +7, and so $HClO_4$ is the strongest oxidizing agent.

15.39 $:\!\overset{..}{Cl}\!-\!\overset{..}{O}\!-\!\overset{..}{Cl}\!:$, AX_2E_2, angular, about 109°. The actual bond angle is 110.9°.

15.41

The thermodynamic stability of the hydrogen halides decreases down the group. The $\Delta G_f°$ values of HCl, HBr, and HI fit nicely on a straight line, whereas HF is anomalous. In other properties, HF is also the anomalous member of the group, in particular, its acidity.

15.43   (a) ICl_5 has too many highly electronegative chlorine atoms present on the central iodine. (b) IF_2 has an odd number of electrons and as a result is a radical and highly reactive. (c) $ClBr_3$ is crowded sterically (too many large atoms on a small central atom).

15.45   (a) IF_3; (b) $3\ XeF_2(g) + I_2(g) \rightarrow 2\ IF_3(s) + 3\ Xe(g)$

15.47   Because $E_{cell}°$ is negative, $Cl_2(g)$ will not oxidize Mn^{2+} to form the permanganate ion in an acidic solution.

15.49 $0.117\ mol\cdot L^{-1}$

15.51 7.59×10^3 kJ of heat released

15.53 Helium occurs as a component of natural gases found under rock formations in certain locations, especially some in Texas. Argon is obtained by distillation of liquid air.

15.55 (a) $+2$; (b) $+6$; (c) $+4$; (d) $+6$

15.57 $XeF_4(aq) + 4\,H^+(aq) + 4\,e^- \rightarrow Xe(g) + 4\,HF(aq)$

15.59 Because H_4XeO_6 has more highly electronegative O atoms bonded to Xe, it should be more acidic than H_2XeO_4.

15.61 In fluorescence, light absorbed by molecules is immediately emitted, whereas, in phosphorescence, molecules remain in an excited state for a period of time before emitting the absorbed light. In both phenomena, the energy of the emitted photon is lower than that of the absorbed photon (emitted photons have a longer wavelength).

15.63 Fluorescent dyes allow much smaller concentrations of biomolecules to be detected than is normally possible.

15.65 The "top down" approach refers to physically assembling the nanoparticles into desired forms; the "bottom up" approach utilizes specific intermolecular interactions to cause the nanomaterials to self-assemble.

15.67 strongest: (a), ion–ion; weakest: (d) dipole–dipole

15.69 (a) sp^3 hybridized

(b) sp^3 hybridization requires bond angles of 109.5°; however, the internal bond angles required by this structure are 60°, which means that the bonds will be strained.

15.71 (a) 14 electrons: CN^-, N_2, and C_2^{2-}; 24 electrons: NO_2^- and O_3; (b) strongest Lewis bases: NO_2^- and C_2^{2-}; (c) strongest reducing agents: NO_2^- and C_2^{2-}

15.73 $I_2 < Br_2 < Cl_2 < F_2$

15.75 (a) $2\,NH_3(aq) + ClO^-(aq) \rightarrow N_2H_4(aq) + Cl^-(aq) + H_2O(l)$;
(b) $Mg_3N_2(s) + 6\,H_2O(l) \rightarrow 3\,Mg(OH)_2(s) + 2\,NH_3(g)$;
(c) $P^{3-}(s) + 3\,H_2O(l) \rightarrow PH_3(g) + 3\,OH^-(aq)$;
(d) $Cl_2(g) + H_2O(l) \rightarrow HClO(aq) + HCl(aq)$;
(a) and (d) are redox reactions. In (a), ClO^- is the oxidizing agent and NH_3 is the reducing agent. In (d), Cl_2 is both oxidized and reduced. (b) is a Brønsted acid–base reaction; H_2O is the acid and N^{3-} is the base. (c) is a Brønsted acid–base reaction; P^{3-} is the base and H_2O is the acid.

15.77 (a) State A: 94.72 kJ·mol^{-1}, State B: 157.85 kJ·mol^{-1}; (b) 1.263 μm

15.79 Due to reduced hydrogen bonding, thiosulfuric acid should be less acidic and have a lower boiling point.

15.81 337.9 g

15.83 Orpiment is As_2S_3 and realgar is As_4S_4. Orpiment is yellow and realgar is orange-red. They are both used as pigments.

15.85 (a) $\ddot{N}=\ddot{N}=\ddot{N}^{-}$; AX_2, linear 180°.
(b) between fluorine and chlorine. (c) HCl, HBr, and HI are all strong acids. For HF, $K_a = 3.5 \times 10^{-4}$, so HF is slightly more acidic than HN_3. The small size of the azide ion suggests that the H—N bond in HN_3 is similar in strength to that of the H—F bond, so it is expected to be a weak acid. (d) ionic: NaN_3, $Pb(N_3)_2$, AgN_3; covalent: HN_3, $B(N_3)_3$, FN_3

15.87 0.156 mol·L^{-1}

15.89 The solubilities of the ionic halides are determined by a variety of factors, especially the lattice enthalpy and enthalpy of hydration. There is a delicate balance between the two factors, with the lattice enthalpy usually being the determining one. Lattice enthalpies decrease from chloride to iodide, so water molecules can more readily separate the ions in the latter. Less ionic halides, such as the silver halides, generally have a much lower solubility, and the trend in solubility is the reverse of the more ionic halides. For the less ionic halides, the covalent character of the bond allows the ion pairs to persist in water. The ions are not easily hydrated, making them less soluble. The polarizability of the halide ions and the covalency of their bonding increases down the group.

15.91 (a) 0.26 V, spontaneous. (b) The formation of AgI precipitate means that the concentration of Ag^+ ions is never high enough to achieve the conditions necessary for the redox reaction to take place. The solubility product, K_{sp}, limits the concentrations in solution, so that the actual redox potential is not the value calculated, which represents the values when $[Ag^+]$ and $[I^-]$ are 1 mol·L^{-1}. If we use the concentrations established by the solubility equilibrium and the Nernst equation, we can calculate the actual redox potential to be -0.68 V.

15.93 (a) The molecular orbital diagram for NO^+ should have the oxygen orbitals slightly lower in energy than the nitrogen orbitals because oxygen is more electronegative. This will cause the bonding to be more ionic than in either N_2 or O_2. There is an ambiguity, however, in that the MO diagram could be similar to either that of N_2 or that of O_2. Refer to Figures 3.31 and 3.32 where you will see that the σ_{2p} and π_{2p} have different relative energies. There are consequently two possibilities for the orbital energy diagram:

NO$^+$
(using N MO diagram)

NO$^+$
(using O MO diagram)

(b) The two orbital diagrams predict the same bond order (3) and same magnetic properties (diamagnetic), and so these properties cannot be used to determine which diagram is the correct one. That must be determined by more complex spectroscopic measurements.

15.95 (a) face-centered cubic. (b) The iron atoms lie at the corners and at the face centers of the unit cell. Eight sulfur atoms lie completely within the unit cell. (c) six. (d) one sulfur atom and three iron atoms. (e) Fe: $+2$, S: -1. An examination of the structure indicates that sulfur atoms are best thought of as S_2^{2-} ions. The locations of the midpoints of the S—S bonds are at the centers of each edge of the unit cell ($12\,S_2^{2-} \times \frac{1}{4}$) plus one S_2^{2-} ion in the center of the unit cell (only half the ions have sulfur atoms within a given cell). This gives a total of four S_2^{2-} ions in the unit cell.

15.97 (a) cloud: $[NO] = 860$ ppt and $[NO_2] = 250$ ppt. clear air: $[NO] = 480$ ppt and $[NO_2] = 260$ ppt. (b) clouds. (c) In addition to jet engines and automobile emissions, NO is also formed when NO_2 dissolves in water. As a result one would expect to find more NO and less NO_2 in clouds, which also contain water vapor. (d) nitrous acid

$$\overset{..}{\underset{..}{O}}=\overset{..}{N}-\overset{..}{\underset{..}{O}}-H$$

(e) Yes. The hydroxyl radical and NO have unpaired electrons; nitrous acid does not. (f) A possible mechanism that fits the observed rate law would be:
step 1 (slow): $NO + O_3 \rightarrow NO_2 + O_2$, step 2 (fast): $NO_2 + O \rightarrow NO + O_2$. (g) 7.56×10^{-19} molecules·cm^{-3}·s^{-1}

Chapter 16

16.1 A strongly negative standard potential indicates a tendency to undergo oxidation, which involves loss of electrons. Elements at the left of the d block have lower ionization energies than those at the right.

16.3 (a) Ti; (b) approximately equal due to lanthanide contraction; (c) Ta; (d) Ir

16.5 (a) Fe; (b) Cu; (c) Pt; (d) Pd; (e) Ta

16.7 Hg is much more dense than Cd, because the decrease in atomic radius that occurs between $Z = 58$ and $Z = 71$ (the lanthanide contraction) causes the atoms following the rare earths to be smaller than might have been expected for their atomic masses and atomic numbers. Zn and Cd have densities that are not too dissimilar because the radius of Cd is subject only to a smaller d-block contraction.

16.9 (a) Proceeding down a group in the d block, there is an increasing probability of finding the elements in higher oxidation states. That is, higher oxidation states become more stable on going down a group. (b) The trend for the p-block elements is reversed. Because of the inert-pair effect, the higher oxidation states tend to be less stable as one descends a group.

16.11 The $+6$ oxidation state is most stable for Cr.

16.13 (a) $TiCl_4(g) + 2\, Mg(l) \rightarrow Ti(s) + 2\, MgCl_2(s)$;
(b) $CoCO_3(s) + HNO_3(aq) \rightarrow Co^{2+}(aq) + HCO_3^-(aq) + NO_3^-(aq)$; (c) $V_2O_5(s) + 5\, Ca(l) \rightarrow 2\, V(s) + 5\, CaO(s)$

16.15 (a) titanium(IV) oxide, TiO_2; (b) iron(III) oxide, Fe_2O_3; (c) manganese(IV) oxide, MnO_2; (d) iron(II) chromite, $FeCr_2O_4$

16.17 (a) yes; (b) no

16.19 (a) $V_2O_5(s) + 2\, H_3O^+(aq) \rightarrow 2\, VO_2^+(aq) + 3\, H_2O(l)$;
(b) $V_2O_5(s) + 6\, OH^-(aq) \rightarrow 2\, VO_4^{3-}(aq) + 3\, H_2O(l)$

16.21 As the value of n increases, d- and f-electrons become less effective at shielding the outermost, highest-energy electron(s) from the attractive charge of the nucleus. This higher effective nuclear charge makes it more difficult to oxidize the metal atom or ion.

16.23 (a) $[Cr(H_2O)_6]^{3+}(aq)$ behaves as a Brønsted acid: $[Cr(H_2O)_6]^{3+}(aq) + H_2O(l) \rightarrow [Cr(H_2O)_5OH]^{2+}(aq) + H_3O^+(aq)$. (b) The gelatinous precipitate is the hydroxide $Cr(OH)_3$. The precipitate dissolves as the $Cr(OH)_4^-$ complex ion is formed: $Cr^{3+}(aq) + 3\, OH^-(aq) \rightarrow Cr(OH)_3(s)$ and $Cr(OH)_3(s) + OH^-(aq) \rightarrow Cr(OH)_4^-(aq)$

16.25 (a) hexacyanoferrate(II) ion, $+2$; (b) exaaminecobalt(III) ion, $+3$; (c) aquapentacyanocobaltate(III) ion, $+3$; (d) pentaamminesulfatocobalt(III) ion, $+3$

16.27 (a) $K_3[Cr(CN)_6]$; (b) $[Co(NH_3)_5(SO_4)]Cl$;
(c) $[Co(NH_3)_4(H_2O)_2]Br_3$; (d) $Na[Fe(H_2O)_2(C_2O_4)_2]$

16.29 (a) 3; (b) 1 or 2; (c) monodentate, 1; (d) 2

16.31 Only (b); in (a) and (c), the amine groups are too far apart to cordinate to the same metal atom.

16.33 (a) 4; (b) 2; (c) 6 (en is bidentate); (d) 6 (EDTA is hexadentate)

16.35 (a) structural isomers, linkage isomers; (b) structural isomers, ionization isomers; (c) structural isomers, linkage isomers; (d) structural isomers, ionization isomers

16.37 (a) yes,

trans-Tetraamminedichlorocobalt(III) chloride monohydrate

cis-Tetraamminedichlorocobalt(III) chloride monohydrate

(b) no;
(c) yes,

cis-Diamminedichloroplatinum(II)

trans-Diamminedichloroplatinum(II)

16.39 (a) chiral; (b) not chiral

16.41 (a) 1; (b) 6; (c) 5; (d) 3; (e) 6; (f) 6

16.43 (a) 2; (b) 5; (c) 8; (d) 10; (e) 0 (or 8); (f) 10

16.45 (a) octahedral, strong-field ligand, 6 e$^-$, no unpaired electrons

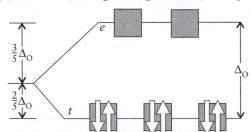

(b) tetrahedral, weak-field ligand, 8 e$^-$, two unpaired electrons,

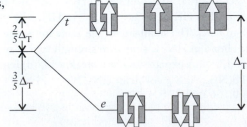

(c) octahedral, weak-field ligand, 5 e⁻, five unpaired electrons

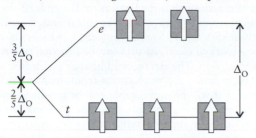

(d) octahedral, strong-field ligand, 5 e⁻, one unpaired electron

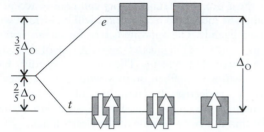

16.47 (a) 0; (b) 2

16.49 Weak-field ligands do not interact strongly with the d-electrons in the metal ion, so they produce only a small crystal field splitting of the d-electron energy states. The opposite is true of strong-field ligands. With weak-field ligands, unpaired electrons remain unpaired if there are unfilled orbitals; hence, a weak-field ligand is likely to lead to a high-spin complex. Strong-field ligands cause electrons to pair up with electrons in lower-energy orbitals. A strong-field ligand is likely to lead to a low-spin complex. The arrangement of ligands in the spectrochemical series helps to distinguish strong-field and weak-field ligands. Measurement of magnetic susceptibility (paramagnetism) can be used to determine the number of unpaired electrons, which, in turn, establishes whether the associated ligand is weak-field or strong-field in nature.

16.51 (a) $[CoF_6]^{3-}$ is blue because F^- is a weak-field ligand. (b) $[Co(en)_3]^{3+}$ is yellow because the en ligand is a strong-field ligand.

16.53 573 nm, yellow

16.55 In Zn^{2+}, the $3d$-orbitals are filled (d^{10}). Therefore, there can be no electronic transitions between the t and e levels; hence, no visible light is absorbed and the aqueous ion is colorless. The d^{10} configuration has no unpaired electrons, so Zn compounds would be diamagnetic, not paramagnetic.

16.57 (a) 162 kJ·mol⁻¹; (b) 260. kJ·mol⁻¹; (c) 208 kJ·mol⁻¹; (d) $Cl^- < H_2O < NH_3$

16.59 the e_g set, which comprises the $d_{x^2-y^2}$ and d_{z^2} orbitals

16.61 (a) The CN^- ion is a π-acid ligand accepting electrons into the empty π^* orbital created by the C–N multiple bond. (b) The Cl^- ion has extra lone pairs in addition to the one that is used to form the σ-bond to the metal, and so it can act as a π-base, donating electrons in a p-orbital to an empty d-orbital on the metal. (c) H_2O, like Cl^-, also has an "extra" lone pair of electrons that can be donated to a metal center, making it a weak π-base; (d) en is neither a π-acid nor a π-base, because it does not have any empty π-type antibonding orbitals, nor does it have any extra lone pairs of electrons to donate. (e) $Cl^- < H_2O < en < CN^-$. Note that the spectrochemical series orders the ligands as π-bases < σ-bond only ligands < π-acceptors.

16.63 nonbonding or slightly antibonding. In a complex that forms only σ-bonds, the t_{2g} set of orbitals is nonbonding. If the ligands can function as weak π-donors (those close to the middle of the spectrochemical series, such as H_2O), the t_{2g} set becomes slightly antibonding by interacting with the filled p-orbitals on the ligands.

16.65 antibonding. The e_g set of orbitals on an octahedral metal ion are always antibonding because of interactions with ligand orbitals that form the σ-bonds. This is true regardless of whether the ligands are π-acceptors, π-donors, or neither.

16.67 Water has two lone pairs of electrons. Once one of these is used to form the σ-bond to the metal ion, the second may be used to form a π-bond. This causes the t_{2g} set of orbitals to move up in energy, making Δ_O smaller; therefore, water is a weak-field ligand. Ammonia does not have this extra lone pair of electrons and consequently cannot function as a π-donor ligand.

16.69 (a) CO. (b) in Zones D & C: $3\ Fe_2O_3(s) + CO(g) \rightarrow 2\ Fe_3O_4(s) + CO_2(g)$; $Fe_3O_4(s) + CO(g) \rightarrow 3\ FeO(s) + CO_2(g)$. These reactions combine to give $Fe_2O_3(s) + CO(g) \rightarrow 2\ FeO(s) + CO_2(g)$. in Zone B: $Fe_2O_3(s) + 3\ CO(g) \rightarrow 2\ Fe(s) + 3\ CO_2(g)$; $FeO(s) + CO(g) \rightarrow Fe(s) + CO_2(g)$. (c) carbon

16.71 The major impurity is carbon; it is removed by oxidation of the carbon to CO_2, followed by capture of the CO_2 by base to form a slag.

16.73 copper and zinc

16.75 Alloys are usually (1) harder and more brittle, and (2) poorer conductors of electricity than the metals from which they are made.

16.77 The compound is ferromagnetic below T_C because the magnetization is higher. Above the Curie temperature, the compound is a simple paramagnet with randomly oriented spins, but below that temperature, the spins align and the magnetization increases.

16.79 The green color suggests chromium or copper but the blue solution formed when the mineral is dissolved in sulfuric acid points to copper as the metal ($CuSO_4$ is blue). The colorless gas that forms either upon heating or on treatment with acid is carbon dioxide; this is proven when the gas is bubbled into limewater (the cloudiness in the solution is due to $CaCO_3$ formation). CO_2 is formed when carbonates are either heated or treated with acid. The mineral is most likely basic copper(II) carbonate ($Cu_2CO_3(OH)_2$).

16.81 (a) +3; (b) $2\ V_2O_5(s) + 8\ C(s) \rightarrow V_4C_3(s) + 5\ CO_2(g)$

16.83 3.6×10^{-31}

16.85 (a)

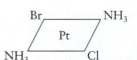

cis-Diamminebromochloroplatinum(II)

Br⟍   ⟋NH₃
        Pt
NH₃⟍   ⟋Cl

trans-Diamminebromochloroplatinum(II)

(b) If $[PtBrCl(NH_3)_2]$ was tetrahedral, a second isomer would not exist.

16.87 (a) The first, $[Ni(SO_4)(en)_2]Cl_2$, will give a precipitate of AgCl when $AgNO_3$ is added; the second will not. (b) The second, $[NiCl_2(en)_2]I_2$, will show free I_2 when mildly oxidized with Br_2, but the first will not.

16.89 (a) $[MnCl_6]^{4-}$:

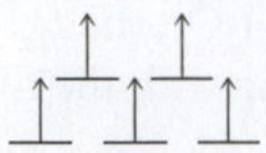

$$\uparrow \quad \uparrow$$
$$\uparrow \quad \uparrow \quad \uparrow$$

$[Mn(CN)_6]^{4-}$:

$$\uparrow\downarrow \quad \uparrow\downarrow \quad \uparrow$$

(b) five in $[MnCl_6]^{4-}$, one in $[Mn(CN)_6]^{4-}$; (c) Complexes with weak-field ligands absorb longer wavelength light; therefore, $[MnCl_6]^{4-}$ absorbs longer wavelengths.

16.91 High-spin Mn^{2+} ions have a d^5 configuration with 5 unpaired electrons. For light to be absorbed in the visible region of the spectrum, an electron from the t_{2g} set has to be moved into the e_g set of orbitals. Because each orbital is already singly occupied by an electron and all five electrons have parallel spins, the electron making the transition must change spin in order to spin-pair in the upper orbital. This sort of transition is called "spin-forbidden" because it has a very low quantum mechanical probability of occurring, and so the complexes usually are only faintly colored.

16.93 If the prismatic structure were correct, six structural isomers should exist, not two, four of which are shown here (X = NH_3):

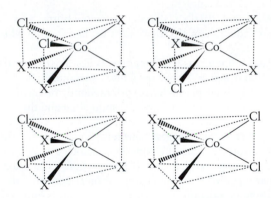

16.95 The correct structure for $[Co(NH_3)_6]Cl_3$ consists of four ions, $Co(NH_3)_6^{3+}$ and 3 Cl^- in aqueous solution. The chloride ions can be easily precipitated as AgCl. This would not be possible if they were bonded to the other (NH_3) ligands. If the structure were $Co(NH_3-NH_3-Cl)_3$, VSEPR theory would predict that the Co^{3+} ion would have a trigonal planar ligand arrangement. The splitting of the d-orbital energies would not be the same as the octahedral arrangement and would lead to different spectroscopic and magnetic properties inconsistent with the experimental evidence. In addition, neither optical nor geometrical isomers would be observed.

16.97 The spectrochemical series given in Figure 16.29 shows the relative ligand field strengths of the halide ions to lie in the order $I^- < Br^- < Cl^- < F^-$. Since their electronegativities follow the same trend, there is a positive correlation between ligand field strength and electronegativity for the halide ions. The value of Δ_O correlates with the ability of the ligand's extra lone pairs of electrons to interact with the t_{2g} set of the octahedral metal ion. Thus, the less electronegative the ligand is, the easier it is for the ligand to donate electrons to the metal ion, the more the energy of the t_{2g} set in the complex is raised, and the smaller Δ_O becomes (see Figure 16.38a).

16.99 To determine these relations, we need to consider the types of interactions that the ligands on the ends of the spectrochemical series will have with the metal ions. Those that are weak-field (form high spin complexes, π-bases) have extra lone pairs of electrons that can be donated to a metal ion in a π fashion. The strong-field ligands (form low spin complexes, π-acids) accept electrons from the metals. The complexes that will be more stable will be produced in general by the match between ligand and metal. Thus, the early transition metals in high oxidation states will have few or no electrons in the d-orbitals. These metal ions will become stabilized by ligands that can donate more electrons to the metal—the d-orbitals that are empty can readily accept electrons. The more stable complexes will be formed with the weak-field ligands. The opposite is true for metals with many electrons, which are the ones at the right side of the periodic table, in low oxidation states. These metals generally have most of the d-orbitals filled, so they would, in fact, be destabilized by π-electron donation. Instead, they form more stable complexes with the π-acceptor ligands (strong-field, π-acids) that can remove some of the electron density from the metal ions.

16.101 (a) $PtCl_2(NH_3)_2$, *cis*-diamminedichloroplatinum(II). (b) The only other isomer is the trans form. Neither the cis nor the trans form is optically active.

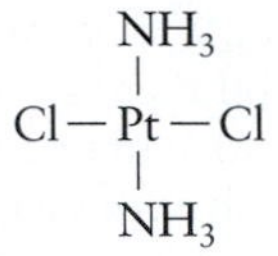

$$
\begin{array}{c}
NH_3 \\
| \\
Cl-Pt-Cl \\
| \\
NH_3
\end{array}
$$

(c) square planar

16.103 $[CrNH_3Cl(H_2O)_2]Cl_2$

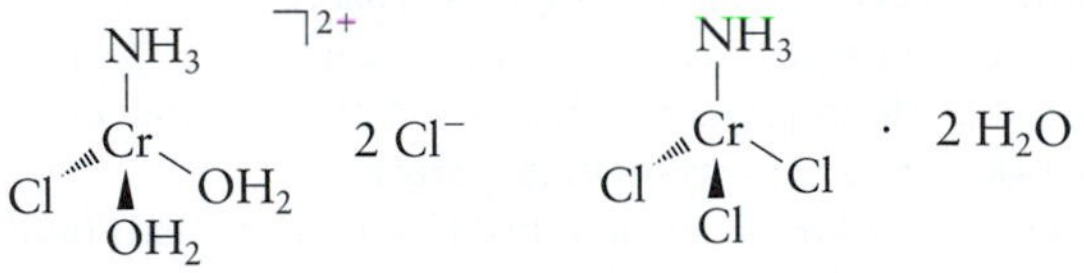

isomers:

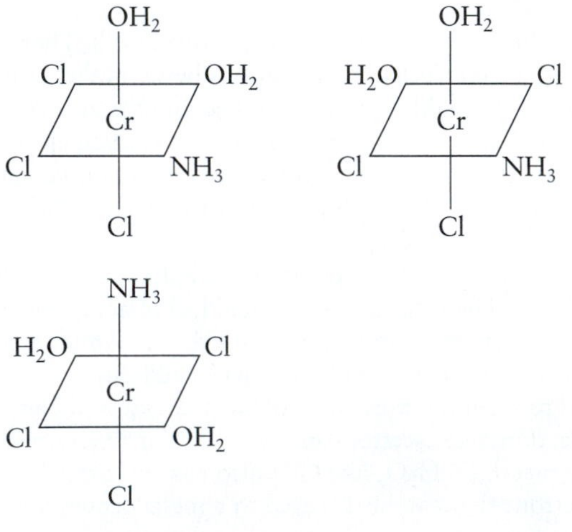

16.105 4.0×10^{-5}

16.107 The electron configuration expected for Ni^{2+} is $[Ar]3d^8$. For this complex to have no unpaired electrons it would have to be (c) square planar in its electronic geometry, as both the octahedral and tetrahedral geometries require a d^8 species to have two unpaired electrons. Square planar does not.

Octahedral d^8 complex   Tetrahedral d^8 complex

Square planar d^8 complex

16.109 (a)

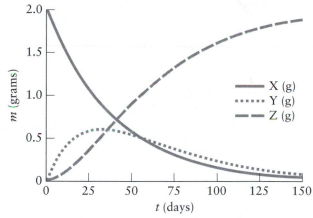

(b) deoxygenated: 4, oxygenated: 0.
(c) For a ligand to bond to the iron atom, it needs to be a Lewis base. Of the species listed, only BF_3 will not be able to do so as it is a Lewis acid.
(d) 86% deprotonated deoxyhemoglobin (Hb^-), 17% deprotonated oxyhemoglobin (HbO_2^-).
(e) Prior to oxygenation, Hb^- is the predominant form of deoxyhemoglobin. Upon oxygenation Hb^- forms HbO_2^-, HbO_2^- then combines with H_3O^+ to form HbO_2H and water. The $[H_3O^+]$ should decrease and the pH of the blood will increase.

Chapter 17
17.1   (a) 3.22×10^{20} s^{-1}, 9.32×10^{-13} m; (b) 3.97×10^{20} s^{-1}, 7.56×10^{-13} m; (c) 2.66×10^{20} s^{-1}, 1.13×10^{-12} m

17.3   argon-39 and calcium-41

17.5   (a) $^3_1T \rightarrow {}^0_{-1}e + {}^3_2He$; (b) $^{83}_{39}Y \rightarrow {}^0_{1}e + {}^{83}_{38}Sr$;
(c) $^{87}_{36}Kr \rightarrow {}^0_{-1}e + {}^{87}_{37}Rb$; (d) $^{225}_{91}Pa \rightarrow {}^4_2\alpha + {}^{221}_{89}Ac$

17.7   (a) $^8_5B \rightarrow {}^0_{1}e + {}^8_4Be$; (b) $^{63}_{28}Ni \rightarrow {}^0_{-1}e + {}^{63}_{29}Cu$;
(c) $^{185}_{79}Au \rightarrow {}^4_2\alpha + {}^{181}_{77}Ir$; (d) $^7_4Be + {}^0_{-1}e \rightarrow {}^7_3Li$

17.9   (a) β, $^{24}_{11}Na \rightarrow {}^{24}_{12}Mg + {}^0_{-1}e$; (b) β,
$^{128}_{50}Sn \rightarrow {}^{128}_{51}Sb + {}^0_{-1}e$; (c) β^+, $^{140}_{57}La \rightarrow {}^{140}_{56}Ba + {}^0_{1}e$;
(d) α, $^{228}_{90}Th \rightarrow {}^{224}_{88}Ra + {}^4_2\alpha$

17.11   (a) $^{11}_5B + {}^4_2\alpha \rightarrow 2{}^1_0n + {}^{13}_7N$;
(b) $^{35}_{17}Cl + {}^2_1D \rightarrow {}^1_0n + {}^{36}_{18}Ar$; (c) $^{96}_{42}Mo + {}^2_1D \rightarrow {}^1_0n + {}^{97}_{43}Tc$;
(d) $^{45}_{21}Sc + {}^1_0n \rightarrow {}^4_2\alpha + {}^{42}_{19}K$

17.13   (a) β decay, $^{68}_{29}Cu \rightarrow {}^0_{-1}e + {}^{68}_{30}Zn$; (b) β^+ decay,
$^{103}_{48}Cd \rightarrow {}^0_{1}e + {}^{103}_{47}Ag$; (c) α decay, $^{243}_{97}Bk \rightarrow {}^4_2\alpha + {}^{239}_{95}Am$;
(d) α decay, $^{260}_{105}Db \rightarrow {}^4_2\alpha + {}^{256}_{103}Lr$

17.15 $^{235}_{92}U \rightarrow {}^4_2\alpha + {}^{231}_{90}Th$, $^{231}_{90}Th \rightarrow {}^0_{-1}e + {}^{231}_{91}Pa$,
$^{231}_{91}Pa \rightarrow {}^4_2\alpha + {}^{227}_{89}Ac$, $^{227}_{89}Ac \rightarrow {}^0_{-1}e + {}^{227}_{90}Th$,

$^{227}_{90}Th \rightarrow {}^4_2\alpha + {}^{223}_{88}Ra$, $^{223}_{88}Ra \rightarrow {}^4_2\alpha + {}^{219}_{86}Rn$,
$^{219}_{86}Rn \rightarrow {}^4_2\alpha + {}^{215}_{84}Po$, $^{215}_{84}Po \rightarrow {}^0_{-1}e + {}^{215}_{85}At$,
$^{215}_{85}At \rightarrow {}^4_2\alpha + {}^{211}_{83}Bi$, $^{211}_{83}Bi \rightarrow {}^0_{-1}e + {}^{211}_{84}Po$,
$^{211}_{84}Po \rightarrow {}^4_2\alpha + {}^{207}_{82}Pb$

17.17   (a) $^{14}_7N + {}^4_2\alpha \rightarrow {}^{17}_8O + {}^1_1p$;
(b) $^{248}_{96}Cm + {}^1_0n \rightarrow {}^{249}_{97}Bk + {}^0_{-1}e$;
(c) $^{243}_{95}Am + {}^1_0n \rightarrow {}^{244}_{96}Cm + {}^0_{-1}e + \gamma$;
(d) $^{13}_6C + {}^1_0n \rightarrow {}^{14}_6C + \gamma$

17.19   (a) $^{20}_{10}Ne + {}^4_2\alpha \rightarrow {}^8_4Be + {}^{16}_8O$;
(b) $^{20}_{10}Ne + {}^{20}_{10}Ne \rightarrow {}^{16}_8O + {}^{24}_{12}Mg$; (c) $^{44}_{20}Ca + {}^4_2\alpha \rightarrow \gamma + {}^{48}_{22}Ti$;
(d) $^{27}_{13}Al + {}^2_1H \rightarrow {}^1_1p + {}^{28}_{13}Al$

17.21   (a) $^{14}_7N + {}^4_2\alpha \rightarrow {}^{17}_8O + {}^1_1p$;
(b) $^{239}_{94}Pu + {}^1_0n \rightarrow {}^{240}_{95}Am + {}^0_{-1}e$

17.23   (a) unbihexium, Ubh; (b) untrihexium, Uth;
(c) binilnilium, Bnn

17.25 1.4×10^{-5} Ci

17.27   (a) 9.2×10^4 Bq; (b) 5.3×10^{12} Bq; (c) 2.7×10^8 Bq

17.29   75 rad, 75 rem, 0.75 Sv

17.31   100 days

17.33   (a) 5.64×10^{-2} a^{-1}; (b) 0.83 s^{-1}; (c) 0.0693 min^{-1}

17.35   8.8 h

17.37   (a) 69.6%; (b) 50.9%

17.39   (a) 55%; (b) 9.29×10^8 a

17.41 3.54×10^3 a

17.43   (a) 9.9×10^{-4} Ci; (b) 0.28 Ci; (c) 0.40 Ci

17.45   362 min or 400 min (1 significant figure)

17.47   26.4%

17.49 $[X] = [X]_0 e^{-k_1 t}$, $[Y] = \dfrac{k_1}{k_2 - k_1}(e^{-k_1 t} - e^{-k_2 t})[X]_0$,

$$[Z] = [X]_0\left(1 + \frac{k_1 e^{-k_2 t} - k_2 e^{-k_1 t}}{k_2 - k_1}\right). \quad k_1 = 0.0253\ d^{-1},$$

$k_2 = 0.0371$ d^{-1}. The plot assume that $[X]_0 = 2.00$ g.

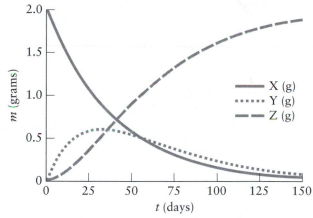

17.51   If $H_2{}^{18}O$ is used in the reaction, the label can be followed. After the products are separated, vibrational spectroscopy or mass spectrometry can be used to determine whether the product has incorporated the ^{18}O. If an O atom came from the water, the molar mass of the methanol produced would be 34 $g\cdot mol^{-1}$, not 32 $g\cdot mol^{-1}$.

17.53   The vibrational frequency will decrease.

17.55   44.4 d

17.57   (a) 9.0×10^{13} J; (b) 8.2×10^{-14} J; (c) 90. J;
(d) 1.51×10^{-10} J

17.59 4.3×10^9 $kg\cdot s^{-1}$

17.61   (a) 1.41×10^{-12} $J\cdot nucleon^{-1}$;
(b) 1.21×10^{-12} $J\cdot nucleon^{-1}$; (c) 1.8×10^{-13} $J\cdot nucleon^{-1}$;
(d) 4.3×10^{-13} $J\cdot nucleon^{-1}$; (e) Ni

17.63 (a) -7.8×10^{10} J·g^{-1}; (b) -3.52×10^{11} J·g^{-1};
(c) -2.09×10^{11} J·g^{-1}; (d) -3.36×10^{11} J·g^{-1}
17.65 (a) $^{24}_{11}$Na $\rightarrow$ $^{24}_{12}$Mg + $^{0}_{-1}$e; (b) -8.85×10^{-13} J;
(c) -3.69×10^{-14} J·nucleon^{-1}
17.67 (a) $^{244}_{95}$Am $\rightarrow$ $^{134}_{53}$I + $^{107}_{42}$Mo + 3$^{1}_{0}$n;
(b) $^{235}_{92}$U + $^{1}_{0}$n $\rightarrow$ $^{96}_{40}$Zr + $^{138}_{52}$Te + 2$^{1}_{0}$n;
(c) $^{235}_{92}$U + $^{1}_{0}$n $\rightarrow$ $^{101}_{42}$Mo + $^{132}_{50}$Sn + 3$^{1}_{0}$n
17.69 (a) False; the dose equivalent is either equal to or larger than the actual dose due to the Q factor. (b) False; 1.8×10^{8} Bq = 0.003 Ci which is smaller than 10 Ci. (c) true (d) true
17.71 (a) 9 dpm; (b) 7×10^{5} decays
17.73 (a) 3.4×10^{8} pCi·L^{-1}; (b) 1.0×10^{19} atoms;
(c) 1×10^{2} days
17.75 (a) 4.5 Ga; (b) 3.8 Ga
17.77 (a) 1.3×10^{-5} s^{-1}, 0.63 d; (b) 0.22 mg
17.79 (a) γ radiation is most effective for diagnostic procedures that detect tumors because it is the least destructive. γ rays pass easily through body tissues and can be counted, whereas α and β particles are stopped by the body tissues. (b) α particles tend to be best for therapy because they cause the most destruction. (c) and (d) ^{131}I, 8d (used to image the thyroid); ^{67}Ga, 78 h (used most often as the citrate complex); ^{99m}Tc, 6 h (used for various body tissues by varying the ligands attached to the Tc atom).
17.81 (a) $^{98}_{42}$Mo + $^{1}_{0}$n $\rightarrow$ $^{99}_{42}$Mo $\rightarrow$ $^{99m}_{43}$Tc + $^{0}_{-1}\beta$;
(b) Tc-99m (N/Z = 1.30 vs. 1.36 for Mo-99
17.83 0.26 L
17.85 1.63742×10^{-13} J
17.87 (a) 262 dpm; (b) 1.18×10^{-4} μCi
17.89 (a) 9%, 13%; (b) 0.47 μCi; (c) 51.4 d; (d) 0.17;
(e) approximately 0%

Chapter 18
18.1 (a)

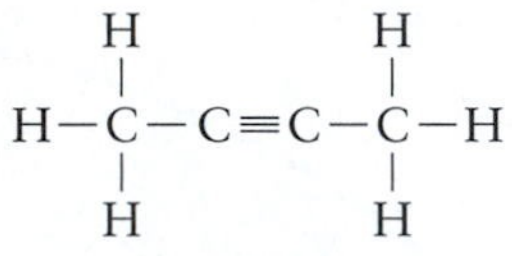

H—C—C≡C—C—H Alkyne

(b) H—C—C—C—C—H Alkane

(c) H—C=C—C—C—H Alkene

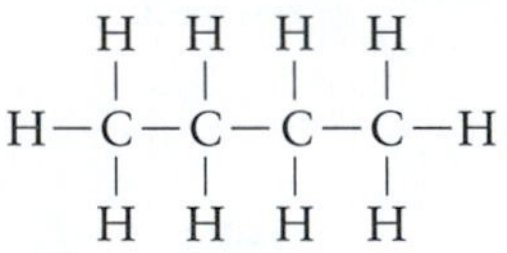

(d) H—C—C=C—C—C≡C—C—H Alkene and alkyne

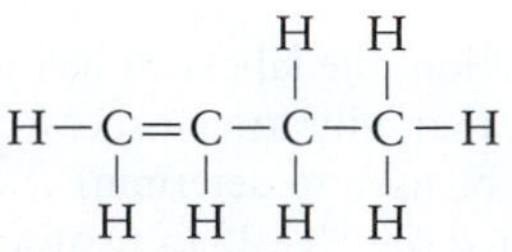

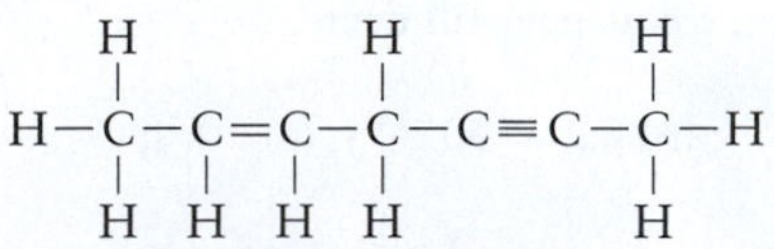

(e) Alkene

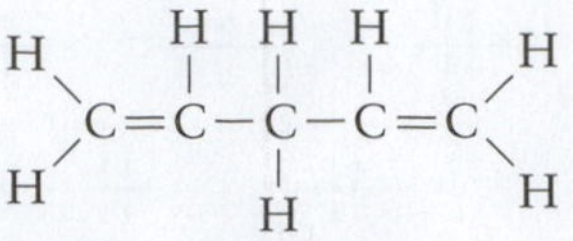

18.3 (a) (CH₃)₃CH or C₄H₁₀, alkane; (b) C₆H₇CH₃ or C₇H₁₀ alkene; (c) C₆H₁₂, alkane; (d) C₆H₁₂ alkane
18.5 (a) $C_{12}H_{26}$, alkane; (b) $C_{13}H_{22}$, alkane; (c) C_7H_{14}, alkane; (d) $C_{14}H_8$, aromatic hydrocarbon
18.7 (a) propane; (b) butane; (c) heptane; (d) decane
18.9 (a) methyl; (b) pentyl; (c) propyl; (d) hexyl
18.11 (a) propane; (b) ethane; (c) pentane;
(d) 2,3-dimethylbutane
18.13 (a) 4-methyl-2-pentene;
(b) 2,3-dimethyl-2-phenylpentane
18.15 (a) $CH_2{=}CHCH(CH_3)CH_2CH_3$;
(b) $CH_3CH_2C(CH_3)_2CH(CH_2CH_3)(CH_2)_2CH_3$;
(c) $HC{\equiv}C(CH_2)_2C(CH_3)_3$;
(d) $CH_3CH(CH_3)CH(CH_2CH_3)CH(CH_3)_2$
18.17 (a)

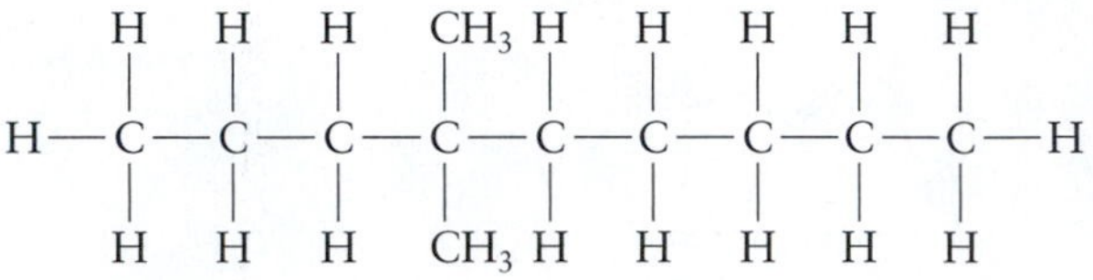

(b)

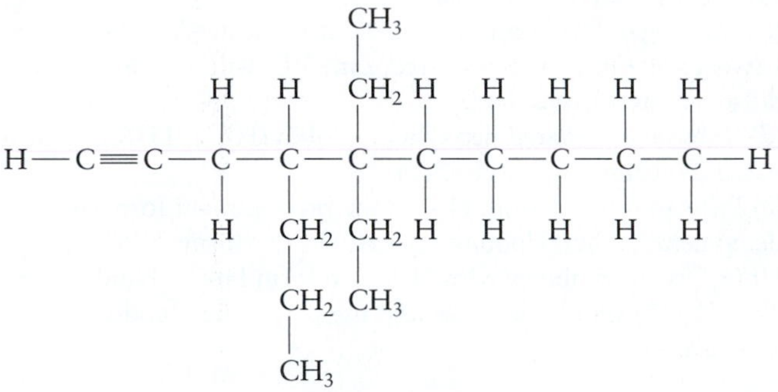

(c)

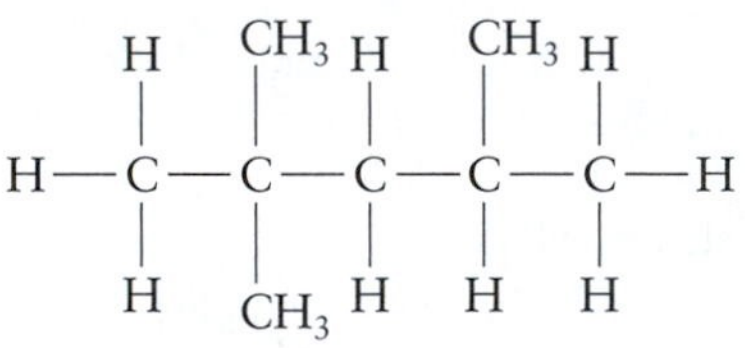

(d) H₃C—CH₂, H / C=C / H, CH₂—CH₃

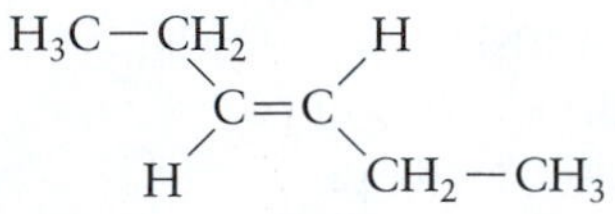

18.19 (a)

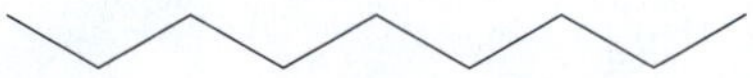

(b)

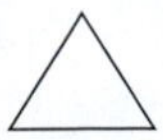

(c)

18.21   (a) hexenes:

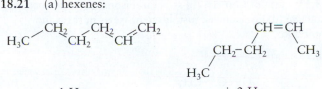

1-Hexene

cis-2-Hexene

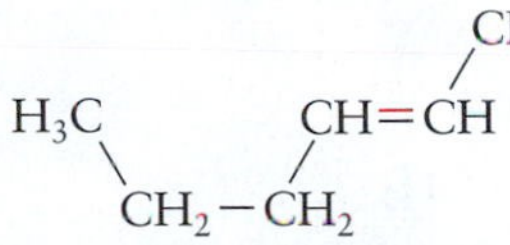

trans-2-Hexene

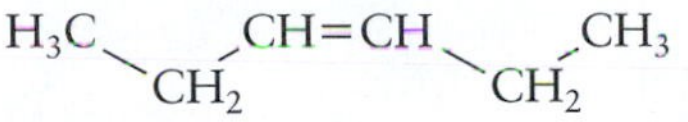

cis-3-Hexene

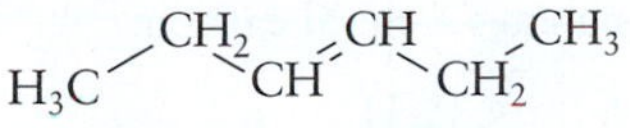

trans-3-Hexene

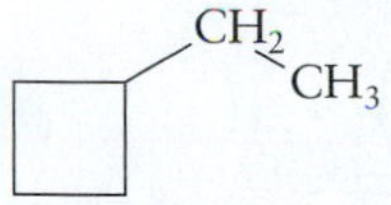

Ethylcyclobutane

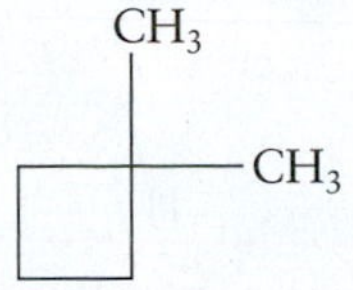

1,1-Dimethylcyclobutane

The following structures are drawn to emphasize the stereochemistry:

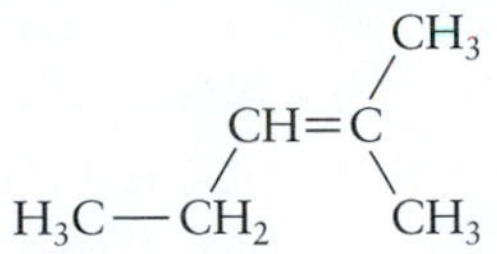

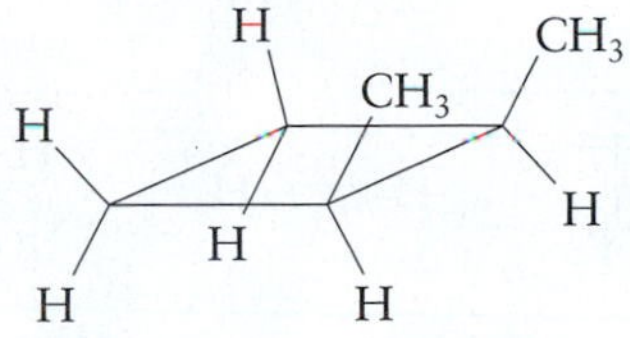

cis-1,2-Dimethylcyclobutane

pentenes:

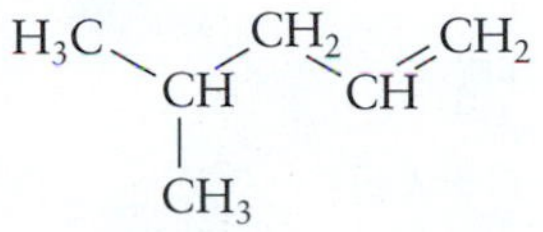

4-Methyl-1-pentene

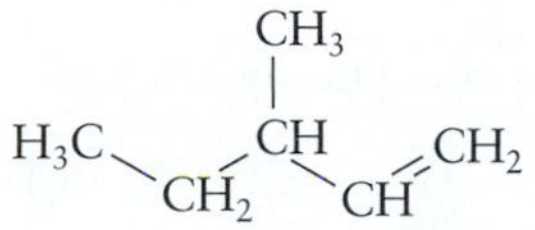

3-Methyl-1-pentene

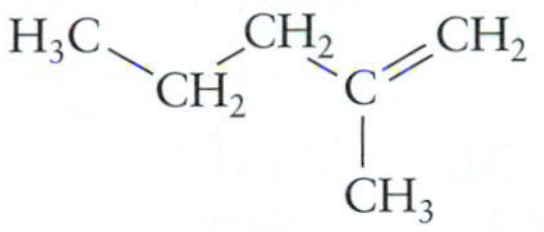

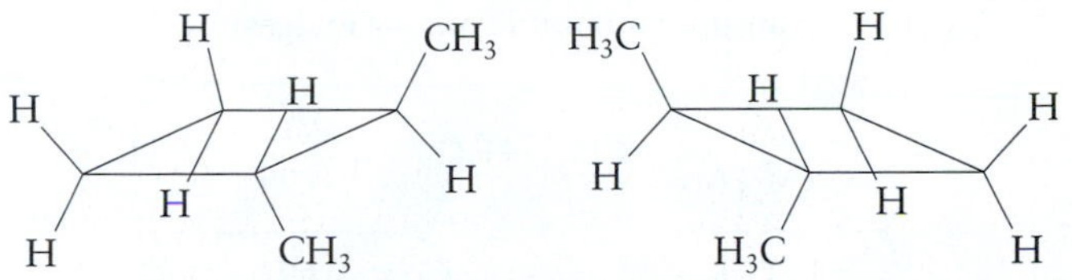

trans-1,2-Dimethylcyclobutane
(nonsuperimposable mirror images)

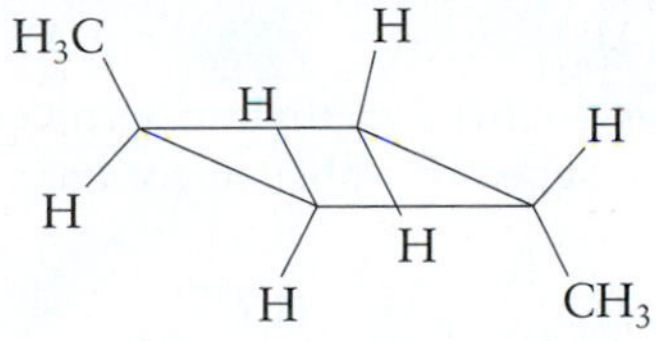

2-Methyl-1-pentene

2-Methyl-2-pentene

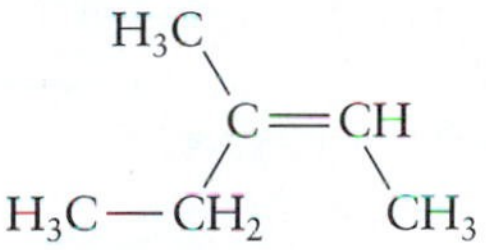

trans-1,3-Dimethylcyclobutane

cis-3-Methyl-2-pentene
(+ trans isomer)

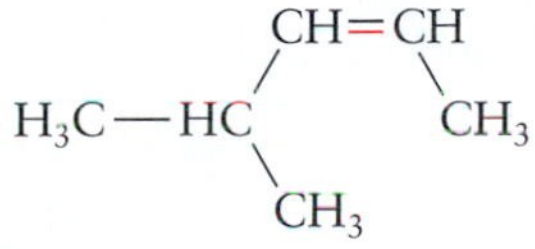

cis-4-Methyl-2-pentene
(+ trans isomer)

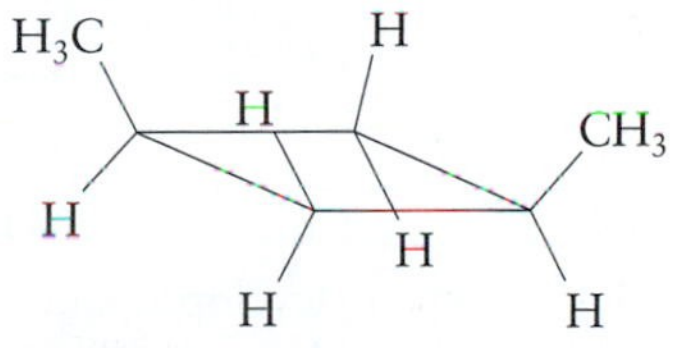

cis-1,3-Dimethylcyclobutane

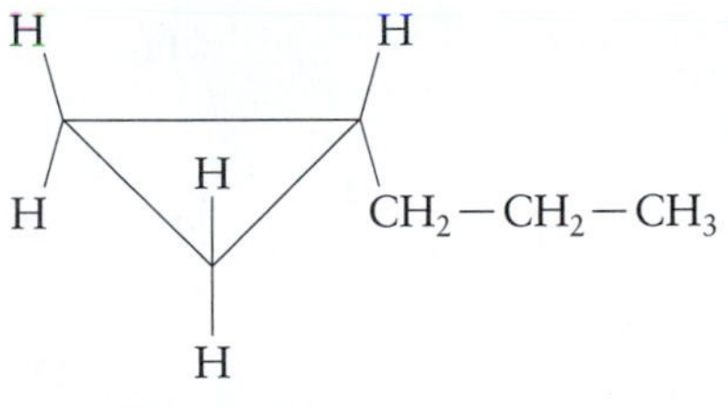

butanes:

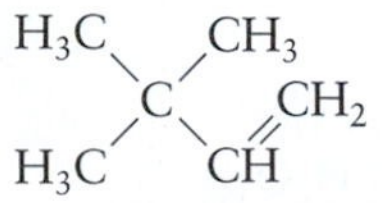

3,3-Dimethyl-1-butene

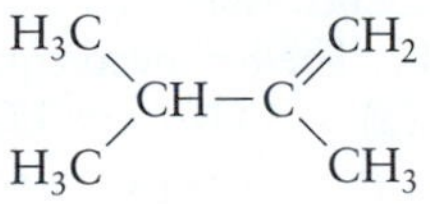

2,3-Dimethyl-1-butene

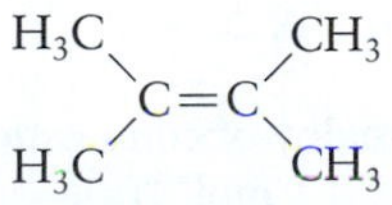

Propylcyclopropane

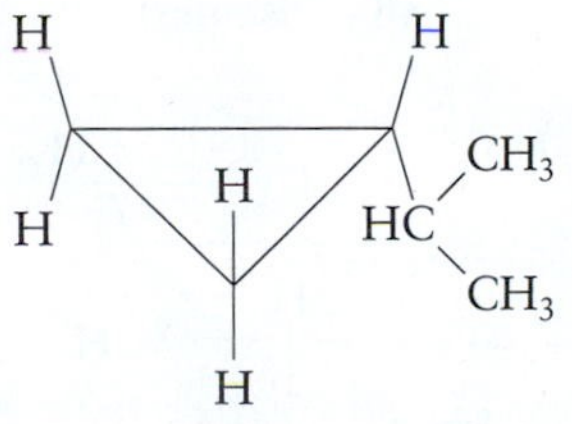

2,3-Dimethyl-2-butene

(b) cyclic molecules:

Cyclohexane

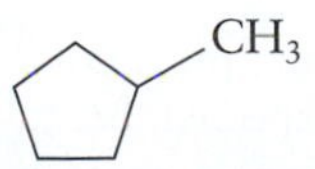

Methylcyclopentane

Isopropylcyclopropane
or 2-cyclopropylpropane

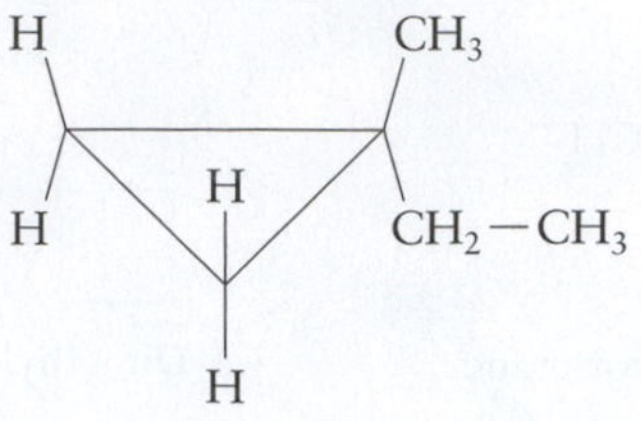

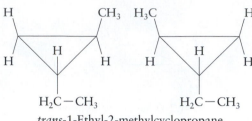

1-Ethyl-1-methylcyclopropane

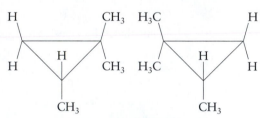

trans-1-Ethyl-2-methylcyclopropane
(nonsuperimposable mirror images)

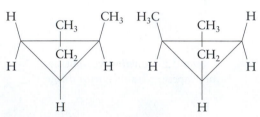

cis-1-Ethyl-2-methylcyclopropane
(nonsuperimposable mirror images)

1,1,2-Trimethylcyclopropane
(nonsuperimposable mirror images)

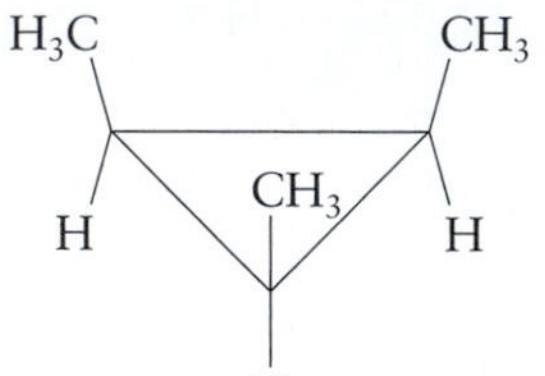

1,2,3-Trimethylcyclopropane
(all cis isomer)

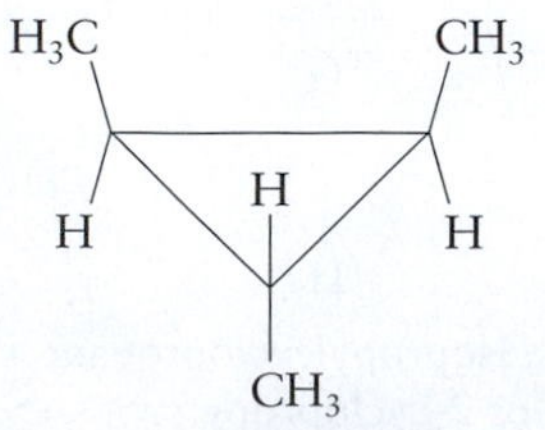

1,2,3-Trimethylcyclopropane
(cis–trans isomer)

18.23 (a) not isomers; (b) structural isomers; (c) geometrical isomers; (d) not isomers, just different views of the same molecule

18.25 (a)

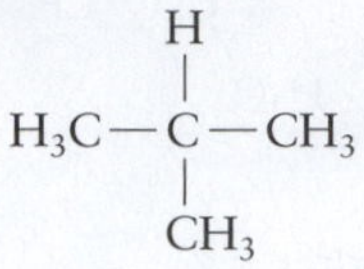

$$H_3C-\underset{\underset{CH_3}{|}}{\overset{\overset{H}{|}}{C}}-CH_3$$

(b)

$$H_3C-\underset{\underset{CH_3}{|}}{\overset{\overset{Cl}{|}}{C}}-CH_3 \qquad H_2C-\underset{\underset{CH_3}{|}}{\overset{\overset{Cl\ H}{|\ |}}{C}}-CH_3$$

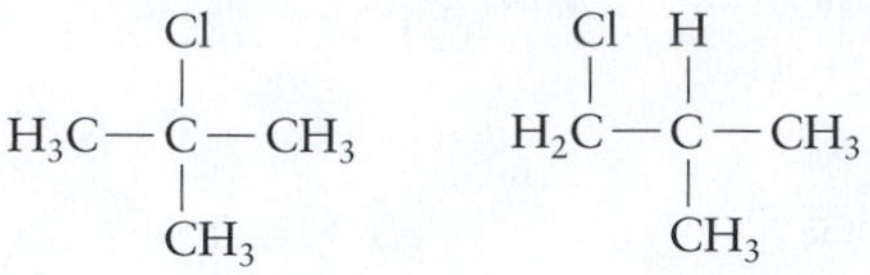

18.27 An * designates a chiral carbon.

(a) optically active,

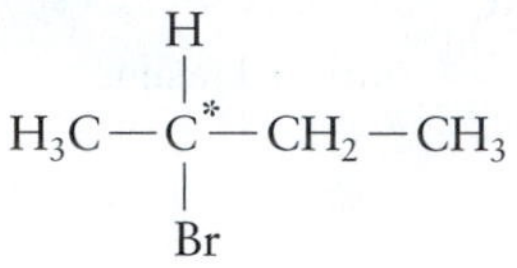

$$H_3C-\overset{\overset{H}{|}}{\underset{\underset{Br}{|}}{C^*}}-CH_2-CH_3$$

(b) not optically active,

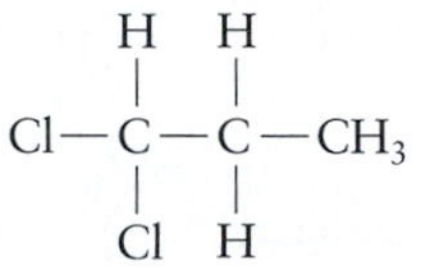

$$Cl-\overset{\overset{H}{|}}{\underset{\underset{Cl}{|}}{C}}-\overset{\overset{H}{|}}{\underset{\underset{H}{|}}{C}}-CH_3$$

(c) optically active,

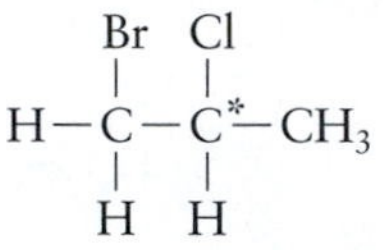

$$H-\overset{\overset{Br}{|}}{\underset{\underset{H}{|}}{C}}-\overset{\overset{Cl}{|}}{\underset{\underset{H}{|}}{C^*}}-CH_3$$

(d) optically active,

$$H_3C-\overset{\overset{Cl}{|}}{\underset{\underset{H}{|}}{C}}-\overset{\overset{Cl}{|}}{\underset{\underset{H}{|}}{C^*}}-CH_2-CH_2-CH_3$$

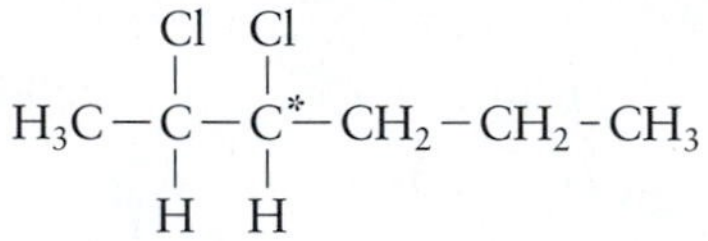

18.29 The difference can be traced to the weaker London forces that exist in branched molecules. Atoms in neighboring branched molecules cannot lie as close together as they can in unbranched isomers.

18.31 The balanced equations are
$C_3H_8(g) + 5\ O_2(g) \rightarrow 3\ CO_2(g) + 4\ H_2O(l)$
$C_4H_{10}(g) + \frac{13}{2}\ O_2(g) \rightarrow 4\ CO_2(g) + 5\ H_2O(l)$
$C_5H_{12}(g) + 8\ O_2(g) \rightarrow 5\ CO_2(g) + 6\ H_2O(l)$
The enthalpies of combustion that correspond to these reactions are listed in Appendix 2:

Compound	(a) Enthalpy of combustion ($kJ \cdot mol^{-1}$)	(b) Heat released ($kJ \cdot g^{-1}$)
propane	−2220.	50.3
butane	−2878	49.5
pentane	−3537	49.0

The molar enthalpy of combustion increases with molar mass as might be expected, because the number of moles of CO_2 and H_2O formed will increase as the number of carbon and hydrogen atoms in the compounds increases. The heat released per gram of these hydrocarbons is essentially the same because the H to C ratio is similar in the three hydrocarbons.

18.33   nine, none

18.35

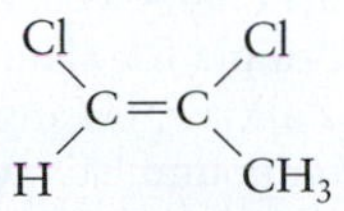

$$\underset{H}{\overset{Cl}{\underset{|}{C}}}=\underset{CH_3}{\overset{Cl}{C}}$$

cis-1,2-Dichloropropene

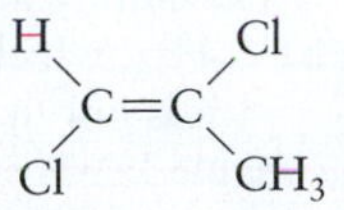

$$\underset{Cl}{\overset{H}{C}}=\underset{CH_3}{\overset{Cl}{C}}$$

trans-1,2-Dichloropropene

cis-1,2-Dichloropropene is polar, although trans-1,2-Dichloropropene is slightly polar also.

18.37   (a)

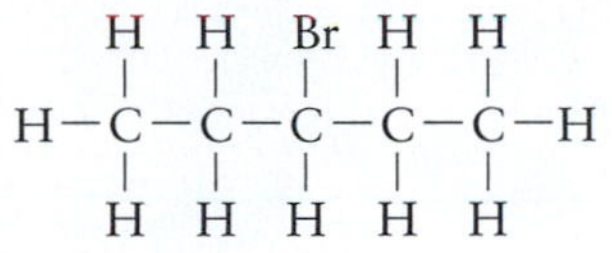

H H Br H H
| | | | |
H—C—C—C—C—C—H
| | | | |
H H H H H

3-Bromopentane

H Br H H H
| | | | |
H—C—C—C—C—C—H
| | | | |
H H H H H

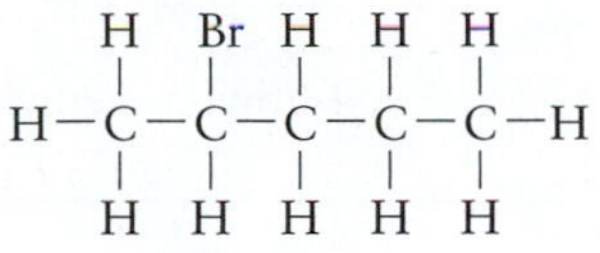

2-Bromopentane

(b) addition reaction

18.39   (a) $C_6H_{11}Br + NaOCH_2CH_3 \rightarrow$
$\qquad\qquad C_6H_{10} + NaBr + HOCH_2CH_3$

(b)

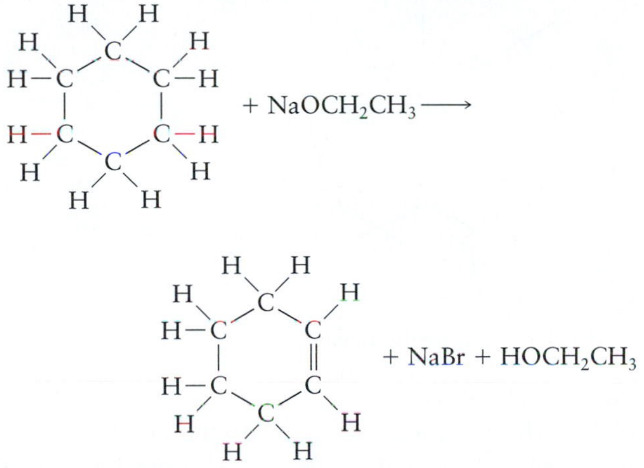

[cyclohexane structure] + NaOCH_2CH_3 ⟶

[cyclohexene structure] + NaBr + HOCH_2CH_3

(c) elimination reaction

18.41   The reaction is less exothermic as the halogen becomes heavier. In general, the reactivity, and also the danger associated with use of the halogens in reactions, decreases as one descends the periodic table.

18.43   (a) 1-ethyl-3-methylbenzene. (b) Pentamethylbenzene (1,2,3,4,5-pentamethylbenzene is also correct, but, because there is only one possible pentamethylbenzene, the use of the numbers is not necessary)

18.45

(a)   CH₃

[benzene with methyl]

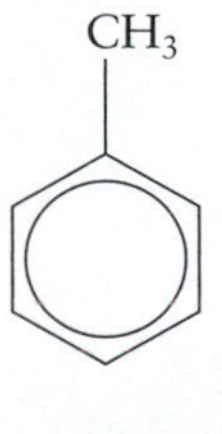

(b)   CH₃

[benzene with methyl and Cl para]
Cl

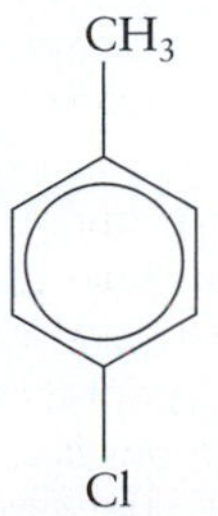

(c)   CH₃

[benzene with methyl and CH₃ meta]
CH₃

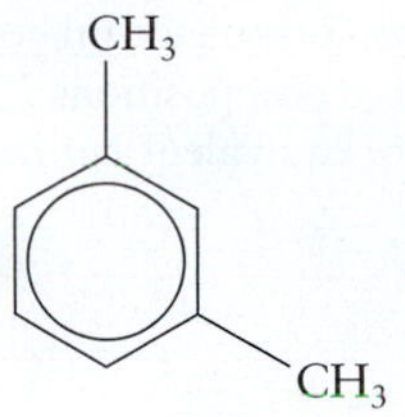

(d)   CH₃

[benzene with methyl and Cl para]
Cl

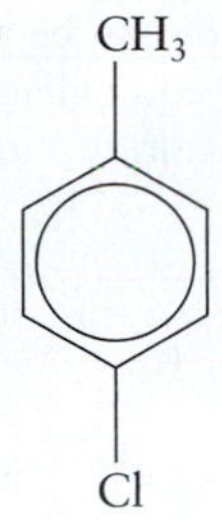

18.47

CH₃
Cl   Cl

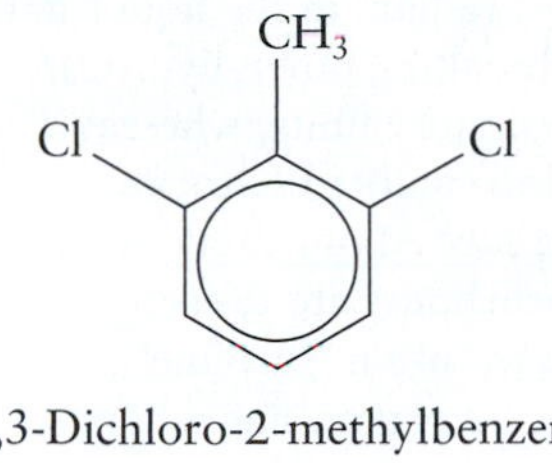

1,3-Dichloro-2-methylbenzene

CH₃
Cl

Cl

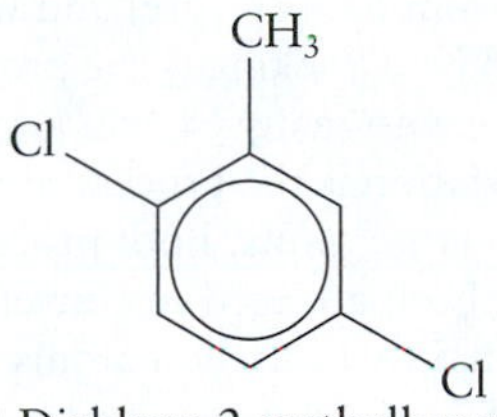

1,4-Dichloro-2-methylbenzene

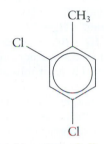

CH₃
Cl

Cl

1,5-Dichloro-2-methylbenzene

CH₃

Cl   Cl

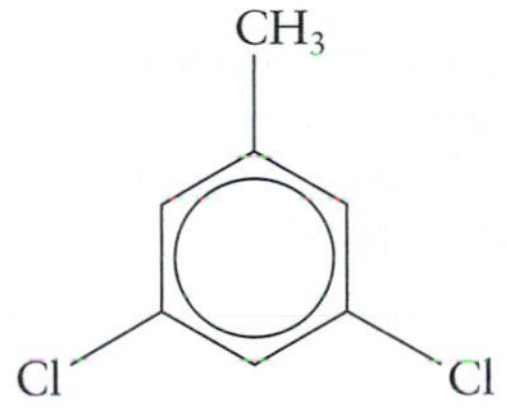

1,3-Dichloro-5-methylbenzene

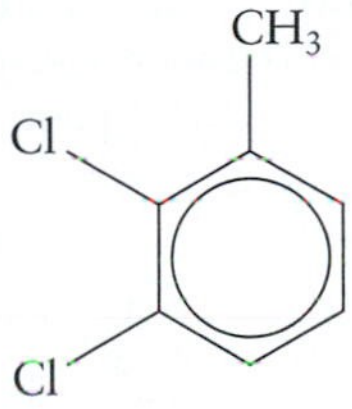

CH₃
Cl

Cl

1,2-Dichloro-3-methylbenzene

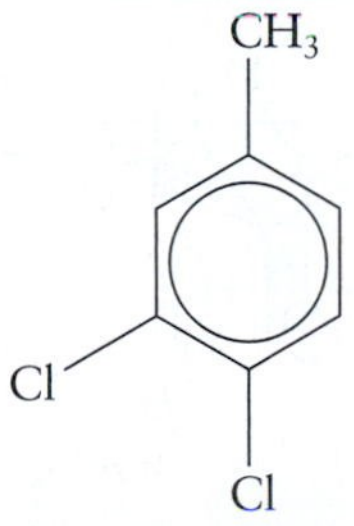

CH₃

Cl

Cl

1,2-Dichloro-4-methylbenzene

(b) All of these molecules are at least slightly polar.

18.49

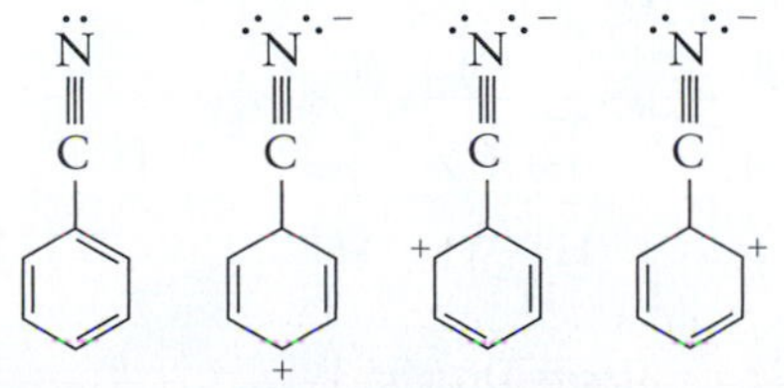

[four resonance structures of N=C attached to benzene ring]

18.51 Two compounds can be produced. Resonance makes positions 1, 4, 6, and 9 equivalent. It also makes positions 2, 3, 7, and 8 equivalent. Positions 5 and 10 are equivalent but have no H atom.

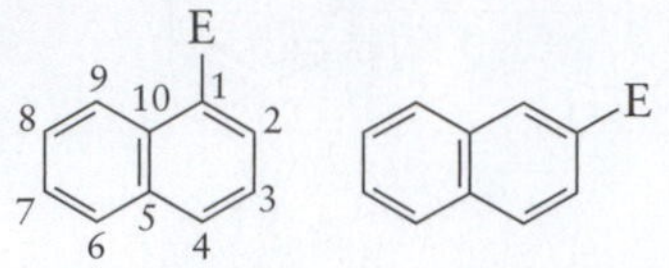

18.53 These hydrocarbons are too volatile (they are all gases at room temperature) and would not remain in the liquid state.

18.55 Cracking is the process of breaking down hydrocarbons with many carbon atoms into smaller units, whereas alkylation is the process of combining smaller hydrocarbons into larger units. Both processes are carried out catalytically, and both are used to convert hydrocarbons into units having from 5 to 11 carbon atoms suitable for use in gasoline.

18.57 (a) 4 σ-type single bonds; (b) 2 σ-type single bonds and 1 double bond with a σ- and a π-bond; (c) 1 σ-type single bond and 1 triple bond with a σ-bond and 2 π-bonds

18.59 (a) substitution, $CH_4 + Cl_2 \rightarrow CH_3Cl + HCl$; (b) addition $CH_2{=}CH_2 + Br_2 \rightarrow CH_2Br{-}CH_2Br$

18.61 Water is not used as the nonpolar reactants will not readily dissolve in a highly polar solvent like water. Also, the ethoxide ion reacts with water.

18.63 The double bond in alkenes makes them more rigid than alkanes. Some of the atoms of alkene molecules are locked into a planar arrangement by the π-bond; hence, they cannot roll up into a ball as compactly as alkanes can. Because they do not pack together as compactly as alkanes do, they have lower boiling and melting points.

18.65 (a) 2-methyl-1-propene, no geometrical isomers; (b) *cis*-3-methyl-2-pentene, *trans*-3-methyl-2-pentene; (c) 1-hexyne, no geometrical isomers; (d) 3-hexyne, no geometrical isomers; (e) 2-hexyne, no geometrical isomers.

18.67 (a) $C_{10}H_{18}$.
(b) naphthalene, 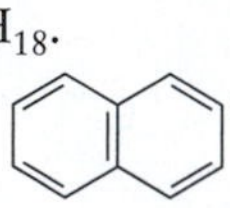 , $C_{10}H_8$.

(c) Yes. Cis and trans forms (relative to the C—C bond common to the two six-membered rings) are possible.

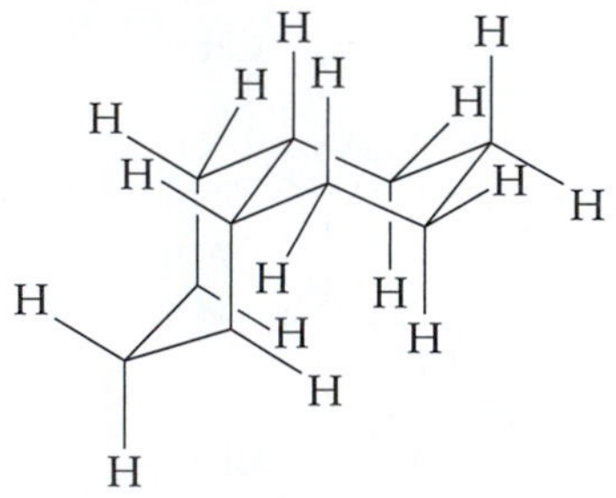

cis-Decalin

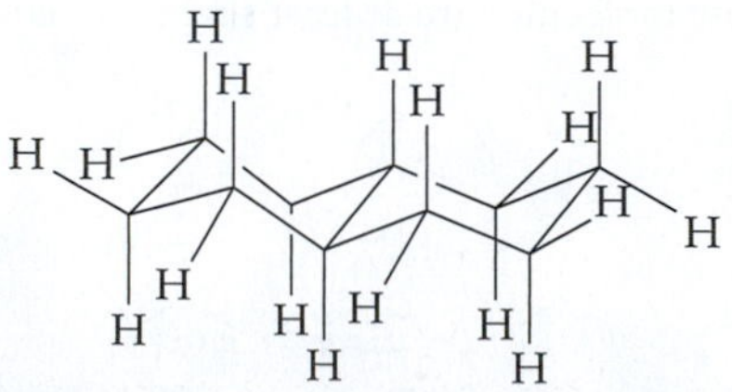

trans-Decalin

18.69 -55 kJ·mol^{-1}

18.71 The empirical formula is C_4H_9; the molecular formula might be C_8H_{18}, which matches the formula for alkanes (C_nH_{2n+2}). It is not likely an alkene or alkyne, because there is no reasonable Lewis structure for a compound having the empirical formula C_4H_9 and multiple bonds.

18.73 (a)

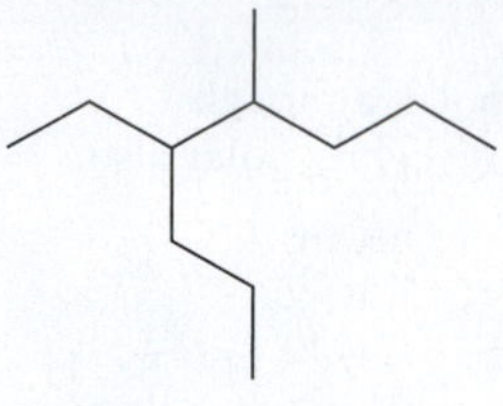

4-Ethyl-5-methyloctane

(b)

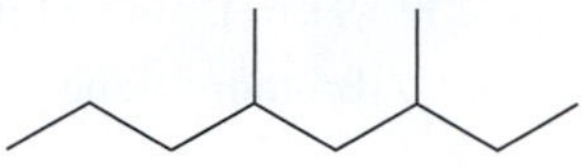

3,5-Dimethyloctane

(c)

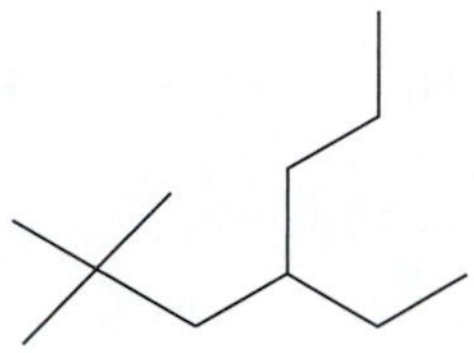

2,2-Dimethyl-4-ethylheptane

(d)

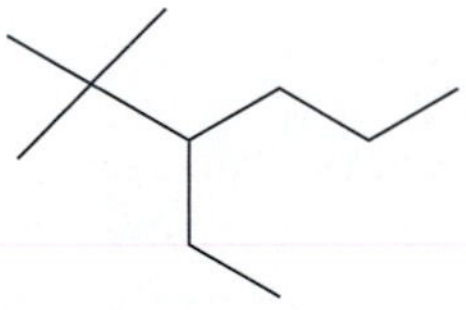

3-Ethyl-2,2-dimethylhexane

18.75 cyclopropane

18.77 The NO_2 group is a meta-directing group and the Br atom is an ortho, para-directing group. Because the position para to Br is already substituted with the NO_2 group, further bromination will not occur there. The resonance forms show that the bromine atom will activate the position ortho to it as expected. The NO_2 group will deactivate the group ortho to itself, thus in essence enhancing the reactivity of the position meta to the NO_2 group. This position is ortho to the Br atom, so the effects of the Br and NO_2 groups reinforce each other. Bromination is thus expected to occur as shown:

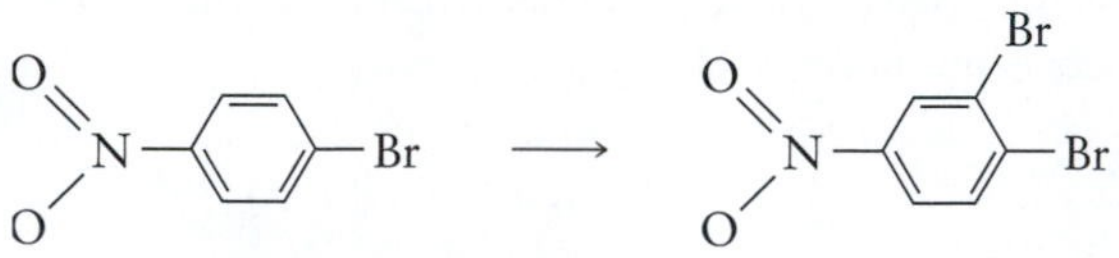

18.79 (a) and (b)

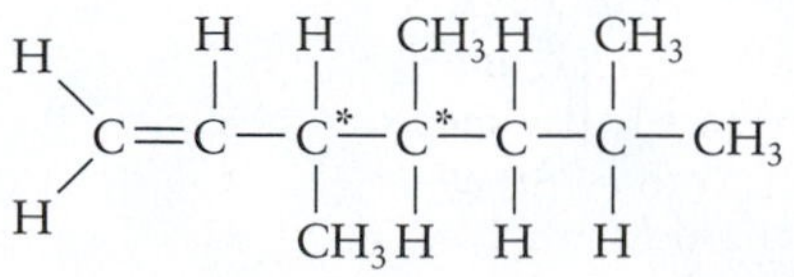

(c) no

18.81 For a molecule such as 1,2-dichloro-4-diethylbenzene, $C_6H_3Cl_2(CH_2CH_3)$, $175.04m_u$, it is relatively easy to lose heavy atoms such as chlorine and groups of atoms such as methyl and ethyl fragments. Molecules can also lose hydrogen atoms. In mass spectrometry, P is used to represent the *parent ion*, which is the ion formed from the molecule without fragmentation. Fragments are then represented as $P - x$, where x is the particular fragment lost from the parent ion to give the observed mass. Because the mass spectrum will measure the masses of individual molecules, the mass of carbon used will be $12.00m_u$ (by definition) because the large majority of the molecules will have all ^{12}C. The mass of H is $1.0078m_u$. Some representative peaks that may be present are listed below.

Fragment formula	Relation to parent ion	Mass (m_u)
$C_6H_3{}^{35}Cl_2(CH_2CH_3)$	P	174.00
$C_6H_3{}^{35}Cl^{37}Cl(CH_2CH_3)$	P	176.00
$C_6H_3{}^{37}Cl_2(CH_2CH_3)$	P	177.99
$C_6H_3{}^{35}Cl(CH_2CH_3)$	P–Cl	139.03
$C_6H_3{}^{37}Cl(CH_2CH_3)$	P–Cl	141.03
$C_6H_3{}^{35}Cl_2(CH_2)$	P–CH_3	158.98
$C_6H_3{}^{35}Cl^{37}Cl(CH_2)$	P–CH_3	160.97
$C_6H_3{}^{37}Cl_2(CH_2)$	P–CH_3	162.97
$C_6H_3{}^{35}Cl_2$	P–CH_2CH_3	144.96
$C_6H_3{}^{35}Cl^{37}Cl$	P–CH_2CH_3	146.96
$C_6H_3{}^{37}Cl_2$	P–CH_2CH_3	148.96
$C_6H_3{}^{35}Cl$	P–CH_2CH_3–Cl	109.99
$C_6H_3{}^{37}Cl$	P–CH_2CH_3–Cl	111.99

etc.

18.83 The presence of one bromine atom will produce in the ions that contain Br companion peaks that are separated by $2m_u$. Any fragment that contains Br will show this "doublet" in which the peaks are nearly but not exactly equal in intensity. Thus, seeing a mass spectrum with such doublets of a compound that is known to have Br or that was involved in a reaction in which Br could have been added or substituted, is almost a sure sign that Br is present in the compound. It is also fairly easy to detect Br atoms in the mass spectrum at $79m_u$ and $81m_u$, confirming their presence. If more than one Br atom is present, then a more complicated pattern is observed for the presence of the two isotopes. The possible combinations for a molecule of unknown formula with two Br atoms is $^{79}Br^{79}Br$, $^{79}Br^{81}Br$, $^{81}Br^{79}Br$, and $^{81}Br^{81}Br$. Thus, a set of three peaks (the two possibilities $^{79}Br^{81}Br$ and $^{81}Br^{79}Br$ have identical masses) will be generated that differ in mass by two units. The center peak, which is produced by the $^{79}Br^{81}Br$ and $^{81}Br^{79}Br$ combinations, will have twice the intensity of the outer two peaks, because statistically there are twice as many combinations that produce this mass. All modern mass spectrometers have spectral simulation programs that can readily calculate and print out the relative isotopic distribution pattern expected for any compound formulation, so that it is possible to easily match the expected pattern for a particular ion with the experimental result.

18.85 C_8H_{10} will have an absorption maximum at a longer wavelength. Molecular orbital theory predicts that in conjugated hydrocarbons (molecules which contain a chain of carbon atoms with alternating single and double bonds) electrons become delocalized and are free to move up and down the chain of carbon atoms. Such electrons may be described using the one dimensional "particle-in-a-box" model introduced in Chapter 1. According to this model, as the box to which electrons are confined lengthens, the quantized energy states avail-

able to the electrons get closer together. As a result, the energy needed to excite an electron from the ground state to the next higher state is lower for electrons confined to longer boxes. Therefore, lower energy photons, that is, photons with longer wavelengths, will be absorbed by the C_8H_{10} molecule because it provides a longer "box" than C_6H_8.

Chapter 19

19.1 (a) RNH_2, R_2NH, R_3N; (b) ROH; (c) RCOOH;
(d) RCHO
19.3 (a) ether; (b) ketone; (c) amine; (d) ester
19.5 (a) 2-iodo-2-butene; (b) 2,4-dichloro-4-methylhexane;
(c) 1,1,1,- triiodoethane; (d) dichloromethane
19.7 (a)

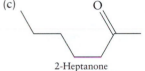

, phenol;

(b) $CH_3CH(CH_3)CH(OH)CH_2CH_3$, secondary alcohol;
(c) $CH_3CH_2CH(CH_3)CH_2CH(CH_3)CH_2OH$, primary alcohol;
(d) $CH_3C(CH_3)(OH)CH_2CH_3$, tertiary alcohol
19.9 (a) $CH_3CH_2OCH_3$; (b) $CH_3CH_2OCH_2CH_2CH_3$;
(c) CH_3OCH_3
19.11 (a) butyl propyl ether; (b) methyl phenyl ether;
(c) pentyl propyl ether
19.13 (a) aldehyde, ethanal; (b) ketone, propanone;
(c) ketone, 3-pentanone
19.15 (a)

Butanal

(b)

3-Hexanone

(c)

2-Heptanone

19.17 (a) ethanoic acid; (b) butanoic acid;
(c) 2-aminoethanoic acid

19.19 (a)

(b)

(c)

(d)

19.21 (a) methylamine; (b) diethylamine;
(c) *o*-methylaniline, 2-methylaniline, *o*-methylphenylamine,
or 1-amino-2-methylbenzene

19.23 (a)

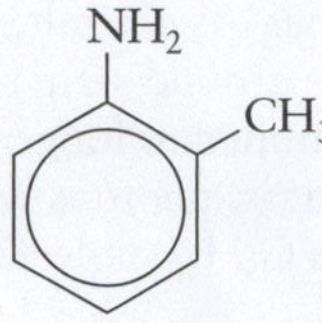

(b) CH₃CH₂ — N — CH₂CH₃ with CH₃CH₂

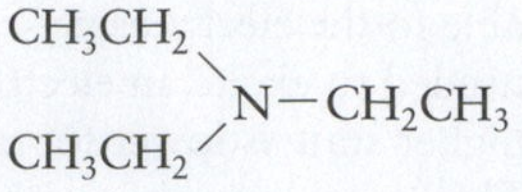

19.23 (a) [aniline with CH₃]; (b) triethylamine structure; (c) tetramethylammonium cation

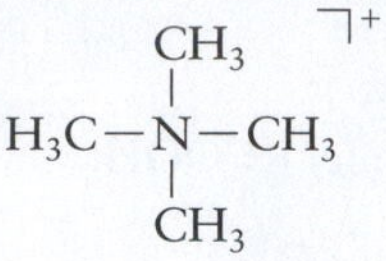

19.25 (a) and (c)

19.27 (a) ethanol. (b) 2-octanol. (c) 5-methyl-1-octanol. These reactions can be accomplished with an oxidizing agent such as acidified sodium dichromate, $Na_2Cr_2O_7$.

19.29 (a) $CH_3CH_2CH_2C(=O)O-CH(CH_3)CH_3$

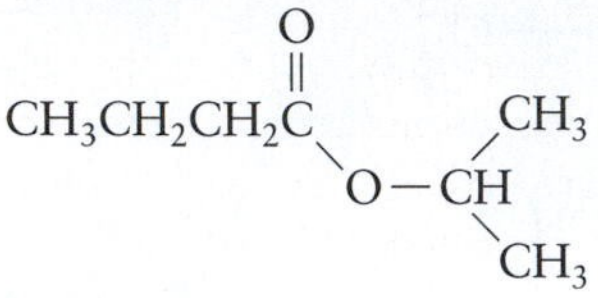

(b) $CH_3C(=O)O-CH_2CH_2CH_2CH_2CH_3$

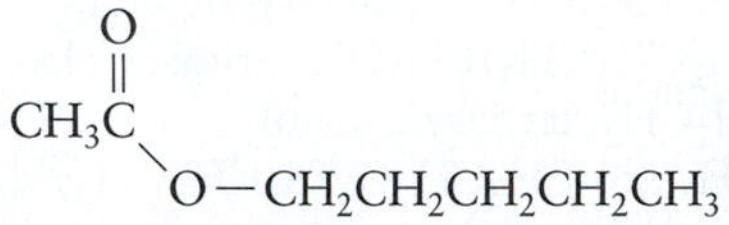

(c) $CH_3CH_2CH_2CH_2CH_2C(=O)N(CH_2CH_3)CH_3$

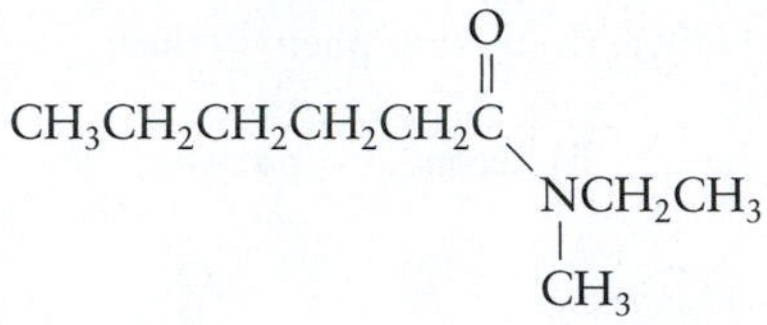

(d) $CH_3C(=O)NHCH_2CH_2CH_3$

19.31 The following procedures can be used:
(1) Dissolve the compounds in water and use an acid–base indicator to look for a color change.

(2) $CH_3CH_2CHO \xrightarrow{\text{Tollens reagent}} CH_3CH_2COOH + Ag(s)$

(3) $CH_3COCH_3 \xrightarrow{\text{Tollens reagent}}$ no reaction

Procedure (1) will distinguish ethanoic acid; (2) and (3) will distinguish propanal from 2-propanone.

19.33 $CH_3CH_2COOH < CH_3COOH < ClCH_2COOH < Cl_3CCOOH$. The greater the electronegativities of the groups attached to the carboxyl group, the stronger the acid (see Chapter 10).

19.35
(a) $-CH_2-C(CH_3)_2-CH_2-C(CH_3)_2-CH_2-C(CH_3)_2-$

(b) $-CH-CH_2-CH-CH_2-CH-CH_2-$ with CN on each CH

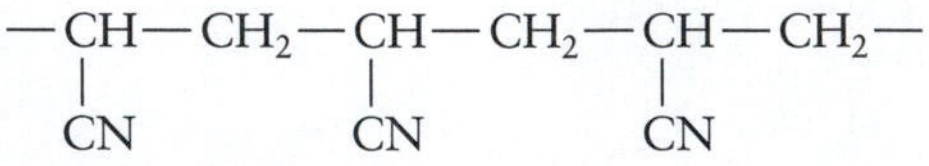

(c)

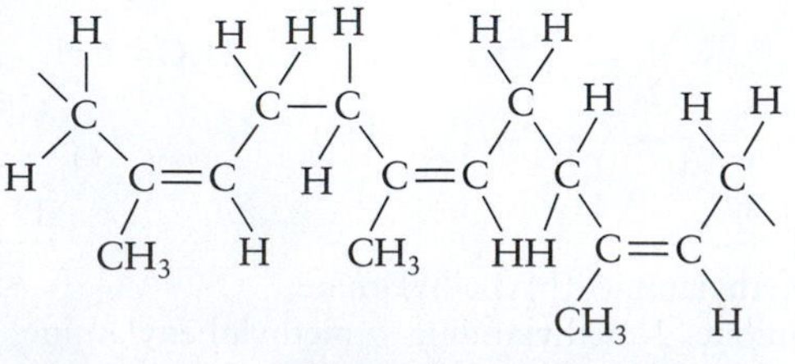

Cis version

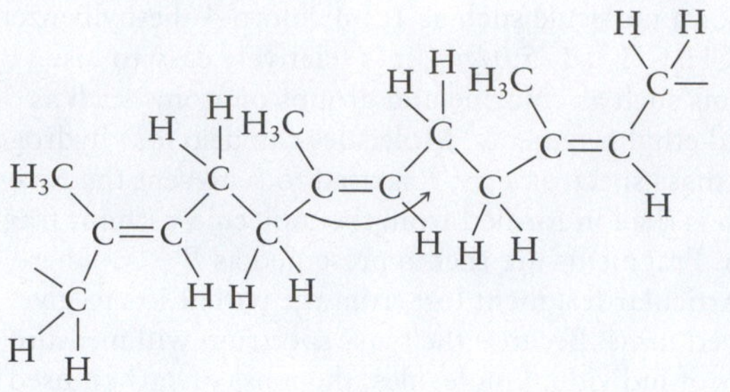

Trans version

19.37 (a) $CHCl=CH_2$; (b) $CFCl=CF_2$
19.39 (a) $-OCCONH(CH_2)_4NHCOCONH(CH_2)_4NH-$;
(b) $-OC-CH(CH_3)-NH-OC-CH(CH_3)-NH-$
19.41 block copolymer
19.43 All increase with increasing molar mass.
19.45 Highly linear, unbranched chains allow for maximum interaction between chains. The greater the intermolecular contact between chains, the stronger the forces between them, and the greater the strength of the material.
19.47 Polymers generally do not have definite molecular masses because there is no fixed point at which the chain-lengthening process will cease. There is no fixed molar mass, only an average molar mass. Because there is no one unique compound, there is no one unique melting point, rather a range of melting points.

19.49 (a) $-C(=O)NH-$

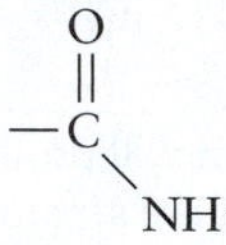

(b) amide;
(c) condensation
19.51 serine, threonine, tyrosine, aspartic acid, glutamic acid, lysine, arginine, histidine, asparagine, and glutamine; Proline and tryptophan generally do not contribute through hydrogen bonding because they are typically found in hydrophobic regions of proteins.

19.53

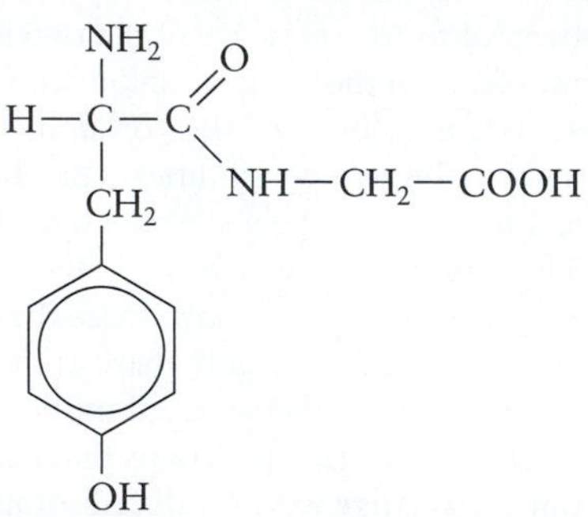

19.55 (a) alcohols and aldehydes.
(b) The chiral carbon atoms are marked with asterisks (*).

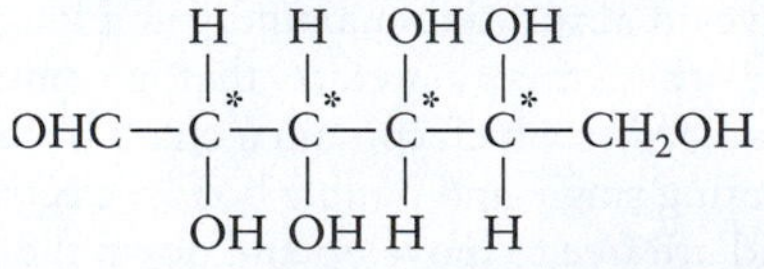

$OHC-\overset{*}{C}-\overset{*}{C}-\overset{*}{C}-\overset{*}{C}-CH_2OH$ with H, H, OH, OH on top and OH, OH, H, H on bottom

19.57 (a) GTACTCAAT; (b) ACTTAACGT
19.59 (a) $C_5H_5N_5O$; (b) $C_6H_{12}O_6$; (c) $C_3H_7NO_2$
19.61 (a) alcohol, ether, aldehyde; (b) ketone, alkene; (c) amine, amide

19.63

(a)

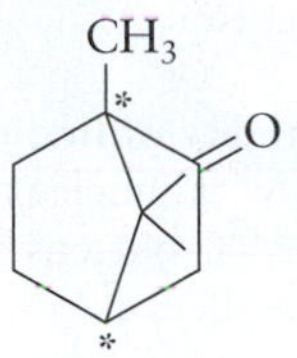

(b)

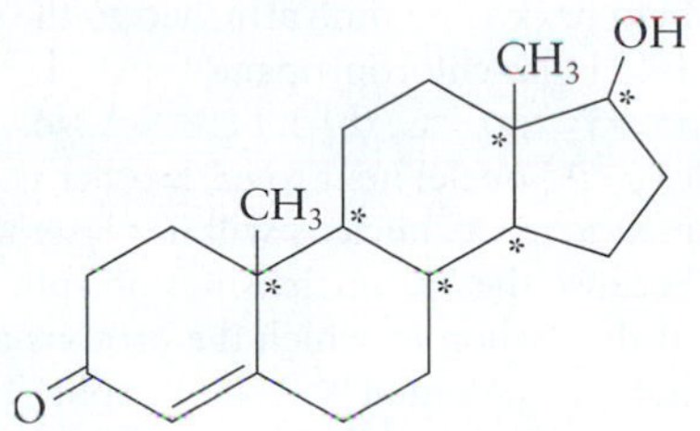

19.65   An asterisk (*) denotes a chiral carbon atom.

(a)

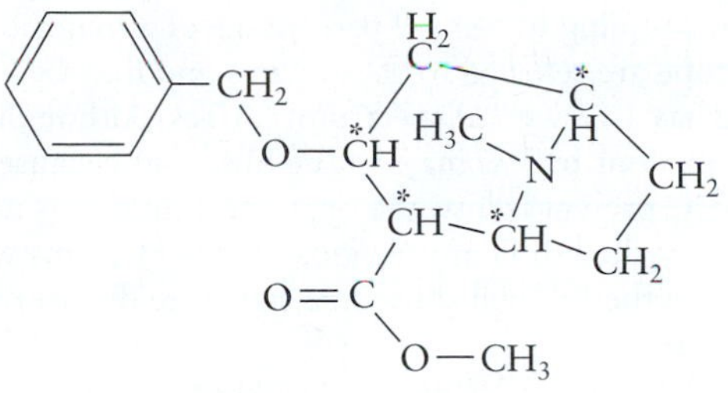

(b)

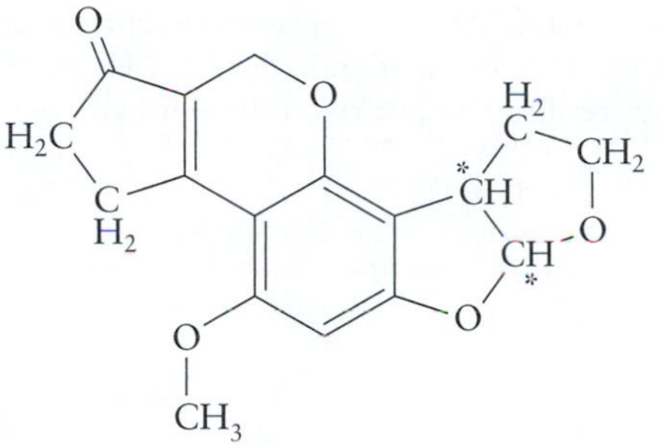

19.67   (a)

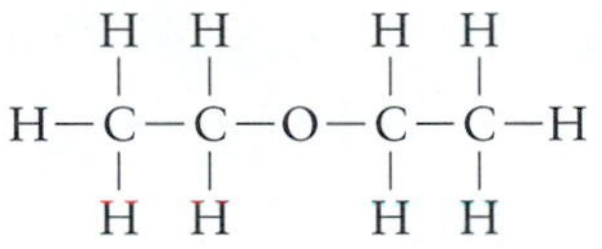

Diethyl ether

(b)

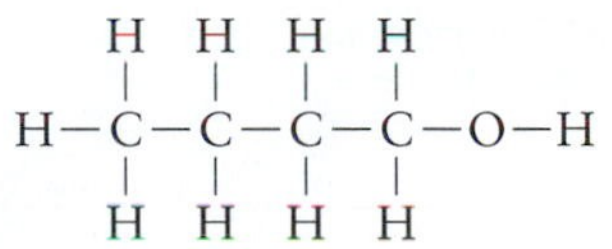

1-Butanol

(b) 1-Butanol can hydrogen bond with itself but diethyl ether cannot, so 1-butanol molecules are held together more strongly in the liquid and therefore 1-butanol has the higher boiling point. Both compounds can form hydrogen bonds with water and therefore have similar solubilities.

19.69   (a)

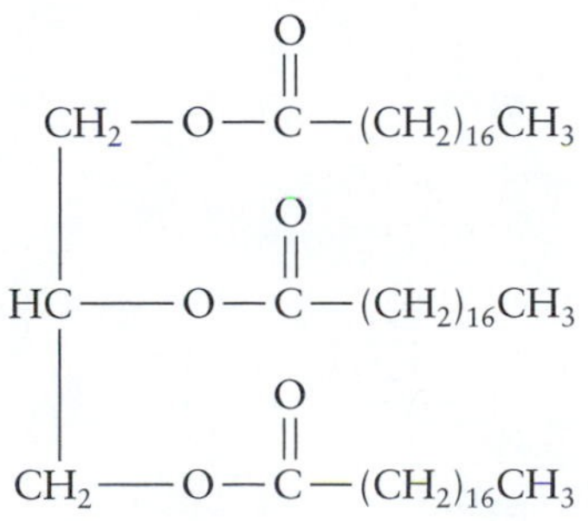

(b)

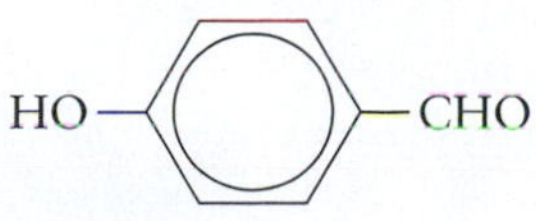

19.71   (a) addition; (b) condensation; (c) addition; (d) addition; (e) condensation

19.73   (a) $HOCH_2CH_2OH + 2\,CH_3(CH_2)_{16}COOH \rightarrow$
$CH_3(CH_2)_{16}COOCH_2CH_2OOC(CH_2)_{16}CH_3 + 2\,H_2O$

(b) $2\,CH_3CH_2OH + HOOCCOOH \rightarrow$
$CH_3CH_2OOCCOOCH_2CH_3 + 2\,H_2O$

(c) $CH_3CH_2CH_2CH_2OH + CH_3CH_2COOH \xrightarrow{\Delta}$
$CH_3CH_2COOCH_2CH_2CH_2CH_3 + H_2O$

19.75   (a) 7189 g·mol^{-1}. (b) 135. (c) 23.75 Torr. (d) The osmotic pressure developed by the resulting polymer solution is readily measured, while the change in partial pressure of $H_2O(g)$ changes by less than 0.1%; therefore, osmometry is the preferred method.

19.77   (a) Primary structure is the sequence of amino acids along a protein chain. Secondary structure is the conformation of the protein, or the manner in which the chain is coiled or layered, as a result of interactions between amide and carboxyl groups. Tertiary structure is the shape into which sections of the proteins twist and intertwine, as a result of interactions between side groups of the amino acids in the protein. If the protein consists of several polypeptide units, then the manner in which the units stick together is the quaternary structure. (b) The primary structure is held together by covalent bonds. Intermolecular forces provide the major stabilizing force of the secondary structure. The tertiary and quaternary structures are maintained by a combination of London forces, hydrogen bonding, and sometimes ion–ion interactions.

19.79   (a) $^{+}H_3NCH_2COOH(aq) + H_2O(l) \rightarrow$
$^{+}H_3NCH_2CO_2^{-}(aq) + H_3O^{+}(aq)$
$^{+}H_3NCH_2CO_2^{-}(aq) + H_2O(l) \rightarrow$
$H_2NCH_2CO_2^{-}(aq) + H_3O^{+}(aq)$

(b) $pK_{a1} = 2.35$, $pK_{a2} = 9.78$
pH = 2, $^{+}H_3NCH_2COOH$
pH = 5, $^{+}H_3NCH_2CO_2^{-}$
pH = 12, $H_2NCH_2CO_2^{-}$

19.81   Condensation polymerization involves the loss of a small molecule, often water or HCl, when monomers are combined. Dacron is more linear than the polymer obtained from benzene-1,2-dicarboxylic acid and ethylene glycol, so Dacron can be more readily spun into yarn.

19.83

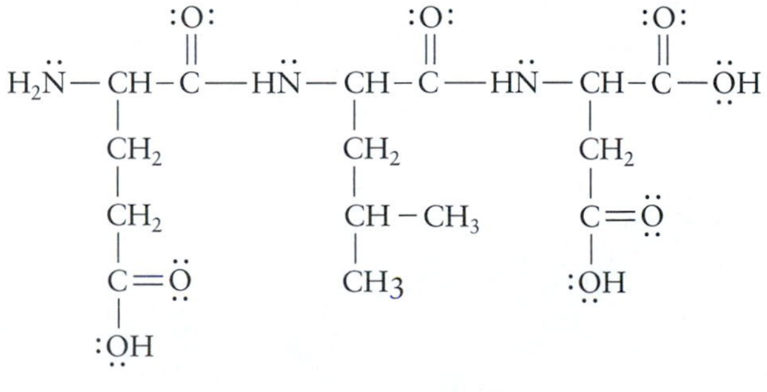

19.85   (a)

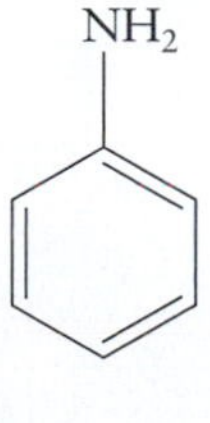

(b) sp^3. (c) sp^2. (d) Each N atom carries one lone pair of electrons. (e) Yes, the N atoms help to carry the current because the unhybridized p-orbital on each N atom is part of the extended π conjugation (delocalized π-bonds) that allows electrons to move freely along the polymer.

19.87 Two peaks are observed with relative overall intensities 3:1. The larger peak is due to the three methyl protons and is split into two lines with equal intensities. The smaller peak is due to the proton on the carbonyl carbon atom and is split into four lines with relative intensities 1:3:3:1.

19.89 The peaks in the spectrum can be assigned on the basis of the intensities and the coupling to other peaks. The hydrogen atoms of the CH_3 unit of the ethyl group will have an intensity of 3 and will be split into a triplet by the two protons on the CH_2 unit. This peak is found at $\delta \approx 1.2$. The CH_2 unit will have an intensity of 2 and will be split into a quartet by the three protons on the methyl group. This peak is found at $\delta = 4.1$. The CH_2 group that is part of the butyl function will have an intensity of two but will appear as a singlet because there are no protons on adjacent carbon atoms. This is the signal found at $\delta = 2.1$. The remaining CH_3 groups are equivalent and also will not show coupling. They can be attributed to the signal at $\delta = 1.0$. Notice that the peak that is most downfield is the one for the CH_2 group attached directly to the electronegative oxygen atom, and that the second most downfield peak is the one attached to the carbonyl group.

19.91 2-chloropropane

19.93 (a) ^{13}C. (b) 1.11%. (c) No. The probability of finding two ^{13}C nuclei next to each other is very low. A ^{12}C nucleus next to a ^{13}C nucleus will not interact with the ^{13}C nucleus because the ^{12}C nucleus has no spin. (d) Maybe. Because most of the carbon to which the protons are attached is ^{12}C, the bulk of the signal will not be split. The protons that are attached to the ^{13}C atoms will be split, but this amounts to only 1.11% of the sample, so the peaks are very small. Peaks that result from coupling to a small percentage of a magnetically active isotope are referred to as satellites and may be observed if one has a very good spectrum. (e) Yes. Although the splitting of protons by ^{13}C may not be observed because the amount of ^{13}C present is low, the opposite situation is not true. If a ^{13}C is attached to H atoms, most of the H atoms will have a spin and so the ^{13}C will show fine structure due to splitting by the H atoms.

19.95 (a) 1. (b) 1. (c) 1. (d) 2. (e) Free rotation around the C—C and C—Cl bonds averages out the environments so that both CH_2 hydrogens are equivalent and all three CH_3 hydrogens are equivalent, resulting in just two different hydrogen signals.

Illustration Credits

13.22, 13.32, and 13.33, W. H. Freeman photos by Ken Karp; Box 13.3, NASA. **Chapter 14** Figs. 14.10 and 14.11, W. H. Freeman photos by Ken Karp; Figs. 14.13a, Richard Megna/Fundamental Photographs; Fig. 14.13c, Charles D. Winters/Photo Researchers; Fig. 14.14, Richard Megna/Fundamental Photographs; Fig. 14.15, W. H. Freeman photo by Ken Karp; Fig. 14.16, Texasgulf; Fig. 14. 19, W. H. Freeman photo by Ken Karp; Fig. 14.20, Houston Museum of Natural Science; Fig. 14.21, Alexander Boden; Fig. 14.22, Aalborg Portland Betonforskningslaboratorium, Karlslunde; Fig. 14.23, L. Linkhart/Visuals Unlimited; Fig 14.25a, Houston Museum of Natural Science; Fig. 14.25b, Lee Boltin; Fig. 14.25c, Houston Museum of Natural Science; Figs. 14.26 and 14.28, Chip Clark; Fig. 14.32, Max Planck/Institut für Kemphysik; Fig. 14.33a, Field Museum of Natural History, Chicago; Fig. 14.36, W. H. Freeman photo by Ken Karp; Figs. 14.39 and 14.40, Chip Clark; Fig. 14.41, Phillip Hayson/Photo Researchers; Fig. 14.43, Chip Clark; Fig. 14.44, Corning Glass; Fig. 14.45, courtesy of R. L. Kugler and J. C. Pashin; Fig. 14.46, courtesy of NASA/JPL/Caltech; Box 14.1, Richard E. Smalley, Rice University, Center for Nanoscale Science and Technology, from *Nature* 361: 297, 1993. **Chapter 15** Fig 15.1, Chip Clark; Fig. 15.2, Photo Researchers; Fig. 15.3, Chip Clark; Fig. 15.4, W. H. Freeman photo by Ken Karp; Fig. 15.5, Chip Clark; Fig. 15.6, W. H. Freeman photo by Ken Karp; Fig. 15.7, Chip Clark; Fig. 15.8, Ross Chapple; Figs. 15.11 and 15.12a, Chip Clark; Fig. 15.12b, National Park Services; Fig. 15.13, Chip Clark; Fig. 15.15, Lee Boltin; Fig. 15.16, W. H. Freeman photo by Ken Karp; Figs. 15.17, 15.20, and 15.21, Chip Clark; Fig. 15.22, W. H. Freeman photo by Ken Karp; Fig. 15.23, Johnson Space Center/NASA; Fig. 15.25, Greater Pittsburgh Neon/Tom Anthony; Fig. 15.27, Argonne National Library; Fig. 15.28, W. H. Freeman photo by Ken Karp; Fig. 15.29, courtesy of S. O. Kim, M. P. Stoykovich, and P. F. Nealey, from S. O. Kim, H. H. Solak, M. P. Stoykovich, N. J. Ferrier, J. J. de Pablo, P. F. Nealey, *Nature* 424: 411, 2003; Fig. 15.30, from M. F. Crommie, C.P. Lutz, D. M. Eigler, *Science* 262: 218–220 1993; Fig. 15.31, Science VU/NASA/ARC/Visuals Unlimited; Box 15.1, F. W. Goro. **Chapter 16** Figs. 16.7 and 16.8, W. H. Freeman photos by Ken Karp; Fig. 16.9, Institute of Oceanographic Sciences/NERC/SPL/Photo Researchers; Fig. 16.10, Chip Clark; Fig. 16.12, Paul Chesley/Tony Stone/Getty Images; Fig. 16.13, Lee Boltin Picture Library; Fig. 16.14, Field Museum of Natural History, Chicago; Figs. 16.16 and 16.17, W. H. Freeman photos by Ken Karp; Figs. 16.31 and 16.32, Carol Moralejo, Concordia University; Fig. 16.33, Richard Megna/Fundamental Photographs; Fig. 16.40, U.S. Steel; Fig. 16.43, S. Odenbach, ZARM, University of Bremen, Germany; Fig. 16.44, from magnets S. K. Ritter, *Chemical & Engineering News* 82(50): 29–32, 2004. **Chapter 17** Figs. 17.2 and 17.3, The Granger Collection; Fig. 17.19, University of Arizona; Fig. 17.26, Tom Carroll/Phototake NYC; Fig. 17.27, Princeton Plasma Physics Laboratory; Fig. 17.28, Chip Clark; Fig. 17.29, Phil Schofield; Fig. 17.30, U.S. Department of Energy; Box 17.1, (left) adapted from S. E. Peterson, P. T. Fox, M. I. Posner, M. A. Mintum, M. E. Raichle, *Nature* 331:585–589, 1998 (right) Marcus E. Raichle and Washington University School of Medicine; Box 17.2 (a), Chip Clark.

Chapter 18 Fig. 18.12, W. H. Freeman photo by Ken Karp; Fig. 18.15, Chevron Corporation. **Chapter 19** Figs. 19.3, 19.7, and 19.8, W. H. Freeman photos by Ken Karp; Fig. 19.10, Fibers Division, Monsanto Chemical Company; Fig. 19.11, Andrew Syred/Science Photo Library/Photo Researchers; Fig. 19.12, Chip Clark; Fig. 19.15, courtesy of Professor Mehmet Sarikaya, University of Washington, Seattle; Fig. 19.16, Gamma Liaison International; Fig. 19.18, A. Marmont and E. Damasio, Division of Hematology, St. Martin's Hospital, Genoa, Italy; Fig. 19.22, Tom Bean/Tony Stone/Getty Images; Fig. 19.23, John P. O'Brien, DuPont; Fig. 19.26b, Institute of Paper Chemistry; Fig. 19.27, Photo Researchers; Box 19.1, James Kilkelly. **Major Technique 7** Fig. 2, Owen Franken/Tony Stone/Getty Images.

INDEX

KEY EQUATIONS

1. General

Roots of the equation $ax^2 + bx + c = 0$:

$$x = \frac{-b \pm (b^2 - 4ac)^{1/2}}{2a}$$

Kinetic energy of a particle:

$$E_K = \tfrac{1}{2}mv^2$$

Gravitational potential energy:

$$E_P = mgh$$

Coulomb potential energy of two charges q_1 and q_2 at a separation r:

$$E_P = q_1q_2/4\pi\varepsilon_0 r$$

2. Structure and spectroscopy

Relation between the wavelength, λ, and frequency, ν, of electromagnetic radiation:

$$\lambda\nu = c$$

Energy of a photon of electromagnetic radiation of frequency ν:

$$E = h\nu$$

de Broglie equation:

$$\lambda = h/p$$

Heisenberg uncertainty principle:

$$\Delta p\Delta x \geq \tfrac{1}{2}\hbar$$

Energy of a particle of mass m in a one-dimensional box of length L:

$$E_n = n^2h^2/8mL^2 \qquad n = 1, 2, \ldots$$

Bohr frequency condition:

$$h\nu = E_{upper} - E_{lower}$$

Energy levels of a hydrogenlike atom of atomic number Z:

$$E_n = -Z^2h\mathcal{R}/n^2, \quad n = 1,2, \ldots$$

Formal charge

$$FC = V - (L + \tfrac{1}{2}S)$$

3. Thermodynamics

Ideal gas law:

$$PV = nRT$$

Expansion work against constant external pressure:

$$w = -P_{ex}\Delta V$$

Work of reversible, isothermal expansion of an ideal gas:

$$w = -nRT \ln (V_{final}/V_{initial})$$

First law of thermodynamics:

$$\Delta U = q + w$$

Definition of entropy change:

$$\Delta S = q_{rev}/T$$

Definition of enthalpy and Gibbs free energy

$$H = U + PV \qquad G = H - TS$$

Change in Gibbs free energy at constant temperature:

$$\Delta G = \Delta H - T\Delta S$$

Relation between the constant-pressure and constant-volume molar heat capacities of an ideal gas:

$$C_{P,m} = C_{V,m} + R$$

Standard reaction enthalpy ($X = H$) and Gibbs free energy ($X = G$) from standard enthalpies of formation:

$$\Delta X^\circ = \sum n\Delta X_f^\circ(\text{products}) - \sum n\Delta X_f^\circ(\text{reactants})$$

Standard reaction entropy:

$$\Delta S^\circ = \sum nS_m^\circ(\text{products}) - \sum nS_m^\circ(\text{reactants})$$

Kirchhoff's law:

$$\Delta H_2^\circ = \Delta H_1^\circ + \Delta C_P(T_2 - T_1)$$

Change in entropy when a substance of constant heat capacity, C, is heated from T_1 to T_2:

$$\Delta S = nR \ln (T_2/T_1)$$

Change in entropy when an ideal gas expands isothermally from V_1 to V_2:

$$\Delta S = nR \ln (V_2/V_1)$$

Boltzmann's formula for the statistical entropy:

$$S = k \ln W$$

Entropy change of the surroundings for a process in a system with enthalpy change ΔH:

$$\Delta S_{surr} = -\Delta H/T$$

4. Equilibrium and electrochemistry

Definition of activity:

For an ideal gas: $\quad a_J = P_J/P^\circ$, $P^\circ = 1$ bar

For a solute in an ideal solution: $\quad a_J = [J]/c^\circ$, $c^\circ = 1 \text{ mol·L}^{-1}$

For a pure liquid or solid: $\quad a_J = 1$

Reaction quotient and equilibrium constant:

For the reaction $a A + b B \longrightarrow c C + d D$, $\quad Q = a_C{}^c a_D{}^d/a_A{}^a a_B{}^b$

For the equilibrium $a A + b B \rightleftharpoons c C + d D$,

$$K = (a_C{}^c a_D{}^d/a_A{}^a a_B{}^b)_{equilibrium}$$

Variation of Gibbs free energy of reaction with composition:

$$\Delta G_r = \Delta G_r^\circ + RT \ln Q$$

Relation between standard reaction Gibbs free energy and equilibrium constant:

$$\Delta G_r^\circ = -RT \ln K$$

van 't Hoff equation:

$$\ln \frac{K_2}{K_1} = \frac{\Delta H_r^\circ}{R}\left(\frac{1}{T_1} - \frac{1}{T_2}\right)$$

Relation between K and K_c:

$$K = (RTc^\circ/P^\circ)^{\Delta n_{gas}}K_c \qquad P^\circ/Rc^\circ = 12.03 \text{ K}$$

Clausius–Clapeyron equation:

$$\ln \frac{P_2}{P_1} = \frac{\Delta H_{vap}^\circ}{R}\left(\frac{1}{T_1} - \frac{1}{T_2}\right)$$

Relation between Gibbs free energy and maximum nonexpansion work:

$$\Delta G = w_{e,max} \text{ at constant temperature and pressure}$$

Relation between pH and pOH:

$$pH + pOH = pK_w$$

Relation between acidity and basicity constants of a conjugate acid–base pair:

$$pK_a + pK_b = pK_w$$

Henderson–Hasselbalch equation:

$$pH = pK_a + \log ([\text{base}]_{initial}/[\text{acid}]_{initial})$$

Relation between the Gibbs free energy of reaction and the emf of a cell:

$$\Delta G = -nFE$$

Relation between the equilibrium constant for a cell reaction and the emf of a cell:

$$\ln K = nFE^\circ/RT$$

Nernst equation:

$$E = E^\circ - (RT/nF) \ln Q$$

5. Kinetics

Average reaction rate:

$$\text{Rate of disappearance of R} = -\frac{\Delta[R]}{\Delta t}$$

$$\text{Rate of formation of P} = \frac{\Delta[P]}{\Delta t}$$

Unique average rate for $a A + b B \longrightarrow c C + d D$,

Unique average reaction rate =

$$-\frac{1}{a}\frac{\Delta[A]}{\Delta t} = -\frac{1}{b}\frac{\Delta[B]}{\Delta t} = \frac{1}{c}\frac{\Delta[C]}{\Delta t} = \frac{1}{d}\frac{\Delta[D]}{\Delta t}$$

Integrated rate laws:

For Rate of disappearance of $A = k[A]$, $\quad \ln \frac{[A]_t}{[A]_0} = -kt$, $\quad [A]_t = [A]_0 e^{-kt}$

For Rate of disappearance of $A = k[A]^2$,

$$\frac{1}{[A]_t} - \frac{1}{[A]_0} = kt, \quad [A]_t = \frac{[A]_0}{1 + [A]_0 kt}, \quad \frac{1}{[A]_t} = kt + \frac{1}{[A]_0}$$

Half-life of a reactant in a first-order reaction:

$$t_{1/2} = (\ln 2)/k$$

Arrhenius equation:

$$\ln k = \ln A - E_a/RT$$

The rate constant at one temperature in terms of its value at another temperature:

$$\ln \frac{k_2}{k_1} = \frac{E_a}{R}\left(\frac{1}{T_1} - \frac{1}{T_2}\right)$$

The equilibrium constant in terms of the rate constants:

$$K = k_{forward}/k_{reverse}$$

THE ELEMENTS

Element	Symbol	Atomic number	Molar mass* (g·mol⁻¹)	Element	Symbol	Atomic number	Molar mass* (g·mol⁻¹)
Actinium	Ac	89	(227)	Mendelevium	Md	101	(258)
Aluminum	Al	13	26.98	Mercury	Hg	80	200.59
Americium	Am	95	(243)	Molybdenum	Mo	42	95.94
Antimony	Sb	51	121.76	Neodymium	Nd	60	144.24
Argon	Ar	18	39.95	Neon	Ne	10	20.18
Arsenic	As	33	74.92	Neptunium	Np	93	(237)
Astatine	At	85	(210)	Nickel	Ni	28	58.69
Barium	Ba	56	137.33	Niobium	Nb	41	92.91
Berkelium	Bk	97	(247)	Nitrogen	N	7	14.01
Beryllium	Be	4	9.01	Nobelium	No	102	(259)
Bismuth	Bi	83	208.98	Osmium	Os	76	190.23
Bohrium	Bh	107	(264)	Oxygen	O	8	16.00
Boron	B	5	10.81	Palladium	Pd	46	106.42
Bromine	Br	35	79.90	Phosphorus	P	15	30.97
Cadmium	Cd	48	112.41	Platinum	Pt	78	195.08
Calcium	Ca	20	40.08	Plutonium	Pu	94	(244)
Californium	Cf	98	(251)	Polonium	Po	84	(209)
Carbon	C	6	12.01	Potassium	K	19	39.10
Cerium	Ce	58	140.12	Praseodymium	Pr	59	140.91
Cesium	Cs	55	132.91	Promethium	Pm	61	(145)
Chlorine	Cl	17	35.45	Protactinium	Pa	91	231.04
Chromium	Cr	24	52.00	Radium	Ra	88	(226)
Cobalt	Co	27	58.93	Radon	Rn	86	(222)
Copper	Cu	29	63.55	Rhenium	Re	75	186.21
Curium	Cm	96	(247)	Rhodium	Rh	45	102.90
Darmstadtium	Ds	110	—	Roentgenium	Rg	111	—
Dubnium	Db	105	(262)	Rubidium	Rb	37	85.47
Dysprosium	Dy	66	162.50	Ruthenium	Ru	44	101.07
Einsteinium	Es	99	(252)	Rutherfordium	Rf	104	(261)
Erbium	Er	68	167.26	Samarium	Sm	62	150.36
Europium	Eu	63	151.96	Scandium	Sc	21	44.96
Fermium	Fm	100	(257)	Seaborgium	Sg	106	(266)
Fluorine	F	9	19.00	Selenium	Se	34	78.96
Francium	Fr	87	(223)	Silicon	Si	14	28.09
Gadolinium	Gd	64	157.25	Silver	Ag	47	107.87
Gallium	Ga	31	69.72	Sodium	Na	11	22.99
Germanium	Ge	32	72.64	Strontium	Sr	38	87.62
Gold	Au	79	196.97	Sulfur	S	16	32.06
Hafnium	Hf	72	178.49	Tantalum	Ta	73	180.95
Hassium	Hs	108	(277)	Technetium	Tc	43	(98)
Helium	He	2	4.00	Tellurium	Te	52	127.60
Holmium	Ho	67	164.93	Terbium	Tb	65	158.93
Hydrogen	H	1	1.0079	Thallium	Tl	81	204.38
Indium	In	49	114.82	Thorium	Th	90	232.04
Iodine	I	53	126.90	Thulium	Tm	69	168.93
Iridium	Ir	77	192.22	Tin	Sn	50	118.71
Iron	Fe	26	55.84	Titanium	Ti	22	47.87
Krypton	Kr	36	83.80	Tungsten	W	74	183.84
Lanthanum	La	57	138.91	Uranium	U	92	238.03
Lawrencium	Lr	103	(262)	Vanadium	V	23	50.94
Lead	Pb	82	207.2	Xenon	Xe	54	131.29
Lithium	Li	3	6.94	Ytterbium	Yb	70	173.04
Lutetium	Lu	71	174.97	Yttrium	Y	39	88.91
Magnesium	Mg	12	24.31	Zinc	Zn	30	65.41
Manganese	Mn	25	54.94	Zirconium	Zr	40	91.22
Meitnerium	Mt	109	(268)				

*Parentheses around molar mass indicate the most stable isotope of a radioactive element.